A POPULAR DICTIONARY OF ZOOLOGY

N WHICH THE CHARACTERISTICS THAT DISTINGUISH THE DIFFERENT CLASSES, GENERAL AND SPECIES, ARE COMBINED WITH A VARIETY OF INTERESTING INFORMATION ILLUSTRATIVE OF THE HABITS, INSTINCTS, AND GENERAL ECONOMY OF THE ANIMAL KINGDOM

BY

SAMUEL MAUNDER

Author of 'The Treasury of Knowledge' 'The Scientific and Literary Treasury' &c.

T. SPENCER COBBOLD, M.D. F.L.S.

Lecturer on Botany, Zoology, and Comparative Anatomy at the Middlesex Hospital Medical College

Embellished with 900 Woodcuts expressly engrabed for this work.

DAYA PUBLISHING HOUSE
DELHI - 110006

First Published 1870

Reprinted 1990

ISBN 81-7035-071-9

Published By

DAYA PUBLISHING HOUSE
1302, Vaidwara, Nai Sarak
Delhi-110006

Printed At

RICHA OFFSET
Delhi

PREFATORY NOTE

BY THE EDITOR

A Sixth Edition of this popular Compendium of Zoological Science having been demanded, I have endeavoured to bring its information up to the present time. This has been accomplished by the introduction of corrections and emendations in the body of the work, and more particularly, by the addition of a Supplement, which had become an imperative necessity. So wide are the recent strides of Natural History Science, that in the brief interval which has elapsed since the last edition of this "Treasury" was issued, a multitude of novel and important facts have been brought to light. Among these may be here specified, in the first place, the re-discovery of that huge living, anthropomorphous ape, the Gorilla, whose position in the scale of animated nature has given origin to a most instructive controversy; secondly, further knowledge of the remarkable genetic changes, or Alternations of Generation, known to occur in many of the lower animals during their passage from the embryonic up to the adult condition; and thirdly, the great variety of extinct monsters, whose skeletal remains have been investigated by the devoted cultivators of palæontological research.

The Supplement, therefore, embracing about one hundred brief articles, will be found to contain not only a record of the above-mentioned discoveries, but also ample notices of remarkable and otherwise interesting animals, such as the Kiang, Eland, Balæniceps, Sieboldia, Euplectella, &c., the previous records of which were either imperfect or altogether wanting.

T. S. C.

INTRODUCTION.

A DICTIONARY of ANIMATED NATURE, upon a comprehensive plan, had long occupied my thoughts, before I first publicly announced my intention of making this addition to my series of popular "TREASURIES."

It will be seen at the first glance, that the whole of the articles are printed in *alphabetical* order; so that the *name* of any animal being previously known, its zoological character and its habits can be instantly ascertained; while those persons who wish to study this branch of Natural History according to the most approved modern *system* will only have to refer to this "INTRODUCTION," and they will find not only an outline of Cuvier's celebrated arrangement, as developed in the last edition of his *Règne Animal* with those alterations and additions required by the present advanced state of the science, but, under each Class and Order, references to the different genera, &c., described in the body of the Work. Thus, this CLASSIFIED INDEX will be the means of supplying the necessary *systematic* information. But whether the articles be so consulted, or merely read in a more desultory way, I believe that a vast fund of instruction and amusement will be found here collected. And so, in truth, there ought. Many of the most celebrated standard zoological works have been put under contribution, and accurate information has been gleaned from all. Nor is it among the least of the advantages which, I presume, this volume will be found to possess over most others on this subject, that, besides numerous entirely new articles, and condensed abridgements of the more elaborate writings of many acknowledged authorities, I have had an opportunity of making myself acquainted with many interesting facts now for the first time recorded in a popular digest of Animated Nature. It will also be apparent that I have not hesitated to make copious extracts from the recent Publications of various living writers who have displayed the wonders of Animated Nature under new aspects, and with increased force, originality, and beauty. In this, I have most scrupulously acknowledged the sources whence my pages have been enriched; and to the many scientific men and pleasing writers to whom I am thus indebted, I beg to return my warmest acknowledgments.

It would be easy to extend my Introductory Remarks to a considerable length by dilating on the uses and advantages to be derived from an acquaintance with Natural History; nor would it be difficult to show how much that is bright and beautiful in Nature is for ever lost to him who has never become conversant with the study. But my inclination is to avoid what some ill-natured critics might term *twaddle*, and my limits forbid me to descant on a theme which others (who are far better qualified than I can ever possibly become) have treated with all the ardent enthusiasm that is inherent in the breast of every *true* votary of Nature. The subject, indeed, presents a wide field for the employment of the mental faculties; and I confess it is difficult to repress some of the thoughts that arise from its contemplation. No part can be viewed as unimportant or uninteresting—none that is unworthy of the most attentive consideration, or that can fail to impress the mind with feelings of profound admiration for the works of NATURE. Marvellous, indeed, as they are all, the most astounding manifestations of Supreme intelligence are unquestionably displayed in his character as "LORD AND GIVER OF LIFE," as the Creator and Preserver of all that "live, move, and have their being." It is therefore that portion of the "wondrous whole" which we term The ANIMAL KINGDOM that demands our especial regard, and is in the highest degree calculated to gratify a laudable curiosity, as well as to reward the labours of the most diligent research.

As to the manner in which this work has been embellished, I can speak with perfect satisfaction. About *Nine hundred accurate Woodcuts* have been given in Mr. R. Branston's best manner, and they will no doubt be properly appreciated.

I shall now offer a *Systematic Classification* of the principal contents of this volume, a mode which, I trust, will be found at once simple and scientific, and calculated to remove any objection that might be urged against my adoption of the *alphabetical arrangement* in the body of the Work.

S. M.

THE ANIMAL KINGDOM,

ARRANGED IN CLASSES, ORDERS, AND GENERA, ACCORDING TO ITS ORGANIZATION.

CLASS IV.

PISCES.

DIVISION II.

ANNULOSA.

CLASS I.

INSECTS.

Class II.

MYRIAPODA.

THE

TREASURY OF NATURAL HISTORY;

OR, A POPULAR

Dictionary of Animated Nature.

AARD-VARK. The name by which the quadruped *Orycteropus Capensis* is known to the Dutch colonists at the Cape of Good Hope. The following cut, which is copied from Daniell's work on the Animals of

AARD-VARK (ORYCTEROPUS CAPENSIS)

Southern Africa, will give a good idea of its form. [For an account of its habits, see ORYCTEROPUS.]

AARD-WOLF. A name given by the European colonists in the neighbourhood of Algoa Bay, in South Africa, to a carnivorous digitigrade animal, which at first sight might be easily mistaken for a young striped hyæna. It is about the size of a full-grown fox, and in habits and manners somewhat resembles it. [See PROTELES.]

ABDOMINALES. An order of fishes in the Linnæan system, consisting of all those species which have the ventral fins placed behind the pectoral, or upon the abdomen, the cartilaginous fishes alone excepted. This arrangement has, however, been departed from by Cuvier as defective; and in its present acceptation the term *Abdominales* denotes a family or subdivision of Malacopterygious or soft-finned fishes only, including therein the greater number of the fresh-water species, and such as periodically migrate from the ocean to deposit their spawn. As familiar instances, we may specify the Salmon and Trout.

ABERDAVINE. A small migratory Passerine bird, more generally termed the SISKIN [which see]. In Sussex it is known as the Barley-bird, because it is usually a visitant of that county about the barley seed-time.

ABOU-HANNES. An African bird, supposed to be the *Ibis religiosa*, or White Ibis, of the ancient Egyptians. [See IBIS.]

ABRAMIS. A genus of Malacopterygious fishes. [See BREAM.]

ABRANCHIATA. An order of the *Annelida*, composed of animals having no branchial appendages. Of these, the Earthworm and the Leech are examples.

ABROCOMA. A genus of small Rodent animals, native of South America, remarkable for the extraordinary fineness of their fur. They have large ears, small claws, and the tail rather long, and not tufted. Their general aspect is intermediate to that of the Chinchillas and Rats or Voles.

ACALEPHÆ. A group of *Cœlenterata*, comprising all those animals which float and swim in the water, by alternate contractions and dilatations of the body, although their substance is merely gelatinous, and without any apparent fibres. They are popularly named *sea-nettles*, from their causing, when touched, a disagreeable sensation, like the sting of a nettle: they are also familiarly known as *jelly-fishes*, *sea-blubbers*, &c., from the extreme softness of their tissues, which

PURPLE OCEANIC JELLY-FISH.
(ÆQUOREA PURPUREA.)

melt away, as it were, when removed from the water. Their form is circular, and there is only one opening into the body, which serves both for the mouth and vent. Although possessed of a certain degree of locomotive power, the movements of the Acalephæ are very feeble; and they are consequently often driven by the winds and rough currents on shore, where they are either beaten to pieces by the waves, or left dry by the tide. The Acalephæ are of various forms: many, indeed, are not yet thoroughly known; but the specimens which are most commonly met with in our climate, when examined in their native element, are seen to be composed of a large mushroom-shaped gelatinous disc, from the lower surface of which various processes and filaments depend. There are, in fact, obvious points of resemblance among them all; but still they admit of division into genera and sub-genera. Thus, we may observe, the genus *Medusa* includes those which have a central disc, more or less convex, on the upper surface, something like the head of a mushroom, and those that have a true mouth on the under-side of the disc; but this mouth is sometimes a simple opening, and at other times placed on a peduncle: while the genus *Æquorea* includes those in which the mouth is simple, and not on a peduncle. When the disc is furnished with tentacula all round, they constitute the genus Æquorea, the species of which are extremely abundant in tropical seas. These animals feed chiefly on Mollusca and Crustacea. [See also JELLY-FISH, P. 355; and for an account of their development, see STROBILA and ALTERNATION OF GENERATION in SUPPLEMENT.]

ACANTHOCEPHALA. An order of Entozoa, which may be exemplified by the *Echinorhynchus gigas*, often found in abundance in the alimentary canal of swine. The form of this parasite is clongated, tapering to the tail: the head consists of a retractile snout or proboscis, armed with four circlets of sharp recurved spines, and it can be withdrawn or protruded at will. These singular parasites have neither mouth nor stomach, their nutriment being imbibed through the general surface of the body. Some of the species are truly formidable to look at, but, fortunately, none of them infest man. [See ECHINORHYNCHUS in SUPPLEMENT.]

ACANTHOCINUS. A genus of Coleopterous insects belonging to the *Longicorn* group. There are but few European species: the antennæ in the male are more than four times the length of the body.

ACANTHOPHIS. A genus of venomous serpents, allied to the Vipers, but distinguished from them in many essential characters. The head of the Acanthophis is broad and compressed, the mouth capable of great extension, and the tail is terminated by a little spur or horny excrescence, whence its name is derived. They are natives of Australia; secrete themselves in holes or beneath the roots of trees, and exhibit an astonishing tenacity of life. The *A. Brownii* is reckoned the most venomous Reptile found near Port-Jackson.

ACANTHOPODA. A tribe of Clavicorn Coleopterous insects (composed of only one genus, *Heterocerus*), distinguished by their flattened feet, which are broad, and armed on the outside with spines; the tarsi short and four-jointed with ordinary sized claws, and the body depressed; the prosternum is dilated; the antennæ are rather longer than the head, eleven-jointed, the last six forming a nearly cylindrical serrated mass.

ACANTHOPTERYGII. One of the three primary grand divisions, or natural orders, of fishes; originally recognized by our countrymen Willoughby and Ray, afterwards systematized by Artedi, and since established by Cuvier. The characteristics of the *Acanthopterygii* are, that they possess bony skeletons, with prickly spinous processes in the dorsal fins. The Stickleback and Perch are familiar examples of this division.

ACANTHURUS, or SURGEON-FISH. A genus of Acanthopterygious fishes, many of which are remarkable for the beauty of their

SURGEON FISH. (ACANTHURUS)

form and the variety of their colours. They are chiefly distinguished by the sharp and lancet-like moveable spines with which they are armed on each side of the tail; hence, as they cannot be handled incautiously with impunity, they have obtained from English sailors, &c., the name of "doctors." They abound in the tropical seas, but are never seen elsewhere.

ACARIDÆ. Of these small spider-like animals, M. Latreille makes four divisions: 1. Mites, (*Trombidites*); 2. Ticks, (*Ricinites*); 3. Water Mites, (*Hydrachnellæ*); and 4. Flesh Worms (*Microphthira*). Some of these exist on the ground, others in the water; some are parasitical, living on the blood and humours of the animals or insects on which they are fixed, while others insinuate themselves under the skin, where they multiply prodigiously. Of these latter, the Itch-insect (*Sarcoptes scabiei*,) is a remarkable example. Their mouths are, in general, formed rather for suction than for mastication; and their extremities are commonly armed with what may be likened to a small pair of pincers. Some have four eyes, some two, and several appear to have none. The common Cheese-mite (*Acarus domesticus*) is familiar to every one. Another has the power of spinning webs, and is well known as the Red Spider, in hot-houses, where it greatly injures the plants by covering the leaves with its webs. There are also Ticks, Harvest-bugs, Water Mites, and many others, which will be noticed in their alphabetical order.

ACARUS DOMESTICUS.

We may here, however, observe, *en passant*, that so widely are the *Acaridæ* disseminated through animate and inanimate objects, that it would be difficult indeed for the most patient naturalist to describe them. Myriads swarm around us: they float in our drink; overspread our food and fruits; and if viewed with a microscopic eye, would make some loathe the choicest viands, and nauseate the most delicious productions of nature. The Mites possess great powers of life, resisting for a time the application of boiling water, and living long in alcohol.—It is a species of Acarus that Mr. Crosse is thought to have produced by galvanic action; but *naturalists* who have attended closely to such matters can readily and rationally account for their production in the usual way.

ACASTA. A genus of Cirrhipedes, found imbedded in sponges. [See BALANUS.]

ACCENTOR. A group of Passerine birds, many of which are peculiar to America; but including also our well-known Hedge-sparrow (*Accentor modularis*).

ACCIPITRES. The first order of birds in the Linnæan system, comprising such as have the beak or upper mandible hooked, and an angular projection on each side near the point; as the Eagles, Falcons, Hawks, and Owls. They are among birds what the *Carnivora* are among quadrupeds.

ACEPHALA. A class of Mollusca, distinguished by having no apparent head, but a mouth only, concealed in the bottom, or between the folds, of their mantle. The testaceous Acephala are by far the most numerous; all Bivalve shells, and some kinds of Multivalves belonging to them. [See LAMELLIBRANCHIATA; and for a familiar example, see OYSTER.]

ACERA. A name applied to a group of Apterous insects, characterized by the absence of antennæ.

ACERÆ. A family of Gasteropodous Mollusca, distinguished by the tentacula being so much shortened, widened, and separated, that there seem to be none at all, or rather, they form together a large, fleshy, and nearly square buckler, under which the eyes are placed. They approximate in many respects to the *Aplysiæ*. The shell, in those which have one, is more or less convolute, without a visible spire, and the mouth has neither sinus nor canal. The genus *Bulla* belongs to this family.

ACHATINA. A genus of terrestrial Pneumonobranchous Gasteropods, popularly known by the name of *agate-snails*. They are characterized by an oval oblong shell, striated longitudinally, with the aperture ovate, and never thickened or reflected, and a smooth, straight columella, truncated at the base. All the species are oviparous; and among them are some which are the largest of all land shells. They always live near water about trees, and are very plentiful in Africa, near the Cape of Good Hope. Some are found in the West Indies; and there are two small species, *Achatina acicula* and *Achatina octona*, found in England, among the roots of trees at the base of limestone

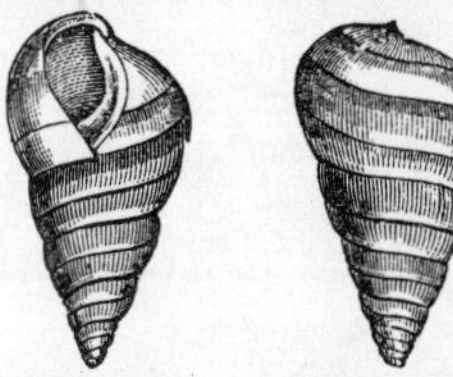

ACHATINA VIRGINIA.

rocks. The *Achatina columnaris* is one of the most remarkable of land shells; it is reversed, and the columella forms a winding pillar, visible within, quite to the summit of the spire. Many are covered with a thick epidermis, as the *Achatina zebra;* but others are destitute.

ACHATINELLA. A small genus of shells, differing from Achatina, in having the inner edge of the outer lip thickened, and a slight groove near the suture of the spire.

ACHERONTIA. A genus of Lepidopterous insects belonging to the family *Sphingidæ*. Of this genus there are two or three species closely resembling each other: one of these is found in this country; and is known as the DEATH'S-HEAD HAWK-MOTH (*Acherontia atropos*). This magnificent insect varies in the expanse of its wings from four to considerably more than five inches. The upper pair are of a very dark brown colour, varied with black, especially near the base, near which is an undulated bar of pale ochre: the disc is varied with deep black

DEATH'S-HEAD HAWK-MOTH (ACHERONTIA ATROPOS.)

undulated lines, and ferruginous patches, minutely irrorated with white, of which colour there is a central spot, and several wavy connected bars beyond the middle. The posterior wings are fulvous orange, with a narrow central and a broader dentated bar running parallel with the hinder margin. The head and thorax are brownish black, the latter with a large pale, skull-like mark on the back: the abdomen is fulvous, with the incisures of the segments black, and a lead-coloured stripe runs down the back. When disturbed or irritated, this insect emits a squeaking sound From this circumstance, as well as from the singular mark just mentioned, its appearance is regarded with much dread by the vulgar in several parts of Europe, as ominous of some approaching calamity. The Caterpillar from which this curious insect proceeds is in the highest degree beautiful, and far surpasses in size every other in this country, measuring sometimes near five inches in length, and being of very considerable thickness. Its colour is a bright yellow; the sides being marked with seven elegant broad stripes or bands, of

CATERPILLAR OF DEATH S-HEAD MOTH.

a mixed violet and sky-blue colour, which meet on the back, and are there varied with jet-black specks: on the last joint of the body is a horn or process, curving over the joint in the manner of a tail. This caterpillar is principally found on the potato and the jessamine, those plants being its favourite food. It usually changes into a chrysalis in the month of September, retiring for that purpose pretty deep under the surface of the earth; the complete insect emerging in the following June or July.

"Another peculiarity connected with the history of this Moth," Mr. Westwood observes, "consists in its attacking bee-hives, ravaging the honey, and dispersing the inhabitants. It is singular that a creature with only the advantage of size should dare, without sting or shield, singly, to attack in their strongholds these well-armed and numerous people; and still more singular, that amongst so many thousands of bees it should always contend victoriously. Huber, who first noticed the fact, asks, 'May not this moth—the dread of superstitious people—also exercise a secret influence over insects, and have the faculty, either by sound or some other means, of paralysing their courage? May not such sounds as inspire the vulgar with dread be also the dread of bees?' He also states that he was witness to the curious fact that some bees, as if expecting their enemy, had barricaded themselves by means of a thick wall of propolis and wax, completely obstructing the entrance of the hive, but penetrated by passages for one or two workers at a time; thus instructing us, that at the period when the moth appears, when also wasps and robber bees attack the hive, it is advantageous to narrow the entrances to it, so as to prevent the depredations of these obnoxious insects. The species appears to be distributed over the greater part of England and Scotland, and many specimens are annually obtained by labourers when employed in getting up potatoes."

ACHETIDÆ. A family of Orthopterous insects, ordinarily called Crickets. [See CRICKET.]

ACHEUS. A name applied by M. F. Cuvier to such of the *Tardigrada*, or Sloths, as have three claws on their fore-feet.

ACHIRUS. A genus of flat-fish, belonging to the order *Malacopterygii;* in form resembling the Sole, but distinguished from all other genera by the total want of pectoral fins; hence their name. The Achiri have no air-bladder, and consequently remain for the most part at the bottom of the sea; yet their motions are there frequently very rapid. They abound mostly in the East and West Indies; and as they keep near the shores, they furnish a plentiful supply of wholesome food to the inhabitants. The most remarkable of this species is the *Achirus marmoratus*, which has the caudal fin distinct from the anal and dorsal, all of which are of a pale bluish-white colour, thickly studded with small black spots. The flesh is of a delicate flavour, and highly esteemed. There is also another of the Achiri deserving notice, the *Achirus pavonicus*, so called from the beauty of its spots, which cover the body of the fish like those on a peacock's tail.

ACIPENSER (sometimes written ACCIPENSER). A genus of fish in the Linnæan system, the distinguishing characteristics of which are, that the mouth is retractile and destitute of teeth, and the gills have only one aperture on each side. [See STURGEON.]

ACONTIAS. A genus of Serpents, formerly confounded with the *Angues*, or common snakes, but differing from those reptiles in certain peculiarities of osteological formation, as well as in their habits; and therefore Cuvier considered it necessary to establish this new genus. They are characterized by the absence of all the bones which represent the extremities of the other angues, while they retain the structure of the head common to those animals and the lizards, and have the body similarly covered with small scales only. The progressive movements of the Acontias are consequently very different from those of common serpents: they carry their heads and breasts erect; and, though by nature harmless and even timid, when pursued they will dart courageously at their assailant. There are few countries in the Old World in which some species of Acontias are not found; but our elder naturalists have generally confounded them with serpents of a dangerously venomous nature: hence the numerous fabulous stories which are related of them by ancient historians.

ACORN-SHELL. The popular name for the *Balanus* and other Cirrhipeds, which inhabit a tubular shell, whose base is usually formed of calcareous laminæ. It is always found attached to some shell or foreign body: it is multivalvular, unequal, and fixed by a stem, or sessile; the valves lie parallel to each other, and in a perpendicular position. The inclosed animal performs its necessary functions by an aperture at the top; for the valves, being destitute of hinges, never open or separate. The tentacula from this animal being feathered, our credulous ancestors conceived the idea that it gave origin to a bird called the barnacle goose; nay, so prevalent was the opinion, that we find inserted in the Philosophical Transactions of this country a grave account of its transformation. [See BARNACLE.]

These curious but common shells are found in all seas, particularlyon the coasts of Africa. They are affixed to marine bodies, generally in numerous groups, and the peduncle is sometimes found a foot long. A large log of timber (as Mr. Broderip remarks) covered with these animals, twisting and diverging in all directions, and so thick as entirely to hide the surface of the log, is a strange sight. They look like an enormous collection of serpents to the ignorant; and a living mass of this description, casually thrown into shallow water and left by the tide, has been so termed. Their growth must be exceeding rapid. A ship going out with a perfectly clean bottom will often return, after a short voyage, covered with them.

ACOUCHI. (*Dasyprocta acuchi.*) A Rodent quadruped, considerably smaller than the Agouti; it is of a deep olive colour, and

ACOUCHI.—(DASYPROCTA ACUCHI.)

has only the rudiments of a tail. It inhabits the woods of Guiana; is of a mild, gentle, and timid disposition; and subsists on nuts, almonds, and other vegetable food.

ACRITA. A term proposed by Professor Owen to be applied to the fourth division of the Animal Kingdom. It is nearly equivalent to the Radiata of Cuvier, in the majority of which there is no distinct nervous system, or separate alimentary canal. In most of the animals composing this sub-kingdom, no muscular fibres are to be perceived, yet of these many contract and expand their bodies, and are furnished with moveable and sensitive tentacles, by which they seize their prey. Many also are capable of locomotion; others, like the plant, are fixed to one spot for life; and some are united together, and form compound animals. There is ordinarily no distinction of sexes; and reproduction takes place either by the simple division of the body, by granular ova, or gemmules which become detached from the parent body, the form of which they ultimately assume.

ACROCHORDUS. A genus of Serpents discovered in Java. They are consid red innoxious, and are distinguished from others by their skin being covered with innumerable small warts or tubercles, which, however, are only apparent when the skin of the living reptile is inflated or in preserved specimens. The only species accurately known at present is the *Acrochordus Javanicus* of Lacepède, chiefly remarkable on account of its diet; which, contrary to the general habit of the order, seems (according to the testimony of Hornstedt) to consist of fruits and other vegetable substances. This animal averages from eight to ten feet in length, the body growing gradually thicker from the head to the vent, and there suddenly contracting so as to form a very short slender tail.

ACROCINUS. A genus of Coleopterous insects belonging to the Longicorn group. The thorax on each side is furnished with a moveable tubercle ending in a spine; the body is depressed; the antennæ very long

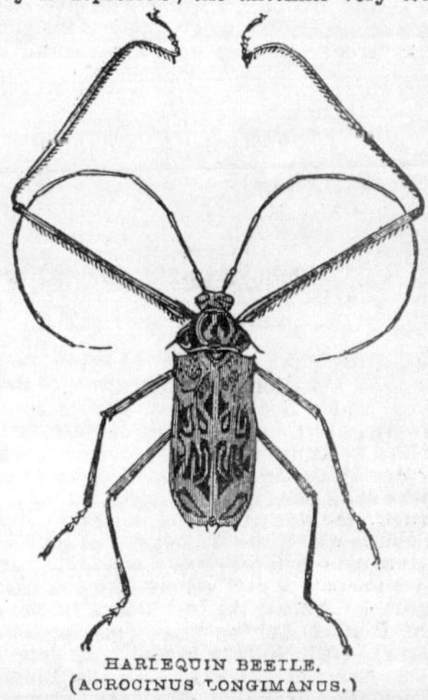

HARLEQUIN BEETLE.
(ACROCINUS LONGIMANUS.)

and slender; the fore-legs much longer than the others; the elytra are truncated at the end and furnished with two teeth. The largest and best known species is the HARLEQUIN BEETLE (*A. longimanus*) of South America, the common name of which is derived from the agreeable mixture of grey, black and red, on the elytra, giving it a resemblance to the garb of a harlequin.

ACRYDIUM. The name applied by Fabricius to a genus of Locusts, characterized by a carinate thorax; filiform antennæ, shorter than the thorax; and equal palpi. [See LOCUST.]

ACTÆON. A genus of Mollusca allied to Doris, a few species of which are found in this country. [For habits, see NUDIBRANCHIATA.]

ACTINIÆ. These curious and interesting marine animals are closely allied to the

Sea-nettles, spoken of under the head ACALEPHÆ. They are distinguished by the form of their body, which is cylindrical, soft, fleshy, and susceptible of contraction and dilatation. They are also furnished with numerous tentacula, which are appended round the margin of the aperture that serves both as the mouth and vent; and these being not only radiated but of various lively colours, have given rise to the popular names of *Animal-flowers* and *Sea-Anemonies*, by which the Actiniæ are familiarly known. They are found on the shores of every sea, often covering the sides of rocks as with a tapestry of flowers. As in other tribes, each species has its peculiar haunt, and they differ from each other in shape, size, and colour; those in tropical regions far surpassing in gorgeous brilliancy such as are met with in the seas of colder latitudes. These creatures may be cut perpendicularly or across, and each cutting will give origin to a new animal. In development the young may be seen issuing from the mouth; and

SEA-ANEMONIES. (ACTINIÆ)

sometimes the base of the old animal is dissevered, a portion remaining attached to the rock, where it continues to live, increasing in size, becoming more and more rounded, while, in a short time, a mouth, stomach, and tentacula are formed, presenting, to the surprise of an observer, a complete Actinia. At length, the side portions of this base give out globules which are detached, and fix themselves upon adjacent rocks, where they grow and produce a new colony like the parent animal. Among the best known species are the Beadlet Anemone (*A. mesembryanthemum*), of which there is also a strawberry-like form or variety, and the Plumose Anemone (*Actinoloba dianthus*). The deep crimson Mediterranean Anemone (*Actinia Jordiaca*) is esteemed a great delicacy by the Italians. All the British species are described and figured in Mr. Gosse's "Actinilogia Britannica."

ACTINOCAMAX. A name given by Miller to the fossil shells of an extinct genus of Cephalopodous Mollusca, apparently connecting the *Belemnites* with the existing *Sepiæ*. They are principally found in the chalk formations of England and Normandy.

ACTINOTROCHA. [See SUPPLEMENT.]

ACULEATA. The name for a section of Hymenopterous insects, whose antennæ are simple, and composed of a constant number of joints, namely, thirteen in the males, and twelve in the females. The larvæ have never any feet, and subsist on food which the females or neuters provide them with: one division of them, the *Prædones*, or predaceous tribes, which do not collect pollen, feeding upon other insects that have been stored up for them; and the other division, the *Melliferæ*, or honey-collectors, feeding upon honey or pollen paste, similarly reserved for their use.

ADAPIS. The name given to a genus of Pachydermatous quadrupeds, now existing only in a fossil state, some imperfect specimens of which were found in the plaster quarries of Montmartre, and described by Cuvier in his great work, *Sur les Ossemens Fossiles*.

ADDA. A small species of Lizard, celebrated throughout the East as being efficacious in the cure of various cutaneous diseases to which the inhabitants of Egypt and Arabia are peculiarly subject. It is about six inches long; the body and tail cylindrical, the latter ending in a very sharp point; the face is covered with black lines, which cross each other; the body is of a light straw colour with black bands; and the scales shine as if they were varnished. It burrows in the sand.

ADDAX. (*Antilope addax.*) A species of Antelope, more heavily formed than the generality of *Antilopidæ*, and having large spiral horns, annulated to within about six inches of the points. It lives solitarily or in pairs on the borders and oases of the Nubian deserts. It has remarkably broad hoofs, provided by nature to enable the animal to move the more easily over the fine loose sand. The general colour of the Addax is a greyish-white; but the head and neck are of a deep reddish-brown with a mark of pure white across the lower part of the forehead.

ADDER. (*Vipera berus.*) The Adder, or Viper, is a venomous reptile of the serpent kind, seldom more than two or three feet long, being considerably shorter than the common snake in proportion to its bulk. It is of a dull yellow colour with black spots, and the abdomen entirely black. [See VIPER.]

ADEPHAGA. A name given to a family of carnivorous and very voracious Coleopterous insects.

ADESMACEA. A family of Lamellibranchiate Mollusca, which either bore tubular dwellings in rocks, wood, &c., or live in testaceous tubes, their shells being consequently destitute of the hinge ligament. The genera *Pholas*, *Teredina*, *Teredo*, *Fistulana*, and *Septaria* belong to this family.

ADJUTANT, or GIGANTIC CRANE (*Leptoptilos Argala*), the *Argala* of India. This remarkable bird is a native of the warmer parts of India, and is of great use in removing noxious animals and carrion, which it devours with avidity. It stands five feet high, measuring from the tip of the bill to the claws seven and a half feet, and from the tip of each outstretched wing not less than fourteen feet. The head and neck are nearly bare; the beak is extremely large,

ADJUTANT (LEPTOPTILOS ARGALA.)

long, and strong; and under it hangs a downy pouch or bag, like a dewlap, which is capable of being inflated; the upper part of the bird is of an ash-grey colour, and the under part white. The voracity of the Adjutant is not more extraordinary than its capacity for swallowing: it makes but one mouthful of a rabbit, a fowl, or even a small leg of mutton; and when domesticated its habits of purloining render it necessary to keep all kinds of provision out of its reach. Dr. Latham observes that these birds in their wild state live in companies, and when seen at a distance, near the mouths of rivers, coming towards an observer, which they often do with their wings extended, "may well be taken for canoes upon the surface of a smooth sea: when on the sand-banks, for men and women picking up shell-fish or other things on the beach."

ADONIS BUTTERFLY. [See POLYOMMATUS.]

ÆGA. A genus of Isopodous Crustacea, parasitic on fish—hence frequently called *Fish-lice*. They are found in all parts of the world. In Newfoundland the fishermen call the species fish-doctors, and believe that the soft matter (mixed with eggs) found on the under side of the body of the females, is very useful in healing wounds. The adjoining cut will give some idea of the form of the species of this genus, which have the eyes generally large and approximating in front.

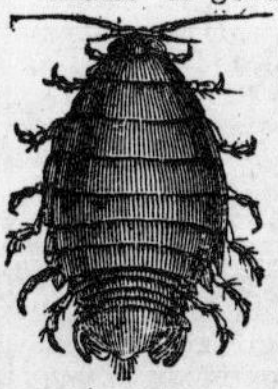

FISH-LOUSE.

ÆGERIIDÆ. A family of Heterocerous Lepidoptera, comprising a moderate number of interesting insects, whose resemblance to various Hymenoptera and Diptera (owing to the elongate form of the body, and the nakedness of the wings, which are more or less transparent in many of the species) is somewhat remarkable. The antennæ are simple, fusiform or thickened towards the tips, and generally terminated by a small pencil of hairs; the ocelli are distinct, and the labial palpi elevated; the abdomen is elongated; the wings have but comparatively few nervures; and the posterior legs are furnished with very long spurs.—The larvæ of these insects are of a cylindrical form, and with naked bodies destitute of a caudal horn: they have six pectoral, eight ventral, and two anal feet. They live in the interior of the branches or roots of trees, where they undergo their transformations to chrysalides, whose abdominal segments are armed with transverse rows of recurved points: these enable the chrysalis to push itself not only through the cocoon which the caterpillar had constructed from the decayed roots or branches, but half out of the hole in the stem previously made, it having had the instinct to turn round in its burrow, so that the head of the pupa should be towards the orifice.

The larvæ of some species, such as the *Ægeria culiciformis* and *Æ. formiciformis*, feed upon the apple, and that of *Ægeria tipuliformis* upon the pith of currant trees; in the neighbourhood of which the perfect insects may be seen flying, in the hottest sunshine, with great activity, or basking upon the leaves, alternately expanding and shutting the fan-like glossy appendages of hair which decorate the end of the body.

ÆGITHALUS. A name given to a genus of birds (the *Pendulous Titmice*) of the order *Passerinæ*. [See TITMOUSE.]

ÆGOTHELES. A genus of Passerine birds, distinguished by long tarsi, and toes apparently fitted for hopping from bough to bough; the wings comparatively short. The only known species inhabits Australia. In the day it resorts to the hollow branches of trees (or *spouts* as they are called), and holes of the gum trees. It feeds on the smaller Coleoptera; lays four or five white eggs, which are nearly round; and has at least two broods a year.

ÆPYORNIS. [See APPENDIX.]

ÆQUOREA. [See ACALEPHA.]

AGAMA: AGAMIDÆ. A genus and family of Saurian Reptiles. They have thick bodies covered with a loose skin, which is capable of being distended with air, at the will of the animal; the head is short, broad, and flat; the neck also is short, and the tail seldom longer than the body. Different species of them are to be met with in every climate; and as some are capable of changing the colours of their skin, they are in some parts of South America called chameleons. They generally lurk among rocks, heaps of stones, and mouldering ruins, their dull and sombre colours protecting them from observation: the more slender and active kinds, however, ascend trees with great facility, sport among the branches, and feed upon the insects which are generally to be found there in abundance. Some of the Agamidæ met with in India, South America, and Australia, are very curious; but so numerous are the species, that we cannot here give more than a description of their general characteristics. [See MOLOCH.]

AGAMI, or GOLD-BREASTED TRUMPETER. (*Psophia crepitans.*) This interesting bird is about the size of a pheasant or large fowl; has long legs, and a long neck, but a very short tail, consisting of twelve black feathers, over which the rump-plumes hang droopingly. It inhabits the forests of South America, where it is found in numerous flocks: it runs swiftly, and when pursued, trusts to its legs rather than its wings. When domesticated, this bird is a pattern of fondness and fidelity; and is so jealous of its master's caresses, that it attacks the dogs and other animals who venture near him. It is sometimes used to protect domestic poultry from the attacks of birds of prey.

AGATHISTEGUES. A family of Cephalapodous Mollusca, in which the cells are gathered together in small numbers, and heaped up in a globular shape.

AGENIOSUS. A genus of Malacopterygious Abdominal fishes, belonging to the *Siluridæ.*

AGGREGATA. The name given to certain shell-less Mollusca, which are collected together in a common enveloping organized substance containing numerous compartments, from each of which a distinct occupant sends forth a circle of organs to collect food, which, after assimilation, is carried by a common and continuous system of vessels for the support and enlargement of the common dwelling.

AGLAURA. A genus of Dorsibranchiate Annelides, distinguished by having numerous jaws, but no tentacles, or which are entirely hidden; and cirrhi, which perform the office of branchiæ.

AGOUTI. (*Dasyprocta.*) A genus of Mammalia belonging to the order *Rodentia*, and classed with the *Cavidæ*, or guinea-pig tribe. It is found in great abundance throughout South America; and as it bears some rude resemblance in its form and manner of living to the hare and rabbit, though it varies from both very essentially, it has sometimes been denominated the rabbit of that continent. It, however, varies still more from that animal in its habitude and disposition, than in its form. It has in a great measure the external covering of a hog; so also has it the hog's voracious appetite: it eats indiscriminately of every thing that comes in its way; and, when satiated, conceals the remainder, like the dog and fox, for a future occasion. The Agouti secretes itself in the holes of trees; its ordinary food consisting of potatoes, yams, and the fruits which fall in autumn. It uses its fore-paws, like the squirrel, to convey the food to its mouth; and as its hind legs are very long, it runs, or rather leaps, with considerable swiftness. The flesh is white and tender, and when fat

AGOUTI. (DASYPROCTA.)

and well dressed it is by no means unpalatable food. Agoutis are particularly destructive to the sugar-cane: the planters consequently use every means to catch them; and although they are still numerous in most places which are not settled and cultivated, their number is not now to be compared with what it was even long after the first colonists took possession of the West India islands. There is one kind of Agouti called the Mara, or Patagonian Cavy, considerably larger and more elegant than any of the others. Differently from most burrowing animals, it wanders, commonly two or three together, to miles or leagues from its home. It feeds and roams about by day; is shy and watchful; and generally produces two young ones at a birth. Naturalists give to this kind and species the name of *Dolichotis Patachonicus.*

AGRAIÆ. An order of quadrupeds, destitute of teeth, but furnished with very long cylindric tongues, which supply that defect. Of this order there are only two distinct genera, the *Myrmecophaga* and the *Manis* [which see].

AGRIOPUS. A genus of Acanthopterygious fishes, particularly distinguished from other genera by having only nine rays in the pectoral fins. The *Agriopus torvus*, or Seahorse, as it is sometimes called, is about two feet long, and is common on the shores of the Cape of Good Hope.

AI. [See SLOTH.]

AIAIA. (*Platalea.*) A bird of the Spoonbill genus, frequently seen in Brazil on the banks of rivers. It is of a pale but very bright and glossy flesh-colour on the back and wings, while the other parts are all beautifully white. Its flesh is considered wholesome and palatable.

AILURUS. A genus of carnivorous quadrupeds belonging to the family *Ursidæ.* The only known species, first found by Maj. Gen. Hardwicke, is the Wah or Panda (*Ailurus fulgens.*) It is about the size of a large cat; the fur soft and thickly set; above of the richest cinnamon-red; behind more fulvous, and deep black beneath.

PANDA (AILURUS FULGENS.)

The head is whitish; the tail annulated with brown; and the soles of the feet are hairy. This elegant animal frequents the vicinity of rivers and mountain torrents, passes much of its time upon trees, and feeds on birds and the smaller quadrupeds.

ALABES. A genus of Malacopterygious apodal fishes, distinguished by having one gill-opening; pectorals well marked, with a disc between them; gill-lids small, with three rays, and pointed teeth. The species inhabit the Indian Ocean.

ALASMADON. A name which has been given to some Bivalve Mollusca, of which the fresh-water Pearl Mussel (*Mya margaritifera*) is an example.

ALATÆ. A family of Mollusca, belonging to the second section of the order Trachelipoda, containing the genera *Rostellaria*, *Pterocera*, *Strombus*, &c. The shells of this family are distinguished by the spreading of the outer lip.

ALAUDA. A genus of granivorous singing-birds, of which there are many species, found in all parts of the globe. They are characterized by a long and straight hind claw, a strong straight bill, and by being able to raise the feathers on the back part of the head into the form of a crest. The greater part of them are migratory: they always build their nests on the ground, and may be considered as peculiarly birds of the fields and meadows. [See LARK.]

ALBATROSS. (*Diomedea.*) A genus of Palmipede birds: the species are the largest of all aquatic birds, the wings of some when extended measuring fifteen feet, and the weight often exceeding twenty pounds. Its plumage is white, with the exception of a few of the wing feathers and some transverse black bands on the back. It has a strong, hard, long beak, of a pale yellow colour: the feet, which are flesh-coloured, are short and webbed; and the wings are long, strong, and narrow. It preys on the wing, and is very voracious; but though formidable from its size and strength, it is not equally courageous, being frequently compelled to yield its prey to the sea-eagles, and sometimes even to the larger species of gulls. These birds are continually met with in the Southern Ocean, and are also seen in immense flocks about Behring's Straits and Kamtschatka in the early part of summer, attracted thither by the vast shoals of fish, whose migrations they follow. Besides the common Albatross, here described, there are two other species of less gigantic proportions, namely, the Albatross of China and the black-beaked Albatross.

When sailors accidentally fall overboard in latitudes where the Albatross abounds, they find it a most formidable enemy, even should only a few minutes elapse before they can be rescued by their comrades. Its powers of flight are prodigious.—Dr. Arnott mentions an instance of one of these birds following a rapidly sailing ship for two or three days. One species is called by sailors the "Cape

WANDERING ALBATROSS (DIOMEDEA EXULANS.)

Sheep," from flocks of them being seen off the Cape of Good Hope.—Captain Sir J. C. Ross, in his voyage to the Southern Seas, mentions that, in one of the islands frequented by seal hunters, the eggs of these birds, each of which averages about a pound in weight, are much esteemed—while the young birds, when first taken from their nest, are described by them as being quite delicious. It is possible, he adds, the sealers may have acquired the Esquimaux taste.

ALBIONES. A genus of Abranchious Annelides, distinguished by having the body bristled with tubercles.

ALBURNUS. A fresh-water fish, a species of the *Cyprinidæ*. [See BLEAK.]

ALCA. [See AUK.]

ALCADÆ. A family of oceanic birds, including the Auks, Puffins, and Guillemots. The power of their wings as organs of flight is generally very circumscribed; but their whole structure is admirably adapted for an aquatic life. The legs are extremely short, but powerful, and placed so far backwards that, in resting on the rocks, the birds appear to stand in an upright position. The toes are usually only three in number, and fully webbed. The bill is generally compressed, and often grooved at the sides, but it varies in form in the different genera. Their food consists of fishes, crustacea, and other marine productions; but they never resort to fresh water. [See AUK, &c.]

ALCEDO: ALCEDINIDÆ. A genus and family of birds, popularly known as *Kingfishers*, of which there are numerous exotic species (all distinguished by the splendid colours of their plumage), but only one kind indigenous to this country. Their principal characteristics are, a long, straight, quadrangular bill, thick and pointed; tongue, short, flat, and fleshy; the nostrils at the side of the base of the bill running obliquely; the tail and legs short. These birds for the most part live on fish, which they transfix with the bill as with a spear: they are solitary in their habits, and build their nests in holes on the banks of rivers. [See KINGFISHER.]

ALCIOPE. A genus of Dorsibranchiate Annelides, distinguished by having two foliated cirrhi, or gills, and a couple of branchial tubercles.

ALCYONEÆ, or ALCYONIAN POLYPES. Under the heads "CORALS," "POLYPES," and "SPONGES," will be found such particulars as are deemed necessary to describe those singular marine productions. It is, therefore, sufficient to observe in this place, that the *Alcyoneæ* are somewhat similar to the last mentioned. They vary much in form, being either lobed, branched, rounded, or existing in a shapeless mass or crust; while the interior substance is of a spongy or cork-like nature, surrounded by tubular rays inclosed in a sort of tough fleshy membrane. The animals are lodged in round cells, separated from each other by thin partitions. They are to be found in all seas, and at various depths, resorting, in general, to sheltered places, or where the water is deep and still.

ALEA. A genus of minute land shells, found in marshy ground, roots of trees, moss, &c.

ALECTOR. (*Crax.*) A large Gallinaceous bird of America, somewhat like a turkey. They have large rounded tails, composed of stiff quills; build their nests in trees; live on buds and fruits; and may be easily domesticated. [See CURASSOW.]

ALECTURA. [See TALEGALLA.]

ALEPOCEPHALUS. A fish belonging to the *Esocidæ*, or Pike family, found in the depths of the Mediterranean. Head naked, body with broad scales, mouth small, teeth minute and crowded, eyes very large, and eight gill-rays.

ALLIGATOR. (*Alligator lucius.*) This very formidable and ferocious Reptile is found in tropical climates, and agrees in every essential property with the Crocodile once so terrible along the banks of the Nile. There are apparently several species belonging to the order *Sauria*, family *Crocodilidæ*, their general plan of structure being the same as that of the lizards. They have a long flat head, thick neck and body, protected by regular transverse rows of square bony plates, raised in the centre into keel-shaped ridges. The mouth is extremely large, extending considerably behind the eyes, and furnished in each jaw with a single row of pointed teeth, all of different sizes, and standing apart from one another. The tongue is short and fleshy, and firmly attached to the under jaw throughout, so as to be incapable of protrusion; the eyes are placed in the upper part of the skull, and provided each with three distinct lids; and beneath the throat are two small glands

ALLIGATOR LUCIUS

which contain a musky substance. They have five toes on the fore-feet and four behind; but only the three inner toes on each foot are provided with claws. But the most remarkable, and, at the same time, most important organ they possess, is their long taper tail, which is strongly compressed on the sides, and surmounted with a double series of strong plates, which, converging towards the middle, there unite and form a single row to the extremity. Their feet are webbed; but it is to the tail they owe most of their progressive power in the water; and although it impedes their motions on dry land, even there it often becomes a powerful weapon of defence.

The Alligator is prodigiously strong; and its arms, both offensive and defensive, are irresistible. Its ordinary length is from fifteen to eighteen feet, though sometimes considerably more. The shortness of its legs, the vertebral conformation of the backbone, the muscles of the legs, and, in short, its whole frame, are calculated for amazing force. Its teeth are sharp, numerous, and formidable; its claws long and tenacious; but its principal instrument of destruction is its tail, with a single blow of which it is capable of overturning a canoe. Its proper element is the water; but it is also very terrible by land: it seldom, however, unless when pressed by hunger, or with a view of depositing its eggs, quits the water: it usually lays between fifty and sixty of these (which are about the same size as those of a goose, but oblong rather than oval), in one place, and covers them up with sand, leaving them to be hatched by the heat of the sun: it generally happens, however, that half of them are devoured by vultures, or fall a prey to various descriptions of ravenous fishes. Both the Alligator and the Crocodile are supposed to be very long-lived, and their growth is extremely slow.

The most extraordinary accounts are related of the ferocity and strength of this terrible destroyer. It usually floats along the surface, and seizes fish, fowl, turtle, or whatever other prey may fall within its reach; but, this method failing, it is then compelled to venture near the shore, where it conceals itself among the sedges in expectation of some land animal coming to drink. As the devoted victim approaches, nothing of its insidious enemy is to be seen; nor is the retreat of the former meditated till it is too late. The voracious reptile instantly springs on its prey with much more agility than might reasonably be expected from such an unwieldy creature; and, having secured it with its teeth and claws, instantly plunges into the water and drags it to the bottom, where it is devoured at its leisure. In its depredations along the banks, however, it sometimes happens that the Alligator seizes on an animal as formidable as itself, and meets with a desperate resistance. With the tiger, in particular, which is in the habit of lurking in the vicinity of great rivers, it has frequent contests; and the instant this animal finds itself assaulted, he turns about with prodigious agility, and forces his claws into the eyes of the assailant, who immediately plunges with its fierce antagonist into the river, where the struggle continues till the tiger submits to a watery death.

As we have spoken at some length of the CROCODILE, and described the different species, it would be inconsistent with our general plan to extend this article much further. We therefore conclude with an anecdote from Waterton's "Wanderings in South America," clearly showing that man is not exempt from the attacks of this ferocious destroyer:—"One Sunday evening, some years ago, as I was walking with Don Felipe de Ynciarte, governor of Angustura on the bank of the Oroonoque, 'Stop here a minute or two, Don Carlos,' said he to me, 'while I recount a sad accident. One fine evening last year, as the people of Angustura, were sauntering up and down here, in the Alameda, I was within twenty yards of this place, when I saw a large Cayman [the common species of Surinam and Guiana] rush out of the river, seize a man, and carry him down, before any body had it in his power to assist him. The screams of the poor fellow were terrible as the Cayman was running off with him. He plunged into the river with his prey: we instantly lost sight of him, and never saw or heard him more."

ALLIGATOR TORTOISE. A genus of the EMYDÆ, or Marsh Tortoises, which are carnivorous in their habits; and some of the species, of which this is one, are formidable from their size and ferocity. It is a native of the lakes, rivers, and morasses of Carolina; and it is remarkable for its activity, darting suddenly upon aquatic birds, fishes, or other animals that come within its reach, and snapping them up: from which habit it is sometimes designated as the "Snapping Turtle." The species is the *Chelydra Serpentina.* [See TORTOISE.]

ALOSA. A genus of Malacopterygious fishes of the *Clupeidæ* or Herring family, greatly resembling the Pilchard and Sardine, *Alosa vulgaris* is the common SHAD (which see].

ALPACA, or PERUVIAN SHEEP. (*Auchenia.*) In form and structure, this animal bears a strong resemblance to the camel; but is greatly inferior in size, and differs from it in the absence of the hump, the want of water-cells in the stomach, and in the conformation of the foot, which con-

PERUVIAN SHEEP (AUCHENIA.)

sists of two toes completely divided, each with a rough cushion beneath, and provided at the end with a strong short hoof. There appear to be three closely allied species of these animals. That which we are now describing is said to be entirely confined to Peru, where the natives keep vast flocks of them for the sake of the silky lustre and fineness of their wool. It inhabits the more elevated parts of the mountain ranges, living almost on the borders of perpetual snow. [See LLAMA, GUANACO, and VICUNA.]

ALUCITIDÆ. A family of small Lepidopterous insects, nearly allied to the TINEIDÆ, but distinguished from that and all others by the wings being singularly divided into narrow feathered rays; the fore wings having two, three, four, or six, and the posterior wings three or six of such rays, which are beautifully feathered on each edge: they are carried horizontally in repose; the antennæ are long, slender, and setaceous; the spiral maxillæ are long; and the legs are long and slender. The larvæ are clothed with very long hairs; they have sixteen feet, and are very inactive; the pupæ are either naked, and enclosed in a transparent silken cocoon; or conical, hairy, and either suspended perpendicularly by a thread, or affixed at the posterior extremity of the body to a layer of silk or leaves. These insects vary in the time of their flight; the *Alucita* frequenting our gardens, and sitting with its beautiful fan-like wings on our hothouses, whilst the *Pterophori*, being crepuscular, fly over low plants. The rays of the wings are composed of the nerves, without any of the intervening membrane, which seems to be transformed into the fringe. In repose the *Pterophori* fold their wings so as to appear to consist of only one broad ray.

ALUTERES. A genus of fishes, belonging to the order *Plectognathi*: they are characterized by a long body, the granulations scarcely visible, and a single spine in the first dorsal; but the pelvis is completely hidden in the skin. For an example, see OSTRACION.

AMADAVADE. A small bird of the Finch tribe (*Fringillidæ*), having a beautiful red bill. The upper part of the body is brown, the rump dark red, and the prime feathers of the wings are black; as are also those of the tail, which are longest in the middle, and gradually slope to the sides: it is frequently kept as a pet in cages, and lives on seeds.

AMBLYRHYNCHUS. The name given to a genus of Lizards, very much resembling the Iguanas, common on all the islands throughout the Galapagos Archipelago. They differ, however, from the Iguana, in having, — instead of the long, pointed, nar-

AMBLYRHYNCHUS CRISTATUS.

row muzzle of that species, — a short, obtusely truncated head, and also in the strength and curvature of the claws. Mr. Darwin (in his "Journal of Researches," &c.) thus speaks of the one which is termed *Amblyrhynchus cristatus*:—"It lives exclusively on the rocky sea-beaches, and is never found, at least I never saw one, even ten yards in-shore. It is a hideous looking creature, of a dirty black colour, stupid and sluggish in its movements. The usual length of a full-grown one is about a yard; but there are some even four feet long. I have seen a large one which weighed twenty pounds. On the island of Albemarle they seem to grow to a greater size than on any other. These Lizards were occasionally seen some hundred yards from the shore, swimming about; and Captain Collnett, in his voyage, says, 'they go out to sea in shoals to fish.' With respect to the object, I believe he is mistaken; but the fact stated on such good authority cannot be doubted. When in the water the animal swims with perfect ease and quickness, by a serpentine movement of its body and flattened tail, — the legs during this time being perfectly motionless and closely collapsed on its sides. A seaman on board sank one, with a heavy weight attached to it, thinking thus to kill it directly; but when, an hour afterwards, he drew up the line, the Lizard was quite active. Their limbs and strong claws are admirably adapted for crawling over the rugged and fissured masses of lava, which every where form the coast. In such situations, a group of six or seven of these hideous reptiles may oftentimes be seen on the black rocks, a few feet above the surf, basking in the sun with outstretched legs." Its habit are entirelyaquatic, as well as its food, which consists of sea-weed.

The species termed *Amblyrhynchus subcristatus* is terrestrial, and is confined to the central islands of the Archipelago. These "inhabit both the higher and damp, as well as the lower and sterile parts; but in the latter they are much the most numerous. Like their aquatic brethren, they are ugly animals; and from their low facial angle have a singularly stupid appearance. In size, perhaps, they are a little inferior to the latter, but several of them weighed between ten and fifteen pounds each. The colour of their belly, front legs, and head (excepting the crown, which is nearly white), is a dirty yellowish-orange: the back is a brownish-red, which in the younger specimens is darker. In their movements they are lazy and half torpid. When not frightened, they slowly crawl along with their tails and bellies dragging on the ground. They often stop and doze for a minute with closed eyes and hind legs spread out on the parched soil. They inhabit burrows, which they sometimes excavate between fragments of lava, but more generally on level patches of the soft volcanic substance. They feed by day, and do not wander far from their burrows. When attentively watching any one they curl their tails, and, raising themselves on their front legs, nod their heads vertically with a quick movement, and try to look very fierce; but in reality they are not at all so: if one just stamps the ground, down go their tails, and off they shuffle as quickly as they can." They live on the leaves of trees and other vegetable productions; and their flesh is considered a delicate kind of food.

AMBLOTIS. [See WOMBAT.]

AMIA. A small Malacopterygious freshwater fish, found in the rivers of South America. It belongs to the *Clupeidæ* family; feeds on Crustacea, and is rarely eaten.

AMMOCETES. A genus of Chondropterygious fishes, allied to the Lampreys, the maxillary ring being without teeth, the fleshy lips semicircular. The common species, *Ammocetes branchialis*, is about the thickness of a goose quill, and is very common in some of the English rivers, where it is known as the Stone Grig. It lodges in the mud, where it preys on worms, insects, &c.; and is of no use but as bait for other fish. It has been accused of sucking the gills of fishes.

AMMODYTE, or LAUNCE. This fish, which is of the Malacopterygious or soft-finned kind, is named Launce from its lance-like shape, and is from eight to ten inches long; its form is slightly square, being rather rounded on the sides, and somewhat flattened above and beneath; the head is small and taper, and the under jaw much longer than the upper; the mouth is destitute of teeth, but at the entrance of the throat are two oblong bones for retaining the prey. The pectoral fins are small, and the tail is slightly forked; the general colour of the body is a greenish-blue on the back, and the belly is either of a silvery white, or of a yellowish hue. These fishes are in England called Sand-eels, being remarkable for their habit of burrowing in the sand, in which they find the worms and insects that constitute their chief food. They are in their turn preyed upon by the larger fishes, par-

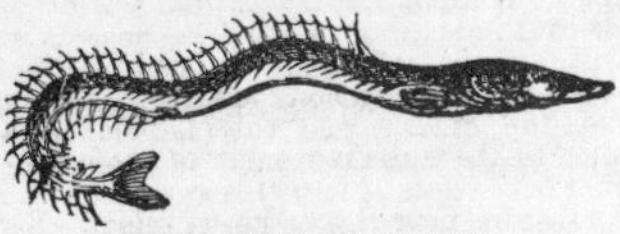

AMMODYTE OR LAUNCE.

ticularly by the mackarel and salmon; to the support of the latter, whilst they are in the estuaries of rivers, the Launces are believed to contribute largely. The Launce spawns in the month of May, depositing its eggs in the mud, near the edges of the coast.

AMMODYTES. A genus of Serpents, nearly the size of the Viper, and allied to it in general appearance, though distinguished from it by an erect pointed process on the tip of the snout: its usual colour is either bluish-grey or brown, with a continued black dorsal band resembling that of the viper. A species of this genus is found in many parts of the East, and is so extremely poisonous as to prove fatal in three or four hours.

AMMONITES, or SNAKE-STONES. Spiral fossil shells, of which there are a great abundance in Europe, Asia, and America, especially in the lias, chalk, and oolite formations. They appear like a snake rolled up: some are very small, but occasionally they are met with upwards of three feet in diameter. In some places they are so numerous, that the rocks seem, as it were, composed of them alone. Upwards of 500

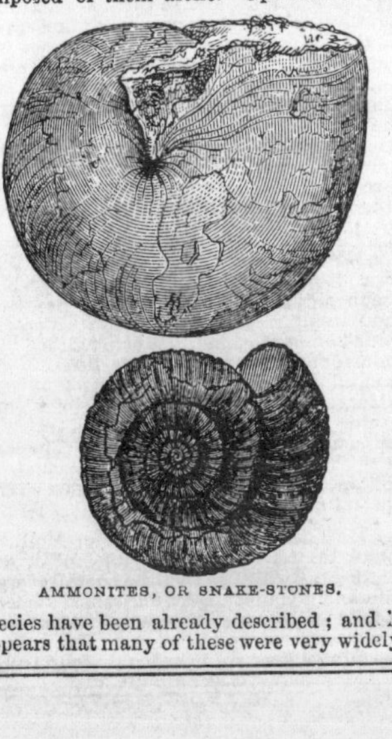

AMMONITES, OR SNAKE-STONES.

species have been already described; and it appears that many of these were very widely distributed; some being found in the Himalaya mountains, at an elevation of 16,000 feet, and others in various parts of Europe. Their numbers must have been very great, as M. Dufresne informed Lamarck that the road from Auxerre to Avalon, in Burgundy, was absolutely paved with them; and we know that it is no uncommon occurrence to find them used in parts of the west of England to pave the roads. It has been suggested that "these animals must have been very important agents, their carnivorous habits duly considered, in keeping the balance among the other tenants of the seas, by preventing the excessive multiplication of crustacea, as well as of other molluscs, and of fishes. The nearest ***recent*** ally is SPIRULA [which see].

AMŒBA. [See PROTEUS, p. 548.]

AMPELIDÆ. [See CHATTERERS.]

AMPHIBIA. This term applies only to animals having the power of living, indifferently, at the same time, either upon land or in water, yet we commonly call Seals, Otters, Beavers, &c., besides many Reptiles, amphibious, because their organization disposes them to resort either to the land or water for procuring food, or whose habits are at once terrestrial and aquatic. But this is by far too comprehensive a sense. Linnæus applied the term generally to the third class of his system of zoology, which comprised not only all the animals since more properly denominated Reptiles, such as the Tortoises, Lizards, Serpents, and Frogs, but likewise the Cartilaginous Fishes. It is now admitted, however, that Linnæus was not correct in this classification, and that a truly amphibious animal should possess lungs and gills at one and the same time, for extracting the principle which supports animal life indifferently from either element.

AMPHICŒLIA. [See SUPPLEMENT.]

AMPHICYON. [See SUPPLEMENT.]

AMPHIDESMA. A genus of small round or rather oblong Shells, slightly gaping and inequilateral, found in the sand on the seacoasts of tropical countries, and also those of England, France, &c. The *Amphidesma variegatum*, described by Lamarck, is a native of the coast of Brazil. "In most bivalve shells," Sowerby observes, "the cartilage and ligament are united in one mass, or placed close to each other; the contrary in this case gives rise to the name, which signifies *double ligament*."

AMPHIOXUS, or LANCELET. A small fish of the Lamprey family. Its form is compressed; the head pointed, without any trace of eyes; a delicate membranous dorsal fin extends the whole length of the back; and the tail is pointed. At one time this was regarded as a mollusc, the best known species being the *Limax lanceolatus* of Pallas. It is found on the coasts of England and Ireland, in the Forth of Clyde, and in the Mediterranean. Mr. Gray has described a second species from the Eastern seas (*A. Belcheri*).

AMPHIPODA. An order of minute Crustaceans, which have the power of swimming and leaping with great facility, but always on one side. Some are found in streams and rivulets, but most in salt water; and their colour is of a uniform pale red or greenish. In this order the eyes are sessile and immoveable; the mandibles are furnished with a palp; the abdominal appendages are always apparent and elongated; and they have ciliæ, which appear to fulfil the office of branchiæ. The antennæ, ordinarily four in number, are composed of peduncle and slender filament; and the body is mostly compressed and bent. The appendages of the tail generally resemble little pointed stylets. Among the Amphipods most common on our shores are the Sandhoppers, (*Talitrus locusta* and *Orchestia littorea*,) found under stones, or under the mass of exuviæ thrown up by the tide on sandy shores, in troops of thousands, all active and leaping when disturbed in their retreats. The following passage from Paley's Natural Theology alludes to these minute crustacea: "Walking by the sea-side, in a calm evening, upon a sandy shore, and with an ebbing tide, I have frequently remarked the appearance of a dark cloud, or rather very thick mist, hanging over the edge of the water, to the height perhaps of half a yard, stretching along the coast as far as the eye could reach, and always retiring with the water. When this cloud came to be examined, it proved to be nothing else than so much space filled with young shrimps in the act of bounding into the air from the shallow margin of the water or the wet sand."

SANDHOPPER.
(TALITRUS LOCUSTA.)

AMPHISBÆNA. A genus of Serpents, natives of South America, distinguished by their bodies having nearly the same uniform thickness throughout, by their small mouths and eyes, short tails, and their numerous rings of small square scales. The two best known species are *Amphisbœna alba* and *Amphisbœna fuliginosa*. They are destitute of fangs, and are consequently harmless and inoffensive; living, for the most part, upon ants and other small insects. The colour of the first mentioned species is white, as its name imports; but in some specimens it is tinged with pale rose colour, while in others the head and back incline to a pale yellow or brownish cast. The *A. fuliginosa* is either black with white variegations, or purple with yellow. The eyes of the Amphisbæna are covered, and almost concealed, by a membrane; which, added to their naturally diminutive size, has given rise to the popular opinion that the animal was destitute of the organs of sight. The head is so small, and the tail so thick and short, that at first sight it is difficult to distinguish one from the other; and this circumstance, united to the animal's habit of proceeding either backwards or forwards as the occasion may require, gave rise to the credulous belief throughout the native regions of the Amphisbæna, that it has two heads, one at each extremity, and that it is impossible to destroy one by simple cutting, as the two heads mutually seek one another, and soon reunite!

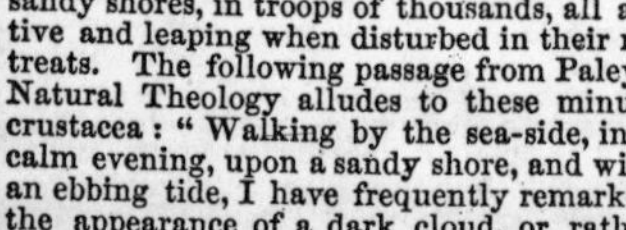

AMPHISBÆNA FULIGINOSA.

AMPHITHERIUM. [See SUPPL.]

AMPHITRITE. A genus of *Annelides* belonging to the division *Tubicolæ*, and easily distinguished by their large golden-coloured setæ, disposed in a comb-like series or in a crown, or in one or several ranges on the front of the head; which may assist them in locomotion, and probably serve them for defence. Around the mouth are very numerous tentacles, and on either side of the commencement of the back are pectinated gills. Some of them compose slight tubes, of a regular conical form, which they carry about with them, when running in search of food; these tubes, which consist of fine grains of sand cemented together with great regularity, are simple cones open at both ends, and not attached; they are usually about two inches long, and may be frequently picked up on our shores. [See TUBICOLÆ.]

AMPHIUMA. A genus of Batrachian reptiles which abound in the lakes and stagnant waters of N. America. They first appear in the tadpole form, respiring by means of gills, and inhabiting the water; they afterwards gradually acquire small legs and feet, and would have an appearance similar to the water-newt, were it not for the extreme length of their body. There are only two known species, one characterised by being three-toed, which is three feet long; the other, a much smaller species, having only two toes.

AMPULLACERA. A genus of Mollusca, allied to the Ampullaria, two species of which are found in great abundance in New Zealand, living in pools of brackish water, and buried in sandy mud. When touched, the animal enters very deeply into its shell, and is at all times much hidden by it. The head is large, flat, divided into two lobes, and having two sessile eyes; no appearance of tentacula; foot short and square. The shell is thick and globular; mouth round, or oblique, having the lips united; spire short; operculum thin and horny.

AMPULLARIA. A genus of Mollusca found in the rivers of Africa, India, and South America. The shell is generally large, thin and globular; spire very short; whorls rapidly enlarging; the operculum thin and horny, and rarely calcareous. The animal has a large head furnished with four tentacula, with eyes at the base of the two longest, and the foot oval and large. Some of the African species have reversed shells, and all that are natives of Africa and America have the operculum horny; while those which come from India usually have it shelly, and are furnished with an internal groove for its reception parallel to the mouth. The animal has a large bag opening beneath, placed on the side of the respiratory organs; this they fill with water, by which means they can exist for a considerable period out of their natural element; and specimens have been brought from Egypt to Paris alive (before steam navigation was common), although packed up in sawdust. The Indian species lay globular eggs, of a pale green colour, about the size of small peas, which are found in clusters attached to sticks or other things in the water; when dry, they have a beautiful appearance.

ANABAS. A genus of Acanthopterygious fishes, whose respiratory organs are so constructed as to enable them to sustain life for a space of time out of water, by having small apertures or some receptacle, where they can preserve sufficient water to moisten their gills. In cold or temperate regions this is not required, but in tropical countries it often happens that many of the rivers and ponds are dried up. At such times no fish but such as, like the Anabas, are furnished with the necessary pharyngeal apparatus for keeping the gills moist, could exist; many of these, however, are able to migrate in search of their natural element, and, it is

CLIMBING PERCH. — (ANABAS SCANDENS)

said, they are guided by a remarkable instinct to travel towards the nearest water. One species is called the Climbing Perch. (*Anabas scandens.*) This species, Mr. Daldorf, a distinguished Danish naturalist, says he observed in the act of ascending palm trees, which it did by means of its fins and tail and the spines of its gill-covers; but other naturalists, who have mentioned its habit of creeping on the ground and living out of water, have not confirmed this Dane's account of its climbing propensities. It is a native of India.

ANABATES. A genus of Passerine birds, distinguished by having the superior ridge of the beak rather convex, like that of a Thrush, without emargination. The tail is long and wedge-shaped, which indicates that it is employed for supporting the bird when in a perpendicular position against the trunks of trees.

ANABLEPS. A viviparous fish belonging to the *Malacopterygii*, remarkable for being apparently possessed of four eyes: this is not, however, really the case; for although the cornea and iris are divided by transverse bands, so that two pupils are observed on each side, yet the other parts of the eye are single. The body is cylindrical, with strong

ANABLEPS TETRAOPHTHALMUS.

scales; the head is flat; the snout blunt, and the mouth across its extremity, with small crowded teeth in both jaws; the intermaxillaries have no peduncle, but are suspended to the nasal bones; the pectorals are in part scaly; the dorsal is small, and nearer the tail than the anal; the pharyngals are large, and covered with small globular teeth. The species here delineated, *Anableps tetraophthalmus*, inhabits the rivers of Guiana.

ANACONDA. (*Boa.*) A Ceylonese serpent, belonging to the Boa family, of enormous magnitude and strength; said to be capable of conquering the largest and fiercest quadruped, and concerning whose actions the most wonderful stories are related. An encounter between one of these serpents and a most powerful tiger is described by an eyewitness in language of fearful interest: "Though unable to get rid of its cruel enemy, the tiger gave it prodigious trouble. A hundred times would it rear up, and run a little way; but soon fell down again, partly oppressed by the weight, and partly by the folds and wreathed twists of the serpent round its body. But though the tiger fell, it was far from being entirely conquered. After some hours it seemed much spent, and lay as if dead; when the serpent, which had many times violently girded itself round the tiger, vainly attempting to break its bones, now quitted its hold, twisting its tail only round the neck of its prey, which was in no condition either to resist or escape. Having by degrees dragged the tiger to a tree, the monster wound its body round the animal and the tree together several times, girding both with such violence that the ribs and other bones began to give way: and, by repeated efforts of this kind, it broke all the ribs, one by one, each of which gave a loud crack in breaking. It next attempted the legs, and broke them severally in the same manner, each in four or five different places. This employed many hours, during all which time the poor tiger remained alive; and at every crack of the bones gave a faint but most piteous howl." A loathsome description of the serpent's "licking the body and covering it with its slaver," preparatory to the act of swallowing, is then given; and the account thus concludes: "Much time was employed in this business; but at length the serpent having prepared the whole to its mind, drew itself up before its prey; and, seizing the head, began to suck that, and afterwards the body, down into its throat." But this, it appears, was the work of some hours; and it had so gorged, that, the next morning, on being attacked by the party who were witnesses to his monster meal, the serpent could neither defend itself nor retreat; and it was dispatched, by repeated heavy blows on the head with large clubs. It was thirty-three feet in length.

ANAMPSES. A genus of Acanthopterygious fishes found in the Indian seas. They are small and beautifully coloured. The head is without scales; and they are distinguished by having two flat teeth, which project from the mouth, and curve upwards.

ANARRICHAS. A genus of Acanthopterygious fishes, bearing great resemblance to the Blennies, except in their being destitute of ventral fins. Their dorsal fin is composed entirely of simple but not stiff rays, and extends, as does also the anal, very close to the base of the caudal, which last, as well as the pectorals, is rounded. The whole body is soft and slimy. Their front teeth are large and conical, and they may be regarded as fierce and dangerous fishes. [See WOLF-FISH.]

ANAS. The name of a large Linnean genus of birds, of the order *Anseres;* whose distinguishing character is, that the beak is convex, terminating in an obtuse point; as the Swan, the Goose, the Duck, Widgeon, &c.

ANASTOMA. A genus of land shells, resembling the other *Helices* in every respect, except in the peculiarity of the last whorl taking a sudden turn and reflecting the aperture upwards, so as to present it on the same plane with the spire; so that the animal must walk with the spire of its shell downwards, resting on its foot.

ANATIFERÆ. A name given to a genus of multivalve Cirrhipeds. [See ACORN-SHELL.]

ANATIDÆ. The *Duck* tribe; a family of web-footed birds; order *Natatores.* They are distinguished by a broad depressed bill, which is covered with a soft skin; and by the hind toe not being included in the web. The bill is furnished with a set of horny laminæ at the edge of each mandible, which serve to filter the fluid taken up by the bill, and retain the solid substances taken up with it: the tongue is large and fleshy, the gizzard strong and muscular, and lined with a tough coat, so as to be capable of grinding down the shells of the mollusca on which they feed. Many are migratory, and fly with great strength at a considerable elevation.

ANCHOVY. (*Engraulis encrasicolus.*) A well known small fish, abounding in many parts of the Mediterranean, particularly on the coasts of Italy, Greece, and Spain: it occurs also, though not in such considerable numbers, on some of our western coasts, as well as on those of France and Holland. It is about four inches long, of a bluish-brown colour on the back, and silvery white on the belly. It is covered with large, thin, and easily deciduous scales, and may be readily distinguished from the Sprat and other kindred species by the anal fin being remarkably short. Mr. Couch, in his Cornish Fauna, says, "this fish abounds towards the end of summer, and if attention were paid to the fishery, enough might be caught to supply the consumption of the British islands;" and he adds, that he has seen it in the Cornish seas of the length of seven inches and a half!

ANCILLA, or ANCILLARIA. A genus of Mollusca, inhabiting a spiral, univalve marine shell, which, when the animal is alive, is so much covered by the foot, that only the middle of the back can be seen. The species are numerous, and they are chiefly confined to tropical climates. The shells are smooth, and appear as if highly polished.

ANCYLUS. A fresh-water Gasteropodous Mollusc, with a shell similar to that of a Patella. They live in stagnant waters and in rivulets, adhering to stones and aquatic plants.

ANDRENIDÆ. A family of solitary Bees, each species consisting only of males and females. The mandibles are simple, or terminated by one or two notches; in which the labium and terminal maxillary lobes do not form an elongated proboscis,—a character which distinguishes them from the APIDÆ [which see]. The antennæ are elbowed; and the hind legs are generally completely clothed with hairs, the trochanters and femora in the females being pollinigerous. The species of the genus Andrena are very numerous; they make their appearance in the early spring and summer months, and have very much the appearance of hive-bees. The females collect pollen from the stamens of flowers, rather by means of the general hairiness of the body than with the posterior tarsi: this they form, by the addition of a little honey, into a paste for the food of their progeny. They burrow in the ground, in sandy situations, especially if exposed to the sun, often to a considerable depth. At the foot of these burrows they deposit an egg, with a sufficient quantity of this prepared food for the supply of the future grub; which they then cover up, and proceed, cell after cell, closing up the hole at the top with earth, to prevent the attacks of parasites, which, notwithstanding, often succeed in entering the hole and depositing their eggs in the cells.

ANDRIAS. [See SUPPLEMENT.]

ANEMONIES, SEA. [See ACTINIA.]

ANGEL-FISH, or MONK-FISH. (*Squatina Angelus.*) This fish, which is more remarkable for its singularity of form than for its beauty, would seem to connect the genus of Rays and Sharks, were it not for the situation of its mouth, which is an exception from each. It is said to have acquired the name of Angel-fish from its extended pectoral fins having the appearance of wings; and it is called Monk-fish, because its rounded head appears as if enveloped in a monk's hood. The head is large, and the mouth very wide; the teeth are broad at the base, but slender and very sharp above, and disposed in five rows round the jaws. By means of muscles uniting them to the jaws, the teeth are capable of being raised and depressed like those of the other shark tribe. The eyes are small, and behind each is an orifice in the shape of a crescent. The back is of a pale ash-colour, and extremely rough, having a prickly tuberculated line

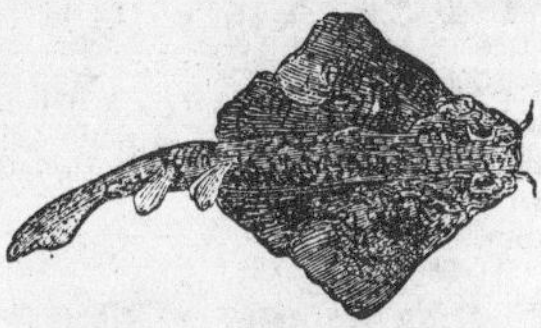

ANGEL-FISH. —(SQUATINA-ANGELUS.)

down the middle; the belly is white and smooth; the pectoral fins are large, and extend horizontally to a considerable distance; the ventral fins are also placed in the same manner, and the tail is bifurcated.

The Angel-fish is met with on many parts of the British coasts, but is most numerous on the Southern. It is very voracious, and feeds on the smaller flat-fishes, which swim close to the bottom; and, like them, it occasionally hides itself in the loose soft soil. It is exceedingly fierce, and dangerous to approach; nor does it look less fierce or malignant than it really is. It sometimes attains the length of seven or eight feet, and weighs nearly a hundred pounds; but instances of this are comparatively rare. Formerly the flesh was held in high estimation, but it is now disregarded as rank and coarse. The skin, being rough, is used to polish wood and ivory, as well as for other uses in the arts.

ANGLER. (*Lophius piscatorius.*) This extraordinary fish is not unfrequently met with on our coasts, and is known also by the names of the Fishing-frog, Toad-fish, and Sea Devil. It is the most uncouth, ill-shapen of the piscatory tribe, resembling the frog in its tadpole state, from which it derives one of its common appellations. The head, which is circumferentially larger than the whole body, is flat on the top; the mouth nearly as wide as the head; the lower jaw is considerably longer than the upper, and bearded all round the edge; both jaws are armed with numerous sharp conical teeth, curving inwards. The nostrils have no external orifice, but there are two internal ones which supply their place; the eyes are large,

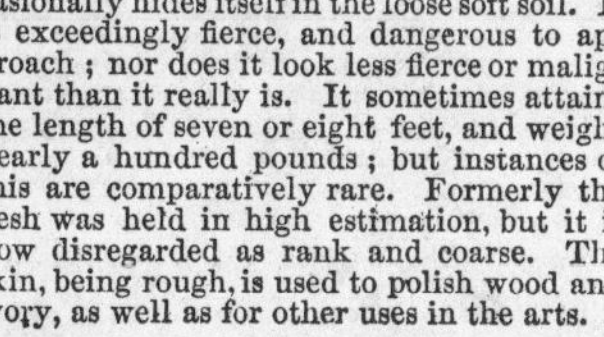

ANGLER. —(LOPHIUS PISCATORIUS.)

the irides brown, and the pupils black; pectoral fins broad, rounded at the edge, and wide at the base; ventral fins broad, thick, and fleshy, jointed like arms, and divided in the insides. The colour of the upper surface of the body is brown, the lower part white, and the skin smooth throughout: ventral and pectoral fins white; tail nearly approaching to black. The Common Angler is usually about three, but sometimes it is six feet in length; lives, as it were, in ambush, at the bottom of the sea; and by means of its fins it stirs up the mud and sand so as to conceal itself from other fishes on whom it preys.

The most interesting circumstances connected with the habits of this animal refer to the mode in which it secures its living dainties, and the singular apparatus specially provided for this purpose. In the annexed cuts it will be noticed that there are two rod-like filamentary appendages affixed to the snout immediately in front of the closely approximated eyes. These filaments are bony, elastic, and comparatively stout below where they are attached; both are moveable, the anterior one being articulated to the snout in a very remarkable manner. This joint, in fact, precisely resembles two links of an ordinary chain, so that the upper link, answering to the hollowed out lower end of the filament, plays freely within the circle of the

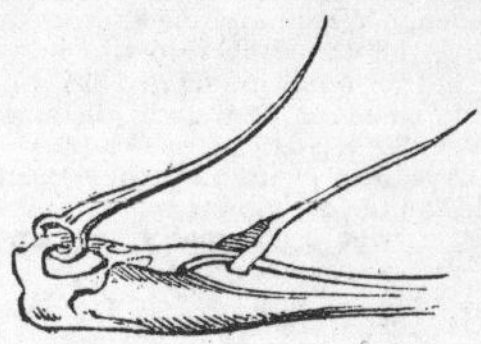

lower link. It will be further observed that the tip of this anterior filament—a veritable Liliputian fishing-rod—is provided with a pendulous flattened membrane, resembling a small flag; this can be hoisted at the animal's will, and, whilst it calls the attention of little fishes in the neighbourhood, at the same time serves as a bait, or, more strictly, a decoy. When about to take a meal, the body of the Fishing-frog is carefully concealed by mud stirred up in the way already mentioned; and the bait being now elevated above the muddy zone, and shaken to and fro with a duly winning or "killing" effect, the prey immediately gather round; whilst, no sooner have a sufficiently numerous group assembled—all being merrily engaged in tugging away at the decoy—than they are remorselessly and by one fell swoop consigned by the Sea-Devil into its capacious stomach. Surely this is a cogent illustration of "final intention."

Allied to the Common Angler are six others:—1. The CORNISH ANGLER (*Lophius Cornubicus*). 2. The MURICATED ANGLER (*Lophius muricatus*). 3. The BEAKED ANGLER (*Lophius rostratus*). 4. The HARLEQUIN ANGLER (*Lophius histrio*). 5. The STRIPED ANGLER (*Lophius striatus*). 6. The MARBLED ANGLER (*Lophius marmoratus*).

ANGUILLA. [See Eel.]

ANHINGA, or White-bellied Darter. (*Plotus anhinga.*) A very elegant species of the *Colymbidæ*, or Diver family, common in some parts of Brazil. Its body is about the size of a tame duck's, but its length, from the tip of the beak to that of the tail, is nearly three feet. It has a long, slender bill, yellowish at the base; a small head; the neck long, round, and slender, and covered with soft downy feathers of a rufous grey colour; while those on the breast, belly, and thighs are of a silvery whiteness. The plumage at the beginning of the back is brown, each feather having an oblong spot of whitish yellow in the centre, so that it appears speckled: the rest of the back is black; and the tail consists of shining black feathers tipped with grey. The legs are remarkably short, the thighs feathered, and the claws very sharp and crooked. The Anhinga builds its nest on trees, on which it roosts at night and when not on the water, being very rarely seen on the ground. It feeds upon fish, which it catches most dexterously, darting upon them with great rapidity.

These birds delight to sit in little communities, on the dry limbs of trees, hanging over still waters, with their wings and tails expanded. When any one approaches, they will drop off the limb into the water as if dead, and for a minute or two are not seen; when on a sudden, at a great distance, their long slender heads and necks only appear, so that whilst swimming they greatly resemble snakes, no other part of them being visible, except occasionally the tip of the tail.

ANI. [See Crotophaga.]

ANIMAL FLOWER. A name given to one species of *Actinia*, the animal bearing some resemblance to a flower with a radiated disc; its tentacula being disposed in regular circles, and tinged with a beautiful variety of bright lively colours, as the marigold, anemone, &c. [See Actinia.]

ANIMALCULÆ, or ANIMALCULES. A term applied to minute animals of various orders, many of which can only be seen by the aid of a microscope. [See Infusoria.]

ANNELIDA. A class of articulate animals, comprising species which may be characterized as possessing an elongated body, divided into numerous segments, marked by transverse lines, and generally furnished with a series of bristly appendages which serve as legs. Many of the Annelida are red blooded, and have a complete apparatus for circulation and respiration. Some live in fresh, others in salt water; and others, like the Hair-worm, are amphibious. In some the bristly appendages are implanted on fleshy tubercles; in others they are only represented by a few short stiff hairs; while in other instances, as in the Leech, there is no trace of any members or appendages to the body. The bristles are usually sharp, and sometimes barbed, serving not only to attach the animals to soft substances, and to hold firmly on to rocks and other solid surfaces, but to aid their movements through the water. When there are no locomotive appendages, the extremities of the body are usually furnished with suckers.

ANNULATA SEDENTARIA. A tribe of soft, elongated, and worm-shaped animals, inhabiting a tube which they never quit. The body has either transverse segments or wrinkles; the head, eyes, and antennæ are furnished with retractile knobs, in lateral rows. They are usually attached to marine substances, and the greater part of them are carnivorous.

ANOA. (*Anoa depressicornis.*) A ruminating animal of Sumatra, at present but imperfectly known to naturalists; by some considered a small species of wild buffalo, and by others a kind of antelope. The living animal has not been brought to this country, but several skulls and horns are deposited in the British Museum, and in that of the Zoological Society. The horns are wrinkled, but perfectly erect and straight, and the head is long and narrow.

ANOBIUM. A genus of Coleopterous insects, some of which inhabit the interior of our houses, where they do much injury in their larva state by gnawing furniture, books, &c., which they pierce with little round holes: others feed upon wafers, preserved specimens of natural history, &c. The curious sound made by one of these has given it the name of Death-watch [which see].

ANODONTA. A fresh-water Molluscous animal, inhabiting a thin, inequivalve, inequilateral shell; hinge straight, with either no teeth or mere rudiments; shell transverse; ligament external. The valves are thin, large, and pearly; and from their shape and lightness they are used in France for skimming milk. The Anodonta is found in every quarter of the world.

ANOLIS. A genus of reptiles peculiar to America, and supplying the place that is occupied by the chameleons in the Old World. Cuvier distinguishes them from the *Iguanas*, by their having teeth in the palate of the mouth as well as in the maxillary bones. The *Anolis* is a small, slender, active animal; frequenting woods and rocky places; and running, leaping, and climbing with singular agility. It is furnished with a loose skin or bag beneath its throat, which, when inflated, frequently changes its colour: in short, whenever these creatures are under the ex-

ANOLIS.

citement of fear, anger, or love, the skin assumes an endless succession of varying hues. They are of more slender proportions than the chameleon, and more agile in their movements; they feed chiefly upon flies and other insects, and inhabit the neighbourhood of marshes and other moist places where insects mostly abound. The head is long, straight, and flattened; the body and tail are long and slender, both being covered with small round scales, which give the skin the appearance of fine shagreen. The hind legs are rather longer than those before; and each foot has five toes. Several species of this genus inhabit the West India Islands; the largest of them not being more than a foot long.

ANOMIA. A genus of marine Mollusca remarkable for the perforation of one of its valves by a large aperture; through which a strong tendinous ligature passes, to be inserted into a third plate, by which the animal adheres to foreign bodies. They are usually found attached to oyster and other shells. This family has long been known in a fossil state, and contains many species. They may be divided into two genera; the inarticulate, and the multarticulate: In the inarticulate Anomia, the hinge of the under valve forms a large cavity, the corners of which make two prominences or joints, and the upper valve is indented into it by corresponding depressions: in the multarticulate Anomia the hinge lies in a long straight line, and is set with many teeth.

ANOMURA. A section of Decapod Crustaceans, consisting of many genera; the habits of some of which, as the Hermit or Soldier Crab (the type of the genus *Pagurus*), are highly curious and interesting. [See HERMIT CRAB.]

ANOPLOTHERIUM. A genus of extinct quadrupeds, found in a fossil state, and which seem to range between the Pachydermata and the Ruminantia. They had six incisor, four canine, and four molar teeth, in each jaw, forming a continued line; and the feet had only two toes, sheathed by separate hoofs; but the toes had separate metacarpal and metatarsal bones, as in the hog, instead of springing from a single canon bone, as they invariably are among the Ru-

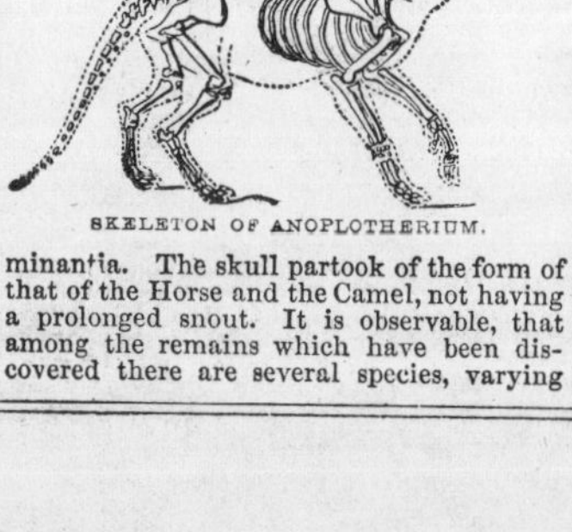

SKELETON OF ANOPLOTHERIUM.

minantia. The skull partook of the form of that of the Horse and the Camel, not having a prolonged snout. It is observable, that among the remains which have been discovered there are several species, varying considerably in their general formation; some presenting a light, slender, and graceful form, probably a fleet and active inhabitant of the dry land, having much of the contour of the gazelle; while another was heavy, bulky, and short-limbed, with a flattened tail, as if aquatic in its habits. But it seems fully demonstrated that these animals were all herbivorous, differing but little in this respect from the Tapirs and Rhinoceroses at present existing.

ANOPLURA. An order of parasitic insects—the *Louse* and its allies; whose presence on the human body is usually regarded as an indication of habitual filthiness. It is to be observed, however, that the inferior animals are subject to them, and that almost every quadruped and bird is infested with some one or other of these parasitic insects.

ANSERES. The third order of the Linnæan class *Aves*, thus characterized: A smooth beak, covered with skin, gibbous at the base, and broader at the point; feet formed for swimming, having palmated toes connected by a membrane; the legs thick and short; and the body bulky, plump, and downy: food fishes, frogs, aquatic plants, worms, &c. The Goose furnishes a ready example.

ANT. (*Formica.*) A well-known genus of Hymenopterous insects, famed from all antiquity for their social and industrious habits, for their love of order and subordination, and for being a pattern of unremitting industry and economy. They are distinguished from other *Hymenoptera* by their habit of residing under ground in numerous societies, and by the existence of *neuters* among them, by which class the labours of the community are chiefly performed. The males have always four wings; the females are larger than the males, and only possess wings during the pairing season; but the neuters have none at any period.

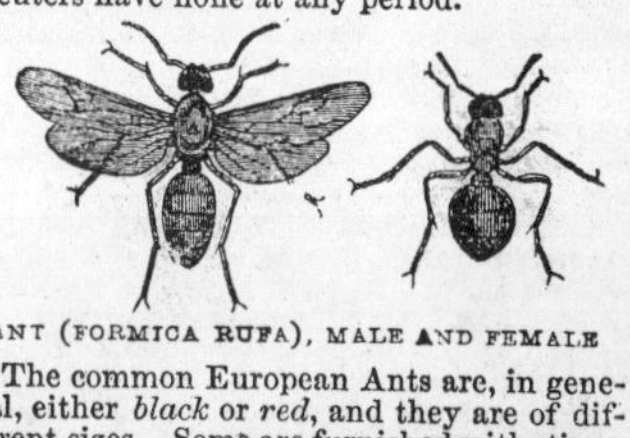

ANT (FORMICA RUFA), MALE AND FEMALE

The common European Ants are, in general, either *black* or *red*, and they are of different sizes. Some are furnished with stings, and others are wholly destitute of them: such as have stings use them for their defence; and such as are unprovided with these weapons have a power of squirting an acid pungent fluid, which inflames and irritates the skin like nettles. The eyes are extremely black; and under them are two small horns or feelers, composed of twelve joints, all covered with fine silky hair. The mouth is composed of two crooked jaws, which project, and in each of which appear incisures resembling teeth. The breast is covered with fine silky hair, from which project six legs, having the extremities of each armed with two small claws, which assist the insect in climbing. The body is of a brown chesnut colour, shining like glass, and covered with extremely fine hair. From this formation, it would appear, the Ant seems bolder and more active than any other creature of the insect tribe of the same size; and, indeed, it possesses sufficient intrepidity to attack an animal many times larger than itself.

The nests of Ants are differently constructed in the different species, but all are very curiously and regularly arranged. "If an Ant-hill," says Mr. Broderip, "be examined any time after Midsummer up to the close of Autumn, there may be seen mixed with the wingless workers a number of both males and females furnished with white glistening wings. These, however, are neither kings nor queens in the state, at least so far as freedom of action is concerned, for they are not allowed to move without a guard of workers to prevent their leaving the boundaries; and if one straggles away unawares, it is for the most part dragged back by the vigilant sentinels, three or four of whom may, in such cases, be seen hauling along a single deserter by the wings and limbs. The workers, so far from ever facilitating the exit, much less the departure of the winged ones, more particularly the females, guard them most assiduously in order to prevent it, and are only forced to acquiesce in it when the winged ones become too numerous to be either guarded or fed. There seems, indeed, to be a uniform disposition in thewinged ones to desert their native colony; and as they never return after pairing, it would soon become depopulated in the absence of females. The actual pairing does not seem to take place within the ant-hill, and we have observed scouts posted all around ready to discover and carry back to the colony as many fertile females as they could meet with. It is probable that, soon after pairing, the males die, as do the males of bees and other insects; for, as the workers never bring any of them back, nor take any notice of them after leaving the ant-hill, they must perish, being entirely defenceless, and destitute both of a sting and of mandibles to provide for their subsistence. The subsequent proceedings of females are very different, and of curious interest. It was supposed by the ancients that all Ants, at a certain age, acquired wings; but it was reserved for the younger Huber, in particular, by means of his artificial formicaries, to trace the development of the wings in the female from the first commencement, till he saw them stript off and laid aside like cast clothes."

"Having directed my close attention to the eggs of ants," says Huber, "I remarked that they were of different sizes, shades, and forms. The smallest were white, opaque, and cylindrical; the largest transparent, and slightly arched at both ends; while those of a middle size were semi-transparent. On holding them up to the light, I observed a sort of white oblong cloud; in some, a transparent point might be remarked at the superior extremity; in others, a clear zone above and underneath the little cloud. There were some whose whole body was so remarkably clear as to allow of my very distinctly observing the rings. On fixing attention more closely upon the latter, I observed the egg open, and the grub appear in its place. Having compared these eggs with those just laid, I constantly found the latter of a milky whiteness, completely opaque, and smaller by one half, so that I had no reason to doubt of the eggs of ants receiving a very considerable increase in size; that in elongating they become transparent, but do not at this time disclose the form of the grub, which is always arched." When the eggs are at length hatched, the young grubs are fed either by the neuters (called also nurse-ants and workers) when any of these are in the colony, and by the mother when she is alone, by a liquid disgorged from the stomach, as is done in a similar way by wasps, humble-bees, and certain birds.

"When the larvæ have attained their full growth they spin a silken covering, called by entomologists a cocoon: in this they completely enclose themselves, and remain perfectly quiescent without receiving any nutriment, awaiting the final change when they are to assume the form of Ants. This stage of its existence is the pupa, but is commonly though very erroneously called the egg. Ants' eggs, as they are vulgarly called, are a favourite food for partridges and pheasants, and are eagerly sought after by persons who rear these birds from the egg. The cocoon containing the pupa is of a long cylindrical form, of a dirty white colour, and perfectly without motion. The pupa within the cocoon has now attained the form which it will finally possess; its limbs are distinct, but want strength and consistence, and are covered by a skin which has yet to be cast. In colour it changes from white to a pale yellow, then to red, and finally becomes almost black; its wings, if a male or female, are distinctly visible, but do not assume the shape, size, or character, they are hereafter destined to bear." — *Newman's Hist. of Insects.*

In England, ant-hills appear formed and arranged with very little regard to order or regularity; but in the more southern parts of Europe they are constructed with amazing ingenuity. They are generally formed in the vicinity of some large tree on the bank of a river; the former for the purpose of securing food, and the latter for supplying them with that abundant moisture which is requisite for the use of these insects. The ant-hill is of a conical shape, and is composed of leaves, bits of wood, sand, earth, stubble, gum, and grains of corn; all united into a compact body, perforated with galleries down to the bottom, and having a variety of tunnels or passages throughout the interior, the number of these avenues depending entirely on the population and extent of the nest. At its commencement the nest is simply an excavation made in the earth; a number of the labourers wander about in quest of materials suitable for the superstructure; others carry out particles of earth from the interior, and these particles, interspersed with the fragments of wood and leaves brought in continually from every quarter, give a kind of stability to the edifice: it daily increases in size, the Ants taking care to leave the spaces required for the galleries which lead to the exterior; while the dome contains a number of spacious chambers or recesses, which communicate with each other by means of galleries constructed in a similar manner. Thus we see that although the exterior of the hillock always presents the appearance of a dome, and appears but a careless heap, it is in reality a most ingenious device for keeping out water, for evading the effects of the wind, and the attacks of enemies; and yet more especially for receiving and husbanding the heat of the sun. There are external apertures, to permit free egress to the multitude of labourers of which the commonwealth is composed; and from the commencement of the warm season they are constantly employed, till the unpropitious winter again suspends their exertions, and terminates their annual industry.

The working Ants are not only employed in sustaining the idlers at home, but in providing sufficient food for themselves. They subsist on various provisions, both animal and vegetable; killing and devouring all weaker insects, as well as in seeking ripe fruits and whatever appears to contain saccharine matter. When they are unable to eat the whole of the substance they have found, they devour what they can; and, tearing the rest in pieces, load themselves with the spoil. When they meet with an insect which they are singly incapable of mastering, several of them join in the attack; its destruction generally follows, each Ant assisting in carrying away a portion of the booty. When a single Ant chances to make a fortunate discovery, it immediately communicates the information to others, and the whole republic soon appear in motion. But while they are thus busied in feeding abroad, and carrying in provisions for the use of those which continue inactive at home, they are by no means unmindful of posterity. The female Ants soon begin to lay their eggs, which are immediately carried to the safest situation, at the bottom of the hill, where they are assiduously defended by the labourers, who always display the fondest attachment to the rising progeny, either attending to the safety of the larvæ, or in feeding the newly born insects. Who, indeed, has not seen them, when the gardener or some formidable enemy has demolished their whole habitation, affectionately solicitous of their offspring, and running wildly about, each loaded with a young one, not unfrequently as large as itself.

For some time the new born Ants remain under the careful superintendence of the labourers: they are attended in all their wanderings about the nest, and are made acquainted with all its galleries and chambers: the wings of the males and females, previously folded together, are extended, and this is always accomplished with such skill and tenderness, that these delicate members are never injured by the operation: in fine, these founders of future colonies are in all respects served with unremitting attention until their final departure from the nest.

In the autumn, says Mr. Newman, we frequently observe one of their hillocks closely covered with a living mass of winged Ants, which continue to promenade, as it were, over its entire surface: they mount on every plant in the vicinity of their nest, and the labourers (for now the entire population of the nest has turned out) accompany them as closely as possible, following them to the extreme tip of every blade of grass; and when at length those possessed of wings spread them in preparation for flight, the labourers will often hold them back, as if loath to trust them alone. . . . When the air is warm and still they rise in thousands, and sailing, or rather floating on the atmosphere, leave for ever the scene of their former existence. Each female, immediately on alighting from her aerial voyage, examines the situation in which chance has placed her, and if she find it adapted to her purpose, she turns her head back over her shoulders, and with her sharp mandibles tears off the wings which bore her from the place of her nativity. Strange as this propensity may seem, it is dictated by an unerring instinct, for the object for which wings were given her is now accomplished, and henceforth they would prove an incumbrance, and retard rather than assist, the performance of her duties. Sometimes a few workers, wandering at this period of excitement far from their home, may happen to meet with her, and if so, they unite their labours with hers in excavating a small and humble dwelling in the earth, which serves as the nucleus of a future colony: in all operations the female, now a queen, takes a most energetic part, and continues to labour until she has laid eggs, when the conduct of the workers undergoes a great change, for they now treat her with the most marked respect, and consider her worthy the honours of a sovereign.

The ingenious author we before quoted gives a very curious account of what he terms the *Slave Ants*, which in substance is as follows: The most remarkable fact connected with the history of Ants, is the propensity possessed by certain species to kidnap the workers of other species, and compel them to labour for the benefit of the community, thus using them completely as slaves; and, as far as we yet know, the kidnappers are red or pale-coloured Ants, and the slaves, like the ill-treated natives of Africa, are of a jet black. The time for capturing slaves extends over a period of about ten weeks, and never commences until the male and female Ants are about emerging from the pupa state, and thus the ruthless marauders never interfere with the continuation of the species. This appears to be a special adaptation of their peculiar instinct; for if the attacks were made on the nests of the Negro Ants, before those by whom the race is propagated are ready to issue forth, it must speedily become extinct. When the Red Ants are about to sally forth on a marauding expedition, they send scouts to ascertain the exact position in which a colony of negroes may be found; these scouts having discovered the object of their search, return to the nest and report their success. Shortly afterwards the army of Red Ants marches forth, headed by a vanguard, consisting of only about eight or ten Ants, which is perpetually being changed,—the individuals which constitute it, when they have advanced a little before the main body, halting, falling into the rear, and being replaced by others. When they have arrived near the Negro colony, they disperse, wandering through the herbage, and hunting about, as if aware that the object of their search was near, though ignorant of its exact position. At last they discover the settlement, and the foremost of the invaders rushing impetuously to the attack, are met, grappled with, and frequently killed by the negroes on guard: the alarm is quickly communicated to the interior of the nest; the negroes sally forth by thousands, and the Red Ants rushing to the rescue, a desperate conflict ensues, which, however, always terminates in the defeat of the negroes, who retire to the innermost recesses of the habitation. Now follows the scene of pillage: the Red Ants with their powerful mandibles tear open the sides of the negro ant-hill, and rush into the heart of the citadel. In a few minutes each of the invaders emerges carrying in its mouth the pupa of a worker negro, which it has obtained in spite of the vigilance and valour of its natural guardians. The Red Ants return in perfect order to their nest, bearing with them their living burdens. On reaching the nest the pupæ appear to be treated precisely as their own, and the workers when they emerge perform the various duties of the community with the greatest energy and apparent good-will. [For an account of the *White Ants*, which belong to a totally different order of insects, see TERMITES. See also DRIVER ANTS.]

The following short passage from Mr. Darwin's Observations on the Natural History of Rio de Janeiro will give the reader a good idea of the magnitude of the Ants' nests there: "Travelling onwards, we passed through tracts of pasturage, much injured by the enormous conical Ants' nests, which were nearly twelve feet high. They gave to the plain exactly the appearance of the mud volcanoes at Jorullo, as figured by Humboldt." And in Gardner's Travels in Brazil we read the following remarks on the immense multitudes of Ants which are found there. "When near Rio de Janeiro," he says, "we passed many habitations belonging to poor people of colour, mostly fishermen. Before reaching the foot of the mountain over which the road leads to Tijuca, we passed a migrating body of small Black Ants. The immense number of individuals composing it may be imagined from the fact, that the column was more than six feet broad, and extended in length to upwards of thirty yards. The ground was completely covered with the little creatures, so closely were they packed together." The species also are more numerous than naturalists are aware of: he says that near Pernambuco he noticed more than 25 different species.

ANT-EATER. (*Myrmecophaga.*) A genus of animals, of the Cuiverian order *Edentata*. Their distinguishing characteristics are, that the body is covered with hair, the mouth is small, and the tongue long and cylindrical, calculated to supply the want of teeth, from being covered with a glutinous saliva, by means of which they entrap and devour the insects upon which they live and from which they derive their name. The head is very long, but the tongue is much longer, and capable of being extended to a surprising distance beyond the snout; the eyes are particularly small, the ears short and round, the legs thick and strong, but most unfavourably formed for locomotion, and consequently their pace is remarkably slow. There are three distinct and well-defined species in South America; and these, with one or two others, we shall briefly describe.

The GREAT ANT-EATER, or ANT-BEAR (*Myrmecophaga jubata*), is by far the largest of the Ant-eaters, and is covered with long, coarse, shaggy hair, except the head, where it is short and close; it has a very long and slender head, and a bushy black tail of

GREAT ANT-EATER.—(MYRMECOPHAGA JUBATA.)

enormous size and length, the whole animal often measuring eight feet from the tip of the snout to the extremity of the tail. Being plantigrade, it stands lower on the hind legs than before, which is the case with bears and other quadrupeds similarly formed. It has four toes on the fore-feet, the second and third being provided with long, sharp-pointed, and trenchant claws; so that nothing upon which it has an opportunity of fastening can escape. The hind feet have five toes, furnished with short weak claws, resembling those of ordinary quadrupeds. The prevailing colour of this animal is a deep grey, with a very broad band of black running from the neck downwards on each side of the body; its habits are slothful and solitary; and it sleeps during the greater part of the day. It lives exclusively on ants, to procure which it opens their hills with its powerful crooked claws, and draws its long flexible tongue, which is covered with glutinous saliva, lightly over the swarms of insects who flock from all quarters to defend their dwellings. It is a native of Brazil and Guiana. It seems almost incredible that so robust and powerful an animal can procure sufficient sustenance from Ants alone; but this circumstance has nothing strange for those who are acquainted with the tropical parts of America, and who have seen the enormous multitudes of these insects, which swarm in all parts of the country to that degree, that their hills often almost touch one another for miles together. The favourite resort of the Great Ant-eaters are the low swampy savannahs, along the banks of rivers and stagnant ponds.

The TAMANDUA (*Myrmecophaga tamandua*), a smaller kind of Ant-eater, is about the size of a full-grown cat; the head not being so disproportionately long as the species above described, though it is of the same general cylindrical form, and equally truncated at the end. The conformation of the extremities, and the number of the toes is in every respect the same as in the Great Ant-eater; but the tail is prehensile, which makes it essentially an arboreal quadruped; while, instead of having long shaggy hair, it is short, shining, and somewhat silky, like the finest wool. There are several varieties of this species, differing chiefly in colour; they reside exclusively on trees, living on termites, honey, and (according to Azara, in his Natural History of the Quadrupeds of Paraguay), bees, which in those countries form their hives among the loftiest branches of the forest, and, having no sting, are readily despoiled of their honied treasure.

The LITTLE ANT-EATER (*Myrmecophaga didactyla*) is an animal of considerable elegance, and not larger than a squirrel: the head is small, the snout sharpened and slightly bent downwards; the fore-feet have only two claws on each, the exterior one much larger and stronger than the interior; the ears are very small, and hid in the fur; the eyes are also small. The whole animal

LITTLE ANT-EATER.—(MYRMECOPHAGA DIDACTYLA)

is covered with a beautiful soft and curled fur of a pale yellow-brown colour: the tail is thick at the base, tapering to the tip; and being prehensile, it greatly assists the Little Ant-eater's operations in its search for insects among the trees, on which it resides.

The STRIPED ANT-EATER. (*Myrmecophaga striata.*) This is a native of Guiana: it is about twenty inches long from the tip of its snout to the end of its tail; the nose is taper, the upper mandible extending very far beyond the lower; the body and tail are of a tawny colour, with the under parts white; the body marked with broad, distant, blackish, transverse stripes, and the tail annulated with similar ones. [For SPINE ANT-EATER, see ECHIDNA.]

ANTELOPE. (*Antilope.*) A genus of hollow-horned Ruminants, of which there are many species, each differing from the other in some important points, but agreeing in the grand leading characteristics. Thus,

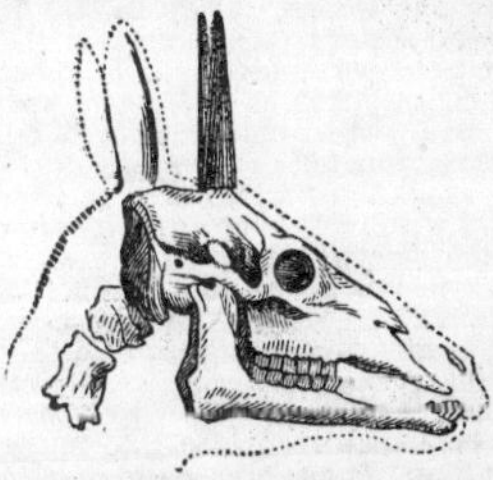

SKULL OF ANTELOPE

speaking generally, it may be said, that Antelopes are of graceful and symmetrical proportions; of a restless and timid disposition, extremely watchful, of great vivacity, remarkably swift and agile, and most of their boundings inconceivably light and elastic. Their horns, whatever shape they assume, are round and annulated; in some species straight, in others curved and spiral; in some the females have no horns, in others they are common to both sexes. They all possess a most delicate sense of smell; their eyes are proverbially bright and beaming; and so fleet are they, that the huntsman is often obliged to call in the aid of the falcon, trained for the purpose, to seize on the animal and arrest its progress, in order to give the greyhounds an opportunity of overtaking it. Their hair is generally short and smooth, and of an equal length over every part of the body: some species, however, have manes on the neck and shoulders; and a few are furnished with long hair on the chin and throat. The ears are long and pointed; the tails short, and tufted at the extremity. For the most part Antelopes are gregarious, some species forming herds of two or three thousand, while others keep in pairs, or in companies of five or six. They often browse like the goat, and feed on the tender shoots of trees; and the flesh of those which are taken in the chase is usually of excellent flavour.

The *Antilopidæ* seem to be a connecting link between the Goat and the Deer. Like the goat, they never shed their horns; but, on the other hand, their size and the delicacy of their conformation, the nature and colour of the hair, their fleetness, &c., are striking points of resemblance to the deer tribe. The hind legs, like those of the hare, being longer than the fore ones, not only give additional swiftness, but greater security, in ascending and descending precipices, a practice in which the Antelope greatly delights. The majority of the species are brown on the back, and white under the belly, with a black stripe separating those colours. The tail is of various lengths, but always covered with pretty long hair; and the ears, which are beautiful and well placed, terminate in a point. The hoof is cloven, like that of the sheep; and the horns are perennial. The length, size, and turn of the horns, the different spots in the skin, or diversities of size, constitute the chief distinctions which mark the several species. They mostly inhabit the torrid regions, or such parts of the temperate zone as are nearly contiguous, frequenting the cliffs and ledges of rocks, or traversing vast untrodden wildernesses. Africa appears to be their great nursery, but many kinds are natives of Asia; very few are met with in Europe; and it is remarkable that, notwithstanding the warmth of South America is well suited to their nature, only a single species of Antelope is to be found in any part of the New World.

Having made these general remarks, it is necessary, for the sake of perspicuity, to consider Antelopes as divided into sub-genera, or families. It has been customary to class them as follows:—1. True Antelopes; 2. Bush Antelopes; 3. Capriform (or goat-like) Antelopes; and 4. Bovine (or ox-like) Antelopes. But some late writers on zoology have rendered the sub-division infinitely more minute; the species in many instances closely bordering on each other, while there are others in which scarcely any corresponding features can be distinctly traced. Thus, as an eminent naturalist has remarked, "the genus Antelope has become a kind of zoological refuge for the destitute, and forms an incongruous assemblage of all the hollow-horned ruminants together. So diversified are its forms, and so incongruous its materials, that it presents not a single character which will either apply to all its species, or suffice to differentiate it from conterminous genera."

The COMMON ANTELOPE, or SASIN. (*Antilope Cervicapra.*) This elegant specimen of the Antelope tribe is a native of many parts of Africa, and also of India. It is somewhat smaller than a fallow deer, and is remarkable for the peculiar beauty of its long spiral horns, which are distinctly marked by numerous prominent rings; its colour is a reddish tawny brown above, and white below; the legs are long and delicate, the body round, but light and well formed; the eyes large and expressive, and their orbits white. They are extremely wary, and when feeding or lying down are guarded by sentinels, who give the alarm on the slightest appear nce of danger; and such is their fleetness and activity, that they often vault over nets ten feet high, and when pursued, will pass over as many yards at a single bound. [See SPRINGBOK, PRONGBUCK, GNU, GAZELLE, KOODO, STEENBOK, NYLGHAU, &c.]

COMMON ANTELOPE.—(ANTILOPE CERVICAPRA.)

ANTHICIDÆ. A tribe of Coleopterous insects, possessing simple or but slightly serrated and filiform antennæ; the maxillary palpi are terminated by a hatchet-shaped joint; and the penultimate joint of the tarsi is bilobed. Some of these species are found upon plants, but the majority live on the ground, and run with great quickness: their larvæ are probably parasites. They compose the genera *Notoxus*, *Anthicus*, &c.

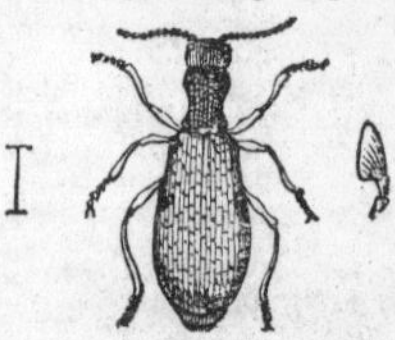

ANTHICUS LATERI PUNCTATUS

ANTHOBII. A section of Coleopterous insects, composed of species inhabiting the southern parts of Europe and the warm parts of both hemispheres. They are distinguished by the two divisions of the lower lip being produced considerably beyond the mentum, and the elytra gaping at the tips, which are rounded; the antennæ have nine or ten joints, the last three composing the club; the terminal lobe of the maxillæ is membranous, silky, and pencil-like, but leathery in others; the upper lip and mandibles are more or less solid, as they are more or less exposed. These insects live upon flowers or leaves.

ANTHOCHÆRA. A genus of birds belonging to the family *Meliphagidæ*, or Honey-eaters, several species of which are found in New Holland. As an example of this interesting genus we give

The ANTHOCHÆRA MELLIVORA, or BUSH WATTLE-BIRD; a bird constantly found where there are Banksias, in New South Wales, South Australia, and Van Diemen's Land. It is bold and spirited, fearlessly attacking and driving away all other birds from the part of the tree on which it is feeding. In spring and summer the male perches on some elevated branch, and screams forth his harsh and peculiar notes,—like a person vomiting,—whence its local

BRUSH WATTLE-BIRD
(ANTHOCHÆRA MELLIVORA.)

name *Goo-gwar-ruck*, in which the natives have tried to imitate it. While thus employed, it frequently jerks its tail, throws back its head, and distends its throat, as if great exertion were required. It breeds in September and three following months: the nest is round, open, and rather small; it is generally placed in the fork of a small branch; and is formed of fine twigs, lined with fibrous roots. Eggs two, and sometimes three.

Banksias are in blossom the greater part of the year; each flower as it expands is diligently examined by the Wattle-bird, which inserts its long feathery tongue into every part, extracting pollen and insects. It is to be observed that Banksias are not a sign of good land, so that the garrulous note of this species may be taken by the settler as an indication of the sterile and unprofitable nature of the soil. (*Gould's Birds of Australia.*)

ANTHOMYZIDÆ. A general division of the *Muscidæ*, composed of species having the appearance of Common Flies; the wings not vibratile; the antennæ inserted near the forehead, always shorter than the head, terminated by a long or linear joint, with the seta mostly plumose; the legs are of moderate size, and the abdomen composed of four joints.

ANTHOPHILA. A name given by Latreille to the fourth family of the Aculeated Hymenoptera (the Bees).

ANTHROCERIDÆ. A family of Lepidopterous insects, of the section *Heterocera;* comprising a rather numerous group of small or moderately sized species, distinguished by their brilliancy of colour and diurnal flight; having the antennæ never terminated by a pencil of hairs, and either simple in both sexes and fusiform, or thickened near the middle: the head is furnished with a pair of ocelli behind the antennæ; the labial palpi are rather small, and the maxillæ greatly elongated: the wings are always deflexed in repose, exhibiting in many species a number of denuded spots; the nervures are very numerous; the legs are long, with the posterior tibiæ furnished with four spurs. The caterpillars are of a cylindrical form, generally clothed with short hairs, and without any spine at the hind part of the body: they feed on various leguminous plants, and considerably resemble those of several of the *Bombycidæ*. The pupæ are of the ordinary conical form, without any angular prominences. The colouring of some of the exotic species of this family is truly beautiful. [See BURNET MOTH.]

ANTHUS. [See PIPIT.]

ANTIPATHES. Black Coral.

ANT-LION. (*Myrmeleon formicaleo.*) A Neuropterous insect which has long been celebrated for its wonderful ingenuity in preparing a kind of pitfall for the destruction of such insects as happen unwarily to enter it. In its complete or fly state it bears no inconsiderable resemblance to a small dragon-fly, from which however it may readily be distinguished by its antennæ, which are hard, and incurvated at the ends. It

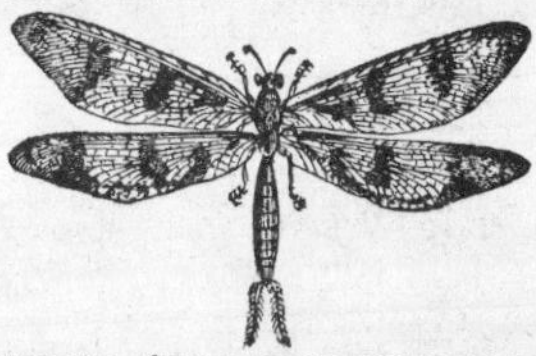

ANT LION.—(MYRMELEON FORMICALEO.)

deposits its eggs in dry sandy situations, and the young larvæ, when hatched, begin separately to exercise their talent of preparing a very small conical cavity in the sand, which they effect by turning themselves rapidly round. Under this cavity it lies concealed, ready to rush forward at a moment, in order to seize any small insect that has been so unfortunate, in approaching the edge, as to fall in; and no sooner has it sucked out the juices of its victim through its tubercular forceps, than it throws it by a sudden exertion to some distance. As the larva increases in size, it enlarges the hole, which at last becomes about two inches in diameter, its own length being when full-grown about half an inch. It is of a flattened figure, broad towards the upper part, and gradually tapering to an obtuse point: the legs are slender; the head and thorax rather small; the tubular jaws long, curved, serrated internally, and very sharp-pointed: it is of a brown colour, beset with numerous tufts of dusky hair; the whole presenting a form bearing some resemblance to a flat-bodied spider. In preparing its pit, it begins by tracing an exterior circle of the intended diameter of the cavity, continuing its motion, in a spiral line, till it gets to the centre, thus making several volutes in the sand, resembling the impression of a large helix or snail-shell; and after having sufficiently deepened the cavity by a repetition of this motion, it smoothes the sides into a regular shape by throwing out the superfluous sand lying on the ridges, which it effects with surprising address and dexterity.

The ingenuity and perseverance of this insect, or rather the admirable instinct it displays, is so amusingly described by Messrs. Kirby and Spence in their "Introduction to

Entomology," that we cannot refrain from indulging in a quotation, the length of which, we trust, its pertinence may well excuse. "In the course of its labours it frequently meets with small stones: these it places upon its head, one by one, and jerks over the margin of the pit. But sometimes, when near the bottom, a pebble presents itself of a size so large that this process is impossible, its head not being sufficiently broad and strong to bear so great a weight, and the height being too considerable to admit of projecting so large a body to the top. A more impatient labourer would despair; but not so our insect. A new plan is adopted. By a manœuvre, not easily described, it lifts the stone upon its back, keeps it in a steady position by an alternate motion of the segments which compose that part; and, carefully walking up the ascent with the burthen, deposits it on the outside of the margin. When, as occasionally happens, the stone is round, the labour becomes most difficult and painful. A spectator watching the motions of the ant-lion feels an inexpressible interest in its behalf. He sees it with vast exertion elevate the stone, and begin its arduous retrograde ascent: at every moment the burthen totters to one side or the other: the adroit porter lifts up the segments of its back to balance it, and has already nearly reached the top of the pit, when a stumble or a jolt mocks all its efforts, and the stone tumbles headlong to the bottom. Mortified, but not despairing, the Ant-Lion returns to the charge; again replaces the stone on its back; again ascends the side, and artfully avails himself, for a road, of the channel formed by the falling stone, against the sides of which he can support his load. This time possibly he succeeds; or it may be, as is often the case, the stone again rolls down. When thus unfortunate, our little Sisyphus has been seen *six* times patiently to renew his attempts, and was at last, as such heroic resolution deserved, successful. It is only after a series of trials have demonstrated the impossibility of succeeding, that our engineer yields to fate, and, quitting his half-excavated pit, begins the formation of another.

"When all obstacles are overcome, and the pit is finished, it presents itself as a conical hole rather more than two inches deep, gradually contracting to a point at the bottom, and about three inches wide at the top. The Ant-lion now takes its station at the bottom of the pit, and, that its gruff appearance may not scare the passengers which approach its den, covers itself with sand, all except the points of its expanded forceps. It is not long before an ant on its travels, fearing no harm, steps upon the margin of the pit, either accidentally or for the purpose of exploring the depth below. Alas! its curiosity is dearly gratified. The faithless sand slides from under its feet; its struggles but hasten its descent; and it is precipitated headlong into the jaws of the concealed devourer. Sometimes, however, it chances that the ant is able to stop itself midway, and with all haste scrambles up again. No sooner does the Ant-lion perceive this (for, being furnished with six eyes on each side of his head, he is sufficiently sharp-sighted), than, shaking off his inactivity, he hastily shovels loads of sand upon his head, and vigorously throws them up in quick succession upon the escaping insect, which, attacked by such a heavy shower from below, and treading upon so unstable a path, is almost inevitably carried to the bottom. The instant his victim is fairly within reach, the Ant-lion seizes him between his jaws, which are admirable instruments, at the same time hooked for holding and grooved on the inner side, so as to form with the adjoining maxillæ, which move up and down in the groove, a tube for sucking, and at his leisure extracting all the juices of the body, regales upon formic acid. The dry carcass he subsequently jerks out of his den, that it may not encumber him in his future contests, or betray the 'horrid secrets of his prison-house:' and if the sides of the pit have received any damage, he leaves his concealment for a while to repair it; which having done, he resumes his station."

Such is the mode of life pursued by the larva of the Ant-lion until nearly two years have elapsed, when, being arrived at its full growth, and ready to change into a chrysalis, it envelopes itself in a round ball of sand, agglutinated and connected by very fine silk, which it draws from a tubular process at the extremity of its body. In this silken cocoon it remains about three weeks; and then bursts forth a four-winged insect, resembling the dragon-fly both in appearance and manners. The Ant-lion is not found in England, but occurs in France, Spain, Germany, &c. [See MYRMELEON.]

APATEON. [See SUPPLEMENT.]

APATURA IRIS, or PURPLE EMPEROR. Of all our native Lepidoptera, there is no Butterfly that is more justly admired than the Purple Emperor. In its

PURPLE EMPEROR. — (APATURA IRIS.)

bold and soaring flight, as it displays its beautiful lines in the effulgence of the meridian sun, or as it settles for repose when the shades of evening approach, it still maintains its acknowledged pre-eminence. The general colour of the wings above is a rich deep brown, changing in the male according to the light, to a lovely purple, or a splendid mazarine blue, and relieved by a triple series of white spots. The posterior wings have a white angular band, placed in continuity with the first series of spots on the anterior wings; and an ocellus at the anal angle with a narrow tawny iris and black pupil: the under surface of the anterior wings is a ferruginous brown, varied with white and black; between the disc and the hinder margin is an ocellus with a black iris and a bluish pupil: body black above, cinereous beneath; antennæ black. The female is considerably larger than the male, but the colours are not so deep, nor are the reflected hues so brilliant. The Caterpillar is a bright green, with greenish-yellow horns, reddish at the tip, and has reddish bristles at the tail. It feeds on the oak, ash, and willow. The Chrysalis is of a pale green hue. The perfect insect seldom makes its appearance before July; is by no means scarce; and in various parts of the South and West of England very beautiful specimens are often taken. — There are other species of the genus *Apatura*, but the above is the only one found in Britain.

APE. (*Pithecus.*) The words APE, MONKEY, and BABOON were formerly applied indiscriminately to any of the Quadrumanous Mammalia; it will therefore be right to state, before we proceed further, that the APES, or SIMIÆ, may be properly divided into four sections; viz.—*Apes*, or such as are destitute of a tail: *Baboons*, or such as have muscular bodies, elongated muzzles, and whose tails are usually short: *Monkeys*, whose tails are in general long: and *Sapajous*, or Monkeys with *prehensile* tails, which can at pleasure be twisted round any object, and thereby in many instances answer the purpose of an additional hand to the animal. It is, however, to the first of these only that our attention is in this place to be directed.

The genus Ape (*Pithecus*) comprises those quadrumanous animals which most closely approach to the human species in anatomical structure, and which, in popular language, are termed monkeys without tails or cheek-pouches. As Buffon justly observes of the whole, they are not quadrupeds, but quadrumana; not four-footed, but four-handed animals. They chiefly inhabit the vast forests of India and Africa, and are numerous in the peninsula of Malacca, and the great islands of the Indian Ocean; living in trees, and feeding on fruits, leaves, and insects; but though frugiverous in a state of nature, yet, from the resemblance of their teeth to those of the human species, it is very evident that their diet may be almost as various as that of Man. They generally live in troops, and some of the species are said to construct a sort of hut of leaves, as a defence against the weather: it is also asserted that they use clubs to defend themselves when attacked.

The Apes are in general fierce and untractable; though some of them appear to be of a grave and gentle disposition; neither petulant nor mischievous, like the monkeys, properly so called. Their arms are so long as almost to touch the ground when the animals stand erect on their hind legs; the fingers and toes are long, flexible, deeply separated from one another, and admirably adapted for prehension: thus they are enabled to spring from tree to tree with surprising agility, even when loaded with their young, who cling closely to them on every appearance of danger. Apes have the power of assuming a nearly erect position; though on the ground this is by no means convenient, as they stand upon the outer edges, being unable to apply the palms of the posterior hands fairly against the soil, and require a staff, or other support, to maintain this attitude, except when they have been taught to stand erect by man. [See GORILLA; CHIMPANZEE; ORANG-OUTANG; SIAMANG; GIBBON, &c.]

APHANIPTERA. An order of Apterous Haustellate insects, having rudimental elytra or wings in the perfect state. It is composed entirely of the different species of Fleas, forming the family PULICIDÆ; the common Flea (*Pulex irritans*) being the type of the order. The legs are long, the posterior formed for leaping; the coxæ are very large; the fore legs are singularly placed, appearing to arise from the front of the head, the coxæ defending the sides of the rostrulum. This peculiarity is caused by the prothoracic epimera being detached from the body, and extended obliquely beneath the head: the femora are short, but strong; the tibiæ very setose; and the tarsi five-jointed, terminated by a pair of strong claws. The female flea deposits a dozen eggs, of a white colour, and rather viscous texture, from which are hatched long worm-like grubs, destitute of feet, which are very active in their motions, winding themselves in a serpentine manner through the substance in which they may be deposited: the head of the larva is protected by a firm skin, and bears two antennæ, but no eyes. The body consists of thirteen segments, bearing little tufts of hair, and the last is armed with a pair of small hooks. When full grown, which occurs in summer in about twelve days, the larvæ enclose themselves in a small cocoon of silk, often covered with dust, and attached to adjoining substances: in this it passes into the pupa state, and in about twelve days more emerges a perfect flea.

In hot countries these insects are exceedingly troublesome: but in the West Indies and South America there is an insect belonging to the family having habits different to those of the common flea, which is even still more obnoxious this is the Chigoe (*Pulex pénetrans*), which lives in the open country, and attacks the naked feet both of men and dogs. [See FLEA and CHIGOE.]

APHIS: APHIDÆ. A genus and family of Homopterous insects, comprising the very numerous and obnoxious species of *Plant-lice*, a tribe of insects analogous, in regard to the vegetable world, to the animal parasites of the order ANOPLURA, or lice. The antennæ are of great length; the ocelli, three in number, form a large triangle; the eyes are entire, prominent, and semiglobose; the abdomen is short and convex, generally furnished with a tubercle on each side near the extremity. Some are winged, and some are wingless, without distinction of sex: the wings are very much deflexed at the side of the body, being almost perpendicular in repose; the fore wings much larger than the posterior, with strong nerves: the legs are very long and slender, formed only for crawling. The species reside in great societies upon almost every species of plant, of which they suck the young shoots, leaves, and stems, by the assistance of their proboscis, producing disease in the plant either by greatly weakening it, or by raising vesicles, or other gall-like excrescences, in which whole generations of Aphides reside. The anal tubercles above mentioned secrete a saccharine fluid of which ants are very fond; and it is this fluid dropped upon the adjacent leaves, or the extravasated sap flowing from the wounds caused by the punctures of the insects, which is known under the name of the *honey-dew*. In the spring they are viviparous, in the autumn and as winter approaches they are oviparous; and by a surprising aberration from the common laws of nature, it appears that one impregnation of the female is sufficient for many generations, without further assistance from the male. All the Aphides which appear in the spring are exclusively females, no males being found till the autumn; and the females are endowed with such astonishing fecundity, that nine generations—each generation averaging 100 individuals—have been produced within three months; "so that from a single Aphis, 10,000 million millions may be generated in that short period!" In some years they are so nume-

PLANT-LICE.—(APHIS PLATANOIDES.)

rous as to cause almost a total failure of the hop plantations; at other times the beans, peas, and potatoes are injured by them to an alarming extent; as also are numerous shrubs, and plants, including the delicate exotics raised in stoves and green-houses. There are numerous species; uniformly deriving their specific name from the tree, shrub, or plant, on which they are commonly found. Of these the *Aphis rapæ*, which has made a great noise as the *Aphis vastator* and feeds on various plants, the *Aphis rosæ* (or rose louse), the *Aphis humuli* (or hop-fly), and the *Aphis vitis* (or vin -fretter), are among the best known and most destructive; but the largest and most remarkable of the British Aphides is the *Aphis salicis*, which is found on the different kinds of willows, and is nearly a quarter of an inch long. Many of the species have the body densely clothed with a white cottony secretion, either in threads or flakes; among these may be particularly mentioned the *Aphis lanigera*, or *American blight*, as it is termed, which infests the stems of Apple-trees, sometimes totally destroying them.

"The injuries occasioned by plant lice," as Dr. Harris very truly observes, "are much greater than would at first sight be expected from the small size and extreme weakness of the insects; but these make up by their numbers what they want in strength individually, and thus become formidable enemies to vegetation. By their punctures, and the quantity of sap which they draw from the leaves, the functions of these important organs are deranged or interrupted, the food of the plant, which is there elaborated to nourish the stem and mature the fruit, is withdrawn before it can reach its proper destination, or is contaminated and left in a state unfitted to supply the wants of vegetation. Plants are differently affected by these insects. Some wither and cease to grow, their leaves and stems put on a sickly appearance, and soon die from exhaustion. Others, though not killed, are greatly impeded in their growth, and their tender parts, which are attacked, become stunted, curled, or warped. The punctures of these lice seem to poison some plants, and affect others in a most singular manner, producing warts or swellings, which are sometimes solid and sometimes hollow, and contain in their interior a swarm of lice, the descendants of a single individual, whose punctures were the original cause of the tumor. I have seen reddish tumors of this kind, as big as a pigeon's egg, growing upon leaves, to which they were attached by a slender neck, and containing thousands of small lice in their interior. Naturalists call these tumors galls, because they seem to be formed in the same way as the oak-galls which are used in the making of ink. The lice which inhabit or produce them generally differ from the others, in having shorter antennæ, being without honey-tubes, and in frequently being clothed with a kind of white down, which, however, disappears when the insects become winged."

Mr. Knapp, in his 'Journal of a Naturalist,' has thus described this species, and its effects. "Our apple-trees here are greatly injured, and some annually destroyed, by the agency of what seems to be a very feeble insect. We call it, from habit, or from some unassigned cause, the 'American blight.' [It seems, however, that it had been noticed in England as early as the year 1787; and there is good reason to believe that in Americ, it is not indigenous, but was introduced there with fruit-trees from Europe.] In the spring of the year a slight hoariness is observed upon the branches of certain species of our orchard fruit. As the season advances this hoariness increases, it becomes cottony; or, in other words, towards the end of summer the under sides of some of the branches are invested with a thick, downy substance, so long as at times to be sensibly agitated by the air. Upon examining this substance, we find that it conceals a multitude of small wingless creatures, which are busily employed in preying upon the limb of the tree beneath. This they are well enabled to do, by means of a beak terminating in a fine bristle; this, being insinuated through the bark and the sappy part of the wood, enables the creature to extract, as with a syringe, the sweet, vital liquor that circulates in the plant. The alburnum, or sap-wood, being thus wounded, rises up in excrescences and nodes all over the branch, and deforms it; the limb, deprived of its nutriment, grows sickly; the leaves turn yellow, and the part perishes. Branch after branch is thus assailed, until they all become leafless, and the tree dies." * * * * "Many remedies have been proposed for removing this evil, efficacious perhaps in some cases upon a small scale; but when the injury has existed for some time, and extended its influence over the parts of a large tree, I apprehend it will take its course, and the tree die."

Upon this part of the subject, Dr. Harris remarks that the application of the spirits of tar, of spirits of turpentine, of oil, urine, and of soft soap, has been recommended; but he is inclined to think that the following mode of treatment will be found the most effectual of any: "Scrape off all the rough bark of the infected trees, and make them perfectly clean and smooth early in the spring; then rub the trunk and limbs with a stiff brush wet with a solution of potash; after which remove the sods and earth around the bottom of the trunk, and with the scraper, brush, and alkaline liquor cleanse that part as far as the roots can conveniently be uncovered. The earth and sods should immediately be carried away, fresh loam should be placed around the roots, and all cracks and wounds should be filled with grafting cement or clay mortar. Small limbs and extremities of branches, if infected, and beyond reach of the applications, should be cut off and burned." He further observes, in reference to remedial measures necessary to counteract the injury done to plants generally by the different species of Aphides, that "solutions of soap, or a mixture of soap-suds and tobacco water, used warm and applied with a watering-pot or with a garden-engine, may be employed for the destruction of these insects. It is said that hot water may also be employed for the same purpose with safety and success. The water, tobacco-tea, or suds should be thrown upon the plants with considerable force, and if they are of the cabbage or lettuce kind, or other plants whose leaves are to be used as food, they should subsequently be drenched thoroughly with pure water. Lice on the extremities of branches may be killed by bending over the branches and holding them for several minutes in warm and strong soap-suds. Lice multiply much faster, and are more injurious to plants, in a dry than in a wet atmosphere; hence in green-houses, attention should be paid to keep the air sufficiently moist; and the lice are readily killed by fumigations with tobacco or with sulphur. To destroy subterranean lice on the roots of plants I have found that watering with salt-water was useful, if the plants were hardy; but tender herbaceous plants cannot be treated in this way, but may sometimes be revived by frequent watering with soap-suds."

The species of this family are greatly subject to the attacks of other insects; the larvæ of the *Hemerobiidæ*, the *Coccinellæ*, and the larvæ of various species of *Syrphidæ* feed upon them, and destroy vast numbers; they are also infested by minute parasitic Hymenoptera belonging to the families *Cynipidæ*, *Ichneumonidæ*, &c. In the papers of Mr. F. Walker, F.L.S., and in his Catalogue of the species in the British Museum, much information on the British Aphides is given.

APHIDIPHAGI. The name of a family of Coleopterous insects, which are for the most part of a hemispherical form, and compose the genus COCCINELLA (or Lady-birds).

APHODIADÆ. A family of minute Lamellicorn beetles, extremely abundant in this and other temperate countries, especially during the spring months, swarming in the dung of the larger herbivorous animals, or hovering over it as soon as it is dropped. The body is of an oblong or oval shape, rounded at the extremity, with the abdomen entirely concealed by the elytra: they are nearly allied to the Scarabæidæ, both in their antennæ, organs of the mouth, and legs, but the body is more elongated.

APHRODITA, or SEA-MOUSE. A small marine annelide, known on our coasts as the Sea-mouse. Its figure is oval and aculeated; and it is covered with a large quantity of silky hairs of a very bright metallic lustre, the colours of which vary with the play of the light. On the back are two rows of large membranous scales, which somewhat resemble the elytra of insects. In many species the lateral setæ or bristles exhibit a beautiful structure, being barbed on each side of the tips, and each of these barbed setæ being inclosed in a smooth horny sheath. It not unfrequently happens that a large number of Aphroditæ are thrown up on the British shores after a gale of wind.

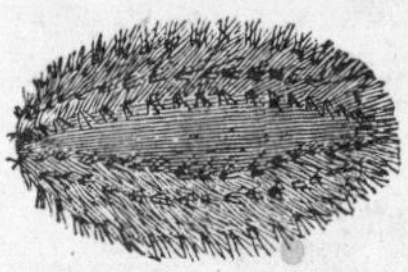

APHROPHORA. A genus of Homopterous insects which in the larva state live on plants enveloped in a saliva-like mass; whence their popular name of *Cuckoo-spits:* the insects in their perfect state are named from their leaping powers, *Frog-hoppers.* [See CICADA.]

APIDÆ. An extensive family of Bees, which may be classed under three heads; namely, 1. Social bees; 2. Solitary working bees; 3. Cuckoo-like parasitic bees. The insects composing this family are distinguished by having the mentum long, with the labium at its extremity, forming an elongated slender seta, with two small lateral filaments, and forming with the maxillæ an elongated proboscis, capable of being porrected in front of the head when in action, or folded up beneath it and the breast when at rest. The antennæ are often elbowed, the basal joint being long. Following the arrangement compiled by Mr. Westwood, we find the APIDÆ are divided into five sub-families:—

1. PANURGIDÆ, consisting of insects nearly allied to the Andrenidæ in the labium being shorter than the mentum, and the structure of the labial palpi, which are composed of continuous linear joints, the two basal ones not being so much elongated as in the following sub-families. The maxillary palpi are six-jointed; the upper lip is short; and the females are destitute of a pollen brush on the under surface of the abdomen. They are, however, furnished with a pollen plate on each side of the metathorax, and another on the posterior femora: the hind legs have also pollen brushes. Nothing is known of their nidification; but Latreille observes that the perfect insects of the genus *Panurgus* are attached to semi-flosculous flowers.

2. DENUDATÆ, or MELECTIDÆ. The insects composing the second sub-family (as well as those of all the following sub-families of bees) have the labial palpi formed of two very long, flattened, scaly basal joints, and two minute apical ones. The abdomen is not provided with a ventral pollen brush, neither do these insects possess any pollen plates, their bodies being in effect naked, whence they are supposed to be parasites. Some of the species resemble small wasps in their colours, whilst in others some parts of their bodies are clothed with small patches of very short hairs. From their evidently parasitic habits they have been termed Cuckoo-bees.

3. LONGILABRES, or MEGACHILIDÆ. The third sub-family of the Apidæ; composed of insects distinguished by the large oblong form of the upper lip, and strong mandibles: the maxillary palpi are but slightly developed; while on the contrary, the labial are very long, with the two last joints obliquely inserted. Nearly all the genera are polliniferous, the pollen brush being very large, and covering the under side of the abdomen: they are, however, destitute of pollen plates. "From their respective economy, they have been termed Mason and Upholsterer bees; the former building their nests of fine moistened earth, whilst the Upholsterers employ i.. the construction of their cells portions of leaves which they have cut from various plants by means of their powerful jaws, which are employed like a pair of scissors." Some of the species of the genus *Osmia* construct their nests of minute grains of sand, cemented together with a glutinous secretion, and which are placed by the insects on the angle of a wall, the crevices between bricks, &c. The genus *Megachile* comprises the Leaf-cutting and some other bees. These form their nests in the trunks of decayed trees, and in old rotten palings. They are lined with pieces of leaves, of a circular form, which the insects have most dexterously clipped off, and afterwards adjusted together so admirably, that, although not covered with any coating of gum, &c., they are honey-tight.

4. SCOPULIPEDES. This sub-family derives its name from the very thick coating of hairs upon the hind legs of the females, which constitute the pollen brushes. The wings have commonly three perfect submarginal cells; the third joint of the antennæ is often long and clavate, and the mouth is occasionally very considerably developed. Notwithstanding the shortness of the wings, and the robustness of the body, these insects fly with great strength and rapidity, making a loud humming noise. They nidificate in the crevices of old walls or in the ground, preferring banks exposed to the sun.

"We are indebted to Reaumur," as this gentleman observes, "for the history of the Carpenter bees, *Xylocopa*, a genus containing the largest species of the family, all of which are exotic. Their wings are often black, with a fine purple or violet gloss, and some of the species are richly coloured. The females of *Xylocopa violacea* appear in the spring, and select posts, palings, espaliers, &c. in gardens, in which they construct their burrows, from twelve to fifteen inches in length, and rather more than half an inch in diameter; the top and bottom of the tunnel is curved, having a passage at each end. When completed, they deposit an egg at the bottom, with a proper supply of pollen paste; the whole is then covered with a layer of agglutinated sawdust, formed during the construction of the burrow: the layer thus formed serves not only as the roof of one cell, but as the floor of another which is placed immediately above it. They thus proceed till about a dozen cells are formed. When the larvæ are full grown, they assume the pupa state, head downward, so as to allow the lowermost and oldest to make its way out of the bottom of the burrow as soon as it becomes winged, and which consequently takes place earlier than in those which occupy the upper cells."

5. SOCIALES. The fifth and last sub-family of the Apidæ. "Here, dependent upon their social habits, we find each species composed of three kinds of individuals; viz. males, females, and neuters, or workers. In addition to their gregarious habits, the circumstances of the larvæ being fed from time to time by the worker bees, and the cells being generally of an hexagonal form, they are also distinguished by their peculiar habit of secreting wax for the manufacture of the cells of their nests. In these insects, the outside of the posterior dilated tibiæ is smooth, and hollowed in the neuters into a shining plate, for the reception and carrying of pollen, which has been accumulated by means of the pollen brushes upon the basal joint of the tarsi of this pair of legs. The maxillary palpi are minute and exarticulate. These bees have the body covered with thick hairs."

"The Humble bees, composing the genus *Bombus*, are at once known by their large and very hairy bodies: they are the largest species of the Mellifera found in England; and they are often of a black colour, with bands of yellow or orange. They form societies consisting of about fifty or sixty individuals, occasionally, however, amounting to two or three hundred. They construct their dwellings under ground, in meadows, pastures, or hedge-rows, generally employing moss for this purpose. Their union, however, lasts only till the cold weather kills the great mass of the inhabitants, a few impregnated females alone surviving to become the foundresses of fresh colonies at the commencement of the following spring. The neuters are late in their appearance, being produced from eggs produced by these foundress bees; and it is not till autumn that the males appear. Unlike the hive-bees, the females take their share in the labours of the community, and they are accordingly furnished with two peculiar organs possessed by the neuters, of which the queen of the hive is destitute, although the neuters of the latter insect possess them; namely, the dense fringe of hairs surrounding the pollen-plate of the posterior tibiæ, and the dilated base of the first tarsal joint. The economy of the humble-bee also, unlike that of the hive, admits of the presence of numerous females in the same nest. The species of Bombus are very difficult to determine, from the colours of the hairs being very liable to fade. It is essential, therefore, to trace the insects from their first leaving the nest."

The Hive-bee, and some other analogous species (forming the second section of the *Sociales*), have the basal joint of the posterior tarsi striated, and the posterior tibiæ have no spurs at the extremity, a character not to be found in any other Hymenopterous group. * * * The principal species of bees kept for domestic purposes are the following:—*Apis mellifica* (Linn.), or the common hive-bee of Europe, and which has also been introduced into the U. S. of America and into New Zealand; *Apis ligustica* (Spinola), kept in some parts of Italy; *Apis fasciata* (Lat.), in Egypt and some parts of Asia Minor; *Apis unicolor* (Fab.), in Madagascar; *Apis Indica* (Linn.), at Bengal; *Apis Adansonii* (Latr.), at Senegal. Lacordaire also observed hives of an undescribed species of Chili; and the Horticultural Society of London, in 1825, received a hive of bees from New Holland, differing materially from the bees of Europe, "being infinitely smaller and wholly without stings." An interesting work on the British Bees, by Mr. Frederick Smith, may be shortly expected: in the "Zoologist," &c. he has published much on the subject.

APION. An extensive genus of Coleopterous insects, deriving the name from their pear-shaped body. The grubs of many kinds of Apions destroy the seeds of plants. In Europe they do much mischief to clover in this way; but in America the species are more numerous and more destructive. *Apion Sayii* is a minute black species, not more than one tenth of an inch long, exclusive of the slender sharp-pointed snout. Its grubs live in the pods of the common wild indigo bush, *Baptisia tinctoria*, devouring the seeds. A smaller kind, somewhat like it, inhabits the pods and eats the seeds of the locust-tree, or *Robinia pseudacacia*.—*Harris.*

APION FRUMENTARIUM.

APLYSIA. A genus of Tectibranchiate Mollusca, of which several species are known. The body of the animal consists of a soft fleshy mass; it has four flattened tentacular appendages; the mouth in the form of a vertical fissure, with two lateral labial plates, and a cordiform tongue beset with denticles; branchiæ covered by a sort of operculum; and shell wanting. From the borders of the mantle is poured out abundantly a deep purple liquor, with which the animal colours the water around to a considerable distance, when it perceives any danger. The *Aplysia depilans*, or Depilatory Aplysia, is found in the European seas adhering to rocks: it is extremely fetid, and it was long supposed that the acrid humour which it exuded occasioned the loss of the hair. Its digestive apparatus consists of a membranous crop, of enormous size, which leads into a muscular gizzard, furnished with pyramidal cartilaginous teeth; and a third stomach beset with pointed hooks, besides a fourth sacculus. Its general colour is blackish, with grey or brown blotches, and tinged with purple. The ova is laid in long glairy entangled filaments, as slender as threads.

APODES or APODA. An order of fishes characterised by Linnæus as being composed of all those which are destitute of ventral fins. According to Cuvier's system, however, they must not only want ventral fins, but be likewise malacopterygious. Of this kind a good and familiar example is seen in the common Eel.

APOLLO [BUTTERFLY]. [See PARNASSIUS.]

APOSURÆ. The name given to a section of the Nocturnal Lepidoptera, differing from all the rest of the order in the caterpillars being destitute of any anal feet, the extremity of the body terminating in a point, which in many is forked, or furnished with two long articulated appendages, forming a kind of tail.

APPLE-MOTH. [See TORTRIX.]

ASPIDOPHORUS. The Armed Bullhead or Pogge. [See BULLHEAD.]

APTENODYTES. The generic appellation of the curiously-formed palmiped birds, known by the name of *Penguins*, a more general and detailed account of which will be found under the letter P. In this place we shall merely make an extract from Capt. Sir J. C. Ross's Voyage to the Antarctic Regions, where he speaks of the Great Penguins: "These enormous birds varied in weight from sixty to seventy-five pounds. The largest was killed by the Terror's people, and weighed seventy-eight pounds. They are remarkably stupid, and you are able to approach them so near as to allow you to strike them on the head with a bludgeon, and sometimes, if knocked off the ice into the water, they will almost immediately leap upon it again as if to attack you, but without the smallest means either of offence or defence. They were first discovered during Capt. Cook's voyage to these regions, and the beautiful unpublished drawing of Forster the naturalist has supplied the only figures and accounts which have been given to the public, both by British and foreign writers on natural history. Mr. G. R. Gray has therefore named it in the zoology of our voyage, *Aptenodytes Forsteri*, of which we were fortunate in bringing the first perfect specimens to England. Some of these were preserved entire in casks of strong pickle, that the physiologist and comparative anatomist might have an opportunity of thoroughly examining the structure of this wonderful creature. Its principal food consists of various species of *cancri* and other crustaceous animals; and in its stomach we frequently found from two to ten pounds weight of pebbles, consisting of granite, quartz, and trappean rocks. Its capture afforded great amusement to our people, for when alarmed and endeavouring to escape, it makes its way over deep snow faster than they could follow it; by lying down on its belly and impelling itself by its powerful feet, it slides along upon the surface of the snow at a great pace, steadying itself by extending its fin-like wings, which alternately touch the ground on the side opposite to the propelling leg."

In No. IV. of the Appendix to the work above quoted, (the Geology of the Southern Islands, by R. Mc Cormick, Esq.,) the writer observes: "As I had no opportunity of landing for specimens, I was in the habit of examining the stomachs of most of the birds which I shot and preserved for the Government Collection; and found the Penguins my best geological collectors, for their crops were frequently filled with pebbles; more especially the large species, *Aptenodytes antarctica*. In one of these individuals I found upwards of a pound of small fragments of rocks; comprising basalt, greenstone, porphyry, granite, vesicular lava, quartz, scoriæ, and pumice; but none of them ever brought me a vestige of aqueous rocks,—all were volcanic,—and such the appearance of the Antarctic lands, even at a distance, would proclaim them to be. We saw three

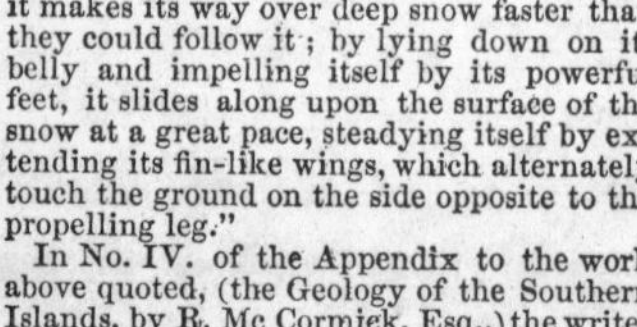

ANTARCTIC PENGUIN.
(APTENODYTES ANTARCTICA.)

species of Penguins within the Antarctic circle. The larger kind, 'Aptenodytes antarctica,' attains a great size. I preserved one, weighing seventy-five pounds. It is a scarce bird, generally met with singly; and I have never seen more than two or three together; whilst the two smaller species congregate in vast numbers. I know not to what cause we can assign this very remarkable paucity of individuals in the larger species."

APTERA. An order of the Linnæan class *Insectæ;* characterized, as the term implies, by having no wings in either sex. It includes the modern orders *Crustacea*, *Arachnida*, and *Myriapoda*.

APTERYX. A bird which in form somewhat resembles a Penguin, and stands about two feet in height. The beak is very long, slender, marked on each side with a longitudinal groove, and furnished with a membrane at its base. Its wings are simple rudiments; a mere stump, terminated by a hook. It has no abdominal air cells, nor are any of its bones hollow. The feathers have no accessory plume, but fall loosely, like those of the emu, and their shafts are prolonged considerably beyond the base.

WINGLESS EMU.—(APTERYX AUSTRALIS.)

The feet have a short and elevated hind-toe, the claw of which is alone externally visible. The eye is small, and a number of bristle-like hairs surround the mouth. Its colour is deep brown; its time of action nocturnal; and it subsists on insects. It runs with rapidity, the limbs are extremely powerful, and it defends itself vigorously with its feet. This bird is chiefly met with in the southern parts of the interior of New Zealand. When chased, it takes refuge in the clefts of rocks, hollow trees, or in deep holes which it excavates in the ground; and it runs with great swiftness, with its head elevated like the ostrich. The natives value it greatly for the sake of its skin, which, prepared with the feathers on, they make into dresses. The name given to this bird by the New Zealanders is *Kiwi*. A second species of this curious genus has been lately received by Mr. Gould from the South Seas.

APUS. A genus of small Crustaceous animals which inhabit ditches, lakes, and standing waters, generally in innumerable quantities. They often swarm in myriads, and, indeed, have been known to be carried up by violent storms of wind, and scattered over the land; hence they often appear suddenly in puddles of rain water where none have been previously, especially in the spring and early in summer. They swim well on the back, and when they burrow in the sand they raise their tails in the

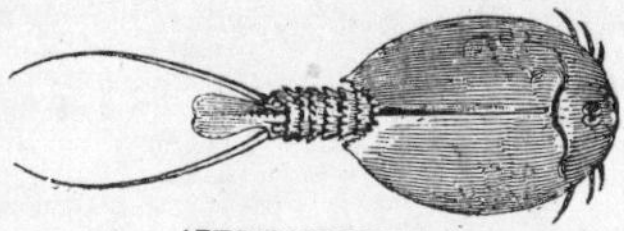

APUS PRODUCTUS

water. Their food principally consists of tadpoles. When first hatched they have only one eye, four oar-like legs, with whorls of hairs, the second pair being the largest: the body has then no tail, and the shell only covers the front half of the body: the other organs are gradually developed during succeeding moultings. These creatures are the common food of the Wagtails. We give the species *Apus productus* as an example.

AQUILA. [See EAGLE.]

ARACHNIDA. A class of Articulated animals, including Spiders, Mites, and Scorpions, all ranked by Linnæus under Insects; but though having a great analogy with them, and being equally fitted to live in the air, are distinguished from them by their number of limbs, their internal structure, and habits. All the Arachnida are destitute of antennæ, and have the head united with the thorax: they have generally eight legs, though some species have six, and others ten; they have no wings; most of them breathe by means of air-sacs, instead of by prolonged tracheæ; and in the greater part there is a complete circulatory system. Most of the Arachnida are carnivorous, and are furnished with appropriate organs for their predatory life; but in general they confine themselves to sucking the juices of insects; and in order to enable them to capture and subdue animals otherwise capable of effectual resistance, Nature has furnished them with a poisonous apparatus. [See SPIDER.]

ARACARI. (*Pteroglossus.*) A genus of birds, which, like the Toucans breed in the hollows of decayed trees, which they enlarge and render commodious by means of the beak; and it is from this habit that the Brazilians give them the name of *Tacataca*, in imitation of the sound made by clipping the decayed wood. We may here mention two species described and figured in Mr. Gould's truly elegant monograph of the Toucans.

ARACARI TOUCAN, (*Pteroglossus pluricinctus.*) This bird, as depicted and described by Mr. Gould, is twenty inches in length, of which the bill is four inches and a half; a broad band of black advances from the nostrils along the whole of the culmen, and forms a narrow belt down the sides of the upper mandible at its base; the elevated basal margin of the bill is yellow; the sides of the upper mandible beautiful orange-yellow, fading into yellowish white towards the tip; under mandible wholly black, with a yellow basal ridge; head, neck, and chest black; the whole of the upper surface, except the rump, which is scarlet, dark olive green; breast marked with two broad bands of black, the upper separated from the throat by an intervening space of yellow dashed with red; a similar but broader space separates the two bands of black, the lower of which is bounded by scarlet, advancing as far as the thighs, which are brownish olive; under the tail coverts light yellow; naked space round the eyes; tarsi and feet dark lead-colour. It is a native of Brazil.

The CURL-CRESTED ARACARI (*Pteroglossus ulocomus*), is one of the most rare and beautiful of its tribe. Its length from the tip of the bill to the end of its tail is eighteen inches: the crown of the head is covered with an elegant crest of curled feathers without barbs, which are of an intense glossy black, but as they approach the occiput they become straight, narrow, and spatulate; the feathers on the cheeks are of a yellowish white colour, tipped with black; the back of the head and upper tail coverts are of a deep blood-red; the breast is a delicate yellow, with slight crescent-shaped bars of red; the back, tail, and thighs are olive green; the quills brown, the tarsi lead-coloured: the beak of this species is lengthened, both mandibles being edged with thickly-set white seratures; the upper has an orange-coloured culmen, bordered by a stripe of dull blue extending nearly to the tip, below which, the sides of the mandibles are fine orange-red; the under mandibles is straw colour, becoming orange at the tip, and a narrow band of rich chesnut encircles both mandibles at the base. During life the colouring of the bills is generally very vivid, but after death the bright hues fade, so as oftentimes to become nearly obsolete.

ARACHNOTHERES, or SPIDER-CATCHERS. Small birds, very similar to the Sun-birds in respect to their long, arcuated beak: they inhabit the Indian Archipelago, and live on spiders.

ARANEA. [See SPIDER.]

ARCA, or ARK SHELL. The *Arcadœ* a family of Bivalve Shells, found in the Atlantic and Pacific Oceans, the Mediterranean, &c., are distinguished by their great number of teeth, resembling those of a fine saw, and forming either a straight or curved continuous line. They bury in the sand near the coasts, and are also sometimes found attached to rocks, coral, &c. The *Arca* is nearly equivalve, inequilateral, heart-shaped, valves ribbed, and in some species gaping at the lower part. A few have one valve larger than the other; and many have a velvety or shelly epidermis, frequently ending in a deep fringe.

ARCHEGOSAURUS. [See APATEON in SUPPLEMENT.]

ARCHER-FISH. [See TOXOTUS.]

ARCHES. A name given to Moths of the genera *Polia* and *Xylophasia*.

ARCTIA CAJA. There are few more striking Insects among the *night-flying Lepidoptera* than *Arctiœ*, or Tiger Moths. The one we have here selected is well known and abundant. It measures from two and a half to three inches in the expanse of the fore wings, which are of a rich brown colour, with numerous irregular spots and streaks of cream white; the hind wings bright red, with blue-black spots; the thorax brown, with a red neck-band, and the abdomen red, with blue-black bars. The insects belonging to this genus are observed to vary consider-

TIGER MOTH.—(ARCTIA CAJA.)

ably in their markings, and the present species is no exception to the general rule; some having the brown and blue-black portions more or less obliterated, whilst in others they are sometimes almost entirely predominant. The Caterpillar is dark brown, and very hairy, the hairs on the back dusky, and those on the neck and sides reddish, the head black: its food is nettles, chickweed, lettuce, strawberries, &c. When full fed it spins itself a web, wherein, at the latter end of April, it changes to the Chrysalis state; and the Moth appears about the end of June or beginning of July.

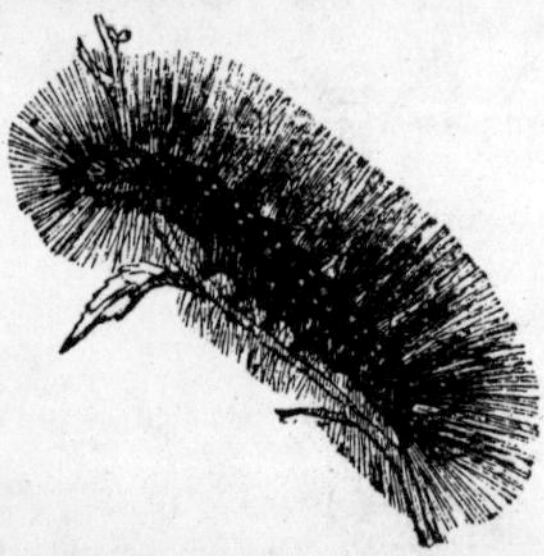

CATERPILLAR OF ARCTIA CAJA

ARCTIC FOX. (*Vulpes lagopus.*) A small species of Fox, inhabiting the high northern latitudes, and justly celebrated for the beauty and fineness of its fur. [See Fox.]

ARCTIIDÆ. A family of Lepidopterous insects, belonging to the general section HETEROCERA, comprising those species which have the wings deflexed in repose, the posterior pair not extending beyond the costa of the anterior. The antennæ of the males are strongly serrated; the spiral tongue is either very small, or obsolete; and the labial palpi are generally short and obtuse at the tip. The caterpillars vary much: in some species they are thickly hairy; some are furnished with long fascicles of hairs; and some are naked, but variously tubercled. They feed upon the external parts of plants, and enclose themselves in cocoons when about to undergo their transformations. The types of the family are distinguished by their larvæ being very thickly clothed with long hairs, whence they have obtained the name of "woolly bears." Such are especially the larvæ of the various species of Tiger Moths, and others nearly allied to them, which are well known, and considered as being amongst the most beautiful of all the species of Moths; their fore wings are ornamented with white, brown, or black, and the hind wings red, with black or blue markings. Some of these caterpillars are extremely destructive, particularly to fruit trees and hedges. Great alarm has been created at times when they were particularly abundant; and, indeed, their polyphagous habits on such occasions may justly be dreaded. The larvæ of some species are furnished, in addition to the long slender hairs all over the body, with several short, thick, truncated tufts of hair on the back as well as at the sides; the majority of these produce species not materially differing in the sexes; but some, forming the genus *Orgyia*, have females with the smallest rudiments of wings, and large swollen abdomens, and which are exceedingly sluggish in their habits, whilst the males are constantly on the wing, flitting about in the hottest weather of autumn. The family likewise comprises several other genera differing widely in the appearance of the sexes, or anomalous as respects their transformations.

ARCTOCEPHALUS URSINUS. The Ursal; a species of Seal, from the north of the Pacific Ocean. It is eight feet long, has no mane, and varies in colour from brown to whitish. [See SEAL.]

ARCTOMYS. [See MARMOT.]

ARCUATA. A section of the genus *Cancer*, of which the true Crabs are the type. [See CRAB.]

ARDEA. The Heron [which see].

ARDEIDÆ. A very extensive family of birds, formed for wading, and generally seeking their food on the margins of rivers and lakes, and in marshes, where they obtain fish, reptiles, and even small mammalia. They are characterized by having very long legs, with a strong, straight, pointed, and compressed bill; in most species finely toothed; the upper mandible usually notched towards the tip; a furrow passing from the nostrils, which are linear, to the apex. They in general build and breed in societies, but always wander alone in search of food, and after the breeding season lead a solitary existence. They have ample wings, and many of them are adorned with elegant plumes and crests. [See HERON, STORK, &c.]

ARENICOLA. A genus of *Dorsibranchiata*, or Cuvier's second order of the class ANNELIDA. The gills are of an arborescent form, on the rings of the middle part of the body. The best known species (*Arenicola marina*) is common on our coasts, where the fishermen, who dig for it as bait, know it by the name of the Lob-worm. It is almost a foot long; the body is of a reddish colour; and on being touched, exudes a quantity of yellow fluid. The animal bores for itself a passage through the sand, and secures the sides of the passage from falling in by applying to them a glutinous cement, which unites the particles of sand into a kind of wall or coating. This covering does not adhere to the body, but forms a detached tube, within which the animal moves with perfect freedom, and which it leaves behind it as it progressively advances; so that the passage is kept pervious throughout its whole length, by means of the lining, which may not inaptly be compared to the brickwork of the shaft of a mine or tunnel.

ARENICOLI. The name given to a section of beetles which live in dung, and form deep burrows in the earth. The elytra entirely cover the abdomen; the mandibles are horny, exposed, and curved; the terminal lobe of the maxillæ is generally straight; and the antennæ are ten or eleven-jointed. They fly about in the twilight after sunset, and counterfeit death when alarmed.

ARGALI. A species of wild Sheep, found on the mountains of Siberia and Kamtschatka. It so closely resembles the MOUFFLON [which see] as to be regarded by many naturalists as the same species.

ARGENTINE. (*Argentina sphyræna.*) A genus of Malacopterygious fishes belonging to the *Salmonidæ;* the mouth of which is small and toothless; the tongue is furnished with strong hooked teeth; and the digestive organs resemble those of the Trout. The well-known species *Argentina sphyræna* is caught in the Mediterranean, and is common in the markets of Rome: it has also, though very rarely, been caught on the British coast. It is about two inches and a half in length; the eyes are large, and the irides silvery; the lower jaw much sloped; the teeth small; the body compressed, and of an equal depth almost to the anal fin; and the tail forked. The back is of a dusky green; the sides and covers of the gills appear as if overlaid with silver: on each side of the belly is a row of circular punctures, and above them another which terminates near the vent. The air-bladder is thick, and loaded with *nacre*, the substance used in making artificial pearls.

ARGONAUT, or PAPER-NAUTILUS. A curious molluscous animal, the shell of which is peculiarly white and delicate; not chambered, as in the true Nautilus, but possessing one spiral cavity, into which the animal can withdraw itself entirely. It has eight arms, two of which expand into wide membranous flaps; and as the animal floats on the surface of the sea, the expanded membranes are spread over the sides of the shell, where, meeting along its keel or edge, they are said to be held in close contact by a double row of suckers, and thus completely inclose it. Such being the structure and action of the Argonauta, it is not surprising that it has had the reputation, from very early times, of using its arms as oars, and spreading these expanded membranes as sails, so as to be wafted along by the wind. [See NAUTILUS.]

The nature and habits of the Argonauta having long been a subject of much controversy, a lady (Madame Jeannette Power) made a series of interesting experiments, in 1836, the result of which she laid before the Academy at Catania. In order to arrive at her conclusions, she had cages constructed, and placed in a shallow part of the sea, near the citadel of Messina, and in these cages she inclosed several of the living animals, which she kept supplied with small molluscs, their natural food. The result of her observations went to prove that the animal is in the habit of sailing on the water, using its dilated tentacula as sails, the remainder as oars, and aiding its movements by means of a kind of proboscis which it employs as a helm. The sail, when spread out, presents a silvery surface, speckled with concentric circles of spots, with a black spot in the middle, surrounded with a beautiful gold colour. The animal is not attached to the shell, but, when under water, it adheres firmly to it by its sail-arms. The shell, which is remarkably brittle when exposed to the air, is quite pliable in water, and thus escapes the destruction to which so thin and tender a fabric would otherwise be liable. The animal at the approach of any object takes in its tentacula, wraps its sails over the shell, and descends, blackening the water at the same time, if hard pressed, by a discharge of inky fluid, to cover its escape.

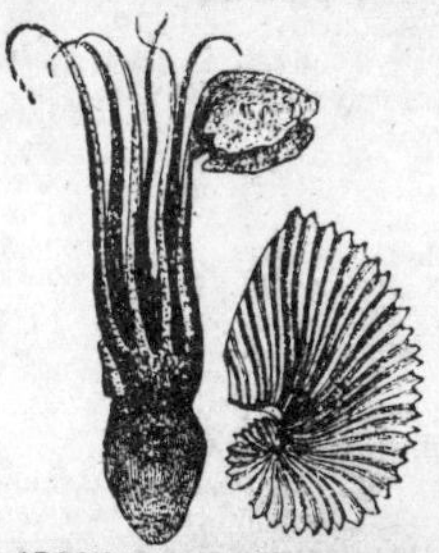

ARGONAUT AND SHELL

ARGULUS. A genus of Crustaceous animals, belonging to the PŒCILOPODA. The best known species, *Argulus foliaceus*, is found in this country. This aquatic parasite attaches itself to the young of Frogs, Sticklebacks, &c., and sucks their blood: it is also found upon the Perch, Pike, Carp, and Trout. The body is flattened; of a greenish-yellow colour; less than a quarter of an inch long; and is divided into five somewhat indistinct segments along the back. The animal turns itself about in the water in a similar manner to the *Gyrini*. The eggs are oval, of a milky white colour, and are attached by gluten to stones or other hard substances; and before the Argulus arrives at the adult state it undergoes several transformations.

ARGUS-PHEASANT. (*Argus giganteus.*) This beautiful but rare bird is a native of many parts of the Indian Islands. The male measures five feet six or eight inches from the beak to the tip of the tail; and the whole of the plumage is remarkable for variety and elegance. The wings consist of very large feathers, nearly three feet long, the outer webs being adorned with a row of large eyes (*ocelli*), arranged parallel to the shaft; the tail is composed of twelve feathers,

ARGUS PHEASANT.—(ARGUS GIGANTEUS.)

the two middle ones being about four feet in length, the next scarcely two, and gradually shortening to the outer ones. The whole plumage is, indeed, so varied, that to attempt to describe it fully in our limits would be vain. Its voice is rather plaintive, and not harsh as in the peacock. It is considered a very shy bird, but one was kept alive a considerable time in the aviary of the Zoological Gardens, where the pleasing variety of its plumage and the beautifully coloured skin of its head were much admired.

ARGUS. [See PECTEN.]

ARGYNNIS. A genus of diurnal Lepidoptera. We here describe two beautiful British species of Butterflies belonging to this division.

ARGYNNIS PAPHIA, or SILVER STREAK BUTTERFLY. There are few of the Lepidoptera more abundant in the woods and meadows of the South of England than the Silver Streak, which is known to

SILVER-STREAK BUTTERFLY. — (A. PAPHIA)

delight in settling on the bramble-blossoms. In the male the wings above are fulvous, in the female virescent, with numerous longitudinal and transverse black lines and bars, and three rows of marginal black spots; anterior wings beneath, paler and less spotted; the posterior wings are greenish beneath, with four irregular narrow pale silvery-waved bands; between the two last is a series of ocelli, with a green iris and pale pupil, and on the margin is a row of green crescents: the cilia of all the wings above are fulvous and black, paler and ferruginous beneath: the body fulvous above, grayish beneath: the antennæ are brownish, with the club black.

ARGYNNIS LATHONIA, or QUEEN OF SPAIN FRITILLARY. This exceedingly beautiful species, though rare in this country, appears to be very common on

QUEEN OF SPAIN BUTTERFLY (ARGYNNIS LATHONIA.)

the Continent. The upper surface in general markings resembles that of the allied species, but it may be at once known by the beautiful and well-defined silver marks on the under surface of the lower wings. British specimens of it are much prized by the collector. Our cuts will give a very good idea of this insect, as we have figured both the upper and under sides.

ARICIA. A genus of Dorsibranchiate Annelidæ. They have neither teeth nor tentacles. The body, which is lengthened, bears two ranges of lamelliform cirrhi along the back; and the anterior feet are furnished with dentelated crests, that do not occur on the other feet.

ARMADILLO. (*Dasypus.*) A genus of mammiferous quadrupeds, belonging to the order *Edentata*, readily distinguished from all others by the singular covering with which Nature has protected them. This is a complete suit of armour; consisting of a triangular or oval plate on the top of the head, a large buckler over the shoulders, and a similar buckler over the haunches, while between these solid portions there intervenes a series of transverse bands or shelly zones, in such a manner as to accommodate this coat of mail to the various postures of the body ; the tail also is covered by a series of calcareous rings ; and the animal altogether exhibits a striking deviation from the usual structure and outward appearance of quadrupeds. Like the hedgehog, it can roll itself up into a ball, thereby offering a uniform, solid surface, impervious to the attacks of birds of prey or small quadrupeds. The interior surface of

POYOU ARMADILLO.
(DASYPUS SEXCINCTUS VAR)

the body, not covered by the shell, is clothed with coarse, scattered hairs, of which some are also seen to issue forth between the joints of the armour. The Armadillos have a rather pointed snout, long ears, short and thick limbs, and stout claws ; all of which are adapted to their habits of burrowing, which they perform with such astonishing rapidity that it is almost impossible to get at them by digging. The hunters are then obliged to smoke them out of their dens ; and as soon as they reach the surface they roll themselves up, and are easily captured. Although they abound in incredible numbers, were it not for their peculiar fecundity they would be speedily exterminated, as they are sought with great avidity on account of their flesh, which is roasted in the shell, and is regarded as a great luxury. Their food consists chiefly of succulent roots, ripe fruits, and other soft vegetable substances ; but they also greedily devour worms, small lizards, ants, and the eggs of birds which build their nests on the ground. The species are distinguished from each other, principally, by the number of bands on the trunk of the body, between the shield on the fore-shoulders and that on the rump. Don F. Azara, however, in his "Essays on the Natural History of the Quadrupeds of Paraguay," showed that the number of these bands is by no means constant in the same species, but that within certain prescribed limits this number varies according to the age and sex of the individual. Baron Cuvier, accordingly, for greater facility of definition, has divided the whole genus into five small groups, principally distinguished from one another by the number and form of their teeth and claws ; and to these sub-divisions he has applied, respectively, the names of Cachicames Apars, Encouberts, Cabassous, and Priodontes.

The CACHICAMES are those which have four toes on each foot, and seven teeth on each side in both the upper and lower jaws. —The APARS have also four toes on each foot, and nine or ten teeth on each side above and below. The Apar has only three moveable bands ; the rest of its tesselated covering being nearly inflexible : it has also the power of rolling itself into a perfect sphere, in which state it is safe from the attack of dogs ; its smooth hard covering offering a better defence than the sharp spines of the hedgehog.—The ENCOUBERTS have five toes on the fore-feet, and in addition to nine or ten teeth on each side in both jaws, have two incisor teeth in the upper.—The CABASSOUS have five toes ; but those of the fore-feet are obliquely placed, so that the thumb and index finger are small, but the middle and fourth claws are armed with immensely large trenchant claws ; on each side above and below are nine or ten teeth.—The PRIODONTES, in addition to the unequal toes and enormous claws of the Cabassous, have, on each side of both jaws, twenty-two or twenty-four small teeth. The GIANT ARMADILLO (*Dasypus gigas*) belongs to this division.

GIANT ARMADILLO.—(DASYPUS GIGAS.)

It is the largest known species of Armadillo; the body, exclusive of the tail, being sometimes three feet in length.

The PICHY (*Dasypus minutus*), as we read in Mr. Darwin's "Researches" in South America, wanders by day over the open plains, feeding on beetles, larvæ, roots, and even small snakes. It prefers a very dry soil ; and the sand-dunes near the coast, where for many months it can never taste water, is its favourite resort. The instant one was perceived, it was necessary in order to catch it, almost to tumble off one's horse ; for if the soil was soft, the animal burrowed so quickly, that its hinder quarters had almost disappeared before one could alight. The Pichy likewise often tries to escape notice by squatting close to the ground.

It is an interesting fact, fully proved by the remains of extinct species discovered by the above-named traveller, that more than one gigantic animal, protected by an armadillo-like covering, were once inhabitants of this earth, but at a period so remote as to render all attempts to ascertain their *exact* nature perfectly unavailing : much, however, has been done towards it by the aid of modern science. [See TOXODON and GLYPTODON.]

ARNEE. (*Bos Arni.*) A large and formidable quadruped, conspicuous for courage, strength, and ferocity; and closely allied to the wild ordinary Buffalo. It inhabits the high lands of Hindostan, and is remarkable for its enormous horns, which often measure from four to six feet in length. They incline outwards and backwards, and then, arching gradually towards each other as they proceed to the points, form a bold crescent: they are rough with numerous ridges and furrows. In Bengal and the neighbouring provinces this animal is known by the name of *Arna.*

ARTAMUS. A genus of birds, one of the species of which was placed by the older writers among the Shrikes. [See WOOD-SWALLOW.]

ARTICULATA. The term applied by Cuvier to a primary division of the animal kingdom. The animals composing it not only present an internal structure which is essentially different from that of the other three divisions — the VERTEBRATA, MOLLUSCA, and RADIATA — but are distinguished by external characters so definite and evident as not to be mistaken. The skeleton is not internal, as in the *Vertebrata*, but is seldom altogether absent, as in the *Mollusca.* Their entire body is divided into segments; the series of articulated rings which encircle the body supplying the place of a skeleton, and being in general hard enough to furnish the necessary resisting fulcra to the muscles of locomotion; whence they are capable of performing the several actions of walking, leaping, swimming, or flying. There are also some which are not furnished with feet, but have only soft and membranous articulated limbs, by which they can merely crawl. In some articulated animals, their ring-like appearance results merely from a certain number of transverse folds, which furrow the skin, and encircle the body; but in the greater number, the animal is enclosed in a kind of case, formed by a series of rings, so united one to another as to allow them a certain degree of movement. In most animals of this sub-kingdom, each ring in its complete state possesses a pair of nervous ganglia, united on the central line; and these ganglia are connected together by a double cord of communication, which runs along the ventral or lower surface of the body. The bulk of the body in the Articulata is made up of the muscles, by which the several segments, and their various appendages, are put in motion; and these muscles are arranged with so much regularity and exactness on the two sides of the central line, that the lateral symmetry of the Articulata is most exact. With the exception of a few of the very lowest species, all the Articulata are furnished with a distinct head, and with jaws for the prehension and reduction of the food: these jaws, however, do not open *vertically*, as in the Vertebrata, but *laterally*, and there are frequently several pairs of them, one behind the other. All the actions of the Articulata are performed with great energy; and at the time of the most rapid increase of the body, the demand for food is so great, that a short suspension of the supply proves fatal.

The members of this great division are distributed into five classes, principally founded on the organs of locomotion. 1. The ANNELIDÆ, or *Red-blooded Worms;* characterized by the presence of a distinct circulating system, and of respiratory organs; the extension of the body into numerous segments; and by the possession of a well-developed nervous system. 2. The CIRRIPEDES, which seem, as it were, to connect the Articulata with the Mollusca. The body is furnished with articulated cirrhi, arranged in pairs, while in many it is provided with a multivalve shell. 3. CRUSTACEA, or *Crabs, Lobsters,* &c. These have articulated limbs, more or less complicated, attached to the sides of the body. Their blood is white, their respiration aquatic, and among them alone, of all the Articulata, do we find a distinct auditory apparatus. They have transverse jaws; two compound eyes; and all are furnished with antennæ or articulated filaments attached to the head, of which there are generally four. 4. ARACHNIDA, or *Spiders, Mites,* &c. In common with a great number of the Crustacea, these have the head and thorax joined into a single piece with articulated limbs on each side: their mouth is armed with jaws, but they have no antennæ. 5. INSECTS; the most numerous in species of any throughout the Animal Kingdom. They are characterized by the division of the body into three distinct portions, — the head, thorax or corselet, and abdomen; by the possession of antennæ on the head; of three pairs of legs, and, in general, of one or two pairs of wings; and by their respiring by means of *tracheæ*, which are elastic vessels that receive the air by orifices termed *stigmata*, pierced in their sides, and which are distributed by minute ramifications over every part of the body.

ARVICOLA. A genus of Rodent Mammalia. [See VOLE.]

ASCARIDÆ. A family of Entozoa, living in the alimentary canal or intestines of various animals. Their bodies are elongated, like the garden worm, round, elastic, and tapering at either extremity. The mouth is provided with three lips or tubercles; but there are no bristles or setæ on the surface of the skin, and they also differ from the earth-worm in having colourless blood, and in exhibiting very slight movements during life. The species are very numerous and common, those most familiar to us being the *Ascaris lumbricoides,* infesting man; the *A. marginata,* found in the Dog; the *A. mystax,* constantly present in the stomach of the Cat; and the *A. suilla* of the Hog or Sow. The Ascaris occupying our bodies frequently gives rise to disagreeable symptoms, and uneducated people sometimes describe these creatures to be serpents, eels, &c., having no knowledge of their true nature. The *A. lumbricoides* varies from four to fourteen inches in length, and from one to four lines in circumference. The body is marked by numerous closely set rings or annulations, and from head to tail by four equidistant longitudinal lines. The worm presents a light yellow colour, the skin itself being transparent and horny (chitinous) in texture. The digestive tube exhibits different degrees of width at those parts of the canal respectively known as the œsophagus, stomach, and intestine, the latter being straight in the main, but slightly undulating; the anus forms a transverse slit near the extremity of the body. This animal is unisexual, the females being furnished with oval-shaped eggs, which, according to the calculations of Eschricht, amount to many thousands in each individual. Of late years numerous experiments have been made with the view of ascertaining how these creatures gain access to the human body; and from the researches of Kitchenmeister, Verloren, Leuckart, Richter, and other distinguished helminthologists, it would appear that we must in the first instance unconsciously swallow the minute eggs which exist in all open waters, and into which the eggs themselves must have been introduced in ways too numerous and unpleasant to be minutely particularised in this place. The young worms develope into active embryos whilst the ova are still in the water, but upon being transferred to the human stomach, it is conjectured that they immediately effect their escape from the shell, attaining maturity in the body of the individual (or *host*, as the Germans call him) who is unluckily destined to entertain their presence.

The Little Maw or Thread-worm of children will be found described under *Oxyurus*. [See SUPPLEMENT.]

ASCIDIA. An order of Acephalous Mollusks answering to the Tunicata of most zoologists. The *simple* Ascidians have the body fixed, roundish, and apparently issuing from a sheath. There are many species, most of which are inhabitants of the European seas, in high latitudes. They adhere by their base to rocks, shells, and other submarine substances; they are more or less gelatinous, and some are esculent; they contract and dilate themselves alternately, and have the power of squirting out the water they have imbibed. This power of ejecting the contents of the branchial sac is, in fact, their principal means of defence: some of the larger species are able to shoot the fluid to a height of three feet. Some of the Ascidiæ are *compound;* different individuals being united together by a common stem; but each having its own heart, respiratory apparatus, and digestive system; and each fixed on a footstalk that branches from a common creeping stem, through which a circulation takes place that connects them all. Both in the solitary and compound Ascidians, the young animal, when it first issues from the egg, has active powers of locomotion, being provided with a large tadpole-like tail, by the aid of which it is propelled through the water. [See SALPA, PYROSOMA, PHALLUSIA, in SUPPLEMENT.]

ASILUS: ASILIDÆ. A genus and family of Dipterous insects; the most common European species of which is the *Asilus crabroniformis*, an insect nearly equalling a hornet in length, but of a much more slender and pointed form; and, though of a somewhat formidable aspect, incapable of piercing with any degree of severity.

ASP. (*Coluber aspis.*) A species of venomous Serpent, often mentioned both by Greek and Roman writers (who, from the discrepancies in their accounts of it, appear to have known several noxious reptiles under this name); but most especially celebrated as the instrument chosen by Cleopatra to put an end to her existence after the defeat of Mark Antony at the battle of Actium. Naturalists now concur in the opinion that the real Asp is the serpent to which the Arabs give the name of *El Haje;* that it is of a green colour, marked obliquely with brown bands, and measures from three to five feet in length. Like the Cobra Capello of India, the Asp has the power of greatly distending the neck when irritated, and of raising itself on its tail to dart forward upon an enemy. The effects of its poison are most deadly, admitting of no remedy where amputation of the part cannot be immediately effected: but Lord Bacon asserts that its bite is the least painful of all the instruments of death, and he supposes its poison to have some affinity to opium, though less disagreeable in its operation.

ASPERGILLUM. A genus of Molluscous animals, furnished with a bivalve shell, inclosed in a tubular calcareous sheath, which is dilated or club-shaped at the lower end, and gradually lessens in diameter to the narrow aperture. The shell, which derives its name from its resemblance to the spout of a *watering-pot* (a name familiarly given to it by collectors), has the form of an elongated cone, terminating at the large end in a disc, which is pierced with a number of small orifices, and bordered by a sort of corolla or frill. By means of two small valves in the tube the water is freely admitted into the interior of the shell. The animals of this genus are borers; some bury themselves in the sand, some in stone, others in wood, and others in thick shells.

ASPIDIPHORA. The name given to a group of Branchiopodous Crustacea, distinguished by having sixty pairs of legs, all furnished on the outside, near the base, with a large oval vesicle, and of which the two anterior, much larger than the rest, resemble antennæ. A large shell, almost entirely disengaged, covers the major part of the upper side of the body. [See BRANCHIOPODA.]

ASS. (*Equus asinus.*) A well-known and most useful domestic quadruped, whose good qualities are too generally undervalued. The Ass is believed to be a descendant of the wild Ass, inhabiting the mountainous deserts of Tartary, &c. (by some naturalists called the *Onager*, and supposed to be identical with the Persian *Koulan*), and celebrated in sacred and profane history, for the fiery activity of its disposition, and the fleetness of its course. But in the state of degradation to which for so many ages successive generations have been doomed, the Ass has long since become proverbial for stolid indifference to suffering and for unconquerable obstinacy and stupidity.

From the general resemblance between the Ass and the Horse, it might naturally enough be supposed that they were very closely allied, and that one had degenerated: they are, however, perfectly distinct; there is that inseparable line drawn, that barrier between them, which Nature provides for the perfection and preservation of her productions — their mutual offspring, the mule, being incapable of reproducing its kind. — The best breed of Asses is that originally derived from the hot and dry regions of Asia; at present, perhaps, the best breed in Europe is the Spanish; and very valuable Asses are still to be had in the southern portion of the American continent, where during the existence of the Spanish dominion the breed was very carefully attended to. In truth, wherever proper attention has been paid to improve the breed by crossing the finest specimens, he is rendered nearly if not quite equal to the horse for most purposes of labour; while on hilly and precipitous roads he is decidedly better adapted from his general habits and formation. The most general colour of the Ass is a mouse-coloured grey, with a black or blackish stripe, extending along the spine to the tail, and crossed by a similar stripe over the shoulders.

The female goes with young eleven months, and seldom produces more than one foal at a time: the teeth follow the same order of appearance and renewal as those of the horse. Asses' milk has long been celebrated for its sanative qualities: invalids suffering from debility of the digestive and assimilative functions make use of it with great advantage; and to those also who are consumptive it is very generally recommended.

The WILD ASS (*Equus hemionus*), [or KOULAN, as it is called by the Persians] stands much higher on its limbs than the

WILD ASS.—(EQUUS HEMIONUS.)

common Ass; its legs are more slender, the forehead is more arched, and it is altogether more symmetrical. The mane is composed of a soft woolly dusky hair, about three or four inches long; the colour of the body is a fine silvery grey; the upper part of the face, the sides of the neck and body, being of a flaxen hue; and a broad brown stripe running down the back, from the mane to the tail, and crossing the shoulders, as in the common Ass. The Koulan inhabits parts of Central Asia, and migrates from north to south, according to the season. Its flesh is held in high esteem by the Tartars and Persians, who hunt it in preference to all kinds of game. We have alluded to the frequent mention of this animal by both sacred and profane writers of antiquity; and we may properly conclude by quoting the book of JOB, xxxix. 5—8: "Who hath sent out the wild ass free? or who hath loosed the bands of the wild ass? Whose house I have made the wilderness, and the barren land his dwellings. He scorneth the multitude of the city, neither regardeth he the crying of the driver. The range of the mountains is his pasture, and he searcheth after every green thing." [See KIANG in SUPPLEMENT.]

ASSERADOR. (The Spanish word for *Sawyer.*) The name applied in Columbia to a remarkable Lamellicorn beetle, which will be better understood by the accompanying wood-cut than by any description. The

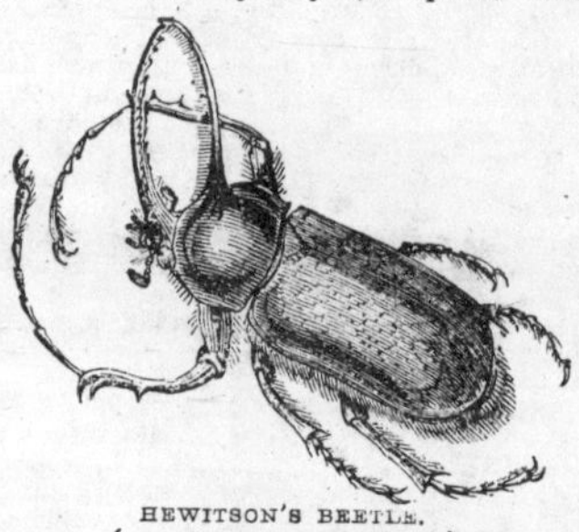

HEWITSON'S BEETLE.
(ASSERADOR HEWITSONI.)

female wants the singular horns on the head and thorax from which the species derives its local name of "The Sawyer;" it being the belief of the country people that the insect saws off the small twigs of trees by means of the friction of the two. Mr. David Dyson informed us that he found it abundantly, and in clusters, on a species of bamboo. Mr. Empson of Bath first discovered this curious insect, and published a figure of it with the name of *Asserador Hewitsoni*, and presented his unique specimen to the British Museum at a time when the insect was very rare. It has also been described by Mr. Hope as the *Golofa Porteri*, and by Erichson as the *Scarabæus Petiverii*; and we see the learned Berlin entomologist now fancies it may be only a variety of the Fabrician species, *S. ægeon.* We give this one example of what naturalists call the *synonymes* of a species, to show the utter impossibility of our attempting to give or to reconcile the different names applied to the same species by different authors.

ASTACUS. A genus of long-tailed Crustaceous animals, whose distinguishing character is derived from the antennæ, the two pairs of which are inserted in the same horizontal line. In it are included those well-known and valuable shell-fish, the Lobster (*Astacus marinus*), and the Crayfish (*Astacus fluviatilis*): the former of these has, however, by recent naturalists been regarded as the type of another genus (*Homarus*). [See LOBSTER and CRAYFISH.]

ASTERIAS. A genus of Radiated animals, of a more or less stellate form, which we find thrown up on every coast, and which are popularly known as *Star-fishes*. They are formed of a semi-transparent and gelatinous substance covered with a thin membrane; and though at first sight they often appear like a lump of inanimate tissue, on minuter inspection they are found possessed of life and motion. "Let a star-fish thus picked up," observes Mr. Rymer Jones, "be placed in some transparent pool left by the tide, within a rocky basin; watch it there, and, doubtless, soon the most incurious looker-on will find himself compelled to gaze in mute astonishment at what he sees. From the inferior surface of each ray, the creature, which before appeared so helpless and inanimate, slowly protrudes numbers of fleshy tubes, which move about in search of firm holding-places, and are soon fixed, by means of little suckers at the end of each, to the smooth surface of a neighbouring stone, or, if the star-fish has been placed in glass filled with salt water, to the inner surface of the glass, where every movement may be plainly seen. When these have laid fast hold, others appear in quick succession, and likewise are attached to the smooth surface, till at last hundreds of little legs, for such these suckers seem, are actively employed, and by their aid the creature glides along with such a gentle motion, that it seems rather to swim than crawl. Thus roused into activity we watch its movements, and perceive that it has appetites and instincts which direct its course. Place within its reach a piece of tainted fish, or other sea-side carrion, and it soon will find it out, and, clasping it between its rays, will swallow and digest it in its ample stomach."

STAR-FISH.—(ASTERIAS.)

ASTERACANTHION. [See SUPPL.]

ASTEROLEPIS. [See SUPPLEMENT.]

ASTRÆA. A genus of fixed Polypi, either incrusting marine bodies, as in the *Astræa rotulosa*, an inhabitant of the West Indian seas, or collected in a hemispherical mass, sometimes though rarely lobated, as in the *Astræa favosa*, commonly found in the seas of the East Indies.

ATELES. [See SPIDER MONKEY.]

ATHERICERA. The fourth section of Dipterous insects, characterised by the antennæ being only two or three-jointed, and the proboscis capable of being withdrawn into the mouth. Few of the Athericeræ are carnivorous in the perfect state. They are found, for the most part, on flowers leaves, and sometimes on human excrement.

ATHERINE. (*Atherina*.) A genus of Acanthopterygious fish, of which there are several species, varying in length from three inches to six. They are abundant on the shores of Italy and Greece, as also on the Peruvian and other coasts of South America, where they are esteemed delicious food. They are likewise taken in considerable numbers on the south-western coasts of England, especially near Southampton, where, from their similarity of appearance, they are called smelts. The Atherine is of a silvery yellow hue, somewhat transparent, and having a well-defined silvery band or stripe running along the sides, from gills to tail.

ATLANTA. (*Atlanta Peronii*.) A small transparent Molluscous animal, found in the seas of all hot climates; it occupies a most delicate shell spirally rolled on itself, having a thin and glassy operculum. The Atlanta belongs to the order *Heteropoda*: it has two tentacula, with large eyes at the base; and the foot large.

ATTAGEN. A local name for the Ptarmigan or White Grouse. [See PTARMIGAN.]

ATTIHAWMEG. [See SALMO ALBUS.]

AUK. (*Alca*.) A genus of aquatic birds of the family *Alcadæ*, consisting of several species; particularly the Great Auk, the Razor-bill, and the Little Auk. They are characterized by having very short wings, and the legs placed so far behind the centre of the body that they stand nearly erect. They are strictly sea birds, and nestle on its borders; breeding in caverns and rocky cliffs, and laying only one large egg. They obtain their food by diving, at which they are very expert; but the power of their wings is very limited; and when they proceed on foot by land, which they do with swiftness, if pursued, their motions are the most awkward imaginable. They all feed on small fishes, crustacea, vermes, mollusca, or marine vegetables.

The GREAT AUK (*Alca impennis*) is three feet long; and has a black bill, four inches

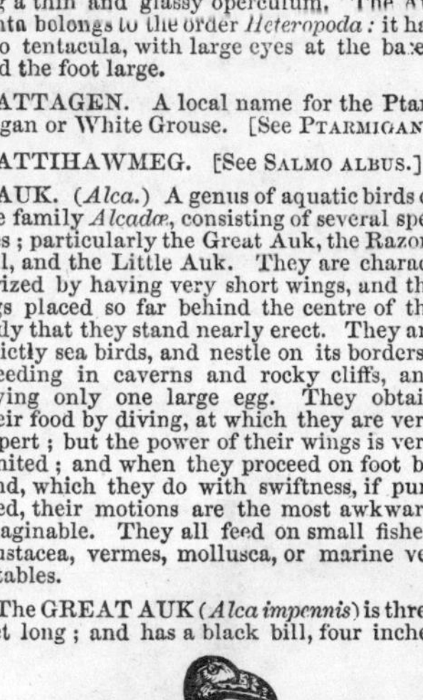
GREAT AUK.—(ALCA IMPENNIS.)

and a quarter long, both mandibles being crossed obliquely with several ridges and furrows. Two oval-shaped white spots occupy nearly the whole space between the bill and the eyes: the head, back part of the neck, and all the upper parts of the body and wings are covered with short, soft, glossy black feathers, excepting a white mark across the wings, formed by the tips of the lesser quills. The wings do not exceed more than four inches and a quarter from the tips of the longest quill-feathers to the first joint: legs black, short, and placed near the vent. This species inhabits Norway, Iceland, Greenland, and the Feroe Islands. They are, sometimes, though very rarely, met with on the northern isles of Britain, but are never observed to be at any great distance from the shore. A British-killed specimen is in the British Museum; it was formerly in the collection of Mr. Bullock.

The **RAZOR-BILL**, or Common AUK, (*Alca torda*). These birds abound in the higher northern latitudes; they are, however, widely diffused; and in England many precipitous cliffs, the Needles, &c., have a fair share of them. The Razor-bill is about eighteen inches long, and the extended wings about twenty-seven inches. They build no nests, but lay their eggs upon the bare edges of lofty rocks hanging over the sea, where they form a very grotesque appearance, from the singular order of the rows in which they sit one above another. Their [one] egg is disproportionately large, being three inches long, the colour a greenish-white irregularly marked with dark spots. Thousands of these birds are killed on the coast of Labrador, for the sake of the breast feathers, which are very warm and elastic; and incredible numbers of eggs are also collected there.

The **LITTLE AUK** (*Mergulus alle*). This is a plump round-shaped little bird, about nine inches long. The crown of the head is

LITTLE AUK.—(MERGULUS ALLE.)

flat and black; nearly all the upper parts of the plumage are of the same colour; the cheeks and under parts white; legs and toes yellowish. These birds inhabit the inhospitable shores of Greenland and Spitzbergen; but their great breeding station is said to be in the northern part of Baffin's Bay. In these dreary regions, we are told, when the ice has been broken up by storms, they watch its motion, and come down in legions to banquet on the various marine animals which lie scattered before them. It is rarely that the Little Auk is seen on our shores, and can hardly be called an occasional visitant. Like the others which have been mentioned, it only lays one egg, which is of a pale bluish-green, and is placed on the most inaccessible ledges of rocks.

Different species of this family of birds are spread over various parts of the northern world; and some of them are met with on almost all the rocky cliffs on the coasts of Britain and Ireland. The female deposits her single egg upon the bare mould, in a hole dug out and formed in the ground by herself and mate, for that purpose, or in one that they find ready made by the rabbits, which they easily dislodge. They assemble early in April, prepare for the business of incubation in May, and hatch their young in the beginning of July; from which time till the middle of August they are employed in nurturing and rearing their brood: they then regularly depart for the southern coasts of France, Spain, and other regions more suited to their exigencies, where they pass the remainder of the year.

AULOSTOMA. A genus of Acanthopterygian fishes, closely allied to *Fistularia*, from which they are distinguished by having

TRUMPET-FISH.—(AULOSTOMA CHINENSIS.)

numerous free spines before the dorsal fin: the jaws are toothless; the tube of the muzzle is shorter, wider, and more compressed than in *Fistularia:* the body is very scaly; the tail is short and slender, ending in a common fin: the air-bladder is also larger than in the true Pipe-fishes. The best known species is a native of the Eastern seas.

AURICULA. A genus of Molluscous animals, having a head furnished with two tentacula, and eyes at their base; foot short and narrow. They inhabit a shell having a fancied resemblance to the ears of certain

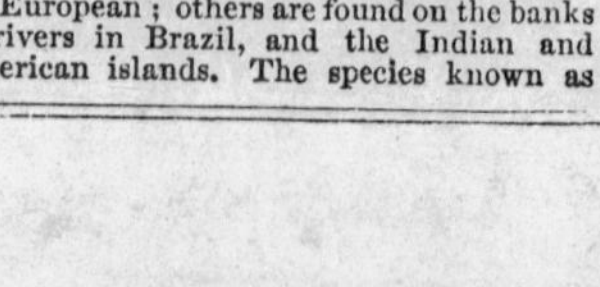

MIDAS' EAR.—(AURICULA MIDÆ.)

animals; hence the name. Several species are European; others are found on the banks of rivers in Brazil, and the Indian and American islands. The species known as

Auricula Midæ, or *Midas' Ear*, is a handsome shell, native of the East Indies: its figure is oval or oblong; the mouth longitudinal, with a reflected lip.

AUXIS. A fish belonging to the *Scomberidæ* or Mackerel family, found in the Mediterranean. It is of a fine blue black on the back, with oblique blackish lines, and the flesh deep red.

AVES. [BIRDS.] The name of a class of vertebrated animals, characterized by oviparous generation, a covering of feathers, and by their anterior extremities being organized as wings, and mostly used for flight. There are six orders, which are distinguished by certain characteristics of the posterior extremities or feet. The FIRST order is termed *Raptores* or *Accipitres;* they have large feet, with three toes before and one behind, all armed with long, strong, sharp, curved, and prehensile talons; this structure is associated with a strong, curved, and sharp-pointed beak; a very muscular body; and capability of rapid and long-continued flight. These are Birds of Prey; the principal of them being Vultures, Eagles, Hawks, Kites, Buzzards, and Owls. The SECOND order is termed *Insessores*, or Perching Birds. The feet of these are all formed for perching, and their power of grasping is very great; the toes are slender, flexible, of moderate length, and provided with long, pointed, and slightly curved claws. It includes the Thrushes, Nightingales, and all the sweetest songsters of our groves; with the Redbreasts, the Sparrows, Larks, Swallows, Crows, Kingfishers, Birds of Paradise, and Humming-birds. From including the smaller tribes of Birds, the term *Passeres* is also given to this order. —The THIRD order is termed *Scansores*, or Climbers. These have the power of throwing one of the fore toes back at pleasure; a construction which enables them to climb the perpendicular trunks of a tree. Of this order the Parrot tribe and the Woodpeckers are the principal members.—The FOURTH order is termed *Rasores*, or Gallinaceous Birds. It is characterized by the hinder toe being raised above the level of the three anterior ones; this reduces the power of perching; but the front toes are united by a slight membrane, and are strong, straight, and terminated by robust, obtuse claws, adapted for scratching up the soil, and for running along the ground; for which purpose they are also furnished with very strong, muscular legs. These birds have the head small in proportion to the body; and the bill generally short, with the upper mandible somewhat curved. In this order are comprised the Peacock, the Turkey, the common Cock and Hen, Partridges, Pheasants, Pigeons, &c.—The FIFTH order is termed *Grallatores*, or Waders. To enable them to wade and seek their food in water, along the margins of rivers, lakes, and estuaries, the birds belonging to this order have long and slender legs, and generally bare thighs. Their three front toes are more or less united at the base by a web, and the central toe is often longer and stronger than the rest; the hind toe is elevated, short, or even sometimes wanting. This order comprises the Ostriches, Cranes, Herons, Storks, Snipes, Woodcocks, Bustards, and Plovers. The SIXTH order is termed *Natatores*, *Palmipedes*, or web-footed Birds; and their whole organization is especially adapted for an aquatic life. Their legs are short, and placed behind the centre of equilibrium; their fore toes are united by a thick and strong web or membrane; and their bodies are covered with a dense layer of down, beneath the outer plumage, which is close, and rendered impervious to the water. The order comprises Swans, Ducks, and Geese; Auks, Penguins, Pelicans, Petrels, Coots, and Grebes. [See the art. BIRDS.]

AVICULA. A genus of Conchiferous Molluscs belonging to the order *Dimyaria*. It is thus defined by Cuvier:—"The shell has the valves equal, with a rectilinear hinge, and is often carried out into wings; the ligament is narrow and elongated; small dentilations often appear on the hinge, on its anterior part; and below the angle on the side near the mouth is the notch for the byssus. The anterior abductor muscle is still extremely minute." The foot of the animal is conical, worm-shaped, and rather long. Some very beautiful species of the Avicula are brought from the Indian Ocean, coast of Brazil, New Holland, the Red Sea, &c. The interior of the shell is pearly in the centre: some species have a broad black border surrounding it, and the margin terminating in a fringe. The *Avicula margaritifera*, or Pearl Oyster, belongs to this genus. [See PEARL OYSTER.]

AVICULARIA. [See SUPPLEMENT.]

AVOSET. (*Recurvirostra avocetta.*) This grallatorial bird, whose great singularity is in the form of its bill, is aquatic, the shores of the ocean and the banks of estuaries being its favourite haunts. On the shores of the Caspian and the salt lakes of Tartary they are abundant; they are widely distributed through the temperate climates of Europe; and on the south-eastern coast of England they are occasionally found. The Avoset is about eighteen inches in length; very erect, and has legs unusually long for its size.

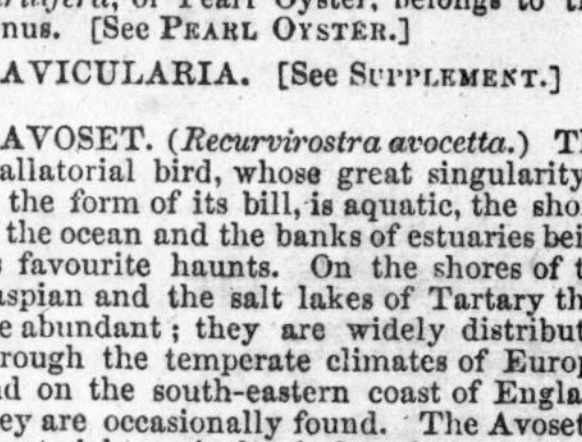

AVOSET.—(RECURVIROSTRA AVOCETTA.)

The bill, which is three inches and a half in length, turns up like a hook, in an opposite direction to that of the hawk or parrot, and is flat, thin, sharp, and flexible. The plumage is black and white, tail consisting of twelve white feathers ; the legs are of a fine blue colour, naked and well calculated for wading ; the feet are palmated, but not so much adapted for swimming as for supporting the bird upon the mud. It feeds on worms, &c., which it scoops out of the mud with its bill ; and it lays two eggs, which are greenish, spotted with brown and black.

AXIS. (*Cervus Axis.*) A species of Indian Deer, of which there are two or three varieties. 1. The *Common Axis* is about the size of a fallow deer, and of a light red colour. The body is beautifully marked with

AXIS DEER. — (CERVUS AXIS.)

white spots, and a line of white runs along the sides, dividing the upper from the under parts of the body. The horns are slender, and tri-forked ; the first ramification being near the base, and the second near the top, each pointing upwards. It is extremely docile, and possesses the sense of smelling in an exquisite degree. Though it is a native of the banks of the Ganges, it appears to bear the climates of Europe without injury. 2. The *Great Axis.* This animal, which is a native of Borneo and Ceylon, is about the height of a horse, and of a reddish-brown colour. The horns are trifurcated, thick, strong, and rugged ; about two feet nine inches long, and two feet four inches between the tips. 3. The *Lesser Axis* is a gregarious animal, inhabiting Java, Ceylon, Borneo, and some other oriental islands. It is hunted with ardour, the sport affording the highest diversion, and the flesh being esteemed excellent.

AXOLOTL. (*Siren pisciformis.*) A singular genus of Batrachian reptiles, being perfectly amphibious, inasmuch as they possess both kinds of respiratory organs at the same period, being furnished alike with gills and lungs ; and they can consequently breathe air and water according to the circumstances in which they happen to be placed. The Axolotl is about eight or nine inches long, the head is broad and flat, the

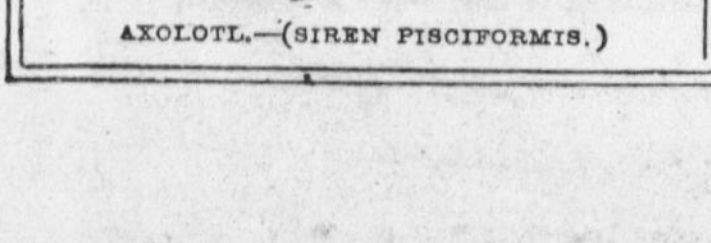

AXOLOTL.—(SIREN PISCIFORMIS.)

nose blunt, the eyes situated near the muzzle, the tail nearly as long as the body, and the toes unconnected by intermediate membranes. The colour is brown, thickly mottled both on the upper and under surfaces of the head and body, as well as on the tail and dorsal fins, with numerous small round black spots. It is commonly sold in the markets of Mexico : it is dressed after the manner of stewed eels, and when served up with a rich and stimulating sauce, is esteemed a great luxury. A second species has been lately discovered and described.

AYE-AYE. (*Cheiromys Madagascariensis.*) A singular quadruped (which in some descriptions has been confounded with the *Ai*, or Sloth, whose habits it somewhat resembles.) It is placed by Cuvier in the order *Rodentia*, but other naturalists have classed it with the Monkey tribe, from the hand-like structure of its hinder feet. It is a native of Madagascar ; it burrows under ground, is very slothful, and is altogether a nocturnal animal. It has large flat ears, like those of a bat, and a tail resembling a squirrel's ; but its most distinguishing peculiarity is the middle toe or finger of the fore-foot, the two last joints of which are very long, slender, and destitute of hair : this, as M. Sonnerat, who describes the one

AYE-AYE.
(CHEIROMYS MADAGASCARIENSIS.)

he had in his possession, remarks, is useful to the animal in drawing worms out of holes in the trees, and in holding on to the branches. It measures about eighteen inches from the nose to the tail ; and its general colour is a pale ferruginous brown, mixed with grey.

BABOON. (*Cynocephalus.*) A genus of Quadrumana, which forms the last link in the chain that unites the Simiæ with quadrupeds ; comprising a large, fierce, and formidable race of animals, who, though they in a slight degree partake of the human conformation, as seen in the Orang-outang, &c., are in their habits, propensities, and dispositions, the very reverse of gentleness and docility. In Apes and other quadrumana which have the head and face round, the nose is flat, and the nostrils are situated about half-way between the mouth and the eyes; but in the Baboon this organ is prolonged uniformly with the jaws, and the nostrils open at the end of it exactly as in the dog. In short, the most distinctive peculiarity of the genus is the marked resemblance which the head and face of these animals bear to a large dog. They have, moreover, long and truncate muzzles, cheek pouches, tails, and sharp claws. Yet, notwithstanding this close approximation to the shape of the dog's head, the form and position of the eyes, combined with the similarity of the arms and hands, give to these creatures a resemblance to humanity as striking as it is humbling and disgusting.

Formed for strength, furnished with dangerous natural weapons, and being wild, restless, and impetuous, this animal, in its native haunts, proves itself to be one of the most formidable of the savage race; nor can it be restrained, even when in confinement, any longer than coercion is continued: allowed to have its own will, its savage nature gains the ascendancy, and its actions are gratuitously cruel, mischievous, and destructive. But there is nothing so revolting as their lascivious habits, which they indulge to such a degree that it is unsafe and highly improper for females to visit exhibitions of animals where these beasts form a part of the number.

In their native haunts they subsist on roots and berries, and partly on eggs, insects, and scorpions; but in cultivated districts they make incursions into the fields and gardens, where they commit the greatest depredations on the fruit and grain. They congregate in troops, and are bold and skilful in their predatory excursions, maintaining their ground even against large parties of men; and it is remarked that "a troop of them will sometimes form a long chain, extending from the vicinity of their ordinary habitation to the garden or field which they happen to be engaged in plundering, and that the produce of their theft is pitched from hand to hand, till it reaches its destination in the mountains."

The Baboon can never be said to be thoroughly tamed, how long soever his confinement may have endured. As he advances in age, all his worst qualities become more strongly developed, and the expression of his physiognomy bears ample testimony to the fierceness and brutality of his disposition.

Having given a general description of these animals, it will be only necessary to particularise a few species where the difference between them seems most to deserve notice.

The DERRIAS. (*Cynocephalus hamadryas.*) This celebrated Baboon inhabits the mountains of Arabia and Abyssinia, and was probably the species known to the ancients, and sculptured in Egyptian monuments. It measures upwards of four feet when standing erect, and about two feet six inches in a sitting posture. The face is extremely long, and of a dirty flesh colour, with a lighter ring surrounding the eyes: the head, neck, shoulders, and all the fore-part of the body is covered with long shaggy hair; that on the hips, thighs, and legs having the appearance of being clipped. The hair of the head and neck forms a long mane, which falls back over the shoulders; and the whiskers are broad, and directed backwards so as to cover the ears. The general colour of the hair is a mixture of light grey and cinereous: a dark brown line passes down the middle of the back; and the tail is terminated by a brown tuft of long hair: the callosities are large, and of a dark flesh colour: the hands are almost jet black; and the feet are rusty brown. The female when full grown is as large as the male, but is destitute of a mane, and the hair of the body is short and of a uniform deep olive-brown colour. The Derrias while young is gentle and playful, but as soon as it has arrived at a mature age it becomes sulky and malicious.

The CHACMA or PIG-FACED BABOON (*Cynocephalus porcarius*) is equal in size, and much superior in strength, to a common English mastiff. It inhabits the mountains in the neighbourhood of the Cape of Good Hope, associates in families more or less numerous, and occasionally levies contributions on the gardens of Cape Town, which it performs in a very adroit and

PIG-FACED BABOON.
(CYNOCEPHALUS PORCARIUS.)

regular manner. The Chacma is of a uniform dark brown colour, mixed throughout with a dark green shade, occasionally relieved by a few hairs of a lighter hue. The hair is long and shaggy, particularly on the neck and shoulders of the males, where it forms a distinct mane; the face and ears are naked, as are likewise the palms of the hands and soles of the feet; and the cheeks of both sexes have small whiskers, directed backwards, of a greyish colour. The hands, face, and ears are of a very dark violet-blue colour; the muzzle is extremely prolonged, and the skull is contracted and flattened. It is no uncommon thing for travellers, while ascending the steep and dangerous mountain passes in South Africa, to meet with troops of these animals, who have been sunning themselves on the rocks: if not attacked, they hasten off, yelling and screaming; but if fired at and wounded, they no sooner get out of the range of the gun than they throw and roll down stones, to resent the injury.

The COMMON BABOON. (*Cynocephalus papio.*) This species is a native of the coast of Guinea, and is the one most commonly exhibited by itinerant showmen. Its appearance is at once grotesque and formidable; its nervous limbs and compressed form indicate great force and agility; the anterior parts especially being extremely strong and muscular. It is of a uniform yellowish-brown colour, with a shade of light red upon the head, shoulders, and extremities; the face, ears, and hands naked, and entirely black. The cheeks are considerably swollen below the eyes; after which the face contracts suddenly, which gives the nose the appearance of having been broken by a violent blow. It is furnished with whiskers, which have a backward direction, but do not conceal the ears. While young, this Baboon is gentle and familiar; but as it approaches adult age, it displays all the repulsive manner, the ferocity and intractability common to the rest of its kind.

The MANDRILL, or VARIEGATED BABOON. (*Cynocephalus maimon.*) The Mandrill is the most remarkable of the whole genus for brilliancy and variety of colour, while for size it is unequalled by any other Baboon, its height when standing upright being upwards of five feet. The limbs

MANDRILL. — (CYNOCEPHALUS MAIMON.)

are large and muscular, the body thick and robust; the head large, face long, scarcely any forehead, and the snout ending abruptly; the eyes small and deeply sunk in the head; the cheek-bones enormously swollen, and marked with several deep furrows of violet-blue, purple, and scarlet; and the muzzle and lips large and protuberant. The hair of the forehead and temples rises, in a remarkable manner, into a pointed form, which gives the head a triangular appearance; and a small pointed orange-yellow beard adorns the chin. Round the back of the neck the hair is long, and inclines forwards, somewhat in the manner of a wreath. On the loins the skin is almost bare and of a violet-blue colour, gradually altering into a bright blood-red, which is more conspicuous on the hinder parts, where it surrounds the tail, which is very short, and generally carried erect. In most of its habits the Mandrill resembles the other Baboons, especially in its growing more morose as it advances in age, and in becoming offensively libidinous. In their wild state they generally march in large bands, and are so formidable that not only are the inhabitants afraid to meet them in the woods, unless they are in considerable companies and well armed, but the beasts of the forest, including even the elephant, quit their respective haunts at the approach of the powerful and savage animals whose habits we have endeavoured to describe. To this truly formidable species belonged "Happy Jerry," long kept in the fine menagerie of Mr. Cross. He was trained to smoke a pipe, and seemed to relish a pot of porter: but he was fierce to most persons who approached him, unless they were his keepers. His stuffed skin and skull may now be seen in the magnificent collection of the British Museum.

There are several other species which our limits forbid us to do more than merely mention; as, the DRILL, the WOOD-BABOON, the PIGTAIL, the CRESTED, the YELLOW, the CINEREOUS, and others.

BABYROUSSA. (*Sus Babirussa.*) This animal is nearly of the size of a common Hog, and has generally been referred to the Hog genus, though in many respects it is essentially different; its form being longer, its limbs more slender, and, instead of bristles, being covered with fine, short, and somewhat woolly hair, of a dark brown colour, interspersed with a few bristles on the upper and hinder part of the back. It is still further distinguished by the very extraordinary position and form of its enormous upper tusks, which, instead of being situated internally, on the edge of the jaw, as in other animals, are placed externally, perforating the upper lip, and turning upwards toward the forehead, like the horns of the Ruminantia: the tusks of the lower jaw are also very long, sharp, and curved; but not of equal magnitude with those of the upper. The tusks are of a very fine ivory, but neither so hard nor so durable as that of the elephant: the eyes are small; the ears erect and pointed; the tail rather long, slender, and tufted at the end with long hairs.

BABYROUSSA. — (SUS BABIRUSSA.)

The Babyroussa is a gregarious animal, inhabiting the woods of Java, Amboyna, the Celebes, and other Indian islands, where large herds are met with. Their food consists chiefly of vegetables, and the leaves of trees. When sleeping or resting themselves in a standing posture, they *are said* often to hook or support themselves by placing the upper tusks across the lower branches of the trees, and, thus suspended, sleep in security. When hunted closely, and in apparent danger, this animal will, if possible, plunge into some great river, or the sea, where it swims with great facility, and by alternate diving and rising, is frequently able to escape from its pursuers. In the gardens of the Zoological Society a fine specimen of this rare animal may now be seen.

BACULITES. A species of Ammonite or Snake-stone. [See AMMONITE.]

BADGER. (*Meles vulgaris.*) The Badger is a carnivorous quadruped, inhabiting most parts of Europe and Asia; and is generally regarded as a solitary, stupid animal, that seeks refuge in the most sequestered places,

BADGER.—(MELES VULGARIS.)

and shuns the light of day. It has very short legs and a broad flat body; the head is long and pointed, the eyes small, the neck short and thick, the tail remarkably short, and the hide thick and tough. The upper parts of the body are covered with long coarse hair, the hue of which is a rusty grey; but on the breast, belly, and limbs it is short and black: the face is white, and along each side of the head runs a long pyramidal band of black, including the eyes and ears.

With its powerful claws it constructs a deep and commodious burrow; and as it continues to bury itself, it throws the earth behind it to a great distance, and thus forms for itself a long winding hole, ending in a round apartment at the bottom, which is well lined with dry grass and hay. This retreat it seldom quits till night, when it steals from its subterraneous abode for the purpose of procuring food. It lives chiefly on roots, fruits, insects, and frogs; but it also robs the bee of his honey, and destroys the eggs of partridges and other birds which build their nests on the ground. It is quiet and inoffensive; but when attacked by dogs it defends itself with great resolution, and seldom dies unrevenged of his enemies. The Badger is about two feet six inches long: his skin is so thick that it resists the impression of the teeth, and so loose, that even when a dog has seized it, he is enabled to turn round easily, and severely bite his assailant. The female produces three or four young at a time. The flesh of the Badger is reckoned a delicacy in Italy, France, and China, and may be made into hams and bacon. The skin, when dressed with the hair on, is impervious to the rain, and consequently makes excellent covers for travelling trunks, &c.; and the hairs or bristles are made into brushes for painters.

The **AMERICAN BADGER,** or **CARCAJOU.** (*Meles Labradorica.*) This animal is rather smaller than the European species; its fore-claws are longer and stronger, and the black bands on the face narrower. Its prevailing colour is a kind of mottled grey, and, with the exception of the head and extremities, which are covered with short coarse hair, it is furnished with a fine, long, silky fur. It is a slow and timid animal, takes to the first earth it meets with when pursued, and, burrowing in the sand, is soon out of the reach of danger. Whilst the ground is covered with snow the American Badger seldom ventures from his hole, but passes the severe winter months in a semi-torpid state. By some naturalists this is regarded as the type of a distinct genus (Taxidea).

The **INDIAN BADGER** (*Mydaus collaris*) is about twenty inches in height and two feet in length, the form of its body and limbs bearing a resemblance to the bear, while its head, eyes, and tail remind us of the hog. The hair is a yellowish white, with black points, which gives the whole a dark brown shade; but the legs and under parts of the body are black. The ears are very small; and on each side of the head are two black bands, which descend down the neck, and enclose the throat. They are exceedingly fierce.

BALÆNA. [See WHALE.]

BALÆNICEPS. [See SUPPLEMENT.]

BALÆNODON. A genus of extinct whales. [See SUPPLEMENT.]

BALANCE-FISH. (*Zygæna.*) A fish, the shape of whose head has been likened to a hammer. It is a native of the Mediterranean Sea. [See ZYGÆNA.]

BALANINUS. A genus of the *Curculionidæ*, furnished with a long slender rostrum, or snout, at the tip of which is a minute pair of sharp horizontal jaws, and by means of which it is enabled to deposit its eggs, which are generally placed in the kernel of some fruit. Of this kind is the *Bala-*

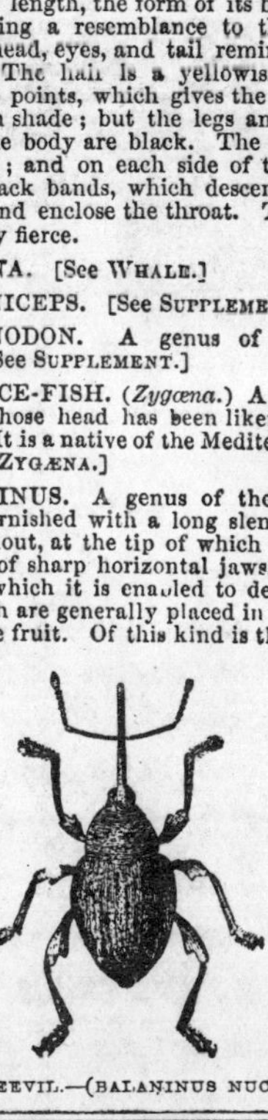

NUT WEEVIL.—(BALANINUS NUCUM.)

ninus Nucum, or Nut-Weevil, whose larva is so commonly found in nuts, filberts, &c. The egg is introduced when the nut is young and soft; and the nut being but slightly injured, continues to grow and ripen, while the larva feeds upon the kernel in which it is imbedded. When about to change its state, it bores through the shell and escapes, leaving a small round orifice: falling on the ground, it then burrows into the earth, where it assumes the pupa state, and in the following summer it comes forth as a perfect insect.

BALANUS. A genus of multivalve Cirripedes, usually found adhering to various submarine productions, whether fixed or moveable; such as the harder sea-plants and all sorts of crustaceous as well as testaceous animals, rocks, ships, timber, &c. The shell shapes itself at the base to the figure of the surface of whatever it adheres to, and from which it is with difficulty removed. It altogether forms a rude hollow cone. The animal inclosed in it is of a very singular structure: it has twelve crooked legs or arms, garnished with a great number of hairs, which it elevates on all occasions; besides eight others, inferior in size and lower in position. In general the *Balani* are considered ineligible as food: but Capt. P. P. King speaks of some large kinds (*Balanus psittacus*) on the southern parts of the South American coast as forming a very common and highly esteemed food of the natives, the flesh equalling in richness and delicacy that of the crab. He also says, it occurs in large b[r]anches, and presents somewhat of a cactus-like appearance. The parent is covered by its progeny, so that large branches are found composed of from fifty to a hundred distinct individuals, each of which becomes in its turn the foundation of another colony.

A GROUP OF BALANUS OVULARIS

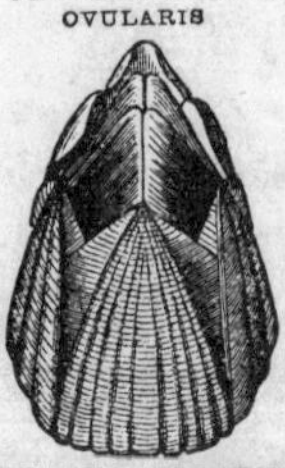

BALANUS OVULARIS.

BALD BUZZARD. [See OSPREY.]

BALEARIC CRANE. [See CRANE.]

BALTIMORE BIRD. [See ORIOLE.]

BAND-FISH. (*Cepola*. This genus of Acanthopterygious fishes is of a form so thin and flat in proportion to its length, as to have obtained among the ancient ichthyologists the name of Tænia or Riband-fish. One species (*C. Mediterranea*) is a native of the Mediterranean, and varies in length from eighteen inches to three feet. The head is short and rather truncated in front; the mouth is wide, and the lower jaw longer than the upper, both being armed with sharp curved teeth, of which there is a double row in the lower jaw. The sides are extremely compressed; and the body, both above and below, sharpens into a kind of carina or ridge. The dorsal fin commences from the back of the head, and is continued as far as the tail; the vent fin also extends nearly throughout the whole length. The colour of the body is bright silver, with a dusky tinge above; the sides are marked with a few large reddish spots; the fins are all of a pale red colour, and the skin is covered with extremely small scales. It is predaceous, and swims with great rapidity.

Another species, found on our coasts, (*Cepola rubescens*) is of a pale carmine colour, and varies from ten to fifteen inches in length. It is very smooth and slender, and tapers very gradually from the head to the tail.

BANDICOOT. (*Perameles.*) A genus of Marsupial animals, indigenous to Australia, and in some respects analogous to the Opossums and Kangaroos; but the disproportion between the fore and hind legs is by no means so great, though sufficient to make their gait rabbit-like, or a succession of leaps, rather than walking or running. Their feet are provided with broad powerful claws, which enable them to burrow with great facility, and to dig up roots, on which they principally feed. The most common species is called the LONG-NOSED BANDICOOT (*Perameles nasuta*): it measures about a foot and a half from the tip of the snout to the origin of the tail; the ears are erect and pointed, the eyes small, and the tail bearing considerable resemblance to that of a large overgrown rat, to which the whole animal, in fact, may be likened as regards its general external appearance, as well as its depredations upon the farm-yards and granaries.

BANXRING. [See TUPAIA.]

BARB. The name given to a fleet and vigorous breed of horses reared by the Moors of Barbary, and introduced into Spain during their dominion in that country, but since their expulsion it has been allowed greatly to degenerate; nor is it much better in their original clime, except among the wild nomadic tribes of the desert, where the breed still ex[ists] in perfection. But the Barb is far [from ex]celling in symmetrical beauty; the true value of these noble animals is to be discovered in their qualities rather than in their appearance. With a large and clumsy head, a short thick neck, and a broad chest, are united a long body and slender legs; but, on the other hand, they are unrivalled in speed, abstinence, docility, patience, and endurance under fatigue. They are sinewy, nervous, and long-winded; they walk well, and stop short, if required, even in full career; walking and galloping, indeed, being the only paces these animals are allowed to practise. It is not customary, except in cavalry exercises, for the Moors to try the powers of their horses very severely; they then, however, gallop them at the height of their speed. The horses are never castrated, and are alone used for the saddle, the mares being kept for breeding. It has been remarked that Barbs grow ripe, but never old, because they retain their vigour to the last; they are also said to be long-lived, and remarkably free from diseases.

BARBARY APE. (*Pithecus inuus.*) This species of Ape, which grows to the height of nearly four feet, is remarkable for docility, and, by force of discipline, is made to exhibit considerable intelligence. Its general colour is a palish olive-brown; the face is a swarthy flesh colour. It is common in Barbary and the lower parts of Africa, and is also found in considerable numbers on the rock of Gibraltar. This species was well known to the ancients, and it has been the "showman's ape" from time immemorial. Though morose and sullen in confinement, it is represented as social, active, and courageous in its wild state, and is particularly distinguished for its attachment to its young.

BARBEL. (*Barbus vulgaris.*) A fresh-water malacopterygious fish, usually frequenting the deep and still parts of rivers,

BARBEL.—(BARBUS VULGARIS.)

swimming with great strength and rapidity, and living not only on aquatic plants, worms, and insects, which it obtains by boring and turning up the loose soil of the banks with its snout, but occasionally by preying on smaller fishes. It is said to receive its name from the barbs or wattles attached about its mouth, by which appendages it is readily distinguished, as well as by the great extension of the upper jaw beyond the lower. It is sometimes found to weigh from fifteen to eighteen pounds, and to measure three feet in length: its more general length, however, is from twelve to eighteen inches. The general colour of the upper part of the head and body is a greenish brown; the scales are small, and in general of a pale gold colour, edged with black on the back and sides, and silvery-white on the belly; the pectoral fins are a pale brown, the ventral and [illegible] fins are tipped with yellow; and the [illegible] is slightly forked, and of a deep purple. The Thames produces Barbel in abundance, and of a large size. "So numerous are they about Shepperton and Walton," says Mr. Yarrell, "that one hundred and fifty pounds weight have been taken in five hours, and on one occasion it is said that two hundred and eighty pounds weight of large sized Barbel were taken in one day." The flesh of the Barbel is very coarse and unsavoury; the fish, consequently, is held in little estimation, except as affording sport for the angler.

BARBET. The Barbets are a family of birds belonging to the order *Scansores*, or Climbers, and are distinguished by their large conical beak, which appears swollen, or, as it were, puffed out at the sides of its base, and by being *bearded* (whence the name) with five tufts of stiff bristles, directed forwards. They inhabit Java, Sumatra, &c., and sport about in all positions on the trunks and among the branches of trees, in search of insects or their larvæ, on which they feed: some of them are said also to devour small birds and fruits; the typical genera, however, appear confined to the former food. The plumage of some of the species is very brilliant.

MANY-COLOURED BARBET.—(BUCCO VERSICOLOR.)

BARIS. A genus of Coleopterous insects, which feed upon the dead parts of trees.

BARKING BIRD. (*Pteroptochos.*) This Tenuirostral bird, which is common in Chiloe and Chonos,—islands in the South American Archipelago,— is called by the natives Guid-guid; "but its English name," says Mr. Darwin, "is well given; for I defy any one at first to feel certain that a small dog is not yelping somewhere in the forest. Just as with the Cheucau, a person will sometimes hear the bark close by, but in vain may endeavour, by watching, and with still less chance by beating the bushes, to see its author; yet at other times the Guid-guid comes fearlessly near." Its manner of feeding and its general habits are very similar to those of the Cheucau. Both species are said to build their nests close to the ground, amongst the rotten branches. [See CHEUCAU.]

BARNACLE. A name given to the cirripedes sometimes found adhering to the bottoms and sides of ships, &c. [See BALANUS.]

BASILISK. (*Basiliscus.*) The Basilisk of modern naturalists has no affinity to the malignant serpent of the poets whose very aspect the ancients believed to be fatal

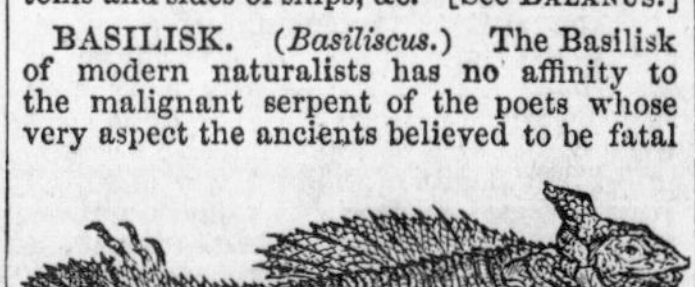

BASILISK.—(BASILISCUS MITRATUS.)

to all who looked upon it. The animal now recognised by the name of Basilisk is a species of lizard, of a very singular shape, being distinguished by a long and broad wing-like process or expansion along the back and upper part of the tail, and furnished at certain distances with interval radii analogous to those in the wings of the *draco*, or flying lizard. This process is capable of being either dilated or contracted at the pleasure of the animal; and the occiput or hind part of the head is elevated into a very conspicuous pointed hood or hollow crest. Notwithstanding its formidable appearance, however, the Basilisk is a perfectly harmless reptile, residing principally among trees, where it feeds on insects, &c. The general colour of this animal is a pale cinereous brown, slightly varied on the back and sides with different shades of brown and blue, and silvery white on the belly. It is possessed of great activity, and from its peculiar structure can adapt itself to the watery element without inconvenience. It is most common in the tropical parts of South America.

BATS. (*Cheiroptera.*) The singular animals which come under this denomination were long considered as partaking so much of the character of birds with that of quadrupeds, that it was thought difficult to assign to them a distinct station in the system of nature. Such doubts, however, have long since yielded to scientific investigation; their anatomical and intes-

BAT.—(VESPERTILIO MURINUS.)

tinal structure, their viviparous nature, their hair, &c., entitling them to be ranked as quadrupeds. Still it is not to be denied that their peculiar conformation is admirably calculated for the exercise of considerable powers of flight. The air, indeed, is their home: through this they move with vast rapidity, and with great apparent ease, wheeling in every direction in search of their insect prey, and performing the most abrupt evolutions to secure it. A remark, indeed, not less true than trite, has been often made that, in their mode of flight, Bats bear a very strong resemblance to swallows; exercising the same purpose in the economy of nature, in restraining the multiplication of the crepuscular and nocturnal insects, as the swallows do in regard to the diurnal.

Their senses of smell, feeling, and hearing are wonderfully acute. In many genera the nose is furnished with a membranous foliation of most delicate structure, by which the sense of smelling is greatly refined; the ears also are in many kinds expanded and capable of being folded down; while their ample wings, and the membranous tissues of the ear and nose, are so abundantly supplied with nerves, as to enable them, even should they be deprived of sight, to pursue uninterruptedly their aerial course, avoiding every obstacle, and passing adroitly through the narrowest apertures.

On the approach of cold weather the Bat hibernates, and in preparing for this state of lifeless inactivity, it seems rather to select a place where it may remain safe from molestation, than where it may be commodiously lodged. "The hibernation of these animals," says Mr. Bell, "is indeed one of the most interesting points in their economy. At an earlier or later period of autumn, according to the species, they retreat, generally in large congregations of various species together, to the most retired places; as under the roofs of houses and churches, in caverns, in the hollows of trees, and similar situations, where they suspend themselves by their hinder claws, with the head downwards. Here they crowd together, holding not only by the surface of the walls of their retreat, but by each other, one crowding over another so closely that it appears scarcely possible for such numbers to occupy so small a space."

Not less than twenty species of Bats are enumerated as known in Great Britain; but these, although differing from each other, either in structure, colour, or habits, can hardly be deemed of sufficient importance to occupy the space that would be necessary to describe them minutely; we shall therefore merely refer to a few of them in the briefest manner possible, and in another place narrate a few particulars relating to two foreign species of a more formidable character.

The COMMON BAT, or FLITTER-MOUSE. (*Vespertilio pipistrellus.*) There are several species known in England, but this is the most common. It is nearly two inches and a half long, or about the size of a mouse. The members usually called wings, are merely the four interior toes of the fore-feet extended to an enormous length, and connected by a thin membrane, reaching also to the hind legs, and from them to the tail: the body is covered with short fur, of a mouse-colour, tinged with red; and the membranes are of a deep dusky hue; the eyes are small, and the ears exactly resemble those of the mouse. This species of Bat makes its appearance in the twilight of fine summer evenings, frequenting the sides of woods, glades, and shady walks, or skimming along the surface of tranquil rivers, where moths, gnats, and other nocturnal insects are most readily to be found; but if the weather be not fine, it remains shut up in the chinks or fissures of crumbling masonry, or lies concealed in the friendly recesses of some hollow tree.

The BARBASTELLE BAT. (*Vespertilio barbastellus.*) This species was originally described by Daubenton, in 1759, but its first detection as a native of Great Britain is due to Mr. Sowerby, who published an account of it with a figure in the British Miscellany. Its general colour is darker than that of any other British species, being nearly black on the back, with here and there a few white hairs, which become more numerous with age; the hinder parts are reddish brown, and the belly a pale grey. The ears, the naked part of the muzzle, and the flying membrane, are of a dusky black. Length of the head and body, two inches; extent of wings, between ten and eleven inches. The muzzle is truncated, and a groove leads on each side upwards to the nostrils. The cheeks are rather tumid, and covered with black hair, which forms a sort of moustache. The ears are about the length of the head, nearly as broad as they are long, and irregularly four-sided; the inner edges are turned back, forming a longitudinal groove just within the margin; the outer and superior angle prominent, rounded, and turned back. The eyes are very small,

BARBASTELLE BAT. —(V. BARBASTELLUS.)

placed close to the base of the auricle, and almost concealed by the hair on the cheek. The fur of the body is long and soft. Mr. Bell having had a Barbastelle Bat in his possession for some time, was able to give a few slight notices of its habits. "He fed readily on small bits of meat, and drank water. He was a timid animal, and did not evince the slightest disposition to become acquainted with me; he would take his food, however, with his companions, and was accustomed to rest with them in a cluster, at the top of the box in which they were placed. The Barbastelle certainly became torpid more readily than any of the others, and more completely so; but when awake, evinced extreme restlessness, and was incessantly biting with great violence at the wires of his box. When suffered to fly about the room, he flew very low, and less actively than any other under similar circumstances; and he was fond of lying before the fire on the hearth-rug, where he appeared quite to luxuriate in the warmth."

The GREAT BAT. (*Vespertilio noctula.*) This was termed *altivolans* by White of Selborne, from its always flying high in the air, in pursuit of its prey. It is gregarious in its habits; the ears are short and rounded; it is about three inches long, and of a reddish ash-colour.

The SEROTINE. (*Vespertilio serotinus.*) This species is somewhat rare in England, but very common in France. It is of a dark chesnut colour; frequents forests; is solitary in its habits; and its flight is slow.

The MOUSE-COLOURED BAT. (*Vespertilio murinus.*) This is the largest of the British Bats, exceeding even the Noctule in its length of body and extent of wings. The head of this Bat is long, with the ears inclining backwards.

The LONG-EARED BAT. (*Plecotus auritus.*) One of the most common of our British Bats, and at the same time one of the most pleasing in its appearance, owing to the extraordinary transparency and beauty of the ears. It is also more familiar and bold when in confinement than any other species.

The HORSE-SHOE BAT. (*Rhinolophus.*) There is a greater and lesser variety of this animal; the former designated *Rhinolophus ferrum-equinum*, the latter *Rhinolophus hipposideros*. This genus is distinguished by a very curious nasal appendage, or foliaceous membrane at the end of the nose, shaped somewhat like a horse-shoe, and supposed to extend in an extraordinary degree the sense of smelling. The upper part of the body is of a deep ash-colour, the lower part inclining to white.

In concluding this article on Bats, we are again tempted to avail ourselves of the intelligent observations of Mr. Bell: "It is perhaps difficult to account for the prejudices which have always existed against these harmless and interesting little animals, which have not only furnished objects of superstitious dread to the ignorant, but have proved to the poet and the painter a fertile source of images of gloom and terror. That the ancient Greek and Roman poets, furnished with exaggerated accounts of the animals infesting the remote regions with which their commerce or their conquests had made them acquainted, should have caught eagerly at those marvellous stories and descriptions, and rendered them subservient to their fabulous but highly imaginative mythology, is not wonderful; and it is more than probable that some of the Indian species of Bats, with their predatory habits, their multitudinous numbers, their obscure and mysterious retreats, and the strange combination of the character of beast and bird which they were believed to possess, gave to Virgil the idea, which he has so poetically worked out, of the Harpies which fell upon the hastily spread tables of his hero and his companions, and polluted, whilst they devoured, the feast from which they had driven the affrighted guests. But that the little harmless Bats of our own climate, whose habits are at once so innocent and so amusing, and whose time of appearance and activity is that when every thing around would lead the mind to tranquillity and peace, should be forced into scenes of mystery and horror, as an almost essential feature in the picture, is an anomaly which cannot be easily explained." [See PTEROPUS and VAMPIRE-BAT.]

BATHYERGUS, or COAST RAT. This Rodent animal is a native of Southern Africa, frequenting sandy tracts along the coast, where it burrows with great rapidity, working out long galleries, and throwing up hil-

COAST RAT. — (BATHYERGUS MARITIMUS.)

locks like the mole. In some districts these are so very numerous, as to render it dangerous to pass over them on horseback, the earth where excavated often suddenly giving way. The Coast Rat is about a foot long, exclusive of the tail, which is three inches; and its general colour is greyish ash. The incisors are of an enormous size; those in the upper jaw having a longitudinal furrow down the front. The whole form and organization of this animal fit it for an underground existence; it is accordingly often called the Sand Mole.

BATRACHOIDEÆ. A family of monstrous-looking Acanthopterygious fishes, whose pectoral fins are supported upon the elongated carpal bones which in some genera perform the functions of hind feet, enabling the fish to creep over sand or mud like small quadrupeds. The ventrals are jugular, and the gill-plates and rays are enveloped in loose skin. Cutaneous appendages or barbels generally fringe the lips, or lower jaw, to the pectorals. In general the skin is destitute of scales, but is sometimes studded with bony tubercles. The skeleton is, for the most part, but imperfectly osseous. Some genera have an air-bladder, and some have not. *Batrachus* has a spiny operculum and suboperculum, and a flat head broader than the body, but not very disproportionate in length. *Lophius* has a depressed form, and *Chironectus* a compressed one, and both have monstrously large heads, with a small hole behind the pectorals for an opening to the gills. In *Malthe* the head is flat, and greatly lengthened laterally by the projection of the large subopercula. The Batrachoideæ can live long out of the water, in consequence of the smallness of their gill-openings. The Chironectes, in particular, are able, even in warm countries, to pass two or three days in creeping over the land. All the Batrachoideæ conceal themselves in the mud or sand, and lie in wait to take their prey by surprise. They exist in the Atlantic, Indian, and Pacific oceans; and several also inhabit the European seas.

BATRACHIANS: BATRACHIA. Frogs, Toads, and all reptiles which, like them, have soft and naked skins (i. e. uncovered with scales), and in the early stage of existence respire by means of gills.

BEAGLE. A small kind of hound, or hunting-dog, formerly much prized for its excellent scent and persevering endurance when employed in hare-hunting. It cannot indeed boast of great speed; but its "slow and sure" qualities are generally rewarded with success, although the chase may be continued for two or three hours. There are several kinds of Beagles; as, the Southern Beagle, the fleet Northern or Cat Beagle, and a very diminutive one called the Lap-dog Beagle.

BEAR. (*Ursus.*) A well-known quadruped, belonging to a family of plantigrades, distinguished by their ponderous bulk, massive limbs, and heavy gait. There are several species of Bears. Of all the Carnivora they are the most omnivorous in their diet, — some of them living almost entirely upon vegetable food, and nearly all being capable of supporting themselves upon it: even the most carnivorous of them, however, will seldom attack man, unless provoked to do so by aggression, or strongly incited by hunger; but when attacked they prove themselves very formidable opponents. They have six incisor and two canine teeth in each jaw, twelve molars in the upper and fourteen in the lower jaw; pendactyle or five-toed feet, armed with strong claws, but which, not being retractile, are more calculated for digging and climbing than for tearing prey. For the most part Bears are unsocial animals, frequenting the recesses of mountains and caverns, and the depths of the forests. During the winter they lay up in caves and hollow trees, passing that inclement season almost without food, and in a comparatively dormant state. In Europe, Asia, and America, Bears are pretty widely diffused, but in Africa they are more rarely found. In the Alpine regions the Bear is brown; in some other parts of Europe, black; and in some parts of Norway it has been seen of a grey colour, and even perfectly white. Bears are reported to be very fond of honey, in search of which they will climb trees, in order to get at the nests of wild bees; for, notwithstanding his awkward form, the Bear is an expert climber. In Russia the skins of Bears are among the most useful as well as comfortable articles of winter apparel; and in many other northern countries they are made into beds, coverlids, caps, and gloves. In England bears' skins are used for the hammer-cloths of carriages, for pistol holsters, &c.; and the leather prepared from them is used for many purposes, as harness, &c., where strength is requisite.

The COMMON or BROWN BEAR. (*Ursus Arctos.*) This species, with some variation as to size and colour, is a native of almost all the northern parts of Europe and Asia. Its usual size is about four feet in length, by about two feet and a half in height. In its nature it is savage and solitary; and though when tamed it appears gentle and placid to its owner, it should be cautiously managed, being often capricious, treacherous, and vindictive. Its retreat, during the period of hybernation, is the natural hollow of a tree, or some cavern; but where these are not conveniently found, it will either form a suitable den for itself by digging, or construct a rude kind of hut with branches of trees, lined with moss. Thus protected, and fat with its summer food, it will remain without further sustenance till the ensuing spring; during which time the female generally produces two cubs, which when first born are not much larger than a mastiff's puppies.

Most writers agree that the Brown Bear was at one time common in the British islands. The Caledonian bears (another name for British with the Romans) were imported to make sport for the Roman people, to whom the excitement of witnessing the suffering of man and beast, in its most distressing shape, seems to have been but too welcome. For many years (says Mr. Broderip) it has been swept away from our islands so completely, that we find it imported for baiting, a sport in which our nobility, as well as the commonalty, of the olden time—nay, even royalty itself—delighted. A bear-bait was one of the recreations offered to Elizabeth at Kenilworth, and in the Earl of Northumberland's Household Book we read of 20*s.* for his bear-ward. In Southwark there was a regular bear-garden, that disputed popularity with the Globe and Swan theatres on the same side of the water. Now, however, so much do tastes alter (in this instance certainly for the better), such barbarous sports are banished from the metropolis."

The AMERICAN BLACK BEAR. (*Ursus Americanus.*) This animal is somewhat smaller than the European Brown Bear just described. It has a long head, pointed nose, small eyes, and short ears rounded at the top; its limbs are strong, thick, and clumsy; its tail is short, its feet large, and the hair on the body and limbs is black, smooth, and glossy. This animal inhabits all the northern parts of America, migrating occasionally from the northern to the more southern parts in quest of food, which consists chiefly of vegetables and grain. So impenetrable are their retreats during the period of gestation, that although immense numbers of Bears are killed annually in America, hardly a single female is ever found among them. The flesh of these Bears in autumn, when they are become exceedingly large by feeding on acorns and other arborescent food, is extremely delicate; the hams, in particular, are much esteemed; and the fat, which preserves a certain degree of fluidity, is remarkably white and sweet. In the *Canadian Naturalist*, by Mr. P. H. Gosse, the following account of this animal forms a portion of the author's interesting 'Conversations':—

"This species appear to be less carnivorous than the *Ursus Arctos* of northern Europe, and less ferocious. His chief food seems to be of a vegetable nature, grain, fruits, and roots. He has an appetite for pork, however, and occasionally makes a visit to the farmer's hog-sty for the purpose of cultivating an acquaintance with the grunting inhabitants. Some years ago, one of our nearest neighbours was aroused in the night by a commotion in his hog-pen; suspecting the cause, he jumped up immediately, took his gun, and saw a bear in the act of getting over the fence with a fine hog, embraced very lovingly in his fore-paws. The man fired (while his wife held a light), and killed the intruder. It is difficult to hurt a bear with any weapon but fire-arms; he fights with his fore-paws like a cat; and so watchful is he, and so expert at warding off every blow that is made at him, that it is next to impossible to strike his head, the only part in which he is vulnerable; for you might almost as well batter a feather-bed as the body of a bear, so encased and shielded by an enormous layer of fat. In our climate he becomes torpid during winter, generally choosing for his hybernaculum some large hollow log, or a cavity beneath the root of an overthrown tree. The species is numerous in all the wooded parts of this continent, even to the shores of the Gulf of Mexico. In the southern states he commits depredations on the farmer's fields of maize: when the corn is in that milky state called 'roasting ears,' so prized for boiling and eating as a table dish, like green peas, or roasting whole on the cob, the bear manifests a singular unity of taste with the farmer, and devours and treads down a large quantity, as he finds no difficulty in climbing over the zig-zag rail fence. I have been told that he repeats his nightly visit to the same field; and, what is singular, always, on such occasions, mounts the fence, night after night, at the same spot where he got over the first time. The planters take advantage of this regular habit, by fastening to the fence a heavily loaded gun at such an angle that it shall point at the bear's breast as he rises on his hind legs. The identical crossing-place is easily known by his great tracks in the soft earth. A stick is attached to the trigger, and this is made fast, at right angles, to a transverse stick resting on two forks about breast high, a few inches outside the fence. The bear rears up to put his fore-paws on the rails, and in getting over presses with his breast against the transverse stick, which drives back the trigger, and poor Bruin instantly receives the reward of his dishonesty."

The GRISLY BEAR. (*Ursus ferox.*) The Grisly Bear is about nine feet long, and is said to attain the weight of eight hundred pounds. The claws are long and very strong, but more adapted for digging than for

GRISLY BEAR.—(URSUS FEROX.)

climbing trees; the muzzle is lengthened, narrowed, and flattened; the canine teeth are highly developed, exhibiting a great increase of size and power; and the eyes are small and sunk in the head. Notwithstanding its bulky and unwieldy form, it is capable of great rapidity of motion; and its strength is so prodigious, that the bison contends with it in vain. Mr. Drummond, in his excursions over the Rocky Mountains, had frequent opportunities of observing the manners of the Grisly Bears, and it often happened that in turning the point of a rock or sharp angle of a valley, he came suddenly upon one or more of them. On such occasions they reared on their hind legs, and made a loud noise like a person breathing quick, but much harsher. He kept his ground, without attempting to molest them; and they on their part, after attentively regarding him for some time, generally wheeled round and galloped off; though, from their known disposition, there is little doubt that he would have been torn in pieces, had he lost his presence of mind and attempted to fly.

The POLAR BEAR. (*Thalassarctos maritimus.*) The accounts given by the early navigators of the size, strength, and ferocity of the Polar Bear are perfectly appalling; but the accuracy of modern investigation has dissipated many of the erroneous ideas which were formerly entertained on the subject, though it is still very clear that this Bear is possessed of immense strength and fierceness. The whole animal is white, except the tip of the nose and the claws, which are jet

POLAR BEAR.
(THALASSARCTOS MARITIMUS.)

black; the ears are small and rounded, the eyes small, the teeth very large, and the limbs extremely large and strong. The shores of Hudson's Bay, Greenland, and Spitzbergen, are its principal places of residence; but it has sometimes been accidentally carried on floating ice as far south as Newfoundland. Their usual food consists of seals, fish, and carcasses of whales; but when on land they prey on various animals, as hares, young birds, &c.: they also eat such roots and berries as they can find. They are said to be frequently seen in Greenland in great droves, allured by the scent of the flesh of seals, and will sometimes surround the habitations of the natives, and attempt to break in. Captain Lyon gives the following account of its hunting the seal: "The bear on seeing his intended prey, gets quietly into the water, and swims until to leeward of him, from whence, by frequent short dives he silently makes his approaches, and so arranges his distance, that, at the last dive, he comes to the spot where the seal is lying. If the poor animal attempts to escape by rolling into the water, he falls into the bear's clutches; if, on the contrary, he lies still, his destroyer makes a powerful spring, kills him on the ice, and devours him at leisure."

During the summer they reside chiefly on the ice-islands, and pass frequently from one to another, being extremely expert swimmers. They lodge in dens, formed in the vast masses of ice, which are piled in a stupendous manner, leaving great caverns beneath: here they breed, and bring forth one or two at a time; and the affection between the parent and the cubs is so great, that they will sooner die than desert each other. During winter they retire, and bed themselves deep beneath the snow, or else beneath the fixed ice of some eminence, where they pass in a state of torpidity the long and dismal arctic night, appearing only with the return of the sun.

JUGGLER BEAR, or JUNGLE BEAR. (*Prochilus ursinus.*) When this uncouth animal was first brought to England (now more than half a century since), it was taken for a Sloth, and obtained the names of *Bradypus pentadactylus.* and *Bradypus ur-*

JUGGLER BEAR.—(PROCHILUS URSINUS.)

sinus, "Five-fingered Sloth," "Ursine Sloth," and "anonymous animal." It is the *Ours jongleur* of the French, who so called it on account of its being a favourite with the Indian mountebanks or jugglers, who rely on the attraction of its ugliness. The Juggler Bear inhabits the mountainous parts of India, its place of retreat being in some cavern. Its short limbs, the depressed air of the head, surmounted by the hillock of a back, and the whole contour of the apparently unwieldy mass, give the idea of deformity. In bulk it is about the size of the Brown Bear. The nasal cartilage is capable of considerable extension, and the lips of protrusion. The muzzle and tips of the paws are a whitish-yellow; and there is a half-collar or Y-like marking on the under side of the neck and breast. With these exceptions, the fur is deep black, with here and there some brown spots, and is rather long, particularly round the head, as the animal grows old. In a state of nature its food consists of fruits, honey, and those destructive insects the white ants. In captivity it appears to be mild, but melancholy.

MALAY BEAR. (*Prochilus Malayanus.*) This animal is jet black, with the muzzle of a yellowish tint, and a crescent-shaped white mark on the breast. Vegetables form its chief diet, but it is said to be extremely fond of delicacies, and in its native forests subsists in a great measure upon the honey which is there found in considerable abundance. It is attracted to the vicinity of man by its fondness for the young shoots of the cocoa-nut trees, to which it is very injurious. It has been frequently taken and domesticated. One which Sir Stamford Raffles possessed when young is thus described by him : "He was brought up in the nursery with the children ; and, when admitted to my table, as was frequently the case, gave a proof of his taste by refusing to eat any fruit

MALAY BEAR. (TROCHILUS MALAYANUS.)

but mangosteens, or to drink any wine but champagne. The only time I ever knew him to be out of humour was on an occasion when no champagne was forthcoming. He was naturally of an affectionate disposition, and it was never found necessary to chain or chastise him. It was usual for this bear, the cat, the dog, and a small blue mountain bird or lory of New Holland, to mess together, and eat out of the same dish. His favourite playfellow was the dog, whose teasing and worrying were always borne and returned with the utmost good humour and playfulness. As he grew up he became a very powerful animal, and in his rambles in the garden he would lay hold of the largest plantains, the stems of which he could scarcely embrace, and tear them up by the roots."

BEAVER. (*Castor fiber.*) The Beaver is a Rodent animal, readily distinguished from every other quadruped by its broad horizontally-flattened tail, which is of a nearly oval form, but rising into a slight convexity on its upper surface, and covered with scales. The hind feet are webbed, and together with the tail, which acts as a rudder, serve to propel it through the water with considerable facility. It is about three feet long, exclusive of the tail, which is one foot more : its colour is a deep chesnut, the hair very fine, smooth, and glossy ; but it occasionally varies, and is sometimes found perfectly black. The incisor teeth are very large and hard ; so hard, indeed, that they were used by the North American Indians to cut bone and to fashion their horn-tipped spears, till they were superseded by the introduction of iron tools from Europe.

Of all quadrupeds the Beaver is considered as possessing the greatest degree of natural or instinctive sagacity in constructing its habitation ; preparing, in concert with others of its own species, a kind of arched caverns or domes, supported by a foundation of strong pillars, and lined or plastered internally with a degree of neatness and accuracy unequalled by the art of any other quadruped. But it should seem, however, that the architecture of the Beaver is nowhere so conspicuous as in the northern parts of America. The favourite resorts of the Beaver are retired, watery, and woody situations. In such places they assemble, to the number of some hundreds ; living, as it were, in families, and building their arched receptacles. From this we may perceive to what a degree animals, unassisted either by language or reason, are capable of concurring for their mutual benefit, and of attaining, by dint of numbers, those advantages which each, in a state of solitude, seems unfitted to possess : for if we view the Beaver only in the light of an individual, and unconnected with others of its kind, we shall find that many other quadrupeds excel it in cunning, and almost in all the powers of annoyance and defence. When kept in a state of solitude or domestic tameness, it appears calm and indifferent to all about it ; without attachments or antipathies ; and never seeking to gain the favour of man, nor aiming to offend him.

BEAVER. — (CASTOR FIBER.)

Few subjects in natural history have more attracted the attention of travellers, or have been more minutely described by naturalists, than the instinctive building operations of the Beaver ; and they have accordingly had attributed to them powers so marvellous, as to render ridiculous that which, if regarded merely as a high species of animal instinct, could not fail to command universal admiration. The account given by Buffon, though graphic and amusing in no ordinary degree, is evidently overcharged : we shall therefore take the more sober narration of Hearne : "The situation of the beaver-houses is various. Where the beavers are numerous they are found to inhabit lakes, ponds, and rivers, as well as those narrow creeks which connect the numerous lakes with which this country abounds ; but the two latter are generally chosen by them when the depth of water and other circumstances are suitable, as they have then the advantage of a current to convey wood and other necessaries to their habitations, and because, in general, they are more difficult to be taken than those that are built in standing water. They always choose those parts that have such a depth of water as will resist the frost in winter, and prevent it from freezing to the bottom. The beavers that build their houses in small rivers, or creeks, in which water is liable to be drained off when the back supplies are dried up by the frost, are wonderfully taught by instinct to provide against that evil by making a dam quite across the river, at a convenient distance from their houses. The beaver-dams differ in shape according to the nature of the place in which they are built. If the water in the river, or creek, have but little motion, the dam is almost straight; but where the current is more rapid, it is always made with a considerable curve, convex toward the stream. The materials made use of are drift-wood, green willows, birch, and poplars if they can be got; also mud and stones intermixed in such a manner as must evidently contribute to the strength of the dam; but there is no other order or method observed in the dams, except that of the work being carried on with a regular sweep, and all the parts being made of equal strength. In places which have been long frequented by beavers undisturbed, their dams, by frequent repairing, become a solid bank, capable of resisting a great force both of water and ice; and as the willow, poplar, and birch generally take root and shoot up, they by degrees form a kind of regular planted hedge, which I have seen in some places so tall that birds have built their nests among the branches.

"The beaver-houses are built of the same materials as their dams, and are always proportioned in size to the number of inhabitants, which seldom exceeds four old and six or eight young ones; though, by chance, I have seen double the number. Instead of order or regulation being observed in rearing their houses, they are of a much ruder structure than their dams; for, notwithstanding the sagacity of these animals, it has never been observed that they aim at any other convenience in their houses than to have a dry place to lie on; and there they usually eat their victuals, which they occasionally take out of the water. It frequently happens that some of the large houses are found to have one or more partitions, if they deserve that appellation, but it is no more than a part of the main building left by the sagacity of the beaver to support the roof. On such occasions it is common for those different apartments, as some are pleased to call them, to have no communication with each other but by water; so that, in fact, they may be called double or treble houses, rather than different apartments of the same house.

"So far are the beavers from driving stakes into the ground when building their houses, that they lay most of the wood crosswise, and nearly horizontal, and without any other order than that of leaving a hollow or cavity in the middle. When any unnecessary branches project inward they cut them off with their teeth, and throw them in among the rest, to prevent the mud from falling through the roof. It is a mistaken notion that the wood-work is first completed and then plastered; for the whole of their houses, as well as their dams, are, from the foundation, one mass of mud and wood mixed with stones, if they can be procured. The mud is always taken from the edge of the bank, or the bottom of the creek or pond near the door of the house; and though their fore-paws are so small, yet it is held close up between them under their throat: thus they carry both mud and stones, while they always drag the wood with their teeth. All their work is executed in the night, and they are so expeditious that, in the course of one night, I have known them to have collected as much as amounted to some thousands of their little handfuls. It is a great piece of policy in these animals to cover the outside of their houses every fall with fresh mud, and as late as possible in the autumn, even when the frost becomes pretty severe, as by this means it soon freezes as hard as a stone, and prevents their common enemy, the wolverene, from disturbing them during the winter; and as they are frequently seen to walk over their work, and sometimes to give a flap with their tail, particularly when plunging into the water, this has, without doubt, given rise to the vulgar opinion that they used their tails as a trowel, with which they plastered their houses; whereas that flapping of the tail is no more than a custom which they always preserve, even when they become tame and domestic, and more particularly so when they are startled."

In the more northern climates the habitations of these animals are finished in August, or early in September, when they begin to lay in their stores. During the summer months they regale themselves on the choicest fruits and plants the country affords; but in winter they subsist principally on the wood of the birch, the plane, and some other trees, which they steep in fresh water from time to time. Those who are accustomed to hunt these animals, being perfectly aware that green wood is much more grateful to them than that which is old and dry, place a considerable quantity round their lodgments; and when they sally forth to seize it, either catch them in snares, or take them by surprise. When the frost is very severe, the hunters sometimes break large holes in the ice; and, on the Beavers resorting to these apertures to breathe the fresh air, they either kill them with their hatchets, or cover the holes with large substantial nets. This being done, they undermine and subvert the whole fabric; when the beavers, expecting to make their escape in the usual way, fly with precipitation to the water, and, rushing to the opening, fall directly into the net.

The Beaver is pursued both for its fur, and for the sake of a peculiar odoriferous secretion, termed *castor*, or *castoreum*, which is contained in two little bags, the inguinal glands, each about the size of a hen's egg. This substance, as we find it in the shops, is of a brownish unctuous consistence, has a disagreeable narcotic smell, and a nauseous acrid taste: it was at one time esteemed as possessing considerable medicinal properties, but is now chiefly employed by perfumers. The fur was formerly a most important article of commerce; but the animals have in recent times been exterminated from so many extensive tracts which they once inhabited, that it is now far less considerable than it was half a century ago. To this may be added, that the present custom of using silk and other materials in lieu of beavers' fur in the manufacture of hats, has wonderfully lessened the demand for it, as well as reduced the price. An idea, however, may be formed of the astonishing number of beavers' skins that were formerly made use of, when we state that in 1808, no less than 126,927 were sent from Quebec alone to this country. The flesh of the Beaver is much prized by the Indians and Canadian voyagers, especially when it is roasted in the skin after the hair has been singed off: and in some districts it requires all the influence of the fur-traders to restrain the hunters from sacrificing a considerable quantity of beaver fur every year to secure the enjoyment of this luxury; and Indians of note have generally one or two feasts in a season, wherein a roasted beaver is the prime dish. It resembles pork in its flavour, but it requires a strong stomach to sustain a full meal of it. (*Richardson's Fauna boreali-Americana.*)

Our readers will see that the foregoing account relates to the American Beaver. The European species does not boast of such architectural habits, but lives in burrows along the banks of the Rhone, the Danube, the Weser, and other large rivers in the north of Europe, yet, from some of the descriptions which have been given of them, we are disposed to believe that, considering the materials within their reach, their instinctive skill is not greatly inferior to those who dwell on the other side of the Atlantic. It is believed that at no period were Beavers common in Britain, though the mention of them by some of our earliest historians is a clear proof of their existence here.

BECCAFIGO, or FIG-EATER. (*Sylvia hortensis.*) A migratory song-bird, about the size of a linnet, but with a remarkably short body. It feeds on fruits and berries, and is highly prized by the Italians for the delicacy of its flesh, particularly in autumn, when it is in excellent condition for the table. It is often seen in England in the summer, where it is called the Pettychaps; but it generally returns to a warmer climate in September. It has a lively, loud, and piercing note; but it is seldom seen, as it usually sings from the midst of some closely embowered covert. Its head, back, neck, wings, and tail are generally of a greenish grey, but some more incline to a greenish brown.

BEE. (*Apis.*) The generic name of a family of Hymenopterous insects, [for the classification of which, see APIDÆ.] Of all the insect tribe none have more justly excited the attention and admiration of mankind than the Bee; and yet, although it has engaged the study of naturalists for two thousand years, we still occasionally find, in the economy of this social and industrious little animal, some obscurely known or unelucidated fact, which is thought worthy of the labours of those who devote their time and abilities in the pursuit and advancement of this interesting branch of natural science.

The most important species is the HONEY-BEE (*Apis mellifica*), so long celebrated for its wonderful polity, the neatness and precision with which it constructs its cells, and the diligence with which it provides during the warmth of summer a supply of food for the support of the hive during the rigours of the succeeding winter. In its natural state the Honey-bee generally constructs its nests in hollow trees; but so universally is it now domesticated that we rarely find it otherwise than hived in any part of Europe.

Peter Kalm, the Swedish traveller, observes, that the people of North America were unanimously of opinion that the

HONEY-BEE.—(APIS MELLIFICA.)

Honey-bee was unknown in that country before the arrival of the Europeans; but that they were first brought over by the English who settled there. The Indians likewise declared that their fathers had never seen any bees either in the woods or elsewhere, before the Europeans had been several years settled there. This, he says, is further confirmed by the name which the Indians gave them: for, having no particular name for them in their language, they call them *English flies*, because the English first brought them over; but at the time he wrote (nearly a century ago) they flew plentifully about the woods of North America.

Honey and wax are the two valuable articles of commerce for which we are indebted to this useful insect. Now, if we examine the structure of the common Bee, the first remarkable part which presents itself is the proboscis, an instrument serving to extract honey from flowers: it is not formed, like that of other flies, in the shape of a tube, by which the fluid is to be sucked up, but rather like a tongue, to lap it up. When thus lapped out of the nectary, it is conveyed to the crop or honey-bag; where it undergoes but little alteration, and is transferred or disgorged into the cells destined to receive it. While the Bee is busy in extracting the sweets of the flowers, it becomes covered with the farina or pollen of the anthers; this pollen it wipes off with the brushes of its legs, collects every particle together, and kneads it into two little masses, which it lodges on the broad surface of the tibia of each hind leg, where a series of elastic hairs over-arches a concavity, and acts as a sort of lid or covering. Thus employed, the Bee flies from flower to flower, increasing its store of honey, and adding to its stock of kneaded pollen, which is called *bee-bread*. The abdomen is divided into six annulations or rings, which are capable of being contracted or extended at pleasure; and the insect is internally furnished with a honey-bag, a venom-bag, and a sting. The honey-bag, which is as transparent as crystal, contains the honey which the Bee has brushed from the flowers, the greatest part of which is carried to the hive, and poured into the cells of the honeycomb, while the remainder serves for the Bee's own nourishment. Wax is a peculiar secretion in little cells beneath the scales of the abdomen. It is from honey that the wax, by some internal process, is elaborated. The wax oozes out between the abdominal rings, in the form of little laminæ; it is then worked with the mouth, and kneaded with saliva that it may acquire the requisite degree of ductility for the construction of the comb, which is finished with a substance called *propolis*, a glutinous or gummy resinous matter procured from the buds of certain trees.

The *sting* is composed of three parts; namely, the sheath, and two extremely small and penetrating darts, each of which is furnished with several points, or barbs, which, rankling in the wound, render the sting more painful. This instrument, however, would prove but a feeble weapon, if the Bee did not poison the wound. The sharp-pointed sheath first enters, and this being followed by the barbed darts, the venomous fluid is speedily injected. Sometimes the sting sticks fast in the flesh, and is left behind; but the death of the Bee invariably follows.

Having examined the Bee singly, we now proceed to an inquiry into its habits as a member of a social community. Viewed in this light, we behold an animal active, vigilant, laborious, and disinterested; subject to regulations, and perfectly submissive. All its provisions are laid up for the community; and all its arts are employed in building a cell, designed for the benefit of posterity. Many interesting accounts of the history and economy of the Bee have been published. We know of none, however, so concise and at the same time so explicit, as that which is given by Mr. Newman, in his "Familiar Introduction to the Study of Insects;" and to that source we are indebted for the following observations, marked with inverted commas:—

"A bee-hive contains three kinds of individuals,—a queen, drones, and workers; the queen is a female, and not only the ruler, but, in great part, the mother of the community; the drones are males, and the workers are abortive females. The sole office of the queen appears to be the laying of eggs, and this occupies her almost incessantly, as a single one only is deposited in each cell, thus causing her to be in continual motion; she is slow and majestic in her movements, and differs from the workers in being larger, having a longer body, shorter wings, and a curved sting. The queen is accompanied by a guard of twelve workers, an office which is taken in turn, but never intermitted; in whatever direction she wishes to travel, these guards clear the way before her, always with the utmost courtesy turning their faces towards her, and when she rests from her labours, approaching her with humility, licking her face, mouth, and eyes, and appearing to fondle her with their antennæ.

"The drones are all males; they are less than the queen, but larger than the workers; they live on the honey of flowers, but bring none home, and are wholly useless, except as being the fathers of the future progeny: when this office is accomplished, they are destroyed by the workers. A buzzing commences in the hive, the drones and the workers sally forth together, grapple each other in the air, hug and scuffle for a minute, during which operation the stings of the workers are plunged into the sides of the drones, who, overpowered by the poison, almost instantly die.

"The workers are the smallest bees in the hive, and by far the most numerous; they have a longer lip for sucking honey than either of the others; their thighs are furnished with a brush for the reception of the pollen of flowers, and their sting is straight. The workers do the entire work of the community; they build the cells, guard the hive and the queen, collect and store the honey, elaborate the wax, feed the young, kill the drones, &c. The average number of these three kinds of bees in a hive is one queen, 2000 drones, and 20,000 workers. The eggs are long, slightly curved, and of a bluish colour; when laid they are covered with a glutinous matter, which instantly dries, attaching them to the bottom of the cell.

"For eleven months the queen lays only workers' eggs; afterwards, those which produce drones: as soon as this change has taken place, the workers begin to construct royal cells, in which, without discontinuing to lay the drones' eggs, the queen deposits here and there, about once in three days, an egg which is destined to produce a queen. The workers' eggs hatch in a few days, and produce little white maggots, which immediately open their mouths to be fed; these the workers attend to with untiring assiduity: in six days each maggot fills up its cell; it is then roofed in by the workers, spins a silken cocoon, and becomes a chrysalis: and on the twenty-first day it comes forth a perfect bee. The drones emerge on the twenty-fifth day, and the queens on the sixteenth."

When the queen-bee has an inclination to deposit her eggs, she goes forth, accompanied by six or eight working bees as a guard, whose stomachs are filled with honey. She is very deliberate in her motions, and seems to proceed with great caution. She first looks into a cell, and if she finds it perfectly empty, she draws up her long body, inserts her tail into the cell, and deposits an egg. In this way she slowly proceeds till she has dropped ten or twelve eggs, when perhaps feeling exhausted, she is fed by one of the attendant bees, who have surrounded her the whole time. This is done by the bee ejecting the honey from its stomach into the mouth of the queen. When this has been done, the bee goes away, and another takes its place. The operation of laying her eggs again goes on, and is succeeded by the same mode of feeding,—the attendant bees frequently touching the antennæ of the queen with their own. When the operation of laying the eggs is completed,—and it generally occupies some time,—the queen retires to that part of the hive which is most filled with bees. During her progress the surface of the comb is very little intruded upon, and the space seems purposely to be left unoccupied. Some few of the cells, however, in a brood comb, are passed over by the queen, and afterwards filled either with honey or farina. These serve as deposits of food, from which the neighbouring brood may be fed more readily, as such cells are never covered with wax. — *Jesse.*

"It has been already stated, that the queen, for nearly a year, lays no eggs that are destined to produce queens; it therefore follows, that if any evil befall her, the hive is left without a queen: it sometimes happens that she dies, or is taken away by the owner of the hive, to observe the result. For twelve hours little notice is taken of the loss; it appears not to be known, and the workers labour as usual: after that period, a hubbub commences; work is abandoned; the whole hive is in an uproar; every bee traverses the hive at random, and with the most evident want of purpose. This state of anarchy sometimes continues for two days; then the bees gather in clusters of a dozen or so, as though engaged in consultation, the result of which seems to be a fixed resolution to supply the loss. A few of the workers repair to the cells in which are deposited the eggs of workers; three of these cells are quickly broken into one, the edges polished, and the sides smoothed and rounded, a single egg being allowed to remain at the bottom. When this egg hatches, the maggot is fed with a peculiarly nutritive food, called royal bee-bread, which is never given to any maggots but such as are to produce queens; work is now resumed over the whole hive, and goes on as briskly as before: on the sixteenth day the egg produces a queen, whose appearance is hailed with every demonstration of delight, and who at once assumes sovereignty over the hive. When, under ordinary circumstances, a young queen emerges from the chrysalis, the old one frequently quits the hive, heading the first swarm for the season, and flying to some neighbouring resting-place, is observed by the owner, captured, placed under a new hive, and a new colony is immediately commenced. Before a swarm leaves the hive, sure indications are given of the intended movement; the workers leave their various occupations and collect in groups, especially near the door of the hive, as though in consultation on the important event about to take place.

"As the summer advances many queens are hatched, but the workers do not allow them instant liberty, as severe battles would take place between them and the reigning queen, in which one would be killed: the workers, therefore, make a small hole in the ceiling of the royal cell, through which the captive queen thrusts her tongue, and receives food from the workers. In this state of confinement the young queen utters a low querulous note, which has been compared to singing. When the reigning, or a newly-created queen, finds one of these captives, she uses every effort to tear open the cell and destroy her rival: to prevent this, the workers often interpose, pulling her away by the legs and wings; to this she submits for a short time, when, uttering a peculiar cry, called her voice of sovereignty, she commands instant attention and obedience, and is at once freed from her assailants. The cocoons spun by the maggots of the workers and drones completely envelope the chrysalis; but that spun by the maggot of the queen appears imperfect, covering only the upper end of the chrysalis: it has been supposed that they are thus designedly exposed to the attacks of other queens, and their destruction, before emerging, facilitated. When the chrysalis of the queen is about to change to a perfect insect, the bees make the cover of the cell thinner by gnawing away part of the wax; and with so much nicety do they perform this operation, that the cover at last becomes pellucid, owing to its extreme thinness.

"The combs of a bee-hive comprise a congeries of hexagonal cells, built by the bees as a receptacle for honey, and for the nurseries of their young: each comb in a hive is composed of two ranges of cells, backed against each other: the base or partition between this double row of cells is so disposed as to form a pyramidal cavity at the bottom of each. There is a continued series of these double combs in every well-filled hive; the spaces between them being just sufficient to allow two bees, one on the surface of each comb, to pass without touching. Each cell is hexagonal, the six sides being perfectly equal. This figure ensures the greatest possible economy of material and space; the outer edges of the cells are slightly thickened, in order to gain strength; the same part is also covered with a beautiful varnish, which is supposed to give additional strength. The construction of several combs is generally going on at the same time: no sooner is the foundation of one laid, with a few rows of cells attached to it, than a second and a third are founded on each side, parallel to the first, and so on till the hive is filled, the combs which were commenced first being always in the most advanced state, and therefore the first completed.

"The design of every comb is sketched out, and the first rudiments laid by a single bee: this foundress-bee forms a block out of a rough mass of wax, drawn partly from its own resources, but principally from those of other bees, which furnish wax from small sacs, in which it has been secreted, that are situated between the segments of the body of the bee; taking out the plates of wax with their hind feet, and carrying it with their fore feet to their mouths, where it is moistened, masticated, and rendered soft and ductile. The foundress-bee determines the relative position of the combs, and their distance from each other: the foundations which she marks serving as guides to the ulterior labours of the wax-working bees, and of those which build the cells, giving them the advantage of the margins and angles already formed. The mass of wax prepared by the assistants is applied by the foundress-bee to the roof or bottom of the hive, and thus a slightly double convex mass is formed: when of sufficient size, a cell is sculptured on one side of it by the bees, who relieve one another in the labour. At the back, and on each side of this first cell, two others are sketched out and excavated: by this proceeding the foundations of two cells are laid, the line betwixt them corresponding with the centre of the opposite cells: as the comb extends, the first excavations are rendered deeper and broader; and when a pyramidal base is finished, the bees build up walls from its edges, so as to complete what may be called the prismatic part of the cell. The cells intended for the drones are considerably larger and more substantial than those for the workers; and being formed subsequently, they usually appear nearer the bottom of the combs: last of all are built the royal cells for the queens: of these there are usually three or four, sometimes ten or twelve, in a hive, attached completely to the central part, but not unfrequently to the edge of the comb. The form of the royal cells is an oblong spheroid, tapering gradually downwards, and having the exterior full of holes: the mouth of the cell, which is always at the bottom, remains open until the maggot is ready for transformation, and it is then closed like the rest.

"When a queen has emerged, the cell in which she was reared is destroyed, and its place is supplied by a range of common cells: the site of this range may always be traced by that part of the comb being thicker than the rest, and forming a kind of knot. The common breeding cells of drones and workers are occasionally made the depositories of honey; but the cells are never sufficiently cleansed to preserve the honey undeteriorated. The finest honey is stored in new cells constructed for the purpose of receiving it, their form precisely resembling that of the common breeding cells: these honey-cells vary in size, being larger or smaller according to the productiveness of the sources from which the bees are collecting, and according to the season."

It is remarkable that all animals which have been long under the protection of man seem to lose a part of their natural sagacity. In those countries where the bees are wild, and unprotected by man, they are always sure to build their waxen cells in the hollows of trees; but with us they appear improvident in their choice; and the first green branch which stops their flight is deemed sufficient for their abode. It does not even appear that the queen chooses the place where they are to alight; for numbers of the swarms when they conceive a predilection for any particular branch, spontaneously settle on it; others follow their example; and at last the queen herself, finding the majority of the swarm convened together, condescends to place herself amongst them. The queen being settled, the rest of the swarm soon flock around her, and in about a quarter of an hour the whole body seems to be perfectly at rest.

When a hive sends out several swarms in a year, the first is always the best as well as the most numerous; for, having the greatest part of the summer before them, they have the more time for making wax and honey, and consequently their labours are the most valuable to their proprietor. Though the swarm is principally made up of the younger Bees, those of all ages generally compose the number of emigrants; and as a single hive sometimes contains upwards of forty thousand inhabitants, such a vast body may well be supposed to work with great expedition.

Among the varied mass of amusing and instructive information with which the volumes of Kirby and Spence abound, we shall make a few condensed extracts ere we close this article:—Bees in their excursions do not confine themselves to the spot immediately contiguous to their dwelling, but, when led by the scent of honey, will go a mile from it, or considerably more; yet from this distance they will discover honey with as much certainty as if it was within their sight. * * * A new-born bee, as soon as it is able to use its wings, seems perfectly aware, without any previous instruction, what are to be its duties and employments for the rest of its life. It appears to know that it is born for society, and not for selfish pursuits; and therefore it invariably devotes itself and its labours to the benefit of the community to which it belongs. Walking upon the combs, it seeks for the door of the hive, that it may sally forth and be useful. Full of life and activity, it then takes its first flight; and, unconducted but by its instinct, visits like the rest the subjects of Flora, absorbs their nectar, covers itself with their ambrosial dust, which it kneads into a mass and packs upon its hind legs; and, if need be, gathers propolis (an unctuous resinous substance, collected from the buds of trees, and used in lining the cells of a new comb, stopping crevices, &c.), and returns unembarrassed to its own hive.

The method of ventilating their hives is thus described:—By means of their marginal hooks, they unite each pair of wings into one plane slightly concave, thus acting upon the air by a surface nearly as large as possible, and forming for them a pair of very ample fans, which in their vibrations describe an arch of 90°. These vibrations are so rapid as to render the wings almost invisible. During the summer a certain number of workers—for it is to the workers solely that this office is committed—may always be observed vibrating their wings before the entrance of their hive; and the observant apiarist will find, upon examination, that a still greater number are engaged within it in the same employment. The station of these ventilators is upon the floor of the hive. They are usually ranged in files that terminate at the entrance; and sometimes, but not constantly, form so many diverging rays, probably to give room for comers and goers to pass. The number of ventilators in action at the same time varies: it seldom much exceeds twenty, and is often more circumscribed. The time also that they devote to this function is longer or shorter, according to circumstances: some have been observed to continue their vibrations for nearly half an hour without resting, suspending the action for not more than an instant, as it should seem to take breath. When one retires, another occupies its place; so that in a hive well peopled there is never any interruption of the sound or humming occasioned by this action, by which it may always be known whether it be going on or not. [For Reproduction of Bees, see PARTHENOGENESIS in SUPPL.]

BEE-EATER. (*Merops.*) There are many species of this genus, all being distinguished by brilliant plumage; they take their prey, consisting of bees, wasps, gnats, &c., on the wing, like the swallow, and without being stung by them. The one we are about to describe is among the most elegant of the European birds, and, next to the Roller and the Kingfisher, may be considered as the most brilliant in point of colour. It is a native of the warmer parts of Europe, particularly among the islands of the Grecian archipelago, and of many parts both of Asia and Africa; but in the north of Europe it is rarely seen. In shape this bird resembles the halcyon tribe, and is about the size of a blackbird. The bill is slightly curved, sharp-pointed, rather long, and black; the irides bright red; the crown

BEE-EATER.—(MEROPS APIASTER.)

of the head and upper parts of the neck and back are of an orange-chesnut colour; the throat yellow, the scapulars, lower part of the back, and wing-coverts pale yellow, more or less shaded with an admixture of red and green. The smaller quill feathers are rufous chesnut, tipped with green; the larger sea-green with dusky tips; the rump and tail sea-green, the latter about three inches long, the two middle feathers projecting, in a pointed form, to some distance beyond the rest. The sides of the head, above the eyes, and the whole under parts are sea-green: from the corners of the bill, on each side of the head, a black streak passes across the eyes, curving downwards, and nearly meeting the tips of a black crescent placed across the snout, and separating the yellow of that part from the sea-green of the under parts. The legs are short, and of a reddish-brown colour. It builds in deep holes in the banks of rivers, forming a nest of moss, and laying from five to seven white eggs.

The INDIAN BEE-EATER (*Merops viridis*) is about half the size of the common or European Bee-eater, but the middle tail-feathers are considerably longer. On the upper part of the breast is a crescent-shaped transverse mark, with the horns pointing upwards; the back and lesser covert-feathers of the wings are of a parrot-green colour; the rump or coverts of the tail of a bluish-green; the breast and belly of a light green, and the tail is green. The greater quills of the wings are dusky at their tips; the centre quills are of an orange colour, bordered with green, and marked with black spots, the extreme tips being orange; the interior quills next the back are wholly green; the first row of coverts above the quills is orange in the centre, and green on the edges. The bill is long and sharp-pointed, having a downward incurvation; the claws are pretty strong; and the legs and feet of a dusky brown colour. This species is a native of Bengal, parts of Madagascar, &c.

BEETLES. The insects composing the order COLEOPTERA, or Beetles, are almost incredible in point of number, as may readily be supposed when it is stated that between 70,000 and 80,000 species at present exist in the cabinets of collectors. The singular forms and brilliant colours of many of them; the size of their bodies; the solid texture of their integuments, which renders their preservation comparatively easy; and the nature of their habits, which affords every facility for their capture; have combined to render them objects of peculiar attention to those who delight in the science of entomology.

Among the beetle tribe some are very remarkable for projections or horns growing from the head and corslet. The species found in warm climates are generally large and of a formidable appearance, though by no means noxious. They are mostly winged, flying with much rapidity and force; but when on the ground their movements are slow and heavy. The wings of beetles are covered and concealed by a pair of horny cases or shells, meeting in a straight line on the top of the back, and usually having a little triangular or semicircular piece, called the scutel, wedged between their bases. Hence the order to which these insects belong is called *Coleoptera*, a word signifying wings in a sheath. They are all produced from eggs; they then become grubs; afterwards they are changed into chrysalides; and lastly, the beetles, leaving their prisons, sally forth as winged insects in full maturity. The conversion of the first pair of wings into *elytra*, or hard wing-cases, and the complete inclosure of the second pair by these, when the insect is at rest, constitute the distinguishing features of the order. The elytra, when expanded, are of little or no use in flight, generally remaining nearly motionless; when closed, they meet along the back in a straight line, which is called the *suture*. The body of the perfect insect is oval, or nearly so, and the head is provided with two *antennæ*, composed of eight or ten pieces; the extremities of the antennæ are club-shaped, and composed of plates or joints, either disposed like the leaves of a book, or arranged perpendicularly to the axis, like the teeth of a comb. The eyes are large and protuberant, especially in the carnivorous species, and in those, the slowness of whose habits makes them need quick powers of sight, for the purpose of avoiding their enemies. Of the three segments of the thorax, the corslet greatly surpasses the two others in size; and the chief movement of the parts of the trunk upon one another, is between the first and second segments of the thorax. The two fore-legs of beetles, and even the others, in some instances, are dentated externally, and suited for burrowing. These are the principal characters which distinguish this numerous family; but it is necessary to observe that nearly all of them are subject to some exceptions.

The *larvæ* are soft, flexible, whitish, semi-cylindric worms, having the body divided into twelve rings, and having a scaly head, armed with strong jaws. They have nine *stigmata*, or breathing-holes, on each side; and the feet, which are six, are scaly. The body is thicker at the posterior than at the anterior extremity, and rounded, almost uniformly curved downwards, so that the larva moves with difficulty over an even surface, and frequently tumbles down. The period during which the larvæ remain in the state of destructive worms, varies in different species; those of some kinds becoming nymphs at the end of several months, and of others not sooner than three or four years. During this period they live in the earth, where they feed upon the roots of vegetables, animal matter in a state of decomposition, &c. It is in this stage of their existence that various species prove exceedingly injurious to farmers and gardeners, from their great numbers and voracity. When about to undergo their change of form, they make an egg-shaped cocoon, from fragments gnawed off wood, &c., which are united by a peculiar glutinous fluid furnished by their bodies. The form of the future beetle is now plainly perceived, the different parts being encased in distinct sheaths. Though the varieties of this genus arising from size and colour are wonderful — some being no larger than a pin's head, while others are several inches in length and circumference, — their most essential difference proceeds from the stages of their existence, some undergoing all their transformations in a few months, and others requiring nearly four years to complete their production.

BELEMNITES. A genus of fossil Cephalopoda, which at different periods have received the names of *Thunderstone*, *Arrowhead*, and *Fingerstone*. The name is derived from *Belemnon* (Gr.), a dart or arrow. They abound in several of the older rocks, especially the lias and oolite; and consist of an interior cone divided into partitions connected by a syphon, as in the Nautilus, and surrounded by a number of concentric layers, made up of fibres radiating from the axis.

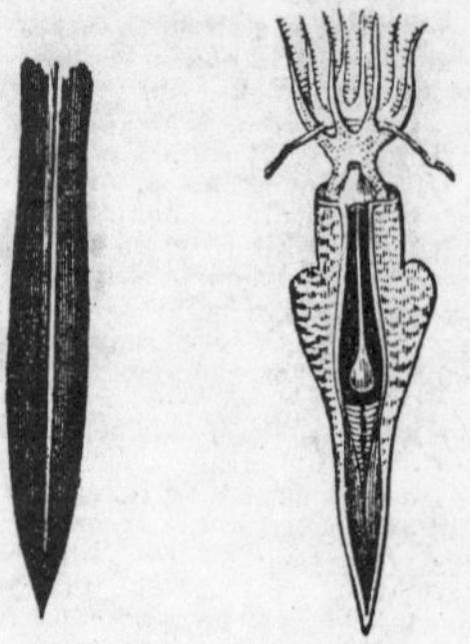

BELEMNITE. (B. ACUTUS.) BELEMNITE [RESTORED.]

These layers are somewhat transparent, and when burnt, rubbed, or scraped, give the odour of rasped horn. From the weight of its dense internal shell the Belemnite may be supposed to have usually maintained a vertical position; and as its chambered portion was provided with a siphuncle analogous to that which we find in the Nautilus, the animal probably had the power of ascending and descending in the water with facility. The animal, of which the Belemnite was the internal "bone," has been proved by Mr. Owen to have been a dibranchiate eight-armed Cuttle-fish, somewhat resembling the recent genus *Onychoteuthis*. This he was enabled to do by access to specimens found near Chippenham, in Wiltshire, during the excavations that were making for the Great Western Railway. The species are now extinct.

BELL-BIRD. (*Procnias carunculata.*) This is a species of Chatterer, distinguished by a long soft caruncle at the base of its beak; it is white when adult, greenish when

BELL-BIRD.—(PROCNIAS CARUNCULATA.)

young. It is a native of South America — the celebrated Campanero or Bell-bird of Guiana — the loud sonorous voice of which, heard from time to time in the depths of the forest, during the stillness of mid-day, exactly resembles the tolling of a bell.

Mr. Waterton, in his hearty "Wanderings in Demerara," often alludes to it. In one passage he says that it "never fails to attract the attention of the passenger; at a distance of nearly three miles you may hear this snow-white bird tolling every four or five minutes like the distant convent bell. From six to nine in the morning the forests resound with the mingled cries and strains of the feathered race, after this they gradually die away. From eleven to three, all nature is hushed as in a midnight silence, and scarce a note is heard, saving that of the *campanero* of the *pi-pi-yo;* it is then that, oppressed by the solar heat, the birds retire to the thickest shade, and wait for the refreshing cool of the evening."

BELLEROPHON. A genus of fossil shells, the animals of which are unknown, but which are now generally supposed to have been allied to *Carinaria*, the structure of whose shell it resembles.

BELODON. A title given by Von Meyer to a genus of extinct reptiles found in the upper white keuper sandstone in the neighbourhood of Wirtemberg. The species appear to be closely allied to the great Dinosaurians of the oolite.

BELONE. A genus of fishes remarkable for the bright green colour of their bones. The jaws are much extended, and furnished with small teeth, without any others in the mouth, except in the pharynx. The body is very long, and covered with scales which are scarcely visible, except one keeled row on each side, near the under edge of the fish. [See GARFISH.]

BELUGA. (*Delphinus leucos.*) A Cetaceous animal, of the size of the Grampus. It chiefly inhabits the seas of the Arctic regions, but is sometimes met with even on the British coasts. [See WHALE.] It is also the Russian name for the largest species of Sturgeon (*Accipenser huso*).

BELYTA. A genus of Hymenopterous insects, being a species of minute four-winged flies, which frequent sandy situations.

BEMBEX: BEMBECIDÆ. A genus and family of Hymenopterous insects, peculiar to hot climates, and, in some instances, very much resembling wasps both in size and colour. *Bembex rostrata*, an insect about the size of a wasp, is the type of this family, and is remarkable for having the lower parts of the mouth produced into a long trunk or proboscis. The female forms oblique cylindrical burrows in sandy banks, with a cell at the end of each, and having collected five or six flies, and placed them in her cell, she deposits a single egg in it; then having carefully closed its mouth, she proceeds in the same manner with another cell. These flies are no sooner hatched than the larva devours them; it then changes into the pupa state, and shortly after to the perfect insect.

BEMBIDIIDÆ. A family of minute carnivorous beetles, which generally frequent damp situations, such as the banks of rivers, ditches, &c. They are usually of a bright blue or green metallic colour, having two or four pale yellow spots on the elytra.

BERNACLE or BARNACLE GOOSE. (*Bernicla leucopsis*). A bird which inhabits the arctic regions, and in its autumnal and brumal migrations visits the more temperate regions of England, France, Germany, Holland, &c. It frequents the north-west coasts of this country, and some parts of Ireland, in large flocks during the winter, but is rarely

BERNACLE GOOSE. (BERNICLA LEUCOPSIS.)

seen in the south except in very severe weather. About February it retires to the north to breed, and is then found in Russia, Lapland, Iceland, Spitzbergen, and other high latitudes.

The length of the Bernacle is rather more than two feet. The bill is black, with a reddish streak on each side, and between it and the eyes is a small black streak; the irides dusky brown; the forehead, sides of the head, and the throat, are of a pure white; the rest of the head, neck, and shoulders black, the upper part of the plumage is marked with blue, grey, black and white; and the legs are black.

The history of this bird has been rendered singularly remarkable by the marvellous accounts which were related in the darker ages concerning its growth; it being a received opinion that the Bernacle was produced in a kind of cirripede, the *lepas anatifera* of Linnæus, growing on rotten ship-timber and other kinds of wood, and trees which lay under water on the coasts! Among these is Gerard, a famous botanist in his day, whose account is too absurd to give in detail, but perhaps a short extract may be tolerated: "When it is perfectly formed, the shell gapeth open, and the first thing that appeareth is the aforesaid lace or string; next cometh the legs of the bird hanging out; and as it groweth greater, it openeth the shell by degrees till at length it has all come forth, and hangeth only by the bill. In short space after it cometh to full maturity, and falleth into the sea, where it gathereth feathers, and groweth to a fowl, bigger than a mallard, and lesser than a goose, having black legs, and bill or beak, and feathers black and white, spotted in such manner as our magpie!" Again, Sir Robert Murray, in his account inserted in the Philosophical Transactions, says that he found "an old fir tree on the coast of Scotland, covered with bernacle shells, and that in every shell that he opened he found a perfect sea-fowl; the little bill, like that of a goose; the eyes marked; the head, neck, breast, wings, tail, and feet formed; the feathers every where perfectly shaped, and blackish coloured; and the feet like those of other water-fowl!" Such are some of the wild chimeras that have been handed down concerning the origin of these birds; such the dangerous contagion of the errors of science, where the imagination is allowed to soar beyond the region of common-sense.

There are several other species, some of which we shall briefly describe:—

The RED-BREASTED BERNACLE. (*Bernicla ruficollis.*) This is a beautiful bird, about twenty-two inches in length; the beak is brown, with its hook black; between the beak and the eye is a white space; behind the eyes and on the sides of the neck it is white; the top of the head, the throat, belly, tail, and all the upper parts are deep black; the vent, under tail-coverts, and rump are pure white; but the breast and fore part of the neck are bright red. A band of black extends the entire length of the hinder part of the neck; the greater wing-coverts are tipped with white; and the legs are black. This beautiful bird inhabits the arctic countries of Asia, living on the borders of the Frozen Ocean: it appears periodically in Russia, and occasionally in Germany; but in England it is very rarely seen. A British-killed specimen, however, has been seen by us in the British Museum.

The WHITE-WINGED BERNACLE. (*Bernicla leucoptera.*) This bird varies in length from about thirty-two to forty inches; the head, neck, lesser wing-coverts, and under parts of the body, white; the lower part of the neck behind, and as far as the middle of the back, crossed with numerous dusky-black lines; the two middle tail-feathers black; the rest white; and the legs black. It stands pretty high upon its legs; walks and flies with great ease; and has not that disagreeable cackling cry peculiar to the rest of its kind. The flesh is wholesome and nourishing. It inhabits the Falkland Islands, where it is called the Bustard Goose.

The ANTARCTIC BERNACLE. (*Bernicla Antarctica.*) This is rather smaller than a tame goose: beak narrow, short, and black; the whole plumage of a dazzling snowy whiteness; on the bend of the wing a blunt knob: legs yellow. It inhabits Christmas Sound, in Terra del Fuego. Its flesh is unfit to be eaten.

BEROE. (*Beröe*, or *Cydippe pileus.*) A small marine animal belonging to the class *Acalepha*, and to which the name *Cydippe* is now very frequently applied. This little animal is nearly of a globular form, somewhat elongated, and about three-fourths of an inch in length. It is composed of a gelatinous substance, strengthened by eight bands of rather firmer texture, which are covered with rows of large vibratile *cilia*, arranged side by side, so as to form narrow plates of a fin-like character. There are, in the most common species, from three to seven cilia in each row, and about twenty rows on each ridge: over these the Beröe has complete control; it can retard or stop their movements at pleasure; and arrest the play of one, two, or more rows, whilst the remainder continue in rapid vibration, and act like so many little paddles. By these means it is capable of swimming through the water with considerable activity, and of changing its course at will. These little animals are of a bright faintly-blue aspect; and the cilia when in motion present vivid iridescent hues. The mouth is situated at one end, which is always directed forward when the animal is in motion, and is then widely dilated. From the stomach, there passes a narrow straight intestine, which terminates at the opposite extremity of the body. When the Beröe is in active movement, therefore, a continual stream of water will enter its mouth, and pass out again behind; and from the minute particles contained in the water, it evidently derives its nourishment; exceedingly minute crustacea may indeed be seen in the transparent stomach for some time after being swallowed. From the posterior part of the body arise two lengthened

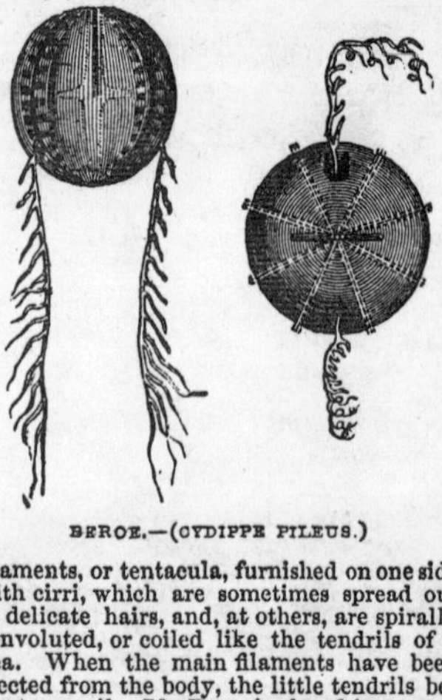

BEROE.—(CYDIPPE PILEUS.)

filaments, or tentacula, furnished on one side with cirri, which are sometimes spread out as delicate hairs, and, at others, are spirally convoluted, or coiled like the tendrils of a pea. When the main filaments have been ejected from the body, the little tendrils begin to uncoil. If a Beröe is placed in a vessel of sea-water, its various movements may be watched with interest: sometimes it remains at the bottom, projecting its long filaments upwards; at others, it darts swiftly upwards, drawing its long filaments after it, and alternately retracting and extending them; not unfrequently it remains for some time at the top of the water, till at length, wishing to descend, it turns over, drawing up its filaments suddenly, and then swims, mouth-downwards, to the bottom.

In a small but intertesing volume on the

Natural History of Arran by the Rev. David Landsborough, the author makes the following remarks on the species *Beröe cucumis*, several specimens of which he had taken during his "Excursions," to that island; the largest being three inches in length, by about one inch and a half in diameter. They varied, he says, from the size of a lemon to that of a lady's thimble, were very beautiful, and in shape resembling an antique pitcher contracted at the neck, with a graceful revolution, or turning back at the brim; but the exact form was difficult to assign, as it varied by partial contractions at the animal's pleasure. "The whole body has a tinge of pink, and the eight ribs closely set with cilia are beautifully adorned, having on each side an edging like fine crimson lace. In the larger specimens, this lace-work was studded with little orange oval-shaped bodies, like little grapes, attached by a capillary peduncle. When the Beröe was at rest, they rested; but when the cilia began rapidly to play, and the current of water, mixed at times with air-bubbles, to rush through the tubes of the ribs, then all the little orange bodies were in quick motion, as if dancing to the music of the spheres; or, believing in fairies as our forefathers did, one might have fancied that they were lace-bobbins, moved by nimble, invisible fairy hands, weaving the beautiful lace edging with which they were intermingled. Professor Forbes, however, says, as I had conjectured, that they are the eggs attached to the placentary membranes; and I doubt not that they are thus shaken by the motion of the cilia, that when fully ripe they may thereby be detached."

Mr. Rymer Jones, in describing the beautiful mechanism of the Beröe, has made some pertinent philosophical reflections on it, in language at once elegant and forcible. "Man," says he, "justly prides himself, among the countless triumphs of his intellect over the stubborn elements, at his success in having found the means of struggling through the opposing surge, propelled by steam revolving wheels whose paddles urge his vessel on with giant force. But man in this contrivance, as in many more, is but a bungling artist when compared with Nature, when he chooses to adopt machinery which she likewise has employed. Examine well the *beröe*, and see if any paddle-wheels can equal hers. Stretching from pole to pole of this translucent little orb, like lines of longitude upon a globe, and placed at equal distances, are eight broad bands of more consistence than the other portions of the body. On these bands are placed thirty or forty paddles, broad flat plates, for such they seem when magnified, with which the little creature rows itself along. But here the difference lies between the art of Man and Nature. Man to move his wheels must have much cumbersome machinery; the furnace, and the boiler, and the Herculean arm that makes the wheel revolve; but here all these may be dispensed with, for the paddles are themselves alive, and move themselves at will with such degree of force as may be needed, either at once, or singly, or in groups, working with mutual consent in any way required. Thus do they all work equally; the *beröe* shoots along meteor-like, or, if a few relax their energy, wheels round in broad gyrations, or revolves on its own axis with an ease and grace inimitable."

BETTONGIA. A genus of Kangaroos, one of the species of which is called "Forest Rat" by the colonists of Van Diemen's Land (*B. cuniculus*): the end of the tail in this species has a white tuft. Another species (*B. fasciata*) was found by M. Peron on the west coast of Australia, at Dirk Hartog. It is very timid, and constructs galleries among the thick brushwood, by cutting away the lower branches and spines. It is of a brown colour, the lower part of the back being banded across with darker lines.

BIBIO: BIBIONIDÆ. A genus and sub-family of Dipterous insects, distinguished from all the other *Tipulidæ* by having the body and legs shorter and more robust; the antennæ cylindric, monliform, or perfoliated; wings large; and the eyes of the males large and generally contiguous. There is great diversity in the sexes of the genus *Bibio;* all the species are of small size; and their flight is slow and heavy. They are found in damp, marshy places, flying in great swarms, and some of the species are amongst the most troublesome pests to our domestic animals.

BIMANA. [Two-handed.] The term applied by Cuvier to the first or highest order of Mammiferous Animals. It contains only one genus, and one species,—MAN; the sole created Being that can be termed truly *bimanous* and truly *biped*. The whole body of Man is adapted for the vertical position: he walks erect; and thus preserves the entire use of his hands for the arts, while his organs of sense are most favourably situated for observation and the great mental purposes assigned to them by the Great Author of Nature. [See the articles MAMMALIA and MAN.]

BIPELTATA. A name given to those Crustacea which have the carapax divided into two shields, the anterior of which is very large, more or less oval, composing the head; and the second, corresponding with the thorax, is transverse and angulated in its outline, and bears the foot-jaws and the ordinary feet. The body is very flat, membranous, and transparent, with the abdomen small, and without spines to the posterior swimmeret. All the species are inhabitants of the Atlantic and Eastern Oceans.

BIPES. A genus of Reptiles in which the hind feet alone are visible, there being externally a total absence of the anterior extremities, though the rudiments of these members are perceptible under the skin. This genus affords an example of one of those beautiful gradations by which Nature glides from one type of form into another, being intermediate between the Saurians (lizards) and the Ophidians (serpents).

BIRDS. In the following observations on the structure, habits, and uses of Birds, we have endeavoured to collect, from the writings of various Ornithologists, such particulars as appeared to be best calculated to illustrate the subject in a manner the most simple, natural, and familiar; and in so doing we have made the just and sensible remarks of the ingenious Thomas Bewick the basis on which to build whatever we have thought necessary to add, or to glean from other sources.

Every part of nature is furnished with its proper inhabitants; the woods, the waters, and the depths of the earth, have their respective tenants; while the passive air and those tracts of seeming space too elevated for man to ascend, are traversed by multitudes of feathered beings, whose buoyancy and beauty are alike the objects of our admiration. But the symmetry and elegance discoverable in their outward appearance, although highly pleasing to the sight, are yet of much greater importance when considered with respect to their peculiar habits and mode of living, to which they are eminently subservient. Instead of the large head and formidable jaws, the deep capacious chest, the brawny shoulders, and the sinewy legs of the quadrupeds; we observe the pointed beak, the long and pliant neck, the gently swelling shoulder, the expansive wings, the tapering tail, the light and bony feet; which are all wisely calculated to assist and accelerate their motion through the yielding air. Every part of their frame is formed for lightness and buoyancy; their bodies are covered with a soft and delicate plumage, so disposed as to protect them from the intense cold of the atmosphere through which they pass; their wings are made of the lightest materials, and yet the force with which they strike the air is so great as to impel their bodies forward with astonishing rapidity, whilst the tail serves the purpose of a rudder to direct them to the different objects of their pursuit. The internal structure of birds is no less wisely adapted to the same purposes; all the bones are light and thin, and all the muscles, except those which are appropriated to the purpose of moving the wings, are extremely delicate and light; the lungs are placed close to the back-bone and ribs; the air entering into them by a communication from the wind-pipe, passes through, and is conveyed into a number of membraneous cells which lie upon the sides of the pericardium, and communicate with those of the sternum. In some birds these cells are continued down the wings, and extended even to the pinions, thigh-bones, and other parts of the body, which can be filled and distended with air at the pleasure of the animal.

All birds are furnished with two very strong pectoral muscles on each side of their breast-bones. In quadrupeds, as well as in men, the pectoral muscles are trifling in comparison with those of birds. In the former, the muscles of the thighs and the hinder parts of the body are by far the strongest; but in birds it is far otherwise; the pectoral muscles which give motion to their wings are amazingly strong, whilst those of their thighs are weak and slender. By means of these a bird can move its wings with a degree of strength which is almost incredible: the flap of a swan's wing would break the leg of a man; and a similar blow from an eagle has been known to cause instant death. Such, consequently, is the force of the wing, and such its lightness, as to be inimitable by human art.

The eyes of birds are admirably adapted to vision, by a particular expansion of their optic nerves, which renders the impression of external objects more vivid and distinct. From this peculiar conformation, it appears that the faculty of sight in birds is infinitely superior to that of other animals, and, indeed, is indispensably necessary to their support and security. Were the eye less perfect, the bird, from the rapidity of its motion, would probably strike against almost every object in its way; as well as be totally incapable of discerning its proper food when soaring in its own element.

In mental capacity birds fully equal quadrupeds, and in some respects surpass them. Parrots, starlings, &c., retain in memory many words and phrases which they have been taught, and many singing-birds whole melodies. Their powers of memory seem also to be evinced by the fact that birds of passage, after an absence of six months, or even a longer time, and after travelling thousands of miles, return to their former home; the swallow to her beam, the finch to the tree where last year she reared her young, or where she herself was hatched. The difference between such birds as love to dwell in uninhabited places, secure from persecution, and such as are found in the neighbourhood of men, surrounded by dangers, is a proof that their prudence, cunning, and docility can be awakened and improved.

The voice is a peculiar gift of Nature, by which the greater part of birds are distinguished from all the rest of the animal world. The wind-pipe of birds is composed of entire rings of cartilage, with an exception in the case of the ostrich. At its bifurcation is a glottis supplied with appropriate muscles, called the lower or inferior larynx. It is here that the voice of birds is formed; the vast body of air contained in the air-cells contributes to the force, and the wind-pipe, by its form and movements, to the modification, of the voice. The superior larynx is very simple and unimportant. The gift of song is given to the male birds only, and their notes are mostly an expression of love. They sing only when they are cheerful; in sadness, during rough weather, and in bodily disorders, they are silent. It is commonly said that the gift of song is confined to the birds in northern climates, and that nature, in the warmer regions, has endowed them, instead, with more brilliant colours; but Foster relates, that in Otaheite the birds sing with charming sweetness; and Cook, on his first voyage, found the forests of Queen Charlotte's Sound, in New Zealand, filled with little birds, whose voices sounded like silver bells. To no other animal have such various tones been granted for giving utterance to different feelings: hunger, fear, the dread of imminent danger, desire for society, or longing for his mate, love, melancholy, &c., are expressed by a variety of notes, which make a language intelligible not only to birds of the same species, but often to the other tribes.

With respect to the feathers of birds, it has been truly said, that every form which the most sportive fancy could create out of the feathery material, and every hue that the warmest imagination could picture to itself, will be found among them. As a general rule, the plumage of the cock bird far exceeds in brilliancy that of the hen; and in many species a striking difference is observable between their plumage in summer and in winter; but in most birds two changes are observable; namely, that in the spring, which indicates the approach of the breeding season, when the bird obtains a certain portion of new feathers without losing any of the old ones; and in autumn, which is the moulting season, when the old feathers are thrown off, and new ones are produced in their places.

Birds may be distinguished, like quadrupeds, into two kinds or classes—granivorous and carnivorous; like quadrupeds too, there are some that hold a middle nature, and partake of both. Granivorous birds are furnished with larger intestines, and proportionally longer, than those of the carnivorous kind. Their food, which consists of grain of various sorts, is conveyed whole and entire into the first stomach or craw, where it undergoes a partial dilution by a liquor secreted from the glands and spread over its surface; it is then received into another species of stomach, where it is further diluted; after which it is transmitted into the gizzard or true stomach, consisting of two very strong muscles, covered externally with a tendinous substance, and lined with a thick membrane of prodigious power and strength: in this place the food is completely triturated and rendered fit for the operation of the gastric juices. The extraordinary powers of the gizzard in comminuting the food, so as to prepare it for digestion, would exceed all credibility, were they not supported by incontrovertible facts founded upon experiments. In order to ascertain the strength of these stomachs, the ingenious Spalanzani made the following among many other curious and interesting experiments:—He fixed twelve small lancets, very sharp, in a ball of lead, which was forced down the throat of a turkey-cock, and left eight hours in the stomach; at the expiration of which the organ was opened, but nothing appeared except the naked ball, the twelve lancets having been broken to pieces, the stomach remaining perfectly sound and entire. We may observe also, that stones taken into the stomach of birds are seldom known to pass with the fæces, but being ground down and separated by the powerful action of the gizzard, are mixed with the food, and, no doubt, contribute essentially to the health of the animal.

Carnivorous Birds are distinguished by those endowments and powers with which they are furnished by nature for the purpose of procuring their food: they are provided with wings of great length, the muscles which move them being proportionally large and strong, whereby they are enabled to keep long upon the wing in search of their prey: they are armed with strong hooked bills, and sharp and formidable claws; they have also large heads, short necks, strong and brawny thighs, and a sight so acute and piercing, as to enable them to view their prey from the greatest heights in the air, upon which they dart with inconceivable swiftness and undeviating aim. The analogy between the structure of rapacious birds and carnivorous quadrupeds is obvious; both of them are provided with weapons which indicate destruction and rapine; their manners are fierce and unsocial; and they seldom live together in flocks, like the inoffensive granivorous tribes. When not on the wing, rapacious birds retire to the tops of sequestered rocks, or to the depths of extensive forests, where they conceal themselves in sullen and gloomy solitude.

Without the means of conveying themselves with great swiftness from one place to another, birds could not easily subsist; the food which Nature has so bountifully provided for them is so irregularly distributed, that they are obliged to take long journeys to distant parts in order to gain the necessary supplies: at one time it is given in great abundance; at another it is administered with a very sparing hand; and this is one cause of those migrations so peculiar to the feathered tribes; the other chief causes are, the want of a proper temperature of air, and a convenient situation for the great work of breeding and rearing their young. Such birds as migrate to great distances are alone denominated "*birds of passage*;" but most birds are, in some measure, birds of passage, although they do not migrate to places very remote from their former habitations. At particular times of the year most birds remove from one country to another, or from the more inland districts towards the shores: the times of these migrations or flittings are observed with astonishing order and punctuality; but the secrecy of their departure, and the suddenness of their re-appearance, have involved the subject of migration in general in great difficulty. Much of this difficulty arises from our not being able to account for their means of subsistence during the long flights of many of those birds which are obliged to cross immense tracts of water before they arrrive at the places of their destination: accustomed to measure distance by the speed of those animals with which we are well acquainted, we are apt to overlook the superior velocity with which birds are carried forward in the air, and the ease with which they continue their exertions for a much longer time than can be done by the strongest quadruped. On this part of the subject we have had occasion to make more particular observations, when speaking of the habits of certain migratory birds; we shall therefore merely add, from Bewick, that from the advantage they possess in being raised to considerable heights in the air, they are enabled, with a sagacity peculiar to instinctive knowledge, to discover the route they are to take, from the appearance of the atmosphere, the clouds, the direction of the winds, and other causes; so that, without having recourse to improbable modes, it is easy to conceive, from the velocity of their speed alone, that most birds may transport themselves to countries lying at great distances, and across vast tracts of ocean.

At the approach of spring, birds begin to pair, and to provide for the support of their future progeny; and the loudest notes, on such occasions, generally proceed from the tuneful throats of the males, while the females express their consent in short interrupted twitterings. The compacts then entered into between the two sexes are, for the season at least, faithfully observed: but many birds live together for years with inviolable fidelity; and when one of them dies, the other does not long survive. We are of course not speaking of the poultry in our yards; but of those denizens of the air where Nature retains her unadulterated simplicity; where the number of males is generally equal to that of the females; and where every little animal seems no less pleased with its progeny than wedded to its mate.

The Nests of Birds now claim our notice; for they are constructed with such exquisite art, as to exceed the utmost exertion of human ingenuity to imitate them with perfect success. Their mode of building, the materials they make use of, as well as the situatious they select, are as various as the different kinds of birds, and are all admirably adapted to their several wants and necessities. Birds of the same species, whatever region of the globe they inhabit, collect the same kind of materials, arrange them in the same manner, and make choice of similar situations for fixing the places of their temporary abodes. Every part of the world furnishes materials for the aerial architects: leaves and small twigs, roots and dried grass mixed with clay, serve for the external; whilst moss, wool, fine hair, and the softest animal and vegetable downs, form the warm internal part of these commodious dwellings. On this subject the author of "The Journal of a Naturalist" thus writes: Birds that build early in the spring seem to require warmth and shelter for their young; and the Blackbird and the Thrush line their nests with a plaster of loam, perfectly excluding, by these cottage-like walls, the keen icy gales of our opening year: yet, should accident bereave the parents of their first hopes, they will construct another, even when summer is far advanced, upon the model of their first erection, and with the precautions against severe weather, when all necessity for such provision has ceased, and the usual temperature of the season rather requires coolness and a free circulation of air. The House-sparrow will commonly build four or five times in the year, and in a variety of situations, under the warm eaves of our houses and our sheds, the branch of the clustered fir, or the thick tall hedge that bounds our garden, &c.; in all which places, and without the least consideration of site or season, it will collect a great mass of straw and hay, and gather a profusion of feathers from the poultry-yard to line its nest. This cradle for its young, whether under our tiles in March or in July, when the parent bird is panting in the common heat of the atmosphere, has the same provisions made to afford warmth to the brood; yet this is a bird that is little affected by any of the extremes of our climate. The Wood-pigeon and the Jay, though they erect their fabrics in the tall underwood in the open air, will construct them so slightly, and with such a scanty provision of materials, that they seem scarcely adequate to support their broods, and even their eggs may almost be seen through the loosely-connected materials: but the Goldfinch, that inimitable spinner, the Arachne of the grove, forms its cradle of fine mosses and lichens, collected from the apple or the pear-tree, compact as a felt, lining it with the down of thistles besides, till it is as warm as any texture of the kind can be, and it becomes a model for beautiful construction. The golden-crested Wren, a minute creature, perfectly unmindful of any severity in our winter, and which hatches its young in June, the warmer portion of our year, yet builds its most beautiful nest with the utmost attention to warmth; and, interweaving small branches of moss with the web of the spider, forms a closely-compacted texture nearly an inch in thickness, lining it with such a profusion of feathers, that sinking deep into this downy accumulation it seems almost lost itself when sitting, and the young, when hatched, appear stifled with the warmth of their bedding and the heat of their apartment; while the White-throat, the Black-cap, and others, which will hatch their young nearly at the same period, or in July, will require nothing of the kind. A few loose bents and goose-grass, rudely entwined, with perhaps the luxury of some scattered hairs, are perfectly sufficient for all the wants of these; yet they are birds that live only in genial temperatures, feel nothing of the icy gales that are natural to our pretty indigenous artists, but flit from sun to sun, and we might suppose would require much warmth in our climate during the season of incubation; but it is not so. The Greenfinch places its nest with little regard to concealment; its fabric is slovenly and rude, and the materials of the coarsest kinds; while the Chaffinch, just above it in the elm, hides its nest with cautious care, and moulds it with the utmost attention to order, neatness, and form. One bird must have a hole in the ground; to another, a crevice in a wall, or a chink in a tree, is indispensable. The Bullfinch requires fine roots for its nest; the grey Fly-catcher will have cobwebs for the outworks of its shed. All the *parus* tribe, except the individual above mentioned, select some hollow in a tree or cranny in a wall; and, sheltered as such places must be, yet will they collect abundance of feathers and warm materials for their infants' bed. Endless examples might be found of the dissimilarity of requirements in these constructions among the several associates of our groves, our hedges and our houses; and yet the supposition cannot be entertained for a moment that they are superfluous, or not essential for some purpose with which we are unacquainted. By how many of the ordinations of Supreme Intelligence is our ignorance made manifest! Even the fabrication of the nests of these little animals exceeds our comprehension — we know none of the causes or motives of that embodied mind that willed them thus."

The difference of climate sometimes occasions vast alterations in the construction of the nests of birds. Some water-fowl strip the down from their own breasts, for the purpose of lining their nests with greater security. In general, however, all birds, when hatching, resort to those climates and places where their food is found in the greatest plenty. Aquatic birds, as well as the largest of the land kinds, select such situations as are most remote from man; their food, in general, being different from that which is cultivated by human labour. Some birds, which regard the serpent as their deadliest foe, build their nests depending from small boughs, and form their entrances from below; thereby equally securing them from the serpent and the monkey tribes: but small birds, which feed upon fruits and corn, make use of every precaution to conceal their nests from man; while the great birds, remote from human society, employ every art to render theirs inaccessible to wild beasts and vermin.

While the female is hatching, nothing can exceed her patience; neither the calls of hunger, nor the near approach of danger, being capable of driving her from her nest. Though fat when she begins to sit, before the time of incubation is expired she is usually reduced to little more than skin and bone. While the young continue in the nest, the old ones provide them with a proper supply of food; and, that no individual may be overlooked, each is served in its turn. If they perceive that man has been busy with their nest, or has handled their little ones, they abandon the place by night, and provide their brood a more secure retreat. When the whole family are fully plumed, and capable of avoiding danger, they are led forth in fine weather, and taught the art not only of providing for their own subsistence, by being conducted to those places where their food is most likely to be found, but of picking it up and carrying it away. After the business of incubation is entirely over, and the young are sufficiently able to provide for themselves, the nests are abandoned by the parents, except by those of the eagle kind.

Most of the smaller birds are supported, especially when young, by a profusion of caterpillars, small worms, and insects; on these they feed, and thus they contribute to preserve the vegetable world from destruction. This is contrary to the commonly-received opinion, that birds, particularly Sparrows, do much mischief in destroying the labours of the gardener and the husbandman. It has been observed, "that a single pair of Sparrows, during the time they are feeding their young, will destroy about four thousand caterpillars weekly; they likewise feed their young with butterflies and other winged insects, each of which, if not destroyed in this manner, would be productive of many thousands of caterpillars." Swallows are almost continually upon the wing, and in their curious winding flights destroy immense numbers of flies and other insects, which are continually floating in the air, and which, if not destroyed by these birds, would render it unfit for the purposes of life and health. That active little bird, the Tom-tit, which is generally supposed hostile to the young and tender buds that appear in the spring, when attentively observed, may be seen running up and down among the branches, and picking up the eggs of insects, or the small maggots or worms that are concealed in the blossoms, and which would effectually destroy the fruit. As the season advances, various other small birds, such as the Redbreast, Wren, Hedge-warbler, White-throat, Redstart, &c., are all engaged in the same useful work, and may be observed examining every leaf, and feeding upon the insects which they find beneath them. These are a few instances of that superintending providential care which is continually exerted in preserving the various ranks and orders of beings in the scale of Animated Nature; and although it is permitted that myriads of individuals should every moment be destroyed, not a single species is lost, but every link of the great chain remains unbroken.

The only disease, if it can be termed one, to which birds are subject, is moulting, or the operation of changing their plumage, during the continuance of which they are sickly and disordered, and many die. This process, which occurs every year, appears to be performed in the following manner: — When the feathers have attained their full size, the quill part, nearest the bird, grows harder, and shrinks in its diameter, thus gradually compressing, and finally obliterating the vessels which supply it with nourishment, and thus becomes an extraneous body which is at last loosened in its socket, and falls off. Whilst these changes are taking place, the rudiments of the new feather are forming beneath, which rapidly attains its natural size, after it has been protruded through the skin, a process which, it will be seen, is very analogous to the annual shedding of the horns in the deer tribe.

Although some birds, by emigrating, make their habitations in different parts of the earth, almost every climate has such as are peculiar to it. Those of the temperate zone are not very remarkable for the beauty of their plumage; but the smaller kinds fully compensate this defect by their melodious notes. Many birds of the torrid zone are resplendent in beauty, but in general they have either harsh and disagreeable voices, or are totally silent: the frigid zone, on the contrary, where the adjacent seas abound with fish, is stocked with birds of the aquatic kind, in much greater plenty than in Europe; and these are generally either clothed with warm coats of feathers, or have large quantities of fat lying beneath the skin, to defend them from the rigours of the climate. In all climates, however, birds are longer lived than quadrupeds of the same climates: indeed, it may be said, that, in proportion to the size of their bodies, birds possess more vitality, and live longer, than either man or quadrupeds.

Naturalists have arranged birds in various orders, founded on the organs of manducation and prehension. The following is that of Cuvier:—1. BIRDS OF PREY (*Accipitres*, Lin.); distinguished by their crooked beak and claws, by means of which they are enabled to overcome and prey upon other birds, and even the weaker quadrupeds. They hold the same rank among birds as the *Carnivora* among quadrupeds. They all have four toes, and the nails of the great and middle toes are the strongest. They form two families, the *diurnal* and *nocturnal*, the first having nostrils inserted in a naked cere, three toes before and one behind, without feathers; eyes directed sideways: the second having nostrils at the anterior edge of the cere, which is more or less covered with stiff hairs; the external toe capable of being turned behind; eyes large, directed forwards.—2. PASSERINE BIRDS (*Passeres*). This is the largest class, and embraces all birds which do not belong to the other five. They present a great resemblance in their structure, and the genera are so closely allied that it is difficult to distinguish between them. They may, however, be separated into two great divisions: 1. Those with the exterior toe united to the middle one, by one or two joints only; and, 2. Exterior toe almost as long as the middle one, and united to it as far as the last joint but one.—3. CLIMBERS (*Scansores*). Birds whose exterior toe directs itself backwards like the great toe, affording a very solid support, by which some of them cling to and climb the trunks of trees.—4. GALLINACEOUS BIRDS. (*Gallinaceæ.*) These have a heavy gait, a short flight, a medium-sized beak, the upper mandible vaulted, nostrils partly covered by a cartilaginous scale, toes generally dentated at the edges, with short membranes between those in front.—5. WADERS (*Grallæ*) may be recognised by the nudity of the lower part of their thighs; very frequently by the length of their legs; and generally by some little web, at least, between the external toes. In flying, they extend their legs behind them, contrary to the habit of other birds, who draw them up close to the body.—6. WEB-FOOTED BIRDS (*Palmipedes*) are strongly characterised by their feet, formed for swimming, being affixed to the hinder part of their body; with very short and compressed tarsi; and palmated between the toes.—For much information on the habits of Birds we would refer to the pages of Loudon's Magazine of Natural History, and to "The Zoologist;" while the great work, "Gray and Mitchell's Genera of Birds," contains an immense supply of matter in illustration of this important and interesting class. In Macgillivray's "British Birds" much valuable information is given by the able professor of Marischal College, Aberdeen.

BIRGUS. A genus of long-tailed Crustaceous animals, of which the PURSE-CRAB (*Birgus latro*) is the largest. This species of land-crab is a native of Amboyna, and other neighbouring islands, where it is said to inhabit the fissures of rocks or holes in the earth by day, and to come forth at night to seek its food on the beach. Some say it climbs cocoa-nut trees in the night to get the cocoa-nuts; and it is certain it can subsist on them, as well as on some other kinds of nuts, when more favourite food is not easily attainable. When properly dressed, this animal is regarded as an excellent dish. [See CRAB.]

BISON. There are two kinds of Bison; one of them European, which is now become very scarce; the other American, which still exists in vast numbers.

The EUROPEAN BISON (*Bos bonassus*) is as large as a bull or ox; and in his native state of wildness, is distinguished not only by his size and the fleshy protuberance on his shoulders, but by the superior depth and shagginess of his hair, which about the head, neck, and shoulders, is sometimes so long as almost to touch the ground. The head is small; the eyes are red and fiery; the forehead is wide; and the horns are short, extremely strong, sharp-pointed, and stand distant from each other at their bases, like those of the common bull. His colour is a dark rufous brown, sometimes nearly black; his limbs remarkably strong; and his whole aspect in the highest degree savage and gloomy. The principal European regions where this animal is at present found, are the marshy forests of Poland, the Carpathian mountains, and Lithuania. Its chief Asiatic residence is the neighbourhood of Mount Caucasus. This animal is very scarce, and would probably soon be extinct but for the strict orders of the emperor of Russia, who will not permit any to be shot in his dominions. This monarch has presented a stuffed specimen and skeleton of one to the British Museum.

The AMERICAN BISON. (*Bos Americanus.*) The American Bison, most frequently called "the Buffalo," differs from the European chiefly in being larger, more shaggy, in having a more protuberant bunch over the shoulders, and by the length and fineness of its woolly hair. The hump is oblong, diminishing in height posteriorly, and gives a considerable obliquity to the outline of the back. The hair over the head, neck, and fore-part of the body is long and shaggy, forming a beard beneath the lower jaw, and descending below the knee in a tuft. The hair on the top of the head rises in a dense mass, nearly to the tips of the horns, and, directly on the front, is curled and matted strongly. The numbers of this species still existing are surprisingly great, when we consider the immense destruction which annually takes place. They were once extensively diffused over the whole of what is now the territory of the United States, but they are no longer found except in the

BISON OR " BUFFALO."—(BOS AMERICANUS.)

remote unsettled regions of the north and west, being rarely seen east of the Mississippi or south of the St. Lawrence.

The Bison, on his native plains, is of a savage and formidable appearance; nevertheless, he is not known to attack man, unless when wounded and at bay. The difference between the summer and winter dress of the Bison consists rather in the length than in other qualities of the hair. In summer, from the shoulders backwards the surface is covered with a very short, fine hair, smooth and soft as velvet. The tail is short, and tufted at the end; and the general colour of the animal is a uniform dun. Varieties of colour are, indeed, so rare among the species, that the hunters and Indians always regard them as matters of special wonder. Herds, consisting of thousands of these fine animals, still roam over the far western prairies, led by the fiercest and most powerful of the bulls. During the sexual season the noise of their roaring is terrific, and the males often fight with all the fury of desperation. While feeding, they are frequently scattered over a vast surface; but when they move onward in a mass, they form a dense, impenetrable column, which, once fairly in motion, is scarcely to be turned. They swim large rivers in nearly the same order in which they traverse the plains; and, when flying from pursuit, it is in vain for those in front to halt suddenly, as the rearward throng dash madly forward, and force their leaders on. The Indians sometimes profit by this habit; they lure a herd to the vicinity of a precipice, and, setting the whole in rapid motion, they terrify them, by shouting and other artifices, to rush on to their inevitable destruction.

There are various modes of capturing or killing these animals; but there are none which require so much dexterity as the hunting them on horseback; which is thus described by Sir John Franklin:—"An expert hunter, when well mounted, dashes at the herd, and chooses an individual which he endeavours to separate from the rest. If he succeeds, he contrives to keep him apart by the proper management of his horse, though going at full speed. Whenever he can get sufficiently near for a ball to penetrate the beast's hide, he fires, and seldom fails of bringing the animal down; though of course he cannot rest the piece against the shoulder, nor take deliberate aim. On this service the hunter is often exposed to considerable danger from the fall of his horse in the numerous holes which the badgers make in these plains, and also from the rage of the buffalo [Bison], which, when closely pursued, often turns suddenly, and, rushing furiously on the horse, frequently succeeds in wounding it, or dismounting the rider." "When the buffaloes are on their guard, horses cannot be used in approaching them; but the hunter dismounts at some distance and crawls in the snow towards the herd, pushing his gun before him. If the buffaloes happen to look towards him he stops, and keeps quite motionless, until their eyes are turned in another direction; by this cautious proceeding a skilful person will be able to get so near as to be able to kill two or three out of the herd." When wounded they are very furious; their hoofs, more than their horns, are their offensive weapons, and whatever opposes them is in no small danger of being trampled into the earth.

The Hon. C. A. Murray, in his Travels in North America, where he had excellent opportunities of studying the habits of this animal in his native haunts, tells us that, "The Buffalo, huge and unwieldy as he is, goes over the ground at a rate which is surprising; he bounds along with large, though clumsy strides; and in a rough country he dashes down the steep sides of the broken ravines, making the dust, the sand, and the stones fly around with a furious rapidity, that defies the pursuit of a rider who has any regard for the neck of his horse or himself. The female, the constant object of the hunter, from the superior quality and tenderness of her flesh, is beyond all comparison swifter than the male; she can run nearly three miles to his two, and gives a very fair chase to a horse of middling speed, fed only on grass, and carrying a man of only ordinary size."

Numerous tribes of Indians are almost wholly dependent on these animals for food, tents, clothing, utensils, &c. The skins, dressed in the Indian fashion, with the hair on, make admirable defences against the cold, and may be used for blankets, &c. They are called *buffalo robes*; the term *Buffalo* being generally, but inaccurately, applied to the Bison. The horns of the Bison are converted into powder-flasks; while their wool has been manufactured into hats, and has also been employed in making coarse cloth. Bison beef is rather coarser grained than that of the domestic ox, but is considered by hunters and travellers as superior in tenderness and flavour. The hump, which is highly celebrated for its richness and delicacy, is said, when properly cooked, to resemble marrow. The Hon. Mr. Murray, in the work from which we have already quoted, says, "I cannot convey any just impression of the total dependence of the remote western tribes on Buffalo for their very existence, without giving a sketch of the various purposes for which that animal is, by their ingenuity, rendered available. First, its flesh is their principal, sometimes their only food; eaten fresh on the prairies during their hunt, and dried in their winter villages. Secondly, the skin is put to various uses; it forms the material of their lodges, of their bales for packing the meat, of their bed by night, and their clothing by day; the coarser parts they make into saddles, or cut into laryettes, or halters; and, more than all, it is now their chief article of trade with the whites, and thus is the source whence they must derive blankets, knives, beads, and every other produce of civilization. Thirdly, they use the sinews as strings to their bows, and the smaller fibres instead of twine or thread; the brains serve to soften and dress the skins, while the hoof, at the end of the shank bone, is made to answer the purpose of a mallet. Fourthly, the bones are not less useful: some of them being serviceable as scrapers or close chisels: others are pointed, and used with the finer fibres as needle and thread; and the ribs, strengthened by some of the stronger fibres, are made to furnish the bow with which other Buffaloes are to be destroyed. This last is the triumph of Indian ingenuity. The first bow that I saw constructed in this manner caused so much surprise and admiration, that I offered nearly the value of a horse for it, but was refused. When I add to the foregoing particulars, that on the barren prairies the Indians frequently depend upon the Buffalo (dung) for their fuel, and on its bladder for the means of carrying water, it will not be denied that the animal is essential to their existence; and when the Buffalo is exterminated, the Indian of the Prairies must perish."—For further particulars we must refer our readers to the delightful pages of Sir John Richardson, M. D. (*Fauna Boreali Americana*), and of Mr. Catlin.

BITTERN. (*Botaurus.*) The Bitterns are a subgenus of the family of Herons, residing in woody swamps and marshy places, and feeding upon aquatic animals, frogs, lizards, insects, &c. The COMMON BITTERN (*Botaurus stellaris*) is about two feet six inches in length, or nearly as large as the common heron, but its legs are stronger; body more plump and fleshy; and its neck is more thickly clothed with feathers. The beak is strong at the base, straight, sharp on the edges, and gradually tapers to an acute point; the upper mandible is brown, the under inclining to green; mouth wide, the gape extending beyond the eyes, with a dusky patch at each angle: irides yellow. The crown of the head is somewhat depressed, and covered with long black feathers; and the neck feathers, which it can raise at pleasure, are long and loose. The general colour of the plumage is dull pale yellow; the back and wings are marked with black zig-zag lines, bars, and streaks, upon a ground shaded with rufous and yellow; an' the greater coverts and quills are regularly barred with black. The tail is very short; the legs are pale green; the toes and claws very long and slender. The female is somewhat smaller than the male, the plumage not quite so bright, and the feather: on the neck shorter. She makes an artless nest, composed chiefly of the withered stalks and leaves of the high coarse herbage, in the midst of which it is placed, and lays from four to six eggs of a greenish white.

The Bittern is a shy solitary bird; it is never seen on the wing in the day-time; but sits generally with the head erect, hid among the reeds and rushes of extensive marshes, from whence it will not stir unless disturbed by the sportsman. When it changes its haunts, it removes in the dusk of the evening, and then, rising in a spiral direction, soars to a vast height. It flies in the same heavy manner as the heron, and might be mistaken for that bird, were it not for the singularly resounding cry which it utters from time to time, while on the wing; but this cry is feeble when compared with the hollow booming noise which it makes during the night, in the breeding season, from its swampy retreats. From the loudness and solemnity of its note, an erroneous notion prevails with the vulgar that it either thrusts its bill into a reed, which serves as a pipe for swelling its note beyond its natural pitch, or that it immerges its head in water, and then produces its boomings by blowing with all its might.

When attacked by the buzzard, or other bird of prey, the Bittern defends itself with great courage, and generally beats off such assailants; neither does it betray any symptoms of fear when wounded by the sportsman, but eyes him with a keen undaunted look, and, when driven to extremity, will attack him with the utmost vigour, wounding his legs, or aiming at his eyes with its sharp and piercing bill. Bitterns reside permanently in England, and in most of the temperate parts of the continent; but in colder climates they are migratory. They were formerly held in great esteem at the tables of the wealthy.

There are several other species of this bird, some of them natives of hot and others of cold climates; but they all resemble the above in its distinguishing characteristics, frequenting the same situations, making their nests on the ground, &c., but differing materially in the colours of their plumage as well as in size.

BIVALVE. The name given to a class of shells composed of two pieces or parts; which, by means of a proper connection by hinges, open and shut, and perform all other functions necessary to the economy or modes of life of the animals included in them. The Mollusca inhabiting them are chiefly distinguished from the other classes by the absence of a visible head or neck, and the consequent deprivation of the organs of sight and hearing: they possess a mouth, but it is a mere opening in the body, with jaws or teeth. The branchiæ are large, placed on each side, between the body and the mantle. The lobes of the mantle are fringed round the edge with numerous filaments, which are very sensitive, and in constant activity. None of the genera are terrestrial, their construction not affording them sufficient powers of locomotion for finding their food on land, and confining them to the water, whether salt or fresh, or to the sands on the coasts. As familiar instances we may mention the Oyster, Mussel, Cockle, &c.

BIZCACHA, or VIZCACHA. (*Calomys bizcacha.*) A Rodent animal, somewhat resembling a rabbit, but with larger gnawing teeth and a long tail: it has, however, only three toes behind, like the Agouti. Near Buenos Ayres they are exceedingly common. They are said to live on roots; which, from the great strength of their gnawing teeth, and the kind of localities they frequent, seems probable. As in the case of the rabbit, a few holes are commonly placed together. In the evening the Bizcachas come out in numbers, and there quietly sit on their haunches. They do not wander far from their burrows: they run very awkwardly, and, when hurrying out of danger, from their elevated tails and short front legs, much resemble great rats. Their flesh, when cooked, is very white and good, but it is seldom used. Of late years the skins of the Bizcacha have found a market in England, on account of the fur.

BLACKBIRD. (*Turdus merula.*) A well-known song-bird, about ten inches long, whose deep-toned warblings are not to be mistaken for those of any other inhabitant of the groves. The plumage of the male bird is altogether black, but that of the female is rather of a brown or dark russet colour: the bill, inside of the mouth, and edges of the eyelids, are yellow, as are also

BLACKBIRD.—(TURDUS MERULA.)

the soles of the feet. The males during the first year resemble the females so much as not easily to be distinguished from them; but after that, they assume the yellow bill and other distinguishing marks of the sex. The Blackbird is a solitary bird, frequenting woods and thickets, chiefly evergreens, especially where there are perennial springs, which together afford it both shelter and subsistence. They feed on berries, fruit, insects, and worms; but never fly in flocks like thrushes. They pair early, and are among the first who render the groves vocal: the note of the Blackbird, indeed, during the spring and summer, when heard at a distance, is rich and enlivening; but when the bird is confined in a cage, its song is too loud and deafening. They build in bushes or low trees, and lay four or five eggs, of a bluish-green, marked irregularly with dusky spots. The young birds are easily tamed, and may be taught to whistle a variety of tunes. They are restless and timorous, easily alarmed, and difficult of access. We occasionally hear of *albinos*, or white blackbirds! but they are so rare as to be regarded in the light of great curiosities.

BLACK-CAP. (*Sylvia atricapilla*). This is a small song-bird, whose notes are so sweet and full that it has obtained the name of the mock-nightingale. The crown of the head, in the male, is black; the hind part of the neck, light ash colour; back and wings, olive grey; throat, breast, and belly, more or less silvery white; legs bluish, and claws black. The Black-cap is migratory, visiting us about the middle of April, and retiring in September. Orchards and gardens are its favourite haunts; and it builds its slightly constructed nest in some low tree or shrub,

BLACK-CAP.—(SYLVIA ATRICAPILLA.)

lining it with the fibres of roots thinly covered with horse hairs: the eggs are reddish brown mottled with a deeper colour, and sprinkled with dark spots. The Black-cap is naturally a very shy bird; and although while banqueting on currants, raspberries, or any of its favourite fruits, it seems to forget its usual timidity, and suffers itself to be looked at, yet at other times it avoids observation as much as possible, and carefully hides itself in the foliage from all familiarity and confidence. Its song, however, never fails to attract attention; for although its modulations are in general short and desultory, yet when this little warbler sits calmly, and is earnestly engaged in singing, it gives utterance to a pleasant and gentle harmony, superior perhaps to any of our other songsters, the nightingale excepted.

BLACK-COCK, and BLACK-GAME. [See GROUSE.]

BLAPS: BLAPSIDÆ. A genus and family of Coleopterous insects; the type of which is the species *Blaps mortisaga:* it is black, but little shining, and the tip of the elytra forms a short obtuse point. It is a very common British insect, found in dark, damp, and dirty places about houses. In Mr. Westwood's "Introduction to the Modern Classification of Insects," the following extraordinary fact is related:—Several instances have been noticed, in which the larvæ of the common species *Blaps mortisaga*, or Church-yard Beetle, has been discharged from the stomach. Of these, the most remarkable account is that published by Dr. Pickells in the *Trans. of Associated*

CHURCH-YARD BEETLE.
(BLAPS MORTISAGA.)

Physicians in Ireland, of a case of a woman, aged twenty-eight, who emitted as many as two thousand larvæ of this insect at various times, as well as one pupa and one imago; and which probably originated in an absurd and superstitious practice, which she had for some time followed, of drinking daily for a certain time a quantity of water mixed with clay, taken from the graves of two Catholic priests, and eating large pieces of chalk. One of these beetles was immersed repeatedly into spirits of wine, but revived after remaining therein all night, and afterwards lived three years."

BLATTIDÆ. A family of voracious insects, of the order *Orthoptera*, of which the troublesome Cockroach (*Blatta orientalis*) is a well-known example. These very destructive and disagreeable insects form one of the principal inconveniences of hot climates. They devour various animal and vegetable substances; and some species have a highly unpleasant smell, which is apt to remain on such articles as they have passed over. The largest of the genus is the

BLATTA GIGANTEA of Linnæus, which is a native of many of the warmer parts of Asia, Africa, and South America. The following description of them is said, by those who have visited the countries where they abound, to be by no means overcharged: "They plunder and erode all kinds of victuals, drest and undrest, and damage all sorts of clothes, especially such as are touched with powder, pomatum, and similar substances; everything made of leather, books, paper, and various other articles, which if they do not destroy, at least they soil, as they frequently deposit a drop of their excrement where they settle, and some way or other by that means damage what they cannot devour. They fly into the flame of candles, and sometimes into the dishes; are very fond of ink and of oil, into which they are apt to fall and perish; in which case they soon turn most offensively putrid: so that a man might as well sit over the cadaverous body of a large animal as write with the ink in which they have died. They often fly into persons' faces or bosoms, and their legs being armed with sharp spines, the pricking excites a sudden horror not easily described. In old houses they swarm by myriads, making every part filthy beyond description wherever they harbour, which in the day-time is in dark corners, behind clothes, in trunks, boxes, and, in short, every place where they can lie concealed. In old timber and deal houses, when the family is retired at night to sleep, this insect, among other disagreeable properties, has the power of making a noise which very much resembles a pretty smart knocking with the knuckle upon the wainscotting. The Blatta Gigantea, in the West Indies, is therefore frequently known by the name of the *drummer*. Three or four of these noisy creatures will sometimes be impelled to answer one another, and cause such a drumming noise that none but those who are very good sleepers can rest for them. What is most disagreeable, those who have not gauze curtains are sometimes attacked by them in their sleep: the sick and dying have their extremities attacked, and the ends of the toes and fingers of the dead are frequently stripped both of the skin and flesh."

The BLATTA ORIENTALIS, or common black Cockroach, which is frequently called in our country by the erroneous name of the *black beetle*, is supposed to have come originally from Asia; but of that there is some little doubt. In its mature state the male has wings extending only half the length of the body; the female has only rudimentary wings; her eggs, which are about sixteen in number, are enclosed in an oblong case, which she carries about with her at first, fixed to the abdomen by a sort of gum. The nocturnal habits and ravages of this species are too well known to require any description.

The BLATTA AMERICANA, or American Cockroach, is of a light chestnut or reddish colour, and is extremely common in the warmer parts of America, and the West India islands. It is somewhat larger than the black or eastern Cockroach. These Blattæ lay their eggs in heaps, and wrap them all round in webs or bags, after the manner of some spiders. When the eggs are hatched, the young ones appear quite perfect, and leave their shells almost instantaneously. Being at first no larger than ants, they are capable of penetrating through the smallest apertures into boxes and chests, where they destroy everything within their reach. When arrived at their full growth, they cast their skins, which burst on their backs; and then the Blattæ, or Cockroaches, are perfectly formed: their wings are at first soft and whitish, and they soon become red; but their heads, horns, and the rest of their bodies, retain the same shapes and colours as they possessed before the exuviæ were shed.

BLEAK. (*Cyprinus alburnus.*) This Malacopterygious fish belongs to the Carp family, and is very common in many of our own rivers: the length about five or six inches; shape slender, with the body much compressed; colour bright silvery, the back olive-green; fins pellucid; scales deciduous; and the tail forked. Bleaks generally keep together in large shoals; and at certain seasons they are observed to tumble about near the surface of the water as if incapable of swimming to

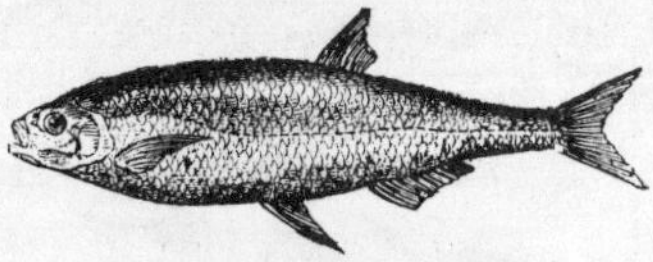

BLEAK.—(CYPRINUS ALBURNUS.)

any considerable distance; but in a short time they recover, and presently disappear. It is from the scales of this fish that the beautiful silvery matter used in the preparation of artificial pearls is chiefly taken; other bright-scaled fishes may, however, be used for the same purpose.

BLENNY. (*Blennius.*) A genus of small Acanthopterygious fishes, living in small shoals, and frequenting rocky coasts, where they may be often found in pools of water left by the tide. The Blennies have one well-marked character in their ventral fins, inserted before the pectorals, and having only two rays each. The stomach is slender, with no cul-de-sac; the intestine large, without cæca, and there is no air-bladder. The form is elongated and compressed, and there is but one dorsal, composed almost entirely of jointless but flexible rays. Their skin is covered with a mucous secretion; they have teeth equal and closely set, forming only a single row in each jaw; their head is blunt, their profile vertical, and their muzzle short. There are several species; a brief description of three, however, will be ample.

The CRESTED BLENNY. (*Blennius galerita.*) This species is about four or five inches in length, and is found about the rocky cóasts of Great Britain. The body is long, compressed, smooth, and slippery; colour yellowish brown, freckled with darker coloured specks; head furnished on the middle with a transverse finny appendage, which can be either raised or depressed at pleasure; and between the eyes is a small triangular prominence, pointing backward, and red about the edges; ventral fins very small and short, dorsal shallow, running from the hind part of the head to the tail, which is of a round shape, and the vent situated under the ends of the pectoral fins.

The OCELLATED BLENNY, or BUTTERFLY FISH. (*Blennius ocellaris.*) This very small species is a native of the Mediterranean, but is occasionally found in the South of England by dredging. It has two lobes in the dorsal, the first marked with a

OCELLATED BLENNY, OR BUTTERFLY-FISH (BLENNIUS OCELLARIS)

round black spot surrounded by a white ring, and then a black one. It lives among the rocks and sea-weed, and is believed to subsist on minute Crustacea and Mollusca.

The GATTORUGINOUS BLENNY (*Blennius Gattorugin*) is about six inches long; the body smooth, and compressed on the sides; the belly rather prominent, and the vent situated as in the crested Blenny. The head is grooved between the eyes, and furnished with two branched membranes, situated just above the eyelids, a distinguishing mark of the species. The pectoral fins, which are broad and rounded, consist of fourteen rays; the dorsal fin has thirty-three; the ventral two; the anal twenty-three; and the tail, which is slightly rounded, has eleven rays. This fish is of a dusky colour, marked across with wavy lines. It has occasionally been found on our western coasts, but is not very common.

BLEPHARIS. A genus of Acanthopterygious fishes, distinguished by their having long filaments to their second dorsal and to their anal fin rays. One species of the *Blepharis*, inhabiting the West Indian seas, is known under the appellation of the *Cobbler-fish*, probably on account of the long thread-like appendages for which it is so conspicuous.

BLEPSIAS. A genus of Acanthopterygious fishes, the generic characters of which are,—compressed head, cheeks mailed, fleshy barbels under the lower jaw, gills with five rays, and one dorsal fin divided into three unequal lobes.

BLETHISA. A genus of carabidous Coleopterous insects, consisting of three known species, only one of which has been found in this country: this is about half an inch long, of a rich bronze or brassy hue, and with numerous indented points on the elytra: it frequents marshy situations, and is often found crawling upon willow trees.

BLIND-WORM, or SLOW-WORM. (*Anguis fragilis.*) A species of viviparous reptile belonging to the third subgenus of the family *Anguidæ*, which may be said to form the connecting link between the lizards and the true serpents. Though somewhat formidable in appearance, the Blind-worm is perfectly innocuous. Its usual length is about eleven inches; the head is small; the eyes are also small, and the irides red; the neck is slender, and thence the body enlarges, continuing of equal bulk to the tip of the tail, which ends bluntly, and is as long as the body. The general colour of the back is cinereous, marked with very small lines of minute black specks; the scales are small, smooth, and shining, of a silvery yellow on the upper parts, and dusky beneath; the tongue is broad and forked; and the teeth are very small and numerous. The Blind-worm feeds on earthworms, insects, &c., and among the uninformed has the character of possessing the most deadly venom. The motion of this reptile is slow; from which circumstance, as well as from the smallness of its eyes, its names are derived. Like all the rest of the kind, in our climate, they lie torpid during the winter, being sometimes found in vast numbers twisted together.

BLOODHOUND. (*Canis [domesticus] sanguinarius.*) A well-marked variety of dog, celebrated for its exquisite scent and unwearied perseverance, qualities which were highly esteemed by our ancestors for tracing and recovering such game as had escaped from the hunters in a wounded state, or had been killed and stolen out of the royal forests. These hounds were also formerly much employed in pursuing criminals escaped from justice, or in tracing out robbers or enemies, whose course was inevitably discovered when once the Bloodhound was placed upon their trail. The genuine Bloodhound breed was large, strong,

BLOODHOUND
(CANIS [DOMESTICUS] SANGUINARIUS.)

muscular, broad-chested, the upper lip large and pendulous; the expression stern and noble; the colour a deep tan, and generally marked with a black spot over each eye; this species, however, seems now to be blended with the other smaller hounds, and the original stock is all but extinct.

Sir Walter Scott and other writers narrate many surprising feats of the "sleut-hound," whose unflinching pertinacity generally overcame all impediments, whether engaged in the usual objects of the chase, or directed against political delinquents. "For such purposes as these," says Mr. Bell, "the Bloodhound has been employed, at various times, in every part of the United Kingdom: in the clan feuds of Scotland, in the border contests of the dĕbatable land of the two kingdoms, and in the unhappy Irish rebellion, its extraordinary powers have been taken advantage of without much regard to the claims either of justice or of mercy. Such scenes, however, have now become mere matter of history and of tradition; for, on the one hand, the improvements which have taken place in the breed of hounds for the purposes of the chase, and on the other, the gradual introduction of a more regular system of police, aided, we may hope, by some amelioration in the feelings of the people, have annihilated the use of the Bloodhound in both the objects for which it was formerly employed."

BLUE-BIRD. (*Sialia.*) This bird is as well known in America as the Redbreast is with us, and its habits of familiarity with man in the summer are on a par with those of our friendly visitor in the winter. It is about seven inches and a half long, and the whole of the upper part of the body is of a rich sky-blue shot with purple. The bill and legs are black; the wings of a

BLUE-BIRD.—(SIALIA SIALIS)

dusky black at the tips, and the shafts of the wings and tail feathers are black; the throat, neck, breast, and sides partially under the wings, reddish chestnut; the belly and vent white. It arrives in the United States early in the spring, and takes its departure in November. Its food consists of large beetles, spiders, and other insects, besides berries, seeds, and fruits. The nest is generally built in holes of trees; and the male is most assiduous in attentions to his mate; the eggs are of a pale blue colour; and it often happens that two or three broods are produced in one season.

BLUE [BUTTERFLY]. A name applied to several species of Butterflies, of the genus *Polyommatus.*

BLUE-BREAST. (*Cyanecula suecica.*) This elegant little bird inhabits different parts of Europe, and is mostly found on the borders of forests. It is five inches and a half in length, of which the tail occupies two and a quarter. The head, back, and wing-coverts are ashy-brown, mottled with a darker tint; a reddish-white line passes above the eyes; a brilliant sky-blue covers the throat and half-way down the breast; this is set off by a spot of the most dazzling white, the size of a pea, placed precisely over the larynx, which, enlarging and diminishing successively by the movement of this part when the bird sings, produces the most beautiful effect. The blue passes into a black band, and the latter into a fine orange; the belly is dusky white; the thighs and sides are reddish; and the quill feathers dark brown. Some males have two little white spots on the throat, and some even three; but some have none. The food of the Blue-breast consists of flies, the larvæ of insects, and worms. The nest is built in bushes and in the holes of trees; and the eggs are of a greenish hue. The females, when young, are of a celestial-blue tint on the sides of the throat; and when very old they have the throat sometimes of a very bright blue.

BOA CONSTRICTOR. Of all the reptiles that exist, none equal in size and power the genus *Boa;* some of them being occasionally met with from thirty to thirty-five feet in length, and of a strength so prodigious as to be able to destroy deer, oxen, and other large and powerful animals, by enveloping

BOA CONSTRICTOR.

them in their ample folds, crushing them to death, and, lubricating the bodies with their saliva, swallowing them at their leisure. In this tribe the branches of the upper and lower jaw, throughout the whole length, as well as the palate bones, are armed with pointed, recurved, solid, and permanent teeth, forming four nearly equal rows above, and two below. They have the tympanic bone or pedicle of the lower jaw moveable, which is itself almost wholly suspended to another bone, analogous to the mastoid, attached to the skull by muscles and ligaments, which contribute to its mobility. The branches of this jaw are not united, and those of the upper jaw are attached to the intermaxillary bone only by ligaments, so that these animals can dilate the mouth sufficiently to swallow bodies much larger than themselves. They are further distinguished by having the *scuta* on the other part of the tail single; a hook on each side of the vent; the tail prehensile; the body compressed, and largest in the middle, and with small scales, at least on the posterior part of the head.

Enormous as the size and power of such animals must be, according to the latest and best authenticated statements of eye-witnesses, yet, *if we may rely* on the accounts of ancient writers, there was a time when serpents far more terrific committed their hideous ravages, and kept whole armies in dismay. One of this kind is described as having had its lair on the banks of the Bagradas, near Utica, and to have swallowed many of the Roman soldiers in the army of Regulus, to have killed others in its folds, and to have kept the army from the river; till at length, being invulnerable by ordinary weapons, it was destroyed by heavy stones slung from the military engines used in sieges: but, according to the historian Livy (quoted by Valerius Maximus), the waters were polluted with its gore, and the air with the steams from its corrupted carcass, to such a degree that the Romans were obliged to remove their camp, taking with them, however, the skin, 120 feet in length, which was sent to Rome. That none of such frightful dimensions now infest the inhabited parts of the earth we have abundant evidence; and there is good reason to believe that as cultivation and population have increased, the larger species of noxious animals have been expelled from the haunts of mankind, and driven into more distant and uncultivated regions.

Some species of the genus Boa are found in the vast marshes and swamps of Guiana, and other hot parts of the American continent: others are natives of India, Africa, and the larger Indian islands. They are at once preeminent from their superior size and their beautiful colours; and though destitute of fangs and venom, nature has endowed them with a degree of muscular power which seems to defy resistance. The ground colour of the whole animal, in the younger specimens, is a yellowish grey, and sometimes even a bright yellow, on which is disposed along the whole length of the back a series of large, chain-like, reddish-brown variegations, leaving large open oval spaces of the ground colour at regular intervals: the largest or principal marks composing the chain-like pattern above mentioned are of a squarish form, accompanied by large triangular and other shaped spots, the exterior of the larger ones being generally of a much darker cast, and the ground colour immediately next to them considerably lighter than on other parts, thus constituting a general richness not easily described.

We cannot reflect upon the history of these great reptiles without being struck with their peculiar adaptation to the situations in which they are commonly most abundant. In regions bordering on great rivers, which annually inundate vast tracts of country, these serpents live securely among the trees with which the soil is covered, and are capable of enduring very protracted hunger, without much apparent suffering, or diminution of vigour. Noxious as such districts are to human life, they teem with a gigantic and luxurious vegetation, and are the favourite haunts of numerous animals, preyed upon, and, to a certain degree, restricted in their increase, by the *boce*. In such situations the Boa Constrictor lurks, or winds itself round the trunk or branches of a tree, until some luckless animal approaches; then, suddenly relinquishing his position, swift as lightning he seizes the victim, and coils his body spirally round its throat and chest, until, after a few ineffectual cries and struggles, the animal is suffocated and expires. The prey is then prepared for being swallowed, which the creature accomplishes by pushing the limbs into the most convenient position, and then covering the surface with a glutinous saliva. The reptile commences the act of deglutition by taking the muzzle of the prey into its mouth, which is capable of vast extension; and, by a succession of wonderful muscular contractions, the rest of the body is gradually drawn in, with a steady and regular motion.

In Mr. M'Leod's narrative of the voyage of H. M. ship *Alceste* to China, is the following characteristic account of the Boa, as observed on shipboard. "Notwithstanding the crowded state of the Cæsar," (the vessel in which the crew of the Alceste returned, their own having been wrecked), "two passengers, of rather a singular nature, were put on board at Batavia for a passage to England: the one, a snake of that species called Boa Constrictor; the other, an Ourang Outang. The former was somewhat small of his kind, being only about sixteen feet long, and of about eighteen inches in circumference; but his stomach was rather disproportionate to his size, as will presently appear. He was a native of Borneo, and was the property of a gentleman (now in England), who had two of the same sort; but, in their passage up to Batavia, one of them broke loose from his confinement, and very soon cleared the decks, as everybody very civilly made way for him, and ran up the rigging, or to some other place of security. Not being used to a ship, however, or taking, perhaps, the sea for a green field, he sprawled overboard, and was drowned. He is said not to have sunk immediately, but to have reared his head several times, and with it a considerable portion of his body, out of the sea. His companion, lately our shipmate, was brought safely on shore, and lodged in the court-yard of Mr. Davidson's house at Ryswick, where he remained for some months, waiting for an opportunity of being conveyed home in some commodious ship sailing directly for England, in which he was likely to be carefully attended to. This opportunity offered in the Cæsar, and he was accordingly embarked on board of that ship with the rest of her numerous passengers. During his stay at Ryswick he is said to have been usually entertained with a goat for dinner, once in every three or four weeks, with occasionally a duck or a fowl, by way of a dessert. He was brought on board shut up in a wooden crib or cage, the bars of which were sufficiently close to prevent his escape; and it had a sliding door, for the purpose of admitting the articles on which he was to subsist; the dimensions of the crib were about four feet high, and five feet square; a space sufficiently large for him to coil himself round with ease. The live stock for his use during the passage, consisting of six goats of the ordinary size, were sent with him on board, five being considered as a fair allowance for as many months. At an early period of the voyage we had an exhibition of his talent in the way of eating, which was publicly performed on the quarter-deck, upon which he was brought. The sliding door being opened, one of the goats was thrust in, and the door of the cage shut. The poor goat, as if instantly aware of all the horrors of its perilous situation, immediately began to utter the most piercing and distressing cries, butting instinctively, at the same time, with its head towards the serpent, in self-defence. The snake, which at first appeared scarcely to notice the poor animal, soon began to stir a little, and, turning his head in the direction of the goat, he at length fixed a deadly and malignant eye on the trembling victim, whose agony and terror seemed to increase; for, previous to the snake seizing its prey, it shook in every limb, but still continued its unavailing show of attack, by butting at the serpent, which now became sufficiently animated to prepare for the banquet. The first operation was that of darting out his forked tongue, and at the same time rearing a little his head; then suddenly seizing the goat by the fore leg with his mouth, and throwing it down, it was encircled in an instant in his horrid folds. So quick, indeed, and so instantaneous was the act, that it was impossible for the eye to follow the rapid convolution of his elongated body. It was not a regular *screw-like* turn that was formed, but resembling rather a knot, one part of the body overlaying the other, as if to add weight to the muscular pressure, the more effectually to crush his object. During this time he continued to grasp with his fangs, though it appeared an unnecessary precaution, that part of the animal which he had first seized. The poor goat, in the mean time, continued its feeble and *half-stifled* cries for some minutes, but they soon became more and more faint, and at last it expired. The snake, however, retained it for a considerable time in his grasp, after it was apparently motionless. He then slowly and cautiously unfolded himself, till the goat fell dead from his monstrous embrace, when he began to prepare himself for swallowing it. Placing his mouth in front of the dead animal, he commenced by lubricating with his saliva that part of the goat; and then taking its muzzle into his mouth, which had, and indeed always has, the appearance of a raw lacerated wound, he *sucked it in*, as far as the horns would allow. These protuberances opposed some little difficulty, not so much from their extent, as from their points; however, they also, in a very short time, disappeared; that is to say, externally; but their progress was still to be traced very distinctly on the outside, threatening every moment to protrude through the skin. The victim had now descended as far as the shoulders; and it was an astonishing sight to observe the extraordinary action of the snake's muscles when stretched to such an unnatural extent—an extent which must have utterly destroyed all muscular power in any animal that was not, like himself, endowed with very peculiar faculties of expansion and action at the same time. When his head and neck had no other appearance than that of a serpent's skin, stuffed almost to bursting, still the workings of the muscles were evident; and his power of suction, as it is erroneously called, unabated; it was, in fact, the effect of a contractile muscular power, assisted by two rows of strong hooked teeth. With all this he must be so formed as to be able to suspend, for a time, his respiration, for it is impossible to conceive that the process of breathing could be carried on while the mouth and throat were so completely stuffed and expanded by the body of the goat, and the lungs themselves (admitting the trachea to be ever so hard) compressed, as they must have been, by its passage downwards.

The whole operation of completely gorging the goat occupied about two hours and twenty minutes: at the end of which time the tumefaction was confined to the middle part of the body, or stomach, the superior parts, which had been so much distended, having resumed their natural dimensions. He now coiled himself up again, and lay quietly in his usual torpid state for about three weeks or a month, when, his last meal appearing to be completely digested and dissolved, he was presented with another goat, which he killed and devoured with equal facility. It would appear that almost all he swallows is converted into nutrition, for a small quantity of calcareous matter (and that, perhaps, not a tenth part of the bones of the animal), with occasionally some of the hairs, seemed to compose his *general* fæces ;— and this may account for these animals being able to remain so long without a supply of food."

There are many other serpents of the species *Boa;* of which a short notice is necessary.—1. The SPOTTED BOA. (*Boa scytale.*) This is sometimes scarcely inferior to the Boa Constrictor, and is of similar habits. It is of a grey colour, marked with large orbicular spots, interspersed with other marks and variegations. It is found in many parts of South America.—2. The RINGED BOA. (*Boa cenchris.*) This also grows to a large size, though considerably smaller than either of the before mentioned ; and may be easily distinguished by the regular distribution of its marks and colours. On the back is a continued series of very large blackish circles from head to tail, while along the sides are interspersed several kidney-shaped spots, with their centres white. It inhabits South America.—3. The EMBROIDERED BOA. (*Boa Phrygia.*) There exists scarcely a more truly elegant species in the whole serpent tribe than this. It is nearly four feet long ; the ground colour white, the back being tinged with a cast of yellowish brown ; while along the whole upper part is a continued series of black variegations, bearing a striking resemblance to embroidery. It is a native of the East Indies.—4. CANINE BOA. (*Boa canina.*) This beautiful snake is about four feet in length : the head is large, and shaped like that of a dog ; the general colour a bright Saxon green, with transverse white bars down the back, the edges of which are of a deeper green than the ground colour of the body : the belly is white. This species belongs to South America.—5. The GARDEN BOA. (*Boa hortulana.*) The ground colour of this species is a light yellowish brown, or sometimes pale violet, variegated with a dark purplish brown pattern resembling rich embroidery. The head is broad, and the neck slender. There are several others, but the foregoing will convey a sufficient idea of them.

BOAR. The male of Swine. [See HOG.]

BOAR-FISH. (*Capros aper.*) An Acanthopterygious fish, resembling the Dory in its general outline, and in having the first

BOAR-FISH.—(CAPROS APER.)

dorsal fin deeply notched ; but it has no spines along the dorsal or anal : The fins are covered with rough scales ; the mouth projects considerably ; and the fins are entirely without filaments. The flesh of the Boar-fish is but in little esteem.

A very few instances of the presence of this rare fish on the British coasts have been recorded. We believe the last was by Waring Kidd, Esq., and the following account of it appeared in "The Zoologist," p. 191 :— "On the 6th of March, 1842, a fish six inches and a half in length and three inches in width, of most brilliant colours, was picked up by a fisherman. It was alive when found : the colours were bright orange and lake. The fish was taken by the person who picked it up to Mr. Griffins, the principal fishmonger of Brighton ; he took it to the Pavilion, where it was presented to her Majesty. It was immediately recognised by His Royal Highness Prince Albert (the Prince being a good naturalist) as the Boar or Hog fish of the ancients, a species very scarce on the British coasts. His Royal Highness, wishing to have it preserved, sent it here ; as it was for Her Majesty, it gave me great pleasure that I succeeded so well, both in preserving the colours and in showing the peculiarity of the mouth, which the fish has the power of extending and contracting at will. When extended, it takes the form of a hog's snout ; hence the name of Boar-fish. I suspended the fish under a glass shade, and placed a few seaweeds, &c. on the stand. When quite completed, I made a painting of it, and succeeded in representing the colours pretty correctly."

BOAT-BILL. (*Cancroma.*) A genus of Grallatorial birds, distinguished by their

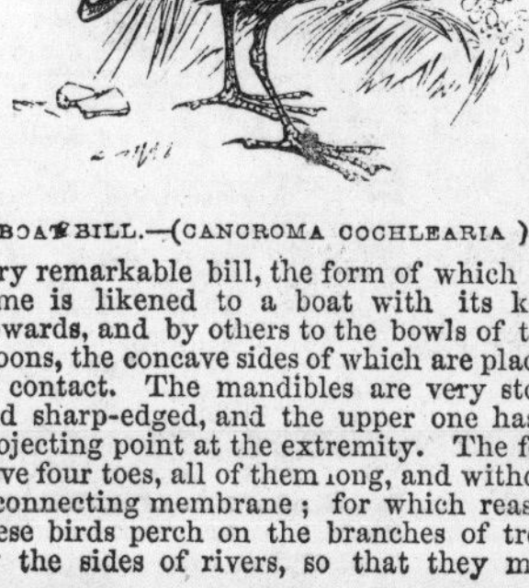

BOAT-BILL.—(CANCROMA COCHLEARIA.)

very remarkable bill, the form of which by some is likened to a boat with its keel upwards, and by others to the bowls of two spoons, the concave sides of which are placed in contact. The mandibles are very stout and sharp-edged, and the upper one has a projecting point at the extremity. The feet have four toes, all of them long, and without a connecting membrane ; for which reason these birds perch on the branches of trees by the sides of rivers, so that they may pounce upon the fish as they swim beneath. The species *Cancroma cochlearia* is the size of a domestic fowl. In the male, the forehead, and upper parts of the neck and breast, are dirty white ; and from the head depends a long crest of black feathers. The female has the top of the head black, without the elongated crest. It inhabits Guiana, Brazil, and other parts of South America. Close to this comes the gigantic Whale-headed Stork, or *Balæniceps*, a bird frequenting the upper waters of the Nile, and subsisting on fishes, young crocodiles, lepidosirens, &c. Two living examples were lately introduced into the Zoological Society's menagerie. [See BALÆNICEPS in SUPPLEMENT.]

BOAT-FLY. (*Notonecta.*) An aquatic Hemipterous insect, the back of which is shaped like the bottom of a boat ; and the hind legs, which are thrice as long as the fore, aptly enough resemble a pair of oars. The legs of the hinder pair have a fringe of bristles along their edge ; by which the surface, with which they strike the water in swimming, is greatly increased. Their general form is well adapted for rapid progression in water ; and it is from the peculiar aspect and movement of the body that they have received their name. They swim on their backs ; and their eyes are so placed that they are able to see both above and below the surface of the water, so that at the approach of danger they instantly descend, and are out of sight.

BOB-O-LINK. [See RICE BUNTING.]

BODIAN. There are several species of fish, of the Carp kind, bearing this name. They are natives of the Indian and American seas ; and vary from one foot to three feet in length.

BOMBUS. The Humble-bee [which see].

BOMBYCIDÆ. A family of Lepidopterous insects, one of the most interesting of which is the *Bombyx Mori*, well known as the Moth to which the Silkworm turns. The caterpillars of most of the species are hairy, and assume the pupa state in a cocoon spun for its protection.

BOMBYCILLA. The name of a genus of birds placed by Cuvier among the Dentirostral genera of his second order of *Passeres*. They may be distinguished at first sight from any other birds by a remarkable appendage on the tips of some of the quills, which has very much the appearance of red sealing-wax. Their principal generic characters are —Bill short, slightly depressed and triangular at the base ; above convex, towards the tip bent down, and emarginate on each side. Nostrils oval, cover d with small feathers. Feet four-toed, with the outer one connected at the base. [See WAXWING.]

BOMBYLIDÆ. A family of insects of the order *Diptera*. They have an appearance somewhat r sembling that of the smaller kinds of Humble-bees, being thickly covered with erect downy hair : their flight is rapid ; and they may be frequently observed to hang, as if suspended, over a flower, sipping its sweets by means of their long proboscis, while their wings vibrate so rapidly as to be scarcely discerned to move ; then darting to another with such rapidity that the eye cannot follow them. They frequent gardens, open parts of woods, and sunny banks ; and are most common in spring.

BOMBYX. [See SILK-WORM.]

BONASSUS. [See BISON.]

BONITO. (*Thynnus.*) A handsome fish of the order *Acanthopterygii*, a native of the Mediterranean, and a rare visitant of our

BONITO.—(THYNNUS PELAMIS.)

shores. It is about three feet long ; has a sharp head, a small mouth, large gills, full silvery eyes, and a crescent-shaped tail. It has no scales except on the middle of the sides, where a line of gold colour runs from the head to the tail. It is greenish on the back and sides, but its belly is of a silvery white ; and it is distinguished by its great activity and voracity, being one of the chief enemies of the flying-fish. It is also called the STRIPED TUNNY.

BOOBY. (*Sula fusca.*) The name given by navigators to a large bird, a species of Gannet, which inhabits the desolate islands

BOOBY.—(SULA FUSCA.)

and coasts of most warm climates. The name was naturally acquired from their apparent stupidity, in quietly sitting on the shore, or perching on the yard of a ship, till knocked on the head, or taken away by any one who may attempt it. [See GANNET.]

BOOK-WORM. A name given to various species of insects in the larva state, in which they destroy books and papers, by boring into them ; such as the *Ptinidæ*, *Anobium*, &c.

BOOPS. A genus of small Acanthopterygious fishes, found mostly in the Mediterranean. The species are generally of brilliant colouring, and characterized by a small mouth, large eyes, and a rounded form.

BOPYRUS. A parasitic Crustacean, of the order *Isopoda*, of which three or four species are known. They fasten on the prawn, hermit crab, and other Crustacea. The sexes differ very much in appearance. The *B. Squillarum* is far from uncommon in this country: it causes swellings on the side of the carapace of the common *Palæmon Squilla.*

BOS. The scientific name for a genus of hollow-horned ruminating quadrupeds, which in their domesticated state contribute most materially to the comforts and conveniences of man. [See OX; BISON, &c.]

BOSTRICHUS: BOSTRICHIDÆ. A genus and family of Coleopterous insects, of the group *Xylophila*, some species of which are highly destructive of timber. These Beetles usually live in wood, which their larvæ

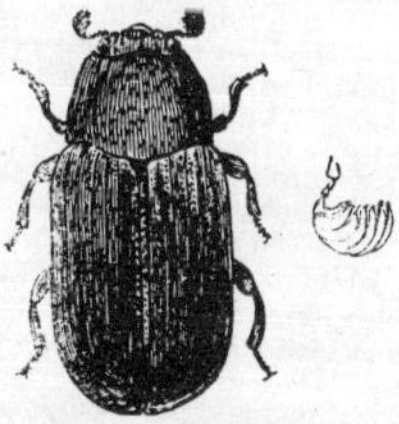

XYLOPHILUS FLABELLICORNIS.—AN ANTENNA MAGNIFIED.

pierce in every direction; and when abundant in forests, especially those of pines and firs, they destroy immense numbers of trees in a few years. One of the most destructive species is the *Bostrichus Typographicus*, or the Typographer Beetle, which has at different times ravaged the forests of Germany. It devours, both in the larva and perfect states, the soft wood beneath the bark, which is most essential to the vegetative process, and thus causes the death of the tree. The females attack the crevices of the bark, and perforate it in diverging lateral channels, in which from sixty to eighty eggs are deposited. At the end of fifteen days the larvæ are hatched, and forthwith commence the work of destruction, each gnawing a serpentine gallery between the bark and the wood, and gradually enlarging its burrow until the period when it is ready to pass into the pupa state; when, having finally become a perfect beetle, it directly bores through the portion of the tree which remains between the wood and the outer bark, and escapes through a small circular aperture in the latter. There appears to be no remedy when the trees are once attacked but to cut down the trees, bark them and burn the bark, and to remove all felled timber without delay.

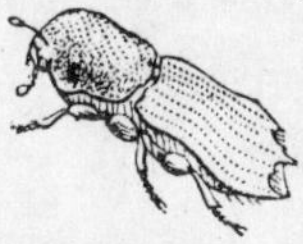

TYPOGRAPHER BEETLE.

BOTATRISSO. The Eel-pout. [See EEL.]

BOTAURUS. [See BITTERN.]

BOTRYLLUS. A genus of Molluscous animals termed *Aggregated Ascidians*, which at first float free and separate, but at a certain period of their existence unite to form one common mass. The aggregated animals thus found together are almost always very small, soft, irritable, and contractile, changing their form with the slightest movement. [See ASCIDIA.]

BOTS. The larvæ or caterpillars of the Gad-fly, belonging to the order *Diptera*, genus *Œstrus*, of which there are numerous species. They infest horses and cattle; and are distinguished by passing the larval state of their existence within some animal, and feeding on the juices or substance of that animal. [See ŒSTRUS; BREEZE-FLY, &c.]

BOTTLE-FISH. [See SACCOPHARYNX.]

BRACHELYTRA. An extensive group of Coleopterous insects, distinguished by the elongate form of the body and the shortness of the wing-cases. They run and fly with equal agility; preying upon decaying animal and vegetable matters, especially fungi, agarics, &c., in which they chiefly reside; they are also found in profusion under heaps of putrescent plants. They are decidedly carnivorous; some species are, however, found in flowers, others upon the margins of running streams, and others under the bark of decaying trees.—One of the commonest, and at the same time most formidable looking members of the family, is a black species, rather more than an inch long, commonly called the "Devil's Coach-horse" (*Goerius olens*). This is frequently to be seen running about garden walks, cellars, and dusty roads. True to the habits of the family, on the least approach of danger it immediately puts itself into a posture of defence, throws the tail over the head like a scorpion, protrudes the anal rings, elevates its head, and opens its long and powerful jaws.—The geographical range of this group of insects is principally confined to the temperate climes of the northern hemisphere; a few species, however, have been received from tropical climates, which are remarkable for the singularity of their forms and the splendour of their colours; but it rarely happens that the exotic species exceed those of our own country in size.

BRACHINUS. A genus of Coleopterous insects, with truncated elytra and a smallish thorax; of which the *Brachinus crepitans* is the most common. This insect, which is found under stones, is about half an inch long; the head, thorax, and legs are of a yellowish red colour; the wing-cases greenish, or blue black; and the antennæ reddish. They possess a remarkable power of violently expelling from the anus a pungent acrid fluid, accompanied by a loud report, considering the size of the insect; whence its common name of *Bombardier Beetle*.

BRACHIONUS. A genus of minute animals, found in stagnant fresh water and in sea water. [See ENTOMOSTRACA.]

BRACHIOPODA. A class of Acephalous or headless molluscous animals, with bivalve shells. They are characterized by having the mantle organized so as to be serviceable for respiration, and by having two long, fleshy, ciliated, spiral arms, but no foot. They have no organs of locomotion, but live fixed to submarine bodies. The species are numerous and widely diffused; and, though comparatively low in the scale of creation, the class is interesting both to the physiologist and the geologist.

BRACHYCERUS. A genus of Coleopterous insects, the species of which are apterous, and very rough. They live upon the ground, and appear to be peculiar to the south of Europe and Africa, particularly abounding in the latter quarter of the globe.

BRACHYPTERES. The name given by Cuvier to that class of birds generally known as "Divers."

BRACHYTELES. A genus of *Quadrumana*, so named on account of the very small development of the thumb.

BRACON. A genus of Hymenopterous insects, allied to the Ichneumons; remarkable for the hiatus which exists between the mandibles and the clypeus.

BRADYPUS. [See SLOTH.]

BRAHMIN BULL, INDIAN OX, or ZEBU. (*Bos Indicus.*) There is a very considerable difference in the various domesticated Asiatic oxen, as to the size and direction of the horns: some are short and suberect; others incline inwards; but they are generally distinguished by a fatty elevated hump upon the withers. The ears are pendulous, and the dewlap is usually very largely developed. Their colour varies from a light ashy grey to a milk white, and their size from the stature of an ordinary bull to

BRAHMIN BULL.—(BOS INDICUS.)

that of a Shetland pony. The limbs of all are light and elegant. The flesh is neither so sweet nor so good as that of the common ox, except the hump, which is allowed on all hands to be delicious when properly cooked. In many parts of India the Zebu is used as an animal of burden, and, when harnessed to a carriage, it will travel, at an easy rate, about thirty miles a day. Antient writers speak of its performing about double that distance; but if that were true, it must have lost much of its fleetness. The Hindoos regard them as animals worthy of veneration, and consequently consider it sinful to slaughter them; they do not, however, generally object to work them. "They are spread," says Mr. Bennett, "over the whole of Southern Asia, the islands of the Indian Archipelago, and the eastern coast of Africa from Abyssinia to the Cape of Good Hope."

BRAMBLING. (*Fringilla montifringilla.*) This bird, which is also called the Mountain-Finch, is larger than the chaffinch. The top of the head and the back are of a glossy black colour, slightly edged with a yellow brown; the throat and breast are orange, as are the lesser coverts of the wings; but those which rest on the quill feathers are barred with black, tipped with orange; and the tail is slightly forked.

BRANCHIOPODA. An order of Crustaceous animals, in which the locomotive extremities fulfil the functions of gills. These Crustaceans, which are for the most part microscopic, are always in motion when in an animated state, and are generally protected by a shell or crust in the shape of a shield, or of a bivalve shell, and are furnished sometimes with four, sometimes with two antennæ. Their feet vary in number, some having not less than a hundred. A great portion have only one eye.

BRANCHIOSTEGI. A tribe of Cartilaginous fishes, comprehending those in which the gills are free, and covered by a membrane; including the Sturgeon and Chimæra [which see].

BREAM. (*Abramis brama.*) A fish of the Carp family, and by anglers often called the Carp-bream; found in lakes, and in the deepest parts of still rivers. The body is extremely deep and thin in proportion to its

BREAM.—(ABRAMIS BRAMA.)

length, and the back much elevated. Length two feet to two feet and a half; colour olive, with a pale or flesh-coloured tinge on the under parts: scales large; dorsal fin rather small, and situated a little beyond the middle of the back; anal fin extending from the vent nearly to the tail, which is pretty deeply forked. Its flesh is generally considered coarse and extremely insipid.

The SEA BREAM (*Pagrus centrodontus*) is a common fish in the Mediterranean, nor is it by any means uncommon on the southern and western coasts of England, especially during summer and autumn. The spawn is shed in the beginning of winter in deep water; and it retreats altogether from our shores in severely cold weather. The young of this fish are commonly known by the name of *Chads*. The Sea bream is not very highly esteemed for the table, either fresh or salted.

BREEZE-FLY. (*Œstrus; Œstridæ.*) The insects we are about to describe are produced from larvæ which when existing in horses are termed *bots;* in sheep, *maggots;* and in cows and oxen, *wornils;* and these three represent three divisions of the family, differing essentially in their history. The perfect insect produced from each kind of larva is properly termed a Breeze-fly. Before we proceed farther, however, we beg to state that the observations which follow are taken from Mr. Newman's History of Insects, who quotes as his authority "An Essay on the Bots of Horses and other animals, by Bracy Clark, F. L. S."

"The opinions of the Breeze-fly of the horse, or *bot*, as it is usually termed, as to the benefit or injury derived from it, are very opposite; some observers go so far as to assert that the larvæ occasionally completely perforate the stomach of the horse, causing disease, pain, and even death; others regard them as perfectly innocuous; and one author [Mr. Bracy Clark], whose careful and laborious investigations entitle his opinions to the greatest respect, believes the effect of bots to be salutiferous rather than otherwise; and from his masterly essay the following particulars are extracted.

"The female fly, in approaching the horse for the purpose of oviposition, carries her body nearly upright in the air, the protruded ovipositor being curved inwards and upwards. Suspending herself for a few seconds before the part of the horse on which she intends to deposit the egg, she suddenly darts upon it, and leaves the egg adhering to the hair: she hardly appears to settle, but merely touches the hair with the egg held out on the extreme point of the ovipositor, the egg adhering by means of a glutinous liquor with which it is covered. She then leaves the horse at a small distance, prepares a second egg, and, poising herself before the part, deposits it in the same way: the liquor dries, and the egg becomes firmly glued to the hair. This is repeated till four or five hundred eggs are sometimes placed on one horse. The skin of the horse is usually thrown into a tremulous motion on the touch of the insect, which merely arises from the very great irritability of the skin and cutaneous muscles at this season of the year, occasioned by the heat and continual teasing of the flies, till at length these muscles appear to act involuntarily on the slightest touch of any body whatever.

"The fly does not deposit her eggs at random on the horse's body, but selects those parts which are most likely to be nibbled by the horse: the inside of the knee is frequently chosen, but all naturalists must have remarked how commonly the eggs of the bot are deposited on that part of a horse's shoulder which he can never reach with his mouth, and thus, to a casual observer, it would seem that they must perish, and fail in the object for which their parent designed them. Now there is a provision of nature which exactly counteracts this difficulty. When horses are together in a pasture, and one of them feels an irritation on any part of the neck or shoulder which he cannot reach with his mouth, he will nibble another horse in the corresponding part of his neck or shoulder, and the horse so nibbled will immediately perform the kind office required, and begin nibbling away in the part indicated. The horses, when they become used to this fly, and find it does them no injury by sucking their blood, hardly regard it, and do not appear at all aware of its object.

"When the eggs have remained on the hairs four or five days, they become mature, after which time the slightest application of warmth and moisture is sufficient to bring forth in an instant the latent larva. At this time, if the lips or tongue of the horse touch the egg, its operculum is thrown open, and the young larva liberated: this readily adheres to the moist surface of the tongue, and is from thence conveyed with the food to the stomach. It is worthy of remark, that it is probable the greater part of the eggs deposited by this fly are taken up in consequence of the irritation of other flies, as the *Tabani* and *Stomoxides*, which, by perpetually settling on the skin, occasion a horse to nibble himself on those parts, and thus receive the larvæ on the tongue and lips, whence they are introduced into the stomach. * * * The larva, when matured, quits the stomach of the animal and falls to the ground, and finding a convenient place of retreat, undergoes its change to a chrysalis, the skin then losing its organization, and changing in colour from a whitish red to a reddish brown. After remaining torpid in the chrysalis state a few weeks, the superfluous moisture being removed and the parts of the future insect hardened by drying, it bursts from its confinement, and the fly makes its exit at the small end of the case. A few hours after quitting their shell they become dry, take wing, and seek their mates.

"A *second* species of Breeze-fly has a still more wonderful history: its eggs are laid in the nostrils of sheep, from one to seven or eight in each individual, and these on becoming larvæ, enter the frontal and maxillary sinuses, and even the horns, and feed on their secretions: when the larvæ are young they are perfectly white and transparent, except two small black horny plates: as they increase in size the upper surface becomes marked with two transverse brown lines on each segment, the anterior being shorter and narrower than the posterior; and some spots are also observable on the sides. The body consists of twelve segments besides the head. These larvæ move with considerable activity, holding with their tentacula to a fixed point and drawing up the body. When full grown the larvæ fall through the nostrils of the sheep, and change to the pupa state lying on the earth or adhering to the side of a blade of grass: in about two months the case of the chrysalis opens, and the fly makes its appearance. Sheep are exceedingly annoyed by these flies, and to avoid them lie down in ruts with their heads close to the ground; at other times we see them huddled together under trees in a dense mass or phalanx, the nose of each being pushed into the fleece of another.

"There is a *third* species of Breeze-fly, far more formidable than either of those previously described: its eggs are laid on the backs and sides of cows and oxen, and the larvæ hatched from them enter the hide, producing tumours as large as pigeons' eggs. The larva itself is of an oblong figure, larger at one extremity than at the other; the body is divided into ten or twelve segments by transverse bands, and these are again intersected by six longitudinal lines, which purse up the skin, and produce along the sides a series of mammiform protuberances, each possessing at its extremity a respiratory pore: on each segment of the body may be observed ridges, or dotted prominent lines, interrupted however by the longitudinal lines already noticed: there are in pairs a narrower and broader line of minute dots or points; the narrower line is found, under a lens, to be formed of hooks bent towards the posterior extremity of the insect; the broader lines consist of smaller hooks bent in an opposite direction, or towards its head: it is probably by the aid of these hooks that the animal raises or depresses itself in the tumour, and finally, when mature, effects its escape.

"The food of the larva appears to be the pus or matter surrounding it in the tumour in which it exists: as regards the period of its continuing to feed we have little satisfactory information. Its colour when young is white, but as it advances towards maturity it becomes browner, and finally of a deep dark brown, approaching to black: having attained its full size, it presses itself against the upper part of the tumour, and by some unknown process makes an aperture in the hide of about sufficient size to admit a pea; through this the larva wriggles itself a segment at a time until it comes quite out, and falling to the ground seeks a convenient retreat in which to become a chrysalis.

"The chrysalis is of a dark brown colour, and in figure somewhat resembles the half of a walnut-shell, being narrower at one end than the other, flat on one side, and very rounded and convex on the other; after lying on the ground for some weeks, a portion of the indurated skin or cover, of a triangular shape, is forced up at the smaller end, and through the aperture thus occasioned the fly emerges. The *fly* is large and handsomely coloured; although the presence of the larvæ in the backs of cattle is frequently observable, the insect in its perfect state is rarely met with, and very few of our cabinets possess good specimens; it flies with rapidity, but apparently without noise, and never ventures over water.

"The act of oviposition appears to be attended with severe suffering, or apprehension at least, which makes the cattle run wild and furious, and gad or stray from the pastures, and hence the ancient epithet of gad-fly. When oxen are yoked to the plough, the attack of this fly is attended with real danger, since they become perfectly uncontrollable, and often run directly forwards through the hedges, or whatever obstructs their way. On this account many ploughs are provided with a contrivance for setting the oxen immediately at liberty. When the cattle are attacked by this fly, it is easily known by the extreme terror and agitation of the whole herd; the unfortunate object of the attack runs bellowing from among them, and seeks a refuge in the nearest water; the tail becomes rigid, and is brandished aloft, or held straight out from the body. Its frightened companions follow in the rear of the animal attacked, and a wild and apparently unmeaning chase takes place, which, from the inelegant gallop of the cows, has often a very ludicrous effect.

BRENT GOOSE. A much smaller bird than the common wild goose, but with longer wings; and it traverses greater distances in its migrations. Its breeding places are in the far north; but it migrates for the winter to France, England, Ireland, &c. The head, neck, bill, and upper part of the breast are black; and on each side of the slenderest part of the neck: the lower part of the breast, the scapulars, and coverts of the wings are ash-coloured; the feathers, both above and below the tail, are white; and the tail, the quill feathers, and the legs are black

BRENTHIDÆ, or BRENTIDÆ. A family of Coleopterous insects, which are among the most remarkable of the beetle tribe, and almost entirely confined to tropical climates. Distinguishing characters:—body much elongated; tarsi with the penultimate joints bilobed; antennæ filiform, or in some with the terminal joint formed into a club; proboscis projecting horizontally; palpi minute. They are found crawling on trees, or under the bark, and sometimes on flowers. Their general colour is black or brown, with red spots or markings.

Dr. Thaddeus Harris, librarian of Harvard College, gives a detailed account of a North American species, in his fine work on the Insects of Massachusetts. We somewhat condense his history of it. It is the *Brenthus septemtrionalis*. The Northern Brenthus, so named because most of the other species are tropical insects, is of a mahogany brown colour; the wing cases are somewhat darker, ornamented with narrow tawny yellow spots, and marked with deep furrows, the sides of which are punctured. Its common length is about six-tenths of an inch, but much larger as well as smaller specimens frequently occur. The Northern Brenthus inhabits the white oak, on the trunks and under the bark of which it may be found in June and July, having then completed its transformations. The female, when about to lay her eggs, punctures the bark with her slender snout, and drops an egg in each hole thus made. The grub, as soon as it is hatched, bores into the solid wood, forming a cylindrical passage, which it keeps clear by pushing its castings out of the orifice of the hole, as fast as they accumulate. These castings or chips are like very fine saw-dust; and the holes made by the insects are easily discovered by the dust around then. The grub is about an inch long and nearly cylindrical; the last segment is of a horny consistence, and is obliquely hollowed at the end, so as to form a kind of gouge or scoop, the edges of which are furnished with little notches or teeth. It is by means of this singular scoop that the grub shovels the minute grains of wood out of its burrow. The pupa, which is of a yellowish white colour, is met with in the burrow formed by the larva; the back is furnished with transverse rows of little thorns or sharp teeth, and there are two larger thorns at the extremity of the body. These minute thorns probably enable the pupa to move towards the mouth of its burrow when it is about to be transformed, and may serve also to keep its body steady during its exertions in casting off its pupa-skin. These insects are most abundant in trees that have been cut down for timber or fuel, which are generally attacked the first summer after they are felled; it has also been ascertained that living trees do not always escape, but those that are in full vigour are rarely perforated by grubs of this kind.

BREVIPENNES. The term given to the first family of *Stilt-birds*, the shortness of whose wings are inadequate to perform the function of flight; the weight of their massive bodies appearing to require more muscular power to support them in the air than nature has furnished them with. The pectoral muscles are reduced to extreme tenuity; but the muscles of the thighs and legs are of an enormous thickness. [See OSTRICH, CASSOWARY, &c.]

BRILL, or PEARL. (*Pleuronectes rhombus.*) In its general form this fish resembles the Turbot, but is inferior to it both in size and quality. It is distinguished from the Turbot by the perfect smoothness of its skin, which is covered with scales of a moderate size, and by its pale brown colour above, marked by scattered yellowish or rufous spots; the lateral line, as in the Turbot, is first arched over the pectoral fins, and from thence runs straight to the tail. The Brill is taken on many parts of our coasts; the principal part of the supply for the London market being derived from the southern coast, where it is most abundant.

BRIMSTONE [BUTTERFLY]. A name applied to the *Gonepteryx Rhamni.*

BROCK. A local name given to the Badger. Burns alludes to a "stinking *brock.*"

BRONTOZOUM. [See SUPPLEMENT.]

BRUCHUS: BRUCHIDÆ. A genus and family of Coleopterous insects, allied to the Weevils, and thus characterized: palpi obvious, filiform, not very minute; rostrum broad; labrum exserted; antennæ eleven-jointed, sub-clavate, with the club formed of distinct joints in some; f[illegible], or gradually thicker towards their points, in others; serrated or pectinated; the anus naked; hind feet generally very large. The female deposits an egg in the young and tender germ of various leguminous or cereal plants, &c., upon which the larva feeds, and within

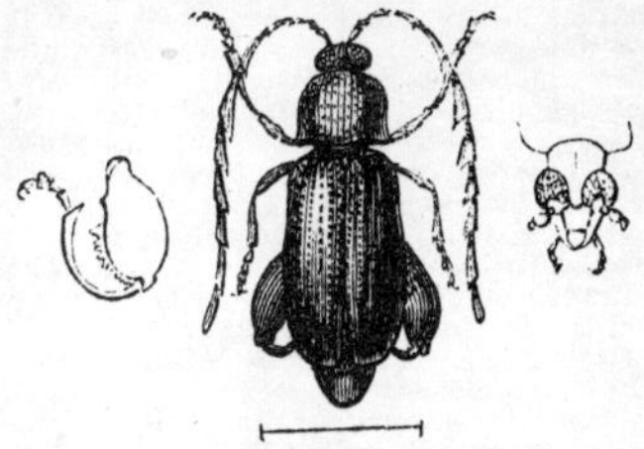

BRUCHUS SERRIPES.

which it undergoes its transformations: the perfect insect, in order to make its escape, detaches a portion of the epidermis like a small cup; hence the small holes often observed in peas, dates, &c. The family is very extensive. *Bruchus pisi*, Linn., which is two lines long, black, with grey spots on the elytra, in some years does great mischief to peas, particularly in North America. *Bruchus serripes*, the figure of which is here given, with the head and posterior limb, is a fine example of this family. By some authors it is placed in a separate genus.

BRUSH TURKEY. The New Holland Tallegalla. [See TALLEGALLA.]

BRYOZOA. [See SUPPLEMENT.]

BUBO. A subgenus of Owls. [See OWL.]

BUCCINUM. A genus of Molluscous animals called Whelks; the general characters of which are, that their mouths are an oblong or very lengthened oval, the upper parts of which are slightly beaked. In the Linnæan system, the Buccina form a distinct genus of the univalve and spiral Testacea. Those species most usually met with on the coasts of the British isles are the brown, massy, waved, striated, reticulated, and small Buccina. The shell of the *Buccinum lapillus* (the common White Buccinum) is one of the shells from which the ancients are supposed to have extracted their indelible purple dye, called the Tyrian purple. The part containing the colouring matter is a longitudinal vein, just under the skin on the back, behind the head. If the vein is laid open with a needle, a tenacious yellow matter will flow, which being applied with a hair pencil to linen, silk, or paper, it will in a short time become of a bright yellow, will soon change to pale green, then assume a bluish cast, and afterwards a deep and brilliant purple.

We learn from Mr. Stevenson's interesting narrative of the erection of the Bell Rock light-house, that the *Buccinum lapillus* preys upon the Mussel (*Mytilus edulis.*) Mr. S. says, "When the workmen first landed upon the Bell Rock, limpets of a very large size were common, but were soon picked up for bait. As the limpets disappeared, we endeavoured to plant a colony of mussels, from beds at the mouth of the river Eden, of a larger kind than those which seem to be natural to the rock. These larger mussels were likely to have been useful to the workmen, and might have been especially so to the light-keepers, the future inhabitants of the rock, to whom that delicate fish would have afforded a fresh meal, as well as a better bait than the limpet; but the mussels were soon observed to open and die in great numbers. For some time this was ascribed to the effects of the violent surge of the sea, but the *Buccinum lapillus* (Purpura) having greatly increased, it was ascertained that it had proved a successful enemy to the mussel. The *Buccinum*, being furnished with a proboscis capable of boring, was observed to perforate a small hole in the shell, and thus to suck out the finer parts of the body of the mussel; the valves of course opened, and the remainder of the fish was washed away by the sea. The perforated hole is generally upon the thinnest part of the shell and is perfectly circular, of a *champhered* form, being wider towards the outward side, and so perfectly smooth and regular as to have all the appearance of the most beautiful work of an expert artist. It became a matter extremely desirable to preserve the mussel, and it seemed practicable to extirpate the *buccinum*. But after we had picked up and destroyed many barrels of them, their extirpation was at length given up as a hopeless task. The mussels were thus abandoned as their prey, and in the course of the third year's operations, so successful had the ravages of the *buccinum* been, that not a single mussel of a large size was to be found upon the rock; and even the small kind which bred there, are now chiefly confined to the extreme points of the rock, where it would seem their enemy cannot so easily follow them."

BUCCO. A genus of birds, called *Barbets*, a name derived from the bristly feathers which surround the base of the bill, and project beneath the chin like a beard. [See BARBET.]

BUCEROS. [See HORNBILL.]

BUCK. The male of the Fallow Deer, the female of which is called a *Doe*. [See DEER.]

BUFFALO. (*Bos bubalus.*) A species of Ox, found in various parts of India; but in America the name of "Buffalo" is universally given to the Bison [which see]. The Buffaloes are of large size, but low in proportion to their bulk; they have no hunch on the back, and only a small dewlap on the breast; the hide is generally black; the tail long and slender. They generally live in small flocks, but sometimes are found in herds of considerable numbers; frequenting moist and marshy situations, and preferring the coarse vegetation of the forest and swampy regions to that of open plains. They swim well, and cross the broadest rivers without hesitation; their gait is heavy, and they run almost always with the nose horizontal, being principally guided by the sense of smelling. They are fierce and stub-

BUFFALO.—(BOS BUBALUS.)

born, and with difficulty subjugated.—The Arnee Buffalo (*Bos arni*) has horns of a prodigious size and length; the horns are turned laterally, flattened in front, and wrinkled on the concave surface. A pair of them are in the British Museum, each of which measures along the curve from base to tip, six feet three inches, and eighteen inches in circumference at the base.

This formidable animal is found wild in many parts of India, and also tame wherever the inhabitants have occasion for its services. Being extremely strong, they are employed in agriculture, and in drawing and carrying burdens, being guided by rings thrust through their noses. All Buffaloes are extremely fearful of fire; and they have a great aversion to red colours. In general, they are very inoffensive, if left undisturbed; but when wounded, or even fired at, their fury becomes ungovernable; they then tear up the ground with their fore-feet, make a horrid bellowing, and pursue the objects of their resentment with determined fury. [See BISON.]

BUFO. [See TOAD.]

BUG. (*Cimex.*) Of the numerous tribe of Hemipterous insects belonging to the genus *Cimex*, we may specify the troublesome and nauseous insect, the *Cimex lectularius*, or common domestic Bug. To give a very particular description of this noxious tormentor would be superfluous: it may be sufficient to observe, that it is of an oval shape, about the sixth of an inch long, of a compressed or

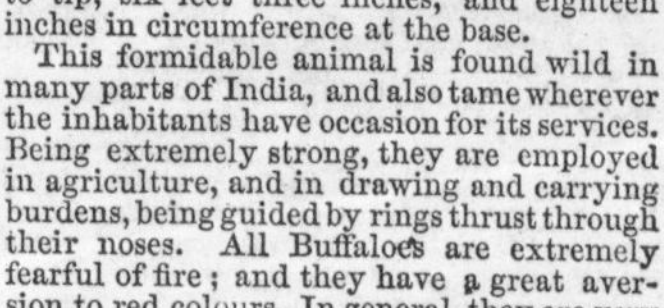

BUG.—(CIMEX LECTULARIUS.)

flat form, and of a reddish-brown colour. It is asserted, though it may be difficult to say how truly, that the Bug was scarcely known in England before the year 1670, having been imported from America among the timber used in rebuilding the city of London after the great fire of 1666; but it appears not to have been an uncommon pest in several countries of Europe before that time. Its blood-sucking properties, and the offensive smell it emits when touched, are too well known to require comment. The female Bug deposits her eggs in the beginning of summer; they are very small, white, and of an oval shape ; each is fixed to a small hair-like stalk, which is glutinous, and readily adheres to any thing it touches. The places in which the eggs are generally deposited are the crevices of bedsteads or other furniture, or the walls of a room. During the winter months these odious insects secrete themselves behind walls, old wainscoting, or any neglected places, where they are capable of bearing the most intense frost without injury, and on the return of warm weather again emerge from their concealment. A Bug always avoids the light, if possible ; and takes advantage of every chink and cranny to make a secure lodgment ; its motion is slow and unwieldy ; but its sight is so exquisite, that although it persecutes its victim with unceasing assiduity in the dark, the moment it perceives the light, it generally makes good its retreat.—The Rev. Leonard Jenyns has described two or three other species found in this country. [See CIMEX.]

BULIMUS. The name of a very extensive genus of terrestrial molluscs, very much resembling the *Helix*. Some attain to great size. [See HELIX.]

BULL-DOG. (*Canis* [*domesticus*] *molossus*). A variety of the Dog, remarkable for its short, broad muzzle, and the projection of its lower jaw. The head is massive and large, and the frontal *sinuses* large ; the lips are thick and pendulous ; the ears pendant at the extremity ; the neck robust and short ; and the legs short and thick. Though inoffensive and harmless when properly domesticated, the Bull-dog presents to the eye a most terrific appearance : the doubtful and designing leer, the tiger-like shortness of the head, the under-hung jaw, the width of the skull, the distension of the nostrils, and the almost constant sight of the teeth, hold forth a very formidable proof of the power he can exert, when that power is angrily brought into action. The breed is by no means so numerous as formerly, in consequence of the abolition of the barbarous sport of bull-baiting ; the butchers, however, use Bull-dogs in catching and throwing down cattle ; and it is surprising to see the apparent ease with which the dog will seize an ox by the nose, and hold him perfectly still, or throw him on his side, at his master's command. They become very vicious, and sometimes extremely dangerous, as they advance in years, inflicting dreadful bites for the slightest provocation; in their unrestrained state, indeed, they are a real nuisance, and therefore ought never to be allowed their full liberty. [See DOG.]

BULLA. A genus of Molluscous animals with univalve shells ; whose general characters are, that the shell is sub-oval, that the aperture is oblong and smooth, and that one end is a little convoluted. The animal breathes by gills, but has no respiratory tube,

and consequently the margin of the aperture of the shell is entire, or without a fissure or canal. There are numerous species, widely diffused ; generally about the size of a bean, tho' sometimes much larger. Most of this genus, especially of the larger sizes, are furnished with an organ exactly resembling the gizzard of a fowl, and which they appear to use for the purpose of masticating their food.

BULLFINCH. (*Loxia pyrrhula* of Linnæus.) A well-known and pretty bird, about the size of a sparrow. Its wild note is a soft low twitter ; but, when tamed, it becomes remarkably docile, and learns with great facility to whistle musical airs, which, if properly taught, it seldom wholly forgets. The bill is strong, short, black, and thick ; the upper part of the head, the ring round the bill, and the origin of the neck, fine glossy black ; the back ash grey ; breast and belly red ; wings and tail black; the upper tail coverts and vent are white ; legs dark brown. The female is very like the male, but the colours are less bright, and the under parts of a reddish brown. These birds are common in every part of our island, as well as in most parts of Europe ; their usual haunts during summer are woods and thickets ; they also frequent our orchards and gardens in the spring, seeking not only the insects which are lodged in the tender buds of fruit-trees, but feeding on the buds ; on which account they are regarded by gardeners as among the most pernicious of the feathered race.

BULLFINCH.—(PYRRHULA PYRRHULA.

The Bullfinch is a native of England, and also of most parts of the European continent. It generally constructs its nest, which is composed of small dry twigs, in the thickest parts of a white or black thorn hedge. The female lays about four or five bluish-white eggs, marked with dark spots at the larger end ; and breeds about the latter end of May. The bird is very common in the mountainous parts of Germany ; from which country the market for piping-bullfinches is principally supplied. Other species are met with in Asia, Africa, and America; but they hardly require a distinct notice here. By many naturalists they are made to constitute a separate genus, called *Pyrrhula*, after the type which we have just described.

BULL-FROG. [See Frog.]

BULL-HEAD, or MILLER'S-THUMB. (*Cottus gobio.*) There are several species of fish, inhabiting different climates, which are denominated Bullheads; but it is the well-known *River Bullhead*, or *Miller's Thumb*, an Acanthopterygious fish, which we are now about to describe. This species is found in clear brooks and rivers, in most parts of

RIVER BULLHEAD.—(COTTUS GOBIO.)

Europe. It is only four or five inches long; the head is of a roundish shape, large, broad, and depressed; the gill-fins are round, and beautifully notched at their circumference; and the rays of all the fins are prettily spotted. The mouth is large, and full of small teeth; the general colour of the body is a dark brownish black; the sides lighter, with small black spots; and the under surface of the head and belly white. It is so remarkably stupid, that whatever number may be together, the most inexpert angler may catch them all. It is generally found among loose stones, under which, from the peculiarly flattened form of its head, it is enabled to thrust itself, and thus to find a hiding place. Mr. Yarrell, in his truly national work "The British Fishes," (the wood-cut illustrations of which are such models of excellency,) so amusingly accounts for the popular names of this fish, that we take the liberty of borrowing his words:— "As the term Bullhead is considered to refer to the large size of the head, so the name of Miller's Thumb given to this species, it has been said, is suggested by, and intended to have reference to, the particular form of the same part. The head of the fish, it will be observed, is smooth, broad, and rounded, and is said to resemble exactly the form of the thumb of a miller, as produced by a peculiar and constant action of the muscles in the exercise of a particular and most important part of his occupation. It is well known that all the science and tact of a miller is directed so as to regulate the machinery of his mill, that the meal produced shall be of the most valuable description that the operation of grinding will permit when performed under the most advantageous circumstances. His profit or his loss, even his fortune or his ruin, depend upon the exact adjustment of all the various parts of the machinery in operation. The miller's ear is constantly directed to the note made by the running-stone in its circular course over the bed-stone, the exact parallelism of their two surfaces, indicated by a particular sound, being a matter of the first consequence: and his hand is as constantly placed under the meal-spout, to ascertain by actual contact the character and qualities of the meal produced. The thumb by a particular movement spreads the sample over the fingers; the thumb is the guage of the value of the produce, and hence has arisen the sayings of "Worth a miller's thumb;" and "An honest miller hath a golden thumb;" in reference to the amount of the profit that is the reward of his skill. By this incessant action of the miller's thumb, a peculiarity in its form is produced, which is said to resemble exactly the shape of the head of the fish constantly found in the mill-stream, and has obtained for it the name of the Miller's Thumb, which occurs in the comedy of "Wit at several Weapons," by Beaumont and Fletcher, act v. scene 1.; and also in Merrett's "Pinax." Although the improved machinery of the present time has diminished the necessity for the miller's skill in the mechanical department, the thumb is still constantly resorted to as the best test for the quality of the flour."

The ARMED BULLHEAD, or POGGE, (*Cottus aspidophorus,*) is found in the Baltic and Northern seas, and is also taken on the British coasts. It seldom exceeds six inches in length; small crustaceous animals and aquatic insects are its food; and its flesh is said to be firm and good. The head is large, bony, and very rugged; the end of the nose is armed with four short upright spines; and the chin is furnished with several minute cirri. The mouth is small, as are the teeth, which are very numerous. The body is divided longitudinally by eight scaly ridges, and is defended by eight rows of strong scaly plates, of which the elevated ridges form the central lines. The pectoral fins are large, with a broad bar of brown across the centre; the general colour of the upper surface of the body brown, with four broad dark brown bands; tail brown; and the under parts of the body nearly white.

The SIX-HORNED BULLHEAD (*Cottus hexacornis*) is a North American species, about seven inches long. The head is large and depressed, and on it are six nail-shaped processes standing erect; the eyes are large; the mouth is capacious, its margins formed by the intermaxillaries and lower jaw; both jaws and the vomer are set with bands of fine teeth; the gill-covers are composed of several bones connected by membrane, and armed on their exterior edges with four or five small spinous teeth: the bones which support the pectoral fins are also armed with small spines, and have sharp rough edges. The body is much narrower than the head, and tapers to the insertion of the caudal fin. The upper aspect of this fish presents a clouded admixture of brown and olive-green tints; the belly white; and the fins streaked with bluish-black. This species is said to be extremely tenacious of life; for, after being drawn from the water, they will leap vigorously over the sands, and inflate the head when touched. In this operation the branchiostegous membrane is distended, and the several pieces composing the gill-covers are separated by the extension of the intervening membranes. Capt. Sir J. C. Ross, who considers it to be the same with the *Cottus scorpioides* of Fabricius, says that, although very abundant on the Greenland coast, it is more rare in the higher latitudes, but several were taken on both sides of the peninsula of Boothia. The natives prize it highly as an article of food, preferring it to cod-fish or salmon.

BUNTING. (*Emberiza.*) The Buntings form a very interesting group of Passerine birds. The general characters of the family are, that the bill is strong and conic, the upper mandible with a strong knob on the "palate," the sides of each mandible bending inwards; they live principally on seeds, for the breaking of which their bill is well adapted. We select a few from among the numerous species as examples.

The COMMON BUNTING. (*Emberiza miliaria.*) The length of this bird is about seven inches and a half; beak brown; head and upper parts light brown, inclining to olive; under parts yellowish white; quills dusky with lighter edges; upper coverts tipped with white; tail slightly forked and dusky; and legs pale brown. These birds

COMMON BUNTING.—(EMBERIZA MILIARIA.)

are common in England, delighting in those parts that abound in corn, and rarely found in uncultivated places: in winter they assemble in vast flocks; and are often taken in nets, and brought to market, where they are sold for larks, but may be easily distinguished by the knob in the roof of the mouth. The female builds her nest on a tuft of dead plants, a few inches from the ground: it is composed externally of grass and a few long hairs. She lays five or six dirty-white eggs, spotted with reddish brown and ash colour.

The BLACK-HEADED, or REED-BUNTING. (*Emberiza schœniclus.*) Birds of this species frequent fens and marshy places, where there is abundance of rushes, among which they nestle. The head, throat, fore part of the neck, and breast, are black; divided by a white line from each corner of the bill, passing downwards and nearly encircling the neck: upper parts of the body and wings reddish brown, each feather with a black streak down the middle; under parts

REED BUNTING.
(EMBERIZA SCHŒNICLUS.)

white, with brownish streaks on the sides; quills dusky, edged with brown; two middle tail-feathers black, the outer ones almost white; legs and feet dusky brown. The head of the female is rust-coloured, spotted with black: it is destitute of the white ring round the neck, but in other respects it resembles the male. This bird was formerly supposed to suspend its nest between four reeds within a few feet of the water; this, however, is not the case, as it generally places it on the ground at a little distance from the water, and occasionally in a bush, in high grass, or in furze, at a great distance from any water: it is composed of stalks of grass, moss, and fibres, lined with fine grass. The eggs are four or five in number, of a dirty bluish white, with many dark-coloured spots and veins.

To the Bunting Family, but by naturalists placed in different genera from the preceding, belong the three following species:—

The SNOW BUNTING. (*Plectrophanes nivalis.*) This hardy bird is an inhabitant of the mountains of Spitzbergen, Greenland, Lapland, Hudson's Bay, and other cold northern countries: in the Highlands of Scotland (where it is known by the name of the Snow-flake) it is said also to be extremely abundant, and is supposed to be the harbinger of severe weather; which drives it from its usual haunts. The Snow Bunting weighs only about an ounce and a half. The bill and legs are black; the forehead and crown are white, with some mixture of black on the hind part of the head; the back is wholly black; the rump is white; the quill-feathers are black, with white bases; and the secondaries are white, with black spots on their interior webs. The inner feathers of the tail are black, the three exterior ones being white, with dusky spots near their ends; and from the chin to the tail is of a delicate white. The claw of the hind toe is very long.

SNOW-BUNTING.
(PLECTROPHANES NIVALIS.)

The nest of this bird is said to be placed in the fissures of the mountain rocks, and to be composed of grass, with a layer of feathers inside, and another of the soft fur of the Arctic Fox within that. The female lays five reddish-white eggs, spotted with brown: on its first arrival in this country it is very lean, but quickly grows fat, and is then excellent eating. It sings very sweetly, sitting on the ground; and does not perch, but runs about like the lark, which at first sight it much resembles.

PAINTED BUNTING. (*Emberiza? ciris.*) A beautiful bird, of the size of a hedge-sparrow, which inhabits various parts of South America: it builds its nest in the orange trees, and will feed on millet, succory, and other seeds. It has a very soft and delicate note; and will live in confinement eight or ten years. The head and neck are of a violet colour; upper part of the back and scapulars yellow-green; lower part and all the under side red; wing coverts and tail of an olive-green, tinged with brown, and edged with red. They seldom obtain their full plumage till the third year, so that they are rarely found quite alike.

The ORANGE-SHOULDERED BUNTING. (*Vidua longicauda.*) This bird, which inhabits the Cape of Good Hope, is the size of a song-thrush; bill strong and dusky, the nostrils almost hid in the feathers; plumage above and below glossy black; lesser wing-coverts crimson, below which is a white spot. The tail consists of twelve feathers, hanging sideways; the two middle ones fifteen inches in length, the rest shortening by degrees, and the outer ones very short; the legs large and brown; claws long and hooked. Of this species M. Vaillant relates some particulars not unworthy of notice in this place. "The female of this beautiful bird," says he, "has the simple colours of the sky-lark, and a short horizontal tail, like that of almost all other birds; the male, on the contrary, is wholly black, except at the shoulder of the wing, where there is a large red patch; and his tail is long, ample, and vertical like that of the common cock. But this brilliant plumage and fine vertical tail subsist only during the season of love, which continues six months. This period over, he lays aside his splendid habiliments, and assumes the more modest dress of his mate. The most extraordinary circumstance is, that the vertical tail also changes to a horizontal one, and the male so exactly resembles the female, that it is not possible to distinguish them from each other. The female has her turn. When she reaches a certain age, and has lost the faculty of propagating the species, she clothes herself for the remainder of her days in the garb which the male had temporarily assumed; her tail, like his at that period, grows long, and like his also, from horizontal becomes vertical. The birds of this species associate together, live in a sort of republic, and build their nests near to each other. The society usually consists of about fourscore females; but whether, by a particular law of nature, more females are produced than males, or for any other reason of which I am ignorant, there are never more than twelve or fifteen males to this number of females, who have them in common." The truth is, that the male, except at the breeding season, when the long-tailed feathers are produced, very nearly resembles the female, and may often be mistaken for it by an inattentive observer.

According to our author, this transmutation is by no means confined to this peculiar species of Bunting. Many females of the feathered creation, when they grow so old as to cease laying eggs, assume the more splendid colours of the male, which they retain during the remainder of their lives.

BUPALUS. A genus of Lepidopterous insects, of which there are many species. The *Bupalus piniarius*, called the Bordered White Moth, may be taken as an example. Its wings on the upper side are of a dusky brown colour, and adorned with numerous pale yellow spots. The Caterpillar is green, with a white stripe down the middle of the back, and two stripes on each side of it.

BUPHAGA. There is but one bird which constitutes this genus of Passerine Conirostres, and that is the AFRICAN OX-PECKER (*Buphaga Africana*). It is said to be frequently found in Senegal, and that its chief food consists in the larvæ of *œstri*, or bot-flies, which it sedulously extracts from the backs of cattle: hence its name. It measures about eight inches and a half in length; is rufous brown above, and of a dull yellowish white beneath. The bill is nearly an inch long, yellowish, with a red tip; the legs and claws are brown. It is extremely wild or shy, and is usually seen in small flocks of six or eight together.

BUPRESTIS: BUPRESTIDÆ. A genus and family of Coleopterous insects, of the family *Serricornes*, distinguished by the toothed or serrated form of the antennæ, and the splendour of its colours; many of its species having spots of golden hue upon an emerald ground, whilst in others azure glitters upon the gold. The subjoined figure shows one of the curious Brazilian species; it is named *B. penicillata* from the pencils of hairs at the tips on the sides of its elytra. The largest and most brilliant of these beetles

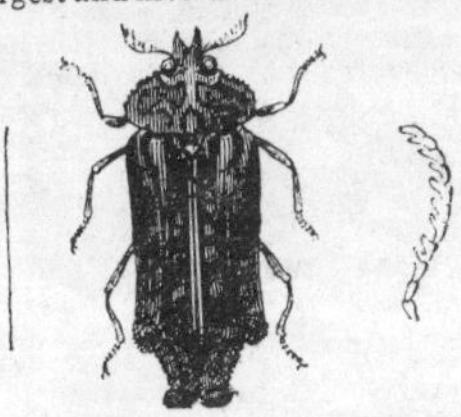

BUPRESTIS PENICILLATA.

are found chiefly in tropical climates. Some of them live for very many years in the larva state. A gentleman in the city of London had a desk that had been brought from India, in which was one of the grubs belonging to this species: several years afterwards the perfect insect made its appearance, and thereby put an end to many surmises of the merchant and his clerks as to certain *scratchings* which they had long heard in silent wonder.

The Buprestians are hard-shelled beetles, often brilliantly coloured, of an elliptical or oblong-oval form, obtuse before, tapering behind, and broader than thick, so that when cut in two transversely, the section is oval; the legs are rather short, and the feet are formed for standing firmly, rather than for rapid motion; the soles being composed of four rather wide joints, covered with little spongy cushions beneath, and terminated by a fifth joint, which is armed with two claws. In the greater number of coleopterous insects the scutel is quite conspicuous, but in the Buprestidæ it is generally very small, and sometimes hardly perceptible. These beetles are frequently seen on the trunks and limbs of trees basking in the sun. They walk slowly, and, at the approach of danger, fold up their legs and antennæ and fall to the ground. Being furnished with ample wings, their flight is swift and attended with a whizzing noise. They keep concealed in the night, and are in motion only during the day.

The larvæ are wood-eaters or borers; and both fruit and forest trees are very subject to their attacks. In the tropical parts of S. America the grub of the *Buprestis gigas*, the perfect insect of which is figured in next column, must be exceedingly destructive. They are in general of a yellowish white colour, very long, narrow, and depressed in form, but abruptly widened near the anterior extremity: the upper jaws are provided with three teeth, and are of a black colour; and the antennæ are very short. There are no legs, nor any organs which can serve as such, except two small warts on the under side of the second segment from the thorax. The motion of the grub appears to be effected by the alternate contractions and elongations of the segments, aided, perhaps, by the tubercular extremity of the body, and by its jaws, with which it takes hold of the sides of its burrow, and thus draws itself along. These grubs are found under the bark and in the solid wood of trees, and sometimes in great numbers. They frequently rest with the body bent sideways, so that the head and tail approach each other; those found under bark usually assuming this posture. The pupa bears a near resemblance to the perfect insect, but is entirely white, until near the time of its last transformation. Its situation is immediately under the bark, the head being directed outwards, so that when the pupa-coat is cast off, the beetle has merely a thin covering of bark to perforate before making its escape from the tree. The form of this perforation is oval, as is also a transverse section of the burrow, that shape being best adapted to the form, motion, and egress of the insect.

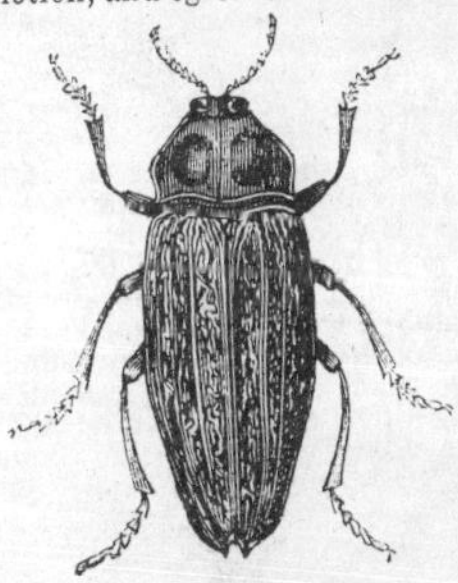

BUPRESTIS GIGAS

Some of these beetles are known to eat leaves and flowers, and of this nature is probably the food of all of them. The injury they may thus commit is not very apparent, and cannot bear any comparison with the extensive ravages of their larvæ. The solid trunks and limbs of sound and vigorous trees are often bored through in various directions by these insects, which, during a long-continued life, derive their only nourishment from the woody fragments they devour. Pines and firs seem particularly subject to their attacks; but other forest-trees do not escape, and even fruit trees are frequently injured by them. We may here remark, that woodpeckers are much more successful in discovering the retreats of these borers, and in dragging out the defenceless culprits from their burrows, than the most skilful gardener or nurseryman.

The wild cherry-tree (*Prunus serotina*), and also the garden cherry and peach trees, suffer severely from the attacks of borers, which are transformed to the beetles called *Buprestis divaricata*, because the wing-covers divaricate or spread apart a little at the tips. These beetles are copper-coloured, sometimes brassy above, and thickly covered with little punctures; the thorax is slightly furrowed in the middle; the wing-covers are marked with numerous fine irregular impressed lines and small oblong square elevated black spots; they taper very much behind, and the long and narrow tips are blunt-pointed: the middle of the breast is furrowed; and the males have a little tooth on the under-side of the shanks of the intermediate legs. They measure from seven to nine tenths of an inch. These beetles may be found sunning themselves upon the limbs of cherry and peach trees during the months of June, July, and August.

Buprestis dentipes, so named from the denticulation on the under-side of the thick fore legs, inhabits the trunks of oak-trees. It completes its transformations and comes out of the trees between the end of May and the 1st of July. It is oblong-oval and flattened, of a bronzed brownish or purplish black colour above, copper-coloured beneath, and rough like shagreen with numerous punctures; on each wing-cover there are three irregular smooth elevated lines, which are divided and interrupted by large thickly punctured impressed spots, two of which are oblique; the tips are rounded. Length rather more than half an inch.

Buprestis Harrisii is a small and broad beetle, of an entirely brilliant blue-green colour, except the sides of the thorax, and the thighs, which, in the male, are copper-coloured: it measures little more than three tenths of an inch in length. The larvæ inhabit the small limbs of the white pine, and young sapling trees of the same kind.

Buprestis Mariana, a species found in the south parts of Europe, is placed along with a closely allied one from America, and two or three other species in the genus *Chalcophora.*

BUPRESTIS MARIANA.

Dr. W. Harris, of Massachusetts, speaking of the great difficulty there is in discovering and dislodging the various grubs of tree-boring beetles, observes:—"When trees are found to be very much infested by them, and are going to decay in consequence of the ravages of these borers, it will be better to cut them down and burn them immediately, than to suffer them to stand until the borers have completed their transformations and made their escape." It is from Dr. Harris's able work on the Insects of Massachusetts that we have derived much of the information in this article.

BURBOT. (*Gadus lota.*) A fish belonging to the order *Malacopterygii;* very highly esteemed for its superior delicacy, and bearing some resemblance to the eel in its body, except that it is shorter and thicker. The head is broad and flat; the eyes small and lateral; the mouth wide; the jaws armed with several rows of sharp teeth; the lower jaw furnished with a beard of considerable length, and two small cirri seated on the top of the nose. The colour of the Burbot varies; some being dusky, and others of a dull green, spotted with black, and often with yellow: the belly in some is white; and the skin is remarkably smooth

BURBOT —(GADUS LOTA

and slippery. The first dorsal fin is short, and the second is placed immediately behind it, extending almost to the tail; the vent is situated near the centre of the belly; the anal fin reaches almost to the tail; and the tail is rather short and rounded. The Burbot is found in several of the English rivers and lakes of the northern countries; but it is said to arrive at its greatest perfection in the lake of Geneva, where it sometimes weighs six pounds, though in this country it seldom exceeds two or three.

BURSATELLA. A genus of marine Mollusca, without shells, found in the Indian seas.

BUSTARDS. (*Otis; Otidæ.*) A genus and family of Cursorial Birds, distinguished for their powers of running and their shyness: some of the Asiatic species, such as the *Florican*, are much sought for by the Indian sportsman as a delicacy for the table. We here mean to confine our attention to the two species indigenous to the British Islands, although now both are very rare birds.

The GREAT BUSTARD (*Otis tarda*) is the largest of European land birds, the male being about four feet long, and measuring nine feet from tip to tip of the wings when extended, while its weight is on an average twenty-five pounds. The head and neck are ash-coloured, and there is a tuft of feathers about five inches long on each side of the lower mandible. The back is transversely barred with black and bright ferruginous colours, and the primaries are black. The tail consists of twenty feathers, broadly barred with red and black; and the legs are naked, dusky, and without a hind toe. The female is not much more than half the size of the male, and has the crown of the head of a deep orange colour, traversed by red lines; the remainder of the head is brown; her colours are not so bright as the male, and she has no tuft on each side of the head. There is likewise another very essential difference between the male and the female; the former being furnished with a sack or pouch, situated in the fore part of the neck, and capable of containing nearly two quarts: the entrance to it is immediately under the tongue. This singular reservoir the bird is supposed to fill with water, as a supply in the midst of those dreary plains where it is accustomed to wander; it is also said to make use of it when attacked by birds of prey, by so violently ejecting it as to baffle their attacks. These birds were formerly seen in considerable flocks on the extensive plains of Wiltshire, Dorsetshire, and in parts of Yorkshire; but as cultivation has advanced, they are become very scarce. They are very shy and vigilant, and

GREAT BUSTARD.—(OTIS TARDA.)

by no means easy to shoot: they run with great speed, and aid their course with their wings, like the ostrich. They feed on grain, seed, worms, &c.; make their nest by merely scraping a hole in the earth; and lay two eggs, as large as those of a goose, of a pale olive tint, with dark spots. They seldom wander far from their accustomed haunts, and have a great unwillingness to rise on the wing; but when once in the air, they can fly several miles without resting.

The LITTLE BUSTARD. (*Otis tetrax.*) This bird is very uncommon in England, but in France it is taken in nets, like the partridge. It is a very shy and cunning bird; if disturbed, it flies two or three hundred paces, not far from the ground, and then runs away much faster than any one can follow on foot. The female lays three or four eggs, of a glossy green colour, in June; and as soon as they are hatched, she leads them about as a hen does her chickens. The length of this bird is seventeen inches: the bill is pale brown; irides red; the top of the head black, spotted with pale rusty; the sides of the head, chin, and throat, reddish with black spots: the whole neck in the male is black, encircled with an irregular band of white near the top and bottom; the back and wings rufous and brown, crossed with fine irregular black lines: the under parts of the body, and outer edges of the wings, are white; the tail tawny and white, with black bands: legs grey. The *thick-kneed Plover* is sometimes also locally named "Bustard," but belongs to another order, the Grallatorial Birds. [See ŒDICNEMUS.]

BUTCHER-BIRD. [See SHRIKE.]

BUTEO. [See BUZZARD.]

BUTTERFLY. (*Papilio.*) The popular English name of an extensive group of beautiful insects, belonging to the order *Lepidoptera*, as they appear in their fully developed state. They are distinguished from other insects by these generical characters: their antennæ are clubbed at the extremities; their wings, when at rest, are closed together over their backs; and they fly only in the day-time. Butterflies are also distinguished from the other Lepidoptera by the superior brilliancy of their colouring, and by the beauty of the under as well as the upper side of the wings. "The Butterfly," as Mr. Knapp observes, "light, airy, joyous, replete with life, sports in the sunshine, wantons on the flower, and trips from bloom to bloom, gay as the brilliant morn, and cheerful as the splendour of heaven: heat and light appear to be the very principle of his being; in a cloudy or a chilly atmosphere his energies become suspended, and, closing his wings, he reposes like a sickly thing upon some drooping flower: but let the cloud disperse, the sun break out, he springs again to active life; associating with the birds of day, and denizen of the same scenes, he only seems of a less elevated order."

Butterflies are very careful in depositing their eggs in places where they are likely to be hatched with the greatest safety and success. They lie dormant through the winter; but when the sun calls forth vegetation, and vivifies the various eggs of insects, caterpillars are seen on various plants, eating their leaves, and preparing for a state of greater perfection. Their form is long and cylindrical, and they consist of thirteen segments, including the head; they have eight feet, and nine spiracles on each side. Those feet which are attached in pairs to the first three segments of the trunk inclose the parts which are developed into the permanent legs of the future Butterfly; the remaining five pairs of feet are membranous, short, and thick, and are finally lost with the moultings of the skin.

The external form of the chrysalids varies according to the species of Butterfly that inhabits them; in all, however, there are apertures opposite to the thorax, by which respiration is carried on during the whole period of their inactive state. After the appointed time, when the creature has acquired sufficient vigour, the shell is broken, which at once constituted "the grave of the caterpillar and the cradle of the butterfly:" the down already grown upon the insect has completely separated it on all sides from the shell, which by the action of the head is broken opposite to that part, and affords free egress to the prisoner it so long confined. The wings of the Butterfly, on its first appearance, are closely folded; but by the help of a fluid constantly circulating through them, they are soon expanded, and sufficiently hardened, by the action of the air, to endure the efforts of flying. It is then that the insect enters upon a more enlarged sphere of action, with increased powers: he ranges from flower to flower, darting his rostrum into their nectaries for the delicious stores they contain. Then, too, in the full possession of every faculty granted to his race, he prepares to multiply and perpetuate it.

This last and most considerable metamorphosis is attended with a greater change in the economy of the insect than of the preceding; for not only the skin, but the teeth, jaws, and even the cranium, are left behind. The work of Herold, on the transformation of the Cabbage Butterfly, may be judiciously consulted by *the student;* while Lyonet's researches and great book on the caterpillar of the Goat Moth, teach many particulars of the transformation of a large moth. There is a paper on the wings of *Lepidoptera* by Bernard Deschamps well worth the attention of persons fond of the microscope (see "Annales des Sciences Naturelles," 2nd ser., t. III. 1835). The quantity of food taken by them in their last state is comparatively small to what they antecedently devoured. For a short time after their appearance on the wing, they discharge some drops of a red-coloured fluid; but its appearance on the surface of the earth has at different times been regarded, by vulgar superstition, as drops of blood fallen from the clouds, and was presumed to be portentous of some heavy calamity.

Various insects prey upon the Butterfly, or hasten the approach of its dissolution. Many species of Ichneumonidæ perforate the body of the insect while a caterpillar, and there deposits its eggs; and although the caterpillar continues to live, and is transformed into a chrysalid, no Butterfly is produced from it, those internal parts that were essential to its perfection being consumed by the larvæ of the ichneumon. From the great fecundity and variety of the insects of this genus, they probably would soon cover the earth, did not nature provide a bar to their increase by multiplying their enemies: hence they are destined to become the food of a great number of animals of various kinds, some of which swallow them entire, others macerate their bodies; while many accomplish their destruction by gradually sucking their juices. It has been calculated that a single pair of sparrows, in order to supply themselves and their young, may destroy three thousand three hundred and sixty Butterflies in one week.

"The clothing of the organs of flight in the Butterfly excites the admiration of the most incurious beholder. The gorgeous wings of these universal favourites owe their beauty to an infinite number of little plumes, thickly planted in their surfaces, and so minute as to seem like powder; but which are in fact an innumerable number of small scales, varying in shape and length in different species, and discoverable only by the assistance of a microscope.

"The Butterfly requires no other food than the nectareous juices which are distilled from flowers, or the saccharine substance which exudes from the leaves of vegetables; it will sometimes alight and suck the sweets of ripe fruit that has been broken by its fall. The skies are its proper habitation—the air is its element; the pageantry of princes cannot equal the ornaments with which it is invested, or the rich colouring that embellishes its wings. There is nothing in the animal creation so beautiful or splendid as many species of these insects; they serve to banish solitude from our walls, and to fill up our idle intervals with the most pleasing speculations.

"Butterflies fly generally only in the day. They accompany the sun in his course, and before he sets disappear. With us, says Mr. Samouelle, many of the species are extremely local; and, from the shortness of their lives, require greater assiduity in the collector, and a wider range of search, than is generally supposed. As an illustration of this fact, we must observe that the number of *Papilionidæ* found in England is about seventy-two. Of this number not more than fifty are to be met with within twenty-five miles of London; and of these several are confined to the vicinity of a chalk-cliff, or are peculiar to a meadow or a certain wood. Even in these situations their appearance in the perfect state is limited but to a few days and at a certain season of the year. Of the remaining number, not found within this distance from London, some are confined to fens, nearly a hundred miles distant from the metropolis, and others to the mountains of Scotland; but they are all equally limited in the times of their appearance and the shortness of their lives. There is also another circumstance in the history of these insects, which must not be passed over in silence; and that is, there are several species which, from some hitherto unknown cause, appear in the proper season, but in certain years only, when they will be found in abundance, and probably extended over a vast tract of the country. These, however, disappear, and not a single specimen is to be found for a period of many years, when they will again be seen as plentiful as before. This is a circumstance that is not confined to England, where it might be attributed to our ever-varying climate, but occurs also in tropical countries."—*Butt. Coll. Vade Mecum.*

"If you denude the wings of any Butterfly, which you may easily do by scraping it lightly on both sides with a penknife," as Messrs. Kirby and Spence observe, "you will be amused to trace the lines in which the scales were planted, consisting of innumerable minute dots: the lines of the under side, in some cases, so cut those of the upper side, as by their intersection to form lozenges. With regard to the position of the scales on the wing, they usually lie flat, but sometimes their extremity is incurved. But though the general clothing of the wings of *Lepidoptera* consists of these little scales, yet in some cases they are either replaced by *hairs* or mixed with them. Thus, in the *clear* parts of the wings of *Heliconians, Attici*, &c., short inconspicuous hairs are planted; in a large number of the Orders the *upper* side of the anal area of the secondary wings is hairy; in several Crepusculars, where there is a double layer, as before mentioned, the upper one consists of dense hairs, except at the apex, and the lower one of scales; and in most of them the scales of the primary wings are piliform, and the secondary are covered by what approach very nearly to real hairs."

The number of exotic Butterflies is very great both in orders and in genera; and students are referred [we limit ourselves to British books] to Dr. Horsfield's elaborate work on those of Java, but especially to the truly admirable work on the Genera of Diurnal Lepidoptera, by Edward Doubleday, F.L.S., illustrated by William Hewitson, and continued by Mr. Westwood since Mr. Doubleday's death. Our space prevents us from even alluding to the numerous genera of those gorgeous insects detailed in this splendid book. It is, however, but bare justice to say that a more beautiful work has rarely been published. We can also heartily recommend Humphrey's British Butterflies, partly edited by Mr. Westwood, and also the more recently published work of the Rev. F. O. Morris, in a single volume, bearing a similar title: the coloured figures of the latter work are admirably given. Butterflies, by their forms, contrasts of colour, and other peculiarities, not only charm the eye, but have afforded valuable information to artists. The great Van Dyck and our countryman Stothard are known to have been indebted to Butterflies for many fine hints on colour, both in harmony and contrast. In the present work we must confine ourselves to the British Genera as much as possible. [See PAPILIO; PONTIA; MELITÆA; ARGYNNIS: LIMENITIS; VANESSA; APATURA LYCÆNA; POLYOMMATUS; THECLA; HIPPARCHIA; HESPERIA; PARNASSUS, &c.]

BUZZARD. (*Falco buteo* of Linnæus.) This bird is supposed to be the most common in England of all the hawk tribe. It has a thick heavy body; measures about twenty-two inches in length, and the full expansion of its wings is about fifty. It is usually of a ferruginous brown above, and yellowish white beneath, with large longitudinal spots and dashes: the tail is barred with black and ash-colour; the tip is dusky white. It breeds in extensive woods, generally fixing on the old nest of a crow, which it enlarges, and lines with wool and other soft materials. It lays two or three eggs, which are sometimes wholly white, and at others spotted with yellow; and when the female happens to be killed during the time of incubation, the cock hatches and rears the brood. The young accompany the old birds for some time after quitting the nest; a circumstance unusual in other birds of prey, which always drive off their young as soon as they can fly. The Buzzard is very sluggish and inactive, remaining perched on the same bough for the greatest part of the day, and always found near the same place. It feeds on birds, frogs, insects, moles, and mice. By modern naturalists it is placed in the genus *Buteo*. [For Honey-Buzzard, see PERNIS.]

BYRRHUS: BYRRHIDÆ. A genus and family of Coleoptera. The insects belonging to this genus have an ovate body, convex or sub-globular in some species, with the elytra covered by a short pile, and the head is retracted under the thorax. *Byrrhus pilula* is about the size of the common Lady-bird: its colour is a dull brown, with a few obscure blackish lines down the wing-shells: it is of an extremely convex shape, and, when disturbed, contracts its limbs, and lies in an inert state, like an oval seed or pill, while thus counterfeiting death as a means of escape from danger. It is found on various plants in gardens and elsewhere.

BYSSOARCA. A sub-genus of Molluscous animals, affixed by byssiform filaments to other bodies, a particular muscle being protruded through the gaping part of the shelly valves.

BYSSOMYA. A genus of Acephalous Molluscs, which live in the fissures of rocks, attached by a byssus: sometimes the animal buries itself in the sand or lodges in small stones, &c.

CACHALOT. (*Physeter macrocephalus.*) The Spermaceti Whale; the head of which nearly equals the rest of the body in length, and surpasses it in bulk. It is an object of great commercial importance on account of the oil and spermaceti which it yields. Mr. Beale has published a most admirable and readable work on it. [See WHALE.]

CACHICAME. [See ARMADILLO.]

CACTORNIS. A subgenus of Passerine birds, closely allied to Geospiza, but differing from it in the beak being elongated, somewhat like that of a Quiscalus, and very sharp-pointed. The typical species is *Cactornis scandens*.

CLIMBING CACTUS BIRD.
(CACTORNIS SCANDENS.)

The male is of a sooty black, the female brownish and spotted. This species was found by Mr. Darwin in the Galapagos; its most favourite resort is the *Opuntia Galapageia*, a species of the Cacti tribe; about the fleshy leaves of which they hop and climb, even with their back downwards, whilst feeding with their sharp beaks, both on the fruit and flowers. They frequently also alight on the ground and search for seeds on the parched volcanic soil.

CADDICE-WORM, or CAD-BAIT. [See PHRYGANEA.]

CÆCILIA. The name of a genus of Serpents, about a foot in length, and having much the appearance of an eel. They are natives of South America, and are said to be innoxious.

CAIMAN. [See ALLIGATOR.]

CALANDRA: CALANDRIDÆ. A genus and family of Coleopterous insects, closely allied to the *Curculionidæ;* some of the minute species of which commit great havoc in granaries, both in their larva and perfect state. The species are very numerous, and among them is the well-known Cornweevil (*Calandra granaria.*) This insect bores a hole into the grain with its proboscis, and there deposits an egg, which turns to a little grub, and devours the whole of the inside of the grain, leaving the husk entire. Another species of Calandra, distinguished by its having four red spots on its elytra, attacks rice in the same way as the one above mentioned does wheat.

These insects must not be confounded with the still more destructive larvæ of the Cornmoth (*Tinea granella*), which also attack stored grain, nor with the orange-coloured maggots of the Wheat-fly (*Cecidomyia tritici*), which are found in the ears of growing wheat. Although the grain-weevils are not actually injurious to vegetation, yet as the name properly belonging to them has often been misapplied, some remarks upon them here may not be inappropriate.

The true Grain-weevil or Wheat-weevil of Europe (*Calandra granaria*), in its perfect state, is a slender beetle of a pitchy red colour, about one-eighth of an inch long, with a slender snout slightly bent downwards, a coarsely punctured and very long thorax, constituting almost one-half the length of the whole body, and wing-covers that are furrowed, and do not entirely cover the tip of the abdomen. This little insect, both in the beetle and grub state, devours stored wheat and other grain, and often commits much havoc in granaries and brewhouses. Its powers of multiplication are very great, for it is stated that a single pair of these destroyers may produce above six thousand descendants in one year. The female deposits her eggs upon the wheat after it is housed, and the young grubs hatched therefrom immediately burrow into the wheat, each individual occupying alone a single grain, the substance of which it devours, so as often to leave nothing but the hull; and this destruction goes on within, while no external appearance leads to its discovery, and the loss of weight is the only evidence of mischief that has been done to the grain. In due time the grubs undergo their transformations, and come out of the hulls in the beetle state, to lay their eggs for another brood. These insects are effectually destroyed by kiln-drying the wheat; and grain that is kept cool, well ventilated, and frequently moved, is said to be exempt from attack.

CALAPPA, or BOX CRAB. A genus of Crustacea, belonging to the family *Calappidæ.* They are named by the French *coqs de mer*, from their crested chelæ, which are large, equal, compressed; with their upper edge, which is notched or crested, very much elevated, and fitting exactly to the external border of the shell or carapace, so as to completely cover the mouth and anterior parts: the rest of the feet short and simple; carapace short and convex, forming, behind, a vaulted shield, under which the posterior legs are hidden when the animal is in a state of repose: eyes mounted on short pedicles, and not far apart. There are several species widely diffused: some inhabit the seas of the Indian Archipelago, and of New Holland: others are met with in the Pacific and Atlantic oceans, the seas of South America, &c.; others, again, inhabit the Mediterranean sea. They frequent the fissures of rocks, some of them at a great depth. The females deposit their eggs in summer.

CALAPPA CRISTATA.

CALATHUS. A genus of Coleopterous insects, belonging to the *Carabidæ.* Several species are found in the British islands, most frequently under stones and house rubbish.

CALLICHTHYS. A genus of abdominal Malacopterygious fishes, family *Siluridæ.* The body and head are protected by large, hard, scaly plates; the mouth is furnished with four long cirri; the teeth are very small; the eyes are also small, and situated on the side of the head. They are natives of South America and other hot climates, where the rivers frequently dry up; and they can not only live for a considerable time out of water, but they are said to perform long journeys over land, directing their course to some other stream.

CALLIDIUM. A genus of Coleopterous insects, belonging to the family *Longicornes;* one species of which (*Callidium bajulus*) in the larva state is particularly destructive to fir timber. This is a flattish rusty black beetle, with some downy whitish spots across the middle of the wing-covers; the thorax is nearly circular, is covered with fine whitish down, and has two elevated polished black points upon it; and the wing covers are very coarsely punctured. It inhabits fir and spruce timber, and may often be seen on wooden buildings and fences in July and August. We are informed by Kirby and Spence that the grubs sometimes greatly injure the wood-work of houses in London, piercing the rafters of the roofs in every direction, and, when arrived at maturity, even penetrating through sheets of lead which covered the place of their exit. One piece of lead, only eight inches long and four broad, contained twelve oval holes made by these insects, and fragments of the lead were found in their stomachs.

The Violet Callidium (*Callidium violaceum*) is of a Prussian blue or violet colour; the thorax is transversely oval, and downy, sometimes having a greenish tinge; and the wing-covers are rough with thick irregular punctures. It is about half an inch in length; may be found in great abundance, in the northern and middle parts of Europe, on piles of pine-wood, from the middle of May to the first of June; and the larvæ and pupæ are often met with in splitting the wood. They live mostly just under the bark, where their broad and winding tracks may be traced by the hardened saw-dust with which they are crowded. Just before they are about to be transformed, they bore into the solid wood to the depth of several inches. In this country it is not so common.

CALLIMORPHA. A genus of Lepidopterous insects, belonging to the family *Bombycidæ*. One of these (*Callimorpha Jacobeæ*) is both a beautiful and common Moth, its wings when expanded measuring about an inch and a half in width: on the upper wings, which are of a greenish black colour, are two round pink spots at the apex, and an oblong pink streak parallel with the outer margin. The under wings are entirely pink, except the margins, which are of the same colour as the ground-colour of the upper wings. The head, body, and legs are quite black. The larva feeds on the common ragwort (*Senecis Jacobœa*); hence the name of the insect.

CALLISTUS. A genus of Coleopterous insects, belonging to the family *Carabidæ*. The species *C. luatus* is found in this country, and is about a quarter of an inch long: the head and under parts of the abdomen are of a greenish black colour, the thorax is reddish-yellow, and the wing-cases are yellow with six black spots: the antennæ and legs are black; the head and thorax are very thickly punctured, and the elytra are punctate-striated.

CALLIONYMUS. [See DRAGONET.]

CALOSOMA. A genus of Coleopterous insects, belonging to the family *Carabidæ*, one of which (*Calosoma sycophanta*) is about an inch long; the head, thorax, and under parts of the body are of a beautiful blue colour, the elytra are green, and the legs and antennæ black. There are about thirty different species of these insects, the prevailing colour being some shade of green with a kind of brassy hue. They are very useful in many places, from the number of noxious caterpillars they destroy. Mr. I. W. Slater (in the Zoologist for 1845) thus speaks of them: "This beautiful beetle is very common in the pine-forests, particularly on the path leading to the Raubchloss, where they spangle the sand and the tree trunks like living gems. The splendour of its elytra, green, gold, scarlet, orange, the rich purple-black of the thorax, the rapidity and ease of its movements, render it a pleasing object even to the most careless, whilst the pungent odour, which it possesses more strongly, I believe, than any other of the Geodephaga, readily betrays its presence. Except from the collector, however, it has nothing to dread, its utility to man being both known and appreciated. The pine-forests, for instance, are exposed to the ravages of various lepidopterous insects, such as *Smerinthus pinastri*, and, in particular, *Gastropacha pini*. Now, a pine-tree, once stripped of its leaves, or *needles*, as the Germans more aptly term them, does not recover like an oak or a sycamore, but dies. Scarcely is vegetation at an end, when the Longicornes seize upon the trunk, and burrow in it; the wood-ants tunnel it in all directions, and it thus becomes worthless. Many hundred acres of the finest timber are thus often destroyed. It is an interesting sight to any but the owner, to visit a forest under the infliction of Gastropacha pini; the thousands of caterpillars eagerly feeding produce a distinct crackling sound, as the hard, dry pine-leaves yield to their persevering jaws. The large moths fluttering lazily about, or perched on the leafless sprays, await the approach of evening, when the gamekeepers kindle large fires in the open spaces. Into these multitudes of moths fall, and are consumed; but this, with all that are destroyed by hand, or devoured by birds, would avail but little, but for the services of certain insects. Our Calosoma is one of the most active; both larva and beetle mount the trees, and slaughter both moths and caterpillars, far more than are requisite to satisfy their appetite. Those seasons in which the pine-moth is most numerous are also remarkably favourable to the Calosoma, and to several kinds of ichneumons, which also prey upon the pine-moth."

CALYMENE. A genus of Trilobite Crustacea, comprising the well-known *C. Blumenbachii*, found in the transition limestone of Dudley. Head deeply divided by two longitudinal grooves, abdominal rings, &c.

CALYPTORHYNCHUS. A genus of birds belonging to the Parrot family, found in New Holland. The plumage is generally black, sometimes of a smoky brown, ornamented occasionally with large spots of a clear red or orange or sulphur colour, forming wide bands on the tail: the beak is short and considerably elevated; by these and other characters the birds contained in it may be distinguished from the light coloured and lively COCKATOO (*Plyctolophus*). One of the best known species is Banks's Cockatoo (*Calyptorhynchus Banksii*), named after Sir Joseph Banks, Bart., who was perhaps the first naturalist that visited Australia. The great Australian Ornithologist, Mr.

Gould, in his very magnificent "Birds of Australia," has figured all the species of this genus; and from his account of the species called *Wy-la* (from its whining call note) by the natives of N. S. Wales, and *Calyptorhynchus funereus* by naturalists, we extract the following observation: — it is usually met with in small companies of from four to eight in number, except during the breeding season, when it is only seen in pairs. Its food is much varied; sometimes the great belts of *Banksia* shrubs are visited, and the seed-covers torn open for the sake of their contents, while at others it searches greedily for the larvæ which are deposited in the wattles and gum trees (*Encalyptus*). Its flight is very heavy, flapping, and laboured; but Mr. Gould informs us that he has sometimes seen it dive between the trees in a most rapid and extraordinary manner. The eggs are white, two in number, and deposited on the rotten wood in the hollow branch of a large gum tree.

CAMBERWELL BEAUTY [BUTTERFLY.] A name given to a species of Butterfly, of the genus VANESSA [which see].

CAMEL. (*Camelus.*) A genus of mammiferous ruminating quadrupeds without horns, further distinguished by the possession of incisive, canine, and molar teeth: the upper lip is divided; the neck long and arched; having one, or two, humps or protuberances upon the back, and naked callosities at the joints of the leg, the lower part of the breast, &c. They have a broad, expanded, elastic foot, terminated in front by two comparatively small hoofs, or toes; the whole structure of it being admirably fitted for enabling the animal to travel with peculiar ease and security over dry, stony, and sandy regions. The native country of this genus is said to extend from Mauritania to China, within a zone of 1000 miles in breadth.

CAMEL.—(CAMELUS BACTRIANUS.)

The common Camel (*Camelus Bactrianus*), having two humps, is only found in the northern part of this region, and exclusively from the ancient Bactria, now Turkestan, to China. It is larger than the Dromedary; the limbs are not so long in proportion to the body; the muzzle is larger and more tumid; the hair of a darker brown, and the usual gait slower: but the most obvious distinction is afforded by the Bactrian Camel having two humps, and the Dromedary or Arabian Camel having but one, which single hump occupies the middle of the back, rising gradually on all sides towards its apex.

The Arabian, or single-hump Camel (*Camelus dromedarius*) is found throughout the entire length of this zone, on its southern side,

DROMEDARY.—(CAMELUS DROMEDARIUS.)

as far as Africa and India. The general height of the Arabian Camel, measured from the top of the dorsal hump to the ground, is about six feet and a half, but from the top of the head when the animal elevates it, not much less than nine feet: the head, however, is generally so carried as to be nearly on a level with the hump, or rather below it, the Camel bending the neck extremely in its general posture. In some particular attitudes, perhaps, the Camel may be said to have an elegant and picturesque appearance, yet its general aspect, and more especially its dorsal hump, at first sight, is apt to impress on the mind the idea of deformity, rather than a truly natural conformation.

It is highly probable that the Camel has long ceased to exist in its wild or natural state, as it has been enslaved by man from the earliest times of which we have record. Unlike the elephant, and other animals which cease to breed in a state of captivity, the Camel is as prolific as if at liberty; and vast numbers are raised and employed throughout the East, especially in the commerce carried on between the people residing in the vicinity of the great deserts. In regions where water is scarce, and wells or springs are several days' journey distant from each other, it would be impossible to traverse the country with the usual beasts of burthen. But the Camel can abstain from drinking for seven or eight days together without injury — an important advantage, which is owing to the possession of an additional cavity in the stomach, destined to receive water, whenever it can be procured, and capable of retaining it unchanged for a long time. "But," as the writer of the zoological articles in Brande's Dictionary observes, "besides a reservoir of water to meet the exigencies of long journeys across the desert, the Dromedary and Camel are provided with a storehouse of solid nutriment, on which they can draw for supplies long after every digestible part has been extracted from the contents of the stomach: this storehouse consists of one or two large collections of fat stored up in ligamentous cells supported by the spines of the dorsal vertebræ, and forming what are called the humps. When the Camel is in a region of fertility, the hump becomes plump and ex-

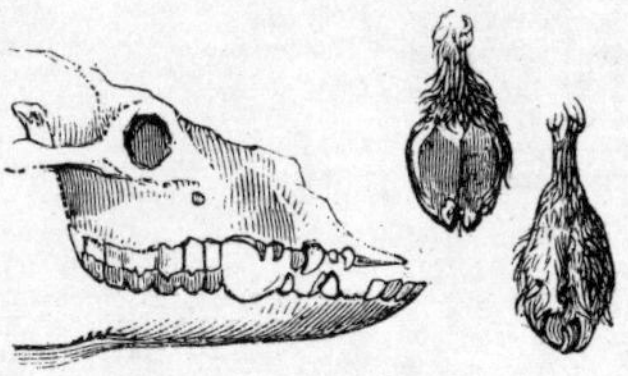

PART OF THE SKULL OF DROMEDARY, WITH ITS FOOT VIEWED FROM BENEATH AND FROM ABOVE.

panded; but after a protracted journey in the wilderness it becomes shrivelled and reduced to its ligamentous constituent, in consequence of the absorption of the fat. Buffon carried his teleological reasoning, or the ascription of design, so far as to assert that the humps on the backs of the Camel were badges of slavery, and intended to adapt them to the burthens of their taskmasters; and he supported this ingenious idea by the unfounded assertion that the dorsal prominences did not belong to the Camels in free nature. But the true uses of the fatty humps, as of the water-cells, relate to the exigencies of the *Camelidæ* of the deserts under every condition."

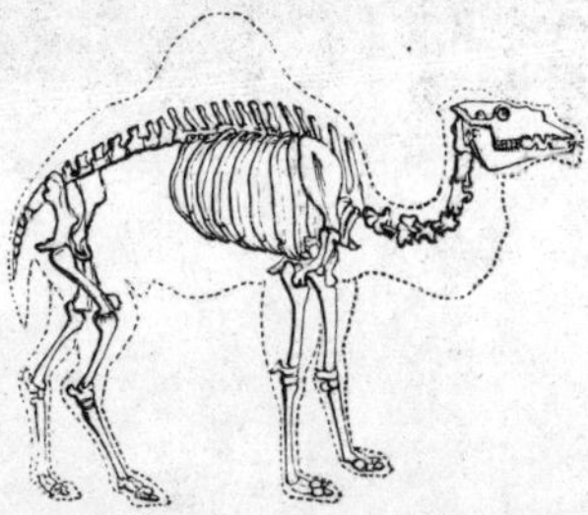

SKELETON OF CAMELUS DROMEDARIUS.

Possessing strength and activity surpassing that of most beasts of burthen, docile, patient of hunger and thirst, and contented with small quantities of the coarsest provender, the Camel is one of the most valuable gifts of Providence. There is nothing, however, in the external appearance of the animal to indicate the existence of any of its excellent qualities. In form and proportions, it is very opposite to our usual ideas of perfection and beauty. A stout body, having the back disfigured by one or two humps; limbs long, slender, and seemingly too weak to support the trunk; a long, thin, crooked neck, surmounted by a heavily-proportioned head, are all ill suited to produce favourable impressions. Nevertheless, there is no creature more excellently adapted to its situation, nor is there one in which more of creative wisdom is displayed in the peculiarities of its organization. To the Arabs and other wanderers of the desert, the Camel is at once wealth, subsistence, and protection. The milk furnishes them with a large part of their nutriment. The flesh of the young animal is one of their greatest luxuries; of the skins they form tents, or manufacture them into saddles, harness, pitchers, shields, and many other articles; the various sorts of hair, or wool, shed by the Camel, are wrought into different fabrics; and its very excrements serve as fuel, or are applied to other useful purposes.

These animals are trained, when extremely young, to the labours which they are afterwards to perform: and with this view, when but a few days old, their limbs are folded under their body, and they are compelled to remain on the ground whilst they are loaded with a weight, which is gradually increased as they increase in strength. The pace of the Camel is a high and swinging trot, which, to persons unaccustomed to it, is at first disagreeable and apparently dangerous, but is afterwards tolerably pleasant and secure. The Arabians in general ride on a saddle that is hollowed in the middle, and has at each bow a piece of wood placed upright, or sometimes horizontally, by which the rider keeps himself in the seat· and the animal is guided, or stopped, by means of a cord that serves as a bridle, and is affixed to a ring which is passed through his nostrils. Small Camels carry from 600 to 800 lbs.; the largest and strongest bear 1000 lbs. or upwards from thirty to thirty-five miles a day; but those which are used for speed alone are capable of travelling from sixty to ninety miles a-day. When a caravan of Camels arrives at a resting or baiting place, they kneel, and, the cords sustaining the load being untied, the bales slip down on each side. They generally sleep on their bellies, crouching between the bales they have carried; the load is, therefore, replaced with great facility. In an abundant pasture they generally browse as much in an hour as serves them for ruminating all night and for their support during the next day. But it is uncommon to find such pasturage, and they are said to prefer nettles, thistles, cassia, and other prickly vegetables, to the softest herbage.

The female goes with young twelve months, and brings forth one at a birth. Her milk is very rich, abundant, and thick, but of rather a strong taste; though when mixed with water it forms a very nutritive article of diet. Breeding and milk-giving Camels are exempted from service, and fed as well as possible, the value of their milk being greater than that of their labour. The young Camel usually sucks for twelve months; but such as are intended for speed are allowed to suck, and exempted from restraint, for two or three years. The Camel arrives at maturity in about five years, and the duration of its life is from forty to fifty years.

There are several races or varieties both of the Arabian and the Bactrian Camel, differing, like those of horses, in strength, size, swiftness, and elegance of form. A breed of peculiar swiftness is said to be reared in China; a white variety occurs in some parts of Siberia; and a hybrid or mixed breed is occasionally obtained between the Bactrian and Arabian Camel.

CAMELOPARD and CAMELEOPARDALIS. [See GIRAFFE.]

CAMELUS. [See CAMEL.]

CAMPAGNOL. A French name sometimes given to the small Rodent animals called *Voles*. [See VOLE.]

CAMPANULARIADÆ. A group of cœlenterate animals, or, more strictly, a family of hydroid polyps, embracing several genera, such as *Campanularia*, *Laomedea*, and *Cymodoçea*, all being zoophytes closely allied to the well-known Sertularias [which see, p. 604]. The Campanularias proper are distinguished by the following characters:—"polypidom, horny, tubular, creeping or erect, plant-like; the main stem giving off branches bearing stalked, campanulate, or bell-shaped cells, the stalks being annulated and either alternately or irregularly dis-

CAMPANULARIA DICHOTOMA

posed." The genus Campanularia contains eleven British species, of which one is here represented. They may be reproduced in one, either, or all of the following ways:—*first*, by a simple process of gemmation or budding, each bud becoming a stalked polyp; *secondly*, by modified polyps, which bud off or pullulate in the form of jelly-fishes (Medusoids), these latter having eggs in them which become transformed into polyps; *thirdly*, by polyp-heads which contain eggs, these latter giving off young ciliated embryos or gemmules, which, after swimming about for a time, settle down and grow up into polyps. [See ALTERNATION OF GENERATION in SUPPLEMENT.]

CAMPULA. [See SUPPLEMENT.]

CANARY-BIRD. (*Carduelis canaria.*) The Canary-bird, or Canary-finch, as it is sometimes called, is a well-known captive songster in this and most other European countries. It is a native of the Canary Islands, but it has continued so long in a domestic state that its native habits seem almost forgotten. In the wild state the prevailing colour is grey or brown, mingled, however, with other colours, but never reaching the brilliancy of plumage exhibited by the bird in captivity. Dr. Heineken, who describes its character and habits in Madeira, where these birds abound, says, "it builds in thick, bushy, high shrubs and trees, with roots, moss, feathers, hair, &c.; pairs in February; lays from four to six pale blue eggs; and hatches five, and often six times in the season. It is a delightful songster, with, beyond doubt, much of the nightingale's and skylark's, but none of the woodlark's song." It was brought into Europe early in the 16th century, and is believed to have spread from the coast of Italy, where a vessel, which was bringing to Leghorn a number of these birds besides its merchandize, was wrecked. The climate being favourable, they increased, and would certainly have become naturalized, had they not been rendered scarce by the desire to possess them, as well as from there being few besides male birds brought over.

In their native islands, a region equally celebrated for the beauty of its landscapes and the harmony of its groves, the Canary-birds are of a dusky-grey colour, and so very different from those usually seen in Europe, that some naturalists have even doubted whether they are of the same species. The original stock has undergone so many changes from its being domesticated, from the climate, and from the union with birds analogous to it, that now we have Canaries of all colours. "Those Canaries that have the upper part of the body of a dusky green or linnet-brown, and the under part the yellowish green of the green-bird, with dark-brown eyes, are the strongest, and most nearly resemble the primitive race. The yellow and white often have red eyes, and are the most tender. The chestnut are the most uncommon, and hold a middle rank for strength and length of life between the two extremes. But as the plumage of the intermediate ones is a mixture of these principal colours, their value depends on the pretty and regular manner in which they are marked. The Canary that is most admired amongst us now is one with the body white or yellow; the head, particularly if crested, wings, and tail, yellowish dun. The second in degree is of a golden yellow, with the head, wings, and tail black, or at least dusky grey. Next follow the grey or blackish, with a yellow head and collar; and the yellow with a blackish or green tuft; both of which are very much valued. As for those that are irregularly spotted, speckled, or variegated, they are much less sought after, and are used to pair with those of one colour, white, yellow, grey, brown-grey, and the like."

In choosing Canary-birds, those are preferable which appear bold and lively. If their eyes look cheerful and bright, it is a sign of health; but, on the contrary, if they hide their heads under their wings, and gather up their bodies, it is symptomatic of their being disordered. The melody of the song should also be regarded in making a selection; for some will open with the notes of the nightingale, and, after running through a variety of modulations, end like the tit-lark; others, again, will begin like the sky-lark, and, by soft melodious turns, fall into the notes of the nightingale. Lessons may be taught this bird in its domestic state; but its native note is loud, shrill, and piercing.

Canary-birds sometimes breed all the year round; but they most usually begin to pair in April, and to breed in June and August. In Germany and the Tyrol, where the breeding of these songsters forms the occupation of numbers, and from whence the rest of Europe is principally supplied, the apparatus for breeding Canaries is both large and expensive. A large building is erected for them, with a square space at each end, and holes communicating with these spaces. In these outlets are planted such trees as the birds prefer. The bottom is strewed with sand, on which is cast rape-seed, chickweed, and such other food as they like. Throughout the inner compartment, which is kept dark, are placed brooms for the birds to build in, care being taken that the breeding-birds are guarded from the intrusions of the rest. With us, however, the apparatus is much less expensive, a breeding-cage often suffices; and, at most, a small room, without any particular preparation.

CANCELLARIA. A genus of Molluscous animals belonging to the *Entomostomata* of De Blainville. There are many species, most of which are found in the Indian and African

CANCELLARIA OBTUSA.

seas, but many are from the warm latitudes of the Pacific side of S. America. There is a monograph of this genus by G. B. Sowerby, Jun., F.L.S., in the "Conchological Illustrations," with figures and descriptions of all the species then known. To this work we refer our readers, as well as to Mr. Sowerby's "Thesaurus Conchyliorum." The genus derives its name from *cancellatus*, cross-barred; and the shell is characterized as oval or turretted; spire generally short, slightly elevated, and pointed; mouth oval, having either a very short canal or a notch only; the outer lip marked within by transverse ridges; inner lip spread over part of the body whorl, terminating in a straight, thick, obtuse columella, with several irregular plaits. The shells are rare, but not remarkable; and are usually rough to the touch, and striped.

CANCER. The name applied by Linnæus to nearly all the species of the class Crustacea. It is now restricted by naturalists in this country to the genus of which the common black-clawed crab (*C. pagurus*) is the type. Other species are found in North America and South America. [See CRAB.]

CANCROMA. [See BOAT-BILL, p. 79.]

CANIS. In the Linnæan system of zoology, a distinct and very large genus of animals of the class *Mammalia*, order *Feræ*, including all the Dog kind. The characters of this genus are,—that the several species (the common dog, the wolf, the jackal, the fox, &c.) have six upper fore-teeth, the lateral ones being longest, and the intermediate ones of a lobated figure; that there are also six fore-teeth in the lower jaw, of which the lateral ones are lobated; that the canine teeth are single and incurvated; and that the grinders are six or seven in number. [See DOG.]

CANTHARIDÆ. A family of Coleopterous insects, the species of which are numerous and widely diffused. They differ from each other in their size, shape, and colour: the largest are about an inch long. Some are of a pure azure, others of a pure gold, some of a mixture of gold and azure, and others scarlet; but all are brilliant, and very beautiful.

The CANTHARIS VESICATORIA, *Spanish Fly*, or *Blister Beetle*, so well known for its medical uses, is the most noted. This insect is about three quarters of an inch in length, and of a beautiful metallic gold green colour, sometimes changing into bluish green: the head is smooth and polished, and in the forehead are two eyes of a golden colour; the

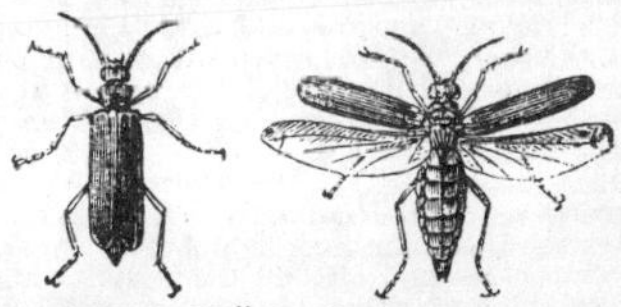

"SPANISH FLY," OR BLISTERING BEETLE (CANTHARIS VESICATORIA.)

wing-cases are membranous, convex above and hollow beneath; thin, but strong, and covering the upper part of the body. The abdomen is composed of eight moveable rings, furrowed from end to end; the legs and antennæ are bluish-black. In Spain, Portugal, and Italy these insects are abundant; in France also they are sometimes found; but are rarely seen in this country. They frequent ash trees, and feed upon its leaves; they are also to be met with on the poplar, the rose, the honey-suckle, and some other sweet-scented shrubs. Although, as we have just said, they but rarely occur in this country; yet at the meeting of the Linnæan Society, Nov. 7. 1837, Mr. Newman exhibited a number of specimens, taken during the preceding summer, near Colchester, where they had appeared by millions, stripping the ash-trees of their leaves. When touched they feign death, and emit a highly offensive odour; which, however, is a guide to those whose business it is to catch them. The most common method of killing them is to expose them to the vapour of hot vinegar: they are then dried on hurdles, and put away in boxes for use.

In Silesia (says Mr. Slater) the *Cantharis vesicatoria* is only a summer guest; it appears there suddenly in June, in rather numerous swarms, which arrive during the night, and are found early in the morning upon the ash, honey-suckle, and some other trees and shrubs, which they soon strip of leaves. Their presence is announced by a most penetrating odour, perceptible at a great distance from the trees on which they sit, and suggesting unpleasant ideas of blistering ointment. Their susceptibility to cold is remarkable; the freshness of early dawn is sufficient to chill and benumb them, and if the trees be then gently agitated, they fall down. In this manner they are collected for sale, and killed by sprinkling with cold water.—(*Zoologist.*)

In North America, according to Dr. Harris, potato-vines are very much infested by two or three kinds of Cantharides, swarms of which attack and destroy the leaves during midsummer. One of these kinds has thereby obtained the name of the *potato-fly*. It is the *Cantharis vittata*, or striped Cantharis. It is of a dull tawny yellow or light yellowish red colour above, with two black spots on the head, and two black stripes on the thorax and on each side of the wing-covers. The under-side of the body, the legs, and the antennæ are black, and covered with a greyish-down. It is more than half an inch long; the thorax is much narrowed before; and the wing-covers are long and narrow, and cover the whole of the back. It does much mischief in potato fields and gardens, eating up not only the leaves of the potato, but those of many other vegetables.—Another species, a jet-black Cantharis (*Cantharis atrata*), measuring nearly half an inch in length, may be seen, about the middle of August, on the potato-vines, and also on the blossoms and leaves of various kinds of golden-rod.—These insects, and others of a similar kind, may be easily taken by brushing or shaking them from the potato-vines into a broad tin pan, and emptied into a covered pail containing a little water, which, by wetting their wings, prevents their flying out when the pail is uncovered; or they may be caught by gently sweeping the plants they frequent with a deep muslin bag-net. They are easily killed by throwing them into scalding water for one or two minutes. (*Ins. of Massach.*)

CAPERCAILZIE. The Scotch name for *Tetrao urogallus*. [See GROUSE.]

CAPITOSAURUS. [See SUPPLEMENT.]

CAPRA. [See GOAT.]

CAPRIMULGUS: CAPRIMULGIDÆ. A genus and family of Passerine birds, popularly termed *Moth-hunters* and *Goat-suckers*. Their habits are nocturnal, and they have the same light soft plumage, minutely mottled with grey and brown, that characterizes other night-birds. Their eyes are large; the beak, very deeply cleft, and generally armed with strong *vibrissæ*, is capable of engulphing the largest insects, which are retained by means of a glutinous saliva; the nostrils, placed at its base, are like small tubes; their wings are lengthened; the feet short, with plumed tarsi, and a membrane connecting the basal portion of the toes: the claw of the middle toe is usually pectinated on its inner edge; and the outer toe has only four phalanges, a conformation extremely rare among birds. They live solitarily, or rather permanently in pairs, and are crepuscular in their time of action, pursuing moths and other nocturnal insects: they deposit their [two] eggs on the bare ground, and have generally singular voices. They bear the same relation to the Swifts that the Owls do to the Hawks; their general anatomy very much resembling that of the Cuckoos. The common European species (*Caprimulgus Europæus*) is remarkable for the loud sound it emits, like the burr of a spinning-wheel. Among the foreign species, a great number have longer tarsi, adapted for running on the ground; and there are some with an appearance of aigrettes on the head. [See GOAT-SUCKER; STEATORNIS; WHIP-POOR-WILL; ÆGOTHELES; LYNCORNIS.]

CAPROMYS. A genus of *Rodentia*, different species of which are found in the West India Islands. They are herbivorous, preferring aromatic plants. In their movements they are slow, somewhat like a bear. One of these was described by Oviedo as the *Chemis*, a name said still to be applied to the *Capromys Fournieri*. Two other species, *C. prehensilis* and *C. Poeyi*, are described. To this genus probably also belongs the "Musk Cavy," described by some authors as almost as large as a rabbit; the upper part of its body is black, and its belly is perfectly white. It inhabits Martinico, and the other Antilles islands; burrows under ground; and smells so strongly of musk, that its retreat may be traced by the perfume.

CAPYBARA. (*Hydrochærus capybara.*) A Rodent animal which has also obtained the name of the Water-hog. It grows to the size of a hog of two years old, and is classed with

CAPYBARA.—(HYDROCHÆRUS CAPYBARA.)

the *Cavidæ*. It inhabits various parts of S. America, but is most common in Brazil. It feeds not only on various vegetables, and particularly on sugar-canes, but also on fish; for which purpose it frequents rivers, swimming with the same facility as the otter, and taking its prey in a similar manner. The Capybara is, in general, considered as of a gentle disposition, and, though shy and timid, is readily tamed and made familiar. It has a very large head, and a thick, divided nose, on each side of which are strong and large whiskers; the ears are small and rounded; the eyes large and black; and the upper jaw longer than the lower. The neck is short; the body thick, and covered with short, coarse, brown hair: the legs short, and the feet long. Like the Peccary, the Capybara is destitute of a tail; and, dissimilar to all others of this kind, instead of a cloven hoof, it is in a manner web-footed, and thus adapted for an aquatic life.

"These great Rodents are generally called 'Carpinchos:' they occasionally frequent the islands in the mouth of the Plata, where the water is quite salt, but are far more abundant on the borders of freshwater lakes and rivers. Near Maldonado three or four generally live together. In the day-time they either lie among the aquatic plants, or openly feed on the turf plain. When viewed at a distance, from their manner of walking and colour, they resemble pigs; but when seated on their haunches, and attentively watching any object with one eye, they reassume the appearance of their congeners, the Cavies. Both the front and side view of their head has quite a ludicrous aspect, from the great depth of their jaw. These animals, at Maldonado, were very tame; by cautiously walking, I approached within three yards of four old ones. As I approached nearer and nearer they frequently made their peculiar noise, which is a low abrupt grunt; not having much actual sound, but rather arising from the sudden expulsion of air: the only noise I know at all like it, is the first hoarse bark of a large dog. Having watched the four from almost within arm's length (and they me) for several minutes, they rushed into the water at full gallop, with the greatest impetuosity, and emitted, at the same time, their bark. After diving a short distance they came again to the surface, but only just showed the upper part of their heads. When the female is swimming in the water and has young ones, they are said to sit on her back."—*Darwin's Journal.*

CARABIDÆ. A very numerous family of Coleopterous insects, containing some of the largest of the carnivorous beetles; many of which are adorned with brilliant metallic colours. The body of these insects is of a very firm consistence, whereby they are enabled to creep about under stones, &c., as well as prevented from falling beneath the power of the insects they attack; most of the species of this family being eminently insectivorous; prowling about, in search of their prey, on the surface of the ground, under stones, &c., or beneath the bark of trees, or in the moss growing at their roots. They are accordingly of essential service in keeping down the numbers of noxious insects with which our gardens and pastures might otherwise be overrun. They are not all, however, exclusively carnivorous; since some of the species generally found in corn-fields are clearly ascertained to feed upon growing grain. Some of the species among the larger Carabidæ exhale a fetid odour, discharging at the same time from the abdomen to a considerable distance a caustic and acrid fluid. Few observations have hitherto been made relative to the larvæ of this family; but those which have been noticed are equally voracious with the perfect insects, and are found in similar situations: they are generally long, with the body of equal breadth throughout. The large work of the Count Dejean is the most complete book yet published on this family. There have been, however, many additions to this family of insects, as to most others. There is a very large collection of them in the British Museum.

CARACARA. A South American Falconidous bird of the genus *Polyborus.* It is of the size of the common kite, and has a tail nine inches long. The beak is black, and hooked; the plumage tawny, with white and yellow specks; the feet are yellow, with semicircular, long, sharp, black talons. In its food the Caracara seems to be content with any animal substance: carrion, reptiles, toads, snails, birds, insects, &c.; whatever, in short, will suit the appetite of other ignoble birds of prey, will content the Caracara. It is by no means shy; and though it ventures to approach inhabited places, it is seldom attacked, as it rarely molests domestic poultry. It builds its nest on the tops of trees where the foliage is close, or in a bushy thicket. It lays two eggs, pointed at one end, and spotted with crimson on a reddish-brown ground.

CARCAJOU. [See BADGER.]

CARACAL. (*Felis caracal.*) This animal, called also the Persian Lynx, is a native both of Asia and Africa. It is about the size of a fox; possesses great strength and fierceness; and is used not only in the chase of the smaller quadrupeds, but of the

CARACAL.—(FELIS CARACAL.)

larger kinds of birds, such as herons, cranes, pelicans, &c., which it is said to surprise with great address. When it has seized its prey, it lies motionless for some time upon it, holding it in its mouth. Its colour is a pale reddish-brown, whitish beneath: the head is small, the face rather long, the ears sharp and slender, of a dark colour, and terminated by a tuft or pencil of long black hairs.

CARCHARIAS. A genus of Chondropterygious fishes, notorious for their bold and predaceous habits, and distinguished by their trenchant-pointed teeth. [See SHARKS, p. 165; and CARCHARODON in SUPPLEMENT.]

CARDIACEÆ. The cockles and their allies; the shells of which are all equivalve, or nearly so. They are furnished with a regularly-toothed hinge, often of great complexity and beauty; and there is always a double abductor muscle: the respiratory orifices are usually prolonged into tubes, which can, however, be drawn within the shell by means of a retractor muscle. There are numerous species, widely diffused; many of them being remarkable for the smallness and delicacy of their shells.

CARDIUM. A genus of Mollusca belonging to the *Cardiaceæ*. The foot is largely developed, and is a most important organ to the animals, it being used by most of them not merely for progression, but in the excavation of hollows in the sand or mud of the shores on which they dwell. As usually seen, the foot of the Cardium, or Cockle, when extended, tapers gradually to a point; and as its diameter is at its largest point much less than the breadth of the shell, it is not apparent by what means the hole that is excavated is made sufficiently large for the reception of the latter: this, however, is accomplished by the distension of the foot with water, through a tube which opens just within the mouth; and thus the size of the borer becomes so nearly equal to that of the shell, that it is enabled, by rotatory motions often repeated, to excavate a burrow large enough to receive the animal with its shell. The shell is generally white, with sometimes a bluish or yellowish cast; it has twenty-six longitudinal ridges, is transversely wrinkled, and has somewhat imbricated striæ. The Cockles, with few exceptions, inhabit the ocean only: they abound most on sandy shores; and are used as a wholesome and nourishing food. The most common species is the Edible Cockle (*Cardium edule*).

CARDIUM FIMBRIA.

Cardium Bechei. This beautiful species (which is a native of the Eastern seas) is dedicated to Sir Henry de la Beche, by its discoverer, Sir E. Belcher, and is described in the "Proceedings of the Zoological Society" (March, 1847), as without exception the most striking and distinct from any hitherto known that can well be imagined. In colour it is of a pure rose tint, with the following singular contrast of character. The middle and anterior portion of the shell is smooth, presenting a peculiar soft velvety appearance, the effect of its being minutely decussated with concentric and radiating striæ, and covered with an exquisite thin shining horny epidermis, disposed in fine concentric cords, abruptly terminating at the posterior area. The posterior portion, accordingly destitute of epidermis, is very thickly rayed with ribs of short compressed spines, as if the delicately clad surface of the shell had been thus far ploughed up, as it were, into furrows.

CARDINAL-BIRD. [See GROSBEAK.]

CARDUELIS. A genus of Passerine birds, of the Finch tribe. [See FRINGILLIDÆ: GOLDFINCH.]

CARIAMA. A Grallatorial bird, of the genus DICHOLOPHUS, the species being *D. cristatus*, —of the size of a heron, inhabiting the great mountain plains of Brazil, "where its sonorous voice often breaks the silence of the desert." Its retired habits are well described by Mr. Broderip: "A tenant of the vast solitudes that form its wide-spreading home, it flies from the face of man; and being almost always on the watch, is very difficult of approach. Stalking slowly on the plain, its eye instantly notes the distant intruder, and, after a moment's hesitation, it decides either to stay or fly, according to the circumstances. Those who have had the best opportunities of observing them in their native wilds, state that the hunters, though surrounded by these birds, cannot, without considerable labour, obtain them. As soon as the bird perceives that it is pursued, it sets off with great rapidity; the pursuer follows on horseback, but it is not till after a sharp and tedious course, with all its turnings and windings, that the Cariama, wearied out, either crouches on the ground, or alights on some bush or tree. Till this happens, the horseman in vain seeks for an opportunity to throw his lasso or pull his trigger. But," adds the writer, "wild as the bird is in its natural state, it is easily domesticated, and will live sociably with the other tenants of the poultry-yard." The Cariama is about two feet eight inches in length: it has an ornamental tuft on the head; the neck covered with long, loose feathers, like those of the bittern; legs long; feet long and slender; and tail rounded. The plumage on the upper part of the bird is brown, and the under parts whitish; the neck feathers are finely rayed with zig-zags of darker brown than the general colour; the wings are dark, traversed with white bands and dotted: the tail feathers are blackish, with white extremities; and the plumage on the front of the neck is prettily variegated with white and brown. The bill is of a bright coral red; and the legs and feet are of an orange red. It feeds chiefly upon lizards and insects. The anatomical structure of the Cariama is interesting to zoologists, on account of the relation it bears both to the waders and gallinaceous birds.

CARINARIA. A genus of Gasteropodous Mollusca, with an elongated, subcylindrical, transparent body, furnished with a sort of fin which performs the part of a rudder. The shells of this genus were formerly known to collectors under the name of "Venus's Slipper," and "Glass Nautilus." A species is found in the Mediterranean, where it is said to feed on small jelly-fish (*Medusæ*), and even true fish, as, for instance, the dwarf Atherine (*Atherina nana.*)

CARNARIA. An immense assemblage of unguiculated quadrupeds, belonging to the third order of Mammalia, and which possess, in common with Man and the Quadrumana, the three sorts of teeth, but have no opposable thumb to the fore-feet.

CARNIVORA. The term applied to the fifth order of Quadrupeds, or Beasts of Prey, which in the structure of their teeth and digestive apparatus, and in their general conformation, show that they are peculiarly adapted for destroying living animals, and for tearing and devouring flesh. In the greater number of the members of this order, the size of the canine teeth is the most obvious mark of distinction; these are large, strong, and pointed, and project somewhat forwards, so as to present themselves rather in front of the line of the other teeth. Between the canines of the two sides are six incisor teeth in each jaw, which are provided

SKULL OF THE LION

with sharp cutting edges. The molar teeth, situated behind the canines, are of three kinds;—those which immediately follow the canines, being more or less pointed, and termed *false molars;* the next being especially adapted for dividing and lacerating animal muscle, by the sharp edge of its summit, and termed *carnivorous teeth;* and the last, or hindmost, being more or less rounded or *tuberculated.* The proportion which these different classes of molar teeth bear to each other in degree and development, accords with the relative carnivorous propensity of the different families; for instance, it may be laid down as a general rule, that those carnivorous animals which have the shortest jaw and the least development of the false molars are those in which the sanguinary propensity and the destructive power co-exist in the highest degree. It should also be remembered that the articulation of the jaw does not permit of horizontal movement, the power being simply that of opening and shutting, like a pair of shears. In these, as indeed in all animals, the structure is admirably adapted to their habits. They feed on living animals and are therefore swift to pursue, and strong to overpower them; they are armed with formidable teeth and claws to tear them in pieces; their sight is keen, and even more so by night than by day; their sense of smell is acute, and their power of hearing delicate; their feet are soft, to enable them to steal silently on their prey; their bodies are long and flexible, so that they may glide unseen; and, finally, their supply of food being uncertain, they are capable of long abstinence.

In every order there is one principal group, which possesses the characteristics of the order in the highest perfection; though the necessary imperfection of all artificial systems of arrangement causes animals to be grouped together, which, although agreeing in very many points, yet differ essentially in others. Thus Lions and Tigers are the principal or tropical group of Carnivora, yet in the order are included Bears, Racoons, &c., which feed principally on vegetables. The CARNIVORA are divided into—1. The *Plantigrades*, or those which walk on the entire soles of their feet: 2. *Digitigrades*, or those which walk on their toes: 3. *Amphibia*, or *Amphibious Carnivora.* The *Plantigrades* comprise Bears, Badgers, Racoons, Gluttons, and Coatimondis. The structure of their feet causes these animals to be slow; but as their food is principally vegetable, speed is not required. The same structure gives them great facility in raising themselves on their hind feet. The *Digitigrades* comprise Lions, Tigers, Cats, Dogs, &c. This structure gives swiftness. This division is also characterized by the claws being retractile: thus preserving them from injury, and keeping them sharp for use when required. The *Amphibia*, or *Amphibious Carnivora*, comprise the Seals and Sea-horses, distinguished by having very short hind legs, and the fore legs formed for swimming.

As we shall have occasion to recur to this subject in describing various animals whose propensities are decidedly carnivorous, though differing in the degree, we shall for the present merely add, that the muscular energy of the Carnivora is very great; their respiration and circulation very active; and the demand for food, as a natural consequence, very constant.

CARP. (*Cyprinus carpio.*) The genus of Malacopterygious abdominal fish, of which this species is the best known, may be easily distinguishable by the small mouth, toothless jaws, and gills of three flat rays. The tongue and palate are smooth, but the gullet is admirably constructed for mastication, having large teeth attached to the inferior pharyngeal bones, which press the food between themselves and a gelatinous knob, connected with a bony plate that is united with the first vertebra, commonly called the carp's tongue. They have but one dorsal fin, and the body is covered with scales, generally of a large size. They frequent fresh and quiet waters, feeding on herbs, grain, and even mud, being, perhaps, the least carnivorous of the finny race. The most noted are the COMMON CARP, and the GOLDEN CARP OR GOLD-FISH (*Cyprinus auratus*).

The COMMON CARP (*Cyprinus carpio*) is found in most of the lakes and smaller rivers of Europe; but those of the southern and temperate parts are most congenial to it, and it is said to decrease in size the farther it is removed to a northern region. It is generally supposed that Carp were introduced into this country about the year 1500; but this is a fact of very little real importance, since they have long become denizens of most of our fish ponds and many of our rivers. They frequent the deepest places, and thrive best in such as have clayey or marly sides, and

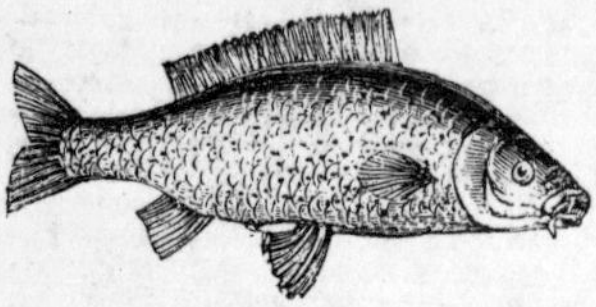

CARP. — (CYPRINUS CARPIO.)

are well provided with aquatic vegetables; Their usual food consists of worms and other insects; but grain of various sorts, and garbage, are frequently thrown into the pond, with a view to aid in fattening them. The Carp is an extremely prolific fish, and the quantity of roe is so great that it is said to have sometimes exceeded the weight of the emptied fish itself when weighed against it.

The age to which the Carp arrives is very great, and several well authenticated instances are adduced of their arriving at that of considerably more than a century; some writers, indeed, affirm that they have been known to live to the age of two hundred years! The usual length of Carp in this country is from about twelve to sixteen inches; but in warmer climates they are frequently more than three feet long, and weigh twenty or thirty pounds. The general colour is a yellowish olive, much deeper or browner on the back, and the sides slightly tinged with a golden hue: the scales large, round, and very distinct; the head large, and the mouth furnished with a moderately long cirrus or beard: above which is a shorter one. The fins are violet brown, except the anal, which has a reddish tinge: the dorsal fin is broad or continued to some distance from the middle of the back towards the tail, which is slightly forked, with rounded lobes.

CARPENTER BEE. (*Xylocopa*). The name given to a very large genus of Hymenopterous insects; not one species of which has yet been found in the British Islands. They are generally of a very dark violet blue, and of considerable size. As an exemplification of this peculiar exotic genus, we may mention the Violet Carpenter Bee (*X. violacea*), a very common insect about Paris. Their bodies are of a very deep blue colour, smooth, and shining; their fore-wings are of a deep violet colour · on their sides, the hinder part of their bodies, and their breasts, there are long black hairs. As we have mentioned, they are not indigenous in this country; but in France and the southern parts of Europe there is scarcely a garden where some of them may not be found at different seasons of the year. They generally form their nests in pieces of half-rotten wood; and the holes are not made directly forward, but inclining to one side, and having apertures large enough to admit a finger; from which run their inner apartments, each generally twelve or fifteen inches long, and diverging into others. In each of these cavities they deposit ten or twelve eggs, which are covered with a sort of paste, serving for the protection of the young insects, as well as for their nourishment. The females perform all the labour; and the males have no stings. In the British Museum cases, may be seen specimens of wood bored by a North American species, or, we should rather say, with holes made by their powerful jaws. [See APIDÆ: BEE.]

CARPET [MOTHS]. A name applied by insect collectors to various species of Moths, of the genera *Harpalyce*, *Cidaria*, *Larentia*, *Cleoria*, and *Alcis*.

CARPINCHO. The Capybara, or Water-hog. From the dung of this, the *Myrapetra scutellaris*, a species of wasp found near Buenos Ayres, constructs its pasteboard nest. [See CAPYBARA.]

CARRIER PIGEON. [See PIGEON.]

CARYOPHYLLÆA. A genus of Zoophytes, belonging to the *Madrephylliæa*, or the first section of the stony *Zoantheria* of De Blainville. The cells in this genus of Zoophytes are furnished with radiating

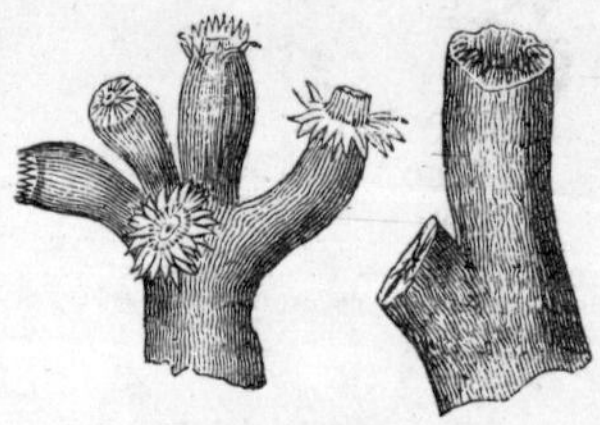

CARYOPHYLLEA.

plates, striated externally, and collected into a solid conical polyparium fixed at the base. In the British Museum are many very fine specimens of this genus, which is found both in a recent and fossil state.

CASHMIRE GOAT. [See GOAT.]

CASSIS. A genus of Molluscous animals inhabiting an oblong shell: found in the seas of warmer climates. [See HELMET SHELL.] The well-known large species of this genus are used as ornaments on chimney pieces, grottos, &c., and are remarkable for the triangular disc, presented by the inner lip, which is thickened and spread over the body whorl, and the angulated outer lip; and as this thickening of the lip takes place at various stages of growth, the same triangular plane is observable at different parts of the spire.

CASSIDA. A genus of Coleopterous insects, of the family *Cassididæ*, or Tortoise-Beetles. They have a flattened body, surrounded by a margin, which is formed by a

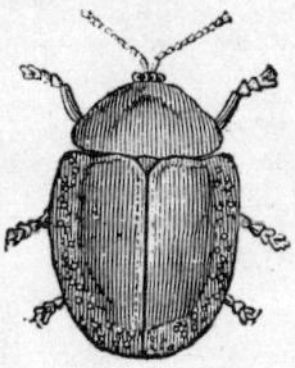

CASSIDA VIRIDIS.

prolongation from the thorax and elytra, and which even conceals the head; and they are able to lie so close upon the surface of the leaves, that no part of the body or limbs can be seen. Their colours are much varied, and often very prettily arranged in spots, points, rays, &c. The family is very numerous in genera and species, and among the exotic species are several of great beauty; but when dead, or taken out of spirits of wine in which they may have been preserved, their metallic brilliancy for the most part disappears.

The COMMON GREEN CASSIDA, or *Cassida viridis*, is often seen during the summer months in gardens on the leaves of mint, &c. Its length is not quite a quarter of an inch; its shape oval, and its colour bright green above, the body or under part being perfectly black. The larva, which is of a highly singular appearance, is oval, of a yellowish brown colour, and has the body edged with a row or fringe of projecting fibres, the two terminal ones being much longer than the rest, and generally carried in an upright position while the insect is in motion. When ready to assume the chrysalid form, it fastens itself to a leaf, and. casting its skin, commences its new state of existence; and from the chrysalis, in the space of three weeks, comes forth the perfect insect.

CASSOWARY. (*Casuarius.*) This large and powerful Struthionidous bird is a native of Java and the adjacent islands of the Indian Archipelago, and is called the *galeated* or *helmeted* Cassowary, from its head being surmounted by a sort of osseous crest or horny helmet. The skin of the head and upper part of the neck are naked, of a deep-blue and fiery-red tint, with pendent caruncles, similar to those

CASSOWARY —(CASUARIUS GALEATUS.)

of the Turkey-cock. It is much inferior in size to the Ostrich, its height when erect being little more than five feet; but it is robustly built, and very strong. From the form of its head, and bright eyes, it would be reasonable to infer that the Cassowary was of a fierce and threatening nature; this, however, is not a true indication of its character, which is rather timorous and shy. The shortness of the wings totally unfits it for flying, and its pectoral or wing-muscles are comparatively slight and weak. All the feathers of the Cassowary are of the same kind, being entirely designed for covering, and externally are all of one colour. They generally grow double, having two long shafts growing out of a short one attached to the skin; yet its whole plumage is so poorly supplied with feathers as to resemble, at a little distance, a coat of coarse or hanging hair. The feathers on the head and neck are so short and scattered, that the skin appears naked, except towards the hind part of the head, where they are somewhat longer.

In many important points of internal structure the Cassowary differs from the Ostrich; particularly in the conformation of its digestive organs. The intestines are short, and the *cæcum* small; there is no stomach intermediate to the crop and gizzard, and the *cloaca* is not larger, in proportion, than that of other birds. It feeds on fruits, eggs of birds, and tender herbage, but not on grain. It eats its food with great voracity, and, like the ostrich, swallows bits of iron, brick, glass, &c., which have the same effect in assisting the digestion of these large birds, that gravel has with ordinary fowls. The Cassowary is an amazingly swift runner; and its mode of progression, being unaided by wings, is as peculiar as it is efficient. It appears to strike out powerfully with one leg, so as to project its body violently forward with a bounding motion, far surpassing the speed of a horse. It also kicks violently when, in a state of captivity, it is provoked to anger, and can inflict a very severe blow. The eggs of the galeated Cassowary are of a greyish-ash colour, verging to green, and are neither as round nor as large as those of the Ostrich: the shell is not very thick, and is marked by numerous little green tubercles.

CASTOR. [See BEAVER.]

CASUARIUS. [See CASSOWARY.]

CAT. (*Felis.*) All animals of the Cat kind, though they may differ greatly in size and colour, are in their wild state equally characterized by fierceness, artfulness, and rapacity. It is not, however, in this place that the habits, propensities, or anatomical structure of the feline race generally are to be discussed: the reader must refer to the article FELIS for such remarks on those heads as we have deemed it essential to introduce.

The Cat, (*Felis catus ferus*), in a state of natural wildness differs in some slight particulars from the domestic animal, having a somewhat shorter tail in proportion, a flatter and larger head, and stronger limbs. The colour of the wild Cat is commonly a pale yellowish-grey, with dusky stripes; those on the back running lengthwise, those on the sides transversely and with a curved direction: the tail is annulated with several alternate circles of blackish-brown and dull white; and the tip of the nose and the lips are black: they are, however, not uniformly alike in every particular. The manners of the wild Cat resemble those of the lynx, and several others of this genus; living in woods, and preying on young hares, birds, and a variety of other animals, which it seizes by surprise. "The wild Cat," says Mr. Pennant, "may be called the British tiger; it is the fiercest and most destructive beast we have; making dreadful havoc amongst our poultry, lambs, and kids. It inhabits the most mountainous and woody parts of these islands, living mostly in trees, and feeding only by night. It multiplies as fast as our common Cats; and often the females of the latter will quit their domestic mates, and return home pregnant by the former."

The varieties of this animal in a domestic state are very numerous: it is either entirely black; black and white; black, fulvous, and white (called the Tortoise-shell or Spanish Cat); entirely white; fulvous and white; dun colour or tawny, either plain or striped; tabby, boldly striped; slate-coloured or blue-grey (called the Chartreux Cat); slate-coloured with very long fur, especially on the neck and tail (the Persian Cat); long hair of silvery whiteness and silky texture (called the Angora Cat); and, lastly, with pencilled or tufted ears, like a lynx, which sometimes though rarely, takes place. Of all the above varieties, the Persian and the Angora are the most remarkable.

Although the Cat is capable of showing considerable fondness for an individual, it seems to be a pretty general opinion that she seldom, if ever, confides fully, even in the warmest demonstrations of kindness; but, being highly sensitive and fond of ease, evinces little anxiety, except for the continuance of her enjoyment. Yet, with all the prejudice that exists against the furtiveness and treachery of the species, no one can deny that, when well educated, the Cat possesses qualities which well entitle her to the regard and protection of mankind; and if she does not exhibit the vivid and animated attachment of the dog, she is still of an affectionate and gentle disposition, and grateful to her benefactors. Nor does any animal, whose habits we have the opportunity of accurately observing, exhibit a greater degree of maternal tenderness; the extreme assiduity with which she attends her young, and the fondness which she shows for them, never fail to attract attention.

At what period Cats became inmates of human habitations, it is scarcely possible, at this period, to determine. Beyond doubt, their usefulness in destroying rats, mice, &c. first introduced them to notice; and there is good reason to believe that they were originally domesticated in Egypt. That Cats, closely allied to the domestic variety, were trained to catch birds, is well known to every one who has seen an Egyptian painting (or a copy from it) in the British Museum, where a cat with a black stripe on the heels (supposed to be the *Felis caligata*) is so represented. The Cat belongs to a genus better armed for the destruction of animal life than all other quadrupeds. The short and powerful jaws, moved by vigorous muscles, are supplied with most formidable trenchant teeth: a cunning disposition, combined with nocturnal habits and much patience in pursuit, gives them great advantages over their prey; and their keen, lacerating claws enable them to inflict a certain death-blow. All animals considerably weaker than themselves prove objects of pursuit: but the mouse is their favourite game; for which they will patiently wait for a whole day till the victim comes within reach, when they seize it with a bound, and after playing with it put it to death.

The pupil of the eye in most animals is capable of but a small degree of contraction and dilatation; it enlarges a little in the dark, and contracts when the light pours upon it too profusely; but in the eyes of Cats, this contraction and dilatation is so considerable, that the pupil, which by day appears narrow and small, by night expands over the whole surface of the eye-ball, and gives the eyes a luminous appearance. By means of this peculiar structure, their eyes are better adapted for vision at night than in the day-time; and they are thus fitted for discovering and surprising their prey.

Cats are extremely fond of strong-smelling plants, and will roll in valerian till they seem almost mad with excitement. Personally, it is a very cleanly animal, avoiding to step in any sort of filth, concealing its excrement in the earth with great care, and preserving its fur in a very neat condition; which being generally clean and dry, readily yields electric sparks when rubbed. The Cat goes with young for sixty-three days, and brings forth from three to six at a litter, which remain blind for nine days.

CATASTOMUS. A sub-genus of Malacopterygious fishes, of the *Cyprinidæ* family, inhabiting the fresh waters of North America. There are many species, the generic description of which is as follows:—Back with a single fin: gill-membrane three-rayed: head and opercula smooth: jaws toothless and retractile: mouth beneath the snout; lips plaited, lobed, or carunculated, suitable for sucking: throat with pectinated teeth. In almost all the species, the scales are marked with radiated lines, and fimbriated on their edges; their form more or less rhomboidal or roundish. In the intestines, river-shells (*Lymnea*, *Bulimus*, &c.), which dwell on aquatic plants and on rocks or bottoms of rivers, are found; the *Catastomi* being enabled to take these shells by means of their lips, which are protruded forwards by their jaws. One species will be sufficient for us to describe.

CATASTOMUS HUDSONIUS; or GREY SUCKING CARP. This is a common fish in all parts of the fur countries, abounding in the rivers, and even in land-locked marshes and ponds, but preferring shallow grassy lakes with muddy bottoms. In the beginning of summer it may be seen in numbers forcing its way up rocky streams, and even breasting strong rapids, to arrive at its proper spawning places in stony rivulets; but it soon afterwards returns to the lakes. Its food appears to consist of soft insects and minute crustacea. In the spawning season (June) it may be readily speared, or even taken by the hand, in shallow streams; but in the winter and autumn it is caught in nets. It is a very soft watery fish, but devoid of any unpleasant flavour, and is excellent for making soup. Like its congeners, it is singularly tenacious of life, and may be frozen and thawed again without being killed. It is about twenty-one inches long; the head is smooth, flattened laterally, with an obtuse snout; the depth of the body exceeds its thickness rather more than one-half. The lateral line runs equidistant from the back and belly, straight till it comes opposite to the anal fin, when it inclines upwards at a very obtuse angle, and passes along the middle of the tail, giving that member a direction slightly different from that of the body. Scales for the most part broadly oval, or nearly orbicular, and of a medium size. Mouth retractile, placed under the snout; lips studded with large soft papillæ; but there are no barbels. The pectoral fins are elliptical; the ventrals obovate; the dorsal fin is nearly quadrangular; the anal extends to within its own length of the caudal, and when it is turned backwards its tip reaches the base of the caudal, which is slightly crescentic. The back and sides of this fish are bluish-grey with considerable lustre, the back being darkest, and the tint of the sides gradually passing into the pearl-white of the belly. Dorsal and caudal fins bluish-grey: pectorals and ventrals ochre-yellow, tinged with red; anal flesh-red. Among the other best-known species are the Red Sucking Carp (*Catastomus Forsterianus*); the Gilt Sucking Carp (*Catastomus aureolus*); and the Black Sucking Carp, or Shoemaker (*Catastomus nigricans*).— These, as well as the preceding, are all described by Sir John Richardson, the most distinguished Ichthyologist of this country, in his Fauna Boreali Americana, and the Supplements to different Arctic Voyages.

CAT-BIRD. (*Turdus* [*Mimus*] *felivox.*) The celebrated American ornithologist, Wilson, has given an account of this bird in a style so amusing, that we are tempted to lay it almost entire before our readers. "In spring or summer," says he, "on approaching thickets or brambles, the first salutation you receive is from the Cat-bird; and a stranger, unacquainted with its note, would instantly conclude that some vagrant orphan kitten had got bewildered among the briers, and wanted assistance; so exactly does the call of the bird resemble the voice of that animal. Unsuspicious, and extremely familiar, he seems less apprehensive of man than almost any other of our summer visitants; for whether in the woods, or in the garden, where he frequently builds his nest, he seldom allows you to pass without approaching to pay his respects in his usual way. This humble familiarity and deference, from a stranger, too, who comes to rear his young, and spend his summer with us, ought to entitle him to a full share of our hospitality. Sorry I am, however, to say, that this, in too many instances, is cruelly the reverse."

The Cat-bird generally succeeds in building his nest about the beginning of May. The place chosen for this purpose is usually a thicket of briers or brambles, a thorn bush, thick vine, or the fork of a small sapling; no great solicitude is shown for concealment, though few birds appear more interested for the safety of their nest and young. The materials are dry leaves and weeds, small twigs, and fine dry grass; the inside is lined with the fine black fibrous roots of some plant. The female lays four, sometimes five, eggs, of a uniform greenish blue colour, without any spots. They generally raise two, and sometimes three, broods in a season.

"In passing through the woods in summer, I have sometimes amused myself with imitating the violent chirping or squeaking of young birds, in order to observe what different species were around me,—for such sounds, at such a season, in the woods, are no less alarming to the feathered tenants of the bushes, than the cry of fire or murder in the streets is to the inhabitants of a large and populous city. On such occasions of alarm and consternation, the Cat-bird is the first to make his appearance, not singly, but sometimes half a dozen at a time, flying from different quarters to the spot. At this time, those who are disposed to play with his feelings may almost throw him into fits, his emotion and agitation are so great, at the distressful cries of what he supposes to be his suffering young. Other birds are variously affected, but none show symptoms of extreme suffering. He hurries backwards and forwards, with hanging wings and open mouth, calling out louder and faster, and actually screaming with distress, till he appears hoarse with his exertions. He attempts no offensive means; but he bewails—he implores—in the most pathetic terms with which nature has supplied him, and with an agony of feeling which is truly affecting. Every feathered neighbour within hearing hastens to the place, to learn the cause of the alarm, peeping about with looks of consternation and sympathy. But their own powerful parental duties and domestic concerns soon oblige each to withdraw. At any other season the most perfect imitations have no effect whatever on him.

"The Cat-bird is one of our earliest morning songsters, beginning generally before break of day, and hovering from bush to bush, with great sprightliness, when there is scarce light sufficient to distinguish him. His notes are more remarkable for singularity than for melody. They consist of short imitations of other birds, and other sounds; but, his pipe being rather deficient in clearness and strength of tone, his imitations fail where these are requisite. Yet he is not easily discouraged, but seems to study certain passages with great perseverance; uttering them at first low, and, as he succeeds, higher and more free, nowise embarrassed by the presence of a spectator even within a few yards of him. On attentively listening for some time to him, one can perceive considerable variety in his performance, in which he seems to introduce all the odd sounds and quaint passages he has been

CAT BIRD.—(TURDUS [MIMU] FELIVOX)

able to collect. Upon the whole, though we cannot arrange him with the grand leaders of our vernal choristers, he well merits a place among the most agreeable *general* performers. In summer, scarcely a thicket in the country is without its Cat-birds; and, were they to fly in flocks, like many other birds, they would darken the air with their numbers. In their migrations they keep pace with the progress of agriculture; and the first settlers in many parts of the Gennesee country have told me, that it was several years after they removed there, before the Cat-bird made its appearance among them. With all these amiable qualities to recommend him, few people in the country respect the Cat-bird; on the contrary, it is generally the object of dislike; and the boys of the United States entertain the same prejudice and contempt for this bird, its nest and young, as those of Britain do for the Yellow-hammer, and its nest, eggs, and young. I am at a loss to account for this cruel prejudice. Even those by whom it is entertained can scarcely tell you why; only they 'hate Cat-birds;' as some persons tell you they hate Frenchmen, they hate Dutchmen, &c.; expressions that bespeak their own narrowness of understanding and want of liberality. Yet, after ruminating over in my own mind all the probable causes, I think I have at last hit on some of them; the prin ipal of which seems to me to be a certain similarity of taste, and clashing of interest, between the Cat-bird and the farmer. The Cat-bird is fond of large ripe garden strawberries; so is the farmer, for the good price they bring at market: the Cat-bird loves the best and richest early cherries; so does the farmer, for they are sometimes the most profitable of his early fruit, &c. Perhaps, too, the common note of the Cat-bird, so like the mewing of the animal whose name it bears, and who itself sustains no small share of prejudice, the homeliness of his plumage, and even his familiarity, so proverbially known to beget contempt, may also contribute to this mean, illiberal, and persecuting prejudice; but, with the generous and the good, the lovers of nature and of rural charms, the confidence which this familiar bird places in man by building in his garden, under his eye, the music of his song, and the interesting playfulness of his manners, will always be more than a recompence for all the little stolen morsels he snatches.

"The Cat-bird measures nine inches in length; at a small distance he appears nearly black; but, on a closer examination, is of deep slate colour above, lightest on the edges of the primaries, and of a considerably lighter slate colour below, except the under tail-coverts, which are very dark red; the tail, which is rounded, and upper part of the head, as well as the legs and bill, are black. The female differs little in colour from the male." The habits, manners, and general appearance of the Cat-bird differ so little from the Thrushes, that the naturalist to whom we are indebted for the foregoing particulars does not hesitate to place him in the genus *Turdus*. He is a great and determined enemy to the common black snake, or horse-runner (*Coluber constrictor*), which rifles its nest whenever an opportunity offers. As the Cat-bird uniformly attacks or pursues this snake, and is frequently seen in the act of hopping eagerly after it, numerous ridiculous stories are related of its being fascinated by the snake; it is, however, well known to naturalists that the bird is almost uniformly the aggressor and victor, driving the reptile to its hiding-place.

CATERPILLAR. The name given to the larvæ of lepidopterous insects; of which we have spoken at some length in the article BUTTERFLY, and to which the following, from "Brande's Dictionary of Science" (art. *Lepidoptera*), may be added. "They have six squamous or hooked feet, which correspond to the legs of the perfect insect, and from four to ten additional membranous ones, or *propedes;* the two last of which are situated at the posterior extremity of the body. Those Caterpillars which have but ten or twelve in all, have been called, from their mode of progression, *Geometræ*. Several of these geometers, when at rest, remain fixed to the branches of plants by the hind feet alone, whence in the form, colour, and directions of their body, they resemble a twig. The body of these larvæ is generally elongated, almost cylindrical, soft, variously coloured; sometimes naked, and sometimes covered with hairs, tubercles, and spines. It is composed of twelve segments or annuli, exclusive of the head, with nine stigmata on each side. Their head is invested with a corneous or squamous dermis, and presents on each side six shining granules, which ap ear to be ocelli: and it is furnished with two very short and conical antennæ, and a mouth composed of strong mandibles; two maxillæ, a labrum, and four small palpi. Most Caterpillars feed on the leaves of plants; some gnaw their flowers, roots, buds, and seeds; others attack the ligneous or hardest parts of trees, softening it by means of a fluid which they disgorge. Certain species attack our woollens and furs, thereby doing us much injury; even our leather, bacon, wax, and lard are not spared by them. Several confine themselves exclusively to a single article of diet; others are less delicate, and devour all sorts of organized matters. Some of them form societies, and frequently live under a silken tent, spun by them in common, which even shelters them in winter. Several construct sheaths for themselves, either fixed or portable; others make their abode in the parenchyma of leaves, where they form galleries. The greater number are diurnal; the others never issue forth but at night."

There are perhaps no insects which are so commonly and so universally destructive as Caterpillars; they are inferior only to locusts in voracity, and equal or exceed them in their powers of increase, and in general are far more widely spread over vegetation. As each female Butterfly or Moth usually lays from two hundred to five hundred eggs, one thousand different kinds of butterflies and moths will produce, on an average, three hundred thousand caterpillars; if one half of this number, when arrived at maturity, are females, they will give forty-five millions of caterpillars in the second, and six thousand seven hundred and fifty millions in the third generation! These data suffice to show that the actual number of these insects, existing at any one time, must be far beyond the limits of calculation.

CATHARTES. [See TURKEY BUZZARD.]

CATTLE. A collective term, denoting all animals of the bovine or ox kind. The domestic cattle of Britain may be divided into two races: those of large size, adapted for the plains; and those of smaller size, adapted for the mountains. Of each of these classes there are several breeds; such as the Highland and the Welch cattle, among the latter; and the Lancashire, the Yorkshire, and the Herefordshire cattle, among the former. There is also an intermediate breed, adapted for moderately hilly countries; such as the Galloway and Fife breeds in Scotland, and the Alderney and Guernsey cattle in England. The best beef brought to the London market is that of cattle of the Highland breed fed in English pastures, or on turnips. The best milk cow for general purposes is the Ayrshire; the best for cream and butter, the Alderney; and the best for immense quantities of milk, the Lancashire. Hence the latter are generally employed in public dairies, the Ayrshire by farmers and cottagers, and the Alderney by the higher classes.

CATURIDÆ. [See SUPPLEMENT.]

CAVY. (*Cavia.*) A genus of Rodents inhabiting tropical America. They are distinguished by the possession of eight grinders on either side of the jaw, and by two wedge-shaped incisive in front. The fore-feet are either three or four-toed, the hind-feet being generally tridactylous and in some cases five-toed, with the two outer digits feebly developed; tail short or absent; and no clavicle. They have generally a slow, and sometimes a leaping pace; they live on vegetable substances, and in their natural state inhabit excavations under ground, or beneath the roots of trees, or other recesses which they either find ready prepared, or form for themselves.

The COMMON CAVY, or GUINEA-PIG. (*Cavia Cobaya.*) From the beauty and variety of its colours, and the neatness of its appearance, this species must have early attracted the attention of those Europeans who first visited South America; but it has

COMMON CAVY, OR GUINEA-PIG.
(CAVIA COBAYA.)

been so long domesticated in this and other countries, as now to have become quite naturalized in the Old World. Its ears are large, broad, and rounded at the sides; its upper lip is half divided; and its hair is erect, somewhat resembling that of a young pig. Its colour is white, varied with orange and black in irregular blotches. It has four toes on the fore-legs, and three on the hind; and is destitute of a tail. In its wild state it lives in societies, inhabiting dry lands covered with low brushwood; and remains concealed during the day, coming forth on the approach of evening to seek its food. It possesses neither cunning to avoid danger, strength to resist, nor swiftness to escape from it; and nothing could save the race from extermination, were it not for its extraordinary rapidity of multiplication. The usual litter consists of six, eight, or ten; and so prolific is it that it breeds almost every two months. The young very soon acquire the necessary degree of strength and perfection of their species, though they continue to grow till seven or eight months. They are very tender animals, and susceptible of cold; and should therefore be provided with warm receptacles to retire into in bad weather. In their habits they are extremely neat, and may be frequently observed in the act of smoothing and dressing their fur. Their general voice is a sort of a grunting squeak, and sometimes a shriller or sharper cry.

The SPOTTED CAVY (*Cœlogenys paca*) is a large species, measuring nearly two feet in length. It is found in Guiana, Brazil, and other parts of South America; inhabiting holes formed underground, and principally near the banks of rivers. Its shape is thick and clumsy, somewhat like that of a pig, for which reason it has been sometimes called the hog-rabbit. It has five toes on each foot, and only the mere rudiment of a tail. The upper jaw is longer than the lower; the ears are short and naked · the lip is divided like that of a hare; and it has long whiskers. The body is covered with coarse, short, thinly-scattered hair of a dusky colour; the throat, breast, and belly are of a dingy white; and on each side the body run five rows of roundish, slightly angular spots. The Spotted Cavy is a nocturnal animal, residing in a solitary manner in his hole nearly the whole day. In a domestic state it readily feeds on almost any kind of vegetable diet, and is particularly fond of sugar and fruits. By the South Americans it is much esteemed as an article of food. [See AGOUTI; CAPYBARA: PACA, &c.]

CEBIDÆ. A term used to include all the Monkeys of the American continent; which differ in several respects from those of the Old World; viz. by a partial or complete absence of the thumb upon the hands; the callosities and cheek-pouches are altogether absent; there is a very considerable space between the nostrils; the tail is usually of great length, never absent, and often prehensile. They are very numerous in those vast forests which occupy the plains between the rivers Oronoko and Amazon. [See MONKEY.]

CEBRIO: CEBRIONIDÆ. A genus and family of Coleopterous insects, of small extent, but comprising several striking peculiarities of structure. The body is of an oblong oval form, of a firm consistence like the Elateridæ, arched above and deflexed in front; the mandibles strong, curved, and entire at the tip; the thorax broadest behind, with the posterior angles acute; and the antennæ generally longer than the head and thorax, and serrated or pectinated in the males. These insects are of moderate size; and their colours generally dull and obscure: for the most part they are inhabitants of the south of Europe, and the north of Africa; and but little is known of their habits.

Some of the genera are most remarkable for their beautiful pectinated antennæ, which in the male sex have the branches often of

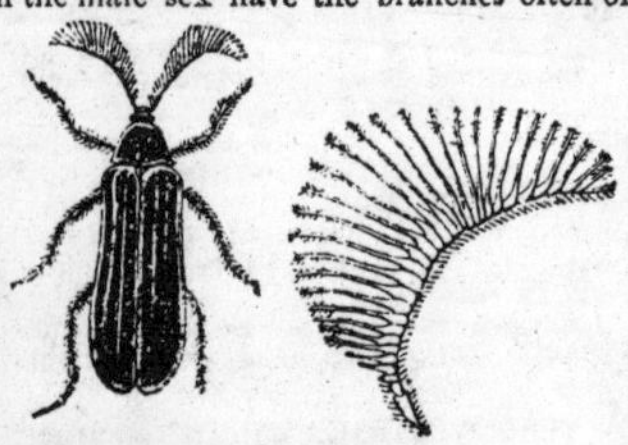

RHIPICERA MARGINATA

very great length. By some authors these are regarded as a separate family, under the name of *Rhipiceridæ*. We figure a beautiful Brazilian species, which is of a blackish green, and pubescent: the anterior and lateral margins of the elytra are yellow: hence it is called *Rhipicera marginata*. The figure in outline represents the beautiful pectinated antenna of the male, considerably magnified.

CECIDOMYIA: CECIDOMYIDÆ. A genus and family of two-winged flies, of which there are many species. They are always of small size; many of them deposit their eggs upon the tender buds of various kinds of plants; others upon the young sprigs, and some upon the flowers. One species (*Cecidomyia salicina*) fixes each of its eggs on a bud of the willow, which becomes enlarged, and ultimately forms a gall in which the larva is lodged and nourished. Another (*Cecidomyia tritici*), known as the Wheat-fly, may sometimes be seen, in great abundance, flying about wheat fields in the month of June. This little fly deposits its eggs in the centre of the corolla, where the larvæ are hatched; and it is probably by devouring the pollen that they are most injurious to the plant. Another species (*Cecidomyia destructor*), known in America under the name of the Hessian-fly, attacks the lower part of the stem of the wheat. Dr. Asa Fitch, an American naturalist, has just published a most admirable and readable account of the *Cecidomyia*, from which we shall make extracts in our article "Wheat-fly." [See WHEAT-FLY and HESSIAN-FLY.]

CECILIANS. A name given to a genus of naked serpents, from their supposed blindness.

CENTIPEDE. (*Scolopendra.*) A genus of carnivorous *annulosa* belonging to the order *Myriapoda* of Cuvier. They are distinguished by having antennæ of fourteen joints or upwards; a mouth composed of two mandibles; a quadrifid lip; two *palpi*, or small feet, united at their base; and a second lip, formed by a second pair of dilated feet, joined at their origin, and terminated by a strong hook, having an opening beneath its point, through which a poisonous fluid is thrown out. The body is long, depressed, and membranous, each ring being covered by a coriaceous or cartilaginous plate, and mostly having one pair of feet: the last is usually thrown backwards, and elongated in form of a tail. These insects conceal themselves under the decayed bark of trees, the decayed timbers of buildings, among stones, lumber, and rubbish, whence they sally forth at night in search of prey. In the West India islands, and throughout the hot parts of America, where they multiply rapidly and grow to a large size, they are very formidable pests. The utmost vigilance is necessary, even in cleanly houses, to prevent these creatures from finding their way into the beds; and although they endeavour to escape as soon as a light is brought into the room, and run with considerable swiftness, they are ready to stand on the defensive, and bite severely: they are accordingly very dangerous when once they have entered a bed; the bite being not only exceedingly painful at the moment, but followed by a high degree of local inflam-

CENTIPEDE.

mation, and a fever of great irritation. This truly noxious Centipede grows to the size of five or six inches in length, and is a formidable inmate of most of the houses in tropical regions. In different countries the species vary: the one common in England is of a reddish-brown colour, about an inch long, with a flat, thin body, and yellowish legs.

Dr. Leach made it the type of his genus *Lithobius*, a word meaning that the Centipede lived under or amongst stones. One species is very common in this country, it is named *L. forcipatus*.

There are other species in the collection of the British Museum, from which they were originally described by the late G. Newport, F.R.S., who in 1844 published an admirable monograph of the *Scolopendridæ* and their allies in the 19th volume of the Linnæan Transactions. From the numerous references to the British Museum, the student will see how rich our great national establishment must be in this important order of the Animal Kingdom.

CENTRISCUS, or SEA-SNIPE. *Centriscus scolopax.*) A genus of Acanthopterygious fishes, principally distinguished by

SEA-SNIPE. — (CENTRISCUS SCOLOPAX.)

their having a long tubular snout; the body compressed, and inclining to an oblong oval form; the abdomen carinated; and the belly-fins united. [See TRUMPET-FISH.]

CENTROLOPHUS. A genus of Acanthopterygious fishes, the technical characters of which are; — body elongate, covered with minute scales; teeth small and numerous; palatine without teeth; one long dorsal fin.

CENTRONOTUS. A genus of Acanthopterygious fishes, family *Scomberidæ;* in which the spines are free or unconnected by membrane, and all have ventral fins.

CENTROPOMUS. A genus of Acanthopterygious fishes; a well known species of which is called the Sea-pike (*Centropomus undecimalis*), and is common throughout South America, where it forms a considerable article of consumption. The Sea-pike sometimes weighs as much as twenty-five pounds: the form of its body is elongate; its colour is greenish-brown above, and silvery beneath.

CENTROPRISTES. A genus of Acanthopterygious fishes; one species of which, (*Centropristes nigricans*), the Black-perch or Black-bass, is of a deep olive-green colour above, and of a pink hue on the under parts; but it is mostly remarkable for having the tail doubly notched, the central and two outer parts projecting.

CENTROPUS. A genus of Scansorial Birds. [See PHEASANT CUCKOO.]

CEPHALOPODA. A class of Molluscous animals, characterized by the possession of locomotive organs (or *feet*) around the head; they are, however, not feet, but prolonged tentacula, or fleshy processes, which project forwards from the head, and more or less conceal the mouth. In the whole range of molluscous animals, the Cephalopods are the most highly organized; they present undoubted rudiments of an internal skeleton, and contain digestive, secretory, respiratory, and generative organs. The nervous system of the Cephalopods approaches that of the lower fishes in many particulars; and they are almost exclusively marine in their habits. The natural division of the class is into those Cephalopods which are naked, and those which are testaceous, (*i. e.* protected by an external shell.) Of the former, the common Cuttle-fish, and of the latter, the Nautilus, may be taken as examples.

CEPHUS. A genus of Hymenopterous insects. The *Cephus pygmæus*, which is common on flowers, particularly buttercups, is about one-third of an inch long; black, with two yellow fasciæ on the abdomen: and its larva is said to live in the stems of wheat.

CEPOLA. A genus of Acanthopterygious fishes, the bodies of which are much compressed and elongated. [See BAND-FISH.]

CERAMBYCIDÆ. A family of Coleopterous insects (*Longicornes*); the most distinguishing feature of which is the very great length of their antennæ. They are found in all parts of the globe, but they abound most in hot climates. They deposit their eggs in old and decaying trees, which the larvæ afterwards feed upon, and thereby assist in removing. The body of these insects is long and subdepressed, occasionally subconvex; the maxillary lobes are distinct and membranaceous; the femora often clavate; and the tarsi short. Mr. Westwood, to whose "Modern Classification of Insects" we are so much indebted, observes that they are "generally of an elegant form, and beautifully variegated in their colours: they are found in forests, hedges, or woods, sitting upon the trunks of trees, or more rarely upon flowers. Some of the exotic species are remarkable for having the antennæ and legs covered with thick pencils of hairs; others are distinguished by the emission of a fragrant odour, not unlike that of attar of roses, which is so powerful, that the insects may be discovered upon trees by passers by, in consequence of the scent diffused through the air, and which is retained for a considerable period after death. Hence the generic names *Callichroma* and *Aromia*, proposed for these insects by Latreille and Serville. The *Cerambyx moschatus*, Linn. (or Musk Beetle, as it is generally but improperly termed, the scent scarcely resembling that of this drug) is the only British species belonging to this scented group: it is more than an inch long, of a fine green colour, and is abundant upon willows in the neighbourhood of London." It has been conjectured that the fragrance, which is always much more powerful in the female, may be intended, like the light of the glowworm, as a guide for the males. The author just quoted informs us, that the larva of *Cerambyx heros*, which is one of the largest European species, is considered by Latreille to have been the *Cossus* of the ancients, by whom it was esteemed a relishing treat.

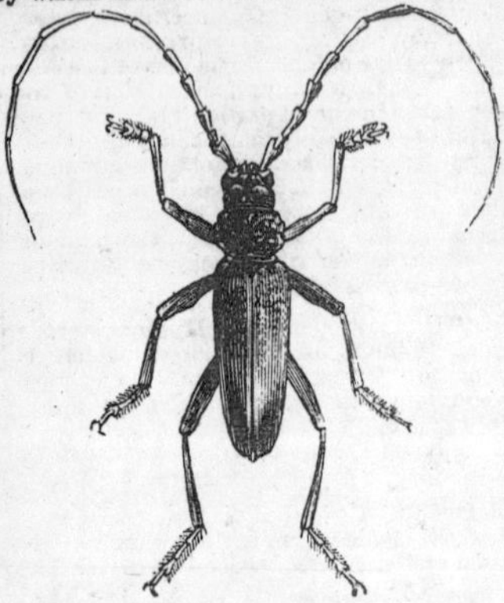

CERAMBYX HEROS.

It resides in the oak, occasioning much injury to the timber, by boring large channels in all directions through the trunk of the tree: this is also the case, as regards young willows, with the Musk beetle; the larva of which is of a thick form and fleshy consistence; the head small; the prothorax large and transverse; the meso and metathorax very short, the former furnished with a pair of spiracles, and the three thoracic segments having three pairs of very short legs. He further says, "The larvæ of the genus *Callidium* are similar to those of *Aromia* (the Musk beetle) both in form and habits. The places where they reside may be known by the long cylindrical burrows which they form, and which are filled with excrement resembling powdered wood. It is not difficult to keep these larvæ alive in the wood in which they are found, and in which they assume the pupa state; it is very rarely, however, that they can be reared to the imago state. Mr. Kirby has given an interesting account of the proceedings of the larva of *Callidium violaceum*, which, in the larva state, feeds principally upon fir timber, upon which the bark has been suffered to remain after it has been felled; residing under the bark, mining its labyrinth-like passages in every direction, and occasioning much damage by means of its powerful jaws, which resemble a large, thick, and solid section of a cone of horn; the whole of their interior flattened surfaces applied together, so as completely to grind the food. It is described as being destitute of feet, pale, folded, somewhat hairy, convex above, and divided into thirteen segments, with the head large and convex. When full grown, it bores down obliquely into the solid wood to the depth of several inches, where it becomes a pupa."

The collection of these insects in the British Museum is very extensive: their form, colour, and appendages make them always pleasing objects to the sight; while to the Natural Theologian, the part they play in the economy of nature is very apparent and easily demonstrable in many striking ways.

CERASTES. A genus of serpent called in England the Horned Snake, having two small protuberances on its forehead. This animal, which partakes of the nature of viviparous serpents, is remarkable for its almost total abstinence from water. It is found in Lybia, Arabia, &c.

CERBERUS. A sub-genus of Ophidians. [See SERPENTS.]

CERCOPIDÆ. An extensive family of Homoptera, comprising several species of singular insects, many of which are tropical. The head is of small or moderate size, with the face broad, the eyes lateral, the antennæ inserted in the middle or lower part of the face; the promuscis short and three-jointed; the prothorax very variable in form and size, and in the sub-family *Cercopidæ* being the portion of the body which assumes the remarkable forms above alluded to. The fore-wings differ in their consistence, but the majority have them strongly veined, forming cells closed before reaching the extremity of the wing. The hind tibiæ vary in structure, being in some nearly simple; in others, furnished with a few strong spurs; and in many, being triangular or quadrangular, each angle throwing out strong spines. The abdomen of the female is furnished with a multivalve ovipositor, variable in its form in the different species. These insects are often beautifully varied in their colours; they are constantly found amongst plants, and on trees, upon the juices of which they subsist, in all their states. One of the best known insects in this family is the *Aphrophora spumaria*, which frequents garden plants, the larva and pupa investing themselves with a frothy excrementitious secretion. [See FROG-HOPPER.]

CERCOPITHECUS. [See MONKEY.]

CEREOPSIS. A genus of Palmipede birds that frequent the coasts in New Holland. The *Cereopsis Novæ Hollandiæ* is about the

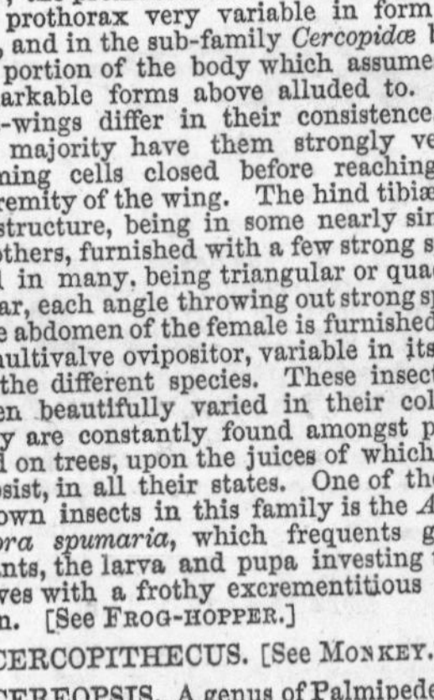

NEW HOLLAND CEREOPSIS.
(CEREOPSIS NOVÆ HOLLANDIÆ.)

size of a common goose, and resembles it in its general appearance, with the exception of the length of the legs, which are from two feet and a half to three feet. Its plumage is of a dingy grey, deeper on the upper than on the under parts. On the top of the head is a large patch of dull white; and the quill-feathers both of the wings and tail are dusky black. The naked extremity of the bill is black; the broadly expanded cere, light straw colour; the naked part of the legs, reddish orange; and the toes, together with their web and claws, black. It has a deep, hoarse, clanging voice; its usual weight is from seven to ten pounds; and its flesh is considered excellent. It is now exceedingly common in aviaries. In the Gardens of the Zoological Society we have been often struck with the grallatorial appearance possessed by this pleasing Australian Cere-faced Goose. Both the genus and species were first described by Dr. Latham.

CERITHIUM. A genus of Mollusca, chiefly inhabiting the Indian and Pacific Oceans. There is a veil on the head of the animal, two distant tentacula, having the eyes at the side, and a round, horny tuberculum. The shell (which is often also found in a fossil state) has a turriculated spire, an oval aperture, and a short but distinct canal curved to the left and backwards. There are very many species, most of which are in the collection of the British Museum.

CERTHIA. [See CREEPER.]

CERTHIADÆ. A family of Tenuirostral or slender-billed Passerine birds, commonly known by the name of *Creepers*. They are birds which for the most part are adapted to live upon the trunks and branches of trees, and to feed upon insects which infest the bark. The form of the bill varies in different species; being long and slender in some; short and stout, and capable of penetrating very hard substances, in others. These birds cling by their feet to the perpendicular surface of trees, resting upon the stiff quills of their tails; and creep from the base to the summit of the stem, with short jerking movements, searching every crevice as they ascend. Several species are described under the word CREEPER.

CERURA. A genus of Bombycidous Moths, of which there are several species; one of the best known is the

CERURA VINULA, or PUSS MOTH. This delicately marked, and at the same time common Moth, varies from two and a half to three and a half inches in the expanse of its wings, which are of a milky-white or pale ash colour; with a transverse row of black spots, near the base, succeeded by a rather more cinereous-coloured bar, edged on both sides with black spots. Then follow in the discoidal cell three curved dusky stripes, which run in arches to the hind margin of the wing. Beyond this are two rows of blackish, very strongly dentate waves; several dark, wedge-like streaks appearing between the veins along the outer margin of the wing. The hind wings are white, but more ashy in the female, with the margin spotted with dusky. Thorax ashy, spotted with black; abdomen white, with dusky marks. There are several varieties, in which the ground colour of the body and the markings of the wings are

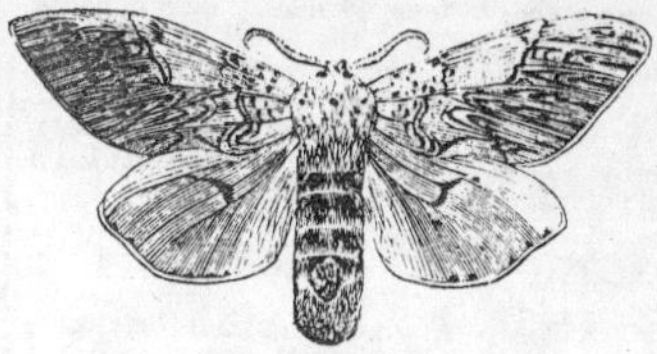

PUSS-MOTH.—(CERURA VINULA.)

more or less intense. The Caterpillar is green, with a reddish head; the back dull lilac, separated from the green colour by an angulated white stripe. Its most striking peculiarity is the possession of two appendages, which, when the creature is disturbed, it pushes out sometimes to a considerable length. The Caterpillar, doubtless, in this way of expressing defiance, alarms some of its enemies; and any one not accustomed to

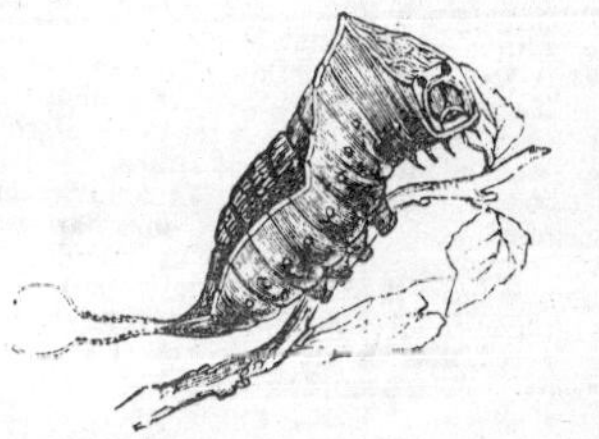

CATERPILLAR OF PUSS-MOTH

insect-studies, would "look twice" before he ventured to touch what *would seem* a beautifully coloured and strange looking "grub," armed with two "weapons" of unknown powers. When full fed it encloses itself in a cocoon formed of chips of wood agglutinated together so firmly that it is difficult to cut it with a knife. It feeds on the willow, poplar, &c. in August, and the moth makes its appearance early in the following summer.

There are other British species of the genus, smaller in size, but more delicately marked; these are all figured in the very elegant British Moths and their transformations of Mr. Humphreys, the descriptions of which were compiled by Mr. Westwood.

CERVIDÆ. The Deer tribe; a group or family of Ruminantia, distinguished by the possession of bony deciduous horns, covered with soft skin, instead of with horny matter, and termed antlers. They are spread very extensively over the globe, each quarter having its own peculiar species, celebrated either for vigour, beauty, or speed, or for all these qualities combined.

CERVUS. [See DEER.]

CESTRACION. A genus of Sharks, found in New Holland; characterized by having two kinds of teeth, arranged in oblique rows; those in front of the mouth being sharp, angular, and pointed; while those in the middle and back part of the jaw are flat and broad; the former evidently adapted for seizing the food, and the latter for crushing and bruising it. Specimens and dissections of *Cestracion Phillipsii* and others of the Fixed-gilled Chondropterygii may be seen in the vast collections of the British Museum and the Royal College of Surgeons.

CETIOSAURUS. [See SUPPLEMENT.]

CESTUM. A marine animal belonging to the *Acalepha ciliograda* and bearing a near resemblance to Beroë. It is a very long gelatinous ribbon, having one of the sides furnished with two rows of ciliæ; and near the sides of the mouth there are two vessels which are probably ovaries.

CETACEA. An order of Mammiferous animals, surpassing in size all others in existence, and inhabiting the sea. Like terrestrial quadrupeds, they are viviparous, suckle their young, have warm blood, and respire through lungs; for which purpose they must frequently come to the surface, to take in fresh supplies of air. But though in their anatomical details they are sufficiently distinguished from *fishes*, it will be seen that these animals have no hind limbs, that the first bones of their anterior extremities are shortened, and the succeeding ones flattened and enveloped in a tendinous membrane, which reduces them to the condition of true fins. The Cetacea are all carnivorous; but the largest species are supported chiefly by minute Mollusca and Medusæ.

J. E. Gray, F. R. S., and Keeper of the Zoological collections in the British Museum, has published, very recently, an elaborate monograph of all the Whales in the Zoology of the Voyage of H. M. SS. Erebus and Terror; and in the Proceedings of the Zoological Society of London for 1847, there are some additional observations and descriptions by this very eminent zoologist. [See WHALE.]

CETONIADÆ. An extensive group of Coleopterous insects, belonging to the family of Lamellicorn beetles, including several distinguished for their brilliant colours. Of these, as an example, the common ROSE CHAFER (*Cetonia aurata*) may be cited. This insect is nearly an inch long, of a shining green colour above, and coppery-red beneath, with white marks on the elytra. It abounds upon roses, and also upon the flowers of the privet. It flies well, with a considerable humming noise, during the hottest part of the day; and although it appears to give the preference to roses, it visits other flowers also, and draws from them their honeyed stores. In its larva state, the Rose-beetle feeds upon moist rotten wood, and is often met with under ground in ants' nests. After remaining about three years in its larva state, it makes a sort of cocoon of chips of wood, glued together by an excretion of its own; in this, as an inactive pupa, it passes the winter, and emerges in the following summer in its perfect form. The insects composing the *Cetoniadæ* are very widely dispersed, but more especially frequent tropical climates.

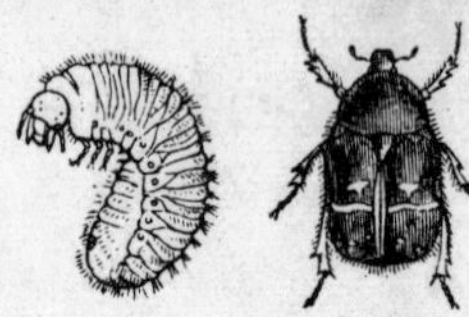

ROSE CHAFER.—(CETONIA AURATA.)

Very few of the flower-beetles are decidedly injurious to vegetation. Some of them are said to eat leaves; but the greater number live on the pollen and the honey of flowers, or upon the sap that oozes from the wounds of plants. In the infant or grub state most of them eat only the crumbled substance of decayed roots and stumps; a few live in the wounds of trees, and by their depredations prevent them from healing, and accelerate the decay of the trunk.

AGESTRATA CHINENSIS.

These beetles (the *Cetoniadæ*) are generally of an oblong oval form, somewhat flattened above, and often brilliantly coloured and highly polished, as in the genus *Agestrata* here figured, species of which are found in Ceylon, India, China, and the Philippine Islands. Mr. Cuming informed the writer that the ladies of Manilla kept a very brilliant metallic green species, *A. luconica*, (preserved in the British Museum collection,) as a pet, in small bamboo cages, which they carried about with them. They are sometimes also covered with hairs. The accompanying cut of the *Cetonia* [*Trichostetha*] *fascicularis*, a native of the Cape of Good Hope, will show another form of this extensive group, which is more or less covered with tufts of hair: the thorax is deep black, with four white longitudinal lines; the elytra are green, their sides being furnished with

C. [TRICHOSTETHA] FASCICULARIS.

several long tufts of yellow hair: the under side of the body is also rather thickly clothed with numerous scattered hairs of the same colour; while the *Cetonia* [*Pachnoda*] *Baxii*, from Senegal and the Gambia, with its harlequin markings, will serve as an illustration of another division of this very extensive family. Most of the bright-coloured kinds are day-fliers; those of dark and plain tints are generally nocturnal beetles. Some of them are of immense size, and have been styled the princes of the beetle tribes; such are the Incas of South America, and the Goliath Beetle (*Goliathus Goliatus*) of Guinea, the latter being more than four inches long, two inches broad, and thick and heavy in proportion. [See GOLIATH.]

C. [PACHNODA] BAXII.

Dr. Thaddeus Harris has so well described a species of this family, that we have made an extract from his valuable work, as follows:—"CETONIA INDA makes its appearance in the United States towards the end of April or the beginning of May, when it may sometimes be seen in considerable numbers around the borders of woods, and in dry open fields, flying just above the grass with a loud humming sound, like a humble-bee, for which, perhaps, it might at first sight be mistaken. Like other insects of the same genus, it has a broad body, very obtuse behind, with a triangular thorax, and a little wedge-shaped piece on each side between the hinder angles of the thorax and the shoulders of the wing-covers; the latter, taken together, form an oblong square, but are somewhat notched or widely scalloped on the middle of the outer edges. The head and thorax of this beetle are dark copper-brown, or almost black, and thickly covered with short greenish yellow hairs; the wing-cases are light yellowish brown; but changeable, with pearly metallic tints, and spattered with numerous irregular black spots; the under side of the body, which is very hairy, is of a black colour, with the edges of the rings and the legs dull red. It measures about six-tenths of an inch in length. During the summer months the Indian Cetonia is not seen; but about the middle of September a new brood comes forth, the beetles appearing fresh and bright, as though they had just completed their last transformation. At this time they may be found on the flowers of the golden-rod, eating the pollen, and also in great numbers on corn-stalks, and on the trunks of the locust tree, feeding upon the sweet sap of these plants. On the approach of cold weather they disappear, and it is conjectured that they get into some warm and sheltered spot, where they pass the winter in a torpid state, and in the spring issue from their retreats, and finish their career by depositing their eggs for another brood. Those that are seen in the spring want the freshness of the autumnal beetles. Their hovering over and occasionally dropping upon the surface of the ground is probably for the purpose of selecting a suitable place to enter the earth and lay their eggs.

Between four and five hundred distinct species are known to exist in collections, and numerous others are yearly added to the list. [See also art. GOLIATHUS.]

CETOTOLITES. Whale ear-bones. [See BALÆNODON in the SUPPLEMENT.]

CHÆTODON: CHÆTODONTIDÆ. A genus and family of Acanthopterygious fishes, abounding in the seas of hot climates, and remarkable for the singularity of their figure and the beauty of their colours. They are, in a general view, distinguished by the great depth and highly compressed form of the body, which, in most species, is beautifully variegated by transverse, oblique, or longitudinal bands, and covered with strong scales; the dorsal and anal fins being remarkably broad. The species are very numerous; but they are rarely, if ever, found in the European seas. It may suffice, perhaps, to describe one species; for which purpose we will take

The IMPERIAL CHÆTODON. This is a magnificent species, growing to the length of twelve or fourteen inches: body oval; dorsal and anal fins broad; and scaled to a considerable distance from the base; gill-covers furnished on each side with a very strong spine; the ground colour a golden yellow, longitudinally but somewhat obliquely striped with very numerous bright-blue parallel rays. It is a native of the seas of Japan, and is in high esteem as an article of food.

CHAFFINCH. (*Fringilla cœlebs.*) A well-known, lively Passerine bird, of elegant plumage, whose short and frequently-repeated song is heard early in spring, but which towards the close of summer becomes a mere chirping note. Its nest is remarkably neat and compact, being constructed of small fibres, roots, and moss, and lined with wool, hair, and feathers. The female generally lays five or six eggs, slightly tinged with red, and sprinkled with dark spots, principally at the larger end; and the male is very assiduous in his attendance during the time of incubation. The bill is pale blue, tipped with black; eyes hazel; forehead black; the crown of the head, and the hinder part and sides of the neck, bluish ash; the cheeks, throat, and fore part of the neck, belly, thighs, and vent, white, slightly tinged with red; the back is reddish-brown, changing to green on the rump; the wing-coverts are dusky, tipped with white, forming two pretty large bars across the wing; the bastard wing and quill feathers are black, edged with yellow; the tail is black, except the outer feather, which is edged with white; legs brown. The plumage of the female is not so vivid, but inclines to a dusky green; and she is destitute of the red on the breast.

Chaffinches subsist chiefly on small seeds; likewise on caterpillars and insects, with which they also feed their young. As they are naturally very hardy, they may be taken from their nests when about ten days old, and brought up with facility; but in England they are seldom kept in cages, as their song is thought to possess no variety, and they are not apt imitators of other song-birds. In Thuringia, however, it is said, there is quite a passion for keeping them, and they accordingly fetch high prices there.

Let us not suppose, however, that the Chaffinch is without a friendly advocate in this country. That he is esteemed by at least one person, and that person a naturalist of no mean calibre, the following extract from the Ornithological Essays of Mr. Waterton afford undoubted evidence:—"Amongst all the pretty warblers," says he, "which flit from bush to bush before me, as I wander through the flowery fields, next to poor cock robin, the chaffinch is my favourite bird. I see him almost at every step. He is in the fruit and forest trees, and in the lowly hawthorn: he is on the house-top, and on the ground close to your feet. You may observe him on the stack-bar, and on the dunghill; on the king's highway, in the fallow field, in the meadow, in the pasture, and by the margin of the stream. If his little pilferings on the beds of early radishes alarm you for the return of the kitchen garden, think, I pray you, how many thousands of seeds he consumes, which otherwise would be carried by the wind into your choicest quarters of cultivation, and would spring up there, most sadly to your cost. Think again of his continual services at your barn door, where he lives throughout the winter, chiefly on the unprofitable seeds, which would cause you endless trouble were they allowed to lie in the straw and to be carried out with it into the land, on the approach of spring.

"His nest is a paragon of perfection. He attaches lichen to the outside of it, by means of the spider's slender web. In the year 1805, when I was on a plantation in Guiana, I saw the humming-bird making use of the spider's web in its nidification; and then the thought struck me that our chaffinch might probably make use of it too. On my return to Europe, I watched a chaffinch busy at its nest: it left it, and flew to an old wall, took a cobweb from it, then conveyed it to its nest, and interwove it with the lichen on the outside of it. Four or five eggs are the usual number which the chaffinch's nest contains; and sometimes only three. The thorn, and most of the evergreen shrubs, the sprouts on the boles of forest trees, the woodbine, the whin, the wild rose, and occasionally the bramble, are this bird's favourite places for nidification. Like all its congeners, it never covers its eggs on retiring from the nest, for its young are hatched blind. There is something peculiarly pleasing to me in the song of this bird. Perhaps association of ideas may add a trifle to the value of its melody; for when I hear the first note of the chaffinch, I know that winter is on the eve of his departure, and that sunshine and fine weather are not far off. * * * The chaffinch never sings when on the wing; but it warbles incessantly on the trees, and on the hedgerows, from the early part of February to the second week in July; and then (if the bird be in a state of freedom) its song entirely ceases. You may hear the thrush, the lark, the robin, and the wren, sing from time to time in the dreary months of winter; but you will never, by any chance, have one single note of melody from the chaffinch. Its powers of song have sunk into a deep and long lasting trance, not to be roused by any casualty whatever. All that remains of its voice, lately so sweet and so exhilarating, is the shrill and well-known monotonous call, which becomes remarkably distinct and frequent whenever the cat, the owl, the weasel, or the fox are seen to be on the move.

"Sad and mournful is the fate which awaits this harmless songster in Belgium and in Holland, and in other kingdoms of the Continent. In your visit to the towns in these countries, you see it outside the window, a lonely prisoner in a wooden cage, which is scarcely large enough to allow it to turn round upon its perch. It no longer enjoys the light of day. Its eyes have been seared with a red-hot iron, in order to increase its powers of song, which, unfortunately for the cause of humanity, are supposed to be heightened and prolonged far beyond their ordinary duration by this barbarous process. Poor chaffinches, poor choristers, poor little sufferers! My heart aches as I pass along the streets, and listen to your plaintive notes. At all hours of the day we may hear these helpless captives singing (as far as we can judge) in apparent ecstasy. I would fain hope that these pretty prisoners, so woe-begone, and so steeped in sorrow, to the eye of him who knows their sad story, may have no recollection of those days when they poured forth their wild notes in the woods, free as air, 'the happiest of the happy.' Did they remember the hour when the hand of man so cruelly deprived them both of liberty and eyesight, we should say that they would pine in anguish, and sink down at last, a certain prey to grief and melancholy. * * * How the song of birds is involved in mystery! mystery probably never to be explained. Whilst sauntering up and down the Continent in the blooming month of May, we hear the frequent warbling of the chaffinch; and then we fancy he is singing solely to beguile the incubation of his female, sitting on her nest in a bush close at hand. But on returning to the town, we notice another little chaffinch, often in some wretched alley, a prisoner with the loss of both its eyes, and singing nevertheless as though its little throat would burst. Does this blind captive pour fourth its melody in order to soothe its sorrows? Has Omnipotence kindly endowed the chaffinch with vocal faculties, which at one time may be employed to support it in distress, and at another time to add to its social enjoyments? What answer shall we make? We know not what to say. But be it as it will, I would not put out the eyes of the poor chaffinch, though by doing so I might render its melody ten times sweeter than that of the sweet nightingale itself. O that the potentate, in whose dominions this little bird is doomed to such a cruel fate, would pass an edict to forbid the perpetration of the barbarous deed! Then would I exclaim, O king of men, thy act is worthy of a royal heart. That kind Being, who is a friend to the friendless, shall recompense thee for this."

CHALCIDÆ. A family of Lizards, long and serpent-like in form, and gradually presenting us with a transition from one to the other class of reptiles. The body is usually cylindrical, and extremely elongated; sometimes destitute of limbs, and mostly with the limbs only a little developed, or merely rudimentary.

For descriptions of the species (eighteen in number, and all in the British Museum), see the very admirable descriptive catalogue of the Lizards in the Museum Collection, by John Edw. Gray, Esq., F. R. S., published in 1845, in 289 closely printed pages, wherein all the species are well characterized.

CHALCIDIDÆ. A family of Hymenopterous insects, composed of a great number of parasitic species, distinguished generally by their very minute size, and many of them displaying splendid metallic colours. So exceedingly minute are some that they are reared within the eggs of other insects, but the majority infest other larvæ or pupæ. Many kinds of insect are subject to them, but they are mostly destructive to the various Lepidoptera; and there are some species, especially those having the ovipositor long and exserted, which deposit their eggs in various kinds of galls, formed by *Cynipidæ*, &c.; their progeny attacking and subsisting upon the larvæ inclosed within.

Mr. Haliday and Mr. F. Walker, F. L. S., have studied this very extensive family of insects. The latter has published a monograph of them; and there are many very beautifully engraved plates, executed by Mr. Ingall, of the Bank of England, in the "Entomological Magazine" and the "Entomologist" of Mr. Newman. There is a large collection of them in the British Museum. Mr. Darwin brought home many from the voyage of H. M. S. Beagle; Mr. E. Doubleday found many new species during his travels in N. America; while, even in high latitudes, Mr. George Barnston, now of Tadousac, found undescribed species of these minute insects—

"The green myriads in the peopled grass."

CHAMA. A genus of large bivalve shells, the characters of which are,—that they are commonly smooth, though in some places a little rugose; and in a few species there are numerous spines. The valves of the shell are equal, elate, and convex; and the mouth gapes, as in the oyster. The *Chama gigas* or Giant Chama, is the largest and heaviest shell yet discovered. It is found in the Indian Ocean. "Many enormous cockles" (*Chama gigas*), Captain Flinders observes, "were scattered upon different parts of the reef. At low water, this cockle seems most commonly to lie half open; but frequently closes with much noise; and the water within the shells then spouts up in a stream, three or four feet high: it was from this noise and the spouting of the water that we discovered them, for in other respects they were scarcely to be distinguished from the coral rock. A number of these cockles were taken on board the ship, and stewed in the coppers; but they were too rank to be agreeable food, and were eaten by few." It is also called *Tridacna*. We have seen an immense pair in the church of St. Sulpice in Paris, where they serve to hold "holy water."

CHAMELEON. A lizard-like Reptile, whose peculiar faculty of changing colour has for ages amused the uninformed, and furnished matter of speculation to the philosopher. The species included in the *Chameleonidæ*, or chameleon-tribe, are distinguished by several very remarkable peculiarities. Their bodies are much compressed, or flattened sideways; and the back is surmounted by a sharp ridge. Two of the toes

CHAMELEON.—(CHAMELEO AFRICANUS.)

are directed backwards, opposing the three anterior ones; and the tail is prehensile. The tongue is a hollow tube, with a swollen fleshy extremity; and it is capable of being darted out instantaneously to a great distance, and of being as rapidly drawn in. This organ is furnished with a glutinous saliva; by which the insect prey that serve for the support of these extraordinary reptiles are attached to it. The eyes of the Chameleon are capable of being moved independently of each other; and they are constantly covered with a sort of eyelid, in which there is a small aperture corresponding with the pupil. It is a creature of a harmless nature, feeding on insects, and is capable of enduring a long abstinence; hence arose the popular idea of the Chameleon being nourished by air alone. It is found in many parts of the world, and particularly in India and Africa: it is also sometimes seen in the warmer parts of Spain and Portugal.

The cause of the different changes of colour which the Chameleon undergoes is not even yet well understood. It is said that "the *rete muscosum*, or coloured layer of the skin, contains two kinds of pigment, situated in different layers; the deeper-seated layer is of a deep green or violet red colour, the superficial pigment is of a greyish colour; the deep-seated pigment is contained in branched cavities, and is moveable, producing by its partial accumulation and varying proportions with the superficial layer the changes of colour for which the Chameleon has in all ages been remarkable." Dr. Shaw thus writes: "The general or usual changes of colour in the Chameleon, so far as I have been able to ascertain from my own observation of such as have been brought into this country in a living state, are from a bluish ash-colour (its natural tinge) to a green and sometimes yellowish colour, spotted unequally with red. If the animal be exposed to a full sunshine, the unilluminated side generally appears, within the space of some minutes, of a pale yellow, with large roundish patches or spots of red-brown. On reversing the situation of the animal the same change takes place in an opposite direction; the side which was before in the shade now becoming either brown or ash-colour, while the other side becomes yellow and red; but these changes are subject to much variety both as to intensity of colours and disposition of spots."

Chameleons are all exceedingly slow, dull, and torpid; often remaining in the same position for many hours together, or traversing the twigs and branches of trees in a slow and cautious manner, with the aid of the grasping powers of the feet and tail. The skin is composed of small granular scales; the lungs are large, and are connected (as in birds) with air-cells that lie among the muscles and beneath the skin; hence the appearance of the animal varies greatly; for, according as these cavities are full or empty, it appears either full and bloated, or lean and shrunken.

CHAMOIS. (*Antilope rupicapra.* Pallas. *Rupicapra tragus.* Gray.) A well-known species of the genus ANTELOPE (to which article we refer the reader); but it being the only animal of Western Europe that partakes in any very considerable degree of the characters belonging to the *Antilopidæ*, we have thought it desirable to describe it separately, under its popular name. The Chamois is found only in high mountainous regions, in small flocks or families, where they feed on the highest cliffs

CHAMOIS.—(ANTILOPE RUPICAPRA.)

and precipices affording vegetation, which are almost inaccessible to man. Their sight, hearing, and smell are so acute, and they are so exceedingly shy, that it is only by the greatest patience and skill that the hunter can approach near enough to shoot them; they are likewise so swift, and leap with such vigour and sureness of foot, that to overtake them in a fair chase is next to impossible; hence the Chamois hunters of the Alps are obliged to encounter the greatest perils in pursuit of this favourite game.

The Chamois is a little more than three feet in length, and two feet in height; its head resembles that of the common goat, but the nostrils are not so large, nor the upper lip so prominent. The whole body is covered with long hair, varying with the seasons, being of a deep brown in winter, of a brown fawn colour in summer, and slightly mixed with grey in spring. The head is of a pale yellow colour, excepting a black brown band, which commences near the nose, and ends at the base of the horns and ears, after encircling the eyes; the tail is short and black; and the edges of the hips and inside of the thighs and ears alone white. The horns are about six or seven inches long, and are nearly parallel throughout: the face is straight; the ears small, erect, and pointed; and there is neither muzzle nor beard. The hoofs are concave beneath, and terminate by a projecting edge, especially on the outside. The colours of both sexes are the same, but the females are rather smaller than the males. The kids are of a deep yellowish brown colour, having the under jaw, both sides of the head, and the throat white; with similar dark bands as the adult, beginning at the corners of the mouth, surrounding the eye, and ending at the forehead. One or two are usually produced at a birth. Their flesh is considered a very superior article of food; and their skin is wrought into a soft, pliable leather, well-known by the name of the animal furnishing it. Their food consists of mountain herbs and flowers, and the tender shoots of shrubs; and it is observed that they seldom drink, and are extremely fond of salt. The father of the present Earl of Derby had several Chamois in his noble menagerie at Knowsley. We saw two young species in London some time ago, which were on their way to his Lordship's. They were sweet, gentle-looking creatures, and were by no means shy.

CHANK SHELLS. The name given to one or more species of shells of the genus *Dolium*. These shells (says Mr. M'Culloch) are fished up by divers in the Gulf of Manar on the coast opposite Jaffnapatam, in Ceylon, in about two fathoms water; and at Travancore, Juticoreen, and other places. Large fossil beds of Chanks have also been found. They are of a spiral shape, and form a considerable article of trade in India, where they are in extensive demand all over the country. They are sawn into narrow rings or bracelets, and are worn as ornaments for the arms, legs, fingers, &c. by the Hindoo women; many of them are also buried with the bodies of opulent and distinguished persons. Those which, from being taken with the fish, are called green Chanks, are most in demand. The white Chank which is the shell thrown upon the beach by strong tides, having lost its gloss and consistency, is not worth the freight up to Calcutta. The value of the green Chank, depends upon its size. A Chank opening to the right, called in Calcutta the right-handed Chank, is so highly prized, as sometimes to sell for 400, or 500, or even 1000 rupees.

CHANNEL-BILL. (*Scythrops.*) A genus of Scansorial birds found in New Holland, &c. [See SCYTHROPS.]

CHARADRIADÆ. A family of Wading birds, or *Grallatores*, including the British Plover and allied species. [See PLOVER.]

CHARADRIUS. [See PLOVER.]

CHARR. (*Salmo salvelinus.*) A fish belonging to the family *Salmonidæ*, which inhabits the lakes of Scotland, Wales, and the north of England, as well as those of the bolder and more mountainous parts of Europe; showing a strong predilection for clear

CHARR.—(SALMO SALVELINUS.)

and pure waters, and being seldom known to wander into running streams, except their bottoms are similar to those of its native lakes. The body is longer and more slender than that of the trout; the back is of an olive colour, speckled with whitish spots; the belly is generally red; the scales are very small, and the lateral lines are straight. The mouth is wide: the jaws are nearly equal; the lower parts of the fins are of a vermillion dye: and the gills are quadruple. Although the Charr is very scarce in this country, it occurs in many of the lakes of Cumberland, Westmoreland, and Lancashire; and its flesh is held in high estimation.

CHATTERERS. (*Ampelidæ.*) The Chatterers are a family of Passerine birds (nearly all of whom are natives of America), subsisting on fruits and berries; but as the generality of them reside far from the habitations of man, few opportunities occur of becoming fully acquainted with their habits, &c. The only species found in Europe is the Bohemian Chatterer, or Waxwing (*Bombycillæ garrula.* [See WAXWING.]

The PURPLE-BREASTED CHATTERER (*Ampelis cotinga.*) inhabits Brazil. It is about nine inches in length; beak black; the head, all the upper parts of the body, and wing-coverts of a splendid glossy blue; the greater coverts, wings, and tail black; the throat and fore part of the neck purple, varied with three or four patches of bright scarlet; breast with a blue, and sometimes also a scarlet band. The female has all the upper parts of the body of a beautiful blue, and the throat, neck, and breast are purple. This and an allied species are often called *Pompadour* Chatterers, from having been introduced into Europe by the extravagant, thoughtless, and ambitious mistress of Louis XV. A fine series of these birds may be seen in the noble collection of birds in the British Museum.

The RED CHATTERER (*Ampelis carnifex*) inhabits Guiana and many other parts of South America. The head is crested, and, with the lower part of the back and belly, rump, thighs, and vent, is of a bright crimson; the rest of the plumage is of a dull red, with the tips of the feathers dusky: the tail is crimson, with the tip black; the legs a dirty yellow. Its length is about seven inches.

CHEGOE, or CHIGOE. (*Pulex penetrans.*) A small and troublesome Apterous insect of the order Aphaniptera, of a black colour, which penetrates the flesh, and will, if neglected, produce malignant ulcers. It is a native of South America and the West India islands. It is, in fact, a very small flea, peculiar to warm climates, and dangerous as well as troublesome to those whom they attack. But our readers shall see what that entertaining naturalist, Waterton, has said upon the subject:—"This apparently insignificant insect far outdoes the bug in the exercise of its noxious qualities. The bug attacks you in an open manner, makes a hearty meal, and then retires to enjoy it: but the Chegoe commences its operations upon you so gently, that they are scarcely felt; and it terminates them in a way that calls for your most serious attention. In a word, it approaches you with such insinuating address, that you absolutely feel a kind of gratification at the very time it is adopting measures which will infallibly end in your certain torment. Soon after the Chegoe has entered your skin, you experience a pleasant itching kind of sensation, by which you begin to suspect that all is not right; and, on taking a nearer view of the part, you perceive that the skin is somewhat discoloured. I know it is supposed by some people, that the accounts concerning the Chegoe have been much exaggerated. I am not of this way of thinking, for I myself have smarted under its attacks; and I have minutely inspected the foot of a Negro, which was a mass of ulcers, formed entirely by the neglected ravages of the Chegoe.

"Not content with merely paying you a visit, and then taking itself off again, as is the custom of most insects, this insidious miner contrives to work its way quite under your skin, and there remains to rear a numerous progeny. I once had the curiosity to watch the movements of a Chegoe on the back of my hand, a part not usually selected by it to form a settlement. It worked its way pretty rapidly for so small an insect. In half an hour it had bored quite through the skin, and was completely out of sight. Not wishful to encourage its intended colony, 'Avast, there! my good little fellow,' said I; we must part company without loss of time. 'I cannot afford to keep you, and a numerous family, for nothing; you would soon eat me out of house and home.' On saying this, I applied the point of my penknife to the place where the Chegoe had entered, and turned it loose upon the world again.

"In the plantations of Guiana there is generally an old negress, known by the name of Granny, a kind of 'Junonis anus,' who loiters about the negro yard, and is supposed to take charge of the little negroes who are too young to work. Towards the close of day, you will sometimes hear some of the most dismal cries of woe coming from that quarter. Old Granny is then at work, grubbing the Chegoe nests out of the feet of the sable urchins, and filling the holes with lime-juice and Cayenne pepper. This searching compound has two duties to perform; first, it causes death to any remaining Chegoe in the hole; and, secondly, it acts as a kind of birch-rod to the unruly brats, by which they are warned, to their cost, not to conceal their Chegoes in future: for, afraid of encountering old Granny's tomahawk, many of them prefer to let the Chegoes riot in their flesh, rather than come under her dissecting hand." In this strain our amusing "Wanderer" continues to recount the Chegoes' annoyances; but our want of space warns us to desist from indulging in a longer extract; and we conclude by observing, that, as these insects have a decided predilection for the toes, the most effectual way to prevent their attacks, is to wear thick stockings, and to bathe the feet often, particularly in sea-water.

CHEIROGALEUS. A genus of Mammalia belonging to the order *Quadrumana*, allied to the *Galagos*. They retain the whole of their inferior incisors during life; the head is round; the nose and muzzle are short; the lips are furnished with *vibrissæ;* the ears are short and oval; the eyes are

HANDED LEMUR.—(CHEIROGALEUS MEDIUS.)

large, and close to each other; the toe-nails are compressed, somewhat clawlike; while the tail is long, bushy, and cylindrical. There are two or three species of these singular Mammalia known, all of which come from Madacascar. The species figured is called *Cheirogaleus medius.*

CHEIROMYS. [See AYE-AYE.]

CHEIROPTERA. The scientific name of an order of Mammalia having the faculty of sustained flight; being characterized by having the anterior extremities so formed as to serve the office of wings, the fingers being extremely long, and connected together by an extended membrane. This power of contiu ued flight, so contrary to the general habits of mammiferous animals, is obtained by the structure of the anterior extremities, the fingers of the fore-hand (or claw) being greatly lengthened; between them is extended a thin membrane, which is continued from the anterior to the hinder extremities, and, in most Bats, is also continued between the hind legs, and it embraces the tail where this member is present. The food of most Bats is insects, which they are incessantly pursuing in their rapid flight; in all of these the membrane is extended between the hind legs, which enables the Bat to turn rapidly in pursuit of its prey. Some Bats, however, feed principally on fruit,

SKELETON OF A BAT

and in these the hind legs are free. They all possess four large canine teeth, but the grinders vary in number, the smallest number being on each side, three in each jaw, and the largest five above and six below, or *vice versâ.* The incisors also vary, the smallest number being two above and two below, and the largest number four above and six below. The order Cheiroptera contains only one division, the *Vespertilionidæ.* The flying Foxes (*Galeopithecus*) being now very properly classed among the Quadrumana. Bats, then, are divided into two families; the first of which, *Istiophori*, are characterized by the peculiar structure of the nose, the skin of which is expanded into leaf-like appendages, which are supposed to increase their power of smell; the second family, *Anistiophori*, have the nose simple. The first family is divided into two subfamilies; the first, *Phyllostomatina*, having the nose-leaf simple, and the second, *Rhinolophina*, in which it is complicated. The second family is divided into three subfamilies; the first, *Vespertilionina*, in which the wings are wide and extended, the head long, and there is only a single phalanx or joint to the fore-finger; the second, *Noctilionina*, having the wings long and straight, head short and obtuse, and there are two phalanges on the fore-finger; and the third, *Pteropina*, in which the wings are rounded, the head long, and having three phalanges on the fore-finger. There are seventeen British species of Bats: two belong to the family *Rhinolophina*, the greater and lesser Horse-shoe Bat; but neither of them are very common. The remaining fifteen belong to the family *Vespertilionina*, twelve being included in the genus *Vespertilio*, the largest of which is the *V. Murinus* or Mouse-coloured Bat, the extent of the wings being fifteen inches; this species is very rare. The common Bat is the Pipistrelle (*V.*

Pipistrellus); the "Common Bat" of the Continent (*V. Murinus*) was long supposed to be our "Common Bat;" but this is now found to be an error. Of the remaining species, two are the most beautiful found in this country, the long-eared Bat, and the lesser long-eared Bat, belonging to the genus *Plecotus*, and the other is the *Barbastelle*. [See BAT and VAMPIRE BAT.]

CHELIFER. A genus of Arachnidæ, belonging to the family called *Pseudo-scorpiones*. Their bodies are oval, and they have the palpi elongated like arms, with a claw-like hand with two fingers; eight legs, all equal, and terminated by two ungues. They resemble small scorpions deprived of tails. The body is flattened, with the thorax nearly square, and having one or two eyes on each side. They run quickly, and often sideways like crabs. Two or three species of this and the closely allied genus *Obisium* are found in this country.

CHELONIA. An order of Reptiles, including the Tortoises and Turtles; characterized by the body being inclosed between a double shield or shell, from which the head, tail, and limbs are protruded. The animals composing this order vary considerably in those details of their structure which adapt them to different habits of life; some of them being adapted to reside exclusively upon the solid ground, and others to dwell amidst marshes, the muddy banks of rivers, &c. The LAND TORTOISES (*Testudinidæ*) have a bulged carapace, sustained by a bony skeleton wholly solid, and anchylosed for the greater part to the lateral edges of the breastplate; their legs are truncated, with very short toes connected almost to the nails, and are capable, together with the head, of being completely withdrawn into the armour. In the MARSH and RIVER TORTOISES (*Emydæ*) the toes are divided and webbed, so as to increase the extent of surface; and in the TURTLES (*Chelonidæ*) they are extended into large undivided paddles, by which they can propel themselves rapidly through the water. J. E. Gray, F.R.S., and Professor Bell have published monographs of this order. [See TORTOISE and TURTLE.]

CHELURA. The name applied to a genus of small Amphipodous Crustacea, first found at Trieste by Dr. Philippi, who has described a species which he calls *C. terebrans*, from its habit of boring into wood-

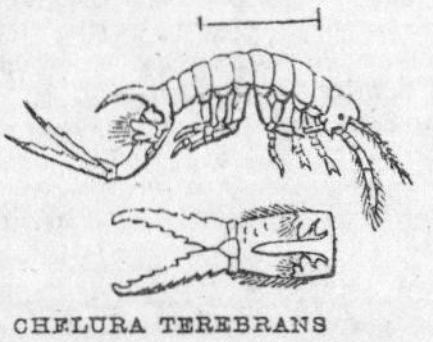

CHELURA TEREBRANS

work in sea water. This species, or a very closely allied one, has been found at Ardrossan, in Ayrshire, by Major Martin, and in Dublin Bay, Ireland, by Dr. Allman and Mr. Thompson. It may prove nearly as destructive as the *Limnoria terebrans* [which see].

CHENALOPEX, or EGYPTIAN GOOSE. A genus of palmiped birds, allied to the Bernacle Geese, but distinguished by the length of its legs, and the small spur on the shoulder of the wing. The only known species (*Chenalopex Ægyptiaca*) is often figured on the

EGYPTIAN GOOSE.
(CHENALOPEX ÆGYPTIACA.)

Egyptian monuments: it is a very common bird in aviaries, where it proves very attractive by its pretty colouring, elegant form, and the ease with which it is kept in confinement. It is a native of the South of Europe, abounding in Sicily, for example; and in N. Africa it is an abundant species, especially in the Valley of the Nile.

CHERMES. A genus of four-winged insects, which, like those of the genus *Aphis*, are found on the leaves, young shoots, and bark of various trees and vegetables. They derive their particular distinctions from the plants or trees on which they feed; as the ash, alder, elm, box, willow, nettle, &c. The abdomen is pointed, and the legs are formed for leaping. In their larva state many of them appear coated, especially on the hind part of the body, with a flocculent or filamentous clammy substance, of a white colour, which exudes from their pores.

CHEUCAU. (*Pteroptochos rubecula.*) This curious bird frequents the most gloomy and retired spots within the damp forests of the islands forming the Chonos archipelago. Sometimes, although its cry may be heard close at hand, let a person watch ever so attentively, he will not see the Cheucau; at other times, let him stand motionless, and the red-breasted little bird will approach within a few feet, in the most familiar manner. It then busily hops about the entangled mass of rotting canes and branches, with its little tail cocked upwards. Mr. Darwin opened the gizzard of some specimens: it was very muscular, and contained hard seeds, buds of plants, and vegetable fibres, mixed with small stones. The Cheucau is held in superstitious fear by the Chilotans, on account of its strange and varied cries.—*Darwin's Journal.* [See BARKING-BIRD.]

CHEVALIER. (*Totanus glottis.*) This Grallatorial bird, which is called by some naturalists the Greenshank, and by others the Green-legged Horseman, is about twelve inches in length, and stands very high on its legs. The bill is long, reddish near the tip, and black near the base; in summer the top of its head and nape are longitudinally rayed with deep black and white; the forehead, throat, fore part of the neck, breast, upper part of the belly and the sides are white, sprinkled with oval dusky spots; the rest of the under parts are pure white, except the under tail-coverts, which have the feathers spotted with black in the direction of the shafts: the greater wing-coverts, and the long feathers which protect the quills, are of a reddish grey, spotted with black; the rest of the coverts are edged with white, which is followed by a band of brown: the two intermediate tail-feathers are ash-coloured, with transverse brown zig-zag stripes. It frequents lakes, meadows, and the margins of rivers; and its flesh is very delicate and well-flavoured.

CHEVROTAIN. [See MUSK DEER.]

CHILOGNATHA. The first division of *Myriapoda.* The body is crustaceous, and often cylindrical, the antennæ rather thickened at the tips: two thick mandibles without palpi, distinctly divided into two portions; legs very short, and always terminated by a single claw. They crawl very slowly, or rather glide along, rolling themselves into a spire or ball. The first segment of the body, and in some also the second, is largest, and represents a corselet or small shield. It is only at the fourth, fifth, or sixth segment, in different species, that the duplication of the legs commences; and the two or three terminal segments are destitute of feet. On this family and the following, the late George Newport, F.R.S., published valuable monographs in the Transactions of the Linnæan Society. [See IULUS and CENTIPEDE.]

CHILOPODA. A division of the class *Myriapoda.* They are characterized by antennæ thick at the base, and gradually growing slender towards the apex; the mouth consists of two mandibles, which are furnished with a palpiform process, and provided at the apex with numerous little denticulations; covering these is an upper and an under lip; above which are two palpi, resembling legs by being terminated by a pointed claw; and covering this under lip is an organ furnished with two lateral processes, each of which is terminated by a large bent claw, through the under part of which a poisonous liquid is said to be ejected. The body is somewhat flattened, composed of numerous segments, defended by plates of a horny substance, and each segment generally furnished with a pair of legs. In hot climates they grow very large, and, from their venomous bite, some of them are truly formidable. They conceal themselves under stones and fallen trees, and are all found in rotten wood. They are nocturnal in their habits, very active, and some emit a phosphoric light. [See SCOLOPENDRA; CENTIPEDE, &c.]

CHIMÆRA. There are two species of this very singular kind of Chondropterygious fish, the Northern and the Southern Chimæra; each named after the ocean it inhabits.

NORTHERN CHIMÆRA.—(C. BOREALIS.)

The NORTHERN CHIMÆRA (*Chimæra Borealis*), generally abides in the deepest recesses of the sea, and is supposed to prey on the smaller fishes, as well as on the various sorts of Mollusca and Testacea. Its usual length is from three to four feet; the body is long, compressed, and gradually tapering towards the tail, which is continued into a long and slender filament: the head is very large and thick, rising up in front into a kind of pyramidal form; and at the top of the head, in the male fish, is a short upright process resembling a tuft. The mouth, placed beneath, is furnished in each jaw with a pair of broad, bony laminæ, notched in the margin into a resemblance of numerous teeth; while in front, both above and below, stand two large sub-triangular teeth: the upper lip is divided into five clefts; the front, from the mouth to the eyes, is marked by transverse undulations and pores; a line runs across the forehead, and is continued in a serpentine course into the lateral line, which is very strongly marked, of a whitish colour, with dark edges, and runs to the tip of the tail: the eyes are very large and bright, of a greenish colour, with silvery irides. The body above the lateral line is of a yellowish brown, and of a bright silver colour beneath it, variegated with numerous irregular spots. The fins are yellowish-brown, varied with darker shades: the first dorsal and the pectoral fins are large and subtriangular; the ventral, similarly shaped, are smaller; and at the base of each, in the males is a lengthened sub-cylindric process, roughened by numerous sharp prominences in a reversed direction. The flesh of the Chimæra is coarse, and unfit to be eaten.—The SOUTHERN CHIMÆRA (*Chimæra Australis*) is nearly of the size of the preceding species, but with the front sloping downwards, and the upper lip extended into a lengthened cartilaginous flap, bending downwards in a reversed direction beneath: general colour of the whole fish silvery, with a yellowish-brown cast on the upper parts: fins pale brown. Its manner of life is similar to that of the *Chimæra Borealis* in the Northern Hemisphere.

CHIMNEY SWEEPER [MOTHS]. A name given by collectors to Moths of the genus *Fumea.*

CHIMPANZEE. (*Pithecus troglodytes.*) Cuvier placed the Oran-Outang foremost in the rank of QUADRUMANA, but later naturalists consider the Gorilla and Chimpanzee to approximate more nearly in their general conformation to man. [See GORILLA in SUPPLEMENT.] The Chimpanzee is an in-

CHIMPANZEE.—(PITHECUS TROGLODYTES.)

habitant of Africa, and especially of the coasts of Congo and Angola; and travellers who have visited those countries assure us that in an adult state the Chimpanzees attain the stature of man, and live in society in the woods; that they construct huts of the leaves and branches of trees, to protect themselves against the extreme heat of the sun and the violence of the rains; that they walk upright, arm themselves with clubs, and make a formidable resistance against the attacks of the largest and most powerful beasts. The body of the Chimpanzee is covered with long black hair on the head, shoulders, and back, but much thinner on the breast and belly; the arms and legs are not so disproportionate as those of the Oran-Outang, the fore-fingers not quite touching the knees when the animal stands upright. The upper part of the head is very flat, with a retiring forehead, and a prominent bony ridge over the eye-brows; the mouth is wide, the ears large, the nose flat, and the face of a blackish brown colour. There was formerly a female Chimpanzee in the Zoological Gardens, Regent's Park, supposed to be about ten years old: she was remarkably docile, and performed certain actions with much apparent rationality.

How truly has it been said, that although the gradations of Nature in the other parts of her works are minute and imperceptible, yet in the ascent from brutes to men the line is strongly drawn, well marked, and decisive! In vain the Chimpanzee or Oran-Outang may resemble Man in form, or may possess the power of imitating his actions, it still continues a wretched helpless creature; and whatever distant resemblance its internal conformation may bear to the human, its whole figure exhibits a picture sufficiently mortifying to those who pride themselves on personal appearances alone. The tongue and all the organs of voice may be the same, yet the animal is dumb; the brain may be formed in the same manner, yet the animal is destitute of reason: an evident proof, as Buffon beautifully observes, that no disposition of matter can constitute a mind; and that the body, how nicely soever constructed, is constructed in vain, when no soul is infused into it for the purpose of directing its operations.

Mr. Newman furnishes the readers of the "Zoologist" (1845) with the subjoined particulars: "A larger, stronger, and more active Chimpanzee than any previously imported, was lately consigned to Messrs. Coleman, Flockhart, and Co., from the river Nunez, near Sierra Leone. On its arrival in the London Docks I paid it a visit, and immediately communicated with Mr. Yarrell, with a view to obtaining it for the Zoological Society: the officials, however, were already on the alert, and the creature has since been purchased by the society for 300*l*. The following paragraph, which has been circulated in the London newspapers, was, I hear, penned by one of the keepers:—'It is singular that she resists every attempt to correct her, fighting with the utmost determination; every other animal, even the Ourang, fears its keeper. The first day of the Chimpanzee's arrival at the Gardens, she tore out three of the strong iron bars of her cage, which have been since strengthened. A temporary nail was driven about half its length, into a piece of wood, about six inches long and three and a half square; she held the wood between her teeth, and doubling the nail backwards and forwards, broke it short off. When in a passion, she tears her hair and rolls herself on the ground violently. Her table is supplied from her keeper's, and she shares in everything and anything he has. She eats her egg with a spoon, takes her grog daily, and, 'tis said, that when on board ship she mixed the latter herself. She will lock and unlock a door or drawer; will thread any needle; she cannot be taken in by the same thing twice, and will imitate almost anything that is done before her. She is considered by Professor Owen to be about nine years old, which well agrees with all accounts of her previous life. She weighs 52 lbs.; measures 2 feet 2 inches round the chest, and is 3 feet 2 inches high; or, as she will not stand upright to be measured, probably her height is nearly 3 feet 6 inches.' On making a more careful examination of this animal in her present abode, I was particularly struck by her want of teeth. Only one incisor and a few imperfect molars appear to remain. I observed her total inability to crack a nut, a feat performed by almost every other monkey with great adroitness. Her manners now are perfectly quiet, and there is no appearance of the ferocity implied in the preceding quotation; she was gentle in the extreme, shaking hands in a very cordial manner with some children who were present, and perfectly on the alert at the sound of her name—"Susan"—whenever it was uttered. I presume the keeper imagined that details of her ferocity would give her an interest in the eyes of the public. I have observed that the captains of Margate steamers always tell their passengers that the present is the roughest passage they ever encountered; so the visitors of this gentle being are assured it is the most savage Chimpanzee. The Captain, to whose care "Susan" was entrusted, told me that in taking her meals on the passage home, she used knife, fork, spoon, and drinking cup, with the same ease as a human being; and with whatever food she was supplied, she preferred using a fork or a spoon to convey it to her mouth, to holding it in her hands. For more than three years she had been in possession of a Mr. Campbell, who left her at perfect liberty, never subjecting her to the slightest confinement. When he received her she was quite young—a mere baby, so that her present age may be supposed four or five years, rather than eight or nine. When on board ship she entertained a great dislike to black men, who used to tease and otherwise misuse her; but with the crew generally she appeared on excellent terms, and exhibited many traits of extreme docility."

CHINA-MARK [MOTHS]. A name given by collectors to Moths of the genera *Hydrocampa* and *Margaritia*.

CHINCHILLA. This little Rodent animal, so highly valued on account of its fur, is a native of South America, inhabiting the valleys in the high mountain districts, where the cold is often very severe. There are several species belonging to the natural family *Chinchillidæ*, of which this animal and the Viscacha of the Pampas are the chief. The colour of the Chinchilla is clear grey above,

CHINCHILLA.—(C. LANIGERA.)

passing into white on the under parts. It associates in numbers, and excavates burrows, in which it resides, feeding chiefly upon roots. In size and general form it much resembles the rabbit, with the exception of the tail, which turns up after the manner of a squirrel's. The fur is of a remarkably close and fine texture, and is, accordingly, much used in muffs, tippets, linings to cloaks, trimmings, &c.

CHIRONOMIDÆ. A sub-family of Dipterous insects, which frequent marshy situations, and very much resemble gnats. The species are of small size and very numerous; they often assemble in immense cloud-like swarms; and the name of *Midge* is given to them.

CHITON. A genus of marine Mollusca, inhabiting multivalve shells, several species of which are found on our own coasts. They adhere to rocks and stones, in general, near low-water mark. The shell is boat-shaped, composed of about eight transverse pieces, folding over each other at their edges, and inserted into a tough ligament. They sometimes attain a large size, but do not usually exceed two inches. They have the power of rolling themselves up into a ball, like the wood-louse. Several new species of these shells were collected in the Eastern Archipelago by Sir Edward Belcher and Mr. Adams during the voyage of H. M. S. Samarang; two of which, viz., *Chiton petasus*, which is described as a beautiful little bright

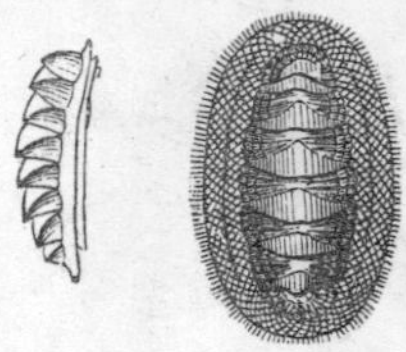

SCALY ARMADILLO-SHELL.
(CHITON SQUAMOSUS.)

scarlet shell enframed within a broad swollen ligament of the same striking colour; and *Chiton formosus*,—a most exquisite little species, of a bright scarlet colour, surrounded with dense tufts of white shining glassy spiculæ. For a revision of this genus, see a paper by Mr. Gray, recently published. Mr. L. Reeve has published figures of many of the species in his "Conchologia Iconica." Mr. Cuming collected a very great number of beautiful species, many of which are in the fine collection of the British Museum; and it is to be hoped that the Government will procure for the National Museum the whole of Mr. Cuming's magnificent collection of shells.

CHLAMYDOSAURUS. A genus of Saurians, described by Mr. Gray, from a specimen discovered in Australia by the late Allan Cunningham, F.L.S., who (between the years 1818 and 1822) accompanied Capt. King's expedition as His Majesty's botanical collector for Kew Gardens. It was taken on the branch of a tree, and sent to Sir Everard Home, by whom it was deposited in the Museum of the Royal College of Surgeons. In Mr. Cunningham's Journal, it is described as a lizard of extraordinary appearance, having a curious crenated membrane, like a ruff or tippet round its neck, covering its shoulders, and when expanded, which it was enabled to do by means of transverse slender cartilages, spreading five inches in the form of an open umbrella. Its head was large, and its eyes, whilst living, rather prominent; its tongue, though bifid, was short, and appeared to be tubular. From Mr. Gray's description of the *Chlamydosaurus Kingii* (the Frilled Lizard), in the Appendix to Capt. King's Voyage, we learn that the animal was scaly; colour yellowish brown, variegated with black; head depressed, with the side erect, leaving a blunt ridge on the upper part wherein the eyes are placed. The frill arises from the hinder part of the head, is attached to the sides of the neck, and extends down to the front part of the chest, supported above by a lunate cartilage arising from the hinder dorsal part of the ear, and in the centre by a bone which extends about half its length. Each frill has four plates which converge on the under part of the chin, and fold it up on the side, and a fifth where the two are united in the centre of the lower part of the neck. The front part of its upper edge is elegantly serrated, and the outer surface is covered with carinated scales; the inner surface being quite smooth. The scales of the back are oval; those of the lower part of the body and upper part of the legs have a short mid-rib, and those of the sides and joints of the limbs are minute. The body is five inches long, the tail twelve, the head nearly six, and the outer edge of the frill ten inches: the toes are long, very unequal, compressed, and scaly: the claws are hooked, and horn-coloured. This frill increases in size more in proportion than the animal's growth; in the young it does not reach to the base of the fore limbs, while in the adult it becomes much fuller, and reaches considerably beyond the axilla.

KING'S FRILLED LIZARD.
(CHLAMYDOSAURUS KINGII.)

It would seem to be not uncommon about Port Essington; and it is found in other parts of Australia. Captain George Grey (now governor of New Zealand) met with it, and gives us the following interesting notice of its habits in the first volume of his Travels. He says, "As we were pursuing our route in the afternoon, we fell in with a specimen of the remarkable Frilled Lizard; this animal measures about twenty-four inches from the tip of the nose to the point of its tail, and lives principally in trees, although it can run very swiftly along the ground: when not provoked or disturbed, it moves quietly about, with its frill lying back in plates upon the body; but it is very irascible, and directly it is frightened, it elevates the frill or ruff, and makes for a tree; where, if overtaken, it throws itself upon its stern, raising its head and chest as high as it can upon the fore-legs, then doubling its tail underneath the body, and displaying a very formidable set of teeth, from the concavity of its large frill, it boldly faces any opponent, biting fiercely whatever is presented to it, and even venturing so far in its rage as to fairly make a fierce charge at its enemy. We repeatedly tried the courage of this lizard, and it certainly fought bravely whenever attacked. From the animal making so much use of this frill as a covering and means of defence for its body, this is most probably one of the uses to which Nature intended the appendage should be applied. The whole animal is fulvous, obscurely varied with brown; the young being more distinctly marked with regularly waved black streaks, forming broad bands across the back, limbs, and tail."

CHLAMYPHORUS. An edentate quadruped, found in South America, in which several characters of different tribes are remarkably blended. Like the Armadillo, it has a tesselated shield, the consistence of which is between horn and leather; but instead of being firmly attached by its whole under surface to the integuments beneath, it is connected with the back only by a ridge of skin along the spine, and with the skull by two bony prominences from the forehead. In the form of its feet, its imperfect eyes, the conical shape of its snout, and its general habits, it resembles the mole. It is a native of Chili, but is so rare even there as to be regarded by the natives as a curiosity. The total length of the entire animal is five inches and a quarter. The shelly covering is composed of a series of plates of a square, rhomboidal, or cubical form, each row separated by a membranous substance, which is reflected above and beneath, over the plates: the rows include from fifteen to twenty-two plates, the shell being broadest at its posterior half, extending about one half round the body. This covering is loose throughout, except along the spine of the back and top of the head. The number of rows of plates on the back, counting from the vertex, where they commence, is twenty-four; the shell then curves suddenly downwards, so as to form a right angle with the body: this truncated surface is composed of plates, nearly similar to those of the back, and are disposed in semicircular rows; the lower margin, somewhat elliptical, has a notch in its centre, in which is attached the free portion of the tail, which curves abruptly, and runs beneath the belly parallel to the axis of the body, the extremity of the tail being depressed, so as to form a paddle. The superior semicircular margin of the truncated surface, together with the lateral margins of the shell, are beautifully fringed with silky hair.

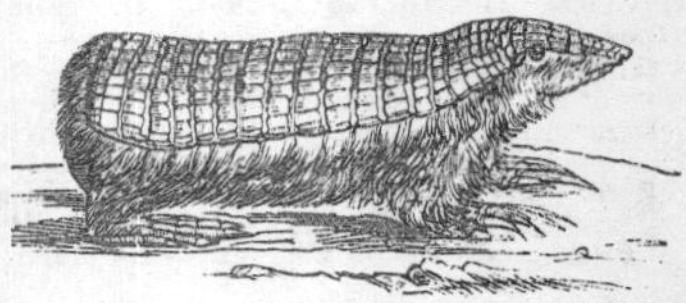

CHLAMYPHORUS TRUNCATUS

The following points of resemblance between the *skeleton* of *Chlamyphorus* and that of other quadrupeds have been noticed by Mr. Yarrell:—1. Beaver (*Castor fiber*), in the form and substance of some of the bones of the limbs, in the flattened and dilated extremity of the tail, and the elongation of the transverse processes of the lower caudal vertebræ. 2. Mole (*Talpa Europea*), in the shortness and great strength of the legs, and

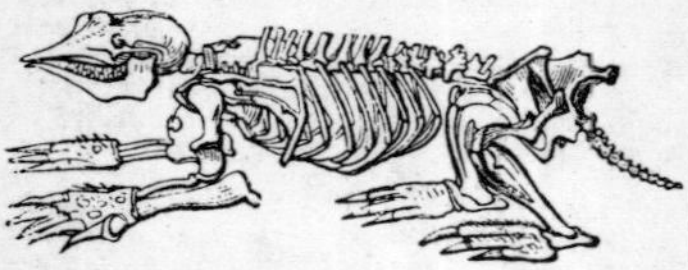

SKELETON OF CHLAMYPHORUS TRUNCATUS.

in the articulation of the claws to the first phalanges of the toes. 3. Sloth (*Bradypus tridactylus*), in the form of the teeth, and in the acute descending process of the zygoma. 4. Armadillo (*Dasypus*), in the coat of mail, in the peculiar ossification of the cervical vertebræ, in possessing the sesamoid bones of the feet, and in the general form of the bones, except those of the pelvis. 5. *Orycteropus Capensis* and *Myrmecophaga jubata*, in some of the bones. 6. *Echidna* and *Ornithorhynchus*, in the form of the first bone of the sternum, and in the bony articulations as well as the dilated connecting plates of the true and false ribs. 7. and 8. *Ruminantia* and *Pachydermata*, in the form of the lower jaw, &c. The unique points in its osteological structure appear to be the form of the head and the open pelvis. Dr. Buckland considers *Chlamyphorus* one of the nearest approximations to *Megatherium*, particularly in regard to its coat of mail, and in the adaptation of the animal for digging.

Dr. Harlan, who first described this remarkable animal, says, "We have been presented in the subject before us with a *new form;* an animal combining in its external configuration a mechanical arrangement of parts which characterizes, respectively, the armadillo, the sloth, and the mole; constituting in themselves, individually and separately, of all other quadrupeds, those which offer the most remarkable anatomical characters. * * * The structure of this animal, Dr. Harlan goes on to say, taken collectively, furnishes us with an example of *organic structure*, if not unparalleled, not surpassed in the history of animals."—*Ann. New York Lyceum*, p. 245.

CHOCOLATE-TIP [MOTHS]. A name given by collectors to Moths of the genus *Clostera*.

CHONDROPTERYGII. The term for one of the great classes or families of fishes; characterized by the cartilaginous nature of the spines and bones. Cuvier divides the *Chondropterygii* into two orders,—those which have their gills free, as in the generality of fishes, and those in which they are fixed.

CHŒROPOTAMUS. [See APPENDIX.]

CHOUGH (CORNISH), or RED-LEGGED CROW. (*Pyrrhocorax graculus.*) A bird somewhat taller and longer than the Jackdaw, whose habits it in many respects resembles. Its colour is a beautiful black, glossed with blue and purple: the bill is long, curved, sharp at the tip, and of a bright orange-red; the legs are of a similar colour, with black claws. It builds on high cliffs, by the sea side, lays four or five eggs, spotted with yellow, and chiefly frequents the coasts of Cornwall, Devonshire, and Wales, though it is sometimes found on the cliffs of Dover, in Scotland, and the Hebrides. In a wild state it feeds principally on insects and berries. It is easily tamed, becomes extremely docile, and is very fond of being caressed by those to whom it shows an attachment, but its shrill notes and mischievous qualities render it sometimes a troublesome inmate. It also becomes bold and pugnacious, and resents an affront with violence and effect.

CHRYSIDIDÆ, or GOLDEN WASPS. A family of Hymenopterous insects, most of which seek the nests of other insects, wherein to deposit their eggs. They are generally distinguished by a peculiar brilliancy of colour, are very active, and are seen flying

CHRYSIS IGNITA.

about in the sunshine, settling upon old walls, palings, &c. The most common, and at the same time most beautiful British species, is the *Chrysis ignita:* it is about the size of the common window fly, and is of a rich deep blue-green colour on the head and thorax, with the abdomen of a burnished golden-copper hue.

CHRYSOCHLORIS, or CAPE MOLE. A Rodent quadruped very much resembling the mole in general structure and habits. There is no external ear, nor any appearance of the eye externally: the body is thick and short; and the claws are particularly well adapted for digging and burrowing in the earth: but it is chiefly distinguished by the splendid colours of its fur, and is the only known quadruped which exhibits anything like the metallic lustre that adorns numerous birds, fishes, and insects. The best known species (*Chrysochloris Capensis*) is, as the name implies, a native of the Cape of Good Hope.

CHRYSOMELA: CHRYSOMELIDÆ. An extensive genus and family of Coleopterous insects, generally of a small or moderate size, and frequently ornamented with the most brilliant colours, amongst which blue, green, and gold are pre-eminently conspicuous. The antennæ are moniliform, thickening towards the tip; the thorax margined; and the body ovate, oblong, or subhemispherical.—*Chrysomela Graminis* is a common but highly elegant insect, of a most vivid, but deep golden-green colour; shape extremely convex.—*Chrysomela Betulæ*, found on birch-trees, is one of the richest of the genus, being entirely of the most brilliant and beautiful grass-green. The species of the genus Chrysomela, and others separated therefrom, are distinguished by the possession of wings, and an oval or rounded body. Among these the *Chrysomela Populi* is one of the most common species. It is of a blue-black colour, with red elytra, tipped with black. It is found upon the willow and poplar. Its larva is of an oblong-ovate form, of a dirty greenish-white colour, with numerous black scaly spots; its meso and metathoracic segments are furnished with two large lateral conical tubercles, and the abdominal segments have also two rows of smaller dorsal and lateral tubercles, from which, as well as from the joints of the legs and mouth, drops of a fetid fluid are emitted when the larva is alarmed. The eggs are deposited upon the leaves in clusters. The pupa is ovate, having the excuviæ of the larvæ collected in a mass at the extremity of the body. The larvæ of some species of this family feed, in society, upon leaves, preserving one or more most orderly rows. Among the most elegant species found in the United States of North America (according to Dr. Harris) is the *Chrysomela scalaris* of Leconte, literally the ladder Chrysomela. The head, thorax, and under side of its body are dark green, the wing-covers silvery white, ornamented with small green spots on the sides, and a broad jagged stripe along the suture or inner edges; the antennæ and legs are rust-red; and the wings are rose-coloured. It is a beautiful object when flying, with its silvery wing-covers embossed with green, raised up, and its rose-red wings spread out beneath them. These beetles inhabit the elm and lime trees, upon which they may be found in April, May, and June, and a second brood of them in September and October. They pass the winter in holes, and under leaves and moss. The trees on which they live are sometimes a good deal injured by them and their larvæ. The latter are hatched from eggs laid by the beetles on the leaves in the spring, and, when full grown, are about half an inch long, of a white colour, with a black line along the top of the back, and a row of small square black spots on each side of the body; the head is horny, and of an ochre-yellow colour; the body is short and very thick, the back arching upwards in the middle.

CHUB. (*Cyprinus cephalus.*) This fish is a native of many parts of Europe, and is

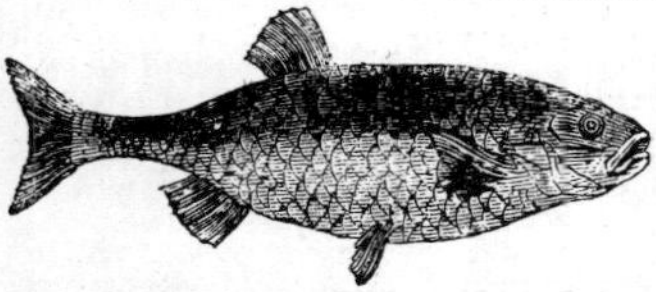

CHUB.—(CYPRINUS CEPHALUS.)

not uncommon in our own island. It frequents the deep holes of rivers, and, during the summer season, commonly lies on the surface of the water, beneath some tree or bush. In shape the Chub rather resembles the Tench, but is of a more lengthened form, and has a larger head in proportion. It is from fourteen to eighteen inches in length; its colour silvery, with a bluish olive cast on its upper parts; the sides bluish white, passing into silvery white on the belly; the scales very large, and the lateral line nearly straight; the dorsal fin is rather small, and situated on the middle of the back; the pectoral fins are of a pale yellow; the ventral and anal fins are red; and the tail is slightly forked, and of a dull bluish-brown colour. It feeds on worms, caterpillars, grasshoppers, beetles, and other coleopterous insects which happen to fall into the water.

CICADA. The family of insects bearing the generic name *Cicadæ*, or *Cicadidæ*, are nearly all inhabitants of tropical or the warmer temperate regions. The most common European species is the *Cicada plebeia* of Linnæus; an insect often commemorated by the ancient poets, but generally confounded by the major part of translators with the Grasshopper. It is a native of the warmer parts of Europe, particularly of Italy and Greece; appearing in the hotter months of summer, and continuing its shrill chirping during the greatest part of the day, generally sitting among the leaves of trees. These insects proceed from eggs deposited by the

MANNAH FLY.—(CICADA ORNI.)

parent in and about the roots of trees, near the ground; and after having remained in the larva state nearly two years, cast their skins, and produce the complete insect.

The male Cicada produces a loud chirping note, and much has been written in praise of it by Anacreon and other ancient authors; it is certain, however, that modern ears are offended rather than pleased with its voice, which is so very strong and stridulous that it fatigues by its incessant repetition. That a sound so piercing should proceed from so small a body may well excite our astonishment; and the curious apparatus by which it is produced has justly claimed the attention of the most celebrated investigators. They have found that it proceeds from a pair of concave membranes, seated on each side the first joints of the abdomen: the large concavities of the abdomen, immediately under the two broad lamellæ in the male insect, are also faced by a thin, pellucid, iridescent membrane, serving to increase and reverberate the sound; and a strong muscular apparatus is exerted for the purpose of moving the necessary organs.

Among the large and elegant insects in this division is the *Cicada hæmatodes*, distinguished by its shining black body, with the divisions of the abdomen marked by numerous scarlet rings or bands; and the *Cicada viridis*, a large species, native of New Holland, of a beautiful green colour, with the transparent wings ornamented by green veins.

Cicada septendecim, or Seventeen-year Cicada. It is well remarked by Dr. Thaddeus Harris, that "the duration of life in winged insects is comparatively very short, seldom exceeding two or three weeks in extent, and in many is limited to the same number of days or hours. To increase and multiply is their principal business in this period of their existence, if not the only one, and the natural term of their life ends when this is accomplished. In their previous states, however, they often pass a much longer time, the length of which depends, in great measure, upon the nature and abundance of their food." The *harvest-flies* continue only a few weeks after their final transformation, and their only nourishment consists of vegetable juices, which they obtain by piercing the bark and leaves of plants with their beaks; and during this period they lay their eggs and then perish. They are, however, amply compensated for the shortness of their life in the winged state by the length of their previous existence, during which they are wingless and grub-like in form, and live under ground, where they obtain their food only by much labour in perforating the soil among the roots of plants, the juices of which they imbibe by suction. To meet the difficulties of their situation and the precarious supply of their food, a remarkable longevity is assigned to them; and one species has obtained the name of *Cicada septendecim*, on account of its life being protracted to the period of seventeen years. This insect, in the perfect state, is of a black colour, with transparent wings and wing-covers, the thick anterior edge and larger veins of which are orange-red, and near the tips of the latter there is a dusky zigzag line in the form of the letter W; the eyes when living are also red; the rings of the body are edged with dull orange; and the legs are of the same colour. The wings expand from two inches and a half to three inches and a quarter.

In those parts of the United States, as we are informed, which are subject to the visitation of this Cicada, it may be seen in forests of oak about the middle of June. Here such immense numbers are sometimes congregated, as to bend and even break down the limbs of the trees by their weight, and the woods resound with the din of their discordant drums from morn to eve. After pairing, the females proceed to prepare a nest for the reception of their eggs. They select, for this purpose, branches of a moderate size, which they clasp on both sides with their legs, and then bending down the piercer at an angle of about forty-five degrees, they repeatedly thrust it obliquely into the bark and wood in the direction of the fibres, at the same time putting in motion the lateral saws, and in this way detach little splinters of the wood at one end, so as to form a kind of fibrous lid or cover to the perforation. The hole is bored obliquely to the pith, and is gradually enlarged by a repetition of the same operation, till a longitudinal fissure is formed of sufficient extent to receive from ten to twenty eggs. The side pieces of the piercer serve as a groove to convey the eggs into the nest, where they are deposited in pairs, side by side, but separated from each other by a portion of woody fibre, and they are implanted into the limb somewhat obliquely, so that one end points upwards. When two eggs have been thus placed, the insect withdraws the piercer for a moment, and then inserts it again and drops two more eggs in a line with the first, and repeats the operation till she has filled the fissure from one end to the other, upon which she removes to a little distance, and begins to make another nest to contain two more rows of eggs. She is about fifteen minutes in preparing a single nest and filling it with eggs; but it is not unusual for her to make fifteen or twenty fissures in the same limb; and one observer counted fifty nests extending along in a line, each containing fifteen or twenty eggs in two rows, and all of them apparently the work of one insect. After one limb is thus sufficiently stocked, the Cicada goes to another, and passes from limb to limb and from tree to tree, till her store, which consists of four or five hundred eggs, is exhausted. At length she becomes so weak by her incessant labours to provide for a succession of her kind, as to falter and fall in attempting to fly, and soon dies.

Although the Cicadas abound most upon the oak, they resort occasionally to other forest-trees, and even to shrubs, when impelled by the necessity for depositing their eggs, and not unfrequently commit them to fruit-trees, when the latter are in their vicinity. Indeed there seem to be no trees or shrubs that are exempted from their attacks, except those of the pine and fir tribes, and of these even the white cedar is sometimes invaded by them. The punctured limbs languish and die soon after the eggs which were placed in them are hatched; they are broken by the winds or by their own weight, and either remain hanging by the bark alone, or fall with their withered foliage to the ground. In this way orchards have suffered severely in consequence of the injurious punctures of these insects. The eggs are one twelfth of an inch long, and one sixteenth of an inch through the middle, but taper at each end to an obtuse point, and are of a pearl-white colour. The shell is so thin and delicate that the form of the included insect can be seen before the egg is hatched.

The young insect when it bursts the shell is one sixteenth of an inch long, and is of a yellowish white colour, except the eyes and the claws of the fore-legs, which are reddish; and it is covered with little hairs. In form it is somewhat grub-like, being longer in proportion than the parent insect, and is furnished with six legs, the first pair of which are very large, shaped almost like lobster-claws, and armed with strong spines beneath. On the shoulders are little prominences in the place of wings; and under the breast is a long beak for suction. These little creatures when liberated from the shell are very lively, and their movements are nearly as quick as those of ants. After a few moments their instincts prompt them to get to the ground, but in order to reach it they do not descend the body of the tree, neither do they cast off themselves precipitately; but running to the side of the limb, they deliberately loosen their hold, and fall to the earth. The instinct which impels them thus fearlessly to precipitate themselves from the trees, from heights of which they can have formed no conception, without any experience or knowledge of the result of their adventurous leap, is still more remarkable than that which carries the gosling to the water as soon as it is hatched. In those actions, that are the result of foresight, of memory, or of experience, animals are controlled by their own reason, as in those to which they are led by the use of their ordinary senses or by the indulgence of their common appetites they may be said to be governed by the laws of their organization; but in such as arise from special and extraordinary instincts, we see the most striking proofs of that creative wisdom which has implanted in them an unerring guide, where reason, the senses, and the appetites would fail to direct them. On reaching the ground the insects immediately bury themselves in the soil, burrowing by means of their broad and strong fore-feet, which, like those of the mole, are admirably adapted for digging. They do not appear ordinarily to descend very deep into the ground, but remain where roots are most abundant. The only alteration to which they are subject during the long period of their subterranean confinement, is an increase of size, and the more complete development of the four small scale-like prominences on their backs, which represent and actually contain their future wings.

As the time of their transformation approaches, they gradually ascend towards the surface, making in their progress cylindrical passages, oftentimes very circuitous, and seldom exactly perpendicular, the sides of which are firmly cemented and varnished so as to be waterproof. When the insect has nearly approached the surface it takes up its temporary habitation till the period for its exit arrives. Here it remains during several days, ascending to the top of the hole in fine weather for the benefit of the warmth and the air, and occasionally peeping forth apparently to reconnoitre, but descending again on the occurrence of cold or wet weather. When at length a favourable moment arrives for them to come forth from their subterranean retreats, they issue from the ground in great numbers in the night, crawl up the trunks of trees, or upon any other object to which they can fasten themselves securely by their claws. After having rested awhile they prepare to cast off their skins, which, in the mean time, have become dry and of an amber colour. By repeated exertions a longitudinal rent is made in the skin of the back, and through this the included Cicada pushes its head and body, and withdraws its wings and limbs from their separate cases, and, crawling to a little distance, it leaves its empty pupa skin, apparently entire, still fastened to the tree. At first the wing-covers and wings are very small and opaque, but, being perfectly soft and flexible, they soon stretch out to their full dimensions, and in the course of a few hours the superfluous moisture of the body evaporates, and the insect becomes strong enough to fly. During several successive nights the pupæ continue to issue from the earth; above 1500 have been found to arise beneath a single apple tree, and in some places the whole surface of the soil, by their successive operations, has appeared as full of holes as a honeycomb. Within about a fortnight after their final transformation they begin to lay their eggs, and in the space of six weeks the whole generation becomes extinct. Fortunately these insects are appointed to return only at periods so distant, that vegetation often has time to recover from the injury inflicted by them. They have also many enemies, which contribute to diminish their numbers. Their eggs are eaten by birds; the young, when they first issue from the shell, are preyed upon by ants, which mount the trees to feed upon them, or destroy them when they are about to enter the ground. Blackbirds eat them when turned up by the plough in fields, and hogs are excessively fond of them, and, when suffered to go at large in the woods, root them up, and devour immense numbers just before the arrival of the period of their final transformation, when they are lodged immediately under the surface of the soil. We may mention that one species has been found in this country, where, however, it is rare; it has been called *Cicada Anglica*, but seems not to be distinct from a common European species.

CICADIDÆ. The first family of Homopterous insects, in the section TRIMERA, and corresponds with the *Cicadæ mannifera* of Linnæus. It embraces the largest insects in the order, one species measuring between six and seven inches in the expanse of its wings. [See CICADA.]

CICINDELA: CICINDELIDÆ; or TIGER BEETLES. A genus and family of Coleopterous insects remarkable for the celerity and vigour of their flight; characterized by the great projection of the eyes, long and sharply pointed jaws; thorax depressed and nearly square; and the legs and antennæ long and slender. They are generally seen on the wing in the hottest part of the day, chiefly frequenting dry meadows, sandy plains or heaths, or the banks of rivers. One of the most striking genera is the *Manticora*, found at the Cape of Good Hope. The common GREEN TIGER-BEETLE (*Cicindela campestris*), one of the most common European species, is a highly beautiful insect, being of a bright grass-green, with the elytra each marked by five small, round, cream-coloured spots: the head, thorax, and limbs are of a rich gilded cast; the eyes black and prominent; the legs long and slender. The larva of this insect lives in cylindrical burrows, excavated by itself, and varying from six inches to a foot in depth. The head is very large, and slightly concave; the jaws are curved and strong; and the

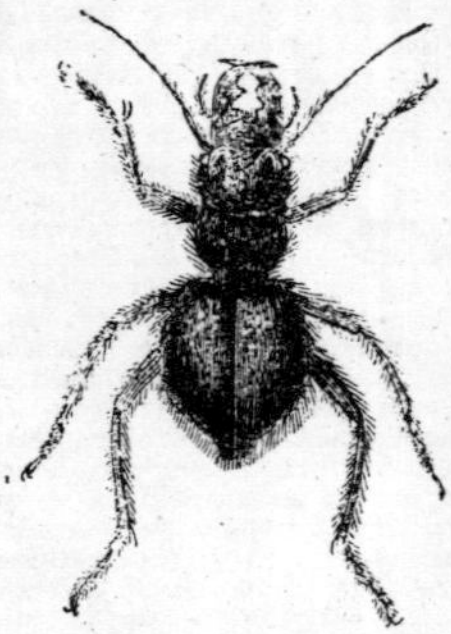

MANTICORA MAXILLOSA

body is humped near the middle of the back, at which part there are two hooked tubercles. In the process of excavation they use their jaws and feet, and load the concave back of their heads with the grains of earth which they have detached; thus loaded, they ascend backwards, resting at intervals, and fixing themselves to the inner walls of their burrow by the assistance of the two hooked tubercles on the back; and when arrived at the orifice, they jerk off their load to a distance. The Cicindelæ are all voracious; and when their prey comes within their reach, they rush upon it with great ferocity.

CILIOGRADA. An order of *Acalephæ*, or gelatinous transparent marine animals, distinguished by their continually agitating the cilia with which their contractile bodies are provided; organs which possess the phosphorescent faculty in a very high degree. [For examples, see BEROE, and MEDUSA.]

CIMBEX: CIMBICIDÆ. A genus and family of Hymenopterous insects, allied to *Tenthredinetæ*, or *Saw-flies* (as they are com-

CIMBEX VARIABILIS AND ITS LARVA.

monly called, from their saw-like ovipositor), comprising those species which have the antennæ alike in both sexes, and terminated by a knob or a reversed cone rounded at the tip, preceded by four or five joints, and the two subcostal nerves being contiguous without a wide intermediate space. The larvæ of these insects greatly resemble the Caterpillars of Lepidopterous insects, but have from eighteen to twenty-two feet, or only six, which distinguishes them from true caterpillars, which have from ten to sixteen feet. In order to undergo their change, they spin, either on the earth or on the plants upon which they have fed, a cocoon, in which, like the rest of the family, they remain unchanged for many months.

CIMEX. A Linnæan genus of Hemipterous insects, now subdivided into several families or sections, according to the general shape or habit of the insects, and severally named *Cimicidæ*, *Pentatomidæ*, *Cydnidæ*, *Coreidæ*, *Lygeidæ*, *Reduviidæ*, *Acanthidæ*, and *Hydrometridæ*: the two terminal joints of the antennæ of hair-like fineness; body much depressed; thorax transverse; antennæ four-jointed; labrum rather long and pointed, and when the proboscis is not in use, recurved under the head. The bed-bug (*Cimex lectularius*) may serve as a general example of this very extensive tribe. [See BUG.]

CIMOLIORNIS. [See APPENDIX.]

CINCLOSOMA. A genus of Passerine birds, belonging to the *Turdidæ* family. The species *Cinclosoma punctatum*, or Spotted Ground Thrush, inhabits Van Diemen's Land and Eastern Australia. It prefers the summits of low stony hills and rocky gullies, particularly those covered with scrubs and bushes. Its flight is very limited; but it readily evades pursuit by running over the stony surface and concealing itself among the underwood: when flushed suddenly, it rises with a loud whirring noise, like a Quail or Partridge. Its note consists of a low piping whistle. It is sold in Hobart-Town market, with Bronzewings, Pigeons, and Wattle-birds, and is known there as the Ground Dove: doubtless from its terrestrial habits and its flesh being excellent eating. To its delicacy, and the large development of the pectoral muscles, and the contour of the body, resembling a Quail, Mr. Gould gives his testimony. It breeds in October and three following months. The nest, which is always placed on the ground, is a slight and rather careless structure, composed of leaves and the inner bark of trees, and is of a round, open form. The stomach of this bird, on dissection, was found to contain seeds and caterpillars, mingled with sand. Another species, *Cinclosoma castanotus*, found near the Swan River, is a much shyer bird than the *C. punctatum*, and runs over the ground faster; its shorter toes considerably assisting its progressive motion.

CINCLUS. The Water-ouzel [which see].

CINNYRIS: CINNYRIDÆ. A genus and family of small birds, remarkable for the splendid metallic lustre of their plumage in which they rival the Humming-birds (*Trochilidæ*). All the species inhabit the

Old World; chiefly Africa and India. [See SUN-BIRD.]

CIRRIPEDIA, or CIRRIPEDES. A class of invertebrated animals, so named from the curled and ciliated branchiæ which protrude from the oval aperture of the shells. They are divided into sessile, that is, either themselves firmly united at their bases to rocks or solid masses; and pedunculated, or attached by a long peduncle or footstalk. They are closely allied to the Crustacea.

CISSITIS. A genus of Coleoptera. [See HORIADÆ.]

CISTELA: CISTELIDÆ. A genus and family of Coleopterous insects, belonging to the section *Heteromera*. They are characterized by antennæ nearly filiform, the

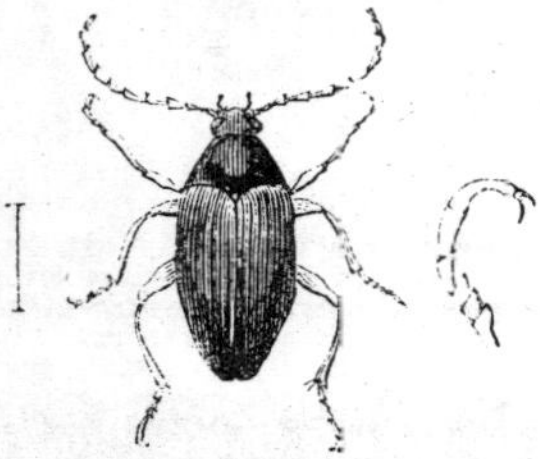

CISTELA SERRICORNIS

joints serrated; body ovoid, arched above; feet long, but none of the legs formed for leaping; penultimate joint of the tarsi bifid; mandibles entire. They generally live amongst leaves and flowers. They are an important group numerically, both as regards genera and species, several being found in this country.

CITILLUS. A small Rodent animal, of the genus *Spermophilus*, with a long thin body, short tail, and of a silvery grey colour. It is a native of the northern parts of Europe, and dwells in communities, great numbers of them being usually found together in the same cave, furnished with a store of nuts, chestnuts, &c. Their flesh is well flavoured, and their skins are much valued. [See SPERMOPHILUS.]

CIVET. (*Viverra civetta.*) This animal, popularly known by the name of the Civet-cat, belongs to a genus of carnivorous, mammiferous quadrupeds, and is a native of several parts of Africa and India. It is particularly distinguished by having a secretory glandular receptacle, situated at some little distance beneath the tail, wherein is formed a powerfully odorous matter called *civet*. In general appearance, this animal reminds one of the fox, which it also resembles in its predatory habits; but the legs are short, the tail is long, hairy, and cylindrical, and the claws, though by no means so acute as those of the cat, are still partially retractile. The ground colour of the body is yellowish-grey, with large dusky spots disposed in longitudinal rows on each side; and a sort of upright mane on the neck and back. The tongue is covered with stout, horny prickles; and the ears are straight and rounded at the tips. The pouch, situated near the genitals, is a deep bag, sometimes divided into two cavities, whence a thick, oily, and strongly musk-like fluid is poured out. When good, this odoriferous substance is of a clear yellowish or brown colour, and of about the consistence of butter; when undiluted, the smell is powerful and very offensive, but when largely diluted with oil or other ingredients, it becomes an agreeable perfume. Important medical virtues were formerly attributed to the civet; it, however, not only no longer forms an article in the Materia Medica, but even as a perfume it has been laid aside. The foregoing description will apply to another species, the *Viverra zibetha*, except that this has no mane: it should be observed also, that the *Viverra civetta* is peculiar to Africa, and the *zibetha* to Asia.

CLADOCERA. An order of minute Crustacea, characterized by the body being inclosed in a bivalve shell, including, among others, the genus Daphnia.

CLAM. The shell of a species of Conchiferous Mollusca. [See TRIDACNA.]

CLAUSILIA. A genus of Mollusca chiefly inhabiting mosses at the foot of trees. The species are very numerous, and they are all small shells, in shape somewhat resembling the pupa or chrysalis of an insect; the largest scarcely exceeding an inch in length. Within the mouth, in the last whorl but one, there is a little elastic shelly plate attached to the shell, and called a *clausium*, from which the genus takes its name. It is used to close up the aperture when the animal has retreated within its shell, and in that respect resembles an operculum, except that the latter is attached to the animal, or is loose and thrown off, whereas the former is fixed permanently to the shell.

CLAVIGER. A genus of Coleopterous insects, of the section *Trimera;* characterized by six-jointed antennæ, the maxillary palpi very short, and the eyes apparently wanting. The species are found under stones, and in the nests of small yellow Ants. One was found a few years ago in a nest of *Formica flava*, by Mr. J. O. Westwood, at Ensham, Oxon, and it was considered one of our rarest insects; but Mr. F. Smith says (in the Zoologist), "I have been an examiner of ants' nests, and an observer of their habits, some years, and have searched in scores of the nests of *Formica flava* for the *Claviger;* and this perhaps is the reason why I have not found it. In the immediate neighbourhood of London there are no stony fields like those in chalky districts; and where the soil is subject to retaining a greater degree of moisture, like the London clay, the ant appears to find it necessary to raise up a hillock like a mole-hill, to the upper chambers of which she conveys her larvæ, eggs, and pupæ, as the atmospheric changes render it necessary; but, on the contrary, at Mickleham I did not observe a single instance of any superstructure being raised, for, in a soil so light as in some places barely to cover the strata of chalk, the ant is glad to find a situation so suited to her purpose as the under side of a large stone, for here the necessary degree of moisture for the development of her progeny is retained in the earth. Now it will be obvious that the difficulty of detecting the Claviger amongst the accumulations of the ant-hill must be very great, but on removing the stones you are at once, as it were, admitted into the channels of the nest, filled with eggs, larvæ, and pupæ, and amongst these it is that Claviger is found. The first question which naturally arises is this:—What is the nature of the connexion between the two insects? P. W. J. Müller, in Germar's Magazin der Entomologie,' informs us that the ants altogether support the Clavigers for the sake of a peculiar secretion which exudes from them, and which the ants suck from the two flocks of hair that terminate the external angles of the elytra,—that the ants occasionally caress the Clavigers, which then give out a fresh supply of the fluid,—that the Claviger is wholly dependent for support on the ants, and that they feed it with juices extracted from flowers, &c. * * * I am inclined to the opinion that the only purpose for which these insects are retained by the ants, is for the sake of the fluid which they extract from them; I feel convinced that there are hundreds of nests without them: they are most numerous during the early summer months, whilst the larvæ are in the nests; and I was at one time inclined, from that circumstance, to think that the fluid extracted from them might serve to nurture

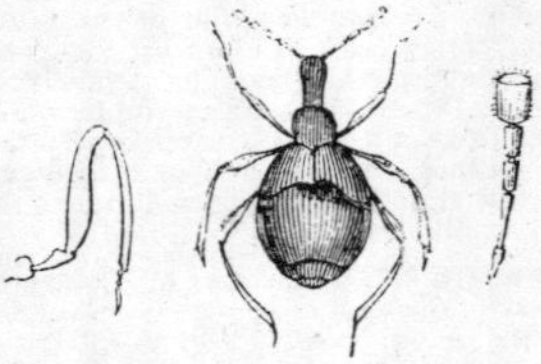

CLAVIGER LONGICORNIS.

particular sexes of ants, but the fact of their not inhabiting every nest at once decides the question." The species figured is the *Claviger longicornis*, which differs considerably from the preceding, but has the same general appearance.

CLAVICORNES. The name given to a family of Pentamerous beetles, whose antennæ end in a club-shaped enlargement: they are partly terrestrial, and partly aquatic.

CLAY [MOTHS]. A name given by collectors to Moths of the genus *Graphiphora*.

CLEAR-WING [HAWK-MOTHS]. A name given to the species of *Sphingidæ*, belonging to the genus *Ægeria*.

CLERUS: CLERIDÆ. A genus and family of Coleopterous insects, of small extent; generally handsomely variegated in their colours, and seldom exceeding an inch in length: the body is firm, long, and often cylindric, with the head and thorax narrower than the elytra; and the antennæ are short, sometimes filiform and serrated. The species of the genus *Clerus* are amongst the largest of the family; having the elytra generally of a bright red colour, ornamented

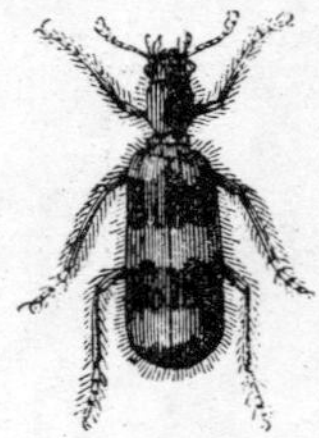

HIVE BEETLE.
(CLERUS [TRICHODES] APIARIUS.)

with purple spots. The perfect insects extract the honey from flowers; but their larvæ, which are of a bright red colour, are very destructive to bees and wasps, in the nests of which the females deposit their eggs during the absence of those insects, upon whose grubs the larvæ of the Clerus prey; they begin in the cell where they were hatched, and proceed from cell to cell, devouring each inhabitant until they arrive at maturity.

CLIO: CLIONIDÆ. A genus and family of naked marine molluscs, belonging to the order *Pteropoda*. They are particularly distinguished by having a pair of fin-like organs, or wings, consisting of an expansion of the mantle on each side of the neck, and furnished with muscular fibres—a peculiarity of structure by which they are enabled to propel themselves rapidly through the water.

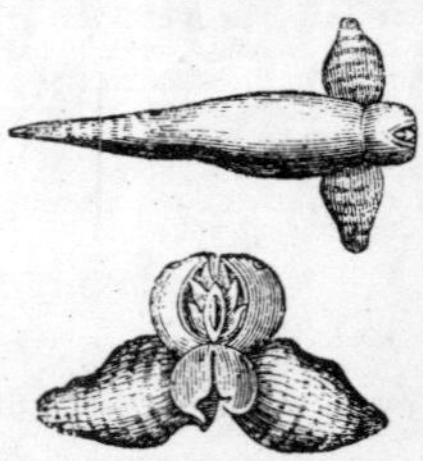

CLIO BOREALIS

So numerous are they in the Northern and Southern oceans, that the water appears literally alive with them; they are called whales' food, and the sea is sometimes so glutted with the Clios, that the whales can scarcely open their mouths without ingulphing thousands of them. The *Clio borealis* of the North and the *C. australis* of the South are equally abundant in the regions to which they belong.

CLIONA. A kind of Sponge. [See SUPPL.]

CLOTHO. A genus of spiders, which inhabit Egypt and the south of Europe, remarkable for the curious nest or habitation which it constructs for its young. This is indeed a singular genus. The best known species (*Clotho Durandii*) is about half an inch long, of a brown maroon colour, with the abdomen black, marked with five yellowish spots. It constructs on the under side of stones, or in crevices of rocks, a cocoon in the shape of a cap or patella, an inch in diameter, its circumference having seven or eight festoons; the points alone being fixed to the stone by means of threads, whilst the edges of the festoons are free. This singular tent, the outer surface resembling the finest taffety, is composed of a number of folds. When young it only constructs two layers, between which it takes its station. But subsequently, perhaps at each moulting, it adds other folds, and when the period of reproduction arrives it weaves another apartment expressly for the reception of the sacs of eggs, and young when hatched, of a softer texture. The inside of its habitation is always remarkably clean. The bags in which the eggs are placed are four, five, or six in number in each habitation; they are about one-third of an inch in diameter, and of a lenticular form. The eggs are not deposited till about the end of December or in January, and they are enveloped in fine down to guard them from the cold. The edges of the festoons not being fastened together, the insect is able to creep in and out at will by lifting them up. When the young are able to dispense with the maternal cares, they quit their common habitation and form separate abodes, and their parent dies in her tent, which is thus its birthplace and its tomb.

CLOUDED YELLOW [BUTTERFLY]. A name applied by insect collectors to Butterflies of the genus *Colias*.

CLUPEA: CLUPEIDÆ. A genus and family of Malacopterygious fishes; distinguished by their wanting the adipose fin, by having the upper jaw composed of the intermaxillary bones in the middle, and the maxillaries at the sides, and by the body being always covered with scales. To this genus belong the Herring, Sprat, Shad, Whitebait, &c. [which see].

CLYTUS. A genus of Longicorn Beetles, abounding in species. A few species (*C. arietis* and *C. arcuatus*) are found in this country; but we prefer quoting, from Dr. Harris's work, his description of two North American species, on account of the interesting notices of the habits of CLYTUS SPECIOSUS:—This beautiful Clytus, like the other beetles of the genus to which it belongs, is a wood-borer; and the noble sugar-maple, which is one of the most beautiful of American forest-trees, is doomed to suffer from its depredations. The Clytus is distinguished from a Callidium by its more convex form, its more nearly globular thorax, which is neither flattened nor indented, and by its more slender thighs. The head is yellow, with the antennæ and the eyes reddish black: the thorax is black, with two transverse yellow spots on each side; the wing-covers, for about two-thirds of their length, are black, the remaining third is yellow, and they are ornamented with bands and spots arranged in the following manner: a yellow spot on each shoulder, a broad yellow curved band or arch, of which the yellow scutel forms the key-stone, on the base of the wing-covers, behind this a zigzag yellow band forming the letter W, across the middle another yellow band arching backwards, and on the yellow tip a curved band and a black spot; the legs are yellow; and the under-side of the body is reddish yellow, variegated with brown. It is the largest known species of Clytus, being from nine to eleven tenths of an inch in length, and three or four tenths in breadth. It lays its eggs on the trunk of the maple in July and August. The grubs burrow into the bark as soon as they are hatched, and are thus protected during the winter. In the spring they penetrate deeper, and form, in the course of the summer, long and winding galleries in the wood, up and down the trunk. In order to check their devastations, they should be sought for in the spring, when they will readily be detected by the sawdust that they cast out of their burrows; and, by a judicious use of a knife and stiff wire, they may be cut out or destroyed before they have gone deeply into the wood. Many kinds of Clytus frequent flowers, for the sake of the pollen which they devour.

CLYTUS PICTUS. This other North American species has the form of the beautiful Maple Clytus. It is velvet black, and ornamented with transverse yellow bands, of which there are three on the head, four on the thorax, and six on the wing-covers, the tips of which are also edged with yellow. The first and second bands on each wing-cover are nearly straight; the third band forms a V, or, united with the opposite one, a W, as in the *C. speciosus;* the fourth is also angled, and runs upwards on the inner margin of the wing-cover towards the scutel; the fifth is broken or interrupted by a longitudinal elevated line; and the sixth is arched, and consists of three little spots. The antennæ are dark brown; and the legs are rust-red. These insects vary from six-tenths to three quarters of an inch in length. We are informed by Dr. Harris, that in the month of September these beetles gather on the locust-trees, where they may be seen glittering in the sunbeams with their gorgeous livery of black velvet and gold, coursing up and down the trunks in pursuit of their mates, or to drive away their rivals, and stopping every now and then to salute those they meet with a rapid bowing of the shoulders, accompanied by a creaking sound, indicative of recognition or defiance. Having paired, the female, attended by her partner, creeps over the bark, searching the crevices with her antennæ, and dropping therein her snow-white eggs, in clusters of seven or eight together, and at intervals of five or six minutes, till her whole stock is safely stored. The eggs are soon hatched, and the grubs immediately burrow into the bark, where they remain during the ensuing winter in a torpid state, but in spring they bore more or less deeply into the trunk, the general course of their winding and irregular passages being in an upward direction from the place of their entrance.

COAL-FISH. (*Gadus carbonarius.*) A Malacopterygious fish, inhabiting the Baltic, the Northern, and the Mediterranean seas: it is common on most of our rocky and deep coasts, but particularly on those of Scotland, the Orkneys, and Yorkshire. The head and body are elegantly shaped; the scales small and oblong: the lateral line silvery white and nearly straight; the under jaw is somewhat longer than the upper; the lips tinged with purple red; the mouth black; the teeth very small; and the irides silvery white. When full grown, it is about two feet and a half long, and weighs thirty pounds; the head, dorsal fins, tail, and upper parts of the body are of a dusky black, which gradually softens into a silvery tinge as it approaches the abdomen; the tail is broad and forked. According to Mr. Pennant, the young begin to appear in vast shoals on the

COAL-FISH. —(GADUS CARBONARIUS.)

coast of Yorkshire, in July, and are at that time about an inch and a half long; in August they are from three to five inches in length, and are taken in great numbers with the rod and line, when they are esteemed a very delicate fish, but when about a year old they are so coarse that few people will eat them. Mr. Couch says, "It is in the highest condition from October to December, at which season it prowls after prey in large companies; so that when met with they prove a valuable capture to the fishermen; for though but coarse food, yet being wholesome, substantial, and cheap, they are eagerly purchased by the poor, either fresh or salted. They swim at no depth, and with great rapidity; but when attracted by bait, will keep near a boat till all are taken; and I have known four men in two boats, two men in each boat, take twenty-four hundred weight with lines in a very few hours. The season for spawning is early in spring; immediately after which this fish becomes so lank as to be worthless, in which state it continues through the summer."

These fish derive their English name from the dusky pigment which tinges their skin, and which, when they are handled, soils the fingers like moist coal. The young resort to the rocky bays of the Orkneys and Hebrides in immense numbers, where, according to the period of their growth, they are known by the names of *cuddy*, *sithe*, and *sillock*. On the Yorkshire coast the young are called *parrs*, and when a year old *billets*.

COATIMONDI. (*Nasua.*) An animal bearing some affinity to the racoon, except that the neck and body are longer, the fur is shorter, and the eyes are smaller; but it is more particularly distinguished by the elongation of its snout, to which its scientific name *nasua* refers. By the assist-

RUFOUS COATIMONDI —(NASUA RUFA.)

ance of this long flexible snout, which is somewhat truncated at the end, it roots up the earth, in the manner of a hog, in quest of earth-worms, &c. It also preys on the smaller quadrupeds; but it lives more upon trees than upon the ground, and is a destructive enemy of birds, their eggs, and unfledged young. It is equal in size to a large cat; its general colour is a cinereous brown; the tail, which is of very considerable length, is annulated with distinct circles of black; the ears are round, like those of a rat, covered with short hair externally, but internally with long whitish hair; the mouth is large, and the under jaw much shorter than the upper. It is a native of Brazil.

COBITIS. [See LOACH.]

COBRA DI CAPELLO. The Portuguese name of the *Vipera naja;* called by the English names of the HOODED SNAKE and the SPECTACLE SNAKE. [See SNAKE.]

COCCINELLA: COCCINELLIDÆ. A genus and family of Coleopterous insects, characterized by their hemispheric form, the upper part being convex, and the lower flat; and further distinguished by the colour and spots of their wings. Among these are included all the Lady-birds; one of which is the *Coccinella septempunctata* of Linnæus, or common seven-spotted Lady-bird, the well-known summer visitant of every field and garden. Though these insects sometimes appear in great numbers, and have occasionally created much alarm, it is erroneous to suppose that they do any injury to vegetation; on the contrary, both in the larva and perfect state, they feed on the *Aphides* which infest plants, and are consequently of great service: its larva has a rather disagreeable appearance; it is of a long oval shape, with a pointed tail, of a black colour,

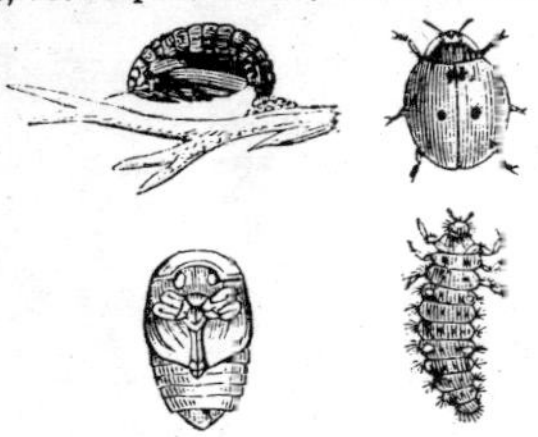

LADY-BIRD, WITH ITS LARVA AND PUPA. (COCCINELLA SEPTEMPUNCTATA.)

with red and white specks, and a rough surface; it changes to a short, blackish, oval chrysalis, spotted with red, and which gives birth to its beautiful inmate in the months of May and June.

The different species of Coccinellæ are very numerous; they are generally divided according to the ground-colour of their elytra, which are either red with black, yellow with black, black with red, or yellow with white spots. One of the most beautiful of the English species is the *Coccinella octodecimpunctata* of Linnæus, or the eighteen-spotted Lady-bird, which is of a bright yellow with numerous black specks, and little more than half the size of the common red kind above described.

Most people who are familiar with our South-eastern coasts have had opportunities of witnessing the flight of extraordinary swarms of Lady-birds during the summer or autumnal months. The most recent instance of this which we have seen publicly noticed is the following:—"On Friday, August 13. 1847, the whole of the coast around Southend was visited by one of the most numerous flights of insects on record. They consisted of at least five species of lady-bird, and they came in such dense numbers, as for miles along the coast to resemble a swarm of bees during hiving. The sea destroyed countless millions of them, the grass and hedge-rows, and every crevice that afforded shelter from the wind, were coloured with their numbers, and for many miles it was impossible to walk, without crushing numbers beneath the tread. The insects evidently came from the east, the wind having veered round to that point during the night. Every true friend of agriculture, however, hails the appearance of these insects, as they are well-known to be the destroyers of *Aphides*, a race of flies the most injurious to vegetation. We found, on inquiry, that this phenomenon was not confined to the above mentioned locality; for on the same day Ramsgate, Margate, Brighton, and the coasts of the adjacent neighbourhoods were similarly visited by swarms of these Aphidivorous insects, which in many places were swept off the public walks, and speedily consigned to a watery tomb."

Dr. Thaddeus Harris has the following sensible remarks on the valuable services of the *Coccinellæ*, when speaking of the "redoubtable enemies" which "seem expressly created to diminish the numbers" of the *Aphides*, or plant-lice. "These lice-destroyers are of three sorts. The first are the young or larvæ of the hemispherical beetles familiarly known by the name of lady-birds, and scientifically by that of *Coccinellæ*. These little beetles are generally yellow or red, with black spots, or black, with white, red, or yellow spots; there are many kinds of them, and they are very common and plentiful insects, and are generally diffused among plants. They live both in the perfect and young state, upon plant-lice, and hence their services are very considerable. Their young are small flattened grubs of a bluish or blue-back colour, spotted usually with red or yellow, and furnished with six legs near the fore-part of the body. They are hatched from little yellow eggs, laid in clusters among the plant-lice, so that they find themselves at once within reach of their prey, which, from their superior strength, they are enabled to seize and slaughter in great numbers. There are some of these lady-birds of a very small size, and blackish colour, sparingly clothed with short hairs, and sometimes with a yellow spot at the end of the wing-covers, whose young are clothed with short tufts or flakes of the most delicate white down. These insects belong to the genus *Scymnus*, which means a lion's whelp, and they well merit such a name, for their young, in proportion to their size, are as sanguinary and ferocious as the most savage beast of prey. I have often seen one of these little tufted animals preying upon the plant-lice, catching and devouring, with the greatest ease, lice nearly as large as its own body, one after another, in rapid succession, without apparently satiating its hunger or diminishing its activity." M. Mulsant, of Lyons, has published a volume on the *Coccinellidæ* of France, most of which are also found in this country: a monograph of the whole group by the same learned entomologist is in the press.

COCCUS. A genus of Hemipterous insects, including the Cochineal insect (*Coccus cacti.*) In this remarkable genus the males are much smaller than the females, and are furnished with wings, of which the females are destitute. The Cocci are found on the leaves and bark of various plants: hence they become injurious to many exotics in our hothouses and conservatories. One of the most common of these is the *Coccus adonidum*, a small oval-shaped insect of a pale rose-colour, slightly convex above, with the body divided into many transverse segments projecting sharply on the sides: it has six short legs, and the whole insect appears more or less covered with a fine white powder. When the female is full of eggs, she ceases to feed, and remaining fixed to one spot, envelops herself in a fine white fibrous cotton-like substance, and soon afterwards dies: the young, which are hatched under the body of the parent insect, proceding from it in great numbers, and dispersing themselves in quest of food. It was originally introduced into Europe along with exotic plants from the warmer regions of Africa and America.

It may be remarked of the *Coccidæ* generally, that they are remarkable for their powers of propagation, and that when they once attack a plant or young tree, the minute size of the larvæ renders their extermination a very difficult task. We were particularly struck with the observations of the President of the Entomological Society (G. Newport, Esq. F. R. S.) in his "Anniversary Address," 1845—that so complete had been the ravages of the Coccus of the orange-trees, that one of the Azores, the island of Fayal, lost its entire produce from this cause alone. The usual exportation of fruit from Fayal had been 12,000 chests annually, but in 1843 not a single chest was exported. This injury had extended to St. Michael's; and the inhabitants of the whole of that group of volcanic islands, depending almost entirely on the produce of their orange-groves, and despairing of retrieving their prospects, were fast turning their attention to the cultivation of other objects of commerce. This amount of injury to a whole population, by a diminutive and apparently contemptible insect, was the result of but three years! It was therefore with great reason that the President laid some stress on the fact, and remarked, that *the effects of this insect on a single article of luxury might fairly be adduced to show that entomological inquiries are deserving of full attention.* They furnish, however, some very important products: the bodies of many species, being deeply coloured through their whole substance, yield dies of great value, the richness of which seems to depend upon the nature of the plant upon which they feed.

By far the most important of all is the *Coccus cacti*, or COCHINEAL CACTUS, so celebrated for the beauty of the colour which it yields. This species is a native of South America, and was for a long time exclusively confined to Mexico, where it feeds on a species of cactus. The female or officinal Cochineal insect, in its full-grown pregnant or torpid state, swells or grows to such a size, in proportion to that of its first or creeping state, that the legs, antennæ, and proboscis, are so small with respect to the rest of the animal as hardly to be discovered by the naked eye; so that on a general view it bears a great resemblance to a seed or berry: hence arose that difference of opinion which at one period subsisted among writers; some maintaining that Cochineal was a berry, while others contended that it was an insect.

When the female insect is arrived at its full size, it fixes itself to the surface of the leaf, and envelopes itself in a kind of white down, which it spins or draws through its proboscis in a continued double filament. The male is a small and rather slender two-winged fly, about the size of a flea, with jointed antennæ and large white wings in proportion to the body, which is of a red colour, with two long filaments proceeding from the tail. When the female insect has discharged all its eggs, it becomes a mere husk, and dies; so that great care is taken to kill the insects before that time, to prevent the young from escaping. The operation of

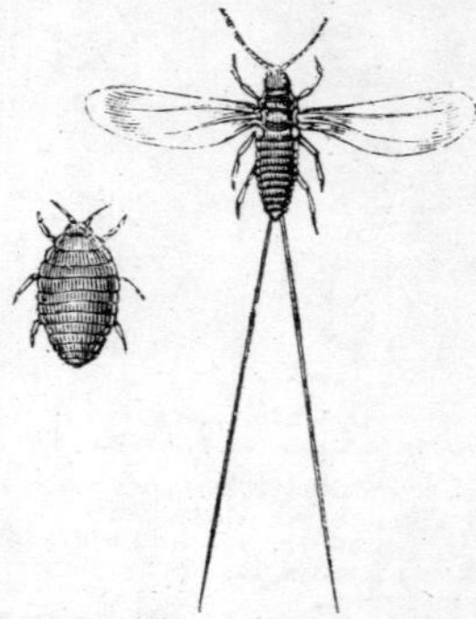

COCHINEAL INSECT - (COCCUS CACTI.)

collecting the insects, which is exceedingly tedious, is performed by the women. "Formerly," says Mr. M'Culloch, "it was in Mexico only that it was reared with care, and formed a valuable article of commerce; but its culture is now more or less attended to in various parts of the West Indies and of the United States. There are two sorts or varieties of Cochineal: the best or domesticated, which the Spaniards call *grana fina*, or fine grain; and the wild, which they call *grana sylvestra*. The former is nearly twice as large as the latter; probably because its size has been improved by the favourable effects of human care, and of a more copious and suitable nourishment, derived solely from the *Cactus cochinellifer*, during many generations. Wild cochineal is collected six times in the year; but that which is cultivated is only collected thrice during the same period. The insects, of which there are about 70,000 in a pound, being detached from the plants on which they feed by a blunt knife, are put into bags, and dipped in boiling water to kill them, after which they are dried in the sun. It is principally used in the dyeing of scarlet, crimson, and other esteemed colours. The watery infusion is of a violet crimson; the alcoholic, of a deep crimson; and the alkaline, of a deep purple, or rather violet hue. It is imported in bags, each containing about 200 lbs.; and has the appearance of small, dry, shrivelled, rugose berries or seeds, of a deep brown, purple, or mulberry colour, with a white matter between the wrinkles."

COCK. (*Gallus domesticus.*) The common domestic Cock, the well-known chieftain of the poultry-yard, is subject to innumerable varieties, scarcely two being found to resemble each other exactly in form and plumage. At what time this valuable bird was brought under the control of man, it is now impossible to determine; but, as the forests of many parts of India still abound with several varieties of the Cock in the wild or natural condition, it is quite reasonable to conclude that the race was first domesticated in the East, and gradually extended thence to the rest of the world. It seems to be generally understood, indeed, that the Cock was first introduced into Europe from Persia: it has, however, been so long established throughout the Western regions, that to attempt to trace its progress from its native wilds would be a useless waste of time. We figure what many naturalists regard as the origin of our domestic poultry, the very handsome Javanese wild fowl; but it is our firm belief that domesticated animals are in general not traceable to any *wild* stock or race.

THE JAVANESE COCK.—(GALLUS BANKIVUS.)

The Domestic Cock has his head surmounted by a notched, crimson, fleshy substance, called a *comb;* and two pendulous fleshy bodies of the same colour, termed *wattles*, hang under his throat. The hen has also a similar, but not so large nor so vividly coloured excrescence on her head. The Cock is provided with a sharp horn or spur on the outside of his tarsus with which he inflicts severe wounds; the hen, instead of a spur, has a mere knot or tubercle. There is, in both sexes, below the ear, an oblong spot, the interior edge of which is reddish, and the remainder white. The feathers arise in pairs from each sheath, touching by their points within the skin, but diverging in their course outwards. On the neck they are long, narrow, and floating; on the rump they are of the same form, but drooping laterally over the extremity of the wings, which are quite short, and terminate at the origin of the tail, the plumes of which are vertical. In the centre of the Cock's tail are two long feathers, which fall backwards in a graceful arch, and add great beauty to the whole aspect of the fowl. It is in vain to offer any description of the colour of the plumage, as it is infinitely varied, being in some breeds of the greatest richness and elegance, and in others of the simplest and plainest hue. Except in the pure white breeds, the plumage of the cock is always more splendid than that of the hen: his apparent consciousness of personal beauty, courage, and gallantry, seem never to forsake him, whether we regard his stately march, at the head of his train of wives and numerous offspring, or watch him as he crows defiance to a rival. His sexual powers are matured when he is about six months old, and his full vigour lasts for about three years.

The hen, if left to herself, forms a very indifferent nest: a simple hole scratched in the ground among a few bushes is the only preparation she usually makes, and she generally lays from twelve to fifteen eggs before she begins to sit upon them for the purpose of hatching. But she now becomes a model of enduring patience, remaining fixed in her place until the urgency of hunger forces her to go in search of food. During the time of her sitting she diligently turns and shifts her eggs, so that each may receive a due degree of genial warmth; and it is not until about three weeks have elapsed that the incubation is completed. The strongest of the progeny then begin to chip the shell with the bill, and are successively enabled to burst their brittle prisons. The whole family being at length emancipated, the parent leads them forth in search of food. In her nature the hen is timid; but in discharging the duties of maternity she becomes bold, and indiscriminately attacks every aggressor, watches over the safety of her young with the utmost jealousy, neglects the demands of her own appetite to divide the food she may obtain among her nurslings, and labours with untiring diligence to provide them sufficient sustenance.

The Cock is very attentive to his females, hardly ever losing sight of them: he leads, defends, and cherishes them; collects them together when they straggle, and seems to eat unwillingly till he sees them feeding around him. Mons. Parrmentier, a celebrated French naturalist, has thus described the Cock:—"He is considered to have every requisite quality when he is of a good middling size; when he carries his head high; has a quick animated look; a strong and shrill voice, short bill, and fine red comb, shining as if varnished; wattles of a large size, and of the same colour as the comb; the breast broad; the wings strong; the plumage black, or of an obscure red: the thighs very muscular; the legs thick, and furnished with strong spurs; the claws rather bent, and sharply pointed. He ought also to be free in his motions, to crow frequently, and to scratch the ground often in search of worms, not so much for himself as to treat his hens. He ought withal to be brisk, spirited, ardent, and ready in caressing the hens; quick in defending them, attentive in soliciting them to eat, in keeping them together, and in assembling them at night."

After the common or dunghill breed which we have described, the principal varieties are—The GAME COCK, which is more distinguished for its unusual length of spur, and its courage, than for any great peculiarity in its plumage: the DORKING fowl, which has two toes behind, and is considerably larger than the other European species; the POLAND breed, which is black-feathered, with white topknots; the BANTAM COCK, a small but most courageous fowl, whose legs are so much feathered as to hinder it greatly in walking; and the DWARF COCK, much smaller than the Bantam, with legs so short that the wings drag on the ground.

COCKCHAFER, or MAY-BUG. (*Melolontha vulgaris.*) This is one of the most common of European beetles, and in this country there is no one with which we are more familiar, the larvæ or caterpillar feeding on the roots of corn, &c., and the complete insect making its appearance during the middle and the decline of summer. It is found on most of the deciduous trees; particularly the oak and willow, and on the hazel and other fruit trees; and often in such numbers that branches bend under their weight. Its duration in the perfect state is very short, each individual living only about a week, and the species entirely disappearing in the course of a month. After the sexes have paired, the males

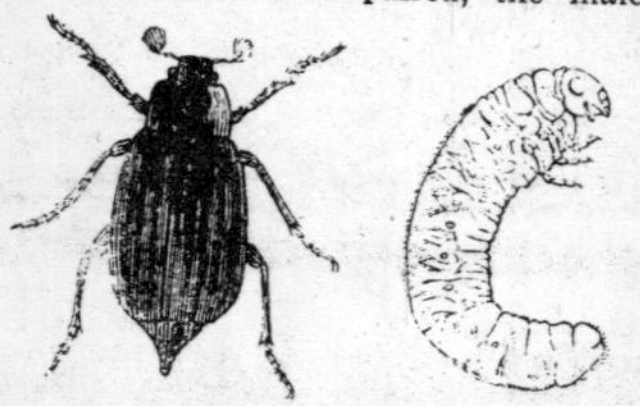

COCKCHAFER AND ITS LARVA (MELOLONTHA VULGARIS.)

perish, and the females enter the earth to the depth of six inches or more, making their way by means of the strong hook which arm the fore-legs; here they deposit their eggs, amounting from one to two hundred from each female, which are abandoned by the parent, who generally ascends again to the surface, and perishes in a short time.

From the eggs are hatched, in the space of fourteen days, little whitish grubs, each provided with six legs near the head, and a mouth furnished with strong jaws. When in a state of rest, these grubs usually curl themselves in the shape of a crescent. They subsist on the tender roots of various plants, committing ravages among these vegetable substances, on some occasions of the most deplorable kind, so as totally to disappoint the best-founded hopes of the husbandman. During the summer they live under the thin coat of vegetable mould near the surface, but as winter approaches they descend below the reach of frost, and remain torpid until the succeeding spring, at which time they change their skins, and re-ascend to the surface for food. At the end of their third summer they have acquired their full growth as larvæ; they then cease eating, and void the residue of their food, preparatory to the metamorphosis which they are about to undergo. As this period approaches they bury themselves deeper in the earth, where they form a rounded cavity, the sides of which are smoothed and consolidated by the application of a fluid disgorged from their mouths. Its abode being thus formed, the larva soon begins to contract in length, swell, and burst its last skin, coming therefrom in the form of a chrysalis, exhibiting the rudiments of elytra, antennæ, &c., and gradually acquiring consistence and colour till it becomes of a brownish hue. In this state it continues about three months, by the end of which time it assumes its rank as a perfect coleopterous insect. During the months of March and April the insect approaches the surface of the earth, and generally bursts from its subterraneous abode during some mild evening about the latter end of May, thus quitting its grovelling mode of life, to soar aloft and disport in the realms of air.

In their winged state, these beetles, with several other species, act as conspicuous a part in injuring the trees, as the grubs do in destroying the herbage. During the month of May they come forth from the ground, whence they have received the name of May-bugs or May-beetles. They pass the greater part of the day upon trees, clinging to the under sides of the leaves, in a state of repose; but as soon as evening approaches, they begin to buzz about among the branches, and continue on the wing till near midnight. In their droning flight they move very irregularly, darting hither and thither with an uncertain aim, hitting against objects in their way with a force that often causes them to fall to the ground. They frequently enter houses in the night, apparently attracted, as well as dazzled and bewildered, by the lights. Their vagaries, in which, without having the power to harm, they seem to threaten an attack, have caused them to be called *dors*, that is, darers; while their seeming blindness and stupidity have become proverbial, in the expressions, "blind as a beetle," and "beetle-headed." Besides the leaves of fruit-trees, they devour those of various forest-trees and shrubs, with an avidity not much less than that of the locust; so that, in certain seasons, and in particular districts, they become an oppressive scourge, and the source of much misery to the inhabitants.

The animals and birds appointed to check the ravages of these insects, are, according to Latreille, the badger, weasel, marten, bats, rats, the common dung-hill fowl, and the goat-sucker or night-hawk. To this list may be added the common crow, which devours not only the perfect insects, but their larvæ, for which purpose it is often observed to follow the plough. In "Anderson's Recreations," it is stated that "a cautious observer, having found a nest of five young jays, remarked that each of these birds, while yet very young, consumed at least fifteen of these full-sized grubs in one day, and of course would require many more of a smaller size. Say, that on an average of sizes, they consumed twenty a-piece, these for the five make one hundred. Each of the parents consumes say fifty; so that the pair and family devour two hundred every day. This, in three months, amounts to twenty thousand in one season. But as the grub continues in that state four seasons, this single pair, with their family alone, without reckoning their descendants after the first year, would destroy eighty thousand grubs. Let us suppose that the half, namely, forty thousand, are females, and it is known that they usually lay about two hundred eggs each; it will appear that no less than eight millions have been destroyed, or prevented from being hatched, by the labours of a single family of jays. It is by reasoning in this way, that we learn to know of what importance it is to attend to the economy of nature, and to be cautious how we derange it by our short-sighted and futile operations.

From Vincent Kollar's useful work on the injuries done to vegetation by various insects, (translated from the German by the Misses Loudon) we derive the following information. "The May-bug is able to do mischief in a double form;—viz. as larva and beetle, in seasons when its increase exceeds the proper limits. The larvæ spare neither meadow nor corn-fields; they often destroy potatoes and other vegetables, and even gnaw the roots of trees and vines, so as to make them sickly. They do particular injury in nurseries, where seeds are raised, to the young plants. By attentively observing the appearance of the young trees, the presence of the larvæ of the May-bug gnawing at the roots may be detected. The plants thus deprived of their roots become yellow and parched, and are easily taken out of the ground. Young fir-trees are not less exposed to the attacks of this insect than deciduous trees. These insects must not be looked for under the already parched-up trees, but under those that are withering; as the former are already deserted from want of nourishment. The fully formed beetle is still more destructive than the larvæ. It attacks cherry, apple, pear, and nut trees, the vine, the oak, and the beech, &c. in multitudes. The leaves and fruit of the trees, when this is the case, are completely destroyed; and the stems, full of sap, become unhealthy, and either recover slowly, or die off. It is worthy of remark, that these insects spare the lime-tree. It is natural that the agriculturist, gardener, and forester should try to discover a method by which so powerful an enemy in their peculiar province may be lessened in number or destroyed. It is impossible to search for the small eggs in the earth; and to dig up the grubs that lie deep in the ground would be attended with an expense which would far exceed that of the ravages they commit, while collecting those which are thrown up by the plough and the spade is not to be taken into consideration. Nothing remains to be done but to catch the fully-formed beetle. Nature, however, as in all other extreme visitations, has provided a more effectual remedy for this evil than can be devised by man. Pigs, moles, field-mice, a multitude of birds (particularly the crow, raven, jackdaw, the woodpecker, and the hawk), and even the large ground-beetles, (*Carabidæ*) instinctively search out the May-bug and its larvæ to feed on. Unfavourable weather often comes on, and if the month of May is wet and cold, the success of the May-bug is at an end; but in order to aid in lessening their too great increase, country magistrates and managers of forests should issue a strict order every spring to the farmers, gardeners, and labourers, to search for and collect these insects as soon as they appear in the gardens, hedges, and forests. For this purpose the children of the peasantry in the country, and those of the lower classes in towns, should be employed and encouraged by rewards. This business should take place in the morning, because the May-bugs, which have been sitting on the blossoms of the trees during the night have become as if torpid, and as long as the branches remain still they do not cling tightly to them with their hooks; in this state they can easily be thrown down in heaps. In shaking the trees, care should be taken that there are no nails or iron on the soles or heels of the shoes of the boys who climb up the branches, so that the abundant sap and tender bark of the trees may not be injured. In order to facilitate the collecting of the fallen beetles, a linen cloth should be spread under each tree, otherwise they will crawl away in the grass. This practice should be continued throughout May, and even to the beginning of June. The collected insects may be killed by pouring boiling water over them, and given as food to fowls and swine; or they may be burnt. It is not advisable either to bury them or to throw them into ponds or rivers, because they would make their way out again, and commit new ravages. Nurseries are best protected by leaves being strewed over the surface of the ground, because (as it is asserted) the beetle never lays its eggs in ground covered with litter.

"Another method of setting a limit to the too great increase of the Cockchafer consists in sparing those birds before named which feed on them, and amongst them the crow undoubtedly claims the first place. These birds follow the plough for the express purpose of consuming worms, the larvæ of insects, and particularly those of the Cockchafer, which are thrown on the surface by the plough. The instinct of the crow to go in quest of this grub, may also be observed in gardens and other places where vegetables are planted. It walks about between the plants, and soon as it sees one that has begun to wither, it approaches it with a joyful spring, digs with its sharp bill deep into the ground near the plant, and knows so well how to seize its prey, that it draws it forth and swallows it almost in the same moment. The crows do the same in meadows, which we sometimes see completely covered with them."

COCKATOO. The Cockatoos belong to the *Psittacidæ*, or Parrot family, but are distinguished from the true parrots, and all others, by a crest, or tuft of elegant feathers, on the head, which they can raise or depress at pleasure. They are in general natives of Australia and the Indian islands, inhabiting the woods, and feeding upon seeds and fruits. They make their nests in decayed trees, and if taken at an early age are easily tamed.

Before we proceed to describe some of the species, we beg to copy from the pages of Capt. Grey (*Travels in Australia*) a most interesting description of "Cockatoo killing." "Perhaps as fine a sight as can be seen in the whole circle of native sports is the killing Cockatoos with the kiley, or boomerang. A native perceives a large flight of Cockatoos in a forest which encircles a lagoon; the expanse of water affords an open clear space above it, unencumbered with trees, but which raise their gigantic forms all around, more vigorous in their growth from the damp soil in which they flourish: and in their leafy summits sit a countless number of Cockatoos, screaming and flying from tree to tree, as they make their arrangements for a night's sound sleep. The native throws aside his cloak, so that he may not even have this slight covering to impede his motions, draws his kiley from his belt, and, with a noiseless, elastic step, approaches the lagoon, creeping from tree to tree, from bush to bush, and disturbing the birds as little as possible; their sentinels, however, take the alarm, the Cockatoos farthest from the water fly to the trees near its edge, and thus they keep concentrating their forces as the native advances; they are aware that danger is at hand, but are ignorant of its nature. At length the pursuer almost reaches the edge of the water, and the scared Cockatoos, with wild cries, spring into the air; at the same instant the native raises his right hand high over his shoulder, and, bounding forward with his utmost speed for a few paces, to give impetus to his blow, the kiley quits his hand as if it would strike the water, but when it has almost touched the unruffled surface of the lake, it spins upwards with inconceivable velocity, and with the strangest contortions. In vain the terrified Cockatoos strive to avoid it: it sweeps wildy and uncertainly through the air, and so eccentric are its motions, that it requires but a slight stretch of the imagination to fancy it endowed with life, and with fell swoops is in rapid pursuit of the devoted birds,—some of whom are almost certain to be brought screaming to the earth. But the wily savage has not yet done with them. He avails himself of the extraordinary attachment which these birds have for one another, and fastening a wounded one to a tree, so that its cries may induce its companions to return, he watches his opportunity by throwing his kiley or spear to add another bird or two to the booty he has already obtained." The preceding animated description refers not only to the species beneath, but also to species of the genus *Calyptorhynchus*, previously described.

Broad-crested Cockatoo. (*Psittacus cristatus.*) This elegant species is about the size of a common fowl; the colour white, with a faint tinge of rose-colour on the head and breast, and of yellow on the inner wing-coverts and tail-feathers: on the head is a very ample crest, consisting of large and long feathers arching over the whole head, which the bird can readily raise or depress: these feathers are white above, but of a fine scarlet hue beneath: the tail is short in proportion to the size of the body, and even at the end; the bill very large, strong, and of a bluish black; the orbits of the eyes bare, and of a deep ash-colour, and the legs deep cinereous. It is of a mild and docile disposition, but can rarely be taught to articulate any other word than its own name, which it pronounces with great distinctness. New Holland is its locality.

Great Sulphur-crested Cockatoo. *Psittacus galeritus.*) This is somewhat larger than the preceding, and measures upwards of two feet in length: its colour is white, slightly tinged with yellow on the sides of the tail, and about the wing-coverts: the head is ornamented with a large, long, and pointed crest, of a fine sulphur colour, slightly reversed at the tip; the bill is black; and the tail longer than in the Broad-crested Cockatoo. Same locality.

Smaller Sulphur-crested Cockatoo. *Psittacus sulphureus*). In almost every respect except in size (being only about fifteen inches long), the description just given would apply to this species. The crest is shaped as in the preceding bird, and is of a fine sulphur-yellow; but it has in addition a large yellow spot beneath each eye. The bill is black; and the legs deep lead-colour. It is a native of the Molucca islands.

Red-vented Cockatoo. (*Psittacus Philippinarum.*) This is not only the smallest of the White Cockatoos, but its crest is smaller in proportion than the rest of the tribe. The bill is of a pale flesh-colour, and the legs cinereous. It is a native of the Philippine isles.

COCKLE. [See Cardium.]

COCK OF THE WOODS. [See Grouse.]

COCKROACH. [See Blatta orientalis.]

COD. [For the generic character of the *Gadidæ*, or Codfish tribe, see Gadus.]—The Common Cod. (*Gadus callarus.*) It is almost impossible to estimate too highly the importance of this truly valuable inhabitant of the deep, whether regarded as a supply of

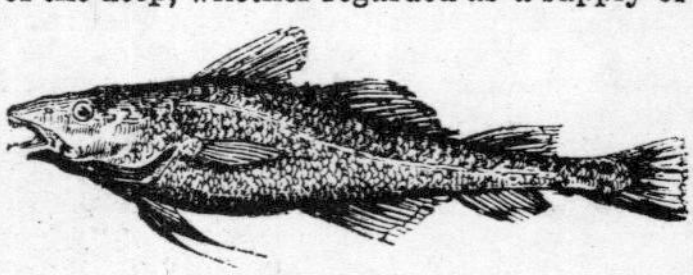

COD —(GADUS CALLARUS.)

food, a source of national industry and commercial wealth, or as a wonder of nature in its astonishing fecundity. It resides in immense shoals in the Northern seas, performing various migrations at stated seasons, and visiting in succession the different coasts of Europe and America. Though found in considerable numbers on the coasts of other northern regions, an extent of about 450 miles of ocean, leaving the chill and rugged shores of Newfoundland, is the favourite annual resort of countless multitudes of Cod, which visit the submarine mountains known as the *Grand Bank*, to feed upon the crustaceous and molluscous animals abundant in such situations. Hither, also, fleets of fishermen regularly adventure, sure of winning a rich freight in return for their toils and exposure. "In this country," Mr. Yarrell observes, "it appears to be taken all round the coast: among the islands to the north and west of Scotland it is abundant: most extensive fisheries are carried on; and it may be traced as occurring also on the shore of almost every county in Ireland. In the United Kingdom alone, this fish, in the catching, the curing, the partial consumption and sale, supplies employment, food, and profit to thousands of the human race."

The Cod is of a moderately long shape, with the abdomen very thick and prominent; the head is large, as also are the eyes; the jaws of equal length, the lower one bearded at the tip by a single cirrus; in the jaws and palate are numerous sharp teeth; the dorsal and anal fins are rather large, the pectoral and ventral rather small; the tail of moderate size, and even at the end; the belly tumid and soft, the body tapering gradually throughout the latter half; the upper part of the head, cheeks, back, and sides, mottled and spotted with dull yellow; the belly white or silvery; the lateral line white; all the fins dusky. The Cod sometimes grows to a very large size. Pennant gives an instance of one taken on the British coasts which weighed seventy-eight pounds, and measured five feet eight inches in length, and five feet in girth round the shoulders; but the general size, at least in the British seas, is far less, and the weight from about fourteen to forty pounds; and such as are of middling size are most esteemed for the table.

Speaking of the localities to which the Cod-fish chiefly resort on our own coasts, Mr. Yarrell says, "A change has lately taken place, from the Cod having shifted their ground. Formerly the Gravesend and Barking fishermen obtained few Cod nearer than the Orkneys or the Dogger Bank; but for the last two or three years the supply for the London market has been obtained by going no farther than the Lincolnshire and Norfolk coasts, and even between that and London, where previously very few fish could be obtained." . . . . . "There appear to be tw well-marked varieties of the Common Cod; one with a sharp nose, elongated before the eye, and the body of a very dark brown colour, which is usually called the Doggerbank Cod. This variety prevails also along our southern coast. The other variety has a round blunt nose, short and wide before the eyes, and the body of light yellowish ash-green colour, and is frequently called the Scotch Cod. Both sorts have the lateral line white. I believe the distinction of more southern and northern Cod to be tenable, and that the blunt-headed lighter-colour fish does not range so far south as the sharper-nosed dark fish. Our fishermen now finding plenty of Cod-fish near home, the London shops for the last year or two have only now and then exhibited specimens of the short-nosed northern Cod: both varieties are equally good in quality, and both are frequently taken on the same ground."

COENURUS. [See SUPPLEMENT.]

COLEOPTERA. [BEETLES.] An order of Insects, having four wings, the external pair of which are not suited for flight, but form a covering or case for the interior pair, and are composed of a hard, tough substance: the inner margins of these wing-cases, or elytra, when closed, touch and form a longitudinal suture; and the inner or true wings, which are large and membranous, when not in use, are folded transversely under them. Under the term Coleoptera, therefore, are included all the beetle tribe; of which naturalists have established a great number of genera, from the different conformations of their antennæ, &c.; presenting among them many that are remarkable for their brilliant colours or singular forms. The larvæ of coleopterous insects undergo a complete transformation: those which burrow in the ground generally prepare for the pupa state by removing the earth which surrounds them so as to form an open oval space; others form a kind of cocoon or web around them; and some assume the perfect state without any preparation.

"Many of these Insects, particularly in the larvæ state, are very injurious to vegetation. The Tiger-beetles (*Cicindelidæ*), the predaceous ground beetles (*Carabidæ*), the diving beetles (*Dytiscidæ*), the Lady-birds (*Coccinellidæ*), and some others, are eminently serviceable by preying upon caterpillars, plant-lice, and other noxious or destructive insects. The water-lovers (*Hydrophilidæ*), rove-beetles (*Staphylinidæ*), carrion-beetles (*Silphidæ*), skin-beetles (*Dermestidæ, Byrrhidæ*, and *Trogidæ*), bone-beetles (some of the *Nitidulidæ* and *Cleridæ*), and various kinds of dung-beetles (*Sphæridiadæ, Histeridæ, Geotrupidæ, Coprididæ*, and *Aphodiadæ*), and the *Pimeliadæ* and *Blaptidæ* act the useful part of scavengers, by removing carrion, dung, and other filth, upon which alone they and their larvæ subsist. Many Coleoptera (some *Staphylinidæ* and *Nitidulidæ, Diaperididæ*, some *Serropalpidæ, Mycetophagidæ, Erotylidæ*, and *Endomychidæ*) live altogether on agarics, mushrooms, and toad-stools, plants of very little use to man, many of them poisonous, and in a state of decay often offensive; these fungus-eaters are therefore to be reckoned among our friends. There are others, such as the stag-beetles (*Lucanidæ*), some spring-beetles (*Elateridæ*), darkling beetles (*Tenebrionidæ*), and many bark-beetles (*Helopidæ, Cistelidæ, Serropalpidæ, Œdemeridæ, Cucujidæ*, and some *Trogositidæ*), which, living under the bark and in the trunks and roots of old trees, though they may occasionally prove injurious, must, on the whole, be considered as serviceable, by contributing to destroy, and reduce to dust, plants that have passed their prime, and are fast going to decay. And, lastly, the blistering-beetles (*Cantharidæ*) have, for a long time, been employed with great benefit in the healing art."

COLIAS. A genus of diurnal Lepidoptera, abounding in species. See Doubleday and Hewitson's Genera of Diurnal Lepidoptera. We here restrict ourselves to the mention of two British species.

COLIAS HYALE, or CLOUDED YELLOW BUTTERFLY. This is a comparatively scarce British butterfly, found chiefly near the sea coast in the counties of Kent, Sussex, and Suffolk. The male is usually of a rich sulphur-yellow, the female nearly white; with a deep black spot in the middle

CLOUDED YELLOW BUTTERFLY (COLIAS HYALE,)

of the anterior wings, and a pale orange spot in the disc of the posterior. The anterior wings have a black border, widest towards the costa, and with a row of yellowish or whitish spots. The under wings have a large orange spot in the centre: beneath, the upper wings are whitish yellow, tipped with orange; having a black ring-spot enclosing a yellow centre near the middle, and with a row of small dusky marks at some distance from the outer margin. The lower wings beneath are entirely orange-yellow, with a row of dusky reddish spots towards the margin, and two silvery spots in the centre. The wings are all ciliated with yellowish red; the body is yellow; the head and the front of the thorax and the legs are ferruginous; the back dusky; the antennæ reddish. The caterpillar is velvety-green, with two yellowish lateral lines, and black spots on the annuli: it feeds on papilionaceous plants. The chrysalis is green, with a yellow lateral line.

COLIAS EDUSA, or CLOUDED SAFFRON BUTTERFLY. The anterior wings of the male insect are of a deep bright fulvous orange above, with a broad black internally-waved band on their outer edge, and a large round central deep black spot: the posterior wings are fulvous above, with a narrow black border on the outer edge, and a greenish tinge on the other; beneath they are greenish, with a sub-ocellated silver spot in the middle, accompanied by a smaller one. The female differs in having a series of irregular yellow spots in the black margin of the anterior wings; but each sex has a row of spots parallel with the edge of the hinder margins of both wings, of which three or four on the anterior ones are deep black, and the rest of a rust-colour: the cilia are yellow and red-brown above, and rose-coloured beneath. The body is yellowish-green, with the back dusky: the antennæ reddish, and the tip of the club inclining to yellow. In some specimens the marginal band is jet black; and the posterior wings are sometimes beautifully iridescent. It is not uncommon during the autumn in the southern counties of England, particularly on the coasts of Kent and Sussex. The caterpillar is deep green with a longitudinal white stripe on each side, spotted with blue and yellow; it feeds on grasses: the chrysalis is green, with a yellow line on each side, and black spots on the wing-cases.

COLIBRI. [See HUMMING-BIRD.]

COLIN. A South American Rasorial bird, by some writers called the Quail, but belonging to the genus ORTYX [which see]. There are several distinct species, all much esteemed for the delicacy of their flesh.

COLOBUS. A genus of quadrumanous animals, of which there are several species. They are natives of Africa, and are in general distinguished by their long, soft, silky hair, which covers the head and upper part of the body. Their "hands" want the thumb; hence their name, derived from the Greek word for *imperfect*. All the species of this genus, most of which are from Western Africa, are in the British Museum. A magnificent species was found by Dr. Ruppell in Abyssinia; it is black, and has long flowing white hair over the sides and back. (*C. Guereza.*) [See MONKEYS.]

COLOSSOCHELYS. (*C. Atlas.*) The name applied by Dr. Falconer and Major Cautley to a gigantic fossil Tortoise discovered by them in India, the remains of which are now in the British Museum.

The first fossil remains of this colossal Tortoise were discovered by the gentlemen above-mentioned in 1835, in the tertiary strata of the Sewalik Hills, or Sub-Himalayahs skirting the southern foot of the great Himalayah chain. They were found associated with the remains of four extinct species of Mastodon and Elephant, species of Rhinoceros, Hippopotamus, Horse, Anoplotherium, Camel, Giraffe, Sivatherium, and a vast number of other Mammalia, &c. The remains of many of the animals associated with the *Colossochelys* in the Sewalik Hills have been discovered along the banks of the Irawaddi in Ava, and in Perim Island in the Gulf of Cambay, showing that the same extinct fauna was formerly spread over the whole continent of India.

"This is not the place (say the discoverers) to enter upon the geological question of the age of the Sewalik strata; suffice it to say, that the general bearing of the evidence is, that they belong to the newer tertiary period. But another question arises: 'Are there any indications as to when this gigantic Tortoise became extinct? or are there grounds for entertaining the opinion that it may have descended to the human period?' Any *à priori* improbability that an animal so hugely disproportionate to existing species should have lived down to be a contemporary with man, is destroyed by the fact that other species of Chelonians which were coeval with the *Colossochelys* in the same fauna, have reached to the present time; and what is true in this respect of one species in a tribe, may be equally true of every other placed under the same circumstances. We have as yet no direct evidence to the point, from remains dug out of recent alluvial deposits; nor is there any historical testimony confirming it; but there are traditions connected with the cosmogonic speculations of almost all Eastern nations having reference to a Tortoise of such gigantic size, as to be associated in their fabulous accounts with the elephant. Was this Tortoise a mere creature of the imagination, or was the idea of it drawn from a reality, like the *Colossochelys?* Without attempting to follow the tortoise tradition through all its ramifications, we may allude to the interesting fact of its existence even among the natives of America. The Iroquois Indians believed that there were originally, before the creation of the globe, six male beings in the air, but subject to mortality. There was no female among them to perpetuate their race; but learning that there was a being of this sort in heaven, one of them undertook the dangerous task of carrying her away. A bird (like the Garuda of Vishnoo, or the Eagle of Jupiter) became the vehicle. He seduced the female by flattery and presents: she was turned out of heaven by the supreme deity, but was fortunately received upon the back of a tortoise, when the otter (an important agent in all the traditions of the American Indians) and the fishes disturbed the mud at the bottom of the ocean, and drawing it up round the tortoise formed a small island, which, increasing gradually, became the earth. We may trace this tradition to an Eastern source, from the circumstance that the female is said to have had two sons, one of whom slew the other; after which she had several children, from whom sprung the human race.

"In this fable we have no comparative data as to the size of the tortoise; but in the Pythagorean cosmogony the infant world is represented as having been placed on the back of *an elephant, which was sustained on a huge tortoise.* It is in the Hindoo accounts, however, that we find the fable most circumstantially told, and especially in what relates to the second Avatar of Vishnoo, when the ocean was churned by means of the mountain Mundar placed on the back of the king of the tortoises, and the serpent Asokee used for the churning-rope. Vishnoo was made to assume the form of the tortoise, and sustain the created world on his back to make it stable. So completely has this fable been impressed on the faith of the country, that the Hindoos to this day even believe that the world rests on the back of a tortoise."

We ought to apologise to our readers, perhaps, for devoting so much space to the "vague and uncertain indications of mythological tradition:" we shall not, however, pursue the subject further, but merely state that the result at which the researches and inquiries of the discoverers arrived was, "that there are fair grounds for entertaining the belief as probable that the *Colossochelys Atlas* may have lived down to an early period of the human epoch and become extinct since:—1st, from the fact that other Chelonian species and crocodiles, contemporaries of the *Colossochelys* in the Sewalik fauna, have survived; 2nd, from the indications of mythology in regard to a gigantic species of tortoise in India."—*Ann. Nat. Hist.* vol. 15.

COLUBER: COLUBRIDÆ. An extensive genus and family of Ophidian reptiles, comprising all serpents, whether venomous or not, whose scales beneath the tail are arranged in pairs; but now, according to Cuvier's arrangement, including only the harmless snakes, many of which habitually reside among trees, and are distinguished by the brilliancy of their colours and the gracefulness of their forms. [See SNAKES.]

COLUGO. The Flying Squirrel. [See GALEOPITHECUS.]

COLUMBIDÆ. A natural family of birds, comprising the pigeons, doves, and turtle-doves. In Britain there are four native species; the Ring-dove or Wood-pigeon; the Rock-pigeon, which is the original of all our domesticated breeds; the Stock-dove, which, like the Ring-dove, chiefly frequents coppices and groves; and the Turtle-dove, which is the smallest, and the most elegant both in form and colour. The *Columbidæ* fly well, and associate invariably in pairs; their nests are constructed in trees, or in the holes of rocks, and both parents sit upon the eggs. They are further remarkable for the peculiar mode in which their young are fed. The crop is furnished with numerous glands, which become developed in both sexes during incubation: these glands secrete a sort of milky substance, with which the food that passes into the crop is moistened; and the food, saturated with this secretion, is regurgitated by the parents for the nourishment of their young. By some naturalists these birds are regarded as forming a distinct order called GYRATORES. [See PIGEON.]

COLUMELLIDÆ. A family of univalve shells, distinguished by their having no canal at the base of the aperture, but a notch, more or less distinct, and plaits on the columella or left lip. Many individuals of this family, as MITRA, MARGINELLA, VOLUTA, &c., are remarkable for their beauty.

COLYMBIDÆ. The *Colymbidæ*, or *Divers*, are a family of birds inhabiting the northern regions, and distinguished by their legs being placed so far back, that they always assume an erect position when standing. Their feet are large and webbed; they are rapid and powerful divers; and they feed both on fish and vegetables. [See DIVER.]

CONCHIFERA. The scientific name given to Bivalve Shells, which are separated into three orders: *Brachiopoda; Dimyaria;* and *Monomyaria:* [which see.] The Mollusca which inhabit them, not having any especial organs for seeing, hearing, or smelling, are limited to the perception of no other impressions but those of immediate contact.

CONDOR. (*Sarcoramphus gryphus.*) A large species of vulture, the most exaggerated descriptions of whose size, as given by the earlier writers and naturalists, caused it to be long regarded as a giant of the feathered race, whose bulk darkened the air, and the rushing of whose mighty wings could only be compared to the roaring of a cataract. But these tales of wonder, like others of a similar nature, have lately given place to the sober reasonings of scientific truth, which, while correcting the extravagance of error, still leave us sufficient room for rational admiration. The Condor is found in the highest and most inaccessible part of the Andes, over the loftiest summits of which it soars, in clear weather, to an amazing height. The elevation it chooses as its breeding-place

CONDOR. — (SARCORAMPHUS GRYPHUS.)

and habitual residence varies from 10,000 to 15,000 feet above the level of the sea; and here, on some isolated pinnacle or jutting ledge, it rears its brood. "The old birds," says Mr. Darwin, "generally live in pairs; but among the inland basaltic cliffs of the Santa Cruz, I found a spot where scores most usually haunt; on coming suddenly to the brow of the precipice, it was a grand spectacle to see between twenty and thirty of these great birds start heavily from their resting-place, and wheel away in majestic circles." "Except when rising from the ground," he adds, "I do not recollect ever having seen one of these birds flap its wings. Near Lima I watched several for nearly half an hour without once taking off my eyes. They moved in large curves, sweeping in circles, descending and ascending without once flapping. As they glided close over my head, I intently watched from an oblique position the outlines of the separate and terminal feathers of the wing; if there had been the least vibratory movement, these would have blended together; but they were seen distinct against the blue sky. The head and neck were moved frequently, and apparently with force; and it appeared as if the extended wings formed the fulcrum on which the movements of the neck, body, and tail acted. If the bird wished to descend, the wings were for a moment collapsed; and then, when again expanded with an altered inclination, the momentum gained by the rapid descent seemed to urge the bird upwards with the even and steady movement of a paper kite." The Condor feeds, like other vultures, chiefly on dead carcasses, but two will frequently unite their forces to overpower and devour the puma, the lama, and other large animals. It occasionally descends to the plains in search of food; but the stories of its attacking children are quite fabulous. It makes no nest, but lays two large white eggs on a shelf of bare rock. The young birds for many months are covered only with a fine thick down, and are said to remain with the parent bird, unable to fly, for an entire year. At mature age the prevailing colour of the male is glossy black, with a tinge of grey. The greater wing-coverts, except at the base and tips, and the secondary quill-feathers, are white; and a white ruff of downy feathers encircles the base of the neck: the crest, or comb, which is fleshy, or rather cartilaginous, occupies the top of the head and about a fourth part of the beak, and is entirely wanting in the female: the skin of the neck is dilated under the throat into a kind of wattle, and along the sides of the neck runs a wrinkled skinny stripe or band, the processes of which are moveable at will. The tail is broad and somewhat wedge-shaped. Length about four feet; expanse of wing about nine feet: tarsi powerful. Various traps and stratagems are made use of to capture the Condors, the *lasso* being among the number. The genus *Sarcoramphus* is peculiar to the New World, and contains, besides the Condor, the King-vulture, and Californian vulture.

CONDYLURA. (*Condylura cristata.*) A mole-like animal of North America, which has the termination of the nostrils surrounded by movable cartilaginous points, that radiate like a star when expanded. The

CRESTED MOLE.—(CONDYLURA CRISTATA.)

head is remarkably large; the body thick and short, growing narrower towards the tail; which is small at the root, large in the middle, and tapering to a fine point at the tip: the fur on the body is very soft, fine, and shining. In Kalm's description of this animal, he says, "It had greater stiffness and strength in its legs than I ever observed in other animals, in proportion to their size. Whenever it intended to dig, it held its legs obliquely like oars. I laid my handkerchief before it, and it began to stir in it with the snout; and taking away the handkerchief to see what it had done to it, I found that in the space of a minute it had made it full of holes, and it looked as if it had been pierced very much by an awl. I was obliged to put some books on the cover of the box in which I kept this animal, or else it was flung off immediately. It was very irascible, and would bite great holes into anything that was put in its way: I held a steel pen-case to it; it at first bit at it with great violence, but having felt its hardness, it would not venture again to bite at anything."

These moles do not make such hills as the European ones, but only little subterraneous walks in the fields, forming banks about four inches broad by two inches thick, and which sink in when trod upon.

CONGER. [See EEL.]

CONIROSTRES. This term is used to denote those Birds which have a strong conical bill, the margin of which is not toothed or indented. The greater part of these are omnivorous; but some are exclusively granivorous. Cuvier observes that they live more or less exclusively upon seeds, in proportion as their bill is more or less thick. Crows, Starlings, and Finches are examples of this class.

CONUS. An extensive genus of univalve Mollusca, the shells of which are thick,

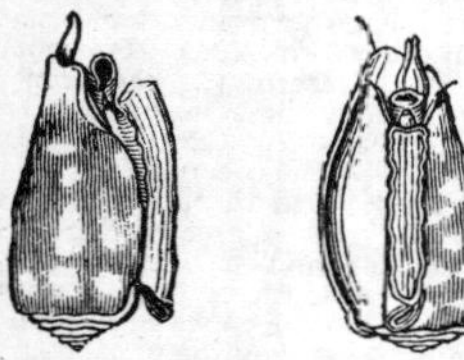

SHELL AND ANIMAL OF THE CONUS BANDANUS.

and rolled up, as it were, in a conical form. They are found principally in the southern and tropical seas; and many of them are very beautiful both in shape and colour. The mollusc is much compressed and involved; the head very distinct, terminated by a trunk capable of great extension; two tentacula, with eyes near the summit; foot oval, and long.

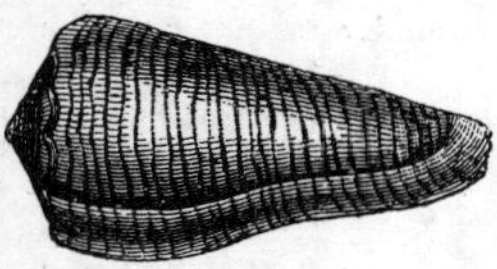

CONUS CALEDONICUS.

Some of the species, such as the *Conus gloria maris*, for example, have fetched enormous prices. The cones are very handsome in shape and agreeable in colour and are consequently much prized by collectors. The accompanying figures will give some idea of their forms. In the British Museum there is a very fine collection of them. The Messrs. Sowerby have monographed the genus and figured all the species.

CONILITES. A genus of fossil Cephalopoda, conical, straight, or slightly curved; having a thin external covering, independent of the alveole. The difference between Belemnites and Conilites, is that the external sheath of the latter is thin, and not filled up with solid matter, from the point of the alveole to the apex, as in the former.

COOT. A genus of birds of the order *Grallatores*. They are distinguished from all other birds by the remarkable structure of the membranes on the toes: the inner toe is furnished with two of these appendages, or rather scallops, the middle one with three, and the outer with four; the hinder toe has a simple membrane only, extending its whole length. They are met with in various parts of Europe, Asia, and America: they delight in marshy and wet places, hiding themselves during the day, and venturing forth in the evening in search of food, which consists of insects and aquatic vegetables. The COMMON COOT (*Fulica atra*) is about sixteen inches in length: its beak is white, slightly tinged with rose-colour; the head and neck deep black; the upper parts of the plumage of a slaty black, and all the under parts of a greyish-blue or lead-colour. The skin is clothed with a thick down, and covered with close fine feathers: thighs placed far behind, fleshy, and strong, bare, and yellow above the knee-joints: legs and toes commonly of a yellowish-green, but sometimes of a lead-colour. From the bill, almost to the crown of the head, there is an excrescence, or fleshy lobe, destitute of feathers, soft, smooth, and round; on which account this bird is sometimes called the *bald* Coot. This species is common in many parts of England, particularly in the Southampton river, and in the Isle of Sheppey; and it is generally believed that it does not migrate to other countries, but changes its stations, and removes in the autumn from ponds and small lakes, where the young have been reared, to the larger lakes, where flocks assemble in the winter. It is usual for them to build their nests in a bush of rushes, surrounded by the water: it is composed of a great quantity of coarse dried weeds, well matted together, and lined within with softer and finer grasses: the female lays from twelve to fifteen eggs, and generally hatches twice in a season; the eggs are about the size of those of a pullet, and are of a pale brownish-white, sprinkled with numerous dark spots, which at the thicker end are like large irregular blotches. A variety, excelling the other in size and the deeper blackness of its plumage, is found in Scotland; also in Lancashire and some of the adjacent counties.

Another species, called WILSON'S COOT (*Fulica Wilsoni*), inhabit various parts of North America, and make their appearance in Pennsylvania in the beginning of October among the muddy flats and islands of the river Delaware, which are overgrown with reeds and rushes, and are periodically overflowed. The chief distinctions between this species and the Common Coot consist in the callous knob on the forehead being of a deep chestnut: the feathers of the vent are quite black, and the under tail-coverts white; and there are a few white feathers on the upper edge of the wing.—There is also a very singular species inhabiting Madagascar, called the CRESTED COOT (*Fulica cristata*). It measures eighteen inches in length; its bill is red at the base, and whitish towards the tip; the crown of the head is bare, of a deep red, and rising into a bifid, detached, crest-like membrane. The entire plumage is blue-black: its legs are dusky, with a tricolor ring or garter above the knee, red green, and yellow.

COPPER [BUTTERFLY]. A name applied by collectors to Butterflies of the genus *Lycæna*. [See LYCÆNA.]

COPRIDÆ. A family of Coleopterous insects allied to the *Scarabæi*. The name *Copris* is from the Greek word for dung, in which the insects are found. Some of them have the head and thorax singularly armed. They are generally of a dull black colour: but some of the species of the American ge-

BLUE AMERICAN DUNG-BEETLE.
(PHANÆUS SAPPHIRINUS.)

nus *Phanæus* perfectly glow with rich green, red, and blue colours. Our figure, derived from Sturm's Catalogue, represents the brilliant blue *Phanæus sapphirinus* of Brazil. There are but few species of this family found in this country.

CORACIAS. A genus of Passerine birds. [See ROLLER.]

CORALLINA. The name given by Linnæus to a genus or group of marine organized bodies, of the class *Vermes*, order *Zoophyta*. The animals of this genus are arborescent or tree-like in form; the stem fixed, with calcareous subdivided branches, mostly jointed. Neither pores nor polypes are distinguishable on the surface of these beings; and they were formerly supposed to be vegetable; but they give the most evident tokens of large portions of ammonia, the common test of animal substance, and have been often traced to spontaneous motion. Every tube, vesicle, or articulation, is probably the enclosure of a distinct animal, so that the entire mass or tree is a family; in this respect resembling the vegetable tree, in which every bud may also be regarded as an individual living plant. [See POLYPES, ACTINIÆ, &c.]

We may in this place very consistently introduce some observations made by late writers on Coral Reefs and Islands, the Coral Fishery, &c. With regard to the growth of coral, it has been observed, that many errors have prevailed upon this subject, both as to the rapidity of their extension, and the depth from which they are built up to the surface of the ocean. It has been commonly stated that many channels and harbours in the Red Sea have been closed up, within the memory of man, by the rapid increase of coral limestone. But Ehrenberg, who carefully examined these localities, attributes the obstruction rather, in some instances, to the quantities of coral sand which have been washed into the harbours, and in others to the accumulation of ballast (generally composed of pieces of coral rock) thrown out from vessels. * * * There can be no doubt that, whether the growth of coral takes place as rapidly as some maintain, or as slowly as it is believed to do by others, it is among the most important of the progressive changes, which have been altering the surface of the globe since it has been tenanted by man. To it is due the existence of a large proportion of the islands of the Polynesian Archipelago, as well as many of those in the Indian Ocean; and the extent of these islands is far less than that of the reefs which are not yet raised above the level of the sea,—some presenting themselves at a distance from any upraised land, others fringing the shores of continents and islands, composed of other formations. It is not correct, however, to affirm (as has been frequently done) that these islands and reefs have been upreared by the Coral-polypes from the depths of the ocean. It is now satisfactorily ascertained that no known species can build from a greater depth than twenty fathoms; and a large proportion seem to prefer a depth of from twenty to thirty feet. As very deep water is found in the immediate neighbourhood of many of these reefs, the question arises, upon what basis they are constructed; and to solve this it is necessary to look at the forms which these massive structures present.

"A large proportion of the Coral Islands of the Polynesian Archipelago," as Dr. Carpenter observes, "are shaped like a crescent, sometimes like a complete ring; and these islands never rise many feet above the surface of the ocean. The highest part is always on the windward [easterly] side, against which the waves are almost constantly dashing. Within the crescent or ring is a basin, termed a *lagoon;* and this usually communicates with the open sea, by a channel, sometimes of considerable width, on the leeward side of the island. Occasionally this channel is completely filled up by the growth of the coral; and the lake, thus inclosed, only communicates with the sea by filtration through the Coral rock. The Coral-polypes never build above low-water mark; and they are not, therefore, immediately concerned in the elevation of the surface from beneath the waves. This is principally accomplished by the action of the sea itself. Large masses are often detached, by the violence of the waves, from the lower part of the structure; and these (sometimes measuring six feet by four) are washed up on the windward side of the reef. Shells, coral-sand, and various other *debris*, accumulate upon it in like manner, until it is at last changed into an island, upon which there is a calcareous soil capable of supporting various kinds of vegetation. When these have once established themselves, the elevation of the surface continues with greater rapidity—successive layers of vegetable mould being deposited by the rapid and luxuriant vegetation of these tropical islands, which are soon tenanted by various forms of animals, and at some subsequent period afford a habitation to Man."

Speaking of an island which was evidently of coral origin, Capt. Flinders thus reasons: "It seems to me, that when the animalcules which form the corals at the bottom of the ocean cease to live, their structures adhere to each other, by virtue either of the glutinous remains within, or of some property in salt water: and the interstices being gradually filled up with sand and broken pieces of coral washed by the sea, which also adhere, a mass of rock is at length formed. Future races of these animalcules erect their habitations upon the rising bank, and die in their turn to increase, but principally to elevate, this monument of their wonderful labours. The care taken to work perpendicularly in the early stages, would mark a surprising instinct in these diminutive creatures. Their wall of coral, for the most part in situations where the winds are constant, being arrived at the surface, affords a shelter, to leeward of which their infant colonies may be safely sent forth: and to this their instinctive foresight it seems to be owing, that the windward side of a reef, exposed to the open sea, is generally, if not always, the highest part, and rises almost perpendicular, sometimes from the depth of 200, and perhaps many more fathoms. To be constantly covered with water seems necessary to the existence of the animalcules, for they do not work, except in holes upon the reef, beyond low-water mark; but the coral sand and other broken remnants thrown up by the sea adhere to the rock, and form a solid mass with it, as high as the common tides reach. That elevation surpassed, the future remnants, being rarely covered, lose their adhesive property; and remaining in a loose state, form what is usually called a *key* upon the top of the reef. The new bank is not long in being visited by sea birds, salt plants take root upon it, and a soil begins to be formed; a cocoa nut, or the drupe of a pandanus is thrown on shore; land birds visit it, and deposit the seeds of shrubs and trees; every high tide, and still more every gale, adds something to the bank; the form of an island is gradually assumed; and last of all comes man to take possession."

A few words in this place respecting the CORAL FISHERY may not be inappropriate. The manner of fishing being nearly the same wherever coral is found, it will suffice to state the method adopted by the French, under the direction of the company established at Marseilles. Seven or eight men go in a boat commanded by the proprietor; and when the net is thrown by the caster, the rest work the vessel, and help to draw the net in. The net is composed of two rafters of wood tied crosswise, with leads fixed to them: to these they fasten a quantity of hemp twisted loosely round, and intermingled with some loose netting. This instrument is let down where they think there is coral, and pulled up again, when the coral is strongly entangled in the hemp and netting. For this, six boats are sometimes required; and if, in hauling in, the rope happens to break, the fishermen run the hazard of being lost. Before the fishers go to sea they agree for the price of the coral; and they engage, on pain of corporal punishment, that neither they nor their crew shall embezzle any, but deliver the whole to the proprietors. Red Coral is found in the Mediterranean, on the shores of Provence, about the isles of Majorca and Minorca, on the south of Sicily; on the coast of Africa; and, lastly, in the Ethiopic Ocean, and about Cape Negro. The divers say that the little branches are found only in the caverns whose situation is parallel to the earth's surface, and open to the south.

CORBULA. A genus of marine Mollusca, some species inhabiting the British coasts. Shell regular, inequivalve, and inequilateral, scarcely gaping; one cardinal spoon-shaped tooth in each valve, but no lateral; ligament interior. These small shells are met with in the seas of New Holland, China, and South America.

COREIDÆ. A family of Hemiptera, of which there are a few brown coloured species in this country; in tropical climates, where there is a luxuriant vegetation, they abound, and from their size, and frequently grotesque shapes, as may be seen in the British Museum collection, are very striking. In the example figured (*Diactor bilineatus*), a native

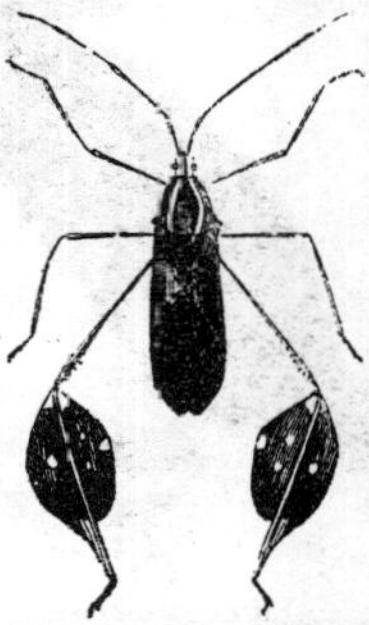

LEAF-LEGGED COREUS
(DIACTOR BILINEATUS.)

of Brazil, the hind legs have singular leaf-like appendages to their tibial joints. This, however, is common to many other species. The smell of these insects is peculiar; the word *cimicine* may be used to express it; it is very far from agreeable, and has associations connected with it by no means pleasing.

COREGONUS. The *Guiniad.* A genus of Malacopterygious fishes, belonging to the *Salmonidæ* family, distinguished by a small trout-like mouth, but with few teeth, and sometimes none; the scales rather large; and the dorsal fin short. There are many species of this genus, some in the sea, others in the fresh waters only. It feeds on insects, and minute fresh-water Crustacea.

They seem to abound in the Arctic parts

of North America; one especially we may mention, the *Coregonus albus*, called the *White-fish* by the fur traders, and *Poisson blanc* by the Canadians. It is from seventeen to twenty inches long. It is bluish-grey on the back, lighter on the sides, and white on the belly; the scales are large and orbicular; there are about eighty scales on the lateral line, and twenty in an oblique series from the dorsal. This species in particular abounds in the lakes of North America. Dr. King, speaking of it, says, "Take, for instance, the *white-fish* only—the *Corregonus albus*, which has never failed to yield to the fisherman's net every demand—the bread of life to the inhabitants of North America, as I have called it, in gratitude for its being the provision which saved my party when in search of Sir John Ross from the death of starvation. This is a food upon which man will not only live for several months together, but actually fatten."

CORMORANT. (*Phalacrocorax.*) Among the whole of the web-footed birds which prey on fish, there are none so voracious as Cormorants. They are most excellent divers, and pursue their prey with astonishing facility beneath the surface of the water, but

COMMON CORMORANT.
(PHALACROCORAX CARBO.)

upon land they are extremely awkward in their movements, owing to their legs being placed so far backwards: they, however, fly with rapidity; and their tail being rather long and furnished with strong feathers, it helps to support their body while walking. As soon as winter approaches, they are seen dispersed along the sea-shores, entering the mouths of fresh-water rivers, and threatening destruction to all the finny tribe. There are several species, but a description of the one common in this country will be sufficient for our purpose. This, which is called by Bewick, the Great Black Cormorant, is said to vary from four to seven pounds in weight, and the size from thirty-two inches to three feet four or five in length, and from four feet to four feet six inches in breadth when the wings are extended. The bill, to the corners of the mouth, measures four inches, and on its ridge two inches and three quarters: it is of a dark horn colour, and the tip of the upper mandible is much hooked and sharp: the lower bill is compressed, and covered about the gape of the mouth with a naked yellowish skin, extended under the chin and throat, where it hangs loose, and forms a kind of pouch, which is capable of distention to a great width: the skin about the eyes is also naked and yellowish, and the eyes have a remarkably wild stare. The crown of the head and the neck are black; and on the former are some loose feathers, which form a sort of short crest; the breast, all the under parts, and the rump, are black glossed with green; the quills and tail-feathers are black; the legs black.

The Cormorant is found in every climate. In Greenland, where it is said they remain throughout the year, the jugular pouch is made use of by the natives as a bladder to float their fishing-darts, after they are thrown; their skins, which are tough, are used by them for garments, and their flesh, which is rank and disagreeable, for food. They usually assemble in flocks on the inaccessible parts of the rocks which overhang or are surrounded by the sea; upon which the female makes her nest of withered sea-weeds, sticks, and grasses: she lays four or more greenish-white eggs, about the size, but somewhat longer, than those of a goose. At sea, or on the inland lakes, they make a terrible havoc. From the greatest height they drop down upon the object of pursuit, dive after it with the rapidity of a dart, and, with an almost unerring certainty, seize the victim; then emerging, with the fish across the bill, with a kind of twirl, throw it up into the air, and dexterously catching it head foremost, swallow it whole.

Notwithstanding the natural wildness of their disposition, it seems that certain species of these birds have formerly been tamed and rendered subservient to the purposes of man, both in this and in other countries. Among the Chinese, it is said, they have frequently been trained to fish, and that some fishermen keep many of them for that purpose, by which they gain a livelihood. In England too, formerly, according to Willoughby, they were hoodwinked in the manner of the falcons, till they were let off to fish, and a leather thong was tied round the lower part of their necks, to prevent their swallowing the fish. The whole deportment of the Cormorant indicates the wary circumspect plunderer, the unrelenting tyrant, and the greedy insatiate glutton, rendered lazy only when the appetite is palled; it ought, however, to be observed, that this bird, like other animals, led only by the cravings of appetite, and directed by instinct, fills the place and pursues the course assigned to it by nature.

It may be thought that we have already dwelt at sufficient length on the nature and habits of the Cormorant; nor would we trespass farther but for the pleasure it affords us to quote from that inimitable writer

Mr. Waterton, whose pleasing descriptions are seldom deficient either in bold originality or graphic power. Walton Hall, the seat of this gentleman, has long been a perfect asylum for such of the feathered tribes as have the good fortune to make it their permanent abode; and he may therefore well say, while pleading their cause [see the Preface to his *Essays*]:—"I possess the very best opportunities of observing the birds whose habits I have described." "The Cormorants," he observes, "often pay me a visit in the winter season; and could they but perceive that there is safety for them here, and great danger elsewhere, they would remain with me while the water is unfrozen. But they wander, unfortunately, through parts where protection is not afforded them; and, being outlandish birds in the eyes of the neighbouring game-keepers, they are immediately shot at. Those which find their way here are so unconscious of danger, that, after they have spent a considerable portion of time in diving for fish, they will come and preen their feathers on the terrace which rises from the water, within ten yards of the drawing-room windows.

"The Cormorant may be justly styled the feathered terror of the finny tribe. His skill in diving is most admirable, and his success beyond belief. You may know him at a distance, among a thousand water-fowl, by his upright neck, by his body being apparently half immersed in the water, and by his being perpetually in motion when not on land. While the ducks and teal and widgeons are stationary on the pool, the Cormorant is seen swimming to and fro, 'as if in quest of something.' First raising his body nearly perpendicular, down he plunges into the deep; and, after staying there a considerable time, he is sure to bring up a fish, which he invariably swallows head foremost. Sometimes half an hour elapses before he can manage to accommodate a large eel quietly in his stomach. You see him straining violently, with repeated efforts to gulp it; and when you fancy that the slippery mouthful is successfully disposed of, all on a sudden the eel retrogrades upwards from its dismal sepulchre, struggling violently to escape. The Cormorant swallows it again; and up again it comes and shows its tail a foot or more out of its destroyer's mouth. At length, worn out with ineffectual writhings, and slidings, the eel is gulped down into the Cormorant's stomach for the last time, there to meet its dreaded and inevitable fate. This gormandizing exhibition was witnessed here by several individuals, both ladies and gentlemen, on Nov. 26. 1832, through an excellent eight and twenty guinea telescope; the Cormorant being, at that time, not more than a hundred yards distant from the observers. I was of the party." [For other species, such as the Chinese Fishing Cormorant and the Australian Cormorant, see PHALACROCORAX.]

CORVIDÆ. The Crow tribe; a family of birds which belong to the *Conirostres*. The *Corvidæ* are very widely diffused over the globe; the general characters are consequently well known. They have a strong bill, compressed at the sides, and covered at the base with stiff feathers, which advance forwards so as to cover the nostrils: the bill is capable of laying hold of almost any kind of food, and the stomach of digesting it. The form of their foot adapts them to traverse the fields and pastures with facility, in search of food; whilst they can also perch with security on trees, the tarsi and toes being moderately long and stout, and the claws arched and acute. Their wings are of that form which ensures a powerful and regular flight; steady without being heavy, and buoyant without wavering; for they are broad and moderately long, and usually rounded at their extremities. The tail, which is chiefly used as a rudder to direct the course of the bird in rapid flight, is short in the species that seek their food entirely on the ground, and long in those which reside chiefly in trees and bushes. Their sight is keen and distant; they often show great sagacity in their natural actions; they possess much docility; and their courage and activity are only equalled by their caution and vigilance. In most of the species inhabiting temperate climates their plumage is rather sombre; but though dark in hue, it is lustrous; while many species in tropical climates exhibit considerable brilliancy and variety in their colouring. [See CROWS.]

CORYPHODON. [See SUPPLEMENT.]

COSSUS. A genus of nocturnal Lepidoptera, the larvæ of which feed on wood.

COSSUS LIGNIPERDA, or GOAT MOTH. This is one of the largest European Moths, being nearly three inches in the expansion of its fore-wings, the colour of which is ashy white, clouded with brown, and marked with an infinite number of short, black, irregular streaks, forming a kind of network: the hind wings are brown, with darker reticulations extending along the margins. The thorax is ochre-coloured in front, pale in the middle, and with a black bar behind: abdomen brown, with the margins of the segments pale yellowish grey.

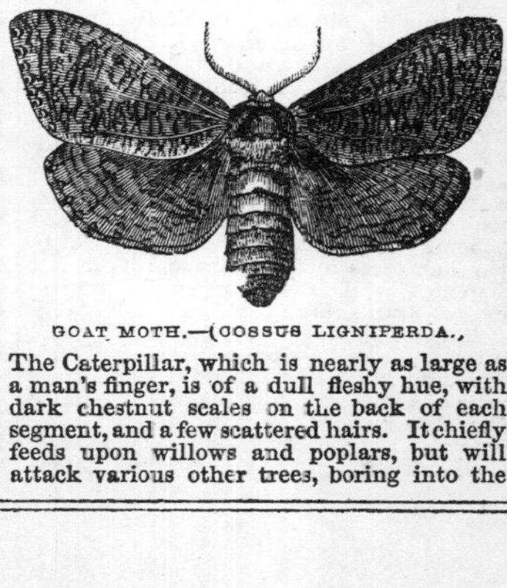

GOAT MOTH.—(COSSUS LIGNIPERDA.)

The Caterpillar, which is nearly as large as a man's finger, is of a dull fleshy hue, with dark chestnut scales on the back of each segment, and a few scattered hairs. It chiefly feeds upon willows and poplars, but will attack various other trees, boring into the timber, and frequently doing very serious damage. It forms a rough cocoon of the chips of wood, which it has bitten to pieces,

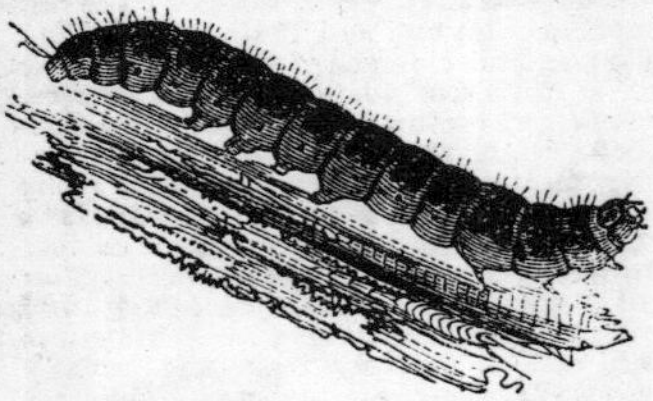

CATERPILLAR OF THE GOAT MOTH

fastening them together with a glutinous secretion, and lining them with its silken web. The pupa has the head-case acute, and each of the abdominal segments is furnished with several rows of reflexed spiny hooks; by the aid of which the pupa, shortly before arriving at the perfect state, is enabled to push itself through its cocoon, and to the surface of the tree; out of the aperture of which the exuviæ may be seen partially sticking after the moth has made its escape.

PUPA AND COCOON OF THE GOAT MOTH

The strength of their jaws is so great that they will very soon destroy any common chip-box in which the larva may be placed, by abrading the edges, to gain its liberty. In breaking up decayed pollards, we not unfrequently find this grub in all the stages of its growth; but more generally observe them without inhabitants, yet perforated with holes large enough to admit the finger. "I suspect," says Mr. Knapp, "that these auger worms are the primary cause of the decay of the tree; having often observed their perforations, and found them, both large and small, in the solid spur or root of the tree, when the upper portion, having been bored, and in a state of decline, is abandoned by them. Those that are full fed appear to form their cases in that part which has lost coherency, while the younger and imperfected creatures mine their way, and obtain nutriment in the solid timber, thus killing the tree by inches; when rain and moisture find lodgment, and complete the dissolution. One year's preparation is the period usually assigned to the larvæ of most insects, before they arrive at their perfect state; but by the Goat-Moth three years are required before it attains its winged state from the egg. Consequently, for the larger portion of its life it is occupied in these destructive operations; and thus this creature becomes a very powerful agent in reducing these Titans of the vegetable world, crumbling them away to their original dust: for what was decreed to be the termination and punishment of Man, is found in active operation throughout the whole chain of Nature's works, which are but dust, and unto dust return, continuing an endless series of production and decay, of restoration and of change."

We may mention, that one of the most extraordinary works on Natural History ever published is devoted to the anatomy of this insect. It is by Lyonet. It will be sufficient to state, in order to give some idea of the careful manner in which the anatomy of this caterpillar has been studied by him, that the author of the "Traité Anatomique" discovered not fewer than 4061 muscles in its body; 228 being attached to the head, 1647 to the body, and 2186 to the intestines, whereas in the human body only 529 have been discovered; so that this caterpillar possesses nearly eight times as many muscles as are contained in the human frame! It has an offensive smell, from which it derives its popular English name.

CORYPHÆNA. A genus of Acanthopterygious fishes, family *Scombridæ;* sometimes called *Dolphins*, but not to be confounded with the Dolphin proper, which belongs to the *Cetacea*. The principal characters by which they are distinguished are as follows:—Body elongated, compressed, covered with small scales; dorsal fin extending nearly the whole length of the back; the tail more or less forked, and the pectoral fin usually arched above and pointed. They have the head much elevated, and the palate and both jaws furnished with teeth. These fishes are very rapid in their motions, generally of large size, and they prey upon the flying-fish. The greater part inhabit the Mediterranean. [See DOLPHIN.]

COTTIDÆ. A family of Acanthopterygious fishes, with hard or mailed cheeks; the sub-orbitals being united to the preoperculum, and so expanded as to cover a large part on the whole of the cheeks. They have many characters in common with the *Percidæ;* in short, a family likeness prevails among the fish possessing this cheek-mail, notwithstanding the various forms of the head that result from its greater or less development. In one group of genera, the head has the form of a cube; in another it is round; in a third it is compressed; and a fourth group is composed of fish of hideous aspect, with a monstrous head and vertical eyes. The only forms among the *Cottidæ* that have anything like a general distribution are the larger genera of *Trigla*, *Cottus*, *Aspidophorus*, *Scorpæna*, *Sebastes*, and *Gasterosteus*, containing the majority of the whole species. The range of individual species is more remarkable in this family than in the more extensive one of *Percidæ;* as is evident when we consider the number of species which cross the Atlantic; and in this respect there is some analogy between the *Cottidæ* and some of the higher classes of animals; it having been observed that the quadrupeds and birds common to the Old and New Worlds are species that have a high northern range.—*Sir John Richardson, M. D., Fauna Bor. Amer.*

COTTUS. A genus of Acanthopterygious fishes, chiefly characterized by having a large head, furnished more or less with spines or tubercles. [See BULL-HEAD.]

COW. The female of the Bovine species, and the most valuable to man of all ruminating quadrupeds. [See OX.]

COW-BUNTING. (*Molothrus pecoris.*) A well-known Passerine bird in North America, the most remarkable trait in the character of which is, the unaccountable practice it has of dropping its eggs into the nests of other birds, instead of building and hatching for itself; and thus entirely abandoning its progeny to the care and mercy of strangers. "About the 25th of March, or early in April," says Wilson, "the cowpen bird makes his first appearance in Pennsylvania from the south, sometimes in company with the red-winged blackbird, more frequently in detached parties, resting early in the morning, an hour at a time, on the tops of trees near streams of water, appearing solitary, silent, and fatigued. They continue to be occasionally seen, in small solitary parties, particularly along creeks and banks of rivers, so late as the middle of

COW-BUNTING.—MOLOTHRUS PECORIS.

June; after which we see no more of them until about the beginning or middle of October, when they re-appear in much larger flocks, generally accompanied by numbers of the redwings; between whom and the present species there is a considerable similarity of manners, dialect, and personal resemblance. In these aerial voyages, like other experienced navigators, they take advantage of the direction of the wind, and always set out with a favourable gale."

"It is well known to those who have paid attention to the manners of birds, that, after their nest is fully finished, a day or two generally elapses before the female begins to lay. This delay is in most cases necessary to give firmness to the yet damp materials, and allow them time to dry. In this state it is sometimes met with, and laid in by the Cow-Bunting, the result of which I have invariably found to be the desertion of the nest by its rightful owner, and the consequent loss of the egg thus dropt in it by the intruder. But when the owner herself has begun to lay, and there are one or more eggs in the nest before the Cow-Bunting deposits hers, the attachment of the proprietor is secured, and remains unshaken until incubation is fully performed, and the little stranger is able to provide for itself. * * * I have never known more than one egg of the Cow-Bunting dropt in the same nest. This egg is somewhat larger than that of the blue-bird, thickly sprinkled with grains of pale brown on a dirty white ground. It is of a size proportionable to that of the bird."

"What reason Nature may have for this extraordinary deviation from her general practice is, I confess, altogether beyond my comprehension. There is nothing singular to be observed in the anatomical structure of the bird that would seem to prevent, or render it incapable of incubation. The extreme heat of our climate is probably one reason why, in the months of July and August, they are not to be seen here. Yet we have many other migratory birds that regularly pass through Pennsylvania to the north, leaving a few residents behind them; who, without exception, build their own nests and rear their own young. This part of the country also abounds with suitable food, such as they usually subsist on. Many conjectures, indeed, might be formed as to the probable cause; but all of them that have occurred to me are unsatisfactory and inconsistent. Future, and more numerous observations, made with care, particularly in those countries where they most usually pass the summer, may throw more light on this matter; till then, we can only rest satisfied with the reality of the fact."

The length of this species is seven inches, breadth eleven inches; the head and neck is of a very deep silky drab; the upper part of the breast a dark changeable violet; the rest of the bird is black, with a considerable gloss of green when exposed to a good light: the tail is slightly forked: legs and claws, glossy black, strong, and muscular; iris of the eye, dark hazel. The young male birds are at first altogether brown, and for a month or more are naked of feathers round the eye and mouth; the breast is also spotted like that of a thrush, with light drab and darker streaks. In about two months after they leave the nest, the black commences at the shoulders of the wings, and gradually increases along each side, as the young feathers come out, until the bird appears mottled on the back and breast with deep black, and light drab. At three months the colours of the plumage are complete, and, except in moulting, they are subject to no periodical change.

COWRIES. A genus of shells used in the East Indies, and many parts of Africa, as the current coin of the natives. [See CYPRÆA.]

COW-FISH. [See MANATUS.]

COYPU. (*Myopotamus coypus.*) A South American rodent animal, resembling the beaver in many respects, though of a smaller size. Its head is large and depressed; ears small and rounded; muzzle pointed, with long stiff whiskers. Its hind feet are webbed, and its habits are aquatic; it swims with

COYPU.—(MYOPOTAMUS COYPUS.)

great ease, lives in the vicinity of water, and burrows in the ground. Its tail is round, instead of being flattened like the beaver, and its scaly covering is partly concealed by scattered hairs. It is easily domesticated, and its manners in captivity are very mild. The Coypu has two kinds of fur: long ruddy hair, which gives the tone of colour; and a brownish ash-coloured fur at its base, which, like that of the beaver, is used largely in the manufacture of hats. It is believed that about 800,000 skins of this animal, under the name of *Neutria* skins, have sometimes been imported into Britain from South America in the course of a year. There is, or was lately, in the gardens of the Zoological Society, a live specimen of this water-loving creature, which enjoyed itself much by diving, while the ease and grace of its manners could not but gratify the visitors.

CRAB. (*Cancer.*) The name of a considerable group of invertebrate animals, whose bodies are covered by an external skeleton, or calcareous crust, having ten articulated limbs, adapted for swimming or walking, and breathing by gills. The head and corselet are united, the latter being broader than it is long: the tail is short in proportion, and concealed by being turned forward beneath the body. They belong to the section of ten-legged, short-tailed Crustacea (*Decapoda brachyura*) of the latest systems, and are of numerous species, exceedingly various in size, colour, and modes of living. The sense of sight, in most of the species, is peculiarly acute, and enables them to distinguish the approach of objects from a very considerable distance. But they are mostly remarkable for a complex and elaborate apparatus for mastication. The mouth is furnished with at least eight pieces or pairs of jaws, which pass the food through an extremely short gullet into a membranous stomach of considerable size. This stomach is rendered curious by having within certain cartilaginous appendages, to which strong grinding teeth are attached. These are five in number, and placed at the pyloric extremity, or outlet of the stomach, so that the aliment, after being subjected to the action of the jaws, is again more perfectly chewed by the stomach-teeth, before entering the digestive tube, where it is exposed to the action of the biliary fluid of the liver. The latter organ is of great size in these creatures, and is all that soft, rich, yellow substance, found immediately beneath the superior shell, called the *fat* of the Crab. A little posterior to the stomach (commonly called *sand-bag*), the heart is situated,—a somewhat globular, whitish body, which propels a colourless lymph to the gills (called *dead men's fingers*) and rest of the body, whence it is brought back to the heart by a hollow vein (*vena cava*) of considerable size.

The process of sloughing, moulting, or throwing off the entire calcareous covering which constitutes their only skeleton, is common to all the *Crustacea*, and is very worthy of attention. As it is obvious that the hard shell, when once perfected, cannot change with the growth of the animal, it becomes necessary that it should be shed entirely; and this shedding takes place at regular periods, at which the increase of size occurs. No one can behold the huge claws or forceps of various species, and the smallness of the joints between them and the body, without feeling some surprise that the creature should be able to extricate them from the old shell, though this is readily accomplished. The aquatic Crabs, when the season of shedding arrives, generally seek the sandy shores of creeks and rivers, and, having selected a situation, they remain at rest, and the change begins. The body of the Crab seems to swell, the large upper shell is gradually detached at the edge, or where it joins the thorax or corselet, and the membrane gradually gives way, and rises up from behind, somewhat like the lid of a chest. The Crab next begins to withdraw the limbs from their cases, and the large muscles of the claws undergo a softening, which allows of their being drawn through the smaller joints. This movement is slowly effected, and, at the time it is accomplished, the parts about the mouth, the antennæ, and eyes, are withdrawn from their old cases, and the animal escapes, retaining its original figure, but soft, helpless, and incapable of exertion or resistance. By a gentle and not very obvious motion, we next observe the sand displaced below the body, and the Crab begins to be covered with it, until, at length, he is sufficiently covered for safety, though still in sight. This is generally in shallow water, where the sun shines freely upon the bottom; and, in the course of twelve hours, the external membrane begins to harden, so as to crackle like paper when pressed upon, and the process of hardening goes on so rapidly, that, by the end of the next forty-eight hours, the Crab regains something of his former solidity and ability to protect himself by flight or resistance.

The habits of Crabs are very various: some are exclusively aquatic, and remain on the sands or rocks, at great depths in the sea; others inhabit excavations formed in the soft coral reefs or bars on certain coasts; some spend their days altogether on shore, living in burrows or dens, formed in a moist or boggy soil; others resort to the rocky flats or beaches, to bask in the sun, where only an occasional wave dashes over them, and seek refuge in the sea when alarmed ; while some species are completely terrestrial, inhabiting holes upon the highest hills and mountains of the West Indies.

Of these LAND CRABS the most remarkable is the species formerly so abundant in the highlands of Jamaica (*Gecarcinus ruricola*), and still common in less densely peopled or uninhabited islands. When the season for

JAMAICA LAND CRAB. (GECARCINUS RURICOLA.)

spawning arrives, vast armies of them set out from the hills, marching in a direct line towards the sea-shore, for the purpose of depositing their eggs in the sand. On this grand expedition nothing is allowed to turn them from their course. With unyielding perseverance they surmount every obstacle which may intervene, whether a house, rock, or other body, not avoiding the labour of climbing by going round, but ascending and passing over it in a straight line. Having reached the destined limit of their journey, they deposit their eggs in the sand, and recommence their toilsome march towards their upland retreats. They set out after nightfall, and steadily advance, until the approach of daylight warns them to seek concealment in the inequalities of the ground, or among any kind of rubbish, where they lie ensconced until the stars again invite them to pursue their undeviating course. On their seaward journey they are in full vigour and fine condition ; and this is the time when they are caught in great numbers for the table. Their flesh, which is of the purest whiteness, is highly esteemed, but, like that of all crustaceous animals, is rather difficult of digestion. Returning from the coast, they are exhausted, poor, and no longer fit for use. They then retire to their burrows, where they slough or shed their shells ; a short time after which operation, and while in their soft state, they are considered by epicures as most delicious, and are consequently sought for with avidity.

Those Crabs which take up their abode in the vicinity of sugar-cane fields are very injurious to the planter ; some of the species being particularly fond of the cane, the juice of which they suck and chiefly subsist on. They are of course narrowly watched, and no opportunity of catching them is lost sight off ; but such is the wonderful facility they have in running, or rather darting in any direction, or with any part of their bodies foremost, that they are almost always enabled to elude capture. It is seldom, however, that they go far from their burrows in the day-time ; and their vigilance is such that they regain them in a moment, and disappear securely, as soon as a man or dog comes near enough to be seen.

Many of the habits of these animals have attracted the notice of travellers. Dr. Gardner, in his "Travels in Brazil," says that while he was near Rio San Francisco,

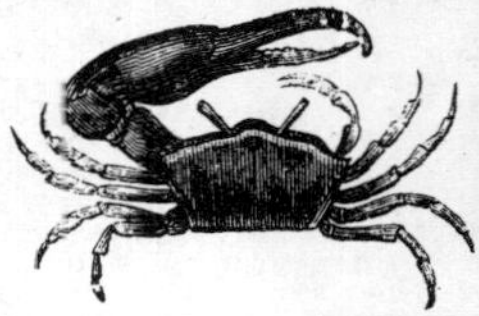

LARGE-CLAWED CALLING CRAB. (GELASIMUS)

he amused himself "by watching the operations of a small species, belonging to the genus *Gelasimus*, that was either making or enlarging its burrow in the sand. About once in every two minutes it came up to the surface with a quantity of sand enclosed in its left claw, which, by a sudden jerk, it ejected to the distance of about six inches, always taking care to vary the direction in which it was thrown, so as to prevent its accumulation in one place."

Another species of Land Crab, apparently belonging to the genus *Thelpheusa*, which inhabits India, is thus noticed by Bishop Heber, in his Journal:—"All the grass through the Deccan generally swarms with a small Land Crab, which burrows in the ground and runs with considerable swiftness, even when encumbered with a bundle of food as big as itself: this food is grass, or the green stalks of rice ; and it is amusing to see the Crabs, sitting, as it were, upright, cut their hay with their sharp pincers, and then waddling off with their sheaf to their holes, as quickly as their sidelong pace will carry them." They have been found on the table-lands, at an elevation of nearly 4000 feet ; but it is believed that they do not perform an annual migration to the sea, for the purpose of depositing their eggs.

CRABRO : CRABRONIDÆ. A genus and family of Hymenopterous insects, popularly known as Wood-wasps. Most of the larger species are marked with yellow rings ; the smaller are generally wholly black. They are extremely active in their movements, and may be seen busily employed, in the hottest sunshine, extracting nectar from the flowers of plants, or running about in search of other insects, on which they prey. They excavate cells in the ground, or in rotten posts, timber, &c., in which they deposit their eggs, together with the flies, &c., which constitute the food of the larvæ when hatched. Many species are found in this country : we refer our readers who may wish further acquaintance with them, to the capital work of Mr. Shuckard on the Indigenous Fossorial Hymenoptera.

CRACIDÆ. A family of Gallinaceous birds, peculiar to tropical America, which approach the turkey in size and grandeur of appearance. They live in the woods, feed on berries, &c., and build on trees; but they are easily domesticated, and their flesh is exceeded by no fowl in delicacy and whiteness. [See CURASSOW.]

CRACTICUS. [See CROW SHRIKE.]

CRAKE. The CORN-CRAKE, or LAND-RAIL, (*Ortygometra crex*), which is very similar to the Water-rail, is fond of woody places, and high herbage or corn-fields in the vicinity of water, or in marshy places, where it breeds; making its nest of a few dry plants, put carelessly together, and laying ten or twelve eggs of a dull white, marked with rust-colour spots. The bill is short, strong, and thick; all the feathers on the upper part of the plumage are of a dark brown, edged with light bay; the wing-coverts and quills are deep chestnut; the fore parts of the neck and breast are pale cinereous; the belly is a yellowish-white; and the legs are a pale flesh-colour. It is much sought after for the delicacy of its flesh, but it is a difficult bird to spring. The legs, which are remarkably long for the size of the bird, hang down while it is on the wing; and, in general, it seems rather inclined to swiftness of foot than rapidity of flight. It is migratory, appearing in England about the beginning of April, and departing in October. At the time of its arrival the bird is extremely lean; but before it quits the island it becomes excessively fat. Its food is chiefly worms, snails, and insects; but it also occasionally feeds on seeds and various vegetables. Its note (crek-crek-crek), rapidly repeated, has been compared to the noise made by drawing a finger along the teeth of a comb.

CRAMP-FISH. A name by which the Torpedo is sometimes called. [See TORPEDO.]

CRANE. Birds of the Crane kind (family *Gruidæ*) subsist on herbs, seeds, worms, frogs, and slugs: they reside in marshy places, rarely visiting the sea shores, and are found in various parts of the world; but only one, the COMMON CRANE, (*Grus cinerea*), is a native of Europe. This bird frequently measures upwards of five feet in length, and weighs about ten pounds; its gait is erect, and its figure tall and slender. The bill is about four inches long, straight, pointed, and compressed at the sides, of a greenish-black, turning lighter towards the point; tongue broad and short, and horny at the top. The forehead, to the middle of the crown, is covered with black hairy down, through which the skin appears red; behind this it is nearly bare to the neck, which is ash grey. The sides of the head behind the eyes, and the hinder part of the neck, are white. The space between the bill and eyes, the cheeks, and fore part of the neck, are a blackish ash; greater wing-coverts also blackish; and those farthest from the body, with the bastard wing and quills, quite black: the rest of the plumage is a fine waved light ash. From the pinion of each wing springs an elegant tuft of loose feathers, curled at the ends, which fall gracefully over the tail, in their flexibility, position, and texture, resembling the plumes of the ostrich. The legs and bare part of the thighs are black.

The Crane is migratory, and, soaring high in the air, performs the boldest and most distant journeys. In summer they spread themselves over the north of Europe and Asia as far as the arctic circle; and in the winter they are met with in India, Syria, Egypt, &c. They formerly visited the fens and marshes of this country in large flocks; but they seem to have been driven away by the advance of cultivation, which has elsewhere, as here, deprived it of many of its most congenial localities. The female lays two greyish-green eggs, spotted with brown.

The SIBERIAN CRANE (*Grus gigantea*) inhabits the great marshes and lakes of Siberia: it builds its nest, of herbs and grass, in almost inaccessible situations amongst the reeds; where the female lays two eggs: both sexes are said to watch the nest alternately; and during the period of incubation, although they are very shy at other times, they will boldly attack any person that approaches their haunts. This species is four feet six inches in height; the bill large and red; the face naked beyond the eyes, and of a red colour; the greater quills and their coverts deep black, the rest of the plumage snowy white; the tail nearly even; the legs red.

The BROWN CRANE (*Grus Canadensis*) is a native of North America, migrating northward in the spring to breed, and returning to the south in autumn. It is three feet three inches long, and its beak about four inches, the tip of the under mandible being of a pale flesh-colour: the top of the head being covered with a red skin, thinly beset with hairs; the hinder part and neck, grey; the scapulars and wing-coverts, pale rufous, margined with brown; the belly, breast, sides, and thighs, ash-colour: the wing-coverts next the body, grey, forming a band on the wing; the greater quills dark brown, with white shafts; the secondaries pale rufous; the tail of a deep ash-colour; the legs and bare part of the thighs, black. The female lays two very large and long eggs at a time; they are much pointed at one end, and freckled with brown. The nest is formed on a tuft upon which much dry grass is accumulated, until it becomes as high as the belly of the bird when standing; this is covered at the top with very fine dried grass, upon which the eggs are laid, and the female stands over them, placing her legs on each side of the heap. [See DEMOISELLE.]

CRAWFISH, or CRAYFISH. (*Astacus fluviatilis.*) A Crustaceous animal of the genus *Astacus*, differing in general appearance but little from the Lobster. They are found in almost every river, and even brook, in England; and their flesh is reckoned cooling and nutritious. Species of this genus are found in all parts of the world. In the mammoth caves of Kentucky, in the United States, a spècies has been discovered; it is the *Astacus pellucidus* of Tellkampf. Mr.

Virtue has written a paper on this subject, and on the other curious animal productions of these caves; to which we refer our readers.

CREEPER. (*Certhia.*) A numerous genus of insectivorous birds, distinguished for the most part by being adapted to live upon the trunks and branches of trees, and to feed upon the insects which infest the bark. The form of the bill is, in some, long and slender; in others, short and stout, and capable of penetrating very hard substances. They have a long, slender, arched bill; wings long and rounded; feet rather slender, but the hinder toe is long and stout; and the tips of the tail-feathers extend beyond the webs. In the splendour and variety of their colours the Creepers rival the Humming-birds, to which they are nearly allied in some of the smaller species. These birds cling by their feet to the perpendicular surface of trees, resting upon the stiff quills of their tails; and they will even pass round a horizontal branch, clinging to its under surface with their backs to the ground.

The COMMON CREEPER (*Certhia familiaris*) weighs only five drams, and next to the Crested Wren is the least of the British birds. The bill is hooked; the legs slender; the toes and claws very long. It breeds in

COMMON CREEPER.—(CERTHIA FAMILIARIS.)

hollow trees; and lays from five to seven spotted ash-coloured eggs. The head and upper part of the neck are brown, streaked with black; the coverts of the wings are variegated with brown and black; the quill-feathers dusky, tipped with white, and barred; the breast and belly white; and the tail very long.

The WALL CREEPER (*Tichodroma muraria*) is considered as one of the rarer European birds, and its principal residence seems to be in Italy and Spain, where it is observed to frequent ruins, creeping about the mutilated walls in quest of spiders and other insects. Its colour is a deep bluish-grey; the wing-coverts and middle quill-feathers black, those nearest the body edged with white; the tail short and black, the two exterior feathers on each side being tipped with white.

CRICKETS. (*Achetidæ.*) A group of Orthopterous insects, belonging to the *grylloid* family, which comprises "the crickets of the hearth," the mole-crickets, and the grasshoppers. The Crickets are distinguished from the other members of this family by their long antennæ, and by the comparative smallness of their thighs. Their bodies are short, thick-set, and soft, with the head, corselet, and abdomen of equal length and breadth: the elytra, which do not completely cover the belly, are curved squarely, and are not roof-shaped, as in the locust and grasshopper. In the winged species the wings exceed the elytra, and project even beyond the abdomen, in the form of a sort of bifid tail.

The Cricket's chirping noise, as it is called, is produced by the friction of the bases of

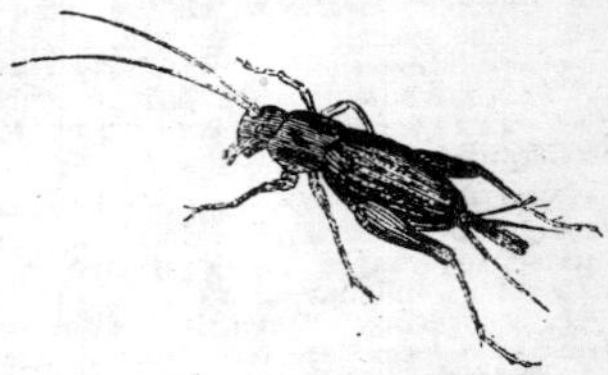

HOUSE CRICKET
(GRYLLUS DOMESTICUS.)

their elytra, or wing-cases, against each other, these parts being curiously adapted to produce this sound. There are some people to whom the chirp of the DOMESTIC CRICKET is not merely an agreeable sound, but who regard the presence of these active insects as a good omen! For our own part, while we are ready to admit that they are perfectly harmless, when, issuing from their warm abodes, they skip round the hearth and join in their monotonous song, we confess that "we would much rather have their room than their company."

The FIELD CRICKET (*Acheta campestris*) is much larger, and also rarer, than the preceding: it is also more noisy. It is of a blackish colour, with a large head in proportion to the body, and full prominent eyes: it frequents hot sandy districts, in which it forms its burrow at the side of footpaths, &c., in situations exposed to the sun, to the depth of from six to twelve inches; and sits at the mouth of it, watching for its prey, which consists of other insects. [See DEINACRIDA: GRYLLUS: MOLE CRICKET.]

CRIMSON UNDERWING [MOTHS]. A name applied by collectors to species of Moths, of the genus *Catocala*.

CREPIDULA. A genus of Molluscous animals, inhabiting an irregularly shaped shell, and often very much flattened; the inside partly covered with a plate, so as to resemble a half-decked boat. There are many recent species, and some fossil. The inside of the *Crepidula onyx* is of the most brilliant black, while the margin of the shell is tinged with a rich brown, and the little half-deck (if such it may be called) is of a beautiful white. These shells are often found upon rocks, where they constantly remain, and form a very irregular outline at the circumference, agreeing with the shape

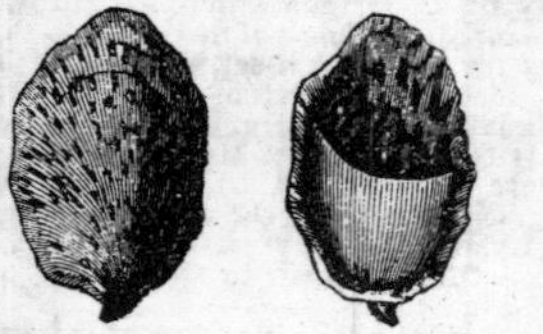

SLIPPER-SHELL.—(CREPIDULA PORCELLANA.)

of the particular part to which they are attached. One species frequently fixes itself upon other living shells, particularly upon the *Purpura*, whose movements it of course follows. The specimen we have here figured is the *Crepidula porcellana*.

CRINOIDEANS. The name of an almost extinct class of invertebrate animals, having a radiated, lily-shaped disc, supported on a jointed stem; and having a crustaceous or coriaceous covering. When this stem is cylindrical, the species are termed *Encrinites;* when it is pentagonal, *Pentacrinites.* [See ENCRINITES, p. 219, and PENTACRINUS, p. 507.]

CRIOCERIS. A genus of Coleopterous insects, belonging to the family *Eupoda*. They live upon aquatic plants, asparagus, &c.; their larvæ feeding upon the same. They have the body soft, short, and swollen; and descend into the earth to become pupæ. One species, *Crioceris Asparagi* (the Asparagus Beetle), is of a blue colour, with the thorax red, and the elytra yellowish-white with blue markings. In its larva state it feeds upon the young sprigs of asparagus, and is sometimes so abundant as to do considerable damage to the plants.

CRIOCERIDIDÆ. A group of oblong leaf-beetles, distinguished by the following characters. The eyes are nearly round and prominent; the antennæ are of moderate length, composed of short, nearly cylindrical or beaded joints, and are implanted before the eyes; the abdomen is narrow and almost cylindrical or square, rounded behind, and much wider than the thorax; and the thighs of the hind legs are often thickened in the middle.

Crioceris trilineata, or Three-lined Leaf-beetle (a North American species), will serve to exemplify the habits of the greater part of the insects of this family. Dr. Harris of Boston, in his truly original work on the Insects of Massachusetts, has described them at length, and it is principally from his work that we are indebted for our notice. This beetle is about one quarter of an inch long, of a rusty buff or nankin-yellow colour, with two black dots on the thorax, and three black stripes on the back, namely, one on the outer side of each wing-cover, and one in the middle on the inner edges of the same; the antennæ (except the first joint) and the feet are dusky; the thorax is abruptly narrowed or pinched in on the middle of each side. When held between the fingers, these insects make a creaking sound like the Capricorn-beetles. They appear early in June on the leaves of the potato-vines, having at that time recently come out of the ground, where they pass the winter in the pupa state. They eat the leaves of the potato, gnawing irregular holes through them; and in the course of a few days begin to lay their oblong oval golden yellow eggs, which are glued to the leaves, in parcels of six or eight together. The grubs, which are hatched in about a fortnight afterwards, are of a dirty yellowish or ashen white colour, with a darker coloured head, and two dark spots on the top of the first ring. They are rather short, approaching to a cylindrical form, but thickest in the middle, and have six legs, arranged in pairs beneath the three first rings. After making a hearty meal upon the leaves of the potato, they cover themselves with their own filth. The vent is situated on the upper side of the last ring, so that their dung falls upon their backs, and, by motions of the body made for this purpose, is pushed forwards, as fast as it accumulates, towards the head, until the whole of the back is entirely coated with it. This covering shelters their soft and tender bodies from the heat of the sun, and probably serves to secure them from the attacks of their enemies. When it becomes too heavy or too dry, it is thrown off, but replaced again by a fresh coat in the course of a few hours. In eating, the grubs move backwards, never devouring the portion of the leaf immediately before the head, but that which lies under it. Their numbers are sometimes very great, and the leaves are then covered and nearly consumed by these filthy insects. When about fifteen days old they throw off their loads, creep down the plant, and bury themselves in the ground. Here each one forms for itself a little cell of earth cemented and varnished within by a gummy fluid discharged from its mouth, and when this is done it changes to a pupa. In about a fortnight more the insect throws off its pupa skin, breaks open its earthen cell, and crawls out of the ground. The beetles come out towards the end of July or early in August, and lay their eggs for a second brood of grubs. The latter come to their growth and go into the ground in the autumn, and remain there in the pupa form during the winter.

CROCODILE. A Saurian reptile of the first magnitude, and celebrated from the remotest antiquity for its terror-striking aspect and destructive power. We of course now allude to the species which inhabits the Nile and other large rivers of Africa; but as we have given the *general* character and habits of Crocodiles under the head "ALLIGATOR," the species peculiar to the American continent, that account should be referred to, and read in connection with what follows:—Crocodiles, like the rest of the Lacertæ, are oviparous: they deposit their eggs in the sand or mud near or on the banks of the rivers they frequent, and the young, when hatched, immediately proceed to the water; but the major part are said to be generally devoured by other animals, as ichneumons, birds, &c. The egg of the common or Nilotic Crocodile is not much larger than that of a goose, but its form is more oblong. When the young are first excluded, the head bears a much larger proportion to the body than when full grown. The Crocodile preys chiefly on fish, but occasionally seizes almost on every animal which happens to be exposed to its rapacity; it is frequently met with twenty feet long, and the armour with which the upper part of the body is covered may be reckoned among the most elaborate specimens of Nature's mechanism. In the full-grown animal it is so strong and thick as easily to repel a musket ball; on the lower parts it is much thinner, and of a more pliable nature; and the whole appears as if covered with the most regular and curious carved-work. The colour of a full-grown Crocodile is blackish-brown above, and yellowish-white beneath; the upper parts of the legs and the sides varied with deep yellow, and in some parts tinged with green: in the younger ones the colour on the upper parts is a mixture of brown and pale yellow, the under parts being nearly white. The eyes are provided with a nictilating membrane, or transparent movable pellicle, as in birds; the mouth is of a vast width, the *rictus* or gape having a somewhat flexuous outline, and both jaws being furnished with very numerous sharp-pointed teeth, of which those about the middle part of each jaw considerably exceed the rest in size, and seem analogous to the canine teeth in the viviparous quadrupeds or mammalia. The tongue is attached by its entire marginal circumference to the lower jaw, and is not extensible, as in all true lizards: the ears are externally closed by two fleshy slips; the nostrils form a long narrow channel, which only opens anteriorly at the back of the throat; and under the throat there are two small pouches, which secrete a strong musky substance. The tail is long, powerful, of a laterally compressed form, and furnished above with an upright process, formed by the gradual approximation of two elevated crests proceeding from the lower part of the back: it accordingly serves as the principal means of propelling the body through the water when in pursuit of fish. The legs are very short, but strong and muscular: the hind feet have only four toes, which are united towards their base by a strong web: the two interior toes on each of the fore-feet, and the interior one on the hind feet, are destitute of claws.

There are also numerous other particulars connected with the anatomy of these beings, which are very curious and interesting. Such are the articulations of the lower jaw with the upper, the joint being so far back as to cause almost every incidental observer to believe that the upper, not the lower, jaw is moved in opening the mouth: the lateral spines on the vertebræ, which prevent the turning of the body, except in a large circle; the curious set of ribs designed exclusively for the protection of the belly, aided by two broad bones standing on the anterior edge of the pelvis; the construction of the external ears; the apparatus for the protection of the eye, &c. &c.

The Crocodile of Egypt is no longer found except in the upper parts of that country, where the heat is greatest, and the population least numerous. Anciently, the species was common nearly to the outlet of the Nile; and it is stated by Pliny, that they used to pass the winter months buried in the mud, or in a state of torpidity. They are still common enough in the river Senegal, Jaire, &c. It is stated by excellent authorities, that they have occasionally been killed in Upper Egypt measuring thirty feet in length; and a very little reflection upon the muscular power of such a reptile will serve to convince us of its ability to commit the most dreadful ravages on the lives of other creatures. Were not such huge and ferocious animals rendered unwieldy by the length of the body and tail, they might become as dreadful on land as in the water; but when on shore, the difficulty they have in turning or of advancing otherwise than directly forward, enables men and animals readily to escape. In the water, the vast force it can exert by means of the long oar-like tail, amply compensates for want of flexibility, and renders the creature more than a match for any of its enemies. Crocodiles are exclusively carnivorous, and they always prefer their food in a certain state of putrefaction. It may be proper to add, that the Crocodile is supposed to be the *Leviathan* of the Scriptures: few persons, indeed, can have read the book of Job without being struck with the magnificent and terrible description of the attributes of Leviathan, to which alone the characters of the Crocodile correspond. [See GAVIAL, and ALLIGATOR.]

CROPPER. A particular species of Pigeon, which receives its name from a large crop under its beak, which it can either raise or depress at pleasure. [See PIGEON.]

CROSSBILL. (*Loxia.*) A genus of Passerine birds, the distinguishing characters of which are—that the tongue is plain, equal, and whole; and that the beak is large, thick,

CROSSBILL.—(LOXIA CURVIROSTRA.)

short, crooked, and convex both ways. This singular structure of the beak was considered as a mere *lusus naturæ* by Buffon; but, notwithstanding the apparently awkward and useless shape, it is found to be most admirably adapted to their particular habits. The two mandibles, instead of lying in a straight direction, cross each other in a similar manner to a pair of scissors, and which enables them to obtain their food with the greatest facility. They live mostly on the seeds of the cones of the fir; in procuring which they exhibit wonderful instinct, as they fix themselves across the cone, then bring the points of their beak immediately over each other, and insinuate them between the scales, when forcing them laterally, the scales open; and then again bringing the points in contact, they pick out the seed with the utmost ease.

The male of the COMMON CROSSBILL (*Loxia curvirostra*) varies from a beautiful red to orange colour on the head, neck, breast, back, and rump; the wing-coverts rufous brown; quills and tail dusky; tail forked; legs short; claws strong. The female in general is of a dull olive-green on those parts that are red in the male; wings and tail similar to the male, but not so dark. This species inhabits Sweden, Germany, and many other parts of Europe, where it breeds and migrates occasionally in vast flocks into the other parts: it is never known to breed in this country, but sometimes appears in immense numbers, fixing on those spots that abound with firs, for the sake of the seeds, which a.e its natural food. It is said to do a deal of mischief to orchards by splitting the apples to get at the seeds; and it is so intent when feeding on the cones of the firs (which it holds in its claws like a parrot), that it may be taken without difficulty. In North America and Greenland this bird is very common, and is said to build its nest in the highest parts of the firs, making use of the resinous matter that exudes from them for fixing it to the trees. It is sometimes called the German Parrot.

Another species, called the WHITE-WINGED CROSSBILL (*Loxia leucoptera*), which is somewhat less than a goldfinch, is common in North America. It is said to make its appearance in March, and to build its nest of mud and fer thers in May, about half-way up a pine-tree, laying five white eggs speckled with yellow: in November both the old and young disappear, and are supposed to retire farther inland.

CROTALUS. [See RATTLESNAKE.]

CROTOPHAGA, or ANI. A genus of Scansorial birds found in the New World. The Crotophaga Americana, or Keel-bill, is principally an inhabitant of the hotter regions of South America, particularly Brazil, though it is met with also in North America, as well as in severa of the West India islands. The general colour of these birds is black, with more or less of metallic reflections; and they have a short, arched bill, very much compressed. They live in flocks; the skirts of woods and the borders of flooded savannahs being their favourite haunts; and their food chiefly consisting of small lizards, insects, and seeds. Their wings are short, and their flight feeble; but they are so bold and fearless as scarcely to be alarmed at the sound of fire-arms; and as they are not reckoned among edible birds, on account of the rank-

KEEL-BILL. — (CROTOPHAGA ANI)

ness of their flesh, they may be said to enjoy a kind of privileged security. Many pairs are said to use the same nest, which is built on the branches of trees, and of a large size; there they lay and hatch in concert. They are observed to breed several times in a year, and their eggs are of a bluish-green colour.

CROW. (*Corvus.*) Under the term CORVIDÆ will be found a brief account of the general characteristics of this gregarious and predatory genus of birds, of which the Raven may be considered the head. We are now about to speak of the COMMON or CARRION CROW (*Corvus corone*), which in form, colour, and appetites, so much resembles it.

The Carrion Crow is similar to the Raven in its habits, colour, and external appearance: length about eighteen inches; breadth three feet. The glossy feathers of the upper plumage have a burnished look, excepting on their edges, which are dull, and form a border to each. They live mostly in woods; build their nests in trees; and lay five or six eggs. They feed on putrid flesh, and garbage of all sorts; likewise on eggs, shell-fish, worms, and insects.

England is said to produce more birds of this kind than any other country of Europe. In the reign of Henry VIII. they were so numerous, and deemed so injurious to the farmer, that they were regarded as an evil worthy of parliamentary redress; and an act was accordingly passed for their destruction, in which rooks and choughs were included. Every hamlet was to provide crownets for ten years: and, during that space, the inhabitants were obliged to assemble at certain times, in order to project the most effectual methods for extirpating them.

The habits of this bird are so amusingly pourtrayed by Mr. Waterton, that we deem it no trespass upon the patience of our readers to quote his observations at considerable length. "This warrior bird," says he, "is always held up to public execration. The very word carrion, attached to his name, carries something disgusting with it, and no one ever shows him any kindness. Though he certainly has his vices, still he has his virtues too; and it would be a pity if the general odium in which he is held should be the means, one day or other, of blotting out his name from the page of our British ornithology. With great propriety he might be styled the lesser raven in our catalogue of native birds; for, to all appearance, he is a raven; and I should wish to see his name changed, were I not devoutly attached to the nomenclature established by the wisdom of our ancestors.

"The Carrion Crow is a very early riser; and, long before the rook is on the wing, you hear this bird announcing the approach of morn, with his loud hollow croaking, from the oak to which he had resorted the night before. He retires to rest later than the rook: indeed, as far as I have been able to observe his motions, I consider him the first bird on wing in the morning, and the last at night, of all our non-migrating diurnal British birds. When the genial voice of spring calls upon him for the continuation of his species, the Carrion Crow, which up to this period has been wary, shy, and cautious, now, all of a sudden, seems to lose these qualities; and, regardless of personal danger, sometimes makes his nest within a hundred yards of the habitation of man, upon a tree, at once the most conspicuous and exposed. To us, who know so little of the economy of birds, this seems a strange phenomenon; nor can any penetration of which we may be possessed enable us to comprehend the true meaning of this change from timidity to boldness, from distance to proximity, from wariness to heedlessness, in so many different species of birds. One would suppose that they would be more shy and distant at this interesting period; and, in imitation of the cat, the rabbit and the fox, conceal as much as possible the place of their retirement. The rook will sometimes build a poor and slovenly nest, but this is never the case with the Carrion Crow; this bird invariably makes its nest firm and compact; it never builds it in hedges, but will construct it in any of our forest trees; and, with me, it seems to give the preference, in general, to the oak, the spruce fir, and the Scotch pine. The young are hatched naked and blind, and remain blind for some days.

"Our ancestors, no doubt, bestowed the epithet *carrion* upon this bird, in order to make a clear and decided distinction between it (whose flesh they probably supposed was rank and bad) and the rook, the flesh of which was well known to be good and wholesome food. Perhaps, too, in those days of plenty, and of less trade, the Carrion Crow had more opportunities of tasting flesh than it has in these our enviable times of divers kinds of improvement. Were a Carrion Crow of the present day to depend upon the finding of a dead cow or horse for its dinner, it would soon become an adept in the art of fasting by actual experiment; for no sooner is one of these animals, in our neighbourhood, struck by the hand of death, than its hide is sent to the tan-pit, and its remains are either made into soup for the hunt, or carefully buried in the dunghill, to increase the farmer's tillage. The poor Crow, in the mean time, despised and persecuted for having an inclination to feed upon that of which, by-the-by, the occupier of the soil takes good care that he shall scarcely have a transient view, is obliged to look out for other kinds of food. Hence you see it regularly examining the meadows, the pastures, and the corn-fields, with an assiduity not even surpassed by that of the rook itself.

"The Carrion Crow will feed voraciously on ripe cherries; and, in the autumn, he will be seen in the walnut-trees, carrying off from time to time, a few of the nuts. With the exception of these two petty acts of depredation, he does very little injury to man during nine or ten months of the year; and if, in this period, he is to be called over the coals for occasionally throttling an unprotected leveret or a stray partridge, he may fairly meet the accusation by a set-off in his account of millions of noxious insects destroyed by him. However, in the spring of the year, when he has a nest full of young to provide for, and when those young begin to give him broad hints that their stomachs would like something of a more solid and substantial nature than mere worms and caterpillars, his attention to game and poultry is enough to alarm the stoutest-hearted squire and henwife. These personages have long sworn an eternal enmity to him; and he now, in his turn, visits, to their sorrow, the rising hopes of the manor with ominous aspect; and he assaults the broods of the duck-pond, in revenge, as it were, for the many attempts which both squire and henwife have made to rob and strangle him.

"In 1815, I fully satisfied myself of his inordinate partiality for young aquatic poultry. The cook had in her custody a brood of ten ducklings, which had been hatched about a fortnight. Unobserved by any body, I put the old duck and her young ones in a pond, nearly 300 yards from a high fir-tree in which a Carrion Crow had built its nest: it contained five young ones almost fledged. I took my station on the bridge, about 100 yards from the tree. Nine times the parent crows flew to the pond, and brought back a duckling each time to their young. I saved a tenth victim by timely interference. When a young brood is attacked by an enemy, the old duck does nothing to defend it. In lieu of putting herself betwixt it and danger, as the dunghill fowl would do, she opens her mouth, and shoots obliquely through the water, beating it with her wings. During these useless movements the invader secures his prey with impunity.

"Let us now examine if the attacks of this bird on domestic poultry cannot be easily counteracted; and whether its assiduous attention to the nests of pheasants and of partridges is of so alarming and so important a nature as to call for its utter extermination from the land. For my own part, I acknowledge that I should lament his final absence from our meadows and our woods.

His loud and varied notes at early dawn, and again at latest eve, are extremely grateful to me; and many an hour of delight do I experience, when, having mounted up to the top of a favourite aged oak which grows on the border of a swamp, I see him chasing the heron and the windhover through the liquid void, till they are lost in the distance. Then, again, how eager is his pursuit!— how loud his croaking!—how inveterate his hostility!—when he has espied a fox stealing away from the hounds, under the covert of some friendly hedge. His compact and well-built figure, too, and the fine jet black of his plumage, are, in my eye, beautifully ornamental to the surrounding sylvan scenery."

The HOODED CROW (*Cervus cornix*) is a bird of passage, which visits England in the beginning of winter, and leaves it with the woodcock. It is found both in the inland and maritime parts of this kingdom; and, in the latter, it feeds on crabs and other shell-fish. It is very common in many parts of the Highlands of Scotland, the Hebrides, Orkneys, &c. They build indifferently in all kinds of trees; lay six eggs; have a shriller voice than the common Crow; and are much more mischievous.

CROW SHRIKE. (*Cracticus.*) A genus of birds found in Madagascar, New Holland, &c., of which there are several species.—The BLACK-THROATED CROW SHRIKE (*Cracticus nigrogularis*, Gould) is a handsome species, with a black head, neck, and breast; the under parts, the hinder part of the neck,

CROW SHRIKE
(CRACTICUS NIGROGULARIUS.)

shoulders, centre of the wing, white; the tail black, the ends of the feathers white, except the two middle feathers, which are black. It is a native of New South Wales, is usually seen in pairs, and from its active habits and conspicuous pied plumage, forms a striking object among the trees. It feeds on insects and small lizards, but is not satisfied with such trifling prey; its powerful and strongly hooked bill makes it a formidable enemy to young birds, mice, and other small animals, which it soon kills, tears to pieces, and devours on the spot. Mr. Gould, in his invaluable work on the "Birds of Australia," from which our figure is copied, tells us that wounded individuals on being handled inflict severe blows. The nest is like that of a jay.

CRUSTACEA, or CRUSTACEANS. The term applied to those animals which are covered with a soft shell or crust. These consist of crabs, lobsters, and many others of a much less complicated structure, and of a different external form. They are called *articulated* animals—that is, those whose members or limbs consist of segments or rings, articulated into each other, to the inside of which their muscles are attached. The tegumentary skeleton of Crustacea generally possesses a considerable degree of stony hardness; and, indeed, contains no small proportion of carbonate of lime. This solid envelope may be looked upon as a kind of epidermis; for beneath it we find a membrane like the true skin of higher animals; and at certain times it detaches itself and falls off, in the same manner as the epidermis of reptiles separates itself from their bodies. The way in which they free themselves from their old shell is exceedingly singular. In general, they manage to get out of it without occasioning the least change in its form. When they are first denuded, the whole surface of their bodies is extremely soft, and it is not for some time that the substance which has been exuded from the pores on the surface of their skin, acquires a hard consistence.

Crustaceous animals present remarkable physiological distinctions. They respire by means of *branchiæ*, or branchial plates, usually attached to their feet or to their jaws; they have from five to seven pairs of feet; their head is frequently not distinct from the trunk, provided with from two to four jointed, setaceous antennæ; and two compound movable eyes seated on peduncles, which are sometimes movable, and at others fixed: they have a distinct heart, and a regular circulating system: and their organs of reproduction are placed either in the feet or tail. In those genera where the head is not separated from the trunk, the shield or covering envelopes the whole thorax. In other genera the head is distinct from the body, which is divided into seven segments, to the lower sides of which the feet are attached; these for the most part have a tail, consisting of many segments. The limbs vary from ten to fourteen, each having six articulations. The two anterior limbs, and sometimes even three on each side, are provided with forceps; at other times they are terminated by simple hooks, and in many instances by appendages which fit them for swimming. The mouth has usually two mandibles, a *labium* or lip below, and from three to five pairs of jaws: these small leg-shaped appendages are not fitted for locomotion, but, being situated near the mouth, assist in the operation of feeding.

Animals of this class live in various situations, suited to their organization: some inhabit considerable depths of the ocean, others are found on rocky shores, or in muddy shallows; some, such as crawfish, inhabit rivers, under stones and banks; while the land-crab takes up its abode in inland situations, making periodical journeys to the coast in vast numbers, for the purpose of depositing its eggs. [See CRAB.] Some of the Crustacea have the power of emitting light in the dark. Others are endowed with the power of not only detaching one of their limbs when seized upon by an enemy, but have also the faculty of reproducing the severed limb, which, however, is always of a less size than the others, until it has once or twice changed its crust.—The reader is referred to Prof. Milne Edwards's *Hist. Nat. des Crustacées*, and to Prof. Bell's *British Crustacea*, for further information. See also numerous scattered articles in this work.

CRUSIAN. (*Cyprinus curassius.*) A fish of the carp kind, which, though common enough in many parts of England, is believed to be not a native fish. It is from eight to ten inches in length; very deep and thick, and the back much arched. The colour is a deep olivaceous yellow, with a slight silvery tinge on the belly; lateral line straight; fins dull violet; the dorsal fin broad, and extending a considerable distance from the middle of the back towards the tail. The flesh is coarse.

CTENOIDEA. [See SUPPLEMENT.]

CTENOMYS. [See TUCUTUCO.]

CUCKOO. (*Cuculus canorus.*) This bird, whose parasitic habits have so long been a subject of popular interest, and regarding whose general economy so much speculation has been indulged in, is about fourteen inches in length, and twenty-five in breadth when extended: the bill is black and somewhat bent; irides yellow; inside of the mouth red; its head, neck, back, and wing-coverts pale blue, darkest on the head and back, and palest on the forepart of the neck and rump; breast and belly white, elegantly crossed with wavy bars of black; quill feathers dusky, the inner webs marked with white oval spots; the tail long, the two middle feathers black, with white spots on each side of the shaft; legs short and yellow; toes, two forward, two backward; the outer one being directed forward or backward at pleasure; claws white. The female is rather less than the male, and

CUCKOO.—(CUCULUS CANORUS.)

somewhat differs in colour; the neck and breast being of a tawny brown, with dusky bars; and the wing-coverts marked with light ferruginous spots. The plumage of the young is very dissimilar to that of the adult bird; it is supposed, indeed, that they do not throw off the nestling feathers till the second year's moulting.

The Cuckoo is a migratory bird, visiting this country early in spring, and generally quitting it at the commencement of July: its well-known note is usually first heard about the middle of April, and ceases at the end of June. Contrary to the general economy of the feathered creation, it constructs no nest, and never hatches its own eggs; but deposits them in the nests of other birds, as the hedge-sparrow, titlark, water-wagtail, &c., preferring, as it would seem, the first-mentioned. During the time the hedge-sparrow is laying her eggs, which generally occupies four or five days, the Cuckoo contrives to deposit her egg among the rest, leaving the future care of it entirely to the hedge-sparrow. This intrusion often occasions discomposure, for the hedge-sparrow, at intervals, whilst sitting, not only throws out some of her own eggs, but injures others in such a way that they become addled, so that not more than two or three of them are hatched along with that of the Cuckoo; and what is very remarkable, she never throws out or injures the egg of the intruder. When she has disengaged the young Cuckoo and her own offspring from the shell, her young ones, and any of the eggs that remain unhatched, are soon turned out by the young Cuckoo, who then remains in full possession of the nest, and becomes the sole object of the care of its foster parents. The young birds are not previously killed, nor the eggs demolished, but all are left to perish together, either entangled in the bush which contains the nest, or lying on the ground near it. The mode of accomplishing the ejectment is curious: The Cuckoo, very soon after being hatched, and consequently while it is yet blind, contrives with its rump and wings to get the hedge-sparrow, or the egg, upon its back, and making a lodgment for its burden by elevating its elbows, clambers backwards with it up the side of the nest, till it reaches the top, where, resting for a moment, it throws off its load with a jerk, and quite disengages it from the nest; after remaining a short time in this situation, and feeling about with the extremities of its wings, as if to be convinced that the business has been properly executed, it drops into the nest again. Nature seems to have provided, even in the formation of the Cuckoo, for the exercise of this peculiar instinct; for, unlike other newly hatched birds, its back, from the scapulæ downwards, is very broad, with a considerable depression in the middle, as if for the purpose of giving a more secure lodgment to the egg, or the young bird, while the intruder is employed in removing either of them from the nest; when about twelve days old, this cavity is filled up, the back assumes the shape of nestling birds in general, and the disposition for turning out any bird or substance placed in the nest entirely ceases. The smallness of the Cuckoo's egg is another circumstance deserving attention in this surprising transaction; in size and appearance it differs little from the egg of the Skylark and Titlark, though the disparity of the bulk of the birds is very great: in short, everything conspires, as might be expected, to render perfect the design which is to be accomplished by the seemingly unnatural propensity of this bird.

The growth of the young Cuckoo is extremely rapid: it has a plaintive chirp which is not learned from its foster-parent; and it never acquires the adult state during its stay here. A fierceness of disposition shows itself long before it leaves the nest; for when irritated it assumes the manners of a bird of prey, often making a chuckling noise like a young hawk. When it is sufficiently fledged, it does not long remain the inmate of its supposed parent's domicile; for as its appetites for insect-food increase, it cannot expect to obtain a supply by imitating its little instructor: it therefore takes a final leave of, and seldom offers any violence to, its nurse. All the little birds, however, who consider the young Cuckoo as their enemy, show an inclination to revenge the general cause, and compose the train of his pursuers; but none of them are so active in the chace as the Wryneck, who, from this circumstance, has been erroneously considered by many as the Cuckoo's attendant and provider. The Cuckoo is said to be a fierce pugnacious bird. Its principal food consists of hairy caterpillars, grasshoppers, snails, moths, cockchafers, &c., of which it disgorges the hard parts after digestion, in the same manner as birds of prey: it is also said to eat the eggs of other birds. Mr. White (of Selborne) remarks, however, that Cuckoos cannot be birds of prey, as they have a weak bill and no talons.

Although we have already extended this article to a greater length than was our intention, we cannot refrain from making room for the following remarks by Mr. Jesse:—"There is still a great mystery attached to the natural history of the Cuckoo, and one would willingly, if possible, rescue it from the charge of a want of that natural affection which has been alleged against it. It has been stated that what has been said by a very ancient and sublime writer, concerning the defect of natural affection in the ostrich, may be applied to the Cuckoo. It is now, however, pretty well ascertained that the ostrich only quits her eggs when the sun is so powerful that the additional warmth from her body would be detrimental to them. She therefore returns to them in the cool of the evening. I am persuaded that the more we inquire and search into the economy of nature, so far from finding any defects, we shall have more and more reason to be convinced that not only every bird, but every animal from the highest to the lowest in the scale of creation, is equally well adapted for the purpose for which it was intended." We should have mentioned that it is to *Dr. Edward Jenner*, who first introduced vaccination, that we are indebted for having given the earliest and fullest account of the habits of this singular bird. Many of our readers are doubtless familiar with Logan's fine address to the Cuckoo, beginning,

"Hail! beauteous stranger of the grove!"

The GREAT SPOTTED CUCKOO. (*Oxylophus glandarius.*) This species inhabits both the south and the north of Europe; and is about the size of a Magpie. The beak is black, and a little bent; head crested; the crest being composed of bluish ash-coloured feathers; from the base of the upper mandible arises a band of black, which passes through the eyes almost to the hinder part of the head, and is broadest in the middle: scapulars, upper wing, and tail-coverts, dark brown, marked with small white and pale cinereous spots; quill-feathers brown; tail wedge-shaped blackish, and all tipped with white except the two middle feathers: legs and claws black.

The ORIENTAL CUCKOO. (*Eudynamys Orientalis.*) There are several varieties of this species. The first is the size of a pigeon: length about sixteen inches; beak grey-brown; plumage nearly black, with a green gloss, which in some parts assumes a sort of violet hue. The tail is eight inches long; the legs are of a dusky grey colour; claws black: it is found in the East Indies.—The next variety inhabits Mindanao: it is fourteen inches long; beak black, yellow at the tip; the plumage a blue-black; and the tail generally carried spread.—A third variety is about nine inches in length: beak bright orange; plumage black, glossed with green and violet; tail wedge-shaped; legs reddish-brown; claws nearly black. This species frequents woods, and for the most part flies in small flocks. It is held in veneration by the Mahometans; but by epicures, who have no religious prejudices in its favour, it is esteemed a great delicacy.

The GILDED CUCKOO. (*Chrysococcyx auratus.*) This beautiful little specimen of the Cuckoo tribe is about seven inches in length: the beak is of a greenish brown colour; and the upper parts of the body are of a rich gilded glossy green; on the head are five stripes of white; nearly all the wing-coverts and the second quills have white tips, as likewise the tail-feathers and the two greater tail-coverts; the throat and breast white; the sides and feathers which fall over the knees marked with a few greenish bars; legs grey, covered with white feathers as far as the middle: tail wedge-shaped, above three inches long, and in its natural state spread out like a fan. Le Vaillant, who discovered this species in Southern Africa, remarks that it is undoubtedly the finest bird of the genus.

There are many genera and species of Cuckoos, it being a very extensive family; and a fine collection of them is to be seen in the British Museum. We find it necessary, however, to give but one more, and that is—

The YELLOW-BILLED AMERICAN CUCKOO (*Coccyzus Americanus*), the description of which we take from Wilson, as follows:—"From the imitative sound of its note, it is known in many parts by the name of the *cow-bird;* it is also called in Virginia the *rain-crow*, being observed to be most clamorous immediately before rain. This species arrives in Pennsylvania, from the south, about the 22nd of April, and spreads over the country, as far at least as Lake Ontario; is numerous in the Chickasaw and Chactau nations; and also breeds in the upper parts of Georgia; preferring, in all these places, the borders of solitary swamps, and apple orchards. It leaves us, on its return southward, about the middle of September.

"The singular, I will not say unnatural, conduct of the European Cuckoo (*Cuculus canorus*), which never constructs a nest for itself, but drops its eggs in those of other birds, and abandons them to their mercy and management, is so universally known, and so proverbial, that the whole tribe of Cuckoos have, by some inconsiderate people, been stigmatized as destitute of all parental care and affection. Without attempting to account for this remarkable habit of the European species, far less to consider as an error what the wisdom of Heaven has imposed as a duty upon the species, I will only remark, that the bird now before us builds its own nest, hatches its own eggs, and rears its own young; and, in conjugal and parental affection, seems nowise behind any of its neighbours of the grove.

"Early in May they begin to pair, when obstinate battles take place among the males. About the 10th of that month they commence building. The nest is usually fixed among the horizontal branches of an apple-tree; sometimes in a solitary thorn, crab, or cedar, in some retired part of the woods. It is constructed, with little art, and scarcely any concavity, of small sticks and twigs, intermixed with green weeds and blossoms of the common maple. On this almost flat bed, the eggs, usually three or four in number, are placed; these are of a uniform greenish blue colour, and of a size proportionable to that of the bird. While the female is sitting, the male is generally not far distant, and gives the alarm, by his notes, when any person is approaching. The female sits so close, that you may almost reach her with your hand, and then precipitates herself to the ground, feigning lameness, to draw you away from the spot, fluttering, trailing her wings, and tumbling over, in the manner of the partridge, woodcock, and many other species. Both parents unite in providing food for the young. This consists, for the most part, of caterpillars, particularly such as infest apple-trees. The same insects constitute the chief part of their own sustenance. They are accused, and with some justice, of sucking the eggs of other birds, like the crow, the blue jay, and other pillagers. They also occasionally eat various kinds of berries. But, from the circumstance of destroying such numbers of very noxious larvæ, they prove themselves the friends of the farmer, and are highly deserving of his protection.

"The Yellow-billed Cuckoo is thirteen inches long, and sixteen inches in extent; the whole upper parts are of a dark glossy drab, or what is usually called a Quaker colour, with greenish silky reflections; from this must, however, be excepted the inner vanes of the wings, which are bright reddish cinnamon; the tail is long, composed of ten feathers, the two middle ones being of the same colour as the back, the others, which gradually shorten to the exterior ones, are black, largely tipt with white; the two outer ones are scarcely half the length of the middle ones. The whole lower parts are pure white; the feathers covering the thighs being large, like those of the hawk tribe. The legs and feet are light blue, the toes placed two before and two behind, as in the rest of the genus: the bill is long, a little bent, very broad at the base, dusky black above, and yellow below; the eye hazel, feathered close to the eyelid, which is yellow. The female differs little from the male; the four middle tail-feathers in her are of the same uniform drab; and the white, with which the others are tipt, not so pure as in the male. In examining this bird by dissection, the inner membrane of the gizzard, which in many other species is so hard and muscular, in this is extremely lax and soft, capable of great distension; and, what is remarkable, is covered with a growth of fine down, or hair, of a light fawn colour." A specimen of this bird is said to have been found in this country.

CUCKOO-SPIT. [See CERCOPIDÆ.]

CUCULIDÆ. An extensive family of Passerine birds, characterized by having the toes situated two before and two behind; and so named from including as the typical species the well-known Cuckoo. These birds are for the most part inhabitants of the warm climates, and none permanently reside in countries subject to severe winter cold. They have a slightly arched compressed beak, and a long rounded tail; their wings are moderately long, and they fly with rapidity. They feed on insects, worms, and soft fruits, which they procure while leaping from branch to branch, or flitting from tree to tree: when on the ground they walk awkwardly, on account of the shortness of their tarsi. [See CUCKOO.]

CUCULINÆ. The name given to denote that sub-family of the CUCULIDÆ which consists of the genuine Cuckoos.

CULEX: CULICIDÆ. A genus and family of Dipterous or two-winged insects, consisting of the various kinds of *Gnats*. They are distinguished by the length of the proboscis, and their beautifully tufted antennæ. They generally abound in damp situations, their larvæ being inhabitants of the water. [See GNAT: MOSQUITO.]

CURASSOW. (*Crax.*) A genus of Gallinaceous birds, inhabiting various parts of South America. They are nearly as large as a turkey.

The CRESTED CURASSOW. (*Crax alector.*) This bird is nearly three feet in length. The crest, which it can elevate or depress at pleasure, is composed of twisted black feathers, narrow at the base and broad at the tip: the whole of the upper part of the plumage is of a deep shining black colour, reflecting purple and green shades; the tail is black, generally tipped with white; the abdomen and the inferior tail-coverts are invariably white. The females have a smaller crest, and their feathers are more dull.

CRESTED CURASSOW. (CRAX ALECTOR.)

They associate in small flocks, and at night roost on high trees: their food consists of maize, rice, bananas, and other fruits. The egg is about the size of that of the turkey, and is of a pure white. Native of Guiana, Mexico, and Brazil.

The RED CURASSOW. (*Crax rubra.*) In size this bird may be compared with the turkey, being about two feet six or eight inches in length. It has a large, strong bill; and a crest composed of twisted and curled feathers, broad at the top, and tipped with black: the front and sides of the head, and the top of the neck, are pure white, the feathers being marked at their tips with a black fringe: the breast and the upper parts of the tail are reddish, the under parts a brighter red than the upper: the feet and the bill are of horn colour. The young of this species are beautifully varied: the sides of the head and top of the neck are barred with black and white; the upper part of the plumage, as well as the tail-feathers, are striated with broad, transverse, red and white bands, margined with a black line: as the bird increases in age these bands gradually disappear, and the feathers of the crest, which are at first straight, begin to twist and curl. In their native country these birds are easily tamed, and readily associate with other poultry; but although they are here common enough in menageries, they have never been known to breed. Temminck, however, says, they have once at least been thoroughly acclimated in Holland, where they were as prolific in their domesticated state as any of our common poultry: and Mr. Bennet, alluding to the same subject, observes, "It may not be too much to expect that the Zoological Society may be successful in perfecting what was then so well begun, and in naturalizing the Curassow as completely as our ancestors have done the equally exotic, and, in their wild state, much less familiar breeds of the Turkey, the Guinea-fowl, and the Peacock." Their flesh is both delicate and nutritious.

CURCULIO: CURCULIONIDÆ. A genus and family of snouted Coleopterous insects, including the diamond beetles and other splendidly coloured species; as well as the corn and nut weevils, and a variety of others scarcely less destructive to grain, fruit, and vegetable products in general; several of which are given under their respective names. M. Schönherr has published a voluminous work which describes the numerous species. Mr. Walton, F. L. S., has studied the British *Curculionidæ*, and published excellent papers on all the species found in this country.

CURLEW. (*Numenius.*) A Grallatorial bird, belonging to the *Scolopacidæ*, or Snipe tribe, all of which inhabit the vicinity of waters and marshes, and feed upon worms, &c. The COMMON CURLEW (*Numenius arquata*) measures about two feet in length; and in breadth, from tip to tip, above three feet. The bill is about seven inches long, of a regular curve, and blunt at the end: the upper mandible is black, gradually softening into brown towards the base; the under one flesh-coloured. The head, neck, and wing-coverts are streaked with brown; the back and scapulars are nearly black in the middle, edged and deeply indented with light grey. The breast, belly, and lower part of the back are of a dull white, spotted with black; the quill-feathers are black, the inner webs crossed with white: tail barred with black, on a white ground tinged with red: thighs bare about half way above the knees, of a bluish-colour: the toes are thick, and slightly membranous. The female makes her nest upon the ground, in a dry tuft of rushes or grass; and lays four eggs of a greenish cast, spotted with brown.

The Curlew is met with in most parts of Europe. In Britain their summer haunts are the large, heathy, and boggy moors, where they breed; their food consisting of worms, flies, and insects, which they pick out of the soft mossy ground by the marshy pools. In autumn and winter they depart to the sea-side in great numbers, and there subsist upon worms, marine insects, small crabs, snails, &c. This bird is extremely common in most parts of Europe, and it occurs also in several parts of Asia. In the winter it is gregarious, and it is at all times very shy and difficult to approach; but it will soon become familiar. In Scotland, from its cry it is called the "Whāāp." [For another species, see WHIMBREL.]

CURSORIUS. This genus of birds inhabits the hot regions of Asia and Africa; one species only, and that very rarely, having been found to visit Europe. The CREAM-COLOURED COURSER (named by Temminck *Cursorius Isabellinus*) is ten inches in length; and has a black, curved beak; the forehead, under parts of the body, back, tail, and wing-coverts of a reddish cream-colour; the latter edged with grey: behind the eyes a double black stripe; the throat and belly whitish; the whole of the lateral tail-feathers black towards the tip, with a small spot of white in the centre of the black: legs yellowish. This rare species is a native of Africa; but with its habits we are unacquainted. Two only are on record as seen in England: one which was shot near St. Albans, in East Kent, the seat of W. Hammond, Esq., Nov. 10. 1785; and another, shot in Chamwood Forest, Leicestershire, Oct. 15. 1827. The former of these was observed to run with incredible swiftness, and at intervals to pick up something from the ground; and was so bold as to render it difficult to make it rise from the ground, in order to take a more secure aim on the wing. The note was unlike that of any known bird. A British-killed specimen of this desert-loving bird is preserved in the fine collection in the British Museum.

CURUCUI. [See TROGON.]

CURVIROSTRA. [See CROSSBILL.]

CUSHAT. The Wood-pigeon [which see].

CUTTLE-FISH. (*Octopus.*) A molluscous animal, belonging to the genus *Sepia*, order *Cephalopoda;* and sometimes called the Ink-fish. It is of an oblong form, about six inches in length, and three and a half in breadth. The body is somewhat oval; but it is broadest near the head, and grows smaller towards the extremity, where it is obtusely pointed. The head is surrounded with eight arms and two feet; the two feet being nearly similar in their structure to the arms, or tentacula, but considerably larger in their dimensions. The head is divided from the sac on all sides by a neck. The sac is furnished on each side throughout its whole length with a narrow fin. The suckers are irregularly scattered on the arms

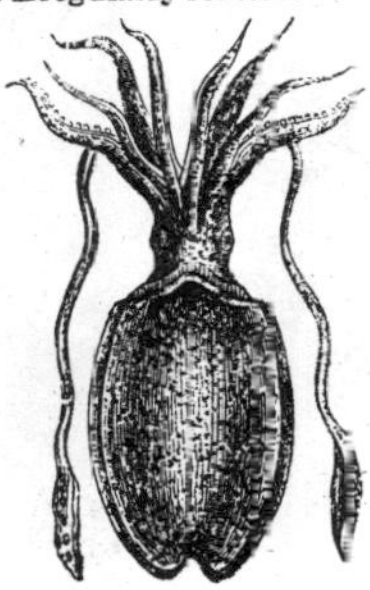

CUTTLE FISH.—(OCTOPUS.)

and feet. The back is strengthened by a complicated calcareous plate, which plate has been long known in the shop of the apothecary under the name *Cuttle-fish bone*, and was formerly much prized in medicine as an absorbent, but is now chiefly sought after for the purpose of polishing the softer metals. The superior half, or the one next the head, is the longest, rounded at the extremity, and thin. The inferior portion becomes suddenly narrow, and ends in a point. It may be considered as consisting of a dermal plate, concave on the central aspect, having its concavity filled up with layers which are convex on their central aspect. The dermal plate consists of three different laminæ, arranged parallel to one another. The external or dorsal layer is rough on the surface, and marked by obscure, concentric arches towards the summit, formed by minute knobs, which become larger towards the base, where they appear in the form of interrupted transverse ridges. It is uniform in its structure, and the tubercles possess a polish and hardness equal to porcellaneous shells, although they blacken speedily when put in the fire, and contain a good deal of animal matter. On the central side of this layer is one flexible and transparent, similar to horn, and smooth on the surface. The third layer is destitute of lustre; and, in hardness and structure, resembles mother-of-pearl shells. The term *bone* has been improperly applied to this complicated plate; for this substance, in composition, is exactly similar to shell, and consists of various membranes, hardened by carbonate of lime, without the smallest mixture of phosphate. Under the throat there is a vessel or bladder containing a fluid blacker than ink, which the Cuttle-fish, when pursued by its enemies, ejects in considerable quantities; and this, darkening the water all around enables it to escape with facility. The most remarkable species of the genus is the *Sepia officinalis*, which is distinguished from the others by its smooth skin. It inhabits the British seas, and although seldom taken, its "bone" is cast ashore on different parts of the coast, from the south of England to the Zetland isles. It is said that the Cuttle-fish is considered a luxury by all classes of the Sandwich islanders, and that when fresh and well cooked it is excellent, being in consistence and flavour not unlike the flesh of a lobster's claw.

The singular habits of the Cuttle-fish did not escape the notice of Mr. C. Darwin, while at the Cape de Verd islands. "I was much interested," says he, "on several occasions, by watching the habits of an Octopus or Cuttle-fish. Although common in the pools of water left by the retiring tide, these animals were not easily caught. By means of their long arms and suckers, they could drag their bodies into very narrow crevices: and when thus fixed, it required great force to remove them. At other times they darted, tail first, with the rapidity of an arrow, from one side of the pool to the other, at the same instant discolouring the water with a dark chestnut-brown ink. These animals also escape detection by a very extraordinary, chameleon-like power of changing their colour" [which Mr. D. minutely describes]. He then adds: "I was much amused by the various arts to escape detection used by one individual, which seemed fully aware that I was watching it. Remaining for a time motionless, it would then stealthily advance an inch or two, like a cat after a mouse; sometimes changing its colour: it thus proceeded, till, having gained a deeper part, it darted away, leaving a dusky train of ink to hide the hole into which it had crawled. . . . That it possesses the power of ejecting water there is no doubt, and it appeared to me certain that it could, moreover, take good aim by directing the tube or siphon on the under side of its body."

CYAMUS, or WHALE LOUSE. A small crustaceous animal belonging to the order *Læmodipoda*. This minute claw-limbed creature, with others allied to it, inhabits the seas of northern and temperate Europe,

WHALE LOUSE.
(CYAMUS CETI.)

and the Southern Seas. As its name indicates, it infests different species of Cetacea; living on their rough skin and gnawing it more or less deeply. Some are found congregated on the heads of the Whale; while others are wanderers, and crawl about various parts of their bodies. It is well worthy of notice that such immense creatures, are subject to parasitic, as well as many other enemies. [See WHALE.]

CYCLOIDEA. [See SUPPLEMENT.]

CYCLOPS. A genus of minute Crustaceans, comprising numerous species, some of which belong to fresh-water, while others are marine. The fresh-water species abound in the muddiest and most stagnant pools, and often too in the clearest springs: the marine species are to be found, often in vast numbers, among the sea-weeds, in small pools on the sea-shore; others there are which inhabit the open ocean, where, by the luminous properties they possess, they contribute to its phosphorescence. They take their name from having but one eye. They have all eight or ten legs, and the abdomen is terminated by a bifid tail adapted for swimming. Dr. Baird has monographed the British species.

CYCLOPTERUS. [See LUMP-FISH.]

CYGNUS. [See SWAN.]

CYNIPS: CYNIPIDÆ. A genus and family of Hymenopterous insects, commonly known by the name of *Gall-flies*. These in-

GALL INSECT,—(CYNIPS QUERCUSFOLII.)

sects puncture, with their ovipositor, the surface of the leaves, buds, and stalks of various plants and trees; and they increase the aperture by means of the toothed edge, forming a kind of saw, with which the extremity of this organ is armed. In this aperture they deposit, with the egg, a drop of fluid, which, from its irritating quality, produces different kinds of gall-nuts, according to the species of *Cynips* by which it has been punctured.

The excrescences on the leaves and buds of trees which are called *Galls* are of various shapes: many are spherical; others are hairy or tomentose, the surface emitting numerous fibrous threads; others resemble buds, flowers, &c.; and there are a few which are flat: in most of the species a single gall supports only a single gall-insect; while others are *polythalamous*, serving for the residence of many. "Probably," says Mr. Westwood, "no insect has been of greater benefit to mankind than the *Cynips Gallæ tinctoriæ*, the galls of which are the common gall-nuts of commerce, growing upon the *Quercus infectoria* in the Levant, and which are employed in the manufacture of ink. The galls are of the size of a boy's marble, very hard and round, with various tubercles on the surface; they contain but a single inhabitant, which may often be found in the interior on breaking the galls. This species resembles some of our English species which reside in globular oak-galls in its habits of undergoing its transformations within the gall, leaving a great portion of the gall unconsumed. Those galls which are gathered before the insect has escaped (and which consequently contain most astringent matter) are known in trade under the name of black or blue galls and green galls; but those from which the insect has escaped are called white galls.

CYNTHIA. A genus of Diurnal Lepidoptera, belonging to the Nymphalidæ: we restrict ourselves to the mention of the British species.

CYNTHIA CARDUI; or PAINTED LADY. This species of Butterfly is noted for the irregularity of its appearance in particular districts. The wings in general

PAINTED LADY BUTTERFLY.
(CYNTHIA CARDUI.)

are of a brownish yellow colour, dappled with black spots or clouds of various shapes; especially those parts of the upper wings

UNDER SIDE OF PAINTED LADY BUTTERFLY.

next the apices, which are all black, except five white spots on each side. On the under side the superior wings are of a fine light orange-colour, but they become of a deep crimson near the body; and the parts towards the tips are of a pale brown. having five white spots, corresponding with those on the upper surface; the remaining parts are dappled with black, with one large white and nearly square spot on the sector edge. The inferior wings are of a pale yellow brown, dappled with dark brown spots nearly equal in size; and near the lower border are five ocelliform spots. The caterpillar, which feeds on thistles, nettles, mallows, &c., is a grayish brown, with yellow lateral lines. The chrysalis is grayish, with golden dots, and whitish brown longitudinal streaks.

"On the blue heads of the pasture scabious (*Scabiosa succisa*,)" says the author of the Journal of a Naturalist, "we occasionally see, toward the end of the summer, the Painted Lady Butterfly (*Cynthia cardui*); but this is a creature that visits us at very uncertain periods, and is vivified by causes infinitely beyond the comprehension of the entomologist, seeming to require a succession and variety of seasons and their change, and then springing into life we know not how. This was particularly obvious in the summer of 1815, and the two following, which were almost unceasingly cold and rainy; scarcely a moth or butterfly appeared. And in the early part of 1818, the season was not less ungenial; a few half-animated creatures alone struggled into being; yet this "painted lady" was fostered into life, and became the commonest butterfly of the year: it has, however, but very partially visited us since that period. The keenest entomologists, perhaps, would not much lament the absence of this beauty, if such cheerless seasons were always requisite to bring it to perfection. Some years ago a quantity of earth was raised in cutting a canal in this county; and in the ensuing summer, on the herbage that sprang up from this new soil on the bank, this butterfly was found in abundance, where it had not been observed for many years before. In some particular seasons we have acres of this scabious in bloom, during the months of September and October, giving a tender shade of lavender colour to the whole field, affording now great pleasure to the entomologist, by reason of the multitude of insects that resort to it for the honey in the tubular florets in the plant. Late as this period is, I have seen, in some bright morning, besides multitudes of bees, flies, and such creatures, eleven different species of lepidopterous insects, feeding and balancing on the blue heads and glancing their gay wings in the sunny beam." This species is, apparently, found every where: and in the Museum collection are specimens from nearly every part of the world.

CYNOCEPHALUS. [See BABOON.]

CYPHONIDÆ. A group of beetles detached from the *Cebrionidæ* on account of their small size; hemispheric, depressed, or ovate, and rather soft bodies, and furcate labial palpi: they are of dull colours, and attached to plants in damp situations; and they fly and run with agility. In some species the hind legs are formed for leaping.

CYPRÆA. A genus of univalve shells, called also *Cowries*, remarkable for the brilliancy of their colours, and for the high polish of which they are susceptible. The animal they contain is a Gasteropodus Mollusc; and the shell of one species, the *Cypræa moneta*, is well-known in commerce as the current coin of the natives of Siam, Bengal,

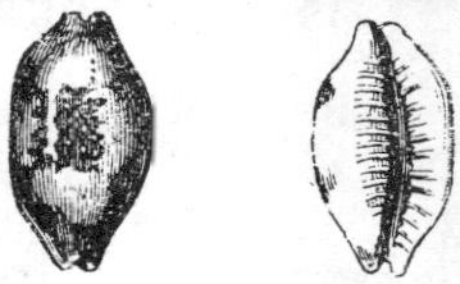

COWRIE.—(CYPRÆA STOLIDA.)

and many parts of Africa: in the latter it is collected by the female negroes, and is thence sent to distant countries. The *Cypræidæ*, or shells of the Cypræa genus, are generally semi-oval, having their mouths placed in their flat part; their spires are not externally visible, the revolutions being performed within the body of the shell; the aperture, or mouth, is a narrow opening, running the entire length of the shell; the lips, which are near each other, are broad, turning inwards, and serrated; and the two ends or extremes on the upper part are very prominent. At one end there appears a wry channel, or opening; the other end has also an opening, but placed perpendicularly. *Cypræidæ* abound both in the old and new world, but their greatest development both in point of size and number of species takes place in warm climates. In the Friendly Islands, permission to wear the *Cypræa aurantia*, or Orange Cowry, as an ornament, is only granted to persons of the highest rank. Mr. Gray, F.R.S., has published an admirable monograph of the Cowries; and the Messrs. Sowerby subsequently figured all the species. They are much prized by collectors.

CYPRINIDÆ. A family of Malacopterygious abdominal fishes, of which there are many genera, the principal being *Cyprinus carpio*, the common Carp. They are for the most part fresh-water fishes; live on

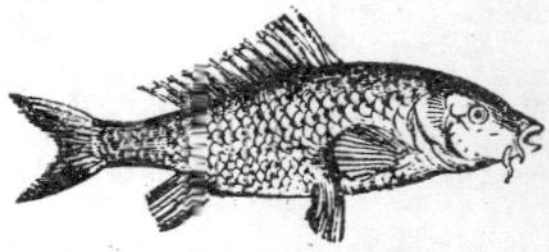

CARP.—(CYPRINUS CARPIO.)

aquatic plants; and are characterized by their small mouth, and by their feeble and generally toothless jaws. They have a scaly body, no adipose fin, a stomach destitute of a *cul de sac*, and no pyloric cæca. The different varieties of Gold and Silver Fish, the Gudgeon, Tench, Bream, Roach, Bleak, Minnow, and many other well-known pond and river fishes, belong to this family.

CYPRIS. A genus of Entomostracous Crustacea, containing numerous species, many of which are British. Their general appearance, to one ignorant of natural his-

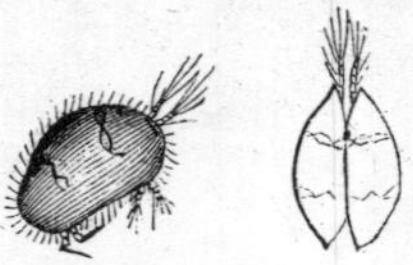

CYPRIS VIDUA

tory, is that of a bivalve shell. They are mostly found in fresh or stagnant water, where they sometimes abound in myriads. Delicate though they are, yet there are abundant indications, in strata of different formations, of their existence in a previous condition of the world, and, like shells and some other fossils, they form curious and interesting medals of creation. [See ENTOMOSTRACA.]

CYPRUS BIRD. The Black-cap (*Sylvia atricapilla*), which has received this name from its frequency in the isle of Cyprus. It is by no means uncommon in this country. [See BLACK-CAP.]

CYTHEREA. A genus of marine Mollusca, of which there are numerous species, inhabiting the Indian and Atlantic Oceans. The shell is equivalve, inequilateral, triangular or transverse; ligament on the longest side; four cardinal teeth in one valve, and three in the other. In their beauty and colouring these shells much resemble the *Venus*. One species, the *Cytherea lusoria*, is found in the Chinese seas: it is used by the Japanese and Chinese in certain games, and the interior is painted by them of various colours.

DAB. (*Pleuronectes limanda.*) This species of flat-fish is of a very broad, ovate shape; generally of a uniform pale brown colour on the upper side, and white on the under side; the lateral line is very much curved at the beginning, but afterwards proceeds straight to the tail. It is usually caught along with Plaice and Flounder, from which it is readily distinguished by the roughness of its scaly surface, and its flesh is considered superior to either. It feeds on small fish and crustacea; and is in highest perfection for the table in February, March, and April. It is caught on various parts of our coasts, averages about eight or nine inches in length, and is well known in the London markets.

The LEMON DAB, SMOOTH DAB, or SMEAR DAB (*Pleuronectes lævis*), is much larger than the preceding, more rare, and its flesh is equally esteemed. It approaches nearer to a rhomboid in form than any of the genus; and is a handsomer fish than the common Dab, on account of the various shades of reddish brown and yellow which are seen on its upper side. The body is smooth, and covered with a mucous secretion: the head is very small; the eyes are placed very near each other; and the mouth is full of small teeth.

The LONG DAB (*Pleuronectes limandoïdes*), as its name imports, greatly exceeds the other species in length, approaching, in fact, much nearer to that of the sole; the form of the body being an elongated oval, almost equally pointed at both ends. It is covered with harsh scales; is an inhabitant of the northern seas; and preys on small crabs and other crustaceous and molluscous animals.

DACE. (*Cyprinus leuciscus.*) The Dace, or, as it is sometimes called, the Dare, or Dart, is a fish of the *Cyprinidæ* family, and is rather like the Roach, both in habits and appearance, but the former is more local and less plentiful than the latter. It is found in clear and quiet streams, and feeds upon worms and other soft substances: it is

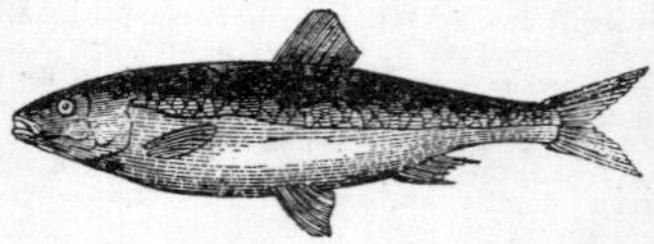

DACE.—(CYPRINUS LEUCISCUS.)

gregarious, extremely vivacious, and, like the Trout, it will occasionally rise either at the day-fly, or at an artificial fly. The head of the Dace is small, and the muzzle pointed; the back is slightly elevated, and the tail much forked; the scales are rather small, the sides and belly silvery, and the general form of the body elegantly shaped. During the months of April and May the Dace is in the highest season. "The Dace," says Mr. Yarrell, "is frequently used as bait for Pike in trolling, on account of its silvery brightness; but where live bait are required, as for night hooks, Roach are preferable, on account of their being more tenacious of life." Although this fish, in warm weather, seldom refuses a fly at the surface of the stream, and thereby affords an expert angler much diversion; during the cold months the bait must be sunk within three inches of the bottom.

DACELO. A genus of Kingfishers, from New Holland. For an account of them we are indebted to Mr. Gould. [See KINGFISHER.]

DACNIS. A genus of small and elegant Passerine birds inhabiting Mexico. The colour of the body is cerulean blue; the forehead, shoulders, wings, and tail are black; and it has a sharp, conical bill.

DACOLYTHUS. An appellation given to a small fish, a species of the Loach, from two to three inches long. The head is broader and flatter than the rest of its body, which is brown with black spots: there are two beards on each side of the upper jaw: and on the gill-covers are two sharp prickles. It is partial to shallow brooks with stony bottoms.

DACTYLOPTERUS. A genus of Acanthopterygious fishes, the generic characters of which are, a large and long flat head, rising suddenly from the muzzle; the pre-operculum furnished with an elongated strong spine; the jaws armed with masses of minute conical teeth; six branchiostegous rays; the sub-pectoral rays numerous, very long, and connected by a membrane; body covered with hard carinated scales. By means of their large fins, these fishes dart out of the water when pursued, and are able to sustain themselves in the air for several seconds. There are only two species; one, the Flying Gurnard (*Trigla volitans* of Linnæus), which inhabits the Mediterranean; the other, the *Dactylopterus orientalis* of Cuvier, inhabits the Indian seas. Neither of these, however, must be confounded with the common Flying-fish, which belongs to the genus *Exocetus*.

DAGGER [MOTHS]. A name given by collectors to Moths of the genera *Acronycta* and *Diurna*.

DALMATIAN, DANISH, SPOTTED, or COACH-DOG. This variety of the canine race, known by each of the foregoing names, is easily distinguished from all other species by its numerous small black spots. Its form is rather elegant, partaking both of the hound

DALMATIAN DOG

and pointer; its limbs are tolerably stout, and its general appearance is showy. The animal has, however, few claims on us for its services; neither its scent nor its sagacity being such as to render it the useful companion of man. It is kept chiefly as an appendage to the carriage, and shows an instinctive fondness for the stable.

DAPHNIA. A genus of minute Crustaceans, belonging to the order *Branchiopoda;* the best known and most interesting of which is the *Daphnia pulex* (sometimes termed *Monoculus*, from having but one eye). This animalcule, which is popularly called the Arborescent Water-flea, is a favourite microscopic object. The head is prolonged into a snout, and provided with a single central compound eye: it is also furnished with antennæ, which serve as oars, to propel it through the water by a series of short springs or jerks. The *Daphnia pulex* is very abundant in many ponds and ditches, being seen on the surface in the mornings and evenings, as well as in cloudy weather; but seeking the depths of the water during the heat of the day. They are extremely prolific; and when, in the summer time, they assume a red colour, the swarms which abound in stagnant water give it the appearance of its being occasioned by blood. In this country, Dr. Baird, F.L.S., and Mr. John Lubbock, F.R.S., have thrown great light upon their structure, habits, and development.

DART [MOTHS]. A name given by collectors to Moths of the genus *Agrotis*.

DART-SNAKE. [See ACONTIAS.]

DARTER. (*Plotus.*) The Darters are a genus of web-footed birds, of the Pelican tribe, found near the eastern coasts of the tropical parts of America, and on the western coast of tropical Africa. The general form of their bodies is like that of the gulls: they have a long slender bill, broad at the base, but compressed and pointed at the tip: they perch on trees by the sides of lakes, lagoons, and rivers; and, after hovering over the water, they suddenly dart at their finny prey with unerring aim. Their movements are alike rapid and graceful.

"The DARTER or SNAKE BIRD, (*Plotus melanogaster*,)" says Wilson, "seems to have derived its name from the singular form of its head and neck, which at a distance might be mistaken for a serpent. In those countries where noxious animals abound, we may readily conceive that the appearance of this bird, extending its slender neck through the foliage of a tree, would tend to startle the wary traveller, whose imagination had pourtrayed objects of danger lurking in every thicket. Its habits, too, while in the water, have not a little contributed to its name

DARTER, OR SNAKE-BIRD.
(PLOTUS MELANOGASTER.)

It generally swims with its body immerged, especially when apprehensive of danger, its long neck extended above the surface, and vibrating in a peculiar manner. The first individual that I saw in Florida was sneaking away, to avoid me, along the shore of a reedy marsh, which was lined with alligators, and the first impression on my mind was, that I beheld a snake, but the recollection of the habits of the bird soon undeceived me. On approaching it, it gradually sank, and my next view of it was at many fathoms distance, its head merely out of the water. To pursue these birds at such times is useless, as they cannot be induced to rise, or even expose their bodies. Wherever the limbs of a tree project over, and dip into the water, there the Darters are sure to be found, these situations being convenient resting-places for the purpose of sunning and preening themselves, and, probably, giving them a better opportunity than when swimming, of observing their finny prey. They crawl from the water upon the limbs, and fix themselves in an upright position, which they maintain in the utmost silence. If there be foliage, or the long moss, they secrete themselves in it in such a manner that they cannot be perceived unless one be close to them. When approached, they drop into the water with such surprising skill, that one is astonished how so large a body can plunge with so little noise, the agitation of the water being apparently not greater than the gliding of an eel.—Formerly the Darter was considered by voyagers as an anomalous production, a monster partaking of the nature of the snake and the duck; and in some ancient charts which I have seen, it is delineated in all the extravagance of fiction."

DARTER-FISH. [See TOXOTUS.]

DASYORNIS. A genus of insectivorous birds, belonging to the great family of Thrushes, and found throughout the greater part of Southern Australia. The BRISTLE-BIRD (*Dasyornis Australis*) inhabits reed-beds and thickets, but owing to its recluse habits is a species familiar to few persons. It carries the tail erect, and threads its way through the thickets with great dexterity; but its powers of flight appear to be very limited. The wings, tail-coverts, and tail, are rufous brown, the latter indistinctly barred with a darker tint; under parts brownish grey; bill brown; legs grayish brown. Another species, of a smaller size, called the LONG-BILLED BRISTLE-BIRD (*Dasyornis longirostris*), is a native of Western Australia, and bears a very close resemblance both in the character and colouring of its plumage to the one above described.

DASYPROCTA. A genus of Rodent Mammalia. In disposition and the nature of their flesh they resemble Hares and Rabbits, which they in some degree represent in the Antilles and hot parts of America. They employ their fore feet to hold up food to their mouth. [See AGOUTI.]

DASYPUS. A genus of Rodent animals, very remarkable among the *Mammalia* for the scaly and hard shell-like armour which, divided into regular compartments, covers their head and body, and often the tail. [See ARMADILLO.]

DAY-FLY. [See EPHEMERA.]

DECAPODA. An order of *Crustacea*, containing those in which we find the highest general organization, the most varied habits, and such as are the most useful to man as food. Their growth is slow, and their habits are mostly aquatic: they are naturally voracious; and they are armed with a pair of powerful claws, by which they seize their food, and convey it to the mouth. In this order are included Crabs, Lobsters, Prawns, Shrimps, &c. [which see]. For the history of the British species, see Dr. Leach's "Malacostraca," or, as more easily accessible, the elegant work on British Crustacea, by Professor Bell, in which are figures and descriptions of all the British species.

DEATH'S-HEAD HAWK-MOTH. A remarkable Lepidopterous insect, belonging to the family *Sphingidæ*. [See ACHERONTIA ATROPOS.]

DEATH-WATCH. (*Anobium tesselatum.*) Among the popular superstitions which the almost general illumination of modern times has not been able to obliterate, as Dr. Shaw very truly observes, the dread of the Death-watch may well be considered as one of the most predominant; yet it must be allowed to be a very singular circumstance that an animal so common should not be more universally known, and the peculiar noise which it occasionally makes be more universally understood. The insect in question is a small beetle belonging to the timber-boring genus *Anobium;* and the popular superstition alluded to is, that when its beating is heard, it is a sign that some one in the house will die before the end of the year. It is chiefly in the advanced state of spring that this little creature commences its sound, which is no other than the call or signal by which the male and female are led to each other, and which may be considered as analagous to the call of birds; though not owing to the voice of the insect, but to its beating on, or striking, any hard substance with the shield or fore-part of its head. The prevailing number of distinct strokes which it beats is from seven to nine or eleven; and this very circumstance may perhaps still add to the ominous character which it bears among the vulgar. These sounds or beats are given in pretty quick succession, and are repeated at uncertain intervals; and in old houses where the insects are numerous, may be heard at almost any hour of the day; especially if the weather be warm. The sound exactly resembles that which may be made by tapping moderately hard with the finger-nail on a table. The insect is of a colour so nearly resembling that of decayed wood, viz. an obscure greyish brown, that it may for a considerable time elude the search of the inquirer. It is about a quarter of an inch in length, and is moderately thick in proportion, and the wing-shells are marked with numerous irregular variegations of a lighter cast than the ground-colour. It is singular that this insect may so far be familiarized as to be made to beat occasionally, by taking it out of its confinement, and beating on a table or board, when it will readily answer the noise, and will continue to beat as often as required. I cannot conclude this slight account of the Death-watch, says our author, without quoting a sentence from that celebrated work the Pseudodoxia Epidemica of the learned Sir Thomas Brown, who on this subject thus expresses himself: "He that could eradicate this error from the minds of the people would save from many a cold sweat the meticulous heads of nurses and grandmothers." In their larva state these insects greatly injure old furniture, by perforating numerous small round holes in it.

DEER. (*Cervus.*) Among the various animals which embellish the forests and animate the solitudes of nature, none are superior to the cervine race. These well-known ruminants are distinguished from the antelopes by their horns, which are composed of a bony substance, caducous, or falling off annually, and again renewed of a larger size than in the preceding year. The form of these is various. Sometimes they

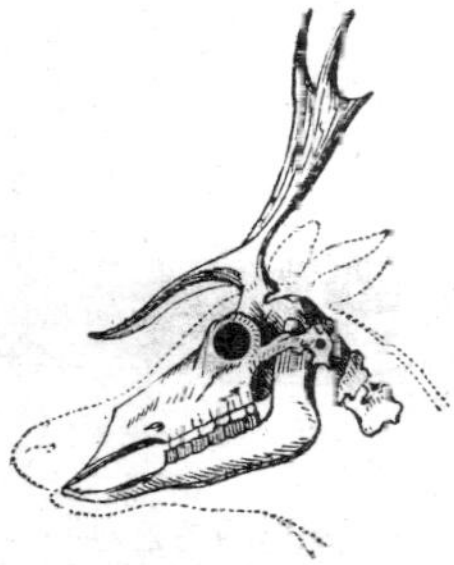

SKULL OF STAG.

spread into broad palms, which send out sharp snags around their outer edges; sometimes they divide fantastically into branches, some of which project over the forehead, whilst others are reared upward in the air, or they may be so reclined backwards, that the animal seems almost forced to carry its head in a stiff, erect posture: yet, in whatever way they grow, they appear to give an air of grandeur to the animal. It may, then, speaking in general terms, be said, that the easy elegance of their form, the lightness of their motions, their size, their strength, their fleetness, and the extraordinary development of those branching horns, which seem fully as much intended for ornament as defence, all contribute towards placing them in the foremost rank of quadrupeds.

RED DEER, or STAG. (*Cervus elaphus.*) Before we speak of the habits &c. of this noble animal, it will be well to enter into a few particulars relative to its distinguishing characteristic, the horns. The first year the stag has properly no horns, but only a kind of corneous excrescence, short, rough, and covered with a thin hairy skin; the second year the horns are single and straight; the third year they have two antlers; the fourth, three; the fifth, four; and the sixth, five. When arrived at the sixth year, the antlers do not always increase; and though the number may amount to six or seven on each side, the Stag's age is then estimated rather from the size and the thickness of the branch that sustains them, than from their number. The proportional length, direction, and curvature of the antlers vary; and it often happens that there is one more or less on the

RED DEER—MALE.—(CERVUS ELAPHUS.)

one side than on the other: the horns also become larger, the superficial furrows more marked, and the burr is more projecting. Notwithstanding their magnitude, these horns are annually shed in the spring of the year, and succeeded by new ones. Of the old horns, which are of a solid, firm texture, a variety of domestic articles are made; but while young they are remarkably soft and

RED DEER.—FEMALE AND YOUNG (CERVUS ELAPHUS.)

tender; and the animal, as if conscious of its want of power, instantly retires from the rest of the herd, and, hiding itself in thickets and unfrequented places, ventures abroad for the sake of pasture only at night. It is nearly three months before the new horns attain their full growth and solidity; and then, by rubbing them against the boughs of trees, they at length clear them of that covering of skin, which had before contributed to their growth and nourishment. "The growth of the horns," says Mr. Bell, "is an astonishing instance of the rapidity of the production of bone under particular circumstances, and is certainly unparalleled in its extent in so short a period of time. A full grown Stag's horn probably weighs twenty-four pounds; and the whole of this immense mass of true bone is produced in about ten weeks. During its growth the branches of the external carotid arteries, which perform the office of secreting this new bone, are considerably enlarged, for the purpose of conveying so large a supply of blood as is necessary for this rapid formation. These vessels extend over the whole surface of the horn as it grows, and the horn itself is at first soft and extremely vascular, so that a slight injury, and even merely pricking it, produces a flow of blood from the wound. It is also protected at this time with a soft, short, hairy or downy coat, which is termed the velvet; and hence the horns are said to be in 'the velvet' during their growth."

The Stag is supposed to have been originally introduced into our own island from France, where it is very common: but it has been in a great degree expelled from most parts of this kingdom to make way for the common, or Fallow Deer, the venison of which is far superior to that of the Red Deer, and the animal itself of a more manageable and placid disposition. The Stag has a fine eye, an acute smell, and a good ear. When listening, he raises his head and erects his ears. When going into a coppice or other half-covered place, he stops to look round him on all sides, and scents the wind, to discover if any object be near that might disturb him. Though a simple, he is a curious and crafty animal. When hissed or called to from a distance, he stops short, and looks steadfastly, and, with a kind of admiration, at horses or men; and if the latter have neither arms nor dogs, he moves on without betraying any symptoms of alarm. He eats slowly; and after his stomach is full, he lies down and ruminates at leisure.

In Dr. A. T. Thomson's notes to an edition of "THE SEASONS," by his celebrated namesake, we find appended to line 454, (*Autumn*,)—

> "The big round tears run down his dappled face;"

the following very apposite remarks:—"This supposed peculiarity of the Stag to shed tears is noticed by several poets, but by none so strikingly as by Shakspere* and our author: but, indeed, it is not wonderful that it was the popular belief before it was noticed by poets, for the eyes of the Stag, and nearly all the deer tribe, display a peculiarly weeping aspect. This is more obviously displayed in the male than in the female. It depends on a remarkable glandular sinus, *crumen*, or *tear-pit*, placed at the inner angle of each eye, close to the nose without having any communication with it, or with what are termed the lachrymal passages. It is composed of a fold of the skin, and is capable of being opened and shut at the pleasure of the animal. It is furnished at the bottom with a gland, which secretes an oily, viscous substance, of the colour and consistence of the wax of the ears, and which hardens and becomes black when exposed to the air. The precise function of this organ is unknown, although many conjectures have been offered in explanation of it: and there can be no doubt that it serves some important purpose in the economy of the animal. The moistened moving edges of the sinus have been mistaken by general observers and the poets, for precious tears let fall over that part of the eyelids; and, in our love for the romantic, we almost regret to be undeceived."

In winter and spring this animal rarely ever drinks, the dews and tender herbage being sufficient to satisfy his thirst; but during the parching heats of summer, he not only frequents the brooks and springs, but searches for deep water wherein to bathe and refresh himself. He swims with great ease and strength, particularly when he is in good condition, his fat contributing to his buoyancy. His voice is stronger, louder, and more tremulous, in proportion as he advances in age; and during the rutting season it is really fearful. The cry of the *hind*, or female, is not so loud as that of the male, and she is never excited but through apprehensions for the safety either of herself or her young. Like all the rest of the Deer tribe, except the Elk, the female is destitute of horns; she is also more feeble and unfit for hunting than the male. The pairing season is in August; the time of gestation is between eight and nine months; and she seldom produces more than one at a time. The usual season of parturition is about May, during which these animals are very assiduous in concealing and tending their young; nor is this a needless precaution, since almost every animal of the canine or feline kind is then an active enemy; nay, unnatural as it may seem, the Stag himself is also one of their most dangerous assailants. At this season, therefore, the courage of the male seems to be transferred to the female: for she resolutely defends her offspring; and if pursued by the hunter, exposes herself to great apparent danger, for the purpose of diverting his attention from the object of her parental regard. The *Calf* (the name by which the young of this animal is called) never quits the dam during the whole summer; and in winter, the hind, together with all the males under a year old, assemble in herds, which are more or less numerous in proportion to the mildness or severity of the season. At the approach of spring they separate, none but those of the age of one year remaining associated. They are, however, generally gregarious; and only danger or necessity can possibly divide them.

"When a Stag stands at bay," says the accomplished Editor of Thomson's Seasons, before quoted, "his instinct leads him to do so in a river or a lake, if either be near; in which case he has a great advantage over the dogs, for he firmly stands and holds his

---

* ———— "the big round tears
Coursed one another down his innocent nose
In piteous chase; and thus the hairy fool,
Much marked of the melancholy Jaques,
Stood on the extremest verge of the swift brook
Augmenting it with tears."
As You Like It, act ii. sc. 1.

position, whilst they swim powerless around him. On land, even, a Stag at bay has great advantage over the hounds, who exhaust themselves with their clamour, whilst he is in a comparative state of rest, and recovers his wind." Powerful as the Stag is, he has never been known to attack a man, unless he has been driven into a corner, and hard pressed, without the means of escape. With regard to hunting the Stag, the pursuit, as carried on in the Highlands, is one capable of rousing all the manly ardour and energy of youth and manhood, whilst all the powers of both body and mind are called into action. The beauty, graceful, and magnificent bearing of the animal, his sagacity in evading the stratagems of the hunter, or deer stalker, and his courage when at bay, add greatly to the pleasure of the chase. In stalking deer, the animal is generally shot; but if he is only wounded, and has power to fly, then the dogs are slipped to the pursuit. But, in olden times, the chief reliance for pulling and killing the deer, was in the dogs; and the fleetness and courage of their hounds were the pride of nobles and kings.'

The food of Stags varies according to the season. In autumn they search for the buds of green shrubs, the flowers of broom or heath, the leaves of brambles, &c. During the snows of winter they feed on the bark and moss of trees; and in mild weather they browse in the fields. In the present cultivated state of this country, Stags are almost unknown in their wild, natural condition; and such as remain amongst us are kept under the name of Red Deer, together with the Fallow Deer; but their excessive ferocity during the rutting season, and the coarseness of their flesh, have contributed in a great measure to effect their almost total extermination. In Scotland, however, they still exist in considerable numbers; and though it was deemed necessary to abolish Stag-hunting by act of parliament, in consequence of the multitudinous gatherings of the clans, upon this pretext, being often made subservient to political purposes, "a Stag-hunt is even in the present day the scene of much of the excitement and profuse hospitality by which this noble sport was characterized in days of yore."

"The Stag is an ancient denizen of the forests of this country. From the most remote periods, it has been the favourite object of the chase; and the severe forest-laws of our earlier Norman monarchs sufficiently attest the importance which they attached to the sport. The afforesting of vast tracts of country, by which not only single cottages were destroyed, but whole villages swept away, and churches desecrated and demolished, was the fertile source of misery to the poorer inhabitants, and of injustice to the ancient proprietors of the soil; and the cruel inflictions of the oppressive laws which were enacted to preserve the Deer, increased tenfold the curse arising from this tyrannical passion for the chase,—for it was a crime less severely penal to kill a man than to destroy or take a Deer."

"The ancient customs and laws of 'Venerie,' that noble science which our simple ancestors looked upon as one of the first accomplishments of the high-bred noble, and a knowledge of which was essential to his education, were formal and technical to a most absurd and ludicrous degree. A few of the terms, betokening the different ages of the Stag and Hind, are still retained, though somewhat altered. The young of either sex is called a CALF; after a few months the male becomes distinguished by the growth of the bossets, or frontal protuberances, on which the horns are afterwards developed, which during the first year are merely rounded knobs, from whence he takes the name of KNOBBER. In the second year they are longer and pointed, and are called dags, and the animal has now the name of BROCKET. In the third year, the first, or *brow antler*, has made its appearance, and the Deer becomes a SPAYAD. In the fourth, the *bez-antler* is added, and he is then termed a STAGGARD. He is a STAG in the fifth year, when the third antler, or *royal*, appears: and in the sixth, the commencement of the *sur-royal*, or crown, is formed; when he takes the name of HART, which name he retains through life. At this time he is called a Hart or Stag of ten, probably because the branches, including the sur-royal, frequently amount to that number. After the seventh year he is said to be *croched*, or *palmed*, or *crowned*, according to the number of branches composing the sur-royal. The female is a *Calf* in the first year, a *Brocket's sister* in the second, and in the third, and ever afterwards, a *Hind*."—*Bell's British Quadrupeds.*

VIRGINIAN DEER. (*Cervus Virginianus.*) This species is found in vast herds in the northern parts of America, and is an animal of great importance to the Indian natives. They are of a light brown colour, and about the size of the Fallow Deer, but their tails are longer. Their horns are slender, bend greatly forwards, and have numerous branches on the interior sides, but no brow antlers. They are of a restless and wandering disposition, and in hard winters are observed to feed much on the different species of *usnea* or string moss, which hangs from the trees: they are also fond of resorting to places impregnated with salt, and vulgarly called *salt-licks*; and it is at these favourite haunts that the hunters generally succeed in killing them. Their flesh, though dry, is very valuable to the Indians, who cure it for their winter provision.

PORCINE DEER. (*Cervus porcinus.*) An Indian species of the cervine genus; about two feet three inches in height; the body clumsy; the legs fine and slender; and the tail about eight inches long. It has slender trifurcated horns, about a foot in length: the colour, on the upper part of the neck, body, and sides is brown; the belly and rump lighter.

[For other species of the genus *Cervus*, or Deer tribe, see AXIS — ELK — FALLOW-DEER — MUNJAC — REINDEER — ROEBUCK — WAPITI.] We may mention that there are many species of Deer, chiefly found in South America and in Asia; specimens of most of the species being in the collection of the British Museum, and many of them are to be seen alive in the Gardens of the Zoological Society, and used to adorn the noble menagerie of the Earl of Derby (President of that Society), at Knowsley, Lancashire.

DEINACRIDA. A genus of Orthoptera belonging to the *Gryllidæ* or Cricket tribe. Our figure is copied from the one that accompanies Mr. White's description in the Zoology of the Voyage of H. M. SS. Erebus and Terror. Mr. W. Stephenson, speaking of it in his remarks on the entomology of New Zealand, says, "It is a peculiarly formidable insect, found in old trees, secreting

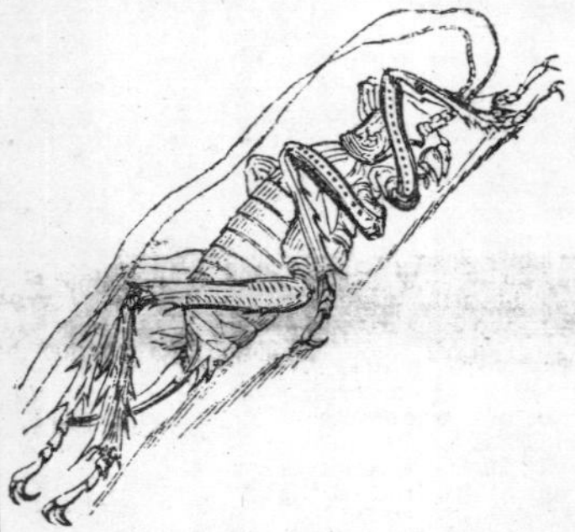

NEW ZEALAND GRAND CRICKET.
(DEINACRIDA HETERACANTHA.)

itself in rents and crevices. It is an abundant species in New Zealand, and is carnivorous. It is called by the Maories *Weta*." The male is distinguished from the other sex (here figured) by its enormous head, the bite of which is very severe. Both sexes are apterous, the female being very prolific in ova. Mr. Stephenson believes there are more than one species of this genus. It is allied to the genus *Anastostomus* of Mr. G. R. Gray.

DELPHINIDÆ. The Dolphin tribe, a family of cetaceous animals, characterized by the moderate size of the head, and usually by the presence of teeth in both jaws. It includes, with the Dolphin and Porpoise, many animals which are ordinarily called Whales; a considerable number of which occasionally visit the northern coasts of Britain. They are in general voracious feeders; and their flesh is for the most part rank, oily, and unwholesome. [See WHALE.]

DEMOISELLE. (*Anthropoïdes Virgo.*) The Demoiselle, or Numidian Crane, is remarkable for the grace and symmetry of its form, and the elegance of its deportment. It measures three feet three inches in length; and has a beak two inches and a half long, the base of which is greenish and the tip red: the irides are crimson: the crown of the head is cinereous; the rest of the head, and neck, black: the feathers of the breast are long and drooping: the under parts of the body, from the breast, the back, and the tail, are bluish ash; the latter and the quills are tipped with black; and the legs are black. This grallatorial bird is a native of many parts of Asia and Africa; and is to be met with along the whole of the southern and eastern shores of the Mediterranean. It delights in damp and marshy places, frequenting those parts in search of small fishes, frogs, &c., which are its favourite food. It is easily domesticated.

There is another species, called the CROWNED DEMOISELLE (*Anthropoïdes Pavonia*), which is less than the one above described, and about the size of the common heron. The crown of the head is covered with soft black feathers, like velvet; on the hind part

CROWNED DEMOISELLE
(ANTHROPOIDES PAVONIA.)

is a tuft of stiff hair, which spreads out on all sides in a globular form; this is four inches in length, and of a reddish brown colour: the sides of the head are bare of feathers; and on each side of the throat hangs a kind of wattle. The general colour of the bird is a bluish ash: the feathers on the fore part of the neck are very long, and hang over the breast; wing-coverts white; the greater ones incline to rufous, and those farthest from the body to black: the greater quills and tail are black, and the secondaries chestnut. The female is black where the male is blue-ash, and the wattles on the throat are wanting. This bird is a tame species, and, like the preceding, is often kept in aviaries: it runs very fast; flies strong, and is able to keep on the wing for a long time together. Another species, the STANLEY DEMOISELLE (*Anthropoïdes paradisea*), is even more elegant than either of the preceding; it is of a light ashy blue colour.

DENDRERPETON. [See SUPPLEMENT.]

DRENDROCOLAPTES, or HOOKED-BILLED CREEPERS. A genus of Tenuirostral Birds, with the bill generally long and curved, the tail feathers stiff and pointed to assist the birds in climbing; the claws are long and curved. There are several species, natives of South America; their general colour is brown, with grey mixtures, and in most of the species there are whitish lines or spots about the head and neck: these birds are marked features in the Fauna of South America. [See FURNARIUS.]

DENDRODUS. [See SUPPLEMENT.]

DENDROLAGUS, or Tree Kangaroo. A

genus of Marsupials. Two species, *D. ursinus* and *D. inustus*, were discovered by M. S. Muller in New Guinea. These were found at Triton Bay, and they also inhabit the interior of the country. They are arboreal in their habits, climbing trees with the utmost facility. The tail is considerably elongated, and in one species (the *D.*

TREE-KANGAROO.—(DENDROLAGUS.)

inustus) of nearly equal thickness throughout. The *D. ursinus* is of a deep blackish brown; the *D. inustus* paler. Now that New Guinea is taken into the possession of the British, specimens and full particulars of this interesting genus may soon be looked for.

DENDROPHYLLIA. The name given to a genus of Polypi, or Madrepores, of coral-like structure. They are of arborescent forms, the stem sending out branches, instead of remaining simply columnar; and these branches again subdividing. The whole structure is covered with a gelatinous or fleshy substance, which, although it has no direct communication with the stomach, seems to constitute the animal, of which the Polypes are only subordinate parts.

DENDROSAURA, or TREE LIZARDS. The name of a tribe of Reptiles containing the Chamæleons, and used by Mr. Gray in his excellent Descriptive Catalogue of the Lizards in the British Museum. The scales of the belly, of the side, and of the back, are granular, and in circular bands: the tongue is worm-like and elongate, and very extensile. The eyes are globular, very mobile, covered with a circular lid pierced with a small central hole. The toes are formed into two grasping opposable groups, which fit them admirably for living on trees. [See CHAMELEON.]

DENTIROSTRES. The name of a tribe of birds, characterized by having a notch and tooth-like process on each side of the margin of the upper mandible. They manifest rapacious habits, and prey on smaller and weaker birds. The *Butcher-bird* will serve as an example of this tribe.

DERMESTES: DERMESTIDÆ. A genus and family of Coleopterous insects, the antennæ of which are elevated and perfoliated transversely. The larvæ or grubs of this tribe devour dead bodies, skins, leather, and almost any animal substance, and are exceedingly destructive to books and furniture. "Although obnoxious in these respects, the insects of this family are of infinite service in the economy of nature, by causing the rapid decomposition of animal matter into a substance fitted for the improvement of the soil, and by their labours, united with those of the Silphæ, Necrophori, &c., destroying such portions of these remains as are left untouched by the Flesh-flies, which only consume the soft portions of the carcasses. Like the perfect insects, their larvæ are seldom observed upon the surface of the matters which they attack." — *Westwood.* This gentleman further observes (in a note), "In some of the Egyptian mummies lately opened, a great number of dead specimens of several species of Dermestes have been discovered in the interior of the body, together with a number of their larvæ, also dead: hence, from the circumstance of these larvæ being found dead in a situation which appears at one time to have been congenial to them, I am induced to think that these insects must have found their way into the body previous to the final operation of embalmment, whereby they were destroyed."

The complete insects are mostly of a lengthened oval shape, and have a habit of withdrawing the head beneath the thorax when handled. — One of the most familiar species is the *Dermestes lardarius*, or Bacon-beetle, which is about a third of an inch in length, and of a dusky brown colour, with the upper half of the wing shells whitish or ash-coloured, and marked with black spots.— Another species, seen in almost every house during the spring and early part of the summer, is the *Attagenus Pellio*. It measures scarcely a quarter of an inch in length, and is of a very dark brown or blackish colour, with a white speck on the middle of each wing-shell.

BACON BEETLE. (DERMESTES LARDARIUS.)

DESMAN, or MUSK-RAT. (*Mygale moschata.*) An insectivorous animal, aquatic in all its habits, and nearly equal in size to

DESMAN, AND FORE-FEET. (MYGALE MOSCHATA)

the Hedgehog. Its muzzle is elongated into a small, very flexible proboscis, which is constantly in motion. It has a long tail, scaly and flattened at the sides; membranous feet; eyes very small; and no external ears. This animal is very common along the rivers and lakes of Southern Russia, where it feeds on worms, the larvæ of insects, and particularly on leeches, which it easily withdraws from the mud by means of its flexible proboscis. It never comes voluntarily on shore, but is often taken in the nets of the fishermen. Its burrow, excavated in a bank, commences under water, and ascends to above the level of the highest floods. Under the tail of the Desman are two small follicles containing a kind of unctuous substance, of a strong musky odour, from which the name of *Musk-rat* is given to it.

DEW [MOTHS]. A name given by collectors to Moths of the genus *Setina*.

DIADEM SPIDER. (*Epeira diadema.*) This spider, so common in the autumn, belongs to Walckenaer's genus Epeira. Its body, when full grown, is nearly as large as a hazel nut, is of a deep chestnut brown colour, and

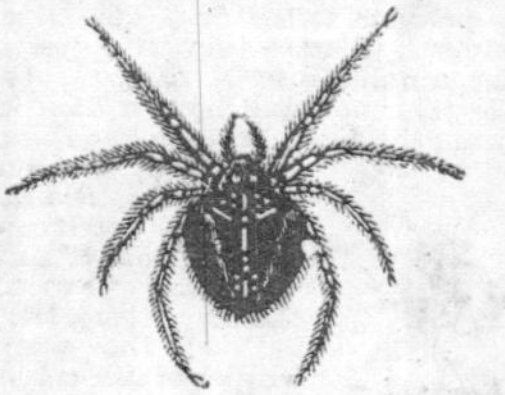

DIADEM SPIDER.—(EPEIRA DIADEMA.)

the abdomen beautifully marked by a longitudinal series of round milk-white spots, crossed by others of a similar appearance, so as to represent in some degree the pattern of a small diadem. It is chiefly seen during the autumnal season in our gardens, where, in some convenient spot or shelter, it forms a large, round, close web of yellow silk, in which it deposits its eggs, guarding this web with a secondary one of a looser texture. The young are hatched in the ensuing May, the parent insects dying towards the close of autumn. At the tip of the abdomen are placed five papillæ or teats, through which the spider draws its thread. The eyes, which are situated on the upper part of the thorax, are eight in number, placed at a small distance from each other. The fangs with which the animal wounds its prey are strong, curved, sharp-pointed, and each furnished on the inside, near the tip, with a small oblong hole or slit, through which is discharged a poisonous fluid into the wound made by the point itself. The feet are of a highly curious structure; the two claws with which each is terminated being furnished on its under side with several parallel processes resembling the teeth of a comb, and enabling the spider to manage with the utmost facility the threads in its web, &c. [See SPIDER.]

DIAMOND BEETLE. (*Entimus.*) This splendid Coleopterous insect belongs to the family *Curculionidæ*, and contains two or three species. It is very abundant in some parts of South America. It is often, with a magnifying glass of no great power, formed

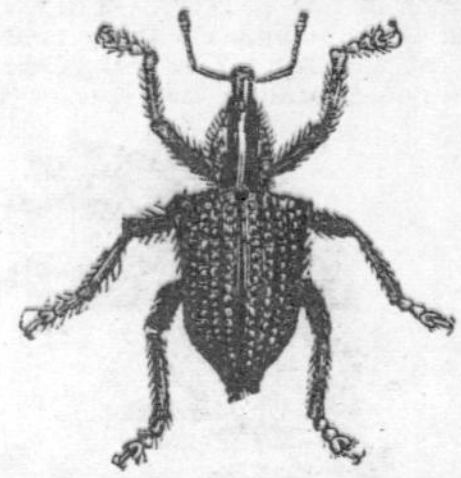

DIAMOND BEETLE.—(ENTIMUS NOBILIS.)

into a very pleasing toy to amuse young people. There are small species of *Curculionidæ* in our own island, however, which are scarcely less brilliant when magnified under a good light, and with sufficient power.

DIAPERIS. [See TAXICORNES.]

DICÆUM. A genus of Tenuirostral Birds, allied to the CREEPERS: they do not use their tails as these birds do; and they are generally brilliant in colour, having more or less of scarlet in their plumage. Species occur in Asia, its islands, and in Australia.

DICHOBUNE. [See SUPPLEMENT.]

DICHODON. [See SUPPLEMENT.]

DICOTYLES. [See PECCARY.]

DICRONOCEPHALUS. A genus of Coleoptera belonging to the family *Cetoniadæ*, described by Mr. Hope. The male, which is the only sex at present known, is remarkable for the two horns on the head, which are bent up. The only known species, *Dicronocephalus Wallichii*, is a native of

DR. WALLICH'S BEETLE, (DICRONOCEPHALUS WALLICHII.)

Northern China, where it has been lately found by Mr. Robert Fortune, the Botanist. It is of a yellowish gray colour, and it is at present very rare in collections (there is one, however, in the British Museum); but, like its congeners, *Narycius* and *Cyphonocephalus*, from India, and *Mycteristes* and *Phædimus*, from the Eastern Islands, the active researches of Indian officers and colonists will make these pretty and singular insects more common.

DICYNODON. [See APPENDIX.]

DIDELPHIDÆ. A family of quadrupeds belonging to the order *Marsupialia*, and consisting of the genus *Didelphis*, or Opossum. They are restricted to America. They are characterized by having ten incisors above and eight below, the canines being one on each side of either jaw, and the molars seven, the four last, or true molars, being crowned with sharp tubercles. The limbs are short; the feet plantigrade; and the toes, which are five on each foot, armed with sharp, strong, curved claws, except the inner toe or thumb on the hinder feet, which is opposable and destitute of a nail. The tail, except at the base, is scaly and naked; and it is usually more or less prehensile. In some species the pouch is entirely wanting, being indicated only by a slight fold of the skin. [See OPOSSUM.]

DIDUNCULUS. A genus of birds found in the South Sea Islands. [See GNATHODON.]

DIDUS. [See DODO, p. 183.]

DIMERA. A section of the order *Homoptera*, comprising much smaller insects than those included in the section *Trimera*, and distinguished from them by having only two joints in the tarsi; with antennæ longer than the head, and composed of from six to ten filiform joints; whilst they differ from the *Monomera* by the winged individuals possessing four wings, the anterior being ordinarily of the same membranous texture as the posterior. The section consists of the families *Psyllidæ*, *Aphidæ*, and *Aleyrodidæ*.

DIMORPHODON. [See SUPPLEMENT.]

DIMYARIA. The name given to the second order of *Conchiferæ*, or Bivalve Shells. It contains a great number of families, which may be grouped into four divisions arising partly from the shape of the foot of its molluscous inhabitant, but chiefly from the more or less perfect manner in which the valves close upon each other.

DINGO, or AUSTRALIAN DOG. This species of the canine race has a very wolf-like appearance. The ears are short and erect; the tail rather bushy; the hair, which

DINGO.—(CANIS FAMILIARIS AUSTRALASIÆ.)

is of a reddish dun colour, is long, thick, and straight. This dog is extremely fierce, and has the same sort of snarling and howling voice as the larger kind of dogs have in general; though by some it has been erroneously said neither to bark nor growl. There is good reason, however, to believe that the Dingo is the descendant of a race once domesticated, which has returned to its wild state.

DINORNIS. A genus of birds allied to the Ostrich tribe, now only found in a fossil state in New Zealand, whence many bones have been sent to this country. One of the species must have been at least fourteen feet high, and it is believed that some specimens may have been still higher. It is known to the natives by the name of *Moa*. For details of its anatomy, the reader should consult Prof. Owen's elaborate Memoir in the third and fourth volumes of the Zoological Society's Transactions.

DINOSAURIA. [See SUPPLEMENT.]

DINOTHERIUM. A genus of extinct herbivorous quadrupeds, of gigantic dimensions; but as only fragments of this huge creature have yet been found, the size of the entire animal cannot be accurately given.

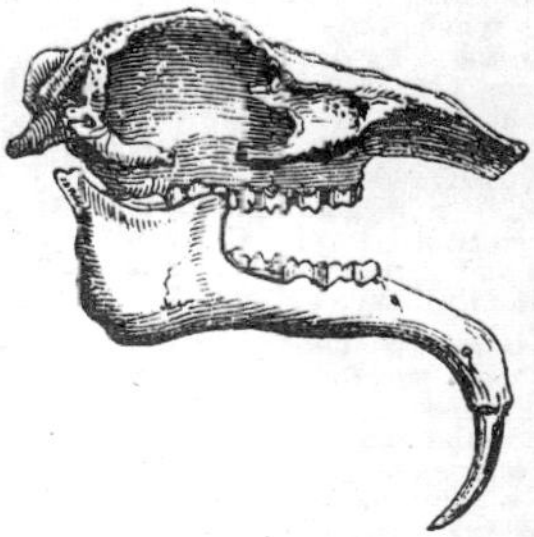

SKULL OF DINOTHERIUM GIGANTEUM.

A skull of one was disinterred at Eppelsheim, in Hesse Darmstadt, in 1836, measuring about four feet in length and three in breadth; from which, according to the calculations of Cuvier and Kaup, the Dinotherium is supposed to have attained the length of eighteen feet. Dr. Buckland, who paid great attention to the remains of this immense specimen of extinct Mammalia, is decidedly of opinion that it was an aquatic animal. "It is mechanically impossible," he observes, "that a lower jaw, nearly four feet long, loaded with such heavy tusks at its extremity could have been otherwise than cumbrous and inconvenient to the quadruped living on dry land. No such disadvantage would have attended this structure in a large animal destined to live in water; and the aquatic habits of the family of Tapirs, to which the Dinotherium was most nearly allied, render it probable that, like them, it was an inhabitant of fresh-water lakes and rivers," &c. The Doctor subsequently says, "Professor Kaup and Dr. Klipstein have recently published a description and figures of this head, in which they state that the very remarkable form and dispositions of the hinder part of the skull show it to have been connected with muscles of extraordinary power, to give that kind of movement to the head which would admit of the peculiar action of the tusks in digging into and tearing up the earth. They further observe that my conjectures respecting the aquatic habits of this animal are confirmed by approximations in the form of the occipital bone to the occiput of *Cetacea;* the Dinotherium, in this structure, affording a new and important link between the *Cetacea* and the *Pachydermata*."

DIODON. A remarkable genus of *Plectognathi*, or bony fishes with soldered jaws.

The DIODON HYSTRIX, commonly termed the Sea-Porcupine, is of a nearly spherical form, sometimes measuring not less than two feet in length; but it possesses the power of inflating or contracting itself at pleasure by means of an internal skin or membrane situated beneath the exterior or spiny covering. Its colour is a pale grey, the back being of a somewhat deeper cast; and the whole body is marked at the base of each spine by a round black spot; the fins being also spotted. When taken by a line and hook, it inflates its body and elevates its spines to the highest possible degree, as if endeavouring to wound in all directions; nor can it be touched without danger until it is dead. It is a native of the Indian and American seas; and its flesh is coarse and worthless.

The OBLONG DIODON (*Diodon atinga*) differs from the former in being of a more lengthened shape, and in having the spines rather round than triangular. Its general colour is grey, deeper on the back, and with a cast of pink on the lower parts like the Diodon Hystrix: it is marked with numerous round black spots; but it is only from twelve to fifteen inches in length. Unless very carefully cleaned, it is dangerous to eat it; for if not absolutely poisonous, the flesh is highly unwholesome. It is a native of the Indian and American seas.

Besides the above, there is the ROUND DIODON (*Diodon orbicularis*), about nine or ten inches in length; which is considered a poisonous fish: PLUMIER'S DIODON (*Diodon Plumieri*), a species very nearly allied to the Oblong Diodon: and the PATCHED DIODON (*Diodon liturosus*), which inclines to a globular shape, and is marked on each side of the body with an oval patch and two transverse ones; and on the back a round spot encircling the dorsal fin: spines white with brown tips, and all the fins greenish yellow.

"One day," says Mr. Darwin (while on the coast of Brazil), "I was amused by watching the habits of a Diodon, which was caught swimming near the shore. This fish is well known to possess the singular power of distending itself into a nearly spherical form. After having been taken out of water for a short time, and then again immersed in it, a considerable quantity both of water and air was absorbed by the mouth, and perhaps likewise by the branchial apertures. The process is effected by two methods; the air is swallowed, and is then forced into the cavity of the body, its return being prevented by a muscular contraction which is externally visible; but the water, I observed, entered in a stream through the mouth, which was wide open and motionless; this latter action must therefore depend on suction. The skin about the abdomen is much looser than that of the back; hence, during the inflation, the lower surface becomes far more distended than the upper; and the fish, in consequence, floats with its back downwards. Cuvier doubts whether the Diodon, in this position, is able to swim; but not only can it thus move forward in a straight line, but likewise it can turn round to either side. This latter movement is effected solely by the aid of the pectoral fins, the tail being collapsed, and not used. From the body being buoyed up with so much air, the branchial openings were out of water: but a stream drawn in by the mouth constantly flowed through them.

"The fish, having remained in this distended state for a short time, generally expelled the air and water with considerable force from the branchial apertures and mouth. It could emit, at will, a portion of the water; and it appears, therefore, probable, that this fluid is taken in partly for the sake of regulating its specific gravity. This Diodon possessed several means of defence. It could give a severe bite, and could eject water from its mouth to some distance, at the same time it made a curious noise by the movement of its jaws. By the inflation of its body, the papillæ, with which the skin is covered, became erect and pointed. But the most curious circumstance was, that it emitted from the skin of its belly, when handled, a most beautiful carmine red and fibrous secretion, which stained ivory and paper in so curious a manner, that the tint is retained to the present day."

DIOMEDEA. [See ALBATROSS.]

DIPHYLLOBOTHRIUM. [See SUPPL.]

DIOPSIS, or TELESCOPE FLY. A very singular genus of Dipterous insects, remarkable for the enormously developed pedicels on which the eyes are situated. They

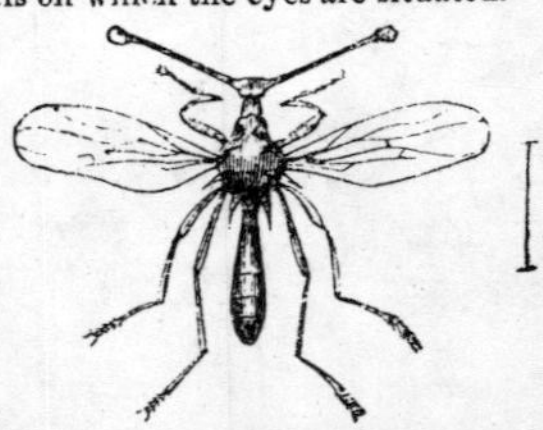

TELESCOPE FLY
(DIOPSIS MACROPTHTHALMA.)

are found in Western Africa, India, and the Indian islands, some of the species being of considerable size. One species, the *Diopsis Sykesii*, was observed by the distinguished Indian statistician and naturalist after whom it is named by Mr. G. R. Gray, in countless multitudes in one of the Indian vallies. Hence it is not improbable that the habits of the numerous species composing the genus are similar. It is one of those well-marked and remarkable groups of insects, all the species of which have been described and figured by Mr. Westwood. They are, however, rare in collections; the British Museum possessing many curious species.

DIPLOPTERA, or DIPLOPTERYGA. A group of Hymenopterous insects, forming the third division of the subsection *Prædones*. These wasps obtain their name from the wings being folded throughout their entire length when at rest. The antennæ are generally elbowed, and either filiform or thickened at the tips; the palpi are short and filiform; the maxillæ are long, coriaceous, and compressed; the thorax is oval and entire; and the collar extends laterally to the base of the wings. The body is generally black, more or less spotted with buff, and either quite naked, or very slightly clothed with hairs: the legs are of moderate length, not furnished with organs fitted for the collection of pollen; and the abdomen is ovate. The sting of the females and neuters is very powerful, and has occasionally caused the death of those persons who have been attacked by these insects. This division forms two families, *Eumenidæ* and *Vespidæ*. [See WASP.]

DIPROTODON. [See SUPPLEMENT.]

DIPSAS. A genus of Snakes [which see].

DIPTERA. An order of two-winged insects; of which the common house-fly and blue-bottle fly afford familiar examples. There are, however, some dipterous insects which are destitute of wings: hence it is necessary to notice other peculiarities belonging to this order. Some possess a proboscis and sucker: others have a proboscis and no sucker. They have six legs, furnished with five-jointed tarsi, two palpi, two antennæ, and three ocelli. The mouth in the insects of this order is formed for suction; but there are considerable varieties in the mode in which this is accomplished. Behind the wings are placed a pair of movable slender bodies, termed *halteres*, or balancers, which are kept in continual motion, and are usually present even when the true wings are not developed. The wings are generally horizontal in their position, and transparent; the eyes are generally large, often occupying nearly the whole head. The Diptera all undergo a complete metamorphosis, as far as respects the comparative structure of the larva and the perfect insect; the former being generally cylindrical footless grubs.

The two-winged insects, though mostly of moderate or small size, are not only very numerous in kinds or species, but also extremely abundant in individuals of the same kind, often appearing in swarms of countless multitudes. Flies are destined to live wholly on liquid food, and are therefore provided with a proboscis, enclosing hard and sharp-pointed darts, instead of jaws, and fitted for piercing and sucking; or ending with soft and fleshy lips, for lapping. In our own persons we suffer much from the sharp suckers and bloodthirsty propensities of gnats and mosquitos (*Culicidæ*), and also from those of certain midges (*Ceratopogon* and *Simulium*), including the tormenting black flies (*Simulium molestum*) of America. The larvæ of these insects live in stagnant water, and subsist on minute aquatic animals. Horse-flies and the golden-eyed forest flies (*Tabanidæ*), whose larvæ live in the ground, and the stinging stable-flies (*Stomoxys*), which closely resemble common house-flies, and in the larva state live in dung, attack both man and animals, goading the latter sometimes almost to madness by their severe and incessant punctures. The winged horse-ticks (*Hippoboscæ*), the bird-flies (*Ornithomyiæ*), the wingless sheep-ticks (*Melophagi*), and the spider-flies (*Nycteribiæ*), and bee-lice (*Braulæ*), which are also destitute of wings, are truly parasitical in their habits, and pass their whole lives upon the skin of animals. Bot-flies or gad-flies (*Œstridæ*), as they are sometimes called, appear to take no food while in the winged state, and are destitute of a proboscis; the nourishment obtained by their larvæ, which, as is well known, live in the bodies of horses, cattle, sheep, and other animals, being sufficient to last these insects during the rest of their lives. Some flies, though apparently harmless in the winged state, deposit their eggs on plants, on the juices of which their young subsist, and are oftentimes productive of immense injury to vegetation; among these the most notorious for their depredations are the gall-gnats (*Cecidomyiæ*), including the wheat-fly and Hessian-fly, the root-eating maggots of some of the long-legged gnats (*Tipulæ*), those of the flower-flies (*Anthomyiæ*), and the two-winged gall-flies and fruit flies (*Ortalides*). To this list of noxious flies, are to be added the common house-flies (*Muscæ*), which pass through the maggot state in dung and other filth, the blue-bottle or blow-flies, and meat-flies (*Luciliæ* and *Calliphoræ*) together with the maggot-producing or viviparous flesh-flies (*Sarcophagæ* and *Cynomyiæ*), whose maggots live in flesh, the cheese-fly (*Piophila*), the parent of the well-known skippers, and a few others that in the larva state attack our household stores. Some are entirely harmless in all their states and many are eminently useful in various ways. Even the common house-flies, and flesh-flies, together with others for which no names exist in our language, render important services by feeding, while larvæ, upon dung, carrion, and all kinds of filth; by which means, and by similar services, rendered by various tribes of scavenger-beetles, these offensive matters speedily disappear, instead of remaining to decay slowly, thereby tainting the air and rendering it unwholesome. Those whose larvæ live in stagnant water, such as gnats (*Culicidæ*), the soldier-flies (*Stratiomyidæ*),

&c., tend to prevent the water from becoming putrid, by devouring the decayed animal and vegetable matter it contains. The maggots of some flies live in toadstools and similar excrescences growing on trees; those of others in rotten wood and bark. And, finally, many lay their eggs on caterpillars, and on various other larvæ, within the bodies of which the maggots hatched from these eggs live till they destroy their victims. Besides performing their various appointed tasks in the economy of nature—flies, and other insects, subserve another highly important purpose, for which an allwise Providence has designed them, namely, that of furnishing food to numerous other animals. Not to mention the various kinds of insectivorous quadrupeds, many birds live partly or entirely on insects. The finest song-birds, nightingales and thrushes, feast with the highest relish on maggots of all kinds, as well as on flies and other insects, while warblers, swallows, &c. &c., devour these two-winged insects in great numbers. —The works of Meigen, Wiedemann, Macquart, and Robineau Desvoidy, are the great authorities on this very numerous and every where distributed order of insects.

DIPUS. [See Jerboa.]

DIRT-DAUBER. The name given in the United States to a species of Hymenopterous insect. [See Pelopeus.]

DISTOMA. [See Fluke, p. 228.]

DIVERS. (*Colymbidæ.*) A genus of aquatic birds, thus described by Bewick:—"The bill is strong, straight, and pointed: the upper mandible longest; the edges of each bending inwards: nostrils linear, the upper part divided by a small cutaneous appendage: tongue long, pointed, and serrated on each side near the base: thighs placed far backward: legs thin and flat, and extended horizontally: toes four in number; the exterior the longest; the back one small, and joined to the interior by a thin membrane: tail short, consisting of twenty feathers. These birds are broad, flat, and long-bodied, and swim in a squat position on the water."

The Great Northern Diver. (*Colymbus glacialis*) measures upwards of three feet in length, and four feet six inches in breadth. The bill is black and strong, and to the corners of the mouth is four inches long; the head and neck are of a deep black, glossed with green and purple; the hind part of the latter being streaked with a large white band shaped like a crescent; exactly under the throat is another band; and both are marked with black oblong strokes pointing downwards. The lower part of the neck is a deep black, tinged with a rich purple gloss; the breast and under side of the body is wholly white; the sides of the breast are marked with black lines; and the back, the coverts of the wings, and the scapulars, are black, thickly marked with white spots. The tail is very short, and almost hid by the scapulars; the legs and feet are black. The female is less than the male, and her whole upper plumage inclines more to brown. This bird inhabits the north of Europe and the Arctic coasts, and is sometimes, though rarely, seen in England. It seldom quits the sea, or retires

GREAT NORTHERN DIVER.
(COLYMBUS GLACIALIS.)

inland, except during the period of incubation, when it repairs to the borders of fresh-water lakes; and the female deposits two large eggs of a pale clear yellowish colour, marked with very large and small spots of ashy-purple. Fish is the principal food of this species, and the herring in particular, the fry of fish, crustaceans, and marine vegetables.

The Red-throated Diver. (*Colymbus septentrionalis.*) This species is about two feet in length, and three feet four inches in breadth. The sides of the head, neck, and throat are mouse colour; the top of the head is spotted with black; the hinder and lower part of the neck are longitudinally rayed with black and white; the upper fore part of the neck, to the throat, is of a deep chestnut-red; the breast and under parts of the plumage are pure white: the sides, the back, and the rest of the upper parts are blackish brown in the very old birds, but in those of the age of three or four years they are slightly sprinkled with small white spots. The male and female are nearly alike in their plumage. This species inhabits the same cold countries as the other Divers, and its manners and habits do not differ from theirs; but it is of a more lively character, and has a more sprightly appearance. They breed and are common in Hudson's Bay, Greenland, Iceland, the Zetland and Orkney Isles, &c. The female makes her nest, which is composed of moss and herbage, lined with a little of her own down, on the very edge of the shore: she lays two eggs, which are somewhat longer than those of a hen, and of a dingy bluish-white, thinly marked with dusky spots. They run swiftly on the surface of the water, but are very awkward on land, from which they rise with difficulty: their flight, however, when once on the wing, is both strong and swift.

The BLACK-THROATED DIVER (*Colymbus arcticus*) differs in plumage from the last described, and is rather larger. The bill and front of the neck are black: the hind part of the head and neck are cinereous; the sides of the neck are marked with black and white lines, and the fore part is of a glossy variable purple, black, and green. The back, the scapulars, and the coverts of the wings are black; the two former being marked with square, and the latter with round white spots: quills dusky; tail black; legs dark, and reddish on the inside. Like the preceding, this bird is common in all the Arctic regions; and in its winter migrations it visits England, Germany, and Holland. Their skins are dressed, and made into caps, hoods, &c., and are much esteemed as a covering for the head and breast in the rigorous climates in which these birds are found, the great thickness of the feathers rendering them very fit for that purpose.

DOBCHICK, or DIDAPPER. (*Podiceps minor.*) A Palmipede bird of the Grebe kind. It seldom exceeds six ounces in weight: the beak is short, large at the base, and tapering to the point: the head is thickly clothed with downy feathers, which it can puff up at pleasure; the eyes are large, the wings small, and it has no tail. Its plumage on the back is of a deep blackish brown colour, and white on the belly. It moves with more facility under the water than on its surface, and raises itself from that element with great difficulty; but when once on the wing, it is capable of continuing its flight for a considerable time. "Ornithologists and sportsmen describe the nest of this bird as being of a large size, and composed of a very great quantity of grass and water-plants at least a foot in thickness, and so placed in the water that the female hatches her eggs amidst the continual wet, in which they were first laid; and it is conjectured that the natural warmth of her body occasions a fermentation of the herbage, which greatly aids the incubation. She lays from four to six eggs, of a yellowish dull white, and is said to cover them up with the surrounding leaves every time she has occasion to stir abroad." (*Bewick.*)—There are several other species of the Dobchick; as the Horned, the Eared, the Black and White Dobchick, &c.; all of which are larger than the one here described. [See GREBE.]

DODO. A large and most unwieldy bird, generally supposed to be extinct, and whose very existence at any period has been doubted. But as there are accounts of it in the works of more than one naturalist, and as it is described most minutely, it behoves us to collect the best information of it we can find. The Dodo is stated to be a native of the Mauritius, or Isle of France; and the Dutch, who first discovered it there, are said to have termed it the nauseous bird, as well from its disgusting figure, as from the disagreeable taste of its flesh. Its appearance, instead of giving one an idea of swiftness, the common attribute of birds in general, seems to strike the imagination as something the most unwieldy and inactive in nature. Its massive and almost globular body, which is covered in general with grey feathers, is barely supported on two clumsy legs; while its head and neck rise from it in a manner truly grotesque. The neck, thick and pursy in itself, is surmounted by a head composed of two enormous mandibles, opening far behind the eyes, which are large, black, and prominent; so that the huge bird, in gaping,

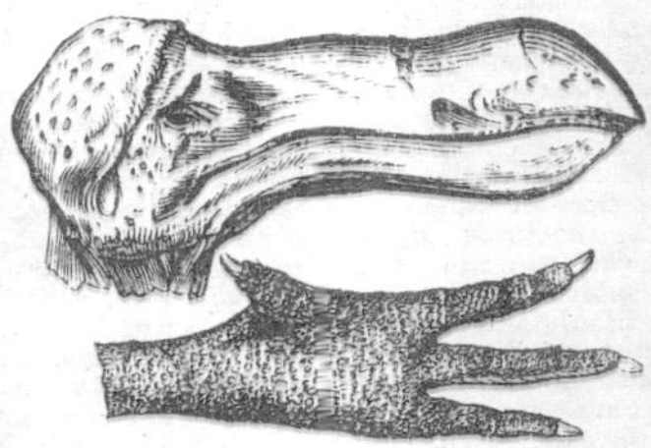

HEAD AND FOOT OF DODO
(DIDUS INEPTUS.)

exhibits a most enormous mouth: hence the bill is of an extraordinary length, thick, sharp at the end, and having each chap crooked in opposite directions; and the two mandibles, which are of a bluish-white colour, in some measure resemble two spoons laid back to back. The Dodo seems to be so weighed down by its own gravity, as scarcely to possess strength sufficient to give energy to its motions; and it appears among the feathered tribe what the Sloth does among quadrupeds, an unresisting creature, equally incapable of flight or defence. Its wings are covered with soft ash-coloured feathers, intermixed with a yellowish-white, but they are too short to render it any essential service in flying: its tail is composed of a few small curled feathers of a light ash-colour; its legs are too short to assist it in running; and its body is exceedingly clumsy. From all that can be gathered concerning this obscurely known bird, it would seem that the species has entirely disappeared; and we now possess no more of it at the present day than a foot preserved in the British Museum, and a head and foot in bad condition at the Ashmolean Museum in Oxford.—Dr. Melville has written, in conjunction with Mr. Strickland, an elaborate memoir on this bird, which they believe to have been *a Pigeon* somewhat allied to the genus TRERON. This memoir, or part of it, which embraces the history of the SOLITAIRE, was read at the meeting of the British Association at Oxford, 184‑. It has been published, with most exquisitely accurate engravings from the pencils of Ford and Dinkel. In the British Museum there is a painting believed to be a representation of this bird; and in the same case are casts from the Oxford remains, and other casts throwing light on its history. It is a bird which would appear to have become extinct within the last 200 years. Mr. Duncan, of the Ashmolean Museum, has published an excellent history of all that was known of it up to his time. [See GNATHODON: SOLITAIRE.]

DOG. (*Canis familiaris.*) This most faithful and valuable domestic — so remarkable for his incorruptible fidelity, his lasting attachment, his inexhaustible diligence, and his ready obedience — deserves all the eulogies that have been bestowed upon him, and all the kindness that can be shown him, by his master and companion, Man. But independent of his being the most sagacious of all known quadrupeds, and the acknowledged friend of mankind, he is possessed of all those native qualities which contribute to the convenience, and generally conciliate the affections, of the human species. A natural courage and ferocious disposition render the Dog in his savage state a formidable enemy to all other animals; but these qualities speedily yield to very different ones in the domestic Dog, whose only ambition seems to be the desire of pleasing: he approaches with a timid respect, and lays his strength, his courage, and all his useful

NEWFOUNDLAND DOG

talents at the feet of his master; he waits his orders, consults his looks, and a single glance is sufficient to put him in motion. Constant in his affections, and grateful for the slightest favours, he is humble and suppliant under his owner's displeasure, and eventually disarms resentment by unwearied submission. When the care of the house is submitted to him, he appears proud of the charge, and, like a faithful sentinel, he goes his rounds, and gives manifest indications that he is intent on his duty. Thus useful in himself. and being admitted, as it were, to a participation of empire, he exerts a degree of superiority over all other animals which stand in need of human protection. The flocks and herds obey his voice more readily even than that of the shepherd or the herdsman; he conducts them, guards them, confines them within their appointed limits, and considers their enemies as his own. Nor are his arts less serviceable in pursuit, or his unflinching courage less valuable to man, than his personal attachment, his obedient watchfulness, and his patient submission, are endearing.

Cuvier observes that the Dog exhibits "the most singular, the most complete, and the most useful conquest that man has ever made. Every species has become our property; each individual is entirely devoted to his master, assumes his manners, distinguishes and defends his property, and remains attached to him even unto death; and all this proceeds neither from mere necessity nor constraint, but solely from true gratitude and real friendship. The swiftness, the strength, and the scent of the Dog have created for man a powerful ally against other animals, and were perhaps necessary to the establishment of society. It is the only animal that has followed man through every region of the earth." What the great French naturalist has here said is strictly true; but every person must agree with Mr. M'Culloch, that "it is singular that neither Cuvier, nor any one of those by whom his statements have been copied, should have mentioned that this account is applicable only to Europe. All Mahommedan nations regard the Dog as impure, and will not touch it without an ablution. The same is also the case with the Hindoos. From the Hellespont to the confines of Cochin-China, dogs are unappropriated, and have no master. They prowl about the towns and villages; and though they are naturally more familiar, they are in no respect more domesticated, than the carrion crows, kites, vultures, &c. which assist them in performing the functions of scavengers."

If we had sufficient space, and it were necessary to the elucidation of the subject, numerous instances might be cited of the sagacity, affection, courage, generous disposition, and other estimable qualities of this animal, which, if such instances were not well authenticated, would appear incredible; but the universality of such cases renders it almost a matter of certainty that there are few of our readers whose own experience will not furnish them with "anecdotes" of this nature, no less wonderful than true. We shall therefore proceed, without further digression, in an attempt to trace, in the pages of the most eminent naturalists, the source or origin of the species, as far as the intermixture of races or the influence of domestication will permit.

Pennant is of opinion that the original stock of Dogs in the Old World is with great reason supposed to be the jackal; that from their tamed offspring, casually crossed with the wolf and the fox, have arisen the numberless forms and sizes of the canine race. Buffon, with much ingenuity, has traced out a genealogical table of all the known Dogs, deducing all the other varieties from the Shepherd's Dog, variously affected by climate, and other casual circumstances. In the Alpine regions, for instance, this Dog is much larger and stronger than in England. From the recent observations of travellers in the high northern parts of the world, where, although Dogs have been employed for an incalculable length of time, they still retain much of the external appearance and general carriage of a wild animal it would seem that Pennant's suggestion is worthy of attention. But at the same time it should be remarked, that the breed of Dogs, produced from the wolf and varieties of the domestic dog, during a long succession of generations, still retains marked characteristics of the predominance of the savage qualities derived from its untamed progenitors, in the keen and vivid expression of the eye, ferocity of disposition, and severity of bite. It is also a singular fact, that the race of European Dogs evince as great an antipathy to the Esquimaux species as they do to a wolf.

In Mr. Bell's History of British Quadrupeds this subject is discussed at considerable length, and with much freedom. "It may not be uninteresting," says our author, "to examine what is the real state of the question, as it regards the original form, from which all the numerous varieties of the DOG have sprung. In order to come to any rational conclusion on this head, it will be necessary to ascertain to what type the animal approaches most nearly, after having for many successive generations existed in a wild state, removed from the influence of domestication, and of association with mankind. Now we find that there are several different instances of the existence of Dogs in such a state of wildness as to have lost even that common character of domestication, variety of colour and marking. Of these, two very remarkable ones are the Dhole of India, and the Dingo of Australia; there is, besides, a half-reclaimed race amongst the Indians of North America, and another, also partially tamed, in South America, which deserve attention: and it is found that these races, in different degrees, and in a greater degree as they are more wild, exhibit the lank and gaunt form, the lengthened limbs, the long and slender muzzle, and the great comparative strength which characterize the Wolf; and that the tail of the Australian Dog, which may be considered as the most remote from domestication, assumes the slightly bushy form of that animal.

"We have here, then, a considerable approximation to a well-known wild animal of the same genus, in races which, though doubtless descended from domesticated ancestors, have gradually assumed the wild condition; and it is worthy of especial remark, that the anatomy of the Wolf, and its osteology in particular, does not differ from the dogs in general, more than the different kinds of dogs do from each other. The cranium is absolutely similar, and so are all, or nearly all, the other essential parts; and to strengthen still further the probability of their identity, the Dog and Wolf will readily breed together, and their progeny is fertile. The obliquity in the position of the eyes in the Wolf is one of the characters in which it differs from the Dogs; and although it is very desirable not to rest too much upon the effects of habit or structure, it is not perhaps straining the point, to attribute the forward direction of the eyes in the Dogs, to the constant habit, for many successive generations, of looking forward to their master, and obeying his voice.

"A point of very considerable importance in the question of the identity of species is the period of gestation. This circumstance is so invariable in individuals of the same species, and so rarely the same in those which are distinct, however nearly they may be allied, that if, in this respect, two animals be found to differ, it would be a strong ground for doubting at least, perhaps even for rejecting, the opinion of their identity; and, on the other hand, their absolute coincidence on this point would afford a collateral argument of equal force in its favour. Buffon indeed relates an instance of the Wolf, in which the period might possibly have been seventy-three days; but even on

GREYHOUND

his own showing, it might have been no more than sixty-three; and certain circumstances detailed in the account afford strong reason for believing this to have been the case. Hunter, who instituted a series of interesting and, as far as they went, important experiments, in order to ascertain whether the Wolf and the Jackal would respectively breed with the Dog, comes to the conclusion, on finding the affirmative to be true in both instances, that the Dog, the Wolf, and the Jackal are of one species. But he found that the period of the Jackal is fifty-nine days, whilst that of the Wolf is sixty-three days, the same as that of the common Dog. Desmarest also gives sixty-three days as the period of the Wolf. As far as this character goes, therefore, it is in favour of the identity of the Wolf and Dog, and of the specific distinctness of the Jackal. The conclusion which Hunter draws from the fact that each of these wild animals will breed with the Dog, and produce young which are fertile again with the Dog, is, however, not yet satisfactory; and the argument would be much stronger were it proved that the progeny would breed with each other, which has not at present been done. It appears that in many other cases, especially amongst birds, the hybrids will breed with either of the parent species; but the more satisfactory experiment just proposed remains to be tried; and until this has been done, the chain of evidence is incomplete, and the validity of the argument derived from the alleged fertility of progeny is inconclusive." [Our author here details some striking facts to prove that the supposed untameable ferocity of the Wolf may not merely be subdued, but that he may become truly attached and friendly to those who treat him with kindness.]

"Upon the whole, the argument in favour of the view which I have taken, that the Wolf is *probably* the originator of all the canine races, may be thus stated: — The structure of the animal is identical, or so nearly so, as to afford the strongest *à priori* evidence in its favour. The Dog must have been derived from an animal susceptible of the highest degree of domestication, and capable of great affection for mankind; which has been abundantly proved of the Wolf. Dogs having returned to a wild state, and continued in that condition through many successive generations, exhibit characters which approximate more and more to those of the Wolf, in proportion as the influence of domestication ceases to act. The two animals will breed together, and produce fertile young. The period of gestation is the same.

"The races of Dogs have at different times been variously classified, according to the views of the respective authors; but, as it appears to me, with very little truth in a zoological point of view, and as little practical advantage. Although it is obvious that certain varieties approach more nearly to each other in habit and conformation than others, there is not sufficient ground for a regular systematic arrangement. Buffon, F. Cuvier, and other authors, have attempted such classifications; but they have been merely artificial, and in many instances have gone upon erroneous suppositions as to the origin of mixed races.

"The food of the Dog is various. It will live on cooked vegetable matters, but prefers animal food, and is particularly fond of it when approaching putrefaction. Its stomach will digest portions of bone. In drinking, it laps with the tongue; it never perspires; but the nose is naked and moist; and when hot, the tongue hangs out of the mouth, and a considerable quantity of water drops from it. It walks round the place it is about to lie down upon, and coils itself up in the same direction. The female goes with young sixty-three days, and usually has about six or eight at a litter; though sometimes as many as twelve or fourteen. These are blind at birth, and do not acquire their sight until the tenth day. It is commonly stated that the male puppies resemble the father, and the female the mother: this, however, if it be true to a certain extent, is not absolutely so; but, like many other animals, the father of the first litter often pro-

SHEPHERD'S DOG.

duces an impression which is scarcely lost in all the subsequent ones. This is a fact worthy of particular attention, as it bears upon a question of as great interest and importance as any in the whole range of animal physiology. Such are some of the general habits of the whole species; and there are many others which are too well known to require repetition. Those which belong to the different races will be briefly mentioned under the separate heads."

Like the young of most animals which bring forth many at a time, the Dog is not perfectly formed when first produced. During their blind state the bones of their skulls are incomplete, their bodies are inflated, their noses are contracted, and their whole figure is but imperfectly represented; but in less than a month the puppy begins to acquire all its senses, and from that time makes hasty advances to perfection. At the end of the fourth month, the Dog, like other animals, sheds some of his teeth, which are renewed by such as are permanent. The teeth of the Dog being his principal, and indeed his only defence, they are formed in such a manner as to render him the most essential services: he cuts with his incisors or foreteeth; he holds with his four great canine ones; and he chews with his grinders, which are fourteen in number, and so placed that when his jaws are shut there remains a distance between them; so that on opening his mouth to the greatest possible stretch, he does not lose the power of his jaws. Though the Dog is a voracious animal, he is capable of enduring hunger for a considerable length of time: it seems, indeed, that water is more necessary to his subsistence than food; and he drinks often, though not abundantly.

We shall conclude this article with some extracts from an ingenious essay "On the Dog, as the Companion of Man in his Geographical Distribution," by Dr. T. Hodgkin, in *The Zoologist*, edited by Mr. Newman: — "The most striking natural group, the most marked in its characters, and the most widely diffused of all the known varieties, is that which we may trace from China, over the northern portion of the old continent, to the islands of the Northern Ocean and the northern part of America. In this wide extent we find, as we should reasonably anticipate, some distinctly marked subdivisions, yet all so evidently maintaining the common type, that the least skilful observer must immediately recognize the family resemblance. The dogs constituting this group may be thus enumerated: 1. Those of China. 2. Those of Kamskatka, and others of the same stock employed in drawing sledges in the northern parts of Asiatic Russia. 3. The very distinctly marked variety of dogs occurring in the northern parts of Europe, and which are called *Spitz* in Germany, but which are known as Pomeranian dogs when introduced into France and England. 4. The dogs of Iceland, with which are probably connected those of Lapland and Greenland. And lastly, those of the Esquimaux. A very remarkable family likeness is to be detected in all this group, of which perhaps the most striking features are the sharpened nose, rather small pointed ears, the approaching eyes but little projecting, the superior length of hair about the neck, with a greater or less tendency to shagginess on the other parts of the body, and, in most instances, an elevated curled tail, with a temper which may be characterized as restless and irritable. We meet with many varieties in stature, colour, and length of hair. Thus it would appear that the dogs of China are often black, the epitelium of the mouth and tongue having the same colour. Those of the north of Europe are almost invariably white or light brown, whilst those of the Esquimaux are often black and white. From China we see specimens both of large and of small size, having the same characteristic form. Those of the Esquimaux and Kamskatkadales are of rather a large size,

ESQUIMAUX DOG.

whilst those of Iceland are small, and probably lower in proportion than any other of the group. The dogs of this group appear to differ as widely in their degree of fidelity and docility. The Pomeranian variety, which is perhaps the most completely domesticated, is faithful and sagacious, and makes an excellent guard, and the smaller specimens become the admired pets of the ladies. From an example which came to my knowledge, I am inclined to believe that the Chinese dogs have the same character. Those of the Esquimaux and the Kamskatkadales are chiefly valuable on account of their strength and endurance of fatigue; but they are often ill-tempered and untractable; and though decidedly sagacious and capable of being trained as retrievers, they are destructive, and cannot be left with safety in the way of live stock, bearing in this, as well as in some other particulars, a strong resemblance to the wolf, with which it is known that their blood is occasionally blended. It may, however, be observed, that independently of such known connexion, the whole group of which we are now speaking has something more of the wolfish expression than any other variety of the *Canis familiaris*. * * *

"Another extensive division of the species, and which appears to me to have been spread over a different portion of the globe, and probably to belong to the western part of Asia, the southern parts of Europe, and north of Africa, may perhaps be regarded as comprising the true hunting dogs. They possess, for the most part, well-developed noses; their ears are large, broad, and pendulous; their proportions rather thick than otherwise; their jaws large as compared with other dogs, and their tails thick. The descriptions of hounds left by Greek authors, seem to have been applied to dogs of this stock, which will also be recognized in the old English hound, and in all the varieties of the modern hound, down to the beagle. The pointer strikingly exhibits the same characters, and all the varieties of spaniel appear to be essentially branches of the same family, though probably modified by a cross, respecting which I shall presently hazard a conjecture. The true smooth terrier appears to be of the same division, though some passing under this name are probably mixed with another stock. In some of the dogs of this group we find probably the most marked effects of culture. Their large pendulous ears, as in some varieties of rabbits, may be referred to this cause. With a greater degree of submission and attention to man, they have also a greater degree of dependence upon him, and some almost resemble the sheep and the cow in this respect, whilst their more artificial faculties, which have been cultivated for many generations, have become innate in the offspring. Thus the pointer's puppy, of a few weeks old, begins to point of his own accord, and anticipates the first lessons of his trainer; just as young horses will frequently adopt the artificial paces which have been taught to their sires.

"A third group is less distinctly marked as a whole, and I am not prepared to lay any great stress on the reasons which have induced me to bring some of its varieties together; but in others we have the strongest evidence of their affinity, both in visible characters and known connection of blood, notwithstanding great apparent differences of figure. In this group I would place the greyhound, and that variety of shepherd's dog which most nearly approaches him in form. It would be quite a mistake to suppose that the shepherd's dog is a single variety, since different kinds of dogs are employed for this purpose in different districts. The transition of the greyhound to one of the shepherd's dogs takes place by almost insensible degrees, and Cowper's description of half lurcher and half cur must be familiar and graphic to almost every one. In the young animals, when no mutilation of the tail has taken place, the resemblance is most striking. Another variety, perhaps, is more related to the greyhound than even any variety of shepherd's dog; I mean the English bull-dog. It was the perception of the striking resemblance in some points exhibited in these animals notwithstanding their general difference of figure, before I was aware of the actual consanguinity which breeders are careful to maintain, which first led me to notice the indications of a natural grouping which would seem to clash with artificial arrangement. Though the bull-dog is short, compact, and heavy, with a proverbially large blunt head and broad face, and the greyhound is the very emblem of lightness, his elongated nose, head, and neck resembling a snake, his back long, curved, and flexible, his body, which, with sufficient room for the organs of circulation and respiration, affords almost none for those of digestion, and supported on long and slender limbs, which seem to render him among quadrupeds what the *hirondelle de mer* is among birds—there are individual points of resemblance between the two dogs which are perhaps more striking than any which can be found among other varieties. The feet and toes are remarkably delicately formed; the ears small and pointed, though generally inclined to be pendulous, capable of being erected; the tail remarkably slender, some of the stoutest bull-dogs having tails which would grace an Italian greyhound. Similar colours also prevail in both varieties, and more especially the brindled, the mottled, and the more or less white. In both, the sense of smell is slow for the dog, whilst the sight is good. Both are ferocious and savage when set on: the ferocity of the greyhound is not unfrequently shown in the destruction of sheep."

The author then gives reasons for presuming that the Newfoundland dog, which he says has been regarded as a large species of water-spaniel, is distinctly traceable to the Esquimaux stock; and he concludes by observing that whatever may be the value and results of inquiries like these, as respects the study of ethnology, the labour need not be in vain as respects the animals themselves, since conclusions of more or less practical value can scarcely fail to be deduced for the guidance of the breeder and the benefit of the public.

Dogs are found in all parts of the world, with the exception of a few groups of islands in the Southern Pacific Ocean. But it is only in temperate climates that they preserve their ardour, courage, sagacity, and other talents. [See BLOODHOUND, BULL-DOG, FOXHOUND, GREYHOUND, HOUND, MASTIFF, POINTER, SPANIEL, SHEPHERD'S DOG, &c.]

DOG-FISH. (*Scyllium catulus.*) This species of Shark, called the Large-spotted Dog-fish, is from two to three feet in length: the head is large; the snout prominent and slightly pointed: the skin rough; body cylindric; the colour a brownish grey, with a slight tinge of pale brick red, and marked with very numerous blackish or dusky spots; the belly whitish, and very smooth. These fish, when at their full growth, weigh about twenty pounds each: they are caught in considerable numbers on our own coasts, where their voracious habits do great injury to the fisheries; and in Scotland they constitute no inconsiderable part of the food of the poor. The rough skin of this fish is used by joiners and other artificers in polishing various substances, particularly wood, and is generally known by the name of "fish-skin."

The SMALL-SPOTTED DOG-FISH (*Scyllium canicula*) is in many respects similar to the preceding, and is one of the most common species on our southern coasts, where, keeping near the bottom of the water, it feeds on small fish and crustacea. The upper part of the body is marked with numerous small, dark, reddish-brown spots, on a pale reddish ground; the spots on the fins rather larger and less numerous than those on the body.

The PICKED DOG-FISH (*Spinax acanthias*) is a species very common on the coasts of Kent and Sussex, where it is almost universally called the Bone Dog; it is also very numerous on the north-eastern and western coasts, and is often seen in shoals among the Scotch islands. This fish is distinguished from the others of this class by having a single spine placed in front of each of its two dorsal fins, and from which it derives its name. We learn from Mr. Yarrell, who quotes Mr. Couch, that "they are sometimes found in incalculable numbers, to the no small annoyance of the fishermen, whose hooks they cut from the lines in a rapid succession. The Picked Dog bends itself into the form of a bow for the purpose of using its spines, and by a sudden motion causes them to spring asunder in opposite directions: and so accurately is this intention effected, that if a finger be placed on its head, it will strike it without piercing its own skin." Length about twenty inches; the upper part of the head and body slate grey; under parts yellowish.

Another species, called the BLACK-MOUTHED DOG-FISH, (from the colour of the inside of its mouth) is well known in the Mediterranean. It runs from two feet to two feet six inches in length; is of a light brown colour on the head and along the back, and on each side are two rows of ocellated spots.

DOLABELLA. A genus of Tectibranchiate Mollusca, closely allied to the Sea-Hares (*Aplysia*), differing from them in having the branchiæ at the posterior part of the body, which looks like a truncated

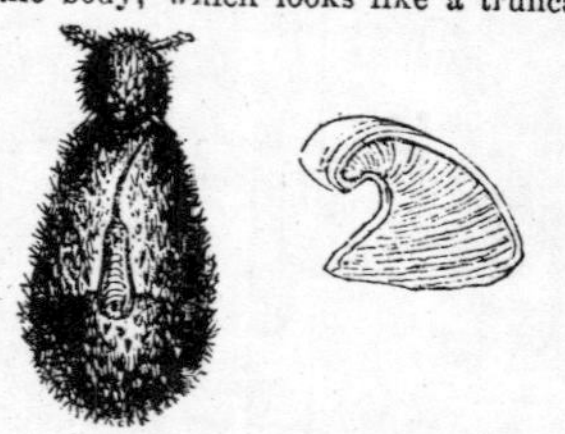

DOLABELLA RUMPHII AND INNER SHELL.

cone. Their lateral crest does not close on branchiæ, leaving a groove. The inner shell is calcareous. There is more than one species; the genus is found both in the Mediterranean and the Eastern seas. Interesting observations on them and allied genera have been recorded by A. Adams, F.L.S.

DOLICHOSAURUS. [See SUPPLEMENT.]

DOLIUM. A genus of Mollusca, inhabiting univalve shells, found, for the most part, in the Indian, African, and South American

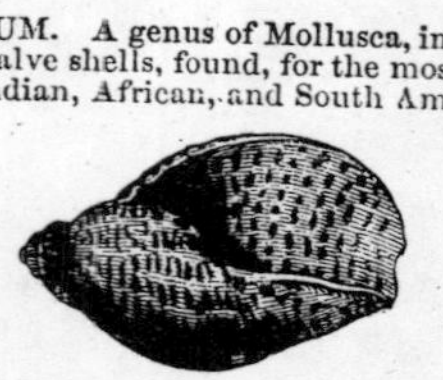

PARTRIDGE SHELL.—(DOLIUM PERDIX.)

seas: the shell is large, light, and oval or globular; the mouth wide and notched, generally transversely banded. The molluscous animal contained in it has a large head with short proboscis, and two tentacula with eyes in the middle. There are several species, most of which may be seen in the fine collection at the British Museum. The foregoing figure of the Partridge Shell (*Dolium perdix*) will give a very good general idea of the form of this genus.

DOLPHIN. (*Delphinus delphis.*) This cetaceous animal bears a great resemblance to the Porpoise, but has a much longer and sharper snout, and the body is of a more slender shape. It often grows to the length of eight or ten feet; the colour on the back and sides is dusky, and the belly whitish; the teeth are very numerous, sharp pointed, and slightly bending forwards; and they are placed so close together, that when the mouth is shut the jaws lock into each other. The Dolphin is found in the Mediterranean and Indian seas, and seems to be generally confounded by navigators with the Porpoise having the general manners and appearance of that animal. It swims very swiftly, and preys on various kinds of fish; and it sometimes happens that either from its impetuosity in the pursuit of prey, or the calls of hunger, it is urged beyond the limits of safety; and the fishermen on the Cornish coasts, who spread their extensive nets for pilchards, sometimes become possessed of a very unwelcome prize.

By ancient writers the Dolphin was celebrated for its supposed affection for the human race, and its appearance was regarded as a favourable omen. Numerous, indeed, are the fables of antiquity in this respect, which could have no better foundation than poetic fiction: its figure is far from prejudicing us in its favour; and its extreme rapacity tends still less to endear it to us.

COMMON DOLPHIN
(DELPHINUS DELPHIS.)

The prejudices of the moderns are of a contrary character; for the appearance both of this species and the porpoise at sea, is generally considered as one of the preludes of an approaching storm. Dolphins inhabit every sea, from the equator to the poles, enduring equally well the extremes of heat and cold. The Dolphin, respiring by lungs, and not in the manner of fishes, is compelled to rise to the surface to breathe, throwing out the water from the blow-hole, or aperture in the head, like a cloud of steam. This hole is of a semilunar form, with a kind of valvular apparatus, and opens nearly over the eyes. The structure of the ear renders the sense of hearing very acute, and the animal is observed to be attracted by regular or harmonious sounds. Compactness and strength are the characteristics of the genus, and the muscular powers of the tail are proverbial. The Dolphin is said to be long-lived, and, like the Whale, seldom brings forth more than one young one at a time, which the parent suckles and watches with great care and anxiety.

It is, perhaps, almost unnecessary to caution the reader not to confound the cetaceous species we have been describing, with the *fish* commonly known as the Dolphin at the present day, and hereunder described.

DOLPHIN. (*Coryphæna hippuris*). This Acanthopterygious fish has a flat and roundish snout, and the body tapers from the head to the tail; but its principal beauty consists in the brilliancy of its colours. The back is

DOLPHIN.—(CORYPHÆNA HIPPURIS)

spangled with bright bluish-green spots; the tail and fins are of a gold colour; and whether viewed alive in its native element, or before it is quite dead, nothing can surpass its lustre. It is about five or six feet long, and nearly as thick as the salmon. A remarkable fin runs from the head, along the back, to the root of the tail, which in the middle is seven inches broad, and consists of a kind of coriaceous membrane with soft spines; opposite to which there is another fin, not more than an inch broad, and extending from the vent to the tail. The tail, which is upwards of two feet and a half long, is divided into two large horns; and the scales are so very minute as to be hardly perceptible. This fish swims with such amazing velocity, as frequently to keep pace with a swift sailing ship for a very considerable time. They abound within the tropics, and are found in all temperate latitudes. In the neighbourhood of the equator, they commit great havoc in the immense shoals of flying-fish which inhabit those regions, and which constitute the principal food of the *Coryphæna*. It is remarkable that, in swallowing their prey, the position of the captured fish is reversed, and it passes down the throat head foremost; by which manœuvre the fins are prevented from impeding the passage.

DONAX. A genus of bivalve shells, the form of which is inequilateral and wedge-shaped. It is found in all parts of the world buried in the sand of the sea-shore. Many of the species are beautiful; but only two, it is said, are found on the British coasts; one called the Yellow Donax, the other the Purple.

DORIPPE. A genus of brachyurous Decapod Crustaceans (comprehended under the general term *Cancer* by Linnæus), found on the sea-coasts of warm climates, where the water is deep; the Mediterranean and Adriatic seas being among the localities given.

They are generically characterized by having rather long external antennæ, inserted above

FLAT-FRONTED DORIPPE,—(DORIPPE SIMA.)

the intermediate ones, which are folded, but not entirely lodged in the cavities where they take their insertion: claws (*chelæ*) small, short, equal; the other feet very long and compressed, the third pair being the greatest; the two last pair elevated upon the back, and terminated by a small hooked nail: carapace slightly depressed, truncated, and spinous before; truncated, sinuous, and bordered behind; the surface marked with small humps or tubercles: inferior and posterior part of the body truncated into a kind of gutter to receive the reflected abdomen, the pieces of which are tuberculous. The eyes are small, lateral, and supported on moderately long peduncles. It is now known, from Mr. Cuming's observations appended to a specimen in the British Museum, that they make use of the feet, elevated on the back to cover themselves, like the *Dromiæ*, with foreign bodies. There are some fossil as well as recent species. One fossil species, brought from New Holland by Peron, is named *Dorippe nodosa*. In the fine collection of James Scott Bowerbank, F. R. S., there are specimens of a species of this genus found on the Isle of Sheppey, in the London clay.

DORIS. A genus of naked Gasteropodous marine Mollusca, which are likewise destitute of any internal testaceous plate. The mantle is covered with retractile papillæ, and separated from the foot by a distinct duplicature. Towards its anterior margin are placed the two superior tentacula: these are retractile, surrounded at the base with a short sheath, and supported on a slender stem, having an enlarged compound plicated

THE GORGEOUS DORIS—(DORIS MAGNIFICA.)

summit. The neck is short, and above the mouth there is a small projecting membrane connected at each side with the oral tentacula, which are in general minute, and of difficult detection. The mouth is in the form of a short trunk, leading to fleshy lips, within which the tongue is placed. The gullet is a simple membranaceous tube, terminating in a stomach. It is obvious, from the structure of the digestive organs, that the species subsist on soft food. The spawn is gelatinous and of a white colour, and is deposited on sea-weed and stones.

Messrs. Alder and Hancock are publishing in one of the works of the Ray Society, descriptions and figures of all the British species of Doris and allied genera, forming the Nudibranchiate Mollusca. It is a truly elegant work, the illustrations in which must arrest the attention of even those who feel but little interest in the subject. Col. Montagu, Dr. Johnston of Berwick, and Messrs. Alder and Hancock, with other naturalists, have shown how rich our own coasts are in these beautifully organized shell-less mollusca.

DORMOUSE. (*Myoxus.*) A genus of mammiferous quadrupeds, of the Linnæan order *Glires*. They appear to be intermediate between the squirrels and mice; inhabit temperate and warm countries, and subsist entirely on vegetable food. They have two cutting teeth in each jaw; four toes before, and five behind; and naked ears. These mice inhabit woods and thick hedges, building their nests, which are lined with moss and dead leaves, either in the hollows of trees, or near the roots of close shrubs. Towards the approach of winter they form little magazines of nuts, beans, acorns, &c., on which to subsist during the inclement season; when they retire to their retreats, roll themselves up, and fall into a torpid or lethargic state, which lasts, with little interruption, till the winter is over. It was formerly believed that their hybernation was a state of continual sleep from the period that they sought their winter quarters until they emerged from them in a more genial season. Buffon, however, very properly exposed the absurdity of the ancient notion; and has observed that these animals occasionally wake, and make use of their stock of provision. They bring forth three or four at a time, which are usually born blind, and remain so for a few days. There are several species.

The COMMON DORMOUSE. (*Myoxus avellanarius.*) The body is about the size of that of the common mouse, but it is of a more plump or rounder form, and the nose is more obtuse: the eyes are large, black, and pro-

COMMON DORMOUSE.
(MYOXUS AVELLANARIUS.)

minent; the ears round and semi-transparent; the tail is two inches and a half long, and more hairy towards the tip than on the other parts: the head, back, sides, belly, and tail are of a tawny red colour, but the throat is white: the fur is remarkably soft, and the animal altogether has a considerable degree of elegance in its appearance. Its habits are similar to those described in the preceding paragraph.

The FAT DORMOUSE. (*Myoxus glis.*) This species is a native of France and the South of Europe. Its body is covered with soft ash-coloured fur; the belly is whitish; the tail is surrounded with very long hair; and the ears are thin and naked. Its length, from the nose to the tail, is nearly six inches, that of the tail being four and a half; and the body is thicker than a squirrel's. Like the last-mentioned animal, although these have not its activity and sprightliness, they can ascend trees in search of their food, which they carefully store up for their winter consumption. During its state of torpidity it is said to grow very fat, contrary to the nature of most of the hybernating or sleeping animals; but there is no doubt that it occasionally wakes, and feeds on its store; in truth, it is at all times fat, and appears as much so in spring as in autumn. Its flesh was esteemed a great delicacy by the Romans, who had their *gliraria*, or places in which they were kept and fattened for the table.

The GARDEN DORMOUSE (*Myoxus nitela*) is a native of the temperate parts of Europe and Asia. It makes its nest, like the rest of this genus, in the hollows of trees, and sometimes in those of walls, where it generally fixes its abode, and remains in a state of torpidity during a great part of the winter, awaking, however, at intervals. Its general length is about four inches and a half, and the tail not quite so much. It is of an elegant rufous colour above, and yellowish white underneath: the eyes are imbedded in a large black patch or spot, which extends to some distance beyond each ear: the tail is rather wide towards the end, sharpening at the extremity, and is marked on that part by a longitudinal black stripe, having the edges white. Delighting in all sorts of fruit, but particularly in wall fruit, these animals prove very destructive in gardens. They produce their young about the middle of summer, which are about five or six in number.

DORSIBRANCHIATA. A name given by Cuvier to an order of *Annelidæ*, or red-blooded worms, which have their organs, and particularly their gills, distributed about equally throughout the middle part of the body. The Nereis, or Sea-centipede, is an example of this order. [See NEREIDA.]

DORY, DOREE, or JOHN DORY. (*Zeus.*) There are several species of this very singular Acanthopterygious fish, which is distinguished by having the spinous portions of the dorsal and anal fins separated by a deep emargination from the soft-rayed portion, and having the base of all the vertical fins, and the carina of the belly anterior to the anal fin, furnished with spines.

The COMMON DORY (*Zeus faber*) is a native of the Mediterranean, Northern, and Atlantic seas; but no locality is more noted for it than Torbay, on our own western coast. It is distinguished by its large and long head, its dusky green colour, accompanied by a strong gilt tinge, and particularly by a large, oval dusky spot on each

DORY — (ZEUS FABER.)

side the body: the mouth is wide, the lower jaw longer than the upper the teeth small and sharp, and the eyes large; the whole body is covered with very small scales, and marked by a curved lateral line, which, descending rather suddenly from the gill-covers, passes on to the tail: the back is much arched, and furnished with a row of strong small prickles, which are also continued along the curve of the abdomen: the tail is of a moderate size, and rounded at the end. The Dory is of an extremely voracious nature, preying on the smaller fishes and their spawn, as well as on various kinds of crustacea and marine insects. The form of the Dory is extremely forbidding, so much so as to deter our ancestors from tasting it; and although its flesh is now esteemed delicious food, its reputation among epicures is but of modern date. The name is said to be derived from the French, *jaune* (yellow) *dorée*, corrupted into John Dory. In general it is from twelve to fifteen inches in length, though it sometimes arrives at a far superior size, and of the weight of ten or twelve pounds.

There are a few other species, but less remarkable than the preceding:—The BRAZILIAN DORY (*Zeus vomer*), which is about six or eight inches long; body very thin, without scales, and of a bright silver colour, tinged with a bluish cast on the upper parts.—The INDIAN DORY (*Zeus Gallus*) is about the same size as the one just mentioned: body very thin, silvery, and without scales: head large, mouth wide. Native of the American and Indian seas.—CILIATED DORY. (*Zeus Ciliaris.*) This species, which is also destitute of scales, is of a bright silver colour, with a cast of bluish-green on the back: head small, and very sloping; lower jaw rather longer than the upper: several of the last rays both of the dorsal and anal fin extend to a vast distance beyond the membrane, reaching farther than the tail itself. It has been supposed that the smaller kind of fishes may be attracted with these long and flexible filaments, and mistake them for worms, while the Dory lies concealed among sea-weeds, &c., waiting for its prey. It is a native of the Indian seas. [See ZEUS.]

DOTTEREL. (*Charadrius morinellus.*) This Grallatorial bird is about ten inches in length: the beak black, slender, and one inch long: forehead mottled with dusky and grey; the hinder part of the head is black; and a broad white line over the eyes surrounds the whole. The back and wings are a light brown; the breast is a pale dull orange; the middle of the belly is black; the edges of the feathers are pale rust colour, and the lower part of the back and rump incline to gray. The tail is composed of twelve brown olive-coloured feathers, barred with black near the ends, and tipped with white: the thighs are a reddish white, and the legs black. The female is rather larger, and the colours more dull; the white line over the eye is smaller; and the crown of the head is mottled with brown and white. Dotterels inhabit the northern parts of Asia and Europe, frequenting the muddy borders of rivers: they are migratory, being seen on our moors and downs in their flights to and from their breeding-places, from April to June, and again from September to November. Being fresh from regions and wilds untrodden by man, and not having experienced persecution, they do not so readily take alarm, as other birds do which have been reared in the vicinity of their general enemy: they have, in consequence, obtained the character of being very stupid birds, and, it is said, may be taken by the most simple artifice; but night-fowling, and all modes of ensnaring them, have yielded to the more certain method of bring them down with a gun.

DOTTEREL, SEA. [See TURNSTONE.]

DOVE. [See PIGEON: RING-DOVE: TURTLE-DOVE.]

DRAGON. (*Draco volans.*) Instead of the formidable monster of this name, which recals to the imagination the wild fictions of romance, the animal we are about to describe is a small and harmless lizard, agreeing in the general form of its body with the rest of that tribe; but furnished with large, expansile, cutaneous processes, which, when expanded, enable it to support itself in the air for a few seconds, in springing from branch to branch, among the lofty trees in which it resides. The total length of this highly curious creature is about ten or twelve inches; the tail being extremely long in proportion to the body, which is not above four inches. The head is of a moderate size, but very singular form, being furnished beneath with a very large triple pouch, one part of which descends beneath the throat, while the two remaining parts project on each side; all being sharp-pointed: the mouth is rather wide; the tongue large, and thick at the base; the teeth small and numerous; the neck, body, and limbs rather slender, and covered with small acuminated and closely-set scales. The colour of this animal on the upper parts is an elegant pale blue, or bluish-grey, the back and tail being marked by several transverse dusky undulations, while the wings are very elegantly spotted with patches of black, brown, and

DRAGON.—(DRACO FIMBRIATUS.)

white: the border of the wings is also white, and the whole under surface of the animal is of a very pale or whitish brown colour. Species of this genus are inhabitants of many parts of Asia, Africa, and South America; they feed on insects; and are in every respect animals of a harmless nature.

All the frightful animals described and figured in the works of some of the older naturalists, under the name of Dragons, are merely fictitious beings, either artificially composed of the skins of different animals, or made by warping some particular species of the ray or skate tribe into a dragon-like shape, by expanding and drying the fins in an elevated position, adding the legs of birds, &c., and otherwise disguising the animals. Such also are the monstrous representations (to be found in Gesner and Aldrovandus) of a seven-headed Dragon, with gaping mouths, long body, snake-like necks and tail, and feet resembling those of birds. Some of the dragons of antiquity are described as having no feet, but as crawling like serpents, and their bodies covered with scales, and so powerful as to crush an elephant with the greatest ease. The animal which gave rise to these is probably no other than the great Boa Constrictor. Again, who has not heard of the fabled Dragon of the middle ages, which had the feet of a lion, the long thick tail of a serpent, and an immense throat, from which streamed flames of fire? This dragon played a distinguished part in the days of chivalry; and was one of those monsters whom it was the business of the heroes of romance to attack and destroy. We have, involuntarily as it were, been led to notice the fabulous history of the Dragon, in order to point out the gross absurdities connected with Natural History, which, though long since exploded, were at one period received as matters of fact.

DOUC. (*Semnopithecus.*) A genus of Monkeys peculiar to Cochin China, the East Indies, and neighbouring islands. They differ from the true Monkeys by having an additional small tubercle on the last of the inferior molars; and are farther distinguished by their lengthened limbs and extremely elongated tail. In their muzzle, as well as in having posterior callosities, they resemble the Gibbons. Though capable of much agility, their movements are staid and deliberate, and their general deportment remarkable for gravity. [See MONKEYS.]

DRAB [MOTHS]. A name given by collectors to Moths of the genus *Orthosia*. They are also called QUAKERS.

DRAGONET. (*Callionymus lyra.*) A beautiful Acanthopterygious fish, inhabiting the Mediterranean and Northern seas, and about a foot in length. The head is large and somewhat depressed; the mouth wide, and the teeth small and numerous; the eyes are placed near each other on the upper part of the head; the body is of a taper form, smooth, and destitute of visible scales. According to Pennant, the pupils of the eyes are of a rich sapphirine blue colour; the irides fine fiery carbuncle, the pectoral fins light brown; the side lines straight; the colours of the fish yellow, blue, and white, making a beautiful appearance when fresh taken.

There are two or three other species, one of which, called the OCELLATED DRAGONET, (*Callionymus ocellatus*) about the size of one's little finger, is a native of the Indian seas. The head is smaller and sharper than in others of the genus, and rather flat at the top; mouth small, with tumid, fleshy lips, the upper one doubled; lateral line straight; tail rounded.

DRAGON-FLY. (*Libellulidæ*). A numerous family of Neuroptera. They are of blue, green, white, crimson, and scarlet colours; in some a variety of the most vivid tints are united; and they are easily distinguished from all other insects by the length of their bodies, the large size of their eyes, and the beautiful transparency of their wings. These

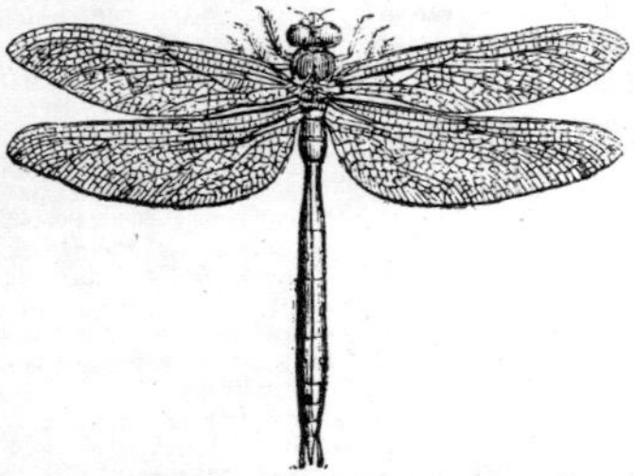

DRAGON FLY.—(AESHNA GRANDIS.)

brilliant and lively insects, which are seen flying with such strength and rapidity round the meadows, and pursuing the smaller insects with the velocity of a hawk, were at one time inhabitants of the water, and resided in that element for a long space of time before they assumed their flying form. The mouth of the Dragon-fly contains a formidable apparatus of mandibles and maxillæ, denticulated at the tip; the antennæ are very short, being merely a pair of small hairs; the wings large and spreading, and the body elongated.

There are many different genera and species of the *Libellulidæ*, both native and exotic. One of the largest English species is the *Æshna varia*, or Great variegated Dragon-fly. This insect makes its appearance principally towards the decline of summer, and is singularly elegant: its general length is about three inches from head to tail, and the wings when expanded measure near four inches from tip to tip: the head is very large; the neck extremely slender; and the eyes, which occupy by far the greatest part of the head, are of a pearly blue-gray cast: the front is greenish yellow; the thorax of the same colour, but marked by longitudinal black streaks; the body, which is very long, slender, and sub-cylindrical, is black, richly variegated with bright blue and deep grass-green: the wings are perfectly transparent, strengthened by very numerous black reticular fibres, and exhibit a strongly iridescent appearance. In the day-time it flies about in pursuit of its prey with astonishing rapidity; but during the early morning hours, and in the evening it is observed to sit with its wings expanded, and may be easily taken.

All the insects of this family are produced from eggs deposited in the water, which, sinking to the bottom, are hatched, after a certain period, into flattish hexapod larvæ: they cast their skins several times before they arrive at their full size, and are of a dusky brown colour: the rudiments of the future wings appear on the back of such as are advanced to the pupa or chrysalis state, in the form of a pair of oblong scales; and the head is armed with a most singular organ for seizing its prey; viz., a kind of flat proboscis, with a joint in the middle, and a pair of strong hooks or prongs at the end. This proboscis, when the Dragon-fly is at rest, is folded or turned up in such a manner as to lap over the face like a mask; but when the creature sees any insect which it means to attack, it springs suddenly forwards, and by stretching forth the jointed proboscis, readily obtains its prey. In this their larva and pupa state they continue for two years, when, having attained their full size, they prepare for their ultimate change; and creeping up the stem of some water-plant, and grasping it with their feet, they make an effort by which the skin of the back and head is forced open, and the enclosed Libellula gradually emerges, its head and wings first appearing. The wings, at this early period of exclusion, are very tender and contracted, all the ramifications or fibres having been compressed within the small compass of the oblong scales on the back of the pupa; but in about half an hour they are fully expanded, and have acquired the strength and solidity necessary for flight. This curious process of the evolution or birth of the Dragon-fly generally takes place on a fine sunny morning; and though for a time it roves the field and forest, or disports itself on the margin of the silvery stream—lightly traversing the air in a thousand directions, and expanding its gossamer wings to the sun—how short is its aerial and terrestrial life, compared with that which it passed in its aquatic state! Scarcely have the frosts of autumn nipped the tenderest plant, ere the whole tribe of Libellulæ perish from the cold.

Among the varieties of the Dragon-fly, many of them may sometimes be observed in the same field, or flying within a small distance of each other on the borders of their natal stream; and though they differ in their size and variegations, their general form and habits correspond too nearly to be mistaken for any other winged insects. A specific notice of each may therefore be thought unnecessary. We shall, however, avail ourselves of Dr. Shaw's description of one species of "exotic" Libellula, and his concluding remarks relative to the extraordinary character of the eyes of these insects in general. "The *Libellula Lucretia* is a native of the Cape of Good Hope (or rather of S. America), and is distinguished by the excessive length of its slender body, which measures not less than five inches and a half in length, though scarcely exceeding the tenth of an inch in diameter: the wings are transparent, of a slender or narrow shape, as in the *L. puella*, to which this species is allied in form, and measures five inches and a half in extent from tip to tip: the colour of the head and thorax is brown, with a yellowish stripe on each side, and the body is of a deep mazarine blue.

"I should not dismiss the genus Libellula without observing that in some species, and particularly in the *L. varia*, *grandis*, &c., the wonderful structure of the cornea or external coat of the eye, which prevails in by far the major part of the insect tribe, is exhibited with peculiar distinctness. Even a common magnifier, of about an inch focus, demonstrates that the cornea is marked by a prodigious number of minute decussating lines, giving a kind of granular appearance to the whole convexity: but when microscopically examined, it exhibits a continued surface of convex hexagons, and if cut from the head, and cleared from its internal pigment, it appears perfectly transparent, and seems to consist of an infinity of hexagonal lenses of equal convexity on both surfaces. This is a subject on which much might be said; but the compass of the present publication forbids too circumstantial a description of minute and disputable particulars. It may be sufficient to observe that on each eye of this animal, according to the computation of Lewenhoek, there are about 12,544 of these lenses."

To those who would study in detail the members of this group, we would recommend the volume of Rambur in the "Suites à Buffon," and the works of Van der Linden, De Selys Longchamps, and especially Charpentier. In this country, J. C. Dale, Esq. F.L.S., has made the group a special subject of study, and Mr. W. Evans has published rough figures of all the British species, which may prove useful in identifying them. Dragon-flies are often found in a fossil state, as early as in the lias formation. [See PETALURA.]

DRILL. [See APE.]

DRIVER ANT. The local name given to a species of Hymenopterous insect belonging to the family of Ants. Its name is *Anomma arcens*. The following very interesting account is derived from a paper by the Rev. Dr. Savage, an American missionary on the coast of West Africa, and published in the "Transactions of the Entomological Society" for 1847.

The writer prefaces his narrative by saying that he is not aware that the insect in question has ever been described, or that it exists in any of the European cabinets; but he thinks it is, without doubt, that of which Mr. Smeathman speaks, when he says, "one species, which seems at times to have no fixed habitation, ranges about in vast armies. By being furnished with very strong jaws, they can attack any animal whatever that impedes their progress, and there is no escape but by immediate flight or instant retreat to the water. The inhabitants of the negro villages are frequently obliged to abandon their dwellings, taking with them their children, &c., and wait till the ants have passed." Dr. Savage says it is evidently closely allied to the *Atta cephalotes* of Fabricius, found in the West Indies and South America, and like that named by the French "*Fourmi de visite*," would be not inappropriately styled the "visiting ant," though he considers the appellation *Driver* more significant of its habits. "Its domicile," he says, "if such it may be called, consists of a shallow excavation under the roots of trees, shelving rocks, and almost any other substance that will afford a shelter; not originating with themselves, but adopted and completed as the wants of their community may require; their mode of life not admitting of cells and magazines, and other interior arrangements, by which the domiciles of other ants more retiring and less aggressive in their habits are characterized.

"Their sallies are made in cloudy days, and in the night, chiefly in the latter. This is owing to the uncongenial influence of the sun, an exposure to the direct rays of which, especially when the power is increased *by reflection, is almost immediately fatal.* If they should be detained abroad till late in the morning of a sunny day by the quantity of their prey, they will construct arches over their path, of dirt agglutinated by a fluid excreted from their mouth. If their way should run under thick grass, sticks, &c., affording sufficient shelter, the arch is dispensed with; if not, so much dirt is added as is necessary to eke out the arch in connection with them. In the rainy season, or in a succession of cloudy days, this arch is seldom visible; their path, however, is very distinct, presenting a beaten appearance, and freedom from every thing movable. They are evidently economists in time and labour; for if a crevice, fissure in the ground, passage under stones, &c., come in their way, they will adopt them as a substitute for the arch. This covered way seems to be designed in part for the protection of workers in transporting prey, pupæ, &c., but chiefly *against the direct rays of the sun, an exposure to which, in places where the* REFLECTON *is strong, is certain death in less than two minutes.* When the sun's rays are intercepted for days, the arch is wanting; and, even with the arch, in a bright strong sunshine, masses of the Drivers are found under the thick grass in holes and other places, regaling themselves in the shade till the decline of the sun, when their work is renewed with their characteristic vigour.

"In cloudy days, when on their predatory excursions, or migrating, an arch for the protection of the workers, &c. is constructed of the bodies of their largest class. Their widely extended jaws, long slender limbs, and projecting antennæ intertwining, form a sort of net-work that seems to answer well their object. Whenever an alarm is given, the arch is instantly broken, and the ants, joining others of the same class on the outside of the line, who seem to be acting as commanders, guides, and scouts, run about in a furious manner in pursuit of the enemy. If the alarm should prove to be without foundation, the victory won, or danger passed, the arch is quickly removed, and the main column marches forward as before in all the order of an intellectual military discipline.

"I will here describe an attempt that I recently made to destroy one of their communities, which, with the facts in the order in which they transpired, and the collateral circumstances attending it, will fairly illustrate many of their habits. My observations were made in part at my former station (Cape Palmas), where I resided nearly eight years. I have been at my present station about eight months. During the first four months of the latter period I was greatly annoyed by the frequent visitations and ravages of these insects; at one time literally *driving* out every member of the female department of the school; at another the male department; then the inmates of my own dwelling; again, attacking my horse, then my pigs, fowls, &c. &c.; nothing, in fine, possessing animal life escaping their assaults. They always pounced upon us at night, and generally when our senses were reposing in sleep. Occasionally we were apprised of their designs at nightfall by a few suspicious individuals lurking in the vicinity in advance of the main body, but mostly they took us by surprise. At last their annoyance seemed to have reached the highest point of our forbearance, and a resolution was forthwith taken to discover their habitation, and, if possible, expel them from the vicinity. Accordingly I commenced cutting over the premises, and had proceeded as far as two-thirds the way down the mount on which my dwellings stand, when, beneath a shelving rock of decomposing granite, their haunt was discovered. They had been roused by the noise and efforts of the workmen, and had come forth in incalculable numbers for defence, literally blackening the surrounding grass and shrubbery. Lines of ants, going and coming agreeably to the rules of their order, were running in opposite directions. Their paths were very distinct and well trodden, of about an inch in width. In other directions were seen covered ways forsaken, the object of their formation no longer existing,—no prey having been discovered, or, if found, being disposed of,—and other regions lying open for exploration. Their numbers could not be computed; millions on millions seemed to be there, besides thousands that were going and coming with astonishing speed and alacrity.

"In attempting their destruction I adopted the mode of the natives, which is, to ignite on the spot a collection of the dried leaves of a species of Corypha (fan palm of this coast), about six feet in diameter, and dried grass, with other combustible matter. A fire of great intensity was thus kindled, which continued to burn for a considerable time. This I supposed would be the last of our troublesome neighbours. Two days after, however, on going to the spot for the purpose of examining into their domicile, I was surprised to see a tree at a short distance, about eighteen inches in diameter, to the height of four feet from the ground, with the adjacent plants and earth, perfectly black with them. From the lower limbs (four feet high) were festoons or lines of the size of a man's thumb, reaching to the plants and ground below, consisting entirely of these insects; others were ascending and descending upon them, thus holding free and ready communication with the lower and upper portion of this dense mass. One of these festoons I saw in the act of formation; it was a good way advanced when first observed: ant after ant coming down from above, extending their long limbs and opening wide their jaws, gradually lengthened out the living chain till it touched the broad leaf of a Canna coccinea below It now swung to and fro in the wind, the terminal ant the meanwhile endeavouring to attach it by his jaws and legs to the leaf; not succeeding, another ant of the same class (the very largest) was seen to ascend the plant, and, fixing his hind legs with the apex of the abdomen firmly to the leaf under the vibrating column, then reaching forth his fore legs and opening wide his jaws, closed in with his companion from above, and thus completed the most curious ladder in the world."

In about two hours Dr. Savage visited the spot again, when the hanging lines or festoons were gone, and about half of the mass also; some below the surface, others on their predatory excursions; and they again underwent the fiery ordeal, which urged them forwards, and they marched on with all their former celerity. Next morning he found them still engaged in removing. Thousands and tens of thousands must have been destroyed by the two fires, and yet apparently their numbers were undiminished. Neither on this nor any other occasion did he detect a winged individual, though it was the season when such are to be found in all communities of ants not apterous.

"Their mode of biting differs from that of the soldiers among the *Termes* The mandibles of the latter are flat and sharp, and move in a cross direction, cutting in the manner of scissors. The mandibles of the Driver of the first class are very prominent and formidable, strongly hooked, having one tooth; those of the second classes are flatter, sharper, and armed with two strong teeth, the edges finely serrated, and admirably calculated for lacerating and cutting muscular fibre. The onset of the former is with a grasp that causes their victim to start and wince as if life were in danger; their mandibles are fixed so strongly into the flesh, and their hold retained with such pertinacity, that a separation is effected often only by a dismemberment of the body. If permitted to retain their hold, the motion of their jaws is alternately from one side to to the other, penetrating deeper and deeper at every stroke. With the second class there is not only this gradual penetration, but at the same time lacerating and cutting of the flesh, with an approximation of the jaws at each effort. This difference in the form and motion in the two classes led me to infer a difference of duties or office in their economy. This impression has been confirmed by repeated observations. To the first class, it would appear, is assigned the defence of the community; it is theirs also to attack and disable their prey. The second lacerate and cut the flesh, and are assisted by the first in tearing it off. Upon the third, who appear to be especially the labourers, devolves the burthen of transportation, whether of prey or pupæ. They are seen to be assisted often by the second class, and, when the prey is too large for either, the first is called in.

They carry their pupæ and prey longitudinally under their bodies, held firmly between their mandibles and legs, the latter of which are admirably calculated by their length and slenderness for this purpose; and the freedom and ease with which they carry their burthen is truly surprising.—Whenever a stream of water intercepts their course in their excursions and migrations, if it should not be extensive they compass it, but if otherwise, they make a line or chain of one another, gradually extending themselves by numbers across, till a connection is formed with the opposite side, and thus a bridge is constructed, over which the main body passes in safety.—Their tenacity of life appears to be truly extraordinary. This was evinced by a series of experiments. An individual of the largest class was submerged to the bottom of a glass of water, where it struggled for about three quarters of an hour, and then apparently expired; but it revived in about ten minutes after it was taken out, exhibiting about as much vitality and ferocity as before. It was re-submerged for five hours, with like results. It was submerged the third time, and kept under water for twelve hours. When taken out it revived, and continued to exhibit signs of life for about twelve hours more, and then expired. Various other experiments were tried. The head of one of the largest class, when dissevered from the body, grasped the finger of an attendant so furiously as to cause an immediate flow of blood; another decapitated head retained its power of biting so as to draw blood, precisely in the manner of the insect in possession of all its parts and powers, twenty-four hours after decapitation; while the body to which it belonged lived more than forty-eight hours!

"I know of no insect," says Dr. Savage, "more ferocious and determined upon victory. They fiercely attack anything that comes in their way,—'conquer or die' is their motto. Yet they are not without their uses in the economy of nature. They keep down the more rapid increase of noxious insects and smaller reptiles; consume much dead animal matter, which is constantly occurring, decaying, becoming offensive, and thus vitiating the atmosphere, and, which is by no means the least important in the Torrid Zone, often compelling the inhabitants to keep their dwellings, towns, and their vicinity, in a state of comparative cleanliness. The dread of them is upon every living thing. It may be literally said that they are against everything, and everything against them. I have known my dog, on meeting them in the road, instead of running any risk by leaping over them, go a great distance round to avoid their well-known bite. My donkey has more than once stopped so suddenly and turned, as to throw me over her head, or to one side, and when urged forward, leaped far over the line.—They will soon kill the largest animal if confined. They attack lizards, guanas, snakes, &c., with complete success. We have lost several animals by them, monkeys, pigs, fowls, &c. The severity of their bite, increased to great intensity by vast numbers, it is impossible to conceive. We may easily believe that it would prove fatal to almost any animal in confinement. They have been known to destroy the *Python natalensis*, our largest serpent. When gorged with prey it lies powerless for days; then, monster as it is, it easily becomes their victim.—Their entrance into a house is soon known by the simultaneous and universal movement of rats, mice, lizards, *Blapsidæ*, *Blattidæ*, and of the numerous vermin that infest our dwellings.—They are decidedly carnivorous in their propensities. Fresh meat of all kinds is their favourite food; fresh oils they also love, especially that of the *Elais guineiensis*, either in the fruit or expressed. It is not true, however, that they devour every thing eatable by us in our houses; there are many articles which form an exception. If a heap of rubbish comes within their route, they invariably explore it, when larvæ and insects of all orders may be seen borne off in triumph,—especially the former."

DROMEDARY. [See CAMEL.]

DROMIA. A genus of Crustacea, somewhat allied to *Dorippe*, of which there are several species. The one figured (*Dromia vulgaris*) is very common in the Mediterranean; its carapace is almost globular; the two posterior pairs of legs are raised above "the plane" of the others; hence the division containing it is named NOTOPODA. They are each furnished with two sharp

SPONGE CRAB.—(DROMIA VULGARIS.)

curved claws, which enable the crab to hold fast by pieces of sponges, medusæ, or other marine productions, under which it conceals itself. In the British Museum collection are some very interesting specimens of a common West Indian species (*Dromia lator*) with pieces of sponge so attached, into which the convexity of the back of the carapace is very nicely fitted.

DRONE. The male honey-bee. [See BEE.]

DRUMMER. [See BLATTA.]

DRYOPITHECUS. [See SUPPLEMENT.]

DUCK. (*Anas.*) A very extensive and natural genus of water-birds, found in all parts of the world. They feed in great part upon animal matter, such as insects and mollusca; as well as upon vegetables and grain: they are generally seen upon the lakes and rivers of the interior, though they occasionally resort to the sea-shore. Ducks can all swim and dive with facility; they can all fly well; and they can all walk, though frequently with difficulty. They feed on soft substances, such as fresh-water insects and tender aquatic plants, which they procure near the surface, or at the bottom in shallow muddy places, and worms and slugs, which they search for among the grass. Their distinguishing characters are these: the beak is shorter in proportion than that of the goose, strong, flat, or depressed and commonly furnished with a nail at the extremity. The feet are proportionably larger than those of the goose kind, the middle toe being the longest; the legs are shorter, and placed farther backwards; the back is flatter; and the body is more compressed. The nostrils are small and oval; and the tongue is broad, the edges near the base being fringed. There are numerous species of this genus, some inhabiting the fresh water, and others the sea.

The COMMON WILD DUCK, or MALLARD (*Anas boschas*), is the original stock of the tame or domesticated Duck, and appears to have been reclaimed at a very early period. This bird measures about twenty-three inches in length, thirty-five in breadth, and is two pounds and a half in weight. The bill is of a yellowish green colour, and the head and neck are a deep shining green: a circle of white surrounds the neck, to about three-fourths of its circumference: the upper part of the breast and shoulders is of a deep vinous chestnut; the breast and belly are gray, marked with transverse speckled lines of a dusky hue; and the scapulars are white, elegantly barred with brown. The spot on the wing is a rich purple; and the tail is composed of twenty-four feathers. The male of this species is distinguished by four middle feathers, which are black, and strongly curled upwards; but of these the female is destitute. Indeed the plumage of the female partakes of none of the male's beauties, except the spot on the wings. She makes her nest, lays from ten to sixteen greenish-white eggs, and rears her young generally in the most sequestered mosses or bogs, far from the haunts of man, and hidden from his sight among high grass, reeds, and bushes. Like the rest of the Duck tribe, the Mallards, in prodigious numbers, quit the north at the end of autumn, and, migrating southward, arrive at the beginning of winter in large flocks, and spread themselves over all the loughs and marshy wastes in the British isles. They pair in the spring, when the greater part of them again retire northward to breed; but many straggling pairs stay with us: they, as well as preceding colonists of their tribes, remain to rear their young, who become natives, and remain with us throughout the year.

WILD DUCK.—(ANAS BOSCHAS.)

The flesh of the Wild Duck is held in general estimation, and various methods are resorted to, in order to obtain these birds in quantities. To describe ever a tithe part of these various contrivances is not our purpose; but it is necessary to state that the *decoy* is by far the most favourite method, and is likely long to continue so, as by that species of stratagem Wild Ducks are taken by thousands at a time; whereas all the other schemes of lying in ambush, shooting, baited hooks, wading in the water with the head covered, &c., are attended with much watching, toil, and fatigue, and are also comparatively trifling in point of success. They abound in Lincolnshire, and are there taken in great numbers. These decoys are usually thus prepared and conducted:—It is generally made where there is a large pond surrounded with wood, and beyond that a marshy and uncultivated country. On the south and north sides of the pond, two or three ditches or channels should be made, broader towards the water, and gradually narrowing till they terminate in a point: these channels should be covered over with nets, supported by hooped sticks, so as to form a vault or arch growing narrower and narrower to the point, where it should be terminated by a tunnel net: along the banks of these netted channels many hedges should be made of reeds slanting to the edges of the gutters, their acute angles being toward the side next the pool; and the whole apparatus should also be concealed from the pool by a marginal hedge of reeds, behind which the operations of the fowler are conducted. Provided with a number of Ducks termed decoys, which are rendered tame by education, and accustomed to attend their master on being summoned by a whistle, the fowler sets them to feed at the mouths of the pipes. No sooner does the evening commence, than the *decoy rises*, to use the language of fowlers, and the wild fowl feed during the night. Should the evening be still, the noise of their wings during their flight is heard at a considerable distance, and produces no unpleasing sensation. The fowler, whenever a fit opportunity offers, and he sees his decoy covered with fowl, walks about the pond, and observes into what pipe or channel the assembled ducks may be enticed or driven with the greatest facility: then, throwing hemp-seed, or some similar allurement which will float on the surface, at the entrance of the pipe, and along its extent, he whistles to his decoy-ducks, which instantly obeying the summons, approach, in expectation of being fed as usual; whither also they are followed by a whole flock of the wild ones, unsuspicious of their meditated ruin. However, their sense of smelling being extremely acute, they would speedily discover the ambuscade, did not the fowler hold a piece of burning turf to his nose, against which he constantly breathes, and thereby prevents the effluvia of his person from affecting their very exquisite senses. The Wild Ducks, therefore, in following the decoy ones, are conducted by them into the broad mouth of the pipe, without the smallest suspicion of danger, the fowler being still hid behind one of the hedges: nevertheless, when they have proceeded a short way up the pipe, and perceive it to grow narrower, they begin to apprehend danger and endeavour to return; but in this attempt they are prevented by the fowler, who now makes his appearance at the broad end below. Thus surprised, intimidated, and utterly unable to rise because of the surrounding net, the only remaining way of escape seems to be through the narrow-funnelled net at the bottom; into which they fly, and are instantly taken.

Pennant had an account sent him of the produce of ten decoys, which, in one winter, amounted to thirty-two thousand two hundred. In Picardy in France, also, vast numbers are taken in decoys, and sold in the Paris market, where, in one season, 30,000 francs have been paid for the produce of the small lake of St. Lambert. Wilson, the celebrated American ornithologist, enumerates several simple and effective contrivances made use of in America for the capture of these wary birds. In some ponds frequented by them, five or six wooden figures, cut and painted to represent ducks, and sunk by pieces of lead nailed to the bottom, so as to float at the usual depth on the surface, are anchored in a favourable position to be raked from a concealment of brush, &c. These attract the passing flock, which alight, and thus expose themselves to certain destruction. In winter, when detached pieces of ice are occasionally floating in the river, some of the sportsmen on the Delaware paint their boats white, and laying themselves flat in the bottom, direct them almost imperceptibly near a flock, before the ducks have distinguished them from a floating piece of ice. On land, another stratagem is sometimes practised with great success. A tight hogshead is sunk in the marsh, or mud, near the place where ducks are accustomed to feed at low water, and where, otherwise, there is no shelter; the edges and top are carefully concealed with tufts of long coarse grass, and reeds or sedge. From within this the sportsman watches his collected prey, and usually commits great havoc. In China, the sportsman covers his head with a calabash, pierced with eye-holes, and, thus equipped, wades into the water, keeping only his head above the surface, and, on arriving amidst a flock, seizes them by the legs, fastens them to his girdle, and takes as many as he wishes, without disturbing the rest.

The TAME DUCK. Some individuals in a domestic state appear in nearly the same plumage as the wild ones; others vary greatly from them, as well as from each other, and are marked with nearly every colour; but all the males or drakes still retain the curled feathers of the tail. The Tame Duck is, however, of a more dull and less elegant form and appearance than the Wild, domestication having deprived it of its lofty gait, long tapering neck, and sprightly eyes. Tame Ducks are reared with more facility than perhaps most other domestic animals. The very instincts of the young direct them to their favourite element; and though they are sometimes hatched and conducted by hens, they seem to contemn the admonitions of their leaders; a circumstance which seems to indicate that all birds receive their manners rather from nature than education, and attain their various perfections without the help of any other guide.

There appears to be good reason for placing duck-eggs under a hen. The Duck generally proves a heedless, inattentive mother; for she frequently leaves her eggs till they become corrupted, and even seems to forget that she is entrusted with the charge: she is also equally regardless of her young brood when they are produced; for she only leads them forth to the water, and then seems to think she has made sufficient provision for them. The hen, on the contrary, who is an indefatigable nurse, broods with unwearied assiduity, and generally hatches a duckling from every egg with which she is entrusted: she does not, indeed, conduct her young to the water, because that is contrary to her nature; but she always keeps a watchful eye over them when they approach the brink. Should the rat, the weasel, or other natural enemy of the feathered tribe, attempt to seize any of them, the hen instantly affords them her best protection; and, leading her supposititious brood to the house when tired with paddling, there nourishes them with all the instinctive ardour of maternal regard. "The village school-boy," as Bewick says, "witnesses with delight the antic movements of the shapeless little brood, sometimes under the charge of a foster-mother, who, with anxious fears, paddles by the brink, and utters her unavailing cries, while the Ducklings, regardless of her warnings, and rejoicing in the element so well adapted to their nature, are splashing over each other beneath the pendent foliage; or, in eager pursuit, snap at their insect prey on the surface, or plunge after them to the bottom: some, meanwhile, are seen perpendicularly suspended, with the tail only above water, engaged in the general search after food."

There are many different varieties of the Tame Duck: the most obvious distinction, however, between the wild and tame species lies in the colour of their feet; those of the tame being black, and the wild yellow. As we before observed, the common species of Tame Ducks derive their origin from the Mallard, and may be traced to that fowl by unerring characters. Tame Ducks are an extremely advantageous kind of poultry; as they subsist on scattered corn, the refuse of vegetable and animal substances, worms, snails, and insects. They lay a great number of eggs annually; require very little attendance when sitting; and, with respect to Ducklings, they may be easily fattened in the course of three or four weeks with any kind of pulse or grain and water.

MUSCOVY DUCK, or MUSK DUCK. (*Cairina moschata.*) This bird, which takes its name from its musky smell, and not from its being originally obtained from Russia, as is supposed, is upwards of two feet in length. In its wild state it is entirely of a black colour, with glosses of blue and green, and white wing-coverts; but when domesticated it varies very considerably: its usual appearance, however, may be thus described. The crown of the head is slightly tufted, and black; the cheeks and fore part of the neck white, irregularly marked with black; the belly chiefly white, and the general colour of the rest of the plumage deep brown, darkest and glossed with green on the back, rump, quills, and tail, the two outer feathers of the latter, and the three first primaries being white: the legs and feet are short, thick, and red. They are more prolific and sit oftener than other ducks; and their eggs, which are frequently tinged with green, are larger and rounder than those of other species.

CANVAS-BACKED DUCK, or POCHARD. (*Aythya vallisneria.*) The zoologist is indebted to the indefatigable Wilson for the first account of this much esteemed species. The Canvas-back is two feet long, and three feet wide, and when in good order weighs three pounds. The beak is large, and of a glossy black; the head and part of the neck of a rich glossy reddish-chestnut hue, ending in a broad space of black that covers the upper part of the breast: back, scapulars,

CANVAS-BACKED DUCK.
(AYTHYA VALLISNERIA.)

lower part of the breast, and belly, white, faintly marked with an infinite number of transverse wavy lines, or points: wing-coverts gray spotted: tail very short, and pointed; legs and feet very pale ash. The female is smaller, and not so brightly coloured as the male. These birds arrive in the United States, from the north, about the middle of October, and, principally, assemble in the numerous rivers in the neighbourhood of the Chesapeake bay. When they first arrive they are very lean; but from the abundance of their favourite food, they become fat about November. From the great demand for these ducks, and the high price they fetch, various methods are employed to decoy them.

Besides the species we have described, there are many, for which we can only find room for very brief notices. The SCAUP DUCK (*Fuligula marila*), somewhat smaller than the common duck. In North America, a variety of this species is better known by the name of the Blue-bill, and is common both to the fresh-water rivers and sea-shores in

SCAUP DUCK.—(FULICULA MARILA: VAR.)

winter, those which frequent the latter being generally much the fattest, on account of the greater abundance of food along the coast. The GOLDEN-EYE (*Clangula glaucion*), the bill of which is black, short, and broad at the base; the head is large, and of a deep black hue, glossed with green; and at each angle of the mouth there is a large white spot. The BLACK DUCK, or SCOTER (*Oidemia nigra*); a bird whose flavour is so rank and fishy, as to be exempted, with a few others, from the interdict which forbids Roman Catholics the use of animal food on certain days, on the supposition of their being cold-blooded, and partaking of the nature of fish. The PIED DUCK (*Anas Labradoria*); a beautiful and rare species, peculiar to America. CHINESE DUCK (*Anas galericulata*); a remarkable bird somewhat less than a widgeon. The SUMMER or WOOD DUCK (*Aix sponsa*); not more remarkable for its

SUMMER DUCK.—(AIX SPONSA.)

great beauty, in which it stands pre-eminent, than for its habits, its migrations being directly opposed to those of the other species. AUTUMNAL DUCK (*Anas autumnalis*); native of the West India islands and South America, where the inhabitants frequently keep them in the farm-yards. TREE DUCK, or WHISTLING DUCK, (*Anas arborea*); this also is an inhabitant of the West India islands and the adjacent continent of America. We learn from Mr. Gosse's "Birds of Jamaica," that its singular whistling note is peculiarly shrill, and is uttered in its crepuscular flights to and from its feeding places, and also when it is alarmed. He also says these birds are much dreaded by those who plant Guinea-corn; and that "numerous flocks of both young and old birds frequent the millet-fields from December till the end of February, when this corn is reaped. They are most busy in their depredations on moonlight nights; and as they sweep round in circles, their remarkable whistle always betrays their movements." Another remarkable peculiarity is thus recorded: "The Whistling Duck endeavours to save her young, when pursued, by throwing herself into the man's way; that is, by rushing up so close to him as to draw his attention, that her young, who are very active, may have an opportunity of escaping. Accordingly, the man, seeing the duck so near him, looking upon her as a much better prize than the young ones, leaves pursuing the ducklings, and endeavours to catch the subtil dame, who runs before, but takes special care to keep out of his reach; yet stopping in front of him occasionally, to make him renew the pursuit, till the young are entirely out of danger; when she flies away, leaving her pursuer to fret at his double disappointment." LONG-TAILED DUCK (*Harelda glacialis*); remarkable for the long and slender feathers of its tail. This Duck is very generally known along the shores of the Chesapeake Bay by the name of the *South-southerly*, from the singularity of its cry, something imitative of the sound of those words; and also, that, when very clamorous, they are supposed to betoken a southerly wind. They inhabit the bays and coasts of North America during the winter only; are rarely found in the marshes, but keep in the channel, diving for small shell-fish, which are their principal food. In passing to and from the bays, sometimes in vast flocks, particularly towards evening, their loud and confused noise may be heard in calm weather at the distance of several miles. They fly very swiftly, take short excursions, and are lively restless birds. Their native regions are in the north, where great numbers of them remain during the whole year; part only of the vast family migrating south to avoid the severest rigours of that climate. They are common to the whole northern hemisphere. In the Orkneys they are met with in considerable flocks from October to April; frequent in Sweden, Lapland, and Russia; and are said to breed at Hudson's Bay, making their nest among the grass near the sea, like the eider duck, and about the middle of June lay from ten to fourteen bluish-white

LONG-TAILED DUCK.—(HARELDA GLACIALIS.)

eggs, the size of those of a pullet. When the young are hatched, the mother carries them to the water in her bill. The nest is lined with the down of her breast, which is accounted equally valuable with that of the eider duck, were it to be had in the same quantity. They come to England only in very severe winters, and then but in small straggling parties. TUFTED DUCK (*Anas cristata*), found in the arctic regions of both continents, and migrating to southern countries in the winter: on the top of the head is a crest consisting of long and slender feathers, which, with the head, neck, and breast, are black, glossed with violet and green. PINTAIL DUCK (*Dafila acuta*); remarkable for the pointed form of its tail: it is abundant in both hemispheres. The NILOTIC MUSK DUCK (*Anas Nilotica*); between the size of the Pintail Duck and the Goose, but stands higher on its legs. It inhabits the Nile, in Upper Egypt; is easily tamed, and lives among other domestic poultry. GREY-HEADED DUCK (*Somateria spectabilis*); with red bill, legs, and feet: native of Hudson's Bay. The GREAT BLACK DUCK (*Oidemia perspicillata*); a species considerably larger than the common Duck. STELLATED DUCK; a species distinguished by its eyes, which are placed higher than usual in an oval black spot; but its principal characteristic is a large white star on its back. The MADAGASCAR DUCK; a large and brilliantly-coloured species: the bill of a yellowish-brown colour; the head and neck of a dusky green; and the back is of a deep purple:

the long feathers of the wings are adorned with red eyes; and the legs and feet are of an orange hue. The Hook-billed Duck (*Anas adunca*), which differs but little from the common Wild Duck except in the bill. The Freckled Duck (*Anas nœvosa*); a very rare species which inhabits the neighbourhood of Swan River, in Australia, but from its scarcity it is little known either to the colonists or the natives. According to Mr. Gould's description of the specimen in his possession, the whole of the plumage is dark brown, minutely freckled and spotted with irregular oblong marks of white in the direction of the feathers; the under surface the same, but lighter and tinged with buff; wings without a speculum; primaries plain brown; irides light brown; bill greenish gray, becoming much darker at the tip; legs bluish green.

DUGONG. (*Halicore.*) A marine animal, herbivorous in regard to its food, and fish-like in its form. It ranks among the *Cetaceæ*; is about seven or eight feet in length; and

DUGONG.—(HALICORE DUGONG.)

has two large permanent incisive teeth in the upper jaw, and four molar teeth above and below. It is a native of the Indian seas, being common among the islands of the Indian Archipelago, and visiting also the coasts of New Holland. Its flesh is said to be tender, and not unlike beef. Professor Owen, in the Appendix to Juke's Voyage of H. M. S. Fly, has described a new species from the Eastern seas.

DUNLIN. (*Pelidna.*) A genus of birds belonging to the *Scolopacidœ* or Snipe tribe. The Dunlins in appearance resemble larks; they fly in troops near the coast; and lay their eggs in the sand. There are several species; but it will be sufficient to describe one of them:—Red Dunlin. (*Pelidna subarcuata.*) This bird is about eight inches in length; top of the head is black, edged with rufous; the forehead and throat are white dotted with brown; the nape is red, with small longitudinal black dashes; the neck, breast, and under parts are red chestnut, sometimes marked with black spots or variegated with white: tail coverts white, transversely rayed with black and red: the back, scapulars, and large wing-coverts of a deep black; on the edge of the feathers is a range of angulated bright red spots, the greater portion of which are terminated with bright ash; the tail is of a dusky ash, bordered with white. The beak is black; and the legs are brown. This bird is a native of most parts of Europe, and is sometimes seen on the British coasts. It rarely appears at any considerable distance inland; but migrates in the spring and autumn. It lays four or five eggs of a dirty white colour, spotted with brown; and its flesh is esteemed a delicacy. The names of the other species are the Pierre Dunlin, (*Pelidna variabilis*); Little Dunlin (*Pelidna pusilla*); Temminck's Dunlin (*Pelidna Temminckii*); Minute Dunlin (*Pelidna minuta*); and the St. Domingo Dunlin, (*Pelidna Dominicensis*).

DYNASTIDÆ. A family of Lamellicorn Coleoptera, comprising several beetles remarkable for their size, strength, and formidable appearance. The males are preeminently distinguished by various singular protuberances, horns, or tubercles, arising from the head or thorax and often from both of these parts of the body. "It must be borne in mind, however," as Mr. Westwood observes, "that these horns are immovable portions of the horny skeleton, and offer no real analogy with the horns of the mammalia; although it is interesting, in respect to the analogies existing in remote tribes of the animal kingdom, that the quadrupeds which are cornuted are herbivorous, and as comparatively harmless as the Dynastidæ." They chiefly inhabit the tropical regions, excavating burrows in the earth,

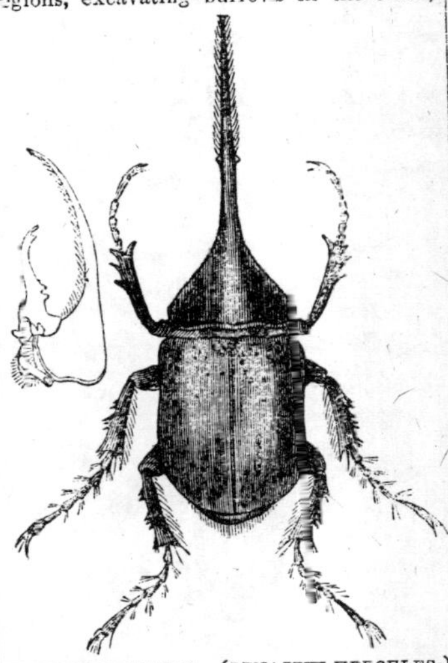

HERCULES BEETLE.—(DYNASTES HERCULES.)

where they conceal themselves during the day, or reside in the decomposed trunks of trees; and they are generally of a dark rich brown or chestnut colour. On the approach of night they leave their retreats, and run about the footpaths in woods, or fly around the trees to a considerable height, with a loud humming noise. It is believed that they subsist principally upon putrescent wood and the detritus of other vegetable matter. Among the most remarkable may be mentioned the *Dynastes Hercules*, or HERCULES BEETLE. It is usually about four, but sometimes measures not less than five, or even six inches in length: the elytra are of a smooth surface, of a bluish or brownish-gray colour, and generally marked with several small round deep-black spots, of different sizes: the head and limbs are jet black; from the upper part of the thorax proceeds a horn or process of enormous length in proportion to the body, sharp at the tip, curving slightly downwards, and furnished throughout its whole length with a fine, short, velvet-like pile, of a brownish orange colour: from the front of the head proceeds also a strong horn, about two-thirds the length of the former, toothed on its upper surface, but not covered with any of the velvet-like pile. This species is a native of several parts of South America, where great numbers are sometimes seen on the tree called the Mammæa Americana, and *have been said by some travellers* to rasp off the rind of the slender branches by working nimbly round them with the horns, till they cause the juice to flow, which they drink to intoxication, and thus fall senseless from the tree. This, however, has a very fabulous air, nor, although the account has been often repeated, do we find it any where sufficiently well authenticated to depend upon; in short, the structure of the horns would render it impossible. The female is destitute both of the frontal and thoracic horn, and but for her large size would hardly be regarded as her lord's mate.

The next species, *Megasoma elephas*, or ELEPHANT BEETLE, is also a native of South

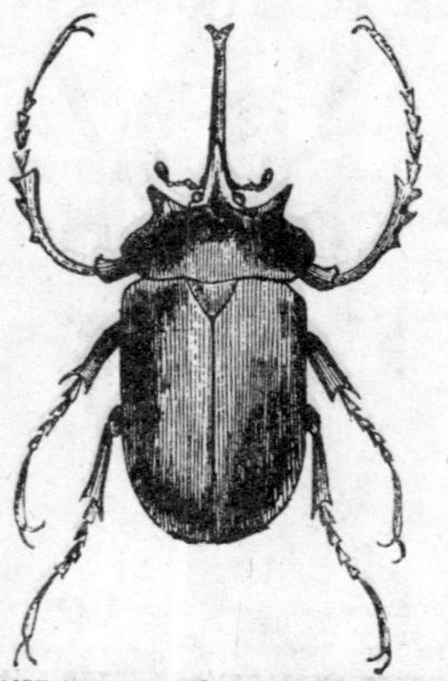

ELEPHANT BEETLE.—(MEGASOMA ELEPHAS.)

America (having been brought from Venezuela by Mr. David Dyson), though for a long time it was erroneously supposed to be a native of Africa. Some specimens are at least three inches long. Our figure will show better than any description its form and general appearance. It is covered with a yellowish gray down, which is very short and thickly set: the head is furnished with a long arched horn, which is bifid at the extremity, of a blackish colour, with a large tubercle at the base, directed forwards: the thorax has two small smooth tubercles in the middle, and a strong somewhat oblique horn on each side: the front legs are considerably arched. The next species figured is from the Eastern Islands. It is the *Chalcosoma Atlas*, or Atlas Beetle. It is of a highly polished metallic surface:

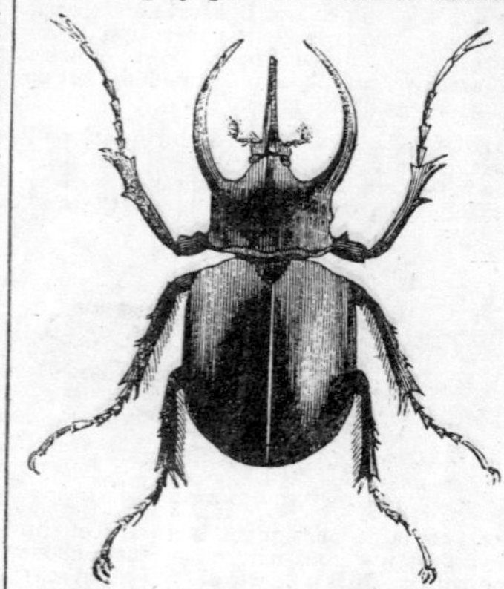

ATLAS BEETLE.—(CHALCOSOMA ATLAS.)

the horns on the head and thorax vary very much in the different specimens; but our figure is taken from a specimen in which these prominences are well developed. It seems to be far from uncommon in the Philippine Islands. There are fine specimens of it and many other species of *Dynastidæ* in the collection of the British Museum. We may here say that the females of the Dynastidæ are devoid of spines or projections on the head and thorax, just like our British *Onthophagi*, and that, like them, the males have the head more or less armed, according to the quality and quantity of the food they have taken in the larva state. This observation, as far as regards European Lamellicorn beetles, is made by M. Mulsant of Lyons, in his excellent and extensive work on the Coleoptera of France.

DYTISCUS: DYTISCIDÆ. A genus and family of insects belonging to the aquatic carnivora; which during their larva and perfect states live in water, but quit that element to undergo their metamorphoses, and to pass the time of their pupa condition. The *Dytiscus marginalis* (one of the largest European species) is common in stagnant waters; it is an inch or rather more in length, and is of a dark olive colour, with the thorax and wing-sheaths bordered with yellow. The larva of this insect in shape much more resembles the shrimp tribe than that of its parent. It measures, when full grown, about two inches and a half in length, and is of a pale yellowish-brown colour, with a high degree of transparency: the head is very large, rather flat, and is furnished with a pair of very strong curved forceps, with which it attacks its prey: its legs are slender, and its abdomen is very considerably lengthened, and ciliated on each side the tail, which terminates in a divided fin or process. The larva is very bold and voracious, committing great ravages not only among water insects, but even among small fishes, and is therefore considered one of the most mischievous animals that can infest a fish-pond. When full-grown, the larva betakes itself to the adjoining banks, where it changes into a chrysalis of a whitish colour. There are numerous species of this genus, but none so large or destructive as the one here described: some of them are, by comparison, extremely minute.

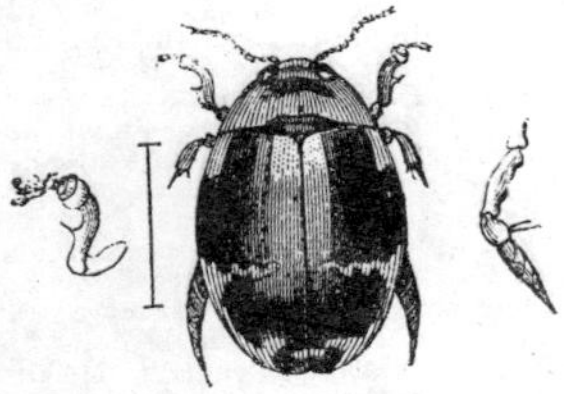

A WATER-BEETLE
(HYDATICUS INTERRUPTUS.)

As an illustration of this family, we have figured (after Sturm) the *Hydaticus interruptus*, belonging to a genus of *Dytiscidæ*, of which some species are found in this country. Dr. Schaum has given in "the Zoologist" a revision of all the British species of this family. In the work of Dr. Aubé of Paris, descriptions of the numerous species and genera of Water Beetles will be found, while in the pages of the "List of Hydrocanthari in the British Museum Collection" will be seen how very extensive and how universally distributed is the family of Water Beetles.

DZERON, or CHINESE ANTELOPE. (*Antilope gutturosa.*) This species of Antelope inhabits China, Thibet, and Tartary, chiefly frequenting the dry and rocky plains, and feeding on the scanty herbage which those barren localities supply. The length of this animal is about four feet and a half: the horns, nine inches long, of an opaque yellow colour, and having a backward direction, are annulated nearly to the tips, and diverge considerably upwards, though the points bend towards each other. The head is rather clumsy; the nose obtuse; the ears small and pointed; and on the middle of the neck grows a large protuberance, occasioned by the peculiar structure of the windpipe. It is of a tawny hue on the upper parts, and white beneath; but in winter the hair grows long, thick, and hoary. This species, which the Chinese call *Whang Yang*, or Yellow Goat, is extremely swift and active, but naturally shy and timid. During the winter season they herd in great numbers, but separate again on the approach of spring. The Tartars hunt them with the utmost eagerness, and esteem their flesh very delicious food: the horns form a considerable article of commerce, and are in great request among the Chinese: the female has no horns. Specimens of this Antelope are in the British Museum collection. It has never, however, been brought alive to this country.

EAGLE. (*Aquila.*) Pre-eminent for courage, strength, and boldness among predaceous birds, is the daring and majestic Eagle. This time-honoured monarch of the feathered tribes, which in the mythology of Greece and Rome was deemed worthy to rank as the chosen associate of Jupiter, was ever regarded as an emblem of dignity and might, and still has the reputation of being equally magnanimous, fierce, and voracious. What the Lion is among quadrupeds, that the Eagle is presumed to be among birds; one who disdains all petty plunder, and pursues only such prey as would seem to be worthy of conquest. This laudatory character of the king of birds, however, though true in the main, and generally acquiesced in, is, it must be confessed, more poetically descriptive than logically accurate; but while, in our zoological character, we are bound to make this admission, far be it from us to disparage the "bird of Jove," or to pluck a single plume from his upsoaring wing. Eagles are distinguished by the feathering of the *tarsi* down to the very base of the claws; and the males are smaller than the females; their plumage varying considerably, according to age and other circumstances. Of all the feathered tribe the Eagle soars the highest; and of all others also it has the strongest and most piercing sight. Though extremely powerful when on the wing, the joints of its legs being rather stiff, it finds some difficulty in rising again after a descent; yet, if not instantly pursued, it will easily carry off a goose or other bird equally large. The Eagle does not rise in the air so much for the purpose of pursuing its prey on the wing, as that it may be enabled to take an extensive survey of the country beneath; for its food does not consist so much of birds that pass most of their time in flight, as of those that live on or near the ground, and of such mammalia as it can vanquish. Buffon remarks, when speaking of the noble nature of this powerful bird, that the Eagle despises small animals, and disregards their insults; that he seldom devours the whole of his prey, but, like the lion, leaves the fragments to other animals; and, except when famishing with hunger, he disdains to feed on carrion. The astonishingly acute sight of the Eagle enables him to discern his prey at an immense distance; and, having perceived it, he darts down upon it with a swoop which there is no resisting.

It is well understood that the Eagle is

able to look stedfastly on the sun, and to sustain his most dazzling rays: which alone must give him a decided superiority over every other denizen of the air: this is accounted for by his being furnished with double eyelids, one of which may be shut while the other is open, so that the glaring light of any dazzling object may be rendered more easily supportable. The nest is composed of sticks, twigs, rushes, heath, &c., and is generally placed upon the jutting ledge of some inaccessible precipice; or in forests, near some lofty tree. The largest species of Eagles seldom lay more than two eggs, and the smallest never more than three.

The IMPERIAL EAGLE. (*Aquila imperialis.*) This is the largest species of Eagle known, measuring three feet and a half from the tip of the bill to the end of the tail; and to it may be referred all the accounts of the ancients respecting the strength, courage, and magnanimity of these birds. Its colour above is rufous gray, barred with black, the black prevailing most on the wings: the head is strongly crested with long gray feathers, the two middle ones being five inches long; the tail is gray, barred and spotted with black, and tipped with rufous: the under parts of the bird are pale cinereous, very soft and downy; the beak and cere black; the feet and legs yellow. It is a native of South America, inhabiting the deep recesses of the forest: and has the reputation of being extremely bold and ferocious.

The GOLDEN EAGLE. (*Aquila chrysaëta.*) This bird is the largest and noblest of the European Eagles; its length being three feet three inches, the extent of its wings seven feet six inches, and its weight from twelve to sixteen pounds. The bill is of a deep lead colour, with a yellow cere; eyes large, deep sunk, and covered by a projecting brow; the irides golden hazel-colour, bright and lustrous. The general colour of the plumage is deep brown, mixed with tawny on the head and neck, and the feathers on the back being finely shaded with a darker hue. The wings, when closed, reach to the end of the tail; the quill-feathers are chocolate-coloured, with white shafts; and the tail brown, the base being generally marked with irregular ash-coloured bars or blotches: the legs are yellow, short, and very strong, being three inches in circumference, and feathered to the very feet, which are covered with large scales, and armed with most formidable claws. It occurs in various mountainous parts of Europe and Asia, and also, though more rarely, in America: in Ireland, Scotland, and Wales, it has also occasionally been found. The eyrie or nest of the Golden Eagle is extremely large and strong, being composed of twigs and branches, interlaced and covered by layers of rushes, heath, &c., and built on the summits of rocks or lofty cliffs. The female lays two or, at most, three eggs, one of which is said to be generally unprolific. The longevity of this species is said to be great indeed, instances being quoted of its having survived more than a century.

The RING-TAILED EAGLE (the *Falco fulvus* of Buffon), though described as a separate species by him, is no other than the *young bird of one and two years* of the preceding. Its colour is a deep brown, the base of the tail being distinguished by a white ring; the bill is of a dark horn-colour, the upper mandible, which is arched, hanging over the lower one about an inch, and having an angle or tooth on each side; and the feet are feathered to the toes. The breast is marked with white triangular spots in the middle of each feather: between the bill and the eyes there are spaces of bare skin of a dirty hue, thinly set with small black hairs: and the tail, which is of an equal length with the wings when closed, is white, except the tips of the feathers, which are black, or dark brown; and the coverts under the tail are a reddish brown or bay. The toes are very thick and strong, and covered with yellowish scales; and the talons, which are black and very powerful, bend almost into semicircular figures, and terminate in very sharp points.

"It is held," says Dr. Richardson, "by the aborigines of America, as it is by almost every other people, to be an emblem of might and courage, and the young Indian warrior glories in his Eagle plume as the most honourable ornament with which he can adorn himself. Its feathers are attached to the calumets, or smoking pipes, used by the Indians in the celebration of their solemn festivals, which has obtained for it the name of the Calumet Eagle. Indeed, so highly are these ornaments prized, that a warrior will often exchange a valuable horse for the tail feathers of a single eagle." He further observes that the mature British Golden Eagle has a darkish brown tail and wings, blackish-brown back, clouded with brownish-black, and a paler and brighter-brown head. The identity of the Ring-tails with the Golden Eagles may now be considered unquestionable, the observations of so many late ornithologists concurring to establish the fact. And though Dr. Richardson says that the Golden Eagle is seldom seen farther to the eastward than the Rocky Mountains, M. Audubon asserts that he has seen it on the coast of Labrador, and various other parts of the United States. It inhabits Russia, Iceland, and Germany, and is said to occur in Northern Africa and Asia Minor. It is also frequently met with in Scotland, and its northern and western islands; in Ireland also, though much less often; and occasionally even in the western counties of England.

The ROYAL EAGLE (*Aquila regalis*) is a bird of great beauty, having an elegantly varied plumage and commanding attitudes; in fact, the account given of it by M. Sonnini, in his edition of Buffon's Natural History, would lead to the belief that it is much larger and more ferocious than any one of which we have a knowledge. His description of it is to this effect:—The head is large, and furnished with a crest in form of a casque: the bill long; the eyes bright and piercing; the claws black, crooked, and of the length of the middle finger: the back, wings, and tail are brown, spotted with black, and variegated with whitish or yellowish streaks; the belly white, the feathers being very soft, and equal in elegance to those of an egret. It flies with majestic rapidity, and such is the expanse of its wings, that it sometimes strikes and kills its prey with them before it touches it with its claws. Its strength is such as to enable it to tear in pieces in an instant the largest sheep; and it pursues, almost indiscriminately, wild animals; but its principal food consists of a particular kind of monkeys, called *Guaribas*, which it instantly kills, and devours with extraordinary voracity. Its general residence is on lofty mountains, and it builds its nest on the highest trees, employing for their construction the bones of the animals it has slaughtered, and some dry branches of trees, which it binds together with the ends of climbers. It is said to lay two or three eggs, which are white, spotted with reddish-brown. It is chiefly found about the borders of the river Amazon. Many virtues are attributed to its burnt feathers. Such is the account, with some slight abridgment, of the description given by Sonnini, and copied by Shaw; the whole resting on the authority of Don Laurent Alvarez Roxo de Postflitz, a Portuguese ecclesiastic at Brazil.

WEDGE-TAILED EAGLE. (*Aquila fucosa.*) This noble bird is the species of Eagle common to Australia, "being of course," as Mr. Gould observes, "more plentiful in such districts as are suited to its habits, and where the character of the country is congenial to the animals upon which it subsists." He further says, that although he has not yet seen it in any collection, either from the

WEDGE-TAILED EAGLE.—(AQUILA FUCOSA.)

northern portion of Australia or any other country, "in all probability it will hereafter be found to extend its range as far towards the tropics in the southern hemisphere as the Golden Eagle (*Aquila chrysaëta*) does in the northern: the two birds being, in fact, beautiful analogues of each other in their respective habitats, and doubtless performing similar offices in the great scheme of creation." In courage, power, rapacity, and size, they are also very similar; but the lengthened and wedge-shaped form of its tail gives to the Australian bird a far more pleasing and elegant contour. One, but by no means the largest, of those which were killed by Mr. Gould, weighed nine pounds, and measured six feet eight inches from tip to tip of the opposite pinions. The Wedge-tailed Eagle frequents the interior portions of the country rather than the neighbourhood of the sea; preying indiscriminately on all the smaller species of Kangaroo which tenant the hills and plains; and whose retreats, from the wonderful acuteness of its vision, it descries while soaring and performing its graceful evolutions in the air. The enterprising ornithologist, from whose splendid work we have derived the foregoing information, goes on to say, that "its tremendous stoop and powerful grasp carry inevitable destruction to its victim, be it ever so large and formidable. The breeders of sheep find in this bird an enemy which commits extensive ravages among their lambs, and consequently in its turn it is persecuted unrelentingly by the shepherds of the stock-owners, who employ every artifice in their power to effect its extirpation, and in Van Diemen's Land considerable rewards are offered for the accomplishment of the same end." He adds, that "the tracts of untrodden ground and the vastness of the impenetrable forests will, however, for a long series of years to come, afford it an asylum, secure from the inroads of the destroying hand of man: still, with every one waging war upon it, its numbers must necessarily be considerably diminished." In the adult bird, the head, throat, and all the upper and under surface of the plumage is blackish brown, stained on the edges and extremities of many of the wing and tail feathers with pale brown; back and sides of the neck rusty-red; irides hazel; cere and space round the eye yellowish white; bill yellowish horn colour, the tip black; feet light yellow. The colour of the young birds is altogether lighter, and the tail is indistinctly barred near the extremity. The nests are of a very large size, built of sticks and boughs, nearly flat, and, placed on trees which from their vast height, are all but inaccessible to man. It appears that although the Wedge-tailed Eagles mostly feed on living prey, they do not scruple to feast on the carcase of a dead bullock when they find one, or refuse to devour carrion, though it may be almost in a state of putridity.

We lately saw three specimens of this very fierce Eagle in the Gardens of the Zoological Society. Their piercing eyes and enormous beaks clearly indicated their "will and power," while their restlessness was a convincing proof that they could ill brook captivity.

VULTURINE EAGLE. (*Aquila vulturina.*) The general colour of this species, which in size is equal to the Golden Eagle, is a deep black, some of the feathers of the back having brownish edges: the bill is very strong, its tip yellow; the legs of a dirty yellow, and feathered for three quarters of their length: the tail rounded, and considerably shorter than the wings. It feeds principally on carrion, but will attack sheep, and devour them on the spot. Native of Caffraria.

CROWNED EAGLE. (*Aquila coronata.*) This species is about one-third less than the generality of Eagles, but of proportionate boldness and strength. It is a native of Africa, and is said to be principally seen in Guinea. The circles round the eyes are of a deep orange colour; the fore part of the head, the space round the eyes, and the throat, are covered with white feathers, with small black spots: the hinder part of the head and neck, the back and wings, are of a dark brown, the outer edges of the feathers being lighter: the ridge in the upper part, and the tips of some of the lesser covert-feathers of the wings, are white: the tail is brown, barred across with black, and on its under side appears of a dark and light ash-colour: the breast is of a reddish-brown, with large transverse black spots on the sides: the thighs and legs, down to the feet, are covered with white feathers, beautifully marked with round black spots. The feet and claws are very strong; the former covered with scales of a bright orange colour, and the latter being black. It takes its name from raising the feathers on the hinder part of the head in the form of a crest or crown.

The SUPERB EAGLE (*Falco superbus*) inhabits the vast forests of Guiana, and is distinguished by a kind of pendent naked craw, like some of the vultures. From the tip of the bill to that of the tail is about twenty-five inches: the upper part of the head and the crest are brown: the back and wings brown, with a few transverse tawny bars: and the tail is alternately barred with black and pale brown: the sides of the neck are tawny; the throat and breast white; the abdomen white, with transverse black stripes, interrupted by the white ground-colour: the feathers of the thighs and legs are white, striped with black.

The CHEELA EAGLE. (*Falco Cheela.*) This species is a native of India, where it is called Cheela. It is of a stout make, two feet long, and of a deep brown colour; but on each side of the head there is a mixture of white: the wing-coverts and thighs are marked with small white spots, and the tail is crossed in the middle by a white band: the bill is blue at the base, and black at the tip.

The WHITE-TAILED EAGLE. (*Haliætus albicilla.*) This bird inhabits all the northern parts of Europe, and is found in Scotland and other parts of Great Britain. The beak, cere, and eyes are of a pale yellow; the sides of the head and neck a pale ash, mixed with reddish-brown: general colour of the plumage brown, darkest on the upper part of the head, neck, and back: quill feathers very dark; breast irregularly marked with white spots; tail white; legs of a bright yellow, and claws black. It is strong, and very ferocious. It usually lays two or three eggs, building its nest upon lofty trees.

The WHITE-HEADED SEA EAGLE. (*Haliætus leucocephalus.*) This distinguished bird is about the size of the Golden Eagle, to which it bears a considerable resemblance; it is, however, of a lighter colour, and the legs are only feathered a little way below the knees. The bill is large, much hooked, and bluish. A row of strong bristly feathers hangs down from under its lower mandible, whence it has sometimes been termed the Bearded

AMERICAN, OR WHITE-HEADED SEA EAGLE (HALIAETUS LEUCOCEPHALUS.)

Eagle. It preys chiefly on fish, which it seizes by darting down upon them while swimming near the surface: it also occasionally preys on birds and other animals. The American variety is superior in size to the European; frequenting the neighbourhood of the sea, and the shores and cliffs of lakes and large rivers, which localities he prefers, from his great partiality for fish. Wilson, the American ornithologist, thus picturesquely describes this powerful bird:—"Elevated upon a high, dead limb of some gigantic tree, that commands a wide view of the neighbouring shore and ocean, he seems calmly to contemplate the motions of the various feathered tribes that pursue their busy avocations below—the snow-white gulls slowly winnowing the air; the busy *tringæ*, coursing along the sands; trains of ducks, streaming over the surface; silent and watchful cranes, intent and wading; clamorous crows, and all the winged multitudes that subsist by the bounty of this vast liquid magazine of nature. High over all these hovers one, whose action instantly arrests all the Eagle's attention. He knows him to be the fish-hawk, settling over some devoted victim of the deep. His eye kindles at the sight, and, balancing himself with half-opened wings on the branch, he watches the result. Down, rapid as an arrow from heaven, descends the distant object of his attention, the roar of its wings reaching the ear as it disappears in the deep, making the surges foam around. At this moment the eager looks of the Eagle are all ardour, and, levelling his neck for flight, he sees the fish-hawk once more emerging, struggling with his prey, and mounting in the air with screams of exultation. These are a signal for our hero, who, launching into the air, instantly gives chase; soon gains on the fish-hawk; each exerts his utmost to mount above the other, displaying in the rencontre the most elegant and sublime aerial evolutions. The unincumbered Eagle rapidly advances, and is just on the point of reaching his opponent, when, with a sudden scream, probably of despair and honest execration, the latter drops his fish: the Eagle, poising himself for a moment, as if to take a more certain aim, descends like a whirlwind, snatches it in his grasp, ere it reaches the water, and bears it silently away to the woods." When this bird has fasted for some time, its appetite is extremely voracious and indiscriminate: even the most putrid carrion, when nothing better can be had, is acceptable. The nest of this species, formed of large sticks, sods, moss, hay, &c., is usually found in a lofty tree, in a swamp or morass; and as it is increased and repaired every season, becomes of great size. Fish are daily carried to the nest in such numbers, that they sometimes lie scattered round the tree; and the odour is very offensive for a considerable distance round it.

EAGLE-OWL. (*Bubo.*) [See OWL.]

EAR-SHELL. (*Haliotis.*) A genus of univalve Mollusca, the shell being of a flattened shape, perforated with small holes on one side, and somewhat resembling the human ear, its base being characterised by a very wide mouth or aperture, the largest in any shell except the Patella or Limpet. This genus sometimes yields small pearls, the rudiments of which are frequently seen in those shells which have not brought them to perfection. The outside is generally rough, worn, or covered with marine substances; the inside presents the same enamelled appearance as mother-of-pearl, and exhibits the most beautiful colours. The holes with which the shell is perforated are for the passage of the lobes of the animal's mantle, and are made at regular intervals as it increases in size: when, however, a new one is formed, the one nearest the spire is closed up. The head of the animal is

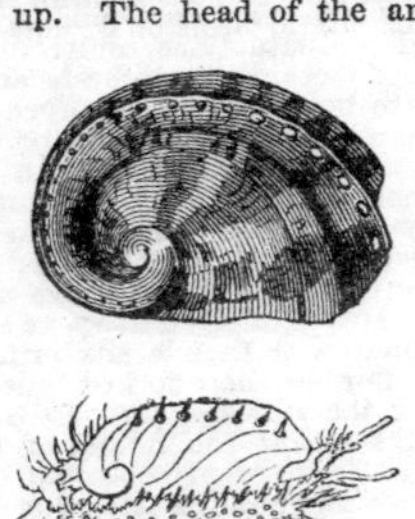

CHANNELLED EAR-SHELL.
(HALIOTIS CANALICULATA.)

large, having two long round tentacula, with eyes at the base on footstalks; foot very large, having the margin fringed all round. In its state of rest, it is able to adhere with such tenacity to the substance it is fixed upon as to be removed with great difficulty, although it can detach itself with ease. It is always found near the surface of the water. There are several species of this shell: the one termed the Great Ear-shell is five inches long and nearly three wide; its shape is an irregular oval, the end where the spiral turn is placed being the largest. It is chiefly found in the East Indies.

EARTHWORM. (*Lumbricus.*) An Annelide, of which there are doubtless many species; characterised by a long cylindrical body, divided by transverse furrows into a great number of rings, and by a mouth without teeth: they have neither eyes, tentacles, gills, nor cirrhi. The common well-known species (*Lumbricus terrestris*) attains nearly a foot in length, and is composed of upwards of one hundred and twenty rings. They are very abundant, piercing and traversing the ground in every direction, subsisting on roots, woody fibres, animal matter, and other organized substances. It is well known that they swallow earthy matter, and that having separated the serviceable portion, they eject at the mouth of their burrows the remainder in little intestine-shaped heaps, or worm-casts.

Though a small and despised creature, the Earthworm is a most important one in the operations of nature. When it is boring, it insinuates its pointed head between the particles of the earth, amongst which it penetrates like a wedge: and in this position the anterior part of the body is fixed by the spines, of which there are four pairs on each segment: the hinder parts are then drawn forwards by a shortening of the body; which swells out the anterior segments, and forcibly dilates the passage into which the head has been already thrust. By the frequent repetition of this process, the little animal more easily permeates hard substances than could be possibly conceived; and by the united labours of myriads, the earth is lightened, and vegetation thereby wonderfully assisted. Mr. Knapp, in his 'Journal of a Naturalist,' thus speaks of the Earthworm:—"This animal, destined to be the natural manurer of the soil, and the ready indicator of an improved staple, consumes on the surface of the ground, where they soon would be injurious, the softer parts of decayed vegetable matters, and conveys with the soil the more woody fibres, where they moulder, and become reduced to a simple nutriment, fitting for living vegetation. The parts consumed by them are soon returned to the surface, whence, dissolved by frosts and scattered by rains, they circulate again in the plants of the soil—'death still producing life.' Thus eminently serviceable as the worm is, it yet becomes the prey of various orders of the animal creation, and perhaps is a solitary example of an individual race being subjected to universal destruction. The very emmet seizes it when disabled, and bears it away as its prize: it constitutes throughout the year the food of many birds; fishes devour it greedily; the hedgehog eats it; the mole pursues it unceasingly in the pastures, along the moist bottoms of ditches, and burrows after it through the banks of hedges, to which it retires in dry seasons. Secured as the worm appears to be by its residence in the earth from the capture of creatures inhabiting a different element, yet many aquatic animals seem well acquainted with it, and prey on it as a natural food, whenever it falls in their way; frogs eat it; and even the great water-beetle (*Dytiscus marginalis*) I have known to seize it when the bait of the angler, and it has been drawn up by the hook. Yet, notwithstanding this prodigious destruction of the animal, its increase is fully commensurate to its consumption, as if ordained the appointed food of all."

In White's History of Selborne the valuable services of the Earthworm are detailed at great length, and with that writer's accustomed perspicacity. Charles Darwin, F. R. S., has made many interesting observations, which have been thus stated from his published researches:—"The burrowing of Earthworms is a process exceedingly useful to the gardener and agriculturist; and these animals are far more beneficial to man in this way, than they are injurious by devouring the vegetables set in the soil. They give a kind of under tillage to the land, performing the same below ground that the spade does above for the garden, and the plough for arable land; and loosening the earth, so as to render it permeable to air and water. It has been shown, too, that they will even add to the depth of soil; covering barren tracts with a layer of productive mould. Thus, in fields which have been overspread with lime, burnt marl, or cinders, these substances are in time covered with finely-divided soil, well adapted to the support of vegetation. That this result, which is commonly attributed by the farmers to the 'working-down' of the materials in question, is really due to the action of the Earthworms, appears from the fact, that, in the soil thus formed, large numbers of worm-casts may be distinguished. These are produced by the digestive process of the worms; which take into their intestinal canal a large quantity of the soil through which they burrow, extract from it the greater part of the vegetable matter it may contain, and reject the rest in a finely divided state. In this manner, a field, manured with marl, has been covered, in the course of eighty years, with a bed of earth averaging thirteen inches in thickness.

"It is commonly supposed," says Dr. Carpenter, "that the Earthworm may be multiplied by the division of its body into two pieces, each of which will continue to live. This does not, however, appear to be the case with the common species. If it is divided across the middle, when in motion, each part will continue to move for a time; but only the piece which bears the head will be found alive after a few hours. This forms a new tail; and soon shows little sign of injury. But if the division be made near the head, the body will remain alive, and will renew the head; and the head, with its few attached segments, will die. There appears, however, to be some species, in which this reproductive power is sufficiently great to produce a new head and body from even a small portion of the original; so that above twenty individuals have been produced in this manner by the division of a single one into as many parts."

EARWIG. (*Forficula.*) A genus of Dermapterous insects. The common Earwig, generally called in Scotland GOLLACH (*Forficula auricularia*), is about three quarters of an inch in length, and has a somewhat flattened body; the wings being folded under very short and truncated elytra or wing-cases, and the extremity of the abdomen armed with a horny forceps. When alarmed, the insect elevates the abdomen, and opens these forceps, in order to defend itself from the attack of its enemies. Though not produced quite perfect from the egg, the Earwig requires but a very small change before it arrives at that state which fits it for flight and

EARWIG.—(FORFICULA AURICULARIA.)

generation. Its natural functions are never suspended; from the instant it leaves the egg, it continues to eat, move, leap, and pursue its prey; and a skin which inclosed a part of its body and limbs bursts behind, and gives full play to a set of wings with which it flies in pursuit of its mate. The places in which they are found are chiefly damp and cool situations, under stones and the bark of trees, among chests and boxes which have been long undisturbed, and in similar haunts. They seem to be as timid as hares, and when disturbed run into the nearest hole, satisfied, like the quadrupeds above named, if they can get their heads under cover, and thus exclude the sight of danger, even when their bodies are fully exposed. Hence, it often happens that they will be found with their heads buried in the bottom of flowers, their forked tails sticking up among the stamens and pistils, so that they might escape the notice of any one but a botanist or an entomologist.

Mr. Newman gives the following interesting description of this insect and its habits:—"The Earwig is one of our most common insects; it is well known to every one, and is very generally an object of unconquerable dislike; the forceps at its tail, and the threatening manner in which these are turned over its back, to pinch anything of which it is afraid, render it peculiarly disgusting. The fore wings of the Earwig are square, short, leathery pieces, which cover but a very small portion of the body: the insect is incapable of bending or folding them in any direction, or of using them as organs of flight. The hind wings are quite different from the fore wings; they are folded into a very small compass, and covered by the fore wings, except a small portion which protrudes from beneath them; and, when examined in this position, appear totally useless as organs of flight. When unfolded, the hind wings are remarkably beautiful; they are of ample size, perfectly transparent, displaying prismatic colours when moved in the light; and are intersected by veins, which radiate from near the centre to the margin. The shape of these wings, when fully opened, is nearly that of the human ear; and from this circumstance it seems highly probable that the original name of this insect was Ear*wing*. [It derives its present name from its supposed habit of insinuating itself into the ears of persons who incautiously lie down and sleep on the grass, &c.: a supposition, if not entirely groundless, unsupported by any well authenticated instances.]

"Earwigs subsist principally on the leaves and flowers of plants, and on fruit; and they are entirely nocturnal insects, retiring by day into dark crevices and corners, where they are screened from observation. The rapidity with which they devour the petals of a flower is remarkable; they clasp the edge of a petal in their fore legs, and then, stretching out their head as far as possible, bite out a mouthful; then another mouthful nearer, and so on till the head is brought to the fore-legs. This mode of eating is exactly that which is practised by the caterpillars of butterflies and moths: the part of a leaf or petal is eaten out in a semicircular form, and the head is thrust out to the extreme part, after a series of mouthfuls. Pinks, carnations, and dahlias, very frequently lose all their beauty from the voracity of these insects. When the time of breeding has arrived, which is generally in the autumn, the female retires for protection to the cracks in the bark of old trees, or the interstices of weather-boarding, or under heavy stones on the ground: here she commences laying her eggs. The eggs are usually from twenty to fifty in number: when the female has finished laying them, she does not forsake them, as is the habit of other insects, but sits on them in the manner of a hen, until they are hatched. When the little ones leave the shell, they are very perceptibly larger than the eggs which contained them. They precisely resemble the parent in structure and habit, except that they are without wings; they also differ in colour, being perfectly white. The care of the mother does not cease with the hatching of the eggs: the young ones run after her wherever she moves, and she continues to sit on them and brood over them with the greatest affection for many days. If the young ones are disturbed or scattered, or if the parent is taken away from them, she will, on the first opportunity, collect them again, and brood over them as carefully as before, allowing them to push her about, and cautiously moving one foot after another, for fear of hurting them. How the young ones are fed until the mother's care for them has ceased, does not appear to have been ascertained; for it is not until they are nearly half grown that they are seen feeding on vegetables with the rest."

A remarkable fact, in relation to the Earwig, is its great abundance at particular times, and its subsequent rarity. From the observations of entomologists, it has been proved that these insects migrate in considerable flocks, selecting the evening for their excursions. It is common with gardeners to hang up, among the flowers and fruit-trees subject to their attacks, and also to place on the ground, pieces of hollow reeds, lobster-claws, and the like, which offer enticing places of retreat for these insects on the approach of daylight, and by means thereof great numbers of them are obtained in the morning. Poultry are very fond of Earwigs. There are many exotic species of this genus, some of them with remarkably elongated forceps.

EBURNA. A genus of marine Mollusca found in the Indian and Chinese seas, inhabiting an oval, thick, smooth, umbilicated shell. The Eburnæ in some respects resemble the Buccina; from which, however, they are essentially different. The head of the animal is furnished with a

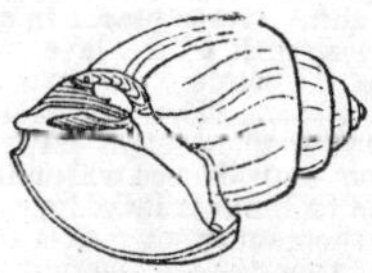

EBURNA SPIRATA.

proboscis, and two tentacula having eyes in the middle; foot short; spire angulated and acute; aperture oval, terminating anteriorly in a canal, posteriorly in a groove; outer lip slightly thickened with an anterior notch, which terminates a spiral fold surrounding the body whorl; umbilicus generally covered by the thickened columellar lip.

ECHIDNA, or PORCUPINE ANT-EATER. (*Echidna hystrix.*) This curious animal is a native of Australia, and is a striking instance of that beautiful gradation, so frequently observed in the animal kingdom, by which creatures of one tribe or genus approach to those of a very different one. It has the external coating and general appearance of the Porcupine, with the mouth and peculiar generic characters of the ant-eaters. It is about a foot in length: the upper parts of the body and tail are thickly coated with strong and very sharp spines, of a yellowish white with black tips, and thicker in proportion to their length than those of a porcupine. The head, legs, and under parts of the body are of a deep brown, and thickly set with bristly hair; the tail is very short, and covered with spines pointing perpendicularly upwards. The snout is long and tubular, the mouth small, and the tongue long and lumbriciform, as in other Ant-eaters. The legs are very short and thick; and are each furnished with five rounded,

PORCUPINE ECHIDNA.—(ECHIDNA HYSTRIX.)

broad toes: on the fore feet are five very strong, long, and blunt claws; but on the hind feet there are only four claws, the thumb being destitute of a claw: the first claw on the hind feet is extremely long, rather curved, and sharp pointed; the next shorter, but of similar appearance; the two remaining ones far shorter, and blunt: it has great strength, and burrows with wonderful celerity.

At a meeting of the Zoological Society, July 22. 1845, Professor Owen communicated his observations on the living *Echidna* exhibited at the Menagerie of the Society in May preceding. The animal when received at the Gardens was active and apparently in sound health. It was placed in a large but shallow box, with a deep layer of sand on one half of the bottom; the top covered with close cross-bars. The animal manifested more vivacity than might have been expected from a quadruped which, in the proportions of its limbs to its body, as well as in its internal organization, makes the nearest approach, after the Ornithorhyncus, to the Reptilia. In the act of walking, which was a kind of waddling gait, the body was alternately bent from one side to the other, the belly was lifted entirely off the ground, and the legs, though not so perpendicular as in higher mammalia, were less bent outwards than in lizards. The broad and short fore paws were turned rather inwards; the hind feet had their claws bent outwards and backwards, resting on the inner border of the sole. The animal was a male; and the tarsal spur, smaller and sharper than in the Ornithorhyncus, projected backwards and outwards, almost hidden by the surrounding coarse and close hair. The small eyes gleamed clear and dark; the ball was sensibly retracted when the animal winked, which it did frequently. It commenced an active exploration of its prison soon after it was encaged: the first instinctive action was to seek its ordinary shelter in the earth, and it turned up the sand rapidly by throwing it aside with strong strokes of its powerful fossorial paws, and repeating the act in many places, until it had assured itself that the same hard impenetrable bottom everywhere opposed its progress downwards. The animal then began to explore every fissure and cranny, poking its long and slender nose into each crevice and hole, and through the interspaces of the cross-bars above. To reach these it had to raise itself almost upright, and often overbalanced itself, falling on its back, and recovering its legs by performing a summerset. I watched these attempts of the animal to escape for more than an hour, and it was not till it had got experience of the strength of its prison, that the Echidna began to notice the food which had been placed there. This consisted of a saucer of bread and milk and some meal-worms. The milk was sucked or rather licked in by rapid protrusion and retraction of the long red cylindrical tongue. The tongue came more than once in contact with the larvæ, which were sometimes rolled over by it, but no attempt was made to swallow them. The Echidna offered little resistance when seized by the hind-leg and lifted off the ground, and made not the slightest demonstration of defending himself by striking with his hind spurs: the only action when irritated was to roll itself into a ball like a hedgehog—the bristles being then erect.—*Ann. Nat. Hist.*

ECHIMYS: ECHIMYNA. The name of a genus and subfamily of Rodents, containing the genus *Echimys* or *Loncheres;* a largish spiny-haired rat-like animal with a long tail; it is a native of South America. The genus *Octodon*, an arboreal type found in Chili; and *Aulacodus* or Ground Pig, from South Africa—also belong to this subfamily.

ECHINEIS. A genus of fish remarkable for a series of suckers on the top of the head. [See REMORA.]

ECHINODERMATA. The name given to an extensive class of Invertebrata of the division RADIATA, comprising all those which have a hard coriaceous integument, which in some species is covered with sharp spines or prickles, like those of the hedgehog; a digestive and vascular system; and a sort of radiating nerves. They are all marine animals, possessing the power of locomotion; the sexes are distinct; and the young are produced from ova. "In this group," as Mr. Patterson observes, "we find animals of extremely dissimilar appearance associated together. One species is attached, for a certain period, to a stem, and resembles a polype, with its waving and sensitive arms. In the common star-fish, or 'five-fingers,' we have the arms radiating from a common centre. In the sea-urchins there are no arms, and the form of the body is globular,

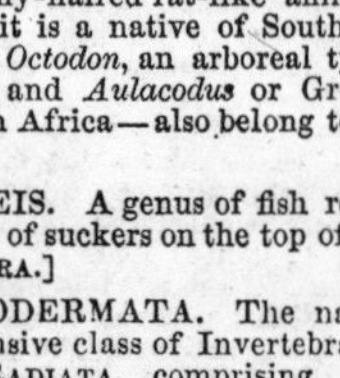

ECHINUS CRENULARIS.

and, passing over some intermediate gradations of figure, we reach creatures which in external aspect, resemble worms, and have even been classed as such. At one extremity of the range, the Echinodermata remind us of polypes—creatures of inferior organization; at the other extremity, they approach the annulose animals, whose structure is of a higher grade. Those occupying the centre of the group may be regarded, therefore, as the types or representatives of the class." [See ASTERIAS; ECHINUS; ENCRINITE; HOLOTHURIA; OPHIURIDÆ; and ALTERNATION OF GENERATION in SUPPLEMENT.]

ECHINOPS. A singular genus of the Shrew tribe, allied to the *Tenrec*, and a native of Madagascar and the Mauritius: one species, *E. Telfairii*, is known.

ECHINORHYNCHUS. [See SUPPL.]

ECHINUS. The generic name of the SEA-URCHINS, or SEA-EGGS, which constitute the type of the class *Echinodermata*, or certain invertebrate animals, which have a crustaceous or coriaceous integument, most commonly armed with tubercles or spines. They are all inhabitants of the sea; and many of them have often been found in a fossil state. The spines are connected with the outer skin by very strong ligaments, and are the instruments of motion. They are generally armed with five sharp teeth; and the pores are furnished with a retractile tentacle or feeler to each, by which the animal affixes itself to any object, and stops its motion. The species most esteemed as an esculent, and thence denominated *Echinus esculentus*, is subglobular; with ten avenues of

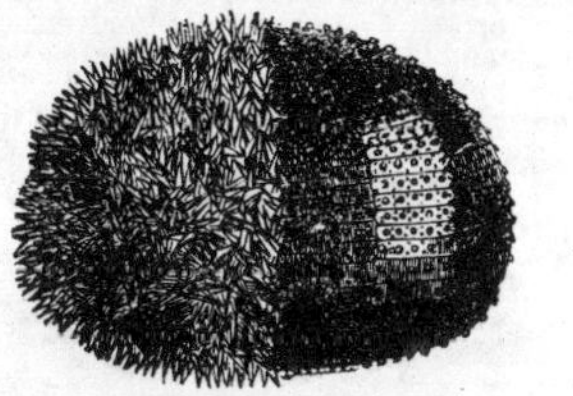

EDIBLE SEA URCHIN (ECHINUS ESCULENTUS.)

pores, the spaces between covered with small tubercles supporting the spines; body reddish or yellowish; spines short, of a violet colour; losing their colour and falling off the dead animal; pores in about three rows; tubercles surrounded with a circle of less ones; vent closed with a coriaceous membrane covered with spines. Echini of this sort constitute no small part of the food of the poor in many countries, and some species are reckoned excellent. In ancient times they were accounted very delicious, being usually dressed with vinegar, honied wine, parsley, and mint; it is recorded that they composed the principal dish at the famous supper of Lentulus, when he was made Flamen Martialis, or Priest of Mars. Our cut exhibits one half of the surface denuded of its spines to show their mode of insertion. *Fossil Echinidæ*, in almost incredible numbers, are to be traced through all the formations, from the epoch of the transition series to the present time; many of them being found in our own chalky or flinty soils. The *Echinus vulgaris*, so perpetually found in a fossil state, is not now traced in a living state: in make it is orbicular; with ten avenues, two of them always near each other.

EDENTATA. The name given to an order of quadrupeds, which although it includes many animals differing from each other widely in habits, and also in certain points of structure, yet agree in so many essential characters, and are connected together by so many intermediate links, as to require being associated in the same group. They all agree in the absence of teeth in the front of the jaws; all resemble each other in the great claws which encompass the ends of their toes; and they are all distinguished by a certain slowness, or want of activity, obviously arising from the peculiar organization of their limbs. The armour-clad, insectivorous *Armadillos*, of South America; the tree-inhabiting *Sloths*, and hairy toothless *Ant-eaters*, of the same continent; the gigantic *Megatherium*, which formerly inhabited it; and the *Manis*, whose lizard-like body, defended by an impenetrable coat of mail, excites our wonder—all belong to the order EDENTATA; which constitute the last group of *unguiculated* animals, and are severally described in the course of this work.

EEL. (*Anguilla.*) The Eel, which in a natural arrangement of the animal world may be considered as in some degree connecting the fish and serpent tribes, is a native of almost all the waters of the ancient continent, frequenting not only rivers but stagnant waters; and occasionally salt marshes and lakes. Its general appearance is so well known, and so unlike most other fishes, as to require but a slight description; we should observe, however, that though the external form of the body resembles the snake, the important internal organs, and the character of the skeleton, are decidedly different. The Eel is distinguished by its uniform colours, but more particularly by the peculiar elongation of the lower jaw, which advances to some distance beyond the

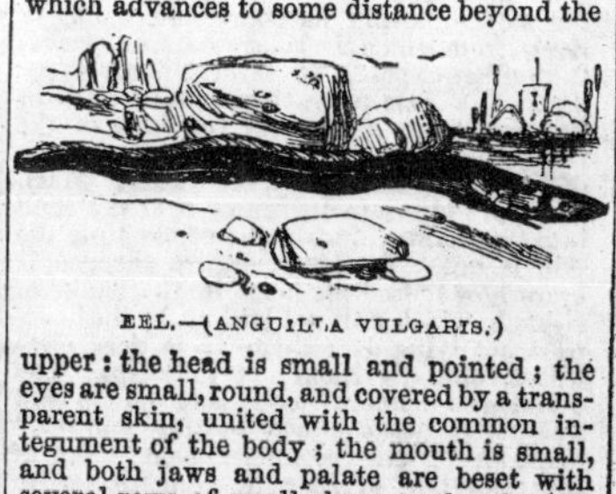

EEL.—(ANGUILLA VULGARIS.)

upper: the head is small and pointed; the eyes are small, round, and covered by a transparent skin, united with the common integument of the body; the mouth is small, and both jaws and palate are beset with several rows of small sharp teeth: the orifices of the gills are very small, of a lunated shape, and are seated close to the pectoral fins, which are small, and of an ovate shape:

the back-fin commences at some distance beyond the head, and is continued into the tail-fin, which is also united with the vent-fin. The general colour of the Eel is olive-brown on the back, and silvery on the sides and beneath; the fins are slightly tinged with violet, and sometimes margined with pale red; it is, however, sometimes seen of a very dark colour, with scarce any silvery tinge, and sometimes yellowish or greenish: those being the most beautiful which inhabit the clearest waters. The skin of the Eel is proverbially slippery, being furnished with a large proportion of mucus: it is also furnished with small deeply-imbedded scales, which are not easily visible in the living animal, but are very conspicuous in the dried skin. Fresh-water eels, inhabiting running streams with gravelly bottoms, are said to be uniformly white upon the belly, and infinitely more delicate than those of muddy waters, which possess a peculiar smell and flavour, by no means agreeable. In the choice of its food the Eel is far from being cleanly, feeding indiscriminately upon all kinds of small fish, and decayed animal matter: they are, however, a most valuable description of fish: their flesh is excellent as food, being highly nutritious, though sometimes too oily for weak stomachs. They are very prolific, hardy, and very easily preserved: they inhabit almost all our rivers, lakes, and ponds; and as they are in great esteem for the table, the consumption in our large cities is very considerable. Few animals are more tenacious of life; they continue to move for a long time even when deprived of the head and skin, preserving the muscular irritability for many hours after death.

The Eel is viviparous, producing its numerous young during the decline of summer: these at their first exclusion are very small. Very gross errors on this subject were formerly indulged in; but it appears that both eggs and ready-formed young are occasionally observed in the same individuals, as is known to be the case also with several other animals. As Mr. Yarrell observes, "during the cold months of the year Eels remain imbedded in mud; and large quantities are frequently taken by eel-spears in the soft soils and harbours and banks of rivers, from which the tide recedes, and leaves the surface exposed for several hours every day. The Eels bury themselves twelve or sixteen inches deep, near the edge of the navigable channel, and generally near some of the many land-drains, the water of which continues to run in its course over the mud into the channel during the whole time the tide is out. In Somersetshire the people know how to find the holes in the banks of rivers in which Eels are laid up, by the hoar-frost not lying over them as it does elsewhere, and dig them out in heaps. The practice of searching for Eels in mud in cold weather is not confined to this country." Some marvellous accounts are on record of the migration of Eels from one river to another, over intervening portions of dry land. On this subject the same author thus expresses himself: "There is no doubt that Eels occasionally quit the water, and when grass meadows are wet from dew, or other causes, travel during the night over the moist surface in search of frogs or other suitable food, or to change their situation. Some ponds continually produce Eels, though the owners of these ponds are most desirous of keeping the water free from Eels, from a knowledge of their destructive habits towards the spawn and fry of other fishes. Other ponds into which Eels have been constantly introduced are obnoxious to them from some quality in the water; and they are known to leave such places during the night, and have been found on their passage to other retreats." The general size of the Eel is from two to three feet, but it is said that it sometimes, though very rarely, attains to the length of six feet, and to the weight of twenty pounds. It is a fish of slow growth, and is supposed to live to a very considerable age; and is attacked by a great many species of intestinal worms.

The CONGER EEL (*Anguilla conger*), in its general appearance is so nearly allied to the common Eel, that on a cursory view it might at first be considered as the same species: it, however, differs materially from it in size, being sometimes ten feet in length, as thick as a man's thigh, and weighing 100 lbs.: it is also in general of a darker colour on the upper part, and of a brighter hue beneath: there is also on the sides a straight, white, broadish line, seemingly composed of a double row of points, which reaches from the head to the tail. The Conger resides generally in the sea, and is only an occasional visitant of fresh waters. In the winter it is supposed to imbed itself under the soft mud, and to lie in an inactive state; but on the approach of spring it emerges from its concealment, and visits the mouths of rivers.

CONGER EEL.—(ANGUILLA CONGER.)

The able naturalist above quoted informs his readers that "the principal fishery for Congers in this country is on the Cornish coast; where, according to Mr. Couch, it is not uncommon for a boat with three men to bring on shore from five hundred weight to two tons, the fishing being performed during the night; for this fish will not readily take a bait by day, and even on moonlight nights it is more shy than when in the dark, except in deep water. The most usual bait with the Cornish fishermen is a pilchard. The Congers that keep among rocks hide themselves in crevices, where they are not unfrequently left by the retiring tide; but in situations free from rocks, Congers hide themselves by burrowing in the ground. The flesh is not in much estimation, but meets a ready sale at a low price among the lower classes. The adult fish is most voracious, not sparing even those of its own species. From the stomach of a specimen weighing twenty-five pounds, I took three common Dabs, and a young Conger of three feet in length. The power of the jaws in this fish is very great: in the stomach of small specimens examined on the coast, I have found the young testaceous coverings of our shell-fish comminuted to fragments. They are often tempted by the crustacea entrapped in the lobster-pots to enter those decoys in order to feed on them, and are thus frequently captured."

EFT. [See NEWT.]

EGG-BIRD. The name given to some species of web-footed birds belonging to the Laridæ family. [See TERN.]

EGGER [MOTH]. A name given by collectors to the species of Moths, of the genera *Lasiocampa* and *Eriogaster*.

EGRET. [See HERON.]

EIDER DUCK. (*Somateria mollissima.*) This valuable species of wild duck is of a size between the tame duck and the goose, measuring about two feet in length. The head is large; the middle of the neck small, with the lower part of it spread out very broad, so as to form a hollow between the shoulders. The bill is of a dirty green or

EIDER DUCK,
(SOMATERIA MOLLISSIMA.)

horn colour, and the upper mandible forked in a singular manner towards each eye, and covered with white feathers on the sides as far forward as the nostrils. The upper part of the head is of a soft velvet black, divided behind by a dull white stroke: the feathers, from the nape of the neck to the throat, are puffed out, and look as if they had been clipped off at the lower ends. The cheeks, chin, upper part of the neck, the back, and lesser wing-coverts, are white; the scapulars, dirty white; bastard wings, and primary quills, brown; secondaries, and greater coverts, darker brown; the front part of the neck, to the breast, is of a buff colour; the breast, belly, rump, and tail-coverts are of a deep sooty black; tail feathers hoary brown; legs short, and yellow; webs and nails dusky. The full-grown male weighs from six to seven pounds; the female only between five and six. Her shape is nearly the same; but her plumage is quite different, the ground colour being of a reddish brown, crossed with waved black lines: the wings are crossed with two bars of white; quills dark; the upper part of the neck marked with dusky streaks; and the belly is deep brown, spotted obscurely with black.

This highly useful and valuable species is a native of the frozen regions of the north: it is extremely abundant in Iceland, Lapland, Greenland, and Spitzbergen, on the shores of Baffin's and Hudson's Bays, &c.; it is also very numerous in the Hebrides and the Orcades, but becomes rare as we advance to the south. The female lays five or six pale greenish-olive eggs in a nest composed of marine plants, and thickly lined with a beautiful down of most exquisite fineness, which is highly esteemed for its excessive lightness, elasticity, and useful qualities. The nest is usually formed on small islands, not far from the shore. As long as the female is sitting, the male continues on watch at no great distance; but as soon as the young are hatched he leaves them: the mother, however, remains with them a considerable time afterwards, and is said to assist them out of the nest almost as soon as they creep from the eggs, and proceeding to the shore, they crawl after her: when she arrives at the water's edge she takes them on her back, and swims a few yards with them; she then dives, and the young being left floating on the surface, they are obliged to take care of themselves.

The manner in which the eider down is taken is as follows:—When the collectors come to the nest, they carefully remove the female, and take away the superfluous down and eggs; after which they replace her. She then begins to lay afresh, and again has recourse to the down on her body to cover her eggs; and in the event of her own stock being exhausted, which is not unfrequently the case, she is now assisted by the male in furnishing the requisite quantity: even this is frequently taken away, when the birds proceed to furnish another supply, both of eggs and down; but if the cruel robbery be repeated again, they immediately abandon the place. One female generally yields about half a pound of down, which is worth about two dollars. This down, from its superior warmth, lightness, and elasticity, is preferred by the luxurious to every other article for beds and coverlets; and from the great demand for it, those districts in Norway and Iceland, where these birds abound, are regarded as the most valuable property, and are guarded with the greatest vigilance. As found in commerce, this down is in balls of the size of a man's fist, and weighing from three to four pounds. It is so fine and elastic, that when a ball is opened, and the down cautiously held near the fire to expand, it will completely fill a quilt five feet square: but it is worthy of observation that although the eider down taken from the nests is so excellent, the down of dead birds is little esteemed, from having lost its elasticity.

Eider Ducks associate in flocks, generally in deep water, diving to great depth for shell-fish, which constitute their principal food. They frequently retire to the rocky shores to rest, particularly on the appearance of an approaching storm. The Greenlanders kill them with darts, pursuing them in their little boats, watching their course by the air bubbles when they dive, and always striking at them when they rise to the surface wearied. Their flesh is eaten by the Greenlanders, but it tastes strongly of fish; the eggs, however, are much esteemed. The female lays from six to eight eggs, in a rock-built nest, lined with her own exquisite down; but the eggs and the down are both frequently obtained at the hazard of life by people let down by ropes from craggy steeps. The skin, taken off, feathers and all, are used by the inhabitants for their under garments. It appears that all the attempts which have been made to domesticate these birds have been unsuccessful.

Another species, called the KING EIDER, (*Somateria spectabilis* of systematic writers), not much unlike the preceding, inhabits the same coasts. Its beak, wattles, and legs are of deep vermilion: a straight band of velvety black surrounds the base of the upper mandible of the beak; and there is a similar double band which becomes spear-shaped on the throat: the top and back of the head are of a fine bluish-grey; the cheeks are a brilliant sea-green; the neck, wing-coverts, and upper part of the back, are pure white: the scapulars, lower part of the back, wings, tail, and all the under parts of the body are deep black. The entire plumage of the female is brown.

ELAND. The largest species of Antelope, found at the Cape of Good Hope, and in the southernmost parts of Africa generally. [See BOSELAPHAS in SUPPLEMENT.]

ELATER: ELATERIDÆ. A genus and family of Coleopterous insects, having setaceous antennæ; but whose leading character is a strong spine situated beneath the thorax, which fits at pleasure into a small cavity on the upper part of the abdomen; thus enabling the insect, when laid on its back, to spring up with great force and agility, in order to regain its natural position. There are various species of these beetles; but few of the European species are comparable in point of size to such as are natives of the tropics.

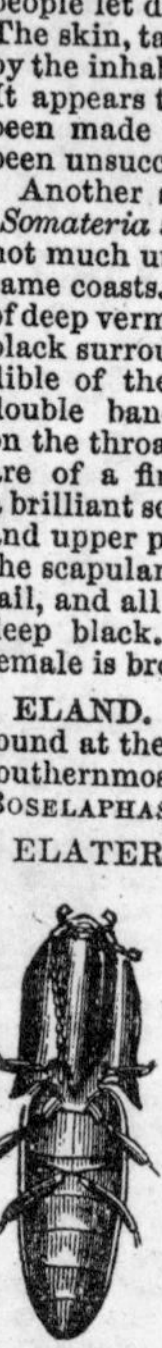

UNDER SIDE OF AN ELATER.

The largest, and one of the most remarkable, is the ELATER FLABELLICORNIS, which is two inches and a half long, and of a uniform brown colour: it differs from the rest in having very strongly pectinated antennæ, the divisions of which, forming a kind of fan on the upper part of each, are nearly a quarter of an inch in length. It is met with in many parts of Asia and Africa.

A species, still more remarkable than the preceding, is the *Pyrophorus noctilucus*, called in South America *Cocujas*. It is about an inch and a half long, of a brown colour, and has a smooth, yellow, semi-transparent spot on each side of the thorax; these spots being, like those on the abdomen of the glowworm, highly luminous in the dark: in short, it is one of the most brilliant of the Fire-flies which inhabit South America and the West India islands. It is asserted that a person may with great ease read the smallest print by the light of one of these insects held between the fingers, and gradually moved along the lines, with the luminous spots above the letters; but if eight or ten of them be put into a phial, the light will be suffi-

FIRE-FLY ELATER.
(PYROPHORUS NOCTILUCUS.)

ciently great to admit of writing by it. Oviedo says, that the Indians travel in the night with these insects fixed to their hands and feet; and that they spin, weave, paint, dance, &c., by their light. In Prescott's "Conquest of Mexico," vol. ii. p. 261, we are told that in 1520, when the Spaniards visited that country, "the air was filled with the 'cocuyos,' a species of large beetle which emits an intense phosphoric light from its body, strong enough to enable one to read by it. These wandering fires, seen in the darkness of the night, were converted, by the excited imaginations of the besieged, into an army with matchlocks!" Such is the report of an eye-witness. (*Bernal Diaz, Hist. de la Conquista*, cap. 122.) Several others might be mentioned of inferior size, one of which it may be necessary to describe; this is the *Elater oculatus*, which is of a dark brown colour, and somewhat smaller than the preceding; distinguished by the thorax being marked on each side by a large, oval, jet black spot, surrounded by a white margin. It is common in North America.

Such species of the Elater as are natives of this country are much smaller than the exotic ones above mentioned, and but rarely distinguished by any peculiar brilliancy.

The larvæ or grubs of the Elaters live upon wood and roots, and are often very injurious to vegetation. Some are confined to old or decaying trees, others devour the roots of herbaceous plants, and are called *wire-worms*, from their slenderness and uncommon hardness. The English wire-worm is said to live, in its feeding or larvæ state, not less than five years; during the greater part of which time it is supported by devouring the roots of wheat, rye, oats, and grass, annually causing a large diminution of the produce, and sometimes destroying whole crops. It is said to be particularly injurious in gardens recently converted from pasture lands; and the method adopted for alluring and capturing these grubs consists in strewing sliced potatoes or turnips in rows through the garden or field; women and boys are employed to examine the slices every morning, and collect the insects which readily come to feed upon the bait. Some of these destructive insects are long, slender, worm-like grubs, closely resembling the common meal-worm; nearly cylindrical, with a hard and smooth skin, of a buff or brownish yellow colour, the head and tail only being a little darker; each of the first three rings provided with a pair of short legs, and a short retractile wart or prop-leg, serving to support the extremity of the body, and prevent it from trailing on the ground. Other grubs of Elaters differ from the foregoing in being proportionally broader, not cylindrical, but somewhat flattened. Such are mostly wood-eaters.

After their last transformation, Elaters or Spring-beetles make their appearance upon trees and fences, and some are found on flowers. They creep slowly, and generally fall to the ground on being touched. They fly both by day and night. Their food, in the beetle state, appears to be chiefly derived from flowers; but some devour the tender leaves of plants.

ELECTRICAL EEL. [See GYMNOTUS.]

ELK, or MOOSE DEER. (*Cervus alces.*) Of all the animals belonging to the genus *Cervus*, none are so large as the Elk, which in size is scarcely inferior to a horse, and its immense horns sometimes weigh near fifty pounds. It is common to both continents, inhabiting only the coldest regions, and is observed to attain larger dimensions in Asia and America than in Europe. It cannot boast of the elegant shape so general in the

ELK. (CERVUS ALCES.)

rest of the deer tribe; the head being disproportionately large, the neck short and thick, and the horns dilating almost immediately from the base into a broad palmated form; while its long legs, high shoulders, and heavy upper lip, hanging very much over the lower, give it an imposing, although an uncouth rather than a majestic appearance. The colour of the Elk is a dark grayish brown, but much paler on the legs and beneath the tail. The hair, which is of a strong, coarse, and elastic nature, is much longer on the top of the shoulders and on the ridge of the neck than on other parts, forming a kind of stiffish mane; under the throat is an excrescence, from which issues a tuft of long hair: the body, which is short and thick, is mounted on tall legs, giving a very ungainly aspect to the animal, which is not diminished when it is in motion, as its gait is a sort of shambling trot.

In Europe the Elk is found chiefly in Sweden, Norway, and some parts of Russia. In Asia it occurs in the woody tracts of the Russian dominions, and in Siberia in particular it is found of a gigantic size. In America it has been found as far north as the country has been explored; its southern range, at former periods, extended to the shores of the great lakes, and throughout the New England states. At present, however, they are seldom heard of to the south of the state of Maine; but in Nova Scotia, around the Bay of Fundy, and in the Hudson's Bay Company's possessions, they are found in considerable numbers. The Elk is a mild and harmless animal, choosing its residence in the midst of forests, and principally supporting itself by browsing the boughs of trees: they feed principally by night; and whenever they graze (which, on account of their short neck and long legs, they do with difficulty), they are observed to choose an ascending ground, for the greater convenience of reaching the surface with their lips.

Though naturally of a peaceable and inoffensive disposition, the Elk displays a high degree of courage, and even ferocity, when suddenly attacked; defending himself with great vigour, not only with his horns, but also by striking violently with his fore-feet, in the use of which he is particularly dextrous. The chase of the Elk or Moose forms an important occupation among the natives of North America, and is performed in different ways, some of which are as remarkable for artfulness as others are for boldness and dexterity: they are also often killed with the gun. Their flesh is more relished by the Indians, and persons resident in the fur countries, than that of any other animal. It bears a greater resemblance, in its flavour, to beef than to venison. It is said that the external fat is soft, like that of a breast of mutton, and when put into a bladder is as fine as marrow. In this it differs from all other species of deer, of which the external fat is hard. Their skins, when properly dressed, make a soft, thick, pliable leather, which the Indians prepare by scraping them to an equal thickness, and removing the hair: they are then smeared with the brains of the animal, until they feel soft and spongy; and, lastly, they are suspended over a fire made of rotten wood, until they are well impregnated with the smoke.

"The Moose," says Mr. Gosse, the author of the "Canadian Naturalist," "is more shy and difficult to take than any other animal. He is more vigilant, and his senses more acute, than those of the buffalo or caribou, while he is more prudent and crafty than the deer. * * * I know not whether the Moose has ever been tamed; but I think it not improbable that it could be trained to harness as well as its congeners, the reindeer and the wapiti: and it would, from its size and strength, be more serviceable than either of them. But in a new country like this, where alone the opportunity for such an experiment is to be found, the inhabitants generally have little time, and less inclination, for innovations."

Many extraordinary accounts have been circulated by travellers who wrote in the 17th century, of Elks or Moose Deer being seen in North America, whose height was twelve feet, and the weight of whose horns was between three and four hundred pounds. Such stories were probably derived from vague and uncertain descriptions furnished by the Indian tribes. That some animal, however, of the deer kind, far superior in size to any at present known, once existed, is sufficiently proved by the enormous fossil horns which have often been found at a considerable depth in the bogs of Ireland and the Isle of Man, as well as in America and other parts of the world. Their appearance, however, differs so considerably from the horns of the Elk, that it seems now pretty generally agreed among naturalists, that they must have belonged to some species either quite extinct or hitherto undiscovered. They are much longer and narrower in proportion than those of the Elk, and are furnished with brow antlers; and the processes or divisions into which the sides and extremities run are much longer, sharper, and more distant in proportion.

ELEPHANT. (*Elephas.*) Largest of all living quadrupeds, and prodigiously strong, the Elephant is not less remarkable for docility and sagacity. Of this we have concurrent testimony from the earliest ages to the present time; yet, were we to form our ideas of its capacities only from the external appearance of this formidable animal, a sagacious character is the last we should be likely to give it credit for. The whole form is awkward; the head is large, the eyes extremely small, and the ears very large and pendulous: the body is huge and thick, and the back much arched; the legs are very clumsy and shapeless, and the feet slightly

AFRICAN ELEPHANT. (ELEPHAS AFRICANUS)

divided into five rounded hoofs: but under this uncouth exterior are qualities which entitle its possessor to the admiration of mankind—a mild and gentle disposition, superior intelligence, great attachment to its master, and invincible perseverance. In point of bulk, the Rhinoceros and the Hippopotamus are the only existing terrestrial animals that can approach the Elephant; though some other species of *Pachydermata* now extinct must have considerably surpassed him. The enormous weight of the body could only be sustained by legs of the most solid construction; and accordingly we find that these have the aspect of straight columns, the joints being so formed that each bone rests vertically upon the one beneath it.

Elephants, of which only two species at present exist, viz. the Asiatic and the African, are distinguished by their extraordinary proboscis or trunk, by the possession of two enormous tusks, which project downwards from the upper jaw, and by the absence of front teeth in the lower. The African and Asiatic species differ from each other in the size of the tusks, which are much longer in the former than in the latter. In the young animals the tusks are not visible; in the more advanced state of growth they are extremely conspicuous; and in a state of maturity they project in some instances six or seven feet; nay, several tusks measured by Eden were nine feet in length; and Hartenfels measured one which exceeded fourteen feet! The largest tusk on record was sold at Amsterdam, and weighed 350 lbs. It is but rarely that the tusks are seen in the females; and when they appear, they are but small, and their direction is rather downwards than upwards. The African Elephant is said to be smaller than the Asiatic; yet the ivory dealers in London affirm that the largest tusks come from Africa, and are of a better texture, and less liable to turn yellow than the Indian ones. The increase of the tusks arises from circular layers of ivory, applied internally, from the core on which they are formed; similar to what happens in the horns of some animals.

But it is the trunk of the Elephant which may justly be considered as one of the miracles of Nature; being, at once, the organ of respiration, as well as the instrument by which the animal supplies itself with food, and sucks up the water it requires to allay its thirst. This wonderful organ is cartilaginous, and composed of numerous rings, divided through its whole length by a septum, and forming a sort of double tube, terminating in a kind of finger-like appendage or movable hook. "Endowed with exquisite sensibility, nearly eight feet in length, and stout in proportion to the massive size of the whole animal, this organ," as is well expressed by Mr. Broderip, "at the volition of the Elephant, will uproot trees or gather grass—raise a piece of artillery or pick up a comfit—kill a man or brush off a fly. It conveys the food to the mouth, and pumps up the enormous draughts of water, which by its recurvature are turned into and driven down the capacious throat, or showered over the body. Its length supplies the place of a long neck, which would have been incompatible with the support of the large head and weighty tusks. A glance at the head of an elephant will show the thickness and strength of the trunk at its insertion; and the massy arched bones of the face and thick muscular neck are admirably adapted for supporting and working this powerful and wonderful instrument."

Elephants are naturally gregarious; large troops assembling together, and living in a kind of society. The skin of the Elephant is of a deep ash-coloured brown; but in some parts of India it is said to be found, though rarely, of a white or cream colour. It sometimes arrives at the height of twelve or fourteen feet, though the more general height seems to be about nine or ten feet. These animals are commonly found in the midst of shady woods, being equally averse to extreme heat as to cold: they delight in cool spots, near rivers, and, as they swim with great ease, they frequently bathe in the water. Their general food consists of the tender branches of various trees, as well as of grains and fruits; on which account it is that their incursions are so much dreaded in plantations of various kinds, where they are said occasionally to commit the most violent depredations; at the same time injuring the crops by trampling the ground with their huge feet.

The wild Elephants of Ceylon, which are much esteemed, live in small groups or families. In wandering from place to place, the males, who are furnished with the largest tusks, put themselves at the head, and are the first to face every danger. In swimming

ASIATIC ELEPHANT.—(ELEPHAS INDICUS.)

over any large river, they lead the van, and seek a proper landing-place: next follow the young Elephants, clinging to each other by means of their trunks, whilst the remainder of the full-grown bring up the rear. In all ages these animals have been eagerly hunted; and some of the arts which have been employed to kill or take them merit attention. The Hottentots in South Africa shoot them with tin balls: this chase is attended with considerable danger; for, with every precaution that can be used, the sagacity of the Elephant often detects the approach of the hunter, who, in this case, will, in all probability, fall a victim to the rage of the animal, unless he can instantly disable him. In the island of Sumatra, the inhabitants split sugar-canes (of which food the Elephant is very fond), and impregnate them with poison. In Abyssinia they are pursued by hunters on horseback, in the following manner:—Two men, perfectly naked, mount the same horse; the hindermost is armed with a broadsword, the lower part of which is covered with cord, and the remainder is exceedingly sharp. In this manner they pursue the Elephants, and, having singled out one, they irritate him to attack them, when they ride up close to him, and the armed man slips from the horse on the off side, and, whilst the Elephant's attention is engaged with the horse, he divides the tendons of his foot with a single blow, and thus disables him, when he is dispatched by lances.

They are also taken alive in pitfalls, or are driven into enclosures; in either case they are fed scantily, though regularly, for a few days, when tame Elephants are employed to engage their attention till they can be tied fast to a tree; after they have become somewhat dispirited, they are led away between two tame ones, and put under the care of keepers, who gradually bring them into subjection,—more, however, by caresses and soothing, than by coercion. When tamed, they become the most gentle and obedient of all domestic animals, and, in most cases, are exceedingly fond of their keepers, and soon learn to distinguish the various tones of the human voice, as expressive of anger, approbation, or command. The domesticated Elephant performs more work than six horses, but at the same time requires much care, and a plentiful supply of food. He is generally fed with rice, either raw or boiled, and mixed with water. To keep him in full vigour, a hundred pounds of this food is said to be required daily, besides fresh herbage to cool him; and he must be led to the water twice or thrice a day to bathe. His daily consumption of water as drink is about forty gallons.

It would be difficult to enumerate *all* the services of these useful animals, so varied are they, and so valuable where strength is necessary. They are employed in carrying burdens on their bodies, necks, and even in their mouths, by means of a rope, the end of which they hold fast with their teeth; they load a boat with amazing dexterity, carefully keeping all the articles dry, and disposing them where they ought to be placed. In propelling wheel carriages heavily laden upon a declivity, they push them forward with their forehead, and support them with their knees. In dragging beams of wood along the ground, they remove obstacles or elevate the ends of the beams so as to clear them. Before the invention of fire-arms, they were used in war by many nations of antiquity; and they are still employed in the East in dragging artillery over mountains. In many parts of India, Elephants are made the executioners of justice; for they will with their trunks either break the limbs of a criminal, trample him to death, or pierce him with their tusks, as they may be directed. The Elephant has been long made the companion of the sports of the Orientalist in the great hunting parties; and from the same early period has been made to minister to the wanton and cruel pleasures of Eastern princes, by being stimulated to combat not only with other Elephants, but with various wild animals: in short, were it consistent with the limits of this work, we might fill many pages with matter, both historical and anecdotical, relating to the uses of this noble animal. We may, however, observe that, its strength being equal to its bulk, it is able to carry on its back three or four thousand weight; on its tusks alone it can support near one thousand; and its ordinary pace is equal to that of the horse at an easy trot.

In the preceding part of this article we have dwelt particularly on the Elephant's trunk and tusks, as deserving especial notice: the organ of hearing would scarcely appear to deserve less. The structure of the Elephant's ear has been investigated with great accuracy by Sir Everard Home. The drum, and every other part of the organ, is much larger in proportion than in other quadrupeds, or in man; and there is a remarkable difference in the arrangement of the muscular fibres of the drum of its ear, when compared with some quadrupeds and the human species. In the human ear, those fibres are radia of a circle; and in the horse, the hare, and the cat, they are of an uniform length; but in the Elephant's ear these fibres are so placed that some are more than double the length of others. Sir E. Home argues, from this remarkable construction, that the Elephant has not a musical ear; but that it has a peculiar compensating power in this length of fibre, as its slower vibrations enable it to hear sounds at a great distance.

The tusks of the Elephant have long been applied, under the denomination of *ivory*, to a variety of important uses in the arts. From the fossil remains which have been discovered, it is apparent that they must have been abundantly distributed over the earth; and some of them appear to have been adapted to a much more northern climate than is now inhabited by the Elephant. It is, indeed, a most curious fact, that skeletons nearly allied to, if not quite resembling, those of Elephants are occasionally found in a fossil state, and in large quantities, at a great depth under the surface, in Russia and Siberia. "All the arctic circle," says Pennant, "is a vast mossy flat, formed of a bed of mud or sand, apparently the effect of the sea, and which gives reason to think that that immense tract was in some distant age won from it. With them are mixed an infinitely greater number of marine bodies than are found in the higher parts of that portion of Asia. I give the fact: let others, more favoured, explain the cause how these animals were transported from their torrid seats to the Arctic regions: I should have recourse to the only one we have authority for; and think that phenomenon sufficient. I mention this, because modern philosophers look out for a later cause: I rest convinced, therefore, to avoid contradicting what can never be proved." Dr. Falconer and Major Cautley have brought from the Sewalik Hills in India numerous fossil remains of Elephants, some of them of enormous size. They are all in the truly magnificent collection of the British Museum, and are described by the donors in their well-known work, the Fauna Antiqua Sivalensis; the illustrations of which by Mr. Ford will hereafter be cited, like the work of Lyonet (alluded to under COSSUS), as a perfect example of excellence in the drawing of Fossils.

ELMIS: ELMIDÆ. A genus and subfamily of aquatic Coleoptera, small in size and of an ovate form, found adhering on the under sides of stones lying at the bottom of running water. They are unable to swim, but are provided with very powerful tarsi and ungues, by which they are enabled to retain firm hold on the stones in the most boisterous currents. Twelve or thirteen species belonging to three genera have been found in this country.

ELOPS. (*Elops saurus.*) A small fish, known in the West Indies by the name of the Sein-fish, or Sea Gally-Wasp. It is about fifteen inches long; in the middle five inches round, and tapering to both ends; the head is smooth, and without scales; the tail much forked, and armed both above and below by a strong spine, forming a first or spiny ray on each side the tail. Its general colour is a silvery gray: dusky on the back, the head slightly tinged with yellow, the fins of a bluish brown, and the belly white.

EMARGINULA. A genus of small Mollusca, inhabiting the seas of all climates, and having two short tentacula, with eyes at the base; foot large and thick. The shell is patelliform, oblong, or oval; anterior margin notched. The Emarginulæ may be known from Patellæ and other approximating genera, by the notch or slit in the anterior edge. Recent species, though widely diffused, yet not numerous; fossil species, rare.

EMBERIZA. The name of a genus of Passerine birds. [See BUNTING.]

EMBLEMA PICTA, or PAINTED FINCH. This Passerine bird is a native of the north-west coast of Australia. It is described and figured by Mr. Gould; and exhibits a singularity in colouring which is rarely ever witnessed among the feathered tribes, the upper parts of the plumage being remarkably plain, while the under surface is extremely beautiful. The face and throat are deep vermillion-red; crown of the head, all the upper surface, and wings, brown; the base of all the feathers of the throat black, giving to that part a mingled appearance of black and red; rump deep vermillion-red; tail dark brown; chest and all the under surface jet-black; the flanks thickly spotted with white; and the centre of the abdomen deep vermillion-red; upper mandible black, under mandible scarlet; feet light red.

EMERALD [MOTHS.] A name given by collectors to Moths of the genus *Hipparchus*.

EMPEROR [MOTH]. The name of a species of Saturnia; a genus of nocturnal Lepidoptera. [See SATURNIA.]

EMU, or EMEU. (*Dromaius Novæ Hollandiæ.*) This singular bird is a native of Australia, and allied to the Cassowary; nearly equalling the Ostrich in bulk; but has shorter legs, a shorter neck, and is thicker in the body. Dr. Latham says, This bird measures more than seven feet in length; the beak is black; the plumage for the most part brown and gray mixed, paler on the under parts: the head differs greatly from that of the common Cassowary, being covered with feathers; nor has it any helmet or rising protuberance whatever, as in that

EMU.—(DROMAIUS NOVÆ HOLLANDIÆ

species; the feathers, however, about the head and neck are of a hairy texture, and the fore part of the chin and throat nearly destitute of any, so that the purple colour of the skin may be seen through them: the long feathers observable in the wings of the Cassowary of the Old Continent are here wanting; but instead of them are real wings, though of so small a size as to be useless for flight; they are covered with feathers like the rest of the body, and when the bird is quite at rest, are scarcely discernible therefrom. The legs are stout, similar to those of the Galeated Cassowary, but greatly indented or jagged at the back part: the three toes placed in the same manner, all forwards. So far the external appearance of the bird: internally it is said to differ from every other species, particularly in having no gizzard, and the liver being so small as not to exceed that of a blackbird. It is shy and timid, trusting to its great speed for safety, except when hard pressed; it then strikes violently with its legs. The flesh of the young is delicate, but that of the full-grown bird is coarse; it is pursued, however, for the oil that is obtained from it, of which the skin produces six or seven quarts.

It is stated by Capt. Grey (*Travels in Australia*), that "Emus are killed in precisely the same manner as Kangaroos, but as they are more prized by the natives, a greater degree of excitement prevails when an Emu is slain; shout succeeds shout, and the distant natives take up the cry until it is sometimes re-echoed for miles: yet the feast which follows the death is a very exclusive one; the flesh is by far too delicious to be made a common article of food,—hence, heavy penalties are pronounced against young men, and unauthorized persons, who venture to touch it; and these are invariably rigidly enforced."

At a meeting of the Zoological Society of London (Feb. 23. 1847), the Earl of Derby observed that it was generally supposed that these birds, like most of the *Rasores*, are polygamous: this, however, was not the fact: the Emu is strictly monogamous; and the male, who attends to the eggs, by no means approves of any other female than the favoured one coming near the nest.

EMU-WREN. The name given in Australia to a singular small species of Malurus, the *Stipiturus malachurus;* so called from the tail feathers being loose webbed and bearing some resemblance to the double feathers of the Emu.

EMYDÆ. A family of Testudinous animals, known as MARSH TORTOISES, They inhabit warm climates both of the Old and New World, and are found also in Australia, where hitherto no land Tortoise has been detected. Their shell is more convex than that of the latter; their feet are webbed, and their toes are armed with sharp claws. Though they seldom venture far from the water, which is their natural element, and which they invariably seek as a refuge from

MARSH TORTOISE.—(EMYS PICTA.)

danger, they are far more active and alert on land than those species which are well-known to us as Land Tortoises. They are carnivorous in their habits; eagerly pursuing frogs, fishes, newts, and insects; and some of them are really formidable from their size and ferocity. [See TORTOISE.]

We refer our readers to the great work of Professor Bell, and to Mr. Gray's very admirable Catalogue of the Tortoises in the British Museum, where all the species are described.

ENALIOSAURI. The name applied to that order of Reptiles which contains the fossil genera ICHTHYOSAURUS and PLESIOSAURUS [which see].

ENCHELIDES. A genus of animalcules, the forms of which are extremely various. In some, scarcely any definite shape can be discovered; their bodies appearing to be composed of a mass of gelatinous matter without any solid support.

ENCRINITES. A genus of petrified radiated animals commonly called *Stone Lilies.* Speaking of their rare occurrence in our modern seas, and of their vast numerical importance among the earliest inhabitants of the ancient deep, Dr. Buckland observes: —"We may judge of the degree to which the individuals of these species multiplied among the first inhabitants of the sea, from the countless myriads of their petrified remains which fill so many limestone-beds of the transition formations, and compose vast strata of entrochal marble, extending over large tracts of country in Northern Europe and North America. The substance of this marble is often almost as entirely made up of the petrified bones of Encrinites as a corn-rick is composed of straws. Man applies it to construct his palace and adorn his sepulchre; but there are few who know, and fewer still who duly appreciate, the surprising fact, that much of this marble is composed of the skeletons of millions of organized beings, once endowed with life, and susceptible of enjoyment, which, after performing the part that was for a while assigned to them in living nature, have contributed their remains

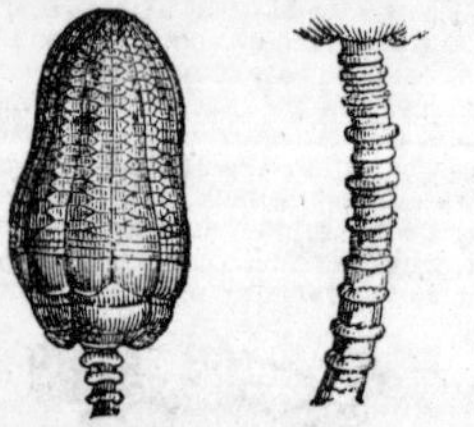

LILY ENCRINITE.—(ENCRINUS LILIIFORMIS.)

towards the composition of the mountain masses of the earth. Of more than thirty species of Crinoïdeans that prevailed to such enormous extent in the transition period, nearly all became extinct before the deposition of the lias, and only one presents the angular column of the Pentacrinite: with this one exception, pentangular columns first began to abound among the Crinoïdeans at the commencement of the lias, and have from thence extended onwards into our present seas. Their several species and even genera are also limited in their extent: *e. g.* the great Lily Encrinite (*E. moniliformis*) is peculiar to the muschel-kalk, and the Pear Encrinite to the middle region of the oolitic formation.

ENHYDRA. The generic name of the Sea Otter of California (*Mustela lutris*, Lin.) See OTTER.]

ENTOMOPHAGA. The name given to a group of Hymenopterous insects, whose larvæ generally feed, parasitically, upon living insects.

ENTOMOSTRACA. A division of the class *Crustacea*, comprising those animals which have only a slight integument in the form of a shell to protect them; of which the genus *Cypris* (many species of which occur in this country) may be given as an example. Their exceeding minuteness and extreme delicacy of structure have deterred most naturalists from examining them and studying them as they require to be studied —fresh from their native habitats. The difficulty of preserving them obliges the naturalist to seek them in their secret lurking places — the fresh-water ponds and ditches, and the little pools in the rocks on the sea-shore, where they are chiefly to be found, and to study them as it were on the spot, with the aid of his microscope. Dr. W. Baird, who has some ingenious remarks on Entomostraca in "The Zoologist" (after alluding to what Latreille had said respecting their organs of mastication being too minute for human observation), says, "The organs of mastication are not however the only organs that are worthy of being noticed. The beautiful and delicate structure of their feet and branchial appendages are worthy of all admiration. These latter organs are almost constantly in motion, and present a most interesting appearance when viewed under the microscope. The extraordinary method they have of reproducing their young, with the transformations which some of them undergo in an early stage, are subjects which might occupy a considerable time in describing, and which cannot be attended to without exciting the greatest interest in the mind of the observer." Dr. Baird has monographed all the British species, and published also descriptions of several exotic species. [See CYPRIS.]

ENTOMYZA. A genus of birds belonging to the *Meliphagidæ*, of which we may mention one of the best known species.

ENTOMYZA CYANOTIS, called by the colonists of New South Wales BLUE-EYE. This bird is found almost exclusively on the Eucalypti, searching among the blossoms and smaller leafy branches for its food, which consists partly of insects and partly of honey, and, as others of the group do, probably, on berries and fruit. They are bold and spirited birds, pugnaciously chasing and drawing about the other species resorting to the same tree. Its cry is loud and monotonous. In every instance that Mr. Gould found its eggs, they were deposited on the deserted, dome-shaped, large nest of the *Pomatorhinus*, never within the dome, but in a neat round depression on the top. It commences breeding early, and has at least two broods in a year

ENTOZOA. A name given to an extensive series of low-organized invertebrate animals, the majority of which are inhabitants of the bodies of other animals during greater part of their existence. They live upon nutriment derived from the tissue or organs in which they dwell; they have colourless blood, no respiratory organs, properly so called, and no articulated members of locomotion. The individual or animal whose body they infest, or who, in other words, entertain their presence, is technically denominated the *host.* (For further particulars of this interesting group, see articles INTESTINALIA, p. 351; FILARIA, p. 230; FASCIOLA, p. 228; TÆNIA, p. 661; and also CÆNURUS and ALTERNATION OF GENERATION in SUPPLEMENT.]

ENTOZOON. A crustaceous parasite, found in the sebaceous follicles of the human skin. It is the *Demodex folliculorum* of Owen.

EPHEMERA: EPHEMERIDÆ. A genus

and family of Neuropterous insects, characterized by the unequal size of the wings; the minute size of the antennæ; the membranous and almost obsolete mouth; and the elongated articulated setæ at the extremity of the body, which is long, soft, and slender: the eyes are large, nearly oval, and three ocelli are placed triangularly between them. The most familiar species is the *Ephemera vulgata*, or common MAY-FLY, so plentiful in the early part of summer about the banks of our rivulets and stagnant waters. It is of a greenish brown colour, with transparent wings, elegantly mottled with brown, and is furnished at the extremity of the body with three very long black bristles. It flutters in the evening about the surface of the water, but during the day is generally seen in a quiescent posture, with the wings closed, and in an upright position. The larva is about an inch in length, having several finny plumes on each side the body, and at the tail three long feathered processes: it has also a pair of moderately long antennæ, though those of the complete insect are extremely short. When arrived at its full size, as above described, it exhibits the rudiments of wings on the back, in the form of a pair of oblong sheaths or scales; its colour being a yellowish or whitish brown. It is supposed to continue in this kind of larva state two years before it assumes the condition of the complete insect. This change takes place in the evening, when the larva rises to the surface of the water, and soon divesting itself of its skin, flies to some neighbouring object, and after having remained some time longer, again casts its pellicle, and appears in its ultimate or perfect form, in which, as well as in its larva state, it is a favourite food of several kinds of fishes, particularly of the Trout. In some seasons it is extremely plentiful, the air in the immediate vicinity of its natal waters being frequently blackened by its numbers during the evening hours.

Swammerdam, the well known writer on Insects, observes, that notwithstanding the dangers to which the eggs, larvæ, and pupæ are constantly exposed, from the attacks of fishes and predaceous aquatic insects, the number of specimens which arrive at the perfect state is sometimes so immense, that the swarms of one species with white wings (*Ephemera albipennis*) has been compared to a fall of snow; whilst, in some parts of Europe where they abound, it is the custom to collect their dead bodies into heaps, and use them for manure. The fishes at such time eagerly wait for them; and so great are the numbers which fall into the water, that the fishermen call them manna.

But the most celebrated of all the Ephemeræ is the species popularly termed the DAY-FLY. It is of a white colour, with the anterior rib of the upper wings black or deep brown, and the tail is furnished with two long bristles. This insect is commemorated as a most remarkable instance of the brevity of animal life; since after its change into the perfect fly it survives but a very few hours, perishing in the course of the same evening that gave it birth. It is to be recollected, however, that its larva lives in its aquatic state two, and even sometimes nearly three years; but when arrived at the proper period, it rises in the evening to the surface of the water; and the skin of the back cracking, and flying off with an elastic motion, the Fly is almost instantly evolved, as in the common species; after which it flies to the nearest convenient spot, and again divesting itself of its pellicle, appears in its last and perfect state. It now flies again to the water, and fluttering over its surface, as if sporting with its innumerable companions, enjoys all the pleasures of its short remainder of existence: the female breeds, deposits her eggs, and, like the male, perishes before or with the dawn of the approaching day. There are several other species of Ephemeræ.

EPIORNIS. [See SUPPLEMENT.]

EPIMACHUS. A genus of birds allied to the Hoopoes, having, like them, a slender beak, but with velvety or scale-like feathers partly covering the nostrils, as in the Birds of Paradise. The plumage in the species figured is of the most gorgeous description. It is of a deep black, with the feathers magnificently glossed with various colours; the

EPIMACHUS MAGNIFICUS

long plumes on the flanks being elongated, turned up, and frizzled: the edges of a burnished steel blue, sometimes inclining to green. It is a native of some of the islands in the Eastern seas, and, like the Birds of Paradise, to which some naturalists have thought it allied, would form a truly beautiful ornament to our aviaries and zoological gardens.

EPOMOPHORUS. A name applied by Mr. Bennett to one or two species of Foxbats (*Pteropidæ*) from W. Africa, which have remarkable tufts of hairs on the sides.

EQUUS. The generic name of certain quadrupeds with solid or undivided hoofs; as the Horse, the Ass, and the Zebra. "This family," says Mr. Gray, speaking of the *Equidæ*, (which is distinguished from all other animals by its undivided hoof, formed of the two anterior toes soldered together, its simple stomach, and its female having the teat placed on the pubes) "may be divided into two very distinct types of form: the one, the Asses and the Zebras, which are always whitish and more or less banded with blackish-brown, and always have a distinct dorsal line, the tail only bristly at the end, and have warts only on the arms and none on the hind legs; and the true Horses, which are not banded, have no dorsal line, are furnished with warts on their arms and legs, and have long hair on the tail, from its insertion to its extremity." Of these species the Horse is by far the most valuable, as well as the most widely distributed over the globe. Dental character of this genus:—six incisors in the front of both the upper and lower jaws, one canine or tusk, and six molars or grinders, on each side of both jaws; in all forty. [See HORSE.]

ERINACEUS. [See HEDGEHOG.]

ERIOMYS. [See CHINCHILLA.]

ERIPHIA. A genus of Decapod short-tailed Crustacea, of which there are several species; one of which, *E. spinifrons*, is com-

GOUTY ERIPHIA.—(ERIPHIA GONAGRA.)

mon in the Mediterranean. The *E. gonagra* is a fine species, with tuberculated fore-legs, found in Brazil. In the Indian Ocean other species are found.

ERMINE. (*Mustela erminea.*) This little digitigrade animal, which is also called the STOAT, resembles in its general appearance the Weasel, but is considerably larger, the Ermine measuring ten inches in length, independent of the tail, whereas the Weasel seldom exceeds six. The colour of the Stoat is a reddish-brown above, white beneath, the tip of the tail being constantly black, whatever may be the cast of colour on the body; for the Stoat, in the northern regions, becomes milk-white during the winter, in which state it is known as the *Ermine:* we may therefore properly say, that an Ermine is a Stoat in its winter dress. Like the Weasel, it lives in hollows under the roots of trees; in banks near rivulets; and it preys on birds, poultry, rats, and all kinds of smaller animals, as well as on rabbits, leverets, &c.; it is also a great devourer of eggs. It is an inhabitant both of the northern parts of Europe and of Asia; and is also found in many parts of North America. Like many other species of this genus, the Ermine has the faculty of ejecting a fluid of a strong musky odour. Its fur is short, soft, and silky; its pure white winter coat being much longer, thicker, and finer than that of summer. The fur of the Ermine is in great request; it was formerly one of the insignia of royalty, and is still used by the judges. When used as linings of cloaks, &c., the black tuft from the tail is sewed to the skin at regular distances.

In the neighbourhood of Hudson's Bay, Ermine are very abundant, particularly in the barren grounds and open plains. In Norway and Siberia also their skins are a great article of commerce. During the winter it is extremely difficult to distinguish them, from their colour so closely resembling that of the snow: and they are generally either taken in traps, or shot with blunt arrows. This animal, which in the pursuit of its prey is one of the boldest of its size, is not readily tamed; but when caught, and kept in a cage, it still exhibits every mark of its ferocious and savage character, by killing or injuring every thing within its reach.

ERNE. A local name for the Sea Eagle, (*Haliæetus albicilla*) which frequents the sea-shores. It is distinguished from the true Eagles, by the absence of feathers at the lower part of the tarsi. [See EAGLES.]

ERODY. The name given to a grallatorial bird (*Dromas ardeola*) allied to the Trumpeter, and found both in India and Abyssinia. It is swift of foot, and was found by Mr. Salt in Abyssinia during Lord Valentia's travels.

EROTYLUS: EROTYLIDÆ. A genus and family of Tetramerous Coleoptera, abounding in South America, where they feed generally on fungi. They form the subject of an admirable monograph by M. Lacordaire, who has published a thick

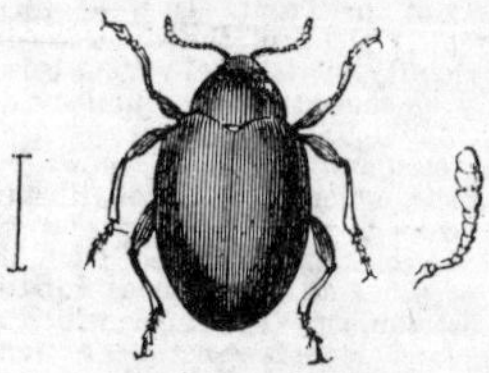

VIOLET-COLOURED EROTYLUS
(EROTYLUS VIOLACEUS.)

volume on the numerous species. A few small British species belonging to the genera *Triplax* and *Tritoma* are found in this country. Most of the South American species have the maxillary palpi terminating in a large crescent-shaped joint; the antennæ end in a very distinct and perfoliated mass.

ERYCINA. A genus of Conchifera, or Bivalves, found in the sand on the shores of New Holland and the Mediterranean. Shell ovate or triangular, transverse, equivalve, smooth; hinge with a ligamentary pit, two diverging cardinal and two lateral teeth in each valve. There are several fossil, and two recent species. Also the name of a genus of Diurnal Lepidoptera.

ERYCINIDÆ. A family of Lepidopterous insects, distinguished by the fore legs of the males being only rudimental: the anal edge of the hind wings is but slightly prominent, and the discoidal cell is either open or closed, partially or entirely, by a false nervure. The caterpillars are very short, pubescent, or hairy; and the chrysalis is short and contracted. These insects are of small size, and generally of very brilliant colours, often varied, and their wings marked with spots. They are almost exclusively confined to South America. Their flight is very rapid, and the majority of them rest with their wings extended on the under side of leaves. Some of the species have the hind wings produced into two or more tails, often of very great length; in this particular resembling the genera Papilio and Thecla; others bear a certain resemblance to the Hipparchiæ, Heliconii, &c. See the fine work of Messrs. Doubleday and Hewitson for the numerous genera and species of this family, where all the leading forms are beautifully figured. In the British Museum there is a very large collection of them.

ESOX: ESOCIDÆ. A genus and family of voracious fishes, many of them inhabitants of rivers. They are destitute of the adipose fin, and the border of their upper jaw is either formed solely by the intermaxillaries, or they have no teeth. The different sub-genera vary greatly in the form of the body, the size of the scales, the length of the jaws, and other striking points. [See PIKE.]

ESQUIMAUX DOG. [See DOG.]

ETHERIA. A genus of Conchifera, or Bivalves, found in the rivers of Africa and Madagascar. Shell inequivalve, very irregular; teeth none; bosses short and indistinct; ligament external, penetrating partly into the shell. In its foliated structure and toothless hinge it resembles *Ostræa*, but differs from it in having two muscular impressions. The exterior is rugged and irregularly shaped, but the interior of the valves is pearly, of a vivid green colour, and raised in small blisters, which are said to be formed by small particles of sand being accidentally introduced during the formation of the nacreous fluid. These shells are abundant in the Nile above the cataracts, where the inhabitants collect them for the purposes of ornamenting their tombs with them. In Lake Tchad, in Central Africa, Major Denham found a fine species of this genus.

EUCHIRUS: EUCHIRIDÆ, or LONG-HANDED BEETLE. A remarkable genus and family of Lamellicorn beetles; the longest known species of which is represented in the two accompanying figures; one of which is the male, and the other the female. The species (*E. longimanus*) is found in the East Indies, where it seems to be by no means common. It is of a rich reddish-brown colour. The two cuts will show the form of the sexes of this species better than the most elaborate description. The insect is most probably a native of one of the islands in the East Indian Archipelago. Another

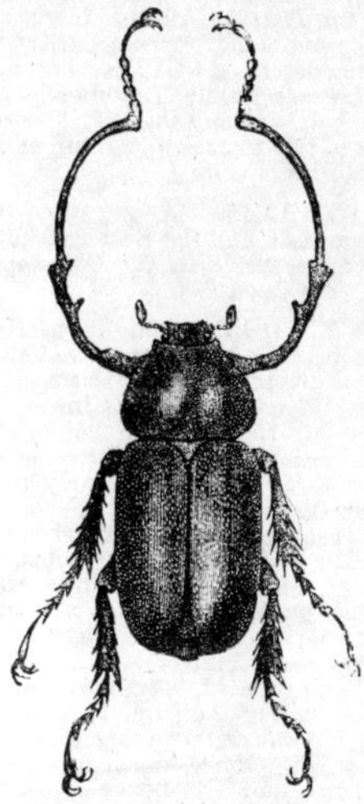

LONG-HANDED BEETLE: MALE
(EUCHIRUS LONGIMANUS.)

species (*E. quadrilineatus*) in the British Museum collection, was found by Mr. Cuming in the Philippine Islands. It is distinguished, among other marks, by four lon-

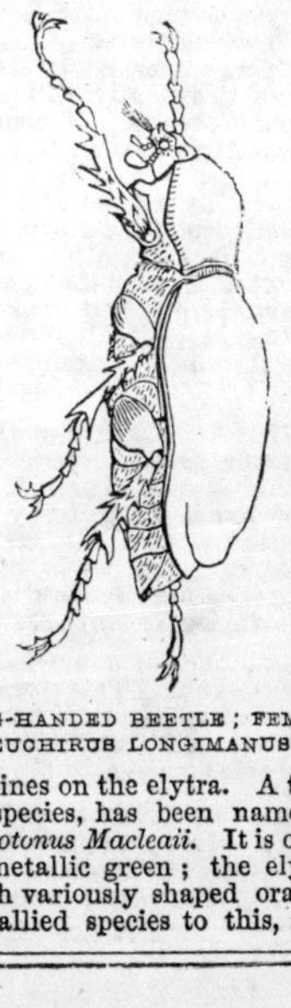

LONG-HANDED BEETLE; FEMALE.
(EUCHIRUS LONGIMANUS.)

gitudinal lines on the elytra. A third most gorgeous species, has been named by Mr. Hope *Chirotonus Macleaii*. It is of the most brilliant metallic green; the elytra being black, with variously shaped orange spots. A closely allied species to this, also in the

British Museum, and described by Mr. Gray, is the *Ch. Parrii*. Another insect belonging to this group is the *Propomacrus Arbaces*, from Smyrna, described by Pallas, and figured by Mr. Newman in the Entomological Magazine. Little, if any thing, is known of the habits of this remarkable group of Lamellicorn beetles.

EUDYNAMIS. A genus of Cuckoos, found in Asia and the Eastern islands. The best known species is the *Eudynamis orientalis*. [See CUCKOO.]

EUMENIDÆ. A family of Hymenopterous insects, of predaceous habits. Reaumur has given the history of a species of this family (*Odynerus*). "This insect, says Mr. Westwood, in his useful compilation, the 'Introduction to the Modern Classification of Insects,' "during the early months of summer, forms a burrow in the sand to the depth of several inches, in which it constructs its cells; besides which it builds, with the grains of sand brought up whilst burrowing, a tubular entrance to the burrow, often more than an inch long, and more or less curved, the grains of sand of which it is formed being agglutinated together; each female forms several of these burrows, and deposits an egg in each cell, together with a number of green caterpillars, which it arranges in a spiral direction, one being applied against the other, and which serve as food for the larvæ when hatched. When the store of food is secured, the insect closes the mouth of the burrow, employing the grains of sand of which the funnel was composed for that purpose. The larvæ of Odynerus are fleshy grubs, destitute of feet, with transverse dorsal tubercles serving in their stead. * * * Geoffroy has described a species of Eumenes which differs somewhat in its habits from the rest of this family. This species constructs, upon the stems of plants, especially heath, small spherical nests, formed of fine earth: at first a hole is left at the top, through which the parent fills the cell with honey, and deposits a single egg therein; the hole is then closed up, and the larva, when hatched, feeds on the honey, undergoes its metamorphosis, and makes its escape through a hole which it forms at the side of the cell, which contains but a single insect."

EUPHEMA. A genus of the *Psittacidæ* or parrot tribe; several species of which are found in Australia. The bill is almost always very much concealed by the long feathers about the face. In Mr. Gould's national work, "The Birds of Australia," several species are described and elegantly figured: of these we may specify—

EUPHEMA AURANTIA, OR ORANGE-BELLIED GRASS PARAKEET. This species is not very abundant in Van Diemen's Land, but in Actæon Islands, in D'Entrecasteau Channel, it is the only bird enlivening the solitary place.

EUPHEMA CHRYSOSTOMA, OR BLUE-BANDED PARAKEET. This beautiful bird is a summer resident of Van Diemen's Land, arriving there in September, and leaving in February and March; running over the ground, and treading its way among the grasses, to feed on the seeds. Its flight is very quick. It can easily be domesticated, and a more elegant or beautiful pet can scarcely be conceived.

EUPHEMA ELEGANS, OR ELEGANT GRASS PARAKEET. This species inhabits South Australia, and is the "Ground Parakeet" of the colonists. It feeds on grass seeds;

ELEGANT GRASS PARAKEET
(EUPHEMA ELEGANS)

congregating in the hot seasons (where there is water in small pools) in almost incredible numbers. Its flight is rapid and even, and frequently at great altitudes. For our figure of this elegant bird we are indebted to the work of Mr. Gould.

EUPHEMA SPLENDIDA, OR SPLENDID GRASS PARAKEET, inhabits the neighbourhood of the Swan River, in Australia. [See MELOPSITTACUS.]

EUPHONIA. A genus of birds allied to the TANAGERS, of which there are many species. We restrict ourselves in this article to the EUPHONIA JAMAICA. This is a small Passerine bird, known in the West Indies as the "Blue Quit," and sometimes also called the Blue Sparrow. It is about four inches and a half long, and rather of an inelegant shape from the abrupt shortness of its tail. The upper parts of the male are of a glossy blue, sometimes tinged with green; throat, breast, and sides pale gray; belly yellow; beak gray, the ridge and tip black. It is common about homesteads, frequenting fruit-trees, busily hopping about the twigs and fruits and picking in any position. It is by no means destitute of musical powers, sometimes delighting in a soft warbling repetition of a single note, and at others treating its hearers with a real song, sweet and musical. Mr. Hill, a gentleman of Jamaica, whose ornithological notes contribute to the entertaining character of Mr. Gosse's work, gives the following description of this little active warbler. "Near the piazza of my house a cotton-bush has flung out its knots of white filaments. Hither come the birds at this season (February) to gather materials for constructing their nests. The Blue Sparrow, a pretty little frugivorous bird that sings in our fruit trees, all the year round, its merry twittering song, has been busily engaged with his mate collecting bill-fulls of cotton. It did not seem to be a thing immediately settled that they should set to work and gather their materials at once. They had alighted on the tree as if they had very unexpectedly found what they were seeking. The male began to twitter a song of joy, dancing and jumping about; and the female, intermingling every now and then a chirp, frisked from stem to stem, and did very little more than survey the riches of the tree: at last she plucked now and then a bill-full of the filaments, and spreading it to flaunt to the wind, tossed it away, as if she had been merely showing that it every way answered the purpose of length and softness, and was in every respect the thing they wanted." We are also told that they build a very snug domed nest, globular in form, and about as large as an infant's head, with an opening in one side, composed of dry grass, the dried stems of the *Tillandsia*, tendrils of passion-flower, bits of rag, &c., profusely intermixed with cotton and the down of plants.

EUPLECTELLA. [See SUPPLEMENT.]

EUPLOCOMUS. An Asiatic genus of Pheasants, of which the males have usually very brilliant plumage. [See PHEASANT.]

EURYALE. A remarkable genus of Radiated animals belonging to the *Asteriadæ*, or Star-fishes, in which each division of the rays is branched again and again, so

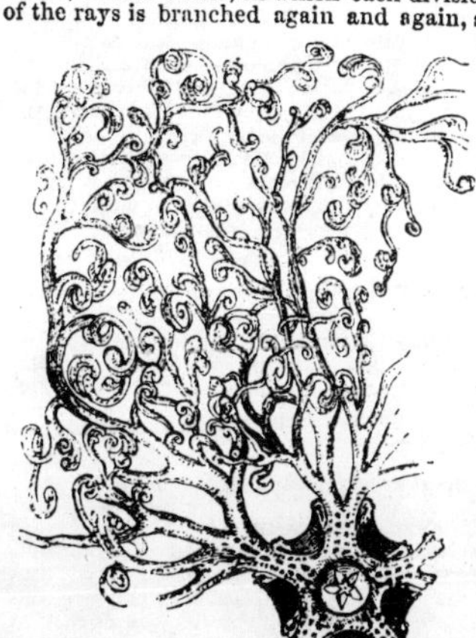

WARTED EURYALE.
(EURYALE VERRUCOSUM.)

that the whole resembles a bunch of serpents' tails. The figure which we subjoin represents the whole of the body, with only two of the rays given in detail, as the cut would otherwise occupy too much space. They are sometimes known by the name of Medusa's heads. These little branches must be of singular use to the animal in securing its prey. In the cases of *Radiata* at the British Museum may be seen some fine examples of these "furies of the deep." [See STAR-FISHES.]

EURYNOME. A genus of Crustacea, belonging to the family *Lambridæ;* of which one species is found in the British seas — the *Eurynome aspera.* It is a pretty little species, rough with projecting knobs; often symmetrically arranged, and of a reddish colour. The fore legs in the male are elongated.

EVANIADÆ. A family of Hymenopterous insects, of small extent, and not possessing any remarkable points of interest. The species are parasitical, the *Evania appendigaster* being attached to the Cockroach (*Blatta orientalis*).

EXOCETUS. [See FLYING-FISH.]

EWE. The female of the Sheep kind. [See SHEEP.]

FAIRY SHRIMP. [See APPENDIX.]

FALCONIDÆ. The genus *Falco* of authors may be considered as constituting five tribes or families of Accipitrine birds, viz. Eagles, Falcons, Kites, Buzzards, and Hawks. They prey, in general, on living animals: the species are extremely numerous; the females are larger than the males; and they vary considerably in their plumage according to age and other circumstances. They are characterized by a powerful beak, which is generally armed with a kind of tooth or process on each side near the apex; and their wings are strong, long, and pointed: they are likewise distinguished by their courage and activity. The true Falcons are symmetrical in their forms; their flight is graceful and vigorous; they possess strength and flexibility; and their sight is so very acute, that they are enabled to discern their prey at a great distance, and to pounce down upon it. The Falcon builds in the hollows of rocks exposed to the south; usually laying its eggs about the close of winter, or very early in the spring: these are often four in number, and are white, spotted vith brown. So rapid is said to be the growth of the young, that in the space of three months they equal the parents in size.

The "noble" breed of Falcons which our ancestors introduced into their service, and so greatly prized, are distinguished from the "ignoble" or baser race of kites, sparrow-hawks, and buzzards, by the peculiar length of their wings, which reach almost as far as their tails; this superiority of wing giving them confidence in the pursuit of the game, and their great power emboldening them to attack it. To train these birds, however, required no small degree of skill and assiduity: but so thoroughly antiquated and obsolete has the once princely sport of Falconry become, that we think the reader will commend us for omitting that which a few centuries ago would have been regarded as indispensable: we mean, a circumstantial account of the training, or education, necessary to teach these magnanimous birds the duties of their office. Numberless indeed are the treatises which have been written on the subject, but in language so fraught with professional technicalities, that at the present day they would be almost unintelligible.

The JERFALCON. (*Falco Gyrfalco.*) This elegant species is generally considered as the boldest and most beautiful of the tribe, approaching in size nearly to that of the Osprey. It is a native of the cold and dreary regions of the north, and is found in Iceland, Russia, Norway, and Baffin's Bay. Its general colour is brown above, in deeper and lighter variegations, and whitish beneath, with brown longitudinal spots; the tail is crossed by numerous deeper and lighter bands, and the bill and legs are generally of a bluish or a pale yellow hue. Buffon mentions three varieties of the Jerfalcon; the first and second very similar to what we have just described; and another which is entirely white. Next to the Eagle, it is the most formidable, active, and intrepid of all rapacious birds, and the most esteemed for falconry. It boldly attacks the largest of the feathered race; and although it is often transported from the coldest regions to some of the warmest, its strength is not diminished by the change of climate, nor its vivacity blunted.

The PEREGRINE FALCON. (*Falco Peregrinus.*) This species is about eighteen inches in length, and three feet six inches wide when its wings are extended; and in its full growth and plumage is a very fine-looking, strong, and bold bird. The bill is pale blue, tipped with black; short, strong, and much hooked. The general colour on the upper parts is a deep bluish lead-colour, barred with black, but the crown of the head and upper part of the neck nearly black: the greater wing-feathers dusky, barred with oval white spots; and the tail of a dark dingy ash, spotted with brownish black,

PEREGRINE FALCON.—(FALCO PEREGRINUS.)

and tipped with pale brown. The under parts, from the chin to the bottom of the breast, are yellowish white, with a deep brown streak down the shaft of each feather; and the remainder are of a dull white, beautifully and distinctly barred with dark brown. The thighs are long, and marked with small heart-shaped spots; legs short, strong, and yellow; claws black, and the toes long. The Peregrine Falcon appears to be a general inhabitant of Europe and Asia: it is common in the north of Scotland, and is known to breed on the rocks of Llandidno, in Caernarvonshire; which have been long celebrated for producing a "generous race."

The BLACK-CHEEKED FALCON. (*Falco melanogenys.*) A noble species of the *Falconidæ*, noted for its bold and rapacious habits, which is universally dispersed over the whole southern portion of Australia, including Van Diemen's Land. Mr. Gould says it gives preference to steep rocky cliffs, and the sides of precipitous gullies, rather than to fertile and woodland districts. It there dwells in pairs throughout the year, much after the manner of the Peregrine Falcon; its nest being placed in the most precipitous and inaccessible parts of the rocks. Their eggs are two in number, the ground colour buff, thickly blotched with deep reddish chestnut. In alluding to the strength and courage of this bird, Mr. Gould has the following comment: "Thus we find in this Falcon a bird well adapted for the sport of Falconry, which, though fallen into disuse in Europe, may at some future time be revived in this new and rising country, since its lagoons and water-courses are well stocked with herons and cranes, and its vast plains are admirably suited to such pastime. The introduction of hounds for the purpose of chasing the native dog (Dingo) and the Kangaroo has already taken place in Australia; and perhaps it is not too much to look forward to the time when the noble science of Falconry shall be resorted to by the colonists. A finer mews of birds could not be formed in any country than in Australia; with such typical Falcons as *F. hypoleucus*, *F. melanogenys*, and *F. frontatus.*"

The WHITE-BREASTED FALCON. (*Falco hypoleucus.*) This fine bird, which greatly resembles the Jerfalcon, belongs to the Australian fauna, and is interesting, as Mr. Gould remarks, "as adding another species to the true or typical Falcons, and as affording another proof of the beautiful analogies which exist between species of certain groups of the southern and northern hemisphere."

The GENTIL, OR GENTLE FALCON. (*Falco Gentilis.*) This is described as somewhat larger than a Goshawk, and of an elegant form. The bill is lead colour; the cere and legs are yellow; and the head is of a light ferruginous colour, with oblong black spots. The whole of the under parts are whitish, with brown spots and dashes; the back is brown; the quill-feathers, which are dusky, are barred on their exterior webs with black, and on the lower parts of their inner ones with white: the wings reach to the middle of the tail, which is alternately banded with black and ash-colour, and tipped with white. The legs are yellow and rather short, and the thighs are well covered with feathers.

There are many other species and varieties; but to give a detailed description of them all would be more monotonous than interesting.

Among the European nations the French and Germans seem to have been the first to encourage the practice of falconry; and the technical terms used by the English are evidently all borrowed or derived from the French. "In our own country," says Mr. Pennant, "I cannot trace the certainty of falconry till the reign of King Ethelbert the Saxon monarch, in the year 760, when he wrote to Germany for a brace of Falcons which would fly at cranes and bring them to the ground, as there were very few such in Kent. It seems highly probable that falconry had its rise in Scythia, and passed from thence to the northern parts of Europe. Tartary is even at present celebrated for its fine breed of Falcons; and the sport is in such general esteem that, according to Olearius, there was no hut but what had its Eagle or Falcon. The boundless plains of that country are as finely adapted to the diversion as the wooded or mountainous nature of most part of Europe is ill calculated for that rapid amusement." In England falconry seems to have continued in full glory till about the time of Cromwell, after which it appears to have gradually declined. With what ardour it was pursued in the reign of James I. may be gathered from the anecdote related by Mr. Pennant, who says that Sir James Monson gave no less a sum than a thousand pounds for a cast of Hawks. [See EAGLE, HAWK, &c.]

FALLOW DEER. (*Cervus dama.*) This animal, so graceful an ornament of our parks, in its general form greatly resembles the Stag, having the same elegance of aspect with a more gentle disposition. It is, however, considerably smaller, being only about

FALLOW DEER.—(CERVUS DAMA.)

three feet, or rather less, to the top of the shoulder. It is generally of a brownish bay colour, more or less beautifully spotted; and it has a longer tail. The horns of the Fallow Deer are broad and palmated at their extremities, pointing a little forward, and branched on their hinder sides; they have two sharp and slender brow-antlers, and, above them, two small slender branches; whereas every branch of a Stag's horn is shaped like the stem that supports it: the form of the horns is, in fact, the *chief* mark of distinction between the two species.

The manners of the Fallow Deer resemble those of the Stag, but it is less delicate in the choice of its food, and browses much closer. It arrives at full growth and perfection in about three years, and is said to live about twenty. The male is called a *buck*. In the first year he is a *fawn*; in his second, a *pricket*; in his third, a *sorel*; in his fourth, a *sore*; in his fifth, a *buck of the first head*; and in his sixth, a *great buck*. The female, or *doe*, in her first year is called a *fawn*; and in her second, a *teg*. The process of acquiring and shedding the horns is in every respect similar to that which takes place in the stag; but the form of them, as we have before described, is very different, and the furrows, &c. are less distinctly marked. Although the males are much less furious in the rutting-season than the Stag, they frequently fight desperately for the possession of the females; and it is not till after repeated conflicts that one buck obtains the sovereignty of the whole. It also often happens that a herd of Fallow-Deer will divide into two parties, and engage each other with great ardour and obstinacy, as if ambitious of securing some favourite spot of pasturage, and of driving the vanquished party into the coarser and more sterile parts.

When closely pursued by the hunters, the buck makes towards some strong hold or thicket with which he is acquainted, either in the more shady parts of a wood, or the steep of some mountain; nor does he fly far before the hounds, nor cross and double like the stag: he will take the water, it is true, when hard run; but in strength, cunning, and courage, he is much inferior to the stag, and, consequently, he affords neither so long nor so various a chace. In England there are two kinds of Fallow Deer: the beautiful dappled kind, supposed to have been brought from the South of Europe, or the Western parts of Asia; and the very deep brown variety, which were brought from Norway by James I., who, while there, noticed that they could endure the cold of that severe climate, and subsist throughout the winter without fodder. Nothing can exceed, in richness and delicacy, the venison of the Fallow Deer. The skins of both the Buck and the Doe are unrivalled for durability and softness; and the horns, like those of the stag, are manufactured into knife handles, &c.; while from the refuse, ammonia (popularly known as *hartshorn*) is extracted.

FANFOOT [MOTHS]. A name given by collectors to Moths of the genus *Polypogon.*

FANTAIL. (*Rhipidura.*) A genus of Birds belonging to the family *Muscicapidæ*, and found in Australia. There are more than one species, but we restrict ourselves here to the

RHIPIDURA ALBISCAPA, or WHITE-SHAFTED FANTAIL. This bird inhabits Van Diemen's Land and South Australia. It is generally seen in pairs, among trees: while in the air it assumes a number of lively and beautiful positions; at one moment mounting almost perpendicularly, spreading out its tail constantly to the full extent, and frequently tumbling over in the descent. It is a very tame bird, allowing near approach without showing the least timidity, and will even enter houses in the bush, in pursuit of gnats and other insects. In the breeding season it is not so familiar. Its nest is

WHITE-SHAFTED FANTAIL. RHIPIDURA ALBISCAPA,)

very elegant, resembling a wine glass in shape; and is generally composed of the inner bark of a Eucalyptus, neatly lined with the down of the tree-fern intermingled with flowering stalks of moss, and outwardly matted together with the webs of spiders, which not only serve to envelope the nest, but also strengthen its attachment to the branch on which it is constructed, which is always within a few feet of the ground. Eggs two in number. Our figure is derived from the beautiful work of Mr. Gould's, and shows the bird flying over its nest.

FASCIOLA, or FLUKE. (*Fasciola [Distoma] hepatica.*) A parasitical animal believed to give rise to that much-dreaded disease called the rot. It is not only found in other ruminants, but likewise infests Man, the Horse, the Hog, and even the Rabbit. In the human subject it is fortunately of rare occurrence. A full-grown example is usually about three-quarters of an inch in length; its form being leaf-like, with a superior projecting portion (the so-called head), resembling a short stalk, the tail end being gradually narrowed and bluntly pointed. These animals are hermaphroditic, in

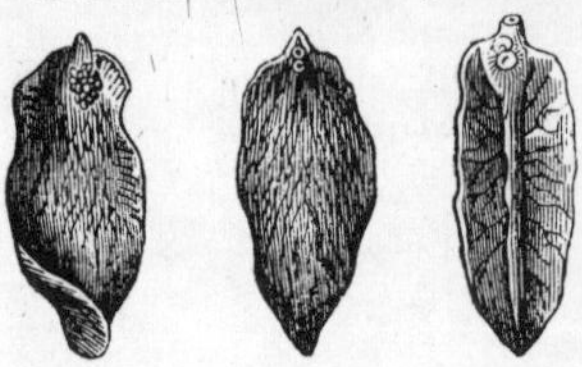

FLUKE. — (FASCIOLA HEPATICA.)

other words, they are bisexual, but, like the Mollusca, they are incapable of fertilising themselves. There are no eyes, or other organs of special sense, although some Entozoologists profess to have discovered a well-marked nervous system. The power of multiplication in these parasites is immense, and the ducts of a single liver have been found to contain upwards of a thousand. From recent investigations and experiments made by Buchenmeister, Leuckart, Siebold, and others, there can be no doubt that these Flukes gain access to the bodies of the creatures they infest whilst they are still in a young condition. The eggs of the parent Fluke having escaped in fields and other open places, the young issue forth from the eggs in the form of ciliated embryos, which swim about in ditches, ponds, &c. After undergoing certain transformations in the water (and probably also in the bodies of aquatic insects, pond-snails, &c.), they are ultimately taken into the bodies of the higher animals along with food or drink. In this way the cycle of their life-development is completed. [See ALTERNATION OF GENERATION in SUPPLEMENT.]

FASCIOLARIA. A genus of Univalves found in the Indian seas, the Antilles, &c., some of which are very beautiful.

FATHER-LASHER. (*Cottus bubalis.*) An Acanthopterygious fish, seldom exceeding eight or ten inches in length, generally found on the rocky coasts of this island, and which is immediately recognized by its large and formidable head, armed with long spines; by

FATHER LASHER.—(COTTUS BUBALIS.)

means of which it immediately combats every enemy that attacks it, inflating its cheeks and gill-covers to a prodigious size. The mouth, which is large, contains two rows of minute teeth, besides others which are in the roof. The back is much elevated; the belly is prominent; the lateral line is rough, but the rest of the body is very smooth, tapering towards the tail. The colour of the body is a dusky brown, marbled with white, and sometimes stained with red; the fins and tail are transparent; and the belly is a silvery white. It feeds on small crustacea. In Greenland this fish attains a much larger size, and forms the principal food of the natives, who make it into soup, which is said to be both wholesome and palatable.

FAWN. An appellation given to a buck or doe of the first year. [See DEER.]

FELIS: FELIDÆ. The name given to animals of the Cat kind, forming a large genus and family of carnivorous quadrupeds, including the lion, tiger, leopard, lynx, domestic cat, &c. They are characterized by having strong, sharp, retractile talons on the feet, and by the teeth being equally fitted for the purposes of destruction. They are all essentially carnivorous; they refuse vegetable food; and in a state of nature they will not, unless pressed by hunger, devour any flesh which they have not themselves killed. They are, consequently, of all Mammalia, the most destructive in their propensities; and their bodily powers are in admirable accordance with their instincts. There are no quadrupeds in which the muscles of the jaws and limbs are more fully developed: their frame is vigorous, but agile; the limbs are well knit, but supple; and every motion is easy, free, and graceful. There is no superfluous flesh; but the whole seems composed of bone, nerve, muscle, and sinew. Though many animals on which they prey excel them in fleetness, in consequence of having longer and more slender limbs, there are none which approach them in the power of leaping and bounding. The under surface of their feet being provided with elastic pads or cushions, their footfall is rendered noiseless; their usual gait is slow, cautious, and stealthy; and when the impetus of the spring is added to the stroke of the paw, their power is almost irresistible. They possess the sense of smell in a very moderate degree, compared with the Canidæ; but their sight is most acute, adapted for vision by night as well as by day; the sense of hearing is also exquisite; and the long whiskers are delicate organs of the sense of feeling. The tongue is furnished with rough horny papillæ, directed backwards; these serve a very important purpose in enabling the animal to scrape off the minute particles of flesh adhering to the bones of its prey.

The different species of this family for the most part bear a very close resemblance to one another in general conformation, though differing widely in size; and it is chiefly by their variations in this respect that their habits are guided. In Brande's Dictionary of Science we find the following judicious observations on the distinguishing characteristics of the different species: "The leopards, panthers, jaguars, are the most typical or truly feline species; in these the beauty of colouring, sleekness of skin, elegance of form, craft, suspicion, bloodthirstiness, agility under excitement, and sloth during repletion, are most strongly manifested. The lion combines more robustness of body with the feline attributes; and his pre-eminent stature receives an air of nobility and grandeur from the mane that decorates his head and neck. He has the credit too of a greater share of boldness and generosity than the other cats. His vocal organs also exhibit a modification of structure not present in the other felines, by which he has the power to utter his tremendous roar—a roar which, when sent forth under the excitement of hunger, scares from their hiding places the timid ruminants which may be lurking within the compass of its fearful reverberations. Among the felines, one group is characterized by the shortness of the tail, and the tuft of hair on the tips of the ears; this includes the lynxes.

"The cheetah, or hunting leopard, deviates most in the half-retractile condition of the talons, and the upright carriage of his body, from the true feline characters; and with these physical modifications is combined so much of the canine disposition, as enables this species to be used in packs for the purposes of the chase.

"The middle-sized cats, which lurk in the branches of trees, as the leopards, ocelots, &c., have a fulvous ground colour, broken by irregular dark spots; a marking which admirably adapts them for concealment amidst foliage. A similar relation of adaptation to the peculiar theatre of their destructive habits may be traced in other species. The tiger, for example, which prowls on the ground, and creeps stealthily towards his victim between the stems of the trees and plants of the jungle, has his bright ground colour interrupted with black vertical stripes. The lion, which traverses the parched deserts of Africa, and lies in wait to intercept the antelopes which bound in troops from one oasis to another, would be rendered too conspicuous if his tawny hide were ornamented with the stripes or spots that characterize the feline livery: these, therefore, which are obvious enough in the earlier periods of his existence, become obliterated as he attains to maturity. A smaller feline species, the puma, or American lion, which plays the predatory character in a corresponding theatre of the New World, presents a similar uniformity of colour. The feline animals bring forth from two to six young ones at a birth." [See CAT: LION: TIGER: &c.]

FENNEC. (*Megalotis.*) This is a beautiful little animal, belonging to the digitigrade Carnivora, closely allied to the Dog, principally found in N. Africa. It is about ten inches in length, five in height, and is of a yellowish-white colour: it has a pointed visage, long whiskers, large bright black eyes, and very large ears, of a bright rose colour, internally lined with long hairs, and the orifice covered with a valve or membrane: the legs and feet are like those of a dog; and it has a taper tail. It inhabits, says Mr. Pennant, the vast deserts of Saara, which extend beyond Mount Atlas; and burrows

NUBIAN FENNEC, (MEGALOTIS NUBIANUS)

in sandy ground, which shows the use of valves to the ears. It is so exceedingly swift that it is very rarely taken alive: feeds on insects, especially locusts; sits on its rump; is very vigilant, and barks like a dog, but much shriller. A fine species of Fennec was lately brought alive from S. Africa, and presented to the Zoological Society of London, by Capt. Sir Edw. Belcher. There seem to be two, if not three, species.

FERÆ. The name of an order of Mammalia, to which the Cats, Dogs, Bears, &c. belong. [See CARNIVORA.]

FERRET. (*Mustela furo.*) This useful but ferocious little animal, of the weasel kind, is kept in a domesticated state in Europe, and is used for rabbit-hunting, as well as for destroying rats. In its general form it resembles the Polecat, but is rather smaller; its usual length being about thirteen inches, exclusive of the tail, which is about

FERRET.—(MUSTELA FURO.)

five. It has a very sharp nose, red and fiery eyes, and round ears. Its colour is a pale yellow, but it also occasionally partakes of all the colours common to the weasel kind, white, black, brown, &c. In the slenderness of its body and the shortness of its legs it also resembles the weasel. In its wild state it is a native of Africa, whence it was originally imported into Spain, and from Spain gradually introduced into other European countries. The cold of our winters is, in fact, too severe for it, so that it becomes necessary to keep it in a warm box, with wool or some other substance in which it may imbed itself. In this state it sleeps almost continually; and when awake, immediately begins to search about for food: that which it is usually given is bread and milk, but its favourite food is the blood of smaller animals. It is by nature an enemy to the rabbit; and Buffon affirms, that whenever a dead rabbit is presented for the first time to a young Ferret, he flies upon it in an instant, and bites it with great fury; but if it be alive, he seizes it by the throat and sucks its blood. When sent into the burrows of rabbits, the Ferret is always muzzled, that he may not kill the rabbits in their holes, but only drive them out to be caught in the nets prepared for them. If the Ferret become unmuzzled he is often lost; for after sucking the blood of his victim, he generally falls asleep in the burrow, from whence he emerges only when by the calls of hunger he goes forth in pursuit of fresh prey; and there, in the midst of abundance, he continues to lead a rapacious life, till the severity of the weather proves fatal to him.

The Ferret, as we have before said, in its nature is ferocious; it is tame without attachment; and such is its appetite for blood, that it has been known to attack and even kill children in their cradles. It is of an irascible nature, and when irritated, the odour it emits is very disagreeable, and its bite not easily cured. The female has two broods in the year, each consisting of from six to nine. She not unfrequently devours her young as soon as they are born; in which case she usually has another brood very soon.

FIBER. A genus of glirine Mammalia, close to the Beaver, the only known species of which is the N. American *Ondrata*, or *Castor zibethicus*, L. [See BEAVER.]

FILANDER. The name given by Brun to the E. India Island Kangaroo, *Halmaturus Asiaticus.*

FINFOOTS. The name applied to two grallatorial birds, allied to the Coots, and closely connected with the web-footed order. One of these, *Heliornis Surinamensis*, is a native of S. America; while the other, *Podica Senegalensis*, or African Finfoot, is, as the name implies, a native of W. Africa.

FILARIA. A genus of Entozoa, having a long, slender, and thread-like body, resembling that of the *Gordii* among the Annelidæ, but with mere marks on the body instead of the rings. These parasitic animals are imbedded in the parenchyma of the cellular tissues, between the coats of the viscera, &c., often existing in numerous bundles, contained in a common cyst or tunic. They are not confined to the larger animals, but are found in insects and their larvæ, and even in various Mollusca. Of these the most common, or at all events the most dreaded by man, is the *Filaria Medinensis*, or Guinea Worm, a most troublesome animal in hot climates, where it insinuates itself under the skin, generally of the leg, and sometimes causes the most excruciating pain. At the seventh anniversary of the Microscopical Society of London, held Feb. 10. 1847, a paper was read, entitled "Observations on the Structure and Nature of the Filaria Medinensis, or Guinea Worm," by G. Busk, Esq. The author, before entering upon the anatomical structure of the worm, premised a short statement of what is known with regard to its habits, and the localities in which it occurs endemically, showing that it is found only in certain portions of the torrid and north temperate zones in Africa and Asia. It is especially frequent on the coast of Africa, and thence derives its name of Guinea Worm. When perfect, its average length is from four to six feet: the body is cylindrical, and of uniform size, or nearly so, throughout; there is no anal or other opening visible at the caudal extremity of the worm, or in any part of its length. The cavity of the worm is occupied by innumerable young. In some worms, or in some parts of them, these occupy the whole space, whilst in others a grumous matter predominates, and only a few young are here and there imbedded in it. The young *Filariæ* differ considerably in their outward form from the parent worm, being furnished with a long tapering tail, which constitutes about two-fifths of the length of the body. The life of *Filaria Medinensis* as a parasite extends over from twelve to eighteen months. When arrived at maturity, it comes to the surface, and is either brought away or comes away piecemeal, thus affording an opportunity for the dispersion of the vivacious young with which its interior is crammed. From these facts, the author suggested that *Filaria Medinensis*, in its parasitic form, presented an instance, among the nematoid Entozoa, of an intermediate or transitory generation, such as have been shown by several naturalists to exist in most of the lower classes of animals.

FIELDFARE. (*Turdus pilaris.*) A bird of the Thrush kind, ten inches in length, and weighing about four ounces. The head is ash-coloured, inclining to olive, and spotted with black; the back and greater coverts of the wings are of a deep chestnut; the throat and breast yellow, regularly spotted with black; the belly and thighs yellowish-white; tail dark brown; legs dusky brown; bill yellow. The Fieldfare is a migratory bird, making its appearance in this country about the beginning of October, in order to avoid the rigorous winters of the north, whence it sometimes comes in great flocks, according to the severity of the season, and leaves us about the latter end of February or the beginning of March. It builds its nest in the loftiest trees; and feeds on hips, haws, and other berries, with various kinds of worms, &c.

Mr. Knapp, speaking of the Fieldfare in his "Journal of a Naturalist," says, "In this county [Gloucestershire], the extensive lowlands of the river Severn in open weather are visited by prodigious flocks of these birds; but as soon as snow falls or hard weather comes on, they leave these marshy lands, because their insect food is covered or become scarce, visit the uplands to feed on the produce of the hedges, and we see them all day long passing over our heads in large flights on some distant progress, in the same manner as our larks, at the commencement of a snowy season, repair to the turnip fields of Somerset and Wiltshire. They remain absent during the continuance of those causes which incited their migration; but, as the frost breaks up, and even before the thaw has actually commenced, we see a large portion of these passengers returning to their worm and insect food in the meadows, attended probably by many that did not take flight with them — though a great many remain in the upland pastures, feeding promiscuously as they can. In my younger days, a keen, unwearied sportsman, it was always observable, that in hard weather these birds increased prodigiously in number in the counties far distant from the meadow lands, though we knew not the reason; and we usually against this time provided tempting bushes of haws, preserved in a barn, to place in frequented hedges, near our secret standings. When the Fieldfare first arrives, its flesh is dark, thin, and scurfy; but, having fed a little time in the hedges, its rump and side veins are covered with fat. This is, in part, attributable to suppression of perspiration by the cold, and partly to a nutritive farinaceous food; its flesh at the time becoming bluish and clean. The upland birds are in this state, from perhaps the end of November till the end of January, according as the hedge fruit has held out; and at this period they are comparatively tame: afterwards, though the flights may be large, they become wild; and the flesh, assuming its darkness, manifests that their food has not been farinaceous. The distant foreign migrations, which have been stated to take place from the meadows of the Severn, I believe to be only these inland trips; and that the supposed migrators returned to those stations fat and in good condition, owing to their having fed during their absence on the nutritious berry of the white thorn. * * * Perfectly gregarious as the Fieldfare is, yet we observe every year, in some tall hedge-row, or little quiet pasture, two or three of them that have withdrawn from the main flocks, and there associate with the blackbird and the thrush. They do not appear to be wounded birds, which from necessity have sought concealment and quiet, but to have retired from inclination; and I have reason to apprehend that these retreats are occasionally made for the purpose of forming nests, though they are afterwards abandoned without incubation. * * * These retiring birds linger with us late in the season, after all the main flights are departed, as if reluctant to leave us; but towards the middle or end of April the stragglers unite, form a small company, and take their flight."

FILE-FISH. (*Balistes.*) There are several species which come under this general denomination; as the Unicorn File-fish, the European File-fish, and many others. The first-named, the UNICORN FILE-FISH (*Balistes monoceros*), grows to a considerable size, often exceeding two feet in length: the body is of an oval shape, and, like most others of this genus, it possesses the power of inflating at pleasure the sides of the abdomen, by means of a pair of bony processes within that part: the skin is everywhere covered with very minute spines, and the general colour is grey, inclining to brown on the upper parts, and varied with irregular, dusky, subtransverse undulations and spots: both fins and tail are of a light brown colour, the latter marked by a few dusky bars. It is a native of the Indian and American seas, and feeds chiefly on crustaceous and testaceous marine animals. The European FILE-FISH (*Balistes capriscus*) is a species well known to the older authors as an inhabitant of the Mediterranean; and instances have occurred, though they are extremely rare, of their having been taken on our own coasts. The shape is ovate; general colour violaceous grey, sometimes variegated both on the body and fins with blue or red spots; first or small dorsal fin furnished with three or four rays, the first of which is very strong; tail rounded. The peculiar structure of the first or dorsal fin is worthy of notice: the bones or rays are so contrived to act in concert with considerable force in suddenly elevating the fin at pleasure; and how hard soever the foremost be pressed, it will not stir; but if the last be only lightly pressed, the other two immediately fall down with it; as a cross-bow is let off by pulling down the trigger. For this reason the fish is called on the Italian shores of the Mediterranean by the name of *Pesce Balestra*.

We shall describe but one more species, which is the singular species named the ACULEATED FILE-FISH (*Balistes aculeatus*). This is twelve or fourteen inches long; of a rufous brown colour, with a few purplish bands across the hinder part of the belly, and sometimes along the middle of the body: from the top of the eye to the crown four bright blue lines diverge, so as to form a blue-striped lozenge on that part; while from the bottom of the eye three or four longer lines of the same colour reach as far as the pectoral fin, the space between the lines being blackish. The skin is rough, and strongly crossed with reticular squares: on each side the end of the body three longitudinal rows of spines: tail rounded. It is a native of the Indian, American, and Red seas; varies in colour, and is sometimes of a bright golden hue.

FINCHES. A numerous group of birds, embracing not only some of the most beautiful, but also some of the most agreeable of the feathered tribe. [See FRINGILLIDÆ.]

FIRE-FLY. The name generally given to any insect which has the singular property of emitting a luminous secretion. [See ELATER: GLOW-WORM: LAMPYRIS.]

FISHING-FROG. [See ANGLER.]

FISSIROSTRES. The name of a tribe of Perching Birds, comprehending those which have a very wide gape, as the Swallow.

FISTULARIA. The name given to a genus of fishes, distinguished by elongated or tube-like noses and cylindrical bodies. [See PIPE-FISH.]

FITCHET. An animal of the weasel kind. [See POLECAT.]

FISH, or FISHES. (*Pisces.*) The name by which we designate the various species of a class of vertebrate animals inhabiting the water; which breathe through the medium of that fluid by means of branchiæ or gills, instead of lungs; which swim by means of fins; and are mostly covered with cartilaginous scales. Though the external form varies, by far the greater part possess considerable similarity of conformation — an elongated oval; a figure which enables them with greater celerity and ease than any other to traverse the aqueous element. They are also, for the most part, furnished with an air bladder in the interior of the body, (an oblong white membranous bag close to the backbone,) by the dilation or compression of which their specific gravity is said to be varied, and by which they are enabled to rise and sink in the water. In most osseous fishes, this organ extends along the back of the abdomen, between the kidneys and the chylopoietic viscera, and sometimes beneath the caudal vertebræ to near the end of the tail. It is seldom bifurcated; still more seldom divided lengthwise into two bladders: it is oftener divided crosswise into two compartments, which intercommunicate by a contracted orifice; or are quite separate. All parts of their bodies seem adapted to accelerate their motion; their fins, their tails, and the undulation of their back bones assist progression — their whole structure, in short, being as evidently adapted for swimming as that of a bird is for flight.

The fins consist of a thin elastic membrane supported by bony rays, and are denominated, according to their position, *dorsal*, *pectoral*, *ventral*, *anal*, or *caudal:* the dorsal and ventral fins apparently serve to balance the fish, and the pectoral to push the creature forward, or to arrest its progress when required; the anal fin occupies that part which lies between the anus and the tail, and this serves to keep the fish in its upright or vertical position: but the tail, which in some fishes is horizontal, and in others perpendicular, seems to be the grand instrument of motion; the fins being all subservient to it, and only giving direction to its powerful impetus. Yet the fins are important, not only as organs of motion, but as affording by their structure, position, and number, materials for distinguishing orders, families, and genera. The surface of the body is termed naked, when destitute of scales; scaly, when furnished with them; smooth, when the scales are without angles; lubicrous, when provided with a mucus; loricate, or mailed, when enclosed in a hard integument; fasciate or banded, when marked with zones from the back to the belly; tuberculate, spinous, striped, reticulate, &c.

Nature appears to have fitted this class of animals with appetites and powers of an inferior kind; and formed them for a sort of passive existence in the heavy element in which they live. To preserve their own existence, and to continue it to their posterity, fill up the whole circle of their pursuits and enjoyments; and to these they seem impelled rather by necessity than choice.

Their senses are incapable of making any nice distinctions; and they move forwards in pursuit of whatever they can swallow, conquer, or enjoy. A craving desire of food seems to give the ruling impulse to all their motions. This appetite impels them to encounter every danger; and to their rapacity no bounds appear prescribed. Even when taken out of the water, and almost expiring, they greedily swallow the very bait which lured them to destruction. Their digestive powers seem, in some measure, to increase with the quantity of food they consume; and a single pike has been known to devour a hundred roaches in three days. The amazing digestive faculties in the cold maws of fishes have justly excited the curiosity of philosophers, and have effectually overturned the system of those who maintain that the heat of the stomach is a sufficient instrument for digestion. The truth seems to be, that there is a power of animal assimilation lodged in the stomachs of all creatures, which we can neither describe nor define, converting the substances they swallow into a fluid adapted for their peculiar support. This is effected neither by trituration, by warmth, by motion, by a dissolving fluid, nor by their united efforts, but by some principle in the stomach yet unknown, which acts in a manner very different from all kinds of artificial maceration. The food taken into the stomach is often seen, though nearly digested, still to retain its original form; and, in fact, is ready for a total dissolution, while to the eye it appears yet untouched by the force of the digestive powers. But though the appetites of Fishes are insatiable, no animals can endure the want of food so long.

Professor Owen, in his 'Lectures on Comparative Anatomy,' observes, "A few species retain the primitive vermiform type, and have no distinct locomotive members; and these members, in the rest of the Piscine class, are small and simple, rarely adapted for any other function than the propulsion or guidance of the body through the water. The form of the body is, for the most part, such as mechanical principles teach to be best adapted for moving with least resistance through a liquid medium. The surface of the body is either smooth and lubricous, or is covered by closely imbricated scales, rarely defended by bony plates or roughened by hard tubercles; still more rarely armed with spines. The central axis of the nervous system presents but one partial enlargement, and that of comparatively small size, at its anterior extremity, forming the brain, which consists of a succession of simple ganglionic masses, most of them exclusively appropriated to the function of a nerve of special sense. The power of touch can be but feebly developed in fishes. The organ of taste is a very inconspicuous one, the chief function of the frame-work supporting it, or the hyoidean apparatus, relating to the mechanism of swallowing and breathing. Of the organ of hearing there is no outward sign; but the essential part, the acoustic labyrinth, is present, and the semicircular canals largely developed within the labyrinth is without cochlea, and is rarely provided with a special chamber, but is lodged, in common with the brain, in the cranial cavity. The eyes are usually large, but are seldom defended by ey lids, and never served by a lachrymal organ. The alimentary canal is commonly short and simple, with its divisions not always clearly marked, the short and wide gullet being hardly distinguishable from the stomach. The pancreas, for the most part, retains its primitive condition of separate cæcal appendages to the duodenum. The heart consists essentially of one auricle, and one ventricle, receiving the venous blood, and propelling it to the gills; whence the circulation is continued over the entire body in vessels only, which are aided by the contraction of the surrounding muscular fibres. The blood of fishes is cold; its temperature being rarely elevated above that of the surrounding medium."

"All writers on animal mechanics," observes the able Professor just quoted, and to whom we are indebted for the following detached extracts, "have shown how admirably the whole form of the fish is adapted to the element in which it lives and moves: the viscera are packed in a small compass, in a cavity brought forwards close to the head, and whilst the consequent abrogation of the neck gives the advantage of a more fixed and resisting connection of the head to the trunk, a greater proportion of the trunk behind is left free for the development and allocation of the muscular masses which are to move the tail. In the caudal, which is usually the longest, portion of the trunk, transverse processes cease to be developed, whilst the dermal and intercalary spines shoot out from the middle line above and below, and give the vertically extended, compressed form, most efficient for the lateral strokes, by the rapid alternation of which the fish is propelled forwards in the diagonal, between the direction of those forces." "You may be reminded that all the vertebræ of the trunk are distinct from one another at one stage of the quadruped's development, as in the fish throughout life; and you might suppose that the absence of that development and confluence of certain vertebræ near the tail, to form a sacrum, was a mark of inferiority in fishes. But note what a hindrance such a fettering of the movements of the caudal vertebræ would be to creatures which progress by alternate vigorous inflections of a muscular tail. A sacrum is a consolidation of a greater or less proportion of the vertebral axis of the body, for the transference of more or less of the weight of the body upon limbs organized for its support on dry land; such a modification would have been useless to the fish, and not only useless, but a hindrance and a defect.

"The pectoral fins, those curtailed prototypes of the fore-limbs of other Vertebrata, with the last segment, or hand, alone projecting freely from the trunk, and swathed in a common undivided tegumentary sheath, present a condition analogous to that of the embryo buds of the homologous members in the higher Vertebrata. But what would have been the effect if both arm and fore-arm had also extended freely from the side of the fish, and dangled as a long flexible many-jointed appendage in the water? This higher development, as it is termed, in relation to the prehensile limb of the denizen of dry land, would have been an imperfection in the structure of the creature which is to cleave the liquid element: in it, therefore, the fore limb is reduced to the smallest proportions consistent with its required functions: the brachial and antibrachial segments are abrogated, or hidden in the trunk: the hand alone projects, and can be applied, when the fish darts forwards, prone and flat, by flexion of the wrist, to the side of the trunk; or it may be extended at right angles, with its flat surfaces turned forwards and backwards, so as to check and arrest more or less suddenly, according to its degree of extension, the progress of the fish; its breadth may also be diminished or increased by approximating or divaricating the rays. In the act of flexion, the fin slightly rotates and gives an oblique stroke to the water. For these functions, however, the hand requires as much extra development in breadth, as reduction in length and thickness; and mark how this is given to the so-called embryo or rudimental fore-limb: it is gained by the addition of ten, twenty, or it may be even a hundred digital rays, beyond the number to which the fingers are restricted, in the hand of the higher classes of Vertebrata. We find, moreover, as numerous and striking modifications of the pectoral fins, in adjustment to the peculiar habits of the species in Fishes, as we do in the fore limbs in any of the higher classes. This fin may wield a formidable and special weapon of offence, as in many Siluroid fishes. But the modified hands have a more constant secondary office, that of touch, and are applied to ascertain the nature of surrounding objects, and particularly the character of the bottom of the water in which the fish may live. You may witness the tactile action of the pectoral fins when gold fish are transferred to a strange vessel: their eyes are so placed as to prevent their seeing what is below them; so they compress their air-bladder, and allow themselves to sink near the bottom, which they sweep, as it were, by rapid and delicate vibrations of the pectoral fins, apparently ascertaining that no sharp stone or stick projects upwards, which might injure them in their rapid movements round their prison." * * * "Everywhere, whatever resemblance or analogy we may perceive in the ichthyic modifications of the Vertebrate skeleton to the lower forms or the embryos of the higher classes, we shall find such analogies to be the result of special adaptations for the purpose or function for which that part of the fish is designed.

"The ventral fins or homologues of the hind legs are still more rudimental—still more embryonic, having in view the comparison with the stages of development in a land animal—than the pectoral fins; and their small proportional size reminds the homologist of the later appearance of the hind limbs, in the development of the land Vertebrate. But the hind limbs more immediately relate to the support and progression of an animal on dry land than the fore limbs: the legs are the sole terrestrial locomotive organs in Birds, whose fore-limbs are exclusively modified, as wings, for motion in another element. The legs are the sole organ of support and progression in Man, whose pectoral members or arms are liberated from that office, and made entirely subservient to the varied purposes to which an inventive faculty and an intelligent will would apply them. To what purpose, then, encumber a creature, always floating in a medium of nearly the same specific gravity as itself, with hind limbs? They could be of no use; nay, to creatures that can only attain their prey, or escape their enemy, by vigorous alternate strokes of the hind part of the trunk, the attachment there of long flexible limbs would be a grievous hindrance, a very monstrosity. So, therefore, we find the All-wise Creator has restricted the development and connections of the hind limbs of Fishes to the dimensions and to the form which, whilst suited to the limited functions they are capable of in this class, would prevent their interfering with the action of more important parts of the locomotive machinery."

"The following short account of some experiments upon fish, made for the purpose of ascertaining the use of their fins, I give (says Mr. Owen) in the words of their gifted describer, PALEY, to whom Comparative Physiology owes many beautiful accessions to its teleological applications. 'In most fish, beside the great fin—the tail, we find two pairs of fins upon the sides, two single fins upon the back, and one upon the belly, or rather between the belly and the tail. The balancing use of these organs is proved in this manner. Of the large-headed fish, if you cut off the pectoral fins, that is, the pair which lies close behind the gills, the head falls prone to the bottom; if the right pectoral fin only be cut off, the fish leans to that side; if the ventral fin on the same side be cut away, then it loses its equilibrium entirely; if the dorsal and anal fins be cut off, the fish reels to the right and left: when the fish dies, that is, when the fins cease to play, the belly turns upwards. The use of the same parts for motion is seen in the following observation upon them when put into action. The pectoral, and more particularly the ventral fins, serve to raise and depress the fish; when the fish desires to have a retrograde motion, a stroke forward with the pectoral fin effectually produces it; if the fish desire to turn either way, a single blow with the tail the opposite way sends it round at once; if the tail strike both ways, the motion produced by the double lash is progressive, and enables the fish to dart forwards with an astonishing velocity. The result is not only in some cases the most rapid, but in all cases the most gentle, pliant, easy animal motion with which we are acquainted. However, when the tail is cut off, the fish loses all motion, and it gives itself up to where the water impels it. The rest of the fins, therefore, so far as respects motion, seem to be merely subsidiary to this. In their mechanical use the anal fin may be reckoned the keel; the ventral fins outriggers; the pectoral fins the oars; and if there be any similitude between these parts of a boat and a fish, observe that it is not the resemblance of imitation, but the likeness which arises from applying similar mechanical means to the same purpose.'"

"Professor Müller concludes, from his experiments, 'that the air-bladder in fishes, in addition to other uses, serves the purpose of increasing by resonance the intensity of the sonorous undulations communicated from water to the body of the fish.' The vibrations thus communicated to the peri- and endo-lymph of the labyrinth are doubtless made to beat more strongly upon the delicate extremities of the acoustic nerve, in osseous fishes, by their effect upon the suspended otolites, also relate to the medium through which the sonorous vibrations are propagated to the fish, and to the mode in which they are transmitted to the organ; in like manner as the eye-balls are expanded in order to take in the utmost possible amount of light. The contracted encephalon harmonises with and suffices for the sensations and volitions, and the simple series of ideas daily repeated in the monotonous existence of the scaled inhabitants of the waters. To say that the fish's ears and eyes were made enormous in order to strike strongly on its dull brain—that the development of the organs of sense has been exaggerated to compensate for the defective size of their nervous centres—implies a want of due appreciation of the beautiful adjustment of the labyrinth and eyeball to the conditions under which the fish receives its impressions of the sonorous and luminous undulations."

It would be impossible, unless we devoted very considerable space to the subject, to enter into all the minutiæ respecting the anatomy, physiology, and habits of Fishes; and sufficient for the purposes of this work, it is hoped, will be found in the descriptions which are given of the various species belonging to this large class of animals. We shall therefore conclude, with a few general observations, derived from different authors. In every point of view Fishes appear inferior to terrestrial animals; in the simplicity of their conformation, in their senses, and in their enjoyments; but theirs is an uniform existence, their movements are without effort, and their lives without labour: their bodies, instead of experiencing the rigidity of age, which is the cause of natural decay in land animals, still continue increasing with fresh supplies; and as their bodies grow, the conduits of life furnish their stores in greater abundance. How long a Fish, which seems to have hardly any bounds prescribed to its growth, continues to live, is not ascertained; but we have sufficient evidence of the extraordinary age of some Fishes. Their fecundity is, however, much more extraordinary than their longevity. Some produce their young alive; others are oviparous: the former are the least prolific, and yet they produce in amazing abundance; the viviparous blenny, for instance, produces two or three hundred at a time. Those which exclude their progeny in eggs, and are obliged to leave them to chance, at the bottom of shallow water, or floating on the surface, where it is deeper, are much more prolific; the stock being in some measure proportioned to the danger there is of its consumption. Mr. Harmer, in the Philosophical Transactions, vol. 57., and recently, Mr. Jesse, have each given a Table, showing the different degrees of fecundity in several kinds of fish: they correspond in almost every instance; it is therefore fair to presume that the one is derived from the other.

Fish	Weight.		Weight of Spawn.	Number of Eggs.
	oz.	dr.	grs.	
Carp	25	5	2,571	205,109
Codfish	-	-	12,540	3,686,760
Flounders	24	4	2,200	1,357,400
Herring	5	10	480	36,960
Mackerel	18	0	1,223½	546,681
Perch	8	9	765½	28,323
Pike	56	4	5,100½	49,304
Roach	10	6½	361	81,586
Smelt	2	0	149½	38,278
Sole	14	8	542½	100,362
Tench	40	0	- -	383,252
Lobster	-	-	1,671	21,699

To which he adds, "The Salmon is far more productive than any of these; the ovarium of one female salmon will produce 20,000,000 eggs.

"That fish have the power of hearing, there can, I think, be no doubt, as I have seen them suddenly move at the report of a gun, though it was impossible for them to see the flash. They also appear to have the sense of smelling, as they will prefer paste and worms that have been prepared with particular perfumes. They have also some curiosity, which I have witnessed by putting some new object into the water, which they have assembled around, and appeared to reconnoitre: carp, especially, would come up to a new fish which was put amongst them. Roach, and other small kinds, are perfectly aware of, and are careful to avoid, those fish which prey upon them. Thus, I have seen large carp swim amongst a shoal of roach without in the least disturbing them, while, if a pike comes near them, they make off in every direction. Fish appear, also, to be capable of entertaining affection for each other. I once caught a female pike during the spawning season, and nothing could drive the male away from the spot at which the female disappeared, whom he had followed to the very edge of the water."

"It may be considered as a law," observes Mr. Yarrell, that those Fish which swim near the surface of the water have a high standard of respiration, a low degree of muscular irritability, great necessity for oxygen, die soon—almost immediately when taken out of the water, and have flesh prone to rapid decomposition: mackerel, salmon trout, and herrings are examples. On the contrary, those Fish that live near the bottom of the water have a low standard of respiration, a high degree of muscular irritability, and less necessity for oxygen; they sustain life long after they are taken out of the water, and their flesh remains good for several days."

In "The Zoologist," (p. 795, *et seq.*) there is an article of considerable interest, entitled "Notes on the Nidification of Fishes," by R. Q. Couch, Esq., from which the following passages are extracted:

"We have been accustomed to look on the inhabitants of the deep as devoid of any thing like intelligence or affection; as beings guided solely by insatiable appetites, which lead them indiscriminately to prey on each other, and to abandon their offspring to the mercy of the sea and their predatory companions, from the instant that the ova are shed. Any attempt to dispel this opinion will probably be received with distrust; for, taken as a whole, fish are certainly the most universally predaceous of any class of animals in existence; being checked only by want of power. But notwithstanding this, some, at least, have a redeeming quality, and show a remarkable care and anxiety for their young. Nests are built in which the ova are deposited, and over which the adult fish will watch till the young make their escape. And where circumstances will not allow of this continued care, as from the reflux of the sea, the old fish will return with the return of the tide, and remain as long as the water will permit.

"During the summers of 1842 and 1843, while searching for the naked molluscs of the county, I occasionally discovered portions of sea-weed, and the common coralline (*C. officinalis*), hanging from the rocks in pear-shaped masses, variously intermingled with each other. On one occasion, having observed that the mass was very curiously bound together by a slender silky-looking thread, it was torn open, and the centre was found to be occupied by a mass of transparent amber-coloured ova, each being about the tenth of an inch in diameter. Though examined on the spot with a lens, nothing could be discovered to indicate their character. They were, however, kept in a basin, and daily supplied with sea-water, and eventually proved to be the young of some fish. The nest varies a great deal in size, but rarely exceeds six inches in length, and four inches in breadth. It is pear-shaped, and composed of sea-weed, or the common coralline, as they hang suspended from the rock. They are brought together, without being detached from their places of growth, by a delicate opaque white thread. This thread is highly elastic, and very much resembles silk, both in appearance and texture: this is brought round the plants, and tightly binds them together, plant after plant, till the ova, which are deposited early, are completely hid from view. This silk-like thread is passed in all directions through and around the mass in a very complicated manner. At first the thread is semi-fluid, but by exposure it solidifies; and hence contracts and bind the substances, forming the nest so closely together, that it is able to withstand the violence of the sea, and may be thrown carelessly about without derangement. In the centre are deposited the ova, very similar to the masses of frog-spawn in ditches.

"It is not necessary to enter into the minute particulars of the development of the young, any further than by observing that they were the subject of observation, till they became excluded from the egg, and that they belonged to the fifteen-spined Stickleback (*Gasterosteus Spinachia*). Some of these nests are formed in pools, and are consequently always in water; others are frequently to be found between tide-marks, in situations where they hang dry for several hours during the day; but whether in the water, or liable to hang dry, they are always carefully watched by the adult animal. On one occasion I repeatedly visited one every day for three weeks, and invariably found it guarded. The old fish would examine it on all sides, and then retire for a short time; but soon return to renew the examination. On several occasions I laid the eggs bare, by removing a portion of the nest; but when this was discovered, great exertions were instantly made to recover them. By the mouth of the fish the edges of the opening were again drawn together, and other portions torn from their attachments, and brought over the orifice till the ova were again hid from view. And as great force was sometimes necessary to effect this, the fish would force its snout into the nest as far as the eyes, and then jerk backwards till the object was effected. While thus engaged, it would suffer itself to be taken in the hand, but repelled any attack made on the nest, and quitted not its post so long as I remained. And to those nests that were left dry between tide-marks, the guarding fish always returned with the returning tide, nor did they quit the post to any great distance till again carried away by the receding tide. * * * But fish vary a great deal in the modes of what may be called their incubation, as much as any other class of animals. Thus, some of the sharks produce their young alive, and in a state quite ready for active life; while others, with the rays, deposit eggs very similar, physiologically, to birds' eggs, which are known as mermaid's purses, being frequently to be found cast on shore on most beaches. Also, among the pipe fishes (*Syngnathi*) of our own seas, we have instances of marsupial fish, as perfect as the kangaroo is marsupial among quadrupeds. But the formation of nests and the watchful attention of fish over their young, which I have repeatedly seen, are unsuspected points of great beauty in their history, and give to them a higher degree of intelligence and interest than we have been accustomed to award. But, from their living in the almost boundless ocean, and wandering where they cannot be observed by man, their habits and economy have been but slightly studied, and they have suffered in reputation accordingly.

But those finer traits of character, which we are so much accustomed to admire in the higher animals, from their being constantly before our eyes, are not found wanting even among fish."

"Aristotle," says Baron Cuvier, in his "Lectures on the History of the Natural Sciences," "in his account of fishes, is truly admirable, giving proof of knowledge on many points superior to our own. Amongst the facts which he relates, many are still in doubt; however, from time to time, new observations teach us the justice of some of his assertions, even of those which seem the most hazardous. He says, for example, that a fish named *Phycis* makes a nest like birds. For a long time the thing was treated as a fable; however, very recently, M. Olivi discovered that a fish named the Goby (*Gobius niger*) has similar habits. The male, in the season of love, makes a hole in the sand, surrounds it with fucus, making a true nest, near which his mate waits, and he never leaves his post till the eggs which have been deposited in it are hatched." [The most extensive general work on Fishes is by Cuvier and Valenciennes, while in this country the works of Sir John Richardson, and Messrs. Yarrell and Lowe, are well worthy of study.]

FLAMINGO. (*Phœnicopterus.*) This is one of the most remarkable of all the aquatic birds for its size, beauty, and, as some say, also for the delicacy of its flesh. The body of the Flamingo is smaller than that of the Stork; but, owing to the great length of the neck and legs, it stands nearly five feet high; and measures six feet from the point of the beak to the tip of the claws. The head is small and round, and furnished with a bill nearly seven inches long, which is higher than it is wide, light and hollow, having a membrane at the base, and suddenly curved downwards from the middle. The long legs and thighs of this bird are extremely slender and delicate, as is also the neck. The plumage is not less remarkable than its figure, being of a bright scarlet. The young differ greatly from the adult, changing their plumage frequently, and which does not become fully coloured till the third year. Flamingoes inhabit the warm climates of Asia, Africa, and America: they live and migrate in large flocks, frequenting desert sea-coasts and salt marshes. They are extremely shy and watchful: while feeding, they keep together, drawn up artificially in lines, which at a distance resemble those of an army; and, like many other gregarious birds, they employ some to act as sentinels, for the security of the rest. On the approach of danger, these give warning by a loud sound, like that of a trumpet, which is the signal for the flock to take wing; and when flying they form a triangle.

Their food appears to be molluscous animals, spawn, and insects, which they fish up by means of their long neck, turning their head in such a manner as to take advantage of the crook in their beak. Their nest is of a singular construction: it is formed of mud in the shape of a hillock, with a cavity at the top, and of such a height as to admit of the bird's sitting on it, or rather standing, her long legs being placed one on each side at full length; thus situated, the female generally lays two or three white eggs somewhat larger than those of a goose. The young do not fly until they have nearly attained their full growth, though they can run very swiftly a few days after their exclusion from the shell. In some parts these birds are tamed, principally for the sake of their skins, which are covered with a very fine down, and applicable to all purposes for which those of the swan are employed. When taken young, they soon grow familiar; but they are not found to thrive in the domesticated state, as they are extremely impatient of cold, and apt to decline from the want of their natural food. They are caught by snares, or by making use of tame ones. There are two species: 1. *Phœnicopterus antiquorum;* which is of a rose colour, with red wings, the quills being black: these inhabit the warm regions of the old continent, migrating in summer to southern, and sometimes to central Europe: these beautiful birds were much esteemed by the Romans, who often used them in their grand sacrifices and sumptuous entertainments; and such of the luxurious emperors as wished to indulge in the very excess of epicurism, were wont to gratify their guests with a dish of Flamingos' tongues! 2. *Phœnicopterus ruber;* deep red; with black quills; which are peculiar to tropical America, migrating in the summer to the southern, but rarely to the middle states.

Some interesting particulars of this species are given by Mr. Gosse's correspondent, Mr. Hill, who observes that when he visited the island of Cuba he had excellent oportunities of noticing their habits — that he was much among the marshes and swamps where the floods of the river and the sea form lakelets, and successively deposit their stores of living atoms, with the rising and falling tides. "Here the Flamingos flock and feed. They arrange themselves in *what seem to be lines,* in consequence of their finding their food along the *edges* of these shallows; and though it is true that whilst their heads are down, and they are cluttering with their bills in the water, they have one of their number on the watch, standing erect, with his long neck turning round to every point, ready to sound the alarm on the apprehension of danger, — what appears to be a studied distribution of themselves back to back, as some observers describe their arrangement, is nothing but their regardlessly turning about in their places, inwardly and outwardly, at a time when all are intent on making the most of the stores which the prolific waters are yielding." Speaking of a pair of Flamingos which had been captured, and were kept on board the vessel he was in on the coast, he says, "I was struck with their attitudes, with the excellent adaptation of their two-fold character of waders and swimmers; to their habits, while standing and feeding in the sort of shoal which we made them in a large tub upon deck. We were here able to observe their natural gait and action. With a firm erectness, like a man treading a wine-press, they trod and stirred the mashed biscuits, and junked fish, with which we fed them; and plied their long lithe necks, scooping with their heads reversed, and bent inwardly towards their trampling feet. The bill being crooked, and flattened for accommodation to this reversed mode of feeding, when the head is thrust down into the mud-shoals and the sand drifts, the upper bill alone touches the ground. The structure of the tongue, of which Professor Owen has given so minute and interesting a description, is admirably adapted for a mode of feeding altogether peculiar. The spines with which the upper surface is armed, are arranged in an irregular and alternate series, and act with the notches on the edge of the upper mandible, on which they press when the bird feeds with the head reversed. In this reversed position, the weight and size of the tongue becomes a very efficient instrument for entrapping the food. The bird muddles, and clutters the bill, and dabbles about, and the tongue receives and holds as a strainer whatever the water offers of food. There is nothing of the Heron character in the Flamingo. Extraordinary length of neck and legs is common to both, but a firm erect posture is its ordinary standing attitude. The neck is never curved inward and outward, convex and concave, like a Crane's, but its movements are in long sweeping curves, which are peculiarly pleasing, when the bird is preening its plumage."

FLEA. (*Pulex irritans.*) The common Flea, a troublesome insect of the order *Aphaniptera*, is well known in every quarter of the globe for its agility, its caution, and its invincible pertinacity in feasting on the blood of man and various animals. Like the major part of the Insect race of other tribes, the Flea is produced from an egg, in the form of a minute worm or larva, which changes to a chrysalis, in order to give birth to the perfect animal. The female drops her eggs, at distant intervals, in any favourable situation: they are very small, of an oval shape, of a white colour, and a polished surface. From these, in the space of six days, are hatched the larvæ, which are destitute of feet, of a lengthened, worm-like shape, beset with distant hairs; the head furnished with a pair of short antennæ, and the tail with a pair of slightly curved forks or holders; their colour is white, with a reddish cast, and their motions quick and tortuous. In the course of ten or twelve days they attain their full growth, and are then nearly a quarter of an inch long: at this period they cease to feed, and, casting their skin, change to an oval-shaped chrysalis, exhibiting the immature limbs of the included insect, which in twelve days emerges in its perfect form: in winter, however, the time required for this evolution is considerably more. It now begins to exert its lively motions, and employs its sharp proboscis in obtaining nourishment from the juices of the first bird or quadruped to which it can gain access. Nothing can exceed the polished elegance of the shelly armour with which the Flea is covered, or the elasticity of its surprising leaps. When examined with a microscope it will be observed to have a small head, large eyes, and a roundish body: it has two short hairy antennæ, composed of five joints; and at a small distance beneath these is the proboscis, which is strong, sharp-pointed, tubular, and placed between a pair of jointed guards or sheaths. Its suit of sable armour appears to be neatly jointed, and beset with a multitude of sharp spines. Its legs are six in number; the joints of which are so adapted, that it can fold them up one within another, and in leaping they all spring out with prodigious force. [See CHEGOE.]

FLITTERMOUSE. [See BAT.]

FLOUNDER. (*Pleuronectes flesus.*) A well-known flat-fish, very similar to the Plaice, but generally smaller and of more obscure colours; the upper side being of a dull brown, and the under of a dull white: the body is covered with very small scales, and along the back runs a row of small sharp spines: the tail is slightly rounded. The Flounder is an inhabitant of the Northern, Baltic, and Mediterranean seas; it is also very common about our own coasts: and it even frequents our rivers at a great distance from the salt water. Though inferior to some others of the genus, its flesh is in considerable esteem.

FLOUNDER.—PLEURONECTES FLESUS.

The ARGUS FLOUNDER (*Pleuronectes Argus*) is a very elegant species, native of the American seas, and of the same general form with the Turbot. It is of a yellowish white colour on the upper side, marked by numerous eye-shaped spots, consisting of bright blue circles with yellow centres: the whole skin is also marked both on the body and fins with small blue and brown specks, and is covered with small scales: the under side is of a whitish or pale gray colour: the lateral line is arched over the pectoral fins, and is thence continued straight to the tail, which is rounded at the tip.

FLUKE. [See FASCIOLA, p. 228.]

FLUSTRA. A genus of *Bryozoa*, or ciliobranchiated Molluscs, having the general aspect of Zoophytes, but lately regarded as belonging to a much higher group. The name is derived from the Saxon *Flustrian*, to weave: hence they are familiarly termed *sea-mats*. They consist of calcareous branches, sometimes forming leaves or stems, with numerous cells, united in clusters like a honeycomb. The aperture of the cells is formed by a semicircular lid, convex externally and concave internally, which folds down when the polypus is about to advance from the cell; and, it is said, the lid of the cells opens and shuts without the slightest perceptible synchronous motion of the polypi. Some species have cells on one side of the leaves only. In the most plant-like of them there is no substance in the least

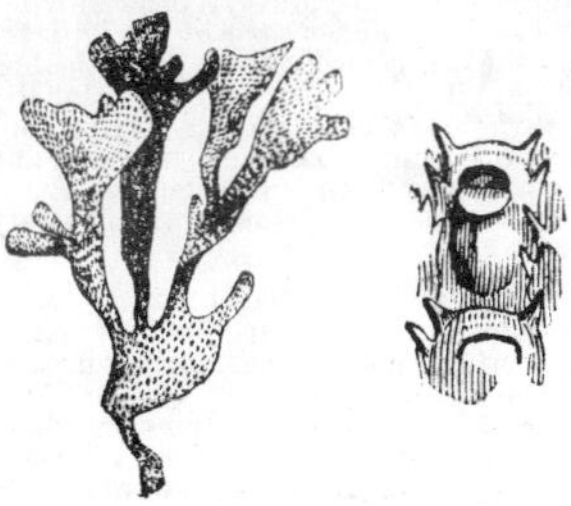

LEAF-LIKE SEA-MAT.—(FLUSTRA FOLIACEA.)

resembling that of the plants with which they agree most in form, nor is there any substance similar to theirs in the most analogous of the true vegetables; they are often, however, called "white sea-weeds." In Dr. George Johnstone's admirable "History of British Zoophytes," we read as follows: "When recent it exhales a pleasant scent, which Pallas compares to that of the orange. Dr. Grant to that of violets, and which a friend tells me smells to him like a mixture of the odour of roses and geranium. On the contrary, Mr. Patterson tells me that the smell is strong, peculiar, and disagreeable. It probably varies, and is often not to be perceived at all."

From the same authentic source we derive the following information respecting another species, *Flustra membranacea;* the cells of which are oblong, with a short blunt spine at each corner. It is thus described:—"Polypidom forming a gauze-like incrustation on the frond of the sea-weed, spreading irregularly to the extent of several square inches, in general thin and closely adherent, but sometimes becoming thickish, and then capable of being detached in considerable portions; cells very obvious to the naked eye, oblong, quadrangular, with a blunt hollow spine at each angle. In many specimens there are some anomalous processes, a quarter of an inch in height, scattered over the surface: they arise from within the cells, are simple, horny, and tubular, but closed at top. When the polypes are all protruded, they form a beautiful object under the microscope, from their numbers, their delicacy, the regularity of their disposition, and the vivacity of their motions, now expanding their tentacula into a beautiful campanulate figure, now contracting the circle, and ever and anon retreating within the shelter of their cells. The tentacula are numerous, filiform, white, and in a single series. The Rev. David Landsborough has seen a specimen (and I have seen its equal, Dr. J. remarks) of *F. membranacea* five feet in length by eight inches in breadth. "As every little cell had been inhabited by a living polype, by counting the cells on a square inch, I calculated that this web of silvery lace had been the work and the habitatior of above two millions of industrious and, we doubt not, happy inmates; so that this single colony on a submarine island was about equal in number to the population of Scotland." [See BRYOZOA in SUPPLEMENT.]

FLY. A name of very general application to insects furnished with wings; but properly restricted to the numerous genus *Musca.* The strong resemblance which exists among all the species of the Fly tribe, together with their small size, makes it difficult to discriminate them readily; but the general and most obvious character of Flies, by which they are distinguished from other winged insects, is in their having transparent and naked wings, totally free from the farina or dust visible on those of butterflies, and in having no cases or covers for them. Thus, by this simple character, they are clearly distinguished from the butterfly, the beetle, the grasshopper, &c. The principal parts or members of which Flies are composed are the head, the thorax, the body, and the wings; from the number of the latter the most obvious distinction for a systematic arrangement of them is drawn. [See MUSCIDÆ.]

FLYCATCHERS. (*Muscicapidæ.*) This very numerous family, which receives its popular name from the expertness of the individuals composing it in catching the flying insects upon which they feed, is found widely diffused throughout both the eastern and western continents; and includes many of the most beautiful of the feathered tribes. The general habits of the Flycatchers are those of the shrikes, and, according to their size, they prey on small birds or insects. They have the beak horizontally depressed and armed with bristles at its base, with the point more or less decurved and emarginated. Those which are called "Tyrant Flycatchers" (*Tyrannus*) are American birds, of a large size and very spirited; they have a long, straight, and very stout bill; the ridge of the upper mandible straight and blunt, its point abruptly hooked: while the species which inhabit Europe, and come under the denomination of "Restricted Flycatchers" (*Muscicapa*), have shorter bristles at the gape, and the bill much more slender, though still depressed, with an acute edge above, and the point a little curved downward. There are, however, only two small species which inhabit this country.

The GREY OR SPOTTED FLYCATCHER. (*Muscicapa grisola.*) This bird is nearly five inches and three quarters in length; bill broad, flattened, and wide at the base, where it is beset with a few short bristles; a ridge runs along the upper mandible; both that and the under one are dusky at the tips, and the latter is yellowish towards the base: all the upper plumage is of a mouse colour, darkest on the wings and tail; head and neck more or less obscurely spotted with dark brown; the wing coverts, secondary quills, and scapulars, also dark brown, edged with dingy white; under parts very pale ash, tinged with rufous on the sides and breast, the latter being marked with

SPOTTED FLYCATCHER.
(MUSCICAPA GRISOLA.)

streaks of brown: the legs are short, and darkish. Of all our summer birds the Flycatcher is the most mute. It visits this island in the spring, and disappears in September. The female builds her nest commonly in gardens, on any projecting stone in a wall, or on the end of a beam, screened by the leaves of a vine, sweet-briar, or woodbine, and sometimes close to the post of a door, where people are going in and out all day long. The nest is rather carelessly made of moss and dried grass, mixed in the inside with some wool and a few hairs. She lays four or five eggs, of a dull white, closely spotted and blotched with rusty red. This bird feeds on insects, for which it sits watching on a branch or a post, suddenly dropping down upon them, and catching them on the wing, and immediately rising, returns again to its station to wait for more. After the young have quitted the nest, the parent birds follow them from tree to tree, and watch them with the most sedulous attention. They feed them with the flies which flutter among the boughs beneath; or, pursuing their insect prey with a quick irregular kind of flight, like that of a butterfly, to a greater distance, they immediately return as before described.

Mr. Knapp says, "We have perhaps no bird more attached to peculiar situations than the Grey Flycatcher (*Muscicapa grisola*); one pair, or their descendants, frequenting year after year the same hole in the wall, or the same branch on the vine or the plum. I once knew a pair of these birds bring off two broods in one season from the same nest. This Flycatcher delights in eminences. The naked spray of a tree, or projecting stone in a building, or even a tall stick in the very middle of the grass-plot, is sure to attract its attention, as affording an uninterrupted view of its winged prey; and from this it will be in constant activity a whole summer's day, capturing its food, and returning to swallow it."

The PIED FLYCATCHER. (*Muscicapa luctuosa.*) This species is found in Sweden, Russia, and sometimes in this country; its distribution here, however, being almost confined to "the lakes" in the north of England. The beak is black; the forehead white; crown of the head, and all the upper parts, black; the lesser wing-coverts and the greater coverts of the primaries are dusky; the first six quills are wholly dusky, the rest white at the base: the under parts of the bird are white; the tail is dusky black; and the legs are black. There is, however, occasionally great variety in their markings. It frequents wild and uncultivated tracts of furze, and open heaths; and constructs its nest in the hole of a tree. The female lays five very pale blue eggs.

The RED-EYED FLYCATCHER. (*Muscicapa olivacea.*) This species is a native of the southern provinces of North America, and is also found in many of the West India islands, particularly Jamaica, where it is called *Whip-Tom-Kelly*, from a fancied resemblance of its note to those words. The head, neck, and back are olive brown; the wing-coverts and quills are edged with green, as is also the tail; the feathers dull brown above and greyish beneath: from the beak passing over the eyes and terminating on the hind head is a reddish white line: the under parts of the body are pale white, irregularly spotted with pale yellow; beak and feet brown.

Mr. Gosse, in describing this species in his "Birds of Jamaica," says he can scarcely understand how the call can be written *Whip-Tom-Kelly*, as the accent is most energetically on the last syllable. The familiar name which he gives to it is *John-to-whit;* and says that sounds closely resembling those words are uttered by this bird with incessant iteration and untiring energy from every grove, nay almost from every tree. Its food, he observes, is both animal and vegetable; for in its stomach he has found seeds of the Tropic birch, and the berries of sweet-wood, and has also observed it jumping out from its umbrageous retreat after stationary, as well as vagrant, prey, "Incubation takes place in June and July. The nest is rather a neat structure, though made of coarse materials. It is a deep cup, about as large as an ordinary tea-cup, narrowed at the mouth; composed of dried grass, intermixed with silk-cotton, and, sparingly, with lichen and spiders' nests, and lined with thatch-threads. It is usually suspended between two twigs, or in the fork of one, the margin being overwoven, so as to embrace the twigs. This is very neatly performed. Specimens vary much in beauty. The eggs, commonly three in number, are delicately white, with a few small red-brown spots thinly scattered over the surface, sometimes very minute and few."

The CAYENNE FLYCATCHER. (*Tityra Cayanensis.*) A beautiful species, above seven inches in length, which inhabits Cayenne and St. Domingo. The crown of the head is a brown yellow; and from the beak, which is dusky, to the hind part of the head, is a white streak: all the upper parts of the body are brown, the feathers lighter on their margins; the wing-coverts and the upper ones of the tail are brown, their edges rufous: the chin is white, and the rest of the under parts bright yellow: quills and tail brown.

PARADISE FLYCATCHER. (*Muscipeta Paradisi.*) A singular bird, measuring upwards of twenty inches long, owing to its disproportioned tail, which is generally about fourteen inches. Its head, hind part of the neck, and throat, are greenish black; the feathers on the former are very long, and form a crest: the back, rump, wing-coverts, and tail-feathers are white; the greater coverts and quills black, fringed with white; the fore part of the neck, and all the under parts of the body, pure white; tail cuneiform; legs ash-coloured. This bird is found in the southern parts of Africa, frequenting the borders of rivers, where its insect food is most abundant.

SWALLOW-TAILED FLYCATCHER. (*Muscivora forficata.*) This bird, whose distinctive appellation is derived from its forked tail, is ten inches in length, of which the tail forms one half. The colour of the beak is black; the head and back are light grey, slightly tinged with red; the under parts of the body white; beneath the wing red; the wing-coverts ash-colour; and the quills black, edged with gray. It inhabits Mexico. [See TYRANNUS: RHIPIDURA: ONYCHORHYNCUS.]

FLYING-FISH. (*Exocetus.*) By the extraordinary length and size of their pectoral fins, the fishes of this genus are enabled to spring occasionally from the water, and to support a kind of temporary flight through the air: hence their name. It is evident, however, that their "flights" are chiefly performed for the purpose of escaping from the jaws of the dolphin, and other predaceous fishes, which are constantly pursuing them. Mr. Gosse, in his admirably graphic "Naturalist's Sojourn in Jamaica," has shown that they can direct and change their flight. The following account cannot, therefore, be entirely depended on: "I have never," observes Mr. G. Bennett, the author of 'Wanderings in New South Wales,' "been able to see any

FLYING-FISH.—(EXOCETUS VOLITANS.)

percussion of the pectoral fins during flight, and the greatest length of time I have seen this volatile fish on the fin has been thirty seconds by the watch, and their longest flight mentioned by Captain Hall has been 200 yards; but he thinks that subsequent observation has extended the space. The most usual height of flight, as seen above the surface of the water, is from two to three feet; but I have known them come on board at a height of fourteen feet and upwards; and they have been well ascertained to come into channels of a line-of-battle ship, which is considered as high as twenty feet and upwards. But it must not be supposed they have the power of elevating themselves in the air after having left their native element; for, on watching them, I have often seen them fall much below the elevation at which they first rose from the water, but never in any one instance could I observe them rise from the height at which they first sprang; for I regard the elevation they first take to depend on the power of the first spring or leap they make on leaving their native element."

In tropical seas the Flying-fish rise from water in flocks, or, more properly, shoals, of hundreds at a time, when disturbed by the passing of a ship, or pursued by their inveterate foe, the dolphin. They spring from the crest of a wave, and, darting forward, plunge into another, to wet the membrane of the fins, and in this manner continue their flights for several hundred yards, often pursued by marine birds in the element to which they are driven for protection against the tyrants of their own.—Gardner, in his 'Travels in Brazil,' confirms Humboldt's assertion, (denied by Cuvier,) that the Flying-fish uses its pectoral fins as wings during the time it remains above water.

The distinguishing characters of the genus are—pectoral fins nearly equal to the body in length; head flattened above and on the sides; the lower part of the body furnished with a longitudinal series of carinated scales on each side; dorsal fin placed above the anal: eyes large; jaws furnished with small pointed teeth. There are but very few of the genus.

The MEDITERRANEAN FLYING-FISH (*Exocetus exiliens*) runs from ten to fifteen inches in length, its general shape resembling that of a herring: the head is rather large, and sloping pretty suddenly in front; the eyes large, and of a silver colour, with a cast of gold; the scales are large, thin, and rounded; and the whole fish is of a bright silvery cast, with a blue or dusky tinge on the upper part. The pectoral fins are of a sharply lanceolate form, and extend as far as the beginning of the tail; the dorsal and anal fins are shallow, and placed opposite each other near the tail, which is deeply forked with sharp-pointed lobes, the lower being nearly twice the length of the upper; the ventral fins, which are rather large and long, are situated behind the middle of the body.

OCEANIC FLYING-FISH. (*Exocetus volitans.*) This species is somewhat more slender, and the head less sloping than the preceding, though from its general resemblance it might be easily mistaken for it; but the principal difference arises from the ventral fins being seated near the pectoral ones, and from their being much smaller and of a slightly lunated form. This species is of a bright silver colour, gradually deepening into purplish brown on the back, the dorsal and anal yellowish, and the ventral fins and tail reddish. It is a native of the Indian and American seas; but it is also sometimes found in the Mediterranean, and some solitary instances occur of its having been seen about our own coasts. In the Gulf of Mexico are found some species with curious appendages or filaments attached to the lower jaw. The air-bladder in this, and doubtless in the rest of the genus, is very large.

FLYING SQUIRREL. (*Pteromys.*) A genus of rodent mammalia, distinguished from the common Squirrels by the extension of the skin of the flank between the fore and hind legs, which gives them the power of supporting themselves a short time in the air, and of making immense leaps. The feet have long bony appendages, which help to support this lateral membrane. To this genus belongs the common Flying Squirrel (*Pteromys volans*), which is chiefly found in the most northern regions, and abounds in the birch and pine woods of Siberia in particular. Its colour on the upper parts is a pale grey, and on the under parts milk-white. It measures about six inches and a quarter in length, from the nose to the tail, the latter being shorter than the body, thickly furred, of a slightly flattened form, and rounded at the extremity. Its manner of flight, or rather springing, is performed by means of an expansile furry membrane, reaching from the fore feet to the hind; and in order the better to manage this part, the thumb of the fore feet is stretched out to a considerable length within the membrane, so as to appear in the skeleton like a long bony process on each side of the fore feet. The Flying Squirrel generally resides in the hollows of trees towards the upper part; preparing its nest of the finer mosses. It is a solitary animal, and is only seen in pairs during the breeding season. It rarely makes its appearance by day, emerging only at the commencement of twilight, when it may be seen climbing about the trees, and darting with great velocity from one to the other. It feeds on the young buds and catkins of the birch and pine, &c.; and in the winter it leaves its nest only in mild weather, but does not become torpid during that season. This animal readily springs, or, as it were, swiftly sails, to the distance of twenty fathoms or more, and thus passes from one tree to another, always directing its flight obliquely downwards. It very rarely descends to the surface, and, when taken and placed on the ground, runs or springs somewhat awkwardly, with its tail elevated, beginning to climb with great activity as soon as it reaches a tree. If thrown from a height, it immediately spreads its membranes, and, balancing itself, endeavours to direct its motion by the assistance of the tail. The young are produced early in May, and are from two to four in number: they are at first blind, and nearly void of hair; and the parent fosters them by covering them with her flying-membrane. In their manner of sitting and feeding, as well as in the action of washing their face with their paws, &c., the Flying Squirrel resembles the common species.

The VIRGINIAN FLYING SQUIRREL. (*Pteromys volucella.*) This species differs from the preceding both in size and colour. Its general length is five inches to the tail, which measures about four inches; the colour being a subferruginous brown above, and yellowish

VIRGINIAN FLYING SQUIRREL.
(PTEROMYS VOLUCELLA.)

white beneath; and the edges of the flying-membrane are of a darker tinge than the rest of the fur, contrasting with the white border of the under part. The tail is of a similar colour to the body, with the hair spreading towards each side, and the extremity somewhat sharpened. The eyes are large, and the ears rather short, almost naked, and slightly rounded. It is a native of the temperate parts of North America; and, being a beautiful little animal and readily tamed, is frequently kept in a state of captivity: it feeds on various fruits, nuts, almonds, &c., and shows a considerable degree of attachment to its possessor. It is naturally of a gregarious disposition, and may be seen flying, to the number of ten or twelve together, from tree to tree. Like the former species, it is chiefly nocturnal in its habits: it prepares its nest in the hollows of trees, with moss, leaves, &c.; several often inhabiting the same retreat. They are capable of swimming, in case of necessity, in the manner of other quadrupeds, and, after leaving the water, can exert their power of flight as before.

FOOTMAN [MOTHS]. A name given by collectors to different species of Moths, of the genera *Eulepia* and *Lithosia*.

FORAMINIFERA. A term equivalent to the *Rhizopoda* of some naturalists, and signifying a class of minute animalcules provided with many-chambered cells. [See RHIZOPODA in SUPPLEMENT.]

FORFICULA. [See EARWIG.]

FORMICA. [See ANT.]

FORMICIDÆ. A family of Hymenopterous insects, composed of the well-known and highly interesting tribes of *Ants;* but not including the still more singular *Termitidæ*, or White Ants (with which they must not be confounded). Both are full of interest, and worthy of the most careful investigation; and to each we have accordingly devoted no inconsiderable space. [See ANT: DRIVER-ANT.]

FOSSANE. (*Viverra fossa.*) An animal of the Weasel tribe, nearly allied to the Genet, which it greatly resembles: its colours, however, are somewhat bolder, and its rows of spots along the sides more regularly disposed: the under part of the body is of a dingy white; and the tail is annulated with black and white. This animal is said to be possessed of considerable fierceness, destroying poultry, &c., in the manner of the common weasel. It is a native of Madagascar, Guinea, Cochin-China, &c.

FOSSORES. An extensive group of Hymenopterous insects, forming a subsection of the *Aculeata*. They are solitary in their habits; and most of the species are organized for excavating cells in earth or wood, in which they bury other insects in a wounded and feeble state, and at the same time deposit their eggs; so that the larvæ, when hatched, find a store of food prepared for their sustenance. The basal joint of the posterior tarsi not being enlarged, the legs are not fitted for carrying pollen, neither is the body clothed with hairs, requisite for its transport. Some species, the structure of whose legs is not adapted for burrowing, are parasitic, and, like the cuckoo among birds, lay their eggs in the nests of other species, at whose expense the young are reared. When full grown, these larvæ spin a cocoon, in which they pass the pupa state. The perfect insects are generally very active, and fond of the nectar of flowers, especially those of the Umbelliferæ. The work of Mr. W. E. Shuckard on the British Fossorial Hymenoptera is very highly esteemed by Entomologists, and we recommend it to those desirous of studying the British species, often endowed with such wonderfully interesting habits.

FOWL. This term, when taken in a general sense, is of similar import with Birds; but, in a limited view, it more peculiarly signifies the larger kind of birds, both wild and domestic, which are either reared or pursued for the purposes of food. In this sense, Fowl includes all the denizens of the poultry yard, with pheasants, partridges, and all other kinds of winged game.

FOX. (*Canis vulpes.*) Of all beasts of prey, the Fox is considered to be the most crafty and sagacious, whether in obtaining food or in eluding pursuit. They appear to be pretty generally diffused throughout all the northern and temperate parts of the

FOX.—(CANIS VULPES.)

globe; occurring with numerous varieties, as to colour and size, in most parts of Europe, the north of Asia, and America. The Fox has a broad head, a sharp snout, a flat forehead, obliquely seated eyes, sharp erect ears, an elongated body well covered with hair, proportionally short limbs, and a straight bushy tail, so long that when pendent it touches the ground. The general colour is a yellow-brown; and on the forehead, shoulders, hind part of the back, as far as the beginning of the tail, and outside of the hind legs, it is a little mixed with white or ash-colour: the lips, cheek, and throat are white, and a white stripe runs along the under side of the legs; the tips of the ears and the feet are black: the tail a reddish-yellow, mixed with a blackish tinge, and internally brownish yellow-white, with a blackish cast; the tip milk white.

The Fox varies considerably in size, but in general measures about three feet six inches from the snout to the end of the tail, of which the latter is sixteen inches; and the height at the shoulders is about fourteen inches. "The general expression of its features," as Mr. Bell remarks, "the obliquity and quickness of the eye, the sharp shrewd-looking muzzle, and the erect ears, afford the most unequivocal indications of that mingled acuteness and fraud which have long rendered it a by-word and a proverb; for it is well-known that this character of its physiognomy is not falsified by the animal's real propensities and habits."

The Fox prepares for himself a convenient den in which he lies concealed during the greater part of the day: this he sometimes obtains by dispossessing the badger of his hole; at other times by forming his own burrow; but it is always so contrived as to afford the best security to the occupant, by being situated under hard ground, the roots of trees, &c., and is furnished with proper outlets through which he may escape when hard pressed by his hunters. Prudent, patient, and vigilant, he waits the opportunity of depredation, and varies his conduct on every occasion. His domicile is generally at the edge of a wood, and yet within a convenient distance of some farm-house: from thence he listens to the crowing of the cock, and the cackling of the domestic fowls; then, concealing his approaches, he creeps stealthily along, attacks his prey, and seldom returns without his booty. Poultry, pheasants, partridges, small birds, leverets, and rabbits are his favourite objects: but he is also fond of certain berries and fruits, and can occasionally make a meal of field-mice, frogs, newts, &c. The Fox seems to be wholly devoid of that instinct of gratitude which characterizes the Dog, and is even found in the Wolf and Jackal; nay, whatever kindness may be shown him when in a state of confinement, he is still sly, timid, and suspicious; insusceptible, as it would seem, of any kind of attachment. His voice is a kind of yelp, or stifled bark, and his bite is very severe and dangerous.

There is no animal that affords more diversion to the huntsman, or that gives him more occupation, than the Fox. When he finds himself pursued, he usually makes for his hole, and, penetrating to the bottom, lies quiet till a terrier is sent in to him; but should his den be under a rock or the roots of trees, he is safe, for the terrier is no match for him there, and he cannot be dug out. When, as is generally practised, the retreat to his den is cut off, his stratagems and shifts to escape are various. He always seeks the most woody parts of the country, and prefers such paths as are most embarrassed by thorns and briars: he runs in a direct line before the hounds, and at no great distance from them; and when overtaken, he defends himself with desperate and silent obstinacy. The fetid odour of the Fox is intolerable: his sight is keen; and he possesses astonishing acuteness of smell. The time of gestation is about sixty-three days; and while the female is suckling her young, nothing can exceed her courage and boldness. The Fox, unmolested, will live twelve or fourteen years. In the first year he is called a *cub;* the second, a *Fox;* and the third, an *old Fox:* he is eighteen months, or nearly two years old, before he arrives at full maturity. The skin makes a warm and soft fur, and is therefore used for muffs, linings, &c.

ARCTIC FOX. (*Canis lagopus.*) This species is smaller than the common Fox, with a sharp nose, and short rounded ears, almost hid in its fur; the legs are short, and the toes are covered both above and below with a very thick soft fur; the tail is shorter than that of the common Fox, but more bushy. It inhabits the countries bordering on the Frozen Ocean in both continents. At the approach of winter their coat of hair becomes thick and ragged; till at length it grows perfectly white, changing colour last on the ridge of the back and tip of the tail.

ARCTIC FOX.—(CANIS [VULPES] LAGOPUS.)

This species preys upon various small quadrupeds, such as hares, marmots, &c., as well as upon all kinds of water-fowl and their eggs; also, when necessity urges, on the carcasses of fish left on shore shell-fish, or whatsoever the sea throws up. Mr. Pennant says, that in Spitzbergen and Greenland, where the ground is eternally frozen, they live in the clefts of rocks, two or three inhabiting the same hole. They swim well, and often cross from island to island in search of prey. They are tame and inoffensive animals; and are killed for the sake of their skins, both in Asia and Hudson's Bay: but though the fur is light and warm, it is not durable. The Greenlanders take them either in pitfalls dug in the snow, and baited with fish; or in springs made with whalebone laid over a hole made in the snow, strewed over at bottom with fish; or in traps similarly baited. The arctic travellers and voyagers, Dr. Sir John Richardson, Captains Parry, Franklin, Ross, Lyon, Back, and Simpson, refer much in their narratives to this inhabitant of snow-covered countries; and those familiar with their writings cannot but sympathize with their *regard* for the limited number of animals and plants which they met with in these dreary wastes. One of the most active, and certainly one of the prettiest, was the White Arctic Fox described above.

ANTARCTIC FOX. (*Canis Antarcticus.*) This species is found in the Falkland Isles, near the extremity of South America, and is about one-third larger than the Arctic Fox; has much the appearance of the wolf in its ears, tail, and the strength of its limbs; whence the French call it *Loup-renard*, or the Wolf-fox. The head and body are of a cinereous brown hue, the hair being more woolly than that of the common Fox; the legs are dashed with rust-colour; the tail dusky, more bushy, and shorter than that of the common Fox, and tipped with white. It resides near the shores, kennels like the rest of its kind, and forms regular paths from one bay to another, probably for the convenience of surprising water-fowl, on which it principally subsists. It is a tame, fetid animal, and barks like a dog.

BLACK OR SILVERY FOX. (*Canis argentatus.*) This species inhabits the northern parts of Asia, Europe, and America, and is only distinguishable from the common Fox

SILVERY FOX.—(CANIS ARGENTATUS.)

by its copious and beautiful fur, which, particularly in the Asiatic one, is of a rich and shining black or deep brown colour, with the longer or exterior hairs of a silvery white, giving a highly elegant appearance to the animal, and rendering its fur more valuable than that of almost any other quadruped.

RED FOX. (*Canis fulvus.*) This species is found throughout North America; its general colour is bright ferruginous on the head, back, and sides: beneath the chin it is white, whilst the throat and neck are of a dark grey: the under parts of the body towards the tail are a very pale red. The skins are much sought for, and employed in various manufactures.

CROSSED FOX. (*Canis decussatus.*) The colour of this animal's fur is a sort of gray, resulting from the mixture of black and white hair. He has a black cross on his shoulders, from which he derives his name. The muzzle, lower parts of the body, and the feet are black; the end of the tail is white. It inhabits the northern parts of America; and it has been suggested as probable that it is only a variety of the Black Fox.

CORSAC FOX. (*Canis Corsac.*) This animal, which inhabits the vast plains of Tartary, is, in summer, of a clear yellow-ferruginous colour: in winter, mixed or shaded with grey, deeper on the back, white on the belly, and reddish on the feet: the eyes are surrounded with a border of white; and a brownish stripe runs from them down the nose: the ears are short; the tail almost as long as the body, both the base and tip being blackish. It commits great ravages among the game; is hunted with falcons and dogs; and it is said that not less than forty or fifty thousand are annually taken, sold to the Russians, and a vast number of them sent into Turkey.

SWIFT FOX. (*Canis velox.*) This beautiful little animal, which is much smaller than any other species, is distinguished by its extraordinary speed, which, it is asserted, surpasses the fleetest antelope, and seems rather to fly than touch the ground in its course. Its body is slender, and the tail rather long, cylindrical, and black: the hair is fine, dense, and soft.

FOX-HOUND. Among those manly and exhilarating field sports for which "Old England" has so long been famous, fox-hunting justly claims pre-eminence; and in the annals of the chase numerous instances of speed, courage, and perseverance are to be found which may well be ranked among the marvellous. With this part of the subject, however, we have no legitimate business, but merely allude to it, in order to account for the extraordinary care and attention which, for centuries, have been bestowed on this peculiar breed of dogs—a breed in which are combined, in the highest possible degree of excellence, fleetness, strength, spirit, fine scent, perseverance, and subordination. The Fox-hound is much smaller than the Stag-hound, his average height being from twenty to twenty-two inches; but in all the requisites for hunting he is unrivalled. To be perfect, we are told, "his legs should be straight as arrows; his feet round and not too large; his shoulders black; his breast rather wide than narrow; his chest deep; his back broad; his head small; his neck thin; his tail thick and bushy, and well carried."

FRANCOLIN. The birds which are thus designated bear so great a resemblance to the Partridge, that many naturalists include them in the genus *Perdix;* but there are others who say that the Francolins are distinguished from the Partridges by the beak being longer and stronger; the tail is also longer, &c. In the manners of the birds also there is a great dissimilarity, the Francolins residing in damp places and perching upon trees, whereas Partridges always rest upon the ground.

The COMMON FRANCOLIN (*Francolinus vulgaris*) is upwards of twelve inches in length: the upper parts of the head, hind part of the neck, back, and wing-coverts are varied with dusky and yellowish rust-colour; the sides of the head, neck, breast, and belly are black; round the neck is a rusty orange collar; the sides of the neck, breast, and body are black, varied with spots of white; the lower part of the belly and thighs striped with black; the lower part of the back and rump crossed with alternate lines of black and yellowish white: the quills dusky, marked with transverse rusty yellow spots: tail rounded, the four middle feathers alternately striped with black and rusty yellow; the others on each side, with black and white for two-thirds of their length; the rest black to the tip: legs reddish, and furnished with a spur. This elegant species is found throughout all the warmer parts of Europe; other allied species are met with in Bengal, and are abundant in Barbary and other parts of Africa. It feeds upon insects and seeds: it has a very loud whistle; and its flesh is greatly esteemed.

The PONDICHERRY FRANCOLIN (*Francolinus Pondicerianus*) is a beautiful species. Its length, including the tail, is fourteen inches: the beak is red at the base and yellow at the tip: the top of the head is gray-brown; the forehead bright red, that colour passing over the eyes like an eyebrow, and ending on the back of the head: the breast is alternately striped with whitish-yellow and bright brown: the back, the greater and lesser wing-coverts, and the rump, gray-brown; the edges of the feathers with black spots, and all of them with three reddish-white stripes: the quills and secondaries are gray, the outer webs striped with yellowish white: the two middle tail-feathers are gray, spotted, and crossed with four yellow-white bands; the belly and abdomen are white, striped with semicircular bands: the legs are red, and armed with a strong spur. It is met with in parts of India, where it frequents gardens and cultivated lands, and is called a partridge.

The PEARLED FRANCOLIN. (*Francolinus perlatus.*) This species is common in China, and is likewise known at Bengal, the Mauritius, and Madagascar. Like the rest of the Francolins, it is a forest bird, and perches upon trees. The male of this beautiful species varies from ten to twelve inches in length: the feathers on the top of the head are black, edged with red; two longitudinal black stripes commence from the beak, and surround the eyes, leaving the space between pure white, of which colour the throat is also: the feathers on the hinder part of the neck are black, marked with four longitudinal white spots; those on the top of the back, the fore part of the neck, the breast, and the lesser wing-coverts, are black, each varied with six rounded white spots: the scapulars are of a reddish chestnut, with whitish spots at their tips: the back, the rump, the upper wing-coverts, and those of the tail at their base, are black, with innumerable white bands; the tip of the tail-feathers is black; the belly is whitish, the sides rather red, both varied with black lines: the under tail-coverts are red; the beak is black, and the feet are bright red: the tarsi are armed with a thick and blunt spur. The female is rather smaller, and differs in several respects from the male.

FRATERCULA. A genus of web-footed birds belonging to the family *Alcadæ*, and containing the common Puffin (*F. arctica.*) [See PUFFIN: AUK.]

FRIAR-BIRD. (*Tropidorhynchus corniculatus.*) This bird is generally dispersed over New South Wales, where it is variously called by the colonists Friar-bird, Monk, and Poor Soldier. It selects the topmost dead branches of the most lofty trees whereon

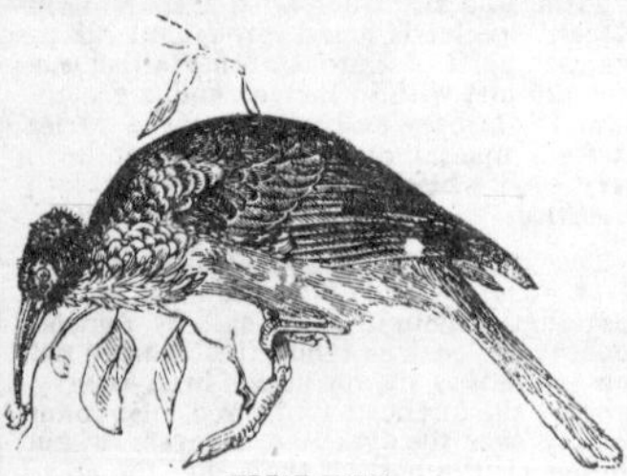

FRIAR-BIRD.
(TROPIDORHYNCHUS CORNICULATUS.)

to perch and pour forth its garrulous and singular notes, and attracts attention more by its loud and singular call than by its appearance. It is called, from some of these notes, Poor Soldier, Pimlico, Four o'clock, &c.: its bare head and neck give it also the appellation of Friar-bird, Monk, and Leather-head. Its flight is undulating and powerful, and it may be seen passing from one part of the forest to another: when among the branches it can cling in every direction; sometimes it hangs by one foot, with its head downwards: if seized when wounded it can inflict with its sharp claws severe wounds on the head of the captor. It feeds on the pollen of Eucalypti, on insects, wild figs, and berries. It begins to breed in November, becoming then animated and fierce, readily attacking hawks, crows, and other birds that may venture near its nest. The nest is cup-shaped, and rather rudely constructed, being composed of the inner rind of the stringy bark and wool, to which succeeds a layer of fine twigs, lined with grasses and fibrous roots; the whole openly suspended to the horizontal branch of an apple (*Angophora*) or gum-tree, frequently within a few feet of the ground. The eggs are generally three in number. The young have merely the rudiment of a knob to the bill.—(*Gould's Birds of Australia.*)

Another species, TROPIDORHYNCHUS ARGENTICEPS, or Silvery-crowned Friar-bird, inhabits the north-west coast of Australia.

FRIGATE-BIRD. (*Tachypetes.*) This is an aquatic bird allied to the Cormorants, from which, however, it differs by having a forked tail, short feet, the membranes of which are very deeply notched, an extraordinary spread of wing (said to be ten or twelve feet in extent), and a beak both man-

FRIGATE-BIRD.—(TACHYPETES AQUILA.)

dibles of which are curved at the tip. The plumage is a richly-empurpled black, the under part of the throat more or less varied with white, and the beak red. In command of wing it is equalled by none of its class; and it is accordingly met with at an immense distance from all land, principally between the tropics, where it is seen darting upon the flying-fish, and attacking the gannets and gulls in order to make them disgorge their prey. It has received from English sailors the names of Frigate-bird and Man-of-war bird. It breeds on trees on uninhabited islands, and lays a single spherical white egg.

Dr. Chamberlaine, in the Jamaica Almanac for 1843, thus writes of the Frigate-bird: "He is almost always a constant attendant upon our fishermen, when pursuing their vocation on the sand-banks in Kingston harbour, or near the Palisados. Over their heads it takes its aerial stand, and watches their motions with a patience and perseverance the most exemplary. It is upon these occasions that the Pelicans, the Gulls, and other sea-birds, become its associates and companions. These are also found watching with equal eagerness and anxiety the issue of the fishermen's progress, attracted to the spot by the sea of living objects immediately beneath them. And then it is, when these men are making their last haul, and the finny tribe are fluttering and panting for life, that this voracious bird exhibits his fierce and voracious propensities. His hungry companions have scarcely secured their prey by the side of the fishermen's canoes, when with the lightning's dart they are pounced upon with such violence, that, to escape its rapacious assaults, they readily in turn yield their hard-earned booty to this formidable opponent. The lightness of its trunk, the short tarsi, and vast spread of wing, together with its long, slender, and forked tail, all conspire to give him a superiority over his tribe, not only in length and rapidity of flight, but also in the power of maintaining itself on outspread pinions in the regions of his aerial habitation amidst the clouds; where, at times, so lofty are its soarings, its figure becomes almost invisible to the spectator in this nether world."

FRINGILLIDÆ. A large family of Passerine birds, known by the general name of *Finches*, and including various minor groups, consisting of several genera, more or less closely related to one another. None of them are of large size; and in their habits and general appearance they bear a very strong relationship. They feed chiefly upon various kinds of grain and seeds; occasionally also upon insects. They are for the most part hardy birds, and do not quit this country during the winter; but some few are driven hither at that season from more northern climates. Many of the *Fringillidæ* are remarkable for their powers of song; others are highly prized for the delicacy of their flesh. They frequent fields, groves, hedgerows, and woodlands; while many, in a state of captivity, are rendered subservient to the amusement and gratification of man. They are severally described in this work, and will be found in their alphabetical order. In this place we shall merely give one species, as an example, which we find among the beautifully coloured specimens in Mr. Gould's superb work. It is called *Estrelda Temporalis*, or the RED-EYEBROWED FINCH. This bird has the crown of the head bluish-gray; wings and tail olive-brown; patch over the eye and rump, crimson; bill red; legs yellowish white. Eggs five or six in number, of a beautiful fleshy white. It is found in the pasture lands of New South Wales and South Australia, and is particularly abundant in the neighbourhood of Sydney. In the autumn it is gregarious, often assembling in very large flocks; but in the spring they are mostly seen in pairs. They build a large nest, formed of dead grass, lined with thistle-down, in any low bush adapted for a site, and in none more frequently than in that beautiful plant, the *Leptospermum squarrosum*. In the extensive and admirable work by Messrs. Gray and Mitchell, "the Genera of Birds," the forms and figures of many of the *Fringillidæ* are described and given. It will be seen by an inspection of that work, or a glance at the large collection of them in the British Museum, or any similar place, how impossible it is for us to enumerate in this place even a tithe of the genera of birds known to naturalists. [See BULLFINCH: BUNTING: GOLDFINCH: LINNET: CANARY-BIRD, &c. &c.]

FRITILLARY [BUTTERFLY]. A name given by insect collectors to various species of Butterflies, of the genera *Nemeobius*, *Melitea*, and *Argynnis* [which see].

FROG. (*Rana.*) Of all the Reptile tribes none are better known than those termed ANUROUS BATRACHIANS, including the genus *Rana*, or common Frog. In colour this animal varies considerably, but its general tinge is olive-brown, variegated on the upper parts of the body and limbs with irregular blackish spots; those on the limbs being mostly disposed in a transverse direction: it has also a long deep brown patch under each eye. The under parts are of a pale greenish-yellow cast, and much more obscurely spotted and variegated than the upper surface. It is not unfrequently seen, however, especially towards the close of summer, of a much brighter cast, and with more vivid variegations; but, like all other species which are in the habit of casting the skin, it differs at intervals as to the brightness or intensity of its colours. The teeth are very small; the eyes large and brilliant, and surrounded with a yellow circle; the ears are placed behind them, and covered with a membrane. Their muscles are considerable in relation to their bulk, and peculiarly elastic, strong, irritable, and sensible to the action of galvanism. The Frog is light, active, and lively; the limbs admirably calculated for the peculiar motions of the animal, and the hind feet strongly webbed, to assist its progress in the water, to which it occasionally retires during the heats of summer, and again during the frosts of winter: for at that time it lies in a torpid state, either deeply plunged in the soft mud at the bottom of stagnant waters, or in the hollows beneath their banks, till it is awakened from its slumber by the return of spring. And here it may be observed, that though the Frog and the Toad have a general similitude, their distinguishing characteristics are very marked and decided. The Frog leaps; the Toad crawls. The Frog is in general the smaller of the two, of a brighter colour, and has a more polished surface; the toad is brown, rough, and dusky. The Frog is light and nimble; the toad slow, corpulent, and heavy. In their internal conformation the difference is not remarkable, except that the Frog has more air-bladders than the toad, by which it is rendered better adapted for an aquatic life.

Some time in March the Frog usually deposits its ova or spawn, consisting of a clustered mass of gelatinous transparent eggs, in each of which is embedded the embryo, or *tadpole*, in the form of a round black globule. In this state it lies for a month or five weeks, before the tadpoles are hatched from it; during which period each egg gradually enlarges in size, and a few days before the time of exclusion the young animals may be perceived to move about in the surrounding gluten. When first hatched, they feed on the remains of the gluten in which they were imbedded, and in the space of a few days, if narrowly examined, they will be found to be furnished, on each side of the head, with a pair of ramified branchiæ, or temporary organs, which again disappear after a certain space. These Tadpoles are so perfectly unlike the animals in their complete state, that a person inconversant in natural history would hardly suppose them to bear any relationship to the Frog, since, on a general view, they appear to consist merely of head and tail; the former large, black, and roundish; the latter slender, and bordered with a very broad transparent finny margin. Their motions are extremely lively, and they are often seen in such vast numbers as to blacken the whole water with their legions. When the tadpoles have arrived at the age of about five or six weeks, the hind legs make their appearance; gradually increasing in length and size; and, in about a fortnight afterwards, are succeeded by the fore legs, which are indeed formed beneath the skin much sooner, and are occasionally protruded and again retracted by the reptile through a small foramen on each side of the breast, and are not completely stretched forth till the time just mentioned. It now partakes of the form both of a frog and a lizard; which figure it retains for about six hours; and then the tail continuing to decrease, it becomes quite obliterated in the space of a day or two afterwards. The Frog, at length arrived at its perfect state, is now seen wandering about the brinks of its parent waters, and sometimes in such vast numbers in particular spots, that not only has their appearance given rise to an absurd belief among the vulgar that it occasionally "rained frogs," but various modes of accounting for so "extraordinary a phenomenon" have presented themselves to the minds of those whose duty it was to dispel the unfounded though popular credulity, by tracing its origin and properly explaining it.

The Frog, no longer of ambiguous form, now feeds on animal food; supporting itself on insects, small snails, worms, &c. It principally lives on insects, for the more readily obtaining of which the structure of its tongue is extremely well calculated; being very long, and so situated that the root is attached to the fore rather than to the hind part of the mouth; and, when at rest, lying backwards, as if the animal were swallowing the tip. By this means it is enabled to throw it some distance from the mouth, which is done with great celerity, the bifid and glutinous tip securing the prey, which is swallowed with an instantaneous motion, so quick as to be scarcely perceptible.

The muscular system of the Frog deserves particular attention. Mr. Broderip observes, "In the Anurous Batrachians, the Frogs especially, the muscles of the abdomen are more developed than in the other Reptiles: offering in this particular some analogy to the abdominal structure of the Mammifers. But it is in the disposition of the muscles of the thigh and leg in the Frogs and other Anurous Batrachians, that the greatest singularity is manifested. These, whether taken conjointly or singly, present the greatest analogy with the muscular arrangement of the same parts in Man. We find the rounded, elongated, conical thigh, the knee extending itself in the same direction with the thigh-bone, and a well-fashioned calf to the leg, formed by the belly of the *gastrocnemii* muscles. It is impossible to watch the horizontal motions of a Frog in the water, as it is impelled by these muscles and its webbed feet, without being struck by the complete resemblance in this portion of its frame to human conformation, and the almost perfect identity of the movements of its lower extremities with those of a man making the same efforts in the same situation. By the aid of these well-developed lower limbs, and the prodigious power of their muscular and bony levers, a Frog can raise itself in the air to twenty times its own height, and traverse at a single bound, a space more than fifty times the length of its own body." The difference of sex in these animals is not perceptible till they have arrived at their fourth year, nor do they begin to propagate till they have reached that period. Hence, on comparing their slow growth with their other habitudes, it would seem that they live about twelve years; but so numerous are their enemies, both by land and water, that it is probable very few arrive at so great an age. The Frog is extremely tenacious of life, and will survive for a considerable space the loss of many of its organs. If confined entirely under water, it is still enabled to support its existence for several days: on the contrary it is not so well able to endure the want of water, nor long exposure to a dry air and a hot sun, though it delights to bask occasionally in a moderate sunshine: it is therefore particularly careful to secure a retreat where it may enjoy the benefit of shade and a sufficient supply of moisture. Frogs are distinguished by a peculiar cry, termed *croaking*, particularly during rain and hot weather, in the morning and evening.— There are several other species of Frogs, a few of which it will be necessary to describe.

The EDIBLE FROG (*Rana esculenta*), so called from its being the kind most approved of for the table by our nearest continental neighbours,—is found plentifully in France, Italy, Germany, and many other parts of Europe, though it is rare in England. It is rather larger than the common Frog, and of an olive-green colour, distinctly and strongly marked on the upper parts of the body with black spots or patches, the limbs being transversely marked with bands of the same colour; and from the tip of the nose down the whole length of the back run three distinct yellow stripes. The under parts of the body and limbs are of a dull white, slightly tinged with green, and variegated with brown spots.

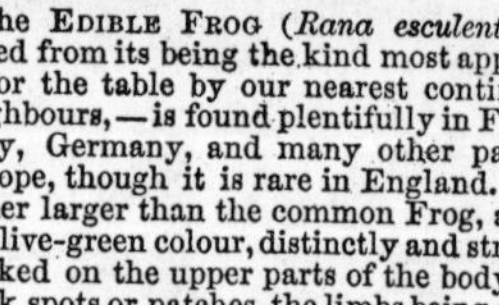

EDIBLE FROG.—(RANA ESCULENTA.)

The proportion of the limbs is nearly the same as in the common Frog, and the hind feet are very strongly palmated; but the head is rather larger and more pointed. The Edible Frog is a very voracious animal, and will occasionally seize on young birds, mice, &c., swallowing them whole, like the rest of its prey. The male of this species, during the breeding season, is observed to protrude from each side of its head a large inflated globular vesicle, and croaks so loud as to be heard at a vast distance: in fact, where these animals assemble in large numbers, their noise is most unpleasant and annoying.

The BULL-FROG (*Rana pipiens*) is the largest species of the genus, being three or four inches broad, and from six to eight in length without including the feet; with the limbs extended it measures about eighteen inches. It inhabits North America, particularly the southern parts of the United States; and has received the name of Bull-Frog from its voice resembling the distant lowing of a bull. Its colour on the upper parts is a dusky olive, somewhat irregularly marked with numerous dark brown spots; the under parts being of a whitish cast tinged with green, and thickly spotted. The fore feet have only four toes, and are unwebbed, but the hind feet, which are large and long, are very widely webbed. The irides of the eyes are red, surrounded with a narrow border of yellow: the external membranes of the ears are large and round, of a reddish brown colour, and surrounded by a pale yellow or whitish margin. In Mr. Catesby's Natural History of Carolina, we are told that this species frequents springs only, where, by the continual running of the water a small pond or hole is usually made before the mouth of the spring, which is rarely without a pair of these Frogs, who when surprised, with a long leap or two enter the mouth of the spring, where they are secure. He adds, that it is commonly believed that they keep the springs clean, and purify the water, and therefore the general prejudice is in their favour, although they are great devourers of young ducks and goslings, which they often swallow whole.

The ARGUS FROG (*Rana ocellata*) is also a native of several parts of North America, choosing moist situations, as the neighbourhood of springs and rivulets. In size it differs but little from the Bull-Frog, except that the limbs are thicker and stouter; but the feet are unwebbed, and are all divided into five toes, each joint being furnished beneath with a kind of tubercle or process. Its colour is a pale reddish brown, with two distinctly marked whitish elevated lines running down the middle of the back, the intervening space being marked with several broad fasciæ of a reddish chestnut colour, while the sides are beautifully ornamented with several ocellated or eye-shaped spots, each being half surrounded by an iris-like paler space or crescent. The limbs are elegantly banded with chestnut-coloured stripes: the under parts are of a dull white. In its general manners it is said to resemble the preceding. [For the Tree Frog, see HYLA.]

FROGHOPPER. (*Aphrophora spumaria.*) The popular name of a small but singular Homopterous insect, belonging to the *Cercopidæ* family. They pass their whole lives on plants, on the stems of which their eggs are laid in the autumn. The following summer they are hatched, and the young immediately perforate the bark with their beaks, and begin to imbibe their sap. Of this they take in such quantities, that it oozes out of their bodies continually, in the form of little bubbles, which soon completely envelope the insects; and from this circumstance the name of *Cuckoo-spits* is also very commonly applied to them. They thus remain entirely buried and concealed in large masses of foam, until they have completed their final transformation. When the pupa, which is of a beautiful green colour, is about to undergo its change into the complete insect, it ceases to absorb any longer the juices of the plant, and to discharge the projecting froth. It then emerges from its concealment. The winged insect is scarely larger than the larva; but its colour is brown, with a pair of broad, irregular, pale bands across the upper wings. It possesses the power of leaping in a remarkable degree; for which purpose, the tips of their hind shanks are surrounded with little spines, and the first two joints of their feet have a similar coronet of spines at their extremities. Their thorax projects somewhat between the basis of the wing-covers; their bodies are rather short, and their wing-covers are almost horizontal and quite broad across the middle, which, with the shortness of their legs, gives them a squat appearance.

FULGORA: FULGORIDÆ. A genus and family of insects bearing great resemblance to the *Cicadidæ*. Many of them are distinguished by a curious prolongation of the forehead, the shape of which varies extremely in the different species, which in tropical regions are numerous. The legs are in general fitted for leaping, with large spurs; and the males are destitute of those organs which are employed in the Cicadæ for the production of sounds. We should observe, that Kirby and Spence, on the authority of Stedman's *Surinam*, assert that *Fulgora laternaria* makes a loud noise in the evening, like that made by a razor-grinder, and that the Dutch in Guiana call it *scare-sleep*. Dr. Hancock, however, states that the razor-grinder, or the [illegible]ria Aria of the natives, is a specis of Cicada. In the typical genus *Fulgora* the head is dilated in front into the most remarkable porrected protuberances, varying in each species, and which is the part of the body asserted by various writers to emit a strong light by night, analogous to that of the fire-flies.

Mr. Westwood alludes to this luminous property at some length. "Much uncertainty (he says) exists as to the real existence of any luminous power possessed by the typical species of this family. This account originated with Madame Merian (*Insecta Surinam*, p. 49.), who asserted it to be possessed by *Fulgora laternaria* in an eminent degree, and her statement long received general assent, and appears to be the only authority for its existence. Olivier appears to be the first author who doubted the luminosity of the *Fulgoræ*, from information given to him by M. Richard, who had reared the *F. laternaria* in Cayenne, and had not found it to be luminous. Hoffmansegg, the Prince Von Nieuwied, and still more recently M. Lacordaire (the two last named authors having been long resident in South America), also concur in this opinion, none of the individuals they had ever seen alive exhibiting the least trace of

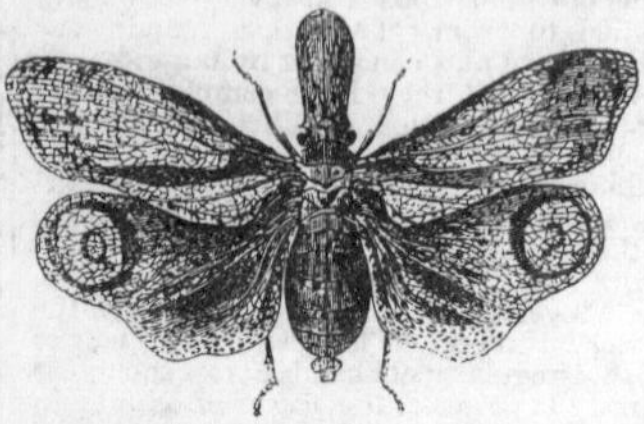

AMERICAN LANTERN-FLY (FULGARIA LATERNARIA.)

luminosity. The majority of the natives also, who had been questioned on the subject, denied the luminous power, although a few affirmed it; hence Lacordaire suggests whether one sex may be luminous and the other not. Dr. Hancock read a memoir on the luminosity of the *Fulg. laternaria* before the Zoological Society, on 24th June, 1834, in which its luminosity is considered entirely fabulous. M. Wesmael has recently reasserted the luminous property of the South American species, on the authority of a friend who had witnessed it alive. And W. Baird, Esq. has informed me of the existence of a Chinese edict, against young ladies keeping lanthorn-flies." Mr. Adam White, in the Annals and Magazine of Natural History, published an extract from a letter of J. Bowring, Esq., of Hong Chong, where the *F. Candelaria* is very abundant, but not known to be luminous. The species are generally very showy, and have been mostly figured by Mr. Westwood. In the British Museum there is a fine collection of them.

FULICA. [See Coot.]

FULMAR. (*Procellaria glacialis.*) A Palmiped bird belonging to the *Petrel* family; abounding in northern latitudes, though rarely seen in warm or temperate climates; in fact, it has been met with not only in arctic and antarctic regions, but even at the foot of those impenetrable barriers, the floating islands and eternal mountains of ice and snow. It measures seventeen inches in length, and weighs twenty-two ounces. The bill is about two inches long, and strongly formed; the hook or nail of the upper mandible, and the truncated termination or tip of the under one, are yellow; the other parts grayish; the nostrils are contained in one sheath, divided into two tubes. The head, neck, all the under parts, and the tail are white; back and wing-coverts blue gray; quills dusky blue; eggs yellowish, sometimes inclining to red. The body is thickly clothed with feathers upon a fine close down.

These birds are extremely greedy and gluttonous, and will devour any floating putrid substances: they feed principally on fish, and on the blubber or fat of whales, and other animals; which being soon convertible into oil, supplies it with provision for its young, and with the constant means of defence; for the Fulmar, like all the Petrels, has a peculiar faculty of spouting from its bill, to a considerable distance, a large quantity of pure oil. Pennant, speaking of those which inhabit the isle of St. Kilda, says—"No bird is of such use to the islanders as this: the Fulmar supplies them with oil for their lamps, down for their beds, a delicacy for their tables, a balm for their wounds, and a medicine for their distempers." The female is said to lay only one white and very brittle egg, which she hatches about the middle of June.

FUNGIA. A genus of Zoophytes, of which there are several species, both recent and fossil, principally from the Indian seas. They belong to the *Madrephylliæa* of De Blainville, and consist of animals in nearly the lowest state of organization; for although they are universally allowed to be animals, they are completely without the power of motion, consisting simply of a living gelatinous film, which is endowed with the capability of constructing for itself a stony support or framework, derived from the surrounding water. In form it is generally orbicular or oval; mouth superior, transverse in a large disc, which is covered by many thick cirriform tentacula; and the disc is solidified internally by a calcareous solid *polyparium*, of a simple figure. We are indebted for the following interesting remarks to the elaborate description given of Fungia by Mr. Rymer Jones. "If we investigate the history of the Fungia a little more closely, it is beautiful to observe in apparently one of the most helpless and useless members of creation, the operations of the same power and foresight that shield and guard the highest and most intelligent. The Fungia, whilst it is alive, lies upon the sand at the bottom of the shallow seas of warm climates, or has its base loosely imbedded in the sand. It is unattached by any pedicle or root, so that a passing wave of any violence might easily take it up and

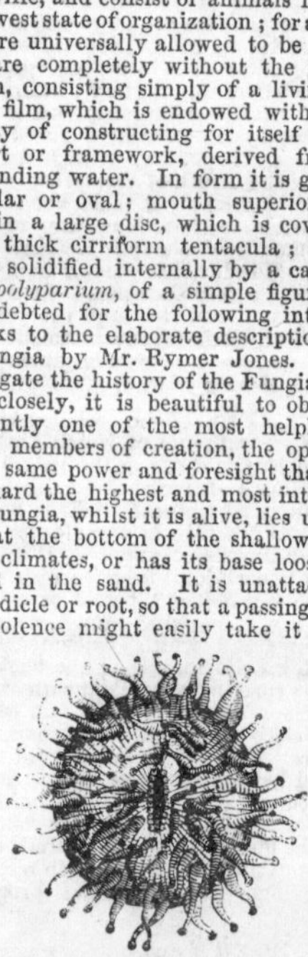

THICK-TENTACLED FUNGIA. (FUNGIA CRASSITENTACULA.)

wash it to a distance from the spot it originally occupied. This being the case, what is to prevent the wave from turning it upside down? It is only upon the upper surface that the living crust is spread, which forms the Fungia, so that should accident reverse its position the creature would inevitably perish. The arrangement adopted to prevent such an occurrence is simple enough, but not on that account less beautiful. The living film that coats its laminated surface has the faculty of secreting little bubbles of air within its substance; the bubbles so produced, although disseminated as it were at random, are sufficiently buoyant to act as floats, and thus provided, let the wave wash it ever so far, still the lightest side keeps uppermost, the floats prevent it from being reversed, and the creature settles down in a right position upon the smooth bottom of the sea." We may mention that our figure of the thick-tentacled Fungia (*F. crasitentaculata*) is derived from one of the French Voyages of Discovery, and shows the animals projecting from their coral home. The collection of corals and zoophytes in the British Museum, now most wondrously increased, contains many fine specimens of this very beautiful and distinct genus. As an ornament on a mantelpiece or on a table, under a bell glass, nothing perhaps is so pleasing as a fine and symmetrical specimen of this coral.

GADFLY. (*Œstrus bovis.*) The Gadfly, or Ox Gadfly, is a Dipterous insect, about the size of a common Bee, with pale brown wings: it is of a pale yellowish brown colour, with the thorax marked by four longitudinal dusky streaks, and the abdomen by a black bar across it, the tip being covered with orange-coloured hairs. The genus is remarkable for its larvæ residing beneath the skin, or in different parts of the bodies of quadrupeds. When the female of this species is ready to deposit her eggs (which chiefly happens in August or December), she fastens on the back of a heifer or cow, and piercing the skin with the tube situated at the top of the abdomen, deposits an egg in the puncture; an operation which she repeats on many parts of the animal's back. Here the several eggs hatch, and the larvæ by their motion and suction cause so many small swellings or abscesses beneath the skin, which growing gradually larger, exhibit tubercles of an inch or more in diameter, with an opening at the top of each, through which may be observed the larva (a whitish oval maggot, which in time becomes brown) imbedded in a purulent fluid. There the larvæ remain till the middle of the next summer, when they force themselves out from their respective cells, and, falling to the ground, each creeps beneath the first convenient shelter, and lying in an inert state becomes contracted into an oval form, but without casting the larva skin, which dries and hardens round it. Having remained in the chrysalis state more than a month, it forces open the top of its coat, or pupa armour, and emerges in its perfect form. [Mr. Bracy Clark, F.L.S., has paid particular attention to the study of the family *Œstridæ:* we refer our readers for further information to the articles BREEZE-FLY: ŒSTRUS.]

GADUS: GADIDÆ. The *Gadidæ*, or Cod tribe, are a family of Fishes belonging to the Malacopterygious (or soft-finned) order. They include the Cod, Haddock, Whiting, Ling, and others; and are distinguished by the following characters:—a smooth, oblong body, covered with small, soft, deciduous scales; head scaleless; eyes lateral; jaws and anterior part of the vomer furnished with several ranges of unequal, pointed teeth; the gills large, seven-rayed, and opening laterally; and a small beard or cirri at the tip of the lower jaw. Almost all the species have two or three dorsal fins, one or two anal, and one distinct caudal fin; and they have a large, strong, swimming-bladder, frequently dentated or lobed at its borders. They live for the most part in the seas of cold or temperate climates; and from their size and their tendency to congregate in particular localities, as well as from the wholesomeness and good flavour of their flesh, they are of first-rate importance to man. [See COD, &c.]

GALAGO. (*Galago* or *Otolicnus.*) A genus of small quadrumanous animals, inhabiting different parts of Africa, and subsisting chiefly on insect food. They have great eyes; large membranous ears, which double down when at rest; hind limbs of a disproportionate length; and a long and tufted tail. The

SENEGAL GALAGO.—(GALAGO SENEGALENSIS.)

best known species are the GREAT GALAGO (*Galago crassicaudatus*), which is as large as a Rabbit; and the SENEGAL GALAGO (*Galago Senegalensis*), or gum animal of Senegal, the size of a Rat. "These pretty animals have at night all the activity of birds, hopping from bough to bough on their hind limbs only. They watch the insects flitting among the leaves, listen to the fluttering of the moth as it darts through the air, lie in wait for it, and spring with the rapidity of an arrow, seldom missing their prize, which is caught by the hands. They make nests in the branches of trees, and cover a bed with grass and leaves for their little ones. They are a favourite article of food in Senegal."

GALATHÆA. A genus of long-tailed Crustacea. In the British seas four species are recorded as native: their porcelain texture, their sculptured carapace and wide tail, joined to their pleasing colours, especially when alive, render them very attractive. Close to this genus is *Grimothea*, one of the species of which (*G. gregaria*) is met with in the Southern seas near the Straits of Magellan, in countless multitudes.

GALECINUS. [See SUPPLEMENT.]

GALEOPITHECUS. An extraordinary quadrumanous animal of the Lemurine tribe, called the *Flying Lemur*, and sometimes termed the *Coluga*; it is a native of the islands of the Indian Archipelago; and its chief peculiarity consists in the extension of its skin between the anterior and posterior limbs on each side, and between the posterior limbs, including also the tail; by which it receives

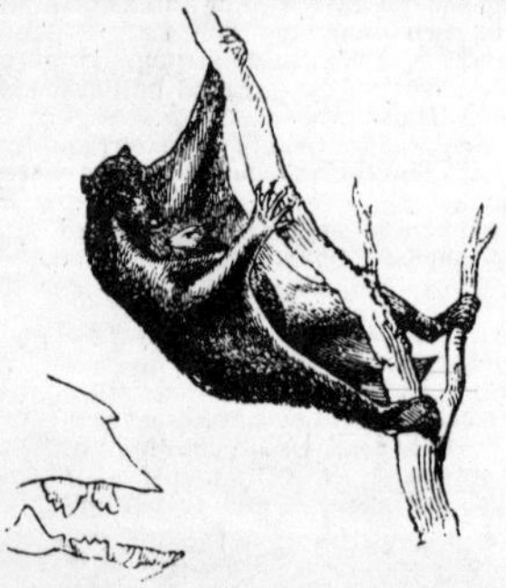

FLYING LEMUR AND YOUNG (GALEOPITHECUS VOLANS), WITH THE FRONT PART OF THE SKULL.

a parachute-like support in the air, and is enabled to take long sweeping leaps from tree to tree, somewhat like flying. They may be considered as connecting the Lemurs with the Bats; differing generically from the latter in having their fingers, which are armed with trenchant nails, no longer than the toes, so that the membrane which occupies their intervals, and extends to the sides of the tail, can only answer the purpose of floating in the air. The general anatomy agrees very closely with that of the Lemurs. They inhabit lofty trees in dark woods; to which they cling with all four extremities, and traverse easily by means of their claws. During the day-time they suspend themselves like Bats from the branches, with the head downwards; but at night they rouse themselves, and make an active search for food, which consists of fruit, insects, eggs, birds, &c. They are very inoffensive; and generally produce two young at a birth.

GALERUCIDÆ. A group of leaf-eating beetles, separated from the *Chrysomelidæ* family, and consisting mostly of dull-coloured beetles: having an oblong oval, slightly convex body; a short and rather narrow thorax; slender antennæ, more than half the length of the body, and implanted close together on the forehead; slender legs, and claws split at the end. They fly mostly by day, and are either very timid or very cunning, for, when we attempt to take hold of them, they draw up their legs, and fall to the ground. They sometimes do great injury to plants, eating large holes in the leaves, or consuming entirely those that are young and tender. The larvæ are rather short cylindrical grubs, generally of a blackish colour, and are provided with six legs. They live and feed together in swarms, and sometimes appear in very great numbers on the leaves of plants, committing ravages, at these times, as extensive as those of the most destructive caterpillars.

The *Galeruca vittata*, or striped Cucumber Beetle, a North American species recorded by Dr. Harris in his "Insects of Massachusetts," is of a light yellow colour above, with a black head, and a broad black stripe on each wing-cover, the inner edge of which is also black, forming a third narrower stripe down the middle of the back; the abdomen, the greater part of the fore-legs, and the knees and feet of the other legs, are black. It is rather less than one-fifth of an inch long. Early in the spring it devours the tender leaves of various plants; and makes its appearance on cucumber, pumpkin, and melon vines, about the end of May or the beginning of June, or as soon as the leaves begin to expand; and as several broods are produced in the course of the summer, it may be found at various times on these plants, till the latter are destroyed by frost. The females lay their eggs in the ground, and the larvæ feed on the roots of plants. Various means have been suggested to prevent the ravages of these striped cucumber beetles; as, wetting the vines with tobacco water, or with infusions of elder, walnut-leaves, or of hops; others recommend the use of soot, sulphur, Scotch snuff, or pepper, to be sifted upon the plants. In this country several species are found, which will be seen referred to in the works of Messrs. Stephens and Curtis.

GALICTIS. A genus of Carnivorous animals allied to the Civets and Genets.

GALLINÆ. The name given to an extensive order of Birds, including all those which constitute what are commonly termed "poultry," and furnishing us with the greater number of our farm-yard fowls, and with much excellent game. The name *Gallinæ* is applied to them from their affinity to the Domestic Cock, in common with which they have generally the upper mandible vaulted, the nostrils pierced in a large membranous space at the base of the beak, and covered by a cartilaginous scale. Their wings are short, their carriage heavy, and their flight laborious. They have an extremely muscular gizzard, and generally a large globular crop. In general they lay and incubate on the ground, on a few carelessly arranged stems of straw or grass. Some species are polygamous, and some monogamous: in the former the male is always larger and more gaily coloured than the female; in the latter the sexes nearly or quite resemble both in size and colour.

GALLINACEÆ. Some of the most

valuable birds we have belong to this order; Peacocks, Turkeys, Fowls, Pheasants, Partridges, &c. being of the number. Their bodies, for the most part, are large and muscular; their wings short; and their toes rough beneath, to enable them to scratch the ground in search of worms, &c. Many feed on grain and seeds, whilst others feed on berries, but the greater portion subsist likewise on insects. They are mostly polygamous, building their rude nests, in retired situations, on the bare ground. The females of several species are extremely prolific, and continue to lay eggs nearly all the year; the young follow the parent mother as soon as hatched, and she continues to protect them till they are fully grown. Some are easily domesticated; others remain in a wild state; but the flesh of nearly all furnish a substantial and wholesome food, while their plumage serves for various domestic and ornamental purposes. In their proper alphabetical order the reader will find them severally described.

GALL-INSECT. (*Gallinsecta.*) A family of insects, of a small size, which live upon trees or plants of various kinds. In the larva state they have the appearance of oval or round scales, closely attached to the plant or bark of the tree they inhabit, and exhibit no distinct external organs. If observed in spring, their bodies are noticed gradually to increase in size, ending in their acquiring the appearance of a gall, being either spherical, kidney-shaped, boat-shaped, &c. The skin in some is entire and very smooth; in others it is incised, or offers traces of segments. It is in this state that the females are impregnated, shortly after which they deposit their eggs, of which the number is very great; these they deposit between the ventral surface of their bodies and a layer of a cottony secretion. Their bodies subsequently dry up and become a solid cocoon, which covers the eggs; others envelope their eggs in a very abundant cottony secretion, which equally defends them. Many of them have been long celebrated for the beautiful dyes they yield. A very curious Gall constructed by a Homopterous insect, of a genus allied to Aphis, has lately been imported from the East by Mr. Morson. This, which is principally composed of gallic acid and tannin, has been particularly described in a late number of the Pharmaceutical Journal. [See Coccus: Kermes.]

GALLINULE. (*Gallinula.*) A genus of birds which frequent fresh waters, swimming and diving about, or running on land with equal ease and swiftness. The common Gallinule (*Gallinula chloropus*), called also the Water-hen or Moor-hen, is about fourteen inches in length, from the tip of the beak to the end of the tail, and weighs from eleven to fourteen ounces. The bill is upwards of an inch long, of a greenish yellow at the tip, and reddish towards the base; whence a kind of horny or membraneous substance shields the forehead as far as the eyes: this appendage to the bill is perfectly red in the breeding season; at other times it varies or fades into white. The head is small and black, except a white spot under each eye: the irides red: all the upper parts of the plumage dark shining olive green, inclining to brown; under parts dark hoary lead gray: vent feathers black; those on the belly and the thighs tipped with dirty white: the long loose feathers on the sides, which hang over the upper part of the thighs, are black, streaked with white; the feathers just beneath the tail are white; and the legs are dusky green. The toes are very long, particularly the middle one; their under sides flat and broad, whereby it is enabled to swim; and, from this part of its conformation, it may be regarded as the bird which connects the web-footed aquatic fowl with the fin-toed. The body is long, and the legs placed far behind; its feathers thickly set, and bedded upon down. It lies concealed during the day among reeds and willows, by the sides of rivulets: it can run over the surface of such waters as are thickly covered with weeds, and it dives and hides itself with equal ease: it flirts up its tail when running, and flies with its legs hanging down. In the evenings, it creeps by the margins of the waters, among the roots of bushes and long loose herbage, in quest of its food, which consists of insects, small fishes, worms, aquatic plants, and seeds: it is also granivorous; and if killed in September or October, after having had the advantage of a neighbouring stubble, its flesh is very good.

The Gallinule, or Moor-hen, makes its nest of reeds and rushes, closely interwoven, choosing for it a very retired spot close by the brink of the water; and it is said the female never quits it without covering her eggs with the leaves of the surrounding herbage. The female lays from five to eight eggs, of a light yellowish brown, marked with rust-coloured spots. Soon after the young are hatched, they take to the water, and shift for themselves. They differ considerably from the adults till after their second autumnal moulting, having till then a much lighter plumage.

"One circumstance respecting this familiar bird," Mr. Gould observes, "appears to have escaped the notice of most ornithologists: we allude to the fact of the female being clothed in a dark and rich plumage, and having the base of the bill and the frontal shield of a bright crimson-red tipped with fine yellow; her superiority in these respects has caused her to be mistaken for the male, which, contrary to the general rule, is at all times clothed in a duller plumage, and has the upper surface more olive than in the female; the bill is also less richly tinted.

There are very few birds of this genus; and most of them inhabit Java; but they are not by any means remarkable.

GALLIWASP. (*Celestus occiduus.*) A reptile of the Saurian order. It is nearly two feet in length from the nose to the tip of the tail, which, like the body, is thick and strong, tapering pretty suddenly towards the tip: the limbs are short, and the animal's whole appearance is remarkably stout and plump: the teeth are small in front, but as they approach the back part of the jaws they increase considerably in size. It is a native of the West India islands, and seems to be particularly common in Jamaica, where it is said to frequent woody and marshy districts. It is usually of a palish brown colour, clouded with spots and bands of deeper cast, but it is reported to change its colour occasionally to a lively golden yellow.

GALLOWAY. A peculiar breed of strong, active, middle-sized horses; so called from the county of Galloway, in Scotland, which was formerly noted for them. Tradition reports, that the stock originated from several Spanish stallions, which swam on shore from some ships wrecked on the coast, belonging to the famous Armada; and, propagating with the mares of the country, furnished the kingdom with their posterity.

GAMBET. (*Totanus.*) A genus of wading birds, allied to the *Scolopacidæ*, and including numerous species. The GREENSHANK GAMBET (*Totanus Glottis*) is the largest European species, being nearly the size of the Godwit, with the beak comparatively stout, and a little recurved; ashy-brown above and on the sides, with the margins of the feathers punctated with brown, the croup and belly white, and tail rayed with narrow irregular bars of gray and white; the feet green: in summer the throat and breast are marked with dusky spots, which disappear after the breeding season. It breeds on the margins of lakes, which it mostly frequents; is very clamorous when on the wing; and in winter resorts to the sea-shore.—The DUSKY GAMBET (*Totanus fuscus*), another European species, but rare in Britain, is more delicately formed, with particularly slender beak and feet, and beautifully barred tail and coverts; it becomes entirely suffused on the under parts with fuliginous black in the spring.—A third, the REDSHANK GAMBET (*Totanus calidris*), is very abundant in this country, breeding also not uncommonly in marshes near the sea-shore, and especially about the estuaries of rivers.—There are others, as the delicate WOOD GAMBET (*Totanus glareola*), remarkable for the extraordinary length of its legs, and its habit of gracefully tripping across the broad floating leaves of aquatic plants when in search of its prey; and the GREEN GAMBET, (*Totanus ochropus*), with shorter legs, and easily known as it flies by its conspicuous white rump.

GAME, BLACK and RED. [See GROUSE.]

GAMMARUS: GAMMARIDÆ. A genus and family of Crustaceans belonging to the order *Amphipoda.* The body of this marine genus is covered with a coriaceous elastic tegument, generally compressed and arched; the posterior extremity of the tail is not furnished with swimmerets, but its appendages are in the form of cylindrical or conical styles. Two at least of the four anterior legs are terminated by claws. The vesicular bags (the use of which has not been ascertained) are situated at the external base of the legs, commencing with the second pair, and accompanied by a small plate. The pectoral scales enclosing the eggs are six in number. There are several species of this family found in the British seas; for an account of which we must refer our readers to the works of Milne Edwards and Kroyer, but especially of the latter. The genera *Talitrus*, *Orchestia*, *Dexamine*, *Amphithoe*, and others recorded in the List of Crustacea in the British Museum, belong to this family. The habits of some of these are very interesting. [See AMPHIPODA, &c.]

GANNET, or SOLAN GOOSE. (*Sula Bassana.*) This Palmipede bird is about the size of the tame goose; its length two feet nine inches, and its weight nearly seven pounds. The bill is six inches long, jagged at the sides, and straight almost to the point, where it inclines downwards; a darkish line passes from the brow over the eyes, which are surrounded with a naked blue skin, and, like those of the Owl, are set in the head so as to look nearly straight forward, and the extreme paleness of the irides gives them a keen wild stare. A loose black bare skin,

COMMON GANNET.—(SULA BASSANA.)

capable of great distension, hung from the blades of the under bill, and extended over the throat, serves it as a pouch to carry provisions in the breeding season to its mate and young. The neck is long; the body flat, and very full of feathers; the crown of the head, and a small space on the hind part of the neck, are buff-coloured; and, with the exception of the quill and bastard-wing feathers, the rest of the plumage is white. The legs and toes are black; but the fore part of both are marked with a pea-green stripe; and the tail is composed of twelve tapering sharp-pointed feathers, the middle ones being the longest. The male and female are nearly alike; but the young birds, during the first year, are of a dusky hue, speckled with numerous triangular white spots; and it is not until the third year that the plumage is perfected.

In the Hebrides, the north of Scotland, and in Norway, this species is very abundant; it is also met with in great numbers on the coasts of Newfoundland and other northern regions, as well as in more temperate climes of both hemispheres. Their food consists chiefly of salt-water fish, the herring and pilchard being their favourites; and they take their prey by darting down upon it from a considerable height. They make their nests, which are composed chiefly of turf and sea-weeds, in the caverns and fissures of rocks, or on their ledges, as well as on the plain surface of the ground. The female (according to Bewick) lays three white eggs, somewhat smaller than those of a goose; but we find it elsewhere stated, that the Gannet, if not disturbed, will lay only one egg throughout the year; but if that be taken away, it will lay another, and in like manner a third, which she is generally permitted to hatch. Their greatest known rendezvous is the Hebrides and other solitary rocky isles of North Britain, such as the Bass in the Firth of Forth, and Ailsa Crag, in the Firth of Clyde, where their nests, in the months of May and June, are described as so closely placed together, that it is difficult to walk without treading upon some of them; and it is said that the swarms of the old birds are so prodigious, that when they rise into the air, they stun the ear with their noise, and overshadow the ground like the clouds. The inhabitants of the islands where these birds breed derive considerable emolument from the produce of their eggs; but to obtain them they encounter the most fearful risks, now climbing rocks which are almost inaccessible, and now clinging to the craggy precipices which, at a prodigious height, overhang a raging sea.

In Mr. Couch's "Cornish Fauna" we are told that "the Gannet takes its prey in a different manner from any other of our aquatic birds; for, traversing the air in all directions, with a heavy and irregular flight, as soon as it discovers the fish it rises to such a height as experience shows best calculated to carry it by a downward motion to the required depth, and then partially closing its wings it falls perpendicularly on the prey, and rarely without success, the time between the plunge and emersion being about fifteen seconds. When pilchards are collected into a narrow space, the number and eagerness of the Gannets are such, that it is surprising they do not fall on and kill each other. Their clamour indeed at such times proves them to be well on their guard, but it is also probable that every one in falling has its eye fixed on the fish it intends to seize, and the well-poised wings direct it unerringly to its prey. The form and setting on of the Gannet's wings well fit it for assuming the perpendicular attitude preparatory to its fall, which is effected with ease, rapidity, and precision. They are attached to the body about the centre of gravity, so that the anterior parts drop as on a pivot, and the elbow being about the middle of the distance between the shoulder and wrist, a slight inclination in any direction is sufficient to regulate the motion." There are also other species bearing the name of Gannet, but the one just described is the best known and the largest. The *White Gannet*, which inhabits China: the *Booby Gannet*, common on the coasts of South America, and described as being a very stupid bird; hence the appellation given to it by sailors: and the *Brown Gannet*, belonging to the West Indies and the western coast of tropical Africa. [See SULA.]

GARFISH. (*Esox belone.*) This fish has a variety of names; as, Garfish, Sea-pike, Sword-fish, Greenbone, Mackerel-Guide, Sea-Needle, &c. It generally precedes the Mackerel in their annual visit to shallow water for the purpose of spawning, and is taken on various parts of the Dutch, English,

GAR-FISH. — (ESOX BELONE.)

Scotch, and Irish coasts. It is from twenty to twenty-four inches in length, with long, narrow, beak-like snout, the under jaw projecting; the teeth are numerous and minute, the eyes large; the dorsal and anal fins opposite each other; pectoral and ventral fins small; and the tail considerably forked. The upper part of the head and back is of a dark green hue, the sides paler, and the belly a silvery white. It is a very vivacious fish, and seizes the bait with avidity. The flesh of the Garfish has somewhat the flavour of Mackerel, but is more dry; and the bones are green.

GARROT. (*Clangula.*) A genus of the Duck family, widely distributed over the colder and temperate regions of both Europe and America. The head is large, compressed, rounded above; bill shorter than the head, higher than broad at the base; neck short and thick; body ovate and depressed; eyes small; legs very short, and placed far behind; hind toe lobed. They breed in the colder regions of Europe and America, returning to more temperate climes in winter. They haunt rivers, lakes, estuaries, and feed chiefly on mollusca, and also on larvæ, crustacea, and sometimes small fish, for which they dive. 1. The GOLDEN-EYED GARROT (*Clangula vulgaris* is a common species in Britain; 2. The HARLEQUIN GARROT (*Clangula histrionica*) occurs as a rare straggler. [See DUCK.]

GASTEROPODA. The name of a class of molluscous animals which move from place to place by means of a fleshy disc, or foot, situated under the abdomen. The greater part of these Mollusca consist of animals inhabiting a univalve shell, which is cone-shaped and rolled into a spiral; and of such the snail is a familiar specimen. Some species, on the contrary, have no shell; of which the slug is an example. The body is elongated, and terminates in front by a head, more or less developed, with a mouth provided with from two to six tentacula; the back is enveloped in a mantle, which secretes the shell; and the belly is covered on its under side by the fleshy mass of the foot. In most aquatic Gasteropods whose shell is spiral, there is a horny or calcareous disc, called the *operculum*, which is attached to the hinder part of the foot, and is used for closing the entrance of the shell when the animal withdraws itself. Some of the Gasteropoda inhabit fresh waters, but most of them are marine animals: some are formed for crawling, as the snail, the whelk, &c.; some are more adapted for swimming; while a few of this class attach themselves to the surface of rocks, scarcely varying their locality, as is the case with the limpet; this attachment being produced by the adhesion of the muscular disc, or foot, which acts like a sucker, and can at any time be detached by the will of the animal.

In the work of Mrs. Gray, of the British Museum, on Mollusca, are figures of the animals of most of the genera of Gasteropoda. To this very carefully executed and authoritative work, we refer our readers, as well as to the various miscellaneous articles in this work; such as ACHATINA, BULIMUS, SNAIL, &c.

Fossil Gasteropoda. Among the numerous organic remains which exist, none are more extensively diffused throughout the globe, occupying the various geological formations, than fossil univalve shells. It is, indeed, asserted by some of the most experienced geologists, that every fossil turbinated univalve of the older beds, from the transition lime to the lias, belongs to the herbivorous genera, which class extends through every stratum in the entire series of geological formations, and still retains its place among the inhabitants of our existing seas: while, on the other hand, the shells of marine carnivorous univalves are very abundant in the tertiary strata above the chalk, but are rare in the secondary strata from the chalk downwards to the inferior oolite; beneath which no trace of them has yet been found.

GASTEROSTEUS. [See STICKLEBACK.]

GASTORNIS. [See SUPPLEMENT.]

GASTROBRANCHUS. (*Gastrobranchus glutinosa*). The HAG. A cartilaginous fish, which in its general appearance bears a near resemblance to the Lampreys, but which in the Systema Naturæ of Linnæus has been considered as belonging to the class *Vermes*. It is of a dusky bluish cast above, and reddish towards the head and tail; is from four to six inches long, and is remarkable for its total want of eyes: the mouth, which is situated beneath, is of an oblong form: on each side are two beards or cirri, and on the upper part four; in front of the top of the head is a small spout-hole, furnished with a valve, by which it can at pleasure be closed: the teeth, which are of an orange-colour, are arranged on each side of the mouth in a double-row, and in the middle of the roof of the mouth is one sharp-pointed and curved tooth. It has no scales, nor any kind of fin but that which forms the tail, the extremity of the body, where it is surrounded by the caudal fin, which is very shallow, being pointed. Beneath the body, from head to tail, runs a double row of equi-distant pores, through which, on pressure, exudes a viscid fluid; and beneath the body are two spiracles having apertures communicating with a series of six globular red cells or vesicles on each side of the body. "The manners of this fish are represented as highly singular: it is said to enter into the bodies of such fishes as it happens to find on the fishermen's hooks, and which consequently have not the power of escaping its attack, and by gnawing its way through the skin, to devour all the internal parts, leaving only the bones and the skin remaining. Another peculiarity in this animal consists in its uncommonly glutinous nature: if put into a large vessel of sea-water it is said in a very short space to render the whole so glutinous as easily to be drawn out into the form of threads. When taken out of water the Gastrobranchus is said to be incapable of living more than three or four hours." The species we have been describing is called the *Blind* Gastrobranchus, and is an inhabitant of the Northern seas. Another, and a much larger one, called the *Dombeyan* Gastrobranchus, from its having been first noticed by M. Dombey, is found in the South American seas.

GASTROCHÆNA. The name given to a genus of Acephalous Molluscs, found on the coasts of Great Britain and America. They inhabit an equivalve, inequilateral shell, united by a ligament, and having in the interior a small spoon-shaped curvature. The Gastrochæna penetrates and makes its abode in hard substances; and seldom exceeds half an inch in length. They are found in the hollows of shells or other marine substances.

GAVIAL, or GARIAL. An enormous Reptile found in India, to which the name of the GANGETIC CROCODILE (*Crocodilus Gangeticus*) is sometimes applied; but the sub-genus termed *Gavial*, by Cuvier, is so strikingly distinguished both from the Crocodile of the Nile and the Alligator by the peculiar form of the mouth, that it is hardly possible, even on a cursory view, to confound it with either of them; the jaws being remarkably long, narrow, and straight, constituting the anterior part or beak, spreading out at its base, and terminating in front, so as to remind the observer of the beak of the Spoonbill. The head, properly so called, has its sides straight and perpendicular, the upper surface being quadrilateral; and the mandible, instead of being continued from the forehead by a gradual slope, sinks suddenly to follow a straight and nearly horizontal direction. In the general form and colour of the body and limbs it resembles the common Crocodile, but the number of transverse zones or bands formed by the rows of scales on the back, is greater than in that species. The teeth are nearly double the number of those of the Nilotic Crocodile, and are of equal size throughout the whole length of the jaws. It is quite as aquatic in its habits as is the African species; its hind feet fully webbed; and the crest on the tail, increasing the surface by which it strikes the water, is much elevated. This powerful animal frequently attains the length of twenty-five feet; and, from its strength and ferocity, is truly formidable. In one respect, however, it is found very serviceable, viz. in devouring the numerous dead bodies of men and animals which are committed to the "sacred river," the effluvia arising from which would otherwise, in all probability, be productive of contagious diseases. Analogous species of Crocodiles have been found in a fossil state in Yorkshire and other places. The fossil group is named *Teleosaurus* [which see].

GAZELLE. (*Antilope Dorcas.*) Of all the Antelopes of the East none are so celebrated for beauty as the Gazelle; and oriental poets, from time immemorial, have thought that the highest compliment they could pay the female sex was to compare the eyes of a lovely woman with the lustrous organs of vision which distinguish that light and graceful animal. This very beautiful species inhabits Arabia and Syria, where they are seen in large groups, bounding across the desert with such amazing fleetness that they seem, bird-like, to skim over the surface. It is so swift that the greyhound is generally unable to overtake it without the assistance of falcons, which fly at its head and thus check its speed till the

GAZELLE.—(ANTILOPE DORCAS)

dogs regain their lost distance. The Ariel Gazelle is about twenty inches high at the shoulder; its limbs are slender, but vigorous; and all its actions are spirited and graceful. It is of a dark fawn colour above, and white below; the upper parts being divided from the lower by a deep dark band along the flanks. On each side of the face a broad stripe of white passes from the horns over the eyes to the nose. Wild and timid as the Gazelle is, when taken young it is readily domesticated; and it is frequently seen at large in the court-yards of houses in Syria, their exquisite form, general beauty, and playfulness rendering them especial favourites.

GECARCINUS. The name given to these Crustaceans which are formed to live at a distance from the sea; some residing in fresh water, and some burrowing in the ground, even at a distance from water. [See *Land Crab*, art. CRAB.]

GECKO. This name is given to a considerable number of Saurian Reptiles, and is said to be taken from the sound of their voice, which resembles the word *gecko* uttered in a shrill tone. Our figure, which represents a common New Holland species, is named by Mr. Gray *White's Phyllure*, or *Gecko*. It was first described by Dr. Shaw in White's Voyage to New South Wales, and is the *Phyllurus platurus* of naturalists; but though very characteristic of the group we prefer giving an account of the COMMON GECKO (*Gecko verus*):—It is of a thicker and

WHITE'S GECKO.—(PHYLLURUS PLATURUS.)

stouter form than most other lizards, having a large and somewhat triangular flattish head, covered with small scales, a wide mouth, large eyes, minute teeth, and a broad flat tongue. The limbs are of moderate length, and the feet are of a broader form than in the rest of the genus *Lacerta*, each toe being dilated on the margins, and divided beneath into a great number of parallel transverse lamellæ, without any longitudinal mark or furrow; all the toes, except the thumbs, are furnished with small claws; the tail, which is generally longer than the body, is marked, more or less, according to the age, into divisions or verticillated rings: the whole animal is covered on the upper parts with numerous, distant, round warts or prominences, approaching more or less to an acute form in different individuals, and sometimes obtuse: beneath each thigh is a row of perforated papillæ, as in the Green Lizard and many others: the under parts of the body are covered with scales of somewhat dissimilar appearance, but all approaching to a round figure."

In describing the habits, food, &c. of the *Geckotidæ*, Mr. Broderip observes that "the greatest number feed on small animals, such as insects, their larvæ and pupæ. These they catch either by lying in ambush, or by pursuing their feeble prey in the holes and dark crevices to which it retires. The structure of their feet enables them to run in every direction over the smoothest surfaces, and they can even remain suspended beneath the large leaves which a luxuriant tropical vegetation so frequently puts forth. The sharp and retractile nails with which the feet of the greater number are armed enable them to cling to and make rapid progress on trees with the smoothest bark, or penetrate the holes of rocks, and to climb walls. Of sombre or varying colours adapted generally to the locality where their lot is cast, they will often remain for hours in positions as extraordinary as the flies and insects for which they watch, the wonderful apparatus with which their feet is furnished enabling them to overcome the general law of gravity, and without which they would instantly fall to the earth. The hues of their skins thus render them less objects of suspicion to the little animals for which they lie in wait, and also serve to dodge even the acute eye of the bird of prey that seeks to destroy them. Their eyes enable them to discern objects in the dark, and are at the same time capable of bearing the rays of a bright sun; for many insects are nocturnal or crepuscular, while the great mass of them are diurnal. The pursuit of their prey leads them near the habitations of man, whose dwelling always attracts certain kinds of insects, and they sometimes fall victims to their appearance, which frequently inspires terror, and often disgust. A Gecko, confident in his powers of flight, appears boldly to await his adversary, and his sudden disappearance at a nearer approach adds to the horror which his uncouth form inspires. The poor Geckos too have a bad name. They are supposed to poison whatsoever they touch, be it animate or inanimate, and their saliva is said to vex the skin of those on whom it falls with foul eruptions. Many of these cuticular irritations, when they have actually existed from the intervention of these animals, may have arisen from the extremely sharp claws of a Gecko running over a sleeping man, or small blisters may have been raised by the adherent apparatus at the bottom of its feet. In each great division of the globe various species of the Geckotidæ are found, though very few of them exist in Europe.

Descriptions of the numerous species will be found in Mr. Gray's catalogue of the reptiles in the British Museum, where there is a large collection of these interesting lizards. By some biblical commentators, "the spider that taketh hold with her hands, and is in king's palaces" is believed to have been a Gecko; Gecko are very common in houses in the East, and may be seen running about the walls.

GENET. (*Viverra genetta.*) This animal belongs to the Weasel tribe; has a very beautiful soft fur; and is about the size of a very small cat, but is of a longer form, with a sharp pointed snout, upright ears, slightly pointed, and very long tail. The colour of the Genet is usually a pale reddish grey, the sides of the body being spotted with black, and a dark line running along the back; where the hair, being longer than on the other parts, resembles a slight mane: the muzzle is dusky; beneath each eye is a white spot; the cheeks, sides of the neck, and the limbs, are spotted in a proportionally smaller pattern than the body, and the tail is marked with black and white rings. Easily tamed, and of a mild disposition, the Genet, at Constantinople, and various other parts of the East, is domesticated like the Cat, and is said to be equally if not more serviceable in clearing houses of rats and other vermin. It is a native of the western parts of Asia, and is also occasionally found in Spain; but though it requires a warm climate for its subsistence and propagation, it has not been discovered in India or any part of Africa. This animal, like the Civet, produces an agreeable perfume; it is, however, less powerful, and its scent much sooner evaporates.

There are two or three other species found in the East; among these may be mentioned

RASSE GENET.—(VIVERRA MALACCENSIS.)

the Rasse (*Viverra Malaccensis*), found in Java by Dr. Horsfield, but also a native of the Indian continent: our figure shows this well-marked species.

GEOMETRIDÆ. A family of Lepidopterous insects, of very considerable extent. It is distinguished from the NOCTUIDÆ by its general weakness of structure and slenderness of body, but still more by the remarkable peculiarities and mode of progression of the caterpillars. The wings are large and of various outlines; in general they are horizontally extended, but in a few species they are carried vertically; the maxillæ short, weak, and nearly membranous; the labial palpi small and cylindrical; the antennæ variable, being in some males strongly bipectinated; the legs are slender, the anterior tibiæ being armed with a spur on the inside, and the posterior with two pairs. From their peculiar mode of progression, the caterpillars are called *Loopers* or *Geometricians:* they have only three pairs of pectoral, and one pair of ventral pro-legs, with a pair of anal feet; they then extend the body to its greatest length, when they put down their fore feet, drawing the hind part of the body as close after them as possible, so as to form an arch, like a pair of compasses, fixing their hind feet, and proceeding again as before. It is evident that they possess great muscular power, and hence their positions during repose are very striking. Fixing themselves by their anal feet alone, they extend their bodies in a straight line, holding it in that position for a long time together. This, together with their obscure colours, and the warts on their bodies, renders it often quite difficult to distinguish them from twigs of trees on which they feed. When alarmed, these caterpillars have the instinct to drop from the leaves, and suspend themselves by a thread, which enables them to remount when the danger is past. The chrysalides are sometimes naked and suspended by the tail, but more frequently enveloped in a slight cocoon, and placed among dry leaves, &c. In their perfect state the Geometridæ fly sluggishly in the twilight, or, if abroad in the day, and are disturbed, they quickly settle again amongst the foliage. Many species have a broad wavy band across the fore wings; these are called Carpet Moths. Figures of all the British species will be found in the very useful work of Mr. Humphreys, "The British Moths."

GEOPHAPS. A genus of birds found in Australia, belonging to a minor group of the *Columbidæ* family, whose habits and economy are very peculiar. Several species are described by Mr. Gould, from whose superb work we glean the following particulars of one, named—

GEOPHAPS SCRIPTA, OR PARTRIDGE BRONZE-WING. This bird is said to be second to none in the world as a delicate viand for the table; while it is equally interesting to the sportsman, no other bird not strictly gallinaceous so closely resembling the genus *Perdix* (Partridges) in many of its habits and manners; in Mr. Gould's opinion, indeed, "in no instance is the theory of the analogical relationship of one group to another more strikingly borne out than in the close resemblance of the members of this group to those of the genus *Perdix*." It is sometimes seen in pairs, but more frequently in small coveys of from four to six in number, which, when approached, generally run off with exceeding rapidity, and crouch down among any scanty herbage, instead of seeking safety by flight; the colouring of the bird assimilating so closely to that of the ground or the herbage, that when crouched down for shelter it is not easily to be seen. When it rises, it does so with great rapidity, making a loud whirring noise with its wings, and generally alighting on the horizontal branch of a large tree. On such plains as are intersected with rivers and pools of water, the Partridge Bronzewing is mostly found; and its principal food is the seeds of various grasses and other small plants, to which are occasionally added insects and berries. The plumage of the head, back, and chest is light, the edges of the primaries and the extremities of the wing-coverts being much paler; a broad stripe of white runs from beneath the mandible to beneath the eye, another stripe from the posterior angle of the eye down the side of the neck, the interspaces being jet black, which colour surrounds the eye, and also forms a crescent across the lower part of the throat; abdomen gray; flanks white; tail grayish brown, tipped with black; naked skin round the eye bluish lead-colour; bill black; feet dark purplish brown. The female lays two eggs on the bare ground, without any nest; and the young birds run and fly strongly when they are only as large as a quail.

GEOPHILUS. [See CENTIPEDE.]

GEOSPIZA. A genus of the Finch tribe, found by Mr. Darwin on the Galapagos islands, and characterised by the species having an enormously thick and well-developed hard bill. They are terrestrial in their habits. The accompanying cut, copied

THICK-BILLED GROUND-FINCH (GEOSPIZA MAGNIROSTRIS.)

from Mr. Gould's figure in the Bird portion of the zoology of the voyage of H.M.S. Beagle, will show the form and appearance of the GEOSPIZA MAGNIROSTRIS, and the accompanying extract from the ever-interesting journal of Mr. Darwin tells us all that is known of their habits. "These birds," he says, "are the most singular of any in the Galapagos archipelago. They all agree in many points, namely, in a peculiar structure of their bill, short tails, general form, and in their plumage. The females are gray or brown, but the old cocks jet-black. All the species, excepting two, feed in flocks on the ground, and have very similar habits. It is very remarkable that a nearly perfect gradation of structure in this one group can be traced in the form of the beak, from one exceeding in dimensions that of the largest grosbeak, to another differing but little from that of a warbler."

GERBILLUS. A genus of Glirine mammalia, chiefly found in South Africa and in India: most of the species are long-tailed, and may be seen in the British Museum collection.

GERBOA. (*Dipus Ægyptus.*) [See JERBOA.]

GHOST-MOTH. (*Hepiolus humuli.*) A nocturnal Lepidopterous insect, which receives this name from the male being of a white colour, and from its habit of hovering with a pendulum-like motion over one spot (often in church-yards), where the female is concealed. Of the singular habits of this insect the following interesting particulars are given in the "Journal of a Naturalist." The larva which produces this creature is hidden in the ground during the season of winter; the fly being formed in the month of May, and soon rising from the soil, then commences its short career. At this time one or more of them may frequently be observed under some hedge in a mead, or some low place in a damp pasture, only a few feet from the ground, persevering for a length of time together in a very irregular flight, and falling, and balancing about in a space not exceeding a few yards in circumference, an action not observable in any other, and fully indicating this moth. This procedure is not the meaningless vagary of the hour, but a frolicsome dance, the wooing of its mate,

LARVA AND CHRYSALIS OF GHOST-MOTH.

which lies concealed in the herbage over which it sports. The two insects are something similar in their general form, but very differently marked. The male exhibitor is known by his four glossy, satiny, white wings, bordered with buff; the lady reposer

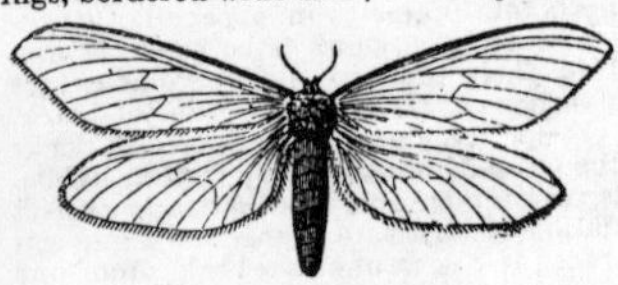

MALE GHOST-MOTH.
(HEPIOLUS HUMULI.)

has her upper wings of a tawny yellow, spotted and banded with deep brown. They are very inert creatures, easily captured; and their existence appears to be of very short duration, as we soon cease to observe them, either in action or at rest. The male probably becomes the prey of every bird that feeds by night; his colour and his actions rendering him particularly obnoxious to dangers of this nature; and the frequency with which we find his wings scattered about, points out the cause of death to most of them. The bat pursues with great avidity all those

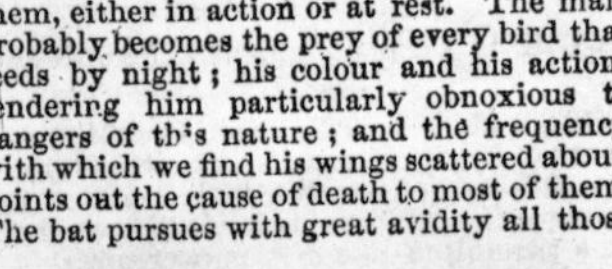

FEMALE GHOST-MOTH.
(HEPIOLUS HUMULI.)

creatures that fly in the evening; and by its actions it seems to meet with constant employment, and has greater probability of success than some insectivorous birds that feed by day, as all the myriads which abound at this time are the sole prey of itself and a few nocturnal ramblers. From this singular flight in the twilight hour, haunting as it were one particular spot, the fancy of some collector, considering it as a spectre-like action, named it the Ghost-moth.

GIBBON. (*Pithecus lar.*) The Gibbon, or Long-armed Ape, is a species of the *Quadrumana*, distinguished from others by the slenderness of its form, but more particularly by the extraordinary length of its arms, which, when the animal is standing erect, reach to the ankle-joints. The hands and feet are even more adapted for climbing than those of the Orang-Outang; their form, in fact, is admirably suited to their arboreal habits; and they are here observed to sweep from the branch of one tree to another with surpassing velocity: suspending themselves by their long anterior limbs, they launch onwards by an energetic muscular movement, seizing with wonderful precision the distant branches, and continuing their progression without any pause or perceptible effort. It is worthy of note, that their feet, which are very long, have the soles turned so much inwards as to afford no support to the erect posture. The colour of the Gibbon is black; but the face is commonly surrounded with a white or grey beard. There is a variety, called the WHITE GIBBON, which

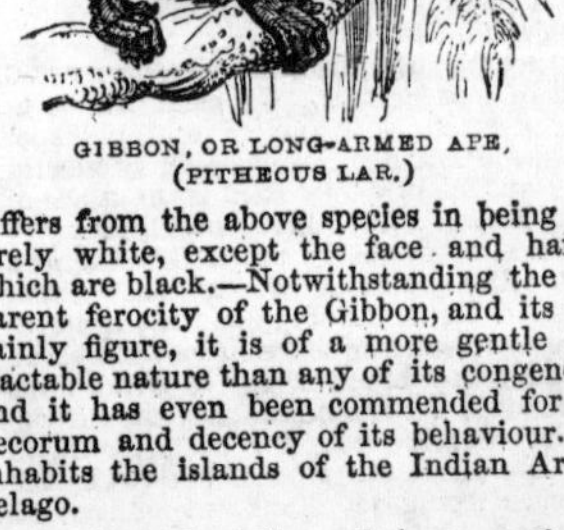

GIBBON, OR LONG-ARMED APE.
(PITHECUS LAR.)

differs from the above species in being entirely white, except the face and hands, which are black.—Notwithstanding the apparent ferocity of the Gibbon, and its ungainly figure, it is of a more gentle and tractable nature than any of its congeners; and it has even been commended for the decorum and decency of its behaviour. It inhabits the islands of the Indian Archipelago.

GILTHEAD. (*Chrysophris auratus.*) An Acanthopterygious fish, of a broad and compressed form, about twelve inches in length, and somewhat resembling the Bream. It is found in abundance in the Mediterranean, and is sometimes taken on the coasts of France and Spain. The back is sharp, and of a dusky green or silvery gray colour; between the eyes there is a gold-coloured crescent-shaped stripe, from which it receives its name: the incisor teeth in each jaw are conical, the molar ones roundish; the tail is very forked, the fins are grayish-blue, the dorsal fin extending almost the whole length of the back. It feeds on various kinds of crustacea and mollusca; and chiefly inhabits deep waters and bold rocky shores.

GIRAFFE, or CAMELOPARD. (*Camelopardalis Giraffa.*) This most remarkable Ruminant, which in its general structure most nearly approaches the Deer, has points of affinity, also, with the Antelopes and Camels, besides very striking peculiarities of its own. If height alone constituted precedency among quadrupeds, the Giraffe, as Le Vaillant justly observes, must hold the first rank. The enormous apparent length of the fore legs and its long tapering neck must strike every one at the first glance: while its small and elevated head, its large and brilliant eyes, its mild aspect, and the whole contour of the animal, differing from all others, cannot fail to excite admiration; for, notwithstanding the unusual proportions of the limbs, its general form is not merely elegant but highly picturesque. The horns of the Giraffe differ both in texture and shape from

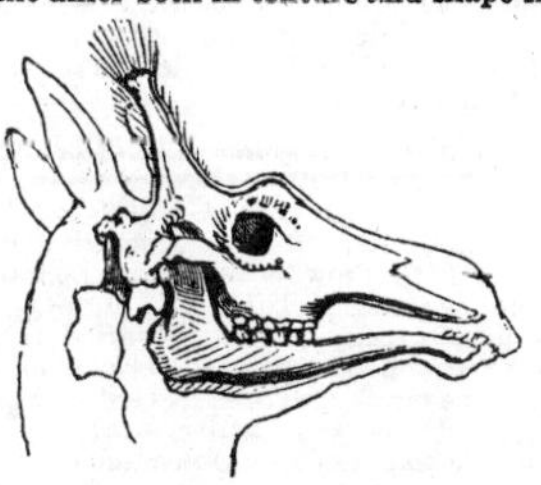

SKULL OF GIRAFFE.

those of all other horned quadrupeds; forming, as it were, a part of the skull, and consisting of two porous bony substances, about three inches long, with which the top of the head is armed, and which are placed just above the ears, and crowned with a thick tuft of stiff upright hairs: a considerable protuberance also rises on the middle of the forehead between the eyes, which appears to be an enlargement of the bony substance, similar to the two horns just mentioned. The neck is furnished with a very short stiff mane. The tail is of moderate length, gradually tapering towards the end, and terminating in a tuft of long hair. The fore part of the body is very thick and muscular; the hind part thin and meagre. The Giraffe, in its wild state, when full grown, measures seventeen feet from the top of the head to the fore feet; the female, however, is not so high; and it must be understood that this measurement is taken at the maximum height, none of those brought to or bred in Europe having reached more than fourteen feet. At first view the fore legs seem twice the length of the hind: but this difference, on accurate examination, appears to result chiefly from the extraordinary height of the shoulders.

The colour of the Giraffe is a light fawn, marked with numerous large spots of a darker hue, less regularly shaped on the sides than on the neck and shoulders. The vertebræ of the neck are slightly curved; but although nothing can exceed the gracefulness of form which this part sometimes presents, the fewness of the joints prevents the neck from being generally bent or arched with swan-like elegance. The peculiarities of conformation which this animal displays are all adapted to the mode of life which is natural to it; for it is destined to browse upon the foliage and young shoots of trees, at a height far greater than that which any other animal can reach, whilst standing on the ground. For this purpose it is furnished with an elongated prehensile tongue, with which it lays hold of the tender branches, and draws them into its mouth; being assisted by its projecting upper lip, which is at once flexible and very muscular. In order to bring its mouth to the ground, which it seldom does except to drink, or to pick up some unusual delicacy, the Giraffe is obliged to stretch its fore legs widely apart, and to bend its neck into a semicircular form. "The head of the Giraffe resembles that of the camel in the absence of a naked muzzle, and in the shape and organization of the nostrils, which are oblique and narrow apertures, defended by the hair which grows from their margins, and surrounded by cutaneous muscular fibres by which the animal can close them at will. This is a beautiful provision for the defence of the air passages, and the irritable membrane lining the olfactory cavities, against the fine particles of sand which the storms of the desert raise in almost suffocating clouds. The large, dark, and lustrous eyes of the Giraffe, which beam with a peculiarly mild but fearless expression, are so placed as to take in a wider range of the horizon than is subject to the vision of any other quadruped. While browsing on his favourite acacia, the Giraffe, by means of his laterally projecting orbits, can direct his sight so as to anticipate a threatened attack in the rear from the stealthy lion, or any other foe of the desert. To an open attack he sometimes makes a successful defence by striking out his powerful and well-armed feet; and the king of beasts is said to be frequently repelled and disabled by the wounds which the Giraffe has thus inflicted with his hoofs. The horns of the Giraffe, small as they are, and muffled with skin and hair, are by no means the insignificant weapons they have been supposed to be. We have seen them wielded by the males against each other with fearful and reckless force; and we know that they are the natural arms of the Giraffe most dreaded by the keeper of the present living Giraffes in the Zoological Gardens, because they are most commonly and suddenly put in use. The Giraffe does not butt by depressing and suddenly elevating the head, like the deer, ox, or sheep; but strikes the callous obtuse extremity of the horns against the object of his attack with a sidelong sweep of the neck. One blow thus delivered at full swing against the head of an unlucky attendant would be fatal:—the female once drove her horns in sport through an inch board. Notwithstanding those natural arms of hoofs and horns, the Giraffe does not turn to do battle except at the last extremity; where escape is possible, it seeks it in flight This is extremely rapid, especially along rising ground; but cannot be maintained for a sufficient period of time to enable it to escape the Arab mounted on his long-winded steed. The paces of the Giraffe, owing to the disproportion between his long legs and short body, are very peculiar: when walking at a brisk rate, it seems to move forward simultaneously the two legs of the same side, as noticed of old by the learned bishop of Sicca, in his account of the presents brought to Hydaspes by the Abyssinian ambassadors." "In the sanded paddock appropriated to the Giraffes in the Zoological Gardens, they exhibit in the warm days of summer all their various and singular paces. In the simple walk, the neck, which is then stretched out in a line with the back, gives them a stiff and awkward appearance; but this is entirely lost when they commence their graceful undulating canter." "The tongue is an organ exquisitely formed for prehension; it is used to hook down the branches which grow beyond the reach of the muzzle of the Giraffe, and the animal in captivity instinctively puts it to use in a variety of ways. We have seen the Giraffe, in the Jardin des Plantes at Paris, stretching upwards its neck and head, and protruding its tongue to the full extent, to hook out single straws, which were platted into the partition, separating it from the contiguous inclosure. In our own menagerie at Regent's Park many a fair lady has been robbed of the artificial flowers which adorned her bonnet, by the nimble filching tongue of the rare object of her admiration. The Giraffe seems, indeed, to be guided more by the eye than the nose in the selection of objects of food; and if we may judge of the apparent satisfaction with which the mock leaves and flowers so obtained are masticated, the tongue would seem by no means to enjoy the sensitive in the same degree as the motive powers. The difference in the size of the nerves of sensation and motion which we observed in the dissection of the tongue accords with these habits of the living animal. From the same dissection it was proved that the movements of the tongue, both those of extension, prehension, and retraction, were due to muscular, and not, as Sir Everard Home supposed, to vascular action. Observations of the living animal, and dissection of the dead, have at length dispelled most of the errors and doubts which obscured the exact knowledge of the nature and zoological affinities of the Giraffe."—"A Giraffe more than two-thirds grown will eat daily in confinement eighteen pounds of clover hay, and eighteen pounds of a mixed vegetable diet, consisting of carrots, mangel-wurzel, barley, split beans, and onions; and will drink four gallons of water. They copulate in March. The female has four inguinal udders: she brings forth one young at a birth; and the period of gestation is fifteen months. The new-born Giraffe measures six feet from the fore-hoofs to the top of the head. In a few hours it is able to follow the dam. It resembles the mature animal in the markings of the hide. The first Giraffe known to have been produced in captivity was brought forth in June, 1839, at the gardens of the Zoological Society of London."—*Brande's Dict.* Two varieties of this curious animal are known; one of them peculiar to Nubia, Abyssinia, and the adjacent districts; the other a native of Southern Africa.

The remains of an animal closely allied to the Giraffe has been found in a fossil state, by Capt. Cautley and Dr. Falconer, in the Seewalik Hills in India. They have described it under the name *Sivatherium.* The head is a gigantic resemblance of that of the Giraffe; as may be seen in the fine specimen preserved in the gallery of the British Museum.

GLASS-SNAKE. The name given in North America to a species of lizard, the OPHISAURUS VENTRALIS. It belongs to the family ZONURIDÆ of Mr. Gray, and has doubtless acquired its name from its "brittleness,"—a habit not uncommon with lizards of allowing their tails to be left in the hands of any who surprise them.

GLAUCOPIS. A genus of birds belonging to the family CORVIDÆ; the only known species being the GLAUCOPIS CINEREA, or NEW ZEALAND CROW. This bird, which has all the habits of a crow, is called by the natives of New Zealand *Kokako.* Its plumage is a very dark green, not much varied in any part of the body; the legs are black and coarse, the claws long. It has a strong black beak, a little curved; and a small brilliant light-blue flap hanging down on each side from the ear: the colour of these flaps fades, however, immediately the bird is dead, and becomes of nearly the same hue as its plumage.

GLAUCUS. A genus of molluscous animals found in the warmer latitudes floating in the open sea, and remarkable for their beautiful azure blue and silvery tints. They are about one inch and three quarters in length, with a subcylindrical body, and the tail terminating in a sharp point, the head furnished with four very short tentacula, and the sides of the body having tufts or branchiæ disposed in pairs, surrounded by digitated appendages, fitted for swimming.

GLIRES. (Lat. *glis*, a dormouse.) The fourth order of Mammalia in the Linnæan system, distinguished by two flat incisors in each jaw. They are also called RODENTIA, or GNAWERS.

GLOBE-FISH. [See DIODON and TETRAODON.]

GLOMERIS. A myriapode bearing a strong resemblance to the woodlouse, in its oval form, and its habit of rolling itself into a ball. [See MYRIAPODA and ZEPHRONIA.]

GLOW-WORM. (*Lampyris noctiluca*). This curious and interesting insect (the female of which, being expressly called by this name), is seen during the summer months, as late as the close of August, on dry banks, about woods, pastures, and hedgerows, exhibiting, as soon as the dusk of the evening commences, the most vivid and beautiful phosphoric splendour. The male insect is

GLOW-WORM.—(LAMPYRIS NOCTILUCA.)

rather more than half an inch in length; the head is of a dun colour, the thorax margined with dusky red, as are also the legs and the edges of the segments of the body; and the wings are shorter than the body. The female is wingless, but in most other respects resembles the male: the thorax is semicircular; the body is very soft, of an oblong form, and pointed at the extremity. It is hardly yet determined with certainty whether the male Glow-worm is at all luminous; but it is universally understood that if it be, it is in a very slight degree. The phosphorescent light emitted by the female, and which can be increased or lessened at will, proceeds from the abdomen, near the tail; it is of a yellow colour, with a very slight cast of green. The larva, pupa, and complete female insect scarcely differ perceptibly from each other in general appearance, but the phosphoric light is strongest in the perfect animal. The general idea among naturalists is, that the light emitted by the female is for the purpose of attracting the other sex; and in numerous instances have poets availed themselves of so pleasing a simile as "the Glow worm's amorous fire," to illustrate the pure intensity of that flame which so often burns in a heroine's breast. Dismissing the poetical metaphor, however, we may observe that the Glow-worm is a slow-moving, inactive insect, and its light not perceptible in the day-time, even if carried into a darkened room, unless the creature is turned on its back, and put in motion; but as night advances, its lamp again begins to burn. On this subject Mr. Knapp remarks, that on a warm dewy evening at the end of September he observed on the house-bank multitudes of these small evanescent sparks in the grass. "The number of them and their actions, creeping away from our sight, contrary to that half-lifeless dulness observed in summer, suggested the idea that the whole body had availed themselves of this warm moist evening to migrate to their winter station. A single spark or so was to be seen some evenings after this, but no such large moving parties were to be discovered again. If we conclude that the summer light of the glow-worm is displayed as a signal taper, the appearance of this autumnal light can have no such object in view, nor can we rationally assign any use of it to the creature itself, unless, indeed, it serves as a point of union in these supposed migrations, like the leading call in the flight of night-moving birds. The activity and numbers of these insects, in the above-mentioned evening, enabled me to observe the frequent presence and disappearance of the light of an individual, which did not seem to be the result of will, but produced by situation." [See LAMPYRIS: ELATER.]

GLUTTON. (*Gulo arcticus.*) A carnivorous quadruped, of a very voracious nature, and about the size of a large Badger; between which animal and the Polecat it seems to be intermediate; nearly resembling the former in its general figure and aspect, and agreeing with the latter as to its dentition. The muzzle, beyond the eyes, is blackish brown, covered with hard shining hair; between the eyes and ears runs a whitish or ash-coloured band or fillet; the top of the head and back are black-brown, the sides inclining to a chestnut colour; and the feet are black. These animals are slow and comparatively deficient in agility; but they are very persevering, determined, and cunning. In the northern regions, both of the Old and New World, they are said to be of

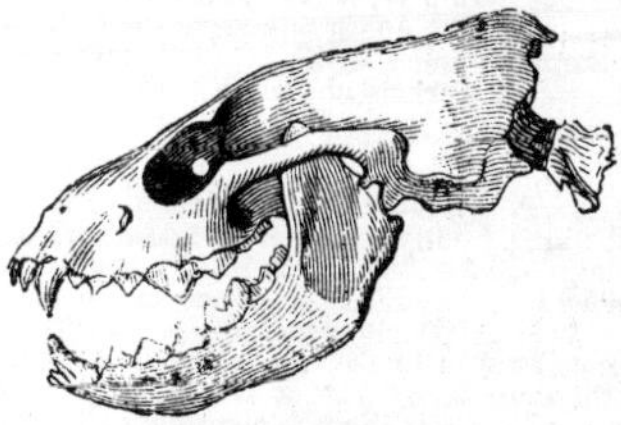

SKULL OF GLUTTON.
(GULO ARCTICUS.)

uncommon fierceness and strength, sometimes even disputing their prey with the Wolf and Bear. They often proceed at a steady pace for miles, hunting out weak or dying animals, and stealing unawares upon hares, marmots, birds, &c. They are said to surprise the larger quadrupeds, such as the Rein-deer and the Elk, as they lie asleep; and to tear the neck and throat in the same manner as the Weasel. What they cannot devour at once they are said to hide under ground or in a hollow tree. They prefer putrid flesh, and are extremely fetid. The female brings forth two young at a litter once a year. The fur is much used for muffs, linings, &c.; and the skins brought from Siberia are much preferred to others, from their being of a more glossy black. This animal is also called the WOLVERENE.

GLYCIPHILA. A genus of Tenuirostral birds belonging to the family *Meliphagidæ*, of which we may mention GLYCIPHILA FULVIFRONS, or the FULVOUS-FRONTED HONEY-EATER. This species, Mr. Gould observes, differs sufficiently from the true *Meliphagi* to fully justify its separation into a distinct genus. It prefers to dwell among the trees that crown the low stony ridges, rather than those growing on the lower lands or the brushes; its flight is rapid, it mounts high in the air, and flies off to a distance with an extremely rapid horizontal and even motion. The song is rather remarkable, being commenced with a single note slowly drawn out, and followed by a quick repetition of a double note, repeated several times in succession, and mostly uttered when the bird is perched on the topmost branch of a tree. It is an exceedingly active bird among the branches, gracefully clinging about and around the flowers of the *Eucalypti* in search of food. It builds in some low bush or scrubby plant, near the ground, the nest being of a compact cup-shaped form, constructed of dried grasses, and lined with soft wool. The eggs are rather large, and often much lengthened; sometimes quite white, but more generally blotched with large marks of chestnut-red. It feeds on the pollen of flowers and insects.

GLYPTODON. The name given to an extinct quadruped, of gigantic dimensions, which, like the Armadilloes of the present day, was covered with a tesselated bony armour. In size it was equal to the Rhinoceros. Prof. Owen has published an elaborate memoir on it, which is beautifully illustrated. The fine specimen in the College of Surgeons must strike every visitor by its dimensions, curious characters, and state of preservation. It was found in South America.

GNAT. (*Culex.*) The *Culicidæ*, or Gnat tribe, are a family of Dipterous insects, whose mouths are furnished with bristly stings, included in flexile sheaths. Some of the species are extremely troublesome, as they pierce the skin to feed upon the blood, and at the same time inject an irritating poisonous fluid. Their flight is accompanied by a humming noise, occasioned by the vibration of their wings: they seldom appear in the day-time, except in thick woods, and they abound in moist situations, which is easily accounted for by their larvæ being inhabitants of the water. In this state they are very active, swimming with great agility, and often descending; but coming to the surface to breathe, which they do head downwards, the respiratory orifice being at the end of a very prolonged spiracle arising from the end of the abdomen. — That well-known insect the COMMON GNAT (*Culex pipiens*) is produced from a singular-looking aquatic larva: it has a large head, furnished on each side with a pair of antennæ-like jointed processes; the thorax large and angular; the body suddenly lessening from this part, and continuing of nearly the same size to the tail, which is abruptly truncated, and tipped with four foliaceous processes. In about fifteen days' time the larvæ are full grown, and arrive at the pupa state; the animal then appears to have a rounded form, is very active, and still inhabits the water; the position of its breathing apparatus, however, is now altered, being situated at the anterior part of the body, and consists of two little tubes, which are applied to the surface of the water for the reception of air. When ready to assume the perfect state, it rises to the surface, and the Gnat quickly emerges from its confinement. A warm, rainy season is most favourable to the evolution of Gnats; and, in such summers, particular districts in most countries are occasionally pestered by them in countless swarms. Those persons who inhabit the more favoured regions of the European continent can hardly conceive what torments are endured from them in some parts of the world; but of all people the Laplanders appear to be the greatest sufferers; for during the heats of their short summer, the Gnats fill the air with such swarming myriads, that the poor inhabitants can hardly venture to walk out of their cabins, without having first smeared their hands and faces with a composition of tar and cream, which is found by experience to prevent their attacks.

A very small black Gnat (*Culex reptans*), with transparent wings, and the legs marked by a white bar, is particularly troublesome in marshy districts during the evening, by its creeping motion on the skin of the face, &c.

To the above we may add, that the MOSQUITO (*Culex mosquito*), so much dreaded by all who visit the West Indies and America, where its bite seems to operate with peculiar malignity, is a species of Gnat which derives additional vigour from the warmer and moister atmosphere. But it is not wonderful that in uncultivated wastes, where the waters stagnate, and the heat of the sun is almost insupportable, that the atmosphere should frequently be filled with clouds of these insects, varying in size from three or four inches in length to a minuteness only discernible by the assistance of a microscope. [See MOSQUITO.]

GNATHODON. A genus of bivalve shells, of which there is one well-known species, (*Gnathodon cuneatus*), from New Orleans. It is ovate, equivalve, and equilateral; and is known from all other shells by the characters of the hinge, having in one valve, a sharp, angular, notched, cardinal tooth, and two lateral teeth, the posterior of which is elongated, and the anterior angulated, tortuous, shaped like a jawbone; in the other valve, two cardinal and two lateral teeth, the interior of which is wedge-shaped. Ligament internal, cuneiform; muscular impressions two. The name has also subsequently been given to a genus of birds. [See next article.]

GNATHODON. A genus of birds described by Sir W. Jardine, from a specimen which belonged to Lady Harvey. From the contour of its beak, which has the upper mandible strongly hooked, as in the Dodo, and the under mandible deeply notched, it is supposed by Mr. Gould to be frugivorous or granivorous; the beak being expressly adapted to denude palm nuts, or other strongly coated seeds, of their hard outer covering. Mr. Gould considers that it is more nearly allied to the Pigeon tribe (*Columbidæ*) than to any other family; the form of the body and wings, and the structure of the feathers, indicating this affinity. The only known species, *Gnathodon strigi-*

rostris, is rather larger than a partridge, and has the head, neck, breast, and belly, of a glossy green black; the back, wings, tail,

GNATHODON STRIGIROSTRIS.

and under tail-coverts, of a deep chestnut red: the beak and naked part round the eye are of a yellowish colour. It is believed to be a native of one of the South Sea Islands; and the *Didunculus*, found by the recent American Voyage of Discovery under C. Wilkes, is thought to be the same bird.

GNU. (*Boselaphus Gnu.*) A very singular species of Antelope, which, at first sight, appears to be a monstrous being, compounded of parts of different animals. Its general colour is a deep umber-brown, approaching to black. It is four feet in height, having the body and crupper of a small horse, and is covered with brown hair; the tail is furnished with long white hairs (like a horse), and on the neck is a beautiful flowing mane, white at the base, and black at the tips. Its horns, approximated and enlarged at the base, descend outwardly, and turn up at the point; the muzzle is large, flat, and surrounded by a circle of projecting hairs; under the throat and dewlap is another black mane; and the legs are as light and slender as those of a stag. The Gnus inhabit the wild karoos of South Africa and

GNU.—(BOSELAPHUS GNU.)

the hilly districts, where they roam mostly in large herds, and migrate according to the season. They are naturally wild and difficult of approach; and when first alarmed, they fling up their heels and plunge about like a restive horse: they soon, however, take to flight, and traverse the desert with such astonishing celerity—not in a tumultuous mass, but in single file, following a leader—that they are quickly out of danger. When wounded they will sometimes turn upon the hunter and pursue him in turn, darting forwards on their assailant with amazing force and impetuosity, so that it requires the utmost coolness on his part to evade the attack. When taken young, this animal is easily domesticated.

GOAT. (*Capra hircus.*) The distinguishing characters in the genus *Capra* in the Linnæan system of Zoology are,—that the horns are hollow, turned upwards, and annulated on their surfaces; that there are eight cutting teeth in the lower jaw, and none in the upper; and that the male is generally bearded. In its domestic state the Goat is found in almost every part of the globe, bearing the extremes of heat and cold, and differing in size and form according to various circumstances; the horns generally having a curvature outwards towards the tips.

Buffon's account of this animal is strikingly descriptive. "The Goat," says he, "is superior to the sheep both in sentiment and dexterity. He approaches man spontaneously, and is easily familiarized. He is sensible of caresses, and capable of a considerable degree of attachment. He is stronger, lighter, more agile, and less timid than the sheep. He is a sprightly, capricious, wandering, wanton animal. It is with much difficulty that he can be confined, and he loves to retire into solitude, and to climb, stand, and even sleep, on rugged and lofty eminences. He is robust and easily nourished, for he eats almost every herb, and is injured by very few. His bodily temperament, which in all animals has a great influence on the natural disposition, is not essentially different from that of the sheep. These two animals, whose internal organization is almost entirely similar, are nourished, grow, and multiply in the same manner; and their diseases are the same, excepting a few, to which the Goat is not subject. The Goat fears not, like the sheep, too great a degree of heat. He cheerfully exposes himself to the sun, and sleeps under his most ardent rays without being affected with the vertigo or any other inconveniency. He is not afraid of rain or storms; but he appears to feel the effects of severe cold. The inconstancy of his disposition is marked by the irregularity of his actions. He walks, stops short, runs, leaps, approaches or retires, shows or conceals himself, or flies off, as if actuated by mere caprice, and without any other cause than what arises from an eccentric vivacity of temper. The suppleness of his organs, and the strength and nervousness of his frame, are hardly sufficient to support the petulance and rapidity of his natural movements."

The original stock of the Common Goat, as of other races of animals early subjugated by Man, cannot be distinctly traced; but it appears to be the same with that of numerous half-domesticated breeds, which abound in Asia. Mr. Bell, in his History of British Quadrupeds, remarks, that "most modern zoologists who have paid much attention to the question, and who have brought to the consideration of it all the helps which recent discoveries in philosophical zoology have furnished, have leaned to the belief that the Ægagrus, or Wild Goat of the mountains of Caucasus and of Persia, is the true original stock. The zoological characters of this animal certainly bear a closer resemblance to those of the domestic breeds; and it is worthy of remark, that the horns of the Persian Domestic Goat, though smaller, are similar in form to those of the *Paseng* or *Ægagrus*. The arguments which have been urged from the intermixture of the *Ibex* with the Common Goat are at present of little value; as the facts recorded are very deficient. The large Goats which are reported to have been brought from the Alps and the Pyrenees to the Garden of Plants in Paris, and which were stated to have been wild, were probably the progeny of the *Ibex* with the Common Goat, as there is no proof of the existence of the true *Ægagrus* in Europe. These were found to be capable of producing offspring, and the details are given by Mr. Fred. Cuvier with great clearness; but the old fault still remains; the question is not set at rest by these observations; for we are only informed that they produce offspring, without any statement whether they will breed *inter se*, or only with the Common Goat. The progeny, however, were either prematurely brought forth, or lived only a short time in a sick and languishing condition."

"The condition of the Goat, in some parts of our own islands," says the same intelligent and accurate writer, "is much more wild than that of any other of our domestic animals. In the mountains of Wales especially the Wild Goats roam over the most inaccessible parts of the mountains and rocks, without the slightest appearance of domestication, or of having been deduced from a domestic stock. It is a hardy, active, powerful animal; capable of maintaining its footing on the smallest point on which its feet can possibly rest, and of taking considerable leaps with the utmost certainty of safely alighting, although the spot which it desires to attain be perhaps but the rugged point or ledge of a precipice. It will thus find its food in places inaccessible to almost all other animals, and live and thrive by cropping the scanty herbage which they furnish. In the mountain ranges of Europe, on the Alps and Pyrenees, the Goat is found at a great elevation, approaching as near the line of perpetual snow as it can find the scanty means of its sustenance; and it feeds on many plants which to other ruminants are distasteful, and even deleterious: thus, hemlock, henbane, and digitalis is eaten by it with impunity, and even the acrid euphorbia is not rejected."

The milk of the Goat is sweet, nutritive, and medicinal; this may be accounted for from the animal's food being chiefly derived from the heathy mountains and shrubby pastures, where sweet and aromatic herbs abound. In ancient times the skin of the Goat was regarded as a most useful article of clothing: it is still manufactured into the best Turkey or Morocco leather; while that of the kid (whose flesh is regarded as a delicacy) forms the softest and most beautiful leather for gloves, &c. The usual colour of the domestic Goat is black and white, or a pale reddish-brown, with a black stripe down the back; but sometimes brown, white, or spotted. In old males the beard is very long; and the horns, which are transversely rugose, bending outwards and falling back, are sometimes nearly three feet long.

We find that the Common Goat inhabits most parts of the world, either native or naturalized. It endures all kinds of weather, being found in Europe as high as Wardhuys, in Norway, where it feeds during the winter season on moss, the bark of fir trees, and even of logs intended for fuel. It is also asserted that they thrive equally well in the hottest parts of Africa and India. The odour of the Goat, which at all times is proverbially strong, is intolerably so in the rutting season (from September to November); but it is commonly believed that horses are refreshed by it, which accounts for this animal being often kept about stables. Upon this subject Mr. Bell observes, "Many persons keep Goats in their stables, from an idea that they contribute to the health of the horses; a fancy not perhaps so far-fetched or absurd as at first sight it might appear; for I believe that all animals are kept in better temper and greater cheerfulness by the presence of a companion than in solitude, and the active and good-humoured Goat may in this way really perform the benefit which has been attributed to it upon mistaken grounds;—indeed, instances of close attachment between the horse and the stable Goat are not unfrequent." The female goes five months with young, and usually produces two kids at a birth; sometimes, however, three, and occasionally but one.

The Angora Goat (*Capra Angorensis*) is by far the most elegant of all the varieties of the Goat, and is a native of Angora, a small district of Asia Minor, and remarkable for producing not only this peculiar race of goats, but also sheep, cats, rabbits, &c., with hair of uncommon fineness. The Goat of Angora is generally of a beautiful milk-white colour, short legged, with black, spreading, spirally-twisted horns, and pendulous ears. Its chief and distinguishing excellence, however, is the wool, which covers the whole body in long pendent spiral ringlets; and it is from the hair of this animal that the finest camlets are made.

The Cashmere Goat, so highly prized for its fleece, is descended from the Goat of Thibet, which pastures on the Himalaya. It is smaller than the common domestic Goat, and has long, fine, silky wool. Thibet is situated at the northern descent of the Himalaya mountains, and Cashmere at the southern; hence there is some difference in the climate; it is observed, also, that the colder the region where the animal pastures, the heavier and finer is its fleece. The Goats which pasture in the highest vales of

Thibet are of a bright ochre colour; in lower grounds, the colour becomes of a yel-

CASHMERE GOAT.

lowish-white, and still farther downwards entirely white. The highest mountains of the Himalaya inhabitable by man contain also a species of black Goats, the wool of which as a material for shawls in India obtains the highest price. The Goats of Thibet and Cashmere have the fine curled wool close to the skin, just as the under hair of our common Goat lies below the coarse upper hair. The flesh of the Cashmere Goat tastes as well as that of the common one; and its milk is as rich; but these animals owe their great celebrity to the extraordinary beauty and costliness of the shawls for which the Asiatics have been so long famous.

The SYRIAN GOAT is remarkable for its large pendulous ears, which are usually from one to two feet in length, and sometimes so troublesome to the animal, that the owners are obliged to trim them, to enable it to feed more at ease. The horns are black, bending a little forwards; and are only about two inches long. The colour of its hair is like that of a fox; and there are two fleshy excrescences under its throat. This variety appears to have been known to Aristotle.

There are several other varieties of the Goat which it is needless to enumerate. But there is one species in North America (the ROCKY MOUNTAIN GOAT), which we should notice, inasmuch as it has given rise to much difference of opinion as to its proper place in a system of arrangement. It has been designated *Ovis montana*. The Rocky Mountain Goat nearly equals in size a common sheep, and has a shaggy appearance, in consequence of the protrusion of the long hair beyond the wool, which is white and soft. Its horns are about five inches long, conical, somewhat curved backwards, and projecting but slightly beyond the wool of the head. They are in great numbers about the head waters of the Columbia, and furnish the principal part of the food of the natives of that district. They appear to be more numerous on the western than on the eastern side of the mountains, and are rarely seen in the plains. The skin is very thick and spongy, and is principally used in making moccasons. Next to the fleece of the Cashmere Goat this is believed to be the finest; and it is prized accordingly.

GOAT-MOTH. [See COSSUS LIGNIPERDA.]

GOATSUCKER, or NIGHT-JAR. (*Caprimulgus Europæus.*) There are many species of Goatsuckers, but this is the only one of the genus that inhabits Europe permanently, the *Caprimulgus ruficollis* being confined to South Western Europe, and appearing there only in the summer. With us it is only a summer visitant, appearing about the middle of May, and retiring in September or October. Before, however, we give a description of the bird, it may be as well to observe that the name *Goatsucker*, although very generally used, and retained in most ornithological works, has no foundation but in the ignorance and superstition of the ancients, who believed it sucked the teats of goats; on which account Bewick suggests the propriety of dropping the

NIGHT-JAR. – (CAPRIMULGUS EUROPŒUS.)

name, and adopting that of NIGHT-JAY, "which, though not universally known, bears some analogy to the nature and qualities of the bird, both in respect to the time of its appearance, which is always the dusk of the evening, as well as to the jarring noise it utters whilst at rest perched on a tree, and by which it is peculiarly distinguished." Like the Owl, it is seldom seen in the daytime, unless disturbed, or on dark and gloomy days, when its eyes are not dazzled by the bright rays of the sun. As moths, gnats, beetles, and other night insects are its food, it is peculiarly formed to enable it to catch them on the wing. For this purpose nature has bestowed on it a mouth of great comparative size, which as the Goatsucker flies, it can quickly [illegible] while the insects are additionally secured by [illegible] bill being surrounded on the inner side with a glutinous substance that prevents their escape. This manner of flying with its mouth open is the cause of that whirring noise which this bird makes while chasing its prey. It arises from the resistance made to the mouth by the air; and is more or less loud according to the velocity with which the bird moves. When perched, it usually sits on a bare twig, with its head lower than its tail, and in this attitude, utters its jarring. It does not perch like other birds, sitting across the branch, but lengthwise, and its hinder toe is capable of being turned forward as well as backward. It is solitary in its habits, and is generally seen alone.

The colours of this bird, though plain, have a beautiful effect from the elegance of their disposition, the plumage being beautifully freckled, barred, and spotted with browns, black, grey, and ferruginous, variously arranged and diversified. The bill is small, flat, and hooked at the tip ; the eyes are large, full, and black ; the legs are short, rough, and scaly, and feathered below the knee : the toes are connected by a membrane as far as the first joint ; the middle one is considerably longer than the rest, and the claw is serrated on one side. The use of this peculiar organ is not clearly ascertained ; by some it is affirmed that the bird employs it to clear away the fragments of insects that may have clogged up the fringe of bristles ; by others, that it strikes its prey with its foot, and that this long serrated claw enables it to hold the insect more securely ; and by others again, that it uses it to clean its plumage. The male is distinguished from the female by an oval white spot, near the end of the first three quill-feathers. These birds frequent moors and wild heathy tracts abounding in fern ; they make no nest, but the female deposits her eggs, which are of a dull-white colour, on the ground. Montbelliard, who wrote this bird's history for Buffon, states, that it no sooner perceives its retreat to be discovered by an enemy, than it carefully rolls its eggs to a more secure situation.

There are other species bearing the same general appellation ; one of which is known in America as *Whip-poor-Will;* another as *Chuck-Will's-Widow*, a third as the *Night-hawk*, and a fourth as the *Rain-bird*. There are also the *Banded Goatsucker*, and *Crested Goatsucker*, natives of New Holland ; besides several inhabiting various parts of India, Africa, &c. These are placed in different genera : for descriptions of figures of which we refer our readers to the works of Mr. Gould, and of Messrs. Gray and Mitchell. The two first-mentioned we shall here describe, from Wilson.

WHIP-POOR-WILL. (*Caprimulgus* [*Antrostomus*] *vociferus.*) The notes of this solitary and celebrated bird, when first heard in the spring, at evening twilight or morning's dawn, seem like the voice of an old friend, and are listened to by almost all with great interest. At first they issue from some retired part of the woods, the glen, or mountain; in a few evenings, perhaps, we hear them from the adjoining coppice, the garden fence, the road before the door, and even from the roof of the dwelling-house, long after the family have retired to rest. He soon becomes a regular acquaintance. Every morning his shrill and rapid repetitions are heard from the adjoining woods ; and when two or more are calling out at the same time, as is often the case in the pairing season, and at no great distance from each other, the noise, mingling with the echoes from the mountains, is really surprising. Their notes seem pretty plainly to articulate the words which have been generally applied to them, *Whip-poor-*

WHIP-POOR-WILL.
CAPRIMULGUS VOCIFERUS.)

Will, the first and last syllables being uttered with great emphasis, and the whole in about a second to each repetition ; but when two or more males meet, their whip-poor-will altercations become much more rapid and incessant, as if each were straining to overpower or silence the other. When near, you often hear an introductory cluck between the notes. At these times, as well as at almost all others, they fly low, not more than a few feet from the surface, skimming about the house and before the door, alighting on the wood-pile, or settling on the roof. Towards midnight they generally become silent, unless in clear moonlight, when they are heard with little intermission till morning. If there be a creek near, with high precipitous bushy banks, they are sure to be found in such situations. During the day they sit in the most retired, solitary, and deep-shaded parts of the woods, generally on high ground, where they repose in silence. When disturbed, they rise within a few feet, sail low and slowly through the woods for thirty or forty yards, and generally settle on a low branch or on the ground. Their sight appears deficient during the day, as, like owls, they seem to want that vivacity for which they are distinguished in the morning and evening twilight. They are rarely shot at or molested ; and from being thus transiently seen in the obscurity of dusk, or in the deep umbrage of the woods, no wonder their particular markings of plumage should be so little known, or that they should be confounded with the Night-hawk, whom in general appearance they so much resemble. The female begins to lay about the second week in May, selecting for this purpose the most unfrequented part of the wood, often where some brush, old logs, heaps of leaves, &c. had been lying, and always on a dry situation. The eggs are deposited on the ground, or on the leaves, not the slightest appearance of a nest being visible. These are usually two in number, in shape much resembling those of the Night-hawk, but having the ground colour much darker, and more thickly marbled with dark olive.

The Whip-poor-Will is nine inches and a half long, and nineteen inches in extent; the bill is blackish, a full quarter of an inch long, and bent a little at the point, the under mandible arched a little upwards; the nostrils are prominent and tubular, their openings directed forward; the mouth is extravagantly large, of a pale flesh colour within, and beset along the sides with a number of long, thick, elastic bristles, which end in fine hair, and curve inwards; these seem to serve as feelers; and prevent the escape of winged insects: the eyes are very large, full, and bluish black; the plumage above is so variegated with black, pale cream, brown, and rust colour, sprinkled and powdered in such minute streaks and spots, as to defy description; the upper part of the head is of a light brownish gray, marked with a longitudinal streak of black, with others radiating from it; the back is darker, finely streaked with a less deep black; the scapulars are very light whitish ochre, beautifully variegated with two or three oblique streaks of very deep black; the tail is rounded, the three outer feathers on each side are blackish brown for half their length, thence pure white to the tips; the deep brown of these feathers is regularly studded with light brown spots; the four middle ones are without the white at the ends, but beautifully marked with herring-bone figures of black and light ochre finely powdered. The cheeks and sides of the head are of a brown orange colour; the wings, when shut, reach scarcely to the middle of the tail, and are elegantly spotted with very light and dark brown; chin black, streaked with brown; a narrow semicircle of white passes across the throat; breast and belly irregularly mottled and streaked with black and yellow ochre, legs and feet light purplish flesh colour, seamed with white; the former feathered before, nearly to the feet; the two exterior toes are joined to the middle one, as far as the first joint, by a broad membrane; and the inner edge of the middle claw is pectinated, and therefore probably employed as a comb to rid the plumage of its vermin. The female is about an inch less in length and in extent, and differs also in being much lighter on the upper parts. Their food appears to be large moths, grasshoppers, ants, and such insects as frequent the bark of old rotten and decayed timber: they are also expert in darting after winged insects. They will sometimes skim in the dusk, within a few feet of a person, uttering a kind of low chatter as they pass. In their migrations north, and on their return, they probably stop a day or two at some of their former stages, and do not advance in one continued flight. This bird, like the owl and other nocturnal flyers, is regarded with a kind of suspicious awe by the weakminded and superstitious. "Night," says Wilson, "to minds of this complexion, brings with it its kindred horrors, its apparitions, strange sounds, and awful sights; and this solitary and inoffensive bird being a frequent wanderer in these hours of ghosts and hobgoblins is considered by the Indians as being, by habit and repute, little better than one of them.

The CHUCK-WILL'S-WIDOW. *Caprimulgus* [*Antrostomus*] *Carolinensis*.) This species is twelve inches long, and twenty-six in extent; bill yellowish, tipt with black; the sides of the mouth are armed with numerous long bristles, strong, tapering, and furnished with finer hairs branching from each; cheeks and chin rust colour, speckled with black; over the eye extends a line of small whitish spots; head and back very deep brown, powdered with rust and cream colour, and marked with long ragged streaks of black; scapulars broadly spotted with deep black, bordered and in-

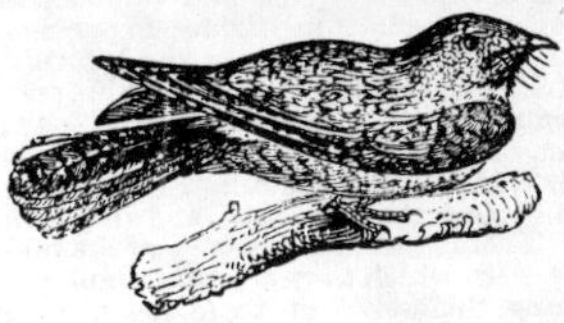

CHUCK-WILL'S-WIDOW
(CAPRIMULGUS CAROLINENSIS.)

terspersed with a creamy white: the plumage of that part of the neck which falls over the back is long, something like that of a cock, and streaked with yellowish brown; wing quills barred with black and bright rust; tail rounded, extending about an inch beyond the tips of the wings; the middle feathers being powdered with various tints of ferruginous, and elegantly marked with fine zig-zag lines, and large herring-bone figures of black; exterior edges of the three outer feathers barred like the wings, their interior vanes being pure snowy white, marbled with black; across the throat is a slight whitish band; breast black, powdered with rust; belly and vent lighter; legs feathered before nearly to the feet, which are of a dirty purplish flesh-colour.

The Chuck-Will's-Widow, whose notes seem exactly to articulate these words, commences its singular call generally in the evening, soon after sunset, and continues it, with short occasional interruptions, for several hours. This note, or call, instantly attracts the attention of a stranger, and is strikingly different from that of the Whip-poor-Will. In sound and articulation it seems plainly to express the words which have been applied to it (*Chuck-Will's-Widow*), pronouncing each syllable leisurely and distinctly, putting the principal emphasis on the last word. In a still evening it may be heard at the distance of nearly a mile, the tones of its voice being stronger and more full than those of the Whip-poor-Will, who utters his wi[illegible] much gre[illegible]ter rapidity. The flight of this bird [illegible] skimming about at a few feet above the surface of the ground, frequently settling on old logs, or on the fences, and from thence sweeping around, in pursuit of various winged insects that fly in the night. Like the Whip-poor-Will, it prefers the declivities of glens and other deeply shaded places, making the surrounding mountains ring with echoes the whole evening. The Chuck-Will's-Widow lays its eggs, two in number, on the ground, in the woods; they are of a dull olive colour, sprinkled with darker specks, and about as large as a pigeon's.

This singular genus of birds, formed to subsist on the superabundance of nocturnal insects, are exactly and surprisingly fitted for their peculiar mode of life. Their flight is low, to accommodate itself to their prey; silent, that they may be the better concealed, and sweep upon it unawares; their sight, most acute in the dusk, when such insects are abroad; their evolutions, something like those of the bat, quick and sudden; their mouths capable of prodigious expansion, to seize with more certainty, and furnished with long branching hairs, or bristles, serving as palisadoes to secure what comes between them. Reposing so much during the heats of day, they are much infested with vermin, particularly about the head, and are provided with a comb on the inner edge of the middle claw, with which they are often employed in ridding themselves of these pests, at least when in a state of captivity. Having no weapons of defence except their wings, their chief security is in the solitude of night, and in their colour and close retreats by day; the former so much resembling that of dead leaves, of various hues, as not to be readily distinguished from them even when close at hand. [See NIGHT-HAWK.]

GOBIOIDEÆ. A family of Acanthopterygious fishes, including the Blennies, Gobies, &c. They may be recognised by the slenderness and flexibility of their dorsal rays. They have an uniformly wide intestinal canal, and no pyloric cæca.

GOBY. (*Gobius.*) A genus of Acanthopterygious fishes, of which there are several species, of a small size, in general varying from three to six inches in length; but none of them are much esteemed for food. They are distinguished by their ventral and thoracic fins being united in their whole length, or at their bases. The spines of the dorsal

RED GOBY.—(GOBIUS CRUENTATUS.)

fins are flexible; the openings of their ears, with four rays. Like the Blenny, they can live a long time out of water. Several species are found in the Mediterranean, American, and Indian seas: some also on our own coasts. Three or four will suffice for examples.

The BLACK GOBY, or ROCK-FISH. (*Gobius niger.*) This is an inhabitant of the Mediterranean and Northern seas, and also of the rocky parts of our own coast: it grows to the length of six inches; the body is soft, slippery, and slender: the head large, the cheeks inflated, and the lips very thick; the mouth is wide, and furnished with numerous small teeth in each jaw, the lower of which is the longest. The ventral fins coalesce, and form a sort of funnel, by which these fish are said to affix themselves immovably to the rocks. The general colour of the fish is a dusky black, and the tail is rounded at the end.

The LANCE-TAILED GOBY (*Gobius lanceolatus*) is distinguished by and named from the peculiar form of its tail, which is large in proportion to the fish, and sharp-pointed at the tip. The body is of a lengthened shape, and nearly of equal diameter throughout: the head is oblong, and truncated in front; the jaws of equal length, and armed with sharp teeth; and the body is covered with scales, those toward the tail being much larger than those on the upper parts. This is a West Indian species.

The BLUE GOBY (*Gobius cœruleus*) is a highly beautiful, though very small species: colour fine blue, rather paler beneath: tail red, with a black border. From the brilliancy of its colours it appears, when swimming in a calm sea, during a bright sunshine, like a small tube of sapphire, tipped with carbuncle. It is found on the eastern coasts of Africa; and the Negroes use it as a bait for other fish.

The SPOTTED GOBY (*Gobius minutus*) is about three inches long; the head is large; the irides blue; the mouth wide, with several rows of small pointed teeth, curving inwards; the dorsal fins distinct, pectoral and ventral fins large; tail a little rounded. The general colour is a pale yellowish-white, freckled with minute light brown specks, and occasionally a row of larger spots along the lateral line. It is frequently taken on our sandy shores in shrimpers' nets; it is also plentiful in the Thames, where it is called by the fishermen Polewig, or Pollybait.

GODWIT. (*Limosa.*) There are several species of these Grallatorial birds. They are a timid, shy, and solitary tribe; characterized by a straight beak, longer than that of the snipes, sometimes slightly bent at the extremity, and by long legs, naked far above the knee. They live amidst the fens, salt marshes, and deep muddy places near the mouths of rivers; seldom remaining above a day or two in the same place, and often removing suddenly in a flock at night, when they fly very high. When pursued, they run with great speed, and scream as they rise. They subsist on worms and larvæ, and their flesh is very excellent. They are migratory, and moult twice in the year.

The COMMON GODWIT (*Limosa Ægocephala*) is sixteen inches in length, and weighs about twelve ounces. The bill is four inches long, bent a little upwards, and black at the point: the head, neck, back, scapulars, and coverts are a dingy reddish pale brown, each feather being marked down the middle with a dark spot. The fore part of the breast is streaked with black; belly, vent, and tail white, the latter barred with black: the webs of the first six quill-feathers black, edged on the interior sides with reddish brown: legs inclining to greenish blue. In the spring and summer the Godwit resides in the fens and marshes, where it rears its young; but when the winter sets in with severity, it seeks the salt-marshes and sea-shores.

The RED GODWIT. (*Limosa rufa.*) This species is not very common in Great Britain, but is found in the north of Europe, and is very plentiful in the fenny parts of North America, about Hudson's Bay, &c. It is larger than the Common Godwit, and is distinguished from it by the redness of its plumage; the head, breast, and sides being a bright ferruginous red, streaked on the head with brown, and marbled on the breast and sides with dusky, cinereous, and white; neck plain dull rusty red. The back, scapulars, greater and lesser coverts, are greyish brown; on the former, some of the feathers are barred and streaked with black and rufous, edged with pale reddish white; and a bar of white is formed across each wing by the tips of the greater coverts. The under parts are white, slightly spotted with brown. The legs are dusky, and bare considerably above the knees. Its flesh is reckoned delicious.

There are also the Great American Godwit, the Cinereous Godwit, the Black-tailed Godwit, the Red-breasted Godwit, &c., all more or less resembling the species above described.

GOLDEN-EYE [BUTTERFLY]. The name given by collectors to Butterflies of the species *Hipparchia pamphilus.*

GOLDEN WASPS, or GOLDEN-TAILED FLIES. (*Chrysis.*) The popular names for a tribe of Hymenopterous insects, which in the richness of their colours are said to "vie with the Humming-birds." They may be observed walking, but in a constant agitation and with great agility, upon walls and palings exposed to the heat of the sun. They are also found upon flowers. The body is elongated and covered with a solid skin; the hind wings are not veined, but the ovipositor is formed by the terminal segments of the abdomen, and terminated by a small sting; the antennæ are filiform, elbowed, and vibratile. The abdomen, which in the female appears to be formed of only three or four segments, is flattened or vaulted beneath, and capable of being folded against the breast, when the insect assumes an orbicular form. They deposit their eggs in the nests of Solitary Mason-bees, or other Hymenoptera, their larvæ destroying those of these insects. [See CHRYSIDIDÆ.]

GOLDFINCH. (*Fringilla carduelis.*) Of all the British Finches, none equal the Goldfinch in brilliant plumage and docility; hence it is one of those most frequently kept in captivity; for though its song is soft and pleasing, it is deficient in power. Its length, from the tip of the bill to the end of the tail, is five inches and a half; and the greatest expansion of its wings is nine inches. The bill is white, tipped with black; the forehead and chin a rich scarlet, which is divided by a black line passing from each corner of the bill to the eyes; the cheeks are white; top of the head black, that colour extending downward from the nape on each side; the back, rump, and breast are of a fine pale tawny brown colour; belly white; the wings and tail are black, but the points of the primaries in both are white; a beautiful yellow stripe runs across the wings: the tail feathers are black, with a white spot on each near the end; legs pale flesh red. The male is distinguished from the female by the feathers on the ridges of the wings, which are of a deep black colour; while those of the hen are a dusky brown; and the black and yellow in the wings of the latter are less brilliant than in those of the male. The nest of the Goldfinch is small, but ex-

GOLDFINCH.—(FRINGILLA CARDUELIS.)

tremely beautiful; the outside consists of very fine moss curiously interwoven with wool, hair, and other materials; and the inside is lined with the down of thistles and other soft and delicate substances. The nest is often found in an orchard, large garden, or plantation, in an apple or pear tree, or carefully placed in some thick evergreen shrub — somewhere in the neighbourhood of Man, but not immediately within his view. The bird lays five or six white eggs, marked with deep purple spots at the larger end. They feed their young with caterpillars and insects; and the old birds feed on various kinds of seeds, particularly those of the thistle, dandelion, and groundsel.

Goldfinches are more easily tamed than other birds; and so reconciled will they in time become to their imprisonment in cages, that they appear as if in reality attached to them. If a young Goldfinch is brought up under a canary, a wood-lark, or any other singing-bird, it will readily catch their song. Goldfinches breed with the Canary; this intermixture, says Bewick, succeeds best between the cock Goldfinch and the hen Canary, whose off[illegible] are productive, and are said to resem[illegible] male in the shape of the bill, and in the colours of the head and wings, and the hen in the rest of the body. Beauty of plumage, observes Buffon, melody of song, sagacity, and docility of disposition, seem all united in this charming little bird, which were it rare, and imported from a foreign country, would be more highly valued.

GOLD-FISH. (*Cyprinus auratus.*) In the fresh waters of China, we are told, certain beautiful species of *Cyprinus*, distinguished for the splendid golden colour of the membrane lying immediately beneath the scales, are as frequent as the most common river fish are here. Nor, indeed, are they at this time either rare or uncommon in our own ponds, being quite naturalized, and breeding freely in open waters.

The colours of Gold-fish are liable to the greatest variations : some are marked with a fine blue, brown, or bright silver hue, but the usually predominant colour is that of a brilliant gold. As an article of food they are not used, and are only valued for their beauty and gentleness.

GOLIATHUS (GOLIATH BEETLES). A group of Lamellicorn Coleoptera, which are chiefly found in Africa, the largest species being indigenous to the western coasts. These insects, which were formerly very scarce, and some of which are still rare in collections, have acquired their name from the large size of some of the species. One of the first specimens was found by Mr. Ogilvie, surgeon of H. M. S. Renown, at least eighty years ago. The specimen was dead, and found floating in the river Gaboon ; it is now in Glasgow, in the Hunterian Museum.

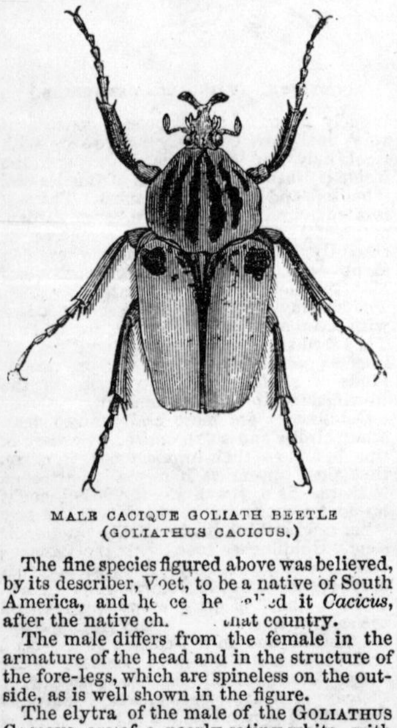

MALE CACIQUE GOLIATH BEETLE
(GOLIATHUS CACICUS.)

The fine species figured above was believed, by its describer, Voet, to be a native of South America, and hence he called it *Cacicus*, after the native chief of that country.

The male differs from the female in the armature of the head and in the structure of the fore-legs, which are spineless on the outside, as is well shown in the figure.

The elytra of the male of the GOLIATHUS CACICUS, are of a pearly satiny white, with a black opaque spot on the shoulder ; the crown of the head and the thorax being of a tawny yellowish brown, with black longitudinal bands on the latter.

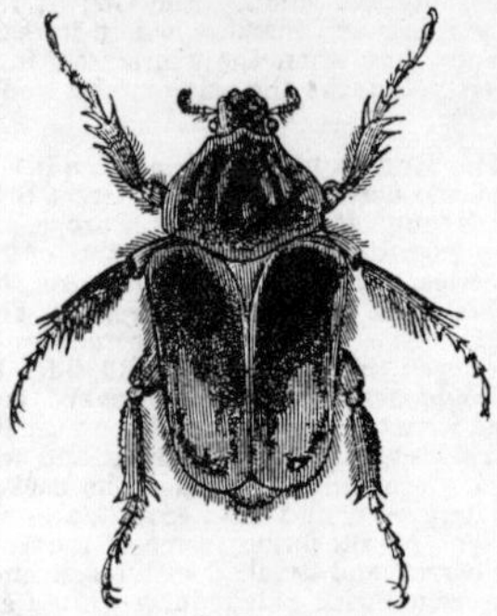

FEMALE CACIQUE GOLIATH BEETLE
(GOLIATHUS CACICUS.)

These insects are said to be roasted and eaten by the natives, who doubtless often make a *bonne bouche* of what would gratify many an entomologist.

The males of these insects, we are informed by Dr. Savage, are much more numerous than the females ; and though the various species of the group cannot be said to be very abundant, yet they are so frequently brought over now, that the large prices of 30*l.* and even 50*l.*, which used to be asked for them, are now very much reduced ; fine specimens, however, of some of the species still fetch 5*l.* or 6*l.*

Separated from them by slight generic characters, are two other groups, a male and female of one of which are figured beneath.

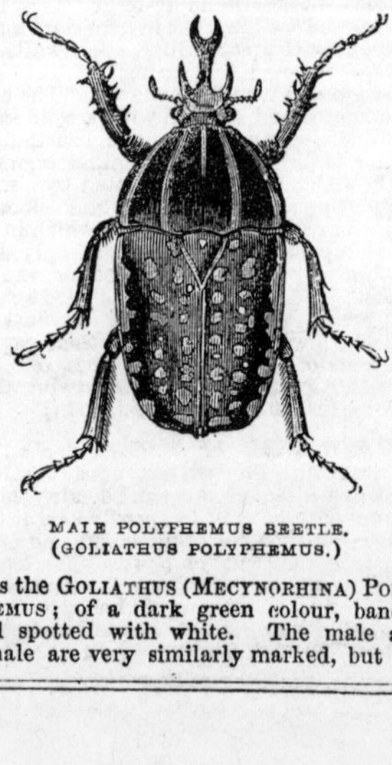

MALE POLYPHEMUS BEETLE.
(GOLIATHUS POLYPHEMUS.)

It is the GOLIATHUS (MECYNORHINA) POLYPHEMUS ; of a dark green colour, banded and spotted with white. The male and female are very similarly marked, but the distinctions in the head and fore leg may be plainly seen in the figures. The female of this species is generally regarded also as

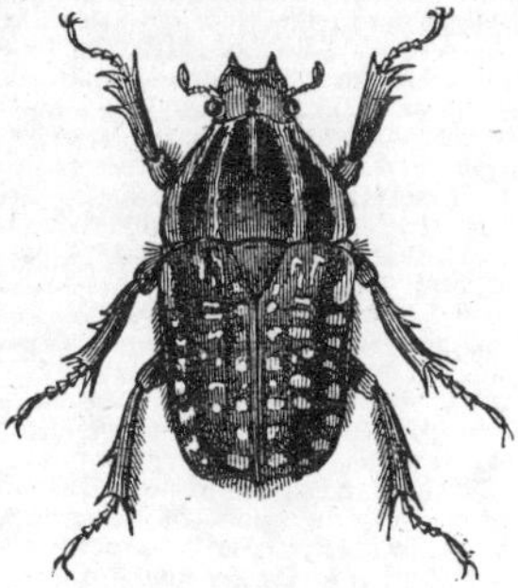

FEMALE POLYPHEMUS BEETLE.
(GOLIATHUS POLYPHEMUS.)

much rarer than the male. This species, as well as one named after Dr. Savage by Dr. Harris of Boston, feeds upon a vine that climbs over very lofty trees. The insects wound the bark of the vine, and extract the juice; the vine being full of a fluid as tasteless and limpid as water.

There are several other genera and species of Goliath Beetles, of most of which there are specimens in the collection of the British Museum; but we must refer our readers to the works of Dr. Burmeister and Mr. Westwood for descriptions and figures of these, it being quite out of the scope of this work to particularize them, splendid as they are. A list of all the species of *Cetonidæ* (including the Goliath), with reference to figures, has been published, and will show how rich the Museum collection is.

The next species (figured beneath) is from West Africa also, and is of a most brilliant green colour; it is the GOLIATHUS (DICRONORHINA) MICANS. The shades on this species vary according as the insect is held to the light.

GLITTERING GOLIATH BEETLE.
(GOLIATHUS MICANS.)

This insect seems to be a native of Senegal and the Calabar coast.

The food of the Goliaths is fluid, like that of the *Cetoniæ* and *Trichii:* the long brushes on their maxillæ, and the diverging rows of hairs that line their lower lips, are admirably fitted for absorbing liquid food, while their horny teeth afford these beetles additional means of obtaining it from the leaves and juicy stems of plants when the blossoms have disappeared. "Thus every new discovery in Natural History, when least expected, serves to increase the evidence of skilful contrivance and perfect adaptation of structure in all organised beings." With this admirable remark of Dr. Harris we conclude this article. [See CETONIADÆ: DICRONOCEPHALUS: INCA.]

GOLLACH. [See EARWIG.]

GONEPTERYX. A genus of diurnal Lepidoptera, so named from its angled wings. The British species is found with very slight variation on the Himalaya mountains; it is the

GONEPTERYX RHAMNI, or BRIMSTONE BUTTERFLY. This gay and lively-coloured insect is one of the earliest among the Papilionidæ that makes its appearance; sometimes, in favourable weather, even as early as the middle of February. Mr. Knapp, in his Journal of a Naturalist, alludes to this butterfly:—

"The very first butterfly that will
—— 'aloft repair,
And sport and flutter in the fields of air,'

is the Sulphur Butterfly (*Gonepteryx rhamni*), which in the bright sunny mornings of March we so often see under the warm hedge, or by the side of some sheltered copse, undulating and vibrating like the petal of a primrose in the breeze." As the spring advances they may be seen on the wing in the woods, meadows, and commons, in tole-

BRIMSTONE BUTTERFLY
(GONEPTERYX RHAMNI.)

rable plenty; and as there is a second brood which comes forth about August, there is no lack of them at any time till autumn summons them away. The male is of a pure sulphur-yellow above, and the female of a greenish-white, and in both sexes a small spot of orange occupies the centre of each wing, and a dusky spot at the base: the abdomen is black above and yellow beneath, its base and the thorax thickly clothed with long glossy white silken hairs: the legs are white; the antennæ reddish. The caterpillar is green, with a paler line on each side of the belly, and very small scale-like black dots on the back. It feeds on the buckthorn (*Rhamnus catharticus*). The pupa, or chrysalis, is green, very gibbous in the middle, and acuminated before: it is vertically suspended on a perpendicular branch, with a loose silken thread round it.

GONOPLAX, or ANGLED CRAB. A genus of decapod short-tailed Crustacea, of which one species is found in this country: it is the *G. angulata* of authors, the young of which has been described as a species of Gelasimus. In the male the fore legs are very long. For figures of this species see Dr. Leach's work, or that of Professor Bell, "The British Crustacea."

GOOSE. A genus of webfooted birds. The Common WILD GOOSE (*Anser ferus*), otherwise called the GREY LAG GOOSE, being the origin of our domestic species, we shall describe it first in order, after having made a few observations on the distinguishing characters of the genus. The bill is the first great distinction of the Goose kind from all the feathered tribes. In other birds it is round and weged-shaped, or crooked at the end; in all the Goose kind it is flat and broad, formed for the purpose of skimming ponds and lakes of the mantling weeds which grow on their surface. The bills of other birds are composed of a horny substance throughout, formed for piercing or tearing; but birds of this genus have their inoffensive beaks sheathed with a skin which entirely covers them; and are only adapted for shovelling up their food, which is chiefly confined to vegetable productions; for though they do not reject animal food when offered to them, they contentedly subsist on vegetable, and seldom seek any other.

The GREY LAG, or COMMON WILD GOOSE, as Pennant remarks, is our largest species; the heaviest weigh ten pounds; the length is two feet nine inches; the extent five feet. The bill is large and elevated, of a flesh colour tinged with yellow; the nail white

WILD GOOSE,—(ANSER FERUS.)

the head and neck cinereous, mixed with ochraceous-yellow; the hind part of the neck very pale, and at the base of a yellowish-brown; the breast and belly whitish, clouded with gray or ash-colour; the back gray, the lesser coverts of the wings almost white, the middle row deep cinereous slightly edged with white; the primaries gray, tipped with black and edged with white; the coverts of the tail and the vent feathers of a pure white; the breast and belly crossed and clouded with dusky and ash on a whitish ground; the middle feathers of the tail dusky, tipped with white, the exterior ones almost wholly white: the legs of a flesh-colour. This species is widely and numerously spread over all the more northerly parts of the globe, whence some flocks of them migrate a long way southward in the winter. Latham says, they seem to be general inhabitants of the world,—are met with from Lapland to the Cape of Good Hope,—are frequent in Arabia, Persia, and China, as well as indigenous to Japan,—and on the American continent from Hudson's Bay to South Carolina. As for their summer residences and breeding-places, the lakes, swamps, and dreary morasses of Siberia, Lapland, Iceland, and the unfrequented northern regions of America seem set apart for that purpose, where, with multitudes of other kinds, in undisturbed security, they rear their young, and are amply provided with a variety of food, a large portion of which must consist of the larvæ of gnats, which swarm in those parts, and the myriads of insects that are fostered by the unsetting sun.

These birds are often seen, in flocks of fifty or a hundred, flying at very great heights, and preserving very great regularity in their motions; sometimes forming a straight line, and at others assuming the shape of a wedge, which is supposed to facilitate their progress. Their cry is frequently heard when they are at an imperceptible distance above us. When on the ground, they range themselves in a line, after the manner of cranes; and seem to have descended rather for the sake of rest than for any other refreshment. Having continued in this situation for an hour or two, one of them, with a long loud note, sounds a kind of signal, to which the rest punctually attend, and rising in a group, they pursue their journey with renewed alacrity.

Their flight is conducted with singular regularity; they always proceed either in a line abreast, or in two lines joining in an angle at the middle, like the letter V. In this order they generally take the lead by turns, the foremost falling back in the rear when tired, and the next in station succeeding to his duty. Their track is generally so high that it is almost impossible to reach them from a fowling-piece; and even when this can be done, they file so equally, that one discharge seldom kills more than a single bird. They are very destructive to the growing corn in the fields where they happen to alight in their migrations. In some countries they are caught at such times in long nets, to which they are decoyed by tame geese placed there for that purpose. Other schemes are contrived to take them; but as they are very vigilant, feed only in the daytime, and betake themselves to the water at night, the fowler must exert his utmost care and ingenuity in order to accomplish his ends; all must be planned in the dark, and every trace of suspicion removed; for nothing can exceed the wary circumspection and acute ear of the sentinel, who, placed on some eminence, with outstretched neck, surveys everything that moves within the circle of his observations, and the instant he sounds the alarm, the whole flock betake themselves to flight.

But though they are seen regularly migrating southward in the autumn, and northward in the spring, they were formerly known to remain and breed in the fens of Lincolnshire and Cambridgeshire, and various other parts of Great Britain; the draining and cultivation of these marshy districts have now, however, nearly depopulated them of their former feathered inhabitants; but in lieu of the wild races, these localities are now teeming with domesticated ones in a highly improved condition. The Wild Goose lays from six to eight, sometimes ten or a dozen eggs, of a dirty greenish colour, the nest being placed among rushes, heaths, &c.

The TAME GOOSE. The wild species we have just described is, as before stated, the original of the domesticated Goose; to describe whose varied plumage, economy, and habits, may to many seem a superfluous task; while others, to whom they are less well known, may deem the account sufficiently interesting. How long they have been reclaimed from their original independence is not easily ascertained: but the time must have been very remote, for from a very distant date they appear to have held their present station, to have been kept for the self-same purposes, and to have been treated in the same manner. Their predominant colours are white and gray, with shades of ash and brown: some of them are yellowish, others dusky, and many are found to differ very little in appearance from the original stock. The only permanent mark, which all the gray ones still retain, like those of the wild kind, is the white ring which surrounds the root of the tail. They are generally furnished with a small tuft on the head; and the most usual colour of the males (the Ganders) is pure white; the bills and feet in both males and females are of an orange red. By studied attention in the breeding, two sorts of these Geese have been obtained—a larger and a smaller sort; the former weighing from ten to upwards of fifteen pounds, and frequently much more. The smaller kind are more delicate eating; delicacy, however, is often not so much regarded as the bountiful appearance and savoury smell of a "fine fat goose" on the festive board. But it is not altogether on account of their use as food that they are valuable; their feathers, their down, and their quills, have long been considered as articles of more importance, and from which their owners reap more advantages. Pennant, in describing the methods used in Lincolnshire, in breeding, rearing, and plucking Geese, says, They are plucked five times in the year: first at Lady-day for the feathers and quills; which business is renewed for the feathers only, four times more between that and Michaelmas: he adds, that he saw the operation performed even upon goslings of six weeks old, from which the tail feathers were plucked; and that numbers of the Geese die when the season afterwards proves cold. But this unfeeling process, as well as the care and attention which are bestowed upon the brood Geese while they are engaged in the work of incubation, is nearly the same everywhere. Wicker pens are provided for them, placed in rows, and tier above tier. Some place water and corn near the nests; others drive them to the water twice a-day, and replace each female upon her own nest as soon as she returns. At length the brood is hatched; and as soon as they are able to follow their parent Geese, they are driven to the neighbouring fens and marshes, on whose grassy-margined pools they feed and thrive without requiring any further attendance until the autumn. In this way immense numbers are reared in many parts of this country; but nowhere are there so many as in the fens of Lincolnshire, where it is said to be no uncommon thing for a single person to keep a thousand old Geese, each of which, on an average, will bring up seven young ones. So far those only are noticed which may properly be called the larger flocks, by which particular watery districts are peopled; but it must be borne in mind that they form only a part of the large family: and when the stock of the various farm-yards throughout the kingdom are added, the immense whole will appear multiplied in a ratio almost incalculable. A great part of those which are left to provide for themselves during the summer, in the solitary distant waters, as well as those which enliven the village green, are put into the stubble fields after harvest, to fatten on the scattered grain, while some are penned up for this purpose; and at length vast numbers are driven in flocks, or otherwise sent, to the great mart and focus of consumption, London; the provincial towns throughout the kingdom being also furnished with an adequate supply.

The Tame Goose lays from seven to twelve eggs, and sometimes more: these are carefully divided among the brood Geese when they begin to sit: those which lay a second time in the course of the summer are seldom, if ever, permitted to have a second hatching; but the eggs are used for household purposes. It is universally believed that the Goose lives to a great age, and particular instances are recorded by ornithologists which confirm the fact—some even emulating the human period of "threescore years and ten."—It has been remarked that none of our domestic birds are so apt to bring forth monstrous productions as Geese—a circumstance which has been attributed to the excessive fatness to which they are liable. The liver of a fat Goose is often larger than all the other viscera, and [illegible] dish i[illegible] so great reputation among the epicu[illegible] Rome, that Pliny thought it deserved a serious discussion, to whom the honour of inventing so excellent a dish was due.

The SNOW GOOSE (*Anser* [*Chen*] *hyperboreus*) is two feet eight inches in length, and its extended wings are five feet. The bill of this bird is very curious, the edges having each twenty-three indentations, or strong teeth, on each side: the inside or concavity of the upper mandible has also seven rows of

SNOW GOOSE.—(ANSER HYPERBOREUS.)

strong projecting teeth; and the tongue, which is horny at the extremity, is armed on each side with thirteen long and sharp bony teeth. The upper mandible of the beak is bright red, the lower whitish; the hooks of both blue. The head, neck, and body are pure white: the quills are white for half their length, the rest black: the legs are of a very deep red. These birds inhabit the regions of the arctic circle, occasionally migrating to the more temperate climates of Prussia, Austria, Hudson's Bay, and the United States of America. They arrive in the River Delaware from the north early in November, sometimes in considerable flocks, and are very noisy: their note is more shrill than that of the Canada Goose. They make but a short stay in winter, proceeding farther south as the severity of the weather increases; and early in the spring they retire to the polar regions, to perform the duties of incubation and rearing their young. Their flesh is esteemed excellent; and in Siberia they form an essential article of subsistence to the natives, each family, it is said, preserving thousands annually.

The method adopted by the Siberians to obtain these Geese is highly curious. According to Pennant's account, they place near the banks of the rivers a great net in a straight line, or else form a hovel of skins sewed together: this done, one of the company dresses himself in the skin of a white rein-deer, advances towards the flock of Geese, and then turns backwards (on all fours) the net or hovel: and his companions go behind the flock. and, by making a noise drive them forwa ds. Th. ·mple birds mistake the man in ·· for their leader, and follow him within reach of the net, which is suddenly pulled down, and thus captures the whole. When he chooses to conduct them even into the hovel, they follow in the same manner; he creeps in at a hole left for that purpose, and out at another on the opposite side, which he closes up. The Geese follow him through the first; and as soon as they are in, he passes round and secures every one of them.

The CANADA GOOSE, or CRAVAT GOOSE. (*Anser Canadensis.*) This is the common Wild Goose of the United States, and is known in every part of the country. It usually weighs about ten pounds. The general colour is a dark ash; head, neck, and tail black; cheeks and throat white; bill and feet black. In their annual migrations to the north, it is the general opinion that they are on their way to the lakes to breed; but, as Wilson observes, it is highly probable that they extend to the utmost polar point, amid the silent desolation of unknown countries, shut out from the prying eye of man by everlasting and insuperable barrriers of ice. After having fulfilled the great law of nature, the approaching rigours of that dreary climate oblige them to return towards the more genial regions of the south; and no sooner do they arrive among men, than an indiscriminate slaughter of them commences. The people at Hudson's Bay greatly depend on these birds, and, in favourable seasons, kill three or four thousand, which are packed up for future use. The autumnal flight lasts from the middle of August to November; the vernal from the middle of April to the middle of May. Their mode of flying, the van being headed by an experienced old Gander, resembles that of the common Wild Goose before described. The Canada Goose generally builds its nest on the ground; but some pairs occasionally breed on the banks of large rivers on trees, depositing their eggs in the deserted nests of ravens or fishing-eagles. The eggs, six or seven in number, are of a greenish white. The bird has been long domiciled in this country, where it breeds freely, and is a great ornament.

CANADA GOOSE.—(ANSER CANADENSIS.)

That most entertaining naturalist, Mr. Waterton, thus speaks of the Canada or Cravat Goose. "The fine proportions of this stately foreigner, its voice, and flavour of its flesh, are strong inducements for us all to hope that, ere long, it will become a naturalised bird throughout the whole of Great Britain. I stop not to give a detailed description of its plumage; that has already been performed by many able hands. Suffice it then to say, that its beautiful black neck and white cheeks render it so particularly conspicuous, that those who have seen it once will never be at a loss to recognise it, when viewed amongst all other species of the Goose tribe. There can be nothing more enlivening to rural solitude than the trumpet-sounding notes of the Canada Goose. They may be heard here at most hours of the day, and often during the night. But spring is the time at which these birds are most vociferous. Then it is that they are on the wing, moving in aërial circles round the mansion,—now rising aloft, now dropping into the water, with such notes of apparent joy and revelry, as cannot fail to attract the attention of those who feel an interest in contemplating Nature's wildest scenery." Mr. W. afterwards relates the following interesting story:—"On my return from Italy in the autumn of 1841, the keeper informed me that, in the preceding spring, one of the little Bernacle ganders, accompanied by an old Canadian Goose, had come on the island where the mansion stands, and formed a kind of nest on the border of a flower-bed near the boat-house; that the female had laid five eggs in it, and that all these eggs had turned out addle. I could easily comprehend the latter part of his information relative to the eggs: but had he told me that the income-tax is a blessing, and that the national debt is an honour to the country, I could more readily have believed him, than that a Canada Goose had been fool enough to unite herself with a Bernacle gander. Nevertheless the man persisted stoutly in what he had affirmed, and I told the story to others, and nobody believed me. In the breeding season, however, of 1842, this diminutive Gander and magnificent Goose appeared on the island; and as the spot which they had occupied on the preceding year was very bleak and quite unsheltered, I thought that I could offer them a more commodious situation. Just opposite the eastern windows of the sitting-room, and two-and-twenty yards distant from them, there is yet alive the remnant of a once superb and fertile black-heart cherry-tree. It was evidently past its prime in the days of my early youth; but I can well remember that it then bore ponderous loads of dainty cherries. This cherry-tree, like the hand that is now writing a description of it, appears the worse for wear; and the wintry blasts of more than half a century have too clearly proved that neither its internal vigour, nor the strength of its gigantic limbs, could make an effectual stand against the attacks of such sturdy antagonists. Its north-western and north-eastern parts have gradually died away, and what remains alive of it to the southward can no longer produce fruit to be compared with that of gone-by periods. The bole, too, which measures full ten feet and five inches in circumference at the graft, seems to show signs of Time's hard usage. Perhaps in a few years more a south-western gale, which often does much damage here, may lay it low in ruins. Close to this venerable tree I made a hollow in the ground, about the size of an ordinary coal-basket, and filled it with hay. The Geese soon took possession of it; and on the third day after they had occupied it, the female laid an egg in it. She ultimately sat on five, and they all proved addle.

"Last year this incongruous though persevering couple visited the island again, and proceeded with the work of incubation in the same place, and upon hay which had been purposely renewed. Nothing could exceed the assiduity with which the little Bernacle stood guard, often on one leg, over his bulky partner, day after day, as she was performing her tedious task. If any body approached the place, his cackling was incessant: he would run at him with the fury of a turkey cock; he would jump up at his knees, and not desist in his aggressions until the intruder had retired. There was something so remarkably disproportionate betwixt this goose and gander, that I gave to this the name of Mopsus, and to that the name of Nisa; and I would sometimes ask the splendid Canadian Nisa, as she sat on her eggs, how she could possibly have lost her heart to so diminutive a little fellow as Bernacle Mopsus, when she had so many of her own comely species present, from which to choose a happy and efficient partner. The whole affair appeared to be one of ridicule and bad taste; and I was quite prepared for a termination of it, similar to that of the two preceding years, when behold! to my utter astonishment, out came two young ones, the remainder of the five eggs being addle. The vociferous gesticulations and strutting of little Mopsus were beyond endurance, when he first got sight of his long-looked-for progeny. He screamed aloud, whilst Nisa helped him to attack me, with their united wings and hissings as I approached the nest in order to convey the little ones to the water; for the place at which the old birds were wont to get upon the island lay at some distance, and I preferred to launch them close to the cherry-tree; which done, the parents immediately jumped down into the water below, and then swam off with them to the opposite shore. This loving couple, apparently so ill assorted and disproportionate, has brought up the progeny with great care and success. It has now arrived at its full growth, and is in mature plumage. The hybrids are elegantly shaped, but are not so large as the mother, nor so small as the father, their plumage partaking in colour with that of both parents. The white on their front is only half as much as that which is seen on the front of the gander, whilst their necks are brown in lieu of the coal-black colour which appears on the neck of the goose. Their breasts, too, are of a dusky colour, whilst the breast of the Bernacle is black, and that of the Canadian white; and throughout the whole of the remaining plumage, there may be seen an altered and modified colour [illegible] to be traced in that of the parent birds.

"I am writing this in the middle of February. In a fortnight or three weeks more, as the breeding season approaches, perhaps my little Mopsus and his beauteous Nisa may try their luck once more, at the bole of the superannuated cherry-tree. I shall have all in readiness, and shall be glad to see them. I certainly acted rashly, notwithstanding appearances, in holding this faithful couple up to the ridicule of visitors who accompanied me to the spot where the novel incubation was going on. I have had a salutary lesson, and shall be more guarded for the future in giving an opinion. Information is always desirable, and is doubly satisfactory when accompanied by a demonstration. In the present instance, my speculation that a progeny could not be produced from the union of a Bernacle gander with a Canada goose has utterly failed. I stand convinced by a hybrid, reprimanded by a gander, and instructed by a Goose." [See BERNACLE.]

The SWAN GOOSE. (*Anser cygnoides.*) This bird is of a size between the Swan and the common Goose, and is distinguished from other species by its upright and stately walk, by having a large knob on the base of the upper mandible, and a sort of wattle under the throat; a white line or fillet runs over the front of the brow, and a black stripe down the hinder part of the neck: the base of the bill is orange; the front of the neck and breast are yellowish-brown; the back, and all the upper parts, darkish-gray; the sides gray edged with white; belly white; and legs orange. It is variously called the Chinese, Spanish, Guinea, Cape, and Swan Goose: and is said to have been originally found only in Guinea, though it is now tolerably common, in a wild as well as in a domesticated state, both in warm and in cold climates. They are more noisy than any other species: nothing can stir by night or day without their sounding the alarm by their hoarse cacklings and shrill cries. They breed with the common Goose, and their offspring are as prolific as others.

The BEAN GOOSE (*Anser segetum*) is a native of the Arctic regions, migrating periodically towards the southern parts of Europe. They arrive in the fens of England in the autumn, sometimes in large flocks, and leave us in April and May for the north; some retiring no farther to breed than the Hebrides. They resort equally to the corn fields and the fens, and are said to show a preference for the green corn as food. The head and top of the neck, back, and wing-coverts are ashy-brown; the base of the neck and under parts of the plumage are bright ash-colour; the rump is nearly black; the vent and under part of the tail are pure white; and the legs are reddish-orange. They lay ten or twelve white eggs, in a nest placed in the marshes, or among the heath.

GORGONIA: GORGONIADÆ. A genus and family of Zoophytes, described in Dr. Johnston's excellent work on the British Zoophytes as "polype-mass rooted, arborescent, consisting of a central axis barked with a polypiferous crust; the crust when recent soft and fleshy, when dried porous and friable." The species here figured (*Gorgonia verrucosa*) is somewhat fan-shaped, much and irregularly branched, the branches cylindrical, flexuous, and barked when dry with a white warted crust. It is found in deep water, and is abundant along the whole of the south coast of England. "The polype-mass is more than twelve inches in height, and fifteen or seventeen in breadth, fixed to rocks by a broad circular fibro-corneous disc, shrub-like, branched from near the

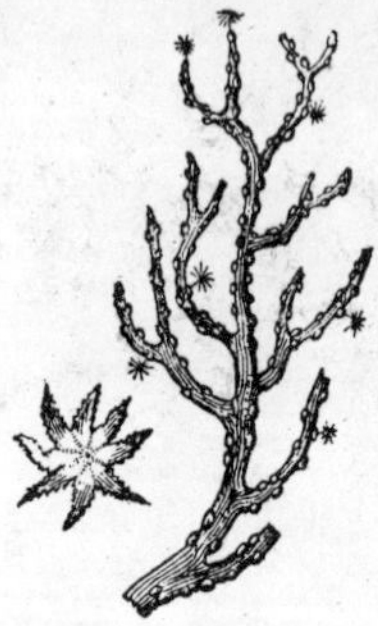

WARTY GORGONIA.—(GORGONIA VERRUCOSA.)

base, the branches expanded laterally, sometimes bushy, cylindrical, erect or erecto-patent, warty. Axis black, smooth, and somewhat glossy, round or a little compressed, compact and corneous, with a snow-white pith in the centre, irregularly cellular and very like the pith of a rush. Crust, in dried specimens, white, cretaceous, friable, warted, with numerous polype-cells and wrinkled in the small spaces between them." Professor E. Forbes, alluding to another species (*Gorgonia flabellum*), which it was supposed had been admitted into the British Fauna on insufficient evidence, says, "The fisherman who brought it described it as being covered with living flesh when taken. On examination we found that it presented the curious appearance of West Indian incrusting shells and British mixed, and the living flesh was doubtless a British sponge, which had grown round the branches in many parts.

GORILLA. [See SUPPLEMENT.]

GOSHAWK. (*Falco palumbarius.*) The Goshawk is twenty-one inches in length; the bill and cere are blue; crown, black, bordered on each side by a line of white, finely speckled with black; upper parts, slate, tinged with brown; legs feathered half way down, and, with the feet, yellow; the breast and belly white, with a number of wavy lines or bars of black; the tail long, of an ash-colour, and crossed with four or five dusky bars: wings much shorter than the tail. The Goshawk frequents the deep solitudes of forests, preying upon hares, squirrels, and the larger ground birds; it also feeds on mice and small birds, and eagerly devours raw flesh. It plucks the birds very neatly, and tears them into pieces before it eats them, but swallows the pieces entire. It is extremely destructive to game, darting through the woods after its prey with great impetuosity : but if the object of its pursuit eludes its first attack, it almost immediately desists, and perches on some bough till new game presents itself. The Goshawk is now rare in the British islands, being chiefly restricted to the Highlands of Scotland ; but it is more abundant in the forest districts of continental Europe, and extends also through the temperate regions of Asia and America. It was formerly used in Europe, in common with the Falcon, Jerfalcon, &c. in the once celebrated royal pastime of falconry ; and it is said to be still used by the emperor of China, in his hunting excursions, when he is usually attended by his grand falconer, and a thousand of inferior rank. Wilson described the American bird under the name of *F. Atricapillus*, but at the same time suspected that it might prove identical with the European, which has since been confirmed.

GRACKLE. (*Gracula.*) The principal species of this genus of birds are natives of Asia and America ; and they chiefly subsist on insects and fruits.

The INDIAN GRACKLE. (*Gracula religiosa.*) Edwards describes two varieties of this species, which resemble each other in every respect except in size ; the one being as large as a Magpie, and the other no larger than a Blackbird. They have round plump bodies, short tails, and legs of moderate length ; the head, neck, whole body, wings, and tail, are covered with glossy black feathers, shining in different lights with green, blue, and purple lustres : a white spot appears in the middle of the wing ; and the legs and feet are of a deep yellow colour. These birds are found in different parts of India and the Indian islands : they are lively, docile, and learn to speak with as much facility as most of the Parrot tribe.

The CRESTED GRACKLE (*Gracula cristatella*) is of a black colour, inclining to a dusky blue ; but the bottoms of some of the first quills are white, which forms a white spot in each wing : though the tail is black, the side feathers are tipped with white : but it is chiefly distinguished for having on the forehead, just at the basis of the bill, a remarkable tuft of feathers, which it can erect at pleasure in form of a crest. It is a native of China.

The PARADISE GRACKLE. (*Gracula tristis.*) This species is rather longer than the Blackbird ; its colour chestnut brown, the head and neck black, but the latter tinged with gray : the plumes on the fore part of the head are fine and narrow, and behind each eye is a triangular bare space of a red colour : the abdomen is white ; the tail dark brown, the lateral feathers tipped with white ; the larger quill-feathers dusky, with white bases, forming an oblong white spot on the upper edge of each wing : the bill and legs are yellow. This bird is a native of India and the Philippine islands, is very voracious, and particularly fond of locusts and grasshoppers ; relative to which Buffon relates the following curious anecdote :— The isle of Bourbon, where these birds were unknown, was overrun with locusts, which had unfortunately been introduced from Madagascar ; their eggs having been imported in the soil with some plants which were brought from that island. In consequence of this, the Governor-general and the Intendant deliberated seriously on the means of extirpating the noxious insects; and for that purpose caused several pair of the Indian Paradise Grackle to be introduced into the island. This plan promised to succeed ; but unfortunately some of the colonists, observing the birds eagerly thrusting their bills into the earth of the new-sown fields, imagined that they were in quest of the grain, and reported that the birds, instead of proving beneficial, would, on the contrary, be highly detrimental to the country. The cause was considered in form. On the part of the birds it was argued, that they raked in the new-ploughed grounds not for the sake of the grain, but the insects ; and were therefore beneficial. They were, however, proscribed by the council ; and in the space of two hours after the sentence was pronounced against them, not a Grackle was to be found in the island. This prompt execution was however followed by a speedy repentance : the locusts gained the ascendancy, and the people, who only viewed the present, regretted the loss of the Paradise Grackles. In a few years afterwards a few pair were again introduced : their preservation and breeding were made a state affair : the laws held out protection to them, and the physicians on their part declared their flesh to be unwholesome : the Grackles accordingly multiplied, and the locusts were destroyed. — The reader will find, under the word "ROOK," this really important subject discussed at some length, in reference to the habits of that well-known insectivorous and granivorous bird.

GRALLÆ. The fourth order of the class *Aves*, comprehending the long-legged wading birds.

GRALLATORES. The fourth order of Birds according to the system of Mr. Vigors, being placed between the *Rasores* and the *Natatores*.

GRAMPUS. (*Delphinus orca.*) A cetaceous animal, from twenty to twenty-five feet long, and of such an extremely fierce and predaceous nature, that it not only destroys the porpoise and dolphin, but it is reported that it will even attack whales. The nose is flat, and reverted at the extremity ; and it has thirty teeth in each jaw, those in front being blunt, round, and slender ; the hinder sharp and thick ; and between each there is a space adapted to receive the teeth of the opposite jaw when the mouth is closed. The body is broad and deep ; the back is black, but on each shoulder there is a large white spot ; the sides are marbled with black and white ; and the belly is perfectly white. The back fin sometimes measures not less than six feet in length from the base to the tip. The Grampus is found in the Mediterranean and Atlantic seas, as well as in both the polar regions; and it occasionally appears on the British coasts.

GRASS-FINCH. A genus of Passerine birds. [See POEPHILA.]

GRASSHOPPER. (*Acrydium.*) This genus of Orthopterous insects is distinguished from the Crickets by the roof-like position of the wing-covers, which in the crickets fold horizontally; and they are distinguished from the Locusts, by the inferior robustness of the body, and the length and slenderness of the legs and antennæ. There are several varieties, but it will be sufficient to give an account of the little Grasshopper that breeds in our meadows, and prolongs its shrill music through the summer, in order to elucidate the history of all.

The general colour of the Grasshopper is green, with a line of brown which streaks the back, and two pale lines under the belly and behind the legs. It may be divided into the head, the corselet, and the abdomen: the head is oblong, prone, and may be likened in shape to that of a horse; the mouth is covered by a kind of buckler, and armed with brownish hooked teeth; the antennæ are long and pointed; and the eyes are black and prominent. The corselet is elevated, narrow, and armed above and below with two serrated spines; the back is covered with a strong buckler, to which the muscles of the legs are firmly bound, and round these muscles the vessels of respiration are seen; the last pair of legs are much longer and stronger than the first two pair, and have muscles extremely well adapted for leaping. There are four wings; the anterior ones springing from the second pair of legs, the posterior from the third pair: the hinder wings are much finer and more expansive than the foremost, and are therefore the principal instruments of flight. The abdomen, which is large, is composed of eight rings, and terminated by a forked tail covered with a kind of down. Towards the latter end of autumn the female is observed to be greatly distended with eggs, and she prepares to deposit her burden. In order to form a proper lodgment for them in the earth, Nature has provided her with an instrument at the end of her body, which she can sheathe and unsheathe at pleasure: with this she pierces the earth to the greatest depths possible; and into the opening thereby made she drops her eggs one after another. Having thus provided for the continuance of her race, she does not long survive: for, as the winter approaches, she gradually withers, and dies through a total decay. In the mean time the deposited eggs continue unaltered, either by the severity of the season or the delay of spring: they are oval, white, and of a horny consistence, and they contain a viscous transparent fluid. When the vernal sun begins to animate all nature, the eggs feel his benign influence; and, generally in the beginning of May, an insect is produced from each about the size of a flea: these are at first of a whitish colour, but at the end of two or three days they turn black; and, soon after, to a reddish brown: from their very origin they exhibit the appearance of Grasshoppers without wings, and hop among the grass, as soon as excluded, with surprising agility. Having continued above twenty days from its exclusion without the use of its wings, which are folded up in its body, at length it prepares for its emancipation; and, in order to make the necessary dispositions for its approaching change, it ceases from its grassy food, and finds some convenient shelter where it may be protected from a passing shower. It then exhibits the same laborious writhings, heavings, and palpitations, which are perceptible in all other insects during their metamorphosis; it struggles hard, in fact, to free itself from prison. At length, the skin which covers the head and breast is observed to divide above the neck; and ere long the little insect extricates itself totally from the old skin, which it leaves adhering to the plant under which the transformation was performed. The Grasshopper, thus disengaged from its exterior skin, appears in its perfect form; but at this period it is extremely feeble, and its body quite soft. It is now of a greenish white colour, which becomes more vivid as the moisture on the surface dries up. Still, however, the insect discovers no signs of life, but appears quite spent, and overcome with its exertions. During this time the body continues drying, and the wings unfolding to their greatest expansion; and a curious observer may perceive them, fold after fold, opening to the sun, till at last they become longer than the two hinder legs: the body of the insect is also lengthened during this operation, and becomes more beautiful than before. These insects are generally vocal in the middle of summer; and, about sunset, their notes are much louder than during the heat of the day. The musical organs of the male consists of what has been termed a pair of taborets. They are formed by a thin and transparent membrane stretched in a strong half-oval frame in the triangular overlapping portion of each wing-cover. During the daytime these insects are silent, and conceal themselves among the leaves of trees; but at night they quit their lurking-places, and the joyous males begin the tell-tale call with which they enliven their silent mates. This proceeds from the friction of the taboret frames against each other when the wing-covers are opened and shut, and consists of two or three distinct notes almost exactly resembling articulated sounds, and corresponding with the number of times that the wing-covers are opened and shut; and the notes are repeated, at intervals of a few minutes, for hours together. Though averse to the exertions of flight, and slow in their aerial excursions, particularly when the weather is moist or cool, they are sometimes seen to fly to considerable distances. When roughly handled they bite sharply; and, in the act of flying, they make a particular noise with their wings. [See LOCUST.]

GRAYLING. (*Thymallus vulgaris.*) A fresh-water fish, of the *Salmonidæ* family, in many respects very similar in its habits to the Trout, delighting in clear rapid streams, and swimming with rapidity. Its figure is elegant, the body, which is longer and flatter than that of the Trout, seldom exceeds eighteen inches: the head is small and pointed, flattened at the top; teeth numerous, small, and incurved; behind the head, the nape and neck rise suddenly; the body deepest at the commencement of the dorsal fin, then tapering off to the tail; the back and sides are a fine silvery gray, but when the fish is just caught they are slightly varied with blue, green, and gold, with a few decided dark spots. The lateral line is straight; the scales are large, their lower edges being dusky, and forming regular rows from head to tail: the top of the back fin is red, the lower part being of a purple hue; the ventral fins

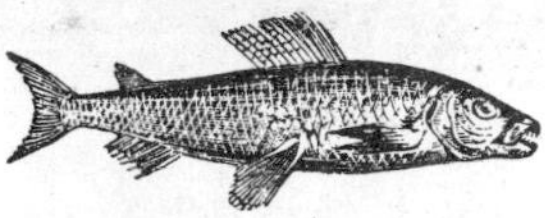

GRAYLING.—(THYMALLUS VULGARIS.)

are bluish, spotted with black; and the tail is considerably forked. The lips are rough; the tongue is smooth; and the gills are quadruple. It is tolerably abundant in several rivers in the north, and also in the north-western counties of Hampshire and Wiltshire, where it is found in the Test and both the Avons. It is known to be plentiful in Sweden, Norway, and Lapland; and it may be generally remarked that it thrives best in rivers with rocky or gravelly bottoms, where stream and pool alternate. The spawning season is in April or May, therein differing from most of the other *Salmonidæ*, which generally spawn late in the autumn: whereas the Grayling is in the finest condition in October and November, when Trout are out of season.

GREBE. (*Podiceps.*) The name given to a natural group of Water birds, allied to the *Divers*. Their distinguishing characters are—a long, straight, and sharp pointed bill; no tail; the toes flattened, separate, but broadly fringed at their edges by a firm membrane. This division of the webbed foot probably assists its action, in waters where there are many aquatic plants. The quickness with which they dive is very remarkable: their progression on land, however, is extremely awkward; for they are obliged to lie upon the whole length of the body, and then to shuffle along like seals, by the action of their feet against the ground. Their flight is very feeble; but in the act of diving, their wings are of great assistance to them.

The GREAT-CRESTED GREBE. (*Podiceps cristatus.*) The length of this bird is about twenty-one inches, and the expansion of its wings thirty. The bill is red at the base and black at the point, and between the bill and the eyes there is a stripe of black naked skin; the irides are pale red, and the head is adorned with a large dusky crest, divided in the middle. The cheeks and throat are surrounded with a long pendent ruff of a bright tawny colour edged with black; the chin is white; the hind part of the neck and the back are of a sooty hue; and the rump is covered with long soft down, which supplies the place of a tail. The covert-feathers on the second and third joints of the wings and the secondaries are white; all the other wing-feathers are dusky: the breast and belly are of a silvery white colour, soft and glossy; the plumage under the wings is dusky; the outsides of the legs are also dusky, but the insides and the toes are a pale green. This bird is found on almost every lake in the north of Europe, and is common in marshes and meres in many parts of England; it breeds among reeds and flags, in a floating nest kept steady by the weeds of the margin; preys on fish; and very rarely quits its watery abode, where by diving and swimming, it is taught to expect food and security. The Grebe is mostly valued for the plumage of its breast, the flesh being rank and nauseous.

The other species of Grebes are the EARED GREBE (*Podiceps auritus*); the RED-NECKED GREBE (*Podiceps rubricollis*); and the LITTLE GREBE (*Podiceps minor*).

GREENFINCH. (*Chlorospiza chloris.*) This bird, which is also known as the Green Linnet and Green Grosbeak, is rather larger than a Sparrow: the beak is thick and whitish: head and back yellowish green; the edges of the feathers grayish inclining to ash-colour about the sides of the head and neck: rump and breast more yellow; greater quills yellow on the outer webs: tail slightly forked; the middle feathers dusky, and the four outer feathers on each side yellow on their exterior webs: legs flesh-colour. Female less bright, and with a brown cast. The Greenfinch is one of the most common birds in this country: it builds its nest in a low and thick bush or hedge, of hay, stubble, grass, and moss, lined with hair, wool, and feathers; laying four or five eggs of a pale green colour, sprinkled with small reddish spots, which are thickest at the larger ends. Its food is principally seed and grain; and it is very easily tamed. Though Greenfinches are frequently caged their note is not to be much admired; but some, if brought up from the nest, will learn to imitate the songs of most other birds. In the winter this bird flocks with the Chaffinches and Yellowhammers; and migrates into warmer districts if the weather be very severe.

GREYHOUND. (*Canis [familiaris] graius.*) This elegant variety of the hound is of no modern origin; for the sport of coursing the hare with Greyhounds was well known in Gaul in the fifth century; and in the annals of our own country it is recorded that among the dogs kept by royal sportsmen of the olden time, this was one; nay, by the forest laws of king Canute it was enacted, that no one under the degree of a gentleman should presume to keep a greyhound; and even he could keep it only if he lived more than two miles beyond a royal forest, unless two of the dog's toes were cut off. "The third Edward," as Mr. Bell writes, "who usually held his Court at Greenwich during the hunting season, in order to be contiguous to his royal forest in Essex, kept his Greyhounds, with his other dogs, in what has from that circumstance been called the Isle of Dogs. In this instance, as in more ancient times, the game coursed by the Greyhounds was principally the Red Deer and the Fallow Deer; and it is clear that the dogs must necessarily have been of a very powerful breed to have pulled down so large and active an animal." The Greyhound is remarkable for the slenderness of its shape, the length and pointed form of its muzzle, and the extreme swiftness of its course; it hunts by sight, and not by scent, the nose

GREYHOUND.—(CANIS [FAMILIARIS] GRAIUS.)

being far from keen; the ears droop at the points, and the eyes are small; the back is broad and muscular; the body is lank, and very much contracted beneath; the limbs combine length with muscular power; the neck is long, the chest is capacious and deep; and the tail is very slender, and curved upwards.—The ITALIAN GREYHOUND is a small and very beautiful variety of the species above described; but in this country it could be but of little value for any kind of hunting, as it is unable to bear even a very moderate degree of cold, and its delicate limbs are unequal to the labour of hard running.—The IRISH GREYHOUND, on the contrary, originally called the Wolf-dog, from its having been used in hunting the Wolf when that animal infested the forests of Ireland, is a large and powerful animal; indicating a considerable approach to the Greyhound in form, and supposed to be a cross of that species with the great Danish Dog.

GRIFFIN. (*Gypäetos.*) A genus of Accipitrine birds, which, though placed by Gmelin in his genus *Falco*, seem more nearly allied in their habits and conformation to the Vultures. [See GYPAETUS.]

GROSBEAK. (*Coccothraustidæ* and *Ploccidæ.*) There are a great variety of birds belonging to this genus; and their general appearance is very similar to birds of the Finch kind. They are distinguished by a strong and thick bill, by means of which they are enabled to break the stones of cherries and other fruit with the greatest facility. In general they are a shy, solitary race, chiefly residing at a distance from the abodes of man; and very few of them are calculated to add much to the harmony of the grove by their "dulcet warblings." Their feet have three toes before and one behind; and their food generally consists of fruits and seeds. Some of the principal species are hereunder described.

The HAWFINCH GROSBEAK. (*Coccothraustes vulgaris.*) This bird is an inhabitant of the milder climates of Europe, visiting this country only occasionally in severe winters, and being nowhere very numerous. The bill is of a horn colour, conical, and prodigiously thick at the base; the space between the bill and the eye, and thence to the chin and throat, is black; the top of the head reddish chestnut; the cheeks somewhat paler, and the back part of the neck grayish ash: the back and smaller wing-coverts chestnut; the greater wing-coverts gray, in some almost white, forming a band across the wing; the quills are all black, excepting some of the secondaries nearest the body, which are brown; and the four outer quills seem as if clipped off at the ends: the breast and belly pale rusty, growing whiter towards the vent; the tail is black, the ends of the middle feathers excepted, which are gray; the outer ones are tipped with white; legs pale brown. These birds vary considerably; in some the head is wholly black: in others the whole upper part of the body is of that colour; while others have been met with entirely white, excepting the wings. The female greatly resembles the male, but her plumage is less vivid. These birds generally inhabit the woods during summer, and in winter resort near the hamlets and farms. The female builds her nest in trees, of small dry roots and grass, lined with wool, feathers, &c. The eggs are of a bluish-green colour, with brown spots.

The PINE GROSBEAK (*Loxia enucleator*) is rather larger than the preceding, being nearly nine inches long. Beak dusky, very thick at the base, and hooked at the tip: head, neck, breast, and rump, rose-coloured crimson; back and lesser wing-coverts black; greater wing-coverts tipped with white, forming two bars on the wing; quills and secondaries black, the latter edged with white; belly and vent straw-coloured. This bird is common in various parts of America, but is found only in this island in the pine forests of Scotland, where it is supposed to breed; its more native habitations are the pine forests of Siberia, Lapland, and the north of Russia. They build on trees, at a small distance from the ground, and there are generally four white eggs, which are hatched early in June.

The GRENADIER GROSBEAK. (*Pyromelana orix.*) This species is gregarious, and builds its nest in large societies, among the reeds, near the rivers and ponds in the vicinity of the Cape of Good Hope. The brilliant plumage of these birds is described as being very striking. The forehead, sides of the head, chin, breast, and belly, are black; wings brown, with pale edges; the rest of the body a most beautiful red; lower part of the thighs brown; legs pale. In size the Grenadier Grosbeak may be compared with the house-sparrow.

The CARDINAL GROSBEAK. (*Cardinalis Virginianus; Loxia cardinalis* of Linnæus.) This species, which is sometimes called the Cardinal-bird, is eight inches in length. The general plumage is a fine red: the bill pale red, and stout: on the head is a crest; and round the bill, and on the

CARDINAL GROSBEAK
(CARDINALIS VIRGINIANUS.)

throat, the colour is black: the quill and tail feathers is not of so bright a red as the body. The song of the Cardinal Grosbeak very much resembles that of the nightingale, and during the spring and summer its sweet notes are heard from the tops of the highest trees. It is met with in several parts of North America: and is said to collect together great quantities of maize and buckwheat, of which it is very fond.

The BLUE GROSBEAK (*Guiraca cœrulea*) is about six inches in length; the bill strong, thick at the base, sharp-pointed, and of a lead colour; surrounded at the base with black feathers: quills and tail brown, with a mixture of green; wing coverts with a red band; all the rest of the plumage blue: legs dusky. It is sometimes found entirely blue, except a black spot between the beak and eye. This species is a native of Brazil.

We might give many more specimens of the Grosbeak genus, if the descriptions were likely to afford matter of an interesting character; and we may also observe that the most important species will be found under other well-known names, as the Bullfinch, Greenfinch, &c. But there is one, called the Sociable Grosbeak, whose habits are worthy of particular notice; and with this species we shall conclude:—

The SOCIABLE GROSBEAK. (*Philetœrus socius.*) This bird, which is about the size of a Bullfinch, and whose prevailing colour is a rufous brown, inhabits the interior country at the Cape of Good Hope, where it was first discovered by Mr. Paterson, who gives the following history of it. "Few species of birds live together in such large societies, or have such an extraordinary mode of nidification as these: they build their nests on the Mimosa trees, which grow to a very large size, and appear to be well calculated for the purpose, as the smoothness of their trunks prevents the birds from being attacked by monkeys and other noxious animals. The method in which their nests are made is very curious. On one tree there could not be less than from eight hundred to a thousand under one general roof. I call it a roof, because it resembles that of a thatched house, and projects over the entrance of the nest below in a very singular manner. The industry of these birds seems almost equal to that of the bee. Throughout the day they appear to be busily employed in carrying a fine species of grass, which is the principal material they employ for the purpose of erecting this extraordinary work, as well as for additions and repairs. Though my short stay in the country was not sufficient to satisfy me by ocular proof that they added to their nest as they annually increased in numbers; still, from the many trees which I have seen borne down by the weight, and others which I have observed with their boughs completely covered over, it would appear that this is really the case. When the tree, which is the support of this aërial city, is obliged to give way to the increase of weight, it is obvious that they are no longer protected, and are under the necessity of rebuilding in other trees. One of these deserted nests I had the curiosity to break down to inform myself of the internal structure of it: and found it equally ingenious with the external. There are many entrances, each of which forms a regular street, with nests on both sides, at about two inches distance from each other. The grass with which they build is called the Boshman's-grass: and I believe the seed of it to be their principal food; though, on examining their nests, I found the wings and legs of different insects. From every appearance the nest which I dissected had been inhabited for many years, and some parts of it were much more complete than others. This, therefore, I conceive to amount nearly to a proof that the animals added to it at different times, as they found necessary from the increase of the family, or rather of the nation or community."

GROUND PIG. (*Aulacodus Swinderianus.*) The name of a South African Rodent belonging to the sub-family *Echimyna:* it gets its name from its burrowing habits.

GROUND SQUIRREL. (*Tamias.*) A genus of Rodent mammalia allied to the true Squirrels, but distinguished from them by the possession of cheek-pouches, and their habit of retreating into subterraneous holes.

The STRIPED GROUND SQUIRREL (*Tamias striatus*) is a very small species, inhabiting the vicinity of the Rocky Mountains. The general colour is reddish above, mixed with black, and whitish beneath, with four broad white lines on the back. It has not been observed to ascend trees, but nestles in holes, or on the edges of rocks; and the nest is composed of a most extraordinary quantity of vegetable substances. Its principal food seems to consist of the seeds of the pine.

LISTER'S GROUND SQUIRREL (*Tamias Listeri*) makes a burrow, generally, about the roots of trees, or along fences and walls, often of considerable extent, and having several branches, and always two openings. On the back are five longitudinal black bands, separated on each side by two white ones. It is a very pretty, lively, and familiar animal, well known in the United States. A closely allied species is said to be extremely common in Siberia, inhabiting the maple and birch woods of that country, and generally forming their nests or burrows near the root of some tree: they are never known to ascend trees in the manner of other Squirrels, unless suddenly surprised or pursued, when they climb with great expedition, and conceal themselves among the branches: they collect their stores during the autumnal season, and on the setting in of winter conceal themselves in their burrows, the entrances of which they stop, and pass the greatest part of the rigorous season in sleep, and in feeding on their collected stores.

GROUSE. Under this general name are comprehended several species of birds classed by Linnæus in the genus *Tetrao*. Their distinguishing characters are, that they have short arched bills; that their exterior and interior toes are connected to the first joint of the middle toe by a small membrane; that their legs are feathered down to the feet; and that they have a broad naked red skin over each eye.

The WOOD GROUSE, called also the COCK OF THE WOOD, and in Scotland CAPERCAILZIE, (*Tetrao urogallus*) is a magnificent species, two feet nine inches in length, nearly four feet in extended breadth, and weighs from eight to fourteen pounds. The bill is very strong, convex, and of a light horn colour; over each eye there is a naked skin, of a bright red; irides hazel; the nostrils small, and almost hid under a covering of short dusky feathers, which extend under the throat, and are there much longer and darker than the rest: the head and neck are elegantly marked with small transverse lines of black and gray, as are also the back and wings, but more irregularly. The upper part of the breast is of a rich glossy green hue; the rest of the breast and belly are black, mixed with a few white feathers: the sides are marked like the neck: the coverts of the wings are crossed with undulated lines of black and reddish brown; the exterior webs of the greater quill feathers are black; the bend of the wing and under tail coverts pure white: the tail consists of eighteen feathers, and is rounded in shape, and black, with a small white spot on the outer feather on each side, near the extremity: the legs are very stout, and covered with brown silky feathers, with loose webs; the feet and claws horn colour, and the toes furnished on each side with a strong pectinated membrane. The female is considerably less than the male, and differs from him greatly in her colours: her throat is red; the transverse bars on the head, neck, and back are red and black; the breast reddish, varied with a few white spots; belly barred with orange and black, the top of each feather white; the back and wings mottled with reddish brown and black; the scapulars tipped with white: the tail is of a deep rust colour, barred with black, and tipped with white.

This fine bird inhabits wooded and mountainous countries, particularly pine forests or plantations of juniper. In Russia, Sweden, and other northern countries, it is very common in the forests of pine, which there abound; and the cones of the fir trees, which it eats, as well as various plants and berries, at some seasons give an unpleasant flavour to its flesh. It was formerly met with in Ireland, the Highlands of Scotland, and parts of North Wales; but it is now very rarely indeed seen in these islands. Early in the spring the season for pairing commences: during this period the cock places himself on an eminence, where he displays a variety of attitudes, appearing unconscious of danger, and insensible to all around him: the feathers on his head stand erect, his neck swells, his tail is expanded, and his wings droop; his eyes sparkle, and the scarlet patch on each side of his head assumes a deeper dye; he at the same time utters his singular cry, which has been compared to the sound produced by the whetting of a scythe: it may be heard at a considerable distance, and never fails to draw to him his faithful mate. The female lays from eight to sixteen eggs, which are white, irregularly spotted with yellow, and larger than those of the common hen: they are generally placed in a dry situation, in an artless nest upon the ground, composed of heath tops; but she covers her eggs carefully with leaves when she is under the necessity of leaving them in search of food. As soon as the young are hatched they follow the mother, who leads them to procure the pupæ of ants and wild mountain berries, which are their first food.

BLACK GROUSE; BLACK GAME; HEATH-COCK, OR BLACK COCK. (*Tetrao tetrix.*) The male bird is about two feet in length, and the expansion of his wings two feet

BLACK GROUSE.—(TETRAO TETRIX.)

nine. The prevailing colour of his plumage is black, richly glossed with blue on the neck, back, and rump; the rest of the body being dull black. The bill is dark; the eyes deep blue; below each eye is a spot of dirty white, and eyebrows formed of a naked space of bright scarlet. The lesser wing-coverts are dusky brown; the greater white, which extends to the ridge of the wing, forming a spot of that colour on the shoulder when the wing is closed; the quills are brown, the lower parts and tips of the secondaries white, forming a bar of white across the wing: the tail is black, changing to deep violet, and when spread out, the feathers form a curve on each side; the under tail-coverts are pure white: the legs and thighs dark brown, mottled with white; the feet brown. Like the former species, these birds are common in Russia, Siberia, and other northern countries, chiefly in wooded and mountainous situations; and in the northern parts of our own island on uncultivated moors. The female is about one third less than the male; her tail is much less forked, and she differs from him considerably in colour; the head, neck, and breast being striped transversely with red and black; the back, wing-coverts, and rump deep red, varied with black lines; and the tail feathers black, with oblique zigzag red stripes, and tipped with white. The males are polygamous, and fight desperately with each other for the females. It is said that when the vanquished are put to flight, and the victors are left in possession of the field, they place themselves on an eminence, clap their wings, and with loud cries give notice to their females, who immediately resort to the spot. The hen makes an artless nest on the heathy ground, and usually lays from six to ten eggs, of a dirty white colour, blotched with spots of rusty brown. The young follow the hen for some time, but quit her at the commencement of the winter, and keep together in flocks of seven or eight till the spring. Their food consists principally of fruits and berries, and in winter, of the tops of the heath and birch; but, though they are particularly fond of wild and mountainous places, in summer they frequently come down from their lofty situations for the sake of feeding on corn.

The Ruffed Grouse (*Bonasia umbellus*) is thus described in Wilson's American Ornithology:—"This elegant species is well known in almost every quarter of the United States, and appears to inhabit a very extensive range of country. It is common at Moose Fort, on Hudson's Bay, in lat. 51°; is frequent in the upper parts of Georgia; very abundant in Kentucky and the Indiana territory; and was found by Captains Lewis and Clarke in crossing the great range of mountains that divide the waters of the Columbia and Missouri, more than three thousand miles, by their measurement, from the mouth of the latter. Its favourite places of resort are high mountains, covered with the balsam pine, hemlock, and such like evergreens. Unlike the pinnated grouse, it always prefers the woods; is seldom or never found in open plains; but loves the pine sheltered declivities of mountains near streams of water. This great difference of disposition in two species, whose food seems to be nearly the same, is very extraordinary. In those open plains called the Barrens of Kentucky, the pinnated grouse was seen in great numbers, but none of the ruffed; while in the high groves with which that singular

RUFFED GROUSE.—(BONASIA UMBELLUS.)

tract of country is interspersed, the latter, or pheasant, was frequently met with; but not a single individual of the former.

The native haunts of the pheasant being a cold, high, mountainous and woody country, it is natural to expect that, as we descend from thence to the sea shores, and the low, flat, and warm climate of the Southern States, these birds should become more rare; and such indeed is the case. In the lower parts of Carolina, Georgia, and Florida, they are very seldom observed: but, as we advance inland to the mountains, they again make their appearance. In the lower parts of New Jersey, we indeed occasionally meet with them; but this is owing to the more northerly situation of the country; for even here they are far less numerous than among the mountains.

"Dr. Turton, and several other English writers, have spoken of a long-tailed grouse, said to inhabit the back parts of Virginia, which can be no other than the present species, there being, as far as I am acquainted, only these two, the ruffed and pinnated grouse, found native within the United States.

"The manners of the pheasant are solitary; they are seldom found in coveys of more than four or five together, and more usually in pairs, or singly. They leave their sequestered haunts in the woods early in the morning, and seek the path or road, to pick up gravel, and glean among the droppings of the horses. In travelling among the mountains that bound the Susquehanna, I was always able to furnish myself with an abundant supply of these birds every morning without leaving the path. If the weather be foggy, or lowering, they are sure to be seen in such situations. They generally move along with great stateliness. The drumming, as it is usually called, of the pheasant, is another singularity of this species. This is performed by the male alone. In walking through solitary woods, frequented by these birds, a stranger is surprised by suddenly hearing a kind of thumping very similar to that produced by striking two full-blown ox-bladders together, but much louder; the strokes at first are slow and distinct, but gradually increase in rapidity, till they run into each other, resembling the rumbling sound of very distant thunder, dying away gradually on the ear. After a few minutes' pause, this is again repeated, and, in a calm day, may be heard nearly half a mile off. This drumming is most common in spring, and is the call of the cock to his favourite female. It is produced in the following manner: The bird, standing on an old prostrate log, generally in a retired and sheltered situation, lowers his wings, erects his expanded tail, contracts his throat, elevates the two tufts of feathers on the neck, and inflates his whole body, something in the manner of the turkey cock, strutting and wheeling about with great stateliness. After a few manœuvres of this kind, he begins to strike with his stiffened wings in short and quick strokes, which become more and more rapid until they run into each other, as has been already described. This is most common in the morning and evening, though I have heard them drumming at all hours of the day. By means of this, the gunner is led to the place of his retreat; though, to those unacquainted with the sound, there is great deception in the supposed distance, it generally appearing to be much nearer than it really is.

"The pheasant begins to pair in April, and builds its nest early in May. This is placed on the ground, at the root of a bush, old log, or other sheltered and solitary situation, well surrounded with withered leaves. Unlike that of the quail, it is open above, and is usually composed of dry leaves and grass. The eggs are from nine to fifteen in number, of a brownish white, without any spots, and nearly as large as those of a pullet. The young leave the nest as soon as hatched, and are directed by the cluck of the mother, very much in the manner of the common hen. On being surprised, she exhibits all the distress and affectionate manœuvres of the quail, and of most other birds, to lead you away from the spot. I once started a hen pheasant with a single young one, seemingly only a few days old; there might have been more, but I observed only this one. The mother fluttered before me for a moment; but, suddenly darting towards the young one, seized it in her bill, and flew off along the surface through the woods, with great steadiness and rapidity, till she was beyond my sight, leaving me in great surprise at the incident. I made a very close and active search around the spot for the rest, but without success. Here was a striking instance of something more than what is termed blind instinct, in this remarkable deviation from her usual manœuvres when she has a numerous brood. It would have been impossible for me to have injured this affectionate mother, who had exhibited such an example of presence of mind, reason, and sound judgment, as must have convinced the most bigoted advocates of mere instinct. To carry off a whole brood in this manner at once would have been impossible, and to attempt to save one at the expense of the rest would be unnatural. She therefore usually takes the only possible mode of saving them in that case, by decoying the person in pursuit of herself, by such a natural imitation of lameness as to impose on most people. But here, in the case of a single solitary young one, she instantly altered her plan, and adopted the most simple and effectual means for its preservation.

"The pheasant generally springs within a few yards, with a loud whirring noise, and flies with great vigour through the woods, beyond reach of view, before it alights. With a good dog, however, they are easily found; and at some times exhibit a singular degree of infatuation, by looking down from the branches where they sit, on the dog below, who, the more noise he keeps up, seems the more to confuse and stupify them, so that they may be shot down, one by one, till the whole are killed, without attempting to fly off. In such cases those on the lower limbs must be taken first; for, should the upper ones be first killed, in their fall they alarm those below, who immediately fly off. In deep snows they are usually taken in traps, commonly dead traps, supported by a figure 4 trigger. At this season, when suddenly alarmed, they frequently dive into the snow, particularly when it has newly fallen, and, coming out at a considerable distance, again take wing. They are pretty hard to kill, and will often carry off a large load to the distance of two hundred yards, and drop down dead. Sometimes, in the depth of winter, they approach the farm house, and lurk near the barn, or about the garden. They have also been often taken young, and tamed, so as to associate with the fowls; and their eggs have frequently been hatched under the common hen; but these rarely survive until full grown. They are exceedingly fond of the seeds of grapes; occasionally eat ants, chestnuts, blackberries, and various vegetables. Formerly they were numerous in the immediate vicinity of Philadelphia: but, as the woods were cleared and population increased, they retreated to the interior. At present there are very few to be found within several miles of the city, and those only singly, in the most solitary and retired woody recesses.

"The pheasant is in best order for the table in September and October. At this season they feed chiefly on whortleberries, and the little red aromatic partridge-berries; the last of which give their flesh a peculiar delicate flavour. With the former our mountains are literally covered from August to November; and these constitute, at that season, the greater part of their food. During the deep snows of winter, they have recourse to the buds of alder, and the tender buds of the laurel. I have frequently found their crops distended with a large handful of these latter alone; and it has been confidently asserted, that, after having fed for some time on the laurel buds, their flesh becomes highly dangerous to eat of, partaking of the poisonous qualities of the plant. The same has been asserted of the flesh of the deer, when, in severe weather and deep snows, they subsist on the leaves and bark of the laurel.

Though I have myself ate freely of the flesh of the pheasant, after emptying it of large quantities of laural buds, without experiencing any bad consequences, yet, from the respectability of those, some of them eminent physicians, who have particularized cases in which it has proved deleterious, and even fatal, I am inclined to believe, that, in certain cases, where this kind of food has been long continued, and the birds allowed to remain undrawn for several days, until the contents of the crop and stomach have had time to diffuse themselves through the flesh, as is too often the case, it may be unwholesome and even dangerous. Great numbers of these birds are brought to our markets, at all times, during fall and winter; some of which are brought from a distance of more than a hundred miles, and have been probably dead a week or two, unpicked and undrawn, before they are purchased for the table. Regulations, prohibiting them from being brought to market unless picked and drawn, would, very probably, be a sufficient security from all danger. At these inclement seasons, however, they are generally lean and dry; and, indeed, at all times, their flesh is far inferior to that of the quail, or of the pinnated grouse. They are usually sold, in Philadelphia market, at from three quarters of a dollar to a dollar and a quarter a pair, and sometimes higher.

"The pheasant, or partridge of New England, is eighteen inches long, and twenty-three inches in extent; bill, a horn colour, paler below; eye, reddish hazel, immediately above which is a small spot of bare skin, of a scarlet colour; crested; head and neck, variegated with black, red brown, white, and pale brown; sides of the neck furnished with a tuft of large black feathers, twenty-nine or thirty in number, which it occasionally raises; this tuft covers a large space of the neck destitute of feathers; body above, a bright rust colour, marked with oval spots of yellowish white, and sprinkled with black; wings, plain olive brown, exteriorly edged with white, spotted with olive; the tail is rounding, extends five inches beyond the tips of the wings, is of a bright reddish brown, beautifully marked with numerous waving transverse bars of black, is also crossed by a broad band of black, within half an inch of the tip, which is bluish white, thickly sprinkled and speckled with black; body below, white, marked with large blotches of pale brown; the legs are covered half way to the feet with hairy down of a brownish white colour; legs and feet, pale ash; toes, pectinated along the sides; the two exterior ones joined at the base, as far as the first joint, by a membrane; vent, yellowish rust colour.

"The female, and young birds, differ in having the ruff or tufts of feathers on the neck of a dark brown colour; as well as the bar of black on the tail inclining much to the same tint."

Red Grouse; Moor Cock, or Gorcock. (*Lagopus Scoticus.*) This species is much smaller than the Black Grouse, its length being only about fifteen inches, and its expanded width twenty-six. The bill is black, and at its base is a white spot on each side: the throat is red; each eye is arched with a large naked spot, of a bright scarlet: the plumage on the head and neck is a light tawny red, each feather being marked with several transverse pars of black; the back and scapulars are a deeper red, and on the middle of each feather is a large black spot; the breast and belly are of a purplish hue, crossed with small dusky lines: the tail consists of sixteen feathers, of equal lengths, the four middlemost barred with red, the others black: the thighs are a pale red, obscurely barred with black; the legs and feet are clothed with soft white feathers down to the claws, which are strong, and of a light colour. This species seems to be peculiar to Britain: it is very plentiful in the Highlands of Scotland, and by no means scarce in any of the wild, heathy, and mountainous tracts in the northern counties of England and Wales. Red Grouse pair in spring, and lay from six to ten eggs: the young brood follows the hen during the whole summer; and, in winter, they unite in flocks of forty or fifty; they are sometimes seen in the valleys, but generally keep on the summits of hills, where they feed on mountain berries, &c., and are exceedingly shy and wild.

White Grouse. [See Ptarmigan.]

Long-tailed Grouse. (*Tetrao Phasianellus.*) This bird, which is about the size of a pheasant, inhabits the mountainous parts of the country about Hudson's Bay, and other northern parts of the American continent. The bill is dusky, the head and neck are of a bright reddish brown, variegated with transverse waved dusky lines; the plumage of the back, wings, and tail is black in the middle, indented with bright brown on the sides, and transversely marked with black and brown at the tips; the outer coverts of the wings, and the quill feathers next the back, have white tips; and the primaries have spots of white along their outer webs. The two middle feathers of the tail are considerably the longest, the rest gradually shortening on each side: the upper part of the breast is brown, but by degrees becomes white; as do the belly, the sides under the wings, and the covert feathers under the tail. The legs are covered with fine filiform feathers of a pale brown colour, transversely variegated with dusky lines. They feed upon juniper berries and buds; associate in small flocks; and lay their eggs, which vary from ten to sixteen, in a nest on the ground, artlessly composed of grass, and lined with a few feathers: the eggs are white, and are hatched about the middle of June, the young immediately following the mother. The flesh of these birds is held in great estimation.

Canada Grouse. (*Tetrao Canadensis.*) This species, which is found in great abundance in the most northerly parts of America, is rather more than thirteen inches in length; the female two inches less. The upper parts of the head, neck, and body of the male bird are transversely barred with dusky and gray brown; over the eyelids is a bare red space; nostrils covered with black, with a small white spot on each side, and one beneath; throat, breast, and belly, black; the latter spotted with white, except the middle: sides of the body barred transversely with grey-brown and dusky; the feathers with a white stripe near the tip: under tail coverts black and white: tail black, tipped with rufous: feathers of the tarsi gray-brown: claws gray: beak black.

PINNATED GROUSE. (*Tetrao Cupido.*) In its voice, manners, and peculiarity of plumage, the Pinnated Grouse is the most singular, and, in its flesh, the most excellent, of all those of its tribe that inhabit the territory of the United States. Though an inhabitant of different and very distant districts of North America, this rare bird is

PINNATED GROUSE.—(TETRAO CUPIDO.)

extremely particular in selecting his place of residence, pitching only upon those tracts whose features and productions correspond with his modes of life, and avoiding immense intermediate regions that he never visits. Open dry plains, thinly interspersed with trees, or partially overgrown with shrub oak, are his favourite haunts: their predilection for such situations being, according to the opinion of Wilson, to be best accounted for by considering the following facts and circumstances:—First, their mode of flight is generally direct, and laborious, and ill calculated for the labyrinth of a high and thick forest, crowded and intersected with trunks and arms of trees, that require continual evolution of wing, or sudden turnings, to which they are by no means accustomed. Secondly, their known dislike of ponds, marshes, or watery places, which they avoid on all occasions, drinking but seldom, and it is believed, never from such places. The last, and probably the strongest inducement to their preferring these plains, is the small acorn of the shrub oak; the strawberries, huckleberries, and partridge-berries, with which they abound, and which constitute the principal part of the food of these birds. These brushy thickets also afford them excellent shelter, being almost impenetrable to dogs or birds of prey.

The Pinnated Grouse is nineteen inches long, twenty-seven inches in extent, and weighs about three pounds; the neck is furnished with supplemental wings, each composed of eighteen feathers, five of which are black, and about three inches long; the rest shorter, also black, streaked laterally with brown, and of unequal lengths; the head is slightly crested; over the eye is an elegant semicircular comb of rich orange, which the bird has the power of raising or relaxing; under the neck wings are two loose, pendulous, and wrinkled skins, extending along the side of the neck for two-thirds of its length, each of which, when inflated with air, resembles, in bulk, colour, and surface, a middle-sized orange; chin, cream-coloured; under the eye runs a dark streak of brown; whole upper parts mottled transversely with black, reddish brown, and white; tail short, very much rounded, and of a plain brownish soot colour; throat elegantly marked with touches of reddish brown, white, and black; lower part of the breast and belly, pale brown, marked transversely with white; legs covered to the toes with hairy down of a dirty drab colour; feet dull yellow; toes pectinated; vent whitish; bill brownish horn colour; eye reddish hazel. The female is considerably less; of a lighter colour; destitute of the neck wings, the naked yellow skin on the neck, and the semicircular comb of yellow over the eye.

The season for pairing is in March, and the breeding time is continued through April and May. Then the male Grouse distinguishes himself by a peculiar sound. When he utters it, the parts about the throat are sensibly inflated and swelled. It may be heard on a still morning for three or more miles. This noise is a sort of ventriloquism. It does not strike the ear of a bystander with much force, but impresses him with the idea, though produced within a few rods of him, of a voice a mile or two distant. This note is highly characteristic. Though very peculiar, it is termed *tooting*, from its resemblance to the blowing of a conch or horn from a remote quarter. The female makes her nest on the ground, in recesses very rarely discovered by men; and she usually lays from ten to twelve brownish-coloured eggs, much resembling those of a guinea-hen. When hatched, the brood is protected by her alone. Surrounded by her young, the mother bird exceedingly resembles a domestic hen with her chickens. When at such times they are surprised, the dam utters a cry of alarm; and while the little ones are hurrying to a place of safety, their anxious parent beguiles the spectator by drooping and fluttering her wings, limping along the path, rolling over in the dirt, and other pretences of inability to walk or fly.

During the period of mating, and while the females are occupied in incubation, the males have a practice of assembling, principally by themselves. To some select and central spot where there is very little underwood, they repair from the adjoining district. From the exercises performed there, this is called a *scratching place*. The time of meeting is the break of day. As soon as the light appears, the company assembles from every side, sometimes to the number of forty or fifty. When the dawn is past, the ceremony begins by a low tooting from one of the cocks. This is answered by another; and they presently come forth one by one from the bushes, strutting about with all the pride and ostentation they can display. Their necks are incurvated; the feathers on them are erected into a sort of ruff; the plumes of their tails are expanded like fans; they strut about in a style resembling the pomp of the turkey-cock. They seem to vie with each other in stateliness; and, as they pass each other, frequently cast looks of insult, and utter notes of defiance. These are the signals for battles. They engage with wonderful spirit and fierceness; and during their contests, they leap a foot or two from the ground, and utter a cackling, screaming, and discordant cry. These places of exhibition have been often discovered by the hunters; and a fatal discovery it has been for the poor Grouse. Their destroyers construct for themselves lurking holes made of pine branches, called "bough houses," within a few yards of the spot. Hither they repair with their fowling-pieces, in the latter part of the night, and wait the appearance of the birds. Watching the moment when two are proudly eyeing each other, or engaged in battle, or when a greater number can be seen in a range, they pour on them a destructive charge of shot. They commonly keep together in coveys of ten or a dozen, or packs, as the phrase is, until the pairing season: and it has been remarked, that when a company of sportsmen have surrounded a pack of Grouse, the birds seldom or never rise upon their pinions while they are encircled; but each runs along till it has passed the person that is nearest, and then flutters off with the utmost expedition.—The interesting facts contained in the foregoing account are derived from the inimitable "American Ornithology" by Alex. Wilson.

GRUB. A name applied more especially to the hexapod worms or maggots hatched from the eggs of beetles.

GRUIDÆ. The name of the family of wading birds represented by the Crane.

GRUS. A genus of Grallatorial birds belonging to the family *Gruidæ*. [See CRANE.]

GRYLLIDÆ. The second family belonging to the *Saltatorial Orthoptera;* containing the Field and House Cricket. [See CRICKET.]

GUACHARO BIRD. (*Steatornis Caripensis.*) A bird of South America, belonging to the family of Gaotsuckers (*Caprimulgidæ*), relative to the locality and habits of which a most interesting account is given by Baron Humboldt, in his "Personal Narrative." This bird is of the size of a common fowl; the plumage sombre, brownish-grey, mixed with small striæ and black dots; large white heart-shaped spots bordered with black on the head, and on the wing and tail feathers; but no spots on the back: the bill is horny, wide, and long; the upper mandible hooked; and the base is furnished with stiff hairs, directed forwards.

The following narrative is derived, in a somewhat abridged form, from an article by the talented author of Zoological Recreations.—"When they (Humboldt and his party) arrived at the foot of the lofty mountain of Guacharo, they were only four hundred paces from the cavern, without yet perceiving the entrance. The torrent runs in a hollow excavated by the waters; and they went on under a ledge or cornice, the projection of which prevented them from seeing the sky. The path winds like the river, and, at the last turning, they suddenly stood before the immense opening of the cave. The *Cueva del Guacharo* is pierced in the vertical profile of a rock, and the entrance is towards the south, forming a vault eighty feet broad and seventy-two feet high, an elevation but a fifth less than that

GUACHARO.—(STEATORNIS CARIPENSIS.

of the Louvre. The rock surmounting the cavern was covered with trees of gigantic height, and all the luxuriant profusion of an intertropical vegetation. The travellers saw with astonishment plantain-leaved heliconias eighteen feet in height, the praga palm, and tree arums, follow the banks of the river, even to the subterranean places. There the vegetation continues, as in the deep crevices of the Andes, half shut out from the light of day, nor does it disappear till a distance of thirty or forty paces from the entrance. The party went forward for about four hundred and thirty feet without being obliged to light their torches. Where the light began to fail, they heard from afar the hoarse cries of the *Guacharo birds.* These birds quit the cavern only at nightfall, especially when there is moonlight; and Humboldt remarks that it is almost the only frugivorous nocturnal bird yet known. It feeds on very hard fruits, and the Indians assured him that it does not pursue either the lamellicorn insects or those *phalænæ* which serve as food to the goatsuckers. He states that it is difficult to form an idea of the horrible noise made by thousands of these birds in the dark recesses of the cavern, whence their shrill and piercing cries strike upon the vaulted rocks, and are repeated by the echo in the depths of the grotto. By fixing torches of copal to the end of a long pole, the Indians showed the nests of these birds fifty or sixty feet above the heads of the explorers, in funnel-shaped holes, with which the cavern roof is pierced like a sieve.

"Once a year, near midsummer, the Guacharo cavern is entered by the Indians. Armed with poles, they ransack the greater part of the nests, while the old birds hover over the heads of the robbers, as if to defend their brood, uttering horrible cries. The young which fall down are opened on the spot. The peritoneum is found loaded with fat, and a layer of the same substance reaches from the abdomen to the vent, forming a kind of cushion between the bird's legs. Humboldt here remarks, that this quantity of fat in frugivorous animals, not exposed to the light, and exerting but little muscular motion, brings to mind what has been long observed in the fattening of geese and oxen. It is well known, he adds, how favourable darkness and repose are to this process. At the period above mentioned, which is generally known at Caripe by the designation of 'the oil harvest,' huts are built by the Indians, with palm leaves, near the entrance, and even in the very porch of the cavern. There the fat of the young birds just killed is melted in clay pots over a brushwood fire; and this fat is named butter or oil of the Guacharo. It is half liquid, transparent, inodorous, and so pure that it will keep above a year without turning rancid. Humboldt observes that the r ce of Guacharo birds would have been extinct long since if several circumstances had not contributed to its preservation. The natives, withheld by superstitious fears, seldom dare to proceed far into the recesses of the cavern. Humboldt had great difficulty in persuading them to pass beyond the outer part of the cave, the only portion of it which they visit annually to collect the oil; and the whole authority of the *Padres* was necessary to make them penetrate as far as the spot where the floor rises abruptly at an inclination of sixty degrees, and where a small subterraneous cascade is formed by the torrent. In the minds of the Indians this cave, inhabited by nocturnal birds, is associated with mystic ideas, and they believe that in the deep recesses of the cavern the souls of their ancestors sojourn. They say that man should avoid places that are enlightened neither by the sun nor the moon; and 'to go and join the Guacharoes' means to rejoin their fathers—in short, to die. At the entrance of the cave the magicians and poisoners perform their exorcisms to conjure the chief of the evil spirits. It appears also, as another cause of preservation, that Guacharo birds inhabit neighbouring caverns too narrow to be accessible to man, and from these perhaps the great cavern is repeopled; for the missionaries declared that no sensible diminution of the birds had been observed. Young birds of this species have been sent to the port of Cumana, and have lived there several days, but without taking any food; the seeds offered to them not suiting them. The crops and gizzards of the young birds opened in the cavern contain all sorts of hard and dry fruits, which are conveyed to them by their parents: these are preserved, and, under the name of *semilla del Guacharo* (Guacharo seed), are considered a celebrated remedy against intermittent fevers, and sent to the sick at Cariaco and other low localities where fever prevails. The *Cueva del Guacharo* is situated nearly in lat. 10° 10′, and consequently in the centre of the torrid zone."

GUAN. A genus of Gallinaceous birds found in the New World. [See PENELOPE.]

GUANA. The name given to several species of Lizards (*Iguana*). The best known species (*Iguana tuberculata*) is found in many parts of America and the West India islands. It inhabits rocky and woody places; feeds on insects and vegetables; and is often seen of the length of from three to even five feet: its general colour is green, shaded with brown: the back is strongly serrated; and this, with its large gular pouch, which it has the power of inflating to a great degree, gives a formidable appearance to an animal otherwise harmless. We learn from Catesby that these reptiles are of various sizes, from two to five feet in length; that their mouths are furnished with exceeding small teeth, but their jaws are armed with a long beak, with which they bite with great strength: and that they inhabit warm countries only. Many of the Bahama islands abound with them, where they nestle in hollow rocks and trees. Their eggs have not a hard shell, like those of alligators, but a skin only, like those of a turtle; and are esteemed a good food: they lay a great number of eggs at a time, in the earth, which are there hatched by the sun's heat. These Guanas are a great part of the subsistence of the inhabitants of the Bahama islands, for which purpose they visit many of the remote *Kayes* and islands in their sloops to catch them, which they do by dogs trained up for that purpose. Their flesh is easy of digestion, delicate, and well-tasted: they are sometimes roasted, but the more common way is to boil them, taking out the leaves of fat, which are melted and clarified, and put into a calabash or dish, into which they dip the flesh of the Guana as they eat it. Though they are not amphibious, they are said to keep under water above an hour. Their pace on land is slow; and when they swim, they do not use their feet, but merely guide themselves with their tails. They are so impatient of cold, that they rarely appear out of their holes but when the sun shines.

GUANACO. The local name of a variety of species of the Llama [which see].

GUDGEON. (*Cyprinus gobio.*) A small Malacopterygious fresh-water fish, generally about five or six inches in length and of a subcylindrical shape; its usual colour is a pale olive brown above, spotted with black; the sides silvery, and the belly white; the scales are small; the tail is forked; and both that and the dorsal fin are spotted with black: the upper jaw is longer than the lower; and furnished with short cirri. Gudgeons appear to delight in slow rivers: they swim together in shoals, feeding on worms, aquatic insects, &c., and affording excellent amusement to anglers from the avidity with

GUDGEON.—(CYPRINUS GOBIO,)

which they seize the bait: they may also be taken in considerable numbers with the casting-net. The Gudgeon spawns in May, generally among stones in shallow water. The flesh is delicate, and easy of digestion.

GUILLEMOT. (*Uria.*) The Guillemots are a genus of sea-birds, having a striking resemblance both to the *Alcidæ* (Auks) and the *Colymbidæ* (Divers). Their bills, though of a slender shape, are firm, strong, and pointed; the upper mandible slightly bending near the end, and the base covered with soft short feathers: tongue long and slender; legs placed far backward; and no hind toe. Some of the species appear to be very stupid, frequent experience not seeming to teach them the danger of fire-arms; while others are sufficiently alert. They are numerously spread over various parts of the northern regions; and, like many others, seek more temperate climes on the approach of winter: thus during that season they are regular visitants of the British coasts.

FOOLISH GUILLEMOT.—(URIA TROILE.)

The Foolish Guillemot. (*Uria troile.*) This bird is about seventeen inches in length, and twenty-seven in breadth. The bill is bluish-black, straight, nearly three inches long, and sharp-pointed: from each eye to the hinder part of the head there is a slight division of the plumage; and the feathers on the upper part of the bill are short, and soft as velvet. The head, neck, back, wings, and tail, are of a deep mouse-colour; the tips of the lesser quill-feathers, the breast, belly, and vent, are white; the entire under side of the body is pure white: legs dusky. Like the Auk, which it greatly resembles, the Guillemot lays but one egg, which is large in proportion to her size: sometimes it is of a pale blue or sea-green colour, and at other times white, spotted, or neatly streaked with intersecting lines. These birds are found in great numbers on the cliffs which encircle several parts of our coasts; and, in the breeding season especially, they will often suffer themselves to be killed or taken, rather than quit the cliff they have chosen for their abode.

The young has been described as a distinct species, by some authors, as the *Lesser Guillemot.* In this state it measures sixteen inches in length, and from tip to tip of its extended wings, twenty-six. The top of the head, the whole upper part of the body, the wings, and the tail, are of a very dark mouse-colour; the cheeks, throat, and lower side of the body, white; from the angle of the eye is a dusky stroke, pointing to the back part of the head; the tips of the secondary feathers are white; the tail is very short; and the legs and feet are dusky.

The Black Guillemot. (*Uria grylle.*) This species, called by seamen the *Dovekey* or *Dovekie*, differs from the preceding principally in the colour of its plumage, which, except a large patch of white on the coverts of each wing, is black, sleek and glossy; its feathers appearing all unwebbed, like silky hair: legs and feet red; claws black. The

BLACK GUILLEMOT.—(URIA GRYLLE.)

nest is made in the deep crevices of the rocks which overhang the sea: the egg (for it is generally said that one egg only is laid) is grey, sometimes spotted with rust-colour. On this much questioned and very questionable fact, the observing and intelligent American ornithologist, Audubon, thus writes:—"Whether European writers have spoken of this species at random, or after due observation, I cannot say. All I know is, that every one of them whose writings I have consulted, says that the Black Guillemot lays only *one egg.* As I have no reason whatever to doubt their assertion, I might be tempted to suppose that our species differs from theirs, were I not perfectly aware that birds in different places will construct different nests, and lay more or fewer eggs Our species always deposits three, unless it may have been disturbed; and this fact I have assured myself of by having caught the birds in more than twenty instances sitting on that number. Nay, on several occasions, at Labrador, some of my party and myself *saw* several Black Guillemots sitting on eggs in the same fissure of a rock, where every bird had three eggs under it; a fact which I communicated to my friend Thomas Nuttall. What was most surprising to me was, that even the fishermen there thought that this bird laid only a single egg; and when I asked them how they knew, they simply and good-naturedly answered that they had heard so."

BLACK GUILLEMOT.—WINTER PLUMAGE. (URIA GRYLLE)

We now turn to the first volume of Mr. Waterton's amusing and characteristic Essays, to extract his *Notes of a visit to the haunts of the Guillemot.* "The immense range of perpendicular rocks, lashed by old ocean's briny surge, offers a choice and favourable retreat to myriads of wild-fowl, from far-famed Flamborough-head to Bempton, and thence to Buckton and Speaton, and outwards to the Bay of Filey. He who wishes to examine the nidification of these birds ought to be at this part of the sea-coast early in the month of May. About five miles from Bridlington Quay is the village of Flamborough, chiefly inhabited by fishermen; and a little farther on is a country inn, called the North Star, which has good accommodation for man and horse; but a lady would feel herself ill at ease in it, on account of the daily visits of the fishermen, those hardy sons of Neptune, who stop at it on their way to the ocean, and again on their return. Here they rendezvous, to fortify their interior with a pint or two of comfort, and to smoke a pipe, by way of compensation for the many buffets which they ever and anon receive in the exercise of their stormy and nocturnal calling.

"On the bare ledges of these stupendous cliffs the Guillemot lays its egg, which is exposed to the face of heaven, without any nest whatever: but the razor-bills and puffins lay theirs in crannies, deep and difficult of access. Here too the peregrine falcon breeds, and here the raven rears its young; while the rock-pigeon and the starling enter the fissures of the precipice, and proceed with their nidification, far removed from the prying eye of man. The Kittiwake makes her nest of dried grass wherever she can find a lodgment, and lays two spotted eggs, very rarely three. The cormorant and shag inhabit that part of the rocks which is opposite to Buckton Hall. You are told that the cormorants had their nests, in former times, near to the Flamborough lighthouse; but now these birds totally abandon the place during the breeding season. The jackdaw is found throughout the whole of this bold and craggy shore: he associates with the sea-fowl, as though he were quite at home amongst his own inland congeners. Towards the top of the cliffs, both rabbits and foxes have descended from the table land above them, and managed to find a shelter among the crevices, in places where you would suppose that no four-footed animal would ever dare to venture. A low mound, half earth, half stone, thrown up by the farmers for the protection of their flocks, skirts the winding summit of the precipice. Cattle have been known to surmount this artificial boundary, and lose their lives in the roaring surge below. This extensive range of rocks, as far as appertains to birds, is not considered private property. Any person who can climb it may carry away what number of eggs he chooses. Still there is a kind of honourable understanding betwixt the different sets of climbers, that they will not trespass over the boundaries which have been marked by mutual consent.

"The usual process of seeking for the eggs is generally carried on by three men, though two will suffice in case of necessity. Having provided themselves with two ropes of sufficient length and strength, they drive an iron bar into the ground, about six inches deep, on the table land at the top of the precipice. To this bar is fastened the thickest of the two ropes, and then it is thrown down the rocks. He who is to descend now puts his legs through a pair of hempen braces, which meet round his middle, and there form a waistband. At each end of this waistband is a loophole, through which they reeve the smaller rope. Sometimes an iron hook and eye are used in lieu of this loop. A man now holds the rope firmly in his hand, and gradually lowers his comrade down the precipice. While he is descending he has hold of the other rope, which was fastened to the iron bar; and, with this assistance, he passes from ledge to ledge, and from rock to rock, picking up the eggs of the Guillemot, and putting them into two bags, which he had slung across his shoulder ere he commenced his arduous undertaking. When he has filled these bags with eggs, he jerks the rope, and the motion informs his friend at the top that it is now time to draw him up. On coming up again to the place from whence he first set out, all the eggs are taken from the bags, and put into a large basket, prior to their being packed in hampers, and carried off in a cart by wholesale dealers, who purchase them from the climbers for sixpence the score. At Bridlington and the neighbouring places the eggs are retailed at a halfpenny a-piece. The rocks are searched for eggs every third day, provided the weather be fair. It requires considerable address on the part of the descending climber to save himself from being hit by fragments of the rock, which are broken off by the rope coming in contact with them. He avoids the danger by moving sidewise when the stone is falling, and by taking care, as he goes down, to clear away with his foot any portion of the rock that seems ready to give way. One of the climbers, while he was imparting to me instructions how to act, grinned purposely, and showed his upper jaw. I learned by his story, that, last year, a falling stone had driven two of his front teeth down his throat; while the poor climber, with all his dexterity, was unable to fend off the blow.

"As I was lowered down, the grandeur and sublimity of the scene beggared all description, and amply repaid any little unpleasant sensations which arose on the score of danger. The sea was roaring at the base of this stupendous wall of rocks; thousands and tens of thousands of wild-fowl were in an instant on the wing: the kittiwakes and jackdaws rose in circling flight; while most of the Guillemots, razorbills, and puffins left the ledges of the rocks in a straight and downward line, with a peculiarly quick motion of the pinions, till they plunged into the ocean. It was easy to distinguish the puffins from the razorbills in their descent: these presented a back of a uniformly dark colour, those had a faint white diagonal line running across the wings. The nests of the kittiwakes were close to each other, on every part of the rocks which was capable of holding them; and they were so numerous as totally to defy any attempt to count them. On the bare and level ledge of the rocks, often not more than six inches wide, lay the eggs of the Guillemots: some were placed parallel with the range of the shelf, others nearly so, and others with their blunt and sharp ends indiscriminately pointing to the sea. By no glutinous matter, nor any foreign body whatever, were they affixed to the rock; bare they lay, and unattached, as on the palm of your outstretched hand. You might see nine or ten, or sometimes twelve old Guillemots in a line, so near to each other that their wings seemed to touch those of their neighbours; and when they flew off at your approach, you would see as many eggs as you had counted birds sitting on the ledge. The eggs vary in size and shape and colour beyond all belief. Some are large, others small, some exceedingly sharp at one end, and others nearly rotund. The rock-climbers assure you that the Guillemot, when undisturbed, never lays more than one egg; but if that be taken away, she will lay another, and so on. They also assure you that when the young Guillemot gets to a certain size, it manages to climb upon the back of the old bird, which conveys it down to the ocean. Having carried a good telescope with me, through it I saw numbers of young Guillemots, diving and sporting on the sea, quite unable to fly; and I observed others on the ledges of the rocks, as I went down among them, in such situations that, had they attempted to fall into the waves beneath, they would have been killed by striking against the projecting points of the intervening sharp and rugged rocks: wherefore I concluded that the information of the rock-climbers was to be depended upon; and I more easily gave credit to it, because I myself have seen an old swan sailing on the water with her young ones upon her back, about a week after they had been hatched.

"He who rejoices when he sees all nature smiling around him, and who takes an interest in contemplating the birds of heaven as they wing their way before him, will feel sad at heart on learning the unmerited persecution to which these harmless sea-fowl are exposed. Parties of sportsmen, from all quarters of the kingdom, visit Flamborough and its vicinity during the summer months, and spread sad devastation all around them. No profit attends the carnage; the poor unfortunate birds serve merely as marks to aim at, and they are generally left where they fall. Did these heartless gunners reflect, but for one moment, how many innocent birds their shot destroys; how many fall disabled on the wave, there to linger for hours, perhaps for days, in torture and in anguish; did they but consider how many helpless young ones will never see again their parents coming to the rock with food; they would, methinks, adopt some other plan to try their skill, or cheat the lingering hour."

GUINEA-FOWL, or PINTADO. (*Numida meleagris.*) The Guinea-fowls are natives of Africa and its adjacent islands: their manners are similar to those of the domestic poultry, and their food the same. This species is bigger than a large cock: the head is bare of feathers, and covered with a naked bluish skin; on the top is a callous conical protuberance: and on each side of the upper mandible, at the base, hangs a loose wattle, which in the female is red, and in the male bluish; the upper part of the neck is almost naked, being very thinly furnished with a few straggling hairy feathers: the skin is of a bluish ash: the lower part of the neck is

GUINEA-FOWL.—(NUMIDA MELEAGRIS.)

covered with feathers of a purple hue; but the general colour of the plumage is dark bluish gray, sprinkled with round white spots of different sizes, on the whole of the feathers, the breast only excepted, which is of a uniform gray blue: the greater quills are white; and the rest are similar to the upper parts of the plumage, spotted and longitudinally barred with white. Its wings are short, and the tail pendulous, or pointing downwards.

This bird is now common in our poultry yards, but from the circumstance of the young ones being difficult to rear, they are not bred in numbers at all equal to those of the domestic poultry. The female lays many eggs in a season, which she frequently secretes till she has produced her young brood. The egg is smaller than that of the common hen, and of a rounder shape; in colour reddish white, obscurely freckled with a darker colour; and is delicious eating. The Guinea-fowl is a restless and clamorous bird: its voice is harsh and unpleasant, consisting chiefly of two notes — *ca-mac, ca-mac* — frequently repeated; which is compared by Latham to a door turning upon its rusty hinges, or to an ungreased axle-tree. During the night it perches on high places, and if disturbed, alarms everything within hearing by its unceasing cry. It scrapes in the ground like the hen, and delights in rolling in the dust to free itself from insects. In a wild state these birds associate in flocks, giving the preference to marshy places, where they subsist almost wholly on insects, worms, and seeds. They formed a part of the Roman banquets; and they are greatly esteemed in this country by many persons, who consider their flavour to resemble that of the pheasant.

In Jamaica and other islands in the West Indies, the Guinea-fowls come in numerous coveys from the woods, and scattering themselves in the provision-grounds at early dawn, commit serious depredations by scratching up and devouring the seed-yams, &c.; and as they are birds of extreme caution and suspicion, it is no easy matter to get at them without the assistance of a dog; but when pursued by an animal whose speed exceeds their own, they instantly betake themselves to a tree, where, their attention being intently fixed upon the dog beneath, they may easily be shot. They are also sometimes caught, Mr. Gosse tells us, by the following stratagem: a small quantity of corn is steeped for a night in proof rum, and is then placed in a shallow vessel, with a little fresh rum, and the water expressed from a bitter cassava, grated: this is deposited within an enclosed ground to which the depredators resort. A small quantity of the grated cassava is then strewed over it, and it is left. The fowls eat the medicated food eagerly, and are soon found reeling about intoxicated, unable to escape, and content with thrusting the head into a corner. Frequently a large part of the flock are found dead from this cause.

The Crested Guinea-fowl. (*Numida cristata.*) This species inhabits the hottest parts of Africa, and is smaller than the one above described. The occiput, upper part of the neck, and the throat, are nearly destitute of feathers: the sides and hinder part of the neck are of a deep blue colour; the space round the ears is blue gray; and the anterior part of the neck is of a crimson red: the head is surmounted by a fine crest, composed of black feathers, with delicate webs, drooping over the hind head and beak. The general plumage is black; the body, with the exception of the neck and breast, being spotted with small points of faint blue, encircled with a brilliant blue: the large wing feathers are dusky brown and spotless; the secondaries are the same, with four longitudinal stripes down the shafts: three or four of these have a large white spot extending the whole length of the under webs; the rest nearer the body having the longitudinal blue rays: legs and feet dusky; hind claw elevated from the ground, and blunt. They live in flocks of many hundreds; and their cry, uttered at the rising and setting of the sun, is very harsh and discordant. In their food and habits they differ very little, if at all, from the other species.

GUINEA-PIG. [See Cavy.]

GUINEA-WORM. [See Filaria Medinensis.]

GULL. (*Larus.*) The birds of this web-footed and well-known marine genus are numerously dispersed over every quarter of the world, and are met with, at certain seasons, in some parts, in prodigious multitudes. They assemble together in a kind of straggling mixed flocks, consisting of various kinds, and greatly enliven the beach and rocky cliffs by their irregular movements, whilst their shrill cries are often deadened by the noise of the waves, or nearly drowned in the roarings of the surge. They occasionally take a wide range over the ocean, and are met with by navigators many leagues distant from the land. They are all greedy and gluttonous, almost indiscriminately devouring whatever comes in their way, whether of fresh or putrid substances, until they are obliged to disgorge their overloaded stomachs; but, at the same time, it appears that they are able to endure hunger a long while. The larger kind of Gulls are most common in cold climates of the north, where they breed and rear their young, feeding chiefly upon the rotting carcasses of dead whales, &c., which they find floating on the sea, among the ice, or driven on shore by the wind and waves. In temperate and cultivated countries some species occasionally leave the shores for the interior, probably to search for a change of food, such as worms, slugs, &c., and of these they find, for a time, an abundant supply on the downs and pastures which they visit. Their general characteristics are — a strong and straight bill, but bent downwards at the point; the lower mandible has an angular prominence on the under side, which tapers towards, and forms its tip; the tongue is a little cloven. The body is clothed with a great quantity of down and feathers, which, together with the large head and long wings, give these birds an appearance of bulk, without a proportionate weight. The legs are small, naked above the knees: feet webbed, and the back toe detached, and very small.

The Common Gull. (*Larus canus.*) This bird, which is one of the most numerous of the genus, breeds on the ledges of cliffs that overhang the sea; and, during the winter season, frequents almost every part of the British coasts where the high bold shores present a favourable situation. Like other rapacious birds, it lays but few eggs. It generally measures about seventeen inches in length, and thirty-six in breadth: the bill is yellow; the head, neck, tail, and the whole under side of the body, are pure white; the back and the coverts of the wings are gray; and the legs are a dull white tinged with green.

The Black-backed Gull. (*Larus marinus.*) This species measures from twenty-six to twenty-nine inches in length, and five feet nine inches in breadth. The bill is pale yellow, very thick and strong; the projecting angle on the lower mandible is light red, with a black spot in the middle, on each side: the irides are yellow, and the edges of the eyelids orange: the upper part of the back and wings black: all the other parts of its plumage, including the tips of the quill feathers, are white: the legs are of a pale flesh-colour. The Black-backed Gull is common in the northern parts of Europe, though but thinly scattered on the coasts of England. In their native haunts, their favourite breeding places are high inaccessible islets, covered with long coarse grass. Their eggs are of a round shape, of a dark olive colour, thinly marked with dusky spots, and quite black at the thicker end. They principally subsist on fish, but when such food is not easily obtainable they will devour carrion. Their cry is hoarse and disagreeable.

The IVORY GULL. (*Larus eburneus.*) A species of Gull, so called from its white plumage, the pureness of which certainly equals in colour new-fallen snow. It is very com-

IVORY GULL.—(LARUS EBURNEUS.)

mon in the arctic regions, especially in Baffin's Bay and the straits leading to it. By our Arctic Voyagers, Captains Parry, Lyons, Ross, and others, it is often mentioned, and is strikingly characteristic of the arctic seas. It is said to have occurred in the Orkney Islands; but in Britain this snow-white bird must be regarded as about as *rara* an *avis* as the black swan was to the ancients.

The HERRING GULL, or SILVERY GULL. (*Larus argentatus.*) This species, which has obtained its name from pursuing and preying upon the shoals of herrings, is met with in the northern seas, and is also well known on our own coasts. In length it is twenty-three inches, and in breadth fifty-two: the bill is yellow, except the spot on the angular knob of the under mandible, which is deep orange: the irides pale yellow, and the edges of the eyelids red: the head, neck, and tail are white; the back and wing-coverts are dark bluish ash; and the legs are of a pale flesh-colour. They make their nests of dry grass, mixed with sea-weed, on the projecting ledges of the rocks, and lay three eggs of a dirty white colour, spotted with black. These Gulls are said to be remarkable for their vigilance; and fishermen describe them as the bold and constant attendants on their nets, from which they find

HERRING-GULL, OR SILVERY GULL
LARUS ARGENTATUS. (ADULT.)

it difficult to drive them. The young, which are ash-coloured spotted with brown, do not assume their mature plumage till they are

HERRING-GULL.
LARUS ARGENTATUS. (YOUNG.)

one year old: a circumstance, indeed, common to others of the genus; and which, not being properly attended to, has occasioned considerable confusion in th descriptions which have sometimes been given of them. [For the Arctic Gull and Skua Gull, see LESTRIS.]

GULO. A genus of carnivorous quadrupeds, the formidably armed skull of which is well shown in the woodcut for the habits

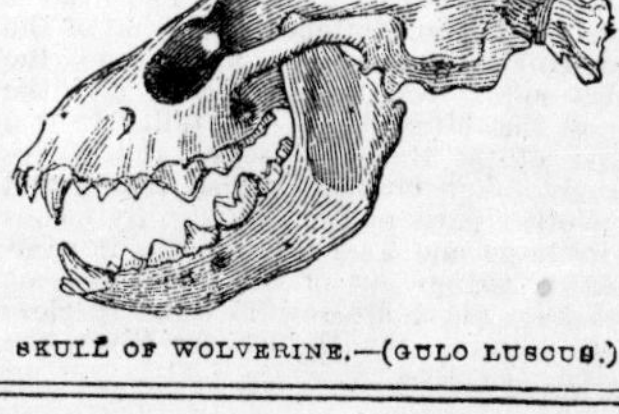

SKULL OF WOLVERINE.—(GULO LUSCUS.)

of this fierce animal. [See GLUTTON and WOLVERINE.]

GURNARD. (*Trigla.*) A genus of Acanthopterygious fishes, of which there are several species. The generic characters are—head nearly square, covered with bony plates; two dorsal fins, the rays of the first spinous, those of the second flexible; teeth in both jaws and on the front of the vomer pointed, small, and numerous; seven branchiostegous rays; and three slender appendages at the base of each pectoral fin.

The GREY GURNARD. (*Trigla gurnardus.*) This fish is distinguished by its elongated body, and varies from one to two feet in length: the back is of a greenish brown colour, marked with black, yellow, and white spots; the lateral line is very prominent, and strongly serrated; and the sides are of a pale hue, variegated with numerous white spots: the belly is white; the nose long, sloping, and bifurcated. The eyes are large: near the extremity of the gill-covers there is a strong, sharp, long spine; and exactly above the pectoral fins there is another. The first dorsal fin consists of eight spiny rays, and the second of nineteen soft rays; the pectoral fins are transparent, and supported by ten rays, bifurcated from the middle; the ventral fins contain six rays, and the anal nineteen. The Grey Gurnard is common on our coasts, feeding on worms, insects, &c. It bites eagerly at a red bait, and is usually taken with a hook in deep water, though in calm weather they may be seen in considerable numbers on the surface. They make a sort of croaking noise, or *croon*, whence probably arises the name of crooner, by which they are called in Ireland.

The RED GURNARD, or CUCKOO GURNARD. (*Trigla cuculus.*) This is an elegant species, about a foot in length, and of a slender form; its colour a beautiful bright red, more or less distinctly marked by whitish transverse bars, the sides and belly silvery white. scales extremely small; lateral line composed of pointed white scales edged with black; a similar row on each side the back: fins transparent; the first dorsal marked on the edge by a black spot; the second tinged near its edge with yellow. It is common on the English coasts; feeds on crustaceous animals; and spawns in May or June.

The SAPPHIRINE GURNARD. (*Trigla hirundo.*) This valuable species is distinguished by the large size of its pectoral fins, which are beautifully edged and spotted with a fine blue colour. It is larger than the preceding, more abundant, and quite equal to any others as food. The head is larger and more flattened than that of the Red Gurnard; the eyes are large; the scales small, oval, and smooth; and the lateral line bifurcates at the tail. It is a native of the European seas; and is frequently taken on the Cornish coasts, and some other parts of this island. By means of its large and long pectoral fins it occasionally springs out of the water to some distance. There are several other species; as the Shining, the Mailed, the Piper, the Japanese, the Carolina, the Lineated, the Flying, &c.; with the last mentioned of which we shall close our account of the Gurnards.

The FLYING GURNARD. (*Trigla volitans.*) This singular and beautiful species is about a foot in length; of a crimson colour above, and pale beneath; the head blunt, and armed on each side with two very strong and large spines pointing backwards. The whole body is covered with strong, sharp-pointed, and closely united scales: the pectoral fins extremely large, transparent, of an olive-green colour, richly marked with numerous bright blue spots: pectoral processes six in number, not separate as in other species, but united, so as to appear like a small fin on each side the thorax: tail pale violet, with the rays crossed by dusky spots, and the base strengthened by two obliquely transverse bony ribs. In the Mediterranean, Atlantic, and Indian seas, the Flying Gurnard swims in shoals; and is often seen darting from the water and sustaining itself for a while in the air, after the manner of the genus Exocetus.

GYMNOPHTHALMIDÆ. The name of a family of Lizards, in which the eyes are distinct and exposed, the eyelids being rudimentary. There are several genera, which will be found described in Mr. Gray's valuable List of the Reptiles in the British Museum; but, important as these are to naturalists, it is quite out of the scope of this work to refer to them.

GYMNOTUS. A genus of Malacopterygious fishes, which contains the well-known GYMNOTUS ELECTRICUS, or ELECTRIC EEL; a fish possessing the extraordinary property of communicating a sensation similar to an electrical shock, when touched with the hand or an electric conductor. The Gymnotus is a fish of a disagreeable appearance, bearing a general resemblance to a large eel, though thicker in proportion, and much darker. It is nearly of equal thickness throughout: the head is broad, depressed, and obtuse: the tail is compressed; and the usual length is from four to five feet, though it is sometimes six, or even eight. It is a native of South America, where it inhabits the larger rivers. The seat of the organs which produce this curious electrical effect is along the under side of

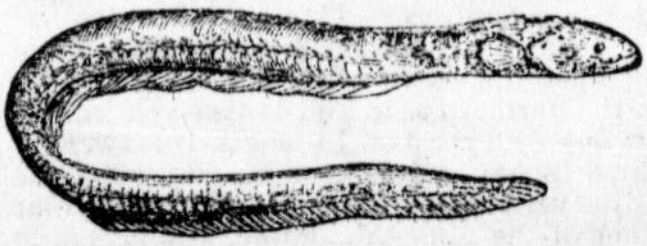

ELECTRIC EEL,—(GYMNOTUS ELECTRICUS.)

the tail. They are composed of four bundles of parallel membranaceous laminæ, placed very near each other, and nearly horizontally, extended from the skin to the central medial plane of the body, connected together by numerous vertical laminæ, arranged transversely. The little cells, or rather the small prismatic and transverse canals, intercepted by these two kinds of laminæ, are, according to Cuvier, filled with a gelatinous substance; and the whole apparatus is abundantly supplied with nerves. It is said to possess power, when in full vigour, to knock down a man, and benumb the limb affected, in the most painful manner, for several hours after communicating the shock; and it is by this extraordinary faculty that the Gymnotus supports its existence: the smaller fishes and other animals which happen to approach it being stupified, and thus falling an easy prey to the electrical tyrant. Those who wish to understand the nature of the organs by which this electrical power is produced may find them minutely described by Hunter in vol. 65. of the Philosophical Transactions. The following observations are given in Brande's Dictionary: "Although to all outward appearance the Gymnotus is nearly allied to the Eel, yet were that part of the body cut off which contains the nutrient, respiratory, and generative organs,—all the parts, in fact, which are essential to the existence of the Gymnotus as a mere fish,—it would present a short and thick-bodied form, very different from that of the eel. The long electric organs are tacked on, as it were, behind the true fish, and thus give the Gymnotus its anguiliform body. The back bone and muscles are of course co-extended with the electric organs for their support and motion; and the air-bladder is continued along the produced electrophorous trunk, to give it convenient specific levity. Two long dorsal nerves are continued from the fifth and eighth cerebral nerves for ordinary sensation and motion. The spinal chord is continued along the vertebral column, for the exclusive supply of the electrical organs. These organs are four in number; two very large above, and two small ones below. The electricity discharged from them decomposes chemical compounds, produces the spark, and magnetizes iron, as does that of the Torpedo. But the magnetizing power seems to be relatively weaker, while the benumbing shock communicated to other animals is stronger than in any other electric fish."

Humboldt has given a lively narrative of the mode of capture of the Gymnoti, employed by the Indians of South America. They rouse the Gymnoti by driving horses and mules into the ponds which those fish inhabit, and harpoon them when they have exhausted their electricity upon the unhappy quadrupeds. "I wished," says Humboldt, "that a clever artist could have depicted the most animated period of the attack: the groups of Indians surrounding the pond, the horses with their manes erect and eyeballs wild with pain and fright, striving to escape from the electric storm which they had roused, and driven back by the shouts and long whips of the excited Indians: the livid yellow eels, like great water-snakes, swimming near the surface and pursuing their enemy: all these objects presented a most picturesque and exciting 'ensemble.' In less than five minutes two horses were killed: the eel, being more than five feet in length, glides beneath the body of the horse and discharges the whole length of its electric organ: it attacks at the same time the heart, the digestive viscera, and, above all, the gastric plexus of nerves. I thought the scene would have a tragic termination, and expected to see most of the quadrupeds killed; but the Indians assured me the fishing would soon be finished, and that only the first attack of the Gymnoti was really formidable. In fact, after the conflict had lasted a quarter of an hour, the mules and horses appeared less alarmed; they no longer erected their manes, and their eyes expressed less pain and terror. One no longer saw them struck down in the water; and the eels, instead of swimming to the attack, retreated from their assailants and approached the shore." The Indians now began to use their missiles; and by means of the long cord attached to the harpoon, jerked the fish out of the water without receiving any shock so long as the cord was dry. All the circumstances narrated by the celebrated philosopher, establish the close analogy between the Gymnotus and Torpedo in the vital phenomena attending the exercise of their extraordinary means of offence. The exercise is voluntary and exhaustive of the nervous energy; like voluntary muscular effort, it needs repose and nourishment to produce a fresh accumulation.

"I was so fortunate (says Professor Owen) as to witness the experiments performed by Professor Faraday on the large Gymnotus which was so long preserved alive at the Adelaide Gallery in London. That the most powerful shocks were received when one hand grasped the head and the other hand the tail of the Gymnotus, I had painful experience; especially at the wrists, the elbow, and across the back. But our distinguished experimenter showed us that the nearer the hands were together within certain limits, the less powerful was the shock. He demonstrated by the galvanometer that the direction of the electric current was always from the anterior parts of the animal to the posterior parts, and that the person touching the fish with both hands received only the discharge of the parts of the organs included between the points of contact. Needles were converted into magnets: iodine was obtained by polar decomposition of iodide of potassium; and, availing himself of this test, Professor Faraday showed that any given part of the organ is negative to other parts before it, and positive to such as are behind it. Finally, heat was evolved, and the electric spark obtained."

There are several other fish belonging to the Gymnotus tribe; but they are much smaller; and whether they possess any electric power is a matter of great doubt: yet the structure of the lower part of their bodies seems to imply a similar contrivance of nature. Most of them are natives of the same climate as the Gymnotus Electricus, and are considered edible food. The principal are the Carapo Gymnote, the Rostrated Gymnote, and the White Gymnote.

GYMNURA. An insectivorous animal belonging to the family *Erinaceadæ*, inhabiting Sumatra. In its dentition and spiny covering it closely resembles the Hedgehog tribe; but it has the long, naked, scaly tail and pointed muzzle of the Shrews. Its generic character has been given by Dr. Horsfield and Mr. Vigors: Head elongated, acuminated, compressed on the sides, flattish above; muzzle obtuse, elongated, and projecting forward considerably beyond the lower jaw; tongue rather smooth, large: auricles rounded, somewhat prominent, naked: eyes small; nostrils lateral, prominent, with the margins convoluted; vibrissæ elongated. Body rather robust; the short fur soft, but with distant, erect, subelongated hairs: tail rather long, smooth, naked, and scaly. Feet plantigrade, pentadactyle, the fore-feet with a rather short thumb. Claws narrow, curved, very acute, and retractile. The body, legs, and first half of the tail are black; the head, the neck, and the shoulders are white; and a black band passes over the eyes. Cuvier, in his "Règne Animal" (1829), observes that the genus *Gymnura* of MM. Vigors and Horsfield appears to approach Cladobates in its teeth, and the Shrews in its pointed muzzle and scaly tail. It has five unguiculated toes on all its feet, and rather stiff bristles projecting forth from the woolly hair. The species is called G. Rafflesii, in compliment to the accomplished founder of Singapore, Sir Stamford Raffles.

GYPAETUS, or BEARDED VULTURE. A genus of birds which may be considered as intermediate between the eagles and vultures. The Bearded Vulture (*Gypaëtus barbatus*), sometimes called the Bearded Griffin or *Lammergeyer*, is the largest bird of prey belonging to the Eastern Continent, and it appears to be the only Vulture which has ever been found in a wild state in Britain. It usually inhabits the high chains of mountains, and nestles in inaccessible accli-

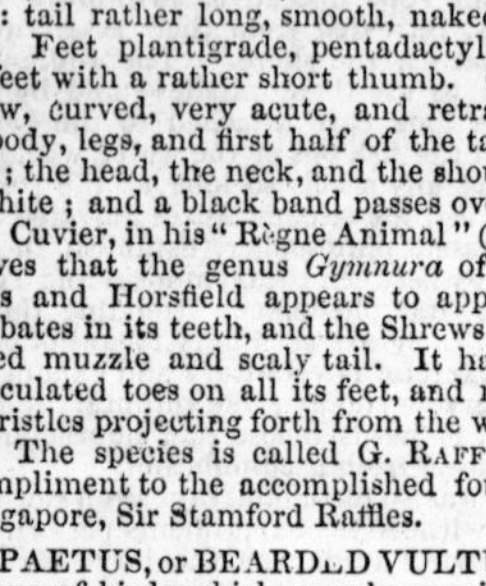

BEARDED VULTURE.—(GYPÆTUS BARBATUS.)

vities. It is found in Europe as far north as Astracan, but is much more common in Spain, on the Pyrenees, Portugal, the isle of Elba, Tuscany, Malta, Turkey, and in the Archipelago; but is nowhere so abundant as in South Africa, in which quarter it attains a larger size than elsewhere. In the adult bird the head and upper part of the neck are of a dirty white colour; a black stripe extends from the base of the beak, and passes above the eyes; another, arising behind the eyes, passes over the ears; lower part of the neck, breast, and belly, orange-red; mantle, back, and wing-coverts, deep grey-brown, but on the centre of each feather is a white longitudinal stripe: wings and tail-feathers ashy-grey, the shafts white; tail long, very much graduated; beak and claws black; feet blue; iris orange, eye surrounded by a red lid. Length about four feet and a half.

"Unlike the typical vultures," says Mr. Gould, "which are distinguished by their bare necks, indicative of their propensity for feeding on carrion, the Lammergeyer has the neck thickly covered with feathers, resembling those of the true eagles, with which it also accords in its bold and predatory habits, pouncing with violent impetuosity on animals exceeding itself in size; hence the young chamois, the wild goat, the mountain hare, and various species of birds find in it a formidable and ferocious enemy. Having seized its prey, the Lammergeyer devours it upon the spot, the straight form of their talons disabling them from carrying it to a distance. It refuses flesh in a state of putrefaction, unless sharply pressed by hunger; hence nature has limited this species as to numbers: while, on the other hand, to the Vultures, who are destined to clear the earth from animal matter in a state of decomposition, and thus render the utmost service to man in the countries where they abound, she has given an almost illimitable increase."

GYRINUS: GYRINIDÆ. A genus and family of aquatic Beetles, the type of which is known under the name of Whirligigs, or Water-flea, from its peculiar motions. They are in general of small or moderate size; and are to be seen, from the first fine days of spring till the end of autumn, on the surface of quiet waters, and even upon that of the sea, often appearing in great numbers, and appearing like brilliant points. They are active swimmers, and curvet about in every direction. Sometimes they remain stationary without the slightest motion; but no sooner are they approached, than they escape by darting under the surface of the water, and swimming off with the greatest agility. The four hind-legs are used as oars, and the anterior ones for seizing the prey: when they dart beneath the surface, a bubble of air like a silvery ball remains attached to the hind part of the body. When seized, they discharge a milky fluid, which spreads over the body, and probably produces the disagreeable odour which they then emit. There are several species found in this country, but it is not necessary to describe them separately. These beetles are almost the only water insects which exhibit a brilliant metallic lustre, a peculiarity dependent upon the habits of the insects which generally swim upon the surface of the water.

HADDOCK. (*Gadus æglefinus.*) This well-known Malacopterygious fish is nearly allied to the cod; and, like it, is a native of the Northern seas, where it assembles in prodigious shoals, visiting particular coasts at stated seasons. Nor is it by any means scarce on the shores of Britain or Ireland; immense quantities, indeed, are taken at different localities, particularly along our eastern coast; and as its flesh is sweet and wholesome, and can be preserved with facility, it is a fish of considerable value. The Haddock is generally about twelve or fourteen inches in length, and weighs from two to three pounds; though, occasionally, they are met with

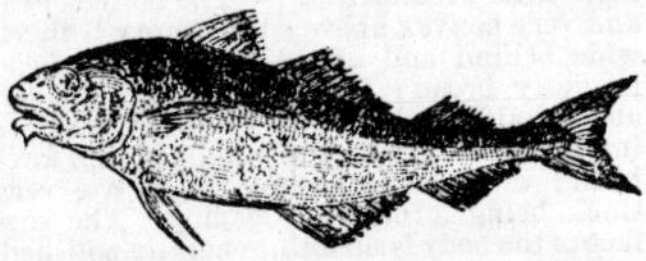

HADDOCK.—(GADUS ÆGLEFINUS.)

nearly three feet long, and weighing ten or twelve pounds: the smaller or moderate sized ones, however, are most esteemed for the table. The body is long and slender; the head slopes suddenly down from the crown to the point of the nose; the lower jaw is longer than the upper, and furnished with a narrow band of teeth: the barbule at the chin is small; the eye is large, and the irides silvery; the head, cheeks, back, and upper part of the sides, are of a dull grayish hue; lower part of the sides and belly, silvery. On each side, is a large black spot, (of which we shall again have to speak.) The lateral line is black: the dorsal fins and tail dusky bluish gray; pectoral, ventral, and anal fins lighter: the tail bifid. Their food is small fish, crustacea, and marine insects: they spawn in February and March; and they are in the best condition for the table from October to January. In stormy weather this fish is said to imbed itself in the ooze at the bottom of the sea; and those which are taken shortly after are observed to have mud on their backs.

We are always loth to make allusion to ignorant superstitions, however popular they may be, unless we can furnish some rational solution for their existence; but they have sometimes taken such deep root, that not to mention, might almost seem to sanction them. We of course allude to the "thumb and finger marks of St. Peter;" and shall therefore extract from Mr. Yarrell's excellent work the following remarks, as supplying additional information of a germane character: "Pennant says, 'Our countryman Turner suggested that the Haddock was the *Opos* or *Asinus* of the ancients. Different reasons have been assigned for giving this name to the species, some imagining it to be from the colour of the fish, others because it used to be carried on the backs of asses to market.' A different reason appears to me more likely to have suggested the name: the dark mark on the shoulder of the Haddock very frequently extends over the back and unites with the patch of the shoulder on the other side, forcibly reminding the observer of the dark stripe over the withers of the ass; and the superstition that assigns the mark in the Haddock to the impression St. Peter left with his finger and thumb when he took the tribute-money out of a fish of this species, which has been continued to the whole race of Haddocks ever since the miracle, may possibly have had reference, or even its origin, in the obvious similarity of this mark on the same part of the body of the Haddock and of the humble animal which had borne the Christian Saviour. That the reference to St. Peter is gratuitous, is shown by the fact that the Haddock does not exist in the sea of the country where the miracle was performed." Independently of which, the Sea of Galilee is a large fresh water lake.

HÆMATOPUS, or OYSTER-CATCHER. A genus of wading birds, the best known species of which is the Common Oyster-Catcher, *H. ostralegus.* [See OYSTER-CATCHER.]

HAG. [See GASTROBRANCHUS.]

HAIR-STREAK [BUTTERFLY]. A name given to various species of Butterflies, of the genus *Thecla.*

HAKE. (*Gadus merlucius* of Linnæus.) This Malacopterygious fish inhabits the seas of the north of Europe and the Mediterranean; it is also found on the western and southern coasts of England, as well as on various parts of the coast of Ireland. It is of a lengthened form, generally from one to two feet, but sometimes more: the head is rather large, broad and flat at the top, but compressed on the sides; wide mouth; lower jaw the longest; teeth slender and sharp, with a single row in each jaw: the colour of the body is a dusky brown above, and lighter beneath; dorsal and caudal fins dark; ventral and anal fins light brown; the pectoral and ventral fins are of moderate size, and of a sharpened shape; and the tail is nearly even at the end. It is salted and dried in the manner of cod, haddock, &c., but is not considered as a delicate fish either in its fresh or salted state, and is rarely admitted to the tables of the affluent: it forms, however, a very useful article of food for the lower orders in many parts both of our own and other countries. It is a very voracious fish; and when pilchards approach the shores, it follows them, continuing in great numbers through the winter; so that when pilchards are taken in a sean, on the Cornish coast, many Hakes are generally found inclosed with them. By Dr. Fleming and other naturalists, this fish is regarded as belonging to a distinct genus, characterized by having one anal and two dorsal fins. (*Merlucius.*)

HALCYON. A genus of the Kingfisher family, of which there are many species:—of these we may specify the SACRED KINGFISHER (*Halcyon sanctus*), which is generally distributed over the Australian continent, and feeds on various insects and reptiles; as Mantidæ, grasshoppers, caterpillars, lizards, and small snakes; and Mr. Gould found that specimens killed in the vicinity of salt marshes had their stomachs literally crammed with crabs and other crustaceous animals. It also excavates holes in the nests of a species of ant which are constructed around the holes and dead branches of the *Eucalypti*, feeding on the larva, a most favourite food.

HALIBUT, or HOLIBUT. (*Hippoglossus.* This is the largest fish belonging to the *Pleuronectidæ* or Flat-fish family, attaining the length of six or seven feet in the northern seas, and weighing from 300 to 400 lbs. In shape and fins the Halibut is like the Flounder; and the lateral line is arched. Its flesh is rather coarse and dry, but it ad-

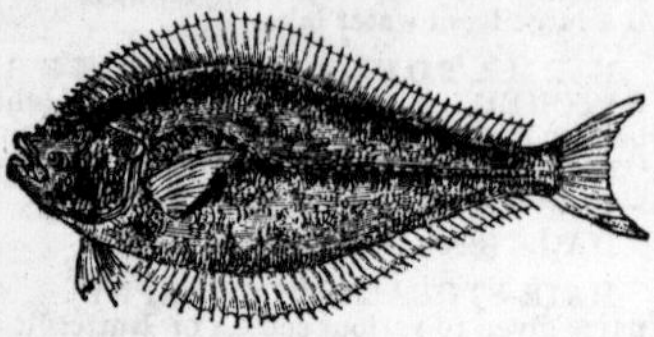

HALIBUT.—(HIPPOGLOSSUS VULGARIS.)

mits of being salted. In some of the smaller species, which are found in the Mediterranean, the eyes look towards the left side, instead of towards the right, the latter being the ordinary rule of the family; and when that happens it is said to be "reversed."

HALICHÆRUS. A genus of Seals. [See SEAL.]

HALICHONDRIA. A genus of Sponges, in which the cartilaginous skeleton is strengthened by siliceous spiculæ. See Dr. Fleming's British Animals, and Dr. Johnston's British Sponges, for an account of the many entirely British species.

HALICORE. A genus of Cetaceous animals, found in the Eastern seas. [See DUGONG.]

HALIOTIS: HALIOTIDÆ. A genus and family of Gasteropodous Mollusca, not widely different from the Limpets, and having the Haliotis or Ear-shell as the type. These splendid shells are remarkable for the pearly iridescence of the inner surface, and the row of holes following the course of the spine. [See EAR-SHELL.]

HALMATURUS. A genus of Marsupialia belonging to the Kangaroo family. As an example we may give Parry's Kangaroo, *Halmaturus Parryi*, a species familiar to the colonists and natives of New South Wales. It is very shy, escapes with great fleetness from its pursuers, and inhabits the mountainous parts. It is easily tamed, becoming very familiar. The male measures five feet from the nose to the end of the tail. The body is bluish gray, whitish beneath; the head brownish; a white streak on the face below the eye, and a short one on each eyebrow. Capt. Sir Edward Parry has given an interesting account of its habits in confinement. Those who wish to get further information on this genus and its allies must consult the noble monagraph of the Kangaroos, by John Gould, F. R. S., where all the species are admirably figured and described. The reader may consult also with profit, Mr. Waterhouse's History of Mammalia, and the volume on Marsupialia in the Naturalist's Library.

HALTICIDÆ. The scientific name *Haltica*, derived from a word signifying *to leap*, has been applied to a family of insects allied to the Chrysomelidæ, and popularly known as *flea-beetles*. The following are their chief peculiarities:—The body is oval and very convex above; the thorax is short, wide behind and narrow before; the head is pretty broad; the antennæ are slender, about half the length of the body, and are implanted nearly on the middle of the forehead; and the hindmost thighs are very thick, being formed for leaping. The surface of the body is smooth, generally polished, and often prettily or brilliantly coloured. The claws are very thick at one end, are deeply notched towards the other, and terminate with a long, curved, and sharp point, which enables the insect to lay hold firmly upon the leaves of the plants on which they live. These beetles eat the leaves of vegetables, preferring especially plants of the cabbage, turnip, mustard, cress, radish, and horse-radish kind, or those which, in botanical language, are called cruciferous plants, to which they are often exceedingly injurious. The flea-beetles conceal themselves, during the winter, in dry places, under stones, in tufts of withered grass, and in chinks of walls. They lay their eggs in the spring, upon the leaves of the plants upon which they feed. The larvæ of the smaller kinds burrow into the leaves, and eat the soft pulpy substance under the skin, forming therein little winding passages, in which they finally complete their transformations. Hence the plants suffer as much from the depredations of the larvæ as from those of the beetles, a fact that has too often been overlooked. The larvæ of the larger kinds live exposed on the surface of the leaves which they devour, till they come to their growth, and go into the ground, where they are changed to pupæ, and soon afterwards to beetles. The mining larvæ are little slender grubs, which arrive at maturity, turn to pupæ, and then to beetles in a few weeks. Hence there is a constant succession of these insects, in their various states, throughout the summer. One of the most destructive species of this family is the Turnip-flea (*Haltica nemorum*), [which see].

HAMSTER. (*Cricetus frumentarius.*) A rodent animal, of the rat tribe, distinguished by two enormous cheek pouches, which will hold a quarter of a pint, and by its remarkable instincts. It inhabits the sandy districts of the north of Europe and Asia, Austria, Silesia, and many parts of Germany, Poland, &c., and is very injurious to the agriculturist, on account of the quantity of grain it devours. The general size of the Hamster is nearly that of a brown or Norway rat, but it is of a much thicker form, and has a short and somewhat hairy tail. Its colour is a pale reddish brown above, and black beneath: the muzzle is whitish, the cheeks reddish, and on each side the body are three white spots, those on the shoulders being the largest: the ears are rather large, and rounded. On the hind feet are five toes, and on the fore feet are four, with a claw in place of a fifth. They sometimes vary in colour; and the male is invariably larger than the female. The quantity of grain which they consume is very great; and what they cannot devour, they carry off in their cheek-pouches, and deposit in their holes for their winter subsistence. Their dwellings are formed under the earth, and consist of more or fewer apartments, according to the age of the animal: a young Hamster makes them hardly a foot deep; an old one sinks them to the depth of four or five feet, and the whole diameter of the residence, taking in all its habitations, is sometimes eight or ten feet. The principal chamber is lined with dried grass, and serves for a lodging: the others are vaults destined for the preservation of provisions, of which he amasses a great quantity during the autumn. Each hole has two apertures; the one descending obliquely, and the other perpendicularly, and it is through the latter that the animal makes its ingress and egress. The holes of the females, who never reside with the males, have more numerous passages. The female breeds two or three times a year, producing from six to ten, and sometimes more: the growth of the young is very rapid, and at about the age of three weeks the old one forces them out of the burrows to shift for themselves.

The Hamster is carnivorous as well as granivorous, for though it feeds on all kinds of herbs as well as corn, it occasionally attacks and devours the smaller kinds of animals. On the approach of winter the Hamster retires into his subterranean abode, the entry of which he closes with great care; and thus remaining tranquil and secure, feeds on his collected store till the frost becomes severe: he then falls into a profound slumber, and in that dormant state continues rolled up, apparently lifeless, his limbs inflexible and his body perfectly cold. This lethargy of the Hamster has been generally ascribed to the effect of cold alone: but more recent observations have proved, that unless at a certain depth beneath the surface, so as to be beyond the access of the external air, the animal does not fall into its torpid state, and that the severest cold on the surface does not affect it. On the contrary, when dug out of its burrow and exposed to the air, it infallibly wakes in a few hours. The waking of the Hamster is a gradual operation: he first loses the rigidity of his limbs, then makes deep inspirations, at long intervals; after this he begins to move his limbs, open his mouth, and utters an unpleasant rattling sound: he at length opens his eyes, and endeavours to rise, but reels about for some time, as if in a state of intoxication, till at length he perfectly recovers his usual powers. When exposed to a cold air he is sometimes two hours in waking; but in a warmer air the transition is effected in half the time. The character which naturalists have given of these animals is very unfavourable. They constantly reject all society with one another, and they will not only destroy every animal which they are capable of conquering but will fight, kill, and devour their own species: yet, fierce as they are, they quail before their deadly enemy the polecat, which chases them into their holes, and unrelentingly destroys them. The fur of the Hamster is said to be valuable; and the peasant who 'goes a *hamster nesting*' in the winter, obtains not only the skin of the animal, but his hoard, which is said commonly to amount to two bushels of good grain in each magazine. Buffon says, that in Gotha, where these animals were proscribed on account of their vast devastations among the corn, 146,132 of their skins were delivered at the Hotel de Ville of the capital in the course of three years.

HARE. (*Lepus.*) A well-known genus of Rodent mammalia, containing several species. We shall first describe—

The COMMON HARE (*Lepus timidus*), which possesses all the characters of the genus *Lepus* in such a degree as to form its most perfect type. Its hearing and sight are most acute; its timidity is unequalled; and its swiftness is surpassed by none. The general length of the Hare is about two feet; the colour a subferruginous gray, with the chin and belly white: the throat and breast ferruginous, and

COMMON HARE.—(LEPUS TIMIDUS.)

the tips of the ears blackish: the tail is blackish above, and white below: the feet are covered beneath as well as above with fur; the upper lip is divided; the eyes are large, prominent, and placed laterally; and they are said to be constantly open even during sleep: the hinder legs are much longer than the fore legs; the feet are hairy; and the tail is short and turned up. Its favourite residence is in rich and somewhat dry and flat grounds, and it is rarely discovered in very hilly or mountainous situations. It feeds principally by night, and remains concealed during the day in its *form*, beneath some bush or slight shelter. To this spot it constantly returns, and becomes so attached to it, that it is with difficulty compelled to abandon it: in choosing its place of rest, however, it is governed by the seasons, and while a cool and shady spot is its resort in summer, it selects for its winter lair a situation where it can best receive the genial warmth of the sun.

The Hare is a very prolific animal, generally producing three or four young at a time, and breeding several times in a year. The eyes of the young are open at birth: the dam suckles them about twenty days, after which they leave her, and procure their own subsistence. Its food consists of various kinds of herbage, but it prefers vegetables of a milky and succulent quality, and is especially fond of parsley. It is at times a very annoying and destructive invader, not only of the field and garden, doing great injury to the young wheat and other grain; but it also frequently commits sad havoc in young plantations, by gnawing off the bark, and feeding on the young shoots of various shrubs. It is proverbially timid, and flies if disturbed when feeding, at the slightest alarm; and, led by a natural instinct, it invariably makes towards the rising ground, the length of its hind legs giving it an advantage in this respect over its pursuers. These animals seldom migrate far from the spot where they are produced; but each makes a form at a small distance, showing a predilection rather for the place of their nativity than the society of their kind. They pair in February; and as they only quit their couch in the night time to obtain food, so they never leave it for the companionship of their mates but at the same silent hour: often, indeed, are they observed by moonlight, playing and skipping about in the most sportive manner; but the slightest breeze, or even the falling of a leaf, is sufficient to disturb their revels; and they instantly fly off, each pursuing a different track.

In order to enable this creature to perceive the most distant approaches of danger, nature has provided it with very long ears, which, like tubes applied to the auditory organs of deaf persons, convey to it such sounds as are remote; and the motions of the Hare are directed accordingly. Its large prominent eyes being placed so far backward as to receive the rays of light on every side, it can almost see distinctly behind while it runs directly forward. The muscles of its body being strong, and unencumbered with fat, it has no superfluous burden of flesh to carry: and to assist it in escaping from its pursuers, the hinder legs are considerably longer than the fore, which adds to the swiftness of its motions. But they generally exhaust their powers by their first efforts, and are consequently much more easily caught than foxes, though these wily creatures are slow when compared with them. When the Hare hears the hounds at a distance, it flies for some time from a natural impulse, till having gained some hill or rising ground, and left the dogs so far behind that their cries no longer reach its ears, it stops, rears on its hinder legs, and looks back, for the purpose of satisfying itself whether its enemies are still in sight or not: but the dogs having once gained the scent, trace it with united and unerring skill; and the poor animal soon again receives indications of their approach. Sometimes, when hard hunted, it will start a fresh Hare, and squat in the same form; at others, it will creep under the door of a sheep-cot, and conceal itself among the sheep; sometimes it will enter a hole, like the rabbit; at others, it will run up one side of a quickset hedge, and down the other; and it has been known to ascend the top of a cut hedge, and run a considerable way, by which stratagem it has effectually evaded the hounds. It is also not unusual for the Hare to betake itself to furze bushes, and leap from one to another, whereby the dogs are frequently misled; and as it swims well, and takes the water readily, it will cross a river with the same intent, if it has the opportunity. It may be observed, however, that the first doubling which a Hare makes generally affords a key to all its future attempts of that kind, the latter exactly resembling the former. The Hare is a short-lived animal, and is supposed rarely to exceed the term of seven or eight years. Its voice, which is seldom heard but in the distress of sudden surprise or when wounded, resembles the sharp cry of an infant. Its enemies are numerous and powerful. Every species of the dog kind pursues it by instinct; the cat and the weasel tribes exercise all their arts to ensnare it; and birds of prey, snakes, adders, &c. drive it from its form, particularly during the summer season: these, with the more destructive pursuits of mankind, contribute to thin the number of these animals, which from their prolific nature would otherwise multiply to an extravagant degree.

The flesh is now much prized for its peculiar flavour, as it was by the Romans; but it was forbidden to be eaten among the Jews, Mahometans, and ancient Britons. The fur, until of late years, when silk became so generally used, was of great importance in the manufacture of hats; and in some parts of the continent it is also woven into cloth.

The IRISH HARE (*Lepus Hibernicus*), usually considered a species of the common Hare of England, is said by Mr. Bell, in his "British Quadrupeds," to be specifically distinct. In support of his opinion, he says, "The characters in which it principally differs from the latter are as follows:—It is somewhat larger; the head is rather shorter; the ears are even shorter than the head, while those of the English Hare are fully an inch longer; the limbs are proportionally rather shorter; and the hinder legs do not so much exceed the fore legs in length. The character of the fur is also remarkably different: it is composed exclusively of the uniform soft and shorter hair which in the English species is mixed with the black-tipped long hairs, which give the peculiar mottled appearance of that animal; it is therefore of a uniform reddish brown colour on the back and sides. The ears are reddish gray, blackish at the tip, with a dark line near the outer margin. The tail is nearly of the same relative length as in the common species. The numerous discrepancies in the colour and texture of the fur, and in the form and proportion of the different parts of the animal, appear to me to be too important to constitute merely the characters of a variety."

The SCOTCH, OR VARYING HARE. (*Lepus variabilis.*) This species, which is intermediate in size between the Common Hare and the Rabbit, differs greatly in its habits from both. Though confined to alpine districts (and therefore sometimes called the Alpine Hare), it is diffused through a wide geographical range; being found on the Alps, in Norway, Sweden, Lapland, Russia, Siberia, and Kamtschatka, and occurring also in our own island on the summits of the Scottish mountains. In summer its colour is a tawny gray, with a slight admixture of black; in winter it is entirely white, except the tips of the ears, which are black. It shelters itself in the cliffs of rocks, is easily tamed, and becomes extremely playful and amusing. Towards the month of September it changes its colour, and resumes its summer dress about April; but in the intensely cold climate of Siberia it continues white all the year round. It has been sometimes found entirely coal black; a variety which is also known to take place occasionally in the common species. When the winter has proved unusually severe, the Varying Hare has been known to migrate from the frozen hills of Siberia, and to descend, in troops of five or six hundred, into the plains and woody districts, where they remained till the returning spring.

In the southern and western provinces of Russia there is a mixed breed of Hares, which sustains only a partial loss of its colours; the sides, and the more exposed parts of the ears and legs, becoming white in the coldest months, while the other parts remain unchanged. This variety is by the Russians called Russak: and prodigious numbers are taken in snares for the sake of their skins only; the Russians and Tartars, like our own druidical ancestors, holding the flesh of Hares in the utmost detestation.

The AMERICAN HARE (*Lepus Americanus*) is not much larger than a rabbit, by which name indeed it is well known throughout the northern parts of that vast continent. The summer hair is dark brown on the upper part of the head, lighter on the sides, and of an ash colour below; the ears are wide, edged with white, tipped with brown, and dark coloured behind; tail dark above, white beneath, having the under surface turned up; the fore legs are shorter and the hinder longer in proportion than those of the European. In the middle and southern states, the change in the colour of the hair is by no means as remarkable as it is farther north, where it becomes nearly white. It is not hunted in America, but is generally roused by a dog, and shot or caught by means of snares or a common box-trap, the latter being the most usual mode. It has the same kind of leaping gait as the European hare; and, like that animal, it breeds several times during the year. It is not of a migratory nature, but always continues to haunt the same places, taking occasional refuge under the roots of trees, or in the hollows near the roots.

In Mr. Gosse's "Canadian Naturalist" we find the following information respecting the American Hare:—"It is found pretty generally over North America, from this province, even to the Gulf of Mexico, where it is more common than it is with us. Here its winter coat is nearly white, but in summer it is of a yellowish brown, with a white tail. It makes a nest or bed of moss and leaves in some hollow tree or old log, whence it issues chiefly by night. Though not so much addicted to gnawing as the squirrels, yet as its teeth are formed in the same manner, it probably resembles them in its food, eating various kinds of nuts and seeds, as well as green herbs. It is said also occasionally to peel off the bark from apple and other trees. A singular mode of taking furred animals out of hollow trees, logs, &c. is practised in the south, called 'twisting.' I once saw it performed on a rabbit (so called); the dogs had tracked him and driven him to his hole in the bottom of a hollow hickory tree. The hole was too small to admit the hunter's hand with convenience, so we made the negroes cut down the tree, which was soon effected. When it fell, we watched the butt, to see that the rabbit did not run out, but he did not make his appearance. The hunter then got some long slender switches, and probing the hollow, found that the rabbit was at the farther end several feet up the trunk. He now commenced turning the switch round in one direction, a great many times, until the tip of it had become so entangled in the animal's fur, as to bear a strong pull. He then began to pull steadily out, but the rabbit held on as well as he could, and made considerable resistance, crying most piteously, like a child; at last the skin gave way, and a great mass of fur and skin came out attached to the switch, pulled off by main force. He now took a new switch, and commenced twisting again, and this time pulled the little thing down, but the skin was torn almost completely off the loins and thighs of the poor little creature, and so tightly twisted about the end of the stick, that we were obliged to cut the skin to get the animal free!" A more cold-blooded or barbarous cruelty, practised on a harmless and defenceless animal, it is scarcely possible to conceive; and were it not for the undoubted veracity of the writer we should reject it as well on the score of its incredibility as of its inhumanity.

The CAPE HARE. (*Lepus Capensis.*) This species, which is about the size of the one last described, inhabits the country near the Cape of Good Hope, frequenting the most rocky and mountainous situations, and taking up its abode in the fissures of the cliffs. The ears are long, broad in the middle, naked, and rose-coloured on the outside, and covered with short grey hairs within: the back and upper parts generally are similar in colour to that of the Common Hare; the cheeks and sides are cinereous; the breast, belly, and legs, ferruginous; and the tail, which is bushy, turns upwards. At the Cape it is called the Mountain Hare, or VLAKTE HAAS. In one of the specimens in the British Museum the nape of the neck has two white streaks.

The BAIKAL HARE (*Lepus Tolai*) is rather larger than the common Hare, and has a longer and smaller head, but in colour and general appearance, pretty much resembles it. This animal is an inhabitant of the open hilly places in Dauria and Mongolia, and is said to extend as far as Tibet. In the colour of its flesh it agrees with the rabbit, but differs both from that animal and the hare in its manners; neither burrowing in the ground, like the former, nor running far when pursued, like the latter; but instantly taking refuge in the holes of rocks. [For Alpine Lagomys, Calling Hare, &c., see LAGOMYS.]

HARELDA. A genus of Ducks, containing the Long-tailed Duck (*H. glacialis*). [See DUCK.]

HARENGUS. [See HERRING.]

HARFANG. The Great Snowy Owl. [See OWL.]

HARLEQUIN BEETLE. [See ACROCINUS.]

HARLEQUIN DUCK. (*Clangula histrionica*). A magnificent species found on

HARLEQUIN DUCK.
(CLANGULA HISTRIONICA.)

both continents; it derives its name from the singularity of its markings. It is seventeen inches in length, and twenty-eight inches in extent: the bill is of a lead colour, tipped with red; upper part of the head black; between the eye and bill a broad space of white, extending over the eye, and ending in reddish; behind the ear, a similar spot; neck black, ending below in a circle of white; breast deep slate; shoulders marked with a semicircle of white; belly black; sides chestnut; body above, black, or deep slate; some of the scapulars white; greater wing-coverts tipt with white; legs and feet deep ash; vent and pointed tail black. It swims and dives well; flies swift, and to a great height; and has a whistling note. The female lays ten white eggs on the grass; the young are prettily speckled. At Hudson's Bay, where it breeds, and is said to frequent the small rivulets inland, it is called the Painted Duck; at Newfoundland and along the coast of New England, the Lord. It is an admirable diver, and is often seen in deep water, considerably out at sea.

HARPA, or HARP-SHELL. A beautiful genus of shells, so regularly marked with parallel longitudinal ribs on the outer surface, as to suggest at the first glance the idea of the stringed instrument to which it owes its name. The upper end of each rib is projected and pointed; spire short, last whorl large and deeply notched; outer lip thickened, and is supposed to have no operculum. The Mollusc which inhabits it has the head large; mouth open below; destitute of a proboscis; but having two ten-

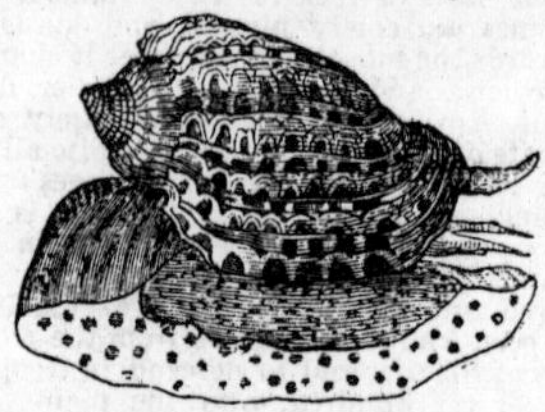

HARP SHELL AND ANIMAL.
(HARPA VENTRICOSA.)

tacula, with eyes in the middle: foot large. It has been asserted by some naturalists that the animal can, when attacked by an enemy, disembarrass itself of part of its foot, and retire entirely within its shell. The principal localities of this genus are the Red Sea and the Indian and South American Oceans. There are several species, all handsome, and some rare; among them the *Harpa multicosta*, which is very rare, and the *Harpa imperialis*, from the Mauritius, the markings of which are very elegant; but perhaps the more abundant species here figured, *Harpa ventricosa*, is as beautiful in form and colouring as any species of this marine carnivorous genus.

HARP-SEAL. The Greenland Seal. [See SEAL.]

HARPY EAGLE. (*Thrasaëtus.*) A genus of Accipitrine Birds found in South America; celebrated for the enormous development of their beak and legs, and the consequent strength and power they evince in mastering their prey. The following short but characteristic notice of this bird occurs

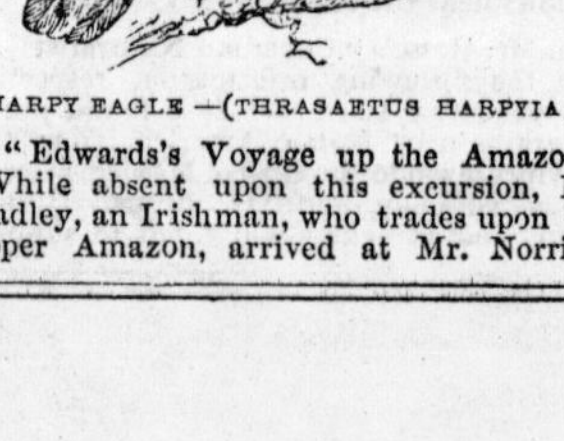

HARPY EAGLE.—(THRASAETUS HARPYIA.)

in "Edwards's Voyage up the Amazon." "While absent upon this excursion, Mr. Bradley, an Irishman, who trades upon the Upper Amazon, arrived at Mr. Norris's, bringing many singular birds and curiosities of various kinds. One of the former was a young Harpy Eagle, a most ferocious looking character, with a harpy's crest and a beak and talons in correspondence. He was turned loose into the garden, and before long gave us a sample of his powers. With erected crest and flashing eyes, uttering a frightful shriek, he pounced upon a young ibis, and quicker than thought had torn his reeking liver from his body. The whole animal world there was wild with fear." No member of the Bird class could look more fierce and indignant than a noble specimen of this formidable Eagle, which we saw some years ago in the Zoological Gardens, Regent's Park. Its whole aspect was that of formidably organized power; and even the appendage of the crest added much to its terrific appearance.

HARRIER. A well-known kind of hound, remarkable for his sagacity in tracing, and boldness in pursuing his game. There are several varieties, but all differing in their services; some being adapted for one sort of game, and some for another. The best breed, and that to which the name is more

HARRIER.

emphatically applied, is the Harrier used for hunting the Hare, which is supposed to have been originally produced by a cross between the Foxhound and the Beagle. The Harrier is generally from sixteen to eighteen inches in height.

HART. The name given to a Stag or male Deer, which has completed his fifth year. [See DEER.]

HARVEST-FLY. [See CICADA *septemdecim.*]

HAWFINCH. [See GROSBEAK.]

HAWK. (*Falconidæ.*) The name by which several birds of prey, closely allied to the Falcons, are designated; as the Goshawk, the Sparrow-hawk, &c., which will be found under their respective names. The beak of the Hawks resembles that of the Falcons in its general form, being curved from the base; but the wings are shorter, and want the pointed tips which are characteristic of that division of the family. The most powerful Hawks are found in cold countries, inhabiting hilly districts where there are woods, and seeking their prey near the ground. Among the whole, none is more bold and pertinacious in pursuit of its prey than the Sparrow-hawk [which see]. In the first volume of Gray and Mitchell's genera of Birds will be found descriptions of the numerous genera, with references to the greater part of the species, and figures of most of the typical forms. In the List of Birds in the British Museum collection, which is exceedingly rich in the Hawk tribe, will be seen how numerous the species are. We refer those desirous of further information to those two works.

HAWK-OWL. [See OWLS.]

HAZEL WORM. [See SLOW-WORM.]

HEATHCOCK. [See GROUSE.]

HECTOCOTYLUS. [See SUPPLEMENT.]

HEDGEHOG. (*Erinaceus Europæus.*) The common Hedgehog is found in most of the temperate parts of Europe and Asia; and though it has a formidable appearance, it is one of the most harmless creatures in existence. It is an insectivorous quadruped, whose generic character may be thus described: — the back covered with sharp strong spines, about an inch long, with the power of rolling itself up in a ball by means of appropriate muscles; muzzle pointed; tail short; and each foot five-toed and armed with robust claws: the head is very conical; the ears short, broad, and rounded; the eyes prominent; the body oblong, and conical above; and the legs short, almost naked, and of a dusky colour. It is about ten inches in length, and its colour is generally a grey-brown. Its close covering of sharp spines, which are firmly fixed in its tough skin, and sufficiently elastic to bear great violence without breaking, protects it from falls or blows, and as effectually secures it from the attacks of an enemy; for when molested, it instantly rolls itself into a kind of ball, and presents nothing but its prickles to the foe; and the more the animal is irritated and alarmed, the more firmly does it contract itself, and the more stiff and strong does its bristly panoply become. Thus rolled up, it patiently waits till the danger is past: the cat, the weasel, the ferret, and the martin soon decline the combat; and though a well-trained wire-haired terrier, or a fox, may now and then be found to open a Hedgehog, it generally remains impenetrable and secure. From this state of security, in fact, it is not easily forced; scarcely anything but cold water obliging it to unfold itself.

The usual food of the Hedgehog is beetles, worms, slugs, and snails; it is also said to devour fruit, the roots of plants, and certain other vegetable substances, while it shows itself not so restricted as has been thought in its choice of animal food; eggs, frogs, toads, mice, and even snakes occasionally, serving for its repast. The Hedgehog is strictly nocturnal, remaining coiled up in its retreat during the day, and wandering about nearly all the night in search of food. It generally resides in small thickets, in hedges, or in ditches covered with bushes, making a hole about six or eight inches deep, which it lines with moss, grass, or leaves. The hibernation of the Hedgehog is undoubted:

although it lays up no store for the winter, it retires to its hole, and in its warm, soft nest of moss and leaves, it lies secure from the rigours of the frost and the violence of the tempest, passing the dreary season in a profoundly torpid state. The female produces from two to four young ones early in the summer, which at their birth are blind, and covered with soft white spines, which in two or three days become hard and elastic. The flesh of these animals, though generally rejected as human food, is said to be very delicate.

Many absurd errors prevail as to the habits of this animal. It is charged with sucking the teats of cows by night, and wounding their udders with its spines, thereby causing those ulcerations which are sometimes observed: from this false accusation, however, the smallness of its mouth is a sufficient exculpation. It is also said to be very destructive to gardens and orchards, by rolling itself among fruit, and thus carrying off a quantity on its spines: but its spines are evidently so disposed, that no fruit would stick on them, even were the experiment attempted. But so far from being mischievous and injurious, the Hedgehog is found to be of real use, and is often kept for the purpose of ridding houses of the numerous cockroaches by which some are infested; and it is well known to devour many destructive insects of the beetle kind and others, which are injurious to the farmer and gardener.

In the "Journal of a Naturalist," this animal is thus noticed:—"Notwithstanding all the persecutions from prejudice and wantonness to which the Hedgehog is exposed, it is yet common with us; sleeping by day in a bed of leaves and moss, under the cover of a very thick bramble or furze-bush, and at times in some hollow stump of a tree. It creeps out in the summer evenings; and, running about with more agility than its dull appearance promises, feeds on dew-worms and beetles, which it finds among the herbage, but retires with trepidation at the approach of man. In the autumn, crabs, fruits, haws, and the common fruits of the hedge, constitute its diet. In the winter, covering itself deeply in moss and leaves, it sleeps during the severe weather; and, when drawn out from its bed, scarcely anything of the creature is to be observed, it exhibiting only a ball of leaves, which it seems to attach to its spines by repeatedly rolling itself round in its nest."

The SIBERIAN OR LONG-EARED HEDGEHOG. (*Erinaceus auritus.*) This species is in general larger than the common or European, and may be easily distinguished by its ears, which are large, oval, open, and naked, with soft whitish hair on the inside, and edged with brown: the upper part of the animal is covered with slender brown spines, with a whitish ring near the base, and another towards the tip, and the legs and belly are clothed with soft white fur. In its general manner and habits this species is said to resemble the common Hedgehog. The EARLESS HEDGEHOG appears to be only a variety of the common species: the head, however, is somewhat shorter and the snout more blunt; there is no appearance of external ears; it is shorter; and the whole animal is of a whitish hue.

HEDGE-SPARROW. [See SPARROW.]

HELAMYS, or JUMPING HARE. This animal constitutes a genus of mammalia, of the order *Rodentia*, allied to the Jerboas. The head is large, the tail long, the fore legs are very short in comparison with the hinder. They have four molars, each composed of two laminæ; their lower incisors are truncated:

AFRICAN JUMPING HARE.
(HELAMYS CAFFER.)

the fore feet have five toes, furnished with long pointed nails; the hind feet have four toes, which are separate as far as the bones of the metatarsus, and furnished with large claws, almost resembling hoofs. The species *Helamys caffer* is pale fulvous, with a long tufted tail, black at the tip. It is as large as a rabbit, and, like it, inhabits deep burrows. Our cut exhibits one about to spring, while another is at the mouth of its burrow.

HELARCTOS. A genus of Bears found in India and the Eastern Islands. The Malay and Java Bears may be given as illustrations. [See BEAR.]

HELICINA. A genus of Mollusca, found in America and the West Indies. Some inhabit the sea, but others are terrestrial, either feeding upon trees or subsisting on the vegetable productions of the fields and gardens. The head of the animal is furnished with a proboscis and two tentacula, with eyes at the base on tubercles; foot short. The shell is of a flattened shape, mouth semicircular, closed by a horny operculum, which is formed of concentric layers, and permanently attached to the foot; outer lip thickened and reflected, inner lip spread over the body-whorl, terminating in a point. There are a great many species.

HELICONIDÆ. A family of Lepidopterous insects; in which the wings vary in shape, but are often very long and narrow, and the discoidal cell of the hind wings is always closed; the antennæ are slightly clavate; the palpi are short, and wide apart at the base, the second joint being generally clothed with hairs directed upwards at its extremity. The caterpillars are cylindrical, and either spinose or furnished with several pairs of long fleshy appendages; and the chrysalides are often ornamented with brilliant golden spots. The species belonging to this family are entirely exotic, of a moderately large size, and of very varied colours. In some of the species the wings are quite denuded of scales and in many they are but slightly covered. One of the species, *Euplœa* (*Danais*) *hamata*, is said to be so abundant in New Holland, that it occasionally darkens the air from the *clouds* of them. By many authors this and the allied genera are placed in the separate family *Danaidæ*. We must refer our readers to Mr. Doubleday's elaborate letter-press to his work on the Diurnal Lepidoptera, so beautifully illustrated by Mr. Hewitson.

HELICTIS. A genus of carnivorous Quadrupeds allied to the Skunks, of which there are at least two species, one found in China, where it was discovered by Mr. Reeves, the other in Nepal, whence it was sent by Mr. Hodgson.

HELIORNIS. A genus of Birds found in South America. [See FINFOOT.]

HELIX: HELICIDÆ. The general name of a large and most extensively diffused class of Molluscous animals with a shelly covering. It is equally adapted to the hottest and the coldest climates, the most cultivated and the most barren situations. In the Cuvierian system this is the type of a family of terrestrial and air-breathing Gasteropods. The common Garden Snail of this country, and the Edible Snail of France and Italy, are well-known examples of this family; but in tropical climates more striking ones are to be found. The work of Dr. Pfeiffer is the latest and the most elaborate on this group. In the works of Wood, Sowerby, Reeve and others, a great number of species are figured. An inspection of the cases containing them in the British Museum will show how varied their forms are, and how beautifully coloured are many of the species. There are some brought from the Philippine Islands by Mr. Cuming, which when wetted lose their colour, but regain it when dry. This is owing to the nature of the epidermis. [See SNAIL.]

HELMET-SHELL. (*Cassis.*) A family of shells, of which there are several species, mostly found on tropical shores, but some

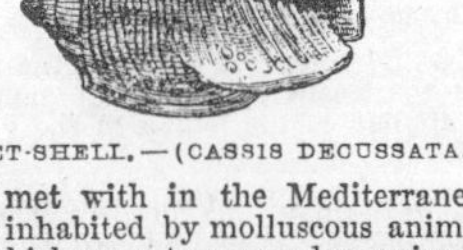

HELMET-SHELL.—(CASSIS DECUSSATA.)

are also met with in the Mediterranean. They are inhabited by molluscous animals, some of which grow to a very large size, requiring of course a corresponding magnitude of shell. They live at some distance from the shore, on the sand, into which they occasionally burrow, so as to hide themselves. The back of the Helmet-shell is convex, and the under part flat: the mouth is long and narrow: the lip is strongly serrated, and rises into a high thick border or ledge on the back; and the pillar is generally strongly toothed, and beset with small asperities. The shells of the *Cassis rufa* and other species are beautifully sculptured by Italian artists in imitation of antique cameos, the different layers of colouring matter resembling the onyx and other precious stones formerly used for this purpose.

Upon this subject some interesting particulars were detailed by Mr. J. E. Gray, at a meeting of the Society of Arts, held April 21. 1847. He observed that numerous attempts have been made to substitute various materials, such as porcelain and glass, for the ancient cameos; but their great inferiority has caused them to be neglected. The best and now most used substitutes are shells; several kinds of which afford the necessary difference of colour, and at the same time are soft enough to be worked with ease and hard enough to resist wear. The shells used are those of the flesh-eating univalves, which are peculiar as being formed of three layers of calcareous matters, each layer being a perpendicular lamina placed side by side. The cameo cutter selects those shells which have the three layers composed of different colours, as they afford him the means of relieving his work; but the kinds now employed, and which experience has taught him are best for his purpose, are the Bull's Mouth (*Cassis rufa*) from the Indian Seas, the Black Helmet (*Cassis Madagascariensis*), a West Indian shell, the Horned Helmet (*Cassis cornuta*), from Madagascar, and the Queen Conch (*Strombus gigas*), a native of the West Indies. The two first are the best shells. After detailing the peculiarities of these shells, Mr. Gray proceeded to give an account of the progress of the art, which was confined to Rome for upwards of forty years, and to Italy until the last twenty years, at which period an Italian commenced the making of them in Paris; and now about three hundred persons are employed in this branch of trade in that city. The number of shells used annually thirty years ago was about three hundred, the whole of which were sent from England; the value of each shell in Rome being 30*s*. To show the increase of this trade, the number of shells used in France last year was nearly as follows:

		Average Price.	Value.
Bull's Mouth -	80,000	1*s*. 8*d*.	£6,400
Black Helmet ·	8,000	5*s*.	1,800
Horned Helmet	500	2*s*. 6*d*.	60
Queen Conch -	1,200	1*s*. 2½*d*.	700
	100,500 shells.		Value 8,960

The average value of the large cameos made in Paris is about six francs each, giving a sterling value of 32,000*l*., and the value of the small cameos is about 8,000*l*., giving a total value of the cameos produced in Paris for the last year of 40,000*l.*, while in England not more than six persons are employed in this trade. *Athenæum*, May 1. 1847.

HELOPIDÆ. A family of insects belonging to the order *Coleoptera*, division *Heteromera*, in which the antennæ are inserted near the eyes, and the terminal joint is always the longest, covered at the base by

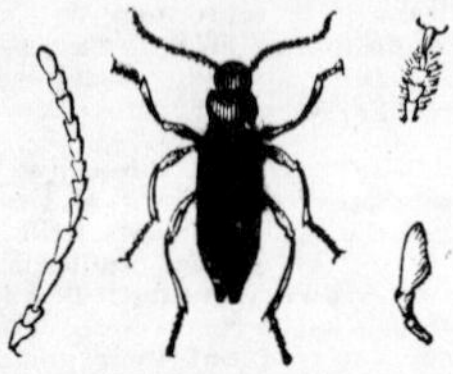

HELOPS PICEUS

the margin of the head, filiform, or slightly thickened at the tip, elytra not soldered together; maxillary palpi, with the last joint largest, hatchet-shaped; eyes generally kidney-shaped. The larvæ generally filiform, with smooth shining bodies and very short feet. They are found in old wood, while the perfect insects are frequently found upon flowers, or below the bark of trees. This family consists of several genera, most of which are exotic. Mr. Paget, of Yarmouth, in his Natural History of that town, mentions that the larva of the common *Helops violaceus* injured the wood of a window-frame very much, in which several of these insects had taken up their abode.

HEMEROBIUS: HEMEROBIIDÆ. LACE-WING FLIES. A genus and family of insects belonging to the order *Neuroptera;* remarkable for the exceeding brilliancy of the eyes in most of the species, and for the

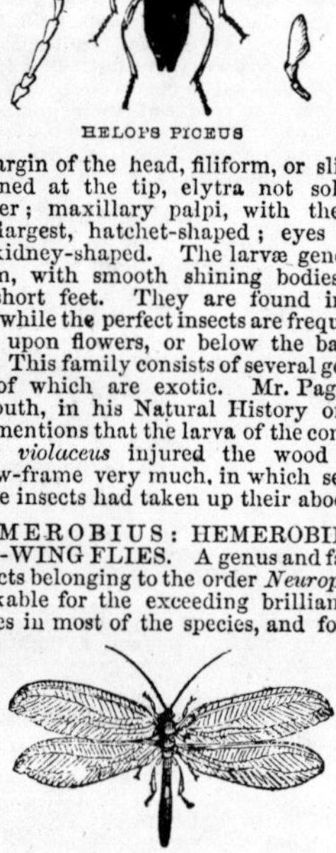

LACE-WING FLY —(HEMEROBIUS.)

delicate structure and varied colours of their long reticulated wings; so that, although of small size, they are very conspicuous. They deposit their eggs upon plants, attaching them at the extremity of a long slender footstalk, the base of which is fastened to the leaf: thus fixed in small clusters, they have the appearance of minute fungi. The larvæ of these insects are extremely ravenous; and, as they feed on the *Aphides*, or plant-lice, are highly beneficial. During the summer they arrive at their full growth in about fifteen days; they then spin a silken cocoon, in which they enter as inactive pupæ, and there remain during the winter.

HEMIDACTYLUS. A genus of Lizards belonging to the Gecko family, in which the tail is depressed, angular above, with cross rows of spines, the toes being free. The species are found in various parts of the world, and will be found described in Mr. Gray's Catalogue of Reptiles; one species seems to be common on the shores of the Mediterranean.

HEMIPODIUS. A genus of Gallinaceous Birds allied to the Quails, of which there are very many species in Africa and Asia chiefly. Colonel Sykes has described many of the East India species. We must refer to Gray's and Mitchell's Genera of Birds for a list of the species and figures of the form, and limit ourselves to the notice of a species figured in the work of Mr. Gould, where it is called the SWIFT-FLYING HEMIPODE. This bird inhabits New South Wales, and is the "Little Quail" of the colonists. The male is little more than half the size of the female. It breeds in September and October: the nest is slightly constructed of grasses, placed in a shallow depression of the ground, under the shelter of a small tuft of grass: eggs four in number. The Hemipodius lies so close as to be nearly trodden on before it will rise, and, when flushed, flies off with such rapidity as to make it very difficult to shoot.

HEMIPTERA. An order of Insects characterized by having a horny beak for suction; four wings, whereof the uppermost are generally thick at the base, with thinner extremities, which lie flat, and cross each other on the top of the back, or are of uniform thickness throughout, and slope at the sides like a roof. Transformation partial. Larvæ and pupæ nearly like the adult insect, but wanting wings.—The various kinds of field and house bugs give out a strong and disagreeable smell. Many of them (some *Pentatomidæ* and *Lygæidæ*, *Cimicidæ*, *Reduviadæ*, *Hydrometridæ*, *Nepidæ*, and *Notonectidæ*) live entirely on the juices of animals, and by this means destroy great numbers of noxious insects; some are of much service in the arts, affording us the costly cochineal, scarlet grain, lac, and manna; but the benefits derived from these are more than counterbalanced by the injuries committed by the domestic kinds, and by the numerous tribes of plant-bugs, locusts or cicadæ, tree-hoppers, plant-lice, bark lice, mealy bugs, and the like, that suck the juices of plants, and require the greatest care and watchfulness on our part to keep them in check. The works of Burmeister, Amyot, and Servilla, Meyer, Hatton and others may be referred to for the species, which are very numerous, and often most beautifully coloured, the colour and odour being by no means in *harmony*.

HEN. The general name of the female among the feathered tribes, but more especially applied to the female of the gallinaceous kind.

HEN-HARRIER. (*Circus cyaneus.*) This bird is a species of hawk; about eighteen inches in length, and three feet in extent from the tips of the wings extended. The bill is black, and covered at the base with long bristly feathers ; cere, irides, and edges of the eyelids, yellow : the upper parts of the plumage bluish gray, mixed with light tinges of rusty ; the breast and under-coverts of the wings white, the former streaked with reddish brown, and the latter marked with bars of the same : the wings and tail are a bluish-gray, variegated with black ; and the legs are long, slender, and yellow. The Hen-Harrier feeds on birds and reptiles ; it flies low, skimming along the surface of the ground in search of prey, and is extremely destructive to young poultry and the feathered game. It makes its nest on the ground, and lays four eggs of a reddish colour, with a few white spots.

HEPATUS. A beautiful genus of Crustacea found in South America, and so named from its liver-coloured marking. The genus is allied to *Calappa*, and belongs to the same family.

HEPIALIDÆ. A family of Lepidopterous insects, in the section HETEROCERA (corresponding with the first group of Latreille's NOCTURNA). It is distinguished by having the antennæ very short and filiform, never feathered to the tip ; the spiral tongue either very short or obsolete ; and the palpi also generally obsolete ; the wings elongated, and deflexed in repose ; the abdomen also elongated, its extremity being attenuated into a long ovipositor, capable of being withdrawn, or introduced into the crevices of the bark of trees, &c. The caterpillars are sixteen-footed fleshy grubs ; and feed upon the roots of vegetables or the wood of standing trees : when full grown, they construct a cocoon of the refuse of what they have been feeding upon. The chrysalis is armed with transverse rows of fine reflexed spines on the abdominal segments, which assist the insect whilst making its efforts to emerge from its confinement and assume the perfect state. The *Hepialidæ* are called *Swifts*, from the rapidity of their flight, which takes place during the twilight. Some of the species are very remarkable, particularly HEPIALUS VIRESCENS, a large species from New Zealand, described by Mr. Doubleday. The caterpillar of this is very frequently attached by a fungus ; which entirely converts it into a vegetable substance, the fructification and its pedicel projecting considerably. This fungus is the *Sphæria Robertsi* of Hooker (*S. erucarum* of Mulsant.) Among the most striking Hepialidæ of this country are *Hepialus Humuli*, or the Ghost Moth, and *Cossus ligniperda*, or the Goat Moth [which see].

HEPIOLUS. [See GHOST-MOTH.]

HERMIT CRAB. The name given to different species of the family *Paguridæ*, which occupy empty shells, in which they protect their soft and otherwise easily injured tails. [See CRAB : PAGURUS.]

HERON. (*Ardea.*) Though birds of the crane, the stork, and the Heron kind, have a strong affinity to each other, the Heron may be distinguished by its smaller size, its longer bill, and particularly by the middle claw on each foot, which is serrated, for the better seizing and securing its slippery prey. Herons reside on the banks of lakes and rivers, or in marshy places : their food consists of fishes and their fry, frogs, and field mice, as well as all sorts of insects, snails, and worms. They build in large societies in the same place ; and when they fly, their neck is contracted and folded over their back, and their legs are extended.

The COMMON HERON (*Ardea cinerea*) is remarkably light in proportion to its bulk, scarcely weighing three pounds and a half, though its length is upwards of three feet, and its extended breadth above five. The bill is six inches long, straight, pointed, and strong ; the upper mandible is of a yellowish horn colour, the under one yellow : the forehead, neck, middle of the belly, edge of the wing, and the thighs, are of a pure white ; the occiput, the sides of the breast, and those of the body, of a deep black : the fore part of the neck is adorned with large longitudinal spots of black and gray ; the back and wings are blue gray. A bare greenish skin is extended from the beak beyond the eyes, the irides of which are yellow, giving them a fierce and piercing aspect. The back part of the head is ornamented with several elongated narrow black feathers, the two middle of which are upwards of eight inches in length ; the whole forming an elegant pendent crest ; the feathers of the scapulars are also elongated, and fall over the back in fine disunited plumes. The tail is composed of twelve short cinereous feathers ; the legs are of a dirty green colour, long, and bare above the knees; and the inner edge of the middle claw is finely serrated. The female is destitute of the long crest of the male, having only a short plume of dusky feathers ; and in general her plumage is gray : the same remarks are also nearly applicable to the young birds. In the breeding season they congregate in large societies, and, like the rooks, build their nests on trees, with sticks, lined with dried grass, wool, and other warm materials. The female lays from four to six eggs, of a pale greenish blue colour.

This bird commits great devastation in ponds and shallow waters. As a proof of its appetite, it is asserted by Willoughby and others, that a single Heron will destroy fifty small roach and dace, one day with another. Though it generally takes its prey by wading into the water, and waiting patiently for its approach, it frequently also catches it whilst on the wing ; but this is only in shallow waters, where it is able to dart with more certainty than in the deep ; for in this case, though the fish does at the first sight of its enemy descend, yet the bird, with its long beak and legs, instantly pins it to the bottom, and there seizes it securely. In general, the Heron is seen taking his gloomy stand by the side of a lake, as if meditating mischief, motionless, and gorged with plunder. His usual attitude on such occasions is that of sinking his long neck between his shoulders, and keeping his head turned on one side, as if viewing the water more intently. When the call of hunger returns, the toil of an hour or two is sufficient to fill his capacious stomach; and, though he often feeds by night, he generally is able to retire long before to his lodging in some wood, which he quits early next morning. But in cold and stormy seasons, when his prey is no longer within his reach — the fish then abiding in the deep as their warmest situation, and frogs, lizards, and other reptiles, also seldom venturing from their retreats during the continuance of such weather — the Heron is obliged to practise abstinence, and to feed on such weeds as the margin of the lake affords: hence he feels the ills both of hunger and repletion, and notwithstanding the amazing quantity he devours, he is always lean and emaciated. While on this subject, it may, however, be well to attend to what Mr. Waterton has written: "I attribute the bad character (says he) which the Heron has with us, for destroying fish, more to erroneous ideas, than to any well-authenticated proofs that it commits extensive depredations on our store-ponds. Under this impression, which certainly hitherto has not been to my disadvantage, I encourage this poor persecuted wader to come and take shelter here; and I am glad to see it build its nest in the trees which overhang the water, though carp, and tench, and many other sorts of fish are there in abundance. Close attention to its habits has convinced me that I have not done wrongly. Let us bear in mind that the Heron can neither swim nor dive; wherefore the range of its depredations on the finny tribe must necessarily be very circumscribed. In the shallow water only can it surprise the fish; and, even there, when we see it standing motionless, and suppose it to be intent on striking some delicious perch or passing tench, it is just as likely that it has waded into the pond to have a better opportunity of transfixing a water-rat lurking at the mouth of its hole, or of gobbling down some unfortunate frog which had taken refuge on the rush-grown margin of the pool. The water-rat may appear a large morsel to be swallowed whole; but so great are the expansive powers of the Heron's throat, that it can gulp down one of these animals without much apparent difficulty. As the ordinary food of this bird consists of reptiles, quadrupeds, and fish, and as the Herons can only catch the fish when they come into shallow water, I think we may fairly consider this wader not very injurious to our property; especially when we reflect for a moment on the prodigious fecundity of fish."

In its aërial journies the Heron soars to a great height, and its harsh cry while on the wing frequently attracts the ear. In flying, it draws the head between the shoulders, and the legs, stretched out, seem, like the longer tails of some birds, to serve as a rudder. The motion of their wings is heavy and flagging, and yet they proceed at a very considerable rate. In England, Heron-hawking was formerly a favourite diversion among the nobility and gentry of the kingdom, at whose tables this bird was a favourite dish, not less esteemed than pheasants and peacocks. It was ranked among the royal game, and protected as such by the laws; and a penalty of twenty shillings was incurred by any person who took or destroyed its eggs. — Dr. Latham says, "In England, and the milder climates, this species of Heron is stationary; migratory in the colder, according to the season; and is rarely seen far north: inhabits Africa and Asia in general, the Cape of Good Hope, Calcutta, and other parts of India; and is found in America from Carolina to New York."

The AGAMI HERON. (*Ardea Agami.*) By general consent, as it were, this bird is allowed to be the most beautiful of the genus. It is a native of Surinam, and is rather more than two feet and a half in length: its beak is about six inches long, and dusky, with the base of the under mandible pale; the crown, the crest, and the hind part of the neck are bluish gray; the upper parts of the body, the wings, and the tail, are a fine glossy green; the quills are black; the sides of the neck bright rufous, with an elegant white and rufous line, bounded by black down the central part: the breast is clothed with long, loose, dark feathers; those on the back of the neck black, with a white streak down the middle of each shaft: the under parts of the body are deep rufous; and the tail is brown.

The GREAT HERON. (*Ardea Herodias.*) This species inhabits North America, and is one of the largest of the genus, measuring upwards of five feet in length: the beak is eight inches long, and of a brown colour, inclining to yellow on the sides: on the back of the head is a long-feathered crest:

GREAT HERON.—(ARDEA HERODIAS.)

the space between the beak and eye is naked, and of a pale yellow: all the upper parts of the body, with the belly, tail, and legs, are brown; the quills black; the neck, breast, and thighs rufous. Like the rest of this genus, the Great Heron frequents the borders of lakes and rivers, and feeds on reptiles and small fishes.

The GREAT WHITE HERON. (*Herodias alba.*) This bird's plumage is wholly white; it may therefore be easily known from the common Heron: it is also rather smaller, the tail and legs are longer, and it has no crest. Its character and manner of living are the same, and it is found in the same countries, though the species is far less numerous, and it is rarely seen in Great Britain. It is found on the shores of the Caspian and Black Seas, the lakes of Great Tartary, and sometimes even much farther north: it is also met with in various parts of Africa and America.

The LITTLE EGRET HERON (*Herodias garzetta*) is one of the most elegant as well as one of the smallest of the Heron tribe. The beak is black, the naked space round the eyes greenish, the legs dusky, and the feet black. Its colour is of the purest white, and it is adorned with soft, silky, flowing plumes on the head, breast, and shoulders, which give the bird a beauty quite peculiar to itself. These delicately-formed feathers are six or eight inches in length, with slender shafts, twisted and bent down towards their tips: they were formerly used to decorate the helmets of warriors, but they now embellish the turbans of Turks and Persians, or are applied to the more consistent purpose of ornamenting the head-dresses of European ladies. The Little Egret is only about eighteen inches in length, and seldom exceeds a pound and a half in weight. These birds are said to have once been plentiful in this country, but they are now nearly extinct here; they are, however, abundant in the south of Europe, and are found in almost every temperate and warm climate. Like the Common Heron, they perch and build on trees, and live on the same kinds of food.

The NIGHT HERON. (*Nycticorax griseus.*) This species, which with its congeners is placed by modern naturalists in a separate genus (*Nycticorax*), is by no means numerous, though widely dispersed over Europe, Asia, and America. It is about twenty inches in length: the bill is slightly arched, strong, and black, inclining to yellow at the base: from the beak round the eyes the skin is bare and of a greenish colour: over each eye is a white line; a black patch, glossed with green, covers the crown of the head and the nape of the neck, from which three long narrow white feathers, tipped with brown, hang loose and waving. The hinder part of the neck, coverts of the wings, the sides, and tail, are ash gray; throat white; fore part of the neck, breast, and belly, yellowish white or buff; the back black; legs greenish yellow. The plumage of the female is considerably less bright and distinct; and she has none of the delicate plumes which flow from the head of the male. She lays three or four white eggs. The Night Heron frequents the sea-shores, rivers, and inland marshes, and lives upon insects, slugs, reptiles, and fish. It remains concealed during the day, and does not roam abroad until the approach of night, when its harsh and disagreeable cry is painfully distinguishable. It builds its nest on trees and on rocky cliffs.

There are numerous other species and varieties of the Heron, differing in their size and plumage, but nearly all having the same habits, and being characterized by similar features with those we have described. Among the most important are the Purple-crested Heron (*Ardea purpurea*), common in the western parts of Asia and the north of Europe; the Violet Heron (*Ardea leucocephala*) of the East Indies; the Cocoi Heron (*Ardea cocoi*), a large species, native of Brazil; the Little White Heron (*Ardea Æquinoctialis*), a native of Carolina and some other parts of North America; besides the Blue, the Brown, the Black, the Ash-coloured Heron, &c.

In Mr. Edwards's narrative of a "Voyage up the Amazon," one cannot but be struck with the multitudes of large birds which almost everywhere met the eye of the voyager; not the least numerous or important among them being various species of Herons. "Upon the trees," says he, "were perched birds of every variety, which flew before our advance at short distances in constantly increasing numbers, or, curving, passed directly over us; in either case affording marks too tempting to be neglected. Upon some topmost limb the great blue Heron, elsewhere shyest of the shy, sat curiously gazing at our approach. Near him, but lower down, herons white as driven snow—some tall and majestic as river naiads, others small and the pictures of grace—were quietly dozing after their morning's meal. Multitudes of night herons, or tacarés, with a loud quack, flew startled by; and now and then, but rarely, a boat-bill with his long-plumed crest would scud before us. The snake-bird peered out his long neck to discover the cause of the general commotion; the cormorant dove, from the dry stick where he had slept away the last hour, into the water below, swimming with head scarcely visible above the surface, and a ready eye to a treacherous shot. Ducks rose hurriedly, and whistled away; curassows flew timidly to the deeper wood; and fearless hawks, of many varieties, looked boldly on the danger."

HERRING. (*Clupea Harengus.*) This Malacopterygious fish, which frequents our coasts in such numbers, and furnishes a large class of persons with an important article of food, is from ten to twelve inches in length. It is principally distinguished by the brilliant silvery colour of its body, the advancement of the lower jaw beyond the upper, and by the number of rays in the anal fin, which are generally found to amount to sixteen: the back and sides are green, varied with

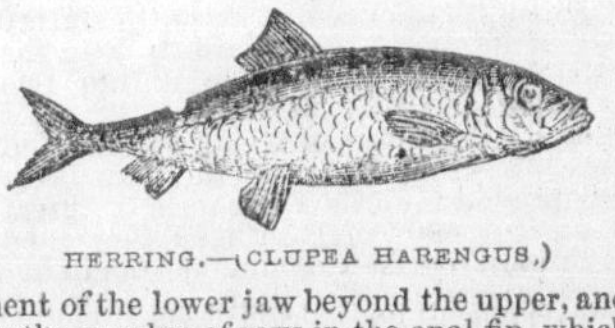

HERRING.—(CLUPEA HARENGUS.)

blue; the eyes are large; the mouth without visible teeth; the openings of the gill-covers very large; the scales moderate in size, oval, and thin; the lateral line not very distinctly visible; the belly carinated, but not serrated: the fins rather small than large for the size of the fish; and the tail considerably forked.

It has long been asserted, and generally believed, that Herrings are found in the greatest abundance in the high northern latitudes; and that the prodigious shoals which at certain seasons fill our seas, are making their migratory excursions from those icy regions. But this "great fact" in natural history has not only been called in question of late, but the migration of the Herring *from one latitude to another* has been denied by men of high scientific attainments who have given the subject great attention, and who assert that the Herring, having passed the winter and spring months in the deep recesses of the ocean, follows the dictates of nature, and at the proper season approaches the shallower water near the coasts to deposit its spawn. We shall therefore lay the statements, *pro* and *con*, before our readers.

Mr. Pennant, in his British Zoology, says, "The great winter rendezvous of the Herring is within the arctic circle: there they continue many months, in order to recruit themselves after the fatigue of spawning, the seas within that space swarming with small crustacea in a far greater degree than in our warmer latitudes." He then thus proceeds: "This mighty army begins to put itself in motion in the spring: we distinguish this vast body by that name, for the word Herring is derived from the German, *Heer*, an army, to express their numbers. They begin to appear off the Shetland isles in April and May: these are only forerunners of the grand shoal which comes in June, and their appearance is marked by certain signs, by the numbers of birds, such as gannets and others, which follow to prey on them: but when the main body approaches, its breadth and depth is such as to alter the very appearance of the ocean. It is divided into distinct columns of five or six miles in lengtl and three or four in breadth, and they drive the water before them with a kind of rippling: sometimes they sink for the space of ten or fifteen minutes; then rise again to the surface, and in bright weather reflect a variety of splendid colours, like a field of the most precious gems, in which, or rather in a much more valuable light, should this stupendous gift of Providence be considered by the inhabitants of the British isles. The first check this army meets in its march southward, is from the Shetland isles, which divide it into two parts; one wing takes to the east, the other to the western shores of Great Britain, and fill every bay and creek with their numbers: others pass on towards Yarmouth, the great and ancient mart of Herrings: they then pass through the British Channel, and after that in a manner disappear: those which take to the west, after offering themselves to the Hebrides, where the great stationary fishery is, proceed towards the north of Ireland, where they meet with a second interruption, and are obliged to make a second division: the one takes to the western side, and is scarce perceived, being soon lost in the immensity of the Atlantic; but the other, which passes into the Irish sea, rejoices and feeds the inhabitants of the coasts that border it. These brigades, as we may call them, which are thus separated from the greater columns, are often capricious in their movements, and do not show an invariable attachment to their haunts."

The foregoing account, so well detailed by Pennant, was until lately, as we have before remarked, the generally received opinion; but it is now supposed that the Herring, like the Mackerel, is in reality at no very great distance during the winter months from the shores which it most frequents at the commencement of the spawning season; and this is thought a sufficient explanation of the glittering myriads which at particular seasons illumine the surface of the ocean for miles together. As a proof of this, Dr. Bloch observes that Herrings are in reality found at almost all seasons of the year about some of the European coasts, and that the northern voyages, supposed by Pennant and others, are impracticable in the short period assigned by naturalists; the fish, in its swiftest progress, being utterly incapable of moving at so rapid a rate as this migration necessarily supposes.

But the subject has been more amply discussed by Mr. Yarrell, who brings forward so many valid and well-supported objections to the theory of the Herring's migration from the arctic seas, that we shall take the liberty of extracting them from his excellent work. "To show that this supposed migration to and from high northern latitudes does not exist, it is only necessary to state, that the Herring has never been noticed, that I am aware, as abounding in the Arctic Ocean: it has not been observed in any number in the proper icy seas; nor have our whale-fishers or arctic voyagers taken any particular notice of them. There is no fishery for them of any consequence either in Greenland or Iceland. On the southern coast of Greenland the Herring is a rare fish; and only a small variety of it, according to Crantz, is found on the northern shore. This small variety or species was found by Sir John Franklin, on the shore of the Polar basin, on his second journey. 'That the Herring is, to a certain degree, a migratory fish,' says Dr. M'Culloch, 'may be true; but even a much more limited migration is far from demonstrable. It is at any rate perfectly certain that there is no such progress along the east and west coasts from a central point.' There can be no doubt that the Herring inhabits the deep water all round our coast, and only approaches the shores for the purpose of depositing its spawn within the immediate influence of the two principal agents in vivification — increased temperature and oxygen; and as soon as that essential operation is effected, the shoals that haunt our coast disappear: but individuals are to be found, and many are caught, throughout the year.

So far are they from being migratory to us from the North only, that Herrings visit the west coast of the county of Cork in August, which is earlier than those which come down the Irish Channel arrive, and long before they make their appearance at other places much farther north. 'In former times, the fishery on the east coast did not commence till that on the west had terminated. It is remarkable also that the eastern fishery has become so abundant as quite to have obscured the western.' And Dr. M'Culloch, from other examples, confirms a statement previously made, that the fishery has commenced soonest on the southern part of the shore; and what is also remarkable, that for some years past it has become later every year. The Herring is in truth a most capricious fish, seldom remaining in one place; and there is scarcely a fishing station round the British islands that has not experienced in the visits of this fish the greatest variations both as to time and quantity, without any accountable reason."

Herrings are full of roe in the end of June, and continue in perfection till the beginning of winter, when they deposit their spawn. The young Herrings begin to approach the shores in July and August, and are then from half an inch to two inches long. The Herring was unknown to the ancients, being rarely, if ever, found within the Mediterranean. The Dutch are said to have engaged in the fishery in 1164. The invention of pickling or salting Herrings is ascribed to one Benkels, or Benkelson, of Biervliet, near Sluys, who died in 1397. The emperor Charles V. visited his grave, and ordered a magnificent tomb to be erected to his memory. Since this early period the Dutch have uniformly maintained their ascendancy in the Herring fishery; but, owing to the Reformation, and the relaxed observance of Lent in Roman Catholic countries, the demand for Herrings upon the Continent is now far less than in the fourteenth and fifteenth centuries. The mode of fishing for Herrings is by drift-nets, very similar to those employed in the pilchard fisheries: the fishing is carried on only in the night; the most favourable time being when it is quite dark, and the surface of the water is ruffled by a breeze.

Though there are some other species of Herrings, none of them are of the same commercial importance as the Common Herring, already described, which so abundantly visits our shores; a slight notice of one of them, therefore, is all that will be necessary.

LEACH'S HERRING. (*Clupea Leachii.*) "The Herring," says Mr. Yarrell, "which I now refer to, is found heavy with roe at the end of January, which it does not deposit till the middle of February. Its length is not more than seven inches and a half, and its depth near two inches. It is known that Dr. Leach had often stated that our coast produced a second species of Herring; but I am not aware that any notice of it has ever appeared in print. In order, however, to identify the name of that distinguished naturalist with a fish of which he was probably the first observer, I proposed for it the name of *Clupea Leachii.*" The flesh of this species is said to differ from that of the Common Herring in flavour, and to be much more mild.

HESPERIIDÆ. A family of Lepidopterous insects, corresponding with the *Plebeii Urbicoli* of Linnæus, and in many respects approaching to the Moths. The six feet are of uniform size in both sexes, the hind tibiæ having a pair of spurs at the apex, and another pair near the middle of the limb; the antennæ are wide apart at the base, and are often terminated in a very strong hook; the maxillæ are very long; and the lower wings are generally horizontal during repose. All the known caterpillars belonging to this family are cylindrical without spines, with the anterior segments narrowed, and the head very large: they roll up leaves, in which they construct a slight silken cocoon, wherein the chrysalis form is assumed; this is entire, without angular prominences, and attached by the tail as well as girt round the middle.

These Butterflies have a peculiar, short, jerking kind of flight, which has obtained for them the name of *Skippers*. The species

HESPERIA [PAMPHILA] SYLVANUS, M.

are of comparatively small size, and of obscure colours, but some are ornamented with bright transparent spots, and others have

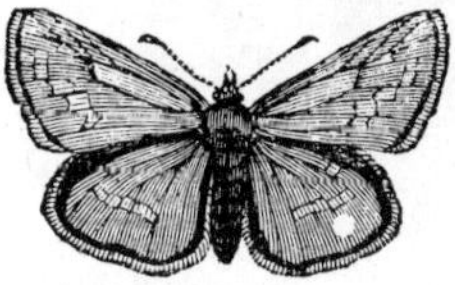

HESPERIA [PAMPHILA] SYLVANUS, F.

the hind wings furnished with long tails. They have a robust body; and frequently settle on flowers, leaves, and branches. There are a few British species, descriptions of which will be found in the works of Stephens and Humphreys. The "Genera of Diurnal Lepidoptera" of Doubleday and Hewitson will be found to contain much information on this family, which in foreign countries abounds in species and genera.

HESSIAN FLY. (*Cecidomyia destructor.*) This far-famed fly, as well as the "wheat-fly, which are common both to Europe and America, are small gnats or midges, and belong to the family called *Cecidomyiadæ*, or gall-gnats. The insects of this family are very numerous, and most of them, in the maggot state, live in galls or unnatural enlargements of the stems, leaves, and buds of plants, caused by the punctures of the winged insects in laying their eggs. The Hessian fly, wheat-fly, and some others differ from the majority in not producing such alterations in plants. The proboscis of these insects is very short, and does not contain the piercing bristles found in the long proboscis of the biting gnats and mosquitoes. Their antennæ are long, composed of many little bead-like joints, which are larger in the males than in the females; and each joint is surrounded with short hairs. Their eyes are kidney-shaped. Their legs are rather long and very slender. Their wings have only two, three, or four veins in them, and are fringed with little hairs around the edges; when not in use, they are generally carried flat on the back. The hind body of the female often ends with a retractile, conical tube, wherewith they deposit their eggs. Their young are little footless maggots, tapering at each end, and generally of a deep yellow or orange colour. They live on the juices of plants, and undergo their transformations either in these plants or in the ground.

The Hessian fly obtained its common name from a supposition that it was imported into England from Germany, and taken to North America in some straw, by the Hessian troops under the command of Sir W. Howe, in the war of the Revolution. This supposition, however, has been thought to be erroneous, because the early inquiries made to discover the Hessian fly in Germany were unsuccessful. Dr. Thaddeus Harris brings together, with much industry, a large amount of information from various sources relative to its economy, its habits, and transformations; and from his statement we shall endeavour to lay the principal facts before our readers. The head and thorax of this fly are black. The hind body is tawny, and covered with fine grayish hairs. The wings are blackish, but are more or less tinged with yellow at the base, where also they are very narrow; they are fringed with short hairs, and are rounded at the end. The body measures about one-tenth of an inch in length, and the wings expand one quarter of an inch or more. Two broods or generations are brought to maturity in the course of a year, and the flies appear in the spring and autumn. It has frequently been asserted that the flies lay their eggs on the grains in the ear; but whether this be true or not, it is certain that they do lay their eggs on the young plants, and long before the grain is ripe. The egg is about the fiftieth of an inch long, and four thousandths of an inch in diameter, cylindrical, translucent, and of a pale red colour. The maggots, when they first come out of the shells, are of a pale red colour. Forthwith they crawl down the leaf, and work their way between it and the main stalk, passing downwards till they come to a joint, just above which they remain, a little below the surface of the ground, with the head towards the root of the plant. Having thus fixed themselves upon the stalk, they become stationary, and never move from the place till their transformations are completed. They do not eat the stalk, neither do they penetrate within it, as some persons have supposed, but they lie lengthwise upon its surface, covered by the lower part of the leaves, and are nourished wholly by the sap, which they appear to take by suction. They soon lose their reddish colour, turn pale, and will be found to be clouded with whitish spots; and through their transparent skins a greenish stripe may be seen in the middle of their bodies. As they increase in size, and grow plump and firm, they become imbedded in the side of the stem, by the pressure of their bodies upon the growing plant; but when two or three are fixed in this manner around the stem, they weaken and impoverish the plant, and cause it to fall down, or to wither and die. They usually come to their full size in five or six weeks, and then measure about three-twentieths of an inch in length. Their skin now gradually hardens, becomes brownish, and soon changes to a bright chestnut colour. This change usually happens about the first of December, when the insect may be said to enter on the pupa state, for after this time it takes no more nourishment. The brown and leathery skin, within which the maggot has changed to a pupa or chrysalis, is long, egg-shaped, smooth, and marked with eleven transverse lines, and measures one-eighth of an inch in length. In this form it has been commonly likened to a flax-seed. It appears, then, from the remarks of the most careful observers that the maggots of the Hessian fly do not cast off their skins in order to become pupæ, wherein they differ from the larvæ of most other gnats, and agree with those of common flies; neither do they spin cocoons, as some of the Cecidomyians are supposed to do. Inclosed within the dried skin of the larva, which thus becomes a kind of cocoon or shell for the pupa, it remains throughout the winter, safely lodged in its bed on the side of the stem, near the root of the plant; and protected from the cold by the dead leaves.

Very soon after the flies come forth in the spring, they are prepared to lay their eggs on the leaves of the wheat sown in the autumn before, and also on the spring-sown wheat, that begins, at this time, to appear above the surface of the ground. They continue to come forth and lay their eggs for the space of three weeks, after which they entirely disappear from the fields. The maggots, hatched from these eggs, pass along the stems of the wheat, nearly to the roots, become stationary, and turn to pupæ in June and July. In this state they are found at the time of harvest, and, when the grain is gathered, they remain in the stubble in the fields. To this there are, however, a few exceptions; for a few of the insects do not pass so far down the side of the stems as to be out of the way of the sickle when the grain is reaped, and consequently will be gathered and carried away with the straw; and from this circumstance it is possible that they might have been imported in straw from a foreign country. In the winged state, these flies, or more properly gnats, are very active, and, though very small and seemingly feeble, are able to fly to a considerable distance in search of fields of young grain.

The best modes of preventing the ravages of the Hessian fly are thus stated by Mr. Herrick, in the 'American Journal of Science,' vol. 41. "The stouter varieties of wheat ought always to be chosen, and the land should be kept in good condition. If full wheat is sown late, some of the eggs will be avoided, but risk of winter killing the plants will be incurred. If cattle are permitted to graze the wheat fields during the fall, they will devour many of the eggs. A large number of the pupæ may be destroyed by burning the wheat stubble immediately after harvest, and then ploughing and harrowing the land. This method will undoubtedly do much good. As the Hessian fly also lays its eggs, to some extent, on rye and barley, these crops should be treated in a similar manner." It is found that luxuriant crops more often escape injury than those that are thin and light. Steeping the grain and rolling it in plaster or lime tends to promote a rapid and vigorous growth, and will therefore prove beneficial. Sowing the fields with wood ashes, in the proportion of two bushels to an acre, in the autumn, and again in the first and last weeks in April, and as late in the month of May as the sower can pass over the wheat without injury to it, has been found useful. Favourable reports have been made upon the practice of allowing sheep to feed off the crop late in the autumn, and it has also been recommended to turn them into the fields again in the spring, in order to retard the growth of the plant till after the fly has disappeared. Too much cannot be said in favour of a judicious management of the soil, feeding off the crop by cattle in the autumn, and burning the stubble after harvest; which will materially lessen the evils arising from the depredations of this noxious insect. Mr. Hardy has described and given the history of the British species.

HETEROCERA. A section of *Lepidoptera*, agreeing with the Linnæan genera *Sphinx* and *Phalæna*. It derives its name from the diversified formation of the antennæ, which are never terminated by a club, like those of the butterflies, but are generally setaceous, filiform, or fusiform, those of the males being moreover often furnished with lateral appendages, forming branches. The caterpillars are much varied, but the pupæ are generally of a conical form, and are ordinarily enclosed in a cocoon, the quiescent state being often undergone in the ground. Modern entomologists have found much difficulty in defining the various groups which compose the *Crepuscularia* and *Nocturna*, and our space precludes us from entering at large upon any subject where much uncertainty exists; nor, indeed, is it essential that we should do so. Mr. Westwood observes, that "Urania, Castnia, Agarista, Sphinx, Ægeria, and Anthrocera are groups of equal value among themselves: and on account of the peculiar conformation of their antennæ, they were united into one group by Linnæus, who, it is well known, considered this character as of the highest importance. Take, for instance, the three English groups, Sphinx, Ægeria, and Anthrocera, and we find the first isolated: the second, in its fenestrated wings, approaches some of the Sphingidæ, but its metamorphoses are totally different, resembling those of Cossus; whilst Anthrocera, on the other hand, is, in its preparatory states, a Bombyx, and in its final one probably intermediate between Macroglossa and Pyralis; Ægeria, nevertheless, is not farther removed from Sphinx than is Castnia or Urania, nor than Hepialus or Lithosia are from Attacus, in the tribe of Bombycidæ. Geometra, Tortrix, Noctua, &c., in their extended state, are groups admirably defined, and yet it is impossible to look at Euclidia, Acosmetia, Nola, or Platypteryx, without perceiving either that we must extend the limits of our families, so as to admit these anomalous groups, or create a far greater number of families than has hitherto been done." * * * "With regard to the primary groups of the *Heterocera*, I candidly admit that I am not able to offer a satisfactory classification, although it seems unquestionable that Sphinx (or the Hawk-moths), Bombyx (or the feather-horned full-bodies), Noctua (or the thread-horned full-bodies), Geometra (or the loopers) Pyralis, Tortrix, and Tinea, are, as Linnæus considered them, amongst the primary types."

HETEROCERIDÆ. A family of Coleopterous insects, of small size and subaquatic habits: body depressed; legs broad, compressed, and serrated, the thorax much narrower than the elytra; the jaws robust; and the antennæ short. These insects burrow in the mud of the banks of ponds or stagnant water, out of which they make their escape when the earth is shaken or stamped upon, and again as quickly bury themselves in the mud. Their bodies are clothed with a fine silky pubescence, whereby the action of the water upon them is prevented. They walk but slowly; yet they are sometimes observed in the hot sunshine to raise their wings, fly off, and again alight, with all the agility of the tiger-beetles. There is every reason to suppose them to be carnivorous.

HETEROMERA. A section of the Coleoptera, comprehending those beetles which have five joints in the tarsus of the first and second pairs of legs, and only four joints in the tarsus of the third pair. This division includes several extensive groups, the majority of the species of which feed upon vegetable substances: some are gaily coloured, and such are generally found in flowers; others, which frequent dark and damp places, are uniformly black: whilst those which inhabit the sandy deserts of tropical regions are of various obscure shades of gray or brown.

HETEROPODA. An order of Molluscous animals, closely allied to the *Gasteropoda*, but distinguished from them and all others by the structure and position of the foot, which is compressed, so as to constitute a vertical muscular paddle, or fin. The gills are external, and form plume-like tufts, situated at the hinder part of the back: the body is gelatinous and transparent; and the mouth is furnished with a kind of muscular tube or proboscis, and a rough tongue. In their general form and structure they correspond very closely with those of the *Carinaria*, which has a thin shell, in form not unlike that of the Argonaut. They all inhabit either the tropical seas, or those of moderately warm climates. The best known genera are *Carinaria*, *Atlanta*, and *Firola*. Figures of the shells and animals of all these will be found in the admirably useful work of Mrs. Gray, "Figures of Mollusca."

HETEROPTERA. The name given to a section of the order of insects called HEMIPTERA, distinguished by the anterior wings being tough at their bases and membranous only towards their points. By far the greater number of them feed upon the juices of plants; some, however, prey upon smaller insects, and others suck the juices of larger animals. They chiefly inhabit tropical regions, and are mostly ornamented with beautiful colours and markings, which

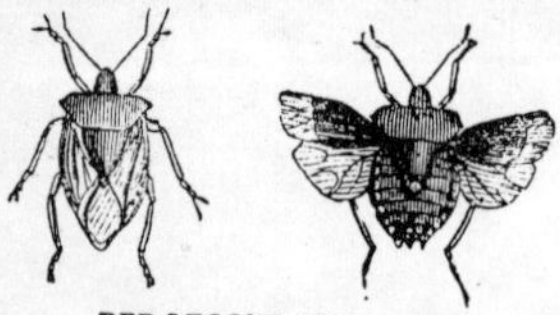

RED-LEGGED PLANT BUG.
(PENTATOMA RUFIPES.)

often vie with the most splendid of the Beetle tribe: those, however, whose habits are aquatic, are of a black or obscure colour. Most of the terrestrial species emit a powerful odour when suddenly alarmed or touched; this is occasionally of an agreeable nature, but more commonly (as in the case of the Bug — *Cimex lectularius*) disgustingly offensive. As an example we give a cut of the *Pentatoma rufipes*, a common species in this country. [See HOMOPTERA.]

HIMANTOPUS, or LONG-LEGGED PLOVER. A genus of Grallatorial birds, distinguished by the great length of their legs; from which circumstance they are sometimes called *Stilt-Birds*.

One species is occasionally found in this country; the LONG-LEGGED PLOVER, (*Himantopus candidus*, or *Charadrius Himantopus* of Linnæus.) White has recorded its appearance in the neighbourhood of his favourite Selbourne; and we believe our readers will be better pleased with the account so graphically pourtrayed by him in a letter to Pennant, than by any other description we perchance might offer. "In the last week of last month (April, 1799), five of these most rare birds, too uncommon to have obtained an English name, but known to naturalists by the terms *Himantopus*, *Loripes*, and *Charadrius Himantopus*, were shot upon the verge of Frinsham pond, a large lake belonging to the Bishop of Winchester, and lying between Wolmer forest and the town of Farnham, in the county of Surrey. The pond-keeper says there were three brace in the flock; but that after he had satisfied his curiosity, he suffered the sixth to remain unmolested. One of these specimens I procured, and found the length of the legs to be so extraordinary, that, at first sight, one might have supposed the shanks had been fastened on to impose on the credulity of the beholder: they were legs in *caricatura;* and had we seen such proportions on a Chinese or Japan screen, we should have made large allowance for the fancy of the draughtsman. These birds are of the *Plover* family, and might with propriety be called *Stilt Plovers*. Brisson, under that idea, gives them the appropriate name of *l'échasse*. My specimen, when drawn and stuffed with pepper, weighed only four ounces and a quarter, though the naked part of the thigh measured three inches and a half. Hence we may safely assert that these birds exhibit, weight for inches, incomparably the greatest length of legs of any known bird. The *flamingo*, for instance, is one of the most long-legged birds, and yet it bears no manner of proportion to the *Himantopus;* for a cock *flamingo* weighs at an average about four pounds avoirdupois: and his legs and thighs measure usually about twenty inches. But four pounds are fifteen times and a fraction more than four ounces and one quarter; and if four ounces and a quarter have eight inches of legs, four pounds must have one hundred and twenty inches and a fraction of legs, viz. somewhat more than ten feet, such a monstrous proportion as the world never saw! If you should try the experiment in still larger birds, the disparity would still increase. It must be matter of great curiosity to see the Stilt Plover move; to observe how it can wield such a length of lever with such feeble muscles as the thighs seem to be furnished with. At best one should expect it to be but a bad walker; but what adds to the wonder is, that it has no back toe. Now without that steady prop to support its steps, it must be liable in speculation to perpetual vacillations, and seldom able to preserve the true centre of gravity. The old name of *Himantopus* is taken from Pliny; and, by an awkward metaphor, implies that the legs are as slender and pliant as if cut out of a thong of leather. Neither Willoughby nor Ray, in all their curious researches, either at home or abroad, ever saw this bird. Mr. Pennant never met with it in all Great Britain, but observed it often in the cabinets of the curious at Paris. Hasselquist says that it migrates to Egypt in the autumn; and a most accurate observer of nature has assured me that he has found it on the banks of the streams in Andalusia. Our writers record it to have been found twice in Great Britain. From all these relations it plainly appears that the Long-legged Plovers are birds of South Europe, and rarely visit our island; and when they do, are wanderers and stragglers, and impelled to make so distant a northern excursion from motives and accidents for which we are not able to account. One thing may fairly be deduced, that these birds come over to us from the

Continent, since nobody can suppose that a species not noticed once in an age, and of such a remarkable make, can constantly breed unobserved in this kingdom.

Mr. Gould observes, in his 'Birds of Europe,' "The Long-legged Plover, as its conformation would lead us to conclude, is a bird whose most congenial habitat is morasses and the low flat shores of lakes, rivers, and seas. Hence in the eastern portions of Europe, where it is said to arrive from Asia in small flocks, it takes up its abode along the lakes and among the vast morasses of Hungary and Russia, where, according to M. Temminck, it rears its progeny, and where it fearlessly wades in search of its food, without much chance of being carried out of its depth; but should such an occurrence happen, or the waves drift it out from the shore, it possesses, like many of the true Wading Birds, the power of swimming with the greatest ease and lightness: in fact, in whatever point of view we consider the Long-legged Plover, we find it adapted in the best possible manner for its habits and modes of life. Few birds exceed it in the powers of flight; its wings far exceed the tail, and it passes through the air with astonishing rapidity. When on firm ground, it appears as if tottering on long and awkward stilts, but firm ground is not its congenial habitat."

An allied species (*H. nigricollis*) is described by Wilson, in his American Ornithology, under the same name as the European, but it is distinct. In 'Gould's Birds of Australia' three species are figured and described; two from Australia, the *H. leucocephalus* and *Cladorhynchus pectoralis*, and one from New Zealand, the *H. Novæ Zealandiæ*, so that this genus and group of Long-legged birds is very widely distributed.

HIND. The female of the Red Deer or Stag. [See DEER.]

HINULIA. A genus of Reptiles closely allied to the officinal *Scincus*, most of the species of which appear to be natives of Australia; a few species are natives of the East. Descriptions of all the species will be found in Mr. Gray's Catalogue of the Reptiles in the British Museum.

HIPPA: HIPPIDÆ. A genus and family of Anomurous Decapod Crustacea, the species of which seem to be fond of working in the sand. One species, the *H. talpoida*, is called sand-bug in North America. To this family belong *Albunea*, *Remipes*, and *Cosmonotus*, which with *Hippa* form very striking and beautiful exotic genera.

HIPPARCHIA, or SATYRUS. A genus of Diurnal Lepidoptera, the species of which are for the most part brown or obscure. In temperate regions and in mountainous districts generally they are numerous; some being found in Lapland (*Chionobas*), and others on elevated ground within the tropics. In this country are several species, some of which are specified beneath; the other British species are recorded in the works of Stephens, Curtis, and Humphreys; while a new species to this country, found lately in Perthshire, is figured and described in "The Zoologist." We must again refer for information to Doubleday and Hewitson's Genera of Diurnal Lepidoptera. The British species we restrict our attention to, are—

The HIPPARCHIA GALATHEA, or MARBLED WHITE BUTTERFLY. This pretty Butterfly is of a yellowish white colour chequered with black, which produces a pleasing effect. In

MARBLED WHITE BUTTERFLY. (HIPPARCHIA GALATHEA.)

some species the black prevails; in others the yellowish-white: the female is larger than the male. It is found abundantly near London, and in most parts of England, chiefly frequenting moist meadows, where it appears in June and July. The caterpillar feeds on grass,

UNDER SIDE OF MARBLED WHITE BUTTERFLY.—(HIPPARCHIA GALATHEA.)

particularly the *Phleum pratense;* is yellowish-green, and strikingly resembles that of some of the moths, and, like most if not all the species of the genus, feeds at night. Our figures, derived from the admirable work of

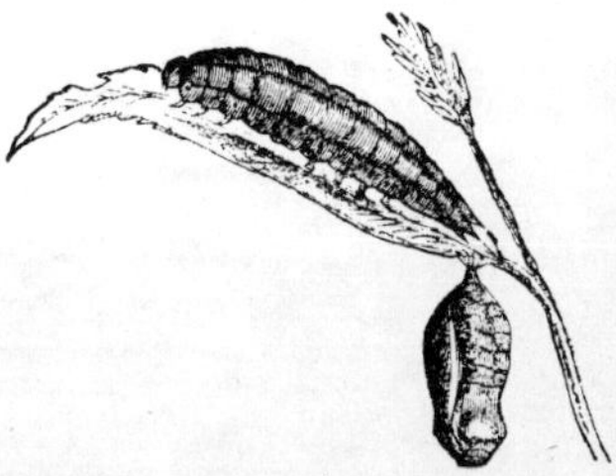

CATERPILLAR AND CHRYSALIS OF MARBLED WHITE BUTTERFLY.—(H. GALATHEA.)

Hübner, show the upper and under sides of the perfect insect, with the caterpillar and the chrysalis.

The HIPPARCHIA SEMELE, or GRAYLING BUTTERFLY. This large, curiously marked species is by no means uncommon on heaths and hilly places, where the caterpillar feeds most probably on grass, forming a cocoon in the earth, according to a French writer. This latter circumstance is very curious if true, and unique amongst British Butterflies; indeed there is only one Butterfly we have heard of, belonging to the genus *Zegris*, which makes a cocoon, although further researches may prove it to be not peculiar to

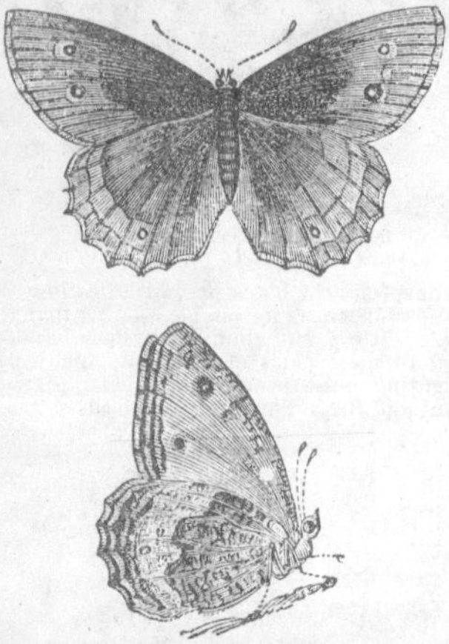

GRAYLING BUTTERFLY, UPPER AND UNDER SIDE.—(HIPPARCHIA SEMELE.)

one or two Butterflies. This species, as well as others of the genus, help to enliven the dullest heaths; and the pedestrian, even though no entomologist, cannot fail to be struck with these brown butterflies and their more gay but smaller comrades, the *Blues* or *Polyommati*. Our figure represents the upper and under sides of this species, and will show, better than any description, its markings and peculiarities.

The HIPPARCHIA PAMPHILUS, or GOLDEN EYE. This species of Butterfly, which makes its appearance in June, and again in September, on every grassy heath and common in the kingdom, has wings of a pale tawny above; the anterior with the margins dusky, and an ocellus near the tip; the posterior nearly resembling them, with an obsolete ocellus near the anal angle: beneath, the anterior wings are cinereous at the base and tip, with a rather large ocellus at the tip, white pupil, and whitish edge: the posterior wings are greenish-brown at the base, with an irregular pale band in the middle, in which are several minute indistinct ocelli; the margin greenish-brown. The body is deep fulvous; the antennæ tawny, with whitish annulations. The colour of the female is scarcely so deep as that of the male, nor are the wings so distinctly edged with dusky above. The Caterpillar has a green hue, and is marked with white dorsal lines: it appears to prefer the dog's-tail grass to other food. Chrysalis green.

The HIPPARCHIA HYPERANTHUS, or RINGLET BUTTERFLY. Of this species of Butterfly there are many varieties, and some of them are scarce; for the most part, however, it is abundant in damp grassy woods and lanes, particularly in the north of Britain. The anterior wings above are plain brown, frequently with one or more black faintly ocellated spots; with three ocelli beneath towards the hinder margin: the posterior wings are also brown above, with two or more obsolete ocelli: beneath, with two approximating ocelli behind the middle of the anterior margin, and three parallel with the hinder margin: all the wings are paler beneath, and edged with a whitish fringe. The body is fuscous, paler beneath: the antennæ brown and lightly annulated. Caterpillar gray or dusky, with a black line behind; it subsists chiefly on the meadow grass, and resides at its roots: the chrysalis is bright brown, obscurely streaked.

The HIPPARCHIA JANIRA, or MEADOW BROWN BUTTERFLY. We know of none among the tribe of papilionaceous insects that is more common than this species; not a meadow or lane in Britain being scarcely to be seen in the month of July where it is wholly absent. The wings are of a dull dark brown or nearly black, the male usually having on the surface of the anterior pair near the tip a black ocellus with a white pupil; beneath fulvous, with the hinder margin grayish-brown: posterior wings beneath tawny-brown, with two or three dusky spots. Female generally with a large irregular tawny orange blotch on the anterior wings above, in which, as in the male, is an ocellus. In some specimens there is a deep black patch on the disc of the anterior wings; while in others irregular and undefined white blotches occur on various parts of the wings. The Caterpillar, which is green, with a white lateral line, and thickly covered with hair, feeds on meadow grass: the Chrysalis is yellowish-green, with dusky streaks on the head and wing-cases.

Knapp, speaking of it in his very interesting Journal of a Naturalist, where he describes the common occurrences of nature as observed near a village in the west of England, says, "Amid the tribes of insects particularly influenced by seasons, there are a few which appear little affected by common events; the brown meadow butterfly, so well known to every one, I have never missed in any year; and in those damp and cheerless summers when even the white cabbage butterfly is scarcely to be found, this creature may be seen in every transient gleam, drying its wings, and tripping from flower to flower with animation and life, nearly the sole possessor of the field and its sweets. Dry and exhausting as the summer may be, yet this dusky butterfly is uninjured by it, and we see it in profusion hovering about the sapless foliage."

The HIPPARCHIA ÆGERIA, or SPECKLED WOOD BUTTERFLY. This species seems to be pretty generally diffused throughout the United Kingdom; and several broods make their appearance between the beginning of April and the end of August. Anterior wings brown on both surfaces, with a number of yellowish spots, and an ocellus towards the tip: posterior wings above brown, with a series of yellowish spots, of which the three inner ones are ocellated, having a white

SPECKLED WOOD BUTTERFLY. UPPER AND UNDER SIDE.—(HIPPARCHIA ÆGERIA.)

pupil with a black iris, and surrounded by a yellowish circle; beneath they are brownish, with irregular angulated brown bands; the hinder margin purplish, with a series of white dots: the cilia are yellowish and brown: the body brown above, pale beneath; the antennæ brownish. The female is generally more spotted than the male, and the spots are larger. Caterpillar green, with white longitudinal lines; it feeds on couch grass. Chrysalis green, and short.

HIPPOCAMPUS. A genus of Lophobranchiate fish of a highly singular appearance, which has obtained the English name of the Sea-horse Pipe-fish. The best known species is the *Syngnathus Hippocampus* of Linnæus, or *Hippocampus Brevirostris* of Cuvier. Its general length is from six to

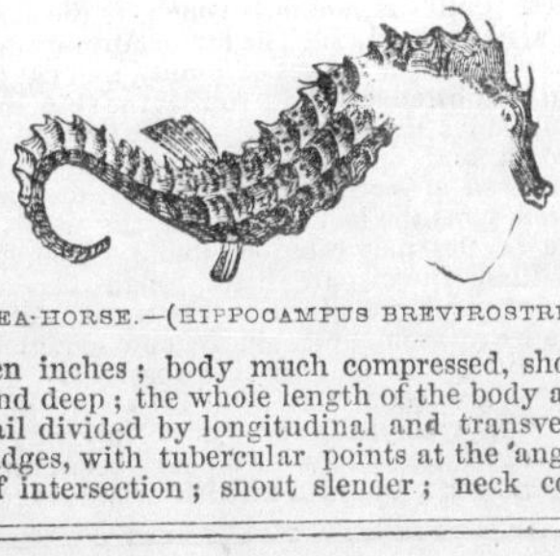

SEA-HORSE.—(HIPPOCAMPUS BREVIROSTRIS.)

ten inches; body much compressed, short, and deep; the whole length of the body and tail divided by longitudinal and transverse ridges, with tubercular points at the angles of intersection; snout slender; neck contracting suddenly beyond the head; and the tail long, quadrangular, and terminating in a naked or finless tip. When swimming about, the Hippocampus maintains a vertical position; but the tail is ready to grasp whatever it meets in the water, and when fixed, the animal darts at its prey with great dexterity. In its dry or contracted state the fancied resemblance from which this fish takes its name is far more apparent than when alive. It is a native of the Mediterranean and Atlantic seas.

HIPPOLYTE. A genus of long-tailed Crustacea allied to the Shrimps, several species of which are found on our coasts. The British Museum contains these: descriptions and figures of them all are given in the works of Dr. Leach and of Prof. Bell on the British Crustacea.

HIPPONYX. A genus of Molluscous animals, of which there are numerous species, though until lately only known in a fossil state. The shell is obliquely cap-shaped; inequivalve, sub-equilateral, and destitute of ligament and hinge teeth; lower valve attached, sub-orbicular, with a muscular impression, composed of two lunulate portions, meeting at one extremity, and presenting the form of a horse-shoe; upper valve conical, with the apex inclined backwards, and the muscular impression marginal. These animals are generally supported on a solid shelly plate, but not always, the same object being gained when a suitable place of attachment can be found by their adhering to other shells.

HIPPOPOTAMUS, or RIVER-HORSE. This gigantic inhabitant of the African rivers is as formidable as the Rhinoceros, and in bulk inferior only to the Elephant, but its limbs are so short that its belly almost touches the ground. Its form is in the highest degree uncouth; the body being

HIPPOPOTAMUS.—(H. AMPHIBIUS.)

extremely bulky, fat, and round; the legs very short and clumsy; the head immensely large; the mouth prodigiously wide, and the teeth of vast strength and size, the incisors and canines of the lower jaw being long, and curved forwards: these canines, or tusks, sometimes measure more than two feet in length, and weigh upwards of six pounds each. Those in the upper jaw are much smaller; and the front teeth are of a moderate size. The lips are very thick and broad, and are beset, here and there, with scattered tufts of short bristles: the nostrils are rather small, and open on the top of the muzzle: the eyes, which are very small, are situated high in the head: the ears are small, slightly pointed, and lined with short soft hair. The tail is short, thick, and sparingly covered with hair. The feet are

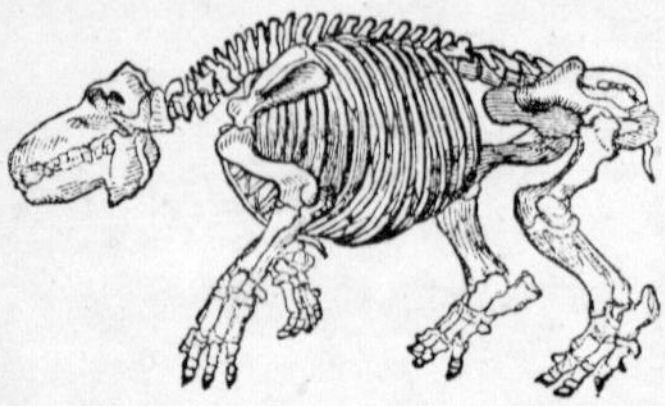

SKELETON OF THE HIPPOPOTAMUS.

large, and have four toes, terminated in separate hoofs. When just emerged from the water, the Hippopotamus appears of a palish brown, or mouse-colour, with a bluish cast on the upper parts; and the belly is flesh-coloured, the skin appearing through the hair: but when perfectly dry, the animal's general colour is an obscure brown.

In the interior of Africa, where the rivers run through countries overshadowed by large forests, the Hippopotamus walks about at the bottom of the stream, raising its head at intervals above the surface, for the purpose of respiration. By night it quits its watery residence in search of its food, which consists of the herbage that grows near the banks of the rivers, and the surrounding pastures. It is not confined to rivers, however; for it also tenants the inland lakes, and is sometimes seen even in the sea, though it will not drink salt water, prey on fish, or live on any kind of animal food. Its voice is described as a peculiar kind of interrupted roar, between that of a bull and the braying of an elephant. When on land it moves in a somewhat slow and awkward manner, but if pursued, can run with considerable speed, and directly plunging into the water sinks to the bottom, and pursues its progress beneath. It i extremely cautious of making its appearance by day, in places much frequented by mankind; but is fearless in rivers which run through unfrequented regions; where it is occasionally seen to rush out of the w ter with sudden impetuosity, trampling down every thing in its way. At such times it is of course highly dangerous; and it sometimes also shows great fury when only slightly provoked: but it is naturally of a harmless disposition; not attacking other animals, but merely committing havoc in plantations of maize, rice, sugar-canes, &c., and destroying trees, by loosening the roots with its vast and powerful teeth.

The Hippopotamus sleeps in the small reedy islets which are here and there found in the rivers it frequents. In such spots it also brings forth its young; having only one at a birth, which it nurses with great care. These animals are occasionally shot, or harpooned; but they are said to be most successfully taken by pitfals, prepared for them near the rivers. But it is chiefly on account of the tusks and teeth that this animal is killed; their hardness being superior to that of ivory, while they are at the same time less liable to turn yellow. The skin, from its great thickness and strength, when dried, is formed into shields, and is said to be bullet-proof; the living animal, indeed, if shot at anywhere but on the head or belly, is scarcely vulnerable; nor is this wonderful when we consider that the hide is two inches deep or more on the back and sides. The flesh of this animal is eaten by the inhabitants of South Africa, and, as would appear from the reports of travellers, with more than usual *gusto* when it is in a half putrid state. The Hippopotamus has properly been considered as the Behemoth of Scripture. Most interesting accounts of the Hippopotamus have recently appeared in the works of Livingstone, Andersson, Cumming, and other African travellers, whilst visitors to the Zoological Gardens, Regent's Park, may now observe the habits of two very fine specimens, male and female. At the Parisian menagerie a similar pair have twice bred, but the young were in both instances destroyed by their mother.

HIPPOPUS. A genus of Conchifera, of which there is but one known recent species, the *Hippopus maculatus* (or Bear's-paw Clam) from the Indian Ocean. The delicate whiteness of the interior, the undulating edge, the radiated fluted columns, and the richness of the variegated colouring are such as to entitle it to the admiration of every one. It is equivalve, regular, and inequilateral; valves closed; transverse; ligament external; shell imbricated with numerous tubercles. It is not nearly so large as the *Tridacna*, but the animal is similar to it. [See TRIDACNA.]

HIRUDO. [See LEECH.]

HIRUNDO: HIRUNDINIDÆ. A genus and family of Fissirostral or wide-gaping birds of the Cuvierian system, embracing the Swallow tribe. Our British species are occasional visitors, and the heralds of summer; but at the approach of winter they resort chiefly, as is supposed, to Africa. [See SWALLOW.]

HISPIDÆ. A family of Coleopterous insects popularly known in the United States as "little leaf-beetles." The upper side of these beetles is generally rough, as the generical name implies. The larvæ burrow under the skin of the leaves of plants, and eat the pulpy substance within, so that the skin, over and under the place of its operations, turns brown and dries, and has somewhat of a blistered appearance, and within these blistered spots the larvæ or grubs, the pupæ, or the beetles may often be found. The eggs of these insects are little rough blackish grains, and are glued to the upper side of the leaves, sometimes singly, and sometimes in clusters of four or five together. The grubs are about one-fifth of an inch in length, when fully grown. The body is oblong, flattened, rather broader before than behind, soft, and of a whitish colour, except the head and the top of the first ring, which are brown, and of a horny consistence. The pupa state lasts only about a week, soon after which the beetles come out of their burrows.

The leaves of the apple-tree in North America are inhabited, according to Dr. Harris, by some of these little mining insects, which, in the beetle state, are probably the *Hispa rosea*, or rosy Hispa. They are of a deep tawny or reddish-yellow colour above, marked with little deep red lines and spots. The head is small, the antennæ are short, and of a black colour; the thorax is narrow before and wide behind, rough above, striped with deep red on each side; the wing-covers taken together form an oblong square; there are three smooth longitudinal lines or ribs on each of them, spotted with blood red, and the spaces between these lines are deeply punctured in double rows; the under side of the body is black, and the legs are short and reddish. They measure about one-fifth of an inch in length. These beetles may be found on the leaves of the apple-tree during the latter part of May and the beginning of June. A small species is found in this country (*H. testacea*).

HISTER: HISTERIDÆ; or MIMIC BEETLES. A genus and family of Coleoptera, which, from the power they possess of contracting their limbs and counterfeiting death, evidently derive their name from the Latin word *Histrio*, a stage mimic. The beetles belonging to this group are distinguished by the very hard consistence of the body, which is generally of an oblong-quadrate form, and of a highly polished surface. The antennæ are short, elbowed, and terminated by a large and solid club; the mandibles very robust, horny, and exserted; the maxillæ elongated and bilobed; the labium bipartite and setose; the palpi filiform; the legs more or less dentate, the two posterior pairs being inserted widely apart; and the elytra generally short and truncate. These insects seldom exceed a third of an inch in length; their colours are generally black and shining; some few have the elytra ornamented with blood-coloured or pale buff spots, or exhibit metallic tints. They creep slowly, but fly well. They feed upon decaying vegetable and animal matter, and are

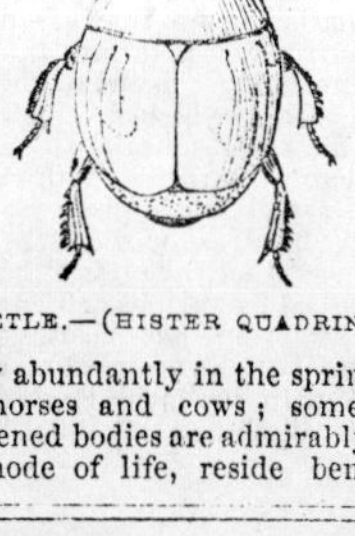

MIMIC BEETLE.—(HISTER QUADRINOTATUS.)

found very abundantly in the spring in the dung of horses and cows; some species, whose flattened bodies are admirably adapted to their mode of life, reside beneath the bark of trees; while some of the more minute species are constantly found as residents in ants' nests. The larvæ are linear, depressed, nearly smooth, of a soft consistence, and white colour; and feed upon the same substance as the perfect insect.

HOBBY. (*Falco subbuteo.*) A bird of the long-winged Hawk kind, formerly used in the humbler walks of falconry, chiefly for larks and other small birds, which were caught in a singular manner: when the Hawk was cast off, the larks, keeping close to the ground through fear, became an easy prey to the fowler, who drew a net over them. The Hobby is about twelve inches in length; has a prominent and crooked bill; the orbits of the eyes are yellow, and over each eye is a light-coloured streak. The crown of the head, the back, and the coverts of the wings, are bluish black; the hinder part of the neck is marked with two pale yellow spots; and each cheek with a large black spot pointing downwards. The breast and belly are pale, marked with dusky streaks; wings brown; the two middle tail feathers deep dove colour, the others barred with rusty, and tipped with white; the legs and feet are yellow. The female is considerably larger than the male; the spots on her breast are more conspicuous; and her legs are greenish. She builds in high trees; and lays three or four bluish white eggs, irregularly spotted with grey and olive: but the Hobby is a bird of passage; and though it breeds here, it migrates from this country in October.

HOG. (*Sus.*) As all the varieties of this useful quadruped are derived from the Wild Boar, we shall proceed to describe that animal before we speak of the domestic species; merely premising that the genus *Sus* is in some points of an ambiguous nature, appearing to form at once a link between the cloven-footed, the whole-hoofed, and the digitated quadrupeds.

The Wild Boar is a native of almost all the temperate parts both of Europe and Asia. In times of yore it was not an unfrequent inhabitant of our own woods and forests; where it served as a beast of chase, as it still does in India, as well as in some parts of Continental Europe; presenting not only the most interesting and exciting sport to the hunters, but at the same time one of the most dangerous in which they can be engaged. This fierce and powerful animal is armed with long, curved, and sharp tusks, capable of inflicting the most severe and fatal wounds; but as he advances in age (after he has passed his fifth year), he becomes less dangerous, on account of the growth of these formidable tusks, which turn up so considerably as often to impede rather than assist his intentions of wounding with them. We learn from Buffon, that wild Boars follow their common parent until they have passed their third year, never wandering alone till they have acquired sufficient strength to resist the attacks of the wolf. "These animals," says he, "when they have young, form a kind of flocks, and it is upon this alone that their safety depends. When attacked, the largest and strongest front the enemy, and by pressing all round against the weaker, force them into the centre. Domestic Hogs are also observed to defend themselves in the same manner. The Wild Boar is hunted with dogs, or killed by surprise during the night, when the moon shines. As he flies slowly, leaves a strong odour behind him, and defends himself against the dogs, and often wounds them dangerously, fine hunting-dogs are unnecessary; and they would have their nose spoiled and acquire a habit of moving slowly by hunting him. Mastiffs, with very little training, are sufficient. The oldest Boars, which are known by the track of their feet, should alone be hunted; a young Boar of three years old is difficult to be attacked; because he runs very far without stopping; but the old Boars do not run far, allow the dogs to come near, and often stop to repel them. During the day the Boar commonly keeps in his soil, which is in the most sequestered part of the woods, and comes out by night in quest of food; and in summer, when the grain is ripe, it is easy to surprise him among the cultivated fields, which he frequents every night."

The Wild Boar is in general more gaunt and bony, the muscular strength much greater, and the temper far more savage, than the domestic Hog. It is of a dark brindled gray colour, or blackish; but when only a year or two old, is of a dull yellowish brown cast; and when quite young, is marked by alternate dusky and pale longitudinal bands along the sides. Between the bristles, next the skin, is a finer or softer hair, of a woolly or curly nature. The snout is somewhat longer in proportion than that of the domestic species; but the principal difference is in the length of the tusks,

WILD BOAR—(SUS SCROFA.)

Though ordinarily timid and inoffensive, it is found that the females show the most determined courage when their young are attacked, and defend them with all imaginable fierceness. If two Boars chance to meet in the early part of the year, at which time the male seeks the female, the most furious encounters ensue. By a forest law of William I. (A. D. 1087), it was ordained that any who were found guilty of killing the Stag, the Roebuck, or the Wild Boar, should have their eyes put out!

The COMMON, or DOMESTIC HOG (*Sus scrofa*) differs from the wild animal principally in having smaller tusks and larger ears, which are also somewhat pendent and of a more pointed form. In colour, as well as size, it varies very considerably, but the prevailing cast is a dull yellowish white, marked or spotted irregularly with black, sometimes perfectly plain or unspotted, sometimes rufous, and sometimes totally black. Of all quadrupeds the Hog is the

BERKSHIRE HOG.

most gross in his manners, and has therefore been generally regarded as the very personification of impurity. The Jews were strictly enjoined not to eat its flesh; and the Mahometans agree in this respect with the Mosaic prohibition. In most parts of Europe, however, it constitutes a very material part of the food of mankind. And we may do well to reflect, while we decry the filthy habits of this animal, that from our own sensations we are often apt to form a partial judgment, and overlook that wise decree of Providence which adapts every part of creation to its respective inhabitants. The Hog is an animal of a remarkably prolific nature; and, as they bring forth from ten to fifteen, and sometimes twenty, at a litter, they would soon become very numerous, were they not diminished for the support of man. Their flesh, says Linnæus, is wholesome food for persons of athletic constitutions, or those who habituate themselves to much exercise, but improper for such as lead sedentary lives. It is, however, an article of general consumption, and one which is of great importance to a naval and commercial nation, as it takes salt better than any other flesh, and consequently is capable of being longer preserved.

The Jews and Mahometans not only abstain from the *flesh* of swine from a religious principle, but even consider themselves defiled by touching it. The Chinese, on the contrary, are so excessively fond of pork, that many, owing to this partiality alone, as it is said, have been prevented from conversion to Mahometanism. The *fat* of swine differs, in its situation, from that of almost every other quadruped, as it forms a thick, distinct, and continued layer betwixt the flesh and the skin. *Lard*, which is chiefly obtained from the fat membranes of the abdomen, is applicable to various uses, both culinary and medicinal; and when good, is white and moderately hard. The *skin*, when properly dressed, is used for the seats of saddles; it is also employed by various artificers.

Great attention has been paid in this country to the improvement of the various breeds; and by judicious crosses much has been effected both as to quality and size.

Some counties in the south-western division of England are considered famous for their breed of Hogs; those of Hampshire, Sussex, Wilts, and Berks being foremost; but since the pains that have of late years been taken by breeders of stock generally throughout the country, and the impetus given to their exertions by the various agricultural associations, we may fairly presume that all manifest the same praiseworthy solicitude in endeavouring to excel in this as in every other branch of rural economy. But this part of the subject, perhaps, does not properly fall within our province; for it has been said, that where art begins, the history of nature ought to end. We shall therefore not notice the different qualities which distinguish one breed from another, but conclude with Mr. Bell's remarks on a well-known variety of the Porcine genus—the Chinese Hog. "The introduction of the Chinese Hog has effected an astonishing change in the native breeds, wherever they have been crossed by it. This very remarkable variety deserves particular mention, not merely as a source of great improvement in an important branch of stock, but also as connected with a zoological question of great interest. M. Frederic Cuvier believes that it is derived from a wild stock specifically distinct from the Wild Boar; and could this be proved, it would go far to settle the long disputed and difficult question of specific distinctions, as connected with the production of infertile progeny: for in that case, as the breed between the Chinese and the Common Hog is perfectly fertile, the argument for specific distinction founded upon that circumstance at once falls to the ground. On the other hand, however, those who contend that the production of fertile young is a proof of specific identity in the parents, would of course hold that the supposition of the celebrated naturalist is erroneous upon this very ground. The Chinese Hog is of small size, short and thick; the belly deep, and when fat, nearly reaching the ground; the legs short and fine; the head very short, and the neck thick. Its influence on the different breeds with which it has intermixed, has been greatly to improve them in the delicacy of the flesh; but while the pork of the Chinese cross is certainly excellent, some of our own breeds are still esteemed as yielding by far the best bacon and hams."

HOG, ETHIOPIAN. (*Phacochœrus Æthiopicus.*) This animal is distinguished from the common Hog by a pair of large, flat, semicircular lobes under the eyes; the snout is also much broader, and is very strong and callous: the ears are large, and slightly pointed: the tusks in the lower jaw are rather small, but those in the upper are large, sharp, and much curved: they have no fore-teeth, their place being supplied by very hard gums: immediately below the eyes the skin is loose and wrinkled, and on each side the corners of the mouth is a callous protuberance. The body is strong, and the limbs muscular; the tail is rather flat, and thinly covered with scattered hairs; the colour of the whole animal, a dusky brown. It is a native of the hotter parts of Africa, residing principally in subterraneous recesses, which it digs with its nose and hoofs. When attacked or pursued, it rushes on its adversary like the Boar, and shows great fierceness.

HOG-LOUSE. [See ONISCUS.]

HOLIBUT, or HALIBUT. (*Pleuronectes hippoglossus.*) This fish not only exceeds in size all the flounder genus, but ranks as one of the largest of fishes; sometimes attaining a length of six or seven feet, and a weight of 300 or 400 lbs. It is a native of the Northern and Mediterranean seas, and appears to arrive at its greatest size in the former. It is considered as the most voracious of its tribe; preying on a variety of other fishes and crustacea. The colour of the Holibut is deep brown above, and white beneath; the body being quite smooth, and the scales moderately small. Its flesh is coarse and dry, but it admits of being salted; and it constitutes no inconsiderable part of the food of the Greenlanders, who cut it into thin slips, and dry it in the sun. In the London markets this fish is usually cut into large pieces when exposed for sale.

HOLOTHURIADÆ. A family of Echinodermata, the distinguishing characters of which are, that the body is of an elongated form, defended by a coriaceous integument; open at both ends, and perforated by numerous small canals, through which suckers are protruded. At the anterior extremity is the mouth, furnished with many retractile tentacula, and at the opposite end is the aperture of the cloaca.

The *Holothuriæ* of the European seas are neither numerous nor brilliantly coloured; but in more tropical seas, where coral reefs rise within a moderate distance of the surface, as in the Red Sea, and the seas to the north and east of Australia, they are exceedingly numerous, and many of them splendidly coloured; so that, together with other Radiata of this and other orders, they make the sea-bottom, when seen by the light of an almost vertical sun, as gay as a tropical garden. The Holothuriæ resemble cucumbers; and various Actiniæ, when their tentacula are expanded, have as gay an appearance as the flowers of almost any plants. Many of this species are esculent, and of a very gelatinous nature. When properly prepared, the Chinese are exceedingly fond of them as a principal ingredient

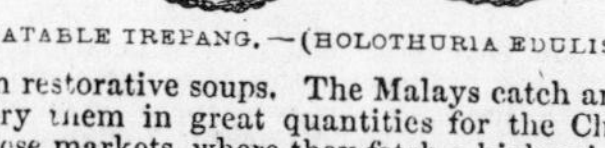

EATABLE TREPANG.—(HOLOTHURIA EDULIS.)

in restorative soups. The Malays catch and dry them in great quantities for the Chinese markets, where they fetch a high price, and are called *Trepang*.

We learn from a paper by Mr. C. W. Peach, read before the Royal Polytechnic Institution of Cornwall, that a species of

Holothuria, called the "Nigger" or "Cotton Spinner" by the Cornish fishermen, is very common in deep water off the Deadman, and is held by them in great detestation, from its throwing out what they call "cotton," and from its slimy nature, and also because where the "Niggers" are numerous and get into the crab-pots, it is very rarely that either crabs or lobsters are caught. Their appearance, when closed up, very much resembles a small cucumber, the back being dark — almost black at times — and the under part light yellowish green, which, with the thorn-like appendages on the back, make the appearance more complete. On being handled, they stain the hand light green. The head is furnished with twenty tentacula, which surround the mouth; the opening is tolerably large, and can be much expanded; and it is amusing to watch the motions of the tentacula acting as feeders. If the tentacula are viewed from the upper part they are club-shaped on the top, this club being placed on a footstalk an inch in length, which is retractile, and is invariably of a lighter colour than the top. When seen from the under side, they appear like the umbels of the elder, and are beautifully branched and tipped much in the manner of the elder flowers. They can completely close in their tentacula, which they do on being disturbed; and they use them at times as organs of locomoti n. Outside the tentacula is a border of spines like processes on a skin, which reaches a short way up the tentacula, and serves as a covering when these are withdrawn. These spines very much resemble the thorns of the brier; the back and sides are covered with similar ones, but not in rows. The under side is furnished with feelers in very great numbers: these feelers they stretch out to a great length, and attach themselves firmly by them; so much so, that in trying to detach them the feelers have been frequently left behind. Each feeler has a small round calcareous plate at the tip, which, under the microscope, shows that it is composed of innumerable plates, an object of great beauty; these plates effervesce with acid and so do the plates of the mouth and tips of the processes. When the softer parts of the feelers are cut transversely, they are composed of fine tubes, and when magnified have very much the appearance of some of the corals. The animal is covered with a dark slimy mucilaginous skin, which peels off freely; underneath this it is light gray, and has a reticulated appearance. They are of various sizes and lengths, often nearly a foot in length, and thick in proportion; they sometimes draw themselves almost into a ball, at others are much inflated in the centre. At times they lie motionless, but generally they are in motion. These animals are enveloped in a film so tenacious that it is a difficult matter to rub them to pieces in the water; on exposure to air they lose their tenacity and crumble to pieces. This species of Holothuria is extremely irritable, and on being touched or disturbed, throws out a bunch of white taper threads, about an inch in length and one-eighth in thickness; these soon become attenuated, either by the agitation of the water or the coming into contact with something, and are drawn into very long threads of great tenacity; they stick to every thing they touch, and from these the animals are called "cotton spinners" by the fishermen. This small bunch is drawn into a large mass of threads, so small that the finest sewing-cotton is not equal to it, and is no doubt one of the means of defence provided for its preservation.

HOMARUS. A genus of long-tailed Crustacea, containing the LOBSTER [which see].

HOMOPTERA. An order of insects, distinguished by their possessing two pairs of wings, usually composed of a firm membrane, and not covered by scales; and by having the anterior pair of the same substance throughout, and roof-like when folded. The mouth is adapted for suction, the tongue being channelled, and surrounded by lancet-like organs, with which the tissues of plants are pierced. All the insects of this group subsist on vegetable juices; and among them there are many which do an incredible amount of damage to the garden crops. This order may be divided into three sections: in the first, *Trimera*, the tarsi have three joints; in the second, *Dimera*, they have two; and in the third, *Monomera*, they have but one joint.

HONEY-GUIDE. (*Cuculus Indicator* — Sparm.) The birds to which this name is given inhabit various parts of Africa, and are closely allied to the Cuckoo tribe, but differ from them in hatching their own eggs. They are celebrated for their curious habit of guiding the natives to the nests of wild bees, enticing them to the spot by flitting before them, and reiterating a peculiar cry. They have a solid, conical, and arched beak; small head; body long and straight; toes strong and short; and wings reaching to the middle of the tail. The feathers are short, hard, and pressed close to the body; and the skin is so thick and tough as to protect them effectually from the stings of bees, unless the enraged insects attack their eyes. The nest of this Honey-guide is composed of slender filaments of bark woven together in the form of a bottle, the neck and opening hanging downwards; and it is said to be constructed in the hollow of trees, which the bird climbs like a woodpecker. The general colour is an olive-green, brownish on the upper parts, and inclining to yellow beneath. One species is described as being about seven inches in length; and another as ten inches: they are called, respectively, the Little and Great Honey-guide (*Indicator minor* and *Indicator major*).

HONEY-SUCKER, and HONEY-EATER. (*Anthochæra* and *Melivora*.) The Honey-suckers are a family of Birds, closely allied to the Humming-birds (*Trochilidæ*), and peculiar to New Holland and the neighbouring islands. The chief difference consists in their bill and legs being stronger, and their powers of flight less conspicuous. Besides the juices of flowers, and the insects obtained with them, many of these birds feed on berries; and one species is said to pick holes in the bark of trees, and to extract insects from them by means of its long

AUSTRALIAN HONEY-EATER.
(MELIPHAGA AUSTRALASIANA.)

tongue, very much after the manner of the Woodpecker. The truly national work of Mr. Gould, "The Birds of Australia," contains figures and descriptions of many species. Our figure is derived from his work. [See ANTHOCHÆRA: MELIPHAGA.]

HOOKTIP [MOTHS]. A name given by collectors to Moths of the genera *Drepana* and *Platypteryx*.

HOOPOE. (*Upupa.*) A genus of birds bearing a close relationship to certain forms of the *Corvidæ* or Crow family. The bill is used in the same manner, and for the same purposes, as in the Hornbills; namely, for seizing insects, &c., squeezing them to death, and throwing them with a jerk into the throat. The tongue is short, and destitute of the power of extension.

The COMMON OR EUROPEAN HOOPOE (*Upupa epops*) is an elegant bird, inhabiting the warmer and temperate parts of the old Continent, and migrating occasionally to the British islands. It is about the size of a thrush, but is easily distinguished from it by its head being ornamented with a handsome crest, composed of cinnamon-coloured feathers of unequal lengths, having a white bar and black tips, which it can expand and depress at pleasure. Its bill is also much longer and more slender, and its feet much shorter. The colour of the head, neck, and body is pale ferruginous, darkest on the back and shoulders: the wings and tail are black, the former crossed by five white bars, the latter crossed in the middle by a white crescent. The Hoopoe feeds on various grubs, worms, &c.; hence it follows the retreat of the Nile in Egypt, whose neighbouring plains swarm with insect life; and it also frequents ploughed lands and pasture grounds, like the crows. It is more abundant on the continent of Europe than in Britain; but its sojourn in temperate climates is but short, its arrival from warmer regions being late in the year, and its departure early. This migration, however, is not universal, since it is observed that great numbers of these

HOOPOE. — (UPUPA EPOPS.)

birds are constantly found about the towns and villages of Egypt, becoming very familiar with man, and building their nests in the immediate vicinity of his habitations. The flight of the Hoopoe is rather slow and undulating; and it seldom perches on trees. The name of the bird seems to be derived from its continually uttering, in soft and rapid tones, a peculiar sound, resembling *hoop, hoop, hoop.*

HOPLOPTERUS. A genus of birds allied to the Plovers: so named from the bony spine or projection on the shoulder of the wing. There are several species, of which the longest known is the SPUR-WINGED PLOVER of Africa (*H. Spinosus*).

HOPPING DICK. The local name given to a species of Thrush (*Merula leucogenys*), common in Jamaica, whose lively and familiar manners, as well as his sable plumage, and clear, rich, and mellow song, greatly resemble the English Blackbird. "The forests skirting the mountain are his favourite haunt. If he frequents the open slopes and crests of the hills, he glides from tree to tree, just above the surface of the grass. If he rises above the lower branches of the pimento, or into some of the loftier shrubs, it is to visit the *Tillandsias*, or parasitical wild pines, to drink from within the heart-leaves at those reservoirs of collected dews, which are the only resource of the birds in these high mountains. His dark sooty plumage, his brilliant orange bill, and his habit, when surprised or disturbed, of escaping by running or flying low, and sounding all the while his alarm scream till he gets away into the thicket, completely identify him with the European Blackbird." — *Gosse.*

HORIADÆ. This family of Coleopterous insects is of very small extent, but the species are comparatively large, handsomely coloured, and principally confined to tropical countries. The larva of one of the species (*Horia maculata*), an inhabitant of South America and the West Indies, is said to destroy the larva of a species of wild carpenter bee (*Xylocopa teredo*), which makes its cells and deposits its eggs in the trunks of

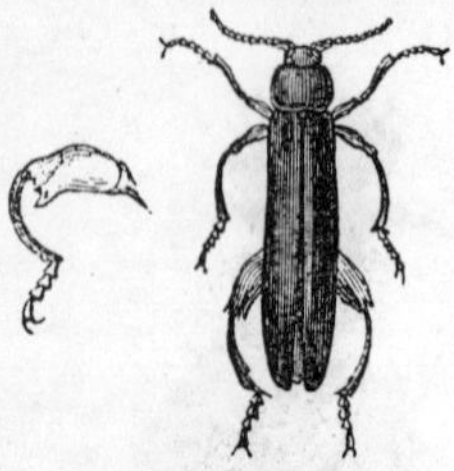

CISSITES TESTACEA.

trees. Our figure represents the *Cissites testacea*, from the East Indies. It is distinguished from *Horia* by the head being narrower than the thorax, and the posterior femora much thickened.

HORNBILL. (*Buceros.*) This family of Conirostral birds is remarkable for the very large size of the beak, and for an extraordinary protuberance with which this is surmounted. They are both carnivorous and frugivorous, feeding not only on various berries, fruits, and other vegetable matter, but also on the smaller kinds of animals, as mice and small birds, as well as on insects and any putrid animal substance. Their large bills are of much less real than apparent strength, and they vary considerably in appearance during the different periods of their age, the upper process or excrescence not exhibiting its genuine form till the full growth of the bird. When cut across, it is found to consist of a very loose bony substance; its interior being traversed in every direction by osseous fibres, the interspaces being quite hollow: all the bones, indeed, of this remarkable bird being more permeated by air than in any other species. They inhabit the warm parts of Asia and Africa; and in their general habits they seem to bear a considerable resemblance to the Crows. The larger species are very difficult of approach; and they perch on the branches of high trees, where their vision can command an extensive range. They may be said to hold the same rank in the old continent that the Toucans do in America; not only from the enormous size of the bill, but also from their habit of swallowing their food whole, throwing it up into the air, and catching it as it falls. There are many species; but two will suffice for our description.

The RHINOCEROS HORNBILL. (*Buceros Rhinoceros.*) This bird is about the size, though rather more slender, than a hen turkey: its colour black, except the lower part of the belly and tip of the tail, which are white: the bill is about ten inches in length, slightly curved, sharp-pointed, irregularly serrated on the edges, and furnished at the base of the upper mandible with an immense appendage in the form of a reverted horn: a longitudinal black line divides this process, the part above it being of a bright red, the part below yellow, and the base of

RHINOCEROS HORNBILL.
(BUCEROS RHINOCEROS.)

it black; the bill itself is black at the base, tinged with bright red, and the remainder is yellow: the legs are short, strong, and of a pale yellow colour.

The UNDULATED HORNBILL. (*Buceros undulatus.*) The beak of this species is more proportioned to the size of the bird, and the colours have more variety and elegance than in any other of the tribe. The length of the bird is about thirty inches, exclusive of the bill, which is only five. The plumage is black, with a strong gloss of blue, and a large patch of red-brown between the shoulders: the chin, the orbits of the eyes, and the space between them and the upper mandible, are covered by a bluish bare skin; the bill is a pale yellow, tinged with brown; and is moderately curved and sharp-pointed.

HORNED OWL. [See OWL.]

HORNET. (*Vespa crabro.*) This insect is of the Wasp kind, but much more formidable, and very considerably larger. It has four wings, the first pair being by far the largest, and with these it flies with great velocity. The head is oblong, and yellowish; the eyes are prominent and semilunated; and between them there are two falciform antennæ. The body is united to the shoulders by a slender filament; the middle of the fore part is of a dark brown hue, marked with a deep yellow belt; and the hinder part is wholly of that colour, except that it is variegated with eight brown spots. The Hornet, like the Wasp, is extremely voracious, and preys on almost any kind of fresh animal substances which it can obtain, as well as on honey, fruit, &c. Its sting is greatly to be dreaded, and is often productive of very serious consequences. The Hornet's nest is generally built in the cavity of some decayed tree, or immediately beneath its roots; and not unfrequently in timber yards and other similar situations. It is of a smaller size than that of the Wasp, and of a somewhat globular form, with the mouths of the cells downwards, which in a great measure preserves them from the rain. In "The Zoologist," p. 162. F. Smith, Esq., Curator of the Entomological Society, thus writes:—"A few years ago, as I was walking by the side of Virginia Water, by moonlight, I heard a loud booming noise, evidently caused by some insect darting rapidly by. I was for some time at a loss to conceive what this could be; at last I succeeded in knocking one down, when I found it was a hornet. By watching the flight of others, I soon discovered the tree containing their nest: they were carrying on their labours by the light of the moon, apparently quite as busily as if it had been open day." [See VESPIDÆ.]

This brief notice of the Hornet leads us to extract from the "Journal of a Naturalist" some interesting remarks relative to this insect:—"Every-day events manifest to very superficial observation, that no created being, from the monster of the ocean, 'that makes the deep boil like a pot of ointment,' to the insect that feebly creeps on the ground, exists free from the persecutions or annoyance of another. Some may be subject to fewer injuries than others, but none are wholly exempt: the strong assail by power, and become assaulted themselves by the minute or weak. This year (1826) the Hornet abounded with us in unusual numbers, and afforded constant evidence of its power and voracity that could not have been exceeded by any ravenous beast. In our gardens the imperious murmur of four or five of them at a time might be frequently heard about our fruit-trees. They would occasionally extract the sweet liquor from the gage, or other rich plums; but the prime object of their visit was to seize the wasps that frequented the same places. This they not only did when the creature was feeding on the fruit, but would hawk after them when on the wing; capture them with a facility to which their heavy flight seemed unequal; bear them to some neighbouring plant, and there feed on the insect, which seemed perfectly overpowered by the might of the Hornet. The first operation was to snip off the head, then to cut away the lower part by the waist; and, when near, we could hear them shearing away the outer coat from the body, and crushing it with their strong mandibles; sometimes devouring it, but generally only sucking the juices it contained. Their avidity for this sort of food is very manifest, when the grape ripens on the wall: being commonly the only remaining fruit, the wasp abounds there; the Hornets flock to the prey, and we may see them in constant progress, bearing their victims from the bunches. The wasp itself seizes the house-fly; but this seems rather the display of wanton power than for food, as it bears the fly about with it for a length of time, and drops it unconsumed. The fly, in its turn, is conducive, after its manner, to the death of many an animal. We know not any insect that destroys the Hornet; but its power and being are terminated by some very effective agent, as in particular years it is almost unknown." To the foregoing the author appends the following note:—"The Hornet is a very pugnacious animal. They will fight desperately with each other at times, when they meet in pursuit of prey, biting each other's body, and trying to get their mandibles under the head of their opponent, to snip it off. I one day confined, under a glass, two of these creatures which had been fighting. One had evidently the mastery; but both had been so injured in the contest, that they soon died; and it is most probable that they fall victims to each other's voracity in the cold damp season that usually terminates the autumn of our year."

HORNET [SPHINX]. A name given by collectors to Hawk-moths of the genus *Trochilium.*

HORSE. (*Equus caballus.*) This most useful and beautiful quadruped demands, perhaps, a more extended notice than the ordinary limits of this work may seem to afford; but we trust we have not omitted any material point, zoological or historical, that is essential to a complete description of an animal, whose services to mankind are everywhere deemed invaluable, and whose noble nature universally excites man's admiration. It has been well observed, that had not custom dignified the Lion with the title of "king of beasts," reason could nowhere confer that honour more deservedly than on the *Horse.* His courage, strength, fleetness, his symmetrical form, and grandeur of deportment, are unalloyed by any quality injurious to other creatures, or calculated to create the aversion of man; whose orders he implicitly obeys, whose severest tasks he undertakes with a cheerful alacrity, and whose pleasures he contributes to with animation and delight. Nor is this all: for, when called to bear our warriors to the battle-field, nothing can excel his resolute fierceness, his courageous ardour! In the poetical language of the Sacred Writings, "His neck is clothed with thunder. The glory of his nostrils is terrible. He paweth the valley, and rejoiceth in his strength. He goeth on to meet the armed men. He mocketh at danger, and is not affrighted; neither turneth he back from the sword." (Job, xxxix.)

What region the Horse originally inhabited, or to what nation we are indebted for his first subjugation, are questions far too remote for history to resolve. That this animal is of Eastern origin, and that the Egyptians were the first to reduce it to obedience, and train it to the various purposes of civilized life, appears highly probable from various passages in the Bible, though no direct testimony of such a fact is to be gathered from that source. The first mention of the Horse occurs during the wise administration of Joseph in Egypt, who, we are told, gave the famishing inhabitants bread "in exchange for horses;" and also when the body of the patriarch Jacob was removed from Egypt to Canaan for burial, we read that "there went up with him both chariots and horsemen." The period when the Horse is thus indicated as a beast both of draught and burden, is 1650 years before the birth of Christ; which is a date anterior to any that profane history affords on the subject.

The generic characteristics of the Horse are a broad undivided hoof; six cutting-teeth or *nippers* in each jaw; two very small tusks or canines; grinders with a flat crown, presenting, when worn, different figures, formed by the enamelled plates of the interior; stomach small and simple, intestines very large. Wild Horses exist in many countries; but Arabia produces the most beautiful breed, and also the most generous, swift, courageous, and persevering. They occur, though not in great numbers, even in the deserts of that country, and the natives make use of every stratagem to take them. They select the most promising for breeding, and, instead of crossing the breed, the utmost care is taken to keep it entire. In other countries it is found necessary to change the races, otherwise, it is said, the Horses would soon degenerate; but in Arabia the same blood has passed down through a long succession, without any diminution either of beauty or strength. A general belief has hence arisen, and been long maintained, that to Arabia are we indebted for the primitive breed of this noble animal, and for its subjugation to man's use. This opinion, however, has been combated by Mr. Bell (in his History of British Quadrupeds) in the following terms: "The long acknowledged superiority of the Horses of Arabia is no proof that they were indigenous to that arid country in a wild state; for there is great reason to conclude that it was only at a comparatively late period that they were employed by that people. Whilst Solomon was receiving from Arabia treasures of various kinds, it was from Egypt only that his Horses were brought: and so highly were they valued by this magnificent and luxurious king, that notwithstanding the Divine prohibition, 'that the king shall not multiply Horses to himself, nor cause his people to return into Egypt, to the end that he should multiply Horses,' it is stated that he had no less than forty thousand stalls of Horses for his chariots, and twelve thousand horsemen. There appears great probability, therefore, in the opinion that Egypt or its neighbourhood is its original country; and still more, that this extraordinary people first rendered it subservient to man, and subsequently distributed it to other countries."

It does not appear, then, that a clue can be obtained to determine, with any degree of precision, in what country the Horse first roamed at large, or where he first submitted to the yoke of man. Those which at present exist in a wild state seem to have been derived from such as had been once domesticated. In the Pampas or plains of South America, on the banks of the river La Plata, there are immense troops of wild horses, which are descended from those of Andalusia, originally carried thither by the Spanish conquerors: and we learn from the accounts given by various travellers, that they not only associate together in herds or troops of several thousands, but that on the appearance of any danger, they evidently put themselves under the direction of a leader, in order the more effectually to resist the enemy's attack. Large herds are sometimes seen in the southern parts of Siberia, in the deserts of the Mongul territory, and among the Kalhas to the north-west of China. Moldavia also abounds with them. At the Cape of Good Hope there are numbers of wild Horses, but they are small and vicious. They are likewise found in other parts of Africa, but the savages there seem ignorant of their value, and also of the methods of taming them.

In Brande's Dictionary of Science, &c. it is remarked, that "wild Horses appear to be free from nearly all those diseases to which the domestic breed are prone. They are generally of a pale or grayish-brown colour, with brown mane and tail, a whitish muzzle changing to black about the mouth. They are less than the domestic breed; with a larger head; longer legs; larger ears, with the apices sub-reflected; the forehead is more convex above the eyes; the hoofs are contracted and sub-cylindrical; mane sub-erect, less lax than in the domestic horse: the coat, in winter, looser and sub-undulated along the back; the tail not very large. They recognize the presence of man at a great distance when he approaches them to windward, and fly from him with wonderful speed; they prefer sunny slopes, and avoid forests and steep places. They do not wander beyond the fiftieth degree of north latitude. Wild stallions attracted by domestic mares are often taken and killed. The first change which domestication works upon the form of the wild Horse is to increase the bulk of his trunk as compared with his head and limbs. This change is beautifully exemplified in the Arabian, which we must regard as an early, if not first remove from his wild neighbours of the more northern deserts, and which the Bedouin still hunts for the sake of their flesh. The head is not only proportionally smaller, but is remarkable for the breadth and squareness of the forehead, the shortness and fineness of the muzzle, the prominence and brilliance of the eye, and the smallness of the ears. The body is still somewhat light, and narrow at the fore part; but the shoulder is superior in its formation to that in any other breed. The Arabian seldom stands more than fourteen hands two inches. The 'Barb,' so called from its native country, Barbary, is somewhat smaller than its near ally the Arabian; it seldom exceeds fourteen hands and an inch; the shoulders are flat, the chest round, the legs rather long, and the head small and very beautiful. The Barb is remarkable for its fine and graceful action; but though it is superior to the Arabian in its general form, it has not its untiring spirit or its speed. Our most valuable English varieties of the Horse date from the introduction of, and interbreeding with, the Barb and Arabian."

The Horse is naturally an herbivorous animal, and is more scrupulous in the choice of his food than most other domestic quadrupeds; in the meadow rejecting several plants which the ox devours with pleasure. His thin and muscular lips, his firm and compressed mouth, and his sharp incisor teeth, are admirably adapted to seize and to crop the grass; and when, free from man's control, he can follow his own propensities, we all know that grass is his chosen food: yet, in order that he may subsist (as in his present state of domestication he necessarily must) on aliment of a much harder kind, he is enabled, by the peculiar structure of some of the bones of his face, so to move his jaws as to comminute and grind down his "corn."—Of the various modes of judging of a Horse's age, the best is from a careful inspection of the teeth. Five days after birth, the four teeth in front, called *nippers*, begin to shoot; these are cast off at the age of two years and a half, but are soon renewed; and in the following year two above and two below, namely one on each side of the nippers, are also thrown off; at four years and a half other four next those last placed fall out, and are succeeded by other four, which grow much more slowly. From these last four corner teeth it is that the animal's age is distinguished, for they are somewhat hollowed in the middle, and have a black mark in the cavities. At five years old these teeth scarcely rise above the gums; at six, their hollow pits begin to fill up, and turn to a brown spot; and before eight years the mark generally disappears. A Horse's age is also indicated by the canine teeth or tusks, for those in the under jaw generally shoot at three years and a half, and the two in the upper at four; till six they continue sharp at the points; but at ten they appear long and blunted. There are, however, many circumstances which render a decision as to the age of the Horse very difficult after the marks are effaced from the lower incisors; and it should be observed, that Horses which are always kept in the stable have the mark much sooner worn out than those that are at grass; to say nothing of the various artful tricks resorted to by dealers and jockeys to deceive the inexperienced and unwary.

The Horse has three natural paces, namely, walking, trotting, and galloping. In the first, he moves off with one of his fore feet, which is immediately followed by the hind leg of the opposite side; and so with the other fore and hind leg. His trot differs from his walk, not only in its greater velocity, but also in this, that he always moves the two opposite legs together. The gallop is a series of leaps, and it is true and regular, when the horse lifts his two feet on one side at the same time, and follows with those on the other side. These three natural paces may be converted into artificial paces by art and skill. But as this is a part of the science of horsemanship, it is not necessary to be more than adverted to in this place: we shall therefore merely observe, that the trot is the pace which enables all quadrupeds to balance and support themselves with ease and firmness; and it is therefore the most proper for ensuring a free determined motion to the Horse.

An old writer, Camerarius, says, a perfect Horse should have the breast broad, the hips round, and the mane long, the countenance fierce like a lion, a nose like a sheep, the head, legs, and skin of a deer, the throat and neck of a wolf, and the ear and tail of a fox. This is as graphic as it is concise; but to be serviceable it is much too general: we therefore turn to the pages of the Penny Cyclopædia for fuller particulars as to the proper conformation of the Horse. "The head should not be disproportionally large, and should be well set on; *i. e.* the lower jawbones should be sufficiently far apart to enable the head to form that angle with the neck which gives free motion and a graceful carriage to it, and prevents its bearing too heavy on the hand. The eye should be large and a little prominent, and the eyelid fine and thin. The ear should be small and erect, and quick in motion. The lop-ear indicates dulness or stubbornness; and when it is habitually laid too far back upon the neck, there is too frequently a disposition to mischief. The nostril in every breed should be somewhat expanded: it can hardly be too much so in the Racer, the Hunter, the Roadster, and the Coach-horse, for this animal breathes only through the nostril, and would be dangerously distressed when much speed is required of him, if the nostril could not dilate to admit and to return the air. The neck should be long rather than short. It then enables the animal to graze with more ease, and to throw his weight more forward, whether he is in harness or galloping at the top of his speed. It should be muscular at its base, and gradually become fine as it approaches the head. The withers should be somewhat high in every Horse, except perhaps that of heavy draught, and it does not harm him, for there is larger surface for the attachment of the muscles of the back, and they act at greater mechanical advantage. A slanting direction of the shoulder gives also much mechanical advantage, as well as an easy and pleasant action, and a greater degree of safety. It must not however exist in any considerable degree in the Horse of draught, and particularly of heavy draught. The chest must be capacious, for it contains the heart and the lungs, the organs on which the speed and endurance of the Horse depend. Capacity of chest is indispensable in every Horse, but the form of the chest admits of variation. In the waggon-horse the circular chest may be admitted, because he seldom goes at any great speed, and there is comparatively little variation in the quantity of air required; but in other Horses the variation is often fearful. The quantity of air expended in the gallop is many times that required in hard work. Here we must have depth of chest not only as giving more room for the insertion of the muscles on the action of which the expansion of the chest depends, but a conformation of the chest which admits of that expansion. That which is somewhat straight may be easily bent into a circle when greater capacity is required; that which is already circular admits of no expansion. A few words more are all that our limits permit us to add, and they contain almost all that is necessary on the conformation of the Horse. The loins should be broad, the quarters long, the thighs muscular, and the hocks well bent and well under the Horse."

Some peculiarity of breed distinguishes the Horses of most civilized countries; or, rather, there is some particular breed for which one country is more celebrated than others. Thus there is the Spanish Genette, a small but fleet and beautiful variety, which is generally ranked next to the Barb: their heads are rather large, their manes thick, their ears long and well pointed, their shoulders somewhat heavy, their chests full and large, and their legs clean and handsome. They move with great ease, and carry themselves very gracefully. They are usually of a black or dark bay colour; and some of them, particularly such as come from the province of Andalusia, are said to possess, in a superior degree, high courage, docility, and other estimable qualities.—France produces a motley breed; adapted rather for the purposes of war than of the chase, and generally considered as heavy-shouldered. But great attention has of late years been paid to the improvement of them by crosses with the best bred English varieties; a remark, by the by, which may in a great measure be applied to the breed of Horses elsewhere throughout the continent. And we may safely assert, that whatever could be gained from long experience and careful assiduity, whatever wealth could procure, or skill effect, in order to arrive at perfection in the various breeds, and in the proper training, of this noble animal, has been fully attained in England.

It is impossible to say at what early period the Horse was first considered an object of interest in Britain; but we know that when our rude and warlike ancestors had to contend with the Roman invaders, they depended much on their cavalry and war-chariots, which they managed with great skill and dexterity. We likewise know that the Saxons paid great attention to the Horse, and took considerable pains to improve the natural breed. King Athelstan obtained several German running-horses from Hugh Capet of France; and William the Conqueror, with his Norman followers, introduced the Spanish horse, in whose veins ran the blood of the swift-footed Barb. When the Crusaders returned from the Holy Land, they brought with them many a noble Eastern steed; and from that time a greater admixture of Arabian blood with the Horses of Europe was a natural consequence. It must however be apparent, when we remember with what a heavy load of armour both horseman and horse were encumbered, that our mail-clad warriors must necessarily have required horses of prodigious strength, and that fleetness was of far less consequence to them than weight and mettle. King John, who appears to have devoted much attention to the breed of Horses, imported a hundred choice stallions of the Flanders kind; to which act may probably be traced the foundation of that character for size, strength, and vigour, which English horses, whether for draught or war, have since maintained. Subsequent monarchs also evinced a strong desire for keeping up, undiminished, a race of Horses which, in a national point of view, had become so valuable, and their exportation was accordingly forbidden. At the period to which we have been alluding, the breeds of Horses most in repute for superior weight and strength were those of Flanders and Normandy. In course of time, the cumbrous armour, the battle-axe and shield, were laid aside; and when the sword and carbine, with the lighter dresses of our cavalry, were introduced, speed and elegance were deemed of more account than size and power. At length the sports of the field engaged the attention and became the amusement of kings and princes; the nobility of the land vied with each other in keeping the choicest studs, the English Hunter was unmatched for ardour in the chase, combined with the most persevering endurance; and the English Race-horse distanced all competition.

RACE-HORSE.

The RACE-HORSE. "Whether or not the blood of our finest Racers be pure Eastern, or a mixture of the Arabian or Barb with the best of our English stock," Mr. Bell observes, "can scarcely, with all the accuracy of our turf genealogy, be positively ascertained: but it is undoubted that the most celebrated Horses that this country has ever produced are traceable from son to sire back to some or other of the well-known Arabian, Barbary, or Turkish stallions which have at different times been imported. The importance of the influence of the sire in breeding Horses is in no point more clearly proved than by the fact that the progeny of the most celebrated Horses have generally sustained the reputation of their sires. Thus the descendants of Eclipse numbered no less than three hundred and sixty-four winners." "The Racer is generally distinguished by his beautiful Arabian head; his fine and finely-set-on neck; his oblique lengthened shoulders; well-bent hinder legs; his ample muscular quarters; his flat legs, rather short from the knee downwards; and his long elastic pastern. From this perfect symmetry, however, many celebrated Race-horses have shown remarkable deviations; and yet they have not failed to enter into the excitation and enjoyment of the sport, straining every muscle, and evincing indescribable energy in their endeavours to outstrip their competitors.

The HUNTER. It is generally allowed that this fine animal, whose spirit is only equalled by his endurance of fatigue, and whose speed is on a par with his beautiful form, presents a happy combination of those

THE HUNTER.

qualities which give swiftness to the racer, vigour to the charger, and muscular power to the draught-horse. "The first property of a good hunter is, that he should be light in hand. For this purpose his head must be small; his neck thin, especially beneath; his crest firm and arched; and his jaws wide. The head will then be well set on. It will form a pleasant angle with the neck, which gives a light and pleasant mouth."

The compact and serviceable ROADSTER, "a hunter in miniature," as a perfect specimen of this truly valuable animal has been called; the splendid CARRIAGE HORSE, with

COACH-HORSE.

his arched crest and high action; the powerful DRAY HORSE, whose united strength and size (derived from the Suffolk Punch and the Flanders breed) are unequalled; the round-chested and long-backed SUFFOLK PUNCH HORSE; and the patient CART HORSE,—have each their peculiar merits, and require careful attention to the breed and management. We have also some smaller varieties, excellent in their kind: as the useful GALLOWAY; the diminutive and hardy SHETLAND PONY; and the sturdy rough Pony bred in the New Forest. But our limits have long warned us to bring this article to a close: we therefore beg to refer our readers to the various works which are exclusively devoted to "the Horse" for whatever further information may be required; and conclude by tritely remarking, that according to the degree of cultivation bestowed on them, Horses improve or degenerate; their qualities of sagacity and docility alone remaining inherent.

CLYDESDALE CART-HORSE.

A curious point, and one of great interest in the investigation of zoological relations, which may be properly introduced in this place, is—"that the characters of the male parent of the mother's first progeny show themselves in her subsequent offspring by other males, however different those males may be in form and colour. Mr. Bell observes that this truth has been illustrated by him when treating on the Dog and on the Hog, and he adds that it receives a remarkable and interesting confirmation from the case of a mare belonging to the Earl of Morton, to which he had before alluded. In that case the mare was young, and after producing the female hybrid by the Quagga, had first a filly, and afterwards a colt, by a fine black Arabian Horse. They both resembled the Quagga in the dark line along the back, the stripes across the forehead, and the bars across the legs: in the filly the mane was short and stiff, like that of the Quagga; in the colt it was long, but so stiff as to arch upwards and hang clear of the sides of the neck; in other respects they were nearly pure Arabian. This and other such cases should not be forgotten by breeders of animals, who are anxious about the perfection of their stock, and should make them particularly careful as to the male influence which first makes its impression on the female.

The mode of catching and taming wild horses in South America is so well described by Mr. Darwin, in his "Researches," and shows so strikingly what mastery over the brute creation man can attain, that we trust it will be considered an appropriate addendum to the foregoing. "A troop of wild young horses is driven into the corral, or large enclosure of stakes, and the door is shut. We will suppose that one man alone has to catch and mount a horse, which as yet had never felt bridle or saddle. I conceive, except by a Gaucho, such a feat would be utterly impracticable. The Gaucho picks out a full-grown colt; and as the beast rushes round the circus, he throws his lazo so as to catch both the front legs. Instantly the horse rolls over with a heavy shock, and, whilst struggling on the ground, the Gaucho, holding the lazo tight, makes a circle, so as to catch one of the hind legs, just beneath the fetlock, and draws it close to the two front. He then hitches the lazo, so that the three legs are bound together. Then sitting on the horse's neck, he fixes a strong bridle, without a bit, to the lower jaw. This he does by passing a narrow thong through the eye-holes, at the end of the reins, and several times round both jaw and tongue. The two front legs are now tied closely together, with a strong leathern thong, fastened by a slip-knot. The lazo, which bound the three together, being then loosed, the horse rises with difficulty. The Gaucho now holding fast the bridle fixed to the lower jaw, leads the horse outside the corral. If a second man is present (otherwise the trouble is much greater) he holds the animal's head, whilst the first puts on the horsecloths and saddle, and girths the whole together. During this operation, the horse, from dread and astonishment at thus being bound round the waist, throws himself, over and over again, on the ground, and, till beaten, is unwilling to rise. At last, when the saddling is finished, the poor animal can hardly breathe from fear, and is white with foam and sweat. The man now prepares to mount, by pressing heavily on the stirrup, so that the horse may not lose its balance; and at the moment he throws his leg over the animal's back he pulls the slip-knot, and the beast is free. Some "domidors" (horse-subduers) pull the knot while the animal is lying on the ground, and, standing over the saddle, allow it to rise beneath them. The horse, wild with dread, gives a few most violent bounds, and then starts off at full gallop: when quite exhausted, the man, by patience, brings him back to the corral, where, reeking hot, and scarcely alive, the poor beast is let free. Those animals which will not gallop away, but obstinately throw themselves on the ground, are by far the most troublesome. This process is tremendously severe, but in two or three trials the horse is tamed. It is not, however, for some weeks that the animal is ridde with the iron bit and solid ring; for it must learn to associate the will of its rider with the feel of the rein, before the most powerful bridle can be of any service."

HORSE-GUARD. This name, we are informed by Mr. Doubleday, is applied in the United States to Hymenopterous insects of the genus *Monedula*, from their habit of capturing Gadflies (*Tabani*).

HOUND. There are several species of Dogs which come under this appellation, as the Foxhound, Greyhound, Bloodhound, &c. which will be found in their proper alphabetical order in this volume. Hounds may be distinguished into such as discover and pursue the game by sight; and those which find and pursue it by the excellence of their scent.

HOUND-FISH. The name applied sometimes to different species of the Shark family. [See DOG-FISH.]

HOWLET. (*Strix aluco.*) A bird of the Owl kind, so called from its mournful, howling voice. It measures eighteen inches in length: the head, back, wings, and tail, are cinereous, with black and white spots; the head is large, round, and full-feathered; and the wings reach to the extremity of the tail.

HUMBLE-BEE. (*Bombus.*) Of the villose or hairy bees popularly called Humble-bees, there are several species. One of the largest and most common is the *Apis lapidarius* of Linnæus, so named from the circumstance of its nest being generally situated in stony or gravelly places. This species is entirely of a deep black colour, except the end of the abdomen, which is red or orange-coloured, more or less deep in different individuals. The female is of large size, measuring nearly an inch in length; the male is considerably smaller, and the labouring bee is still smaller than the male.

Humble-bees are the only tribe besides the hive-bees that in this part of the world construct nests by the united labour of the society. Their habitations are sometimes excavated at a considerable depth in the ground, and sometimes built upon its surface, beneath stones, &c. The societies consist, in some species, of about fifty or sixty individuals; in others of as many as two or three hundred. They contain males, females, and workers or neuters. The females alone survive the winter; and they employ the first fine days in spring to commence their nests, which they very quickly excavate, and supply with a mixture of honey and pollen for the nourishment of the first brood, which consists exclusively of workers. These, after having undergone their transformations, assist in the construction of new cells, the collection of the food, and the rearing of the larvæ. In autumn the males and females are produced; and at the commencement of winter all but the larger females die; these remain in a sort of chamber distinct from the rest, but, as it would appear, without any supply of food. It should be observed that though the Humble-bees collect honey as well as the common ones, it is neither so fine nor so good: nor is their wax so clean, or so capable of fusion.

HUMBLE-BEE FLY. (*Bombylius.*) A name very usually given to a species of the large order Diptera, comprehending species of different sizes, but all agreeing in the great resemblance they bear, at first sight, to the Humble-bees of the smaller or middle-sized kinds; but on examination, it will appear that they are destitute of trunks, and have but one pair of wings. Nature has assigned for the larvæ of some of the species a very singular habitation—the intestines of horses, or under the thick skins of oxen. In the latter case, the worm hatched from the egg of its parent fly, deposited there, forms a tumour which furnishes it with food and lodging, and in the middle there is an aperture for the purpose of respiration. Some, however, feed on vegetable substances, and one species in particular shows a strong predilection for the bulbous roots of flowers.

HUMMING-BIRD. (*Trochilus.*) The birds included in the family of *Trochilidæ*, or Humming-birds, are at once the most diminutive and the most brilliantly coloured of the whole feathered race. Their vivacity, swiftness, and singular appearance, unite in rendering them the admiration of mankind; while their colours are so radiant that we can only compare their peculiar splendour with the brilliancy of polished metals and the superior lustre of the ruby, the sapphire, or the emerald. This is their general character; but there are some species whose plumage is comparatively obscure, exhibiting only a golden-green tinge, diffused over the brown or purplish colour of the back and wings. In size they vary from that of a wren to a humble-bee; the muscles of their wings are very strong, and their plumage dense and compact: they are almost ever in motion; and the velocity with which they dart through the air, and the rapidly-vibrating motion of their wings, are quite inconceivable. These lovely gems of animated nature are peculiar to America, and almost exclusively tropical; some species, however, migrate into the temperate regions on either side of the equator during the warm season; and stragglers have occasionally been met with even in cold situations. They are characterized by a long and extremely slender bill, inclosing an extensile and retractile tongue, which is divided into two filaments from the middle to the tip, by which they extract the nectar and the small insects which may lurk in the recesses of flowers. Their feet are very small, their wings long and narrow; the mechanism of their whole form being, in fact, like that of the Swift, formed for rapid and powerful flight. When hovering before a flower, they seem suspended in the air, rather than sustained by the vigorous movement of their pinions; and it is to the constant murmur or buzzing sound, caused by the rapid vibration of them, that these beautiful little creatures derive their name. How greatly they must add to the richness of a Transatlantic landscape, when fluttering from flower to flower in the morning sunbeams!

"Wherever a creeping vine opens its fragrant clusters, or wherever a tree-flower blooms, may these little things be seen. In the garden or in the woods, over the water, every where they are darting about; of all sizes, from one that might easily be mistaken for a different variety of bird, to the tiny Hermit (*T. rufigaster*), whose body is not half the size of the bees buzzing about the same sweets. Sometimes they are seen chasing each other in sport with a rapidity of flight and intricacy of path the eye is puzzled to follow. Again circling round and round, they rise high in mid air, then dart off like light to some distant attraction. Perched upon a little limb, they smooth their plumes and seem to delight in their dazzling hues; then starting off leisurely, they skim along, stopping capriciously to kiss the coquetting flowerets. Often two meet in mid air and furiously fight, their crests and the feathers upon their throats all erected and blazing, and altogether pictures of the most violent rage. Several times we saw them battling with large black bees, who frequent the same flowers, and may be supposed often to interfere provokingly. Like lightning our little heroes would come down, but the coat of shining mail would ward their furious strokes. Again and again would they renew the attack, until their anger had expended itself by its own fury, or until the apathetic bee, once roused, had put forth powers that drove the invader from the field."—*Edwards's Voyage up the Amazon.*

The Humming-birds are generally divided into two classes—those with curved bills, and those whose bills are straight. We shall endeavour to give a description of the most remarkable species; reserving some of them for insertion under the article *Trochilidæ.* Their nests are very beautifully constructed, being usually composed of vegetable down, such as that of the cotton-plant or silk-cotton tree; and being covered on the outside with bits of lichen, leaves, moss, &c. Sometimes they are suspended from the extremities of twigs of the orange, the pomegranate, or the citron-tree; and sometimes from a house, provided they can find convenient twigs for the purpose: for it is to be observed, that although these birds are most numerous in the dense forests, where the wild blossoms almost vie with themselves in splendour, they are also seen in the gardens of cultivated districts, and do not appear at all disinclined to the society of Man, though it is very difficult to keep them in a state of domestication. The Humming-bird is very irascible, two males scarcely ever meeting without a contest ensuing: they will also attack birds of a much larger size, as wrens or king-birds, and they sometimes even have contests for a flower with the humble-bee.

TOPAZ-THROATED HUMMING-BIRD. (*Trochilus pella.*) Both in size and colours this species is decidedly superior to any others of the curve-billed kind. Its body is as large as that of a wren; and from the tip of the bill to the end of the two long-tailed feathers, it measures from eight to ten inches. The upper part of the head and neck are of a glossy black, the back and smaller wing-coverts being of a fine deep orange-purple colour; the throat and part of the neck is of the most splendid topaz yellow, changing from the lustre of polished gold to deep emerald green, according to the situations in which it is viewed: the topaz-coloured plumage is separated from the breast and sides of the neck by a black line, beneath which the whole breast and sides are of a deep but shining purple rose-colour: the wings are of a purplish brown; the rump of a bright grass-green; and the tail orange-purple, except the two middle feathers, which are purple-brown, of a narrow shape, and pointed at the tips, and exceed the rest in length by about four inches. The bill is moderately long, curved, and black; the legs are also black. The female is far less brilliant than the male, being of a dark coppery-green colour, with dusky wings, and the two middle feathers of the tail no longer than the rest. This species is said to be principally found in Surinam and Guiana, where it frequents the banks of rivers and brooks, the surface of which they skim after the manner of swallows.

FORK-TAILED HUMMING-BIRD. (*Trochilus forficatus.*) This species is chiefly noticeable for the shining beauty of its tail-feathers, which appear of a brilliant blue, green, or golden colour, according to the lights in which they are seen, and form a very long and broad tail in proportion to the body of the bird: the crown of the head is blue, and a shining golden lustre pervades the rest of the plumage, but it is trifling in comparison with the beauty of the tail: the legs, feet, and claws are black.

BAR-TAILED HUMMING-BIRD. (*Trochilus sparganurus.*) This elegant bird is nearly eight inches long: its colour is green-gold, but not very bright, except on the throat, where it is rich and brilliant: the tail is long and strongly forked, and the feathers are velvet-black, each being crossed by a broad golden crimson bar, and rounded at the end: bill and legs black. Native of Peru.

HARLEQUIN HUMMING-BIRD. (*Trochilus multicolor.*) A highly elegant species, remarkable for the variety of its colours. Its length is about four inches; the bill long, slightly bent, and of a pale yellow hue: the crown of the head, throat, neck, breast, upper part of the back, rump, and wing-coverts, fine gilded grass-green: the whole upper part of the neck, ultramarine blue, divided from the green of the back by a narrow black bar; the wings and tail light brown; belly and vent-feathers red; wings long in proportion to the bird; tail rounded at the tip.

CRESTED HUMMING-BIRD. (*Trochilus cristatus.*) This bird is a native of the West Indies: the bill is slender, sharp-pointed, incurvated, and blackish; the top of the head, from the bill to the hinder part, which terminates in a crest, is partly green and partly blue, and shines with a most brilliant metallic lustre: the plumage on the upper part of the body and wings is dark green intermixed with gold colour; the breast and belly are of a dingy grey; the tail is a bluish-black, glossy on the upper surface; and the legs and feet, which are very small, are blackish.

SAPPHIRE AND EMERALD HUMMING-BIRD. (*Trochilus bicolor.*) The two brilliant colours with which this bird is invested, not only merit the title of the gems by which they are called, but possess a vivid metallic splendour not exhibited by the gems themselves. The sapphire colour covers the head and throat, beyond which it blends with the lucid golden emerald colour of the breast, belly, and back: the wings are brown; the tail glossy bluish-black; and the belly white: the upper mandible is black, the lower whitish. Native of South America and the West Indies.

RUBY-THROATED HUMMING-BIRD. (*Trochilus colubris.*) This beautiful species is about three inches and a half in length from the tip of the bill to that of the tail: the bill is black; the crown, upper part of the neck, back, and coverts of the wings are of a most resplendent and variable green and gold colour; the chin and throat rival the ruby in brilliancy, changing, according to the light, either into a burnished gold colour, or a deep brown tint: the breast and belly are white; the wings and tail purplish-brown, but the two middle tail-feathers green. It is a native of, and continues in the southern parts of the American continent during the whole year, but appears in North America only in summer. It breeds in Florida, Carolina, and some of the West India islands; and is even seen in Canada. In that entertaining book, "The Canadian Naturalist," this bird and its habits are thus noticed in one of the "Conversations:"—"*C.* Ha! there is what I have long wished to see, a Humming-bird sucking the flowers. There are two of them: let us take a closer view of them.—*F.* No, no: stay where you are, and remain quite still, and talk in a low voice; for on the slightest alarm, and their brilliant little eyes are glancing in every direction, they shoot off with the straightness and speed of an arrow. See how they hover on the wing, in front of the blossoms, quite

N. AMERICAN HUMMING-BIRD (TROCHILUS COLIBRIS.)

stationary, while their long tongue is inserted, but their wings vibrating so rapidly as to be only visible as an indistinct cloud on each side.—*C.* One of them has suddenly vanished, but I did not see him fly, though I was watching him.—*F.* He has gone only about a yard: you may see him stationary again to the right of where he was before. These starts are so sudden and so rapid, that they are often lost to the sight.—*C.* How very little and how very beautiful! the body glitters in the sun with green and gold, and the throat is just like a glowing coal of fire. Now they rest on a twig; one of them I perceive has not the brilliant throat of the other.—*F.* That is the female; in other respects her plumage is like that of the male. It is the Ruby-throated Humming-bird (*Trochilus colibris*), and is scattered over the whole of this continent, at least to the latitude of 57 degrees north."

Least Humming-bird. (*Trochilus minimus.*) This is the smallest of the whole feathered tribe; being about an inch and a quarter in length, and weighing only about twenty grains: the general colour on the upper parts is green gold; the quill and tail feathers glossy violet-brown, and the exterior tail-feathers edged and tipped with white; the under parts of the body are of a dull white; and the legs and feet black. Native of several parts of South America, and of some of the West India islands.

The **Long-tailed Humming-bird** (*Trochilus polytmus*) is called by Mr. Gosse "the gem of American ornithology;" its slender form, velvet crest, emerald bosom, and lengthened tail-plumes, rendering it one of the most elegant even of this most brilliant family. The length of the male is ten inches and a quarter; wings expanded, six and three-eighths, and longest tail-feather seven inches and a half. Irides black; beak coral red, the tip black; feet purplish-brown, soles paler. Crown, hind head, and nape deep

LONG-TAILED HUMMING-BIRD: MALE.
(TROCHILUS POLYTMUS.)

velvety-black, very slightly glossed; back, rump, wing, and tail-coverts, rich golden green; wings purplish-black; tail deep black, with bluish gloss, the uropygials, and the outer edges of the others, glossed with golden green, varying in intensity. The tail is slightly forked, the feathers regularly graduating from the uropygials outwards, save that the outmost but one is exceedingly lengthened. Throat, breast, and belly, gorgeous emerald-green, extending to the thighs; vent and under tail-coverts, purpled black. The plumage of the hind head long and loose, descending in two lateral tufts upon the nape, which are to some extent erectile. The whole upper plumage of the female, from the hind head, is rich golden green; tail blue black, the exterior two feathers on each side broadly tipped with white: uropygials golden-green. Wings as in the male. Under parts white, the feathers having round tips of metallic green on the sides of the neck, and being mingled with green ones on the sides of the body.

"The Long-tail is a permanent resident in Jamaica, and is not uncommonly seen at all seasons and in all situations. It loves to frequent the margins of woods and roadsides, where it sucks the blossoms of the trees, occasionally descending, however, to the low shrubs. There is one locality where it is abundant, the summit of that range of mountains just behind Bluefields, and known as the Bluefields Ridge. Behind the peaks which are visible from the sea, at an elevation of about half a mile, there runs through the dense woods a narrow path, just passable for a horse, overrun with beautiful ferns of many graceful forms, and always damp and cool. No habitation occurs within several miles, and no cultivation, save the isolated provision grounds of the negroes, which are teeming with enormous arums, and these are hidden from view far up in the thick woods. The refreshing coolness of this road, its unbroken solitude, combined with the peculiarity and luxuriance of the vegetation, made it one of my favourite resorts. Not a tree, from the thickness of one's wrist up to the giant magnitude of the hoary figs and cotton trees, but is clothed with fantastic parasites; begonias with waxen flowers, and ferns with hirsate stems, climb up the trunks; enormous bromelias spring from the greater forks, and fringe the horizontal limbs; various orchideæ with matted roots and grotesque blossoms droop from every bough, and long lianes, like the cordage of a ship, depend from the loftiest branches, or stretch from tree to tree. Elegant tree-ferns and towering palms are numerous; here and there the wild plantain or heliconia waves its long flag-like leaves from amidst the humbler bushes, and in the most obscure corners over some decaying log, nods the noble spike of a magnificent limodorum. Nothing is flaunting or showy; all is solemn and subdued; but all is exquisitely beautiful. Now and then the ear is startled by the long-drawn measured notes, most richly sweet, of the Solitaire, itself mysteriously unseen, like the hymn of praise of an angel. It is so in keeping with the solitude, and with the scene, that we are unconsciously arrested to admire and listen. The smaller wood consists largely of the plant called Glass-eye berry, a Scrophularious shrub, the blossoms of which, though presenting little beauty in form or hue, are pre-eminently attractive to the Long-tailed Humming-bird. These bushes are at no part of the year out of blossom, the scarlet berries appearing at all seasons on the same stalk as the flowers. And here at any time one may with tolerable certainty calculate on finding these very lovely birds. But it is in March, April, and May that they abound: I suppose I have sometimes seen not fewer than a hundred come successively to rifle the blossoms within the space of half as many yards in the course of a forenoon. They are, however, in no respect gregarious; though three or four may be at one moment hovering round the blossoms of the same bush, there is no association; each is governed by his individual preference, and each attends to his own affairs."

"The Humming-birds in Jamaica do not confine themselves to any particular season for nidification. In almost every month of the year I have either found, or had brought to me, the nests of Polytmus in occupation. Still, as far as my experience goes, they are most numerous in June; while Mr. Hill considers January as the most normal period. It is not improbable that two broods are reared in a season. In the latter part of February, a friend showed me a nest of this species in a singular situation, but which I afterwards found to be quite in accordance with its usual habits: it was composed wholly of moss, and suspended to one of the fibres, not thicker than whipcord, belonging to the root of a tree, and contained two eggs. Mr. Gosse goes on describing, in his peculiarly pleasant manner, his further operations in endeavouring to become acquainted with every particular respecting the nidification and general habits of this interesting species. We select one example. "On the 12th of November, we took, in Bluefields morass, the nest of a Polytmus, containing two eggs, one of which had the chick considerably advanced, the other was freshly laid. The nest was placed on a hanging twig of a black-mangrove tree, the twig passing perpendicularly through the side, and out at the bottom. It is now before me. It is a very compact cup, one inch and three quarters deep without, and one inch deep within; the sides about a quarter of an inch thick, the inner margin a little over-arching, so as to narrow the opening: the total diameter at top, one inch and a half; one inch in the clear. It is mainly composed of silk cotton very closely pressed, mixed with the still more glossy cotton of an *asclepias*, particularly round the edge; the seed remaining attached to some of the filaments. On the outside the whole structure is quite covered with spider's web, crossed and recrossed in every direction, and made to adhere by some viscous substance, evidently applied after the web was placed, probably saliva. Little bits of pale green lichen, and fragments of thin laminated bark, are stuck here and there on the outside, by means of the webs having been passed over them. The eggs are long-oval, pure white, save that, when fresh, the contents produce a reddish tinge, from the thinness of the shell."

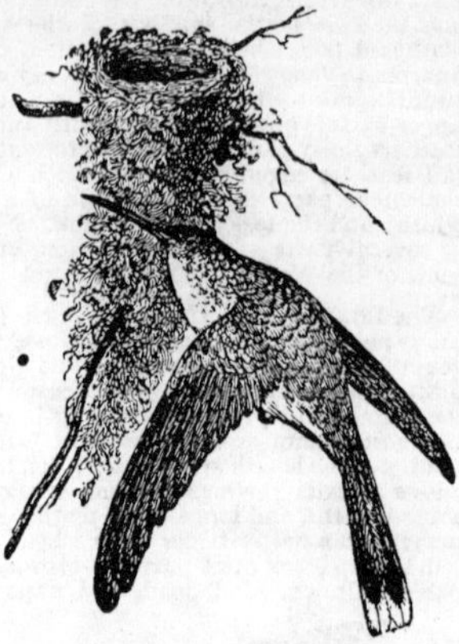

FEMALE LONG-TAILED HUMMING-BIRD, AND NEST.

"All the Humming-birds have more or less the habit when in flight of pausing in the air, and throwing the body and tail into rapid and odd contortions; this seems to be most the case with Mango, but perhaps is more observable in Polytmus from the effect that such motions have on the beautiful long feathers of the tail. That the object of these quick turns is the capture of insects I am sure, having watched one thus engaged pretty close to me; I drew up and observed it carefully, and distinctly saw the minute flies in the air, which it pursued and caught, and heard repeatedly the snapping of the beak. My presence scarcely disturbed it, if at all. * * * When I left England, I had laid myself out for the attempt to bring these radiant creatures alive to this country: and after a little acquaintance with the Jamaican species, Polytmus seemed, from its beauty, its abundance, its size, its docility, and its mountain habitat, to be the species at once most likely to succeed, and most worthy of the effort. My expectations were disappointed: yet as the efforts themselves made me more familiar with their habits, the reader, I trust, will pardon some prolixity of detail in the narration of these attempts."

[We have already so fully availed ourselves of Mr. Gosse's labours, that we beg to refer, for further information, to the work itself; and we take the opportunity of assuring him, at the same time, that his readers will be far more inclined to applaud than to condemn what he is pleased to call his "prolixity." We have personally derived both pleasure and instruction from its perusal; and we trust that many who consult our volume will be induced, from the extracts they have seen, to become possessed of "The Birds of Jamaica;" for a more delightful specimen of descriptive ornithology never came from the pen of a naturalist thoroughly imbued with his subject, and perfectly competent to impart his knowledge to the world in the most enticing form.] The works of M. Bourcier and Mr. Gould on Humming-birds are full of figures of these living jewels. In 1851, the visitors to the Zoological Gardens were charmed with Mr. Gould's display of so many gorgeous groupes of these birds.

HURON. (*Perca nigricans.*) An Acanthopterygian fish belonging to the family *Percidæ*, known to the English settlers on the borders of Lake Huron by the name of "Black Bass"—the word *bass* being almost synonymous with *perch*. It haunts deep holes at the mouths of rivers or edges of banks, and readily takes a hook baited with a small fish, or a piece of white rag trailed after a boat, as in fishing for mackerel. The flesh is firm, white, and well-flavoured; and it is, accordingly, in high estimation as an article of food.

HYÆNA. A well-known genus of digitigrade and carnivorous quadrupeds, distinguished by having no tuberculous teeth or small teeth behind the carnivorous, which, from their peculiar conformation, aided by the enormous strength of their jaws, are adapted for crushing the hardest substances. The skull of the Hyæna is short, and re-

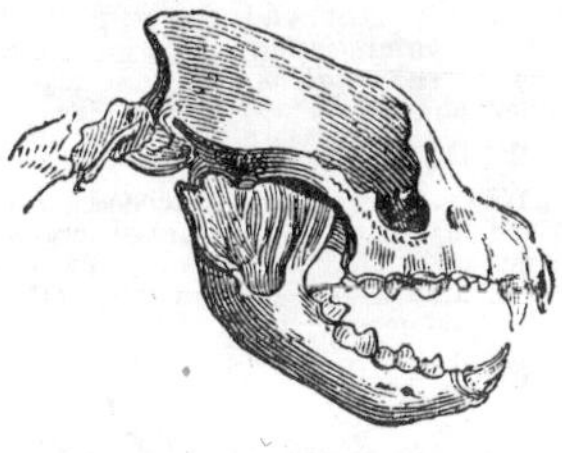

SKULL OF HYÆNA.

markable for its solidity; the muzzle also is short; and the temporal muscles, which raise the lower jaw, together with those of the neck, are very fully developed. The tongue is rough, the eyes are projecting, and the ears are large. The neck, chest, and shoulders are extremely powerful; but the hind-quarters are low, and the hind-legs seem comparatively feeble. It has four toes on each foot, furnished with blunt, stout, unretractile claws. Beneath the tail is a glandulous pouch, analogous to that of the Civets, but not secreting a similar odorous substance.

The common or STRIPED HYÆNA (*Hyæna vulgaris*) is a native of Asiatic Turkey, Syria, Abyssinia, &c. It is of a brownish gray colour, marked by several transverse dark brown bands on the body, which are more numerous as well as of a deeper colour on the legs: from the neck along the upper part of the back runs a strong bristly mane; the nose is black; the ears are rather long, sharp-pointed, and nearly naked; the tail is short rather than long, and very full of hair. Many absurd notions respecting the Hyæna were entertained by the ancients—its annual change of sex, its imitation of the human voice, its power of charming or fascinating shepherds, &c., subjects which at

STRIPED HYÆNA.—(HYÆNA VULGARIS.)

the present day scarcely deserve to be mentioned. Hyænas generally inhabit caverns and rocky places, prowling about at night to feed on the remains of dead animals, or whatever living prey they can seize; but they seldom attack man, except in self-defence. As carrion-feeders they seem destined to fill up an important station in the economy of nature, by cleansing the earth of the decaying carcasses of the larger beasts, whose remains might otherwise infect the atmosphere with pestilential effluvia. Though not gregarious from any social principle, they sometimes assemble in troops, and follow the movements of an army in order to feast on the bodies of those who perish on the field of battle: nay, it is asserted—nor is it inconsistent with their insatiable voracity and the peculiar strength of their claws—that they have been often known to tear newly-buried corpses out of their graves.

The aspect of the Hyæna seems to indicate a gloominess and malignity of disposition, with which its manners in a state of captivity appear in general to correspond: savageness and intractability mark its every look and movement; and it is said that its courage is equal to its rapacity. It was formerly supposed, and universally believed, that the Hyæna was untameable; but that it is possible, however difficult it may be, to tame it, there now exists not the shadow of a doubt. A remarkable peculiarity in this animal is, that when he is first obliged to run, he always appears lame for a considerable distance, and that, in some cases, to such a degree as to induce a belief that one of his legs is broken; but after running for some time, this halting disappears, and he proceeds on his course very swiftly. Mr. Bruce, the persevering and entertaining Abyssinian traveller, says, "I do not think there is any one that has hitherto written of this animal who ever saw the thousandth part of them that I have. They were a plague in Abyssinia in every situation, both in the city and in the field, and, I think, surpassed the sheep in number. Gondar was full of them from the time it turned dark till the dawn of day, seeking the different pieces of slaughtered carcasses which this cruel and unclean people expose in the streets without burial, and who firmly believe that these animals are Falasha from the neighbouring mountains, transformed by magic, and come down to eat human flesh in the dark in safety." "One night in Maitsha, being very intent on observation, I heard something pass behind me towards the bed, but upon looking round could perceive nothing. Having finished what I was then about, I went out of my tent, intending directly to return, which I immediately did, when I perceived large blue eyes glaring at me in the dark. I called upon my servant with a light, and there was the Hyæna standing nigh the head of the bed, with two or three large bunches of candles in his mouth. To have fired at him I was in danger of breaking my quadrant or other furniture, and he seemed, by keeping the candles steadily in his mouth, to wish for no other prey at that time. As his mouth was full, and he had no claws to tear with, I was not afraid of him, but with a pike struck him as near the heart as I could judge. It was not till then he showed any sign of fierceness; but, upon feeling his wound, he let drop the candles, and endeavoured to run up the shaft of the spear to arrive at me, so that, in self-defence, I was obliged to draw a pistol from my girdle and shoot him, and nearly at the same time my servant cleft his skull with a battle-axe. In a word, the Hyæna was the plague of our lives, the terror of our night-walks, the destruction of our mules and asses, which above all others are his favourite food."

The SPOTTED HYÆNA bears a considerable resemblance to the former species, but is marked with numerous round blackish-brown spots instead of stripes, and the mane is much less. Its habits are similar to the Striped Hyæna, and it commits equal ravages

SPOTTED HYÆNA.—(HYÆNA CROCUTA.)

amongst the cattle of the districts in which it resides. This species inhabits many parts of Africa, but is numerous round the Cape of Good Hope, where it is much dreaded. It rarely, however, moves abroad during the day, but passes that period in a state of repose, either in holes in the ground, or in retired situations densely covered with bush. Till lately, it seems, Hyænas were in the habit of paying nightly visits to the streets of Cape Town, and were regarded as very useful in carrying away the animal refuse; but partly from better regulations now existing in the town, and partly from the number of these animals having greatly decreased, this no longer occurs. In the interior of Southern Africa the ravages of this beast are still frightful, and it is no uncommon thing to find that they have entered the hut of a native in the night, and devoured or dreadfully mangled some of the younger branches of the family. And yet, notwithstanding this ferocity, the Spotted Hyæna is sometimes domiciliated in the houses of the peasantry, among whom, we are told, "he is preferred to the dog himself for attachment to his master, for general sagacity, and even, it is said, for his qualifications for the chase."

Another species (*Hyæna villosa*) is mentioned by Cuvier, as differing from the preceding by having stripes on the legs, the rest of the body being of a dark grayish-brown It inhabits the south of Africa, and is known there under the name of the *sea-shore wolf*.

HYÆNA-DOG. A name given to the Wild Dog of the settlers at the Cape. It is smaller and more slender than either the Hyæna or the Wolf. In its general osteological structure it agrees with the dogs, and it has no mane; but the head resembles the Hyæna's, and it has only four toes to each foot. Its colour is a reddish-brown, with patches of black and white intermixed: nose and muzzle black, with a strong black line passing from them up the centre of the forehead to between the ears. It is fierce, swift, and active; is very destructive to animals which are less fleet and powerful than itself; and commits great havoc on sheep. It is the *Hyæna venatica* of Dr. Burchell's "Travels," and the *Lycaon* of Mr. Gray.

HYÆNODON. [See SUPPLEMENT.]

HYALÆA. A genus of *Pteropoda*, distinguished by their wing-shaped organs of locomotion. They are found in the Atlantic and Mediterranean, and the shell is known in this country by the trivial name

HYALEA GLOBULOSA.

of Venus' Chariot. It is globose, glassy, and transparent, with a triangular opening at the upper part where the dorsal portion advances beyond the ventral, which is vaulted; dorsal more flat; lower extremity tridendate. The head of the animal is very indistinct, and it has no eyes.

HYALONEMA. A genus of siliceous sponges, by some considered a true zoophyte.

HYDATIDS. A term applied to nearly all kinds of cyst-like productions or cystic parasites invading the bodies of men and animals. They are now known to be only the young of higher forms of Entozoa.

HYDRA. The name given to a genus of minute polypi found in stagnant pools of water, where numbers are often seen clustering upon aquatic plants, &c. These animals present us with the simplest kind of structure which has yet been ascertained. The Hydra consists simply of a fleshy tube, open at both extremities, and the aperture of the tube serving as a mouth, which is situate in the more dilated end, and this mouth is provided at its margin with a single row of *tentacula*, or long flexible arms, which diverge from each other like the spokes of a wheel. Looking to this animal, we may suppose that nature has formed it to prove that animal life may be carried

HYDRA FUSCA.

on without the aid of the complicated machinery which she has given to the higher orders of creation. The *Hydra viridis*, or Green Polype, has the power of fixing itself in an erect position by the foot, and if it wishes to change place, it slowly bends till its head touches the plane on which it is moving, and adheres to it by the mouth, or one or two of its tentacula; the foot is then detached, and by a curve of the body placed close to the head, where it is again fixed, preparatory to a new step, which it performs by a repetition of the same movements. When in search of prey, the Hydra permits its arms to float loosely through the water; by which means it succeeds in obtaining a supply of food; for if, in their active course, any of the minute crustacea and aquatic worms should but touch one of the tentacula, it is immediately seized, other arms are soon coiled round it, and the unfortunate victim is speedily conveyed to the mouth.

With regard to the powers of reproduction possessed by these simple animals, it is to be observed that, when mature and well supplied with food, minute gemmules or buds are seen to become developed from the common substance of the body: these gemmæ appear at first like delicate gelatinous tubercles upon the exterior of the parent polype; but, as they increase in size, they gradually assume a similar form, and become perforated at their unattached extremity. During the first period of the formation of these sprouts, they are evidently continuous with the general substance from which they arise; but, at length, when the young is fully formed and ripe for independent existence, the point of union between the two becomes more and more slender, until a slight effort on the part of either is sufficient to detach them, and the process is completed.

But among the many remarkable features in the history of the Hydra, that which appears the most so, is its capability of reproducing the whole structure from separate portions of it. New tentacula will replace any which have been accidentally lost or removed. If the body is divided transversely, each segment will become a new animal; the upper one closing the aperture at its base, and the lower one speedily developing tentacula around the newly-formed mouth. If divided longitudinally, each half will, in a very short space of time, begin to ply its tentacula; nay, if cut transversely into several segments, each will in time become a perfect animal.—Trembly was the first man who gave a detailed account of this curious polype. [See POLYPI.]

HYDRACHNA. A genus of aquatic insects closley allied to the *Acaridæ* [which see]. One of the largest and by far the most elegant of the genus is the *Hydrachna geographica*, so named from the fancied map-like distribution of its variegations. Its shape is globular, and its colour a polished black, ornamented with red spots, which in a certain light have a kind of gilded lustre. The legs of this insect, as in the rest of the genus, are hairy; it swims with great celerity, and appears in almost constant motion. The eggs of the Hydrachna are small and round: the young, when first excluded, are furnished with six legs only; but they acquire two more legs after the first or second change of their skin.

HYDROBRANCHIATA. The first section of the order *Gasteropoda*, containing Mollusca which breathe water only.

HYDROCANTHARI, or WATER BEETLES. The name of a great group of Coleoptera, containing *Dytiscus*, *Colymbetes*, and numerous other genera.

HYDROCHŒRUS. [See CAPYBARA.]

HYDROMETRIDÆ. A family of insects, some species of which may be met with in almost every pond or stream, skimming along the surface, and turning about with the greatest rapidity. The body is boat-shaped, the hind feet serving as a rudder, while the two middle feet brush along the surface of the water, and give the required motion: the under side of the body is clothed with a thick coating of fine hairs, evidently intended to prevent the insect from coming in contact with the water.

HYDROPHILUS. A remarkable genus of aquatic insects, differing from that of Dytiscus only in the structure of the antennæ, which, instead of being setaceous, are short, and furnished with a clavated and perfoliated tip or knob. One large species, common in our ponds and ditches, is an inch and a half long, oval, and of a deep brown colour, highly polished. The eggs are laid in a sort of cocoon, spun by the female, and coated with a gummy matter that is impervious to the water on which it floats. The larvæ are observed to prey on the smaller kinds of water-snails, tadpoles, &c., and appear very voracious; and they remain about two years before they change into pupæ or chrysalides. When the larva is arrived at

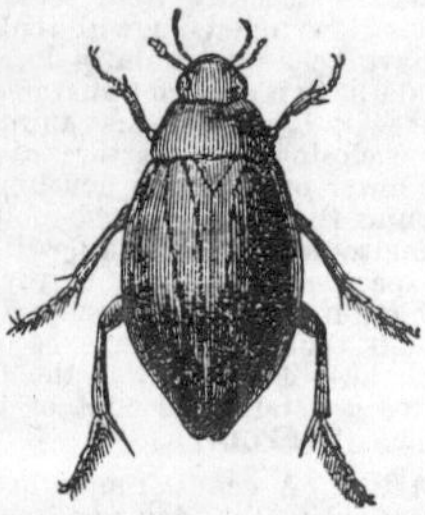

HYDROPHILUS PICEUS.

its full growth, it secretes itself in the bank of the water it inhabits, and having formed a convenient cell, lies dormant for some time; after which it divests itself of its skin, and appears in the form of a chrysalis; in this state it remains some time longer, when it again releases itself of its exuviæ, appears in its complete or beetle form, and as soon as the elytra or wing-cases acquire a sufficient degree of strength and colour; it comes forth from its retreat, and commits itself in its new form to its native element. It is a curious circumstance that some of the species of *Hydrophilidæ* found in this country exceed in size those from tropical climates; many of the species are, however, very minute. [See DYTISCUS.]

HYDROPHIS, or WATER-SNAKE. This genus of reptiles is very common in certain parts of the Indian Seas, where it feeds on fishes, and is considered excessively venomous. They have the back part of the body and tail very much compressed and

WATER SNAKE.—(HYDROPHIS.)

raised vertically, which, imparting to them the power of swimming, renders them aquatic animals. They have a range of scales a little broader than the rest under the belly; the head small, not bulged, obtuse, and covered with large plates. Several species are found in the salt water of Bengal, and others in the Indian Ocean.

HYDRUS. A species of small aquatic serpents, having the extremity of their tails enlarged, and very much compressed; which conformation gives them greater facility in moving through the water. They inhabit the intertropical parts of Asia, and the neighbouring islands, and in some situations are very abundant.

HYLA. A genus of Batrachian reptiles, known as *Tree Frogs*, and generically differing from the common Frogs in no respect, excepting that the extremity of each of their toes is widened and rounded into a sort of viscous palette, which enables them to adhere to the surfaces of bodies, and to climb trees, to which last they resort during the summer, in pursuit of insects; but they deposit their eggs in water, and penetrate into the mud in winter, like other Frogs.

The TREE FROG (*Hyla arborea*) exceeds all other European species in the beauty of its colours, the elegance of its form, and the agility of its movements; while its size is smaller than any of the tribe. It is a native of France, Germany, Italy, and many other countries of Europe, but is not found in the British islands. During the summer months its principal residence is on the upper parts of trees, where it wanders among the foliage in quest of insects, which it catches with extreme celerity, either stealing softly towards its prey, or springing upon it with a sudden leap; and it is often seen suspending itself by its feet to the under parts of the leaves, to enjoy their shade. Its colour on the upper parts is green, more or less bright: the abdomen is whitish, and marked by numerous granules: on each side of the body is a dark violet-coloured streak, tinged underneath with yellow, separating the green of the upper parts from the white colour of the lower. The body is short, plump, and smooth: the hind legs are very long and slender; the fore feet have four and the hind feet five toes, all of which are terminated by round, flat, and dilated tips, the under surface of which, being soft and glutinous, enables the animal to hang with perfect security from the leaves of trees, &c.; it can also adhere to any substance by its abdomen, which is covered with small glandular granules, by merely pressing itself against it. Though the Tree Frog inhabits the woods during the summer months, yet on the approach of winter it retires to the waters, and there submerging itself in the soft mud, or concealing itself beneath the banks, remains in a state of torpidity, and again emerges in the spring, at which period it deposits its spawn in the water, like the rest, in small clustered masses. The male at this period inflates its throat in a surprising manner, and croaks in so loud and sharp a key as to be heard at an immense distance. During their residence among the trees they are observed to be particularly noisy on the approach of rain.

HYLÆOSAURUS. [See SUPPLEMENT.]

HYLOBATES. [See APE, Long-armed.]

HYMENOPTERA. An order of insects having four naked membranous wings, as seen in Bees, Wasps, Ants, Ichneumons, &c. The anterior wings are usually much larger than the posterior; and the nervures or hard framework on which the membrane of the wing is extended are but few. The mouth is furnished with mandibles and maxillæ, and the abdomen is terminated either by an ovipositor or a sting. The larvæ of some resemble those of the order Lepidoptera, but differ in the number of their legs, &c. Most hymenopterous insects, when in the perfect state, are constantly resorting to flowers, either for the purpose of gathering honey, or of preying upon the less powerful species of their own class. Hymenopterous insects love the light of the sun; they take wing only during the day-time, remaining at rest in the night, and in dull and wet weather; and it must be admitted that they excel all other insects in the number and variety of their instincts, which are wonderfully displayed in the methods employed by them in providing not merely for their own welfare, but for the comfort and future wants of their offspring.

In the adult state these insects live chiefly on the honey and pollen of flowers, and the juices of fruits. The larvæ of the Saw-flies (*Tenthredinidæ*), under the form of false caterpillars and slugs, are leaf-eaters, and are oftentimes productive of much injury to plants. The larvæ of the *Xiphydriadæ*, and of the Horn-tails (*Uroceridæ*), are borers and wood-eaters, and consequently injurious to the plants inhabited by them. Pines and firs suffer most from their attacks. Some of the warty excrescences on the leaves and stems of plants, such as oak-apples, gall-nuts, and the like, arise from the punctures of four-winged gall-flies (*Diplolepididæ*), and the irritation produced by their larvæ, which reside in these swellings. The injury caused by them is, comparatively, of very little importance, while, on the other hand, we are greatly indebted to these insects for the gall-nuts that are extensively used in colouring, and in medicine, and form the chief ingredient in ink. We may, therefore, write down these insects among the benefactors of the human race. Immense numbers of caterpillars and other noxious insects are preyed upon by internal enemies, the larvæ of the ichneumon-flies (*Evaniadæ*, *Ichneumonidæ*, and *Chalcididæ*), which live upon the fat of their victims, and finally destroy them. Some of these ichneumon-flies (*Ichneumones ovulorum*) are extremely small, and confine their attacks to the eggs of other insects, which they puncture, and the little creatures produced from the latter find a sufficient quantity of food to supply all their wants within the larger eggs they occupy. The ruby-tails (*Chrysididæ*), and the cuckoo-bees (*Hylæus*, *Sphecodes*, *Nomada*, *Melecta*, *Epeolus*, *Cœlioxys*, and *Stelis*), lay their eggs in the provisioned nests of other insects, whose young are robbed of their food by the earlier hatched intruders, and are consequently starved to death. The wood-wasps (*Crabronidæ*), and numerous kinds of sand-wasps (*Larradæ*, *Bembicidæ*, *Sphegidæ*, *Pompilidæ*, and *Scoliadæ*), mud-wasps (*Pelopeus*), the stinging velvet-ants (*Mutilladæ*), and the solitary wasps (*Odynerus* and *Eumenes*), are predaceous in their habits, and provision their nests with other insects, which serve for food to their young. The food of ants consists of animal and vegetable juices; and though these industrious little animals sometimes prove troublesome by their fondness for sweets, yet, as they seize and destroy many insects also, their occasional trespasses may well be forgiven. Even the proverbially irritable paper-making wasps and hornets (*Polistes* and *Vespa*) are not without their use in the economy of nature; for they feed their tender offspring not only with vegetable juices, but with the soft parts of other insects, great numbers of which they seize and destroy for this purpose. The solitary and social bees (*Andrenadæ* and *Apidæ*) live wholly on the honey and pollen of flowers, and feed their young with a mixture of the same, called bee-bread. Various kinds of bees are domesticated for the sake of their stores of wax and honey, and are thus made to contribute directly to the comfort and convenience of man, in return for the care and attention afforded them. Honey and wax are also obtained from several species of wild bees (*Melipona*, *Trigona*, and *Tetragona*), essentially different from the domesticated kinds. While bees and other hymenopterous insects seek only the gratification of their own inclinations, in their frequent visits to flowers, they carry on their bodies the yellow dust or pollen from one blossom to another, and scatter it over the parts prepared to receive and be fertilized by it, whereby they render an important service to vegetation.

HYRAX. A curious genus of small rabbit-like animals, inhabiting rocky and mountainous districts in Africa and Syria. The best known species are the Cape Hyrax,

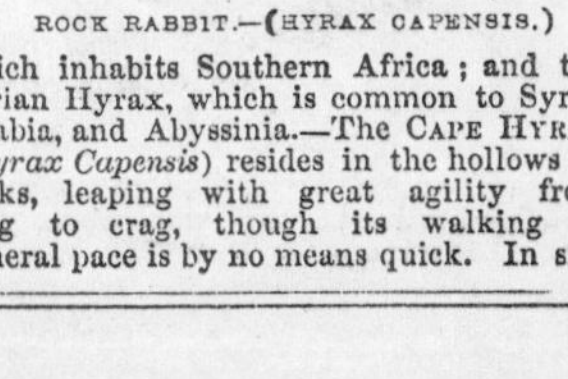

ROCK RABBIT.—(HYRAX CAPENSIS.)

which inhabits Southern Africa; and the Syrian Hyrax, which is common to Syria, Arabia, and Abyssinia.—The CAPE HYRAX (*Hyrax Capensis*) resides in the hollows of rocks, leaping with great agility from crag to crag, though its walking or general pace is by no means quick. In size and colour it greatly resembles the rabbits: it is of a thick form, with short limbs, the hinder being the longest, and it is destitute of a tail. The head is rather small; the ears short and rounded; the eyes large and black; the fore feet have each four soft pulpy toes, with flattish, rounded nails; the hind feet have only three, the inner one of

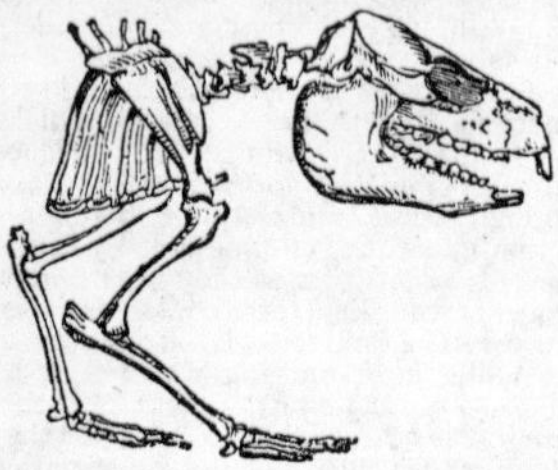

SKULL AND PART OF SKELETON OF HYRAX.

which is furnished with a sharp crooked claw. Both this and the Syrian Hyrax live in families, and take up their abode in caves or crevices in the sides of rocks. They subsist on grain, fruit, roots, the young shoots of shrubs, herbs, and grass: they are easily tamed, and are lively, active, docile, and cleanly when domesticated. Although the external appearance and the habits of the Hyrax appear to point it out as being a rodent quadruped, Cuvier says that its osteological structure shows it to belong to the *Pachydermata*, and that, notwithstanding the smallness of its proportions, it must be regarded as intermediate between the Rhinoceros and the Tapir. The Syrian species is doubtless "the Coney" of the Scriptures.

HYSTRIX. [See PORCUPINE.]

IBEX. A quadruped of the Goat kind, several distinct species of which are said to exist among the mountain ranges of Europe, Asia, and Africa, most of them resembling

IBEX.

each other in structure and habits. Those best known are the IBEX CAPRA, and the IBEX ÆGAGRUS, or Caucasian Ibex: they are each much larger and stronger than the common domestic Goat; and to the one or the other of these, that animal is believed to owe its origin. The Ibex Capra inhabits the Carpathian and Pyrenean mountains, various parts of the Alps, &c. Its colour is a deep hoary brown; the under parts of the body and insides of the limbs being of a much paler or whitish hue: its body is thick, short, and strong; it has a small head; large eyes; strong legs; very short hoofs; and a short tail. The horns, which are extremely large and long, and of a deep brown colour, are marked on the upper surface with protuberant transverse knots or half circles: the hair is harsh; and the male is furnished with a beard. The female is smaller than the male, with smaller horns in proportion, much less boldly knotted. These animals usually resort to the most precipitous heights of lofty mountains, where they assemble in small flocks, sometimes consisting of ten or fifteen individuals. They are remarkably swift, and display amazing agility and dexterity in leaping. They are objects of the chase, but the danger attendant on the pursuit of them is great indeed; for not only are strength, address, and activity necessary to the hunter when following the Ibex from one precipice to another, or in tracking him among difficult passes; but, when close pressed, he will sometimes turn on his pursuer with impetuous rapidity, and hurl him down the most frightful declivity. The fore legs being considerably shorter than the hinder, enables these animals to ascend with more facility than to descend, and hence, when pursued, they always attempt to gain the summits of the mountains. The season for hunting them is during August and September, when they are usually in good condition. The voice of the Ibex is a sharp, short whistle, not unlike that of the chamois, but of shorter duration; sometimes, and especially when irritated, they make a snorting noise. The female has seldom more than one young one at a time; to this she pays great attention, defending it with courage and obstinacy.

The CAUCASIAN IBEX (*Ibex Ægagrus*) is considerably larger than the Common Goat, and in form bears considerable resemblance to the animals of the cervine genus. It inhabits the loftiest rocky points about Mount Caucasus. Its general colour is a brownish-gray above, and white beneath: the forehead is nearly black: and a black stripe is continued down the back; the horns, which are very large, and bend far backwards, are smooth, black, sharply ridged near the top, and hollowed on their exterior side, but have no appearance of either knots or rings; they are about three feet long, close at the base, about a foot apart in the middle, and eight or nine inches at the tips. The male has a large brownish beard: the female has neither horns nor beard.

One of the handsomest of these animals is the JEMLAH IBEX, an inhabitant of the Himalaya Mountains. Its head is finely formed, full of beauty and expression: it has no beard; and its horns are remarkably massive at the base. It lives solitarily or in small herds; and though bold and pugnacious, it is easily tamed.

IBIS. A genus of birds which in their general habits and conformation closely approach the Storks: they chiefly inhabit warm countries, but, except in very cold regions,

SACRED IBIS.—(IBIS RELIGIOSA.)

they are to be found in all parts of the world. Generic characters:—beak arched, long, slender, thick at the base, and quandrangular, rounded at the tip, which is obtuse; nostrils linear, extending from the root to the tip of the beak, and dividing it into three portions, of which the upper is the broadest, and flattened; head and throat bare; legs long, and four-toed, the front webbed at their base as far as the first joint, the hind toe very long, all provided with claws. They frequent the borders of rivers and lakes, feeding on insects, worms, mollusca, and occasionally on vegetable matter. They perform powerful and elevated flights, extending their neck and legs, and uttering a hoarse croak.

THE GLOSSY IBIS (*Ibis falcinellus*) is nearly two feet in length. In the adult bird, the neck, breast, top of the back, and all the inferior parts of the body, are of a bright red chestnut; the wing-coverts, quills, tail-feathers, and the rest of the back, of a dusky green, glossed with bronze and purple; but it varies much in its plumage at different ages. This species builds in Asia, and is found on the streams and lakes, in flocks of thirty or forty. They migrate perodically to Egypt; and in their passage they are numerous in Poland, Hungary, Turkey, and the Grecian Archipelago. They occasionally visit the banks of the Danube, Switzerland, and, more rarely, England and Holland.

The WHITE IBIS (*Ibis religiosa*) arrives in Egypt about the time that the inundation of the Nile commences, its numbers increasing or diminishing with the increase or diminution of the waters: and it migrates about the end of June, at which time it is first noticed in Ethiopia. This species does not collect in large flights, more than eight or ten seldom being seen together. They

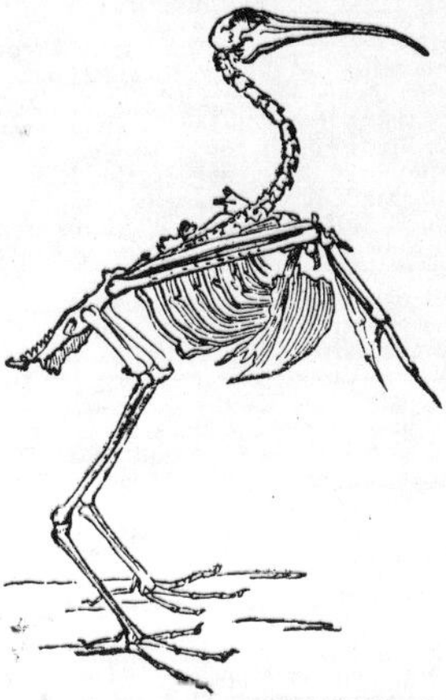

SKELETON OF WHITE IBIS.

are about the size of a fowl; the head and neck bare; the body white; the primaries of the wings tipped with shining, ashy black, among which the white forms oblique notches; the secondaries bright black, glossed with green and violet; the quill feathers of the tail white. This and the above described were the two species of birds adored by the ancient Egyptians, who used to rear them in their temples, and after death embalm them. Their mummies are found to this day in numbers, in the vast catacombs of ancient Memphis.

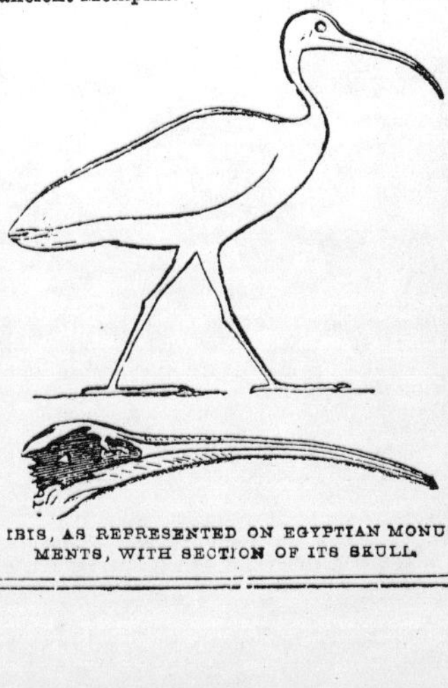

IBIS, AS REPRESENTED ON EGYPTIAN MONUMENTS, WITH SECTION OF ITS SKULL.

"This," says Cuvier, "is the most celebrated species: it was reared in the temples of ancient Egypt, with veneration which approached to worship; and it was embalmed after its death, as some said, because it devoured the serpents which would otherwise have become dangerous to the country: according to others, because there was a resemblance between its plumage and some of the phases of the moon: finally, according to other some, because its advent announced the rising of the Nile. For a long time it was thought that this Ibis of the Egyptians was the Tantalus of Africa: we now know it belongs to the genus of which we are treating. It is as large as a hen, with white plumage, except the end of the wing-feathers, which is black; the last coverts have their barbs elongated, loose, black, with violet reflections, and thus covering the end of the wings and tail. The bill and the feet are black, as well as all the naked part of the head and neck: this part is covered in youth, at least on its upper surface, with small blackish feathers. The species is found throughout the extent of Africa."

The SCARLET IBIS (*Ibis rubra*) is a very splendid bird, and is found in the hottest parts of America in large flocks. They fly rapidly, but rarely, except at morning and evening, in search of food. The plumage is scarlet; beak naked; part of the cheeks, legs, and feet, pale red. Before the Scarlet Ibis reaches its full age, its plumage varies considerably.—Other species are found in India, Madagascar, the Cape of Good Hope, and Mexico.

ICHNEUMON. (*Herpestes.*) An animal bearing a very close resemblance to the weasel tribe both in form and habits. From the snout to the root of the tail it is about eighteen inches long: it has a long, agile body, short limbs, semi-plantigrade feet, small glowing eyes, and a pointed nose. It glides towards its prey with a snake-like movement, and then darts suddenly upon it. These animals feed upon birds, reptiles,

ICHNEUMON.—(HERPESTES ICHNEUMON.)

rats, mice, &c. Their disposition is as sanguinary as their habits are predatory; but though the destruction they cause among the poultry is very annoying, it is well compensated by the incessant war they wage against reptiles, the eggs of which they devour with the greatest avidity. The most celebrated species inhabits Egypt and the adjacent countries, where it is called "Pharaoh's rat." It is larger than a cat, but formed like a weasel; it is of a gray colour, and has a long tail, terminated by a black tuft. This species was ranked by the ancient Egyptians among their numerous divinities on account, it is supposed, of the benefits it confers on man by the destruction of crocodiles, whose eggs it digs out of the sand, and sucks. It is also a natural enemy of the whole serpent race, and so exceedingly expert in seizing them by the neck, as to avoid any injury to itself. The Ichneumon is easily domesticated, seeming to form an attachment to its place of residence; and it is not unfrequently kept tame both in India and Egypt, for the purpose of clearing the houses of mice and rats. Ichneumons are sometimes seen to squat on their haunches, and feed themselves with their fore paws, like the squirrel. When they sleep, they bring their head and tail under their belly, and appear like a round ball. In a wild state they generally reside along the banks of rivers; and they swim and dive like the otter, being able to continue under water for a great length of time.

ICHNEUMONIDÆ, or ICHNEUMON-FLIES. A family of hymenopterous insects, the genera and species of which are very numerous, and their manners extremely diversified, but all agreeing in this characteristic — that they deposit their eggs in the bodies of other living insects, and generally in those of caterpillars. The

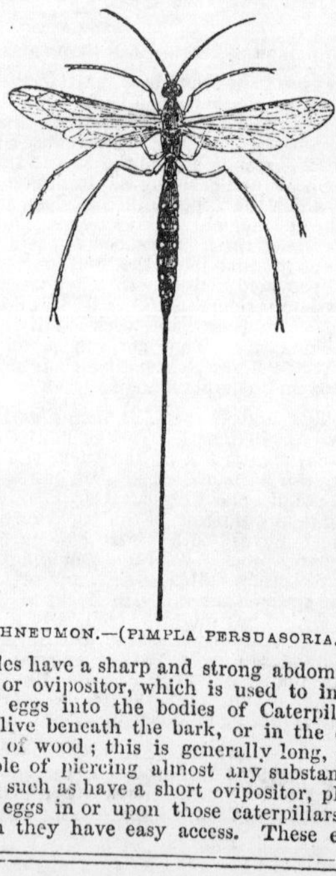

ICHNEUMON.—(PIMPLA PERSUASORIA.)

females have a sharp and strong abdominal tube, or ovipositor, which is used to insert their eggs into the bodies of Caterpillars that live beneath the bark, or in the crevices of wood; this is generally long, and capable of piercing almost any substance; while such as have a short ovipositor, place their eggs in or upon those caterpillars to which they have easy access. These eggs are in a few days hatched, and the young larvæ, which resemble minute white maggots, subsist on the juices of their victim, but without absolutely destroying it: in fact, the animal they infest may continue to exist for some time, thus affording them a continued supply of nutriment; but when the Ichneumons are ready to undergo their last metamorphosis, they pierce the skin, and each spinning itself up in a small oval silken case, changes into chrysalis, the whole number forming a group on the shrivelled body of the caterpillar; and, after a certain period, they emerge in the state of complete Ichneumons. One of the most familiar examples of this process is afforded by the caterpillar of the common white or cabbage butterfly, which in autumn may be frequently observed to creep up some wall or other convenient surface, in order to undergo its own change into chrysalis; but in the space of a day or two a numerous tribe of small maggots will be seen to emerge from it, and immediately proceed to envelope themselves in distinct yellow silken cases; the whole forming a group around the caterpillar. The perfect Ichneumons feed solely upon the juices of flowers, and fly about with considerable agility in search of their food, or of proper situations for the deposition of their eggs.

These carnivorous insects are of various sizes; some are so small, that the Aphis, or plant-louse, serves as a cradle for their young; others again, from their size and strength, are formidable even to the spider, destroying them with their powerful stings: some place their eggs within the aurelia of a nascent insect; others deposit them within the nest, which the wasp has curiously contrived for her young; and, as both are produced at the same time, the offspring of the Ichneumon not only devour the young wasps, but the whole supply of larvæ which the parent had carefully provided for their support. The best known, and perhaps the most formidable of this genus, is the common Ichneumon with four wings like the bee; a long slender black body; and a three-forked tail, consisting of bristles, of which the two exterior are black, and the central one is red. But when we read that "probably more than 3000 species exist in Europe alone, and the number peculiar to other parts of the globe may fairly be reckoned as at least equal," it would seem to be of little use to attempt to give more than this general description: we shall therefore conclude by remarking, that however terrible to other insect tribes the Ichneumon-flies may be, their destruction of countless myriads, which would otherwise be left to banquet on the fruits of the earth, must be of the most essential service to mankind.

ICHTHYIAËTUS. A sub-genus of the Falcon family of birds, so named from their living principally on fish. From the account given in Mr. Gould's work, some members of the genus would seem to partake of the habits of the vulture family: among these is the

ICHTHYIAËTUS LEUCOGASTER, or WHITE-BELLIED SEA EAGLE. This is a fearless and familiar bird, found throughout the whole of South Australia. It is distinguished by its never plunging beneath the surface of the water, but living on dead cetacea, fish, &c., left on the shore by the tide. In Bass's Straits it subsists principally on Petrels and Penguins, which are easily captured. On the main land it builds a large flat nest on a fork of the loftiest trees, on islands, on the flat surface of a large stone, and sometimes on the twigs and branches of barilla, a low shrub. One nest was observed on a tree 200 feet high and 41 feet round, where it probably had its nest for several years.

ICHTHYOSAURUS, or FISH-LIZARD. A genus of extinct marine animals which combined the characters of saurian reptiles and fishes, with some peculiar to cetaceous mammalia. This extraordinary creature, whose fossil remains discover its anatomical conformation, has been the subject of much learned investigation; and the anatomy and animal economy of it are in a manner established. Some of the largest exceeded thirty feet in length; and, from their structure, it is easy to conceive that they must have been very formidable enemies to the other inhabitants of the deep. The spinal column was formed like that of a fish, the vertebræ

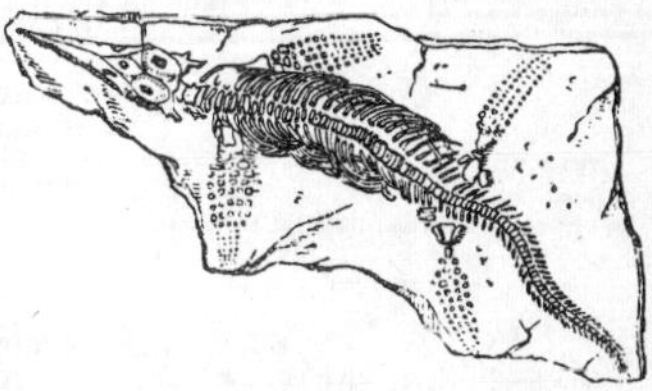

ICHTHYOSAURUS COMMUNIS.

being concave on both surfaces, and the arches which enclosed the spinal chord always remained distinct from the bodies as in reptiles; hence the body must have had great flexibility; but the progression of these animals through the water was chiefly by means of the anterior and posterior extremities (of which there were four), formed very much upon the plan of the feet or paddles of the whale. The general form of the head was not unlike that of the porpoise; and it had an elongated and pointed muzzle, the mouth being armed with numerous crocodile-like teeth; and its body terminated in a long and powerful tail. From the absence of any remains of scales or plates, it may be concluded that the skin was naked, like that of the whales and their allies; and that it was an air-breathing animal, coming to the surface of the water occasionally, no doubt exists. From the remains of crushed and partially-digested fish bones and scales which are found with their bones, it appears that the Ichthyosauri principally preyed upon fishes. Dr. Buckland states that the fossil remains of these animals abound along the whole extent of the lias formation, from the coasts of Dorset, through Somerset and Leicestershire, to the coast of Yorkshire. The lias of Germany and France also contains them.

Mr. Pearce found, within a specimen of the Ichthyosaurus, what he thinks may have been an embryo; and although the Ichthyosaurus, by analogy, might have been supposed to be oviparous in its generation, yet Dr. Buckland and Professor Owen think there is no reason why it should not have been viviparous; and from the evidence of Mr. Pearce's specimen it appears fair to suppose that they really were so. The collection of remains of Ichthyosauri in the British Museum is very perfect and great; and with the allied Plesiosauri forms quite a feature in the room devoted to their exhibition.

ICTERIA, or CHATTERING FLY-CATCHER. (*Icteria viridis.*) This bird abounds in most parts of North America during the summer months, and is remarkable both for its colouring and its habits. It is about six inches in length: nearly the whole of the upper parts of its body are of a rich deep olive green, the tips of the wings excepted, which are of a dusky brown: the throat, breast, and sides of the body are of a bright yellow; the belly and vent white; the forehead pale ash; from the nostrils a line of white extends to the upper part of the eyes, which it nearly surrounds; another white spot is situated at the base of the under mandible: beak strong and black; legs and feet pale blue. It has the faculty of mimicking almost any noise that it hears, which it will repeat during the whole night if the weather be fine. Its favourite residence is in close hazel or bramble thickets, and its nest is composed of dry leaves with layers of grape vine bark, lined with fibrous roots and dry grass. The female lays four flesh-coloured eggs, sprinkled with brown and dull red spots.

IGUANA. A genus of Reptiles of which there are several species; the genus including several of large dimensions, common in the tropical parts of America, some of which feed on vegetable substances, and are esteemed delicious food; while there are others which appear to be omnivorous. They are thus characterized by Cuvier:

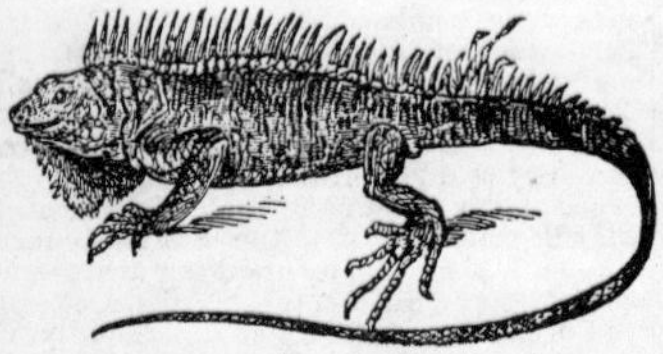

IGUANA.—(I. TUBERCULATA.)

body and tail covered with small imbricated scales: the ridge of the back garnished with a row of spines, or rather of elevated, compressed, and pointed scales; under the throat a depressed and depending dewlap, the edge of which is attached to a cartilaginous appendage of the hyoid bone. Their thighs are provided with a similar arrangement of porous tubercles with the true lizards, and their head is covered with scaly plates. Each jaw is furnished with a row of compressed triangular teeth, having their cutting edges serrated: there are also two small rows on the posterior part of the palate. They live for the most part on trees, but sometimes take to the water, and swim with ease. They attain a great size, being sometimes found five feet in length, though they are much more often from two to three: their upper parts are bluish-green, and sometimes slate colour; the under parts yellowish-green; in general, on the sides of the body are brown stripes or zigzags edged with yellow; and the tail is surrounded with large brown or yellow rings. The Common Iguanas (*Iguana tuberculata*) are eagerly sought, especially in the spring, being esteemed a great delicacy. They are caught by means of a noose attached to the end of a stick: for the animal, though formidable in appearance, is timid and defenceless. It is very active; but when it has taken refuge in a tree, it appears to depend on the security of its situation, and never offers to stir unless roused; hence it is easily taken. The female deposits her eggs, which are about the size of a pigeon's egg, in the sand, where they are left to be hatched by the genial warmth of the sun.

IGUANIDÆ. The family of Saurians, or lizard-like reptiles, of which the animal just described is the type, and of which there are several sub-genera: descriptions of which are given in Mr. Gray's elaborate work, the Catalogue of Lizards in the British Museum. [See LIZARD.]

IGUANODON. The name given to an extinct gigantic reptile, closely resembling the Iguana in osseous structure, whose remains were discovered by Dr. Mantell in the wealden formation of the South of England, in the localities of Purbeck, the Isle of Wight, and Maidstone. From its dentition there seems to be no doubt that it was herbivorous; the form of the teeth, considered with relation to the demands made by the habits of the animal, being well adapted for cropping tough vegetable food, such as the *Clathraria* and similar plants which are found buried with the Iguanodon. From the proportions which the bones of the Iguanodon bear to those of the Iguana, this extinct monster is calculated to have been 70 feet in length from the snout to the end of the tail; the length of the tail alone 52½ feet, and the circumference of the body 14½ feet. The thigh bone of the full-sized Iguanodon is twenty times the size of that of the Iguana; and on the snout of this prodigious reptile was a short but strong horn: its whole appearance, indeed, must have realized the wildest poetical fictions of the dragons of old. In the British Museum are contained all the specimens of Iguanodon obtained by Dr. Mantell; and a comparison of the teeth and bones, with those of its recent comparatively Lilliputian analogue, is a most interesting and curious study.

INCA. A genus of Lamellicorn Beetles, by many authors placed among the Goliath beetles, but whose situation in the system, according to more modern views, is nearer *Trichius*. They are natives of South America. The species figured here is *Inca Weberi*. It is of a violet black; the thorax edged with white; three-banded, the outer bands connected with the white edge of the thorax: the elytra have a reddish tinge, spotted with

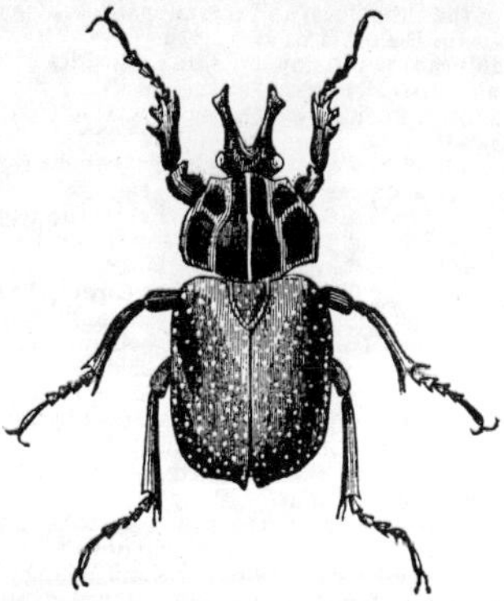

WEBER'S INCA BEETLE.—(INCA WEBERI.)

small palish marks. It is a native of South America: and the accompanying figure will show its form and appearance. [See GOLIATH.]

INDICATOR. [See HONEY-GUIDE.]

INDRI. The name of a quadruped belonging to the family *Lemuridæ*. It is a native of Madagascar, and from its fine long hair is called *Indris laniger*.

INFEROBRANCHIATA. An order of molluscous animals (Gasteropods), characterized by the position of the gills, which are situated beneath the produced margin of the mantle. They are incapable of swimming, and are therefore confined to the sea-shore, where they subsist upon sea-weeds and other aquatic plants.

INFUSORIA. A term applied by naturalists to the numerous minute animals found in water, which are commonly called animalcules. Had the microscope never been invented, the existence of myriads of living creatures whose forms and properties are now in some measure revealed to us, would have been wholly unknown. Ehrenberg, who by means of a most powerful microscope, was enabled to describe species which are not larger than from one-thousandth to two-thousandth of a line in diameter, infers, that a single drop of water may hold 500 millions of these animalculæ. By what arithmetical power, then, shall the numbers that swarm in every stagnant pool or lake be calculated? "All true Infusoria," says he, "even the smallest monads, are organized animal bodies (none consisting of a homogeneous jelly), and distinctly provided with at least a mouth and internal nutritive apparatus." They are found equally abundant as fossils. The Norwegian earth, called *Beargmehl*, or Mountain meal, is principally composed of fossil animalcules. Professor Bailey tells us that the town of Charleston, in the United States, is built upon a bed of animalculæ several hundred feet in thickness, every cubic inch of which is filled with myriads of perfectly preserved microscopic shells. He says also, that these *polythalamia*, or many-chambered shells, to whose labour South Carolina owes so large a portion of her territory, are still at work, in countless thousands, upon her coasts, filling up harbours, forming shoals, and depositing their shells to record the present state of the sea-shore, as their predecessors, now entombed beneath Charleston, have done with regard to ancient oceans. The most highly organized Infusoria are called by Cuvier *Rotifera* [which see].

The immense importance of the Infusoria in the scale of animal existence is chiefly seen by those who visit the Arctic and Antarctic seas. Although remotely supporting the higher animals, yet the want of them would be materially felt. This is well stated by Capt. Sir James Clark Ross, who, in speaking of a small fish found by him in the South Seas, and described by Dr. Sir John Richardson, under the name of *Notothenia phocæ*, says, "They occupy the place of the *Merlangus polaris* and *Ophidium Parryii*, of the Arctic seas, the latter of which they much resemble; like them, they conceal themselves from the persecutions of their enemies in the small cracks and cavities of the pack ice, and may be seen when driven from shelter by the ships striking and passing over their protecting pieces of ice. The seals and petrels are their chief enemies, whilst they, in their turn, live upon the smaller *Cancri* and *Limacinæ*. Thus we behold in these regions, where the vegetable kingdom, which constitutes the support of animal life in milder climates, has no representative, a chain of animal existences, maintained by each preying upon that next below it in the order of created beings, and all eventually nourished and sustained by the minute infusorial animalculæ which we found filling the ocean with an inconceivable multitude of the minutest forms of organic life."—*Antarctic Voyage*, vol. ii. p. 161.

INSECTS. (*Insecta.*) A class of invertebrate animals, to which the term *insecta* has been applied, in reference to the insected, or divided, appearance of the body, which is not only composed of a continuous series of segments, articulating with each other, but is also often divided or cut into three very marked portions, to which the names *head*, *thorax*, and *abdomen* have been applied. There is no class of the animal kingdom which has been the subject of more numerous and various attempts at classification than that of Insects: nor is it at all surprising; since it is pre-eminent in regard to the number of distinct species which it includes, and unsurpassed by any, save the

Infusory Animalcules, in regard to the number of individuals at any time existing on the earth's surface, which belong to the numerous and diversified races comprehended in it. In ordinary phraseology, an Insect may be defined as a little animal without bones or cartilages; furnished with a trunk, or else a mouth opening lengthwise; and with eyes destitute of coverings. This definition will comprehend the whole class of Insects, either with or without wings; either in their caterpillar or perfect state. Every year adds to the difficulty of defining in a brief manner the characters of the great and smaller divisions of the animal kingdom. Hence it will appear, that in this class of animals there are numerous distinctions, and that no general description will serve for all: so various are the appetites, manners, and modes of propagation, that every species requires its distinct history. Though so far inferior in point of magnitude, Insects, it must he confessed, surpass in variety of structure and singularity of appearance all the larger branches of the animal world. The general characters by which they are distinguished from other animals are these:—First, they are furnished with several feet: secondly, the muscles are affixed to the internal surface of the skin, which, though hard, sometimes preserves a certain degree of flexibility: thirdly, they breathe, not like the generality of larger animals, by lungs or gills, but by spiracles or breathing holes, distributed in a series or row on each side the whole length of the abdomen, and communicating with two long air-pipes within their bodies, and a number of smaller ones, to carry the air to every part. The head is furnished with a pair of *antennæ*, or horns, which are extremely various in the different tribes, and which, by their differences of structure, form a leading character in the institution of the genera into which Insects are distributed.

Insects have a very small brain, and instead of a spinal marrow, a kind of knotted cord, extending from the brain to the hinder extremity; and numerous small whitish threads, which are the nerves, spread from the brain and knots, in various directions. The heart is a long tube, lying under the skin of the back, having little holes on each side for the admission of the juices of the body, which are prevented from escaping again by valves or clappers, formed to close the holes within. Moreover, this tubular heart is divided into several chambers, by transverse partitions, in each of which there is a hole shut by a valve, which allows the blood to flow only from the hinder to the fore part of the heart, and prevents it from passing in the contrary direction.

The ancients entertained an idea that Insects were destitute of blood; hence they called them *animalia exsanguinea:* but now they are well known to be so far from bloodless animals, that in many of them the circulation itself of the blood is clearly and distinctly perceived. The blood of Insects differs from that of the larger animals chiefly in colour, since in most insects it wants redness, being generally of a clear or watery aspect, and sometimes of a yellowish hue. The circulation of the blood is particularly conspicuous in Spiders, and in some species of Cimex or bug, especially the *Cimex lectularius;* it is to be observed, however, that it does not circulate in proper arteries and veins; but is driven from the fore part of the heart into the head, and thence escapes into the body, where it is mingled with the nutritive juices that filter through the sides of the intestines, and the mingled fluid penetrates the crevices among the flesh and other internal parts, flowing along the sides of the air-pipes, whereby it receives from the air that influence which renders it fitted to nourish the frame and maintain life.

The first state in which the generality of Insects appear is that of an egg; some few, however, are viviparous. From the egg is hatched the Insect in its Larva state; the Larvæ or Caterpillars of Insects differing materially from each other, according to the different tribes to which they belong. There are some Insects, however, which undergo no change of shape, but are hatched from the egg complete in all their parts, and undergo no farther alteration than that of casting their skin from time to time, till at length they acquire the complete resemblance of the parent animal.

Most insects, in the course of their lives, are subject to very great changes of form, attended by equally remarkable changes in their habits and propensities. These changes, transformations, or *metamorphoses*, as they are called, might cause the same insect, at different ages, to be mistaken for as many different animals. For example, a caterpillar, after feeding upon leaves till it is fully grown, retires into some place of concealment, casts off its caterpillar-skin, and presents itself in an entirely different form, one wherein it has neither the power of moving about, nor of taking food; in fact, in this, its second or chrysalis state, the insect seems to be a lifeless oblong oval or conical body, without a distinct head, or movable limbs; after resting awhile, an inward struggle begins, the chrysalis-skin bursts open, and from the rent issues a butterfly, or a moth, whose small and flabby wings soon extend and harden, and become fitted to bear away the insect in search of the honeyed juice of flowers and other liquids that suffice for its nourishment.

In the different tribes of Insects the Pupa or Chrysalis differs almost as much as the Larva. In most of the Beetle tribe it is furnished with short legs: in the Butterfly tribe it is perfectly destitute of all appearance of legs, and has no other motion than a mere writhing when touched: in the Locust tribe it differs very little from the perfect Insect, except in not having the wings complete: and in most of the Fly tribe it is perfectly oval, without any apparent motion, or distinction of parts. The Pupæ of the Bee tribe, and other Insects of a similar cast, are less shapeless than those of Flies, exhibiting the faint appearance of the limbs: while those of the Libellulæ or Dragon-flies are locomotive, as in the Locust tribe, but differ most widely from the appearance of the complete

Insect, and may be numbered among the most singular in the whole class of Insects. From the Pupa or Chrysalis at length emerges the Insect in its complete or ultimate form, from which it can never change, nor can it receive any further increase of growth.

Hence there are three periods in the life of an insect, more or less distinctly marked by corresponding changes in the form, power, and habits. In the first, or period of infancy, an insect is technically called a *larva*, a word signifying a mask, because therein its future form is more or less masked or concealed. This name is not only applied to grubs, caterpillars, and maggots, and to other insects that undergo a complete transformation, but also to young and wingless grasshoppers, and bugs, and indeed to all young insects before the wings begin to appear. In this first period, which is generally much the longest, insects are always wingless, pass most of their time in eating, grow rapidly, and usually cast off their skins repeatedly. The second period, wherein those Insects that undergo a partial transformation, retain their activity and their appetites for food, continue to grow, and acquire the rudiments of wings, while others, at this age, entirely lose their larva form, take no food, and remain at rest in a deathlike sleep,—is called the *pupa* state, from a slight resemblance that some of the latter present to an infant trussed in bandages, as was the fashion among the Romans. The pupæ from caterpillars, however, are more commonly called *chrysalids*, because some of them, as the name implies, are gilt or adorned with golden spots; and grubs, after their first transformation, are often named *nymphs;* the reason for which is not very obvious. At the end of the second period Insects again shed their skins, and come forth fully grown, and (with few exceptions) provided with wings. They thus enter upon their last or adult state, wherein they no longer increase in size, and during which they provide for a continuation of their kind. This period usually lasts only a short time, for most Insects die immediately after their eggs are laid. Bees, wasps, and ants, however, which live in society, and labour together for the common good of their communities, continue much longer in the adult state.

Insects possess some particular parts which are not to be found in any of the larger animals: among these are the *antennæ* before mentioned, which are those processes or jointed bodies situated on each side the head. They differ extremely in the different tribes of Insects, and are found to constitute one of the most convenient parts to fix upon in the distribution of Insects into genera and species. It is therefore necessary slightly to enumerate some of them:—*Antenna setacea*, or setaceous antenna; bristle-shaped, or growing fine and sharp at its termination; *antenna filiformis*, or thread-shaped, being of equal size throughout: *antenna moniliformis*, or moniliform; each joint being globular, or nearly so: *antenna clavata*, club-shaped; having a knob at the top, as in the major part of Butterflies: *antenna fissilis*, or fissile; one which is split or divided at the tip into several lamellæ or flat separations: *antenna pectinata*, or pectinated; one which is divided along each side into numerous processes in such a manner as to resemble the teeth of a comb: *antenna barbata*, or bearded; one which is slightly feathered, either on one or both sides, with fine lateral fibres or hairs: *antenna perfoliata*, perfoliate; the joints of a flattened and circular shape, with the stem or body of the antenna passing through them, as in the leaves of some plants, in which the stem seems to pass through them. Another part peculiar to Insects consists in a pair or two of short jointed processes proceeding from the mouth: these are termed *palpi*, or feelers, which in some Insects are very conspicuous, but not in all. The mouth in Insects is generally situated at the lower part of the front, and varies much in structure in the different orders. In some it is furnished with very strong jaws, often notched or serrated on the inner side into the appearance of teeth, and which always meet horizontally; in others the mouth consists of a tube or instrument for suction, either simple, or guarded by various kinds of appendages. The eyes in Insects are commonly situated on each side of the head, and are two in number; but in some Insects, as in Spiders, there are six or eight. In most of the Insect tribes the eyes may be considered as compound, the cornea presenting when viewed with a microscope the appearance of an infinite number of separate convexities, like so many real convex lenses. There are also on the heads of many Insects three small, smooth, lucid globules resembling so many separate eyes, placed on the top of the head, between or above the lateral ones: these Linnæus distinguishes by the title of *stemmata;* they are also called *ocelli.* The body in the major part of Insects is divided into the thorax or upper part, and the abdomen or lower part. In many of the Beetle tribe the back of the thorax is distinguished by a small triangular piece or division, situated at its lower part, between the juncture of the wing-sheaths: this triangular part is called *scutellum*, or the escutcheon. The under part of the thorax is called the breast, or *pectus*, and in this the sternum is frequently distinguishable. The abdomen is marked into transverse sections, and the last joint terminates in the tail. The wing-sheaths or shelly coverings, in the Beetle tribe and some others, are termed *elytra.*—The name of the *orders* into which Insects are divided, as *Coleoptera, Hymenoptera, Diptera, Neuroptera, &c.*, have reference chiefly to the number and nature of their wings; but as the definitions will be found under their respective names, we need not here repeat them.

In the Introductory Letter to "Kirby and Spence's Entomology," the beauties of the Insect world are thus graphically pourtrayed:—"Insects, indeed, appear to have been Nature's favourite productions, in which, to manifest her power and skill, she has combined and concentrated almost all that is either beautiful and graceful, interesting and alluring, or curious and singular, in every other class and order of her children. To these her valued miniatures she has given the most delicate touch and highest finish of her pencil. Numbers she has armed with glittering mail, which reflects a lustre like that of burnished metals; in others she lights up the dazzling radiance of polished gems: some she has decked with what looks like liquid drops, or plates of gold and silver; or with scales or pile, which mimic the colour and emit the ray of the same precious metals. Some exhibit a rude exterior, like stones in their native state; while others represent their smooth and shining face after they have been submitted to the tool of the polisher: others, again, like so many pigmy Atlases bearing on their backs a microcosm, by the rugged and various elevations and depressions of their tuberculated crust, present to the eye of the beholder no unapt imitation of the unequal surface of the earth, now horrid with misshapen rocks, ridges, and precipices — now swelling into hills and mountains, and now sinking into valleys, glens, and caves; while not a few are covered with branching spines, which fancy may form into a forest of trees.... The sight indeed of a well-stored cabinet of Insects will bring before every beholder not conversant with them, forms in endless variety, which before he would not have thought it possible could exist in nature, resembling nothing that the other departments of the animal kingdom exhibit, and exceeding even the wildest fictions of the most fertile imagination."

Before we close this article, we beg leave to quote from Mr. Newman's work on "The History of Insects," the following brief but admirable summary:—"The *senses of insects* are, properly speaking, seven: love, touch, taste, smell, hearing, sight, and the commanding and governing sense, called volition, mind, thought, or instinct. *Love* is that sense which ensures obedience to the great command, 'Increase and multiply:' its gratification seems the great object of an insect's life, after having arrived at maturity: its seat is in the organs of generation. *Touch* is a most invaluable sense to insects; they have two antennæ and four feelers attached to the mouth, which appear provided purposely for the exercise of this sense: the *tarsi* are also employed to ascertain qualities by touch; but the other parts of the body appear insensible to feeling, either as regards the ascertaining of qualities or the sensation of pain. *Taste* is undoubtedly possessed by insects in an eminent degree; and they seem to have the same preferences for animal or vegetable food which are evinced by vertebrated animals. *Smell* appears to be the sense by which insects are led to discover strongly-scented substances at a great distance, where it is quite impossible that sight should aid them; its seat, however, is wholly unknown. *Hearing* seems also to be possessed by insects, or to what purpose would the merry cricket sing his evening song, if there were none of his kind to listen to and admire it? The seat of this sense is also wholly unknown. *Sight* is a sense of which we have abundant evidence; it is seated in two large compound eyes, often occupying nearly the whole head, and also occasionally in three minute simple eyes, situated in a triangle on the crown of the head. The *mind* of insects is more wonderful than our own: it has neither speculation, retention, judgment, nor power; it is, in fact, an existence which comes perfect from the Creator: the new-born bee is perfectly mistress of architecture; she is heaven-instructed: the mind is not only the ruling sense, but is a distinct immaterial element."

INSECTIVORA. The fourth Order of of Mammiferous animals, comprising the Shrews, Hedgehogs, Moles, &c. As the name denotes, they subsist principally on insects, worms, &c. In general they lead a nocturnal and subterranean life; and in cold countries most of them pass the winter in a torpid state. Their legs are short, and in running they place the entire sole of the foot upon the ground. There is great variety in the front teeth of the animals belonging to this Order; in some the canine teeth are longer than the incisors, while in others the canines are very small or entirely wanting. The Order is naturally divided into those with simple fore legs, and those in which the fore legs are peculiarly fitted for digging. The first division comprises the Hedgehogs, the Tenrecs or Madagascar Hedgehogs (which have the muzzle very long, and have not the power of rolling themselves into a ball), the Shrews, and the Desmans (*Sorex moschatus*), a curious tribe of aquatic animals, with webbed feet, and the nose so much prolonged as to look like a proboscis. The second division comprises the Moles, the Shining or Cape Moles (the only quadrupeds whose fur has a metallic gloss), the Radiated Moles, and the *Scalops* or Shrew Mole of the United States. [See MOLE: HEDGEHOG: SHREW, &c.

INSESSORES. The name given by Mr. Vigors to a most extensive order of *Perching* Birds; in which are comprehended all those tribes which live habitually among trees, with the exception of the birds of prey and the climbing birds. In all the true Insessorial Birds, the toes are three before and one behind. The adaptation of the foot of this order to grasping or perching is evident from the situation of the hinder toe; which is invariably placed on the same level with those in front; and by which they are distinguished from the Gallinaceous and Wading Birds. The toes are slender, flexible, and of moderate length, with long, slender, and slightly curved claws; of which the foot of the Canary affords a very good example. The birds of this order are generally on the wing; and we accordingly find that, in proportion as the legs are small and weak, the wings are highly developed. The male is nearly always larger than the female, and is more distinguished for the brilliancy of his plumage. The Perchers live in pairs, and construct their nests, usually in trees, bushes, &c., with great art. This order is divided, by the form of the beak, into four subordinate groups; namely — 1. *Conirostres*, or conical-billed birds;

the greater part of which are omnivorous, though some are exclusively granivorous, 2. *Dentirostres*, or tooth-billed birds; which are characterized by a tooth or notch near the extremity of the upper mandible: these feed on insects, small birds, &c. 3. *Tenuirostres*, or slender-billed birds: these have a long slender bill, adapted for sucking up vegetable juices, &c.; and to this group belong also many whose principal food consists of insects. 4. *Fissirostres*, or gaping-billed birds; in which the beak is very much flattened, in order to afford them greater facility for capturing insects when on the wing, as is seen in the swallow and others of that kind.

INTESTINALIA. Another name for those parasites, worms, or *Entozoa*, which infest the intestinal canal of the higher animals. There are four orders; viz. 1. *Nematoidea* (Round-worms); 2. *Acanthocephala* (Hooked-worms); 3. *Trematoda* (Fluke-worms); 4. *Cestoidea* (Tape-worms). The hydatids or Cystic-worms are only the young of the last-named order.

IRIDINA. A genus of Conchifera, consisting of one species only, the *Iridina exotica*, which is found in the Nile, and in many other rivers of warm climates. The shell is equivalve and inequilateral; teeth very small and numerous; inside very iridescent and of a red cast; ligament external; the hinge lamina crenulated in its whole length. It is used by the natives of Egypt as spoons in measuring oil butter, and different kinds of provisions.

ISOCARDIA. A genus of Conchifera, the shells of which are remarkable for the beautiful curvature of the diverging umbones.

ISOPODA. The name of an order of aquatic Crustacea, many of which are parasitic upon other animals, very frequently upon larger crustacea.

IULUS, or JULUS. The *Julidæ* are a family of Myriapoda, very nearly allied to the Centipedes (Scolopendræ); but their body, instead of being flattened, as in that genus, is nearly cylindrical. Each of the numerous segments of the body is furnished with two pair of feet or legs, which are scarcely large or strong enough to support its weight; so that the animal, instead of appearing to walk, seems to have a sort of undulatory motion, like a serpent or worm. They roll themselves up in a spiral form; and the firmness of the rings of the body enables them to resist considerable pressure. The eyes of the Iulidæ are composed of nu-

GIGANTIC MILLIPEDE.—(JULUS MAXIMUS.)

merous hexagonal convexities, as in the greater part of the insect tribes; and the mouth resembles that of the larvæ of many insects by being furnished with a pair of denticulated jaws; by means of which they are enabled to divide with facility the portions of decaying vegetable matter on which they usually feed. Some are found under stones, others in the earth, and some inhabit nuts. The most common species is the *Iulus sabulosus*, about an inch and a quarter in length: its colour is a polished brownish black, with whitish legs: it is oviparous, and the young, when first hatched, have only three pair of legs, which are situated near the head; the remainder being gradually acquired till the number is complete, which usually amounts to a hundred and twenty on each side. In its young or growing state, it is of a pale colour, with a dark red spot on each side of every segment; and in this state it may sometimes be found in the soft mould of hollow trees.—The largest species known is the *Iulus Indus*, or *Iulus maximus*, which in its conformation resembles the species above described, but is from six to seven inches long. This is found in South America and the warmer parts of Asia, inhabiting woods and other retired places. The *Iulidæ* have no poisonous organs, and are perfectly innoxious to man: indeed, by their consuming vegetable substances that are in a state of decomposition, they may be considered beneficial. Mr. Newport, F.R.S., has made them and the Scolopendridæ a special object of study, and has published the results in the Linnæan Transactions. In the British Museum there is a very extensive collection of these interesting Myriapoda. [See CHILOGNATHA.]

JABIRU. A large aquatic bird, allied to the stork, three species of which are known, respectively inhabiting America, Western Africa, and Australasia. It is the *Mycteria* of Linnæus. It is somewhat larger than the swan; the head is large; the neck thick;

SENEGAL JABIRU.
(MYCTERIA SENEGALENSIS.)

and the bill is long, conical, smooth, and pointed. The body is entirely white; the head and neck are very bare of feathers, and covered with a thick black skin; and the tail is broad and short: the legs, which are more than two feet long, are thick and scaly; and the bill and feet are black. Our figure represents a gigantic species from the west coast of Africa, the Mycteria Senegalensis, on the shore of which it must form a striking featnre. In the enormous size of the beak this species resembles the Adjutant of India, and like that bird, we suppose this species to be somewhat of a carrion eater.

JACAMAR. (*Galbula.*) The birds belonging to this genus, of which there are but a few known species, are very much allied to the Kingfishers, except that their feet are quite different, and they inhabit moist woods; whereas the Kingfishers are only found on or near the banks of rivers. Their plumage has a metallic lustre, which it is extremely difficult to imitate.

The GREEN JACAMAR (*Galbula viridis*) is a splendid bird, about the size of a lark; its prevailing colour on the upper parts of the body being a most brilliant, changeable green, glossed with copper and gold. The beak is about two inches in length, black, slightly incurved, and sharp-pointed: the legs, which are short and weak, are a greenish yellow; and the claws black. Its breast and belly are of a dusky yellow hue, and the chin white. The tail is wedge-shaped; the two middle feathers very long, and the outer ones much shorter. The habits of this

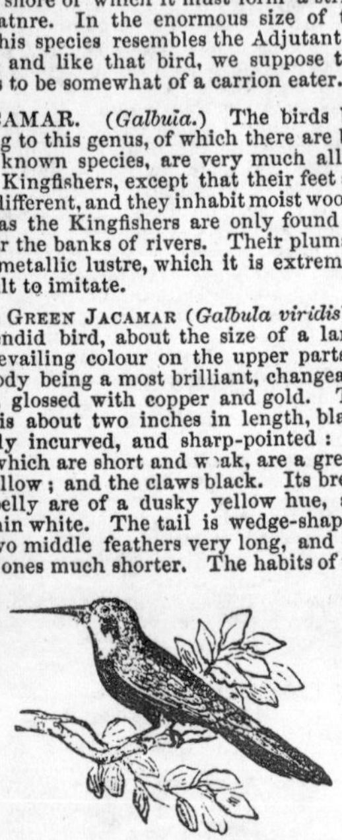

GREEN JACAMAR.—(GALBULA VIRIDIS.)

bird are very solitary; it resorts to the thickest parts of the woods, where it can obtain plenty of insects, and is seldom seen in company with another. Its flight is short and quick; and it is said to have an agreeable note.

The PARADISE JACAMAR (*Galbula paradisea*) is a less solitary bird than the others of this genus; being found in pairs, and frequenting the more open parts of the woods. It is nearly a foot in length: beak two and a half inches long, black, pointed, square, and compressed on the sides: head of a dull violet green; throat, fore part of the neck, and under wing-coverts white; the rest of the plumage green, varying in shades and glosses according to the lights in which it is viewed; the two middle tail-feathers six inches long, and the outer one only an inch: legs black. Native of Surinam and Cayenne.

JACANA. A genus of wading birds, distinguished by the extraordinary length of their toes and their spine-like claws, especially that of the hinder toe. They are very light birds; and the wide surface over which their toes extend, enables them the more easily to procure their food, consisting of worms, small fishes, and insects, by walking on the leaves of aquatic plants which float on the water. Various species of the Jacana, which in contour and habit resemble our moor-hen, are spread over the tropical regions both of the Old and New World.

The COMMON JACANA (*Parra Jacana*) is about ten inches long, the beak being upwards of one inch, and orange-coloured: the head, throat, neck, breast, and under parts are black; the back, wing-coverts, and scapulars bright chestnut; spur on the wing yellow, and the bend of the wing varied with black: the quills olive-yellow, tipped and partly edged with dusky; tail rounded, chestnut tipped with black; legs greenish ash. These birds inhabit Brazil, Surinam, and other parts of South America and the West Indies: they are very shy and noisy, and their note is very shrill.

The INDIAN JACANA (*Parra Indica*) is a shy bird, frequenting stagnant lakes, and building its nest upon floating materials, among weeds, near the banks. It has a yellow beak, with the base of its upper mandible dusky blue; and near the gape a red spot: over the eye is a white streak, which reaches some distance down the side of the neck: the head, neck, and upper parts of the body are deep blue-black: the back and wings are ashy-brown: legs dirty yellowish-brown.

The BRONZED JACANA (*Parra œnea*) is a highly elegant species inhabiting Brazil. The prevailing colour of the body is black, brilliantly glossed with blue and violet reflections: its head and neck are of a brilliant bronzed-green colour: behind the eye is a white streak: the quills are black; the wing-coverts dull yellow; and the rump and tail are blood-red.

JACCHUS, or MARMOZET. A genus of Monkeys, of a small size, with short muzzle, flesh-coloured face, and round head. The five fingers are armed with claws, except the thumbs of the posterior extremities, which have nails: fur very soft; tail full and handsome. Length of body about eight inches; tail eleven. General colour olive-gray; head and shoulders nearly black; the tail and lower part of the back are annulated with pale gray; and two tufts of pale hair grow round the ears. They are squirrel-like in their habits, and omnivorous; feeding on roots, seeds, fruits, insects, snails, and young birds. Native of Guyana and Brazil.

JACK. [See PIKE.]

JACKAL. (*Canis aureus.*) This animal in its appearance somewhat resembles the fox, while its habits are more like those of the wolf. It is a native of India, Persia, and various other of the warmer parts of Asia, as also of Barbary, and the interior of South

Africa. Its size is about that of a fox, but longer in the legs; its colour a light orange-yellow or yellowish gray above, and whitish below, with dark shades about the back: the

JACKAL.—(CANIS AUREUS.)

tail hangs straight, is rather bushy, and generally black at the tip: the ears are very ruddy, and the muzzle pointed. The voice of the Jackal is described as peculiarly hideous, consisting of an indistinct bark and a piteous howl. It resides in woods, holes, and rocky places; and preys indiscriminately on all the weaker animals, committing ravages among flocks, in the poultry-yard, &c., though it seldom ventures abroad till nightfall. Jackals frequently go in great troops to hunt their prey, and by their dreadful yellings alarm and put to flight deer, antelopes, and other timid quadrupeds; while the Lion, instinctively attending to the clamour, is said to follow till the Jackals have hunted down the prey, and, having satiated himself on the spoil, leaves only the scanty remains to the famished hunters. Hence the Jackal has been popularly called "the lion's provider." Some say that the Jackal has a natural propensity to follow mankind, instead of flying from him, like the Wolf and the Fox: also that the whelp is readily tamed, and, when grown up, assumes all the habits of the domestic Dog: nay, it is well known that the Jackal interbreeds with the common dog; its period of gestation is the same, and the hybrid progeny is fertile. We should, however, observe, that between the Jackal and the Dog there exists such an irreconcileable antipathy, that they never meet without a combat.

JACKDAW, or DAW. (*Corvus monedula.*) A well-known English bird, considerably less than the Rook, being about thirteen inches in length, and twenty-eight in breadth. The bill and legs are black; the claws strong and hooked; eyes white; the hinder part of the head and neck is silvery gray; the rest of the plumage is of a fine glossy blue-black above, beneath dusky. Jackdaws frequent church steeples, old towers, and ruins, in flocks, where they build their nests: the female lays five or six eggs, paler and smaller than those of the crow. They are easily tamed, and may be taught, like the magpie, &c., to imitate human articulation: they have also the mischievous faculty of stealing and hiding money, spoons, or other glittering and metallic substances. They feed on insects, grain, fruit, small pieces of flesh, eggs, &c. They remain in this country during the whole year; but in France, Germany, and other parts of the Continent, they are migratory. From an article, headed "Habits of the Jackdaw," in Mr. Waterton's Essays, we glean the following observations: "Though the Jackdaw makes use of the same kind of materials for building as those which are found in the nest of the rook; though it is, to all appearance, quite as hardy a bird; and though it passes the night, exposed to the chilling cold and rains of winter, on the leafless branches of the lofty elm; still, when the period for incubation arrives, it bids farewell to those exposed heights, where the rook remains to hatch its young, and betakes itself to the shelter which is afforded in the holes of steeples, towers, and trees. Perhaps there is no instance in the annals of ornithology which tells of the Jackdaw ever building its nest in the open air. Wishing to try whether these two congeners could not be induced to continue the year throughout in that bond of society which, I had observed, was only broken during incubation, I made a commodious cavity in an aged elm, just at the place where it had lost a mighty limb, some forty years ago, in a tremendous gale of wind which laid prostrate some of the finest trees in this part of Yorkshire. At the approach of breeding-time, a pair of Jackdaws took possession of it, and reared their young in shelter; while the rooks performed a similar duty on the top of the same tree, exposed to all the rigours of an English spring. This success induced me to appropriate other conveniences for the incubation of the Jackdaw; and I have now the satisfaction to see an uninterrupted fellowship exist, the year throughout, between the Jackdaw and the Rook."

JAGUAR, or OUNCE. (*Felis onca.*) A fierce and destructive animal of the feline kind, partaking of the qualities and habits of the Tiger: it is a native of the hotter parts of South America, and from its being the most formidable quadruped there, is sometimes called the tiger or panther of the New World. It is as large as a wolf, and lives solely on prey. Its ground colour is a pale brownish yellow, variegated on the upper parts of the

JAGUAR.—(FELIS ONCA.)

body with streaks and irregular oblong spots of black; the top of the back being marked with long uninterrupted stripes, and the sides with rows of regular open marks: the thighs and legs are marked with full black spots; the breast and belly are whitish; the tail not so long as the body; the upper part irregularly marked with large black spots, the lower with smaller ones. It swims and climbs with ease; and preys not only on the larger domestic quadrupeds, and on smaller that are wild, but also on birds, fish, tortoises, turtles' eggs, &c. It must, however, be very hard pressed before it will attack man.

JANTHINA. A Molluscous animal, belonging to the *Pectinibranchiata*. The shell has some resemblance to our land snails, but the aperture is angular at its lower part and at its outer side, where, however, the angle formed by the union of the upper and lower halves of the outer lip is much rounded in most of the species; the columella straight

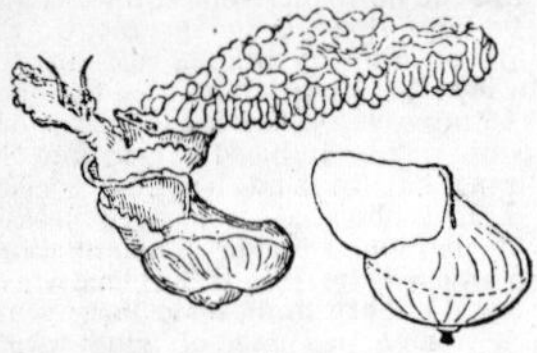

PURPLE SEA-SNAIL.—(JANTHINA COMMUNIS.)

and elongated, the inner lip turned back over it. The animal has no operculum, but carries under its foot a vesicular organ, like a congeries of foam-bubbles, of solid consistence, that prevents creeping, but serves as a buoy to support it at the surface of the water. The head is a cylindrical proboscis; and is terminated with a mouth cleft vertically, and armed with little curved spines: on each side of it is a forked tentaculum. The shells are of a violet colour; and when the animal is irritated it pours forth an excretion of deeper blue to tinge the sea around it.

"The method in which this animal fills its float," says Capt. Grey, "is curious: it throws it back, and gradually lifts the lip of the valve out of water, until the valve stands vertical; it then closes the valve tightly round a globule of air, around which it folds, by means of the most complex and delicate machinery. The valve is then bent over until it touches the edge of the float nearest the head, and when it is in this position, the portion of it which is inflated with air looks like a bladder, the air gradually is expelled into the float, and as this process takes place the bladder in the valve diminishes, and the valve becomes by degrees like a lip pushed forwards until it lies flat on the float: the valve is composed of two portions, a cup and a lip. The time occupied from first removing the valve from the float, until the inflation, and the expulsion of air into the float being completed, so that the valve begins to move again, is sixty-one seconds, from the mean of several experiments. These animals have also the power of compressing the valve into a hollow tube, which they elevate above the water like a funnel, and draw down air through it. The colouring matter which they emit has no stinging, electric, or deleterious properties whatever, that I could discover. I found that when this colouring matter was mixed with water, it became of a deep blue. In those which I caught in Nov. 1837, I may have been deceived, and the colouring matter might also possibly have been scarlet directly it was emitted. It is difficult to conceive what use this liquid can be to the fish against its foes, yet it certainly uses it as a means of defence. To one of these shells, the fish in which was alive and well, we found attached a number of barnacles, some of which were of large size." —*Narrative of Expedition in South Australia.*

JAY. (*Garrulus glandarius.*) The Jay is the most elegant bird of the Corvine genus in Britain, and is about thirteen inches in length. Its general colour is a light purplish buff, paler on the under parts; the wings black, with a large white spot in the middle: its bill and tail are black; the former notched on each side near the tip, and the latter rather rounded at the end: the feathers on the forehead are white, streaked with black, and form a tuft which it can erect or depress at pleasure: the greater wing-coverts are elegantly barred with black, fine pale blue, and white alternately; the lesser wing-coverts bay; the belly and

JAY.—(GARRULUS GLANDARIUS.)

vent almost white: the greater quills are black, with light edges; the bases of some of them white; lesser quills black; those next the body chestnut: legs of a dirty flesh colour. The Jay is very common in this country, and is found in most of the temperate parts of Europe, frequenting woods, and feeding on acorns, beech-mast, berries, and fruits of various kinds, insects, and sometimes young birds in the absence of the old ones. The Jays are distinguished as well for the beautiful arrangement of their colours, as for their harsh, grating voice, and petulant, restless disposition. In confinement, however, it loses the beauty of its plumage, and becomes of a dull or brownish tinge. When an owl or other bird of prey appears in the woods, they utter piercing cries, and assemble in great numbers to attack the common enemy: the same thing takes place when they see a sportsman, whom they often frustrate by their vociferous noise. Like their kindred, the magpie and jackdaw, they can be taught a variety of words and sounds, particularly those of a harsh and grating character, as that of a saw, &c. They sometimes assemble in great numbers in the spring, and seem to hold a conference, (as Bewick says) probably for the purpose of pairing and of fixing upon the districts they are to occupy; and the noise made on these occasions may be aptly compared to that of a distant meeting of disorderly drunken persons. The Jay builds in woods, and makes an artless nest, composed of sticks, fibres, and slender twigs; lays five or six eggs, ash-gray, mixed with green and faintly spotted with brown.

In the 'Journal of a Naturalist' we find, in reference to the love of offspring, as being particularly manifested in birds, the following remarks on the Jay. "This bird is always extremely timid and cautious, when its own interest or safety is solely concerned; but no sooner does its hungry brood clamour for supply, than it loses all this wary character, and it becomes a bold and impudent thief. At this period it will visit our gardens, which it rarely approaches at other times, plunder them of every raspberry, cherry, or bean, that it can obtain, and will not cease from rapine as long as any of the brood or the crop remains. We see all the nestlings approach, and, settling near some meditated scene of plunder, quietly await a summons to commence. A parent bird from some tree surveys the ground, then descends upon the cherry, or into the rows, immediately announces a discovery by a low but particular call, and all the family flock into the banquet, which having finished by repeated visits, the old birds return to the woods, with all their chattering children, and become the same wild cautious creatures they were before."

The BLUE JAY. (*Garrulus cristatus.*) This elegant species is a native of North America, considerably smaller than the European Jay, with a tail much longer in proportion: the head is handsomely crested, with loose silky plumes; bill black; legs brown: the whole bird is of a fine blue colour on the upper parts, with the wings and tail marked by numerous black bars; neck encircled with a

BLUE JAY.—(GARRULUS CRISTATUS.)

black collar; under parts blossom-colour, with a slight cast of blue; tail tipped with white; legs, feet, and thighs of a dusky brown. Its note is less discordant than the European Jay; but its manners are very similar. It is said to be a great destroyer of maize or Indian corn, often assembling in large flocks to devour it.

Mr. Gosse, in his 'Canadian Naturalist,' thus speaks of this bird, in his observations made during the month of December. "The Blue Jay continues as numerous and as noisy as ever. His harsh screaming voice may be heard above that of all the other feathered inhabitants of our groves, all the year through. A beautiful bird he is, with his bright violet, white, and sky-blue coat, long tail, and pointed crest; and by his airs and grimaces he appears to have no mean idea of his own personal attractions, and probably he may think his voice as charming as his plumage, as he so continually gives us the benefit of his music. He appears to tyrannize over his brethren occasionally. I once saw, in the south, a Blue Jay in close and hot pursuit of a summer Red-bird (*Tanagra æstiva*), and Wilson records a parallel incident. He has other notes, besides his common loud squall, some of which are difficult to recognize. In the clearing, the parties of these birds, for they are hardly numerous enough to be called flocks, generally fly high, and alight about the summits of lofty trees; but in the woods, particularly in spring, they as frequently choose a lower altitude. They are wary, and rather difficult of approach.

JELLY-FISH. Under the heads "Acalephæ," "Beröe," and "Medusa," will be found various information applicable to the present article, the popular name of "Jelly-fishes" being very generally used (by the unscientific) to denote the different marine substances forming that branch of the division RADIATA which is comprised in the class *Acalephæ.* Extreme delicacy of structure is common to the whole group; most of them have no hard support whatever, and the animals when removed from their natural element wholly lose their form; but there are a few species which have a very thin cartilaginous covering, and these retain a semblance of the animal as it appeared when alive. We find that in every climate the ocean swarms with infinite multitudes of animals, which, from their minuteness and transparency, would be almost imperceptible, were it not for the phosphorescent properties of some of them being rendered evident on the slightest agitation. All, however, are not equally minute: some grow to a large size, and their forms are perfectly well known to the casual observers of marine substances which lie on every beach. Most of these are highly phosphorescent; and in tropical regions, more particularly, where they exist in the greatest abundance, the path of a vessel is marked by a brilliant line of glowing light, and the whole surface of the ocean often displays a beautiful luminosity. Even on our own coasts a similar effect is very frequently observed, though the luminous appearance is vastly less brilliant.

In Patterson's "Introduction to Zoology" we find the following practical observations: —"Our admiration for the various functions performed by the *Acalephæ* is much increased when we reflect upon the extremely small quantity of solid matter which enters into their composition. This fact admits of easy illustration, both in the Beroes and in the Medusæ. On one occasion, we took a dead Cydippe, and placing it on a piece of glass, exposed it to the sun. As the moisture evaporated, the different parts appeared as if confusedly painted on the glass; and when it was become perfectly dry, a touch removed the only vestiges of what had been so lately a graceful and animated being. With regard to the Medusæ, we may mention an anecdote which we learned from an eminent zoologist [the late Edward Forbes, F.R.S., Professor of Natural History in the University of Edinburgh]. He had, a few years ago, been delivering some zoological lectures in a seaport town in Scotland, in the course of which he had reverted to some of the most remarkable points in the economy of the Acalephæ. After the lecture, a farmer who had been present came forward, and inquired if he had understood him correctly, as having stated that the Medusæ contained so little of solid material that they might be regarded as little else than a mass of animated sea-water? On being answered in the affirmative, he remarked, that it would have saved him many a pound had he known that sooner, for he had been in the habit of employing his men and horses in carting away large quantities of jelly-fish from the shore, and using them as manure on his farm, and he now believed they could have been of little more real use than an equal weight of sea-water. Assuming that so much as one ton weight of Medusæ recently thrown on the beach had been carted away in one load, it will be found that, according to the experiments of Professor Owen, the entire quantity of solid material would be only about four pounds of avoirdupois weight, an amount of solid material which, if compressed, the farmer might, with ease, have carried home in one of his coat-pockets." [For development, &c., see STROBILA in SUPPLEMENT.]

JERBOA. (*Dipus.*) A genus of Rodents between the Squirrels and the Rats, but agreeing with the latter rather than the former; while the enormous development of its hind legs and tail cause it to bear considerable resemblance in form to the Kangaroo. One species is a native of Egypt, Syria, &c.; and was known to the ancients under the name of *Dipus*, (two-footed,) which is still its scientific appellation. The most common species is the *Dipus sagitta*. It is of a pale yellowish fawn-colour on the upper parts, and white beneath; the length of the body is about eight inches, and of the tail ten, being terminated by a tuft of black hair, the tip of which is white, but the rest short and rough. The head is short; the ears thin, broad, upright, and rounded; the eyes large, round, and dark coloured; the fore-legs about an inch long, with five toes to each foot, the inner toe very small, but furnished with a sharp, crooked claw like the rest; the hind-legs are extremely long, thin, sparingly covered with short hair, and very much resemble those of a bird: the hind feet have three toes on each, the middle of which is somewhat larger than the rest, and all are

JERBOA.—(DIPUS SAGITTA.)

furnished with sharp and strong claws: there is also a very small spur or back toe, with its corresponding claw. On each side the nose are several long hairs or whiskers; and the cutting teeth are sharp and strong, resembling those of a rat. In its attitudes and manner of progression this animal much resembles a bird; generally standing, like the Kangaroo, on its hind feet, and leaping with much celerity, and to a great distance. It principally uses the fore legs in feeding, putting to his mouth the ears of corn, and various other vegetable substances on which it feeds.

The Jerboas inhabit dry, hard, and clayey ground, in which they make their burrows. These are of considerable length, and run obliquely and winding; at about half a yard below the surface of the ground, they terminate in large excavations or nests; they are usually provided but with one opening, though the animals are provident enough to make another passage, to within a short distance from the surface, through which they rapidly penetrate in case of necessity. They keep within their holes during the day, sleeping rolled up, with their head between their thighs: at sunset they come out, and remain abroad till morning. From the rapidity with which they take their leaps (of six or seven feet at a time), it is nearly impossible to overtake them. In leaping, they carry their tails stretched out; but in standing or walking, they carry them in a curved form, the lower curve touching the ground. In their wild state these animals are very fond of bulbous roots; but, when confined, they will feed on raw meat. They are tamed without much difficulty, but require to be kept warm.

There are some other species of the Jerboa; by far the largest of which is the CAPE JERBOA, a native of the mountainous country to the north of the Cape of Good Hope. Its length from the nose to the tail is fourteen inches, and the tail itself somewhat more. The head is broad, the muzzle sharp, and the upper jaw longer than the lower: the ears are large, the whiskers long and black, and the tail is extremely full of hair. It is an animal of great strength and activity, and will spring to the distance of twenty or thirty feet at once. When eating, it sits upright in the manner of a squirrel; and it burrows in the ground, like the smaller kind of Jerboas, with great ease and expedition; having five very strong and long claws on each of its fore feet: those on its hind feet are short, and four in number. [See HELAMYS.]

JERFALCON. [See FALCON.]

JOHN CROW VULTURE. The local name in Jamaica for the Turkey Buzzard. [See TURKEY BUZZARD.]

JOHN DORY. [See DORY.]

JUMNOS. A singular genus of Coleoptera belonging to the family *Cetoniadæ*, one species of which, described by Mr. W. W. Saunders, is still very rare in collections; this is the *J. Ruckeri*; it is of a brilliant green with large yellow marks on the elytra, and the male has long fore legs. It is a native of Northern India.

JUMPING HARE. [See HELAMYS.]

JUNGLE-FOWL. (*Megapodius tumulus.*) Mr. Gould, in his able work on the "Birds of Australia," gives an interesting account of this bird, which in size is about that of a common Fowl, and must not be confounded with the Jungle Cock of India, a very different bird. Its mode of constructing its mound-like nest, and its manner of depositing the eggs, &c., very much resemble those described under TALEGALLA [which see]. "The Jungle-fowl," we learn, "is almost exclusively confined to the dense thickets immediately adjacent to the sea-beach: it appears never to go far inland, except along the banks of creeks. It is always met with in pairs or quite solitary, and feeds on the ground, its food consisting of roots which its powerful claws enable it to scratch up with the utmost facility, and also of seeds, berries, and insects, particularly the larger species of Coleoptera. It is at all times a very difficult bird to procure; for although the rustling noise produced by its stiff pinions when flying away be frequently heard, the bird itself is seldom to be seen. Its flight is heavy and unsustained in the extreme; when first disturbed it invariably flies to a tree, and on alighting stretches out its head and neck in a straight line with its body, remaining in this position as stationary and motionless as the branch upon which it is perched: if, however, it becomes fairly alarmed, it takes a horizontal but laborious flight for about a hundred yards, with its legs hanging down as if broken. I did not myself detect any note or cry, but from the natives' description and imitation of it, it much resembles the clucking of the domestic fowl, ending with a scream like that of the peacock." The head and crest of this bird is of a very deep cinnamon brown; back of the neck and all the under surface very dark gray; back and wings cinnamon brown; upper and under tail coverts dark chestnut brown; tail blackish brown; bill reddish brown, with yellow edges; tarsi and feet bright orange. It appears that on Mr. Gilbert's arrival at Port Essington his attention was attracted to numerous great mounds of earth which were pointed out to him by some of the residents as being the tumuli of the aborigines. The natives, on the other hand, assured him that they were formed by the Jungle-fowl for the purpose of hatching its eggs: and so it afterwards proved. One of these mounds is described as fifteen feet high, and sixty in circumference at the base, and so enveloped in thickly foliaged trees as to preclude the possibility of the sun's rays reaching any part of it.

KAHAU. The Proboscis Monkey. [See MONKEYS.]

KAKAPO. A New Zealand parrot. [See STRIGOPS.]

KALONG. The name given to several species of Fox-bats (*Pteropidæ*). [See PTEROPUS.]

KANGAROO. (*Macropus.*) This extraordinary animal is peculiar to Australasia, and belongs to the marsupial order of quadrupeds; but it receives its scientific name from the enormous length of the hind feet, which is the distinguishing characteristic in all the animals included in the family *Macropopidæ*, or Kangaroo tribe. But before we proceed to describe the form and habits of this singular quadruped, we shall mention the circumstances (as detailed by Dr. Shaw) attending its first discovery. This was in 1770, when the celebrated navigator Captain Cook was stationed for a short time on that part of the coast of New Holland which is now called New South Wales. On Friday, June 22, says Captain Cook, a party who were engaged in shooting pigeons for the use of the sick of the ship, saw an animal which they described to be as large as a greyhound, of a slender make, and extremely swift. The following day the same kind of animal was again seen by a great many other people. On the 24th it was seen by Captain Cook himself, who, walking at a little distance from the shore, observed a quadruped, which he thought bore some resemblance to a greyhound, and was of a light mouse-colour, with a long tail, and which he should have taken for a kind of wild dog, had not its extraordinary manner of leaping, instead of running, convinced him of the contrary. Mr. Banks also obtained a view of it, and immediately concluded it to be an animal perfectly new and undescribed. Some time after, this gentleman, accompanied by a small party, had an opportunity of chasing two with his greyhound, which the Kangaroo, by its bounding leaps over the high grass, soon outstripped. It was not long, however, before one was shot; and the scientific associates in this expedition of discovery were then fully gratified.

The upper parts of the Kangaroo are small, while the lower are remarkably large in proportion; yet its general appearance is decidedly picturesque. The head bears some resemblance to that of a deer, and the visage is mild and placid: the ears are moderately large, rather pointed, and upright; the eyes large, and the mouth rather small; the neck thin and finely proportioned; the are legs extremely short, with the feet divided into five toes, each furnished with a short and somewhat hooked claw; the hinder feet, on the contrary, are provided with only four toes, the middle one of which is long, of great strength, and terminated by a large and powerful hoof-like nail or claw: so that the head and upper parts seem strangely disproportioned to the posterior parts of the animal, which are robust and powerful. The tail, which is very long, is extremely thick at the base, gradually tapering, and appears to act as a supplemental limb, when the animal assumes its erect or sitting posture. When feeding, it is seen in a crouching position, resting on its fore paws, as well as on the hinder extremities, whilst it browses on the herbage; and in this attitude it hops gently along, deriving some assistance from its tail. On the least alarm, however, it raises itself on its hind legs, and bounds away to a distance with great rapidity. The leap is of very great length; and is accomplished by the muscular action of the tail, almost as much as by that of the limbs. They use their tails and hinder feet also as weapons of defence: for when pursued and overtaken by dogs, they turn, and seizing them with their fore feet, strike them with their hinder ones, sometimes causing death by a single blow. The under side of the hind foot has a callous sole along its whole length; and its great length is chiefly given by the elongation of the metatarsal bones. Kangaroos have no canine teeth: their incisors are six in the upper jaw, and but two in the lower; the former short, and the latter long: the molars, which are separated from the incisors by a large vacant space, are ten in number in each jaw. They are exclusively herbivorous in their diet, feeding chiefly on grass: and they associate in small herds, under the guidance of the older males. The ventral pouch, or receptacle for the young, with which the female Kangaroo is furnished, is indeed a most curious provision of nature. Being situated just below her breasts, there the young ones sit to suck; and even when they are old enough to leave the pouch, for exercise or amusement, they immediately seek refuge in it on the least alarm.

The number of species which are now known are very considerable: they vary in size, from that of a rat to the Great Kangaroo, the male of which has been known to measure nearly eight feet from the nose to the tip of the tail, and to weigh 220 lbs.; but in form and habits they bear a strong resemblance to each other. The young are produced in an extremely imperfect state, and are even disproportionately small; not exceeding an inch in length. These animals are easily tamed; and when in a state of domestication, they are harmless and timid. Their flesh is eaten in Australia, and is said to be nutritious. Some persons are loud in their commendations of it; Colonel Light, indeed, goes so far as to recommend all who are fond of ox-tail soup (and they are not a few), to take a trip to South Australia, and eat Kangaroo-tail soup; which, he says, if made with the skill that soups in England are, would as far surpass the ox as turtle does the French *potage*.

Mr. Gould's great work on the Kangaroo Family is a most noble contribution to Natural History: in it all the species are figured and described with the hand of a master. We must also refer to the work of Mr. Waterhouse, who has devoted a thick octavo volume to their history. Both these works are indispensable to those who would desire to study this important family.

KERMES. (*Coccus ilicis.*) An insect produced in the excrescences of a small oak, the *Quercus coccifera*, and found in many parts of Asia and the South of Europe. The body of this insect is full of reddish juice, and when dead, and transformed into an apparent grain or berry, it is used for the purpose of dyeing a brilliant red colour. They were long taken for the seeds of the tree on which they live, and hence called *grains of Kermes*. Kermes is now nearly superseded by the use of cochineal, but though much inferior in brilliancy to the scarlet cloths dyed with real Mexican cochineal, they retain the colour better, and are less liable to stain. This is said to have been the celebrated Phœnician dye. [See Cochineal.]

KESTREL. (*Falco Tinnunculus.*) A beautiful bird of the Hawk kind, known also as the Stannel Hawk, and Windhover. The male is about fourteen inches in length, and in breadth two feet three inches. Its colours, at first sight, distinguish it from all other hawks: the crown of the head, and the greater part of the tail, are of a fine light gray hue; and on the lower part of the latter there is a broad black bar, succeeded by white tips. The back and coverts of the wings bright cinnamon brown, spotted with black; quill feathers dusky, with light edges; inside of the wings white, beautifully spotted with brown on the under coverts, and barred on all the quills with pale ash. The whole under side of the bird is of a pale rust colour, streaked and spotted with black. The bill blue; cere and eyelids yellow; legs yellow; claws black. The colours of the female are less vivid than those of the male: the back and wing-coverts are rusty brown, and elegantly marked with numerous undulated bars of black; the breast, belly, and thighs are of a pale reddish buff, with dusky streaks pointing downwards; and the tail is marked by a pretty broad dark ash-coloured bar near the end.

The Kestrel is widely diffused throughout Europe, and is by no means rare in the more temperate parts of North America. It breeds in the hollows of decayed trees, and in the holes of rocks, towers, and ruined buildings; and lays four or five pale reddish eggs. It feeds on small birds, field mice, reptiles, and insects: after securing its prey, it plucks the feathers very dexterously from birds, but swallows mice entire, and discharges the hair, in the form of round balls, from its mouth. This bird, when in quest of food, "glides softly through the air, at a moderate height, now poised in the breeze on fluttering pinion, now resting in the void apparently without motion; till, at last, down he comes, like a falling stone, upon the unconscious prey below." That discerning friend of the feathered tribes, Mr. Waterton, whose words we have just quoted, thus apostrophizes in this bird's favour: "Did the nurseryman, the farmer, and the country gentleman, know the value of the Windhover's services, they would vie with each other in offering him a safe retreat. He may be said to live almost entirely on mice; and mice, you know, are not the friends of man; for they bring desolation to the bee-hive, destruction to the pea-bed, and spoliation to the corn-stack. Add to this, they are extremely injurious to the planter of trees." Again, "I prize the services of the Windhover Hawk, which are manifest by the quantity of mice he destroys; and I do all in my power to put this pretty bird on a good footing with the gamekeepers and sportsmen of our neighbourhood. Were this bird properly protected, it would repay our kindness with interest; and we should then have the Windhover by day, and the owls by night, to thin the swarms of mice which overrun the land." "The Windhover," he further observes, "is a social bird, and, unlike most other hawks, it seems fond of taking up its abode near the haunts of men. What heartfelt pleasure I often experience in watching the evolutions of this handsome little falcon! and with what content I see the crow and the magpie forming their own nests; as I know that, on the return of another spring, these very nests will afford shelter to the Windhover! Were I to allow the crow and the magpie to be persecuted, there would be no chance for the Windhover to rear its progeny here; for Nature has not taught this bird the art of making its nest in a tree. How astonishing, and how diversified, are the habits of birds! The Windhover is never known to make use of a nest until it has been abandoned for good and all by the rightful owner; whilst, on the contrary, the cuckoo lays her egg in one of which the original framer still retains possession."

KIANG. [See SUPPLEMENT.]

KING-BIRD. [See TYRANT FLYCATCHER.]

KING-FISH. [See OPAH.]

KINGFISHER. (*Alcedo.*) A rather numerous genus of birds, and widely diffused in warm climates, although there is but one species occurring in Europe. They are, in general, birds of an inelegant shape, the head being large in proportion to the size of the body, and the legs and feet very small; but they are of singular brilliancy of plumage, in which blue, green, and orange are the prevailing colours. They are distinguished by having a long, straight, strong, and acute bill; wings rather short; body thick and compact; head large and elongated; plumage thick and glossy. In some of the larger species, however, the colours are more obscure, exhibiting a mixture of brown, black, and white, variously modified in the differen birds. In their manners they all seem to agree; frequenting the banks of rivers, &c., where, perched on a branch of a tree, or other projecting object, they will remain motionless for hours, watching till some fish comes under its station, when the bird dives perpendicularly down into the water, and brings up its prey with its feet, carries it to land, kills it by repeated strokes of the bill, and immediately swallows it: afterwards casting up the scales and other indigestible parts, in pellets, like birds of prey.

The COMMON OR EUROPEAN KINGFISHER. (*Alcedo ispida.*) This retired and solitary bird, which is only to be found near rivers, brooks, or stagnant waters, subsisting entirely on the smaller kinds of fish, is only seven inches in length, and eleven in breadth: its bill is nearly two inches long, the upper mandible being black, and rather red at the base; the under one, as well as the inside of the mouth, orange-coloured: the throat is white: the crown of the head is a deep shining green, with numerous transverse bright blue streaks: the shoulders and whole wings dark green, but the edges of the quill feathers are glossed with pale blue, and the shoulders marked by numerous small blue spots. The middle of the back, the rump, and coverts of the tail are of a most resplendent azure: the tail is very short, and of a deep rich blue colour; and the whole under part of the body is of a bright orange: legs red; claws black. The female commonly deposits her eggs (which are from five to eight in number, and perfectly white) in a hole in the river's banks, which has probably been made by the mole or the water-rat. If the nest be robbed, the bird returns and lays in the same situation. "I have had," says Reaumur, "one of these females brought me, taken from her nest about three leagues from my house. After admiring the beauty of her colours, I permitted her to fly; when the fond creature was instantly seen to repair to the nest where she had just before been made a captive: there joining the male, she again began to lay, though it was for the third time, and the season was very far advanced. At each time she had seven eggs." In this country the Kingfisher begins to lay early in the season, and excludes her first brood about the beginning of April. The fidelity of the male exceeds that of the turtle: he brings the female large supplies of fish during the season of incubation; and she, contrary to most other birds, is always plump and fat at that time. The male, who on other occasions always makes a twittering noise, now enters the nest with all the silence and circumspection imaginable. The young are hatched at the expiration of twenty days; but they do not acquire the beauty of their plumage till after the first moulting season. This bird is usually seen flying rapidly near the surface of the stream; and the velocity with which it maintains its flight, considering the shortness of its wings, is really surprising.

The ancients attributed to the Kingfisher innumerable habits and properties equally improbable. They supposed that it built its nest upon the ocean; but as this floating cradle would be likely to be destroyed by storms, they endowed the bird with powers to lull the raging of the waves during the period of incubation: hence those tranquil days near the solstice were termed *halcyon* days; and that the feathered voyager might want no accomplishment, they attributed to it the charm of song. They also kept the dead body of the bird as a safeguard against thunder, and as a relic by which the peace of families would be preserved. But it is not to the fanciful genius of the ancients alone that this bird is indebted for wonderful attributes. The Tartars and Ostiaks preserve the skin about their persons as an amulet against every ill; and they consider that the feathers have magic influence, when properly used, in securing a female's love: nor are such superstitions entirely confined to barbarous nations; for there are persons, it is said, who believe that if the body of a Kingfisher be suspended by a thread, its breast, by some magnetic influence, will invariably turn to the north.

We shall now endeavour to point out, in the briefest manner possible, some of the other most important species.—The GIANT KINGFISHER. (*Dacelo gigantea.*) This is the largest species known, measuring eighteen inches from the tip of the bill to the end of the tail: the colour of the plumage chiefly composed of olive-brown and a pale blue-green. Native of Australia.—PIED KINGFISHER. (*Alcedo rudis.*) Size of the song-thrush. The plumage chiefly party-coloured of black and white. Native of various parts of Asia and Africa.—SMYRNA KINGFISHER. (*Alcedo Smyrnensis*). Size of the missel-thrush. A most brilliantly coloured bird; the bright blue of the wings yielding in lustre to none of the feathered tribes. Native of the hotter parts of both Africa and Asia.—SACRED KINGFISHER. (*Alcedo sacra.*) Crown of the head and upper parts blue-green; the throat white; the under parts pale ferruginous, passing upwards like a collar round the neck. Native of the South Sea Islands.—CRESTED KINGFISHER. (*Alcedo cristata.*) A singularly brilliant and elegant species. The crown of the head covered with long blue-green feathers, barred with black, form a crest; the back, wings, and tail are of an exceeding fine ultramarine blue; the breast, belly, thighs, and covert-feathers under the tail are of a bright orange-colour; and the legs scarlet. Native of Madagascar. The next species demands a more lengthened notice.

The AMERICAN or BELTED KINGFISHER. (*Alcedo alcyon.*) This species is distinguished by being of a bluish slate-colour, with a ferruginous band on the breast; having a large collar of pure white round the neck; and an elevated crest on the head: legs extremely short. It inhabits all parts of the North American continent, and is the only species of its tribe found within the United States. "Like the love-lorn swains, of whom poets tell us," says Wilson, "he delights in murmuring streams and falling waters; not, however, merely that they may soothe his ear, but for a gratification somewhat more substantial. Amidst the roar of the cataract, or over the foam of a torrent, he sits perched upon an overhanging bough, glancing his piercing eye in every direction below for his

BELTED KINGFISHER.—(ALCEDO ALCYON.)

scaly prey, which, with a sudden circular plunge, he sweeps from their native element, and swallows in an instant. His voice, which is not unlike the twirling of a watchman's rattle, is naturally loud, harsh, and sudden; but is softened by the sound of the brawling streams and cascades among which he generally rambles. He courses along the windings of the brook or river, at a small height above the surface, sometimes suspending himself by the rapid action of his wings, like certain species of hawks, ready to pounce on the fry below; now and then settling on an old dead overhanging limb to reconnoitre. Mill-dams are particularly visited by this feathered fisher; and the sound of his pipe is as well known to the miller, as the rattling of his own hopper. Rapid streams, with high perpendicular banks, particularly if they be of a hard clayey or sandy nature, are also favourite places of resort for this bird; not only because in such places the small fish are more exposed to view, but because those steep and dry banks are the chosen situations for his nest. Into these he digs with bill and claws horizontally, sometimes to the extent of four or five feet, at the distance of a foot or two from the surface. The few materials he takes in are not always placed at the extremity of the hole, that he and his mate may have room to turn with convenience. The eggs are five, pure white, and the first brood usually comes out about the beginning of June, and sometimes sooner, according to the part of the country where they reside. They are very tenacious of their haunts, breeding for several successive years in the same hole, and do not readily forsake it, even though it be visited."

It is this species that Mr. Gosse, in his "Birds of Jamaica," thus prettily describes:—"Where the mangrove or the sea-grape stretches its branches down to the water's edge, stopping the way along the yellow beach, the Kingfisher delights to resort, sitting on a projecting twig; here he waits patiently for the approach of some small fish, on which he drops perpendicularly, and having seized it in his powerful beak, emerges from the wave, and returns to his former station to swallow it. It is a very shy and recluse bird; I have found scarcely any more difficult of approach: the posts of observation which he chooses are mostly such as command a wide view; and it was very wary; long before the gunner can creep within shot, the bird takes alarm, and darts away to a distant tree. Often as it sits watching, and sometimes at the moment of flying, it utters a loud rattling *churr*." "The form of the body of this bird, in conjunction with the head and beak, is wedge-shaped, the tip of the latter being the point. This form is admirably suited for its sudden and impetuous plunges upon its fishy prey: as the powerful texture, great size, sharp point, and cutting edges of the beak, are for holding it. The feathers of the throat and breast are of the closest texture, and lie on each other like scales, preventing the access of any water to the body, while, from their glossy, satiny surface, the water is thrown off instantly on emersion, as from the plumage of a duck. The feet again, though small, are muscular, the tarsus very short, the toes united into a broad, flat palm, and the claws unusually strong, short, and sharp. When one remembers that the Kingfisher digs his own cave out of the clayey or gravelly cliffs to the depth of several feet, we shall see the use of his strong and broad feet, as we may see it also in the Mole."

Mr. Waterton, in his 'Essays,' has furnished some interesting notes on the habits of the Kingfisher, with a selection from which we will conclude the article: "Modern ornithologists," says this gentleman, "have thought fit to remove the Kingfisher from the land birds, and assign it a place amongst the water-fowl. To me the change appears a bad one; and I could wish to see it brought back again to the original situation in which our ancestors had placed it: for there seems to be nothing in its external formation which can warrant this arbitrary transposition, The plumage of the Kingfisher is precisely that of the land bird, and, of course, some parts of the skin are bare of feathers; while the whole body is deprived of that thick coat of down so remarkable in those birds which are classed under the denomination of water-fowl. Its feet are not webbed, its breast-bone is formed like that of land birds; and its legs are ill calculated to enable it to walk into the water. Thus we see that it can neither swim with the duck, nor dive with the merganser, nor wade with the heron. Its act of immersion in the water is quite momentary, and bears no similarity to the immersion of those water-fowl which can pursue their prey under the surface, and persevere for a certain length of time, till they lay hold of it. Still the mode of taking its food is similar to that of the gulls, which first see the fish, and then plunge into the deep to obtain it; but this bird differs from the gull in every other habit." "If the Kingfisher is to be considered a water-bird merely because it draws its sustenance from the water, then our modern innovators ought to consider the osprey in the same light: and even the barn owl might give them a hint that she feels inclined to seek a new acquaintance; for I myself have seen her plunge into the water, bring out a fish, and convey it to her nest. Indeed, the swallow, with a still better grace, might ask permission to form a new division, distant both from land and water-birds, and call it ethereal; because it procures the whole of its sustenance from insects in the circumambient air." "I love to take my stand behind a large tree, and watch the Kingfisher as he hovers over the water, and at last plunges into it, with a velocity like that of an arrow from a bow. How we are lost in astonishment when we reflect that instinct forces this little bird to seek its sustenance underneath the water; and that it can emerge from it in perfect safety; though it possesses none of the faculties (save that of plunging) which have been so liberally granted to most other birds which frequent the deep! I sometimes fancy that it is all over with it, when I see it plunge into a pond, which I know to be well stocked with ravenous pike; still it invariably returns uninjured, and prepares to take another dip." "There are people who imagine that the brilliancy of the plumage of birds has some connection with a tropical sun. Here, however, in our own native bird, we have an instance that the glowing sun of the tropics is not required to produce a splendid plumage. The hottest parts of Asia and of Africa do not present us with an azure more rich and lovely than that which adorns the back of this charming little bird; while throughout the whole of America, from Hudson's Bay to Tierra del Fuego, there has not been discovered a Kingfisher with colours half so rich or beautiful. Asia, Africa, and America offer to the naturalist a vast abundance of different species of the Kingfisher. Europe presents only one; but that one is like a gem of the finest lustre."

KINKAJOU. (*Cercoleptes.*) A genus of Plantigrade Carnivora allied to the Coatimondis. It has a very long tail, which is prehensile at the end: the muzzle is short, the

KINKAJOU.—(CERCOLEPTES CAUDIVOLVULUS)

tongue slender and extensile; with two pointed molars before, and three tubercular ones behind. One species only is known, (*Cercoleptes caudivolvulus.*) [See POTTO.]

KITE. (*Falco milvus.*) This well-known bird may be distinguished from all the rest of the hawk kind by its forked tail. Its length is a little more than two feet, and its breadth five: the bill is two inches long, very much curved at the end, and of horn colour: the feathers on the head and neck are long and narrow, of a hoary colour, streaked with brown; legs yellow; claws black. It is almost perpetually on the wing; and appears to repose on the bosom of the air without making the least effort to support itself, so easy and elegant is its

KITE.- (FALCO MILVUS.)

motion there. It is, however, intent on its prey beneath; and as the young chicken, ducks, goslings, &c. suffer by the Kite's depredations, it is proscribed by the universal voice of every rural district. Were it not for this, its appearance would be welcomed as the harbinger of clear skies and fine weather; for it is in such that it makes its principal excursions. It breeds in large forests, or wooded hilly countries; and lays two or three eggs, of a whitish colour, spotted with pale yellow, and of a roundish form. In the breeding season it will at times approach near the outskirts of villages, seeking materials for its nest; but in general it avoids the haunts of man. The nest is usually in the fork of a thick tree, where it is concealed by the branches: the external part is formed of twigs, thickly matted together; and the interior is lined with wool, or some other soft and warm substance. The young remain a long time in the nest, and are extremely voracious in their appetite; so that to provide them with food requires considerable labour, and greatly heightens the parent bird's audacity.

There was a time when the Kite appears to have been of as much service in London, as the Vulture still is in some of the crowded cities of the East; for we read that in the reign of Henry VIII. the British metropolis swarmed with Kites, attracted hither by the various kinds of offal thrown into the streets, and that these birds fearlessly descended, and fearlessly performed the scavenger's office in the midst of the people, it being forbidden to kill them. When such a fact as this is brought before our eyes, the "street nuisances" of the present day appear like a comparative luxury; and we are apt to think that "metropolitan improvements" must have since gone on at such a rate that there can no longer be any room for them.

The MISSISSIPPI KITE. (*Elanus Mississippiensis.*) The celebrated American ornithologist, Wilson, thus introduces this species: "In my perambulations I frequently remarked this hawk sailing about in easy circles, and at a considerable height in the air, generally in company with the turkey buzzards, whose manner of flight it so exactly imitates as to seem the same species, only in minature, or seen at a more immense height. Why these two birds, whose food and manners, in other respects, are so different, should so frequently associate together in air, I am at a loss to comprehend. We cannot for a moment suppose them mutually deceived by the similarity of each other's flight: the keenness of their vision forbids all suspicion of this kind. They may perhaps be engaged, at such times, in mere amusement, as they are observed to soar to great heights previous to a storm; or, what is more probable, they may both be in pursuit of their respective food. One, that he may reconnoitre a vast extent of surface below, and trace the tainted atmosphere to his favourite carrion; the other, in search of those large beetles, or coleopterous insects, that are known often to wing the higher regions of the air; and which, in the three individuals of this species of hawk which I examined by dissection, were the only substances found in their stomachs. For several miles, as I passed near Bayo Manchak, the trees were swarming with a kind of cicada, or locust, that made a deafening noise; and here I observed numbers of the hawk now before us sweeping about among the trees like swallows, evidently in pursuit of these locusts; so that insects, it would appear, are the principal food of this species. Yet when we contemplate the beak and talons of this bird, both so sharp and powerful, it is difficult to believe that they were not intended by nature for some more formidable prey than beetles, locusts, or grasshoppers; and I doubt not but mice, lizards, snakes, and small birds, furnish him with an occasional repast.

"This hawk, which proved to be a male, though wounded and precipitated from a vast height, exhibited, in his distress, symptoms of great strength, and an almost unconquerable spirit. I no sooner approached to pick him up than he instantly gave battle, striking rapidly with his claws, wheeling round and round as he lay partly on his rump; and defending himself with great vigilance and dexterity; while his dark red eye sparkled with rage. Nowithstanding all my caution in seizing him to carry him home, he struck his hind claw into my hand with such force as to penetrate into the bone. The Mississippi Kite measures fourteen inches in length, and three feet in extent. The head and neck of a hoary white; the lower parts a whitish ash; bill, cere, lores, and narrow line round the eye, black; back, rump, scapulars, and wing-coverts, dark blackish ash; wings very long and pointed; the primaries are black, marked down each side of the shaft with reddish sorrel: all the upper plumage at the roots is white; the scapulars are also spotted with white; tail slightly forked, and, as well as the rump, jet black: legs vermillion, tinged with orange, and becoming blackish towards the toes; claws black; iris of the eye dark red; pupil, black. The long pointed wings and forked tail point out the affinity of this bird to that family or subdivison of the *falco* genus, distinguished by the name of Kites, which sail without flapping the wings, and eat from their talons as they glide along."

KITTEN [MOTHS]. A name given by collectors to Moths of the genus *Cerura*.

KIWI. A remarkable and curious bird of New Zealand, which we have described under APTERYX, (*Apteryx Australis*) or Wingless Emu. "These birds," the Rev. W. Yate observes, "hide themselves during the day; and come out of their retreats, which are generally small holes in the earth, or under stones, at night, to seek for their food. They run very fast, and are only to be caught by dogs, by torch-light, which they sometimes kick and bruise severely. They are highly prized, when taken, which is very rarely, by the natives; and their skins are kept till a sufficient number are collected to make into a garment. I have only seen one garment made of skins of this bird, during my six years and a half residence in New Zealand: and no consideration would induce the man to whom it belonged to part with it." The flesh is black, sinewy, tough, and tasteless. [See APTERYX.]

KNOTHORN [MOTHS]. A name given by collectors to Moths of the genus *Phycita*.

KOKAKO. The name given by the natives of New Zealand to a corvine bird, called, by some, the New Zealand Crow. [See GLAUCOPIS CINEREA.]

KOODOO, or STRIPED ANTELOPE. (*Antilope strepsiceros.*) This magnificent animal has no rival among the Antelope genus for beauty and height, or for bold and widely-spreading horns. It is eight feet in length and four feet in height at the shoulder; with ponderous horns beautifully twisted, having a prominent spiral ridge running obliquely from the base to the point, and extending to the length of about four feet. The colour of the back and sides is a light brown, with a narrow white band along the spine, and several similar stripes descending obliquely down the sides and hips; the belly and under parts being of a pale hue. The head is large, the ears broad, and the limbs thick and robust; yet, notwithstanding its heavy make, it takes long bounding leaps with surprising agility. It inhabits the woody parts of Caffraria, along the banks of the rivers; and when pursued takes to the water.

KOODOO, OR STRIPED ANTELOPE.
(ANTILOPE STREPSICEROS.)

KUKUPA. A beautiful species of Wood-pigeon known by this name in New Zealand, where it is very plentiful. It is described by the Rev. W. Yate as "much larger than the largest wild or tame pigeons in England, and has a plumage unrivalled among the extensive family of doves for splendour and variety: green, purple, and gold are, however, the prevailing colours. It is a heavy-flying bird, which makes it an easy prey to the hawks, with which the woods abound. They are easily killed with a spear or a musket; and if two birds are found upon the same tree, they are either so sluggish or stupid as not to fly when one is either killed or wounded. They feed upon the berries of the *Miro;* are most delicious eating; and are in season from January to June. The natives destroy vast numbers of these birds, and value them much, on account of both the quantity and the quality of their flesh.

LABRUS: LABRIDÆ. A genus and family of Acanthopterygious fishes, the species of which are very numerous in tropical seas; and even on our own shores they are abundant. The *Labridæ* family (Wrasses or Rock-fish, as they are also called) are chiefly remarkable for their thick fleshy lips, their large and strong conical teeth, their oblong scaly body, and their brilliant colours. They are further generically distinguished by a single dorsal fin, extending nearly the whole length of the back. [See WRASSE.]

LABYRINTHODON. [See SUPPLEMENT.]

LACERTA: LACERTIDÆ. A genus and family of reptiles. [See LIZARD.]

LACKEY [MOTHS]. A name given by collectors to species of Moths of the genus *Clisiocampa*.

LADY-BIRD. The popular name given to a well-known genus of coleopterous in-

LADY-BIRD AND LARVA.

sects, which are sometimes seen, in vast numbers, in hop plantations, &c., where they are of infinite service in destroying the various species of *Aphides*, which are so prejudicial to certain plants and fruit-trees. [See COCCINELLA.]

LÆMODIPODA. The name of an order of marine Crustaceans, with sessile eyes, and in which the posterior extremity of the body exhibits no distinct branchiæ. The body is almost linear or filiform, and with the head consists of eight or nine segments, with some small tubercle-like appendages at its posterior and inferior extremity: the limbs are terminated by a strong hook. The females carry their ova beneath the second and third segments of the body in a pouch formed of approximated scales. The *Cyamus Ceti*, or Whale louse, is an example of this order.

LAGOMYS. A genus of Rodent Quadrupeds, separated from the Hares. The ALPINE LAGOMYS (*Lagomys Alpinus*) has sometimes been confounded with the Varying Hare, in consequence of the latter having also obtained the name of *Alpine;* but is a far smaller animal, scarce exceeding a Guinea-pig (*Cavia copayba*) in size, and measuring only nine inches in length; while it has a long head, and the ears are short, broad, and rounded. It is a native of the Altaic mountains, extending even as far as Kamtschatka; inhabiting woody tracts amidst rocks and cataracts, and forming burrows beneath the rocks, or lodging in their fissures. In fair weather they seldom leave their holes in the day-time; but when the weather is dull they are seen running about among the rocks, and frequently uttering a sort of whistle or chirping bird-like sound. During the autumn they prepare for their winter support, by collecting a plentiful assortment of the finest herbs and grasses; which, after drying in the sun, they dispose into heaps of various sizes, according to the number of animals employed in forming them: these are easily distinguishable even through the deep snow, being often several feet in height and breadth. These little hayricks, raised by their industrious labours, are often found of great service to the adventurous sable-hunters, whose horses would perish were it not for the supplies which they thus occasionally discover. For this reason the Alpine Hare has a name among every Siberian and Tartar nation where it is found: a circumstance which marks its importance to society; for few animals, so diminutive, are noticed in those regions, unless possessed of some valuable or attractive qualities.

The OGOTONA HARE. (*Lagomys Ogotona.*) This little animal, whose length is only six inches, inhabits the vast deserts of Mongolia, and the frontiers of Chinese Tartary, living in sandy plains or on rocky mountains. It sometimes burrows under the soil, or conceals itself under heaps of stones, and forms a soft nest at no great depth from the surface. Before the approach of winter these animals collect large quantities of herbs, with which they fill their holes; and, directed by the same instinct as the Alpine Lagomys, they also form hemispherical ricks of hay, about a foot high, for their support during the inclement season. The colour of the Ogotona Hare is a pale brown above, and white beneath: on the nose is a yellowish spot, which colour is seen on the outsides of the limbs and the space about the rump. Hawks, magpies, and owls indiscriminately prey on them; but their most formidable enemies are the cat, the fitchet, and the ermine.

The CALLING HARE (*Lagomys pusillus*) extremely resembles the Ogotona Hare, just described, but is rather smaller. The head is long, and covered with fur to the very tip of the nose; the ears are large and rounded; and the legs very short. The whole body is covered with very soft, long, smooth fur, of a brownish lead colour, with the hairs tipped with black; but on the sides a yellowish tinge prevails. It is an inhabitant of the south-east parts of Russia, and is an animal of so solitary a nature, that it is very rarely to be seen even in places it most frequents. It commonly chooses its residence

CALLING HARE.—(LAGOMYS PUSILLUS.)

in some dry gentle declivity, where the turf is firm and covered with bushes: it there forms an obliquely descending burrow, the entrance of which is scarcely more than two inches in diameter; and so numerous and intricate are the avenues which lead to their retreats, that they would with great difficulty be discovered, did not their voice betray them. This voice resembles the piping of a quail, but is so loud that it may be heard at a surprising distance, particularly as there is nothing in the structure of its organs which can account for so powerful a tone. These little animals are of an extremely gentle disposition, and easily tamed. Their pace is a kind of leaping motion, but not very quick, nor do they run well, on account of the shortness of their legs.

LAGOPUS. [See PTARMIGAN.]

LAGOSTOMUS. A genus of Rodent Mammalia, in which the fore feet are furnished with four toes, the hinder with three only, as in the Cavies, all of them armed with stout claws adapted for digging. The ears are of moderate size, and the tail comparatively short. Their three anterior molars of the upper jaw consist each of two double layers, and the last of three. The only known species (*Lagostomus trichodactylus*) is about the size of a Hare, and inhabits Chili and Brazil: its general colour is grayish; the fur of two sorts, one entirely white, and the other, which is coarser, black, except at the base; the under parts white. Its motions are quick, and resemble those of a Rabbit; and it seeks its food by night,

LAGOSTOMOS VISCACCIA.

subsisting wholly on vegetables; inhabits the level country; and is not esteemed as food. It has very generally obtained the name of *Viscacha;* and it has also been figured in Griffith's edition of Cuvier's Regne Animal under the name of the *Diana Marmot.*

LAGRIIDÆ. A family of small Coleopterous insects, found in woods and hedges, and upon plants, counterfeiting death when alarmed, like the Cantharidæ. The head and thorax are narrow; the elytra soft and flexible; and the antennæ filiform. Their bodies are soft, and although they creep but awkwardly, they are active on the wing. The larvæ are found in the winter, under dead oak leaves, upon which they feed: when disturbed, they roll themselves up, with the head bent towards the tail; and they assume the pupa state without forming any cocoon. The species are few in number, but widely dispersed.

LAMB. The young of the SHEEP.

LAMBRUS. A genus of short-tailed Crustacea, most of the species of which are tropical: many of them have very long fore-legs, and are curiously covered with knobs and spines.

The *Eurynome aspera* is the only member of this group found in the British seas.

LAMELLIBRANCHIATA. An order of acephalous (headless) molluscs, in bivalve shells; all the species being aquatic. In these the mouth is not situated upon a prominent part of the body, nor assisted in its choice of food by organs of special sensation in its neighbourhood; but the entrance to the stomach is buried between the folds of the mantle. The shell of these animals is composed of particles of carbonate of lime, exuded from the surface of the mantle, and contained in the cavities of cells, or between layers of membrane; and a constant relation is preserved between the size of the animal and that of its shell. The valves are connected together in various ways. In the first place, they are jointed by a hinge; which is sometimes formed by the locking of a continuous ridge on one valve into a groove in the other, and sometimes by little projections which fit into corresponding hollows in the opposite valve. Near the hinge is fixed the *ligament;* which is composed of an elastic animal substance, and answers the purpose of binding the valves together, and at the same time keeping them a little apart, which may be regarded as their natural position. The Lamellibranchiata have usually more power of motion than the other Acephala; but they do not in general attain any great size. They are distributed over the whole globe, principally frequenting the shores or shallows; but the largest kinds are only found in warm latitudes.

LAMELLICORNES. The name by which an extensive section of Coleoptera tribe is distinguished. With respect to the size of the body, and the variety of forms exhibited in the head and thorax, it is one of the most beautiful of the coleopterous order; while those species which in their perfect state live upon fresh vegetable substances display metallic colours of great brilliancy: the majority, however, are of an uniform black or brown colour. All have wings; and they crawl but slowly on the ground. They feed on manure and other decomposed substances; but some species subsist on the roots of vegetables, and in their larva state do great injury to the cultivator. The antennæ are always short; they usually consist of nine or ten joints, and are terminated in a club, generally composed of the three last, which are lamellar, and are either arranged like a fan, a comb, or the leaves of a book. The larvæ have the body long, nearly semi-cylindrical, soft, often transversely wrinkled, whitish-coloured, twelve-jointed, with the head scaly, armed with strong jaws and six scaly feet. A general idea of their form may be obtained from that of the grub which produces the common Cockchafer. Some species do not change to pupæ until they have passed three or four years as larvæ; they form for themselves in their retreats, with the earth or the debris of the materials they have gnawed, a cocoon of an ovoid form, or in the shape of an elongated ball, of which the particles are fastened together with a glutinous secretion.

LAMELLIPEDES. The term applied to the third section of the order *Conchifera Dimyaria*, containing Bivalves, with the foot of the animal broad and thin; as in the genus *Cardiacea*, &c.

LAMPREY. (*Petromyzon marinus.*) This fish has a long and slender body, nearly cylindrical, resembling an eel; and its skin, which has no scales, is covered with a glutinous mucus. The Marine or Sea Lamprey sometimes grows to a very large size (three feet in length); the British specimens, however, are generally far inferior in magnitude. The usual colour of the Lamprey is a dull brownish olive, clouded with yellowish-white variegations; the back darker than the other parts, and the abdomen paler: the fins are tinged with dull orange, and the tail with blue. The mouth is of a round form, resembling that of a leech, and, like it, possesses the power of sucking and adhering to stones or other substances with extraordinary tenacity. The tongue, which moves to and fro like a piston, and which is the principal instrument in the act of suction, is furnished with two longitudinal rows of small teeth, and the mouth is lined with several circular rows. On the top of the head is a small orifice or spout-hole, through which is discharged the superfluous water taken in at the mouth and gills; and on each side the neck is a row of seven equidistant spiracles, or breathing-holes. In reference to this respiratory apparatus Mr. Owen has remarked, that "when the Lamprey is firmly attached, as is commonly the case, to foreign bodies, by means of its suctorial mouth, it is obvious that no water can pass by that aperture from the pharynx to the gills; it is therefore alternately received and expelled by the external apertures." The first dorsal fin, which is rather shallow, with a rounded outline, commences towards the lower part of the back; the second is nearly of the same extent, but with a subtriangular outline: the tail is short, and slightly rounded.

The Lamprey usually quits the sea in the spring for the purpose of spawning, and after an absence of a few months returns to its original marine element. When in motion this fish is observed to swim with considerable vigour and rapidity, but it is more commonly seen attached by the mouth to some large stone or other substance, the body hanging at rest, or obeying the motion of the current, so strong is its power of suction. Its general habits seem pretty much to resemble those of the eel; and, like the eel, it is remarkably tenacious of life. The Lamprey, though its ancient repute no longer remains, is still considered as a delicacy at certain seasons of the year; and the potted Lampreys and Lamperns of Worcester are in high estimation; those taken in the Severn being preferred to all others. During cold weather, this fish conceals itself in the crevices of rocks; and it is a usual expedient with anglers to form pits extending to the water-side in the vicinity of its haunts; into these a little blood is thrown, to induce the Lamprey to come forth, when it is readily taken.

The RIVER LAMPREY, OR LAMPERN. (*Petromyzon fluviatilis.*) This well-known species inhabits fresh waters, and is common in the Thames, the Severn, the Dee, the Tweed, &c. It is from twelve to fifteen inches in length; has a rounded head, a slender cylindrical body for about two-thirds of its length, and then compressed to the end of

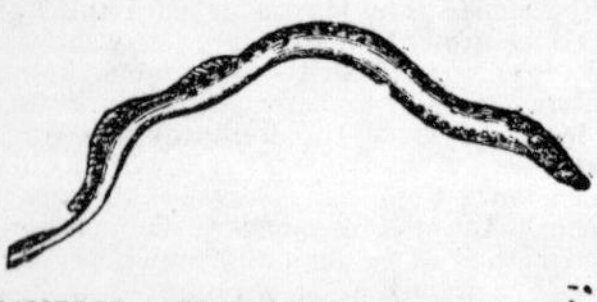

LAMPREY,—(PETROMYZON FLUVIATILIS.)

the tail. "Formerly," says Mr. Yarrel, "the Lampern was considered a fish of considerable importance. It was taken in great quantities in the Thames from Battersea Reach to Taplow Mills, and was sold to the Dutch as bait for the Turbot, Cod, and other fisheries. Four hundred thousand have been sold in one season for this purpose, at the rate of forty shillings per thousand. From five pounds to eight pounds a thousand have been given; but a comparative scarcity of late years, and consequent increase in price, has obliged the line fishermen to adopt other substances for bait. Formerly the Thames alone supplied from one million to twelve hundred thousand Lamperns annually. They are very tenacious of life, and the Dutch fishermen managed to keep them alive at sea for many weeks." Great quantities are also taken in the rivers of Germany: after being fried, they are packed in barrels by layers, between each of which is a layer of bay leaves and spices, sprinkled over with vinegar; and in this state they are sent to other countries. This species spawns in April and May. It feeds on insects, worms, &c., and is a prolific fish.—There are a few more species, of a smaller size; but in all the main characteristics they correspond with the foregoing.

LAMPYRIDÆ. A family of Coleopterous insects, having for its type the genus *Lampyris*. The *Lampyridæ* are distinguished by having five joints to all the tarsi; flexible elytra; and the body usually elongated and somewhat depressed; by the thorax projecting more or less over the head; small mandibles, terminated by a sharp point; the penultimate joint of the tarsi always bilobate; the terminal claws simple; and the antennæ approximated at the base. In some species the females are apterous, and in others furnished only with short elytra. They are voracious in their habits; preying in the larva state upon the bodies of snails, and not upon plants. The species are, for the most part, exotic, and are often ornamented with red or yellow and black colours. Scarcely any exceed an inch in length. When alarmed, they fold their antennæ and legs against the body, and remain motionless, as though dead; many, also, at such times, bend their abdomen downwards. The three most important genera are *Lycus*, *Omalisus*, and *Dictyoptera*. [See GLOW-WORM.]

LANNER. (*Falco lannarius.*) A bird of the long-winged Hawk kind, rather less than the buzzard. It breeds in France, where it continues the whole year, is very docile, and seems well adapted to all the purposes of hawking. It is also met with in Ireland, and is thus described by Pennant: The ear is a palish blue; the crown of the head, brown and yellow clay-colour; above each eye a broad white line passes to the hind part of the head; and beneath each a black mark points downwards. The throat is white; the breast is tinged with dull yellow, and marked with brown spots pointing downwards; the thighs and vent are spotted in a similar manner; the back and coverts of the wings are a deep brown, edged with a paler tinge; the quill-feathers are dusky; the inner webs are marked with oval rust-coloured spots; and the tail is spotted in the same manner as the wings.

LANTERN-FLY. (*Fulgora lanternaria.*) This insect is curious both on account of its size and its singular properties. It is nearly three inches and a half in length from the tip of the front to that of the tail, and about five inches and a half broad with its wings expanded: the body is of a lengthened oval shape, sub-cylindric, and divided into several rings or segments; while the head is distinguished by a singular prolongation, which sometimes equals the rest of the body in size. In this projection the luminous property of the Lantern-fly is said to exist; but the luminosity of this insect — of which there are several species — is doubted by most naturalists; who say, that if it really exists, it is only at particular seasons. It is therefore but right that we should give our authority: — Madame Merian, in her work on the Insects of Surinam, says, "The Indians once brought me, before I knew that they shone by night, a number of these Lantern-flies, which I shut up in a large wooden box. In the night they made such a noise that I awoke in a fright, and ordered a light to be brought; not knowing from whence the noise proceeded. As soon as we found that it came from the box, we opened it; but were still much more alarmed, and let it fall to the ground in a fright, at seeing a flame of fire come out of it; and as many animals as came out, so many flames of fire appeared. When we found this to be the case, we recovered from our fright, and again collected the insects, highly admiring their splendid appearance."

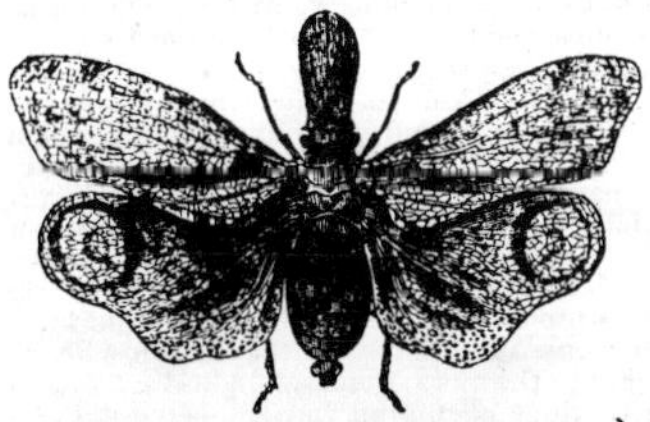

LANTERN-FLY.—(FULGORA LANTERNARIA.)

The ground-colour is an elegant yellow, with a strong tinge of green, and marked with numerous bright brown stripes and spots: the wings are very large, and the lower pair are decorated with a large eye-shaped spot on the middle of each, the iris or border of the spot being red, and the centre half red and half semitransparent white; the head or lantern is pale yellow, with longitudinal red stripes. This beautiful insect is a native of several parts of South America. [See FULGORA.]

The *Fulgora candelaria*, a native of China, is a much smaller species; measuring about two inches in length, and two inches and a half in breadth. The body is oval, and the head produced into a long horn-shaped process: the colours are very elegant; the head and horn being of a fine reddish brown, and covered with numerous white specks: the thorax is of a deep yellow, and the body black above, but deep yellow beneath: the wings are oval; the upper pair blackish, with numerous green reticulations, dividing the whole surface into innumerable squares, and farther decorated by several yellow spots: the under wings are orange-coloured, with broad black tips.

LAP-DOG. The little pets of the drawing-room and boudoir who bear the enviable appellation of Ladies' Lap-dogs, and who for years past have been growing "small by degrees and beautifully less," belong to that race of Dogs which have been described as "timid, fond, and affectionate — the most grateful for kindness, the most patient under ill-treatment;" — of course we mean the SPANIELS; and, therefore, under that word will the Lap-dog's zoological character be found. Nature originally, without doubt, had some hand in the production of these highly-favoured diminutives; but her empire over them has long been usurped by Fancy and Fashion, who have agreed— that the ears of these companions of female loveliness should be remarkably long and full, and the hair (of the ears more especially) plentiful and beautifully waved; that "liver-colour-and-white," though its pretensions to beauty are but moderate, is not to be despised; that "black-and-white" is entitled to our particular regard; but that the dear little "black-and-tan" variety is vastly to be preferred to either; while it is absolute treason to honour any with the title of "King Charles's breed" which do not possess certain indubitable signs of royal descent, as a black-roofed mouth, &c. There is also a variety of the Spaniel, generally of a white colour, and the hair of which is extremely long: it is called the Maltese dog, and is said to be one of the most elegant of the Lap-dog tribe.

LAPPET [MOTH]. A name given by collectors to species of Moths, of the genera *Gastropacha* and *Eutricha*.

LAPWING or PEEWIT. (*Vanellus cristatus.*) This bird, which is about the size of a pigeon, belongs to the snipe and plover tribe. It is found in this country in large flocks, except during the pairing season, when it separates for the purposes of incubation. It builds a slight inartificial nest on the ground, and lays four eggs of an olive cast spotted with black. The Lapwing's bill is black; the crown of the head and the crest are of a shining black; the cheeks and sides of the neck, white: the throat and forepart of the neck are black; the hind part, a mixture of red, white, and cinereous. The back and scapulars are of a glossy green colour, the latter variegated with purple: the small wing-coverts are of a resplendent black blue and green hue; the greater quill-feathers are black; and the breast and belly are white. The vent and coverts of the tail are orange-coloured; the tail is black and white; and the legs are red. The young birds run about very soon after they are hatched. During this period the old ones are very assiduous in their attention to their charge : on the approach of any person, they flutter round his head with great inquietude, and if he persists in advancing they will endeavour to draw him away, by running along the ground as if lame, and thereby inviting pursuit. It subsists chiefly on worms and the animalcula of the sea-shore. These birds are very lively and active during their love season, being almost continually in motion, sporting and frolicking in the air in all directions, or springing and bounding from spot to spot with great agility.

"Far from her nest, the lapwing cries 'away.'"—*Shaks.*

In the month of October they are in good condition for the table, and their eggs are considered a delicacy. [See TERUTERO.]

LARIDÆ. Birds of the Gull tribe, all of which are oceanic in their habits, and distinguished for great powers of flight. [See GULL and LESTRIS.]

LARK. (*Alauda.*) There are many species of this bird, and their great characteristic distinction from other birds consists in the extreme elongation in an almost straight line of their hinder claws ; by this formation the prehensile faculty is nearly destroyed, and consequently, with the exception of a few species with shorter claws, they are incapable of perching upon trees. The bill is straight, slender, bending a little towards the end, and sharp-pointed : the toes are all divided to their origin, the nostrils are covered with feathers, and the tongue is bifid. These birds are famed for singing during flight ; and there is something very delightful in listening to their melodious strains when the performers are soaring aloft, beyond the reach of human ken. From the situation of their nests they are greatly exposed to the attacks of predaceous animals of the weasel kind, which destroy great numbers of the eggs and young. The species which first claims our notice is

The SKY-LARK. (*Alauda arvensis.*) This delightful songster, the most harmonious of the whole family, is universally diffused throughout Europe, and is everywhere extremely prolific. It is about seven inches in

SKY-LARK.—(ALAUDA ARVENSIS.)

length : bill dusky, the base of the under mandible yellowish : the feathers on the top of the head are dusky, edged with rufous brown ; they are rather elongated, and may be set up as a crest : the plumage on the upper part of the body is reddish-brown, with the middle darkest, and the edges rather pale : the upper part of the breast is yellow, spotted with black ; and the lower part of the body is a pale yellow. The tail is dusky brown ; legs dusky ; claws dusky ; the hind one being very long, straight, and strong. The male is of a deeper colour, and larger than the female ; and is further distinguished by having the hind claw longer. The species is subject, however, to considerable variety ; and has even been found of a pure white colour. The Sky-lark commences his song early in the spring, continuing it during the whole summer, and is one of the few birds that chant whilst on the wing. When it first rises from the earth, its notes are feeble and interrupted ; as it ascends, however, they gradually swell to their full tone, and long after the bird has reached a height where it is lost to the eye, it still continues to charm the ear with its melody. It mounts almost perpendicularly, and by successive springs, and descends in an oblique direction, unless when threatened with danger, when it drops like a stone. The female forms her nest on the ground, close to some turf, which serves at once to hide and shelter it ; sometimes in corn-fields ; and, at others, in various sorts of pasturage. She lays four or five dirty white eggs, blotched and spotted with brown ; and she generally produces two broods in a year. These prolific birds live on seeds and insects, they are most abundant in the more open and highest cultivated situations abounding in corn, being but seldom seen in extensive moors at a distance from arable land. In winter they assemble in vast flocks, grow very fat, and are taken in great numbers for the table.

The WOOD-LARK (*Alauda arborea*) greatly resembles the Sky-lark, though it is much smaller, and the colours are less distinct. The feathers on the crown and upper parts of the body are marked with dusky spots edged with light reddish brown : from the beak over the eye is a narrow yellowish white band surrounding the crown of the head ; the feathers over the ears are brown, beneath which is another light band : quills dusky ; neck and breast yellowish white, tinged with brown, and marked with dusky spots : tail short ; the four outer feathers on each side black, with dirty white tips ; tail-coverts very long and brown : legs yellowish flesh-colour : hind claws long, and slightly bent. It is generally found near the borders of woods, perches on trees, and sings during the night, so as sometimes to be mistaken for the nightingale. When kept in a cage, near one of the latter birds, it often strives to excel it, and, if not speedily removed, will fall a victim to emulation. This species can be easily distinguished from the Sky-lark during flight, as it does not mount in the air in a perpendicular manner, and continue hovering and singing in the same spot like that bird; but will often rise to a great height, and keep flying in large irregular circles, singing the whole time with little intermission, sometimes for an hour together.

WOOD-LARK.—(ALAUDA ARBOREA.)

Its song, though not so loud as that of the Sky-lark, is more melodious, and may sometimes be heard in fine weather even in the depth of winter. This bird feeds on grain, seeds, and insects: its nest is placed under a tuft of high grass or furze, or in a low bush; and is made of dry grass, lined with finer grass and hair. The female lays four or five eggs, of a dusky colour, interspersed with deep brown spots, and, like the former species, often raises two broods in the year. It is a general inhabitant of Europe, but not so plentiful as the Sky-lark. It is more abundant in Devonshire than in any other part of England. These birds get very fat after harvest, and are taken in great numbers.

The CRESTED LARK (*Alauda cristata*) is distinguished from other species by the feathers on the crown of the head being much elongated and forming a crest, which is darker than the rest of the plumage. The back is ash-coloured, spotted with brown; the breast and belly yellowish white; and the throat is beautifully spotted. The tail is rather short; the two outer feathers, with their exterior edges, white, tinged with red. The song of this bird is fine, but not equal to the Skylark: its aërial excursions are likewise shorter. Though found in many parts of Europe, frequenting the banks of lakes and rivers, it does not appear to be known in England.

The TIT-LARK. (*Anthus pratensis.*) This bird, which by the older writers was classed with the larks, belongs to a different genus and family; but may be described here. It is of an elegant and slender shape, five inches and a half in length, and nine in breadth. The bill is black; the back and head are of a greenish brown colour, spotted with black; the throat and lower part of the belly are white; the breast is yellow, spotted with black; the tail is dusky; and the feet are of a pale yellow colour. In many parts of England this is a very common bird; and is met with in marshes, barren moors, and mountainous heaths: its nest, made of dry grass and stalks of plants, lined with fine grass and horse-hair, is placed on the ground amongst furze and long grass. The eggs are generally six in number, but vary considerably in size and colour: and the Cuckoo is said to deposit its eggs among them. During the period of incubation, *the male* pours forth its short but pleasing song; it generally springs into the air, increasing its song as it descends to the ground, from a height of 30 or 40 feet. In winter many Tit-larks betake themselves to the sea-shore.

The TREE TIT-LARK. (*Anthus arboreus.*) This bird greatly resembles the Tit-lark; but may be readily distinguished from it, by the bill being much broader at the base, and the legs being yellowish-brown instead of dusky. It is a solitary species, never associating in flocks, nor seen on the moors and downs, where the Titlark is most abundant. The nest of this species is placed only amongst high grass in the most cultivated parts, where there are plenty of trees. Its eggs, of which there are four, are of a dirty bluish white, blotched and spotted with purplish brown. Its flight is very peculiar, mounting up in a fluttering manner, and after some time descending to a neighbouring tree with motionless wing and expanded tail, and then alighting on the ground, warbling during its descent. It is chiefly found in the western and south-western counties of England.

There are other species enumerated by ornithologists; as the MEADOW LARK, a species common in many parts of Italy: the SHORE LARK, known as an inhabitant both of Europe and America, and very abundant in the latter continent: the BROWN LARK: the ROCK LARK, found at the Cape of Good Hope: the MARSH LARK, native of Germany: the SIBERIAN LARK: the RED LARK: the BLACK LARK, &c.

LARRIDÆ. A family of Hymenopterous insects, small in extent, and the species of which it is composed are but of moderate size. They are distinguished by the labrum being entirely or partially concealed, and the mandibles deeply notched on the inner side near the base. They are ordinarily found in sandy situations, and are fossorial in their habits. One species the beautiful and rare *Dinetus pictus*, is remarkable for the convoluted antennæ of the males; and the exotic genus *Palarus* is not less distinguished by the constricted segments of its abdomen.

LATHAMUS. A genus of Parrakeets found in Australia; as an example we may mention

LATHAMUS DISCOLOR, termed by the colonists of Van Diemen's Land the "Swift Parrakeet." During September and the four following months this migratory species is abundant in the gum forests, and very common in the shrubberies and gardens at Hobart Town; small flights of them continually flying up and down the streets and over the houses. They gather a fine clear honey from the fresh-blown flowers of the *Eucalypti* (especially *E. gibbosus*), which daily expand. They are quite fearless, and allow the inhabitants to pass within a few feet of their heads. Their eggs are laid in holes in the loftiest and most inaccessible trees. For other species see Mr. Gould's Birds of Australia.

LEAF-CUTTING BEES. [See MEGACHILE.]

LEECH. (*Hirudo.*) A genus of suctorial animals, or red-blooded worms, of aquatic habits, provided with a sucker at both ends of the body: the greater part are inhabitants of fresh water; some, however, are only found in the sea; while others live in moist situations near stagnant water, pursuing earth-worms, &c. Many of them accumulate their eggs into cocoons, enveloped by a fibrous excretion, at first sight so closely resembling sponge in structure as to have been once mistaken by a distinguished naturalist for a new genus of that family. The species which principally deserves our attention is

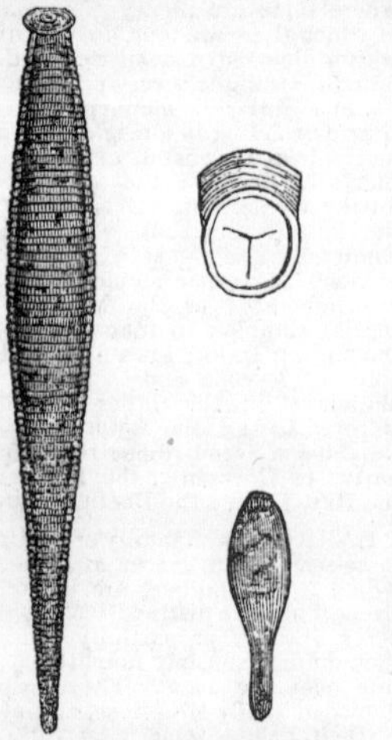

LEECHES.—(HIRUDO MEDICINALIS.)

the common Leech (the *Hirudo medicinalis* of Linnæus). This species, which is usually about the length of the middle finger, bears a considerable resemblance to the earthworm in its general structure, but differs as to the conformation of its mouth and digestive apparatus. Its skin is composed of from ninety to a hundred or more soft rings, by means of which it acquires its agility, and swims in the water. It has a small head; a black skin, edged with a yellow line on each side, and some yellowish spots on the back; and the belly, which is of a reddish colour, is marked with pale yellow spots. But the most remarkable part is the mouth, which is situated in the middle of the cavity of the anterior sucker; and three little cartilaginous bodies, or jaws, are seen to be disposed around it, in such a manner that the three edges form three radii of a circle. Each of these has two rows of minute teeth at its edge, so that it resembles a small semicircular saw. It is imbedded at its base in a bed of muscle, by the action of which it is worked, in such a manner as to cut into the skin, — a sawing movement being given to each piece separately. It is in this manner that the tri-radiate form of the leech-bite is occasioned; each ray being a separate little saw, this apparatus enabling the leech to penetrate the skin without causing a dangerous wound. The lacerated character of the wound is very favourable to the flow of blood; which is further promoted by the vacuum created by the action of the sucker. The alimentary canal consists of an œsophagus, a long stomach, with cæcal sacs, and an intestine. The operation of digestion is extremely slow, notwithstanding the rapid and excessive manner in which the Leech fills its stomach; a single meal of blood will suffice for many months; nay, more than a year will sometimes elapse, before the blood has passed through the alimentary canal in the ordinary manner, during all which period so much of the blood as remains undigested in the stomach continues in a fluid state. This accounts for the reluctance of the Leech, after being used to abstract blood, to repeat the operation; it not only being gorged at the time, but provided with a sufficient supply for so much longer. Indeed, the true medicinal Leech does not seem to take any solid aliment, but subsists on the fluids of frogs, fish, &c. Leeches are furnished with eight or ten simple eyes, which may be detected with a magnifying glass as a semi-circular row of black points, situated above the mouth upon the sucking surface of the oral disc; and to these visual specks it is supposed they are indebted for whatever sight they possess.

Leeches derive their principal interest from the use that is made of them as a remedial agent; but it should be observed that there are only two species so employed, and these are principally derived from the South of France, Sweden, Poland, and Hungary. It is common for the leech-dealers to drive horses and cows into the ponds, that the Leeches may fatten and propagate more abundantly by sucking their blood. Children are also employed to catch them by the hand; and grown persons wade into the shallow waters in the spring of the year, and catch the Leeches that adhere to their naked legs. In summer, when they have retired to deeper waters, a sort of raft is constructed of twigs and rushes, by which a few are entangled. They are also taken by laying baits of liver, to which the Leeches resort, and are then caught; but this last method is thought to make them sickly. A Leech may be known to be in good health if it be active in the water, and plump when taken out. The most certain method of inducing Leeches to bite, is to cleanse the skin thoroughly; and they should be exposed to the air for a short time previous to their application, as by this means they will bite more freely. If they are voracious, they may be applied to the part by being held lightly in the fingers, or they may be placed in a leech-glass, which is a preferable mode.

They should not be disturbed whilst sucking, nor the patient be exposed to too great warmth, or they will fall off; this they should always be permitted to do of their own accord. When the Leech has dropped off, it should be seized by the tail, and drawn between the finger and thumb, so as to cause it to disgorge most of the blood; or this may be effected by putting it in a week solution of common salt. It should then be placed in many successive fresh waters, and if not injured, it may be used again at a future time.

"The increasing scarcity of Leeches," as is remarked by Mr. Broderip, "renders their preservation and propagation objects of primary importance. The death of a vast number of Leeches is occasioned by errors in the method of keeping them. Though aquatic animals, it is not enough that they be supplied with water. They breathe by their entire surface, and are accustomed to change their skins every four or five days. Their body is covered, like that of all animals and plants which inhabit the water, by a shiny or mucilaginous fluid, which not only enables them to glide through the water, but keeps an aërial stratum in close contact with their respiring surface. When present in a limited degree, this mucous secretion is highly serviceable to them; in excess it is destructive. It is impossible for them to diminish it when it has accumulated, or to denude themselves entirely of their old skin, in water only. They must have some resisting body to creep over or through in order to accomplish this object." The most effectual method of preserving them appears to be that recommended by Fee; which is as follows:—"Into a marble or stone trough a layer of seven inches of a mixture of moss, turf, and charcoal of wood is to be put, and some small pebbles placed above it; at one extremity of the trough, and midway between the bottom and the top, place a thin plate of marble pierced with numerous small holes, upon which there should rest a stratum of moss, or portions of the equisetum palustre, or horse-tail, firmly compressed by a stratum of pebbles. The trough to be replenished with water only so high that the moss and pebbles should be but slightly moistened. A cloth is to be kept over the mouth of the trough. This is imitating as near as possible their natural condition, and the charcoal not only aids in keeping the water sweet, but appears to prevent the Leeches being attacked by parasitic animals, to which they are very liable. The water should be changed about once a week, and more frequently in warm weather." To judge of the vast numbers of Leeches that are required for medical uses, and of the great importance it is to ascertain the best method of preserving them, it is only necessary to state, that four only of the principal dealers in London import between seven and eight millions annually!

The Horse-Leech. (*Hirudo sanguisuga*, Linn.) The body of this species is depressed; and in the bottom of the mouth there are certain sharp tubercles. The mouth and tail are slender; the body is pretty thick; the belly is of a yellowish green colour, and the back is dusky. It is very common in shallow pools and stagnant waters.

The Mechanical Leech. (*Hirudo geometra*, Linn.) This species is found adhering to the trout and some other fishes after the spawning season. Its motions are performed by a particular expansion of the head and tail, as if measuring like a compass; and hence it receives its name. The body is greenish, spotted with white; and both ends are dilatable and equally tenacious.

The Tuberculated Leech. (*Pontobdella muricata.*) A marine species, which adheres strongly to fish, and leaves a black impression on the place. The body, which is taper and rounded at the greater extremity, is furnished with two small horns, strongly annulated, and tuberculated on the rings; and the tail is dilatable.

LEIODON. [See Supplement.]

LEIPOA OCELLATA. The "Native Pheasant" of the colonists of Western Australia; which in its habits is very like the domestic fowl. It deposits its eggs in a mound of sand, about three feet high, the inside being lined with layers of dried leaves,

OCELLATED LEIPOA.—(L. OCELLATA.)

grasses, &c. The bird never sits on the eggs, but leaves them to be hatched by the heat of the sun's rays. The natives are very fond of the eggs, and rob the mounds twice or thrice in a season. These mounds resemble ant-hills; and, indeed, ants often abound in them. Captain Grey observes that the nests are at least nine feet in diameter and three feet high. By the natives this bird is named *Ngowoo*.

LEMMING. (*Georychus lemmus.*) There are several species of this animal, varying in size and colour according to the regions they inhabit. They are found in Norway, Lapland, Siberia, and the northern parts of America; those of Norway being nearly the size of a water rat, and of a tawny colour, variegated with black, the sides of the head and the under parts being white; while those of Lapland and Siberia are scarcely larger than a field mouse, and much less distinctly marked. The head of the Lemming is large, short, thick, and well furred; the eyes and ears small; the body thick; and the limbs short and stout, especially the f..e legs: they have five toes on each foot, and the claws on the fore feet are strong, compressed, and rather crooked: the tail is very short, thick, cylindrical, and covered with strong hairs, disposed like those of a

LEMMING.—(GEORYCHUS LEMMUS.)

pencil at the tip. They subsist entirely on vegetable food. They form shallow burrows, in summer time, under the ground, and in winter make long passages under the snow in search of food. The most extraordinary characteristic of these animals is their migrations, which they undertake at irregular epochs — upon an average about once in ten years: these migrations are supposed to arise from an unusual multiplication of the animals in the mountainous parts they inhabit, together with a deficiency of food; and, perhaps, a kind of instinctive prescience of the severity of the approaching winter. They descend from the mountains in incredible numbers, and assemble in the plains; and then, as it were with one consent, they march on in a direct course, no obstacle deterring them, and nothing seeming to make them turn aside. If they are disturbed while swimming over a lake or river, they will not recede, but swim on, and soon re-assume their former order. They chiefly move at night, or early in the morning; so completely devouring the herbage as they pass, that the ground has the appearance of having been burnt. Exposed as they are to every attack from owls, hawks, weasels, &c., and so many being destroyed in attempting to cross rivers and lakes, the diminution of their numbers is very great; so that comparatively very few return to their native haunts. When enraged, they raise themselves on their hind feet, and utter a barking sound. They breed several times in the year, producing five or six at a birth. Formerly, so gross was the superstition of the common people of Norway, and so great their terror at these devastating marches, that they believed the Lemmings fell from the clouds; and they were actually exorcised by the clergy.

The HUDSON'S BAY LEMMING is of an ash colour, with a tinge of tawny on the back, having a dusky stripe along its middle, and a pale line on each side. The hair is very fine, soft, and long. It is known that they migrate like the foregoing species. It occurs in Labrador, and all parts of Northern America bordering on the Polar Sea. It has the character of being very inoffensive, and so easily tamed that, when caught, it will become not only reconciled to its situation in a day or two, but show a fondness for the caresses of its master.

LEMUR. A genus of Quadrumanous animals which approximate to the Monkey tribe in having opposable thumbs on both pairs of extremities, and to the ordinary quadruped in their elongated pointed head and sharp projecting muzzle: they are also void of that mischievous and petulant disposition which so much distinguishes the monkey tribe; and at the same time they differ from them in their dentition. They are all natives of Madagascar and of some of the smaller islands in its neighbourhood. The general form of the body is slender and elongated; the head shaped somewhat like that of a fox; and the eyes large, as in the generality of nocturnal animals. A long curved claw on the first fingers of the hind feet distinguishes them from all other quadrumana. Their hind legs are much longer than their anterior limbs, and for the most part they are excellent leapers. Gentle and harmless as these animals in general are, they will defend themselves with great resolution when attacked. In their natural haunts they associate in troops; but they are seldom seen abroad in the day-time, always as much as possible secluding themselves from the light. They subsist on fruit, insects, and small birds. Their fur is usually very fine and silky; and the tail long and bushy: there are some species, however, which are wholly destitute of a tail, and others where that member is merely rudimentary. [See LORI: for the *Flying Lemur*, see GALEOPITHECUS.]

LEO. The classical appellation for the Lion. [See LION.]

LEOPARD. (*Felis leopardus.*) A graceful and active animal belonging to the feline tribe, but so like the Panther as to be frequently taken for it by the mere casual observer. The principal difference is in size; the Leopard being considerably the smaller of the two, and of a paler yellow colour; while the ocelli or rounded marks on the Panther are larger, and more distinctly formed. Both animals are widely diffused through the tropical regions of the Old World; being natives of Africa, Persia, China, India, and many of the Indian islands. The general length of the Leopard, from

LEOPARD.—(FELIS LEOPARDUS.)

nose to tail, is four feet; and of the tail, two and a half; and so great is its flexibility of body, that it is able to take surprising leaps, to swim, climb trees, or crawl like a snake upon the ground, with nearly equal facility. When pursued, they often take refuge in trees, and occasionally spring upon their prey from the branches. In speaking of the Leopard, Mr. Swainson observes, "Although the names of Leopard and Panther have been long familiar in common language, and have conveyed the idea of two distinct species, yet it is perfectly clear that no scientific writer of the last generation either described, or indeed appeared to know, in what respects the animals differed. It seems that numerous specimens of what are called the Leopard are in the Zoological Gardens, and one has been figured in the book so entitled; but Mr. Bennett has not made the slightest attempt to investigate the subject, or to throw any light upon this difficult question. In this dilemma we shall therefore repose on the opinions of Major Hamilton Smith, whose long experience and accuracy of observation are well known, and whose authority in this department of nature deservedly ranks above that of any other naturalist of this country. The Leopard, as defined by Major Smith, when compared with the Jaguar and Panther of naturalists, is uniformly of a paler yellowish colour, rather smaller, and the dots rose-formed, or consisting of several dots partially united into a circular figure in some instances, and into a quadrangular, triangular, or other less determinate forms: there are also several single isolated black spots, which more especially occur on the outside of the limbs." Mr. Swainson then proceeds to say, "Our own opinion of the specific dissimilarity between the Leopard and the Panther, judging from what has been written on the subject, is in perfect unison with that of Major Smith; while the following remark of that observing naturalist, incidentally inserted in his account of the Panther of antiquity, seems to us almost conclusive:—'The open spots which mark all the Panthers have the inner surface of the annuli or rings more fulvous (in other words darker) than the general colour of the sides; but in the Leopard no such distinction appears, nor is there room, as the small and more congregated dots are too small to admit it.' In truth, if there is any reliance to be placed in the most accurate figures hitherto published, the small spots of the Leopard and the large ones of the Panther must strike even a casual observer, and lead him to believe that the two animals were called by different names." Like most feline animals, Leopards are fierce and rapacious; and, it is remarked, that though they are ever devouring, they always appear lean and emaciated. They are taken in pitfalls, covered over with slight hurdles, on which a bait of flesh is placed. Their skins are very valuable.

The HUNTING LEOPARD, or CHEETAH. (*Gueparda jubata.*) This species exhibits in its form and habits a mixture of the feline and canine tribes; so much so, indeed, as to have induced some naturalists to designate it as a distinct genus under the name of *Cynailurus*, or *Gueparda*. "Intermediate in size and shape between the Leopard and the hound, (observes Mr. Bennett, in the *Tower Menagerie*) he is slenderer in his body, more elevated on his legs, and less flattened on the fore part of his head than the former, while he is deficient in the peculiarly graceful form, both of head and body, which characterizes the latter. His tail is entirely that of a cat, and his limbs, although more elongated than in any other species of that group, seem

HUNTING LEOPARD.—(GUEPARDA JUBATA.)

to be better fitted for strong muscular exertion than for active and long-continued speed." Though the Hunting Leopard possesses much of the sagacity and fidelity of the dog, its anatomical structure and general habits are undoubtedly feline. The general ground-colour is a bright yellowish brown above, lighter on the sides, and nearly white beneath; marked with numerous small black spots on the back, sides, and limbs; and which are continued along the tail, so closely set as to appear like rings; the tip of the tail is white, as is also the whole of its under surface, with the exception of the rings just mentioned. The ears are short and rounded, marked with a broad black spot at the base, the tip and inside being whitish. The upper part of the head is of a deeper tinge than the rest; from each eye is a blackish line running down to the corners of the mouth, and the extremity of the nose, like that of a dog, is black. The fur does not possess that sleekness which distinguishes the feline race in general, but has a peculiar kind of crispness; and there is very little appearance of a mane, except that the hair is somewhat longer and more crisp along the back of the neck.

This useful and docile species, which it is believed might be reduced to a state of perfect domestication, inhabits the greater part both of Asia and Africa. In India and Persia, where they are employed in the chase, they are carried, chained and hoodwinked, to the field in low cars. When the hunters come within view of a herd of antelopes, the Leopard is liberated, and the game is pointed out to him: he does not, however, immediately dash forward in pursuit, but steals along cautiously till he has nearly approached the herd unseen, when with a few rapid and vigorous bounds he darts on the timid game, and strangles it almost instantaneously. Should he, however, fail in his first efforts and miss his prey, he attempts no pursuit, but returns to the call of his master, evidently disappointed, and generally almost breathless.

LEPADOGASTER. A genus of small Malacopterygious fishes, which have the power of attaching themselves to rocks and other hard substances, by means of the disc, whereby they are enabled to remain and find their food in situations where every other species of fish would be swept away by the current of the water. They have large pectorals reaching to the under side of the body; head broad and depressed; snout curved and protractile; body without scales; gills with little opening, and four or five rays: they have no air-bladder, but they swim briskly.—In the Rev. David Landsborough's 'Excursions to Arran,' the little two-spotted sucker-fish, (*Lepadogaster bimaculatus*), whose fry he found on opening a scallop, furnishes him with a subject which he treats in a very pleasant and edifying manner: "How wonderfully the Lord teaches the feeblest of his creatures to provide for their own safety and that of their offspring! What a charming nursery this little sucker-fish selects for itself! It is rather nice in its choice. It is not an old, weather-beaten scallop that it takes possession of, but one that is fresh without, and smooth and pure within. After it has entered, it certainly has some way of gluing the valves together, for it is not without difficulty that they can be torn asunder. Neither is it imprisoned, though the apartment is thus shut against intruders; for, closely as the valves cohere, there are some little apertures about the ears of the shell through which it can make its exit with its numerous family, or by which such little creatures as they feed on may, in their simplicity, enter."

LEPAS. A genus of Cirrhipedous animals, of which the Barnacle (*Lepas anatifera*) is a specimen. They adhere in clusters to rocks, shells, floating wood, and other extraneous marine substances, and, being incapable of changing place, are supposed to be true hermaphrodites. The word *Lepas*, in the Linnæan system, contains all the Cirrhipeds, or Multivalves. It was formerly applied to Limpets or Patella: in short, the ancient definition, "Concha petræ adhærens," would apply to any shells attached to rocks. The scientific world is much indebted to that distinguished naturalist, Charles Darwin, Esq., F.R.S., for an elaborate account of the structure, habits, and form of all the known species, which has been published in two volumes of the Ray Society's works. [See BALANUS.]

LEPIDOPTERA. An order of four-winged insects, containing some of the largest and most beautiful in nature, and comprehending all those usually ranked as Butterflies, Moths, and Sphinxes. The wings, which vary in size, figure, and position, are covered with a multitude of minute scales, which when rubbed appear like powder or farinaceous dust; the nervures of the wings being disposed chiefly in a longitudinal direction. The antennæ are composed of numerous minute joints, and are generally distinct. They are also furnished with a proboscis, composed of two sub-cylindrical tubes, between which there is an intermediate one, or sucker; and by means of it they are enabled to extract the nectar from flowers, that being the only aliment on which they subsist. The head, thorax, and abdomen are always more or less covered with hair.

In the Linnæan system, this order is composed of three genera: 1. *Papilio* (Butterflies); which in the Cuvierian system is represented by the *Diurna*: 2. *Sphinx* (the Hawk Moths); viz. the *Crepuscularia* of Cuvier, which mostly fly in the morning or evening twilight: 3. *Phalæna* (or Moths); called by Cuvier the *Nocturna*, or those which in general fly only during the night. Some of these are domestic pests, and devour cloth, wool, furs, feathers, wax, lard, flour, and the like; but by far the greatest number live wholly on vegetable food, certain kinds being exclusively leaf-eaters, while others attack the buds, fruits, seeds, bark, pith, stems, and roots of plants. The larvæ of Lepidopterous insects are well known by the name of *Caterpillars*. [See BUTTERFLY and CATERPILLAR.]

So great is the number of insects belonging to this order, that Dr. Burmeister supposes them to amount to nearly one sixth of the whole of the insect tribes. The imago state is characterized by several peculiarities not occurring in any of the other orders. "The body is compact, and densely clothed with hairs or scales: the head is free, not being received into a frontal prothoracic cavity, but attached by a narrow ligament; it is furnished at the sides with a pair of large granulated eyes, and its hinder part often with a pair of ocelli, which are generally hidden by the thick covering of hairs or scales: the antennæ are inserted on the upper part of the head, and are generally long and multiarticulate, very variable in form, and often very complicated in the males; the mouth, at first sight, appears to consist of a long and delicate spirally convoluted organ, which, when examined, is found to consist of two pieces, each of which is sometimes provided with a small jointed appendage or palpus at its base. This very slender proboscis (or antlia as it is called by Kirby and Spence) is employed to pump up the nectar of flowers, upon which alone it subsists, into the mouth and stomach of the insect, and which, from its peculiar construction, is admirably adapted for penetrating to the depths of the narrowest blossoms. When at rest, it is coiled up, and defended by two large and compressed palpi, composed of three joints inserted upon a fleshy piece, soldered to the front of the head." * * * * "The thorax is robust and compact, the prothoracic portion minute, owing to the fore-legs performing no supplemental functions, whilst the mesothorax, to which is attached the anterior pair of large wings, is greatly enlarged, the metathorax being again reduced in size. The prothorax bears upon its upper side a pair of organs, especially characteristic of the order, namely, a pair of scales covered with hair quite distinct from the wing-covers (tegulæ), which Kirby and Spence call patagia or tippets, but which have been overlooked by all other authors except Chabrier, who first discovered them." * * * * "The wings are attached to the lateral and superior parts of the meso- and meta-thorax, and are always present, except in a few species, of which the females alone are apterous, or have the wings reduced to small and useless appendages: these wings are of large size, and are not folded up; the two fine layers of membrane of which the wings are composed, like the upper and lower surface of a leaf, are kept expanded by a number of longitudinal corneous veins, or nerves, as they have been called." * * * * "The wings in this order offer another peculiarity, since, instead of being naked and transparent, they are clothed with a double layer of minute scales, somewhat resembling those of fishes. These scales, upon which the beauty of these insects so entirely depends, are easily detached in the form of a fine dust, and, when examined under the microscope, are exceedingly variable in their form, but generally more or less wedge-shaped, or oval; sometimes toothed or notched at the broadest end, and having a slender point at the base, by which they are attached upon the membranous surface of the wing, which, when denuded, presents the appearance of numerous minute impressions arranged in lines, in which the base of the scales are planted, being laid upon each other like the tiles on the roof of a house. The number of these scales is very great, there being more than 400,000 on the wings of the silkworm moth, according to Leuwenhoeck: in some species, however, the wings are partially, or even entirely, denuded of scales; and in others, small patches only are thus denuded, as in the great Atlas Moth. In many species, these scales exhibit the most brilliant metallic tints, so that in the bright light of the sun it is almost impossible to look upon them." * * * * "The variations in the colours and markings of the wings are almost as numerous as the species themselves: the sexes also often differ materially from each other; still some general principles are evident in the distribution of these colours and markings. Thus the *Pontiæ* and *Pierides* are almost uniformly white; *Colias* and its allies yellow; the *Fritillaries* rich brown, spotted with black and with silvery spots on the under side; *Hipparchia* and its allies ornamented with eye-like marks; the *Lycenæ* copper-coloured; the *Polyommati* fine blue, with small eyes on the under side: the *Zyganæ* with red under wings; the *Noctuidæ* with an ear-like mark in the middle of the fore wings; the *Geometridæ* with waved carpet-like marks."—We are indebted for the previous extracts to Mr. Westwood's excellent "Introduction to the Classification of Insects." We must refer our readers, who wish to study the subject more particularly, to the works of Dr. Boisduval, Messrs. Doubleday and Hewitson, and others. The recently published Lists of British Species, drawn up with so much study and care by Mr. H. Doubleday of Epping, and by Mr. Stephens (Cat. Lep. in Brit. Mus.) are indispensable to all who study the British species; as British authors, up to his time, have been apt to multiply species, and occasionally to misapply the names, from the want of authentic specimens to compare with their species. More particular information will be found under the different species of Lepidoptera described in the course of the work.

LEPIDOSIREN. A genus placed by some authors among the Fish, by others among the Amphibia: of late it has been the subject of many learned papers, abroad and at home; the best known species is named *Lepidosiren annectens*, and is a native of Africa.

Dr. Melville differs from Professor Owen with regard to the position of this remarkable genus in the Animal Kingdom, as he regards it as a true Amphibian. He rests its character on the absence of the super-occipital bone, the presence of the large epi- and basi-cranial bones, the non-development of the maxillary and intermaxillary bones; and especially the enormous magnitude of the Wernerian bones, which become subservient to mastication, and anchylosed to the expanded pterygoids; on the nostril being doubled; on the existence of external cutaneous gills during the adult condition, which did not occur in any fish; and on the co-existence of external and internal gills, with lungs: in other words, on its exhibiting the different modes of circulation, respiration, &c., in the produce—second stage of the larva of the frog and Amphiuma, or Menopoma, and other characters.

One species (about a foot long) inhabits the upper part of the river Gambia; and another (between two and three feet in length) is a native of the large rivers of South America. In its respiratory apparatus, it bears the closest correspondence with the Perennibranchiate Batrachia; but in many other points of its internal structure, it more resembles certain species of fishes. The African species is said to pass nine months out of the twelve in a state of torpidity; burying itself in the mud during the dry season, and again reviving when the sources of the river are swollen by the rains.

LEPIDOSTEUS. A genus of fishes with very bony polished scales, one species of which is found in the United States. Many allied genera are found in a fossil state.

LEPISMA. A Linnæan genus of Apterous insects; distinguished by an elongated body, covered with small scales, frequently silvery and brilliant. They have six feet, run with great velocity, and some of them by means of their caudal appendage are

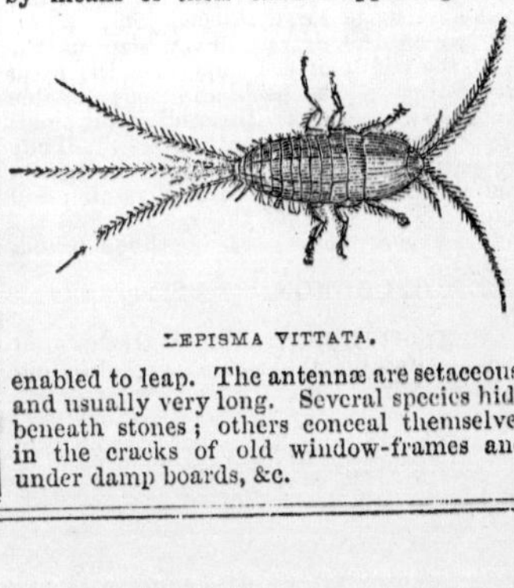

LEPISMA VITTATA.

enabled to leap. The antennæ are setaceous, and usually very long. Several species hide beneath stones; others conceal themselves in the cracks of old window-frames and under damp boards, &c.

LEPTIDÆ. A subfamily of Dipterous insects, distinguished by the proboscis being short and membranous; the lips terminal and thick; and the abdomen usually with five distinct segments.

LEPTOCEPHALUS, or ANGLESEA MORRIS. A Malacopterygious fish, characterized by a very small and short head, and a remarkably compressed body. It is common in the seas of hot climates. One specimen of it was taken on the coast of Anglesea by a gentleman named Morris, and is described by Pennant; but since that time many others have been found on our coasts. It is four inches long; head very small; the eyes large; lower jaw slender; numerous small teeth in each jaw; the body compressed sideways; extremely thin, and almost transparent: the bones forming the vertebræ have no spinous processes whatever; the dorsal and abdominal margins, as well as the lateral line, exhibit a series of small black specks; and its general opal-like hue and graceful motions give it a very pleasing appearance. It is usually found among sea-weed.

LEPTOCONCHUS. A genus of Mollusca, found in the Red Sea, where it is imbedded in calcareous masses of Polyperia. The head of the animal is furnished with a proboscis; two tentacula, with eyes in the middle; foot of moderate size, and no operculum. The shell is of a dirty white colour, subglobular, delicate, fragile, and translucent; spire low; aperture large, and furrowed externally.

LEPTOPHINA. The name given to a subfamily of serpents belonging to the family *Colubridæ*. They are characterized by a long and very slender body, slightly depressed: the head elongated, and narrowed before; and a very long, slender, and acutely pointed tail. "The whole of the serpents composing these genera live," Mr. Bell observes, "in woods, entwining themselves amongst the branches, and gliding from one to another. Their habits, combined with the graceful slenderness of their form, the beautiful metallic reflection from the surface in some species, and the bright and changeable hues in others, place them among the most interesting of the serpent tribe. Their food consists of large insects, young birds, &c., which the extraordinary size of the head, the width of the gape, and the great dilatability of the neck and body, enable them to swallow, notwithstanding the small size of these parts in a state of rest." They are perfectly harmless; and it is even said that children are in the habit of taming and playing with some of the species, and the snakes appear pleased at being thus caressed.

LEPTOPLEURON. [See SUPPLEMENT.]

LEPTOPTILUS. A genus of Grallatorial birds, containing the well known *Adjutant* of India [which see].

LEPTURIDÆ. The third family of Longicorn beetles, comprising such as have the eyes rounded, or very slightly emarginate; the antennæ of moderate length, inserted before the eyes; the head is inclined downwards, and narrowed into a neck at its union with the thorax, which is conical or trapezoid, and narrower in front than the head; the mandibles are acute at the tips; the elytra are narrowed to the tips, so as to give the terminal part of the body the appearance of an elongated but reversed triangle. These insects are of moderate size, active, and generally gaily coloured, being often ornamented with yellow markings; they are found either upon umbelliferous flowers in the hot sunshine, or on the trunks of trees, where they usually reside in their previous states.

One of the largest and finest of these beetles is a North American species, the *Desmocerus palliatus*, which appears on the flowers and leaves of the common elder towards the end of June and until the middle of July. It is of a deep violet or Prussian blue colour, sometimes glossed with green, and nearly one half of the fore part of the wing-covers is orange-yellow, suggesting the idea of a short cloak of this colour thrown over the shoulders, which the name *palliatus*, that is, cloaked, was designed to express. The head is narrow; the thorax is narrow before and wide behind, and has a little sharp projecting point on each side of the base. The larvæ live in the lower part of the stems of the elder, and devour the pith, as Dr. Harris informs us. In this country are many species, some of which are rather large and handsome. They are described in the works of Mr. Stephens.

LEPUS. [See HARE.]

LERNÆADÆ. A group of parasitic crustacea; one species of which infests the Sun-fish (*Orthogariscus*). The fish and its parasite are thus described in Capt. Grey's Travels in Australia: "We caught also a fish (*Orthogariscus*), which the seamen called a devil-fish. The length of it was six feet two inches; breadth from fin to fin, three feet six inches; length from tip of nose to pectoral fin, two feet; thickness through the breast, one foot six inches. This fish was infested about its nose with a kind of parasite (*Lernea*), having two long thin tails. The sailors stated that these animals frequently cause large sores about the nose of the fish, and that when suffering from this, it will allow the sea birds to sit on it, and peck away at the affected part. The habit of the fish is to swim during calms, with one of the hind fins out of water, and it is then harpooned from a boat. I have myself seen petrels perched upon them; and directly one of these fish was hoisted on board, the sailors looked for the parasites and found them. They were an inch long, and covered with a transparent shell marked with gray spots and lines; the hind part of the body, near the tail, being darker than the fore part, as if the intestines were seated there. These little creatures adhered strongly to any substance that they were laid on, and caused an irritating feeling to the skin, if placed on it; they swam with great rapidity when put into sea water, and in their movements in swimming much resembled a tadpole ; their tails were merely long transparent fibres."

ANTHOSOMA SMITHII

Our figure represents a species, *Anthosoma Smithii*, which derives its generic name from its body resembling the blossom of a flower. In Dr. Baird's British Entomostraca figures and descriptions of the British species are given.

LESTRIS. A genus of Palmipede Birds distinguished from the true Gulls by their membranous nostrils being larger and opening nearer to the point and edge of the beak; the tail also is pointed. The females are larger than the males, which is the reverse of what is observable in the genus *Larus;* and they lay but two eggs, of a dark colour.

LESTRIS PARASITICUS ; the ARCTIC GULL. This species is common in the northern parts of Europe, Asia, and America. Numbers of them frequent the Hebrides in the breeding season ; and they are also to be seen in the Orkneys, and on the coast of Yorkshire. They make their nest of moss, on the dry grassy tufts in boggy places, and lay two eggs of an ash colour, spotted with black. The length of the Arctic Gull is twenty-one inches : the bill is dusky, pretty much hooked at the end, but the straight part is covered with a sort of cere : the nostrils are narrow, and placed near the end. In the male the crown of the head is black ; the back, wings, and tail are dusky ; and the whole under side of the body is white : the legs are small, scaly, and black. The female is entirely brown. They are ravenous and ferocious to such a degree, that they pursue other gulls of a less vigorous and determined nature, whenever they observe them to have a prize worth contending for, and compel them to drop or disgorge their prey ; which the pursuer usually catches as it falls. Mr. Fisher, in his Journal of a Voyage to the Arctic regions, in H. M. SS. Hecla and Griper (1820), gives the following information on this subject : "Several Arctic Gulls were seen to-day for the first time. This bird is commonly called by our Greenland seamen the Boatswain, and sometimes Dirty Allen, a name somewhat analogous to that by which it is characterized by the Danes, viz. Stroudtjager, or Dung-bird. All these names have had their origin from a mistaken notion that these birds lived on the excrements of the lesser gulls, which, on being pursued, either from fear, or to relieve themselves from the persecution of fierce enemies, voided something to satiate the voracious appetites of their pursuers, and by that means escape from further molestation. The fallacy of this opinion is now, however, pretty generally known. That the Arctic Gulls do pursue those of their own genus which they can master (particularly the Kittiwakes) is an incontestable fact ; but the object of their pursuit is not the excrement, but the prey that the pursued is at that time possessed of, and which at length they are forced to drop, to secure their own safety ; which they effect during the time that their enemy is employed picking it up, although that is done in a very short period, for they manage the business with such dexterity, that the object dropped is generally caught before it reaches the water."

LESTRIS CATARACTES ; the SKUA GULL. This is the most formidable of all the Gull kind, preying not only on fishes, but also on the smaller kinds of water-fowl, and, as some assert, even on young lambs. It is a stout bird, two feet in length, and between four and five from tip to tip of the extended wings. The bill is dark, strong, much hooked, sharp at the tip, and covered to the nostrils with a kind of cere. The whole upper plumage is of a deep brown, edged with a dull rust colour ; the under parts being considerably lighter : the tail is white at the root, the shafts are of the same colour, and the webs of deep brown : the legs and toes are covered with coarse black scales ; and the claws are strong and hooked. This fierce species is met with in the high latitudes of both hemispheres, where they are much more common than in the warm or temperate parts of the globe. They are uncommonly courageous in defence of their young, and attack, with eagle-like courage and ferocity, any animal that dares to disturb them ; nay, those persons who are about to rob their nests, aware of the reception they are likely to meet with, hold a knife or other sharp instrument over their heads, upon which the enraged bird rushes, to its own destruction. By many people their feathers are preferred to those of the goose ; and in some parts they are killed in great numbers merely for the sake of them.

LEUCISCUS. A genus of Malacopterygious fishes, of the genus *Cyprinidæ.* It contains many species, chiefly distinguished from others of the Carp tribe by the comparative shortness of the dorsal and anal fins, and a deficiency of barbules about the mouth. [For examples of this genus, see BLEAK : CHUB : DACE : ROACH, &c.]

LEUCOPHASIA ; called by Hübner *Leptoria.* A genus of Butterflies distinguished from the other "*Whites*" by the narrow elongated wings, rounded at the end. There are few species in this genus ; we particularize the British

LEUCOPHASIA SINAPIS ; or WOOD WHITE BUTTERFLY. In certain woods and copses this insect is to be met with at the end of

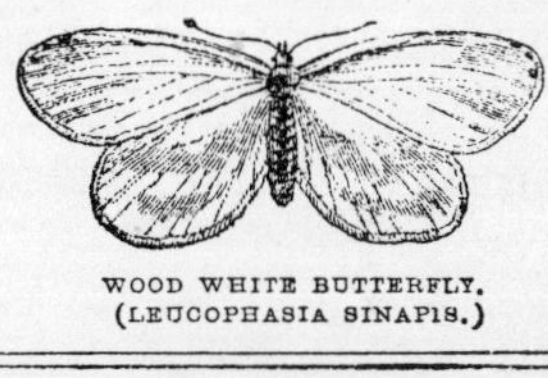

WOOD WHITE BUTTERFLY.
(LEUCOPHASIA SINAPIS.)

May and beginning of August. Its wings above are milk-white, with a dusky rounded spot at the tip of the anterior, and the base sprinkled with dusky; beneath, the tip and base are yellowish tipped with green: the posterior wings are faintly tinged with yellow and sprinkled with dusky clouds: body cinereous above, white beneath; antennæ white, with black rings. In the female the wings are more rounded. Caterpillar green, with a deep yellow lateral line: it feeds on the *lotus corniculatus.* The Chrysalis is fusiform; greenish with a yellow streak on the sides, and white spots on the stigmata.

LEUCOSIADÆ. A family of Decapod Crustacea, containing many fine round porcellane exotic crabs; the genus *Ebalia* of the British seas belongs to this family.

LEVERET. The young of the Hare during the first year of its age.

LIBELLULA: LIBELLULIDÆ. A genus and family of Neuropterous insects; the distinguishing characters of which are: that the mouth is furnished with jaws; that the antennæ are shorter than the thorax; that the wings are extended; and that the tail is terminated by a kind of forceps. [See DRAGON-FLY: PETALURA.]

LICMETIS. A genus of Scansorial birds found in New Holland: it contains the LICMETIS NASICUS, or LONG-BILLED COCKATOO. This species of the *Psittacidæ* or Parrot tribe, like the common *Cacatua galerita*, assembles in large flocks and spends much of its time on the ground, where it grubs up the roots of orchids and other bulbous plants, upon which it mainly subsists. It not unfrequently makes inroads to the newly-sown fields of corn, where its attacks are most destructive. In confinement they appear dull and morose, and show a very irritable temper. The general plumage is white, washed with pale brimstone-yellow on the under surface of the wing, and with bright brimstone-yellow on the under surface of the tail; line across the forehead and lores scarlet; the feathers of the head, neck, and breast are also scarlet at the base, showing through the white, particularly on the breast; irides light brown; bill white; naked skin round the eye greenish blue; legs and feet dull olive gray. The sexes are alike in size and colour. The female deposits two white eggs on a layer of rotten wood at the bottom of holes in the larger gum trees.

LIMA. A genus of Conchifera, inhabiting a longitudinal shell, almost always white, nearly equivalve, obliquely fan-shaped, and slightly eared; valves gaping near the bosses, which are distant; hinge with a triangular disc between the umbones, divided in the centre by a triangular ligamentary pit, without teeth. The animal makes use of the valves of his shell as natatory organs, working them like fins or paddles, and by this means proceeding at a rapid rate through the water. Two or three species are found on our coasts, and fossil species occurring in lias, inferior oolite, &c.

LIMACINA. A genus of Mollusca belonging to the order *Pteropoda.* It inhabits the northern seas; and is said to be devoured by whales in vast quantities. The shell is

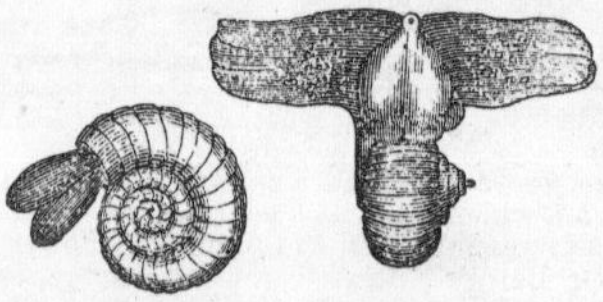

LIMACINA ARCTICA.

thin, fragile, papyraceous, spiral, and obliquely convolute; spiral side rather prominent, the other side umbilicated; aperture large. The body of the animal is long; and it can retire completely into its shell.

LIMAX: LIMACINÆ. A genus and family of voracious naked Molluscs, commonly called Slugs. In most of the terrestrial species of this order there is a prominent head, with four retractile tentacula; and at the end of the longest pair the eyes are situated. The figure of the Limax is oblong, approaching to cylindric. On the back there is a kind of shield or disc, formed by the mantle; and this shield covers the pulmonary sac, the orifice of which is on the right side. They are diffused throughout all climates, particular species being restricted to each; and they are every where regarded as inveterate destroyers of garden produce. [See SLUG.]

LIMENITIS. A genus of Butterflies, one species of which is found in this country.

LIMENITIS CAMILLA; or HONEYSUCKLE BUTTERFLY. This somewhat rare species, which on the Continent is known as *Sybilla*, is noted for the graceful elegance with which it floats along with outstretched wing. Its general colour is a dark brown, spotted with black, the anterior wings having a curved central white band, intersected with black veins, a grayish crescent and three or four small white dots; the posterior wings are very similar, but the white band in the

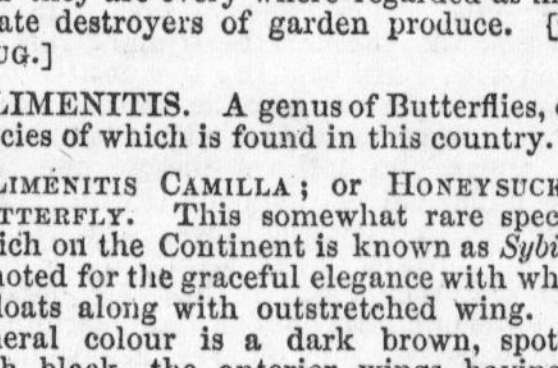

HONEYSUCKLE BUTTERFLY
(LIMENITIS CAMILLA.)

centre is oblique and straight; between the fascia and the margin is a double parallel series of obscure black spots: beneath, the anterior wings are brown, clouded with fulvous, and there are several white spots: the posterior wings at the base are a mixture of tawny-orange and bluish-gray, with several black zig-zag lines and dots; then brownish orange, a white band, a double series of black spots, and a few white dots. The body is dusky black above, white beneath; antennæ black above, tawny beneath and at

LIMENITIS CAMILLA—UNDER SIDE.

the tip. Caterpillar green, with the head and legs reddish; it feeds on the various species of honeysuckle: the chrysalis is green spotted with gold, forked in front. The Honeysuckle Butterfly appears to delight in settling on the blossoms of the bramble.

LIMOSA. A genus of Wading Birds, frequenting marshes and the sea-shore. They are characterized by a long straight beak, slightly bent at the extremity; and by the external toes, which are long and slender, being palmated at the base. [See GODWIT.]

LIMPET. (*Patella.*) A genus of marine Molluscous animals; the distinguishing characters of which are: that the shell is univalve, of a gibbous shape, almost conical, always fixed to a rock or some hard body; and having its apex sometimes sharp-pointed, at others obtuse: straight, or crooked; whole, or perforated: these variations occasioning so many specific distinctions. The means by which the Limpet affixes

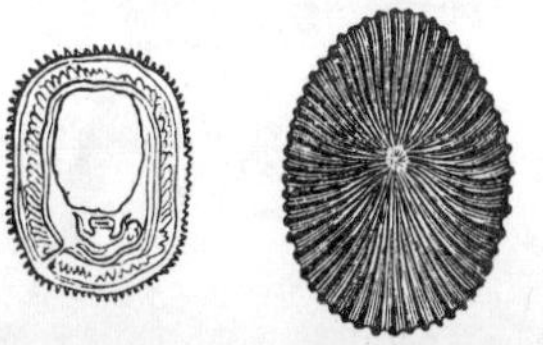

LIMPET AND SHELL.—(PATELLA LUGUBRIS.)

itself to a rock were first clearly explained by Reaumur. The shell approaches to a conic figure; the base of which is occupied by a large muscle, which alone contains nearly as much flesh as the whole body of the fish: this muscle is not confined within the shell, but assists the creature in its progressive motion, or in fixing itself at pleasure. When in a quiescent state, which is commonly the case, it applies this muscle every way round to the surface of some stone, and so firmly attaches itself to it that it is not easily separated even with the assistance of a knife. It is said that crows and other birds which endeavour to detach them for food, are sometimes caught by the points of their bills, and are held there until drowned by the advancing tide.

The Common Limpet (*P. vulgaris*), which is very numerous on the British coasts, has rough prominent striæ, with edges sharply crenated; and the vertex is near the centre. Another species, frequent on the Cornish coast, is called the Transparent Limpet; it has a pellucid shell, longitudinally marked with rows of rich blue spots; and the vertex is placed near one of the edges. But the most beautiful varieties are found on the shores of the Oriental seas and the coasts of the Mediterranean. Limpets are herbivorous, feeding upon sea-weeds, which they reduce with their long riband-shaped tongues.

Many and very opposite opinions have been given to account for the extraordinary tenacity with which this animal adheres to the rock: that which to us appears the most feasible, ascribes the true cause to a viscous juice emitted from the muscle of which we have spoken, which, though imperceptible to the eye, is nevertheless capable of producing these surprising effects. This, it is observed, may be perceived by the touch; for if the finger be applied to the place immediately after the removal of the Limpet from a stone, the tenacity of this juice will be extremely strong; but if any wet touches the stone after the removal of the fish, no viscosity will be perceptible, the whole substance of the glue being instantly dissolved, and its effects totally lost. Water, therefore, is a sufficient solvent for this glue; but the close adhesion of the outer rim of the great circular muscle prevents the external water from acting on it, otherwise it must always be destroyed as soon as discharged. However, the under surface of the body of the animal is entirely covered with small tubercles, containing water, which the creature discharges whenever inclined to liberate itself, and the whole cement immediately dissolves before it.

LIMULUS, or KING-CRAB. A genus of large Crustacea, belonging to the group *Xyphosura* or *Sword-tails*, sometimes attaining the length of two feet. The Limuli are of a very singular form and structure: their bodies are divided into two parts; of which the anterior, covered by a large semicircular shield, bears the eyes, the antennæ, and six pairs of legs, which surround the mouth, and are used both for walking and for mastication; whilst another shield of a somewhat triangular shape covers the posterior portion of the body, which supports five pairs of swimming legs, and terminates in a long pointed process. The Limuli are confined to the shores of tropical Asia, the Asiatic Archipelago, and tropical America. The best known species comes from the Molucca islands: hence they are sometimes termed Molucca crabs. Their habits do not appear to be very well understood: it seems, however, that they prefer the neighbourhood of sandy shores; and it is said that, in order to avoid the violent heat of the sun, which becomes fatal to their existence, they bury themselves in the sand. The long horny process is used by some of the Malays as a

KING-CRAB. (LIMULUS MOLUCCANUS.)

point for their arrows; the wounds it makes being dangerous, like those made by the jagged spines of certain fishes.

LING. (*Gadus molva.*) This is a valuable fish of the *Gadidæ* family, or Cod tribe. The body is very long and slender, usually from three to four feet; the head is flat; the teeth in the upper jaw are numerous and very small, while those in the lower are few, long, and sharp; and the lower jaw is

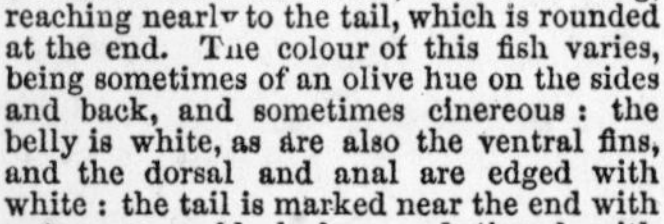

LING.—(GADUS MOLVA.)

shorter than the upper, with a single barbule at its extremity: lateral line straight; scales small; two dorsal fins of equal height; one short near the head; the other long, reaching nearly to the tail, which is rounded at the end. The colour of this fish varies, being sometimes of an olive hue on the sides and back, and sometimes cinereous: the belly is white, as are also the ventral fins, and the dorsal and anal are edged with white: the tail is marked near the end with a transverse black bar, and tipped with white. The Ling is an inhabitant of the Northern seas, and forms in many places a considerable article of commerce. Large quantities are taken among the Western Islands, in the Orkneys, on the Yorkshire and Cornish coasts, and, generally speaking, all round the Irish coast. They spawn in June, depositing their eggs in the soft oozy ground at the mouths of rivers; at which period the males separate from the females. While the Ling continues in season, its liver is very white, and abounds with a fine-flavoured oil; but no sooner does it cease to be in season, than its liver becomes red, and destitute of oil. The same, indeed, happens to the Cod and some other fishes, in a certain degree, but not so remarkably as in the Ling. Besides a certain portion which are consumed fresh, considerable quantities are cured for exportation. The young are called *drizzles*.

LINGULA. A genus of Conchifera, found in the Philippine Islands, &c., and constituting a singular anomaly, as being the only bivalve shell that is pedunculated. The animal has two long ciliated arms, which are curled up during repose. Shell thin, either horny or calcareous, equivalve, equilateral, peaked at the apex, and generally open at the base. There are several recent species found in the Moluccas, and some fossils in sandy indurated marl, limestone, &c. *Lingula anatina* is so named from its resemblance to a duck's bill.

LINNET. (*Fringilla linota.*) The Brown or Gray Linnet is a well-known song-bird, being common in every part of Europe. Its length is about five inches and a half, including the bill and tail: the bill is bluish gray; the neck, back, and upper parts of the head, dark reddish brown, the edges of the feathers being pale; under parts dirty reddish white; breast deeper than the rest, sides streaked with brown; quills dusky, edged with white; the tail, which is a little forked, is of a brown colour, edged with white, except the two middle feathers, which are bordered with a dull red: legs short and brown. The female is marked on the breast with streaks of brown; her wings have less of white; and her colours are in general less bright. The Linnet usually

LINNET.—(FRINGILLA LINOTA.)

builds in some thick bush or hedge, preferring the white-thorn and furze; the outside is composed of moss, dry grass, and roots; and the inside of fine soft wool and hair. The female lays four or five eggs, which are white, spotted with blue, and irregularly spotted with brown at the larger end. The young are hatched towards the end of April or beginning of May. The song of the Linnet is lively and sweetly varied; its manners are gentle, and its disposition is docile. When confined with other birds it easily adopts their song, and when taken young it may be readily taught to modulate its voice to any sound to which it is accustomed. But those persons who have paid most attention to the natural note of this bird must be well aware that its native strains are more delightful than any in which it is capable of being instructed. Linnets, says Bewick, are frequently seen in flocks during the winter; and their assembling with other kinds of small birds is the sure presage of a coming storm. They may be caught in clap-nets during the summer months; but flight-birds are most plentiful about the beginning of October. They feed on various seeds, and are particularly fond of linseed; from which circumstance, it is said, they derive their name.

In alluding to the domestic attachments of some species of birds, the *Journal of a Naturalist* thus speaks of the Linnet: "This songster is no solitary visitor of our dwellings: it delights and lives in society, frequenting open commons and gorsy fields, where several pairs, without the least rivalry or contention, will build their nests and rear their offspring in the same neighbourhood, twittering and warbling all the day long. This duty over, the families unite, and form large associations, feeding and moving in company, as one united household; and, resorting to the head of some sunny tree, they will pass hours in the enjoyment of the warmth, chattering with each other in a low and gentle note; and they will thus regularly assemble during any occasional bright gleam throughout all the winter season,—'and still their voice is song,' which, heard at some little distance, forms a very pleasing concert, innocent and joyous. The Linnet is the cleanliest of birds, delighting to dabble in the water, and dress its plumage in every little rill that runs by. The extent of voice in a single bird is not remarkable, being more pleasing than powerful; yet a large field of furze, in a mild sunny April morning, animated with the actions and cheering music of these harmless little creatures, united with the bright glow and odour of this early blossom, is not visited without gratification and pleasure."

LION. (*Felis leo.*) This most noble as well as most formidable of all carnivorous animals is chiefly distinguished by the presence of a full flowing mane in the male, and by a tufted tail and the disappearance of the feline markings in both sexes before they arrive at maturity. The Lion is principally an inhabitant of the interior wilds of Africa, but is also found, though far less plentifully, in the hotter regions of Asia.

LION.—(FELIS LEO.)

It is in Africa, however, that he reigns supreme among the weaker quadrupeds and exerts his power to the greatest extent. A Lion of the largest size has been found to measure about eight feet from the nose to the tail, and the tail itself about four feet: the general colour is a pale tawny, still paler or more inclining to white beneath: the head is very large, the ears rounded, the face covered with short or close hair; the upper part of the head, the neck, and shoulders coated with long shaggy hair, forming a pendent mane; on the body the hair is short and smooth; and the tail is terminated by a tuft of blackish hair. The Lioness is smaller than the Lion, has no mane, and is of a whiter cast beneath. During the day the Lion usually slumbers in his retreat; and as night sets in, he rouses from his lair, and begins his prowl. Being of the cat tribe, his eyes are incapable of bearing a strong light; the night is therefore his proper time for action.

Much has been written respecting a sharp prickle, or corneous process, concealed in the tuft of hair at the extremity of the Lion's tail, with which he was said to lash himself when angry, or to arouse his dormant rage. Homer, Lucan, and Pliny had so described him: but though they appeared to have no doubt of his lashing his sides with that object, they did not advert to this peculiarity of caudal structure. Didymus Alexandrinus,

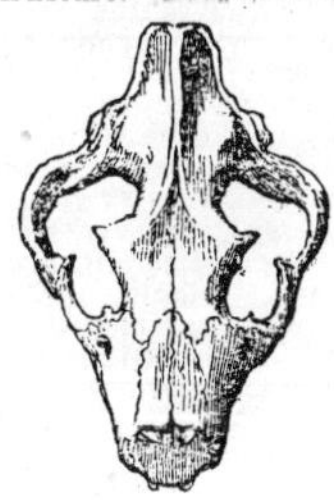

SKULL OF LION

a commentator on the 'Iliad,' however, having found a black horny prickle among the hair of the tail, immediatey conjectured that he had ascertained the true cause of the stimulus when the animal flourishes that member in defiance of his enemies. The subject afterwards remained unnoticed for centuries, till at length Blumenbach verified the fact of its existence, although he did not admit that it could produce the effect attributed to it by the ancient scholiast. He remarked, indeed, that the tail was terminated by a horny prickle, surrounded at its base by an annular fold of the skin, and so buried in the tuft of hair that its use for the purpose stated could only be imaginary. Since that time it has been clearly proved, by the examination of Lions, both living and dead, that there is occasionally present at the extreme tip of the tail, a horny prickle, scarcely three-eighths of an inch in length, which is altogether unconnected with the caudal vertebræ, and easily detached from the skin; what its real use may be is purely conjectural, but that the animal is furnished with it in order to incite him to anger cannot for a moment be entertained. We should here observe, that in one of the *bas reliefs* discovered, through the laudable zeal of Sir

FOOT OF LION, DISSECTED, TO SHOW THE MUSCLES WHICH MOVE THE RETRACTILE CLAWS.

Stratford Canning, in the excavations of Nimroud (the supposed site of the ancient city of Nineveh), and now in the British Museum, an *exaggerated* representation of this "prickle" is very apparent. From this it is certain that the fact of its existence was perfectly established in the time of the Assyrians, or it would not have been prominently introduced in the figure of the sculptured Lion.

When in quest of prey his roaring resembles the sound of distant thunder, and, being re-echoed by the rocks and mountains, appals and puts to flight every animal within hearing. In general, however, he waits in ambush, or creeps insidiously towards his victim; and then, springing on it with a tremendous bound, he seizes it with his powerful claws. His strength is prodigious: a single stroke of his paw, it is affirmed, is sufficient to break the back of a horse; and his strength is such as to enable him to carry off a buffalo or antelope, with as much apparent ease as a cat carries off a rat. The Lion is supposed to be destitute of a fine scent, and to hunt by the eye alone: he will devour as much at one time as will serve him for two or three days; and, when satiated with food, he is said to retire to his den, which he seldom quits, except for the purpose of prowling about for his prey. His teeth are so strong, that he breaks the bones with perfect ease, and often swallows them together with the flesh: his tongue, as in other feline animals, is furnished with reversed prickles, but they are so large and strong in the Lion as to be capable of lacerating the skin: the muscles which raise the jaw are of enormous size; and those which support the head, as well as the *ligamentum nuchæ* which runs along the spinous processes of the vertebræ to the occiput, are very highly developed. The Lioness is said to go with young five months, and to produce but one brood in the year: the young are generally from two to four in number, which the parent nurses with great assiduity, and attends in their first excursions for prey; and it is remarked that in a state of captivity she usually becomes very savage as soon as she becomes a mother.

From the writings of ancient historians it appears very clear that Lions were at one time found in Europe, but they have long since totally disappeared. They are also no longer seen in Egypt, Palestine, or Syria, where they once were evidently far from uncommon; and, as Cuvier remarks, even in Asia generally, with the exception of some countries between India and Persia, and some districts of Arabia, they have become comparatively rare. Nor is this to be wondered at, when we reflect on the constantly increasing numbers of the human race, the superior advantages given to man by the arts of civilization, and, above all, the destruction which is caused by using fire-arms against them, instead of the spear and the arrow. "His true country," as Mr. Bennett observes, "is Africa, in the vast and untrodden wilds of which, from the immense deserts of the north to the trackless forests of the south, he reigns supreme and uncontrolled. In the sandy deserts of Arabia, in some of the wild districts of Persia, and in the vast jungles of Hindostan, he still maintains a precarious footing; but from the classic soil of Greece, as well as from the whole of Asia Minor, both of which were once exposed to his ravages, he has been utterly dislodged and extirpated." How different was it in the time of the Romans! Struck with the magnificent appearance of these animals, they imported them in vast numbers from Africa, for their public spectacles. Quintus Scævola, according to Pliny, was the first in Rome who exhibited a combat of Lions; but Sylla the dictator, during his prætorship, exhibited a hundred Lions; after him, Pompey the Great produced no less than six hundred in the grand circus; and Cæsar the dictator four hundred. Mark Antony appeared in the streets of Rome in a chariot drawn by these noble animals, accompanied by his mistress Cytheris, an actress from the theatre: a sight, says Pliny, that surpassed in enormity even all the calamities of the times.

"The general prey of the African Lion," Mr. Broderip observes, "consists of the larger herbivorous quadrupeds, very few of which it is unable to master; and it is a severe scourge to the farmer, who is consequently ever on the look-out for Lions, and generally a most imperturbable and unerring shot. Though mortal accidents frequently happen in these huntings, the cool sportsman seldom fails of using his rifle with effect. Lions when roused, it seems, walk off quietly at first, and if no cover is near, and they are not pursued, they gradually mend their pace to a trot, till they have reached a good distance, and then they bound away. Their demeanour upon these occasions has been described to us by eye-witnesses to be of a careless description, as if they did not want a fray, but, if pressed, were ready to fight it out. If they are pursued closely, they turn and couch, generally with their faces to the adversary; then the nerves of the sportsman are tried. If he is collected and master of his craft, the well-directed rifle ends the scene at once ; but if, in the flutter of the moment, the vital parts are missed, or the ball passes by, leaving the Lion unhurt, the infuriated beast frequently charges on his enemies, dealing destruction around him. This, however, is not always the case ; and a steady unshrinking deportment has, in more instances than one, saved the life of the hunter."

The distinctions which some naturalists have pointed out as existing between the African and Asiatic Lions have been altogether denied by Buffon, with whom Cuvier appears to coincide. On the other hand, modern writers, who have lately paid great attention to the subject, state that the African Lion is larger, has a more regular and graceful form, is generally of a darker colour, and has a less extensive mane. The African varieties are, 1. The *Barbary Lion*, which is described as having a deep yellowish brown fur and a full flowing mane : 2. The *Senegal Lion*, the fur of which is of a brighter yellow tint, and the mane thinner : 3. The *Cape Lion*, of which there are two varieties, one brown, the other yellowish ; the former being the most powerful and ferocious. The Asiatic varieties are generally distinguished as the *Bengal Lion;* the *Persian* or *Arabian Lion;* and the *Maneless Lion* of Guzerat ; the last of which appears to be limited to a comparatively small district. There is also the *Puma*, or *American Lion;* a description of which will be found in its proper alphabetical position : but that animal, it should be observed, is destitute of several of the distinguishing characters of the true Lion, and is not entitled to the appellation.

We would willingly, if space permitted, insert some of the stirring narratives which recent travellers have given of Oriental Lion hunts : our readers must, however, be content with a brief notice, which we copy from the excellent publication last quoted : —"The habits of the Asiatic Lions do not differ much from those of Africa, excepting that the former, from the state of the country, frequent the jungles. In India the elephant is generally employed in the chase which is even now conducted with more pomp and circumstance than in Africa. The grand Asiatic huntings of former times, those of Genghis Khan for instance, will occur to many of our readers. The accounts of most Asiatic modern sportsmen give a most courageous bearing to the Lions in these encounters. One of these states that the Lions in India, instead of running away when pursued through a jungle, seldom take to cover as a refuge at all. On the approach of their enemies, they spring out to meet them open-mouthed in the plain. They are thus easily shot ; but if they are missed or only slightly wounded, they are most formidable adversaries. They are even said to have sprung on the heads of the largest elephants, and to have fairly pulled them to the ground, riders and all."

The mane and tuft on the tail of a Lion are not fully developed till the animal is six or seven years old ; and the natural period of a Lion's life is generally supposed to be about twenty-two years ; but instances are on record which show they have sometimes attained the "age of man."

LION-LIZARD. The name applied by Catesby in his Natural History of Carolina to the Basilisk (*Basiliscus Americanus.*)

LITHODOMUS. A Molluscous animal inhabiting a bivalve shell, oblong, almost equally rounded at both ends, and the summits very near the anterior. They at first suspend themselves to stones, like the common Mussels, but then they perforate them,

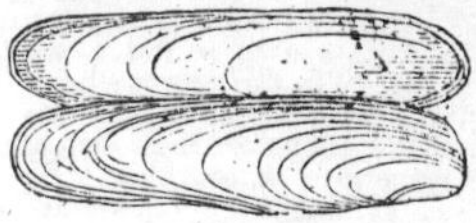

STONE-BORER.
(LITHODOMUS LITHOPHAGUS)

and bury themselves in the excavations, whence they cannot issue. Cuvier says, that when young, the Lithodomus suspends itself to rocks by a byssus, but, as it grows, it pierces a hole, and introduces itself, forming a cavity which thenceforward it never leaves: indeed, after a short time, as it merely enlarges the interior, without making the entrance any wider, its increasing bulk renders it unable to quit its cell, and in such cases the byssus dies away.

LITHOPHAGIDÆ. A family of *Mollusca*, consisting of terebrating bivalves, without accessory valves.

LITHORNIS. [See SUPPLEMENT.]

LITHOTRYA. A genus of Molluscous animals, allied to the family of Pedunculated Cirripedes, inhabiting an irregularly shaped pyramidal shell, consisting of eight unequal pieces ; having at the base an irregularly cap-shaped appendage, like the inverted shell of a Patella, and to which the lower part of the peduncle is affixed. The genus derives its name from the power possessed by the animal of making dwelling holes in stones or pieces of rock.

LITTORINA. A genus of Mollusca found on the sea-shores in all parts of the world, feeding upon sea-weed. They inhabit a turbinated, thick shell, consisting of few whorls ; spire acuminated ; columella rather flattened ; operculum, horny, spiral, with rapidly increasing volutions. The common Periwinkle is a specimen of this genus.

LIZARD. (*Lacerta.*) A group of Reptiles, which not only differ from every other class of animals, but they also vary widely from each other. With respect to size, the ranks of no class of beings are so opposite: contrast the gigantic and ferocious Crocodile with the inoffensive Chameleon ; or the monstrous Saurian reptiles, whose fossil remains excite the wonder of all beholders, with the harmless little Lizard of our walls and copses ! They v·y too in colour greatly, and they differ con·'derably in form. But the principal disti·'tion between the Lizard species arises from t·: manner of bringing forth their young. Some are viviparous; others emit their spawn like fishes. The Crocodile, the Iguana, and all the larger kinds, produce eggs, which are hatched by the vivifying heat of the sun: the animals that issue from them are complete on leaving their shells, and their first efforts to run are in order to procure subsistence in their native element. The viviparous kinds, in which are all the Salamanders, are produced from the bodies of the females perfect and active, and undergo no future change: but those which are bred in the water, and, as is generally supposed, from spawn, suffer a very considerable change in their form; being generated with external skins or coverings, which sometimes enclose their feet, and give them a serpentine appearance. To these adscititious skins fins are superadded above and below their tails, which assist the animals in swimming; but when the false skins drop off, these likewise disappear; and then the Lizards, with their four feet, are completely formed, and exchange the water for the land. The most important of all these will be found described in other parts of this volume; and we shall therefore have to consider in this place those only which are denominated *Lacertidæ*, or True Lizards, which are bright-eyed, active, slender little animals, adorned with brilliant colours, and whose aspect and manners have nothing repulsive about them.

The GREEN LIZARD. (*Lacerta agilis.*) This elegant species, which is found in all the warmer parts of Europe, and seems pretty generally diffused throughout the Old Continent, is from ten to fifteen inches in length; exhibiting a rich and varied mixture of darker and lighter green, interspersed with spots and marks of yellow, brown, &c. The head is green, covered with large angular scales; the rest of the upper parts with very small ovate ones: the tail, which is commonly much longer than the body, is marked into very numerous scaly rings;

GREEN LIZARD.—(LACERTA AGILIS.)

and the under part of the animal, both on the body and limbs, is of a pale blue-green cast: beneath the throat is a kind of collar, formed by a row of scales much larger than the rest; the abdomen and under surface of the limbs is likewise covered with scales: the tongue is moderately long, broad at the base, bifid towards the tip, and covered on its broad part with numerous rows of minute sharp papillæ pointing backwards, and thus the better enabling the animal to retain and swallow its prey, which consists chiefly of insects, small worms, &c. The Green Lizard is found in various situations, in gardens, about warm walls, buildings, &c. It is extremely active, pursuing its insect prey with great celerity, and readily escaping from pursuit when disturbed. If taken, however, it soon becomes familiar, and to a certain degree may even be tamed; for which reason it is regarded with favour in many countries. It appears to run into many varieties both as to size and colour, but in all these states the particular characteristics of the species are easily ascertained.

The VARIEGATED LIZARD. (*Teius teguixin.*) The colour of this large species is highly beautiful, consisting of an elegant variegation of brown, blackish, and purple spots, on a pale bluish-white, and, in some parts, yellowish ground. The head is covered, as in the Green Lizard, with large scales or plates; the body with small scales, so disposed as to mark the sides into numerous tapering annuli or striæ; and the tail, which is very long, is surrounded by extremely numerous rings of small square scales, and tapers to a slender point. The head is rather longer and more tapering than that of the Green Lizard: the tongue is broad, flat, long, forked at the tip, and curiously striated on each side. Native of South America.

The VIVIPAROUS LIZARD. (*Zootoca vivipara.*) As its name imports, this reptile is produced alive. It frequents thickets, heaths, and sunny banks; and several are often seen in such situations basking in the summer sun, and watching for their insect prey. They burrow in the ground, and retreat to their hiding-places on the slightest alarm. The average length of this species is about six inches.

We have specified three species of this group, and others will be found scattered throughout this work. Lizards are so numerous in genera and species that we must refer our readers who are desirous of making further acquaintance with them to the elaborate and admirable descriptive Catalogue of Lizards in the British Museum, by J. E. Gray, F.R.S., a goodly volume of nearly 300 pages.

LIMNORIA. A genus of Isopodous Crustacea, in which the head is as broad as the first segment of the body, and the eyes gra-

LIMNORIA LIGNORUM.

nulated. The only known species (*L. lignorum*) is like a small wood-louse in general appearance. It is of an ash-colour, with black eyes: it was first discovered in Britain by Mr. Stevenson, the builder of the Bell Rock Lighthouse, who found it exceedingly destructive to the wood-work necessary in laying the foundations of that useful structure, which it perforated in every direction. It is found in other parts of the British and Irish coasts, and has even attracted notice in France by its perforating ravages into wooden piles, piers, jetties, and other structures embedded in the sea. The small line by the side of the figure denotes the natural size of this Lilliputian but destructive Isopode. [See CHELURA.]

LITHOSIIDÆ. This family of Heterocerous Lepidoptera is of small extent, and the species are weak and inactive: the body is slender; the antennæ are slender and setaceous; the mouth considerably developed, the maxillæ being long and spiral, and the labial palpi of moderate size, and three-jointed; the thorax is not crested; and the wings are comparatively of delicate structure, and elongated. The brilliant colours of some of these insects would seem to indicate that they flew by day; but the contrary is the case, and their flight is short and feeble. The larvæ are cylindrical, often somewhat hairy, with six pectoral, eight ventral, and two anal feet: they are solitary in their habits, and never reside either in a case or in a general tent-like web. There are several exotic species of this family which are very splendid. The species of the genus *Lithosia* found in this country are very sombre in colour.

LLAMA, or GUANACO. (*Auchenia glama.*) This animal bears a strong resemblance to the Camel in form and structure, but is much inferior in size. It is a native of South America, and is particularly plentiful in Peru, where it inhabits, in a wild

LLAMA.—(AUCHENIA GLAMA.)

state, the highest and coldest parts of mountains, feeding in numerous herds, and flying with great rapidity at the sight of man. The ancient Peruvians, however, completely subdued and domesticated it as a beast of burthen, and to them it answered the same purposes as the camel and dromedary of the old continent. The general size of the Llama is nearly that of a stag, or about four feet and a half in height, and six feet in length: the neck is very long, and habitually upright; the head is small; the eyes large and brilliant; the lips thick; and the ears long and movable: the haunches are slightly elevated; and on the breast is a bunch which constantly exudes a yellowish oily matter. Its general colour is a light brown, the under parts being whitish; and sometimes it is said to be varied or patched with darker and lighter shades on different parts, and to have a black stripe running down the back. The tail is about five inches long, small, straight, and slightly curved downwards. The hoofs are divided, and terminated by small horny appendages, rounded above, and on either side somewhat curved. It has no upper cutting teeth. In the wild animal the hair is long and shaggy; in the domesticated smoother and closer. It requires no care or expense with respect to attendance or provision for its sustenance; it is satisfied with vegetables, requiring neither corn nor hay; and it even exceeds the camel in its abstinence and endurance of thirst. The voice of the Llama resembles the shrill neighing of the horse. It is naturally patient and enduring; but when angry or attacked, it strikes with its feet, and ejects from its mouth a quantity of saliva, which is said to be of so caustic a nature as to inflame the skin and produce slight eruptions. When the Spaniards invaded South America, it was kept in immense numbers for the purposes of traffic, and also for food; its skin, also, was prepared as leather, and its wool spun and manufactured into cloth. Immense numbers were constantly employed in the transport of ore from the mines; the ordinary load of each Llama was about 100 lbs., and its rate of travelling with this burthen over rugged mountain passes was from twelve to fifteen miles a day. At the present time, however, the horse, the ass, and especially the mule, which have been introduced from Europe, have very generally superseded the Llama as beasts of burthen; whilst the introduction of the sheep, the goat, and the ox, has rendered it less necessary as affording either food, leather, or wool. The fleece of the *Guanaco*, the name usually given to the *wild* Llama, is longer than that of the domesticated animal, and is in much request for the manufacture of many woollen cloths of a delicate texture.

LOACH, or LOCHE. (*Cobitis barbatula.*) A small fish, often found secreted under stones in small, shallow, clear streams, and which swims rapidly away when disturbed

LOACH.—(COBITIS BARBATULA.)

by moving the stone. It seldom exceeds four inches in length; has six barbules about the mouth; feeds on worms and aquatic insects; and the flesh is accounted excellent. The head, back, and sides are clouded and spotted with brown on a yellowish white ground; the fins spotted with dark brown; and the belly and under surface white.

LOBIVANELLUS. A genus of Birds allied to the Lapwings, of which we may particularize the LOBIVANELLUS LOBATUS, or WATTLED PEWIT. This is an attractive and showy bird, of the Plover kind, common in most parts of New South Wales, and when unmolested approaching sufficiently close to the dwellings of the settlers to permit its habits, &c. to be minutely observed. In some districts, however, it has been much persecuted, and has become so shy and distrustful as to obtain the name of the Alarm Bird, from its rising high in the air and screaming at the approach of every intruder. It is distinguished by a beautiful primrose-coloured wattle, with which the colouring of the bill and the bold eye closely assimilate; the head, back of the neck, and sides of the chest, are black; back, wing-coverts, and scapularies, dark grayish-brown; primaries black; tail white, crossed near the extremity by a broad band of black; tarsi purplish red; scales black; spur yellow. The colours of the plumage are strongly contrasted; and, taken altogether, it is one of the most beautiful of the Plovers yet discovered. "While on the wing," Mr. Gould observes, "it has much of the carriage of the common European Pewit (*Vanellus cristatus*), but a decided difference is observable in its mode of running, and in its more bold and attractive manners."

LOBSTER. (*Homarus vulgaris*). A crustaceous animal, belonging to the sub-order MACROURA, or *long-tailed Decapods* (but constituting a species of *Cancer*, or crab, in the Linnæan system). Lobsters are found in great plenty about many of the European shores; their general habitation being in the clearest waters, about the foot of such rocks as impend over the sea. The colour of this animal alive is a fine bluish black, beautifully variegated with paler spots and clouds: it has a smooth thorax; a short serrated snout; very long antennæ, and between them two shorter bifid ones. The claws and fangs are large, the greater being tuberculated, and the lesser serrated on their interior edges: it has four pair of legs; the tail has six joints; and the caudal fin is rounded. The two great claws of the Lobster constitute its instruments of provision and defence: they open like a pair of nippers, possess great strength, are notched like a saw, and take a firm hold. Besides these powerful members, which may be considered as arms, the Lobster has eight legs and a tail; the latter, expanded laterally, being a very powerful instrument for motion in water. Between the two claws is placed the head, very small, and furnished with eyes, which are projectile or retractile at pleasure. The mouth, like that of an insect, opens longitudinally, and is furnished with two teeth for the comminution of its food; and between them there is a fleshy substance shaped like a tongue. The intestines consist of one long canal; and the spinal marrow is lodged in the breast-bone. The ovary, or place where the spawn is first produced, is situated backward towards the tail, where a red substance is always found, composed of a number of small spawns, too minute for exclusion: from this receptacle proceed two canals, which open on each side of the junctures of the shell, towards the belly; and through these passages the small round particles, destined for the future young, descend to be excluded, and arranged under the tail. No sooner do the young quit the parent Lobster than they seek refuge in the minute crevices of the rocks and other secure apertures; and in a few weeks they acquire hard, firm shells, which furnish them both with defensive and offensive armour.

Like the crabs, they change their shelly covering annually; previous to which process they appear sick, languid, and restless; no longer laboriously harrowing up the sand, or hunting for their prey, but lying torpid and motionless, as if in anxious expectation of their approaching fate. They acquire the new shell in about three or four days, during which time, being perfectly defenceless, they become the prey, not only of fish, but also of such of their own species as are not in the same condition. It is difficult to conceive how they are able to draw the muscles of their claws out of their hard covering; but persons who have paid particular attention to the subject say, that during the pining state of the animal, before casting its shell, the limb becomes contracted to such a degree as to be capable of being withdrawn through the joints and narrow passage near the body. Like all other crustaceous animals, they only increase in size whilst in a soft state; and on comparing the dimensions of the old shell with that of the new, the latter is frequently found to be one-third larger—an amazing addition in such a short interval, and which cannot be explained on any known principle of animal vegetation.

These animals are very sensible to the shock communicated to the fluid in which they live, by the firing of cannon; and the circumstance of Lobsters losing their claws from this cause, or from thunder-claps, is well authenticated. The restoration of claws lost thus, or from their frequent combats with each other, in which the vanquished party generally leaves one of his limbs in his adversary's grasp, may be readily observed, as the new limb seldom, if ever, attains the size of the former one. In the water they are very rapid in their motions, and, when suddenly alarmed, can spring to a great distance. They effect their retreat in a rock with surprising dexterity, throwing themselves into a passage barely sufficient for their bodies to pass. Lobsters begin to breed in the spring, and continue breeding during part of the summer. In the months of July and August the young may be observed in great numbers in the little pools left by the tide among the rocks. In some places Lobsters are caught with the hand; but they are generally taken by means of pots or traps, constructed of osier twigs, and baited with garbage; they are then attached to a cord thrown into the sea, and their stations marked by means of buoys. Lobsters are esteemed a very rich and nourishing aliment; and they are generally in their best season from the middle of October till the beginning of May. There are several varieties; with some differences in the claws, the size, and the places of resort, but few in the habits or conformations.

LOBSTER MOTH. [See STAUROPUS.]

LOCUSTS. (*Locustidæ.*) These noxious insects, whose numbers and voracity constitute one of the severest pests of the hotter regions of the globe, are classed with the Grasshoppers by Linnæus, under the genus *Gryllus;* but more modern entomologists have applied the term *Saltatoria* to them, on account of the power of leaping which the species possess; and in this instance, as in many others where the scientific names of genera and subgenera (of insects in particular) differ, some unavoidable confusion exists. They have coloured elytra, and large wings, disposed when at rest in straight folds, covered by the long narrow wing-cases, and frequently exhibiting blue, green, or red colours: the antennæ are short; the feet have only three joints; and the hind legs are long, strong, and formed for leaping.

The most celebrated species is the MIGRATORY LOCUST (*Gryllus migratorius*), which, of all the animals capable of injuring mankind, seem to possess the most dreadful powers of destruction. In Syria, Egypt, and almost all the south of Asia, these insects make their appearance in legions, and carry

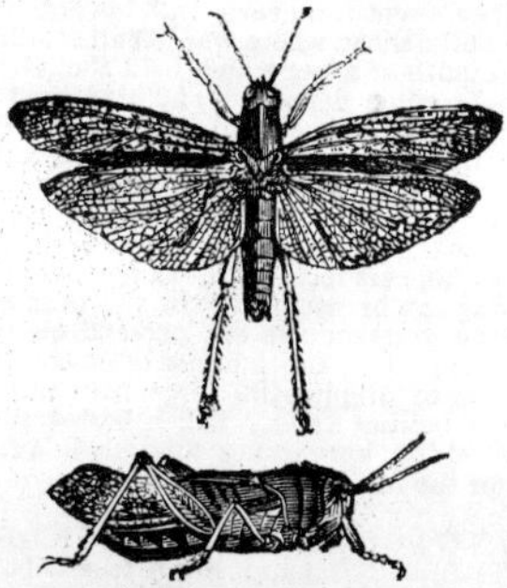

MIGRATORY LOCUST.
(GRYLLUS MIGRATORIUS.)

desolation with them, in a few hours changing the most fertile provinces into barren deserts, and darkening the air by their numbers. This formidable Locust is generally of a brownish colour, varied with pale red, and the legs are of a bluish cast. Happily for mankind, this awful visitation is not frequently repeated; for they are often not only the precursors of famine, but, when they die, the putrefaction which arises from their inconceivable number is so great, that it is justly regarded as the cause of some of those desolating pestilences which almost depopulate whole districts of country. Mr. Barrow, in his "Travels," states, that in the southern parts of Africa the whole surface of the ground might literally be said to be covered with them for an area of nearly 2000 square miles. When driven into the sea by a north-west wind, they formed upon the shore for fifty miles a bank three or four feet high, and when the wind was south-east the stench was so powerful as to be smelt at the distance of 150 miles: the air, in short, became poisoned by their fetid exhalations. Mr. Darwin, in his "Researches," has the following graphical description of a swarm of Locusts, closely resembling the species (*Gryllus migratorius*) which he saw in South America, in 1835. It was at the passage of the Cordilleras, near the village and river of Luxan. "Shortly before we arrived at this place, we observed to the southward a ragged cloud of a dark reddish-brown colour. For some time, we had no doubt but that it was thick smoke proceeding from some great fire on the plains. Soon afterwards we found it was a pest of locusts. The insects overtook us, as they were travelling northward, by the aid of a light breeze, at the rate, I should suppose, of ten or fifteen miles an hour. The main body filled the air from a height of twenty feet, to that, as it appeared, of two or three thousand above the ground. The noise of their approach was that of a strong breeze passing through the rigging of a ship. The sky seen through the advanced guard appeared like a mezzotinto engraving, but the main body was impervious to sight; they were not, however, so thick but that they could escape from a stick moved backward and forward. When they alighted they were more numerous than the leaves in a field, and changed the green into a reddish colour: the swarm having once alighted, the individuals flew from side to side in any direction. Of course this swarm cannot even be compared to those of the Eastern world, yet it was sufficient to make the well-known descriptions of their ravages more intelligible."

But to recount the various devastations which these famished insects have at different times occasioned, would be endless. They have several times visited Poland and the south of Europe in amazing numbers; and instances have been known of their reaching our own coasts: happily for us, however, the cold and humidity of the climate are by no means favourable to their production.

One of the largest Locusts known is the *Gryllus cristatus* of Linnæus, a highly beautiful species; being of a bright red, with the body annulated with black, and the legs varied with yellow: the upper wings marked alternately with dark and pale green; the lower with transverse wavy streaks: its length is about four inches; and the expanse of wings when fully extended about seven and a half. These, with other large kinds, are made use of in some parts of the world as an article of food; and sold, both fresh and salted, in the markets of some parts of the Levant. Hasselquist, alluding to the passage in the New Testament in which John the Baptist is said to have fed on Locusts and wild honey, thus expresses himself: "They who deny insects to have been the food of this holy man, urge that this insect is an unaccustomary and unnatural food; but they would soon be convinced to the contrary, if they would travel hither, to Egypt, Arabia, or Syria, and take a meal with the Arabs. Roasted Locusts are at this time eaten by the Arabs, at the proper season, when they can procure them; so that in all probability this dish had been used in the time of St. John." He further says, that when corn is scarce the Arabians grind the Locusts in handmills, or pound them in stone mortars, and bake them as bread; that he has frequently seen Locusts used by the Arabians, even when there was no scarcity of corn; but then they stew them with butter, and make them into a kind of fricasee, the flavour of which is by no means disagreeable. Later travellers have fully confirmed these remarks. We may accordingly see the folly of that dispute among Divines about the nature of St. John's food in the wilderness — some maintaining that the word in the original text means the fruit of a certain tree; others that a species of bird is intended, &c.; while those who adhered to the *literal* meaning were the only ones who were both consistent and orthodox.

To give a description of the various species of Locusts, would extend this article to an unreasonable length, and at the same time afford but little of useful information: it may be necessary, however, to show how the three large groups or genera may be distinguished from each other:— *Acrydium.* (Spine-breasted Locusts.) The thorax and wing-covers of ordinary dimensions; a projecting spine in the middle of the breast; and a little projecting cushion between the nails of all the feet. 2. *Locusta.* (Locusts proper.) The thorax, and usually the wing-covers also, of ordinary dimensions; no projecting spine in the middle of the breasts; cushions between the nails of the feet. 3. *Tetrix.* (Grouse Locust.) The thorax greatly prolonged, tapering to a point behind and covering the whole of the back to the extremity of the abdomen; wing-covers exceedingly minute, consisting only of a little scale on each side of the body; forepart of the breast forming a projection like a cravat or stock, to receive the lower part of the head: no spine in the middle of the breast; no cushions between the nails.

"In the South of France," says Dr. Thaddeus Harris, "the people make a business, at certain seasons of the year, of collecting Locusts and their eggs, the latter being turned out of the ground in little masses cemented and covered with a sort of gum in which they are enveloped by the insects. Rewards are offered and paid for their collection, half a franc being given for a kilogramme (about 2 lb. 3¼ oz. avoirdupois) of the insects, and a quarter of a franc for the same weight of their eggs. At this rate twenty thousand francs were paid in Marseilles, and twenty-five thousand in Arles, in the year 1613; in 1824, five thousand five hundred and forty-two, and in 1825, six thousand two hundred francs were paid in Marseilles. It is stated that an active boy can collect from six to seven kilogrammes (or from 13 lb. 3 oz. to 15 lb. 7 oz.) of eggs in one day. The Locusts are taken by means of a piece of stout cloth, carried by four persons, two of whom draw it rapidly along, so that the edge may sweep over the surface of the soil, and the two others hold up the cloth behind at an angle of forty-five degrees. This contrivance seems to operate somewhat like a horse-rake, in gathering the insects into winrows or heaps, from which they are speedily transferred to large sacks. A somewhat similar plan has been successfully tried in this country (United States of America), as appears by an account published in the "New England Farmer." It is there stated that, in July, 1826, Mr. Arnold Thompson, of Epsom, New Hampshire, caught, in one evening, between the hours of eight and twelve, in his own and his neighbour's grain-fields, five bushels and three pecks of grasshoppers, or more properly locusts. His mode of catching them was by attaching two sheets together, and fastening them to a pole, which was used as the front part of the drag. The pole extended beyond the width of the sheets, so as to admit persons at both sides to draw it forward. At the sides of the drag, braces extended from the pole to raise the back part considerably from the ground, so that the grasshoppers could not escape. After running the drag about a dozen rods with rapidity, the braces were taken out, and the sheets doubled over; the insects were then swept from each end towards the centre of the sheet, where was left an opening to the mouth of a bag which held about half a bushel; when deposited and tied up, the drag was again opened and ready to proceed. When this bag was filled so as to become burthensome, (their weight is about the same as that of the same measure of corn,) the bag was opened into a larger one, and the grasshoppers received into a new deposit. The drag can be used only in the evening, when the grasshoppers are perched on the top of the grain. His manner of destroying them was by dipping the large bags into a kettle of boiling water. When boiled, they had a reddish appearance, and made a fine feast for the farmer's hogs."

LONGICORNES; or LONGICORN BEETLES. The name given to a tribe of coleopterous insects, or beetles, which are readily distinguished by the great length of the antennæ, and by the first three joints of the tarsi being furnished with a brush. The larvæ mostly reside in the interior of trees, or under the bark; and are either wholly destitute of feet, or have them very small. Both in their larva and perfect state, but particularly in the former, they do much injury to vegetation. Some of the tropical species are brilliantly coloured; and some are remarkable for exhaling an agreeable musky odour.

Mr. Westwood observes: "From the habits of these insects, in burrowing into the very heart of solid timber, there can be no doubt that the marvellous accounts which we constantly meet with in the journals of the discovery of insects, in cutting up logs of wood (especially foreign timber), relate to the larvæ, or perfect states, of these insects; and it is owing to the same circumstance that our English catalogues have been swelled by the introduction of numerous species, which have, indeed, been captured alive in this country, but which have no legitimate claim to be regarded as natives, having been entirely produced from larvæ imported in timber from abroad. * * * From the large size of many of these larvæ, and the long period during which they remain in that state, it may easily be conceived that they do much damage to trees, boring very deeply, and cutting channels into them. A few species appear to subsist in the larva state upon the roots of plants. Another peculiarity resulting from their lignivorous habits is exhibited in their geographical distribution; the tropical and thickly wooded districts of South America possessing a far greater number of species (and these, too, of the largest size) than are to be found in corresponding latitudes in Africa; the speedy decay of vegetable matter requiring the presence of great quantities of such insects. In India but very few gigantic species of Longicornes are to be found."

LONGIPENNES. Cuvier's name for a family of aquatic birds, whose wings are remarkably long, their powers of flight proportionally great, and their habits entirely marine. The beak is hooked at the top, and the hind toe is wanting. The *Albatross* furnishes an example.

LONGIROSTRES. The name given by Cuvier to a tribe of wading birds, divided into families and genera, and distinguished principally by the length and tenuity of their bills.

LOPHIADÆ. A tribe of spiny-finned fishes, distinguished by the bones of the carpus being so elongated as to form a sort of arm, by which the pectoral fins are supported. The Angler is a type of this family. [See ANGLER, p. 17.]

LOPHIODON. [See SUPPLEMENT.]

LOPHOBRANCHII. An order of osseous fishes, distinguished by the structure of their gills, which are in the form of small round tufts, disposed in pairs, and arranged along the branchial arches. They are also further distinguished by having their body covered with shields or small plates, which often give it an angular form. [See PIPE-FISH.]

LOPHOPHORUS, or IMPEYAN PHEASANT; in India also called MONAUL. A genus of gallinaceous birds, belonging to the Pheasant tribe, having the head surmounted by an egret, the feathers in the male being very much elongated. The tail is large and flat, the tail-coverts short; the male is of the most brilliant coloured plumage. The circumference of the eye and the cheeks are destitute of feathers: the upper mandible overhanging the under one very much, a structure which is very important to this bird, as it enables it to root up bulbs, upon which it chiefly feeds. The best known species, which was named after Sir Elijah Impey by Dr. Latham (*L. Impeyanus or refulgens*), seems to be common in the Himalaya mountains; and a pair, in May, 1847, were brought alive to this country. The crest and the greater part of the plumage of the back in the male is composed of the most beautiful and resplendent colours, reflecting various hues of gold, copper, sapphire, and emerald. The tail is of a reddish chestnut; the rump white. The female and young are brown, varied with gray and tawny yellow. It is to be hoped that this fine species may be domiciled in this country. It can easily be brought down to the plains of India, but, from the great heat, it seldom long survives.

LORIS. (*Loris* or *Stenops.*) A genus of Quadrumanous animals, allied to the Lemurs. They have a short muzzle, slender

SLOW-PACED LEMUR.
(LORIS TARDIGRADUS)

body, no tail, large approximating eyes, and rough tongue. Two species only are known, both of which are natives of the East Indies, the SHORT-LIMBED LORIS (*Lemur tardigradus*), and the SLENDER LORIS (*Lemur gracilis*): the latter is remarkable for the disproportionate length of its limbs, and especially of its fore-arms. They are nocturnal and arboreal in their habits; they subsist on insects, occasionally on small birds or quadrupeds, and have an excessively slow gait. During the day they sleep clinging to a branch: at night they prowl among the forest boughs in quest of food. Nothing can escape the scrutiny of their large glaring orbs, or the tenacity of their grasp; and when they have marked their victim, they cautiously and noiselessly approach it till it is within their reach.

"The genus *Loris*," Mr. Bennett observes, in his 'Gardens and Menagerie of the Zoological Society,' "forms part of that division of the Quadrumanous order which is essentially distinguished by an unequal number or irregular disposition of the incisor teeth in the two jaws; terminal nostrils with sinuous openings; and a long subulate or sickle-shaped claw upon the fore-finger of the hinder hands, all the rest of the nails being flat and rounded like those of the greater part of the monkeys and of man. The Loris differs from the other genera of this family in having four incisors in the upper jaw, placed in pairs with a vacant space between, and six in the lower, directed obliquely forwards; canines of moderate

SLENDER LORIS AND PART OF SKULL.
(LORIS GRACILIS.)

size; twelve molars above and ten below; a short rounded head, and little or no tail. * * * * In addition to these primary characters, the Loris are distinguished by large prominent eyes, placed in front of the head and at no great distance from each other; short ears, scarcely rising through the hair with which they are invested, a rough tongue; nostrils projecting beyond the mouth and surrounded by a naked muzzle; and thumbs widely separated from the fingers, both on the fore and hinder hands."

Little is known of the habits of the Loris in a state of nature; but the following description of one in confinement is from the pen of Sir W. Jones: "In his manners he was for the most part gentle, except in the cold season, when his temper seemed wholly changed: and his Creator, who made him so sensible of cold, to which he must often have been exposed even in his native forests, gave him, probably for that reason, his thick fur, which we rarely see on animals in these tropical climates: to me, who not only constantly fed him, but bathed him twice a week in water accommodated to the seasons, and whom he clearly distinguished from others, he was at all times grateful; but when I disturbed him in winter, he was usually indignant, and seemed to reproach me with the uneasiness which he felt, though no possible precautions had been omitted to keep him in a proper degree of warmth. * * * * From half an hour after sunrise to half an hour before sunset he slept without intermission, rolled up like a hedgehog; and, as soon as he awoke, he began to prepare himself for the labours of *his* approaching day, licking and dressing himself like a cat, an operation which the flexibility of his neck and limbs enabled him to perform very completely: he was then ready for a slight breakfast, after which he commonly took a short nap; but when the sun was quite set, he recovered all his vivacity. His ordinary food was the sweet fruit of this country; plantains always, and mangoes during the season; but he refused peaches, and was not fond of mulberries, or even of guaiavas: milk he lapped eagerly, but was contented with plain water. In general he was not voracious, but never appeared satiated with grasshoppers, and passed the whole night, while the hot season lasted, in prowling for them.

LORY. A name given to several birds of the Parrot tribe, from their frequently repeating the word. They are remarkable for their brilliant colours, dense soft plumage, and comparatively feeble beaks. They are

BLACK-CAPPED LORY.
(LORIUS PHILIPPINENSIS.)

very active and gay, even in captivity. They are found for the most part in the Moluccas, and are held in great estimation in some parts of the East. Many of the species are very docile and familiar. The following are of great beauty.

The COLLARED LORY. (*Lorius domicella.*) This species is about the size of a common pigeon; general colour of the body scarlet; the wings grass green, with the ridge of the shoulders blue, and the tops of the quill feathers rather dusky: across the breast is a moderately broad yellow bar, sometimes waved or intermixed with a portion of red; thighs violet-blue; crown of the head violet-black; bill deep yellow; under coverts of the wings violet-blue; and the under surface of the tail inclining to purple. It is lively, gay, and remarkable for its distinctness of utterance.

CERAM LORY. (*Lorius garrulus.*) Size of the preceding; colour scarlet, with deep grass-green wings and thighs: shoulder tips yellow: tips of the wings inclining to violet-brown: tail generally of the same scarlet colour with the rest of the plumage for about half its length, the remainder blue, but the two middle tail-feathers of a green hue.

SCARLET LORY. (*Lorius cæruleatus.*) The head, neck, body, and coverts of the tail are of a shining scarlet hue, except the feathers on the lower part of the neck behind, which are tipped with yellow. The greater quills of the wings are a dark green, and those which fall over them are a lighter green. The upper part of the tail is of a bright blue colour, the central feathers being slightly tinted with green. The crown of the head is red; and the legs and feet are of a blue-black.

RAJAH LORY. (*Lorius rajah.*) The colour of this splendid bird is a vivid scarlet, with the wings entirely golden yellow: on the top of the head is a broad spot of the same colour, and across the breast a broad bar: the thighs are yellow; the bill yellowish white; and the legs blackish.

KING LORY. (*Aprosmictus scapulatus.*) The habitat of this showy and noble species is New South Wales, where it is said to be almost wholly confined to the brushes, as it there finds a plentiful supply of seeds, fruit, and berries; but we find in Mr. Gould's description, that "when the Indian corn is becoming ripe it leaves its umbrageous abode and sallies forth in large flocks, which commit great devastation on the ripening grain." The sexes differ very considerably in the colouring of the plumage: the male has the head, neck, and all the under surface scarlet; back and wings green, the inner webs of the primaries and secondaries being black; along the scapularies a broad line of pale verdigris green; the rump and upper tail-coverts rich deep blue; tail black; bill scarlet; legs nearly brown. The female has the head and all the upper surface green; throat and chest green tinged with red; abdomen and under tail-coverts scarlet; rump dull blue; two centre tail feathers green; the remainder green, passing into bluish black; and with a rose-coloured spot at the extremity of the under surface. — Another species, the RED-WINGED LORY (*Aprosmictus erythropterus*), is said by Mr. Gould to have much of the character of the King Lory, being morose, indocile, shy, and wary; and is as exclusively an inhabitant of the interior of Australia as its near ally the King Lory is a denizen of the thick brushes which extend along the coast. He further tells his readers, that the extensive belts of *Acacia pendula* which stretch over and diversify the arid plains of the great Australian basin, are tenanted with thousands of this bird, besides numerous other species, roaming about either in small companies of six or eight, or in flocks of a much greater number. It is beyond the power of my pen (says Mr. Gould) to describe or give a just idea of the extreme beauty of the appearance of the Red-winged Lory when seen among the silvery branches of the acacia, particularly when the flocks comprise a large number of adult males, the gorgeous scarlet of whose shoulders offers so striking a contrast to the surrounding objects.

LOTTIA. A genus of Mollusca, closely resembling Patella; but the shells are generally rather flatter, and have the apex placed somewhat nearer the posterior margin.

LOUSE. (*Pediculus.*) A genus of parasitic aptera, most disagreeable and unseemly to us, from the idea that invariably accompanies their presence — viz. that they are seldom prevalent where cleanliness is not wholly neglected. They are characterized by having six feet formed for walking, a mouth furnished with a proboscis, antennæ as long as the thorax, and the abdomen depressed, and formed of several segments. They undergo no metamorphosis, they are very prolific, and their generations succeed each other very rapidly. The number of species is very considerable; for not only are the human race, but many animals also, subject to the intrusive visits of its peculiar parasite.

The *Pediculus humanus*, or common louse, is distinguished by its pale and livid colour, and lobated, oval abdomen. It is produced from a small oval egg, popularly called a *nit*, fastened or agglutinated by its smaller end to the hair on which it is deposited. From this egg proceeds the insect, complete in all its parts, and differing only from the parent animal in its smaller size. When examined by the microscope the principal appearances are as follow: the trunk or proboscis, which is generally concealed in its sheath or tube, is of a very sharp form, and is furnished, towards its upper part, with a few reversed aculei or prickles: the eyes are large, smooth, and black; the stomach and intestines, which possess the greater part of the abdominal cavity, afford an extremely distinct and curious view of the peristaltic motion; while the ramifications of the tracheæ or respiratory tubes appear dispersed throughout various parts of the animal: the legs are short, and terminated by a sharp-pointed double claw; and the insect is everywhere covered by a strong granulated skin. It would be as unnecessary as disgusting to dwell on the habits of this insect, or on the dreadful and loathsome disease by which, in ancient times, the human race was visited; and from which Herod, Antiochus, Callisthenes, Sylla, and many others, are said to have perished. Those who would study the history, scientific and popular, of these parasites, must take advantage of Mr. Denny's elaborate work: the number of species found on Birds, &c., in this and other countries, is v ry great.

LOVE-BIRD. (*Psittacula.*) The name given to a beautiful and diminutive group of birds belonging to the *Psittacidæ.* They are distinguished by the tail being slightly graduated: they are found in both continents; and are remarkable for having no furcula.

LOXIA. A genus of Conirostral passerine birds, remarkable on account of the peculiar conformation of the bill, which is compressed, and the two mandibles so strongly curved, that their points cross each other. [See CROSSBILL.]

LUCANIDÆ. [STAG BEETLES.] An important family of Coleopterous insects, comprising some of very large dimensions. The *Lucanidæ* are distinguished by having the antennæ terminated by a large club, composed of several of the apical joints; by the legs being robust, the anterior tibiæ being generally dilated and toothed; by the males of many species having singular horns affixed to the head and thorax; and by the great size of the mandibles. The larvæ are large fleshy grubs, having the extremity of the

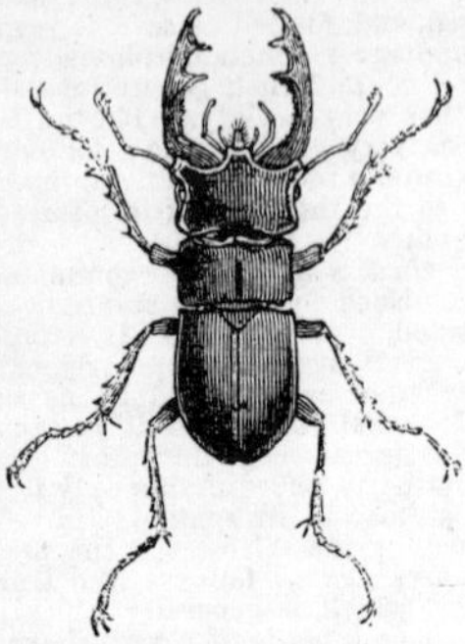

STAG BEETLE. – (LUCANUS CERVUS)

body curved towards the breast, so that it is not able to creep upon a flat surface, but compelled to lie on its side. Both in their larva and perfect states these insects are herbivorous, their habits, however, varying in the different families, according to their several structures. The family is of moderate extent, and but sparingly scattered over the globe. — Amongst the exotic genera, the beautiful Australian genus *Lamprima* is distinguished by its splendid metallic colouring, the remarkable porrected and villose mandibles, and the large plate which arms the extremity of the anterior tibiæ of the males. — Another most remarkable genus is *Chiasognathus*, in which the mandibles are longer than the body, rather slender, bent down towards the tip, where they are suddenly reflexed; they are also furnished on the under side at the base with a long horn: the colours of this genus are exceedingly splendid and metallic.

The beetles of this family fly abroad during the night, and frequently enter houses at that time, somewhat to the alarm of the occupants; but they are not venomous, and never attempt to bite without provocation. They pass the day on the trunks of trees, and live upon the sap, for procuring which the brushes of their jaws and lip seem to be designed. They are said also occasionally to bite and seize caterpillars and other soft-bodied insects, for the purpose of sucking out their juices. They lay their eggs in crevices of the bark of trees, especially near the roots, where they may sometimes be seen thus employed. The grubs of the large kinds are said to be six years in coming to their growth, living all this time in the trunks and roots of trees, boring into the solid wood, and reducing it to a substance resembling very coarse sawdust; and the injury thus caused by them is sometimes very considerable. When they have arrived at their full size, they enclose themselves in egg-shaped pods, composed of gnawed particles of wood and bark stuck together and lined with a kind of glue: within these pods they are transformed to pupæ, of a yellowish white colour, having the body and all the limbs of the future beetle encased in a whitish film, which being thrown off in due time, the insects appear in the beetle form, burst the walls of their prison, crawl through the passages the larvæ had gnawed, and come forth on the outside of the trees. Our figure represents the *Lu-*

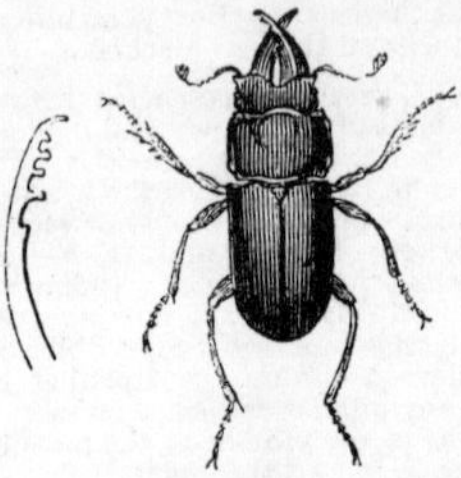

LUCANUS IBEX.

canus Ibex, a very common Brazilian species; but we may rather refer to the common Stag-beetle (*Lucanus cervus*), a highly characteristic species of the group, which is seen flying about in the evening, in the middle of summer, especially round the oaks, upon the wood of which the larva feeds; remaining in that state for several years, before undergoing its final transformation.

LUCERNARIA. A genus of Polypi belonging to the *Radiata*. They affix themselves by a slender peduncle to sea-weeds and other substances. The upper part expands like an inverted parasol, and is sur-

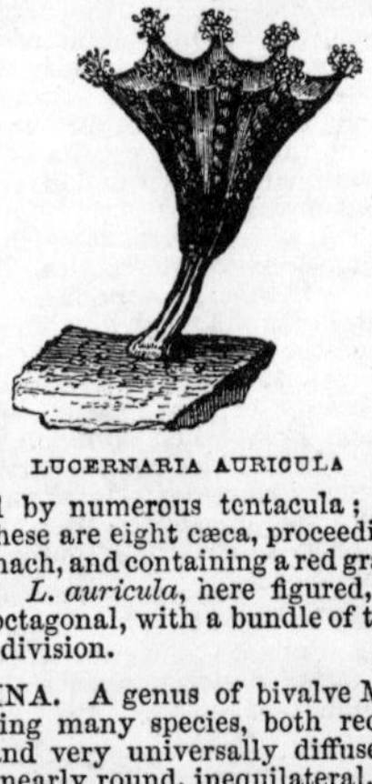

LUCERNARIA AURICULA

rounded by numerous tentacula; and between these are eight cæca, proceeding from the stomach, and containing a red granulated matter. *L. auricula*, here figured, has the border octagonal, with a bundle of tentacula in each division.

LUCINA. A genus of bivalve Mollusca, comprising many species, both recent and fossil, and very universally diffused. The shell is nearly round, inequilateral, and radiately striated; bosses small and pointed; the outer surface sculptured, the interior often punctured with small holes; cardinal and lateral teeth distinct, but variable in number. The foot of the animal is long and cylindrical.

LUMBRICUS. A genus of worms in the Linnæan system, of which the common Earth-worm is the type. They generally live beneath the surface of the ground, either perforating the dry soil, or burying themselves in mud, where many of them lead a semi-aquatic life. [See EARTH-WORM.]

LUMP-FISH. (*Cyclopterus lumpus.*) A Malacopterygious fish, deriving its name from the clumsiness of its form: its height being about half its length, and its thickness about half its height. The names LUMP-SUCKER and COCK PADDLE are also given to it. These fish are very remarkable for the manner in which their ventral fins are arranged. They are united by a membrane so as to form a kind of oval and concave disc; by means of which they are enabled to adhere with great force to any substance to which they apply themselves. This, the largest of the genus, sometimes weighs seven pounds. The back is arched and sharp, of a blackish colour, variegated with brown; the body is

LUMP-SUCKER.—(CYCLOPTERUS LUMPUS.)

covered over with sharp, black tubercles; and on each side there are three rows of large bony scales, and another on the back. The great resort of this species is on the Northern seas, about the coast of Greenland: it is also caught in many parts of the British seas, during the spring season; when it approaches the shore for the purpose of depositing its spawn. In the Northern seas great numbers of them are devoured by the seals, who swallow all but the skins, quantities of which, thus emptied, are seen floating about in the spring months; and it is said that the spots where the seals carry on their depredations can be readily distinguished by the smoothness of the water. Its power of adhesion is truly wonderful. Pennant says, "that on placing a fish of this species, just caught, into a pail of water, it fixed itself so firmly to the bottom, that on taking it by the tail, the whole pail by that means was lifted, though it held some gallons, and that without removing the fish from its hold." The colours of the Lump-fish, when in the highest perfection, combine various shades of blue, purple, and rich orange; and in the month of March it may be frequently seen in the shops of London fishmongers, suspended by the middle of the back, its singular form and brilliant colours being sure to attract the attention of the public. The flesh is soft and insipid; but the Greenlanders gladly avail themselves of the arrival of the species.

LURCHER. A species of Dog whose principal use is to assist the poacher in his nefarious and demoralizing nocturnal trade. It is supposed to be descended from the Shepherd's Dog and the Greyhound, exhibiting the stout, rough, homely character of the former, combined with the long muzzle and limbs of the latter. It is not so tall as the Greyhound; its hair is rough and wiry; the ears are half erect; and the tail is short and pendent. None of the canine species evince more sagacity, or serve their masters with more fidelity. Whether it be required to drive partridges into the net, to run down a hare, to seize a fallow-deer, or to start a rabbit, the Lurcher pursues his object in silence, and with so much skill as to render almost useless to the owner of him any other description of sporting dog.

LUTRARIA. A genus of Conchifera, found in the sand at the mouth of rivers in temperate climates. Foot of the animal sharp, oval, and long. The shell is inequilateral, oblong or ovate, gaping at both extremities; hinge with two cardinal teeth in one valve, and a triangular pit; no lateral teeth; in which respect it differs from the genus *Mactra*, which it otherwise much resembles.

LYCÆNA. A genus of Butterflies closely allied to *Polyommatus*. Referring the student to the work of Messrs. Doubleday and Hewitson, we here restrict ourselves to the notice of two British species.

The LYCÆNA DISPAR, or LARGE COPPER BUTTERFLY. It is generally remarked that this splendid insect is chiefly confined to the fenny counties of Cambridge and Huntingdon, and the neighbouring ones of Suffolk and Norfolk. The upper surface of the wings of the male are a brilliant copper colour, with an obscure row of spots towards the tip; the costal and posterior margins, and a patch at the base, black; the posterior, with a slender oblong discoidal line, and the margins black; beneath, the anterior wings are pale fulvous orange, with ten distinct ocelli, with a large black pupil and slender white iris: posterior wings bluish, with an elongate discoidal streak, and numerous rather obsolete ocelli, with a black pupil and pale blue iris: the hinder margin is deep orange, except where it unites with the anterior, margined internally and externally with a series of black spots. In the female the anterior wings above are divested of the gloss so conspicuous in the male, and have nine or ten black spots, two or three of which are placed near the base of the costal margin, the rest in an arcuated band near the tip: the posterior wings are dusky brown, with the nervures and a denticulated hinder band copper-coloured. The ocellated spots vary considerably in both sexes. Caterpillar bright green, and somewhat hairy, with innumerable white dots: it feeds upon a kind of dock. The chrysalis is at first green; afterwards pale ash, with a dark dorsal line, and two abbreviated white ones on each side.

The LYCÆNA PHLÆAS, or SMALL COPPER BUTTERFLY. One very part of our island, as well as on the adjacent continent, this pretty Butterfly is tolerably abundant on commons, roadsides, pastures, and heaths. The an-

SMALL COPPER BUTTERFLY. (LYCÆNA PHLÆAS.)

terior wings above are of a brilliant copper colour, with the posterior margin and eight discoidal spots black; the hinder wings are brownish black, with a copper band on the

UNDER SIDE OF LYCÆNA PHLÆAS.

hinder margin, which is externally denticulated, and has a black line and some dots on the disc: beneath, the colour is paler and not glossy, and there are ten distinct black spots on the disc; the posterior wings are drab-coloured, tinged with copper, and

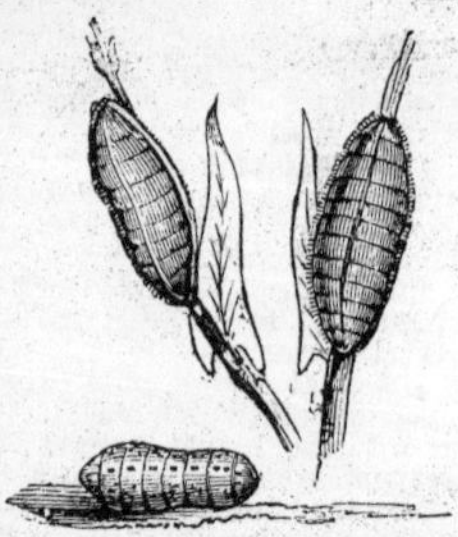

CATERPILLAR AND CHRYSALIS OF L. PHLÆAS.

sprinkled with numerous blackish dots: the cilia are rose-coloured at the tip, and black at the base: the body is black with tawny hairs above; the antennæ black, annulated with white. Caterpillar green, with a yellow dorsal stripe. It is observed to feed much on the sorrel. Mr. Knapp, in his attractive work, the 'Journal of a Naturalist,' speaking of this pretty little butterfly, says, "We shall see these diminutive creatures, whenever they come near each other, dart into action, and continue buffeting one another about till one retires from the contest; when the victor returns in triumph to the station he had left. Should the enemy again advance, the combat is again renewed; but should a cloud obscure the sun, or a breeze chill the air, their ardour becomes abated, and contention ceases. The *papilio phlœas* enjoys a combat even with its kindred. Two of them are seldom disturbed, when basking on a knot of asters in September, without mutual strife ensuing."

LYCÆNIDÆ. A family of lepidopterous insects, comprising several distinct groups of small, but beautiful Butterflies, including *Polyommati*, or the Blues; *Lycœnœ*, or the Coppers; and *Theclœ*, or the Hair-streaks. The majority of these have at least the anal angle, if not the entire under surface of the wing, ornamented with eye-like spots of various colours. The flight of these insects is feeble and slow. The caterpillars have a great resemblance to wood-lice; and the chrysalis is short, obtuse at each end, and girt round the middle as well as attached by the tail. "They have hitherto been observed to feed only upon the leaves of different trees and plants in the larva state; but a beautiful Indian species (*Thecla Isocrates*) resides within the fruit of the pomegranate, several being found within one fruit, in which, after consuming the interior, they assume the pupa state, having first eaten as many holes as there are insects through the rind of the fruit, and carefully attached its footstalk to the branch, by a coating of silk, in order to prevent its falling."—*Westwood.*

LYMEXYLON: LYMEXYLONIDÆ. A genus and family of Serricorn Beetles; having the antennæ simple and sub-moniliform, and the thorax nearly cylindrical. They are nearly allied to the *Elateridœ* and *Buprestidœ*. From the latter, however, the insects of this small group are distinguished by having the head broad before, narrowed behind, and not sunk into the thorax; they have not the breast-spine of the Elaters, and their legs are close together, and not separated from each other by a broad breast-bone, as in the Buprestians; and the hip-joints are long, and not sunk into the breast. In the principal insects of this family the antennæ are short, and from the third joint, flattened, widened, and saw-toothed on the inside; and the jaw-feelers of the males have a singular fringed piece attached to them. The body is long, narrow, nearly cylindrical, and not so firm and hard as in the Elaters. The feet are five-jointed, long, and slender. The larvæ of *Lymexylon* and *Hylecœtus* are very odd-looking, long, and slender grubs. The head is small: the first ring is very much hunched; and on the top of the last ring there is a fleshy appendage, resembling a leaf in *Lymexylon*, and like a straight horn in *Hylecœtus*. They have six short legs near the head. These grubs inhabit oak-trees, and make long cylindrical burrows in the solid wood. The generical name *Hylecœtus* means a sleeper in the woods, or one who makes his bed in the forest. One species of these insects (*Lymexylon navale*) is very common in the oak forests of the north of Europe, but rare in England. Its larva is very long. At one time it multiplied to such an extent in the dock-yards at Toulon, that the injuries it committed in the wood-works were very serious. It is recorded that Linnæus was once consulted by the King of Sweden upon the cause of the decay and destruction of the ship-timber in the royal dock-yards, and having

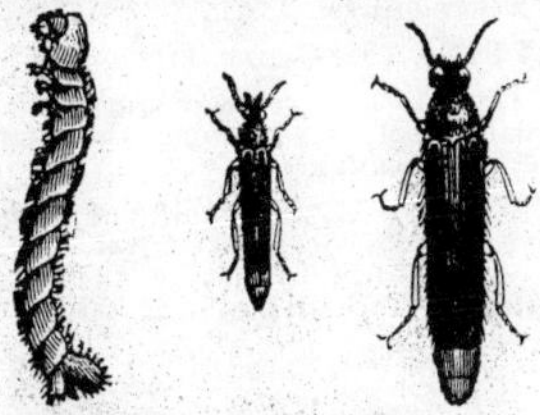

SHIP-TIMBER BEETLE. (LYMEXYLON NAVALE.)

traced it to the depredations of insects, and ascertained the history of the depredations, by directing the timber to be sunk under water during the season when these insects made their appearance in the winged-state, and were busied in laying their eggs, he effectually secured it from future attacks.

LYMNÆA. A genus of Mollusca, inhabiting a thin, oval or oblong shell; and having two triangular tentacula, with eyes at the base; foot oval and thin. Like the

LYMNÆA STAGNALIS.

Physæ, which they much resemble in appearance, they are abundantly found in our rivers and ponds, particularly the latter. They feed on aquatic plants, to the under side of the leaves of which they adhere, and come to the surface of the water for air; the number of their eggs is very great, and they are deposited on stones, stems of vegetables, &c., in long masses enveloped in a glairy substance.

LYNX. (*Felis lynx.*) The name given to certain species of feline animals, which differ slightly from others of the cat tribe, in having the ears tufted with hair, in the greater elevation of the body at the haunches, and in having a shorter tail. They are less courageous than the other felines, and show a sullen and suspicious disposition: they live upon small quadrupeds and birds, pursuing the latter to the tops of trees; and some of them also resort to the water, to feed on fishes. With some slight varieties as to size and colour, the Lynx appears to be found in all the colder regions of Europe, Asia, and America, residing in thick woods, and preying on hares, deer, birds, and almost every kind of defenceless animal. Its average length is about three feet. In colour the Lynx varies, but is generally of a pale gray, with a slight reddish tinge: the back and whole upper parts are obscurely spotted with small dusky or blackish marks; the throat, breast, and belly are white; the tail white, with a black tip; and the ears tipped with pencils of long black hair. Its eyes are brilliant and penetrating, its aspect mild, and its general air sprightly and agreeable. Though possessing nothing in common with the wolf but a kind of howl, it is often mistaken for that animal when heard at a distance. The female produces two or three young at a birth, and carefully secretes them in the recesses of the woods. The Lynx is clothed with a very thick soft fur; and the colder the climate, the more valuable it generally is: those skins which approach to a pale or whitish colour, and on which the spots are most distinct, are the most valued. The skin of the Canada Lynx forms a considerable article in the fur trade; the Hudson's Bay Company alone annually importing from seven to nine thousand skins. The fur is close and fine on the back, longer and paler on the belly. When blown aside it shows on the middle of the back a dark liver-brown colour from the roots to near the tip, but on the sides it is for the greatest part of its length of a pale yellowish brown, being merely a little darker near the roots.

LYRE-BIRD OF AUSTRALIA. (*Menura superba.*) Among the many curious and beautiful genera and species of the feathered tribes which Mr. Gould has delineated and described in his elegant work, 'The Birds of Australia,' no one seems to deserve more attention than the Lyre-bird; for, independently of its remarkable form, and the opposite opinions entertained by ornithologists as to the situation it should occupy in the natural system; "and although," as Mr. Gould observes, "more than fifty years have now elapsed since the bird was first discovered, little or no information has been hitherto published respecting its economy and habits." After paying considerable attention to the subject, while in Australia, this gentleman is decidedly of opinion that it has not, as has been generally considered, the most remote relationship to the *Gallinaceæ*, but that it forms, with certain American genera, a family of the Insessorial order. "Notwithstanding the great size of *Menura*, and the extraordinary form of its tail, in almost every other point it presents a striking resemblance to its minute congeners: like them, it possesses the bristles at the base of the bill, but to a less extent, the same unusual mass of loose, flowing, hair-like feathers on the back and rump, the same extraordinary power of running, and the like feebleness of flight." The great stronghold of the Lyre-bird is the colony of

New South Wales: it inhabits equally the brushes on the coast, and those that clothe the sides of the mountains in the interior. "Of all the birds I have ever met with," says Mr. Gould, "the Menura is by far the most shy and difficult to procure. While

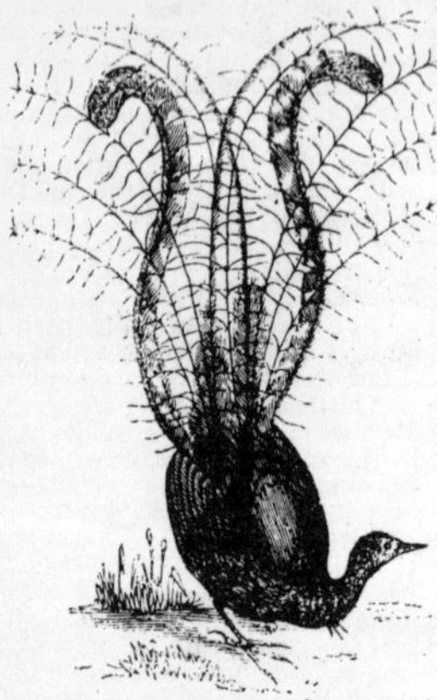

LYRE BIRD.—(MENURA SUPERBA.)

among the brushes I have been surrounded by these birds, pouring forth their loud and liquid calls, for days together, without being able to get a sight of them; and it was only by the most determined perseverance and extreme caution that I was enabled to effect this desirable object." The Lyre-bird is constantly engaged in traversing the brush from one end to the other, from mountain-top to the bottom of the gullies whose steep and rugged sides present no obstacle to its long legs and powerful muscular thighs. When running quickly through the bush they carry the tail horizontally, that being the only position in which it could be borne at such times.

Besides its loud full call, which may be heard at a great distance, it has an inward and varied song, the lower notes of which can only be heard when you have stealthily approached to within a few yards of the bird while it is singing. Its habits appear to be solitary, seldom more than a pair being seen together. It constructs a large nest, formed on the outside of sticks and twigs, like that of a magpie, and lined with the inner bark of trees and fibrous roots. The eggs are two in number, of a light colour, freckled with spots of red. The general colour of the plumage is brown; the secondary wing-feathers nearest the body, and the outer webs of the remainder, rich rufous brown; upper tail-coverts tinged with rufous; chin and front of the throat rufous, all the under surface brownish ash-colour; upper surface of the tail blackish brown; under surface silvery gray, becoming very dark on the external web of the outer feather; the inner webs fine rufous, crossed by numerous transparent bands; the margin of the inner web and tips black; bare space round the eye of a dark lead colour; legs and feet black. The female is destitute of this singularly formed tail, and in having the bare space round the eye less extensive.

LYTTA. [See CANTHARIDES.]

MACAUCO. A genus of quadrumanous animals nearly approaching the Monkey tribe. [See LEMUR.]

MACAW. These magnificent birds belong to the *Psittacidæ*, or Parrot tribe, and are distinguished by having their cheeks destitute of feathers, and their tail-feathers

RED AND YELLOW MACAW.
(MACROCERCUS ARACANGA.)

long. They are all natives of the tropical regions of South America; and abound in the swampy grounds which are covered with palm-trees, the fruit of which they are particularly fond of. They generally appear in pairs, and are always observed to perch on the summits of trees, or on the highest branch. During the day they wander to the

BLUE MACAW
(MACROCERCUS HYACINTHINUS.)

distance of about a league from their favourite spot or home, but always return in the evening. They build their nests in the hollow of decayed trees; and lay twice in the year, generally two eggs at a time. The male and female share alternately in the labour of incubation, &c. When young they are easily tamed, and soon grow familiar with persons they are accustomed to see; but, like all the Parrot tribe, they show an aversion to strangers. They are particularly fond of fruits, but in a domesticated state they will feed on almost every article, more especially sugar, bread, and fruits. Like other Parrots, they use their claws with great dexterity, though in climbing they always begin by taking hold with their bill in the first instance, using their feet only as a second point of their motion. They may be taught to speak, but their articulation is hoarse and unpleasant. Rarely, however, are those which are brought to Europe known to articulate more than a word or two, and their general voice is a loud and piercing scream. We have figured a lovely Brazilian species called, from its fine hyacinthine plumage, the MACROCERCUS HYACINTHINUS. It is not so common in aviaries as the other species.

The SCARLET MACAW. (*Macrocercus macao.*) This bird is allowed to be the most splendid with regard to colour, as well as one of the largest of all the *Psittacidæ.* From the tip of the bill to the extremity of the tail some of them measure thirty-six inches. The arch of the upper mandible, from the forehead to the point of the bill, is nearly three inches; the upper mandible is whitish, the lower black or dusky. The nostrils are placed in the upper part of the bill, just within the feathers. The sides of the head are destitute of feathers, and covered with a whitish, wrinkled skin: the head, neck, breast, belly, thighs, upper part of the back, and lesser covert-feathers of the wings, are of a very fine bright red or scarlet colour; the quill-feathers of the wings are externally of a fine blue, and on their under sides of a faint red: the first feathers next above the quills are a bright yellow, some of the feathers being tipped with green; the blue quills which fall next the back are tinged with green; and the hinder part of the thigh has some green intermixed with the red. The lower belly and covert-feathers under the tail, as also the lower part of the back and coverts on the upper side of the tail, are of a very fine blue colour: the tail-feathers gradually shorten towards the sides; some of the longest or middle-feathers are wholly red; the shorter, or side-feathers, are partly red and partly blue; the legs and feet are covered with dusky scales; and the toes are disposed two forwards and two backwards, as in others of the parrot tribe, all armed with strong claws. This noble bird, which occasionally varies in some degree in point of size and colours, was justly considered at its first introduction into Europe as a present fit for royalty, and was one of the principal ornaments in the halls of palaces.

BLUE AND YELLOW MACAW. (*Macrocer araraunа.*) This species is less common than the Scarlet Macaw, and but little inferior in point of size. The bill is arched and of a black colour: the nostrils are placed at the base of the upper mandible, in a white bare skin, which extends all round the eyes, this skin being variegated with fine lines of small black feathers: immediately under the bill is a large black spot, which encompasses part of the bare white space on the sides of the head: the feathers on the top of the head are green, gradually becoming blue on the neck: the upper side of the neck, the back, and upper sides of the wings and tail are of an exceeding fine blue colour, the lesser wing-coverts and the rump being a little tinged with green, and the tail and upper sides of the quill-feathers with purple: all the blue feathers of the back, wings, and tail are of a reddish yellow on their under sides: the fore part of the neck, the breast, belly, thighs, and covert-feathers under the tail, are of a fine yellow orange-colour, except the hinder parts of the thighs, where there is a little blue intermixed: the covert-feathers withinside the wings are yellow, which appears outwardly on the ridge or joint in the upper part of the wing: the legs and feet are nearly black.

BRAZILIAN GREEN MACAW. (*Macrocercus severus.*) The colour is a fine green; the bend of the shoulders and whole under side of both wings and tail red: quill-feathers and some of the larger coverts fine blue: tail green above, but growing blue at the tips; the two middl. feathers blue throughout their whole length on the outer edges: bill black, with flesh-coloured cere; legs black, with a feathery red zone round the bottom of the thighs. It is said to be common in Brazil, appearing in innumerable flocks, and committing great devastation among the coffee plantations, by devouring the ripe berries.

MACHAIRODUS. [See SUPPLEMENT.]

MACKEREL. (*Scomber scomber.*) This well-known fish is one of the most beautiful as regards the brilliancy of its colours, and at the same time one of the most useful as regards the food of man, among the inhabitants of the watery element. It is a native

MACKEREL. —(SCOMBER SCOMBER.)

of the European and American seas, generally appearing at stated seasons, in vast shoals, round particular coasts. The periodical appearance of these large shoals was formerly imputed to its migration from north to south: but many facts are opposed to this idea; and there is abundant reason to believe that it inhabits the deeper parts of the seas around our island through the whole year, and that its periodical appearance on our coasts, in such vast numbers, is solely due to its seeking the shore, for the purpose of depositing its spawn. The observations on this subject, which were made when speaking of the Herring, are equally applicable here; and, to the able zoologist (Mr. Yarrell) whom on that occasion we quoted, we are now further indebted for the following sensible arguments in support of this theory. He says, "It does not appear to have been sufficiently considered, that, inhabiting a medium which varied but little either in its temperature or productions, locally, fishes are removed beyond the influence of the two principal causes which make a temporary change of situation necessary. Independently of the difficulty of tracing the course pursued through so vast an expanse of water, the order of the appearance of the fish at different places on the shores of the temperate and southern parts of Europe is the reverse of that which, according to their theory, ought to have happened. It is known that this fish is now taken, even on some parts of our own coast, in every month of the year. It is probable that the Mackerel inhabits almost the whole of the European seas: and the law of nature which obliges them and many others to visit the shallower water of the shores at a particular season, appears to be one of those wise and bountiful provisions of the Creator, by which not only is the species perpetuated with the greatest certainty, but a large portion of the parent animals are thus brought within the reach of man; who, but for the action of this law, would be deprived of many of those species most valuable to him as food. For the Mackerel dispersed over the immense surface of the deep, no effective fishery could be carried on: but, approaching the shore as they do from all directions, and roving along the coast collected in immense shoals, millions are caught, which yet form but a very small portion compared with the myriads that escape."

The usual length of the Mackerel is about fourteen inches, or varying from twelve to sixteen; but in the northern seas it is occasionally found of greater size. Its colour on the upper parts, as far as the lateral line, is a rich, deep blue, accompanied by a varying tinge of green, and marked by numerous black transverse streaks, which in the male are nearly straight, but in the female beautifully undulated: the jaws, gill-covers, and abdomen are of a bright silvery hue, with a slight varying cast of gold-green along the sides. The scales are small, oval, and transparent; the pinnules or spurious fins are small, and five in number both above and below: the nose is pointed; the under jaw the longest; the teeth are alike in both jaws, curving slightly inward; and the tail is crescent-shaped. Beautiful as are the colours of the Mackerel when alive, no sooner is it caught than its lustre begins to disappear. It is a voracious feeder, and its growth is rapid; but it is not the largest fish that are accounted the best for the table. Those taken in May or June are considered superior in flavour to such as are caught either in the spring or autumn. Most Mackerel are taken by drift-nets.

MACRAUCHENIA. [See SUPPLEMENT.]

MACROTHERIUM. [See SUPPLEMENT.]

MACROURA. The Long-tailed Decapods, such as Lobsters, Prawns, Shrimps, &c. At the end of the tail is a sort of fin, expanded laterally, which serves, by its vertical strokes, to propel the animals through the water.

MACTRA: MACTRADÆ. A genus and family of Molluscous animals of the order Conchifera Dimyaria. Shell oval, transverse, with thin cardinal and lateral teeth; valves slightly inequilateral, and gaping a little on each side; bosses protuberant. Animal, foot sharp, oval, and long. The Mactræ live in the sand, and are universally diffused. The genus includes many rare and beautiful species; though the shells exhibit rather a diversity of form, they are generally more or less triangular.

MADREPHYLLIÆA. The name given to an extensive group of Zoophytes, forming part of the MADREPORES. [See next *Art.*]

MADREPORE. A submarine substance, resembling coral, and consisting of carbonate of lime with some animal matter. It is of a white colour, wrinkled on the surface, and full of cavities or cells, inhabited by a small animal, which discharges a liquid from which the stony substance is formed. "Those beautiful rocky masses," observes Mr. Rymer Jones, "for such they appear to the vulgar

MADREPORA ABRATANOIDES.

eye, called *Madrepores*, which, branching into countless varieties of arborescent forms, are abundantly met with in the ocean, and so frequently ornament the cabinets of the curious, are merely fabrics constructed by compound Polyps, and owe their growth to the accumulation of earthy particles deposited within a fleshy substance that is nourished by countless Polyps, more or less resembling *Hydræ*, diffused over all its external surface. * * * Every one of the branchy stems of the Madrepore is seen, upon a cursory survey, to be covered with multitudes of minute pits or depressions, although these, from the smallness of their size, are scarcely visible to an inattentive observer. Examined with a magnifying glass, however, each of these multitudinous orifices is found to be a cell of beautiful construction, equally remarkable for the mathematical regularity with which it is formed and the exquisite fineness of the materials composing it. * * * Let us endeavour to picture to ourselves an extent of the bed of the ocean, spacious as these realms that we inhabit, carpeted with living plants; every blade of grass and every flower instinct with life, and all the vast expanse busily engaged in deriving from the surrounding water materials for subsistence: let us consider that from age to age, the wide-spread scene is building up, by constant precipitation from the sea, a rocky territory, co-extensive with itself, and then we shall perceive that, in the course of time, even these almost unknown members of the animal creation may perform achievements at which the boldest mind is startled when it comes to survey what they have accomplished."

MAGILUS. A genus of Mollusca, inhabiting a thick, tubular, irregularly contorted shell; spire short, consisting of three or four whorls; aperture longer than wide, without any notch, but an angle at the base. When in a young state," observes Miss Catlow, "this curious shell presents all the character of a regular spiral univalve. This animal

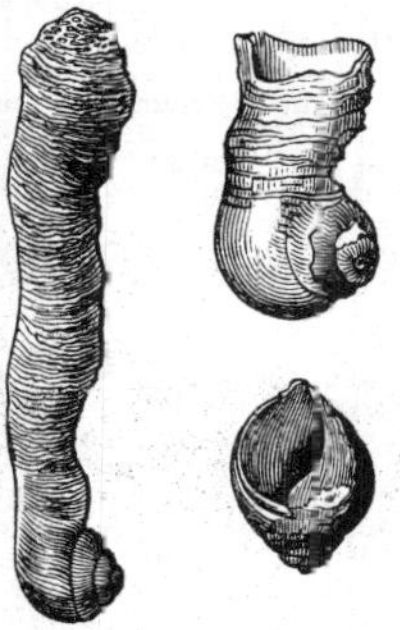

MAGILUS ANTIQUUS.

establishes itself in the excavations of Madrepores; and as the coral increases around it, the Magilus is obliged, in order to have its aperture on a level with the surrounding surface, or near it, to construct a tube, the growth of the coral determining its length. As this tube goes on increasing, the animal abandons the spiral for the tubular part of the shell; and in the operation it leaves behind no partitions, but secretes a compact calcareous matter, which reaches to the very summit of the spiral part; so that in an old specimen the posterior part of the shell presents a solid mass. One species only, *Magilus antiquus*, is known. The colour is white, more or less pure."

MAGOT. The Barbary Ape. (*Pithecus Inuus.*) [See APE.]

MAGPIE. (*Pica caudata.*) A crafty and familiar bird of the corvine family, whose plumage of black and white, green and purple, with the rich and gilded variegations of its tail, may be safely pronounced beautiful; yet its propensity for mischief, its noise, and its restless and quarrelsome disposition, render it every where an unwelcome intruder. In length it is about eighteen inches; its bill is strong and black; eyes hazel; head, neck, back, breast, and tail-coverts deep black, forming a fine contrast with the snowy whiteness of the under parts and scapulars. The plumage is in general glossed with green, purple, and blue, which catch the eye in different lights, and are particularly resplendent on the tail, which is very long, and rather wedge-shaped: vent, under tail-coverts, thighs, and legs black: on the throat and part of the neck the feathers are mixed with others, resembling strong whitish hairs. It feeds both on animal and vegetable substances; and when satisfied with its present meal, it will hide the remainder of its provision for a future occasion. It builds its nest of sticks and clay, with great art and sagacity; defending it on all sides with sharp thorny twigs, leaving only a hole for entrance, and furnishing the inside with a lining of fibrous roots and other soft materials. The female lays seven or eight eggs, pale green, spotted with black. The Magpie may easily be tamed and taught to pronounce words and short sentences, but

MAGPIE.—(PICA CAUDATA.)

its tones are too shrill and sharp to be a perfect imitation of the human voice. Like other birds of its kind, it is addicted to stealing and hoarding. It will occasionally plunder the nests of some other birds, and even carry off whole broods of stray ducklings when its young demand more food than is easily obtained: but it has its good qualities also; for it frees our pastures of an incredible number of grubs and slugs, and often performs a friendly office for sheep and oxen, by getting on their backs and freeing them from troublesome vermin. Magpies may be said to be social, though not actually gregarious.

MALACODERMATA. A section of Pentamerous Coleoptera: for a familiar example of which we must refer to the Glow-worm (*Lampyris*), and Soldier-beetle (*Telephorus*). The antennæ differ in the two sexes. The accompanying figure represents the female of the *Lamprocera Latreillii*, also called

Homalisus grandis, a native of Brazil. The pectinated figure on the one side shows the

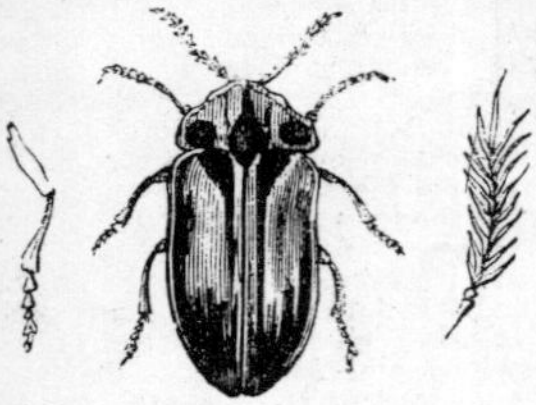

LAMPROCERA LATREILLII.

antennæ of the male; the other figure representing the leg, with its five-jointed or pentamerous tarsus.

MALLEUS, or HAMMER-HEADED OYSTER. (*Malleus vulgaris*). A genus allied to *Ostrea*, chiefly remarkable for its singular form; the two sides of the hinge being extended so as to resemble in some measure the head of a hammer, while the valves, elongated nearly at right angles to these,

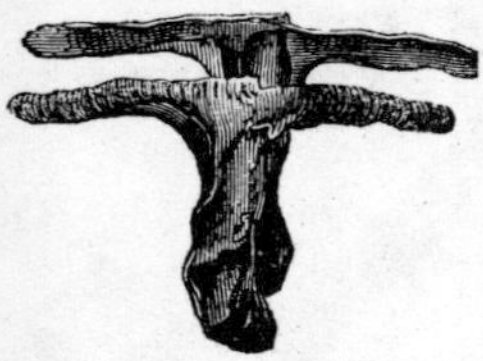

MALLEUS VULGARIS

represent the handle. It inhabits the Indian archipelago, attaching itself by a byssus to submarine rocks. The shape of the shells are so very various, that scarcely two of a species can be found alike; externally their appearance is very rude and irregular, but the interior is extremely beautiful, being lined with the most brilliant mother-of-pearl; hence, as they are rather rare also, they generally obtain a good price.

MALURUS. A genus of Passerine birds, abundantly dispersed throughout New South Wales, containing several species, one of which,

MALURUS CYANEUS, named by the colonists the SUPERB WARBLER, BLUE WREN, &c., is the oldest known species of the whole of the lovely group forming the genus; and its favourite haunts are localities of a wild and sterile character, thinly covered with low scrubby brushwood, near the borders of rivers and ravines. The male in summer has the crown of the head, ear-coverts, and a lunar-shaped mark on the upper part of the back light metallic blue; lores, line over the eye, occiput, scapularies, back, rump, and upper tail-coverts velvety black; throat and chest bluish black; tail deep blue, indistinctly barred with a darker hue, and finely tipped with white; wings brown; under surface buffy white, tinged with blue on the flanks; bill black; feet brown. The female has the lores and a circle surrounding the eye reddish brown; wings and tail brown; under surface brownish white; bill reddish brown; feet pale brown.

The Malurus Cyaneus is of a very wandering disposition, but seldom travels far beyond the district in which it was bred. During the winter they associate in small flocks; but as spring advances they separate into pairs, the male undergoing a most surprising change of plumage, which for a few months is as resplendent as it is possible to conceive: indeed, its whole character and nature appear also to have received a new impulse; the little creature now displaying great vivacity, proudly showing off its gorgeous attire, and pouring out its animated song almost unceasingly, until the female has completed her task of incubation. In the winter no bird can be more tame and familiar, seeming to court, rather than shun, the presence of man. Its mode of progression is a succession of bounding hops, performed with great rapidity, its short and rounded wing incapacitating it for protracted flight. Two, if not three, broods are reared in a season; and, independently of her own young, the female is the foster-parent of the Bronze Cuckoo, a single egg of which species is frequently found deposited in her dome-shaped nest, which has a small hole at the side for an entrance, and is usually placed near the ground, in a secluded bush, tuft of grass, or under the shelter of a bank. The song is a hurried strain, somewhat resembling that of the European Wren.

MAMMALIA. That class which is placed at the head of the Animal Kingdom, because it is composed of the beings whose faculties are the most numerous, whose structure is the most perfect, whose movements are the most various, and whose intelligence is the most developed. The term is derived from *mammæ* [*breasts*], and the class contains all those animals which suckle their young by means of breasts. Most mammiferous animals are formed for walking; a few, however, can sustain themselves in the air; and a limited number are destined to live in the water. From Man, who, from his most perfect organization, stands at the head of the system, to Whales and other cetaceous animals, which are classed at the end of Mammalia, the skeleton is formed upon the same general principles, and its parts are only altered and modified to suit the station which the animal is destined to fill. All Mammalia are viviparous; the fœtus derives its nourishment direct from the blood of the mother, and, after birth, she supports it, for a longer or shorter time, by her milk, a nutritious liquid secreted by particular glands, called *mammary*. Sometimes the young are born with their eyes open, and can immediately run about, and procure their own food; but many come into the world with their eyes closed, and in a state of utter helplessness.

Linnæus was the first to bring under review the whole animal, vegetable, and mineral kingdoms, wherein he described and named every natural object which had been discovered up to his time, and introduced into his writings a language fitted to supply all the wants of the age: and not long after his death, Gmelin edited a new edition of the "Systema Naturæ," with additions up to that date (1788). Various scientific men subsequently attempted to improve the arrangement of Linnæus; and at length appeared the "Regne Animal," by Cuvier, who, having shown that there are "immutable laws prescribed to living beings," divides his class Mammalia into the following orders:—1. BIMANA; with two hands, of which Man is the only species. He has three kinds of teeth.—2. QUADRUMANA; animals with four hands, and having three kinds of teeth: Monkeys, &c.—3. CARNARIA. These have three kinds of teeth, which are more or less of a carnivorous character. Thumb of the anterior extremities never opposable to the other fingers or toes. It is divided into three families:—*Cheiroptera*, or bats; *Insectivora*, or such animals as feed much on insects, as the Hedgehog, &c.; *Carnivora*, animals which subsist on flesh; Cats, &c.—4. MARSUPIALIA; animals provided with a pouch for the protection of their young after birth, as the Kangaroo, &c.—5. RODENTIA, or Gnawers; animals with two large incisors in each jaw, separated from the molars by a void space. The molars in most genera with flat or riggled crowns, and in others blunt tubercles: Hares, Squirrels, &c.—6. EDENTATA; generally destitute of teeth, some genera with molars only; their toes varying in number, and provided with large hoof-like nails: Ant-eaters, &c.—7. PACHYDERMATA, or thick-skinned animals; it includes all the hoofed quadrupeds, except the ruminants: Horses, &c.—8. RUMINANTIA; animals which ruminate or chew the cud, with cloven feet, and provided with four stomachs: Deer, &c.—9. CETACEA; Whales and their congeners.

The essential characters of the Mammalia are taken from the number and structure of their teeth, and the construction of their hands and feet: on the perfection of the organs of touch the expertness of the animal depends; and from their dental formula may, in a great measure, be deduced the nature of their food and digestive functions. Living for the most part on the earth's surface, the Mammalia are exposed to the transitions of heat and cold: hence the bodies of most of them are covered with a coating of hair, varying in thickness. As their habitation approaches the northern regions, it is more dense, and thinner towards the equator. And it is to be observed, that the cetaceous animals which inhabit the sea are totally divested of hair.

The Mammalia are, of all animals, those which approach the nearest to Man, in regard to their intellectual powers; yet in this respect they present the greatest differences amongst themselves. This the reader will observe, as he turns to the various articles in this volume, where the instincts and habits of each species are described.

MAMMOTH. (*Elephas primogenius.*) A term employed to designate an extinct species of elephant, the fossil remains of which have been at various times discovered embedded in the newer tertiary deposits both in Europe and Asia. A great quantity of fossil ivory is obtained from Siberia; and even whole carcasses, covered with flesh and skin, preserved by the eternal frost of those regions, have been found in the northern parts of that country. It is not to be confounded with the *Mastodon*, a gigantic fossil animal of North America.

Some authors derive the name "Mammoth" from the word *Behemoth*, used in the book of Job to designate an immensely large animal, or from *Mehemoth*, an Arab term applied to elephants of extraordinary size; while others are of opinion that it is merely an adoption of the word *Mammouth*, given by the Siberians to a huge animal, which they (in order to account for the quantity of Mammoth-horns, or fossil ivory) pretend lived underground in the manner of moles, and could not bear the light of day. This story is in a manner corroborated by the Chinese account of a subterranean animal, which in their great work on Natural History is thus described: "The animal called *tien-schu*, *tyn-schu*, or *yn-schu* (signifying the mouse that conceals itself), lives entirely in subterranean caverns; in form it resembles a mouse, but is equal to an ox or a buffalo in size. It has no tail, and is of a dark colour; it is exceedingly strong, and digs caverns in which it lives, in rocky and woody places." It is the universal opinion throughout Siberia, that Mammoths have been found with the flesh quite fresh and filled with blood; this, although an exaggeration, is founded on the fact that entire bodies have been discovered, preserved in ice, with the flesh comparatively in a state of freshness. The best authenticated instance of this was that of the Mammoth brought to St. Petersburg by Mr. Adams, and first recorded in Oct. 1807 in the 'Journal du Nord.' The account is related in 'The Zoologist' as follows:—

"In 1799 a Tungusian fisherman observed, in a bank on the shore of the Frozen Ocean, at the mouth of the river Lena, a shapeless mass, almost enveloped in ice, and he was quite unable to make out what it could be. The year following, a larger portion of this mass became visible, but the fisherman was still unable to ascertain its nature. Towards the end of the following summer one of the tusks and an entire side of the animal were exposed. It was not, however, until the fifth year from its discovery, when the ice having melted sooner than usual, that the enormous animal became entirely detached from the bank or cliff in which it was first observed, and came thundering down on to a sand-bank below. In the month of March, 1804, the fisherman extracted the tusks, which were 9 feet 6 inches long, and together weighed 360 lbs., and sold them at Jakutsk for fifty rubles. Two years afterwards Mr. Adams visited the animal, and found it much mutilated. The Jakoutes residing in the neighbourhood had cut away the flesh to feed their dogs; wild beasts, especially white bears, foxes, &c., had also eaten a great quantity of it. Nevertheless, the skeleton was entire, with the exception of a fore leg; the other bones being still held together by ligaments and portions of skin. The head was covered with dried skin; one of the ears was entire, and furnished with a tuft of hairs: the pupil of the eye was still to be distinguished; the brain was in the skull, but somewhat dried; the lower lip had been gnawed by animals, the upper one was entirely gone, and the teeth consequently exposed; the neck was furnished with a long mane; the skin was covered with long hair and a reddish wool; the portion of skin still remaining was so heavy, that ten men could scarcely carry it: according to Mr. Adams, more than thirty pounds weight of hair and wool was collected from the wet sand into which it had been trodden by the white bears while devouring the flesh. Mr. Adams took the greatest pains in collecting what remained of this unique specimen of an ancient creation, and procured the tusks from Jakutsk. The Emperor of Russia purchased the skeleton, which is now in the Museum of the Academy of St. Petersburg. The height of the creature is about nine feet, and its extreme length to the tip of the tail about sixteen feet. Portions of the skin and hair were presented to most of the continental museums, as well as to the College of Surgeons in London."

"The Mammoth seems a link connecting the past and the present worlds—a being whose body has outlived its destination. All the arguments which have been used to prove that the earth has undergone some great convulsion since this huge animal was endowed with life, appear perfectly untenable. In the first place, it is evident that its life became a sacrifice to a sudden snow-storm, by which it was overtaken, overwhelmed, and suffocated. The suddenness of the storm might have been accidental; the winter might have set in earlier, it might have been more severe than usual: but the animal was well adapted for such winters; its long, warm, and shaggy coat proclaim it a denizen of arctic countries, and is admirably adapted to exclude the severest cold: such a clothing would have been intolerable in tropical regions, where elephants now abound. We learn from Bishop Heber that in some of the colder and mountainous districts of northern India, hairy elephants still exist, thus showing that this clothing is provided as an especial protection against the climate; and at the same time leading to the obvious conclusion, that the well-clad Mammoth, like the Polar Bear, was the destined denizen of still severer climes. Nature ever adapts her creatures to the circumstances under which she has chosen to place them."

Dr. Falconer and Major Cautley found numerous species of fossil elephants in the Sewalik Hills, which are described in their beautifully illustrated work, and are now in the British Museum.

MAN. Linnæus was the first who ventured to class Man in a scientific system with the rest of animated nature; nor did he wholly escape censure for degrading the dignity of the human race by such an approximation: but whether considered as the head of the animal creation, and a part of it; or as a sole genus and sole species, distinct from others, and lord of all; it is not merely correct, but absolutely necessary, to define Man—if viewed solely in a physical light, and setting aside his divine reason, and his immortal nature—as a being provided with two hands, designed for prehension, and having fingers protected by flat nails; with two feet, destined for walking; with a single stomach; and with three kinds of teeth,—incisive, canine, and molar. His position is upright, his food both vegetable and animal, his body naked.

Man is the only animal truly *bimanous* and *biped*. His whole body is modified for the vertical position. His feet furnish him with a larger base than those of other mammalians; the muscles which retain the foot and thigh in the state of extension are more vigorous, whence results the swelling of the calf and buttock; the flexors of the leg are attached higher up, which permits of complete extension of the knee, and renders the calf more apparent. The pelvis is larger, which separates the thighs and feet, and gives to the trunk that pyramidal form favourable to equilibrium: the necks of the thigh-bones form an angle with the body of the bone, which increases still more the separation of the feet, and augments the basis of the body. And the head, in this vertical position, is in equilibrium with the trunk, because its articulation is exactly under the middle of its mass. Man thus preserves the entire use of his hands for the arts, while his organs of sense are most favourably situated for observation. His two eyes are directed forwards; which produces more unity in the result of his vision, and concentrates his attention more closely on objects of this kind. He has a particular pre-eminence in his organ of voice: of all mammalians, he can alone articulate sounds. Hence results his most invaluable mode of communication; for of all the signs which can be conveniently employed for the transmission of ideas, variations of sound are those which can be perceived at the greatest distance, and in the most various directions simultaneously.

The ordinary produce of the human species is but one child at a birth; the period of gestation, nine months. The fœtus grows more rapidly as it approaches the time of birth. The infant, on the contrary, increases always more and more slowly. It has reached upwards of a fourth of its height when born; attains the half of it at two years and a half; and the three-fourths at nine or ten years. By the eighteenth year the growth almost entirely ceases. Man rarely exceeds six feet, and seldom remains under five. Woman is ordinarily some inches shorter. Scarcely has the body attained its full growth in height, before it commences to increase in bulk; fat accumulates in the cellular tissue. The different vessels become gradually obstructed; the solids become rigid; decrepitude and decay follow in their turn; and most of the species, either from disease, accidents, or merely old age, perish ere they are "threescore years and ten." Occasionally one lives upwards of a hundred years; but long before that patriarchal age is reached, the survivor needs no monitor to tell him that "all is labour and sorrow."

It has been made a subject of dispute, whether there is more than one species in the human race; but it is merely a dispute of words; and if the term *species* is used in its common scientific sense, it cannot be denied that there is but one species. There are, however, certain and constant differences of stature, physiognomy, colour, nature of the hair, or form of the skull, which have given rise to subdivisions of this species. Blumenbach reduces these varieties to five:—

The first variety, usually called the *Caucasian*, from its supposed origin in the Caucasus, occupies the central parts of the old continent, namely, Western Asia, Eastern and Northern Africa, Hindostan, and Europe. Its characters are the colour of the skin, more or less white or brown; the cheeks tinged with red; long hair, either brown or light; the head almost spherical; the face oval and narrow; the features moderately marked, the nose slightly arched; the mouth small; the front teeth placed perpendicularly in the jaws; the chin full and round. The regularity of the features of such a countenance, which is that of the European, causes it to be generally considered (by them at least) as the most agreeable.—2. The second variety has been called the *Eastern* variety. The colour in this race is yellow; the hair black, stiff, straight, and rather thin; the head almost square; the face large, flat, and depressed; the features indistinctly marked; the nose small and flat; the cheeks round and prominent; the chin pointed; the eyes small. This variety comprises the Asiatics to the east of the Ganges and of Mount Beloor, except the Malays.—3. The *American* variety resembles that last described in several points. Its principal characters are the copper colour; stiff, thin, straight black hair; low forehead; eyes sunk; the nose somewhat projecting; cheek bones prominent; the face large. This variety comprises all the Americans except the Esquimaux. There are several branches, however, which differ considerably.—4. The fourth variety is called by Blumenbach the *Malay*, and described as of a tawny colour; the hair black, soft, thick, and curled; the forehead a little projecting; the nose thick, wide, and flattened; the mouth large; the upper jaw projecting. This variety comprehends the islanders of the Pacific Ocean.—5. The remaining variety is the *Negro*. Its characters are: colour black; hair black and woolly; head narrow; forehead convex and arched; cheek-bones projecting; nose large, and almost confounded with the upper jaw; the upper front teeth obliquely placed; the lips thick; the chin drawn in; the legs crooked. This race is found in Western and Southern Africa, and the great islands of the Pacific, generally in the interior. There are very great differences in the tribes included in this variety; the Negro, with the complexion of jet, and wool; the Caffre, with a copper complexion, and long hair; the sooty Papous, or New Guineamen; the native of Van Diemen's Land, &c. "Within each of these varieties are included numerous smaller divisions, which are certainly, though less prominently, distinct in their features. The varieties of national appearance between the Scotch, English, French, and Germans, for example, are in general distinguishable, though it would be difficult to define their differences. Similar subdivisions of character exist among all the varieties, and so fill up the intervals between the extreme specimens of each as to form a regular and nearly perfect series, of which the Esquimaux and Negro might occupy the extremities, and the European the middle place, between the broad and high features of the one, and the narrow, elongated, and depressed skull and face of the other."

Those writers who have gone deeply into the subject, and attempted to account for *all* the causes which have contributed to the diversity of the human species, have generally been led into a more discursive field than they had anticipated; while the result, perhaps, has been both inconclusive and unsatisfactory. In such a compendium as this, where brevity is scarcely less essential than precision, we are constantly warned not to exceed our limits. We shall therefore not pretend to describe minutely the anatomical structure of Man, neither shall we attempt to follow him from his entrance into life to his mortal exit; but shall endeavour to lay before the reader such of our "gleanings" as we conceive will best illustrate the subject, without extending the article to an unwarrantable length.

If Man be compared with the other classes of animated nature, we shall find that he possesses most of those advantages united, which the rest only partially enjoy. Infinitely superior to all others in the mental powers, he is also superior to them in the aptness and proportion of his form. He would indeed be one of the most wretched beings on earth, if, with a sentient mind, he was so constructed as to be incapable of obeying its impulses. In the lectures of Professor Green, this subject has been handled with philosophical acuteness and masterly power. He says, "In a comparison of the frame and capabilities of Man with those of the inferior animals, if we take the human frame as the ideal standard of form, it will be found that all others present many declensions from the idea by exaggeration or defect; and it will be found from this survey that Man is unquestionably endowed with that structure, the perfection of which is revealed in such a balanced relation of the parts to a whole as may best fit it for a being exercising intelligent choice, and destined for moral freedom. It is not, therefore, an absolute perfection of the constituents singly, but the proportional development of all, and their harmonious constitution to One, for which we contend;—a constitution which implies in a far higher degree than in any other animal a balanced relation of the living powers and faculties, and which requires, therefore, in Man pre-eminently, the endowment of rational will as necessary for the control and adjustment of the balance. Man has not the quick hearing of the timid herbivorous animals; but it was not intended that he should catch the sound of distant danger, and be governed by his fears: he has not the piercing sight of the eagle, nor the keen scent of the beast of prey; but neither was Man intended to be the fellow of the tiger, or a denizen of the forest. Hence the departure from the perfect proportion of Man which we observe in the inferior animals may be regarded as deformities by exaggeration or defect, dependent upon a preponderance of a part that necessitates a particular use, or the absence of a part that deprives the animal of a power, and in both instances alike abrogates that freedom for which provision is made in the balanced relation of the constituents of the human fabric, which permits the free choice of means, and the adaptation to any purpose determined by an intelligent free-will. Dilate the head, and you have a symptom of disease; protrude the jaws, you have a voracious animal; lengthen the ears, timidity is expressed; let the nose project, and the animal is governed by its scent; enlarge the belly, and you are reminded of the animal appetites: long arms may fit him for an inhabitant of the trees, and a fit companion for the ape; and predominant length of legs are infallibly associated with the habits of the wading or leaping animals. In all, regarding Man's form with reference to his destination as the ideal standard, the means become ends; deformity prevails, and becomes the badge of unintelligent slavery to the mere animal nature."

"In the contemplation of the human skeleton, its most striking characteristic, and that which contradistinguishes it from the bony fabric of all other animals, is its adaptation to the erect position; an attribute not only peculiar to Man, but without which his structure could not correspond with his spiritual endowments, since it is at once the need and symbol of a being raised above the servile condition of the mere animal nature. Thus the skull is poised with a slight preponderance anteriorly, at the top of the vertebral column; and a plumb-line dropped from the point of its support falls through the centre of gravity between the feet, which present the base of support to the whole towering fabric. We remark, however, that the supporting parts do not range with this line. The spine is bent like an italic *S*: it recedes at the chest, in order to give room to its cavity; and at the same time is harmoniously inflected forwards at the loins and neck, in order to facilitate its balance over the points of support: and it cannot be doubted that these curves contribute to the capability of bending and changing the position of the trunk, without endangering the loss of balance. But the balance of the body is also greatly aided by the breadth of the human pelvis, which, supplying a broad base of support, permits the inclinations of the trunk without the necessity of altering the position of the lower limbs. The lateral breadth of the pelvis, however, throws the heads of the thigh-bones, upon which the weight of the body is transmitted, to some distance on each side of the line that falls through the centre of gravity: and in order to provide a compensating adjustment, the thigh-bones are placed obliquely, inclining towards each other; so that in the upright posture with the feet together they touch at the knees, and the weight is then received upon the heads of the leg-bones or *tibiæ*, which stand perpendicularly under the centre of gravity: and these again are planted upon the arch of the foot or instep, on which the whole weight of the body securely rests. Then, in order to secure in the foot the requisite firmness in standing, we find that it is articulated with the leg at right angles, so that both the heel and toes touch the ground; and the joint is placed nearer the posterior than the anterior part of the foot, so as to increase the base of support in that direction towards which the body tends most to fall: besides which, the weight is here received on the inner side of the foot, where it is most arched, thereby offering not only the advantage of a strong support, but one which is highly elastic, yielding without injury in alighting upon the feet, and acting as a spring in progression. Thus the majestic column of the human form is raised and built up upon its pedestal; and the living pillar, readily maintaining its equipoise, bears aloft its capital, whilst the upper limbs are left free to adlibitive motion. Thus the place of the head, as the corporeal representative of that which perceives and wills; the disposition of the senses therein as the *media* of intelligence, and of the organs of speech as the interpreters of thought; and the arrangement of the upper limbs as the instruments of volition, no longer subservient to mere animal needs, all impress us with the conviction that even the skeleton cannot be intelligible to us without admitting that the human bodily frame was designed for the instrument and dwelling of a being contradistinguished from, and elevated above, all other animals."

It has been well argued by a popular writer of the present day, that, "destitute of either projecting teeth or strong claws, covered neither with hard scales nor with bristles, nor with a thick hide, and surpassed in speed by many of his more powerful antagonists, Man's condition would seem most pitiable, and inferior to that of any other animal; for on all the rest of those to whom she has denied the weapons of attack, Nature has bestowed the means either of defence, or of concealment, or of flight. But Man, by his superior reason, has subdued all other animals. His intellect can scarcely suggest the mechanism which his hands cannot frame; and he has made for himself arms more powerful and destructive than any other creature wields; he has clothed himself in armour and built walls of defence with which he can defy the attacks of any but his fellow-men. Naturally unarmed, Man has conquered the whole armed creation: some he has driven from their abodes, and almost exterminated; others he has forced to share his labour; and others he uses for his food, his clothing, or his pleasure. The only other part of the human structure which it is now necessary to notice is the brain, whose size in proportion to the rest of the nervous system far surpasses that of any other animal. This may be at once seen by observing the proportion which the cranium, or rather the cavity containing the brain, and the face, bear to each other. In many cases also it may be estimated by what is called the facial angle of Camper, which is found by drawing a line from the most prominent part of the forehead to that of the upper jaw-bone, and observe the angle which it forms with another line drawn through the *meatus auditorius externus* to the base of the nose, or (the head being held in a vertical position) with a horizontal line. In Man the facial angle is in the average of Europeans 80°; in some children it is a right angle, but in some negroes is not more than 70°. In the adult chimpanzee (which approaches in this respect nearest to Man) the facial angle is only 35°, and in the orang or satyr 30°. In other animals it is still less, except when it is increased by the prominence of large frontal sinuses, or by the comparative shortness of the jaws. In regard to its structure the human brain exceeds all other in development of its cerebral hemispheres, in the number and development of parts, in the depth and number of its convolutions, and in the quantity of its medullary matter in proportion to the cortical.

"In the economy of the human body there are peculiarities not less marked than those in its structure. Perhaps the most characteristic is the ability which Man enjoys of living on almost any part of the globe, and of thriving alike in either extreme of natural temperature. Thus the Greenlanders and Esquimaux have reached between 70° and 80° of north latitude, while the negro of Africa and the red man of America live under the equator. But even Europeans, accustomed to a temperate climate, can bear either of these extremes of cold and heat, as has been sufficiently proved by the numerous instances in which those who have gone on the Arctic expeditions have been obliged to winter in high northern latitudes; and, on the other hand, by the slight degree in which European settlers in the hottest parts of Africa are influenced by the temperature.

"In adaptation with his ability to inhabit almost every climate, Man can subsist on the most varied food. In the northern regions, where the earth is covered through the greater part of the year with snow, and vegetables of any kind can be procured only in the smallest quantity, the Esquimaux and Samoides subsist as well on animal food alone as the European does on the most carefully mixed diet; and on the other hand, the inhabitant of the torrid zone is not more inconvenienced by his daily subsistence on the cocoa-nut, banana, yam, rice, and other farinaceous and acid vegetables. In the temperate climates, where animal and vegetable food can be procured with equal facility, Man is truly omnivorous; towards the poles animal food or fish becomes more exclusively his diet; and towards the equator his food is chiefly composed of vegetables: and there is no doubt that in each case that food which is most universally adopted is that which is best adapted for the health of the inhabitants.

"There is not a proof in the whole history of animals that any species or individual has ever made an advance towards an improvement, or an alteration in its condition: whether solitary or living in herds, the habits of all remain the same; all of the same species appear endowed with the same faculties and dispositions, and each is in mental power the same throughout his life. Contrast with these the progress of Man. In his origin weak, naked, and defenceless, he has not only obtained dominion over all the animate creation, but the very elements are made to serve his purpose. Of the earth he has built his houses, and constructed weapons and the implements of art; he uses the wind to carry him in ships, and to prepare his food; and when the wind will not suit him, he employs fire and water to replace or to resist it. By artificial light he has prevented the inconveniences of darkness; he has stopped and made rivers, and has forced deserts, marshes, and forests alike to bear his food; he has marked out and measured the course of the celestial bodies, till he has discovered from them the size and form of the earth that he himself inhabits."

With regard to the proportions of the human figure, we have no exact knowledge; for the beauty of the best statues is better conceived by observing than by measuring them. Those of antiquity, which were at first copied after the human form, are now become the models of it; nor is there one Man found whose person approaches to those inimitable performances that have thus, in one figure, united the perfections of numbers. It is sufficient to say that, from being at first models, they are now become originals; and are used to correct deviations in that form from whence they are taken. We must not, however, pretend to give the proportions of the human body as taken from these, there being nothing more arbitrary. Some, for instance, who have studied after models, divide the body into ten times the length of the face, and others into eight. Some even pretend to assert that there is a similitude of proportion in different parts of the body: thus, that the head is the length of the face; the thumb the length of the nose; the space between the eyes the breadth of an eye; the breadth of the thigh, where thickest, double that of the thickest part of the leg, and treble the smallest; that the arms when extended are equal to the length of the figure; and that the legs and thighs are half the length of the body. All this, however, is extremely arbitrary; and the excellence of a shape, or the beauty of a statue, results from the attitude and position of the whole, rather than from any determined measurements, begun without experience, and sanctioned by caprice. It may in general be remarked, that the proportions alter in every age, and are obviously different in the two sexes. In Women the shoulders are narrower, and the neck is proportionally longer, than in Men; the hips are also considerably larger, and the thighs shorter. These proportions, however, vary greatly at different stages of life: in infancy the upper parts of the body are much larger than the lower; and the legs and thighs do not nearly constitute half the height of the whole figure. In proportion as the child increases in age, the inferior parts lengthen, so that the body is not equally divided till it arrives at its full stature.

There is a striking difference in the size of Men. Those are said to be tall who measure from five feet eight to six feet in height; the middle stature is from five feet five to five feet eight; and such as fall short of these proportions are said to be of a diminutive size. It should be observed, however, that the same person is always taller in the morning than on going to bed at night; there being sometimes the difference of an inch. The reason of this is obvious. Between all the joints of the back-bone a glutinous liquor, styled *synovia*, is deposited, which serves, like oil in a machine, to give the parts an easy play on each other: this lubricating liquor, or synovia, according to anatomists, is poured in during the season of repose, and is consumed by exercise and employment; so that after hard labour scarcely any of it remains, but the joints grow stiff, and their motion is painful and uneasy. Hence, therefore, the body diminishes in stature: for this moisture being drained away from between the numerous joints of the back-bone, they lie close on each other, and their entire length is thus very sensibly diminished; but sleep, by restoring the fluid, again swells the spaces between the vertebræ, and the whole is extended to its former dimensions.

A comparison between the strength of Men and other animals may be estimated by various modes. First, by the weight they are able to carry. It is affirmed that the porters of Constantinople carry burdens of nine hundred pounds weight: and Desguliers tells us that, by means of a certain harness, by which every part of a Man's body was proportionably loaded, the person he employed in this experiment was able to support in an erect posture, a weight not less than two thousand pounds. A horse, about seven times our bulk, would be thus able to raise a weight of fourteen thousand pounds, if his strength were in the same proportion. But the fact is, a horse cannot carry on his back above two or three hundred weight; while a Man can support two thousand pounds. But if we reflect for a moment, the reason of this will be apparent: a load on a Man's shoulders is placed to the greatest advantage; while, on the contrary, on the back of a horse it is placed to the greatest disadvantage. Suppose a Man to be standing as upright as possible under this before mentioned enormous weight; then all the bones may be compared to pillars supporting a building, and his muscles will have very little employment in this dangerous duty: however, they are not absolutely inactive; as Man, let him stand ever so upright, will have some bending in different parts of his body. The muscles therefore give the bones a partial assistance, and that with the greatest possible advantage. The greatest force of a horse, and of other quadrupeds, is exerted when the load is placed in such a position that the column of the bones can be properly applied, which is lengthwise. When, therefore, we estimate the comparative strength of a horse, we must not regard what he can carry, but what he can draw: and in this case his amazing superiority over Man is easily discovered; for one horse can draw a load which ten Men would be unable to move.

Among the ancients, strength was a quality of much greater use than at present; as, in time of war, the same Man who had strength enough to carry the heaviest armour, had also ability sufficient to strike the most fatal blow. In this case, his strength was at once his protection and his power. We should not, therefore, be surprised when we read of one Man whose personal prowess rendered him terrible in war, and irresistible, though we may fairly make allowances for its being greatly exaggerated by flattery, or magnified by terror. And, in an age of ignorance, which is ever an age of wonder, mankind, having no just idea of the human powers, were pleased rather to represent what they wished than what they knew; and exalted human strength, to fill up the whole sphere of their limited conceptions. Great strength is an accidental endowment; two or three persons in a country may possess it, and these may institute a claim to heroism; but prodigious strength is not hereditary, like family honours; and when we contemplate the splendid characters of Homer's heroes, who are all represented as the descendants of heroes, we may well believe that they are more indebted to their princely titles, than to their bodily strength and indomitable vigour, for their splendid attributes and their herculean achievements.

There are indeed, in later ages, some instances of amazing strength, which cannot be questioned; but in these Nature is found to pursue her ordinary course. These strong men have originated from the lowest ranks, and gradually risen into notice as their adventitious superiority had more opportunities of being displayed. Among this number may be ranked the Roman tribune who obtained the name of the second Achilles, and who is said to have killed, with his own hand, at different times, three hundred of the enemy; and, when insidiously attacked by twenty-five of his own countrymen, though past his sixtieth year, to have killed fourteen of them before he himself was slain. Of this number, too, was Milo, who, when he stood upright, could not be moved from his place. Pliny also mentions one named Athenatus, who walked across the stage at Rome loaded with a breast-plate which weighed five hundred pounds, and buskins of the same weight. But of all the prodigies of strength recorded in authentic history,

Maximinius, the Roman emperor, may be reckoned the chief. Whatever we are told of him is well attested: his character was too exalted not to be perfectly known; and that very strength for which he was celebrated, at last procured him no less a reward than the empire of the world. Maximinius was upwards of nine feet high, and one of the best-proportioned men in the whole empire. He was a Thracian by birth; and, from being a simple herdsman, rose, through the several gradations of office, till he became Emperor of Rome. The first opportunity which offered of exerting his strength, was in the presence of a numerous assembly of citizens in the theatre, where he overthrew twelve of the strongest men in wrestling, and outstripped two of the fleetest horses in running, on the same day. He could draw a loaded chariot, which two strong horses were unable to move; and could break the jaw of a horse with one blow of his fist, and his thigh with a kick. In war he was always engaged in the foremost ranks, where he displayed feats of activity that could only be equalled by his success; and happy had it been for him and his people, if, from being formidable to his enemies, he had not become still more so to his subjects. He reigned for some time at enmity with all the world; all mankind wishing for his death, yet none daring to strike the blow; and, as if Fortune had resolved that through life he should continue unconquerable, he was killed at last by his own subjects while asleep.

In more modern times we have several instances of bodily strength, and not a few of amazing swiftness; but these merely corporeal perfections are now considered as of small advantage, either in peace or war. The invention of gunpowder in some measure levelled all flesh to one standard, and wrought a total change in martial education through all parts of the world. In peace also, the discovery of new machines almost every day, and the application of the strength of irrational animals to the purposes of life, and, above all, the wondrous uses of the steam-engine, have rendered human strength of less value. The boast of corporeal strength is therefore consigned to barbarous nations, where, from the deficiency of art, its value is still felt; but in more civilized countries, its proud pre-eminence has fallen in a ratio commensurate with the progress of art, and the advancement of intellectual superiority.

But Man, though invested with superior powers, and possessed of more numerous privileges, with respect to his necessities seems to be inferior to the meanest animals. Nature has introduced him into life with a greater variety of wants and infirmities than the rest of her creatures, unarmed in the midst of enemies. Among the many thousand imaginary wants peculiar to Man, he has two in common with all other animals, which nevertheless he feels in a greater degree than they: these are the want of sleep, and hunger. The latter is a more destructive foe to mankind than watchfulness: but, though fatal without its proper antidote, it may always be removed by food; and to acquire this, Men have been known to encounter certain death. Hunger, however, terrible as it is in its approaches, is said to be not proportionately so in its duration; for the pain occasioned by famine decreases as the strength fails, and a total insensibility at length comes to the relief of the wretched sufferer. It is, however, incontestably certain that Man is less able to support hunger than any other animal: nor is he better qualified to bear a state of watchfulness. Sleep, indeed, seems much more necessary to him than to any other creature; as, when awake, he may be said to exhaust a greater proportion of the nervous fluid, and consequently to stand in need of an adequate supply. Other animals, when most awake, are but little removed from a state of slumber: their inert faculties, imprisoned in matter, and rather exerted by impulse than deliberation, require sleep more as a cessation from motion than from thought. But with respect to Man it is far otherwise; his ideas, fatigued with their various excursions, demand a cessation, not less than the body from toil. Fortunately for mankind, sleep generally arrives in time to relieve the mental powers, as well as the bodily frame: but it is often in vain that all light is excluded, all noise removed, and warmth and softness conspire, as it were, to invite sleep; the restless and active mind still retains its former vigilance; and reason, that wishes to resign the reins, is obliged, in spite of herself, to maintain them. In this disagreeable state, the mind ranges from thought to thought, willing to lose the distinctness of perception, by increasing the multitude of images. At last, when sleep makes nearer approaches, every object of the imagination begins to blend with that which lies next to it; a part of their distinction fades away; and ensuing sleep fashions out dreams for the remainder.

In sleep the whole nervous frame is relaxed, while the heart and lungs seem more forcibly exerted. This fuller circulation produces also a tension of the muscles: it may be considered as a kind of exercise, continued through the whole frame; and by this the perspiration becomes more copious, though the appetite for food is entirely removed. Too much sleep dulls the apprehension, weakens the memory, and unfits the body for supporting fatigue: too little sleep, on the contrary, emaciates the frame, produces melancholy, and wastes the constitution. A life of study, it is well known, unfits the body for receiving this grateful refreshment; and the approaches of sleep are averted by intense reflection: when, therefore, it comes at last, its continuance should not be hastily interrupted. Sleep is, indeed, by some pronounced to be a very agreeable period of Man's existence, in consequence of the pleasurable dreams which sometimes attend it. This, however, is rather fanciful than just; the pleasure which dreams are capable of conveying seldom reaching to our waking pitch of felicity: the mind often, in the midst of its visionary satisfactions, demands of itself, whether it does not owe them to an illusion? and not unfrequently awakes with the reply.

But it is seldom, except in cases of the highest delight or the deepest distress, that the mind has power thus to disengage itself from the empire of fancy: in the common course of its operations, it submits to those numerous fantastic images which succeed each other, and which, like many of our waking thoughts, are generally forgotten. There are others on whom dreams appear to have a very different effect; and who, without seeming to remember their impressions the succeeding morning, have yet evidenced, by their actions during sleep, that they were very powerfully impelled by their dominion; performing many of the ordinary duties to which they have been accustomed when awake; and, with a ridiculous industry, completing by night what they had failed to accomplish by day. Numerous instances might indeed be cited to show that the imagination is equally active by night as by day, and that it often involuntarily intrudes where it is least commanded or desired. While awake, and in health, this busy principle cannot much deceive us: it may raise a thousand phantoms before us, build schemes of happiness, or shudder at ideal misery; but the senses are all alive and sound to evince its falsity. Our eyes show us that the prospect is not present: our hearing and our touch depose against its reality; and our taste and smelling are equally vigilant in detecting the imposition. Reason, therefore, at once determines on the cause; and the fleeting intruder, Imagination, is restrained or banished from the mind. But it is otherwise in sleep: the senses being as much as possible at rest, having lost their peculiar functions, the imagination is then left to riot at large, and to lead the understanding captive. Every incursive idea then becomes a reality; and the mind, being destitute of every power that can correct the illusion, receives them for truths.

But we fear we have trespassed too long on this part of our subject; we therefore hasten from the consideration of what may be thought ideal and imaginary, to that which is actual and manifest. Every object in nature has its improvement and decay. The human form no sooner arrives at maturity, than it instantly begins to decline. The waste is at first insensible, and frequently several years revolve before we perceive any considerable alteration: but we ought to feel the weight of our years better than their number can be estimated by strangers; and as those are seldom deceived who judge of our age by external signs, we might be more sensible of the truth, were we more attentive to our feelings, and did not suffer ourselves to be deceived by vanity and fallacious hopes. When the body has acquired its full stature, and is extended to its just dimensions, it begins to increase in thickness; and this augmentation is the first step towards a decay, being merely an addition of superfluous matter, which inflates the body, and loads it with an useless weight: this matter, which is denominated *fat*, about the age of thirty-five or forty, begins to cover the muscles and interrupt their activity: every action then requires a greater exertion to perform it; and the increase of size is at the expense of ease, activity, and strength. The bones also become every day more solid. In the embryo they are almost as soft as the muscles and the flesh; by degrees they harden and acquire their natural vigour; but the circulation is still carried on through them; and how hard soever the bones may seem, the blood holds its current through them, as through all other parts of the body. Like the softer parts, they are furnished, through all their substance, with their proper canals, although in the different stages of existence they are of very different capacities. In infancy they are capacious, and the blood flows through the bones with almost the same facility as through the other channels. In manhood their size is greatly diminished; the vessels are almost imperceptible, and the circulation through them is proportionably slow. But in the decline of life, the blood which meanders through the bones no longer contributing to their growth, of necessity tends to increase their rigidity. In proportion as we advance in years, the bones, the cartilages, the membranes, the flesh, the skin, and every fibre of the body, become more solid, hard, and dry: every part shrinks, every motion becomes more slow; the circulation of the fluids is performed with less freedom; perspiration diminishes; the secretions alter; the digestion becomes slow and laborious; and the juices no longer serving to convey their accustomed nutriment, those parts may be said to live no longer when the circulation ceases. Thus the body dies by little and little; all its functions are weakened by degrees; life is driven from one part of the frame to another; universal rigidity prevails; and death at last closes the scene. When the natural stamina are good, life may perhaps be prolonged for a few years, by moderating the passions, by temperance, and by abstemiousness: but no human art can prolong the period of life to any considerable extent. It is apparent, indeed, that the duration of life has no absolute dependence either on manners, customs, or the qualities of particular food: much, it is true, is to be ascribed to the quality of the air; but we may rely upon it that, if luxury and intemperance be excepted, nothing can alter those laws of mechanism which regulate the number of our years.

Well may it be said, that Man is a compound being — the link between spiritual and animal existence; partaking of both their natures, but having also something peculiar to himself. His intellectual faculties prove his alliance to a superior class of beings: his sensual appetites and passions show his affinity to the brute creation.

We cannot close this article without referring to Dr. Prichard's admirable Researches into the Physical History of Man, — a work which, although we have not here quoted it, we recommend to the attention of our readers as one which discusses a most important subject with consummate ability.

MANAKIN. [See PARDALOTUS.]

MANATUS. A genus of herbivorous marine animals, familiarly called Sea Cows, and usually associated with the order *Cetacea*. The body of the *Manatus* is of an oblong

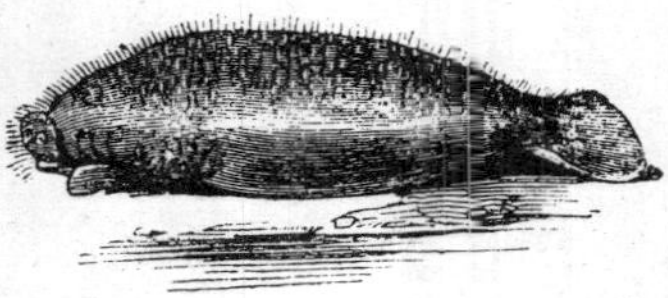

THE MANATEE. — (MANATUS AUSTRALIS.)

shape, terminated by a lengthened oval fin: it generally measures six or seven feet in length, but sometimes grows to an enormous size; and its paddles or fins exhibit rudiments of nails, by the aid of which the unwieldy animal drags its body along on the shore, to browse on the herbage that grows on and near the banks of the great rivers to which it resorts. The skin of the Manatus is of a blackish colour, very tough and hard, and full of inequalities, like the bark of an oak; and on it are sprinkled a few bristly hairs, about an inch in length. The eyes are exceedingly small in proportion to the size of the animal. It has no external ears, but only two orifices, scarcely large enough to admit a quill; the tongue is pointed, and extremely small; the mouth is destitute of teeth, but furnished with two solid white bones, extending the entire length of both

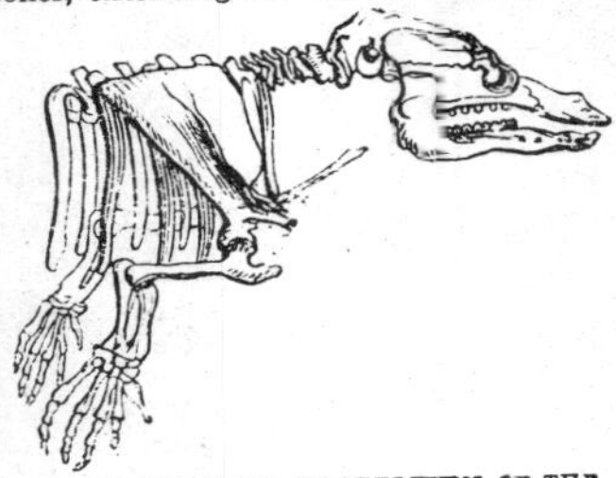

SKULL AND PART OF SKELETON OF THE MANATEE.

jaws, which serve instead of grinders; the lips are double; and near the junction of the two jaws the mouth is full of white tubular bristles, answering the same purpose as the laminæ in whales, to prevent the food from issuing out with the water. The lips are also thick-set with bristles, serving, instead of teeth, to cut the strong roots of the marine plants, which, floating ashore, point out the vicinity of these animals.

MANDRILL. The great blue-faced Baboon. [See BABOON.]

MANGO-FISH. [See POLYNEMUS,]

MANIS; PANGOLIN; or SCALY ANT-EATER. The Linnæan genus *Manis* consists of certain singular animals, known also by the name of Pangolins and Scaly Ant-eaters; and are limited to the warmest parts of Asia and Africa. They resemble the *Myr-*

SCALY ANT-EATER.
(MANIS CRASSICAUDATA.

mecophaga, or Hairy Ant-eaters, in having a very long extensible tongue, furnished with a glutinous mucus for securing their insect food, and in being destitute of teeth; but differing wholly from them in the body, limbs, and tail being covered with a panoply of large, imbricated scales, overlapping each other, after the manner of lacertine reptiles; and also in being able to roll themselves up when in danger, by which their trenchant scales become erect, and present a defensive armour sufficient to repel the assaults of the most ferocious of their enemies. They are quite harmless in their nature, entirely subsist on ants, termites, &c., and differ from the true Ant-eaters of South America in little else than in being provided with this scaly integument. They are remarkable for the strength and number of their caudal vertebræ; and in a general view of the animal kingdom, they may be considered as having the appearance of forming a kind of link between the proper viviparous quadrupeds and the Lizards.

The LONG-TAILED MANIS. (*Manis tetradactyla.*) This species is generally upwards of two feet in length, and the tail is more than twice as long as the body: the head is small, the snout narrow; the whole body, except the under part, covered with broad but sharp-pointed scales, which are striated throughout their whole length. The legs are very short; scaled like the body; and on each of the feet are four claws, those on the fore feet being stronger than those on the hind. The colour of the whole animal is an uniform deep brown, with a cast of yellow, and a glossy surface. It is a native of Africa.

The SHORT-TAILED MANIS. (*Manis pentadactyla.*) In this species the head is small as in the former, but the tail is much thicker and shorter, being not so long as the body, wide at the base, gradually tapering, but terminating very obtusely. The feet are furnished with five toes each, those on the fore feet, except the exterior one, which is very small, being extremely strong. The scales differ in shape from those of the preceding, being much larger and wider in proportion to the body and tail: they are also much harder, and so impenetrable when the animal rolls itself up, that when the tiger, panther, or hyæna attempts to force it, the Manis remains perfectly secure, and the assailant suffers for his temerity. The Manis chiefly inhabits the most obscure parts of the forest, and digs itself a retreat in the cleft of some rock, where it brings forth its young. It is a native of India, in many parts of which it is called the *Bajjerkeit.*

MANTICORA. [See CICINDELIDÆ.]

MANTIS: MANTIDÆ. A genus and family of Orthopterous insects, whose singular appearance, and the grotesque forms they usually assume when lying in wait for their prey, have not only attracted great attention, but have given rise to the most superstitious notions among the vulgar. The *Mantidæ* are characterized by having a narrow and elongated body; the anterior legs of enormous length; short palpi, terminating in a point; the tarsi five-jointed, and the wings plaited longitudinally.—These insects frequent trees and plants; and the forms and colours of their wings and bodies are so like the leaves and twigs which surround them as to give them remarkable power to elude observation.

The PRAYING MANTIS (*Mantis religiosa*) is of a beautiful green colour, nearly three inches in length, of a slender shape, and in its general sitting posture holds up the two fore-legs, slightly bent, in an attitude resembling that of a person when at prayer; in which position it will some-

PRAYING MANTIS.—(MANTIS RELIGIOSA.)

times remain motionless for several hours. It is termed by the French *prie-Dieu.* Its food consists of flies and other insects, which it is exceedingly dexterous in catching and retaining. "The monkish legends tell us that St. Francis Xavier, seeing a Mantis moving along in its solemn way, holding up its two fore-legs as in the act of devotion, desired it to sing the praises of God; whereupon the insect carolled forth a fine canticle! (*Ins. Arch.*, p. 63.) Mouffet, also, informs us, that 'so divine a creature is this esteemed, that if a childe ask the way to such a place, she will stretch out one of her feet, and shew him the right way, and seldom or never misse. As she resembleth those diviners in the elevation of her hands, so also in likenesse of motion; for they do not sport themselves as others do, nor leap, nor play; but, walking softly, she retains her modesty, and shews forth a mature kind of gravity!' But this gravity (as Mr. Westwood aptly says) has an object of a very different kind to that of the sorcerer. It is thus, after exhibiting a wonderful degree of patience, that, like a cat approaching a mouse, the Mantis moves almost imperceptibly along, and steals towards its prey, fearful of putting it to flight. When sufficiently near, the fore leg is extended to its full length, and the insect seized, being immediately secured between the tibia and femur, where it is held by the numerous teeth with which those parts are armed." These insects are of a very voracious and pugnacious nature; and when kept with others of their own species in a state of captivity, will fight till one or the other is destroyed in the contest.—Very similar to the foregoing is the *Mantis precaria.* It is of a beautiful green colour, with the thorax ciliated on each side, and the upper wings each marked in the middle by a transparent spot. This species is held in the highest veneration by some of the ignorant African tribes.—But of all the Mantes, perhaps the most singular in its appearance is the *Empusa gongylodes*, which, from its thin limbs and the grotesque form of its body, especially in its dried state, seems to resemble the conjunction of several fragments of withered stalks, &c.

MANTISPA: MANTISPIDÆ. A genus and family of insects belonging to the order Neuroptera. They appear to be very closely allied to the *Hemerobiidæ* in the general character of the veins of the wings. The species are but of small size, of dull colours, and widely dispersed throughout the globe. They chiefly reside upon oaks, and the structure of the fore legs and mouth seems to indicate that their habits are predaceous.

MARECA. A genus of Palmipede birds, containing the Widgeon (*Mareca Penelope*), [which see.]

MARGARITACEÆ. An order of Mollusca, containing several interesting genera; among which is the *Avicula margaritifera*, the shell of which produces the most valued Pearls, as well as the greatest quantity of Mother-of-Pearl; the latter being simply the nacreous interior of the shell. The pearls are separate formations of a similar substance, deposited by the mantle. The best Pearls are generally produced at the point, where the attachment of the adductor muscle causes a roughness in the shell. The gradual change which takes place in the position of this muscle, in accordance with the growth of the animal, causes the detachment of the pearl; and it is generally found imbedded in the substance of the muscle, by the motion of whose fibres its spherical form seems chiefly occasioned. But the formation of Pearls is by no means confined to this species; for any shell, univalve or bivalve, with a nacreous interior, may produce them.

MARGAY. (*Felis tigrina.*) A species of wild cat, native of South America. It is about the size of the common cat; and is very fierce and untameable. The ground-colour is a bright tawny: the face striped downwards with black; the shoulders and body marked both with stripes and large oblong black spots; small spots on the legs; the breast, belly, and insides of the limbs, whitish: long tail, marked with black and gray. It resides principally on trees, preying on birds; and seldom brings forth more than two young ones at a birth.

MARGINELLA. A genus of Molluscous animals, inhabiting an oval, smooth, shining shell, often handsomely coloured; the spire exceedingly short; the right lip having a thick margin; plaits nearly equal in size; and no operculum. The head of the animal is very distinct, with a small proboscis, and two tentacula having eyes at the base. It covers the greater part of the shell with the mantle, and by continually depositing vitreous matter gives it a bright polish, which, together with the delicately neat arrangement of colours in most species, renders them very beautiful.

MARIKINA. An appellation given to a Brazilian species of Monkey, furnished with a mane, and having a tuft of hair at the end of its tail. It is the *Jacchus Rosclia* of naturalists.

MARMOT. (*Arctomys.*) A genus of Rodent animals of which there are several species. The Marmots in their dentition are nearly allied to the squirrels, though in their general form they are very dissimilar to those active little quadrupeds, and have been generally classed with the rats. They have five molar teeth on each side of the lower jaw; short legs; a rather short tail; heavy body; and a short flat head: four toes on the fore feet, and five on the hinder. They live in communities; have extensive burrows on the sides of high and cold mountains; and pass the winter in a dormant state.

The Alpine Marmot (*Arctomys Alpinus*) is about the size of a rabbit; of a grayish yellow colour, approaching to brown towards the head; and has a short tail. This

MARMOT.—(ARCTOMYS ALPINUS.)

species inhabits the mountains of Europe (particularly those of the Alps and Pyrenees), just below the region of perpetual snow; and feeds on insects, roots, and vegetables. They live in large societies; and when they are eating, they post a sentinel, who on the approach of any danger gives a shrill whistle, when they all retire into their burrows, which are contrived with great art, and are well lined with moss and hay. In these retreats they remain in a torpid state from the autumn till April. In fine weather they are seen sporting about the neighbourhood of their burrows; they delight in basking in the sunshine, and frequently assume an upright posture, sitting on their hind feet. Before they retire to their winter quarters they are observed to grow excessively fat; and, on the contrary, appear greatly emaciated on first emerging from them. In a domestic state the Marmot will eat almost any kind of animal or vegetable food.

There are many Marmots inhabiting North America which have been considered as belonging to the sub-genus *Spermophilus.* The most celebrated of these is the Prairie Dog. (*Arctomys ludovicianus.*) The name of Prairie Dog has been given to it from a supposed similarity between its warning cry and the barking of a small dog. They live in large communities; their villages, as they are termed by the hunters, sometimes being many miles in extent. The entrance to each burrow is at the summit of the mound of earth thrown up during the progress of the excavation below. The hole descends vertically to the depth of one or two feet, after which it continues in an oblique direction. This Marmot, like the rest of the species, becomes torpid during the winter, and, to protect itself against the rigour of the season, stops the mouth of its hole, and constructs a neat globular cell at the bottom of it, of fine dry grass, so compactly put together as almost to form a solid mass. In the "Travels in North America" by the Hon. C. A. Murray, we find an account of this animal. Speaking of an extensive and desolate prairie through which he was passing, is the following description of the "Prairie Dog." "In this waste there was not either bird or beast to be seen, except Prairie Dogs. I do not know how these little animals obtained this absurd appellation, as they do not bear the slightest resemblance to the canine species, either in formation or habits. In size they vary extremely, but in general they are not larger than a squirrel, and not unlike one in appearance, except that they want his bushy tail; the head is also somewhat rounder. They burrow under the light soil, and throw it up round the entrance to their dwelling like the English rabbit: on this little mound they generally sit, chirping and chattering to one another, like two neighbour gossips in a village. Their number is incredible, and their cities (for they deserve no less a name) full of activity and bustle. I do not know what their occupations are; but I have seen them constantly running from one hole to another, although they do not ever pay any distant visits. They seem on the approach of danger always to retire to their own homes: but their great delight apparently consists in braving it, with the usual insolence of cowardice when secure from punishment; for, as you approach, they wag their little tails, elevate their heads, and chatter at you like a monkey, louder and louder the nearer you come: but no sooner is the hand raised to any missile, whether gun, arrow, stick, or stone, than they pop into the hole with a rapidity only equalled by that sudden disappearance of Punch, with which, when a child, I have been so much delighted in the streets and squares of London." Their holes seem to be tenanted also by a species of owls (*Strix cunicularia*); and this apparently discrepant couple live together united not in the bonds of matrimony but of friendship.

There are several other American species. The Quebec Marmot (*Arctomys empetra*), a solitary animal, whose burrows are almost perpendicular, and situated in dry spots, at some distance from the water. The Wood-chuck (*Arctomys monax*); they make their burrows in the sides of hills, which extend a considerable distance, and terminate in chambers lined with dry grass, leaves, &c. They are easily tamed, and are very cleanly.

Besides the foregoing, many species of the Marmot are found in the north of Europe and Asia: they swarm in the Ukraine, about the Boristhenes, in the southern desert of Great Tartary, and in the Alcaic mountains south of the Irtis. They burrow, and form magazines of corn and nuts; sit like squirrels while they eat; and generally bring forth from five to eight young. They are both herbivorous and carnivorous.

MARMOZET, or OUISTITIS. (*Jacchus.*) A genus of American monkeys distinguished from the rest by the absence of the additional molar, and by the sharpness and crookedness of their nails. The thumb is not opposable, being placed in the same line with the other fingers; and that of the hind feet is very short. The tail is large, and thickly covered with hair; but it is not prehensile; and in many species it is marked by transverse bars, giving it a very elegant appearance: several are also distinguished by tufts of hair projecting from the sides of the head. They are very agile in their movements, and extremely cautious and wary; exhibiting a degree of wildness and distrust even when in confinement. They show much instinctive sagacity in their search for insect food.

MARSUPIALIA, or MARSUPIALS. A singular family of the order *Carnivora*, in the class *Mammalia;* and so called from the females having a pouch (*marsupium*), or temporary abode for the young immediately after birth, and into which they retreat long after they can walk, whenever they are apprehensive of danger. Two particular bones, called the marsupial bones, attached to the pubis, and placed amidst the abdominal muscles, support this pouch. Professor Owen says, "they assist in producing a compression of the mammary gland, necessary for the alimentation of a peculiarly feeble offspring, and they defend the abdominal viscera from the pressure of the young as they increase in size, during their mammary or marsupial existence, and still more when they return to the pouch for temporary shelter." It should moreover be observed, that these marsupial bones are found likewise in the male, and even in species where the pouch-formed fold of skin is scarcely perceptible. New South Wales abounds in marsupial animals, but they are found also in America and the Asiatic islands. [See Kangaroo: Opossum.]

MARTEN. (*Mustela foina.*) This elegant and lively animal, whose agile and graceful motions are not excelled by any of the weasel tribe, resides in woods, and preys chiefly on birds and small animals. Its general length from nose to tail is about a foot and a half, and the tail is ten inches long. The female breeds in hollow trees, produces from three to seven young at a time, and has at least two litters in a year. They are very destructive to game of every kind, and to all sorts of domestic poultry, eggs, &c.: they will also feed on rats, mice, and moles; are very fond of honey, and will sometimes eat seeds and grain. The Marten is of a dark tawny colour, with a white throat; and the belly is of a dusky brown: the tail is bushy, and of a darker colour than the other parts; the ears are moderately large and rounded; muzzle pointed; and the eyes bright and lively. It is very wild and untameable if captured when full grown, but if taken young is susceptible of great docility. It has two sorts of fur; the outer, which is very long, and brown of different shades in different parts of the body; and the inner, which is extremely soft, short, and of a light yellowish gray colour.

The Pine Marten (*Mustela martes*) is an inhabitant of the woody districts in the northern parts of America, from the Atlantic to the Pacific; it is also found about the region of Mount Caucasus, as well as in Sweden, Norway, &c. It very closely resembles the preceding, but may be distinguished by its smaller size, longer legs, finer, thicker, and more glossy fur, and from the throat being marked with a broad yellow spot. The Pine Marten preys on mice, rabbits, partridges, &c. It never frequents

PINE MARTEN.—(MUSTELA MARTES.)

houses, as the common Marten occasionally does; but confines itself altogether to the woods and fields. Its fur is far superior in quality to that of the former species, and the skins form a great article of commerce. When this animal is pursued, and its retreat cut off, it shows its teeth, erects its hair, arches its back, and hisses like a cat. It burrows in the ground, carries its young about six weeks, and brings forth from four to seven in a litter about the latter end of April. Both this and the former species have a kind of musky smell.

Pennant's Marten. (*Mustela Canadensis.*) This is also a native of the northern parts of America. It is a larger and stronger animal than the Pine Marten; lives in the woods, preferring damp places to dry; and climbs with facility. It brings forth once a year, from two to four young. It is sought for its skin, of which considerable numbers are every year exported by the fur traders.

MARTIN. [See Swallow.]

MASON-BEE. A species of the genus *Osmia*, remarkable for constructing its nest of agglutinated sand, fixing it on the sides of walls, &c., or availing itself of some cavity or suitable projection for that purpose. This species constructs six or eight cells near each other, though irregularly placed; and the female, having deposited an egg, with a supply of honey and pollen in each, covers the whole and fills the spaces between the cells with the same kind of material she had used in constructing them; the whole having the appearance of a dab of mud, which might have been placed there by accident. This viscid mud, or mortar, which is at first soft, soon becomes as hard as stone; and the eggs being laid in it, undergo the same metamorphosis as those of the common bees.—Several species select the deserted shells of snails, in the spiral tubes of which they construct their nests. The bee having found a shell suitable to her purpose, deposits an egg, together with a suitable supply of pollen and honey, at the extremity of the tube; the space occupied thereby being not quite half an inch in length: this space she closes by a thin partition, which is composed of abraded leaves or moss, repeating the operation until she has constructed the required number of shells; she next closes up the entrance to the tube, for which purpose she collects pellets of earth, small pieces of stick, pebbles, &c., which, being mixed with some liquid secreted by the animal, form a secure protection to her works. The larva having consumed the store laid up by the provident parent, spins a cocoon of a toughish texture and of a dark brown colour; and in due time the perfect insect makes its appearance.

The genus *Osmia* contains many species, each having a favourite locality for its nest-building operations, but all of them varying their economy in accordance with accidental circumstances. Some of these bees are red, and others black; but they are all nearly of the same size, being about the length of drones, though not so thick. The black Mason-bees have stings; but the red, being males, have none.

MASON-WASP. (*Odynerus murarius, parietinus, &c.*) Hymenopterous insects, whose nests may be found in this country in most sandy banks exposed to the sun,

MASON WASP.—(ODYNERUS PARIETINUS.)

and who received the name from the ingenuity with which they construct their habitation. An account of this is so pleasingly given by Messrs. Kirby and Spence in their "Introduction to Entomology," that we take the liberty of extracting it. "This insect (*O. murarius*) bores a cylindrical cavity from two to three inches deep, in hard sand which its mandibles alone would be scarcely capable of penetrating, were it not provided with a slightly glutinous liquor which it pours out of its mouth, that, like the vinegar with which Hannibal softened the Alps, acts upon the cement of the sand, and renders the separation of the grains easy to the double pickaxe with which our little pioneer is furnished. But the most remarkable circumstance is the mode in which it disposes of the excavated materials. Instead of throwing them at random on a heap, it carefully forms them into little oblong pellets, and arranges them round the entrance of the hole so as to form a tunnel, which, when the excavation is completed, is often not less than two or three inches in length. For the greater part of its height this tunnel is upright, but towards the top it bends into a curve; always, however, retaining its cylindrical form. The little masses are so attached to each other in this cylinder as to leave numerous vacuities between them, which give it the appearance of filagree-work. You will readily divine that the excavated hole is intended for the reception of an egg, but for what purpose the external tunnel is meant is not so apparent. One use, and perhaps the most important, would seem to be to prevent the incursions of the artful Ichneumons, Chrysidæ, &c., which are ever on the watch to insinuate their parasitic young into the nests of other insects: it may render their access to the nest more difficult; they may dread to enter into so long and dark a defile. I have seen, however, more than once a Chrysis come out of these tunnels. That its use is only temporary is plain from the circumstance that the insect employs the whole fabric, when its egg is laid and store of fruit procured, in filling up the remaining vacuity of the hole; taking down the pellets, which are very conveniently at hand, and placing them in it until the entrance is filled." Speaking of the care which Mason-wasps take for their young, the same authors say: "One species not only incloses a living caterpillar along with its eggs in the cell, which it carefully closes, but at the expiration of a few days, when the young grub has appeared and has consumed its provision, re-opens the nest, incloses a second caterpillar, and again shuts the mouth: and this operation it repeats until the young one has attained its full growth."

MASTIFF. (*Canis molossus.*) This noble and powerful variety of the Canine race is distinguished by a large head and broad muzzle, very thick pendulous lips, moderate sized dependent ears, heavy brow, a strong and well-proportioned body, and the tail rather full. Like most of the larger kinds of dogs, although extremely vigilant over any thing committed to his charge, he will not abuse the power with which he is intrusted, nor call it into action unless provoked by injuries. In this he shows a disposition the very reverse of that of the Bull-dog, who seldom waits for aggression, but savagely and insidiously makes the first

THIBET DOG.

attack. So famous was great Britain for its Mastiffs when the Romans were its masters, and in such high estimation were their strength, courage, and sagacity held by the Roman emperors, that a resident officer was appointed, for the purpose of breeding them, and transmitting to the imperial city such as he thought capable of sustaining the combats in the amphitheatre. Strabo says that the Gauls trained British mastiffs for war, and used them in their battles. According to Dr. Caius, three were a match for a bear, and four for a lion.

A remarkable variety, if not a distinct species of this animal, is the Thibet Dog.

MASTODON. A genus of extinct quadrupeds, the remains of which in a fossil state show that it was a pachydermatous animal allied to the elephants. It has received its name from the conical projections on the surfaces of the molar teeth. Some of these were natives of the Old World; but by far the largest in size have been found on the American continent. The skeleton of one, termed the *Mastodon giganteus*, which was lately exhibited in London, under the name of the Missouri Leviathan, and is now in the British Museum, must have considerably exceeded in its dimensions the largest elephants now existing. In some parts of North America the fossil remains of this stupendous animal are abundant, particularly in the saline morass popularly termed the Big-bone Lick, in the northern part of Kentucky. There are no traces within the period of tradition or history of the existence of these animals as a living genus. When and how they perished, if ascertained at all, must be revealed by geological data. It is worthy of remark, that the skeletons seem to have been unmoved since the death of the animal; some, in fact, which were found near the banks of the great rivers, appearing in a vertical position, as if they had sunk down or been imbedded in the mud.

Among many curious traditions which were believed by the native Indians concerning this gigantic animal and its destruction, the following may be noticed: The Shawnee Indians believed that with these stupendous quadrupeds there existed men of proportionate dimensions, and that the Great Being destroyed both with thunderbolts. Those of Virginia state that as a troop of these terrible quadrupeds were destroying the deer, the bisons, and the other animals created for the use of the Indians, the Great Man slew them all with his thunder, except the big Bull, who, nothing daunted, presented his enormous forehead to the bolts, and shook them off as they fell, till, being at last wounded in the side, he fled towards the great lakes, where he is to this day.

MAY-FLY. [See EPHEMERA.]

MEADOW BROWN [BUTTERFLY]. A name given by collectors to Butterflies of the species *Hipparchia janira*.

MEAL [MOTH]. The name given to the *Pyralis farinalis*.

MEDUSA. The name given to a genus of marine animals, in the class *Acalepha*, which present to the eye, when floating in their native element, an umbrella-shaped disc, from beneath which a number of tentacula or filaments depend. In the central part of the concave side of this disc is the stomach, in the middle of which is the mouth, opening downwards, and surrounded by four leaf-like tentacula. The *Medusæ* are commonly known by the name of "sea-blubber," "jelly-fish," &c. They receive nutriment by means of innumerable minute pores; and in their stomachs are found small crustacea, mollusca, and even fishes. At certain seasons many of them sting and inflame the hand that touches them; and their tentacula seem to possess considerable muscular power, capable of drawing towards the mouth almost any thing that comes within their reach. They swim by muscular contraction of the margins of the disc. Many of the Medusæ are phosphorescent, and give that luminous appearance to the sea which has been so often described and variously accounted for. [See ACALEPHÆ, p. 1, and STROBILA in SUPPLEMENT.]

MEDUSA'S HEAD. A name sometimes applied to those species of Star-fishes which have the rays very much branched. [See EURYALE.]

MEGACEPHALON. The name of a singular genus of birds allied to the *Talegalla* and *Leipoa*, and doubtless resembling these genera in habits.

MEGACEROS. [See SUPPLEMENT.]

MEGACHILE. A genus of bees, popularly named *leaf-cutters*, from their habit of cutting off pieces of the leaves of the rose, elm, and other trees, and using them in the construction of the cases in which they deposit the pollen and honey necessary for the food of the larvæ. There are several species; but one of them will be amply sufficient for us to describe. *Megachile Willughbiella*: the WILLOW BEE. The male is about half an inch long; colour, black: the face densely clothed with bright yellow, the vertex with pale ferruginous hair: the antennæ have the apical segment compressed, and when viewed in front broader than the rest: the cheeks and under side of the thorax are clothed with an ashy pubescence; above with yellow ferruginous hair: the femora are yellow, with three black stripes in front: the tibiæ are black above, yellow at their extreme apex: tarsi palmated, and all the joints fringed with white silvery hairs. These insects exhibit wonderful mechanical ingenuity in the construction of their pollen-cases; the same species sometimes choosing trees, posts, or rails in a decaying state, at other times burrowing in banks, or in the mortar of old walls, or availing itself of the interstices from which the mortar has fallen out. Mr. F. Smith tells us that on one occasion he split off a large portion of an old willow tree, which was perforated in all directions by the bees, and in doing so, laid open to view a channel, about eight inches long, containing seven cells, constructed of rose-leaves. These he preserved for some weeks; at length a male bee made its escape, and on examination, it proved to have quitted the upper cell. The rest followed in regular succession, three other males, and three females. Mr. Smith observes, that he is not acquainted with any species of this genus which continues its burrow to the outside of the substance in which it is constructed, as a means of escape for its young brood. The Leaf-cutter Bees are subject to the intrusion of parasites, belonging to the genus *Cœlioxys*.

"The process which one of these bees employs in cutting the pieces of leaf that compose her nest is worthy of attention. Nothing can be more expeditious: she is not longer about it than we should be with a pair of scissars. After hovering for some moments over a rose-bush, as if to reconnoitre the ground, the bee alights upon the leaf she has selected, usually taking her station upon its edge, so that the margin passes between her legs. With her strong mandibles she cuts without intermission in a curve line so as to detach a triangular portion. When this hangs by the last fibre, lest its weight should carry her to the ground, she balances her little wings for flight, and the very moment it parts from the leaf flies off with it in triumph; the detached portion remaining bent between her legs in a direction perpendicular to her body. Thus without rule or compasses do these diminutive creatures mete out the materials of their work into portions of an ellipse, into ovals or circles, accurately accommodating the dimensions of the several pieces of each figure to each other. What other architect could carry impressed upon the tablet of his memory the entire idea of the edifice which he has to erect, and, destitute of square and plumb-line, cut out his materials in their exact dimensions without making a single mistake? Yet this is what our little bee invariably does. So far are human art and reason excelled by the teaching of the Almighty."—*Kirby and Spence's Entomology*.

MEGALOSAURUS. The name given to an extinct genus of lizard-like reptiles, of gigantic size, discovered in the oolitic slate of Stonesfield, near Oxford. Some of them measured from forty to fifty feet in length; but no *perfect* skeleton has been found. The generic character of this animal is founded by Dr. Buckland chiefly on the structure of the teeth, which he describes as presenting "a combination of mechanical contrivances analogous to those which are adopted in the construction of the knife, the sabre, and the saw." These teeth were arranged in a pretty close series, in sockets, along the alveolar border of the jaws; and when it is remembered that, according to the measurement of the imperfect remains which have been discovered, the Megalosaurus was about seventy feet in length, the predaceous powers of this carnivorous extinct monster must have been truly apalling.

MEGALOTIS. A genus of Mammalia allied to the family *Canidæ*. [See FENNEC.]

MEGATHERIUM. This name has been given by Cuvier to an extinct genus of gigantic quadrupeds, whose structure bears a great resemblance to that of the Bradypus or Sloth family. Several remains of the Megatherium have been found in South America: the one described by Cuvier was in a fossil state, and found a hundred feet below the surface of a sandy soil, in the vicinity of the river La Plata; other specimens, however, have since been found on the same continent, but not in so complete a state. The skeleton was twelve feet (French) long, by six feet in height; the thigh-bones excessively thick, and the leg-bones still more

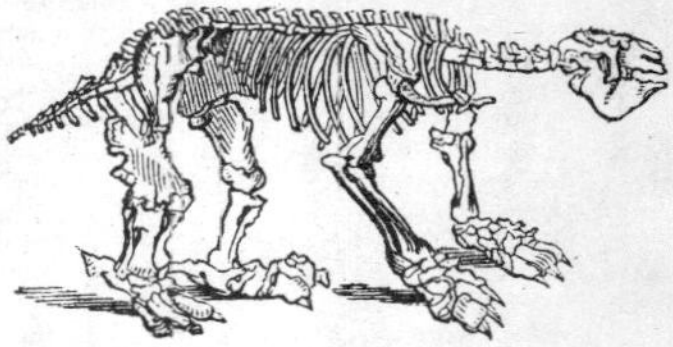

SKELETON OF THE MEGATHERIUM.

so in proportion: the fore limbs were longer than the hind, and there were three enormous claws on the fore feet, but only a single one on the hinder. The head was relatively small: in the upper jaw were five teeth on each side, and in the under jaw four—all molars. "As to its place in the system of quadrupeds," Cuvier observes, "it is perfectly marked by the sole inspection of the ordinary indicatory characters, that is, the claws and teeth. These show that it must be classed in the family of unguiculated quadrupeds destitute of cutting teeth; and, in fact, it has striking relations with these animals in all parts of its body. The great thickness of the branches of the lower jaw, surpassing even that of the elephant, seems to prove that the vast animal was not content with leaves, but, like the elephant and rhinoceros, broke and ground the branches themselves, its close and flat-crowned teeth appearing very proper for that purpose. The position of the bones of the nose, having some analogy with that of the elephant and tapir, would induce a suspicion that the animal wore a trunk, but it must have been very short, since the length of the head and neck together equals that of the fore legs. However this be, we find in the absence of canine teeth and the shortness of the muzzle, sufficient characters to constitute a new genus in the family of the *edentated*, which ought to be placed between the Sloth and the Armadillo;

MEGATHERIUM RESTORED.

since to the shape of the head of the former, it joins the teeth of the latter. It would be necessary to know particulars of which a skeleton cannot inform us, such as the nature of the teguments, the form of the tongue, the position of the mammæ, &c., in order to determine to which of these it approached the most. In the mean time, I thought I might give it the generic name of *Megatherium*, and the trivial one of *Americanum*. It adds to the numerous facts which apprise us that the animals of the ancient world were all different from those we now see on the earth; for it is scarcely probable that, if this animal still existed, so remarkable a species could have hitherto escaped the researches of naturalists. It is also a new and very strong proof of the invariable laws of the subordination of characters, and the justness of the consequences thence deduced for the classification of organized bodies; and under both these views it is one of the most valuable discoveries which have for a long time been made in Natural History."

Remains of a similar animal were collected by Sir Woodbine Parish, in the river Salado, which runs through the flat alluvial plains to the south of the city of Buenos Ayres. It was found there after a succession of three unusually dry seasons, which lowered the waters in an extraordinary degree, end exposed part of the pelvis to view, as it stood upright in the bottom of the river. This animal apears to have been larger than the one described by Cuvier. The thigh bone was twice the thickness of that of the largest elephant; the fore foot measured more than a yard in length, and more than twelve inches in width, and was terminated by an enormous claw; and the upper part of the tail was two feet wide. [See the articles SLOTH and MYLODON.]

MEGAPODIUS. A genus of Rasorial birds; so called from their large feet, which serve an important part in their economy. The eggs of these birds are very large; we may mention

The DUPERREY'S MEGAPODIUS (*Megapodius Duperreyii*), which inhabits the umbrageous forests of New Guinea. In size it is rather less than the partridge: the neck is well clothed with feathers; and a very thick crest, raised towards the occiput, covers the head: the wings are concave, an inch longer than the tail, and terminated in a point; tail sub-oval, pointed, and very short: legs grayish, and feathered down to the tarsi. The neck, throat, belly, and lateral parts, are of a gray slate-colour: the feathers of the back and the wing-coverts are large, and of a ruddy yellowish brown: rump, upper part of the tail, and vent-feathers ochreous red. The bird is timid, runs very fast among the bushes, like a partridge in standing corn, and utters a feeble cluck.

The MEGAPODIUS TUMULUS. [See JUNGLE-FOWL.]

MELANDRYIDÆ. A family of Coleopterous insects, specially distinguished by the large size of the three terminal joints of the maxillary palpi: the body is generally elongate and sub-cylindric or depressed; the mandibles are short and often bifid at the tips; and the tarsal claws are entire: the penultimate joint of the tarsi is generally bilobed in the two anterior pair of legs; in those species in which it is entire, the hind legs are formed for leaping, being long and compressed with slender tarsi. These insects chiefly reside beneath the bark of trees.

MELEAGRIS. A genus of Rasorial birds, which contains two species, the COMMON TURKEY (*M. gallopavo*) and the still more splendid HONDURAS TURKEY (*M. ocellata.*) [See TURKEY.]

MELITÆA. A genus of Butterflies belonging to the family *Nymphalidæ*, and distinguished by their antennæ, which have a wide flat club; the eyes are naked. There are several British species, for which we must refer to such works as Stephens, Wood, and Humphreys and Westwood: we particularize two.

MELITEA SELENE, APRIL FRITILLARY, or SILVER-SPOT BUTTERFLY. This is a well-known and beautiful insect, occurring on heaths and in woods throughout the south and west of England; two broods being produced, one in May, and another in August. The wings above are pale fulvous, spotted with black, and a marginal series of dusky spots, bounded by a slender black line: the ground colour of the posterior wings is ferru-

SILVER-SPOT BUTTERFLY.
(MELITEA SELENE.)

ginous, with a brighter band at the base, which is bordered on each side with a row of irregular silver and yellow spots, and having a large black ocellus in the centre, with a rufous pupil; the rest of the wing is varied with ferruginous and yellowish, with three silvery spots, placed transversely; on the inner and anterior margins a striga

MELITEA SELENE—UNDER SIDE.

composed of black dots, and six silver spots, edged internally with black: the anterior wings are distinctly varied with black, the hinder margin being strongly tipped with deep brown, and having a distinct row of conical black spots. Caterpillar black, with a clear lateral stripe; spines half-yellow.

The MELITÆA ARTEMIS, GREASY FRITILLARY, or SCABIOUS BUTTERFLY. This insect makes its appearance towards the end of May: it is more local than most of its kind; rare in the neighbourhood of London,

GREASY FRITILLARY.—(MELITEA ARTEMIS.)

but particularly abundant near Brighton; occurring plentifully also in various other parts of the south and west of England, but being in some places rarely seen. The wings above are reddish-fulvous, undulated with black, and spotted with yellow; the posterior marked with three distinct bands, the middle one bearing a striga composed of from four to seven black dots: the under

MELITEA ARTEMIS—UNDER SIDE.

surface of the anterior wings is glossy, with some ochraceous dashes at the tip: the posterior wings beneath are fulvous, with three transverse yellow bands, slightly edged with black: between the outer bands is a row of seven black dots, edged with ochraceous; and the basal band is broken and irregular: the cilia are yellowish: the body and antennæ dusky. The Caterpillar is black above and yellowish beneath, with a row of white dots down the back and on each side; head and spines black; legs red-brown. It feeds on the *Scabiosa succisa*, plantain, &c., and appears in September: about the end of April it changes to a pale green chrysalis, spotted with black, and having yellow tubercles at the extremity of the body. In about fifteen days the butterfly is produced.

MELIPHAGA. A genus of Tenuirostral birds belonging to the *Meliphagidæ* family, very many species of which will be found described in the great work of Mr. Gould on the Birds of Australia, the country where they abound; of these we may specify

The MELIPHAGA NOVÆ-HOLLANDIÆ, or NEW HOLLAND HONEY-EATER. This is one of the most abundant and familiar birds inhabiting the colonies of New South Wales, Van Diemen's Land, and South Australia; breeding among shrubs and flowering plants, and being common, in fact, on the sandy districts wherever the Banksias abound. "Nor is it the least attractive of the Australian Fauna; the strikingly contrasted markings of its plumage, and the beautiful appearance of its golden-edged wings, when passing with its quick, devious, and jumping flight from shrub to shrub, rendering it a conspicuous and pleasing object."—*Gould.* It usually rears two or three broods during the course of the season, which lasts from August to January. The nest is composed of small wiry sticks, coarse grasses, and strips of bark; the inside lined with the soft woolly portion of the blossoms of small ground plants. It lays two, and sometimes three eggs, of a pale buff colour, spotted with deep chestnut-brown at the larger end. Its food principally consists of the juices and pollen of flowers; but it also feeds on fruit and insects.

The MELIPHAGA SERICEA, or WHITE-CHEEKED HONEY-EATER. This species appears to be more confined to an eastern locality in Australia than the one above described, found in more open districts, and less seen in the interior of the country. When perched on the trees it is a most showy bird, its white cheek-feathers and contrasted tints of colouring rendering it very conspicuous. It is readily known from the *Meliphaga Novæ-Hollandiæ* by its white cheeks and the absence of white tips to the tail-feathers.

The MELIPHAGA AUSTRALASIANA, or TASMANIAN HONEY-EATER. This species, which is smaller than either of the preceding, and less brilliantly marked, is abundantly dispersed over every part of Van Diemen's Land, preferring such parts of the forests as are clothed with a thick brush of dwarf shrubby trees, growing beneath the more lofty gums, where numbers of these birds may be heard pouring forth their loud, shrill, and liquid notes in quick succession. It also resorts to the more open hills, where it finds thick beds of the *Epacris impressa*, whose bright red and white heath-like blossoms afford it an abundant supply of food. But, independently of the honey it obtains from the tube of every floret, which it rifles by means of its slender brush-like tongue, it feeds on various kinds of insects. The nest is placed on a low shrub near the ground; it is of a circular form, outwardly constructed of the inner rind of the stringy bark gum-tree, generally lined with fine grasses. The male has a black stripe passing from the base of the bill through the eye, and a lunar-shaped black mark down each side of the breast; a narrow stripe above the eye and one behind the lunar marks on the breast white; all the upper

TASMANIAN HONEY-EATER.
(MELIPHAGA AUSTRALASIANA.)

surface dusky black; wings blackish brown, the primaries and secondaries margined externally with golden yellow; tail feathers brownish black, fringed with golden yellow at the base, the two lateral feathers having a long oval spot of white on their inner webs at the tip; throat and chest white, flanks and under tail-coverts sooty gray: bill and feet black. The female is of a nearly dusky brown above and beneath; and has only a faint tinge of the golden yellow on the wings and tail.

MELLIFERA. A very extensive and interesting group of aculeated Hymenoptera, comprising th various species of Bees, which, from their peculiar construction and admirable economy, may be considered as the types of the order. These insects are characterized by having the basal joint of the posterior tarsi dilated into an oblong or subtriangular plate, which is hirsute on the inside, and provided with instruments for collecting and carrying pollen; the jaws are strong, and varied in the different species; the maxillæ and labium are elongated, and often transformed into a proboscis capable of being folded up many times beneath the head. The larvæ feed exclusively upon pollen or honey. Some of the species live in society, residing in dwellings of such regular construction, that the beauties of insect architecture may be said to rival the skill of the mechanic, while insect industry, order and good government may well command the admiration of mankind, and furnish them with lessons worthy of their imitation. It is not necessary, however, to do more in this place than to refer to the articles on the various species of Bees, which will be found, at considerable length, arranged according to their respective alphabetical situations.

MELOE, OIL, or MAY-BEETLE. A genus of Coleopterous insects belonging to the *Cantharidæ;* "now confined," as Mr. Westwood informs us, "to those apterous species, which have the body large and distended, with the elytra short, oval, and lapping over each other at the base of the suture. These insects crawl slowly along upon the ground, or amongst low herbage, upon which they feed, especially relishing the wild buttercups (*Ranunculus bulbosus* and *R. acris*).

MAY-BEETLE.—(MELOE PROSCARABŒUS.)

Mr. Jeffreys also found them very abundant on *Arum maculatum*, near Cromlyn Burrows. When alarmed, they emit from the joints of the legs an oily yellowish liquor, whence they have obtained the name of Oil Beetles. In some parts of Spain they are used instead of the blister-fly, or are mixed with it. They are also said by Latreille to be employed by farriers; and Hoppe tells us that they were, when he wrote (1795) in use as a specific against hydrophobia in Germany; and the oil which is expressed from these insects is used in Sweden with the greatest success, in the cure of rheumatism, by bathing the affected part. (*Drury's Insects.*) General Hardwick has also described a species of Meloë, found in all parts of Bengal, Bahar, and Oude, possessing all the properties of the Spanish blistering-fly. From the medicinal properties of these insects, Latreille has surmised in his ingenious memoirs upon the Buprestis of the ancients, that that noxious animal must have been a Meloë. M. Blot, however, contends, on the contrary, that the Meloë is not serviceable in medicine. The preparatory states of these insects have been the subject of much controversy. According to Gœdart, Linnæus, Frisch, and De Geer, the females burrow into the earth, and there deposit a large mass of yellow eggs, agglutinated together, which produce minute larvæ of a long narrow flattened form, with thirteen jointed bodies, six short legs, and two long anal setæ. They are exceedingly active in their movements, attaching themselves to flies, bees, &c., which it is said that they suck." Mr. Newport has lately proved the accuracy of these statements in most particulars, and in his admirable memoir on the Natural History of the Oil Beetle, in the twentieth volume of the Transactions of the Linnæan Society, has settled this hitherto much "vexed question," and traced the Meloe from the egg to the perfect insect. [See OIL BEETLE.]

MELOLONTHIDÆ. A very extensive and widely distributed group of Coleoptera; of which the well-known and destructive Cockchafer (*Melolontha vulgaris*) is the type. [See COCKCHAFER.]

MELOPSITTACUS. A sub-genus of the Parrot family, found in Australia, which contains

The MELOPSITTACUS UNDULATUS, or WARBLING GRASS-PARRAKEET. We learn from Mr. Gould that this lovely little bird is pre-eminent among the numerous members of the Parrot family in Australia, both for beauty of plumage and elegance of form:

WARBLING GRASS PARRAKEET.
(MELOPSITTACUS UNDULATUS.)

it is also remarkable for its sprightly and animated manners. It is believed to be generally dispersed over the central parts of Australia, but so exclusively an inhabitant of the vast inland plains that it is rarely seen between the mountain ranges and the coast. They breed in the hollow spouts of the large *Eucalypti*, and may be seen in flocks of many hundreds feeding upon the grass-seeds that are found in abundance on the plains. The nature of their food and the excessive heat of these plains compel them frequently to seek the water; but before going to drink, they settle together in clusters on the neighbouring trees. Their flight is remarkably straight and rapid, and is generally accompanied by a screeching noise. During the heat of the day when sitting motionless among the leaves of the gum-tree, they so closely assimilate in colour as to be detected with difficulty. The breeding season is at its height in December, and by the end of the month the young are generally capable of providing for themselves; they then assemble in vast flights, preparatory to their great migratory movement. The eggs are pure white, in number three or four, and are deposited in the holes and spouts of the gum-trees without any nest. They are particularly interesting as cage-birds; for, independently of their highly ornamental appearance, they have a most animated and pleasing song; besides which, they are continually billing, cooing, and feeding each other; and their inward warbling is constantly heard from morning to night.

The young gain their full livery in about eight months, the sexes being precisely alike in the colouring and marking of their plumage. Forehead and crown straw yellow; the remainder of the head, ear-coverts, nape, upper part of the back, scapularies, and wing-coverts pale greenish yellow, each feather having a crescent-shaped mark of blackish brown near the extremity; wings brown; the outer webs of the feathers deep green, margined with greenish yellow: face and throat yellow, with a patch of rich blue on each cheek, below which are three circular spots of bluish black; rump, upper tail-coverts, and all the under surface bright green; two centre tail-feathers blue, the remainder green, crossed in the middle by an oblique band of yellow; irides straw white; nostrils bright blue or greenish blue and brown; legs pale bluish lead colour. In a state of nature they feed exclusively upon grass-seeds; but in confinement they thrive equally well on canary-seed.

MELYRIDÆ. A family of Coleopterous insects, having an oblong or ovate body, soft, and but slightly convex: the palpi are short, filiform, and pointed at the tip; the thorax rather convex; and the antennæ moderately long, serrated, nodose, or pectinated in the males of some of the species. These insects are generally of small size, and very gaily coloured, green and red being most conspicuous. They may be ordinarily found upon flowers, as they frequent them for the sake of the insects which they find there to feed on. Some of the species of the British genus *Malachius* have the anterior angles of the thorax and the base of the abdomen furnished with several red bladder-like appendages, which the insect is able to contract or dilate at will; it may therefore be provided for the purpose of increasing or decreasing its gravity during flight, or be used as a portion of an apparatus for emitting an offensive effluvium. The exotic genera are few, and exhibit no remarkable features.

MEMBRACIS: MEMBRACIDÆ. (Tree-hoppers.) A genus and family of Hemipterous insects, in many respects resembling the *Cicadidæ*, but they enjoy the faculty of leaping, which the Cicadas do not. This faculty does not, as in the grasshoppers and other leaping insects, result from an enlargement of their hindmost thighs, which do not differ much in thickness from the others; but is owing to the length of their hinder shanks, or to the bristles and spines with which these parts are clothed and tipped. These spines serve to fix the hind legs securely to the surface, and when the insect suddenly unbends its legs, its body is launched forward in the air. Some of them, when assisted by their wings, will leap to the distance of five or six feet, which is more than two hundred and fifty times their own length; in the same proportion, "a man of ordinary stature should be able at once to vault through the air to the distance of a quarter of a mile." Some of these "leaping harvest-flies" have the face nearly vertical, and the thorax very large, tapering to a point behind, covering the whole of the upper side of the body, and overtopping even the head, which is not visible from above. In others the face slopes downwards towards the breast, the thorax is of moderate size, and does not extend much, if at all, beyond the base of the wing-covers, and does not conceal the head when viewed from above.

The habits of some of the "Tree-hoppers" are presumed to be much the same as those of the musical harvest-flies [See CICADIDÆ], for they are found on the limbs of trees, where they deposit their eggs, only during the adult state, and probably pass the early period of their existence in the ground. Others, however, are known to live and undergo all their changes on the stems of plants. Among the former is the American Two-spotted Tree-hopper, or *Membracis bimaculata* of Fabricius, which may be found in great abundance on the locust-tree (*Robinia pseudacacia*) during the months of September and October. These, as well as other tree-hoppers, show but little activity when undisturbed, remaining without motion for hours together on the limbs of the trees; but, on the approach of the fingers, they leap vigorously, and, spreading their wings at the same time, fly to another limb and settle there, in the same position as before. They never sit across the limbs, but always in the direction of their length, with the head or fore part of the body towards the extremity of the branches. On account of their peculiar form, which is that of a thick cone with a very oblique direction, their dark colour, and their fixed posture while perching, they would readily be mistaken for the thorns of the tree, a circumstance undoubtedly intended for their preservation. This insect measures about half an inch from the tip of the horn to the end of the body; the male is blackish above, with a long yellow spot on each side of the back; the female is ash-coloured, and without spots. While on the trees, these insects, though perfectly still, are not unemployed; but puncture the bark with their sharp and slender beaks, and imbibe the sap for nourishment. The female also appears to commit her eggs to the protection of the tree, being furnished with a piercer beneath the extremity of her body, with which to make suitable perforations in the branches. Another species, the White-lined Tree-hopper (*Membracis univittata*), which may be found upon the oak-tree in the U. States during the month of July. It is about four-tenths of an inch in length; the thorax is brown, has a short obtuse horn extending obliquely upwards from its fore part, and there is a white line on the back, extending from the top of the horn to the hinder extremity. Tree-hoppers are often surrounded by ants, for the sake of their castings, and for the sap which oozes from the punctures made by the former, of which the ants are very fond.

MENOBRANCHUS. A genus of Reptiles belonging to the Salamander group, distinguished from the allied genera by its persistent branchiæ; the head having two rows of teeth in the upper and one row in the lower jaw. There are four toes to each foot, the toes being destitute of claws. There are at least two species of this genus found in North America.

MENURA. [See LYRE-BIRD.]

MEPHITIS. A genus of carnivorous animals, notorious for their disagreeable smell. [See SKUNK.]

MERGUS. A genus of Palmipede Birds; three species of which are found in this country.

The MERGANSER, or GOOSANDER. (*Mergus merganser.*) These birds are nearly allied to the Duck and Diver tribes. They are inhabitants of the arctic regions, breeding very far north in summer, and migrating southwards in winter; in severe seasons occasionally frequenting the lakes and rivers of Britain, but leaving this country early in the spring. Their food consists principally of fish, which they take by rapid diving: crustaceans, mollusca, and insects are also devoured by them; but they seem to be incapable of digesting vegetable matter of any kind. The male weighs about four pounds, and measures in length two feet three inches, and across the wings three feet two inches. The bill is three inches long, narrow, and finely serrated, the tip being armed with a hooked horny tip: both mandibles are black on the upper and under parts, and crimson on the sides: the head is large, and crowned with a great quantity of long loose feathers, which, when erected, form a crest; these feathers are of a glossy bottle-green; the cheeks and upper part of the neck are a dull black; the lower part, breast, belly, vent, and inner wing-coverts of a fine cream colour: the upper part of the back, and the lower scapulars, are black; the lower part of the back and the tail are ash-coloured, the latter consisting of eighteen feathers. The legs and feet are very deep orange-colour. The flesh of this aquatic bird is accounted rank and fishy.

The RED-BREASTED MERGANSER or GOOSANDER. (*Mergus serrator.*) This species measures one foot nine inches in length, and weighs about two pounds: the bill is long, hooked at the tip, and toothed at the edges; the upper mandible is dark brown, tinged with green, and edged with red; the lower one wholly red: the irides are a purplish red: the head, long pendent crest, and upper part of the neck, are of a glossy violet black, varying in different lights to a golden-green; the neck and belly white; the breast

RED-BREASTED MERGANSER.
(MERGUS SERRATOR.)

rusty red, spotted with black on the front, and bordered on each side with five or six white feathers, edged with black; the upper part of the back glossy black; the lower, the rump, and sides being marked with transverse zigzag lines of brown and gray: the feathers nearest to the wings are white; the greater coverts, some of the secondary quills, and the scapulars, black and white: the primary quills are black; some tipped with white, and others white on the upper half and black to their points. The tail is short, its colour brown: the legs and feet of a deep orange-colour. It is remarked, however, that these birds, both male and female, differ much in their plumage; some being whiter, brighter, and more distinctly marked than others. They are met with in great flocks at Newfoundland, Hudson's Bay, &c.

The HOODED MERGANSER. (*Mergus cucullatus.*) This species is peculiar to America, and is usually found along the lakes and fresh water rivers rather than near the sea; tracing up creeks, and visiting millponds, diving perpetually for their food. Like the Red-breasted, they are migratory, the manners, food, and places of resort of both being very much alike. On the seacoast this species is very commonly called the *hairy-head*. It is eighteen inches in length, and two feet in extent; bill blackish red, narrow, thickly toothed, and furnished with a projecting nail at the extremity; the head is ornamented with a large circular crest, which the bird has the faculty of raising or depressing at pleasure; the fore part of this, as far as the eye, is black, thence to the hind head white, and elegantly tipped with black; it is composed

HOODED MERGANSER. (MERGUS CUCULLATUS.)

of two separate rows of feathers, radiating from each side of the head; irides golden; eye very small; neck black; part of the lesser wing-coverts very pale ash, under which the greater wing-coverts and secondaries form four alternate bars of black and white; tertials long, black, and streaked down the middle with white; the black on the back curves handsomely round in two points on the breast, which, with the whole lower parts, are pure white; sides, under the wings and flanks, reddish brown, beautifully crossed with parallel lines of black; tail pointed; legs and feet, flesh-coloured; claws, large and stout. The female is rather less than the male; the crest is smaller; and the plumage in general is less decided and handsome in its markings. Her nest is composed of grass, lined with feathers from the breast; and she lays six white eggs.

MERIONES. A genus of Mammalia, belonging to the order *Rodentia*, distinguished from *Gerbillus*, to which they are closely allied, by their hind feet being much longer, the tail nearly naked, and the existence of a small tooth before the superior molars. There are two species found in

MERIONES LABRADORIUS.

North America; one is the *Meriones Canadensis*, well known to the inhabitants of Canada for its extreme agility. It closes itself up in its burrow, and passes the winter, like many of its congeners, in a state of lethargy. The *Meriones Labradorius*, figured by Sir John Richardson in his 'Fauna Boreali Americana,' is another interesting but closely allied species.

MERLIN. (*Falco œsalon.*) The smallest bird of the Falcon tribe, scarcely exceeding a Blackbird in size; but, though small, not inferior in courage to any of its more powerful congeners. It flies low, and with great

MERLIN.—(FALCO ÆSALON.)

celerity. Small birds are its natural prey; and in the palmy days of falconry it was used for taking quails and partridges, which it would strike on the head, breast, or neck, and kill with a single blow. The bill is of a bluish lead colour; head ferruginous, streaked with black; back and wings of a dark brown, tinged with bluish ash colour, streaked down the shafts with black, and edged with ferruginous spots: quill feathers dusky, marked with reddish oval spots; the under coverts of the wings brown, beautifully marked with round white spots: the tail is five inches long, crossed with alternate bars of dusky and reddish clay-colour: the breast and belly are of a yellowish white, with oblong brown spots pointing downwards: the legs yellow. It breeds in woods; and lays from four to six white eggs, mottled at the end with brown.

MEROPIDÆ. [See BEE-EATER.]

MERULIDÆ. [See THRUSH.]

MICROLESTES. [See SUPPLEMENT.]

MICROTHERIUM. [See SUPPLEMENT.]

MILLEPEDE. [See IULUS.]

MILLER'S THUMB. [See BULLHEAD.]

MILVUS; or KITE. A genus of *Falconidæ*, with long wings and generally forked tail; it contains the COMMON KITE [see KITE] and other species; of these Mr. Gould found a square-tailed species in Australia—the MILVUS ISURUS, or SQUARE-TAILED KITE. This true species of the Kite tribe inhabits South and West Australia; and may at one time be seen soaring high above the trees, and at others hunting over the open wastes in search of food.

MINNOW. (*Cyprinus phoxinus.*) This active and elegant little fish, the length of which seldom exceeds three inches, is commonly found, swimming in shoals, in some gravelly rivulets and trout-streams. The top of the head and back are of a dusky olive, mottled, and lighter on the sides; the belly white and silvery, with a tinge of yellow, and sometimes in summer of a bright rosy red colour; the scales very small; the lateral line straight and of a golden yellow; and the tail forked. It bites readily at a small

MINNOW.—(CYPRINUS PHOXINUS.)

red worm; and we know of no fish that affords more amusement to the youthful angler. When they are in abundance, a small casting-net may be used with advantage; and they make an excellent fry; but the Minnow is principally used as a bait for pike and large trout.

MINOR [MOTHS]. A name given by collectors to Moths of the genera *Miana* and *Celœna*.

MIRAFRA. A genus of Larks found on the plains and open districts of New South Wales. The species MIRAFRA HORSFIELDII, which is larger, redder in colour, and has a stouter bill than others found in the same region, is more terrestrial in its habits than arboreal; and, when it rises, very commonly flies merely to a short distance and descends again: it may often be seen perched upon the strong blades of grass, and occasionally on the trees; it frequently mounts high in the air after the manner of the well-known Sky-lark of Europe, singing all the time very melodiously, but with a weaker strain than that favourite bird; it also occasionally utters its pleasing song while perched on the branches of the trees. The general plumage is ashy brown, the centre of the feathers dark brown, the latter colour predominating on the head, lower part of the back and tertiaries; wings brown, margined with rufous; over the eye a stripe of buff; chin white; under surface pale buff; throat crossed by a series of dark brown spots arranged in a crescentic form; under surface of the wing rufous; bill dark brown at the tip; feet fleshy brown.

MITE. By this name several minute insects, of different species, are known. Some have six legs, others eight; each leg being furnished with two small claws at the extremity, surrounded with hairs. Many resemble the *Cheese Mite* in structure and habits; others are parasitic, &c. [See ACARIDÆ.]

MITRA. A genus of Mollusca, inhabiting a small and pretty turreted shell; spire long and pointed at the end; columella with several oblique thick plates. They exhibit a great variety of patterns; some are smooth, others grooved, some are angulated, some coronated; and they are variegated with every kind of hue. They abound in the seas of hot cli-

MITRA VULPECULA.

mates, the greater number being found in the Pacific Ocean, generally in shallow water, near coral reefs, but sometimes at great depths. It has been asserted that the animal is of a poisonous nature, and to wound, with its pointed trunk, those who touch it; but this wants confirmation. The species are very numerous, both recent and fossil.

MOA. The name by which the *Dinornis*, a gigantic fossil bird, is known to the natives of New Zealand. [See DINORNIS.]

MOCHA [MOTHS]. A name given by collectors to Moths of the genus *Cyclophora*.

MOCKING-BIRD. (*Mimus polyglottus.*) This remarkable bird, sometimes called the Mocking Thrush, receives its name from its amazing powers of voice, being able to imitate that of almost every species of animal, as well as many noises that are produced artificially. But its notes are not entirely imitative: its own song is bold, full, and exceedingly varied; and in confinement it loses little of its energy. It inhabits most parts of America, and the West Indies. Its general colour is cinereous, paler beneath: but though it cannot vie with most of the American birds in brilliancy of plumage, its own sweet and varied notes, no less than its peculiar faculty of imitation, render it an especial favourite, and a large price is often obtained for it. To use the words of

MOCKING-BIRD.—(MIMUS POLYGLOTTUS.)

Wilson, "He whistles for the dog; Cæsar starts up, wags his tail, and runs to meet his master. He squeaks out like a hurt chicken; and the hen hurries about, with hanging wings and bristled feathers, clucking to protect her injured brood. The barking of the dog, the mewing of the cat, the creaking of the passing wheelbarrow, follow with great truth and rapidity. He repeats the tune taught him by his master, though of considerable length, fully and faithfully. He runs over the quiverings of the canary, or the clear whistlings of the Virginia nightingale or red-bird, with such superior execution and effect, that the mortified songsters feel their own inferiority, and become altogether silent; while he seems to triumph in their defeat, by redoubling his exertions." It builds its nest in fruit-trees, feeds on berries and other fruits, and is easily tamed. The female lays from four to five eggs, of an ash-blue colour, marked with patches of brown; she incubates fourteen days, and is extremely jealous of her nest, being very apt to desert it if much disturbed.

The observant author of the 'Birds of Jamaica' remarks, that at this time the old birds are watchful and courageous, and that any winged intruder, though ever so unconscious of evil intent, or ever so large, is driven away with fearless pertinacity. But the hogs are the creatures that give him the most annoyance. They are ordinarily fed upon the inferior oranges, the fruit being shaken down to them in the evenings; hence they acquire the habit of resorting to the orange trees, to wait for a lucky windfall. The Mocking-bird, says he, feeling nettled at the intrusion, flies down and begins to peck the hog with all his might: — Piggy, not understanding the matter, but pleased with the titillation, gently lies down and turns up his broad side to enjoy it; the poor bird gets into an agony of distress, pecks and pecks again; but only increases the enjoyment of the luxurious intruder, and is at last compelled to give up the effort in despair.

MOLE. (*Talpa Europæa.*) A quadruped of the genus *Talpa*, whose structure admirably fits it for a subterranean life. It is from five to six inches in length: the body is thick and cylindrical; the head is much prolonged, especially the muzzle, which projects far beyond the jaws, and is very flexible, strong, and tendinous, serving to convey food to the mouth: it has no external ears, but the auricular apparatus is highly developed, and the sense is very acute: its eyes are so very minute, and concealed by its fur, that it is a vulgar opinion that it is deficient in these important organs. The

MOLE.—(TALPA EUROPÆA.)

head is not distinguished from the body by any appearance of neck; the legs are so short as scarcely to project perceptibly from the body; the fore feet, situated obliquely outwards, are excessively strong and broad, and furnished with very large and stout claws, so as to give the animal the power of working under the surface with the utmost rapidity; the hind feet are small in proportion to the fore feet; and are calculated for throwing back with ease the mould from behind, during the animal's subterranean labours. The rapidity with which the Mole can make its way through a favourable soil would be quite astonishing, did not its whole conformation and great muscular strength account for it. The tail is short and small: the skin is much thicker and tougher in proportion than in other quadrupeds, and the fur with which it is covered is close set and soft as the finest velvet. The food of the Mole consists chiefly of earthworms and the larvæ of insects; but it is not confined to these; for during the summer months it not unfrequently leaves its subterranean retreat, and wanders upon the surface in quest of prey, such as birds, mice, frogs, snails, &c.; and during these nocturnal excursions, it often meets with a vigilant and successful enemy in the owl. Moles are extremely voracious. We are told, that if two are shut up together without food, the strongest will devour the weakest, even to the bones: nothing but the skin is left, which they never eat, and which, when one has killed the other, is always seen to be ripped up along the belly. They are incapable of long fasting; if kept ten or twelve hours without food, it is said they die of starvation.

"The farmer views the operations of the Mole as destructive to his crops by exposing and destroying their roots, or by overthrowing the plants in the construction of the mole-hills; his burrows, moreover, become the haunts and hiding-places of the field-mouse and other noxious animals. The Mole is also accused of carrying off quantities of young corn to form its nest. Hence every means are devised to capture and destroy it, and men gain a livelihood exclusively by this occupation. Some naturalists, however, plead that the injury which it perpetrates is slight, and that it is more than counterbalanced by the benefit which it produces by turning up and lightening the soil, and especially by its immense destruction of earth-worms, and many other noxious animals which inhabit the superficial layer of the ground, and occasion great injury to the roots of grass, corn, and many other plants. The soundest practical conclusion lies probably in the mean of these opinions; and the enlightened agriculturist, while he takes prompt measures to prevent the undue increase of the Mole, would do well to reflect on the disadvantages which might follow its total extermination." — *Brande's Dict. of Science.*

From a mass of interesting information relative to the habits of this animal, in Mr. Bell's History of British Quadrupeds, we select the following: — "Every one is aware of the fact that the Mole burrows for its food, that its nest is formed under ground, that a larger hillock than the rest is raised for the reception of its young; but it is not so generally known that its subterranean excavations are of the most distinct and determinate character; that there are permanent passages or high roads for its ordinary travels from one part of its domain to another; that into these roads open the excavations in which it follows its daily labour in search of food; that its fortress — the house in which it resides from the autumn to the spring — is of a complex and most ingenious structure, and that this domicile is always a distinct and even remote building from that in which the nest is formed." After stating that we are principally indebted to the researches of Henri le Court, a French gentleman who devoted many years to the study of the habits, &c. of the Mole, he thus proceeds: "The district or domain to which an individual Mole confines himself may be termed its encampment. Within its limits, or at least in immediate communication with the district, all the labours of the animal are pursued. It consists of the habitation or fortress, from which extends the high road by which the animal reaches the opposite extremity of the encampment, and of various galleries or excavations opening into this road, which it is continually extending in search of food, and which constitute, in fact, its hunting-ground. The fortress is formed under a large hillock, which is always raised in a situation of safety and protection; either under a bank, against the foundation of a wall, at the root of a tree, or in some similar locality. The earth, of which the dome covering this curious habitation is composed, is rendered exceedingly strong and solid, by being pressed and beaten by the Mole in forming it. It contains a circular gallery within the base, which communicates with a smaller one above by five nearly equi-distant passages; and the domicile or chamber is placed within the lower and beneath the upper circular gallery, to which last it has access by three similar passages. From the chamber extends another road, the direction of which is at first downwards for several inches; it then rises again to open into the high road of the encampment. From the external circular gallery open about nine other passages, the orifices of which are never formed opposite to those which connect the outer with the inner and upper gallery: these extend to a greater or less distance, and return, each taking an irregular semicircular route, and opening into the high road at various distances from the fortress. Such is a very hasty description of this most singular structure; and nothing surely can be imagined more admirably calculated to ensure the security or the retreat of the inhabitant than such an arrangement of internal routes of communication as this. The chamber communicating beneath directly with the road, and above with the upper gallery, — this with the lower by five passages, and the latter again with the road by no less than nine, — exhibit altogether a complication of architecture, which may rival the more celebrated erections of the Beaver." "The nest is always distinct, and frequently remote from the fortress, and is usually, but not always, covered by a hillock; which, when it exists, is much larger than an ordinary mole-hill. It is formed simply by excavating and enlarging the point of intersection of three or four passages. The bed of the nest is composed of a mass of herbage, grass, roots, or leaves: in one which was examined by Geoffrey and Le Court, no less than two hundred and four blades of young wheat were counted. This, however, can scarcely be considered as an ordinary occurrence, as they generally prefer dry and soft substances. The period of gestation is supposed to be about two months or upwards: and the young are brought forth in April, — sometimes earlier, at others later, according to the season: indeed, young Moles have been found at all times from the beginning of April till August, which has led some persons to believe that there are more than one brood in the year. There are generally four or five, sometimes as few as three, rarely six."

"That Moles were intended to be beneficial to mankind," observes Mr. Jesse, "there can, I think, be no doubt. I have been assured that where old Mole-hills are most abundant on sheep pastures, the latter animal is generally in a healthy state, as it feeds on the wild thyme, and other salubrious herbs, which grow on these heaps of earth. Where these have been levelled and cleared away, sheep are not found to thrive as well as they did previously. This fact was confirmed to me by the Ettrick Shepherd, who deprecated the practice of removing Mole-hills. On the fine and extensive pastures in Leicestershire, where old Mole-hills are extremely abundant, sheep thrive well, and are generally healthy: and I have been assured that after the mole-hills had been destroyed in a park which belonged to the Earl of Essex, in Herefordshire, the deer in it never throve."

The Rev. C. A. Bury, who has published some very White-of-Selborne-like notes on the "Mammalia of the Isle of Wight," in the pages of "The Zoologist," observes that "On some lands the drainage is effected wholly or in part by the Moles. So far, then (he says), I think the farmer might spare the Moles to his own advantage, and save some shillings, perhaps pounds, to the mole-catcher. Man is too fond of meddling, and often blunders to his own cost. In his attempts at improvement, he only disturbs the balance of creation. Granted that occasionally some species of animal, favoured by circumstances, either the scarcity of its appointed check (occasioned, perhaps, by the meddling hand of man), or a superabundance of its natural food, may increase beyond due bounds, and so require the interposition of human force or skill, let that force and skill be then exerted ; but I believe that this would be seldom necessary ; things would right themselves. They have been generally found to do so, unless man has carried his meddling propensities to the extent of utterly extirpating the appointed check ; for it seems to be a law of creation, that where there is food, there will be provided that which feeds on it, and that in just proportion. The Mole is evidently an appointed check to the undue increase of the earthworm : it not only devours numbers itself, but by its burrowing drives to the surface many more, which, in their attempt to escape the Mole, fall a prey to the robin and the thrush. The earthworm, unquestionably, has its uses, in drawing vegetable substances beneath the surface, and so the gases that are released in the process of decomposition, and which would otherwise be lost, are preserved for the nutriment of the growing plant, while the portion devoured by the worm is again thrown to the surface in the form best adapted for the nutriment of the plant above ground. But worms devour the roots of plants ; and were there no checks to their increase, vegetation would be seriously injured, instead of benefited, by their existence : so long, however, as they are kept in check by the mole beneath, and the birds above ground, perhaps even their destruction of some plants is benefical in preventing a too crowded herbage. Thus, then, all is well arranged by Divine Wisdom ; but if man steps in, throttles the mole, and shoots or snares the birds, he must, if he carry his interference far, produce a disturbance among God's works, to his own detriment."

MOLE CRICKET. (*Gryllotalpa vulgaris.*) Of all the British Orthopterous insects, the Mole Cricket is by far the most curious. It derives its name from the peculiar formation of its anterior extremities, and its resemblance in its habits to those of the Mole. It is about two inches long, and of a broad shape. In making its burrows, it cuts through or detaches all the roots of plants that lie in its way. It is readily distinguished by the extraordinary structure of its fore legs, which are excessively strong, and furnished with very broad feet divided into several sharp, claw-shaped segments. The Mole Cricket emerges from its subterraneous retreats only by night, when it creeps about the surface and occasionally

MOLE CRICKET.
(GRYLLOTALPA VULGARIS.)

employs its wings in flight : it is at that time also that it exercises its chirping call. It lives entirely on vegetables, devouring the young roots of grasses, corn, and various esculent plants, and commits great devastation in gardens.

When the female is fecundated, she forms a cell of clammy earth, in which she deposits about a hundred and fifty eggs : this nest, which is about the size of a common hen's egg, is carefully closed up on every side, as well to defend its contents from the injuries of the weather, as from the attacks of carnivorous beetles ; which, being themselves underground inhabitants, would certainly, but for this precaution, either devour or destroy them. Nothing, indeed, can exceed the care and assiduity of the Mole Cricket in the preservation of its young. Wherever a nest is situated, fortifications, avenues, and entrenchments surround it : there are also numerous winding by-ways which lead to it ; and a ditch encompasses the whole, which few insects are capable of passing. But the diligence of these little animals does not end here : at the approach of winter they move their nests entirely away, and sink them deeper in the ground ; so that the influence of the frost cannot retard the young brood in their progress to maturity. When the weather grows milder, they raise their habitations, &c. in proportion ; till at last they are brought as near the surface as possible, without being wholly exposed to view, in order to receive the genial influence of the sun ; but should the frost unexpectedly return, they again sink them to their former depth.

MOLLUSCA, or MOLLUSCS. The term applied to that large division or class of animals which inhabit and form shells. Their bodies are soft, and destitute of an articulated skeleton or vertebral column : and instead of the nervous system being developed in the form of a spinal chord, it is simply dispersed more or less irregularly in different parts of the body. Some species are terrestrial, and breathe air ; but the greater part live entirely in the water, from which they derive their nutriment, and in which they breathe by the aid of *branchiæ*, or certain gill-like appendages. Those which are terrestrial are seen in our gardens, pastures, and plantations; on the trunks and stems of trees, and in moist and shady places: while multitudes of aquatic species are to be found in the seas of tropical and arctic regions, as well as in those which environ our own islands. Others dwell on the margin of fresh-water lakes and ponds, or live at the bottom of rapid streams. In short, they may be said to be universally diffused, and produced in every variety of form and colour.

The organs of sensation and motion have not the same uniformity in point of number and position as in the vertebrate animals; and a greater aberration is observable in the position of the heart and organs of respiration, as well as in the structure of the latter. The body of the Mollusca is almost entirely occupied by the organs of nutrition; and the organs of sensation and locomotion are entirely subservient to the supply of these.

The motions of Molluscous animals consist of different contractions, varying in their direction, producing inflections and prolongations, together with relaxations of their several parts, by means of which they creep, swim, and seize upon such objects as the formation of these parts are adapted to: they are, however, incapable of rapid progress, their limbs not being supported by articulated and solid levers. The name they have received designates them as *soft* animals; and this they are pre-eminently. Almost all of them have a peculiar development of the skin, which covers their body like a *mantle*, and has received that appellation. This process, however, is sometimes narrowed in a simple disc, or is tubiform, or is hollowed into a sac, and in some cases it is divided and extended in the shape of fins.

There are two distinct kinds of molluscous animals, namely, *cephalous*, or such as are provided with a head; and *acephalous*, or headless. Those which have heads are usually provided with *tentacula*, by which they feel their way, and which they have the power of easily retracing when in danger: some have also the organs of sight and hearing; whilst others, lestitute of these, only enjoy the sense of touch. This they possess in the organ of motion, to which the name of *foot* has been given, and which is a member of considerable importance. Many move along the surface of the ground or bottom of the sea, by means of their foot, which they thrust forward and fix to some solid object, and then by a strong muscular contraction they draw their body to it; and by a repetition of this action the animal continues to make progressive motion. Others swim, by using their foot as a fin; while others, again, permanently attach themselves to a rock or other substance.

Mollusca are again distinguished into such as are *naked*, and such as are *testaceous*, viz. furnished with a shelly covering. The naked Mollusca have a membranous or fleshy mantle (as before noticed), which, however, has frequently one or more hard laminæ in its texture. Shells are formed, like bones, of a combination of earthy and animal matter. The former consists entirely of carbonate of lime; the latter is composed of layers of membrane, alternating with the mineral matter. The shell is most solid and massive in those species which lead an inactive life. The variety of form, surface, colour, brilliancy, and substance is almost infinite. They are nearly all calcareous, although some are simply of a horny consistence; but in both cases they consist of matter deposited in layers, or exuded from the skin under the epidermis, in the same manner as nails, hairs, horns, scales, &c. The shelly covering differs according as its transudation is deposited either in parallel laminæ or in close-set vertical filaments. And it is worthy of observation, that the Mollusc always appears inclined to adapt its shell to the form of the body, by reducing its cavity if necessary, as well as by extending it.

The Mollusca are, for the most part, extremely voracious; and are not particular in their selection of food. Their digestive apparatus is always highly developed; in fact, every mode of mastication and deglutition is met with: their stomachs are simple, complicated, and frequently provided with a peculiar armature: most species have salivary glands, and always a liver, but neither pancreas nor mesentery; and the intestinal tube is often of considerable length, and much convoluted. The blood is either colourless, or tinged with a bluish cast; and circulates, in all Molluscs, in a regular system of arteries and veins, issuing from a heart, which is either muscular or nearly so; and seems to contain a smaller proportion of fibrin than that of vertebrate animals. Several of the Mollusca are bisexual some produce their young alive, while others are oviparous: the eggs in some are covered by a shelly envelope, and others only by a simple viscosity. Some genera of Mollusca inhabit the sea and fresh waters, while others are entirely terrestrial; and a few species are amphibious. They feed on all substances, both animal and vegetable. Many of them are taken and used as food for man; others supply nutritious prey for birds and fishes; and their shelly coverings are converted into many useful articles of commerce. [See CEPHALOPODA; PTEROPODA; GASTEROPODA; HETEROPODA; LAMELLIBRANCHIATA; PALLIOBRANCHIATA; TUNICATA: also SHELLS.]

MOLOCH. A singular genus of Reptiles, established by Mr. Gray, and thus described in the Appendix to Capt. Grey's Travels in Australia:—"Body depressed, covered with irregular, unequal, small, granular plates, each furnished with a more or less prominent central spine, and with a series of large, conical, convex, acute spines; head and limbs covered with similar scales and spines; head small, with very large spines over the eyebrows; tail with irregular rings of very large acute spines; femoral and subanal pores none; teeth small, subequal; toes 5·5, short, covered above and below with keeled scales; claws long, acute. The external appearance of this lizard is the most ferocious of any that I know, the horns of the head and the numerous spines on the body giving it a most formidable aspect. The scales of the back are small and unequal; they gra-

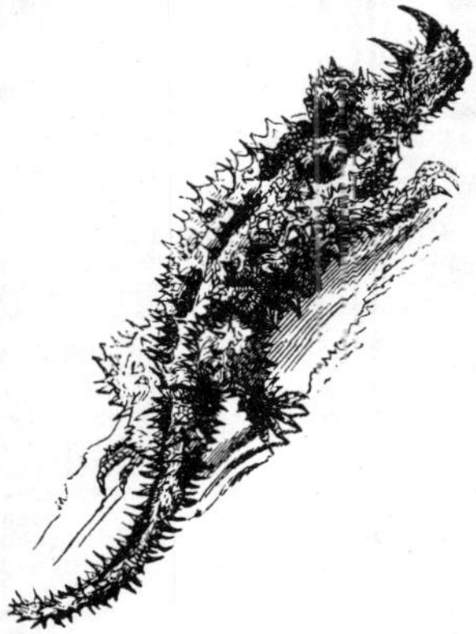

MOLOCH LIZARD.—(MOLOCH HORRIDUS.)

dually increase in size as they approach the base of the conical spines, which is surrounded with a ring of larger scales with longer spines. The large spines are conical, rather compressed, spinulose below, smooth and acute at the tip, and are usually furnished with a sharp-toothed ridge on the front edge, and sometimes on both. These spines only consist of a horny sheath, placed on a fleshy process of the exact form and appearance of the spines they bear. The scales of the under side of the body are of the same form as those of the back, and are furnished with similar but smaller and less produced spines. The back of the neck of the two specimens I have seen is furnished with a large rounded protuberance like a cherry, covered with large granular spinous scales, and armed on each side with a large conical spine; but I do not know if this is common to the species, or merely accidental in these individuals; at any rate it adds considerably to the singularity of their appearance. I have named this genus, from its appearance, after *Moloch*, " horrid king."

MONAD. The name given to the smallest creature that exists among the Infusorial Animalculæ; a mere atom or point, so small indeed as sometimes almost to elude microscopical examination; yet we are told by those who have devoted great attention to this branch of Natural History, that indescribably minute as these Monads are, they present a distinct organization, and are capable of locomotion; and have, moreover, senses sufficient for their guidance.

MONITOR. A name given to certain large Saurian reptiles, belonging to the Lizard tribe. They have teeth in both jaws, but none on the palate, and the greater number have the tail laterally compressed, as more adapted to their aquatic habits. They are divided into two distinct groups; (the first, or Nilotic Monitors, are known by their numerons small scales upon the head and limbs, the belly, and around the tail, which latter has a keel above, composed of a double range of projecting scales. The other group of Monitors has angular plates upon the

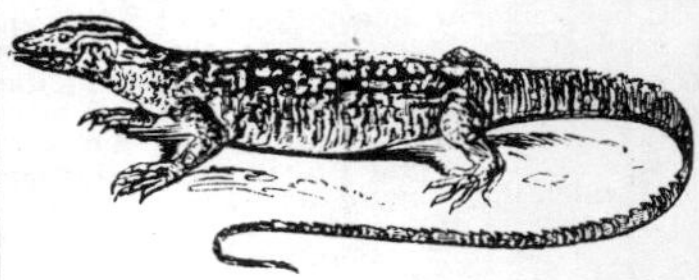

MONITOR OF THE NILE.
(MONITOR NILOTICUS.)

head, and great rectangular scales upon the belly and around the tail. The skin of their throat is invested with small scales, and forms two transverse folds. The Monitors frequent the haunts of Crocodiles and Alligators, and are said to receive the name from their giving warning, by a whistling sound, of the approach of those dangerous reptiles.

MONKEY. (*Simiadæ.*) Under the word APE will be found a general as well as a particular description of the higher *Quadrumana*. Then come the BABOONS, which the reader will find similarly arranged in alphabetical order. To them succeed the MONKEYS, which, for the most part, are distinguished by their having cheek-pouches for the temporary reception of their food, a long muscular tail, and callosities on each side of it. The species are very numerous; many inhabit India and the Malay Archipelago; but Africa may be regarded as the head-quarters of the Monkey tribe; for there they literally swarm. A great variety of species are spread over the face of the country; each species being said to be restricted to a certain extent of territory, and violently resisting any intrusion upon it. The Monkeys being the smallest of the quadrumana, are endued with less power for doing mischief than the Ape and Baboon: their ferocity, indeed, appears to diminish with their size; and, when taken wild in the woods, they are tamed with more facility, as well as sooner taught to imitate human actions, than the larger kinds. Most of the species are gregarious, associating in large troops; but each troop is invariably formed of the same species.

It has been well observed that the Monkey tribes are in reality the masters of the forests: for their dominion is not disputed either by the tiger or the lion, since they easily escape them by their nimbleness, and live on the tops of trees beyond their reach. The only animals they have to dread are serpents, who make perpetual war on them. Some of these serpents are of prodigious size, and swallow a Monkey with as much ease as it can swallow a bird. Others are smaller, but more agile, and go in quest of Monkeys along the branches of trees; and, the more effectually to secure their prey, watch the time when they are asleep. Thus creatures that are the objects of our terror, prey upon others that are objects of our disgust.

Monkeys subsist principally on fruits, the buds of trees, or succulent roots and plants. They are all fond of sweets, and show a particular predilection for the pleasant juice of the palm-tree and the sugar-cane; but when it happens that these fail, or that a different kind of food becomes more agreeable, they have recourse to insects and worms; and sometimes such as inhabit the coasts descend to the sea-shores, where they feast on oysters, crabs, &c. The crafty and ingenious manner in which they obtain these is thus effected. The oysters of the tropical climates being larger than ours, the Monkeys, when they reach the sea-side, pick up stones, and thrust them between the opening shells, which being thus prevented from closing, the cunning animals then eat the fish at their ease. In order to attract the crabs, they put their tails to the holes in which they have taken refuge; and when the crabs have fastened on the lure, the Monkeys suddenly withdraw their tails, and thus drag their prey on shore. The females generally bring forth one at a time, and sometimes two. They but rarely breed when brought to Europe; but such as do, exhibit great parental affection. Both the male and female seem indefatigable in nurturing, fondling, and caressing their young; nor do they instruct it with less assiduity; often severely correcting it, if stubborn or disinclined to profit by their example. We all know that these animals, when domesticated, are highly amusing; and there are few persons who have not laughed at their droll mimicries and capricious feats of activity. But it is generally when in company with other animals of a more simple nature, whom they appear to delight in tormenting, that their tricks and superior instincts are most effectively displayed.

We shall now give a few specimens, beginning with Monkeys that belong to the Old Continent.

The SPOTTED or DIANA MONKEY. (*Cercopithecus Diana.*) This species has a long white beard; the upper parts of the body are of a reddish colour, marked with white specks; the belly and chin are whitish; it has a crescent of white hair on the brow; and the tail, which is very long, is of the same colour as the body. It is a native of Congo and Guinea; and is one of the most lively and playful of the whole tribe.

DIANA MONKEY.—(CERCOPITHECUS DIANA.)

The GREEN MONKEY. (*Cercopithecus Sabæus.*) The prevailing colour of this species is a fine olive, a little varied with gray. The under parts of the animal and the insides of the limbs are of a light silvery gray. The face is of a swarthy flesh-colour; the nose black; the cheeks furnished with thick and long pale-yellow hair, falling back on each side the face, and almost covering the ears. It is a native of several parts of Africa.

The MOUSTACHE MONKEY. (*Cercopithecus cephus.*) On the cheeks of this Monkey there are two large tufts of yellow hair, from which it derives its name. It is about a foot in length, and the tail a foot and a half. The face is bare, and of a bluish-black; the nose blunt, with a dilated, transverse white patch immediately below it; the edges of both lips and the space round the eyes black; the ears are round, and tufted with whitish hair: the hair on the head is yellow mixed with black; that on the body and limbs is a mixture of red and ash-colour: the under part of the body is somewhat paler than the upper; and the feet are black. It is a native of Western Africa.

The PATAS, or RED MONKEY. *Cercopithecus ruber.*) The upper parts of this animal are of a vividly bright bay colour, so as almost to appear red; but the under parts and insides of the limbs are ash-coloured: the cheeks are bearded, as is also the chin, with whitish hairs, and across the forehead runs a black band. The body is about eighteen inches long; the tail somewhat shorter. It is a native of Senegal.

The PROBOSCIS MONKEY. (*Nasalis larvatus.*) This is the most grotesque in appearance of all the different species: the nose being of such a length and form, as to present to the mind no other idea than that of caricature. It is a large species, measuring two feet from the tip of the nose to the tail, which is more than two feet long. The face is of a brown colour, marked with blue and red; the ears broad, thin, naked, and hid within the hair. The head is large, and covered with chestnut-coloured hair; the whole body is also of a similar colour, except that on the breast it approaches to orange. It is chiefly found in Cochin-China and Borneo; and is sometimes seen in large troops. It feeds only on fruits.

The FULL-BOTTOM or KING MONKEY. (*Colobus polycomus.*) This species is distinguished by its head and shoulders being covered with long, coarse, flowing hair, like a full-bottomed perriwig, and of a dirty yellow colour mixed with black. Its body, arms, and legs are of a glossy black; hands naked, and furnished with no more than four fingers; on each foot five very long toes. The tail is very long, and of a snowy whiteness, with very long hair at the end, forming a tuft: body and legs very slender; and the length of the former is about three feet. It is a native of Sierra Leone.

The COCHINCHINA MONKEY, or DOUC, is a very large species, distinguished by the singular variety and brilliancy of its colours. The face is rather flat, and of a yellowish bay colour; and across the forehead runs a narrow dusky band: the sides of the face are bounded by long yellowish hairs; round the neck is a collar of purplish-brown; the upper part of the arms and thighs are black; and the legs and knees are of a chestnut-colour. The back, belly, and sides are of a yellowish gray; the lower part of the arms and tail are white; the feet black; and the rump (like the American Monkeys) is covered with hair. In an upright position this animal measures three and a half or four feet in height, being nearly as large as the Barbary ape.

The Monkeys that follow belong to the American continent, all of which differ from those of Asia and Africa by having neither cheek-pouches, posterior callosities, nor opposable thumbs, and being generally furnished with prehensile tails; while the nostrils are separated by a broad space in front: they constitute the family CEBIDÆ.

The PREACHER MONKEY. (*Mycetes Beelzebub.*) This animal is about the size of a fox; with long, black, glossy hair; a round beard beneath the chin and throat; black shining eyes; short round ears; and a long tail. It is a native of Brazil and Guiana, inhabiting the woods in great numbers, which resound with its dreadful howlings. It receives its name from the following circumstance, the authenticity of which is abundantly verified by different writers. It is common for one of these creatures to ascend a lofty tree; while numbers of them assemble on the lower branches: the Monkey who is elevated above the rest then sets up a howl, so loud and shrill that it may be heard at an immense distance: after a certain space he stops and gives a signal with his hand, when the whole assembly join in chorus; but on another signal a sudden silence prevails, and the orator concludes his harangue. The clamour on such occasions is most astounding and disagreeable. This howling faculty is accounted for by the peculiar conformation of the *os hyoides*, or throat bone, which, communicating with the larynx, gives great additional resonance to the voice. These howlings are usually sent forth in the morning, at sunset, and in the darkness of night; they are also heard when the over-clouded sky threatens an approaching storm. [See MYCETES.]

The FOX-TAILED MONKEY. *Pithecia leucocephala.*) This animal, which is about the size of a large cat, has a very singular aspect; the middle of the face being black, bare, and surrounded by white downy hair on the cheeks and forehead, and which, gradually expanding on the top and sides of the head, forms a very thick and full kind of beard, which divides under the chin, so as to leave a bare space there. Its general colour is a dusky brown: the eyes are large, and the ears round and flat: the feet and hands are furnished with sharpish claws; and the tail is long and very full of hair. It is a native of South America.

The FOUR-FINGERED, or SPIDER MONKEY. (*Ateles paniscus.*) A species of Monkey, distinguished no less by its active, lively, and tractable disposition, than by the slenderness of its body and limbs, and the absence of thumbs on its fore-paws. Its colour is uniformly black, except on the face, which is of a dark flesh-colour; and it has a long prehensile tail, which more than compensates for the defects of the hand. It inhabits the woods of South America, associating in great multitudes, and assailing such travellers as pass through their haunts by throwing dry and withered sticks at them, and by numberless sportive and mischievous gambols. In order to pass from one lofty tree to another, whose branches are too distant for a leap, they form a kind of chain, by hanging down, linked to each other by their tails; and swinging in that manner till the lowermost catches hold of a bough of the next tree, and draws up the rest.

The SQUIRREL MONKEY. (*Callithrix sciureus.*) This species, which scarcely exceeds in size the animal whose name it bears, is of a bright golden yellow colour, with orange-yellow feet and hands: the nails of the hands are flat, and those of the feet resemble claws. The head is round; the nose blackish; the orbits of the eyes flesh-colour; and the ears hairy: under parts whitish: tail very long, with a black tip. The specimens usually brought to Europe are rather of a yellowish brown or greenish cast.

MONODON. [See NARWHAL.

MONODONTA. A genus of Mollusca, inhabiting a pyramidal shell; the lips disunited at the upper part, the left having a tooth-like process, from which the name is derived; and it is on account of this tooth

MONODONTA PHARAONIS.

or notch, with which the columella terminates, that the genus is divided from *Trochus*. The animal is characterized as—head distinct, having two tentacula, with eyes at the base; foot short. They are found in most seas, and recent species are rather numerous.

MONOMYARIA. The name of an order of *Conchifera*, consisting of those bivalve shells which have but one principal muscular impression in each valve, and which includes several well-known useful Mollusca;

as the Oyster, Mussel, and Pearl Oyster. Many of the genera arranged under this order attach themselves to rocks, &c.

MONOPTERUS. A genus of Malacopterygious fishes; characterized by their having the gill-openings united, but with a partition; and the dorsal and anal apparent only from the middle of the tail backwards. The known species is from the Moluccas; it is green above and fawn-coloured below.

MOOR-HEN. [See GALLINULE.]

MOOSE-DEER. [See ELK.]

MORDELLIDÆ. A family of Coleopterous insects, distinguished by the peculiar structure of their body, and their extreme activity both in flying and leaping. The body is elevated and arched, with the head inserted very low; the thorax is trapezoid or semicircular; the elytra either very short or acuminated at the extremity, as well as the abdomen: the antennæ rather short.

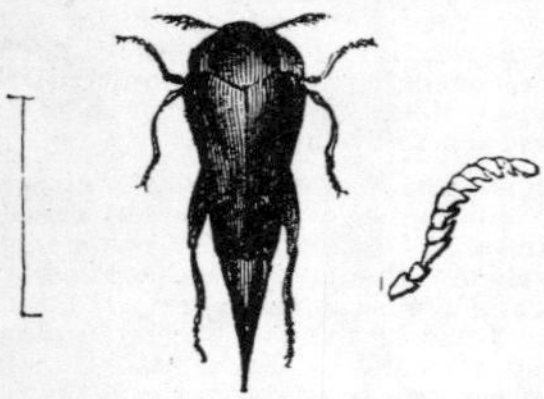

LUNATED POINT-TAIL BEETLE. (MORDELLA LUNATA.)

The smaller typical species frequent flowers, especially those of the white-thorn and *umbelliferæ*. Some of the species are parasitic upon other insects. *Ripiphorus paradoxus*, for instance, inhabits in the perfect state the nests of the common wasp; "whence Latreille observes, that it has been inferred that it subsists in that situation in the larva state, and is probably nourished by the wasps as their own offspring. On arriving at the perfect state, it emerges from the nest, and seeks the flowers; and it is probable that the female deposits her eggs in the already formed cells of the wasps, her abdomen being well adapted for such purposes, being long and acuminated."—*Westwood.*

MORMOLYCE. A singular genus of Coleopterous insects found in Java, one species of which has been described by M. Hagenbach: our figure gives a very good idea of its form, which is remarkable for its extreme flatness, the elongation of the head, and the very great leaf-like dilatation of the elytra: it was first found by Kuhl and Van Hasselt. The larva has only lately been described and figured. M. Van Ovendyk found the larva and pupa in the *Polyporus fomentarius*, or an allied species of fungus growing on the trunks and roots of trees: the larva closely resembles that of *Carabus* and *Calosoma*. Naturalists generally place this curious Carabidous insect near the South American genus *Agra*.

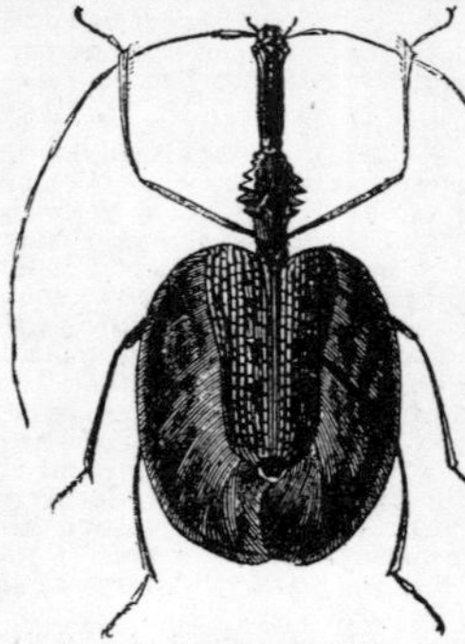

JAVANESE MORMOLYCE. (MORMOLYCE PHYLLODES.)

MORMYRUS. A genus of Malacopterygious fishes, nearly allied to the *Esocidæ* or Pike family. The body is compressed, oblong, and scaly; tail thin at the base, but swelling near the fin; skin of the head naked, covering the operculum and gill-rays, and leaving no opening for the latter but a vertical fissure. The gape is small, the angles being formed by the maxillaries: the teeth are small, notched at the extremities, and occupy the intermaxillaries and

SHARP-NOSED NILE MORMYRUS (MORMYRUS OXYRHYNCHUS.)

lower jaw; and there are bands of small crowded ones on the vomer and tongue. The stomach is a roundish sac, followed by a slender intestine with two cæca, almost always covered with fat; and the air-bladder is long, large, and simple. Two species have a cylindrical muzzle,—the one having a long dorsal fin, and the other a short one; a third has both the snout and dorsal short; and in a fourth, the forehead forms a protuberance advancing in front of the mouth. The species here figured is the sharp-nosed Mormyrus (*M. oxyrhynchus*), which is regarded as one of the best fishes found in the river Nile.

MORSE. [See WALRUS.]

MOSASAURUS. [See SUPPLEMENT.]

MOSCHIDÆ. [See MUSK-DEER.]

MOSCHUS. A genus of Ruminants allied to the antelopes, most of them being delicately graceful in form. They are found in Western Africa, in India, and the Indian islands. The accompanying figure of the Moschus Kanchil will give a good idea of

MOSCHUS KANCHIL.

the form of this genus, some of which we have seen alive in the gardens of the Zoological Society. They are very delicate animals. [See MUSK.]

MOSQUITO. (*Culex.*) A gnat-like insect common in America and the West India islands, whose stinging qualities are most annoying. These insects, of which there are many species, are furnished with a proboscis for piercing the flesh, and at the same time forming a kind of siphon through which the blood flows; but that which renders the Mosquito so dangerous as well as troublesome is, that the proboscis not only makes a wound, but injects into it a poison which causes inflammation. Mr. Edwards, in his 'Voyage up the Amazon,' has the following notice of these troublesome pests: "Soon

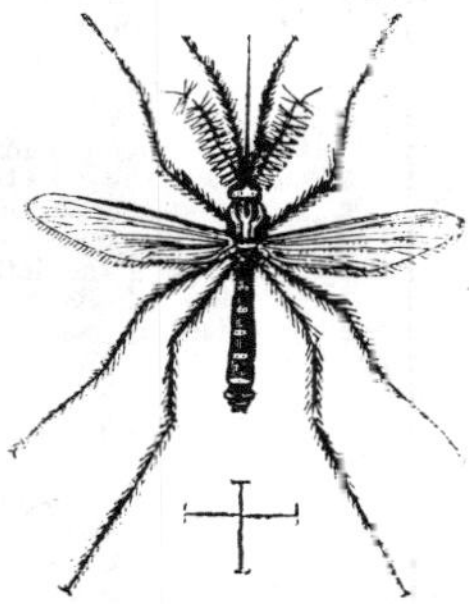

MOSQUITO. — (CULEX MOSQUITO.)

after dark we crossed the mouth of the Xingu (Shingu), much to the displeasure of the Indians, who wished to stop upon the lower side. And they were very right; for scarcely had we crossed, when we were beset by such swarms of cárapanás, or mosquitoes, as put all sleep at defiance. Nets were of no avail, even if the oppressive heat would have allowed them, for those which could not creep through the meshes would in some other way find entrance, in spite of every precaution. Thick breeches they laughed at, and the cabin seemed the interior of a bee-hive. This would not do; so we tried the deck; but fresh swarms continually poured over us, and all night long we were foaming with vexation and rage."

MOTACILLA. [See WAGTAIL.]

MOTH. (*Phalæna* of Linnæus.) The name of Moths is given to a numerous and beautiful divison of Lepidopterous insects, readily distinguished from Butterflies by their antennæ tapering to a point, instead of being terminated by a knob, and by their being seldom seen on the wing except in the evening or night. It should also be observed that the antennæ are often feathery, or comb-shaped; and that the legs have two spiny processes or thorn-like points at the middle joints of each.

The diurnal Lepidoptera are all provided with a tongue for gathering their food; but a great proportion of the Moths are destitute of that organ, whilst in others it is exceedingly small: a considerable number of them, therefore, must pass the whole of their winged state without food. The larvæ or caterpillars from which the various Moths are produced, exhibit nearly the same variety of appearance as the winged insects which spring from them. Some are large, while others are extremely minute; many are provided with ten, others twelve or fourteen feet, and the largest have sixteen. Some of the caterpillars are smooth, others are covered with hairs; but all of them, after having several times cast their skin, spin for themselves the materials of a habitation, in which they are to be transformed into chrysalids.

All the nocturnal Lepidoptera were included in the genus *Phalæna* by Linnæus; but since the time of that great naturalist they have been divided by Cuvier, Latreille, and others, into a number of different groups, the classification of which is too complicated and embarrassing to be thoroughly explained in this work. A few of the species are here appended as examples of the group. We may remark that there are several thousand species of Moths, varying in size from a line in breadth to eleven inches, and even more. The variety of form and colour is endless. We limit ourselves to the notice of two or three species which are *notorious* for their depredations; readers who wish to see how useful some members of the group are to mankind, must consult the articles SILKWORM MOTH and SATURNIA, though there is not a Moth that is not more or less useful in many ways, to Birds and Bats, if not to us. Referring to the article SPHINGIDÆ for an account of the HAWK-MOTHS, and to the article HEPIALUS for a short description of another important subdivision; to the word COSSUS for the GOAT-MOTH; and to other words scattered over the work, — we begin with noticing the BOMBYCIDÆ, which contains the largest of all the Moths yet known, — the *Saturnia Atlas*, — the extent of whose wings measures between eight and nine inches. The ground colour is a fine deep orange-brown, and in the middle of each wing is a large sub-triangular transparent spot; each of these transparent parts is succeeded by a black border; and across all the wings run lighter and darker bars, exhibiting a very fine assortment of varying shades: the upper wings are slightly curved downwards at their tips, and the lower wings are edged with a border of black spots on a pale buff-coloured ground: the antennæ are widely pectinated with a quadruple series of fibres, which have a very elegant appearance. This species, or at least a closely allied one, is found both in the East and West Indies.

The RYE-GRASS MOTH. (*Penthophora morio.*) This Moth is of a middling size; the male, with extended wings, is nearly an inch broad, and black. The antennæ are strongly pectinated in two rows: and the head, back, and abdomen are black, the latter with yellow notches posteriorly. The wings are very thin, membranous, transparent, and black, with fringes of the same colour, or sometimes brownish. The female is distinguished by a proportionally thick, long abdomen, which is whitish gray, and woolly at its exterior; and by small, slender, brownish gray wings, which are not adapted for flying. The caterpillar is found in April and May, living on rye-grass (*Lolium perenne*), and many other plants in meadows: its ground colour is velvety black, yellow at the incisions and sides, with a black head and small reddish yellow warts, having ash-gray hairs on them. The cocoon consists of only a few threads; the pupa is yellow, streaked with black lengthways, blackish brown on the wing-covers, and beset with whitish gray tufts of hair. After pairing, the female lays her eggs at the end of May and beginning of June, round the stems of the grass, and covers them with the down from her abdomen, to secure them from the weather. Two generations appear in long, warm summers; but in general the caterpillar passes the winter at the roots of the grass. The destruction of this caterpillar when in great abundance is very difficult, as it prefers living in long grass in the day time, or in the ground. Breaking up the meadows in autumn appears to be the best method of destroying the pupæ concealed there; they will thus be exposed to the enemies appointed by Nature herself, such as *Ichneumonidæ*, &c.

The BROWN TAIL MOTH (*Porthesia auriflua*) is remarkable for the ravages which its caterpillar commits by destroying the foliage of trees and hedges. The Moth is of a fine satiny white, except the hinder part of the body, which is of a deep brown. The caterpillar is brown and hairy, having a row of white spots along each side, and two red spots on the lower part of the back. It is of a gregarious nature, vast numbers residing under one common web: they are hatched early in autumn, and immediately form for themselves a small web, and begin feeding on the foliage of the tree or shrub on which they were placed: they marshal themselves with great regularity for this purpose in rows, and at first devour only the upper pellicle and the green parenchyma of the leaves, and in the evening retire to their web. In about three weeks they cast their skin, and proceed to feed as before, enlarging their web from time to time, and forming it on all sides as strong and secure as possible. Here they remain during the whole winter in a state of torpidity, till, being enlivened by the warmth of the returning spring, they issue from their covering with increased strength, and devour the whole substance of the leaves. When full grown, which is usually about the beginning of June, each spins itself a separate web, in which it changes to a dark brown chrysalis, out of which in about a month the Moth issues. The ravages of these caterpillars have in some years been so great as to cause the most serious apprehensions. In 1782, so numerous were they in many parts of England, and particularly in the neighbourhood of London, that subscriptions were opened and poor people employed to cut off and collect the webs at one shilling per bushel; they were then burned, under the inspection of the parochial officers: and it is asserted, that in one day, in the parish of Clapham alone, eighty bushels were thus collected and destroyed.

The larvæ of Psyche and other allied genera of Moths inhabit a case constructed by themselves. In some species found in New Holland and South Africa, these cases are often beautifully ornamented with straws, spines, or little bits of wood. It is an example of this kind that Capt. Sir James Alexander describes, in his 'Excursions in Western Africa,' under the name of the *Lictor*. "That strange insect, the Lictor, or bundle of sticks, occurs here (Kaffir-land), and, with its caterpillar-like tunic, on which are stuck the ends of little sticks, all raking aft like the quills of a porcupine, it may be seen walking along by projecting its head and six legs from its case. In some of these insects the sticks are irregular, the longest being near the tail. In others again there are three sets of regular fasces connected by a 'diarthrodial articulation;' which makes the ingenuity of this insect the more remarkable. All the fasces are about the same length; but the set about the head are thick; the middle are less so; and the tail fasces taper nearly to a point. This variety is found suspended to dried rhenaster bushes. A third species, more delicate than the other two, feeds on yellow everlasting flowers; and has one set of regular fasces about its body."

Of the family NOCTUIDÆ, we may specify the CABBAGE MOTH. (*Mamestria brassicæ.*) This Moth is about an inch and a half broad, when the wings are extended; its head, collar, and back, are blackish-gray, intermixed with whitish and yellowish hairs. The back has a thick double crest; the abdomen is dark ash gray, the upper half beset in the middle with black tufts. The upper wings are gray, with a mixture of yellow and white. The anterior border is very light to beyond the middle, with dark spots; on the watered band are two or three yellowish spots: the cross lines are distinct, the first is rather broad, and the next double the width, with a dark edge; the usual middle spots surrounded with white, the kidney-shaped one in the middle, with a whitish-gray lunule, surrounded with a blackish colour; the usual conical spot is dark, and surrounded with brown. The watered band is extremely light, and terminates at the white notched line, marked with a W. At the first end of the above line is a softened-off rusty spot. Near the border of the gray, yellowish-striped, and toothed fringes, is a row of small, black, triangular marks. The under wings are light gray, with dark veins, and central spots; blackish towards the outer edge The Moth appears in May and June, sits in the daytime on hedges, the stems of trees, or on the earth, and only flies at night.

The caterpillar is green, more or less covered with gray or black. It has a dark stripe on the back, on which there is a pale indistinct line. Above, it is sometimes furnished with dark or pale spots placed lengthwise. At the sides is a dirty yellow stripe, which becomes reddish above: and close above this spot are two white spiracles, surrounded with black, each in a small black spot. When this caterpillar appears in great numbers, it does considerable damage to several vegetables, such as cabbages, lettuces, &c., by eating out the heart. It appears in July, August, and September. To look for and kill them, although troublesome, is the only sure way of getting rid of them.

The ANTLER, or GRASS MOTH (*Cerapteryx graminis*) is another species injurious to meadows. This Moth is of a moderate size; its head and back are yellowish brown, the collar almost yellow; the brown antennæ are covered with yellow scales; and the abdomen and the legs are brownish gray, the latter with darker joints. The upper wings are usually brownish gray, with a darker mixture in the middle. The ordinary middle spots are whitish, yellowish, or bluish; the first round, the second half-moon shaped. A strong narrow line runs from the root lengthwise through the middle of the wings into the half-moon spot, which it intersects in such a manner as to give it the appearance of a three-pronged fork, or horn, whence the common English name of the Moth. The upper border of the wings is lighter; the fringes brownish yellow. The under wings are yellowish gray, nearly black towards the outer border, with yellowish fringes. It flies in the latter end of July and beginning of August.—The caterpillar is brown or blackish, with five lighter stripes along the back; the first and last sections are covered with a hard, smooth scale. The stripes meet at the edge of the anus: the abdomen is blackish: the hind feet project beyond the anal point. The larvæ are an inch long; and they undergo their transformation about Midsummer, within a light cocoon, under moss, stones, and such like materials, changing into a blackish-brown shining pupa. The food of the caterpillar consists of all the soft sorts of grasses. It lives at the roots, and eats all the germs. Although it is in existence in autumn, it lies benumbed in the earth in winter, and begins to eat again in the spring; yet the effects of its devastations appear chiefly in the beginning of June, when it has changed its skin for the last time. This insect appears only to be injurious in dry situations, particularly in mountain pastures; the caterpillar has never been met with on low, wet, and marshy meadows. The only means of extirpating or diminishing this caterpillar consists in surrounding the attacked places, as the ground permits, with shallow ditches, or by means of a plough with deep furrows, as broad as possible, and turning pigs into these places to eat the caterpillars. Crows are also among the natural enemies of this insect, and should be encouraged.—For the information contained in this article we are indebted to 'Köllar's Treatise on Insects injurious to Vegetation,' &c.

The GAMMA MOTH. (*Plusia gamma.*) The ground colour of this beautiful Moth is light and dark gray mixed with rust colour. The head and collar are of a brownish hue, edged with light gray lines, as well as the crested back and shoulders: the abdomen is yellowish gray, with elevated brown tufts of hair. The upper wings are marbled, and have a metallic lustre: the inner edge is wavy, and toothed near the fringes: the notched cross-lines are silvery: towards the inner border is situated a silvery or gold-coloured shining mark, which resembles the Greek letter Gamma (γ). The under wings are yellowish brown at the base above the fringes, with black bands. The blackish-brown pupa is enclosed in a white cocoon.—The caterpillar is green, beset with single hairs, having only twelve feet, and a brownish green head. On the back are four very small yellowish or whitish lines: the feet have a yellow stripe: the spiracles are blackish green. These caterpillars are found from spring to autumn in a variety of generations, and are so plentiful in some years, that they do great damage to vegetables, peas, and various sorts of fodder-herbage. The only possible means of destroying them is by shaking them off and hand-picking.

The RED UNDERWING. (*Catocala.*) The antennæ of this handsome Moth resemble threads: the thorax is crested, and of a brownish-gray colour, as is the abdomen and superior wings; the latter having double lines and zig-zag bars crossing them in several places, and a remarkable spot in the middle. The under wings are of a fine scarlet colour, having two broad bands of black; the tongue is spiral; and all the wings are dentated. The caterpillar is about two inches and a half in length; feeds on the willow; and is in colour so like the bark, as not to be easily seen. About the latter end of June it changes to a red chrysalis: and the Moth appears in August; flies in the day; and is very fond of settling against barns, or the sides of such houses as are boarded.

Amongst the family TINEIDÆ, so numerous in genera and species, generally of small size, may be mentioned the HONEY-COMB MOTH. (*Galleria cereella.*) This Moth is one of the larger species of the family of *Tineidæ;* the male being considerably smaller than the female, and the sexes differing much in size, colour, and in the form of the upper wings. The caterpillar of this Moth feeds on the *wax* of the honey-bee, and not unfrequently destroys a hive by the filth and stench which it occasions. Hence, though it does not destroy the honey, it is a most formidable enemy to apiarians. *Male:* Antennæ, head, and back clay-yellow; on the back behind the scutellum rises a small blackish-brown tuft of hair, the point of which is white; abdomen yellowish brown. The upper wings are broad, short, and obtuse; the fore border slightly curved, the fringe border lunated, notched inwardly, the inner border rather waving, with a small hollow exactly opposite the corner of the inner angle. The colour is dusty ash gray. From the base to the middle there is a white slender band, on which are scattered single dark brown minute spots. Upon the fore-border, and along the fringe-border, are lines in the same direction, proceeding from a spotted band, which is angularly bent, and faint. The inner border is light yellowish for a considerable breadth, from the base to the inner angle, with many purplish-brown, short, elevated, wool-like lines in the same direction; so that when the wings are closed above, an apparently furrowed surface will be formed. The fringes are jagged, brown and white at the points, and surrounded inwardly with a darker hair-like line. The under wings are light ash gray, and sometimes brownish gray, with lighter fringes edged with white, and a yellowish line for their inner boundary. The female is not only much larger than the male, but distinguished from him by a darker rusty brown head and back. The abdomen is thick and club-shaped, furnished with a brownish gray ovipositor, and the feet of the same colour. The upper wings are darker, obtuse, straight, almost rectangular, and a trace of the faint spotted band is rarely perceptible. The under wings are much lighter, white, but with a dark gray dusty border, and darker veins of the same colour, as far as the white fringes, which are surrounded with a yellowish colour.

The caterpillar is cylindrically spindle-shaped, when fully grown from ten to twelve lines long, and two lines thick, dirty white, with scarcely visible brown single tubercles, emitting slender hairs. The head is chestnut brown, the back of the following segment rather darker, divided lengthwise by a whitish line which line is sometimes continued indistinctly along the back; the belly and sixteen feet are bone-coloured. It prepares for itself, immediately on issuing from the egg, a web, or covered passage, with thick, strong threads, in which it lodges by day safe from the attacks of the bees, and only seeks its food, which consists of wax, at night, when the bees are at rest. At first, these caterpillars only live in the lower cells, but when they are bigger they ascend higher, lengthening their passage as they proceed; so that when there are many of them, in a hive, it is entirely filled with these webs. The bees which are entangled in them and cannot get away, die. Three hundred caterpillars have been found in a hive. They attain their full size within three weeks, and are then ready for entering the pupa state. When this is the case, they make for themselves a much firmer and entirely closed web, either in the above-named passages, or in a concealed corner of the hive. In this web the caterpillar lives from ten to twenty-eight days unchanged, but is finally transformed into a brown pupa, out of which the moth appears in fourteen days. Those which become pupæ in autumn lie the whole winter in that state. There are two generations of them in a year. The moth of the first generation appears in spring, and that of the second in the beginning of July. The female lays her eggs at night, in the cracks of the lower part of the hive, from which the young caterpillars find their way to the honeycomb.

There is but one sure method of clearing the bee-hives of this moth, and this consists in looking for and destroying the larvæ and pupæ. If the hives are examined only once a week for this purpose, any traces of covered passages will easily be perceived, and must be immediately removed, and destroyed with the caterpillars in them. The corners of the hive must also be closely examined, in case of cocoons being there, which must also be destroyed. A lighted candle is also recommended to be held before the hole of the bee-hive, that the moth, flying out to the light, may be burnt. But this is labour in vain, for the female does not leave the hive till she has laid her eggs; and it is only supernumerary males that perish in the flame.—*Köllar.*

The CLOTHES-MOTH. (*Tinea vestianella.*) As soon as the caterpillar quits the egg, it begins to form a nest. For this purpose, having spun a thin coating of silk round its body, it cuts filaments of wool or fur close to the thread of the cloth, and applies the pieces to the outside of its case; which covering it never leaves except in cases of urgent necessity. When it wishes to feed, it puts out its head at either end of the case, as may be most convenient. When inclined to change its position, it protrudes its head and about half its body, and thus moves forward, dragging its case by fixing its hinder legs firmly in it: and when, from its increase in size, the case becomes too small, it makes an addition to it at each end. This operation can be readily traced by transferring it from cloth of one colour to another, when each addition will be conspicuous from the difference of colour. After changing into a chrysalis in April, or May, it remains quiescent for about three weeks, when a small nocturnal Moth, of a silvery gray colour, comes forth. It is said that Moths never attack *unwashed* wool—that is, wool as it comes from the sheep's back, before any cleansing process has been employed that will deprive it of its natural oil or smell. It is therefore recommended to be placed in layers between clothes, or kept in small parcels in the corners of shelves or drawers. The most important, by far, of all the species is the Silkworm Moth. [See SILKWORM. See also ARCTIA: STAUROPUS: COSSUS: HEPIALUS, &c.—Such of our readers as may wish for further information respecting the genera and species of this most extensive class of insects, are referred to the works of Stephens, Curtis, Wood, Humphreys, and Westwood's British Moths; also to Mr. Henry Doubleday's (of Epping), and to the papers of Messrs. Stainton and Douglas.

MOTMOT. (*Motmotus*, or *Prionites Braziliensis.*) A curious and handsome bird, inhabiting many parts of South America. Its back is of a dark rich green colour, and it has a long wedge-shaped tail, two feathers of which extend some inches beyond the others. The shafts of these are stripped of their webs near the extremities, giving the bird a very singular appearance. One would suppose that these birds trimmed their feathers thus themselves, for many are found with quills perfect, and others partly denuded. The Motmots are generally in pairs in the deep woods, and are easily recognized by their note, *mot-mot*, slowly repeated. There are several species of Motmots: the edge of the beak in these birds is serrated both in the upper and lower mandibles.

MOUFFLON. (*Ovis Musmon.*) An animal of the sheep kind, called also the Musmon, inhabiting the mountainous parts of Corsica, Sardinia, Greece, &c., and which, though by nature extremely wild, retains all the characteristic marks of the primitive race. The general size of the Moufflon is that of a small fallow deer; but, though covered with hair, it bears a stronger resemblance to the ram than to any other animal, both in regard to its horns, which sometimes grow to a vast size, as well as in its general conformation. The muzzle and the inside of the ears are of a whitish hue, tinged with yellow; but the other parts of the face are of a brownish gray. The body is covered with hair instead of wool; in which particular consists its chief difference from the general aspect of the sheep: the upper part of it is brown, but the under part and the insides of the limbs are whitish. In summer its hair is close, like that of a deer; in winter it becomes rough, wavy, and a little curled, concealing at its roots a fine white woolly down. About the neck and shoulders

MOUFFLON. — (OVIS MUSMON.)

as well as under the throat, the hair is considerably longer than elsewhere.

From spring to autumn the Moufflons feed in the little vallies among the upper regions of the mountains, on the young shoots of the Alpine plants, and are said to grow very fat. As winter approaches, they descend lower, and eat grass and other vegetables. The whole form of this animal seems better calculated for agility and strength than that of the common sheep; but still it is very timid, and, when closely pursued, does not run in a directly progressive course, but obliquely, from side to side, in the manner of other sheep; ascending the rocky mountains with great agility, and, like the wild goat, going over the narrowest and most dangerous passes with perfect safety. Their chace is dangerous and difficult; and they are so wild as to be seldom taken until shot by the hunters, who lie in wait for them among the mountains. The female is less than the male, and her horns never grow to the same magnitude as in the ram. These have sometimes been found to measure, in their convolutions, above two ells in length: with these they often maintain very furious battles among their own kind; and sometimes they are broken in the conflict. The young, when first born, are covered with a soft, gray, curling fleece, which gradually changes into hair towards the end of summer. Such is the sheep in its wild state; by no means that seemingly helpless animal which we view it under the shepherd's care; but in the highest degree active and vigorous.

MOUSE. (*Mus.*) A genus of Rodent quadrupeds, including not only what are usually termed *Mice*, but also the *Rats*. [The latter are described at page 562.]

The Common Mouse. (*Mus musculus.*) This little animal is a general inhabitant of almost every country in the world; for though it is said to be not a native of America, but taken there by European settlers, it is now found in every part of that continent. There are several varieties, distinguished by their colour; but the fur is usually of a brownish ash-colour above, and light beneath; the tail not quite so long as the body; and the ears about half the length of the head. As Mr. Bell observes, "there are few animals more generally associated with mankind, or whose very existence appears to be more essentially dependent upon human arts and human civilization, than this pretty, but annoying little pest. Domestic in its habits, nourished by almost every article of human food, and obtaining effectual shelter in the secret recesses of the habitations which human art has raised, it has accompanied man in all his adventures for colonization, and identified itself with every new territorial occupation of our race." All its actions appear to be regulated by fear and necessity. It seldom leaves its hole but when impelled thereto by the want of food; and then, unlike the rat, who travels from one house to another, it seldom quits the spot where it has once taken up its residence. The Mouse makes a nest not unlike that of a bird, and brings forth several times in a year, generally having from six to ten at a litter; when first born, mice are naked and helpless, but in about fifteen days they are able to shift for themselves. No animal has more enemies than the Mouse, and few are so incapable of resistance. Cats, snakes, hawks, owls, weasels, and rats are their incessant destroyers; and but for their amazing fecundity, the extirpation of the whole race would seem to be a natural consequence. The Mouse is capable of being tamed, and will sometimes show considerable attachment to its keeper: the *albino*, or white variety, which may be perpetuated by breeding, and is frequently kept as a pet, is particularly so.

The HARVEST MOUSE. (*Mus messorius.*) This is the smallest, and one of the prettiest, of all quadrupeds that exist in Britain. The upper part of the body is of an iron colour, the lower being white; a straight line runs along the sides, dividing the colours; and the tail is somewhat hairy: the length of the body, from the nose to the tail, is two inches and a half; and that of the tail is about two inches. These little animals never enter houses, but during the harvest are carried into ricks and barns with the sheaves; and there they live and multiply. They build a curious nest amidst the straws of standing corn, and sometimes in thistles. In the winter months they appear to retire to burrows, where they hybernate; but their grand rendezvous seems to be in corn-ricks. They are very common in Hampshire, Wiltshire, and some of the neighbouring counties; yet they almost escaped the notice of naturalists till the appearance of White's History of Selborne, where its ingenious author thus describes the nest, &c. "They breed as many as eight at a litter, in a little round nest composed of the blades of grass or wheat. One of these nests I procured this autumn (1767), most artificially platted, and composed of the blades of wheat; perfectly round, and about the size of a cricket ball; with the aperture so ingeniously closed, that there was no discovering to what part it belonged. It was so compact and well filled, that it would roll across the table without being discomposed, though it contained eight little mice that were naked and blind. As this nest was perfectly full, how could the dam come at her litter respectively, so as to administer a teat to each? Perhaps she opens different places for that purpose, adjusting them again when the business is over; but she could not possibly be contained herself in the ball with her young, which moreover would be daily increasing in bulk. This wonderful procreant cradle, an elegant instance of the effects of instinct, was found in a wheat field, suspended in the head of a thistle."

The LONG-TAILED FIELD-MOUSE, or WOOD-MOUSE. (*Mus sylvaticus.*) This species is found in fields and gardens, widely diffused throughout Europe, and is everywhere considered among the minor pests of the farmer and gardener. The head is long, the muzzle tapering; the eyes are black, large, and prominent; the ears large, and of an oblong oval shape; and the legs long. From the tip of the nose to the end of the body is about four inches and a half; and the length of the tail four inches: the head, back, and sides are of a yellowish brown hue, mixed with some dusky hairs; a yellowish gray patch on the breast; belly white; and the tail slightly covered with short hair. These animals retire to holes among brushwood, and under the trunks of trees, where they form large magazines of grain, acorns, nuts, &c., for their winter provision; but the injury done to the farmer consists less in the quantity the Field Mice collect, than by the hogs rooting up the ground in their search after their hoards. The nests which they provide for their young are generally very near the surface of the heart, and frequently in thick tufts of grass.

The BARBARY MOUSE. (*Mus barbarus.*) This African species is intermediate in size between the common Rat and common Mouse. It is of a darkish brown colour, with five or six longitudinal stripes on each side, about half as wide as the intervening spaces, and becoming more indistinct to-

BARBARY MOUSE.—(MUS BARBARUS.)

wards the under parts, which are nearly white. Mr. Bennett observes, that on the fore feet only three of the toes are at first visible; and that this circumstance, mentioned in the specific character given by Linnæus, has led many subsequent naturalists to doubt whether the Barbary Mouse really belonged to the genus *Mus*. Linnæus himself, however, states, that rudiments of a thumb, and also of a fifth toe, were observable on a closer inspection; which fact subsequent examination of living specimens has fully confirmed.

Amongst the Mice, though it belongs to a distinct genus, is generally placed the RUSTIC MOUSE, or FIELD VOLE. (*Arvicola agrestis.*) This species, like the common Field Mouse, inhabits corn-fields and frequents granaries, but is chiefly confined to the northern parts of Europe, as the temperate tracts of Russia, Siberia, &c., where in particular seasons it appears in great multitudes, and devours great quantities of grain. It has a sharp nose, an oblong head, and small ears lined with fur; the colour of the body and head are ferruginous, with a dusky line along the back; the belly and limbs are whitish; above each hind foot there is a dusky circle; the body is somewhat less than that of the Field Mouse; and the tail is only half the length of the body. It burrows at a small distance from the surface, each retreat having a long gallery, with a chamber at the end, in which the winter food is deposited.

MUGIL: MUGILIDÆ. A genus and family of Acanthopterygious fishes, distinguished by a nearly cylindrical form, large scales, and two distinct dorsals; the head is protected by large scales or polygonal plates; the snout is very short, and the orifice of the mouth is transverse. The species are found not only in the European seas, the North

Atlantic, but ranging through the African Atlantic to the Cape of Good Hope. They enter bays and the mouths of rivers in large shoals, and have the habit of leaping high out of the water. [See MULLET.]

MULE. A hybrid animal between the horse and the ass, differing in size, strength, and beauty, according to the predominance of its parental species; those between a male ass and a mare being far superior to the progeny of a she-ass with a horse. In mountainous countries Mules are highly serviceable; no beast of burden being either so sure-footed, or so capable of enduring fatigue; but in beauty of form they fall very short of that noble quadruped the Horse: the Mule having a large, clumsy head, long erect ears, a short mane, and a thin tail. In Spain, Portugal, Italy, the East, and in South America, this animal is much valued for the saddle, and for drawing carriages. Savoy produces very large ones, but the finest are bred in Spain. They are sometimes fifteen or sixteen hands high, thick set, and capable of travelling, for months together, with six or eight hundred weight on their backs. It has been thought that they are altogether incapable of producing their kind: but some few instances have occurred in which female Mules have had foals, and in which the male has impregnated females, both of the horse and ass species. Such instances are, however, very rare.

MULLET. The name given to two genera of Acanthopterygious fishes, viz. the gray Mullets (*Mugil*), and the red Mullets (*Mullus*). The former are distinguished by a nearly cylindrical body covered with large scales; six branchiostegous rays; head somewhat depressed, the scales large; the muzzle short; an angular rise in the middle

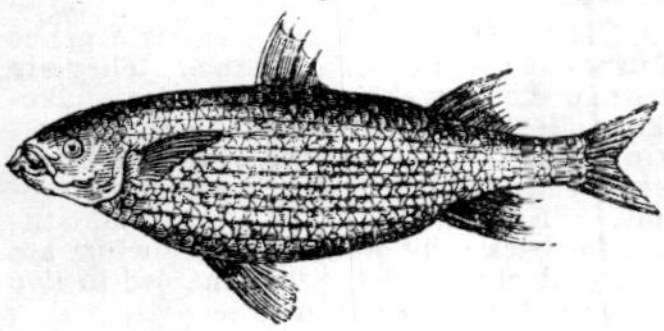

MUGIL CEPHALUS.

of the lower jaw which fits into a corresponding hollow in the upper; and very minute teeth.

The GRAY MULLET. (*Mugil capito.*) This is a common inhabitant of the Mediterranean and Northern seas, where it is chiefly found frequenting the shallow water near the shores; nor is it by any means uncommon on our own western and southern coasts. Its general length is from twelve to fifteen inches; its colour bluish-gray, darker on the back, and silvery underneath: the sides are marked with several dusky stripes; and the fins have a tinge of bluish-white: the head is rather large, pointed in front, and somewhat flattened at the top; cheeks and operculum silvery white; the mouth small, and the tongue rough: the first dorsal fin, which is situated on the middle of the back, consists of four very strong rays; the second dorsal fin, placed opposite the anal, has only soft rays; and the tail is considerably forked. In the spring and early summer months this fish, like the salmon, ascends rivers to a considerable distance, and when preparing for these expeditions is observed in shoals near the surface of the water; but they are so extremely wary and active, that when surrounded by a net, the whole shoal frequently escapes by leaping over it. They rise freely at the flies used for trout, and when hooked, require great care in the management of them, as they are strong in the water, and plunge violently. They are fond of rooting in the sand or soft mud in search of food, leaving ample evidence of their having been so employed, in the round holes made by them in the operation.

The THICK-LIPPED GRAY MULLET. (*Mugil chelo.*) This species is distinguished from the foregoing chiefly by its large and fleshy lips, the edges of which are ciliated, and through their thickness the teeth penetrate like so many hairs. Body solid; scales large; tail broad and concave: colour of the head and back greenish; all besides silvery, with six or seven parallel lines along the sides, of the same colour as the back. This species is common on many parts of the Devonshire coast in the autumn, and is also occasionally seen in larger shoals on the east coast of Scotland.

The SHORT GRAY MULLET. (*Mugil curtus.*) This is a smaller and less well-known species than either of the preceding: the body is also deeper in proportion; the head wider, more triangular and pointed; the eye larger in proportion; and the fin-rays longer, particularly those of the tail. In colour it nearly resembles the species first described.

The AMERICAN MULLET. (*Mugil albula.*) A fish greatly allied to the common Mullet, but of a more slender form: mouth small and toothless; tail large and forked. It inhabits the American seas, and is abundant about the Bahama Islands. Its flesh is considered excellent.

Fish of the genus *Mullus* are distinguished by the oblique form of their head; by two long appendages under the chin, and large scales on the head and body, which are easily detached; the body oblong, and generally of a red or yellow colour; and the eyes situated very close to each other.

The PLAIN RED MULLET. (*Mullus barbatus.*) This fish is caught in the Mediterranean; and its usual length is about six inches. The head is remarkable for its almost vertical profile: the body is pretty thick; the back is rather flat; and from the head to the tail the fish gradually diminishes in thickness. In colour and general appearance it very much resembles the species just described: the scales are thin and easily separated, and when rubbed off, the skin itself appears of a brighter red. It is generally considered as a very delicate fish; and is celebrated for having been the fashionable object of Roman luxury, and for which such enormous sums were paid; though it is probable that the high estimation in which it was held by those ancient epicures was more

RED MULLET.—(MULLUS BARBATUS.)

owing to its elegant appearance than to its real merit as a food. To such a pitch, indeed, was their "refinement in luxury" carried, that before the fish was dressed, it was brought alive into the apartment in a glass vessel, in order that the guests might enjoy the pleasure of contemplating the beautiful changes of its evanescent colours during its expiring agonies; immediately after which it was prepared for the repast.

MUNTJAK. (*Cervus muntjac.*) An animal of the Deer tribe, about one-fifth larger than the Roebuck, being about two feet two inches high at the shoulders; head large; ears rather large; eyes large, with lachrymal sinuses; tail short and flattened: general colour reddish-brown above; belly and front of the thighs pure white. The male has large canines in the upper jaw; the female has none, nor has she horns. The Cervus Muntjak is a native of Java; and is described by those who are fully acquainted with its character as possessing a great portion of craftiness, combined with much indolence. It has a strong scent, and is easily tracked by dogs. When pursued, it does not go off, like the stag, in any accidental direction: its flight indeed is very swift at first, but it soon relaxes, and, taking a circular course, returns to the spot from which it was started. After several circular returns, if the pursuit be continued, the Muntjak thrusts its head into a thicket, and in this situation remains fixed and motionless, as in a place of security, and regardless of the approach of the sportsmen. If it remains unobserved, it is still unwilling to quit its haunts: and experienced hunters, acquainted with its natural disposition, after an unsuccessful pursuit, return the following day, and in many cases find the object of their previous chase in the same spot. The native dogs, vulgarly denominated pariahs, are regularly trained to the sport: and many of them are extremely ardent and courageous in the pursuit. The male animal possesses a great share of courage, and when the dogs are at bay with him, with his tusks he makes a most vigorous defence; and many dogs are wounded in the attack. But although possessed of great courage, the life of the Muntjak is not tenacious, and the sportsman, on arriving at the spot, generally dispatches it with small shot. The natives of the most distant districts are in the habit of bringing their best dogs to the capitals, on occasion of their half-yearly attendance, to perform the feudal services to the sovereigns, when their sports afford an occasional amusement to the European inhabitants. Other modes of chase are also employed by the natives of rank. One in particular is common in the western parts of the island. A district is surrounded by a line of hunters, and the Muntjak is driven in towards a central spot: forty or fifty animals are in this manner often obtained at a single pursuit. Many of the hunters are mounted, and the horses are trained to the chase. The sportsman endeavours to overtake the animal, and to kill it by a stroke with a sword. The inhabitants of Pugar and Blambaugen, two provinces at the eastern extremity of Java, possessing a small population, but abounding in extensive plains and acclivities, which afford an ample range and abundant pasture, are particularly skilled in this sport. The best horses are trained for it: the sportsman, without a saddle, mounts on the naked back, and carries on the pursuit with a frantic impetuosity, at the risk of his limbs and neck. In Banka a less arduous, but more destructive method is employed to take this animal: a long rope of rattan is suspended, at a proper height above the ground, between two trees; numerous nooses, of the same material, hang from this, in a close and continued series, and the Muntjak, driven towards it, pursued by dogs and blinded by fear, does not perceive the slender rattan, and thrusting his head into a noose, is strangled on the spot. The Muntjak is impatient of confinement, and requires a considerable range to live comfortably: it is cleanly in its habits, and delicate in its choice of food. The flesh, it is said, affords an excellent venison, which is often found on the tables of Europeans. We are indebted to Dr. Horsfield's admirable Zoological Researches in Java for the foregoing particulars.

MURÆNIDÆ. By this term is a genus of fishes of the Eel tribe known. They are distinguished by their long, slender, snake-like bodies, covered with a thick and soft skin, and having the scales very minute, deeply imbedded, and often scarcely perceptible. The gill-opening is small, and situated far back; by which the branchiæ are more protected, and the fish enabled to live a considerable time out of water.

MUREX A genus of Mollusca, containing numerous species, many of them remarkable for the form and beautiful colouring of the shells, particularly those which come from the seas of tropical climates. The shell is oblong; spire short, with three or more longitudinal, continuous, branched, spinose, or fringed varices; channel generally long, and sometimes closed; inner lip smooth; mouth round and small; operculum horny. The head of the inclosed animal has too long tentacula furnished with eyes, foot round, and generally short. The general character of the genus may be seen in the accompanying figure. The *Murex tenuispina*, or Venus's Comb, is entirely beset with long sharp spines, which the animal has the power of dissolving and replacing by a smooth and even surface, whenever it finds it necessary, in enlarging its shell, to remove them. Others also

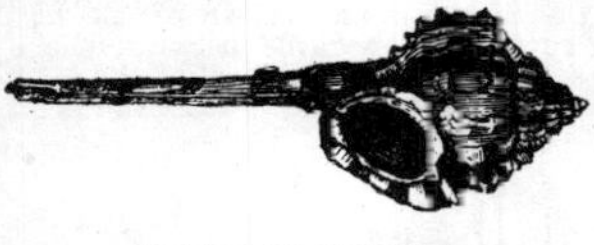

MUREX HAUSTELLUM.

have their distinguishing peculiarities, such as the Rosebud Murex, with its pink-tipped fringes; the Ducal Murex, the Royal Murex, &c.; and are much sought after by collectors.

MURIDÆ. (*Mus*, a mouse.) The name of an extensive group of Rodent animals (of the Linnæan order *Glires*) consisting of *Mice*, *Rats*, and other animals allied to them; which, though none of them attain to any considerable size, become worthy of serious notice from their prodigious multiplication, and the destructive influence they exert over vegetation and the products of the husbandman's toil. They are distinguished by their long, round, scaly tails, and by the presence of only three molars in each jaw. [See MOUSE: RAT.]

MUSCICAPIDÆ. A family of insectivorous birds which take their prey as they fly. There are a multitude of species, diffused over every quarter of the globe, which, although differing in many points of generic distinction, may be all known by their essential characteristics—a notched, depressed, and angular bill, with strong hairs surrounding the base. [See FLY-CATCHER: RHIPIDURA: TODY.]

MUSCIDÆ. A most extensive family of Dipterous insects, distinguished by having a proboscis distinct, short, thick, membranaceous, terminated by two large labial lobes, and entirely retractile within the oral cavity: the antennæ are triarticulate; the body is short and robust; the legs and wings are of moderate length; and the nerves extend to the posterior extremity of the wings. The largest known species (*Musca grossa*) is nearly as large as a Humble-bee. It is black, very bristly, with the head buff, eyes brown, and base of the wings reddish. It makes a loud buzzing, settles upon flowers in woods, and often upon cow-dung, on which its larva resides: the body of the larva is yellowish, shining, conical, with a single hook, and two fleshy horns at its anterior extremity; the other being terminated by a circular plate, upon which are two spiracles, each placed upon a reticular lobe, elevated in the middle: the segment after the head is also furnished on each side with a spiracle. —*Musca vomitoria*, the common Meat-fly, has the forehead fulvous, the thorax black and abdomen blue, with black marks. It possesses a remarkably fine sense of smelling, and makes a loud buzzing noise, when it enters our houses in order to deposit its eggs on meat. When ready to assume the pupa state, it quits its food and descends into the earth, or else undergoes its change in some dry and retired situation.—The species of *Musca domestica*, the small common house-fly, are found more especially in houses, settling upon and sipping at almost every article of food. Their larva is elongated, slightly attenuated in front, with a small head; and the terminal segment bearing a pair of spiracles, entire, and without any radiating points.—Amongst the exotic species of this family many exhibit remarkable variations of structure; as the splendid *Rutiliæ* from New Holland; the Indian genus *Celyphus*, and others.

Vincent Kollar, in his "Treatise on Insects," observes, that "the species of true flies (*Muscidæ*) deserve to be mentioned as particularly troublesome guests in our houses. Although their bites do not cause pain, still it is extremely disagreeable to feel them crawling over our faces, particularly when we are in a state of repose. But they may even be dangerous, particularly in their larva state. The larvæ or maggots feed upon animal as well as vegetable matter, particularly when it begins to decay. Open wounds, when they begin to suppurate, attract flies, and they deposit their eggs in them. In a very short time the maggots are hatched, and increase, by their sucking, the malignity and pain of the wound. They will even deposit their eggs on sound parts if they happen to be smeared with matter fit for the nourishment of their progeny." And he gives instances of cases in which much pain and suffering were occasioned by the eggs of flies having been thus introduced into the ears and nostrils of females.

MUSK, or MUSK-DEER. (*Moschus moschiferus*.) These animals, which give name to the well-known perfume, inhabit the great extent of elevated country which occupies a large part of central Asia, and are principally found in Thibet, Nepaul, Tonquin, and the districts adjacent to the north of

MUSK.—(MOSCHUS MOSCHIFERUS.)

India and China. Their favourite haunts are the tops of mountains covered with pines, where they roam in places most difficult of access, resembling in their manners the Chamois and other mountain quadrupeds. In size and general appearance the Musk-deer is not very unlike a small roebuck, the length of the body being about three feet four inches. The upper jaw is considerably longer than the lower, and is furnished on each side with a curved tusk, about two inches long, the inner edges of which are quite sharp. The general colour of the body is a kind of deep iron-gray: the ears are erect, about two inches long, of a deep brown colour externally, and pale yellow within; the hoofs long, and much divided; and the tail extremely short. These animals are hunted for the sake of their musk; which is contained in an oval receptacle, or small glandular pouch, situate at the hinder part of the abdomen, and peculiar to the male. The unctuous secretion contained in this receptacle is of the most powerful and penetrating nature; but from the ease with which it can be adulterated, very little of it reaches Europe in a pure state. The follicle containing the musk is covered with short brown hair, and is more or less full according to the age, health, &c. of the animal: the musk, when dry, is of a dark reddish brown colour; has a bitterish sub-acrid taste; and a fragrant smell, agreeable at a distance, but so strong and pungent as to be highly unpleasant when quite near. It is held in high estimation as a medicine among oriental nations.

The JAVANESE MUSK DEER. (*Moschus Javanicus.*) This animal is rather larger than a full-sized hare: body heavy; limbs very delicate: head arched and long; eyes

JAVANESE MUSK DEER.
(MOSCHUS JAVANICUS.)

large, but not expressive. Its general colour is brown mixed with gray or yellowish reflections, the yellow predominating along the back and tail, on the legs, the neck, and head.

There are other Musk-deer, which are very small, and to which the general term of *Chevrotains* is given: they are inhabitants of Java, Sumatra, Ceylon, and Southern India; and are adapted to a forest rather than a mountain life. They are timid and wild in their native haunts, but gentle and mild in captivity, and particularly elegant in their appearance and movements.

MUSK OX. (*Ovibos moschatus.*) This animal, which by some naturalists has been considered as intermediate between the sheep and ox, inhabits the more northern parts of America, where the country is mostly rocky and barren, except on the banks of the larger rivers. When they are fat the flesh is well-flavoured, but smells strongly of musk. They herd together in flocks of twenty or thirty. The Musk Ox is about the height of a deer, but of much stouter proportions. The horns are very broad at the base, covering the forehead and crown of the head; and curving downwards between the eye and ear, until about the level of the mouth, when they turn upwards. The head is large

MUSK OX.—(OVIBOS MOSCHATUS.)

and broad, and the nose very obtuse: the ears are short, and not very conspicuous. The hair of the body is in general brown: on the neck and between the shoulders it is long, matted, and somewhat curled; and this bushy state of the hair on those parts causes the animal to appear humped. On the back and hips the hair is also long, but lies even and smooth; and on the shoulders, thighs, and sides, it is so long as to hang down below the middle of the leg. The tail is so short as to be concealed in the fur. Beneath the long hair, on all parts of the animal, is a fine kind of soft ash-coloured wool, which, if it could be procured in sufficient quantity, would be highly useful to the manufacturer. The legs of the Musk Ox are short and thick, and furnished with narrow hoofs, resembling those of the Moose. The female is smaller than the male, and has also smaller horns, whose bases do not meet. Her general colour is black, except that the legs are whitish, and between the horns there is a bed of white hair intermixed with rust colour: an elevated ridge or mane of dusky hair runs along the back, and on the middle of the back is an oblong patch or bed of white hair, shorter than the rest, and which has been termed the saddle. The Musk Ox runs nimbly, and climbs hills and rocks with great ease. When pursued by the hunter, they seek for safety by instant flight; but the bulls are sometimes dangerously irascible when closely pressed. Perhaps the only specimen now in Europe is that preserved in the noble collection at the British Museum. This individual was brought by Capt. Parry from Melville Island.

MUSK-RAT, called in Canada, where it abounds, the MUSQUASH. (*Fiber zibethicus.*) This animal is about the size of a small rabbit, and of a reddish-brown colour; its feet are partly webbed; and its tail somewhat flattened. It has four very strong cutting teeth, of which those in the lower jaw are nearly an inch long: the fur on the whole body is soft and glossy, and beneath is a fine fur or thick down, as in the beaver. It has also similar instincts and dispositions;

living in a social state in the winter, in curiously constructed huts, built near the edge of some lake or river. These huts are about two feet and a half or three feet in diameter, plastered with great neatness in the inside, and covered externally with a kind of basket-work, of rushes, &c., carefully interlaced together so as to form a compact and secure guard, impermeable by water. The entrance to them is under water, for the purpose of procuring food, which consists entirely of roots and vegetables. In summer these creatures wander about in pairs, feeding voraciously on herbs and roots: at this season they become extremely fat, and are much sought after, partly for their flesh, but chiefly for their skins, which are valuable. Their odour resembles that of musk; and the skin, when taken from the body, still retains the scent. This musky odour is owing to a whitish fluid deposited in certain glands near the origin of the tail. The fur is used in hat-making.

Dr. Richardson states that the Musquashes vary considerably in size, and that though they have a strong musky smell, particularly the males, in spring, their flesh is eaten by the Indians, who prize it for a time when it is fat, but soon tire of it. They generally have three litters in a year, and from three to seven in a litter. Great numbers are destroyed by the inundations which cover the low grounds where they haunt, and in severe winters they are almost extirpated from some localities by the freezing of the swamps inhabited by them. Famine in such cases prompts them to destroy each other; and they are subject to some disease which occasionally proves fatal to vast numbers. The principal seasons for taking the Musquash are, the autumn, before the snow falls, and the spring, after it has disappeared, but while the ice is still entire. In the winter time the depth of snow prevents the houses and breathing-holes from being seen. One of the first operations of the hunter is to stop all the holes with the exception of one, at which he stations himself to spear the animals that have escaped being speared through the walls of their houses, and come hither to breathe. In the summer the Musquash burrows in the banks of the lakes, making branched canals many yards in extent; and forming its nest in a chamber at the extremity, in which the young are brought forth. When its house is attacked in the autumn, it retreats to these passages, but in the spring they are frozen up. It is, a watchful but not a very shy animal. It will approach very near a boat or canoe, but dives instantly on perceiving the flash of a gun. It may be frequently seen sitting on the shore of small muddy islands in a rounded form, and not easily to be distinguished from a piece of earth, until, on the approach of danger, it suddenly plunges into the water. There are several varieties of this animal.

MUSLIN [MOTHS]. A name applied by collectors to Moths of the genera *Psyche*, *Penthophora*, *Nudaria*, &c.

MUSOPHAGA: MUSOPHAGIDÆ; or PLANTAIN-EATERS. A genus of Scansorial birds, evidently allied to the Insessorial or Perchers. The base of the bill is enormously dilated, so as to spread like a casque or helmet over the fore part of the head as far as the crown, where its thickened sides form a semicircle. Nostrils oval, open, placed nearer to the tip than to the eyes, and pierced in the substance of the bill. The species *Musophaga violacea* here figured is a very magnificent bird. Bill rich yellow, passing into crimson; orbits naked, and, like the compact velvety feathers of the crown, glossy crimson; a white stripe

PLANTAIN-EATER.
(MUSOPHAGA VIOLACEA.)

beginning below the eye and extending above the ear; secondary and part of the primary quills carmine, margined and tipped with blackish violet, which is the general colour of the plumage, changing into a very deep green on the under parts, which is very rich on the tail; legs strong and black; gape wide. The Gold Coast and Senegal, in Africa, are its localities.

MUSSEL. (*Mytilus*.) A genus of Molluscous animals, the characters of which are, that the shell is bivalve, of an oblong triangular form, terminating in a point, and having its two extremities equal. The head of the animal is in the acute angle. The COMMON SALT-WATER MUSSEL (*Mytilus edulis*) is distinguished by a strong shell, slightly incurvated on one side, and angulated on the other; the end near the hinge being pointed, and the other rounded. Mussels abound on the rocks of our own coasts, to which they are fixed by their byssus. From the circumstance of their being always found attached to rocks, stones, or to the shells of each other, they have been supposed by many to be incapable of progressive motion; but although they have no tendency to change of place, they seem possessed of a certain degree of locomotive power; and their manner of exerting it has been examined and well explained by Reaumur. He discovered that their mode of progression consisted in thrusting their tongue-like foot out of the shell, curving it, hooking it to some adjacent body, and thus drawing themselves forward to the point of attachment.

Although Mussels commonly afford a supply of wholesome food, they sometimes (in spring) acquire very poisonous properties;

and many persons have been suddenly attacked with violent symptoms after eating them. It frequently happens, indeed, with some constitutions, after partaking of certain kinds of shell-fish, that intolerable itchings all over the body take place, accompanied by great restlessness and agitation, and followed by cutaneous eruptions.

MUSTELA: MUSTELIDÆ. A genus and family of carnivorous Mammalia, distinguished by the length and slenderness of their bodies. The characters of this genus are: six cutting-teeth in each jaw, the upper being erect, acute, and separate; the lower more obtuse; the tongue smooth. [See WEASEL.]

MUTILLIDÆ. A family of Hymenopterous insects, generally found in hot sandy situations, and bearing considerable resemblance to other sand-wasps. They are solitary in their habits: the males occasionally frequent flowers; but the females are always found on the ground, and they run with great quickness, secreting themselves, on the approach of danger, amongst grass and under stones. The antennæ are filiform or setaceous, the first and third joints being elongated; the labrum is transverse and ciliated; the mandibles notched; and the body often very much clothed with hair. The females are destitute of wings and ocelli, but they are provided with a powerful sting.

MYA. A genus of Molluscous animals, inclosed in a bivalve shell. The Myæ are to be found both in the ocean and in rivers: the marine kinds generally burrow in the sand, and those which inhabit rivers are generally found in the mud. They are of considerable importance, in consequence of the shell sometimes producing a quantity of pearls; and the shell is well known by the name of the Pearl Mussel. One end of this shell gapes considerably, and at the aperture the thick epidermis is lengthened out into a tube, which can be withdrawn by the animal at pleasure. It is found on the shores of the European, Asiatic, and African seas; and in several places it is used as food; it is also devoured by various aquatic birds. According to Camden, Sir John Hawkins had a patent for fishing for it in the river Irt, in Cumberland. There was also a great fishery for pearls in the river Tay, which extended from Perth to Loch Tay; and it is said that the pearls sent from thence, from the year 1761 to 1764, were worth 10,000*l.* At the present day it is not uncommon to find pearls in these shells which bring from 1*l.* to 2*l.*

MYCETES. A genus of *Quadrumana* inhabiting the American continent, and popularly called *Howling Monkeys.* They are distinguished by a pyramidal head, the upper jaw of which descends much below the cranium, while the branches of the lower one ascend very high, for the purpose of lodging a bony drum, formed by a vesicular inflation of the hyoid bone, which communicates with their larynx, and imparts to their voice prodigious volume and a most frightful sound. Hence the appellation bestowed on them. They are shaggy animals, about the size of a fox, of different shades of brown or blackish; the females carry their young upon their shoulders, and some of them are differently coloured from the males: these

HOWLING MONKEY.—(MYCETES URSINUS.)

are of a social disposition and grave deportment; and most of them have thick beards. They utter their hideous yells and howling by night; and subsist on fruits and foliage.

MYCETOPHILIDÆ. A subfamily of Dipterous insects, of small and active habits. They are found in damp situations, amongst various plants; and many of the species enter our houses, and are found on the windows. They are capable of leaping by means of their hind legs; and are distinguished by having two or three unequal-sized ocelli; eyes generally round; head not rostrated; the antennæ slender, and never fasciculated.

MYCTERIA. A genus of Grallatorial Birds allied to the Storks, of which there are several species: the best known is the *M. Americana*, or common Jabiru. [See JABIRU.]

MYLABRIS. A genus of Vesicatory Beetles (*Cantharidæ*). The head is large, broad, and rounded behind; the thorax nearly orbicular; and the elytra slightly inclined at the sides. They have long an-

MYLABRIS CICHORII.

tennæ, with eleven distinct joints in both sexes. This genus abounds in species, Africa, and Asia being the chief countries where they are found. *Mylabris cichorii*, the species here figured, inhabits the south of Europe;

and its vesicatory properties are as powerful as the *Cantharis* of the shops, with which it is said to be mixed in Italy.

MYLODON. A gigantic animal which has long since become extinct.

The teeth of the Mylodon are eighteen in number, five on each side above and four below: they are simple, long, fangless, of uniform substance and nearly straight, with the exception of the first tooth in the upper jaw, which is slightly curved. From its dentition, therefore, and the peculiar conformation of the jaws, it is concluded that it fed on the leaves or slender terminal twigs of trees, in this respect resembling the giraffe, the elephant, and the sloth. The extraordinary stature of the giraffe raises its mouth with ease to its food; the trunk of the elephant conveys the food to its mouth; and the comparatively light weight of the sloth enables him to run along the under side of the boughs till he has reached a commodious feeding-place: but the Mylodon and his congeners had short and massive necks, and were as bulky as the Rhinoceros; so that it is apparently impossible they could obtain their food in the same manner as either of the animals we have mentioned. In his analysis of the osteological structure of the Mylodon, Professor Owen, after alluding to its very perfect clavicles, which have been alternately received as evidence of the burrowing and climbing hypothesis, does not admit them to be necessarily essential to those qualities, since the bear and the badger, the one a climbing and the other a burrowing animal, are perfectly destitute of them: but from a comparison of the hand of the Mylodon with that of certain ant-eaters, he infers that it was an instrument employed in digging or removing the earth. The great bulk of the posterior extremities, and the corresponding excess of muscular power, as shown by the spinal crest of the sacrum, he regards as farther evidence against the climbing theory; while he believes that the enormous tail formed a tripod with the hind legs, which could well support the weight of the animal, and leave the anterior limbs at liberty. "If the foregoing physiological interpretation of the osseous frame-work of the gigantic extinct sloths be the true one," says Mr. Owen, "they may be supposed to have commenced the process of prostrating the chosen tree by scratching away the soil from the roots; for which office we find in the Mylodon the modern scansorial fore-feet of the sloth modified after the type of that of the partially fossorial ant-eater. The compressed or subcompressed form of the claws, which detracts from their power as burrowing instruments, adds to their fitness for penetrating the interspaces of roots, and for exposing and liberating them from the attached soil. This operation having been duly effected by the alternate action of the fore-feet, aided probably by the unguiculate digits of the hind feet, the long and curved fore-claws, which are habitually flexed and fettered in the movements of extension, would next be applied to the opposite sides of the loosened trunk of the tree: and now the Mylodon would derive the full advantage of those modifications of its fore-feet by which it resembles the Bradypus; the correspondence in the structure of the prehensile instruments of the existing and extinct sloths, extending as far as was compatible with the different degrees of resistance to be overcome. In the small climbing sloth the claws are long and slender, having only to bear the weight of the animal's light body, which is approximated by the action of the muscles towards the grasped branch, as to a fixed point. The stouter proportions of the prehensile hooks of the Mylodon accord with the harder task of overcoming the resistance of the part seized and bringing it down to the body. For the long and slender branchial and anti-branchial bones of the climbing sloth we find substituted in its gigantic predecessor a humerus, radius, and ulna of more robust proportions, of such proportions, indeed, in the *Mylodon robustus* as are unequalled in any other known existing or extinct animal. The tree being thus partly undermined and firmly grappled with, the muscles of the trunk, the pelvis, and hind limbs, animated by the nervous influence of the unusually large spinal chord, would combine their forces with those of the anterior members in the efforts at prostration. And now let us picture to ourselves the massive frame of the Megatherium, convulsed with the mighty wrestling, every vibrating fibre reacting upon its bony attachment with a force which the sharp and strong crests and apophyses loudly bespeak: —extraordinary must have been the strength and proportions of that tree, which, rocked to and fro, to right and left, in such an embrace, could long withstand the efforts of its ponderous assailant.

[For some additional particulars, accompanied by a figure, see MYLODON ROBUSTUS in SUPPLEMENT.]

MYOCHAMA. A genus of Mollusca, of which only one species is known (the *M. anomioides* of New South Wales), described by Mr. Sowerby as — "inequivalve, irregular, attached, sub-equilateral; attached valve flat, with two marginal, diverging teeth, and one end of a little testaceous appendage fixed between them by a horny cartilage; free valve convex, with umbo incurved, and two very minute diverging teeth, between which the other end of the testaceous appendage is placed; external surface of both valves conforming to the grooves or undulations in the shell to which the specimen is attached: muscular impressions two in each valve; palleal impression with a short sinus."

MYOXUS. [See DORMOUSE.]

MYRAPETRA. A genus of Hymenoptera, which constructs a singular nest. [See WASP.]

MYRIAPODA. The name given to the lowest class of articulated animals; included by some naturalists among the insects, and bearing considerable affinity to them; but differing from that large class in the absence of wings, and in the body being composed of an extensive series of segments, each provided with a pair of legs. [See IULUS: CENTIPEDE: CHILOPODA: CHILOGNATHA: SCOLOPENDRA.]

MYRMECOBIUS. A genus of marsupial animals which feed on ants. *Myrmecobius fasciatus*, the only known species, is a native of Australia: it is formed like a squirrel, and is of the size of a rat: has fifty-two teeth; and is marked on the lower part of the back with white bands on a reddish ground tint.

MYRMECOPHAGA. The name of a genus of edentate quadrupeds. [See ANT-EATER.]

MYRMELEON; MYRMELEONIDÆ. A genus and family of Neuropterous insects, one species of which, the *Myrmeleon formicaleo*, or ANT-LION, has been described under its well-known English name.

MYSIS, or OPOSSUM SHRIMP. (*Mysis vulgaris.*) This curious little Crustacean bears, in its general form, a strong resemblance to the ordinary Shrimps; it is, however, distinguished from the true Decapoda by the external position of its branchiæ, &c. In regard to the number of the feet, it holds an intermediate place between the Decapods and ordinary Stomapods. Each of the legs has a natural appendage, so much developed as to make the limb appear bifid: and thus, including the feet-jaws, which also possess similar appendages, the Opossum Shrimp may be said to have thirty-two legs. The

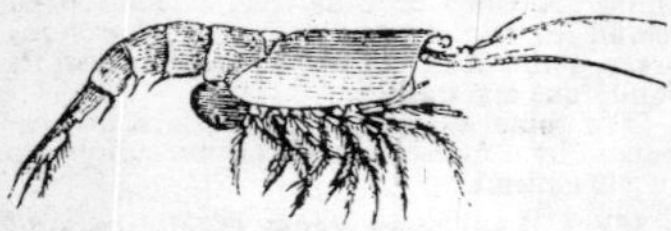

OPOSSUM SHRIMP.—(MYSIS VULGARIS.)

common name of this Crustacean is derived from the peculiar conformation which enables it to afford a special protection to the eggs. Attached to the inner division of each of the posterior legs the female has a large concave scale: and thus a pouch is formed, which is capable of considerable extension. Here the eggs are received when they quit the ovarium, and here the young remain till their form is fully developed; when the parent opens the valves of the pouch, and liberates the whole brood at once. These are the chief crustacea which inhabit the Arctic Ocean in such amazing numbers as to constitute the principal food of the Whalebone Whale, and to support the prodigious shoals of Salmon which resort thither in the months of July and August.

MYTILACEÆ. An order of Mollusca, of which the common *Mussel* furnishes an example.

MYTILUS. [See MUSSEL.]

MYZANTHA. A genus of Australian birds, belonging to the family *Meliphagidæ*. One species is the *M. Garrula*, or GARRULOUS HONEY EATER, called the *Miner* by the colonists in Van Diemen's Land: it moves about in small flocks of from four to ten in number. Mr. Gould tells us that it is very restless and inquisitive if its haunts be intruded upon; "no sooner does the hunter come within the precincts of its abode than the whole tribe assemble round him and perform the most grotesque actions, spreading out their wings and tail, hanging from the branches, and keeping up all the time

AUSTRALIAN BELL-BIRD.
(MYZANTHA MELANOPHRYS.)

one incessant babbling note:" by following up the intruder in this way, "they become very troublesome and annoying, awaking as they do the suspicions of the other animals of which you are in pursuit." It feeds among the branches of the Eucalypti, from the pollen of the flowers of which it obtains abundance of genial food; but it also preys with avidity on insects. Its nest is cup-shaped, and very neatly built of fine twigs and coarse grass, and lined with feathers, and it is about the size of that of the common Thrush (*Turdus musicus*). Another pretty yellow-olive species peculiar to New South Wales is the AUSTRALIAN BELL-BIRD (*Myzantha melanophrys*), figured in our cut from Mr. Gould's fine work. The note of this is peculiar, and from it the colonists have given the species the name of *Bell-bird;* the sound having been compared, and justly, to the sound of distant sheep-bells: and when this is poured forth from a hundred throats it produces a most singular effect. The Bell-bird of Demerara is quite another bird. [See CAMPANERO (*Procnias carunculata*).] By some naturalists the generic name given to these birds is *Manorhina*. Other species will be found figured in the great work of Mr. Gould.

NAIDES. (*Nais.*) These are small semi-aquatic worms, of the order *Terricolæ*, closely allied to the Earthworms, but having the elongated body, and the rings less marked. They live in holes which they perforate in mud at the bottom of water, and from which they protrude the anterior portion of the body, incessantly moving it. Some have black points upon the head, which have been regarded as eyes. Many species exist in our fresh waters; and their reproductive power is not less astonishing than that of the Hydra or Polypus. Some have very long bristles; others a long protrusile trunk; and several have small tentacles at the hind extremity.

NAKOO. One of the native names of the Narrow-beaked Crocodile of India. (*Gavialis gangeticus.*)

NARWHAL. (*Monodon monoceros.*) This extraordinary marine animal, which is also known under the name of the SEA UNICORN, belongs to the *Cetacea*, but differs from every other kind of Whale by having no teeth, *properly so called*, and in being armed with a formidable horn, projecting directly forward from the upper jaw, in a straight line with the body. This horn is from six to ten feet long, spirally striated throughout its whole length, and tapering to a point: it is harder and whiter than ivory, for which article it was at one time not only substituted, but was also in high repute for its *supposed* medicinal powers. The Narwhal is generally from twenty to thirty feet in length from the mouth to the tail; sometimes much more; and it is occasionally, though not very often, found with two of these horns, or tusks, sometimes of equal length, and sometimes very unequal. The head of this animal is short, and convex above; the mouth small; the spiracle or breathing-hole duplicated within; the tongue long; the pectoral fins small; the back finless, convex, and rather wide; becoming gradually accuminated towards the tail, which, as in other Whales, is horizontal. The skin is darkly marbled on the back, lighter on the sides, and nearly white on the belly: it is quite smooth, and there is a considerable depth of oil or blubber beneath. The Narwhal chiefly inhabits the arctic seas; and its food is said to consist of the smaller kinds of flat-fish, medusæ, and other marine animals. It is taken by means of the harpoon; and its flesh is eaten by the Greenlanders. Although both swift and strong, as well as being armed with such a prodigious weapon, the Narwhal is one of the most peaceable inhabitants of the ocean.

We have the authority of Mr. Bell for stating that in the general form of the body, in the obtuse and rounded head, its small gape, its finless back, and in the form and structure of the cranium — the Narwhal approaches very near to the Beluga. Of its *tooth* or *horn*, this gentleman thus speaks: "To what extent the ascribed power of the tooth may be true, we have but little means of ascertaining; but there is the structural evidence of its form, and its extraordinary development, to indicate that there must be some especial use for so long and sharp and powerful a weapon; and really there seems no reasonable ground for assigning to it any other object than that formerly attributed to it by the ignorant, — namely, that of defence. In this respect, it forms, indeed, an additional instance to numerous others, of gregarious animals, to the males of which alone belongs such a development of the teeth or the horns as shall constitute them the natural defenders of the herd. The elephant, the wild boar, and even the horse, offer examples of the former, and the antelopes and deer of the latter; and there can be no doubt that the restriction of this weapon to the males in the Narwhal has a similar object." "It would be a strange anomaly were the apparent singleness of this weapon real; but the truth is, that both the teeth are invariably formed in the jaw, not only of the male, but of the female also — but that in ordinary cases one only, and this in the former sex, is fully developed, the other remaining in a rudimentary condition, as is the case with both in the female."

NASALIS. A genus of monkeys, containing the curious Bornean long-nosed Monkey. [See PROBOSCIS MONKEY.]

NASSA. A genus of Mollusca, inhabiting a small globular or oval shell, according to the spire, which in some is short, and in others long; mouth oblong, notched; inner lip thickened, and spread out, occasionally very large; right lip often wrinkled; operculum horny. By some authors this genus is united to *Buccinum*, on account of the little tooth-like projection terminating the columella. The head of the animal is large, the proboscis short, or altogether wanting; two tentacula, with eyes in the middle; foot very large. They abound in the South of Europe, and some are occasionally seen on our own coasts. They may sometimes be seen feeding on the *Mactra*, which they effect by piercing the shell with their proboscis, and extracting the contents through the small round aperture which by this means they have formed.

NASUA. A genus of Plantigrade quadrupeds, distinguished by the elongation and upward curve of the snout, which the animals belonging to this genus have the power of turning about, and uprooting the earth, when in search of worms, &c. [See COATIMONDI.]

NATATORES. The name given to an Order of birds, viz. those which are *web-footed*, and otherwise adapted for an aquatic life. This order includes five families; the *Anatidæ*, or Ducks; the *Colymbidæ*, or Divers; the *Alcidæ*, or Auks; the *Laridæ*, or Gulls; and the *Pelicanidæ*, or Pelicans. The form and size of the wings, and the powers of flight, vary greatly in the different tribes of this order; but they are all most easily distinguishable from the rest of the feathered race on account of the peculiar structure and position of their feet; the toes being invariably connected together by a membrane, and the legs placed behind the equilibrium of the body, so as to be more efficient instruments for its propulsion in the water. The body is also covered with a thick coat of down beneath the feathers; and the plumage is oiled by a secretion of certain glands near the tail, so that the water runs off without scarcely wetting the surface. Their food consists chiefly of fish, mollusca, and insects. They live much more upon the water than on land; and they resort to the shore chiefly for the purpose of building their nests and rearing their young.

NATICA. A genus of Mollusca, the shell of which is globose, thick, and generally smooth; spire short, pointed, and with few volutions; outer lip thin; inner lip and inside smooth; operculum shelly in some species, horny in others; epidermis thin, light, and transparent. The head of the

NATICA PLUMBEA.

animal is very large, having two tentacula with eyes at the base; foot large and thin. The straight, callous, smooth edge of the columella serves to distinguish this genus from Nerita, Helix, &c. There are very many recent marine species, and not a few fossil.

NATRIX. A genus of *Colubridæ*, a family of snakes destitute of poison-fangs; of which our common harmless snake (*Coluber natrix*) is a type. [See SNAKE.]

NATTER-JACK. (*Bufo calamita.*) The English name of a species of Toad, of a lightish yellow colour, inclining to brown, and clouded with dull olive; but its most distinguishing mark is a bright yellow line running down the middle of the back. It never leaps, nor does it crawl with the slow pace of a toad, but its motion is more like running. They are found in considerable numbers near stagnant pools and ditches, where they congregate for the purpose of breeding; and their hoarse voices are heard at a great distance.

NAULTINUS. A genus of Lizards, containing four or more generally green coloured species, natives of New Zealand. They are allied to the Geckos.

NAUTILITES. The name given to numerous chambered shells existing in a fossil state, nearly resembling the Nautilus, above described, and which are found in almost all marine strata, from the oldest limestones and sandstones of the Silurian system, down to those overlying the chalk.

NAUTILUS. (*Nautilus Pompilius.*) The Pearl Nautilus, so named from the nacreous lining of its shell, belongs to a genus of *Tetrabranchiate Cephalopods;* but though the shell of this animal is well known, being found in the seas of most tropical latitudes, the most vague and incorrect ideas were, until lately, formed of its living inhabitant:

NAUTILUS AND SECTION OF SHELL.

we believe, indeed, it was only in 1829 that this animal was known with any certainty, one having been caught alive by Mr. G. Bennett, near the New Hebrides Islands; which, preserved in spirits, is now in the museum of the College of Surgeons. The Nautilus is very rarely met with in the living state, owing to its being an inhabitant of the open sea, and possessing the power of sinking at the slightest alarm. Externally the shell presents nothing remarkable, being a flattened spiral; but on examining its interior, we find it divided into chambers, by a large number of transverse partitions of shelly matter; sometimes as many as thirty or forty separate chambers or divisions, each communicating with the rest by a small tubular hole near the centre. The opening or mouth of the shell therefore presents a large but shallow concavity, pierced with a central or nearly central hole; while beyond it lie all the divisions adverted to. The outer chamber is by far the largest, and to this the body of the animal is restricted; but it maintains a connection with the rest by means of a membranous tube, called the *siphuncle*, which passes through the centre of each partition, and thus penetrates even to the innermost and smallest chamber. These animals are furnished with numerous tentacula, short, slender, and unprovided with suckers. They usually remain at the bottom of the water, and are able to creep along rather quickly, supporting themselves upon their tentacula, with the head downwards, and the shell raised above. After stormy weather, as it becomes more calm, they may be seen, in great numbers, floating upon the surface of the waves, with the head put out, and the tentacula resting upon the water, the shell at the same time being undermost: they remain, however, but a short time sailing in this manner, as they have the power of easily returning to their situation at the bottom of the sea, by merely drawing in their tentacula and upsetting the shell.

NAVICELLA. A genus of fluviatile Mollusca, inhabiting the clear rivers of India, the Isle of France, &c. The shell is transversely oval; dorsal surface convex; with the apex straight and bent down to the edge, not spiral; operculum testaceous, flat, subquadrate, with a lateral articulation; the shell, indeed, altogether much resembling a *Patella*. The animal is distinguished by a large head, having two tentacula, the eyes placed on the summit of two small protuberances at their base; foot large: they creep well on the rocks, and do not continue fixed to one spot.

NAXIA. A genus of short-tailed Decapod Crustacea, containing some singular spine-fronted species of Crabs, found in the Eastern Seas.

NEBALIA. A genus of singular Crustacea belonging to the Entomostraca, order *Phyllopoda*, and containing two or more interesting British species.

NECROPHAGA. The appellation given by Latreille to an extensive group of Coleopterous insects, highly serviceable in removing the decaying remains of animal matter and such kinds of impurities. According to Mr. Westwood's definition, they are "chiefly distinguished by having the antennæ gradually or suddenly thickened at the tips; the mandibles generally robust and exserted; the maxillæ with the outer lobe large, but not palpiform nor articulated; the maxillary palpi with the basal joint often small; the body often oval or oblong, with the prosternum not anteriorly produced; the elytra sometimes shorter than the abdomen; the legs formed for running, and not contractile." Though the Necrophaga in general subsist on the decaying remains of animal matter, some of the species feed upon decaying fungi and other vegetable remains; while in others are to be traced vestiges of those predaceous habits which characterize another group of beetles. [See next *Art.*]

NECROPHORUS; or SEXTON-BEETLES. A genus of Coleoptera belonging to the preceding group, and containing several species found in Europe and North America principally; though some are found also on the mountains of South America and Asia. The habits of all the species are believed to be similar to the example referred to beneath. Our figure represents the *Necrophorus vespillo*, perhaps the first species on which observations were made. It has

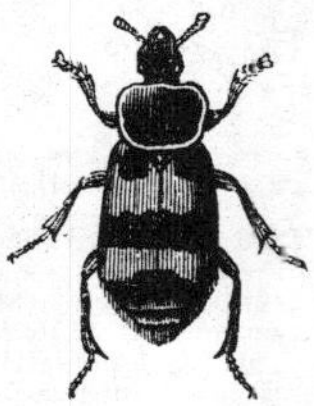

THE SEXTON OR BURYING BEETLE. (NECROPHORUS VESPILLO.)

the elytra red, and banded with black. From Mr. Newman's interesting History of Insects we extract, as a good summary of the habits of the genus, his account of the habits of the GREAT BLACK SEXTON BEETLE (*Necrophorus germanus*).—"It is about an inch in length, of a black colour, and so fetid that the hands smell for hours after handling it; and if it crawl on woollen clothes which are not washed, the smell continues for several days. The Sexton Beetle lays its eggs in the bodies of putrefying dead animals, which, when practicable, it buries in the ground. In Russia, where the poor people are buried but a few inches below the surface of the ground, the Sexton Beetles avail themselves of the bodies for this purpose, and the graves are pierced with their holes in every direction; at evening, hundreds of these beetles may be seen in the church-yards, either buzzing over recent graves, or emerging from them. The Sexton Beetle in this country seldom finds so convenient a provision for him, and he is under the necessity of taking much more trouble; he sometimes avails himself of dead dogs and horses, but these are too great rarities to be his constant resort; the usual objects of his search are dead mice, rats, birds, frogs, and moles; of these a bird is most commonly obtained. In the neighbourhood of towns, every kind of garbage that is thrown out attracts these beetles as soon as it begins to smell, and it is not unusual to see them settling in our streets, enticed by the grateful odour of such substances. The Sexton Beetles hunt in couples, male and female; and where six or eight are found in a large animal, they are almost sure to be males and females, in equal numbers; they hunt by scent only, the chase being mostly performed when no other sense would be very available, viz. in the night. When they have found a bird, great comfort is expressed by the male, who wheels round and round above it, like a vulture over the putrefying carcass of some giant of the forest,—the female settles on it at once, without this testimonial of satisfaction; the male at last settles also, and a savoury and ample meal is made before the great work is begun. After the beetles have appeased the calls of hunger, the bird is abandoned for a while; they both leave it to explore the earth in the neighbourhood, and ascertain whether there is a place suitable for interment: if on a ploughed field there is no difficulty; but if on grass, or among stones, much labour is required to draw it to a more suitable place. The operation of burying is performed almost entirely by the male beetle, the female mostly hiding herself in the body of the bird about to be buried, or sitting quietly upon it, and allowing herself to be buried with it: the male begins by digging a furrow all round the bird, at the distance of about half an inch, turning the earth outside; his head is the only tool used in this operation; it is held sloping outwards, and is exceedingly powerful. After the first furrow is completed, another is made within it, and the earth is thrown into the first furrow; then a third furrow is made, and this is completely under the bird, so that the beetle whilst working at it is out of sight: now, the operation can only be traced by the heaving of the earth, which soon forms a little rampart round the bird: as the earth is moved from beneath, and the surrounding rampart increases in height, the bird sinks. After incessant labour for about three hours the beetle emerges, crawls upon the bird, and takes a survey of his work. If the female is on the bird, she is driven away by the male, who does not choose to be intruded on during the important business. The male beetle then remains for about an hour perfectly still; he then dismounts, dives again into the grave, and pulls the bird, down by the feathers for half an hour; its own weight appears to sink it but very little. At last, after two or three hours' more labour, the beetle comes up, again gets on the bird, and again takes a survey, and then drops down as though dead, or fallen suddenly fast asleep. When sufficiently rested, he rouses himself, treads the bird firmly into its grave, pulls it by the feathers this way and that way, and having settled it to his mind, begins to shovel in the earth; this is done in a very short time, by means of his broad head. He goes behind the rampart of earth, and pushes it into the grave with amazing strength and dexterity; the head being bent directly downward at first, and then the nose elevated with a kind of jerk, which sends the earth forwards. After the grave is thus filled up, the earth is trodden in, and undergoes another keen scrutiny all round, the bird being completely hidden; the beetle then makes a hole in the still loose earth, and having buried the bird and his own bride, next buries himself. The female having laid her eggs in the carcass of the bird, in number proportioned to its size, and the pair having eaten as much of the savoury viand as they please, they make their way out, and fly away." The eggs are quickly hatched, and when the grubs become perfect insects, they make holes in the ground, and come forth.

NECTARINIADÆ. A family of Passerine birds, comprising the Honey-suckers, all of which are foreign. They are distinguished by a beak of medium length, arched, pointed, and compressed; but they neither use the tail, nor climb. Some of the smaller species have a very vivid plumage. They are natives of Africa and Asia for the most part.

NEGRO-FLY. (*Psila rosæ.*) This Hemipterous insect, which is sometimes called the Carrot-fly, in its perfect state is slightly haired, shining black, rather of a metallic green. The head is reddish yellow, antennæ and palpi with black tips. Legs light yellow; balancers white; and wings clear like glass. It is found throughout the summer. The larva lives in the carrot, where it eats passages; it is found particularly near the extremity of the main root. The carrots die off by degrees, as they cannot draw sufficient nourishment from the fibrous roots. When carrots have been attacked by this insect, they lose their sweet taste, and become rusty, so called from the rusty colour assumed by the passages of the maggots. The larva of the Negro-fly is cylindrical, pointed anteriorly, like parchment, shining, smooth, bare, pale yellow; the anal joint is rounded, having posteriorly above two black, rather elevated spiracular plates, the latter having a sharp point at the end. Leaving the carrot, the larva is transformed in the earth into a small light brown, obliquely impressed, little oval mass; the short, roundish head end of which is obliquely truncated, and rather hollowed out above. At the anal end, the two spiracular plates of the larva form two small tail points. The only way to diminish their numbers is to pull up the sickly infested carrots, which are distinguishable by their yellow outer leaves, and early withering; and to destroy the insects contained in them before they change into pupæ.

NEMATURA. A genus of Mollusca belonging to the family *Turbinacea*. The shell is thin and nearly oval, somewhat compressed from back to front; spire acute, consisting of few rounded whorls, the last being large, but contracted near the aperture; operculum spiral, horny, with few volutions.

NEMEOBIUS. A genus of Diurnal Lepidoptera, which contains one British species, the NEMEOBIUS LUCINA; or DUKE OF BURGUNDY BUTTERFLY. This small indigenous species is somewhat local in its haunts, though not by any means rare. The upper surface of all the wings is obscure brown, irregularly spotted with fulvous, disposed transversely, the base of the wings being immaculate, and a central black dot being surrounded by an outer row of spots: the anterior

DUKE OF BURGUNDY BUTTERFLY
(NEMEOBIUS LUCINA.)

wings beneath are paler than the upper surface, with two ranges of fuscous spots towards the tip: the posterior wings beneath are deep fulvous, with two rows of white spots, and a marginal striga of black dots: the cilia on both upper and lower surfaces are white, interrupted with fuscous: the antennæ and upper part of the body dusky. The Caterpillar is said to feed on grasses; but neither this nor its chrysalis appears to be well known.

NEOMORPHA. A genus of birds allied to Epimachus, of which the only known species is NEOMORPHA GOULDII, a native of New Zealand, which, according to Dr. Dieffenbach, is confined to the hills near Port Nicholson, whence the feathers of the tail are in great request among the natives, who send them to all parts of the island. The straight and stout-beaked bird is regarded as the male; the slender curved-billed as the female. The natives entice them by a shrill and long-continued whistle. Their food consists of seeds and insects. James Pomare, the New Zealand boy who accompanied Mr. Angus, had a tail of this bird in his hair. The plumage is deep black; the tip of the tail white; the beak horn coloured; wattle rich orange.

NEPHROPS. A genus of long-tailed Crustacea allied to the lobster, and containing a species (*N. Norwegicus*), occasionally brought to the London markets.

NEREIDEÆ. A family of *Dorsibranchiate Anellidæ*, of which the genus *Nereis* is the type. They have an even number of tentacula attached to the sides of the base of the head, two other biarticulated ones a little more forward, and between these two simple ones. Their branchiæ consist of little laminæ, traversed by a network of vessels; each foot is furnished with two tubercles, two bundles of bristles, and a cirrhus above and beneath. A great number of species inhabit our coasts.

One species, the *Nereis prolifera*, exhibits, a singular peculiarity in its mode of propagation, merely by spontaneous division, the hind part of the body being gradually transformed into an additional animal, the head and tentacular cirrhi being already developed. [See DORSIBRANCHIATA.]

NERINEA. A genus of Mollusca, family *Canalifera*, only found in a fossil state, and not resembled by any other. It is oblong, turreted, and consists of numerous whorls; aperture with a strong fold on the columella, one on the outer lip, and one on the inner lip at the edge of the body whorl.

NERITACEA, or NERITIDÆ. A family of the first order of *Trachelipoda*, containing the genera Navicella, Natica, Nerita, and Neritina. The shells constituting this family are chiefly distinguished by the smallness and flatness of the spire, which appears placed on one side; the mouth is generally semicircular, being half closed by a flat partition, which forms the left lip.

NERITA. A genus of marine Mollusca, inhabiting the Eastern and American seas, the West Indies, Moluccas, &c. The shell is thick, smooth or ribbed, semiglobose; spire short, consisting of few volutions; aperture large, semicircular; inner lip flattened, and frequently toothed, as well as the outer, the operculum horny, covered with shelly laminæ. One species (*Nerita pelodonta*) is called the Bleeding Tooth, from the red appearance of the teeth on the inner lip. The head of the animal is furnished with two pointed tentacula having eyes at the base; foot large. There are about thirty species recent, and ten fossil.

NERITINA. A genus of fresh-water Mollusca, found in the East and West Indies, the Isle of France, &c. The shells are prettily marked, and are considered sufficiently handsome to be often worn as ornaments by the Indians. The shell is thin, semi-globose, obliquely oval, smooth, and rather flat in front; spire somewhat depressed, and consisting of few rapidly increasing whorls; aperture semicircular; columellar lip broad, flat, its inner edge straight, denticulated; right lip destitute of teeth; animal, head large, having two tentacula, with eyes at the base; foot short. Many of the species are covered with an epidermis; and some of the genus are found in the rivers of England adhering to stones.

NESTOR. A genus of the Parrot family, containing the NESTOR PRODUCTUS, or PHILIP ISLAND PARROT. It appears that this species of the genus *Nestor* has a very limited habitat, the entire race, as Mr. Gould was credibly informed, being confined to Philip Island, whose whole circumference is not more than five miles in extent. In consequence, therefore, of the war of extermination that has been carried on against it since the settlement of Norfolk Island, it would seem that the time is not far distant, when, like the Dodo, its skin and bones will become the only mementos of its existence. It is found among the rocks and upon the loftiest trees of the island; is easily taken; and, like the rest of the *Psittacidæ*, bears captivity remarkably well. In its wild state it feeds upon the blossoms of the white-wood tree, or white *Hibiscus*, sucking the honey of the flowers. A knowledge of this circumstance induced Mr. Gould "to examine the tongue of the bird, which presented a very peculiar structure, not, like that of the true honey-feeding Parrakeets (the *Trichoglossi*), furnished with a brush-like termination, but with a narrow horny scoop on the under side, which, together with the extremity of the tongue, resembled the end of a finger with the nail beneath instead of above; this peculiarity in the structure of the organ is doubtless indicative of a corresponding peculiarity in the nature of the food upon which the bird subsists." The general colour of the plumage is brown above, the head and back of the neck tinged with gray; cheeks yellow, tinged with red; throat and chest yellow; tail feathers banded at the base with orange-yellow and brown; rump, belly, and under tail-coverts deep red: bill and feet dark olive-brown.

Notwithstanding what we have stated above, on the authority of Mr. Gould, as to the habitat of this bird being confined to Philip Island, we believe that the *Kaka* of New Zealand, described by the Rev. W. Yate, is the identical species. He says, "This bird feeds upon all kinds of fruit, berries, and farinaceous roots. It bites holes in trees, in which it makes its nest; laying four, and sometimes five, eggs, perfectly white. Generally, three of these birds are found together in the same hole, one male and two females; and during the season of incubation, the nests, though separated, are so close together, that either of the mother-birds can sit upon the eggs, feed their neighbour's young, and cover them with one of her wings, without leaving her own nest, or neglecting her own offspring." It is much larger than any other New Zealand Parrot; but possessing all their mischievous qualities, and capable of learning to imitate the human voice to an astonishing degree; but when ranging at large in the woods, its cry is harsh and disagreeable in the extreme.

NETTAPUS. A genus of web-footed birds, allied to the Bernacle Geese, but of small size; it contains the COROMANDEL TEAL (*N. Coromandelianus*), a well-known native of India, and the MADAGASCAR TEAL (*N. auritus*), a native of Africa: we may particularly refer to the NETTAPUS PULCHELLUS, or BEAUTIFUL PIGMY GOOSE. The male of this small and elegant species of the genus

Nettapus has the head brownish-green, indistinctly barred with light brown; beneath the eye an oval spot of white; neck, back, and wings, deep glossy green; primaries black; outer webs of the secondaries snow white; feathers of the chest, sides, and back of the neck white, with a number of greenish black circles one within the other, so numerous that the white is nearly lost; flanks similarly marked, but bolder; tail black, glossed with green; abdomen white; under tail-coverts black; bill dark greenish gray; legs and feet blackish brown. The female differs from the male in being destitute of the white spot beneath the eye; in having the crown, occiput, and a stripe down the back of the neck dark brown; in having the chin and upper part of the throat white, speckled with brown. Mr. Gould, to whose 'Birds of Australia' we are indebted for the description, says that it is an extremely shy species, and at the slightest movement of anything near it, dives and remains under water a long time. His specimens were shot at Port Essington.

NETTLE [BUTTERFLY]. A name given by collectors to Butterflies of the species *Vanessa urticæ*.

NETTLETAP [MOTHS]. A name given by collectors to Moths of the genus *Simæthis*.

NEUROPTERA. [Dragon-flies, Lace-winged Flies; May-flies, Ant-lion, Day-fly, White Ants, &c.] One of the Orders into which the class *Insecta* is divided. The *Neuroptera* are distinguished by having four wings, each pair being membranous and transparent; the nervures forming a very beautiful and minute network, subdividing and uniting again, so as to divide the whole surface into a great number of minute cells. The antennæ are in general setaceous; the mouth is usually furnished with mandibles and maxillæ; and the abdomen is unprovided with a sting. The larvæ have six legs, and are very active. Some insects of this order merely pass through a semi-metamorphosis; others a complete one. Although the posterior wings are usually as large as the anterior, they are occasionally much smaller, and may even be altogether wanting. The Dragon-fly and May-fly are familiar examples of this order. — The White Ants, Wood-lice, and Wood-ticks (*Termitidæ* and *Psocidæ*), the latter including also the *Anobium* or Death-watch, are almost the only noxious insects in this order, and even these do not injure living plants. The Dragon-flies (*Libellulidæ*) prey upon gnats and mosquitos; and their larvæ and pupæ, as well as those of the Day-flies (*Ephemeridæ*), and those of some of the May-flies, called Caddis-worms (*Phryganeidæ*), all of which live in the water, devour aquatic insects. The predaceous habits of the Ant-lion are so well-known as to be almost proverbial. The Lace-winged flies (*Hemerobiidæ*), in the larva state, live wholly on plant-lice, great numbers of which they destroy; and the Scorpion-flies (*Panorpidæ*) are also predaceous insects. The particular history of several of the more interesting members of this class is given in separate articles under some of the above names.

NEWFOUNDLAND DOG. This noble species of the canine race is justly entitled to the peculiar regard of man. For faithful attachment to his master, great strength, sagacity, and perseverance; for good temper, patience, and quiet fondness to all who belong to the household; as well as for being the fearless protector of whatever may be entrusted to his vigilant care, the genuine Newfoundland Dog has no superior. In his

NEWFOUNDLAND DOG.

native country he serves to convey light loads of wood or provision, on sledges, over many a rugged track; nor is he a contemptible assistant to the aquatic sportsman, either there or here, in rescuing his birds from the water. With so many excellent qualities, we may well excuse him if he sometimes shows impatience of restraint at being kept chained up, or if, apparently unprovoked, he should bite the hand that has been accustomed to caress him. There are several varieties of the Newfoundland species, differing in size, and in the character and colour of the fur. In general the muzzle is broad, the head raised, and the carriage majestic: the colour is black and white, the latter generally predominating; the hair waved or curly; the tail thick, bushy, and the end of it curled upwards. There is also a smaller kind, at present very common, not much larger than a Water Spaniel; the hair of which is almost wholly black, and whose general appearance is much less noble than the one we have described.

NEWT. There are several species of these small reptiles, the greater part of them aquatic. The principal one is called the GREAT WATER-NEWT (*Triton palustris*). When full grown this species measures about six inches in length, and is greatly allied to the Salamander in its general appearance. Its colour on the upper parts is an extremely dark brown; the sides being marked with numerous small whitish specks; and the under parts are of a bright orange-colour, variegated with large and irregular patches of black. The tail is of a flattened form, with thin edges, and pointed at the extremity: on each side the tail, in the male, is a silvery-white broad band or stripe, tinged with blue. The eyes are of a bright gold-colour; the head rather small; the limbs short; the fore-feet divided into four, and the hind into five toes, all destitute of claws. It frequents shady places and stagnant waters; lives principally on insects; and is perfectly innoxious.

The COMMON WATER-NEWT (*Triton aquaticus*) is much smaller than the preceding, being only about three inches and a half in length. The dorsal crest of this animal is remarkably transparent, so that when viewed with a lens of even moderately magnifying power, it exhibits very distinctly the ramifications of the blood-vessels dispersed through it; but if examined by the microscope, it shows, in the most distinct and beautiful manner, the rapid circulation of the blood, the particles of which, in this animal, as well as in the rest of the Amphibia, are of an oval form, not round, as in the Mammalia. The general colour is not very different from that of the preceding; varying, however, sometimes in the course of the same day, according to the temperature of the weather, &c. The Water-Newt breeds in the early part of the spring, depositing small clusters of spawn, from which are soon hatched the larvæ or young, which, for a considerable period, are furnished with a triple pair of ramified branchial fins or processes on each side the neck. These parts, after having served their temporary purpose of assisting the respiration of the animal during its growing state, are gradually obliterated. The Water-Newts frequently cast their skins; and are remarkable for a high degree of reproductive power.

NIGGER. A name given by the Cornish fishermen to a species of *Holothuria*. It is sometimes also called "Cotton Spinner." [See HOLOTHURIA.] The word *Nigger* is also a local name for the larva of the Saw-fly, (*Tenthredo*) so destructive to the turnip crop.

NIGHT-HAWK. (*Chordeiles Virginianus.*) This beautiful Passerine bird, belonging to the family *Caprimulgidæ*, is eight inches and a half long, the expanded wings being twenty inches. It appears in Jamaica about the beginning of April, and is supposed by Mr. P. H. Gosse to winter in Central America. This gentleman informs his readers, in his valuable and interesting work, entitled 'The Birds of Jamaica,' that the manners and voice of this species are so superior as to force themselves upon our attention. "About an hour before the sun sets," he observes, we hear a loud, abrupt, and rapid repetition of four or five syllables in the air above our heads, resembling the sounds, *piramidig*, or *gi' me a bit*, or perhaps still more, *wittawitawit*. On looking up we see some two or three birds, exceedingly like swallows in figure and flight, but considerably larger, with a conspicuous white spot on each wing. * * * Like them the Piramidig is pursuing flying insects; and though the prey, from its great height, and probably its minute size, is invisible from the earth, we may very often observe that it is captured, by a sudden arresting of the career, and by the swift zigzag dodgings, or almost stationary flutterings that ensue. * * * It is when the afternoon rains of the season have descended plentifully, that these birds are most numerous and most vociferous; and they continue to fly till the twilight is beginning to fade into darkness. After this, they appear for the most part to retire, and the strange and startling voices, that before were sounding all around and above us, are rarely heard by the most attentive listening. Early in the morning, before the gray dawn has peeped over the mountain, I have heard great numbers of these birds flying low, and hawking to and fro. Their cries were uttered in rapid succession, and resounded from all parts of the air, though it was too dark to distinguish even such as were apparently in near proximity. Now and again, the hollow booming sound, like blowing into the bunghole of a barrel, produced at the moment of perpendicular descent, as described by Wilson, fell on my ear."

Whither the Piramidig retires after its twilight evolutions are performed, or where it dwells by day, Mr. Gosse says he has little evidence. He remarks that "these birds are usually solitary, except inasmuch that several hawking over the same circumscribed region, must often come into close proximity; but this seems, in general, neither sought nor avoided; each swoops on its own course, regardless of its momentary neighbour. Yet the tender passion sets aside even the most recluse solitariness in any animal; and to this I attribute it that now and then I have seen one Pirimadig following another in close and pertinacious pursuit, ever and anon uttering its singular cry, and evidently desiring to come into contact with, but not to strike or hurt its coy companion. I would not assert from hence that the nuptials of this species are performed upon the wing, because the premises are too slight to decide so important a fact; but it is known that it is so with the European Swift, a bird whose manners greatly resemble those of our Night Hawk."

In some parts of Jamaica this bird bears the appellation of "Turtle-dove;" but more often, and with more propriety, that of "Mosquito-hawk." In one which Mr. Gosse shot in its evening career, and afterwards dissected, the stomach was stuffed with an amazing number of insects, consisting chiefly of small beetles of the genus *Bostrichus*, of which alone there were about two hundred.

NIGHTINGALE. (*Philomela luscinia.*) Whether poets have contributed most to the popular celebrity of the Nightingale, or the aspirants to poetic fame have been most indebted to this delightful songster for affording them an inexhaustible theme for their laudations, is not exactly within the province of natural history to determine: we will therefore not trespass on a subject so puzzling and profound, but at once proceed to describe this "tenant of the grove," which Milton apostrophised as

"Sweet bird, that shunn'st the noise of folly,
Most musical, most melancholy!"

Though so universally esteemed for its vocal powers, the Nightingale cannot boast of the variety or the richness of its plumage.

It is about six inches in length: the upper part of its body is of a rusty brown, tinged with olive; the under parts pale ash-colour, almost white at the throat and belly: the bill is brown, yellow on the edges at the base; eyes hazel; legs pale brown. It is

NIGHTINGALE.—(PHILOMELA LUSCINIA.)

common in the southern counties of England, but never visits the northern parts of the island, and is but seldom seen so far west as Devon and Cornwall. Montague informs us that it is said to be found only as far north as Yorkshire, and certainly not farther west than the eastern borders of Devonshire; although they are plentiful both in Somersetshire and Dorsetshire. "Why (he adds) they should not be found in all the wooded parts of Devonshire and Cornwall, which appear equally calculated for their residence, both from the mildness of the air and variety of ground, is beyond the naturalist's penetration. The bounds prescribed to all animals, and even plants, is a curious and important fact in the great works of nature. It has been observed, that the Nightingale may possibly not be found in any part but where cowslips grow plentifully; certainly, with respect to Devon and Cornwall this coincidence is just." It is a bird of passage, appearing in this country, and the rest of Europe, about the beginning of April, and returning, as it is supposed, to the distant regions of Asia, by the end of the summer. They neither appear to winter on the European continent, nor to stay in Africa; but are at all times seen in India, Persia, China, and Japan, where they are even more esteemed for their song, and sell for higher prices, than here.

Mr. Gould (in his 'Birds of Europe') remarks that the Nightingale appears to be confined to particular districts; remarking that Devonshire appears to be its limit westward, and Doncaster in Yorkshire in a northern direction, few if any authenticated instances being on record of its occurrence beyond that town, which is the more singular as Nightingales are common in Sweden and other countries situated farther north than England. "Our own observation," continues Mr. Gould, "respecting the migrations of the Nightingale, is, that after leaving our island it proceeds to the opposite shores of the Continent, and gradually makes its way southward, until it arrives in Africa, which is its ultimate resting-place during our winter months. We have ourselves received specimens killed in the northern districts of Africa, but have never obtained any from the central or southern parts of that portion of the globe; it would appear, therefore, that its distribution over that vast continent is comparatively limited. In no part of Europe is it more abundant than in Spain and Italy; from whence, however, equally as from our own, it regularly migrates on the approach of winter.

These birds are solitary in their habits, never associating in flocks, like most of the smaller birds. They make their nest in the lower part of a thick bush or hedge, where it is well sheltered and secure; and the female lays four or five eggs, of a greenish brown colour. The nest is composed of dry grass, moss, and leaves, and lined with hair, down, and other soft substances. Whilst the business of incubation is performed by the female, her mate, at no great distance, entertains her with his delightful melody: as soon, however, as the young are hatched, he leaves off singing, and joins her in the care of providing for them. A second and sometimes a third hatch takes place; and in hot countries they are said to have four. The note of the Nightingale is soft, various, and interrupted; frequently pausing, but more pleasing than the warbling of any other bird; the more so because it is heard at a time when all the rest are silent — when every melodious sound is heard to advantage, and has a powerful effect on the imagination. Its food consists principally of insects, small worms, eggs of ants, and sometimes berries of various kinds.

It has been frequently remarked that the Nightingale is not only famous among the moderns for its singing, but almost every one of the ancients who undertook to describe the beauties of nature, has contributed to raise its reputation. "The Nightingale," says Pliny, "that for fifteen days and nights, hid in the thickest shades, continues her note without intermission, deserves our attention and wonder. How surprising that so great a voice can reside in so small a body! such perseverance in so minute an animal! With what a musical propriety are the sounds it produces modulated! The note at one time drawn out with a long breath, now stealing off into a different cadence, now interrupted by a break, then changing into a new note by an unexpected transition; now seeming to renew the same strain, then deceiving expectation! She sometimes seems to murmur within herself; full, deep, sharp, swift, drawling, trembling; now at the top, the middle, and the bottom of the scale! In short, in that little bill seems to reside all the melody which man has vainly laboured to bring from a variety of musical instruments. Some even seem to be possessed of a different song from the rest, and contend with each other with great ardour. The bird overcome is then seen only to discontinue its song with its life." From Pliny's description, we should be led to believe this bird possessed of a persevering strain; but, though it is in fact so with the Nightingale in Italy, yet in our hedges in England the little songstress is by no means so liberal of her music. It is true that for weeks together, if undisturbed, Nightingales will sit on the same tree, begin their song in the evening, and, with short interruptions, continue it throughout the night. It is therefore by no means wonderful that their sweet notes and unceasing perseverance in pouring forth such a volume of rich melody, when all else is hushed in the silence of night, should have been the theme for poets in all ages to descant on; but that the philosophic Gessner should gravely relate a long story respecting this bird's oratorical talents, and describe the conversation which a friend of his heard between two of them while passing a sleepless night at an inn in Ratisbon, is not only too much for human credibility, but almost too much for human patience.

We conclude with a passage from Sturm: "When we listen to the brilliant sounds of that voice, we are apt to conclude that the bird must be large, that the throat must have great strength; and the inimitable charm of her melodious notes makes us presume she surpasses all others in the beauty of her form. But it would be to no purpose to seek these advantages in the Nightingale: it is a bird of poor appearance, whose colour, form, and the whole of its exterior, is void of anything attractive or majestic. Nature has, however, compensated for its plainness, by giving it a voice irresistibly charming. Listen to its fine long quivering notes: what variety, sweetness, and brilliancy in them! When she begins her song, she seems to study and compose beforehand the melodious notes she wishes to be heard. She begins softly: then the notes swell gradually, till they run with the rapidity of a torrent: she goes from serious to gay, from simple notes to the wildest warblings; from the lightest turns and shakes to languishing sighs; and has, throughout the whole, the art to please the nicest ear. This bird may give rise to many useful and edifying reflections: for example, we learn this truth from it, that homeliness of body is sometimes united with very estimable qualities, and does not exclude beauty from the soul. When we hear the skilful harmony of the Nightingale, does it not naturally lead us to the Creator, from whom she has this talent? What wisdom must there be in the formation of this bird, to make it capable of giving utterance to such sounds! Lungs so delicate as those of the Nightingale, the motions of which are so violent, must be easily wounded, if they had not the singular advantage of being fastened to the backbone by a number of little sinews. The orifice of the windpipe is very large, and that is certainly what most contributes to the variety of those sounds, which, in charming the ear, fill the soul with sweet and pious joy. Is it possible not to trace a divine wisdom and providence in this? and will not even the song of the Nightingale lead us to glorify the Author of all nature? Lovely songstress! I will not leave thee till I have learned from thee the art of praising my Creator and thine. O pour, with thy song, gratitude and joy into the hearts of the many insensible mortals who contemplate the beauties of the creation with indifference."

NIGHT-JAR. [See GOATSUCKER.]

NOCTILUCA. A minute genus of *Acalephœ*, often seen on our own coasts, which in size and appearance much resembles a grain of boiled sago, or a little granule of jelly with a long stalk, the stalk appearing to be a trunk or sucking-tube. The luminous property of these minute Acalephæ always appears to become more vivid when the animals are alarmed or stimulated in any way: hence the curling of the waves, and their ripple on the shore, the movement of a boat, or the stroke of the oars, is marked by lines of increased brilliancy. Nay, if the hands be dipped in the water thus phosphorescent, and then rubbed together, they will be covered with luminous spots, occasioned by these delicately-formed little animals, the bodies of which are often so transparent, that they can scarcely be distinguished from the water, except when displaying their phosphorescence. When we consider that the whole surface of the ocean, as far as the eye can reach, is sometimes seen to exhibit a uniform luminosity, and it is ascertained to be due to these otherwise almost invisible atoms, the vast amount of organic life that ordinarily escapes our notice must strike the most inattentive observer of the works of Nature with astonishment and admiration.

NOCTUIDÆ. An extensive family of Lepidopterous insects, corresponding with the Linnæan section *Phalœna Noctua*. The body is robust, and clothed with scales; the antennæ almost always simple, or but rarely pectinated or ciliated in the males; the thorax stout, and often crested; and the mouth well developed, the maxillæ being greatly elongated. The wings are of moderate size, with strong nervures, and ear-shaped spots on the disc of the anterior pair; and when in repose the wings are ordinarily deflexed at the sides of the body. The larvæ, for the most part, are naked, with sixteen feet; and they in general undergo their transformations underground in cocoons, often formed of particles of earth mixed in with the silk. The typical groups of this family, as their name imports, fly only by night, and repose during the day in the crevices of the bark of trees, old walls, palings, &c.: there are others, however, which fly also during the afternoon and at twilight. The generality of these insects appear in very sombre colours; but in some species, more accustomed to be abroad in the day-time, the wings, especially the posterior ones, are occasionally more gaudy: this is the case with the *Catocalœ* or Scarlet Underwing Moths; whilst the *Plusiœ* are bedecked with spots and patches of silver or gold. There is a considerable diversity in the form of the wings; in general the anterior ones are elongate-triangular, and the posterior somewhat triangular-orbiculate; and it is further to be observed that the anterior wings are mostly adorned with two stigmata, one round or nearly so, and the other reniform. The larvæ are usually solitary; and they neither reside in a web, nor are they subcutaneous.

NOCTURNA; or NOCTURNAL LEPIDOPTERA. [See MOTHS.]

NODDY. A bird of the Tern genus (*Sterna stolida*), well known to seamen for the stupidity with which it throws itself on vessels, and allows itself to be taken. [See BOOBY.]

NOTACANTHA. The name of a family of Dipterous insects, mostly small and gaily coloured. Some of the larvæ are completely aquatic; and respire like the larvæ of the Gnats, by extending their tails to the surface, the spiracle being in that situation.

NOTHOSAURUS. [See SUPPLEMENT.]

NOTONECTA. [See BOAT-FLY.]

NOTORNIS. [See SUPPLEMENT.]

NOTOTHERIUM. [See SUPPLEMENT.]

NOTOXIDÆ. A family of Coleopterous insects, of small extent, and composed of species minute in size. The majority of them are found upon the ground, and at the roots of grass in sandy situations; some frequent flowers, and others evidently prefer the neighbourhood of decayed vegetable matter. They are active in their motions, and fly well. In the genus *Notoxus* the front of the thorax is produced into a long horn extending over the head.

NUCIFRAGA. A genus of birds. [See NUTCRACKER.]

NUCULA. A genus of Conchifera, found in the Baltic and Mediterranean, the Indian seas, the English Channel, &c. These shells are small, and vary in shape, but are generally pearly inside; they are equivalve, inequilateral, and covered with a green or dark brown epidermis; hinge linear; bosses contiguous and curved; teeth small, numerous, and prominent, with a large one in the middle; muscular impressions two, simple. The row of teeth on each side of the umbones, and the ligamentary pit in the centre of the hinge, are the distinguishing characteristics of this genus. Foot of the animal large, but thin. They are chiefly found on the sand and mud, either on the open coast or at the mouth of rivers. The species are both recent and fossil.

NUDIBRANCHIATA. A numerous order of marine molluscous animals, which are often found at a great distance from land. The British Nudibranchiates, published by the Ray Society, gives figures and descriptions of all the species, beautifully drawn from the life. [See DORIS.]

NUMENIUS. A genus of Grallatorial birds, containing the well-known CURLEW and WHIMBREL [which see].

NUMIDA. A genus of Rasorial birds, containing the well-known GUINEA-FOWL or PINTADO (*Numida meleagris*), and five other species, like it, natives of Africa.

NUMIDIAN CRANE. [See DEMOISELLE.]

NUMMULITES. Small round fossil shells, which in various parts of the world are found in immense numbers, and which receive their name from their external resemblance to battered coins. They are orbicular, convolute, and show no trace of spire externally; whorls contiguous, and not apparent; cells numerous and small; partitions transverse, and not perforated. Some are very minute, and scarcely any are more tha[n] an inch in diameter. It is said that they are in

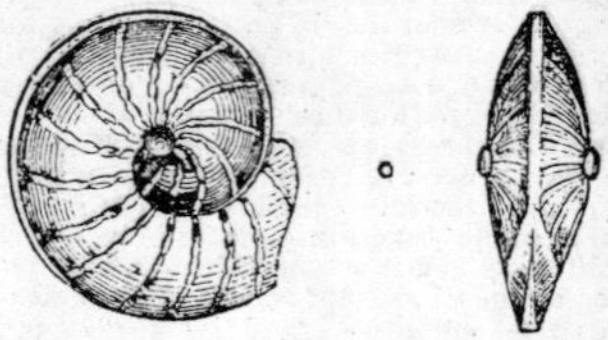

NUMMULINA DISCOIDALIS

some places accumulated in such vast masses as to form entire mountains, and that many buildings have been constructed of limestone crowded with them. The pyramids of Egypt, for example, are built of stone composed of the *Nummulina discoidalis*, and perhaps other species.

NUTCRACKER. (*Nucifraga caryocatactes.*) An Insessorial bird, resembling in its manners and habits both the Jay and the Woodpecker. It is about the size of a Jackdaw; its wings, when closed, measuring near seven inches. The nostrils are covered with whitish feathers, which point forwards; the plumage of the head, neck, and body is of a dark brown hue, a little inclining to red; and the feathers on the lower side of the head and neck have each a triangular white spot at their tips: the wings are black, with triangular white spots on the lesser coverts; the tail is composed of black feathers, tipped with white; and the legs, feet, and claws are black. It feeds on nuts, berries, and insects; climbing the trees and tapping the bark with its bill to get at the larvæ beneath. It lays five or six yellowish-white eggs.

NUTHATCH. (*Sitta Europæa.*) A Scansorial bird which frequents woods, and, like the Woodpecker, moves up and down the trunks of trees with great facility, in search of food. It is near six inches in length; bill strong; black above, beneath almost white; and the eyes hazel. A black stroke passes over each eye, from the bill, extending down the side of the neck; all the upper part of the body is a fine blue gray; breast and belly of a pale orange, sides marked with streaks of chestnut; quills dusky; tail short, the two middlemost gray, and the three outermost feathers spotted with white; legs pale yellow; claws large, sharp, and much bent, the back claw very strong. The female lays her eggs, which are white with a few pale brown spots, in holes of trees, frequently in those which have been deserted by the Woodpecker; and when driven from her nest, on being disturbed, hisses like a snake. The Nuthatch, like the Woodpeckers, runs with facility upon and about the trunks and branches of trees; but the tail, which is short and rounded, is of no assistance to the bird in its progress. Unlike the Woodpecker, however, the Nuthatch runs with the head downwards as well as upwards, and indeed the former position of the head appears to be the favourite one; it generally alights on a branch with the head in a down-

NUTHATCH.- (SITTA EUROPÆA.)

ward position, and sleeps in that posture. The Nuthatch feeds on caterpillars, beetles, and various kinds of insects: it also eats nuts, of which it lays up considerable hoards in the holes of trees. Its mode of fastening the nut in a chink, perforating the shell, and extracting the kernel, is as ingenious as it is amusing to witness: when disturbed at its work, it very readily removes the nut, and flies away with it. These birds are found in all cold and temperate climates.

The courage and perseverance of the Nuthatch, when made captive, are notorious. It is related in the Magazine of Natural History, that one of these birds had been winged by a sportsman, and was put into a small cage made of oak and wire. During a night and day, the period of his confinement, his tapping labour was incessant, and at the end of that time the wood-work of his prison was pierced and worn like worm-eaten timber. His impatience of his situation was excessive; his efforts to escape were unceasing, and displayed a degree of shrewd intelligence perfectly surprising. He was fierce and fearlessly familiar, and voraciously devoured the food placed before him. His hammering is described as having been peculiarly laborious, for he did not peck as other birds do, but taking a firm grasp with his great feet, he turned upon them as upon a pivot, striking with his whole weight, and thus assuming with his body the appearance of the head of a hammer in motion. But all his energy was fruitless; his liberation was beyond his own power to effect; and the unfortunate bird expired at the close of the second day under the combined effects of his vexation and assiduity.

NUT-WEEVIL. [See Weevil.]

NYCTIBIUS. A genus of birds belonging to the *Caprimulgidæ* family.

NYCTICORAX. A genus of the Heron tribe, containing the Night Herons. [See Heron.]

NYLGHAU, or White-footed Antelope. (*Antilope* [*Portax*] *picta.*) This animal, which inhabits various parts of India, is one of the largest and finest Antelopes known. Its face is long and narrow; its horns are black, round, pointed, and slightly curved forwards, though only about seven inches long; the ears broad and fringed with white hairs; the neck deep and compressed: along the top of the neck runs a slight mane of black hair, which is continued to some distance down the back; and on the breast is a long hanging tuft of a similar colour. The general colour of the Nylghau is a fine dark gray or slaty blue on the upper parts, and white underneath. The female resembles the male in general appearance, but is considerably smaller, of a pale brown colour, and has no horns. There is a large white

NYLGHAU.—(ANTILOPE [PORTAX] PICTA.)

spot on the throat, and a smaller one on each cheek; and the pastern joints are marked in front with one, and behind with two white spots or bars. The native haunts of this powerful animal are the dense forests of India. It is said that in the days of Aurungzebe they abounded between Delhi and Lahore, and formed one of the objects of the chase with that mighty monarch during his journey to Cashmire; his army of hunters inclosing them within a limited space by means of nets. The king and his omrahs, attended by their huntsmen, then entered, and, somewhat after the manner of a modern *battu*, dispatched them with their arrows, spears, &c.

NYMPHALIDÆ. The third family of *Lepidoptera*. They are distinguished by the rudimental structure of the fore legs, which are thickly covered with hair; the labial palpi are proportionably longer; the wings more robust; the posterior grooved to receive the abdomen; and the discoidal cell either open, or closed by a slender nerve. The caterpillars are variable in form; and the chrysalis is simply suspended by the tail. Among the species belonging to this family are many of the most beautifully varied in their markings and colours. The well-known species bearing the English names of the Peacock, Painted Lady, Camberwell Beauty, and Red Admiral, are included in it; also the Fritillaries (so called from the spotted flower of that name), the under side of whose wings is delicately ornamented with pearl or silvery spots. Others, belonging to the genus *Morpho*, which comprises also some of the largest known Butterflies, have the upper surface of the wings adorned with the most splendid silvery blue; while others, as the males of *Apatura Iris*, or Purple Emperor, present the eye with a changeable gloss of intense purple. The numerous species forming the genus *Hipparchia* are of feeble construction in the imago state, and cannot bear comparison with those before mentioned, which are the most robust and active of lepidopterous insects. The species of this family are extremely liable to sport into varieties, which is especially the case with the *Hipparchiæ;* the caterpillars of which, it is to be observed, confine themselves to the different grasses, and feed only in the night. The caterpillars of *Vanessa* are armed with long and rough spines, arranged in transverse whorls upon the segments, except the first. Those of the *Fritillaries* are similarly armed, but have two long spines on the neck.

NYROCA. A genus of Ducks, containing the POCHARD (*N. ferina*), and CANVAS-BACKED DUCK (*N. valisneria*). [See DUCK and POCHARD.]

OCELOT. (*Felis pardalis.*) An animal of the feline tribe, less than the Ounce, but its skin is more beautifully variegated. The ground colour of the male is a bright reddish tawny above, and nearly white on the lower part of the sides, breast, limbs, and belly. Several large, long, broad stripes, of a deeper

OCELOT.—(FELIS PARDALIS.)

tinge, and edged with black, are variously disposed over the upper parts of the body; the head is streaked and spotted with black, and the limbs and belly are beautifully marked with numerous small round spots; the tail is spotted or marked with patches also. The colours of the female are less vivid, and more inclining to ash-colour. The Ocelot inhabits the hotter parts of South America; is extremely ferocious; and preys upon various kinds of game.

OCTOCERA. The first family of the order *Cryptodibranchiata* of Blainville, in the class Mollusca, containing the genus Octopus, — a species of which being found in the Argonauta, or Paper Sailor, has given rise to the long-continued controversy as to whether it is really the constructor of the shell, or whether it is a mere pirate, which, having destroyed the true animal of the Argonaut, has possessed itself of the habitation."

OCTODON. A genus of small Rodent Mammalia, inhabiting Chili. They have large ears, and a long and tufted tail, and are somewhat allied to the Chinchilla group. The only known species is the *Octodon Degus*, which is often seen traversing the branches of low underwood. In size and shape this species generally resembles the Water Rat, with which, indeed, it appears to be connected systematically. "These animals," Mr. Bennett observes, "burrow in the ground, but always under brushwood fences or low thickets. They are so abundant in the neighbourhood of Valparaiso, that in the high road between that place and St. Jago more than a hundred may frequently be seen at one time in search of food. Sometimes, but not often, they are observed on the lower branches of the shrubs, and on those which form the fences. They fly at the least alarm, and in running carry their tufted tails like a bent bow." Mr. Bennett adds that two living specimens brought by Mr. Cuming from Chili, were placed by him, in 1831, in the menagerie of the Zoological Society: one of them escaped, but the other was alive when Mr. B. wrote (Dec. 1835), and was as active and lively as it was on its first arrival. They were rather shy, and not very playful. They leaped readily, and without any exertion, from the floor of their cage to a narrow perch placed at the height of nearly a foot, and there remained seated at their ease. They lived on vegetable food.

OCTOPODA. The name of a tribe of Dibranchiate Cephalopods, with eight feet or tentacular appendages.

OCTOPUS. The common Octopus or Poulp is the *Polypus* of ancient naturalists. It has eight arms, each of which is six times the length of its body, and furnished with 120 pairs of suckers. Every sucker is composed of a circular adhesive disc, which has a thick fleshy circumference, and presents a number of lines radiating towards the circular orifice of an inner cavity. In this cavity is a moveable circular piston, which in its operation forms an air-pump of the most precise and beautiful construction. When a fish becomes infolded within the tenacious grasp of its arms, resistance is vain; for with such tenacity do the suckers adhere, that they may be sooner wrenched off than unfixed. Some of these *Octopi* measure four feet between the ends of the arms; and it is said that much larger ones are sometimes met with in the warmer regions of the globe. It has been justly remarked that "there is something strange and uncouth in the aspect of this creature; its long flexible arms moving and curling in all directions; and its large eyes, which stare with fixed gaze, rendering it really repulsive."

Mr. Adams, in his Natural History of the

countries visited by H. M. S. Samarang, says, "*Octopi*, of enormous size, are occasionally met with among the islands of the Meïa-co-shima group. I measured one, which two men were bearing on their shoulders across a pole, and found each brachium rather more than two feet long, giving the creature the power of exploring a space of about twelve feet, without moving, taking the mouth for a central point, and the ends of the arms for the periphery." "On moonlight nights among these islands, I have frequently observed the *Sepiæ* and *Octopi* in full predatory activity, and have had considerable trouble and difficulty in securing them; so great is their restless vivacity at this time, and so vigorous their endeavours to escape. They dart from side to side of the pools, or fix themselves so tenaciously to the surface of the stones, by means of their sucker-like *acetabula* that it requires great force and strength to detach them. Even when removed, and thrown upon the sand, they progress rapidly, in a sidelong shuffling manner, throwing about their long arms, ejecting their ink-like fluid in sudden violent jets, and staring about with their big, shining eyes (which at night appear luminous like a cat's), in a very grotesque and hideous manner."

OCYPODA. A genus of Brachyurous Crustaceans, inhabiting the sea-shores of warm climates in both hemispheres. They derive their name from the rapidity of their motions; those who have observed these

AMERICAN SAND CRAB
(OCYPODA ARENARIA.)

animals in their native haunts declaring that they run so fast that a man can hardly overtake them. They form holes for themselves in the sand immediately above the level of the wash of the sea, and in these they reside during the summer, but they pass the winter in a state of hybernation. There are several species, differing but little from each other: the one here figured is *Ocypoda arenaria*, or SAND-CRAB: length about two inches; colour yellowish. In the summer their general time of quitting the burrow to seek their food is the night; but towards the end of October they retire inland to hybernate in the earth; and when they have found a suitable place, they dig a hole like that which they had occupied on the edge of the sea, enter it, and close up the entrance so thoroughly that no trace of it can be seen. There they remain until they repair to their marine residences.

ODONTOSAURUS. [See SUPPLEMENT.]

ŒDEMERIDÆ. A family of Coleopterous insects, of a moderate size, and generally of lively colours. In the perfect state they frequent flowers and hedges: they fly with agility, but walk slow: they are, however, enabled to retain firm hold upon the leaves and stems of plants, by means of their dilated tarsi. The body is long and narrow, with the elytra broader than the head and thorax; the antennæ moderately long and filiform; the head elongated in front, and inserted deeply in the thorax, without any distinct neck.

ŒDICNEMUS. A genus of Grallatorial birds, having the tip of the bill inflated both above and beneath; the groove of the nostrils half the length of the beak: legs reticulated, with a short membrane at the base of the three toes. Mr. Gould, in his 'Birds of Europe,' considers the genus as connecting the Bustards and Plovers, and observes that while the normal or typical groups are abundant in species, the aberrant forms, which appear to be created for the purpose of filling up the intervening chasms, are restricted for the most part to a limited number of species: thus, while the Bustards and Plovers comprise a vast multitude of species, the genus *Œdicnemus* contains at most but five or six, and these confined entirely to the Eastern hemisphere. Their English name is derived from the usual habitat being arid and stony districts, where they pick up slugs and insects. [See STONE CURLEW.]

ŒSTRUS. A family of Dipterous insects, or flies, whose larvæ are known by the name of *bots*. The perfect insects resemble large meat-flies in form, are very hairy, and have these hairs coloured in rings, like Humble-bees; but the duration of their lives is so short in this condition that they are seldom seen. They deposit their eggs on the body of various herbivorous quadrupeds; each species almost invariably confining its attacks to a certain species of animal. The egg is, in some cases, deposited by the parent in situations where the larva may burrow into the flesh, where it finds its nutriment in the inflammatory tumours it occasions. In other instances, the eggs or larvæ, deposited upon spots which the animal is in the habit of licking, are taken up by the tongue, conveyed to the mouth, and thus pass into the stomach. And the species which inhabits the Sheep, are found in the frontal sinuses of the skull. Hence they are called *cutaneous*, *gastric*, or *cervical*, according to the locality in which they are bred. When full grown they quit the body, and fall to the ground; beneath the surface of which they undergo their transformations. [See GADFLY.]

ŒTHRA. The name of a genus of *Crustacea* whose general organization nearly approaches that of the Crabs. They are from two to three inches in length, and the whole surface of the body is extremely rugged. The species *Œthra scruposa* is a native of the Indian Archipelago.

OIDEMIA. A genus of Wading Birds, containing the Scoter Duck and others. [See DUCK: SCOTER.]

OIL BEETLE. (*Meloë.*) A genus of Coleopterous insects, belonging to the tribe of Vesicatory Beetles, whose economy until lately has remained one of the most difficult unsolved problems in the natural history of the *Articulata*. At a meeting of the Linnæan Society of London, Nov. 18. 1845, the history, development, and general economy of this insect, formed the subject of a memoir by G. Newport, Esq., F. R. S., and is reported in their 'Proceedings.' The writer observes that many naturalists, more particularly Gœdart, Frisch, and De Geer, have well described the perfect insect, and have even given detailed observations on the oviposition of the female and the early stage of the larva, but they have invariably failed to carry their inquiries further, and have been quite unacquainted with the adult larva and the nymph, as well as with the early stage of the imago. This deficiency in our knowledge of the history of these common insects is to be attributed principally to the anomalous habits of the insect in its earliest stages, and to the little credit that has been given to the statements of former observers.

Mr. Newport commenced his observations on the habits of *Meloë* about fifteen years ago; but although he succeeded at that time in rearing the larva from the egg, as had been done by Gœdart and De Geer, and soon afterwards obtained the full-grown larva, the nymph, and the imago, before it left its cell, he has never been able to obtain the larva in a stage intermediate between its earliest and its full-grown condition. The species on which Mr. Newport made his investigations are *Meloë violaceus*, *Meloë proscarabæus*, and *Meloë cicatricosus*, all which he procured at Richborough, near Sandwich, in Kent. The first two of these species come forth about the middle of March, and the latter from ten days to a fortnight later in the season. They feed chiefly on the buttercup (*Ranunculus acris*), and one species, *M. cicatricosus*, also on the dandelion.

When the *Meloës* first appear they are feeble, and have the body very small and contracted. In the course of a few days they become more active and are increased in size. They expose themselves much to the sun, and pair in the middle and warmest part of the day. On the 8th of April, 1830, the author first observed a female preparing to deposit her eggs, and he has since had numerous opportunities of observing her thus occupied. She excavates a burrow, to the depth of about two inches, beneath the roots of grass, in a dry soil exposed to the sun, usually at the side of a foot-path. Into this burrow she passes her body backwards, and having deposited a large packet of yellow-coloured cylindrical eggs, she closes up the burrow with earth and begins again to feed. Each female deposits eggs from three to four times during the season, at intervals of from one to two or three weeks. The greatest number are deposited at the first laying. In order to ascertain the number deposited at the first laying by *Meloë proscarabæus*, Mr. Newport removed the ovaries from a specimen that had recently been impregnated, and having divided one ovary into pieces, counted the number of eggs in each under the microscope, and found that one ovary contained 2109 eggs ready for deposition; so that the two ovaries contained the astonishing number of 4218 mature eggs, besides an almost equal number in the course of formation.

The larva of *Meloë*, as it comes from the egg, is a yellow, slender, active little hexapod, scarcely one-twelfth of an inch in length. It attaches itself with great readiness to bees and flies, and clings so securely to them, that the insects are not able to remove it from their bodies, as was noticed in several experiments. These facts confirm the observations of Gœdart and De Geer, who first bred the larva from eggs deposited by *Meloë*. The structure of the larva is next described, and compared with that of the *Pediculus apis* of Linnæus, as found on Hymenopterous insects, and the two are shown to be identical in every particular. The *Meloë* larva is also compared with the *Pediculus Melittæ* of Mr. Kirby, with which also it agrees exactly in form and general structure, but differs in colour, that of the latter insect being always black, while the larva of *Meloë* is yellow. From this circumstance the author concludes that Mr. Kirby's insect is the larva of another genus of the same family.

The habits of the larva of *Meloë* are then investigated, and the effects produced on it by exposure to light are minutely detailed. When light was totally excluded, the larvæ remained perfectly quiet for several days; but the instant light was admitted they were in motion, travelling rapidly in a direction towards it. The experiments were made by enclosing larvæ in a phial, which was inverted and turned in opposite directions. When the phial was placed perpendicularly they invariably ascended to the top, and when placed in a horizontal direction they always ran to that end which was nearest the light, even when the stopper around which they had been lying was removed to allow of their escape. This influence of light Mr. Newport conceives may be that which induces them to ascend the yellow flowers of the dandelion and buttercup preparatory to their attaching themselves to bees that alight on the flowers to collect pollen, and which then carry them into their nests. This seems to be the object of their attacking the bees, to be carried to the nest, where they are to reside as parasites, and subsist on the food stored up for the bee-larva, and not to prey on the bee itself.

The full-grown larva of *Meloë cicatricosus* is then described, and also the nymph and the imago. The author had found the insect in those stages in the nests of *Anthophora retusa*; but he had not succeeded in his attempts to rear the young larva of *M. violaceus* and *M. proscarabæus* in the nests of that insect. He concludes, therefore, that these species inhabit the nests of some other bees. In the stage between the very young and the full-grown period the larva is believed to be active and retain its six scaly feet, and to feed on the food prepared for the young bee. In its full-grown state the legs of the larva are reduced to six short tubercles. The insect is then very fat, inanimate, and of an orange-yellow colour, has ten pair of spiracles and greatly resembles the full-grown Hymenopterous larva. It remains but a short time in this condition before it changes to a nymph, and soon afterwards to an imago, in which form it passes the winter in a state of hybernation, and comes forth in the spring.

In the course of this paper, while detailing the influence of light on the larva of Meloë, Mr. Newport stated that he had been led by these and other facts, which showed the great influence of light on the instincts of the young animal, "to regard light as the primary source of all vital and instinctive power, the degrees and variations of which may, perhaps, be referred to modifications of this influence on the special organization of each animal body." This view had suggested itself to him in connection with the discovery recently made by Mr. Faraday of the analogy of light with magnetism and electricity, and the close relation previously shown by Matteucci to exist between electricity and nervous power, on which not only all the vital actions, but also the instinctive faculties, seem to depend.

In another paper, read on the 19th of January, 1847, in which this subject is resumed by Mr. Newport, he entered on an examination of the habits of the entire group of insects allied to Meloë, and showed that the whole of them in their larva state bear a general resemblance to the larva Meloë, not only in their organization, but also in their habits; and that the more closely the larvæ of different genera approach in structure, the more nearly also are they allied in instinct and economy. This accordance between structure and instinct he regards as universal throughout nature, and as particularly marked in the *Articulata*: and he believes that, by carefully comparing our observations on the natural history of animals with their peculiarities of structure, and these on the other hand with their instincts, what might otherwise remain useless and isolated facts, may be rendered truly important to science, "as data on which a correct knowledge of the laws of creation and life may be established." In this way, he states, "natural history may be made to occupy its proper position as an important branch of useful knowledge, and mainly help to demonstrate the connection which subsists between structure and function, and function and the habits of animals." In pursuing this view, he showed that the organization and instinct of the larva *Meloë* closely agree. At the moment of birth, when the larva is destined to attach itself parasitically to the *Hymenoptera* which alight on flowers to collect pollen, and which are to convey it to their nests, its organs of vision are largely developed, and those of locomotion are elongated, powerful, and constructed like those of the parasitic *Anoplura*; and it is extremely active and sensitive of light. But when, at the period of full growth, it is found in the cell of *Anthophora*, it is a fattened, yellow-coloured, almost motionless larva, with its legs atrophied and reduced to mere pedal tubercles previous to a further change in their structure when the larva passes to the state of nymph.

It was further observed, that the limbs of this beetle, although strong, are unguiculated, like those of the *Anoplura*, and fitted for clinging rather than for regular progression; and its mandibles, retaining the jointed pediform structure of the corresponding organs in the carnivorous *Chilopoda*, are fitted for piercing soft structures, rather than for triturating or for incising their food. This fact, overlooked by the author in his former memoir, now induced him to believe that the young *Meloë* pierces and preys on the bee larva rather than that it subsists on its food.

OLIVA. A genus of Mollusca, common in the seas of warm climates. The species are very numerous; some of the shells being large, and ornamented with a great variety of rich markings and brilliant colours. The animal has a small head, terminated by a proboscis; two tentacula enlarged at the base, and having the eyes situated in the middle; foot very large, as is also the mantle. The shell is oblong, cylindrical, smooth, and shining; spire short, with sutures dis-

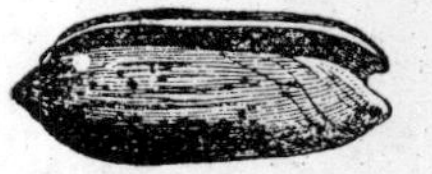

OLIVA ISPIDULA.

tinctly grooved; aperture narrow and long, and notched at both extremities; outer lip generally thick; columella obliquely striated; operculum horny and small in some species, in others not existing. They are brought principally from Asia, but some are also met with on the coasts of Africa and America.

ONCHIDIUM. A genus of Mollusca, belonging to the *Aquatic Pulmonea* (a class remarkable for their coming frequently to the surface to breathe, and which in consequence can only inhabit waters of inconsiderable depth). The *Onchidium* has a large fleshy buckler-shaped mantle, which overlaps the foot on every side, and even covers the head when this is contracted. It has two long retractile tentacula, and over the mouth a veil, sinuated, or formed of two triangular compressed lobes. The anus and air-passage are under the hinder margin of the mantle, where, a little deeper, is also the pulmonary sac. Destitute of jaws, they have a muscular gizzard, succeeded by two membranous stomachs. Several species inhabit the coasts of the sea, but always in such a situation that they are uncovered at ebb tide, when they obtain the air necessary to respiration.

We have the following account of a species of *Onchidium* in Mr. Arthur Adams's 'Notes on the Natural History of the Countries visited during the Voyage of H. M. S. Samarang:' "Among molluscous animals, the Onchidium of Singapore offers a curious instance of what may be termed an Arboreal Slug. It is a limaciform animal, which is found crawling among the foliage of the trees in the woods, and appearing more particularly after heavy showers. During the heat of the day it collapses its broad, flattened body, and retires under the shade of large leaves, where it remains apparently in a half torpid condition. It leaves no shiny trail behind, when it crawls, as the limax and snail do. It is of a chestnut brown colour, minutely tuberculated, with numerous small, dark, scattered spots, and with the raised middle line of a pale brown; the eyes are terminal on the long superior pair of tentacles."

ONISCIA. A genus of Mollusca, littoral in its habits, and occupying an oblong, subovate, and slightly turbinated shell; spire short, base rather pointed; aperture elongated, terminating anteriorly in a short, scarcely recurved canal; outer lip thickened, denticulated within; inner lip expanded and granulated: outside ribbed.

ONISCIDÆ. There are several species of Crustacea, of the order *Isopoda*, thus designated, some of which are terrestrial, and some aquatic. The type of the group, *Lygia oceanica*, is about an inch long, of a gray

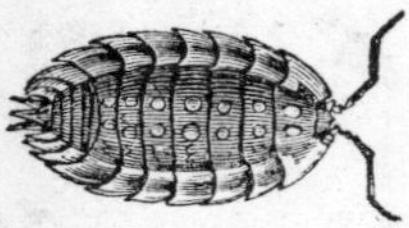

GRANULATED HOG-LOUSE.
(PORCELLIO GRANULATUS.)

colour, with two large yellow patches on the back. It is very common on the coast, clinging to the rocks and to the parapets of maritime erections. When it is attempted to be seized it immediately folds up its legs, and drops.—The *terrestrial Oniscus* frequents dark and concealed places, such as cellars, caves, holes in walls, under stones, &c. They feed upon decaying vegetables and animal matter, and only come forth from their retreat in wet and moist weather. They are popularly known by the name of Wood-lice and Slaters.

ONTHOPHAGUS. A genus of Lamellicorn Beetles living amongst dung. There are very many species.

ONYCHORHYNCUS, or KING TODY. [See TODY.]

OPAH, or KING-FISH. (*Lampris luna.*) This large and beautiful fish, though a native of the Eastern seas, has sometimes, though very rarely, been met with on our own coasts. It is about four feet and a half in length, and weighs from 140 to 150 lbs.: the body is of an oval form; the mouth small, without teeth; tongue thick, with rough papillæ pointing backwards. The dorsal, pectoral, and ventral fins very long, and falciform; and the shape of the tail lunate. The colours of the Opah are particularly rich and showy; the back and sides are green, reflecting both purple and gold in different lights, and passing into yellowish green below. Above and beneath

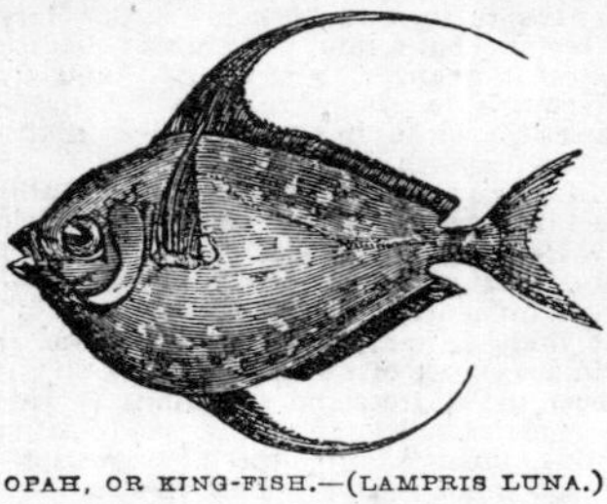

OPAH, OR KING-FISH.—(LAMPRIS LUNA.)

the lateral line are numerous round, yellowish-white spots; and all the fins are bright vermillion. This fish is held sacred by the Japanese, who regard it as the peculiar emblem of happiness.

OPEN-BILL. (*Anastomus.*) A genus of Wading Birds, allied to the Storks and Jabirus. The mandibles of their beak come in contact only at the base and tips, leaving a wide interval between their edges, at the medial portion; the fibres of the horny substance of the bill in this part appearing as if worn away. One species (*Anastomus oscitans*) is whitish, with black tail-feathers:

PONDICHERRY OPEN-BILL.
(ANASTOMUS OSCITANS.)

another (*A. lamelliger*) is of a shining black, and remarkable for the stem of each of its feathers terminating in a narrow horny disc, which passes beyond the vane. They are natives of India.

OPHISAURUS. A snake met with in the Southern United States; about eighteen inches in length, and of a yellowish green colour, with black spots on the upper part: the head is very small, and the tail is longer than the body. So great is its fragility, that, according to Catesby, a small blow with a stick will cause the body to separate, not only at the place struck, but at two or three other places; the muscles being articulated in a singular manner, quite through to the vertebræ. Hence this reptile has obtained the name of the *Glass Snake*.

OPHIDIA. The name given to an order of Reptiles which includes the Serpent tribes. The species belonging to this order may be naturally grouped in three sections: — 1. Harmless Snakes; — 2. Venomous Snakes; — 3. Water Snakes. The Harmless Snakes are divided into two families, the *Colubridæ* and the *Boidæ*: many of them being very large, and possessed of enormous muscular strength. The Section of Venomous Snakes also contains two families, — the *Crotalidæ*, or Rattle-snakes, and the *Viperidæ*, or Vipers. The Water Snakes belong to the family *Hydridæ*, and are characterized by the compressed form of their bodies, and by the vertical flattening of the tail, which enables them to swim with facility. [See SERPENTS: SNAKES: BOA CONSTRICTOR: RATTLE-SNAKE: VIPER, &c.

OPHIDIUM. There are three or four species of fish bearing this name; all small and anguiliform. One called the BEARDED OPHIDIUM, is about eight or ten inches in length, and has two long bifid cirri or barbules beneath the chin. It is found in the Mediterranean sea, and is in great plenty in the Adriatic. It feeds upon small crabs and fishes, but the flesh is considered rather coarse. The BEARDLESS OPHIDIUM is only about three inches long; the head is very obtuse, and the body is ensiform, considerably compressed towards the tail.

OPHIOPS. A genus of Lizards, principally distinguished by the absence of eyelids. *Ophiops elegans*, a species found in Smyrna, is of an olive colour above: two yellowish lines extend along each side of the trunk; each of these lines separates two rows of black spots, which are small and very distinct when the reptile is young, but more or less dilated and confused in adults; they are white beneath.

OPHIURA. Lamarck's name for a genus of Star-fishes.

OPOSSUM. (*Didelphis.*) The name of a family of marsupial quadrupeds (of the genus *Didelphis*), peculiar to the American continent; and of which about twenty species are known; some of them being scarcely larger

OPOSSUM.—(DIDELPHIS CANCRIVORA.)

than a mouse. They are characterized by the number of the incisor teeth, — which are ten above and eight below; three anterior compressed molars, and four sharply tuberculated back molars, the superior of which are triangular, the inferior oblong: so that, with the four canines, they have in all fifty teeth, a number greater than has as yet been observed in any other quadruped, except the newly-discovered *Myrmecobius*. The limbs are short; the feet plantigrade; and the toes, which are five on each foot, armed with sharp, strong, curved claws, except the inner toe or thumb on the hinder feet, which is opposable and destitute of a nail. The tail is scaly and naked, except at its base; and is usually more or less prehensile. In some of the smaller species the pouch is almost entirely wanting, being indicated only by a slight fold of skin; and in these the young adhere to the mother by entwining their little prehensile tails around hers, clinging to the fur of her back.

When, on the discovery of the Western Continent, this singular genus first became known, this hitherto unheard-of contrivance of nature for the protection and preservation of the young justly excited the admiration of naturalists; nor can any one, indeed, who for the first time witnesses this marsupial wonder, withhold the expressions or conceal the signs of his astonishment.

The VIRGINIAN OPOSSUM (*Didelphis Virginiana*), being one of the largest and most robust of the genus, and at the same time common in many parts of the southern states of North America, we shall take it as the best species to describe. This animal is about the size of a cat, but appears thicker owing to the length and upright growth of the fur. It has a long sharpened face, and very wide mouth, armed with numerous

VIRGINIAN OPOSSUM.
(DIDELPHIS VIRGINIANA.)

sharp teeth; the ears are thin, naked, round, and blackish, edged with a border of white: the legs are short; the feet armed with short claws, but the interior toes of the hind feet are flat and rounded. The whole hair is of a wool-like softness, short on the face and body, but long on the legs; and the general colour is a light gray. The tail is thick and black for upwards of three inches at the base, and is covered with small scales. The Opossum is nocturnal and timid animal, residing in the day-time in the hollows of trees, or among the branches, and prowling at night in search of its food, which consists of insects, eggs, birds, small reptiles, &c., as also fruits and roots; sometimes even invading the precincts of the farm-house, and killing the poultry. Its movements on the ground are slow and awkward; but it climbs trees with great facility, and uses its prehensile tail with great effect in suspending itself from the branches. When alarmed or irritated, it emits a most disgusting odour. In captivity it is slothful, and becomes inordinately fat, eating both animal and vegetable food with voracity. The flesh resembles in flavour that of a young pig. The wool, especially of those killed during the winter, is very long and fine, and might be advantageously employed in many manufactures.

The places in which the Opossum is usually found are thick woods, where they generally dwell in the hollow of decayed trees. They are usually hunted in the autumn, after the first frosts. Instead of taking to flight as soon as they perceive the approach of danger, they lie close to the branch on which they were clinging; and when they are discovered, the hunters take them by shaking the branch violently, when they fall to the ground: if, however, the hunter is unaccompanied by dogs, they either steal quietly away, or assume a death-like position, in which they will persevere even if taken up and handled. The female has ten to fifteen young, and she conceals herself in a thick nest of dry grass, in some obscure retreat. When first born, the young are in a very undeveloped state, being minute, blind, naked, and shapeless; but they are always found adhering to the teats of the mother, protected by her pouch. There they remain for fifty days, until they have attained the size of a mouse, at which period their eyes are opened, and their bodies are covered with hair. They may now be seen occasionally venturing from their hiding-place, but return to it on the least appearance of danger: nor do they absolutely withdraw from the care of the parent for a long time after; for when they no longer resort to her pouch for protection, it is said that they jump on her back, and twine their tails securely in hers, so that she may carry them out of the reach of danger.

ORANGE-TIP [BUTTERFLIES]. A name applied by insect collectors to Butterflies of the genus *Mancipium*.

ORANG-OUTANG. (*Simia Satyrus.*) Wonderful are the accounts which some of the earlier travellers have related of this quadrumanous animal, — the far-famed "Wild Man of the Woods,"—his size, swiftness, address, and ferocity. Persons, however, who have viewed these creatures only in a state of captivity have been particularly struck with their patient and docile dispositions, and their comparative helplessness: but the fact is, that the specimens seen in Europe have all been very young; and it is well-known that in their adult state, when their muscular power is more fully developed, their disposition alters, and they become as dangerously mischievous as they are then formidable.

The Orang-Outang is a native of the most unfrequented forests in the interior of Sumatra, Borneo, Malacca, &c.; living chiefly on fruits, but occasionally eating eggs, insects, and reptiles. In early youth it is remarkable for its rotundity of cranium and height of forehead; but these outward marks of superior mental power disappear as the animal advances in age. They have arms so long that the tops of the fingers can touch the ground when they stand upright; the body is covered with coarse reddish hair; on the head, shoulders, and back it is thick, but on the fore parts of the body rather thin; the neck is short and thick; the voice has a peculiarly shrill and hollow tone; the lips are thin and protuberant, the ears small, the nose particularly flat, and the face has a bluish cast.

ORANG-OUTANG. — (SIMIA SATYRUS.)

One of the most authentic accounts of this animal in its wild state, and which at the same time conveys a good idea of its powerful frame and arboreal habits, is given by Dr. Clarke Abel, in the 'Asiatic Researches,' who describes the capture of an Orang-Outang on the north-west coast of Sumatra. He was discovered by the company of a merchant's ship at a place called Ramboon; and on the approach of the boat's crew he came down from a tree, and made for a clump at some distance, "walking erect with a waddling gait, but sometimes accelerating his motion with his hands, and occasionally impelling himself forward by the bough of a tree. On being driven to a small clump, he gained by one spring a very lofty branch, and bounded from one branch to another with the swiftness of a common monkey, his progress being as rapid as that of a swift horse. After receiving five balls his exertions relaxed, and, reclining exhausted against a branch, he vomited a quantity of blood. The ammunition of the hunters being by this time exhausted, they were obliged to fell the tree in order to obtain him. But what was their surprise, to see him, as the tree was falling, effect his retreat to another, with seemingly undiminished vigour! In fact, they were forced to cut down all the trees before they could force him to combat his enemies on the ground, and when finally overpowered by numbers, and nearly in a dying state, he seized a spear made of a supple wood, which would have withstood the strength of the stoutest man, and broke it like a reed. It was stated by those who aided in his death, that the human-like expression of his countenance, and his piteous manner of placing his hands over his wounds, distressed their feelings so as almost to make them question the nature of the act they were committing. He was seven feet high, with a broad expanded chest, and narrow waist. His chin was fringed with a beard that curled neatly on each side, and formed an ornamental rather than a frightful appendage to his visage. His arms were long even in proportion to his height, but his legs were much shorter. Upon the whole he was a wonderful beast to behold, and there was more about him to excite amazement than fear. His hair was smooth and glossy, and his whole appearance showed him to be in the full vigour of youth and strength."

ORBICULA.' A genus of Conchifera, found in large masses on the coasts of Peru and Chili, and also in the Northern seas. The shells of these bivalves are horny, sub-orbicular, rather compressed, the upper valve patelliform, the lower flat. In the centre of the latter is a small oval depression, with an oblique fissure in it for the passage of a tendon; four muscular impressions in each valve; no hinge. The animal has two short ciliated arms.

ORGYIA, or VAPOURER MOTH. The genus *Orgyia* comprises those species of Moths which fly by day, with a vapouring kind of motion (whence their English name), which have unwieldy partners, furnished with slight rudiments of wings, and therefore incapable of flight. The male of the common

VAPOURER MOTH — MALE.
(ORGYIA ANTIQUA.)

species of this genus (*O. antiqua*), which we have here selected, varies from one inch and a sixth to one inch and a half in the expanse of the fore wings, which are of a red brown, with dusky clouds and two undulated strigæ,

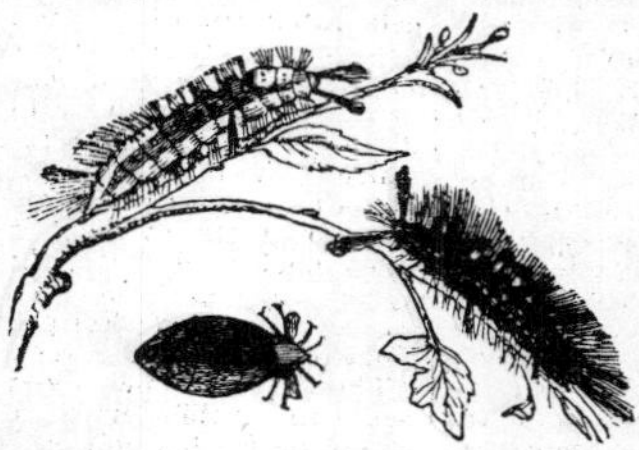

TWO CATERPILLARS, AND THE ALMOST WINGLESS FEMALE OF THE VAPOURER MOTH.

the second of which terminates in a kidney-shaped white spot near the anal angle of the fore wings, and with a pale clay-coloured, crescent-shaped, discoidal spot. The hind wings are dark orange-brown. The female is dull ash-coloured, with the rudiments of wings very pale. The caterpillar is very handsome, being spotted with red, and with four thick whitish tufts of hair on the back, and with long pencils of clavate hairs on the sides of the head, at the sides of the body, and over the tail. The ground colour of the body of the male is darker than in the female, which is varied with pale yellow and gray. They feed on a great variety of trees, and are found throughout the summer. The Moth appears in the autumn, and is seen flying during the day-time, even in the streets of London.

ORIOLE. A name applied to birds of different groups,—the European Oriole being allied to the Thrushes, the American Oriole to the Starlings. The first species we describe

GOLDEN ORIOLE. — (ORIOLUS GALBULA.)

is the GOLDEN ORIOLE (*Oriolus galbula*), or, as it is sometimes called, the GOLDEN THRUSH, is during the summer months an inhabitant of many of the temperate and warmer parts of Europe, though its presence in this country is very rare. It is about the size of a blackbird, but its bill is larger, arched, and slightly notched at the tip. Its colour is a very fine bright golden yellow, except the wings and tail, which are black; but the quill feathers and some of the larger coverts are tipped with yellow, the latter forming a small yellow spot on the edge of the wing: all the tail feathers, except the two middle ones, are tipped with yellow: the bill is brownish-red, and between the bill and eye is a black stripe; the legs are brown. The female differs widely from the male in colour: where he is yellow, she is of a dull olive green; her wing-coverts, secondary quills, and upper parts of the tail feathers, partake of the same colour, but are much darker; the quills and lower ends of the tail feathers are dusky, and, as well as the former, are all tipped, more or less, with pale dull yellow. This bird is of a migratory nature, and is supposed to spend the winter in Asia and Africa, and to pass the summer in Europe. When about to construct its nest, the Golden Oriole selects the forked extremity of some slender branch (usually in the lower part of a high tree), and wreathing the two forks round with straws, grasses, or other vegetable fibres proper for the purpose, at length connects the two ends in order to form the verge of the nest; then continuing the straws from the one side to the other, giving the whole a proper depth, and crossing and interweaving them as the work proceeds, forms the concavity or basket, which is afterwards thickened with the stems of the finer grasses, intermixed with mosses and lichens, and lined with feathers and still finer materials. It usually lays four or five eggs, which are of a dull white colour, speckled with black.

The name Oriole, as we have said, is also applied to birds allied to the Starlings, to which naturalists have applied the names of *Icterus* and *Agelaius.* Nearly all the birds which belong to this numerous and beautiful genus are natives of the American continent. Some of the species are gregarious, others solitary in their habits; but all are of a noisy and restless disposition, and feed on various kinds of fruit, grain, and insects. They are remarkable for the structure of their nest, which in some species hangs from the branch to which it is attached, and in others is sewed or fastened with peculiar art beneath the surface of some very large leaf. The bill of this genus is straight, conical, and very sharp-pointed; mandibles of equal length; nostrils small, placed at the base of the bill, and partly covered; tongue divided at the end; toes three forward and one backward; the middle joined near the base to the outermost one.

The Baltimore Oriole (*Icterus Baltimore*) takes its name (according to Catesby) from its colours, which are black and orange, being those of the arms or livery of Lord Baltimore, formerly proprietary of Maryland. This bird is seven inches in length; bill almost straight, strong, tapering to a sharp point, black, and sometimes lead-coloured, above, the lower mandible light blue towards the base. Head, throat, upper part of the back and wings, black; lower part of the back, rump, and whole under parts, a bright orange, deepening into vermillion on the breast; the black on the shoulders is also divided by a band of orange; exterior edges of the greater wing-coverts, as well as the

BALTIMORE BIRD.—(ICTERUS BALTIMORE.)

edges of the secondaries, and part of those of the primaries, white; the tail feathers under the coverts, orange; the two middle ones, from thence to the tips, are black, the next five, on each side, black near the coverts, and orange towards the extremities, so disposed, that when the tail is expanded, and the coverts removed, the black appears in the form of a pyramid, supported on an arch of orange. Tail slightly forked, the exterior feather on each side a quarter of an inch shorter than the others: legs and feet light blue, or lead colour: iris of the eye hazel.

Almost the whole genus of Orioles build pensile nests. In Wilson's American Ornithology we read, that "so solicitous is the Baltimore to procure proper materials for his nest, that, in the season of building, the women in the country are under the necessity of narrowly watching their thread that may chance to be out bleaching, and the farmer to secure his young grafts; as the Baltimore, finding the former, and the strings which secure the latter, so well adapted for his purpose, frequently carries off both; or, should the one be over heavy, and the other too firmly tied, he will tug at them a considerable time before he gives up the attempt. Skeins of silk and hanks of thread have been often found, after the leaves were fallen, hanging round the Baltimore's nest; but so woven up, and entangled, as to be entirely irreclaimable. Before the introduction of Europeans, no such material could have been obtained here; but, with the sagacity of a good architect, he has improved this circumstance to his advantage; and the strongest and best materials are uniformly found in those parts by which the whole is supported. Their principal food consists of caterpillars, beetles, and bugs, particularly of one of a brilliant glossy green, fragments of which I have almost always found in their stomach, and sometimes these only. The song of the Baltimore is a clear mellow whistle, repeated at short intervals as he gleans among the branches. There is in it a certain wild plaintiveness and *naïveté* extremely interesting. It is not uttered with the rapidity of the ferruginous thrush (*Turdus rufus*), and some other eminent songsters; but with the pleasing tranquillity of a careless ploughboy, whistling merely for his own amusement. When alarmed by an approach to his nest, or any such circumstance, he makes a kind of rapid chirruping, very different from his usual note. This, however, is always succeeded by those mellow tones which seem so congenial to his nature.

"The Baltimore inhabits North America, from Canada to Mexico, and is even found as far south as Brazil. Since the streets of our cities have been planted with that beautiful and stately tree, the Lombardy poplar, these birds are our constant visitors during the early part of summer; and, amid the noise and tumult of coaches, drays, wheelbarrows, and the din of the multitude, they are heard chanting "their native wood-notes wild;" sometimes, too, within a few yards of an oyster-man, who stands bellowing, with the lungs of a Stentor, under the shade of the same tree; so much will habit reconcile even birds to the roar of the city, and to sounds and noises, that, in other circumstances, would put a whole grove of them to flight. These birds are several years in receiving their complete plumage. Sometimes the whole tail of a male individual in spring is yellow, sometimes only the two middle feathers are black, and frequently the black on the back is skirted with orange, and the tail tipped with the same colour. Three years, I have reason to believe, are necessary to fix the full tint of the plumage, and then the male bird appears as already described. The chief difference between the male and female Baltimore Oriole is the superior brightness of the orange colour of the former to that of the latter. The black on the head, upper part of the back and throat of the female, is intermixed with dull orange; whereas, in the male, those parts are of a deep shining black; the tail of the female also wants the greater part of the black, and the whole lower parts are of a much duskier orange."

The RED-WINGED STARLING (*Agelaius phœniceus*) — the *Sturnus predatorius* of Wilson — is thus described by that observant and industrious ornithologist: — "This notorious and celebrated corn-thief, the long reputed plunderer and pest of our honest and laborious farmers, now presents himself before us, with his female copartner in iniquity, to receive the character due for their very active and distinguished services. In investigating the nature of these, I shall endeavour to render strict historical justice to this noted pair; adhering to the honest injunction of the poet,

Nothing extenuate,
Nor set down aught in malice.

Let the reader divest himself equally of prejudice, and we shall be at no loss to ascertain accurately their true character.

"The Red-winged Starlings, though generally migratory in the states north of Maryland, are found during winter in immense flocks, sometimes associated with the purple grakles, and often by themselves,

RED-WINGED STARLING.
(AGELAIUS PHŒNICEUS.)

along the whole lower parts of Virginia, both Carolinas, Georgia, and Louisiana, particularly near the sea coast, and in the vicinity of large rice and corn fields. In the months of January and February, while passing through the former of these countries, I was frequently entertained with the aerial evolutions of these great bodies of Starlings. Sometimes they appeared driving about like an enormous black cloud carried before the wind, varying its shape every moment. Sometimes suddenly rising from the fields around me with a noise like thunder; while the glittering of innumerable wings of the brightest vermillion amid the black cloud they formed, produced on these occasions a very striking and splendid effect. Then descending like a torrent, and covering the branches of some detached grove, or clump of trees, the whole congregated multitude commenced one general concert or chorus, that I have plainly distinguished at the distance of more than two miles; and, when listened to at the intermediate space of about a quarter of a mile, with a slight breeze of wind to swell and soften the flow of its cadences, was to me grand, and even sublime. The whole season of winter, that, with most birds, is passed in struggling to sustain life in silent melancholy, is, with the Red-wings, one continued carnival. The profuse gleanings of the old rice, corn, and buckwheat fields, supply them with abundant food, at once ready and nutritious; and the intermediate time is spent either in aerial manœuvres, or in grand vocal performances, as if solicitous to supply the absence of all the tuneful summer tribes, and to cheer the dejected face of nature with their whole combined powers of harmony.

"About the 20th of March, or earlier, if the season be open, they begin to enter Pennsylvania in numerous, though small parties. These migrating flocks are usually observed from daybreak to eight or nine in the morning, passing to the north, chattering to each other as they fly along; and, in spite of all our antipathy, their well-known notes and appearance, after the long and dreary solitude of winter, inspire cheerful and pleasing ideas of returning spring, warmth, and verdure. Selecting their old haunts, every meadow is soon enlivened by their presence. They continue in small parties to frequent the low borders of creeks, swamps, and ponds, till about the middle of April, when they separate in pairs to breed; and, about the last week in April or first in May, begin to construct their nest. The place chosen for this is generally within the precincts of a marsh or swamp, meadow, or other like watery situation, — the spot, usually a thicket of alder bushes, at the height of six or seven feet from the ground; sometimes in a detached bush, in a meadow of high grass; often in a tussock of rushes or coarse rank grass; and not unfrequently on the ground: in all which situations, I have repeatedly found them. When in a bush, they are generally composed outwardly of wet rushes, picked from the swamp, and long tough grass in large quantity, and well lined with very fine bent. The rushes, forming the exterior, are generally extended to several of the adjoining twigs, round which they are repeatedly and securely twisted; a precaution absolutely necessary for its preservation, on account of the flexible nature of the bushes in which it is placed. The same caution is observed when a tussock is chosen, by fastening the tops together, and intertwining the materials of which the nest is formed with the stalks of rushes around. When placed on the ground, less care and fewer materials being necessary, the nest is much simpler and slighter than before. The female lays five eggs, of a very pale light blue, marked with faint tinges of light purple and long straggling lines and dashes of black. It is not uncommon to find several nests in the same thicket, within a few feet of each other.

"During the time the female is sitting, and still more particularly after the young are hatched, the male, like most other birds that build in low situations, exhibits the most violent symptoms of apprehension and alarm on the approach of any person to its near neighbourhood. Like the lapwing of Europe, he flies to meet the intruder, hovers at a short height over-head, uttering loud notes of distress; and, while in this situation, displays to great advantage the rich glowing scarlet of his wings, heightened by the jetty black of his general plumage. As the danger increases, his cries become more shrill and incessant, and his motions rapid and restless; the whole meadow is alarmed, and a collected crowd of his fellows hover around, and mingle their notes of alarm and agitation with his. When the young are taken away, or destroyed, he continues for several days near the place, restless and dejected, and generally recommences building soon after, in the same meadow. Towards the beginning or middle of August, the young birds begin to fly in flocks, and at that age nearly resemble the female, with the exception of some reddish or orange, that marks the shoulders of the males, and which increases in space and brilliancy as winter approaches. It has been frequently remarked, that, at this time, the young birds chiefly associate by themselves, there being sometimes not more than two or three old males observed in a flock of many thousands. These, from the superior blackness and rich red of their plumage, are very conspicuous.

"Before the beginning of September, these flocks have become numerous and formidable; and the young ears of maize, or Indian corn, being then in their soft, succulent, milky state, present a temptation that cannot be resisted. Reinforced by numerous and daily flocks from all parts of the interior, they pour down on the low countries in prodigious multitudes. Here they are seen, like vast clouds, wheeling and driving over the meadows and devoted corn fields, darkening the air with their numbers. Then commences the work of destruction on the corn, the husks of which, though composed of numerous envelopments of closely wrapt leaves, are soon completely or partially torn off; while from all quarters myriads continue to pour down like a tempest, blackening half an acre at a time; and, if not disturbed, repeat their depredations till little remains but the cob and the shrivelled skins of the grain; what little is left of the tender ear, being exposed to the rains and weather, is generally much injured. All the attacks and havoc made at this time among them with the gun, and by the hawks,—several species of which are their constant attendants,—has little effect on the remainder. When the hawks make a sweep among them, they suddenly open on all sides, but rarely in time to disappoint them of their victims; and, though repeatedly fired at, with mortal effect, they only remove from one field to an adjoining one, or to another quarter of the same enclosure. From dawn to nearly sunset, this open and daring devastation is carried on, under the eye of the proprietor; and a farmer, who has any considerable extent of corn, would require half-a-dozen men at least, with guns, to guard it; and even then, all their vigilance and activity would not prevent a good tithe of it from becoming the prey of the blackbirds. The Indians, who usually plant their corn in one general field, keep the whole young boys of the village all day patrolling round and among it; and each being furnished with bow and arrows, with which they are very expert, they generally contrive to destroy great numbers of them.

"It must, however, be observed, that this scene of pillage is principally carried on in the low countries, not far from the sea-coast, or near the extensive flats that border our large rivers; and is also chiefly confined to the months of August and September. After this period, the corn having acquired its hard shelly coat, and the seeds of the reeds or wild oats, with a profusion of other plants, that abound along the river shores, being now ripe, and in great abundance, they present a new and more extensive field for these marauding multitudes. The reeds also supply them with convenient roosting places, being often in almost unapproachable morasses; and thither they repair every evening from all quarters of the country. In some places, however, when the reeds become dry, advantage is taken of this circumstance, to destroy these birds, by a party secretly approaching the place, under cover of a dark night, setting fire to the reeds in several places at once, which, being soon enveloped in one general flame, the uproar among the blackbirds becomes universal; and, by the light of the conflagration, they are shot down in vast numbers while hovering and screaming over the place. Sometimes straw is used for the same purpose, being previously strewed near the reeds and alder bushes, where they are known to roost, which being instantly set on fire, the consternation and havoc is prodigious; and the party return by day to pick up the slaughtered game. About the first of November, they begin to move off towards the south; though, near the sea coast, in the states of New Jersey and Delaware, they continue long after that period.

"Such are the general manners and character of the Red-winged Starling; but there remain some facts to be mentioned, no less authentic, and well deserving the consideration of its enemies, more especially, of those whose detestation of this species would stop at nothing short of total extirpation.

"It has been already stated, that they arrive in Pennsylvania late in March. Their general food at this season, as well as during the early part of summer (for the crows and purple grakles are the principal pests in planting time), consists of grub-worms, caterpillars, and various other larvæ, the silent, but deadly enemies of all vegetation, and whose secret and insidious attacks are more to be dreaded by the husbandman than the combined forces of the whole feathered tribes together. For these vermin, the Starlings search with great diligence; in the ground, at the roots of plants, in orchards, and meadows, as well as among buds, leaves, and blossoms; and, from their known voracity, the multitudes of these insects which they destroy must be immense. Let me illustrate this by a short computation: if we suppose each bird, on an average, to devour fifty of these larvæ in a day (a very moderate allowance), a single pair, in four months, the usual time such food is sought after, will consume upwards of twelve thousand. It is believed, that not less than a million pair of these birds are distributed over the whole extent of the United States in summer; whose food, being nearly the same, would swell the amount of vermin destroyed to twelve thousand millions. But the number of young birds may be fairly estimated at double that of their parents; and, as these are constantly fed on larvæ, for at least three weeks, making only the same allowance for them as for the old ones, their share would amount to four thousand two hundred millions; making a grand total of sixteen thousand two hundred millions of noxious insects destroyed in the space of four months by this single species! The combined ravages of such a hideous host of vermin would be sufficient to spread famine and desolation over a wide extent of the richest and best cultivated country on earth. All this, it may be said, is mere supposition. It is, however, supposition founded on known and acknowledged facts. I have never dissected any of these birds in spring without receiving the most striking and satisfactory proofs of these facts; and though, in a matter of this kind, it is impossible to ascertain precisely the amount of the benefits derived by agriculture from this and many other species of our birds, yet, in the present case, I cannot resist the belief, that the services of this species, in spring, are far more important aad beneficial than the value of all that portion of corn which a careful and active farmer permits himself to lose by it.

"The Red-winged Starling is nine inches long, and fourteen inches in extent; the general colour is a glossy black, with the exception of the whole lesser wing-coverts, the first, or lower row of which is of a reddish cream colour, the rest a rich and splendid scarlet; legs and bill, glossy brownish black; irides, hazel; bill, cylindrical above, compressed at the sides, straight, running considerably up the forehead, where it is prominent, rounding and flattish towards the tip, though sharp-pointed; tongue, nearly as long as the bill, tapering and lacerated at the end; tail, rounded, the two middle feathers also somewhat shorter than those immediately adjoining.

"The female is seven inches and a quarter in length, and twelve inches in extent; chin, a pale reddish cream; from the nostril over the eye, and from the lower mandible, run two stripes of the same, speckled with black; from the posterior angle of the eye backwards, a streak of brownish black covers the auriculars; throat, and whole lower parts, thickly streaked with black and white, the latter inclining to cream on the breast; whole plumage above, black, each feather bordered with pale brown, white, or bay, giving the bird a very mottled appearance; lesser coverts, the same; bill and legs as in the male."

We observe that Mr. Darwin, in his 'Researches,' speaking of the various birds which abound on the undulating grassy plains of Maldonado, says, "Several species of the genus Cassicus, allied to our Starlings in habits and structure, and of Tyrant Flycatchers, and a Mocking-bird, from their numbers, give a character to the ornithology. Some of the Cassici are very beautiful, black and yellow being the prevailing colours; but *Oriolus ruber* offers an exception, in having its head, shoulders, and thighs of the most splendid scarlet. This bird differs from its congeners in being solitary. It frequents marshes; and, seated on the summit of a low bush, with its mouth wide open, utters a plaintive agreeable cry, which can be heard at a long distance."

ORNITHICNITES. [See SUPPLEMENT.]

ORNITHORHYNCHUS. The Duck-billed Platypus or Mullingong is peculiar to New Holland and Van Diemen's Land. When first sent to this country, the abnormal character of its beak excited the suspicion of naturalists that some trick had been at-

DUCK-BILLED MULLINGONG.
(ORNITHORHYNCHUS PARADOXUS.)

tempted to be played off upon them; nor was it until one or two more arrived, that they were disposed to believe it was a *bonâ fide* appendage to the animal's body. The Ornithorhynchus is about twenty inches long, having a long and flattened body like that of the Otter, covered with a thick soft fur, moderately dark brown above, and whitish beneath. The muzzle is elongated, enlarged, and flattened, resembling the beak of a duck, like which, its edges are armed with transversal plates. The teeth are situate in the back part of the mouth, two on each side, with flat tops and no roots. The feet are furnished with a membrane uniting the toes, and in the anterior feet extending beyond the nails. The tail is flat and obtuse. From the form of this animal, it is fitted to reside in the water; and it must feed on soft food, as the structure of the beak will not enable it to grasp anything firmly. The central portion of the mandibles is a bony continuation from the skull and anteriorly and laterally a cartilaginous substance, perfectly moveable, extends some little distance from the bony portion. Feet five-toed and webbed. In the fore-feet the web extends a short distance beyond the claws, is loose, and falls back when the animal burrows: claws strong, blunt, the two lateral shorter than the three middle ones. Hind feet short, narrow, turned backwards, and, when the animal is at rest, somewhat resembling a fin. The male Ornithorhyncus is armed with a spur on each hind leg, having a canal in it similar to that in the poison-fang of venomous serpents, and, like this, also furnished with a gland at the base, secreting a fluid: hence it has been thought likely, though there is no evidence of the fact, that wounds produced by them would be dangerous. They have no external ear, and their eyes are very small, but brilliant. The motions of the mandibles in this animal, when seeking its food in the mud and water, are the same as those of a *duck* when feeding in similar situations. Their burrows are excavated in the banks of the streams they inhabit, and are of very curious construction. The entrance is situated near the water's edge, on a steep part of the bank, and is concealed amongst the herbage. The young are produced in a very imperfect state, and are very unlike the full-grown animal. The skin is entirely destitute of fur; the eyes are not formed, and their place is merely indicated by the presence of a few wrinkles on the skin. The margin of the bill is at that time soft, and the tongue advances to its front edge; so that the young animal can obtain nourishment by sucking, which was at first thought impossible. The mammary gland is very simple in structure, and is divided into a large number of separate lobes. The Ornithorhyncus, when asleep, rolls itself up like a hedgehog, or curls itself like a dog, keeping its back warm by bringing over it the flattened tail. It dresses its fur, combing it with its feet, and pecking at it with its beak; and seems to take great delight in keeping it smooth and clean.

In conclusion, we cannot but join in the remark of Dr. Shaw, who was the first to describe it. "Of all the Mammalia yet known it seems the most extraordinary in its conformation; exhibiting the perfect resemblance of the beak of a duck engrafted on the head of a quadruped."

ORTHOCERAS. A genus of fossil shells, found in strata mostly below the carboniferous or mountain limestone. These shells are straight, with septa regularly concave towards the aperture, perforated by a nearly cylindrical siphuncle near the centre of the disc.

ORTHOPTERA. An order of insects, distinguished by the following characters: The body generally less firm in texture than the *Coleoptera*, and covered by soft semi-membranous elytra furnished with nervures. The superior wings often overlap horizontally, as in the Cockroaches, but in many species they meet at an angle, as in the Grasshoppers and Locusts. The legs of some are formed for running, others for leaping. The antennæ are usually filiform, and sometimes extremely long and slender, in which case they are composed of innumerable minute joints. The parts of the mouth are well developed, and approach in structure those of the order Coleoptera. The Orthoptera undergo a semi-metamorphosis, of which all the mutations are reduced to the growth and development of the elytra and wings that are always visible in a rudimental state in the nymph. All the insects of this order, without exception, are terrestrial, even in the first two states of their existence. Some are carnivorous, or omnivorous, but the greater part feed on living plants. The Order comprises numerous well-known insects, often of large size and splendid colours; such as Grasshoppers, Locusts, &c.: nay, some of the largest known insects belong to it; a few species attaining the length of eight or nine inches. Comparatively few are found in temperate regions; the tropics claiming the largest and most splendidly coloured among them. All the insects belonging to this order, except the *Mantidæ*, which prey on other insects, are destructive to vegetation, or injurious to our household possessions.

ORTOLAN. (*Emberiza hortulana.*) This bird, so much esteemed for the delicacy of its flesh, is a native of the southern parts of Europe, and a summer visitor also of the central and northern parts. It is a species of *Fringillidæ*, rather more than six inches in length, and to a cursory observer might be easily mistaken for the yellow-hammer. It is yellow on the throat and around the eyes; the breast and belly are a reddish bay; the rump red; and the upper part of the body brown, varied with black; bill and feet inclining to flesh-colour. No bird what-

ORTOLAN. (EMBERIZA HORTULANA.)

ever has been so highly celebrated in the annals of gastronomy as the Ortolan, whether we consider the practices resorted to at the present day to fit them for the tables of the wealthy, or refer to the enormous prices paid for them by the epicures of ancient Rome. The manner in which they are artificially brought to the highest degree of perfection, in Italy and the south of France, is by confining them in a room from which the rays of the sun are excluded, and which is lighted by lamps kept constantly burning. There the birds are kept plentifully supplied with millet seed and other food of the most nutritive kind, till they become mere lumps of fat; in which state they are regarded as most delicious, although so rich as soon to satiate the appetite of even a professed gourmand. A great traffic was formerly carried on from the island of Cyprus in these birds. They are caught in vast numbers there, and pickled in casks, each containing from three to four hundred, prepared with spice and vinegar. In some years the number of casks exported has amounted to 400, or, upon an average, 14,000 of these highly-prized morsels. — The Ortolan frequents bushy places, but sometimes makes its nest on the ground in corn-fields; and lays four or five dull white or grayish eggs, speckled and spotted with black.

ORTYGOMETRA. A genus of Grallatorial birds, containing the COMMON CRAKE [which see].

ORTYX. A genus of Gallinaceous birds, which may be regarded as the Partridges

CALIFORNIAN QUAIL.—(O. CALIFORNICUS.)

and Quails of America, but differing from those of the Eastern hemisphere in some striking features. They have a shorter and stouter beak, more convex above; and some have remarkable recurved top-knots. They perch on branches, and when disturbed, even on trees.

Mr. Gould has published a beautifully illustrated monograph of them. Our first figure represents the finely crested and gracefully shaped and coloured Californian Quail, which, as the name implies, is a native of California: our second represents a species which has been long known and described as the AMERICAN QUAIL, or PARTRIDGE, as it is termed (*Ortyx Virginanus*), and is found from New England to Honduras. It is about nine inches in length; the bill is black; eye dark hazel; crown, neck, and upper part of the breast, red brown; sides of the neck spotted with white and black on a reddish brown ground; line over the eye, down the neck, and whole chin pure white, bordered by a band of black, which descends and forms a crescent on the throat: the back, scapulars, and lesser coverts, red brown, mixed with ash, and minutely marked with black; wings plain and dusky; lower part of the breast and belly whitish, marked with black arrow-heads; tail ash, spotted with yellow brown. Notwithstanding there is some resemblance in form and general appearance between the Quails of the two continents, they differ very widely in their habits. Instead of being a bird of passage, scarcely any of the feathered tribe appear to have such strong local attachments as the American Quail. The female constructs her simple nest, in May, generally at the foot of a thick tuft of grass, that shelters and conceals it; and lays from fifteen to twenty eggs, which

AMERICAN QUAIL.—(ORTYX VIRGINIANUS.)

are perfectly white. Wilson is of opinion that the common idea, that Quails occasionally lay in each other's nests, is correct. About the beginning of September the young birds nearly attain their full growth, and associate in flocks or coveys of various sizes; at which time also their untiring persecution by sportsmen and trappers begins. During the end of the summer and the beginning of the autumn, the note of the male is everywhere heard; and by the commencement of October they enter on what is termed their *running* season, when they are to be met with in swamps and thickets, instead of the open fields. They are particularly fond of buck-wheat and Indian corn; but grain of all kinds, seeds, and insects, supply them with food. Like the rest of the gallinaceous tribe, they fly with a loud whirring sound, occasioned by the shortness of their wings and the rapidity with which they move them. During the winter they often suffer severely from the inclemency of the weather and whole coveys are found frozen in spots where they had endeavoured to shelter themselves.

ORYCTEROPUS, or AARD-VARK. (*Orycteropus Capensis.*) This insectivorous animal partakes of the nature both of the Ant-eater and the Armadillo; agreeing with the former in its general habits, but, although entirely destitute of scaly armour, more resembling the latter as to its anatomical structure. The skin is thick, coarse, and covered with stiff hair; the limbs short, thick, and very muscular. It is of a deep gray colour, tinged with reddish brown on the sides, and blackish legs: the fore-feet have each four stout toes, armed with large solid nails, the hind ones five; and the nails or claws on all the feet are remarkably strong. This animal is very common in some parts of South Africa, and has received the name of Aard-vark [earth hog] from the Dutch colonists at the Cape of Good Hope, from its habit of burrowing (which its taper head and powerful claws are admirably adapted for), as well as from its fancied resemblance to a small short-legged pig, which at first sight it conveys a tolerable notion of. When full grown the Aard-vark measures about five feet from the tip of the

ORYCTEROPUS CAPENSIS.

snout to the end of the tail, the latter being nearly half the length of the body. By means of its long glutinous tongue it feeds principally on ants, which, after it has effected an entrance into their dome-like habitations, it literally devours by thousands; and as these insects in tropical climates are not only very large, but of a fat and unctuous nature, and found in immense abundance, the animal is generally in good condition. The flesh is allowed to be wholesome and palatable food, and the hind quarters especially, when cured as hams, are much esteemed.

ORYX; or EGYPTIAN ANTELOPE. (*Oryx gazella.*) The size of the Egyptian Antelope, or Pasan, is somewhat superior to that of a deer, and it is more easily distinguished than many others in this extensive race; the horns affording a character perfectly clear and constant, being three feet long, nearly straight, annulated half way up, and gradually tapering to the point. The head is white, with triangular patches of black on the forehead and under the eyes: the neck and upper part of the body are of a pale bluish gray; the belly and insides of the limbs are white; and a dark stripe runs along the back to the tail, which much resembles that of a horse. The hoofs and horns are black: the hair under the throat, along the ridge of the back, and over the shoulders, is long and rough. It inhabits different parts of Africa, and is met with also in Persia, India, and Arabia. It is resolute and dangerous when hard pressed, its long sharp horns being used with amazing energy and address.

OSPHRANTER. A genus of Kangaroos figured in Mr. Gould's fine work. One species, *O. Antilopinus*, or Red Wallaroo, is from Port Essington. Capt. Chambers informed Mr. Gould, that, when hard pressed, this robust-formed animal becomes exceedingly fierce and bold, and while among the rocks and at bay, a most dangerous antagonist, one of his finest dogs being tumbled over a precipice and killed by an old male. The female is much smaller than the male, the former being but five feet six inches from the snout to the end of the tail; while the latter is at least seven feet three inches.

OSPREY. (*Pandion haliœtus.*) This is one of the most numerous of all the large birds of prey, and is found scattered over the whole of Europe. Its haunts are on the sea-shore, and on the borders of rivers and lakes: its principal food is fish, upon which it darts with great rapidity and undeviating aim. It is nearly two feet in length: bill black, eye yellow; the head is small and flat, the crown white, marked with oblong dusky spots; the cheeks, and all the under parts of the body, are white, slightly spotted with brown on the breast; from the corner of each

OSPREY.—(PANDION HALIÆTUS.)

eye a streak of dark brown extends down the side of the neck towards the wing; the upper part of the body is brown; the legs are very short, thick, strong, and of a pale blue colour, and the claws black: the outer toe is larger than the inner one, and easily turns backwards, by which means this bird can more readily secure its slippery prey. The Osprey builds its nest among rocks or on fir-trees, and lays three or four eggs, of an elliptical form, rather less than those of a hen.

OSTEOLEPIS, or BONY-SCALE FISH. An ichthyolite of very singular structure, discovered by Mr. Hugh Miller, and described by him in his work, entitled 'The Old Red Sandstone,' &c. We shall give the account of it in his own words: "We are accustomed to see vertebrated animals with the bone uncovered in one part only,—that part the teeth,—and with the rest of the skeleton wrapped up in flesh and skin. Among the reptiles we find a few exceptions; but a creature with a skull as naked as its teeth,—the bone being merely covered, as in these, by a hard shining enamel, and with toes also of bare enamelled bone,—would be deemed an anomaly in creation. And yet such was the condition of the *Osteolepis* and many of its contemporaries. The enamelled teeth were placed in jaws which presented outside a surface as naked and as finely enamelled as their own. The entire head was covered with enamelled osseous plates, furnished inside like other bones, as shown by their cellular construction, with their nourishing blood-vessels, and perhaps their oil, and which rested apparently on the cartilaginous box, which must have enclosed the brain, and connected it with the vertebral column. I cannot better illustrate the peculiar condition of the fins of this ichthyolite, than by the webbed foot of a water-fowl. The web or membrane in all the aquatic birds with which we are acquainted, not only connects, but also covers the toes. The web or membrane in the fins of existing fishes accomplishes a similar purpose; it both connects and covers the supporting bones or rays. Imagine, however, a webbed foot in which the toes—connected but not covered—present, as in skeletons, an upper and under surface of naked bone; and a very correct idea may be formed from such a foot, of the condition of fin which obtained among at least one-half the ichthyolites of the Lower Old Red Sandstone. The supporting bones or rays seem to have been connected laterally by the membrane, but on both sides they presented bony and finely-enamelled surfaces. In this singular class of fish, all was bone without, and all was cartilage within; and the bone in every instance, whether in the form of jaws or of plates, of scales or of rays, presented an external surface of enamel." "The *Osteolepis* was cased from head to tail in complete armour. The head had its plaited mail, the body its scaly mail, the fins their mail of parallel and jointed bars; the entire suit glittered with enamel; and every plate, bar, and scale was dotted with microscopic points. Every ray had its double or treble punctulated group; the markings lie as thickly in proportion to the fields they cover, as the circular perforations in a lace veil; and the effect, viewed through the glass, is one of lightness and beauty."

OSTRACEA: OSTREA. [See OYSTER.]

OSTRACION. A singular genus of fishes, distinguished from all others by the bony crust or covering in which they are enveloped, and the species differing also from each other by certain peculiarities of form. They are termed Ostracions or Trunk fishes. The head and body are covered with plates of bone, so united as to form an inflexible cuirass: leaving only the tail, fins, mouth, and a small portion of the gill-opening, capable of motion,—all of which movable parts pass through openings of the armadillo-like defensive coat of mail. The vertebræ are also compactly fixed together. There are no ventral fins, and the dorsal and anal are small and placed far back. There is little flesh; but the liver is large, and abounds in oil. The surface is often armed with spines. Nearly all the species are natives of the Indian and American seas; and some are considered excellent fish for the table. None are known in the British seas.

OSTRICH. (*Struthio.*) The Ostrich and its allies, belonging to the order *Cursores*, are distinguished by having their wings but little developed; and accordingly, instead of being denizens of the air, they may be considered as exclusively terrestrial. They have wings, it is true, admirably adapted to assist them in running; but they are totally incapable, by their most energetic action, of raising the birds from the ground. Nor is it only in the absence of perfect wings, but in the character of the plumage, that the want of adaptation of these birds to flight in the air is manifested: for the barbs of the feathers have so little adhesion to each other, that the air can pass readily between them. It may indeed be said, that while the Ostrich has the general outline and properties of a bird, it still retains many of the traits of a quadruped, and appears to fill up the chasm in nature which separates one class of beings from another. No bird, however, is more justly celebrated, not only from the beauty and value of its plumage, but also from its great size and peculiar habits.

The AFRICAN OR TRUE OSTRICH (*Struthio camelus*) is from seven to eight feet high from the top of its head to the ground: much of this, however, is made up by the great length of its neck. Its head is small, and both it and the neck are destitute of feathers, having only a few scattered hairs. The feathers on the body of the male bird are black; but on the female dusky; those of the wings and tail are white, sometimes marked with black; and on each of the wings are two spurs, about an inch long. The thighs are naked, and the legs hard and scaly. It has two very

OSTRICH.—(STRUTHIO CAMELUS.)

large toes, of unequal size; the largest, which is on the inside, is seven inches long, including the claw; the other, about four inches, is destitute of a claw. It inhabits the sandy deserts of Arabia and Africa, in large flocks; everywhere avoiding the presence of Man, but not disliking the society of other animals. The wings are furnished with loose and flexible plumes. The elegance of these feathers, arising from their slender stems and the disunited barbs, has occasioned them to be prized in all ages; and as they still constitute a valuable article of commerce, there is no chance of the Ostrich being allowed to remain undisturbed, even in the desolate regions which he inhabits. The hunting of this bird is extremely laborious, as he is far swifter than the fleetest horse. The mode adopted by the Arabians and Moors is to continue the pursuit as long as possible, when the chase is taken up by another on a fresh horse, till the bird is worn down; which is the more readily done, as the Ostrich, instead of pursuing a straight course, runs in a circuitous direction. The European sportsman, we are told, after riding so that the bird shall pass within shot, dismounts and brings it down with the rifle. In Sir James Alexander's Travels we read that the Kaffirs nimbly pursue the fleet and powerful Ostrich, and enclose him; when he makes a rush at a part of the circle, kicking out furiously, and clearing all before him, if not mortally and speedily *assegaied*.

The Ostrich has a capacious crop, strong gizzard, and voluminous intestines; feeds voraciously on grain, grass, &c., and so obtuse is its taste that it will swallow pieces of leather, metal, wood, or any hard substances. In this it is probably guided by the same instinct that leads the fowl to swallow gravel: for they are probably of use in assisting the action of the gizzard in the reduction of the food. Dr. Shaw asserts that he saw one at Oran that swallowed, without any seeming inconvenience, several leaden bullets, as they were thrown upon the floor, scorching hot from the mould. The female lays from ten to twelve eggs in a hole in the sand; and, although she does not incubate them continually, no bird has a stronger affection for its offspring, or watches its nest with more assiduity. Contrary to the general opinion, she always broods over her eggs at night, only leaving them during the hottest part of the day. In procuring the eggs from the nest, the natives are very careful not to touch any with their hands, as the parent birds are sure to discover it on their return, and not only desist from laying any more in the same place, but trample to pieces all those that have been left; therefore a long stick is always used to push them out of the nest. The eggs, which weigh about three pounds each, are said to be a great delicacy, and are prepared for the table in various ways. Ostriches are polygamous birds; one male being generally seen with two or three females, and sometimes with more. In a tame state they are tractable and familiar towards persons whom they know, but are often fierce towards strangers, whom they will attempt to push down by running furiously upon them; and on succeeding in this effort, they not only peck at their fallen foe with their beak, but strike at him with their feet with the utmost violence. When thus engaged, they make a fierce hissing noise, and have their throats inflated and mouths open, but at other times they have a kind of cackling voice.

The species *Rhea Americana*, which by some is called the American Ostrich, inhabits various parts of South America to the southward of the equator, but is principally found on the great plains in Buenos Ayres and the adjoining states. It differs essentially, however, from the true Ostrich, having three toes instead of two; is much smaller; and is of a uniform gray colour, except on the back, which has a brown tint. The back and rump are furnished with long feathers, but not of the same rich and costly kind as those of the former species. It is capable of great speed, and its running is accompanied with a singular motion of its wings; each being alternately raised and outstretched, and then depressed. It is taken by being chased on horseback, and catching it with the *lasso*, or by means of balls connected by a strip of hide, and thrown in such a way as to entangle its legs.

In describing the habits of this bird Mr. Darwin tells us that "When several horsemen appear in a semicircle, it becomes confounded, and does not know which way to escape. They generally prefer running against the wind; yet at the first start they expand their wings, and, like a vessel, make all sail. On one fine hot day I saw several Ostriches enter a bed of tall rushes, where they squatted concealed, till quite closely approached. It is not generally known that Ostriches readily take to the water. Mr. King informs me that at the bay of San Blas, and at Port Valdes in Patagonia, he saw these birds swimming several times from island to island. They ran into the water both when driven down to a point, and likewise of their own accord when not frightened: the distance crossed was about 200 yards. When swimming, very little of their bodies appear above water, and their necks are extended a little forward: their progress is slow. On two occasions, I saw some Ostriches swimming across the Cruz river, where its course was about 400 yards wide, and the stream rapid.

"The inhabitants who live in the country readily distinguish, even at a distance, the cock bird from the hen. The former is larger and darker-coloured, and has a bigger head. The Ostrich, I believe the cock, emits a singular deep-toned hissing note. When first I heard it, standing in the midst of some sand-hillocks, I thought it was made by some wild beast, for it is a sound that one cannot tell whence it comes, or from how far distant. When we were at Bahia Blanca in the months of September and October, the eggs, in extraordinary numbers, were found all over the country. They either lie scattered single, in which case they are never hatched, and are called by the Spaniards *huachos;* or they are collected together into a shallow excavation, which forms the nest. Out of the four nests which I saw, three contained twenty-two eggs each, and the fourth twenty-seven. In one day's hunting on horseback sixty-four eggs were found; forty-four of these were in two nests, and the remaining twenty scattered huachos. The Gauchos unanimously affirm, and there is no reason to doubt their statement, that the male bird alone hatches the eggs, and for some time afterwards accompanies the young. The cock when on the nest lies very close; I have myself almost ridden over one. It is asserted that at such times they are occasionally fierce, and even dangerous, and that they have been known to attack a man on horseback, trying to kick and leap on him. The Gauchos unanimously affirm that several families lay in one nest. I have been positively told, that four or five hen birds have been seen to go, in the middle of the day, one after the other, to the same nest. I may add, also, that it is believed in Africa, that two females lay in one nest."

At a meeting of the Zoological Society of London (Feb. 23, 1847) the Earl of Derby took an opportunity of noticing some of the differences which appear to characterize the Struthious tribe in their breeding, and which he believed were not generally known. Having shown that the Emu is strictly monogamous, he observed that the *Rheas*, on the contrary, are clearly polygamous; and with them the male not only selects the place for and forms the nest, but actually collects together in it the eggs (which are frequently laid at random about the enclosure), and roll them along by inserting his beak between the egg and the ground, as a boy would roll a cricket-ball along by the aid of a long stick with a hooked end to it. He does this in order that he may incubate them; and it has been observed that he shows no signs of anger when the females approach the nest.

OTIDÆ. The name given to a family of birds (the *Bustards*). Those which are peculiar to the Eastern Hemisphere and to Australia, have the long neck and legs, stout body, and strong limbs of the Ostrich. [For the European species, see BUSTARD.]

OTION. A genus of Pedunculated Cirripedes, found on the Indian coast, commonly attached to buildings covered by the sea. The body is sub-quadrate, supported on a fleshy pedicle with a gaping aperture and two posterior auricular tubes; five small testaceous valves, adhering near the sides of the aperture.

OTOLITHUS. A sub-genus of fishes belonging to the family *Sciænidæ*, inhabiting the Indian Ocean and Atlantic coasts of America. The *Otolithus regalis*, or SQUETAGUE, is commonly from a foot to fifteen inches long, but it often grows much larger. The head and back are brown, with frequently a tinge of greenish; faintly silvery with dusky specks above the lateral line, which gradually disappear on the sides; and the under part is wholly of a clear white. The eyes are large and pale yellow. There are two strong canine teeth in the upper jaw, which is also armed with a single row of very small pointed teeth; and the under jaw is furnished with a row of small teeth which is doubled anteriorly. The two dorsals are well separated, and the second, as well as the caudal and anal, is in a great part covered with small scales. Dr. Mitchell, describing this species, observes that it is "a fish of a goodly appearance, wholesome and well-tasted, though rather soft. He is taken both by the line and sean, and is brought to the New York market in great numbers during the summer months. He is called *weak-fish*, as some say, because he does not pull very hard after he is hooked; or, as others allege, because labouring men, who are fed upon him, are weak by reason of the deficient nourishment in that kind of food. Certain peculiar noises under water, of a low rumbling or drumming kind, are ascribed by the fishermen to the Squetague. Whether the sound came from these fishes or not, it is certain, that during their season, only, they may be heard from the bottom of the water, in places frequented by the weak-fish, and not elsewhere. The swimming-bladder is convertible into good glue. I have eaten as fine blanc-mange made from it as from the isinglass of the sturgeon."

OTTER. (*Lutra vulgaris.*) This aquatic quadruped is about two feet long from the nose to the insertion of the tail, which is fifteen inches more; its body is elongated and much flattened; the tail is flat and broad; the legs are short and strong, but so loosely articulated as to turn in every direction while swimming; the feet broad, and the toes connected by a complete web: its structure is consequently well adapted for an aquatic life: and it feeds almost entirely on fish. It swims and dives with great readiness, and with peculiar ease and elegance of movement. Its teeth are sharp and strong, and the tubercles of the molars very pointed, a modification necessary to secure the prehension and speedy destruction of their agile and slippery prey. It has a black nose and long whiskers; the eyes are

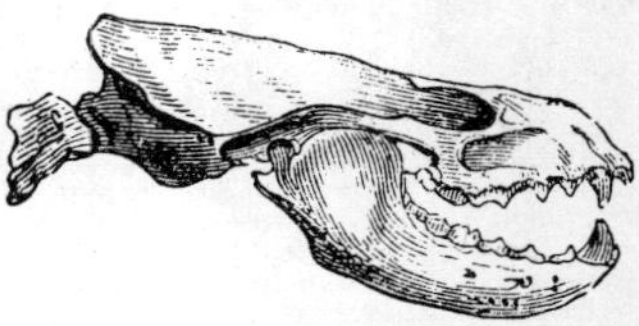

SKULL OF OTTER.

very small, and placed nearer the nose than in most other animals; the upper jaw is longer and broader than the lower; the ears small and erect; and the skin is protected by a compact fur, which consists of two kinds of hair, the longer and stiffer shining hairs, which are grayish at the base and a rich brown at the point, concealing an extremely fine and soft fur of a light gray colour, brown at the tip: the under parts of the body, inner parts of the limbs, and the cheeks, are of a brownish-gray throughout.

The Otter can be domesticated, though, from its ferocious disposition, this is a task of much difficulty. In order to do it effectually, so that the animal might be trained to catch fish or assist in fishing, it is recommended that they should be procured as young as possible, and be first fed with small fish and water. Then bread and milk is to be alternated with the fish, and the proportion of the former gradually increased till they are led to live entirely on bread and milk. They are then taught to fetch and carry, as dogs are trained, and when they are brought to do this well, a leather fish stuffed with wool is employed as the thing to be fetched: they are afterwards exercised with a dead fish, and chastised if they attempt to tear it. Finally they are sent into the water after living fish. Otters generally bring forth their young under hollow banks, on a bed of rushes, flags, or such weeds as the place affords in greatest quantities. They are always found at the edge of the water; and when under the protection of the dam, she teaches them instantly to plunge into the deep and escape from their pursuers among the rushes or weeds that fringe the stream; and, except in the absence of the parent, they are not to be easily taken. When the Otter, in its wild state, has taken a fish, it carries it on shore, and devours the head and upper parts, rejecting the remainder. The female produces four or five young in the spring of the year.

There are numerous instances on record of their being tamed and educated; one of which we copy, as peculiarly interesting, from the Journal of the late Bishop Heber: "We passed," says this exemplary prelate, "to my surprise, a row of no less than nine or ten large and very beautiful Otters, [we presume, of the species *Lutra nair*, F. Cuv.] tethered with straw collars and long strings to bamboo stakes on the banks (of the Matta Colly). Some were swimming about at the full extent of their strings, or lying half in and half out of the water; others were rolling themselves in the sun on the sandy bank, uttering a shrill, whistling noise, as if in play. I was told that most of the fishermen in this neighbourhood kept one or more of these animals, who were almost as tame as dogs, and of great use in fishing: sometimes driving the shoals into the nets, sometimes bringing out the larger fish with their teeth. I was much pleased and interested with the sight. It has always been a fancy of mine that the poor creatures whom we waste and persecute to death, for no cause but the gratification of our cruelty, might by reasonable treatment be made the sources of abundant amusement and advantage to us."

In the older annals of sporting in this country, Otter-hunting holds no inconsiderable place; and it is even still practised occasionally with dogs especially trained for the sport. "When the Otter is found," says Mr. Bell (*Hist. of Brit. Quad.*), "the scene becomes exceedingly animated. He instantly takes the water and dives, remaining a long time underneath it, and rising at a considerable distance from the place at which he dived. Then the anxious watch that is kept for his rising to 'vent,' the steady purpose with which the dogs follow and bait him as he swims, the attempts of the cunning beast to drown his assailants by diving whilst they have fastened on him, the baying of the hounds, the cries of the hunters, and the fierce and dogged resolution with which the poor hopeless quarry holds his pursuers at bay, inflicting severe, sometimes fatal wounds, and holding on with unflinching pertinacity even to the last, must altogether form a scene as animated and exciting as the veriest epicure in hunting could desire."

The following interesting paper on the Breeding of the Otter in confinement in the Zoological Gardens, Regent's Park, in 1846; by James Hunt, Head Keeper; was read at a meeting of the Society, March 23. 1847:—

"The female Otter was presented to the Society by Lady Rolle on the 4th of February, 1840, being apparently at that time about three months old. She remained without a male till the 11th of March, 1846, when a large male was presented to the Society by the Rev. P. M. Brunwin, of Braintree, Essex, in whose possession it had been for some months, and had been kept in a cellar. His weight when first taken was twenty-one pounds, but he was not above half that weight when he arrived at the Gardens, having wasted much in confinement, and become very weak in the loins, from which he soon recovered after his arrival. About a month after his arrival there was a continual chattering between him and the female during the night, which lasted for four or five nights; but they did not appear to be quarrelling. Nothing further was observed in their manners or in the appearance of the female to make me think she was with young, until the morning of the 13th of August, when the keeper that has the charge of them went to give them a fresh bed, which he does once a week; while in the act of pulling out the old bed he observed two young ones, apparently five or six days old, and about the size of a full-grown rat: he immediately put back the bed, with the young on it, and left them. On the 21st the mother removed them to the second sleeping-den, at the other end of their enclosure, and several times after she was observed to remove them from one end of the house to the other, by pushing them before her on a little straw: her object in removing them appeared to be to let them have a dry bed: on the 9th of September they were first seen out of the house; they did not go into the water, but crawled about, and appeared very feeble.

"On the 26th of September they were first seen to eat fish, and follow the mother into the water: they did not dive into the water like the mother, but went into it like a dog, with their head above water, and it was not until the middle of October that they were observed to plunge into the water like the old ones. On the 22nd of December the water was let out of the pond for the purpose of cleaning it, which is done once a week: the animals were shut up in their sleeping-den, but they let themselves out when the pond was but half full of water, and the young ones got into it and were not able to get out without assistance; after they had been in the water some minutes the mother appeared very anxious to get them out, and made several attemps to reach them from the side of the pond where she was standing; but this she was not able to do, as they were not within her reach. After making several attempts in this manner without success, she plunged into the water to them, and began to play with one of them for a short time, and put her head close to its ears, as if she was making it understand what she meant; the next moment she made a spring out of the pond, with the young one holding on by the fur at the root of the tail with its teeth; having safely landed it, she got the other out in the same manner: this she did several times during a quarter of an hour, as the young ones kept going into the water as fast as she got them out. Sometimes the young held on by the fur at her sides, at others by that at the tail. As soon as there was sufficient water for her to reach them from the side of the pond, she took hold of them by the ears with her mouth, and drew them out of the pond, and led them round the pond close to the fence, and kept chattering to them, as if she was telling them not to go into the pond again."

The AMERICAN or CANADA OTTER (*Lutra lataxina*) resembles the European species, but is considerably larger. Both have a habit peculiar to these animals: this is sliding or climbing to the top of a ridge of snow in winter, or a sloping moist bank in summer, where, lying on the belly, with the fore legs bent backwards, they give themselves an impulse with the hind legs that enables them to glide swiftly down the eminence. This sport they continue for a long time. The colour of the whole body, except the chin and throat, which are a dusky white, is a glossy brown. Many are caught for the sake of their skins, which are very dense and fine, and are much esteemed. The common mode of taking them is by sinking a steel trap near the mouth of their burrow.

The SEA-OTTER (*Enhydra lutris*) is a much larger species than the preceding, and presents such modifications of its palmated feet, and of its teeth, as to form the type of a distinct sub-genus (*Enhydra*), which connects the Otter with the Seal. It weighs from seventy to eighty pounds. Its colour when in full season is perfectly black; at other times of a dark brown. It has six incisors in the upper, and four in the lower jaw; the grinders being broad, and well adapted for breaking crustaceous animals. It runs very swiftly, and swims with extreme celerity, either on its back, sides, or sometimes as if upright in the water. The hind feet resemble those of a seal, and have a membrane skirting the outside of the exterior toe, like that of a goose. It is exclusively found between the forty-ninth and sixtieth degrees north latitude, on the north-western coasts of North America, and the shores of Kamtschatka and the adjoining islands. It is caught by placing a net among the sea-weed, or by chasing it in boats. The female brings forth but one at a birth, and is extremely sedulous in her attention to her offspring, playing with and fondling it in various ways. The Sea-Otter haunts sea-washed rocks, lives mostly in the water, and approximates to the seals more than to the Otters in its habits. The flesh of the young Sea-Otters is said to be very delicate food, not unlike lamb. The Kamtschatdales, on whose coasts these animals are chiefly killed, exchange the skins with the Russians for those of the fox and sable, and the Russian merchants sell them principally to the Chinese, with whom they are in great request, and who pay for them at the rate of from seventy to a hundred roubles each. This great price, and the great distance from where they are obtained, are the obvious reasons why so few are seen in the European market. Sir George Simpson tells us that since 1814 the Russians have sent to market from California the enormous number of 80,000 Sea-Otters, besides Fur Seals.

OUDENODON. [See SUPPLEMENT.]

OUNCE. [See JAGUAR.]

OUZEL, or RING OUZEL. (*Turdus torquatus.*) This bird is somewhat larger than the Blackbird, which it much resembles in its general habits. Its general colour is dull black; each feather margined with ash

RING OUZEL.—(TURDUS TORQUATUS.)

gray; the bill is dusky; corners of the mouth and inside yellow; eyes hazel; and the legs dusky brown. The breast of the male is distinguished by a crescent of pure white, which almost surrounds the neck: on the female this crescent is much less conspicuous, and in some birds it is wholly wanting. Ring Ouzels are found in various parts of this kingdom, chiefly in the wilder and more mountainous districts. The female builds her nest in the same manner and in the same situations as the Blackbird, and lays four or five eggs of the same colour. Their food consists of insects and berries.

The WATER OUZEL, or DIPPER, (*Cinclus aquaticus*), is a bird of a very retired nature, resorting to small brooks and rivulets which flow rapidly among stones and fragments of rocks in hilly countries. There it may be seen perched on the top of a stone in the midst of a torrent, in a continual dipping motion, whilst watching for its food, which consists of small fishes and insects. The Rev. George Gordon mentions that in some of the rivers of the north of Scotland it is very partial to and destroys the spawn of the Sea Trout (*Salmo trutta*), from which it most probably has obtained its no enviable place in the following ancient distich:

> "The Gordon, the guile, and the *water-craw*
> Are the three worst ills that the Moray ever saw."

The *Gordons* being one of the Highland clans, famed for their incursions in former times; and the *guile* being a weed, very destructive in corn-fields. The upper parts of the head and neck are deepish rusty brown; the back, rump, scapulars, wing-coverts, belly, and tail are black; but each feather on these parts is distinctly edged with hoary gray. The breast, fore part of the neck, and throat are of a snowy white, and the black and

WATER OUZEL, OR DIPPER.
(CINCLUS AQUATICUS.)

white on the belly are separated by a rusty brown. The legs are short and strong; the claws curved; and the toes are distinctly parted, without any membrane between to join them. It forms its nest in the holes of banks; and lays five eggs of a whitish colour, slightly tinged with red. "The most singular trait in its character," observes Bewick, "is that of its possessing the power of walking, in quest of its prey, on the pebbly bottom of a river, and with the same ease as on dry land." Upon this "trait" we find Mr. Waterton thus commenting: "This is the bird whose supposed sub-aquatic pranks have set the laws of gravity at defiance, by breaking through the general mandate which has ordained that things lighter than water shall rise towards its surface, and that things that are heavier shall sink beneath it." "If the Water-Ouzel, which is specifically lighter than water, can manage, by some inherent power, to walk on the ground at the bottom of a rivulet, then there is great reason to hope that we, who are heavier than air, may, any day, rise up into it, unassisted by artificial apparatus, such as wings, gas, steam, or broom-staff."

OVIS. [See Sheep.]

OVULA. A genus of Mollusca, inhabiting the Indian and Chinese seas. Shell oblong, with elongated aperture, the ends of which in some species are so much lengthened as to make it fusiform or spindle-shaped; outer lip crenulated, inner lip smooth. The animal is furnished with two tentacula, having eyes at the base on small projections, like the *Cypræa*; mantle and foot large; the former however, having only one lobe.

OWL. (*Strigidæ.*) It is a common remark, that Owls may be considered as a kind of nocturnal hawks, differing, as Linnæus has observed, from those birds in the same manner as Moths differ from Butterflies; the one being chiefly nocturnal, and the other diurnal. They are distinguished by having a large head; great projecting eyes directed forwards, and surrounded with a circle or disc of loose and delicate feathers, covering the base of the beak and the opening of the ear; a strong hooked bill; crooked claws; and a downy plumage, generally spotted or barred with different shades of brown or yellow. The feet are chiefly remarkable for the power possessed by the external toe, of being turned either backwards or forwards. Unable to bear the brighter light of the sun, the Owl retires to some lonely retreat, where it passes the day in silence and obscurity; but at the approach of evening, when all nature is desirous of repose, and the smaller animals, which are its principal food, are seeking their nestling places, the Owl comes forth from its lurking holes in quest of prey. Its eyes are admirably adapted for this purpose, being so formed as to distinguish objects with greater facility in the dusk than in broad daylight. Its flight is low and silent during its nocturnal excursions, and when it rests, it is then only known by the frightful and reiterated cries with which it interrupts the silence of night. If forced from his retreat during the day, his flight is broken and interrupted, and he is sometimes attended by numbers of small birds, who, seeing his embarrassment, pursue him with incessant cries, tormenting him with their movements; while the Owl remains perched upon the branch of a tree, and regards the assembled group with all the appearance of mockery and affectation. There are *some* species of Owls, however, able to fly, and see distinctly in open day. And we may remark further, that although the *Strigidæ* are dazzled by too refulgent a light, they do not, as some have imagined, see best in the darkest nights. Their vision, generally speaking, is clearest in the dusk of the evening, at the dawning of the morning, or by moonlight, when they are not incommoded either by too much or too little light: their faculty of nocturnal vision differing considerably, however, in different species; some seeing with exquisite acuteness in the gloom of night, while others invariably roam abroad at early morn or in the shades of evening. Their hearing is very acute, and their plumage soft and loose, enabling them to fly without noise, and thus to come on their prey in an unexpected manner. They feed on small birds, mice, bats, and moths, swallowing them entire, and casting up the indigestible parts in the form of small balls. They breed in fissures of rocks, in old buildings, or in holes of trees, the female laying from two to six eggs; and they are found in every part of the globe.

Mr. Hewitson, in his 'Illustrations of the Eggs of Birds,' remarks that there is a strong and perfect similarity amongst the eggs of the different species of Owls which we could scarcely expect to find in the eggs of birds which differ from each other so much in their mode of breeding. The eggs of those species which are deposited in the hollows of old trees, and deserted ruins, and those which are found on the bare sod, and exposed to the broad light of day and the pelting storm, are alike without colour.

The Owl family is very numerous, and may be subdivided into the three following groups:—1. The *Typical Owls* (whose adaptation to nocturnal habits is most complete, and who during the day, with their eyes half shut, present a great appearance of gravity) have a large external ear, and large and complete discs around the eyes. 2. The *Horned Owls*, in which the external ear is smaller, but the discs around the eyes still large; and in which the head is furnished with two feathery tufts, resembling horns. 3. The *Hawk Owls*, in which the external ear is very small, the facial discs are wanting, and the feathery tufts absent.

The BARN OWL. (*Strix flammea.*) The Common White, Barn, or Screech Owl is so well-known an inhabitant of this country, that every village is acquainted with its history. It is also spread through the temperate and warmer regions of Europe, but

BARN OWL — (STRIX FLAMMEA.)

is not found in the higher latitudes. It is a most beautiful species, though, from the frequency of its appearance, but little attended to. The Barn Owl is about fourteen inches in length. The head and upper parts of the body are of a fine pale orange-colour, slightly marked by small scattered chestnut-coloured spots: the feathers of the upper parts of the back and the wing-coverts are gray towards their tips, finely sprinkled with blackish transverse specks; while down the shaft of each runs a short series of alternate black and white oblong spots: the face is white, but the ruff elegantly edged by a rufous verge intermixed with white: the quill-feathers barred with pale brown, and the tail slightly crossed by brownish freckles. The whole under parts are white, sometimes marked by a few small dusky spots. Occasionally in this species the under parts are yellowish. The legs are feathered or plumed to the toes, which are covered with fine hair. It conceals itself by day in deep recesses among ivy-clad ruins, in antique church towers, in the hollow of old trees, in barns, hay-lofts, and other out-houses. Towards twilight it quits its perch, and takes a regular circuit round the fields, skimming along the ground in quest of field mice, rats, moles, shrews, and large insects. During the time the young are in the nest, the male and female alternately sally out and beat the fields with the regularity of a spaniel. As soon as they have seized their prey they return with it in their claws; but as it is necessary to shift it into their bill, they always alight for that purpose on the roof, before they attempt to enter their nest.

Mr. Waterton (to whose intelligent remarks we are so much indebted, and who never fails to put the services of the feathered tribes in their proper light) tells his readers that "if this useful bird caught its food by day, instead of hunting for it by night, mankind would have ocular demonstration of its utility in thinning the country of mice; and it would be protected, and encouraged, everywhere. It would be with us what the Ibis was with the Egyptians. When it has young, it will bring a mouse to the nest every twelve or fifteen minutes. But, in order to have a proper idea of the enormous quantity of mice which this bird destroys, we must examine the pellets which it ejects from its stomach in the place of its retreat. Every pellet contains from four to seven skeletons of mice. In sixteen months from the time that the apartment of the owl on the old gateway was cleaned out, there has been a deposit of above a bushel of pellets. . . . When farmers complain that the Barn Owl destroys the eggs of their pigeons, they lay the saddle on the wrong horse. They ought to put it on the rat. Formerly I could get very few young pigeons till the rats were excluded effectually from the dovecot. Since that took place, it has produced a great abundance every year, though the Barn Owls frequent it, and are encouraged all around it. The Barn Owl merely resorts to it for repose and concealment. If it were really an enemy to the dovecot, we should see the pigeons in commotion as soon as it begins its evening flight; but the pigeons heed it not: whereas, if the sparrow-hawk or hobby should make its appearance, the whole community would be up at once; proof sufficient that the Barn Owl is not looked upon as a bad, or even a suspicious, character by the inhabitants of the dovecot."

Many other species, more or less resembling the Barn Owl, are found in different parts of the temperate regions of the globe.

The TAWNY OWL. (*Syrnium aluco.*) This bird is about the size of the Barn Owl. Its bill is white; its eyes dark blue: the radiated feathers round the eyes are white, finely streaked with brown; the head, neck, back, wing-coverts and scapulars are tawny brown, finely dotted with dark brown and black: on the wing-coverts and scapulars are several large white spots, regularly placed, so as to form three rows; the quill-feathers are marked with alternate bars of light and dark brown; the breast and belly are pale yellow, with narrow dark streaks pointing downwards, and crossed with similar ones: the legs are feathered down to the toes; the claws large, much hooked, and white. — This is the Owl that hoots by night, and sharply gives out the repeated cry of *tee-whit*, particularly in cold frosty nights. When these birds are slightly disturbed amid their slumbers in the vast and solitary woods, they will utter an inward tremulous hooting of *too-who*, the subdued and gloomy shivering of which is peculiarly horrific.

There are *some* lovers of nature, it seems, who are of a different opinion; or Mr. Hewitson would not thus express himself: "This is the Owl from which issues forth that loud melancholy sound at night, which, however much it may be associated with goblins in the minds of others, is *extremely agreeable* to the ear that is fond of nature's sylvan sounds." The same writer tells us that "the Tawny Owl usually lays its eggs in a hollow tree, sometimes in the holes of rocks, and occasionally in the deserted nest of some other bird; they are round, large, bright, and glossy, from three to five in number, and are deposited at irregular intervals, the first being sat upon as soon as laid; the young of the same nest differ in consequence very considerably in their size."

The GREAT HORNED OWL. (*Bubo Virginianus.*) The Great Horned or Eagle Owl is but little inferior in size to the Golden Eagle; and is very destructive to young fawns, hares, rabbits, rats, moles, reptiles, partridges, grouse, and other game. It is found in the mountainous parts of Central Europe, and in almost every quarter of the United States; frequenting deep forest glens, and making its nest in the fissures of rocks, ruined castles, &c.; but in Britain it has been very rarely seen. "Along the mountainous shores of the Ohio, and amidst the deep forests of Indiana," says Wilson, "this ghostly watchman has frequently warned me of the approach of morning, and amused me with his singular exclamations, sometimes sweeping down and around my fire, uttering a loud and sudden *Waugh O! Waugh O!* sufficient to have alarmed a whole garrison. He has other nocturnal solos, no less melodious, one of which very strikingly resembles the half-suppressed screams of a person suffocating, or throttled, and cannot fail of being exceedingly entertaining to a lonely benighted traveller, in the midst of an Indian wilderness!" "There

GREAT HORNED OWL.—(BUBO VIRGINIANUS.)

is something in the character of the Owl so recluse, solitary, and mysterious, something so discordant in the tones of its voice, heard only amid the silence and gloom of night, and in the most lonely and sequestered situations, as to have strongly impressed the minds of mankind in general with sensations of awe and abhorrence of the whole tribe. The poets have indulged freely in this general prejudice; and in their descriptions and delineations of midnight storms and gloomy scenes of nature, the Owl is generally introduced to heighten the horror of the picture. Ignorance and superstition, in all ages and in all countries, listen to the voice of the Owl, and even contemplate its physiognomy, with feelings of disgust and a kind of fearful awe." "Nothing is a more effectual cure for superstition than a knowledge of the general laws and productions of nature; nor more forcibly leads our reflections to the first, great, self-existent CAUSE of all, to whom our reverential awe is then humbly devoted, and not to any of his dependent creatures. With all the gloomy habits and ungracious tones of the Owl, there is nothing in this bird supernatural or mysterious, or more than that of a simple bird of prey, formed for feeding by night, like many other animals, and of reposing by day. The harshness of its voice, occasioned by the width and capacity of its throat, may be intended by Heaven as an alarm and warning to the birds and animals on which it preys to secure itself from danger. The voices of all carnivorous birds and animals are also observed to be harsh and hideous, probably for this very purpose." Its general colour is ferruginous, varied with larger and smaller spots and markings of brown, black, and gray; together with innumerable minute specks. The larger wing and tail-feathers are obscurely varied by dusky transverse bars: the bill is black; the eyes very large, and of a golden-orange colour: the legs are short and strong, thickly clothed down to the very claws with fine downy plumes; and the claws are extremely large, strong, and black. It rarely lays more than two eggs, which are larger and rounder than those of a hen, and of a reddish-brown colour, with darker blotches and variegations.

The GREAT SNOWY OWL. (*Surnia nyctea.*) This is one of the most beautiful of all the species, on account of its snowy whiteness; and in size it nearly equals the Eagle Owl, which it also resembles in its general habits. It is one of the hardiest of all birds, and is

OWL.—(SURNIA NYCTEA.)

found in very high northern latitudes of both the Old and New World; obtaining its food and rearing its young among rocky mountains and islands, in spite of all the vicissitudes of temperature and season. The bill is hooked, like a hawk's, with stiff feathers like hairs round its base, reflected forward; and bright yellow irides. The head, whole body, wings, and tail, are of a pure white:

on the top of the head are numerous small brown spots; the upper part of the back is marked with transverse lines of dusky brown; a few dusky spots are on the covert-feathers of the wings; but withinside they are purely white; and the lower part of the back is spotless: the middle feathers of the tail have a few spots on each side the shafts of the feathers: the legs and feet are covered with white feathers; and the claws are long, strong, black, and sharp pointed. The Snowy Owl, it is supposed, used to breed on some of the Shetland Islands, but is now a very rare visitor to any part of the British Islands.

The BURROWING OWL. (*Athene cunicularia.*) This singular species is widely spread through the American continent, and is peculiar to it. It inhabits the burrows of the marmot, viscacha, and other small rodent animals; and when these do not present themselves, it makes excavations for itself. This is a small bird, its length not exceeding ten inches.

OX: OXEN. The general designation for the different species and varieties of the ruminant quadrupeds belonging to the genus *Bos;* generically distinguished by having smooth hollow horns, directed sideways, and their curving upwards or forwards in a semilunar form; body thick and heavy; tail long, terminated by a tuft of hair; and four inguinal mammæ. The male of this genus is called a BULL; the female, a COW; and the young, a CALF. The name of *Ox* is given to the castrated male, and he is called an *Ox-calf* or *Bull-calf* until he is a twelvemonth old; a *Steer* until he is four years old, and after that an *Ox* or *Bullock.*

Truly does Mr. Bell, in his 'History of British Quadrupeds,' say, "Of all the animals which have been reduced into the immediate service of man, the Ox is without exception that to which he is most indebted for the variety and extent of its means of usefulness. If the qualities of the Dog are of a higher and more intellectual character, and bring it into closer communication with man as a social being; and if the Horse, as a beast of burden and of draught, serve more to his immediate personal assistance; the Ox surpasses these and all others in the devotion of its powers while living, and the appropriation of every part of the body when dead, to the wants, the comforts, and the luxuries of his owner." "This universal utility of the animal," he adds, "appears to have been very soon detected, and we find consequently that its domestication constituted one of the earliest triumphs of human authority over the natural instincts and habits of the brute creation. That this event took place before the Flood, and induced even then that propensity to a pastoral life which has ever been characteristic of man in his less cultivated state, wherever the climate was such as to encourage or permit it, we have the Sacred Writings to attest; for we are told that Jubal, the son of Lamech, was the father or ancestor 'of such as live in tents, and of such as have cattle.' From the time when the family of Noah issued from the Ark, in every quarter of the earth which his varied and multitudinous descendants have cultivated, the Ox has been reared as the most useful and important aid to the necessities of mankind. In Egypt it was the object of worship; and after the Israelites had left that seat of idolatry, when they themselves were disposed to lapse into that high and rebellious offence against the Majesty of Heaven, it was in the form of a golden calf that they modelled the object of their absurd and impious rites."

It has been the general opinion that the domestic races of our cattle are originally sprung from the *Bos bubalus*, the Indian and European Buffalo; but some treat of them as arising from the aurochs or wild cattle of Germany and Poland. Baron Cuvier, however, differs from both these suppositions, and considers our present cattle identical with a species no longer existing in a wild state, but which has, by the exertions of man, as in the instance of the camel and dromedary, been for ages entirely subjected to his power. The remains of this animal have been found in a fossil state, and it is upon the comparison of these remains with the skeleton of the aurochs, the buffalo, and our domestic races, that Cuvier founded his opinion.

The COMMON OX (*Bos taurus*) has a flat forehead, longer than it is broad, and round horns placed at the two extremities of a projecting line which separates the front from the occiput: the horns, however, differ so much in their form and direction in the numerous varieties which domestication has produced in this species, that no specific character can be based upon them. The colours of these animals are extremely variable, being reddish, white, gray, brown, black, &c. From what species the present useful and valuable domesticated breeds owe their origin, it would be very difficult to determine; but it is certain that their utility was well known to mankind in the very infancy of society, and that they still form the basis of the wealth of many countries, where the people subsist and flourish in proportion to the cultivation of their lands and the number of their cattle. Throughout a great part of the world, the flesh of the Ox is the principal article of animal food; while from the milk of the Cow, of itself an almost indispensable part of our diet, are manufactured cheese and butter. There is, indeed, scarcely any part of this animal that is not useful to mankind: the skin, the horns, the bones, the blood, the hair,—nay, the very refuse of all these,—each and all have their separate uses. Though at the present day, in this country, the Ox is less used for the purposes of agricultural labour than it was formerly, in many parts of the world the practice still remains; and wherever it prevails to any extent, its excellence is universally felt and acknowledged. The period of gestation of the Cow is nine months; and the young, like that of the horse, is very perfect and vigorous soon after birth, though it needs the care of the mother for a considerable time. It attains its full vigour in three years, and the term of its natural life is abc ıt fourteen.

The climate as well as the pasture of Great Britain is excellently adapted to the moderate nature of these animals; and the verdure and fertility of our plains are perfectly suited to their manner of feeding; for, being destitute of the superior fore-teeth, they love to graze in high and rich pastures; nor do they seem to be very choice as to the quality of their food, provided they have always an abundant supply of herbage. For this reason, in our English pastures, where the grass is rather high and flourishing, than succulent and nutritious, the Cow thrives admirably; and there is no part of Europe in which this animal grows larger, yields more milk, or fattens sooner.

In the Islands and Highlands of Scotland the breed of Oxen is very small, and the majority of them black. They are very light,

KYLOE OX.

and traverse with great ease the boggy ground which abounds in these parts. They are said to derive their name of *Kyloe* oxen from the Islanders having to cross the *kyles* or ferries on their way to the market. Thousands of these are annually driven to market, and the ferries have frequently very strong currents through which the animals are made to swim.

The varieties produced by domestication and climate are almost innumerable: but the principal kinds in this country are thus described by Mr. Youatt. "The breeds of cattle, as they are now found in Great Britain, are almost as various as the soil of the different districts, or the fancies of the breeders. They have, however, been very conveniently classed according to the comparative size of

LONG-HORNED BULL.

the horns: the long horns originally, so far as our country is concerned, from Lancashire, much improved by Mr. Bakewell of Leicestershire, and established through the greater part of the midland counties; — the short horns, originally from East York, improved in Durham, mostly cultivated in the northern counties and in Lincolnshire, and many of them found in every part of the kingdom, where the farmer attends much

SHORT-HORNED BULL.

to his dairy, or a large supply of milk is wanted; — and the middle horns, not derived from a mixture of the two preceding, but a distinct and valuable and beautiful breed, inhabiting principally the north of

SHORT-HORNED COW.

Devon, the east of Sussex, Herefordshire, Gloucestershire; and of diminished bulk, and with somewhat different character, the cattle of the Scottish and Welsh mountains. The Alderney, with her crumpled horn, is

ALDERNEY COW.

found on the southern coast, and, in smaller numbers, in gentlemen's parks and pleasure-grounds every where; while the polled or hornless cattle prevail in Suffolk and Norfolk, and in Galloway, whence they were first derived. These, however, have been intermingled in every possible way. They are found pure only in their native districts, or on the estates of some opulent and spirited individuals. Each county has its own mongrel breed, often difficult to be described, and not always to be traced, — neglected enough, yet suited to the soil and to the climate; and, among little farmers, maintaining their station, and advantageously maintaining it, in spite of attempts at supposed improvements by the intermixture or substitution of foreign varieties."

"It does not appear," says Mr. Bell, "that any very decided steps were taken towards the improvement of the English breeds of cattle until within the last half century, or rather more. They were, it is true, bred in great numbers to supply the profuse hospitality of the ancient nobility; but there is no reason to believe that any particular care was taken to procure the best breeds, or to increase their size by a particular mode of feeding. The establishment of prizes has led, perhaps, as much as the real advantage of the pursuit, to that great interest which has of late years been taken in the breeding and fattening of cattle. The result has been the establishment of numerous distinct breeds, of which some are particularly advantageous for particular districts."—That the encouragement given to agricultural pursuits in general by associations expressly devoted to that object has been attended with many beneficial results, cannot be denied; but as regards the fattening of cattle for public exhibition, we consider that it has been carried to a most ridiculous excess. It is, however, no part of *our* duty to describe the *fat beauties* which are exhibited at monster cattle-shows; nor are we disposed to eulogise those patriots who think they are entitled to the gratitude of their country, for spending their time and money in heaping up mountains of fat on the carcases of animals which, in our humble opinion, are quite as likely to die of repletion as they are to grace the shops of aristocratic butchers.

Volumes have been written on the different breeds of cattle, but this is not the place for discussing the comparative merits of long-horned and short-horned bulls, or whether the milk of Alderney cows be superior to that supplied in the dairies of Somersetshire, Cheshire, or Ayreshire; and as for the "vexed question" of the size of the carcase, or the smallness of the bone, of the rival breeds, and the relative qualities of the hide, we must leave our readers to consult the many well-known elaborate works on these subjects. Our purpose in this book being more zoological than economical, let us pass to a short notice of a variety or species of Ox which is believed to be nearly in its primitive state.

The wild cattle which anciently inhabited the Great Caledonian Forest (the *Bos Scoticus* of some authors) are now restricted to a few individuals preserved by noblemen at Chillingham Park, Cadzowe, near Hamilton &c. A specimen of a bull from the first mentioned of these places is preserved in the British Museum, to which it was presented by the Earl of Tankerville. This variety is thus described by Leslie: — Their colour is invariably of a creamy white. muzzle black: the whole of the inside of the ear, and about one-third of the outside, from the tips downwards, red; horns white with black tips, very fine and bent upwards; some of the bulls have a thin upright mane, about an inch and a half or two inches long. At the first appearance of any person they set off in full gallop, and at the distance of two or three hundred yards make a wheel round, and come boldly up again, tossing their heads in a menacing manner: on a sudden they make a full stop, at the distance of forty or fifty yards, looking wildly at the object of their surprise; but upon the least motion being made, they all again turn round and fly off with equal speed, but not to the same distance; forming a shorter circle, and again returning with a bolder and more threatening aspect than before, they approach much

WILD BULL OF THIS COUNTRY.

nearer, probably within thirty yards, when they make another stand, and again fly off; this they do several times, shortening their distance, and advancing nearer, till they come within ten yards; when most people think it prudent to leave them, not choosing to provoke them further; for there is little doubt but, in two or three turns more, they would make an attack."

WILD COW.

We might make elaborate extracts from a paper read before the British Association by Mr. Hindmarsh (in 1838), in which a good account is given of specimens preserved by the Earl of Tankerville, in his park at Chillingham; but it would occupy too much of our space to do so. We therefore refer our readers to the second volume of the Annals of Natural History, in which are some notes from Lord Tankerville himself; and we agree in his conclusion, that the same species of wild cattle prevalent in Scotland had extended to the northern districts of England; that in proportion as population and culture advanced, they became here, as in Scotland, the subjects of almost universal slaughter; and that a few of those that escaped had found sanctuary in the great wood at Chillingham (as well as in some other ancient forests), where they escaped the fury of their destroyers. The only other tenable hypothesis is, that after the enclosure of the park at Chillingham, they had been brought from Scotland and located there as a relic of the ancient Caledonian cattle; but the absence of all tradition and record upon the subject, and the circumstance of a similar breed having been found in places far removed from the Borders, render this supposition less probable than the former.

The CAPE OX, or CAPE BUFFALO. (*Bos [Bubalus] Caffer.*) This species of the Bovine genus is superior in size to the largest English Ox, is very strong and muscular, and has a most fierce and malevolent aspect. It inhabits the interior parts of Africa, north of the Cape of Good Hope, where these animals are found in large herds. Its colour is a deep cinereous brown: the hair on the body is rather short, but that on the head and breast very long, coarse, and black, hanging down the dewlap, like that of a Bison: from the hind part of the head to the middle of the back is also a loose black mane: the tail is nearly naked at the root, the remainder being covered with long loose hair. The horns are black, and extremely broad at their base: they are transversely wrinkled above, and are very large and long, spreading far over the head towards the eyes, then growing taper, and bending down on each side of the neck; the ends inclining backwards and upwards: the space between the tips is sometimes five feet. The ears are a foot long, and half-pendulous. These powerful animals are described as sometimes rushing suddenly on the African traveller, goring and trampling both man and horse under foot. The skin is excessively strong, and is, on this account, in high estimation with the colonists at the Cape, for its superior excellence in making harness, &c.

The GRUNTING OX. (*Bos [Poephagus] grunniens.*) Respecting the size of this animal (which is also called the Horse-tailed Ox, or Yack) there is much dispute, some travellers describing it as smaller than the domestic breeds of Oxen, while others contend that it is much larger; but from the accounts of Russian naturalists, it appears probable that there are two varieties, differing materially in size, but in other respects corresponding. It has a short head, broad nose, and large ears: the horns are short, slender, round, upright, sharp-pointed, and bent inwards. The whole body is covered with long hair, and is entirely black, except the front, ridge of the back, and tail, which are white. One peculiarity belongs to this species, which is, that instead of lowing, like others of the genus, they utter a sound resembling the grunting of a hog. In Thibet and other parts of central Asia, where they exist in a wild state, they are very dangerous, fighting, desperately when attacked; and though they are susceptible of domestication, they always retain some of their natural ferocity. The tails of these animals are very valuable: they form the standards designating the rank of superior officers in the Turkish army; they are extensively used in India as brushes to drive away insects; and the Chinese adorn their caps with them.

The JUNGLE OX, or GYALL (*Bos frontalis*) resembles the domestic Ox in most of its characters, but has horns flattened from before backwards, and no angular ridges. They are directed laterally, and more or less upward, but not backward. It is a domestic race in the mountain districts of the north-east of India, and although it has been suspected by some persons to be derived from the intermixture of the Buffalo with the common species, is quite a distinct species from either. [See BISON: BUFFALO: URUS.]

OX-BIRD. [See SANDERLING.]

OX-PECKER. [See BUPHAGA.]

OXYLOPHUS. A genus of Cuckoos; the best known species of which is the *Oxylophus glandarius*. [See CUCKOO.]

OYSTER. (*Ostrea edulis.*) A well-known edible Mollusc, the shell of which is formed of two unequal valves, connected together by a hinge of the simplest character. Externally the shell has a coarse and dirty appearance; each shell being composed of a great number of laminæ irregularly closed down on each other. In some species it is smooth; in others striated, tuberous, or prickly; the lower shell being always the deepest. The animal itself is also of very simple structure: no vestige of a foot can be seen; and the ligament which unites the valves is of small size. On separating the valves, four rows of gills, or what is called the *beard*, are observed at a little distance from the fringed edge of the mantle. The abductor muscle is situated at about the centre of the body, near which the heart is to be distinguished; and the mouth may be seen beneath a kind of hood, formed by the union of the two edges of the mantle near the hinge. Many curious discussions have arisen as to whether Oysters possessed the faculty of locomotion. It is well known that, in general, they are firmly attached to stones, or to each other; and it has been stated, and generally believed, that they are not endowed with any powers of changing their position. This much, indeed, is certain, that it is one of the most inanimate of the Mollusca; remaining fixed upon some submarine substance, enjoying only the nourishment brought it by the waves, and giving scarcely a sign of life, except the opening and shutting of its valves. In the British Museum there is a large specimen of a crab, to the back and claws of which a number of good-sized oysters have attached themselves. From the observations and experiments of naturalists, it appears, however, they can move from place to place by suddenly closing their shells, and thus ejecting the water contained between them with sufficient force to throw themselves backward, or in a lateral direction.

The principal breeding time of the common Oyster is in April or May, when their spawn is usually cast: this appears at first like little spots of grease, which fasten upon rocks, stones, or other hard substances that happen to be near. Very commonly they adhere to adult shells; and thus are formed the large masses termed oyster-banks. In about a year and a half they attain a size fit for the table; and they are taken by dredging, and stored in pits formed for the purpose, furnished with sluices, through which at spring tides the water is suffered to flow. In these receptacles they acquire a green tinge, which arises from the *confervæ*, and other marine vegetable matter, on which they feed. The powers of multiplication which Oysters possess are so wonderful, that the *banks* or *beds* which they form occupy portions of the sea, in shallow parts, extending for miles; and in some places (particularly along the alluvial shores of Georgia, in North America) walls of living Oysters literally counteract the otherwise resistless force of the tide. Oysters are particularly plentiful on the British coasts, and form a most important article of commerce. The breeding and fattening of them for the London market forms a considerable branch of business, which is principally carried on in Essex and Kent; but exclusive of the Oysters bred there, vast numbers are found on the coasts of Hants and Dorset; and they are also exceedingly abundant in the Jersey fishery, employing in it, during the season, about 1500 men, 1000 women and children, and 250 boats.

From the spawning time till about the end of July, the Oysters are said to be *sick*, but by the end of August they become perfectly recovered. Our Oyster fisheries are regulated by a court of admiralty; and after the month of May it is felony to carry away the *cultch*, (which means any substance the Oysters adhere to), and otherwise punishable to take any Oyster between whose shells, when closed, a shilling will rattle.

Oysters form the basis of many culinary preparations, but are much more digestible in their raw state than after any mode of cooking them, as this process in a great measure deprives them of the nourishing animal jelly which forms so large a portion of their substance. The shell of the Oyster is composed of carbonate of lime and animal matter, and was, at one time, supposed to possess peculiar medical properties; but analysis has shown that the only advantage of these animal carbonates of lime over those from the mineral kingdom arises from their containing no admixture of any metallic substance.

The Oyster is a very entertaining object to those who are fond of microscopic investigation. In the clear liquid around the animal, many minute, round, living animalcules have been found, whose bodies being conjoined, form spherical figures with tails, not changing their place otherwise than by sinking to the bottom, being heavier than the fluid; these have been frequently seen separating, and coming together again. In other Oysters, animalcules of the same kind were found not conjoined, but swimming by one another, where they seemed in a more perfect state, and were judged by Lewenhoek to be the animalcules in the roe or melt of the Oyster.

All bivalves which adhere by the shell are covered at their birth with a mucilaginous liquid which attaches them to the surface of any object on which they rest. The animal strengthens this first adhesion in the same manner that it increases the size of its shell. At the mouths of several American, African, and Indian rivers, great quantities of Oysters are found attached to the roots of trees, and even to their branches, where they are so situated as to be covered by the tide. Mrs. Lee, in her 'Stories of Strange Lands,' says, "The flavour of the tree-oyster is delicious; they are small, and cover the lower branches of the mangroves. Two or three of these branches form an ample luncheon, and in the river Gaboon we had a daily supply."

OYSTER-CATCHER. (*Hæmatopus.*) A wading bird which resides on the sea shore, where it feeds on marine animals. Its feet are strong and muscular; and it both runs and flies swiftly. In the British species (*Hæmatopus ostralegus*) the head, neck, upper part of the breast, back, wings, and tip of the tail, are of a deep black; the rump, base of the tail, a transverse stripe on the wings, and the whole of the under parts of the body, of a pure white: the beak and circle round the eyes of a bright orange; the irides crimson; and the feet red. Oyster-catchers abound on the western coasts of England: feeding on limpets and oysters; and from their dexterity in procuring the latter their name is derived. It does not construct any nest; but deposits its eggs on the bare ground, above high-water mark: they are from two to four in number, of an olivaceous brown, blotched with black. During the period of incubation the male is very watchful, and upon the least alarm utters a loud scream and flies off, as does also the female, after running some distance from the place where the eggs are deposited. If taken young they may be easily reared, and will frequent the ponds and ditches during the day, keeping in company with domestic poultry. They are abundant throughout all the coasts of Europe, and other species occur on many of those of Asia and America.

PACA. (*Cœlogenys.*) A genus of Rodent animals, allied in many points to the Capybaras and Agoutis, but presenting also considerable differences, particularly in the complex structure of the molar teeth. They inhabit the woods of South America, and are generally found in the vicinity of water,

PACA.—(CŒLOGENYS SUBNIGER.)

concealing themselves in burrows so near the surface that the foot of the pedestrian often breaks through. There are generally three openings to a burrow, which the animal takes care to cover with dry leaves and branches. They are of a thick and clumsy forn and, when full grown, measure about two feet in length from the tip of the nose to the extremity of the body, and about one foot in height, the hinder limbs being much longer (but considerably bent) than the anterior ones. The claws are conical, thick and strong, and proper for digging. Their eyes are large, prominent, and of a brownish hue; their ears are nearly naked, and their whiskers rigid. They swim and dive remarkably well; and, although heavy and corpulent, they run and jump with activity. Their cry resembles the grunting of a young pig. The food of the Paca consists of fruits and tender plants, which it seeks in the night, seldom quitting its burrow in the day. Its flesh is said to be very savoury, and forms a staple article of food in many parts of South America. The mode of taking these animals is by closing two of the apertures of their burrows, and digging up the third; and it often happens that they show a very determined resistance, biting their pursuers very severely. When undisturbed, the Paca often sits up and cleans its head and whiskers with its two fore paws, which it moistens with its saliva like a cat. It is readily tamed, very cleanly, and shows a quiet and contented disposition in captivity. The fur is composed of silky hairs, very short, thin, and stiff; of a blackish-brown on all the upper parts of the body, excepting four rows of parallel spots, from the shoulders to the rump, which, viewed in some situations, appear to form an almost uninterrupted line.

PACHYDERMATA. An order of Mammiferous Quadrupeds distinguished by the thickness of their skins, including various animals that in other respects are by no means closely allied; as, for instance, the Elephant, the Horse, and the Hog. The order is, however, subdivided into—1. The *Proboscidea;* or those possessing a prolonged snout or proboscis, and having five toes on each foot, included in a very firm horny skin; as the Elephant, and certain extinct gigantic species. 2. The *Pachydermata ordinaria;* in which the feet have four, three, or two toes on each foot. Among these are the Rhinoceros, Hippopotamus, Tapir, Wild Boar, &c. 3. The *Solidungula;* or quadrupeds with only one apparent toe and a single hoof to each foot, although beneath the skin, on each side of their metacarpus and metatarsus, there are bony points or processes which represent two lateral toes; as the Horse and its congeners.

Mr. Darwin has many excellent remarks on the extinct Pachydermata, which, he says, appear formerly to have had a range over the world, like that which deer and antelopes now hold. "If Buffon had known of these gigantic Armadilloes, Llamas, great Rodents, and lost Pachydermata, he would have said with a greater semblance of truth, that the creative force in America had lost its vigour, rather than that it had never possessed such powers. It is impossible to reflect without the deepest astonishment, on the changed state of this continent. Formerly it must have swarmed with great monsters, like the southern parts of Africa, but now we find only the tapir, guanaco, armadillo, and capabyra; mere pigmies compared to the antecedent races. The greater number, if not all, of these extinct quadrupeds lived at a very recent period; and many of them were contemporaries of the existing molluscs. Since their loss, no very great physical changes can have taken place in the nature of the country. What then has exterminated so many living creatures? In the Pampas, the great sepulchre of such remains, there are no signs of violence, but, on the contrary, of the most quiet and scarcely sensible changes." "That large animals require a luxuriant vegetation, has been a general assumption, which has passed from one work to another. I do not hesitate, however, to say that it is completely false; and that it has vitiated the reasoning of geologists, on some points of great interest in the ancient history of the world. The prejudice has probably been derived from India, and the Indian islands, where troops of elephants, noble forests, and impenetrable jungles are associated together in every account. If, on the other hand, we refer to any work of travels through the southern parts of Africa, we shall find allusions in almost every page either to the desert character of the country, or to the numbers of large animals inhabiting it. . . . . Dr. Andrew Smith, who, at the head of his adventurous party, has so lately succeeded in passing the Tropic of Capricorn, informs me that, taking into consideration the whole of the southern part of Africa, there can be no doubt of its being a sterile country. On the southern and south-eastern coasts there are some fine forests; but with these exceptions, the traveller may pass, for days together, through open plains, covered by a poor and scanty vegetation. It is difficult to convey any accurate ideas of degrees of comparative fertility; but it may be safely said, that the amount of vegetation supported at one time by Great Britain, exceeds, perhaps, even tenfold, the quantity on an equal area, in the interior parts of southern Africa." After remarking on the improbable effects which have been attributed to the variation of climate and food, the introduction of enemies, or the increased numbers of other species, to account for the succession of races, he observes, in conclusion, "We see that whole series of animals, which have been created with peculiar kinds of organization, are confined to certain areas; and we can hardly suppose these structures are only adaptations to peculiarities of climate or country; for otherwise, animals belonging to a distinct type, and introduced by man, would not succeed so admirably, even to the extermination of aborigines. On such grounds it does not seem a necessary conclusion, that the extinction of species, more than their creation, should exclusively depend on the nature (altered by physical changes) of their country. All that at present can be said with certainty, is, that as with the individual, so with the species, the hour of life has run its course, and is spent.

PACHYPTILA; or WHALE BIRD. A genus of web-footed birds, allied to the Petrels, but distinguished from them by having the nostrils separate, and the beak widened at the base, the edges of it furnished in the inside with fine, pointed, vertical laminæ. There are two species of this genus, which occur frequently in the seas of the Southern hemisphere. They are often called Blue Petrels, from their ashy-gray colour above, while the under parts are white. The *Pachyptila vittata* is very numerous in certain parts. Capt. George Grey tells us that "their flight much resembles that of a snipe. The name by which they are known to the sailors is the Whale Bird: they appear to take their food upon the wing; for I have never seen them sit upon the water even for a single second, although I have observed them frequently and at all hours; but night and day they hurry on with the same restless, rapid flight, sometimes going in large flocks. I never heard them utter any cry or sound."

PAGURUS: PAGURIDÆ. A genus and family of anomourous Crustacea; called also *Soldier* and *Hermit Crabs.* They are very peculiar as to both their conformation and their habits. The tail, or post abdomen, is of large size, but its envelope is little else than a membranous bag, entirely destitute of the usual hardness of the Crustaceous integument, and presenting no division into segments. The thorax itself is not very firm; and it is only on the claws, which are of large size, that we find the true calcareous envelope. For the protection of their soft tails, the *Paguridæ* resort to various artificial methods. Many of them seek univalve shells, in which they take up their abodes; attaching themselves to their interior by a sucker with which the tail is furnished at its extremity, and also holding by its six false legs which it bears at its hinder portion. When they are feeding or walking, the head and thorax project beyond the mouth of the shell; but when they are alarmed they draw themselves in, closing the mouth with one of the claws, which is much larger than the other, and holding to the interior so firmly, that they will rather be torn asunder than quit their attachment. As they increase in size, they are obliged to change their habitation for a more commodious one; and the way in which they accomplish this is very amusing. They may be frequently observed crawling slowly along the line of empty shells, &c., left by the last wave; and as if unwilling to part with their old domicile till a new one has been obtained, they slip their tails out of the old house into the new one, again betaking themselves to the former, if the latter is not found suitable. In this manner they not unfrequently try a large number of shells before they find one to their liking. If it happens that two hermit-crabs stop before the same shell, a dispute arises, and the weakest yields to the strongest. There are several species of various sizes, some of which may be found on our own coasts, but the greater part belong to tropical shores. For the most part they feed upon dead fish, but it does not appear that they are very nice in their food, as all kinds of garbage that may be thrown on the shore are devoured by them.

Sloane, in describing the species which is most common in Jamaica, thus writes:—"This small lobster or crab differs in very little from the European soldier or hermit-crab. It hath two large forked claws like those of an ordinary lobster, one of which is bigger than the other, both rounded, more tumid, less prickly, and of a paler red than that of Europe. They fit themselves with any shell they find empty, whether it be of the land or sea, and cover themselves almost over in it, carrying it on their backs wherever they go, like a snail. It is not possible to believe how quick the land-crabs and this crab will run, upon the least appearance of danger. Till they are turned up, nothing appears but a dead shell, the mouth of which lies undermost, out of which some little part of the crab appears after it is taken up." The species we have figured as an example is the *Cenobita Diogenes*, which is thus described by Catesby:—"They crawl very fast with their shell on their back; and at the approach of danger draw themselves within the shell; and thrusting out the

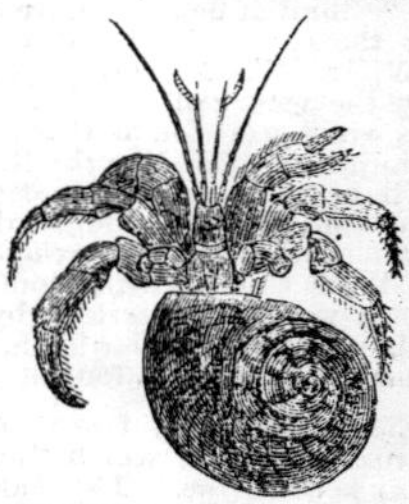

DIOGENES HERMIT CRAB.
(CENOBITA DIOGENES.)

larger claw in a defensive posture, will pinch very hard whatever molests them. They frequent most those parts of the sea-shores which are covered with trees and shrub, producing various wild fruits on which they subsist; though I have seen them feed on the fragments of fish and other animal substances cast on shore. They being roasted in the shell are esteemed delicate." The species are very numerous.

PALAPTERIX. [See SUPPLEMENT.]

PALÆOTHERIUM. A genus of extinct Pachydermatous animals, discovered (in company with *Anoplotherium*) in the gypsum beds of Paris; and of which discovery Cuvier thus speaks. "I found myself, as if placed in a charnel-house, surrounded by mutilated fragments of many hundred skeletons of more than twenty kinds of animals piled confusedly around me; the task assigned to me was to restore them all to their original position. At the voice of Comparative Anatomy, every bone and fragment of bone resumed its place. I cannot find words to express the pleasure I experienced in seeing, when I discovered one character, how all the consequences which I predicted from it were successively confirmed. The feet accorded with the characters announced

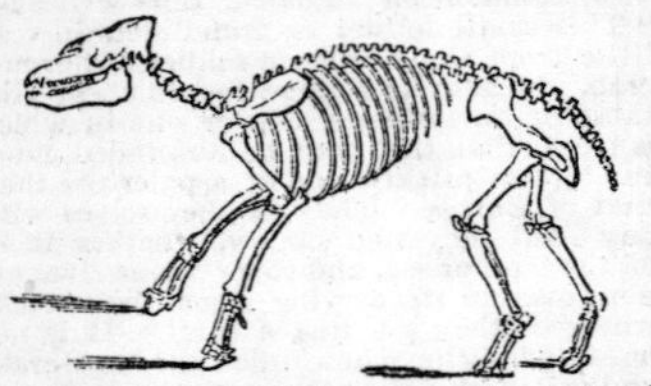

SKELETON OF PALÆOTHERIUM (RESTORED.)

by the teeth; the teeth were in harmony with those previously indicated by the feet. The bones of the legs and thighs, and every connecting portion of the extremities, were found to be joined together, precisely as I had arranged them before my conjectures were verified by the discovery of the parts entire. Each species was, in fact, reconstructed from a single unit of its component elements." Similar deposits have also been found in the corresponding strata in the Isle of Wight. That these deposits were formed by the agency of fresh water, or that the bones which were found there were the relics of animals which, like the Rhinoceros and Tapir of the present day, frequented the borders of lakes and large rivers, by whose waters they were occasionally ingulphed, there can be little doubt. The Palæotheria were characterized by having twenty-eight complex molar teeth, four canines, and twelve incisors, four in each jaw.

PALÆMONIDÆ. A family of Long-tailed Crustaceans, of which the Prawn (*Palæmon*) is the type. There are several species; among them some are extremely small, and their habits curious. As an example of this family we figure the beautifully

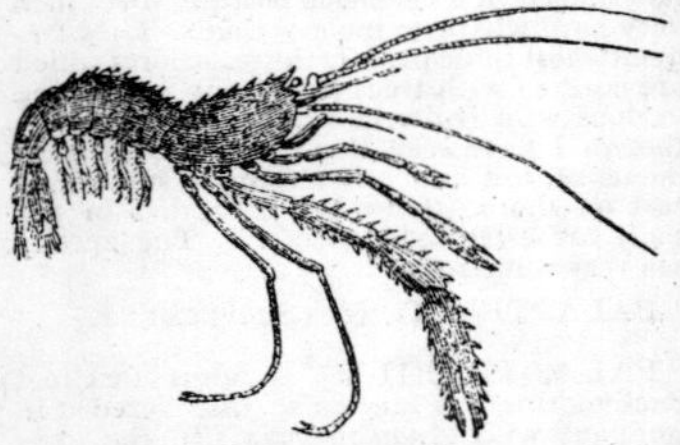

STENOPUS HISPIDUS.

marked STENOPUS HISPIDUS, found in the Eastern seas; when alive this species, as seen by Mr. Arthur Adams, is most delicately marked with red and blue colours, which may be looked for in vain in the dried specimens. Many species of *Palæmonidæ* are excellent to eat; of which we may specify the Prawn. In Kalm's Travels in America, we find a species of minute shrimp (*Palæmon fuci*) and a small crab (*Cancer minutus*) thus spoken of: — "Of the latter I collected eight, of the former three, all of which I put in a glass with water; the little shrimp moved as swift as an arrow round the glass, but sometimes its motion was slow, and sometimes it stood still on one side, or at the bottom of the glass. If one of the little crabs approached, it was seized by its fore paws, killed, and sucked; for which reason they were careful to avoid their fate. It was quite of the shape of a shrimp; in swimming it moved always on one side, the sides and the tail moving alternately. It was capable of putting its fore paws entirely into its mouth: its antennæ were in continual motion. Having left these little shrimps together with the crabs during night, I found in the morning all the crabs killed and eaten by the shrimps."

PALÆOPHIS. [See SUPPLEMENT.]

PALÆOPHRYNOS. [See SUPPLEMENT.]

PALÆOSAURUS. [See SUPPLEMENT.]

PALAMEDEA. The *Anhima* of the Brazilians. A genus of aquatic Grallatorial birds inhabiting the marshy or inundated places in South America, somewhat resembling a crane, and as large as a swan. The most distinguishing peculiarity of this bird is a long pointed horn which grows from the fore part of the head, and is surrounded by small black and white feathers. Its tail is about eight inches long; and its wings, when folded, reach more than half the length of the tail. The head and neck are of a greenish-brown colour, and covered with very soft feathers; the breast, belly, and thighs are of a silvery white; and the back is black, except the upper part, which is brown with yellow spots. Its food consists of grain and aquatic herbs: and it has a loud and wild cry.

PALINURUS. A genus of long-tailed Crustacea, containing many of the largest species. It is popularly known as the Sea-crawfish, or Spiny Lobster; and is distinguished by the very large size of its lateral antennæ, which are beset, like the body, with sharp points. The legs are all single-fingered; not even the first pair being furnished with pincers. The *Palinurus vulgaris* frequents deep waters, especially off rocky shores; and is common in such situations off the British coasts, especially in the south, and on the like coasts of France. They not unfrequently weigh ten or twelve pounds each, and are in general use when in season as an article of food. There are many other fine species in the West Indies and Indian ocean.

PALLIOBRANCHIATA. The name of an order of Acephalous Molluscs; very limited, both as to the number of the existing species it includes, and the small number of these which seem to be distributed through the ocean. It includes those in which the gills are situated on the internal surface of the lobes of the mantle. They are usually furnished with numerous vibratory filaments; and are attached, in some way or other, to solid bodies.

PALMER-WORM. An appellation given to larvæ of very different species and genera of Coleoptera. [See CALANDRA.]

PALOLO. A genus of *Annelides* apparently allied to *Arenicola*. By the 'Proceedings of the Zoological Society' (March 9. 1847), we learn that numerous specimens of this Sea Worm were presented to the British Museum by the Rev. J. B. Stair, of the London Missionary Society, and which has been described by J. E. Gray, Esq. as follows:— Body cylindrical, separated into equal joints, each joint with a small tuft of three or four spicula on the middle of each side. Head, ——? Last joint ending in a couple of tentacles. Eggs globular.

Most of the specimens, unfortunately, were broken into short pieces, and Mr. Gray was unable to discover any specimen with a head.

Palolo viridis, n. s. Green, with a row of round black spots down the middle of the dorsal? surface; one spot on the middle of each joint. *Habitat*. Navigator Islands.

The following is Mr. Stair's account:— "*Palolo* is the native name for a species of Sea Worm which is found in some parts of Samoa (the Navigator Islands) in the South Pacific Ocean. They come regularly in the months of October and November, during portions of two days in each month, viz. the day before and the day on which the moon is in her last quarter. They appear in much greater numbers on the second than on the first day of their rising, and are only observed for two or three hours in the early part of each morning of their appearance. At the first dawn of day they may be felt by the hand swimming on the surface of the water; and as the day advances their numbers increase, so that by the time the sun has risen, thousands may be observed in a very small space, sporting merrily during their short visit to the surface of the ocean. On the second day they appear at the same time and in a similar manner, but in such countless myriads that the surface of the ocean is covered with them for a considerable extent. On each day, after sporting for an hour or two, they disappear until the next season, and not one is ever observed during the intervening time. Sometimes, when plentiful at one island in one month, scarcely any are observed the next; but they always appear with great regularity at the times mentioned, and these are the only times at which they are observed throughout the whole year. They are found only in certain parts of the islands, generally near the openings of the reefs on portions of the coast on which much fresh water is found; but this is not always the case.

"In size they may be compared to a very fine straw, and are of various colours and lengths, green, brown, white, and speckled, and in appearance and mode of swimming resemble very small snakes. They are exceedingly brittle, and if broken into many pieces, each piece swims off as though it were an entire worm. No particular direction appeared to be taken by them in swimming. I observed carefully to see whether they came from sea-ward or rose from the reef, and feel assured they come from the latter place. The natives are exceedingly fond of them, and calculate with great exactness the time of their appearance, which is looked forward to with great interest. The worms are caught in small baskets, beautifully made, and when taken on shore are tied up in leaves in small bundles, and baked. Great quantities are eaten undressed, but either dressed or undressed are esteemed a great delicacy. Such is the desire to eat Palolo by all classes, that immediately the fishing parties reach the shore, messengers are dispatched in all directions with large quantities to parts of the island on which none appear."

PALUDINA. A genus of fluviatile Mollusca, very widely diffused in rivers and ponds, and occasionally found in salt marshes, but not in the sea. The shell is cone-shaped, varying in form from oval to globose, and having the whorls rounded; aperture roundish, angulated above; margins of the inner and outer lip united; operculum horny; shell covered by a greenish epidermis. The head of the animal is furnished with a proboscis, and two tentacula, having eyes at the base; foot somewhat triangular. The Paludinæ are viviparous.

PAMPHILA. A genus of diurnal Lepidoptera; two species of which are hereunder described.

PAMPHILA SYLVANUS; OR CLOUDED SKIPPER BUTTERFLY. This well-known insect is commonly found on the borders of woods and in woody lanes, about the end of May and in July. The wings above are a bright fulvous, with the hinder margin and the nervures brown, the margin itself marked with a strong black line: the anterior wings faintly spotted with fulvous: the posterior tawny ash-colour; beneath fulvous, with the tip of the anterior wings slightly tipped with greenish, and a black patch at the base; posterior wings obscure greenish, faintly spotted with yellowish-white, with a very slender marginal line: cilia fulvous. The male has a black line on the disc of the superior wings, and the nervures and marginal streak are broader and of a deeper black than in the female; in which sex the spots are more distinct on both surfaces of the wings.

PAMPHILA PANISCUS; OR CHEQUERED BUTTERFLY. A somewhat scarce and very local species, which makes its appearance about the end of May. Its wings above are black brown, spotted with tawny: anterior with a central blotch, followed by an interrupted band, intersected with black veins, with two smaller posterior spots, and a marginal band of tawny dots: posterior wings with three discoidal spots, and a row of dots, all tawny; fringe of the same colour, but black at the base: beneath the anterior wings are yellowish, with three discoidal spots, and four or five smaller posterior ones: posterior wings yellowish-brown, with seven larger spots, and five smaller and paler on the hinder margin, where there is also a pale yellowish streak. Caterpillar dark brown on the back, sides paler, with two yellow longitudinal stripes; black head, and an orange-coloured ring round the neck. It feeds on the Great plantain (*Plantago major*).

PANDA. (*Ailurus.*) See AILURUS.

PANDORA. A well-known genus of Conchiferous Mollusca, found in the sandy shores of Europe at a considerable depth; also in the Persian Gulf and Pacific Ocean. Shell regular, inequivalve, the upper one flat, and the lower convex; an obtuse, oblong tooth in one valve, and a receptacle for it in the other; ligament internal. The foot of the animal is large and triangular. The shells are small, and pearly inside.

PANGOLIN. [See MANIS.]

PANOPEA. A genus of Conchiferous Molluscs, belonging to the *Solenidæ* family; found in the Mediterranean and Australia. The shell is large and handsome; equivalve, transverse, and gaping at both extremities; one conical tooth in each valve, and a thick callosity on the side; two oval muscular impressions, and one deep palleal impression.

PANORPIDÆ. A family of insects belonging to the order *Neuroptera;* distinguished by the front of the head (which is vertical) being produced into an elongated slender deflexed rostrum; the eyes prominent and semiglobose; the antennæ long, slender, and multi-articulate; the body moderately long and slender; the maxillæ bilobed at the extremity, membranous, and pilose; the wings of moderate and equal size, numerously reticulated, the posterior not being folded when at rest; the legs long and slender; and the tarsi five-jointed, simple, with two tibial spurs, and denticulated ungues, and a large pulvillus. The type of this family is the *Panorpa communis*, an abundant species, ordinarily known as the SCORPION-FLY [which see].

PANTHER (*Felis Pardus.*) A feline quadruped, measuring about six feet and a half from nose to tail, which is itself about three feet long. Its colour is a bright tawny-yellow, thickly marked all over the upper parts of the body, shoulders, and thighs, with roundish black spots, disposed into circles consisting of four or five separate spots; and there is commonly, but not always, a central spot in each circle; in which, as well as in its superior size and deeper colour, the Panther differs from the Leopard. On the face and legs the spots are single, and along the top of the back is a row of oblong spots, which are still longer as they approach the tail. The breast and belly are white; the former marked with transverse dusky stripes; the latter and the tail with large irregular black spots. The Panther is principally found in Africa, and is to that country what the Tiger is to Asia, but is less to be dreaded, inasmuch as it prefers the flesh of brutes to that of human beings. The manner it seizes its prey—lurking near the sides of woods, &c., and darting forward with a sudden spring—resembles that of the Tiger. These animals and the Leopard were the *Varii* and *Pardi* of the ancients. The Romans drew immense numbers from the deserts of Africa for their public spectacles. Scaurus exhibited at one time a hundred and fifty Panthers; Pompey, four hundred and ten; and Augustus, four hundred and twenty. It is one of the wildest of the feline tribe, always retaining its fierce aspect and perpetual muttering growl. The female is pregnant nine weeks, and the young are born blind, continuing so for about nine days. In America, other species are called "Panthers."

PAPILIO: PAPILIONIDÆ. A genus and family of Lepidopterous insects, comprising numerous and distinct species of the diurnal tribes: it is distinguished by the perfectly ambulatory structure of the fore legs; the ungues distinct and simple, or bifid; antennæ having a distinct club, but never hooked at the tips; the hinder wings entire, and the discoidal cell of the hind wing closed; the body small, slender, and compressed; and the proboscis short, or moderately long. This family comprises two very distinct sub-families, namely, the PAPILIONIDÆ and PIERIDÆ.

In the PAPILIONIDÆ the anal edge of the hind wings is concave or folded; the palpi are very short; the club of the antennæ forms an elongated mass; the ungues are entire and simple; the wings are broad, with the discoidal cell always closed; the abdomen free. The caterpillars are slow, cylindrical, thickened, never villose nor hairy, with two retractile tentacles placed on the neck, in the shape of a fork, arising from a common tubercle, and which the insect throws out when alarmed, emitting at the same time a disagreeable odour. The species of Papilionidæ are for the most part tropical; but one has been found in England, *P. machaon.* Many of the species have the hind wings produced into a pair of tails, whence they have obtained the name of Swallow-tails. From the beauty of their colours and large size, these insects were by Linnæus styled *Equites.* Their flight is rapid.

The sub-family PIERIDÆ, comprising the *Danii candidi* of Linnæus, is distinguished from the preceding by the hind wings forming a groove for the reception of the abdomen; the palpi are porrected, with distinct joints; the minute labrum and mandibles are perceived above the base of the spiral maxillæ; the fore legs are long and perfect, without the dilated spine; the ungues are bifid, often with a long pulvillus and a narrow hirsute appendage on each side. The caterpillars are finely pubescent and attenuated at each end, without any nuchal tentacle; the chrysalides angular, slightly compressed, and terminated in a point at each extremity, sometimes assuming the appearance of a curved canoe. These insects, which include our common well-known white garden butterflies, are not equal, either in size or beauty, to the preceding sub-family; white, orange, and brimstone being their prevailing tints. The last-named are, however, occasionally very destructive, the larvæ feeding for the most part upon the cabbages and other vegetable produce of our gardens.—Such of our readers as wish to have additional information on the Butterflies must consult Doubleday and Hewitson's Genera of Diurnal Lepidoptera, where they will find much valuable information and accurate description, accompanied with most admirable coloured figures of the principal forms; it is a book quite indispensable to any one who wishes to *study* the subject; in our popular sketch any great detail would be misplaced. [See LEPIDOPTERA: BUTTERFLY.]

PAPILIO MACHAON, or SWALLOW-TAIL BUTTERFLY. This very elegant and conspicuous Butterfly is of all our indigenous species the largest; the female, which, as usual, exceeds the male in size, frequently measuring in expanse of wing considerably more than three inches. The general colour of the wings is black, powdered with yellow, and relieved by bold yellow markings.

SWALLOW-TAILED BUTTERFLY.
(PAPILIO MACHAON.)

The basal half of the hinder wings is also yellow; and from the posterior margin of them an acute "tail" projects, which may be fancifully compared to the outer tail-feathers of the swallow—hence its name: at each inner corner is an ocellated spot of red, with an anterior crescent of light blue; the whole nearly surrounded by a ring of black. The body is yellow, with two lines

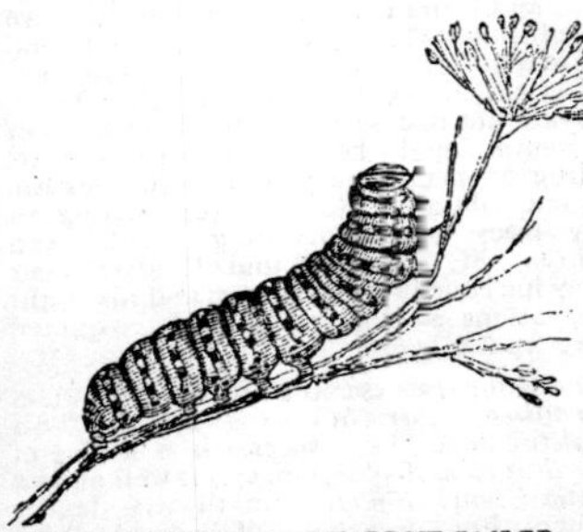

CATERPILLAR OF SWALLOW-TAILED BUTTERFLY.

beneath, and the back black; the antennæ and legs black. Though this species does not appear on the wing in our island till the beginning of June, and is rarely seen at all in the northern counties, it is by no means rare in the south and west of England. It is common in several parts of France and Italy, and abundant in Syria and Egypt. It flies with rapidity, and is difficult to catch. The caterpillar is smooth, green, with velvety black rings: the organ with which it is armed on the top of the neck is red; and it secretes an acrid liquor, which emits an unpleasant smell. It feeds solitarily on umbelliferous plants; and about July it changes to the chrysalis, which is greenish, with a longitudinal black band on each side.

PARADISEIDÆ, or BIRDS OF PARADISE. The genus *Paradisea*, distinguished in most species by a peculiar union of splendour and elegance, appears to be confined to the regions of Papua or New Guinea, and the small isles in the immediate vicinity; extending only a few degrees on each side the Equator. For a long time the most absurd fables and traditions were current respecting these magnificent specimens of the feathered tribes: namely, that they passed their whole existence in sailing in the air, the dew of heaven being their only food; that they were destitute of legs; that they never took rest except by suspending themselves from the branches of trees by the shafts of the two elongated feathers which form a characteristic of this beautiful race, and that they never touched the earth till the moment of their death. From such a tissue of absurdity and error the world has, however, long been free; and time has discovered that these birds have not only legs, but that they are both large and strong.

Birds of Paradise, which are allowed to exceed all others in the beauty, variety, and peculiar construction of their plumage, associate in large flocks in the delightful aromatic woods and groves of their native islands: and the inhabitants themselves, not insensible to their charms, give them the name of God's Birds. From the rapidity of their flight, as well as their being continually on the wing in pursuit of insects, their usual prey, they are sometimes called the swallows of Ternate. However, as the country where they breed is visited with tempestuous seasons, these birds are seldom seen at such times; and it is supposed that they then migrate to countries where their food is to be found in greater abundance; for, like swallows, they have their stated periods of return. There are several species of this beautiful group; but as it would be impossible to convey a perfect idea of the originals, unless we could represent their vivid and ever-changing tints, descriptions of two or three will suffice.

The GREAT EMERALD PARADISE BIRD. (*Paradisea apoda.*) The general length of this most elegant bird, from the tip of the bill to the end of the long side-feathers, is about two feet, but to the end of the real tail about twelve inches, the size of the bird being that of a thrush. The bill is slightly bent, and of a greenish colour; the base being surrounded, for the distance of half an inch, with close-set, velvet-like black plumes, with a varying lustre of gold-green: the head, together with the back part of the neck, is of a pale gold-colour, the throat and fore part of the neck of the richest changeable gold-green: the whole remainder of the plumage on the body and tail is of a fine deep chestnut, except on the breast, which is of a deep purple colour. From the upper

GREAT BIRD OF PARADISE.
(PARADISEA APODA.)

part of each side of the body, beneath the wings, springs a vast assemblage of extremely long, loose, broad floating plumes, of the most delicate texture and appearance; in some specimens of a bright deep yellow in others of a paler hue, but most of them marked by a few longitudinal dark red spots: and from the middle of the rump spring a pair of naked shafts, considerably exceeding in length even the long loose plumes of the sides. This bird is a native of the Molucca Islands and the islands around New Guinea, particularly in the Island of Aroo. Latham mentions that a specimen was once brought alive to England, and it is occasionally brought to Macao in China.

The chiefs of the countries where they are found use them in their turbans; and in many parts of the East, as well as in this country, parts of the birds are used by the fair sex as ornaments in their head-dress.

The ROYAL OR KING PARADISE BIRD. (*Paradisea Regia.*) This is supposed to be the smallest of all the Birds of Paradise, measuring only five inches and a half in length, without reckoning the two tail feathers, which are about six inches long. The colour of this bird on the upper parts is a most intense and beautiful red or purplish chestnut; the bill of a brownish yellow; the base, as well as the fore part of the head, being surrounded with velvet-like plumes: the throat and upper part of the breast are of a deep purple red, and across the lower part of the breast runs a broad gold-green zone, separated from the red above by a line of yellow: from the green zone or crescent downwards, the body and under wing-coverts are white:

KING BIRD OF PARADISE — MALE.
(PARADISEA [CICINNURUS] REGIA.)

beneath the wings, on each side the body, is a set of feathers of a dusky brown colour, with tips of the richest golden-green, each tip separated from the brown by a bounding line of white. The quill feathers are of a bright orange-brown beneath; and from

KING BIRD OF PARADISE—FEMALE.

the upper part of the rump, over the middle of the tail, extend two very long naked shafts, each terminating, in the most beautiful manner, in a moderately broad gold-green web, rising from one side only of the shaft, and forming a flat spiral of nearly two convolutions. The legs are moderately stout, and of a yellowish brown colour. This species is called the King-bird by the Dutch, and said not to associate with other birds of the genus, but to be of a solitary nature, feeding on berries, particularly such as are of a red colour; seldom, if ever, settling on lofty trees, but frequenting shrubs and bushes. M. Lesson found it alive near Dorey harbour in New Guinea, and his slight observations confirm what we have quoted above from a Dutch author.

The GOLD-BREASTED BIRD OF PARADISE. (*Paradisea* [*Parotia*] *sex-setacea.*) This beautiful deep black species is a native of New Guinea and Waigiou, and is well named by the French *Sifilet*, from the six slender feathers, three on each side of the head, which want webs, except at the end, where they spread into an oval. The breast has a rich gilded changeable green gorget, which is very brilliant. Our figure, which is copied from the work of Lesson, will show the form and

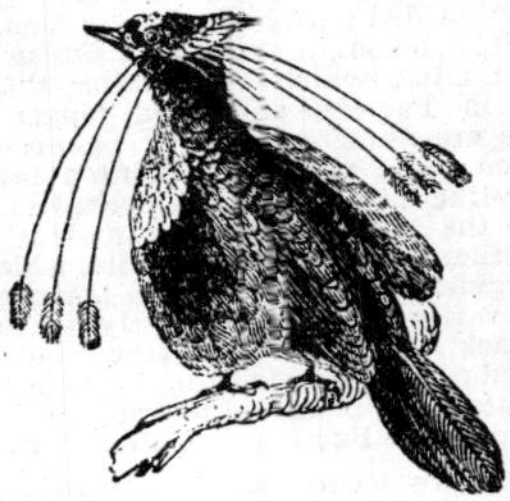

GOLD-BREASTED BIRD OF PARADISE—MALE. (PARADISEA [PAROTIA] SEX SETACEA.)

general appearance of such specimens as are preserved in museums. The female, which is also figured here, wants the six long-shafted feathers and the gorgeous breast of

GOLD-BREASTED BIRD OF PARADISE—FEMALE

the male, but instead, the feathers on the neck and side and under parts of the body are of a very light brown colour, transversely marked with rather wide deep brown bars. It is to be hoped that in a short time this, as well as the other superb Paradiseæ, will be found alive in our aviaries and Zoological Gardens.

The Superb Paradise Bird. *Paradisea* [*Lophorhina*] *superba.*) According to Mr. Forster, this magnificent native of that ornithological paradise the island of New Guinea, is brought down to Salawat by the inhabitants, in the shape of skins dried in the smoke, and deprived of the legs and wings. M. Lesson obtained his specimens at Dorey, and from his figure the cut which accompanies this is copied Nothing but a Humming-bird can exceed in splendour of colour some parts of the breast of this bird; the closely imbricated feathers on the throat and breast are of a bronzed green, with iridescence and corruscations of violet. The crest at the base of the beak, the long feathers on the side of the neck, looking like a second pair of wings, and the brilliant

SUPERB BIRD OF PARADISE—MALE. (PARADISEA [LOPHORHINA] SUPERBA.)

deeply-notched projecting green shield on its breast, are indicated in the figure: no description can give an adequate idea of the splendour of this or any other Bird of Paradise; we must refer our readers to the cases in the British Museum, or to other collections which contain these "children of the sun."

PARANDRA. A genus of Longicorn beetles belonging to the *Prionidæ* family, the species of which, as yet, have only been found in America. Their form and general appearance will be better indicated to our readers by the accompanying figure than by any description. We may only observe that the body is parallelopiped and very glossy; that the antennæ are simple, somewhat moniliform, and rather short; that the ligula has the form of a short transverse segment of a circle, not lobed in front; and that the

PARANDRA FERRUGINEA.

penultimate joint of the tarsi is hardly bilobed. Like most of the members of this family, in the larva state they feed upon timber. There are several species found in both North and South America.

PARDALOTUS. A genus of Australian birds, which in affinity of manners and general appearance seem to be allied to the Titmice and Wrens. We give a figure of the Spotted Manakin, as characteristic of the genus, and, as an example, may allude to the Pardalotus Affinis, or Striped-

HEADED MANAKIN. This bird inhabits Van Diemen's Land, and is the commonest of the isla d; wherever, indeed, the gum and wattle trees are, there may the bird be found, creeping about in the most easy and elegant manner, examining the upper and under sides of leaves for insects. It is found in

SPOTTED MANAKIN.
(PARDALOTUS PUNCTATUS.)

the gardens and shrubberies even in towns; where its sprightly action, and piping though monotonous note, are thought pleasing. It breeds in September and four following months, and has two or three broods in a year. The nest is of a round domed shape, like that of a Wren, with a small hole for an entrance: it is outwardly composed of grasses, and warmly lined with feathers.

PARMOPHORUS, or DUCK'S BILL LIMPET. A genus of Mollusca, found in New Holland and New Zealand; the shell of which is oblong, slightly depressed, and convex on the outside; the interior exhibiting very strong muscular impressions, which in some species are marked with a blood red colour. The head of the animal is rather indistinct, with two tentacula, having eyes at the base; foot very large.

PARNASSIUS. A very beautiful genus of Butterflies found on mountains in Europe and Asia, and lately ascertained by the most

APOLLO BUTTERFLY.
(PARNASSIUS APOLLO.)

profound Lepidopterist of this country, Mr. Edw. Doubleday, to be indigenous to North America, on the Rocky Mountains. The best known species, which is here figured, is the Apollo Butterfly (*Parnassius Apollo*), found in Norway, Sweden, and Switzerland. It is white, with various black markings; and these colours, with the beautiful crimson spots on the wings and the elegant shape of this pretty species, combine to make a most pleasing object to look at. The species has been reported to be found in Scotland, but the statement does not rest on good authority.

PAROQUET, or PARRAKEET. (*Palæornis.*) A distinctive appellation for a group of birds belonging to the *Psittacidæ*, or Parrot tribe, which are smaller than the common Parrots, and have longer tails. There are numerous species; some, distinguished by a very long pointed tail and collar-like mark round the neck, which inhabit the Asiatic continent and islands; and others, natives of Australia, which are distinguished by their colours being gorgeously variegated, and peculiarly mottled on the back; by their tail-feathers not being pointed; and by their being furnished with elongated tarsi, adapted for running on the ground. [See PEZOPORUS: PLATYCERCUS.]

The RING PAROQUET. (*Palæornis Alexandri.*) This beautiful species, no less remarkable for its symmetrical form and graceful movements, than for its docility and imitative powers, is supposed to have been the first bird of the parrot kind known to the ancient Greeks and Romans, having been brought from the island of Ceylon, after the Indian expeditions of Alexander the Great. They afterwards obtained other species from Africa.—The size of the Alexandrine or Ring Paroquet is that of a common pigeon; its general length about fifteen inches, and its colour an elegant bright green above, paler or yellower beneath; across each shoulder, on the smaller coverts, is a lengthened purplish-red patch or spot; and from the base of the

ROSE-RINGED PARRAKEET.
(PALÆORNIS TORQUATUS.)

lower mandible, on each side, proceeds a moderately broad black band or stripe, which, after descending a little way, passes backwards so as almost to encircle the neck, growing very narrow as it approaches the back part, which is marked by a red collar, near half an inch wide, but narrowing as it passes forwards immediately beneath the black one, almost reaching the front of the neck: the back part of the head, towards the commencement of the red collar, has a slight bluish tinge, and the edges of the tail-feathers are often of a similar cast: the bill is of a bright orange-red; the legs ash-coloured; and the under surface of the tail, which is strongly and regularly cuneated, is of a yellowish hue.

The GRASS PARRAKEET. [See EUPHEMIA.]

PARRA. A genus of Grallatorial birds, the species of which have very long toes, which enable them to support themselves on aquatic plants. They are often named JACANAS, and are found chiefly in the warm parts of America, Africa, and Asia. We may particularize PARRA GALLINACEA, a species inhabiting Australia, one of the most typical members of this genus; its hind toe and claw being so largely developed as to expressly adapt it for traversing those floating leaves and herbage that merely rise to the level of the water. Mr. Gould thus describes it:—Back of the head, line down the back of the neck, tips of the shoulders, under surface of the wing, flanks, and a broad band crossing the chest and abdomen, deep bluish-black; chin and throat white; orbits, ear-coverts, sides of the neck and breast, pale glossy orange, the white and the orange gradually blending into each other; back and scapularies bronzy olive-green, becoming nearly black at the base of the neck and on the rump; wing-coverts olive-brown; the remainder of the wing and tail greenish black; vent and under tail-coverts buffy white; irides light sulphur yellow; eyelash light ash-gray; bill greenish-gray at the extreme tip, then black to near the nostrils; the basal portion of the upper mandible and the helmet aurora-red; base of the lower mandible light primrose-yellow; fore-part of the tibia red, with a mixture in patches of yellow and greenish-gray; hinder part of the tibia, tarsi, and toes dark greenish-gray. Their powers of diving and of remaining under water are very great indeed, but their powers of flight are inconsiderable. At the slightest alarm they dive down at once or take to flight.

PARROTS. (*Psittacidæ.*) The Parrot family is a very numerous and splendid one; and is subdivided, chiefly according to the form of the bill and tail, into several groups; as the Macaws, Cockatoos, Lories, Paroquets, &c., which are each inserted in their alphabetical order. Under the word PSITTACIDÆ will be found a few general observations relating to the distinguishing characters of the genus, &c.—The *True Parrots*, which we are now to consider, have the upper mandible toothed, and longer than it is high; and the tail is short, or even and rounded at the end. They unite great beauty with great docility; and their faculty of imitating the human voice is superior to that of any other bird. The luxuriant tracts of the torrid zone seem to be the favourite residence of these richly-plumaged tribes: they are not, however, confined to that zone, as Buffon imagined, but are found in latitudes as far as forty or forty-five degrees on each side the equator. The tongue is fleshy, obtuse, and entire: their feet are formed for climbing, in which they assist themselves with their bill: they feed on the seeds and fruits of various plants; and often attain to a very great age.

The GREY PARROT. (*Psittacus erithacus.*) This species is remarkable for its loquacity, docility, and distinctness of articulation; and appears to have been one of the earliest imported species from Africa, in many parts of which it is common. It is about the size of a small pigeon, and in length about twelve inches. Its colour is an elegant ash-gray, deeper on the upper parts, and more inclining to white beneath; the whole, though of a sober colour, having a very elegant appearance, from the deeper and lighter undulations formed by the edges of the feathers, as well as from a kind of efflorescence resembling fine powder, which, in a healthy state, is perpetually diffused over the plumage. The whole tail is of the brightest crimson; the temples or orbits of the eyes bare and white; the bill black, and the legs cinereous. It is extremely long-lived; there are well-recorded instances of their having attained the age of seventy years; and some authors speak positively of individuals living to the age of 100. The surprising facility with which they repeat sentences has been often noticed; sometimes too ludicrously apposite, we should imagine, to obtain perfect credence. It was one of this species to which the memorable anecdote, first related by Gesner, and often referred to by succeeding writers, refers: "A Parrot belonging to King Henry VIII., who then resided at Westminster, in his palace by the river Thames, had picked up many words from hearing the passengers talk as they happened to take water. One day, sporting on its perch, the poor bird fell into the river; and then very seasonably remembering the words it had often heard some, whether in danger or in jest, use, cried out amain, 'A boat! a boat! twenty pound for a boat!' A waterman, who happened to be near, hearing the cry, made to the place where the Parrot was floating, and knowing to whom it belonged, restored it to its royal master, in the full expectation, as the bird was a great favourite, of receiving the promised reward. The king, however, preferred appealing to the Parrot himself to determine the sum, which being consented to by the boatman, the bird immediately cried out, 'Give the knave a groat!'"

The BRAZILIAN GREEN PARROT. (*Psittacus Braziliensis.*) This beautiful bird is rather larger than the Common Gray Parrot. Its plumage is fine grass-green, rather paler beneath; the feathers edged with purplish-brown: the front, all round the base of the bill, is bright red; the cheeks deep blue, and the top of the head yellow: the edge of the wing, at some distance beyond the shoulders, is red; all the wing-coverts and the shorter quill feathers deep or dusky blue; the outside feather on each side the tail deep blue, tipped with yellow; the next feather red, with a similar yellow tip, and all the remaining ones green with yellow tips: the bill pale, and the legs and feet dusky.

The AMAZON PARROT. (*Psittacus Amazonius.*) There are several varieties found on each side of a great extent of the river Amazon to whom the general appellation of *Amazonian* Parrots is given. Their usual length is about fourteen inches; the bills varying in colour: the irides yellow or orange; and the plumage bright green, with the feathers marked by dusky or blackish margins: the tops or edges of the shoulders, and a conspicuous patch on the middle of the wings, bright red; the red wing patch is usually bounded by shades of blue, green, and yellow, which colours are only completely visible in the expanded state of the wings: the tail-feathers are green, but appear red beneath the base when expanded. A bright blue band generally reaches from eye to eye, beyond which the feathers of the crown, cheeks, and throat are of a jonquil yellow: the legs and feet are either dusky or of a pale grayish brown. The species we have selected for description corresponds in its main features with what we have above stated; and, that the general colour of the plumage is a bright and beautiful green, deepest on the back and wings, and lighter beneath, a yellowish garter encircling the bottom of the thighs. The smallest of the wing-coverts, forming the ridge of the shoulders, are of a splendid red colour; the larger wing-feathers are externally of a deep blue with a cast of violet; the middle ones of the same colour at their tips, but red on their outward edges. The tail is deep green above, and yellowish beneath, and has some red on the upper part of each feather, which, however, is not seen when the tail is closed: the bill is dark brown, and the legs light gray. The Amazon Parrot abounds in Guiana and Surinam, where it causes great injury to the plantations. It builds in the midst of impenetrable forests, the female laying four white eggs in the hollow of a tree.

CAROLINA PARROT. (*Psittacus Carolinensis.*) The only species found native in the United States is the Carolina or Illinois Parrot, which is resident from the Gulf of Mexico to the neighbourhood of Lake Michigan, and on the east of the Alleghanies to Maryland. Their favourite food is the seeds of the cockle-bur, which grows in great abundance along the shores of the Mississippi and the Ohio; where they are seen in large flocks, screaming round the salt-licks. They are very sociable in their dispositions, extremely fond of each other, and showing the greatest grief for the loss of their companions. The plumage is very beautiful, the general colour being a bright yellowish silky green, with light blue reflections.

Wilson's American Ornithology furnishes us with the following particulars of this bird:—"In descending the river Ohio, by myself, in the month of February, I met with the first flock of paroquets, at the mouth of the Little Sioto. I had been informed, by an old and respectable inhabitant of Marietta, that they were sometimes, though rarely, seen there. I observed flocks of them, afterwards, at the mouth of the Great and Little Miami, and in the neighbourhood of numerous creeks that discharge themselves into the Ohio. At Big Bone lick, thirty miles above the mouth of Kentucky river, I saw them in great numbers. They

CAROLINA PARROT.
(PSITTACUS CAROLINENSIS.)

came screaming through the woods in the morning, about an hour after sunrise, to drink the salt water, of which they, as well as the pigeons, are remarkably fond. When they alighted on the ground, it appeared at a distance as if covered with a carpet of the richest green, orange, and yellow: they afterwards settled, in one body, on a neighbouring tree, which stood detached from any other, covering almost every twig of it, and the sun, shining strongly on their gay and glossy plumage, produced a very beautiful and splendid appearance. Here I had an opportunity of observing some very particular traits of their character: having shot down a number, some of which were only wounded, the whole flock swept repeatedly around their prostrate companions, and again settled on a low tree, within twenty yards of the spot where I stood. At each successive discharge, though showers of them fell, yet the affection of the survivors seemed rather to increase; for, after a few circuits around the place, they again alighted near me, looking down on their slaughtered companions with such manifest symptoms of sympathy and concern, as entirely disarmed me. I could not but take notice of the remarkable contrast between their elegant manner of flight, and their lame crawling gait among the branches. They fly very much like the wild pigeon, in close compact bodies, and with great rapidity, making a loud and outrageous screaming, not unlike that of the red-headed woodpecker. Their flight is sometimes in a direct line; but most usually circuitous, making a great variety of elegant and easy serpentine meanders, as if for pleasure. They are particularly attached to the large sycamores, in the hollow of the trunks and branches of which they generally roost, thirty or forty, and sometimes more, entering at the same hole. Here they cling close to the sides of the tree, holding fast by the claws and also by the bills. They appear to be fond of sleep, and often retire to their holes during the day. probably to take their regular *siesta*. They are extremely sociable with and fond of each other, often scratching each other's heads and necks, and always, at night, nestling as close as possible to each other, preferring, at that time, a perpendicular position, supported by their bill and claws. In the fall, when their favourite cockle-burs are ripe, they swarm along the coast, or high grounds of the Mississippi, above New Orleans, for a great extent. At such times, they are killed and eaten by many of the inhabitants; though, I confess, I think their flesh very indifferent. I have several times dined on it from necessity, in the woods: but found it merely passable, with all the sauce of a keen appetite to recommend it.

"The Carolina or Illinois parrot (for it has been described under both these appellations) is thirteen inches long, and twenty-one in extent; forehead and cheeks, orange red; beyond this, for an inch and a half, down and round the neck, a rich and pure yellow; shoulder and bend of the wing, also edged with rich orange red. The general colour of the rest of the plumage is a bright yellowish silky green, with light blue reflections, lightest and most diluted with yellow below; greater wing-coverts and roots of the primaries, yellow, slightly tinged with green; interior webs of the primaries, deep dusky purple, almost black, exterior ones, bluish green; tail, long, cuneiform, consisting of twelve feathers, the exterior one only half the length, the others increasing to the middle ones, which are streaked along the middle with light blue; shafts of all the larger feathers, and of most part of the green plumage, black; knees and vent, orange yellow; feet, a pale whitish flesh colour; claws, black; bill, white, or slightly tinged with pale cream; iris of the eye, hazel; round the eye is a small space without feathers, covered with a whitish skin; nostrils placed in an elevated membrane at the base of the bill, and covered with feathers; chin, wholly bare of feathers, but concealed by those descending on each side; from each side of the palate hangs a lobe or skin of a blackish colour; tongue, thick and fleshy; inside of the upper mandible near the point, grooved exactly like a file, that it may hold with more security. The female differs very little in her colours and markings from the male. After examining numerous specimens, the following appear to be the principal differences. The yellow on the neck of the female does not descend quite so far; the interior vanes of the primaries are brownish, instead of black, and the orange red on the bend and edges of the wing is considerably narrower; in other respects, the colours and markings are nearly the same."

PARROT-FISH. (*Scarus.*) This fish obtains its name from the peculiar hooked conformation of its mouth, or the brillancy of its colours, or perhaps from both. It has large, convex, rounded jaws, covered with hard, scale-like teeth, which succeed each other from the rear to the front in such a manner, that the bases of the newest form a cutting edge. It is about a foot long; the colour is greenish, variegated near the head with yellow; the fins are blue, and the scales are very large. Numerous species of this genus inhabit tropical seas, some of them being remarkably brilliant; but they are mostly noticeable for the immense strength of their jaws and teeth enabling them to browse without difficulty on the newest layers of the stony corals, digesting the animal matter therein contained, and setting free the carbonate of lime in a chalky state. The flesh of the Parrot-fish is firm and well-tasted.

PARTHENOGENESIS. [See Suppl.]

PARTHENOPE. A genus of short-tailed Crustacea, the rugosities on the back of one of which present the appearance of a fragment of eroded rock; this crab is the *P. horrida*, and is found in the Indian Ocean.

PARTRIDGE. (*Perdix cinereus.*) This well-known bird is about thirteen inches in length. The general colour of its plumage is brown and ash, beautifully mixed with black, and each feather streaked down the middle with buff: the upper part of the neck is transversely varied with dusky gray, and a tinge of red: the sides of the head are tawny; under each eye is a small saffron-coloured spot, which has a granulated appearance, and between the eye and the ear a naked skin of bright scarlet, which is not very conspicuous but in old birds: the under part of the neck and breast are bluish gray, marked with transverse black lines, and

COMMON PARTRIDGE (PERDIX CINEREUS.)

sprinkled with small reddish spots: on the lower part of the breast is a rich gorget of deep chestnut, in form of a horse-shoe: the tail is short and drooping; the legs are greenish white, and furnished with a small knob behind. The female has no crescent on the breast; and her colours in general are not so distinct and bright as those of the male. Partridges pair early in the spring:

the female lays from fourteen to eighteen or twenty eggs, of a greenish colour, making her nest of withered leaves and grass upon the ground. The young birds run as soon as hatched, frequently encumbered with part of the shell. The affection of the Partridge for her young is peculiarly strong; and she is greatly assisted by her mate in the care of rearing them: they lead them out in common, call them together, gather for them their proper food, and assist in finding it by scratching the ground — at first furnishing them with the larvæ of ants, on which they principally feed while very young. It is no unusual thing to introduce Partridges' eggs under the domestic hen, who watches and rears them as her own; in which case the young birds require to be fed with ants' eggs, which are their favourite food, and without which it is almost impossible to bring them up. Care should be taken to supply them well with fresh water: it is also recommended to give them a mixture of wood-lice and earwigs: and occasionally fresh curds, mixed with lettuce, chickweed, or groundsel. They likewise eat insects, and when full grown, all kinds of grain and young plants.

Whenever a dog or other formidable animal approaches the nest of a Partridge, the hen practises every art to allure him from the site: she keeps at a little distance before him, feigning to be incapable of flight, and just hopping up and falling down before him, but never advancing to such a distance as to discourage her pursuer: at length, having successfully misled him, she at once takes wing and disappears. The danger being over, and the dog withdrawn, she returns and finds her scattered brood, who immediately assemble at her call, and follow her. Corn fields are the places that Partridges most delight in, especially while the corn is growing; for that is a safe retreat, where they remain undisturbed, and under which they usually breed. They frequent the same fields after the corn is cut down, but with a different intent; for they then feed on such corn as has dropped from the ears; and find a sufficient shelter under cover of the stalks, especially in wheat stubble. When the winter comes on, and the stubble fields are trodden down or ploughed up, they sometimes retire to the upland meadows, where they lodge in the high grass and among rushes: they also sometimes resort to the low coppice-woods, especially if they are contiguous to corn lands. The eggs of these birds are frequently destroyed by weasels, foxes, &c., but still they are in general sufficiently numerous to furnish the sportsman with employment enough in the "shooting season." The sexual ardour of the male has been the theme of many writers on natural history; and there are instances out of number in which the parental solicitude of the female has justly called forth their eulogistic admiration. Partridges are found throughout nearly the whole of Europe, and nowhere in greater plenty than in this island, the north of France, Holland, and Germany.

The RED-LEGGED PARTRIDGE. (*Perdix rufus.*) A very beautiful and delicate bird, common in Barbary and South Europe; while it has been introduced into Norfolk and other parts of England: the bill is of a fine scarlet colour; the top of the head is a bright chestnut, becoming more dusky as it reaches the back part, and forming a ring round the neck, beautifully varied with small white spots. The sides of the head and throat are of a light bluish ash-colour, which gradually changes on the breast to a

RED-LEGGED PARTRIDGE.
(PERDIX RUFUS.)

faint rose-colour: the belly, thighs, and tail-coverts are light brown; the upper side of the neck, back, and wings are of a darkish hue; the prime quills of the wings are tipped with a light yellowish brown colour; and the scapulars are a bright blue, bordered with a dark red. The sides are covered with beautiful feathers, transversely variegated; the tips are orange, within which there are bars of black, succeeded by others of white; the rump is ash-coloured; the middle feathers of the tail are rather darker, and transversely barred; the side feathers of the tail are ash colour towards their roots, and their upper parts of a dirty orange. The legs and feet are red; and the claws are brown. — In South America the name of Partridge is applied to species of the genus TINAMUS [which see].

The PARTRIDGE BRONZEWING. [See GEOPHAPS.]

PARUS. A genus of Conirostral passerine birds; characterized by a conical beak, straight, and rather slender, with few hairs at its base, and a strong hind toe, armed with a long hooked claw. They are active little birds, continually flitting from spray to spray, suspending themselves in all sorts of attitudes, rending apart the seeds on which they feed, devouring insects, &c. They build their nests in the holes of trees, and store up provisions of grain. [See TOMTIT.]

PASAN. A species of Egyptian Antelope.

PASSENGER PIGEON. (*Columba* [*Ectopistes*] *migratoria.*) This species abounds in America most prolifically; but their numbers can scarcely be conceived without seeing the account given of them by the graphic pen of Wilson, the celebrated American ornithologist. The Passenger Pigeon is of a bluish-slate-colour, white underneath; wings long, and acuminated; the throat, breast, and sides vinaceous; tail, of twelve feathers, the two middle ones black, the lateral ones whitish; bill black; iris bright orange-red;

the naked orbit purplish-red. The female is paler, and her breast of a cinereous brown. These birds visit the states in prodigious numbers, but are more abundant in the Western States, where they breed, and which abound in beech mast, their favourite food.

PASSENGER PIGEON.
(COLUMBA [ECTOPISTES] MIGRATORIA.)

"The roosting-places are always in the woods, and sometimes occupy a large extent of forest. When they have frequented one of those places for some time, the appearance it exhibits is surprising. The ground is covered to the depth of several inches with their dung; all the tender grass and underwood destroyed; the surface strewed with large limbs of trees, broken down by the weight of the birds collecting one above another; and the trees themselves, for thousands of acres, killed as completely as if girdled with an axe. The marks of their desolation remain for many years on the spot; and numerous places could be pointed out where, for several years after, scarcely a single vegetable made its appearance. When these roosts are first discovered, the inhabitants, from considerable distances, visit them in the night with guns, clubs, long poles, pots of sulphur, and various other engines of destruction. In a few hours they fill many sacks, and load horses with them. By the Indians, a pigeon-roost or breeding-place is considered an important source of national profit and dependence for that season, and all their active ingenuity is exercised on the occasion. In the western countries, viz. the states of Ohio, Kentucky, and Indiana, these are generally in back woods, and often extend in nearly a straight line across the country for a great way. Not far from Shelbyville, in the state of Kentucky, about five years ago, there was one of these breeding-places, which stretched through the woods in nearly a north and south direction, was several miles in breadth, and was said to be upwards of forty miles in extent. In this tract almost every tree was furnished with nests wherever the branches could accommodate them. The Pigeons made their first appearance there about the 10th of April, and left it altogether with their young before the 25th of May. As soon as the young were fully grown, and before they left the nests, numerous parties of the inhabitants, from all parts of the adjacent country, came with waggons, axes, beds, cooking utensils, many of them accompanied by the greater part of their families, and encamped for several days at this immense nursery. Several of them informed me that the noise was so great as to terrify their horses, and that it was difficult for one person to hear another speak without bawling in his ear. The ground was strewed with broken limbs of trees, eggs, and young squab pigeons, which had been precipitated from above, and on which herds of Hogs were fattening. Hawks, Buzzards, and Eagles were sailing about in great numbers, and seizing the squabs from the nest at pleasure, while, from twenty feet upwards to the top of the trees, the view through the woods presented a perpetual tumult of crowding and fluttering Pigeons, their wings roaring like thunder, mingled with the frequent crash of falling timber; for now the axe-men were at work, cutting down those trees that seemed to be most crowded with nests, and contrived to fell them in such a manner, that in their descent they might bring down several others; by which means the falling of one large tree sometimes produced 200 squabs, little inferior in size to the old ones, and almost one heap of fat. On some single trees, upwards of 100 nests were found, each containing one squab only; a circumstance in the history of this bird not generally known to naturalists.* It was dangerous to walk under these flying and fluttering millions, from the frequent fall of large branches, broken down by the weight of the multitudes above, and which, in their descent, often destroyed numbers of the birds themselves; while the clothes of those engaged in traversing the woods were completely covered with the excrements of Pigeons.

"These circumstances were related to me by many of the most respectable part of the community in that quarter, and were confirmed in part by what I myself witnessed. I passed for several miles through this same breeding-place, where every tree was spotted with nests, the remains of those above described. In many instances I counted upwards of ninety nests on a single tree; but the Pigeons had abandoned this place for another, sixty or eighty miles off, towards Green River, where they were said at that time to be equally numerous. From the great numbers that were constantly passing over our heads, to or from that quarter, I had no doubt of the truth of this statement. The mast had been chiefly consumed in Kentucky; and the Pigeons, every morning a little before sunrise, set out for the Indiana territory, the nearest part of which was about sixty miles distant. Many of these returned before ten o'clock, and the great body generally appeared on their return a little after noon. I had left the public road to visit the

* The reader would naturally infer from this, that the Passenger Pigeon hatches a single young one only at a time; but Mr. Audubon observes, that the bird lays *two eggs* of a pure white, and that, as is the case with the rest of the genus, each brood generally consists of a male and a female. Every other part of Wilson's account he confirms.

remains of the breeding-place near Shelbyville, and was traversing the woods with my gun, on my way to Frankfort, when, about ten o'clock, the Pigeons which I had observed flying the greater part of the morning northerly, began to return in such immense numbers as I had never before witnessed. Coming to an opening by the side of a creek called the Benson, where I had a more uninterrupted view, I was astonished at their appearance: they were flying with great steadiness and rapidity, at a height beyond gunshot, in several strata deep, and so close together that, could shot have reached them, one discharge could not have failed of bringing down several individuals. From right to left, as far as the eye could reach, the breadth of this vast procession extended, seeming every where equally crowded. Curious to determine how long this appearance would continue, I took out my watch to note the time, and sat down to observe them. It was then half-past one; I sat for more than an hour, but instead of a diminution of this prodigious procession, it seemed rather to increase both in numbers and rapidity; and, anxious to reach Frankfort before night, I rose and went on. About four o'clock in the afternoon, I crossed Kentucky river, at the town of Frankfort, at which time the living torrent above my head seemed as numerous and as extensive as ever. Long after this I observed them in large bodies, all moving in the same south-east direction, till after six o'clock in the evening. The great breadth of front which this mighty multitude preserved would seem to intimate a corresponding breadth of their breeding-place, which, by several gentlemen who had lately passed through part of it, was stated to me at several miles."

Having endeavoured to make a rough calculation of the numbers composing this mass, he believes that, at the lowest estimate, there were 2,230,272,000 Pigeons; that they extended full 240 miles in length; and allowing each bird to consume half a pint of food daily, it would amount to 17,424,000 bushels per day!

PASSERINÆ, or PASSERINE BIRDS. The name of a most extensive and varied order of birds, which feed on insects, fruit, or grain, according to the slenderness or strength of their beak. They have all short and slender legs, with three toes before and one behind; the two external toes being united by a very short membrane: all the toes are slender, flexible, and moderately elongated, with long, pointed, and slightly curved claws. [See INSESSORES.]

PATELLA. [See LIMPET.]

PAUSSIDÆ. A family of Coleopterous insects, the various species of which are small in size, varying from a quarter to half an inch in length; but containing amongst them some of most remarkable form. The body is of a firm consistence, and of an oblong, quadrate, subdepressed form, narrowed in front; the head small, and generally narrowed behind into a neck; the antennæ, which are the most singular parts of these insects, are of a very large size, composed of two or more joints, of a very irregular construction; the elytra are broader than the

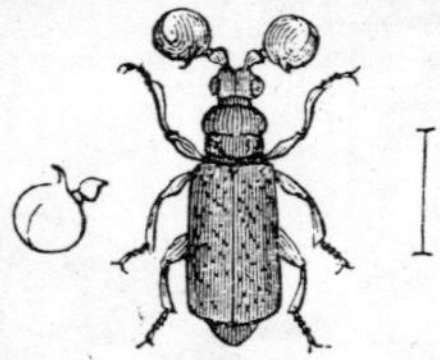

GLOBE HORNED ANT-BEETLE.
(PAUSSUS SPHÆROCERUS.)

rest of the body; the legs short, strong, and compressed. These extraordinary insects appear almost exclusively to inhabit the Old World; but lately a species has been described which was found by Mr. Miers, the eminent traveller and botanist, in South America. They are rare, and little is known of their habits; but they are believed to be nocturnal, and are said to crepitate like the Bombardier Beetles (*Brachinidæ*), while recent dissections have shown that in other respects they are not very distantly removed from that tribe. They are found about the nests of ants; and those who are desirous of *studying* the many curious species of these insects must consult the monographs of them given by Mr. Westwood in the Linnæan Transactions, in the Arcana Entomologica, and in the Transactions of the Entomological Society of London.

PAVONIA. A genus of Zoophytes, containing many foliated species of great beauty; specimens of which may be seen in the fine collection at the British Museum. Our figure

PAVONIA LACTUCA WITH POLYPES IN NATURAL POSITION.

represents a portion of a specimen of the *Pavonia lactuca*, or *Lettuce Coral*, in which the Zoophyte as well as the Polypidom are exhibited. It is copied from one of the finely illustrated recently published French works.

PEACOCK. (*Pavo.*) A genus of splendid Gallinaceous birds, of which but two species are recorded, viz. the *Common Peacock* and the *Javanese Peacock*. The COMMON PEACOCK (*Pavo cristatus*) is universally well known; and, as Buffon truly says, "its matchless plumage seems to combine all that delights the eye in the soft and delicate tints of the finest flowers, all that dazzles it in the sparkling lustre of the gems, and all that astonishes it in the grand display of the

COMMON PEACOCK.—(PAVO CRISTATUS.)

rainbow." Though long naturalized in Europe, it is of Eastern origin, occurring in the greatest profusion in the neighbourhood of the Ganges, and in the extensive plains of India, the kingdom of Siam, &c. As early as the days of Solomon they were imported into Judea by the fleets which that monarch equipped on the Red Sea. From India they were brought into Greece about the time of Alexander; and towards the decline of the Roman republic they were introduced into Rome, and were esteemed as one of the choicest luxuries of the table. They are still found wild in many parts of Asia and Africa, but more particularly in the fertile plains of India, where they attain a great size, and exhibit colours which seem to vie with the glittering gems and precious stones produced in those luxurious regions. Of the exact period when it was introduced into England we have no authentic record; but we learn from good authority that it long formed one of the dishes in the second course of every great feast; being usually baked in a pie, made in the form of the bird, with the head raised above the crust, the beak richly gilt, and the tail expanded. In the days of chivalry it was common for the knights to make their vows of enterprise at a solemn feast, on the presentation to each knight, in turn, of a roasted peacock in a golden dish.

The ordinary length of the Peacock, from the tip of the bill to that of the tail, is about four feet. Its finely-shaped head is adorned with a tuft, consisting of twenty-four feathers, whose slender shafts are furnished with webs only at the ends, painted with the most exquisite green, edged with gold: the head, throat, neck, and breast are of a deep blue, glossed with green and gold; the back of the same, tinged with bronze; the scapulars and smaller wing-coverts, reddish cream colour, variegated with black; the middle coverts deep blue, glossed with green and gold; and the belly and vent are dusky, with a greenish hue. The tail, which is of a gray-brown, is hidden beneath that which constitutes the distinguishing character of this beautiful bird—its magnificent *train*, which rises above it, and, when expanded, forms a superb fan of the most resplendent hues: the shafts are white, and are furnished from their origin, nearly to the end, with divided irridescent barbs; at the extremity of these feathers the barbs unite, and form a flat extended vane, decorated with what is called "the eye." This is a brilliant spot, or circlet, enamelled with the most enchanting colours; yellow, gilded with various shades; green, running into blue and bright violet, varying according to its different positions; the whole receiving additional lustre from the colour of the centre, which is a fine velvet black. "When pleased or delighted, and in the sight of his females, the Peacock erects his train, and displays the majesty of his beauty: all his movements are full of dignity; his head and neck bend nobly back, his pace is slow and solemn, and he frequently turns slowly and gracefully round, as if to catch the sunbeams in every direction and produce new colours of inconceivable richness."—These gorgeous plumes, however, whose versatile hues he has so often displayed with all the pride of conscious superiority, are shed every year; and then, as if sensible of his loss, he seeks the most obscure retreats to conceal himself, till the returning spring restores him to his accustomed beauty. The cry of the Peacock, especially on a summer evening and at night, is often repeated, and his loud and discordant screams are generally considered as the sure prognostic of bad weather. The legs are gray-brown, those of the male being furnished with a strong spur; and the feet are clumsy in the extreme.

The female (called the Pea-hen) is rather less than the male; and her train is not only very short, but destitute of those dazzling colours by which he is distinguished: her whole plumage, in fact, partakes of a light brown or cinereous hue. She seldom lays more than four or five eggs at a time, and always chooses some sequestered or secret spot, where she can conceal them from the male, who is apt to break them. The eggs are white and spotted; and she sits from twenty-five to thirty days, according to the temperature of the climate or the warmth of the season. The young birds do not acquire their perfect brilliancy till the third year. Occasionally the Peacock has the whole of the plumage of a pure white colour, the eyes of the train not excepted, but they may be traced by a different undulation of shade upon that part. There is also a variegated or mixed breed, between the common and the white variety; in which every proportion of colour between the two is at different times observed. Sometimes the female assumes the plumage of the male, which is said to take place only after she has done laying; but instances of this, we believe, are very rare.—The Peacock formerly graced the tables of the magnates of the land, and was served up with the feathers of the neck and tail preserved: its flesh, however, is far less delicious than that of the turkey, and it now rarely appears on the festive board.

In size and proportions the two species are nearly similar, but the crest of *Pavo Javanicus* is much longer than that of *P. cristatus*, and the feathers of which it is composed are regularly barbed from the base upwards in the adult bird, and of equal breadth throughout. Head and crest interchangeably blue and green. A naked space on the cheeks, including the eyes and ears, is coloured of a

JAVANESE PEACOCK.
(PAVO JAVANICUS.)

light yellow behind, and bluish-green towards its fore part. The feathers of the neck and breast, which are broad, short, rounded, and imbricated like the scales of a fish, are at their base of the same brilliant hue as the head, and have a broad, lighter, somewhat metallic margin; those of the back have still more of the metallic lustre. The wing-coverts have a deeper tinge of blue. The tail-feathers and their coverts (the train) are of a splendid metallic brown, changing into green; their barbs very long, loose, and silky; and the latter are almost all terminated by ocellated spots similar to those which mark the train of the common species, and of nearly the same size. Like it, also, they are of a beautiful deep purple in the centre, which is surrounded by a band of green, becoming narrow behind, but widening in front and filling up a kind of notch that occurs in the blue; then comes a broad brownish band; and lastly, a narrow black ring, edged with chestnut, all beautifully iridescent. Bill of a grayish horn-colour; iris deep hazel. Legs strong, naked, reticulated, dusky black.

The beauty of the Peacock's plumage was a theme of admiration in the remotest times; and the bird was sought after as capable of adding splendour to the magnificence of Solomon. The chief display of this beauty arises from that arrangement of long and gorgeous feathers which spring from the space between the region behind the wings and the origin of the tail; but the use of this to the bird itself has been a subject of doubt. At first sight it seems to be no better than a luxuriance of nature, and an encumbrance rather than a benefit. The action by which their splendour is outspread has also been deemed an absurd manifestation of pride. But men are imperfect interpreters of the actions of animals; and a closer examination of the habits of this bird affords a different explanation.

PEACOCK [BUTTERFLY]. A name given by insect collectors to Butterflies of the species *Vanessa Io.*

PEARL [MOTHS]. A name given by collectors to Moths of the genus *Margaritia.*

PEARL OYSTER. (*Avicula Margaritifera.*) A bivalve Mollusc, celebrated for the valuable nacreous substance, called *mother-of-pearl*, with which the inside of the shell is lined, but still more for the little globular, oval, or pear-shaped concretions, called *pearls*, which are sometimes found free and detached within the lobes of the mantle. They are thus described by an eminent conchologist: "Pearls are small nacreous balls, that become formed and hardened within the body of the animal; they are found deposited in the most fleshy parts, particularly within and around the adductor muscle, and are said to be occasioned by the overcharge of those glands whose function it is to secrete the nacreous fluid destined for the internal lining of the shell. When the animal is thus diseased, this beautiful irridescent fluid

PEARL-OYSTER.—(AVICULA MARGARITIFERA.)

is very irregularly discharged, being also deposited upon the inner surface of the shell in little excrescences; these are often detached, and form articles of commerce as pearls of inferior value, the former being considered more precious, both on account of their rotundity of form and the clearness and beauty of their complexion."—*Reeve's Con. Sys.*

From an interesting article on Pearls and Pearl Fisheries, by Dr. Baird (in Chambers's Miscellany, No. 167.), we glean the following:—"Substances so unlike the composition of the shells in which they are found must naturally give rise to speculations respecting their origin; and thus we find, in times ere science had determined their real nature, various amusing hypotheses to account for their existence. Pliny, the celebrated Roman naturalist, gravely tells us that the oyster which produces pearls does so from feeding upon heavenly dew. Our own early writers entertained the same notion; and Boethius, speaking of the pearl-mussel of the Scottish rivers, remarks, that 'these mussels, early in the morning, when the sky is clear and temperate, open their mouths a little above the water, and most greedily swallow the dew of heaven; and after the measure and quantity of the dew which they swallow, they conceive and breed the pearl. These mussels,' he continues, 'are so exceedingly quick of touch and hearing, that, however faint the noise that may be made on the bank beside them, or however small the stone that may be thrown into the water, they sink at once to the bottom, knowing well in what estimation the fruit of their womb is to all people." In the East, the belief is equally common that these precious gems are

'Rain from the sky,
Which turns into pearls as it falls in the sea.'

But, alas for poesy and romance! the science of chemistry — which has, with its sledge-hammer of matter-of-fact, converted the all-glorious diamond into vulgar charcoal — has also pronounced the precious pearl to be composed of 'concentric layers of membrane and carbonate of lime!' Admitting its composition, the question still remains as to the cause of a substance so dissimilar in appearance to the shell in which it exists, and why it should be present in some shells, and absent in others.

"In all cases, it appears that the ultimate cause of the animal's forming this beautiful substance is to get rid of a source of irritation. Sometimes this happens to be a grain of sand, or some such small foreign body, which has insinuated itself between the mantle of the oyster and the shell, and which, proving a great annoyance, the animal covers with a smooth coat of membrane, over which it spreads a layer of nacre. At other times, it is caused by some enemy of the inhabitant of the shell perforating it from the outside to get within reach of its prey. With a plug of this same matter, the oyster immediately fills up the opening made, and shutting out the intruder, balks it of its nefarious design. In both these cases, we find the pearl usually adhering to the internal surface of the shell. The best, however, and the most valuable specimens, are generally found in the body itself of the animal; and the source of irritation here is proved, according to the observations of Sir Everard Home, who has paid great attention to this subject, to be an ovum or egg of the animal, which, instead of becoming ripe, proves abortive, and is not thrown out by the mother along with the others, but remains behind in the capsule in which the ova are originally contained. This capsule, being still supplied with blood-vessels from the parent animal, goes on increasing in size for another year, and then receives a covering of nacre, the same as the animal spreads over the internal surface of the shell.

"Sir Everard Home does not appear to have been aware that Sandius, as long ago as 1673, communicated the same fact to the Royal Society of London; but was led to it when investigating the mode of breeding of the fresh-water mussel, by generally finding in the ovarium round hard bodies, too small to be noticed by the naked eye, having exactly the appearance of seed-pearls, as they are called. Sometimes he found these bodies connected with the surface of the shell, in contact with the membrane covering it. In further examining into the structure of pearls, he ascertained that all split pearls upon which he could lay his hands universally possessed a small central cell, which surprised him by its extreme brightness of polish; and in comparing the size of this cell with that of the ovum when ready to drop off from its pedicle, he found it sufficiently large to enclose it. He came thus to the conclusion that these abortive eggs are the commencement or nuclei of the pearl. Being once formed, the animal continues to increase its size by the addition of fresh coats, adding, it is said, a fresh layer every year. It is extremely probable, however, that its presence being still a source of irritation to the creature, the nacral covering is more rapidly deposited upon the pearl than upon the shell itself. Those pearls found in the substance of the animal are generally round, but occasionally we find them of a pyramidal form, the pedicle by which the egg is attached appearing to have received a coat of nacre as well as itself. People conversant with the pearl-fishery assert that they do not appear till the animal has reached its fourth year, and that it takes from seven to nine years for the oyster to reach maturity.

"The true pearl is remarkable, as is well known, for its beautiful lustre — a lustre which cannot altogether be given to artificial ones. According to Sir Everard Home, this peculiar lustre arises from the central cell, which is lined with a highly-polished coat of nacre; and the substance of the pearl itself being diaphanous, the rays of light easily pervade it. Previous to Sir Everard's theory, it was supposed by opticians that the peculiar splendour was the effect of light reflected from the external surface. They took for granted that pearls were solid bodies, denied them to be diaphanous, and, therefore, considering the subject mathematically, they contended that their brilliancy must be produced by the reflection from the nacral surface. In the Edinburgh Encyclopædia, we are told by Sir David Brewster that the fine pearly lustre and iridescence of the inside of the pearl-oyster arises from the circumstance, that we find in all 'mother-of-pearl a grooved structure upon its surface, resembling very closely the delicate texture of the skin at the top of an infant's finger, or the minute corrugations which are often seen on surfaces covered with varnish or with oil paint.' Similar appearances, we are told, are to be seen in the structure of pearls. 'The direction of the grooves,' says Sir David, 'is in every case at right angles to the line joining the common image and the coloured image; hence in irregularly-formed mother-of-pearl, where the grooves are often circular, and have every possible direction, the coloured images appear irregularly scattered round the ordinary image. In the real pearl these coloured images are crowded into a small space round the common image, partly on account of the spherical form of the pearl; and the various hues are thus blended into a white unformed light, which gives to this substance its high value as an ornament.' Pearls, however, at least the most valuable, are not perfectly solid, and are certainly translucent. In fact, in a split pearl we find the transparency to be considerable. 'Upon taking a split pearl,' says Sir Everard Home, 'and putting a candle behind the cell, the surface of the pearl became immediately illuminated; and upon mounting one with coloured foil behind the cell, and by putting a candle behind the foil, the outer convex surface became universally of a beautiful pink colour.' If we take a split pearl and set it in a ring with the divided surface outwards, and look at this through a magnifying glass, this central cell becomes very conspicuous, and the different layers of which the pearl is composed are also beautifully displayed. It is the brilliancy above described that distinguishes the real from the factitious pearl—a lustre which no art can altogether give, though often attempted with considerable success."

Much valuable information on the subject of pearls and pearl fisheries is also given in Mr. M'Culloch's Commercial Dictionary, to which work we are indebted for most of the following particulars. Pearls should be chosen round, of a bright translucent silvery whiteness, free from stains and roughness. Having these qualities, the largest are of course the most valuable. The larger ones have frequently the shape of a pear; and when these are otherwise perfect, they are in great demand for ear-rings. Pearls were in the highest possible estimation in ancient Rome; but, owing partly to the changes of manners and fashions, and still more, probably, from the admirable imitations that may be obtained at a very low price, they are now less esteemed, and comparatively cheap. When the pearls dwindle to the size of small shot, they are denominated *seed* pearls, and are of little value. One of the most remarkable pearls of which we have any authentic account was bought by Tavernier, at Catifa, in Arabia, a fishery famous in the days of Pliny, for the enormous sum of 110,000*l.*! It is pear-shaped, regular, and without blemish. The diameter is ·63 inch at the largest part, and the length from two to three inches.—Very good imitations of pearls have been made with hollow glass globules, the inside of which is covered with a liquid called pearl essence, and then filled with white wax: the essence is composed of the silver-coloured particles which adhere to the scales of the Bleak (*Cyprinus alburnus*).

The Pearl Oyster is fished in various parts of the world, particularly on the west coast of Ceylon; at Tuticoreen, in the province of Tinnevelly, on the coast of Coromandel; at the Bahrein Islands, in the gulf of Persia; at the Soloo Islands; off the coast of Algiers; off St. Margarita, or Pearl Islands, in the West Indies, and other places on the coast of Colombia; and in the Bay of Panama, in the South Sea. Pearls have sometimes been found on the Scotch coast, and in various other places. The most extensive pearl fisheries are those on the several banks not far distant from the island of Bahrein, on the west side of the Persian Gulf; but Pearl Oysters are found along the whole of the Arabian Coast. The fishing season is divided into two portions—the one called the short and cold, the other the long and hot. In the cooler weather of the month of June, diving is practised along the coast in shallow water; but it is not until the intensely hot months of July, August, and September, that the Bahrein banks are much frequented. The water on them is about seven fathoms deep, and the divers are much inconvenienced when it is cold; indeed, they can do little when it is not as warm as the air, and it frequently becomes even more so in the hottest months of the summer. When they dive, they compress the nostrils tightly with a small piece of horn, which keeps the water out, and stuff their ears with bees' wax for the same purpose. They attach a net to their waists, to contain the oysters; and aid their descent by means of a stone, which they hold by a rope attached to a boat, and shake it when they wish to be drawn up. A person usually dives from twelve to fifteen times a day in favourable weather; but when otherwise, three or four times only. They continue under water from a minute to a minute and a half, or at most two minutes. The exertion is extremely violent; and the divers are unhealthy and short-lived.

PECCARY. (*Dicotyles tajacu.*) This Pachydermatous animal, which at first view has very much the appearance of a small Hog, is a native of South America. It is of a short compact form, thickly covered on the

WHITE-LIPPED PECCARY.—(D. LABIATUS.)

upper parts of the body with thick and strong dark-coloured bristles, each marked by yellowish-white rings; and round the neck is generally a whitish band or collar. The head is rather large; the snout long; the ears short and upright; and the under part of the body nearly naked. Instead of a tail, it has merely a fleshy protuberance ; and at the lower part of the back is a glandular orifice, from which exudes a strong-scented fluid, and which is surrounded by strong bristles. The Peccary is a gregarious animal, and in its wild state is fierce and dangerous ; sometimes attacking the hunters with great vigour, and often killing the dogs. It is useful in destroying several reptiles, particularly the rattle-snake, which it does without the least dread or inconvenience. It is capable of being tamed like the hog, lives on the same kind of food, and has nearly the same habits and natural inclinations. The flesh of the Peccary is tolerable food, but, to prevent an unpleasant flavour, the dorsal gland must be cut away as soon as the animal is killed. Our figure represents what some of the older naturalists regarded as a variety of the Tajacu ; but modern writers have proved its distinctness as a species, and from its white lips have named it *Dicotyles labiatus*. It is also a native of South America.

PECTEN. A Molluscous animal, whose testaceous covering has a hinge like that of the Oysters; but they are easily distinguished from the *Ostrea* family, by their inequivalve semicircular shell being almost always regularly marked with ribs, which radiate from the summit of each valve to the circumference, and are furnished with two angular productions called *ears*, that widen the sides of the hinge. The animal has a small oval foot supported on a cylindrical peduncle, in front of an abdomen in form of a sac hanging between the branchiæ. In some species, known by the strong sinus under their anterior ear, there is a byssus. The others are not adherent. and can even swim with considerable velocity, by flapping their valves together. The mantle is surrounded with two rows of filaments, several

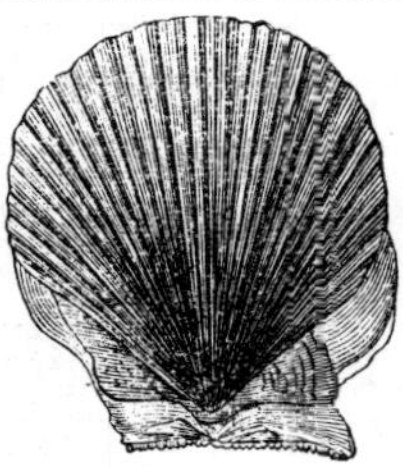

GIBBOUS SCALLOP-SHELL.
(PECTEN GIBBOSUS.)

of those of the exterior row being terminated by a little shining green globule. The mouth is garnished with many branched tentacula instead of the four usual labial laminæ. The Clam-shells are often coloured in a lively manner, and many species are remarkable for the difference in colouring observable in the two valves. The well-known large species found on our coasts (the *Pecten jacobæus* of authors) is the Scallop or Pilgrim's shell, worn in front of the hat by those who had visited the shrine of St. James, in the Holy Land. There are numerous species, some of which are found in the British seas.

PECTENIBRANCHIATA. The name given by Cuvier to an order of Gasteropods. It includes almost all the spiral univalve shells, as well as several which are merely conical. The animals of this order are so named from the comb-like form of the gills, which are usually situated in a cavity behind the head.

PECTUNCULUS. A genus of Conchiferous Mollusca, found in the Atlantic Ocean, the Mediterranean Sea, and in the West Indies, where it lives on the sandy or muddy coasts, and moves by the aid of its foot, which is large. The shell is orbicular, equivalve, sub-equilateral, thick, striated longitudinally ; and many of the species covered with a soft downy epidermis : hinge curved, with a line of teeth diverging on each side, those in the middle being incompletely formed ; ligament external. No byssus.

PEDICULUS. [See LOUSE.]

PEDIONOMUS. A genus of Gallinaceous birds, allied to the Partridges and Quails, which contains the PEDIONOMUS TORQUATUS, or COLLARED PLAIN WANDERER. This is a small quail-like bird, with lengthened bustard-like legs, admirably suited for running, and a small hind toe. It is a native of South Australia, on the desert plains of which it is not unfrequently found.

PEDUM. A singular genus of Conchifera, only one species of which is known, and that is found in the Indian seas, at great depths, and is rare. The shell is hatchet-shaped, inequivalve, and slightly eared ; attached by a byssus passing through a sinus in the lower valve ; hinge toothless, with a triangular area in each valve, separating the umbones ; ligament contained in a groove running across the area ; bosses unequal and distant, the lower valve rather convex, with the sides reflected over the upper. This rare shell is white, slightly tinged with purple near the bosses ; and buries itself partially in madrepores, in crevices of its own boring.

PEEWIT. [See LAPWING.]

PEGASUS. A genus of Lophobranchiate fishes, native of the Indian Seas, and in some degree allied to the genus *Syngnathus*. They have a snout, with the mouth under it, and movable, like that of a sturgeon, only composed of the same bones as in other osseous fishes. The body is armed as in Hippocampus, but their thorax is broad, depressed, and with the gill openings in the sides. They have two distinct ventrals in rear of the pectorals, which are often large, and have procured these fishes the name of Pegasus, or Flying Horses. The dorsal and anal fins are opposite each other; the abdominal cavity is wider and shorter than in Syngnathus, and the intestine has two or three flexures.

The principal species, the DRAGON PEGASUS (*Pegasus draco*) is a small fish, three or four inches in length, and is remarkable for the size of its pectoral fins, which are supposed to enable it, like the Exoceti and some other fishes, to support itself for a few moments in air, while it springs occasionally over the surface of the water. The thorax or superior part of the body is of a broad, slightly flattened, squarish form, and is marked both above and beneath by several radiated shields or bony tubercles of considerable size: from each side the abdomen springs a lengthened cirrus, which supplies the place of a ventral fin: from the thorax the body decreases suddenly in diameter, and is marked into several divisions or transverse segments; the tail is small and slightly rounded; and the pectoral fins are large, of a rounded shape, with an indented outline: the eyes are large and protuberant, and the snout of a sub-conical form, but with a slight dilatation towards the tip. The colour of this fish is whitish, with a cast of pale brown.

The FLYING PEGASUS (*Pegasus volans*) is somewhat smaller than the preceding: snout much elongated, flattened, rounded, and slightly dilated at the tip; marked by a longitudinal channel, and denticulated on the edges: on the head a rhomboidal depression, and behind it two deep sub-pentagonal cavities: last joints of the body, next the tail, pointed on each side.

The SWIMMING PEGASUS (*Pegasus natans*) is of a much more slender shape than that of the *P. volans:* colour yellowish brown,

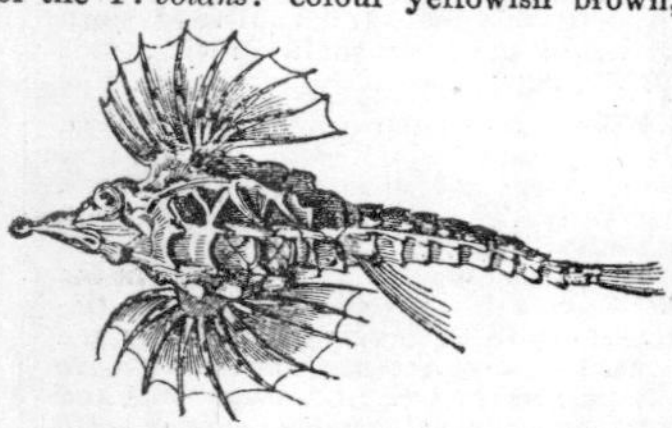

SWIMMING PEGASUS.—(P. NATANS.)

whitish beneath; snout slender, slightly dilated and rounded at the tip: pectoral fins rounded, and of moderate size; dorsal situated on the middle of the back; tail small, and slightly rounded: ventral cirri slender and flexible.

PEKAN, or WOOD-SHOCK. The name given to a species of Marten (the *Martes Canadensis*) found in North America.

PELECANIDÆ. The name given to a family of *Natatores* or Swimming Birds. They have the hind toe united with the others by a single membrane; they are excellent swimmers, often perch on trees, and have short legs: their beak is long, the edge of it generally toothed; and the skin of the throat is more or less extensible, forming a bag in which they keep the fish as they catch them, to feed their young. They are a large, voracious, and wandering tribe, living for the most part on the ocean, flying with ease and swiftness, and never visiting the land for any length of time but at the season of incubation.

PELICAN. (*Pelecanus.*) This is a genus containing several large web-footed species of birds, residing on rivers, lakes, or along the sea-coast, and preying on fish. They have a long, straight, broad, and much depressed bill; upper mandibles flattened, terminated by a nail, or very strong hook, the lower formed by two bony branches, which are depressed, flexible, and united at the tip; and from these branches is suspended a naked skin in form of a pouch; face and throat naked; nostrils basal, in the form of narrow longitudinal slits; legs short and strong; all the four toes connected by a web; wings of moderate dimensions. "The expansive pouch, whose elasticity is well known to all who have witnessed the shapes into which it is stretched and formed by the itinerant showmen, will hold a considerable number of fish, and thus enables the bird to dispose of the superfluous quantity which may be taken during fishing excursions, either for its own consumption or for the nourishment of its young. In feeding the nestlings — and the male is said to supply the wants of the female when sitting in the same manner — the under mandible is pressed against the neck and breast, to assist the bird in disgorging the contents of the capacious pouch, and during this action the red nail of the upper mandible would appear to come in contact with the breast, thus laying the foundation, in all probability, for the *fable* that the Pelican nourishes her young with her blood, and for the attitude in which the imagination of painters has placed the bird in books of emblems, &c., with the blood spirting from the wounds made by the terminating nail of the upper mandible into the gaping mouths of her offspring." — *Broderip.* Pelicans are gregarious, and fish is their favourite food: they store up their prey in their gular pouch, from which it is gradually transferred to the *œsophagus*, as the process of digestion goes on; but when harassed or pursued, they readily reject the contents of the stomach, like the Gull tribe. Though remarkable for their voracity, some of the species have been trained to fish in the service of man. The species are widely spread throughout the world, but are not numerous. In external appearance the sexes very nearly resemble each other.

The COMMON PELICAN. (*Pelecanus onocrotalus.*) The colour of the Pelican is white, faintly tinged with flesh or light rose colour, which is brightest in the breeding season; gullet with a bright yellow pouch. The first quill-feathers and spurious wings are black; the bag at the throat is flaccid, membranous, and capable of great distension; naked space round the eyes and at the base of the bill, where the frontal feathers form a point, flesh-colour; the upper mandible bluish, with a crimson line running along the top, reddish at the base, yellowish at the tip, and the terminal nail red; irides hazel; feet livid; tail short. Length between five and six feet; expanse of wings twelve or thirteen feet. The young are distinguished by the prevalence of ash-colour in their plumage. About the middle of September, flocks of this species repair to Egypt, in regular bands, terminating in an obtuse angle. During the summer months they take up their abode on the borders of the Black Sea and the shores of Greece. In France they are very rare; in Great Britain unknown. They generally take their prey in the morning and evening, when the fish are most in motion.

COMMON PELICAN,
(PELECANUS ONOCROTALUS.)

At night the Pelican retires a little way on the shore to rest, with its head leaning against its breast; and in this attitude it remains almost motionless, till hunger calls it to break off its repose. It then flies from its resting-place, and, raising itself thirty or forty feet above the surface of the sea, turns its head, with one eye downwards, and continues on wing till it sees a fish sufficiently near the surface, when it darts down with astonishing swiftness, seizes it with unerring certainty, and stores it away in its pouch; it then rises again, and continues the same manœuvres till it has procured a competent stock. The female feeds her young with fish that have been macerated for some time in her pouch. The Pelican generally breeds in marshy and uncultivated places, particularly about islands and lakes, making its nest, which is a foot and a half in diameter, and proportionably deep, of sedges and aquatic plants, and lining it with soft grass. It lays two or more white eggs, of equal roundness at the two ends, and which, when persecuted, it sometimes hides in the water. When it nestles in dry and desert places, it brings water to its young in its bag, which is capable of containing nearly twenty pints. Pelicans are rarely seen farther than twenty miles from the land. To a certain extent, they appear to be gregarious.

The account which Capt. Flinders gives of the Pelicans which he saw while on his voyage of Discovery at '*Terra Australis*' is almost as pathetic as it is descriptive: "Flocks of the old birds were sitting upon the beaches of the lagoon, and it appeared that the islands were their breeding places: not only so, but from the number of bones and skeletons there scattered, it should seem that they had for ages been selected for the closing scene of their existence. Certainly none more likely to be free from disturbance of every kind could have been chosen. than these islets in a hidden lagoon of an uninhabited island, situate upon an unknown coast near the antipodes of Europe: nor can anything be more consonant to the feelings, if Pelicans have any, than quietly to resign their breath, whilst surrounded by their progeny, and in the same spot where they first drew it." It was on this passage that Mr. James Montgomery founded his beautiful poem, 'The Pelican Island.'

In many places the Pelicans are almost regarded as sacred birds: for instance, a correspondent of the Athenæum, when travelling in Persia, speaks of "an immense flock of Pelicans which got up out of the reeds, and flew across our course, many passing quite close to the peak of our sail; one of our Greek servants, Yanni, a Cypriote, drew his pistol to fire at them: but his arm was caught by an Arnout, who told him the bird was sacred, Pelicans having brought water in their bill-pouches to Ali after a battle, when he lay on the desert faint with extreme thirst and toil."

PELIDNOTA. A genus of Lamellicorn Beetles of an elongated shape, somewhat related to the Cockchafer. It contains many Brazilian species, some of them with brilliant metallic green and copper reflections. In this genus we may specify, from Dr. Harris's work, the common North American species.

The PELIDNOTA PUNCTATA, or SPOTTED PELIDNOTA. A large beetle, arranged among the *Rutelidæ*, which is found on the cultivated and wild grape-vine, sometimes in great abundance, during the months of July and August. It is of an oblong oval shape, and about an inch long. The wing-covers are tile-coloured, or dull brownish yellow, with three distinct black spots on each; the thorax is darker, and slightly bronzed, with a black dot on each side; the body beneath, and the legs, are of a deep bronzed green colour. These beetles fly by day; but may also be seen at the same time on the leaves of the grape, which are their only food. They sometimes prove very injurious to the vine. The only method of destroying them is to pick them off by hand, and crush them under foot. The larvæ live in rotten wood.

PELOPEUS, or DIRT-DAUBER. [See WASP.]

PENELOPE, or GUAN. (*Penelope cristata.*) This bird resembles, both in appearance and manners, the Curassows, and seems, like them, to be capable, with proper care and attention, of being added to our stock of domesticated poultry. In a wild state they inhabit Guiana and Brazil, and are said to furnish an excellent dish for the table. They are about thirty inches in length, the tail being about thirteen. Upper parts dusky black or bronze, glossed with green and olive; fore part of neck and breast spotted with white; belly and legs, lower part of the back, and under tail-coverts, reddish. Cheeks naked, and of a purple violet colour. Bill dusky. On the head a thick tufted crest, which the bird can raise or depress at pleasure. Naked part of the throat scarlet, with an extensile fold of depending skin. Their food consists principally of seeds and fruits,

GUAN.—(PENELOPE CRISTATA.)

which they search for and eat upon the ground; but they build their nests and perch on trees. The females lay from two to five eggs. From the shortness of their wings their flight is low and heavy. Their note is so extremely loud, that when any number are collected near the same spot, they make the woods resound with their clamorous cries.

PENGUIN, or PINGUIN. (*Spheniscus aptenodytes.*) The name of a remarkable group of aquatic birds, exclusively found in the Antarctic seas, and deriving their name from their *pinguidity*, or excessive fatness. Their feet are placed so far back, that the body is quite upright when the bird is standing on the ground, for which purpose the tarsus is enlarged like the sole of the foot of a quadruped. The wings are very small, and lose altogether the power of raising the body in the air, being covered with short, rigid, scale-like feathers, disposed in regular order, instead of having their surface extended by prolonged feathers. While in the water, which is their natural element, they move with great alertness and rapidity; but on the land their motions are slow and awkward, and, from the form of their wings, they cannot fly. The female lays from one to three eggs, forming a rude excavation or burrow in the sand, instead of a nest, and it is only during the period of incubation that they are to be found on shore. The largest species is the GREAT MAGELLANIC PENGUIN (*Spheniscus Magellanicus*), which, although not more than two feet in length, is sometimes so bulky as to weigh from thirty to forty pounds.

The KING PENGUIN (*Aptenodytes Patachonica*), as described by Mr. G. Bennett, who saw a colony of these birds which covered an extent of thirty or forty acres, "are arranged, when on shore, in as compact a manner and in as regular ranks as a regiment of soldiers, and are classed with the greatest order, the young birds being in one situation, the moulting birds in another, the sitting hens in a third, the clean birds in a fourth, &c.; and so strictly do birds in similar condition congregate, that should a bird that is moulting intrude itself among those which are clean, it is immediately ejected from them. The females hatch the eggs by keeping them close between their thighs; and if approached during the time of incubation, move away, carrying their eggs with them. At this time the male bird goes to sea and collects food for the female, which becomes very fat. After the young are hatched, both parents go to sea, and bring home food for it; it soon becomes so fat as scarcely to be able to walk, the old birds getting very thin. They sit quite upright in their roosting-places, and walk in the erect position until they arrive at the beach, when they throw themselves on their breasts, in order to encounter the very heavy sea met with at their landing-place." Two species have been confounded under this name: the species of Forster is named by Mr. G. R. Gray *Aptenodytes Forsteri*, while Pennant's, which would seem the species described by Mr. Bennet, is called by Mr. Gray *Aptenodytes Pennantii.*

Of the habits of another species, called the JACKASS PENGUIN (*Eudyptes demersa*), Mr. Darwin gives the following account: "One day, having placed myself between a Penguin and the water, I was much amused by watching its habits. It was a brave bird: and, till reaching the sea, it regularly fought and drove me backwards. Nothing less than heavy blows would have stopped him; every inch gained he firmly kept, standing close before me, erect and determined. When thus opposed, he continually rolled his head from side to side, in a very odd manner, as if the power of vision only lay in the anterior and basal part of each. This bird is commonly called the Jackass Penguin, from its habit, while on shore, of throwing its head backwards, and making a loud strange noise, very like the braying of that animal; but while at sea and undisturbed, its note is very deep and solemn, and is often heard in the night-time. In diving, its little plumeless wings are used as fins; but on the land, as front legs. When crawling (it may be said on four legs) through the tussocks, or on the side of a grassy cliff, it moved so very quickly that it might readily have been mistaken for a quadruped. When at sea, and fishing, it comes to the surface, for the purpose of breathing, with such a spring, and dives again so instantaneously, that I defy any one at first sight to be sure that it is not a fish leaping for sport."

We are told by Sir James Clark Ross, in his Voyage of Discovery in the Southern and Antarctic Regions, that when he was performing the ceremony of taking possession of the newly-discovered lands, since called Victoria Land, in the name of Her Majesty, he was surrounded by Penguins in countless multitudes. These are his words: "Possession Island is situated in lat. 71° 56′, and long. 71° 7′ E., composed entirely of igneous rocks, and only accessible on its western side. We saw not the smallest appearance of vegetation, but inconceivable myriads of Penguins completely and densely covered the whole surface of the island, along the ledges of the precipices, and even to the summits of the hills, attacking us vigorously as we waded through their ranks, and pecking at us with their sharp beaks, disputing possession; which, together with their loud coarse notes, and the insupportable stench from the deep bed of guano, which had been forming for ages, and which may at some period be valuable to the agriculturists of our Australasian colonies, made us glad to get away again, after having loaded our boats with geological specimens and penguins." The *Auks*, *Razor-bills*, and *Puffins* [which see] are birds of the northern hemisphere, and belong to the genus *Alca*.

We must refer our readers to Mr. G. R. Gray's account of the Penguins in the Zoology of the Voyage of H.M. Ships Erebus and Terror. The galleries of the British Museum contain a very fine collection of these singular birds.

PENTACRINUS. A genus of stalked Echinoderms, mostly fossil. Their stems are pentangular; hence they are sometimes called "Five-angled Lily-shaped animals." Mr. Thompson found a living species (*Pentacrinus Europæus*) in the Cove of Cork and elsewhere on the Irish coast; but this turns out to be only the young pentacrinoid condi-

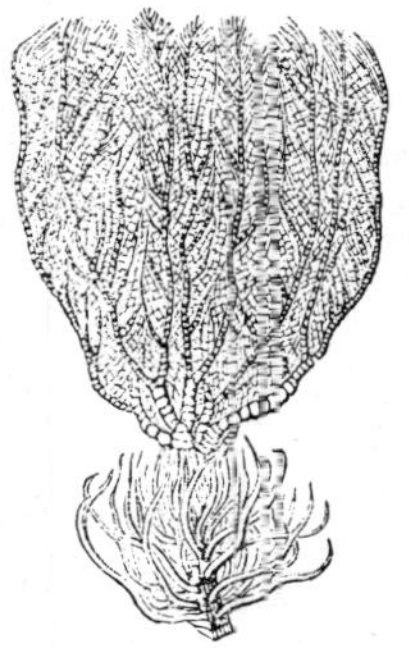

PORTION OF THE PENTACRINUS BRIAREUS. (FOSSIL.)

tion of the Star-fish known as *Comatula rosacea*. There is but one living species of the Pentacrinus, and that is the *P. Caput-Medusæ*, found in the West Indian seas. Numerous fossil species are found in Lias formation, associated with other allied Crinoids.

PENTALASMIS. A genus of Pedunculated Cirripedes. [See ANATIFA.]

PENTAMERA. A family of carnivorous beetles; some terrestrial, others aquatic. They have five joints to the tarsi of all the legs; hence the name.

PERAMELES, or PURSED BANDICOOT. A genus of Marsupial animals, of which several species are found in Australia.

PERCH. (*Perca fluviatilis.*) This well-known fish is to be found in clear rivers and lakes throughout nearly the whole of the temperate parts of Europe; and in England there is scarcely one of either in which it is not common. Its general size varies from ten to eighteen inches in length, and its weight from one to three pounds. Occasionally it is much larger. The body of the Perch is compressed, and its height is about

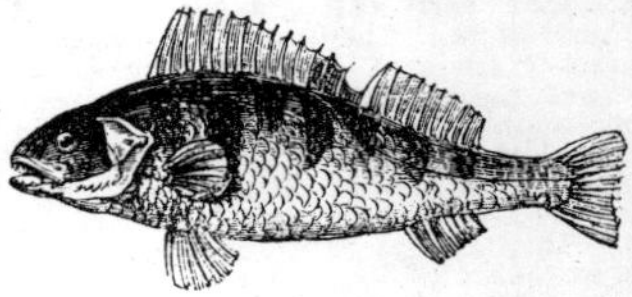

COMMON PERCH.—(PERCA FLUVIATILIS.)

one-third of its length; the upper part is a rich olive brown, passing into golden yellowish white below; and the sides have usually five or six dark transverse bands: the first dorsal fin is a pale violet-brown, marked at the back of the spiny part by one large and one small black spot; the second dorsal and pectoral fins pale brown; ventral, anal, and caudal fins, bright vermillion; the irides golden yellow. The Perch usually spawns in the early part of the spring; is of a gregarious disposition, and is fond of frequenting deep holes in rivers which flow with a gentle current: it is extremely voracious; bites eagerly at a bait; and is very tenacious of life. Its flesh is firm and delicate.

PERCHING BIRDS. [See INSESSORES.]

PERCIDÆ. A family of Acanthopterygious fishes; of which the *Perch* furnishes an example.

PERDICIDÆ. The name given to a family of birds which includes the Partridges, Quails, Francolins, &c. [See PARTRIDGE.]

PEREGRINE FALCON. [See FALCON.]

PERISTERA. A genus of the family *Columbidæ*, containing the Partridge Pigeon (*Peristera montana*), and the White-bellied Pigeon (*Peristera Jamaicensis*) and many other species.

PERISTERA HISTRIONICA, or HARLEQUIN BRONZEWING. A fine Pigeon found on the Mokai, a river falling into the Namoi, in Australia. Mr. Gould saw two or three immense flocks, and supposes it must be a bird belonging to the interior of Australia: its wings are long, and it has great power of flight. [See PIGEON.]

PERIWINKLE. (*Turbo littoreus.*) A well-known species of Mollusca, more extensively used as food than any of the other testaceous univalves. This shell is easily gathered, as it is found on all our rocks which are left uncovered by the ebbing of the tide. Children are principally employed in the fishery, and they are sold by measure. They are in general used after being plainly boiled, and are consumed in great quantities by the poor inhabitants on the coast.

PERLIDÆ. A family of Neuropterous insects, comprising a few species of moderate size; distinguished by the large size of the posterior pair of wings; the body oblong, depressed, and of equal breadth throughout; the prothorax large, flat, and quadrate; the eyes prominent and globose, and between them three ocelli in a triangle; the mandibles small, flat, and membranous; and the antennæ nearly as long as the body, and multi-articulate. The insects belonging to this family frequent damp marshy situations, and the borders of lakes and rivers, resting upon stones, palings, and plants growing close to the water's edge; they are sluggish in their movements, and the larger species are esteemed an excellent bait for trout. In their preparatory states they reside in the water, the larvæ being naked, not enclosed in a case, and in general form resembling the imago, except in wanting wings. These insects have been studied much in this country by Mr. Newman, and abroad by M. Pictet of Geneva; the latter has published their history with much detail.

PERN, or HONEY-BUZZARD. (*Pernis apivorus.*) This is one of the most elegant of the British birds of prey, or rather of such migratory species as become occasional visitants here. It is a trifle longer than the common Buzzard, and rather more slender: the bill is black, the irides yellow, the crown of the head is ash-coloured, and the cheeks are covered with small feathers. The neck, back, scapulars, and covert feathers of the wings, are of a deep brown; the breast and belly are white, marked with dusky spots pointing downwards; the tail is long, of a dull brown colour, and marked with three broad dusky bars, between each of which there are two or three of the same colour, but narrower: the legs are short, strong, and thick; and the claws are large and black. The Honey Buzzard generally lays two eggs, blotched over with a fainter and a deeper red: it builds its nest on small twigs, which it covers with wool. It feeds on bees, wasps, &c.

In an interesting article on the changes which take place in the plumage of this bird, communicated by W. R. Fisher, Esq., of Great Yarmouth, to the "Zoologist," the writer says, "As the Honey Buzzard has, I believe, never, except in the instance recorded by White of Selbourne, in the year 1780, been satisfactorily ascertained to have bred in this country, British ornithologists are deprived of this means of watching the changes by which it ultimately assumes the adult dress. For even if it were possible to procure the eggs or young from those countries of the East to which this species is said to be indigenous, the process would be so tedious and expensive, that few naturalists would be willing to undertake it; and the difficulty of rearing young birds, and the many casualties to which they are subject during the process of moulting, are well known."—Six specimens are delineated; and the gradations from a dark clove brown in the plumage of one, to the almost pure white (except of the wings and tail) in another, are truly remarkable: but that these striking differences are partly to be attributed to certain periodical changes, and partly arising from the difference of age and sex, there can be no doubt. Four of the birds there figured were taken in the county of Norfolk in the month of September, 1841. It is an error, however, to imagine that the Honey-buzzard does not breed in this country, or that it so rarely happens as to render it necessary for naturalists to refer to so distant a date as the year 1780: several recent instances of its nidification in different parts of this country could easily be given; the usual season for it being about the beginning of June. The nests are chiefly composed of sticks and twigs, and made very shallow, except just where the eggs are deposited; and the situation chosen for the nest is generally on one of the largest branches of an oak.

PERNA. A genus of Conchiferous Mollusca, the shell of which is sub-equivalve, irregular, compressed, and foliaceous; hinge straight and broad, divided into parallel grooves; bosses small; margins very brittle. They are mostly from India, the Cape Verde Islands, &c., and generally found adhering to rocks deep in the sea; considerable clusters being frequently found attached firmly to each other by the byssus. The genus is chiefly distinguished by the straightness, number, and regularity of the grooves in the hinge and the sinus, for the passage of the byssus. Its shape recals to mind that of a gammon of bacon: hence the name.

PETALOCERA. A tribe of Coleopterous insects, comprising those which have antennæ terminated by a foliated mass. The mandibles are very variable in their structure, corresponding with the habits of the various groups; and the head and thorax of the males are armed with strange horns or protuberances, of which it is difficult to conceive the uses. The body is generally more or less oval and convex, the legs robust, and the anterior tibiæ dentated on the outside. In many of the insects, especially those which feed upon leaves, the internal edge of the mandibles is formed into a broad horny plate, with various transverse channels, well formed for masticating. These insects subsist on vegetable substances, some while in a state of decay, and others upon fresh leaves and flowers, their larvæ devouring the roots of grass, &c., and often doing much damage. The head of the larva is generally large and horny, convex in front, with the top curved; the mandibles are strong, flat on the anterior surface, concave on the posterior; the legs are robust, with four joints, terminated by a strong hook. Some of these larvæ are several years in attaining their full size; they then form, in situations where they reside, an oval cocoon, composed of earth, excrements, and morsels of gnawed wood, &c. agglutinated together. The pupæ are of the ordinary form; but the sheaths of the lower wings are rather longer than those of the elytra. As examples of this tribe we may cite the COCKCHAFER, STAG-BEETLE, ROSE-BEETLE, SCARABÆUS, &c.

PETALURA. A genus of gigantic Dragon-flies, found in New Holland and New Zealand; a closely allied species of which seems at one time to have lived in this country, although now alone known by fossil remains, figured by Mr. Strickland, under the name of *Æshna liassina.* There are two species known — the *Petalura gigantea* of New Holland, and the *Petalura Carovei* of New Zealand; both of which may be at once known by the largely developed appendages at the end of the abdomen. The accompanying figure, copied

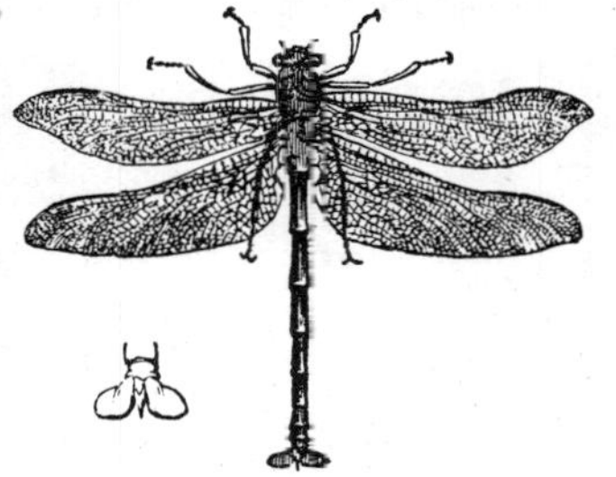

CAROVE'S DRAGON-FLY.
(PETALURA CAROVEI)

from the Zoology of H. M. SS. Erebus and Terror, will give a tolerably accurate idea of the form of this curious genus of Neuroptera. The name, we may remark, was given to it in compliment to the author of 'The Story without an End,' in which a Dragon-fly is made to act an important part.

PETAURUS. The Flying Phalanger: a Marsupial animal which bears the same relationship to the true Phalanger, as the flying squirrel does to the ordinary squirrel. By means of the skin which is extended between the fore and hind limbs, the animal can partially sustain itself in the air; and its aerial evolutions, when favoured by the shades of evening, are considered peculiarly graceful. It is destitute of the prehensile tail of the true Phalangers.

PETREL. (*Thalassidroma.*) A genus of celebrated oceanic birds.

The STORMY PETREL, so well known and much dreaded by sailors as the harbingers of a storm,—and to whom the *soubriquet* of Mother Carey's Chicken has been given,—are the least of all the web-footed birds, being only about six inches in length. The bill is half an inch long, hooked at the tip; the nostrils tubular. The upper parts of the plumage are black, sleek, and glossed with bluish reflections; the brow, cheeks, and under parts, sooty brown; the rump, and some feathers on the sides of the tail, white: legs slender, black, and scarcely an inch and three quarters in length, from the knee joint to the end of the toes. In the length of its wings, and the swiftness of its flight, it resembles the Chimney Swallow. It is met with on every part of the ocean, diving, or swimming over the surface of the heavy rolling waves of the most tempestuous sea, quite at ease, and in security; and yet it seems to foresee and fear the coming storm before the seaman can discover any appearance of its approach; flocking together, and making a clamorous piercing cry, as if to warn the mariner of his danger. They feed on small marine animals and seeds of seaweeds, and appear very fond of fat or grease, for which, and for the animals put in motion, they will follow in the wake of ships for great distances. They breed in the fissures of rocks, and the female lays two eggs. They fly rapidly, and generally close to the water; and, when in pursuit of food, they suspend themselves by extending their wings, and appear to run on the surface of the water. There are four species, which are so closely allied as to be often confounded. C. Buonaparté, who paid much attention to this genus of oceanic birds, designates them as follows: — *Thalassidroma Wilsonii* (Stormy Petrel); deep sooty black; tail even; wings reaching a little beyond its tip; tube of the nostrils recurved; tarsus one and a half inch long. —*T. Leachii* (Forked-tailed Petrel); brownish black; tail forked; wings

FORK-TAILED PETREL.
(THALASSIDROMA LEACHII.)

not reaching beyond the tip; tube of the nostrils straight; tarsus one inch long. Our figure represents this species it is copied from the great work of Audubon 'The Birds of America.'—*T. pelagica;* sooty black; tail even; wings reaching a little beyond it; tube of the nostrils almost straight; tarsus seven-eighths of an inch long.—*T. oceanica;* brownish black; tail slightly emarginate; wings reaching more than an inch beyond it; tube of the nostrils recurved; tarsus nearly one and three-fourths of an inch long. —"Whisking with the celerity of an arrow through the deep valleys of the abyss, and darting away over the foaming crest of some mountain wave, they attend the labouring bark in all her perilous course. When the storm subsides they retire to rest, and are no more seen." [For Sooty Petrel, see PUFFINUS.]

PETRICOLA. A genus of Conchifera, found in various parts of the world, in rocks, corals, &c., but most abundant in America. They are delicate, white, and radiated; and contain a tongue-shaped mollusc, the foot of which is small. The shell is equivalve, inequilateral, transverse, and variously oblong; anterior side rounded, posterior side more or less attenuated, slightly gaping; hinge with two cardinal teeth in each valve, and two muscular impressions in each; ligament external.

PETROGALE. A well characterized genus of the Kangaroo family, first described by Mr. Gray. The species frequent rocky mountains, preferring in some instances those that are most precipitous. The Brush-tailed Rock Wallaby (*P. penicillata*) has a harshish long fur, of a dusky brown hue, tinged with red and grey: a white streak passes down the middle of the throat; the tail is black, very long, and furnished with long hairs which form a brush. The length of the male is about three feet and a half. It is a strictly gregarious species, assembling in such numbers (Mr. Gould informs us) as to form well-beaten paths along the sides of the mountains: their agility is very great, leaping from rock to rock, and, like the chamois or goat, alighting on perilously narrow ledges—a habit which protects them from the aborigines and the native Australian dog. The species is strictly nocturnal in its habits. It occasionally ascends trees, not using the tail as a help. We are assured by Mr. Gould that the flesh is excellent. Captain Grey, in his Travels in South Australia, has described the habits of one of these, and as they are believed to be all somewhat similar, we cannot do better than quote him. He is speaking of the species called the Short-eared Rock Kangaroo (*P. brachiotis*), which is found in North-western Australia. He says, "This graceful little animal is excessively wild and shy in its habits, frequenting in the day-time the highest and most inaccessible rocks, and only descending into the valleys to feed early in the morning and late in the evening. When disturbed in the day-time, among the roughest and most precipitous rocks, it bounds along from one to the other with the greatest apparent facility, and is so watchful and wary in its habits that it is by no means easy to get a shot at it. One very surprising thing is, how it can support the temperature to which it is exposed in the situations it always frequents amongst the burning sandstone rocks, the mercury there during the heat of the day being frequently 136°. I have never seen these animals in the plains or lowlands, and believe that they frequent mountains alone."

PETROICA. A genus of interesting Passerine birds, found in Australia, of which Mr. Gould has described several species.

The PETROICA MULTICOLOR, or SCARLET-BREASTED ROBIN. This elegant species is a native of New Holland and all the small islands lying off the southern coast, the low bushes and woods skirting the open plains and sterile districts being its favourite places of resort. We learn from Mr. Gould, that, like our own Redbreast, the familiarity with which this beautiful Robin enters the gardens and dwellings, necessarily makes it a great favourite; "its attractiveness is moreover much enhanced by its more gay attire, the strong contrasts of scarlet, jet-black, and white rendering it one of the most beautiful to behold of any of the birds of Australia;" but its song and call-note, though resembling the European bird, are much more feeble. The head, throat, and upper surface of the male are black; forehead snowy-white; a longitudinal and two oblique bands of white on the wings; breast and upper part of the belly scarlet; lower part of the belly white; bill and feet black. The breast of the female is strongly tinged with red, but the chief part of her plumage is brown. The nest, which is compactly made and securely placed, is composed of dried grasses, narrow strips of bark, mosses and lichens, all bound firmly together with cobwebs and the finest fibres of vegetables, and lined with feathers, wool, or soft hair. The eggs are generally three or four in number; greenish-white, slightly tinged with flesh colour, and rather minutely freckled all over with olive-brown. Two or three broods are generally reared in the year, the period of nidification commencing in August and ending in February.

The PETROICA SUPERCILIOSA, or WHITE-EYEBROWED ROBIN. This species was discovered in the neighbourhood of the Burdekin Lakes, by Mr. Gilbert, while in company with Dr. Leichardt, during his adventurous expedition from Moreton Bay to Port Essington; and it is thus noticed in that gentleman's journal:—"May 14th. In a ramble with my gun I shot a new bird, the actions of which assimilate to those of the *Petroicæ* and the *Eopsaltriæ;* like the former, it carries its tail very erect, but is more retiring in its habits than those birds; on the other hand, its notes resemble those of the latter. It inhabits the dense jungle-like vegetation growing beneath the shade of the fig-trees on the banks of the Burdekin." Over the eye is a long white stripe, and the throat, abdomen, under surface of the shoulder, and the bases of the primaries and secondaries are white; lores, ear-coverts, wing-coverts, and the primaries and secondaries, for some distance beyond the white, deep black; all the upper surface, wings, and tail, sooty brown; all but the two central tail-feathers largely tipped with white; bill and feet black.

Another species, the PETROICA ERYTHROGASTRA, or NORFOLK ISLAND ROBIN, to which locality it is believed to be strictly confined, is thus described:—The male has the forehead silvery white; a small patch on the wings near the shoulder, under wing-coverts, the flanks and under tail-coverts white; chest and abdomen very rich scarlet; the remainder of the plumage deep black; bill black; feet brown. The female has the crown of the head, all the upper surface, wings, and tail reddish brown; throat white, tinged with brown; chest and centre of the abdomen washed with scarlet; lower part of the abdomen and under tail-coverts white; flanks brown; bill blackish brown; feet yellowish brown.

The other species are *Petroica Goodenovii*, or Red-capped Robin; *Petroica phœnicea*, or Flame-breasted Robin; *Petroica bicolor*, or Pied-Robin; and *Petroica fusca*, or Dusky Robin:—all closely assimilating in manners and economy to the one above described, but differing from that and each other in specific characteristics.

PEZOPORUS, or GROUND PARRAKEETS. A genus of the family *Psittacidæ*; so called from their terrestrial habits. To this genus belongs the beautiful green and black-marked New Holland Parrot—the PEZOPORUS FORMOSUS, to which Mr. Gould, the historiographer of the Birds of Australia, thus alludes in his large and noble work: "Unlike some of the African members of its family, who are inelegant in form and slow and ungraceful in their actions, the *P. formosus* is as active and graceful as can be well imagined; and although in its colouring it cannot vie with some of its more gaudily attired brethren, it possesses a style of plumage and diversity of markings far from unpleasing. Having very frequently encountered it in a state of nature, I am enabled to state that in its action it differs from every other known species of its race, as it does also in its habits and economy, which I shall now attempt to describe. Whether the power of perching is entirely denied to it or not I am uncertain, but I never saw it fly into a tree, nor could I ever force it to take shelter on the branches. It usually frequents either sandy sterile districts covered with tufts of rank grass and herbage, or low swampy flats abounding with rushes and the other kinds of vegetation peculiar to such situations. It is generally observed either singly or in pairs, but from its very recluse habits, and great powers of running, it is seldom or ever seen until it is flushed, and then only for a short time, as it soon pitches again and runs off to a place of seclusion, often under the covert of the grass-tree (*Xanthorrhœa*) which abounds in the districts it frequents." * * * It flies near the ground with great rapidity, frequently making several zigzag turns in the short distance of a hundred yards, beyond which it seldom passes without again resting on the ground. Its flesh is excellent, being much more delicate in flavour than that of the snipe, and equalling, if not surpassing, that of the quail. Its white eggs are deposited on the ground. It is a native of South Australia, and is found also in Van Diemen's Land.

PHACOCHŒRUS. A genus of Pachydermata allied to Swine, and from the projecting appendages about the head called *Warty Hogs*. They are natives of Africa.

PHÆTON. A genus of web-footed birds. [See TROPIC BIRD.]

PHALACROCORAX. A genus of Palmipede birds; for the characters and various species of which see the article CORMORANT. In addition to what is there given, we think that the following extract, although long, is sufficiently interesting to warrant the introduction of it here: it is taken from Mr. Robert Fortune's "Three Years' Wanderings in the Northern Provinces of China:"—

"The most singular of all the methods of catching fish in China is that of training and employing a large species of cormorant for this purpose, generally called the fishing-cormorant. These are certainly wonderful birds. I have frequently met with them on the canals and lakes in the interior, and, had I not seen with my own eyes their extraordinary docility, I should have had great difficulty in bringing my mind to believe what authors have said about them. The first time I saw them was on a canal a few miles from Ning-po. I was then on my way to a celebrated temple in that quarter, where I intended to remain for some time, in order to make collections of objects of natural history in the neighbourhood. When the birds came in sight I immediately made my boatmen take in our sail, and we remained stationary for some time to observe their proceedings. There were two small boats, containing one man and about ten or twelve birds in each. The birds were standing perched on the sides of the little boat, and apparently had just arrived at the fishing-ground, and were about to commence operations. They were now ordered out of the boats by their masters; and so well trained were they, that they went on the water immediately, scattered themselves over the canal, and began to look for fish. They have a beautiful sea-green eye, and, quick as lightning, they see and dive upon the finny tribe, which, once caught in the sharp-notched bill of the bird, never by any possibility can escape. The cormorant now rises to the surface with the fish in its bill, and the moment he is seen by the Chinaman he is called back to the boat. As docile as a dog, he swims after his master, and allows himself to be pulled into the san-pan, where he disgorges his prey, and again resumes his labours. And what is more wonderful still, if one of the cormorants gets hold of a fish of large size, so large that he would have some difficulty in taking it to the boat, some of the others, seeing his dilemma, hasten to his assistance, and with their efforts united capture the animal and haul him off to the boat. Sometimes a bird seemed to get lazy or playful, and swam about without attending to his business; and then the Chinaman, with a long bamboo, which he also used for propelling the boat, struck the water near where the bird was, without, however, hurting him, calling out to him at the same time in an angry tone. Immediately, like the truant schoolboy who neglects his lessons and is found out, the cormorant gives up his play and resumes his labours. A small string is put round the neck of the bird, to prevent him from swallowing the fish which he catches; and great care is taken that this string is placed and fastened so that it will not slip farther down upon his neck and choke him, which otherwise it would be very apt to do.

"Since I first saw these birds on the Ning-po Canal, I have had opportunities of inspecting them and their operations in many other parts of China, more particularly in the country between the towns of Hang-chow-foo and Shang-hae. I also saw great numbers of them on the river Min, near Foo-chow-foo. I was most anxious to get some living specimens, that I might take them home to England. Having great difficulty in inducing the Chinese to part with them, or, indeed, to speak at all on the subject, when I met them in the country, owing to our place of meeting being generally in those parts of the interior where the English are never seen, I applied to her Majesty's consul at Shang-hae (Captain Balfour), who very kindly sent one of the Chinese connected with the consulate into the country, and procured two pairs for me. The difficulty now was to provide food for them on the voyage from Shang-hae to Hong-Kong. We procured a large quantity of live eels, this being a principal part of their food, and put them into a jar of mud and fresh water. These they eat in a most voracious manner, swallowing them whole, and, in many instances, vomiting them afterwards. If one bird was unlucky enough to vomit his eel, he was fortunate indeed if he caught it again, for another, as voracious as himself, would instantly seize it, and swallow it in a moment. Often they would fight stoutly for the fish, and then it either became the property of one, or, as often happened, their sharp bills divided the prey, and each ran off and devoured the half which fell to his share. During the passage down we encountered a heavy gale at sea; and as the vessel was one of those small clipper schooners, she pitched and rolled very much, shipping seas from bow to stern, which set everything on her decks swimming. I put my head out of the cabin door when the gale was at its height, and the first thing I saw was the cormorants devouring the eels, which were floating all over the decks. I then knew that the jar must have been turned over or smashed to pieces, and that of course all the eels which escaped the bills of the cormorants were now swimming in the ocean. After this I was obliged to feed them upon anything on board which I could find; but when I arrived at Hong-Kong they were not in very good condition: two of them died soon after: and as there was no hope of taking the others home alive, I was obliged to kill them and preserve their skins.

"The Chinaman from whom I bought these birds has a large establishment for fishing and breeding the birds about thirty or forty miles from Shanghae, and between that town and Chapoo. They sell at a high price even amongst the Chinese themselves; I believe from six to eight dollars per pair, that is, from thirty shillings to forty shillings. As I was anxious to learn something of their food and habits, Mr. Medhurst, junior, the interpreter to the British consulate at Shang-hae, kindly undertook to put some questions to the man who brought them, and sent me the following notes connected with this subject:—'The fish-catching birds eat small fish, yellow eels, and pulse-jelly. At 5 P. M. every day each bird will eat six taels (eight ounces) of eels or fish, and a catty of pulse-jelly. They lay eggs after three years, and in the fourth or fifth month. Hens are used to incubate the eggs. When about to lay, their faces turn red, and then a good hen must be prepared. The date must be clearly written upon the shells of the eggs laid, and they will hatch in less than twenty-five days. When hatched, take the young and put them upon cotton, spread upon some warm water, and feed them with eel's blood for five days. After five days they can be fed with eel's flesh chopped fine, and great care must be taken in watching them. When fishing, a straw tie must be put upon their necks, to prevent them from swallowing the fish when they catch them. In the eighth or ninth month of the year they will daily descend into the water at ten o'clock in the morning, and catch fish until five in the afternoon, when they will come on shore. They will continue to go on in this way until the third month, after which time they cannot fish until the eighth month comes round again. The male is easily known from the female, it being generally a larger bird, and in having a darker and more glossy feather, but more particularly in the size of the head, the head of the male being large, and that of the female small.' Such are the habits of this extraordinary bird. As the months named in the note just quoted refer to the Chinese calendar, it follows that these birds do not fish in the summer months, but commence in autumn, about October, and end about May — periods agreeing nearly with the eighth and third month of the Chinese year." [See CORMORANT.]

The SPOTTED CORMORANT. (*Phalacrocorax punctatus.*) This beautiful species of Cormorant is a native of New Zealand, where it is said to be abundant, although it is extremely rare in ornithological collections. It builds among rocks, and also on trees which grow near the water. It is described in Mr. Gould's splendid "Birds of Australia" as follows:—"Vertical and occipital crest, crown of the head and throat sooty black; back of the neck, lower part of the back, and rump glossy green; a white stripe commencing above the eye passes down each side of the neck to the flanks; lower part of the neck, chest, and abdomen, beautiful leaden gray; under tail-coverts and tail black; mantle, scapularies, and wings brownish ash, all the feathers except the secondaries and primaries having a small spot of black at their tip; from the throat, sides, and back of the neck and thighs, arise numerous plume-like white feathers of a soft loose texture; those on the sides and back of the neck are very numerous, but on the other parts they are few and thinly scattered."

PHALÆNA. [See MOTH.]

PHALANGER. (*Phalangista.*) A genus of Marsupial animals, distinguished by having the second and third toes of the hind feet united as far as the last phalanx in a common cutaneous sheath. The *Phalangista Cuvieri* may be taken as an example.

There are several species of Phalangers in Australia, belonging to the genera *Phalangista*, *Dromicia*, and *Hepoona*. They are

LONG-EARED PHALANGER.
(PHALANGISTA CUVIERI.)

particularly organized for living in trees. In Mr. Gould's works on the Quadrupeds of Australia, descriptions and figures of all the species will be found.

PHALANGIDÆ. The name of a family of *Arachnidæ*, called Harvest-men, or Shepherd-Spiders. They have two thread-like palpi, terminated by a small hook; the legs are long and slender, the tarsi consisting of more than fifty joints. The majority of them live upon the ground, upon plants, or at the roots of trees, and are very active: others, less agile, hide themselves between stones, in moss, &c.

PHALAROPE. (*Phalaropus.*) A genus of birds, belonging to the Cuvierian family *Longirostres*. They live in small flocks on the sea-coasts, and feed on aquatic and molluscous animals. They fly well, and swim expertly, resisting the heaviest waves, but never dive. They inhabit far north, migrating in the autumn and winter to the temperate regions of both continents. The female builds on the shore among the grass, laying from four to six eggs. Both sexes incubate, and attend on the young.

PHASCOLOTHERIUM. [See SUPPLEMENT.]

PHASCOLARCTOS, or KOALA. (*Phascolarctos cinereus.*) A Marsupial animal, closely allied to the Phalangers. It is stoutly made, has robust limbs and powerful claws, but is entirely destitute of tail. It lives chiefly on fruits, and its habits are arboreal: as it passes along the branches of trees, it suspends itself by its claws, after the manner of a Sloth. It also visits the ground, however; burrows with facility: and there remains in a dormant state during the cold season. The fore feet of this animal have each five toes, of which *two* are opposed to the other three; a zoological fact worthy of note, as it is the only instance among Mammalia: in the hind feet this power does not exist. When the young one leaves the pouch, it clings to the back of the parent for some time. In New South Wales, where they are common, they are often called "Monkeys," and sometimes "Bears." We extract a short account of them which appeared in the Saturday Magazine for Dec. 31, 1836, and was written by one who has shot them, and also kept them in a state of confinement for some time. "They have four hands, having naked palms, which are armed with crooked pointed nails, exceedingly sharp, and rather long. They are covered with fur of a bluish-gay colour, very thick, and extremely soft. It is darker on the back, and paler under the throat and belly, but slightly tinged with a reddish-brown about the rump. The nose is somewhat elongated, and appears as if it was tipped with black leather. The ears are almost concealed in the thickness of the fur, but have inwardly long whitish hairs. The eyes are round and dark, sometimes expressive and interesting. The mouth is small, and they have no tail. Their countenance altogether is by no means disagreeable, but harmless-looking and pitiful. They seemed formed for climbing trees, but they are rather slow in motion, and but moderately active. Like many other animals of the colony, they are drowsy and stupid by day, but become more animated at night, and when disturbed they make a melancholy cry, exciting pity. They feed upon the tops of trees, selecting the blossoms and young shoots; and they are also said to eat some particular kinds of bark. When full-grown, they appear about the size of a small Chinese pig. They are certainly formed differently from every other species of the quadrumana, and it is probable they possess different enjoyments. They are very inoffensive and gentle in manners, if not irritated. The first I ever saw of these animals was caught in a particular manner by a native; and as we witnessed his manœuvres with considerable curiosity, it may afford some interest to relate the anecdote.

"We were ascending very early in the morning Mount Tourang, one of the trigonometrical stations in Argyle, when the native perceived a very large monkey in the act of ascending a tree: he caught it, and being desirous of preserving the animal, we tied it with some silk kerchiefs to the trunk of a small tree, intending to take it to the camp on our return. About sunset we were descending the mountain, and did not forget the prisoner; but, lo! on arriving at the spot, the creature was gone. The native shook his head, whistled, and commenced examining the neighbouring trees, when presently he espied the animal perched upon the top of a high tree, quite at home. "Me catch the rascal directly," said the black, and proceeded first to cut a thin pole about ten feet in length. He next tore a long strip of ropy bark, which he fastened to one end of the pole, in the form of a loop or noose; after which he commenced climbing the tree in good spirits, and confident of success. The animal, on observing the approach of his enemy, ascended higher and higher till he reached the very extremity of the leafy bough on the top of the tree: while the native, mounting as high as he could safely go, could but scarcely reach him with his pole. For a long time he tried to get the noose over the head of the monkey, and several times, when the native imagined he had succeeded, the monkey, at work with his fore-hand, would repeatedly tear it off and disengage himself. The poor animal, as he looked down upon his perplexing adversary, looked truly piteous and ridiculous, and we began to think that the black would fail in his attempt.

"The native, however, growing impatient and angry, ascended a step higher, till the very tree bent with his weight. He tried again, and having succeeded in slipping the noose over the monkey's head, immediately twisted the pole so as to tighten the cord. "Me got him rascal," he exclaimed, as he looked downward to see the best way of descending. "Come along, you rascal, come come, come!" he cried, tugging away at the monkey, who seemed unwilling to quit his post. Down they came by degrees, the black cautiously managing his prisoner, every now and then making faces at him and teasing him, with great apparent delight and satisfaction to himself. We could not but observe the cautious manner in which he appeared at times to treat the monkey; but this caution we soon perceived was very necessary, for when they had descended to where the tree divided into two branches, the black endeavoured to make the animal pass him, so that he might have better command over him. In so doing the monkey made a spiteful catch or spring at the native, but which he cleverly avoided by shifting himself to the other branch with great dexterity. At length, however, both the man and the monkey arrived nearly to the bottom of the tree, when the latter, being lowermost, jumped upon the ground, got loose, and having crawled to the nearest tree, commenced ascending again. We seized him by the rump, thoughtless of danger, but soon thought it advisable to quit our hold, when the native, now enraged, sprung to his tomahawk, and threw it with such force at the unlucky monkey as to knock him clean off the tree. We took the animal to the camp, where it was soon despatched, as we thought, from its pitiful cries, that it was suffering torture from the blow of the tomahawk."

PHASCOLOMYS. [See WOMBAT.]

PHASIANELLA. A genus of Mollusca found in South America, India, New Holland, the Mediterranean, &c. The head of the animal has two long and round tentacula, with eyes on two footstalks; foot oblong. The shell is smooth, oval, variegated; aperture entire, oval; outer lip thin; inner lip thin, spread over a portion of the body

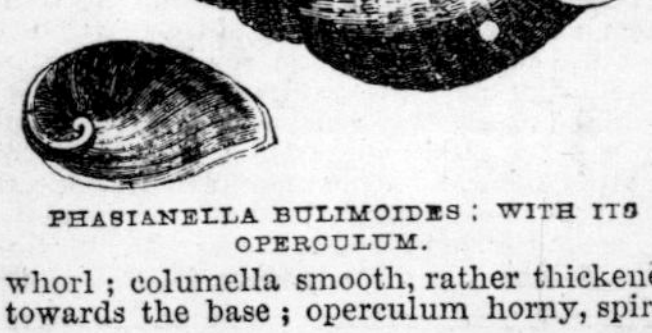

PHASIANELLA BULIMOIDES; WITH ITS OPERCULUM.

whorl; columella smooth, rather thickened towards the base; operculum horny, spiral within. The shells composing this genus are richly marked with lines and waves of various and delicate colours.

PHASIANIDÆ. The name of a family of Gallinaceous birds, of which the genus *Phasianus* is the type. [See PHEASANT.]

PHASMIDÆ, or SPECTRE INSECTS. A family of Orthopterous insects, allied to the *Mantidæ*, peculiar to warm climates, and remarkable for their very close resemblance to the objects in the midst of which they live. Some of them are destitute of wings, and have the appearance of dead twigs; their legs being extended, and the absence of all motion for a considerable time, favouring the deception: others appear like leaves, &c. Their larvæ differ but little from the perfect insects, except in their colours, and the absence of wings; and there are several species in which these are never developed. It not unfrequently happens that they lose a limb by violence; and this is reproduced, provided the complete growth of the animal has not been attained. A species found in the Navigators' Islands, and described by the late Mr. Williams in his admirable 'Missionary Enterprises'—destroys the top of the Cocoa-nut tree, and has been named *P. cocophaga* from this circumstance. We must refer our readers to the works of Mr. G. R. Gray and of M. De Haan on this singular group of Insects; we may remark that their eggs are solitary and not enclosed in a case, and that they often resemble small beans or other seeds. [See PHYLLIUM.]

PHEASANT. (*Phasianus colchicus.*) This beautiful Gallinaceous bird derives its origin from Eastern climes, and is said to have been first imported into Europe from the banks of the Phasis, a river of Colchis, in Asia Minor (as its name imports); but it has now become so thoroughly naturalized in this country, and indeed in most others where the temperature is not too low for its constitution, and where ordinary care is taken for its preservation, that both here and in many parts of the south of Europe it is well known and highly appreciated. Of all birds, except, perhaps, the Peacock, the Pheasant has the most beautiful and finely variegated plumage. In size the male may be compared with the domestic Cock. The irides are yellow; round the eyes is a naked skin, of a beautiful scarlet, with small black specks; and under each eye is a small patch of short feathers, of a dark glossy purple colour: the upper parts of the head and neck are of a deep purple, with green and blue reflections; the lower parts of the neck and breast are of reddish chestnut, edged with black, under which appears a transverse golden streak; the whole body, indeed, uniting the finest tints of golden yellow and green with the richest ruby and purple, set off with spots of glossy black. The legs, feet, and toes are horn-coloured. The tail, which is very long and regularly wedge-shaped, partakes of the beautiful colouring above described; and the whole bird has an air of great elegance. This brilliant plumage is, however, denied to the female, though she is by no means uncomely in form or colour. The natural home of the Pheasant is in the woods, which he leaves at the close of day, to perambulate the corn-fields and pastures, accompanied by his females, in search of food. When young, however, they principally subsist on insects, and are exceedingly fond of ants' eggs. The female constructs her nest, of leaves, in some retired spot; and lays from ten to twenty eggs; but in a state of captivity she seldom produces above ten. In the wild state she hatches her brood with patience, vigilance, and courage; but when kept tame, she becomes so very remiss in her duty, that a common hen is generally made her substitute. The males and females only associate together in the first spring months. When disturbed, they make a whirring noise, like the Partridge, and, from being a large mark and flying slowly, they are readily brought down by the sportsman. There are several varieties, produced by climate and domestication, among which are the following:—The *White Pheasant*, marked with a few small black spots on the neck, and rufous ones on the shoulders; the *Pied Pheasant*, the tail feathers of which are black edged with white, and the upper part of the body reddish brown and white; the *Variegated Pheasant*, which is white and rufous; and the *Ringed Pheasant*, which has a white collar. Foxes and Polecats destroy many Pheasants; and as these are commonly females engaged in incubation, the tendency to diminution of the race from this cause is increased: but the chief loss of the Pheasant-breeder is caused by the mortality of the young birds, about the time of changing their nestling feathers, produced by a convulsive attempt to gasp the air, or expel the worms, (a peculiar species of Entozoon) that have occasioned a disease known by the name of "the gapes." In their wild state Pheasants feed upon all kinds of grain and herbage, like the rest of the gallinaceous tribes. From their size, their beauty, and the delicacy of their flesh, they are every where considered by the sportsman as excellent game; and there is, accordingly, no bird upon which such pains have been taken in its propagation in parks and preserves.

Among the various pleasing and edifying observations to be found in Mr. Waterton's 'Essays,' there are some on the habits of the Pheasant, and (intimately connected with the subject) on that most exciting topic, the game laws. The following are extracts. "By the laws of England, the Pheasant is considered game; and the sportsman is under the necessity of taking out a licence from government, in order to qualify himself to shoot it. When we consider the habits of this bird, we are apt to doubt the propriety of placing it under the denomination of *feræ naturâ*, and I am one of those who think it would be a better plan to put it on the same footing with the barn-door fowl, by making it private property; that is, by considering it the property of the person in whose field or wood it may be found. The Pheasant is a more than half-retained bird. While the Hare and the Partridge wander in wildest freedom through the land, heedless of the fostering care of man, the bird in question will come to us, at all hours of the day, to be fed. It will even sometimes associate with the poultry on the farm; and, where it is not disturbed, it will roost in trees close to our habitations. Its produce with the barn-door fowl is unprolific, and seems to have nothing to recommend it to our notice on the score of brilliancy of plumage, or of fineness of shape. The Pheasant crows at all seasons, on retiring to roost. It repeats the call often during the night, and again at early dawn; and frequently in the day-time, on the appearance of an enemy, or at the report of a gun, or during a thunder-storm. I am of opinion that it does not pair. The female lays from seven to eighteen eggs; but in general the nest contains about twelve.

"Notwithstanding the proximity of the Pheasant to the nature of the barn-door fowl, still it has that within it which baffles every attempt on our part to render its domestication complete. What I allude to is, a most singular innate timidity, which never fails to show itself on the sudden and abrupt appearance of an object. I spent some months in trying to overcome this timorous propensity in the Pheasant, but I failed completely in the attempt. The young birds, which had been hatched under a domestic hen, soon became very tame, and would even receive food from the hand, when it was offered cautiously to them. They would fly up to the window, and would feed in company with the common poultry. But if any body approached them unawares, off they went to the nearest cover with surprising velocity. They remained in it till all was quiet, and then returned with their usual confidence. Two of them lost their lives in the water by the unexpected appearance of a pointer, while the barn-door fowls seemed scarcely to notice the presence of the intruder. The rest took finally to the woods at the commencement of the breeding season. This particular kind of timidity, which does not appear in our domestic fowls, seems to me to oppose the only, though at the same time an insurmountable, bar to our final triumph over the Pheasant. After attentive observation, I can perceive nothing else in the habits of the bird, to serve as a clue by which we may be enabled to trace the cause of failure in the many attempts which have been made to invite it to breed in our yards, and retire to rest with the barn-door fowl and turkey.

"Though a preserve of Pheasants is an unpopular thing, still I am satisfied in my own mind that the bird cannot exist in this country without one; at the same time, I am aware that a preserve may be overdone. Thus, when Pheasants are reserved for a day of slaughter, under the appellation of a *battu*, the regular supply of the market is endangered, the diversion has the appearance of cruelty, and no good end seems to be answered. It exposes the preservers of Pheasants in general to the animadversions of an angry press, which are greedily read, and long remembered, by those whose situation in life precludes them from joining in the supposed diversion. However ardently I may wish to protect the Pheasant in an ornithological point of view, — I say ornithological, for its flesh I heed not, — still, I am fully aware that the danger to be incurred and the odium to be borne are mighty objections. We read, that the ancients sacrificed a cock to Æsculapius: perhaps the day is at no great distance when it will be considered an indispensable act of prudence for the country gentleman to offer up his last hecatomb of Pheasants at the shrine of public opinion.

"To the illegal possession of the Pheasant alone may be traced the sanguinary nocturnal conflicts between the poachers and those who are appointed to watch for its safety. The poacher is well aware that he cannot procure Pheasants without the aid of a gun; and he knows, at the same time, that the report of that gun will betray him, and bring up the watchers, against whom he would have no chance single-handed. Wherefore, in order that he may come off victorious, he musters an overwhelming force of tinkers, cobblers, masons, smiths, and labourers, armed with bludgeons, and, perhaps, here and there a rusty gun. Taking the precaution to get well primed with beer, off they go, fully bent on having every thing their own way. The Pheasants fall; the watchers come up; oaths and curses are poured out, and a desperate fray commences. Here are furnished, work important for the nearest magistrate, profit to his clerk, expense to the county, and practice for Mr. Ketch.* Let it be also observed, that the unlawful capture of the Hare and the Partridge (which are really *feræ naturâ*) does not produce similar work of mischief. These are taken with nets and snares. The fewer poachers employed, the more certain is their success. A number of men would only do harm, and mar the plan of capture. So silently is this mode of poaching carried on, that the owner of the soil is not aware of the loss he is about to sustain in the plunder of his game. When his Hares and Partridges are actually on their way to the dealer's shop, he, 'good easy man,' may fancy that they are merely on a visit to his neighbour's manor, or that the Fox and the Polecat may have made free with them. Not so with regard to the capture of the Pheasant. The mansion is sometimes beset; guns are fired close to the windows; females are frightened into hysterics; and, if the owner sallies forth to meet the marauders, his reception is often the most untoward and disagreeable that can well be imagined.

"Pheasants would certainly be delightful ornaments to the lawn of the country gentleman, were it not for the annoying idea, that any night, from November to May, he runs the risk of getting a broken head, if he ventures out to disturb the sport of those who have assembled to destroy them. There must be something radically wrong in the game laws. How or when these laws are to be amended, is an affair of the legislature. The ornithologist can do no more than point out the grievance which they inflict upon society, and hope that there will soon be a change in them for the better. But to the point. Food and a quiet retreat are the two best offers that man can make to the feathered race, to induce them to take up their abode on his domain; and they are absolutely necessary to the successful propagation of the Pheasant. This bird has a capacious stomach, and requires much nutriment; while its timidity soon causes it to abandon those places which are disturbed. It is fond of acorns, beech mast, the berries of the hawthorn, the seeds of the wild rose, and the tubers of the Jerusalem artichoke. As long as these, and the corn dropped in the harvest, can be procured, the Pheasant will do very well. In the spring it finds abundance of nourishment in the sprouting leaves of young clover; but, from the commencement of the new year till the vernal period, their wild food affords a very scanty supply; and the bird will be exposed to all the evils of the vagrant act, unless you can contrive to keep it at home by an artificial supply of food. Boiled potatoes (which the Pheasant prefers much to those in the raw state) and beans are, perhaps, the two most nourishing things that can be offered in the depth of winter. Beans, in the end, are cheaper than all the smaller kinds of grain; because the little birds, which usually swarm at the place where Pheasants are fed, cannot swallow them; and, if you conceal the beans under yew or holly bushes, or under the lower branches of the spruce fir tree, they will be out of the way of the rooks and ring-doves. About two roods of the thousand-headed cabbage are a most valuable acquisition to the Pheasant preserve. You sow a few ounces of seed in April, and transplant the young plants, two feet asunder, in the month of June. By the time that the harvest is all in, these cabbages will afford a most excellent aliment to the Pheasants, and are particularly serviceable when the ground is deeply covered with snow. I often think that Pheasants are unintentionally destroyed by farmers during the autumnal seed-time. They have a custom of steeping the wheat in arsenic water. This must be injurious to birds which pick up the corn remaining on the surface of the mould. I sometimes find Pheasants, at this period, dead in the plantations, and now and then take them up, weak and languid, and quite unable to fly."

We must now briefly describe some of the rarer species, viz. the GOLDEN PHEASANT (*Phasianus pictus*), a native of China, remarkable for the beauty of its plumage: the prevailing colours are red, yellow, and blue, and it is distinguished by a crest upon the head, which can be raised at pleasure. The iris, bill, and legs are yellow. The tail is longer and more richly tinted than that of the European species; and from above it arise a number of long, straight feathers, of a scarlet hue, mixed with yellow. Cuvier is

* This was written before the abrogation of *capital punishment* for this offence — and others much more heinous — became the law of the land.

of opinion that the description given by Pliny of the Phœnix is meant for this bird.

GOLDEN PHEASANT.—(PHASIANUS PICTUS.)

Another fine species found in China is the SILVER PHEASANT (*Phasianus nycthemerus*). This is of a silvery white colour, with very delicate black lines on each feather and

SILVER PHEASANT.
(PHASIANUS NYCHTHEMERUS.

black abdomen.—But the most splendid of all is the so-called ARGUS PHEASANT (*Argus giganteus*). This species, which is as large as a turkey, is an inhabitant of the mountains of Sumatra, and of some other of the Indian islands. The male has a very long tail; the feathers of the wings are large and long; and both are thickly covered with ocellate spots. [See ARGUS.]

PHEASANT CUCKOO. (*Centropus.*) A genus of Scansorial birds belonging to the Cuckoo family.

PHILETAERUS. A genus of Grosbeaks, remarkable for building their nests in society. [See GROSBEAK.]

PHOCÆNA. A sub-genus of Dolphins, distinguished by the absence of the beak-like prolongation of the jaws. [See PORPOISE.]

PHOCIDÆ. The name of the family of carnivorous and amphibious Mammalia, of which the Seal (*Phoca*) is the type.

PHŒNICOPTERUS. The generic name of the *Flamingo* [which see]. The term is also applied to other birds which have red wings, as the *Bombycilla phœnicoptera*.

PHOLADOMYA. A genus of Conchiferous Mollusca, one species only of which (*P. candida*) is known to be in existence at the present time, and that is from the island of Tortola, where it is frequently found after hurricanes; but there are several fossil species occurring in rocks of the oolitic series. The shell is thin, equivalve, ventricose, elongated, and gaping, transparent, white or yellowish, hinge having a long narrow hollow or pit; ligament external; bosses worn by being placed near each other; muscular impressions two in each valve, rather indistinct. Professor Owen has given a description of the animal of this curious genus.

PHOLAS: PHOLADES. A genus and family of Conchiferous Mollusca, protected by a testaceous bivalve shell; and it is worthy of notice that the Pholas is the only testaceous mollusca which has the property of evolving a phosphorescent light. This quality in the Pholas was first observed by Pliny, and has since been confirmed by Reaumur. Pliny says that the whole substance of the animal is charged with a fluid that has the

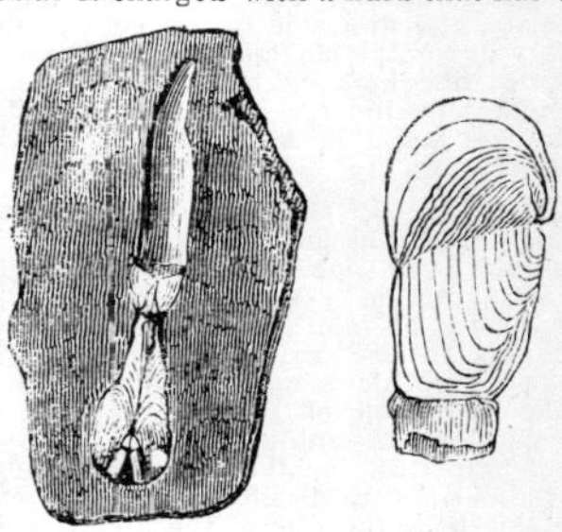

PHOLAS STRIATA

property of emitting a phosphorescent light; and that it will illuminate any substance which it touches. Dr. Priestley says, "This fish illuminates the mouth of the person who eats it: and it is remarkable, that, contrary to the nature of other fish, which give light when they tend to putrescence, this is more luminous the fresher it is: when dried, its light will revive on being moistened either with salt water or fresh; brandy, however, immediately extinguishes it."

The geographical distribution of the *Pholades* is very wide, and their habit of boring hard substances, such as indurated mud or clay, wood, and stone, renders them, as well as other terebrating testaceans, an object of anxious interest to those who construct submarine works. The species are numerous, and some are very abundant on our own coasts. "Of these," says Mr. G. B. Sowerby, "*Pholas crispata, Dactylus, candida,* and *parva*, are the most common; several others are described by Turton, in his 'British Bivalves,' of which we are quite convinced the *P. lamellata* is only the young of *P. papyracea:* we are not acquainted with his *P. tuberculata*. Much confusion appears to prevail in regard to several very distinct species." We refer our readers to the newly published work of Messrs. Forbes and Hanley on the British Mollusca and their shells, in which these points and many others are dwelt on and settled.

The *Pholas* has a delicate, milky white, rather transparent shell, covered sometimes with a thin epidermis, oval, elongated, inequilateral, gaping posteriorly, and especially at the antero-inferior part; umbones hidden by a callosity; hinge toothless; a flat, recurved, spoon-shaped process enlarged at its extremity, elevating itself within each valve below the umbo; muscular impressions very distant, the posterior one large, oblong, elongated, always very visible, the anterior one small and rounded, both more or less approximated to the edge of the shell, and joined by a pallial impression, which is long, narrow, and deeply excavated backwards. The animal is thick and somewhat elongated; mantle reflected on the dorsal part; anterior aperture rather small; foot short, oblong, and flattened; siphons often elongated and united into a single, very extensible, and dilatable tube; mouth small, with very small labial appendages; branchiæ narrow, unequal, and greatly elongated.—Some interesting specimens of *fossil* Pholades are found in Italy and France, but they are rare.

PHORUS: PHORIDÆ. A genus and family of Molluscous animals which have generally been placed with the *Trochi;* but Mr. Gray, in his systematic arrangement of the genera of Mollusca published in the Synopsis to the British Museum (1840), formed for this genus a peculiar family under the name of *Phoridæ;* having observed that the animal, though a Phytophagous mollusc, had the annular operculum of the *zoophagous* division. These animals are small for the size of the mouth of the shell, and have much the general appearance of the animal of *Strombus;* but their eyes are sessile. The foot is small, and divided into two parts, the front rather expanded, the hind part small and tapering. In colour they are dull opaque white; the proboscis pinkish, and the eyes black. They crawl like a tortoise by lifting and throwing forward the shell with the long tapering tentacles stretched out, the proboscis bent down, and the operculum trailing behind. They are numerous in the Javan and China seas, preferring deep water, and a bottom composed of detritus of dead shells and sand mixed with mud.

The most noted species of this family of turbinated Gasteropods is the PHORUS AGGLUTINANS, or the CARRIER SHELL. The shell is thick and conical; ordinarily nacreous; the spire sometimes lowered, and at others rather lofty and pointed at the summit; trenchant or carinated on its circumference; aperture transversely depressed, angular or sub-angular; edge of the outer lip disunited from the inner at the top; inner lip curved, rather oblique at the base; the columella bent, twisted, and often projecting in front; operculum generally horny, thin, and spiral, with numerous narrow whorls. *Phorus agglutinans*, figured below, is remarkable for the singular habit of accumulating, during its formation, different substances, as stones, corals, small shells, &c., which adhere to its shell. From this circumstance it has received the name of the *Carrier Shell.* Some of the foreign species are peculiarly distinguished by their bright colouring, but those which are common on our own coasts are not. The animal has a distinct head, with two tentacula, and eyes at the base; foot short and round.

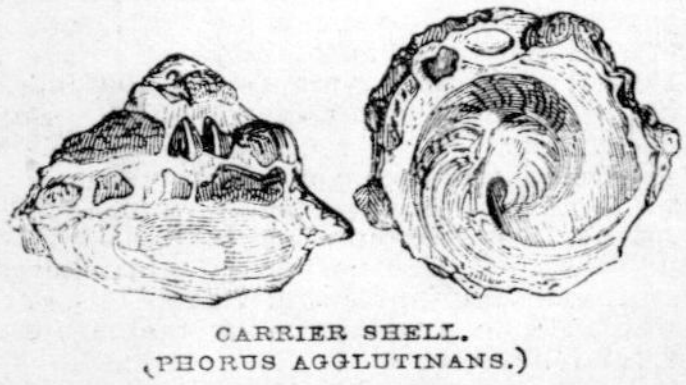

CARRIER SHELL.
(PHORUS AGGLUTINANS.)

PHOSPHODES. A genus of birds allied to the Honey-eaters, of which one species is recorded by Mr. Gould: it is

The PHOSPHODES CREPITANS, or COACH-WHIP BIRD. Like the *Menura* and *Wattled Talegalla*, this bird, which is abundant in many parts of New South Wales, frequents the dense brushes so common on the Australian continent, threading its way with the utmost ease through the matted foliage and thick climbing plants which it meets with in its arboreal retreats. It is a shy and recluse bird; but its loud full note, ending sharply like the cracking of a whip, reverberating through the woods, indicates the locality where it is to be found. It is extremely animated and sprightly in all its actions, raising its crest and spreading its tail in the most elegant manner. This is most observable in the spring, when the males may be often seen chasing each other, while they occasionally stop to pour out their full tide of song; but independently of the Coach-whip Bird's shrill whistle, it possesses a low inward song of considerable melody. The male has the head, ear-covers, chin, and breast, black; a large patch of white on each side of the neck, all the upper surface, wings, flanks, and base of the tail-feathers olive-green; the remaining portion of the tail-feathers black, the three lateral feathers on each side tipped with white; under surface olive-brown, some of the feathers on the centre of the abdomen tipped with white, and forming a conspicuous irregular patch: bill, inside and out, base of the tongue, black; feet reddish brown. The sexes are much alike in colour, the plumage of the female being more obscure, and her size rather less than that of the male. The food consists of various kinds of coleopterous and other insects.

PHRYGANEA: PHRYGANEIDÆ. A genus and family of Trichopterous insects, comprising the well-known Caddice-flies, or Water Moths of the angler; their larvæ being called *cad-bait*, and residing in portable tubes, composed of various extraneous materials. The type of the family (*Phryganea grandis*), in its perfect state, has a body of a leathery consistence, and thickly clothed with hair; the head small, with prominent semiglobular eyes, and three ocelli. The antennæ are often much longer than the body, slender, setaceous, and multi-articulate; the mouth consists of an elongated upper lip; the prothorax forms a very short collar; and the mesa- and meta-thorax are dilated into an oval or orbicular mass. The anterior wings are elongated and lanceolate in the females, but rather more obtuse in the males; deflexed at the sides of the body during repose, and are furnished with numerous branching veins; the posterior pair are shorter, but considerably broader, than the anterior, and are folded when at rest. The larvæ ordinarily live in cylindrical cases, open at each end, some composed of fine sand, and others formed of bits of stick and various other light materials, which they attach to it by the assistance of silken threads spun from the mouth in the same

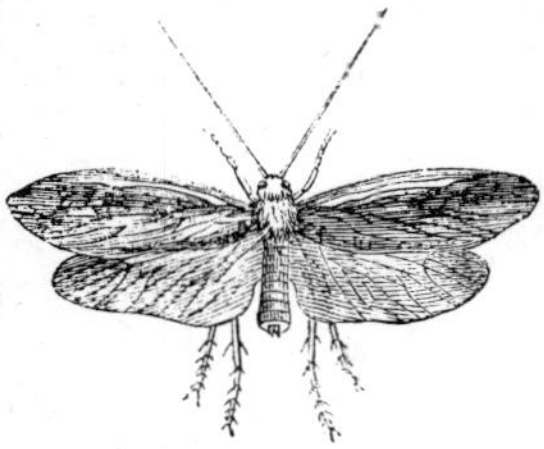

GREAT CADDIS-FLY.—(PHRYGANEA GRANDIS.)

manner as caterpillars. Here the larva remains, exposing only its head and three anterior segments of the body, and which on the slightest alarm it suddenly withdraws. The food of the larva in some species consists of minute aquatic larvæ, but the greater number are purely herbivorous.

"When the period for assuming the pupa state is arrived, the larvæ, which reside in movable cases, fasten them to some fixed substance beneath the water, and close the two extremities with an open-work fence, which varies in form in the different species, and which, by admitting a current of water, permits the respiration of the pupa; indeed, Reaumur states, that he actually saw this grate-work in alternate motion from convex to concave, as the water passed out and in. Within this retreat they then become inactive pupæ, in which they bear a considerable resemblance to the imago, except that the antennæ, palpi, wings, and legs are shorter, enclosed in separate sheaths, and arranged upon the breast; the antennæ, in the species which have those organs, greatly exceeding the length of the body; being extended beyond the abdomen, with the extremities curled up." * * * * "The perfect insects are of small or moderate size, seldom reaching a couple of inches in the expanse of the wings. They are very active, running with agility, with a kind of gliding motion, not unlike that of certain Tipulidæ, and other insects with long tibial spurs; but their flight is awkward, except in some of the smaller species, which assemble in troops, and fly over the surface of the water towards sunset: they frequent damp, marshy situations. From the weak structure of the mouth, it is evident they can live but a very short time in the perfect state, taking no nourishment, and only anxious to continue their species. Their colours are obscure, being ordinarily brown or gray; when handled, they emit a very disagreeable odour. A very few exotic species are ornamented with spots and markings. Few only have been brought from extra-European countries."—*Westwood.*

PHRYNISCUS. A genus of Batrachian Reptiles, containing the *Phryniscus nigricans*, which is the toad so graphically described by Mr. Darwin, who noticed it at Bahia Blanca. "Amongst the Batrachian reptiles," he remarks, "I found only one little Toad, which was most singular from its colour. If we imagine, first, that it had been steeped in the blackest ink, and then when dry, allowed to crawl over a board freshly painted with the brightest vermillion, so as to colour the soles of its feet and parts of its stomach, a good idea of its appearance will be gained. If it is an unnamed species, surely it ought to be called *diabolicus*, for it is a fit Toad to preach in the

BAHIA TOAD.
(PHRYNISCUS NIGRICANS.)

ear of Eve. Instead of being nocturnal in its habits, as other Toads are, and living in damp and obscure recesses, it crawls during the heat of the day about the dry sand-hillocks and arid plains, where not a single drop of water can be found. It must neces-

UNDER SIDE OF BAHIA TOAD.

sarily depend on the dew for its moisture; and this probably is absorbed by the skin, for it is known that these reptiles possess great powers of cutaneous absorption. At Maldonado I found one in a situation nearly as dry as at Bahia Blanca, and thinking to give it a great treat, carried it to a pool of water; not only was the little animal unable to swim, but, I think, without help would soon have been drowned."

PHYLLIDEA: PHYLLIDIDÆ. A genus and family of Mollusca, generally found adhering to rocks, buildings washed by the sea, or other marine objects; and which, though of a dull colour outside, are often very beautiful in the interior. Some of the species differ from all other univalves, in being composed of movable pieces at the back, formed to facilitate the motion of the animal. Some are particularly simple in form, the shape being no more than that of a conical cup or deep dish; others are boat-shaped; and, when fixed, so fast do they retain their situation, by exhausting the air beneath their bodies, that it is very difficult to force the animal from its position without breaking the shell.

PHYLLIUM. A genus of insects belonging to the family *Phasmidæ*, and popularly known as *Walking-leaves;* some of which have wing-covers so closely resembling the leaves of plants, that the insects are easily mistaken for the vegetable productions around them. They are for the most part natives of the East Indies, Australia, and South America. G. R. Gray Esq., of the British Museum, in a communication to 'The Zoologist,' observes that "in the time of Linnæus only one species was known as the *Mantis siccifolium*, which is figured by Rösel. And it was the general opinion of authors long after that great man's time, that there existed but one species of these remarkable insects, until Stoll gave many figures of them, one of which he considered to differ in some points, and gave to it the name of *Phasma chlorophyllium*. The general opinion having been thus broken in

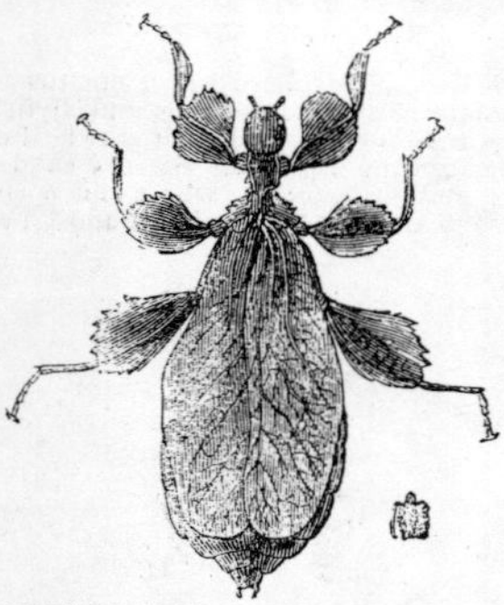

WALKING LEAF-INSECT, AND EGG.
(PHYLLIUM SICCIFOLIUM.)

upon, other species have since been added [five of them are described in the notice to which we refer]. * * * These extraordinarily formed insects were, at one time, supposed to partake both of insect and vegetable life; and not only has the perfect insect such similarity to portions of vegetables, but even their eggs might at first sight be mistaken for the deeply ribbed fruits of various umbelliferous plants." The species *Phyllium bilobatum* is thus described by Mr. Gray:— Abdomen narrow at the base, enlarging on each side to the middle of the third segment, and then gradually decreasing to the end of the fifth; the outer margins of the sixth and seventh are lobed, with the remaining segments suddenly lessened to the tip. Femora of the fore legs dilated; the inner dilation has the margin inwardly entire and outwardly much dentated, the outer dilation oval and entire. Length of body two inches nine lines. Inhabits the Philippine islands.

PHYLLOSOMA, or GLASS CRAB. A striking genus of Crustacea, belonging to the order *Stomapoda*. Our figure will illustrate, better than any description we might give, the form and general character of this genus. There are many species, found for

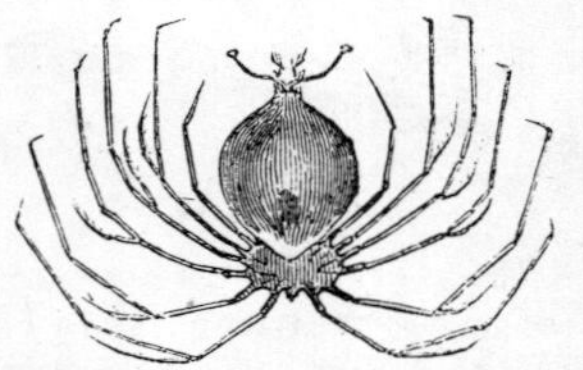

GLASS-CRAB.—(PHYLLOSOMA STYLICORNIS.)

the most part in the tropical parts of the Atlantic and Indian oceans: they are highly transparent. Captain Grey, who had many opportunities of observing them, in speaking of one, says, "When it was taken out of the water, it stood upright on its legs, and crawled a little like a large beetle, but soon died. In the water it swam with the legs, the large joint of which appeared to be feathered. It was not thicker than the thinnest wafer; the back was marked with curved lines; and it shrank instantly when touched. The species have a horny feel to the touch, are destitute of smell, and look like a transparent scale when they lie in your hand."

PHYSA. A genus of fresh-water Mollusca occupying a small oval or oblong, smooth, thin shell, generally sinistral or reversed; and no operculum. The animal has two long tentacula, with eyes at the base: foot long; mantle large, so as to cover part of the shell, and very transparent. These animals are most frequently found on the under side of the leaves of aquatic plants: they have a very singular way of adhering to the surface of the water with the shell downwards, and crawl in that direction with as much apparent ease as on a solid surface, and they will occasionally let themselves down gradually by a thread.

PHYSALIA. A genus of Acalepha, remarkable for its size, the brilliancy of its hues, and the severe burning pain produced by its contact. [See PORTUGUESE MAN-OF-WAR.]

PICIDÆ. The name given to a family of Scansorial birds. [See WOODPECKER.]

PICHIAGO. The *Chlamyphorus truncatus* [which see].

PIERIS. A genus of diurnal Lepidoptera which, amongst numerous exotic species, contains our native PIERIS CRATÆGI, the BLACK-VEINED WHITE OR HAWTHORN BUTTERFLY. This is an elegant insect: both surfaces of the wings are white, with black nervures; above, the anterior wings are margined on their outer edge with irregular dusky spots, transparent and triangular: the posterior ones are similarly bordered on their outer edges, but the nervures are less expanded on the disc; beneath, the anterior wings resemble the upper surfaces, but the

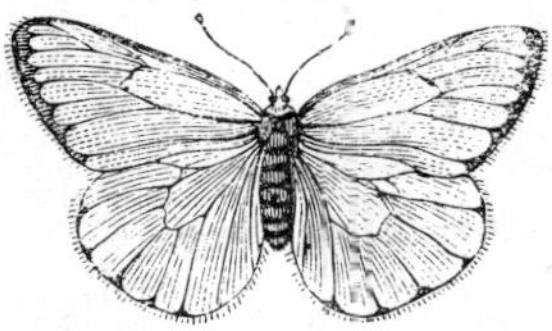

BLACK VEINED WHITE BUTTERFLY (PIERIS CRATÆGI.)

nervures are more dilated; the posterior ones, on the contrary, have the nervures much stronger on the upper surface, and very thickly irrorated with dusky: in both sexes the wings are very transparent, the female more especially. During the earlier periods of its existence the caterpillar lives beneath a silken web: it is at first black,

PIERIS CRATÆGI—UNDER SIDE.

but is afterwards furnished with short yellow and white hairs, and is marked with three black longitudinal lines: it feeds on the whitethorn. The chrysalis is either yellow or white, with small black stripes and spots. In about three weeks the perfect insect makes its appearance. It is by no means scarce, but periodical in its visits rather than generally abounding.

PIGEON. As the Ring-Dove, Stock-Dove, and Turtle-Dove will be found described under those words respectively, we shall devote this article almost exclusively to the tame or domesticated Pigeons, the tenants of the dove-cot. These are the willing attendants on man, and depend on his bounty, seldom leaving the dwellings provided for them, and only roaming abroad to seek amusement, or to procure subsistence; but when, as Bewick observes, we consider the lightness of their bodies, the great strength of their wing, and the amazing rapidity of their flight, it is a matter of wonder that they should submit even to a partial domestication, or occupy those tenements fitted up for the purpose of breeding and rearing their young. Pigeons occur in every climate, and although they thrive best in warm countries, yet with care they succeed also in very northern latitudes. Their manners are gentle and lively; they are fond of society, and have always been held emblematic of peace and innocence; they are faithful to their mates, whom they solicit with the softest cooings, the tenderest caresses, and the most graceful movements. The exterior form of the Pigeon is elegant: the bill is weak, straight, slender, somewhat curved at the point, and has a soft protuberance at the base, in which the nostrils are placed: the legs are short and red, and the toes divided to the origin. They moult once, and the sexes do not differ in plumage.

It would be as fruitless as unnecessary to attempt to describe *all* the varieties of the tame Pigeon; for human art has so much altered the colour and figure of this bird, that pigeon-fanciers, by pairing a male and female of different sorts, can, as they express it, "breed them to a feather." Hence we have the various names of Carriers, Tumblers, Jacobins, Croppers, Pouters, Runts, Turbits, Shakers, Fantails, Owls, Nuns, &c., all birds that at first may have accidentally varied from the Stock-dove, and, by having these varieties still improved by pairing, food, and climate, the different kinds have been propagated. The Dove-house Pigeon breeds every month; but when the weather is severe, and the fields are covered with snow, it must be supplied with food: at other times it may be left to itself, and generally repays the owner for his protection, The Pigeon lays two white eggs, which produce young ones of different sexes. When the eggs are laid, the female sits fifteen days, exclusive of the three days she is employed in laying, and is relieved at intervals by the male; the female performing her share of the duty by night, and the male during the day. When hatched, the young only require warmth for the first three days; a task which the female takes entirely upon herself, and never leaves them except for a few minutes to take a little food. After this they are fed for about ten days, at first with a milky secretion prepared from the glandular coat of the crop, and regurgitated; and afterwards with what the old ones have picked up in the fields, and kept treasured in their crops. This way of supplying the young with food from the crop, in birds of the Pigeon kind, differs from all others. They have the largest crops, for their size, of any birds; and they have the power of distending the crop with air in such a manner, that, in one species in particular, (the

Croppers), the bird's breast appears larger than its body. The numerous glands, assisted by air, and the heat of the bird's body, are the necessary apparatus for secreting the milky fluid before mentioned: but as the food is macerated, that also swells, and becomes considerably dilated.

Though the constancy of the Turtle-dove is proverbial, the Pigeon of the dove-house is not so faithful, and having become subject to man, puts on incontinence among its other domestic qualities. Two males are often seen quarrelling for the same mistress; and when the female encourages the freedoms of a new gallant, her old companion shows visible marks of his displeasure, quits her company, or if he approach, it is only to chastise her. Many instances have been known where two males, being dissatisfied with their respective mates, have thought fit to make an exchange, and have lived in peace and friendship with the new objects of their choice. The dove-cot Pigeons, like the rest of the genus, retire to their roost at a very early hour; but they leave it unusually late in the morning; and though they will perch on trees in the day-time, nothing will induce them to roost there at night. They are greatly attached to the cot of their choice; so much so, that they are scarcely to be driven from it but by fire-arms. Pigeons have sharp sight, and are quick of hearing; and when pursued by a hawk they show that they can fly with great velocity. It is their nature to congregate together, to bill in courtship, and to utter a plaintive note.

The CARRIER PIGEON. Of all the varieties, the most remarkable for its attachment to its native place is the CARRIER PIGEON, or MESSENGER; so called from its being used to convey letters from one place to another. These birds are rather larger than most of the common-sized Pigeons; their feathers lie very close and even, and their necks are long and straight; so that when they stand upright on their legs, they show more gentility of shape than most other Pigeons, From the lower part of the head to the middle of the lower chap there grows out a white, naked, fungous flesh, which is called the wattle, and is generally met by two small protuberances of the same luxuriant flesh, rising on each side of the under chap. The eyes are surrounded with the same sort of corrugated flesh; and the circle round the black pupil of their eyes is commonly of a red brick-dust colour, though more esteemed when it is of a brilliant red. When the luxuriant flesh round the eye is thick and broad, it is considered that the Carrier will be a good breeder, and rear very fine young ones. Extraordinary attention was formerly paid to the training of these Pigeons. An actual post system, in which Pigeons were the messengers, was established by the Sultan Noureddin Mahmoud, who died in 1174; which flying post lasted till 1258, when Bagdad fell into the hands of the Mongols, and was destroyed by them. At present they are kept only by a few wealthy individuals in the East, much time and attention being required to train them properly. As soon as the young are fledged, a cock and a hen bird are made as tame as possible, and accustomed to each other's society. They are then sent, in an uncovered cage, to the place whither they are usually to carry messages. If one of them should be lost, or carried away, after having been well treated for some time, it will certainly return to its mate. A small letter is written on the finest kind of thin paper; then placed lengthwise under one wing, and closely fastened with a pin (the point being turned from the body) to a feather. The custom, however, was not altogether confined to the East; nor is it obsolete; for although we no longer hear of Pigeons conveying tidings of distress from a besieged town, or of promised deliverance from an army advancing to its relief, we know that they are frequently employed with effect in "stock-jobbing *transactions*," or in enabling an adept in the mysteries of betting to pocket a few "cool hundreds," whether it be from his friend on the turf or a fellow-patron of the more ignoble "ring." Nay, we imagine they are not even now likely to be wholly superseded,—wonderful as are the powers of steam and locomotion—if it be true, as stated, that a Carrier Pigeon will perform the distance of forty miles in half an hour!

Having dwelt so fully on the qualities of this *serviceable* Pigeon, it may be proper to give some instructions for its education. "In order to train a Pigeon for this purpose," says our authority, "take a strong, full-fledged young Carrier, and convey it in a basket or bag about half a mile from home, and there turn it loose; having repeated this two or three times, then take it two, four, eight, ten, or twenty miles, and so on till they will return from the remote parts of the kingdom. For if they be not practised when young, the best of them will fly but insecurely, and stand a chance of being lost. Be careful that the Pigeon intended to be sent with the letter is kept in the dark, and without food, for about eight hours before it is let loose, [rather a long abstinence, it would seem, but "use is second nature,"] when it will immediately rise to a great height, and turning round, as is their custom, will continue on the wing till it has reached its home." By what chart it is guided in its unerring flight is among the wonders of instinctive reasoning.

The TUMBLER. These birds, which are of various colours, receive their name from their extraordinary motions in flying, frequently turning themselves in the air, and proceeding with an undulating and irregular motion. They will also frequently rise to such an amazing height in the air as to be almost imperceptible to the keenest eye. They all keep quite close together while flying, and in fair weather they will continue their aerial evolutions for many hours at a time.

The JACOBIN, or JACK. This kind has a range of inverted feathers on the back part of the head, which turns towards the neck, like the cap or cowl of a monk: thus deriving their name from the religious of that order, who wear cowls.

The CROPPER. The body of this variety is thick, short, and clumsy; as are also the legs, which are feathered down to the feet: they have a large pouch or bag hanging under their beak, which they can inflate with wind or depress at pleasure: their crop hangs low, but is very large; and they are so loose-feathered on their thighs, as to be styled flag-thighed.

The NUN. The head of this bird is almost covered with a veil of feathers, whence its name. Its body is chiefly white; its head, tail, and the six flight-feathers of its wings should be entirely red, yellow, or black; that is, when its head is red, the tail and flight-feathers should be red; and when its head is yellow or black, the tail and flight-feathers should invariably correspond with it.

The BALD-PATE, or WHITE-CROWNED PIGEON. (*Columba leucocephala.*) We derive our information respecting this and the succeeding species from Mr. Gosse's 'Birds of Jamaica.' The author tells us that "this fine dove is common in almost all situations, but chiefly affects the groves of pimento, which generally adorn the mountain pens. The sweet aromatic berries afford him abundant and delicious food during the pimento season; the umbrageous trees afford him a concealment suited to his shy and suspicious character; and on them his mate prefers to build her rude platform-nest, and rear her tender progeny. Wary exceedingly, the Bald-pate, from his seat among the topmost twigs, discerns the gunner, himself unseen, and intimates his vicinity only by the rushing of his strong wings, as he shoots off to some distant part of the grove. In the breeding season, however, when alarmed from the nesting tree, he does not fly far, and soon returns; so that the sportsman, by concealing himself, and watching the bird's return, may bring him down. When the pimento is out of season, he seeks other food; the berries of the sweet-wood, the larger ones of the bread-nut, and burn-wood, of the bastard cedar, and the fig, and the little ruddy clusters of the fiddle-wood, attract him. He feeds early in the morning, and late in the afternoon: large numbers resort to a single tree (though not strictly gregarious), and when this is observed, the sportsman, by going thither before dawn, and lying in wait, may shoot them one by one, as they arrive. In September and October they are in fine condition, often exceedingly fat and juicy, and of exquisite flavour. In March the clammy-cherry displays its showy scarlet racemes, to which the Bald-pates flock." "Late in the year they resort to the saline morasses, to feed on the seeds of the black mangrove, which I have repeatedly found in the craw; I have even seen one descend to the ground beneath a mangrove, doubtless in search of the fallen seeds. In general, however, the Bald-pate is an arboreal pigeon, his visits to the earth being very rare. He often feeds at a distance from home; so that it is a common thing to observe, just before nightfall, straggling parties of two or three, or individuals, rushing along with arrowy swiftness in a straight line to some distant wood. The Bald-pate is a noble bird; plump, yet of a graceful form; the iridescent scale-like feathers of his neck, with their black borders, are very striking: he is staid and sedate in manners, when sitting, and there is something of supercilious sternness in his countenance, which, combined with his snow-white head, always reminds me, strange as the comparison may appear, of the grand Bald Eagle. His coo is *Sary-coat-blue*, uttered with much energy, the second syllable short and suddenly elevated, the last a little protracted and descending. Incubation takes place chiefly in the months of June and July. The nest is merely a very slight platform of dry twigs, rudely attached, on which two eggs are laid. They are of delicate whiteness, in form very regularly oval, and in dimensions an inch and a half by one-tenth. The length of the Bald-pate is sixteen inches, expanse twenty-three inches and a half. Irides cream-white; eyelids purplish flesh-colour.

The WHITE-BELLIED PIGEON (*Peristera Jamaicensis*) is chiefly confined to the upland districts; where its loud and plaintive cooing makes the woods resound. It is one of those species which habitually live on the ground: in unfrequented woods, as well those which are open as those which are choked with underwood, the White-belly walks about singly or in pairs, picking up various seeds. Its flesh is generally esteemed; it is white, juicy, and well-flavoured, without being liable to bitterness. "If flushed, it betakes itself to a low tree not far off, whence, if unmolested, it is soon down again. Often, when seen in the woods, it runs a few yards, and then rises to fly, but as if trusting less to its powers of flight than to those of running, alights again immediately, and runs swiftly off among the bushes. It has no regular roosting-place, often spending the night on a stone, or a log, or a low bush that happens to be near the spot where it was feeding at nightfall. This is not the case with the other Doves. The aspect and air of the White-belly are unlike those of its kindred. Its round head, the prevalence of light hues, and its height upon the legs, contribute to this peculiarity. Essentially a ground-pigeon, its length of tarsus enables it to run with ease and celerity; perhaps more rapidly than any other of the family." It is nearly thirteen inches in length, and nineteen in expanse: feet crimson: beak black: forehead pure white, becoming slate blue on crown; hind-head delicate gray-blue; neck reddish-brown, changing to amethyst, the lowest feathers brilliant green and purple. Back, wing-coverts, and uropygials dusky-brown, with slight reflexions. Wing-quills deep brown, the outer edge narrowly white, the basal part of inner webs chestnut; true tail-feathers blue-gray, with white tips. Under parts pure white, tinged with flesh-colour on breast: inner surface of wings chestnut. Eyelids bluish, the edges and angles dark lake. The White-belly usually builds in rather a low situation; the nest consisting of a few loose sticks, with some leaves in the centre: the eggs are white.

PIKE. (*Esox.*) A genus of Malacopterygious fishes, of the family *Esocidæ.* These fish are distinguished by having only one dorsal fin near the tail; a long slender body, compressed laterally; and the lower jaw projecting beyond the upper. They are extremely voracious and destructive, and their digestive powers are as remarkable as their voracity.

The COMMON PIKE (*Esox lucius*) is found in the fresh waters of most parts of Europe. The body is elongated, and the surface covered with small scales. When in season, it is beautifully marked with a mixture of green and bright yellow spots, passing into white on the abdomen; when out of season, however, these colours become dull. Pikes grow to a large size, occasionally attaining a weight of thirty or forty pounds; and are taken in great numbers as an article of food; their flesh being white, firm, and well tasted. They are strong, fierce, active, and particularly bold; swim rapidly, and occasionally dart along with inconceivable velocity.

PIKE.—(ESOX LUCIUS.)

They are caught either in what are called crown nets, or by the hook; when the latter mode is used, the line must be very strong, and the hook fastened with wire. The bait generally used is a small fish. They attain a great longevity: Pennant speaks of one that was ninety years old; but Gesner relates that, in the year 1497, a Pike was taken at Halibrun in Suabia, with a brazen ring attached to it, on which were these words in Greek characters: "I am the fish which was first of all put into this lake by the hands of the Governor of the Universe, Frederick the Second, the 5th of October, 1230." This fish was therefore two hundred and sixty-seven years old, and was said to have weighed three hundred and fifty pounds! The skeleton, nineteen feet in length, was long preserved at Manheim as a great curiosity in Natural History. Several instances have, indeed, occurred in the lakes of Scotland where Pike of seventy or eighty pounds each have been caught; but nothing like the patriarchal age of the Halibrun Pike was ever heard of elsewhere. Rapid growth requires to be sustained by a corresponding proportion of food; and there can be no fear of a Pike starving while any thing eatable is in the way. Mr. Jesse mentions, that eight Pike, of about five pounds weight each, consumed nearly 800 gudgeons in three weeks; and that one of these devoured five roach, each about four inches long, within a quarter of an hour. The Pike not only makes sad havoc among other fish; but, in default of a sufficient quantity, it will devour frogs, water-rats, field mice, small aquatic birds, and other animals, whether alive or dead. In short, so great is its rapacity, that it has been known to contend with the Otter for his prey.

The SEA PIKE. [See GARFISH.]

PILCHARD. (*Clupea pilchardus.*) This fish, which resembles the Herring, not only in general appearance but in its habits, is about nine inches in length, and of a somewhat less compressed and rounder form than the Herring; the scales considerably larger. The head is rather flat, and the mouth is destitute of teeth; the back is of a bluish cast, the belly and sides are silvery, and the upper angle of each of the gills is marked with a large black spot. They feed on minute Crustacea and other marine insects found at the bottom of the water. These fish annually appear on the English coast, and are taken in immense quantities; the same reason having until late years been

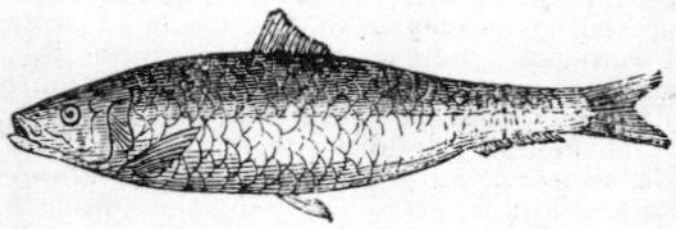

PILCHARD.—(CLUPEA PILCHARDUS.)

universally assigned for such vast shoals of Pilchards appearing periodically, as for the Herrings, namely, their presumed migration from the Arctic regions to warmer latitudes for the purpose of spawning. This theory, however, is now, with sufficient reason, abandoned; and it is established, almost beyond a doubt, that they inhabit our own seas, merely forsaking the deep waters and coming towards the shore to deposit their spawn; thus fulfilling a great law of nature in the propagation of their species, and at the same time providing multitudes of human beings with food. On this head we have already spoken in the articles HERRING and MACKEREL, to which we beg to refer. And we shall now avail ourselves of Mr. M'Culloch's account of the *Pilchard fishery*, which has evidently been obtained from the most authentic sources.

"It is carried on along the coasts of Cornwall and Devon, from the Bolt Head in the latter, round by the Land's End to Padstow and Bossiney in the former. Its principal seats are St. Ives, Mount's Bay, and Mevagissey. The fish usually make their appearance in vast shoals in the early part of July, and disappear about the middle of October; but they sometimes reappear in large numbers in November and December. They are taken either by *seans* or by drift-nets, but principally, perhaps, by the former. A sean is a net, varying from 200 to 300 fathoms in length, and from 10 to 14½ ditto in depth, having cork buoys on one edge and lead weights on the other. Three boats are attached to each sean, viz., a boat (*sean-boat*), of about 15 tons burden, for carrying the sean; another (*follower*), of about the same size, to assist in mooring it; and a smaller boat (*lurker*), for general purposes. The number of hands employed in these three boats varies from 13 to 18, but may be taken, at an average, at about 16. When the shoals of fish come so near the shore that the water is about the depth of the sean, it is employed to encircle them; the fishermen being directed to proper places for casting or shooting the nets by persons (*huers*) stationed for that purpose on the cliffs and in the boats. The practice is to row the boat with the sean on board gently round the shoal; and the sean being, at the same time, thrown gradually into the water, assumes, by means of its buoys and weights, a vertical position, its loaded edge being at the bottom, and the other floating on the surface. Its two ends are then fastened together; and, being brought into a convenient situation, it is moored by small anchors or grapnels; sometimes, however, one or two smaller seans are employed to assist in securing the fish. At low water, the enclosed fish are taken out by a *tuck net*, and carried to the shore. A single sean has been known to enclose at once as many as 4,200 hogsheads (1,200 tons) of fish! But this was the greatest quantity ever taken, and it is but seldom that as many as 1,200 hogsheads are caught at a time. The "take," in fact, depends upon so many accidental circumstances, that while one sean may catch and cure in a season from 1,000 to 2,000 hogsheads, others in the neighbourhood may not get a single fish. In some places the tides are so strong as to break the seans, and set the fish at liberty. When the quantity enclosed is large, it requires several days to take them out, as they must not be removed in greater numbers than those who salt them can conveniently manage.

"As soon as the fish are brought on shore, they are carried to cellars or warehouses, where they are piled in large heaps, having a sufficient quantity of salt interspersed between the layers. Having remained in this state for about 35 days, they are, after being carefully washed and cleaned, packed in hogsheads, each containing, at an average, about 2,600 fish: they are then subject to a pressure sufficient to extract the oil, of which each hogshead yields, provided the fish be caught in summer, about three gallons; but those that are taken late in the season do not yield above half this quantity. This oil usually sells for from 12 to 15 per cent. under the price of brown seal oil. The broken and refuse fish and salt are sold to the farmers, and are used as manure with excellent effect. The skimmings which float on the water in which the Pilchards are washed are called *dregs*, and are chiefly sold as grease for machinery. The sean fishery employs about 1,500 hands regularly throughout the season, and a vast number more when any considerable shoals are inclosed. Four fifths of the persons employed on shore in the salting, curing, packing, &c. of the fish, are women."—*Commercial Dictionary* (where further statistics may be seen).

To the foregoing account we may add, that the Cornish Pilchard Fisheries produce, upon an average, 60,000,000 per annum, or 21,000 hogsheads of Pilchards; and that the season of 1845 produced 100,000,000.

Pilchards frequent both the French and Spanish coasts, but not in very considerable numbers, or with much regularity: the coast of Cornwall seems to be their native home; for there they are found through all the seasons of the year.

PILOT-FISH. (*Naucrates ductor.*) This fish is in size and shape like the mackerel, and may be immediately recognized by certain conspicuous bands which surround its body. Its general colour is a silvery grayish blue, darkest on the back; five dark blue transverse bands pass round the body, and both on the head and tail are slight indications of another band. The head is small, the under jaw rather the longer, and the nose rounded; the scales are small and oval; the ventral fins are attached to the abdomen

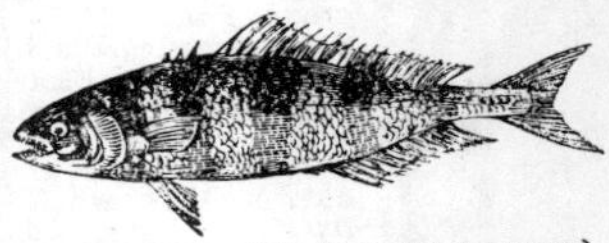

PILOT FISH.—(NAUCRATES DUCTOR.)

by a membrane through one-third of their length; the pectoral fins are clouded with white and blue, the ventrals nearly black. The Pilot-fish will frequently attend a ship during its course at sea for weeks or even months together; and there are many curious stories told respecting its habits, in occasionally directing a Shark where to find a good meal, and also in warning him when to avoid a dangerous bait. We shall, however, leave the relation of such wonders to others; and be content with observing that the Pilot-fish is frequently found in company with the Shark, and is well rewarded for his attendance in being able to snatch up the morsels which are overlooked by his companion.

PILUMNUS. A genus of short-tailed Decapod Crustacea: so called from most of the species being more or less covered with long hairs: one species (*P. hirtellus*) is found on the British coasts.

PIMELIIDÆ. An extensive family of Coleopterous insects, which, although little known in this country, are abundant in Southern and Eastern Europe, and in the

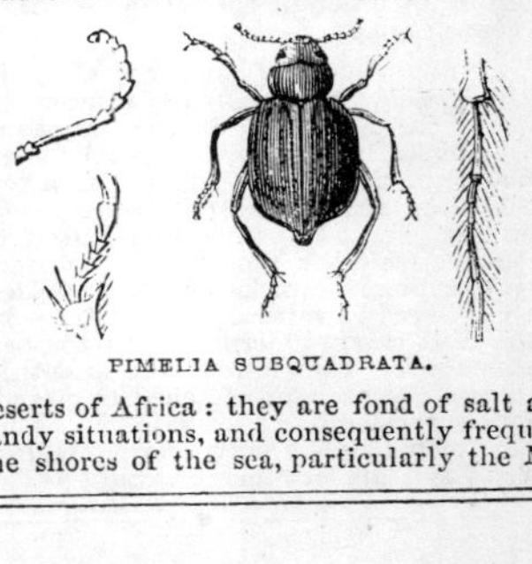

PIMELIA SUBQUADRATA.

deserts of Africa: they are fond of salt and sandy situations, and consequently frequent the shores of the sea, particularly the Mediterranean. They are distinguished by having the elytra soldered together; the wings rudimental or obsolete; the palpi filiform, the mandibles bifid at the tips; and the maxillæ concealed by the mentum, which is very large, and not narrowed at the base. Their colours are black or obscure; they are exceedingly sluggish, and on being alarmed they emit a disagreeably fetid fluid.

PIMELODUS. A genus of malacopterygious abdominal fishes, separated by Lacepede from the genus *Silurus* of Linnæus: by modern Ichthyologists again this genus is subdivided. The head is depressed; there are two dorsal fins, the second adipose. There are very many species of this group, found chiefly in South America, the Nile, and some of the Eastern rivers. We figure a small species discovered by M. Humboldt in the kingdom of Quito, where it lives in streams, and is only occasionally eaten by the very poorest of the Indians. This species has two cirri on the head; it is of an olive colour, sprinkled with small black spots, and is about four inches long; but the striking fact whence it derives its specific name, (*P. Cyclopum*) is its being found sometimes in thousands, ejected from the crater or the apertures on the sides of volcanoes. The inhabitants know them well, and call them *prenadillas*. They are believed to abound in subterraneous lakes, and only to be found by accident in the streams. On emerging from the crater they are found so little changed, that they can always be recognized: a proof that the heat of the water thrown up from the volcano can have little effect on them: this may in part be owing to the mucilage wit' which they are covered.

PIMELODUS CYCLOPUM

PINE MARTEN. [See MARTEN.]

PINION [MOTHS]. A name applied by collectors to different species of Moths, of the genus *Cosmia*.

PINNA. A genus of Mollusca, called also the *Wing-shell*, which in many respects approaches the Mussels. It has two equal wedge-shaped valves, united by a ligament along one of their sides; and attains a very considerable size, sometimes being nearly three feet long. The animal fi..es itself, by its *byssus*, which is remarkably long and silky, to submarine rocks and other bodies; where it lives in a vertical position, the point of the shell being undermost, and the base or edge above. Sometimes large quantities of them are even found attached to a sandy bottom at the depth of a few fathoms. They are common in some parts of the Mediterranean; and are not merely sought as food by the inhabitants on the coasts, but they gather the byssus, of which a stuff may be formed that is remarkable for its warmth and suppleness. The filaments are extremely

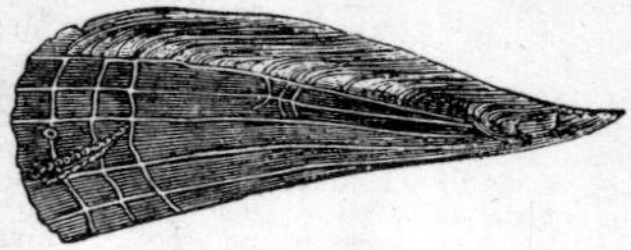

PINNA ANGUSTANA.

fine and strong, and the colour, which is a reddish-brown, never fades. The finest byssus of the ancients was fabricated from these filaments; and in Sicily they are still sometimes manufactured into gloves and other articles of dress, though, it must be confessed, more as an object of curiosity than for use.

PINNIPEDES. A group of Decapodous Crustacea, met with at a distance from the coasts. They are characterized by having the hind pair of legs terminated by a flattened plate for swimming. The most noticeable of these swimming or shuttle-crabs, as they are termed, are the exotic species composing the genus *Matuta*, which have the carapax nearly circular, and armed on each side with a strong spine, and with the four posterior pairs of legs terminated by a dilated plate for swimming. Some of the smaller species, found on our own coasts, are exceedingly abundant, and furnish the lower orders in London and elsewhere with an article of food.

PINNOTHERES, or OYSTER CRAB. A genus of Decapod Crustaceans, of very small size (some of them called Pea-crabs), which reside, during a portion of the year at least, inside various bivalve shells, such as

PEA-CRAB,—(PINNOTHERES VETERUM.)

mussels, &c. The carapace of the females is suborbicular, very thin and soft; whilst that of the males is firmer and nearly globular, and rather pointed in front; the legs are of moderate length, and the claws of the ordinary form; the tail of the female is very ample, and covers the whole of the under side of the body. The ancients believed that the Pea-crab lived upon the best terms with the inhabitant of the shell in which it was found; and that they not only warned them of danger, but went abroad to cater for them.

PINTAIL DUCK. (*Dafila acuta.*) This is an elegantly formed, long-bodied Duck, the neck longer and more slender than most others. It is a shy and cautious bird, feeding in the mud flats and shallow freshwater marshes, but rarely resides on the sea coast.

They inhabit the whole northern parts of Europe, Asia, and America. Great flocks of them are sometimes spread along the isles and shores of Scotland and Ireland, as well as on the interior lakes of both those countries. The male Pintail Duck is twenty-six inches in length, and two feet ten inches in extent: the bill is a dusky lead colour; head and half of the neck pale brown, each side of the neck marked with a band of purple violet, bordering the white; hind part of the upper half of the neck black, bordered on each side by a stripe of white, which spreads over the lower part of the neck before; sides of the breast and upper part of the back white, thickly and elegantly marked with transverse undulating lines of black, here and there tinged with pale buff; throat and middle of the belly whitish; flanks finely pencilled with waving lines; vent white; under tail-coverts black; lesser wing-coverts brown ash; greater wing-coverts black, tipped with orange; below which

PINTAIL DUCK.—(DAFILA ACUTA.)

is the speculum of rich golden green, bordered below with a band of black, and another of white; primaries dusky brown; tertials long, black, edged with white, and tinged with rust; rump and tail-coverts pale ash, centred with dark brown; tail greatly pointed, the two middle tapering feathers being full five inches longer than the others, and black; the rest brown ash, edged with white; legs, a pale lead colour. The female has the crown of a dark brown colour; back, and root of the neck above, black, each feather elegantly waved with broad lines of brownish white, these wavings becoming rufous on the scapulars; but the general plumage is a dull brownish white, speckled with dark brown.

PIPA. A genus of Batrachian reptiles, closely allied to the common Toad, but distinguished by the body being horizontally flattened, the head large and triangular, tongue wanting, tympanum concealed beneath the skin, the eyes small, placed near the margin of the upper jaw. The best known species is the SURINAM TOAD, PIPA SURINAMENSIS (the *Bufo pipa* of Linnæus).

This species considerably exceeds in size the Common Toad. It is one of those animals which, at first view, every one pronounces deformed and hideous; the general uncouthness of its shape being often aggravated by a phenomenon unexampled in the rest of the animal world, namely, the young in various stages of exclusion, proceeding from cells dispersed over the back of the parent. It was for a long time supposed that the ova of this extraordinary reptile were produced in the dorsal cells, without having been first excluded in the form of spawn; but it is

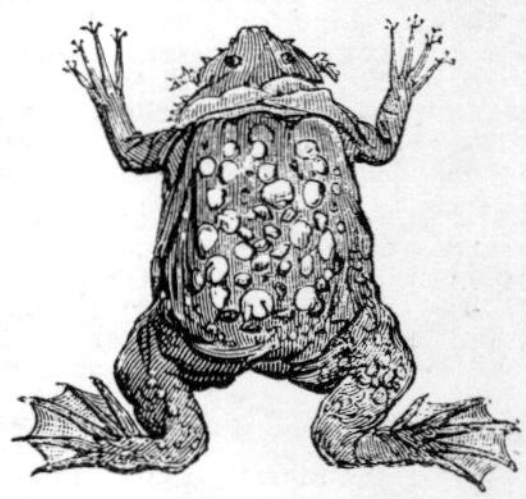

SURINAM TOAD.—(PIPA SURINAMENSIS.)

now thoroughly ascertained that the female Pipa deposits her eggs or spawn at the brink of some stagnant water; and that the male collects or amasses the heap of ova, and deposits them with great care on the back of the female, where, after impregnation, they are pressed into the cellules, which are at that period open for their reception, and afterwards close over them; thus retaining them till the period of their second birth; which happens in somewhat less than three months, when they emerge from the back of the parent in their complete state. During the time of their concealment, however, they undergo the usual change of the rest of this genus, being first hatched from the egg in the form of a Tadpole; and gradually acquire their complete shape, some time before their exclusion. This species inhabits the obscure nooks of houses in Cayenne and Surinam, and has a granulated back, with three longitudinal ranges of larger granules.

PIPE-FISH. (*Syngnathus.*) There are several species of this genus, the distinguishing characters of which are, that the body is greatly elongated, slender, and covered with a series of indurated plates, arranged in parallel lines; that the nose is long and tubular; that the gills are arranged in small round tufts along the branchial arches; and that there are no ventral fins.

PIPE-FISH.—(SYNGNATHUS.)

The GREAT PIPE-FISH (*Syngnathus acus*) is one of the most common species found on our coasts, sometimes among sea-weed at low water, and at other times in deep water. It is usually seen of the length of twelve or fifteen inches, but is sometimes found, especially in the northern seas, measuring from two to three feet. Its form is extremely slender, gradually tapering towards the extremity; of a palish brown colour, varied throughout its whole length with broad alternate zones of a deeper hue, slightly variegated: the laminæ with which the joints of the body are covered, appear to be finely radiated from the centre by numerous streaks: the dorsal fin is thin, shallow, and small, the pectorals are small and slightly rounded; and the tail is of a corresponding shape and size. In spring, the ova of this genus are found lying in a longitudinal division at the lower part of the abdomen; and from these are hatched the young, completely formed.

The LITTLE PIPE-FISH (*Syngnathus ophidion*) is about five or six inches long, slender, nearly cylindrical, and tapering off to a point. It wants both the pectoral and caudal fins; and is covered with a smooth skin, whereas the other kinds are covered with a sort of crust. They are either olive green, or tinged with yellowish-brown.

But by far the most extraordinary species is the FOLIATED PIPE-FISH (*Syngnathus foliatus*). In its general shape it is greatly allied to the *Hippocampus*, but is considerably longer: its great singularity, however, consists in certain large leaf-shaped appendages with which the back, tail, and abdomen are furnished; these appendages are situated on very strong, rough, square spines or processes, and, were it not for the regularity of their respective proportions, might be mistaken for the leaves of some kind of fucus adhering to the spines. The colour of the whole animal is a dusky olive, thickly sprinkled on all parts, except on these appendages, with small, round, whitish specks, and accompanied by a kind of metallic gloss on the abdomen: the fins are soft, tender, and transparent. This remarkable species is a native of the Indian seas.

PIPING CROW. (*Barita tibicen.*) A striking bird, by some placed among the Shrikes, by others among the Crows. As the crow in the fable was proved unable to sing or chant, and as our present bird is decidedly most musical, his *talents would* remove him from that despised group, even if his characters were not somewhat different. It is a common species in New South Wales, whence it is not unfrequently brought alive to this country. The visitors to the Zoological Gardens in the Regent's Park cannot have failed to be amused with his peculiar musical pipe, as well as his pleasant look: black is the most prevalent colour of his plumage; the hinder part of the neck, and the top of the back, and the base of the wing-coverts are white, tinged with grayish blue: by some authors this genus is named *Cracticus*.—For another species see CROW-SHRIKE.

PIPIT. (*Anthus.*) The Pipits are birds very much resembling the Larks, both in regard to their generally having a long hind claw and in the colour of their plumage. The TREE PIPIT (*Anthus arboreus*), a migratory species, and very sweet songster, is of common occurrence in Britain. This bird generally rises singing from the ground, and after attaining a certain height, descends and rests on the summit of a tree; from which it again rises and descends singing to the ground. Its colour is a streaked olive-brown above, paler underneath, with longitudinal dark spots on the breast, and two pale transversal bands on each wing. The COMMON PIPIT (*Anthus pratensis*) is extremely common throughout Europe, inhabiting mountain moors, and lowland heaths and marshes. It is a more slender bird than the preceding. The SHORE PIPIT (*Anthus aquaticus*) abounds on the sea coast, and is very rarely met with inland. It is larger and darker coloured, and is a superior songster to the last named.

PIPRA. A genus of Dentirostral birds, comprehending the different species of Manakins. They are for the most part natives of the warmer regions of America, and noted for the brilliancy of their colours. They have a compressed bill, higher than broad, emarginate, with great nasal fossæ. Their tail and limbs are short; and their general proportions occasioned them to be long regarded as allied to the Tits. They frequent woods, are very active, and their flight is short, but quick.

In Mr. Edwards's narrative of a 'Voyage up the Amazon,' he says, the Manakins, in their different varieties, form a beautiful family, the most numerous of any, and corresponding much in their habits to our Warblers. "They are tiny things, generally having black bodies, and heads of yellow, red, white, and other colours. Like perpetual motion personified, they move about the branches and low shrubs, always piping their sharp notes; and, unless upon a feeding-tree, almost defying shot."

PISTOSAURUS. [See SUPPLEMENT.]

PITHECIA. [See MONKEY.]

PLACODUS. [See SUPPLEMENT.]

PLACUNA. A genus of Conchiferous Mollusca, family *Ostracea*. The shell is compressed, thin, equivalve, and nearly equilateral; planorbicular, fibrous, foliaceous, and nearly transparent: hinge flat. The most noted species is the *Placuna placentia*, or Chinese Window Oyster, which is used for windows, lanthorns, &c., in China, as horn is used here. The valves, when closed, are so thin as to appear to touch; the animal is consequently exceedingly flat. The Chinese also use the powder of this shell for silver in their water-colour drawings.

PLACUNANOMIA. A genus of Conchiferous Mollusca; the shell of which is thin, smooth, inequivalve, plaited round the edge; attached by a bony substance passing through a fissure in the lower valve. It partakes, as its name denotes, of the characters both of *Placuna* and *Anomia;* the hinge resembling the former, and the opening in the lower valve for the passage of the tendon being like the latter.

PLAGIAULAX. [See SUPPLEMENT.]

PLAICE. (*Pleuronectes.*) This common Flat-fish is easily distinguished by its being very broad, flat, and of a fine pale brown above, marked both on the body and fins by numerous moderately large orange-coloured spots; while the whole of the under part is perfectly white: behind the left eye is a row of six tubercles, reaching as far as the commencement of the lateral line; the mouth is rather small, the lower jaw longer than the upper, and both furnished with a row of small teeth. When near the ground they swim slowly and horizontally; but if suddenly disturbed, they sometimes change

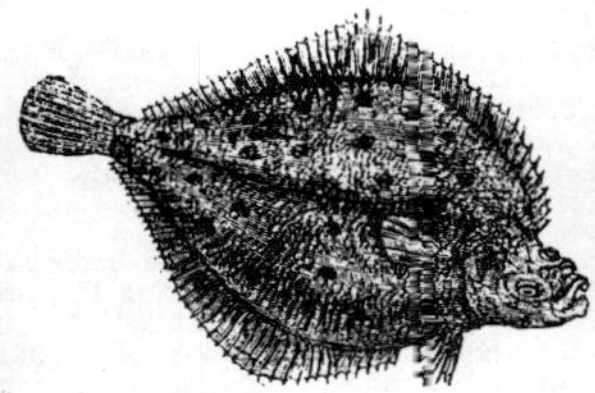

PLAICE.—(PLEURONECTES PLATESSA.)

the horizontal to the vertical position, darting along with meteor-like rapidity, and then again quickly resuming their inactive habits at the bottom of the water. Plaice feed on small fish and young crustacea, and have sometimes been taken on our coasts weighing fifteen pounds, but a fish one-half that weight is considered very large. The finest kind, called Diamond Plaice, are caught on the Sussex coast. These fish are in considerable esteem as food, though by no means equal to the Turbot and Sole. Those of a moderate size are reckoned the best eating.

PLANAXIS. A genus of Mollusca, resembling the *Phasianellæ*, very abundantly found in India, South America, and the Isle of France. The shell is small and oval, the spire consisting of few whorls; outer lip thickened and denticulated within; operculum thin and horny, with a terminal nucleus.

PLANIPENNES. The name given to a tribe of Neuropterous insects, comprehending those in which the inferior pair of wings almost equal the superior ones, and are simply folded underneath at their anterior margin. The antennæ are multi-articulate, and much longer than the head; the maxillary palpi are shorter than the head, and are composed of four or five joints. The Ant-Lions (*Myrmeleon*) and Termites are examples of this tribe.

PLANORBIS. A genus of snails, chiefly inhabiting ponds or the banks of rivers, and deriving their name from the form of the shell, which is that of a flattened orb, occasioned by the volutions being coiled on the same plane. Many of the species are common in Great Britain; and fossil species are found in the freshwater strata of the Isle of Wight, and in the neighbourhood of Paris.

WEST INDIA PLANORBIS.
(P. GUADALOUPENSIS.)

PLANTIGRADA. (Lat. *planta*, the sole of the foot; *gradior*, I march.) The name of a tribe of carnivorous Mammalia, which apply the whole or part of the sole of the foot to the ground in walking, &c. The Bears, Racoons, Badgers, &c. are examples of Plantigrade Carnivora.

PLANT-LICE. [See APHIS.]

PLATESSA. A sub-genus of the *Pleuronectidæ*, or flat-fish family, comprising the Flounders, Plaice, &c.

PLATYCERCUS. A genus of the Parrot tribe, which derives its name from its fine wide tail: there are many species, most of which are natives of Australia. As an example, we may cite the PENNANTIAN or BLUE-CHEEKED PARRAKEET (*Platycercus Pennantii*). This beautiful Parakeet is very generally dispersed over New South Wales, its true habitat, and is chiefly found on the ranges of grassy hills and brushes. Although much variation exists between the plumage of these birds in youth and maturity, the colouring of the sexes when fully adult is alike. "The head, neck, all the under surface, the rump and upper tail-coverts, are of a rich deep crimson-red; the feathers of the back and scapularies black, broadly margined with rich crimson-red; the cheeks and shoulders cerulean blue; the greater wing-coverts pale blue; the primaries and secondaries black, with the basal half of their external webs margined with deep blue; the two centre tail-feathers green, passing into blue on their margins and at the tip; the remainder black on the inner webs for three-fourths of their length; deep blue for nearly the same length on their outer webs, and largely tipped on both webs with pale blue, which becomes still paler to the tips of the feathers; bill horn-colour; irides very dark brown; feet blackish brown." It breeds in the holes of the large gum-trees; the months of September, October, and November constituting the breeding season. It makes no nest, but deposits from four to seven white eggs on the rotten wood at the bottom of the hole. In disposition this species is tame and familiar; few can exceed it in interest or beauty; and consequently it is one of the commonest living Parakeets sent from Australia to this country. The plumage of the young birds during the first autumn is a nearly uniform green, which is gradually changing to a party-coloured livery of scarlet, blue, and green, till it assumes the rich and well-defined colours of the adult.

PLATYPUS. [See ORNITHORHYNCUS.]

PLATYRRHINI. The name given to a division of the *Quadrumana*, comprehending all the large species of Monkey-like animals belonging to the New World. They are characterized by having thirty-six grinders (being four more than the others); the tail, in general, long; and in some species prehensile; no cheek-pouches; posteriors hairy and without callosities; nostrils opening on the sides of the nose, and not underneath. [See MONKEY.]

PLECTOGNATHI. The name of an order of fishes, in some measure connecting the osseous with the cartilaginous kinds; comprehending those which have the jaws formed by the maxillary bones being anchylosed to the sides of the intermaxillaries.

PLESIOSAURUS. The name of a genus of extinct marine Saurians, of gigantic dimensions, which may be thus described: the head short, somewhat oblong, and obtuse; the neck extremely long, consisting of about thirty-three vertebræ; body elongated; tail short; nostrils small; teeth numerous, lodged in small alveoli; ribs composed of two parts, the one vertebral and the other ventral, the vertebral column consisting of about ninety joints.—That in the earlier periods of animal existence reptiles were created of much greater dimensions, and were far more numerous in proportion, than at present, seems evident from the discovery and examination of the organic remains which from time to time have come under the observation of men of science; and there is scarcely any one more entitled to our notice, on account of its extraordinary form than the Plesiosaurus. Its neck is five times the length of its head; the trunk of the body four times the length of the head; and the tail three times; while the head itself is only a thirtieth part of the whole body. From the whole physiology of the animal, Mr. Conybeare says, that it was aquatic is evident from the form of its paddles; that it was marine, is almost equally so, from the remains with which it is universally associated; that it may have occasionally visited the shore, the resemblance of its extremities to those of the Turtle may lead us to conjecture; its motion, however, must have been very awkward on land: its long neck must have impeded its progress through the water, presenting a striking contrast to the organization which so admirably fits the Ichthyosaurus to cut through the waves. May it not, therefore, be concluded, (since, in addition to these circumstances, its respiration must have required frequent access of air,) that it swam upon or near the surface; arching back its long neck like the swan, and occasionally darting it down at the fish which happened to float within its reach. It may perhaps have lurked in shoal water along the coast, concealed among the sea-weed, and, raising its nostrils to a level with the surface from a considerable depth, may have found a secure retreat from the assaults of dangerous enemies; while the length and flexibility of its neck may have compensated for the want of strength in its jaws, and its incapacity for swift motion through the water by the suddenness and agility of the attack which they enabled it to make on every animal fitted for its prey. The remains of the Plesiosauri occur in the formations from the muschel-chalk to the chalk inclusive; but are most common in the lias and Kimmeridge clay beds. They were discovered in England, and have since been found in France and Germany.

"It is of the Plesiosaurus," says Dr. Buckland, "that Cuvier asserts the structure to have been the most heteroclite, and its characters altogether the most monstrous that have been yet found amid the ruins of a former world. To the head of a lizard it united the teeth of a crocodile; a neck of enormous length, resembling the body of a serpent; a trunk and tail having the proportions of an ordinary quadruped, the ribs of a chameleon, and the paddles of a whale. Such are the strange combinations of form and structure in the *Plesiosaurus*—a genus, the remains of which, after interment for thousands of years amidst the wreck of millions of extinct inhabitants of the ancient earth, are at length recalled to light by the researches of the geologist, and submitted to our examination in nearly as perfect a state as the bones of species that are now existing upon the earth." The finest collection of remains of the Plesiosaurus is in the British Museum. [See ICHTHYOSAURUS.]

PLEUROBRANCHUS. A genus of marine Mollusca, having a very light thin internal shell; nearly flat, and obliquely oval; slightly convex towards the spiral apex. It is found in the Indian seas and the Mediterranean.

PLEURONECTIDÆ. The name of a family of Malacopterygious fishes, commonly known by the appellation of *Flat-fish*. They are distinguished not only from all other Fishes, but even from all other vertebrated animals, by several peculiarities of structure. Their body is extremely compressed, or flattened at the sides. Both eyes are on one side, and this side always remains uppermost when the animal is swimming. The upper side is in general deeply coloured, while the other side is whitish. The two sides of the mouth are not equal, and the pectoral fins are rarely so. The body is depressed, and elevated in the direction of the spinous processes; the dorsal extends along the whole back; the anal occupies the lower edge of the body, and the ventrals are sometimes united with it. They have six gill-rays; the abdominal cavity is small, but extends in a cavity imbedded in the flesh on the two sides of the tail, for the purpose of containing some of the viscera: they have no air-bladder, and they seldom rise far from the bottom; but when disturbed, they will raise themselves into a vertical position, so as to show their white sides, and they then dart along with great rapidity; but they soon return to their usual posture, and glide along with a sort of undulating motion near the bottom. They are found along the shores of almost all countries; and are, generally speaking, wholesome and agreeable food. The Sole, Plaice, Turbot, Flounder, &c. are examples of *Pleuronectidæ*.

PLEUROSTERNON. [See SUPPLEMENT.]

PLEUROTOMA. A genus of Mollusca inhabiting eastern and tropical seas. The shell is turreted or fusiform; generally

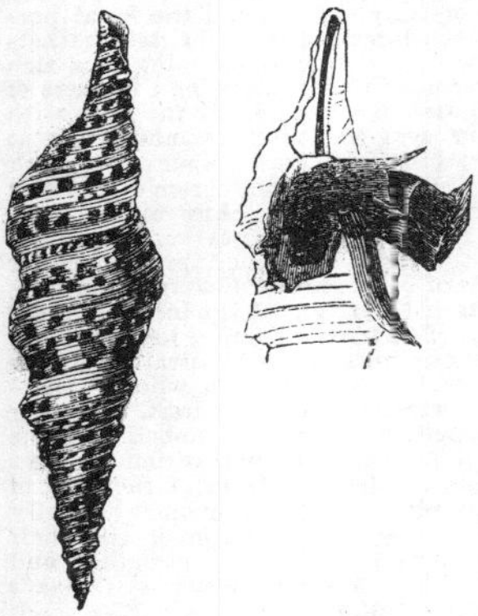

PLEUROTOMA BABYLONICA.

ribbed or striated transversely; aperture oval, terminating anteriorly in an elongated canal; outer lip thin, with a fissure near its union with the spire; operculum small and horny. The species *Pleurotoma gracilis* is found on the British shores. Our figure shows the shell and the mouth of another species, the *Pleurotoma Babylonica*, from which the carnivorous mollusc protrudes. There are many fossil species.

PLICATULA. A genus of Conchiferous Mollusca, found both in a recent and fossil state.

PLIOLOPHUS. [See SUPPLEMENT.]

PLIOPITHECUS. [See SUPPLEMENT.]

PLIOSAURUS. [See SUPPLEMENT.]

PLOTUS. A genus of swimming birds, belonging to the *Pelecanidæ*. Mr. Gould has described the species PLOTUS NOVÆ-HOLLANDIÆ as follows:—"*Male:* An arrow-head-shaped mark of white on the throat; a broad stripe of the same colour commences at the base of the mandibles, extends for about four inches down the sides of the neck, and terminates in a point; head, neck, and all the upper surface of the body greenish black, stained with brown and with a patch of deep rusty red in the centre of the under side of the throat; under surface deep glossy greenish black; wings and tail shining black; all the coverts with a broad stripe of dull white, occupying nearly the whole of the outer and a part of the inner web, and terminating in a point; scapularies lanceolate in form, with a similar shaped mark of white down the centre, and with black shafts, the scapular nearest the body being nearly as large as the secondaries, and having the outer web crimped and the inner web with a broad stripe of dull white close to the stem; the secondaries nearest the body with a similar white stripe on the outer web, close to the stem; centre tail-feathers strongly and the lateral ones slightly crimped; orbits naked, fleshy, protuberant, and of a yellowish olive, mottled over with brown specks; irides of three colours, the ring nearest the pupil being dull orange-buff; to this succeeds another of marbled buff and brown, and to this an outer one of orange-buff; naked skin at the base of the lower mandible wrinkled and yellow; upper mandible olive, under mandible dull yellow, both becoming brighter at the base; feet yellowish flesh-colour, becoming brown on the upper part of the outer toes.—*Female:* Upper surface blackish brown, each feather margined with grayish white; under surface, buffy white. In other respects similar to the male. Total length, 36 inches; bill, 4; wing, 13½; tail, 9; tarsi, 2. Inhabits the rivers of the whole of the southern coast of Australia. [For habits of Plotus, see DARTER.]

PLOVER. A genus of birds, in many respects allied to the Wader tribes, but generally partaking of the nature of land birds, and therefore more properly classed with them. Many, as Bewick remarks, breed upon our loftiest mountains, and though they are sometimes seen feeding upon the sea shores, yet they are no more water birds, on that account, than many of our small birds which repair thither for the same purpose. They are gregarious, and are generally seen in meadows or on the sea shore, in search of food, which they procure by stirring the earth or mud with their feet, and thus inviting worms and aquatic insects to the surface. They are generically distinguished by a large full eye; the bill is straight, short, and rather swollen towards the tip; the head large; legs naked above the knee; and most of the species are without the hind toe.

The RING PLOVER (*Charadrius hiaticula*, Linn.) is very abundant on the sea-coasts of Great Britain. Its plumage is grayish-brown above, white beneath, with a black or dark brown collar on the lower part of the neck, very broad anteriorly; the head marked with black and white, and the beak yellow, tipped with black; orange-coloured legs. It generally breeds on heaths not far from the coast.—Another British species is the KENTISH PLOVER (*Ch. Cantianus*), less deeply coloured, with longer and black legs, and a rufous occiput. It is almost always to be seen as a frequenter of shingle-beaches.

GOLDEN PLOVER. (*Charadrius pluvialis.*) The length of this bird is about ten inches.

On all the upper parts of the plumage the feathers are indented on the edge with bright yellow spots upon a dark brown ground; the front of the neck and the breast are the same, but much paler; the belly is almost white; the quills are dusky; the tail is marked with dusky and yellow indentings and bars; the legs are black; and the bill is dusky. The Golden Plover is common in this country and all the northern parts of Europe; it is also very numerous in various parts of America, migrating from one place to another according to the seasons. It breeds on high and heathy mountains; and the female lays four eggs of a pale olive colour, variegated with irregular umber-brown blotches. The young, when excluded, are covered with a beautiful particoloured down of bright yellow and brown; they quit the nest as soon as hatched, and follow their parents till able to supply and support themselves, which is in the course of a month or five weeks. The old birds display great anxiety in protecting their young brood, using various stratagems to divert the attention of the enemy. When aware of an intruder near, the female invariably runs to some distance from her nest before she takes wing, a manœuvre tending to conceal its true situation; and the discovery of it is rendered still more difficult by the colour and markings of the eggs assimilating so closely to that of the ground and surrounding herbage. The usual call-note of the Plover is a plaintive monotonous whistle, by imitating which it may frequently be enticed within a very short distance. In the breeding season a more varied call is used, during which it flies at a great elevation, and continues soaring round for a considerable time. Towards the end of August the Plover leaves the moors, and descending to the cultivated vales, gets fat by picking up the larvæ and worms in the newly-sown wheat fields; but as the winter draws on it moves to the coast, where it remains until the approach of spring. In autumn the flesh of the Plover is scarcely inferior to the woodcock; but it was more esteemed formerly than at present. The "Plover's eggs" frequently seen at the tables of the opulent and luxurious, are not those of the Golden Plover, but of the Lapwing. Plovers fly in small flocks, and make a shrill whistling noise, by an imitation of which they are sometimes enticed within gun-shot. When merely wounded they run so fast that they often escape. While tending the brood, the old birds employ a number of stratagems to divert the attention of any one approaching them. Like the Lapwing, they feign lameness, tumble over as if unable to fly; and then, after running for some distance, they take wing and perform many gyrations in the air before they again alight. Scarcely any difference is observable between the male and the female. In young birds the plumage inclines more to gray, and the yellow spots are not very distinguishable.

There are several other species of Plovers, some of which are peculiar to America, and others common to both continents. [See DOTTEREL; LAPWING, &c.]

PLUME [MOTHS]. A name given to Moths of the genus *Alucita*.

PLUSIA. [See MOTH: GAMMA MOTH.]

PLUTEUS. [See SUPPLEMENT.]

PLYCTOLOPHUS. A genus of birds belonging to the *Psittacidæ*.

The LEADBEATER'S COCKATOO. (*Plyctolophus* or *Cacatua Leadbeateri.*) Of all the Cockatoos yet discovered, this species is at once the most beautiful and elegant of the genus. Its general plumage is white; the forehead, front and sides of the neck, centre of the under surface of the wing, middle of the abdomen, and the basal portion of the inner webs of the tail-feathers tinged with rose colour, becoming of a rich salmon-colour under the wing; feathers of the occipital crest crimson at the base, with a yellow spot in the centre and white at the tip; bill light horn-colour; feet dark brown. It enjoys a wide range over the southern portion of the Australian continent; it never approaches very near the sea, but evinces a decided preference for the belts of lofty gums and scrubs clothing the sides of the rivers of the interior of the country. Few birds tend more to enliven the monotonous hues of the Australian forests than this beautiful species, whose "pink-coloured wings and glowing crest," says Sir T. Mitchell, "might have embellished the air of a more voluptuous region." Two examples, in the possession of the Earl of Derby, appear to bear confinement equally as well as any of their congeners: in their disposition they are not so sprightly and animated, but they are less noisy. (*Gould's Birds of Australia.*)

PNEUMORA. A genus of Orthopterous insects, remarkable for the *blown-up* appearance of their abdomen, which seems to resemble an inflated balloon. The Dutch at the Cape of Good Hope, where some of the species are common, call them *Blos op*, from

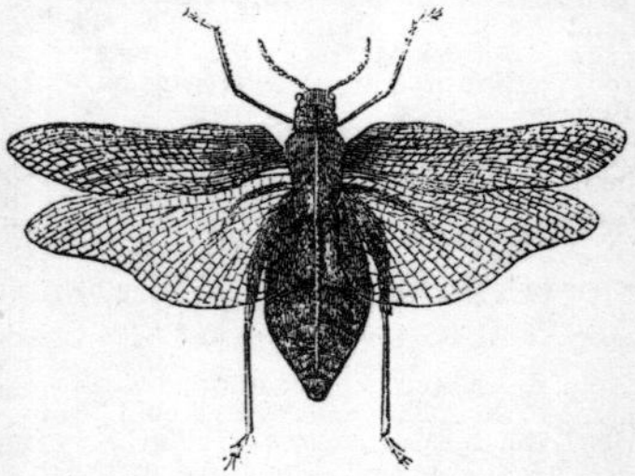

PNEUMORA VARIOLARIS

their swollen appearance. The noise they make is very great. The species are of delicate green or rose tints, some of them spotted with silver.

PODARGUS. A genus of insectivorous birds, natives of Australia, whose habits are strictly nocturnal. During the day the Po-

DARGUS HUMERALIS sleeps so soundly on the dead branch of a tree that it is almost impossible to arouse it; and Mr. Gould remarks that he has frequently shot one without disturbing the mate close by. It does not appear to take its prey on the wing, but creeps about the trees in search of it. It has the power of shifting the outer toe backwards; and the wing is short and concave. The nest is flat, carelessly interwoven, and placed in the fork of a branch. The female generally lays two eggs, which are white, and the male assists in incubation.

Another species, the PODARGUS CUVIERI, which is readily distinguished from the preceding by the bill being much less robust, and the bird itself smaller in size and altogether more slender, is almost exclusively found in Van Diemen's Land. Like the other members of the genus, it feeds principally upon coleopterous and other insects, and is nocturnal in its habits. It displays considerable alertness in the capture of its food; but never flies by day, its whole diurnal existence being passed in a sitting posture across a dead branch, perfectly motionless; and it is not easily to be roused, so as to take wing, either by the discharge of a gun or any other noise. "Like the owl," says Mr. Gould, "it is considered by some a bird of ill omen, principally from the extraordinary sound of its hoarse, unearthly cry, which resembles the words *more-pork* (the name given to it by the colonists); it not only approaches the immediate vicinity of the houses, but emits this sound while perched in their verandahs and on the buildings themselves; and it is often to be seen perched on the tombstones of the churchyard." Considerable variation appears to occur in the colour of the plumage; the prevailing tints in some being a dull ashy gray, while in others they are a rich chestnut-brown; but altogether it may be characterized as striped and minutely freckled with grayish white and dark brown. The nest is rather neatly formed and flat; and the female lays two white eggs.

PODURA. The Poduræ are small insects which, in general, are found in damp places, under stones, on the bark of trees, &c. When disturbed, they suddenly spring to a small distance by the help of a long forked process, or tail, which is bent forwards beneath the abdomen; and it is by the sudden extension of it that the leap is produced. Hence these insects are commonly known under the name of Spring-tails. One of the most common of this genus is the *Podura aquatica* of Linnæus, a minute black insect occasionally seen in vast numbers, particularly near the brinks of ponds, and sometimes even on the surface of the water itself.

POE-BIRD. (*Prosthemadera cincinnata.*) This elegant species of the family of the Honey-eaters is about the size of a blackbird, and is a native of New Zealand, and of some of the scattered islands in the South Sea. The general colour is a glossy greenish black, with strong varying glosses of green accompanying the general plumage of the back and wings; the rump a rich deep blue; and the larger wing-coverts white, forming a bar of white across the wings. The feathers of the neck are of a loose silky texture, rather long, and curving slightly upwards at the tips; but the principal mark of distinction in this elegant bird is a moderately large and lengthened pendent tuft of broad white feathers curving upwards at the tips, and situated on each side the neck: the bill is black and slightly curved; and the legs are black. This bird is greatly valued by the natives of the Southern islands; its glossy plumage often contributing to the ornaments of the feathered mantles worn by their chiefs. As a song-bird also it has considerable merit; and it is said that as a delicious food it is one of the greatest luxuries afforded by the woods of New Zealand.

[For further interesting particulars from the pages of the Rev. Mr. Yate, an accurate observer, who resided long in New Zealand, see PROSTHEMADERA.]

POEPHAGOMYS. A genus of Rodent animals found in South America. They have narrow incisors; the auditory couch

POEPHAGOMYS ATER

small, but distinct; and claws adapted for burrowing. The only well-ascertained species is *Phoephagomys ater*, which is a native of Chili. It has also been described under the name SPALACOPUS.

POËPHILA. A genus of Passerine Birds, belonging to the Finch family, and deriving their name from their fondness for grass-seeds, on which they feed. We may specify, from Mr. Gould's Birds of Australia,

The POEPHILA LEUCOTIS, or WHITE-EARED GRASS FINCH. It is a native of Australia, and has a band crossing the forehead, lores, throat, and a large patch on each flank, deep velvety black; ear-coverts, a narrow line beneath the black of the throat, and a space surrounding the black patch on the flanks, white; crown of the head deep reddish chestnut; all the upper surface and wings deep cinnamon-brown; chest and abdomen pale vinous brown; upper and under tail-coverts white, the former margined externally with deep black; tail black; irides dark brown; feet red; bill yellowish horn-colour. Like the other members of the gends, it inhabits the open spots of country, and feeds on grass-seeds.

POINTER. (*Canis familiaris avicularis.*) The Pointers are a breed of valuable sporting dogs. They are used in finding feathered game of various sorts, partridges, pheasants, &c. When they scent their game, they suddenly stop, and remain motionless as a statue, until the sportsman comes near enough, and is prepared to take his shot; he then gives the word, and the Dog immediately springs the game. So admirably have these Dogs been trained, that their acquired propensities seem almost as inherent as a natural instinct, and appear to be transmitted from parent to progeny: at least, they now require but very little *breaking* to stand at any kind of game. Their scent and sight are equally acute. In all probability Spain is the native country of this valuable Dog, which is found there and also in France with very slight difference of form; but the English breed is much to be preferred, for good temper, beauty of appearance, docility, patience, and activity. "Those Pointers," says Johnson, in his Shooter's Companion, "which I have seen direct from Spain, are heavy and clumsily formed; those from Portugal are somewhat lighter; while the French breed is remarkable for a wide furrow which runs between the nostrils, and give to the animal's countenance a very grotesque appearance. They are all thick and heavy, with large chubby heads, long pendent ears, and short smooth hair; they are often ill-tempered and snappish, and, in fact, are good for little in this country till they have been crossed with the more generous blood of these islands. Yet the conjunction of the Setter and Pointer is by no means advisable. Excellent Pointers have been produced by the Foxhound and the Spaniard. In crossing with the Spanish Pointer, the deep-flewed Hound is to be preferred, and from judicious crossing excellent Pointers are to be met with in most parts of England. They differ from the Setter, as, when they have approached sufficiently near the game, they stand erect, whereas the true-bred Setter will either sit upon his haunches, or lie close to the ground, generally the latter. Pointers often suffer much from sore feet. I have generally found white-footed Dogs much more tender in this respect than those whose feet are of a dark colour. Pointers are sometimes used with bells round their necks in cover-shooting. When the Dog sets, the ringing ceases, and the shooter proceeds to the spot. Pointers are very susceptible of education, and not so apt to forget their lessons as the Setter; and their speed, strength, and persevering spirit, enable them to continue the chase for a length of time almost incredible." "I have heard my father, a man of close observation, and an enthusiastic sportsman," observes Mr. Bell, "offer the opinion that the stand of the Pointer and the crouching of the Setter are but the natural start of surprise or interest, which all dogs give when coming suddenly upon the scent or sight of their natural prey; modified of course by cultivation, and by transmission through many generations, each, by education, improving upon the capabilities of the former."

POLECAT, FITCHET WEASEL, or FOUMART. (*Mustela putorius.*) This animal is known by each of the names here given, but most frequently by the first. It is one of the most remarkable European species of the Weasel tribe, and is found in most parts of Europe, as well as in some of the Asiatic regions. Its colour is a deep blackish-brown, with a tawny cast slightly intermixed; the ears are edged with white, and the space round the muzzle is also whitish. It is about seventeen inches in length, exclusive of the tail, which is about six inches. In its habits it greatly resembles the other Weasels; it preys indiscrimately on the smaller animals, is very destructive to poultry, and most inimical to rabbits, which it destroys like the ferret, by sucking their blood, instead of immediately tearing them to pieces, so that, it is said, a single Polecat is often sufficient to clear a whole warren; and twenty rabbits have been found dead, which one Polecat had destroyed, and that by a wound which was hardly perceptible. It steals into barns, pigeon-houses, &c., where it occasionally makes great havoc, biting off the heads of fowls and pigeons, and then

POLECAT.—(MUSTELA PUTORIUS.)

carrying them away to its retreat. It is also a great lover of milk, and often robs the dairy. During the summer, however, it principally frequents rabbit-warrens, or the hollow trunks of trees, &c., and prowls about in quest of young birds, rats, and field-mice. Sometimes it forsakes the field, the wood, and poultry-yard, to roam by the rivulet's side, and indulge in its propensity for fish. The Polecat is a strong and active creature, and will spring with great vigour and celerity when preparing to attack its prey, or to escape from pursuit; at which time it arches its back considerably to assist its effort. It is of a smell proverbially fetid, being furnished, like several others of the Weasel tribe, with a pouch or follicle beneath the tail, which secretes a thickish fluid of a peculiarly strong and offensive odour. The fur of the body is of two sorts; the shorter being woolly, of a pale yellowish or fulvous colour; the longer, shining, and of a rich black or brownish black; which, though far less valuable than either that of the Sable or the Marten, is still much esteemed; and numbers of the skins are annually imported here from the north of Europe, under the name of *Fitch*. The spring is the season in which the Polecat breeds, generally producing three or four at a birth, which the parent is said to suckle but a short time, accustoming them early to suck the blood of the animals which she brings to them, as well as eggs, &c. The Polecat has been known to breed with the Ferret; nay, it is asserted to be a practice with warreners, in order to improve the breed of the latter, to procure a mixed breed from time to time, which are of a colour between the Ferret and the Polecat, or of a dingy yellowish-brown.

POLLACK. (*Gadus pollachius.*) This fish, sometimes called the Whiting Pollack, is common on many of the rocky coasts of this island; and during summer large shoals of them are seen sporting on the surface of the water, ready to bite at any bait that may be thrown to them. The under jaw is longer than the upper; the head and body rise pretty high; and the lateral line is incurvated, rising towards the middle of the back, then sinking, and running straight to the tail, which is broad and of a brownish colour: the back is dusky, inclining to green: the sides are marked with yellow streaks; and the tail is slightly forked. Fine specimens of the Pollack are taken at Scarborough, where it has the name of Leet. It is also caught at Hastings, Weymouth, and on the Devonshire coast, and bought by the inexperienced as Whiting. Hand-line fishing for Pollacks, Mackerel, &c is called *whiffing.*

POLYBORUS. A genus of rapacious birds which frequent the extratropical parts of South America, and in their habits (according to Mr. Darwin) well supply the place of our carrion-crows, magpies, and ravens; a tribe of birds not known there. The reader will observe that we have frequently availed ourselves of valuable zoological information contained in that gentleman's 'Journal of Researches;' and in this instance we are largely indebted to the same source for the following graphic ornithological observations, which we have only slightly abridged.

"To begin with the *Polyborus Braziliensis.* This is a common bird, and has a wide geographical range; it is most numerous on the grassy savannahs of La Plata (where it goes by the name of Carrancha), and is far from unfrequent throughout the sterile plains of Patagonia. The Carranchas, together with the *Polyborus Chimango*, constantly attend in numbers the estancias and slaughtering-houses. If an animal dies on the plain the Gallinazo commences the feast, and then the two Caracaras pick the bones clean. These birds, although thus commonly feeding together, are far from being friends. When the Carrancha is quietly seated on the branch of a tree, or on the ground, the Chimango often continues for a long time flying backwards and forwards, up and down, in a semicircle, trying each time, at the bottom of the curve, to strike its larger relative. The Carrancha takes little notice, except by bobbing its head. Although the Carranchas frequently assemble in numbers, they are not gregarious: for in desert places they may be seen solitary, or more commonly by pairs. Besides the carrion of large animals, these birds frequent the borders of streams and sea beaches, to pick up whatever the waters may cast on shore." * * * * "A person will discover the *Necrophagous* habits of the Carrancha, by walking out on one of the desolate plains, and there lying down to sleep. When he awakes, he will see, on each surrounding hillock, one of these birds patiently watching him with an evil eye. It is a feature in the landscape of these countries, which will be recognized by every one who has wandered over them. If a party goes out hunting with dogs and horses, it will be accompanied during the day by several of these attendants. After feeding, the uncovered craw protrudes; at such times, and indeed generally, the Carrancha is an inactive, tame, and cowardly bird. Its flight is heavy and slow, like that of an English rook. It seldom soars; but I have twice seen one at a great height gliding through the air with much ease. It runs (in contradistinction to hopping), but not quite so quickly as some of its congeners. At times the Carrancha is noisy, but is not generally so: its cry is loud, very harsh, and peculiar, and may be likened to the sound of the Spanish guttural *g*, followed by a rough double *r r.* Perhaps the Gauchos, from this cause, have called it Carrancha. Molina, who says it is called Charu in Chile, states, that when uttering this cry, it elevates its head higher and higher, till at last, with its beak wide open, the crown almost touches the lower part of the back. This fact, which has been doubted, is quite true. I have seen them several times with their heads backwards in a completely inverted position. The Carrancha builds a large coarse nest, either in a low cliff, or in a bush or lofty tree. To these observations I may add, on the high authority of Azara, that the Carrancha feeds on worms, shells, slugs, grasshoppers, and frogs; that it destroys young lambs by tearing the umbilical cord; and that it pursues the Gallinazo, till that bird is compelled to vomit up the carrion it may have recently gorged. Lastly, Azara states that several Carranchas, five or six together, will unite in chase of large birds, such as herons. All these facts show that it is a bird of very versatile habits and considerable ingenuity.

"The *Polyborus Chimango* is considerably smaller than the last species. It is common on both sides of the continent, but does not appear to extend so far northward as the last species. We have already remarked that it feeds on carrion, in common with the Carrancha. It is generally the last bird which leaves the skeleton; and may often be seen within the ribs of a cow or horse, like a bird in a cage. The Chimango often frequents the sea-coast and the borders of lakes and swamps, where it picks up small fish. It is truly omnivorous, and will eat even bread, when thrown out of a house with other offal. They are more active than the Carranchas, but their flight is heavy: they are very tame; not gregarious; and frequently utter a gentle, shrill scream.

"The *Polyborus Novæ Zelandiæ* is exceedingly numerous over the whole of the Falkland Islands. In many respects these hawks resemble in their habits the Carranchas. They live on the flesh of dead animals and on marine productions. They are extraordinarily tame and fearless, and haunt the neighbourhood of houses for offal. If a hunting party kills an animal, a number soon collect, and patiently await, standing on the ground on all sides. After eating, their uncovered craws are largely protruded, giving them a disgusting appearance. They readily attack wounded birds: a cormorant in this state having taken to the shore, was immediately seized on by several, and its death hastened by their blows. The Beagle was at the Falklands only during the summer, but the officers of the Adventure, who were there in the winter, mention many extraordinary instances of the boldness and rapacity of these birds. They actually pounced on a dog that was lying fast asleep close by one of the party; and the sportsmen had difficulty in preventing the wounded geese from being seized before their eyes. It is said that several together (in this respect resembling the Carranchas) wait at the mouth of a rabbit-hole, and together seize on the animal when it comes out. They were constantly flying on board the vessel when in the harbour; and it was necessary to keep a good look out to prevent the leather being torn from the rigging, and the meat or game from the stern. These birds are very mischievous and inquisitive; they will pick up almost anything from the ground; a large black glazed hat was carried nearly a mile, as was a pair of the heavy balls used in catching cattle. Mr. Usborne experienced during the survey a more severe loss, in their stealing a small Kater's compass in a red morocco leather case, which was never recovered. These birds are, moreover, quarrelsome, and very passionate; tearing up the grass with their bills from rage. They are not truly gregarious; do not soar; their flight is heavy and clumsy; on the ground they run with extreme quickness, very much like pheasants. They are noisy, uttering several harsh cries; one of which is like that of the English rook; hence the sealers always so call them. It is a curious circumstance that, when crying out, they throw their heads upwards and backwards, after the same manner as the Carrancha. They build on the rocky cliffs of the sea-coast, but only in the small islets, and not in the two main islands. This is a singular precaution in so tame and fearless a bird. The sealers say that the flesh of these birds, when cooked, is quite white, and very good eating."

POLYGASTRICA. The name given by Ehrenberg to the most minute and simple kinds of Infusorial Animalculæ that exist. They occur in all parts of the world, and differ according to diversity of climate, region, kind of water, &c.; and though they are invisible to the naked eye, they are all endowed with an organization characteristic of the animal kingdom; most of them having a distinct mouth, and internal cavities for the reception of food; and they enjoy the most extensive powers of reproduction. They are not confined to infusions of organized matter; they are found in the stagnant waters around our cities; in the waters of rivers, harbours, and lakes; and even, it is believed, in every fluid drop of the ocean. Their forms are extremely various: some appear composed of a mass of gelatinous matter that may assume almost any shape; others seem to undergo various forms according as they are differently situated; while other species remain unchanged, their soft bodies being enclosed in a delicate but firm integument, strengthened by an envelope formed of siliceous matter, and termed the sheath. Most of the Polygastrica have the power of freely moving through their native element; but others attach themselves to a solid base, like Polypes. In almost all, we find the body furnished more or less abundantly with *cilia*, usually disposed around the mouth, towards which they produce a vortex of fluid, that brings a supply of alimentary particles. But it is necessary to state, that, notwithstanding this subject has occupied the attention of many learned naturalists, and many extraordinary discoveries have been made of late years, the nature of the organization and life of these Animalcules is still involved in great mystery; and the question is infinitely too comprehensive for us to attempt to enter into any of the details upon which the various scientific opinions have been formed.

POLYNEMUS, or MANGO-FISH. A group of Abdominal Fishes, chiefly confined to the warmer latitudes; and distinguished by the rays of the pectoral fins being extended into long filaments, which hang loosely on each side of the body, giving it a singular and beautiful appearance. The fishes of this genus are usually very brilliant in their colours; and are reckoned very delicious as articles of food. The general form of the body somewhat resembles that of the Perch.

"Considerable interest is attached to the *Polynemi* on account of some recent discoveries, which tend to show that they produce isinglass in considerable abundance. The attention of the members of the Zoological Society (says Mr. Broderip) was first directed to this subject by Dr. Cantor. 'In the December number (1838) of Parbury's Oriental Herald, says this naturalist, 'appears a letter on the Suleah fish of Bengal, and the isinglass it affords:' this fish, says the anonymous writer, 'when at its full size, attains about four feet in length, and is squaliform, resembling the Shark species in appearance, but exhibiting a more delicate structure. The meat of this fish is exceedingly coarse, and is converted by the natives, when salted and spiced, into "burtah," a piquant relish, well known at the breakfast tables of Bengal. The bladder of the *Suleah* may be considered the most valuable part of it; this, when exposed to the sun, and suffered to dry, becomes purely pellucid, and so hard, that it will repel the edge of a sharp knife when applied to it. These bladders vary in weight from half a pound to three quarters of a pound avoirdupois when perfectly dry. This fish abounds in Channel Creek, off Sangor, and in the mouths of all the rivers which intersect the Sunderbuns, and are exceedingly plentiful in certain seasons.

"The discovery of isinglass as a product of India was so important, that Dr. Cantor determined to investigate the subject, and to ascertain, if possible, what the Suleah might be; when, quite unexpectedly, he received a letter from Mr. M'Clelland, in which that naturalist stated that he had examined this fish, and found it to be the *Polynemus Sele* of Hamilton's 'Fishes of the Ganges;' he moreover discovered that an individual of that species weighing two pounds would yield sixty-five grains of pure isinglass, an article which in India sells at sixteen rupees (1*l.* 12*s.*) per pound. Thinking it highly probable that other species of *Polynemus* besides the *P. Sele* will yield isinglass, Dr. Cantor proceeds to give a short account of those species which came under his observation while attached as surgeon to the Honourable Company's Survey of the sea-face of the Gangetic delta.

"'The species best known,' says the author, 'is the *Polynemus Risua* of Hamilton (*Pol. longifilis*, Cuvier; the TUPSEE, or MANGO-FISH, of the Anglo-Indians): this inhabits the Bay of Bengal and the estuaries of the Ganges, but enters the mouths of the rivers even higher up than Calcutta during the breeding season (April and May), when the fish is considered in its highest perfection, and is generally sought as a great delicacy. This species is the smallest, for its length seldom exceeds eight or nine inches, and one and a half or two inches in depth.' It is remarkable for the great length of filaments, or free rays, of the pectoral fins, these being about twice the length of the body, and seven in number on each side."

In Dr. Shaw's Zoology is a curious and interesting account (taken from Bruce's Travels) of a species called *Polynemus Niloticus*. "This, according to Mr. Bruce, who describes and figures it in the appendix to his Travels, is a large species, and may vie, both for the elegance of its form and its taste, with any fish inhabiting the rivers running either into the Mediterranean or the Ocean. The specimen from which Mr. Bruce's figure was taken weighed thirty-two pounds, but it is said often to arrive at the weight of seventy pounds or more. It is an inhabitant of the river Nile, where it is by no means uncommon, as far up the river as Syene and the first cataract: the whole body is covered with scales of a brillant silver colour, so as to resemble spangles lying close together; and there is no variety of tinge on the fish, except a shade of red on the end of the nose, which is fat and fleshy.

"We are informed by Mr. Bruce, that in order to take this fish the Egyptian peasants prepare a pretty large mass or cake, consisting of oil, clay, flour, honey, and straw, kneading it with their feet till it is well incorporated: they then take two handfuls of dates, and break them into pieces about the size of the point of a finger, and stick them in different parts of the mass, into the heart of which they put seven or eight hooks with dates upon them, and a string of strong whipcord to each: this mass of paste is then conveyed by the fisherman or shepherd into the stream, the man sitting for this purpose on a blown-up goat-skin. When arrived at the middle, he drops the mass in the deepest art of the stream, and cautiously holding ie ends of each of the strings slack, so as not to pull the dates and hooks out of the middle of the composition, he makes to shore again, a little below the spot where he has sunk the mass, and separating the ends of the strings, ties each of them, without straining, to a palm-branch fastened on the shore, to the end of which is fastened a small bell. He then goes and feeds his cattle, or digs his trenches, or lies down to sleep: in the mean time the cake beginning to dissolve, the small pieces of date fall off, and, flowing down the stream, are eagerly seized on by the fishes as they pass; they rush up the stream, picking up the floating pieces as they go, till at length they arrive at the cake itself, and voraciously falling to work at the dates which are buried in it, each fish in swallowing a date, swallows also the hook in it, and feeling himself fast, makes off as speedily as possible: the consequence is, that in endeavouring to escape from the line by which he is held, he pulls the palm-branch to which it is fastened, and thus gives notice of his capture by ringing the bell. The fisherman runs, and having secured the fish puts a strong iron ring through his jaw, ties a few yards of cord to it, and again commits him to the water, fastening the cord well to the shore. This is practised in order to preserve the fish ready for sale, since fish in general, when dead, will not keep long in these regions. It is rarely that on those occasions a single hook is found empty."

There are several other species found in the Indian, African, and American seas, bearing a tolerably near resemblance to the Mango-fish before described.

POLYODONTA. A name applied by Lamarck and De Blainville to the *Ark-shells*, &c. of collectors, comprehending the forms collected by Linnæus under the genus Arca. The word signifies "many-toothed;" and the family is defined by Lamarck:—"cardinal teeth small, numerous, entering, and disposed in each valve in either a straight, a curved, or a broken line."

POLYOMMATUS. A genus of diurnal Lepidoptera, so called from many of the species having numerous eye-like marks on the under side. There are many British species.

POLYOMMATUS ARGUS; or LEAD BLUE BUTTERFLY. The male of this insect has the wings above deep blue, tinged with lilac, the hinder margin broad and black, the costa white; beneath grayish-blue: anterior wings with a central ocellus, behind which is a bent series of six ocelli; and the hinder margin with a double band of black spots: posterior wings with three ocelli at the base placed obliquely; a triangular dis oidal spot, with a large black dot in the centre; behind this a waved series of ocelli, and a band of orange tawny, containing several brilliant silvery blue spots on a black ground, and bordered internally with a series of black crescents, and externally with whitish: the outer margin of all the wings black; cilia white. The female is brown above, the disc sometimes bright blue, with or without a marginal tawny band; beneath grayish, with the ocelli larger and more distinct than in the male, and a dentated white band traversing all the wings, between the ocelli, and the fulvous band: cilia brown. Caterpillar green, hairy, with whitish tubercles; a reddish-brown dorsal line, edged with white, another on the sides, and some oblique stripes of the same; head and true legs brown: it feeds on common food and saintfoin. Chrysalis at first green, afterwards brown.

Polyommatus Arion; or Arion Butterfly. This insect is considered one of great rarity, and is usually found on commons and pastures early in July. Wings above brown, with a blue disc, or blue with a brown margin posteriorly; anterior with a central transverse black spot, behind which is an undulated row of black bars, disposed longitudinally; the posterior wings have some obsolete ocelli towards the hinder margin: beneath, dusky ash-colour; the anterior wings with about eight ocelli, forming an undulated band near the hinder margin, all with a black pupil and white

ARION BUTTERFLY.—(POLYOMMATUS ARION.)

iris: on the hinder margin are two rows of black wedge-shaped spots, with a pale dot attached to each; the cilia white, with brown bars beneath: the posterior wings with the base blue-green, and having an angulated row of four ocellated dots, followed by a transverse discoidal crescent, and then by

P. ARION. — UNDER SIDE.

an interrupted angulated and waved band, consisting of eight ocelli, the inner but one being frequently double; beyond this, on the margin, are two rows of lunulated dots: cilia as in the anterior wings. Body dusky, with bluish hairs above, hoary beneath: antennæ black, annulated with white. It is occasionally caught in the vicinities of Dover, Winchester, and Bath.

Polyommatus Artaxerxes; or Scotch Argus Butterfly. This unassuming species of the papilionaceous tribe was until lately supposed to be peculiar to Scotland; but it is no longer so, as instances are given of its having been met with both in the north and west of England. It frequents meadows and grassy places, like its congeners, and makes its appearance first in June, and again in August. The wings above are in both sexes black brown, with a discoidal white spot on the anterior, and sometimes on the posterior; they have also an orange-coloured band; fringe white, brown at the base: beneath, the anterior wings have a central white spot, between which and the posterior margin are five similar spots, followed by a broad orange-coloured band, terminating externally in a white spot with a black pupil, and interiorly in a series of black and white crescents: on the margin of the posterior wings this band is continued; there is a large white blotch on its interior edge, and between it and the base of the wing are several scattered white spots. Like its congeners, however, it is subject to considerable variations.

Polyommatus Corydon; or Chalk Hill Blue Butterfly. In all chalky districts this pretty butterfly abounds, especially on the downs, and under the cliffs, near Dover; in various parts of the Isle of Wight, on the beds of chalk round Winchester; and in many other similar situations. The wings above are of a rich pale silvery-blue, with the hinder margin and nervures dusky, and cilia white: the posterior wings with five sub-ocellated spots in the hinder margin: beneath, the anterior wings are whitish, ocellated, and usually with two or three spots towards the base of the wing: beyond the undulated band of ocelli is an interrupted brown streak, between which and the hinder margin is a series of sub-ocellated dots, with a whitish circle: posterior wings cinereous, greenish-blue at the base, with four ocellated spots at the base, and eight forming an angulated band behind the middle; the pupils black, with a white iris; in the centre of the wing, between the bands, is a white sub-triangular spot; and on the hinder margin is a series of ocellated black spots, with a white iris, marked internally with orange; which series is connected by a pure white oblong patch to the external ocellated band: cilia white. In the female, the fulvous-orange spots on the hinder margin of all the wings are more distinct than in the male, and the cilia are browner.

Polyommatus Adonis; or Clifden Blue Butterfly. This truly beautiful insect is extremely local, but still very plentiful on all the Sussex downs and Kentish

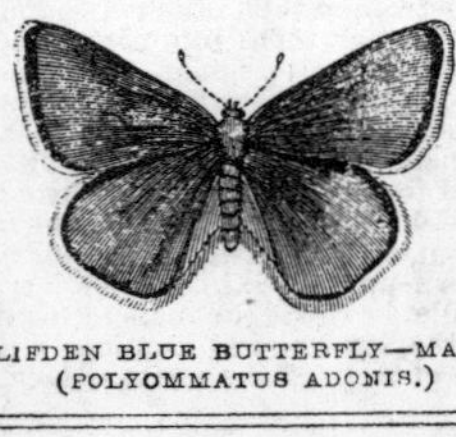

CLIFDEN BLUE BUTTERFLY—MALE.
(POLYOMMATUS ADONIS.)

coast. The male is of a most lovely azure or silvery blue, varying in lustre; now taking a tinge of green, and now of lilac, according to the light in which it is presented to the eye; the hinder margin of all the wings marked with a slender black line,

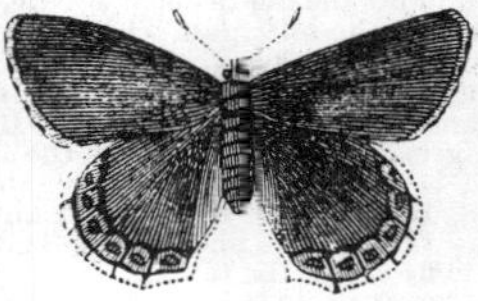

P. ADONIS — FEMALE.

the cilia white, interrupted by brown: beneath, the anterior wings are whitish, with spots distinctly ocellated; the margin with the fulvous ocelli of a deep hue, and the ground colour deep. The female is of a deep brown, with a black discoidal spot, the

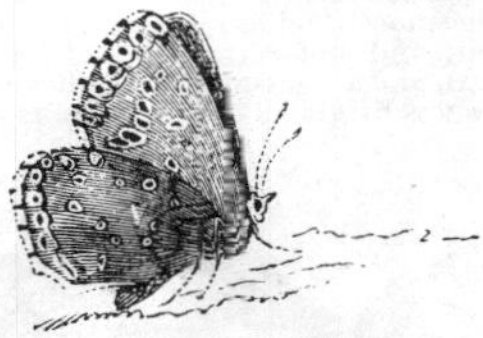

P. ADONIS — UNDER SIDE

disc frequently bluish; the hinder margin of the posterior wings with a slightly ocellated fulvous streak; and the colour beneath much darker than the male, and the ocelli more distinct. Caterpillar green, with dorsal rows of fulvous spots: it feeds on clover. Chrysalis green, or brown.

POLYOMMATUS ALEXIS; or ALEXIS BUTTERFLY. This, the commonest of our blue butterflies, is seen disporting itself by the sides of grassy lanes, in meadows, and in marshy places, wherever we go. Two broods make their appearance, the first in May, the last in August. Male above of a bright lilac blue, with the costa of the anterior

ALEXIS BUTTERFLY
(POLYOMMATUS ALEXIS.)

wings white, and a slender marginal black line to all the wings; the fringe usually white: the anterior wings have two ocelli placed transversely towards the base of the wing, then an ovate central spot with a transverse black streak, followed by a regularly curved transverse series of ocelli, seven in number; between which and the outer margin are a row of dusky lunules, edged with fulvous, and a series of dusky spots on a whitish ground; the extreme margin is

P. ALEXIS — UNDER SIDE.

black: the posterior wings are usually bluish at the base, with four ocelli placed obliquely towards the inner margin; the disc has a triangular white spot, with black centre, behind which is a waved series of eight or nine ocelli, externally bordered with a fulvous patch, the extreme edge of which is black: and a white blotch connects the

P. ALEXIS — CATERPILLAR AND CHRYSALIS.

fulvous band with the waved series of ocelli. Female, above brown, with the disc more or less blue: beneath, all the wings are deep ash-colour or drab, with all the ocelli very distinct and large. Caterpillar bright green, slightly hairy, with a dark dorsal line, and triangular yellow spots: it feeds upon the wild strawberry, and grasses. Chrysalis dark brown.

POLYPI: POLYPIARIA. The animals belonging to this extensive and remarkable class possess an organization so low in the scale of being (by which we mean, that the distinctive characters of animal life are so slightly developed), that there is very considerable difficulty in distinguishing many of them from the cryptogamic families of the vegetable kingdom; and, accordingly, we find in the works of the older botanists that the *Zoophytes*, generally, were arranged with the Sea-weeds and Mosses; nor was any idea entertained of their possessing a different character. That such should have been the case can excite the wonder of no person who merely regards the apparent *structure* of these plant-like animals. They see that a bulb is formed, which shoots up into a stem, and sends off branches; that there is also a root is evident, which, however we now know is merely the organ of attachment, affording no nourishment to the animal. Most of the Polypi form compound animals, attached to one another by lateral appendages, or by their posterior extremity, participating in a common life, while at the same time they enjoy their individual and independent existence. In reference to the different views which have been entertained on this once questionable subject, Mr. Broderip makes the following just observation: "Borrowing from Aristotle and Pliny the term Polypus, by them applied to a cephalapod, the systematic naturalists who followed Linnæus collected under this title many really animalized masses in the form of plants, and after abundant examinations by Ellis and others, of the membranous, horny, or stony 'fulcra,' bases, or axes, which remain after desiccation or decay of the softer parts, generally agreed in opinion that to all these plant-like bodies were associated active living animals like the Hydræ described by Trembley. As in a tree the flowering and reproductive organs manifest more active and varied functions than the general mass of bark and wood which serves to unite them in one common life, so in these Zoophyta the little Polypi expanding from their cells for food, light, or aëration, and shrinking back upon the agitation of the water, or withdrawal of the light, seemed like so many animal flowers, which might be studied apart from the Polypiaria which they adorned. They were in fact studied apart, and unfortunately attention was more directed to the wonderful permanent fabric, or 'Polypidom,' as Lamouroux calls it, than to the Polypi themselves. Hence the imperfection of all the schemes of classification for this portion of the Zoophytic division of the animal kingdom, and many of the erroneous generalizations and hypotheses regarding the lower forms of animal life."

POLYPTERUS. This is a fish, which is usually about eighteen inches in length, and partakes in some degree both of the osseous and cartilaginous kinds, but seems most nearly allied to those species of the genus *Esox* which are furnished with large, strong, and bony scales. Its shape is elongated and nearly cylindrical: its head is defended by large bony pieces or plates, and the body covered with large and strong scales, very closely affixed to the skin: the pectoral fins are placed immediately beyond the head; the ventral at a vast distance beyond it, the abdomen in this fish being of a very unusual length; the anal fin is seated very near the tail; and the tail is of a rounded or ovate form. At a small distance beyond the head, along the whole length of the back, runs a continued series of small dorsal fins, varying in different individuals from sixteen to eighteen: each of these fins is of an ovate shape, upright, or but very slightly inclining backwards, and is furnished with a very strong spine at its base, while the remaining part consists of four or five soft and branched rays, connected by their united membrane. The lateral line runs nearly straight from the gill-covers to the tail: the eyes are small and round; the mouth of moderate width; a row of small and sharp teeth in each jaw; and the upper lip furnished with a pair of small and short tentacula at its tip. This fish inhabits the depths of the Nile, remaining among the soft mud, which it is thought to quit only at particular seasons, and is sometimes taken in the fishermen's nets at the time of the river's decrease. Its colour is sea-green, paler or whitish on the abdomen, which is marked by some irregular black spots. It is called by the Egyptians *Bichir*, and is said to be one of the best of the Nilotic fishes for the table.

POLYTHALAMIA. Ehrenberg's name for the Rhizopoda or *Foraminifera* of D'Orbigny. [See INFUSORIA.]

POLYPTYCHODON and POLYZOA. [See SUPPLEMENT.]

POMFRET. (*Stromateus.*) A genus of Acanthopterygious fishes, having the same compressed form as the Dory (*Zeus*), and the same smooth epidermis; but the muzzle is blunt, and not retractile. It has a single dorsal, and a few concealed spines anteriorly, but no ventrals. The vertical fins are thick-

POMFRET. — (STROMATEUS NIGER.)

ened as in the scaly-finned fishes; the gullet has a number of spines attached to the membrane. They are found in the Mediterranean, the Indian Ocean, and Pacific. Some of the species differ considerably in form. The one here represented is the Black Pomfret (*Stromateus niger*).

POMOTIS. A genus of fish, belonging to the *Percidæ* family. The Northern Pomotis (*Pomotis vulgaris*) frequents the sheltered inlets of Lake Huron and the ponds in that vicinity, concealing itself, in the summer time, beneath the broad leaves of the nuphar and water-lily, where it may be readily taken with a hook baited with a small fish or worm. The form of this fish is a broad oval, the anterior apex rather acute, and formed by the lower jaw, which projects slightly beyond the upper one. It is about eight or nine inches long; the body is much compressed; the scales adhere firmly to the skin and are rather large; the head is small, and the opposing surfaces of both mandibles are covered with small teeth crowded closely together. The branchiostegous rays are considerably curved; and the caudal fin is slightly sinuated at the extremity with rounded lobes. Its principal food is small crustacea.

PONGO. A quadrumanous animal, being a species of the Orang-Outang, found in Borneo; characterized by the extraordinary size of its canine and incisor teeth, and by its black hair being relieved with hair of a dark red colour. [See ORANG-OUTANG.]

PONTIA. A genus of diurnal Lepidoptera, containing numerous species, a few of which are British. Some of the exotic species have the under side beautifully marked with red and yellow. Among the British species we may mention the common White Butterfly:—

PONTIA BRASSICÆ, or CABBAGE BUTTERFLY. This common and destructive insect makes its appearance in our gardens about the middle of May, and lays its eggs on the under side of cabbage leaves. Both sexes have the upper surface of all the wings white, with the tip of the anterior wings above black, the patch on its inner edge being indented, and the extreme tip being slightly irrorated with white: beneath, the under surface of the anterior wings is yellowish, the base slightly irrorated

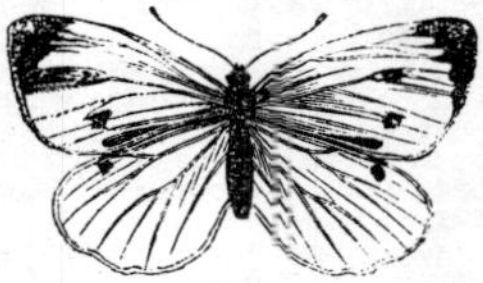

CABBAGE BUTTERFLY.—(P. BRASSICÆ.)

with dusky, and two transverse spots adorning the disc. The posterior wings are pale yellowish, rather sprinkled with dusky. The body and antennæ are black above and white beneath. There are, however, several varieties of this Butterfly; slightly differing

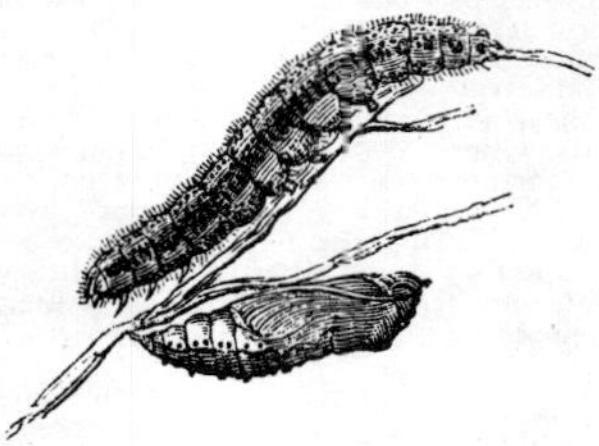

CATERPILLAR AND CHRYSALIS OF THE CABBAGE BUTTERFLY.—(PONTIA BRASSICÆ.)

from each other. The Caterpillar is bluish-green, with three yellow longitudinal lines, one on the back, the others on the sides; between these are several tubercular spots, each bearing a small hair; the tail is black. They are hatched in a few days, and continue to feed together till the end of June: when they have found a convenient place to attach themselves, they fasten their tail by a web, and carry a strong thread of silk round the upper part of their body; after hanging a few hours the chrysalis (which is greenish, spotted with black, with three yellow stripes) is perfectly formed; and in about six days the butterfly appears. The eggs laid by the second brood produce caterpillars which feed during the remainder of the summer, and remain in the pupa state during the winter, to be hatched in the succeeding spring. So prolific is this destructive species, that were it not for the ichneumon fly, which deposits her eggs within the body of the caterpillar, and in the larva state continues to prey on its vitals, the ravages of this insect would be of the most serious consequence to our vegetable productions.

PONTIA (*Anthocharis*) CARDAMINES; ORANGE-TIP, or WOOD LADY BUTTERFLY. This beautiful species is commonly seen during the month of May, whether we walk in the garden or stroll through the green lanes. The upper surface of all the wings in both sexes is white, with the base dusky; in the male the anterior wing is marked with a small black lunule, from which a deep fulvous spot extends to the tip; the tip itself being black or dusky: the posterior wings have a few dusky spots on the edge: beneath, anterior wings white, with a white lunule in the centre, the costa marked with a few black dots, and the tip varied with a greenish yellow: the posterior wings in both sexes are alike beneath; they are white, prettily marbled with green and yellow. The body is black above and white beneath; antennæ white, annulated with light brown. The caterpillar is green, with a longitudinal stripe placed above the legs; it feeds on the *Cardamine impatiens*, *Brassica campestris*, &c. The Chrysalis is either green or brown, with a fulvous spot on the wing-cases in the male; the middle is swollen and conical, with the extremities much produced and fusiform.

PORCELLANA. A genus of *Anomurous Crustacea*, in general form resembling that of the *Brachyura*, and distinguished by a fan-like caudal fin. The carapace is suborbicular and depressed above. The pincers

FLAT-CLAWED PORCELAIN CRAB. (PORCELLANA PLATYCHELES.)

are strong, and little or not at all dentated: the three succeeding pairs of feet are nearly cylindrical, and terminated by a conical tarsus. The species here figured, *Porcellana platycheles*, is found on the coasts of England and France, and is about seven lines long, and of a brownish colour.

PORCELLIO. A genus of *Isopoda*, distinguished from the *Onisci* (true Wood-lice) by the number of joints of their lateral antennæ, which are only seven. They are

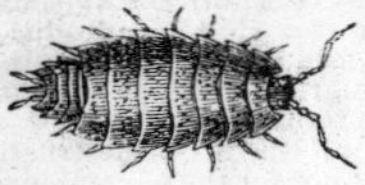

PORCELLIO GRANULATUS.

found under stones, old logs of wood, &c. Their food consists of decayed vegetable and animal substances; they move slowly when in danger; and they seldom come forth from their retreats except in damp weather.

PORCUPINE. (*Hystrix.*) A genus of Rodent quadrupeds, characterized by having the clavicles imperfect, two incisor teeth in each jaw, and four molars, both above and below, on each side: these have flat crowns, surrounded by a line of enamel, which enters into both edges, and appears to divide the tooth into two portions; the muzzle is thick and truncated; the lip divided; the tongue furnished with spiny scales; the ears short and rounded; the fore feet furnished with four toes; and the hind ones with five, all armed with thick nails. Many of them live in burrows, and have much the habits of rabbits; but their grunting voice, joined to their large and truncated muzzle, has caused them to be compared to the hog. The singular appearance of this animal, so different from that of the generality of quadrupeds, must in the earliest ages have attracted the attention of even the most incurious; the variegated spines or quills with which it is covered naturally suggesting the idea of a fierce and formidable animal: it is, however, of a harmless nature, and the quills are merely defensive weapons, which, when disturbed or attacked, the animal erects, and thus endeavours to repel his adversary.

The COMMON PORCUPINE (*Hystrix cristata*) is a native of Africa, India, and the Indian islands; and is also found in some of the warmer parts of Europe. When full grown, it measures about two feet in length, independent of the tail, which is five or six inches. The upper parts of the animal are covered with long, hard, and sharp quills; those towards the middle and hind part of the body being longer than the rest, very

COMMON PORCUPINE.—(HYSTRIX CRISTATA.)

sharp-pointed, and measuring from ten to twelve or fifteen inches in length: they are variegated with several alternate black and white rings; and their root, or point of attachment, is small. In their usual position they lie nearly flat upon the body, with their points directed backwards; but when the animal is excited, they are capable of being raised. The head, belly, and legs are covered with strong dusky bristles, intermixed with softer hairs; and on the top of the head the hair is very long, and curved backwards. The Common Porcupine, though known from the earliest ages, has given rise to numberless fables, among which that most commonly received is, that it possesses the power of darting its quills with great violence to a considerable distance when irritated or pursued. Perhaps in shaking the general skin of its body, like other quadrupeds, it may sometimes cast off a few of its loose quills to some distance, and thus slightly wound any animal that may happen to stand in its way: and this may have given rise to the popular idea of its darting them at pleasure against its enemies. In Bewick's Quadrupeds, the subject is thus mentioned: "Upon the smallest irritation it raises its quills, and shakes them with great violence, directing them to that quarter from whence it is in danger of being attacked, and striking at the object of its resentment with its quills at the same time. We have observed, on an occasion of this sort, at a time when the animal was moulting or casting its quills, that they would fly out to the distance of a few yards, with such force as to bend the points of them against the board where they struck; and it is not improbable that a circumstance of this kind may have given rise to an opinion of its power to use them in a more effectual manner." The use of this armature does not appear even now to be well understood: the most probable supposition, however, is, that it is merely for defence, as, like the hedgehog, it has the power of rolling itself up in a ball, and thus presenting a phalanx of spears on every side, that renders the attack of most animals fruitless. The Porcupine feeds principally on roots, fruit, bark, and other vegetable substances: it inhabits holes or subterraneous retreats, which it is said to form into several compartments or divisions, leaving only a single hole or entrance. It seldom leaves its burrow during the day, but makes its excursions for food by night. It is a solitary animal, and becomes torpid during winter. The female produces two young at a birth.

The CANADA PORCUPINE. (*Hystrix dorsata.*) This is a very unsightly and sluggish animal, approaching somewhat to the form of a Beaver, and principally found in the Northern States of the Union and in Canada. It is not provided with the long quills so remarkable in the common species, its armature consisting of short, sharp spines, almost concealed by the hair with which they are intermingled. It is about two feet long, and is remarkable for the length and fulness of its fur, which is soft, of a dusky brown colour, and intermixed with longer and coarser hairs with whitish tips: the head is short, the nose blunt, the ears small and rounded, the teeth very strong; the limbs short; the feet armed with strong, crooked claws.

Small and insignificant as their spines may appear, yet they are capable of seriously injuring dogs and other animals that incautiously attempt to seize the Porcupine. This animal makes its retreat amongst the roots of an old tree, and when not occupied in search of fruit, roots, and other vegetables, is said to pass most of its time in sleeping.

The PREHENSILE PORCUPINE (*Hystrix prehensilis*) is found in Brazil and other parts of South America; where it inhabits woods, and occasionally clings to the branches of the trees by its tail, in the manner of some of the Opossum and Monkey tribes. Its general length is about a foot, and the tail about eighteen inches. The whole animal, except on the belly and insides of the limbs, is covered with short, strong, and very sharp spines, of which the longest measure three inches, and are white with black tips. The colour of the hair with which the under parts are covered is a dusky brown. The head is small; the nose extremely blunt; the teeth very large and strong; and the ears short and rounded: the feet have four toes each, with strong claws, and a tubercle in place of a fifth toe: the tail is covered with spines for about a third part of its length; the remainder being naked, and strongly prehensile.

PORIFERA. The name given to the lowest of the classes of organized beings in the Animal Kingdom, including the marine and fresh-water Sponges; in which the absence of characteristic structure does not extend to external form alone, but is equally remarkable in the internal arrangement of the parts of which these beings are composed; They possess nothing, in fact, beyond the very simplest apparatus for reproduction. nothing distinctly characteristic of an animal nature; the only *obvious* vital action which can be observed in their ordinary state being a rapid movement of fluid through their channels. [See SPONGE.]

PORPOISE. (*Phocœna vulgaris.*) Of all the Cetacea, this and the closely allied varieties may be considered the most common, being found in almost all the European seas and on the American coasts. It greatly resembles the Dolphin (*Delphinus delphis*), and is frequently confounded with it; but may be readily distinguished by its shorter snout, thicker head, and smaller size. It rarely exceeds the length of six feet; is of a thick form on the fore parts, and gradually tapers towards the tail, which is horizontal and crescent-shaped, like that of other Cetacea. Its colour is a bluish black or a very dark brown above, and nearly white underneath. The back fin, situated rather nearer the tail than the head, is somewhat of a triangular shape, and placed nearly upright. The spiracle or spout-hole is upon the crown of the head, of a semilunar form, and divided internally by a cartilaginous membrane: the mouth is of moderate width; the teeth small, sharp, and numerous. The whole body is covered with a coat of fat, nearly an inch in thickness, beneath which the flesh appears red and muscular, resembling that of the hog. The Porpoise feeds on small fish, such as the Herring and Mackerel, of which they destroy great numbers. they root about the shores with their snout in quest of food, like hogs, and are believed to act in concert when in pursuit of their prey, urging them from one bay or estuary to another, deterring them from the shallow water, and driving them towards each other's ambush, with all the art of a well-trained dog. Before a storm, they may be seen gamboling and *tumbling* about (as it is termed) in the ocean, and they are occasionally observed to congregate together in large numbers. Their flesh was formerly considered a great delicacy; but is now seldom eaten. The term *Porpoise*, *Porpesse*, or *Porpus*, is said to be derived from the Italian *Porcopesce*, or hog-fish, from the supposed resemblance of its projecting snout to that of the Hog.

PORTUGUESE MAN-OF-WAR. (*Physalia atlantica.*) The name given by early English voyagers to a species of *Physalia*, belonging to the group of *Hydrostatic Acalephæ*. It is an inhabitant of the seas of warm climates, but a shoal of them are sometimes driven into our bays, particularly on the south-west coast; and it has been very probably mistaken for *Argonauta Argo*, the Paper Nautilus, by those who have declared that they have seen fleets of the latter sailing. These Acalephæ are characterized by the presence of one or more large air-sacs, by which great buoyancy is given to them; and it would appear that they have considerable power over these organs. The species we are now describing possesses a

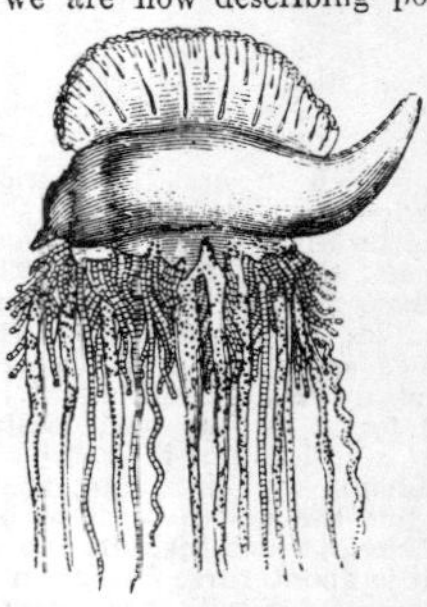

PORTUGUESE MAN-OF-WAR.
(PHYSALIA ATLANTICA.)

single large air-sac, beneath which the digestive apparatus is disposed; and the sac is surmounted by a sort of crest, which possesses considerable muscular power, and is elevated entirely above the water, when the animal is floating at the surface. The air-sac is provided with two orifices, one at each extremity, through both of which air is forced out when the bag is compressed by the hand; each of these orifices is provided with a little circular muscle, which usually keeps them closed, but which allows of their dilatation during the continuance of the outward flow of air. By means of this organ, it appears, they either force out the air, or compress it into a much smaller compass, when they wish to sink; and distend the sac when they desire to rise. From the under side of the air-sac there is a mass of short flask-shaped appendages hanging down, which are terminated by suckers, with an orifice in each. Whilst the lower surface of the air-sac is not itself above six inches from one end to the other, the tentacula sometimes hang down like fishing lines, to an extent of fifteen or sixteen feet. They generally possess an active stinging power, and are also very contractile, so that they are able to draw up the prey which they have attacked. It would seem that the short suckers are attached to the bodies of animals thus entrapped; and that the *Physalia* derives its nourishment by imbibing their juices through the pores of these numerous cirrhi.

PORTUNIDÆ, or FIN-FOOTED CRABS. A family of Brachyurous Crustacea mostly found in the seas of warm climates, and vulgarly called *Paddling Crabs.* They are in general remarkable for the flatness and great transversed extent of their

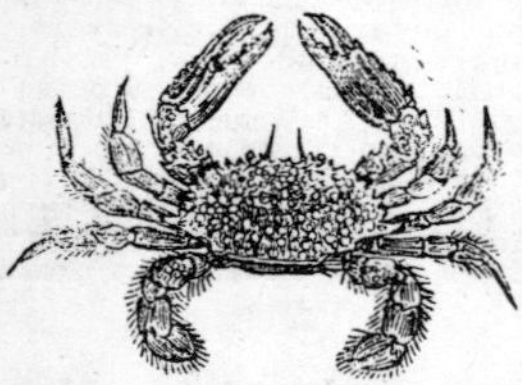

SPOTTED-FIN CRAB.—(LUPA CRIBRARIA.)

carapace, which in breadth is more than double its length. Their general form does not ordinarily differ much from that of the greater part of the Cancerians. The last pair of legs is flat and oar-like; a structure which enables these crabs to swim with great ease: hence some of the species are found at a great distance from land. Some of these are found on our own coasts, one of which, the small common Crab, is hawked about London, and eaten by the poorer classes; but the one here figured is the *Lupa cribraria*, which inhabits the coasts of Brazil: it is about three inches in length; of a yellow colour, with numerous whitish spots.

PORZANA. A genus of birds belonging to the family *Rallidæ;* by most authors it is included in the genus *Ortygometra*, the type of which is our COMMON CRAKE (*O. crex*). To it belongs the SPOTTED GALLINULE (*O. Porzana*), a species which is not very common in Britain; on the upper side it is of an olive-brown colour, with dusky streaks and white spots; beneath, it is of an ashy-olive, with white spots. Mr. Gould describes a fine species in his Australian Birds. This is the PORZANA (*Ortyg.*) FLUMINEA, or SPOTTED WATER-CRAKE. This Grallatorial bird inhabits various parts of Australia; and, like its European ally (*Rallus porzana* of Linnæus) frequents morasses, reed-beds, and the neighbourhood of rivers clothed with dense herbage; but the uniform gray tint of its breast and under surface, and its smaller size, are characters by which it is readily distinguished from it. The whole of the upper surface is of an olive colour, with a broad stripe of blackish brown down the centre, and two oval spots of white, bounded above and below with black on the margin of each web of every feather; primaries and secondaries brown; tail dark brown, margined with lighter brown, and with an indication of white spots on the extreme edge; face, throat, chest, and upper part of the abdomen dark slate-gray; lower part of the abdomen and flanks grayish-black, crossed by narrow irregular bars of white; under tail-coverts white; bill orange-red at the base, and dark olive-green for the remainder of its length; feet dark olive-green.

POTAMOCHŒRUS. [See SUPPL.]

POTOO. (*Nyctibius Jamaicensis.*) This bird is a native of Jamaica, and belongs to the *Caprimulgidæ* family. It is sixteen inches long, and in expanse from the tip of each wing thirty-three inches and a half. Plumage mottled with black, brown, gray, and white; the white prevailing on the tertiaries, tertiary-coverts, and scapulars, the black upon the primaries and their coverts; the tail-feathers barred transversely with black on a gray ground, and delicately mottled; tail broad, very slightly rounded: inner surface of the wings black, spotted with white. On each side of the throat is a black streak; a bay tint prevails on the breast; and some of the feathers there have broad terminal spots of black. Under parts pale gray; but every feather of the whole plumage is marked with a black stripe down the centre. The beak is black; the tongue sagittiform, slender towards the tip, reverted barbs along the edges. Irides orange-coloured or brilliant straw-colour. Feet whitish, and scurfy.

"The Potoo is not unfrequently seen in the evening, taking its station soon after sunset on some dead tree or fence-post, or floating by on noiseless wing, like an owl, which the common people suppose it to be. Its plumage has the soft, puffy, unwebbed character which marks that of the owls, and which prevents the impact of its wings upon the air from being audible, notwithstanding the power and length of those organs. Now and then it is seen by day; but it is half concealed in the bushy foliage of some thick tree, distrustful of its powers by day. It ever and anon utters a loud and hoarse *ho-hoo*, and sometimes the same syllables are heard, in a much lower tone, as if proceeding from the depth of the throat. * * * If I may judge of the habits of the Potoo (observes Mr. Gosse) from what little I have observed of it when at liberty, and from the manners of my captive specimen, I presume that, notwithstanding the powerful wings, it flies but little; but that, sitting on some post of observation, it watches there till some crepuscular beetle wings by, on which it sallies out, and having captured it with its cavernous and viscid mouth, returns immediately to its station. Mr. Swainson appears to consider that the stiff bristles with which many *Caprimulgidæ* are armed have a manifest relation to the size and power of their prey, beetles and large moths, while these appendages are not needed in the swallows, their prey consisting of 'little soft insects.' But here is a species, whose prey is the hardest and most rigid beetles, of large size, and often set with formidable horns, —which has no true rictal bristles at all. * * * I have seen that which serves this bird for a nest: it is simply a round flat mat, about five inches wide, and little more than one thick, composed of the fibrous plant called Old Man's Beard (*Tillandsia usneoides.*) It was found on the ground on a spot whence the Potoo bird had just risen. This bird is a permanent inhabitant of Jamaica; it is common in the lowlands of the south side, and probably is generally distributed in the island: it is found also in Brazil."—*Gosse's Birds of Jamaica.*

POTOROO. [See KANGAROO RAT.]

POTTO. (*Cercoleptes caudivolvus.*) A singular quadruped of South America, resembling the Lemurs somewhat in its structure and aspect, but closely allied to the COATIMONDI (*Nasua*), and, like it, consequently belonging to the order *Carnivora.* It has short round ears; short nose; a tongue of great length; a large prehensile tail; and eats like a squirrel, holding the food in its hands. It is a nocturnal animal; climbs like a Lemur, with agility; and is said to be a great destroyer of wild bees' nests. In captivity it is very mild, and climbs about the chairs, &c. in a room, if suffered to go at large. [See KINKAJOU.]

POUNDSTONE [also called QUOITSTONE.] A local name, in Oxfordshire and the adjacent counties, for a fossil found in the Oolite, belonging to the Sea-eggs or *Echinidæ.* The dairy-women in these counties frequently use them as pound weights: hence the name.

PRAIRIE DOG. (*Arctomys Ludovicianus.*) [See MARMOT.]

PRATINCOLE. (*Glareola.*) A genus of birds allied to the Plovers. They are characterized by a short, hard, convex bill, curved for upwards of half its length, and compressed towards the point; legs feathered nearly to the knee; toes, three before and one behind; claws long, and drawn to a fine point; wings very large, the first quill-feather the longest; tail more or less forked. Length upwards of nine inches. In Mr. Gould's 'Birds of Europe,' he observes that the genus *Glarcola* appears to be strictly confined to the Old World, no Transatlantic example having ever been discovered. It may be said to be truly a native of the eastern provinces of Europe on the Asiatic borders, and especially Hungary, where wide tracts of morass and flat lands, abounding in lakes both fresh and saline, and traversed by mighty rivers, afford it food and security. It is also abundant in Western Tartary. In England it is only an occasional visitor, but in Germany, France, and Italy, it is a bird of periodical occurrence. "With the long wings and forked tail of the swallow," says Mr. Gould, "the Pratincole possesses that rapidity and power of flight for which the bird is so remarkable. It takes its food, which consists of insects, and especially such as frequent marshes and the borders of rivers, while on the wing, darting along in the chase with the rapidity of an arrow; nor is it less distinguishable for celerity on the ground, and often catches its prey as it nimbly runs along. This elegant and graceful bird incubates in the concealment afforded by reeds, osiers, and tall herbage, laying three or four white eggs." A few months ago we had the pleasure of seeing specimens of this curious bird, brought alive to the Zoological Gardens by Mr. Fraser. They seemed to be moping and unhappy.

PRAWN. (*Palæmon serratus.*) A crustaceous animal; a species of *Macroura*, or Long-tailed Decapod, well known, and esteemed as an agreeable article of food. The species ordinarily sold in the fish-shops is the *Palæmon serratus.* It is generally about three inches long, and of a pale red colour, which is brightest in the antennæ, and especially in the swimmeret of the tail. Its frontal spine extends beyond the peduncle of the middle antennæ: it is curved upwards at the tip, with seven or eight spines above, and five beneath. They are taken on many parts of the British coasts, but are by no means so abundant as Shrimps. Some of the exotic species acquire a very large size. [See PALÆMONIDÆ.]

PRION. A genus of oceanic birds, belonging to the *Procellaridæ* or Petrel kind. They are distinguished by a strong, stout, and wide bill, very much depressed, the upper mandible convex on the sides, terminated by a compressed hook; the edges furnished internally with cartilaginous lamellæ; nostrils opening by two distinct orifices, and disposed in the form of a short tube. No hind toe, but in place of it a very small claw. In a letter addressed by Mr. Gould to the Zoological Society, dated Van Diemen's Land, May 10. 1839, several interesting particulars are detailed relative to oceanic birds observed by him on his voyage.

BROAD-BILLED PETREL.—(PRION VITTATUS.)

Mr. Gould crossed the equator on the 7th of July, having been more than twenty days within the tropics, part of which time the vessel lay becalmed. On the 23rd July (lat. 31° 10′ S., long. 24° W.) they were surrounded by the feathered race. Independently of an abundance of Cape Petrels, two other species and three kinds of Albatrosses were observed. A few days after this, they commenced running down their longitude, and from that time until they reached the shores of Van Diemen's Land, several species of *Procellaridæ* accompanied the ship. Mr. Gould found the Australian seas inhabited by their peculiar Storm Petrels (*Thalassidroma*), fourdistinct speciesof which he had already observed since leaving the Cape. "From the westerly winds which prevail in the southern hemisphere," adds Mr. Gould, "between the latitudes 35° and 55°, I am induced to believe that a perpetual migration is carried on by several members of the oceanic family continually passing from west to east, and circumnavigating this portion of the globe. This remark more particularly refers to the Albatrosses, Prions, and other large kinds of Petrels; the same individuals of several of these species having been observed to follow our ship for some thousands of miles. Until I had ascertained that they were nocturnal, it was a matter of surprise to me how the birds which were seen around the vessel at nightfall were to be observed crossing our wake at daybreak on the following morning, the ship having frequently run a distance of nearly a hundred miles during the night."

PRIONODON. A genus of quadrupeds, of which the type described by Dr. Horsfield is *P. gracilis*, a native of Java, partaking of the structural character both of the *Felidæ* and *Mustelidæ;* though in its general economy and habits it resembles the former only. It has a long, annulated, and cylindrical tail; light brown body, with four very wide dorsal bands and two narrow anal bands; two broad lateral striæ, the narrow cervical striæ, the numerous humeral and femoral spots, and the seven caudal rings, very deep brown. Mr. B. H. Hodgson has lately described other species of this genus from India.

PRIONUS: PRIONIDÆ. A genus and family of Longicorn Coleoptera. These insects only fly in the evening or during the night, and always settle upon trees. They are known by the following characters:— Eyes emarginate; head not narrowed behind into a neck; mandibles very large; palpi moderately long; labium small; the antennæ inserted between the base of the mandibles and the eyes; and the thorax generally square or transverse, and denticulated at the sides. The perfect insects are generally of dark colours, and are usually found on the trunks of trees: they are very inactive during the day, but take flight in the twilight. Several curious species are found in South America and in India; but not many inhabit Europe. The transformations of *Prionus coriarius* are thus described:—The larva, a broad, flattish, white grub, with the body gradually narrowed towards the posterior extremity, and divided into a head of moderate size, thirteen segments and an anal lobe; the mandibles are very powerful, but small and triangular, and are employed in gnawing the wood,

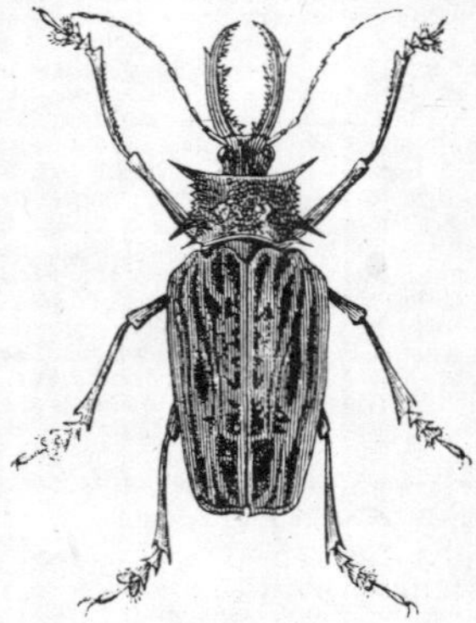

STAG-HORN BEETLE.
(PRIONUS CERVICORNIS.)

upon which the insect feeds. When full fed, it forms a large cocoon, chiefly composed of chips of gnawed wood, wherein it passes its pupa state; the antennæ at that time lying along the sides of the body, over the elytra: before the larva undergoes its change to a pupa, it instinctively bores a hole close to the outer surface of the tree, in order that the escape of the perfect insect may be the more readily effected. The genus comprises a very great number of species, which, from the variety in the form and size of their mandibles, antennæ, thorax, and abdomen, are divisible into many smaller subgenera. Some (chiefly exotic species) have the body elongated, straight, with the thorax much shorter than the abdomen, and greatly curved at the sides, and the mandibles of large size in the males. Others have the body not so oblong, somewhat depressed in front, and with moderate-sized mandibles in both sexes, and the antennæ strongly serrated in the males. The one here figured is the *Prionus cervicornis;* the larvæ live in the wood of the Gossampinus tree, and are eaten by the natives of South America. The largest of the tribe is also a native of South America; it is called *Titanus giganteus*, and well merits the name.

PRIVET HAWK MOTH. [See SPHINX LIGUSTRI.]

PROBOSCIDEÆ, or PROBOSCIDIANS. A term applied to those Pachydermatous animals which are distinguished as possessing a prolonged prehensile snout or proboscis, and having five toes on each foot, included in a very firm, horny skin; as the Elephant.

PROBOSCIS MONKEY. [See MONKEYS.]

PROCELLARIA. [See PETREL.]

PROCNIAS. A genus of Brazilian birds remarkable for the enormous width of their mouths, which enables them to swallow the large tropical berries, on which, as Mr. Swainson says, they wholly subsist, and not on insects, as Cuvier asserts. Although, he adds, they perfectly resemble the swallows in the construction of their bills, their wings are not formed for rapid flight; and their feet are much stronger, and calculated for searching among branches for their food, in which situations Mr. Swainson frequently saw them.

PROCTOTRUPIDÆ. A family of Hymenopterous insects, consisting of numerous minute species, distinguished by having the wings entirely destitute of, or with but very few, veins; and the body being extremely long and slender. Their colours are generally black, varied with brown: some frequent aquatic plants, whilst others are found in hot sandy situations, and the greater part run and fly with great agility. They are parasitic; and some are so extremely small as to be visible only to the naked eye when creeping up the glass of windows opposed to the light.

PROMEROPS. A genus of birds, many of which are remarkable for the beauty of their plumage, and its singular arrangement. They have an extensible tongue; and feed upon insects, soft fruits, and the saccharine juices of plants.

The SUPERB PROMEROPS (*Promerops superba*) is four feet in length from the tip of the bill to the end of the tail; the tail being extremely long in proportion to the body, which is delicate and slender. In this respect it resembles the Birds of Paradise, which are inhabitants of the same region; as it also does in the metallic lustre of its plumage. The feathers of the head, neck, and under part of the body, are of a glittering green, and soft as the finest velvet. The back is of a purple or violet hue; the wings, which also possess a velvety texture, appear blue, violet, or black, according to the light in which they are held; and the brilliancy of the tail and wing-coverts may well be likened to polished steel. On each side the lower part of the body beneath the wings is a thick and moderately long group of loose-webbed, pendent, brownish feathers; in which, as well as in some other points, it may be likened to the *Paradiseæ*. The legs are of moderate length, strong, and black. Native of New Guinea.

The RED-BILLED PROMEROPS. (*P. erythrorhynchus.*) This elegant species, which is a native of Africa, is about fifteen inches in length. Its general colour is black, with varying glosses of red, violet, and golden green: the red cast predominates on the head, the green on the wing-coverts, and the violet on the back and tail: the latter is very long and cuneated, the outside feathers measuring about three inches long, and the rest gradually lengthening to the two middlemost, which measure about eight inches: all the tail-feathers, except the two middle ones, are marked near the tip by an oval white spot on each side the web; the first six quill-feathers of the wings have also a white spot on the inner web near the tip: the bill is rather long, slender, moderately curved, and of a red or orange colour, as are likewise the legs, which are rather disproportionately short.

PROMINENT [MOTHS]. A name applied by collectors to different species of Moths, of the genera *Notodonta*, *Leiocampa*, *Ptilodontis*, and *Lophopteryx*.

PRONGBUCK, or PRONG-HORNED ANTELOPE. (*Antilocapra furcifera.*) A species of Antelope, inhabiting the extensive plains of the centre and west of North America in vast herds. It is about four feet four inches long, and three feet high; the whole form of the animal being peculiarly graceful and elegant. The horns rise perpendicularly from the front of the skull, and are perfectly straight till within two or three inches of the tips, when they curve suddenly inwards in the form of a hook: the horns below the

PRONGBUCK.—(ANTILOCAPRA FURCIFERA.)

prong are like the antler of a deer, but above they are round, black, and polished. The ears are long and pointed, the eyes large and animated, the tail short and bushy. The hair, which in the summer season has the ordinary texture and appearance of other Antelopes, becomes as the winter approaches long and tubular, and so inelastic that when pressed it crushes like a dry reed: on the head, ears, and legs the fur is close and smooth, but down the back of the neck it is six inches in length, and forms a mane. The general colour is pale fawn, the under parts being white; a broad disc of white surrounds the tail; and there are two transverse white bands on the throat. Like some other species, it migrates from north to south, according to the season, but is never found to inhabit forests or closely-wooded districts. It is an active and vigorous animal, though less enduring in its speed than most other Antelopes.

PROPITHECUS. A genus of quadrupeds allied to the *Lemurs*, but distinguished from them by its shorter muzzle and its rounded ears, as well as by the marked disproportion in length between its hinder and anterior extremities, the greater length of its hands, and the shortness of its anterior thumb. *Propithecus diadema:* Length of body and head, twenty-one inches; tail, seventeen inches. Face nearly naked. Above the eyes, the long, silky, waved hairs which cover the body commence by a band of yellowish white crossing the front and passing beneath the ears to the throat; the back of the head and neck clothed with black hair, which is freely intermingled with white on the shoulders and sides, the white gradually increasing towards the posterior portion of the body: the under surface white throughout.

PROSCOPIA. A genus of Locusts peculiar to South America; which have a membranous pellet between the terminal hooks of the tarsi, the antennæ filiform, and the posterior legs long and approximated to the intermediate pair, which are remote from the anterior pair. [See LOCUST.]

PROSTHEMADERA. A genus of birds belonging to the family of the *Honey-eaters*. It contains the POE-BIRD (*Prosthemadera cincinnata*), a Passerine bird of New Zealand, the native name of which is *Tui*. It is thus described by the Rev. W. Yate, a missionary

POE BIRD.—(PROSTHEMADERA CINCINNATA.)

there, belonging to the Church Missionary Society. "This remarkable bird," says the writer, "from the versatility of its talents for imitation, has, by some, been called 'the Mocking-bird;' and, from its peculiar plumage, has by others been denominated 'the Parson Bird.' It is so restless in its disposition, as to seem incapable of remaining in one situation, or unemployed, for a single moment. There is not a note of any bird of the woods but what it exactly imitates; and when confined in a cage, it learns with great ease and correctness to speak long sentences. It imitates dogs, cats, turkeys, geese, and, in fact, every sound which is repeated a few times in its hearing. Its size is that of the thrush; and its plumage a beautiful glossy black, with a few very fine white hairy feathers scattered about the head and breast, a few stronger ones about the nostrils, and two small clusters of long white feathers hanging down from the neck upon the breast, resembling a pair of clerical bands. Its eye is penetrating, and its voice peculiarly mellow. Its general food is flies and small insects, which it is very expert in catching; supplying itself in a very short time with great abundance. It also feeds upon the berries of various plants, and will not reject earthworms. This bird seems to associate with every other warbler of the wood; and, next to the ground-lark, is found in the greatest number of all the birds of New Zealand. It is delicious eating. It seems to be of a tender constitution, short-lived, and not able to bear the extremes of either heat or cold."

PROTELES, or AARD-WOLF. (*Proteles Lalandii.*) A carnivorous animal, about the size of a full-grown fox, inhabiting the southern parts of Africa. The genus Proteles, of which, as far as is at present known, this is the only species, resembles both the Civets and Hyænas; the teeth and pointed head resembling the former, while its striped fur, and the stiff bristly hair which runs along the neck and back, give it the appearance of the latter. The body is covered with coarse woolly fur, the general colour being a yellowish gray, radiated with transverse stripes of dusky black; and the tail is short and bushy. It has five toes on the fore-feet, but only four on the hinder ones; the claws on all being large and strong. It burrows like a fox, and, like that predatory animal, it ventures abroad at night only in search of its food, which consists chiefly of carrion and the smaller kinds of vermin.

PROTEUS. A very singular amphibious reptile, peculiar to certain subterranean waters, or underground lakes, of the Tyrol. It is very eel-like in its appearance and movements, but has four short limbs. The waters in which it dwells are sometimes dried up; and when this happens, it buries itself in the mud. They retain their external gills through life, the lungs not being developed sufficiently to maintain respiration by themselves. It is particularly found in the great Cave of Adelsberg, and

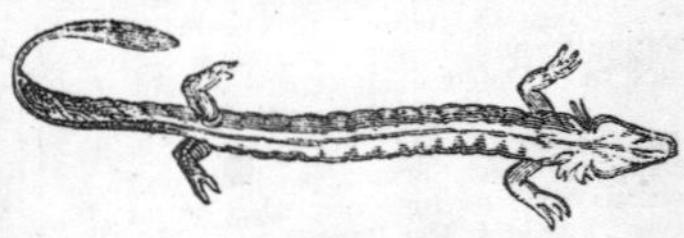

(PROTEUS ANGUINUS.)

is known to the inhabitants of the country *Bela Riba*, while the Germans call it *Weiss Fish*. A live specimen was exhibited at the Linnæan Society in June, 1847, by a gentleman who had it in his possession for eighteen months. The water in which it resides is strongly impregnated with carbonate of lime; but the party was not aware on what it fed.

The name PROTEUS is also given to an infusorial animalcule (*Amœba diffluens*) which is often met with in some vegetable infusions; and, under the microscope, appears to consist of a mass of gray-looking jelly, a film that can change its form at will, and assume every diversity of outline. Sometimes it shrinks up into a gelatinous ball, then shoots out rays in every direction, which appear like limbs; or it moulds itself into any form adapted to the shape of the animalcules it swallows for food.

PROTOSAURUS. [See SUPPLEMENT.]

PSEUDOTETRAMERA. The third general section of the Coleoptera, comprising those beetles which have the tarsi apparently four-jointed, although in reality consisting of five joints, the fourth being so exceedingly minute as to have escaped the notice of the tarsal systematists, who gave to these insects the sectional name of Tetramera. The whole of them feed upon vegetable matter, and are found in their perfect state upon flowers, leaves, or the bark of trees: the larvæ are fleshy grubs. Latreille divides this section into seven groups or families—the Rhyncophora (Curculio); Xylophaga (Scolytus, &c); Platysoma (Cucujus); Longicornes (Cerambyx); Eupoda (Crioceres); Cyclica (Chrysomela); and Clavipalpi (Erotylidæ).

PSITHYRUS. A genus of Hymenopterous insects belonging to the family *Apidæ*. Until lately the insects of this genus were confounded with the Humble-bees (*Bombus*), which in many respects they nearly resemble, but differ widely from them in others; viz. they make no nests of their own, neither do they collect food for their young, but, like the cuckoo among birds, they deposit their eggs in the nests of others, and leave their young to be hatched and reared by them. They may be distinguished also from the *Bombi* by the structure of their hinder legs, the tibia being narrower and covered throughout with hair.

PSITTACIDÆ. The name of a tribe of Scansorial Birds, of which the Parrot is the type. They are characterized by their short, hard beaks, which are generally highly arched, and surrounded at the base by a naked skin, in which are the orifices of the nostrils. They are natives of tropical and the warmer temperate regions; and they subsist for the most part upon fruits, seeds, honey, &c. Parrots, Macaws, Cockatoos, &c., are included in this numerous family. [See PARROTS, &c.]

PSOCIDÆ. A family of minute Neuropterous insects, which frequent the trunks of trees, palings, old walls, moss-covered stones, old books, &c. for the purpose of feeding, either upon the still more minute animalculæ which inhabit those situations or upon the decaying vegetable matter to be there met with. The eyes are semiglobose, lateral, and prominent; the ocelli are three in number, and placed triangularly between the eyes; the antennæ are slender and setaceous; the body is gibbose, ovate, and short; the meso- and meta-thorax larger and deeply impressed; the wings are hyaline, deflexed, with conspicuous veins, the anterior larger than the posterior, and often variegated and coloured; legs long and slender. These insects are remarkably active, and when approached they quickly endeavour to hide themselves by running to some obscure place. Towards the end of summer the perfect insects sometimes appear in great numbers. The larva differs from the imago in being apterous, while the pupa has rudimental wings.

PSOPHODES. A genus of birds belonging to the family *Meliphagidæ*; it contains the PSOPHODES CREPITANS, or COACH-WHIP BIRD. This is a shy and recluse bird, peculiar to South Australia, and renowned for the singularity of its note, which is loud and full, ending sharply like the cracking of a whip—whence its name. It rarely exposes itself to view, but keeps in the midst of the densest foliage and among the thickest climbing plants, through which it threads its way with the utmost ease. Its actions are always animated and sprightly, but particularly so in the spring, when the males may often be seen chasing each other, while they make the brushes ring with their clear and voluble song, or rather whistle. The male has the head, ear-coverts, chin, and breast black; on each side of the neck is a large patch of white; all the upper surface, wings, flanks, and base of the tail-feathers olive-green; the remaining portion of the tail-feathers black, the three lateral feathers on each side tipped with white; under surface olive brown; bill black; feet reddish-brown. The female may be distinguished by her more obscure plumage and smaller size. The food consists of various kinds of insects, mostly obtained from the ground by scratching up the leaves and turning over the small stones. Besides its peculiar shrill song, an idea of which it would be difficult to convey in words, it possesses a low inward song of considerable melody.

PSYLLA: PSYLLIDÆ. A genus and family of Homopterous insects, similar in their general habit, as well as in their saltatorial powers and deflexed wings, with some of the *Cicadæ*. They subsist in all their states upon plants, and have received specific names from the various trees and vegetables which they frequent. Their larvæ have the body very flat, the head broad, and the abdomen rounded behind; the pupæ are distinguished by having four large and broad scales on the back, which are the rudimental wings. Many species in the preparatory stages are covered with a white cottony secretion, and their excrement forms threads or masses of a gummy sucreous nature. Some species also, by puncturing vegetables, in order to suck the sap, produce gall-like monstrosities, especially upon the leaves and buds. Two species, *Psylla pyri* and *Chermes mali*, are very injurious in orchards, the former to the young shoots and leaves of the pear, and the latter to the apple.

PTARMIGAN. (*Lagopus mutus.*) This bird, which is also called White Grouse, is about fifteen inches long, or nearly the size of the Red Grouse. The bill is black; orbits bright red; the upper parts of the body pale brown or ash, mottled with small dusky spots and bars; the head and neck with broad bars of black, rust-colour, and white: the under parts are white, as are also the wings, excepting the shafts of the quills, which are black. In winter this plumage is changed to a pure white, except that in the male there is a black line between the bill

PTARMIGAN.—(LAGOPUS MUTUS.)

and the eye. The tail consists of sixteen feathers; the two middle ones ash-coloured in summer, and white in winter; the next two are slightly marked with white near the ends, the rest are wholly black; and the feathers incumbent on the tail, and nearly covering it, are white. The Ptarmigan, or White Grouse, is fond of lofty situations, and is found in most of the northern parts of Europe, even as far as Greenland: in this country it is only to be met with on the summits of some of our highest hills, chiefly in the Highlands of Scotland, in the Hebrides and Orkneys, and sometimes, but rarely, on the lofty hills of Cumberland and Wales. The female lays eight or ten eggs, which are white, spotted with brown: she makes no nest, but deposits them on the ground. These birds fly in small flocks, and feed on the wild productions of the hills: their flesh is dark-coloured, and has somewhat the flavour of the hare.

PTERICHTHYS, or WINGED FISH. A fossil genus of fish found in the Old Red Sandstone by Mr. Hugh Miller, and described by him in his interesting geological work. "Imagine," says he, "the figure of a man rudely drawn in black on a gray ground, the head cut off by the shoulders, the arms spread at full, as in the attitude of swimming, the body rather long than otherwise, and narrowing from the chest downwards; one of the legs cut away at the hip joint, and the other, as if to preserve the balance, placed directly under the centre of the figure, which it seems to support. Such, at a first glance, is the appearance of the fossil. The body was of very considerable depth, perhaps little less deep proportionally from back to breast than the body of the tortoise; the under part was flat, the upper rose towards the centre into a roof-like ridge, and both under and upper were covered with a strong armour of bony plates, which, resembling more the plates of the tortoise than those of the crustacean, received their accessions of growth at the edges or sutures. The plates on the under side are divided by two lines of suture, which run, the one longitudinally through the centre of the body, the other transversely, also through the centre of it; and they cut one another at right angles, were there not a lozenge-shaped plate inserted at the point where they would otherwise meet. There are thus five plates at the lower or belly part of the animal. They are all thickly tuberculated outside with wart-like prominences; the inner present appearances indicative of a bony structure. The plates on the upper side are more numerous and more difficult to describe, just as it would be difficult to describe the forms of the various stones which compose the ribbed and pointed roof of a Gothic cathedral, the arched ridge or hump of the back requiring, in a somewhat similar way, a peculiar form and arrangement of plates. The apex of the ridge is covered by a strong hexagonal plate, fitted upon it like a cap or helmet, and which nearly corresponds in place to the flat central part of the under side. There runs around it a border of variously-formed plates, that diminish in size and increase in number towards the head, and which are separated, like the pieces of a dissected map, by deep sutures. They all present the tuberculated surface. The eyes are placed in front, on a prominence much lower than the roof-like ridge of the back; the mouth seems to have opened, as in many fishes, in the edge of the creature's snout, where a line running along the back would bisect a line running along the belly; but this part is less perfectly shown by my specimens than any other. The two arms or paddles are placed so far forward as to give the body a disproportionate and decapitated appearance. From the shoulder to the elbow, if I may employ the terms, there is a swelling muscular appearance, as in the human arm; the part below is flattened so as to resemble the blade of an oar, and it terminates in a strong sharp point. The tail — the one leg on which, as exhibited in one of my specimens, the creature seems to stand — is of considerable length, more than equal to a third of the entire figure, and of an angular form, the base representing the part attached to the body, and the apex its termination. It was covered with small tuberculated rhomboidal plates, like scales; and where the internal structure is shown, there are appearances of a vertebrated bone, with rib-like processes standing out at a sharp angle." The species has been named by Agassiz, *P. Milleri* in honour of the author of "The Old Red Sandstone" and the "Asterolepis."

PTEROCERAS. A genus of Molluscous animals, inhabiting the Indian Ocean. The head is furnished with a proboscis and two tentacula, which are short; the eyes are situated on foot-stalks longer than the tentacula: foot small. The shell is oblong; spire small; mouth terminated by a rather long canal; right lip dilated into several claws, and having a sinus near the canal;

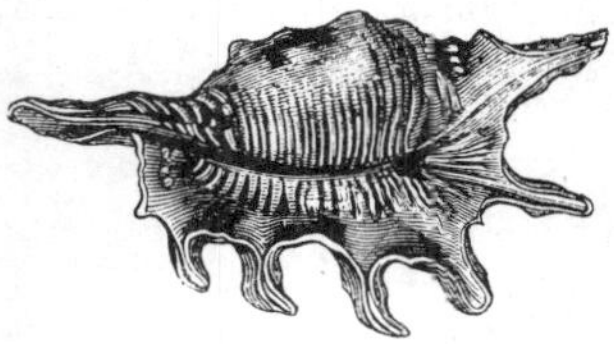

DEVIL'S CLAW.—(PTEROCERAS SCORPIO.)

operculum horny. The appearance these shells present at various periods is strikingly different. When the animal is young the shell has no claws; but they gradually make their appearance, at first in the form of short and open canals, which by degrees assume the length and curve of the adult and completed shell, and ultimately are closed up with shelly matter and become solidified. The number of claws varies in different species; in some they are straight and smooth; in others they are numerous, but small; whilst many have these appendages very much curved. Some of them exhibit the most beautiful colours on their internal surface. Our cut represents *Pteroceras scorpio:* the DEVIL'S CLAW.

PTERODACTYLUS. The name given to a genus of extinct Reptiles, which are supposed from their structure to have occupied that share in the economy of nature which is at present assigned to the Bats and Insectivorous Birds. From the size and form of the posterior extremities, the Pterodactylus seems to have been able to walk and perch upon them, after the manner of birds; and by using both its anterior and posterior limbs, it could probably walk and climb on rocks and cliffs, like Bats and Lizards. They have been found in the lias and oolite formations, greatly varying in size, and generally mingled with the remains of Dragon-flies, Beetles, and other insects.

It appears that the opinions of philosophers with regard to the true nature of this extinct animal were various and contradictory, until the reasonings of the great French Naturalist solved this zoological puzzle. "Behold," says Cuvier, "after having built, as it were, the animal before our eyes, an animal which, in its osteology, from its teeth to the end of its claws, offers all the characters of the Saurians; nor can we doubt that those characters existed in its integuments and soft parts — in its scales, its circulation, its general organs. But it was at the same time an animal provided with the means of flight, — which, when stationary, could not have made much use of its anterior extremities, even if it did not keep them always folded, as birds keep their wings, — which nevertheless might use its small anterior fingers to suspend itself from the branches of trees, but when at rest must have been ordinarily on its hind feet, like the birds again; and, also like them, must have carried its neck sub-erect and curved backwards, so that its enormous head should not interrupt its equilibrium." Dr. Buckland, whose attention has been especially directed to the examination of extinct animals, dwells at considerable length on the presumed habits and character of the Pterodactylus; and exclaims, "Thus, like Milton's fiend, all-qualified for all services and all elements, the creature was a fit companion for the kindred reptiles that swarmed in the seas or crawled on the shores of a turbulent planet. With flocks of such-like creatures flying in the air, and shoals of no less monstrous Ichthyosauri and Plesiosauri swarming in the ocean, and gigantic crocodiles and tortoises crawling on the shores of the primeval lakes and rivers, air, sea, and land must have been strangely tenanted in these early periods of our infant world."

PTEROGLOSSUS. [See ARACARI.]

PTEROMYS, or FLYING SQUIRREL. [See SQUIRREL.]

PTERONARCYS. A genus of Neuroptera first described by Mr. Newman. It is allied to PERLA; the finest species, *Pteronarcys regalis*, is a native of Canada and other more northern parts of North America. Mr. Barnston, a gentleman belonging to the Hudson's Bay Company's service, and who has paid much attention to Natural History, especially to that part of it which regards insects, mentions in his notes that the species of this genus shun light, concealing themselves during the day under stones in damp places, and appear on the wing at nightfall, when the air is charged with moisture. But the most remarkable fact connected with this insect is the discovery by Mr. Newport, of persistent branchiæ in the perfect state; in the larva, and, it is believed, in the pupa state, the insect lives constantly in the water; and in ordinary states of the atmosphere, such branchiæ would be no longer necessary, but in this case their continuance would seem a peculiar provision of Nature suited to the damp atmosphere in which it lives, — Mr. Newport observing, that "the function of branchiæ, or aquatic organs, is equally well performed in the open air as in water, so long as the air is charged with a sufficiency of fluid to preserve these organs in a healthy state." We eagerly look for an elaborate memoir on this anomalous occurrence from the pen of our most talented comparative anatomist in the field of articulated animals.

PTEROPODA. The name of a class of Molluscous animals, particularly distinguished by the possession of a pair of fin-like organs, or wings, consisting of a natatory expansion of the mantle on each side of the neck, by the aid of which they are rapidly propelled through the water. Some of them have a shelly covering; others are unprovided with such a protection; but wherever it exists, it seldom covers more than the posterior half of the body, and is extremely light and delicate. The head of these animals is generally prominent, possessing eyes and sensory tentacula; and their internal organization is of a very complex nature. These animals abound in the seas of warm climates; to which, however, the species are not restricted; for some of them, as *Clio borealis*, are so numerous in the Arctic Seas, as at certain seasons to furnish whales with their ordinary food. Mr. Arthur Adams calls these little active and energetic mollusca "the very butterflies of the deep;" "insatiate and greedy, they are ever on the move, spinning, diving, and whirling in every direction." They are stated to be all hermaphrodites.

PTEROPTOCHOS. [See BARKING BIRD and CHEUCAN.]

PTEROPUS: PTEROPIDÆ. A genus and family of Mammalia, belonging to the *Cheiroptera*, and distinguished as Frugiverous Bats. The species are very numerous; they produce early; and the sexes are separately gregarious. They are found in the south of Asia, the Indian Archipelago, Japan, Madagascar, and Australia. The KALONG, or FOX BAT. (*Pteropus Javanicus.*) This species, which is a native of Java, measures in the spread of the wings about five feet. They congregate in companies, and, selecting a large tree for their resort, suspend themselves by the claws of their hind limbs to the naked branches, affording to the stranger a very singular spectacle: in short, to a person unaccustomed to their habits, they might be readily mistaken for fruit of a large size suspended from the branches. They thus pass the greater portion of the day in sleep; but soon after sunset they gradually quit their hold, and pursue their nocturnal flight in quest of food. They direct their course, says Dr. Horsfield, by an unerring instinct, to the forests, villages, and plantations, occasioning incalculable mischief, attacking and devouring indiscriminately every kind of fruit, from the abundant and useful cocoa-nut, which surrounds the dwelling of the meanest peasantry, to the rare and most delicate productions which are cultivated with care by princes and chiefs of distinction. The flight of the Kalong is slow and steady, pursued in a straight line, and capable of long continuance.

PTILINOPUS. A genus of beautiful birds belonging to the *Columbidæ* family, some species of which are natives of Australia, and others distributed over the Indian and Polynesian islands. The one we are about to describe has been named by Mr. Gould, in compliment to Mr. Swainson, the celebrated naturalist.

PTILINOPUS SWAINSONII, or SWAINSON'S FRUIT-PIGEON. This bird has by many authors been considered as identical with or as a mere variety of the *Columba purpurata;* but Mr. Gould was convinced, by comparing them, that they possessed characters sufficiently different to constitute a distinct genus. The forehead and crown deep crimson-red, surrounded except in front with a ring of light yellow; back of the neck grayish green; all the upper surface bright green tinged with yellow, the green becoming deep blue towards the extremities of the tertiaries, which are broadly margined with yellow; tail-feathers deep green, tipped with rich yellow; throat greenish gray; breast green, each feather forked at the end, and with a triangular silvery-gray spot at each point; flanks and abdomen green, with a large patch of orange-red in the centre of the latter; under tail-coverts orange-yellow; thighs green; bill greenish black; feet olive brown. The sexes are so nearly alike as to render them scarcely distinguishable.

PTILOCERCUS. A genus of mammalia allied to *Tupaia*, and remarkable for its tail being fringed on each side, at the end, like a quill. This remarkable genus was described in February, 1848, by Mr. Gray, in a paper read at the Zoological Society. The only species, PTILOCERCUS LOWII, was brought by Mr. Low from Borneo. As that gentleman has lately returned to Borneo, we may expect from him shortly, an account of the habits of this very curious animal.

PTILOGONYS. A genus of Passerine birds, the best known species of which is PTILOGONYS ARMILLATUS, found in Hayti, Jamaica, &c., and there called the Solitaire; remarkable for its singularly clear, slow, and melodious notes. It is eight inches in length, and its wings expanded rather less than a foot: the upper part of the plumage is blue-gray; wing-quills black with gray edges, the bases of the interior primaries white; breast ashy-gray, paler beneath; tail black; vent and under tail-coverts rusty orange: bill black; feet bright fulvous, claws black. The following description, which is taken from Mr. Gosse's charming work, conveys a lively idea of this sweet vocalist. "As soon as the first indications of day-light are perceived, even while the mists hang over the forests, these minstrels are heard pouring forth their wild notes in a concert of many voices, sweet and lengthened like those of the harmonica or musical glasses. It is the sweetest, the most solemn, and most unearthly of all the woodland singing I have ever heard. The lofty locality, the cloud-capt heights, to which alone the eagle soars in other countries,—so different from ordinary singing-birds in gardens and cultivated fields,—combine with the solemnity of the music to excite something like devotional associations. The notes are uttered slowly and distinctly, with a strangely-measured exactness. Though it is seldom that the bird is seen, it can scarcely be said to be solitary, since it rarely sings alone, but in harmony or concert with some half-dozen others chanting in the same glen. Occasionally it strikes out into such an adventitious combination of notes, as to form a perfect tune. The time of enunciating a single note, is that of the semi-breve. The quaver is executed with the most perfect trill. It regards the major and minor cadences, and observes the harmony of counterpoint, with all the preciseness of a perfect musician. Its melodies, from the length and distinctness of each note, are more hymns than songs. Though the concert of singers will keep to the same melody for an hour, each little coterie of birds chants a different song, and the traveller by no accident ever hears the same tune. * * *. Its plumage being blue passing into violet, it has hence obtained the name of Bishop. It is so sweet-throated, so flexible in its tones, and so soft in its warblings, that those who once hear it become somewhat measured in their praises of the Nightingale. The notes of its song are lengthened out like those of a *miserere.* Whilst it sings it does not seem to draw breath; but it rests a double time before it recommences, and this alternation of singing and resting will be continued for two hours." The foregoing account Mr. Gosse derived from his ornithological friend Mr. Hill; and after he had proved their general correctness by auricular observation, he hazards the very probable conjecture, that "these *true melodies* are peculiar to the nuptial season, and indicate that the period of incubation is either begun or near." In the specimens which he dissected he found no insects; they were evidently baccivorous, their stomachs being full of the green berries of the pimento.

PTILONORHYNCHUS. [See SATIN-BOWER-BIRD.

PTILORIS PARADISEUS. [See RIFLE BIRD.]

PTILOTIS. A genus of birds found in Australia. Among the species particularised by Mr. Gould, we may allude to the PTILOTIS ORNATUS, a bird of Western Australia. It is found among gum trees, searching for insects, pollen, and saccharine juices. It has a loud, ringing, and not unpleasing sound, constantly poured forth. Its nest, which is neat, small, open, and cup-shaped, is generally suspended from a horizontal forked branch; and is composed of fine vegetable fibres and grasses matted together, with spiders' webs, and sometimes wool.

PTILOTIS PLUMULUS. This bird, which is also an inhabitant of Western Australia, is distinguished by its note — loud and shrill, like the sportsman's pea-whistle, continued without intermission for a great length of time.

PTINIDÆ. A family of Coleopterous insects, comprising a rather extensive group, which, though small in size, are of very destructive habits. The body is of an oval or subcylindric form, generally short and obtuse at each end; head small; antennæ long, and filiform or serrated; mandibles small; palpi short; tarsi five-jointed, and occasionally very broad. When touched, they counterfeit death by withdrawing their head and antennæ, and contracting their legs. Some species are found in old houses, rotten palings, stumps of decayed trees, &c., which their larvæ perforate in every direction; others feed upon collections of dried plants, skins of insects, &c.; whilst others attack our household furniture, books, &c.; in short, there are some which will devour almost any substance they come in contact with, whether it be ship-biscuit or Cayenne pepper, old woollen clothes or rhubarb, the wheat deposited in our granaries, or the timbers with which they are constructed. That *alarming* insect, *Anobium tesselatum*, or the Death-watch, is the largest British species belonging to the family; and to it we refer our readers for further particulars.

PTINUS. A genus of Coleoptera belonging to the family *Ptinidæ*. The body of these insects is of rather solid consistence, sometimes ovoid or oval, and sometimes cylindrical, but generally short, and rounded at each end: the head is almost orbicular, and received in the thorax, which is swollen, or hood-shaped; the antennæ of some are filiform, or become gradually slender to the

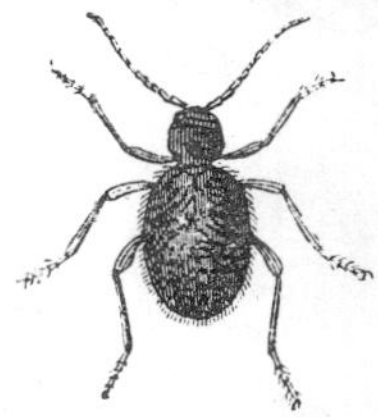

PTINUS FUR.

tip, while others terminate in three joints, abruptly thicker and longer than the preceding joints; the mandibles are short, thick, and toothed. All these insects are of small size; and their colours are always obscure, and but slightly variegated. *Ptinus fur*, the species here figured, has the antennæ inserted below the eyes, and the body is oblong. They frequent houses, and especially granaries. Their larvæ devour dried plants, and the prepared dry skins of animals. The antennæ of the males are longer than those of the females, and in many species the latter are wingless.

PUFF-ADDER. [See SUPPLEMENT.]

PUFFIN. (*Fratercula Arctica.*) The Puffin belongs to the sub-genus *Fratercula;* and is about twelve inches long. It has a very large singular looking bill, which has the appearance of a sheath slipped over both mandibles; it is curved towards the point, compressed vertically, and transversely furrowed on the sides: the chin and cheeks are white, bordered with gray, the latter much puffed up with feathers, which make the head look large and round. The crown of the head and upper part of the plumage are black, and a collar of the same colour encircles the neck: the under parts are white, and the legs are orange. The Puffin can fly with great rapidity when once upon the wing. In tempestuous weather it takes shelter in the holes of caverns and rocks, or in those made by rabbits; for they are unable to brave the storm. They live chiefly upon small crustaceans, sea-weed, &c., as it is said; but it is evident, from the structure and great strength of their bill, that they are able to crush and pluck out other kinds of shell-fish. The female deposits her single whitish-coloured egg in a hole dug out and formed in the ground, by her mate and herself, or in one ready made by the rabbits,

COMMON PUFFIN.—(FRATERCULA ARCTICA.)

which they easily dislodge. Puffins are met with on almost all the rocky cliffs on the coasts of Great Britain and Ireland, and on many of the surrounding islands, in immense numbers. They are gregarious and migratory. They hatch their young early in July; from which time till about the middle of August they are employed in nurturing and rearing their brood; which being done, the whole company leaves the breeding-place, and pursues its route to other regions, more suited to their future exigencies.

PUFFINUS. A genus of web-footed birds allied to the Petrels. The nostrils have separate openings, and the end of the lower mandible is bent downwards.

Our British species, which is called the MANKS PETREL, and sometimes the SHEARWATER or SCRABE (*Puffinus anglorum*), has the wings longer than the tail; it is of a black colour above, and is white beneath, the sides of the neck being freckled with black and white. It arrives at its breeding places in March, and generally leaves in August. They breed on the Isle of Man in rabbit-holes, in the Scilly Islands, and in different parts of Scotland. The young are fat, and sought after by the inhabitants, who salt them and eat them with potatoes and cabbage: the feathers also are collected. Another closely allied and widely distributed species is

The SOOTY PETREL (*Puffinus major*) is mentioned by one of our voyagers as frequenting some of the tufted, grassy parts of the South Sea islands in astonishing numbers. It is known that these birds make burrows in the ground, like rabbits; that they lay one or two enormous eggs in these holes, and bring up their young there. In the evening they come in from sea, having their stomachs filled with a gelatinous substance gathered from the waves: and this they eject into the throats of their offspring, or retain for their own nourishment, according to circumstances. A little after sunset, the air at Preservation Island used to be darkened with their numbers; and it was generally an hour before their squabblings ceased, and every one had found its own retreat. These birds are about the size of a pigeon, and when skinned and dried in smoke we thought them passable food. Any quantity could be procured, by sending people on shore in the evening. The sole process was to thrust in the arm up to the shoulder, and seize them briskly; but there was some danger of grasping a snake at the bottom of the burrow, instead of a Petrel.

Capt. Sir James Clark Ross, in his Voyage &c. to the Antarctic Regions, observes that when in lat. 47° 17′ S. long. 58° 50′ E. "we were accompanied on our course by many of the great Albatross, and the large dark Petrel, and still more numerously by the speckled Cape Pigeon (*Daption capensis*) and Stormy Petrel, of two or three different kinds. These birds added a degree of cheerfulness to our solitary wanderings, which contrasted strongly with the dreary and unvarying stillness of the tropical region, where not a sea-bird is to be seen, except only in the vicinity of its few scattered islets, which is the more remarkable where the ocean abounds so plentifully with creatures fit for their food. [See THALASSIDROMA.]

PUG [MOTHS]. A name applied by collectors to various species of Moths of the genus *Eupithecia*.

PULEX. [See FLEA.]

PULMOGRADA. [See ACALEPHA.]

PULMONARIA. The name of an order of the *Arachnida*, or Spiders, having small foot-like palpi, not terminating in pincers; and the *Pedipalpi*, or Scorpions and their allies, having very large palpi, which terminate in pincers. [See ARACHNIDA.]

PULMONEA; or PULMONATA. The name of an order of Gasteropodous Molluscs, comprehending those which breathe air, to which the blood is exposed while circulating through a vascular network which lines the internal surface of the bronchial cavity. Although the greater part of the Molluscs of this order live on land, some are aquatic; but these are obliged to come occasionally to the surface to breathe. They all feed upon vegetables, and many of them do so exclusively; but some are extremely voracious. Those without a shell, commonly known as Slugs, constitute the family *Limacinæ*. Those which have a shell, viz. the Snails and their allies, constitute the family *Helicinæ*.

PUMA. (*Felis concolor*.) This animal, which is the largest of the feline species found in America, and has sometimes been termed the American Lion, is about five feet from nose to tail; the tail itself measuring somewhat more than two feet and a half. The Puma is of a brownish red colour, with small patches of rather a deeper tint, which are only observable in certain lights, and disappear entirely as the animal advances in age: the breast, belly, and insides of the thighs are of a reddish-ash colour; the lower jaw and throat entirely white;

and the tail of a dusky ferruginous tinge, with a black tip. When at a mature age, however, its general colour is a silvery fawn. The Puma was formerly found in most parts of the American continent, and is still numerous in South America; but

PUMA.—(FELIS CONCOLOR.)

the advance of population in the north has rendered it scarce. It is a savage and destructive animal, possessing all the watchful caution of the cat kind; and although it generally confines its attacks to the smaller quadrupeds, it will sometimes attack those of large size and strength. When domesticated, (as it is occasionally,) its manners closely resemble those of the common cat, showing its fondness at being caressed by the same kind of gentle purring. It can climb trees with great facility, and will watch the opportunity of springing on such animals as happen to pass beneath. In the day-time, however, it is seldom seen, the night being the time it selects for committing its depredations. It is asserted that the Puma always kills its prey by springing on the shoulders, and then drawing back the head with one of its paws, until the vertebræ break.

PUNDIB. The local name in Oxfordshire and the adjacent counties for an oolite fossil belonging to the genus *Terebratula*.

PUPA. A genus of Mollusca, which derive their name from the resemblance of the

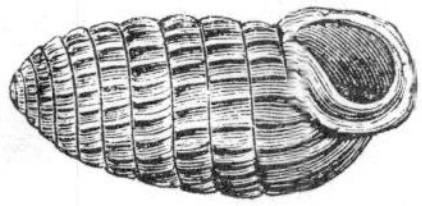

PUPA UVA.

shell in shape to the pupa or chrysalis of an insect. The shell is cylindrical; spire long. Animal like the Helix.

PURPLE EMPEROR [BUTTERFLY]. A name given by insect collectors to Butterflies of the species *Apatura Iris*.

PURPLE GRACKLE. [See QUISCALUS.]

PURPURA. A genus of Mollusca, found most abundantly in the seas of warm climates, where the shells attain a very large size. They are thick and oval, either smooth or tuberculated; spire short. A few species are met with in Europe, chiefly found on or near the sea shores. It was from the *Purpura patula*, as is supposed, that the Roman purple dye was obtained. There are very many recent species, and a few fossil.

PUSS MOTH. [See CERURA VINULA.]

PYCNOGONUM. A genus of Crustacea belonging to the group *Podosomata*, and forming as it were a connecting link with the *Arachnida* or Spiders, with which some naturalists used to class them. There are several genera belonging to the same group, all of which are marine. These animals conceal themselves among sea-weeds and corallines, and under stones; and they are not unfrequently dredged in deep water. Their motions are very slow, so that their prey must be either dead animal matter or

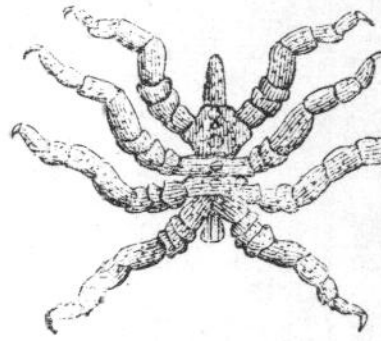

PYCNOGONUM LITTORALE.

living animals as sluggish as themselves. They are said to live chiefly on the animals of bivalve shells, and on minute insects and worms. The species here figured, *P. littorale*, is not uncommon on our coasts; by Linnæus it was believed to be parasitic on whales. The female, Dr. Johnston informs us, carries her innumerable ova, enveloped in a broad square gelatinous membrane or apron, under the body between the legs, where they are attached in front to a pair of filiform jointed organs. M. Kroyer has lately published interesting descriptions and figures of the metamorphoses of this and the allied genera. They would seem to be softer and larger-bodied proportionally than in the perfect state, in which it is represented in the above figure.

PYRALIDÆ. A family of Lepidopterous insects, moderate in extent, belonging to the general section HETEROCERA. The species are of a small size, having a slender and elongated body; the antennæ are simple, or but slightly ciliated in the males; the labrum and mandibles small; the labial palpi often greatly elongated and porrected, but occasionally recurved; the head is sometimes furnished with a pair of ocelli; and the thorax never crested. The wings are of moderate size, and generally placed in a triangular form during repose, the anterior ones being slightly angulated at the tip: the legs are ordinarily very long, especially the fore pair, the coxæ of which are nearly as long as the tibiæ, thereby indicating the great activity of movement so frequently exhibited by these insects. Owing to the fore legs of some of the species being ornamented with fascicles of hairs capable of expansion, they have received the name of "fan-footed" moths. The caterpillars are, in general, long and slightly hairy. For the most part they have only three, but sometimes four pairs of ventral feet.

The genus *Hypena* and its allies are the largest in the family; the species are found in hedges, and amongst low herbage; the larvæ are well distinguished by having only three pairs of ventral feet; and the chrysalis is enclosed in a slight cocoon in a leaf rolled up by the larva. The species of *Pyrausta* (*P. purpuralis*) are gaily coloured insects, which frequent hedges, and revel in the sunshine, hovering over grassy spots, but immediately settling as soon as the sun is overclouded; whilst those of *Hydrocampa* and its allies frequent aquatic plants, upon which the larvæ feed.

PYRALIS. A genus of Pyralidous Moths, one species only of which deserves notice, as its caterpillar sometimes greatly injures several different sorts of vegetables. This is the CABBAGE-GARDEN PEBBLE MOTH (*Pyralis forficalis*). The head, back, and upper wings of the Moth are hazel brown, and brownish gold; the antennæ light brown; the abdomen and under wings whitish. On the upper wings are two distinct and two faint deep rusty brown stripes. The under wings have a brownish-yellow curved line, towards the outer edge. Breadth, one inch. The first brood flies in May, and the second in August. The caterpillar is found in May and June, and the second generation in September and October. It has a light brown head, and a yellowish green body, with blackish stripes running lengthwise, and blackish dots, having fine white lines between, and white incisions and spiracles. Its length is about eight lines. When these caterpillars are numerous, they do important damage to the cabbage tribe and horse-radish. There is scarcely any other means of destroying them, than that of shaking them off, and burying them immediately.

PYRAMIDELLA. A genus of Mollusca, with pretty little spiral shells, found in the Indian and American seas; and of which there are both recent and fossil species. The shell is pyramidal, smooth, and polished; spire long, pointed, and composed of numerous whorls; outer lip somewhat expanded; columella tortuous, with several folds.

PYROCHROIDÆ. A family of small Coleopterous insects, found in the spring and early part of the summer. The front part of the body is narrow and flat, with the neck distinct, and the thorax sub-orbicular: they are generally gaily coloured and active in their flight; they frequent leaves and flowers, but the larvæ are found under the bark of trees and in rotten wood. The only British genus is *Pyrochroa*, distinguished by its pure red colour.

PYROMELANA. A genus of Grosbeaks, distinguished by the fine red and black colour of the plumage. [See GROSBEAK.]

PYROSOMA. The name given to certain compound Ascidians, remarkable for their brilliant phosphoric luminosity. [For further details see SUPPLEMENT.]

PYRULA. A common and numerous genus of Mollusca, chiefly found in the Indian Ocean and Red Sea. The shell is large and pear or fig-shaped; the spire short, and sometimes flattened; aperture wide, terminating in a long, open canal; outer lip thin; columella smooth; operculum horny. In the

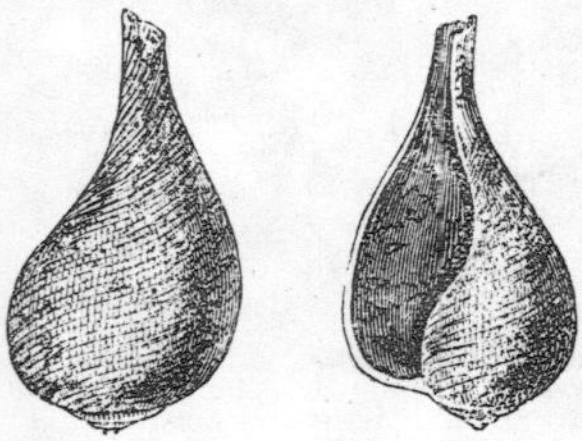

PYRULA FICUS

British Museum is a specimen of a *Pyrula bezoar* that appears to have grown with perfect regularity until the formation of its last half whorl, which is thrown considerably more than half an inch out of its proper position by a group of barnacles. These shells had probably attached themselves to the back of the *Pyrula* at an earlier stage, and as the latter had increased in size at length filled the place that should have been occupied by the inner lip, which, on meeting with this interruption, diverged from its course, and was thrown over the barnacles. Had the shell not been taken until a later period, there can be little doubt that the animal would have at length destroyed the barnacles, and completely hidden them from view, although it would appear that it had not the power to remove them by absorption while they retained their vitality.

PYTHON. A name given to the great constricting serpents of the Old World. The size to which the Pythons grow is fully equal to that attained by the *Boœ*, if it does not exceed it: some have been seen upwards of thirty feet long; and their strength is proportionate to their gigantic size. Indeed, a good idea of these reptiles may be gathered from the article BOA [which see]. At the same time it may not be amiss to give the reader an example of the Python's nature, as related by Mr. A. Adams, of H. M. S. Samarang, while in the Philippine Archipelago (March, 1844): "While lying in the truly delightful bay at this place [Manilla], a trifling incident occurred, showing the extreme vivacity, and rapidity of movement, in the larger serpents, even in those of the Boa tribe, especially when first captured. They are, indeed, then very different from those apathetic listless monsters one sees coiled up in blankets, at Zoological Gardens and in menageries. Sir Edward Belcher had a very beautiful specimen of the *Python Schneideri* presented to him, about twelve feet long, and having one day given it a chicken, the reptile, as usual, compressed it nearly to death, within the muscular folds of its body, when one of the bystanders, more tender-hearted than the rest, begged the life of the fowl. I had no sooner, however, introduced my arm with that benevolent intention, than throwing back its head, and unwinding its body from its prey, 'the spirited, sly snake,' as Milton would have termed it, darted at my hand with the greatest velocity, and held me fast with its teeth, by the ball of the thumb, nor was it without some trouble that I was able to extricate myself, owing to the fact that the long, sharp, curved teeth of a serpent all point backwards. Some time after this event the death-warrant of the poor reptile was sealed, and I appointed myself his executioner."

QUADRUMANA. The name of an order of Mammalia, characterized by the four limbs being each terminated by a *hand;* as the Ape, Baboon, &c. The term *Quadrumana*, or *four-handed*, does not, however, correctly apply to all the animals thus designated; the thumb on the hands, or fore extremities, being wholly or in part wanting in the monkeys of the Western Hemisphere. There are very marked differences among the tribes of this order, as to the degree in which they approach Man in their general conformation; some of them bearing a strong resemblance to him in structure, aspect, and gait; whilst others are but little removed from the ordinary Mammalia. In their food and habits of life, also, there is great variety. Some live solitarily or in pairs; but the larger proportion congregate together: some dwell on the ground or inhabit rocky heights, while others are altogether arboreal, and spend their lives among the branchy foliage of the forest. The three families or tribes which this order includes are thus distinguished: 1. SIMIADÆ, *Monkeys of the Old World;* 2. CEBIDÆ, *Monkeys of the New World;* and 3. LEMURIDÆ, *the Lemur tribe.* But the reader is referred to the words, APE, ORANG-OUTANG, BABOON, MONKEY, &c., for further information applicable to the particular species.

QUADRUPEDS. In this familiar term is comprehended a large and most important class of terrestrial animals; whose essential characters are—that their bodies are covered with hair; that they have four feet; that the females are viviparous, or bring forth their young alive; and that they suckle them.

Although the word Quadruped is not used, in a strict zoological sense, as indicative of a particular group of animals, yet a few observations under this head, though trite, appear to be not wholly uncalled for. Widely different from each other as many of the genera are, no one can fail to notice how admirably adapted they all are by Nature to fill their respective stations. Some have limbs formed to support a vast unwieldy frame, and possess neither flexibility nor beauty. The Elephant and the Rhinoceros have legs resembling pillars: they are not destined to pursue other animals for their support; and, conscious of their own superior strength, there are none which they study to avoid. Deer, Hares, and other animals whose safety depends on flight, have slender legs, and are so formed as to escape from their pursuers by superior swiftness. Beasts of prey have their feet armed with sharp and powerful claws; while, on the contrary, animals of peaceful habits generally have hoofs. It is also ordained that those animals which are large and formidable produce but few at a time, while such as are small are extremely prolific: were it otherwise, many of the former would perish for want of food, and life would be indulged without the necessary means of subsistence. Besides, did the Elephant or the Rhinoceros, the Tiger or the Lion, possess the same degree of fecundity with the Rabbit, all the arts of Man would be unequal to the contest; and he who now styles himself "lord of the creation," would soon become its most abject slave.

QUAGGA. (*Asinus Quagga.*) This animal is an inhabitant of the southern parts of Africa, and bears a great resemblance to the Zebra. It is less, however, than the Zebra, with the hinder parts higher, and the ears shorter. The head, neck, mane, and shoulders are blackish brown, banded with white; the ground colour gradually becoming paler, and the bands less distinct and diffused, as we proceed along the back towards the rump, which is grayish; the hind parts being rather spotted than striped. The dorsal line is black, margined on each side with a white

QUAGGA.—(ASINUS QUAGGA.)

line: belly, tail, and legs whitish; ears with two irregular black bands and white tip. The Quagga is a social animal, living in large troops, is much more tractable than the Zebra, and is said to be occasionally used at the Cape of Good Hope for domestic purposes. Notwithstanding this mildness of character when domesticated, it is exceedingly fearless in its native plains, and is even said to be more than a match for the Hyæna, fighting desperately both with its hoofs and teeth. Though it inhabits the same parts of Africa, it never associates with the Zebra. The Quagga has received a variety of names from authors; thus Pennant terms it the *Quacha*, Masson the *Opeagha*, and Sparrman the *Quagga;* a name, Mr. Gray observes, derived from its voice, which resembles the barking of a dog.

QUAIL. (*Coturnix vulgaris.*) This bird greatly resembles the Partridge, but is smaller, has a more delicate beak, and no spur on the legs. It is about seven inches and a half in length. Bill dusky, eyes hazel: the feathers of the head, neck, and back are a mixture of brown, ash, and black; the hinder part of the neck and crown of the head are divided by a long pale yellow line; the chin and throat are white, bounded by a black crescent, which is deepest in front; the breast is of a yellowish-red, spotted with black; the scapular feathers are marked by a light yellowish streak down the middle of each; quills lightish brown, with small rust-coloured bands on the exterior edges of the feathers; the breast is pale rusty, spotted with black, and streaked with pale yellow; the tail barred like the wings; belly and thighs yellowish white; legs pale brown. The female's plumage is less vivid, and the wing-coverts are barred with yellowish white. She lays from eight to a dozen eggs, of a yellowish colour, with dusky spots and blotches.

Quails are very generally diffused throughout Asia, Africa, and the southern parts of Europe, but are more rare in northern and temperate climates. In the British islands they are never abundant. They are migratory, and are seen in immense flocks flying across the Mediterranean, from Europe to the shores of Africa, in the autumn, and returning again in the spring, frequently alighting in their passage on the intervening islands. Such prodigious numbers have appeared on the western coasts of the kingdom of Naples, that a hundred thousand have been taken in a day within the space of four or five miles. From these circumstances (observes Bewick) it appears highly probable that the Quails which supplied the Israelites with food, during their journey through the wilderness, were driven thither on their passage to the north, by a wind from the south-west, sweeping over Ethiopia and Egypt towards the shores of the Red Sea. Quails are said to be very indolent birds; they usually sleep through great part of the day, concealed among the high grass, lying on their sides, with their legs extended, and should a dog approach, he must absolutely run in upon them before they are flushed. The males are birds of great courage, and their quarrels frequently terminate in mutual destruction. Quail-fighting was practised by the Greeks and Romans; and we are informed that Agustus punished a prefect of Egypt with death, for bringing to his table one of these birds which had acquired celebrity by its victories! The Chinese are much addicted to the *sport;* and it is said to be also a common practice in some parts of Italy.

The CHINESE QUAIL (*Coturnix excalfactoria*) is an elegant little species, measuring only four inches in length. The male has a triangular deep black spot on the throat: from the base of the beak extends a white whisker, surrounded by black; below which is a pure white gorget, bordered with black: the forehead, breast, and sides are of a lead colour, the latter marked with black bands: the middle of the belly, the thighs, and the abdomen are red chestnut; the upper parts of the body and the tail-coverts are a grayish-brown, varied with black spots; and most of the feathers have whitish shafts: the wings are gray-brown, the greater coverts lead colour, and the whole tipped and fringed with chestnut: the beak is black; the feet and the claws are yellow. In the female, the cheeks, the forehead, and a stripe above the eyes, are of a bright red; the throat pure white; the feathers of the head are dusky, tipped with gray; and a narrow longitudinal band extends over the middle of the crown from the forehead. The plumage of the back and rump is red, with black spots, and longitudinal reddish-white dashes: the scapulars and wing-coverts are gray-brown, marked with delicate black undulated lines, and many black spots on their inner webs; the breast, sides, thighs, and abdomen, are bright gray, striped transversely with black; the wings gray-brown; the beak brown; and the legs yellow. This bird is abundant in the Manilla and Philippine Islands, and in China it is amazingly numerous. There they are kept in cages, for the singular purpose of warming the hands of their owners in winter: they also rear them for the purpose of fighting.

Several other species, in appearance and habits not greatly differing from the common Quail, are known; as the New Holland Quail (*Coturnix Australis*); the White-throated Quail (*Coturnix torquata*), &c.

QUERQUEDULA, or **TEAL**. A genus of web-footed birds, containing the COMMON TEAL (*Querquedula crecca*), the GARGANEY (*Q. circia*), and other species. [See DUCK: TEAL.]

QUISCALUS. A genus of birds allied to the Starlings, and indigenous to America. Of these we may particularly describe the QUISCALUS VERSICOLOR, or PURPLE GRACKLE. We are told by Wilson that this "noted depredator" is well known to every careful farmer of the northern and middle states. About the 20th of March (he says) the Purple Grackles visit Pennsylvania from the south, fly in loose flocks, frequent swamps and meadows and follow in the furrows after the plough; their food at this season consisting of worms, grubs, and caterpillars, of which they destroy prodigious numbers, as if to recompense the husbandman beforehand for the havoc they intend to make among his crops of Indian corn. Towards evening they retire to the nearest cedars and pine trees to roost, making a continued chattering as they fly along. On the tallest of these trees they generally build

PURPLE GRACKLE.
(QUISCALUS VERSICOLOR.)

their nests in company, about the beginning or middle of April; sometimes ten or fifteen nests being on the same tree. One of these nests, taken from a high pine tree, is now before me. It measures full five inches in diameter within, and four in depth; is composed outwardly of mud, mixed with long stalks and roots of a knotty kind of grass, and lined with fine bent and horse hair. The eggs are five, of a bluish olive colour, marked with large spots and straggling streaks of black and dark brown, also with others of a fainter tinge. They rarely produce more than one brood in a season. The trees where these birds build are often at no great distance from the farm-house, and overlook the plantations. From thence they issue in all directions, and with as much confidence, to make their depredations among the surrounding fields, as if the whole were intended for their use alone. Their chief attention, however, is directed to the Indian corn in all its progressive stages. As soon as the infant blade of this grain begins to make its appearance above ground, the Grackles hail the welcome signal with screams of peculiar satisfaction, and, without waiting for a formal invitation from the proprietor, descend on the fields and begin to pull up and regale themselves on the seed, scattering the green blades around. While thus eagerly employed, the vengeance of the gun sometimes overtakes them; but these disasters are soon forgotten, and those

—— who live to get away,
Return to steal another day.

About the beginning of August, when the young ears are in their milky state, they are attacked with redoubled eagerness by the Grackles and Redwings, in formidable and combined bodies. They descend like a blackening, sweeping tempest on the corn, dig off the external covering of twelve or fifteen coats of leaves, as dextrously as if done by the hand of man, and, having laid bare the ear, leave little behind to the farmer but the cobs, and shrivelled skins, that contained their favourite fare." About the middle of November, it appears, they move off towards the south, their winter residences being North and South Carolina, Georgia, &c. "Here numerous bodies, collecting together from all quarters of the interior and northern districts, and darkening the air with their numbers, sometimes from one congregated multitude of many hundred thousands. A few miles from the banks of the Roanoke, on the 20th of January, I met with one of those prodigious armies of Grackles. They rose from the surrounding fields with a noise like thunder, and, descending on the length of road before me, covered it and the fences completely with black, and when they again rose, and, after a few evolutions, descended on the skirts of the high-timbered woods, at that time destitute of leaves, they produced a most singular and striking effect; the whole trees for a considerable extent, from the top to the lowest branches, seeming as if hung in mourning; their notes and screaming the meanwhile resembling the distant sound of a great cataract, but in more musical cadence, swelling and dying away on the ear, according to the fluctuation of the breeze. In Kentucky, and all along the Mississippi, from its juncture with the Ohio to the Balize, I found numbers of these birds, so that the Purple Grackle may be considered as a very general inhabitant of the territory of the United States." That they are great destroyers of corn, there can be no doubt; but it must not be forgotton that they are also particularly destructive to almost all the noxious worms, grubs, and caterpillars, that infest the fields, which, were they allowed to multiply unmolested, would soon consume nine-tenths of all the production of the cultivator's labour, and desolate the country with the miseries of famine. The Purple Grackle is twelve inches long and eighteen in extent; on a slight view seems wholly black, but placed near, in a good light, the whole head, neck and breast, appear of a rich glossy steel blue, dark violet, and silky green; the wings and other parts of the plumage reflecting these and various other glosses in a greater or less degree.

RABBIT. (*Lepus cuniculus.*) This animal belongs to the *Leporidæ*, or Hare tribe, and is a native of most of the temperate and warmer parts of the old continent, but is not found very far north; neither was it originally a native of Britain, but is said to have been introduced from Spain. In structure the Rabbit very much resembles the Hare, but may be readily distinguished from it by its smaller size, its shorter ears and hind legs, and the absence of the black tip to the ears. In its habits it is extremely different from that animal; being unable to outstrip its enemies in the chase, it seeks its safety and

RABBIT.—(LEPUS CUNICULUS.)

finds shelter by burrowing in the ground; and, instead of leading a solitary life, its manners are eminently social. The fecundity of Rabbits is truly astonishing: they will breed seven times in one year, and perhaps bring forth eight each time; and, on a supposition that this happens regularly for four years, a single pair would in that time multiply to 1,274,840. We should, however, add that although this is *possible*, such extraordinary fertility is not very probable. When the time of parturition draws near, the female forms a separate burrow, more intricate than the ordinary one, and lines it at the bottom with a part of her own fur: the young are born blind, and very scantily covered with hair; and for nearly six weeks she continues to suckle them. During this period the female is seldom visited by the male; but as soon as the little progeny are capable of coming abroad he seems anxious to acknowledge and caress them.

"In sandy heaths, coveredwith large bushes of furze," says Mr. Bell, "Rabbits often multiply to a great extent; as the soil is easily removed, and the dense furze affords a secure cover to their retreat, and a wholesome, ready, and never-failing food; for the young tops of the plants are found constantly eaten down, and the bushes present the appearance of a solid mass, with the surface even and rounded, as far as the Rabbits can reach them standing on their hinder legs. They make extensive inroads, however, upon corn-fields and plantations, in which they do considerable mischief by devouring the newly-sprung corn, and barking the young trees. They generally retire within their burrows during the day, coming abroad about twilight to feed. . . . . The rapid multiplication of the Rabbit would soon render it one of the greatest scourges of our agriculture, were it not, on the one hand, destroyed by numerous birds and beasts of prey, and on the other, sought by man as an article of food, and on account of its fur, which is used for various purposes. The supply for this latter object would, however, be wholly unequal to the demand, were our furriers dependent upon the produce of our country only. Hundreds of thousands of Rabbit-skins are annually imported here from Germany, and other parts of the northern and middle districts of Europe, where myriads of Rabbits are bred for this purpose."

In its wild state the colour of the Rabbit's fur is grayish brown, paler or whitish on the under parts; its tail black above, and white beneath; but when domesticated, as every one knows, it varies greatly in colour; being gray, reddish-brown, or black, more or less mixed with white; and often perfectly white. In England, Rabbits are reared either in warrens or in hutches: the best situations for the former are sandy hills, on which the juniper is thickly planted, as the leaves of this shrub are eagerly eaten by these animals, and impart a delicate and aromatic flavour to the flesh. The cleanliness of hutches should be particularly attended to; otherwise their inhabitants will be sickly, and Rabbit-breeding turn out a losing speculation. The ingenuity of Rabbit-fanciers has been shown in the production of various breeds, chiefly remarkable for the excessive length of their ears; and we occasionally see them exhibited of such an enormous size and fatness, as to be well entitled to the appellation of "prize cattle." The Half-lop, the Oar-lop, the Perfect-lop, &c., are names by which these varieties are distinguished.

RABBIT-FISH. A local name for the Northern Chimæra, or King of the Herrings. [See CHIMÆRA.]

RACOON. (*Procyon lotor.*) This Plantigrade carnivorous animal is a native of America, and chiefly found in the northern parts of that continent: it is also met with in some of the West Indian islands. Its average length is about two feet from the nose to the tail, and the tail about ten inches. The head somewhat resembles that of the Fox, the forehead being broad and the nose sharp, but the ears are short and slightly rounded: the body is broad, the back arched, the limbs rather short, and the fore legs shorter than the hinder. Its colour is grayish-brown, with a dusky line running from the top of the head down the middle of the face, ending below the eyes. The tail is very thickly covered with hair, and is annulated with several black bars, on a yellowish-white ground. There are, however, several varieties as regards colour. In the wild state the Racoon is savage and sanguinary, committing great slaughter among both wild and domesticated birds, as it always destroys a great number without consuming any part of them except the head, or the blood which flows from their wounds; in this particular resembling the Polecat. It will also occasionally commit ravages in plantations of sugar-cane or of Indian corn, especially while the latter is young: it also feeds on various kinds of fruit, and is said to devour birds and their eggs, on which account it has the reputation of being destructive to poultry. This animal is a good climber, and the form of its claws enables it to adhere so firmly to a branch of a tree, that it requires no slight exertion of strength to disengage it. It chiefly feeds by night, keeping in its hole during the day, except in dull weather: it has a kind of oblique gait in walking; can leap and climb with great ease, and is very frequently seen on trees. In the domesticated state it is extremely restless and inquisitive, examining everything; will live on bread, milk, fish, eggs, &c.; is particularly fond of sweets of every kind, and has as great a dislike to acids. Captivity, however, produces considerable changes in the habits of the Racoon; for instead, as in a state of nature, of sleeping during the day, and roaming about at night in search of food, it will learn to be active during the day, and to remain quiet at night. In eating, it commonly sits on its hind legs, and uses its fore feet like a squirrel. One of its most marked peculiarities, and on which its specific name of *lotor*, or the washer, is founded, is its habit of plunging its dry food into water before eating it. It is extremely expert in opening oysters, on which, as well as on crabs and other crustacea, it frequently feeds. Although when tamed it is noted for its active and playful habits, it is capricious, and not easily reconciled when offended. In its wild state it generally inhabits the hollows of trees; but when domesticated, it shows no particular inclination for warmth. When inclined to sleep, it rolls itself up into a kind of ball; and in this position it sleeps so profoundly as not to be easily disturbed. The female has from two to three young at a birth; which usually takes place in May. The fur of the Racoon is valuable, particularly in the manufacture of hats, and forms no inconsiderable article of commerce.

RADIATA. A term given to a subdivision of the Animal Kingdom which includes all those animals in which, as in the Star-fish, Sea Anemone, &c., there is a regular disposition of similar parts around a common centre. Their organs of motion, when they have any, are movable spines attached to the skin, or flexible papillæ, capable of inflation. They have no true system of circulation, and their nervous system is always obscure, and sometimes cannot be traced. Some have a mouth and vent, others only one opening, and others appear to be nourished through pores. Some are of distinct sexes; some bisexual, and some are produced by buds or division. Many grow in clusters upon stalks, or Polypidoms — dwellings of polypi, — which are sometimes leathery or horny, and sometimes calcareous.

RAIL. (*Rallus.*) A genus of Wading-birds, of which there are many species. They are distinguished by a very compressed form of body, with wings of a middling length, rounded, and the first quill shorter than the second, third, or fourth. They seldom fly, but run or swim with celerity: they frequent large ponds or lakes, the borders of which are well clothed with plants; and they subsist on vegetables and seeds, as well as on insects, snails, and worms.

The Common European Water Rail (*Rallus aquaticus*) is nearly twelve inches long. It has a red beak shaded with brown at the tip; irides orange; throat whitish: the sides of the head, neck, breast, and belly are of an ashy lead-colour: all the feathers on the upper parts of the body are reddish-brown, with a deep black mark in the centre of each; the flanks are deep black, transversely rayed with white bars; the under tail-coverts are white; the legs lead-coloured. The young of the year have the middle of the belly of a brown-red, and are destitute of the white band on the sides.

This bird is not very common in Britain, though it is found throughout the country, and continues with us all the year: it is said to be very numerous in the northern countries of Europe, migrating southward during the severity of winter: it is very abundant also in Germany, France, and Holland. It is shy and solitary in its habits, resorting to low damp situations overgrown with sedges, reeds, and coarse herbage, among which it shelters, and is seldom put to flight unless pressed by the dogs, rather depending on its legs for safety; but when once flushed it is easily shot, as it flies in a heavy and awkward manner, with its legs hanging down: it runs, however, very fast, and frequently flits up its tail. Though it swims, and even dives well occasionally, it delights most in shallow water, where it can wade through without swimming. Its nest is constructed of sedges and coarse grass amongst the thickest aquatic plants; and it lays from six to ten eggs, of a yellowish colour, spotted with red-brown.

The Virginian Rail (*Rallus Virginianus*) very much resembles the European Water Rail described above; but it is smaller, and has none of the slate or lead colour on the breast. It feeds more on animal and less on vegetable food than the common and more numerous species known as the Clapper Rail. During the months of September and October, when the reeds and wild oats swarm with the latter, feeding on their nutritious seeds, there are but few of the Virginian Rail to be met with. The food of this species consists of small snails, worms, and the larvæ of insects, which it extracts from the mud; hence the cause of its greater length of bill, to enable it the more readily to search its food. In most of its habits, its thin compressed form of body, its aversion to take wing, and the dexterity with which it runs or conceals itself among the grass and sedge, are exactly similar to those of the common Rail. The Virginian Rail is migratory, never wintering in the northern or middle States. It makes its first appearance in Pennsylvania early in May, and leaves the country on the first smart frosts, generally in November. They frequent those parts of the salt marshes only where fresh-water springs rise through the bogs, and in these places the female usually constructs her nest. The usual number of eggs is from six to ten: they are shaped like those of the

AMERICAN RAIL.—(RALLUS VIRGINIANUS.)

domestic hen, and are of a dirty white or pale cream-colour, sprinkled with specks of reddish or pale purple, most numerous near the great end. This species is ten inches in length: bill, dusky red; cheeks and stripe over the eye, ash; over the lores and at the lower eyelid, white; crown and whole upper parts, black, streaked with brown, the centre of each feather being black; wing-coverts, hazel-brown; quills, plain deep dusky; chin, white; throat, breast, and belly, orange-brown; sides and vent black, tipped with white; leggs and feet, dull red-brown; edge of the bend of the wing, white.

RALLIDÆ. A family of birds (the Rails, Gallinules, Water-hens, &c.), chiefly distinguished by their long and slender toes, often with a membranous margin along their sides; by means of which, and their generally compressed bodies, they are not only enabled to support themselves on the aquatic herbage which is seen floating on the surface of the water, but to move with great facility through high grass, bulrushes, and other closely-set herbage. Mr. Swainson describes them "for the most part as solitary and timid birds, hiding themselves at the least approach of danger, but quitting their semi-aquatic retreats in the morning and evening, to feed in more open spots: their flight, from the shortness of their wings, is very feeble, but they run with swiftness; and by the peculiarly compressed form of their body, are able to make their way through dense masses of reeds and high grass with so much facility as to escape even after being desperately wounded. The flesh of all these birds is delicate; and from living chiefly upon aquatic seeds and vegetable aliment, they may be considered as aquatic *Gallinacea*." The Jacanas and Screamers of tropical climates are often placed in this family; their general structure and habits rendering such an arrangement quite natural.

RAMPHASTIDÆ. A family of birds found in tropical America. [See TOUCAN.]

RAMPHORHYNCHUS. [See SUPPLEMENT.]

RANA. [See FROG.]

RANELLA. A genus of Mollusca, whose shells are for the most part covered with tubercles and granulations, and which, from the colour and squat shape of some of the species, have been likened to frogs (*rana*), whence their name. The shell is ovate or oblong, depressed, and thick, with two rows of varices situated at the distance of half a whorl from each other, and longitudinally united, forming a continued ridge on each side of the shell. They are mostly from the Indian seas. A few fossil species occur in the London clay.

RAPHIDIA: RAPHIDIIDÆ. A genus and family of Neuropterous insects, which are of comparatively small size, and of active habits; the structure of the head and neck, powerful jaws, and the elongated coxæ of the legs, as well as the membranous attachment of the segments of the body, indicating predaceous habits. They are chiefly found in the neighbourhood of woods and streams; and from the form of the head and neck, and the facility with which they turn the front of the body in different directions, they have received the English name of Snake-flies. The wings are moderately large, strongly veined, and of nearly equal size, the posterior not folded when at rest, when they are deflexed at the sides of the body; the maxillæ and labium are distinct, with short palpi; the legs are short; the abdomen is unarmed; the eyes resemble ocelli, and are situated near the base of the antennæ. The pupa resembles the perfect insect in general form, but is furnished only with short rudiments of wings, lying at the sides of the body, and is not inclosed in a cocoon.

RAPTORES. The name given to an order of Birds—*Birds of prey*. They are characterized by a strong, curved, sharp-edged, and sharp-pointed beak, suitable for tearing the flesh of the animals they devour; their legs are short and robust; and their toes, equally vigorous, are furnished with strong hooked talons, by which they seize their prey. Considerable strength is also indicated by the general formation of the body; and their whole appearance bespeaks a ferocious character. Some are distinguished by their dense plumage, and by the lateral direction of their eyes; as the Vultures, Falcons, Eagles, Hawks, Buzzards, &c. Others are characterized by their loose plumage, and by the anterior direction of their full round eyes: these are nocturnal, and constitute the family of Owls; differing

A RAPTORIAL BIRD.

from the former kinds by their obtaining their food rather by the stealthiness with which they approach it, than by the vigour with which they attack it. The *Raptores* always associate in pairs, the same males and females continuing to live together—not pairing anew every season, as is the case with many of the feathered tribe belonging to other Orders. They generally build their nests in the loftiest situations, and are totally destitute of the power of song. The young birds are long dependent upon their parents for support, which chiefly devolves upon the female.

RASORES. An Order of Birds (called also *Gallinaceæ*). They have strong feet, provided with obtuse claws for scratching up grains, seeds, &c., of which their food principally consists. Their bodies are for the most part bulky, and their legs strong: but their wings not being of a size proportionate to their bulk, their powers of flight are inconsiderable. The beak is usually arched, and surrounded at the base with a soft skin, in which the nostrils are pierced. They are polygamous, the male taking no part in the construction of the nest, or in the nurture of the young. Generally speaking, the Birds of this order are easily domesticated; they multiply with great rapidity; and as they furnish Man with a large quantity of wholesome and delicate food, they are justly entitled to his especial regard. Most of them fly badly, do not perch on trees, and seek their food on the ground. [See GALLINACEÆ.]

RAT. (*Mus decumanus.*) The BROWN RAT is a large, destructive, and very prolific species of the genus *Mus*, originally brought to Europe from Asia, and not, as is commonly though erroneously supposed, imported into this country from Norway. But from whatever country it might have originally come, it is now generally distributed throughout every quarter of the globe. The length of the head and body is about ten inches, and of the tail eight; the head, back, and sides are of a light brown colour, mixed with tawny and ash; the breast and belly are a dirty white; the feet are naked, and of a dull flesh colour, the fore ones being furnished with four toes, and a claw instead of the fifth. Whenever it conveniently can, the Rat forms its hole very near the edge of the water, where it chiefly resides during the summer, feeding on small animals, fish, and grain: it also haunts the corn-fields, where it makes burrows, and breeds. When winter approaches, it draws near some farm-house, and burrows in the corn, where it consumes much, but wastes more. It destroys rabbits, poultry, and all sorts of game; and scarcely any of the feebler animals can escape its rapacity. "Its astonishing fecundity," Mr. Bell observes, "its omnivorous habits, the secresy of its retreats, and the ingenious devices to which it has recourse, either to retain its existing place of abode, or to migrate to a more favourable situation, all conduce to keep up its almost overwhelming numbers. It digs with great facility and vigour, making its way with rapidity beneath the floors of our houses, between the stones and bricks of walls, and often excavating the foundations of a dwelling to a dangerous extent. There are many instances of their fatally undermining the most solid mason-work, or burrowing through dams which had for ages served to confine the waters of rivers and canals."

Mr. Waterton, to whom in the course of this volume we have so often been indebted both for zoological facts and interesting anecdotes, has given his readers some "Notes on the History and Habits of the Brown or Grey Rat," which are not the less *genuine* for being rather whimsical. "It is known to naturalists," says he, "sometimes by the name of the Norwegian, sometimes by that of the Hanoverian, Rat. Though I am not aware that there are any minutes in the zoological archives of this country which point out to us the precise time at which this insatiate and mischievous little brute first appeared among us; still, there is a tradition current in this part of the country, that it actually came over in the same ship which conveyed the new dynasty to these shores. [By the way, Mr. W., like a true Jacobite, as he professes himself to be, can never forgive the *new dynasty*, or forget the *old*,—but surely neither we nor our readers have any right to quarrel with him for the consistency with which he espousesthe claims of the Stuarts, or the heartiness with which he anathematizes those who upset them; nay, his indignation at times appears to us so honest and original, as to heighten rather than to detract from his merits as a popular writer.] My father," continues he, "who was of the first order of field naturalists, was always positive on this point; and he maintained firmly, that it *did* accompany the House of Hanover in its emigration from Germany to England. Be this as it may, it is certain that the stranger Rat has now punished us severely for more than a century and a quarter. Its rapacity knows no bounds, while its increase is prodigious beyond all belief. But the most singular part of its history is, that it has nearly worried every individual of the original Rat of Great Britain. So scarce have these last-mentioned animals become, that in all my life I have never seen but one single solitary specimen; it was sent, some few years ago, to Nostell Priory, in a cage, from Bristol; and I received an invitation from Mr. Arthur Strickland, who was on a visit there, to go and see it. Whilst I was looking at the little native prisoner in its cage, I could not help exclaiming—'Poor injured Briton! hard indeed has been the fate of thy family! in another generation at farthest it will probably sink down to the dust for ever. Vain would be an attempt to trace the progress of the stranger Rat through England's wide domain, as the old people now alive can tell nothing of its coming amongst them. No part of the country is free from its baleful presence: the fold and the field, the street and the stable, the ground and the garret, all bear undoubted testimony to its ubiquity and to its forbidding habits. After dining on carrion in the filthiest sink, it will often manage to sup on the choicest dainties of the larder, where, like Celæno of old, 'vestigia fœda relinquit.' We may consider it saddled upon us for ever. Hercules himself, could he return to earth, would have his hands full, were he to attempt to drive this harpy back again to Stymphalus. It were loss of time to dwell on its fecundity. Let any body trace its movements in the cellar, the dairy, the outhouse, and the barn, and he will be able to form some notion of the number of hungry mouths which we have to fill. Nine or ten young ones at a time, twice or thrice during the year, are an enormous increase, and must naturally recall to our minds one of the many plagues which formerly desolated the fertile land of Egypt. In the summer months it will take off to the fields, and rear its young amongst the weeds which grow in the hedgerows; plundering, for their support, the birds' nests with a ferocity scarcely conceivable in so small an animal. Man has invented various instruments for its destruction; and what with these, and with poison, added to the occasional assistance which he receives from his auxiliaries, the cat, the dog, the owl, the weasel, the ferret, and the foumart, he is enabled, in some degree, to thin its numbers, and to check its depredations."

The Black Rat. (*Mus rattus.*) The Old English or Black Rat was, previously to the introduction of the Brown Rat, just described, as numerous and perhaps as extensively distributed as that species has since become; it is, however, smaller and weaker; and hence we may account for its almost total extinction by its more powerful enemy. Its length from the nose to the tail is about seven inches, and the tail nearly eight, almost bare, and covered with numerous rings and scales. The nose, which is sharp-pointed, is furnished with long whiskers; the colour of the head, and the whole upper part of the body, is a deep iron gray, bordering on black; the throat a·d belly are of a dirty white; and the feet and legs are of a dirty pale flesh-

BLACK RAT.—(MUS RATTUS.)

colour, almost destitute of hair. The ears are rounded, long, and naked; the eyes large; the feet plantigrade, with five toes on each; but the thumb on the anterior pair is concealed within the skin, except the terminal joint, with its claw. In its habits it resembles the Brown Rat, both in respect to its destructive propensities and its amazing fecundity; and in warmer climates, where there is no winter to interrupt their breeding, or to diminish their supply of food, the multiplication of this species, as well as of the Brown Rat, is enormous. Like most of the genus, it can hold its food in the fore paws whilst eating, and drinks by lapping. The Rat is a cleanly animal; and as it occupies the greater part of its time in cleaning itself, its skin is ordinarily kept in excellent order.

The WATER RAT, or WATER VOLE (*Arvicola amphibia*) belongs to another group of *Rodentia*, but may be as well described here. This animal is found in most parts of Europe; frequenting the banks of rivers, ditches, and ponds; excavating its habitations to a considerable distance, and breeding in the burrows it has thus formed. It

WATER RAT.—(ARVICOLA AMPHIBIUS.)

is not at all carnivorous, its food consisting entirely of roots, subaquatic plants, and other vegetable substances; yet, from its being confounded with the common Brown Rat, it is by no means unusual to hear it asserted that it destroys young ducks, small fish, frogs, &c. It is an expert swimmer and diver, instantly seeking the water upon every alarm, and plunging at once to the bottom; where, however, it can remain only for about a minute at a time without coming to the surface for respiration. This animal is nearly as large as the Brown Rat, but has a larger head, a nose more blunt, and smaller eyes: its ears are very short, and almost hid in the fur; and the tip of its tail is whitish: the cutting teeth are of a deep yellow colour in front, very strong, and much resembling those of the Beaver. Its head and back are covered with long black hair, and its belly with iron gray. Tail more than half the length of the body, covered with hairs. Fur thick and shining; of a rich reddish brown, mixed with gray above, yellowish gray beneath. The female produces a brood of five or six young once (and sometimes twice) a year.

RATEL. (*Ratellus mellivorus.*) The name given by the Hottentots to an animal of the weasel-kind which inhabits the country near the Cape of Good Hope, and is celebrated for the destruction it makes among the nests of the wild bee, to the honey of which it is very partial, and in the discovery of these nests it is said to be assisted by the actions and voice of a bird, called the Honey-guide. It has a blunt black nose; no external ears,

RATEL.—(RATELLUS MELLIVORUS.)

but a small rim round the orifice; a rough tongue; short legs, and very long claws. The colour of the forehead, crown, and whole upper part of the body, is a cinereous gray; the cheeks, throat, breast, belly, and limbs are black; and a dusky line extends from each ear to the tail along the sides, beneath which there is another of gray. It has a remarkably tough and loose skin, with thick hair. Its length from the nose to the tail is forty inches, and the tail is twelve.

RATTLESNAKE. (*Crotalus horridus.*) One of the most deadly of poisonous serpents, sometimes found as thick as a man's leg, and six feet in length; but more usually from four to five feet long. Till the discovery of the Western Hemisphere the knowledge of these Serpents was concealed from the rest of the world, and naturalists then first beheld with amazement a reptile of the most fatal nature, furnished, as if by a peculiar institution of Providence, with an instrument capable, in general, of warning mankind of their danger in too near an approach. There are several species, two of which are well distinguished, viz. the *Crotalus horridus* (or Banded Rattlesnake) of the United States, and the *Crotalus durissus* of Guiana. The former is of a yellowish brown colour, marked throughout its whole length with several transverse and somewhat irregular fasciæ of deep brown, and from the head to some distance down the neck run two or three longitudinal stripes of the same colour: the head is large, flat, and covered with small scales; the rest of the upper parts with moderately large oval ones, all furnished with a prominent line down the middle: the under parts are of a dingy yellowish brown colour, marked here and there with numerous dusky variegations and freckles: at the extremity of the tail is situated the *rattle*, consisting of several hard, dry, bony processes. It consists, in fact, of a number of hollow, hard, dry, and semitransparent bones, nearly of the same size and figure; resembling in some degree the shape of the human *os sacrum;* for although only the last or terminal one seems to have a rigid epiphysis joined to it, yet have every one of them the like; so that the tip of every uppermost bone runs within two of the bones below it; by which they have not only a movable coherence, but also make a more multiplied sound: each bone hitting against two others at the same time. The rattle is placed with the broad part perpendicular to the body, and not horizontal; and the first joint is fastened to the last vertebra of the tail by means of a thick muscle under it, as well as by the membranes which unite it to the skin: all the remaining

RATTLESNAKE.—(CROTALUS HORRIDUS.)

joints are so many extraneous bodies, as it were, or perfectly unconnected to the tail by any other means than their curious insertions into each other. These bony rings increase in number with the age of the animal, and it is said that it acquires an additional one at each casting of the skin.

The habits of the Rattlesnake are sluggish; they move slowly, and only bite when provoked, or for the purpose of killing their prey. They have two kinds of teeth, viz. the smaller, which are seated in each jaw, and serve to catch and retain their food; and secondly, the fangs or poisonous teeth, which kill the prey, and are placed *without* the upper jaw. They feed principally upon birds, squirrels, and other small animals, which it is believed they have the power of fascinating. Whatever may be the nature of this power, it is certain that its effects on the little animals are irresistible When the piercing eye of the Rattlesnake is fixed on them, terror and amazement render them incapable of escaping; and, while involuntarily keeping their eyes fixed on those of the reptile, birds have been seen to drop into its mouth, as if paralyzed, squirrels descend from their trees, and leverets run into the jaws of the expecting devourer. They love to reside in woods and on lofty hills, especially where the strata is rocky or chalky. Being slow of motion, they also frequent the sides of rills, where frogs, &c. resort. They are generally found during summer in pairs; in winter collecting in multitudes, and retiring under ground, beyond the reach of frost. The Rattlesnake is viviparous, producing its young (generally about twelve in number) in the month of June; and it is said to practise the same extraordinary mode of preserving them from danger which is attributed to the Viper in Europe, viz., of receiving them into its mouth, and retaining them in its stomach till the danger is past, when they issue forth again uninjured. It is well known that in the Western States of North America, where Rattlesnakes are plentiful, the hogs kill and eat them; nor is their bite formidable to their swinish enemy, on whom its venomous fangs seem to produce no effect. It is owing to this well-known fact, that families resident in those districts conceive that hog's lard must be a kind of antidote to their poison, and frequently use it, I believe, successfully, as a remedy.—*Murray.*

The STRIPED RATTLESNAKE (*Crotalus durissus*) may be distinguished from the preceding by the different disposition of its colours, being of a deep brown above, with pale yellow streaks, forming a continued series of large rhombs or lozenges down the back, the stripes growing less distinct as they descend on the sides. The neck is marked by a longitudinal streak on each side, and the under parts of the body are of a dusky yellowish brown, with numerous small dark spots and patches. It is a native of the same parts of America as the one already described; resembling it also in size and general proportions, as well as in the fatal effects of its bite.

There is also the WOOD RATTLESNAKE (*Crotalus dryinus*), which is of a paler colour than either of those previously mentioned, and more particularly distinguished by its having a much longer rattle. And the GROUND RATTLESNAKE (*Crotalus miliarius*), a small species, inhabiting the Southern and Western States of America. It has but two or three rattles on the tail, and is much dreaded, as its small size, and the slight noise of its rattle, render it more liable to be overlooked.

RAVEN. (*Corvus corax.*) Of all the corvine birds this is the largest European species; its general length being about two feet two inches. The bill is strong and black, covered with hairs or bristles, and the upper mandible is convex: colour of the whole bird is black, finely glossed with blue, except on the belly which is of a dusky hue. In times of ignorance and superstition the Raven was regarded as a bird of ill omen, announcing, by its croaking, impending calamities; and of such vast importance was it considered, that the various modulations of its voice were studied with the most careful attention.* It is proverbially long-lived, and is supposed sometimes to attain the age of a hundred years. Its favourite food is carrion, which it scents at a great distance; it will also destroy rabbits, young ducks, and chickens; nay, it has been known to seize on young lambs, and even sheep when sick and weak, and pick out their eyes while yet alive; in short, the Raven is a most voracious plunderer; and whether his prey be living or dead he greedily devours it. "Considered as a domestic bird, the Raven possesses many qualities which render him extremely amusing: active, curious, and impudent, he goes every where; pries into every thing; runs after dogs; plays tricks with poultry; and with great skill and address insinuates himself into the favour of the cook-maid, sensible of her ability to reward him for his attachment and attention." It has often been taught to pronounce a variety of words; and, being a crafty bird, it will frequently pick up things of value, and carry them to its hiding place. They build early in the spring, in trees and the holes of rocks, laying five or six eggs, of a pale bluish green, spotted with brown. The female sits about twenty days, her mate not only providing her with abundance of food during the time, but relieving her in turn, and taking her place in the nest.

Upon the fate of the Raven in modern times our old and oft-quoted friend thus feelingly apostrophises:—"Pity it is that the Raven, a bird of such note and consequence in times gone by, should be exposed to unrelenting persecution in our days of professed philanthropy. His noble aspect, his aërial evolutions, and his wonderful modulations of voice, all contribute to render him an ornament to any gentleman's park. He can scarcely be styled a bird of rapine, in the strict sense of the word; for, in the few inland parts of this country where he is still protected, we hear of no very alarming acts of depredation on his part. A stray chicken or so, during the time that he is obliged to feed his young—a rickety lamb which would never make mutton—a leveret started from her seat by the village mole-catcher—make up nearly the whole amount of a Raven's plunder." Again he says, "No bird in the creation exhibits finer symmetry than the Raven. His beautiful proportions, and his glossy plumage, are calculated to strike the eye of every beholder with admiration. He is by far the largest of the pie tribe in Europe; and, according to our notion of things, no bird can be better provided with the means of making his way through the world; for his armour is solid, his spirit unconquerable, and his strength surprising."

RAY. (*Raia.*) A genus of Cartilaginous fishes, distinguished by the remarkable breadth and thinness of their disc-shaped body, the pectoral fins appearing like a continuation of the sides themselves, being covered with the common skin: their rays are cartilaginous, straight, and furnished with numerous swellings or knots; the teeth are very numerous, small, and placed in ranges over the lips or edges of the mouth; the eyes are furnished with a nictitating membrane or skin, which can at pleasure be drawn over them like an eyelid; and at some distance above the eyes are situated the nostrils, each appearing like a large and somewhat semilunar opening edged with a reticulated skin: behind the eyes are likewise a pair of holes communicating with the mouth and gills. But the most distinguishing peculiarity of the Ray kind is their prickles, which the different species have on different parts of their bodies: some are armed with spines both above and below; others have them on their upper parts only; some have their spines at their tails; some have triple rows of them; while others have them single: in some species the spines are comparatively soft and feeble; but in others they are strong and piercing: and it is by these spines that the different species are distinguished. They in general feed on the smaller crustacea, testacea, marine insects, and fishes; lying concealed during part of the winter among the mud or sand, from which they occasionally emerge. When disturbed, they glide along in an undulating manner, with a slight motion of the pectoral fins; and if attacked, they defend themselves by lashing violently with the tail, which is often furnished with sharp spines. After these general observations on the genus, it will be necessary only to describe a few of the species.

The PAINTED RAY (*Raia microocellata.*) The Painted, or Small-eyed Ray, is described by Mr. Yarrell as the most beautiful of the British Rays in regard to the distribution of its colours. "The upper surface is a light gray, with a lighter line running along the back and middle of the tail, enclosing the central row of spines. The disc is regularly and beautifully quartered, first by three white lines enclosing each other, and passing from near the eye circularly to near the extremity of the expansion, the convexity of the arch inwards, and consequently the shorter line nearer the margin; on the hinder edge of the disc, formed by the pec-

* "I am no friend," observes the author of The Journal of a Naturalist, "to the superstition of converting natural transactions, or occasional events, into signs and indications of coming things; superstitions are wearing out, and shortly will waste away, and be no more heard of; but, I fear, in their place, deism, infidelity, impiety, have started up, the offspring of intuitive wisdom: the first belief arises from weakness and ignorance; the latter disbelief is ingratitude, pride, wickedness."

torals, are two other lines passing from behind the expansion circularly to the neighbourhood of the abdominal fins, the convexity of the arch inwards; on the more central part of the disc are a few whitish spots, those of both sides answering to each other; the extreme edge of the disc posterior to its greatest expansion, and also the abdominals, as well as the fin-like margin of the tail, are edged with white." Length thirty-three inches; breadth across the fins twenty-four: the eyes very small, three inches apart, and five inches and a half from the snout: the body covered with rough granulations, but altogether without spines, except a row that runs along two-thirds of the back, and down the middle of the tail to the fins; and an irregular row of similar hooked spines, extending along each side of the tail.

The STING RAY. (*Raia pastinaca.*) The shape of this fish is subrhomboidal, but somewhat approaching to ovate, snout pointed, and body rather convex: colour yellowish olive above, and whitish beneath: tail without fin, of considerable length, very thick at the base, and gradually tapering to the extremity, which is very slender: near the middle, on the upper part, it is armed with a very long, flattened, and sharp-pointed bone or spine, finely serrated in a reversed direction on both sides: with this the animal is capable of inflicting very severe wounds on such as incautiously attempt to handle it; and it answers the purpose both of an offensive and defensive weapon. It is annually cast; and as it frequently happens that the new spine has arrived at a considerable size before the old one has been cast, the fish is occasionally found with two, in which state it has sometimes been erroneously considered as a distinct species. This species, which is numbered among the edible Rays, is an inhabitant of the Mediterranean, Atlantic, and Indian seas. On account of the danger attending the wounds inflicted by the spine, it is usual with the fishermen to cut off the tail as soon as the fish is taken, and in some countries it is illegal to sell it before this has been done. The spine was formerly supposed to contain a most active poison; but that notion, like many others in zoology, equally erroneous, has long since been exploded.

The general habits of the Sting Ray are similar to those of the rest of the genus, often lying flat on the soft mud at the bottom of the shores which it frequents, and there seizing its prey by surprise; while at other times it pursues it through the depths of the ocean.

RAZOR-BILL. (*Alca torda.*) [See AUK.]

RAZOR-SHELL. [See SOLEN.]

RECURVIROSTRA. [See AVOSET.]

RED ADMIRAL [BUTTERFLY]. A name given by collectors to Butterflies of the species *Vanessa Atalanta.*

RED-BIRD, of SURINAM. (*Ampelis carnifex.*) [See CHATTERER, RED.]

RED-BIRD of CAROLINA. (*Cardinalis rubra.*) This bird is of the size of a Skylark: the bill is thick, strong, and of a palish red colour, with a black ring round the base; on the head is a crest, which it can raise and depress at pleasure; and the whole body is of a fine scarlet colour, except the back and tail, which are of a dark red. The hen is brown, with a reddish hue on the wings, &c. In America this bird is caged for its song as well as for its beauty.

REDBREAST. (*Rubecula familiaris.*) This well-known favourite song-bird, called also the Robin-Redbreast or simply the Robin, has a slender and delicate bill; large, black, and expressive eyes; and a mild familiar aspect: the head and all the upper parts are brown, tinged with greenish olive; the forehead, throat, and breast are of a fine deep reddish orange colour; the belly and vent dull white; and the legs dusky.

REDBREAST.—(RUBECULA FAMILIARIS.)

In spring the Redbreast retires to woods and thickets, where, with its mate, it prepares for the accommodation of its future family. The nest, constructed of moss and dried leaves, intermixed with hair and lined with feathers, is placed near the ground, by the roots of trees, and sometimes in old buildings, but always artfully concealed as much as possible. The female lays from four to eight eggs, of a dull white, with reddish spots. During the time of incubation, the male sits at no great distance, and makes the woods resound with his enlivening strains; while he exerts no common watchfulness in driving all intruders from his little settlement. As soon as the business of incubation is over, and the young are able to provide for themselves, he leaves his retirement, and again draws near the habitations of mankind: when the frost grows severe, and the snow covers the ground, he approaches the house, taps at the window with his bill, and solicits an admission, which is always cheerfully granted.

> "Half afraid, he first
> Against the window beats; then brisk alights
> On the warm hearth; then, hopping o'er the floor
> Eyes all the smiling family askance,
> And pecks, and starts, and wonders where he is:
> Till, more familiar grown, the table crumbs
> Attract his slender feet." THOMSON.

Most of the soft-billed birds, such as the famed Nightingale, and its congeners, leave us in the winter, when their insect food is no longer found in abundance; but the Redbreast continues with us the whole year; and endeavours to support himself in the dead of winter by entering those places from which the inclemency of the season is artificially expelled, and where insects, themselves attracted by a similar cause, are the most numerous.

Redbreasts are never seen in flocks, but always singly; and when all other birds associate together, they still retain their solitary habits. As soon as the young birds have attained their full plumage, they prepare for their departure from woods and thickets; but in thus changing their situation, they do not gather in flocks, but perform their journey singly, one after another; which, as Bewick has observed, is a singular circumstance in the history of this bird. It is worthy of note, also, that social as it is with the human race, it lives in a state of continued hostility with its own tribe, and has acquired a character for petulance and pugnacity which it well deserves. But where such universal favour is shown, there must surely be some cause for it; and whether its domestic qualities entitle it to our regard, or our kindlier sympathies have been first awakened by a legendary tale of the nursery, is of little importance, so long as either the one or the other serves to implant in the youthful breast a single humane or generous sentiment. "A favourite by commiseration, the Redbreast seeks an asylum with us; by supplication and importunity it becomes a partaker of our bounty in a season of severity and want; and its seeming humbleness and necessities obtain our pity; but it slights and forgets our kindnesses the moment it can provide for itself, and is away to its woods and its shades. Yet it has some little coaxing ways, and such fearless confidence, that it wins our regard; and its late autumnal song, in evening's dusky hour, as a monologue, is pleasing, and redeems much of its character."

To one of the poet Wordsworth's Sonnets, addressed to a "Wild Redbreast" which had pecked at his lip in the woods of Rydal, the author has appended the following characteristic note "The scene of the incident having been a wild wood, it may be doubted, as a point of natural history, whether the bird was aware that his attentions were bestowed upon a human, or even a living, creature. But a Redbreast will perch upon the foot of a gardener at work, and alight on the handle of a spade when his hand was half upon it — this I have seen. And under my own roof I have witnessed affecting instances of the creature's friendly visits to the chambers of sick persons, as described in the verses to the Redbreast, vol. i. p. 253. One of these welcome intruders used frequently to roost upon a nail in the wall, from which a picture had hung, and was ready, as morning came, to pipe his song in the hearing of the invalid, who had long been confined to her room. These attachments to a particular person, when marked and continued, used to be reckoned ominous; but the superstition is passing away."

REDBREAST, BLUE. (*Sialia Wilsoni.*) This bird, which is the *Motacilla Sialis* of Linnæus, is migratory, and makes its appearance in Carolina and Virginia very early in the spring, in flocks. It is about six inches long; beak dusky: the whole of the upper parts of its plumage are of a fine blue colour; the throat, fore part of the neck, breast, and sides, rufous; the belly and under tail-coverts white. The female is less brilliant in colour, and has the upper parts varied with brown. It feeds on insects; to procure which it frequents fields of maize and marshy places. It has only a slight plaintive note; and its nest is placed in the hole of a tree or wall.

RED DEER. [See Deer.]

RED-POLE, or RED-HEADED WARBLER. (*Sylvicola petechia*, or *æstiva.*) This bird inhabits Pennsylvania, where it makes its first appearance in March, and retires in the autumn. It has a black, slender, sharp-pointed bill; the top of the head is red; the upper parts of the body, from the head to the tail, olive green; the wings and tail dusky, with yellow edges: the under parts of the plumage are bright yellow, sprinkled on the breast and belly with red: the legs dusky. It frequents bushy places, and is a solitary species.

REDSHANK. (*Totanus calidris.*) This is an aquatic bird, about the size of the common Plover: the back is of a grayish or greenish brown colour, spotted with black; the neck is gray; the throat is variegated with black and white, with a few loose streaks of black; and the wing-feathers are a mixture of black, brown, and white. The bill is long, slender, and shaped like that of a woodcock, reddish at the base, and blacker lower down, and the legs are of a bright red. This bird breeds in fens and marshes, and is generally observed singly, or at most in pairs. When disturbed, it flies round its nest, making a noise like the Lapwing. It lays four eggs, of a whitish colour tinged with olive, and marked with irregular spots of black.

REED-BUNTING. (*Emberiza schœniclus.*) [See Bunting.]

REGENT BIRD. (*Sericulus chrysocephalus.*) A very beautiful bird belonging to the *Meliphagidæ* or Honey-eaters, found in the eastern portion of Australia, figured and described by Mr. Gould, in his celebrated work as one of the finest birds of the Australian Fauna, "which, when adorned in its gorgeous livery of golden yellow and deep velvety black, exhibits an extreme shyness of disposition, as if conscious that its beauty, rendering it a conspicuous object, might lead to its destruction." The plumage of the male bird is exceedingly rich and brilliant, but is not acquired until the second or third year. It is thus described: — Head and back of the neck, running in a rounded point towards the breast, rich bright gamboge yellow, tinged with orange, particularly on the centre of the forehead; the remainder of the plumage, with the exception of the secondaries and inner webs of all but the first primary, deep velvety black; the secondaries bright gamboge-yellow, with a narrow edging of black along the inner webs; the first primary is entirely black, the next have the tips and outer webs black —the half of the inner web and that part of the shaft not running through the black tip are yellow; as the primaries approach the secondaries, the yellow of the inner web extends across the shaft, leaving only a black edge on the outer web, which gradually narrows until the tips only of both webs remain black; bill yellow; irides pale yellow; legs and feet black. The plumage of the female is totally different, the prevailing colour being a dull brownish white on the head, throat, and breast, with the wings and tail pale olive-brown: the young males resembling them until they gradually change to the livery of the adult. Their food consists of ripe fruits, berries, and seeds. A few specimens of this bird were lately brought alive to this country, and were in the possession of Mr. Warwick.

REGULUS. A genus of Passerine birds, sometimes called *Kinglets*, and identical with or closely allied to the Wrens. The GOLDEN-CRESTED WREN (*Regulus cristatus*) is supposed to be the least of all European birds; being less than three inches and a half in length, and when stripped of its feathers the body is only about an inch

GOLDEN-CRESTED WREN.
(REGULUS CRISTATUS.)

long. The bill is slender and dark; eyes hazel; on the top of its head the feathers are of a bright orange colour, bordered on each side with black, which forms an arch above the eyes, and with which it sometimes conceals the crown, by contracting the muscles of the head; the upper part of the body is yellowish olive green; all the under parts pale reddish white, tinged with green on the sides; the greater coverts of the wings are dusky brown, edged with yellow, and tipped with white: legs yellowish brown. The female is distinguished by a pale yellow crown; and her whole plumage is less than that of the male. This delightful little fairy bird frequents the largest trees, such as oaks, elms, tall pines, and firs, particularly the first, in which it finds both food and shelter; in these it builds its nest, which is suspended like a hammock, from a branch by a kind of cordage made of the materials of which the nest is chiefly composed; it is of an oblong form, having an aperture on one side, and is made principally of moss, lined with the softest down, mixed with slender filaments: the female lays from six to a dozen eggs, scarcely larger than peas, which are white, sprinkled with very small dull-coloured spots. These birds are very agile, and are almost continually in motion, fluttering from branch to branch, creeping on all sides of the trees, clinging to them in every situation, and often hanging like the Titmouse. Their food consists chiefly of the smallest insects, which they find in the crevices of the bark of trees, or catch on the wing; they also eat the eggs of insects, small worms, and various sorts of seeds. The song of the Golden-crested Wren is said to be very melodious, but weaker than that of the common Wren; and it has besides a sharp shrill cry, somewhat like that of the Grasshopper. The species is diffused throughout Europe; it stays with us the whole year; and is able to bear great extremes of temperature. There are two other European species, the *R. ignicapillus* and *R. modestus*, the former of these is occasionally found in this country. Three closely allied species are found in North America. These are the *R. Satrapa*, the species regarded by Wilson as similar to the European *R. cristatus*, the *R. Cuvieri*, and the *R. Calendula*.

REDSTART. (*Ruticilla phœnicurus.*) This is a beautiful little bird belonging to the family *Sylviadæ*, rather more than five inches in length. The bill, legs, and claws are black; the forehead is white; the crown of the head, hind part of the neck, and the back are of a deep blue gray colour; the cheeks and throat are black; the breast, rump, and sides are red; and the two middle feathers of the tail are brown; the belly is white. The female differs considerably from the male: her colours are less vivid; the top of the head and back are ash gray; chin white. The Redstart visits us about the

REDSTART.—(RUTICILLA PHŒNICURUS.)

middle of April, and takes its departure at the end of September or the beginning of October. Though wild and timorous, it is frequently found in the midst of cities, always choosing the most inaccessible places for its residence: it likewise builds in the holes of forest trees, or in high and dangerous precipices. Its nest is chiefly composed of moss, lined with hair and feathers. It is distinguished by a peculiar quick shake of its tail from side to side, when it alights These birds feed on flies, spiders, ants' eggs, soft fruits, berries, &c.

The AMERICAN REDSTART (*Setophaga ruticilla*), which belongs to the family *Muscicapidæ*, is described by Wilson as one of the most expert flycatchers of its tribe. He says, "it is almost perpetually in motion; and will pursue a retreating party of flies from the tops of the tallest trees, in an almost perpendicular, but zig-zag direction, to the ground, where the clicking of its bill is distinctly heard; and I doubt not but it often secures ten or twelve of these in a descent of three or four seconds." Wherever flying insects abound, there this little bird is sure to be seen. It builds frequently in low bushes, or on the drooping branches of the elm, within a few feet of the ground, fastening its nest to two twigs; outwardly it is formed of flax, well wound together, and moistened with its saliva, interspersed here and there with pieces of lichen, and lined with a very soft downy substance. The female lays five white eggs, sprinkled with gray and blackish specks. The general colour of the plumage above is black, which covers the whole head and neck, and spreads on the upper parts of the breast in a rounding form; where, as well as on the head and neck, it is glossed with steel blue; sides of the breast below this, black; the inside of the wings, and upper half of the wing-quills, are of a fine orange colour; belly and vent, white, slightly streaked with pale orange; legs black. This species has the constant habit of flirting its extended tail from side to side, as it runs along the branches, occasionally shooting off after winged insects. Its notes are few and feeble, repeated at short intervals, as it darts among the foliage.

REIN-DEER. (*Cervus tarandus.*) The Rein-deer is an inhabitant of the most northerly regions. In Europe its chief residence is in Lapland and Norway; in Asia it frequents the north coast as far as Kamtschatka, and the inland parts as far as Siberia; and in America it is common in

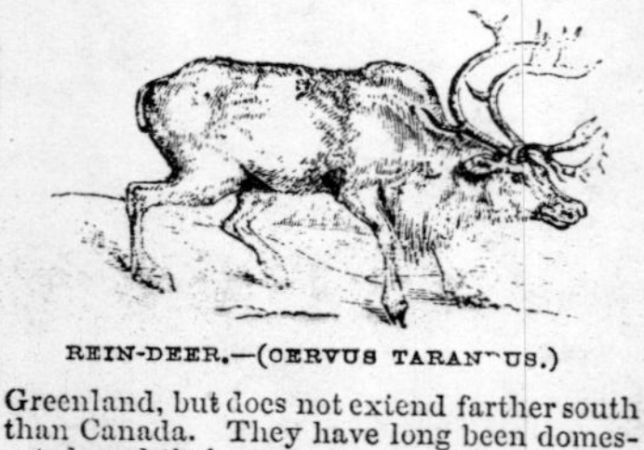

REIN-DEER.—(CERVUS TARANDUS.)

Greenland, but does not extend farther south than Canada. They have long been domesticated, and their appearance and habits have been minutely described by many travellers and naturalists. They are about four feet six inches in height. Their horns are remarkably long and slender, and they have branched, recurved, round antlers, the summits of which are palmated. The body is of a thick and square form; and the legs shorter in proportion than those of the stag; but it should be observed that the size varies according to the climate, those in the Arctic regions being the largest. The colour of the Rein-deer is brown above and white beneath; but as it advances in age, it often becomes of a grayish-white, and sometimes almost entirely white: the space about the eyes is always black: the under part of the neck is much longer than the rest, and forms a kind of hanging beard. Both sexes have horns, but those of the male are much larger and longer than those of the female. The hoofs are long, large, and black, as are also the false or secondary hoofs behind; and these latter, while the animal is running, make by their collision a remarkable clattering sound, which may be heard at a considerable distance.

It is an observation no less true than trite, that to the Laplanders this animal is the substitute for the horse, the cow, and the sheep: harnessed to the sledge, the Rein-deer bounds over the frozen lakes and rivers, or the equally hardened surface of the snow; of its milk they make their cheese; its flesh supplies them with food; and the skins furnish them not only with clothing, but with their tents and bedding. In short, this animal is deservedly celebrated for its services to the simple and harmless inhabitants of Lapland, who, undisturbed by the sound of war, or the anxieties which commerce brings, lead a kind of pastoral life, even within the frozen limits of the arctic circle, and have no other cares than those of providing for the rigours of their long winter, and of rearing and supporting their numerous herds of Rein-deer, which may be said to constitute almost their whole wealth. Some writers, indeed, have eulogized the great happiness of the Laplanders in terms too extravagant for the sober pen of truth, and, quitting descriptive prose, have indulged in the pleasant reveries of poetical fiction: thus—

> "Here stands secure, beneath the northern zone,
> O sacred Innocence, thy turf-built throne:
> 'Tis here thou wav'st aloft thy snowy wings,
> Far from the pride of courts and pomp of kings."

But, in fact, if the poor Laplander be really so happy, it arises from his being ignorant of the wants of luxury, occasioned by the sterility of his native land, and his non-intercourse with highly civilized and polished nations. Their state of felicity has, indeed, some serious drawbacks. The winter may be said to continue nearly nine months, and is of a rigour unknown in the more southern regions of the world; the sun is invisible for a certain period, and the moon and stars, with the frequent coruscations of the aurora borealis, and the reflection from the snow, constitute the only light afforded by nature. The short summer, on the contrary, when once fairly commenced, is scarcely less oppressive, from the innumerable legions of musquitoes, which abound to such a degree in the marshy districts, as to oblige the inhabitants, in order to walk abroad with common comfort, to anoint their faces with a mixture of tar and milk!

The chief food of the Rein-deer is a species of lichen, or moss, which covers large tracts of the northern regions, and on which these animals delight to browse. "Lapland," we are told, "is divided into two tracts, called the alpine and the woodland country. Those immense mountains, called in Sweden Tjellen, divide that country from Norway, extending towards the White Sea as far as Russia, and are frequently more than twelve miles in breadth. The other, called the woodland division, lies to the east of this, and differs from the neighbouring provinces of Norway by its soil, which is exceedingly stony and barren, being covered with one continued tract of wood, of old pine-trees. This tract has a very singular appearance. The trees above are covered over with great quantities of a black hanging lichen, growing in filaments resembling locks of hair, while the ground beneath appears like snow, being totally covered with white lichens. Between this wood and the Alps lies a region called the Woodland, or Desert Lapmark, of thirty or forty miles in breadth, of the most savage and horrid appearance, consisting of scattered and uncultivated woods, and continued plains of dry barren sand, mixed with vast lakes and mountains. When the mosses on part of this desert tract have been burnt, either by lightning or any accidental fire, the barren soil immediately produces the white lichen which covers the lower parts of the Alps. The Rein-deer in summer seek their highest parts, and there dwell amidst their storms and snows, not to fly the heat of the lower regions, but to avoid the gnat and gad-fly. In winter these intensely cold mountains, whose tops reach high into the atmosphere, can no longer support them, and they are obliged to return to the desert and subsist upon the lichens."

"To the natives of North America," to use the words of a contemporary writer, "the Reindeer is only known as a beast of chace, but it is a most important one: there is hardly a part of the animal which is not made available to some useful purpose. Clothing made of the skin is, according to Dr. Richardson, so impervious to the cold, that, with the addition of a blanket of the same material, any one so clothed may bivouack on the snow with safety in the most intense cold of an arctic winter's night. The venison, when in high condition, has several inches of fat on the haunches, and is said to equal that of the fallow-deer in our best English parks; the tongue and some of the tripe are reckoned most delicious morsels. Pemmican is formed by pouring one-third part of melted fat over the pounded meat, and incorporating them well together. The Esquimaux and Greenlanders consider the stomach or paunch with its contents a great delicacy; and Captain James Ross says that those contents form the only vegetable food which the natives of Boothia ever taste. For further particulars, and there are many and interesting, we must refer to Dr. Richardson's Fauna Boreali-Americana, and the works of our gallant northern voyagers generally."

REMORA, or SUCKING-FISH. (*Echineis.*) This fish, which in form bears some resemblance to the herring, and is from fifteen to eighteen inches in length, is the *echeneis* of the Greeks, and has been celebrated from remote antiquity for its power of adhesion to any other animal or inanimate substance: in short, the most incredible stories are related

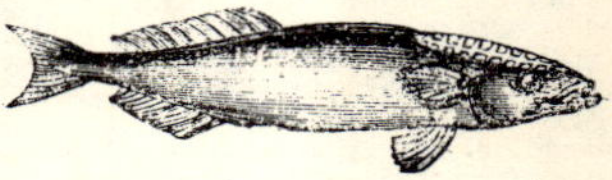

REMORA — (ECHINEIS REMORA.)

by Pliny and other ancient naturalists with all possible gravity and good faith; among others, that Antony's ship, at the battle of Actium, was kept motionless by the exertions of the Remora, notwithstanding the efforts of several hundred sailors; and that the vessel of Caligula was detained between Astura and Actium by another of these fish found sticking to the helm, and whose solitary efforts could not be countervailed by a crew of four hundred able seamen, till several of the latter, on examining into the cause of the detention, perceived the impediment, and detached the Remora from its hold. The real fact is, that the fins of this fish are particularly weak, on which account it attaches itself to various bodies, and is found not only fastened to ships, but to whales, sharks, and other fishes; and with such extreme tenacity is this hold maintained, that, unless the effort of separation be applied in a particular direction, it is impossible to effect the disunion without the destruction of the fish itself. In stormy and boisterous weather, the Remora, like the lumpfish and some others, will also often adhere to rocks.

The Remora is principally an inhabitant of the Mediterranean and Atlantic seas. Its general colour is an uniform brown; the skin smooth and destitute of scales, but marked with numerous impressed points or pores: the mouth is large, and furnished with very numerous small teeth; and the lower jaw is rather longer than the upper: the eyes are small, with yellow irides: the lateral line commences above the pectoral fins, and from thence, pretty suddenly descending, runs straight in the tail, which is of a slightly forked, or rather lunated form. Another species, the INDIAN REMORA (*Echeneis neucrates*), which is of a more slender or lengthened shape, is said to be employed by the natives of the coast of Mozambique in their pursuit of turtles, with great success. A ring is fastened round the tail of the fish in such a manner as to prevent its escape, and a long cord fastened to the ring. When the boat has arrived as near as it well can to a turtle that is sleeping on the surface of the water, as is the custom of these animals, the boatmen throw the Remora into the sea, and giving it the proper length of cord, it soon attaches itself to the breast of the sleeping turtle, and both are then drawn into the boat with ease. The apparatus by which this adhesion is accomplished by the Remora consists of an oval area on the top of the head, traversed by numerous partitions, each of which is fringed at the end by a row of very num rous perpendicular teeth, or filaments, while the whole oval space is strengthened by a longitudinal septum.

REPTILES. (*Reptilia.*) The name given to a class of cold-blooded vertebrated animals, whose movements are usually confined to crawling and swimming, and whose respiration is aerial and incomplete. They have the heart so constructed that at each contraction a portion only of the blood received from the various parts of the system is sent into the lungs, the remainder of this fluid returning into the general circulation without having passed through the lungs, and consequently without having been subjected there to respiration. Hence it results that the action of oxygen upon the blood is less than in the Mammalia; and though several of them leap and run with celerity on certain occasions, their habits are generally sluggish, their digestion excessively slow, their sensations obtuse, and, in cold or temperate climates, they pass nearly the whole winter in a state of lethargy. In their general form, Reptiles approach Mammalia nearer than Birds; but they offer in this respect many variations, as may be seen by comparing together, a Tortoise, a Crocodile, a Serpent, and a Frog. Their head is almost always small, and their body very much lengthened out; some, as Serpents, are entirely destitute of members, or have only traces of them; but the greater number of these animals, the Lizard and Frog for instance, have four limbs, formed so as to serve for walking or swimming. The skeleton in this class presents much greater variations in its structure, than in warm-blooded Vertebrata. All the parts of which it is composed are wanting in one or another group, excepting the head and the vertebral column; but the bones of which these are composed always preserve a great resemblance to those of Mammalia and Birds, and are easily recognized as being analagous to them. Their brain, which is proportionally very small, is not so essentially requisite to the exercise of their animal and vital faculties as in the Mammalia; for they continue to live and to execute voluntary movements for a considerable time after being deprived of the brain, and even after the loss of the head: their muscles also preserve their irritability for a considerable time after being severed from the body; and their heart continues to pulsate for hours after it has been torn from the body. Reptiles dive with more facility, and remain longer under water than either the Mammalia or Birds, the smallness of the pulmonary vessels permitting them to suspend the process of respiration, without arresting the course of the blood. No Reptile hatches its eggs. Some on quitting the egg have the form and gills of fishes; and certain genera retain these organs even after the development of their lungs. In other Reptiles which produce eggs, the young is already formed and considerably advanced within the egg at the time the parent deposits it. Reptiles also present more varied forms, characters, and modes of gait, than the other classes of animals; and it is in their production more especially, that Nature seems to have tried to imagine grotesque forms, and to have modified in every possible way the general plan adopted for all vertebrated animals, and for the oviparous classes in particular.

Reptiles are endowed with five senses, but none of them in great perfection. In those species which are covered with scales or plates, the sense of touch is very obtuse; and in the species which have a naked skin, such as the Frog, it is also weak, in consequence of not being adherent to the body, but envelopes it like a bag. In the Serpents the eyes are immovable, destitute of eyelids, and covered with a corneous substance; in some genera three eyelids are distinguishable, while others are destitute of sight. Their nostrils are small, and they appear to have a very weak sense of smell. They have no delicacy of taste, for almost all the species swallow their food entire; and those in which the tongue is soft and flexible, this organ serves chiefly as an instrument for the seizure of their food. None of them have true fleshy lips; and some, such as the Tortoises, are provided with a horny bill, like that of a parrot; others have teeth of various forms, which are not, however, formed for mastication, but to assist in holding their prey. Various serpents have hollow fangs, which they can erect at pleasure, when they open their mouths to bite, and these fangs have apertures, from which they inject into the wounds made by them an active and deadly poison.

From the earliest times the forms and habits of the reptile world attracted attention, and appear to have been pretty well understood. The ancient monuments of the Egyptians show this; and numerous passages in the Old Testament prove that a similar knowledge existed when the Scriptures were written. Nor must it be forgotten that among the Organic Remains which the industry and science of inquiring minds have lately brought to light, none present forms more wonderful, or proportions more gigantic, than some of the Fossil Reptilia. [See ICHTHYOSAURUS: IGUANODON: PLESIOSAURUS: NOTHOSAURUS, &c.]

RETEPORA. A genus of the Polypiferous corallines which is allied to *Eschara*, and has the leaf-like expansion pierced like net-

NEPTUNE'S RUFFLES.
(RETEPORA CELLULOSA.)

work; our figure will show the appearance of this genus. The species represented is often called Neptune's Ruffles, and is the *Retepora cellulosa* of naturalists. It is some of the recent species; there are others found in a fossil state.

RHEA. [See OSTRICH, AMERICAN.]

RHINOCEROS. (*Rhinoceros.*) This large and uncouth-looking Pachydermatous genus inhabits the hotter regions of Asia and Africa, and, next to the Elephant contains the most powerful of quadrupeds. The common INDIAN RHINOCEROS (*R. unicornis*) is usually about twelve feet long from the tip of the nose to the insertion of the tail; its height is about seven feet; and the circumference of its body is nearly equal to its length. The back, instead of rising, as in the Elephant, sinks in considerably; the head is moderately large and long; the upper lip protrudes considerably, and being extremely pliable, answers the end of a small proboscis: but its most distinguishing mark is the possession of a solid, slightly curved, sharp-pointed horn, which rests on a strong arch

INDIAN RHINOCEROS.
(RHINOCEROS UNICORNIS.)

formed by the nasal bones. This horn is sometimes (but not generally) as much as three feet in length, and eighteen inches in circumference at its base, and is used as a most powerful and effective weapon. The animal is also characterized by having seven molars on each side above and below, with only four incisors, and no canine teeth. The ears are moderately large, upright, and pointed; the eyes small and half closed. The skin is thick and coarse, with a knotty or granulated surface; and so impenetrable on the body and limbs, as to resist either the claws of the lion or the tiger, the sword or the shot of the hunter. About the neck the skin is disposed in several large plaits or folds; another fold passes from the shoulders to the fore legs, and another from the hind part of the back to the thighs. The tail is slender, flattened at the end, and covered on the sides with very stiff and thick black hairs: the belly is somewhat pendulous; the legs very short, strong, and thick; and the feet divided into three large hoofs, all standing forwards. In India the Rhinoceros leads a tranquil indolent life, wallowing on the marshy borders of lakes and rivers, and occasionally bathing itself in their waters. Its movements are usually slow; and it carries its head low, like the Hog, ploughing up the ground with its horn, and making its way by sheer force through the jungle. It is naturally of a quiet and inoffensive disposition, but very furious and dangerous when provoked or attacked; charging with great impetuosity, and trampling down, or ripping up with its horn, any animal which opposes it. The bones of the Rhinoceros, like those of the Elephant, are often found in a fossil state in various parts of the world; and in the year 1772 an entire Rhinoceros was found buried in the banks of a Siberian river, in the ancient frozen soil, with the skin, tendons, and some of the flesh, in the highest state of preservation.

The TWO-HORNED RHINOCEROS. (*Rhinoceros bicornis.*) This species is found in various parts of Africa, and seems to have been the kind known to the ancient Romans, and by them exhibited in their public shows and combats of animals. In size it equals the common or single-horned species; and its habits and manner of feeding are the same: but it differs greatly in the appearance of its skin, which, instead of the vast and regularly marked armour-like folds of the former, has merely a slight wrinkle across the shoulders, and on the hinder parts, with

AFRICAN RHINOCEROS.

a few fainter wrinkles on the sides; so that, in comparison with the common Rhinoceros, it appears almost smooth: the skin, however, is rough or tuberculated: but what constitutes the specific or principal distinction is, that the nose is furnished with *two* horns, one of which is smaller than the other, and situated higher up; and that they are fixed to the nose by a strong apparatus of muscles and tendons, so that they are loose when the animal is in a quiescent state, but become firm and immovable when he is enraged. His manner of feeding, with some other particulars, is thus given by Mr. Bruce, the Abyssinian traveller. He informs us, that, "besides the trees capable of most resistance, there are, in the vast forests within the rains, trees of a softer consistence, and of a very succulent quality, which seem to be destined for his principal food. For the purpose of gaining the highest branches of these, his upper lip is capable of being lengthened out so as to increase his power of laying hold with this in the same manner as the Elephant does with his trunk. With this lip, and the assistance of his tongue, he pulls down the upper branches which have most leaves, and these he devours first; having stripped the tree of its branches, he does not therefore abandon it, but, placing his snout as low in the trunk as he finds his horns will enter, he rips up the body of the tree, and reduces it to thin pieces, like so many laths; and when he has thus prepared it, he embraces as much of it as he can in his monstrous jaws, and twists it round with as much ease as an ox would do a root of celery. When pursued, and in fear, he possesses an astonishing degree of swiftness, considering his size, the apparent unwieldiness of his body, his great weight before, and the shortness of his legs. He is long, and has a kind of trot, which, after a few minutes, increases in a great proportion, and takes in a great distance; but this is to be understood with a degree of moderation. It is not true that in a plain he beats the horse in swiftness. I have passed him with ease, and seen many worse mounted do the same; and though it is true that a horse can seldom come up with him, this is owing to his cunning, and not his swiftness. He makes constantly from wood to wood, and forces himself into the thickest part of them. The trees that are frush, or dry, are broke down, like as with a cannon shot, and fall behind him and on his sides in all directions. Others that are more pliable, greener, or fuller of sap, are bent back by his weight and the velocity of his motions. And, after he has passed, restoring themselves like a green branch to their natural position, they sweep the incautious pursuer and his horse from the the ground, and dash them in pieces against the surrounding trees. The eyes of the Rhinoceros are very small, and he seldom turns his head, and therefore sees nothing but what is before him. To this he owes his death, and never escapes if there is so much plain as to enable the horse to get before him. His pride and fury, then, make him lay aside all thoughts of escaping, but by victory over his enemy. He stands for a moment at bay, then, at a start, runs forward at the horse, like the wild boar, whom, in his manner of action, he very much resembles. The horse easily avoids him, by turning short aside; and this is the fatal instant: the naked man, with the sword, drops from behind the principal horseman, and, unseen by the Rhinoceros, who is seeking his enemy, the horse, he gives him a stroke across the tendon of the heel, which renders him incapable of further flight or resistance."

Another species of Rhinoceros, less powerful and savage, is found in Java; of this we figure the skull, which will serve also to illustrate the structure of the head; a third, which possesses two horns, in Sumatra; and

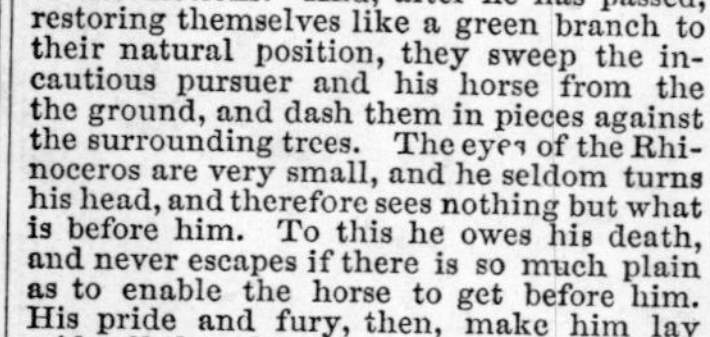

SKULL OF RHINOCEROS JAVANUS.

three species are said to be known in Africa: but the most formidable are those we have described. The *skin* of the Rhinoceros is an article in great demand in several countries of Asia and Africa. It is manufactured into the best and hardest leather that can be imagined; and targets and shields are made of it, that are proof against even the stroke of a scimitar. When polished, the skin is very similar in appearance to tortoise shell. Their *horns* are manufactured into drinking cups, the hilts of swords, and snuff-boxes, by several oriental nations; and in the palmy days of ancient Rome, we are told, the ladies of fashion used them in their baths, to hold their essence bottles and oils.

In M. de Blainville's great work on the Osteography of the Vertebratæ, he admits five living species as indisputable; two of which are African — the black rhinoceros of the Cape (*R. bicornis*), and the white rhinoceros of Southern Africa, first distinguished by Dr. Burchell (*R. simus*); three are Asiatic — the Rhinoceros of India (*R. unicornis*), the rhinoceros of Java, with one horn (*R. Javanus*), and that of Sumatra, with two horns (*R. Sumatranus*). Dr. Andrew Smith discovered a third species, distinguished, among other peculiarities, by the great length of the second horn. This is the *Rhinoceros Ketloa*, described by that distinguished naturalist; a fine specimen of it exists in the collection of the British Museum. Some accounts would likewise lead us to believe in the existence of a rhinoceros in Africa with one horn, which would form another species to be added to the preceding.

Among the fossil rhinoceroses, M. de Blainville admits but three European species as certain. The first is the rhinoceros with partitioned nostrils (*R. tichorhinus*). This species, destitute of incisors, had three toes on each foot, the cranium elongated, the nostrils separated by a bony partition; its nose was provided with two horns; its molars approached those of the *Rhinoceros camus;* its bones were short and strong, and its body covered with hair. On this subject he remarks, that these hairs have sometimes been erroneously described as forming a long and thick fur, but at most they did not exceed three lines in length. *R. tichorhinus* is found in the deposits formed during the diluvian epoch. It is probable that it inhabited Siberia, and the greater part of Europe. This is the species which has been found preserved in the ice of the North of Asia. The second species is the rhinoceros with nostrils not partitioned (*R. leptorhinus*), which had persistent incisors, but concealed in the gums, three toes on each foot, two horns, an elongated cranium, and slender bones. This species, which is not so well characterized as the preceding, has been found chiefly in the superior tertiaries of Italy and the south of France. M. de Blainville likewise refers the bones found in caverns in the south of France to *R. tichorhinus*, while those of the north and of Belgium contain only the remains of the preceding species. — The third species is the rhinoceros with incisors (*R. incisivus*), characterized by half salient incisors in the two jaws, four toes on the anterior feet, flat metatarsi, &c. It would appear that the male bore two horns, and that the female was destitute of these appendages. The latter, for this reason, has been made the type of the genus *Accrotherium* of M. Kaup. The *R. incisivus* is found in the middle tertiary formations, and has been described under many names. In the Sewalik Hills, in India, Dr. Falconer and Major Cautley have discovered remains of other fossil species; figures of these are given in their *Fauna antiqua Sivalensis:* the originals are preserved in the British Museum.

It appears that rhinoceroses have not existed during the whole commencement of the tertiary epoch, for the eocene formations yield no trace of them. They have appeared, for the first time, in the middle or miocene period, during which the *R. incisivus* has inhabited the greater part of Europe. Towards the close of the tertiary epoch this species has been replaced by the *R. leptorhinus*, and during the diluvian epoch, it is the *R. tichorhinus* which has been the most abundant and most widely diffused. In the present day rhinoceroses do not exist in Europe, and are only found in the warmest countries. We find three species in Africa, one species in Continental Asia, and two in the Sunda Islands. America and New Holland have not any at present, and do not appear to have possessed any in the epoch anterior to our own.

RHIPIDURA. [See FANTAIL.]

RHIZOPODA. [See POLYTHALAMIA, p. 540.]

RHIZOSTOMA. A genus of Acalephæ.

RHOPALOCERA. The first section of the Lepidoptera, in the recent Classification of Insects, corresponding with the genus *Papilio* (Linn.), and deriving its name from generally having the antennæ, which are thin and elongated, terminated by a knob. This section comprises the well-known tribes of Butterflies, whose elegant forms and beautiful colours may be mistaken for "winged flowers or flying gems." They vary greatly in size, as well as in the diversity of their colours: here, in our native fields we have some species not an inch across the wings, while in India and South America are to be seen fluttering in the sun's warm rays, gorgeous specimens nine or ten inches in expanse. Their flight is also as varied as that of the feathered tribes, and can as readily be distinguished by the skilful collector. Some skim along the plain with graceful elegance; others fly more slowly, and with an undulating motion; while others, again, rise high into the air, and sail over the topmost branches of the sturdy oak. The prevalence of particular colours in certain groups also deserves mention: thus the *Polyommati* are chiefly blue; among the *Picrides* the colour is either white or orange tipped with black; in the *Hipparchiæ*, dull brown; in *Lycæna*, bright copper colour; while the *Nymphalidæ* have their wings varied with beautiful eyes or spots; and the *Fritillaries* are fulvous, varied on the under side with pearly patches.—We might extend this article to an indefinite length were we to attempt to describe the various habits, the distinctive characters, and the transformations, &c. of these beautiful insects; but we trust the reader will excuse us if we at once refer him, for such additional information as our space would allow, to the articles PAPILIO and LEPIDOPTERA.

RHYNCHŒA. A genus of Grallatorial birds allied to the Snipes. The species RHYNCHŒA AUSTRALIS, which is a summer visitant of New South Wales, in its habits and disposition partakes both of the true Snipes and the Sandpipers; running about, like the latter, among the rushes or on the bare ground at the edge of the water. Olive-green, with narrow bars and marblings of dark brown, is the prevailing colour; and a pale buff stripe runs from the bill down the centre of the head to the nape; breast and all the under surface white; legs pale green. The male is much smaller than the female and has the sides, back, and front of the neck much lighter and mingled with patches of white; wings more olive, the coverts ornamented with numerous large irregular patches of buff, encircled with a narrow line of black; the buff bands on the primaries richer and more distinct; the scapularies speckled with white; the patch on each side of the chest dark olive, with large patches of white surrounded by a line of black. The plumage of the female, contrary to the general rule, is darker, richer, and more distinctly defined. Mr. Gould says that on dissection be observed an anatomical peculiarity of a very extraordinary nature, the more so as it exists in the female alone; namely, the great elongation of the trachea, which passes down between the skin and the muscles forming the breast for the whole length of the body, making four distinct convolutions before entering the lungs. This was afterwards examined by Mr. Yarrell, who states that the form and position of the trachea in the *Rhynchœa Australis* is similar to that of the Semipalmated Goose, figured in the 15th volume of the Trans. Linn. Soc. Tab. 14.

RHYNCHOPHORA. An extensive group or subsection of Coleopterous insects, distinguished by the front of the head being produced into a long snout or rostrum, at the extremity of which is the mouth. The body is oval or rounded; the antennæ are inserted at the sides of the rostrum, and are short, elbowed, and often terminated in an oval club; the mandibles are small but robust; the palpi short and conical; and the third tarsal joint deeply bilobed. The majority of the species are of small or moderate size; but the elytra of some of them are most brilliantly coloured: they are widely distributed, but abound chiefly in hot countries, and all are herbivorous. The larvæ are white and fleshy grubs, with strong and horny jaws, whereby they are enabled to gnaw the harder parts of vegetable food, on which they subsist.

These beetles are often very hurtful to plants, by boring into the leaves, bark, buds, fruit, and seeds, and feeding upon the soft substance therein contained. They are diurnal insects, and love to come out of their retreats and enjoy the sunshine. Some of them fly well; but others have no wings, or only very short ones, under the wing-cases, and are therefore unable to fly. They walk slowly, and being of a timid nature, and without the means of defence, when alarmed they turn back their antennæ under the snout, fold up their legs, and fall from the plants on which they live. The larvæ have strong and horny jaws, with which they gnaw those parts of plants which serve for their food. Some of them bore into and spoil fruits, grain, and seeds; some attack the leaves and stems of plants, causing them to swell and become cankered; while others penetrate into the solid wood, interrupt the course of the sap, and occasion the branch above the seat of attack to wither and die. Most of these grubs are transformed within the vegetable substances upon which they have lived; some, however, when fully grown, go into the ground, where they are changed to pupæ, and afterwards to beetles. This subsection corresponds with the Linnæan genera Bruchus, Attelabus, and Curculio. Some of the most extraordinary species of Rhyncophorous insects compose the South African genus Antliarhinus, in one sex of which the rostrum is nearly three times as long as the body, porrected, and as thin as a fine needle. In the great work of Schœnherr on these insects at least 7000 species are described or alluded to.

RHYNCHOPS. We learn from Wilson's American Ornithology, that this truly singular bird was the only species of its tribe discovered at the time he wrote. Another species at least, the *R. orientalis* has been since found in Africa. The species whose history we here subjoin is the BLACK SKIMMER; or SHEERWATER. (*Rhynchops nigra.*) It is a bird of passage in the United States, and makes its first appearance on the shores of New Jersey early in May. It resides there, as well as along the whole Atlantic coast, during the summer, and retires early in September. Its favourite haunts

BLACK SKIMMER.—(RHYNCHOPS NIGRA.)

are low sand bars, raised above the reach of the summer tides; and also dry flat sands on the beach in front of the ocean. Early in June these birds form themselves into small societies, fifteen or twenty pair frequently breeding within a few yards of each other. The nest is a mere hollow formed in the sand; and the female lays three eggs, almost perfectly oval, of a clear white, marked with large round spots of brownish black, and intermixed with others of a paler dusky hue. The female sits on them only during the night, or in wet and stormy weather. The young remain for several weeks before they are able to fly; are fed with great assiduity by both parents; and seem to delight in lying with loosened wings, flat on the sand, enjoying its invigorating warmth. They breed but once in the season.

The Sheerwater is formed for skimming, while on wing, the surface of the sea for its food, which consists of small fish, shrimps, young fry, &c., whose usual haunts are near the shore, and towards the surface. That the lower mandible, when dipt into and cleaving the water, might not retard the bird's way, it is thinned and sharpened like the blade of a knife; the upper mandible being, at such times, elevated above water, is curtailed in its length, as being less necessary, but tapering gradually to a point, that, on shutting, it may offer less opposition. To prevent inconvenience from the rushing of the water, the mouth is confined to the mere opening of the gullet, which, indeed, prevents mastication taking place there; but the stomach, or gizzard, to which this business is solely allotted, is of uncommon hardness, strength, and muscularity, far surpassing in these respects any other water bird with which I am acquainted. To all these is added a vast expansion of wing, to enable the bird to sail with sufficient celerity while dipping in the water. The general proportion of the length of our swiftest hawks and swallows, to their breadth, is as one to two; but in the present case, as there is not only the resistance of the air, but also that of the water, to overcome, a still greater volume of wing is given, the Sheerwater measuring nineteen inches in length, and upwards of forty-four in extent. In short (says Wilson), whoever has attentively examined this curious apparatus, and observed the possessor, with his ample wings, long bending neck, and lower mandible, occasionally dipt into and ploughing the surface, and the facility with which he procures his food, cannot but consider it a mere playful amusement, when compared with the dashing immersions of the tern, the gull, or the fish-hawk, who, to the superficial observer, appear so superiorly accommodated.

The voice of the Sheerwater is harsh and screaming, resembling that of the tern, but stronger. It flies with a slowly flapping flight, dipping occasionally, with steady expanded wings and banded neck, its lower mandible into the sea, and, with open mouth, receiving its food as it ploughs along the surface. It is rarely seen swimming on the water, but frequently rests in large parties on the sand bars at low water. The length and breadth of this bird we before noticed as nineteen inches by forty-four: the length of the lower mandible is four inches and a half; of the upper, three inches and a half; both of a scarlet red, tinged with orange, and ending in black; the lower extremely thin; the upper grooved, so as to receive the edge of the lower: the nostril is large and pervious, placed in a hollow near the base and edge of the upper mandible, where it projects greatly over the lower; upper part of the head, neck, back, and scapulars, deep black; wings the same, except the secondaries, which are white on the inner vanes, and also tipt with white; tail forked, the two middle feathers being about an inch and a half shorter than the exterior ones, all black, broadly edged on both sides with white; tail-coverts, white on the outer side, black in the middle; front, passing down the neck, below the eye, throat, breast, and whole lower parts, pure white; legs and webbed feet, bright scarlet. The female is less than the male, but the colours and markings are very similar. The Sheerwater is found on various coasts of Asia, as well as America, residing principally near the tropics, and migrating into the temperate regions of the globe for the purpose of rearing his young. He is rarely or never seen far out at sea, and must not be mistaken for the Sheerwater Petrel (a species of *Puffinus*).

RHYNCHOSAURUS. [See SUPPL.]

RIBBON-FISH. (*Cepola.*) A genus of Acanthopterygious fishes belonging to the *Tæniadæ* family. The peculiar characters of this genus are indicated by the name; the species being distinguished by their lengthened bodies, much flattened at the sides, and having very small scales. In this family are three tribes; one having the muzzle

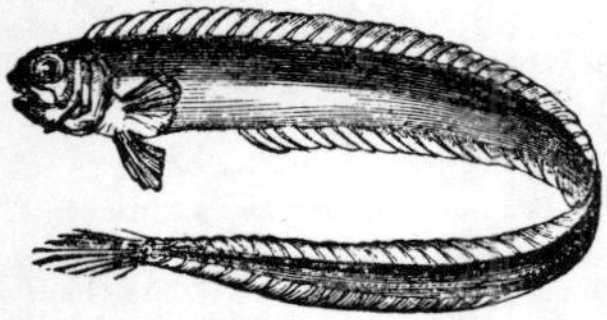

RIBBON FISH.—(CEPOLA RUBESCENS.)

elongated, the mouth deeply cleft, with strong trenchant teeth, and the lower jaw projecting beyond the upper: the other tribe comprehending genera which have the mouth small and little cleft. These occur in the Mediterranean, the Indian, the Atlantic, and the Arctic Seas; and some of them are ten feet in length. A third tribe has the muzzle short, and the mouth cleft obliquely.

RICE-BIRD. (*Loxia oryzivora.*) This bird, which is about the size of a Greenfinch, is a native of Java, and is sometimes called the Java Sparrow. The bill is extremely thick, and of a fine red colour above and beneath, except towards the point, where there is a little space of white. The eyes are dark, and the irides red. The whole head is black, except a white oval spot on each cheek: the neck, breast, back, and coverts of the wings, are of a fine bluish ash-colour, the rump being somewhat lighter than the back; while the ash-colour on the breast changes gradually towards the belly into a blossom-colour, terminating in a dirty white. The greater quill-feathers, as well as the whole tail, are black; the legs and feet are of a faint red hue; and the claws are of a dingy white. From the general

RICE-BIRD.—(LOXIA ORYZIVORA.)

plumage being remarkably smooth and even, this bird derives a peculiar beauty.

RICE-BUNTING, RICE-BIRD, or BOB-O-LINK. (*Dolichonyx oryzivora.*) The specific characters of this bird are — tail-feathers very acute; adult male, in spring dress, black, the hind head yellowish white; scapulars, rump, and tail-coverts, white, tinged with ash. The Rice-Bunting migrates over the continent of America from Labrador to Mexico, and over the Great Antilles, appearing in the southern extremity of the United States about the end of March. Their food is insects and worms, and the seeds of the grassy meadows. In the autumn they sometimes attack the crops of oats and barley. The song of the male continues, with little interruption, as long as the female is sitting, and is singular and pleasant; it consists of a jingling medley of short, variable notes, confused, rapid, and continuous. The relish for song and merriment is confined to the male, but he

RICE-BUNTING — MALE.
(DOLICHONYX ORYZIVORUS.)

generally loses his musical talent about the end of the first week in July, from which time, or somewhat earlier, his plumage begins to lose its gay colours, and to assume the humble hue of that of the female. About the middle of August they enter New York and Pennsylvania, on their way to the south. There, along the shores of the large rivers lined with floating fields of wild rice, they find abundant subsistence, grow fat, and their flesh becomes little inferior in flavour to that of the European Ortolan; on which account the Reed or Rice-birds, as they are then called, are shot in great numbers. When the cool nights in October

RICE-BUNTING — FEMALE.
(DOLICHONYX ORYZIVORUS.)

commence, they move still farther south, till they reach the islands of Cuba and Jamaica.

RIFLE BIRD. (*Ptiloris Paradiseus.*) This magnificent bird which Mr. Gould considers is without exception the most gorgeously plumaged one yet discovered in Australia, is found in the south-eastern portion of that country, inhabiting the "brushes." The general colour of the male is a rich velvety black, glossed on the upper surface with brownish lilac; under surface similar, but all the feathers of the abdomen and flanks broadly margined with rich olive-green; feathers of the head and throat small, scale-like, and of a shining metallic blue-green; two centre tail-feathers rich shining metallic green, the remainder deep black; bill and feet black. But while the male is adorned with hues only equalled by some species of the *Trochilidæ* or Humming-birds, the dress of the female is remarkably plain and sombre. The Rifle Bird's powers of flight are very limited, arising from the shortness and truncate form of the wing; but owing to that structure it ascends the upright boles of trees precisely after the manner of the *Climacteri*, many of whose habits it possesses.

RING-DOVE. (*Columba palumbus.*) This is the largest of all the Pigeon tribe, measuring above seventeen inches in length. Its bill is pale red; the eyes pale yellow; the upper parts of the body bluish ash, deepest on the upper parts of the back; the head and fore part of the neck, pale ash gray; the lower part of the neck and breast, vinous ash; the belly, thighs, and vent, dull white. It receives its name from having a semicircular line of white on the hinder part of the neck, above and beneath which the feathers are glossy, and of a varying hue according to the light in which they are seen: the greater quills are dusky, and all of them excepting the outermost, edged with white: from the point of the wing a white line extends downwards, passing above the bastard wing: the tail is ash gray, tipped with black: legs red, and partly covered with feathers; claws black. The Ring-dove is very generally diffused throughout Europe; and is said to be migratory; though it is certain that there are many which remain and regularly breed here. They seldom fly singly, but in large flocks; and they subsist on grain, acorns, ivy-berries, and other wild fruits. Their cooing is louder and more plaintive than that of the common Pigeon, but is not heard except in pairing time, or during fine weather.

"The Ring-dove," as Mr. Waterton observes, "lays two snow-white eggs on a nest which may be termed a platform of sticks, so sparingly put together, that the eggs are easily seen through it by an eye habituated to look for them. On inspecting this apparent commencement or remnant of a nest, one is led to surmise, at the first glance, that the young are necessarily exposed to many

RING DOVE.—(COLUMBA PALUMBUS.)

a cold and bitter blast during the spring of this ever-changing climate. 'But God tempers the wind,' said Maria, 'to the shorn lamb;' and in the case before us, instinct teaches the parent bird to sit upon its offspring for a longer period after they are hatched than, perhaps, any other of the feathered tribe. In the mean time, the droppings of the young, which the old birds of some species carefully convey away, are allowed to remain in the nest of the Ring-dove. They soon form a kind of plaster, strong and scentless. This adds consistency to the nest, producing, at the same time, a defence against the cold. The ornithologist, while going his autumnal beats, in quest of knowledge, on seeing this, will know immediately that the nest has contained young: should this be wanting, he may conclude that the nest has been abandoned at an early period. As he will find but very few nests with this species of plaster in them, he may conclude, to a certainty, that the Ring-dove has a host of enemies in this country, and that it is seldom fortunate enough to rear its young to that state in which the faculty of flying saves them from destruction. No bird in the British dominions seems to resort to so many trees and shrubs for the purpose of incubation as the Ring-dove. Not a tree, from the towering pine to the lowly thorn, ever comes amiss to it. * * * During the winter months they are exceedingly shy and timorous, seeking for safety in lofty flight, the moment they see you approach. They become quite silent towards the last week in October, and their notes are reduced to half their number for some days before they cease to coo entirely. At this period they discontinue those graceful risings and sinkings in the air, in which they appear to so much advantage during the whole of the breeding season. * * * As yet, all attempts to reclaim this Pigeon have been of no avail. I should suppose it is not in the power of man to make it breed within the walls of a dovecot."

RINGLET [BUTTERFLY]. A name given by collectors to the species *Hipparchia Hyperanthus*.

RIVULET [MOTHS]. A name given by collectors to species of Moths of the genus *Emmelesia*.

ROACH. (*Cyprinus rutilus.*) This fish inhabits deep, still, and clear rivers; is considered coarse rather than delicate; and in general weighs from about a pound to a pound and a half, though it is occasionally larger. In shape it is deep, but rather thin; the back much arched; the scales large and

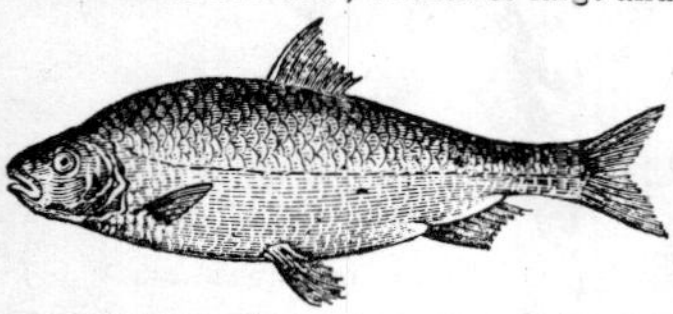

ROACH.—(CYPRINUS RUTILUS.)

easily deciduous; and the lateral line is considerably incurvated towards the abdomen. Its general colour is silvery, with a cast of dull yellow, growing more dusky on the upper parts: fins red; dorsal fin rather small, and situated on the middle of the back: tail slightly forked. The Roach is a gregarious fish, always swimming in large shoals, and feeding on worms and herbs. It generally spawns about the middle of May, and is very prolific. Although in no great esteem in this country, it is considered in many parts of Europe as an excellent fish for the table, its flesh being white and firm.

ROBIN. The REDBREAST [which see]. Also the name given in America to the *Turdus migratorius*. Nearly every country has its "Robin;" colonists assigning the name to the most familiar Red-breasted bird which occurs in the land of their adoption. Thus the Robins of Great Britain, America, New Holland, or other countries, belong to very different genera.

RODENTIA. RODENT or GNAWING ANIMALS. This name is given to an Order of mammiferous quadrupeds, occupying, in many respects, an intermediate place between the purely carnivorous and purely herbivorous Mammalia, and so forming the connecting link between them. Like the Carnivora, they are *unguiculated*, or furnished with claws; but the chief peculiarity of this order is seen in the remarkable conformation of the teeth. They have two long chisel-shaped incisors in each jaw, by some zoologists said to be canines, and a vacant space between the incisors and the molars. The conformation of the gnawing teeth is beautifully adapted to the purpose they have to fulfil: they are required to have a sharp edge, in order to make their way through tough vegetable substances, and they must at the same time be very strong and firm; this is effected by the principal substance of the tooth being composed of very tough *ivory*, with a plate of hard *enamel* in front only, which latter, wearing most slowly, is always left as a sharp projecting edge. The molar teeth, which are separated from the canines by a wide interval, are composed of alternate plates of enamel and ivory, which, wearing unequally, stand up in ridges, and give them a rasp-like surface. The ridges are always transverse, or in a direction from side to side of the head; and as the lower jaw has considerable facility of moving backwards and forwards, it greatly increases the power of trituration. In the frugivorous species of the Order, however, the surface of the molar teeth is raised into rounded tubercles, as is the case with the Squirrel, for instance; whilst in those animals which have any carnivorous tendency, as in the Rat, they are raised into sharp points, thus bearing some resemblance to those quadrupeds which are wholly carnivorous. At the same time, it should not be forgotten that there are some animals belonging to the Order *Rodentia*, whose propensities to devour almost anything that falls in their way, are such as to be entitled to the term omnivorous. The animals composing this order are mostly of small size; some are docile and gentle, whilst others are savage and untameable; their instinctive powers are great, but they possess not much sagacity. In form they may be said to be disproportionate, the posterior limbs being generally much larger than the anterior; they rather leap than walk; and most of them have the habit of sitting upon their haunches, and of using their fore paws for the prehension of food, &c. The brain of the Rodents is, as Cuvier remarks, nearly smooth and without convolutions; the orbits are not separated from the temporal fossæ, which have but little depth; the eyes are entirely directed laterally; the zygomatic arches, delicate and curved below, plainly indicate the weakness of their jaws; the anterior limbs have scarcely any rotary motion, and their two bones are nearly united; in short, the inferiority of these animals shows itself in the greater part of the details of their organization. Nevertheless, the genera which have the strongest clavicles enjoy a certain dexterity, and use their fore feet for carrying their food to their mouth; while others (the squirrels for instance) climb trees with the utmost facility.

ROEBUCK, or ROE DEER. (*Cervus Capreolus.*) Although there are very few, if any, of this light and agile species of the Deer tribe in England, they are still to be met with in the mountainous parts of Scotland, making their couches, like hares, among the heather. In size they are far inferior to the Fallow Deer, being only about two feet four inches in height, and three feet six inches in length. They are of a reddish brown colour on the back, the chest and belly yellowish, and the rump white: the horns are about nine inches long, round, and divided into three branches. The Roebuck seems naturally attached to shady thickets and rising slopes. All its motions are elegant

ROEBUCK.—(CERVUS CAPREOLUS.)

and easy; bounding with the utmost facility, and continuing the course with little apparent fatigue. In many essential particulars, this animal differs from its congeners; not merely in its lightness of figure and limb, but in its appetites, inclinations, and general habitudes. Instead of herding together in large communities, these Deer live in separate families; the sire, the dam, and the young associating together, to the exclusion of all others. Every other species of Deer is inconstant in affection; but the Roebuck never forsakes its mate.

The female goes with young about five months; when she forms a retreat in the thickest part of some wood, and generally produces two at a birth, which she carefully conceals from the buck. The fawns continue to follow the dam eight or nine months; and, on separating, their horns begin to appear, simple and without ramification the first year, as in those of the stag: these they shed at the end of autumn, and renew them during the winter; differing from the stag in this particular, the latter shedding his horns in the spring, and renewing them in the summer. The Roebuck can easily be subdued, but never perfectly tamed. No arts can teach it familiarity with its feeder, much less to show any attachment to him; but it always retains some portion of its natural wildness.

ROLLER. (*Coracias.*) A genus of birds allied to the Crows and Jays, in general distinguished by peculiar elegance and splendour of colours. They are very shy, inhabiting the thickest and most unfrequented woods, though, like the crows, they are frequently seen in newly ploughed fields, searching for worms and larvæ. They are more wild and untractable than their congeners, and do not appear to possess the imitative faculty of the Jay or Magpie. These birds, of which there are several species, are found in Asia, Africa, and the hotter parts of America; but only one is a native of Europe. This is

The COMMON OR GARRULOUS ROLLER. (*Coracias garrula.*) This bird is frequently met with in Italy, France, Spain, and many other parts of Europe, but is seldom seen in England. It is about the size of a Jay, and of an elegant shape. The bill is black, straight, and hooked at the point; the orbits of the eyes bare, and beyond each eye is a small bare spot or protuberance. The head, neck, breast, and under parts are of a light bluish or sea-green colour; the back, and

ROLLER.—(CORACIAS GARRULA.)

feathers of the wings next to it, are of a reddish brown; the shoulders, or smaller wing-coverts, are of a rich ultra-marine blue; the larger coverts bright sea-green; the lower part of the smaller wing-feathers ultra-marine blue, forming a tolerably large patch of that colour on the middle of the wing; the remainder of the wing black: the back and scapular feathers pale chestnut; the rump a fine deep blue, as are also the under parts; and the tail is of a dull green, the two outer feathers, which are rather longer than the rest, being tipped with black. The legs, which are short, are of a dirty yellow hue. Its note is loud and chattering; hence its specific name.

The INDIAN ROLLER. (*Coracias Indica.*) In size this bird is nearly equal to the Common Roller, and is remarkable for the rich and vivid appearance of its colours. The crown of the head is greenish blue; the remainder, with the back, scapulars, neck, and breast, pale ferruginous; the feathers of the throat and upper part of the breast each marked by a paler streak down the shaft: the belly, thighs, and vent are sea-green: the whole wing is varied with deep and light blue, so as to divide it into five alternate bands; the deep or predominating colour being the richest smalt blue, while the paler or middle part is a most brilliant blue-green, changing, according to the direction of the light, into pure pale blue, and forming a delightful contrast with the other. The tail is of the same vivid appearance as the wings; the base and the tips being deep blue, and the middle part pale or greenish-blue. In this species the two exterior tail-fathers do not project beyond the rest. The bill is black; and the legs reddish brown. Native of many parts of India and the Indian islands.

ROOK. (*Corvus frugilegus.*) Both in size and figure the Rook greatly resembles the Carrion Crow; their colour is also similar, the plumage of each being glossed with a rich purple; the principal distinction between the two species is found in the bill of the Rook, the base of which is covered with a rough scabrous skin. These birds are very sociable, and build close to each other in trees, frequently in the midst of large and populous towns. In these aerial abodes they establish a kind of legal constitution, and suffer none to fix their domicile among them but acknowledged natives of the place. Rookeries are, in consequence, often the scenes of bitter contests; the half-built nests of new comers torn in pieces, and the unfortunate couple forced to begin anew in some more undisturbed situation. At the commencement of spring the Rooks begin to build their nests; and one of them brings materials, while the other watches the building, lest it should be plundered by its brethren. All the old inhabitants, however, are already provided with nests; those which served them in former years requiring only a little trimming and dressing to render them equally commodious with new habita-

ROOK.—(CORVUS FRUGILEGUS.)

tions. The young Rooks indeed are unprovided, but they do not long remain so. As soon as the male and female have fixed on an eligible branch, they begin to collect suitable materials; the outside of the nest consisting chiefly of sticks, and the inside usually lined with fibrous roots; the whole regularly and substantially arranged. It not unfrequently happens that some of the old birds take umbrage at the young ones for making choice of a spot too near the abode of the former; hostilities ensue, and a speedy removal is the consequence; but all opposition ceases whenever the female begins to lay, and not one of the whole colony will afterwards molest her. They are gregarious, and fly in immense flocks morning and evening to and from their roosting-places in quest of food.

Rooks are often accused of feeding on the corn just after it has been sown, and various methods have been contrived both to kill and frighten them away; but persons who have paid the most attention to this subject are of opinion that the advantages derived from the destruction which they make upon grubs, larvæ, worms, and noxious insects, greatly overpay the injury done to the future harvest, by the small quantity of corn they may destroy in searching after their favourite food. But as much has been written on this oft-disputed question, as well as on the instincts of Rooks, a few extracts in this place may not be deemed inappropriate.

"When the waters retire from meadows and low lands, where they have remained any time, a luxurious banquet is provided for this *corvus*, in the multitude of worms which it finds drowned on them. But its jubilee is the season of the cockchafer (*melolontha vulgaris*), when every little copse, every oak, becomes animated with it and all its noisy, joyful family, feeding and scrambling for the insect food. The power or faculty, be it by the scent, or by other means that rooks possess of discovering their food, is very remarkable. I have often observed them alight on a pasture of uniform verdure, and exhibiting no sensible appearance of withering or decay, and immediately commence stocking up the ground. Upon investigating the object of their operations, I have found many heads of plantains, the little autumnal dandelions, and other plants, drawn out of the ground and scattered about, their roots having been eaten off by a grub, leaving only a crown of leaves upon the surface. This grub beneath, in the earth, the Rooks had detected in their flight, and descended to feed on it, first pulling up the plant which concealed it, and then drawing the larvæ from their holes. By what intimation this bird had discovered its hidden food we are at a loss to conjecture; but the Rook has always been supposed to scent matters with great discrimination."—*Journ. of a Naturalist.*

"The Rook entices its young from the breeding trees as soon as they can flutter to any other. These young, for a few evenings after flight, will return with their parents, and roost where they were bred; but they soon quit their abode, and remain absent the whole of the summer months. As soon, however, as the heat of summer is subdued, and the air of autumn felt, they return and visit their forsaken habitations, and some few of them even commence the repair of their shattered nests; but this meeting is very differently conducted from that in the spring; their voices have now a mellowness approaching to musical, with little mixture of that harsh and noisy contention, so distracting at the former season, and seems more like a grave consultation upon future procedure; and as winter approaches they depart for some other place. The object of this meeting is unknown; nor are we aware that any other bird revisits the nest it has once forsaken." —*Ibid.*

"There is no wild bird in England so completely gregarious as the Rook, or so regular in its daily movements. The Ring-doves will assemble in countless multitudes, the Finches will unite in vast assemblies, and Waterfowl will flock in thousands to the protected lake, during the dreary months of winter: but when the returning sun spreads joy and consolation over the face of nature, their congregated numbers are dissolved, and the individuals retire in pairs to propagate their respective species. The Rook, however, remains in society the year throughout. In flocks it builds its nest, in flocks it seeks for food, and in flocks it retires for food." . . . . "Sometimes these birds perform an evolution, which is, in this part of the country, usually called the shooting of the Rooks. Farmers tell you, that this shooting portends a coming wind. He who pays attention to the flight of birds has, no doubt, observed this downward movement. When Rooks have risen to an immense height in the air, so that, in appearance, they are scarcely larger than the lark, they suddenly descend to the ground, or to the tops of trees exactly under them. To effect this, they come headlong down, on pinion a little raised, but not expanded, in a zig-zag direction (presenting alternately their back and breast to you), through the resisting air, which causes a noise similar to that of a rushing wind. This is a magnificent and beautiful sight to the eye of an ornithologist. It is idle for a moment to suppose that it portends wind. It is merely the ordinary descent of the birds to an inviting spot beneath them, where, in general, some of their associates are already assembled, or where there is food to be procured. When we consider the prodigious height of the Rooks at the time they begin to descend, we conclude that they cannot effect their arrival at a spot perpendicular under them by any other process so short and rapid." "Rooks remain with us the year throughout. If there were a deficiency of food, this would not be the case; for, when birds can no longer support themselves in the place which they have chosen for their residence, they leave it, and go in quest of nutriment elsewhere. Thus, for want of food, myriads of wild fowl leave the frozen north, and repair to milder climates; and in this immediate district, when there is but a scanty sprinkling of seeds on the whitethorn bush, our flocks of Fieldfares and of Redwings bear no proportion to those in times of a plentiful supply of their favourite food. But the number of Rooks never visibly diminishes; and on this account we may safely conclude that, one way or other, they always find a sufficiency of food. Now, if we bring, as a charge against them, their feeding upon the industry of man, as, for example, during the time of a hard frost, or at seed-time, or a harvest, at which periods they will commit depredations, if not narrowly watched; we ought, in justice, to put down in their favour the rest of the year, when they feed entirely upon insects." — *Waterton's Essays.*

But while admitting the truth of many of the foregoing remarks, in regard to the meritorious services of Rooks, so ably contended for by their protectors and defenders, it is impossible to overlook the fact that they consume an enormous quantity of grain, thereby occasioning great loss to the husbandman, unless they are watched at certain seasons with unremitting assiduity. It was stated at a meeting of Scotch agriculturists, held no longer ago than April, 1847, that there were no less than 2663 Rooks' nests in *one rookery* at Newliston, near Edinburgh; and that, attracted by so numerous a colony, it had become a kind of rendezvous for the species from all parts of the surrounding country, insomuch that the flocks of Rooks almost darkened the air. A calculation had been made, by which it appeared that, allowing their numbers to be 30,000, it would require 30 bolls (or 180 bushels) of wheat to furnish them with *one meal a day!*

RORQUAL. (*Balænoptera.*) A genus of Cetaceous Mammalia, closely allied to the common Whales, but distinguished by having a dorsal fin, with the throat and under parts wrinkled with deep longitudinal folds,

RORQUAL.—(BALÆNOPTERA RORQUAL.)

which are supposed to be susceptible of great dilatation; the use of which in their economy is as yet unknown. Two or three species are known, but they are rather avoided, on account of their ferocity, and the small quantity of oil they produce.

ROSE-BEETLE, or ROSE-FLY. (*Cetonia aurata.*) A well-known Coleopterous insect; about an inch long, of a shining green colour above, coppery red underneath, with white marks on the elytra. In its larva state, it frequents rotten timber, and is often met with underground in ants' nests, where it appears to feed upon the bits of wood of which they are composed. In consequence of this the larva of the Rose-beetle is sometimes called the "king of the ants." Having remained about three years in the larva state, it makes a sort of cocoon of chips of wood, glued together by an excretion of its own; here, in an inactive state, it passes the winter, and emerges in the following summer as a perfect insect. In the heat of the day the Rose-beetle is seen flying from flower to flower, sucking their honey, but evidently preferring the rose to all others.

ROSE CHAFER. The name commonly given in this country to a Coleopterous insect (*Cetonia aurata*) found on the rose. [See CETONIA]. In the United States, according to Dr. Harris, this name is applied to an insect belonging to a different family, which is known as the *Macrodactylus subspinosus.* It is about one third of an inch in length; the body slender, tapering before and behind, and is entirely covered with very short and close ashen-yellow down; the thorax is long and narrow; the legs are slender, and of a pale red colour; the joints of the feet are tipped with black and are very long, which caused Latreille to call the genus *Macrodactylus*, that is, long toe or long foot. The natural history of the Rose Chafer, according to this very observant and intelligent writer, shows it to be one of the greatest scourges with which the gardens and nurseries in the "States" are afflicted, and was for a long time involved in mystery. "For some time after they were first noticed, rose-bugs appeared to be confined to their favourite, the blossoms of the rose; but within thirty years they have prodigiously increased in number, have attacked at random various kinds of plants in swarms, and have become notorious for their extensive and deplorable ravages. The grapevine in particular, the cherry, plum, and apple trees, have annually suffered by their depredations; many other fruit-trees and shrubs, garden vegetables and corn and even the trees of the forest and the grass of the fields, have been laid under contribution by these indiscriminate feeders, by whom leaves, flowers, and fruits are alike consumed. The unexpected arrival of these insects in swarms, at their first coming, and their sudden disappearance, at the close of their career, are remarkable facts in their history. They come forth from the ground during the second week in June, or about the time of the blossoming of the damask rose, and remain from thirty to forty days. At the end of this period the males become exhausted, fall to the ground, and perish, while the females enter the earth, lay their eggs, return to the surface, and, after lingering a few days, die also. The eggs are hatched about twenty days after they are laid; and the young larvæ begin to feed on such roots as are within their reach. They attain their full size in the autumn, being then nearly three quarters of an inch long, and about an eighth of an inch in diameter. They are of a yellowish white colour, with a tinge of blue towards the hinder extremity, which is thick and obtuse or rounded. In October they descend below the reach of frost, and pass the winter in a torpid state. In the spring they approach towards the surface, and each one forms for itself a little cell of an oval shape, by turning round a great many times, so as to compress the earth and render the inside of the cavity hard and smooth. Within this cell the grub is transformed to a pupa, during the month of May, by casting off its skin, which is pushed downwards in folds from the head to the tail. The pupa has somewhat the form of the perfected beetle; but it is of a yellowish white colour, and its short stump-like wings, its antennæ, and its legs are folded upon the breast, and its whole body is enclosed in a thin film, that wraps each part separately. During the month of June, this filmy skin is rent, the included beetle withdraws from it its body and its limbs, bursts open its earthen cell, and digs its way to the surface of the ground. Thus the various changes, from the egg to the full development of the perfected beetle, are completed within the space of one year.

Such being the metamorphoses and habits of these insects, it is evident that we cannot attack them in the egg, the grub, or the pupa state; the enemy, in these stages, is beyond our reach, and is subject to the control only of the natural but unknown means appointed by the Author of Nature to keep the insect tribes in check. When they have issued from their subterranean retreats, and have congregated upon our vines, trees, and other vegetable productions, in the complete enjoyment of their propensities, we must unite our efforts to seize and crush the invaders. They must indeed be crushed, scalded, or burned, to deprive them of life, for they are not affected by any of the applications usually found destructive to other insects. Our insect-eating birds undoubtedly devour many of them, and deserve to be cherished and protected for their services. They are also eaten greedily by domesticated fowls; and when they become exhausted and fall to the ground, or when they are about to lay their eggs, they are destroyed by moles, insects, and other animals, which lie in wait to seize them."

ROSTELLARIA. A genus of Molluscous animals, inhabiting the seas of hot climates, or rather the muddy sand on their coasts. The body is subcylindrical, marbled with rich brown on the outer side, and white on the inner and front side: the trunk is subcylindrical, and annulated with a central broad line of deep bronze-black: the margins yellow with a narrow vermillion line externally. The eyes are on long cylindrical peduncles, of a deep blue with a black pupil: the tentacula are subulate, elongate, arising from the peduncle rather below the eye. The foot is narrow, rather dilated in front and small behind: the operculum is ovate, triangular, annular, semi-transparent, and horny. Like the *Strombidæ*, it progresses by means of its powerful and elastic foot, which it places under the shell in a bent position, when suddenly by a muscular effort it straightens that organ and rolls and leaps over and over. The shell is oblong, turreted, and acuminated; the spire long, consisting of numerous whorls.

ROTELLA. A genus of Mollusca, inhabiting a smooth, shining, orbicular shell, with a conical spire, and horny operculum; left lip very thick, and spreading over the under surface so as to form a callosity. The animal has two very long and pointed tentacula, with eyes at the base; foot short.

ROTIFERA. The name of a class of highly organised Infusorial Crustaceans, commonly called WHEEL-ANIMALCULES (*Rotifer vulgaris*), &c. These wonderfully minute objects possessing life and motion (some of them less than the 500th part of an inch in length!) are of course wholly invisible to the naked eye, but their structure is beautifully revealed to us by the powers of the microscope. Nearly all of them are aquatic in their habits; their bodies are transparent, and consequently their general structure is to be easily recognized. They have usually an elongated form, similar on the two sides; and at the anterior extremity are one or more rows of vibratile *cilia*, usually arranged in a circular manner, which, when in motion, appear like revolving wheels (as in the Wheel Animalcule, which has a circular row of cilia on each side). The posterior extremity is prolonged into a tail, possessing three joints, each of which has a pair of prongs or points. The *cilia* are disposed in two circles, forming what are called the *wheels*. By the successive vibration of these, the appearance of a continual rotation is produced; and their action creates rapid currents in the surrounding fluid, by which the supply of food is obtained. Between the wheels the head is occasionally protruded, bearing two red spots, supposed to be eyes; and on its under surface there is a projecting tubular spike, which is believed to act as a syphon for the introduction of water into the general cavity for the purpose of respiration. The microscope alone reveals the wonderful structure and variety of the animals of this class. Mr. Brightwell of Norwich was the first who discovered the males, while Mr. Gosse has confirmed and extended our knowledge of these interesting Crustaceous animals. He is preparing a fine work on the subject, with admirable drawings.

RUDD, or RED-EYE. (*Cyprinus erythrophthalmus.*) This Acanthopterygious fish, which is from eight to ten inches long, is very common in many of the lakes and rivers of the European continent, and is found in the Thames and various other rivers in the British islands. It has a small head, blunt nose, and orange-coloured irides; back arched, and sloping rather suddenly towards the head and tail; scales large: general colour pale gilded olive, the back being browner, and the whole varying when viewed in different positions in reference to the light; the fins more or less bright red or reddish-brown; dorsal fin rather small, and placed beyond the middle of the back: tail forked. It is said to be a better fish to eat than the Roach, with which it corresponds in size and weight.

RUFF. (*Machetes.*) A genus of Wading birds belonging to the *Scolopacidæ* family. The Ruff (*Machetes pugnax*) is a bird of a very pugnacious character; the female of which is called the REEVE. It is about a foot in length; and is principally distinguished by a very remarkable circle of long feathers round the neck, whence it receives its name: in some birds these feathers are black, in others white, yellow, or ferruginous; and even in the same bird they frequently differ in colour. It is only the male, however, that is furnished with this appendage, which he does not gain till the second year. These birds are migratory, appearing at certain seasons of the year, in great numbers, in the north of Europe. They arrive in this country early in the spring, take up their abode in Lincolnshire, Yorkshire, &c., and disappear about Michaelmas. Soon after their arrival, the males begin to *hill*, as it is termed; that is, to assemble on some dry bank, near a pool of water, in expectation of the females, which there resort to them. Each male takes possession of a small spot of ground, round which he runs so often as to make a bare circular path; and as soon as a female, alights, all the males within a certain distance commence a general fight, placing their bills to the ground, spreading their ruff, and using the same action as the common cock. They are generally taken in large nets. When fattened, they are dressed like the Woodcock, without withdrawing the intestines; and when killed at the proper season, are reckoned a most delicious treat for an epicure. The pugnacious disposition of these birds is so strong, that when they are kept for the purpose of fattening, their place of confinement is obliged to be dark, as, the moment any light is admitted, they attack each other with such fury as to occasion a great slaughter. The female lays four white eggs, marked with large rusty spots, in a tuft of grass, during the first week in May, and sits on them about a month.

RUMINANTIA. An order of herbivorous Mammalia, which not only feed exclusively on vegetable matter, but which ruminate, or 'chew the cud,' (thereby meaning, that they possess the faculty of masticating a second time their food, which they return into the mouth after a previous deglutition;) as Oxen, Sheep, Deer, Goats, Camels, &c. "The stomach of the Ruminants is especially organized for rumination, consisting of four distinct cavities, all of which communicate with a muscular canal, at the termination of the œsophagus. Hard, solid, or coarsely masticated food passes from the beginning of the muscular canal into the first cavity of the stomach, called the *rumen*, or paunch. Water is received into the second cavity, called the *reticulum*, and almost exclusively occupies the honeycomb cells of that cavity; it is gradually mixed with the coarsely divided food which is undergoing mastication in the rumen. When this is sufficiently advanced, a portion of the mass is received into the muscular canal at the termination of the œsophagus: it is there moulded into a ball, and propelled by a rapid and inverted action of the muscles of the gullet into the mouth, where it is more perfectly masticated, mixed with fluid, and again swallowed. It now passes directly into the third stomach, called the *psalterium*, from the broad leaf-like plates of membrane with which it is occupied; here the superfluous fluid, which otherwise might have too much diluted the gastric juice, is absorbed, and the subdivided cud passes gradually into the fourth or true digesting stomach, called the *abomasus*."—*Brande's Dict.* The senses of the Ruminantia are extremely acute, and serve to indicate to them the approach of danger, as well as to direct them in their choice of food. Their eyes are placed at the side of the head, so that their range of vision is greatly extended. The ears also are placed far back, and are very movable; so that they can be turned to catch sounds in any direction; and their sense of smell is particularly acute. Of all animals, Ruminants are the most useful to Man. In the first place, they furnish him with nearly all the animal flesh which he consumes. Some of them serve him as beasts of burden; and others supply him with milk, tallow, hides, horns, and other products most important to his comfort, and even to his subsistence. Many of them have from the earliest periods been domesticated, and have accompanied Man in his gradual diffusion over the globe; while some, as the Rein-deer and Camel, are invaluable in certain localities, to which they are expressly and admirably adapted.

RUSTIC [MOTHS]. A name given by collectors to species of Moths, of the genera *Charœas*, *Rusina*, *Caradrina*, and *Segetia*.

RUTELIDÆ. A group of Coleopterous insects, in some respects allied to the Melolonthidæ and Cetoniadæ. The body is shorter, rounder, and more polished than in the Scarabæidæ, and ornamented with brilliant colours. The head and thorax are identical, and not cornuted in either sex; the maxillæ are scaly, truncated at the tip, with five or six strong teeth. The mesosternum is often porrected, the scutellum large, and the tarsal claws unequal-sized. With few exceptions, they are confined to the warmer parts of America. Dr. Thaddeus Harris has described a well-known American species. He says, "One of the most common and the most beautiful of the Tree-beetles of this country is the *Areoda lanigera*, or woolly Areoda, sometimes also called the GOLDSMITH-BEETLE. It is about nine-tenths of an inch in length, broad oval in shape, of a lemon-yellow colour above, glittering like burnished gold on the top of the head and thorax; the under side of the body is copper-coloured, and thickly covered with whitish wool; and the legs are brownish-yellow, or brassy, shaded with green. These fine beetles begin to appear in Massachusetts about the middle of May, and continue generally till the 20th of June. In the morning and evening twilight they come forth from their retreats, and fly about with a humming and rustling sound among the branches of trees, the tender leaves of which they devour. Pear-trees are particularly subject to their attacks, but the elm, hickory, poplar, oak, and probably also other kinds of trees, are frequented and injured by them. During the middle of the day they remain at rest upon the trees, clinging to the under sides of the leaves; and endeavour to conceal themselves by drawing two or three leaves together, and holding them in this position with their long unequal claws. In some seasons they occur in profusion, and then may be observed in great quantities by shaking the young trees on which they are lodged in the day-time, as they do not attempt to fly when thus disturbed, but fall at once to the ground. The larvæ of these insects are not known; probably they live in the ground upon the roots of plants."

SABELLA. A marine animal, belonging to the second order of *Annelida*, which forms its tube or shell partly by a calcareous exudation from its own body and partly by granules of clay or fine mud. The species are rather large, and their branchial tufts extremely delicate.

SABLE. (*Mustela zibellina.*) Of all the Weasel tribe this is the most celebrated, not only on account of the richness of its fur, but from the horrors of the chace, which is carried on in the depth of winter among mountains covered with ice and snow, in the coldest and most desolate regions into which man has yet penetrated. The Sable has long whiskers, rounded ears, large feet, the soles of which are covered with fur, white claws, and a long bushy tail. The general colour of the fur is brown, more or less brilliant, with the lower parts of the throat and neck grayish. They resemble the rest of the weasel kind in vivacity and agility; in sleeping by day, and hunting their prey by night: they usually live in the depths of the forest, in holes of the earth, or beneath the roots of trees; and sometimes, like the marten, they form their nests in the boughs of trees. The females bring forth from three to five young at a time, which they suckle for a month or five weeks. They inhabit all the northern parts of Europe and Asia; and as prodigious numbers are killed in Siberia, their skins form a very considerable article of commerce with the Russians.

Sables' skins are in the highest perfection betwixt the months of November and January; accordingly, at the commencement of the winter, the Sable hunters assemble in very considerable companies, and proceed along the great rivers in boats, taking with them provisions for three or four months. When they arrive at their place of rendezvous, the different parties, each under the direction of a leader, take up their respective quarters, where they form huts of trees, and bank up the snow round them: near these they lay their snares; and then advancing farther, they set more, still building new huts in every quarter, and returning successively to every old one, to visit the traps, from which they take the game and skin it. Their snares or traps are generally a sort of pitfalls, with loose boards placed over them, baited with fish or flesh: but when Sables grow scarce, the hunters trace them to their holes through the new-fallen snow, place nets at their entrances, and frequently watch two or three days for the appearance of the animals. Other modes of taking them are also resorted to; sometimes fire-arms are used, and sometimes cross-bows. When the nature of the employment, and the intense cold which the Sable hunters must endure in the depth of a Siberian winter, are considered, we think there are few persons dwelling in more congenial climes who are likely to envy them the *sport*.

SACCOPHARYNX AMPULLACEUS, or BOTTLE-FISH. This anguilliform fish belongs to the singular genus *Saccopharynx*, in which the body, capable of being inflated like a sack or leathern bottle, is terminated by a very long and slender whip-like tail, edged above and below by the narrow dorsal and anal which unite at its tip. It is thus described by Dr. Richardson in his 'Fauna Boreali-Americana:'—the mouth, armed with long sharp teeth, is cleft far past the eyes, which are close to the very short pointed snout. The gill-openings, having the form of irregular slits, and large enough to permit the three branchiæ to be seen, are under the very small pectorals. The skin is soft, slimy, loose, and slightly granular in appearance. The extensibility of the jaws and throat is extraordinary, being even greater than that exhibited by the serpent tribe. Only two examples of the genus are known to have been taken, and, with the exception of dimensions, they realise many of the popular accounts of the great American sea-serpent. They are voracious fish, with a capacious stomach and short straight gut. One of the specimens had recently before its capture swallowed a fish longer than its own body, and the other had apparently exhausted itself in vain attempts to gorge a sea-perch thicker than itself. The individual described by Dr. Harwood (*Saccopharynx ampullaceus*), measuring four feet and a half in length, was captured in the entrance of Davis' Strait, by Capt. Sawyer, of the ship Harmony; the other (*Saccopharynx chordatus*), which was six feet long, was taken by Capt. Hector Coffin, about midway between the Labrador coast and Ireland, in the fifty-second parallel of latitude.

SAGOIN, or SQUIRREL MONKEY. The little animals belonging to this group are extremely light, active, and graceful in their movements, as well as elegant in their forms. They use their tail as a protection against cold, to which they are acutely sensitive. Their food chiefly consists of insects, eggs, and small birds.

SAJOU. A lively and active Monkey, of the genus *Cebus;* docile, but somewhat capricious. It has a prehensile tail, though it is not so delicate an organ of touch as in some other species. In their native forests they live in troops; feeding on fruits, grain, eggs, &c. [See MONKEYS.]

SAKI. A monkey belonging to the genus *Pithecia*, and called the Fox-tailed Monkey. These animals usually reside in the outskirts of forests, in small societies of ten or twelve individuals. Upon the slightest provocation they display a morose and savage temper; and, like the Howlers, they utter loud cries before sunrise and after sunset.

SALAMANDER. (*Salamandra.*) A genus of reptiles, closely allied to the frog, but differing from it in having an elongated body, a long tail, and four feet of equal length. They have the general form of lizards, but have all the characters of *Batrachians*, and have therefore been removed from the genus *Lacerta*, where Linnæus had placed them. The head is flattened; the jaws are armed with numerous small teeth; and there are two longitudinal rows on the palate. The young are born in the shape of tadpoles, are provided with gills, and have their tails vertically compressed. In the adult state they respire in the same manner as frogs and tortoises. The terrestrial Salamanders inhabit the water only during the tadpole state, or during the time that they are laying their eggs: they are distinguished by a rounded tail. The aquatic species remain during life in water, and are enabled to swim with considerable briskness

SALAMANDER.—(SALAMANDRA MACULATA.)

by means of their compressed tails. They possess the most extraordinary powers of reproducing their parts; renewing, many times successively (according to the experiments of Spallanzani), the same member after it had been severed, and this with all its bones, muscles, vessels, &c. Another faculty, not less singular, consists (as shown by Dufoy) in their recovering after having been long frozen up in ice.

The COMMON SALAMANDER of Europe (*Salamandra vulgaris*) is a sluggish, clumsy reptile, six or eight inches long, of a blackish colour, with large, irregular, rounded spots of bright yellow. It is found in moist places, under stones or the roots of trees, near the borders of springs, in deep woods, &c., and passes its life under ground, except during rains or at night, when it comes out, but does not wander far from its place of residence. It lives on slugs, insects, worms, &c.; does not appear to shun the presence of man or other animals; is oviparous; and exudes a mucous and acrid secretion in great abundance. Among the most absurdly ignorant of all popular superstitions, was that which ascribed to this poor reptile the power of subsisting in the fire: and how the idea could ever have originated appears truly wonderful, when all the haunts and habits of the animal are connected with cold and moisture. There are many species in North and South America.

The GIGANTIC SALAMANDER. [See SIEBOLDIA in SUPPLEMENT.]

SALLOW [MOTHS]. A name given by collectors to Moths of the genus *Xanthia*.

SALMO. A numerous genus of Malacopterygious fishes, most of which are highly prized as food; among the species are the

SALMON. (*Salmo salar.*) This well-known fish, so highly esteemed for its delicacy of flavour, and so important in a commercial sense, is one of the largest and most plentiful species of the *Salmonidæ*, or Salmon and Trout tribe, a family of fishes belonging to the Malacopterygii Abdominales. They have the body covered with scales, and are characterized by having all the rays of the first dorsal fin soft or jointed, and the second dorsal entirely adipose : they are generally very muscular, and possessed of great strength ; and they are voracious in their habits, feeding rather upon insects and small crustacea than upon other fishes. The common Salmon (*Salmo salar*) is chiefly an inhabitant of the northern temperate regions, where it occurs at different periods both in salt and fresh waters ; quitting the

SALMON.—(SALMO SALAR.)

sea at certain seasons to deposit its spawn in the gravelly beds of rivers, at a great distance from their mouths. It grows to the length of three, four, or five feet, and is usually about ten or twelve pounds when taken ; but the full-grown Salmon averages a weight of between twenty and thirty pounds. Enormous specimens, however, are now and then captured : sometimes weighing forty or fifty pounds ; and it is a fact that, in 1821, a Salmon was exhibited in a fishmonger's shop in London (Mr. Grove's of Bond Street), weighing eighty-three pounds. It was a female fish, of extraordinary thickness, good colour, and excellent quality. The body of the Salmon is elongated and compressed ; the colour a dark blue, dotted with black spots on the back ; silvery gray on the sides with spots, and white with a faint shade of pink below ; the head of moderate size, and the upper jaw rather the longest. Almost all parts of the mouth are furnished with pointed teeth. The usual time at which the Salmon leaves the sea, is the autumn ; it remains in the rivers during the winter ; and returns to the sea after having deposited the spawn, in the spring. In ascending rivers there are scarcely any obstacles which these fish will not surmount : they will force themselves against the most rapid streams, and spring with amazing agility over cataracts of ten or twelve feet in height. On this account, small cascades on the Tweed, the Severn, and other rivers where they resort, are called Salmon-leaps. If alarmed, they dart away with such velocity that the eye can scarcely follow them. They penetrate far into the interior of the continents, and deposit their spawn near the head-waters of the longest rivers ; but before depositing it, the Salmon makes a furrow in the gravelly bed of the river ; and its eggs, when deposited in this, are carefully covered up. When the young are about a foot in length, they descend the rivers, and take refuge in the ocean. Late in the following spring, or the beginning of summer, and after the old ones have ascended, the young again enter the rivers, and are then about eighteen inches in length. They again seek the ocean on the return of frosts. At two years old the Salmon weighs six or eight pounds, and generally requires five or six years to attain the weight of ten or twelve.

In Mr. Yarrell's excellent work on British Fishes, much information is given, both with respect to the habits of those which resort to our rivers and the various modes of taking them. "The adult fish having spawned, being out of condition, and unfit for food, are considered as unclean fish. They are usually called *Kelts;* the male fish is also called a *Kipper*, the female a *Shedder*, or *Baggit*. With the floods of the end of winter and the commencement of spring they descend the river from pool to pool ; and ultimately gain the sea, where they quickly recover their condition, to ascend again in autumn for the same purpose as before ; but always remaining for a time in the brackish water of the tide-way before making either decided change ; obtaining, it has been said, a release from certain parasitic animals, either external or internal, by each seasonal change ; those of the salt water being destroyed by contact with the fresh, and *vice versâ*." "The Salmon fry at first keep in the slack water by the sides of the river ; after a time, as they become stronger, they go more towards the mid-stream ; and when the water is increased by rain, they move gradually down the river. On meeting the tide, they remain for two or three days in that part where the water becomes a little brackish from the mixture of salt water, till they are inured to the change, when they go off to the sea all at once. There, their growth appears to be very rapid, and many return to the brackish water, increased in size, in proportion to the time they have been absent." "It has been a constantly received opinion, that all the young fish after their first visit to the sea return to the rivers in which they had been bred ; and numbers of marked fish are stated to have been retaken in their native rivers : but it is equally certain that some have been taken in other rivers not far off. The difficulty of supposing that they could find and return to the same spot after roving for miles along the coast remains to be solved. That they do thus rove for miles is proved by the thousands that are taken in nets placed in the bays along the coast." The flesh of the Salmon is of a bright orange colour when raw, redder when salted, and a little paler when boiled ; as a food it is rich, tender, and sweet ; it is, however, considered to be difficult of digestion, and should be eaten as early as possible after its capture, it being very unwholesome when stale. This, indeed, may be remarked of all the *Salmonidæ*. The principal Salmon fisheries in Europe are in the rivers, or on the sea-coasts adjoining the mouths of the large rivers of England, Scotland, and Ireland. The Tweed is the most famous river for Salmon fishing, and prodigious quantities are caught there :

in several other large streams also very considerable quantities are taken; as the Severn, the Mersey, the Thames, the Tyne, the Trent, the Medway, &c. A young Salmon under two pounds in weight is called a *Salmon Peel*, and a larger one a *Grilse*.

In the Transactions of the Royal Society of Edinburgh is an account of repeated observations and experiments by Mr. Shaw of Drumlanrig, clearly proving that the small Salmonoid fish, called the *Parr*, is, as many naturalists had suspected, the young of the Salmon.

How far the legitimate province of a writer on zoology may extend, when describing the habits and instincts of animals, we are unable exactly to define, or what bounds are to be prescribed to his fancy (if he happen to possess any); but we would rather incur the charge of supererogation, justly founded or not, than forego the strong inclination we feel for the adoption of an apposite passage — particularly one so graphic and spirited as the following description of the capture of a salmon, from the vigorous pen of the well-known Christopher North:— "She is a salmon, therefore to be viewed — she is a salmon, therefore to be won; but shy, timid, capricious, headstrong, now wrathful, and now full of fear: the cruel artist has hooked her, and in spite of all her struggling, will bring her to the gasp at last." — "But the salmon has grown sulky, and must be made to spring to the plunging stone. There, suddenly instinct with new passion, she shoots out of the foam like a bar of silver bullion; and relapsing into the flood is in another moment at the very head of the waterfall! Give her the butt—give her the butt — or she is gone for ever with the thunder into ten fathoms deep! Now comes the trial of your tackle — and when was Phin ever known to fail at the edge of cliff or cataract? Her snout is southwards — right up to the middle of the hill-born river, as if she would seek its very source where she was spawned. She still swims swift and strong, and the deep, and the line goes steady. There is yet an hour's play in her dorsal fin — danger in the flap of her tail — and yet may her silver shoulder shatter the gut against a rock." — "What another mad leap! yet another sullen plunge! Ha, ha, my beauty! Methinks we could fain fond and kiss thy silver side, languidly lying afloat on the foam, as if all further resistance now were vain, and gracefully thou wert surrendering thyself to death! No — she trusts to the last trial of her tail — sweetly workest thou, O reel of reels! and on thy smooth axle spinning sleeps't, even as Milton describes her, like our own worthy planet." — "The gaff! the gaff! Into the eddy she sails, sick and slow, and almost with a swirl — whitening as she nears the sand — there she has it — stuck right in the shoulder — and lies at last in all the glorious length and breadth of beaming beauty, fit prey for giant or demigod angling before the Flood!" —*Chris. North's Recreations.*

With another picturesque *morceau*, from the 'Days and Nights,' &c. of Mr. Scrope, we will take our leave of this noble sport-creating fish. The author is remarking on the difference between fishing for Salmon in the briny tide and in its favourite rivers, and exclaims: "No, the wild main I trust not. Rather let me wander beside the banks of the tranquil streams of the warm South, 'in the yellow meads of Asphodel,' when the young spring comes forth, and all nature is glad; or if a wilder mood comes over me, let me clamber among the steeps of the North, beneath the shaggy mountains, where the river comes foaming and raging everlastingly, wedging its way through the secret glen, whilst the eagle, but dimly seen, cleaves the winds and the clouds, and the dun deer gaze from the mosses above. There, amongst gigantic rocks, and the din of mountain torrents, let me do battle with the lusty Salmon, till I drag him into day, rejoicing in his bulk, voluminous and vast."

SALMO ROSSII; or ROSS'S ARCTIC SALMON. This species was named by Dr. Richardson in honour of Capt. Sir James Clark Ross, "whose scientific acquirements and contributions to Natural History" are equalled only by the "professional skill,

SALMO ROSSII.

exertions, and perseverance" he exhibited during his various expeditions of discovery in the Arctic seas. The *Salmo Rossii* is of a more slender form than the common Salmon, with a straighter back, much less arched forehead and shoulders, and slightly larger head. The remarkable length of the under jaw, and the truncated snout, give a peculiar appearance to the fish, and, in conjunction with the nature of the scales (which are small, and each surrounded by a distinct space of smooth skin), and the colour of the skin, readily characterize the species. In regard to colour, the back, top of the head, dorsal and caudal fins, have a hue intermediate between oil-green and hair-brown; the cheeks are nacry, and the sides pearl-gray, with a blush of lilac and a silvery lustre; near the lateral line are numerous scattered dots of carmine; and the colour of the belly varies in different individuals from faded orange to deep red. — "The *Salmo Rossii* is so extremely abundant in the sea, near the mouths of the rivers of Boothia Felix, at certain seasons, that 3378 were obtained at one haul of a small-sized sean. They varied in weight from two to fourteen pounds, and rather exceeded, in the aggregate, six tons. In some the colour of the flesh was of a dark red, in others it was very pale, the dark ones being the firmest and best flavoured." Dr. R. adds, that the *malma*, or *golet* of the Russians, which enters the rivers of Kamtschatka, agrees with the *Salmo Rossii* in its comparatively slender cylindrical form, scarlet spots on the sides, and the colours of some other parts; but that the habits of the two are evidently unlike, if it be true, as is asserted, that the malma never congregates in shoals.

SALMO ALBUS. This fish, which bears the name of *Attihawmeg* by the native Indians, many of whom mainly subsist upon it, is an inhabitant of all the interior lakes of America, from Erie to the Arctic Sea. It belongs to the sub-genus *Coregonus*, family *Salmonideæ*. The Attihawmeg has some resemblance to the Herring in the structure of its jaws and gill-covers, and, like that fish, it dies speedily when taken out of the water. It measures about twenty inches in length; its usual weight is from two to three pounds, and when very fat it attains to seven or eight pounds, and occasionally more; but these large fish are confined to particular localities. The form of the Attihawmeg is ovate, more or less gibbous before the dorsal fin, with a slightly-tapering tail inclining a little upwards. The body is compressed; the upper surface of the head is smooth and even; the eyes are large, and situated a little more than a diameter of the orbit from the tip of the snout, which projects a little from the shut mouth. The jaws and tongue are furnished with a few teeth, which are too minute to be readily seen by the naked eye, and too slender to be very perceptible to the finger: the palate and vomer are quite smooth. The scales are about half an inch in diameter; they have a bright pearly lustre, and are thin and very deciduous. The caudal fin is forked, and spreads widely. The colour of the Attihawmeg, in the shade, is a bluish-gray on the back, lighter on the sides, and white on the belly; but when in a full light, it assumes a nacry and iridescent pearly lustre.

In certain lakes, and in some seasons, this fish is loaded with fat, particularly about the shoulders, where it produces a hump; but though it is rich and fat, instead of producing satiety it daily becomes more agreeable to the palate; and it is confidently asserted, that, though deprived of bread and vegetables, one may live wholly upon this fish for months, or even years, without tiring. After the spawning season its flesh becomes lean and rather watery, but not unwholesome, and it may be improved by being hung in the open air for a month or six weeks; at least it is allowed by the ichthyophagists of the fur countries to be richer, firmer, and altogether more agreeable to their palates. It is a gregarious fish, and resorts to different parts of a lake according to the season of the year, its movements being in all probability regulated by its supply of food. In winter the fisheries are generally established in deep water, remote from the shore; after the spawning period, the full-fishery, as it is termed, is more productive in shallow bays and on banks near the shore. The Attihawmeg feeds on soft insects and small shelly mollusca; and it is worthy of observation that it differs from the other known *Coregoni* in the extraordinary thickness of its stomach, which has been thought to bear some resemblance to the gizzard of a fowl.

SALMON TROUT. (*Salmo Trutta.*) This fish, which in Scotland is called the Sea Trout, is next in value to the Salmon, and in its habits exceedingly similar. It has a large smooth head, of a dusky colour, with a gloss of blue and green; the back is of the same colour, except that it becomes fainter towards the lateral line: the sides, as far as the lateral line, are marked with large, irregular-shaped spots of black; and the abdomen is white. Like the Salmon, this fish migrates to and from the sea, and consequently, when it has entered the rivers in order to deposit its roe, it is occasionally found in lakes and streams at a great distance from the sea. They continue in season during the whole summer; and may be angled for either in the mornings or evenings. They are usually from about two to four pounds weight; and great quantities are sent from Scotland to the London market. The flesh is much esteemed, but it ought to be dressed as soon as possible.

The 'Fordwich Trout' of Izaak Walton, as we are informed by Mr. Yarrell, is the Salmon Trout; "and its character for affording 'rare good meat,' besides the circumstance of its being really an excellent fish, second only to the Salmon, was greatly enhanced, no doubt, by the opportunity of eating it very fresh. Fordwich is about two miles east-north-east of Canterbury. The stream called the Stour was formerly very considerable; it communicates with the sea opposite the back of the Isle of Sheppy, and from Fordwich, one branch, going eastward, again enters the sea at Sandwich. The ancient right to the fishery at Fordwich was enjoyed jointly by two religious establishments: it is now vested in six or seven individuals, who receive a consideration for their several interests. It was formerly the custom to visit the nets at Fordwich every morning to purchase the fish caught during the night. I have seen specimens of the Salmon Trout from the Sandwich river exposed for sale in the fishmongers' shops at Ramsgate, during the season for visiting that watering-place; and the Salmon Trout is also occasionally taken in the Medway by fishermen who work long nets for smelts during the autumn and winter." The same writer also says, "This fish is the White Trout of Devonshire, Wales, and Ireland; it is found in the Severn, in the rivers of Cornwall, and is plentiful in the Esk and the Eden, which communicate with the Solway, where it is called Sea Trout." "Great quantities of it are sent to the London market; those from Perth, Dundee, Montrose, and Aberdeen appear, from their comparative depth of body, to be better fed, are higher in colour, and are considered to be finer in flavour than from some other localities."

SALPA. [See SUPPLEMENT.]

SALPINGIDÆ. A family of Coleoptera, having the head produced in front into a flattened rostrum; the antennæ inserted in front of the eyes; the body generally oval, or oblong and depressed. The species are of small size, sometimes brightly coloured, and are found either beneath the bark of trees or in flowers.

SALTATORES, or SALTIGRADES. A tribe of Spiders, so named in consequence of their legs being fitted rather for *leaping* than for running. Many species of this group construct, amongst leaves, under stones, &c., silken nests, open at each end, into which they retire; but if menaced with danger, they make a precipitate retreat. One species (*Salticus scenicus*) is very commonly seen in summer upon walls and windows exposed to the sun, moving about in short leaps. When it discovers a small fly or a gnat, it cautiously approaches till within leaping distance, when it suddenly darts upon it; not fearing to take even a perpendicular leap, because it always at the same time suspends itself by a thread, which it winds off as it advances. By this thread it can also suspend itself in the air, and is enabled to mount up again to the spot from which it leaped.

SALTATORIA. A section of Orthopterous insects, corresponding with the Linnæan genus *Gryllus*, and consisting of all those species which have the four anterior legs simple and short, and the two hind legs long, and formed for leaping. The body is generally compressed; the tarsi vary in the number of their joints, as well as the antennæ, which are also greatly variable in length, being in some species several times longer than the body. The males are enabled to make a peculiar shrill noise, which is produced in different ways in different groups; being in some caused by the friction of the posterior femora against the wing-covers, and in the others by the friction of the strong veins enclosing a talc-like spot at the base of the wing-covers. In general, the females deposit their eggs by the assistance of a horny ovipositor, in the earth; and the species are almost exclusively herbivorous. [See CRICKETS and LOCUSTS.]

SANDERLING. (*Arenaria vulgaris.*) A small wading bird which frequents many of our shores, and is a pretty general inhabitant of the globe. It is about eight inches in length. Its autumnal and winter plumage differs considerably from that which it assumes in spring: the face, throat, neck, and the whole of the under parts of the body, being of a pure white in winter; whereas in spring the face and top of the head are marked with large black spots, and the feathers are bordered with red; and the neck, breast, and upper parts of the sides, are gray-red, with the middle of each feather spotted with black, and their tips whitish: the back and scapulars are deep rufous, with large black spots, and the whole of the feathers edged and tipped with white: beak, irides, and feet, black. It feeds on small marine insects; breeds in the north; and is sometimes called the Ox-bird.

SANDPIPER. (*Totanus.*) This name is applied to different species of wading birds of the genus *Tringa*, but properly restricted to the sub-genus *Totanus*. The Sandpipers chiefly frequent saline marshes and the seashore; but they are also found on the banks of inland lakes and rivers, and even in damp meadows. They fly in flocks, and perform periodical migrations in large bodies. Their food consists of worms, crustacea, and small mollusca, and they also occasionally subsist upon small fish and their fry. They have the tip of the beak depressed, and the nasal furrow very long, as in the Godwits, but the mandibles in general are not longer than the head; their toes are not palmated at the base, and the back toe hardly reaches to the ground. Many of the species are very widely diffused, and several are found, more or less regularly, on the British shores.

The COMMON SANDPIPER. (*Totanus hypoleucos.*) This species, which is less than eight inches long, visits England in the spring, and leaves it in the autumn. All the upper parts of the body are brown, glossed with an olive hue, and marked with a blackish ray in the direction of the shafts: the feathers of the wings and back are transversely streaked with narrow zigzag dusky bands; the throat, breast, and under parts are pure white, the sides of the neck and breast being varied with streaks of brown: the middle tail feathers are marked with transverse dusky lines, and the outer one on each side white with brown bars; legs greenish ash. The nest of this bird is composed of moss and dry leaves, and generally placed in a hole on a river's bank; and the eggs, usually five in number, are of a reddish-green, with dark spots mostly at the larger end. When disturbed they make a clear piping note, by which they are easily recognized.

There are several species of these birds, differing but slightly from each other, and we may remark of them, generally, that their legs are destitute of feathers for some distance above the knee, and the toes are short and incapable of grasping; hence they do not perch, but frequent the borders of ponds, rivers, and marshes, especially in the vicinity of the ocean, and are often seen rapidly coursing along the strand, following the flux and reflux of the waves. Their wings are long, and their flight powerful.

SAND-WASP. (*Ammophila.*) A genus of Hymenopterous insects, which, together with several other genera, form a group that from their peculiar habits are termed *Fossores*, or diggers, and commonly known as

SAND-WASP.—(AMMOPHILA ARENARIA.)

Sand and Wood Wasps. In general the females excavate cells in the ground, or in posts, timbers, &c.; in which they deposit—together with their eggs—various larvæ or perfect insects, and (in some species) even spiders, which are destined for the support of their progeny when hatched. It happens that the insects composing this store are sometimes first stung to death; but more frequently they are only slightly stung, and finally killed by the larvæ when they come forth from their eggs,—being in this manner rendered powerless, whilst their bodies are prevented from decomposing. The antennæ have about thirteen joints, attenuated exteriorly, and mostly recurved; mandibles long, and dentate at the apex; labium short, with its ligula short and trilobed; ocelli three, distinct; wings alike in both sexes; legs long, spiny; female armed with a sting. The Sand-Wasp inhabits sunny banks in sandy situations, running among grass, &c. with great activity, and continually vibrating its antennæ and wings. It feeds on insects.

SAPAJOU. A small species of Monkey, of the genus *Cebus*. [See MONKEYS.]

SARCIOPHORUS. A genus of Grallatorial birds closely allied to the Pewits, which derive their name from the wattles or fleshy appendages about the neck. The species SARCIOPHORUS PECTORALIS, or BLACK-BREASTED PEWIT, inhabits South Australia, Van Diemen's Land, &c., its favourite localities being open sterile downs, thinly covered with vegetation, and occasionally to be met with on the grassy flats in the neighbourhood of rivers. It trips very quickly over the ground, much after the manner of the true Pewits, and when flushed generally flies off in a straight line, very near the ground. Crown of the head, line running from the angle of the mouth beneath the eye, and down the sides of the neck, and a broad crescent-shaped band across the breast, jet-black; line from the eye to near the occiput, chin, throat, flanks, abdomen, upper and under tail-coverts, white; back light brown; primaries brownish black; scapularies and lower part of the back bronzy brown, passing into black towards the tip of each feather, and tipped with white; tail white, crossed near the tip by a broad irregular band of black; bill primrose yellow, the tip horn-colour; naked parts of the thigh and knees dark pink; tarsi and toes blackish brown, the latter inclining to pink-red.

SARCORAMPHUS, or KING OF THE VULTURES. This powerful species of the *Vulturidæ* family is about two feet and a half in length, and upwards of five feet across the expanded wings. The naked skin of the head and neck is brilliantly coloured; the beak reddish, with a shade of black; cere bright orange, prolonged between the nostrils into a comb, loose in texture, and falling on each side of the bill when the head is erect. The back of the head is covered with a short blackish down, and the side of the head is purplish black. A scarlet circle surrounds the eye; and on each side behind the eye are several broad and deep wrinkles, whence rises a thick fold extending obliquely downwards along the neck: from the bright red upper part of the neck the colour gradually lessens in intensity, fading into orange and yellow lower down. Round the bottom of the neck is a broad ruff of soft, downy, ash-gray feathers: the back and tail-coverts bright fawn;

KING OF THE VULTURES.
(SARCORAMPHUS PAPA.)

greater wing-coverts and tail-feathers glossy black; legs and claws dusky, or dirty yellow.

In the central parts of America the *Sarcoramphus papa* is frequently to be seen, alone or in pairs, perched on the highest trees; though it is said that considerable flocks may be occasionally met with. The expanse and strength of wing of the King Vulture enables it to reach a lofty height, where its piercing sight brings under its observation a wide tract of country, while its exquisite sense of smell detects the effluvia arising from putrid fish which during the summer perish in the lakes. The story that the other vultures stand patiently by till this, their monarch, has finished his repast, may be accounted for by the superior strength and courage of this species; and it is, in fact, confirmed by the testimony of many travellers; among others, by Mr. Edwards, in his 'Voyage up the Amazon;' who observes—"If a King Vulture makes its appearance where a number of the other species are collected about carrion, the latter instinctively give way and stand meekly around while their sovereign leisurely gorges himself. These birds are not very common upon the Amazon, and we never had an opportunity of shooting them, but several times we observed them circling in pairs over the forest. Senhor Henriquez informed us at the Barra that they were not unfrequently taken alive, particularly if a putrid snake, of which they are fond, be exposed to them. A noose is arranged to fall over their heads, and the caught bird is transformed from a wild marauder into a peaceable citizen. At Para they are highly valued. We saw a pair in perfect plumage which were presented to Mr. Norris, and felt nothing of the disgust inspired by the other common species. Their bare necks were beautifully marked with red and black, orange and yellow, and were surrounded near the base by a ruffle of feathers. Their breasts were white, and the general colour of the upper parts was a light ashy gray. These birds were very active, moving about the yard with a leap rather than a step." They are said to make their nests in the hollows of trees, and to lay two eggs.

SARDINE. (*Clupea sardina.*) A fish closely allied to the Pilchard, though smaller. It is found in the Mediterranean, and its flesh is highly esteemed.

SATIN BOWER BIRD. (*Ptilonorhynchus holosericeus.*) Mr. Gould, to whose splendid work on the 'Birds of Australia' we are so much indebted, observes, that although this species has been long known to ornithologists, and is familiar to the colonists of New South Wales, its habits have never been brought before the scientific world; and he very naturally congratulates himself on being the first to place them on record. One point to which he more particularly alludes — a point, as he truly says, "of no ordinary interest, both to the naturalist and the general admirer of nature — is the formation of a bower-like structure by this bird for the purpose of a playing ground or hall of assembly,— a circumstance in its economy which adds another to the many anomalies connected with the fauna of Australia.' It appears to be altogether granivorous and frugivorous. "Independently of numerous berry-bearing plants and shrubs, the brushes it inhabits are studded with enormous fig-trees, some of them towering to the height of two hundred feet; among the lofty branches of these giants of the forest, the Satin Bower-bird and several species of Pigeons find in the small wild fig, with which the branches are loaded, an abundant supply of a favourite food: this species also commits considerable depredation on any ripening corn near the localities it frequents." The extraordinary bower-like structures above alluded to are usually placed under the shelter of the branches of some overhanging tree in the most retired part of the forest: "the base consists of an extensive and rather convex platform of sticks firmly interwoven, on the centre of which the bower itself is built: this, like the platform on which it is placed and with which it is interwoven, is formed of sticks and twigs, but of a more slender and flexible description, the tips of the twigs being so arranged as to curve inwards and nearly meet at the top: in the interior of the bower the materials are so placed that the forks of the twigs are always presented outwards, by which arrangement not the slightest obstruction is offered to the passage of the birds. For what purpose these curious bowers are made, is not yet, perhaps, fully understood; they are certainly not used as a nest, but as a place of resort for many individuals of both sexes, which, when there assembled, run through and around the bower in a sportive and playful manner, and that so frequently that it is seldom entirely deserted. The proceedings of these birds have not been sufficiently watched, to render it certain whether the runs are frequented throughout the whole year or not; but it is highly probable that they are merely resorted to as a rendezvous, or playing ground, at the pairing time and during the period of incubation." In the British Museum specimens of these Bowers may be seen.

The whole plumage of the male is of a deep shining blue-black, closely resembling satin, with the exception of the primary wing-feathers, the wing-coverts, and the secondaries and tail-feathers, which are of a deep velvety black, tipped with the shining blue-black lustre; irides light blue, with a circle of red round the pupil; bill bluish horn, with yellow tip; legs and feet yellowish white. The female has the head and all the upper surface grayish green; wings and tail dark sulphur brown; under surface much lighter, and yellowish, each feather having a crescent-shaped mark of dark brown near the extremity. Besides the loud liquid call peculiar to the male, both sexes frequently utter a harsh, unpleasant, guttural note, indicative of surprise or displeasure.

The SPOTTED BOWER-BIRD. (*Chlamydera maculata.*) The able ornithologist from whom we derived the information given in the preceding article, observes, that this species is as exclusively an inhabitant of the interior of the country as the Satin Bower-bird is of the brushes between the mountain-ranges and the coast, and is especially interesting as the constructor of a bower, even more extraordinary than that of the latter. But to describe them fully we must borrow the author's own words:— "They are considerably longer and more avenue-like than those of the Satin Bower-bird, being in many instances three feet in length. They are outwardly built of twigs, and beautifully lined with tall grasses, so disposed that their heads nearly meet; the decorations are very profuse, and consist of bivalve shells, crania of small mammalia, and other bones. Evident and beautiful indications of design are manifest throughout the whole of the bower and decorations formed by this species, particularly in the manner in which the stones are placed within the bower, apparently to keep the grasses with which it is lined fixed firmly in their places: these stones diverge from the mouth of the run on each side, so as to form little paths, while the immense collection of decorative materials, bones, shells, &c., are placed in a heap before the entrance of the avenue, this arrangement being the same at both ends. . . . I frequently found these structures at a considerable distance from the rivers, from the borders of which they could alone have procured the shells and small round pebbly stones; their collection and transportation must therefore be a task of great labour and difficulty. As these birds feed almost entirely upon seeds and fruits, the shells and bones cannot have been collected for any other purpose than ornament; besides, it is only those that have been bleached perfectly white in the sun, or such as have been roasted by the natives, and by this means whitened, that attract their attention. I fully ascertained that these runs, like those of the Satin Bower-bird, formed the rendezvous of many individuals; for, after secreting myself for a short space of time near one of them, I killed two males which I had previously seen running through the avenue."

The Spotted Bower-bird has the crown of the head, ear-coverts, and throat, of a rich brown, each feather surrounded with a narrow line of black; a beautiful band of elongated rose-pink feathers crosses the back of the neck, forming a broad, fan-like occipital crest; all the upper surface, wings, and tail, of a deep brown; every feather of the back, rump, scapularies, and secondaries, tipped with a large round spot of rich buff; primaries slightly tipped with white; all the tail-feathers terminated with buffy white; feathers of the flanks marked with faint, transverse, zigzag lines of light brown; bill and feet dusky brown; bare skin at the corner of the mouth thick, prominent, and of a deep flesh-colour.

SATYRUS. A genus of Diurnal Lepidoptera, also called HIPPARCHIA; it contains several British species: amongst these is

SATYRUS GALATHEA; or MARBLE BUTTERFLY. This delicate and rather singular species of Butterfly, is known by its yellowish and black-spotted wings, but though pretty general, it is so decidedly local and limited in its particular habitat as to be very rarely seen at all in many places. The anterior wings have a series of yellowish dots parallel with the hinder margin above, and the under surface nearly similar, with a small ocellus near the tip: the posterior wings have also a series of marginal dots or lunules, with a yellowish broad patch in the centre, and another at the base. Body black above, yellowish beneath: antennæ black, with white rings, and tips reddish. Caterpillar bright green, with obscure lines on the back and sides: head rather brown. It feeds on the cat's tail grass. Chrysalis of yellowish colour. [See HIPPARCHIA.]

SAURIA, or SAURIANS. The name of an order of Reptiles, including all those which, like the Crocodile and Lizard, are covered with scales and have four legs. The most gigantic and remarkable specimens of Saurian reptiles are now extinct, but their fossil remains, immense in size and wonderful as they appear, afford incontestable evidence of their similarity in structure to the harmless little Lizard of the present day. The diversity in the habits of the existing Saurians is very considerable; some being more or less aquatic; others strictly terrestrial; while others are essentially arboreal. The greater part feed on animal substances; some of them preferring flesh, and others attacking small animals; while some are entirely insectivorous, and a few are herbivorous. They are all furnished with teeth, which are of a simple conical form and adapted rather for securing and tearing their prey, than for masticating it: their toes are generally furnished with claws, and they have all a tail more or less strong, and generally very thick at the base. A few species, exceptions to the general character, have only two legs. [The distinguishing characteristics of different Saurian Reptiles will be found under the words CROCODILE, ALLIGATOR, CHAMELEON, AGAMA, LIZARD, IGUANA, GECKO, PLESIOSAURUS, ICHTHYOSAURUS, &c.]

SAWFISH. (*Pristis antiquorum.*) A fish belonging to the family of the *Squalidæ*, or Shark tribe; and which receives its name from the extension of its snout into a long flat blade, furnished with a row of sharp spines, on each side, so as to resemble a large toothed saw. With this formidable weapon the Sawfish attacks the largest Whales, and

SAWFISH.—(PRISTIS ANTIQUORUM.)

inflicts very severe wounds. It sometimes attains the length of twelve or even fifteen feet. The back is ash-coloured, and the belly white: the head is cordiform, and flattened, the mouth is placed far below the end of the snout; and the lips are rough and sharp like a file, supplying the place of teeth. This fish is very widely distributed, being found in the arctic, antarctic, and tropical seas; but it seldom approaches the shore.

SAW-FLY. [See TENTHREDINIDÆ.]

SAXICAVA. A genus of Conchiferous Molluscs, (family *Lithophagidæ*); often found in the hollows of rocks, in cavities on the backs of oysters, and among the roots of sea-weed, &c. The foot of the animal is thin and pointed; and in its habits it appears to resemble the *Pholas*, masses of rock being to be seen on different parts of the coast of England, which are pierced with innumerable small holes, the entrance to the habitations of these animals. The shell is transverse, irregular, generally oblong, and gaping externally; teeth and bosses obsolete or indistinct.

SCALARIA, or WENTLE TRAP. A genus of Pectinibranchiate Gasteropodous Mollusca, allied to the Periwinkles (*Turbo*), but distinguished from them by the turreted spire being covered with longitudinal, elevated, rather sharp ribs, and the mouth being encircled by a varix. The finest species (*Scalaria pretiosa*) was long famous for the rarity and high prices given for a single specimen. It is now found to be not an uncommon shell in the Eastern seas. It is known by the whorls being separated from each other.

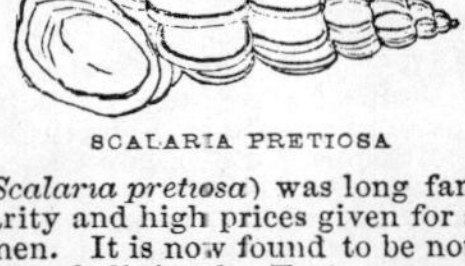

SCALARIA PRETIOSA

SCALE INSECTS. A name given to insects belonging to the family *Coccidæ*, many species of which live as parasites on various plants, particularly on hot-house plants, and do them considerable injury. They belong to the order *Hemiptera*, in which the bugs, plant-lice, and *Cicadæ* are included; although the main characteristic of the order corresponds only with the males, as they are winged. The females are shaped like a scale or shield, convex above, flat or concave below, provided with six very delicate feet, which sometimes, chiefly when the female has grown old, merge into the substance of the body. Anteriorly, at about the third part of the length of the insect, is situated a short or long rostrum on the under side, which it inserts into the epidermis of plants, and sucks out their juices. After pairing, when the eggs begin to develop themselves, the female dies, and her body serves as a protection to her posterity, by covering the eggs till the young are hatched, when they crawl away. Almost all sorts of plants suffer from the attacks of some species or other of Scale Insects, but chiefly in warm weather, and more especially at all times are those affected which are reared and kept in hot-houses. The Scale Insects are much more difficult to destroy than the Aphides; as they do not die from the effects of tobacco: the best remedy is to brush off the insects from the twigs and stems, and to wipe them off with a cloth or sponge from the leaves of more tender plants; and it is advisable to cleanse plants in pots at a distance from the greenhouse, as the insects are apt to creep up again and renew their depredations. The trees mostly infested with them are the peach and nectarine, the plum and damson, the wild chestnut and the vine. [See COCCUS.]

SCANSORES. The name of an order of birds, whose feet are peculiarly adapted for climbing. It comprehends the families of *Psittacidæ*, or Parrots; *Rhamphastidæ*, *Picidæ*, or Woodpeckers; and *Cuculidæ*, or Cuckoos. That which particularly distinguishes this order is the power of turning one of the front toes backwards, so as to oppose two hind toes to the two front ones. In their food, habits, outward appearance, and structure, the above-named families are very dissimilar; and therefore no general statement will be applicable to all of them: but it will be seen that the form of their feet, which gives them great power of prehension, and thus enables them to cling with firmness to their perch, renders walking more difficult; and that, as they pass most of their time in trees, their powers of flight are usually moderate.

SCARABÆIDÆ. An extensive and important group of Coleoptera, including the numerous dung-feeding Lamellicorns, of which the majority are inhabitants of tropical countries: some of these are among the most bulky species of beetles, but such as our own country produces are of small size. "From the great similarity in the structure of the mouth of all these insects," says Mr. Westwood, "a great uniformity of habits is evident. But a more remarkable peculiarity exists in the structure and situation of the hind legs, which are placed so near the extremity of the body, and so far from each other, as to give the insect a most extraordinary appearance whilst walking. This peculiar formation is, nevertheless (as Mr. McLeay observes), particularly serviceable

SACRED BEETLE.—(SCARABÆUS SACER.)

to its possessors in rolling the balls of excrementitious matter in which they enclose their eggs; whence these insects were named by the first naturalists *Pilulariæ*. These balls are at first irregular and soft, but, by degrees, and during the process of rolling along, become rounded and harder: they are propelled by means of the hind legs; and the insects occasionally mount to the top, when they find a difficulty in urging them along; probably in order to destroy the equilibrium. Sometimes these balls are an inch and a half or two inches in diameter; and in rolling them along the beetles stand almost upon their heads, with their heads turned from the balls. These manœuvres have for their object the burying of the balls in holes, which the insects have previously dug for their reception; and it is upon the dung, thus deposited, that the larvæ, when hatched, feed. It does not appear that these beetles have the instinct to distinguish their own balls, as they will seize upon those belonging to another, in case they have lost their own; and, indeed, it is said that several of them occasionally assist in rolling the same ball. They fly during the hottest part of the day."

"The type of this family is the renowned 'Sacred Beetle' of the Egyptians, of which so many models, carvings, amulets, &c. are discovered, occasionally of a gigantic size, in sarcophagi, and rolled up in the mummies and relics of that remarkable people, by whom its appearance in great numbers on the sandy margins of the Nile, after the annual rising and falling of the river, together with its extraordinary motions whilst rolling along its little globular balls of dung, were regarded as mystically representing the motions of the earth, the sun and planetary bodies. It was also regarded as the emblem of fertility; and, even at the present day, we are informed by Dr. Clarke that it is eaten by the women of Egypt. The various species of Sacred Beetles, whereof Dejean enumerates twenty-six, are distinguished by their flattened form, radiated clypeus, long hind legs, clothed with hairs, with the posterior tarsi obliquely inserted; head and thorax unarmed, and elytra with the margins not sinuated."

SCARUS. [See PARROT-FISH.]

SCELIDOSAURUS and SCELIDOTHERIUM. [See SUPPLEMENT.]

SCIÆNIDÆ. A family of fishes, of which there are several subgenera. The general characters are, — the head inflated, and supported by cavernous bones; the body compressed and broad; only one dorsal fin, but it is bifid, and so deeply divided at the middle that it seems to form two and the soft part is much longer than the spinous; the anal short, the pre-operculum toothed,

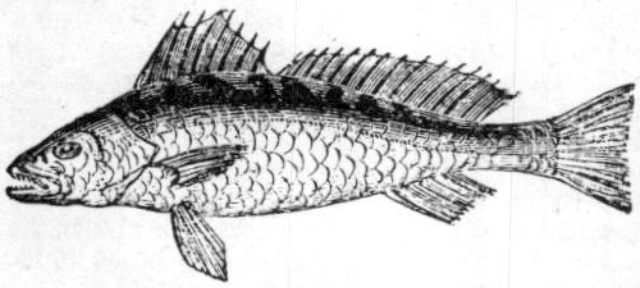

MAIGRE.—(SCIÆNA AQUILA.)

and the operculum divided into points at its extremity; seven arches in the gills. They resemble the Perches, except that they have no teeth in the palate. — The *Sciænidæ* with less than seven gill-rays, and the lateral line interrupted, form several genera of small oval fishes, generally finely coloured, and distinguished by the armature of their heads. The Sciæna Aquila, or Maigre, may be given as an example. It has occasionally been found on the English coast.

SCINQUE, or SKINK. (*Scincus.*) The name given to a family of lizard-like reptiles, in which there appears to be a gradual transition from the form of the Lizards to that of the Serpents. They are all natives of warm climates; and one species, common in Arabia, Northern Africa, &c., was long held in repute on account of its supposed medicinal virtues. They are recognized by the shortness of their feet, the non-extensibility of the tongue, and the tile-like scales which cover the whole body and tail, presenting almost the appearance of a coat of mail.

SCISSOR-BILL. (*Rhynchops.*) A genus of palmipede birds closely allied to the Terns, but easily distinguished by the singular bill, which is compressed like a knife, and has the lower mandible longer than the upper, and broken off (as it were) at the tip. At least two species are known; one of these is peculiar to the New World (*R. nigra*), while the other (*R. orientalis*) is found in the Eastern hemisphere.

In our article "Rhynchops," we gave some interesting extracts from Wilson's Ornithology respecting the habits of the species found in North America. We may here add a short account of the same birds seen in more southern latitudes, by Mr. Darwin, and described by that most accurate and intelligent observer in his 'Journal.' He is in the vicinity of the Rio Parana, and thus writes: — "I here saw a very extraordinary bird, called the Scissor-beak (*Rhynchops nigra*). It has short legs, web feet, extremely long pointed wings, and is of about the size of a tern. The beak is

SCISSOR-BILL.—(RHYNCHOPS NIGRA.)

flattened laterally, that is, in a plane at right angles to that of a spoonbill or duck. It is as flat and elastic as an ivory paper-cutter, and the lower mandible, differently from every other bird, is an inch and a half longer than the upper. I will here detail all I know of the habits of the Scissor-beak. It is found both on the east and west coasts, between lat. 30° and 45°, and frequents either salt or fresh water. The specimen now at the Zoological Society was shot at a lake near Maldonado, from which the water had been nearly drained, and which, in consequence, swarmed with small fry. I there saw several of these birds, generally in small flocks, flying backwards and forwards, close to the surface of the lake. They kept their bills wide open, and with the lower mandible half buried in the water. Thus skimming the surface, they ploughed it in their course: the water was quite smooth, and it formed a most curious spectacle to behold a flock, each bird leaving its narrow wake on the mirror-like surface. In their flight they frequently twist about with extreme rapidity, and so dexterously manage, that with their projecting lower mandible they plough up small fish, which are secured by the upper half of their scissor-like bills. This fact I repeatedly saw, as, like swallows, they continued to fly backwards and forwards, close before me. Occasionally when leaving the surface of the water their flight was wild, irregular, and rapid; they then also uttered loud harsh cries. When these birds are fishing, the length of the primary feathers of the wings is seen to be quite necessary, in order to keep the latter dry. When thus employed, their forms resemble the symbol by which many artists represent marine birds. The tail is much used in steering their irregular course.

"These birds are common far inland along the course of the Rio Parana; it is said they remain during the whole year, and breed in the marshes. During the day they rest in flocks on the grassy plains, at some distance from the water. Being at anchor, as I have said, in one of the deep creeks between the islands of the Parana, as the evening drew to a close, one of these Scissor-beaks suddenly appeared. The water was quite still, and many little fish were rising. The bird continued for a long time to skim the surface, flying in its wild and irregular manner up and down the narrow canal, now dark with the growing night and the shadows of the overhanging trees. At Monte Video I observed that some large flocks during the day remained on the mud-banks at the head of the harbour, in the same manner as on the grassy plains near the Parana; and every evening they took flight direct to seaward. From these facts, I suspect that the Rhyncops generally fishes by night, at which time many of the lower animals come most abundantly to the surface. M. Lesson states that he has seen these birds opening the shells of the *mactræ*, buried in the sand-banks on the coast of Chile."

SCISSOR-TAIL. (*Milvulus forficatus.*) "A bird with a forked tail, terminated by two long feathers, and named by the Spaniards *Scissor-tail*, is very common near Buenos Ayres. It belongs to the family *Laniidæ* or Butcher-birds. It commonly sits on a branch of the *ombu* tree, near the house, and thence takes a short flight in pursuit of insects, and returns to the same spot. When on the wing, it presents in its manner of flight and general appearance a caricature-likeness of the Common Swallow. It has the power in the air of turning very shortly, and in so doing, opens and shuts its tail sometimes in a horizontal or lateral, and sometimes in a vertical direction, just like a pair of scissors. In structure this bird is a true tyrant-flycatcher, although in its habits certainly allied to the Swallows."—*Darwin.*

SCIURIDÆ. The name given to the Squirrel tribe.

SCIURUS. [See SQUIRREL.]

SCOLIADÆ. A family of Hymenopterous insects, distinguished by having the collar laterally extending to the base of the wings; the legs short and robust, the tibiæ being thick, spinose, or denticulate. The antennæ are generally short, thick, and more or less serrated: the abdomen is elongate-ovate, and attached by a short peduncle: both sexes are winged; and the body is often very hirsute. The greater part of the species are exotic, and some are of a very large size. They abound in the very hottest situations, and make deep burrows in sand banks; and they are said to be particularly fond of revelling in strong-scented flowers, such as rue, &c.

SCOLOPACIDÆ. The name of a numerous family of Wading Birds, the greater part of which were comprehended by Linnæus in his genus *Scolopax*, consisting of the different species of the Snipe tribe, the Sandpipers, Curlews, Godwits, &c.; all of which inhabit marshy lands, the borders of lakes, rivers, and the sea-shore. They are all more or less migratory in their habits; breeding in high latitudes; and their powers of flight are considerable. They are characterized by a long, slender, and feeble bill, provided with a very peculiar distribution of nerves, which render its exterior sensitive, especially towards the tip, the membrane of which is fleshy; and in many species there is a peculiar muscle, that enables the bird to separate the flexible points of the mandible, so as to seize its prey the moment it is felt, while the bill is still buried in the ground. When it is considered that their food consists of insects, worms, slugs, &c., which they find in the mud or soft earth, it will be seen how admirably adapted the bill we have described is for the purpose of obtaining it. [See SNIPE; WOODCOCK; SANDPIPER; GODWIT; RUFF; &c.]

SCOLOPENDRA. A genus of annulosa of the class *Myriapoda*. They inhabit the southern parts of Europe, and all the tropical portions of the globe, where they are much dreaded for their venomous properties: their mandibles are terminated by a sharp hook, which is pierced for the transmission of a venomous fluid. The *Scolopendræ* have the body long, slender, and depressed, and protected by coriaceous plates: they run very fast, and shun the light, living for the most part under logs of wood and the loose bark of decayed trees. [See CENTIPEDE.]

SCOLYTIDÆ. A family of Coleopterous insects, the type of which is named *Scolytus destructor*. The body is oblong or cylindric, convex and rounded above, with the head globular; and the antennæ have the basal joint elongated, and the terminal joints form a more or less solid oval mass. We learn from Mr. Westwood, that of all the species, "the *Scolytus destructor* is the most obnoxious in this country, annually destroying a great number of elm trees in the neighbourhood of London; and the injury is gradually spreading into the provinces, owing to the inattention or ignorance of those whose duty it is, or ought to be, to adopt decided measures for stopping the mischief. The parks and public gardens and walks around London have been especially subject to the attacks of these insects. It has, indeed, been a question whether the insects were the *primary* cause of the mischief, or whether the trees were not previously infected in some way or other, and thus rendered an agreeable nidus for the insect. From the recent observations of Messrs. Audouin and Spence, it, however, appears evident that, in the first instance, both the male and female insects attack the trees for the purpose of obtaining food, burrowing into the trunk. This brings the trees into a state of ill health, which is adapted for the reception of the eggs and food of the larva. The female insect then burrows deeper into the trunk, and there deposits her eggs; and the larvæ, when hatched, form cylindrical galleries, diverging at right angles from the track of the parent, and parallel to each other; within which they also become pupæ; and so great is the fecundity of these insects, that their countless numbers are soon sufficient to destroy the largest tree." "The larvæ of *Scolytus destructor* are thick, fleshy, curved, apod grubs, of a whitish colour, the back much wrinkled, armed with a scaly head and powerful horny jaws."

"The sudden change in the temperature that generally occurs in the early part of May, brings out great numbers of insects, from their winter quarters, to enjoy the sunshine and the ardent heat which are congenial to their natures. While a continued hum is heard among the branches of the trees, from thousands of bees and flies, drawn thither by the fragrance of the bursting buds; swarms of little beetles of various kinds come forth to try their wings, and,

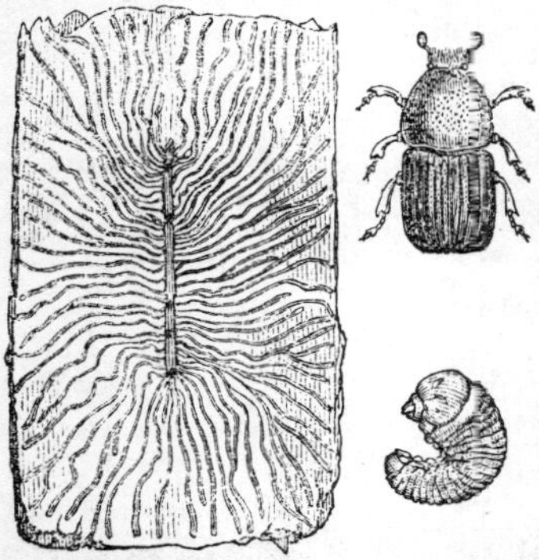

SCOLYTUS DESTRUCTOR AND LARVA, WITH A PIECE OF WOOD TO SHOW THE RAVAGES OF THE INSECT.

with an uncertain and heavy flight, launch into the air. Among these beetles there are many of a dull red or fox colour, nearly cylindrical in form, tapering a very little before, obtusely rounded at both extremities, and about one quarter of an inch in length. They are seen slowly creeping upon the sides of wooden buildings, resting on the tops of fences, or wheeling about in the air, and every now and then suddenly alighting on some tree or wall, or dropping to the ground. If we go to an old pine-tree, we may discover from whence they have come, and what they have been about during the past period of their lives. Here they will be found creeping out of thousands of small round holes which they have made through the bark for their escape. Upon raising a piece of the bark, already loosened by the undermining of these insects, we find it pierced with holes in every direction, and even the surface of the wood will be seen to have been gnawed by these little miners. After enjoying themselves abroad for a few days, they pair, and begin to lay their eggs. They gnaw little holes here and there through the rough bark of the trunk and limbs, drop their eggs therein, and, after this labour is finished, they become exhausted and die. In the autumn the grubs hatched from these eggs will be found fully grown. They devour the soft inner substance of the bark, boring through it in various directions for this purpose; and they gnaw a passage to the surface, for their escape after they have completed their transformations. Their depredations interrupt the descent of the sap; the bark becomes loosened from the wood, and the tree decays.

SCOMBERESOX. A fish, called also the Mackerel Pike, or Saury Pike, found in the Mediterranean, and remarkable for its long, beak-like jaws. They are gregarious fishes; and are followed and preyed upon by Porpoises, and also by the Tunny and other large members of the Mackerel family.

SCOMBERIDÆ. A family of Acanthopterygious fishes, of which the genus *Scomber*, or Mackerel, is the type. They are characterized by having a smooth body covered with small scales, and a very powerful tail and caudal fin; in most of the species the pectoral fins are long, narrow, and pointed. This family includes species of the greatest utility to mankind, in consequence of their abundance and their wholesomeness as food. [See MACKEREL.]

SCORPÆNA. A genus of Acanthopterygious marine fish, associating in shoals, and haunting rocky shores. The head is tuberculated and compressed laterally, but in other respects they much resemble a Perch. The species are popularly termed Hog-fish.

SCORPION. (*Scorpio.*) A genus of the class *Arachnida*, distinguished from other groups of Spiders by having the abdomen articulated, and its hinder part, or tail, terminated by a curved spur or sting, beneath the extremity of which are two small orifices, by which a venomous fluid is discharged: the stigmata are eight in number, and situated along the inferior and lateral part of the abdomen. Between the thorax and abdomen there is no distinct division; and the latter is composed of twelve segments, six of which are broad, forming the posterior part of the body, whilst the other six form the tail. The palpi are very large, resembling extended arms, and the terminal segment assumes the form of the lobster's claw, being in like manner provided with pincers. The *Scorpionidæ* inhabit the warm countries of both hemispheres, living on the ground, hiding themselves under stones or other bodies, generally amongst ruins or other dark and cool places, and sometimes taking up their abode in houses. They run with considerable swiftness, curving the tail over the back: they can turn it in all directions, and employ it as a weapon of offence or defence. With their forceps they seize wood-lice and various other ground insects, on which they feed, after having pierced them with their sting; and it is remarkable

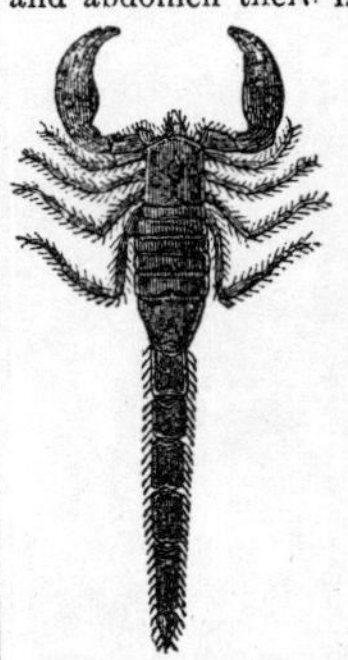

SCORPION. (SCORPIO AFER.)

that they are particularly fond of the eggs of spiders and insects. The larger species of Scorpions, of whose malignity and venom so much has been related, are five or six inches long, but they are confined to tropical climates; those of the south of Europe are very troublesome pests also; but their sting, though painful, is seldom productive of serious mischief to man. The generality of this tribe (*Scorpio Europœus*) have six eyes; but there are some of the most formidable kind (*Scorpio afer*) which have eight. The female pays great attention to her young; carrying them upon her back for several days, at first not quitting her abode; and she afterwards takes care of them for the space of a month, by which time they are enabled to shift for themselves. Messrs Klug, Ehrenberg, Koch, and Gervais have described numerous new species of the family *Scorpionidœ*.

SCORPION-FLY. (*Panorpa.*) A winged insect thus designated on account of the remarkable conformation of the posterior extremity of the abdomen in the male, which is turned up like a Scorpion's sting. The abdomen of the female is also prolonged into an ovipositor; by which she can deposit her eggs in deep holes or crevices. The Scorpion-flies, or *Panorpidæ*, are very active, and prey upon other insects in the perfect state. There are two beautiful species of this insect; the one has silvery wings, variegated with three transverse streaks of black towards the ends; the head is black; the breast, shoulders, and feet are whitish; and the rest of the body is black. The tail, which represents a sting, has five joints, three red and two black; and the extremity of the tail is forked and reverted. This insect may be commonly seen frequenting our hedges and woods.

SCOTER. (*Oidemia.*) The Black Diver, an aquatic bird about the size of the common Duck; but the hind toe has a widish membrane, and the beak is high at the base and

BLACK SCOTER.—(OIDEMIA NIGRA.)

suddenly flattened; while the body is more round, and of a deep shining and beautiful black colour. It is very common on the shores of Lancashire, and some other maritime counties. There are at least four species of Scoter-Ducks; the whole of these go to the sea chiefly in quest of their food, and one of them (*O. perspicillata*) has acquired its English name of Surf-Duck, from being found for the most part on its edge.

SCREAMER. (*Palamedea.*) A genus of Wading Birds, natives of South America. The chief characteristics are, that the bill is conical, the upper mandible being hooked; and the feet are cloven, having each four toes. They are remarkable for their harsh and discordant voices, and for the sharp hard spurs with which the wings are armed at the shoulder-joint. These are very efficient weapons of defence, and enable the birds to resist the attacks of the snakes which infest the places they inhabit. One species is also remarkable for having on the top of the head a slender pointed horn, three or four inches long, which curves gently forwards, but the use of it does not appear to be known.

SCULPIN. (*Cottus octodecimspinosus.*) An Acanthopterygious fish, of the genus *Cottus*, found on the American coasts, and which has often been confounded with the European *Cottus scorpius:* it is, however, quite distinct from it, and considerably larger. The Sculpin abounds on the coasts of the United States, and is also plentiful at Newfoundland. Sir John Richardson says it is a pity that Cuvier did not retain the original specific name (*Scorpius Virginianus*) given by Willoughby, who figured it correctly; it being preferable to *octodecimspinosus*, which may lead to error, their being, in fact, twenty spines on the head. *Cottus scorpius* has exactly the same number, viz., ten on each side.

SCUTIBRANCHIATA. The name given to an order of Molluscous animals, including those Gasteropods which have the gills covered with a shell in the form of a shield. The *Haliotis* is a beautiful example [which see].

SCYDMÆNUS: SCYDMÆNIDÆ. A genus and family of Coleoptera, belonging to the *Palpatores*. They are generally of very minute size, some of the species of which are found, according to Mr. Doubleday, under planks placed upon hotbeds, where they prey upon minute Thysanura, carrying them about in their jaws. Others

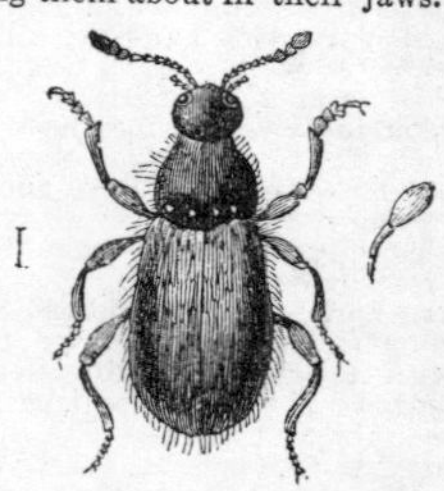

SCYDMÆNUS TARSATUS.

are found under stones and among moss at the roots of trees, and also in ants' nests. Our figure, copied from Mr. Sturm's catalogue, gives a highly magnified representation of a species of this interesting family; but the nature of this work precludes us from entering either into generic or specific characters. We must refer to the works of

Mr. Denny, of Leeds, and Dr. Schaum, of Stettin, for descriptions of the various species.

SCYLLÆA. A genus of Gasteropodous Nudibranchiate Mollusca. In this genus the body is compressed; the foot narrow and furrowed, to enable it to embrace the stems of sea-weed; the mouth forming a small pro-

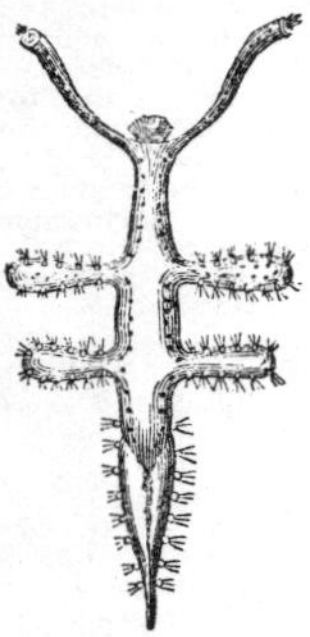

SCYLLÆA PELAGICA.

boscis; the tentacula compressed, terminating in a cavity from which a little point, with an unequal surface, can be protruded: and upon the back are two pairs of membranous crests, carrying, on their inner aspect, some pencils of branched filaments. The middle of the stomach is covered with a fleshy ring, armed with very sharp horny laminæ. The common species is found on *Fucus natans*, or gulf-weed, wherever this appears.

SCYLLARUS: SCYLLARIDÆ. A genus and family of Macrurous Decapods, or Long-tailed Crabs. They are distinguished by a very wide carapace, and but little elevated; its anterior border nearly straight, and presenting a horizontal prolongation which advances between the base of the external antennæ, which are foliaceous and

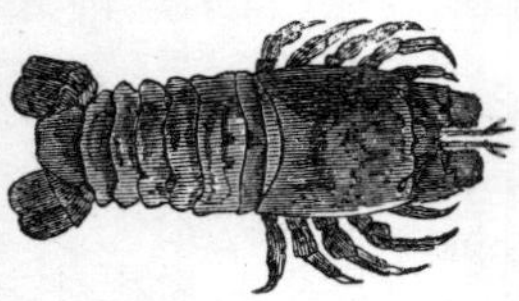

SCYLLARUS EQUINOXIALIS.

extremely wide. The buccal frame is small and the jaw-feet are moderate and nearly pediform. There are several species, differing considerably from each other. The one here figured is *Scyllarus Equinoxialis;* the body of which is very much depressed, and much narrowed from before backwards. It is of a yellowish colour mingled with red, and about a foot in length: its locality the Antilles.

SCYTHROPS or CHANNEL-BILL. A remarkable genus of Birds, by some naturalists considered as allied to the Horn-bills, by others to the Toucans, but in reality belonging to the family of the Cuckoos. The Bill has two narrow longitudinal grooves; and the space round the eyes and nostrils is void of feathers. But one species of this genus is known, the S. NOVÆ-HOLLANDIÆ or CHANNEL-BILL. It is a migratory bird in New South Wales, arriving in October, and leaving in January; it is chiefly seen in the morning and evening, sometimes in small parties of seven or eight, but more frequently in pairs. It makes a loud screaming noise when a hawk is in sight. The tail, which is nearly as long as the body, and has most of the feathers tipped with white with a black band before the tip, is occasionally displayed like a fan, and gives the bird a majestic appearance. Mr. Gould informs us that it feeds on the larger kinds of *Phasmidæ* and *Coleoptera;* but he could not ascertain whether the species was parasitic or not, like the other cuckoos.

SEA-DEVIL. The Lophius Piscatorius. [See ANGLER.]

SEA-EAGLE. The Erne: also the name sometimes given to the Osprey. [See EAGLE.]

SEA-EGG: SEA-URCHIN. Names frequently given to different species of the family *Echinidæ*. We give a representation of the half of a beautiful species of this

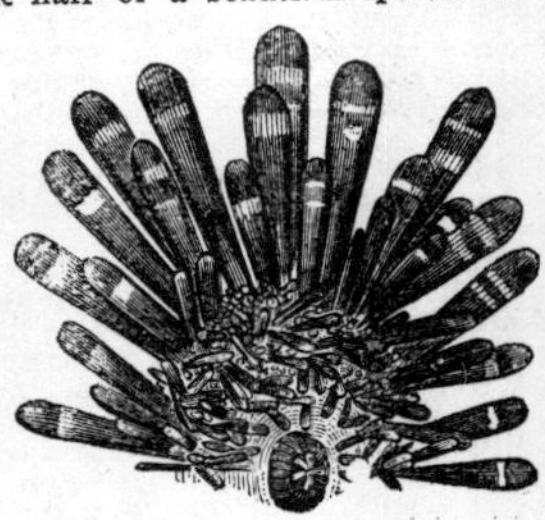

ECHINUS MAMMILLATUS.

group, the above figure showing it in its natural state, covered with elegant knobbed spines, which Mr. Williams, the missionary to the South Sea Islands, tells us, in his interesting Narrative, are very often used as slate pencils in those distant islands. The

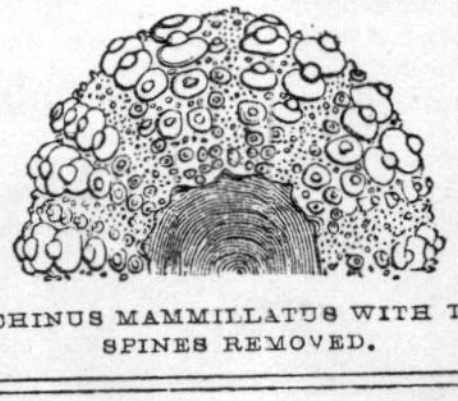

ECHINUS MAMMILLATUS WITH THE SPINES REMOVED.

other figure represents the half of the same animal divested of its spines, to show the bases of their attachment. [See ECHINUS and ECHINODERMATA.]

SEA-HORSE. [See HIPPOCAMPUS.]

SEA-OWL. [See LUMP-FISH.]

SEA-PIE. One of the names of the Oyster-Catcher (*Hœmatopus ostralegus.*) [See OYSTER-CATCHER.]

SEA-SERPENT. [or the KRAKEN.] The appearance of this *fabulous* monster is thus accounted for by Mr. A. Adams: "In the Sooloo seas I have often witnessed the phenomenon which first gave origin to the marvellous stories of the great Sea-serpent, namely, lines of rolling porpoises, resembling a long string of buoys, oftentimes extending seventy, eighty, or a hundred yards. These constitute the so-named protuberances of the monster's back, keep in close single file, progressing rapidly along the calm surface of the water, by a succession of leaps or demi-vaults forwards, part only of their uncouth forms appearing to the eye. At the same moment of time, I have seen beautifully-banded Water-snakes, of the thickness of a man's leg, lying extended supinely along the glassy surface, or diving and swimming gracefully, with slow undulating lateral movements of their vertically-compressed bodies."

Shortly after the appearance of the first edition of "The Treasury of Natural History," the public were equally surprised and amused with a marvellous account of a marine monster which, on the authority of Capt. M'Quhae of Her Majesty's ship Dædalus, was called a *Sea Serpent*, and asserted to have been seen by him and part of his crew while sailing in lat. 24° 44′ S., long. 9° 20′ E. Had the said account appeared before the publication of our work we should have treated its existence as doubtful (although vouched for by a captain of the royal navy in a letter to the admiralty), no record of any thing like it being to be found in the works of zoologists, nor any fragment of such a skeleton having ever been seen in any part of the world. Capt. M'Quhae's Sea Serpent, however, made some considerable stir in the newspaper world; at length Professor Owen thought proper to publish a long letter in "The Times" of Nov. 14, containing a dissertation on the subject, giving his reasons for presuming that the animal was a species of seal, and grounding his disbelief as to its reality on various well-established zoological principles (but too long for quotation here), and concluding as follows: "I regard the negative evidence from the utter absence of any of the recent remains of great sea serpents, krakens, or Enaliosauria, as stronger against their actual existence than the positive statements which have hitherto weighed with the public mind in favour of their existence. A larger body of evidence from eye-witnesses might be got together in proof of ghosts than of the sea serpent."

It is right, indeed, to mention that in our volume entitled "The Scientific and Literary Treasury" (first published in 1840) we showed that we had paid due attention to reports of a similar nature, as will be seen from the following notice we therein took of the "SEA SERPENT. At various times within the last quarter of a century the public have been entertained with marvellous stories respecting an enormous marine animal seen on the coasts of America, of a size and length varying according to the opinions of those who assert that they have witnessed it, some declaring it 100 feet long, while others describe it as nearly as many yards. All accounts, however, agree in regard to the protuberances on its back, its vertebral sinuosities, and its serpent-shaped head. As there is no absolute reliance to be placed on any of the descriptions of this marine monster, we think it merely necessary thus to notice it; and we beg to refer our readers to the article KRAKEN in this volume."

SEA-SNIPE. [See TRUMPET-FISH.]

SEA-SWALLOW. A species of Tern [which see].

SEA-UNICORN. [See NARWHAL.]

SEAL. (*Phoca.*) The family *Phocidœ*, or Seal tribe, are, of all four-limbed Mammiferous animals, those which display the most complete adaptation to residence in the water. The Seal (*Phoca vitulina*) resembles a quadruped in some respects, and a fish in others. The head is round, and the nose, which is broad, resembles that of a dog, with the same look of intelligence and mild and expressive physiognomy. It has large whiskers, oblong nostrils, and great black sparkling eyes. It has no external ears, but a valve exists in the orifices, which can be closed at will, so as to keep out the water; the nostrils have a similar valve; and the clothing of the body consists of stiff glossy hairs, very closely set against the skin. The body is elongated and conical, gradually tapering from the shoulders to the tail. The spine is provided with strong muscles, which bend it with considerable force; and this movement is of great assistance to the propulsion of the body. Although in most of the foregoing particulars the Seal resembles the quadruped kind, it greatly differs from all of them with respect

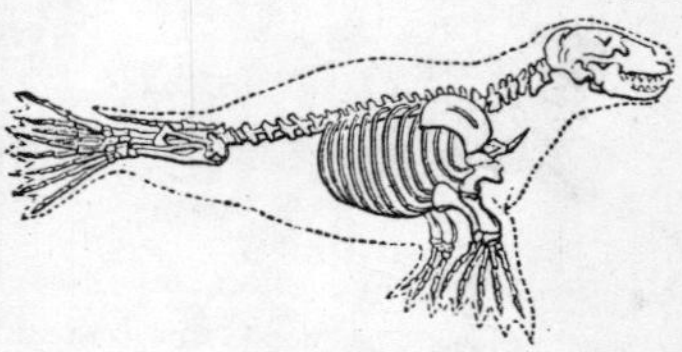

SKELETON OF SEAL.

to its feet; for, though furnished with the same number of bones as in quadrupeds, they are united to the body in such a singular manner, and so covered with a membrane, that they would rather resemble fins than feet, did not the sharp strong claws with which they are pointed show their proper analogy. The limbs, in fact, are converted into oars and paddles. The anterior pair have the arm and fore-arm so short, that little more than the paw advances from the body. The hinder limbs are directed backwards, so as almost to seem like a continuation of the body; the thigh and leg are very short, and the foot is formed on the same plan as the fore-paw,—the toes being in contact, however, and the web folded, when it is not in use as a paddle, but being spread out when the animal is swimming. When on land, or on masses of ice, the movements of the Seal are particularly awkward, its body being forced onward by the action of the fore-limbs only, and the wriggling motion of the abdominal muscles; they accordingly seldom venture from the shore, but usually bask on the rocks; and when disturbed, plunge immediately to the bottom of the water.

The Seals live in herds, more or less numerous, along the shores of the sea; and upon uninhabited coasts they bring forth and suckle their young, and exhibit the most tender solicitude for their welfare. They are easily tamed, become strongly attached to their keepers, recognise them at a distance, and seem to be endowed with a very considerable share of intelligence. The form of their teeth and jaws shows them to be carnivorous; and their food consists of fish, crabs, and sea-birds, which they are enabled to surprise while swimming.

SEAL.—(PHOCA VITULINA.)

Seals swim with great rapidity and ease; and by a peculiar arrangement of their bloodvessels, nearly similar to that which exists in the whale tribe, they can remain under water for a considerable time. There are many species of these animals; some are found in almost every quarter of the globe, but chiefly in the frigid or temperate regions; and they exist in vast numbers in the seas round Spitzbergen, and on the coasts of Labrador and Newfoundland. Their habits are migratory; and it is known that at least four species visit the shores of Britain. Quadrupeds are in general contented with their native plains and forests; seldom wandering far from those situations where they were produced, unless compelled by necessity or fear; but Seals frequently shift their places of abode, and are seen in myriads directing their course from one continent to another. On the northern coasts of Greenland they are observed to retire in July, and to return again in September. These animals produce two or three young at a time; and they suckle them for six or seven weeks, generally in the cavernous recesses of rocks; after which they take to the sea. The young are remarkably docile: they recognise and are obedient to the voice of their dams amidst the numerous clamours of the flock, and mutually assist each other when in danger or distress. Thus early accustomed to subjection, they continue to live in society, hunt and herd together; and have a variety of cries by which they encourage or pursue, express apprehension or success. When incited by natural desire, however, their social spirit seems to forsake them; they then fight most desperately; and the victorious male always keeps a watchful eye over those females whom his prowess has secured. In some of the species there is a remarkable disproportion in the sexes; and some also are far more pugnacious than others.

The species to which the foregoing observations more particularly refer is the COMMON SEAL (*Phoca vitulina*), which is from four to five feet in length. The Greenland or HARP SEAL (*Phoca Greenlandica*) is about six feet in length, and is remarkable for the changes of colour it undergoes in the course of its advance towards maturity. The BEARDED SEAL (*Phoca barbata*), another northern species, is from seven to ten feet long; and is distinguished from others by having thicker and stronger moustaches. The HOODED SEAL (*Stemmatopus cristatus*) is remarkable for a globular sac, susceptible of inflation, which is situated upon the summit of the head of the males. It grows to the length of seven or eight feet, and inhabits the seas about Greenland and Newfoundland. But by far the largest known species is the ELEPHANT SEAL, OR SEA-ELEPHANT (*Macrorhinus proboscideus*); its length being from twenty to twenty-five or thirty feet, and its girth at the largest part of the body being from fifteen to eighteen feet. It is said that a full-grown male will yield seventy gallons of oil. These animals inhabit the Antarctic seas, and are found upon the southern coasts of Australasia, Juan Fernandez, and the neighbouring parts of South America. Their voice resembles the lowing of cattle: and they migrate towards the tropic in winter, and return southwards in summer. They are very inert, not easily alarmed, and make little defence when attacked. The name of Elephant Seal is given to them partly from the large size of their tusk-like canines, and partly from the faculty which the male possesses of elongating the upper lip into a kind of proboscis: they are much sought after, on account of the large quantity of oil they yield; as well as for the skin, which, being of great strength and thickness, is much used in harness-making. —Two more species must be noticed; one called the Sea-Lion, the other the Sea-Bear. The SEA-LION (*Platyrhynchus leoninus*) grows from the length of from six to ten feet, and is said to inhabit both the northern and southern coasts of the Pacific. The colour is yellowish brown; and the males have a large mane upon their necks, which partly covers their head and shoulders. The nails of the fore-feet are very small, and in part wanting. The voice of the males is very powerful.—The SEA-BEAR (*Arctocephalus ursinus*), so named from the fur and shape of the head, grows to the length of five or six feet, and has small external ears. The membrane of the hinder feet is prolonged into as many lobes as there are toes, and the fore-feet are placed very far back. The colour of the fur is brown, and when old takes a grayish tint. This species inhabits the coasts of the North Pacific, and is also said to be found in the northern hemisphere.

Sir George Simpson, who has had so many excellent opportunities of studying the manners of the North American animals in their native haunts, speaking of the Fur Seals, says—"Some twenty or thirty years ago there was a most wasteful destruction of the Fur Seal, when young and old, male and female, were indiscriminately knocked on the head. This improvidence, as every one might have expected, proved detrimental in two ways. The race was almost extirpated; and the market was glutted to such a degree, at the rate, for some time, of 200,000 skins a year, that the prices did not even pay the expenses of carriage. The Russians, however, have now adopted nearly the same plan which the Hudson's Bay Company pursues in recruiting any of its exhausted districts, killing only a limited number of such males as have attained their full growth—a plan peculiarly applicable to the Fur Seal, inasmuch as its habits render the system of husbanding the stock as easy and certain as that of destroying it. In the month of May, with something of the regularity of the almanack, the Fur Seals make their appearance at the island of St. Paul, one of the Aleutian group. Each old male brings a herd of females under its protection, varying in number according to his size and strength; the weaker brethren are obliged to content themselves with half a dozen wives, while some of the sturdier and fiercer fellows preside over harems that are 200 strong. From the date of their arrival in May to that of their departure in October, the whole of them are principally ashore on the beach. The females go down to the sea once or twice a day, while the male, morning, noon, and night, watches his charge with the utmost jealousy, postponing even the pleasures of eating, and drinking, and sleeping, to the duty of keeping his favourites together. If any young gallant venture by stealth to approach any senior chief's bevy of beauty, he generally atones for his imprudence with his life, being torn to pieces by the old fellow; and such of the fair ones as may have given the intruders any encouragement are pretty sure to catch it in the shape of some secondary punishment. The ladies are in the straw about a fortnight after they arrive at St. Paul's; about two or three weeks afterwards, they lay the single foundation, being all that is necessary, of next season's proceedings; and the remainder of the sojourn they devote exclusively to the rearing of their young. At last the whole band departs, no one knows whither. The mode of capture is this: at the proper time, the whole are driven, like a flock of sheep, to the establishment, which is about a mile distant from the sea; and there the males of four years, with the exception of a few that are left to keep up the breed, are separated from the rest and killed. In the days of promiscuous massacre, such of the mothers as have lost their pups would ever and anon return to the establishment, absolutely harrowing up the sympathies of the wives and daughters of the hunters, accustomed as they were to the scene, with their doleful lamentations." —*Narrative of a Journey round the World in* 1841 *and* 1842.

SEBASTES. (*Sebastes Norvegica.*) The Northern Sebastes, or Norway Haddock, is an Acanthopterygious fish, of the family *Cottidæ* (genus *Scorpæna*, Linn.) It inhabits the Icy Sea and Northern Ocean; is plentiful on the coast of Norway, and is found at Iceland, Greenland, off Newfoundland, &c. It inhabits the deepest bays of South Greenland, and does not approach the shore, except when driven thither by tempests. Its colour, when quite fresh, is a bright carmine, which is paler towards the belly, and mixed with brown on the back; there is likewise a blackish mark on the top of the gill-cover. It resembles the perch in form, its body being somewhat compressed, its profile oblong, and the dorsal and ventral curves being slightly convex: the mouth is oblique, and the lower jaw projects a little. The whole fish is clothed with small rough scales. Its flesh is dry, but much esteemed by the Greenlanders, who eat its lips raw, and were formerly accustomed to use its spines as sewing needles.—There are several other species of Sebastes: one at the Cape of Good Hope, which very nearly resembles the above-described; and another which differs from it in a few characters, and is more like one found in the Mediterranean: there are also two or three in the Indian and Polynesian seas; several in the sea of Japan; and one in the sea of Kamtschatka (*Sebastes variabilis*), which has the head less armed than any other species.

SECRETARY. [See Serpent-eater.]

SEDGE-WARBLER, or SEDGE-BIRD; sometimes called also the Willow-Lark. (*Sylvia salicaria.*) This is a smaller kind of Reed-sparrow; like the *Emberiza schoeniculus*, it frequents reedy and marshy places. It is a bird of a slender, elegant figure: it frequents low, wet grounds; sitting on the top of some spray, with its wings dishevelled; while it utters a loud and somewhat discordant song of only two notes.

SEMNOPITHECUS. A genus of Monkeys, bearing many points of resemblance to the Gibbons. They are, however, readily distinguished by their having a very long, slender, and powerfully muscular tail, which is cylindrical for the greatest portion of its length, and terminated by a close tuft of long hairs. The colour of the adult animal is intensely black, except the breast, the abdomen, and the root of the tail, which are gray. On the crown of the head the black hairs are tipped with gray: and as age advances the latter colour becomes more extensive, showing itself on the upper parts of the body; but the extremities externally, and the tail, retain their blackness to the last. The hair is long, soft, and silky. There are many species, one of the most famous being the *S. Entellus*, a species of Monkey venerated by the Hindoo. [See MONKEY.]

SEPIA: SEPIADÆ. A genus and family of Cephalopods belonging to the Cuttlefish tribe. The best known species is that figured in our article, the Common or officinal Cuttlefish (*Sepia officinalis*); the little figure at the side representing the shell, which is often found cast ashore, and is used in medical purposes from the pureness of the calcareous

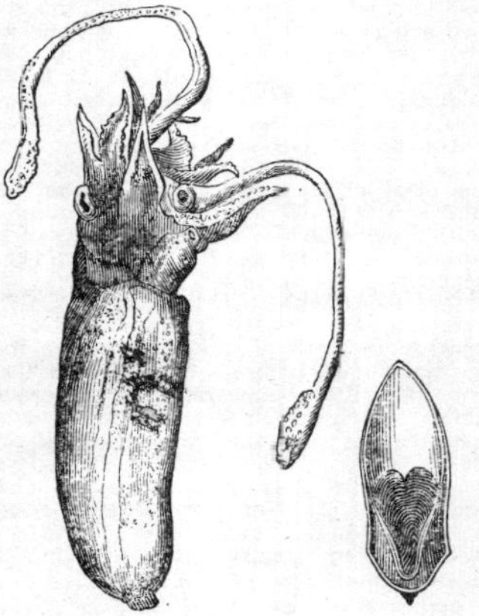

CUTTLE-FISH.—(SEPIA OFFICINALIS.)

matter of which it is composed. In ancient times, and in some part of the Levant even now, as we learn from Forbes and Spratt's Lycia, the Cuttle-fish of different species were used as articles of food; and we know, from the works of travellers, that in other parts of the world, when cooked, they are esteemed luxuries. Cuttle-fish are furnished with a curious receptacle for a fluid, which they use not only as a direct means of annoyance, but also for the sake of making the water turbid, and thus eluding pursuit. [See CUTTLE-FISH: CEPHALOPODA.]

SEPS. The name of a genus of Saurian reptiles, which have a long serpentiform body, and four very short legs, each terminated by only three toes. They differ from the Skinks by having the body still more elongated.

SERICULUS. A genus of birds found in Australia. [See REGENT-BIRD.]

SERPENTS. (*Ophidia.*) The general name of the third order of Reptiles, according to the arrangement of Cuvier. We have described several of the most remarkable of these Reptiles in separate articles, [see BOA CONSTRICTOR, RATTLESNAKE, &c.] Some observations, however, on their general character will in this place be necessary. These Reptiles are in general easily distinguished from others by the total absence of external feet, hardly a vestige of which is discoverable on the most minute dissection. Their motion is, notwithstanding, very rapid in some species, and is accomplished by means of the sinuosities or folds which they form with their bodies. When in a state of repose they usually dispose themselves in coils, with the head in the centre; and many are enabled to spring to a certain distance by the sudden unfolding of these coils. Serpents are destitute of movable eyelids, or distinct tympanums. All have teeth, but they serve only to retain their food, and are not adapted to mastication. The venomous species have the maxillaries very small and movable, and in them are implanted two teeth much longer than the rest, and traversed by a canal for the purpose of transmitting the poison. These fangs are projected forward in the action of biting, but at other times are disposed along the roof of the mouth, in such a manner as hardly to be discoverable at first sight. The jaws of Serpents are united by ligaments so as to admit of great extension, which enables them to swallow animals of much greater diameter than their own bodies. The tongue is remarkably extensible, and terminates in two long cartilaginous points. They have only one lung. The skin in different genera is annulated, coriacious, or granulated, or, most frequently, covered with scales. They feed on quadrupeds, reptiles, insects, or worms, and swallow their prey entire. They do not drink, and the power of digestion is slow, one meal serving them for weeks, or even months; but when an opportunity offers, they take an enormous quantity of food. The ribs are very numerous, and surround a great portion of the trunk. The muscles, even in the smaller species, are endowed with an astonishing power of constriction; and those species which attain the enormous dimensions of thirty feet or more, are enabled to destroy the larger quadrupeds by involving them in their folds. The Serpent tribes are almost universally regarded with feelings of horror and aversion, which doubtless originate in the venomous qualities of some, and the terrific strength that characterizes others; and also in the insidious manner in which they usually approach their victims. Natural as these prejudices are, it is certain, however, that by far the greater part are perfectly harmless. In northern climates they pass the winter in a torpid state, and change the epidermis in the spring. The eggs are rounded, and agglutinated in bead-like rows by a mucous substance, and, in the venomous species, hatch before they are excluded from the oviduct, and the young are born alive. The females often take care of their young for a time, and, on the approach of danger, have been seen to receive the whole family in their throats, and, when it has passed, to restore them again to the open air. More than three hundred species are enumerated, most of which, including all the gigantic species, inhabit tropical climates. South America, in particular, abounds with them. The venomous species compose about a fifth of the whole number; and among these are some whose bite is fatal in a few hours, and even minutes. [See SNAKES.]

SERPENT-EATER. (*Gypogeranus.*) A South African bird of prey, often called the Secretary-bird, or Secretary Falcon; agreeing in its general character and some of its habits with the *Falconidæ*, especially those which prey on reptiles, while it differs from the Hawks and Owls in having feet incapable of grasping, and very long legs. Accordingly, it keeps constantly on the ground, in sandy and open places, and wages continual war against reptiles, especially Serpents, which

SECRETARY BIRD.
(GYPOGERANUS SERPENTARIUS.)

it pursues on foot. When this bird attacks a serpent, it covers its breast with one wing (the wings being armed with spurs on the elbow-joints,) to protect itself from the bite, and with the other strikes violent blows, until it has stunned its prey. It then breaks the cranium with its beak, and tears the reptile in pieces, or, if small, swallows it entire. In its wild state the Serpent-eater is shy and difficult of approach; but it is easily tamed, and is often kept in poultry-yards by the inhabitants of the Cape of Good Hope, for the purpose of destroying lizards, snakes, rats, &c. It soon becomes habituated to the poultry; but if left too long fasting, it does not scruple to satisfy its hunger with the young chickens. It runs with great rapidity. Le Vaillant mentions, that having killed one of these birds, which he had seen to vanquish a serpent, he found in its crop eleven rather large lizards, three serpents of an arm's length, and eleven small tortoises very entire, — all of which had received the stroke on the head; as well as a number of locusts, beetles, and other insects, very little injured. The colour of this bird in its perfect plumage is a bluish gray on the head, neck, breast, back, and wing-coverts; the throat white; abdomen black, streaked with rufous; thighs black, streaked with brown; tail feathers black and gray, tipped with white. The skin of the throat and neck are capable of great extension. It builds its nest on high trees, or dense thickets; and is not at all disposed to associate with its fellows. The name of *Secretary* was given to it by the Dutch settlers at the Cape, from a pendent crest on the back of the head reminding them of the pen stuck behind the ear, according to the custom of writing-clerks.

SERPULA. The name of a genus of Anellidans inhabiting cylindrical and tortuous calcareous tubes; generally parasitic on testaceous Mollusca. The tubes of the Serpulæ are found clustering in masses, attached to the surface of stones, shells, or other bodies, which have been immersed for any length of time in the sea: they are usually more or less contorted in form, varying in this respect according to the position in which they grow; but they are always closed at one end, which tapers to a point, the wide end being open to give exit to the head and mouth of the inhabitant. The animal which forms this shell, and resides in it, has its branchial filaments or gill-tufts all assembled round the head; where they form a pair of elegant fan-like appendages. At the base of each series there is a fleshy filament, one of which fits to the mouth of the shell, and serves to close it when the animal is withdrawn into the tube. The body of the animal is composed of a great number of segments; but these are for the most part unprovided with any appendages. The largest species of Serpulæ are found in tropical regions, where they usually form their habitations in the midst of corals, lengthening their tubes as the coral is built up around them. Their length is sometimes as much as three feet; and their expanded gill-tufts are of extremely vivid colours, strongly resembling the most brilliant carnations in general aspect. Numerous smaller species are found on our own coasts; the gills of some of them being remarkable for their brilliant hues.

SERRICORNES. A family of Coleopterous insects, distinguished by the toothed or serrated form of the antennæ. It includes many of the Beetle tribe, which are distinguished for the splendour of their colours, the largest and most brilliant of which are found chiefly in tropical climates. [See BUPRESTIS and ELATER.]

SERTULARIA. A genus of compound tubular Polypes; consisting of those species in which the cells are arranged on two sides of the stem, either opposite or alternate; of these there are many British species, often taken by the ignorant for Sea-weeds. Some of these are most beautiful objects, being finely branded: those indigenous to our coast are described and figured in Johnston's admirable 'British Zoophytes.'

SERVAL. (*Felis serval.*) This fierce and rapacious animal is a native of India and Thibet. It resides principally among trees; leaping with great agility from one to another, and pursuing birds. It resembles the Panther in its spots, but the Lynx in its size, the robustness of its make, and the shortness of its tail.

SESIA. A genus of Lepidopterous insects, comprising those with the antennæ always simple, elongate-fusiform, and often terminated by a small bundle of scales. Many of the species resemble Wasps and other Hymenopterous and Dipterous insects, and fly about in the hottest sunshine.

SETTER. (*Canis familiaris index.*) This variety of the Dog is little inferior in point of sagacity to any of the species, and surpassed by none in docility or grateful attachment, while its excellent nose and endurance

ENGLISH SETTER.

of fatigue in the field render it of great service to the sportsman. In figure it partakes of the characters of the Pointer and Spaniel, the hair having much of the wavy appearance of the latter, and also the ears. Its general colour is white, with large liver-coloured patches.

SHAD. (*Clupea alosa.*) This fish resembles the Pilchard in general appearance, but is much larger, and flatter in proportion. The colour of the body is bright silver, inclining to dusky on the back, and marked on each side, at a small distance from the gills, by four or more rounded black spots: the scales are rather large: the fins are of a bluish tinge; and the tail is forked. It is a native of the Mediterranean and Northern seas, and, like the Salmon, ascends rivers, at particular seasons, for the purpose of depositing its spawn. Like the Herring, it dies almost immediately after being taken out of the water, and is supposed to feed chiefly on worms, insects, and young fish. It is found in greater perfection in the Severn than in any other British river. The Thames Shad is comparatively a very coarse fish.—The Shad which frequents American waters is probably a different species. It usually weighs four or five pounds, but sometimes twelve. It is highly esteemed for food, and is consumed in great quantities in the fresh state: great quantities are salted, but they are then less esteemed than when eaten fresh. During the season they are an important source of wealth to the inhabitants of the borders of the Hudson, Delaware, and Chesapeake rivers.

SHARKS. (*Squalidæ.*) A family of Cartilaginous fishes, allied to the Rays, and celebrated for the size and voracity of some of the species. The form of the body is elongated, and the tail is thick and fleshy. The mouth is large, generally situated beneath the snout, and is armed with several rows of compressed, sharp-edged, and sometimes serrated teeth; these are movable at the will of the animal, and are usually laid down and directed backwards, but become erect at the moment he is seizing his prey. The skin is usually very rough, covered with a multitude of little osseous tubercles; and that of some species forms the substance called *shagreen.* They devour with indiscriminating voracity almost every animal substance, whether living or dead. They often follow vessels for the sake of picking up any offal that may be thrown overboard; and, in hot climates especially, man himself frequently becomes a victim to their rapacity. No fish can swim with such velocity as the Shark, nor is any so constantly engaged in that exercise: he outstrips the swiftest ships, and plays round them, without exhibiting a symptom of strong exertion or uneasy apprehension; and the depredations he commits on the other inhabitants of the deep are truly formidable. The eggs of Sharks are few and large, in comparison of those of bony fishes; they are enveloped in a hard, horny, semi-transparent shell, terminated at the four angles with long filaments. Messrs. Müller and Henlé have described many new genera and species of this family.

The WHITE SHARK. (*Squalus carcharias.*) The White Shark, in size and voracity the most formidable of all the species, is an inhabitant of most parts of the globe, though much more frequently seen in the warmer than in the colder latitudes. It is believed to reside principally in the depths of the ocean, rising at intervals in order to pursue its prey. It sometimes attains the length of from twenty to thirty feet, and its mouth is sufficiently wide to enable it to receive the thigh, or even the body of a man. The head is of a depressed shape and broad, terminating in an obtusely pointed snout; the margin of each jaw is furnished with from three to six rows of strong, flat, triangular, sharp-pointed, and finely serrated teeth; the tongue is broad, thick, and cartilaginous, and the throat extremely wide; the eyes, as in most of the genus, of a bluish or greenish cast. The pectoral fins are large, strong, broad, and pointed; the first dorsal fin falcated behind, and pointed; the second is situated near the origin of the tail, which is slightly lengthened, and of a bilobate shape. The general colour is a pale or whitish ash, but darker on the upper parts. The internal parts of the Shark present many remarkable particulars: the brain is small; the throat is very short, and of a diameter not greatly inferior to that of the beginning of the stomach, which is of vast size, and dilatable to a great degree: the intestinal canal, instead of forming a mere continued tube, consists rather of a large series of meshes or divisions, placed in a spiral direction throughout its length. During the breeding season, which takes place at different periods in different climates, the Sharks are observed to approach the shores, in order to deposit their young in the most favourable situations. The length of the newly-hatched Shark does not exceed a few inches.

The BASKING SHARK. (*Selachias maximus.*) This species is scarcely, if at all, inferior in size to the White Shark. They generally appear in the Firth of Clyde and among the

Hebrides in June, in pairs, or in small shoals of seven or eight; and depart again in July. They are said to have nothing of the fierce and voracious nature of other Sharks; but are seen sometimes lying quietly near the surface, and at other times leaping with vast agility several feet out of the water. The tail is very large, and the upper part of it remarkable for its extreme length. The upper part of the body is of a deep leaden colour, the belly white: on the back the skin is granulated, like shagreen; and within the mouth, towards the throat, is a very short sort of whalebone. They are viviparous. They are killed by harpooning, which, owing to their strength, is often a long and difficult operation. When killed, they are either hauled on shore, or, if at a distance from land, to the vessel's side: the liver, which is the only part of any value, is then taken out, and melted into oil; of which a large fish will yield eight barrels.

The BLUE SHARK. (*Squalus glaucus.*) This fish is of a more slender and elegant shape, as well as the most beautiful in point of colour, of all the Sharks. The colour above is blue-green, beneath white: head rather large, with the snout very long and pointed; and the mouth wide, and placed very far under: teeth nearly triangular, sharp, and disposed in three or four rows: eyes large: the tail deeply bilobate, with the lower lobe much larger and longer than the upper. It grows to the length of eight feet, and is an inhabitant of most parts of the globe. It is a very voracious and bold fish, and is scarcely less dreaded by sailors than the Common or White Shark. It is said principally to prey on herrings, shads, and tunnies. It frequents several of the British coasts, particularly that of Cornwall, during the pilchard season, when it is extremely troublesome to the fishermen, by cutting their lines and nets, and devouring the fish. It is taken with large iron hooks prepared for the purpose.

The FOX SHARK. (*Squalus vulpes.*) The Fox Shark, or Thresher, is distinguished for the great development of the upper lobe of the caudal fin, or tail, into which the vertebral column is prolonged; this being nearly as long as the entire body, which is plump and sub-ovate. The first dorsal fin is triangular, and placed on the middle of the back; and the pectoral fins are of considerable size: the eyes are large; the mouth small; the teeth triangular, small, and in three rows. Colour, dusky ash above, and whitish beneath. It inhabits the Mediterranean and other seas, and is occasionally met with on our own coasts: it grows to the length of twelve or fourteen feet; and is considered as a voracious and artful fish.

The HAMMER-HEADED SHARK. (*Zygæna vulgaris.*) Of all marine animals this is perhaps the most deformed. It resembles the ordinary Sharks in the form of its body, which is sub-cylindric and rather slender; but its head is dilated on each side to a great extent, in the form of a double-headed hammer; the eyes, which are very large, being placed at each extremity: mouth beneath, as in other Sharks. It is a native of the Mediterranean and Indian seas, where it is scarcely less voracious and formidable than even the White Shark itself; attacking

HAMMER-HEADED SHARK (ZYGÆNA VULGARIS.)

such as are accidentally bathing in its neighbourhood. It is observed about the coasts of the South Sea islands, and particularly of Otaheite, where the natives, trusting to their dexterity in swimming, appear to hold it in but little dread.

The PICKED SHARK. (*Galeus acanthias.*) This species is from three to four feet in length; and is readily distinguished by a very strong bony spine, situated before each dorsal fin, and connected at its base with

PICKED SHARK.– GALEUS ACANTHIAS.)

the fin itself: teeth small and sharp, and disposed in rows along the jaws; upper lobe of the tail longer or more projecting than the lower, which is continued to some distance beneath. It inhabits the European seas, and is very common about the coasts of Scotland, where it is taken in order to be prepared for sale by splitting and drying, and is then much used as a food among the poorer classes.

SHARK [MOTHS]. A name given by collectors to Moths of the genus *Cucullia*.

SHEARS [MOTHS]. A name given by collectors to Moths of the genus *Hadena*.

SHEATH-BILL. (*Chionis.*) A genus of Shore-birds, or Waders, which have short toes, nearly as in domestic poultry, the tarsi scutellated, the beak thick and conical, and enveloped at the base by a hard substance, which, it appears, the bird has the power of raising and depressing. This remarkable bird approaches very near to the Oyster-catchers in its whole anatomy, and the affinity is discernible on comparison of their external characters. The species *Chionis necrophaga*, which is from New Holland, is the size of a large Partridge, and entirely white. It frequents the sea shore, and feeds on dead animal matter thrown up by the tide.

SHEATHBILL.—(CHIONIS ALBA.)

SHEEP. (*Ovis aries.*) A genus of Ruminant quadrupeds, belonging to the class Mammalia; and differing so slightly in the anatomical structure from the Goat, that both genera are by some naturalists united. The principal distinctive characters consist in the Sheep having no beard; in the horns being directed backwards, and then inclining spirally more or less forwards; in having a convex forehead; and in the existence of a sac or fossa, situated at the base of the toes, lined with hair, and furnished with sebaceous follicles. It is generally imagined that the primitive stock may be traced to the Wild Sheep of Sardinia and Corsica

BLACK-FACED RAM

[see MOUFFLON], or to the Argali of Asia; but whether either of these are to be regarded as the parent stock, or as the descendants of those which have escaped from the dominion of man (as some have suggested), is of little importance; but this is certain, that although the coat of these wild sheep consists of coarse, stiff, and long hairs, they possess the essential character of wool—an imbricated scaly surface—which gives to the shorter and finer wool of the domestic races that remarkable *felting* * property upon which its peculiar utility depends.

It is universally allowed, that, with the exception of horses, and perhaps cattle, Sheep are by far the most important of all the domestic animals we have. They not only afford a large supply of food, and furnish one of the principal materials of clothing, in the manufacture of which an immense number of people are employed; but it should be remembered that they can be reared in situations and upon soils where other animals could not find sufficient food for their support. "The dressed skin," says Mr. Pennant, "forms different parts of our apparel; and is used for covers of books. The entrails, properly prepared and twisted, serve for strings of various musical instruments. The bones, calcined (like other bones in general), form materials for tests for the refiner. The milk is thicker than that of cows, and consequently yields a greater quantity of butter and cheese; and in some places is so rich, that it will not produce the cheese without a mixture of water to make it part from the whey. The dung is a remarkably rich manure; insomuch that the folding of sheep is become too useful a branch of industry for the farmer to neglect. To conclude: whether we consider the advantages that result from this animal to individuals in particular, or to these kingdoms in general, we may, with Columella, consider this, in one sense, as the first of the domestic quadrupeds."

Many persons are accustomed to consider the Sheep as the most stupid of all domestic quadrupeds, and as the only one which is probably incapable of returning to a state of nature; that it neither knows how to avoid danger, nor to seek shelter from the changes of the atmosphere, nor even to procure nourishment, except in abundant pasturage. To a certain extent this may be true; but those who have witnessed the boldness and agility with which the Sheep of the Welsh mountains leap from crag to crag,—or the safety with which others descend the rocky precipices of the south-western coasts of the Isle of Wight, to graze on the sweet but scanty herbage which occasionally shows itself among the chalk, and then re-ascend till they reach the summit, bounding upwards with a sureness of foot and strength of spring that seem to rival the goat,—would be disposed to consider that their instincts were neither so obtuse, nor their return to a state of nature, under favourable circumstances, by any means so difficult as they had imagined.

The history of the Sheep may be traced to the remotest antiquity; for we read that "Abel was a keeper of Sheep," and that "Abel brought as an offering to the Lord the firstlings of his flock, and the fat thereof." "There probably is not a species amongst all our domesticated animals," observes Mr. Bell, "which in its historical relations is so interesting as the Sheep. Its early domestication, its employment as the subject of the first sacrifices, its typical character as an offering of atonement, its importance as forming the principal wealth of the early patriarchs—its various connection, in short,

* "Cloth and woollen goods are made from wool possessing this property; the wool is carded, spun, and woven, and then, being put into the fulling mill, the process of felting takes place. The strokes of the mill make the fibres cohere; the piece subjected to the operation contracts in length and breadth, and its texture becomes more compact and uniform. This process is essential to the beauty and strength of woollen cloth. But the long wool of which stuffs and worsted goods are made is deprived of its felting properties. This is done by passing the wool through heated iron combs, which takes away the laminæ or feathery part of the wool, and approximates it to the nature of silk and cotton."—*M'Culloch.*

with the political, the religious, and the domestic customs of those primitive magnates of the Jewish nation, are all of them subjects forming ample food for deep and delightful reflection. The relation which existed between the patriarchal shepherds and their flocks was indeed of so intimate, and even affectionate a nature, as to have afforded the subject of many of the most beautiful and touching parables and moral illustrations in the Sacred Writings. It is scarcely necessary to refer to the unequalled appeal of Nathan to David, to the still higher and prophetic allusion to the character of the Messiah, or to the sublime illustration of the beneficence of the "great Shepherd of Israel," in the beautiful and well-known pastoral psalm. These are subjects which cannot be discussed here; but it is impossible to pass them wholly without notice. But the historical interest attached to this animal does not stop here. The customs observed in the treatment of their flocks by the shepherds of the Eastern nations in the present day, offer numerous and highly important coincidences with those incidentally alluded to or more distinctly described in the Scriptures."

The habits of the Sheep in its domestic state are too well known to render a detailed account necessary, or to speak of the methods which have been adopted to improve the breed. We shall, however, glean from different sources many particulars relating to them, which are too important to be omitted. We know that the products of this animal are the flesh, milk, skin, and especially the wool, which employs a vast capital in the manufacture of clothing. The fineness of the wool is considerably influenced by the temperature: in a hot climate Sheep yield a comparatively coarse wool; in a cold climate they have a closer but a warmer fleece. The filaments of wool taken from a healthy Sheep present a beautiful polished and even glittering appearance; while that of the neglected or half starved animal exhibits a paler hue. "As for the carcase of the Sheep, it is comparatively lately that even in Great Britain it has been regarded in the light that it deserves. In many foreign countries it is disliked, or at least rarely eaten. The Calmucks and Cossacks seldom touch it. Even in some parts of America there is a prejudice against it. It is an object of little or no value in Spain; and, except among the poorest, it is not there considered fit for food. Since the British sheep-master has begun, and judiciously, to look more to the profit to be derived from the carcase — since the system of artificial feeding has been brought to so great perfection, and a far greater number of sheep can be fed and perfected on the same number of acres, perhaps the wool may have somewhat altered in character. It has grown in length, and it has increased in bulk of fibre. It has not deteriorated, but changed. If no longer fit for the purposes to which it was once devoted, it has become suited to others. The increase of the number of fleeces and the increase of weight in each fleece go far to compensate for the diminution of price, while the improvement of the carcase more than supplies the deficiency, if in truth there was any deficiency to supply. It has been proved by authentic documents, that the number of sheep in the United Kingdom has been more than trebled in the last 150 years. How was this managed? for it was supposed that no more than a given number of Sheep could be kept on a certain space of ground. The quantity of ground was rapidly increased, and much that was formerly deemed unworthy of cultivation was rendered productive; but more effective than this was the new system of husbandry that was introduced — the artificial or turnip husbandry, by which a regular supply of food could be provided for every season. With this was connected the attempt to fatten Sheep still more expeditiously than could be accomplished by any former method. This succeeding beyond his most sanguine expectations, the sheep-master next attempted to increase the size of the breed. He had not, however, sufficiently taken into account a consequence of this. As the carcase increased in size, the wool became longer, heavier, and coarser. The breeder would not see this at first; but he soon began to find that the manufacturer would not purchase it, for it could not be used for the purposes to which it was formerly applied. His stock accumulated. It weighed heavily on his hand. Still he would not believe that his once favourite and yet valuable wool was deteriorated, although he was compelled to sell it at a diminished price. And what was the consequence? Why, that he had no just reason to complain; for the early maturity of the Sheep, and the continued value of the wool for many important purposes rendered his profits greater than they were before he had begun to alter his system." — *Penny Cyclopœdia.*

The varieties of the Sheep are very numerous, differing in size, the length of their legs, and the size of their horns: some are covered with hair instead of wool; others have enormous tails; and others, again, pendent ears. The variety most celebrated for the fineness of the wool is the Spanish Merino, as improved in Germany, where both governments and individuals have paid great attention to the improvement of the wool, and in some parts of that country it has been brought to such perfection as to surpass that of any other part of the world. When we look for the origin of the improvements which have been made in the breeding of this animal, which has become so important an element of national wealth, and the source of so much manufacturing and commercial industry, we are obliged to go back to the Romans. Columella, who lived under the Emperor Claudius, gives us interesting information on this point. Among other things, he says that his uncle, who lived in Bœtica (which comprehends the present province of Estremadura), procured some wild African rams at Cadiz, of a coarse fleece, but of an admirable colour. He put them to some fine-woolled ewes, and, the male progeny being again put to Tarentine ewes, the offspring, with their descendants, united the colour of the sire with the dam's softness of fleece. Other agriculturists undoubtedly imitated him, and thus the purest white was communited to the black or particoloured native flocks, which, according to Pliny, were common in Spain. The attention paid by the ancients to the Sheep was excessive, and the animal was extremely tender; so that we must account for the transition from the ancient Sheep to the Merino, which is a hardy animal, thriving almost in any climate, by supposing that other agriculturists imitated Columella, and by crossing the breed imparted a stronger constitution to the fine-fleeced but delicate Sheep of ancient Italy. Strabo, indeed, describes the beginning of this improvement as having taken place in the reign of Tiberius. Five rams were at that time sold at Truditania, part of Bœtica, for a talent, or about 250*l*., a price which, considering the value at that period, is immense. When the Roman empire was overrun, and almost all traces of civilization swept away the Tarentine stock in Greece and Italy, being very tender and requiring the greatest care, became extinct; but the regenerated stock of Bœtica — the Merinos — being able to live on the mountains, survived the conquest of

MERINO RAM.

Spain by the Goths and Vandals; and from these Merinos are descended those animals which supply all the manufactories of fine cloth in Europe. Care was early taken in Spain that the improved Sheep should not mix with the coarse native Sheep. The government soon took this important branch of national industry under its protection, and established particular courts to have jurisdiction over all subjects connected with Sheep, wool, shepherds, pastures, &c.

It does not appear that in this country the Sheep was an object of much attention prior to the invasion of the Romans; but they established a woollen manufactory at Winchester, which soon acquired so much celebrity that it supplied the finest and most expensive woollen cloths for the Roman market, such as were employed for the festival dresses of the patricians. Surrounded by downs and grazing land, and the valley in which that ancient city is situated being plentifully supplied with streams of the purest water, the site was well chose[illegible]nd for many centuries, after England [illegible]b-mitted to less civilized conquerors [illegible]e haughty Roman, it continued to be th[illegible] emporium of the woollen cloth tra[illegible] well as for the raw material.

The Spanish breed of Sheep was first [illegible]troduced into Great Britain in 1787. So[illegible] individuals of the black and spotted kinds had indeed been procured and kept in the parks of noblemen previously, but without any regard to the wool; nor was much interest awakened by the flock imported in 1787. Subsequently, great attention was paid to the improvement of English wool; but it was ascertained that though the fleece of the Merino did not degenerate in England, it did not improve; and the farmers accordingly found it for their interest to return to the native breeds, and to give up the Spanish Sheep. The breed of Sheep that was carried out to New Holland and Van Diemen's Land has succeeded remarkably well; and the former promises, at no distant day, to be one of the principal wool-growing countries in the world.

It appears that a great deterioration in the quality of British wool has taken place in the present century. The great object of the agriculturists has been to increase the weight of the carcase and the quantity of the wool, and it seems very difficult, if not impossible, to accomplish this withoat injuring the fineness of the fleece. We now have to speak of some of the principal breeds of Sheep of which Great Britain can boast.

The *Southdown Sheep* takes its name from an immense tract of downs, formed by a long range of chalk hills extending more than sixty miles in length, through part of the counties of Sussex, Surrey, and Kent. Its head is small and hornless; the face dun, black, or speckled; the ears tolerably wide, and well covered with wool, as is also the forehead, and the space between the ears; the eyes full and bright, but not prominent: the chest wide, deep, and projecting between the fore-legs: the shoulders on a level with the back, and not too wide above, but bow-

SOUTHDOWN SHEEP.

ing outward from the top to the breast: the back flat from the shoulders to the setting on of the tail; the loin broad and flat; the rump long and broad; the hips wide; and the ribs presenting a circular form, like a barrel: the belly as straight as the back: the fore legs straight from the breast to the foot, not bending inward at the knee, and standing far apart both before and behind; the hocks having a direction outwards, and the meeting of the thighs being particularly full: the bones fine, but having no appearance of weakness; and the legs of a speckled dark colour. The wool short, close, curled, and free from spiry, projecting hairs. The Southdown is adapted to almost any situation in the midland part of England, but the northern hills are occasionally too cold. It is capable of enduring occasional short keep and hard stocking equal to any other Sheep; and the flesh is finely grained and of good flavour. The figure of this Sheep was formerly inferior to that of many others, but very great improvement in them has arisen during the last few years (in the figure and size of the animal, as well as in the wool), not from crossing with other breeds, but from the system of sorting the flocks. These Sheep occupy the whole of the upper and under hill-grounds of Sussex; and they have also succeeded well in all the southern districts of the kingdom.

The *Dorset* breed. The Dorset Sheep has a strong well-formed body and limbs, a clear white fleece, and finely-curved horns; altogether presenting to the eye an animal which, whatever its intrinsic merits may be, must be considered handsome. The face is long and broad, with a tuft of wool on the forehead; the shoulders low, but broad; the chest deep; the loins broad; and the bone small. Their chief peculiarity is the forwardness of the ewes, which supply the market with lamb when it fetches the highest price. A very profitable variety is found in a cross between the Southdown and the Dorset Sheep; the carcase being increased, and the wool rendered more valuable. In Hampshire, Berkshire, Wiltshire, and Somersetshire, the old breeds, for which these counties were once famous, have generally yielded to cross breeds with the Southdown, or been supplanted by the true Southdowns. In short, we find the same breed either pure, or materially improving the breeds of many other counties, both westward and midland.

The small hardy Sheep, called the *Ryelands*, are still met with in Herefordshire. They are small, polled, with white faces, the wool almost covering the eyes, and the carcase round and compact: they have a tendency to fatten quickly; and they are particularly distinguished by the fineness of their wool. — The *Cheviot* breed, so called from the Cheviot Hills, in Northumberland, have no horns, and are mostly white-faced and white-legged; the body is long, with fine, small, clean-boned legs; the fleece fine, short, close, and thick set. — Wales, both North and South, is celebrated for a small and valuable breed of Sheep, principally used for the supply of the London market, where the Welsh mutton is always in request.— The *Lincolnshire* Sheep are of a large size, and afford a great quantity of wool, owing to the rich marshes where they graze; but their flesh is coarser, leaner, and less finely flavoured than that of the smaller breeds. The old Lincolnshire Sheep was, however, unrivalled in its wool, both as to quality and quantity; and since they have been crossed with the Leicesters, which were always remarked for their disposition to fatten well, the value of the carcase has increased, though, in some measure at the expense of the fleece. The *Cotswold* Sheep, so called from the cots or sheds in which they were housed, formerly inhabited the counties of Gloucester, Hereford, and Worcester. They were a long-woolled breed, yielding in the 15th century a description of wool much valued on account of the fabrics in the construction of which it was employed.

But of all the various breeds of Sheep, it must be confessed that none have attained such deserved celebrity as the *New Leicester*, a breed brought to the highest state of perfection by the skill and perseverance of Mr. Bakewell, of Dishley, Leicestershire, — the eminent agriculturist and improver of live stock. It would not be consistent with the nature of this work (even if our space would permit) to describe the various means he made use of; but his principle was, to select such kinds of Sheep as his experience told him had an aptitude to fatten, and with little bone and offal: he cared not about near or distant affinities, but his object was to increase every good point, and get rid of every bad one. They were not different sorts of Sheep that he selected, but the best of the breed to which he had been accustomed. He also introduced the practice of *letting* some of his rams, which extended the benefit of his system far and wide: and so great was the desire for improvement when the *Dishley* or *New Leicester* breed became known, that it was calculated that 100,000*l.* was annually spent by the midland farmers in the hiring of rams. There are few other varieties of long-woolled Sheep which do not owe much of their excellence to the new Leicesters, and even some of the short-woolled flocks are deeply indebted to the breed. The deficiency of the fleece was formerly objected to; but

LEICESTERSHIRE

it has now not only considerably increased in length, but improved in fineness and strength of fibre, and averages between six and seven pounds the fleece. In short, it has been truly said, that it is difficult to select any part of the kingdom into which the Leicester and the Southdown Sheep have not penetrated, and where they have not materially improved the native breed

"Different names are given to the Sheep according to its sex and age. The male is called a *ram* or *tup*. After weaning he is said to be a *hog*, a *hoggett*, a *tup-hog*, or a *teg;* and if castrated, a *wether hog*. After shearing, and when he is probably a year or a year and a half old, he is called a *shear-hog*, or *shearling*, or *dinmont*, or *tup;* and when castrated, a *shearing-wether*. After a second shearing, he is a *two-shear ram*, or *tup*, or *wether*. At the expiration of another year, he is a *three-shear ram*, &c.—The female is a *ewe* or *gimmer lamb* until weaned; and then a *gimmer*, or *ewe-hog*, or *teg*. After being shorn she is a *shearing ewe* or *gimmer*, or *sheave* or *double-toothed ewe;* and after that, a *two*, or *three*, or *four-sheared ewe* or *sheave*. The age of the Sheep is reckoned, not from the period of their being dropped, but from the first shearing.

There are several remarkable varieties of the genus *Ovis* in different parts of the globe, which must here be noticed.

The MANY-HORNED SHEEP. (*Ovis polycerata.*) This variety, which is found in Iceland and the most northern parts of the Russian dominions, resembles the domestic breed in the shape of its body and tail, though it has three, four, five, or more horns, sometimes placed with great regularity, and sometimes differing in proportion and situation. This animal is large and formidable in appearance: but in its nature it is timid and gentle. The wool, which is long, smooth, hairy, and very different from that of the common Sheep, is of a dark brown colour; and under its exterior coat there is a fine, short, and soft kind of wool or fur.

The BROAD-TAILED SHEEP. (*Ovis laticauda.*) This variety is very common in Tartary, Arabia, Persia, Barbary, Syria, and Egypt; and is principally remarkable for its large heavy tail, often so loaded with a mass of fat as to weigh from ten to twenty

AILED SHEEP.

pounds; nay, some writers assert that the tails are occasionally double that weight, a foot broad, and supported by a small board, which runs on wheels. The upper part is covered with wool, but it is bare underneath; and the fat or marrow of which it consists is reckoned a great delic .

The CRETAN SHEEP trepsiceros.) This animal is principally found in the island of Crete, and is kept in several parts of Europe for the singularity of its appearance; the horns being very large, long, and spiral: those of the male are upright; those of the female at right angles to the head. By Buffon this variety is termed Wallachian Sheep.

The FAT-RUMPED, TAILLESS SHEEP Sheep are met with in all the deserts of Tartary, from the Wolga to the Irtis and the Altaic chain of mountains. They have long legs, a somewhat arched visage, horns in the male, like those of the domestic Sheep, black heads, and large pendent ears. The tail is sometimes so enveloped in fat as to be scarcely visible, the parts on each side swelling out into two naked hemispheres.

The AFRICAN SHEEP. (*Ovis Guineensis.*) The African or Guinea Sheep is a native of all the tropical climates, both of Africa and the East. It is large, with rough hairy skin, short horns, and pendulous ears. Under its chin there is a kind of dewlap; and it has a long mane, which reaches below the neck. They are stronger, larger, and more fleet than other Sheep, and therefore better adapted to a precarious forest life; but their flesh is very indifferent food.

SHEEP-TICK. (*Hippobosca Ovis.*) A well-known insect, extremely common in pasture grounds, about the commencement of summer. The body is very compressed and smooth, the shape triangular, and the colour a blackish brown. It fixes its head in the skin of the animal, and extracts the blood, leaving a large round tumour. This singular animal has no wings, nor does it ever attain them; yet it evidently belongs, from the conformation of its body, to the family *Hippoboscidæ*, as the bed-bug belongs to the tribe of the winged bugs. The fore part of its body is uncommonly small; the thick roundish abdomen, however, is proportionally very large, and generally in circumference about the size of a middling-sized pea. Its colour is pale red, the abdomen lighter, with an irregular white line on each side, and a red spot on the back. The Sheep-tick lays only one egg, which is the nymph or pupa, as in the forest-fly, and is fastened to the wool of the sheep. At first it is white, then brown, and finally the perfect insect escapes from it. As a remedy for this insect, Bock advises that the infested sheep should be washed with a decoction of the crushed or bruised leaves of the common maple. Another method of diminishing or destroying this troublesome insect is given in the Farmer's Magazine for Nov. 1828, by a farmer in Suffolk. He advises the lambs to be put into a bath, by which the production of the sheep-tick will be prevented. The best time for this is July or August. Should it, however, have been neglected, then it is still time, if the weather permits, till Christmas. A pound of arsenic is boiled with a pound of soft soap and a pound of purified potash, in four gallons of water. The arsenic will be perfectly dissolved by the other ingredients. As soon as this is the case, the solution is thrown into a bathing tub sufficiently large to dip a sheep in, and forty gallons more water added to it. In order to dip the sheep, its fore legs must be held by one man and its hind legs by another, so that the feet are held upwards. A man must also stand at the tub, to prevent the head being dipped, so that no poison may get into the ears, which would do it an injury. This man is provided with a sort of tressel, which he holds under the lamb as soon as it is withdrawn from the bath. He then squeezes the fleece with his hands, so that the greater part of the water sucked up by the fleece runs again into the tub. In this way the above-named quantity may serve to dip one hundred moderate-sized lambs in. The author adds, that the arsenic has no injurious effects, if carefully used;

and that one essential advantage of this proceeding consists in its protecting the lambs from the sheep bot fly, and consequently their larvæ, if it is done early enough.

SHELDRAKE, or SHIELDRAKE. (*Tadorna vulpanser.*) An elegant species of Duck, belonging to the genus *Tadorna*, upwards of two feet in length, which frequents many parts of our coast, and remains throughout the year. The head and neck of the male is of a dark green; lower part of the neck, coverts of the wings, back, sides, rump, and base of the tail, pure white; the scapulars, abdomen, quills, a large band across the belly, and tips of the tail feathers, deep black. A large bay-coloured gorget adorns the breast, and the wing is ornamented with a spot of

SHELDRAKE.—(TADORNA VULPANSER.)

purple-green. The bill, and the fleshy knob at its base, deep red. Feet, flesh-colour. It may often be seen about our largest rivers. Its food consists of small testaceous mollusca, small fish, small crustacea, and marine plants. The female commonly selects a rabbit-hole in which to deposit her eggs; these are ten or twelve in number, and of a pure white: when there is not the convenience of a burrow for nidification, she resorts to the fissures and cavities of rocks. Sheldrakes are very abundant in Holland and on the coasts of France. They may be domesticated, and are handsome ornaments in poultry-yards; but their flesh is rank and bad.

SHELL. The hard calcareous substance which either protects the testaceous Mollusca externally, or supports certain species of them internally. It has been truly said, that he who would know the nature of Shells, must know first the nature of the animals of which Shells form a part; and to this end we at once refer the reader to the article MOLLUSCA. Although Shells, properly so called, which form the habitation of testaceous animals, are sometimes confounded with the shelly coverings which protect the Crustacea (Crabs and their numerous allies), a very obvious and striking difference exists between them, as well as between the kinds of animals which respectively inhabit them. The Shells of Testacea are composed of carbonate of lime, combined with a small portion of gelatinous matter; they are, in general, permanent coverings for their inhabitants; and the animal is of a soft substance, without bones of any kind, and attached to its domicile by a certain adhesive property. On the other hand, those animals which are defended by a crustaceous covering cast their shells, and renew them annually; while the animals themselves are of a fibrous texture, with articulated limbs, and protected, as it were, by a coat of mail.

Shells are divided into *Multivalves*, *Bivalves*, and *Univalves*. The first order, MULTIVALVE, is made up of Shells consisting of more shelly parts or pieces than two. Every part of a Shell which is connected with a corresponding part by a cartilage, ligament, hinge, or tooth, is called a *valve* of such shell. — The second order, BIVALVE, is made up of Shells having two parts or valves, generally connected by a cartilage or hinge; as in the Cockle and Mussel. The hinge is entirely formed by the inner layer of shell, and consists of either a simple cardinal process, or a serrated edge, or of projections called teeth, and corresponding cavities into which they are inserted. To this hinge is superadded a highly elastic ligament, composed of a number of fibres parallel to each other and perpendicular to the valves which they connect; which is a beautiful contrivance for the necessities of the animal; for by means of it, while undisturbed, the valves are kept open, and the animal functions are carried on without effort; whereas, if danger be apprehended, or any circumstances require it, the adductor muscles contract, overcome the resistance of the hinge, and shut the valves close until they may be opened in safety. The valves of some bivalve Shells are formed exactly alike; others are very different; one may be smooth, and the other rough; one flat, the other convex; and one is sometimes shorter than the other. —The Third Order, UNIVALVE, is made up of Shells complete in one piece, as in the Periwinkle and the Whelk: and they are subdivided into Shells with a regular spire, and those without a spire. The shells composing this order are far more numerous than those of the two preceding, both in genera and species. The spire is a prominent feature of the Univalve; and upon its being lengthened or elevated, shortened or depressed, &c., depends much of the generic and specific definition. Shells increase in size by the deposition of new layers internally upon those already formed. Each new layer extends more or less beyond the margins of the layer to which it is applied, so that as the animal becomes older, its Shell becomes larger and thicker. The outer surface is generally covered by a thin layer of membranous or horny matter named the *epidermis*, and the inner surface is often covered with a layer of a pearly nature.— It is universally found that the Marine Shells of warm climates exceed all others in beauty of colouring. and in taking a fine polish. Several of the Land Shells also that are met with in tropical countries are remarkable for their bright colours and elegant forms. River and Land Shells, with very few exceptions, are thinner than those of the sea.

The following observations, which we abridge from an article in 'Brande's Dictionary of Science,' &c.. are intended to supply the best additional information we can give on the subject, consistent with our confined limits. — Shells, instead of consisting, like bones, of living organized substance permeated by blood-vessels, absorbents, and nerves, are mere inorganic laminated, concretionary, or crystalline deposits of calcareous earth, more or less combined with albuminous matter: they are also formed in the skin, and are appendages to the dermal system, which in all cases of animals is the principal seat of variety. In many cases, therefore, there exists very little correspondence between the structure or even the presence of a shell and the general character of the organization of a mollusc; and the absence of uniformity between the condition of the shell in closely allied species is exemplified in the highest as well as the lowest classes of the molluscous sub-kingdom.—The formation of a shell commences with the exudation of layers of albumen from the outer surface of the mantle or skin of the embryo mollusc, which is generally followed by the admixture of rhombic or prismatic crystalline particles of the carbonate of lime: and this first-formed shell of the embryo constitutes the nucleus of the shell of the mature mollusc. The nucleus is developed in most cases before the embryo quits the egg-coverings; but it is never coequal with the first formation of the animal; it is preceded by several distinct stages in the development of the embryo. The subsequent growth of the shell depends upon the deposition of fresh layers to the inner surface of the circumference of those previously formed; beyond which the new-formed layers extend in proportions which determine the figure of the future shell.—In many Univalves, the aperture of the shell is entire; in others, it is broken by a notch, or perforated by one or more holes; or a portion of it is produced into a canal or siphon; or it may present a pallial notch opposite to the siphon. These modifications are important, on account of their relation to certain conditions of the respiratory organs: thus the conchologist, in grouping together all the spiral univalve shells of which a part of the margin was either notched or produced into a grooved siphon, would really indicate a very natural tribe of Mollusca, every species of which he might be assured was aquatic and marine, and breathed by means of two gills having a pectinated structure, to which the water is conducted by a fleshy tube. Were a like correlation between the shell and its inhabitant to hold good in other families of Mollusca, the classification of Shells would then be a subject of much importance, and worthy the attention of the scientific naturalist: unfortunately, the reverse of this is frequently the case.—True bivalve Shells are peculiar to the Acephalous Mollusca; and their presence is constant, although they are in a few instances too small to cover the whole of the body, and in the ship-borers (*Teredo*) exist only as small instruments, limited to the function of excavating the burrows inhabited by these Molluscs. But all the species in which the bivalve shell is inadequate to the protection of the whole of the body derive extrinsic defence by burrowing in sand, stone, or wood; and they also commonly line their burrows with a layer of smooth and compact calcareous matter, forming a tube.—In all the Lamellibranchiate Bivalves which are free, the two valves are symmetrical, and the shell is termed *equivalve;* in all those which adhere by one of their valves to foreign bodies, this valve is deeper and larger than the unattached valve: such shells are termed *inequivalve.* If the shell of the common Cockle (*Cardium edule*) be examined, each valve will be seen to be produced into a conical prominence, bent towards, and nearly meeting at, that part by which the valves are joined together. These prominences are termed the *umbones.* The apex, or beak of the umbo, corresponds to the apex of the univalve shell, and is the point at which the development of the bivalve commences. When the apex is directed in the transverse plane of the shell, and so placed that a bisection of the shell in that plane through the apices shall divide the valve into two equal parts, the shell is termed *equilateral:* of this form the common scallop (*Pecten*) is an example. When, upon a similar division, a slight difference is observed in the two valves, the shell is termed *sub-equilateral;* but where the difference is well marked, it is an *inequilateral* bivalve. When the circumference or margin of one valve fits exactly at every part to that of its fellow, it is said to be "regular," or entire; but if it be notched at any part, so as not to come into contact with the corresponding part of the opposite valve, it is "irregular," or emarginate. The most important part of the margin is that which is modified to form the joint or hinge upon which the two valves open and shut. This part is called the "cardinal edge," and generally presents certain prominences and depressions, the projections of one valve interlocking with the depressions of the other. The projections or "teeth," together with the cavities or "cardinal pits," are very regular in their formation in each genus and species of bivalve. What is of more importance is, that every modification in the structure of the hinge is generally found to coincide with some recognizable and more or less important difference in the organization of the soft parts; so that conchologists have justly attached great value to the characters derivable from the hinge, especially for the purpose of generic distinctions. When the teeth are situate beneath the apex or centre of the hinge, they are called *cardinal;* when they are removed from the centre of the hinge, they are named *lateral teeth;* when two only are present, one is called anterior, the other posterior; when there are three, they are distinguished respectively as the anterior, median, and posterior teeth; but when the hinge is composed of a great number of teeth, it is said to be "serial," as in *Arca.* The direct medium of union of the two valves is a dense fasciculus of elastic albuminous fibres, generally of a brown colour, called the "ligament," or "elastic ligament."

SHEPHERD'S DOG. (*Canis* [*familiaris*] *domesticus*.) This variety of the canine tribe stands at the head of the class of farm Dogs, and is said to be preserved in the greatest purity in the northern part of Scotland, where its aid is highly necessary in managing the numerous herds of Sheep bred in those extensive wilds. It is distinguished by its upright ears; the hair soft, long, shaggy, and somewhat waved; and its remarkably bushy tail slightly pendulous: the same variety is diffused over most parts of Europe. The Dog prevents the Sheep from straggling; conducts them from one part of the pasture to another; and will not suffer any strange Sheep to mix with them. In driving a number of well-trained Sheep to a distance, a well-trained Dog always confines them to the road, watching every avenue that leads from it, and pursuing every straggler; and at the Shepherd's signal, this faithful assistant will conduct the Sheep to him from a considerable distance. "In temper and disposition," Mr. Bell observes, "the Sheep Dog is calm, serene, and quiet; but under a thoughtful and almost heavy aspect, there sparkles an expression of readiness and inquiry in his eye, as it peers out from under his shaggy brow, which betokens a spirit always on the alert, and prepared for instant obedience to the commands or wishes of his master. He has not, it is true, the noble port of the Newfoundland Dog, the affectionate fondling of the Spaniel, nor the fierce attachment which renders the Mastiff so efficient a guard; but he exceeds them all in readiness and extent of intelligence, combined with a degree of docility unequalled, perhaps, by any other animal in existence."

Numerous well-authenticated instances of the watchful fidelity, patient care, and instinctive sagacity of the Shepherd's Dog might be adduced; but nothing, perhaps, more interesting than the account which Mr. Darwin gives of the Dogs which are trained to this employment in Banda Oriental, in South America. "While staying at this estancia," he observes, "I was amused with what I saw and heard of the Shepherd Dogs of the country. When riding, it is a common thing to meet a large flock of Sheep guarded by one or two dogs, at the distance of some miles from any house or man. I often wondered how so firm a friendship had been established. The method of education consists in separating the puppy, while very young, from the bitch, and in accustoming it to its future companion. An ewe is held three or four times a day for the little thing to suck; and a nest of wool is made for it in the sheep-pen; at no time is it allowed to associate with other dogs, or with the children of the family. The puppy is, moreover, generally castrated; so that, when grown up, it can scarcely have any feelings in common with the rest of its kind. From this education it has no wish to leave the flock, and just as another dog will defend its master, man, so will these, the sheep. It is amusing to observe, when approaching a flock, how the Dog immediately advances barking, and the Sheep all close in his rear, as if round the oldest ram. These Dogs are also easily taught to bring home the flock, at a certain hour in the evening. Their most troublesome fault, when young, is their desire of playing with the Sheep; for in their sport they sometimes gallop their poor subjects most unmercifully. The Shepherd Dog comes to the house every day for some meat, and immediately it is given him, he skulks away as if ashamed of himself. On these occasions the house-dogs are very tyrannical, and the least of them will attack and pursue the stranger. The minute, however, the latter has reached the flock, he turns round, and begins to bark, and then all the house-dogs take very quickly to their heels. In a similar manner a whole pack of the hungry wild dogs will scarcely ever (and I was told by some, never) venture to attack a flock guarded by one of these faithful shepherds. The whole account appears to me a curious instance of the pliability of the affections in the dog race; and yet, whether wild, or however educated, with a mutual feeling of respect or fear for those that are fulfilling their instinct of association. For we can understand on no principle the wild dogs being driven away by the single one with its flock, except that they consider, from some confused notion, that the one thus associated gains power, as if in company with its own kind. F. Cuvier has observed, that all animals that readily enter into domestication, consider man as a member of their society, and thus fulfil their instinct of association. In the above case the Shepherd Dogs rank the Sheep as their fellow-brethren; and the wild dogs, though knowing that the individual Sheep are not Dogs, but are good to eat, yet partly consent to this view, when seeing them in a flock with a Shepherd Dog at their head."

SHOVELLER. (*Rhynchaspis*.) A genus of aquatic birds, of which there are several species. They are distinguished from the rest of the group by the singular form of the beak, which is larger than the duck's: at its origin the upper mandible is semicylindric; it then becomes depressed, and at the tip is greatly expanded on the sides, the tip itself being furnished with a very small incurved nail. The lamellæ at the edges of the mandibles are very long and fine; and those of the opposite mandibles fit into each other in such a manner that very little food can escape the bird while in search of it.

THE COMMON SHOVELLER. (*Rhynchaspis clypeata*.) This is a beautiful species: length upwards of eighteen inches: beak broad and black, but yellowish beneath: the head and neck deep glossy green; breast pure white; belly and sides of a chestnut red; the back a blackish brown; the wing-coverts clear blue; scapulars white, and dotted; the spot or speculum on the wing deep green; the legs reddish-orange. The female has a head of a clear red, marked with small streaks. This species inhabits various countries of the north of Europe and of America, frequenting the marshes, lakes, and rivers, and occasionally visiting the sea-coasts. They are not unfrequent in France, and are sometimes also met with in England, but they are by no means common. They are of a wild, shy, and solitary disposition. The female makes her nest on the ground, with withered grass, and lays ten or twelve rust-coloured eggs. Their food consists of worms and the larvæ of insects.

SHREW. (*Sorex araneus.*) The Shrew is a small insectivorous animal, covered with short velvety fur, and having much of the general form and aspect of the Mouse. It may, however, be easily distinguished from the mouse by its long, taper, cartilaginous snout; the eyes, too, are very minute, and almost hidden in the surrounding hairs; and the ears are round and close. The Shrew is usually of a reddish mouse colour above, grayish beneath, and sometimes tinged with yellow. The whole structure of this animal seems especially adapted to facilitate his progress under the earth; though it is to be observed that he is not only able to make his way rapidly under ground, but can run quite fast when on the surface. The total length, from the point of the snout to the beginning of the tail is under five inches, and the tail is one inch long. The Shrew frequents dry situations, feeding upon insects, worms, and grubs, for the pursuit of which its thin pointed snout is admirably fitted, either among the closest herbage, or under the surface of the soil. The body

SHREW (SOREX ARANEUS.)

exhales a rank musky odour, which renders them distasteful to cats, though they will readily kill them; but its flesh does not seem to be disliked by weasels, hawks, and owls, which destroy these little nocturnal insectivora in great numbers. They are common in hedge-rows, thickets, gardens, &c.; and make long superficial burrows in banks, among the roots of trees and brushwood. These animals show much of the pugnacity and voracious- ness of the Mole. The female makes a of soft herbage, in any hole of a bank, covered over at the top, and entered a side; and she brings forth in the spring five to seven young ones.

Among the superstitions of olden was one, that the Shrew Mouse had of inflicting serious injury upon cattle mere contact of its body. That enterta... naturalist Gilbert White, in his History of Selborne, thus alludes to it and its supposed remedy. "At the south corner of the plestor, or area, near the church, there stood, about twenty years ago, a very old, grotesque, pollard-ash, which for ages had been looked upon with no small veneration as a Shrew-ash. Now, a Shrew-ash is an ash whose twigs or branches, when applied to the limbs of cattle, will immediately relieve the pains which a beast suffers from the running of a Shrew Mouse over the part affected; for it is supposed that a Shrew Mouse is of so baneful and deleterious a nature, that wherever it creeps over a beast, be it horse, cow, or sheep, the suffering animal is afflicted with cruel anguish, and threatened with the loss of the use of the limb. Against this accident, to which they were continually liable, our provident forefathers always kept a Shrew-ash at hand, which, when once medicated, would maintain its virtue for ever. A Shrew-ash was made thus: into the body of the tree a hole was bored with an auger, and a poor devoted Shrew Mouse was thrust in alive, and plugged in, no doubt with several quaint incantations long since forgotten!"

There are two other British species, the Water Shrew and the Oared Shrew, the habits of both of which are aquatic, as their names import. Their burrows are formed in the banks of rivers, and their food consists of aquatic insects and larvæ, in pursuit of which they dive with great facility. The WATER-SHREW (*Sorex fodiens*) possesses the same general conformation as the Common Shrew—a body equally slender; a snout nearly as thin and pointed; and its fur has the same soft and silky kind of texture. Its feet are rather broad and formed for swimming, having a lash of stiff white hairs on the edge of the toes; the tail rather slender, compressed at the tip, and fringed with stiff hairs beneath. The head, back, and flanks, a rich brownish black; the under parts nearly pure white. The author of the "Journal of a Naturalist" thus speaks of these pretty little animals: "It is very amusing to observe the actions of these creatures, all life and animation in an element they could not be thought any way calculated for enjoying; but they swim admirably, frolicking over the floating leaves of the pondweed, and up the foliage of the flags, which, bending with their weight, will at times souse them in the pool, and away they scramble to another, searching apparently for the insects that frequent such places, and feeding on drowned moths and similar insects. They run along the margin of the water, rooting amid the leaves and mud with their long noses for food, like little ducks, with great earnestness perseverance. Their power of vision is limited to a confined circumference. smallness of their eyes, and the growth ur about them, are convenient for the its of the animal, but impediments to ex- led vision; so that, with caution, we can roach them in their gambols, and observe their actions. The general blackness of body, and the triangular spot beneath tail, as mentioned by Pennant, afford best ready distinction of this mouse from the Common Shrew." "Its swimming," says Mr. Bell, "is principally effected by the alternate action of its hinder feet, which produces an unequal or wriggling motion: it makes its way, however, with great velocity; and as it swims rather superficially, with the belly flattened, the sides as it were spread out, and the tail extended backwards as a rudder, it forms a very beautiful and pleasing object, moving on the calm surface of a quiet brook, or diving, in an instant, after its food, its black velvety coat becoming beautifully silvered, with the innumerable bubbles of air that cover it when submerged; and on rising again, the fur is observed to be perfectly dry, repelling the water as completely as the feathers of a Water-fowl."

SHREW MOLE. (*Scalops aquaticus.*) This little insectivorous quadruped inhabits a great part of North America, along the rivers; and so nearly resembles the European Mole externally, as to be readily mistaken for it; by Pennant it is described as the "Brown Mole." The muzzle is elongated as in the Shrews, and their limbs are adapted for digging into the ground precisely as in the Moles, which they entirely resemble in their mode of life. Their eyes are exceedingly small, and so completely concealed by the hair, as to require the closest attention for their detection. The auricle is entirely wanting, and the integument of the head nearly covers the tube leading to the internal ear. The feet are very short, and five-toed; the fore-feet terminate in a remarkably large hand, of which the fingers are armed with long, flat, and linear nails. The hind feet are very delicate, and the toes are provided with small hooked nails. This animal burrows with great quickness: his soft and polished fur, preventing friction, tends to facilitate his subterranean march; which is generally straight forward, or in gentle curvatures, at a very little distance from the surface; though sometimes numerous galleries are formed, communicating with each other, through which he is enabled to travel in various directions. Shrew Moles are most active in the morning and at mid-day; and it is observed that their daily appearance above ground at twelve o'clock is truly remarkable. The Shrew Mole is covered with a bright glossy fur, about half an inch in length, and of a very bright lead colour, very closely set, and in all parts directed backwards.

SHRIKE. (*Lanius.*) There are many species of these birds; and it is in this family (the *Laniadæ*, or Shrikes) that we find the largest and most rapacious of the Dentirostral tribe. In their general habits many of the Laniadæ resemble the Raptorial birds; for they sit motionless upon their perch, watching for their prey, and then suddenly dart upon it. They live in families for a few weeks after the breeding season; fly irregularly and precipitately, uttering shrill cries; nestle on trees or in bushes; lay five or six eggs, and take great care of their young. Some have the upper mandible arched: those in which its point is strong and much hooked, and in which the notch forms a small tooth on each side, manifest a degree of courage and cruelty which has led to their association with the Birds of Prey by many naturalists. Many of them have the curious habit of impaling the animals they have caught upon a large thorn; and then pulling them to pieces, and devouring them at their leisure. Hence they have derived the name of *Butcher-birds*. The Shrikes have great power of clutching with their toes, and always hold their prey in one foot, resting on the tarsal joint of that foot, unless when they have fastened it upon a thorn, when they pull it to pieces in a contrary direction. They exhibit great courage in defending themselves and their nests from more powerful enemies; and the parents show great attachment to each other and to their young.

Of this genus there are three British species, two only of which are commonly met with; these are

The RED-BACKED SHRIKE (*Lanius collurio*), which has derived its English name from the back, scapulars, and wing-coverts being of a rusty red colour. This species arrives in England in May, breeds in the southern counties, and departs in September.

RED-BACKED SHRIKE.—(LANIUS COLLURIO.)

Its nest, which is formed of moss and lined with hair, is placed in hedges. It is considerably smaller and sca[illegible] species.

The SENTINEL SHRIKE, or GREAT GRAY SHRIKE. (*Lanius excubitor.*) This species is as large as a Thrush. Its bill is black, and furnished with bristles at the base; the upper parts of its plumage pale blue ash; white underneath: the wings, tail, and a band crossing the eyes, black; some white on the scapulars and tail. It is common all the year in France, and is known in this country chiefly as a somewhat rare winter visitant. "It is one of our late birds of passage, but its arrival is soon made known to us by its croaking, unmusical voice from the summit of some tree. Its nest is large and ill-concealed; and during the season of incubation the male bird is particularly vigilant and uneasy at any approach towards his sitting mate, though often by his clamorous anxiety he betrays it and her to every bird-nesting boy. The female, when the eggs are hatched, unites her vociferations with those of the male, and facilitates the detection of the brood. Both parents are very assiduous in their attentions to their offspring, feeding them long after they have left the nest, for the young appear to be heavy, inactive birds, and little able to capture the winged insects that constitute their principal food. I could never observe that this bird destroyed others smaller than itself, or even fed upon flesh. I have hung up dead young birds, and even parts of them, near their nests, but never found that they were touched by the Shrike. Yet it appears that it must be a butcher too, and that the name "*lanius*," bestowed on it by Gesner two hundred and fifty years ago, was not lightly given. My neighbour's gamekeeper kills it as a bird of prey, and tells me he has known it draw the weak young pheasants through the bars of the breeding-coops; and others have assured me that they have killed them when banqueting on the carcase of some little bird they had captured. All small birds have an antipathy to the Shrike, betray anger, and utter the moan of danger, when it approaches their nests. I have often heard this signal of distress, and, cautiously approaching to learn the cause, have frequently found that this Butcher-bird occasioned it. They will mob, attack, and drive it away, as they do the Owl, as if fully acquainted with its plundering propensities. Linnæus attached to it the trivial name "*excubitor*," a sentinel; a very apposite appellation, as this bird seldom conceals itself in a bush, but sits perched upon some upper spray, or in an open situation, heedful of danger, or watching its prey." This species of the Shrike tribe feeds upon mice, shrews, small birds, frogs, lizards, and large insects. The nest is generally built on trees, and is framed of grass-stalks, roots, and moss, with a lining of down or wool. The eggs, from five to seven, are grayish white, spotted on the larger end with light brown and ash. Wilson, speaking of the American Shrike (*Lanius septentrionalis*, a species closely allied to the *L. excubitor*, says, "The character of the Butcher-bird is entitled to no common degree of respect. His activity is visible in all his motions; his courage and intrepidity beyond every other bird of his size (one of his own tribe only excepted, *L. tyrannus*, or King-bird); and in affection for his young, he is surpassed by no other. He associates with them in the latter part of summer, the whole family hunting in company. He attacks the largest hawk or eagle in their defence, with a resolution truly astonishing; so that all of them respect him, and, on every occasion, decline the contest. As the snows of winter approach, he descends from the mountainous forests, and from the regions of the north, to the more cultivated parts of the country, hovering about our hedgerows, orchards, and meadows, and disappears again [illegible] April."

There are numerous ex[illegible] with arcuated beaks, the points of which diminish by degrees. Other Shrikes have the superior mandible straight, and abruptly hooked at the tip. Others again, with a straight and slender bill, are remarkable for their crests of vertical feathers. Some species have the beak conical and rounded, without any ridge, somewhat arched towards the tip, with a very fine point, slightly emarginated on each side. Their feet are very short, and the wings in particular reach beyond the tail, which renders their flight similar to that of a Swallow; but they have the courage of the Shrike family, and do not fear to attack even the Crow. Numerous species inhabit the coasts and islands of the Indian Ocean, where they are continually seen on the wing, flying swiftly in pursuit of insects.

SHRIMP. (*Crangon vulgaris.*) A small crustaceous Decapod, allied to the Lobster and Crawfish, which frequents shallow waters along the sea-coast. It does not exceed two inches in length, and is of a pale glaucous green colour, dotted with grey. In shape it resembles the larger crustacea just mentioned, but it is more elongated in proportion, and is destitute of the large anterior claws; and it is distinguished from the Prawn by the absence of the long, anterior, serrated spine. The Shrimp has ten feet; the tail is as long as the body, and terminated at the extremity with scale-like appendages, which unfold somewhat in the manner of a fan. During life the body is semi-transparent, and so much resembles sea-water that the animal is distinguished with difficulty. Its ordinary motion consists of leaps. It is abundant in sandy places on the coast; and besides furnishing nutriment to great numbers of fishes, aquatic birds, &c., it is in great request for the table.

Shrimp-catching, or *Shrimping*, as it is termed, affords abundant employment on the flat sandy parts of our coast to boys and women, who wade up to their knees, pushing a sort of dredge-net at the end of a long pole before them; but a more wholesale way of collecting them is by means of sweep-nets, dragged over the fishing ground by men in boats.

SIALIDÆ. The name given to a small group of Neuropterous insects, having very large anterior wings. They frequent the neighbourhood of water, and pass their larva state in that element. The ordinary species (*Sialis lutaria*) is of a dull brown colour, and is a well-known bait with the angler, being produced in the spring months in great quantities, and may be found upon walls or palings near the water. The female attaches her numerous eggs, with the greatest regularity, to rushes or other aquatic plants. The larva swims well by the assistance of several pairs of articulated setose filaments attached at the sides of the abdominal segments. When full grown, this larva quits the water, and burrows into the adjoining bank, in which it forms a cell, wherein it is transformed into an inactive pupa, with the limbs laid along the breast. The insect assumes its perfect form in its cell.

SIAMANG. (*Hylobates syndactylus.*) The Siamang is a quadrumanous animal, inferior to the Chimpanzee and Orang-Outang both in structure and intelligence; and belongs to that division of Apes called *Gibbons*. These animals have long, thick, glossy black hair over the whole body, but particularly on the shoulders, back, and limbs: they are distinguished by the possession of small rudimentary callosities; and they derive their specific appellation of *syndactylus* from having the second and third toes of the hind foot united by a narrow membrane the whole length of the first joint. They are slow and heavy in their gait, but so vigilant as not to be easily surprised; when it does happen, however, they are so conscious of their inability to make effectual resistance, that overwhelmed with fear, they quickly fall into the hands of their pursuers. They live in numerous troops, which, it is said, are conducted by vigilant and courageous chiefs, and at sunrise and sunset they make the forests resound with frightful cries, which may be heard at a prodigious distance. From the accounts given by M. Duvaucel, who had numerous opportunities of observing the Siamang, in Sumatra, both in his wild state and in bondage, we learn that while dwelling in his native woods he exhibits an absence of all intellectual faculty, hunger itself being insufficient to excite, or divest him of his natural apathy; and that confinement, however long, seems to have no effect in modifying his characteristic stupidity and sluggishness; in short, he never acquires the familiarity of other apes; and even his submission appears to be rather the result of extreme apathy, than of any degree of confidence or affection.

SIBERIAN DOG. This useful variety of the canine race is distinguished by having its ears erect, and the hair of its body and tail very long; it is also distinguished for its steadiness, docility, and endurance of fatigue when used for the purpose of draught. In many northern countries these dogs are employed in drawing sledges over the frozen snow, five of them being yoked to each sledge, two and two, with the fifth in front as a leader. In general only one person rides in a sledge, who sits sideways, and guides the animals by reins fastened to their collars Such is their fleetness, that they have been known to perform a journey of 270 miles in three days and a half, and such their strength that they will convey a sledge containing three persons and their luggage sixty miles in a day.

SILIQUARIA. A molluscous animal, very long and spiral, inhabiting an irregularly twisted tube, tapering towards one end; the other end open; and a longitudinal fissure throughout its whole length, corresponding with a similar cleft in that part of the mantle which covers the branchial cavity. Along the whole side of this cleft is a branchial comb, composed of numerous delicate and tubular-like leaflets. It has a distinct head, and two small tentacula, with eyes at the base. Found in the Mediterranean and the Indian Seas.

SILKWORM. The SILKWORM MO' (*Bombyx mori*) is a whitish moth, wit' broad pale brown bar across each of upper wings. The larva or caterpillar, - phatically styled the Silkworm, is of a yellowish gray colour, and, when full grown, nearly three inches long: on the upper part of the last joint of the body is a horn-like process, similar to that on several of the Sphinx Moths. It feeds, as every one knows, on the leaves of the white mulberry, or, when they cannot be obtained, on those of the black mulberry or lettuce. The Silkworm remains in its larva state about six weeks, changing its skin four times during that period, and, like other caterpillars, abstaining from food for some time before each change. When full grown it entirely ceases to feed, and begins to form itself a loose envelopment of silken fibres in some convenient spot which it has chosen for that purpose, and afterwards proceeds to enwrap itself in a much closer covering, forming an oval yellow silken case or ball about the size of a pigeon's egg, in which it changes to a chrysalis, and after lying thus enclosed about fifteen days, gives birth to the Moth. This, however, is always carefully prevented when these insects are reared for the purpose of commerce, the Moth greatly injuring the silk of the ball by discharging a quantity of coloured fluid before it leaves the cell.

The Silkworm, when first hatched, is black, and does not exceed in length one fourth of an inch. The desire for food is the first symptom it exhibits of life, and at this period it is more active than at any other. When about eight days have elapsed after its hatching, its head becomes considerably enlarged, and it turns sick, refuses food, and remains in a state of lethargy for about three days. This sickness would appear to arise from the pressure of the animal's skin, which has become too tight for the increased bulk of its body. Indeed, the very great difference in the size of the worm, from the beginning to the end of its caterpillar state, is so great, that nature has furnished it with several skins, each of which it casts in succession. The body is begirt by twelve rings, which approach to or recede from each other, according to its motions: there are nine breathing holes on each side of the body; seven eyes on each side of the head; and two small orifices below the jaw, through which the worm ejects its silken filament.

The art of making the filamentous substance available for the use of man, seems to have originated with the Chinese, and to have been discovered at a very early period; but although the propagation of the Silkworm was confined to that country, the raw material was purchased and manufactured by the Persians, Tyrians, Indians, &c. for a long time before any attempt was made to establish it in Europe. For many ages silk bore an enormous price at Rome; but about the middle of the sixth century, during the reign of Justinian, two monks arrived at Constantinople from India, bringing with em .he white mulberry, and the eggs of Silkworm. This, however, is not the e for pursuing the history of the silk acture, or we might trace its progress .he East to Greece, and thence through , Spain, and France, where the culture e mulberry-tree, and the attention paid to the rearing of Silkworms, still form a most important feature in the industrial resources of the country.

"I was occupied the other day," says Mr. Jesse, in his 'Gleanings,' "in reflecting on the benefits accruing to mankind from a remarkable instinct impressed by the great Creator on that insignificant insect the Silkworm. What warmth and comfort does it afford to us! How useful, convenient, and elegant, is the clothing we derive from it! But this is not all. Let us, for one moment, consider how many thousands of persons are indebted to it for almost their very existence, in consequence of the employment it affords to man in nearly every country of the known world. There is, however, another striking and interesting peculiarity attending the Silkworm. It is this; that while the caterpillars of all the other tribes of moths and butterflies, when they have arrived at a certain state of maturity, show a restless disposition, and wander about and hide themselves in a variety of places in order to spin their cocoons, preparatory to their making their escape as Moths, &c., the Caterpillar of the Silkworm, on the contrary, is content to remain stationary in the open tray, or box, in which it may be placed. After consuming its immediate supply of mulberry leaves, it waits for a further quantity; and when the period arrives for spinning its cocoon, instead of showing any migratory disposition, it seems to place itself with confidence under the care of man to provide it with a suitable place for its convenience and protection. In the fly or moth state, the female is quite incapable of flight; and the male, although of a much lighter make, and more active, can fly but very imperfectly. This latter circumstance insures to us the eggs for the following season, thus completing the adaptation of the insect in its different stages to the purposes it is destined to fulfil for our advantage. To my mind this striking peculiarity in the habits of the Silkworm, illustrates the care and kindness of the Almighty, in thus making an apparently insignificant insect the means of so many important benefits to man.

SILURIDÆ. A family of Malacopterygious fishes, of which the genus *Silurus* is the type. They are chiefly distinguished by the want of true scales, having merely a naked skin, or large osseous plates. The species included in this group are mostly

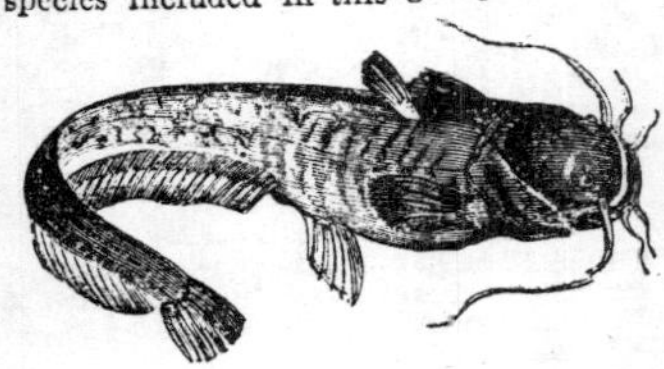

SILURUS GLANIS.

river-fish, of considerable size, inhabiting warm climates. Many of them have the first ray of the pectoral fin very strong and bony; and the fish can, at pleasure, lay it flat on the body, or keep it fixed in a perpendicular direction, in which case it becomes a formidable weapon, capable of inflicting very severe wounds.

The only known European species of the *Siluridæ* is the SILURUS GLANIS, a fish of very large size, found in the lakes of Switzerland, the Danube, the Wolga, the Elbe, and other large rivers in the north of Europe; as also in many of the fresh waters of Asia and Africa. It sometimes grows to the length of from six to eight feet, and to the weight of 300lbs. The head is broad and flat; the body thick and of a lengthened form, with the abdomen very thick and short; the mouth very large and wide, and on each side of the upper lip is a long barbule; the jaws are circular, the lower one the longest, and both furnished with numerous small incurved teeth. The back is round, of a dark green; paler below; and the whole body covered with dark irregulary-formed spots. Mr. Yarrell observes, that "the Silurus is represented as sluggish in its habits, and a slow swimmer, taking its prey by lying in wait for it, in a manner somewhat similar to the Angler (*Lophius*); hiding itself in holes or soft mud, and apparently depending upon the accidental approach of fishes or other animals, of which its long and numerous barbules may be at the same time the source of attraction to the victims, and the means of warning to the devourer. From its own formidable size, it can have but few enemies in the fresh water; and from them its dark colour, in addition to its habit of secreting itself either in holes or soft mud, would be a sufficient security. In spring the male and female may be seen together, about the middle of the day, near the banks or edges of the water, but soon return to their usual retreats. The ova when deposited are green; and the young are excluded between the sixteenth and nineteenth days. The flesh of the Silurus is white, fat, and agreeable to many persons as food, particularly the part of the fish near the tail; but on account of its being luscious, soft, and difficult to digest, it is not recommended to those who have weak stomachs. In the northern countries of Europe, the flesh is preserved by drying, and the fat is used as lard."

The ELECTRICAL SILURUS, or MALAPTERURUS, which inhabits the Nile, the Senegal, and other African rivers, is from ten to fifteen inches in length: the head very broad and depressed; on the upper lip two cirri, on the lower four; teeth small and numerous. It appears to derive the power of giving electrical shocks from a particular tissue situated between the skin of the sides and the muscles. It possesses this electric or galvanic power, however, in a much slighter degree than the Torpedo.

SILVER-FISH. A well-known small species of the Carp tribe. [See GOLD-FISH.]

SILVER-LINE [MOTHS]. A name applied by collectors to Moths of the genus [illegible]LOEOPHORA].

[illegible]GLET [BUTTERFLY]. A [illegible]llectors to Butterflies of [illegible] *hero*.

[illegible] name applied by [illegible]pecies of quad- [illegible]mals, except the Lemurs. They are divided into numerous sub-genera;

but the term *Simia* is no longer used, except by some modern naturalists to the Orang-outang. [See APE, MONKEY, &c.]

SIMULIUM, or SAND-FLY. An extremely troublesome Dipterous insect, respecting which, in its different stages, Mr. Newman furnishes the following information:—"The eggs of the *Simulia* or *Sand-fly* appear to be at present unknown; there is, however, little doubt that, like those of other gnats, they are deposited on the surface of the water, and in that situation are hatched by the warmth of the sun combined with the moisture of the water. The larva is found on the stems of water-plants (*Phellandrium*, &c.), on those portions which are always covered by the water. It is long, cylindrical, considerably thickened posteriorly, and nearly transparent; its head is distinctly separated from the body, and is of an oblong form; it has four jaws moving horizontally, each bifid at the tip, and two little horns in the usual place of antennæ, inserted in the front of the head, rather towards each side: each of these is composed of two joints, the first or basal joint stout, the second or apical one divided into many rays, which fold back on the first joint: there are two very small eyes on each side of the head. The body of the larva is divided into twelve segments, besides the head; of these, the second is incrassated, and furnished below with a retractile conical foot; the last segment is very minute, and furnished with two small prehensile feet: the air-tubes, so very plainly seen in other aquatic larvæ, are totally wanting; neither is there the least appearance of spiracles or breathing-holes in the sides.

"The motion of the larva in the water is tolerably brisk; but on any object coming in contact with it, it instantly becomes motionless, attaches itself by the anterior prehensile foot, and remains for a long time perfectly still and immovable. When it moves from one place to another, its progression is undulating, somewhat like that of a leech, being performed in this manner:—the anterior foot is firmly attached to some object, then the posterior pair of feet are brought up to it, the back arching up during the operation; the anterior foot then releases its hold; the body is again elongated, the foot attached further on, and the posterior feet again brought up to it. The food of the larva is unknown: when full grown, it spins a little silken sheath, in shape like a watch-pocket, which is attached to the plant frequented by the larva, and in this it shortly changes to a pupa in an upright position: the case being always open at top, the head and shoulders of the pupa are seen projecting above it. The pupa much resembles that of a moth: it is perfectly motionless, of a brown colour, and exhibits very distinctly the parts of the perfect insect through its skin: from the back of its head arise, on each side, four hair-like appendages; these are tubular, and appear to be designed for breathing. About the 6th of July the little creature bursts from its sheath; the case of the chrysalis opens in a right line down the back, and the perfect insect emerges through the opening, surrounded by a bubble of air, and slowly begins to unfold its wings under the water; finally, its skin being cast, and maturity attained, the imago disengages itself from its former habitation, and mounts within its bubble to the surface of the water, when the bubble bursts, and the creature, with its new organs, has acquired a new element. The imago is a small black fly, with two large transparent wings, which, when at rest, repose horizontally on its back; moderately long legs, and short stout antennæ: it flies with ease, and somewhat sportively, rising and falling. In this country it is found in the damp parts of woods, and other similar situations; but, happily, in very limited numbers."—"The *Simulium* seems to have adopted the world for its country: no known land appears to be without it; all temperatures suit it—the polar snows and the blaze of tropical sands. Yet *all* the flies of which travellers complain as so dreadfully annoying, are not *Simulia;*—many of our commonest gnats have a similar taste for blood. Although, from what is related, there can be no doubt that the blood of man is an acceptable food to the *Simulia*, yet it is remarkable that the greatest multitudes of these creatures inhabit those bleak, inhospitable, and almost inaccessible regions where the foot of man seldom treads, and where other warm-blooded animals are scarcely known to exist. It is clearly ascertained that the female *Simuliæ* alone suck the blood of man; the males spend their lives among the leaves of trees, or settle on flowers, from which they appear to derive nutriment; it is therefore far from impossible that, on the failure of animal, the females may also have recourse to vegetable food."—*Hist. of Insects.*

SIPHONAPTERA. A name given by Latreille to an order of insects, including those Apterous species which have a mouth in the form of a siphon.

SIPHONARIA. A genus of Mollusca, the shell of which greatly resembles the Patella in shape. The animal has no tentacula or visible eyes. They are found on the coasts of South America, Australia, and in the Mediterranean.

SIPHONOBRANCHIATA. The name of an order of Gasteropodous Mollusca, including those in which the branchial cavity terminates in a tube or siphon, by which the respiratory current of water is received and expelled.

SIPHONOSTOMA. An order of Crustacea, all of which are parasitic upon Fishes, aquatic Batrachia, &c., comprehending those which have a siphon-shaped mouth for suction.

SIPUNCULUS. The name of a genus of worms which conceal themselves in the sands of the sea-shore, and occasionally protrude their heads from the orifice. They are much sought after by the fishermen, who use them, like the Common Lob-worm (*Arenicola Piscatorum*), as baits for their hooks. Some of them attach stony particles to their skin, by a glutinous exudation, so as to cover it with a hard crust, resembling that formed by some Annelida.

SIREN. A genus of remarkable Batrachian reptiles, peculiar to the Southern provinces of the United States. They have an elongated form, nearly like that of eels, three branchial tufts on each side; only one pair of feet; a flattened head, and obtuse muzzle; eye very small; the ear concealed; lower jaw armed with a horny sheath and several rows of small teeth; the upper jaw toothless; but numerous small, retroverted teeth occur on the palatal region. The anomalous organization of this reptile, and its apparent relationship with different families, rendered it for a long time doubtful to which it belonged. At length Cuvier satisfactorily established, that the Siren was a reptile *sui generis*, which never could have hind feet, and whose bony framework differed especially from that of the Salamanders; that there was no probability that it ever changed its form or lost its branchiæ; and that the Siren is consequently a true amphibian, which respires at will throughout its life, either in the water by means of branchiæ, or in the air by means of lungs. The same naturalist adds, that it is to the Salamanders that the Sirens approach most nearly by the structure of the head, although neither the general form nor the proportions of the parts have so near a similarity. The Axolotl belongs to a closely allied genus. [See AXOLOTL.]

SIREX: SIRICIDÆ. A genus and family of Hymenopterous insects, of which the *Sirex gigas* may be taken as a type. They have the antennæ jointed, and inserted near the forehead; the mandibles toothed internally; the maxillary palpi very small, nearly conical, and two-jointed, with the extremity of the abdomen prolonged into a horn, and the ovipositor exserted and formed of three threads. These insects are of large size, and

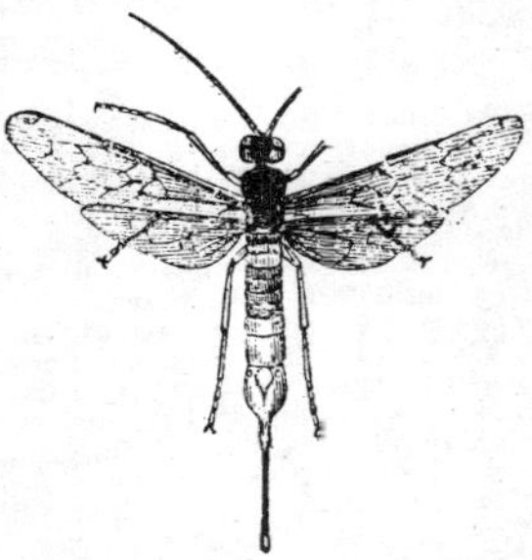

SIREX GIGAS.

generally inhabit pine forests in cold and mountainous countries, and produce during flight a buzzing noise like that of the Humble-bees. In those countries they appear, in certain seasons, in such abundance that they become objects of popular dread. The larvæ have six feet, with the posterior extremity of the body terminated in a point; they live in wood, where they spin a cocoon and undergo their transformations. The *Sirex gigas* has sometimes, though rarely, occurred in this country, and is as large as a Hornet.

SISKIN, or ABERDEVINE. (*Carduelis spinus.*) A song-bird, very similar in colour and general appearance to the green variety of the canary, though somewhat more dusky on the back and head. It is a lively and persevering songster. It breeds in Sweden,

SISKIN OR ABERDEVINE, MALE AND FEMALE. (CARDUELIS SPINUS.)

Norway, the north of Germany, and sometimes in the Highlands of Scotland, visiting England only in the autumn and winter. In most places they are migratory, but do not seem to observe regular periods, as they are sometimes seen in large, and at other times in very small numbers. Buffon observes that these immense flights happen only once in the course of three or four years. They conceal their nest with much art. In some parts of the south of England it is called the Barley-bird, being seen about that seed-time; and in the neighbourhood of London it is known by the name of the Aberdevine.

SITTA. The Linnæan name of a genus of birds, of which the Nuthatch is the type. [See NUTHATCH.]

IVATHERIUM. [See SUPPLEMENT.]

SKATE. (*Raia batis.*) This fish, the true Skate, in proportion to its bulk, is the thinnest of any of the *Raiadæ* as well as the largest, some being known to weigh near two hundred pounds. The breadth of the body is to its length nearly as four to three. The nose is conical; and above the eyes there is a set of sharp spines. The whole upper part is of a dull brown colour, sometimes streaked with black; the lower part is of a dusky white, marked with many small black spots; and the jaws are covered with small granulated but sharp-pointed teeth. The tail is of a moderate length, and two fins near its extremity; along the top of it there is one row of spines, and on the edges a few more are irregulatly dispersed. In the males of this species the fins are full of spines. The females are generally called Maids; and fishermen distinguish the females of the three species of most frequent occurrence by the names of Skate Maid, Thornback Maid, and Homelyn Maid,—frequently calling the old male of the Skate with his two long appendages the Three-tailed Skate. It is a very voracious fish, and commits great havoc among numbers of the finny tribe and crustacea. It is found on the coast of Scotland, among the Orkneys, in many parts of the Irish coast, and on the British coasts from Cornwall to Kent.

The FLAPPER SKATE. (*Raia intermedia.*) This species is distinguished from *Raia batis*, in the upper surface of the body being perfectly smooth, without granulations, and of a dark olive colour spotted with white; in the dorsal fins being more remote from each other, and in the anterior margins of the pectorals being rather more concave, giving the snout a sharper appearance.

SKIPPER. A name commonly applied to the Mackerel Pike, or Saury Pike (*Scomberesox saurus*). They are gregarious fishes; and are followed and preyed upon by Porpoises, and also by the Tunny, and other large members of the *Esocidæ* or Mackerel family.

SKIPPER [BUTTERFLIES]. A name applied to several species of Butterflies, of the genera *Thymele* and *Pamphila*.

SKUA GULL. [See GULL.]

SKUNK. (*Mephitis Americana.*) A carnivorous animal of the genus Mephitis inhabiting both North and South America. It has short round ears, black cheeks, and a white stripe extending from the nose to the back. The upper part of the neck and the whole back are white, divided at the bottom by a black line, commencing at the tail, and passing a little way up the back. The belly

SKUNK.—(MEPHITIS AMERICAN

and legs are black; the tail is very full of long coarse hair, generally black, sometimes tipt with white; and the claws are long, like those on the fore feet of the badger. This animal is remarkable for the intolerable odour of the secretion from its glandular pouches, which it has the power of ejecting on its pursuers, and serves as a most complete means of defence; the least quantity of it being enough to produce nausea and a sense of suffocation. Clothes that are infected with this smell retain it for many weeks; and nothing, it is said, will render them sweet, but burying them for a time in the fresh earth. As soon as the animals are dead, the glands from which this vapour issues are cut away, and the flesh, then untainted, is eaten by the American Indians, who say the flavour much resembles that of a young pig. There are several species of this genus, all of them American.

SKY-LARK. [See LARK.]

SLOTH, or AI. (*Bradypus torquatus.*) An herbivorous Edentate quadruped, of most uncouth appearance, treated by Buffon as one whose existence must be a burthen to it, from its imperfect formation; but though uncouth and apparently disproportioned, it is found on examination that the organization and habits of the Sloth are as completely adapted to each other, as are those of any other animal. It is true that the arms or fore legs are nearly twice as long as the hinder pair; and that when it attempts to walk on the ground, the action is most awkward and laborious: but when we consider that the Sloth is formed to live not on the ground but in trees, and not *on* the branches of trees, like the squirrel, but *under* them, the complete adaptation of its whole structure to its mode of life becomes apparent. No man had a better opportunity of observing this animal than Mr. Waterton, during his long residence in the wilds of South America; and he, a close observer and just reasoner, thus writes: "He moves suspended from the branch, he rests suspended from the branch, and he sleeps suspended from the branch. Hence his seemingly bungled composition is at once accounted for; and in lieu of the Sloth leading a painful life, and entailing a miserable existence upon its progeny, it is but fair to conclude that it just enjoys life as much as any other animal, and that its extraordinary formation and singular habits are but further proofs to engage us to admire the wonderful works of Omnipotence." They bring forth and suckle their young like ordinary quadrupeds; and the young Sloth, from the moment of its birth, adheres to the body of its parent till it acquires sufficient size and strength to shift for itself. The head of the Sloth is short, the face small and round, the hair coarse and shaggy, differing considerably in colour in different individuals, but resembling, in general, dry withered grass or moss. Its powerful claws, and the peculiarly enduring strength of its long arms, make very efficient weapons of defence against the large snakes by whom it is often attacked. It has sometimes been brought to this country; a specimen was in the Zoological Gardens, Regent's Park, in 1846.

The following is Dr. Lund's account of the THREE-TOED SLOTH (*Bradypus torquatus*), which he kept in his house for a considerable time. "This animal climbs with remarkable sureness and aptitude, although, as is well known, with a degree of slowness which, however, may be called rapidity in comparison with its terrestrial movements. The manner in which it moves is this:—Lying on its belly with all its four extremities stretched out from its body, it first presses one of its hind feet with all its might against the ground, whereby the corresponding side of the body is a little raised. The fore leg

THREE-TOED SLOTH.
(BRADYPUS TORQUATUS.)

on the same side thus becomes sufficiently free for the animal to advance it a trifle forward. It then hooks its powerful claws fast in the earth, and so drags its body a little onwards. The same manœuvre is next repeated on the opposite side; and thus the poor creature progresses in the slowest and most laborious manner possible. But in proportion as the Sloth's organization unfits it for terrestrial progression, is it wonderfully adapted to climbing trees. With its long arms it reaches high up, aud clings fast to the branches with its strong crooked claws. The *inverted* position of the soles of its hind feet gives it a power of *grasping* the trunk of the tree which no other mammal possesses. So that truly when we see it climbing a tree, we can scarcely believe it to be the same animal that lies so helpless on the ground. Hence we see that the Sloth's organization is entirely adapted for living in trees. Compared with the slowness of its motions, it is the best climber among mammals, while it is the worst walker; or rather, it is the only mammal that can neither walk nor stand

SLOW-WORM. [See BLIND-WORM.]

SLUG. (*Limax.*) A naked mollusc, the order *Pulmonea*, family *Limacinæ*. The common small gray Slug (*Limax cinereus*) is too well known as a destructive pest in our gardens to need much describing. It has a prominent head, with four tentacula; and at the end of the longer pair the eyes are situated. These tentacula can be drawn inwards, by a process resembling the inversion of the finger of a glove. On the back there is a kind of shield or disc, formed by the mantle, and which covers the pulmonary sac; and the head can be partially con-

VARIEGATED SLUG.—(LIMAX VARIEGATUS.)

tracted or withdrawn beneath it. In the mouth is an upper jaw only, of a crescent form, and toothed, which enables it to devour with voracity herbs and fruits. The stomach is elongated, simple, and membranous. Its progress on the ground may easily be traced by the slime which it leaves in its track.

The BLACK SLUG, whose appearance in our fields and meadows in the summer season is considered as an indication of approaching rain, feeds on the leaves of different kinds of plants, and is in all respects except its size and colour, similar to the preceding.

Another species, called the TESTACELLA, (*T. haliotidea*), which feeds largely on earthworms, has the respiratory aperture, and the anus, near the posterior extremity; where their mantle, which is very small, is also placed, and contains a small ear-shaped shell which does not equal one tenth the length of the body. This animal is abundant in the south of France, and has been lately introduced into the gardens of this country, where it is said to be rapidly multiplying.

SMELT. (*Osmerus eperlanus.*) A small but delicious Malacopterygious fish, inhabiting the salt water about the mouths of rivers, and in its habits resembling the salmon. All parts of the mouth are armed with long and pointed teeth, and there are four or five upon the tongue. The body is long and somewhat compressed; the eyes large and round; and the under jaw longest. The European Smelt is from four to eight inches long; the head and body are semi-transparent, with the most brilliant tints of green, and silvery: all the fins pale yellowish white; the ends of the caudal rays tipped with black. The

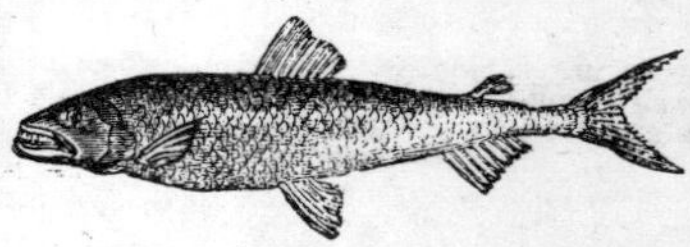

SMELT.—(OSMERUS EPERLANUS.)

Smelt inhabits fresh water from August to May. After spawning in the beginning of April, they return to the sea. In August the fry are found about three inches long, swimming near the surface in shoals in the rivers, ascending and descending with the tide, when the adult fish are again visiting the fresh water. The Smelt is generally in great request from its delicate and peculiar flavour. Its well-known cucumber-like smell is very powerful when they are first taken out of the water. They are taken both on our eastern and western coasts, and are abundant in the Thames and Medway.

The AMERICAN SMELT (*Osmerus viridescens*) is considered a different species. The body is long, green on the back, and silvery-white on the sides. It inhabits the coasts of New England, and as far as the Hudson, but is unknown farther south.

SMEW. (*Mergus albellus.*) This is a web-footed bird, about the size of a Wigeon, which seldom visits this country except in very severe winters. It has a bill nearly two inches long, of a dusky blue, thickest at the base, and tapering into a more slender and narrow shape towards the point. On each side of the head is an oval-shaped black patch, glossed with green; under side of the crest black; the other parts of the head and neck white: the breast, belly, and vent are also white, excepting a curved black line on each side of the upper part of the breast, and similar marks on the lower part: the back, the coverts on the ridge of the wings, and the primary quills are black; the secondaries and greater coverts tipped with white; the middle coverts and scapulars white; and the sides, under the wings to the tail, are variegated and crossed with dark wa ed lines. The legs and feet are of a bluish lead colour. This species is easily distinguished from its congeners by its black and white piebald

HOODED SMEW.—(MERGUS CUCULLATUS.)

appearance. Our figure represents a most beautiful species, the HOODED SMEW (*Mergus cucullatus*), which is common in North America, but only accidentally found in Europe. [See MERGANSER.]

SNAIL. The Garden Snail (*Helix a... sa*), and its allies, constituting the ... *Helicidæ*, are closely allied to the Slugs ... organization, and differ from them in little else than in their being inclosed in a shell, which is univalve, spiral, sub-pellucid, and brittle, and has a semilunar aperture. Its head is furnished with four tentacula; on the superior pair the eyes are placed; while the inferior pair have no visual organs, but seem more exclusively adapted to the perception of tactile impressions. Both the upper and lower tentacula are retractile, and can be completely inverted so as to be withdrawn into the interior of the body. Each tentacle is a hollow flexible cylinder. When partially retracted, the extremity of the organ is drawn inwards, and two cylinders are thus formed, one within the other: if the outer cylinder is elongated, as in protruding the tentacle, it is at the expense of the inner one; and, on the contrary, the inner cylinder, when the organ is retracted, is lengthened as the other becomes shorter. Snails lay eggs, and carefully bury them in the ground. These eggs are very numerous, round, semi-transparent, about the size of a small pea, and covered with soft shells: they are also united to each other by an imperceptible slime. When the Snail leaves the egg, it is observed with a very small shell on its back, having only one whorl; but, in proportion as it grows, the shell increases in the number of its spiral turns. The addition is always at the mouth, the first centre still remaining; the animal sending forth from its body that slime which hardens into a calcareous substance, and is still fashioned into similar convolutions. Thus fitted with its covering, which is light and firm, the Snail finds itself well defended from external injury; and it has only to retire into its fortress to escape impending danger. It derives its chief subsistence from the leaves of plants and trees, and, although very voracious, is extremely delicate in its choice. When in quest of food, it moves forward by means of that broad muscular skin, which is sometimes seen projecting beyond the mouth of the shell: this is expanded before, and then contracted with a kind of undulating motion. It is also able to ascend in a perpendicular direction, and has its progress facilitated by means of that viscous excretion which it emits whenever it moves. On this glutinous matter it can proceed slowly and in safety along a rugged path, or ascend trees and fences for the purpose of feeding; and it also descends by the same aid, without danger of falling and injuring its she...

At t... nter the Snail buries itself in the earth, or retires to some hole, where it continues in a torpid state during the severity of the season: thus it sometimes lies torpid for six or seven months, till the genial warmth of spring awakens it to a state of activity; when it quickly makes ...nds for its long abstinence by feasting ...ry vegetable substance that falls in its ... Before, however, they commence this ...tive state of existence, Snails close the ...uth of their shells with an epiphragma ...r covering, not attached to or forming a part of the animal), which, stopping it up entirely, protects it from every external injury: it is composed of a whitish substance somewhat resembling plaster. In the centre is an exceedingly minute orifice, communicating with the lungs; and this minute hole, though not large enough to admit a drop of water, is of sufficient capacity for the passage of air. The multiplication of Snails is at times prodigious; and it is uniformly observed that a rainy season contributes much to their increase. It has been asserted, and on apparently good authority, that Snails have been known to revive after remaining in torpidity a number of years; and they also possess extraordinary powers of reproduction, being able to renew almost any part of the body that has been amputated, or of the shell that has been broken. This species of Mollusca is universally diffused: throughout the continents of Europe, Asia, and Africa; in the hottest and coldest climates; in the most cultivated as well as in the most barren situations; in the forests of Guiana and Brazil, at the foot of Chimborazo, and even in the great desert of Zahara, the common Garden Snail will be found.

The GREAT VINE SNAIL, or EDIBLE SNAIL. (*Helix pomatia.*) This species was considered by the ancient Romans one of their table luxuries, and such great attention was paid to the mode of feeding them, that they frequently attained an immense size. On the shores of the Mediterranean they are still regarded as a valuable article of food, when boiled in the shell, and eaten with rice: and in some countries, as Switzerland and parts of France, they form a considerable article of commerce. They are fed by

EDIBLE SNAIL.—(HELIX POMATIA.)

thousands in places, called *escargatoires*, which are made on purpose for them. They are used, boiled in milk, for diseases of the lungs; and are also sent to America from this country as a delicacy. Some authors tell us that this species has been introduced into this country from abroad; while others suppose it to be indigenous. It is almost peculiar to chalky and gravelly soils.

Among the members of the family *Helicidæ* one genus deserves especial notice from its structure. There are only two species known, *Anastoma depressa* and *Anastoma globulosa*. "The peculiarity," says Mr. Sowerby, "which distinguishes this genus from all the other Heliciform Univalves is so extraordinary, that it appears to us to be deserving of particular notice, inasmuch as it evidences a considerable alteration in the habit and economy of the animal which produces it, at the time of its arrival at the last period of growth, when it forms the reflected outer lip, and the teeth in the aperture. Until then, the animal must crawl about like other Snails, with the spire of its shell uppermost; but as soon as it arrives at maturity, and is about to form its complete aperture, it takes a reverse position, and afterwards constantly carries its spire downwards." It is very rare, and is brought from the East Indies.

SNAKES. Under the words SERPENTS, RATTLESNAKE, BOA, HYDROPHIS, &c., will be found descriptions of many of the most formidable among the *venomous* species: we shall therefore in this article notice a few of the *Colubridæ*, all of which are perfectly innoxious. We commence, then, with the Common or RINGED SNAKE (*Coluber natrix.*) This species is very common in all parts of England; frequenting low moist woods, damp meadows, and hedgerows in the vicinity of water; feeding upon young birds, mice, and other small quadrupeds, and lizards; but, in preference to all these, upon frogs. The Ringed Snake grows to the length of more than three feet. The head is of an elegant ovate form, and considerably depressed, the back part broader than the neck. The teeth are small, curved backwards, as in all the other innocuous Snakes, arranged in two series on each side of the jaw both above and below. Tongue long and flexible, and bifid to about one third of its length. The back and sides are covered with small scales; and the belly with oblong, narrow, transverse plates. The colour of the back and sides is dusky or brown; the upper parts of the body and head being of a light brownish gray with a green tinge, sometimes approaching to a dull pale olive: the middle of the back is marked with two rows of small black spots, running from head to tail; and from them proceed numerous lines of spots crossing the sides. The plates on the abdomen are dusky; and the scales on the sides are a bluish white colour, sometimes marbled with black. On each side of the neck there is a pale yellow spot; and the base of each has a triangular black spot, one angle of which points downwards. It lays its eggs in dunghills and hotbeds, whose heat, aided by that of the sun, promotes the exclusion of its young. During the winter these reptiles resort to the banks of hedges, the hollow roots of old trees, or some sequestered and sheltered corner, where they remain, coiled together, sometimes in considerable numbers, till, like the other tribes wh h hybernate, a warmer season calls them forth to resume their natural functions.

Mr. Bell remarks, that "Snakes, like most other Reptilia, shed their cuticle or outer skin at greater or less intervals. It is a mistake to assign a particular period to this process; some have stated it to occur once, some twice in the summer; but I have found it to depend upon the temperature of the atmosphere, and on the state of health, and the more or less frequent feeding of the animal. I have known the skin shed four or five times during the year. It is always thrown off by reversing it; so that the transparent covering of the eyes, and that of the scales also, are always found concave in the exuviæ. Previously to this curious circumstance taking place, the whole cuticle becomes somewhat opaque, the eyes are dim, and the animal is evidently blind. It also becomes more or less inactive; until at length when the skin is ready to be removed, being every where detached, and the new skin perfectly hard underneath, the animal bursts it at the neck, and creeping through some dense herbage, or low brushwood, leaves it attached, and comes forth in far brighter and clearer colours than before." At times a strong fetor proceeds from it; but this appears to be sexual, or made use of as the means of annoying its enemies.

The JAVA SNAKE. (*Coluber Javanicus.*) This Snake grows to the length of nine feet, and is principally seen in the rice fields of Java. The head is large and flat, and covered with large scaly plates: the mouth is furnished with double rows of teeth; but not being of a poisonous nature, it is destitute of fangs. From behind the eyes pass two deep-blue stripes to the upper part of the neck, where they unite: a third stripe of the same colour proceeds from the snout to the occiput, where it divides into two, and surrounds a yellow spot, marked with a few blue specks. The upper part of the body is divided, as it were, into squares resembling a kind of lattice-work, formed by stripes of bright blue with gold-coloured edges; the middle parts of the squares exhibiting changeable hues of gray, yellow, blue, and green: each side of the body is also marked with a row of white spots situated at the crossings of the blue stripes. It is altogether a superb species. It devours rats and other small quadrupeds, birds, &c.

The ÆSCULAPIAN SNAKE. (*Coluber Æsculapii.*) This is common in most of the warm parts of Europe, and is nowhere more frequent than in the neighbourhood of Rome: it is therefore not improbable to be the species peculiarly consecrated by the ancient Romans to the benevolent deity whose name it bears. It is nearly four feet in length, of a rufous colour on the upper parts, and marked on each side by a blackish longitudinal band: the scales on the sides, nearest the scuta, are white bordered beneath with black, thus forming a range of small whitish triangles along each side of the body. In its general habits it much resembles the *Coluber natrix* or Ringed Snake. — The following species are all natives of North America.

The BLACK SNAKE (*Coluber constrictor*) is found throughout the United States. The colour is black, inclining to slate colour beneath, with the throat and lips white. It grows to the length of six feet; the scales are smooth; and its motions are rapid. — The CHAIN SNAKE (*Coluber getulus*) is of a black and white colour, the black predominating. The white often forms transverse lines on the back, which unite on the sides, thus forming the semblance of a chain. The markings are, however, extremely variable; some being thickly sprinkled all over with irregularly oval white specks. — The WATER SNAKE (*Coluber sipedon*), which is found in all parts of the United States, is generally brown on the back, beneath pale, with indistinct dark spots; but the markings vary exceedingly, and it is often found transversely banded with white. It sometimes grows to the length of five feet. It frequents exclusively the borders of streams, and, when disturbed, often takes refuge in them, and conceals itself at the bottom. — The SCARLET SNAKE (*Coluber coccineus*), which is beautifully marked with scarlet, black, and yellow, inhabits the Southern States. — The PINE SNAKE. (*Coluber melanoleucus.*) This species sometimes attains the length of eight feet: the colour whitish, with large blackish spots. It is common in all the more southern and western parts of the United States: is of a gentle disposition, and is sometimes tamed and kept about houses. — The CHICKEN SNAKE, or HOUSE SNAKE, (*Coluber guttatus*) is a beautiful species. The body is elongated, somewhat flattened on the back, with smooth scales; the colour whitish; a row of large brownish spots, bordered with black, upon the back; a second series of smaller and darker ones on each side, alternating with the former; beneath, with small, square, black specks. The abdominal plates and sub-caudal scales are very numerous. It attains a large size, and inhabits all parts of the United States. Some of the Snakes here described belong to different subgenera of *Colubridæ*. We must refer our readers to the works of Schlegel and Gunther.

SNAKE-FLY. The Snake-flies, or *Raphidiadæ*, are a group of Neuroptera which receive their common name from the elongated form of the head and neck, and the facility with which they move the front of the body in different directions. They are mostly to be found in the neighbourhood of woods and streams; they are of comparatively small size, very active in their motions, and possess very powerful jaws: they prey upon other insects inhabiting the same situations.

SNIPE. (*Scolopax gallinago.*) The common Snipe is eleven or twelve inches long, and weighs about four ounces. The bill is nearly three inches long; pale brown or greenish yellow, rather flat and dark at the tip, and very smooth in the living bird, but it soon becomes dimpled when the bird is dead: the head is divided lengthwise by three reddish or rusty white lines and two of

COMMON SNIPE. — (SCOLOPAX GALLINAGO.)

black: the chin under the bill is white; the neck is a mixture of brown and red; the breast and abdomen are white. The scapulars are elegantly striped lengthwise on one web, and barred on the other, with black and yellow: quills dusky, the edge of the primaries, and tips of the secondaries, white, those next to the back barred with black, and pale rufous: the tip of the tail is commonly of a pale reddish yellow; and the legs pale green. The Snipe frequents marshy places and wet meadows, and in frosty weather, the edges of rushy hills, where it is almost constantly digging and nibbling in the soft mud. Their food consists of worms, insects, slugs, &c., which abound in such places. In these retreats, when undisturbed, the Snipe walks leisurely, with its head erect, and at intervals moving its tail. When disturbed, it usually springs, and takes flight beyond the reach of the gun, turning nimbly in a zigzag direction for two or three hundred paces, and sometimes soaring almost out of sight.

The Snipe, like the Woodcock, shuns the extremes of heat and cold, by keeping upon the bleak moors in summer, and seeking the shelter of the valleys in winter. In severe frosts and storms of snow, driven by the extremity of the weather, they seek the unfrozen boggy places, springy rills, or any open streamlet of water, and there they will sometimes sit till nearly trodden upon before they will take to flight. Although it is well known that numbers of Snipes leave Great Britain in the spring, and return in the autumn, yet it is equally well ascertained that many constantly remain and breed in various parts of the country; for their nests and young ones have been so often found as to leave no doubt of the fact. The female makes her nest (which is very inartificially composed of withered grasses and a few feathers) in some retired spot, generally under the stump of an alder or willow. The eggs, which are large and generally four in number, are pale-yellowish or greenish-white with rather elongated rusty spots at the big end. Sir Humphrey Davy describes the parent birds as exceedingly attached to their young, and says that if any one approach their nest, they make a loud and drumming noise above the head of the intruder, as if to divert his attention. The young birds run off soon after they leave the shell, but they are attended by their parents until their bills have acquired a sufficient firmness to enable them to provide for themselves. The Snipe is a very fat bird, but its fat does not cloy, and very rarely disagrees even with the weakest stomachs. It is much esteemed as a delicious and well-flavoured dish.

The JACK-SNIPE, or JUDCOCK, (*Scolopax gallinula*), in its figure and plumage very much resembles the Snipe; but it seldom exceeds two ounces in weight, or is above eight inches and a half in length. The bill is black at the tip, and light towards the base. A black streak passes over the head lengthwise; and another of a yellowish colour over each eye. The neck is white, spotted with brown and pale red. The scapulars and tertials are very long and beautiful; being bordered on their exterior edges with a stripe of yellow, and the inner webs streaked with bright rust colour on a bronze ground, reflecting shades of purple and green. The rump is glossy violet; the abdomen and vent white; the tail dark brown, edged with rust colour; legs dull green In its general habits this bird resembles the common Snipe: it feeds upon the same kinds of food, lives and breeds in the same swamps and marshes, and conceals itself from the sportsman with as great circumspection, among the rushes or tufts of coarse grass. It differs, however, in this, that it seldom rises from its lurking place until it is almost trampled upon, and, when flushed, does not fly to so great a distance. It seldom abandons for any length of time the place it has once fixed upon; and though roused from it, and fired at repeatedly, perhaps, through the day, neither the noise nor any sense of danger seems to alarm it; and if we should seek for the little Judcock on the following morning, in all likelihood we should find it at its spring again.

SNOUT [MOTHS]. A name applied by collectors to various Moths, of the genera *Hypena*, *Crambus*, and *Cledeobia*.

SNOW-BUNTING. The *Emberiza Nivalis*. [See BUNTING.]

SNOW-GOOSE. [See GOOSE.]

SOLAN-GOOSE. [See GANNET.]

SOLDIER BEETLE. [See TELEPHORUS.]

SOLE. (*Pleuronectes solea.*) This well-known and much esteemed fish is most abundant on the sandy shores all round our coast, where it keeps close to the bottom, preying on the smaller testaceous animals, and the spawn and fry of other fishes. It is also an inhabitant of the Northern Baltic, Mediterranean, and American seas. The form of the body is a long oval, widest at a short distance behind the head, becoming gradually narrower and rather pointed towards the tail. It sometimes grows to the

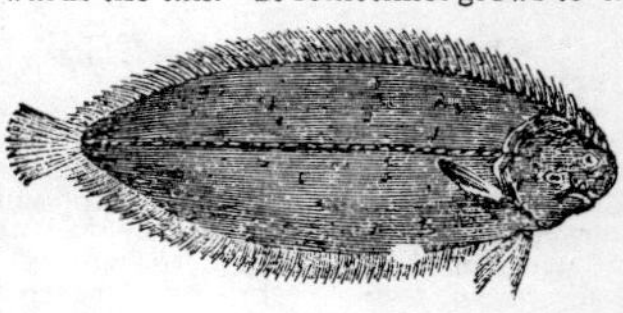

SOLE.—(PLEURONECTES SOLEA.)

length of two feet, and to the weight of six or eight pounds: its general size, however, is much smaller. Its colour is obscure brown above, and white beneath; it is covered with small rough scales of an oblong form, each terminated by numerous spines, and very strongly fastened to the skin. The head is small; the eyes and mouth of moderate size; both jaws furnished with minute teeth on the under or white side of the fish only; the eyes small. Soles seldom take any bait, but are almost entirely taken by *trawling*. At Hastings, Brighton, and the great fishing station at Brixham in Torbay, and, indeed, nearly all along the southern and western coast of England, they are taken in great numbers. They are also caught on various parts of the Irish coast: and Mr.

Yarrel asserts, that eighty-six thousand bushels of Soles have been received at Billingsgate market only within twelve months! Next to the Turbot this fish is considered as the most delicate of the genus, and is by many even preferred to the former; the flesh being remarkably firm, white, and well-flavoured: those of moderate size are in general most esteemed.

There are several varieties, as the Lemon Sole, the Variegated Sole, the Zebra Sole, the Silver Sole, &c., none of which are by any means so abundant as the common species just described, nor differing from it in any very important point. There is also the SOLENETTE or LITTLE SOLE (*Pleuronectes lingula*), considerable numbers of which are taken in the trawl nets off Brixham throughout the whole year; but from their diminutive size, they are generally thrown back into the sea.

SOLENIDÆ. The name given to a family of Mollusca, distinguished by the great length of their respiratory tubes. The *Solen*, or *Razor-shell*, is a well-known example. It has an elongated shell, the hinge being furnished with distinct teeth, and the ligament altogether external. The animal burrows in the sand (sometimes to the depth of nearly two feet), into which it sinks rapidly on the approach of danger; and as it very rarely quits its hole, its movements are nearly limited to an ascent or descent in it. This it accomplishes by means of its foot, which it attenuates into a point when it is about to bore, and afterwards contracts into a rounded form, so as to fix it by its enlargement when it desires to rise. In places where they

RAZOR-SHELL.—(SOLEN VAGINA.)

abound, they are sought after as bait for fish, and are taken in the following manner. Although the Solen is an inhabitant of salt water, yet salt in its pure state appears to have an irritating effect upon the animal: the fisherman, therefore, having discovered its retreat, throws into the hole a small quantity of salt, which generally brings the creature to the surface, when he endeavours to grasp it firmly; to do which some address and quickness are required; but should he fail, and the animal make good its retreat, there is no other way to capture it than to dig it out of the sand: for it is either insensible to the additional irritation, or its instinct of self-preservation teaches it to remain beneath. When the tide is low the burrow of the Solen is often recognized by the little jet of water which the animal throws out, when alarmed by the shaking of the sand occasioned by the motion of the fisherman above. Some species are common on the English coast; others come from India, America, &c. One of the Indian varieties is remarkable for its beautiful colour; the shell under the epidermis being of a delicate violet, striped with white.

SOLITAIRE. The name given to an extinct species of Dodo: also the name applied in Jamaica to a species of Thrush. [See PTILOGONYS.]

SOREX: SORICIDÆ. The name given by Cuvier to a genus and family of nocturnal insectivorous quadrupeds, of which the Shrews or Shrew-mice are the type. [See SHREW.]

SOUSLIK. (*Spermophilus citillus.*) A pretty little Rodent quadruped allied to the Marmots, but distinguished by having cheek-pouches in which it stores away seeds and nuts. It is not uncommon in different parts of Germany and Russia, and seems to vary considerably in markings. This species and its congeners lay up, for the winter, seeds,

SOUSLIK.—(SPERMOPHILUS CITILLUS.)

acorns, nuts, and beech-mast, which they carry to their burrows. Pennant informs us that in the more primitive times, when foreign furs were scarcer than they are now, the ladies in Bohemia made cloaks of the skins of Sousliks; and they are sometimes used to line articles of dress even at the present time. Numerous species of this genus occur in North America.

SPALACOTHERIUM. [See SUPPL.]

SPANIEL. (*Canis [familiaris] avicularius.*) The name given to several varieties or distinct breeds of the canine race, all more or less elegant; the distinguishing characters of which are,—that the muzzle is rather broad; the ears remarkably long and full; the hair plentiful, and beautifully waved, particularly that of the ears, tail, and hinder parts of the thighs and legs. The prevailing colour is liver and white; sometimes red and white, or black and white; and sometimes deep brown, or black on the face and breast, with a tan spot over each eye. England has been famous for producing dogs of this sort, particular care having been taken to preserve the breed in its utmost purity; so that notwithstanding the name Spaniel is supposed to be derived from Spain, it is more than probable that the English Spaniel (the most common and useful breed) is indigenous. The fond attachment and timid submission of the Spaniel are proverbial; there are few persons, indeed, who could not bear witness to the truth of the following description given by Mr. Bell: "If punished, it receives the chastisement with submission, and looks in the face of its offended master with an expression of humble sorrow for having been the cause of his anger; and the instant that the punishment is over, it comes courting the caresses of the hand that had inflicted the stripes, and asking him again to be received into favour. At the slightest look of encouragement, its joy at the reconciliation seems to know no bounds, and is expressed by the liveliest indications of delight, jumping and fawning upon the person of him who had just before been inflicting bodily pain and mental distress — capering round him, and barking loudly with ecstasy."

The SPRINGER is a small and elegant breed, generally red and white, with black nose and palate. In this elegant variety length of ears and a small head are essential points. —The WATER SPANIELS, large and small, differ only from the common Spaniel in the roughness of their coats, and in uniting the aquatic propensities of the Newfoundland Dog with the fine hunting qualities of their own race. — The beautiful breed known as King Charles's are highly prized for their diminutive size, length of ears, &c. [See LAP-DOG.]

SPARROW. The Common or HOUSE-SPARROW (*Pyrgita domestica*) the most familiar representative of the Finch tribe (*Fringillidæ*) is so constantly seen in the vicinity of our habitations, even in the midst of populous cities, that no person can be ignorant of its appearance or habits: although it must be admitted that, as seen in smoky towns, it is difficult to trace that agreeable variety in the plumage which distinguishes the male bird as it hops about among the ricks and mingles with the poultry in the farm-yard. This bird is nearly six inches in length, and of a robust form: bill dusky, eyes hazel; the top of the head and back part of the neck ash gray; the throat, fore part of the neck, and space round the eyes black; the cheeks whitish; the breast and all the under parts pale ash; the back, scapulars, and wing-coverts are reddish brown, mixed with black — the latter tipped with white, forming a light bar across the wing; tail brown, edged with gray, and rather forked; legs pale brown. The plumage of the female is plainer and duller than that of the male;

SPARROW.—(PYRGITA DOMESTICA.)

beyond each eye there is a line of white, and she has no black patch on the throat. Sparrows are bold and crafty; and their partiality to the vicinage of man does not originate from any social affection on their part, but because their chief subsistence is there most abundantly to be found. They follow society, and live at its expense: granaries, barns, court-yards, pigeon-houses, and all places, in short, where grain is scattered, being their favourite resorts. Their voracity is extreme; they are inconveniently familiar, and their incessant and monotonous note is fatiguing to the ear. But if Buffon's estimate be true that a pair of Sparrows will destroy about 4000 caterpillars weekly in feeding their young, there is good reason to suppose that they sufficiently repay the trivial damage they may cause either in the garden or the field. The Sparrow builds under the eaves of houses, in holes of walls, &c.; the nest being made of hay, and lined with feathers. The female lays five or six eggs of a reddish white, spotted with brown; and has generally three broods in the year.

The following characteristic observations on the habits of this well-known bird are from the pen of Mr. Knapp:—"A dispensation that exists throughout creation is brought more immediately to our notice by the domestic habits of this bird. The natural tendency that the Sparrow has to increase, will often enable one pair of birds to bring up fourteen or more young ones in the season. They build in places of perfect security from the plunder of larger birds and vermin. Their art and ingenuity in commonly attaching their nests beneath that of the rook, high in the elm, a bird whose habits are perfectly dissimilar, and with which they have no association whatever, making use of their structure only for a defence to which no other bird resorts, manifest their anxiety and contrivance for the safety of their broods. With peculiar perseverance and boldness they forage and provide for themselves and their offspring; will filch grain from the trough of the pig, or contend for its food with the gigantic turkey; and, if scared away, their fears are those of a moment, as they quickly return to their plunder; and they roost protected from all the injuries of weather. These circumstances tend greatly to increase the race, and in some seasons their numbers in our corn-fields towards autumn are prodigious; and did not events counteract the increase of this army of plunderers, the larger portion of our bread corn would be consumed by them. But their reduction is as rapidly accomplished as their increase, their love of association bringing upon them a destruction, which a contrary habit would not tempt." The common Sparrow is found in all parts of Europe, and almost throughout the eastern continent, supporting equally well severe cold and extreme heats. America is, however, free from it; but they have, in its place, the CHIRPING SPARROW,—a delicate bird, almost as familiar, but nowise intrusive.

The TREE SPARROW, or MOUNTAIN SPARROW (*Pyrgita montana*), is somewhat less than the common Sparrow: the bill is thick and black; the crown of the head and hinder part of the neck chestnut brown; sides of the head white; throat and auriculars black; the greater quills are black, bordered with rust-colour; the lesser coverts of the wings of a bright bay colour, spotted with black, and crossed with two white bars; breast and under parts dirty white. Just above the greater coverts there is a row of black feathers edged with white; the lower part of the back is of an olive-brown hue; the tail is reddish-brown, and even at the end; legs pale yellow. This species, though plentiful on the continent, and even in some of our southern and eastern counties, is seldom seen in the north of England. It differs from the House-Sparrow in making its nest in the holes of trees far from towns or villages. It feeds on fruits, seeds, and insects. It is a lively, active bird, and, when it alights, has a variety of motions, whirling about, and jerking its tail upwards and downwards, like the Wagtail.

The WHITE-THROATED SPARROW. (*Fringilla albicollis.*) Of all the Sparrows known in North America, Wilson says this species is the largest as well as the handsomest. From Connecticut to Savannah he found these birds numerous, particularly in the neighbourhood of the Roanoke river, and among the rice plantations. In summer they retire to the higher inland parts of the country, and also farther north, to breed; but during their residence in the above-mentioned localities, they collect together in flocks, always preferring the borders of swampy thickets, creeks, and mill-ponds, skirted with alder bushes and long rank weeds, the seeds of which form their principal food. The length of the White-throated Sparrow is six inches and a half, breadth nine inches; the upper part of the back and the lesser wing-coverts are beautifully variegated with black, bay, ash, and light brown; a stripe of white passes from the base of the upper mandible to the hind head; this is bordered on each side with a stripe of black; below this again is another of white passing over each eye, and deepening into orange yellow between that and the nostril; this is again bordered by a stripe of black proceeding from the hind part of the eye; breast, ash; chin, belly, and vent, white; tail, somewhat wedged; legs, flesh-coloured; bill, a bluish horn colour; eye, hazel. All the parts that are white in the male are in the female of a light drab colour.

The HEDGE-SPARROW (*Accentor modularis*) is about the size of the Redbreast, and belongs to the family *Sylviadæ*. The beak is black, and rather long and slender; the head is of a deep brown hue, mixed with ash-colour; and the cheeks are marked with oblong spots of dirty white: the back and coverts of the wings are dusky, edged with reddish-brown; the quill-feathers and the tail are also dusky; the rump brown, tinged with green; the throat and breast are of a dull ash-colour; the sides, thighs, and vent feathers, pale tawny brown; and the legs are of a dull flesh-colour. This bird frequents low hedges, particularly those of gardens; making its nest in some small bush, where it lays four or five pale blue eggs; and, during the season of incubation, it has a remarkable flirt with its wings. The male utters a short, but very sweet plaintive note, which it begins about the commencement of the first frosty mornings and continues till the melody of the returning spring drowns its voice.

The Hedge-Sparrow appears to be a prime favourite with the entertaining author of the 'Journal of a Naturalist,' who thus speaks of it. "Not influenced by season or caprice to desert us, it lives in our homesteads and our orchards through all the year, our most domestic bird. It is nearly the first bird that forms a nest; and this being placed in an almost leafless hedge, with little art displayed in its concealment, generally becomes the booty of every prying boy; and the blue eggs of the Hedge-Sparrow are always found in such numbers on his string, that it is surprising how any of the race are remaining, especially when we consider the many casualties to which the old birds are obnoxious from their tameness, and the young that are hatched from their situation. The plumage of this *motacilla* is remarkably sober and grave, and all its actions are quiet and conformable to its appearance. Its song is short, sweet, and gentle. Sometimes it is prolonged, but generally the bird perches on the summit of some bush, utters its brief modulation, and seeks retirement again. Its chief habitation is some hedge in the rick-yard, some cottage-garden, or near society with man. Unobtrusive, it does not enter our dwellings like the Redbreast, but picks minute insects from the edges of drains and ditches, or morsels from the door of the poorest dwelling in the village. As an example of a household or domestic bird, none can be found with better pretensions to such a character than the Hedge-Sparrow."

The REED-SPARROW. [See BUNTING, REED.]

The SOLITARY SPARROW. ("*Passer solitarius.*") This beautiful bird, which seems to be a species of Thrush, may be described here. It is a native of the southern parts of Europe. In shape it resembles the blackbird, but is rather smaller: the bill is straight, and of a dusky brown colour, the upper mandible bending a little downwards at the point; the eyes dark hazel, and the eyelids yellowish. The entire plumage, except the quills and tail, is blue, darker on the back and lighter on the breast: the feathers on the breast and abdomen being transversely barred with a lighter colour: the quills and tail-feathers are of a dusky brown hue, except that there is a small portion of blue on their exterior webs. The legs, feet, and claws are black. It feeds on insects, grapes, and other fruit.

The following pleasing observations relative to this bird are given by Mr. Waterton in his Essays. "Would my readers," says he, "lend a patient ear for a short time, they shall have both the history and the true name of this bird placed in a proper light. The royal psalmist, whilst bending down in penitential prayer before his offended Maker, exclaims, 'I have watched, and am become as a Sparrow all alone upon the house-top.' I have often wondered what bird this could be; knowing, by daily experience, that it could not actually be the house-sparrow; for the house-sparrow is not solitary in its habits. I despaired of being able to trace its character satisfactorily and I should probably have long remained in ignorance of it, had I not visited the southern parts of Europe. My arrival at Rome let me at once into the secret. The bird to which the repentant king of Israel compared himself in the seven penitential psalms is a real thrush in size, in shape, in habits, and in song; with this difference from the rest of the tribe, that it is remarkable throughout all the East for sitting solitary on the habitations of man. The first time I ever saw this lonely plaintive songster was in going to hear mass in the magnificent church of the Jesuits at Rome. The dawn was just appearing, and the bird passed over my head, in its transit from the roof of the palace Odescalchi to the belfry of the church of the Twelve Apostles, singing as it flew. I thought it had been the Italian blackbird, with notes somewhat different from those of our own; for its song was partly that of the blackbird, and partly that of the stormcock, but not so loud as the last, nor so varied as the first. I found out my mistake in due time; and, on seeing that the bird was the true Solitary Thrush, I paid particular attention to its habits. It is indeed a solitary bird. for it never associates with any other, and only with its own mate at breeding time; and even then it is often seen quite alone upon the house-top, where it warbles in sweet and plaintive strains, and continues its song as it moves in easy flight from roof to roof. It lays five eggs of a very pale blue. They much resemble those of our Starling. The bird itself is blue, with black wings and tail; the blue of the body becoming lighter when placed in different attitudes."

SPARROW-HAWK. (*Falco* [*Accipiter*] *nisus.*) This is a bold and spirited bird, but the most pernicious of the Hawk kind that inhabits Britain, making great depredations among pigeons, partridges, and the young of domestic poultry. The difference of size between the male and female is very disproportionate; the former measuring about twelve, and the latter fifteen inches. Individuals of this species also vary considerably in their colours: in some, the back, head, coverts of the wings, and tail, are of a deep bluish-gray, edged with a rusty red. The quill-feathers are dusky, barred with black on their exterior webs, and spotted with white on the lower part of the interior webs. On the tail, which is of a deep ash-colour, there are fine broad black bars, and the tip is white. The breast and belly are of a cream-colour, with transverse bars at the base, of a deep brown in some, and orange-colour in others; and the skin at the base of the bill, the irides, and the legs, are yellow. The colours of the female are different from those of the male: the head, back, and coverts of the wings being browner, and the tail of a brighter dove colour; the waved lines on the breast more numerous, and the breast containing a greater portion of white. She builds her nest in hollow trees, high rocks, or lofty ruins; sometimes in the old nest of a crow; and generally lays four or five whitish eggs, spotted with red at the thicker end. Mr. Selby says that it occasionally makes its nest in low trees or thorn-bushes, that it is flat and shallow, and very similar to that of the ring-dove, but rather

SPARROW-HAWK.
(FALCO [ACCIPITER] NISUS.)

larger, and is composed of tender twigs. The Sparrow-hawk is found, in considerable numbers, in various parts of the world, from Russia to the Cape of Good Hope. This bird was held in great veneration among the ancient Egyptians, because it was made the emblem of their god Osiris. Among the Greeks it was consecrated to Apollo.

The AMERICAN SPARROW-HAWK (*Falco sparverius*) is a beautifully marked bird, belonging to the same subdivison which contains the *Kestrel*, and appears to reside principally in the warmer parts of the United States: they are found also in the West Indies, south of the Equator. The female is eleven inches long; the male not quite ten. The cere and legs are yellow; the head bluish ash; crown rufous. The upper parts are reddish-bay, striped transversely with dusky brown; the lower parts pale yellowish white, marked with longitudinal spots of brown: the claws black. The nest is built in a hollow, shattered, or decayed tree, at a considerable elevation. It lays four or five eggs, of a light brownish colour, and spotted with brown. It preys upon sparrows and other small birds, also mice, grasshoppers, and lizards; but it has been remarked that it will very seldom, if ever, eat of any thing which it has not itself killed.

Another species, called the COLLARED SPARROW-HAWK, (*Accipiter torquatus*), which is well known in Van Diemen's Land and New South Wales, has all the bold and daring characteristics of its European ally. The head and all the upper part of the plumage is a deep brownish gray, the tail indistinctly barred with deep brown, and on the back of the neck an obscure collar of reddish brown; the throat, breast, and thighs, rufous, crossed by numerous bars of white; under surface of the wings and tail gray, barred with brown; irides and eyelash yellow; cere green; bill lead-colour, the tip black; legs yellow, slightly tinged with green.

SPARUS. The name given to a genus of Acanthopterygian fishes in the Linnæan system, the characteristics of which are—that the gill-openings are scaly; the mouth is furnished with strong cutting teeth; the grinders are obtuse, close set, and covered with lips; the branchiostegous membrane consists of five rays; the body is compressed; the lateral line is curved behind; and the pectoral fins are rounded. For an example of this genus, see GILTHEAD.

SPATANGUS, or HEART URCHIN. A genus of *Echinidæ*, common on many of our sandy shores. In this species the radiated form is considerably departed from, the shell being oval instead of round, and often much prolonged in one direction. Little is known of the habits of the *Spatangi*. They are almost always found buried in the sand, in which substance they seem to find enough nutriment (composed probably of the minute animals mingled therewith), their alimentary canal being filled with it. As they seem to be unable to bring their suckers into proximity with the mouth, they must derive their nourishment from the chance-supplies which the substances in contact with their mouths may furnish. Their whole organization is, certainly, adapted to this mode of existence; yet it is difficult to conceive how they can obtain the necessary amount of aliment, with so little power of either locomotion or prehension.

SPERMACETI WHALE. The common Cachalot. [See WHALE.]

SPHÆRIDIADÆ. A small group of Coleopterous insects, very similar in general structure to the *Hydrophilidæ*, but in their habits very different; since they frequent putrescent vegetable matter which has passed through the bodies of animals, the excrement of horses and cows being their chief abode, over which, when recently ejected, they may be seen hovering, and in which they burrow. The species of *Sphæridium* are the largest in

SPHÆRIDIUM SCUTELLATUM.

the family, not, however, exceeding a quarter of an inch in length; they are generally of a shining black colour, with the elytra variegated with large patches of red or dingy yellow. Seventy species or upwards are believed to occur in this country; most of these belong to the genus Cercyon, the characters distinguishing which are most frequently very obscure and unsatisfactory.

SPHEGIDÆ. A family of Hymenopterous insects, some inhabiting tropical climates, which are the largest belonging to the Order, and others noted for their varied and splendidly metallic colours. The body is long, with the abdomen often attached to the thorax by a peduncle; the collar laterally dilated, and extending as far as the wings; the antennæ long, and filiform or subsetaceous; the legs long, and in general fossorial; the mandibles are long, curved, and acute at the tips; and their sting is very powerful. They are exceedingly active and very restless in their motions, and may often be seen upon sand-banks, &c., running along with their wings in constant vibration.

SPHINGIDÆ. A family of *Lepidoptera*, called by the English name of Hawk-Moths; comprising the most robust and powerful insects in the order, and generally distinguished by their strength of flight and large size. The antennæ are prismatic, and terminated by a little feather or thread; the tongue is often extremely long, in some species even exceeding the whole body in length; the labial palpi are broad and compressed, and closely covered with scales; the labrum and mandibles are minute; the body is long, and acute behind; and the wings, especially the hinder pair, small. The caterpillars are naked, cylindrical, and sixteen-footed; they are ornamented with pale oblique stripes upon the sides of the body, and are usually furnished with a short horn on the back of the eleventh segment. They descend into the earth to become pupæ, which are naked and conical. Various modifications occur in the character of the imago in this family. The maxillæ vary considerably in length, exceeding that of the entire body in Sphinx, but scarcely exceeding that of the head in the Death's-head Hawk-moth (*Acherontia Atropos*), [See ACHERONTIA], and in *Smerinthus* not longer than the labial palpi; this variation in length corresponds with the rapidity of flight, and the habit of the insects of extracting the nectareous juices of tube-bearing flowers by means of their elongated tongue. The caterpillars of the typical species are remarkable for the attitude in which they are usually seen, and from which they have obtained the genuine name of *Sphinx*, from their supposed resemblance to the figures of that fabulous creature. Some of them are also remarkable for the faculty they possess of elongating and contracting the three anterior segments of the body, giving them somewhat of a proboscis-like appearance; whence they have been termed Elephant Sphinxes.

Although the Sphinxes in general are only seen on the wing in the twilight hour, this is not absolutely the case with all. Mr. Knapp, in his 'Journal of a Naturalist,' thus speaks of the HUMMING-BIRD HAWK-MOTH (*Macroglossa stellatarum*). "It frisks about all the summer long, and in very fine seasons continues with us as late as the second week in October. The vigilance and animation of this creature are surprising, and seem to equal those of its namesake, that splendid meteoric bird of the tropics, 'that winged thought, as some one has called it; though our plain and dusky insect can boast none of its glorious hues. Our little Sphinx appears chiefly in the mornings and evenings of the day, rather avoiding the heat of the midday sun, possibly roused from its rest by the scent, that 'aromatic soul of flowers' which is principally exhaled at these periods; delighting in the jasmine, marvel of Peru, phlox, and such tubular flowers; and it will even insert its long, flexible tube into every petal of the carnation, to extract the honey-like liquor it contains. Nature seems to have given this creature some essential requisites for its safety: its activity, when on the wing, renders its capture difficult; and when it rests it is on a wall, the bark of a tree, or some dusky body, that assimilates so nearly to its own colour as to render it almost invisible, though watched to its settlement. We sometimes see it enter our rooms, attracted by flowers in the open windows; but it seems to be immediately aware of its danger, disappears in an instant, and is safe from capture. Wild and fearful as this creature is by nature, yet continued gentle treatment will remove much of its timidity, and render it familiar to our presence. Perfectly free from any annoyance as they are when ranging from sweet to sweet on my borders, and accustomed to a close inspection of all their operations, I have frequently touched their wings with my fingers, while hovering over a flower, and dipping their long tubes into the corolla of a geranium: they would retire a little, confused with such freedoms and interruptions, but, experiencing no harm, they would return and finish their meal, unmindful of such petty annoyances. I have known this creature, like some other insects, counterfeit death when apprehensive of danger, fall on its back, and appear in all respects devoid of life when in a box; and, as soon as a fit opportunity arrived, dart away with its usual celerity."

In some species the extremity of the abdomen is elongated, and very acute, and in others broader, and furnished on each side with a brush. Some, again, have scaleless wings (*Sesia*), whence the smaller species have received the names of *Sesia fuciformis* and *S. bombyliformis*, in reference to their analogical resemblance to drones or *Bombylii*. These last mentioned species, as well as those with tufted abdomens, fly during the day, the latter thence obtaining the name of Humming-bird Hawk-moths; whereas the others fly during the twilight, darting about with the greatest rapidity, or hovering, hawk-like, in front of the flowers, from which they extract the nectar with the assistance of their elongated tongue.

As a striking and yet common example of this family we figure and describe

The Sphinx Ligustri, or Privet Hawk-Moth. Among the numerous Moths which make their appearance on fine summer evenings, we have no one that is more handsome, and scarcely one that is more common, than the *Sphinx ligustri*. It varies in the expansion of its wings from three and a half to nearly five inches. The fore wings are of an ashy colour, with the base pale, and slightly tinted with rose colour, and having a large dark patch along the inner margin, extending nearly from the base to the tip; while slender black lines run longitudinally between the veins of the wings; along the extremity of this dark patch runs a weaved ashy and black stripe, and a slender wavy white line running parallel with the outer margin. The hind wings are of a pale rosy colour, with three black bands, two of them long and broad, running parallel with the extremity of the wing: the fringe is uniformly grayish-brown. The hind part of the head and the sides of the thorax are of an ashy-white, but the back of the latter is

PRIVET HAWK-MOTH.—(SPHINX LIGUSTRI.)

black, posteriorly irrorated with gray: the sides of the abdomen are of a rich pink-red colour, interrupted by black bars, and with a broad dorsal ashy bar, along the middle of which runs a darker line. The under side of the body is a light dun colour, with a black line down the centre. The Caterpillar is green, with the caudal horn black above, and yellow beneath, and seven oblique stripes

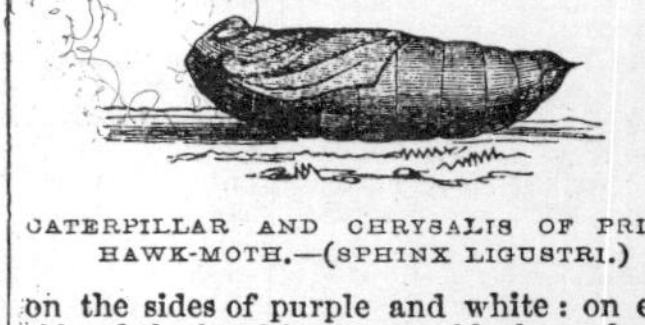

CATERPILLAR AND CHRYSALIS OF PRIVET HAWK-MOTH.—(SPHINX LIGUSTRI.)

on the sides of purple and white: on each side of the head is a strong black mark, and the spiracles are orange. When first hatched the young Caterpillars have the tails remarkably long, and the bodies very rugose, but they become smooth at the final moulting. By the end of August or the middle of September they are full grown, and become of a dirty-red colour, when they descend into the earth, where they change into a dark brown chrysalis, with the extremity slightly bifid, and the tongue-case straight. The Moth appears in the following June and July. Sometimes, however, it will remain two and even three years in the chrysalis state, and then become winged as perfectly as if it had appeared at the ordinary period.

SPIDERS. (*Arachnida.*) These well-known animals, if not among the most admired, are undoubtedly among the most interesting, of the annulose world, from their habits and mode of life. They differ essentially in their internal structure, from insects proper; and their external form is so peculiar that they are easily recognized. The body is composed of two pieces only, the head being united with the thorax; and the feet are always eight in number. Their cephalothorax appears as if composed of but a single segment, and is covered with a sort of horny buckler, generally oval, to which the abdomen, consisting of a soft and tumid mass, is appended. Generally they have eight eyes, though sometimes only six, variously disposed in the different genera, but always simple. The mandibles terminate in a very short movable hook, having near its extremity a small aperture, which serves as a passage for the poison. The legs are inserted almost in a circular manner around the cephalothorax; they are all nearly of the same form; and each is composed of seven joints, the last being armed with two hooks. The pulmonary sacs are placed near the base of the abdomen, and indicated externally by a brownish or whitish spot. They are now divided into groups or families, according to the arrangement of the mandibles and eyes, which corresponds very remarkably with their respective modes of life.

The Spider being formed for a life of rapacity, and incapable of living on any other than insect food, all its habits are calculated to deceive and surprise: it spreads toils to entangle its prey; it is endued with patience to expect its approach; and possesses power sufficient to destroy it when captured. For the purpose of constructing its web, Nature has supplied the Spider with a large quantity of glutinous matter within its body, and with five papillæ, or teats, for spinning it into thread. This substance is contained in a little bag, and at first sight resembles soft glue; but when more accurately examined, is found twisted into many coils of an agate colour; and, on breaking it, the contents may easily be extended into threads, from the tenacity of the substance — not from those threads being already formed. The machine by which wire is drawn will furnish us with some idea of the manner in which this creature forms the threads of its little net; the orifices of the five teats, through which the thread is drawn, contracting or dilating at pleasure. The threads which we see, and which appear so fine, are, notwithstanding, composed of five joined together; and these are repeatedly doubled as the work proceeds. When a house or common Spider is about to form a web, it first selects some commodious and secure spot, where insects appear to be in sufficient abundance. It then distils a small drop of its glutinous liquor, which is very tenacious; and, creeping up the wall, and joining its thread as it proceeds, darts itself in a very surprising manner to the opposite station where the other end of the web is to be fastened. The first thread thus spun, drawn tight, and fixed at each end, the Spider runs on it, to and fro, still assiduously employed in doubling and strengthening it, as on its force depends the strength and stability of the whole. The scaffolding being thus completed, the Spider draws a number of threads parallel to the first, in the same manner, and then crosses them with others; the adhesive substance of which they are formed serving to bind them together when newly spun. After this operation the wary architect doubles and trebles the thread that borders its web, by opening all its papillæ at once; and so secures the edges as to prevent the wind from displacing the work. The edges being thus fortified, the retreat is next to be attended to; and this is formed like a funnel, where the little workman lies concealed. To this there are two passages or outlets, one above and the other below, very artfully contrived, to allow the animal an opportunity of making excursions at proper seasons, of examining every corner, and clearing those parts which become foul or encumbered. It often happens also, that from the main web there are several webs extended at some distance on each side: these may be considered as the outworks of the fortification, which, whenever touched from without, the Spider prepares for attack or self-defence. If the insect impinging happens to be a fly, it springs forward with great agility; but if, on the contrary, some enemy stronger than itself, it then keeps within its fortress, and never ventures out till the danger is past.

The Garden Spider (*Epeira*) appears to work in a different manner. It spins a large quantity of thread, which, floating in the air in various directions, happens, from its glutinous quality, at last to adhere to some object near it — a lofty plant, or the branch of a tree. The Spider is anxious to have one end of the line fixed, that it may be enabled to secure and tighten the other: it accordingly draws the line when thus fixed; and then, by passing and repassing on it, strengthens the thread in such a manner as to answer all its intentions. The first cord being thus stretched, the Spider walks along a part of it, and there fastens another; and dropping from thence, affixes the thread to some solid body below; then climbs up again, and begins a third, which it fastens by a similar contrivance. When three threads are thus fixed, it forms a figure somewhat resembling a square; and in this the animal is generally found to reside. It often happens, however, when the young Spider begins spinning, that its web becomes too buoyant; and not only the web floats in the air, but the spinner also. The struggles of an entangled insect communicate an undulatory motion to the whole web, which gives notice to the Spider, who immediately sallies forth, and, if his victim be small, seizes it at once, and sucks its blood: if, however, it be too large to be thus disposed of, the Spider rolls it with his hinder feet, encircling it with a new thread at every turn, until, sometimes, the insect is completely coated, and it may be devoured at pleasure. Some Spiders spin an irregular web, consisting of threads intersecting each other at every angle: others, again, make a horizontal, closely-matted web, having a funnel-shaped retreat, into which they convey their prey: while others make only a retreat by binding a few leaves together, from which they sally forth and seize insects which approach them. Some of these

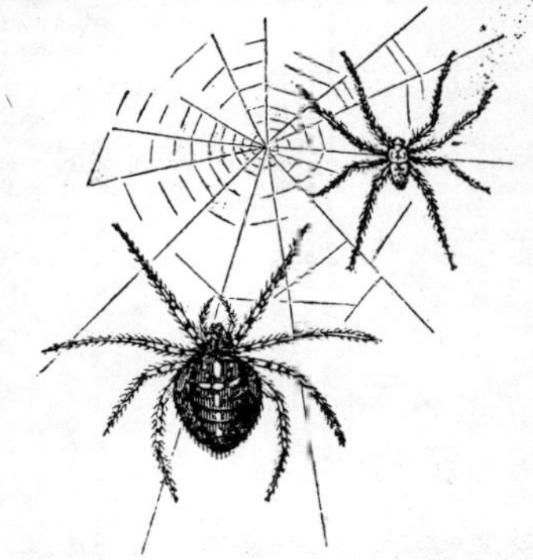

GARDEN SPIDER.—(EPEIRA DIADEMA.)

seem to be extremely venomous; for it is observed that no insect that has been once bitten by them, ever recovers, even though it be many times larger and more powerful than its adversary. Some are aquatic, and spin a cup-like web, which answers the purpose of a diving-bell, under which they disengage the air they bring down from the surface, and pass their lives feeding on aquatic insects. Some Spiders spin no web, but take their prey by running; others by approaching quietly till within a certain distance, when they suddenly leap upon their prey: other Spiders form perpendicular and cylindrical holes in the ground, into which they retreat on the approach of danger.

The female Spider generally lays nearly a thousand eggs in a season; which are separated from each other by a glutinous substance. These eggs are small or large in proportion to the size of the animal that produces them. In some they are as large as a grain of mustard-seed; but in others they are too minute to be distinctly visible. The female never begins to lay till she is two years old; and her first brood is never so numerous as when she arrives at full maturity. When the eggs have continued to dry for an hour or two after exclusion, the Spider prepares a bag for their reception, where they remain to be hatched till they leave the shell. For this purpose she spins a web four or five times stronger than that intended for the catching of flies. This bag, when completed, is as thick as paper, smooth on the inside, but somewhat rough without: in this the eggs are deposited; and nothing can exceed the concern and industry which the parent manifests in the preservation of it: by means of the glutinous fluid, it is stuck to the extremity of her body; so that, when thus loaded, she appears as if double. If the bag should happen by any accident to be separated from her, all her assiduity is employed to fix it again in its former situation; and this precious treasure she seldom abandons but with her life. When the young are excluded from their shells within the bag, they remain for some time in their confinement; till the female, instinctively knowing their maturity, bites open their prison, and sets them at liberty. But her parental care does not terminate with their exclusion: she receives them on her back from time to time; till having acquired sufficient strength to provide for themselves, they leave her to return no more, and each commences a web for itself. The young ones begin to spin when they are scarcely large enough to be discerned; and discover their propensity to a life of plunder before Nature has conferred on them strength for the conquest.

In Mr. Low's 'Sarawak' it is said that, "the Spiders, so disgusting in their appearance in many other countries, are in Borneo of quite a different nature, and are the most beautiful of the insect tribe. They have a skin of a shell-like texture, furnished with curious processes, in some long, in others short, in some few, in others numerous; but are found of this description only in thick woods and shady places. Their colours are of every hue, brilliant and metallic as the feathers of the humming-bird, but are, unlike the bright colours of the beetle, totally dependent on the life of the insect which they beautify; so that it is impossible to preserve them."

In the 'Excursions to Arran,' by the Rev. David Landsborough, we find an account of the persevering labours of an *Epeira*, "who had pitched his tent by the way-side," which, in our opinion, is sufficiently interesting to warrant us in extracting nearly the whole of it. "The Spider is in kings' palaces;" and kings and queens too may learn a lesson from it, and so, surely, may we. Spiders have not got justice done to them: they are a much more interesting race than many suppose. They improve on acquaintance: the better they are known, the more they are admired. At that time a whole colony of them were encamped by the road-side, within the compass of half a mile. "As he was rather a gigantic Spider, his tent, instead of being on the ground, was elevated, like the house of a giant of whom in early life we have all read. It was built on the tops of the common grass, *Holcus lanatus*, more than a foot above the ground. Had he built his house on the top of one stalk of grass, the house and its inhabitant might have borne down a single slender stalk. But he had contrived to bring together several heads whose roots stood apart, and, with cordage which he could furnish at will, had bound them firmly together, so that his elevated habitation was anchored on all sides. From whatever *airt* the wind blew, it had at once halser and stay. Not only did he bind the heads together, but he bent, doubled, and fastened them down as a thatch roof, under which his habitation was suspended. As he was a larger Spider than usual, his house was large; the more capacious apartment, which I believe was the nursery, being below; and the smaller one, which was his observatory or watch-tower, being above, from which he could pounce on his prey, or, in case of hostile attack, could make his escape by a postern gate, so as to conceal himself among the grass.

"During my visit in June last, I was anxious, as we returned from Whiting Bay, to ascertain whether this interesting colony of tent-makers was still in a thriving state; and not seeing any at first, I began to fear that a Highland clearance had taken place. When I at last discovered a few of them, I saw that, as there are times of low trade among our industrious two-footed artizans in towns, so are there occasionally hard times among our six-footed operatives in the country. The field in which they encamped had, I suppose, been overstocked. The stately *Holcus* had been eaten down; but these shifty children of the mist had availed themselves of the heather—doubling down the tops of some of the heath-sprigs, and under this thatched canopy forming their suspension-tabernacles. As yet, however, it was too early in the season. The house had only one apartment; the web of which it was formed was as yet thin, so that through it I could see the Spider, which being but half grown, had not yet got in perfection its fine tiger-like markings. 'Go to the ant, thou sluggard;' go also to the Spider. He who taught the one taught the other; and, learning humility, let both teach thee.

"I said that kings might learn of the Spider; and one of the greatest of our Scottish kings, some five hundred years ago, disdained not to learn of an Arran Spider in the very district in which this Spider is found. The tradition still lingers in Arran, that King's-cross-point was so named, because from this point in Arran, King Robert the Bruce sailed for C rrick, his own district in Ayrshire. When he was, by a train of adverse circumstances, almost driven to despair, it is said that after a sleepless night, in a humble cot on this rocky point, he in the morning observed from his lowly bed a Spider actively employed in weaving her silken web. To make it firm and extensive, she endeavoured to fasten her filmy threads on a beam projecting from the roof, but in attempting to reach this beam she fell down to the ground. Six times she repeated the attempt with no better success, but instead of being discouraged, she made a seventh attempt—reached the wished-for point, fastened her adhesive cords, and went triumphantly on with her work. On observing this, the King sprang up with reviving hopes and fresh resolution. 'Shall I,' said he, 'be more easily discouraged than this reptile? Shall she, in spite of repeated failures, persevere till crowned with success, though her object is to enslave and destroy? and shall I leave anything untried that I may deliver from thraldom my oppressed subjects?' He hastened to the beach, launched a fishing-boat, sailed from *King's-cross-point* for Ayrshire, which he reached in safety—secretly assembled his liege men in Carrick—made a bold, and sudden, and successful attack on his own castle of Turnberry, which he took from the vanquished English garrison; and, following up this auspicious blow, he advanced on the side of victory, till, at Bannockburn, he drove the cruel invaders from the land, and set once more our beloved Scotland free."

As we have already seen, the species are very numerous; some differing widely from others; but the space we have already occupied compels us to confine ourselves, in the present instance, to the general description we have already given of their structure, habits, &c. One particularity, however, in the history of Spiders remains to be noticed, which is their power of flight. This is chiefly exercised by those of minute size. It is principally in the autumnal season that these diminutive adventurers ascend the air, and contribute to fill it with that infinity of floating cobwebs which are so peculiarly conspicuous at that period of the year. When inclined to make these aërial excursions, the Spider ascends some slight eminence, as the top of a wall, or the branch of a tree, and, turning itself with its head towards the wind, darts out several threads from its papillæ, and, rising from its station commits itself to the gale, and is thus carried far beyond the height of the loftiest towers. During their flight it is probable that Spiders employ themselves in catching such minute winged insects as may happen to occur in their progress; and when satisfied with their journey and their prey, they suffer themselves to fall, by contracting their limbs, and gradually disengaging themselves from the thread that supports them.

"We read in various works," says Vincent Köllar, "that Spiders often eject a corrosive poisonous juice, in consequence of which the joints become inflamed and swelled; and even that the crawling of a Spider is sufficient to cause inflammation in the parts which it touches. It might perhaps be too rash to contradict the assertions of many writers, but I have never found these observations adduced by men who have been exclusively occupied with the study of Spiders, nor have I ever experienced any thing of this kind myself throughout the many years in which I have been engaged in studying insects and spiders. All Spiders are, however, insects of prey, and feed on other insects, which they catch alive, kill, and then suck out their fluids. For this end they are mostly provided with very strong *chelæ* or mandibles. These *chelæ* are of a horny substance, bent inwards, hollow, and provided with an opening at the top, and are connected with glands, which secrete a corrosive juice. They discharge this juice into the captured insects they have wounded, apparently to kill them sooner. The same thing happens when they wound a person who has caught one, and gives it pain. Pain will naturally be the consequence of the wound, and the corrosive juice communicated to it; the wounded part becoming inflamed, and swelling. The larger the Spider, the warmer the climate or season of the year, and the more susceptible the wounded individual, so much worse will the effects be; and it is therefore no wonder that people who would have a fester from a simple prick with a needle, should feel more violent effects from the bite of a Spider. Thus the bite of the Tarantula in southern Italy, namely Apulia, according to late observations, is said not to be nearly so dangerous as it was considered formerly, and the disease attributed to the bite of the Tarantula is said to be more the consequence of the climate and manner of life of the people. It is, however, an indisputable fact that Spiders defend themselves when they are persecuted and captured, bite with their *chelæ*, and drop into the wound a more or less poisonous juice, although the consequences are very seldom dangerous."

RED SPIDER. There is a small Tick, so commonly called the Red Spider (*Acarus telarius*), that it may be described here. It is scarcely visible to the naked eye, and does considerable injury to various plants in warm dry summers. It is also called the Plant Mite. Like most of the *Arachnidæ*, it has eight legs; its colour changes from yellowish to brown and reddish, and on each side of the back is a blackish spot. In the open air it usually attacks kidney-beans. Among trees, the young limes principally suffer, and the Mites are found in thousands on the under side of the leaves. These leaves assume a dirty yellow or brownish appearance, and in the middle of summer the trees acquire an autumnal hue. In hot-houses the Red Spider feeds during the whole year, and is a great pest to nurserymen and gardeners. It spins a sort of web over the leaves, particularly on the under surface, and sucks the juice of the plants with its rostrum, which completely enfeebles and defoliates them. Vincent Köllar says that frequently sprinkling the plants with cold water has been found efficient as a means of destroying these insects: fumigating the hot-house repeatedly with strong tobacco smoke also injures them in some degree. They are most abundant when the plants are kept too warm in summer; and as most hot-house plants thrive well when placed in the open air in July and August, placing them out will almost entirely free them from these insects. When hot-house plants are placed in the open air, the precaution must be taken of sinking the pots in a warm dung or tan-bed to keep the roots warm. The roots being preserved in this way, the plants will defy the coldest weather they are ever likely to be exposed to in summer. For kidney-beans that are trained on sticks in the open air, it is necessary in autumn and winter to cleanse the sticks from all loose rind, as the Mites take up their winter quarters within it, in whole families, and if they are not destroyed, proceed from it to the young plants the ensuing spring. The Ray Society has recently published a beautiful work on Spiders by J. Blackwall, Esq., F.L.S.

SPIDER MONKEY. (*Ateles.*) A genus of Quadrumana, the species being thus called from their long slender limbs, and sprawling

BLACK SPIDER MONKEY.—(ATELES ATER.)

movements. They exhibit some remarkable resemblances to the human conformation in their muscles, and, of all animals, alone have the biceps of the thigh like that of a man. They accordingly make little use of their fore-hands in progression. Their colours are chiefly or wholly black, or fulvous-gray; face black, or flesh-coloured. They are gentle and confiding, and capable of much attachment.

SPIRULA: SPIRULIDÆ. A genus and family of Dibranchiate Cephalopods. According to the judgment of J. E. Gray, Esq., of the British Museum, "there is every reason to believe that the Spirula is the nearest recent ally of the *Ammonites*, so abundant and so numerous in kinds, found in the different fossiliferous strata." "The animal," Mr. Gray observes, "has all the general external characters of the cuttle-fish; that is

SPIRULA; WITH THE ANIMAL.

to say, it has a large distinct head with eyes on each side, eight short conical arms with series of small discs on the inner side, two long arms with elongated peduncles, and a bag-like mantle with a process in the middle above, and one on each side of the anal tube below; but it differs from the cuttle-fish in being entirely destitute of any fins, being rather compressed behind, and showing in the specimen under examination a part of the whorls of the shell above and below; but from the ragged edges of the skin it appears as if this shell was covered with a skin when the animal is alive, and that the exposure of the surface of the shell has only been caused by the contraction of the animal, and especially of the skin over the shell, from the animal having been placed in very strong spirits when caught. * * * The mantle is free from the body on all sides at its oral edge, and without any cartilaginous ridges; this edge is formed into a point on the centre of the dorsal aspect, and into two mesial processes, one situated on each side of the anal funnel on the ventral side; the funnel is quite free from the mantle. The part of the shell which is exposed is covered with minute rugosities and indistinct reticulations, somewhat like the surface of a cuttle-fish bone.

"The examination of this animal (continues Mr. G.) confirms me in the opinion which I expressed in the 'Synopsis of the British Museum' (1840, p. 149.), that the *Ammonites*, from their texture and the small size of the last chamber, are internal shells, and should be arranged with the Decapodous Cephalopods, being chiefly distinguished from the *Spirulæ* by the siphon being always on the dorsal margin of the whorls, and the septa being foliated on the edge. I am aware that this opinion is not in conformity with the ideas of many zoologists and comparative anatomists, for Mr. Owen, in the last arrangement of these animals (Todd, Ency. Comp. Anat.), though he places the *Spirulæ* with the *Dibranchiate* Cephalopods, places the *Ammonites* with *Tetrabranchiata* next to *Nautilus*, with the following character, "animal unknown, presumed to resemble the *Nautilus*."

SPONDYLUS. A genus of Mollusca, for the most part inhabiting a rough and foliated bivalve shell, with spines and plaits; the valves closely united by two strong teeth. Like the Pectens, the margins of the mantle of the animal are garnished with two rows of tentacula, and in the outer row there are

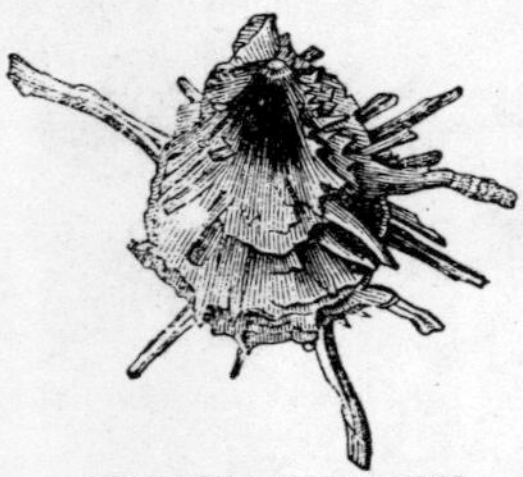

SPONDYLUS AMERICANUS.

several terminated with coloured tubercles: in front of the abdomen is a broad radiated disc with a short pedicle, capable of contraction and elongation; and from its centre there hangs a thread terminated with an oval mass, the use of which is unknown. They live at great depths in the sea, and attach themselves to corals, &c. They are also frequently found adhering to anchors, cannons, and other iron articles that have been for some time at the bottom of the sea. Many of the species are very beautiful, and of very vivid colours, such as bright red, pink, and yellow, or orange. Some of the species of Spondylus, as the water-clam (*Spondylus varius*), form a series of chambers by secreting successive layers of nacreous shell at a distance from each other. The genus is included in the Ostracean family of the *Acephalous Testacea*, by Cuvier. The Spondyli are eaten like Oysters; and the form of their shells is generally modified by the surface of the objects on which they grow. They are found in the Indian Ocean, the Mediterranean, and on the American coasts.

SPONGE. (*Spongia.*) A cellular fibrous tissue, or reticulated porous substance, found adhering to rocks, and produced by minute Polypi, — animals almost imperceptibly small, — which live in the sea. This tissue is covered in its recent state with a kind of thin coat of animal jelly, susceptible of a slight contraction or trembling on being touched — its only symptom of vitality. After death this soft gelatinous matter disappears. Every coast, from the Equator to the highest Polar regions, furnishes some species of Sponge; but they exist in much

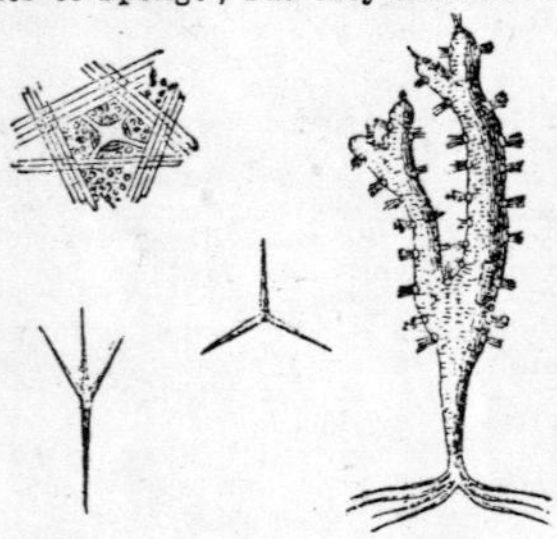

SPONGIA OCULATA; WITH ONE OF THE PORES AND TWO SPICULÆ MAGNIFIED.

greater abundance in warm latitudes than in cold, and they attain also a much greater size. "There are certain forms of organization," observes Mr. Rymer Jones, "so closely allied to both the animal and the vegetable kingdom, that it is difficult to say precisely in which they ought to be included. Such are the Sponges, which, although by common consent admitted into the animal series, will be found to be excluded, by almost every point of their structure, from all the definitions of an animal hitherto devised. What is an animal? How are we to distinguish it as contrasted with a mineral or a vegetable? The concise axiom of Linnæus upon this subject is well known, — 'Stones grow; vegetables grow and live; animals grow, live, and feel.' The capability of feeling, therefore, formed, in the opinion of Linnæus, the great characteristic separating the animal from the vegetable kingdom; yet, in the class before us, no indication of sensation has been witnessed; contact, however rude, excites no movement or contraction which might indicate its being perceived;

no torture has ever elicited from them an intimation of suffering; they have been pinched with forceps, lacerated in all directions, bored with hot irons, and attacked with the most energetic chemical stimuli, without shrinking or exhibiting the remotest appearance of sensibility. On the other hand, in the vegetable world we have plants which apparently *feel* in this sense of the word. The sensitive plant, for example, which droops its leaves upon the slightest touch, would have far greater claims to be considered as being an animal than the sponges, of which we are speaking."

We have thought proper to introduce the foregoing quotation, there being no point of dispute in natural history which has been more often or more fiercely contested than the *true nature of sponges*. That the animal and vegetable organizations both terminate obscurely toward the inorganic structures of creation, and that in this approach to their common boundary they touch and melt into each other at more than one point, must be evident to all who have given the subject the slightest consideration; and it cannot be wondered at that in this instance, where the lines of demarcation are so indistinct, different reasoners have come to different conclusions. Thus we find Dr. Johnston, who omits them in his work on British Zoophyta, asserting that they have no animal structure or individual organs, and exhibit no one function usually supposed to be characteristic of the animal kingdom. "Like vegetables," he says, "they are permanently fixed; like vegetables, they are non-irritable; their movements, like those of vegetables, are extrinsical and involuntary; their nutriment is elaborated in no appropriated digestive sac; and, like cryptogamous vegetables or algæ, they usually grow and ramify in forms determined by local circumstances, and if they present some peculiarities in the mode of the imbibition of their food, and in their secretions, yet even in these they evince a nearer affinity to plants than to any animal whatever."

We all know that the common Sponge is made up of horny, elastic fibres of great delicacy, united with each other in every possible direction, so as to form innumerable canals, which traverse its substance in all directions; and to this structure it owes its useful properties, the resiliency of the fibres composing it making them, after compression, return to their former state. But it is principally to the observations of Dr. Grant (which have been confirmed by other naturalists) that we owe the elucidation of the real character of the spongy structure, and of its vital action The dried sponge is only the skeleton of the living animal: in its original state, before it was withdrawn from its native element, every filament of its substance was coated over with a thin film of glairy semifluid matter, composed of aggregated transparent globules, which was the living part of the sponge, secreting, as it extended itself, the horny fibres which are imbedded in it. When Sponges are examined in their living state and natural condition, a constant and rapid stream of water is seen to issue from the larger orifices or vents. This stream is made apparent by the movement of the minute particles contained in it, and by the disturbance of those which may be floating in the surrounding fluid. On the other hand, it is easily made apparent that water is as constantly being imbibed through the minute pores; and that, after traversing the smaller cavities of the spongy structure, it finds its way into the canals through which it is expelled. No cause has been assigned for this movement, nor is it easy to assign any: no cilia have been discovered; and the tissues are possessed of so little contractility, that it is difficult to suppose the fluid propelled through the tubes by any mechanical influence on their part. That the nutrition and growth of the Sponges depend on the water which enters the pores, and on the minute substances it holds in solution, appears evident. And not only does this circulation of fluid answer the purposes of nutrition, but it is subservient also to the process of excretion. On watching the currents of water that issue from the vents, it is observed that minute flocculent particles are incessantly detached and carried out by them. "The growth of the Sponge is thus provided for; the living gelatinous matter continually accumulates, and, as it spreads in every direction, secretes and deposits, in the form peculiar to its species, the fibrous material and earthy spicula which characterize the skeleton." From this description of the structure of a Sponge it will be apparent that all parts of the mass are similarly organized: a necessary consequence will be, that each part is able to carry on, independently of the rest, those functions needful for existence. If, therefore, a Sponge be mechanically divided into several pieces, every portion becomes a distinct animal.

"The multiplication of Sponges, however," as is observed by the author before quoted, and to whose able work we again refer, "is effected in another manner, which is the ordinary mode of their reproduction, and forms a very interesting portion of their history. At certain seasons of the year, if a living Sponge be cut to pieces, the channels in its interior are found to have their walls studded with yellowish gelatinous granules, developed in the living parenchyma which lines them; these granules are the germs or gemmules from which a future race will spring; they seem to be formed indifferently in all parts of the mass, sprouting as it were from the albuminous crust which coats the skeleton, without the appearance of any organs appropriated to their development. As they increase in size, they are found to project more and more into the canals which ramify through the Sponge, and to be provided with an apparatus of locomotion of a description which we shall frequently have occasion to mention. The gemmule assumes an ovoid form, and a large portion of its surface becomes covered with innumerable vibrating hairs or cilia, as they are denominated, which are of inconceivable minuteness, yet individually capable of exercising rapid movements, which produce currents in the surrounding fluid. As soon, therefore, as a gemmule is sufficiently mature, it becomes detached from the nidus where it was formed, and whirled along by the issuing streams which are expelled through the fecal orifices of the parent, it escapes into the water around. Instead, however, of falling to the bottom, as so apparently helpless a particle of jelly might be expected to do, the ceaseless vibration of the cilia upon its surface propels it rapidly along, until, being removed to a considerable distance from its original, it attaches itself to a proper object, and, losing the locomotive cilia which it at first possessed, it becomes fixed and motionless, and developes within its substance the skeleton peculiar to its species, exhibiting by degrees the form of the individual from which it sprung." The uses to which the Sponges of commerce are applied are various and well known. They are usually prepared before they come to the market, by being beaten and soaked in dilute muriatic acid, with a view to bleach them, and to dissolve any adherent portions of carbonated lime. The late Dr. Johnston's work on British Sponges is an admirable book; but naturalists are looking for a more extended treatise by Dr. Bowerbank, F.R.S., F.L.S.

SPONGE-CRAB. [See DROMIA.]

SPONGILLA. The generic name for our well-known fresh-water Sponge, which is found growing in canals, especi lly in the neighbourhood of locks.

SPOONBILL. (*Platalea.*) A genus of Grallatorial birds, of which the two best known species are the *White* and the *Roseate Spoonbill;* the former appearing to be a general inhabitant of the Old Continent, and the latter of the New. In their general structure and habits they are allied to the Storks and Herons; but their beak, from which their name is derived, is long, flat, and broad throughout its length, widening and flattening more particularly at the end, so as to form a round spatula-like disc.

The WHITE SPOONBILL (*Platalea leucorodia*) is about two feet six inches in length; its beak is eight inches and a half, and dusky, with several undulated transverse ridges of black, and the tip of an orange-yellow. The feathers at the back of the head form a beautiful crest, which is of a yellowish colour. The whole of the plumage is of a pure white, except the lower part of the neck, which is yellowish buff: the naked space round the eyes and on the throat pale yellow; the base of the latter part slightly tinged with rufous. The Spoonbill frequents the borders of rivers and sea-coasts, migrating with the Herons and Storks; but in England it is now scarcely ever seen. The nest is placed on a high tree near the sea-side, where the female lays three or four white eggs, generally marked with a few pale red spots: during breeding-time this bird is very clamorous. It feeds upon small fish, frogs, snails, and aquatic insects: the flesh is high-coloured when dressed, and is said to have the flavour of a goose. The trachea is curiously formed, having a double flexure, like

WHITE SPOONBILL.
(PLATALEA LEUCORODIA.)

the figure 8, but the convolutions do not cross each other, the points of contact being united by a fine membrane.

The ROSEATE SPOONBILL (*Platalea Ajaja*) is a most elegant species, two feet three inches in length: the beak, six inches in length, and marked all round with a groove parallel to the margin, is of a grayish white, and slightly transparent, showing the ramifications of its blood-vessels: the forehead and throat are naked, and whitish. The plumage is of a fine rose-colour, deepest on the wings; the tail coverts crimson: the legs gray; the claws dusky. It inhabits Guiana, Mexico, and other parts of America; and its habits are very similar to those of the white species.

Mr. Edwards, in his 'Voyage up the Amazon,' alludes to this species as being very abundant in that wildly magnificent region, and as excelling (with the scarlet ibis, which he names with it) all the water-birds in gorgeousness and delicate colouring. "The roseate Spoonbills," he adds, "do not migrate, as do the ibises, being quite common upon the whole coast, and sometimes being seen far up the Amazon in summer. The delicate roseate of their general colouring, with the rich lustrous carmine of their shoulders and breast-tufts, as well as the singular formation of their bills, render them objects of great interest as well as beauty. They are seen fishing for shrimps and other small matters along the edges of the water, or in the mud left exposed by the ebbing tide, and, as they eat, grind the food in their mandibles moved laterally. As well as the ibis, they are exceedingly shy at every season except when breeding. They breed in the same places with the scarlet ibises and wood ibises, and the nests of the three resemble each other in every respect but in size. The eggs of the Spoonbill are from three to four, large, white, and much spotted with brown."

SPRAT. (*Clupea sprattus.*) This fish is so much allied to the Herring in all points except in size, as to be frequently mistaken for its young: its distinctive characters are, however, sufficiently evident on examination. The chief difference consists in the abdomen, which in the Sprat is more strongly serrated: the back fin is also placed farther from the head than in the Herring, and the under jaw is longer than the upper: the tail is deeply forked; the scales are large, round, and deciduous; the upper part of the head and back dark blue, with green reflections passing into silvery white on the sides and belly. In length a full-sized Sprat measures six inches, and nearly an inch and a quarter in depth. These fish are taken in considerable numbers on our coasts, particularly the southern and western; and though not so valuable as Herrings, their coming into the market in immense quantities and at a very moderate price immediately after the Herring season is over, they prove very useful as a cheap and agreeable food during the winter months. The fishing season begins early in November; and in dark, foggy nights, especially, large shoals are often taken. So great, indeed, is the supply occasionally, that many thousand tons are in some seasons sold to the Kentish hop-growers and farmers for the express purpose of being used as manure.

SPRINGBOK, or SPRINGBUCK. (*Antilope Euchore.*) Few Antelopes are more entitled to our notice than the graceful Springbok, whose name is derived from the extraordinary perpendicular leaps it makes when alarmed, or as it scours the plain. These animals are gregarious, and nothing is more remarkable than their habits of mi-

SPRINGBOK.—(ANTILOPE EUCHORE.)

gration. The vast wilds in the interior of South Africa, which they inhabit in almost incredible multitudes, are subject to seasons of such excessive drought that not a green leaf or a blade of grass is visible. When this scene of barrenness occurs, myriads of Springbucks make their way towards the fertile cultivated districts, which they literally inundate, to the great dismay of the colonist, who is compelled to drive his flocks and herds to a distant pasturage while the work of desolation is going on. They continue in the neighbourhood of the Cape for two or three months; when the rainy season has set in they return in troops of many thousands, covering the extensive plains for several hours in their passage. But their migrations are not made with impunity: lions, hyænas, and other beasts of prey make great havoc in their ranks; and the gun of the colonist is used with unerring aim. Several English travellers have witnessed and described these extraordinary marches, among whom is Mr. Pringle, who says that he once passed through one of these migratory hordes, near the Little Fish river, which whitened, or rather speckled, the country as far as the eye could reach; and he estimates the numbers at one time in view, at not less than 25,000 or 30,000. The general colour of the Springbuck is a light yellowish brown; the sides and belly divided by a broad band of chestnut, which runs down part of the shoulders: the upper part of the tail is white, the lower black; and from the tail some way up the back is a stripe of white, expansible at pleasure. When taken young the Springbuck is easily tamed, and displays the confident sportiveness of a goat.

SQUALUS: SQUALIDÆ. A genus and family of cartilaginous voracious fishes. [See SHARK.]

SQUATINA. [See ANGEL-FISH.]

SQUILLA; or MANTIS CRABS. A genus of Crustacea, belonging to the order *Stomapoda*. Its carapace only covers the anterior half of the thorax; the hinder being formed of rings like those of the abdomen. It is provided with enormous claws, terminating in a sharp hook; the last joint furnished with six sharp projecting spines, and the preceding joint with three, and so hollowed

SPOTTED MANTIS-CRAB.
(SQUILLA MACULATA.)

as to render this claw a most efficient instrument of prehension; bearing a considerable resemblance to the fore legs of the orthopterous genus *Mantis:* hence the more popular name applied to the species of this genus. The other foot-jaws, and the three anterior pairs of thoracic members, share in this conformation, and serve to hold the prey against the mouth. The three posterior pairs of legs, which are attached to the annulated portion of the thorax, are furnished with a brush instead of a hook at their extremities, and more resemble the abdominal swimming-legs. The tail is expanded into a broad fin. By the nature of its conformation we see that this animal is adapted both for seizing and holding its prey, as well as for swimming, but not at all for walking.

SQUIRREL. (*Sciurus.*) A genus of Rodent mammalia, characterized by the lower incisors being very compressed, and the tail long and bushy. From this latter member being turneu over its back when the animal is in a state of rest, the genus has derived its scientific name, *sciurus* (*skia*, a shade, and *oura*, a tail), of which the English is only a corruption. The fore feet have four toes, with a trace of an anterior thumb; the hinder feet have five distinct toes: there are four molars to each jaw, and a very small additional one in front, which soon falls. The head is large, the eyes prominent: they are active animals, ascending trees with facility. Their beauty and extreme neatness combined with their light and graceful motions have made them general favourites.

The COMMON SQUIRREL (*Sciurus vulgaris*) is completely formed for an arboreal life; and its tail is extremely long, beautiful, and spreading. Its general length from the nose to the tip of the tail is about fifteen inches. The ears are terminated by long tufts of hair; the colour of the head, body, tail, and legs, is a bright reddish brown; the belly and breast are white; the eyes large, black, and sparkling; the fore feet strong, sharp, and well adapted to hold its food; the legs short and muscular; the toes long; and the nails sharp and strong. The upper lip is cleft; the fur short and silky; there are four molar teeth on each side of the lower jaw, and five in the upper, the first of which is only a small tubercle; and the incisors are two in each jaw. When on the ground, they move by successive leaps, with the tail extended and undulating; when sitting, the tail is elevated over their backs like a plume; but the forest is their home, and they display wonderful agility in leaping from bough to bough. The Squirrel lives upon nuts, acorns, beech-mast, the bark of young trees, leaf-buds, and tender shoots. Like the hare and rabbit, it generally sits on its hinder legs, using its fore paws to convey its food to the mouth. It is most provident in laying up its winter stores, not merely in one place of safety, but in several holes of trees, in the immediate neighbourhood of its own retreat; and there vast magazines of nuts and acorns are to be found in that dreary season when the trees are divested both of their fruits and foliage. The Squirrel's nest is constructed with great art, of moss, twigs, and dry leaves, curiously interlaced, and usually placed either in the hollow of a tree, or in the fork between two branches, where it is well defended from the weather, and can be least easily discovered. Here, generally in May or June, the little animal brings forth its young, the young family rarely consisting of more than four or five. The Squirrel never appears in the open fields, but keeps among the tallest trees, and avoids as much as possible the habitations of men. It is so extremely vigilant, that if the tree in which it resides be only touched at the bottom, it instantly takes the alarm, quits its nest, leaps to another and another tree, and thus travels on till it finds itself in perfect security; and it returns to its home by similar arboreal paths, unattainable by any other quadruped. Their agility is such that it is a very difficult thing to shoot a Squirrel in motion. "They have been seen, when hard pressed, and when the distance to the next tree has been beyond their most extravagant leaps, to throw themselves off, spreading abroad their limbs so as to make their body as parachute-like as possible to break their fall; and on reaching the ground without harm, bound along for the few intervening paces, and ascend the tree with a celerity almost too quick for the eye to follow." The Common Squirrel inhabits Europe, North America, and the northern and temperate parts of Asia. In Sweden and Lapland the colour changes to gray in the winter season; in Siberia it is often seen entirely white; and even in this country some slight variation in the colour is observable; sometimes, indeed, it is found with the tail milk-white, all the other parts being of the usual colour. When in captivity the Squirrel may be said to be always in motion.

The GREY SQUIRREL (*Sciurus Carolinensis*) is extremely common in North America, especially in oak, hickory, and chestnut forests. Formerly it was so abundant in many districts as to become a scourge to the inhabitants. Its colour is usually a fine bluish-gray, mixed with a slight tinge of orange, and the tail is edged with white. It is a small species, remarkable for its beauty and activity, and when kept in confinement is exceedingly playful and mischievous.

The FOX SQUIRREL (*Sciurus vulpinus*) is a large species, and inhabits, exclusively, the pine forests of the Southern States of America. The body is fourteen inches in length, and the tail sixteen. The colour is gray and black, or mottled.

The CAT SQUIRREL (*Sciurus cinereus*) is distinguished by the fineness of the texture of its fur. The length of the body is twelve inches, and of the tail fourteen: the colour cinereous above, and white beneath: the tail is less distichous than in the others, and striped with black. It has four molar teeth only on each side of the upper jaw.

The RED SQUIRREL, or HUDSON'S BAY SQUIRREL, (*Sciurus Hudsonius*), is a beautiful species, inhabiting the pine-forests of Hudson's Bay and the Northern States of America. It is marked along the middle of the back with a ferruginous line from head to tail; the sides are paler; and the belly is of a pale ash-colour, mottled with black. The tail, which is neither so long nor so bushy as that of the common kind, is of a ferruginous colour, barred with black; and towards the tip has a broad belt of the same colour. In size it is somewhat less than the European Squirrel.

The BARBARY SQUIRREL. (*Sciurus getulus.*) This is a native of Barbary and other parts of Africa, living usually in palm trees. It has full black eyes, with white orbits; the head, feet, body, and tail are cinereous, inclining to red; the sides are longitudinally marked with two white stripes; the belly is white; and the tail, which is bushy, is regularly marked with shades of black. It is about the size of the common Squirrel.

The PALM SQUIRREL. (*Sciurus palmarum.*) This species has acquired its name from its being commonly seen frisking about palm trees in the East Indies. It principally feeds on fruit, and is said to be very

PALM SQUIRREL.—(SCIURUS PALMARUM.)

fond of the palm wine, which is extracted from the cocoa trees. The female lays her young in the holes of old walls. This species is not unfrequently brought alive to this country.

The PLANTAIN SQUIRREL. (*Sciurus bilineatus.*) This pretty species is a native of Java, and is constantly kept by the Javanese as a pet. One which Mr. Adams obtained when visiting that country in the Samarang, he describes as "an amusing little animal, full of frolic, and as playful as a kitten. He never carried his tail over his back, like the greater number of his consimilars, but would trail it gracefully along the ground. When angry, he would dilate this ornamental appendage, and bristle up the hairs, like an irritated cat. His natural cry was a weak chirping sound, but when teazed beyond his powers of endurance, he would make a sharp, low, and passionate noise. He seemed to court caresses, and would receive them with pleasure. His food consisted of bananas and cocoa-nuts, which he would usually nibble like a rat, though sometimes he would place it between his paws. He was a remarkably cleanly little creature, continually dressing his fur in the manner of the *Felinæ.* When he slept, he rolled himself up like the dormouse, with his tail encircling his body. Always active and blithe, he would sometimes perform feats of extraordinary agility, bounding to great distances, and clinging to every object within his reach."

STAG. [See DEER, RED.]

STAG-BEETLE. In the article LUCANUS we have entered at some length on a description of the most common genus of the *Stag-beetle.* We may here refer our readers to two or three of the most remarkable genera of exotic *Lucanidæ.* In Australia we find the genus *Lamprima*, in which the prevalent colouring is metallic green; the mandibles are short, and clothed with hairs on the inside. In New Zealand occurs the genus *Dendroblax*, which at first sight resembles a small species of Dynastes. In South America we meet with the genus *Pholidotus*, in which the body is covered with scales. On the island of Chiloë the truly remarkable genus *Chiasognathus* occurs, in the male of which the mandibles are longer than the body, bent down towards the tip, where they are reflexed; on their under side, at the base, there is a long horn. Our figure will illustrate the form of this curious insect better than the longest description; that with the long jaws shows the

CHIASOGNATHUS GRANTII—MALE.

male, while the figure of the thorax and head represent those of the female, in which the

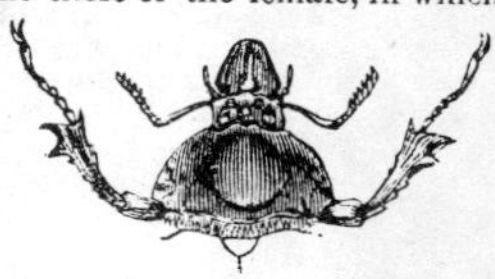

HEAD AND THORAX OF FEMALE (C GRANTII.)

mandibles are very short. Another curious genus allied to this is *Sphenognathus*, a native of Columbia.

STAPHYLINUS: STAPHYLINIDÆ. A genus and family of Coleopterous insects. They have in general the head large and flat, strong mandibles, antennæ short, the thorax as broad as the abdomen, the elytra truncate at the tip, but still covering the wings, which are of ordinary size. The species are usually found under dead leaves, stones, dung, &c. Our figure represents the *Staphylinus* (*Ocypus*) *olens*, a large and characteristic species of the family, which is very common in this country, and by many known familiarly as the "DEVIL'S COACH HORSE."

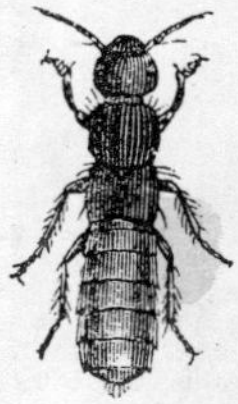

STAPHYLINUS OLENS.

Dr. Erichson, of Berlin, has published, in one volume, a monograph of all the species. Many of them are found in Great Britain.

STAR-FISHES. [See ASTERIAS, p. 41.]

STARLING. (*Sturnus.*) A genus of Passerine birds, having the beak compressed, particularly towards the point, which is blunt and nail-like. Of these, the best known species is the COMMON STARLING (*Sturnus vulgaris*), which is about the size of the blackbird: the bill is straight, sharp-pointed, and of a yellowish brown; in old birds deep yellow: the whole plumage dark, glossed with green, blue, purple, and copper, each feather being marked at the end with a pale yellow spot: the wing-coverts are edged with yellowish-brown; the quill and tail-feathers dusky, with light edges: the legs are reddish brown. The Starling is an inhabitant of almost every climate; and as it is a familiar bird, and easily trained in a state of captivity, its habits have been more frequently observed than those of most other birds. They make an artless nest in hollow trees, the eaves of old houses, towers, and cliffs overhanging the sea. In the autumn they fly in vast flocks, and may be known at a great distance by their whirling mode of flight. So attached are they to society, that they not only join those of their own species, but also birds of a different kind, and are frequently seen in company with Redwings, Fieldfares, Jackdaws, &c. Their principal food consists of worms, snails, and caterpillars; they also eat various kinds of grain, seeds, and berries; are said to be particularly fond of cherries; and are accused of breaking and sucking the eggs of other birds. They are very docile in confinement, and may be easily taught to repeat short phrases, or whistle tunes, with great exactness, — their powers of imitation being considerable.

STARLING.—(STURNUS VULGARIS.)

"The Starling shall always have a friend in me," says that genuine defender of the feathered race, Charles Waterton, Esq. "I admire it for its fine shape and lovely plumage; I protect it for its wild and varied song; and I defend it for its innocence." "There is not a bird in all Great Britain more harmless than the Starling: still it has to suffer persecution, and is too often doomed to see its numbers thinned by the hand of wantonness or error. The farmer complains that it sucks his pigeons' eggs, and, when the gunner and his assembled party wish to try their new percussion cap, the keeper is ordered to close the holes of entrance into the dovecot overnight; and the next morning three or four dozen of Starlings are captured to be shot: while the keeper, that slave of Nimrod, receives thanks, and often a boon, from the surrounding sportsmen, for having freed the dovecot from such a pest. Alas! these poor Starlings had merely resorted to it for shelter and protection, and were in no way responsible for the fragments of egg-shells which were strewed upon the floor. These fragments were the work of deep-designing knaves, and not of the harmless starling. The rat and the weasel were the real destroyers; but they had done the deed of mischief in the dark, unseen and unsuspected; while the stranger Starlings were taken, condemned, and executed, for having been found in a place built for other tenants of a more profitable description."

We take leave also to add a few lines respecting this bird from the 'Journal of a Naturalist.' "They vastly delight, in a bright autumnal morning, to sit basking and preening themselves on the summit of a tree, chattering all together in a low song-like note. There is something singularly curious and mysterious in the conduct of these birds previous to their nightly retirement, by the variety and intricacy of the evolutions they execute at that time. They will form themselves, perhaps, into a triangle, then shoot into a long, pear-shaped figure, expand like a sheet, wheel into a ball, as Pliny observes, each individual striving to get into the centre, &c., with a promptitude more like parade movements than the actions of birds. As the breeding season advances, these prodigious flights divide, and finally separate into pairs, and form their summer settlements; but probably the vast body of them leaves the kingdom."

A second species is found in the south of Europe, and is distinguished from the former by its uniform colour, wanting the whitish spots, and having the feathers longer and more pointed. This is the *Sturnus unicolor*.

STAUROPUS. A genus of nocturnal Lepidoptera, containing the

STAUROPUS FAGI, or LOBSTER MOTH. This Moth is found in various parts of the south of England, but is comparatively rare. It varies from two inches and

LOBSTER-MOTH.—(STAUROPUS FAGI.)

a quarter to three inches in expanse; its colour a dull grayish-brown, with the fore-wings varied towards the base and hind margin with reddish: the base is pale, succeeded by a broad dark bar, with several paler patches occupying the middle of the wing, followed by a waved and toothed pale stripe. The apical portion of the wing is paler, with a row of small black sub-marginal dots, preceded on the costa by a luteous stripe: the hind wings are brown, with an angulated pale stripe running from the outer margin half-way across the middle of the wings. The Caterpillar is rusty gray, or

CATERPILLAR OF LOBSTER-MOTH. (STAUROPUS FAGI.)

fawn colour: the cocoon is closely woven, and more resembles silver paper than any other material. The perfect insect appears in June and July, and the caterpillar in the autumn. It feeds on the hazel, alder, sloe, &c.

STEENBOK. (*Antilope tragulus.*) Few of the Antelope tribe are more graceful than the Steenbok. Its body is well made and compact; its legs long and slender; its head small and well formed; and its tail scarcely perceptible. The length of this animal is about three feet six inches, and the height at the shoulder under twenty inches. The upper parts of the body are of a reddish-fawn colour; the hair on shoulders, back, and sides appearing to be tipped with a silvery hue: the nose and legs are dark brown; the breast, belly, and inner parts of the limbs white: but what most conspicuously marks this species is a black line which passes from the root of each horn backwards, uniting between the ears, and forming an obtuse angle. The horns are smooth, polished, and finely pointed; the ears very long and broad. It inhabits the stony plains and rocky hills of South Africa; is particularly shy, and runs with remarkable swiftness. It is much hunted on account of the delicacy of its flesh, which is esteemed excellent venison.

STELLERIDÆ, or STAR-FISHES. [See ASTERIAS, p. 41.]

STELLIO. The name of a genus of Saurians belonging to the Iguana family. They are characterised chiefly by having the tail encircled with rings of large scales that are often spinous.

STEREOGNATHUS. [See SUPPLEMENT.]

STERLET. (*Accipenser ruthenus.*) The smallest species of Sturgeon, being from two to three feet in length: it is found in the Volga and some other Russian rivers, and is considered a great delicacy. The caviar made from this fish is confined almost exclusively to the use of the royal table.

STICKLEBACK. (*Gasterosteus.*) A genus of Acanthopterygious fishes, comprising several species, which differ principally in their number of spines, and are named accordingly. The THREE-SPINED STICKLEBACK (*Gasterosteus aculeatus*) is found in almost every river, brook, and pond. It seldom grows to the length of two inches and a half: the eyes are large; the belly is prominent; the body near the tail is square; and the sides are covered with large bony plates, placed transversely. On the back there are three sharp spines, capable of elevation or depression at pleasure; the dorsal fin is placed near the tail; the pectoral fins are broad; the ventral spine triangular at the base; and a small fold of skin forms a horizontal crest on each side of the tail. The colour of the back is green; the cheeks, sides, and belly silvery white; but in some the lower jaw and belly are of a bright crimson. These fishes are sometimes so plentiful in the fens of Lincolnshire and Cambridgeshire as to be collected and used for manuring the land in their vicinity. The males are exceedingly pugnacious, and they use their spines with such fatal effect, that one occasionally rips up and kills the other.

THREE-SPINED STICKLEBACK. (GASTEROSTEUS ACULEATUS.)

The other species deserving mention is the FIFTEEN-SPINED STICKLEBACK (*Gasterosteus spinachia*), which is of a more elongated form than the others, and is common around our coast, and in the Baltic; seldom, however, ascending rivers. Though less active than its brethren of the fresh water, it is scarcely less voracious, devouring the fry of other fishes, crustaceans, &c. It keeps near rocks and stones covered with sea-weeds, among which it takes refuge on any alarm; but is very pugnacious, and seldom loses any opportunity that presents itself of displaying its natural ferocity. It spawns in spring, and the young, less than half an inch long, are seen in considerable numbers in summer at the margin of the sea.

It has been satisfactorily ascertained that this species of Stickleback constructs a nest wherein to deposit its spawn, and guards it with watchful care till the young fry make their appearance. In our article "FISHES" a circumstantial account of this fact, as given by Mr. Couch, appears; and in the 'Transactions of the Berwickshire Naturalists' Club' the fact is further confirmed by Mr. Duncan and the Rev. Mr. Turnbull. They say, "These nests are to be found in spring and summer, on several parts of our coasts, in rocky and weedy pools between tide-marks. They are about eight inches in length, and of an elliptical form or pear-shaped, formed by matting together the branches of some common *fucus*, as, for example, the *fucus nodosus*, with various confervæ, alvæ, the smaller Florideæ, and corallines. These are all tied together in one confused compact mass, by means of a thread run through, and around, and amongst them in every conceivable direction. The thread is of great length, as fine as ordinary silk, tough, and somewhat elastic; whitish and formed of some albuminous secretion. It is evident that the fish must first deposit its spawn amid the growing fucus, and afterwards gather its branches together around the eggs, weaving and incorporating at the same time all the rubbish that is lying or floating around the nucleus. They were narrowly watched for some weeks, and it was observed that the same fish was always in attendance upon its own nest. During the time of hope and expectation, they become fearless, and will allow themselves to be taken up by the hand repeatedly. There can be no doubt that their object in remaining near the nest is to guard it against the attacks of such animals as might feel inclined to prey upon its contents.

STILIFER. A genus of Molluscous animals, one of which was discovered by Arthur Adams, Esq., of H. M. S. Samarang, living on the body of a starfish (*Asterias*) on the coast of Borneo. It has two elongate subulate tentacles, with the eyes sessile near the outer side of their base, and a small rounded head. The mantle is entirely enclosed and covered by the thin shell, and the foot is narrow, slender, very much produced beyond the head in front, and scarcely extended at all behind.

STIPITURUS, or SOFT-TAILED FLY-CATCHER. (*Stipiturus malachurus.*) This curious species of bird inhabits Australia. The beak, which is dark brown, is furnished with strong bristles: the general colour of the plumage is ferruginous, but the feathers of the upper parts of the body and wings are streaked down the middle with brownish black: over the eyes, arising at the base of the beak, is a pale blue streak; throat and fore part of the neck of the same blue colour: the feathers of the rump are soft, long, and silky; wings short, nearly reaching to the base of the tail, which is upwards of four inches long; the shafts very slender and black, the webs on each side consisting of minute slender hairy black filaments, placed at distances, and distinct from each other, as in the feathers of the Cassowary. It is fond of marshy places, abounding with long grass and rushes. When disturbed, its flight is very short, and it runs on the ground with great swiftness.

STOAT. [See ERMINE.]

STILT, or STILT PLOVER. (*Himantopus.*) A genus of wading birds, remarkable for the extreme length and slenderness of their legs, and for the peculiar form of the bill, which is round, slender, and pointed. Stilts, though not numerous, are found in every quarter of the globe; the species which occasionally visits England and Western Europe being spread throughout Asia and Africa — two others being met with in America, and one in Australia. The European species is white, with a black calotte and mantle, and red legs. They are destitute of a hind toe, and their long legs are so feeble as to render walking a painful effort to them; but they fly with great swiftness, and swim well. They frequent marshes, shallow lakes, &c.; and feed upon minute shell-fish, insects, crustacea, &c. In constructing their nests, six or eight pairs of birds unite to build a platform, by which the nests may be raised above the middle of the water. [See HIMANTOPUS.]

STOCK-DOVE. (*Columba œnas.*) From the Stock-dove, or common WILD-PIGEON, it was once supposed that most of the beautiful varieties of the *Columbidæ*, which in a state of domestication are dependent upon man, derived their origin: hence the name *Stock*-dove. It is, however, now believed that the Rock-dove (*Col. livia*) is the parent stock. It is fourteen inches in length: the bill is pale red; the head, neck, and upper part of the back are of a deep bluish gray, reflected on the sides of the neck with glossy green and gold; breast pale reddish purple; the lower part of the back and the rump light gray or ash, as are also the belly, thighs, and under tail-coverts; the primary quill feathers are dusky, edged with white, the others gray, marked with two black spots on the outer webs; tail ash-gray, tipped with black: legs and feet red; claws black. The pigeon-fancier's art has been carried so far as to produce an almost endless variety of tints among the various do-

STOCK-DOVE.—(COLUMBA ŒNAS.)

mesticated species. Wild Pigeons are said to migrate into England at the approach of winter, from the northern regions, and return in the spring; many of them, however, remain in this country. They build in the hollows of decayed trees, and usually have two broods in the year; but in a state of domestication they generally breed every month; and although they only lay two eggs at a time, if all were suffered to live their increase in a few years would become enormous. The male and female perform the office of incubation by turns, and feed their young by casting up the provisions which they have treasured up in their capacious crop. At first the young are served with food considerably macerated; but as they grow older, the parents gradually diminish the trouble of preparing it; and at length send forth the young birds to provide for themselves. However, when they have plenty of provisions, it is not uncommon to see young ones almost fit for flight, and eggs hatching at the same time in the identical nest.

STOMAPODA. An order of the class *Crustacea*, all the species of which are marine, and the largest only found in tropical climates. In many of the animals composing this Order the feet approach the mouth. The general form of the body bears considerable resemblance to that of the Crayfish and its allies; the abdomen being much prolonged, the tail-fin much expanded laterally, and the appendages beneath the abdomen being developed and used as fin-feet. As they inhabit the deep parts of the sea, their habits are not well known, but they are supposed to be voracious. [See SQUILLA, and OPOSSUM SHRIMP.]

STOMATIA. A genus of Mollusca, the shells of which are auriform, but distinguished from *Haliotis* by being destitute of the series of holes; mouth large, oblong, interior pearly. They are found in the East Indian seas, and in those of Australia.

STONECHAT. (*Saxicola rubicola.*) This bird, which belongs to the same family as the Robin Red-breast, is nearly five inches in length; and is chiefly found on wild heaths and commons, where it feeds on small worms and all kinds of insects. The bill is black; the head, neck, and throat black, faintly mixed with brown; on each side of the neck, immediately above the wings, there is a large white spot; the back and wing-coverts are of a fine velvet black, edged with brown; the quills next the body are white at the bottom, forming a spot of that colour on the wings; the breast is a bright bay; the rump white; tail and legs black. This solitary bird builds at the roots of bushes, or underneath stones, carefully concealing the entrance of the nest, and using a variety of arts to prevent any one from tracing it to its retreat. It is almost continually on the wing, flying from bush to bush, and resting but for a few seconds at a time. The sound of its note has been thought to resemble the clicking of two stones together, which circumstance has been given as the origin of its name.

STONE CURLEW. (*Œdicnemus crepitans.*) This bird is also called the Whistling or Norfolk Plover, and belongs to the order *Grallatores*. It is larger than the Woodcock, the expansion of the wings being three feet. It has a straight bill, two inches long, black towards the base, and yellow at the tip. Under each of the eyes there is a bare space, of a yellowish green: the breast and thighs are a yellowish white; the middle of the back, the head, and the neck are black, edged with a reddish ash-colour; on the quill-feathers there are transverse white spots; and some of the wing-feathers, which are tipped with white, appear beautifully mottled. The tail is about six inches long, and variegated like the wings; the legs are long and yellowish; the claws small and black. This bird has no hind toe, and those before are united by a small membrane. It is a native of several English counties, particularly Norfolk. It is rapid on foot, and powerful in flight, which it executes in wide circles; and it is remarkable for its piercing shrill note, which it sends forth in the evening. It lays two eggs of a dirty white, marked with spots and streaks of a deep reddish colour; feeds on slugs, worms, and caterpillars; and its flesh, when young, is considered delicious.

White of Selborne observes in a letter to Pennant, "I wonder that the Stone Curlew, *Charadrius Œdicnemus*, should be mentioned by writers as a rare kind: it abounds in all the campaign parts of Hampshire and Sussex, and breeds, I think, all the summer, having young ones, I know, very late in the autumn. Already (March 30.) they begin clamouring in the evening. They cannot, I think, with any propriety, be called, as they are by Mr. Ray, 'circa aquas versantes;' for with us, by day at least, they haunt only the most dry, open, upland fields and sheep-walks, far removed from water: what they may do in the night I cannot say. Worms are their usual food, but they also eat toads and frogs. It lays its eggs, usually two, never more than three, on the bare ground, without any nest, in the field; so that the countryman, in stirring his fallows, often destroys them. The young run immediately from the egg, like partridges, &c., and are withdrawn to some flinty field by the dam, where they skulk among the stones, which are their best security; for their feathers are so exactly of the colour of our gray-spotted flints, that the most exact observer, unless he catches the eye of the young bird, may be eluded. . . . *Œdicnemus* is a most apt and expressive name for them, since their legs seem swollen like those of a gouty man. After harvest I have shot them before the pointers in turnip-fields." This bird appears to be pretty generally distributed throughout Europe; in the south of France and in Italy it is abundant: and in many parts, as in Britain and Germany, it is migratory; but it is seldom met with in the northern counties of England, and scarcely ever in Scotland.

STORK. (*Ciconia.*) A genus of large Grallatorial birds. In most countries Storks are held in great esteem by the inhabitants, as they tend to prevent the increase of noxious vermin by destroying great numbers, all the species being extremely voracious. They reside in marshy places, where their chief food (reptiles, worms, and insects) is found; and they migrate in large flocks to immense distances, returning regularly to their former habitations. They have no voice, but produce a clattering with their bills, by striking the mandibles together. Among the ancients, to kill them was considered a crime, which, in some places, was punished even with death; and, like the Ibis, this bird became the object of worship. The Stork is remarkable for its great affection towards its young, but more especially for its attention to its parents in old age.

The WHITE STORK (*Ciconia alba*) is upwards of three feet six inches long. The head, the neck, and the whole of the body are of a pure white; the scapulars and wings black: the bill, seven inches in length, is of a fine red colour; and the legs and bare part

WHITE STORK.—(CICONIA ALBA.)

of the thighs are also red. The neck is long and arched; and the feathers near the breast are long and pendulous. The Stork inhabits various parts of the temperate regions of the Old Continent, though it rarely visits England. The nest is made of dry sticks, twigs, and aquatic plants, sometimes on large trees, or the summits of high rocky cliffs: this, however, says Bewick, seldom happens, for the Stork prefers the neighbourhood of populous places, where it finds protection from the inhabitants; who, for ages, have regarded both the bird and its nest as sacred, and commonly place boxes for them on the tops of the houses wherein to make their nests; to which they return, after the most distant journeys, and every Stork takes possession of his own box. When these are not provided for them, they build on the tops of chimneys, steeples, and lofty ruins. The Stork lays from two to four eggs, the size and colour of those of a goose, and the male and female sit by turns. They are singularly attentive to their young, both together never quitting the nest, but each by turns bringing provisions for them. Their food consists of serpents, lizards, frogs, small fish, &c. In their migrations these birds avoid the extremes of heat and cold; never being seen in summer farther north than Russia or Sweden, nor in winter farther south than Egypt, where it is constantly seen during that season. Before they take their departure they assemble in large flocks, making a clattering noise, and appearing to be all bustle and consultation; but when they are actually about to leave, the whole body become silent, and move at once, generally in the night. The flesh of this bird is very rank, and not fit for food.

The BLACK STORK. (*Ciconia nigra.*) This species is not so large as the preceding, being about three feet in length. Its head, neck, the whole of the upper parts of its body, wings, and tail are dusky, with green and purple hues; the under parts of the breast and belly are pure white; the beak, the naked skin about the eyes, and throat, are deep red; as also are the legs. The Black Stork inhabits many parts of Europe, but is not so common as the white. It is said to be a solitary bird, frequenting the most sequestered places to breed: it builds on trees, laying two or three eggs, of a dull white, shaded with green, and slightly marked with brown spots. Like the White Stork, its flesh is wholly unfit for food.

The AMERICAN STORK. (*Ciconia maguari.*) There is little difference in size between this species and the common White Stork: the head, neck, back, tail, and all the under parts of the body are of a pure white: the feathers at the base of the neck are long and pendent; the wings and upper tail-coverts are dusky, glossed with green; a large naked space on the upper part of the throat, which is capable of dilatation, is of a fine vermillion hue, as is also the skin which surrounds the eyes. The beak is greenish-yellow; the feet red, and the claws brown. This bird inhabits various parts of America, and is said to be good food.

STRATIOMIDÆ. A family of Dipterous insects, which in the perfect state are generally found, in damp situations, upon flowers, sucking their sweets. They are mostly prettily coloured, and some of the species have beautiful metallic tints. The larvæ of some are aquatic, whilst others are found under ground, in dung, or the rotten detritus of wood; but they all agree in retaining the larva skin in its original form during their existence in the pupa state.

STORMY PETREL. [See PETREL.]

STREPSIPTERA. The name given by Kirby to an order of insects (consisting only of a single family, *Stylopidæ*) which possess rudimental elytra in the form of linear and spirally twisted scales. The species composing this order are all of small size, none of them reaching a quarter of an inch in length. The body is long and narrow; the thorax large and singularly developed; mouth with two slender acute jaws, wide apart, and two biarticulate palpi; anterior wings transformed into a pair of short, slender, contorted appendages; posterior wings very large, folding longitudinally like a fan. The head is distinct and exposed: it is transverse, with the eyes very large, lateral, and prominent, being placed upon the contracted sides of the head, which gives them the appearance of being inserted upon short footstalks. The number of hexagonal facets is small, and they are singularly separated from each other by a septum or partition, which, being elevated above the lenses, gives the eyes a cellular surface. In *Elenchus tenuicornis* Mr. Templeton could detect only about fifteen lenses in the eyes, which are quite sessile. The antennæ are of singular construction, although consisting of but a few joints. Mr. Kirby (says Mr. Westwood) noticed the analogy which existed between the antennæ of some of these insects and many Coleoptera and Hymenoptera which have branching or furcate antennæ. But it is to be observed, that, with the exception of a very few, antennæ thus constructed are found only amongst male insects; and hence it appears not improbable that all the winged individuals of this order yet discovered are males, all exhibiting a complicated structure in their antennæ. The true wings, which are very large and membranous, are attached at the anterior lateral angles of the metathorax, and can be folded up longitudinally at the sides of the body. The legs are moderately long, and rather weak, but the coxæ of the two anterior pairs are elongated, giving them considerable powers of motion. The femora are simple; the tibiæ not furnished with spurs; and the tarsal joints are furnished beneath with large fleshy cushions, without any terminal ungues. These insects, in their early states, are parasitic in the bodies of various bees and wasps; the larva, when full-grown, protruding its head between the abdominal segments of these insects, appearing, at first sight like a small flattened acarus.

Mr. Kirby's account of the discovery of these insects, and of the bursting forth of the imago, is, in Mr. Westwood's opinion, so interesting, that he gladly avails himself of the following extract. After mentioning that he had repeatedly observed something upon the abdomen of various Andrenæ, which he had at first regarded as a kind of acarus, he at length determined to examine and describe one of them: "But what was my astonishment when, upon attempting to disengage it with a pin, I drew forth from the body of the bee a white fleshy larva a quarter of an inch long, the head of which I had mistaken for an acarus. How this animal receives its nourishment seems a mystery. Upon examining the head under a strong magnifier, I could not discover any mouth or proboscis with which it might perforate the corneous covering of the abdomen, and so support itself by suction: on the under side of the head, at its junction with the body, there was a concavity; but I could observe nothing in this but a uniform unbroken surface. As the body of the animal is inserted in the body of the bee, does that part receive its nutriment from it by absorption? After I had examined one specimen, I attempted to extract a second; and the reader may imagine how greatly my astonishment was increased when, after I had drawn it out but a little way, I saw its skin burst, and a head as black as ink, with large staring eyes and antennæ, consisting of two branches, break forth, and move itself briskly from side to side. It looked like a little imp of darkness just emerging from the infernal regions. I was impatient to become better acquainted with so singular a creature. When it was completely disengaged, and I had secured it from making its escape, I set myself to examine it as accurately as possible; and I found, after a careful inquiry, that I had not only got a nondescript, but also an insect of a new genus, whose very class [order] seemed dubious."

"In the perfect state, these insects are but short-lived, delicate creatures. Mr. Dale, who has been very fortunate in his discoveries of this order, thus describes the proceedings of one which he caught flying, on the 7th of May, over a quickset hedge of a garden. 'It looked milk-white on the wing, with a jet-black body, and totally unlike any thing else; it flew with an undulating or vacillating motion amongst the young shoots, and I could not catch it till it settled on one, when it ran up and down, its wings in motion, and making a considerable buzz or hum, as loud as a Sesia; it twisted about its rather long tail, and turned it up like a Staphylinus. I put it under a glass, and placed it in the sun; it became quite furious in its confinement, and never ceased running about for two hours. The elytra, or processes were kept in quick vibration as well as the wings; it buzzed against the sides of the glass, with its head touching it, and tumbled about on its back. By putting two bees (*Andrena labialis*) under a glass in the sun, two Stylops were produced: the bees seemed uneasy, and went up towards them, but evidently with caution, as if to fight; and moving their antennæ towards them, retreated. I once thought the bee attempted to seize it but the oddest thing was to see the Stylops get on the body of the bee and ride about, the latter using every effort to throw his rider." These insects appear at different times of the year, and seem widely distributed.

To the foregoing we should add, that in the 'Anniversary Address delivered at the Entomological Society, Feb. 10. 1845, by the President, G. Newport, Esq., F. R. S.,' it was shown, from the discoveries of Dr. Siebold of Erlangen, that the *Strepsiptera* undergo a singular metamorphosis; that the males and females differ from each other, the metamorphosis of the males being complete, they alone being furnished with wings: the females, on the contrary, have neither legs, wings, nor eyes, and greatly resemble larvæ. These females are viviparous, and never quit the bodies of the Hymenoptera in which they live as parasites. The young Strepsiptera, at the moment that they burst the eggs in which they are developed within the body of the parent, have six legs, and are furnished with organs of manducation.

STRIGOPS. A remarkable genus of Scansorial birds belonging to the Parrot family, which at first sight has a strong resemblance to an Owl. It was established by Mr. G. Gray for the reception of a species in the British Museum, to which he has given the name of *Strigops habroptilus*. It is of a greenish colour, mottled and streaked with black; and from a letter of Mr. Strange, read at the Zoological Society, it appears that in New Zealand, where it is very rare, it is called *Kakapo*, and is nocturnal in its habits; resorting in the day-time to burrows formed under the roots of trees, or to large masses of rock. It feeds on the roots of the fern, and on the outer covering of the *Phormium tenax*, or New Zealand flax. The French Museum has subsequently obtained a specimen from Stewart Island, to the south of New Zealand. A figure of it is given in 'L'Illustration,' December 4. 1847.

STRIGIDÆ. The name of the family of Nocturnal birds of which the Owl (*Strix*) is the type. [See OWL.]

STROMBUS: STROMBIDÆ. A genus and family of Mollusca, for the most part found in the seas of tropical countries, inhabiting large and thick oval shells. The head of the animal is furnishedwith a proboscis and two short tentacula; and the eyes are situated on a lateral peduncle longer than the tentacula itself. Spire of the shell moderate; mouth long, and rather narrow, terminated by a canal more or less long and recurved; right lip dilated in the adult, and having a small notch or sinus near the canal; left lip sometimes thickened; operculum horny, long, and narrow. In many species the spire is quite hidden by the expansion of the outer lip. In the *Strombus gigas*, a very large species, which is caught for the table, pearls are said to be occasionally, though very rarely, found. Mr. Wood, in

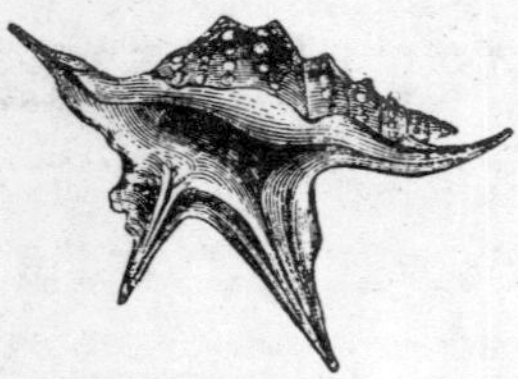

PELECAN'S-FOOT SHELL.
(STROMBUS [APORRHAIS] PES-PELECANI.)

his 'Zoography,' relates that he saw a pink pearl, weighing twenty-four grains, taken from the body of one of this species that was caught off the island of Barbadoes. As an example of this group of shells we have subjoined a figure of the *Strombus* (*Aporrhais*) *Pes-Pelecani*, or Pelecan's-foot shell, which has received its name from the processes round the mouth being arranged and connected much as in the foot of that well-known bird. Some of the Strombs are used to make artificial cameos. [See HELMET SHELL.]

STRUTHIO. [See OSTRICH.]

STRUTHIOLARIA. A genus of marine Mollusca, found in New Holland and New Zealand. The shells are oval, in shape like a *Buccinum*, but differ in having a thickened lip; the spire is elevated; mouth oval, terminated by a very short straight canal; no varices; operculum horny: they are both rare and singular.

STRUTHIONIDÆ. The name of a family of large birds, incapable of flight, having mere rudimentary wings, but long and strong legs; including the Ostrich, the Cassowary, and other congeneric species.

STURGEON. (*Accipenser.*) A genus of large Cartilaginous fish, allied somewhat to the Shark and Ray, but differing essentially in structure, as well as in habits. There are several species.

The COMMON STURGEON (*Accipenser sturio*) is generally about six feet long, but sometimes attains to the length of eighteen. It inhabits the Northern, European, and American seas, migrating during the early summer months into the larger rivers and lakes, and returning to the sea again in autumn, after having

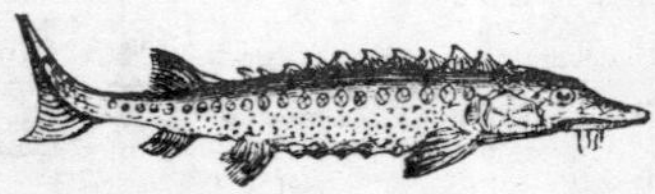

COMMON STURGEON.—(ACCIPENSER STURIO.)

deposited its spawn. Its form is long and slender, gradually tapering towards the tail, and covered throughout the whole length by five rows of strong, large, bony tubercles, rounded at the base, and terminated above by a sharp curved point in a reversed direction. The body of the Sturgeon is more or less covered with bony plates, arranged in longitudinal rows; and the head is armed in a similar manner: the snout is long and slender, obtuse at the tip, and furnished beneath, at some distance from the end, with four long worm-shaped cirri: the mouth, placed under the elongated muzzle, is small and toothless; and the palatal bones form the upper jaw; the air-bladder is very large, and communicates by a wide opening with the gullet. The pectoral fins are oval, and middle-sized; the dorsal small, and situated very near the tail; the ventral and anal fins are also small, and placed nearly opposite the dorsal. The tail is lobed or slightly forked, the upper lobe extending far beyond the lower. The general colour is cinereous above, with dusky specks, and yellowish-white beneath; and the tops of the tubercles are of a similar cast. Though generally considered as a fish of slow motion, it is sometimes seen to swim with great rapidity, and also to spring out of the water with great force at intervals. It is rarely taken at any great distance from shore, but frequents such parts of the sea as are not remote from the estuaries of large rivers. In North America they appear in great abundance during the early summer months. The flesh of the Sturgeon is white, delicate, and firm: it is said to resemble veal, when roasted; but it is generally eaten pickled, and the major part of what we receive in that state comes either from the Baltic rivers or those of North America. It annually ascends the large rivers in our country, but not in any quantities, and is occasionally taken in the salmon-nets. From the roe, when properly salted and dried, is prepared the substance known by the name of *caviar;* but a very superior sort is made from a smaller species, called the *Sterlet*.

The Sturgeon was a fish in high repute

among the Greeks and Romans, and, according to Pliny, was brought to table with much pomp, and ornamented with flowers, the slaves who carried it being also adorned with garlands, and accompanied by music. Its flesh has, indeed been esteemed in all ages; but modern nations do not consider it so great a luxury as the ancients. Its fishery, however, is an object of importance.

The largest species of Sturgeon, called the ISINGLASS STURGEON (*Accipenser huso*), is chiefly found in the Black and Caspian seas, ascending the tributary streams in immense multitudes. It frequently attains the length of twenty or twenty-five feet; and some have been taken weighing nearly 3000 lbs. It enters the rivers in the middle of winter, while they are still covered with ice, is very voracious, and pursues all the smaller fishes, but feeds likewise on vegetables. The fishery of this species is vastly important in the south of Russia; upwards of a hundred thousand being taken yearly. The caviar of commerce is chiefly made from its eggs, which exist in such abundance as to constitute nearly one-third of the total weight. This is a very common aliment in Turkey, Russia, Germany, Italy, and especially in Greece, and forms an important article of commerce, very profitable to Russia. The flesh is nutritious, wholesome, and of an agreeable flavour. The isinglass of commerce is prepared from the air-bladder; and the fat may be used as a substitute for butter or oil.

STURIO: STURIONIDÆ. A genus and family of Cartilaginous fishes, of which the Sturgeon is the type. [See STURGEON.]

STURNUS: STURNIDÆ. A genus and family of Passerine birds, of which the common Starling is a familiar example. [See STARLING.]

SUCKER. [See LUMPFISH.]

SUIDÆ: SWINE. (*Sus*, Linn.) A family of *Pachydermata*, highly important to man as food. The animals composing this family are characterized by having on each foot two large principal toes shod with stout hoofs, and two lateral toes which are much shorter and hardly touch the earth. The incisor teeth are variable in number, but the lower incisors are all levelled forwards; the canines are projected from the mouth and recurved upwards. The muzzle is terminated by a truncated snout fitted for turning up the ground. The Babyroussa, Peccary, and other allied genera, are included in the family *Suidæ*. [See HOG.]

SUN-BIRDS. (*Cinnyridæ*.) A family of Tenuirostral birds, of the most brilliant plumage, living upon the juices of flowers. Cuvier defines the genus *Cinnyris* as being distinguished by a long and slender bill, with the edge of the two mandibles finely serrated; and the tongue, which can be protruded from the bill, terminating in a fork. They are, he observes, small birds, the plumage of whose males glitters in the season of love with metallic colours, approaching in splendour that of the Humming-birds, which they represent in this respect in the Old Continent, where they are found principally in Africa and the Indian Archipelago. Their subsistence for the most part is drawn from the nectar of flowers; their nature is gay, their song agreeable, and their beauty makes them much sought after in our cabinets; but as the tribe is confined exclusively to the torrid zone and the southern hemisphere, the naturalists of our northern latitudes have little opportunity of observing their manners or of inspecting their internal construction. There is an obvious affinity between the *Cinnyridæ*, the *Trochilidæ*, and the *Meliphagidæ*. One species will be sufficient to describe.

The SUN-BIRD. (*Cinnyris splendida*.) The length of this beautiful bird is rather more than five inches. The bill and legs are black; the head and throat deep violet-blue, with a gloss of gold on the crown; upper part of the neck, back, wing and tail-coverts, of a deep but very brilliant golden green, and stretching to a considerable extent over the tail: across the middle of the breast runs a bright red bar, beyond which the abdomen and thighs are of the same deep violet-blue colour as the breast: the wings and tail are black.

SUN-FISH. (*Orthagoriscus*.) A genus of Cartilaginous fish, remarkable for its peculiarity of form: the body is compressed, broad, abruptly truncated, resembling, in fact, the head of a large fish separated from the body. Its nearly circular form, and the silvery whiteness of the sides, together with their brilliant phosphorescence during the night, have obtained for it very generally the appellations of sun or moon-fish. While

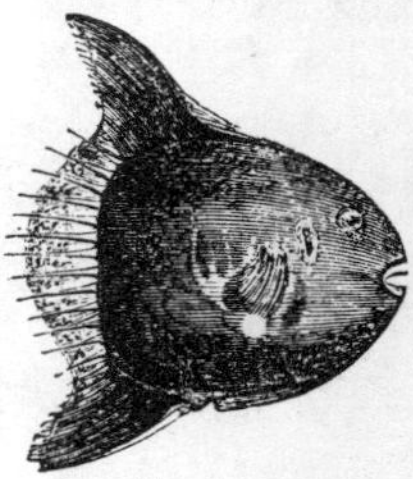

SHORT SUN-FISH.—(ORTHAGORISCUS MOLA.)

swimming, it turns round like a wheel: it has also the power of floating with its head and eyes above water, but not of inflating or distending itself with air; in this state it moves along sideways, very slowly, however; and appears like a dead or dying fish. It grows to an immense size, often attaining the diameter of four feet, sometimes even double that size, and occasionally weighing from three to five hundred pounds. It is very fat, and yields a great quantity of oil; but the flesh is ill tasted, and exhales a disagreeable odour. It is found in almost all seas, from the arctic to the antarctic circle. There are three or four species; two of which, the Short Sun-fish (*Orthagoriscus mola*), and the Oblong Sun-fish (*O. oblongus*), are found in the British seas.

SURGEON-FISH. [See ACANTHURUS.]

SURINAM TOAD. [See PIPA.]

SURMULLET, or STRIPED RED MULLET. (*Mullus surmuletus.*) This fish, which is a native of the Mediterranean, and found there in abundance, is also of frequent occurrence on the southern and western coasts of England. It seldom exceeds fourteen inches in length, and even that is accounted very large. Its colour is an elegant rose-red, tinged with olive on the back, and

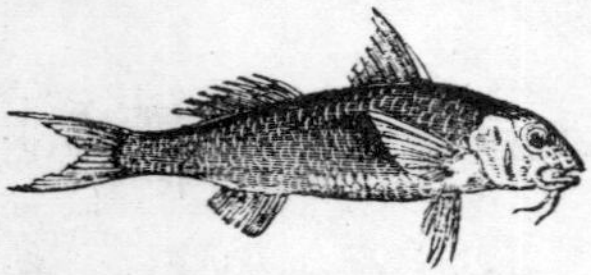

SURMULLET.—(MULLUS SURMULETUS.)

of a silvery cast towards the abdomen; marked on each side by two, and sometimes three, longitudinal lines. In the Mackerel season they are often taken with a draught of those fishes; and so abundant are they occasionally, that in August 1819, five thousand were taken in one night in Weymouth Bay. Mr. Yarrell observes that "the Striped Red Mullet has been considered migratory; but it appears in the shops of the London fishmongers throughout the year, though in much greater plenty during the summer, at which time their colours are most vivid, and the fish, as food, in the best condition. The food appears to be selected from among the softer crustaceous and molluscous animals." [See MULLET.]

SWALLOW. (*Hirundo.*) This well-known group of birds has often been eulogised by the lovers of nature; but no one, perhaps, has expressed his admiration with more truth and fervour than our own philosophic countryman, Sir Humphry Davy. "The Swallow," he says, "is one of my favourite birds, and a rival of the nightingale, for he cheers my sense of seeing as much as the other does my sense of hearing. He is the glad prophet of the year, the harbinger of the best season — he lives a life of enjoyment amongst the loveliest forms of nature — winter is unknown to him; and he leaves the green meadows of England in autumn for the myrtle and orange groves of Italy, and for the palms of Africa; he has always objects of pursuit, and his success is secure. Even the beings selected for his prey are poetical, beautiful, and transient. The ephemeræ are saved by his means from a slow and lingering death in the evening, and killed in a moment when they have known nothing but pleasure. He is the constant destroyer of insects, the friend of man, and may be regarded as a sacred bird. His instinct, which gives him his appointed season, and teaches him when and where to move, may be regarded as flowing from a Divine source; and he belongs to the oracles of nature, which speak the awful and intelligible language of a present Deity."

The habits and modes of living of the Swallow tribe are perhaps more conspicuous, and consequently more noticed by us, than any other. Their arrival has ever been associated in our minds with the idea of spring; and till the time of their departure they seem continually before our eyes. The air seems to be truly their home: they eat, drink, sometimes even feed their young, on the wing, and surpass all other birds in the untiring rapidity of their flight and evolutions. The beak is very short, broad at the base, much flattened, and very deeply cleft, forming a large mouth, well adapted to the purpose of seizing winged insects, which constitute their accustomed food. The feet are very short, and the wings remarkably long. In winter they migrate to tropical climates, a few days being sufficient for them to pass from the arctic to the torrid zone. In the spring they return; and it has been found by experiment that individual birds always come back to their former haunts. They usually have two broods in the year; some will occasionally have three; their nests are made of mud, rendered firm by a mixture of hair, twigs, and such kinds of materials. They are fond of flying over the surface of rivers and brooks, and sipping the water, without staying their flight. They are found in every country of the world.

Few subjects in natural history have given rise to more discussion than to determine the winter retreat of Swallows. It has long been clearly ascertained, that they migrate to warmer climates when they disappear in northern countries; and that they also creep into hollow trees and holes in the clefts of rocks, where they lie all the winter in a torpid state; but at one time it was firmly believed that they also retreated into water, and revived again in spring. Upon this subject, however, we will quote some of Wilson's graphic and, we think, conclusive remarks. After stating that the Swallow flies, in his usual way, at the rate of one mile in a minute; that he is so engaged for ten hours every day; and that his active life is, on an average, extended to ten years—which would give us two million one hundred and ninety thousand miles; upwards of eighty-seven times the circumference of the globe! "Yet," says he, "this little winged seraph, if I may so speak, who, in a few days, and at will, can pass from the borders of the arctic regions to the torrid zone, is forced, when winter approaches, to descend to the bottoms of lakes, rivers, and mill-ponds, to bury itself in the mud with eels and snapping turtles; or to creep ingloriously into a cavern, a rat-hole, or a hollow tree, there to doze, with snakes, toads, and other reptiles, until the return of spring! Is not this true, ye wise men of Europe and America, who have published so many *credible* narratives on this subject? The geese, the ducks, the cat-bird, and even the wren, which creeps about our outhouses in summer like a mouse, are all acknowledged to be migratory, and to pass to southern regions at the approach of winter: the Swallow alone, on whom Heaven has conferred superior powers of wing, must sink in torpidity at the bottom of our rivers, or doze all winter in the caverns of the earth. I am myself something of a traveller, and foreign countries afford many novel sights: should I assert, that in some of my peregrinations I had met with a nation of Indians, all of whom, old and young, at the commencement of cold weather, descend to the bottom of their lakes and rivers, and there remain until the breaking up of frost; nay, should I affirm, that thousands of people in the neighbourhood of this city, regularly undergo the same semi-annual submersion,—that I myself had fished up a whole family of these from the bottom of Schuylkill, where they had lain *torpid* all winter, carried them home, and brought them all comfortably to themselves again—should I even publish this in the learned pages of the 'Transactions' of our Philosophical Society, who would believe me? Is, then, the organization of a Swallow less delicate than that of a man? Can a bird, whose vital functions are destroyed by a short privation of pure air and its usual food, sustain, for six months, a situation where the most robust man would perish in a few hours, or minutes? Away with such absurdities! they are unworthy of a serious refutation."

English Chimney or House Swallow. (*Hirundo rustica.*) In length this bird is rather more than six inches: the bill is black; forehead and chin chestnut red; top of the head and all the upper parts of the body black, glossed with purplish hue; the quills of the wings, according as they are seen in different positions, are bluish black or greenish brown; while those of the tail are black, with green reflections: upper part of the breast black; lower part and belly white: the inside and corners of the mouth yellow: tail very long and much forked; and each feather, except the two middle ones, marked with an oval white spot on the inner web: legs short, delicately fine, and dusky. Every person must have observed the elegant and

ENGLISH CHIMNEY SWALLOW.
(HIRUNDO RUSTICA.)

varied flight of this bird, during the summer months, when it is almost continually on the wing, performing its various evolutions, and searching for its insect food, which it takes flying, and by suddenly opening its mouth. The nest of this bird is composed of mud, rendered tough by a mixture of hair and straw, lined with feathers, and fixed firmly about three or four feet from the top of the inside of a chimney. The female lays five or six eggs, white, speckled with red; and it generally has two broods in the year. The nestlings are sometimes dislodged from their nest, and fall down; and when that is the case, the old ones will frequently continue to supply them with provisions until they are able to shift for themselves. They generally make their first appearance in this country in the early part of April, and retire from us to the south on the approach of winter. For some time before they quit this part of the world they forsake houses, and roost on trees, preferring the dead, leafless branches; and within a day or two of retiring, they assemble in vast flocks on house-tops, the leads of churches, and on trees, especially by the water side, from which circumstance it has been erroneously supposed that they retired into the water.

The Barn Swallow (*Hirundo rufa*) inhabits America, and receives its name from its frequently attaching its nest to the rafters in barns, &c. The upper parts are steel blue, the lower light chestnut, and the wings and tail brownish black: the tail is greatly forked, and each feather, except the two middle ones, is (like the Chimney Swallow) marked on the inner vane with an oval white spot. The Barn Swallow's nest is in the shape of an inverted cone, with a perpendicular section cut off on that side by which it adheres to the wood: it is formed of mud, mixed with fine hay, and disposed in regular strata from side to side; within there is a quantity of hay, which is profusely lined with goose feathers. The eggs are extremely transparent; white, sprinkled with reddish-brown, and are five in number. When the young birds first leave the nest they are observed to fly about within doors, for some days before they venture out; which when they do they are conducted by the old ones to the sides of rivers, &c., where the food is most abundant, and they are fed by them in the same manner as the European Swallow does its young. These birds are easily tamed, and soon become very gentle and familiar. Their song is a sprightly warble, and is sometimes continued for a length of time.

The Cliff Swallow (*Hirundo fulva*) is easily distinguished by its even tail. The upper parts of the body are black, glossed with violaceous; the under parts whitish, tinged with ferruginous brown; the throat and cheeks dark ferruginous; and the front pale rufous. It lives in communities, building in unsettled places, under projecting ledges of rocks. The nests are formed of mud, are very friable, and somewhat resemble in form a chemist's retort. Their note is very singular, and may be imitated by rubbing moistened cork round the neck of a bottle. It is a native of North America.

The Martin Martlet, or Window Swallow. (*Hirundo* [*Chelidon*] *urbica.*) This species of Swallow, with which all persons are familiar, and which Shakspeare terms "the temple-haunting Martlet," is found throughout Europe and Asia, and is much more abundant in England than the Swallow, which generally arrives here about ten days previous to this bird. It is about five inches and a half in length; bill black; upper parts of the body and tail of a glossy blue black; rump and all the under parts of the body white; ends of the secondary quill-feathers finely edged with white; and the legs covered with white downy feathers down to the claws, which are white also, very sharp and much hooked. Should the weather prove favourable, it begins to build early in May; placing its nest generally beneath the eaves of a house or building against rocks and cliffs by the sea-side. The nest is composed externally of mud and straw, and lined with feathers. The first hatch consists of five eggs, which are white inclining to dusky

MARTIN – (HIRUNDO URBICA.)

at the thicker end: the second of three or four; and if a third, of only two or three. While the young birds are confined to the nest, the parents feed them, adhering by the claws to the outside; but as soon as they are able to fly, they receive their nourishment on the wing, by a quick and almost imperceptible motion. As the season advances the flocks increase in number daily, from the addition of the second and third broods; and during the month of October they generally migrate, continuing to depart till about the 6th of November, by which time they have generally all disappeared.

The SAND MARTIN, or BANK SWALLOW (*Hirundo* [*Cotyle*] *riparia*), is the smallest as well as the least numerous of our Swallows. It has no partiality for the society of man, but dwells in communities along steep gravelly and sandy banks, in which it makes deep holes for breeding places. Several of these holes are often within a few inches of each other, and extend in various strata along the front of the precipice, sometimes for eighty or a hundred yards. At the end of the hole is placed the nest, which is carefully constructed of straw, dry grass, and feathers. The female lays five or six white eggs, almost transparent, and has commonly two broods in the year. The young are hatched late in May; and Wilson tells us that he has taken notice of the common crow, in parties of four or five, watching at the entrance of these holes, to seize the first straggling young that should make its appearance. He also observes, that "from the clouds of Swallows that usually play round these breeding-places, they remind one at a

SAND MARTIN.—(HIRUNDO RIPARIA.)

distance of a swarm of bees." This species is common to Europe and America, arriving in this country first of the Swallow tribe: it is the scarcest and most local with us, but is extremely abundant in America. "They are particularly fond of the shores of rivers, and, in several places along the Ohio (says he), they congregate in immense multitudes. We have sometimes several days of cold rain and severe weather after their arrival in spring, from which they take refuge in their holes, clustering together for warmth, and have been frequently found at such times in almost a lifeless state with the cold; which circumstance has contributed to the belief that they lie torpid all winter in these recesses. I have searched hundreds of these holes in the months of December and January, but never found a single Swallow, dead, living, or torpid. I met with this bird in considerable numbers on the shores of the Kentucky river, between Lexington and Danville. They likewise visit the sea-shore in great numbers, previous to their departure, which continues from the end of September to the middle of October." The plumage is mouse-colour above; the throat, fore part of the neck, belly, and vent, white; wings and tail brown, the outer feather slightly margined with white: legs dusky, slightly feathered behind; feet smooth and dark brown. The manners of this species are similar to those of the Common Martin, with which bird it often associates, and flies over the water in pursuit of insects.

The PURPLE MARTIN (*Progne purpurea*) is a native of America, inhabiting all parts of the United States and Canada to Hudson's Bay. It is a general favourite, and takes up its abode among the habitations of men. The Indians and Negroes hang up gourds, properly hollowed for its convenience; and in some parts of the Union, considerable expense is sometimes incurred in preparing for it a suitable residence. In the country it renders essential service, by worrying and driving away crows, hawks, and other large birds. To observe with what spirit and audacity this bird dives and sweeps upon and around the hawk or the eagle is astonishing. He also bestows an occasional bastinading on the King-bird when he finds him too near his premises; though he will, at any time, instantly co-operate with him in attacking the common enemy. Its note is loud and musical. The colour of the male is a rich and deep purplish blue, with the wings and tail brownish-black: the female is more plainly attired, and has the under parts whitish, with dusky and yellowish

PURPLE MARTIN.—(PROGNE PURPUREA.)

stains. The food of the Purple Martin is usually the larger winged insects; as wasps, bees, large beetles, &c. In flight it possesses all the swiftness, ease, and grace of the tribe; sometimes sailing among the clouds at a dizzy height, at others darting through the crowded streets with the rapidity of thought. It lays from four to six eggs, which are pure white. About the middle of April these Martins first begin to prepare their nest, which is formed of dry leaves, slender straws, hay, and feathers. The first brood appears in May, the second late in July. During the period in which the female is laying, and before she commences incubation, they are both from home the greater part of the day. When the female is sitting, she is frequently visited by the male, who also occupies her place while she takes a short recreation abroad. He often passes a quarter of an hour in the apartment beside her; and, when not thus engaged, sits on the outside dressing and arranging his plumage. His notes, at this time, seem to have assumed a peculiar softness, and his gratulations are expressive of much tenderness. Conjugal fidelity, even where there is a number together, seems to be faithfully preserved by these birds. For ESCULENT SWALLOW and SWIFT, see SWIFT.

SWALLOW-TAIL [BUTTERFLIES]. A name given by insect collectors to some species of Butterflies of the genus *Papilio*.

SWAN. (*Cygnus.*) A genus of web-footed birds, distinguished by their graceful and majestic appearance, their muscular power, and superior size. The generic character of *Cygnus* is thus given: beak of equal breadth throughout; higher than wide at the base, and depressed at the point; both mandibles furnished along the sides with transverse serrated lamellæ the nostrils placed about midway; and the neck very long, and slender: legs short, the hind toe small and free. They feed chiefly on the seeds and roots of aquatic plants, and on the grass which grows near the brink of the water. The plumage of Swans, as in Geese, is similar in both sexes, is moulted only once in the year, and undergoes no seasonal variation of colour: like Geese, also, they attack with the same hissing note, strike similarly with their wings; and the male guards the female during incubation, and accompanies her while followed by her brood. In their anatomical structure, although infinitely superior in size and beauty, and easily recognized, they are so closely allied to the Duck and Goose, that it is difficult to point out distinctive characters.

The WILD SWAN, or WHISTLING SWAN. (*Cygnus ferus.*) This noble bird is nearly five feet in length, above seven in breadth with its wings extended, and weighs about fifteen pounds. Its bill is black, covered at the base with a yellowish white cere, the bare space over the eye being yellow: the entire plumage in adult birds is of a pure white, and, next to the skin, they are clothed with a thick fine down: the legs are black. "They generally," says Bewick, "keep together in small flocks, or families, except in the pairing season, and at the setting in of winter. At the latter period they assemble in multitudes, particularly on the large rivers and lakes of the thinly inhabited

WHISTLING SWAN.—(CYGNUS FERUS.)

northern parts of Europe, Asia, and America; but when the extremity of the weather threatens to become insupportable, in order to shun the gathering storm, they shape their course, high in air, in divided and diminished numbers, in search of milder climates. In such seasons they are most commonly seen in various parts of the British isles, and in other more southern countries of Europe: the same is observed of them in the North American states. They do not, however, remain longer than till the approach of spring, when they again retire northward to breed." The female makes her nest of the withered leaves and stalks of reeds and rushes, and usually lays six or seven thick-shelled eggs, which in about six weeks are hatched; when both parents unremittingly watch and guard them. Much has been said in ancient times of the singing of the Swan, and many beautiful and poetical descriptions have been given of its dying song. No fiction of natural history, no fable of antiquity, was ever more celebrated, oftener repeated, or better received: it occupied the soft and lively imagination of the Greeks; poets, orators, and even philosophers, adopted it as a truth too pleasing to be doubted. The truth, however, is very different from such amiable and affecting fables; for the voice of the Swan is very loud, shrill, and harsh; though, when high in the air, and modulated by the winds, the note, or *hoop*, of an assemblage of them is not unpleasant. Equally absurd stories are current of their great strength of wing, and how dangerous it is to approach their nests, it being asserted that a blow from the wing of a Swan is capable of breaking a man's thigh. "It is high time," as Montagu observes, "such absurdities should be erased in this philosophic age, and that the mind of man should reason before he continues to relate such accounts, only calculated to frighten children." In Iceland, we are told, Swans are an object of chase. In the month of August they lose their feathers to such a degree as to be incapable of flying. The natives, at that season, resort in great numbers to the places where they most abound, and are accompanied with dogs and horses trained to the sport; by which means they take great numbers. But when in full plumage, Swans are so extremely swift on the wing as to make it very difficult to shoot them.

The TAME SWAN, OR MUTE SWAN. (*Cygnus olor.*) Our half-domesticated Swan is very properly entitled the peaceful monarch of the lake; conscious of his superior strength, he fears no enemy, nor suffers any bird, however powerful, to molest him. The plumage of this species is of the same snowy whiteness as that of the Wild Swan, and the bird is covered next the body with the same kind of fine close down; but it greatly exceeds the Wild Swan in size, weighing about twenty-five pounds, and measuring more in the length of the body and extent of the wings. It also differs in being furnished with a projecting, callous, black tubercle or knob on the base of the upper mandible, and in the colour of the bill, which is red, with black edges and t[illegible]; the naked skin between the bill and the eyes is also of the latter colour; legs black. This species cannot with strict propriety be called domesticated; they are only, as it were, partly reclaimed from a state of nature, and invited by the friendly and protecting hand of man to decorate and embellish the artificial lakes and pools which ornament his pleasure grounds. On these the Swan cannot be accounted a captive, for he enjoys all the sweets of liberty. Placed there, as he is the largest of all British birds, so is he to the eye the most pleasing and elegant. "What in nature," exclaims Bewick, "can be more beautiful than the grassy margined lake, hung round with the varied foliage of the grove, when contrasted with the pure resplendent whiteness of the majestic Swan, wafted along with erected plumes by the gentle breeze, or floating, reflected on the glassy surface of the water, while he throws himself into numberless graceful attitudes, as if desirous of attracting the admiration of the spectator." The Tame Swan is found, in its wild state, in the eastern countries of Europe and Asia; and, domesticated, it occurs in almost every European country. Swans are supposed to live to a great age, but no satisfactory evidence has yet been brought forward to prove the assertion. The young do not acquire their full plumage till the second year: during this period they are called *cygnets*, and in former times were much esteemed as food, though they are not at present.

The BLACK SWAN (*Cygnus atratus*) is nearly the size of the Tame Swan. Its beak is large and red, the tip being rather paler; at the base of its upper mandible, near the nostrils, is a bifid protuberance; its under mandible is red on the sides and white beneath: the irides are red. The prevailing colour of the plumage is of a deep black,

BLACK SWAN.—(CYGNUS ATRATUS.)

with all the primary quills, the greater part of the secondaries, and part of the wing-coverts white: the belly and thighs are cinereous: the legs brownish flesh-colour. The female is destitute of the nasal protuberance on the beak. These birds inhabit various parts of Australia, and are generally seen floating on some lake in small flocks of eight or nine. The Swan River, in Western Australia, derives its name from the abundance of them found there. Their habits are but little known; but in a domesticated state their food is similar to the common species. When disturbed, they generally fly off in line or single file, and are so shy that it is difficult to get within gunshot. Their note is rather melodious than harsh, though not of long continuance. This species, like the Tame White Swan, is frequently kept as an ornament in parks in this country, and is now by no means the "rara avis in terris" of antiquity.

SWIFT, or BLACK MARTIN. (*Cypselus apus.*) This species of the *Hirundinæ*, or Swallow tribe, arrives later in this country, and departs sooner than any of its congeners. It is larger, stronger, and its flight is more rapid than that of any other of the tribe. Its length is nearly eight inches: general colour a sooty black, with a greenish tinge; the wings very long in proportion to the size of the body; tail much forked; bill black; chin white; legs dark brown, and very short; toes stand two and two on each side of the foot, and consist of two phalanges or joints only. The female is less than the male, and the general colour of her plumage more inclined to brown. The Swift builds its nest in the holes and crevices of high towers or lofty steeples; it is constructed of dried grass, silk or linen threads, pieces of muslin, feathers, and such kind of materials, which the bird collects on the wing, picking them up from the ground with great dexterity. It lays only two white, oblong eggs; and during the period of incubation the male is continually flying to and fro, uttering its loud screaming note. It has but one brood in the year, so that the young ones have time to gain strength enough to accompany the parent birds in their distant excursions. They have been noticed at the Cape of Good Hope, and probably visit the more remote regions of Asia. Swifts fly higher, and wheel with bolder wing than the Swallows, with which they never intermingle. Their life seems to be divided into two extremes; the one of the most violent exertion, the other of perfect inaction; they must either shoot through the air, or remain close in their holes. They are seldom seen to alight; but if by any accident they should fall upon a piece of even ground, it is with difficulty they can recover themselves, owing to the shortness of their feet and the great length of their wings. They are said to avoid heat, and for this reason pass the middle of the day in their holes: in the morning and evening they go out in quest of provision; they then are seen in flocks, describing an endless series of circles upon circles, sometimes in close ranks, pursuing the direction of a street, and sometimes whirling round a large edifice, all screaming together: they often glide along without stirring their wings, and on a sudden they move them with frequent and quickly repeated strokes They arrive about the beginning of May, and depart in August.

American Chimney Swallow. *Acanthylis pelasgia.*) "This species," says the great American ornithologist, Wilson, "is peculiarly our own; and strongly distinguished from all the rest of our Swallows by its figure, flight, and manners. This Swallow, like all the rest of its tribe in the United States, is migratory, arriving in Pennsylvania late in April or early in May, and dispersing themselves over the whole country wherever there are vacant chimneys in summer sufficiently high and convenient for their accommodation. In no other situation with us are they observed at present to build. This circumstance naturally suggests the query, Where did these birds construct their nests before the arrival of Europeans in this country, when there were no such places for their accommodation? I would answer, Probably in the same situations in which they still continue to build in the remote regions of our western forests, where European improvements of this kind are scarcely to be found, namely, in the hollow of a tree, which, in some cases, has the nearest resemblance to their present choice of any other." "The present site which they have chosen must hold out many more advantages than the former, since we see that, in the whole thickly settled parts of the United States, these birds have uniformly adopted this new convenience, not a single pair being observed to prefer the woods. Security from birds of prey and other animals, from storms that frequently overthrow the timber, and the numerous ready conveniences which these new situations afford, are doubtless some of the advantages. The choice they have made certainly bespeaks something more than mere unreasoning instinct, and does honour to their discernment.

AMERICAN CHIMNEY SWALLOW
(ACANTHYLIS PELASGIA.)

"The nest of this bird is of singular construction, being formed of very small twigs, fastened together with a strong adhesive glue or gum, which is secreted by two glands, one on each side of the hind head, and mixes with the saliva. With this glue, which becomes hard as the twigs themselves, the whole nest is thickly besmeared. The nest itself is small and shallow, and attached by one side or edge to the wall, and is totally destitute of the soft lining with which the others are so plentifully supplied. The eggs are generally four and white, and they have generally two broods in a season. The young are fed at intervals during the greater part of the night,—a fact which I have had frequent opportunities of remarking both here and in the Mississippi territory. The noise which the old ones make in passing up and down the funnel has some resemblance to distant thunder. When heavy and long continued rains occur, the nest, losing its hold, is precipitated to the bottom. This disaster frequently happens. The eggs are destroyed; but the young, though blind, (which they are for a considerable time,) sometimes scramble up along the vent, to which they cling like squirrels, the muscularity of their feet, and the sharpness of their claws, at this tender age, being remarkable. In this situation they continue to be fed for perhaps a week or more. When these birds first arrive in spring, and for a considerable time after, they associate together every evening in one general rendezvous; those of a whole district roosting together. This place of repose, in the more unsettled parts of the country, is usually a large hollow tree, open at top; trees of that kind, or *swallow trees*, as they are usually called, having been noticed in various parts of the country, and generally believed to be the winter quarters of these birds, where, heaps upon heaps, they dozed away the winter in a state of torpidity. Here they have been seen on their resurrection in spring, and here they have again been remarked descending to the death-like sleep in autumn."

"The Chimney Swallow is easily distinguished in air from the rest of its tribe here, by its long wings, its short body, the quick and slight vibrations of its wings, and its wide unexpected diving rapidity of flight; shooting swiftly in various directions without any apparent motion of the wings, and uttering the sounds *tsip tsip tsip tsee tsee* in a hurried manner. In roosting, the thorny extremities of its tail are thrown in for its support. It is never seen to alight but in hollow trees or chimneys; it is always most gay and active in wet and gloomy weather; and is the earliest abroad in the morning, and latest out in evening, of all our Swallows. About the first or second week in September they move off to the south, being often observed on their route, accompanied by the purple martins." This species is four inches and a half in length, and twelve inches in extent; of a deep sooty brown, except the chin and line over the eye, which are of a dull white.

The Esculent Swallow. (*Collocallia esculenta.*) This bird is four inches and a half in length, and eleven in expanse: its beak is black: the upper parts of the plumage shining dusky black; under parts pale ash-colour; wings, when closed, one inch longer than the tail, which is slightly forked, and has all the feathers of an uniform black colour, and rounded at the end. The nest of this bird is exceedingly curious, and is composed of such materials that it is not only edible, but is accounted among the greatest dainties by the Asiatic epicures. It generally weighs about half an ounce, and is shaped like a common Swallow's nest, the flat side adhering to the rock. They are found in vast numbers in caves of various islands in the Soolo Archipelago, and are particularly abundant in Sumatra, about Croe, near the south end of the island: they have the appearance of fibrous, imperfectly concocted isinglass. More or less of this substance is contained in the nests of all Swallows in that region. The manner in which the substance is procured is not ascertained: the most probable suppositions are, that it is the spawn of fish gathered by the bird, or a secretion elaborated in the bird's body. The birds, after having spent nearly two months in preparing their nests, lay each two eggs, which are hatched in about fifteen days: when the young birds become fledged, it is thought the proper time to seize upon their nests, which is done regularly three times a year, and is effected by means of ladders of bamboo and reeds, by which the people descend into the caves; but when these are very deep, rope ladders are used. It is attended with considerable danger, and many perish in the attempt. The Javanese and Chinese collect the nests, and make of them a profitable article of commerce. Dissolved in broths, &c., they make a delicious jelly. The finest are those obtained before the nest has been contaminated by the young birds; they are pure white, and are scarce and valuable. The inferior ones are dark, streaked with blood, or mixed with feathers: they are chiefly converted into glue. The only preparation which the birds' nests undergo is that of simple drying, without direct exposure to the sun, after which they are packed in small boxes, usually of half a picul. They are assorted for the Chinese market into three kinds, according to their qualities, distinguished into *first* or *best*, *second*, and *third* qualities. Caverns that are regularly managed, will afford in one hundred parts, fifty-three three-tenth parts of those of the first quality, thirty-five parts of those of the second, and eleven-seventeenth parts of those of the third. They are regarded only as an article of expensive luxury, and are sold at the most extraordinary prices; they are, consequently, consumed only by the great; and, indeed, the best part is sent to the capital for the use of the court. The sensual Chinese use them under an idea that they are powerfully stimulating and tonic; but it is probable that their most valuable quality is their being perfectly harmless.

The Fairy Martin. (*Collocalia Ariel.*) This curious and beautiful species is numerously dispersed over all the southern portions of Australia, where it usually arrives in the month of August, and departs again in February or March; during which interval it rears two or three broods. It is seldom seen within a few miles of the sea-coasts, but wherever suitable situations for breeding present themselves in the interior, it abounds. The nest, which is bottle-shaped with a long neck, is composed of mud or clay, and, like that of our Common Martin, is only constructed in the morning and evening, unless the day be wet or lowering. In the construction of the nests they appear to work in small companies, six or seven assisting in the formation of each nest, one remaining within and receiving the mud brought by the others in their mouths: in shape they are nearly round, but vary in size from four to six or seven inches in diameter; the spouts being eight, nine, or ten inches in length. Sometimes they are built in low decayed trees; sometimes under verandahs or in the corners of windows; and not unfrequently clusters of them are attached to the perpendicular banks of rivers, the sides of rocks, &c.; but always in the vicinity of water. They are lined with feathers and fine grasses. Eggs four or five in number, sometimes white, at others blotched with red. The Fairy Martin has the crown of the head rust-red; back, scapularies, and wing-coverts deep steel-blue; wings and tail dark brown; rump buffy white; upper tail-coverts brown; under surface white, tinged with rust-red, particularly on the sides of the neck and flanks; the feathers of the throat with a fine line of dark brown down the centre ; irides blackish brown ; bill blackish gray ; legs and feet olive-gray. — *Gould's Birds of Australia.*

The PALM SWIFT. (*Tachornis phœnicobia.*) We are told by Mr. Gosse, in his interesting work on the 'Birds of Jamaica,' that this delicately formed little Swift, conspicuous even in flight, from the broad belt of white across its black body, is a very common species in Jamaica, where it resides all the year. It is thus described:—Irides dark hazel ; beak black ; feet purplish flesh-colour ; claws horn-colour ; inside of mouth flesh-colour, tinged in parts with bluish. Head smoke brown, paling on the sides ; back, wings, tail-coverts, and tail sooty-black, unglossed, or with slight greenish reflections on the tail. Across the rump a broad band of pure white, the black descending into it from the back, in form of a point; sometimes dividing it. Chin and throat silky white, the feathers brown at the base ; sides smoky black, meeting in a narrow, ill-defined line across the breast : medial belly white. Thighs, under tail-coverts, and inner surface of wings smoky black. "Over the grass-pieces and savannas of the lowlands, the marshy flats at the seaward mouths of the valleys, as well as the pens of the mountain slopes, this swift-winged sylph daily urges its rushing course in parties of half a dozen to fifty or a hundred, often mingled with other Swallows, performing mazy evolutions, circling and turning, crossing and recrossing, now darting aloft, now sweeping over the grass, till the eye is wearied with attempting to follow them. The length of its wings, which is scarcely less than that of the whole bird, renders it a fleet and powerful flyer ; an attentive observer will be able to indentify it, when mingling in aerial career, by a more frequent recurrence of the rapid vibration of the wings, the momentary winnowing, by which a fresh impetus is gained. There is a very interesting structure in the sternum of this bird, which, as far as I know, is unprecedented. The sternum, though void of emarginations, possesses two oblong foramina of large size, one on each side of the middle of the ridge, and a round one perforating the ridge itself near the front margin. As all three are closed by the usual membrane, the object may be, the decrease of weight by the abstraction of bone, while the surface for the attachment of the muscles of flight remains undiminished."

Our author then proceeds with an interesting description of their nests. "I observed," says he, "several small Swallows flying above some cocoa-nut palms ; they uttered, as they flew, a continued twittering warble, shrill but sweet, which attracted my attention. I commenced a careful search, with my eye, of the under surface of the fronds and spadices of one, and at length discerned some masses of cotton projecting from some of the spathes, which I concluded to be their nests. This conjecture proved correct ; for presently I discovered a bird clinging to one of these masses, which I shot, and found to be this white-rumped Swift. On my lad's attempt to climb the tree, eight or ten birds flew in succession from various parts, where they had been concealed before. The tree, however, was too smooth to be climbed, and as we watched beneath for the birds to return, one and another came, but charily, and entered their respective nests. Although several other cocoa-nuts were close by, I could not discern that any one of them was tenanted but this, and this so numerously ; whence I inferred the social disposition of the bird. At some distance we found another tree, at the foot of which lay the dried fronds, spadices, and spathes, which had been, in the course of growth, thrown off, and in these were many nests. They were formed chiefly in the hollow spathes, and were placed in a series of three or four in a spathe, one above another, and agglutinated together, but with a kind of gallery along the side, communicating with each. The materials seemed only feathers and silk-cotton (the down of the *Bombax*) ; the former very largely used, the most downy placed within, the cotton principally without ; the whole felted closely, and cemented together by some slimy fluid, now dry, probably the saliva. * * * All the nests were evidently old ones, for the *Bombax* had not yet perfected its cotton, and hence I infer that these birds continue from year to year to occupy the same nests, until they are thrown off by the growth of the tree. The entrance to the nests, which were subglobular, was near the bottom." Another opportunity afterwards presented itself, and Mr. Gosse became better acquainted with the habitations of the Palm Swift ; and he thus describes the nests he had in his possession : "They have a singularly hairy appearance, being composed almost exclusively of the flax-like cotton of the Bombax, and when separated, are not unlike a doll's wig. They are in the form of those watch-fobs which are hung at beds' heads, the backs being firmly glued by the saliva to the under surface of the fronds, the impressions of the plaits of which are conspicuous on the nest when separated. The thickness is slight in the upper part, but in the lower it is much increased, the depth of the cup descending very little below the opening. The cotton is cemented firmly together as in the case of the others, but externally it is allowed to hang in filamentous locks, having a woolly but not altogether a ragged appearance. A few feathers are intermixed, but only singly, and not in any part specially. One specimen is double, two nests having been constructed so close side by side, that there is but a partition wall between them. Many nests had eggs, but in throwing down the fronds all were broken but one, which I now have. It is pure white, unspotted, larger at one end, measuring 13-20ths of an inch by 9-20ths. The average dimensions of the nests were about five inches high, and three and a half wide."

SWIFT [MOTHS]. A name applied by collectors to Moths of the genus *Hepialus*.

SWORD-FISH. (*Xiphias.*) A genus of Acanthopterygian fishes, the distinguishing characteristic of which is a long pointed beak, constituting one third of its whole length, and shaped like a straight sword; being a most powerful offensive weapon. They are placed by Cuvier among the *Scomberidæ*, or Mackerel family. The common Sword-fish (*Xiphias gladius*) is sometimes more than twenty feet long, the beak included. It swims with greater swiftness than almost any inhabitant of the deep, and is possessed of vast muscular strength. It attacks, and generally puts to flight, the smaller cetaceous animals, notwithstanding its food is usually vegetable. Its flesh is good; and in some countries the fishery is an object of importance. It is taken with the harpoon, and

COMMON SWORD-FISH — (XIPHIAS GLADIUS)

usually tears the net, if enclosed. It has not unfrequently happened that a Sword-fish has struck a ship, and driven its sharp weapon through the planking. It is very abundant in the Mediterranean, but less so in the Atlantic. Notwithstanding its formidable weapon, its great strength, and its almost incredible celerity, a small crustaceous animal penetrates the flesh of the Swordfish, and sometimes so torments it, that it dashes itself on the shore with mortal violence. In the Mediterranean it is regularly pursued by the fishermen; and its flesh is much esteemed in some places as an article of food. The female approaches the shores in the latter part of spring or beginning of summer. Mr. Gray has described a fine species of Sword-fish from the Cape of Good Hope, in which the skin is strengthened with bony spicula. It is nearly eleven feet long, and having been found in Table Bay during the visit of Sir John Herschel (the astronomer), has been named *Tetrapturus Herschelii*, in compliment to him. It belongs to a genus subdivided from *Xiphias* by its possessing ventral fin ; the caudal fin is furnished on each side with two small prominent crests. The specimen is in the collection of the British Museum.

SYLVIA: SYLVIADÆ. The *Sylviadæ*, or *Warblers*, are a family of small birds, with rather long and slender bills, with the tip slightly curved and toothed; and it contains a large proportion of the species which are most remarkable for their power of song. "The chief peculiarity," observes Mr. Swainson, "which runs through this numerous family, is the very small size and delicate structure of its individuals. Excepting the Humming-birds, we find among these elegant little creatures the smallest birds in the creation. The diminutive Golden-crests, the Nightingale, the Whitethroat, and the Wood-wren, are all well-known examples of genuine Warblers, familiar to the British naturalist. The groups of this extensive family, spread over all the habitable regions of the globe, are destined to perform an important part in the economy of nature: to them appears intrusted the subjugation of those innumerable minute insects which lurk within the buds, the foliage, or the flowers of plants; and, thus protected, escape that destruction from swallows, to which they are only ex-

SYLVIA HORTENSIS.

posed during flight. The diminutive size of such insects renders them unfit for the nourishment of the thrushes and the larger insectivorous birds, while their number and variety only become apparent when the boughs are shaken and their retreat disturbed. How enormous then would be their multiplication, had not nature provided other races of beings to check their increase! No birds appear more perfectly adapted for this purpose than the Warblers." The same writer then notices their arrival, for the most part, on the first appearance of spring, when the insect world is called into life and activity by the renewal of vegetation; and their departure towards autumn, when the insect hosts diminish, and consequently no longer require the agency of these little birds to keep their numbers within due bounds. He remarks also, that as different localities are assigned to different tribes of insects, so a similar diversity of haunts is allotted to different groups of Warblers. [See WARBLER.]

SYNALLAXIS. The name given to a genus of birds by Vieillot, placed by Mr. Swainson in the family of *Certhiadæ*. Their generic character is thus described:—Bill short, strong, and straight; both mandibles of equal thickness, and much compressed: wings short, and much rounded; the primaries scarcely exceeding the tertials: tail broad and long, either granulated or cuneated; the webs loose, the shafts rather rigid, the tips lanceolate: feet very large: tarsus lengthened: the claws slender, acute, and slightly curved. The SYNALLAXIS GARRULUS is given as an example of the genus. Colour of the plumage: brown; beneath whitish; feathers on the front of the head rigid, pointed, and rufous; lines before and behind the eye whitish; tail rounded. This bird is remarkable for its very singular nest, which in the woodland scenery of Bahia (Brazil) forms a striking object. It is built in low trees, formed externally of dried sticks, usually three or four feet long, and resembling at a distance a thick twist of bean-stalks thrown in the branches by accident. Sometimes two of these nests appear as if joined together, and there is an opening on the side, besides one on the top. Both male and female are generally seen near the nest, uttering a shrill, incessant, monotonous chirp, particularly in the morning and evening.

SYNBRANCHUS. The name of a subdivision of the *Murænidæ*, or Eel-shaped fishes; characterized by having the gill-opening entirely single, no pectorals, fins fatty, head thick, snout rounded, operculum cartilaginous, with six rays, stomach and anal perfectly straight, and bladder long and narrow. They are found chiefly in tropical seas.

SYNDACTYLI. The name given to a tribe of Perching Birds, including those which have the external and middle toe united as far as the second joint; the word *Syndactyli* indicating the adhesion of the fingers. The plumage is generally of a brilliant blue or green colour; and very smooth and glossy. [For examples, see BEE-EATER and KINGFISHER.]

SYNGNATHUS. [See PIPE-FISH.]

SYRPHIDÆ. A family of Dipterous insects, generally of a moderate or large size, and of variegated colours. Many of the species resemble humble-bees, wasps, &c., and are frequently mistaken for them by the inexperienced. The proboscis is long, membranous, elbowed near the base, terminated by two large lobes, and the sucker enclosed in an upper canal; a long horny upper lip, hollow, and notched at the tip; a pair of slender acute maxillæ, and a slender pointed tongue; the head is hemispherical, and covered for the most part by the eyes, especially in the males: the front of the head is often produced into a kind of beak, receiving the proboscis when it is folded in inaction. They are all fond of flowers: they fly with amazing swiftness, and many of them, if disturbed from their favourite haunts even for a number of times, will return and continue to hover there again. The larvæ of the typical genus *Syrphus* feed upon all kinds of Aphides, which they often hold up in the air, and suck very quickly: the body of these larvæ is of an elongate-conic form, uneven, and sometimes spinose. When ready to metamorphose, they fix themselves to leaves or other substances by a glutinous secretion; the body shortens, and its anterior end, which was the slenderest, becomes the thickest.—The larvæ of the genus *Volucella* are also insectivorous, but reside in the nests of Humble-bees and Wasps, upon the larvæ of which they subsist.

TABANUS: TABANIDÆ. A genus and family of Dipterous insects, comprising various large flies, pre-eminently distinguished for the tormenting powers which different species possess; piercing the skin, in order to suck the blood, of various quadrupeds, wild and domesticated. The TABANUS BOVINUS of Linnæus is the largest of the British species. It has the appearance of a very large pale brown fly, marked on the back by a series of large, whitish, triangular spots. This insect, like the rest of its genus, is seen during the middle and the decline of summer, generally in the hottest part of the day, and chiefly abounding in woods and pastures. It is extremely troublesome to cattle, piercing their skin with the lancets of its proboscis, and sucking the blood in such a manner as to cause considerable pain. It proceeds from a large, dusky-yellowish larva, marked by transverse blackish rings: it resides under ground, in moist meadows, &c., and changes to a cylindric, brownish chrysalis; out of which, in about a month, proceeds the perfect insect.

TABBY [MOTHS]. A name applied by collectors to Moths of the genus *Aglossa*.

TACHYPETES. [See FRIGATE-BIRD.]

TADORNA. A genus of web-footed birds, founded on the *Anas Tadorna* of Linnæus. [See SHELDRAKE.]

TADPOLE. The Frog in its nascent state. [See FROG.]

TADPOLE-FISH, or LESSER FORKED BEARD. (*Barbus minor.*) A somewhat rare fish of the *Gadidæ* family, measuring about a foot in length, and in its general form and colour bearing some resemblance to the imperfect animal from whom the name is derived. The head is very large, obtuse, and flattened on the crown; the mouth is wide; under the chin there is a small conical barb or feeler; and the lips are rounded and white. Tail wedge-shaped; scales small. It has been taken on the Scottish coast: it spawns in April, and feeds on small insects; but it is too scarce for naturalists to be much acquainted with its history.

TÆNIA. An intestinal worm, belonging to a numerous and, unfortunately, but too well-known a genus. *Tœnia solium* is characterized by an extremely long body, flat, and composed of a number of joints or articulations, which sometimes amount to several hundred; the whole animal occasionally attaining the length of five yards or more. They are thinner anteriorly, and generally have a square head, with four small suckers. Their numerous segments are all connected by the nutritive canal, which runs from one end to the other; but the reproductive apparatus is repeated in each division. That only one can exist in one human body at the same time is a vulgar error, since as many as forty have been found in one person. [For development see STROBILA in SUPPLEMENT.]

TÆNIDÆ. The name given to a family of Acanthopterygious fishes, distinguished by their lengthened and flattened bodies, and having very small scales. [See RIBBON-FISH.]

TAGUAN. A species of *Pteromys*, or Flying Squirrel.

TAILOR BIRD. A name applied to more than one species of soft-billed Indian Birds, allied to the Warblers. Some of them, if not all, belong to Dr. Horsfield's genera

Orthotomus and *Prinia*. The first described Tailor Bird (*Sylvia sutoria*, Latham) is a native of Ceylon, whence its curious nest is very frequently brought. It is for the most part composed of two leaves, one of them being dead; the latter is fixed by the ingenious bird to the living leaf as it hangs from the tree, by sewing both together, like a pouch or purse; this is open at the top, the cavity being filled with fine down; it is suspended from the branch, so as in great measure to secure it from the attacks of Reptiles and monkeys. Col. Sykes has described another interesting Tailor Bird, from the East Indies. This is the *Orthotomus Bennettii*. It constructs its nest by sewing together the leaves with threads of cotton and with fibres; in some cases, this naturalist found these threads actually knotted at the end. Professor Savi has described the habits of a species of Sylvia (*S. cysticola*), common in various parts of Italy, which constructs its nest among sedges and reeds which it unites together by real stitches; and the edge of each leaf is pierced by this bird with minute holes, through which it passes threads formed of spiders' web, particularly from the silk of their egg-pouches. These threads, as described by the Rev. Mr. Kirby, are not very long, and are sufficient only to pass two or three times from one leaf to another; there are knots scattered here and there, which in some places divide into two or three branches.

TALAPOIN. The name given by Buffon to a species of Monkey, distinguished by its beautiful variety of green, white, and yellow hair. It is the *Cercopithecus Talapoin* of zoologists.

TALBOT. A species of Dog, remarkable for its quick scent, and for its eagerness in pursuit of game.

TALEGALLA. A large gregarious Rasorial bird, which, according to Mr. Gould, may be considered, in a degree, as the representative of the Turkey in Australia. The plumage of the upper parts of the body, wings, and tail, blackish-brown; the feathers of the under parts blackish-brown at the base, becoming silvery-gray at the tip; skin of the head and neck deep pink-red, thinly sprinkled with short hair-like feathers; wattle bright yellow, tinged with red where it unites with the red of the neck; bill black: feet brown. It is about the size of a Turkey; and moves about in small companies. When it is disturbed, it readily eludes pursuit by the facility with which it runs through the tangled bush. If hard pressed, or rushed upon by their great enemy, the native dog, the whole company spring upon the lowermost bough of some neighbouring tree, and, by a succession of leaps from branch to branch, ascend to the top, and either perch there or fly off to another part of the brush. It is remarkable that this bird does not hatch its eggs by incubation. It collects together a great heap of decaying vegetables as the place of deposit of its eggs, thus making a hot-bed, arising from the decomposition of the collected matter, by the heat of which the young are hatched. This mound varies in quantity from two to four cart-loads, and is of a perfectly pyramidical form: it is not, however, the work of a single pair of birds, but is the result of the united labour of many; and the same site appears to be resorted to for several years in succession. "The mode," says Mr. Gould, "in which the materials

BRUSH TURKEY.—(TALEGALLA LATHAMI.)

composing these mounds are accumulated is equally singular, the bird never using its bill, but always grasping a quantity in its foot, throwing it backwards to one common centre, and thus clearing the surface of the ground to a considerable distance so completely, that scarcely a leaf or blade of grass is left. The heap being accumulated, and time allowed for a sufficient heat to be engendered, the eggs are deposited, not side by side, as is ordinarily the case, but planted at the distance of nine or twelve inches from each other, and buried at nearly an arm's depth, perfectly upright, with the large end upwards: they are covered up as they are laid, and allowed to remain until hatched. I have been credibly informed, both by natives and settlers living near their haunts, that it is not an unusual event to obtain nearly a bushel of eggs at one time from a single heap; and as they are delicious eating, they are eagerly sought after. Some of the natives state, that the females are constantly in the neighbourhood of the heap about the time the young are likely to be hatched, and frequently uncover and cover them up again, apparently for the purpose of assisting those that may have appeared; while others have informed me that the eggs are merely deposited, and the young allowed to force their way unassisted. In all probability, as Nature has adopted this mode of reproduction, she has also furnished the tender birds with the power of sustaining themselves from the earliest period; and the great size of the egg would equally lead to this conclusion, since in so large a space it is reasonable to suppose that the bird would be much more developed than is usually found in eggs of smaller dimensions. The eggs are perfectly white, of a long, oval form, three inches and three-quarters long by two inches and a half in diameter." It was originally described by Dr. Latham as a vulture under the name of "the New Holland Vulture," and at first sight a dried skin has considerable resemblance to that of some species of the group. In Australia it is called the BUSH TURKEY, and, as we remarked at the beginning, it is to the Rasorial order, and not the Raptorial, that this singularly interesting genus belongs: in the same family with it are birds with similar habits. [See MEGAPODIUS and LEIPOA.]

TALPA: TALPIDÆ. [See MOLE.]

TAMANDUA. [See ANT-EATER.]

TANAGRA. A group of birds of which there are several genera, and numerous species, all peculiar to America, and which are conspicuous for their brilliant colours. They have a conical beak and short wings; representing the Finches, &c. of Europe and Asia in their conformation and habits, and in the nature of their food.

TANTALUS: TANTALIDÆ. The *Tantalidæ* are a family of Wading Birds, the chief of which inhabit tropical latitudes, living almost entirely on the swampy banks of rivers and lakes. The genus *Tantalus* greatly partakes of the character of the Storks and Herons, and is characterized by Cuvier as having the feet, the nostrils and the bill of a stork; but the back of the bill, he observes, is rounded, and its point curved downwards and slightly notched on each side; a portion of the head, and sometimes of the neck, being denuded of feathers. It includes the American Scarlet *Ibis* (*Ibis rubra*), of which the following is a description. Length twenty-three inches: bill five inches long, thick, and of a somewhat square form at the base, gradually bent downwards, and sharply ridged; black, except near the base, where it inclines to red.

SCARLET IBIS.—(IBIS RUBRA.)

Iris dark hazel. The face naked, slightly wrinkled, pale red. Chin bare, wrinkled also. Plumage rich, glowing scarlet, except about three inches of the extremities of the four outer quill-feathers, which are deep steel blue. Legs pale red; the three anterior toes united by a membrane as far as the first joint. "This brilliant and exclusively American species," says Nuttall, in his 'Ornithology of the United States,' &c., inhabits chiefly within the tropics, abounding in the West India and Bahama Islands, and south of the equator, at least as far as Brazil. They migrate in the course of the summer (about July and August) into Florida, Alabama, Georgia, and South Carolina; but retire into Mexico, or the Caribbean islands, at the approach of cool weather. They generally associate in numbers, frequenting the borders of the sea, and the banks and æstuaries of neighbouring rivers, feeding on small fry, shell-fish, crustacea, worms, and insects, which they collect at the ebbing of the tide. They are said to be in the habit of perching on trees in companies; but they lay their eggs, which are greenish, on the ground, amidst the tall grass of the marshes, on a slight nest of leaves. When just hatched, the young are black, soon changing to gray, but are nearly white before they are able to fly; by degrees they attain their red plumage, which is not complete until the third year. The young and old associate in distinct bands. In the countries where they abound, they are sometimes domesticated, and accompany the poultry. The Ibis shows great courage in attacking the fowls, and will even defend itself from the insidious attacks of the cat. It is generally esteemed as good food; and its rich and gaudy plumage is used by the Brazilians for various ornaments." [See IBIS.]

TANYSTOMA. The name of a group of Dipterous insects, comprehending those which have a projecting proboscis, with the last joint of the antennæ undivided.

TAPE-WORM. [See TÆNIA.]

TAPIR. The name of a genus of Pachydermatous quadrupeds, of which three species are at present known; two of them being natives of South America, whilst the other inhabits Sumatra and Malacca. In its general form and contour, the Tapir reminds us of the Hog; but it is sufficiently distinguished from that animal by its snout, which is lengthened into a flexible proboscis, that looks like the rudiment of the trunk of the elephant, and partly serves the same purpose.

AMERICAN TAPIR.—(TAPIRUS AMERICANUS.)

The anterior feet have four toes, but the posterior only three; and these have only their tips cased in small hoofs. The eyes are small and lateral, and the ears long and pointed. The incisor teeth are six in number; the canines small; and the molars are seven on each side of the upper jaw, and six in the lower. The common AMERICAN TAPIR (*Tapir Americanus*) is the largest animal of South America, and is found in all parts of that continent, though most abundant in Guiana, Brazil, and Paraguay.

It is of a deep brown colour throughout, approaching to black; between three and four feet in height, and from five to six in length. The hair of the body is scanty, very short, and closely depressed to the surface; scarcely distinguishable at a short distance. The inmost recesses of deep forests are the chosen haunts of this species, which is not gregarious, and shuns the society of man. It is for the most part nocturnal in its habits, sleeping or remaining quiet during the day, and at night seeking its food, which, in its natural state, consists of shoots of trees, buds, wild fruits, &c. It is, however, when in confinement, an indiscriminate swallower of every thing, filthy or clean. Its enormous muscular power, and the tough thick hide which defends its body, enable it to tear its way through the underwood in whatever direction it pleases. Its ordinary pace is a sort of trot; but it sometimes gallops, though awkwardly, and with the head down. It is very fond of the water, and frequently resorts to it. Its disposition is peaceful and quiet; and though it will defend itself vigorously, and in so doing inflict severe wounds with its teeth, it never attempts to attack either man or beast, unless hard pressed. The flesh is dry, and has a disagreeable flavour.

The MALAY TAPIR (*Tapirus Malayanus*) in its general form resembles the American, and has a similar flexible proboscis, which is six or eight inches in length. Its general appearance is heavy and massive: the skin is thick and firm, thinly covered with short hair; the eyes are small; the ears roundish, and bordered with white. The tail is very short, and almost destitute of hair; and it has no mane on the neck. Legs short and stout; the fore feet furnished with four toes, the hind feet with three. The general colour is glossy black, with the exception of the back, rump, and sides of the belly, which are white, and separated by a defined line from the parts that are black. It is a native of Sumatra, and was first described by Sir Stamford Raffles.

TAPIRIDÆ. The first family of pachydermatous quadrupeds, including the Rhinoceros, Tapir, Hyrax and several extinct genera occasionally found in a fossil state.

TARANDUS. [See REINDEER.]

TARANTULA. (*Lycosa tarantula.*) A species of Spider found in some of the warmer parts of Italy, whose bite produces a train of symptoms long believed to be only curable by music (and still exercising the faith and ignorance of the vulgar in some countries), is the largest of all the European Spiders, and is generally found in dry and sunny plains. When full grown, it is as large as a chestnut; and, like all the Spiders, has a poison-gland in its mandibles. It is of a brown colour, with the back of the abdomen marked by a row of trigonal black spots with whitish edges, and the legs marked beneath by black and white bars. In the present enlightened period it may be sufficient to observe that the extraordinary symptoms supposed to ensue from the bite of this insect, as well as their supposed cure, are entirely without foundation. We may, however, be expected to give some account of the *nature* of the symptoms &c. formerly so generally attributed to it: we therefore extract from the pages of an old popular writer the following particulars: "The bite of this creature oc-

TARANTULA.—(LYCOSA TARANTULA.)

casions a pain which at first resembles that of the sting of a bee or an ant. In a few hours the patient feels a numbness; and the part affected is marked with a small livid circle, which soon after rises into a very painful swelling: shortly after this he falls into a profound sadness, breathes with much difficulty, his pulse grows feeble, and his senses fail. At length he loses all sense and motion; and, according to some naturalists, expires, unless speedily relieved. But these symptoms come on somewhat differently, according to the nature of the Tarantula, and the disposition of the patient. An aversion for black and blue; and, on the contrary, a predilection for white and red; are among the unaccountable symptoms of this disease. All the medical assistance hitherto discovered, consists in some chirurgical applications on the wound, and in cordials and sudorifics which are of little service; but music, which reason perhaps never could have pointed out, is said to be infinitely more efficacious. No sooner has the person affected lost his sense and motion, than a musician tries several tunes on an instrument; and when he has hit on one whose tones and modulations suit the patient, he is immediately observed to make a faint motion; his fingers begin to move in cadence, then his arms, next his legs, and by degrees his whole body: then he rises on his feet and begins to dance, his strength and activity still increasing. Some will continue to dance for six hours without intermission. After this the patient is put to bed; and when he is judged to be sufficiently recruited from his first dance, he is allured out of bed by the first tune, in order to a second. This exercise is reiterated for several days successively; seven or eight at least; in which time the patient finds himself excessively fatigued, and unable to dance any longer, the characteristic proof of his being cured; for, as long as the poison acts on him, he would dance, if encouraged, till he fainted through extreme lassitude. Perceiving himself thus tired, he begins to recover his reason; and awakes, as out of a profound sleep, without the smallest recollection of what had passed in his paroxysm, or even in his dancing."

TARSIPES. A singular genus of Marsupial animals, found at King George's Sound, in Australia, only one species of which is as yet known. This has a longish muzzle, and is not much bigger than a mouse. It derives its name from the length of part of its hind legs.

TARSIUS. A genus of Quadrumanous Mammalia, inhabiting the Moluccas. They have the teeth and insectivorous regimen of the Loris; the tarsi elongated, which gives to their hinder limbs a disproportionate extent; tail very long and tufted; large membranous ears; and great eyes, which indicate a nocturnal life. Two species are known, *Tarsius fuscomanus* of Fischer, and

TARSIUS BANCANUS.

T. bancanus of Horsfield. These animals have an aversion to light, and retire by day under the roots of trees; feed chiefly on lizards, and leap about two feet at a spring; are easily tamed, and capable of some attachment. They hold their prey in their fore hands, while they rest on their haunches; produce one young at a birth, and live in pairs.

TASMANIAN CROW SHRIKE. (*Gymnorhina organicum.*) This animated and elegant bird is a native of Van Diemen's Land, inhabiting and enlivening by its presence the interior of the country. Mr. Gould tells us that "it runs, and occasionally hops, over the surface with great quickness, but flies rather slowly, and upon alighting on a branch raises and closes one wing several times in quick succession, and in a very peculiar manner. When on the plains it utters a loud ringing call, but when perched on the dead branches of the trees soon after day-break, it pours forth a succession of notes of the strangest description that can be imagined, much resembling the sounds of a hand-organ out of tune, which has obtained for it the colonial name of the Organ Bird. It is very easily tamed; and as it possesses the power of imitation in an extraordinary degree, it may be readily taught to whistle tunes as well as to articulate words; it consequently soon becomes a most amusing as well as ornamental bird for the aviary or cage." The male has the crown of the head, cheeks, throat, all the under surface, scapularies, primaries, and tips of the tail jet black; nape of the neck, back, upper and under tail-coverts, and base of the tail-feathers white; bill dark lead colour at the base, passing into black at the tip; legs black; irides bright hazel. In the female the nape of the neck and back are gray. It builds a round cup-shaped nest on the topmost branches of the gum-tree, constructing it awkwardly of sticks interspersed with strips of bark, &c., and lining it with coarse grass, sheep's wool, and a few feathers, felted together, and forming a dense and warm receptacle for the eggs, which are of a greenish ashy gray colour, spotted and blotched with umber-brown and bluish-gray.

TASMANIAN HONEY-EATER. [See Meliphaga Australasiana.]

TATOU. The native name for the giant armadillo of South America (*Priodonta gigas*). [See Armadillo.]

TAXICORNES. An extensive group of Heterimerous Coleoptera: two or three genera of which are natives of this country. The greater part of the beetles composing this family live on *fungi*, and are either found upon them, or beneath the bark of trees which produces them. A few live on the ground under stones. They are dis-

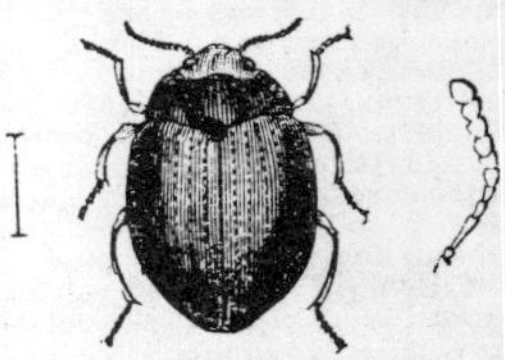

DIAPERIS RUFICORNIS.

tinguished by having no corneous hook on the inner edge of the maxillæ: they are generally furnished with wings: the antennæ are usually inserted beneath the margin of the sides of the head, and more or less perfoliated, and gradually thickened or ending in a club. We figure a species of the genus *Diaperis* as an example of this group. Most of the species are of a small size.

TAXUS. [See Badger.]

TEAL. (*Querquedula crecca.*) The common Teal is a small species of duck that frequents ponds, marshes, and the reedy shores of creeks, inlets, and rivers, but rarely visits the sea-shore. It is about fifteen inches in length: the beak is dusky; the top of the head, cheeks, and neck are chestnut-red ; the throat is black ; a broad green band extends from the eyes to the nape ; the lower part of the neck, back, scapulars, and sides are alternately striped with zigzag lines of white and black ; the breast is reddish, and spotted ; the belly a yellowish white ; the speculum of the wings is half white, half black, and edged with two white bands : the legs are ash-coloured. The female is smaller than the male, and has a

COMMON TEAL.—(QUERQUEDULA CRECCA.)

reddish-white band, spotted with brown, behind and beneath the eyes ; the throat is white ; the plumage above is blackish-brown, edged with a broad band of clear brown ; and the under parts are whitish. This species is a native of the north, occurring equally in Europe and America : it is very abundant in England during its migration ; but it does not appear usually to breed here, although its nest is sometimes met with, and is said to be not uncommon in France. The nest is large, and is composed of soft dried grasses, lined with feathers, and generally concealed in a hole among the roots of reeds and rushes near the water's edge. The female lays about a dozen reddish-white eggs, which are indistinctly sprinkled with brown dots, and in size about those of a pigeon. The Teal is widely and numerously dispersed over the whole of Norway, Sweden, and Lapland : it is abundant in Germany, Holland, France, Spain, and Italy ; it is also found in the winter in considerable numbers in Ireland ; and sometimes it inhabits the edges of the Scottish lakes. The flesh is dry and difficult of digestion, but, notwithstanding, is in great request. In the reign of Henry VIII. it held a high place among the luxuries of a royal banquet.

The BLUE-WINGED TEAL (*Querquedula discors*), says Wilson, in his 'American Ornithology,' is the first of its tribe that returns to us in the autumn from its breeding place in the north. They are usually seen early in September, along the shores of the Delaware, where they sit on the mud close to the edge of the water, so crowded together that the gunners often kill great numbers at a single discharge. When a flock is discovered thus sitting and sunning themselves, the experienced gunner runs his batteau on shore at some distance above or below them, and, getting out, pushes her before him over the slippery mud, concealing himself all the while behind her ; by this method he can sometimes approach within twenty yards of the flock, among which he generally makes great slaughter. They fly rapidly, and, when they alight, drop down suddenly, like the snipe or woodcock, among the reeds or on the mud. They feed chiefly on vegetable food, and are eagerly fond of the seeds of the reeds or wild oats. Their flesh is excellent, and, after a residence for a short time among the reeds, they become very fat. As the first frosts come on, they proceed to the south, being a delicate bird, very susceptible of cold. They abound in the inundated rice fields in the southern States, where vast numbers are taken in traps placed on small dry eminences that here and there rise above the water. These places are strewed with rice, and they are caught alive in hollow traps. This species is a trifle smaller than the preceding : the bill is long, and of a dark dusky slate colour ; the front and upper part of the head are black ; from the eye to the chin is a large crescent of white, the rest of the head and half the neck are of a dark slate, richly glossed with green and violet ; remainder of the neck and breast is black or dusky, thickly marked with semicircles of brownish white, elegantly intersecting each other ; belly, pale brown, barred with dusky, in narrow lines ; back, deep brownish-black, each feather waved with large semi-ovals of brownish-white ; lesser wing-coverts, a bright light blue ; primaries, dusky brown ; speculum, or beauty-spot, rich green ; tertials edged with black or light blue, and streaked down their middle with white : the tail pointed : legs and feet yellow.

TELEOSAURUS. The name given to a group of extinct Reptiles allied to the Crocodiles, the fossil remains of which have been found in this country. They had a long muzzle, somewhat resembling that of the Gavial, or Gangetic Crocodile.

TELEPHORUS : TELEPHORIDÆ. A genus and family of Coleopterous insects ; of a long and narrow form, with perfect wings and elytra ; head broad and not concealed under the thorax ; mandibles acute and curved ; and the antennæ simple, moderately

SOLDIER BEETLE.—(TELEPHORUS FUSCUS.)

long, and inserted closely together. These insects, which are known by the name of *Soldiers*, *Sailors*, or *Doctors*, are found in the spring in considerable numbers upon flowers, especially those of the *Umbelliferæ*. So voracious are they, that they not only feed upon other insects, but the weaker of their own species fall a prey to the stronger. They walk awkwardly, and their flight is heavy.

TELESCOPE FLY. [See DIOPSIS.]

TELLINA : TELLINIDÆ. A genus and family of bivalve Mollusca, which have in the centre of the hinge a tooth on the left and two teeth on the right, often bifid, and at some distance in front and behind; on the right valve a lateral tooth or plate, which does not penetrate into a cavity of the opposite one. There is a slight fold near the extremity of both valves, which renders them unequal in that part, where they gape a little. The animal has two long tubes, respiratory and excrementitial, which can be withdrawn into the shell, and concealed in a duplicature of the mantle. The shells are generally transversely striated, and beautifully coloured: some are oval and thickish; others are oblong and much compressed; others lenticular. Instead of a fold, we

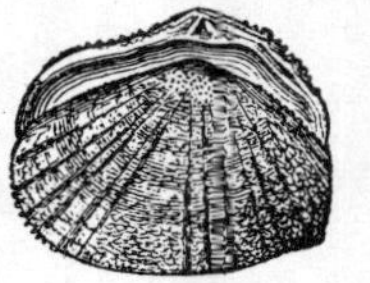

CAT'S-TONGUE TELLINA.
(TELLINA LINGUA-FELIS.)

often find in the latter merely a deviation in the course of the transverse striæ. Sowerby says, "The irregular flexuosity of the anterior ventral margin appears to have been constantly regarded as the principal distinguishing character of this beautiful genus; and when we consider the number of species possessing this character, and agreeing also in other general circumstances, it may perhaps be still considered the essential character of the genus."

TENCH. (*Tinca vulgaris.*) A fish belonging to the Cyprinidæ family, or Carp tribe; common in most of the lakes of the European continent, and more or less abundant in ornamental waters and ponds in this country, but is seldom found in our rivers, being more fond of still and muddy waters. Its general length is about twelve inches: its usual colour a deep olive, accompanied

TENCH.—(TINCA VULGARIS.)

by a slight golden tinge; the abdomen being paler or yellower than the other parts, and the fins, which are thick and opaque, of a dull violet colour. The body is short and thick; and the skin is covered, like that of an eel, with a tenacious mucus or slime, beneath which its small and numerous scales appear: the head is rather large, the eyes small, the lips thick, and on each side of the mouth is a small cirrus. It is considered as a very prolific fish, and of quick growth. It deposits its spawn, consisting of very small greenish ova, among aquatic plants, &c.; this takes place about the middle of June, when the female is attended by two males. By some it is supposed to lie during the winter in a torpid state, concealed beneath the mud of the waters it inhabits, being rarely taken during that season. We glean from Mr. Yarrell the following observations: "As the Tench is one of our most useful fresh-water fishes, from the ease with which it may be preserved and its increase promoted, the facility of transportation from its great tenacity of life, and the goodness of its flesh, —which is not, however, generally held in the estimation which I think it deserves,— as the Tench is also, like the Carp, one of those species first selected as stock for ornamental waters, I venture to recommend that large and fine fish be chosen as breeders, as the most certain mode of obtaining sizeable fish for table in the shortest space of time. Two males to one female, or not less than three to two, should be the proportion of the sexes; and from the pond which is found by experiment favourable for breeding, the small fish should be in part withdrawn from time to time, and deposited elsewhere, to afford more space for all. The male of the Tench is recognized by the large size of the ventral fins, which reach far enough to cover the vent, and are deeply concave internally: in the females the ventral fins are smaller, shorter, and less powerful."

A most beautiful variety, called the GOLDEN TENCH, is found in some parts of Germany, differing from the common Tench in being of the richest orange-yellow, variegated with small black spots, while the fins are thin, transparent, and of a bright red colour. It is said to be of slow growth, and to live, like most others of this genus, on worms, water insects, &c. These Tenches are delighted with warmth, and in bright weather are observed to swim in small shoals near the surface.

TENEBRIONIDÆ. A family of Coleopterous insects, distinguished by having the elytra not soldered together, with the wings fitted for flight. The body is generally oblong or ovate; depressed, or but slightly elevated; the thorax square or trapeziform; and the palpi enlarged at the tip, the last joint being generally hatchet-shaped. One of the most familiar of these is the *Tenebrio molitor*, the larva of which is commonly called the MEAL-WORM, and may be regarded as the type of the family. This insect frequents corn-mills, granaries, bakehouses, &c., doing much damage by devouring flour, meal, bran, &c. It is also very destructive to ship-biscuits packed in casks, which when opened are found eaten through in holes by these insects and their larvæ. The latter are about an inch long, of a cylindrical and linear form, very smooth and glossy, of a fulvous colour, consisting of twelve segments, exclusive of the head, which is provided with short trophi, and a pair of very small antennæ. This larva changes its skin several times, avoids the light, and at length assumes the pupa state, without forming any cocoon; the imago appearing at the expiration of about six weeks,

MEAL-WORM BEETLE.
(TENEBRIO MOLITOR.)

at first being o. a reddish-colour, but soon assuming its black hue. The larvæ of these insects are greedily devoured by nightingales and other insectivorous birds, and are accordingly bred by bird-fanciers; it is necessary to keep the vessels in which they are placed firmly closed, or they make their escape, neither wood nor cloth being sufficiently strong to prevent them from boring their way through.

TENREC. (*Centetes.*) A genus of small insectivorous quadrupeds found in Madagascar, closely resembling the Hedgehog in their general character, but differing in their dentition, the feebleness of their spines, and in their being much less able to roll themselves into a ball. The Tenrec is known also by the name of the ASIATIC or STRIPED HEDGEHOG. It is seven inches in length; and is characterized by a long, pointed muzzle; short legs, with five toes on each foot, separated and armed with crooked claws; and no tail. It is of a black colour; with five longitudinal bands on the body: all the black parts are covered with hard hair; the white bands with small prickles. From the black bands on the back spring long scattered hairs which reach to the ground; the head is covered with short black hairs or prickles; the snout is white; the eye surrounded by a white circle; and the feet are reddish. The Tenrecs move slowly; grunt like pigs, (for which reason they are sometimes called Ground-hogs or Pig-porcupines): they grow extremely fat; multiply prodigiously; and burrow under ground: they are nocturnal in their habits; and remain torpid during great part of the summer.

TENTHREDINIDÆ. The name given to a family of Hymenopterous insects, popularly known as *Saw-flies*, from the saw-like character and action of the ovipositor. With this they make a number of small holes in the branches of trees, inserting an egg in each hole, and closing the hole with a drop of frothy fluid. The wound thus made becomes more and more convex by the increase in size of the egg, and sometimes these parts assume the size of a gall, either woody or pulpy, according to the parts injured: these tumours form the abode of the larvæ which reside within them, and the insect makes with its teeth a circular hole for its escape. They greatly resemble the Caterpillars of Lepidopterous insects; but usually differ from them as to the number of their feet, which are either restricted to six, answering to those of the perfect insect, or amount to eighteen or twenty-two. In order to undergo their change into the pupa state, they spin a cocoon, either on the earth or on the plants on which they have fed; but they do not become pupæ until they have been inclosed in this for many months, and only a few days before they come forth as perfect Saw-

CIMBEX VARIABILIS

flies. Our figure represents the *Cimbex variabilis*, a member of this large family. It belongs to a section which by some naturalists has been raised into a distinct family, from the species having clubbed antennæ.

TENUIROSTRES. The name of a tribe of Insessorial birds, comprehending those which are distinguished by a long and slender bill. Mr. Swainson, in his 'Classification of Birds,' remarks that the most aberrant division of the insessorial order is that of the *Tenuirostres*, or honey-suckers, so called from the great majority deriving their subsistence both from insects and the nectar of plants, which they suck up by means of a long or filamentous tongue adapted for the purpose. [See NUTHATCH: CREEPER: SUNBIRD, &c.]

TEREBELLUM. A genus of Mollusca, whose shells are oblong, subcylindrical, and very smooth; spire pointed; sutures not channelled; aperture narrow and long, wider anteriorly; outer lip slightly thickened, truncated; inner lip thin, smooth, nearly straight, and spread over a portion of the body whorl. These shells are brought from the Indian seas; they are thin, delicate, and prettily marked with bands and cloudy spots. In its habits the animal of the *Terebellum* is exceedingly shy and timid, retracting its body into the shell on the slightest alarm. It will remain stationary for a long time, moving its tentacles about cautiously in every direction, when, suddenly, it will roll over with its shell, and continue again perfectly quiet.

TEREBRANTIA. The name of a section of Hymenopterous insects, characterized by the possession of an anal instrument organized for the perforation of the bodies of animals, or the substance of plants. The borer (*terebra*) is peculiar to the female, and is composed of three long and slender pieces, of which two serve as a sheath for the third; it is placed at the anal extremity of the abdomen, and the oviduct is continued into it. The females instinctively use this weapon to prepare a place for the deposition of their eggs, where the maggot may be incubated in safety, and upon its exclusion be surrounded by already organized matter adapted for its sustenance. Some genera select vegetables for the parasitic support of their young, as *Sirex* (Linn.), which infests the pine-tree; and *Cephus* (Latr.), which perforates the stalks of corn for the purpose of oviposition. Others, as the ichneumons, pierce the skins of insects, and deposit their eggs in the subcutaneous fatty and nutrient material.—[See ICHNEUMONIDÆ.]

TEREBRATULA. A genus of Conchiferous Mollusca, found at great depths in the Southern Ocean, and also in the European seas. The animals have a curious kind of internal skeleton, as it may be termed, consisting of a slender, flattened, calcareous loop, with other pieces diverging from it; and a ciliated appendage on each side of the body. The shell is inequivalve, equilateral, oval or sub-trigonal, ventricose or compressed, adhering by a short gelatinous tendon. Our figure, which exhibits a species of this extensive genus, shows the peculiarity

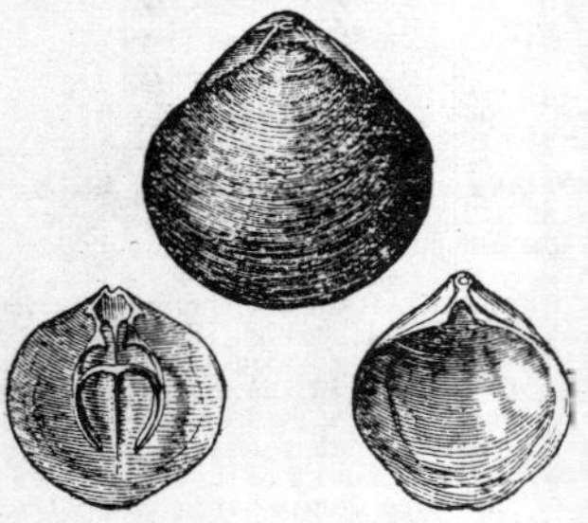

TEREBRATULA GAUDICHAUDII.

of structure above alluded to; the upper figure representing the whole shell and the two lower cuts, the insides of each valve. There are numerous species of this Molluscous genus, of various forms, and some of them curiously ribbed, found in a fossil state.

TEREDO. The name given to a genus of testaceous molluscs, which form their habitations by boring holes in submerged timber, and thereby occasion destructive ravages in ships' bottoms, sunken piles, &c. The *Teredo navalis* is worm-shaped, and about six inches long. In making its excavations into the wood, which it does by boring into the substance in the direction of the grain, each individual is careful to avoid the tube formed by its neighbour, and often a very thin leaf alone of wood is left between; it also, when a knot occurs in its path, makes a turn to avoid it. It is commonly supposed that this animal, so injurious to mankind, was introduced into Europe from warmer climates; but however that may be, it now unfortunately swarms in our seas. The rapidity of its growth, and the destructive celerity with which it works, are hardly credible. In Holland, in order to prevent the irruption of the sea, where the land is below the level of high water, immense dykes are constructed along the shore, formed of large masses of sand, and strengthened by large piles driven

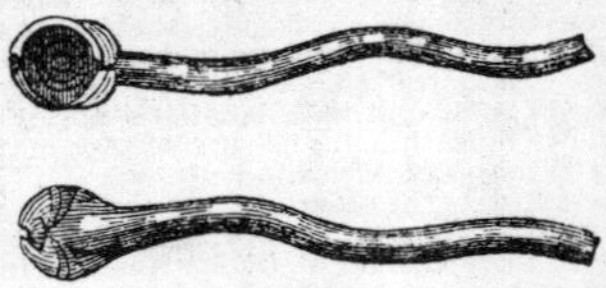

SHIP-WORM.—(TEREDO NAVALIS.)

into the ground. In the year 1730 it was discovered that these piles were attacked by the Teredo, and, on examination, were found to be pierced in all directions to such an extent, that, had it not been for a timely discovery of the mischief, the whole of that part of the country might have been overwhelmed by the sea breaking through the worm-eaten defences. The only effectual way of preventing the attacks of the Teredo upon piles is said to be by covering all that part which is continually beneath the surface with short broad-headed nails: the action of the sea-water on the nails producing a strong coating of rust, said to be superior to a copper sheathing.—Another species, the *Teredo gigantea*, is described by Sir Everard Home as sometimes exceeding four feet in length and several inches in circumference.

TERMITIDÆ. An extensive and important family of the Neuropterous order of Insects, to which the name of *White Ants* is very commonly given. Some few species are found in temperate regions, but they are chiefly confined to the tropics, where they perform a considerable share in the necessary operation of completing the comminution and destruction of dead and decomposing organized matter. Next to Locusts, they may be reckoned the most destructive insects known to Man. They are characterized by four-jointed tarsi; but the wings are carried horizontally on the body, and very long; the head rounded, and the prothorax short and square. The body is depressed, with the antennæ short; the mouth very similar to that of the *Orthoptera*, with the four-cleft lower lip; three ocelli; the wings slightly transparent, coloured, with the nervures forming a close network; and the legs short. They live in societies, often prodigiously numerous, and, like the Bee and the Ant, are composed of three sorts of individuals. In all the stages of their existence, save that of the ovum, they are active, carnivorous or omnivorous; and are, beyond all doubt, the greatest pest of tropical climates: destroying all articles of furniture made of wood, cloths, &c., and even entering the foundations of houses, and eating out the whole interior of the timbers, so that while they appear perfectly sound externally, they will fall to pieces under the slightest blow. One species is celebrated for the edifices it rears, in the form of a sugar-loaf, ten or twelve feet in height, and so solid that the wild cattle mount upon them without breaking through. Internally they are divided into numerous apartments, and have subterranean galleries connected with them, from the extremities of which the insects issue. But, so extraordinary is the whole history and economy of these insects—so wonderful their habits and instincts—that, in order to do justice to the subject, we feel ourselves under the necessity of inserting, with but little abridgment, Mr. Smeathman's celebrated 'Account of the Termites of Africa.'

These insects (he observes) have generally obtained the name of Ants from the similarity in their manner of living, which is in large communities, that erect very extraordinary nests, for the most part on the surface of the ground, from whence their excursions are made through subterraneous passages or covered galleries, which they build whenever necessity obliges, or plunder induces, them to march above ground, and at a great distance from their habitations carry on a business of depredation and destruction, scarce credible but to those who have seen it. But notwithstanding they live in communities, and are, like the Ants, omnivorous; though, like them, at a certain period they are furnished with four wings, and emigrate or colonize at the same season; they are by no means the same kind of insects, nor does their form correspond with that of Ants in any one state of their existence, which, like most other insects, is changed several times. They resemble the Ants also in their provident and diligent labour, but surpass them as well as the Bees, Wasps, Beavers, and all other animals which I have ever heard of, in the arts of building, as much as the Europeans excel the least cultivated savages. It is more than probable they excel them as much in sagacity and the arts of government; it is certain they show more substantial instances of their ingenuity and industry than any other animals; and do, in fact, lay up vast magazines of provisions and other stores.

The different species of this genus resemble each other in form, in their manner of living, and in their good and bad qualities; but differ as much as birds in the manner of building their habitations or nests, and in the choice of the materials of which they compose them. There are some species which build upon the surface of the ground, or part above and part beneath, and one or two species, perhaps more, that build on the stems or branches of trees, sometimes aloft at a vast height. Of every species there are three orders; first, the working insects, or *labourers;* next, the fighting ones, or *soldiers*, which do no kind of labour; and, last of all, the winged ones, or *perfect insects* (called kings and queens), which are male and female, and capable of propagation. These neither labour, or toil, or fight, being quite incapable of either, and almost of self-defence; and nature has so ordered it, that they emigrate within a few weeks after they have arrived at this state, and either establish new kingdoms, or perish within a day or two.

The *Termes bellicosus* is the largest and best known species on the coast of Africa; this account of the Termites is therefore taken from observations made thereon. The nests of this species are so numerous all over the island of Barranas, and the adjacent continent of Africa, that it is scarcely possible to stand upon any open place, such as a rice plantation or other clear spot, where one or more of these buildings is not to be seen within fifty paces. In some parts near Senegal, as mentioned by Adanson, their number, magnitude, and closeness of situation, make them appear like the villages of the natives. These buildings are usually termed *hills*,

ANT-HILL OF TERMES BELLICOSUS.

from their outward appearance, which is that of little hills, generally pretty much in the form of sugar-loaves, and about ten or twelve feet in height. These hills continue quite bare until they are six or eight feet high; but in time become, like the rest of the earth, almost covered with grass and other plants; and in the dry season, when the herbage is burnt up by the rays of the sun, they somewhat resemble very large haycocks. The exterior of the building consists of one large dome-shaped shell; large and strong enough to enclose and shelter the interior from the weather, and to protect the inhabitants from the attacks of most of their enemies. It also serves to collect and preserve a regular degree of genial warmth and moisture; which in all probability is quite necessary for hatching the eggs. The interior is divided, with great regularity and contrivance, into a great number of apartments; some of which are intended for the residence of the kings and queens, and for the rearing of their numerous progeny; whilst others serve as magazines, and are always well filled with stores and provisions. These hills make their first appearance above ground by a little turret or two in the shape of sugar-loaves; which only rise to the height of a foot, or a little more. Soon afterwards, at some little distance, while the former are increasing in height and size, the Termites raise others, and so go on increasing the number, and widening them at the base, till their works below are covered with these turrets, which they always raise the highest and largest in the middle, and by filling up the intervals between each turret, collect them, as it were, into one dome.

They are not very curious or exact about these turrets, except in making them very solid and strong; and when, by the junction of them, the dome is completed (for which purpose the turrets answer as scaffolds), they take away the middle ones entirely, except the tops, which, joined together, form the crown of the cupola; and they apply the clay to the building of the works within, or to erecting fresh turrets for the purpose of raising the hillock still higher.

The *royal chamber*, so called on account of its being adapted for, and occupied by, the king and queen, appears to be thought of the most consequence, being always situated as near as possible to the centre of the hillock. It resembles the shape of half an egg cut lengthways, or an obtuse oval, and is at first not above an inch long; it is afterwards, however, increased to six or eight inches, or even more, being always in proportion to the size of the queen, who, increasing in bulk as in age, at length requires a chamber of such dimensions. The floor and roof of this chamber are very solid, and are composed of hardened clay. Its walls are pierced by several door-ways or entrances, at pretty equal distances from each other, and of sufficient size to admit the soldiers and labourers, but not large enough to allow the king and the queen (the latter being, at full size, a thousand times the weight of a king) to pass out. Surrounding the royal chamber are a number of others, of different shapes and sizes, but all of them arched: these are occupied by the soldiers and labourers that guard the pair, on whose safety depends the happiness, and probably even the existence, of the whole community. These apartments, being connected together by openings and passages, form an intricate labyrinth, which extends a foot or more in diameter from the royal chamber on every side; and they are surrounded by the magazines and nurseries. The former are chambers of clay; and are always well fitted with a kind of provisions, which appear to consist of the gums or other thick juices of plants. The nurseries, which are so called because they are invariably found to contain eggs and young ones, are entirely composed of wooden materials, seemingly joined together with gums. These nurseries are exceedingly compact, and divided into very small irregularly-shaped chambers, not one of which is to be found half an inch in width. They are placed as near as possible to the royal apartments. When the nest is in the infant state, they are close to the royal chamber; but as, in process of time, the queen enlarges, it becomes necessary to enlarge this chamber for her accommodation; and as she then lays a greater number of eggs, and requires a greater number of attendants, so is it necessary to enlarge and increase the number of the adjacent apartments; for which purpose, the small nurseries that were first built are taken to pieces, and are rebuilt a little farther off. The nurseries are always found slightly overgrown with mould, and plentifully sprinkled with white globules, about the size of a small pin's head. These may at first be mistaken for eggs; but on being examined under a microscope, they evidently appear to be a species of fungus, in shape like a young mushroom. The nurseries are enclosed in chambers of clay, like those which contain the provisions, but much larger. In the early state of the nest they are not larger than a hazel nut; but in old hills are often as large as the head of a child a year old. Under the dome is a large open space, which is surrounded by three or four large gothic-shaped arches, which are sometimes two or three feet high in front of the area, but diminish very rapidly as they recede from thence, and are soon lost among the innumerable chambers and nurseries behind them. There are, comparatively speaking, few openings into the great area, and they, for the most part, seem intended only to admit into the nurseries that genial warmth which the dome collects.

The subterraneous passages which run under the lowest apartments in the hill, in various directions, are of an astonishing size, being wider than the bore of a large cannon. These passages or galleries, which are very thickly lined with the same kind of clay of which the hill is composed, ascend the inside of the external shell in a spiral manner, winding round the whole building up to the top, and intersecting each other at different heights, opening either immediately into the dome in various places, or into the interior buildings, the new turrets, &c., and sometimes communicating therewith by other galleries of different diameters, either circular or oval. Under the ground there are a great many which lead downwards by sloping descents, three and four feet perpendicularly among the gravel; from this the labouring Termites cull the finer parts, which being worked up in their mouths to the consistence of mortar, form that solid clay or stony substance of which all their hills and buildings, except the nurseries, are composed. Other galleries again ascend, leading out horizontally on every side, and are carried under ground, near to the surface, a vast distance; for if you destroy all the nests within a hundred yards of your house, the inhabitants of those which are unmolested further off will nevertheless carry on their subterraneous galleries, and invade the goods and merchandize contained in it, by undermining them, and do great mischief if you are not very circumspect. Sometimes their passages cannot be continued under ground in the required direction; and the Termites then make pipes or covered ways along its surface, composed of the same materials as the nests. These they continue, with many windings and ramifications, for great lengths; and they construct, where it is possible, subterranean pipes running parallel with them, into which they may sink and save themselves, if their galleries above ground are destroyed by violence, or the tread of men or animals alarm them.

As we before observed, each community of Termites consists of a king and queen, soldiers, and labourers. The labourers are the most numerous, there being at least a hundred of them to one soldier; they are about a quarter of an inch long, run extremely fast, and appear to be incessantly occupied. The second order, or soldiers, have a very different form from the labourers, but they are in fact the same insects, only they have undergone their first metamorphosis, and approached one degree nearer to the perfect state. They are now much larger, being half an inch long, and equal in bulk to fifteen of the labourers. The third order, or insect in its perfect state, varies in form still more than ever. The head, thorax, and abdomen differ almost entirely from the same parts in the labourers and soldiers; and besides this, the animal is now furnished with four fine, large, brownish, transparent wings, with which, at the time of emigration, it is to wing its way in search of a new settlement. In their winged state they are also much altered in size as well as form. Their bodies now measure between six and seven tenths of an inch in length, and their wings above two inches and a half from tip to tip, and they are equal in bulk to about thirty labourers or two soldiers. They are now also furnished with two large eyes, one on each side of the head; if they have any before, they are so small as not easily to be distinguished; and as they live, like moles, always under ground, they have as little occasion for these organs; but the case is widely different when they arrive at the winged state, in which they are to roam, though but for a brief space, through the air, and explore new and distant regions. In this form the animal comes abroad during or soon after the first tornado, which at the latter end of the dry season proclaims the approach of the ensuing rains, and seldom waits for a second or third shower, if the first, as is generally the case, happens in the night and brings much wet after it. The numbers that are to be found next morning all over the surface of the earth, but particularly on the water, is astonishing; for their wings are only calculated to carry them a few hours; and after the rising of the sun not one in a thousand is to be found with four wings, unless the morning continues rainy, when here and there a solitary being is seen winging its way from one place to another, as if solicitous only to avoid its numerous enemies, particularly various species of ants, which are hunting on every spray, on every leaf, and in every possible place, for this unhappy race, of which probably not a pair in many millions get into a place of safety, fulfil the first law of nature, and lay the foundation of a new community.— Not only do ants, birds, and reptiles destroy them, but even the inhabitants of the country eagerly seek after these wingless creatures and devour them with the greatest avidity. It is, indeed, wonderful that a pair should ever escape so many dangers and get into a place of security. Some, however, are so fortunate; and being found by some of the labouring insects that are continually running about the surface of the ground under their covered galleries, are elected kings and queens of new states; all those which are not so elected and preserved certainly perish, and most probably in the course of the following day. The manner in which these labourers protect the happy pair from their innumerable enemies, not only on the day of the massacre of almost all their race, but for a long time afterwards, seems to justify the use of the term *election*. The little industrious creatures immediately enclose them in a small chamber of clay suitable to their size, into which at first they leave but one small entrance, large enough for themselves and the soldiers to go in and out, but much too small for either of the royal pair to make use of; and when necessity obliges them to make more entrances, they are never larger: so that, of course, the voluntary subjects charge themselves with the task of providing for the offspring of their sovereigns, as well as to work and fight for them, until they shall have raised a progeny capable at least of dividing the task with them.

About this time a most extraordinary change begins to take place in the queen, to which we have nothing similar, except in the Jigger of the West Indies (*Pulex penetrans* of Linnæus), and in the different species of *Coccus*. The abdomen of this female begins gradually to extend and enlarge to such an enormous size, that in an old queen it will increase so as to become fifteen hundred or two thousand times the bulk of the rest of her body, and twenty or thirty thousand times the bulk of a labourer! The skin between the segments of the abdomen extends in every direction; and at last the segments are removed to half an inch distance from each other, although at first the length of the whole abdomen is not half an inch. They preserve their dark brown colour, and the upper part of the abdomen is marked with a regular series of brown bars throughout its entire length, while the intervals between them are covered with a thin, delicate, transparent skin, and appear of a fine cream colour, a little shaded by the dark colour of the intestines and watery fluid seen here and there beneath. The animal is supposed to be upwards of two years old when the abdomen is increased to three inches in length; and they are sometimes found nearly twice that size. The abdomen is now of an irregular oblong shape, being contracted by the muscles of every segment, and is become one vast matrix full of eggs, which make long circumvolutions through an immense number of very minute vessels that circulate round the inside in a serpentine manner, which would exercise the ingenuity of a skilful anatomist to dissect and develope. This singular matrix is not more remarkable for its amazing extension and size, than for its peristaltic motion, which resembles the undulation of waves, and continues incessantly without any apparent effort of the animal; so that one part or other is alternately rising and falling in perpetual succession, and the matrix seems never at rest, but is always protruding eggs, to the number of sixty in a minute in old queens, or eighty thousand and upwards in one day of twenty-four hours. These eggs are instantly taken from the body of the queen by her attendants (of whom there always are, in the royal chamber, and galleries adjacent, a sufficient number in waiting), and carried to the nurseries, some of which in a large nest may be four or five feet distant, in a straight line, and consequently much farther by their winding galleries. Here, after they are hatched, the young are attended and provided with every thing necessary, until they are able to shift for themselves, and take their share of the labours of the community.

When a person accidentally enters any solitary grove, where the ground is pretty well covered with their arched galleries, the Termites give the alarm by loud hissings, which may be distinctly heard at every step: soon after this, their galleries may be searched in vain for the insects; but little holes are found, just large enough to admit of their escape into the subterraneous roads. These galleries are of sufficient size to allow the Termites to pass and repass without stopping each other (though there are always numerous passengers), and to shelter them equally from light and air, as well as from their enemies,—of which the Ants, from being the most numerous, are the most formidable. If the Termites are dislodged from their covered ways, the various species of Ants (which are probably as numerous above ground as the Termites are in their subterranean passages) instantly seize and drag them away to their nests, to feed their young brood. The Termites are, therefore, exceedingly solicitous about preserving their covered ways in good repair; and if one of these be demolished for a few inches in length, it is wonderful how soon they will rebuild it. At first, in their hurry, they run into the open part an inch or two, but stop so suddenly that it is evident they are surprised; for though some will run straight on, and get under the further part of the arch as quickly as possible, most of them run back as fast, and very few will venture through that part of the gallery which is left uncovered. In a few minutes they may be seen rebuilding the arch; and even if three or four yards of their gallery have been destroyed, it will be restored by the next morning, and will be found to contain numerous Termites passing along in both directions. If the gallery be several times destroyed, they will at length seem to give up the point, and build another in a different direction; but if the old one led to some favourite plunder, they will rebuild it again in a few days; and unless the nest be destroyed, they will never totally abandon their gallery.

The Termites generally make their approaches to the nest under ground, descending below the foundations of houses and stores at several feet from the surface, and rising again either in the floors, or entering at the bottoms of the posts of which the sides of the buildings are composed, following the course of the fibres to the top, and having lateral perforations or cavities here and there. While some of them are employed in gutting the posts, others ascend from them, entering a rafter or some other part of the roof, in search, as it would seem, of the thatch, which appears to be their favourite food; and if they find it, they bring up wet clay, and build galleries through the roof in various directions, as long as it will support them. In this manner a wooden house is speedily destroyed; and all that it contains is, at the same time, subjected to the ravages of these destructive insects. In carrying on this business, they sometimes find, by some means or other, that the post has a certain weight to support, and then, if it is a convenient track to the roof, or is itself a kind of wood agreeable to them, they bring their mortar; and, as fast as they take away the wood, replace the vacancy with that material, which they work together more closely and compactly than human strength or art could ram it. Hence, when the house is taken to pieces, in order to examine if any of the posts are fit to be used again, those made of the softer kinds of wood are often found reduced almost to a shell; and almost all of them are transformed from wood to clay, as solid and as hard as many kinds of stone that is used for the purposes of building.

Another African species (*Termes arborum*) builds its nest among the branches of trees, sometimes at the height of sixty or eighty feet from the ground. They also frequently establish their nests within the roofs and other parts of houses, to which they do considerable damage if not timely extirpated. They are not, however, so destructive or so difficult to be guarded against as the species we have been so minutely describing.

TERN. (*Sterna.*) A genus of web-footed birds readily distinguished by the great length of their wings and their forked tail, as well as by the form of their beak. Their nourishment consists almost exclusively of small live fishes, which they seize upon while on the wing, descending like a shot to the water, and capturing their prey. They are pretty generally diffused over the globe, and are abundant in the temperate regions. There are several species.

The COMMON TERN, GREAT TERN, or SEA-SWALLOW. (*Sterna hirundo.*) This bird is upwards of fourteen inches in length: the bill is crimson, tipped with black, and about two inches and a quarter long; the forehead, top of the head, and the long occipital feathers are deep black; the hinder part of the neck, the back, and wings, are bluish-ash: the under parts are pure white, the breast excepted, which is slightly shaded: the tail, which is long and greatly forked, is white, except the two outside feathers, which are black on their exterior webs; the legs and feet are red. This clean-looking pretty bird is common in the summer months on the sea-coasts, rivers, and lakes of the British isles, and is also met with in various parts of Europe and Asia. The female forms her nest in the moss or long coarse grass, near the lake, and lays three or four eggs of a dull olive, marked with different sized black spots at the thicker end. It is a bold bird, and during the period of incubation will attack any person approaching too near its nest.

"The flight of the Great Tern," says Wilson, "and, indeed, of the whole tribe, is not in the sweeping, shooting manner of the land swallows, notwithstanding their name; the motions of their long wings are slower, and more in the manner of the gull. They have, however, great powers of wing and strength in the muscl s of the neck, which enable them to make such sudden and violent plunges, and that from a considerable height too, headlong on their prey, which they never seize but with their bills." Mr. Gould, in his great work on the 'Birds of Europe,' says, "The Common Tern does not confine itself entirely to the sea, but frequently resorts to inland streams, &c.; and when thus ascending our creeks and rivers these little fairies of the ocean fearlessly fish around our boats, nothing can be more pleasing than to observe their poise and dip. When with their scrutinizing eyes they have observed a fish sufficiently near the surface, they precipitate themselves upon it with unnerring certainty, and a rapidity that is truly astonishing: this mode of capture strongly reminds us of the fissirostral tribe among the land birds; and they may indeed be truly termed the swallows of the ocean, their long and pointed wings, and small but muscular bodies, being admirably adapted for rapid and sustained flight, and affording the means by which they are enabled to traverse the surface of the deep with never-tiring wings."

The LITTLE or LESSER TERN (*Sterna minuta*), which is only about nine inches in length, looks like the preceding in miniature; is equally if not more delicately elegant in its plumage and general appearance, and its manners and habits are very similar; but it is not nearly so numerous, or so widely dispersed. It differs from the Common Tern in having the black patch on its head bounded by a white line on the front of the brow, and over each eye; in the tail being wholly white; and, in proportion to the size of the bird, much shorter or less forked; while the bill and the feet are more inclined to orange or yellow. Nothing can exceed the clean, clear, and glossy whiteness of its close-set feathers on the under parts of the body; but the upper plumage is of a plain lead-coloured gray. The Lesser Tern feeds on beetles, crickets, spiders, and other insects, which it picks up from the marshes, as well as on small fish, on which it plunges at sea. Like the former, it also makes extensive incursions inland along the river courses, and has frequently been shot several hundred miles from the sea. It is extremely tame and unsuspicious, often passing you on its flight, and within a few yards, as it traces the windings and indentations of the shore in search of the various small crustacea on which it delights to feed. Indeed, at such times it appears either altogether heedless of man, or its eagerness for food overcomes its apprehensions for its own safety. The eggs, which are generally four in number, are dropt on the dry and warm sand, the heat of which, during the day, is fully sufficient for the purpose of incubation; but the parent sits upon them during the night: they are of a yellowish brown colour, and nearly an inch and three-quarters long. This bird is met with in the south of Russia, and about the Black and Caspian Sea. It also inhabits the shores of England during the summer, where it breeds, and migrates to the south as the cold of autumn approaches.

TERRAPIN, or BOX-TORTOISE. (*Terrapene.*) A genus of fresh-water Tortoise; the breastplate of which is divided into two pieces by a movable articulation; and they have the power of closing their carapace when the head and limbs are withdrawn into it.

TERRICOLÆ. An order of vermiform animals, of the class *Annelida*, including two principal groups, the *Earth-worms* and the *Naiads;* the former being terrestrial, and the latter semi-aquatic. The Annelidæ of this order have a cylindrical body, tapering to a point at its extremities, and furnished only with several rows of bristles; which, although frequently invisible to the naked eye, may be plainly felt by passing the finger along the body from behind forwards; their points being directed backwards, in order to give the animal a firm hold of the earth through which it is boring. They have neither eyes, antennæ, mandibles, cirrhi, nor external gills: their bodies, however, are distinctly divided into segments; and these are marked by minute spots on each side, which are apertures leading to small respiratory sacs. [See EARTH-WORM.]

TERRIER. (*Canis familiaris terrarius.*) There are two varieties of this breed of Dogs; the one smooth, sleek, and of rather slender form; colour bright black and tan: the other, a hardy and fierce animal, known as the White-haired or Scotch Terrier; whose rough harsh hair, short muzzle, stout and short-limbs, and dirty white colour, sufficiently distinguish it from the former. The English or Common Terrier carries his head high, has a sharp muzzle, quick and bright eye, neat and compact body, erect ears, with the tips sometimes pendulous, legs slender but strong, and the tail erect and stiff. It should be observed, however, that although both these varieties of a bold, active, and useful animal are highly valued, and often preserved in all their purity, mongrel breeds are common; and therefore very many serviceable dogs, usually called Terriers, are every where to be found. Mr. Bell thus speaks of the species: "The Terrier is applied to several purposes in which its diminutive size, its strength, courage, activity, and perseverance are all called into action. In the office of unearthing the fox it is an essential addition to the pack, and a good kennel can scarcely be without them; and it takes the earth with much eagerness, from which it has received its name. But if the Terrier contribute so much to the enjoyment of the regular sportsman, it offers no less amusement to those of a less dignified character, by the feats it displays in the destruction of minor vermin,—the Badger, the Polecat, and the whole tribe of *Mustelidæ*, and particularly the Rat. The clever manner in which it deals with the largest and boldest of these savage creatures, and the rapidity with which it kills them, can scarcely be described. The celebrated dog 'Billy' was turned into a room where there were one hundred rats; the object being to decide a wager that he would kill that number within a given time. It was done in less than seven minutes. A large breed crossed with the Bull-dog, and termed the Bull-Terrier, constitutes one of the most savage and determined races of Dogs known."

TERU-TERO. (*Vanellus cayanensis.*) This is a bird of the Plover kind, which Mr. Darwin speaks of as "disturbing the stillness of the night," in the Pampas of South America. "In appearance and habits," he says, "it resembles in many respects our Peewits; its wings, however, are armed with short spurs, like those on the legs of the common cock. As our peewit takes its name from the sound of its voice, so does the Teru-tero. While riding over the grassy plains, one is constantly pursued by these birds, which appear to hate mankind, and I am sure deserve to be hated, for their never-ceasing, unvarying, harsh screams. To the sportsman they are most annoying, by telling every other bird and animal of his approach: to the traveller in the country, they may possibly, as Molina says, do good, by warning him of the midnight robber. During the breeding season, they attempt, like our peewits, by feigning to be wounded, to draw away from their nests dogs and other enemies. The eggs of this bird are esteemed a great delicacy."

TESTACELLA. [See SLUG.]

TESTUDINATA. The name of a tribe of Chelonian reptiles, of which the Tortoise (*Testudo*) is the type.

TESTUDO. [See TORTOISE.]

TETRABRANCHIATA. The name of an order of *Cephalopods*, which are nearly extinct; the only remaining representative of it being the Pearly Nautilus (*Nautilus pompilius*).

TETRAMERA. The name given to the third general section of the *Coleoptera*, comprising exclusively those species which have four distinct joints to all the tarsi. All these insects feed upon vegetable substances: their larvæ have generally short feet, or they are wanting and replaced by fleshy lobes. The perfect insect is found upon the flowers and leaves of plants. The larvæ of many of these beetles live mostly hidden in the interior of vegetables, and are generally deprived of feet, or have them very minute. Some of the larger kinds devour the hard and ligneous particles.

TETRAO: TETRAONIDÆ. A genus and family of Rasorial Birds. [See GROUSE: PTARMIGAN.]

TETRAODON. A genus of bony fishes belonging to the order *Plectognathi*. Like the Diodon, they have the faculty of inflating themselves, by filling with air a thin and extensile membranous sac, which adheres to the peritoneum the whole length of the abdomen. When thus inflated, they roll over, and float with the belly uppermost, without any power of directing their course. Each jaw of the Tetraodon is marked with a suture, so as to give the appearance of four teeth. The spines are small and low; and

GLOBE FISH.—(TETRAODON HISPIDUS.)

some species are reckoned poisonous. One is electrical (*Tetraodon lineatus*), straight, brown and whitish: it is found in the Nile, cast on shore by the inundations, and collected by the children as a plaything.

TETRAPTURUS. A genus of Acanthopterygious fishes, nearly allied to the *Xiphias*, or Sword-fish, inhabiting the Mediterranean. The beak is shaped like a stiletto; each ventral fin consists of one jointless blade; and there are two small crests on each side of the base of the tail, as in the Mackerel, which appear to steady that powerful organ. [See SWORD-FISH.]

TETTIGONIA: TETTIGONIADÆ. A genus and family of Hemipterous insects, to which the name of "leaf-hoppers" has been applied. They have the head and thorax somewhat like those of the Frog-hoppers, but their bodies are, in general, proportionally longer, not so broad across the middle, and not so much flattened. The thorax is wider than long, with the front margin curving forwards, the hind margin transverse, or not extended between the wing-covers, which space is filled by a pretty large triangular scutel or escutcheon. The wing-covers are generally opaque, and moulded somewhat to the form of the body. The eyes, which are placed at the sides of the head are pretty large, but flattish, and not globular, as in the Cicadas. Notwithstanding the small size of most of these insects, they are deserving our attention on account of their beauty, delicacy, and surprising agility, as well as for the injury sustained by vegetation from them.

Tettigonia Vitis, which for many years was supposed to be the common European "vine-fretter," is a small insect, as Dr. Harris informs us, abundant in Massachusetts, United States, and in its perfect state measuring only one-tenth of an inch in length. It is of a pale yellow or straw-colour; there are two little red lines on the head; the back part of the thorax, the scutel, the base of the wing-covers, and a broad band across their middle, are scarlet; the tips of the wing-covers are blackish, and there are some little red lines between the broad band and the tips. The head is crescent-shaped above, and the eyelets are situated just below the ridge of the front. The vine-hoppers, as they may be called, inhabit the foreign and the native grape-vines, on the under surface of the leaves of which they may be found during the greater part of the summer; for they pass through all their changes on the vines. They make their first appearance on the leaves in June, when they are very small and not provided with wings, being then in the larva state. During most of the time they remain perfectly quiet, with their beaks thrust into the leaves, from which they derive their nourishment by suction. If disturbed, however, they leap from one leaf to another with great agility. As they increase in size they have occasion frequently to change their skins; and great numbers of their empty cast-skins, of a white colour, will be found, throughout the summer, adhering to the under sides of the leaves, and upon the ground beneath the vines. When arrived at maturity, which generally occurs during the month of August, they are still more agile than before, making use of their delicate wings as well as their legs in their motions from place to place; and, when the leaves are agitated, they leap and fly from them in swarms, but soon alight and begin again their destructive operations. The infested leaves at length become yellow, sickly, and prematurely dry, and give to the vine at Midsummer the aspect it naturally assumes on the approach of winter. But this is not the only injury arising from the exhausting punctures of the vine-hoppers; the plant languishes, and, if the evil be allowed to go on unchecked, in a few years the vines become exhausted, barren, and worthless. In the autumn the vine-hoppers desert the vines, and retire for shelter during the coming winter beneath fallen leaves and among the decayed tufts and roots of grass, where they remain till the following spring, when they emerge from their winter quarters, and in due time deposit their eggs upon the leaves of the vine, and then perish.

THALARCTOS. [See BEAR, POLAR.]

THALASSIDROMA. A genus of web-footed Birds closely allied to Procellaria, and commonly called Stormy Petrel, under which word, two or three species are described. We may here describe another oceanic species, the

THALASSIDROMA LEUCOGASTER, or WHITE-BELLIED STORMY PETREL. This is a fine and powerful species of the Petrel family of birds; easily distinguished from all others by the total absence of black down the centre of the abdomen, and the shortness of its toes. It is seen (says Mr. Gould) fluttering over the glassy surface of the ocean during calms with an easy butterfly-like motion of the wings, and buffeting with equal vigour the crests of the loftiest waves of the storm; at one moment descending into their deep troughs, and at the next rising with the utmost alertness to their highest points, apparently from an impulse communicated as much by striking the surface of the water with its webbed feet, as by the action of the wings. The head and neck is of a deep sooty black; back grayish black, each feather margined with white; wings and tail black; chest, all the under surface, and the upper tail-coverts, white; bill and feet jet-black. Like the other members of the genus, it feeds on mollusca, the spawn of fish, and any kind of fatty matter that may be floating on the surface of the ocean. [See PETREL.]

THECLA. A genus of diurnal Lepidoptera, abounding in exotic species, but of which only six or seven are met with in this country. They are called by collectors "*Hair-streaks*," from the under-side of the wings being frequently ornamented with two or three delicate, straight, or zigzag pale lines on a dark ground. We particularize

The THECLA QUERCUS, or PURPLE HAIR-STREAK BUTTERFLY. About the middle of July this species of the Papilionaceous tribe is seen frequenting the tops of lofty oaks and ash trees. Its wings are dusky black above: the male with the disc of the anterior deep glossy blue, formed of an oblong patch, and extending towards the anal angle; the female with the entire disc purple, and a dusky margin: beneath,

PURPLE HAIR-STREAK BUTTERFLY.
(THECLA QUERCUS.)

both sexes are similar; the anterior wings are cinereous, with a short white streak on the costa towards the apex; between which and the posterior margin the wing is paler, with a few whitish spots: the posterior wings are similar at the base, and have an undulated white streak, slightly edged internally with dusky; beyond this are two rows of whitish crescents, with a

PURPLE HAIR-STREAK BUTTERFLY.
UNDER SIDE.

fulvous spot at the anal angle. Body black above, cinereous beneath; tail black; antennæ black, faintly annulated with cinereous. The purple blotch on the anterior wings of the male varies greatly in size; and the wings of the female are sometimes only slightly purpurescent. Caterpillar lightish brown, with three rows of green dots; it feeds on the oak. Chrysalis rust-coloured, with three rows of brown dots.

THECLA PRUNI; or, PLUME HAIR-STREAK BUTTERFLY. This insect is in many parts of England considered scarce, while in others it frequently abounds; myriads, indeed, may sometimes be seen hovering over the flowers and bramble blossoms in one district, though in another part, not very remote perhaps, hardly one is to be seen. From the beginning to the middle of July is the usual time of its appearing. Wings above deep black or brown, immaculate; beneath paler; anterior with a transverse abbreviated white streak on the costa towards the apex: posterior wings with a similar streak, which becomes of a zig-zag form, gradually lessening towards the inner margin: beyond this is an irregular deep-rufous orange marginal band, edged internally with black, and sometimes accompanied by a narrow white streak, and spotted with black externally; the tailed appendages are black, tipped with white, those of the females being the longest. Body black above, drab-colour beneath; the legs bluish; antennæ black, with white rings, and an orange tip. Caterpillar dusky-green, with whitish lateral lines; the back dentated: it feeds on the blackthorn. Chrysalis dusky-brown, with a white head.

THECLA RUBI; or GREEN HAIR-STREAK BUTTERFLY. This is a pretty but not very abundant species; it frequents hedges and bramble bushes, upon the buds

GREEN HAIR-STREAK BUTTERFLY.
(THECLA RUBI.)

of which shrub its larvæ feed. Colour of the wings above dusky brown, with the nervures blackish: beneath green, the anterior wings

THECLA RUBI—CATERPILLAR AND CHRYSALIS

usually immaculate, with the thinner margin pale dusky-brown: the posterior wings not tailed, but denticulated on the hinder margin with an interrupted series of white dots: the cilia, both above and below, are brown, dotted with black on the posterior wings: body deep brown above, pale beneath. Caterpillar green and yellow, with black head: it feeds on the bramble, saintfoin, and broom. Chrysalis brown.

THECLA BETULÆ; or BROWN HAIR-STREAK BUTTERFLY. This insect resorts chiefly to birch woods, but cannot be considered a very common species any where. Wings above dark brown: the anterior with a transverse black streak at the apex of the

BROWN HAIR-STREAK BUTTERFLY.
(THECLA BETULÆ—MALE.)

basal areolet; beyond which, in the female, is a large kidney-shaped orange spot, and in the male a slight fulvous cloud: the pos-

THECLA BETULÆ—FEMALE,

terior wings have a tawny spot on the inner angle, and a streak of the same on the tail: beneath, the sexes resemble each other, but the colours are more vivid in the female: all the wings are orange-tawny, with a bright orange margin; the posterior ones

THECLA BETULÆ—CATERPILLAR
AND CHRYSALIS.

have an oblique slightly-waved elongate orange band, with black inner and white outer margin: the anal angle is spotted with black, and the cilia on the anal areolets have a fuscous stripe: the body is brown above, cinereous beneath; the antennæ black, annulated with white. Caterpillar green, with oblique yellowish streaks on the sides, and two yellow dorsal lines: it feeds on the birch and blackthorn: the chrysalis is reddish-brown, with paler streaks.

THECODONTOSAURUS. [See SUPPL.]

THORNBACK. (*Raia clavata.*) A well-known fish of the *Raiadæ* family, rarely equal to the Skate in point of size. It is an inhabitant of the Mediterranean; and is taken in great abundance in the spring and summer (when it visits the shallows for the deposition of its eggs) on the Cornish coast, and also on the coasts of Scotland and Ireland. The colour of the skin is brownish

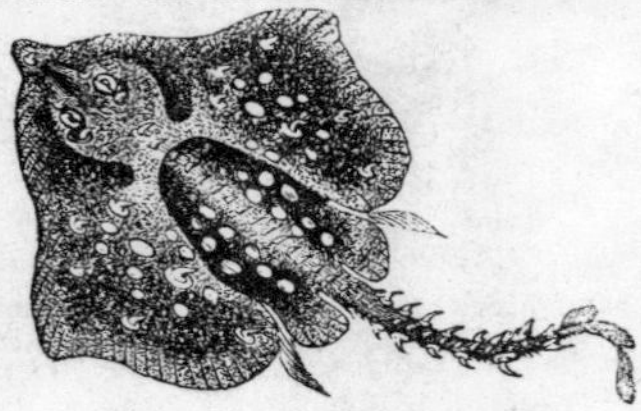

THORNBACK.—(RAIA CLAVATA.)

gray, with irregular dusky variegations; and of a rough or shagreen-like surface: the under part is white, with a slight cast of flesh colour. Its whole upper surface is covered with strong curved spines, which are most conspicuous down the middle and on each side of the back, where four or six of much larger size than the rest are generally seen. The back is marked with a number of pale round spots, of different sizes, and which are commonly surrounded with a dark-coloured edge. Along the middle of the back runs a single row of strong spines, continued to the tip of the tail; and it often happens that there are three, or even five rows of spines on this part. The tail is furnished with two membranous fins on the upper central ridge, and ends with a small dilatation. The Thornback is in the best condition for table about November. They feed on various other fish, particularly flat-fish, testaceous mollusca, and crustacea.

THORN [MOTHS]. A name given by collectors to Moths of the genus *Geometra*.

THRUSH. (*Turdus.*) Birds of the family *Turdidæ*, or Thrushes, are extremely numerous, and are found in nearly every part of the world; the several species being adapted to almost every climate. They generally frequent the fields and pastures for their food, which usually consists of soft animal and vegetable substances, as berries and other fruits, worms, and snails. Some species are remarkable for their power and variety of song, and others for their powers of imitation. Their nests are generally constructed on the branches of trees, and most of the species lay from four to six eggs. They have the beak arcuated and compressed, but its point is not hooked. Their habits are in general solitary; but the majority, however, are gregarious during the winter. The females of all the Thrush kind, though somewhat less brilliant, are very similar in plumage to the males.

The SONG-THRUSH, MAVIS, or THROSTLE (*Turdus musicus*), is a well-known and much admired bird in this country, charming us not only with the sweetness, but the variety of its song, which it commences early in the spring, and continues to the beginning of autumn. It measures nine inches in length: its beak is dusky, the under mandible yellowish at the base: head, and upper parts of the body, yellowish brown, with a few obscure dusky lines on the former: the throat, neck, and sides are yellowish; the breast white, spotted with dusky; and the abdomen white: the under wing-coverts dull orange yellow: legs light brown. The food of the Song-thrush consists of insects and berries; it is also particularly fond of snails, the shells of which it breaks against stones. The female builds her nest generally in bushes; it is composed of dried grass and green moss, with a little earth or clay intermixed, and lined with rotten wood: she lays four or five eggs, of a pale blue colour, marked with dusky spots on the larger end.

Bewick says, that although this species is not considered migratory with us, it has, nevertheless, been observed in some places in great numbers during the spring and summer where not one was to be seen in the winter; which has induced an opinion that they either shift their quarters entirely, or take shelter in the more retired parts of the woods. They have been observed to pass through Courland, Prussia, &c. in great numbers, in their way to the Alps: and in France they are migratory, visiting Burgundy when the grapes are ripe, and committing great ravages among the vineyards.

In the Journal of a Naturalist, the habits of this bird are thus pleasantly noticed: "The Throstle is a bird of great utility in a garden where wall-fruit is grown, by reason of the peculiar inclination which it has for feeding upon snails, and very many of them he does dislodge in the course of a day. When the female is sitting, the male bird seems to be particularly assiduous in searching them out, and I believe he feeds his mate during that period, having frequently seen him flying to the nest with food, long before the eggs were hatched: after this time the united labours of the pair destroy numbers of these injurious creatures. That he will regale himself frequently with a tempting gooseberry or bunch of currants, is well known; but his services entitle him to a very ample reward. The Blackbird associates with these Thrushes in our gardens, but makes no compensation for our indulgences after his song ceases, as he does not feed upon the snail; but the Thrush benefits us through the year, by his properties for this particular food, and every grove resounds with his harmony in the season; and probably if this race suffered less from the gun of the Christmas popper, the gardener might find much benefit, in his ensuing crop of fruit, from the forbearance."

The MISSEL THRUSH, or STORMCOCK, (*Turdus viscivorus*), is the largest of the European Thrushes, being nearly twelve inches in length. Bill dusky; eyes hazel; the head, back, and lesser coverts of the wings olive brown, the latter tipped with dull brownish white; the lower part of the back and rump tinged with yellowish brown and ash; sides of the head and throat yellowish white, spotted with brown; from thence to the vent white, with dusky spots; those of the breast triangular in shape, and of the belly and sides roundish: tail feathers brown, the three outermost tipped with white: legs yellow; claws black. Its food consists principally of berries, those of the misletoe being its favourite; from whence the ancients erroneously believed that the plant could not vegetate without having passed through its body,—hence the proverb, "*Turdus malum sibi cacat.*" It will also eat insects, and their larvæ, with which it feeds its young. It builds its nest in the fork of low trees, particularly those that are covered with moss; coarse grass, woven together with wool, and a lining of fine dry grass, being the materials. The eggs are four or five in number, of a flesh-colour, varied with deep and light rust-coloured spots. This bird is common throughout all Britain, and resident at all seasons. It is very wild and distrustful, except at the season of propagation, when it approaches the vicinity of human habitation, and is remarkable for the spirit with which it attacks and drives away Magpies, &c. from near its nest, uttering a loud harsh shriek. Its song is powerful and monotonous; and if the weather be mild, it will begin to sing at the commencement of the year.

That most entertaining naturalist, Charles Waterton, Esq., whose art of story-telling is only excelled by the soundness of his ornithological observations, concludes his remarks on this bird as follows:—"The Stormcock surpasses all other Thrushes in size, and is decidedly the largest songster of the European birds. He remains with us the whole of the year; and he is one of three birds which charm us with their melody during the dreary months of winter, when the Throstle and the Lark are silent, and all the migratory birds have left us, to sojourn in warmer climates. On this account I prize him doubly. He appears to be gregarious in the months of August and September. I have occasionally counted from forty to fifty of these birds in a flock; and I suspect they are sometimes mistaken for an early arrival of fieldfares, by those who pay attention to the migration of birds. The Stormcock is remarkably fond of the berries of the mountain-ash. He who loves to see this pretty songster near his dwelling would do well to plant a number of mountain-ashes near his pleasure-grounds: they are of quick growth, and they soon produce an abundance of berries.

"Whilst the fruit of these trees affords a delicious repast to the Stormcock, the branches that bear the berries are well known to be an effectual preservative against the devilish spells of witchcraft. In the village of Watton I have two small tenants: the name of one is James Simpson, that of the other Sally Holloway; and Sally's house stands a little before the house of Simpson. Some three months ago I overtook Simpson on the turnpike road, and I asked him if his cow were getting better, for his son had told me she had fallen sick. 'She's coming on surprisingly, Sir,' quoth he. 'The last time that the cow-doctor came to see her, 'Jem,' said he to me, looking earnestly at old Sally's house, 'Jem,' said he, 'mind and keep your cow-house door shut before the sun goes down, otherwise I won't answer what may happen to the cow.'—'Ay, ay, my lad,' said I, 'I understand your meaning; but I am up to the old slut, and I defy her to do me any harm now.'—'And what has old Sally been doing to you, James?' said I.—'Why, Sir,' replied he, "we all know too well what she can do. She has long owed me a grudge; and my cow, which was in very good health, fell sick immediately after Sally had been seen to look in at the door of the cow-house, just as night was coming on. The cow grew worse and worse; and so I went and cut a bundle of wiggin (mountain-ash), and I nailed the branches all up and down the cow-house; and, Sir, you may see them there if you will take the trouble to step in. I am a match for old Sally now, and she can't do me any more harm, so long as the wiggin branches hang in the place where I have nailed them. My poor cow will get well in spite of her.' Alas! thought I to myself, as the deluded man was finishing his story, how much there is yet to be done in our part of the country by the schoolmaster of the nineteenth century!"

The HERMIT THRUSH. (*Turdus solitarius.*) The favourite native haunts of this silent and recluse species, according to Wilson, are the dark solitary cane and myrtle swamps of the southern States of America. It has been supposed, he says, to be only a variety of the Wood-thrush; but it is considerably less, being only about seven inches in length, and altogether destitute of the clear voice and musical powers of that charming minstrel. Its upper parts are a plain, deep olive-brown; lower, dull white; upper part of the breast and throat, dull cream colour, deepest where the plumage falls over the shoulders of the wing, and marked with large dark brown pointed spots; ear, feathers, and line over the eye, cream, the former mottled with olive; edges of the wings lighter, tips dusky, tail-coverts and tail, inclining to a reddish fox-colour. Tail slightly forked; legs dusky; bill black above, and at the tip, whitish below.

The WOOD THRUSH (*Turdus melodus*) is about eight inches in length: the whole of the upper parts of the body are fulvous brown, brightest on the head, and inclining to olive on the rump and tail: throat and breast white, tinged with buff, and sprinkled all over with dusky spots: belly and vent pure white: eyes surrounded with a white circle: legs and claws flesh colour. Very little difference in the colour of the sexes. This species inhabits the whole of North America, from Hudson's Bay to Florida: its song is heard every morning and evening during the months of May and June, and is greatly admired; but during the day it is silent: its favourite haunts are thick shaded hollows by the sides of brooks or rivulets: its nest, made of withered beech leaves with layers of dry grass mixed with mud, and lined with dry fibrous roots, is often placed in an alder bush. Its eggs are four or five in number, and of a light blue colour.

The REDWING THRUSH, (*Turdus iliacus*), like the Fieldfare, which it much resembles, is migratory, generally arriving in Britain about the latter end of September, and departing gradually, not in flocks, in the spring. It is about eight inches and a half in length: the flanks and beneath the wings are deep rufous; the back brown, inclining to olive green; a conspicuous pale streak over the eye; and longitudinal markings on the under parts. It is abundant in Norway, Sweden, and Prussia. Its nest is placed in a low bush or shrub; and it lays five or six blue-green eggs; spotted with black. Its song is not very attractive.

In Mr. Hewitson's elegant work on the Eggs of Birds, is the following interesting account of the Redwing: "In our long rambles through the boundless forest scenery of Norway, or during our visits to some of its thousand isles, whether by night or by day, the loud, wild, and most delicious song of the Redwing seldom failed to cheer us. Unlike its neighbour the fieldfare, it was solitary and shy, and on our approach to the tree on the top of which it was perched, would drop down and hide itself in the thick of the brushwood. Throughout that part of the country which we visited, it is known by the name of Nightingale, and well it deserves to be so; to a sweeter songster I have never listened. Like the nightingale of more southern skies, its clear sweet song would occasionally delight us during the hours of night, if the two or three delightful hours of twilight which succeed the long day of a Norwegian summer can be called night. The birds, like the other inhabitants of the country, seem loth to lose in sleep a portion of this short-lived season.

"Anxious to extend our researches onwards, in the hope that as we proceeded north we should prove more successful, we had lingered but little to search for the nest and eggs of the Redwing, and our inquiries with regard to them had been unavailing. One afternoon, as we approached the seacoast, and at the same time the northern limit of a beaten road, we discovered a nest of the Redwing, but to our great disappointment it had young ones. Having almost reached the boundary of our woodland rambles for the present, we spent the whole of the following day in exploring the beautiful woods by which we were on all sides surrounded. We found a second nest of the Redwing, but the eggs were again hatched. The nest of the Redwing is placed, like those of the thrush and blackbird, in the centre of a thorn or other thick bush. It is similar to those of the blackbird, fieldfare, and ring ouzel. Outwardly, it is formed of moss, roots, and dry grass; inwardly, cemented with clay, and again lined with finer grass."

The author of 'The Journal of a Naturalist' says it is well known to every sportsman that the Redwing and the Fieldfare feed chiefly upon "heps and haws," the fruit of the white thorn and the wild rose. Yet he admits that "these birds, generally speaking, give the preference to insect food and worms; and when flights of them have taken their station near the banks of large rivers, margined by lowlands, we shall find that the bulk of them will remain there, and feed in those places; and, in the uplands, we shall observe small restless parties only. But in the midland and some other counties, the flocks that are resident have not always these meadows to resort to, and they then feed on the haws as long as they remain. In this county, the extensive lowlands of the river Severn in open weather are visited by prodigious flocks of these birds; but as soon as snow falls, or hard weather comes on, they leave these marshy lands, because their insect food is covered or become scarce, visit the uplands to feed on the produce of the hedges, and we see them all day long passing over our heads in large flights on some distant progress, in the same manner as our larks, at the commencement of a snowy season, repair to the turnip fields of Somerset and Wiltshire. They remain absent during the continuance of those causes which incited their migration; but, as the frost breaks up, and even before the thaw has actually commenced, we see a large portion of these passengers returning to their worm and insect food in the meadows, attended probably by many that did not take flight with them—though a great number remain in the upland pastures, feeding promiscuously as they can."

The RED-BREASTED THRUSH. (*Turdus migratorius.*) This species of the Thrush, to which the name of the ROBIN is also commonly applied, is one of the loudest and most delightful songsters of the North American continent. "His notes," as Dr. Richardson truly remarks, "resemble those of the common Thrush, but are not so loud. Within the arctic circle the woods are silent in the bright light of noon-day, but towards midnight, when the sun travels near the horizon, and the shades of the forest are lengthened, the concert commences, and continues till six or seven in the morning. Even in these remote regions, the assertion of those naturalists who have declared that the feathered tribes of America are void of harmony, might be fully disproved. Indeed, the transition is so sudden from the perfect repose, the deathlike silence of an arctic winter, to the animated bustle of summer; the trees spread their foliage with such magical rapidity, and every succeeding morning opens with such agreeable accessions of feathered songsters to swell the chorus — their plumage as gay and unimpaired as when they enlivened the deep-green forests of tropical climes, that the return of a northern spring excites in the mind a deep feeling of the beauties of the

RED-BREASTED THRUSH.
(TURDUS MIGRATORIUS.)

season, a sense of the bounty and providence of the Supreme Being, which is cheaply purchased by the tedium of nine months of winter. The most verdant lawns and cultivated glades of Europe, the most beautiful productions of art, fail in producing that exhilaration and joyous buoyancy of mind which we have experienced in treading the wilds of Arctic America, when their snowy covering has been just replaced by an infant but vigorous vegetation. It is impossible for the traveller to refrain, at such moments, from joining his aspirations to the song which every creature around is pouring forth to the great Creator."

The RING THRUSH. (*Turdus torquatus.*) This species is migratory, and is found throughout the greatest part of Europe, Asia, and Africa. It is eleven inches in length: the beak is partly orange-yellow: the whole upper part of the plumage is black, with scarcely any gray on the margins of the feathers; the quills and wing-coverts dusky, bordered with pale gray: a gorget of pure white: under wing-coverts pale brown, with broad gray margins: legs dusky brown. It breeds in Wales and many of the mountainous parts of Britain and Ireland; and it is very abundant in the isle of Portland, upon their arrival and departure, every spring and autumn. Its nest is generally placed on the ground, under some small bush: it is formed like that of the Blackbird; and the eggs in size and colour are very like that bird's. During the breeding season it is a rare occurrence to observe a second pair in the same neighbourhood. When they have young, they are very clamorous if disturbed. Their food consists of snails, insects, and berries, particularly those of the juniper.

The WATER THRUSH. (*Seiurus aquaticus.*) This bird, which is called in America a Thrush, belongs, however, to a different sub-family, but may be described here: it is remarkable for its partiality to brooks, rivers, shores, and ponds; wading in the shallows in search of aquatic insects, chattering as it flies. It is only about six inches in length: the whole upper parts are of a uniform and very dark olive, with a line of white extending over the eye, and along the sides of the neck; the lower parts are white, tinged with an ochreous yellow; the breast and sides marked with pointed spots or streaks of black or deep brown: bill brown: legs flesh-colour. Wilson remarks that the cane-brakes, swamps, river shores, and deep watery solitudes of Louisiana, Tennessee, and the Mississippi territory, possess them in abundance; there they are eminently distinguished by the loudness, sweetness, and expressive variety of their notes, which begin very high and clear, falling with an almost imperceptible gradation till they are scarcely articulated. At these times the musician is perched on the middle branches of a tree over the brook or river bank, pouring out its charming melody, that may be distinctly heard for nearly half a mile. The voice of this little bird, says he, appeared so exquisitely sweet and expressive that I was never tired of listening to it, while traversing the deep shaded hollows of those cane-brakes where it usually resorts.

We can afford no more space for the description of other species; but may remark that there are foreign species of this extensive genus intermediate, in every possible way, to all those of Europe. In a group inhabiting Australia, the Indian Archipelago, and slopes of the Asiatic mountains, the dorsal plumage is mottled at all ages; a character peculiar to the nestling dresses of the others. One species belonging to it (*Turdus Whitii*) the largest of all the Thrushes, resembles the Missel Thrush in its form and proportions, and occasionally strays to the west of Europe, having been met with even in Britain. Other Thrushes, peculiar to America, and breeding in the northern division of that continent, are solitary in habit, and successively diminish in size; having the bill weaker and tarsi more elongated, assuming the russet tint of the Nightingale, and gradually losing the breast-spots, &c. In short, the Thrushes form a great centre of radiation, which ramifies in every direction, till the normal generic features disappear.

THYLACINUS. A genus of Marsupial animals. The *Thylacines* are distinguished from the Opossums by the hind feet having no thumb, by a hairy and not prehensile tail, and two incisors less to each jaw. There is only one *existing* species known, a native of Australia. It is smaller than a wolf, and lower on the legs; of a grayish colour, barred with black across its hinder limbs. In its native island (Van Diemen's Land) it is called both *Tiger* and *Hyæna*.

THYLACOLEO. [See SUPPLEMENT.]

THYLACOTHERIUM. [See SUPPL.]

THYMALLUS. [See GRAYLING.]

THYMELE. A genus of Diurnal Lepidoptera belonging to the family *Hesperiadæ*, or "Skippers" as they are called in this country. Of the British species we may particularize the THYMELE MALVÆ; or THE GRIZZLED SKIPPER BUTTERFLY. This elegant and variable insect is distinguished by its numerous white or cream-coloured quad-

GRIZZLED SKIPPER BUTTERFLY (THYMELE MALVÆ.)

rangular notched spots on a dusky ground; the posterior wings with the white spots in the centre forming an interrupted band: all the

THYMELE MALVÆ—UNDER IDE.

wings have a white or cream-coloured fringe barred with black: beneath, the anterior wings are pale greenish-gray, with white spots, as above; the posterior wings are grayish-green

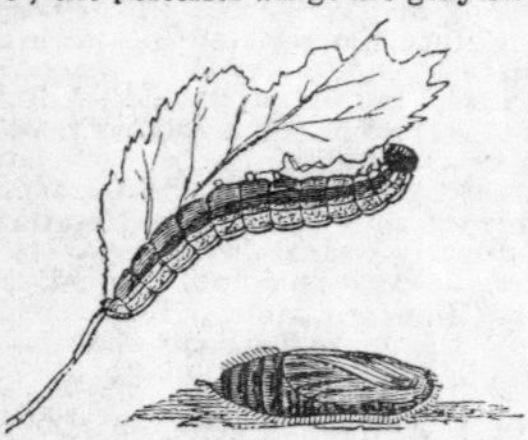

CATERPILLAR AND CHRYSALIS OF THYMELE MALVÆ.

also spotted as above: fringe with black bars narrower than on the upper surface. The Grizzle frequents woods, commons, dry banks, and meadows about the end of May.

THYNNUS. [See TUNNY.]

THYSANOPTERA. The name given to an order of insects of a very minu e size, scarcely exceeding a line in length; characterized by long, narrow, membranous wings, neither folded nor reticulated, with long ciliæ, laid horizontally along the back when at rest; mouth with two setiform mandibles; two triangular flat palpigerous maxillæ, and a palpigerous labium; tarsi, with two joints, vesiculose at the tip: pupa active, semi-complete. The order comprises but a single family, *Thripidæ*, the species of which, however, are rather numerous. These insects are found upon various plants, sometimes swarming in immense profusion in various kinds of flowers, especially the large white hedge-convolvulus: they feed upon the juices of plants, and are often extremely injurious, especially in hot-houses, vine-houses, melon and cucumber beds, &c., the leaves upon which they reside being marked all over with small decayed patches. One species (*Thrips cerealium*) infests the wheat, sometimes to a mischievous extent. It takes its station in the furrow of the seed, in the bottom of which it fixes its rostrum, and by depriving the seed of its moisture, causes it to shrink up. One sex of this species is apterous; the larva is yellow and very nimble, and the pupa is whitish, with black eyes, and very sluggish. This species also gnaws the stems above the knots, and causes the abortion of the ear. It is said that in 1805 the wheat crops in England suffered materially from this minute insect.

THYSANOURA. The name given to an order of apterous insects, comprising many species, none of which undergo a metamorphosis. They are furnished with six legs, and have at the sides of the body, or its extremity, peculiar organs of locomotion. The order contains two families. In the first, the LEPISMADÆ, the abdomen is furnished on each side with a row of movable appendages, like false legs; and is terminated by long pointed bristles, of which three are usually most remarkable. In the second family (the PODURIDÆ), the appendages to the sides of the abdomen are wanting; but the extremity of it is prolonged into a forked tail, by which these insects can take very surprising leaps. [See PODURA.]

TICK. The *Ricinia*, commonly known as *Ticks*, belong to the ACARIDÆ [which see].

They are small, disagreeable animals, usually of a flattened, round, or oval form; generally destitute of eyes, but have the mouth provided with lancets, that enable them to penetrate more readily the skins of animals whose blood they suck. They fasten upon horses, cows, sheep, dogs, and other quadrupeds; and they bury their suckers (which are often furnished with minute recurved hooks) so firmly in the skin, that they can scarcely be detached without a portion of it coming away with them. They acquire a very considerable size by suction, being frequently distended like a blown bladder, and full of blood. It is common to find them in thick woods, abounding in brushwood, briars, &c., and attaching themselves to plants with the two anterior legs. [For an account of a curious species of Tick commonly called *Red-Spider*, see SPIDER.]

TIGER. (*Felis tigris.*) This most beautiful, but most destructive of quadrupeds, is undoubtedly, next to the Lion, also the most powerful animal of the feline species. It is a native of the warmer parts of Asia, and is principally found in India and the Indian islands; though the species extends as far as China, Chinese Tartary, and the Altaic mountains. It has all the zoological characters, prowling habits, and sanguinary propensities common to the rest of the genus; but it is distinguished from them all by the peculiar markings of its coat. The ground colour is a bright orange-yellow; the face, throat, and under side of the belly being nearly white; the whole elegantly striped by a series of transverse black bands or bars, which form a bold and striking contrast with the ground-colour. About the face and breast the stripes are proportionally smaller than on other parts; and the markings are continued, in an annular form, upon the tail, the tip of which is black.

TIGER.—(FELIS TIGRIS.)

The Tigers exhibited in our menageries seldom fail to engage the spectator's especial notice; but a wide difference is observable between such animals as by long confinement, and an alteration of climate, have lost the native brilliancy of their colours, and those which roam the forest, or lurk in the jungle and morass. When seen in perfection, and before its health has been impaired by confinement, it is scarcely possible to conceive a more elegantly variegated animal: the bright and intense orange-yellow; the deep and well-defined stripes of black, in some parts double, in others single; the pure white of the cheeks and lower part of the sides, over which a part of the black striping is continued—form, altogether, an appearance superior in beauty to that of any other regularly marked quadruped. Although the Tiger is generally inferior in size to the Lion, it has sometimes been seen even larger, viz. of the length of fifteen feet from the nose to the tip of the tail. The largest are those of India which are termed *Royal Tigers*.

As this animal is said to surpass in ferocity every other, it is accordingly considered as the most dreadful scourge of the hotter regions of Asia. It has been common to represent it as quite untameable even when in confinement; but many instances might be given to show that such an assertion is without foundation; though no one can deny that it is extremely difficult to overcome their naturally cruel and ferocious nature. The Tiger's method of seizing his prey is by concealing himself from view, and springing with a horrible roar on his victim, which he carries off, and tears to pieces, after having first partly satiated himself by sucking the blood; and such is his strength, that he is able to carry off a buffalo with seeming ease. The Tigress, like the Lioness, produces four or five young at a litter: she is at all times furious, but when robbed of her young her rage exceeds all bounds. Braving every danger, she then pursues her plunderers, who are often glad to release a cub in order to retard her while they make their escape: she stops, takes it up, and carries it to the nearest cover, but instantly returns, and renews her pursuit, even to the very gates of buildings, or the edge of the sea; and when her hope of recovering them is lost, she expresses her agony by hideous and terrific howlings.

The following observations on the habits, chase, &c. of the Tiger are much to the purpose: "The bound with which the ambushed Tiger throws himself upon his prey is as wonderful in its extent as it is terrible in its effects. Pennant justly observes that the distance which it clears in this deadly leap is scarcely credible. Man is a mere puppet in his gripe; and the Indian buffalo is not only borne down by the ferocious beast, but carried off by his enormous strength. If he fails, it has been said that he makes off. This may be true in certain instances, but in general he does not slink away, but pursues the affrighted prey with a speedy activity which is seldom exerted in vain. This leads us to the observation of Pliny celebrating its swiftness, for which the Roman zoologist has been censured, most unjustly, apparently; nor is he the only author among the ancients who notices its speed. Oppian *Cyneg.*, i. 323.) speaks of the swift Tigers as being the offspring of the Zephyr. Pliny, says Pennant, has been frequently taken to task by the moderns for calling the Tiger 'animal tremendæ velocitatis;' they allow it great agility in its bounds, but deny it swiftness in pursuit. Two travellers of authority, both eye-witnesses, confirm what Pliny says; the one indeed only mentions in general vast fleetness; the other saw a trial between one and a swift horse, whose rider escaped merely by getting in time amidst a circle of armed men. The chase of this animal was a favourite diversion with the great Cam-Hi, the Chinese monarch, in whose company our countryman, Mr. Bell, that faithful traveller, and the Père Gerbillon, saw these proofs of the Tiger's speed."

Numerous are the instances which might be given of the Tiger's ferocity, and contempt of danger. The following fatal event, as described by an eye-witness, though frequently related, possesses such an unusual degree of fearful interest, that we are tempted to repeat it here. It took place in 1792; the unfortunate victim was the son of Sir Hector Monro, Bart. "We went," says the narrator, "on shore on Saugar Island, to shoot deer, of which we saw innumerable tracks, as well as of Tigers; notwithstanding which, we continued our diversion till near three o'clock, when, sitting down by the side of a jungle to refresh ourselves, a roar like thunder was heard, and an immense Tiger seized on our unfortunate friend, and rushed again into the jungle, dragging him through the thickest bushes and trees, every thing giving way to his monstrous strength: a Tigress accompanied his progress. The united agonies of horror, regret, and fear, rushed at once upon us. I fired on the Tiger: he seemed agitated: my companion fired also, and in a few minutes after this my unfortunate friend came up to us, bathed in blood. Every medical assistance was vain, and he expired in the space of twenty-four hours, having received such deep wounds from the claws and teeth of the animal as rendered his recovery hopeless. A large fire, consisting of ten or twelve whole trees, was blazing by us at the time this accident took place; and ten or more of the natives with us. The human mind can scarce form any idea of this scene of horror. We had hardly pushed our boat from that accursed shore, when the Tigress made her appearance, almost raging mad, and remained on the sand all the while we continued in sight."

The following narrative of the almost miraculous escape of an European soldier from the grasp of a Tiger is given by an officer who some years since was in command of a party in India, and may therefore be regarded as authentic: —

"It was after a long day's march of fifteen miles, across a country where with difficulty a road could be traced, and that made by deep ravines cut by the rains, with here and there upright stones, that we arrived at a jungle unusually swampy, which, from its size, and the fatigued state of the soldiers and cattle, I thought it prudent to defer passing until the following morning, when, probably, we should fall in with an enemy about three thousand strong, with several pieces of cannon, under the command of Ally Newas Khan, with whom we were not over anxious to hazard an engagement, from the inferiority of our force; which consisted only of six hundred Europeans and two small field pieces.

"I had observed several flocks of wild peacocks and turkeys while the tents were pitching, which always frequent the same dry heathy ground adjacent to jungles, that tigers do; but from our numbers, the compactness of our encampment, and the precaution I had taken to order fires to be kindled in various directions, I conceived we had little to fear from any visit those gentlemen might think proper to pay us. I had but just entered my tent, and wrapped myself in my boat cloak, with a view to doze away the remaining hour or two, before we broke up for another day's march to join General R——y, when the report of a musket roused me. I instantly started to the entrance of my tent, and was questioning the sentinel who stood there, as to the direction of the sound, when a huge tiger, with monstrous bounds, passed within a few yards of the spot where I was standing, with one of our brave fellows struggling in his jaws. My sentinel immediately fired at him; but the agitation of the moment prevented his taking a deadly aim. The ball, to all appearance, struck him, from the enormous bound he immediately made, but only to increase his speed. We were, however, enabled to follow him, by some blood that now fell from him, or his unhappy prey, and had already entered the jungle several hundred yards, before we began to despair of finding the latter alive, and of aiding his rescue. Judge of our horror, on hearing on a sudden a kind of sullen growl, or roar, which made the hills echo a still more dreadful sound; and the next moment, of our joy, on being greeted with a halloo from our lost companion, about fifty yards farther in the jungle than we had penetrated, which was heartily returned by those who joined me in the pursuit, and in a few moments more we met him limping towards us, with as joyous a face as ever I witnessed, even after the most flattering success.

"The following account of his escape he afterwards committed to paper: —'I was just returning, at a good brisk pace, from one of the posts down the jungle, where I had been taking some victuals to my bedfellow, when I heard a kind of rustling noise in some bushes, about six or seven yards behind me; and, before I could turn round to ascertain the cause, I was pounced upon, and knocked down with such force, as to deprive me of my senses, till I arrived opposite your tent; when the sudden report of a musket, together with a kind of twitching in my thigh, brought me to myself, and to a sense of the great danger in which I was; but, nevertheless, I did not despair. I now began to think of some plan of saving myself; and, though carried away very rapidly, I felt, as well as saw, that your sentinel's ball had, instead of hitting the Tiger, struck me, and that I was losing blood very fast. I remembered that the bayonet was in my belt, and reflected, that if it was possible for me to draw it, I might yet escape the horrible death that awaited me. I with difficulty put my arm back, and found it, and several times attempted to draw it from its sheath; but, from my position, I was unable. To describe the fears I now felt would be impossible; I thought it was all over. At last, thank Heaven! after another attempt with my utmost force, I drew it out, and instantly plunged it into his shoulder. He bounded aside, and his eyes flashed frightfully; he let me down, but instantly seized me again above the hip, which at first prevented me from drawing my breath. I now had, from the change of position, a fair opportunity of killing the monster and saving my life. I stabbed him behind the shoulder several times as deeply as the bayonet would enter; he staggered, and fell, and again letting me go, rolled several yards beyond me. I now thought myself safe; and was getting up, when he rose, and, with a dreadful roar, again attempted to seize me, but again fell down, and rolled close to my feet. I now had the advantage of a fallen enemy, which I forgot not to turn to the best account, and again plunged my bayonet into his side, which I suppose, from his struggles, pierced his heart. I then fell upon my knees, and endeavoured, but from the fulness of my heart I was unable, to return thanks aloud to Almighty God for his gracious goodness in delivering me from so terrible a death. I rose, and hallooed; my halloo was returned, and just afterwards I met you, or perhaps I might have been lost from my weakness.'

"It would appear that the Tiger, either from the distance of his leap or the hardness of the soldier's cartouch box, fortunately missed his hold, and seized him after he had knocked him down, by his clothes, the cartouch box saving him from being bitten. But I am convinced that never did any man, if we take into consideration the distance he was carried before he released himself, and the circumstance of his being wounded by the ball intended for the tiger, which directed us what road to follow, more providentially escape to all appearance an inevitable death."

The annexed "Tiger adventure" some years since appeared in the *Literary Gazette*, to which journal it was sent by an Indian correspondent. As it contains a mixture of the marvellous with what, at a distance, appears more ludicrous than tragical, it may serve, perhaps, to allay any agitation of the nerves which the dreadful catastrophe above related may have caused:—

"Our annual supply of good things having reached us this morning, we were enjoying a bottle of some delicious Burgundy and 'La Rose' after dinner, when we were roused by violent screams in the direction of the village. We were all up in an instant, and several men directed to the spot. Our speculations on the cause were soon set at rest by the appearance of two *hircarras* (messengers), and a lad with a vessel of milk on his head. For this daily supply they had gone several miles, and had nearly reached the camp, when, having outwalked the boy, they were alarmed by his vociferations, 'Oh, uncle, let go, let go—I am your child, uncle—let me go!' They thought the boy mad, and, it being very dark, cursed his uncle, and desired him to make haste; but the same wild exclamations continuing, they ran back, and found a huge tiger hanging on his tattered cold-weather doublet. The *hircarras* attacked the beast most manfully with their javelin-headed sticks, and adding their screams to his, soon brought the whole village, men, women, and children, armed with all sorts of missiles, to the rescue; and it was their discordant yells that made us exchange our good fare for the jungles of Morwun. The 'lord of the black rock,' for such is the designation of the Tiger, was one of the most ancient *bourgeois* of Morwun: his freehold is Kálá-pahár, between this and Mugurwar, and his reign for a long series of years had been unmolested, notwithstanding his numerous acts of aggression on his bovine subjects: indeed, only two nights before, he was disturbed gorging on a buffalo belonging to a poor oilman of Morwun. Whether this Tiger was an incarnation of one of the Mori lords of Morwun, tradition does not say; but neither gun, bow, nor spear, had ever been raised against him. In return for this forbearance, it is said, he never preyed upon man, or if he seized one, would, upon being entreated with the endearing epithet of *mamoo*, or uncle, let go his hold; and this accounted for the little ragged urchin using a phrase which almost prevented the *hircarras* returning to his rescue."

Of all the grand and exciting field-sports of the East, there is none, it is said, that equals a Tiger-hunt. When it is remembered, however, that from ten to thirty well-trained elephants, each carrying sportsmen, armed with rifles, not unfrequently join in the chase, it will readily be conceived how great must be both the excitement and the danger.

Tiger's skins are occasionally imported into Europe, as objects of curiosity rather than of use, except as hammer-cloths for carriages. In China they are used by the mandarins as covers for their seats of justice, as well as for cushions, pillows, &c., in the winter: the more intense the yellow, and the better defined the stripes, the more valuable are the skins.

TIGER BEETLES. [See CICINDELIDÆ.]

TIGER [MOTHS]. A name given by collectors to different species of Moths, of the genera *Arctia*, *Hypercampa*, and *Nemeophila*.

TIGER-WOLF. [See APPENDIX.]

TIGRISOMA. A sub-genus of Bitterns found in South America, and so called from the markings on their bodies somewhat resembling those of the Tiger.

TIMALIA. A genus of birds found in the groves and small woods which abound throughout Java. The species (*T. pileata*) described by Dr. Horsfield is six inches and a half in length; and having a body rather stout, and ovate. General colour above, brown with an olivaceous tint; underneath, dull testaceous, inclining to gray; crown of the head, chestnut; throat and cheeks white: breast white, inclining to gray, marked with intensely black stripes by the shafts of the plumes. A narrow white band commences at the forehead, near the base of the bill, encircles the eye, and unites with the white plumes of the cheeks. Quills brown, tinged with chestnut on the edges: lesser wing-coverts, as well as the plumes which cover the nape and back, grayish-blue at the base: bill black and shining: feet brown. It constructs its nest in hedges, and is a bird of social habits, delighting to dwell in the neighbourhood of plantations and human dwellings. Its flight is low and interrupted; and it is generally a welcome neighbour wherever it resides, in consequence of the peculiarity and pleasantness of its note, which is remarkably slow and regular.

TIMARCHA. A genus of Coleopterous insects, allied to Chrysomela. The *Timarcha lævigata* is a common British species, between half and three quarters of an inch long. It frequents woods, turf, and low herbage; crawls slowly, and emits a reddish-yellow fluid from the joints when disturbed: from which circumstance it is vulgarly known as the Bloody-nose Beetle. The larvæ bear a strong resemblance to the perfect insect, both in appearance and general habits: when disturbed, they roll themselves up after the manner of a wood-louse.

TINAMOU. (*Tinamus*.) A genus of Gallinaceous birds, consisting of several species, all natives of South America. Their flight is low, heavy, and of little duration, but they run swiftly. They live in small scattered coveys; some species residing in the open fields, while others prefer the borders of woods. Their eggs are deposited in a hole or furrow ready formed on the ground; and two broods are usually produced in the year. Birds of this genus are remarkable for a long and slender neck, covered with feathers, the tips of the barbs of which are slender and slightly curled, which imparts a peculiar air to that part of their plumage. The beak is long, slender, and blunt at the end; somewhat vaulted, with a small groove at each side. Their wings are short, and they have scarcely any tail. The membrane between the base of their toes is very short; and their hind claw, reduced to a spur, cannot touch the ground. Their size varies from that of a Pheasant down to that of a Quail.

The GREAT TINAMOU (*Tinamus Brasiliensis*) is eighteen inches long; and the general colour of the plumage grayish-brown, inclining to olive, with a mixture of white underneath and on the sides, and of greenish on the neck: upper part of the back, wing-coverts, and tail, marked with dusky transverse spots: sides of the head, throat, and fore-part of the neck, not well clothed with feathers. The tail is short; the beak black; and the legs yellowish-brown, the hind part of them very rough and scaly. This species inhabits the great forests of Guiana, and roosts upon the lower branches of trees, two or three feet from the ground. The female lays from twelve to fifteen eggs, the size of those of a hen, and of a beautiful green colour, in a nest formed of moss and dried leaves, and placed on the ground, among the thick herbage, near the root of some large tree. The young run after the mother almost as soon as hatched, and hide themselves on the least appearance of danger. Their cry, which is a kind of dull whistle, is heard every evening at sunset, and again at sunrise; and so well do the natives imitate it, that the birds are easily decoyed within reach of the gun: they also take many during the night, while roosting on the trees. Their food consists of various fruits and grain, worms, and insects; and their flesh is highly esteemed.

The RUFESCENT TINAMOU. (*Tinamus rufescens.*) This bird, the most beautiful of the genus, is fifteen inches and a half in length. The top of the head is spotted with black, and bordered with rufous: the shoulders, back, wing-coverts, and rump are gray with a reddish shade, and transversely striped with black and white: the quills the outer border of the wing, and the spurious wing are rusty red: the throat is white; the neck, breast, and belly are rufous, the last slightly striped transversely with fuscous: the abdomen and sides are of a gray hue, varied with stripes of rufous and black. The beak is long, strongly curved, and of a brown blue: the feet are pale red. It resides among thick herbage, and feeds night and morning, when it regularly utters its melancholy and feeble cry. The female deposits seven eggs of a fine bright violet hue, in a hollow, situated beneath tufts of grass; and the young reside within a short distance of each other, and not in families. This is the species of which Mr. Darwin speaks, in his description of the country around Maldonado: "We every where saw great numbers of partridges (*Tinamus rufescens*). These birds do not go in coveys, nor do they conceal themselves like the English kind. It appears a very silly bird. A man on horseback, by riding round and round in a circle, or rather in a spire, so as to approach closer each time, may knock on the head as many as he pleases. The more common method is to catch them with a running noose or little lazo, made of the stem of an ostrich's feather, fastened to the end of a long stick. A boy on a quiet old horse will frequently thus catch thirty or forty in a day. The flesh of this bird, when cooked, is delicately white."

TINCA. [See TENCH.]

TINEIDÆ. A family of Lepidoptera, comprising an extensive series of minute insects, distinguished by their narrow wings and the slenderness of their palpi; the head is often densely clothed with scales in front, and the body is generally long and slender; the antennæ are of moderate length, either simple in both sexes, or pubescent beneath in the males; the maxillary palpi are well developed, and, although occasionally short, are sometimes extraordinarily developed; the wings are entire, often very narrow, and mostly convoluted in repose; and when at rest the posterior pair are much folded. The larvæ are generally naked or slightly hairy; many undergoing their transformations in portable cases formed of various materials; whilst others reside either within the stalks or upon the leaves of plants. In the perfect state, they are of a sombre hue rather than of a bright metallic appearance, their longitudinal markings or streaks being conspicuous. In the larva state they are notoriously destructive to woollen materials of every description, feathers, furs, skins, &c., upon which they feed; using the material also for the construction of their cases; in which, when full grown, they become chrysalides. The species included in the genus *Galleria* inhabit the nests of bees, the larvæ feeding upon honey, and forming galleries in the honeycomb. Others make great havoc in granaries and malthouses: and one, *Diatræa sacchari*, is a most destructive pest of the sugar-cane in the West Indies, the larva burrowing into the centre of the stems, and often destroying whole acres.

TIPULA: TIPULIDÆ. A genus and family of Dipterous insects, distinguished by the proboscis being very short, its internal organs slightly developed, and terminated by two large fleshy lips; the palpi longer than the proboscis, four-jointed, and generally folded back. The body is long and slender, as also are the legs; the head is rather small, the antennæ are very variable in length; and the alulets are mostly obsolete. The larger species appear to be the types of the family; such as *Ctenophora*, *Pedicia*, and the true *Tipulæ*, which are vulgarly termed Daddy-long-legs.

The *Tipulidæ Culiciformes* resemble Gnats, having the antennæ entirely pilose, but with the hairs much longer in the males than in the females. Their larvæ live in the water, and resemble those of Gnats. Some of them have false feet; others have arm-like appendages at the posterior extremity of the body; and they are generally of a red colour. The pupæ are also aquatic, and respire by two outer appendages placed at the anterior extremity of the body. Some have the power of swimming.

The *Tipulidæ Terricolæ* comprise the largest species in the family, with the antennæ longer than the head, and slender; destitute of ocelli; the eyes round and entire; the wings, extended in many, have always membranous nerves, united together transversely, and closed discoidal cells. The front of the head is narrowed, and prolonged into a muzzle, with a basal prominence; the palpi generally long, and the extremity of the tibiæ spinose. The larvæ of many species live in the earth, the decayed parts of trees, &c. The thorax is not distinct, and they have no false feet. The pupæ are naked, with two respiratory tubes near the head; and the edges of the abdominal segments spinose.

TITLARK; or TITLING. The English name of birds of the genus *Anthus*. [See LARK.]

TITMOUSE. (*Parus*.) A genus of active little birds, continually flitting from spray to spray, and suspending themselves in all kinds of attitudes. They are noted for the peculiarly elegant construction of their nests, which are composed of the softest materials; and many of them are fastened to the extreme end of a small branch of a tree that projects over the water — a contrivance by which they are well secured from the attacks of quadrupeds and reptiles. They are extremely prolific, and provide for their numerous young ones with the most indefatigable industry. Such is their strength and courage that they will venture to attack birds above three times their own size; and when they kill an opponent (or even if they find one that has recently died) they always pierce a hole in the skull and eat the brains. Their principal food consists of insects, which they obtain in the spring by biting off the opening buds, and in the summer by searching in cracks and crevices of trees. The Titmice have short conical bills, with the tips not dentated, and a few bristles at the base. Though essentially insectivorous, many of them also feed on fruit and seeds of various kinds, and show great fondness for animal fat. A writer in the Quarterly Review (Dec. 1842) tells us that "Tom-tits are called '*Bee-biters*' in Hampshire. They are said to tap at the hives of the bees, and then snap up the testy inmates, who come out to see what it is all about: if birds chuckle as well as chirp, we can fancy the delight of this little mischievous ne'er-do-good at the success of his lark." Our figure represents a characteristic species of the group, the COLE TITMOUSE. (*Parus ater*.) This species is not so common in England as it appears

COLE TITMOUSE.—(PARUS ATER.)

to be in Scotland, where it abounds in the woods. The head, neck, and upper part of the breast black; the cheeks and nape white. This species makes its nest in holes of old trees near the ground, forming it of moss lined with hair; its eggs are from six to eight, white with reddish spots.

The BLACK-CAPPED TITMOUSE (*Parus atricapillus*), which Wilson, the American ornithologist, suspects to be identical with the *Parus Hudsonicus* of Latham, is thus described by him: — "This is one of our resident birds, active, noisy, and restless; hardy beyond any of his size, braving the severest cold of our continent as far north as the country round Hudson's Bay, and always appearing most lively in the coldest weather. The males have a variety of very sprightly notes, which cannot, indeed, be called a song, but rather a lively, frequently repeated, and often varied twitter. They are most usually seen during the fall and winter, when they leave the depths of the woods, and approach nearer to the scenes of cultivation. At such seasons they abound among evergreens, feeding on the seeds of the pine tree; they are also fond of sunflower seeds, and associate in parties of six, eight, or more, attended by the Carolina Nuthatch, the Crested Titmouse, Brown Creeper, and small Spotted Woodpecker; the whole forming a very nimble and restless company, whose food, manners, and dispositions are very much alike. About the middle of April they begin to build, choosing the deserted hole of a squirrel or woodpecker, and sometimes, with incredible labour, digging one out for themselves. The female lays six white eggs, marked with minute specks of red; the first brood appear about the beginning of June, and the second towards the end of July; the whole of the family continue to associate together during winter. They traverse the woods in regular progression, from tree to tree, tumbling, chattering, and hanging from the extremities of the branches, examining about the roots of the leaves, buds, and crevices of the bark, for insects and their larvæ. They also frequently visit the orchards, particularly in the fall of the year, the sides of the barn and barn-yard, in the same pursuit, trees in such situations being generally much infested with insects. We, therefore, with pleasure, rank this little bird among the farmer's friends, and trust our rural citizens will always recognize him as such. This species has a very extensive range; it has been found on the western coast of America as far north as lat. 62°; it is common at Hudson's Bay, and most plentiful there during winter, as it then approaches the settlements in quest of food. Protected by a remarkably thick covering of long, soft, downy plumage, it braves the severest cold of those northern regions.—The Black-capped Titmouse is five inches and a half in length; the throat, and whole upper part of the head and ridge of the neck, black; between these lies a triangular edge of white, ending at the nostril; bill, black and short; tongue truncate; rest of the upper parts, lead coloured or cinereous, slightly tinged with brown; wings edged with white; breast, belly, and vent, yellowish white; legs light blue; eyes dark hazel. The male and female are nearly alike.

The BLUE TITMOUSE. (*Parus cœruleus.*) The length of this elegant little bird is four inches and a half; its beak is dusky; forehead and cheeks white, that on the forehead forming a line round the crown of the head, which is of a clear blue; behind this there is a circle of blue, surrounding the head, and joining at the base of the under mandible, where it is nearly black: from the beak, through the eyes, is a narrow black line. The back is yellowish-green: quills black, with bluish edges; wing-coverts blue, edged with white; under parts of the body yellow: tail blue, the two middle feathers longest. The female is rather smaller than the male, has less blue on the head, and the colours in general are not so bright. This bird is an inhabitant of Europe, and in no country more common than in our own. It has long had the unenviable reputation of being very destructive to gardens and orchards, by plucking off the buds in search of insects and their larvæ that are lodged within; but whether as their destroyer it does more good, than as the horticultural depredator it does harm, is a question not thoroughly ascertained. It is fond of flesh of any description, either fresh or putrid; and it displays its pugnacious and predaceous disposition whenever it has a fair chance of coming off conqueror. The nest is made in the hollows of trees, of moss lined with feathers and hair. The female lays seven or eight eggs, white, speckled with rust colour: she is very tenacious of her nest, and will suffer herself to be taken rather than quit it; nay, upon that occasion she will erect all her feathers, utter a noise like the spitting of a cat, and if handled, will bite very sharply. The note of this bird consists only of a disagreeable shriek.

Another of the *Parus* tribe is thus pleasingly described by the author of the 'Journal of a Naturalist.' "Our tall hedgerows and copses are frequented by a very amusing little bird, the LONG-TAILED TITMOUSE (*Parus caudatus*). Our boys call it the Long-tailed Tom-tit, Long Tom, Poke-pudding, and various other names. It seems the most restless of little creatures, and is all day long in a state of progression from tree to tree, from hedge to hedge, jerking through the air with its long tail like a ball of feathers, or threading the branches of a tree, several following each other in a little stream; the leading bird uttering a shrill cry of *twit, twit, twit,* and away they all scuttle to be first, stop for a second, and then are away again, observing the same order and precipitation the whole day long. The space travelled by these diminutive creatures in the course of their progresses from the first move till the evening roost must be considerable; yet, by their constant alacrity and animation, they appear fully equal to their daily task. We have no bird more remarkable for its family association than this Parus. It is never seen alone, the young ones continuing to accompany each other from the period of their hatching until their pairing in spring. Its food is entirely insects, which it seeks among mosses and lichens, the very smallest being captured by the diminutive bill of this creature. Its nest is as singular in construction as the bird itself. Even in years long passed away, when, a nesting boy, I strung my plunder on the benty grass: it was my admiration; and I never see it now without secretly lauding the industry of these tiny architects. It is shaped like a bag, and externally fabricated of moss and different herbaceous lichens, collected chiefly from the sloe and the maple; but the inside contains such a profusion of feathers, that it seems rather filled than lined with them—a perfect feather-bed! I remember finding fourteen or sixteen pea-like eggs within this downy covert, and many more were reported to have been found. The excessive labour of the parent birds in the construction and collection of this mass of materials is exceeded by none that I know of; and the exertions of two little creatures in providing for, and feeding, with all the incumbrances of feathers and tails, fourteen young ones, in such a situation, surpass in diligence and ingenuity the efforts of any other birds, persevering as they are, that I am acquainted with." Modern naturalists place it in a separate genus which from the great length of the tail they call *Mecistura.*

PENDULOUS TITMOUSE. (*Ægithalus pendulinus.*) This species derives its name from its pensile purse-like or flask-like nest, generally suspended at the end of some willow twig or other flexible branch of a tree that overhangs the water. This skilfully wrought habitation is woven from the cotton-like wool or down of the willow or poplar, with an opening in the side for the ingress and the egress of the artificers and their young; and it is generally so placed as to droop over the brink of a rivulet or pond. This bird is four inches in length: the bill is black, straight, and a little pointed; forehead, top of the head, and nape, pure ash-colour; feathers round the eyes and ears deep black; back and scapulars reddish gray; throat white; the lower parts generally whitish with rosy tints; coverts of the wings chestnut, bordered with light rusty and white; wings and tail blackish, bordered with whitish rusty; tail feathers tipt with white. The female is rather less than the male; the black on the forehead not so large nor so deep; and the

PENDULOUS TITMOUSE.
(ÆGITHALUS PENDULINUS.)

upper parts more clouded with rust-colour. They are found in Russia, Poland, along the banks of the Danube, where it breeds, and in the south of France and Italy. It frequents the reedy banks of rivers and lakes; and its food consists of the seeds of the reeds, and of molluscs and aquatic insects.

TOAD. A tailless Batrachian Reptile belonging to the genus *Bufo;* of which there are several species. They are characterized by a thick and squat body, much swollen, and covered with warts or tubercles; the head large, flat on the top, with a protuberance studded with pores behind each eye, from which a fetid milky secretion is expressed; no teeth in either jaw, the hind limbs but little elongated; and the toes very slightly webbed. The COMMON TOAD (*Bufo vulgaris*) is found in gardens, woods, and fields, and frequently makes its way into cellars, or any obscure recesses where it may find a supply of food and security from too great a degree of cold. Early in spring, it retires to the

TOAD.—(BUFO VULGARIS.)

waters, where it continues during the breeding season, and deposits its ova or spawn in the form of double necklace-like chains or strings of beautifully transparent gluten, of the length of three or four feet, the ova throughout the whole length having the appearance of so many small jet-black globules or beads: these are in reality no other than the tadpoles or larvæ convoluted into a globular form, and waiting for the period of their evolution or hatching, which takes place in the space of about fourteen or fifteen days, when they break from the surrounding gluten, and, like the tadpoles of Frogs, swim about in the water, imbibing nourishment from various animalcules, &c., till their legs are formed, the tail gradually becomes obliterated, and the animals quit the water for the surface of the ground, which generally happens early in the autumn. The prevailing colour of the Common Toad is an obscure brown above, much paler and irregularly spotted beneath. It is, however, occasionally found of an olive cast, with darker variegations; and sometimes, particularly in the earlier part of summer, the limbs are marked with reddish spots; while a tinge of yellow often pervades the under parts both of the limbs and body. Much has been said by the older writers with respect to the Toads's supposed venom, but it appears to be perfectly free from any poisonous properties. It is true that dogs, on seizing a Toad, and carrying it for some little time in their mouth, will appear to be affected with a very slight swelling of the lips, accompanied by an increased discharge of saliva — the mere effect of the slightly acrimonious fluid which the Toad on irritation exudes from its skin, and which seems, in this country at least, to produce no dangerous symptoms in such animals as happen to taste or swallow it. The limpid fluid also, which this reptile suddenly discharges when disturbed, is a mere watery liquor, perfectly free from any acrimonious or noxious qualities. Its usual pace is a kind of crawl; and on being alarmed or threatened with danger, it stops, swells its body, and, on its being handled, a portion of the cutaneous secretion, just mentioned, exudes from the follicles.

It is well known that the Toad, like many other Amphibia, can support a long abstinence, and requires but a small quantity of air: but in the accounts generally given of Toads discovered in stones, wood, &c., the animals are said to have been completely impacted or imbedded, and without any space for air. In confirmation of this doctrine, Mr. Jesse relates "the following fact. A gentleman put a Toad into a small flower-pot, and secured it so that no insect could penetrate into it, and then buried it in the ground at a sufficient depth to protect it from the influence of frost. At the end of twenty years he took it up, and found the Toad increased in size, and apparently healthy." He then informs his readers that "Dr. Townson, in his tracts on the respiration of the Amphibia, proves from actual experiment, that, while those animals, with whose economy we are best acquainted receive their principal supply of liquids by the mouth, the frog and salamander tribes take in theirs through the skin alone; all the aqueous particles being absorbed by the skin, and all they reject being transpired through it. He found that a frog, when placed on blotting-paper well soaked with water, absorbed nearly its own weight of the fluid in the short time of an hour and a half; and it is believed that they never discharge it, except when they are disturbed or pursued, and then only to lighten their bodies, and facilitate their escape. That the moisture thus imbibed is sufficient to enable some of the Amphibia to exist without any other food, cannot (he thinks) be reasonably doubted; and if this is admitted, the circumstance of Toads being found alive in the centre of trees is fully accounted for."

We are quite ready to admit that many very extraordinary cases of this animal having lived for years embedded in stone, wood, or otherwise in a state of total exclusion from the air, and also without the means of obtaining a particle of food during the whole time, are to be met with, supported by most respectable authority; and yet on this oft-discussed question we still confess to a degree of scepticism, the grounds of which we cannot better explain than Dr. Shaw has already done for us: We suspect "that proper attention, in such cases, was not paid to the real situation of the animal. That a Toad may have occasionally '*latibulized*' in some part of a tree, and have been in some degree overtaken or enclosed by the growth of the wood, so as to be obliged to continue in that situation, without being able to effect its escape, may perhaps be granted: but it would probably continue to live so long only as there remained a passage for air, and for the ingress of insects, &c., on which it might occasionally feed; but that it should be completely blocked up in any kind of stone or marble, without either food or air, appears entirely incredible, and the general run of such accounts must be received with a great many grains of allowance for the natural love of the marvellous, the surprise excited by the sudden appearance of the animal in an unsuspected place, and the consequent neglect of minute attention at the moment, to the surrounding parts of the spot where it was discovered." Well, indeed, may Mr. Bell exclaim, "To believe that a Toad enclosed within a mass of clay, or other similar substance, shall exist wholly without air or food, for hundreds of years, and at length be liberated alive, and capable of crawling, on the breaking up of the matrix, now become a solid rock, is certainly a demand upon our credulity which few would be ready to answer!"

"Like the other Amphibia, and the Reptilia generally," observes this gentleman, "the Toad sheds its skin at certain intervals, the old cuticle coming off, and leaving a new one which had been formed underneath in its stead. There are some very remarkable circumstances connected with this process, which I detailed many years ago to the Zoological Club of the Linnæan Society, and of which the following is the substance. Having often found, amongst several Toads which I was then keeping for the purpose of observing their habits, some of brighter colours than usual, and with the surface moist and very smooth, I had supposed that this appearance might have depended upon the state of the animal's health, or the influence of some peculiarity in one or other of its functions: on watching carefully, however, I one day observed a large one, the skin of which was particularly dry and dull in its colours, with a bright streak down the mesial line of the back; and on examining further I discovered a corresponding line along the belly. This proved to arise from an entire slit in the old cuticle, which exposed to view the new and brighter skin underneath. Finding, therefore, what was about to happen, I watched the whole detail of this curious process. I soon observed that the two halves of the skin, thus completely divided, continued to recede further and further from the centre, and become folded and rugose; and after a short space, by means of the continued twitching of the animal's body, it was brought down in folds on the sides. The hinder leg, first on one side and then on the other, was brought forward under the arm, which was pressed down upon it, and on the hinder limb being withdrawn, its cuticle was left inverted under the arm; and that of the anterior extremity was now loosened, and at length drawn off by the assistance of the mouth. The whole cuticle was thus detached, and was now pushed by the two hands into the mouth in a little ball, and swallowed at a single gulp. I afterwards had repeated opportunities of watching this curious process, which did not materially vary in any instance."

"The Toad becomes torpid during the winter, and chooses for its retreat some retired and sheltered hole, a hollow tree, or a space amongst large stones, or some such place, and there remains until the return of spring calls it again into a state of life and activity. Its food consists of insects and worms, of almost every kind. It refuses food which is not living, and, indeed, will only take it at the moment when it is in motion. The Toad, when about to feed, remains motionless, with its eyes turned directly forwards upon the object, and the head a little inclined towards it, and in this attitude it remains until the insect moves; when, by a stroke like lightning, the tongue is thrown forward upon the victim, which is instantly drawn into the mouth. So rapid is this movement that it requires some little practice as well as close observation to distinguish the different motions of the tongue. This organ is constructed as in the Frog, being folded back upon itself; and the under surface of the tip being imbued with a viscid mucous secretion, the insect is secured by its adhesive quality. When the prey is taken it is slightly pressed by the margins of the jaw; but as this seldom kills it, unless it be a soft tender larva, it is generally swallowed alive; and I have often seen the muscles of the Toad's sides twitch in a very curious manner, from the tickling movements of a hard coleopterous insect in the stomach." [For the *Surinam Toad*, see PIPA: see also PHRYNISCUS.]

TODY. (*Todus.*) A genus of Scansorial birds, principally natives of the warmer parts of America, and nearly similar to the Kingfishers in their general form. They are characterized by a peculiar flatness or depression of the beak, which is blunt at the end, and has a gape extending as far back as the eyes. They are birds of gaudy plumage and rapid flight; and they feed on insects, worms, small reptiles, &c. The most elegant species is the ROYAL or KING TODY (*Onychorhynchus* or *Todus regius*), described by Buffon, who considered it as belonging to the *Muscicapidæ* or Flycatchers. Its bill is somewhat disproportionably broad, very much flattened, and beset with numerous strong bristles at the base: the colour of the plumage on the upper parts is a deep yellowish brown or chestnut, passing round the fore part of the neck like a collar: the throat, and all the under parts, are whitish yellow, the breast being crossed by numerous dusky undulations: the tail is bright ferruginous; and on the head is a most beautiful, broad, transverse crest, consisting of very numerous feathers, disposed in several series, lengthening as they recede from the front or base. These feathers are of a bright or red-ferruginous colour, and are each terminated by a black tip, so that the crest resembles that of a Hoopoe, placed in a transverse direction. Over each eye is a narrow white streak; the bill is dark brown; and the legs flesh-colour. This curious bird is a native of Cayenne and Brazil.

GREEN TODY. (*Todus viridis.*) This elegant little bird is about the size of a Wren, and has a bill long, like that of a Kingfisher, and ridged along the top of the upper mandible, which is of a dusky brown, the lower being of an orange or yellow colour; at the base of the bill are several stiff black hairs or bristles, standing forwards. The whole upper side of the bird is of a fine vivid green; the inner coverts of the wings are white; the inside of the quills and the under side of the tail are of a brownish-ash colour; and a few of the prime quills are black at their tips: the throat is of a very fine red; the breast, belly, thighs, and covert feathers under the tail are white, a little shaded with pale green: the legs and feet are dusky; and the toes are united, as in the Kingfisher.

Mr. Gosse tells us, that in all parts of Jamaica which he visited, the Tody is a very common bird. On the summit of Bluefields mountain, about three thousand feet from the level of the sea, and particularly where the deserted provision-grounds are overgrown with thicket, almost impenetrable, of jointer, or joint-wood (*Piper geniculatum*), it is especially abundant. Always conspicuous from its bright grass-green coat and crimson velvet gorget, it is still a very tame bird; yet this seems rather the tameness of indifference than of confidence; it will allow a person to approach very near, and, if disturbed, alight on another twig a few yards distant. It hops about the twigs of low trees, searching for minute insects, occasionally uttering a querulous, sibilant note: but more commonly it is seen sitting patiently on a twig, with the head drawn in, the beak pointing upwards, the loose plumage puffed out, when it appears much larger than it is. It certainly has an air of stupidity when thus seen. But this abstraction is more apparent than real; if we watch it, we shall see that the odd-looking gray eyes are glancing hither and thither, and that, ever and anon, the bird sallies out upon a short feeble flight, snaps at something in the air, and returns to his twig to swallow it. The following details are so pleasingly characteristic of the bird's habits, that we cannot refrain from making the extract: "One captured with a net in April, on being turned into a room, began immediately to catch flies, and other minute insects that flitted about, particularly little destructive *Tineadæ* that infested my dried birds. At this employment he continued incessantly, and most successfully, all that evening, and all the next day from earliest dawn till dusk. He would sit on the edge of the tables, on the lines, on shelves, or on the floor, ever glancing about, now and then flitting up into the air, when the snap of his beak announced a capture, and he returned to some station to eat it. He would peep into the lowest and darkest corners, even under the tables, for the little globose, long-legged spiders, which he would drag from their webs, and swallow. He sought also about the ceiling and walls, and found very many. I have said that he continued all day at this employment without intermission, and though I took no account, I judged that, on an average, he made a capture per minute. We may thus form some idea of the immense number of insects destroyed by these and similar birds; bearing in mind that this was in a room, where the human eye scarcely recognized a dozen insects altogether; and that in the free air insects would be much more numerous. Water in a basin was in the room, but I did not see him drink, though occasionally he perched on the brim; and when I inserted his beak into the water, he would not drink. Though so actively engaged in his own occupation, he cared nothing for the presence of man; he sometimes alighted voluntarily on our heads, shoulders, or fingers; and when sitting, would permit me at any time to put my hand over him and take him up; though when in the hand he would struggle to get out. He seemed likely to thrive, but incautiously settling in front of a dove cage, a surly Baldpate poked his head through the wires, and with his beak aimed a cruel blow at the pretty green head of the unoffending and unsuspecting Tody. He appeared not to mind it at first, but did not again fly; and about an hour afterward, on my taking him into my hand, and throwing him up, he could only flutter to the ground, and on laying him on the table, he stretched out his little feet, shivered, and died."

The Green Tody is exclusively an insect feeder, and burrows in the earth to breed. The banks of ravines, and the scarps of dry ditches, are excavated by its feeble feet, in which two out of three of its front toes are united together, having only the terminal joint free, and hence the feet of this kind of birds are called *syndactylous*. The hole runs into the banks some eight inches or a foot: at the extremity of this subterranean lodging it nestles in secrecy and security. The excavation is made by means of the beak and claws. It is a winding gallery, rounded at the bottom, and terminating in a sufficiently wide lodging, lined with pliant fibres, and dry moss and cotton. Four or five gray, brown-spotted eggs are laid, and the young are fed within the cave till they are full-fledged.

GREEN TODY.—(TODUS VIRIDIS.)

Mr. Gosse remarks that the inhabitants of Jamaica are not in the habit of domesticating many of the native birds; else this is one of the species which would become a favourite pet. In a state of liberty, however, it attracts the admiration even of the most unobservant, and an European is charmed with it. As it sits on a twig in the verdure of spring, its grass-green coat is sometimes undistinguishable from the leaves in which it is embowered, itself looking like a leaf, but a little change of position bringing its throat into the sun's rays, the light suddenly gleams as from a glowing coal. Occasionally, too, this crimson plumage is puffed out into a globose form, when its appearance is particularly beautiful.

TORNATELLA. A genus of Mollusca found on the shores of the Indian Ocean and Senegal; and one species (*Tornatella fasciata*) inhabits our own coasts. The shell is oval, spirally grooved; spire short, consisting of few whorls, and usually striped transversely; aperture long, narrow, rounded anteriorly; outer lip simple, inner lip slightly spread; columella spiral; no epidermis. Several fossil species occur in the London clay and inferior oolite.

TORPEDO. A genus of fishes belonging to the *Raiidæ* family; distinguished for their powers of imparting electric shocks to whatever animals they may come in contact with.

The TORPEDO (*Torpedo Vulgaris*), called also the Cramp-fish and the Electric Ray, is thus described by Pennant:—head and body indistinct, and nearly round; greatest breadth two-thirds of the entire length; thickness in the middle about one-sixth of the breadth, attenuating to extreme thinness on the edges; mouth small; teeth minute, spicular; eyes small, placed near each other; behind each a round spiracle, with six small cutaneous appendages on their inner circumference; branchial openings five in number; skin everywhere smooth; two dorsal fins on the trunk of the tail; tail one-third of the entire length, tolerably thick and round; the caudal fin broad and abrupt; ventrals below the body, forming on each side a quarter of a circle; colours, cinereous-brown above, whitish beneath. Mr. Yarrell (who calls this the Old British Torpedo) says, "The electrical powers of the Torpedo are so well understood by the different names that have been applied to it, as well as by the various and voluminous accounts that have been published, that it is unnecessary to repeat what has already appeared so often in print elsewhere. The situation of the apparatus or structure from which these species derive their extraordinary power is indicated by the two elevations, one of which is placed on each outside of the eyes and temporal orifices, and extending to the lateral external rounded edges. The apparatus occupies the whole of the space between the upper and under surface of the body, and is composed, as shown by the figures of Walsh and Pennant, of a great number of tubes arranged perpendicular to the plane of the upper and under surfaces, which, when exposed by a transverse section, have very much the appearance of a portion of honeycomb. The tubes contain a mucous secretion, and the structure is largely provided with nerves derived from the eighth pair. It is said that when the shock is given, the convex part of the upper surface is gradually depressed, the sensation is then felt, and the convexity suddenly returns. The whole use of the electrical apparatus and power to the fish can only be conjectured. That it serves as a means of defence is very probable; that it also enables a slow, inactive fish to arrest and obtain as food some of the more active inhabitants of the deep, is also probable."

At the meeting of the British Association, held at Oxford in June, 1847, Sir R. H. Inglis, the president, in referring to the experiments of the Tuscan philosopher, Matteucci, on the existence of electrical currents in all living animals, made this important observation:—"The delicate experiments of Matteucci on the *Torpedo* agree with those made by our own Faraday upon the *Gymnotus electricus*, in proving that the shocks communicated by those fishes are due to electric currents generated by peculiar electric organs, which owe their most immediate and powerful stimulus to the action of the nerves. In both species of fishes the electricity generated by the action of their peculiar organized batteries—besides its benumbing and stunning effects on living animals, renders the needle magnetic, decomposes chemical compounds, emits the spark, and, in short, exercises all the other known powers of the ordinary electricity developed in inorganic matter, or by the artificial apparatus of the laboratory."

The electric powers of the Torpedo, it is said, by no means affect its flesh; for it is frequently eaten in the south of Europe. It is a native of the Mediterranean and many other seas, and sometimes, though rarely, found on the British coasts. [See GYMNOTUS.]

TORSK. The name of a fine fish (*Brosme vulgaris*) closely allied to the Cod, and found abundantly in the Shetland Islands, where it is much prized.

TORTOISES. (*Testudinata.*) Under this general appellation is included a numerous and interesting order of Reptiles, which are distinguished, at the first glance by their body being inclosed in a double buckler, which only allows the head and neck, the tail, and the four limbs to be protruded. They are slow, quiet, and inoffensive animals; extremely tenacious of life; suffering the severest mutilations for days or even weeks; and for longevity they are unequalled, actual proof having been given of some which were known to have lived upwards of two hundred years. Their motions are usually slow and awkward, and their sensations appear to be very obtuse. They feed upon vegetable substances, and seldom wander far from their usual haunts; but they require very little nourishment, and can even remain for months without taking any. For the most part they inhabit the warmer regions of the globe; though many species will bear removal to colder climates, where, however, they pass the winter in a torpid state. The upper buckler, termed the *carapace* or shield, is highly arched and very strong in the Land Tortoises; but more flattened in the aquatic families, for the better adaptation of their form to motion in a liquid. The inferior buckler, named the *plastron* or breast-plate, differs considerably in degree of development, and in the relative consolidation of its different parts. It is most complete in the Land Tortoises; in many of

TORTOISE.—(TESTUDO RADIATA.)

which the anterior and posterior portions of it are so jointed to the centre-piece, as to be able to close the orifices before and behind, after the head, tail, and legs have been drawn in. Notwithstanding this unusual arrangement of the osseous frame-work of the Tortoise tribe, we find the same *constituent parts* (though greatly modified) as in ordinary vertebrata. We see that the carapace on its upper surface is formed by a great number of bony plates, united together by sutures; of these plates, eight occupy the median line sixteen constitute a longitudinal range on each side of these, and twenty-five or twenty-six surround the whole like an oval frame. But if we examine the carapace by its lower surface, we find that the central pieces are appendages to the dorsal vertebræ. On the under side, the body of each of these bones is found, in fact, to present its ordinary form; as is also the vertebral canal, which serves to lodge the spinal cord; but the upper portion of the walls of the ring which constitute this canal is here spread out sideways as a disc, and is continuous without interruption with the corresponding plates, belonging to the vertebra which precedes, and to that which follows. These dorsal vertebræ, thus become immovable, have attached to each a pair of ribs, as in most vertebrated animals: but these ribs are so much widened as to touch each other along the whole or nearly the whole of their length, and are connected together by sutures. Lastly, the marginal pieces, which are articulated with the extremities of the ribs, and which form a kind of border to the carapace, represent the sternal portions of these bones; which in Mammalia remain in a cartilaginous state, as, in fact, they do, in some Tortoises.

COMMON OR GREEK TORTOISE. (*Testudo Græca.*) This well-known species is supposed to be a native of almost all the countries bordering on the Mediterranean sea, and is thought to be more frequent in Greece than elsewhere. It is found in the scattered islands of the Archipelago, and in Corsica and Sardinia; occurring likewise in many parts of Africa. The general length of the shell of this species is from six to eight inches; rarely exceeding the latter measure; and the weight of the full-grown animal is about forty-eight ounces. The shell is of an oval form, extremely convex on the upper part, and composed of thirteen middle pieces, and about twenty-five marginal ones: the middle pieces, or those constituting the disc of the shield, are mostly of an oblong square form, and of a blackish or dark brown colour, varied by a broad yellow or citron band running along one side of each, and continued about half way along the upper part: there is also an oblong patch of a similar colour running down the lower part or side of each: and on the top or centre of each piece is an obscurely square or oblong space; rather more depressed than the rest, and marked, as in many other tortoise-shells, with roughish spots or granules: several furrows more or less distinct in different individuals, appear traced round the sides of each piece, becoming gradually less distinct as they approach the upper part or space just mentioned. The colours of the shell are more or less bright in the different specimens, and are subject to some occasional variations, as well as sometimes in the shape of the pieces themselves. The under part of the shell is of a citron or pale yellow colour, with a broad blackish or deep brown zone down on each side, leaving the middle part plain. The head is rather small; the eye small and black; the mouth not extending beyond the eyes; the upper part of the head covered with somewhat irregular, tough scales, and the neck with smaller granulations so as to be flexible at the pleasure of the animal. The legs are short, and the feet moderately broad, covered with strong ovate scales, and commonly furnished with four tolerably stout claws on each; but these parts are found to vary in number, there being sometimes five claws instead of four on the fore feet. The tail is still shorter than the legs, is covered with small scales, and terminates in a naked horny pointed tip or process. This animal lives to a most extraordinary age, of which fact several well-attested instances are on record. One, whose shell is still preserved in the library of Lambeth Palace, was introduced into the archiepiscopal garden in the time of Archbishop Laud, about the year 1633, and continued to live there till 1753, when it was supposed to have perished rather from accidental neglect on the part of the gardener, than from the mere effect of age.

The general manners of the Tortoise, in a state of domestication in this country, are very agreeably detailed by the Rev. Gilbert White, in his History of Selborne. He thus writes to the Hon. Daines Barrington, from Ringmer, near Lewes, in October, 1770:—"A land-tortoise, which has been kept for thirty years in a little walled court belonging to the house where I am now visiting, retires under ground about the middle of November, and comes forth again about the middle of April. When it first appears in the spring, it discovers very little inclination towards food, but, in the height of summer, grows voracious, and then, as the summer declines, its appetite declines; so that, for the last six weeks in autumn, it hardly eats at all. Milky plants, such as lettuces, dandelions, sow-thistles, are its favourite dish. In a neighbouring village, one was kept, till, by tradition, it was supposed to be an hundred years old—an instance of vast longevity in such a poor reptile."—In April, 1772, he again writes to the same corespondent:—"While I was in Sussex last autumn, my residence was at the village near Lewes, from whence I had formerly the pleasure of writing to you. On the 1st of November, I remarked that the old tortoise, formerly mentioned, began first to dig the ground, in order to the forming of its hybernaculum, which it had fixed on just beside a great tuft of hepaticas. It scrapes out the ground with its fore-feet, and throws it up over its back with its hind; but the motion of its legs is ridiculously slow, little exceeding the hour hand of a clock; and suitable to the composure of an animal said to be a whole month in performing one feat of copulation. Nothing can be more assiduous than this creature, night and day, in scooping the earth, and forcing its great body into the cavity; but, as the noons of that season proved unusually warm and sunny, it was continually interrupted and called forth by the heat, in the middle of the day; and though I continued there till the 13th of November, yet the work remained unfinished. Harsher weather and frosty mornings would have quickened its operations. No part of its behaviour ever struck me more than the extreme timidity it always expresses with regard to rain; for though it has a shell that would secure it against the wheel of a loaded cart, yet does it discover as much solicitude about rain as a lady dressed in all her best attire, shuffling away on the first sprinklings, and running its head up in a corner. If attended to, it becomes an excellent weather-glass; for as sure as it walks elate, and, as it were, on tiptoe, feeding with great earnestness in a morning, so sure will it rain before night. It is totally a diurnal animal, and never pretends to stir after it becomes dark. The Tortoise, like other reptiles, has an arbitrary stomach, as well as lungs; and can refrain from eating as well as breathing for a great part of the year. When first awakened, it eats nothing; nor again in the autumn, before it retires: through the height of the summer it feeds voraciously, devouring all the food that comes in its way. I was much taken with its sagacity in discerning those that do it kind offices: for, as soon as the good old lady comes in sight, who has waited on it for more than thirty years, it hobbles towards its benefactress with awkward alacrity; but remains inattentive to strangers. Thus not only 'the ox knoweth his owner, and the ass his master's crib,' but the most abject reptile and torpid of beings distinguishes the hand that feeds it, and is touched with feelings of gratitude."

There are several species of Land Tortoises, which differ from the foregoing both with regard to their size, and the colour, form, &c. of their buckler. Some are remarkable for the pleasing distribution of their colours, as the GEOMETRICAL TORTOISE (*Testudo geometrica*), a small species with a black carapace, each scale of which is regularly adorned with yellow lines radiating from a disc of the same colour.—Another, the CLOSE TORTOISE (*Testudo clausa*), obtains its name from the unusual manner in which the under part of the shell is applied to the upper; being continued in such a manner round the margin, that when the animal withdraws its head and legs, it is enabled accurately to close all parts of the shell entirely together, so as to be in a complete state of security; and so strong is the defence of this little animal, that it is not only uninjured by having a weight of five or six hundred pounds laid upon it, but can walk in its usual manner beneath the load. Its length rarely exceeds four or five inches. It is a native of many parts of North America; and is principally sought for on account of its eggs, which are reckoned a delicacy, and are about the size of pigeons' eggs.

We shall conclude our account of Land Tortoises with the following from 'Darwin's Researches,' &c. In describing the reptiles common in the Galapagos Archipelago, that gentleman particularly notices the habits of the large Tortoise (*Testudo Indicus*). "These animals," says he, "are found, I believe, in all the islands of the Archipelago; certainly in the greatest number. They frequent in preference the high damp parts, but likewise inhabit the lower and arid districts. He then quotes Dampier, in proof of their number, who says, 'They are here so numerous, that five or six hundred men might subsist on them for several months without any other sort of provisions; and they are so extraordinarily large and fat, and so sweet, that no pullet eats more pleasantly.' "The Tortoise is very fond of water, drinking large quantities, and wallowing in the mud. The larger islands alone possess springs, and these are always situated towards the central parts, and at a considerable elevation. The Tortoises, therefore, which frequent the lower districts when thirsty, are obliged to travel from a long distance. Hence broad and well-beaten paths radiate off in every direction from the wells even down to the sea-coast; and the Spaniards, by following them up, first discovered the watering-places. When I landed at Chatham Island, I could not imagine what animal travelled so methodically along the well-chosen tracks. Near the springs it was a curious spectacle to behold many of these great monsters; one set eagerly travelling onward with outstretched necks, and another set returning, after having drunk their fill. When the Tortoise arrives at the spring, quite regardless of any spectator, it buries its head in the water above its eyes, and greedily swallows great mouthfuls, at the rate of about ten in a minute. The inhabitants say each animal stays three or four days in the neighbourhood of the water, and then returns to the lower country; but they differed in their accounts respecting the frequency of these visits. The animal probably regulates them according to the nature of the food which it has consumed. It is, however, certain, that Tortoises can subsist even on those islands where there is no other water than what falls during a few rainy days in the year. I believe it is well ascertained, that the bladder of the Frog acts as a reservoir for the moisture necessary to its existence: such seems to be the case with the Tortoise. For some time after a visit to the springs, the urinary bladder of these animals is distended with fluid, which is said gradually to decrease in volume, and to become less pure. The inhabitants, when walking in the lower district, and overcome with thirst, often take advantage of this circumstance, by killing a Tortoise, and if the bladder is full, drinking its contents. In one I saw killed, the fluid was quite limpid, and had only a *very slightly* bitter taste. The inhabitants, however, always drink first the water in the pericardium, which is described as being best. The Tortoises, when moving towards any definite point, travel by night and day, and arrive at their journey's end much sooner than would be expected. The inhabitants, from observations on marked individuals, consider that they can move a distance of about eight miles in two or three days. One large Tortoise, which I watched, I found walked at the rate of sixty yards in ten minutes, that is, three hundred and sixty in the hour, or four miles a day,—allowing also a little time for it to eat on the road. The flesh of this animal is largely employed, both fresh and salted; and a beautifully clear oil is prepared from the fat. When a Tortoise is caught, the man makes a slit in the skin near its tail, so as to see inside its body, whether the fat under the dorsal plate is thick. If it is not, the animal is liberated; and it is said to recover soon from this strange operation. In order to secure the Tortoises, it is not sufficient to turn them like Turtle, for they are often able to regain their upright position. [See Turtle.]

TORTOISE-SHELL [BUTTERFLY]. A name given by insect collectors to different Butterflies, of the species *Vanessa polychloros* and *V. urticæ*.

TORTRICIDÆ. A family of Heterocerous Lepidoptera, comprising an extensive group of minute, generally dull-coloured moths, distinguished by their broad entire fore wings, which form a triangle with the body when at rest. The labial palpi are broad and very compressed; the spiral tongue is generally short; the thorax rarely crested;

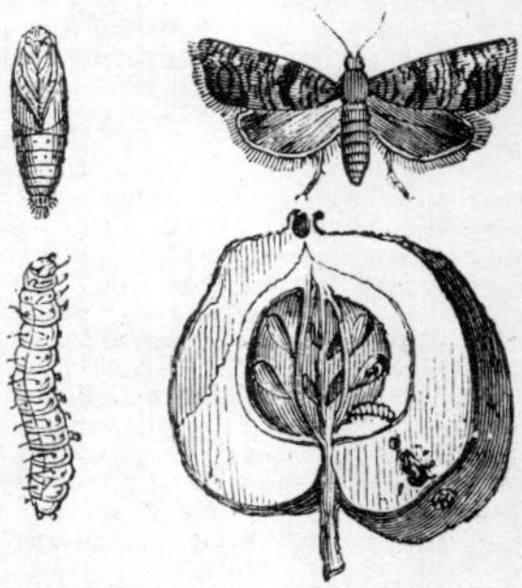

APPLE MOTH, WITH THE CATERPILLAR AND CHRYSALIS. (TORTRIX POMONANA.)

and the antennæ simple. The wings in some species are ornamented with small tufts of scales. The larvæ are naked fleshy grubs, which, for the most part, take up their abode in a leaf, curled up by the insect itself, and fastened with silken threads, forming a cylindrical tube, open at each end, which thus serves them for abode and food; others frequent the young buds and shoots of various plants, fastening several of the leaves together so firmly as to impede its growth; others, again, find their home in the pulpy substance of various fruits, particularly the apple and plum. Another insect of this family (*Carcocapsa Pomonella*), the Codling Moth, is one of the most destructive enemies to the apple crops in this country, laying its eggs in the eyes of the newly-formed fruit, within which the larva feeds, its presence being only indicated by the premature falling of the fruit. Another species (*Tortrix viridana*) feeds upon the oak, which, in certain years, it totally strips of its foliage, its numbers being so great, that when the branches of that tree are sharply beaten, a complete shower of these moths is dislodged. But there is no species of the family so truly injurious as the *Tortrix vitana*, a species which, in the larva state, attacks the leaves of the vines in France, rolling them up, and fastening them together with threads.

TOTANUS. Cuvier's name for a genus of Wading birds, comprehending many species, which, under different names, are found in nearly all parts of the world. They are characterized by a slender, round, pointed, and solid beak, the nasal groove of which only extends half its length, and the upper mandible is slightly arcuated towards the tip. Their form is slight, and the legs very long.

There are four or five British species; among these are *Totanus ochropus*, the GREEN SANDPIPER, called by sportsmen the Whistling Snipe from the shrill note it utters when first flushed: *Totanus glareola*, the WOOD SANDPIPER, which sometimes visits us in winter: *Totanus calidris*, the REDSHANK, which is resident in this country: and *Totanus fuscus*, the SPOTTED SNIPE of Montagu, which is found on our coasts during winter.

One of the most singular species, which is described by Wilson as a native of America, is his *Scolopax vociferus*, but it belongs properly to the genus Totanus, and is the *Totanus melanoleucus* of modern authors. He tells us that this species and the *Totanus flavipes* are "both well known to our duck-gunners along the sea coast and marshes, by whom they are detested, and stigmatized with the names of the greater and lesser tell-tale, for their faithful vigilance in alarming the ducks with their loud and shrill whistle, on the first glimpse of the gunner's approach. Of the two, the present species is by far the most watchful; and its whistle, which consists of four notes rapidly repeated, is so loud, shrill, and alarming, as instantly to arouse every duck within its hearing, and thus disappoints the eager expectations of the marksman. Yet the cunning and experience of the latter are frequently more than a match for all of them; and before the poor tell-tale is aware, his warning voice is hushed for ever, and his dead body mingled with those of his associates.

"The tell-tale seldom flies in large flocks, at least during summer. It delights in watery bogs, and the muddy margins of creeks and inlets; is either seen searching about for food, or standing in a watchful posture, alternately raising and lowering the head, and, on the least appearance of danger, utters its shrill whistle, and mounts on wing; generally accompanied by all the feathered tribes that are near. It occasionally penetrates inland along the muddy shores of our large rivers, seldom higher than tide water, and then singly and solitarily. They sometimes rise to a great height in the air, and can be distinctly heard when beyond the reach of the eye. In the fall, when they are fat, their flesh is highly esteemed, and many of them are brought to our markets." [See GAMBET.]

TOUCAN. (*Ramphastos: Rhamphastidæ.*) A genus and family of Scansorial birds, distinguished by the enormous size of the bill, which in some of the species is nearly as long and as large as the body itself but which is light, cellular, and irregularly notched at the edge, having both mandibles arched towards the tip. The tongue is also of a highly singular form, being narrow and elongated, and laterally barbed like a feather. The structure of the bill renders it necessary for these birds to throw each morsel of their food up into the air, and catch it as it descends, in the throat; a habit observed in many others whose tongue is of a form unfavourable to assist in deglutition. The Toucans are only found in tropical America, where they live in small flocks, in the recesses of the forests. They subsist on fruit and insects, and during the nesting season on the eggs and young of other birds. Their feet are rather short, their wings but moderate, and a rather long tail, which, when the bird is at rest, it commonly holds erect. They nestle in the trunks of trees, and uniformly produce two delicately white eggs, of a rotund form. Their flight is straight, but laborious; among the branches of trees, however, their movements are easy and active; with such lightsome agility, indeed, do they leap from bough to bough, that the beak has then no appearance of being disproportionately large.

In Mr. Swainson's 'Classification of Birds' he states that the fourth family of the *Scansores*, or *Climbing Birds*, is represented by the *Toucans*, whose enormous bills give to these birds a most singular and uncouth appearance. He remarks that their feet are formed, like those of the Parrots, more for grasping than climbing, and that they do not appear to possess the latter faculty; but as they always live among trees, and proceed by hopping from branch to branch, their grasping feet are peculiarly adapted to such habits. He also observes, that the apparent disproportion of the bill is one of the innumerable instances of that beautiful adaptation of structure to use which the book of Nature every where reveals. It is now universally believed that the *Ramphastidæ* are decidedly omnivorous; and although, as Mr. Gould remarks, their elastic bill and delicately feathered tongue would lead us to conclude that fruits constituted the greatest proportion of their diet, we have abundant testimony that they as readily devour flesh, fish, eggs, and small birds; to which, in all probability, are added the smaller kinds of reptiles, caterpillars, and the larvæ of insects in general. — We shall now briefly describe a few species.

The RED-BREASTED TOUCAN. (*Ramphastos dicolorus.*) This bird is a native of Brazil and other parts of South America. Its length is about eighteen inches: colour black, with a gloss of green: cheeks, throat, and fore part of the breast, in some sulphur-yellow, in others orange-yellow: across the lower part of the breast is a broad crimson bar, sometimes extending nearly to the thighs, and sometimes falling far short of those parts; so that, according to this variation, the belly appears either black or crimson: thighs black; vent feathers crimson; rump either crimson or orange-yellow; bill darkish olive-green, with pale yellow base, bounded by a black bar: legs dusky.

We are told in Mr. Edwards's entertaining 'Voyage up the Amazon,' that there are many varieties of Toucans, appearing there at different seasons; but the Red-billed (*R. erythrorynchos*), and the Ariel (*R. ariel*), are the largest and most abundant, seen at every season, but towards autumn particularly in vast numbers throughout the forest.

Their large beaks give them a very awkward appearance, more especially when flying; yet in the trees they use them with as much apparent ease as though they were to our eyes of a more convenient form. Alighted

TOUCAN.—(RAMPHASTOS)

on a tree, one usually acts the part of a sentinel, uttering constantly the loud cry *Tucáno*, whence they derive their name. The others disperse over the branches, climbing about by aid of their beaks and seize the fruit. We had been told that these birds were in the habit of tossing up their food to a considerable distance, and catching it as it fell; but, as far as we could observe, they merely threw back the head, allowing the fruit to fall down the throat. We saw at different times tamed Toucans, and they never were seen to toss their food, although almost invariably throwing back the head. This habit is rendered necessary by the length of the bill and the stiffness of the tongue, which prevents their eating as do other birds. All the time while feeding, a hoarse chattering is kept up, and at intervals they unite with the noisy sentry, and scream a concert that may be heard a mile. Having appeased their appetites, they fly towards the deeper forest, and quietly doze away the noon. Often in the very early morning a few of them may be seen sitting silently upon the branches of some dead tree, apparently awaiting the coming sunlight before starting for their feeding-trees. Toucans, when tamed, are exceedingly familiar, playful birds, capable of learning as many feats as any of the parrots, with the exception of talking. When turning about on their perch, they effect their object by one sudden jump. They eat anything, but are particularly fond of meat. When roosting they have a habit of elevating their tails over their backs. The beaks of the red-billed Toucans are richly marked with red, yellow, and black; but preserved specimens soon lose their beauty."

The COLLARED TOUCAN. (*Ramphastos torquatus.*) Total length eighteen inches; of the bill seven: upper mandible whitish; lower, black: general colour of the plumage black, with the back of the neck crossed by a red collar or bar: fore part of the neck whitish, spotted with red, and streaked with black: belly green; vent feathers red; thighs purple, and legs greenish. This bird is a native of those parts of Mexico which border on the sea, and is supposed to feed on fish.

Illiger separated from *Ramphastos*, under the name of *Pteroglossus*, those species which have the beak not so thick as the head, and are of inferior size, the tail being graduated.

TOURACO. (*Corythaix.*) A genus of birds allied to the *Scansores*. They are natives of Africa. Their generic character may be thus stated:—Bill short, rather small, high, and greatly compressed: the frontal feathers lying upon and concealing the nostrils: culmen high, curved to the tip: lower mandible narrow, both mandibles being distinctly notched at the tip and finely serrated: wings short, and rounded: tail long, broad, and rounded: feet short and strong: claws short, thick, and much com-

TOURACO.—(CORYTHAIX PERSA.)

pressed. The prevailing colour of these elegant birds is green, varied in some species with purple on the wings and tail. They are natives of Africa, where they perch on the highest branches of the forest trees; and feed principally on soft fruits. The most delicate species is thought to be *Corythaix erythrolophus* of Swainson: its crest is red, erect, and compressed; sides of the head, ears and chin, and patch round the eye (which is large, red, and brilliant) white; general plumage green, inclining to bluish on the body and belly; quills rich purple violet; tail rounded; bill yellow; feet grayish black. When the bird is excited or in action, the crest is elevated in a compressed subconical shape; and when thus erected it gives the head a helmeted air.

TOXODON. The name given to an extinct genus of gigantic mammiferous animals, discovered by Mr. Darwin during his sojourn in Banda Oriental, and thus named by Professor Owen, whose notice of this interesting discovery appears in the 'Proceedings of the Geological Society of London,' in 1837. The following clear and concise account, which we extract from Mr. Darwin's Journal, will give the reader a good idea of this wonderful genus of extinct animals. "Having heard of some giant's bones at a neighbouring farm-house on the Sarandis, a small stream entering the Rio Negro, I rode there accompanied by my host, and purchased for the value of eighteen pence, the head of an animal equalling in size that of the Hippopotamus. Mr. Owen, in a paper read before the Geological Society, has called this very extraordinary animal, Toxodon, from the curvature of its teeth. The following notice is taken from the proceedings of that society: Mr. Owen says, judging from the portion of the skeleton preserved, the Toxodon, as far as dental characters have weight, must be referred to the Rodent order. But from that order it deviates in the relative position of its supernumerary incisors, in the number and direction of the curvature of its molars, and in some other respects. It again deviates, in several parts of its structure which Mr. Owen enumerated, both from the *Rodentia* and the existing *Pachydermata*, and it manifests an affinity to the *Dinotherium* and the *Cetaceous* order. Mr. Owen, however, observed, that 'the development of the nasal cavity and the presence of frontal sinuses, renders it extremely improbable that the habits of the Toxodon were so exclusively aquatic as would result from the total absence of hinder extremities; and concludes, therefore, that it was a quadruped, and not a Cetacean; and that it manifested an additional step in the gradation of mammiferous forms leading from the *Rodentia*, through the *Pachydermata* to the *Cetacea;* a gradation of which the Water-hog of South America (*Hydrochœrus capybara*) already indicates the commencement amongst existing *Rodentia*, of which order it is interesting to observe this species is the largest, while at the same time it is peculiar to the continent in which the remains of the gigantic *Toxodon* were discovered.'

"The people at the farm-house told me that the remains were exposed, by a flood having washed down part of a bank of earth. When found, the head was quite perfect; but the boys knocked the teeth out with stones, and then set up the head as a mark to throw at. By a most fortunate chance, I found a perfect tooth, which exactly fits one of the sockets in this skull, embedded by itself on the banks of the Ri Tercero, at the distance of about 180 miles from this place. Near the Toxodon I found the fragments of the head of an animal, rather larger than the horse, which has some points of resemblance with the Toxodon, and others perhaps with the Edentata. The head of this animal, as well as that of the Toxodon, and especially the former, appear so fresh, that it is difficult to believe they have lain buried for ages under ground. The bone contains so much animal matter, that when heated in the flame of a spirit-lamp, it not only exhales a very strong animal odour, but likewise burns with a slight flame.

"At the distance of a few leagues I visited a place where the remains of another great animal, associated with large pieces of armadillo-like covering, had been found. Similar pieces were likewise lying in the bed of the stream, close to the spot where the skeleton of the Toxodon had been exposed. These portions are dissimilar from those mentioned at Bahia Blanca. It is a most interesting fact thus to discover that more than one gigantic animal in former ages was protected by a coat of mail, very similar to the kind now found on the numerous species of armadillo, and exclusively confined to that South American genus. — I may here just mention that I saw in the possession of a clergyman near Monte Video, the terminal portion of a tail, which precisely resembled, but on a gigantic scale, that of the common armadillo. The fragment was 17 inches long, 11½ in circumference at the upper end, and 8½ at the *extreme* point. As we do not know what proportion the tail bore to the body of the animal, we cannot compare it with that of any living species. But at the same time we may conjecture that, in all probability, this extinct monster was from six to ten feet long."

TOXOTES, or ARCHER-FISH. A genus of Acanthopterygious fishes, belonging to Cuvier's sixth family of *Squamipennes*, distinguished from its congeners by the body being short and compressed, the dorsal fin situate far back, the snout short, the

TOXOTES JACULATOR

lower jaw projecting beyond the upper; the mouth is crowded with small teeth, and the opercula are finely toothed. The species obtains its name from projecting drops of water at insects three or four feet above the surface of the water, which it seldom fails in bringing down.

TRACHELIDES. A family of Coleopterous insects, obtaining this name from having the head, which is triangular or heart-shaped, carried on a kind of neck, which separates it from the thorax. The body is soft, and the elytra are flexible. The majority of this group live in the perfect state upon different vegetables, devouring the leaves, or sucking the juices of the flowers. Many of them, when seized, depress the head and contract the feet, as if they were dead. Their colours are often very brilliant. We refer for an example of them to the *Cantharis vesicatoria*, commonly known as the Blistering-fly; an insect of a shining green metallic hue, mostly abundant in Spain.

TRAP-DOOR SPIDER. The name applied to Spiders of the genera *Cteniza* and *Actinopus*, separated by modern authors from the genus *Mygale* of Walckenaer, and remarkable for forming in the ground a long cylindrical tube, protected at the top by a circular door, which is connected to the tube by a hinge. Mr. Westwood remarks:—"Of all the habitations constructed by annulose animals for their own abodes, those cylindrical retreats lined with silk and fitted to the size of the creature's body, are amongst the most ingenious. These are of two kinds: 1st, those which are movable, the creature generally weaving various extraneous materials into the texture of the web, and often with the greatest regularity (amongst which I may particularly mention the nests made by the caddice-worms and the caterpillars of various *Lepidoptera*); and 2ndly, those which are fixed, being formed either in wood or the earth. Instances of the latter are afforded by various species of wild bees and wasps, but they are of a comparatively rude construction compared with the cells of the Trap-door Spider. The interest excited by the accounts of these Spiders has been kept alive since the middle of the last century, when M. Sauvages published his account of an "*Araignée maconne* (*Mygale cœmentaria*) in the Memoirs de l'Academie des Sciences, for 1758." The writer then gives several instances of specimens having been described and published in various scientific works since that date; and proceeds to quote the description of one first given in Brown's History of Jamaica. "*Tarantula* 2. The black *Tarantula* (*Cteniza nidulans*). The valves of the nest are so well contrived, and so strongly connected, that whenever they are forced open, the native elasticity of the ligaments that fix them restore them immediately to their usual position. It is most frequent in the loose rocky soils, and nestles under ground." Mr. W. adds, by way of a note, "Brown's figure represents the regular trap-door partly opened, having a larger and looser flap attached to its base at the hinge above, and falling backwards; and a specimen of the nest in the Linnæan Society's collection is furnished with a *short* lax membranous appendage on the outside of the trap-door immediately behind the hinge."

Another species (*Mygale Ionica*) is described by Sydney Smith Saunders, Esq., who noticed a number of nests during a short excursion to Zante. These nests were found close round the roots of the olive-trees in a somewhat elevated situation, and were generally observed two or three together about the same tree: the soil a sort of sandy clay, of a light ochraceous colour. "The upper portion of the nests was slightly raised above the surface of the ground; but this may have arisen from the washing away of the surrounding earth during the heavy autumnal rains, the more especially as from the coating of moss which showed itself in many cases upon the upper surface of the operculum, they could not have been of very recent construction. The form and structure of the opercula were also peculiar, all of them being more or less provided with an elevation of the posterior margin directly above the hinge, to the extent in some instances of one-third of the diameter of the lid. The object of this projection could not be mistaken, for, acting as a lever, the slightest pressure upon it would suffice to raise the operculum, and afford the readiest ingress. This elevation appears to be produced by a gradual lengthening in the direction of the hinge of the respective layers of which the lid is composed. * * * The interior lining of the tube of *M. Ionica* appears, from all the nests which I have seen, to be of a less perfect consistency than that of *M. fodiens*, and divested of that circumference of macerated earth, or exterior wall, of a more solid consistency than the surrounding mass, which in those of the last-mentioned species give strength to the work, and facilitate the separation of the tubes from the mass in which they are imbedded. In attempting such separation, the tubes of the Zante *Mygale* invariably broke asunder, although this effect may be in some measure attributed to the excessive dryness of the earth at the time of excavation. The length of these tubes was about four and five inches."

We now return to Mr. Westwood's observations on the species of Trap-door Spiders, to notice one which he names *Actinopus ædificatorius*. "This Spider is of a pitchy black colour, and (with the exception of the abdomen) very shining and polished; the abdomen (which is considerably larger than the cephalo-thorax and greatly elevated and gibbose) is obscure, very finely sericeous, and of an uniform dull brown black colour: the legs are clothed with hair and fine bristles of various lengths, and the various joints are connected together by a very pale whitish membrane, which gives them the appearance of being annulated; these limbs are nearly of equal size, but variable in thickness; the palpi are also of considerable length, and have all the appearance of a pair of feet, at least in the female, which is the only sex I have seen either of this or the Jamaica species. This species is a native of North Africa, where it was discovered by Mr. Drummond Hay. The nests are about four inches deep, slightly curved within, and three-quarters of an inch in diameter; the valve at the mouth not being circular, but rather of an oval form, one side, where the hinge is placed, being straighter than the other. The valve is formed of a number of layers of coarse silk, in the upper layers of which are imbedded particles of the earth, so as to give the cover the exact appearance of the surrounding soil, the several successive layers causing it, when more closely inspected, to resemble a small flattened oyster-shell. The mouth of the nest is shelved off at the edge, so that the valve, which is also shelved off at the edge, falls into and upon the orifice, and shuts it far more completely than if the edges of the valve had been cut straight. The inner lining of the nest and of the valve is pure white.

TREE-FROG. [See HYLA.]

TREPANG. (*Holothuria edulis.*) A marine Radiated animal, belonging to the genus *Holothuria*; sometimes called the Sea Cucumber, which is said to be so abundant in certain parts of the Australian coasts, that by diving for them, in from three to eight fathoms water, a man will bring up eight or ten at a time. The mode of preserving it is this: the animal is split down one side, boiled, and pressed with a weight of stones; then stretched open by slips of bamboo, dried in the sun, and afterwards in smoke, when it is fit to be put away in bags, but requires frequent exposure to the sun. [See HOLOTHURIA.]

TRERON. A genus of Pigeons with thickish bills, to which by some writers the DODO is considered to have been nearly allied.

TRICHECHUS. [See WALRUS.]

TRICHOGLOSSUS. A genus of the Parrot family.

TRICHOGLOSSUS SWAINSONII, or SWAINSON'S LORIKEET. This bird, whose habitat is the south-eastern portion of the Australian continent, is thus described in Mr. Gould's splendid work:—Head, sides of the face and throat blue, with a lighter stripe down the centre of each feather; across the occiput a narrow band of greenish yellow; all the upper surface green, blotched at the base of the neck with scarlet and yellow; wings dark green on their outer webs; their inner webs black, crossed by a broad oblique band of bright yellow; tail green above, passing into blue on the tips of the two central feathers; under surface of the tail greenish yellow; chest crossed by a broad band, the centre of which is rich scarlet, with a few feathers fringed with deep blue, and the sides being rich orange-yellow margined with scarlet; under surface of the shoulder and sides of the chest deep blood-red; abdomen rich deep blue, blotched on each side with scarlet and yellow; under tail-coverts rich yellow, with an oblong patch of green at the extremity of each feather; bill blood-red, with the extreme tip yellow; nostrils and bare space round the eye brownish black; irides reddish orange, with a narrow ring of dark brown near the pupil; feet olive. The flowers of the various species of *Eucalypti* furnish this bird with an abundant supply of food; and as those trees which are covered with newly expanded blossoms afford the greatest quantity of nectarine juice and pollen, to them they chiefly resort for their subsistence. Three or four species, indeed, are often seen on the same tree, and often simultaneously attacking the pendent blossoms of the same branch.

TRICHODON. (*Trichodon Stelleri.*) The only species belonging to the genus *Trichodon* (which stands among the *Thoracic Percideæ* in Cuvier's system) inhabits the most northern part of the Pacific, being found both on the American and Kamtschatdale coasts, and abounding particularly at Unalaschka. It buries itself in the sands at low water, and is dug up by the natives with their hands. "The females deposit their roes in holes in the sand, where the males fecundate them, and it would appear that the parents look after their offspring, as they are often dug up in the same pits with their little ones."

TRICHOPTERA. The name of an order of insects specially founded by Kirby for the case-worm flies; which are characterized by four hairy membranous wings, bearing considerable resemblance in their nervures to the Lepidoptera; the under ones folding longitudinally. [See PHRYGANEA.]

TRIDACNA, or CLAMS. A genus of Conchiferous Mollusca, some of the species of which are of gigantic size, and all are more or less beautiful, of a delicate white colour tinged with buff. They are equivalve, radiately ribbed, the ribs adorned with vaulted foliations, waved at the margins, with a large anterior hiatus close to the umbones, for the passage of a large byssus, by which the animal fixes itself on marine substances, rocks, and with the most extraordinary tenacity; hinge with a ligament partly external; two laminar teeth in one valve, one in the other. The shells of some of the *Tridacna gigas* weigh 500lbs., and are used in some Catholic countries as receptacles for the holy water used in churches. The animal is correspondingly large.

TRIGLA. A genus of fishes belonging to the second family of *Acanthopterygii*, which in Cuvier's system comprehends a number of fishes of which the appearance of the head is singular, being variously mailed,

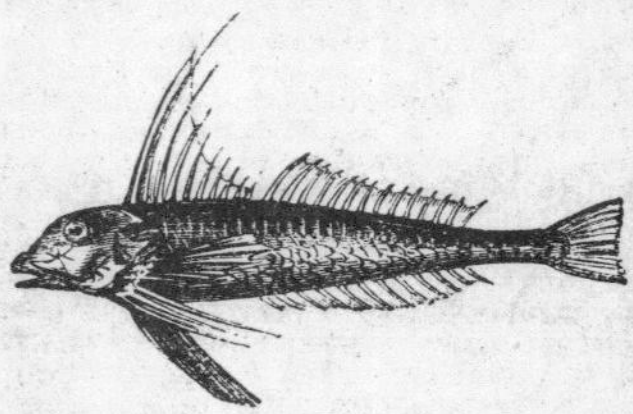

GURNARD.—(TRIGLA.)

or defended by spines or scaly plates of hard matter; but they have many characters in common with the *Percidæ*. Their principal distinction consists in the suborbital bone being more or less extended over the cheek and articulated with the operculum. Our figure represents one of the principal genera, the *Gurnards*, so called from the sounds which they utter with their gill-lids when taken out of the water. They have an immense suborbital plate, to which the operculum or gill-lid is articulated by an immovable suture, so as to be incapable of separate motion. They have the head vertical in the sides, hard and rough bones, two distinct dorsals, three free rays under the pectorals, twelve cæca, and an air-bladder of two lobes. Their pectorals are very large, but not sufficiently so for raising them out of the water, like those of the Flying-fishes. There are many species found in the temperate seas. [See GURNARD.]

TRIGONIA. A genus of Conchiferous Mollusca, of which there is only one species recent, which is found in the deep seas of New Holland; but many fossil. The animal is characterized as having the mantle open along its length; no posterior tubes; foot powerful and trenchant. The shell is equivalve, inequilateral, transversely furrowed denticulated on the inner margin, rounded anteriorly, truncated posteriorly; hinge with four oblong, compressed, diverging teeth in one valve, receiving between their grooved sides two similar teeth in the other; in each valve two muscular impressions. The inside is of a brilliant pearly texture, tinged with purple or golden brown. The *Trigonia pectinata* was formerly so very rare, that even an old worn-out valve would fetch a high price. The fossil species are numerous, and occur in the upper and lower oolites, the lias, and in the beds of green sand.

TRIGONOCEPHALUS. A genus of poisonous serpents, characterized by having the tail terminated by a horny conical process or spur. They are closely allied to the Rattlesnakes.

TRILOBITES. These Crustacea, which, as Cuvier tells us, appear to have been annihilated during the ancient revolutions of our planet, are defined in that most scientific work of Dr. Burmeister's, devoted to their history, were a peculiar family of Crustacea nearly allied to the existing *Phyllopoda*, approaching this family most nearly in the genus *Branchipus*, and forming a link connecting the *Phyllopoda* with the *Peciopoda*. We need not add that they are only found in a fossil state, and that our figure, which represents the *Asaphus caudatus*, a very typical form of the family, will give a general idea of the appearance of this group of animals.

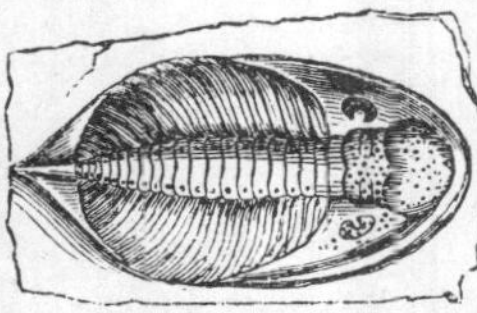

ASAPHUS CAUDATUS.

Our countryman, Edward Lhwyd, curator of the Ashmolean Museum, Oxford, considerably more than a century ago, was the first author who wrote on them. Dr. Burmeister, whose work was translated and published by the Ray Society in 1846, believes, from a comparison of their structure with recent analogues, that these animals moved only by swimming, and remained close beneath the surface of the water; they swam in an inverted position, the belly upwards, and made use of their power of rolling themselves into a ball as a defence against attacks from above. Their food was the smaller water-animals. Their habitat was not the open sea, but the vicinity of coasts in shallow water, where they lived gregariously in vast numbers. Dr. Burmeister believes that the numbers of species could never have been very great, and thinks that some geologists, by judging of size and such characters, have multiplied the species too much.

TRIMERA. The first section of the order HOMOPTERA, which corresponds with the Linnæan genus *Cicada*, and comprises the most numerous portion of the order, consisting of the largest and most beautiful of the species. They are generally saltatorial, but the hind legs are never disproportionably incrassated: they have ordinarily three joints in the tarsi, and very small antennæ: the wings are varied in their consistence in different species, but the upper pair never exhibit two different textures, so remarkable in the Heteroptera.

TRINGA. A genus of Grallatorial birds, having the bill generally not longer than the head, with its tip depressed, and the nasal groove very long. Their toes have no web at the base, and the back toe scarcely reaches the ground. Their legs are shortish, and in general the birds are of small size. In this genus is the Sandpiper (*Tringa canutus*), and the Purple Sandpiper (*Tringa maritinia.*) The Ruffs are closely allied to them. [See RUFF and SANDPIPER.]

TRITON. A genus of Mollusca, found in the Mediterranean, Indian, and South Seas. The animal has two long tentacula furnished with eyes; foot round, and generally short. The shell is oblong, ribbed or tuberculated, with continuous varices placed alternately on each whorl: spire prominent; right lip often wrinkled, and left occasionally thickened, generally denticulated within; epidermis rough; operculum horny. By some uncivilized nations inhabiting the countries near which it abounds, this shell, often from one to two feet long, is used as a military horn; the apex having a hole bored in it, notes can be produced by blowing through the aperture, and thus it becomes a rude instrument of music.

TROCHILIDÆ. A family of extremely diminutive Tenuirostral birds, celebrated alike for the brilliant lustre of their plumage and the rapidity of their flight. They have a long slender beak, and a tongue split, almost to its base, into two filaments, which, being capable of protrusion upon the same principle as that of the Woodpeckers, they are said to employ it in sucking up the nectar of flowers: they, however, also feed on insects. They have very small feet, a great tail, and excessively elongated and narrow wings; balancing themselves in the air by a rapid motion of the latter, now hovering and humming round flowering shrubs and plants, and now darting through the air with almost incredible swiftness. They fight desperately with each other, and defend their nests with courage. Two of the same species can rarely suck flowers from the same bush without a rencontre: this is abundantly confirmed by Mr. Gosse, when describing the species *Trochilus mango*. "In the garden were two trees, of the kind called the Malay apple (*Eugenia Malaccensis*), one of which was but a yard or two from my window. The genial influence of the spring rains had covered them with a profusion of beautiful blossoms, each consisting of a multitude of crimson stamens, with very minute petals, like bunches of crimson tassels; but the leaf-buds were but just beginning to open.

A Mango Humming-Bird had, every day, and all day long, been paying his devoirs to these charming blossoms. On the morning to which I allude, another came, and the manœuvres of these two tiny creatures became highly interesting. They chased each other through the labyrinth of twigs and flowers, till, an opportunity occurring, the one would dart with seeming fury upon the other, and then, with a loud rustling of their wings, they would twirl together, round and round, until they nearly came to the earth. It was some time before I could see, with any distinctness, what took place in these tussles; their twirlings were so rapid as to baffle all discrimination. At length an encounter took place pretty close to me, and I perceived that the beak of the one grasped the beak of the other, and thus fastened, both whirled round and round in their perpendicular descent, the point of contact being the centre of the gyrations, till, when another second would have brought them both on the ground, they separated, and the one chased the other for about a hundred yards, and then returned in triumph to the tree, where, perched on a lofty twig, he chirped monotonously and pertinaciously for some time; — I could not help thinking, in defiance. In a few minutes, however, the banished one returned, and began chirping no less provokingly, which soon brought on another chase and other tussle. I am persuaded that these were hostile encounters, for one seemed evidently afraid of the other, fleeing when the other pursued, though his indomitable spirit would prompt the chirp of defiance; and, when resting after a battle, I noticed that this one held his beak open as if panting. Sometimes they would suspend hostilities to suck a few blossoms, but mutual proximity was sure to bring them on again, with the same result. In their tortuous and rapid evolutions, the light from their ruby necks would now and then flash in the sun with gem-like radiance; and as they now and then hovered motionless, the broadly-expanded tail, whose outer feathers are crimson-purple, but when intercepting the sun's rays transmit orange-coloured light, added much to their beauty. A little Banana Quit, that was peeping among the blossoms in his own quiet way, seemed now and then to look with surprise on the combatants; but when the one had driven his rival to a longer distance than usual, the victor set upon the unoffending Quit, who soon yielded the point, and retired, humbly enough, to a neighbouring tree. The war, for it was a thorough campaign, a regular succession of battles, lasted fully an hour, and then I was called away from the post of observation. Both of the Humming-birds appeared to be adult males."

In an earlier part of the same article, our author observes that the interior of flowers is almost always inhabited by very small insects, and that he believes it is principally to pick out these that the Humming-birds probe the tubular nectaries of blossoms. That they also pursue flies on the wing seems, however, no less certain; for he has often seen the Mango, just before night-fall, fluttering round the top of a tree on which were no blossoms, and from the manner in which it turned hither and thither, while hovering in a perpendicular position, it was manifest that it was catching minute insects. This species (he says) when flying often flirts and flutters the tail in a peculiar manner, throwing it in as he hangs perpendicularly in mid air, when the appearance of the broad lustrous feathers, expanded like a fan, is particularly beautiful. The length of the Mango Humming-bird is rather more than five inches; and in expanse it somewhat exceeds seven inches.

The Vervain Humming-bird. (*Mellisuga humilis*). The male of this species is about two inches and a half in length, its wings expanded being three and a half. The whole upper parts of the plumage metallic-green; wings purplish black, tail deep black; chin and throat, white speckled with black; breast white; sides metallic green; belly whitish, each feather tipped with green; under tail-coverts white, faintly tipped with green. The female is rather less than the male; and of a yellower green above, which descends half way down upon the tail: whole under parts pure white, unspotted, and untinged with green; tail-feathers, except the uropygials, tipped with white. Irides, beak, and feet black.

"The West Indian Vervain (*Stachytarpheta*)," as Mr. Gosse informs us," is one of the most common weeds in neglected pastures, shooting up everywhere its slender columns, set around with blue flowers, to the height of a foot. About these our little Humming-bird is abundant during the summer months, probing the azure blossoms a few inches from the ground. It visits the spikes in succession, flitting from one to another, exactly in the manner of the honey-bee, and with the same business-like industry and application. In the winter, the abundance of other flowers and the paucity of vervain-blossoms, induce its attention to the hedgerows and woods. I have sometimes watched, with much delight, the evolutions of this little species at the moringa tree. When only one is present, he pursues the round of the blossoms soberly enough, sucking as he goes, and now and anon sitting quietly on a twig. But if two are about the tree, one will fly off, and, suspending himself in the air a few yards distant, the other presently shoots off to him, and then, without touching each other, they mount upward with a strong rushing of wings, perhaps for five hundred feet. Then they separate, and each shoots diagonally towards the ground, like a ball from a rifle, and wheeling round, comes up to the blossoms again, and sucks, and sucks, as if it had not moved away at all. Frequently one alone will mount in this manner, or dart on invisible wing diagonally upward, looking exactly like a humble-bee. Indeed, the figure of the smaller Humming-birds on the wing, their rapidity, their arrowy course, and their whole manner of flight, are entirely those of an insect; and one who has watched the flight of a large beetle or bee, will have a very good idea of the form of one of these tropic gems, painted against the sky. I have observed all our three species at one time engaged in sucking the blossoms of the moringa at Content; and have noticed that whereas Polytmus and Mango expand and depress the tail, when hovering before flowers, Humilis, on the contrary, for the most part *erects* the tail, but not invariably. The present is the only Humming-bird that I am acquainted with, that has a real song. Soon after sunrise in the spring months, it is fond of sitting on the topmost twig of some mango or orange tree, where it warbles, in a very weak but very sweet tone, a continuous melody, for ten minutes at a time: it has little variety. The others have only a pertinacious chirping."

"One day in June," observes Mr. Gosse, while speaking of their mode of nidification in the zig-zag terraces cut in the mountain roads of Jamaica, "I found two nests attached to twigs of bamboo and one just commenced. Two parallel twigs were connected together by spiders' webs, profusely but irregularly stretched across, and these held a layer of silk-cotton, which just filled up the space (about an inch square) between them. This was the base. The others were complete cups of silk-cotton exceedingly compact and neat, ornamented outside with bits of gray lichen, stuck about. Usually the nest is placed on a joint of a bamboo branch, and the diverging twigs are embraced by the base. The nest is about the size of half a walnut-shell, if divided not lengthwise, but transversely. To see the bird sitting in this tiny structure is amusing. The head and tail are both excluded, the latter erect like a wren's: and the bright eyes glance in every direction. One of these contained two eggs, the other a single young one nearly fledged." * * * "Several times I have enclosed a nest of eggs in a gauzed cage, with the dam, taken in the act of sitting; but in no case did she survive twenty-four hours' confinement, or take the slightest notice of her nest. When engaged in the attempt to domesticate a colony of Polytnus, an opportunity offered to add this minute species to my aviary. For at that time two large tamarind-trees very near the house were in full blossom, and round them the Vervain Humming-bird was swarming. They flocked together like bees, and the air resounded with their humming, as if in the neighbourhood of a hive. We caught several of them with the net, but could make nothing of them; they were indomitably timid. When turned into the room, they shot away into the loftiest angle of the ceiling, and there hovered motionless, or sometimes slowly turning as if on a pivot, their wings all the time vibrating with such extraordinary velocity as to be visible only as a semicircular film on each side." * * * "The spirit of curiosity is manifested by this little bird as well as by the larger species. When struck at, it will return in a moment, and peep into the net, or hover just in front of one's face." The magnificent work of Mr. Gould on this family, and the no less elegant and admirable book of M. Bourcier, will leave little to be done in the illustration of the members of this "angelically graceful group."

TROGON. (*Curucui.*) A genus of Scansorial birds, mostly inhabitants of South America. They differ so much in the various stages of growth, that it has caused considerable confusion of species; but as they all agree in their general habits of life, the description of one will suffice. — TROGON CURUCUI. This bird is ten inches and a half in length: beak pale yellow, the under mandible armed with stiff black bristles; head, neck, back, rump, and upper tail-coverts shining green, with a blue gloss in certain lights; wing-coverts bluish-gray, marked with many undulating black lines; quills black, with part of the shafts white; the breast, belly, sides, and under tail-coverts of a beautiful red; thighs dusky; upper surface of the tail green, except the three outer feathers, which are blackish, and crossed with narrow transverse lines of gray; tail wedge-shaped; legs brown. It is a very solitary bird, being found only in the thickest forests; and in the pairing time the male has a very melancholy note (by which his haunts are discovered), which is never uttered at any other time than while the female is sitting, for as soon as the young make their appearance he becomes again perfectly mute. They begin to pair in April, and build in the hole of a rotten tree, laying three or four white eggs, about the size of a pigeon's, on the decayed dust, or if there be no dust, they bruise the sound wood into powder by means of their strong bill. The young when first hatched are quite destitute of feathers; the head disproportionately large, and the legs very long: the old birds feed them with small worms, caterpillars, and insects; and when able to shift for themselves, desert them and return to their solitary haunts till August or September; when they are again instinctively prompted to produce another brood. To this genus belongs the gorgeous long-tailed Trogon or *Quezal*, the feathers of which were allowed to be worn only by Mexicans of the highest rank in former times.

Mr. Edwards thus speaks of those he saw while pursuing his voyage up the Amazon. "There were half a dozen varieties, differing in size — from the *T. viridis*, a small species whose body was scarcely larger than many of our sparrows, to the Curucua grande (*Calurus auriceps*), twice the size of a jay. All have long spreading tails, and their dense plumage makes them appear of greater size than the reality. They are solitary birds; and early in the morning, or late in the afternoon, may be observed sitting, singly or in pairs, some species upon the tallest trees, and others but a few feet above the ground, with tails outspread and drooping, watching for passing insects. Their appetites appeased, they spend the remainder of the day in the shade, uttering at intervals a mournful note, well imitated by their common name, *curuqua*. This would betray them to the hunter, but they are great ventriloquists, and it is often impossible to discover them, although they are directly above one's head. The species vary in colouring as in size, but the backs of all are of a lustrous green or blue, and bellies of red, or pink, or yellow. The *curucua grande* is occasionally seen at Barra; but, frequenting the tallest forests, it is exceedingly difficult to be obtained. We offered a high price for a specimen, and employed half the garrison for this single bird without success. They reported that they every day saw them, and frequently shot at them; but that they never would come down. Their feathers were so loose, that, in falling when shot, they almost invariably lost many; and thus, together with the tenderness of their skins, made them the most difficult of birds to preserve."

TROGONOTHERIUM. [See SUPPL.]

TROOPIALS. (*Cassicus: Icterus: Xanthornus.*) Passerine birds, in which the beak is large, conical, thick at the base, and very sharp at the point. Their manners somewhat approach the Starlings: they frequently construct their nests close together, and feed on insects and grain; and when in numerous flocks they commit great ravages in cultivated districts, especially in maize plantations. In his voyage up the river Amazon, in 1846, Mr. Edwards was much struck with their nests, and from his lately published narrative we make the following extract:—"The most singular nests, and most worthy of description, were those of the Troopials (*Cassicus icteronotus*, Swain.), a large black bird, much marked with yellow, and frequently seen in cages. Their native name is Japim. They build in colonies pensile nests of grass, nearly two feet in length, having an opening for entrance near the top. Upon one tree standing in the middle of the lake, not more than ten feet high, and the thickness of a man's arm, were forty-five nests of these birds, built one upon another, often one depending from another, and completely concealing all the tree-top except a few outermost leaves; at a distance the whole resembled a huge basket. Part of these nests belonged to the Red-rumped Troopial (*C. hæmorrhous*); and a singular variety of Oriole, the Ruff-necked of Latham, called Araona or Rice-bird, after the fashion of our cow-bird, deposits its eggs in the Troopials' nests, leaving the young to the care of their foster-mothers. Usually Troopials build nearer houses, and are always welcome, being friendly sociable birds, ever ready to repay man's protection by a song. Often in such situations large trees are seen with hundreds of these nests dependent from the limbs and swaying in the wind. A colony which had settled upon a tall palm near the mill was one night entirely robbed of eggs by a lizard. Snakes are sometimes the depredators, and, between all their enemies, the poor birds of every species are robbed repeatedly. Probably owing to this cause it is very unusual to find more than two eggs in one nest. The Red-rumped Troopials shot in this place were of different sizes, some being several inches longer than others, although all were in mature plumage. Their nests were perhaps larger than those of the Japims, but differed in no other respect. The eggs were white, spotted with brown, and particularly on the larger end. The Japim's eggs were cream-coloured, and similarly spotted; and the eggs of the ruff-necked orioles were large in proportion to the size of the bird, bluish in colour, and much spotted, and lined with dark brown."

TROPIC-BIRD. (*Phaëton.*) A genus of Palmipede birds, distinguished by two long slender tail-feathers, and well known to navigators as the harbingers of the tropics. They are characterized by extraordinary length of wing and feeble feet; they are accordingly well formed for flight, and disport in the air far at sea: when on land, to which they seldom resort for any length of time together, except at the period of nidification, they are seen perching on rocks and trees. Two species only are known: *Phaëton æthereus* and *Phaëton phœnicurus.*

The COMMON TROPIC-BIRD (*Phaëton æthereus*) is about the size of a Partridge, and has very long wings: the bill is red, with an angle under the lower mandible, like those of the Gull kind, of which it is a species: the eyes are surrounded with black, which ends in a point towards the back of the head: three or four of the larger quill-feathers towards their ends are black tipped with white: all the rest of the bird is white, except the back, which is variegated with curved lines of black: the legs and feet are of a vermillion red: the toes are webbed: the tail consists of two long straight narrow feathers almost of equal breadth from their quills to their points. These birds are rarely seen but between the tropics, at the remotest distance from land. Their name seems to imply the limit of their abode; and, indeed they are seldom seen but a few degrees north or south of either tropic.

Nothing, says Lesson, who had good opportunities of observing both species of the Tropic-bird, can be more graceful than their flight. They glide along, most frequently without any motion of the wing, on the sustaining air, but at times this smooth progression is interrupted by sudden jerks. When they perceive a ship, they never fail to sail round it, as if to reconnoitre. They ordinarily return every evening to the land, to roost in the midst of the rocks where they place their nests. Their food appears to consist entirely of fish. The long feathers of the tail are employed by the natives of the greater part of the South Sea Islands as ornaments of dress.

TROPIDORHYNCHUS. [See FRIAR-BIRD.]

TROUT. (*Salmo fario.*) The common name of Trout is given to several species of the genus *Salmo*. The one we are about to describe in this place is the well-known RIVER TROUT, a valuable fish, which frequents most of the rivers and lakes of Great Britain, affording much diversion to the angler, and, from its vigilance and caution, combined with its boldness and activity, requiring all his patience and no little skill. The colours of the Trout, and its spots, vary greatly in different waters, and in different seasons; it being remarked that those that inhabit clear, swift, and shallow streams, and live mostly on insect food, have the most brilliant red spots on the sides, and their flesh is of the finest quality; whilst those which are obliged to live chiefly on aquatic vegetables are dull in colour, and their flesh is less deli-

TROUT.—(SALMO FARIO.

cate. The Common River Trout is generally from twelve to fifteen inches long, and from three quarters to a pound and a half in weight; sometimes, however, but not often, considerably exceeding it: the form of the head is blunt; the eye large, the irides silvery, with a tinge of pink; the teeth numerous, strong, and curving inwards, extending along the whole length of the vomer: the convexity of the dorsal and ventral outline nearly similar; and the scales small. The colour of the back and upper part of the sides is made up of numerous dark reddish-brown spots on a yellow-brown ground; about a dozen bright red spots along the lateral line, with a few other red spots above and below it; the lower part of the sides golden yellow; belly and under surface silvery white; dorsal fin and tail light brown, with numerous darker brown spots; the adipose fin brown, frequently with one or two darker brown spots, and edged with red; the pectoral, ventral, and anal fins uniform pale orange-brown. The female fish is of a brighter and more beautiful appearance than the male.

In streams that flow rapidly over gravelly or rocky bottoms, the Trout are generally remarkable for the brilliancy and beauty of their spots and colours: and they are finest in appearance and flavour from the end of May till towards the end of September; an effect produced by the greater quantity and variety of nutritious food obtained during that period. Dr. A. T. Thomson remarks that "each species of Trout has its peculiarities of colour: but the common Trout is the most beautiful of its class: the variations of its tints and spots, from golden-yellow to crimson and greenish-black, are almost infinite, and depend, in a great measure, on the nature of its food; for the colours are always the most brilliant in those fish that feed on the water-shrimp; and those are, also, the most highly prized for the table. It is a curious fact that the brightness of the colours is not diminished when the fish dies; for, even after he has been played with for an hour or longer by the practised angler, and at length is brought floating upon his side to the margin of the stream, and thrown upon the bank floundering, till, gasping with distant and feeble motions, he is either knocked on the head, or dies from exhaustion, his scaly splendour is as bright as before."

It is observed that during the day the larger-sized fish move but little from their accustomed haunts; but towards evening and during the night they rove in search of small fish, insects, and their various larvæ, upon which they feed with eagerness. With no food, however, do they seem so delighted as with the May-fly. The young Trout fry may be seen throughout the day sporting on the shallow gravelly scours of the stream, where the want of sufficient depth of water, or the greater caution of larger and older fish, prevents their appearance. The season of spawning with the Trout is generally in October, at which time the under jaw of the old male exhibits in a smaller degree the elongation and curvature observed in the male Salmon. The stomach of this fish is uncommonly thick and strong; but this circumstance is observed to be nowhere so remarkable as in those found in some of the Irish lakes, and particularly in those of the county of Galway. These are called Gillaroo Trouts: on the most accurate examination, however, it does not appear that they are specifically different from the common Trout, but by living much on different kinds of Crustacea, and swallowing small stones at the same time, their stomachs acquire a much greater degree of thickness, and a kind of muscular appearance, so as to resemble a sort of gizzard. [See SALMON-TROUT.]

TRUMPET-FISH. (*Centriscus scolopax.*) This is a singular looking small Acanthopterygious fish, sometimes called the Sea Snipe. Its body is of an oval shape, and it is distinguished by its long tubular beak, which seems well adapted for drawing from among the sea-weed and mud the minute Crustacea upon which it is supposed to feed. On the back is a slight ridge; and the first dorsal fin is armed with a strong, pointed spine, movable and serrated, constituting a formidable weapon of defence. The colour of the back is red, the sides lighter red; the sides of the head and belly silvery, tinged with a golden hue; the scales on the body hard and rough; and the fins of a grayish white. The Trumpet-fish is found in the Mediterranean, and the flesh is reckoned good. [See AULOSTOMA.]

TRUNK-FISH. [See OSTRACION.]

TRYGON. [See RAY: STING-RAY.]

TUBICOLÆ. The name of an order of *Anellidæ*, comprehending those which live in tubes. One of the commonest of these is the *Serpula*, the shell of which is formed of calcareous matter, resembling that of the shells of Mollusca, and apparently secreted from the surface of the body in a similar manner. They are generally found clustering in masses, attached to the surface of stones, shells, or other bodies, which have been for any length of time immersed in the sea, and more or less contorted according to the position in which they grow. The animal residing in this shell has its branchial filaments or gill-tufts all assembled round the head; where they form a pair of fan-like appendages, usually possessing very brilliant colours. At the base of each series is a fleshy filament, one of which is prolonged and dilated at its extremity into a flat disc, which fits to the mouth of the shell, and serves to close it when the animal is withdrawn into the tube. These groups are found in tropical regions, where they usually form their habitations in the midst of corals, and lengthen their tubes as the coral is built up around them. Numerous smaller species are also found on our own coasts, some of which are remarkable for the brilliant hues of their expanded gills. Others there are which do not form their tubes by a calcareous exudation from their own bodies, but by cementing together particles of shell, sand, &c., by means of a glutinous secretion.

TUBIFERA. The name given by Lamarck to an order of the class *Polypi*, comprising those which are united upon a common substance fixed at its base, and whose surface is wholly or partially covered with retractile hollow tubes.

TUBULARIA. The name of a genus of Corallines which have tubes of a horny substance, simple or branched, from the extremities of which the polypes are protruded. Many of them are found in stagnant fresh water; but the *Tubularia marina* have two ranges of tentacula, the exterior as rays the interior as a tuft.

TUBULIBRANCHIATA. An order of hermaphrodite Gasteropodous Molluscs, comprehending those which have the shell in the form of a more or less irregular tube in which the branchiæ are lodged.

TUCUTUCO. (*Ctenomys Braziliensis.*) A curious small animal, native of South America, described by Mr. Darwin as a rodent, with the habits of a mole. It is extremely abundant in some parts of the country, but is difficult to be procured, and still more difficult to be seen when at liberty. It lives almost entirely under ground, and prefers a sandy soil with a gentle inclination. The burrows are said not to be deep, but of great length. They are nocturnal in their habits; and their principal food is afforded by the roots of plants, which is the object of their extensive and superficial burrows. This animal is universally known by a very peculiar noise, which it makes when beneath the ground. A person, the first time he hears it, is much surprised; for it is not easy to tell whence it comes, nor is it possible to guess what kind of creature utters it. The noise consists of a short, but not rough, nasal grunt, which is repeated about four times in quick succession; the first grunt is not so loud, but a little longer, and more distinct than the three following: the musical time of the whole is constant, as often as it is uttered. The name Tucutuco is given in imitation of the sound. In all times of the day, where this animal is abundant, the noise may be heard, and sometimes directly beneath one's feet. When kept in a room the Tucutucos move about slowly and clumsily, which appears owing to the outward action of their hind legs; and they are likewise quite incapable of jumping the smallest vertical height, which is accounted for by the socket of the thigh-bone not being attached by a *ligamentum teres*. When eating, they rest on their hind legs and hold the piece in their fore paws.—Mr. Darwin observes, that the wide plains north of the Rio Colorado are undermined by these animals; and near the Strait of Magellan, where Patagonia blends with Terra del Fuego, the whole sandy country forms a great warren for the Tucutuco.

TUI. The native name of a Passerine bird of New Zealand; it is called by some the "Parson Bird," and by others the "Mocking Bird." It is the *Prosthemadera (merops) cincinnata* [which see].

TUNICATA. An order of Acephalous Mollusca; for a lucid and interesting description of which, we are indebted to the 'History of British Mollusca and their Shells, by Prof. E. Forbes, F.R.S., and Sylv. Hanley, F.L.S.' "The Tunicata are Mollusca which have no true shell, but are enveloped in a coriaceous tunic or mantle; whence their name. This is constructed in the form of a sac with two openings, or else is shaped like a tube, of greater or less dimensions, open at both ends. Within the tunic we find the viscera, consisting of well-defined organs of respiration, circulation, and digestion, and a muscular and a nervous system. The branchial organ is usually in the form of a sac, placed at the commencement of the alimentary canal, of which it forms, as it were, the antechamber, and is never arranged in distinct leaflets, as it is in the lamellibranchiate conchifera. The circulation of their blood is remarkable, on account of its fluctuations and periodical changes of direction. They have no distinct head, and no organs serving as arms or feet. Sometimes they are free, more usually fixed; but in all cases free during some portion of their existence. Some are simple, some present various degrees of combination; some are simple in one generation, combined in another. They are all dwellers in the sea. Their various states and structures enable naturalists to group them under several well-marked tribes, of most of which we have examples in the British seas. The best classification of them is that proposed by Professor Milne Edwards. He divides them into three suborders, of which the Salpa, the Ascidia, and the Pyrosoma are the types, and subdivides the Ascidians proper into simple, social, and compound. Of all, except the Pyrosoma, we have British examples.

"These animals attracted the notice of the all-observing Aristotle. Like most philosophic naturalists, the question of the distinction between the animal and vegetable kingdoms had for him great attractions. The Ascidia was one of the many creatures which he examined, in the hopes of gaining definite information respecting such distinction. Its inert and sponge-like form, rooted to the ground, seemed to indicate a vegetable nature; but Aristotle was not content with a mere external survey—he explored its internal structure, and soon perceived its highly animal condition," &c. &c. "It is worthy of remark, that very lately the Ascidians have again played a part in that much-vexed question of the distinction between animals and vegetables.' This part of the subject is pursued, with much ability, in the work from which the foregoing is extracted, and to which we beg to refer our readers. [See SALPA in SUPPLEMENT.]

TUNNY. (*Thynnus vulgaris.*) This Acanthopterygious fish has been known and celebrated from a very remote period, and at present forms a valuable source of profit to the inhabitants of the northern coasts of the Mediterranean and the island of Sicily, where in the summer season they resort in vast shoals, and are taken in large nets, or by means of what the Italians call the *tonnaro.* Though bearing a general resemblance to the Mackerel in form, it is a far larger and stouter fish. Each jaw is furnished with a row of small sharp-pointed teeth, slightly curved inwards; the tongue and inside of the mouth very dark coloured; the cheeks are covered with long, narrow, pointed scales; the operculum smooth: the dorsal and anal fins are each followed by nine small finlets; and the tail is crescent-shaped. The upper part of the body is very dark blue; the belly a light gray, spotted with silvery white: the first dorsal fin, pectorals, and ventrals, black; the second dorsal and anal, nearly flesh-colour, with a silvery tint; the finlets above and below yellowish, tipped with black. Mr. Yarrell, quoting the MS. of Mr. Couch, states that "the Tunny

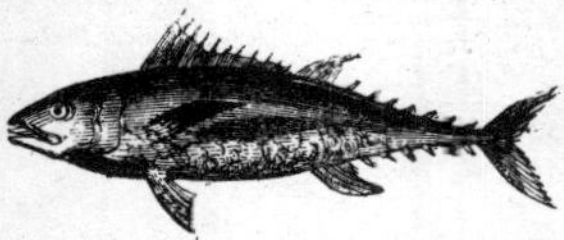

TUNNY.—(THYNNUS VULGARIS.)

appears on the Cornish coast in summer and autumn; but is not often taken, because it does not swallow a bait, or at least the fishermen use no bait that is acceptable to it; and its size and strength seldom suffer it to become entangled in their nets. It feeds on Pilchards, Herrings, and perhaps most other small fishes; but the Skipper (*Esox saurus*) seems to be a favourite prey; for it not only compels it to seek another element for safety, but will also spring to a considerable height after it, usually across its course, at the same time attempting to strike down its prey with its tail."

The Tunny sometimes acquires an immense size, some having been taken which measured nine feet in length, and weighed five hundred pounds: the specimens, however, do not usually exceed from three to four feet. The flesh is considered very delicious, though very solid food: as firm as Sturgeon, but of a finer flavour. It is dressed in a variety of ways in France; served as a ragout, as soup, plain broiled or fried, made into pies, or pickled and eaten cold, as we eat pickled salmon. [See BONITO.]

TUPAIA. A genus of remarkable insectivorous animals, of which there are only three species at present known, and these are found in Sumatra and Java. Their habits are diurnal, and they feed on fruit and insects; but instead of being decidedly terrestrial, they lead the life of Squirrels, whose appearance they greatly resemble, and whose sprightliness and activity they almost rival. They have soft glistening hair and a long bushy tail; and were it not for their long, pointed muzzle, they could not be easily distinguished at a distance from Squirrels. The name of *Banxring* is as often given to this animal as *Tupaia.*

TURBINELLA. A genus of marine Mollusca, inhabiting a large, heavy, and rather fusiform shell; turbinated; thick and wide near the apex; spire short; aperture rather narrow, terminating anteriorly in an open canal; left lip with from three to five distinct plaits, which are at right angles with the axis of the shell. They are found in the Indian and African oceans.

TURBO: TURBINIDÆ. A genus and family of Gasteropodous Mollusca, characterized by having a shell of a regular turbinated form, with an entire and rounded mouth. The largest and perhaps the best-known species is the *Turbo marmoratus;* but there are numerous others; and we cannot, perhaps, give a clearer or more interesting account of the genus than is to be found in the "Popular Conchology" of Agnes Catlow: "*Shell* rather turretted, base not flattened; mouth round; lips not united; outer lip thin; an operculum, shelly and solid. *Animal,* head having two pointed tentacula, with eyes at the base; foot short. Thirty-four species recent, and four fossil. The shells of this genus, if placed upon their mouths, will stand steadily in that position, with the axis very much inclined. They are brought from China, India, Africa, &c. The *Turbo littoreus,* or common Periwinkle, is used by mankind as an article of food, and is found on the shores of England in great numbers. In Sweden, where they also abound, they serve to prognosticate the approaching state of the weather; the peasants having observed that whenever the periwinkles ascend the rocks it is a sure sign of a storm being near, as their instinct teaches them to place themselves out of the reach of the dashing of the waves; on the contrary, when they make a descent upon the sand it is an indication of a calm. In hot countries some species are often seen on the trees near the coast, and on the rocks elevated above the surface of the water; they remain stationary on the latter during the hottest hours, even when it is painful to walk on them from their great heat: they leave the water early in the morning, but return at night. These circumstances prove that, although marine, many species are amphibious. These shells are often highly iridescent; and the mouth in some species, as the *Turbo chrysostomus,* is of a deep and beautiful golden colour."

TURBOT. (*Pleuronectes maximus.*) Of all our Flatfishes this is both the best and one of the largest; and when we consider that the number brought to Billingsgate market alone amounts to about 90,000 in one year, it will be seen that, although they are sometimes scarce and dear, the piscivorous epicures of the metropolis need be under no apprehension of being deprived of such delicious fare. The Turbot is an inhabitant of the Northern and Mediterranean seas, where it often arrives at a very large size.

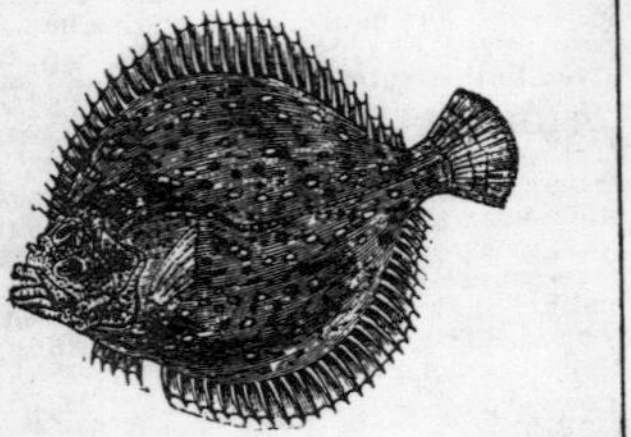

TURBOT.—(PLEURONECTES MAXIMUS.)

It has a broader and squarer form than any others of the genus; and is of a dark brown on the upper surface, marbled with blackish spots of different sizes; and white beneath: the scales are so small as to be scarcely observable, but the skin is of a wrinkled appearance, and covered with pretty numerous and moderately large pointed tubercles or abrupt spines, those on the upper or coloured side being far larger than those on the under side: the lateral line forms an arch over the pectoral fins, and from thence runs straight to the tail. It generally lies in deep water, preying on worms, crustacea, and marine insects, as well as on small fishes: it is taken in great quantities about the northern coasts of England, as well as those of France, Holland, &c., and is baited for with pieces of herring, haddock, &c., but more particularly with the river lamprey, vast quantities of which are said to be purchased of our fishermen by the Dutch.

It is stated in the Encyclopædia Britannica, that "The only fishery, perhaps, which neither the Scotch nor the English follow up with the same success as the Dutch, is that of the Turbot; the finest of which are supposed to be taken upon the Flemish banks. The Turbot fishery begins about the end of March, when the Dutch fishermen assemble a few leagues to the south of Scheveling. As the warm weather approaches, the fish gradually advance to the northward, and during the months of April and May they are found in great shoals on the banks called the Broad Forties. Early in June they have proceeded to the banks which surround the small island of Heligoland, off the mouth of the Elbe, where the fishery continues to the middle of August, when it terminates for the year. The mode of taking the Turbot is as follows:—At the beginning of the season the trawl-net is used; which being drawn along the banks, brings up various kinds of Flatfish, as Soles, Plaice, Thornbacks, and Turbots; but when the warm weather has driven the fish into deeper water, and upon banks of a rougher surface, where trawling is no longer practicable, the fishermen have then recourse to their many-hooked lines. The hooks are baited with the common Smelt, and a small fish resembling an eel, called the Gore-bill [Garfish]. Though very considerable quantities of this fish are now taken on various parts of our own coasts, from the Orkneys to the Land's End, yet a preference is given in the London market to those caught by the Dutch, who are supposed to have drawn not less than 80,000*l.* a year for the supply of this market alone; and the Danes from 12,000*l.* to 15,000*l.* a year for sauce to this luxury of the table, extracted from one million of lobsters, taken on the rocky shores of Norway,—though our own shores are in many parts plentifully supplied with this crustacean, equal in goodness to those of Norway."

TURDUS: TURDIDÆ. A genus and family of Passerine birds, embracing the various species of Thrushes, &c. [See THRUSH.]

TURKEY. (*Meleagris gallipavo.*) The Turkey came originally from North America, where it still associates in large flocks, though it is fast decreasing in its wild state, being only found in remote and unfrequented spots. It is about three feet and a half in length; being somewhat larger than the domestic variety. The general colour of the plumage is black, variegated with bronze and bright glossy green, in some parts changing to purple; the quills are green gold, black towards the end, and tipped with white; the tail consists of eighteen feathers, brown, mottled, and tipped with black; the tail-coverts are waved with black and white; on the breast is a tuft of black hairs, eight

TURKEY.—(MELEAGRIS GALLIPAVO.)

inches in length: in other respects it resembles the domestic bird, especially in having a bare carunculated head and neck, a fleshy dilatable appendage hanging over the bill, and a short blunt spur or knob at the back part of the leg.

Tame Turkeys, like every other animal in a state of domestication, vary considerably in colour, but the prevailing one is dark gray, inclining to black, with a little white towar's the end of the feathers; some are black and white; others perfectly white; there is also a beautiful variety of a fine deep copper colour, with the greater quills pure white, and the tail of a dirty white; but in all of them the tuft of black hair on the breast is prevalent. The young males do not put out the tuft on the breast till they are about three years old. Great numbers are bred in Norfolk, Suffolk, and other counties, whence they are driven to the London markets in flocks of several hundreds. The drivers manage them with facility, by means of a bit of red rag tied to a long rod, which, from the antipathy these birds bear to that colour, effectually drives them forward. The females lay their eggs in the spring, generally in a retired and obscure place, as the male will often break them. They are usually from fourteen to eighteen in number, white mixed with reddish or yellow freckles: the female sits with so much perseverance, that if fresh eggs be introduced into the nest immediately upon the young being hatched, she will long continue the business of incubation; but in this climate she has seldom more than one hatch in a season. Young Turkeys require great care in rearing, being subject to a variety of diseases from cold, rain, and dews; but as soon as they are sufficiently strong, the hen abandons them entirely, and they are capable of enduring the utmost rigour of our winters.

The motions of the male, when agitated with desire, or inflamed with rage, are very similar to those of the Peacock: he erects his tail, and spreads it like a fan, whilst his wings droop and trail on the ground, and he utters at the same time a dull, hollow sound; he struts round and round with a solemn pace, assumes all the dignity of the most majestic of birds, and every now and then bursts out abruptly into a most unmusical gurgle. The familiar name of this bird, it is said, originated in an erroneous idea that it originally came from Turkey.

The OCELLATED TURKEY. (*Meleagris ocellata.*) This magnificent species is a native of Honduras, whence it has been brought alive to this country and preserved in the aviary of the Earl of Derby. It is a much more splendid bird than the common Turkey, and among other characters may be distinguished by the eye-like marks on the tail and upper coverts.

TURKEY BUZZARD. (*Cathartes aura.*) This is a rapacious bird belonging to the *Vulturidæ* family, and often called the TURKEY VULTURE. It inhabits a vast range of territory in the warmer parts of the American continent, but in the northern and middle states of the Union it is partially migratory, the greater part retiring to the south on the approach of cold weather. The Turkey Buzzard is two feet and a half in length, and with wings extended upwards of six feet in breadth. The bill from the corner of the mouth is almost two inches and a half long, of a dark horn colour for upwards of an inch from the tip, the nostril a remarkably wide slit or opening through it; the tongue is concave, cartilaginous, and finely serrated on its edges; eyes dark, and bright; the head and neck are furnished with a reddish wrinkled skin, beset with short black hairs; from the hind head to the neck feathers the space is covered with down of a sooty black colour; the fore part of the neck is bare as far as the breast-bone, the skin on the lower part or pouch very much wrinkled, but is not discernible without removing the plumage which arches over it; the whole lower parts, lining of the wings, rump, and tail-coverts, are of a sooty brown; the plumage of the neck is large and tumid, and, with that of the back and shoulders, black; the scapulars and secondaries are black on their outer webs, skirted with brown, and the latter slightly tipped with white; primaries plain brown; coverts of the secondaries tawny brown, centred with

TURKEY BUZZARD. — (CATHARTES AURA.)

black. The tail is twelve inches long, rounded, and of a brownish black; inside of wings and tail, light ash. The whole body and neck beneath the plumage are thickly clothed with a white down; the plumage of the neck, back, shoulders, scapulars, and secondaries, is glossed with green and bronze, and has purple reflections; the thighs are feathered to the knees; feet considerably webbed; middle toe three inches and a half in length; claws dark horn colour; legs pale flesh colour.

Much contention has arisen between certain naturalists with respect to the olfactory powers possessed by this bird; and there are some very amusing strictures on this subject in Mr. Waterton's Essays, in which the writer (who is a warm advocate for its existence in a high degree) seems to have by far the best of the argument. It appears, however, that their food is carrion, in search after which they are always soaring in the air. They continue a long time on the wing, and with an easy swimming motion mount and fall, without any visible motion of their wings. They have a wonderful sagacity (says Catesby) in smelling; no sooner is there a dead beast, but they are seen approaching from all quarters of the air, wheeling about and gradually descending and drawing nigh their prey, till at length they fall upon it.

They are generally thought not to prey on any thing living, though I have known them kill lambs, and snakes are their usual food. Their custom is to roost many of them together on tall dead pine or cypress trees, and in the morning continue several hours on their roost, with their wings spread open, that the air, as I believe, may have the greater influence to purify their filthy carcasses. They are little apprehensive of danger, and will suffer a near approach, especially when eating.

In Mr. Darwin's Journal we read that "the Turkey-buzzard is a solitary bird, or, at most, goes in pairs. It may at once be recognized from a long distance by its lofty, soaring, and most elegant flight. It is well known to be a true carrion-feeder. On the west coast of Patagonia, among the thickly-wooded islets and broken land, it lives exclusively on what the sea throws up, and on the carcasses of dead seals. Wherever these animals are congregated on the rocks, there the Vultures may be seen. . . . They certainly may be called gregarious, for they seem to have pleasure in society, and are not solely brought together by the attraction of a common prey. On a fine day a flock may be observed at a great height, each bird wheeling round and round without closing its wings, in the most graceful evolutions. This is clearly done for sport sake, or perhaps is connected with their matrimonial alliances."

This bird is also abundantly found in Jamaica, where it goes by the name of the John Crow Vulture. Its history is given in Mr. Gosse's entertaining volume, from which we shall make a few extracts. The first relates to the disputed question of *scent*. "Those who ascribe the power which the Vulture possesses of discerning from a distance its carrion food, to the sense of seeing or the sense of smelling, *exclusively*, appear to me to be both in error. It is the two senses, exerted sometimes singly, but generally unitedly, which give the facility which it possesses of tracing its appropriate food from far distances. I shall relate one or two occurrences which seem to me to be instances in which the sense of seeing and the sense of smelling were sometimes separately and sometimes unitedly exerted by the Vulture in quest for food.

"A poor German immigrant, who lived alone in a detached cottage in this town, rose from his bed after a two days' confinement by fever, to purchase in the market some fresh meat for a little soup. Before he could do more than prepare the several ingredients of herbs and roots, and put his meat in water for the preparation of his pottage, the paroxysm of fever had returned, and he laid himself upon his bed exhausted. Two days elapsed in this state of helplessness and inanition; by which time the mass of meat and pot-herbs had putrefied. The stench becoming very perceptible in the neighbourhood, Vulture after Vulture as they sailed past were observed always to descend to the cottage of the German, and to sweep round, as if they had tracked some putrid carcass, but failed to find exactly where it was. This led the neighbours to apprehend that the poor man lay dead in his cottage, as no one had seen him for the two days last past. His door was broken open; he was found in a state of helpless feebleness, but the room was most insufferably offensive from something putrefying, which could not immediately be found, for the fever having deprived the German of his wits, he had no recollection of his uncooked mess of meat and herbs. No one imagining that the kitchen pot could contain any thing offensive, search was made every where but in the right place: at last the pot-lid was lifted, and the cause of the insupportable stench discovered in the corrupted soup-meat. Here we have the sense of smelling directing the Vultures, without any assistance from the sense of sight, and discovering unerringly the locality of the putrid animal matter, when even the neighbours were at fault in their patient search."

The next instance is one in which the senses of hearing, seeing, and smelling were all exercised; but not under the influence of the usual appetite for carrion food, but where the object was a living, though a wounded animal.

"A person in the neighbourhood of the town, having his pastures much trespassed on by vagrant hogs, resorted to his gun to rid himself of the annoyance. A pig which had been mortally wounded, and had run squealing and trailing his blood through the grass, had not gone far before it fell in the agonies of death. At the moment the animal was perceived to be unable to rise, three Vultures at the same instant descended upon it, attracted no doubt by the cries of the dying pig, and by the scent of its reeking blood; and while it was yet struggling for life, began to tear open its wounds and devour it." Mr. Gosse further says, that "the common opinion is erroneous, which attributes to the Vulture a confinement of appetite to flesh in a state of decomposition. Flesh is his food; and that he does not pounce upon living prey like the falcons, is because his structure is not adapted for predatory warfare, and not because he refuses recent, and even living flesh when in his power. If the John Crow Vulture discovers a weakling new-born pig apart from the rest, he will descend, and seizing it with his beak, will endeavour to drag it away; its cries of course bring the mother, but before she can come, the Vulture gives it a severe nip across the back, which soon ensures the pig for his own maw. If a large hog be lying in a sick condition beneath a tree, the Vulture will not hesitate to pick out its eyes, having first muted upon the body, that it may discover whether the animal be able to rise; the contact of the hot fæces arousing the hog if he be not too far gone. Cattle also he will attack under similar circumstances."

"The Aura Vultures are often to be observed soaring in companies, particularly previous to a thunder-storm. This occurrence is commonly remarked, because at almost all other times this species is seen solitary, or, at most, scouring the country in pairs. They appear to delight in the hurly-burly of transient squalls, gathering together, and sweeping round in oblique circles, as the fitful gust favours them with an opportunity of rising through the blast, or winging onwards through the misty darkness of the storm. The effect which this imparts to a tropical landscape at a time when thick clouds are upon the mountains, and all vegetation is bending beneath the sudden rush of the tempest, as gust gathers louder and louder, is particularly wild and exciting. Ordinarily, however, in the evening, when the sea-breeze is lulling, and the fading day-beam is changing like the hues of the dying dolphin, they delight to congregate, and career at an immense height. At this time they soar so loftily, that they are scarcely discernible as they change their position in wheeling from shade into light, and from light into shade. They seem as if they rose upward to follow the fading daylight, and to revel in the departing sunbeams, as one after the other, the varying lines are withdrawn, or irradiate only the upper heavens."

TURNIP-FLEA. (*Haltica nemorum.*) The generic description of this small Coleopterous insect will be found under the article HALTICIDÆ, which refers to the "flea-beetles" generally. The Turnip-flea, or more properly Turnip Flea-beetle, is one of these HALTICÆ, which lays waste our turnip-fields, devouring the seed-leaves of the plants as soon as they appear above the ground, and continuing their ravages upon new crops throughout the summer. It is stated in 'Young's Annals of Agriculture,' that the loss in Devonshire alone, in one season, from the destruction of the turnip crops by this little insect, was estimated at one hundred thousand pounds sterling. We could scarcely believe that so small a creature was capable of causing perceptible injury to vegetation; but what these beetles want in size, is made up by their numbers and voracity; the extent of the injury is also much increased by the circumstance of their attacking, when young, many vegetables, and not gnawing the young leaves, like most other insects, only on the edge, but eating their surface, piercing them like a sieve, and disturbing the cellular tissue; thus preventing their growth, and finally causing the total destruction of the plant.

The ravages of the Turnip Flea-beetle have naturally attracted great attention, and have caused many and various experiments to be tried with a view of checking them. The chief object of the farmer should be to accelerate as much as possible the growth of the turnip as soon as it appears above ground, and to keep the insects from the crop until the plants are in the rough leaf, when they are secure from danger. Many practical men consider that the careful and systematic use of lime will, in a great degree, obviate the evil, and indeed there is good reason to expect that it will effectually protect plants from the various kinds of flea-beetles, if dusted over them, when wet with dew, in proper season. Watering plants with alkaline solutions, it is said, will kill the insects without injuring the plants. The solution may be made by dissolving one pound of hard soap in twelve gallons of the soap-suds left after washing. Köllar very highly recommends watering the leaves of plants with an infusion of wormwood, which prevents the flea-beetles from touching them. Sprinkling with road-dust also, while the young plants are still wet with dew, is also strongly recommended.

TURNIX. A genus of Gallinaceous birds closely allied to the Quails, containing several species, one of which (*T. Andalusica*), the Andalusian Quail, has been shot in this country; to which it is a very rare and stray visitor.

TURNSTONE. (*Strepsilas interpres.*) A small Grallatorial bird, met with in almost every part both of the northern and southern hemispheres. They reside on the sea shores, and on the gravelly borders of lakes and rivers; are most abundant in the northern parts of Europe, less frequent in the temperate regions, and extremely rare to the south. The Turnstones have a short bill, thick at the base, and narrowing gradually to the point; and with this they turn over the stones on the sea-shore, in quest of the small molluscous and crustaceous animals on which they feed. They breed in high latitudes, and migrate towards the tropics for the winter season; visiting our shores in August, and departing towards the north in the spring. They lay four eggs of an olive colour, spotted with black.

TURRILITES. A genus of fossil shells, occurring only in the chalk marl. They are spiral and turreted; whorls contiguous and apparent; septa sinuous and lobate, perforated by a siphon; aperture rounded, with the outer lip expanded.

TURRITELLA. A genus of Mollusca, the shell of which is very long, and pointed at the apex, with numerous whorls, usually transversely striped; aperture round; lips thin, and disunited at the upper part; operculum horny. The animal is furnished with two long tentacula, with eyes at the base. The shells of this genus are all marine, and many of the species found in the Eastern seas attain a very large size, but none are known to possess vertical ribs or thickened bands. There are about a dozen species recent, and as many fossil.

TURTLE. (*Chelonia.*) The Marine Tortoises, or Turtles (*Chelonidæ*), as they are usually called, differ from the *Testitudinata*, or Land Tortoises, in many essential points, although their exterior, like that of the latter, is composed of a strong bony covering or shield, in which are imbedded the ribs, and which is coated externally by hard horny plates. Their distinguishing characteristics are the compressed and paddle-like form of the feet, particularly the anterior pair, which they use as oars, and by their means can move through the water in any direction, with considerable rapidity. Their progression on land is however, by this conformation, rendered much more difficult, so that it is only with laborious efforts they are enabled to shuffle slowly along; while, from the flattened form of the carapace, they are unable to recover their natural position when turned upon their backs.

The EDIBLE or GREEN TURTLE (*Chelonia midas*) is one of the largest of this genus, often measuring above five feet in length, and weighing above five or six hundred pounds. Its shell consists of thirteen dorsal segments or divisions, surrounded by twenty-five marginal pieces; and its form is somewhat heart-shaped, or pointed at the extremity: its colour is a dull palish brown, more or less variegated with deeper undulations, but not exhibiting those strong and beautiful colours which so peculiarly distinguish that of the Imbricated Turtle. But so much is the flesh esteemed, that it not only furnishes an agreeable viand to those navigators who traverse the torrid zone, and is eaten by the inhabitants of our West India islands, but is in such high estimation in this country as a delicious luxury, that large quantities are continually imported for the supply of the London taverns alone. The eggs of this species are very fine.

"Of the Sea Turtles," says Catesby, "the most in request is the *Green Turtle*, which is esteemed a most wholesome and delicious food. It receives its name from the fat, which is of a green colour. Sir Hans Sloane informs us, in his History of Jamaica, that forty sloops are employed by the inhabitants of Port Royal, in Jamaica, for the catching them. The markets are there supplied with Turtle as ours are with butchers' meat. The Bahamians carry many of them to Carolina, where they turn to good account; not because that plentiful country wants provisions, but they are esteemed there as a rarity, and for the delicacy of their flesh. They feed on a kind of grass, growing at the bottom of the sea, commonly called turtle-grass. The inhabitants of the Bahama islands, by frequent practice, are very expert at catching Turtles, particularly the Green Turtle. In April they go, in little boats, to Cuba and other neighbouring islands, where, in the evening, especially in moonlight nights, they watch the going and returning of the Turtle to and from their nests, at which time they turn them on their backs, where they leave them, and proceed on, turning all they meet; for they cannot get on their feet again when once turned. Some are so large that it requires three men to turn one of them. The way by which the Turtle are most commonly taken at the Bahama islands is by striking them with a small iron peg of two inches long, put in a socket, at the end of a staff of twelve feet long. Two men usually set out for this work in a little light boat or canoe, one to row and gently steer the boat, while the other stands at the end of it with his weapon. The Turtle are sometimes discovered by their swimming with the head and back out of the water, but they are more often discovered lying at the bottom, a fathom or more deep. If a Turtle perceives he is discovered, he starts up to make his escape; the men in the boat pursuing him, endeavour to keep sight of him, which they often lose, and recover again by the Turtle putting his nose out of the water to breathe."

The Isle of Ascension is called by Sir J. E. Alexander "the head quarters of the finest Turtle in the world," and his account of it in that locality, which we subjoin, is really interesting: "We walked down to the Turtle ponds, two large enclosures near the sea, which flowed in and out through a breakwater of large stones. A gallows was erected between the two ponds, where the Turtle are slaughtered for shipping, by suspending them by the hind flippers, and then cutting their throats. About three hundred Turtle, of four and five hundred pounds each, lay on the sand, or swam about in the ponds: a sight to set an alderman mad with delight!

"In the hot months of January, February, March, and April, the females land at night; and waddling over the sands in the various bays of the island far above high-water mark,—for by a pole in the ponds the tide only rises here two feet,—they scrape up, by alternate scoops of their flippers, a hole deep enough to cover their bodies. Into this they get, sighing heavily, and deposit from one hundred and fifty to two hundred eggs; cover them up; leave them to the sun to hatch; and then waddle again towards the sea. Two stout hands are, meanwhile, on the look-out, watching the movements of the unfortunate Turtle; and running up to her after the completion of her task, one seizes a fore-flipper, and dexterously shoves it under her belly, to serve as a purchase; whilst the other, avoiding a stroke which might lame him, cants the Turtle over on her back, where she lies helpless. From fifteen to thirty are thus turned in a night; and six hundred had been so captured in the season of 1834. In the bays, where the surf, or heavy rollers, prevent the boats being beached to take on board the Turtle when caught, they are hauled out to them by ropes.

"No ships' crews are now allowed to turn Turtle, which is converted into a government monopoly; and two pounds ten shillings is the fixed price for each. Strange to say, from the time that the young Turtle, the size of a dollar, are observed scuttling down to the water, they are never seen again here until they are four or five hundred pounds weight; and how long they take to attain this great size, and where they spend the intermediate time, is as yet a mystery. I was surprised to hear that Turtle are kept in the ponds for a year and upwards without a morsel of food of any kind. They sometimes deposit their eggs in the sand, on the sides of the ponds; and in due time the little animals are allowed to make their escape to sea. One old female, called 'Nelson,' because one of her flippers had been carried off by a shark, was kept, out of respect, for two or three years in the ponds. She contrived, however, one night to crawl round the enclosure, and make her escape; but she was turned next year in Clarence Bay. Another Turtle was also turned there, a short time since, on the back shell of which was carved the name of a mate of a British vessel, who had bought it and sailed with it three weeks before: it is probable that, imagining it to be dead, he had thrown it overboard. The best way to send home Turtle from Ascension, is to 'head them up' in a *sealed* cask, and have the water changed daily by the bunghole and a cock. Turtle, though the extremes of heat and cold are equally injurious to them, should always arrive in hot weather in England. Thus, an unfortunate captain, on one occasion, took from Ascension two hundred Turtle; and timing his arrival badly, brought only four alive to Bristol!"

Mr. Darwin, in his Journal, when describing Keeling Island, gives an account of another method of catching Turtle. He says, "I accompanied Captain Fitzroy to an island at the head of the lagoon: the channel was exceedingly intricate, winding through fields of delicately branched corals. We saw several Turtles, and two boats were then employed in catching them. The method is rather curious: the water is so clear and shallow, that although at first a Turtle quickly dives out of sight, yet in a canoe or boat under sail, the pursuers, after no very long chase, come up to it. A man standing ready in the bows at this moment dashes through the water upon the Turtle's back; then, clinging with both hands by the shell of the neck, he is carried away till the animal becomes exhausted, and is secured. It was quite an interesting chase to see the two boats thus doubling about, and the men dashing into the water, trying to seize their prey."

The IMBRICATED TURTLE (*Chelonia imbricata*) is so named from its scales overlapping each other at their extremities, in the manner of tiles on the roof of a building. The outline of the shell is more heart-shaped than any other species, and terminates more acutely: each of the middle row of scales on the back is also of an acute form at the tip, and has a ridge or carina down the middle: the head is smaller than in other Turtles; the neck longer, and the beak narrower,

HAWKSBILL TURTLE.
(CHELONIA IMBRICATA.)

sharper, and more curved so as to bear no inconsiderable resemblance to the bill of a hawk: hence its common or popular name—*Hawksbill Turtle.* The fore legs are longer than in the rest of the tribe, and it is said that when turned or laid on its back, the animal is enabled by their assistance to recover its former position, which no other Turtle can do. It is a native of the Asiatic and American seas, and is occasionally also found in the Mediterranean. Its general length is about three feet, though it is sometimes much larger, and in the Indian ocean in particular, specimens are said to have occurred of more than twice that size. The flesh is in no estimation as a food: but the lamellæ or plates of the shell are stronger, thicker, and clearer than in any other kind, afford the valuable substance called *tortoise-shell:* they are semi-transparent, and most elegantly variegated with whitish, yellowish, reddish, and dark brown clouds and undulations, so as to constitute, when properly prepared and polished, one of the most elegant articles for various ornamental purposes. "The goodness of tortoise-shell depends mainly on the thickness and size of the scales, and in a smaller degree on the clearness and brilliancy of the colours. The best is that of the Indian Archipelago; and the finest of this quarter is obtained on the shores of the Spice Islands and New Guinea."—*M'Culloch.*

The natural or general number of the dorsal pieces is thirteen; the marginal row consisting of twenty-five smaller pieces. This external coating is raised or separated from the bony part, which it covers, by placing fire beneath the shell; the heat soon causing the plates to start, so as to be easily detached from the bone. These plates vary in thickness, according to the age and size of the animal, and measure from an eighth to a quarter of an inch in thickness.

The CORIACEOUS TURTLE (*Sphargis coriacea*) differs from the rest of its tribe, not only in the form of its body, which is longer in proportion, but still more in its external covering, which is of a substance resembling strong leather, marked over the whole surface into small, obscurely subhexagonal and pentagonal subdivisions or lineations, which do not, however, detract from its general smoothness. Along the whole length of this leathery shield run five distinct, strongly prominent, tuberculated ridges, besides those which border the sides. There is no under or thoracic shell; and the general colour of the whole animal is dusky brown, paler beneath. The head is large, and the upper mandible notched at the tip in such a manner as to give the appearance of two large teeth or processes, between which, when the mouth is closed, is received the tip of the lower mandible. The fins or legs are large and long, and covered with a tough leathery skin; the tail is rather short and sharp pointed. This singular animal is a native of the Mediterranean sea; it is occasionally seen both on the coasts of South America and Africa; and has been taken at different periods both on the coasts of France and England. Instances have been known of their being eight feet long, and weighing a thousand pounds.

The LOGGERHEAD TURTLE (*Testudo caretta*) is of extraordinary size, and the boldest and most voracious of any; but, considered in a commercial view, it is of little or no value, except that it affords some oil, which may be used for lamps, &c. It is distinguished by having fifteen, instead of thirteen, dorsal segments, or scutella; each of the scutella in the middle dorsal range being extremely protuberant at the end, rising into a subacute prominence, and thus forming a row of tubercles along the back of the shield. The fore feet are very large and long; the hind feet short but broad.

It is well known that the different kinds of great Marine Tortoises, or Turtles, at their appointed times every year, deposit their eggs in the sand, on the shores of the sea, and banks of rivers where the strand is gently declivous. There the females hollow out a strong vaulted nest, wherein the eggs (amounting to a hundred laid at one time) may have the benefit of the concentrated rays of the sun, so as to enjoy an equable heat, as in the case of eggs under a sitting hen. The shell of these eggs is generally solid, and their form globular, or nearly so.

TURTLE-DOVE. (*Columba Turtur.*) This species of the *Columbidæ* family — whose gentle and soothing accents when "cooing" to its mate, combined with its general deportment, have caused it to be regarded as the most perfect emblem of connubial attachment — arrives in this country late in the spring, and departs about the latter end of August; during which time

TURTLE DOVE.—(COLUMBA TURTUR.)

the birds pair, breed, and rear their young. In warmer climates, however, they are supposed to breed several times in the year. In length the Turtle-dove is rather more than twelve inches: bill brown, eyes yellow, encompassed with a crimson circle; top of the head ash gray, mixed with olive; each side of the neck is marked with a spot of black feathers, tipped with white; the back is ash gray, each feather margined with reddish brown; wing coverts and scapulars reddish brown, spotted with black; quill feathers dusky, edges pale; the fore part of the neck and the breast are light purplish red; the belly, thighs, and vent white; the two middle feathers of the tail brown, the others dusky, tipped with white, the two outermost also edged with the same: legs red. The Turtle-dove frequents the thickest and most sheltered parts of the woods, where it builds on the highest trees; and the female generally lays two eggs.

The AMERICAN TURTLE-DOVE, or CAROLINA PIGEON (*Columba Carolinensis*), is thus spoken of by Wilson: — "This is a favourite bird with all those who love to wander among our woods in spring, and listen to their varied harmony. They will there hear many a sprightly performer; but none so mournful as this. The hopeless woe of settled sorrow, swelling the heart of female innocence itself, could not assume tones more sad, more tender and affecting. Its notes are four [Mr. Gosse says *five*]; the first is somewhat the highest and preparatory, seeming to be uttered with an inspiration of the breath, as if the afflicted creature were just recovering its voice from the last convulsive sobs of distress; this is followed by three long, deep, and mournful moanings, that no person of sensibility can listen to without sympathy. A pause of a few minutes ensues, and again the solemn voice of sorrow is renewed as before. This is generally heard in the deepest shaded parts of the woods, frequently about noon, and towards the evening. There is, however, nothing of real distress in all this; quite the reverse. The bird who utters it wantons by the side of his beloved partner, or invites her by his call to some favourite retired and shady retreat. It is the voice of love, of faithful connubial affection, for which the whole family of doves are so celebrated; and, among them all, none more deservingly so than the species now before us." Our author then describes it as a general inhabitant, in summer, of the United States, from Canada to Florida, and from the seacoast to the Mississippi, and far to the westward. Their flight, he observes, is quick, vigorous, and always accompanied by a peculiar whistling of the wings, by which they can easily be distinguished from the wild pigeon. The nest is very rudely constructed, generally in an evergreen, among the thick foliage of a vine, in an orchard, on the horizontal branches of an apple-tree, and, in some cases, on the ground. It is composed of a handful of small twigs, laid with little art, on which are scattered dry fibrous roots of plants; and in this almost flat bed are deposited two eggs of a snowy whiteness. The male and female unite in feeding the young, and they have rarely more than two broods in the same season.

The American Turtle-dove is twelve inches long, and seventeen inches in extent; bill black; eye of a glossy blackness, surrounded with a pale greenish-blue skin; crown, upper part of the neck and wings, a fine silky slate blue; back, scapulars, and lesser wing-coverts, ashy brown; tertials, spotted with black; primaries, edged and tipped with white; forehead, sides of the neck, and breast, a pale brown vinous orange; under the ear-feathers, a spot or drop of deep black; immediately below which the plumage reflects the most vivid tints of green, gold, and crimson; chin, pale yellow ochre; belly and vent, whitish; legs and feet, coral red, seamed with white; the tail is long and cuneiform, consisting of fourteen feathers; the four exterior ones, on each side, are marked with black, about an inch from the tips, and white thence to the extremity; the next has less of the white at the tip; these gradually lengthen to the four middle ones, which are wholly dark slate; all of them taper towards the points, the middle ones most so. The female is an inch shorter, wants the rich silky blue on the crown, and has altogether less brilliancy of colour. The flesh of this bird is considered much superior to that of the wild pigeon; but its seeming confidence in man, the tenderness of its notes, and the innocency attached to its character, are, with many, its security and protection.

TUSSOCK [MOTHS]. A name given by collectors to Moths of the genera *Dasychira* and *Demas*.

TUYUYU. A local name for the *Mycteria Americana*, a Grallatorial bird which when full-grown is upwards of six feet in height. Its general plumage is white its neck is bare of feathers, and, for two-thirds of its length from above, black; the remainder is of a dark red. Its bill is about fifteen inches long, and by its habit of striking the mandibles together a loud clattering noise is produced. Though shy and difficult to be got at, they are occasionally domesticated.

TYRANT FLYCATCHER, or KING-BIRD. (*Tyrannus intrepidus.*) This very singular species of a group of Passerine birds, known as the Flycatchers, has received its trivial names of *tyrant* and *king*, from its extraordinary behaviour, and the authority it assumes over all others, during the time of breeding. It is eight inches long, and fourteen in extent; the general colour above is a dark slaty ash; the head and tail are nearly black; the latter *even* at the end, and tipt with white; the wings are of a brownish cast; the quills and wing-coverts are edged with dull white; the throat, and all the rest of the lower parts, are pure white; the plumage on the crown (though not forming a crest) is frequently erected, and discovers a rich bed of brilliant orange, which when the feathers lie close, is altogether concealed. The bill is very broad at the base, overhanging at the point, and notched, of a glossy black colour, and furnished with bristles at the base; the legs and feet are black, seamed with gray. The female differs chiefly in being of a browner cast on the upper parts, and having a narrower border of duller white on the tail. In the breeding season, as we are told by Wilson, in his 'American Ornithology,' the Tyrant Flycatcher's extreme affection for his mate, and for his nest and young, makes him suspicious of every bird that happens to pass near his residence, so that he attacks, without discrimination, every intruder. In the months of May, June, and part of July, his life is one continued scene of broils and battles; in which, however, he generally comes off conqueror. Hawks and crows, the bald eagle, and the great black eagle, all equally dread a rencounter with this dauntless little champion, who, as soon as he perceives one of these last approaching, launches into the air to meet him, mounts to a considerable height above him, and darts down on his back, sometimes fixing there to the great annoyance of his sovereign, who, if no convenient retreat or resting-place be near, endeavours by various evolutions to rid himself of his merciless adversary. But the King-bird is not so easily dismounted. He teazes the eagle incessantly, sweeps upon him from right and left, re-

TYRANT FLYCATCHER.
(TYRANNUS INTREPIDUS.)

mounts, that he may descend on his back with the greater violence; all the while keeping up a shrill and rapid twittering; and continuing the attack sometimes for more than a mile, till he is relieved by some other of his tribe equally eager for the contest. . . . . All his turbulence, however, vanishes as soon as his young are able to shift for themselves; and he is then as mild and peaceable as any other bird."

His usual mode of flight is singular. The vibrations of his broad wings (says this observing writer, whom we quote with slight deviations), as he moves slowly over the fields, resemble those of a hawk hovering and settling in the air to reconnoitre the ground below; and the object of the King-bird is no doubt something similar, viz. to look out for passing insects, either in the air, or among the flowers and blossoms below him. In fields of pasture he often takes his stand on the tops of the mullein, and other rank weeds, near the cattle, and makes occasional sweeps after passing insects, particularly the large black gadfly, so terrifying to horses and cattle. His eye moves restlessly around him, traces the flight of an insect for a moment or two, then that of a second, and even a third, until he perceives one to his liking, when, with a shrill sweep, he pursues, seizes it, and returns to the same spot again, to look out for more. This habit is so conspicuous when he is watching the bee-hive, that many intelligent persons are of opinion that he picks out only the drones, and never injures the working bees. Be this as it may, he certainly gives a preference to one bee, and one species of insect, over another. . . . Whatever antipathy may prevail against him for depredations on the drones, or, if you will, on the bees, this bird is greatly the farmer's friend, in destroying multitudes of insects, whose larvæ prey on the harvests of his fields. These noxious insects are the daily food of this bird: and he destroys, upon a very moderate average, some hundreds of them daily. The death of every King-bird is therefore an actual loss to the farmer, by multiplying the numbers of destructive insects, and encouraging the depredations of crows, hawks, and eagles, who avoid as much as possible his immediate vicinity. "For myself," says Wilson, "I must say, that the King-bird possesses no common share of my regard. I honour this little bird for his extreme affection for his young; for his contempt of danger, and unexampled intrepidity; for his meekness of behaviour when there are no calls on his courage, a quality which even in the human race is justly considered so noble: but above all, I honour and esteem this bird for the millions of ruinous vermin which he rids us of; whose depredations, in *one* season, but for the services of this and other friendly birds, would far overbalance all the produce of the bee-hives in fifty." The nest is large, remarkably firm and compact, consisting of small slender twigs on the outside, and usually lined with fine dry fibrous grass, and horse-hair. The eggs are five, of a very pale cream colour or dull white, marked with a few large spots of deep purple, and other smaller ones of light brown, chiefly, though not altogether, towards the great end. They generally build twice in the season.

UMBRE. A genus of Grallatorial birds, of which there is but one known species, namely, the CRESTED UMBRE (*Scopus umbretta*), which is as large as a Crow, of an umber colour, and the male is crested. It is diffused throughout all Africa. The Umbres are only distinguished from the Storks by their compressed beak, the trenchant ridge of which is inflated towards the base, and the nostrils are prolonged by a groove that runs parallel with the ridge to its tip, which is slightly hooked.

UMBRELLA. A genus of Mollusca, one species of which inhabits the Indian Ocean, and, from the shape of the shell, is very commonly called the *Chinese Umbrella*. It is sub-orbicular, slightly convex on the outside, with central apex slightly raised; margin sharp; internal surface with a central, callous, coloured disc, surrounded by a continuous, irregular, muscular impression. Another species, much smaller, named the *Umbrella Mediterranea*, is from the Gulf of Tarento, and differs from the former also in not being marked with rays.

UMBRINA. An Acanthopterygious fish, belonging to the *Sciænidæ* family. It is remarkably beautiful, the ground colour being golden, with bright bands of steel blue. It is sometimes forty pounds in weight, but is by no means a long fish. In the Mediterranean it is plentiful, and occasionally some are met with on the southern coasts of Britain: the flesh is highly esteemed.

UNAU. The two-toed Sloth, which, like the *Ai*, or common three-toed Sloth, is an inhabitant of the dense forests of the tropical portion of South America, and has all the singularities of conformation and habits which distinguish that species. [See SLOTH: BRADYPUS.]

UNGKA-PUTI. (*Hylobates agilis.*) A species of Gibbon, found in Borneo, Java, &c. They are arboreal in their habits, and are distinguished by astonishing activity. Their mode is to suspend themselves by their long arms to the forest branches, and by an energetic muscular movement to spring forward from one tree to another, although the distance may be from thirty to forty feet; which they are enabled to accomplish, when required, with apparent ease and precision. This animal is further remarkable for a curious call-note, which it frequently utters during its most active movements, especially in the morning. It is of a timid and gentle disposition, and in confinement shows attachment to those who have the care of it. [See GIBBON.]

UNGULINA. A genus of Conchiferous Molluscs, the shell of which is equivalve, sub-orbicular, and rayed; valves nearly equilateral, with margins entire, simple, closed all round; hinge with one short, subdivided cardinal tooth in each valve, and at the side an oblong ligamentary pit, divided into two portions, one of which receives the cartilage: two muscular impressions in each valve, and the impression of the mantle entire. The animal perforates rocks, &c.; and the shells are small, thin, and transparent.

UNIO, or FRESH-WATER MUSSEL. This fluviatile genus of Mollusca is found in the rivers of Europe and America, the East and West Indies, &c. The shell is thick and solid, transverse, equivalve, inequilateral; cardinal teeth solid, short, and oblique; umbones prominent, and generally corroded. The hinge is somewhat complicated: there is a short plate in the left valve, received into a cavity in the right; and behind this a longer plate closing between two others of the opposite side. The interior of the valves are sometimes pink, sometimes white, often highly iridescent; and they occasionally contain tolerably large pearls. Several are natives of this country; but they more especially abound in the rivers and lakes of North America. The animal is of no value as food, from the insipidity of its taste.

There is a large family of fresh-water conchifers to which the genus *Unio* belongs, abounding in the North American rivers, and comprising the genera *Unio*, *Hyria*, *Anodonta*, and *Iridina*. Among the observations made on them by Mr. Lea, of Philadelphia, who paid much attention to their classification, &c., and who has described their habits with great minuteness, we learn that the animal of *Anodonta*, which is essentially the same with that of *Unio*, is hermaphrodite, and seems viviparous; for the eggs pass into the oviduct placed along the superior branchiæ, where the young are found with their shells complete. He dissected a specimen of *Anodonta undulata* nearly three inches long, and found the oviducts charged with about 600,000 (as nearly as he could calculate) young shells perfectly formed, both valves being distinctly visible with the microscope. Whilst engaged in this investigation, Dr. Kirtland, of Portland, Ohio, informed Mr. Lea of his ability to distinguish the female and male shells of the same species, without having recourse to the included animal; and he says that a very short series of examination satisfied him fully as to the establishment of the difference of sexes. The female, sustaining her very large burthen, naturally requires, he observes, more space within the valves; hence an enlargement of the posterior portion of the shell is generally found, differing in its form in various species.

It seems to be a matter of doubt, according to Mr. Lea, on what these animals subsist, but he had strong reasons for believing that they feed on animalcules which are ever found to exist in water, and which they might separate from the constant stream which they pass from the posterior part of the shell, and which must be taken in at another part. This operation he witnessed frequently in a vessel in which he kept the *Naiadæ* for some months. If the water was not changed for twenty-four hours, he uniformly found the animals quiet, but within a few minutes after it was changed they as uniformly commenced the passage of this constant stream, which he considers to be the result of the action of the separation of the animalcules from the water. Referring to the fact of pearls being found in other freshwater bivalves, Mr. Broderip observes that the brilliant and variously-coloured nacre with which many of the species are lined, and the extreme thickness of some of the shells, are very remarkable. That pearls should be found in them will not surprise those whose attention has been drawn to their internal surface. Pennant remarks that *Mya Margaritifera* of Linnæus (*Unio elongatus*) is noted for producing quantities of pearls, and formerly there were regular fisheries in many of our rivers to obtain them. As many as sixteen have been taken from one shell. The Esk and the Conway were famous in this way. The latter river, in the days of Camden, was noted for them. Sir Richard Wynn, of Gwidir, chamberlain to Catherine, queen to Charles II., is said to have presented her Majesty with a Conway pearl which is to this day honoured with a place in the regal crown. Pennant, who states this, adds, that the shells are called by the Welsh, *Crigen Dilune*, or Deluge Shells, as if left there by the deluge. The river Irt, in Cumberland, also produced them; and Sir John Hawkins, the circumnavigator [as mentioned in the article MYA], had a patent for fishing that river. Britain, indeed, had early acquired a reputation for its pearls; for, according to Suetonius, they were Cæsar's inducement for undertaking his British expedition. This, however, does not seem very probable. Pliny, indeed, speaks of the pearls of our island as small and ill-coloured, and refers to the breastplate which Cæsar himself had brought home and dedicated to Venus Genetrix in her temple, adding that he wished it to be understood that the offering was formed of British pearls."

UNIPELTATA. A family of Crustacea, belonging to the order *Stomapoda*, and comprising one genus only, *Squilla* [which see].

UNOGATA. The name given by Fabricius to a part of the Arachnida, order *Pulmonaria*, and comprehending the *Scorpions* [which see].

UPHOLSTERER BEE. (*Osmia papaveris.*) This name is given to a species of wild bee, found in France, belonging to the genus *Osmia*. These ingenious artificers excavate holes in the earth for the reception of their young, and line them with an elegant coating of flowers or leaves; an operation which is so pleasingly described in Messrs. Kirby and Spence's Introduction to Entomology, that we beg to transfer the account they give of it, from their pages to our own. "This little bee, as though fascinated with the colour most attractive to our eyes, invariably chooses for the hangings of her apartments the most brilliant scarlet, selecting for its material the petals of the wild poppy, which she dexterously cuts into the proper form. Her first process is to excavate in some pathway a burrow, cylindrical at the entrance, but swelled out below to the depth of about three inches. Having polished the walls of this little apartment, she next flies to a neighbouring field, cuts out oval portions of the flowers of poppies, seizes them between her legs, and returns with them to her cell; and though separated from the wrinkled petal of a half-expanded flower, she knows how to straighten their folds, and, if too large, to fit them for her purpose by cutting off the superfluous parts. Beginning at the bottom, she overlays the walls of her mansion with this brilliant tapestry, extending it also on the surface of the ground round the margin of the orifice. The bottom is rendered warm by three or four coats, and the sides have never less than two. The little Upholsterer, having completed the hangings of her apartment, next fills it with pollen and honey to the height of about half an inch; then, after committing an egg to it, she wraps over the poppy lining so that even the roof may be of this material, and lastly closes its mouth with a small hillock of earth. The great depth of the cell compared with the space which the single egg and the accompanying food deposited in it occupy, deserves particular notice. This is not more than half an inch at the bottom, the remaining two inches and a half being subsequently filled with earth."

UPUPA. [See HOOPOE.]

URAN-UTAN. [See ORANG.]

URANIIDÆ. A family of Lepidopterous insects, belonging to the HETEROCERA; comprising several very anomalous exotic genera, which, from their apparently occupying a station between the Hesperiæ and Sphinges, Latreille named *Hesperi-Sphinges*. Since the discovery of its preparatory states, however, it is allowed to belong to the Heterocerous section of the order. Mr. Westwood tells us, that "the splendid colours of the typical *Uraniæ* are, it is true, indicative of diurnal flight, and give them, in conjunction with their form, all the appearance of a butterfly, to which the tailed hind wings add considerably; but there are other species (*Nyctalemon Orontes* and *Patroclus* and *Sematura Lunus*, &c.), which in their more sober colouring would be considered as moths, and some of these seem so nearly related to *Coronis*, whilst *Urania* is in several respects so close to *Agarista* (in its larva, palpi, and antennæ), that I am induced to unite them into one family, a step which seems to be supported by the neuration of the wings." . . . "The flight of *Urania Fernandinæ* is diurnal, and exceedingly swift, somewhat like that of *Apatura Iris*, sporting about the topmost branches of forest trees; and when it alights, its four wings are expanded horizontally. *Agarista* consists of New Holland insects, having much more the appearance of moths, but with filiform antennæ slightly thickened in the middle, and terminated in a point."

URANOSCOPUS. A very remarkable but repulsive-looking genus of the *Percidæ* family (of Acanthopterygious fishes); one species of which, *Uranoscopus scaber*, familiarly called the STAR-GAZER, inhabits the Mediterranean. This name has been given to them on account of the eyes being placed on the upper surface of the nearly cubical head, and directed towards the heavens. Their pre-operculum is toothed on the lower part; their mouth is cleft vertically; they have a strong spine on each shoulder, and only six rays on each gill. Behind the tongue is a narrow slip which they can protrude, and with which they attract small fishes, while the mud effectually conceals them from their prey. They have an immensely large gall-bladder.

URCHIN. The Hedgehog [which see].

URIA. [See GUILLEMOT.]

UROCERATA. The name given to a tribe of Hymenopterous insects, comprising the genus *Sirex*, which deposit their eggs in old fir trees, &c. [See SIREX.]

UROMASTIX. A genus of Saurian reptiles belonging to the Iguana group, and distinguished from others of the same family by all the body-scales being small, uniform, and smooth; but those of the upper surface of the tail are large and spinous, though there are none underneath it.

UROPELTIS. A genus of Serpents, found in America, distinguished externally by a very small head and pointed muzzle; the tail short, and obliquely truncated above, is flat and beset with little scales at the truncation; and they have a range of scales under the tail, a little larger than the rest, with a double range beneath its truncate portion.

UROPTERA. A subsection of minute Crustaceans, of the order *Amphipoda*, which reside in the bodies of various Acalephæ and some other zoophytes. They have the head generally large, the antennæ often short, and the body soft; all the legs except the fifth pair simple, the anterior either short or small, and the tail either furnished at the tip with lateral swimmerets, or terminated by appendages or dilated points, bidentate or forked at the extremity.

URSAL. A species of Seal, about eight feet in length, inhabiting the shores of the North Pacific Ocean. It is said to be one of the most pugnacious and ferocious of the whole tribe. There is a remarkable disproportion in the number of the sexes in this species; each family consisting of but one male with a crowd of females; and if one family encroaches on the station of another, a desperate fight generally ensues. [See SEAL.]

URSIDÆ. A family of Plantigrade Mammalia; comprising the true Bears, the Badgers, the Racoons, and the Wolverines. They are characterized by a plantigrade walk; grinders more or less tuberculated; stature generally large; carnivorous and frugivorous; claws formed for digging; tail generally short. [The reader is referred to the articles above-named, as BEAR, BADGER, &c., for particulars of the various genera belonging to the *Ursidæ*.]

URSUS. [See BEAR.]

URUS. (*Bos Urus.*) The Aurochs, a species of Bovine animals still existing in Lithuania, though till recently supposed, by most naturalists, to have become extinct. The distinction between the species *Bos taurus* and *Bos urus* is thus carefully marked by Cuvier: "The forehead of the ox is flat, and a little concave; that of the aurochs protuberant, although less so than the buffalo's; the forehead is square in the ox, its height, taking its base between the orbits, being very nearly equal to its breadth; in the aurochs it is much wider than high, in the proportion of three to two. The horns are attached in the ox to the extremities of a salient line, the most elevated of the head, that which separates the occiput from the forehead; in the aurochs this line is placed two inches farther backward than the roots of the horns: in the ox the plane of the occiput makes an acute angle with that of the forehead; in the aurochs this angle is obtuse: finally, the plane of the occiput is square in the ox, but represents a half circle in the aurochs."

A noble stuffed specimen of the Aurochs, and a skeleton of the same animal, were some years ago presented to the British Museum by the Emperor Nicholas of Russia, and subsequently also he forwarded to the Gardens of the Zoological Society in the Regent's Park, young specimens of the male and female. These fine animals died, but had they attained maturity they would have proved a most attractive addition; they were taken in July, 1846, in the forest of Bieloviege, in Grodnau, where a herd of about a thousand head is preserved with great difficulty by 300 families, who are stationed there by the Emperor to take charge of them. They feed on grass, and on the bark of trees, in gnawing which, however, they frequently destroy their teeth.

Cuvier considers the Aurochs to be a species which man has never subdued; and observes, in his *Ossemens Fossiles*, that if Europe possessed a *Urus*, a *Thur* of the Poles, different from the *Bison* or the *Aurochs* of the Germans, it is only in its remains that the species can be traced; such remains are found, in the skulls of a species of ox different from the Aurochs, in the superficial beds of certain districts. This, Cuvier thinks, must be the true Urus of the ancients, the original of our domestic ox, the stock perhaps whence our wild cattle descended; while the Aurochs of the present day is nothing more than the Bison or Bonasus of the ancients, a species which has never been brought under the yoke. [See OX: BISON.]

VAMPIRE-BAT. (*Vampyrus spectrum.*) This bat is a native of South America, of a reddish-brown colour, and as large as a magpie. It is said, by Piso, to "seek out every kind of animal and suck their blood." This fact has often been most circumstantially related, and as often positively denied; but if we compare the accounts of many highly respectable modern travellers, the truth of the statement will appear to be fully established. Captain Stedman, who had himself been bitten, thus describes the operation. "Knowing by instinct that the person they intend to attack is in a sound slumber, they generally alight near the feet, where, while the creature continues fanning with its enormous wings which keeps one cool, he bites a piece out of the tip of the great toe, so very small, indeed, that the head of a pin could be scarcely received into the wound, which is consequently not painful; yet through this orifice he continues to suck the blood until he is obliged to disgorge. He then begins again, and thus continues sucking and disgorging till he is scarce able to fly; and the sufferer has often been known to sleep from time into eternity." To the same effect is the testimony of several other naturalists who have paid attention to the subject, among whom may be named Messrs. Darwin, Swainson, and Waterton; the last of whom observes, that "Europeans may consider as fabulous the stories related of the Vampire; but, for my own part, I must believe in its powers of sucking blood from living animals, as I have repeatedly seen both men and beasts which had been sucked, and, moreover, I have examined very minutely their bleeding wounds." But he admits that he could never find out how the Vampires actually draw the blood; and that he continued as ignorant of the real process as though he had never been in the Vampire's country. "For the space of eleven months," adds this most amusing writer, "I slept alone in the loft of a woodcutter's abandoned house in the forest; and though the Vampire came in and out every night, and I had the finest opportunity of seeing him, as the moon shone through apertures where windows had once been, I never could be certain that I saw him make a positive attempt to quench his thirst from my veins, though he often hovered over the hammock."

VANESSA. A genus of Diurnal Lepidoptera belonging to the family *Nymphalidæ*, in most of the species of which the wings are angulated. The caterpillar has numerous bristly spines, and the pupa is much angulated and suspended by the tail. In Doubleday and Hewitson's work, descriptions and figures of the various forms will be found: we limit our notice to the British species, which are all eminently handsome.

VANESSA C. ALBUM; or COMMA BUTTERFLY. Of late years this insect appears to have become much more scarce than formerly, or it may have forsaken its old localities and found new ones: it frequents woods, thickets and gardens; and there are two broods in the year, one towards the end of June, the other in September. Wings above dark orange, with black or

COMMA BUTTERFLY.—(VANESSA C ALBUM.)

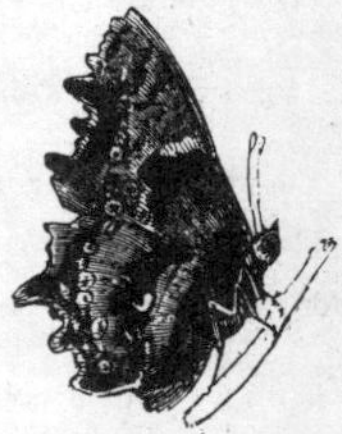

UNDER-SIDE OF COMMA BUTTERFLY.

brown spots, and a brown posterior margin; on the disc of the anterior wings are two roundish spots, and near the interior margin two other larger spots; beneath, the anterior wings are dusky-brown, with a broad, irregular, green-marbled pale band near the posterior margin: posterior wings very similar, with a pure white crescent in the centre: near the posterior margin of all the wings is an irregular series of spurious ocelli. Body above dusky, with greenish hairs on the thorax: antennæ black above, brown annulated with white beneath. Caterpillar red-brown and yellow: it feeds on the hop, nettle, elm, gooseberry, and honey-suckle. The chrysalis is flesh-coloured, spotted with gold.

VANESSA POLYCHLOROS; or GREAT TORTOISE-SHELL BUTTERFLY. This insect frequents woody places and lanes where elms abound, and in some seasons it is particularly abundant in some situations. Wings above dark orange, with the base dusky, and furnished with greenish hairs: the anterior with two transverse costal fasciæ; between which and the base is a somewhat ovate black spot; on the disc, and also near the interior margin, are two other spots: the hinder margin is black, with a series of pale crescents: on the posterior wings is a large black costal spot, with a yellowish patch adjoining; and the margin is black, with obscure bluish crescents; interiorly the wings are furnished with long tawny or greenish hairs: beneath, all the wings are clouded with black, with a broad ash-coloured fascia behind, in which is a series of bluish lunules: the anterior wings have three pale equidistant spots on the costa, and the posterior a white discoidal spot; the body is dusky, with tawny hairs; and the antennæ black. The caterpillar is brownish, with a yellow lateral stripe, and the spines slightly branched. It feeds chiefly on the elm; and while young, the brood continues under a silken web. The chrysalis is flesh-coloured, with golden spots on the neck.

VANESSA URTICÆ; or SMALL TORTOISE-SHELL BUTTERFLY. This elegant and very prevalent British species has the wings above of a rich reddish orange, with the base and the hinder margin black, the latter with a series of blue crescents: the anterior wings above have the costal areolet mottled with black and tawny: on the costa are two large sub-quadrate black spots; at the base two others, placed obliquely; and posteriorly, on the disc, two small round ones: between the two large costal spots and the anterior basal one are two yellow spots, and towards the tip of the wing adjoining the posterior costal spot is a light one. The posterior wings are black at the base, powdered with tawny, and covered with long hairs: beneath, the anterior wings are pale, variegated with black, with a pale band marbled with brown, in which is a series of angular black spots. The body is dusky, with a greenish pubescence: the antennæ are marked with black and white rings, and the tip of the club is ochraceous. The Caterpillar of this species feeds chiefly on the nettle, and is found in lanes, gardens, &c.; it is about an inch in length, covered with bristles, and of a reddish-brown colour, marked with two greenish-yellow lines on the back, and one on each side. The chrysalis is grayish, with golden spots on the neck: sometimes the whole body is entirely golden; from which the words *chrysalis* and *aurelia* are supposed to have suggested themselves to entomologists to denote the pupa state of insects. Two broods occur every year — one early in spring, the other in autumn; and in Italy it continues on the wing in fine weather even in winter.

We may here take an opportunity of referring to a valuable paper in the proceedings of the Royal Society, recorded in vol. 15. of the *Annals of Natural History*, "On the Reproduction of lost parts in Myriapoda and Insects," by G. Newport, Esq. F.R.S., Pres. Ent. Soc., &c. (communicated by Dr. Roget). It has long been known that the limbs of *Crustacea* and *Arachnida*, accidentally lost or designedly removed, are, in course of time, replaced by the growth of new limbs; but whether such a power exists in those insects, such as *Lepidoptera*, which undergo a complete metamorphosis, changing not only their form, but also their food and mode of life, in passing from the larva to the adult state, has been considered doubtful. "The first observation which led the author to believe that true insects might possess the power of reproducing lost parts, was that of a specimen of *Phasma* in the collection of the British Museum, in which the right anterior leg had evidently been reproduced. He then instituted a series of experiments on the larva of the *Vanessa urticæ*, or common nettle butterfly, which belongs to the order *Lepidoptera*, and undergoes complete metamorphosis. He removed some of the true legs of the larva, sometimes in their tibial portion, and sometimes at their base: in the first case, parts similar to those removed were *invariably reproduced* in different states of development, and in the latter *entire new limbs* were formed; in some instances, at the second change of the larva, when it passed into the pupa state; but in two or three instances no reproduction took place. At first view, this difference in the results might appear to favour the opinion that this reproduction of limbs depends on the existence of parts especially adapted to perform this function, and which, in those experiments that had failed to exhibit the phenomenon, had been themselves removed. But the author found that in every instance of the mutilations thus practised, the perfect insect possessed a coxa, or basilar part of the limb; and this was the case even in those in which a new organ was not reproduced. From this fact, taken in conjunction with the formation of new entire limbs in the *Iulidæ* after the removal of every portion of the previous ones, the author infers that the power of reproduction resides in the whole of the organized tissues.

VANESSA ANTIOPA; WILLOW BUTTERFLY, or CAMBERWELL BEAUTY. The wings of this insect are of a reddish black or purplish hue above, with a broad, velvety-black posterior band, in which on each wing are seven or eight violet-blue spots; followed by a broad straw-coloured border, waved internally, and minutely speckled with black dots, particularly on the prominent angles of the wing. The anterior wings above have the costal areolet marked with white, and two large white spots near the tip. Beneath, all the wings are obscure black, with darker waves, and a broad white border on the outer margin. Body and antennæ dark brown. The Caterpillar is black, with a row of square dorsal spots, and the eight anterior prolegs red: it feeds on the willow, birch, and poplar. The chrysalis is dusky, with bluish and tawny spots. This species is remarkably irregular in its appearance, scarcely any being met with in some seasons, and then again appearing perhaps in immense numbers.

VANESSA IO, or PEACOCK BUTTERFLY. This highly beautiful species of Butterfly occurr pretty abundantly in lanes, woods, and commons where nettles and thistles abound. The wings above are of a purplish hue, with the base and hinder margin dusky ash, and a large ocelliform spot

PEACOCK BUTTERFLY.—(VANESSA IO.)

on each wing, the posterior wings having towards the margin a large ocellus with a large black pupil spotted with blue, and a gray iris, terminated anteriorly with a black crescent. The under side of the wings are glossy brown, marbled and spotted with black: the body is dusky, with rusty down; the antennæ blackish, the tip yellow. The Caterpillar is glossy black, spotted with white: the chrysalis green, dotted with gold.

VANESSA ATALANTA; or RED ADMIRAL BUTTERFLY. Common as this species is, it is one of the most splendid of the British butterflies; the intense black of its wings being so beautifully relieved by the red fascia and pure white spots, and the marbled veinings of its posterior wings beneath, defying the utmost efforts of the painter's skill. The wings above are deep silken black; the anterior with a central-bent orange-red band, sometimes bearing a round white spot towards the anal angle of the wing: between this and the tip are six white spots, the largest on the costa; and between them and the margin is a slight bluish wave: the posterior

RED ADMIRAL BUTTERFLY.
(VANESSA ATALANTA.)

wings have a broad orange-red border, with a transverse series of black triangular dots, and some black spots on the cilia; the tip of the inner areolet is varied with bluish, and the black dot in the following areolet is also sometimes externally edged with blue: beneath, the anterior wings have the central band interrupted with white, and blue streaks; beyond these are seen the three larger spots of the upper surface; two imperfect ocelli occupy the place of two others; and the costal areolet is black, marbled with blue. Nothing can excel the beautiful variegations of the posterior wings, mottled with black, brown, and pale fulvous; in the middle of the anterior margin is a pale triangular spot, a band of obscure ocelli parallel with the hinder margin, and a streak of silken blue between this and the margin, all the wings, both above and below, are fringed with white, interrupted at the nervures with black. Body black above, grayish beneath; antennæ black, annulated with white, the tip rather yellow: palpi black above, white sides, and yellowish beneath. The caterpillar is greenish, or dusky, with a yellowish spotted line on each side: it feeds on the *Urtica urens* and *U. dioica*. The chrysalis is dusky, or gray, with golden spots.

VANELLUS. [See LAPWING.]

VANGA. A genus of Passerine birds, indigenous to South America, and allied to the Shrikes and Flycatchers. They are distinguished by a large beak, very much compressed throughout, its tip strongly hooked, and that of the lower mandible bent downwards.

VELIA. A genus of Hemipterous insects; belonging to which is a British species (*Velia currens*), commonly seen running on the surface of brooks. The antennæ are filiform, with the sheath of the sucker only two-jointed; the legs moderately long, and placed at equal distances apart.

VENEER [MOTHS]. A name given by collectors to Moths of the genus *Chilo*.

VENERICARDIA. A genus of Acephalous Testacea, inhabiting an almost round shell, the muscular impressions in which indicating that the animal has a resemblance to that of the Carditæ and Unio, both of which approach the Cardia in general form and in the direction of their ribs.

VENUS. A genus of Conchiferous Mollusca, which are found buried in the sand, at a short distance from the shore, particularly in hot climates. The recent species are very numerous; most of the animals serving as food for man; while amongst the shells are some so beautiful as to fully justify the name given to the genus. They are equivalve, inequilateral, nearly round or oval, transverse, externally rugose, striated, ribbed, cancellated or smooth; margins entire, simple, close; bosses slightly turned on one side; ligament external, and on the longest side.—"The species *Venus mercenaria* is cut by the North American Indians into beads, of which they construct their Wampum or treaty belts, and the shells are also used amongst them as money, and are made into ornaments for their dresses."

VERMES. The name by which ancient naturalists designated a class of all the lower animals resembling the earthworm, but considered obsolete since Cuvier, in 1798, limited the term to the animals now known as *Annelides* and *Entozoa*. In Mr. Broderip's observations on this subject he makes the following sensible remarks: "The history of the now obsolete class of *Vermes* is interesting to the philosophical naturalist, since to him it symbolises the progress of zoological science. At first, with few materials on which to build his arrangement, the zoologist was guided solely by a vaque perception of analogy. Similarities of external form were made the basis of classification. The distinction between the resemblance of animals adapted for existence under similar conditions of the earth's surface and their relations to each other according to their organization, correspondent with their position in the series, could not be expected to strike the naturalist when his data were as yet so scanty. But as the discovery of species, the observation of their distribution and habits, and the anatomical investigation of their structure progressed, a new light opened on his mind, and he learned to separate forms merely analogous, and to combine such as had a true affinity of structure in well-defined divisions."

VERMETUS. A genus of Mollusca, consisting of only one species, *Vermetus lumbricalis*, which may be found in groups, twisted together in great numbers, in the seas near Senegal. The animal has two tentacula, with eyes at the base; foot cylindrical. The shell is thin, tubular, irregularly and slightly twisted; aperture round; apex pointed.

VERMILIA. A genus of *Annulata* composed of species of *Serpulæ*, and found on stones, shells, fuci, &c. They are attached by the whole length of their shell, no part being free. The tube is testaceous, cylindrical, gradually lessening at one end, and more or less twisted.

VESICULOSA. A group of Dipterous insects, nearly allied to *Bombylius;* with the wings deflexed at each side of the body; the alulets very large, and covering the halteres; the head small and globular; the thorax very gibbose; the abdomen vesiculose; and the proboscis directed backwards, or wanting.

VESPA: VESPIDÆ. A family of aculeated Hymenopterous insects, (including the common Wasp and Hornet,) which live in temporary societies, consisting of males, females, and workers or neuters. They are characterized by their geniculate antennæ, composed in the males of thirteen joints, and sometimes, in this sex, hooked at the extremity. Mandibles strong and dentated; clypeus large; ligula plumose or bilobed. The sting of the females and neuters long, powerful, and highly venomous. The economy of these insects is scarcely less interesting than that of the hive bee (with which they agree in their habit of constructing hexagonal cells arranged in combs of different size.) [See BEE.]

The societies are, however, annual, being dissolved at the approach of winter. The nests are of varied size, according to the number of the society by which they are inhabited, being from time to time enlarged during the summer, as the community becomes more and more extensive. Previous to the setting in of the winter, the females, which have been but recently developed, are impregnated by the males, which soon afterwards die; the females then disperse, seeking winter quarters, in sheltered situations; and those which survive the rigours of winter commence the building of a new nest at the return of the spring, in which they deposit eggs and tend their young themselves; these at first consisting entirely of neuters, which assist their parent in the duties of the nest. The nests are either built underground in holes, in banks, or are attached to the branches of trees, or the woodwork of outhouses. They are composed of a paper-like substance formed of finely-gnawed wood, or the bark of trees, reduced to a kind of paste by the action of the jaws, and contain a variable number of cells, which are of an hexagonal form, arranged in tiers with the mouth downwards, or opening sideways, in which the larvæ and pupæ are contained. The larvæ of the wasp tribe are vermiform and without feet: those of the solitary species are enclosed separately in a cell, in which the mother deposits, with singular apparent foresight, at the same time with the egg, the bodies of insects, killed for the purpose, and upon which the larva feeds. The nest is generally surrounded by an envelope, pierced with a common central opening. The larvæ are nourished with the juices or pulp of fruit provided for them by the neuters; they are shut up, and spin for themselves a cocoon, when about to become nymphs.

These insects are very voracious, preying upon other insects, sugar, meat, fruit, honey, &c., which, after being properly prepared in the stomach of the winged insects, is disgorged, and serves as food for the young, which are fed therewith daily; the females as well as neuters assisting in this task. The males, as in all other social insects, are drones performing no kind of labour. Notwithstanding the powerful sting of the Wasp, it is liable to the attacks of other insects. The Hornet (*Vespa crabro*) builds its nest in decaying hollow trees, under the eaves of barns, &c. [See WASP and HORNET.]

PASTE-BOARD WASPS. (*Chartergus.*) We shall conclude the article *Vespidæ* with an account of a South American Wasp which collects honey; as described by Mr. Adam White. "Some of the Wasp tribe of the New World form their nests of a solid and rather thick pasteboard. Such structures have been met with in Pennsylvania, while they occur frequently in the more tropical parts of South America as far as Buenos Ayres, and very probably much to the south of that point: in the description of the Isthmus of Darien, Wafer mentions "the bird's nest bee, the hives of which are black and hard, hanging from the trees like birds' nests." The best known is that of the *Chartergus nidulans*, which is formed of a beautifully polished white and solid pasteboard, impenetrable by the weather. It has been fully described by Reaumur in the sixth volume of his 'Memoires:' in the British Museum there are two specimens of this nest. They are securely attached to the branch of a tree by their upper end, and vary much in length, from a few inches, as in the Museum specimens, to two feet or even more. In the former case they are more or less round, and have but four or five combs, while in the latter they are of a long cylindrical shape, and have a corresponding number of partitions; additional combs are added to the lower part as the occupants increase in number. These combs are horizontal, convex on the under side, and fixed to the walls of the nest by their whole circumference. The cells are hexagonal and open downwards, as in most other nests constructed by the *Vespidæ*. Each of the combs has a hole near the middle, through which access is obtained to the uppermost apartments. The upper entrance is by a small round orifice near the middle of the under side, which is more or less funnel-shaped.

"The insects which form these curious habitations have been observed by Lacordaire in their native country. Their societies are not dissolved each year, as happens with the wasps of our climates, which, on the approach of cold weather, are nearly all cut off. The nests are found in copse-wood, principally near plantations (at least in Guiana), and are generally suspended at a height of three or four feet from the ground. During the rainy season, from January to the middle of June, only perfect nests are to be met with; in January and February the cells are in great measure filled with larvæ; in March and April these decrease in number, and by the end of May scarcely any are to be found. These are thought to turn into females, which, not finding room in their old nursery, emigrate and form new colonies, as when the fine season returns, which is about the middle of June, nests are to be found in progress; but instead of only one female being at work, as is the case with our wasps, Lacordaire has observed as many as a dozen busily engaged in constructing their new abode. As soon as a series of cells is completed larva may be found in them, and the nest is gradually increased by the addition of new combs. In September the structure is half finished, and towards the end of November it is most frequently completed. The old nests of the preceding year continue peopled as before, but new larvæ were only observed in them in abundance in September or October; these are believed to turn into neuters: if this is the case, the reverse takes place with the European wasps, the neuters of which are first excluded.

Mr. Walter Hawkins has presented to the collection of the British Museum a pasteboard nest discovered in June, 1837, in the woods situated along the banks of the Yancay, a tributary stream of the Uruguay, and takes its rise in the province of Entrerios: it was about seven feet from the ground. Viewed sideways, it is of an oblong form, rounded at the base: the orifices at the side, near the bottom, bulge out considerably. When viewed from beneath it is somewhat ovate. It is very generally covered with conical knobs of various shapes, nearly all of which are more or less rubbed at the end, but in some places, less exposed, they are pointed, and in many instances nearly three quarters of an inch long. At the very top, and on the side above the entrance, there are but few of these projections; in two or three places the surface is very distinctly contracted; and in the concavities are no projecting points; the knobs seem to run in

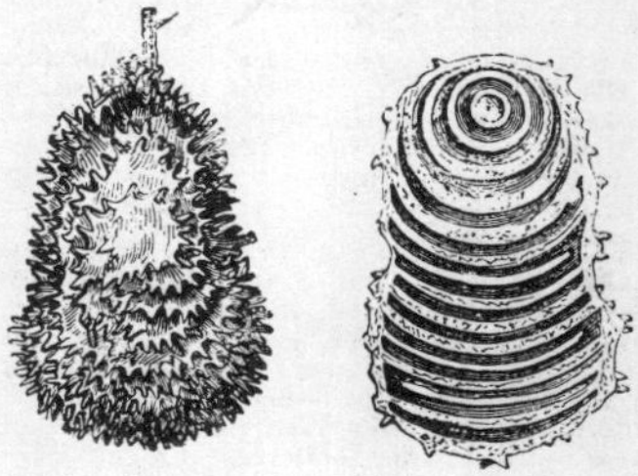

NEST OF HONEY WASP.—(MYRAPETRA.)

irregular, generally transverse, ridges. The entrances are artfully protected by pent roofs from the weather, which, in the rainy season, is sometimes very violent; they are also so intricately twisted, as to prevent the ingress of any moth or other enemy, at least of any size. The hardness of the whole mass must tend very much to protect its constructor from the attacks of insect or honey-seeking animals; and the natives, with some degree of probability, believe, that feline and other animals are deterred from taking the nest by the pointed knobs with which it is covered. The substance is hard, the texture close, and, when seen with a slight magnifying power, seems curiously matted. The natives say that it is principally formed of the dried dung of the Carpincho and dried rushes and underwood. The Carpincho is a species of Tapir or Water-hog, and is amphibious.

Many of the uppermost combs have the cells, in the middle, filled with a brownish red honey, which, in its present state, possesses scarcely any smell or taste. Azara,

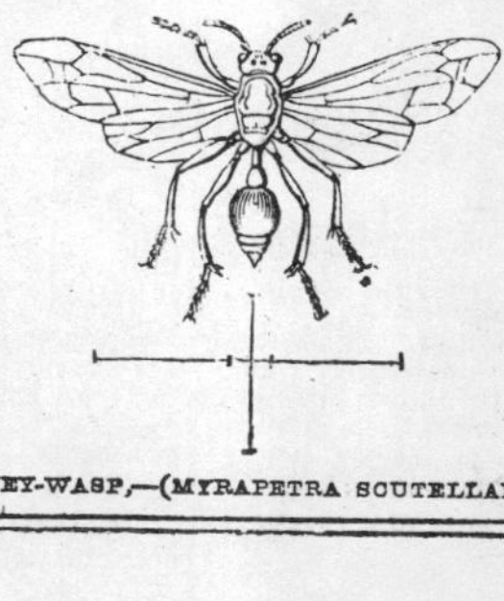

HONEY-WASP.—(MYRAPETRA SCUTELLARIS.)

in the account of his residence in various parts of South America, mentioned the fact of several *Wasps* of these countries collecting honey. The occurrence of honey in the combs of these *Myrapetra* confirms the accuracy of Azara' observation, and is made by a Vespidous insect having the first joint of the abdomen elongated into a pedicel.

VESPERTILIONIDÆ. The name given to a family of Bats, including most of those belonging to temperate climates. [See BAT.]

VICUGNA. A Ruminant quadruped of South America, belonging to the *Camelidæ* family, and bearing considerable resemblance to the Alpaca. They inhabit the mountain ranges, and are remarkable for the fineness of their wool, which has a texture that may be termed silken; and they are accordingly much in request.

VIPERIDÆ. A group of venomous snakes, of which there are many species and varieties diffused almost everywhere throughout the habitable globe, Africa and America being, however, far more infested by them than Europe. A description of the different kinds, after what we have said of these reptiles under their respective appellations, would be neither interesting nor instructive; we therefore pass on to the common Viper, occasionally met with by those whose occupations take them to our heaths, woods, and water meadows.

The COMMON VIPER OR ADDER (*Vipera berus*) is the only poisonous reptile indigenous to this country; and is abundantly found in many parts of Scotland, England, and Wales, particularly in chalky and stony districts, frequenting heaths, dry woods, and banks. In Ireland it certainly does not exist. On the continent of Europe it is extensively distributed, being found from the northern parts of Russia to the south of Italy and Spain, and its presence is everywhere dreaded on account of its venomous qualities. It seldom arrives at a greater length than two feet, though it is occasionally met with above three. The ground colour of the male is a dirty yellow; that of the female is deeper. The back is marked throughout its whole length with a series of rhomboidal black spots, touching each other at the points; the sides are marked with triangular ones; and the belly is entirely black. It is chiefly distinguished from the common black snake by the colour which in the latter is more beautifully mottled; as well as by the head, which is thicker than the body; but particularly by the tail, which in the Viper, though it terminates in a point, does not run tapering to so great a length as in the other: when, therefore, other distinctions fail, the difference of the tail can be discerned with ease. The venom of the Viper is less virulent than that of many of the poisonous serpents, but still sufficiently severe, in the warmer climates, to produce even the most fatal results. The remedies usually employed are the external application of oil and the internal administration of ammonia.

VIPER.—(VIPERA BERUS.)

The apparatus by which the poison wounds are inflicted, which render these and so many other serpents so formidable, is described by Mr. Bell, in his work on 'British Reptiles,' as follows:—On each side of the upper jaw, instead of the outer row of teeth which are found in non-venomous serpents, there exist two or three, or more, long, curved, and tubular teeth, the first of which is larger than the others, and is attached to a small movable bone, articulated to the maxillary bone, and moved by a muscular apparatus, by which the animal has the power of erecting it. In a state of rest the fang reclines backwards along the margin of the jaw, and is covered by a fold of skin; but when about to be called into use, it is erected by means of a small muscle, and brought to stand perpendicular to the bone. The tooth itself is it as were perforated by a tube. This tube, although completely enclosed, excepting at its basal and apical orifices, must be considered as formed merely by the closing round of a groove in the external part of the tooth itself, and hence not in any way connected with the inner cavity of the tooth, in which exists the pulp upon which the substance of the tooth is formed. The base of the tooth, and consequently the basal orifice of the tube just described, is embedded in a sac, into which the poison is poured from the ducts of the glandular structure by which it is secreted, and which is believed to represent the parotid gland of the higher vertebrata. The poisonous fluid itself is inodorous, tasteless, and of a yellow colour. It is secreted in greater quantities, and its qualities are more virulent in a high temperature than in cold. . . . When the animal inflicts the wound, the pressure on the tooth forces a small drop of the poison through the tube; it passes through the external orifice, which is situated on the concave side of the curved tooth, and is in the form of a slit. The manner in which the blow is inflicted is as follows. The animal generally throws itself in the first place into a coil more or less close, and the anterior part of the body is raised. The neck is bent somewhat abruptly backwards, and the head fixed almost horizontally. In an instant the head is, as it were, launched by a sudden effort towards the object of its anger, and the erected tooth struck into it and with the velocity of thought. It is found by experiment that the effect of subsequent wounds is greatly diminished either by the diminution of the quantity of venom, or by some deterioration of its strength; so that if a venomous Serpent be made repeatedly to inflict wounds, without allowing sufficiently long intervals for it to recover its powers, each successive bite becomes less and less effective."

The Viper, like many other of the poisonous groups of Serpents, is ovo-viviparous. If a female Viper about to bring forth her young be killed, and the young ones set at liberty by opening the abdomen, they will immediately crawl about, and on being irritated will throw themselves into an attitude of defence. The number of young produced at each birth varies from about twelve to twenty. During the cold months of the year the Viper, like the other *reptilia*, finds a secure retreat in which to hibernate. Shrews, field-mice, and other small animals are the Viper's food. There are two or three varieties, as the Red Viper, the Black Viper, &c.

VIRGULARIA. A genus of Coralliferous Polypi, closely allied to the genus Pennatula, but having the laminæ between which the polypi are situated much shorter. Like that

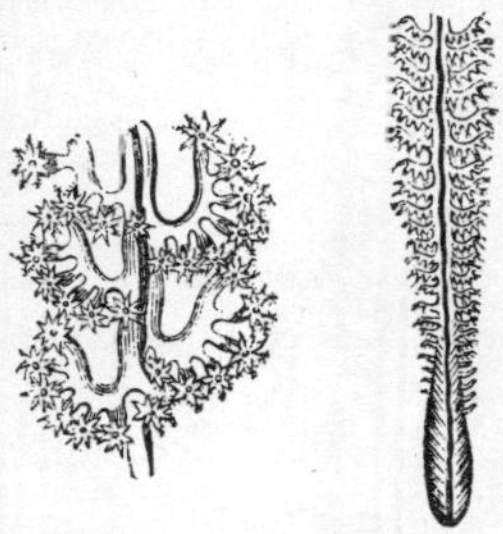

VIRGULARIA MIRABILIS.

genus, one extremity of it is always without polypi, and somewhat resembles the barrel of a feather. It is believed to be phosphorescent, like many of the other allied genera. Our figure shows, better than a description, the form of this singularly beautiful genus.

VITRINA. A genus of small land shells, ovate, thin, glossy, and fragile; spire short, the last whorl large; aperture oval. The body of the animal is long, with four tentacula, two of which have eyes at the summit. The species are all recent, and found among moss and grass in damp situations. They greatly resemble young specimens of the genus *Helix*, from which they are distinguished by their never being umbilicated or perforated.

VIVERRA: VIVERRIDÆ. A genus and family of carnivorous quadrupeds, which in the Linnæan arrangement included (besides the true Civets, to which the genus *Viverra* is now restricted) various animals differing remarkably in form, in structure, and in habits; as ichneumons, coatimondis, genets, weasels, &c. The true Civets, as Mr. Bennett, in his 'Tower Menagerie,' observes, yield in the extent of their carnivorous propensities to the cats alone, whom they approach very closely in many points of their zoological character, as well as in their predatory, sanguinary, and nocturnal habits. In addition to the six incisors and two canines which are common to the whole of the true *Carnivora*, they have on each side and in each jaw six molars, one of which is peculiarly adapted for lacerating flesh, while the rest are more or less of the ordinary form. Their tongues are furnished with the same elevated and pointed papillæ which give so remarkable an asperity to those of the cats, and their claws are half retractile. The toes are five in number on each of their feet, and their extremities alone are applied to the ground in walking; the animals are consequently completely digitigrade. But the most distinctive character of the group consists in an opening near the tail, leading into a double cavity of considerable size, furnished with glands and follicles for the secretion of the peculiar odoriferous substance so well known as the produce of the Civet, and from which the animal derives its name. [See CIVET.]

VIZCACHA. [See BIZCACHA.]

VOLUTA: VOLUTIDÆ. A genus and family of testaceous gasteropodous Mollusca, principally found in tropical seas, and whose shells are prized above most others for their beauty and rarity. The animals inhabiting them have the head distinct, and two short triangular tentacula, with eyes at the base, and a long thick proboscis or trunk; foot very large. The *Volutidæ* comprise numerous species, both recent and fossil, and may be regarded as one of the most interesting and beautiful families of the spiral *Testacea*, whether in regard to the elegance of the shells themselves, or as exhibiting a principle of variation in their structure hardly to be excelled. They are generally smooth, shining, and the colours bright and varied; they differ exceedingly in form and size, some being globular, others oval, some turreted, and others with only a very small spire; but though they vary in the figure of the shell and of the aperture, they are recognized by the emargination without a canal which terminates it, and by the oblique plaits of the columella. Some of them have spines at the upper part of each whorl, which form a kind of thorny crown; many are curiously marked with lines and spots, so as to form some resemblance to a line of printed music; and one very scarce species is marked with five or six transverse milk-white bands upon a dark ground, and spotted with reddish brown, forming a beautiful contrast of colours. Many of them attain a very large size; but the fossil species are generally smaller than the recent.

VOLVARIA. A genus of Univalve Mollusca, found on the coasts of Africa and

China. The shell is oval, cylindrical, and spirially striated; spire very short; aperture narrow, and as long as the shell; columella with three oblique plaits; outer lip thin.

VOLE. (*Arvicola.*) Under the word RAT will be found a description of the Bank Vole or Water Rat. The species we have now to describe is called the FIELD VOLE or SHORT-TAILED FIELD MOUSE (*Arvicola agrestis*); a small Rodent animal, which is exceedingly prolific, and whose depredations in the field, the rick-yard, and the granary are highly injurious to the agriculturist. This little creature is of a reddish-brown colour, mixed with grey, on its upper parts, and ash-colour beneath; feet and tail dusky. Length of the head and body, four inches; tail about one inch and a quarter. The head is large; muzzle very obtuse; the body thick; the tail not more than one-third the length of the body, sparingly covered with hair. The female forms her nest of dried grass, and produces six or seven young at a time. The nature of the Field Vole's food is decidedly vegetable, as we might indeed infer from the following interesting facts, related by Mr. Jesse in the first series of his 'Gleanings:'—"An extraordinary instance of the rapid increase of Mice, and of the injury they sometimes do, occurred a few years ago in the new plantations made by order of the Crown in Dean Forest, Gloucestershire, and in the New Forest, Hampshire. Soon after the formation of these plantations, a sudden and rapid increase of Mice took place in them, which threatened destruction to the whole of the young plants. Vast numbers of these were killed; the Mice having eaten through the roots of five-year-old oaks and chestnuts, generally just below the surface of the ground. Hollies also, which were five or six feet high, were barked round the bottom; and in some instances the Mice had crawled up the tree, and were seen feeding on the bark of the upper branches. In the reports made to Government on the subject, it appeared that the roots had been eaten through wherever they obstructed the runs of the Mice. Various plans were devised for their destruction; traps were set, poison laid, and cats turned out; but nothing appeared to lessen their number. It was at last suggested, that if holes were dug, into which the Mice might be enticed or fall, their destruction might be effected." Holes, it appears, were accordingly made in Dean Forest, about twenty yards asunder, and from eighteen to twenty inches in depth, hollowed out much wider at bottom than at the top; so that the animal, when once in, could not easily get out again. In these holes at least thirty thousand Mice were found in the course of three or four months; and it was calculated that a much greater number than these were taken out of the holes, after being caught, by stoats, weasels, kites, hawks, owls, crows, magpies, &c. The Field Vole either burrows itself, or takes possession of the excavations made by the mole and other burrowing animals.

VOLVOX. The name given to certain infusorial animalcules which swarm in our stagnant waters. They are globular masses, and are made up of smaller, minute oval bodies. The Volvox is now more commonly regarded as a low vegetable organism.

VULPES. [See FOX.]

VULSELLA. A genus of Conchiferous Mollusca, the shells of which are brought from the Indian Ocean and the seas of New Holland, and are generally found buried in sponge. They are oblong, longitudinal, nearly equivalve, and irregular; hinge with a prominent callosity in each valve, showing an impression of a conical and arched pit for the ligament: the interior is iridescent.

VULTURIDÆ. A family of diurnal Accipitrine birds, characterized by an elongated beak, curved only at the tip, and by having a greater or less proportion of the head, and sometimes of the neck, denuded of feathers. In general, the birds belonging to this family are of a cowardly nature, living on dead carcases and offal; their gullet dilates into a considerable crop, which, when distended with garbage, projects above the furcular bone. When gorged with food the bird is reduced to a state of stupidity, and a fetid humour is discharged from the nostrils.

"The Vultures," as Mr. Swainson has remarked, "are the great scavengers of nature in hot latitudes, where putrefaction is most rapid, and most injurious to health; and the disposition of their numbers is regulated by an all-wise Creator according to their usefulness. They are sparingly scattered over the south of Europe; in Egypt they are more numerous; but in tropical America, although the species are fewer, the individuals are much more plentiful. No sooner is an animal dead than its carcase is surrounded by numbers of these birds, who suddenly appear, coming from all quarters, in situations where not one had just before been seen. The nakedness of the head, and frequently of the neck, is most apparent in those whose geographic range is limited to the New World, at the head of which division stand two remarkable species, the celebrated Condor of the Andes, and the *Papa*, or King Vulture, of the Brazilian forests. The first is well known for the loftiness of its flight and its amazing strength, while the latter is the only species whose colouring is not dark or sombre." We shall now describe a European species.

The GRIFFON VULTURE. (*Vultur fulvus.*) This bird inhabits the mountainous parts of the north of Europe, Silesia, Dalmatia, the Tyrol, Spain (where, near Gibraltar, it is abundant), the Alps, the Pyrenees, Turkey, and the Grecian Archipelago. Its nest is usually formed upon the most elevated and inaccessible rocks, or upon the loftiest trees of the forest. Its eggs, generally two or three in number, are of a dull greenish or grayish white, slightly marked with pale reddish spots, and with a rough surface. "Like all the other birds of its tribe," says Mr. Bennet, "it feeds principally upon dead carcases, to which it is frequently attracted in very considerable numbers. When it has once made a lodgment upon its prey, it rarely quits the banquet while a morsel of flesh remains, so that it is not uncommon to see it perched

GRIFFON VULTURE.—(VULTUR FULVUS.)

upon a putrefying body for several successive days. It never attempts to carry off a portion even to satisfy its young, but feeds them by disgorging the half-digested morsel from its maw.

The SOCIABLE VULTURE. (*Vultur auricularis.*) This is a gigantic species inhabiting the greater part of Africa, and said by some naturalists to be also found in Greece. Its head and greater portion of the neck are red and naked, the folds of red naked skin originating behind the ears, and surrounding the upper part of them: the throat is covered with blackish hairs, and the lower and back part of the neck clothed with a ruff of blackish curling feathers. The plumage of the

SOCIABLE VULTURE.
(VULTUR AURICULARIS.)

body, wings, and tail are of a blackish-brown colour, rather lighter beneath than above; feathers of the breast, belly, and sides beneath, narrow, long, pointed, projecting from the body so as to discover the nearly pure white down which everywhere closely covers it, and extends beyond the feathers on the lower and anterior parts of the neck. Legs brownish; claws light brown. In size the Sociable Vulture is equal to the Condor, measuring upwards of ten feet across the wings expanded. The nest is built in the fissures of rocks, and the female generally lays two, sometimes three eggs. During the period of incubation the male keeps watch at the entrance of the cave.

It has been observed of this gigantic species, that it is "a fit machine for assisting in the clearance of the soil of Africa from the putrid bodies of elephants, hippopotami, rhinoceroses, and giraffes, that it haunts the caverns of rocks, and is altogether a mountain bird. There its night is passed, and there, among the lofty crags, it retires to repose when it has sated its appetite. Le Vaillant saw large flocks of them perched at sunrise on the precipitous entrances to their abodes, and sometimes the extent of the rocky region was marked by a continued chain of these birds. Their tails are worn down by friction against their craggy haunts and by the soil of the plains, in consequence of the laborious efforts which they make to raise themselves into the air: when once on the wing, however, their flight is grand and powerful. They rise higher and higher, till their enormous bulk is lost to human ken; but though beyond the sphere of man's vision, the telescopic eye of the bird is at work. The moment any animal sinks to the earth in death, the imperceptible Vulture detects it. Does the hunter bring down some large quadruped beyond his powers to remove, and leave it to obtain assistance?—on his return, however speedy, he finds it surrounded by a band of Vultures, where not one was to be seen a quarter of an hour before."

The EGYPTIAN VULTURE. (*Neophron percnopterus.*) The Egyptian Neophron, which has also been denominated Pharaoh's Chicken, is the smallest of the Vulture tribe; its natural habitation, the shores of the Mediterranean Sea. The adult has the front of the head, the upper part of the throat and cere naked, and of a bright yellow. The plumage is altogether of a pure white, with the exception of the quill feathers, which are black: legs, feet, and base of the bill yellow; point of the bill, black. There is scarcely any difference in the colouring and plumage in the adults of both sexes. The young of the year are of a deep brown, slightly spotted with lighter brown and white, and do not attain their adult plumage for two or three years.

In our description of the Bearded Vulture we entered rather fully on the often-discussed question of the very acute sense of smelling which has been attributed to birds of the Vulture tribe. Before we quit the subject, it may be proper to mention that Audubon, in his 'Birds of America,' insists on it that it is the organs of sight, and not those of smelling, that enables Vultures and other birds of prey to discover carcases at such immense distances as they are said to do. We quote from him the following passage: "We were led to call in question the accuracy of this opinion, on recollecting the observations of some travellers, who have remarked birds of prey directing their course towards dead animals floating in the rivers in India, where the wind blows steadily from one point in the compass for many months in succession. It is not easy to conceive that the effluvia from a dead carcase in the water should proceed in direct opposition to the current of air, and affect the olfactory nerves of birds at so many miles distant." In order to satisfy himself on this point, Audubon made several experiments, one of which was as follows. Having stuffed and dried the skin of a deer, he retired from it: a Vulture soon approached, attacked its eyes, which were made of painted clay, then walked to the other extremity, tore some of the stitches, until much of the fodder and hay with which it was stuffed was pulled out, and after reiterated attempts to discover flesh, took flight. Afterwards, he had a large dead hog put into a ravine and concealed in the briars; he saw many Vultures pass over it, but some approached it, although several dogs had made a meal on it. He then tried to approach it himself, but found the stench too intolerable. This species has great power of wing, and specimens have sometimes been killed in the British isles.

The BLACK VULTURE or GALLINAGO. (*Cathartes atratus.*) In our article Turkey Buzzard we have described a species closely allied to this in appearance and habits. We introduce this species as a well-marked form of the family Vulturidæ. It is a native of the United States, and is found in South America, as Darwin informs us, as far south as Lat. 41°. It prefers a humid climate, or

BLACK VULTURE.—(CATHARTES ATRATUS.)

rather the neighbourhood of fresh water, and in Peru is protected as a scavenger. These Vultures may be called gregarious, and, as Darwin observes, are not solely brought together by the attraction of a common prey, but seem to have pleasure in society. He has observed a flock of them on a fine day at a great height, each bird wheeling round and round without closing its wings, in the most graceful evolutions. It is clearly done for sport-sake, or may be connected with their matrimonial alliances.

For *King Vulture*, see SARCORAMPHUS.

For *Bearded Vulture*, see GYPAETUS.

For *Turkey Vulture* and *John Crow Vulture*, see TURKEY BUZZARD.

WAGTAIL. The species of Wagtails, which are few, are chiefly confined to the European continent, where the individuals are numerous. Bewick remarks that "in almost all languages the name of this bird is descriptive of its peculiar habits. In Latin, Motacilla; in French, Moteux, La Lavandière, or Washer; in England they are sometimes called Washers, from their peculiar motion; in German their name signifies Brook-stilts; and in Italian, Shake-tail," &c. They are easily distinguished by their brisk and lively motions, as well as by the great length of their tails, which they jerk up and down incessantly, from which circumstance they derive their name. They do not hop, but run along the ground very nimbly after flies and other insects, on which they feed: they likewise feed on small worms, in search of which they frequently flutter round the ploughman, and follow the flocks in search of the flies which generally surround them. Their flight is weak and undulating, during which they make a twittering noise; and they seldom perch. As the species do not differ in their habits, and are not very dissimilar in appearance, it will be sufficient that we describe the one most common with us, viz.—

The PIED WAGTAIL, or BLACK and WHITE WATER WAGTAIL. (*Motacilla alba.*) Its length is about seven inches: the bill is black; eyes hazel; hinder part of the head and neck black; forehead, cheeks, and sides of the neck white; the fore part of the neck and part of the breast are black, bordered by a whitish line, form-

PIED WAGTAIL.—(MOTACILLA ALBA.)

ing a gorget; the back and rump are dark ash; wing-coverts and secondary quills dusky, edged with light gray; prime quills black, with pale edges; lower part of the breast and belly white; the middle feathers of the tail are black, the outermost white, except at the base and tips of the inner webs, which are black: legs black. These birds are to be seen wherever there are shallow springs and running waters. They make their nest on the ground, of dry grass, moss, and small roots, lined with hair and feathers; the female lays five white eggs, spotted with brown; and both parents continue to feed and train their young for three or four weeks after they are able to fly. As the winter approaches they migrate from north to south.

WAINSCOT [MOTHS]. A name given by collectors to different species of Moths, of the genera *Nonagria* and *Leucasia*.

WALKING-LEAF. [See PHYLLIUM.]

WALRUS or MORSE. (*Trichecus.*) A genus of the *Phocidæ* or Seal family, though differing greatly from them in the cranium and the teeth. The head is well proportioned, round, obtuse, eyes small and brilliant, upper lip remarkably thick, co-

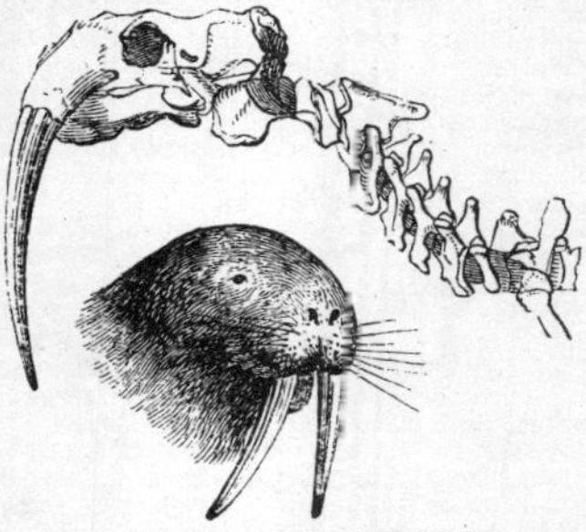

SKULL AND HEAD OF WALRUS.

vered with large pellucid whiskers or bristles. Nostrils large, rounded, placed on the upper part of the snout: no external ears. In the adult lower jaw there are neither incisors nor canines, and the lower jaw itself is compressed anteriorly so as to fit between the two enormous tusks (canines) of the upper jaw, which are directed downwards, and are sometimes two feet long.

WALRUS OR MORSE.
(TRICHECUS ROSMARUS.)

The great alveoli, or sockets for containing these formidable teeth, produce the characteristic form of the skull of the Walrus, and make the anterior part of the upper jaw present an immense convex muzzle, the nostrils having an upward direction, and not terminating at the snout. It is evident that there is a general resemblance between the organization of the Walrus and that of the Seal; but the development of the brain is not so great in the former as it is in the latter, and the Walrus appears to be gifted with less intelligence.

It is the opinion of most naturalists that Walruses feed on shell-fish and marine vegetables which adhere to the bottom of the sea, and that one of the uses of their tusks is to root up their food from the spot to which it is fixed; and the probability is, that though the Walrus does not abstain entirely from carnivorous habits, marine plants form the bulk of its food. They swim rapidly, but their progress on land is awkward and tedious. They appear to be monogamous, and the female is said to bring forth her young, one only at a birth, either on shore or on the ice. The flesh is highly valued by the inhabitants of the arctic regions, and our own northern voyagers have often found it a most acceptable repast. According to Professor Macgillivray a small specimen was shot on the East coast of Harris, one of the Western Isles, December 1817. It was formerly abundant in the Norwegian seas, but is now driven further north.

WANDEROO MONKEY. (*Macacus Silenus.*) A fine species of monkey, native of Ceylon, which is of a deep black colour, excepting the long hairs about the head, which are more or less of an ash colour, and sometimes almost white. This mane, as it may be called, descends on each side of the face like a ruff. The tail ends in a brush of tufted hair. It is occasionally brought to this country, but is by no means common in a state of confinement. Father Maria has given the following account, which we quote from Mr. Bennett. "There are found four sorts of monkeys on the coast of Malabar; the

WANDEROO MONKEY.—(MACACUS SILENUS.)

first is quite black, with glossy hair and a white beard round the chin measuring rather more than a palm in length. The other monkeys pay to this so profound a respect that they are humble in his presence, as though they appreciated his superiority. The princes and mighty lords hold him in much estimation for his endowments of gravity, capacity, and the appearance of wisdom above every other monkey. He is readily trained to enact a variety of ceremonies and affected courtesies, which he goes through with so grave a face, and so perfectly, that it is a most wonderful thing to see them so exactly performed by an irrational creature." We need hardly add that this monkey is not endowed with more capacity than his congeners, but from his lion-like mane and aspect as well as his strongly marked features and colour, looks peculiar among his allies.

WAPITI. (*Cervus Canadensis.*) This animal, which is frequently called the Canada Stag, more nearly resembles the European red deer, in colour, shape, and form, than it does any other of the cervine race, though it is much larger and of a stronger make. It is, in fact, one of the most gigantic of the deer tribe, frequently growing to the height of our tallest oxen, and possessing great activity as well as strength. His horns, which he sheds annually, are very large, branching in serpentine curves, and measuring, from tip to tip, upwards of six feet. Most of the upper parts of the Wapiti

WAPITI DEER.
(CERVUS CANADENSIS.)

are of a lively yellowish brown colour; the neck, mixed red and black, with coarse black hairs descending from it like a dewlap; from the shoulders to the hips French gray; a pale yellowish patch on the buttocks, bounded on the thighs by a black line. They are considered more stupid than the rest of the deer kind; and they frequently make a shrill quivering noise, which is "not very unlike the braying of an ass." The flesh is coarse, and but little prized by the natives; but its hide, when made into leather after the Indian fashion, is said not to turn hard in drying after being wet—a quality which justly entitles it to a preference over almost every other kind of leather.

WARBLERS. (*Sylvia. Sylviadæ.*) The small singing-birds comprised under this general name form an interesting and comprehensive group, spread over the whole globe, and were arranged by Linnæus under his genus *Motacilla*. Their bill, as Nuttall remarks, is slender, straight, awl-shaped, higher than it is wide at the base, and furnished with scattered bristles; the lower mandible straight. Nostrils basal, lateral, oval, half closed by a membrane. Tongue lacerated at the tip. Tarsus longer than the middle toe; inner toe free. Wings moderate or short; scapulars considerably shorter than the quill-feathers. The same author further observes, that they are generally small, sprightly, and endowed with an incessant activity, in accordance with the subtleness of their flying insect prey: they therefore approach, both in habit and character, the Flycatchers, Thrushes, Saxicolas, and Wrens so nearly, that it is rendered at times doubtful to which of these several genera they ought to be referred. They principally inhabit forests or thickets, and some affect watery situations or reed-marshes. Many are remarkable for the melody of their song and the sprightliness of their airs, which in the period of incubation they almost incessantly pour forth. The Nightingale, so celebrated for his powerful, varied, and pathetic lay, as well as the humble but tuneful Robin Redbreast, belong to this highly vocal genus (*Sylvia*); and though many species seek out the arctic solitudes in which to waste their melody or soothe alone their mates, yet other species may be numbered among the more familiar tenants of our gardens, groves, and orchards. Living almost exclusively on the winged insects of summer, which they dexterously catch in the air or pick from off the leaves, they migrate to the south in autumn, and pass their winter in the warm or tropical regions. The greater part of the group we have described under their several names; we shall therefore now only select the undermentioned:—

The DARTFORD WARBLER. (*Melizophilus provincialis.*) Fond of retirement and seclusion, this pretty little Warbler secretes itself in the thickest parts of the bushes, where it may be heard but not seen. Though not by any means abundant in this country, they are met with in the neighbourhood of London, and also in several of the south-western counties. Mr. Gould observes, with reference to its secluded habits, that in the spring it becomes more lively and more frequently visible, "rising on quivering wing above the tops of the furze, and uttering a hurried babbling song, much after the manner of the Whitethroat; at these times it erects the feathers of the head into a crest, and distends the throat, exhibiting many attitudes and gesticulations." It is truly a mocking-bird, imitating the notes of various kinds, generally beginning with those of the Swallow, and ending with the song of the Blackbird. It is between five and six inches long; the whole upper part of the body is darkish brown, tinged very slightly with olive; throat, breast, and sides, reddish purple; tail very long, blackish brown, the external feather only terminated with white; wings very short; legs dark brown, and feet yellowish; bill black, but yellowish white at its base. Its food consists of insects, and also of such berries as it can obtain near its retreats. The nest, which is composed of dry stalks and grass intertwined with fibres of plants and roots, is generally securely placed in the middle of a furze-bush, not far from the ground. Eggs greenish white, speckled with brown.

The ORANGE-CROWNED WARBLER. This is one of the American Warblers, of which there are numerous species, but none of them much distinguished as vocalists. It is five inches long and seven in extent. The general plumage above is dull greenish olive, the rump and tail coverts being bright yellowish olive. The head is slightly crested, the feathers of the crest are orange at base, constituting a spot on the crown, visible only when they are elevated, being tipped with the common colour. The whole bird beneath is dull olive yellow; the inferior tail-coverts pure yellow. The tail is even, the feathers being dark brown, edged with olive green on the outer, and with white on the inner web. The manners of the orange-crowned Warbler resemble those of the kindred species, though, as Wilson observes, they have a remarkable habit of inflecting the tail.

Among the *Australian Warblers*, we select one described by Mr. Gould, in his magnificent work, as the WHITE-FRONTED EPTHIANURA (*Epthianura albifrons*). It is described as a most active and sprightly little bird, particularly the male. It gives a decided preference to spots of a sterile character, and is in the habit of frequently perching on the summit of a stone, or on the extremity of a dead and leafless branch. It is rather shy in its disposition, and when disturbed flies off with considerable rapidity to a distance of two or three hundred yards before it alights again. The forehead, face, throat, and all the under surface of the male is pure white; occiput black; chest crossed by a broad crescent of deep black, the points of which run up the sides of the neck, and join the black of the occiput upper surface dark gray with a patch of dark brown in the centre of each feather; wings dark brown; upper tail-coverts black; two centre tail-feathers dark brown, the remainder dark brown, with a large oblong patch of white on the inner web at the tip: bill and feet black.

WART-HOG. [See PHACOCHÆRUS.]

WASP. A name given to many Hymenopterous insects, but more properly applied to the species of the genus *Vespa*. Under the article VESPIDÆ we have described the habits, &c. of the family of Hymenopterous insects which compose it, namely, Wasps and Hornets. We shall therefore in this place introduce the genus *Pelopœus*, or DIRT-DAUBERS, which by accident was omitted in its proper place. These curious insects belong to the *Sphegidæ* family. For the interesting particulars respecting them the public are indebted to the pages of the Zoologist for 1844, the following account having been communicated to it by P. H. Gosse, Esq.—"One of the many things that struck my attention on first going into the Southern United States, was, in most of the farm-houses, lumps of yellowish mud stuck on the walls and rafters, and particularly the large projecting chimneys. Some of these were of irregular shape, nearly as large as one's fist, and others were cylindrical, as thick as one's thumb, and three or four inches long. The little boys (and boys in the back-woods know a good deal about natural history) informed me that these were the nests of the Dirt-daubers: and on taking down one of the shapeless lumps, which had been fixed right over my bed, and carefully opening it, I found within, many long-oval cells lined with a thin coat of brittle shelly substance. These were arranged side by side, in two rows: each contained the slough of a perfected insect. In a much smaller nest I found but one cell, and no exuviæ, but six spiders, all dried. The long thimble-like nests were divided into cells, in a single series, by transverse partitions of mud. The children soon showed me the insects to which the nests belonged, although, as the season was spring, they were not then building. By and by, in the summer, I cultivated an acquaintance with these funny little architects, and had opportunities of watching the whole process of building; and thus of setting at rest, to my own satisfaction, the disputed point of ownership to these nests, which some entomologists have attributed to Eumenes, supposing the Pelopæus to be parasitical. The following observations will show that sometimes, at least, the latter builds. I transcribe now from my journal.

"*June* 30.—I watched with much interest the proceedings of a Dauber, in building her mud cells: it is a pretty species, *Pelopœus flavipes*. She has chosen the ceiling of a cupboard in my sitting-room, where, previously to my observing her, she had made one cell, and half another parallel to it; the former was closed, the latter had got its contents of spiders and only wanted closing. Such was the *statu quo*. I had not seen the Dauber go in for some time, so that when she did go in, I watched her from her recommencement. She came empty, and having for some moments peeped in, and examined the contents to see that all was right, she suddenly flew out at the door (which as well as the window was almost constantly open), and returned in about a minute with a lump of soft wet mud in her jaws, about twice as large as her head. Where she got it in so short a time, I don't know; it was perfectly kneaded, and free from all lumps, or grit, and was worked, when laid on, as freely as butter. I suspect that it was formed of dry dust, on which she had poured a drop of fluid from her mouth. She laid the substance on the open end of the unfinished cell, and spread it about with her jaws very expeditiously and skilfully, till the orifice was quite closed up. She then flew off, and returned with a similar load, which she applied upon the last to make it thicker. When she was gone the third time, to observe her behaviour, I thrust the head of a pin through the newly laid mortar, opening a hole into the cell. On her return, she at once perceived the hole, and deposited her lump upon it, spreading it about as before. I played her the same trick several times, at all of which her proceedings were the same, save that at length she seemed to become very angry, and endeavoured to catch the house-flies that were flying and crawling near. I have no doubt that she suspected them of having a hand in it. At all events, she jumped at them very snappishly whenever they came near, and sometimes even with the load in her mouth, but I did not see that she caught one. Once too, a large Ichneumon was lurking about, at whom she fiercely flew, and I think they had a short struggle. At times she would linger at a little distance after depositing her load, apparently hoping to catch the insidious housebreaker, 'in the manner,' as lawyers say.

"At length I broke off a large piece from the side and bottom, exposing the spiders to view; this, however, she speedily built up as before, at two or three loads, adding to the standing part all round the hole, and not at one side only. After this I did not put her industry to the task any more, but suffered her to finish her work, which she did by adding another layer or two to the end. I, however, made a hole in the first cell, which was quite hard and dry, to see if she would observe it, which she did at once, and clapped her load of mortar on it. I noticed, that while working, though the wings were closed incumbently, she kept up a shrill buzz, like that of a bee when held in the fingers: her antennæ, which were usually carried nearly straight, were, during the plastering, curled up, and continually vibrating, and moving on the surface of the work, evidently trying it by touch, which I could not see without rejecting the theory that calls the antennæ 'ears.' In seeking her materials, she was gone never more, often less, than a minute, and always brought a similar lump in appearance, which was invariably carried in the jaws, without any aid from the feet.

"*July* 1.—The Dauber built another cell to-day, on the other side of the first, which is now therefore in the middle. I again pestered her, by sticking a small tin tack in the newly laid mud, just where she would have to deposit the next load. When she came she appeared quite 'bothered;' she ran backward and forward, and round and round, over the cells for some time, with the mud in her jaws, as if at a loss what to do in so novel an exigency. It was a different case from the former; a hole could be stopped up, but here was an intruding substance just where she wanted to deposit; should she lay it on, the incumbrance would be more firmly imbedded; should she place it elsewhere, it would be wasted, not being needed, or perhaps be positively injurious; should she attempt to remove the evil, her mouth was occupied, and she was unwilling to lose her burden. At length, however, as the least of the evils, she seized the tack with her jaws and drew it out, dropping her mud in the effort. When away the next time, I bundled up a worsted thread, and pressed it on the soft work, which presented a still more serious obstacle, as she could seize only a small part of it which would yield without coming away; however, by taking hold of several parts successively, and tugging at them a long time, and by walking round and round with it in her mouth, she at length got it out. These instances of sagacity and perseverance greatly pleased me. After laying on a load, she always cleans her antennæ with her fore-feet, and her feet with her jaws: on arriving she never alights at the nest, but always on the inside of the cup-board-front, and crawls along the ceiling to it.

"*Aug.* 6.—I pulled down the nest of the yellow-footed Dauber, to which other cells had been added in succession after the last record. On examining them now, I find three perfected insects have made their exit, one has died in making its way out, two are in pupa, one black and near perfection, the other white and nearly turned, and two are in larva, one large, the other very small, making eight originally in the nest. Many of the spiders remained uneaten, most of them were handsomely studded with scarlet spots on a black ground. It was in looking at these pupæ that I first was aware how a difficulty of no ordinary magnitude was got over. How do insects, whose abdomen is peduncled, draw it out of the pupa skin, seeing the peduncle is so slender? I should have guessed that the skin would be ruptured, but it is not so. These Daubers have a very long and slender peduncle, but the skin of the pupa, close in every other part, is as wide around the peduncle as around the abdomen, stretching across from the thorax to the summit of the abdomen, like a loose garment. What a beautiful example of Divine foresight in creation!"

WATER-HEN. [See GALLINULE.]

WATER-OUZEL. [See OUZEL.]

WATER-SNAKE. [See HYDROPHIS.]

WATTLE-BIRD. [See TALEGALLA.]

WAVE [MOTHS]. A name given by collectors to different species of Moths, of the genera *Ptychopoda*, *Emmelesia*, *Cabera*, &c.

WAXWING. (*Bombycilla.*) We learn from Bonaparte's supplement to Wilson's entertaining 'American Ornitholgy,' that the Waxwings, "having no other representative in Europe or North America are easily recognized by their short turgid bill, trigonal at base, somewhat compressed and curved at tip, where both mandibles are strongly notched; their short feet, and rather long subacute wings. But their most curious trait consists in the small, flat, oblong appendages, resembling in colour and substance red sealing-wax, found at the tips of the secondaries in the adult. These appendages are merely the coloured corneous prolongation of the shafts beyond the webs of the feathers." "The Waxwings," he adds, "live in numerous flocks, keeping by pairs only in the breeding season; and so social is their disposition, that, as soon as the young are ably to fly, they collect in large bands from the whole neighbourhood. They perform extensive journeys, and are great and irregular wanderers. Far from being shy, they are simple and easily tamed, but generally soon die in confinement. Their food consists chiefly of juicy fruits, on which they fatten, but to the great detriment of the orchard, where they commit extensive ravages. When fruits are scarce they seize upon insects, catching them dexterously in the same manner as their distant relatives the fly-catchers. No name could be more inappropriate for these birds than that of chatterers, as there are few less noisy, and they might even be called mute with much better reason. They build in trees, and lay twice in a year, about five eggs."

The BOHEMIAN WAXWING. (*Bombycilla garrula.*) "Whence," exclaims C. Bonaparte, "does the Bohemian Waxwing come at the long and irregular periods of its migrations? Whither does it retire to pass its existence and give birth to its progeny? These are circumstances involved in darkness, and which it has not been given to any naturalist to ascertain. It has been stated, and with much appearance of probability, that these birds retire during summer within the arctic circle: but the fact is otherwise, naturalists who have explored these regions asserting that they are rarer and more accidental there than in temperate climates. It seems probable that their chief place of abode is in the oriental parts of the old continent, and, if we may hazard an opinion, we should not be surprised if the extensive and elevated table land of Central Asia was found to be their principal rendezvous, whence, like the Tartars in former times, they make their irregular excursions."

It seems that in Northern Russia, and the extreme north of Norway, they are seen in great numbers every winter; and, notwithstanding they at times invade peculiar districts in vast numbers, so remarkable was their appearance in former times considered, that they have alarmed whole regions, and been looked upon as the precursors of war, pestilence, and other public calamities. "In 1552, Gesner informs us, they appeared along the Rhine, near Mentz in Germany, in such numbers as to obscure the sun. They have, however, of late years, in Italy and Germany, and in France especially, at all times, been extremely rare, being seen only in small companies or singly, appearing as if they had strayed from their way. In England, the Bohemian Waxwing has always been a rare visitant, coming only at long and uncertain intervals. In the winter of 1810 large flocks were dispersed through various parts of that kingdom, from which period we do not find it recorded by English writers till the month of February, 1822, when a few came under Mr. Selby's inspection, and several were again observed during the severe storm in the winter of 1823. Upon the Continent, its returns are subject to similar uncertainty. In M. Necker's very interesting memoir lately published on the Birds of Geneva, we read, that from the beginning of this century only two considerable flights have been observed in that canton, one in January, 1807, and the other in January, 1814, when they were very numerous, and spent the winter there, all departing in March. In 1807 they were dispersed over a great portion of western Europe, and were seen near Edinburgh in the first days of that year."

WEASELS. A genus of digitigrade Carnivora, belonging to the *Mustelidæ* family, many of which are described in this volume under their several well-known names, as MARTEN, ERMINE, &c. We shall therefore now only give the COMMON WEASEL (*Mustela vulgaris*), a species which inhabits many countries of Europe, and, in much greater abundance, North America. In Mr. Bell's excellent work on the British Quadrupeds, he makes the following accurate remarks on the resemblance that exists between the Weasel and the Stoat: "The Stoat is brown above, dirty white beneath; the tail always black at the tip, longer and more bushy than that of the Weasel, and the former animal is twice as large as its elegant little congener. The Weasel, on the other hand, is red above, pure white beneath; the tail red and uniform. Their habits also, though generally similar, are, in many of their details, considerably distinct, and we are fully borne out by observation in saying that the accusations against the Weasel, of the mischief which he is said to perpetrate in the farm-yard and the hen-roost, as well as amongst game of every description—on hares and rabbits, no less than on the feathered tribes—are principally due to the Stoat. It is not meant to be asserted that the Weasel will not, when driven by hunger, boldly attack the stock of the poultry-yard, or occasionally make free with a young rabbit or a sleeping partridge; but that its usual prey is of a much more ignoble character is proved by daily observation. Mice of every description, the field and the water-vole, rats, moles, and small birds, are their ordinary food; and from the report of unprejudiced observers, it would appear that this pretty animal ought rather to be fostered as a destroyer

COMMON WEASEL.—(MUSTELA VULGARIS.)

of vermin, than extirpated as a noxious depredator. Above all, it should not be molested in barns, ricks, or granaries, in which situations it is of great service in destroying the colonies of mice which infest them. Those only who have witnessed the multitudinous numbers in which these little pests are found, in wheat-ricks especially, and have seen the manner in which the interior is sometimes drilled, as it were, in every direction by their runs, can at all appreciate the amount of their depredations; and surely the occasional abduction of a chicken or a duckling, supposing it to be much more frequently chargeable against the Weasel than it really is, would be but a trifling set-off against the benefit produced by the destruction of those swarms of little thieves."

The Weasel's courage in defending itself when attacked by birds of prey, is universally admitted; nor is it deficient in fierce opposition to dogs and even men, when its nest is invaded by either. The nest is constructed of dry leaves and herbage, and is generally lodged in some snug locality, as a dry ditch, the hollow of a tree, &c. It produces four or five young at a birth, and generally has two or three litters in the year.

WEAVER. (*Ploceus.*) The Weavers, of which there are several species, belong to the *Fringillidæ*, have a conical beak, more or less stout at its base, and the upper mandible slightly bulging. These birds are found in both continents, and the greater number of those of the eastern hemisphere are remarkably skilful nest-builders, in which they interweave blades of grass, from which circumstance they derive their name. They generally build their nests independently of each other, as the Philippine Weaver-bird, whose spherical pensile nest is entered by a vertical canal, which communicates with a lateral opening of the cavity wherein the eggs are deposited; but some of them build a vast number of contiguous nests, which form a single mass divided into numerous compartments.

WEEVER. (*Trachinus.*) There are two Acanthopterygious fishes of this name, not uncommon in the British Seas: the GREAT and the LESSER WEEVER, but the last-mentioned is the one most frequently met with on different parts of our coast. It is seldom more than five or six inches long; the GREAT WEEVER OR STING BULL is, however, double that length. Their most dis-

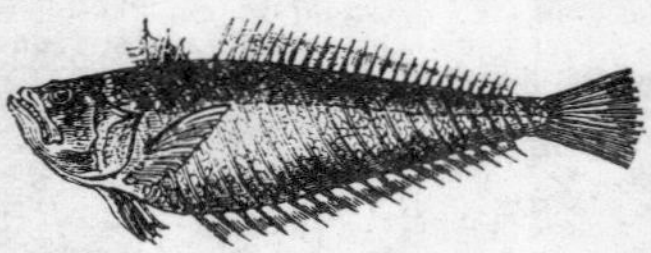

COMMON WEEVER.—(TRACHINUS DRACO.)

tinguishing characteristic is the power they have of inflicting wounds by means of their spinous fins; and fishermen almost invariably cut off the first dorsal fin, and both opercular spines, before they bring them on shore. The LESSER WEEVER OR STING-FISH is much quicker in its motions, and is even more difficult to handle with security, than the larger species. "In its habits," Mr. Yarrell tells us, "it is actve and subtle, burying itself in the loose soil at the bottom of the water, the head only being exposed; it thus waits for its prey — aquatic insects, or minute crustaceous animals, which the ascending position of its mouth enables it to seize with certainty. If trod upon or only touched while thus on the watch, it strikes with force either upwards or sideways; and Pennant states, that he had seen it direct its blows with as much judgment as a fighting-cock." Whether the supposed venomous quality of the sharp spines is justly founded, or not, is difficult to determine, but it appears that the wounds inflicted by these offensive weapons usually exhibit spmptoms of great inflammation and pain. The back is reddish-gray; lower part of the sides and belly silvery white, membrane of the first dorsal fin black; caudal fin tipped with black, the other fins pale brown.

WEEVIL. The name applied to Coleopterous insects of the family CURCULIONIDÆ. The CORN-WEEVIL (*Calandra granaria*) in its larva state is exceedingly destructive to grain; the female perfect insect lays a single egg in each grain, which when hatched turns into a grub which eats away the interior of the grain and perfectly destroys it: in granaries the perfect insect may be destroyed by sorting the grain into conical heaps, when the beetles cluster at the top and may be taken away in great quantities. (See BALANINUS.)

WHALES. (*Cetacea.*) An order of aquatic Mammalia, characterized by having fin-like anterior extremities, and the posterior extremities having their place supplied by a large horizonal caudal fin or tail, and the cervical bones so compressed as to leave the animal without any outward appearance of a neck. In this order are comprised the largest animated forms in existence. Their abode is in the sea or the great rivers, and they resemble the *Fishes* so closely in external appearance, that they are not only so regarded by the vulgar, but even many of the earlier zoologists considered them as belonging to that class. Nay, to the present day, when the capture of Whales is spoken of in the public papers, we read that one ship has returned from the *Whale Fishery* with two *fish*, another with three *fish*, &c. Mr. Bell, indeed, in his description of the *Cetacea*, says, "The outward form of the cetaceous animals, organized as they are for a permanent residence in the ocean, resembles so nearly that of the fishes, that the ancients were wont to arrange them together. Ray himself was not prepared to separate them from the fishes; and even the example of the great Linné, who with his wonted correctness and judgment placed them in their true position, was not sufficient to counterbalance the prejudices of Pennant, whose knowledge of the true principles of zoological science was too limited to enable him to look beneath the surface. Hence he follows Ray, and considers the Cetacea as forming a division of the class of fishes; and this notwithstanding he was well aware that they bring forth their young alive, and nourish them by means of mammary organs, similarly constructed to those of the whole class of mammalia. This fact, however, being established, it becomes a matter of great interest to ascertain what relation the other organs of the body bear to the corresponding ones in the other groups of this class, and by what modifications of structure they are rendered subservient to a mode of life so different from that of the more typical forms. These huge beings, then, have all the essential characters of mammiferous animals:

they have warm blood, a complete double circulation; they breathe the atmosphere by means of true lungs; and their reproduction and the nourishment of their offspring associate them with the true mammiferous type."

Mr. Bell then proceeds to say, that "the general form of the Cetacea is similar to that of fishes, in the horizontal elongation of the body, the rounded and smooth surface, the gradual attenuation of the extremities of the trunk, and the development of fins and especially of the tail as means of progression. The arrangement of the bones composing the anterior limb is one of the most important and curious parts of this subject. The whole of the fin consists of exactly the same elements as those which compose the arm and hand of man; but so concealed underneath the thick skin which envelopes it, that not a trace of these bones is to be seen externally. In this respect an intermediate structure is exhibited by the anterior extremities in the Seals."

"The posterior extremity is in the whole order, either absolutely wanting, or merely rudimentary. In the latter case, its only vestige consists of certain small bones, the imperfect representative of a pelvis, suspended, as it were, in the flesh and having no connexion with the spinal column. In this respect a striking difference is observed between these animals and the Seals: in the latter, the posterior extremities are carried backwards, and perform the office of a true caudal fin; but in the Whales this most important organ of progression consists of an extremely broad and powerful horizontal disc, varying in figure in the different genera, but in all constituting the principal instrument of locomotion. This extraordinary organ is not placed vertically as in fishes, but horizontally; and the admirable adaptation of such a peculiarity in its position to the requirements of the animal forms a fresh and beautiful illustration of the perfection of Creative Wisdom. The fishes, respiring only the air contained in the dense medium in which they live, do not require any access to the atmosphere; and their progression therefore is principally confined to the same place: but the Whales, breathing the atmosphere, are necessitated to come to the surface for each respiration, and hence require an oar of inconceivable power, the position of which applies its impulse in a vertical direction, so as to impel their ponderous bodies from the lowest depths of the ocean to the surface, every time the lungs require to be replenished with fresh air. The greatest rapidity of motion is produced by alternate strokes of the tail against the water, upwards and downwards; but their more ordinary progression is effected by an oblique lateral and downward impulse, first on one side and then on the other, as a boat is impelled forwards by a single oar in the act of sculling. The extent of the tail in some of the larger species is enormous; its superficies being no less than about a hundred square feet, and its breadth considerably upwards of twenty feet."

The respiration of these animals is another important part of their physiology. It appears that often, when the blow-holes are far out of the water, a jet of water of considerable size is thrown up with great force and to a considerable height;—a circumstance which can only be accounted for by supposing that the water taken into the mouth, and carried back into the pharynx, is then regurgitated by the blow-holes. "Let us suppose," says Cuvier, "the Cetacea to have taken into its mouth some water which it wishes to eject. It moves its tongue and jaws as if it were about to swallow it; but, closing the pharynx, it forces the water to mount into the nasal passages, where its progress is accelerated by annular muscular fibres, until it raises the valve (between the nasal passage and two pouches or reservoirs) and distends the membranous pouches above. The water once received into these pouches can be retained there until the animal wishes to spout. For that purpose it closes the valve to prevent the descent of the water again into the nasal passages below; and forcibly compresses the pouches by means of the fleshy expansions which cover them: thus compelled to escape by the narrow crescentic aperture or blow-hole, it is projected to a height corresponding with the force of the pressure."

With the unflinching firmness of a master mind, relying upon philosophical principles, and not yielding to popular prejudice by calling that a fish which he knew to be a mammiferous animal, Linnæus separated these cetaceans from the fishes, and associated them with the mammalia, on account of their warm bilocular heart, their lungs, their movable eyelids, their viviparous generation, the teats by means of which they suckle their young, and the other details of their anatomy which he, Cuvier, and all succeeding naturalists of note, allow to sufficiently distinguish them.

The Cetaceans are divided by Cuvier into two great tribes or families, one of which he terms *Herbivorous Cetacea;* the other, *Ordinary Cetacea.* And M. F. Cuvier thus arranges the order: Tribe 1. PHYTOPHAGA. These are characterized by having teeth of different kinds; the *molars* with flattened crowns, corresponding to the vegetable nature of their food. *Mammæ* two, pectoral. *Lips* provided with stiff bristles. *External nostrils* two, situated at the extremity or upper part of the rostrum, which is obtuse. Genera: *Manatus*, Cuv.; *Halicore*, Cuv.; *Rytina*, Ill. Tribe 2. ZOOPHAGA. Teeth of one kind or wanting, not adapted for mastication. *Mammæ* two, pudendal. *External nostrils* double or single, situated on the top of the head. And he further divides them into—1. *Those which have the head of moderate size:* family *Delphinidæ;* characterized by having teeth in both jaws, all of simple structure, and, generally, conical form: and—2. *Those with the head of immoderate size, equalling one-third the length of the body;* forming two families, 1. *Catodontidæ;* with numerous conical *teeth* in the lower jaw, and *blow-holes confluent.* 2. *Balænidæ;* which have *no teeth*, but their place supplied by the plates of baleen or whalebone attached to the upper jaw : *blow-holes distinct.*

The following simple and natural arrangement is by Dr. J. E. Gray. He describes the Whales (*Cete*) as the third order of *Mammalia*, stating that they are peculiar for their fish-shaped, nearly bald body; that their hinder limbs are united, forming an horizontal tail; and that they have simply conical rootless teeth or whalebone in the jaws. — The family of the Whales (*Balænidæ*), he observes, have a very large head, at least one-third the length of the body, as the tribe of Whales (*Balænina*), which have whalebones in the jaws, and the Catodons or *Physeterina*, which have simple conical teeth, as the Spermaceti Whale (*Catodonta*) and Cachalot (*Physeter*). — The family of Porpoises (*Delphinidæ*), which have a moderate or small head and an elongated or smooth body, as the Dolphins (*Delphinus*), which have conical jaws and teeth, the Porpoises (*Phocæna*), which have a shorter head and compressed teeth, the *Hyperoodons*, which only have a few teeth, — all these, Dr. Gray observes, have tapering front limbs, while the Susuk (*Platanista*) has triangular truncated limbs, an elongated beak with compressed teeth, and the bones of the skull bent over the forehead, so as to form an arched cavity. — In the other families, Dr. Gray remarks, the skin is more or less horny, and the lips always furnished with rigid whiskers; the teeth are flat-topped. The Manatees (*Manatidæ*) have eight grinders in each jaw, and the tail rounded at the end. The Dugongs (*Halicoridæ*) have only three or five grinders in each jaw, and the end of the tail truncated or two-lobed.

The COMMON, TRUE, or GREENLAND WHALE (*Balæna Mysticetus*), is principally met with in the northern arctic circle, but it is also to be found, in considerable numbers, in many other parts of the world. Although not the largest of the tribe, it is, on many accounts, the most valuable in a commercial point of view, being, like several other genera and species, pursued by man for the sake of oil and other valuable products. Its size is usually, in length, about sixty feet; its greatest circumference from thirty to forty feet. The body is bulky forwards, largest about the middle, and tapers rather suddenly towards the tail. The head

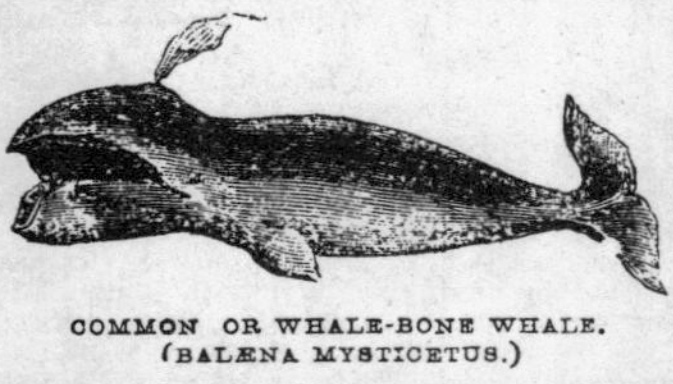

COMMON OR WHALE-BONE WHALE.
(BALÆNA MYSTICETUS.)

is very large, narrow above; very broad, flat, and rounded beneath; it occupies about one-third of the entire length, being about sixteen or twenty feet long, and ten or twelve broad: the lips are five or six feet high; and the upper jaw bends down at the extremity to close the cavity of the mouth. There are no teeth: the laminæ of whalebone which fill the cavity of the mouth are ranged in two series, consisting of about three hundred in each: the eyes are remarkably small; the external opening of the ears scarcely perceptible; the pectoral fins are of moderate size, and placed about two feet behind the angle of the lips. The tail is of great breadth, semi-lunate on its anterior margin deeply divided in the middle; the posterior outline sinuous, and the termination of the lobes pointed: the anterior and middle parts of the body nearly cylindrical; the posterior part rhomboid, the highest ridge or angle being upwards. General colour blackish gray; the anterior part of the lower jaw, and part of the throat and belly, white.

"The family of the *Balænidæ*," says Mr. Bell, "consisting, as there is now reason to believe, but of two known generic forms, are distinguished by the following generic characters: — Rivalling the *Physeters* in their huge general dimensions, the head is proportionally much smaller, and the whole form less clumsy. They have no teeth in either jaw; but the upper, which is extremely narrow, is furnished with numerous horny laminæ, — the whalebone of commerce, — descending perpendicularly from the palate, and varying in proportional breadth and length in the different species. . . . The

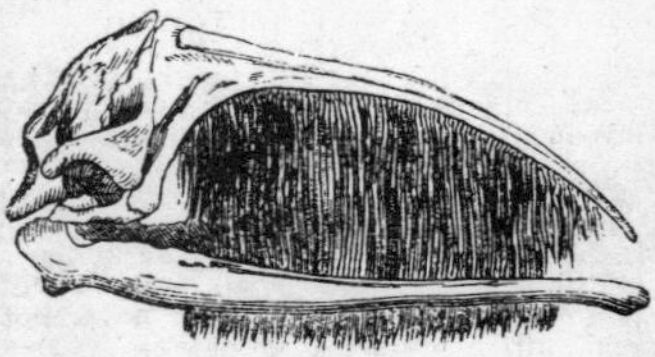

SKULL OF WHALE, WITH THE BALEEN.

whalebone, or baleen, as it has been called, consists of numerous parallel laminæ, each of which is formed of a central coarse fibrous layer lying between two which are compact and externally polished. The external part does not cover the internal to its extreme edge; the latter appears therefore beyond the former, and terminates in a loose fringed or fibrous extremity. The base of each plate of baleen has a conical cavity, covering a pulp which corresponds with it, and which is embedded within the substance of the gum or buccal membrane which covers the palate and upper jaw. The outer compact layers of each baleen plate, which have been described, are continuous with a white horny layer of the gum, which passes on to the surface of each plate; and the pulp appears therefore to be the secreting organ of the internal coarse structure only. The filaments of the fringe are very numerous, and fill up the cavity of the mouth sufficiently to form a most complete and efficient strainer; and as the swallow is extremely small, not being large enough to admit even the smaller fish, and the food of these Whales being consequently restricted to very small animals, such a structure is necessary in order to retain the whole of those which are taken into the mouth. The manner in which the food is taken, then, is as follows: — The whole of the seas of the Arctic regions, no less than those of the more southern climates, abound in innumerable shoals of molluscous, radiate, and crustaceous animals, which swarm in such hosts as often to colour the surface of the sea. When a Whale, therefore, is taking its food, the immense mouth being opened, a large number are as it were shovelled up by the great expanse of the lower jaw, and as the mouth is closed the water is regurgitated, and the numerous captives are retained by the apparatus just described. When the number of Whales which are found in the Northern Seas and the immense bulk of each individual are considered, imagination itself must fail to appreciate the countless myriads of small beings which are consumed for the nourishment of these stupendous bodies." [See CLIO.]

But although this baleen, or whalebone, which the Greenland Whale yields in such large quantities, is a product of such value as to render it an object of eager pursuit to those engaged in the Whale Fishery, the principal reward arising from the perilous employment is to be found in the large quantities of oil which are obtained from its thick cutaneous layer of fat, or blubber, as it is usually termed. A Whale sixty feet in length will frequently yield more than twenty tons of pure oil, and some of the pieces of baleen are twelve feet long. It is for these prizes that men willingly expose themselves to the rigours of an Arctic winter, the chance of falling victims to the united effects of cold and hunger, or shipwreck in its most horrid form, occasioned by the irresistible crush of icebergs. And should the hardy mariner escape from dangers such as these, the harpooner not unfrequently perishes from the upsetting of the boat owing to the violent plunges which the wounded animal makes in the water, or the whirlpool produced by his rapidly rushing down into the deep.

Who can read the following passage, which we extract from the able author before quoted, without a feeling of remorse and shame? "The female of this species, like most others of the Cetacea, is extremely attached to her young, and often rushes into the most imminent danger, and even upon certain death, to rescue or defend it. The whalers take advantage of this affectionate attachment, and strike with the harpoon the young Whale, quite sure that the mother will before long approach for the purpose of saving her offspring, but too frequently, in fact, to perish with it!" The Whale has usually but one young one, and brings forth in the early spring; at birth it is about ten or twelve feet long.

The NORTHERN RORQUAL, OR RAZOR-BACKED WHALE. (*Balænoptera physalis.*) This Whale, which is probably the longest of the animal creation, is so named from its having a prominent ridge, or spine, on its back. It is about a hundred feet long, and from thirty to thirty-five feet in circumference; but in proportion to its size, and the difficulty of killing it, its value in oil and whalebone is far less considerable than that of the preceding; and on that account it is not sought after by whalers, and not always attacked when met with. It is less quiet and tranquil in its general movements than the Common Whale, seldom lying motionless on the surface of the water whilst blowing, but making way at the rate of about five miles an hour. When struck, the velocity of its descent is such as very frequently to break the line, of which the Rev. Dr. Scoresby mentions several instances. The food of this enormous animal consists not only of the mollusca and smaller crustacea which constitute the aliment of the *Balæna*, but also of fish of considerable size. One of this species was some years since towed into the harbour of Ostend; and its magnificent skeleton, ninety-five feet in length, was exhibited at Charing-cross (near the King's Mews, the present site of Trafalgar-square). When taken, this specimen of the Whale weighed 249 tons; and 4000 gallons of oil were extracted from the blubber.

The BROAD-NOSED WHALE (*Balæna musculus*) in many respects much resembles the preceding, except in its never attaining so gigantic a size: its length being from fifty to eighty feet.

The smallest of the Whales is called the BEAKED WHALE (*Balæna rostrata*); its length being about twenty-five feet.

The SPERM WHALE, OR SPERMACETI WHALE (*Physeter macrocephalus*) now demands our attention. Mr. T. Beale, surgeon, to whom we are indebted for a circumstantial account of the natural history of this species, says: — "In length it comes next to the Balæna Physalis, and in bulk, probably, generally exceeds it, and in commercial value, perhaps, equals the Balæna Mysticetus; for although it does not possess the valuable whalebone of this animal, it furnishes us with the beautiful substance spercameti, and is rich in abundance of the finest oil: it is also the source of the perfume termed ambergris: its length is about eighty feet; circumference about thirty or thirty-five."

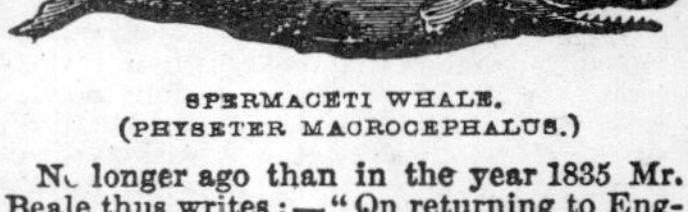

SPERMACETI WHALE.
(PHYSETER MACROCEPHALUS.)

No longer ago than in the year 1835 Mr. Beale thus writes: — "On returning to England, after completing an engagement which occupied upwards of two years, in the South Sea Whale Fishery, I was surprised to find, that, when the knowledge of every useful and interesting subject is so widely diffused, so little should be generally known of the natural history of almost the largest inhabitant of our planet, the great Sperm Whale;

in fact, till the appearance of Mr. Huggins' admirable print [published by that gentleman about six months before] few, with the exception of those immediately engaged in the fishery, had the most distant idea even of its external form. Of its manners and habits, people in general seem to know as little as if the capture of this valuable animal had never given employment to British capital, or encouragement to the daring courage of our hardy seamen. The very term whale fishery seems associated with the coast of Greenland, or ice-bound Spitzbergen, and the stern magnificence of Arctic scenery; few connect the pursuit of this 'sea beast' with the smiling latitudes of the South Pacific and the Coral Islands of the Torrid Zone; and fewer still have any more distinct conception of the object of this pursuit, than that it is a whale producing the substance called spermaceti, and the animal oil best adapted to the purpose of illumination.

"The head of the Sperm Whale presents, in front, a very thick blunt extremity, called the snout, or nose, and constitutes about one-third of the whole length of the animal; at its junction with the body, is a large protuberance on the back, called by whalers the 'bunch of the neck;' immediately behind this, or at what might be termed the shoulder, is the thickest part of the body, which from this point gradually tapers off to the tail, but it does not become much smaller for about another third of the whole length, when the 'small,' as it is called, or tail commences; and at this point also, on the back, is a large prominence, of a pyramidal form, called the 'hump,' from which a series of smaller processes run half way down the 'small,' or tail, constituting what is called the ridge. The body then contracts so much as to become finally not thicker than the body of a man, and terminates by becoming expanded on the sides into the 'flukes,' or tail, properly speaking. The two 'flukes' constitute a large triangular fin, resembling, in some respects, the tail of fishes, but differing in being placed horizontally; there is a slight notch, or depression, between the flukes posteriorly: they are about 6 or 8 feet in length, and from 12 to 14 in breadth. The chest and belly are narrower than the broadest part of the back, and taper off evenly and beautifully towards the tail, giving what by sailors is termed a clear run: the depth of the head and body is in all parts, except the tail, greater than the width.

"In the right side of the nose and head is a large almost triangular-shaped cavity, called by whalers the 'case,' which is lined with a beautiful glistening membrane, and covered by a thick layer of muscular fibres, and small tendons running in various directions, and finally by the common integuments. This cavity is for the purpose of secreting and containing an oily fluid, which after death concretes into a granulated substance of a yellowish colour, the spermaceti. The size of the case may be estimated, when it is stated that in a large whale it not unfrequently contains upwards of a ton, or more than ten large barrels of spermaceti. Beneath the case and nostril, and projecting beyond the lower jaw, is a thick mass of elastic substance, called the 'junk;' it is formed of a dense cellular tissue, strengthened by numerous strong tendinous fibres, and infiltrated with very fine sperm oil and spermaceti. The enormous mouth extends nearly the whole length of the head; both the jaws, but especially the lower, are in front contracted to a very narrow point; and, when the mouth is closed, the lower jaw is received within a sort of cartilaginous lip or projection of the upper one, but principally in front; for further back, at the sides and towards the angle of the mouth, both jaws are furnished with tolerably well developed lips. In the lower jaw are forty-two teeth of a formidable size and conical shape, but none in the upper, which instead presents depressions corresponding to, and for the reception of the crowns of those in the lower jaw. The tongue is small, and does not appear to possess the power of very extended motion. The throat is capacious enough to give passage to the body of a man, in this respect presenting a strong contrast with the contracted gullet of the Greenland Whale. The mouth is lined throughout with a pearly white membrane, which becomes continuous at the lips, and borders with the common integuments. The eyes are small in comparison with the size of the animal, and are furnished with eyelids, the lower of which is the more movable; they are placed immediately above the angle of the mouth, at the widest part of the head. At a short distance behind the eyes, are the external openings of the ears, of size sufficient to admit a small quill, and unprovided with any external auricular appendage. Behind, and not far from the posterior termination of the mouth, are placed the swimming paws, or fins, which are analogous in formation to the anterior extremities of other animals, or the arms of Man: they are not used as instruments of progression, but probably in giving a direction to that motion, in balancing the body, in sinking suddenly, and occasionally in supporting their young.

"A peculiarity of the Sperm Whale, which strikes at first sight every beholder, is the apparently disproportionate and unwieldy bulk of the head; but this peculiarity, instead of being, as might be supposed, an impediment to the freedom of the animal's motions in his native element, is, in fact, on the contrary, in some respects very conducive to his lightness and agility, if such a term can with propriety be applied to such an enormous creature; for a great part of this bulk of the head is made up of a large thin membranous case, containing, during life, a thin oil of much less specific gravity than water, below which again is the junk, which, although heavier than the spermaceti, is still lighter than the element in which the whale moves; consequently the head, taken as a whole, is lighter, specifically, than any other part of the body, and will always have a tendency to rise, at least, so far above the surface as to elevate the nostril, or 'blow hole,' sufficiently for all purposes of respiration; and more than this, a very slight effort on the part of the fish would only be necessary to raise the whole of the anterior flat surface of the nose out of the water; in case the animal should wish to increase its speed to the utmost, the narrow inferior surface which has been before stated, to bear some resemblance to the cutwater of a ship, and which would in fact answer the same purpose to the whale, would be the only part exposed to the pressure of the water in front, enabling him thus to pass with the greatest celerity and ease through the boundless tracks of his wide domain. It is in this shape of the head that the Sperm Whale differs in the most remarkable degree from the Greenland Whale, the shape of whose head more resembles that of the porpoise, and in it the nostril is situated much further back, rendering it seldom or never necessary for the nose to be elevated above the surface of the water, and when swimming even at the greatest speed, the Greenland Whale keeps nearly the whole of the head under it, but as his head tapers off evenly in front, this circumstance does not much impede his motion, the rate of which is, however, never equal to that of a Sperm Whale. It seems, indeed, in point of fact, that this purpose of rendering the head of light specific gravity, is the only use of this mass of oil and fat, although many have supposed, and not without some degree of probability, that the 'junk' especially may be serviceable in obviating the injurious effects of concussion, should the Whale happen to meet with any obstacle when in full career; this supposition, however, would appear hardly tenable when we consider the Greenland Whale, although living among the rock-like icebergs of the Arctic Seas, has no such convenient provision, and with senses probably in all, and certainly in one respect, less acute than those of the Sperm Whale, on which account it would seem requisite for him to possess this defence rather than the Sperm Whale, whose habitation is, for the most part, in the smiling latitude of the Southern Seas.

"The several humps and ridges on the back of the Sperm Whale constitute another difference in their external aspect; these prominences, however, are by no means peculiar only to the Sperm Whale, as they are possessed also by several other species of Whales, as the Razor-back and Broad-nosed Whales, and some others; and it would seem that the possession of these parts marks those Whales which are noted for their swiftness in flight, and their activity in endeavouring to defend themselves when attacked, which may be explained in this way, or it may be considered probable, that these prominences result from a greater development, in the situations where they are placed, of those processes of the vertebræ or bones composing the spine, called the spinal processes, and to which the muscles principally used in progression and other motions are attached, as well as those muscles and ligaments which support the long and bulky head; they consequently must indicate an increase in the size and strength of these muscles and ligaments, &c., and on this account constitute a very remarkable difference between those Whales possessed of them, and those not so furnished. This distinction is so great, that it induced Lacépede to divide the genus Balæna into those with a hump and those without, employing the name Balæna for the latter, and styling the others Balænoptera.

"The skin of the Sperm Whale, as of all other cetaceous animals, is without scales, smooth, but occasionally, in old whales, wrinkled, and frequently marked on the sides by linear impressions, appearing as if rubbed against some angular body. The colour of the skin, over the greater part of its extent, is very dark—most so on the upper part of the head, the back, and on the flukes, in which situations it is in fact sometimes black; on the sides it gradually assumes a lighter tint, till on the breast it becomes silvery gray. In different individuals there is, however, considerable variety of shade, and some are even piebald. Old 'Bulls,' as full-grown males are called by whalers, have generally a portion of gray on the nose, immediately above the fore part of the upper jaw, and they are then said to be gray-headed. In young whales the skin is about three-eighths of an inch thick, but in old ones it is not more than one-eighth. Immediately beneath the 'black skin' lies the blubber, or fat, which on the breast of a large whale acquires the thickness of 14 inches, and on most other parts of the body, it measures from 8 to 11. This covering is called, by South Sea whalers, the blanket; it is of a light yellow colour, and when melted down furnishes the Sperm oil. The blubber serves two excellent purposes to the whale, in rendering it buoyant, and in furnishing it with a warm protection from the coldness of the surrounding element, in this last respect answering well to the name bestowed upon it by the sailors."

The ingenious and intelligent author, from whose pamphlet we have made the preceding extracts, gives an account of the Rise and Progress of the Fishery, and of the modes of pursuing, killing, and "cutting in" the Sperm Whale. To the Pamphlet itself, as well as to Mr. Beale's more elaborate history of this important Whale, therefore, we beg to refer such of our readers as would wish for a more detailed narrative. We shall conclude with one short extract more from Mr. Beale's able "Observations."

"In calm weather great difficulty is sometimes experienced in approaching the Whale on account of the quickness of his sight and hearing. Under these circumstances the fishers have recourse to paddles instead of oars, and by this means can quietly get near enough to make use of the harpoon. When first struck, the Whale generally 'sounds,' or descends perpendicularly to an amazing depth, taking out perhaps the lines belonging to the four boats, 800 fathoms! afterwards, when weakened by loss of blood and fatigue, he becomes unable to 'sound,' but passes rapidly along the surface, towing after him perhaps three or four boats. If he does not turn, the people in the boats draw in the line by which they are attached to the Whale, and thus easily come up with him, even when going with great velocity; he is then easily lanced, and soon killed."

Dr. J. E. Gray has published an elaborate monograph of the Whales in one of the parts of the Zoology of H. M. SS. Erebus and Terror, and also a Catalogue of the Cetacea in the British Museum. The works of Scoresby and Beale give us large details of what is known about the history and capture of the two most important species in a commercial point of view. [See DOLPHIN, NARWHAL, PORPOISE, &c.]

WHEATEAR. (*Saxicola œnanthe.*) This Passerine bird is very generally diffused over the globe, and visits us early in the spring. It frequents new-tilled grounds, and is a close attendant on the plough, in search of insects and small worms, which are its principal food. In length the Wheatear is about five inches and a half. The Bill is black; eyes

WHEATEAR.—(SAXICOLA ŒNANTHE.)

hazel; over the eyes, cheek, and ears is a broad black streak, and above it a line of white; the top of the head, hinder part of the neck, and the back are bluish gray; the wing-coverts and quills are dusky, edged with rusty white; the rump is perfectly white, as is also part of the tail; the rest black; the under parts are pale buff, tinged with red on the breast: legs and feet black. The Wheatear breeds under shelter of a tuft or clod, in newly-ploughed lands, or under stones, and sometimes in old rabbit burrows: its nest, which is constructed with great care, is composed of dry grass or moss, mixed with wool, and is lined with feathers, and defended by a sort of covert fixed to the stone or clod under which it is formed: the female generally lays five or six light blue eggs, the larger end encompassed with a circle of a somewhat deeper hue. In some parts of England great numbers are taken in snares made of horse-hair, placed beneath a turf. They leave us about the latter end of August and September, and about that time are seen in great numbers by the sea shore, where, probably, they subsist some little time before they take their departure.

WHEAT-FLY. (*Cecidomyia tritici.*) The European Wheat-fly is a two-winged gnat, somewhat resembling a musquito in form, but is very small, being only about one tenth of an inch long. Its body is orange-coloured. Its two wings are transparent, and changeable in colour; they are narrow at the base, rounded at the tip, and are fringed with little hairs on the edges. Its long antennæ are composed of twelve little bead-like joints, each encircled with minute hairs. Towards the end of June, or when the wheat is in blossom, these flies appear in swarms in the wheat-fields during the evening, at which time they are very active. The females generally lay their eggs before nine o'clock at night, thrusting them, by means of a long retractile tube in the end of their bodies, within the chaffy scales of the flowers, in clusters of from two to fifteen, or more. By day they remain at rest on the stems and leaves of the plants, where they are shaded from the heat of the sun. They continue to appear and lay their eggs throughout a period of thirty-nine days. The eggs are oblong, transparent, and of a pale buff colour, and hatch in eight or ten days after they are laid. The young insects, produced from them, are little footless maggots, tapering towards the head, and blunt at the hinder extremity, with the rings of the body somewhat wrinkled and bulging at the sides. They are at first perfectly transparent and colourless, but soon take a deep yellow or orange colour. They do not travel from one floret to another, but move in a wriggling manner, and by sudden jerks of the body, when disturbed. As many as forty-seven have been counted in a single floret. It is supposed that they live at first upon the pollen, and thereby prevent the fertilization of the grain. They are soon seen, however, to crowd around the lower part of the germ, and there appear to subsist on the matter destined to have formed the grain. The latter, in consequence of their depredations, becomes shrivelled and abortive; and, in some seasons, a considerable part of the crop is thereby rendered worthless. The maggots, when fully grown, are nearly one eighth of an inch long. It is said that the maggots quit the ears of the wheat by the first of August, descend to the ground, and go into it to the depth of half an inch. It is probable that there they remain unchanged through the winter, and having finished their transformations, come out of the ground in the winged form in the spring, when the wheat is about to blossom. Dr. Asa Fitch has entered into the history, transformation, and habits of this insect with great detail in the sixth volume of the Transactions of the New York State Agricultural Society (1847). Jealous for the honour of his country, he has tried to prove that it is not a native of North America, and was unknown there anterior to the revolutionary war; but there is some doubt whether the Wheat-fly of North America is not a distinct species from the European one. Mr. Say has named it *Cecidomyia destructor*. [See HESSIAN-FLY.]

WHIDAH FINCH. (*Vidua.*) A genus of beautiful birds, inhabiting Western Africa, and particularly abundant in the kingdom of Whidah, — whence their name; but which has been corrupted, and is frequently written

WIDOW BIRD. The body of the Whidah-finch is generally about the size of a canary-bird, but the male is remakable for an astonishing development of plumage during the breeding season, after which its splendid tail drops off, and the sexes are then barely distinguishable. There are several species, one of which, VIDUA PARADISEA, will be sufficient to describe. The upper part of the plumage is of a faded or deep brownish-

WHIDAH FINCH.—(VIDUA PARADISEA.)

black; but this colour becomes of a paler hue on the wings and lateral tail-feathers. The head, chin, and throat are of this faded black, which extends downwards narrowing as it descends, to the middle of the breast. A broad rich orange rufous colour proceeds from the upper part of the back of the neck and unites with a tinge of the same colour on the sides of the neck and breast; this last hue passes into the pale buff of the body, but leaves the under tail-covers black, like the upper ones.

WHIMBREL. (*Numenius Phæopus.*) A species of grallatorial bird closely allied to the Curlew, but considerably smaller in size, being not above eighteen inches long. The plumage is of a grayish white, the feathers being streaked with brown; the scapulars are brown, with pale edges; the upper part of the head is divided longitudinally by a white line, bounded on each side by a black one; the bill is at least three inches long; the upper mandible is blackish brown, the lower one pale red. Dr. Fleming, in his British Animals, informs us that it forms its nest on exposed heaths in Zetland, and lays four or five eggs. After the breeding season it nearly disappears from the northern islands, but, during winter, frequents the English shores, associating in small flocks. [See CURLEW.]

WHINCHAT. (*Saxicola rubetra.*) A species of Passerine bird which is not unfrequent in the British islands, and may be commonly found on broom and furze, on the highest twigs of which it perches, and occasionally sings very sweetly. It builds its nest on the ground, forming it of dried sticks, and lining it with fine grass. The female lays six eggs of a uniform blue. It is rather larger than the Stonechat, to which it is closely allied. [See STONECHAT.]

WHIP-POOR-WILL. The American name of a species of Goatsucker (*Caprimulgus vociferus.*) Wilson tells us, in his interesting work on the Ornithology of America, that "on or about the 25th of April, if the season be not uncommonly cold, the whip-poor-will is first heard in Pennsylvania, in the evening, as the dusk of twilight commences, or in the morning as soon as dawn has broke. In the state of Kentucky I first heard this bird on the 14th of April, near the town of Danville. The notes of this solitary bird, from the ideas which are naturally associated with them, seem like the voice of an old friend, and are listened to by almost all with great interest. At first they issue from some retired part of the woods, the glen, or mountain; in a few evenings, perhaps, we hear them from the adjoining coppice, the garden fence, the road before the door, and even from the roof of the dwelling-house, long after the family have retired to rest. Some of the more ignorant and superstitious considered this near approach as foreboding no good to the family, nothing less than sickness, misfortune, or death, to some of its members; these visits, however, so often occur without any bad consequences, that this superstitious dread seems on the decline.

"He is now a regular acquaintance. Every morning and evening his shrill and rapid repetitions are heard from the adjoining woods, and when two or more are calling out at the same time, as is often the case in the pairing season, and at no great distance from each other, the noise, mingling with the echoes from the mountains, is really surprising. Strangers, in parts of the country where these birds are numerous, find it almost impossible for some time to sleep; while to those long acquainted with them, the sound often serves as a lullaby to assist their repose.

"These notes seem pretty plainly to articulate the words which have been generally applied to them, *whip-poor-will*, the first and last syllables being uttered with great emphasis, and the whole in about a second to each repetition; but when two or more males meet, their whip-poor-will altercations become much more rapid and incessant, as if each were straining to overpower or silence the other. When near, you often hear an introductory cluck between the notes. At these times, as well as at almost all others, they fly low, not more than a few feet from the surface, skimming about the house and before the door, alighting on the wood pile, or settling on the roof. Towards midnight they generally become silent, unless in clear moonlight, when they are heard with little intermission till morning. If there be a creek near, with high precipitous bushy banks, they are sure to be found in such situations. During the day they sit in the most retired, solitary, and deep shaded parts of the woods, generally on high ground, where they repose in silence. When disturbed, they rise within a few feet, sail low and slowly through the woods for thirty or forty yards, and generally settle on a low branch or on the ground. Their sight appears deficient during the day, as, like owls, they seem then to want that vivacity for which they are distinguished in the morning and evening twilight. They are rarely shot at or molested; and from being thus transiently seen in the obscurity of dusk, or in the deep umbrage of the woods, no wonder their particular markings of plumage should be so little known, or that they should be confounded with the night hawk, whom in general appearance they so much resemble. The female begins to lay about the second week in May, selecting for this purpose the most unfrequented part of the wood, often where some brush, old logs, heaps of leaves, &c. had been lying, and always on a dry situation. The eggs are deposited on the ground, or on the leaves, not the slightest appearance of a nest being visible. These are usually two in number, in shape much resembling those of the night hawk, but having the ground colour much darker, and more thickly marbled with dark olive.

"Early in June, as soon as the young appear, the notes of the male usually cease, or are heard but rarely. Towards the latter part of summer, a short time before these birds leave us, they are again occasionally heard; but their call is then not so loud, much less emphatical, and more interrupted than in spring. Early in September they move off towards the south.

"The Whip-poor-will is nine inches and a half long, and nineteen inches in extent; the bill is blackish, a full quarter of an inch long, much stronger than that of the night hawk, and bent a little at the point, the under mandible arched a little upwards, following the curvature of the upper; the nostrils are prominent and tubular, their openings directed forward; the mouth is extravagantly large, of a pale flesh-colour within, and beset along the sides with a number of long, thick, elastic bristles, the longest of which extends more than half an inch beyond the point of the bill, end in fine hair, and curve inwards; these seem to serve as feelers, and prevent the escape of winged insects: the eyes are very large, full, and bluish black; the plumage above is so variegated with black, pale cream, brown, and rust-colour, sprinkled and powdered in such minute streaks and spots, as to defy description."

WHITE ADMIRAL [BUTTERFLY]. A name given by collectors to Butterflies of the genus *Limenitis*.

WHITE ANT. [See TERMES.]

WHITEBAIT. (*Clupea alba.*) This small fish, which of late years has gained a sort of tavern celebrity as a dish suited to the epicurism of certain "diners-out," appears in the Thames about the beginning of April, and becomes abundant during the summer months till September. It grows to the length of six inches, and its sides are uniformly of a white colour, whence its name. For a long time it was denied to be a distinct species, and supposed to be the fry of other members of the Herring tribe, and there are legislative enactments (now rarely if ever enforced) against Whitebait fishing, on account of the necessity of using nets with small meshes. It is now, however, a well-established fact, that no fry of valuable fishes swim along with them; and those who are prone to indulge in the luxury of a Whitebait dinner on the banks of Father Thames need be under no apprehension of having gratified an epicurean taste at the expense of piscatorial impropriety.

WHITE BORDER [BUTTERFLY]. A name given by collectors to a species of Butterfly, *Vanessa Antiopa*.

WHITE [BUTTERFLIES]. A name applied by collectors to species of Butterflies, of the genera *Pieris*, *Pontia*, and *Leucophasia*.

WHITE SHARK. [See SHARK.]

WHITING. (*Merlangus vulgaris.*) A well-known fish belonging to the *Gadidæ* or cod tribe, and valuable on account of its delicacy and lightness as an article of food. It does not usually exceed a pound and a half in weight; abounds on all the British

WHITING.—(MERLANGUS VULGARIS.)

coasts, and comes in large shoals towards the shore in the months of January and February, for the purpose of depositing its spawn. It is easily distinguished from the haddock by the absence of the barbule on the chin; and from the pollack and coal-fish by having the under jaw shorter than the upper, and the tail even at the end.

WIDGEON. (*Anas [Mareca] Penelope.*) A species of migratory birds, bred in the morasses of the north, which they quit on the approach of winter, and as they advance towards the end of their southern journey, they spread themselves along the shores, and over the marshes and lakes in various parts of the Continent, as well as those of the British Isles. Here they remain during the winter, at the end of which the old birds pair: and the whole tribe, in full plumage, take their departure northward about the end of March. They commonly fly, in small flocks, during the night, and may be known from their congeners by their whistling note while they are on the wing. They are easily domesticated in places where there is plenty of water, and are much admired for their beauty and sprightliness. The bill is an inch and a half long, narrow, and serrated on the inner edges, the upper mandible being of a dark lead colour, tipped with black. The crown of the head, which is very high and narrow, is of a cream colour, with a small spot of the same under each eye: the rest of the head, the neck, and the breast, are bright rufous chestnut, obscurely freckled on the head with black spots, and darkest on the chin and throat, which are tinged with a vinous colour: a band of waved or indented narrow ash brown and white lines separates the breast and neck: the back and scapulars are marked with similar feathers, as are also the sides of the body under the wings: the belly, to the vent, is white: the great wing-coverts are brown, edged with white and tipped with black, which forms an upper border to the changeable green beauty-spot of the wings, which is also bordered on the under side by another stripe formed by the

WIDGEON.—(MARECA PENELOPE.)

deep velvet black tips of the secondary quills: the exterior webs of the adjoining quills are white, and those next the back are of a deep brown, edged with yellowish white: the vent and upper tail-coverts are black. The tail is of a brownish ash colour, edged with yellowish white; the two middle feathers being sharp-pointed, darker and longer than the rest. The legs and toes are of a dull lead colour, faintly tinged with green: the middle of the webs and nails black. The female is of a sober brown; the fore part of the neck and breast paler; scapulars dark brown, edges paler; wings and belly as in the male. The young of both sexes are gray, and continue so till February, when the plumage of the male begins gradually to assume its rich colourings; but after July the feathers become dark and gray, so that he is hardly to be distinguished from his mate.

WIDOW-BIRD. [See WHIDAH-FINCH.]

WILLOW WREN. (*Sylvia trochilus.*) For a most pleasing description of this "fairy bird," we turn to Mr. W. C. Hewitson's elegant 'Illustrations,' &c.: and, with his consent, we copy the greater part of it. "Much as I love all the dear birds of summer," says this gentleman, "there is not one the return of which I have yearly witnessed with so much pleasure as that of the Willow Wren; and however more highly the rich melody of some of the other warblers may be prized, there is a simplicity and a sweet cadence about the note of this species, which never fails to excite within me feelings of pleasure, which none but the lover of nature can either appreciate or understand, but which are to him amongst the chief enjoyments of his life. The Willow Wren is one of the most abundant of the warblers, and almost every wood and copse is enlivened by its beautiful form and graceful motions. It is, too, an inhabitant of more northern countries; and I shall not readily forget the delight I experienced on hearing its soft sweet note, whilst seated within the Arctic Circle, upon one of the bleak isles of Norway.

"The Willow Wren builds its nest upon the ground, sometimes in the midst of woods, when not thick, but more commonly near their margin, or in open places, or by the side of those grassy drives which are cut through them. It may be found in most of those grassy banks where brushwood occurs. In shape the nest resembles that of the common wren, being arched over, and entered from the side; it is, however, much more fragile, and not easily moved entire; it is composed of dry grass and moss, with dead leaves, warmly lined with feathers. . . . Mr. Neville Wood, in his British Song Birds, quotes a letter from Dr. Liverpool, describing the readiness with which the Willow Wren becomes sociable. To this I can add a most interesting instance. To ascertain beyond doubt the identity of the two varieties of the eggs figured, I had captured, on their nests, several of the birds. Amongst these was one which I had carried home and confined during the night in a large box, and such was its tameness, that when I took it out the following morning, and would have set it at liberty, it seemed to have no wish to leave my hand, and would hop about the table at which I was sitting, picking up flies which I caught for it. In the autumn, previous to their departure, the Willow Wrens frequent our gardens and orchards, where they may be seen busily picking insects from the pea-straw, and other vegetables, the young ones easily distinguished by their brighter yellow colouring; sometimes warbling a farewell song, but in a tone far different to their joyous carol in the spring, and so subdued that it is scarcely audible."

We are also indebted to the kindness of Mr. A. Hepburn, of Whittingham, for the following interesting notes on the WILLOW WREN. This plainly coloured but elegantly shaped species is a summer visitant in Britain, arriving in April and departing in September, and is abundantly distributed over the whole wooded parts of the country. The male announces his presence by a simple song, composed of a few notes, on a descending scale, but the tone is so silvery that it seems to tell of all the sweet influences of spring, the April shower and sunshine, the bursting bud and the opening flower; and what eye for the beautiful can fail to mark the elegance of his form as he nimbly glides amongst the young leaves, springs into the air after an insect, or flits from tree to tree? By and by, when mated, a snug arched nest is built on the ground, in a tuft of grass or amongst other rank vegetation, and six or seven little white eggs spotted with red are deposited: the young are fed on a variety of caterpillars and insects, by the destruction of which, great benefit is conferred on the labours of the husbandman and gardener. There are often two broods in the season.

WINDHOVER. [See KESTREL.]

WOLF. (*Canis Lupus.*) A ferocious quadruped belonging to the Digitigrade Carnivora, in habits and physical development closely related to the Dog. The Common

E ropean Wolf is yellowish or fulvous gray: hair harsh and strong, longest below the ears and on the neck, shoulders, and haunches: muzzle black; cheeks and parts above the eyes ochreo s or gray: upper lip and chin white: eyes oblique: tail straight or nearly so; and a blackish streak or band on the fore-legs about the carpus. Cuvier states that this Wolf, which more commonly infests the western countries of Europe, is found from Egypt to Lapland, and seems to have passed over into America. The French wolves are generally browner and somewhat

WOLF.—(CANIS LUPUS.)

smaller than those of Germany; while those of Russia are longer, and appear more bulky and formidable from the great quantity of long coarse hair on the cheeks, throat, and neck. In Sweden and Norway the Wolves are very similar to the Russian race, but are lighter in colour, and in winter totally white. The Alpine Wolves are brownish-gray, and smaller than the French; those of Italy and to the eastward towards Turkey, fulvous. There is no doubt whatever that Wolves formerly lurked in the uncleared woody districts of Britain; and there is sufficient historical evidence to prove that the Romans endeavoured to extirpate them; but although they considerably thinned these ferocious and cowardly beasts of prey, enough was left for their Saxon and Norman successors to do; and notwithstanding the laws of Edgar were specially directed to their extirpation, by liberating the Welsh from the tax of gold and silver on condition of an annual tribute of three hundred Wolves, and the punishment awarded to English criminals was commuted to a delivery of a certain number of Wolves' tongues, yet the vast wild tracts and extensive forests of ancient Britain were holds too strong even for his wise and vigorous measures.

There are several species of this animal, the chief of which is the BLACK WOLF, frequent in the Pyrenees and to the south of those mountains, where it is more numerous than the Common Wolf above described, and exceeds it in strength and stature. "The Spanish Wolves," says Col. Hamilton Smith, "congregated formerly in the passes of the Pyrenees in large troops, and even now the *lobo* will accompany strings of mules as soon as it becomes dusky. They are seen bounding from bush to bush by the side of travellers, and keeping parallel with them as they proceed, waiting an opportunity to select a victim; and often succeeding, unless the muleteers can reach some place of safety before dark."

WOLF-FISH. (*Anarrhicas lupus.*) An Acanthopterygious fish, belonging to the *Gobioideæ* family, generally of a large size, and furnished with jaws so well-armed as to render it a dangerous inhabitant of the deep. The whole body is smooth and slimy: the jaws, vomer, and palate-bones are armed with large bony tubercles which support on their summits little enamelled teeth, but the anterior teeth are conical and longer. There are six gill-rays, and neither cæca nor air-bladder. This fish inhabits the North Sea, being common enough as low as the French coast. They sometimes attain the length of six or seven feet, but their more common size is from eighteen inches to three feet, the latter of which will weigh about twenty pounds. It has a hoary colour, with a whitish belly, dark head with white specks, and two rows of large blackish lateral spots. It feeds upon crustacea and shell-fish, which it breaks in pieces with its teeth. Its motion is serpentine, like that of an eel and when

WOLF-FISH.—(ANARRHICAS LUPUS.)

it is seen reposing in the cleft of a rock its body is undulated. Fabricius says, that on the Greenland coast it associates itself with the common Lump-fish, migrating along with it; that is, retiring from the deep sea in autumn, and returning again in spring. Its great size and formidable teeth do not protect it from the assaults of the Lump-fish, for the latter, when alarmed for the safety of its offspring, pursues the Wolf-fish, and fastening upon its neck persecutes it to death.

WOLVERINE. [See GULO.]

WOMBAT. (*Phascolomys Wombat.*) This little bear-like Marsupial quadruped is known in New South Wales, and called by the natives *Womat*, *Wombat*, or *Wombach*, according to the different dialects, or perhaps to the different rendering of the wood-rangers who brought the information. It burrows like the badger, and on the con-

WOMBAT.—(PHASCOLOMYS WOMBAT.)

tinent does not quit its retreat till dark; but it feeds at all times on the uninhabited islands, and was commonly seen foraging amongst the sea refuse on the shore, though the coarse grass seemed to be its usual nourishment. Bass, the intrepid surgeon and voyager, whose name is commemorated in the Strait separating Van Dieman's Land and Australia, first found this curious animal. The natives eat its flesh, and, as is usual with them, roast it. It is sometimes brought to this country; the only specimen we have seen was very dull and unintelligent. It has a clumsy body, and a large flattish head; fore feet with five toes, armed with crooked nails, hind feet with four, and a little tubercle without a nail in place of the great toe. The hair is coarse, thinly set upon the belly, thicker on the back and head, and thickest upon the loins and rump; the colour of it a light sandy brown of varying shades, but darkest along the back. According to the natives, the Wombat of the mountains is never seen during the day, but lives retired in his burrow, which is long and winding; it feeds only during the night; but that of the islands is said to feed in all parts of the day. [For note on a fossil species see PHASCOLOMYS in SUPPLEMENT.]

WOODCOCK. (*Scolopax rusticola.*) A species of migratory birds of the Snipe tribe; measuring fourteen inches in length, twenty-six in breadth, and weighing about twelve ounces. The shape of the head is remarkable, being rather obtusely triangular than round, with the eyes placed near the top, and the ears very forward. The upper mandible, which measures about three inches, is furrowed nearly its whole length, and at the tip it projects beyond, and hangs over, the under one, ending in a kind of knob, which is susceptible of the finest feeling, and calculated, by that means, to find the small worms in the soft moist grounds, from whence it extracts them with its sharp-pointed tongue. The crown of the head is ash colour; the nape and back part of the neck are black, marbled with three bars of rusty red: a black line extends from the corners of the mouth to the eyes, the orbits of which are pale buff: the whole of the under parts are yellowish white, numerously barred with dark waved lines. The tail is black, indented across with reddish spots on the edges: the tip is ash above, and glossy white below. The legs are short, feathered to the knees, and are either of a sallow flesh-colour or a bluish hue. The upper parts of the plumage are so mottled, barred, streaked, and variegated, as would render a minute description both difficult and tedious. The black, white, red, gray, brown, rufous, and yellow, are so disposed in rows, crossed and broken at intervals by lines and marks of different shapes, that the whole, seen at a little distance, appear to be undistinguishably blended together and confused; the sportsman, however, by being accustomed to it, is enabled to discover it (among the withered stalks and leaves of ferns, sticks, moss, and grasses, by which it is generally sheltered in its moist and solitary retreats) by its full dark eye and glossy silver-white tipped tail. The Woodcock leaves the countries bordering upon the Baltic in the autumn and setting in of winter, on its route to this country. They neither come in flocks, nor remain near the shores to take their rest longer than a day. In temperate weather, they retire to the mossy moors and bleak mountainous parts of the country; but as soon as the frost sets in, and the snow begins to fall, they return to lower and warmer situations, where they meet with boggy grounds and springs, and little oozing mossy rills which are rarely frozen, and seek the shelter of close bushes of holly, furze, &c. in the woody glens by day, and remove to different haunts and feed only in the night. The female makes her nest on the ground, generally at the root or stump of a decayed tree; it is carelessly formed of dry fibres and leaves, upon which she lays four or five rusty gray eggs, blotched and marked with dusky spots. The flight of the Woodcock is rapid when pursued by the sportsman. Its flesh is highly esteemed.

WOOD-CRACKER. A name not uncommonly applied to the Nuthatch (*Sitta Europœa*). [See NUTHATCH.]

WOODLARK. [See LARK.]

WOOD-LEOPARD. The name applied to a beautiful species of Moth (*Zeuzera Æsculi*). [See ZEUZERA.]

WOODPECKERS. A great group of Zygodactylous Birds, well characterized by their striking and singular habits, to which their whole structure is singularly adapted. Mr. Bewick has described the family as having the bill large, strong, and fitted for its employment: the end of it is sharp and formed like a wedge, with which it pierces the bark of trees, and penetrates through the outside sound wood of the tree to the inside decayed part, where its food is lodged. Its neck is short and thick, and furnished with powerful muscles, which enable it to strike with such force as to be heard at a considerable distance: the noise thus occasioned is not by vibration round a hole, as some authors assert, but by a succession of strokes repeated with surprising rapidity. Its tongue is long and taper, and capable of great elongation; at the end of it there is in most of the species a hard horny substance, curving slightly downwards, which penetrates into the crevices of trees, and extracts the insects and their eggs which are lodged there: it is also lubricated by a glutinous secretion. The tail consists of ten stiff, sharp-pointed feathers, rough on the under sides, and bent inwards, by which it supports itself on the trunks of trees while in search of food: for this purpose its feet are short and thick, and its toes, which are placed two forward and two backward, are armed with strong hooked claws, by which it clings firmly, and creeps up and down in all directions.

Mr. Yarrell observes that another anatomical peculiarity remarkable in the skeleton of the Woodpecker, but admirably adapted to its habits, is the small size of the keel of the breast-bone. "Moderate powers of flight," he says, "sufficient to transport the bird from tree to tree, are all that it seems to require; large pectoral muscles with a deep keel to the breast-bone would to this bird be an inconvenience. The descending position of the bones of the tail indicate the mode by which the stiff points of the tail feathers ar brought into contact with the surface of the bark of the tree to form an accessory prop."

The GREEN WOODPECKER. (*Picus viridis.*) The bill of this bird, which is the second in size of the British kinds, is two inches long, triangular, and of a dark horn colour; the tongue towards the tip is furnished with numerous fibres, projecting transversely, of the size of minute hairs; the outer circle of the eye is white, surrounding another of red; top of the head bright crimson, which extends down the hinder part of the neck, ending in a point behind; the eye is surrounded by a black space; and from each corner of the bill runs a crimson streak pointing downwards; the back and coverts

GREEN WOODPECKER.- PICUS VIRIDIS.)

olive green; rump yellow; the quill feathers are dusky, barred on the outer web with black and white; under parts of the body white, slightly tinged with green: and the tail is marked with bars like the wings. This species obtains its food both upon trees and on the ground: its flight is short, undulating, and rather laborious. "When seen moving upon a tree," says Mr. Yarrell, "the bird is mostly ascending in a direction more or less oblique, and is believed to be incapable of descending unless this action is performed backwards. On flying to a tree to make a new search, the bird settles low down on the bole or body of the tree, but a few feet above the ground, and generally below the lowest large branch, as if to have all its work above it, and proceeds from thence upwards, alternately tapping to induce any hidden insect to change its place, pecking holes in a decayed branch, that it may be able to reach any insects that are lodged within, or producing its long extensible tongue to take up any insect on the surface; but the summit of the tree once obtained, the bird does not descend over the examined part, but flies off to another tree, or to another part of the same tree, to recommence its search lower down nearer the ground." The female differs from the male in not having the red mark from the corner of the mouth: she makes her nest in the hollow of a tree, fifteen or twenty feet from the ground. Both male and female labour by turns in boring through the sound part of the wood until they penetrate to that which is decayed and rotten, where she lays five or six eggs, of a greenish colour, marked with small black spots. The Green Woodpecker is frequently seen on the ground where there are ant-hills. It inserts its long tongue into the holes through which the ants issue, and draws out those insects in abundance. Sometimes, with its feet and bill, it makes a breach in the nest, and devours them at its ease, together with their eggs.

The IVORY-BILLED WOODPECKER. (*Picus principalis.*) This fine species of Woodpecker is a native of Brazil, Mexico, and the Southern States of North America. This bird is about twenty inches in length, and thirty in extent. "He is never found in cultivated tracts," says Nuttall; "the scene of his dominion is the lonely forest, amidst trees of the greatest magnitude. His reiterated trumpeting note, somewhat similar to the high tones of the clarionet, is heard soon after day, and until a late morning hour,

IVORY BILLED WOODPECKER.
(PICUS PRINCIPALIS.)

echoing loudly from the recesses of the dark cypress swamps, where he dwells in domestic security, without showing any impertinent or necessary desire to quit his native solitary abodes. Upon the giant trunk and moss-grown arms of this colossus of the forest, and, amidst inaccessible and almost ruinous piles of mouldering logs, the high rattling clarion and repeated strokes of this princely Woodpecker are often the only sounds which vibrate through and communicate an air of life to these dismal wilds. His stridulous interrupted call, and loud industrious blows, may often be heard for more than half a mile, and become audible at various distances, as the elevated mechanic raises or depresses his voice, or as he flags or exerts himself in his laborious employment. His retiring habits, loud notes, and singular occupation, amidst scenes so savage yet majestic, afford withal a peculiar scene of solemn grandeur, on which the mind dwells for a moment with sublime contemplation, convinced that there is no scene in nature devoid of harmonious consistence. Nor is the performance of this industrious hermit less remarkable than the peals of his sonorous voice, or the loud choppings of his powerful bill. He is soon surrounded with striking monuments of his industry: like a real carpenter (a nickname given him by the Spaniards), he is seen surrounded with cartloads of chips and broad flakes of bark, which rapidly accumulate round the roots of the tall pine and cypress where he has been a few hours employed; the work of half a dozen men, felling trees for a whole morning, would scarcely exceed the pile he has produced in quest of a single breakfast upon those insect larvæ which have already, perhaps, succeeded in deadening the tree preparatory to his repast. The plumage of this bird is black with a gloss of green: forepart of the head black, the rest of the crest crimson, with some white at the base: a stripe of white proceeding from a little below the eye, down each side of the neck, and along the back nearly to the rump. Tail black, tapering from the two exterior feathers, legs lead colour. Bill an inch broad at the base, channelled, and of the colour and consistence of ivory. Tongue white: iris vivid yellow. The female lays four or five white eggs, which are generally deposited in a hole in the trunk of a cypress tree.

The BLACK WOODPECKER, (*Picus* [*Dryocopus*] *martius*.) Of all the species of Woodpeckers known in Britain this is the largest and the scarcest. It is about sixteen inches in length; bill nearly two and a half, of a horn colour, and pale yellow on the sides; the top of the head, occiput and moustaches brilliant red; face black, upper parts a beautiful green; tail shaded with brown and striped transversely; rump tinged with yellowish; quills brown, and all the rest of the plumage dull black. The legs are lead gray, having the fore part covered with feathers half their length. The female differs from the male, the hinder part of her head only being red, and in some specimens the red is entirely wanting; the black parts of her plumage are also duller. They form their nest in the deep hollows of old trees, and lay two or three white eggs.

We have given descriptions of only three species of Woodpeckers, although the number is very considerable, and they are to be met with in each quarter of the globe. Among the Asiatic Woodpeckers may be named the species *Picus squamatus* and *Picus occipitalis*, described by Mr. Gould; among those of Africa, *Picus cafer*, the head, belly, and rump of which are yellow, and the upper coverts of the tail orange; and among those of America is the *Gold-winged Woodpecker* (*Colaptes auratus*), at once distinguished by the comparative slightness and length of its bill and its beautifully varied plumage, part of the quills being of a yellow colour, whence its name; another species is black and white speckled or mottled—"the finest," says Lawson, "I ever saw. The cock has a red crown. He is not very wild, but will let one come up to him; then shifts on the

GOLD-WINGED WOODPECKER.
(COLAPTES AURATUS.)

other side of the tree from your sight; and so dodges you for a long time together. This would seem to be the RED-HEADED WOODPECKER (*Picus erythrocephalus*), of

RED-HEADED WOODPECKER.
(PICUS ERYTHROCEPHALUS.)

which the subjoined cut gives a very good representation; M. Malherbe of Metz has made the extensive family of Woodpeckers a particular object of study, and has described many new species.

WOOD-SWALLOW. (*Artamus.*) Several species of this genus of birds are described by Mr. Gould, in that magnificent work, 'The Birds of Australia;' from his account of one of which (*Artamus sordidus*) we take the liberty of making the following extract: "This Wood-Swallow, besides being the commonest species of the genus, must I think be rendered a general favourite with the Australians, not only from its singular and pleasing actions, but by its often taking up its abode and incubating near the houses; particularly such as are surrounded by paddocks and open pasture-lands skirted by large trees. It was in such situations as these in Van Diemen's Land, at the commencement of spring, that I first had an opportunity of observing this species: it was then very numerous on all cleared estates on the north side of the Derwent, about eight or ten being seen on a single tree, and half as many crowding against one another on the same dead branch, but never in such numbers as to deserve the appellation of flocks: each bird appeared to act independently of the other; each, as the desire for food prompted it, sallying from the branch to capture a passing insect, or to soar round the tree and return again to the same spot; on alighting it repeatedly throws up and closes one wing at a time, and spreads the tail obliquely prior to settling. At other times a few were seen perched on the fence surrounding the paddock, on which they frequently descended, like starlings in search of coleoptera and other insects. It is not, however, in this state of comparative quiescence that this graceful bird is seen to the greatest advantage, neither is it that kind of existence for which its form is especially adapted; for although its structure is more equally suited for terrestrial, arboreal, and aerial habits than that of any other species I have examined, the form of its wing at once points out the air as its peculiar province: hence it is, that when engaged in pursuit of the insects which the serene and warm weather has enticed from their lurking-places among the foliage to sport in higher regions, this beautiful species in these aërial flights displays its greatest beauty, while soaring above, in a variety of easy positions, with white-tipped tail widely spread." Another very extraordinary and singular habit of this bird is its manner of suspending itself in perfect clusters, like a swarm of bees; a few birds suspending themselves on the under side of a dead branch, while others of the flock attach themselves one to the other, in such numbers, we are told, that they have been observed nearly of the size of a bushel measure.

The head, neck, and whole of the body fuliginous gray; wings and tail dark bluish black; the external edges of some of the primaries white, and the tail feathers tipped with white; bill blue with a black tip; feet lead colour. The nest, which is variously placed, sometimes in the naked fork, and at others in a thickly foliated bough near the ground, is about five inches in diameter, round, and rather shallow.

The other species described by Mr. Gould are the GRAY-BREASTED WOOD-SWALLOW, (*Artamus cinereus*), which is the largest of the genus: the LITTLE WOOD-SWALLOW (*Artamus minor*), which in colouring bears considerable resemblance to the one above described; the WHITE EYE-BROWED WOOD-SWALLOW (*Artamus superciliosus*), yielding to none in the variety and beauty of its plumage; the MASKED WOOD-SWALLOW, (*Artamus personatus*), a species that is more shy and retired than the others, never being seen but in the most secluded parts of the bush; and the WHITE-RUMPED WOOD-SWALLOW (*Artamus leucopygialis*), which, as it flies near the ground, "reminds one of the House Marten of our own country."

WOU-WOU. The native name of the Silvery Gibbon (*Hylobates leuciscus*), a pretty species of long armed Ape found in the Malay peninsula and other parts of the Asiatic continent.

WRASSE. (*Labrus.*) There are several species of this Acanthopterygious fish, viz. the BALLAN WRASSE (*Labrus tinca*), the GREEN-STREAKED WRASSE (*L. lineatus*), the COOK WRASSE OR BLUE-STRIPED WRASSE (*L. variegatus*), the COMBER WRASSE (*L. comber*), and the RAINBOW WRASSE (*Julis vulgaris*), all of which are more or less plentiful on the eastern, southern, and western coasts of England; it is hardly necessary, however, to describe more than one, and we take the first mentioned as an example. The BALLAN WRASSE, called also the ANCIENT WRASSE OR OLD WIFE (*Labrus tinca*). The Ballan Wrasse frequents deep gullies among rocks, where it shelters itself among the larger kinds of sea-weeds, and feeds upon crabs and other crustaceous animals. It takes a bait freely, and fishermen remark that when they first fish in a place, they take but few, and those of large size; but on trying the same spot a few days after, they catch a greater number, and those smaller;

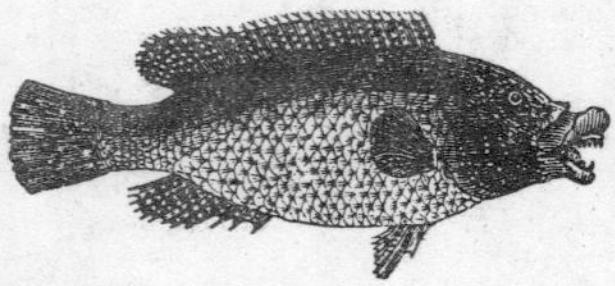

WRASSE. OLD WIFE.—(LABRUS TINCA.)

from whence they conclude that the large fish assume the dominion of a district, and keep the younger at a distance. The genus is distinguished by an elongated body, covered with large thin scales; a single dorsal fin, extending nearly the whole length of the back, part of the rays spinous, the others flexible; behind the point of each spinous ray a short membranous filament; lips large and fleshy; teeth conspicuous, conical, sharp; cheek and operculum covered with scales. The flesh is soft, and they are not in much estimation as food. A fine specimen, eighteen inches long, and weighing three pounds seven ounces, Mr. Yarrell observes, was taken in January 1831, in Swansea Bay, of which a notice and short description was furnished him by L. W. Dillwyn, Esq. The colour was red, becoming pale orange on the belly; the body ornamented with bluish green oval spots: the dorsal fin had spots along the base only. This fish spawns in April, and the young, scarcely more than an inch in length, are seen about the margin of the rocks in shallow water through the summer.

WREN. (*Troglodytes vulgaris.*) This active little Passerine bird, whose length is but three inches and a half, is very common in England, braving our severest winters, which it helps to cheer by its sprightly note. The bill is slender, and a little curved; upper mandible and tips of a brownish horn colour, the under one, and edges of both, dull yellow; a whitish line extends from the bill over the eyes, which are dark hazel; the upper parts of the plumage are clear brown, obscurely marked on the back and rump with narrow double wavy lines of pale and

WREN.—(TROGLODYTES VULGARIS.)

dark brown colours; the belly, sides, and thighs are marked with the same colours, but more distinctly; the throat is dingy white; cheeks and breast the same, faintly dappled with brown; the quills and tail are marked with alternate bars of a reddish brown and black; legs pale olive brown. During the winter season this brisk little warbler approaches near the dwellings of man, and takes shelter in the roofs of houses, barns, and in hay-stacks; it sings till late in the evening, and not unfrequently during a fall of snow. In the spring it betakes itself to the woods, where it builds on the ground, or in a low bush, and sometimes on the turf, beneath the trunk of a tree, or in a hole in a wall; its nest is constructed with much art, of an oval shape, with one small aperture in the side for an entrance; it is composed chiefly of moss, or other surrounding materials, so as not to be easily distinguished from them, and lined with feathers: the female lays from ten to sixteen or eighteen eggs, which are white, thinly sprinkled with small reddish spots, mostly at the larger end.

[For GOLDEN-CRESTED WREN, see REGULUS.]

The AMERICAN HOUSE WREN. (*Troglodytes domestica.*) We copy the following amusing account, *verbatim*, from 'Wilson's American Ornithology.'

"This well-known and familiar bird arrives in Pennsylvania about the middle of April; and, about the 8th or 10th of May, begins to build its nest, sometimes in the wooden cornishing under the eaves, or in a hollow cherry tree; but most commonly in small boxes, fixed on the top of a pole, in or near the garden, to which he is extremely partial, for the great number of caterpillars and other larvæ with which it constantly supplies him. If all these conveniences are wanting, he will even put up with an old hat, nailed on the weather boards, with a small hole for entrance; and, if even this be denied him, he will find some hole, corner, or crevice, about the house, barn, or stables, rather than abandon the dwellings of man. In the month of June, a mower hung up his coat, under a shed, near the barn; two or three days elapsed before he had occasion to put it on again; thrusting his arm up the sleeve, he found it completely filled with some rubbish, as he expressed it, and, on extracting the whole mass, found it to be the nest of a wren completely finished, and lined with a large quantity of feathers. In his retreat, he was followed by the little forlorn proprietors, who scolded him with great vehemence, for thus ruining the whole

AMERICAN HOUSE WREN.
(TROGLODYTES DOMESTICA.)

economy of their household affairs. The twigs with which the outward parts of the nest are constructed are short and crooked, that they may the better hook in with one another, and the hole or entrance is so much shut up, to prevent the intrusion of snakes or cats, that it appears almost impossible the body of the bird could be admitted; within this, is a layer of fine dried stalks of grass, and lastly feathers. The eggs are six or seven, and sometimes nine, of a red purplish flesh colour, innumerable fine grains of that tint being thickly sprinkled over the whole egg. They generally raise two broods in a season; the first about the beginning of June, the second in July.

"This little bird has a strong antipathy to cats; for, having frequent occasion to glean among the currant bushes, and other shrubbery in the garden, those lurking enemies of the feathered race often prove fatal to him. A box fixed up in the window of the room where I slept, was taken possession of by a pair of wrens. Already the nest was built, and two eggs laid, when one day, the window being open, as well as the room door, the female wren, venturing too far into the room to reconnoitre, was sprung upon by grimalkin, who had planted herself there for the purpose; and, before relief could be given, was destroyed. Curious to see how the survivor would demean himself, I watched him carefully for several days. At first he sung with great vivacity for an hour or so, but, becoming uneasy, went off for half an hour; on his return, he chanted again as before, went to the top of the house, stable, and weeping willow, that she might hear him; but seeing no appearance of her, he returned once more, visited the nest, ventured cautiously into the window, gazed about with suspicious looks, his voice sinking to a low melancholy note, as he stretched his little neck about in every direction. Returning to the box, he seemed for some minutes at a loss what to do, and soon after went off, as I thought, altogether, for I saw him no more that day. Towards the afternoon of the second day, he again made his appearance, accompanied with a new female, who seemed exceedingly timorous and shy, and who, after great hesitation, entered the box; at this moment the little widower or bridegroom seemed as if he would warble out his very life with ecstacy of joy. After remaining about half a minute in, they both flew off, but returned in a few minutes, and instantly began to carry out the eggs, feathers, and some of the sticks, supplying the place of the two latter with materials of the same sort; and ultimately succeeded in raising a brood of seven young, all of which escaped in safety.

"The immense number of insects which this sociable little bird removes from the garden and fruit trees, ought to endear him to every cultivator, even if he had nothing else to recommend him: but his notes, loud, sprightly, tremulous, and repeated every few seconds with great animation, are extremely agreeable. In the heat of summer, families in the country often dine in the piazza adjoining green canopies of mantling grape vines, gourds, &c., while overhead the trilling vivacity of the wren, mingled with the warbling mimicry of the cat-bird, and the distant softened sounds of numerous other songsters, form a soul-soothing and almost heavenly music, breathing peace, innocence, and rural repose. The European who judges of the song of this species by that of his own wren (*m. troglodytes*) will do injustice to the former, as in strength of tone, and execution, it is far superior, as well as the bird is in size, figure, and elegance of markings, to the European one. Its manners are also different; its sociability greater. It is no underground inhabitant; its nest is differently constructed, the number of its eggs fewer; it is also migratory; and has the tail and bill much longer. Its food is insects and caterpillars, and, while supplying the wants of its young, it destroys, on a moderate calculation, many hundreds a day, and greatly circumscribes the ravages of these vermin. It is a bold and insolent bird against those of the titmouse or woodpecker kind that venture to build within its jurisdiction; attacking them without hesitation, though twice its size, and generally forcing them to decamp. Even the bluebird, who claims an equal and sort of hereditary right to the box in the garden, when attacked by this little impertinent, soon relinquishes the contest, the mild placidness of its disposition not being a match for the fiery impetuosity of his little antagonist. With those of his own species who settle and build near him he has frequent squabbles; and when their respective females are sitting, each strains his whole powers of song to excel the other. When the young are hatched, the hurry and press of business leave no time for disputing, so true it is that idleness is the mother of mischief. These birds are not confined to the country; they are to be heard on the tops of houses in the most central parts of our cities, singing with great energy. Scarce a house or cottage in the country is without at least a pair of them, and sometimes two; but unless where there is a large garden, orchard, and numerous outhouses, it is not often the case that more than one pair reside near the same spot, owing to their party disputes and jealousies. It has been said, by a friend to this little bird, that "the esculent vegetables of a whole garden may, perhaps, be preserved from the depredations of different species of insects by ten or fifteen pair of these small birds;"* and probably they might, were the combination practicable; but such a congregation of wrens about one garden is a phenomenon not to be expected but from a total change in the very nature and disposition of the species.

"Though Europeans are not ignorant of the existence of this bird, they have considered it, as usual, merely as a slight variation from the original stock (*m. troglodytes*), their own wren: in which they are, as usual, mistaken; the length and bent form of the bill, its notes, migratory habits, long tail, and red eggs, are sufficient specific differences.

"The house wren inhabits the whole of the United States, in all of which it is migratory. It leaves Pennsylvania in September; I have sometimes, though rarely, seen it in the beginning of October. It is four inches and a half long, and five and three quarters in extent, the whole upper parts of a deep brown, transversely crossed with black, except the head and neck, which is plain; throat, breast, and cheeks, light clay colour; belly and vent, mottled with black, brown, and white; tail, long, cuneiform, crossed with black; legs and feet, light clay colour; bill, black, long, slightly curved, sharp pointed, and resembling that of the genus *certhia*, considerably; the whole plumage below the surface is bluish ash; that on the rump having large round spots of white, not perceivable unless separated with the hand. The female differs very little in plumage from the male."

WRYNECK. (*Yunx torquilla.*) This bird, though in many aspects nearly allied to the Woodpeckers, being similar to that tribe in the formation of its bill and feet, never associates with them, and constitutes a genus of itself. Its principal colours consist of different shades of brown, exquisitely arranged. The larger quill feathers are marked on the outer webs with alternate spots of dark brown and rust colour, which, when the wing is closed, give it the appearance of chequered work; the rest of the wing and the scapulars are nicely freckled, and shaded with brown spots of different sizes; the tail-feathers are irregularly barred with black, the intervening spaces being finely freckled, and powdered with dark brown spots. The bill is rather long, sharp pointed, and pale gray; the eyes light brown; but what chiefly distinguishes this bird is the structure of its tongue, which is of considerable length, of a cylindrical form, and capable of being pushed forward and drawn into its bill again. Legs short and slender; toes long, two before and

* Barton's Fragments, part i. p. 22.

two behind; the claws sharp, much hooked, and formed for climbing branches of trees, on which it can run with the utmost facility. The Wryneck is found in various parts of

WRYNECK.—(YUNX TORQUILLA.)

Europe, and generally precedes the Cuckoo a few days. Its food consists chiefly of ants and other insects, of which it finds great abundance lodged in the bark and crevices of trees.

XANTHO. A genus of Brachyurous Crustaceans, of which there are numerous species, extensively distributed. The carapace is very wide, but never regularly ovoid, and not very convex. They are arranged by Milne Edwards into those species whose carapace is granulous or tuberculous above—and those species whose carrapace is not covered either with granulations or tubercles. One species, *Xantho floridus*, about two inches in length, of a reddish brown colour, with black claws, is common on the English and French coasts.

XANTHORNUS. The generic name used by Brisson for certain American birds. [See ORIOLE, BALTIMORE.]

XENOPS. The name used by Illiger for a genus of Fissirostral birds of South America; one species of which (*Xenops genibarbis*) is thus described by Mr. Swainson: above reddish, beneath gray-brown; chin, eyebrows, and spots on the throat and breast whitish; beneath the ears a snowy spot; lesser quills blackish, the base fulvous, the tips and margins rufous. Mr. Swainson remarks that this extraordinary and not inelegant little creature has a bill totally different from that of any other bird. Its general habit, he states, evinces a close connection with the *Sittæ*, particularly those of New Holland; some of which have their bills (which are slender) slightly inclining upwards, thus forming a connection between *Xenops* and the straight-billed *Sittæ* of the Old World.

XIPHIAS. [See SWORD-FISH.]

XIPHODON. [See SUPPLEMENT.]

XYLOCOPA. A genus of Hymenoptera, from their habits termed *Carpenter Bees.* They have very thick covering of hairs upon the hind legs of the females, which are used by them as pollen-brushes. They form their nests in crevices of old walls or in sunny banks; their cells are composed of earth, and are very smooth in the inside, and the mouth of the nest is closed with the same material. Their wings are most commonly black, with a fine purple or violet gloss.

XYLOPHAGA. A genus of small Conchiferous Molluscs, very similar to those of *Teredo*, and which are found in light wood that the animals have penetrated to the depth of about an inch. The valves are equal, globose, inequilateral, and closed at the back; they have no calcareous tube, but two small accessory testaceous pieces placed near the hinge, and one small tooth in each valve.

XYLOPHAGI. A family of insects of the order Coleoptera Tetramera, distinguished from the Weevils by the absence of a proboscis. These insects generally live in wood, which is perforated and channelled in various directions by their larvæ. The different species commit their ravages on various kinds of trees, some feeding on pines and firs, others on olives, and some restricting themselves to fungi. [See SCOLYTIDÆ.]

XYLOPHILI. An extensive series of gigantic Coleopterous insects, the males of which are particularly distinguished by various singular protuberances, horns or tubercles, arising from the head and thorax. They reside for the most part in tropical regions, and some of them acquire an immense size. [See DYNASTIDÆ and RUTELIDÆ.]

XYPHOSURA. A sub-class of Crustacea, so called from the long tail-like spine, so characteristic a mark of the King-Crab. There is only one well-marked genus of this group, which will be found described under the head of LIMULUS.

YAK. (*Poephagus grunniens.*) A species of Ox found in Thibet, among the mountains; the bushy white tail is much prized in the East, where it is used to brush away flies, and also as an emblem of authority.

YARKE. The native name of different South American monkeys of the genus *Pithecia.*

YELLOWHAMMER. (*Emberiza citrinella.*) This Passerine bird, which is about seven inches in length, is found a resident in this country, and generally throughout Europe. The male is known by the head, cheeks, front of the neck, belly, and tail-coverts being of a bright yellow; on the breast and sides reddish spots, which on the sides leave a black streak in the centre. Feathers of the top of the back, blackish in the middle, and reddish-brown on the sides; those on the rump bright chesnut, terminated with grayish; tail-feathers blackish, the two lateral ones with a conical white spot on the inner barbs. Feet yellowish. The female is smaller than the male; and the yellow of the head, throat, and neck more thickly marked with the brown and olive spots with which those parts are sprinkled. Their food consists of grain, seeds, and insects. In summer the well-known notes of the male are almost incessantly heard from the roadside hedge. In winter the yellowhammer joins the flocks of greenfinches, chaffinches, &c., which congregate in the fields and farmyards. The nest, made on or near the ground,

YELLOWHAMMER.
(EMBERIZA CITRINELLA.)

is composed of moss, roots, and hair, well interwoven. The female lays four or five pale purplish white eggs, streaked and speckled with dark reddish-brown, and the male takes his turn with her in the business of incubation.

YPONOMEUTIDÆ. A family of Heterocerous Lepidoptera, comprising an extensive collection of minute Moths. The body is ordinarily slender and elongated; the head is small and occasionally clothed with long scales in front; the antennæ long, slender, and generally simple in both sexes; the wings are entire, and often long, and more or less convoluted; the legs are of moderate length and spurred; the anterior tibiæ having one, the intermediate two, and the posterior four spurs; the palpi are generally long and slender, and mostly recurved. Some of the species reside in the larva state on flowers, upon which they subsist; others are found within the surfaces of leaves, devouring only the parenchyma; some form extensive webs, and live in society; others are solitary. Some species are remarkably brilliant; their wings being ornamented with highly polished metallic scales, and some of them being extremely varied in the number of their tints.

"The typical insects of this family, forming Latreille's genus *Yponomeuta*, are amongst the largest in the family, having the fore wings long, and convoluted when at rest, and the posterior large, and with moderate ciliæ. They are generally of white or slate colours with black spots, whence their names of small Ermine Moths: the larvæ reside in large societies under a common web on various fruit-trees, and especially on white-thorn hedges, which are sometimes entirely defoliated by them. I have also seen the apple-trees, along the sides of the roads in France, equally deprived of their leaves by these insects, and festoons of their webs suspended from the trees, and clothing the surface of the ground beneath the trees. These larvæ are of a slate colour with black dots, and let themselves down to the ground when alarmed. They form their cocoons in company together; in the midst of their webs. The elegant species of *Œcophora* fly during the day, frequenting gardens and hedges. *Adela* also comprises day-flying species, known under the name of "Japan Moths," from their polished metallic wings, and sometimes called "Long-horns," from the great length of the antennæ. They frequent woods, and fly in troops, like gnats, over the bushes in the sunshine. But the most beautiful species in the family are those minute moths with metallic spotted wings, the majority of which in the larvæ state are leaf-miners.

ZEBRA. The name given to at least two species of South African mammalia, belonging to the family that contains the Horse and the Ass. They are beautifully banded, and have never yet been thoroughly domesticated. The Zebras are closely allied to the common ass, the gradations, as it were, being the Quagga (See QUAGGA), and Dzigetai (*Equus hemionus*). Two species of Zebra are known, both natives of the Southern parts of Africa, where they are frequently found in large herds: the one frequents mountainous districts, while the other only occurs in the plains, where it associates with some of the antelopes, and even with the Ostrich.

The COMMON ZEBRA (*Equus Zebra*, L.) is found in South Africa, both within and beyond the Cape Colony, but is confined to the mountainous regions. Dr. Burchell, not knowing that it was *the* Zebra of the older naturalists, has very characteristically named it *Equus montanus*. It may be at once known from the following species by the pure white ground-colour of its coat, and the numerous glossy jet-black bands with which it is striped, except on the belly; the legs also are striped from the top to the bottom; the ears are longer than in the following species, while the tail is tufted, like that of the ass, the tuft being of a black colour. There are other characters, but these may suffice to distinguish it. Major Harris as well as other travellers tell us that it seeks the wildest and most sequestered spots, so that it is exceedingly difficult of approach, not only from its watchful habits and very great agility of foot, but also from the inaccessible nature of its highland abode. The herds graze on the steep hill side, with a sentinel posted on some adjacent crag, ready to sound the alarm in case of any suspicious approach to their feeding quarters, and no sooner is the alarm given than away they scamper with pricked ears, and whisking their tails aloft, to places where few, if any, would venture to pursue. It is the *wilde Paard* (wild horse) of the Cape Colonists, and the *Daaw* of the Hottentot.

BURCHELL'S ZEBRA (*Equus Burchellii*, Gray). This beautiful species inhabits the plains of South Africa beyond the Gariep or Orange river, but is never, according to Major Harris, found to the southward of that stream. The ears and tail more resemble those of the horse than the preceding species, which approaches the ass in these particulars. The back, neck, and head are tinged with brown, harmoniously banded with black and deep brown transverse stripes; the belly and legs are pure white: there are obscure traces of black transverse markings on the arm. Major Harris, who had so many opportunities of seeing this fine species in a state of nature, remarks that, "beautifully clad by the hand of nature, possessing much of the graceful symmetry of the horse, with great bone and muscular power, united to easy and stylish action — thus combining comeliness of figure with solidity of form, this species, if subjugated and domesticated, would assuredly make the best pony in the world. Although it admits of being tamed to a certain extent with considerable facility, —a half-domesticated specimen with a jockey on its brindled back being occasionally exposed in Cape Town for sale, — it has hitherto contrived to evade the yoke of servitude.

BURCHELL'S ZEBRA.—(EQUUS BURCHELLII.)

. . . The voice of this free-born son of the desert has no analogy to the discordant braying of the ass, but consists of a shrill abrupt neigh, which may be likened to the barking of a dog, as heard by a passer by, from the interior of a house. The senses of sight, hearing, and smell are extremely delicate. The slightest noise or motion, no less than the appearance of any object that is unfamiliar, at once rivets their gaze, and causes them to stop and listen with the utmost attention; any taint in the air equally attracting their olfactory organs. Instinct having taught these beautiful animals that in union consists their strength, they combine in a compact body when menaced by an attack either from man or beast; and, if overtaken by the foe, they unite for mutual defence, with their heads together in a close circular band, presenting their heels to the enemy, and dealing out kicks in equal force and abundance. Beset on all sides, or partially crippled, they rear on their hinder legs, fly at the adversary with jaws distended, and use both teeth and heels with the greatest freedom." — *Harris*, '*Game and Wild Animals of South Africa*,' p. 19. It is called *Bonte Quagga* by the Cape Colonists, and *Peetsey* by the Bechuanas.

ZEBU. The name given to the humped varieties of oxen which are found in India and the Asiatic Islands, and extend along the eastern coast of Africa to the Cape of Good Hope. They are used as beasts of burden, and serve as articles of food, though in this respect its flesh is by no means equal to that of our domestic breeds. The hump, which is chiefly composed of fat, is regarded as a great delicacy. Zebus differ greatly in dimensions; some are of large size, while others, of which we have figured a female and young, are not much larger than a sheep:

ZEBU.

they vary in colour; the most common variety is of a light gray, passing into cream-colour. The Hindoos treat the larger breed with superstitious veneration. [See BRAHMIN BULL.]

ZEE-KOE. The name given by the Dutch colonists in South Africa to the Hippopotamus. [See HIPPOPOTAMUS.]

ZERDA. [See FENNEC.]

ZEUGLODON. [See SUPPLEMENT.]

ZEUS, ZEIDÆ. A genus and family of Acanthopterygious fishes, remarkable for their compressed form; to this group belongs the JOHN DORY and the OPAH DORY (*Zeus Opah*) which latter is a very superb species, and inhabits the seas of warm regions, being only an occasional visitant of the Mediterranean and Northern seas. In size it exceeds every other species, measuring between four and five feet in length; in colour it appears to vary, the ground being sometimes a brilliant silvery green, and sometimes a bright gold-colour; but in either case the body is variegated on the sides with pretty numerous oval white spots, while the fins and tail are bright scarlet. The skin is apparently destitute of scales, and perfectly smooth. Two or three instances have occurred of this very beautifully coloured species having been taken on the British coast; one, which weighed between seventy and eighty pounds, was thrown upon the sands at Blyth, near Newcastle, in 1769; the colours and beauty of which are stated to "beggar all description; the upper part being of a bright green, variegated with whitish spots, and enriched with a shining golden hue, like the splendour of a peacock's feather." Another specimen was caught at Brixham, in Torbay, in 1772, which "weighed a hundred and forty pounds, measuring four feet and a half in length, and two feet and a quarter in breath: its greatest thickness was only four inches, and the general colour was a vivid transparent scarlet varnish over burnished gold, bespangled with oval silver spots of various sizes." [See DORY.]

ZEUZERA. A genus of nocturnal Lepidoptera, two species of which are found in this country, one of these, however, is extremely rare; the more common species, the Wood-leopard, (*Zeuzera Æsculi*) is white and spotted with black, whence it has derived its English name; the antennæ in the male are beautifully bipectinated for half their length; the larva, which is yellow and spotted, feeds like that of the Goat-moth, in the interior of trees, and as well as it, forms a cocoon of chips of wood agglutinated together; it feeds on various trees, but seems particularly fond of the elm.

ZIBET. A species of carnivorous mammalia belonging to the genus *Viverra* and the family *Viverridæ*. It is found on the Asiatic coast, and in some of the larger islands of the Indian Archipelago. It has a short and thick neck, the breast being full and somewhat distended, and differs considerably in its markings from its African congener, the Civet. Dr. Horsfield, in his Zoological Researches, informs us that it is of a comparatively mild disposition, and is sometimes found in a state of partial do-

ZIBET.—(VIVERRA.)

mestication. The substance secreted by an opening near the tail resembles that of the Civet, and is, perhaps, equally prized. [See CIVET.]

ZIMB. A fly, supposed to be a species of Tabanus, described by Bruce, the Abyssinian traveller, but not previously referred to by any naturalist. From Bruce's account we learn that it is in size very little larger than a bee, of a thicker proportion, and has wings, which are broader than those of a bee, placed separate, like those of a fly: they are of a fine gauze, without colour or spot upon them. The head is large; the upper jaw or lip is sharp, and has at the end of it a strong pointed hair, of about a quarter of an inch long; the lower jaw has two of these pointed hairs, and the pencil of hairs when joined together makes a resistance to the finger nearly equal to that of a hog's bristle. Its legs are serrated on the inside, and the whole covered with brown hair or down. He has no sting, though he seems to me rather of the bee kind; but his motion is more rapid and sudden than that of the bee, and resembles that of the gadfly in England. There is something peculiar in the sound or buzzing. It is a jarring noise, together with a humming, which induces me to believe it proceeds, at least in part, from a vibration made with the three hairs at its snout. As soon as this plague appears, and this buzzing is heard, all the cattle forsake their food, and run wildly about the plain till they die, worn out with fatigue, fright, and hunger. No remedy remains but to leave the black earth, and hasten down to the sands of Atbara, and there they remain while the rains last, this cruel enemy never daring to pursue them farther. Though his size is a immense as is his strength, and his body covered with a thick skin defended with strong hair, yet even the camel is not unable to sustain the violent punctures the fly makes with his pointed proboscis. He must lose no time in removing to the sands of Atbara, for, when once attacked by this fly, his body, head, and legs break out in large bosses, which swell, break, and putrefy, to the certain destruction of the creature. Even the elephant and rhinoceros, which, by reason of their enormous bulk and the vast quantity of food and water which they require daily, cannot shift to desert and dry places as the season may require, are obliged to roll themselves in mud and mire, which when dry, coats them over like armour, and enables them to stand their ground against this winged assassin; yet I have seen some of these tubercles upon almost every elephant and rhinoceros that I have seen, and attribute them to this cause. All the inhabitants of the sea-coast of the Melinda, down to Cape Gardefui, to Saba, and the south coast of the Red Sea, are obliged to put themselves in motion and remove to the next sand in the beginning of the rainy season, to prevent all their stock of cattle being destroyed. This is not a partial emigration, the inhabitants of all the countries, from the mountains of Abyssinia to the confluence of the Nile and Cestaboras northwards, are once a year compelled to change their abode and seek protection in the sands of Beja; nor is there any alternative or means of avoiding this.

Providence from the beginning, it would seem, had fixed its habitation to one species of soil, being a black fat earth, extraordinarily fruitful; and, small and inconceivable as it was, it seems from the first to have given law to the settlement of the country. It prohibited absolutely those inhabitants of the fat earth called Mazaga, domiciled in caves and mountains, from enjoying the help or labour of any beasts of carriage. It deprived them of their flesh and milk for food, and gave rise to another nation whose manners were just the reverse of the first. These were the shepherds, leading a wandering life, and preserving their immense herds of cattle by conducting them into the sands beyond the limits of the black earth, and bringing them back again when the danger from this insect was over."

"We cannot read the history of the plagues which God brought upon Pharaoh by the hands of Moses," observes our author, "without stopping a moment to consider a singularity — a very principal one — which attended this plague of the fly. It was not till this time, and by means of this insect, that God said he would separate his people from the Egyptians. And it would seem that then a law was given to them that fixed the limits of their habitation. It is well known that the land of Goshen or Geshen, the possession of the Israelites, was a land of pasture, which was not tilled or sown, because it was not overflowed by the Nile. But the land overflowed by the Nile was the *black earth* of the valley of Egypt, and it was here that God confined the flies; for he says, it shall be a sign of this separation of the people, that not one fly should be seen in the sand or pasture-ground, the land of Goshen; and this kind of soil has ever since been the refuge of all cattle emigrating from the black earth to the lower part of Atbara."

To the foregoing graphic narrative by Bruce we shall only add, that, much as this, as well as other particulars on subjects equally extraordinary, were at one time ridiculed and regarded as unworthy of belief, strong corroborative testimony may be found in the works of modern naturalists, as well as of recent African travellers (Denham and Clapperton among others), whose veracity has never been called in question.

ZOANTHUS. A genus of Zoophytes established by Cuvier, and giving its name to a division of the great group of animals to which it belongs (Zoantharia); in this genus the body is elongated, conic and pedunculated, and springs from a base common to several individuals; as the name implies, the species of the genus resemble flowers, such as an expanded daisy.

ZOËA. The name given by Bosc to what he regarded as a distinct genus of decapod Crustacea, different species of which are found in the ocean; Mr. Thompson discovered that these curious-looking spined creatures were the larvæ of long and short-tailed Crustacea, immediately after their exclusion from the egg. Mr. Arthur Adams was much struck with their curious and fantastic shapes; one form, he observes, would serve as an excellent model for a grotesque monster in a pantomime; in fact they all more resemble phantasms than the ordinary organizations we are in the habit of contemplating. He doubted the accuracy of Mr. Thompson's opinion, that these whimsical-looking creatures are merely the larvæ of different kinds of crabs, particularly as they are found in the high seas, where few of the larger crustacea are ever discovered. However in many cases Mr. Thompson has observed the metamorphosis take place, especially on the Irish coast. We must refer to his memoirs in the third volume of the Entomological Magazine, as well as to his Memoirs on Crustacea.

ZONITIS. A genus of Coleopterous insects belonging to the family Cantharidæ, the species of which are found on flowers.

ZONURIDÆ. A name given by Dr. Gray to a family of Saurian reptiles.

ZOOARCES. [See Viviparous Blenny.]

ZOOPHYTES. A great division of the Animal Kingdom, containing beings which are always evidently more simple in organization than in the other divisions, and which have their parts more or less distinctly arranged round an axis, a dispositon which frequently gives them the shape of flowers, and hence the name, which means *living plants*, or *plant-like animals*. The name *Radiata*, or radiated animals, is also applied to this division. It contains the Star-fishes and Sea-eggs, as well as the Actiniæ, Corals, and Corallines. For the history of the two first of these, so far as they are found in the British Islands, we must refer our readers to the work of Professor Forbes, which is devoted to them, while Dr. Johnston's History of British Zoophytes will give ample and interesting information, as well as admirable figures of all the genera and species belonging to the last mentioned. There are none who have opportunities of visiting the sea-coast who should neglect to examine and study these animals. The Reverend David Landsborough, in his Excursions to Arran, has well alluded to one of these Coralline Zoophytes, which he had taken from a scallop-shell to which it was attached. When out of the water, the *Plumularia pinnata* looks like a dirty and worn white feather. He says, you would not think that that feather had life, but, place it in water, it immediately recovers from its state of collapse, and, though still a feather, has become one of great beauty and elegance. "But it is only the habitations that you see; the alarmed inhabitants have fled into their houses. But place the polypidom, as it is called, in a tumbler of sea-water, and, when the alarm is over, the inhabitants will again appear. The polypes are hydra-form, and spread forth many tentacula in search of food, which they greedily grasp. The feather is formed of calcareous matter, mixed with gelatine, to give it flexibility, so that it may the better stand the buffeting of the waves. Observe the stem or quill of the feather, and you will see that it is full of red matter. That is the medullary pulp. Every plumule of the feather is a street. Even with the naked eye you may observe on each plumule about a dozen notches or denticles. Each of these is the house or *cell*, as it is called, of a polype: so that, in a good specimen, we see a kind of marine village, which, under the teaching of God, has been beautifully constructed by the thousand inhabitants which it contains." Many of the more transparent Zoophytes are highly luminous, and, in some cases, as Mr. Landsborough mentions in the Edinburgh New Philosophical Journal, vol. xxxii. p. 170., each polype seems as if it had a will of its own, for when agitated, after being taken from the water, "they lighted and extinguished their little lamps, not simultaneously, but with rapid irregularity, so that this running fire had a very lively appearance." Mr. Darwin, in the admirable journal to which we have referred so often, speaks of a Zoophyte closely allied to Clytia, of which he put a large tuft in a basin of salt-water. "When it was dark," he adds, "I found that as often as I rubbed any part of a branch, the whole became strongly phosphorescent with a green light; I do not think I ever saw any object more beautifully so. But the remarkable circumstance was, that the flashes of light always proceeded up the branches, from the base towards the extremity." This luminosity would seem to be chiefly produced by irritation, for living specimens have been kept for days in sea-water, and observed at all hours, and no appearance of light was perceptible.

ZOOTOCA. A genus of small Saurian reptiles, in which is placed our pretty little olive-coloured Lizard, *Zootoca vivipara*. [See LIZARD.]

ZORILLA. A genus of carnivorous quadrupeds, closely allied to the weasels, of which a species (*Zorilla striata*) is found in South Africa.

ZOSTEROPS. A genus of Birds closely allied to the Warblers, and seemingly intermediate between them and the Titmice. A marked peculiarity of the species belonging to the genus is that their eyelids are surrounded by a narrow ring of snow-white feathers. The birds are all small, and

WHITE EYE. (ZOSTEROPS DORSALIS.)

generally of a yellowish green or brown colour. They are found principally in Africa, Asia, and Australia. Our figure, copied from Mr. Gould s truly elegant work, represents the ZOSTEROPS DORSALIS or White-eye of the colonists of New South Wales ; Mr. Gould informs us that in South Australia, New South Wales, and Van Diemen's Land this is the bird which is seen more frequently than any other species. In the forests and thickets it abounds, and is far from a welcome visitor in gardens, where it does great damage to buds and fruits of every kind, though it is upon insects that it principally feeds ; in its disposition it is very familiar, often building its nest and rearing its young in shrubs and rose-trees bordering on the garden walks. This nest, which is also figured in the cut, is a very beautiful structure, being of a round deep cup-shaped form and composed of fine grasses, moss, and wool, and most carefully lined with fibrous roots and grasses ; the eggs are of a beautiful pale blue colour. The song of this bird is very pretty and lively, and there is no perceptible difference in the plumage of the sexes.

Another species, ZOSTEROPS CHLORONOTUS, also described by Mr. Gould, was found by Mr. Gilbert in Western Australia ; it is particularly fond of figs and grapes, and is often to be seen in gardens where these fruits are grown, in flocks as numerous as sparrows in this country. It takes flies on the wing like the true fly-catchers.

ZYGÆNA. A genus of Chondropterygious fishes belonging to the Shark family, and at once distinguished from all its members by the horizontally flattened head, truncated in front, its sides extending transversely like the head of a hammer, whence the common name of the species Hammer-headed Sharks. Mr. Arthur Adams when on the east coast of Borneo mentions a circumstance which shows the extreme voracity of a species of Zygæna. One of these fish sprang from the water, seized a bullock's hide which was drying at the bows of the ship, (H.M.S. Samarang) and succeeded in tearing a portion of it off. He also mentions that when one hundred miles from Batan, a shark was caught with a partially digested pig in his stomach, which had been thrown overboard at the anchorage of San Domingo in that island. [See SHARK.]

The name ZYGÆNA is also applied by some naturalists to the pretty black and red sphingidous insects called Burnet-moths ; the word Anthrocera however is now generally substituted for it. [See ANTHROCERIDÆ.]

ZYGODACTYLI. The name given by some ornithologists to that order of birds in which two of the toes are directed forwards and two backwards, the term Scansores howhowever is more generally used ; it contains the Parrots, Woodpeckers, Cuckoos, &c. [See SCANSORES.]

SUPPLEMENT.

ACTINOTROCHA. A generic name given by J. Müller to a marine animalcule, which is in all probability the larva of an Echinoderm allied to the common Sea-urchin. The creature is about the 1-40th of an inch in length, highly transparent, furnished with numerous tentacula, and clothed with a series of active vibratile cilia. Individuals were first discovered on our own shores by Dr. Cobbold, F.L.S., by whom also they were carefully figured and described in the sixth volume of the Transactions of the London Microscopical Society, to which we must refer our readers who desire further information.

ALTERNATION OF GENERATION. As the term *Parthenogenesis* is now generally employed in a more restricted sense than originally intended by its distinguished framer [see PARTHENOGENESIS, below], it is here proposed to explain in detail the phenomena of non-sexual reproduction generally. The above title does not correctly embrace all the phenomena of reproduction without the direct influence of the male, but until a more comprehensive general term be employed, it is certainly most convenient to describe these changes under the present head.

It is in the lower animals only that we find the ordinary sexual reproductive process superseded by the non-sexual production of individuals. Even in these the phenomena are comparatively rare. Nevertheless they are by no means accidental, but, as Von Siebold remarks, have a definite position in the history of the development of organic beings, being especially manifested in the Cœlenterata, the cestode and trematode Entoza, and in certain families of Insecta.

Perhaps the true relation of the *direct* and *indirect* processes of generation will be better understood by presenting all the phenomena of development in a tabulated form, as has been done by Prof. Huxley, thus :—

Development.	Continuous		Growth. Metamorphosis. Gemmation (without fission).
	Discontinuous (Gemmation with fission).	Agamogenesis	Metagenesis. Parthenogenesis.
		Gamogenesis.	

By gamogenesis is understood "sexual reproduction;" by agamogenesis, the non-sexual process. When the producing individual (or protozooid) has no sexual organs, Prof. Owen's term *metagenesis* may be employed; but when there are sexual organs, and the buds resemble ova, then Prof. Huxley adopts the term Parthenogenesis in its restricted sense.

The essential nature of the phenomena of alternate generation has been most ably described by Prof. Allen Thomson, of Glasgow, who observes that it consists in this, namely, that in some animals "the body or individual which is developed immediately from the ovum is not, in general, itself the bearer of the sexual organs, but nevertheless maintains for a time an independent existence, or presents the structural and functional characters of a separate or distinct individual, these characters often differing remarkably from those of the sexual individuals from which the ovum derived its origin; and that subsequently this individual, or one or other of its successors, has formed in connection with it, either internally or externally, and without sexual organs, a new progeny, which may consist of one or of many individuals, which have each of them more of the structure and properties of independent animals, and which, however variable their organisation may be, present *this* in common, that they are sexually complete and renew the true generative act by the formation of fecundated ova. In some animals it is the immediate offspring of the individual developed from the ovum which resumes the sexual functions; in other animals this offspring bears a second brood, or a third, and even more successive generations, before the return is made to sexual reproduction."

Such being a general statement of the facts representing so many links, as it were, in the complicated chain of phenomena of non-sexual reproduction, we now proceed to adduce a selection of illustrations by which this interesting law of alternate generation may be clearly understood.

Probably the most instructive and at the same time the most practically useful exemplification which can be brought forward, is that which we derive from a consideration of the development of a cestode parasite or Entozoon which, unfortunately, infests the human body. In this view, therefore, we particularly invite attention to the natural history of the common Tapeworm, or *Tœnia solium.*

The Solitary Tapeworm (so misnamed from the false notion that only one lives in

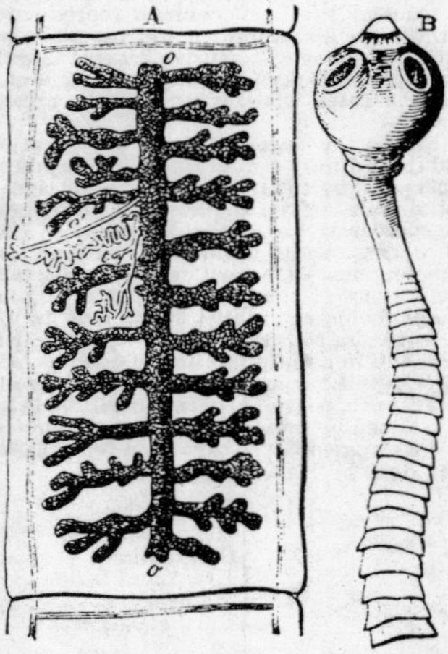

TAPEWORM: A, SINGLE JOINT OR PROGLOTTIS; B, HEAD OF THE COLONY, OR STROBILA

the same person at once), in the full-grown condition, is not, strictly speaking, a creature or animal, but rather a great many creatures attached to one another, so as to form a colony, or, more scientifically, the Strobila. [See STROBILA, below.] This colony is usually composed of several hundred joints, and each of these joints represents an individual worm (*proglottis*); those which are nearest to the lower end (or so-called tail of the Tapeworm) being sexually mature. They are indeed hermaphroditic, *i. e.* having both male and female reproductive organs. Those feebly developed joints which form the so-called neck of the worm are imperfect or immature individuals; whilst the little head is neither more nor less than a single individual (equivalent to a joint or proglottis) singularly modified, and furnished with an apparatus by which the strobila or colony is, as it were, securely anchored to the walls of the bowel of the unhappy person which the Tapeworm infests. The man, woman, or child, thus infested, or harbouring the parasite, is technically said to be the *host*, because he or she entertains its presence.

Looking, therefore, at the mature proglottis as the adult individual worm, we have now to consider the manner in which it reproduces itself.

After the proglottis (which is furnished with male and female reproductive organs) has undergone impregnation by contact with another proglottis, there results from this the formation of eggs within it, which eggs, whilst still within the body of the parent, develope into embryos, the latter still retaining the egg coverings. At this time the proglottis is about to undergo a passive migration, for having detached itself from the strobila, it is soon expelled from the bowel of the host, and therefore finds its way into some cesspool, or it may be into the open fields. The proglottides move about for a time, but the growth of the multitudes of embryos within causes the proglottis sooner or later to burst, and the embryos thus become dispersed: some are thus conveyed down drains and sewers, others are lodged by the roadsides in ditches and waste places, whilst multitudes are scattered far and near by winds or insects in every conceivable direction. Each embryo is furnished with a special boring apparatus, having at its anterior end three pairs of hooks: the entire group or family, therefore, of any single proglottis is called the "six-hooked brood." After a while, by accident, as it were, a pig coming in the way, either of these embryos or of the proglottides, swallows them along with other matters taken in as food. The embryos, immediately being transferred to the digestive canal, escape from the eggshells and bore their way through the living tissues of the animal, and having lodged themselves in the fatty parts of the flesh, they there rest to await their further transformations or destiny. The animal thus infested becomes measled, and thus it is that we are acquainted with measly pork. In this situation the embryos drop their hooks or boring apparatus, and become transformed into the *Cysticercus cellulosœ.* A portion of this measled meat being eaten by ourselves, either in a raw or imperfectly cooked condition, transfers the Cysticercus to our own alimentary canal, in which situation the Cysticercus attaches itself to the wall of the human intestine, and, having secured a good anchorage, begins to grow at the lower or caudal extremity, producing numerous joints or buds to form the strobila or Tapeworm colony.

Thus the cycle of life-development is completed, and we have a simple alternation of generation in which the immediate product of the proglottis (or sexually mature individual) is a six-hooked brood; by metamorphosis the latter becomes transformed into the Cysticercus, having a head with four suckers, and a double crown of hooks; and by gemmation the latter gives rise to a whole colony (strobila) of individuals, the greater part of which are destined to become sexually mature individuals, or proglottides. It will be observed, therefore, that the product of a single ovum is in the first instance a single non-sexual embryo (or protozooid); in the second phase it becomes a non-sexual Cysticercus (or deutozooid); in the third change it gives off, by budding, numerous gemmules (or tritozooids), most of them destined to be sexually mature individuals, and in this way to resemble their original parents.

A more complicated alternation of generation occurs in the Cœlenterata, especially in that division which we call the Hydrozoa. Thus the common zoophyte *Campanularia* may be taken as an example.

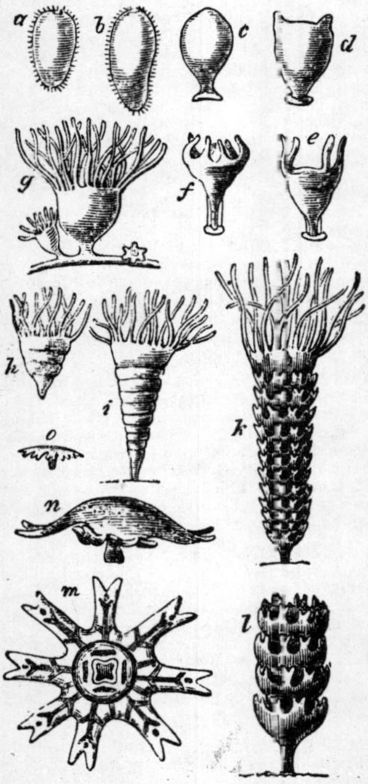

DEVELOPMENT OF CHRYSAORA

Certain polyp-like cells or gonoblastidia (*f*, in the right half of the diagram given at next page) of the polyp colony contain ova (*g*), which latter, after contact with spermatozoa developed from other cells, form ciliated embryos (*i* and *c*). These, having escaped from their gonoblastidium, swim about for a time, and, losing their cilia, ultimately settle down on some weed or rock, where undergoing a change of form, they sprout upwards to form a young polypidome (or cœnosarc). By a process of gemmation numerous polyp-heads (polypites) are produced, and also at intervals other modified polypites, which are contained in similar capsules or hydrothecæ (*k* and *f*). Some of these capsules give rise to medusoids (*l*) by a process of budding, and these latter are capable of producing ova by an ordinary sexual process (gamogenesis). Others form the gonoblastidia above mentioned, whilst the embryonic products of both these modified individuals form new polyp colonies in the way we have just described. If space admitted, we should be glad to enter into minuter details of this curiously complicated process, a modification of which is seen in the development of certain Medusæ, such as Chrysaora. From these sexually mature forms Prof. Thomson has described the process as follows:—"The fecundated ova which they produce are first developed into a ciliated moving animalcule (*a* and *b*, opposite), somewhat like a polygastrican. This creature, after undergoing a slight change of form, fixes itself by the narrowest end, and acquires tentacles like a polype at the other (*c*, *d*, and *e*) amounting for some time to eight (*f*). In this condition it appears to be capable of multiplying itself, or producing other similar attached polyps by gemmation from its side or base, or from a running stolon below it (*g*). The subsequent change of each of these polypoids is remarkable. It has been described by Sars and Dalyell as follows:—The body undergoing some elongation becomes partially divided by transverse grooves (*h*, *i*) into a range or column of imperfect Medusæ, attached still to each other by their adjacent surfaces, but presenting at their borders, in various degrees of advancement, the division into rays or lobes which belong to the Medusa, the upper or terminal one having developed upon it a set of radiated processes distinct from the tentacles of the polype, and much longer than those of the rest (*k*). The young Medusæ are successively separated from the stock by the deepening of the transverse clefts between them (*l*). They then move about as independent animals, and proceed in their further growth and development to sexual and other completeness (*m*, *n*). These bodies, therefore, are subject to two kinds of multiplication, which are very different: by simple gemmation a number or a colony of strobilæ may be produced, and by transverse fission and development a number of Medusæ may be thrown off from each strobila." In the Tapeworm colony, as we have seen, the individuals of the strobila attain their sexual maturity whilst still associated together; but here, in the Medusæ, the Medusæ are very small and immature at the time of their separation.

A still more startling modification of the non-sexual process of reproduction is seen in the Plant-lice or Aphides. In these tin, though highly organised insect members of the animal kingdom, whole generations, sometimes eleven in succession, are reproduced from sexual parents, all of them being the product of a single ovum. This is brought about by a process of internal budding in the parents, the extruded viviparous young being developed internally from egg-like bodies, which are not true ova and have never undergone impregnation. The distinguished author above quoted describes the course of the generative process as follows:—"Perfect male and female winged insects are observed only towards the autumn season. These fly about in great quantities; the impregnated females deposit their eggs, covered with a protecting case of mucus, in the axils and other recesses of plants where they remain during the winter. In spring there are developed from these ova a brood of larvæ, or imperfect female Aphides, which soon produce, by an act of viviparous generation, and without any concurrence of the male sex, a progeny of a similar kind; and this is repeated in successive generations for nine or ten times in the common species, or for ten or twelve weeks during the summer, at the end of which time the last brood brings forth male and perfect female individuals, both of which die after having provided, by the production of fecundated ova, for the continued generation during the next season."

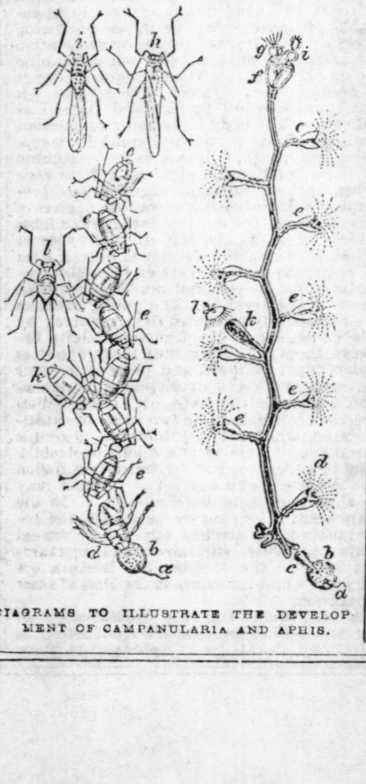

DIAGRAMS TO ILLUSTRATE THE DEVELOPMENT OF CAMPANULARIA AND APHIS.

In the accompanying diagram *a* represents the spermatazoon fertilising the ovum (*b*), which gives rise to *d*, the first embryo or viviparous larva. This larva gives origin to eight somewhat differently formed larvæ (*e*), whilst each one of these in their turn produces eight others, and so on, until the successive progenies are completed; the last giving birth to true sexual individuals, a pair of which are here drawn, *h* being the male and *i* the female. The multitudes of larval Plant-lice thus produced from a single ovum, though not defying calculation on paper, yet almost defies the imagination to conceive. This process, while it constitutes a true parthenogenesis, is not only remarkable as occurring in creatures so high in the scale of animal organisation, but also on account of the very close resemblance which obtains in the mode of the formation of the young of these viviparous larvæ as compared with the mode of formation of the eggs within the body of the true sexually mature female. In both creatures there are special organs which give rise to the germs; but in the non-fertilised broods, or viviparous larvæ, they may be compared to a multitude of buds capable of attaining the development and appearance of full-grown insects, which latter differ from the perfect insect chiefly in their usually not possessing any wings, and having imperfect reproductive organs.

Those who desire to make themselves acquainted with the intimate structural or histological changes which the germs undergo during their development within the body of the viviparous larvæ, should consult Prof. Huxley's elaborate memoir, "On the Reproduction of the Aphides," in the 22nd volume of the Linnæan Society's Transactions.

AMPHICŒLIA. A term applied by Prof. Owen to a group of extinct Crocodiles, whose backbone consists of vertebræ which are concave or hollowed out at both ends. They were furnished with long slender jaws like the existing "Gavial" of the Ganges; the earliest known fossils being found in the lias formation The *Teleosaurus* may be taken as a type [which see, at p. 666].

AMPHICYON. A genus of extinct mammals, whose fossil remains occur in the miocene deposits of Southern France. They appear to have belonged to the plantigrade carnivora, *i. e.* to those flesh-eating quadrupeds which walk on the soles of their feet, such as the Bears. Two species are known, the *A. giganteus*, and *A. diaphorus*. The latter is found at Eppelsheim, and has been described by Kaup as a species of *Gulo*.

AMPHITHERIUM. A highly interesting genus of Insectivorous mammalia only known in a fossil state. The animals of this genus seem to be the most ancient representatives of their order on the globe; the remains of them were by some naturalists believed to have belonged to extinct species of reptiles or other cold-blooded oviparous animals. Cuvier at first sight determined the jaw of one of these to have belonged to a marsupial animal, and was followed by another French naturalist, who called the creature in consequence *Thylacotherium*, from two Greek words signifying "pouched wild beast." Agassiz gave it another name, while a writer in the Athenæum, reviewing the disputation of the savans about its place in the system, jocularly called it *Botheratiotherium*, a title gravely objected to by a learned Frenchman in a paper read at the Academy of Sciences, and published in their "Comptes Rendus."

The name here given is that of Blainville, adopted by Professor Owen, and therefore, it is to be hoped, settled. The quadrupeds, to which these remains belonged, seem by their teeth to have been particularly adapted for living upon insects, like our shrews, and we may mention that remains of beetles occur in the oolitic slate of Stonesfield, where they were found. The teeth have many sharp points, well fitted to crush the hard bodies of the beetles on which they probably depended for their chief supply of food.

ANDRIAS SCHEUCHZERI. The scientific appellations of an extinct Salamander whose fossil remains were discovered in the tertiary strata at Œningen, so early as the year 1726. The bones were first considered to be human, and were accordingly described by Scheuchzer under the title of *Homo Diluvii testis*. The animal, however, does not appear to have been much larger than the Great Salamander now living in Japan. [See SIEBOLDIA, in this SUPPLEMENT.]

APATEON. The generic name given by Hermann von Meyer of Frankfort to those species of gigantic reptiles to which the better-known name of Archegosaurus has been applied by Prof. Goldfuss of Bonn. Their fossil remains occur in the coal measures. At first it was supposed that the animals thus represented belonged to the fishes, but the last-named palæontologist demonstrated that they were more properly to be regarded as primæval lizards, in which were strangely blended characters common alike to frogs and toads in their early condition, to crocodiles, and to lizards properly so called. From the investigations of Prof. Owen, and those of the authors above named, it would seem that we have in *Apateon* a remarkable transition between the reptilian and piscine classes; and so equally is the balance of characters manifested, that it has been seriously proposed to consider these two vertebrate divisions as only one great natural group. For further particulars consult Owen's Palæontology.

AQUARIUM or AQUA-VIVARIUM. A receptacle of water, either fresh or salt, where, by preserving the fluid in a pure state by artificial means, or by the action of plants growing in it, animals are kept alive. During the last two or three years the visitors to the Zoological Gardens have been much gratified with the tanks of water, both fresh and salt, in which are displayed the inhabitants of our streams, lakes, and seas, in a living and thriving state. From the work of one who has contributed much in promoting the introduction of marine aquaria, we derive the information which follows.

Mr. Gosse published in 1854, a most beautifully illustrated and excellent book, "The Aquarium; an Unveiling of the Wonders of the Deep Seas." He shows that the animals cannot be preserved in health for a single day except by the help of plants to produce oxygen, the principle in water which supports life. "By their means, however, nothing is easier than to have an Aquarium on almost as small a scale as we please; and any visitor to the sea-side, though there for ever so brief a stay, may enjoy with the least possible trouble the amenities of zoological study in a soup-plate, or even in a tumbler. It is easy to knock off with a hammer, or even to dislodge with a strong clasp-knife, a fragment of rock on which a minute sea-weed is growing, proportioning the surfaces of leaf to the volumes of water, —and you have an Aquarium. A wide-mouthed phial affords a capital opportunity for studying minute Zoophytes, Bryozoa, Nudibranch Mollusca, &c., as they may be examined through the clear glass sides with perfect ease, by the aid of a pocket lens. The influence of light should be allowed to operate on the sea-weed, to promote the elaboration of oxygen, but at the same time, if the weather be warm, care must be taken that the subjects be not killed by the sun's heat.

Mr. Warrington, an able chemist and naturalist, has done much by his experiments to promote the keeping of Aquaria. This gentleman found that the decay of the older leaves of the plants produced turbidity in the water, and that a green scum began to grow on the sides of the vessel and on the surface of the water. When the water was fresh he introduced a few common pond snails (*Lymnæa*); these greedily fed on the decaying vegetable matter and on the green scum, and quickly restored the water to a healthy state. When the water is salt, some of the little winkles (*Littorinæ*) answer the same purpose, and effectually act the part of scavengers by clearing away the decaying matter, as well as the young algæ which are produced in plenty, and which, but for the winkle *mowing* them down with its mouth, would soon cover the inside of the vessel, and prevent its contents from being seen. It is necessary that the Aquarium be placed where the sun's rays can fall most freely on the leaves, as the free access of light to the plants is indispensable for the production of oxygen. Mr. Gosse remarks, that "the difference between the profusion of oxygen-bubbles produced on a sunny day, and the paucity of those seen in a dark cloudy day, or in a northern aspect, is very marked." Pieces of rock, and corals or shells, should be placed in the Aquarium, to which the plants and animals may attach themselves, and into their crevices fish and crustacea may also take refuge. Artificial rocks may be made of Roman, Portland, or some other cement which hardens in the water, and into this, when soft, pieces of branching corals may be fixed, or fragments of stone with sea-weeds attached to them, so that the tufts may droop elegantly. When cement is used, it must be allowed to remain in water for at least a month, before animals are introduced, in order that the free lime may soak out, the lime being very injurious to animal life.

The bottom of the Aquarium must be covered with gravel and sand, not only from its pleasant look, but that the burrowing animals may have a field for their labours; the yellow earthy sand must be avoided, as it tinges the water with ochre.

The sea-water must be got from the clear open sea, out of the reach of rivers, and the cask to contain it should be either new, or at least nothing deleterious to animal life should have been held in it previously, such as spirits, wine, chemicals, or acids. A cask made of fir is preferable; the casks in which wine is kept are generally made of oak, the tannin and gallic acid of which, by their astringency, turn the animal integuments into leather; if nothing but an oak cask can be obtained, it should be well seasoned for two or three weeks before it is used, by filling it with water, which should be changed every day. If only a small quantity of water be required, stone-ware jars are best. Mr. Gosse finds that the best plants for a marine Aquarium are sea-weeds of the red and green orders. He particularly specifies the *Phyllophora rubens*, the various species of the elegant genus *Griffithsia*, and the no less beautiful *Bryopsis*. The *Ulva latissima*, a sea-weed easily procured on every shore, he finds particularly adapted for the purpose. Care should be taken in seeing that the pieces of rock to which the plants are attached should be as clean as possible: any adhering sponges should be carefully scraped off, as they are sure to die, and the sulphuretted hydrogen, which they then emit, corrupts the water and turns everything black.

If the plants assume an orange hue, it is a sure sign that they are not in a healthy state; they must be taken up, and the diseased parts cut away. Any of the animals which die should also be as soon as possible taken out.

It is necessary occasionally, too, that the water should be artificially aerated. The simplest way is by pouring the water from a little height, so that the particles of water may imbibe air and get rid of some of "the animal excretions which they hold in suspension, which thus become chemically changed and deprived of their putrescent qualities. This is what takes place in nature. By the perpetual dashing of the waves against the shore, and especially against the rugged rocks, an immense quantity of air becomes entangled in the form of minute bubbles, which by the various currents are diffused through the sea, and even carried to considerable depths, before they rise to the surface and become dissipated. Thus the violent agitation of the sea is a powerful agent in its purification. One of the simplest modes by which this object can be effected is the drip-glass. I have been accustomed to suspend over the Aquarium a perforated bell-glass of suitable size, into the orifice of which a bit of sponge may be pushed, or a cork drilled with small holes. The cord which suspends the drip-glass passes over a pulley at the top of the window, so as to be raised or lowered at pleasure. Every morning sufficient water from the tank is drawn or dipped off to fill the drip-glass, which is then hoisted to its full height. The contents run out in slender streams, or in a rapid succession of drops, which, passing through some four or five feet of air before they reach the tank, become effectually purified." (GOSSE's *Aquarium*, p. 272.)

As a great deal of water daily evaporates, and as only the pure water rises in vapour, it is necessary every now and then to add pure *fresh* water, to replace what is lost. A glass cover, however, greatly prevents loss from evaporation.

By means of the Aquarium, the singular habits of many water animals may be studied. Some naturalists had invented an apparatus, by which they could go down into the sea, and attempt to notice the ways of the curious creatures which there take up their abode. We never, however, heard of any practical results arising from this awkward and dangerous as well as difficult mode of observation. The visitors to the Zoological Gardens, whether scientific or not, cannot fail to be struck with the beauties of the numerous Actiniæ which there display their charms like so many opening flowers. The hermit crabs crawl about and vibrate their singular pedicelled eyes, and beautifully jointed antennæ and feelers. Shrimps and curious fish swim about, quite apparently at home, and feed among the gracefully drooping sea-weeds, which hang in tufts from the rocks. Even dead shells are covered with innumerable living beings, which open and shut their wonderful ciliary processes, as they take their microscopical food. Unless a person lived on the sea-shore, or had good correspondents there, it would be difficult, and expensive besides, to keep a marine Aquarium; but any one with a simple glass dish, and a few fresh-water fishes and plants, may keep, and with a little care keep long and in health, a most pleasing fresh-water Aquarium, the objects in which would afford a constant source of amusement. To the work from which we have quoted, we refer the reader who desires to form an Aquarium, or to see the valuable and interesting researches which may be made by one who possesses a well-stocked tank, even when far removed from the sounding sea.

ARCHEGOSAURUS. [See APATEON, above.]

ASTERACANTHION. A genus of Starfishes, several species of which are tolerably abundant on our own shores, but they are generally described under the better-known generic title of *Uraster*. The development of the young has been especially investigated by Sars and Müller; and still more recently by Prof. Wyville Thomson, who has published a lengthened paper on this subject in the first volume of the new series of the London Journal of Microscopical Science.

ASTEROLEPIS. A genus of Ganoid

fishes, found in a fossil state in the old red sandstone. The head of this was covered with bony plates, which were fretted with star-like tubercles, whence the name (*asterolepis*, i. e. star-scales); the body was covered with bony scales which were beautifully sculptured. Mr. Hugh Miller, who has taken the Asterolopis as the subject of an admirable volume, in which he controverts the views of creation propounded in the "Vestiges of Creation," remarks on these scales: "I have seen a richly inlaid coat of mail, which was once worn by the puissant Charles the Fifth; but its elaborate carvings, though they belonged to the age of Benvenuto Cellini, were rude and unfinished, compared with those which fretted the armour of the Asterolepis." The fish was of large size, equalling in that respect a large Porpoise, and sometimes attaining much greater dimensions. Mr. Miller believed that its vertebral column was cartilaginous, like that of the sturgeon, while its outer covering was composed, as in the Lepidosteus of America, and the Polypterus of the Nile, of strong plates and scales of solid bone. Its teeth partook of the characters of both the fish and reptile classes; the outer row being thickly set as in the fish, while the inner row was thinly set as in the reptiles.

Remains of this curious fish occur at Stromness, and at other places in the Orkney Islands.

AVICULARIA. A term applied to certain peculiar developments generally attached to the cells of the polypidoms, in that class of zoophyte-like molluscs termed *Polyzoa* or *Bryozoa*. These appendages are of two kinds, namely, the bird's-head processes or *Avicularia*, properly so called, and the vibrating filaments, or *Vibracula*. The former bear a striking resemblance to a bird's head, having parts severally called the body, hinge, and stalk, and being also furnished with a special moveable piece, which exhibits the characteristic shape as well as action of the lower jaw of the wide-gaping Goatsuckers (*Caprimulgidæ*). The Vibracula display a less complicated structure, consisting of a simple filament which is set in constant motion by contractile substance lodged within a tubular portion on which it rests. [See BRYOZOA, in this SUPPLEMENT.]

BALÆNICEPS REX. The scientific names given by Mr. John Gould, F.R.S., to a remarkable bird closely allied to the Storks and Herons. It is better known by the English titles of the Shoebird and Whale-headed Stork. In the year 1860 an indefatigable naturalist, Mr. Petherick, H.M. Vice-Consul at Khartoum, attempted to transmit to this country from the Nile a most valuable collection of rare animals, consisting of three African Elephants, a young male Hippopotamus, a Monkey (*Colobus*), two Rhinoceroses, and eleven birds; but out of all these only the Hippopotamus and two Whale-headed Storks reached England alive, and both the birds have since died. The Balæniceps, dead specimens of which were first brought to Europe by Mr. Mansfield Parkyns in 1850, is an inhabitant of the Upper or White Nile. The most extraordinary feature of this bird has reference, as the generic name implies, to the singular form and bulk of the bill, which is much longer than the head, and shaped like an inverted shoe or boat. By the Arabs the Balæniceps is known by the name of "Abou makoub," which means literally the "father of the shoe." At first Mr. Gould was inclined to regard this bird as a kind of Pelican, whilst other ornithologists thought it most closely allied to the Boatbill properly so called, a South American bird, which exhibits a similar form of beak. The latter, however, is a much smaller bird, and is moreover furnished with a curiously pectinated claw attached to the middle toe, besides exhibiting other characters which render it distinct. [See the art. BOATBILL, p. 79.] The living specimens formerly in the Zoological Gardens were hatched under hens at Khartoum, a town situate at the junction of the White and Blue Niles, the eggs having been obtained by Mr. Petherick's hunters "from nests of this bird found in the reedy marshes of the regions traversed by the upper branches of the White Nile." As already hinted, the affinities of the Balæniceps have been made the subject of considerable discussion; but the question seems now pretty well set at rest, especially since the reading of a paper before the Zoological Society by Mr. A. D. Bartlett, who finds a curious confirmation of the view above advocated in the circumstance that the so-called "powder-down patches" are less complicated in Balæniceps than in Cancroma, in which respect the Herons form an intermediate type. The Shoebirds have only one pair of these "patches," as also occurs in *Eurypyga* (an aberrant form of Heron); the Bitterns have two pairs, the true Herons three pairs, and the *Cancroma*, or Boatbill proper, four pairs. Seeing the curious results thus obtained, ornithologists would do well to follow up Mr. Bartlett's interesting observations, and, in all future determinations of affinity, pay due regard to the form, number, and arrangement of the hitherto little-regarded "powder-down patches."

BALÆNODON. A genus of extinct whales whose fossil remains are found in a very fragmentary condition, in the red crag deposits of Suffolk. Portions of ribs, entire teeth, and ear-bones (cetotolites), all bearing evidences of being much water-worn, are the principal indications of the former existence of these huge leviathans of the deep. They are constantly associated with those phosphatic nodules (cops, or coprolites as they are erroneously called) which occur so abundantly in these strata; and one not unfrequently finds them imbedded in a pseudo-coprilitic mass, forming a kind of nucleus for the concretion. The whale-ear-stones, to which Prof. Owen appropriately gave the name of "Cetotolites," form the harder portion of the entire ear-bone (petro-tympanic), the two elements having always been rudely torn from each other by the rubbing, rolling, and erosion of oceanic currents. They exhibit the characteristic conchoidal form, and have an extremely dense texture. There appears to have been a great variety of these tertiary whales, but we have no accurate data by which to determine their several skeletal peculiarities.

BRONTOZOUM GIGAS. The name given by Prof. Hitchcock to an extinct bird, the impressions of whose feet are extremely numerous in the red sandstone deposits of the valley of Connecticut. In the first instance, these appearances were made known to science under the more simple title of *Ornithicnites giganteus*. By whatever name they are called, there is every reason to believe that these struthious birds were four times the size of our living African Ostrich, and, in all probability, they closely resembled it in form, structure, and mode of living.

BRYOZOA. This term is synonymous with POLYZOA, and is employed to denote a class of molluscous animals, the species of which were until late years invariably associated with the Polyps. The Sea-mat (*Flustra foliacea*) so common on our coasts, may be taken as a type of the group, and of this zoophyte we have already given a description at p. 238. The *Bryozoa* are polyp-like animals, forming colonies, as it were, and depending for their mutual support on the presence of numerous horny, closely-connected cells, the aggregate of which constitutes the so-called polypidom. This support or framework is usually more or less branched, sometimes in a flat leafy manner, at other times in a tubular and finely dendritic form. The mouths of these minute creatures are surrounded by non-contractile ciliated tentacles, and the anal opening lies in their immediate neighbourhood. Some species occur in fresh waters, the majority being marine. The former have been carefully figured and described by Prof. Allman, F.R.S., in a publication of the Ray Society; the latter by Mr. Busk, F.R.S., in the British Museum Catalogue, and in successive numbers of the Microscopical Journal.

CAMPULA OBLONGA. The names given by Dr. Cobbold, F.L.S., to a peculiar entozoon found infesting the liver-ducts of the common Porpoise. This little fluke-worm is distinguished from those of the genera *Distoma* and *Fasciola*, by its possessing a zigzag-shaped alimentary canal. For further particulars see the above-mentioned author's various papers in the Linnean Society's Transactions.

CAPITOSAURUS. A generic name applied by Münster to a certain species of the labyrinthodont reptiles. It is the *L. arenaceus* of Owen. [See LABYRINTHODON, below.]

CARCHARODON. Under this generic title are included those large extinct Sharks, whose teeth very closely resemble those of the existing White Sharks. These fossil remains occur only in a very fragmentary condition, consisting for the most part of teeth and the bodies of vertebræ, all being very much water-worn. These organic débris are found in the red crag deposits of the Suffolk coast; and from the large size of some of the teeth, which occasionally measure as much as six inches in longitudinal diameter, it has been conjectured that some of the species of this genus attained a length of no less than sixty feet.

CATURIDÆ. A family of extinct fishes, whose fossil remains occur abundantly in the lias formation. They were ganoids possessing a homocercal or equally developed tail; but the spines supporting the caudal fin were unsymmetrically developed. The family comprises a large number of genera, but none of them are represented by species now living. The North American Lepidosteus is the nearest approach. One species of Caturus (*C. similis*) is found in the chalk.

CETIOSAURUS. Under this generic title Prof. Owen long ago indicated evidences of the existence of ancient crocodiles, whose length and bulk must have been something prodigious. Their fossil remains consisting of vertebræ occur in the Oolitic and Wealden formation, these bones being what is termed opisthocœlian, that is, having their centra or bodies convex or flat anteriorly and concave behind. For minute details the reader should consult Owen's Report on British Fossils in the British Association's volume for 1841.

CETOTOLITES. Fossil whale ear-bones. [See BALÆNODON, above.]

CHŒROPOTAMUS. A genus of Pachyderm quadrupeds, founded by Cuvier on fossil remains associated at Montmartre with those of the Palæotherium and Anoplotherium. In the Isle of Wight, near Ryde, the Rev. D. Fox discovered some remains of this interesting animal. The quadruped seems to have been closely allied to the Peccaries, at present abundant in the woods of South America. Professor Owen has shown that the structure of some of the teeth, and the form of part of the lower jaw, manifest a marked approximation to the feline type. He alludes to the occasional carnivorous propensities of our common hog, and shows that the extinct Chœropotamus was still better adapted by its dentition for predaceous habits, and "presents an interesting example of one of those links, completing the chain of affinities, which the revolutions of the earth's surface have interrupted, as it were, and for a time concealed from our view." He shows, too, that it is curious to notice that the living sub-genus of the family to which the Chœropotamus is most nearly allied, is an inhabitant only of South America, "where the Llama and the Tapir, the nearest living analogues of the Anoplotherian and Palæotherian associates of the Chœropotamus now exist, and which was formerly inhabited by a genus, *Macrauchenia*, which connects the Llama with the Palæothere."

CIMOLIORNIS. In the chalk near Maidstone, the Earl of Enniskillen found fragments of fossil bones, which Professor Owen believes to have belonged to the shaft of the humerus of a long-winged, web-footed bird, equalling the gigantic Albatross in size. To these remains the learned Professor has given the name of *Cimoliornis diomedeus*, or Long-winged Bird of the chalk. It is singular how few really fossil remains of birds have been found in England, while the remains of the other classes of vertebrata are far from uncommon. The slight texture and sponginess of the bones of birds do not admit of their preservation. The late Keeper of the Mineralogical Department in the British Museum figured the fossil cranium of a bird which was found in the London clay at the Island of Sheppy. Mr. König regarded this as a species of Gull, but Professor Owen has shown that it must have belonged with much more probability to a bird of the Kingfisher family, and has named it *Halcyornis*.

CLIONA. A curious genus of sponges, which have the singular power of boring into shells, and even the hardest rocks. This tunnelling process they appear to effect by means of minute spicules, which project from the surface. The species are all marine, the most common being that which is found attached to old oyster-shells, giving to the valves a honeycomb-like appearance. The above generic title was first employed by Dr. Grant, F.R.S., whose earlier researches on the subject of sponges obtained for him a world-wide repute.

CŒNURUS. An old name given by Rudolphi to a group of cystic worms, which take up their abode in the brains of the horse, sheep, and rabbit, but most commonly in the sheep. Their presence is indicated by the formidable cerebral symptoms to which they give rise. The disease thus produced is termed the "Staggers" in England and the "Sturdy" in Scotland. The cysts are now thoroughly well known to be neither more nor less than the young, or, rather, larval forms of tape-worms. Hitherto they have been supposed to occupy only the brains of animals, but Dr. Cobbold, F.L.S., has discovered a remarkable colony of these larvæ in the lungs of a Madagascar lemur, which died at the Zoological Society's Gardens, Regent's Park, and whose death was entirely due to their presence. The cysts in all cases have the form of vesicles, supporting a multitude of small processes, which, when unfolded and subjected to microscopic examination, are each found to contain the head of a tape-worm. Thus a single cœnurus lodged in the brain of a sheep will display two or three hundred heads, and (as the reader will find more fully explained under the article ALTERNATION OF GENERATION in the present SUPPLEMENT) each one of those heads is destined to become a tape-worm, or more strictly a strobila or tape-worm colony, in the intestines of some unfortunate dog which may chance to have devoured the brain of a sheep thus affected.

CORYPHODON. A genus of extinct pachyderms, whose fossil remains, few and very fragmentary as regards the single British species, are found in eocene formations. With rare acumen, Prof. Owen established this genus from the examination of a small portion of the right side of the lower jaw, which supported an entire molar tooth, and half of an adjoining tooth. This imperfect fossil was dredged up off the coast of Essex, and has thus supplied science with the means of determining the existence of an animal (*Coryphodon eocænus*) which, in a zoological point of view, forms as it were a link between the Elephants and Tapirs. A second species from France has recently been described by Prof. Hebert, under the title of *C. Oweni;* this may, however, on further evidence, turn out to be identical with the eocene coryphodon, above mentioned.

CTENOIDEA and CYCLOIDEA. These terms severally represent orders of fishes, thus named on account of certain peculiarities in the form of their scales. In the cycloids the scales have their posterior border spined, dentated, or toothed like a comb, as, for example, in the common fresh-water Perch, whilst the cycloids have oval or circular scales, with uniformly even margins, as in the Roach and most of our fresh-water fishes. Very few representations of either of these orders occur in the fossil state. The two great orders of fossil fishes (in Agassiz's Classification) are denominated *Ganoidea* and *Placoidea*. The scales of the former are osseous, thick, angular, and regularly arranged, the latter occurring in the form of irregularly disposed tubercles or plates. In both these orders the scales are enamelled. This mode of classifying the fishes into four orders, though very simple, and in some respects convenient, is now partly abandoned, as a few fishes are found to combine the characters of different orders.

CYCLOIDEA. [See CTENOIDEA, above.]

DENDRERPETON. A genus of batrachian reptiles, established by Prof. Owen from the examination of a few small bones and teeth, discovered by Sir Charles Lyell and Mr. Dawson in the coal strata of Nova Scotia. These remains were curiously enough found in the cavity of a fossil tree (*Sigillaria*), the trunk of which was about two feet in diameter. The only species at present known is described in the ninth volume of the "Quarterly Journal of the Geological Society," under the title of *D. Acadianum*.

DENDRODUS. A genus of extinct fishes, whose fossil teeth occur in the old red sandstone beds of Scotland in the neighbourhood of Elgin. On transverse section (as seen in the subjoined figure, p. 766) these teeth exhibit under the microscope a beautiful and singularly complex structure, due chiefly to the presence of radiating lines, which, proceeding from the centre, branch out and enclose leaf-like bundles of dentinal tubes. This labyrinthic or dendritic character suggested the generic name employed by Prof. Owen.

DICHOBUNE. An interesting genus of extinct quadrupeds, presenting marked affinity to the Ruminants, and coming, as it were, between them and the *Anoplotherium*, under which latter title two members of the genus were originally described by Cuvier. These species are *Anoplotherium minus* and *Anoplotherium leporinum*. Prof. Owen, in his recent systematic summary of extinct animals, speaks of a new species (*Dichobune ovina*), which he has "founded upon an almost entire lower jaw, with the permanent dental series (wanting only the four middle incisors), which now forms part of the palæontological collection in the British Museum." These fossils occur in eocene formations, the last-named having been found in Hampshire.

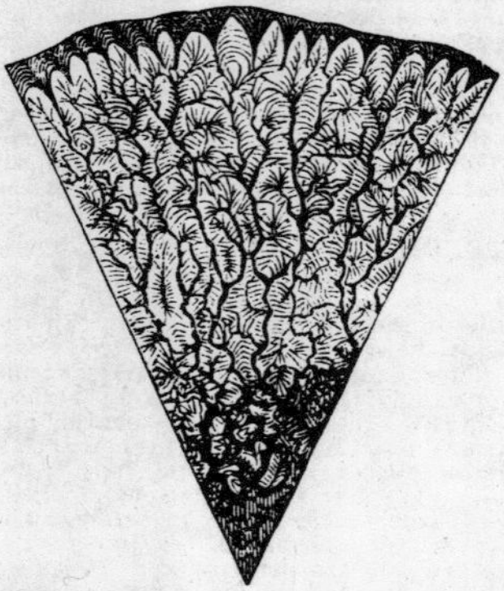

TRANSVERSE SECTION OF THE TOOTH OF DENDRODUS.

DICHODON. A genus of extinct animals closely allied to Dichobune, and whose fossil remains also occur in the upper eocene strata of Hampshire. At present only one species has been determined, and this the distinguished palæontologist above mentioned judges to have been about the size of our existing fallow-deer. It is called *Dichodon cuspidatus*, its specific nomenclature having reference to the number of points or cusps surmounting the crowns of the unworn molar teeth. Prof. Owen supposes that its food did not entirely consist of vegetable matters.

DICYNODON. A curious reptile, the remains of which were discovered by Mr. Bain in a fossil state in the south-eastern extremity of Africa. This singular animal was described in 1845 by Professor Owen. He named it Dicynodon from the most important character of the genus, the possession of two large tusks, like those of the walrus, and which were probably used as weapons of offence and defence. This creature seems to have united in its structure the characters of the lizards, crocodiles, and tortoises. It appears from the geological observations of Mr. Bain and others, that there formerly existed in Southern Africa a great salt-water lake or inland sea, now dried up and converted into dry land, but which, when covered with water, was tenanted by these anomalous creatures. Professor Owen described three species: the first he named *D. lacerticeps*, from its analogies with the lizards: a restoration of this by Mr. B. Waterhouse Hawkins is now exhibited in the Crystal Palace, where it is shown to have been as big as the Walrus or Morse, that characteristic denizen of the Arctic Seas. In the remains of this there is an exhibition of unusual strength in the bones of the face; there is no mark of any other teeth than the two large tusks which give the peculiar character to the animal. The whole of the anterior part of the jaws seems to have been sheathed with horn, as in the Tortoises, that horn which makes a bite from some of the larger specimens so formidable.

Professor Owen gave the name of *D. testudiniformis* to a second species from its having considerable resemblance to a tortoise. To a third species, with the tusks placed far behind the orbit of the eye, and with the head apparently having had a form somewhat resembling that which gives its peculiar physiognomy to the owl, the professor applied the name *D. strigiceps*. The nearest analogue to this genus is a reptile named Rhynchosaurus, found in the new red sandstone of England.

Professor Owen has remarked that "some bones of the back, or vertebræ, by the hollowness of the co-adapted articular surfaces, indicate these reptiles to have been good swimmers, and probably to have habitually existed in water; but the construction of the bony passages of the nostrils proves that they must have come to the surface to breathe air."

DIMORPHODON. A genus of Pterodactyles, whose fossil remains are found in the lower liassic deposits of Lyme Regis.

DINOSAURIA. An order of gigantic reptiles, whose fossil remains abound in the Oolite and Wealden formations. It includes such genera as *Iguanodon*, *Megalosaurus*, and *Hylæosaurus*, whose restored forms are so admirably represented in the grounds attached to the Crystal Palace at Sydenham.

DIPHYLLOBOTHRIUM. The generic name given by Dr. Cobbold, F.L.S., to a remarkable form of Tape-worm which dwells in the intestinal canal of the common Porpoise. At present only one species is known (*D. stemmacephalum*); this attains a length of about ten feet. The head is so small as to be barely visible to the naked eye, but under low magnifying powers it is seen to support two elegant leaf-like appendages or suckers, one on either side, their combined outline presenting a festooned appearance; hence the generic and specific name adopted.

DIPROTODON. Sir Thomas Mitchell, during his expeditions into the interior of the Australian continent, obtained a large number of fossil mammalian remains from old caverns, similar to the well-known bone caves of this country. Among these fragments was a large tooth, which Prof. Owen, in the year 1838, determined to be that of a gigantic marsupial, allied to the Kangaroos, but having remarkable affinities to our large non-marsupial pachyderms. Subsequent explorations brought to light other portions of the skeleton, and at length an entire skull was safely deposited in the palæontological collection of the British Museum. The head of this kangaroo-like quadruped (*Diprotodon Australis*) attained a length of fully three feet, its width being correspondingly great. In regard to the teeth and other skeletal peculiarities, Prof. Owen observes, that "like the contemporary gigantic sloth in South America, the Diprotodon of Australia, while retaining the dental formula of its homologue, shows great and remarkable modifications of its limbs. The hind pair were much shortened and strengthened compared with those of the kangaroo; the fore pair were lengthened as well as strengthened; yet, as in the case of the Megatherium, the ulna and radius were maintained free, and so articulated as to give the fore paw the rotatory actions. These in Diprotodon would be needed, as in the herbivorous kangaroo, by the economy of the marsupial pouch." At present only one species of this genus is known, and from the circumstance of the skull just mentioned having been found in a lacustrine formation, it is reasonably conjectured that the animal lived, geologically speaking, in comparatively recent times, at least not earlier than the pleistocene period.

DOLICHOSAURUS LONGICOLLIS. The scientific names applied by Prof. Owen to a small extinct lizard, whose fossil skeletal remains are found in the Wealden formation of Sussex. It appears to have displayed an almost serpent-like form, as is conjectured from the comparatively elongated vertebral column in the regions of the body and neck.

DRYOPITHECUS FONTANI. Terms proposed by Lartet to characterise a large Ape, whose fossil remains occur in a very incomplete state in miocene formations. In regard to the question of its affinities, Prof. Owen remarks, that "from the portion of humerus, associated with the jaw of *Dryopithecus*, the arm would seem to have been proportionally longer and more slender than in the chimpanzee and gorilla, with a cylindrical shaft, more like that in the long-armed apes (*Hylobates*), and less like the arm of the human subject."

ECHINORHYNCHUS. A genus of the *Acanthocephala*, or Thorn-headed family of worms. The species of this group of Entozoa are very numerous in fishes, being likewise tolerably abundant in reptiles and mammals; in birds, however, they are less frequent, and, so far as is at present known, they have never been found in man. On this score we may congratulate ourselves, seeing that the Echinorhynchi are all provided with a proboscis or head, in most cases armed with numerous hooks, by means of which they fasten themselves into the walls of the intestine of the animal they inhabit. On the species infesting water-newts (*E. anthurus*) Dr. Cobbold, F.L.S., has recorded many particulars in the 22nd vol. of the "Linnæan Transactions."

ELAND. (*Boselaphus oreas.*) This magnificent animal, the largest of the Antelopes, is worthy of an extended notice. It is also known by the names of the Cape Elk, Ganna, and *Impoofoo*, the latter term being employed by the Bechuanas and Matabili. The importance of this ruminant will be readily understood when it is mentioned, that not only is its flesh of the most palatable and nutritive character, but breeding experiments have recently established the fact that it may be thoroughly acclimatised in this country. Elands are now flourishing not only in the Zoological Society's Menagerie, Regent's Park, but also in the more extended demesnes of Viscount Hill at Hawkstone Park in Shropshire, and in the grounds of the Marquis of Bredalbane at Taymouth. Elands were first imported into England by the late Earl of Derby in 1840, several being afterwards bequeathed by him to the Zoological Society. A full-grown Eland is capable of attaining a weight of from fifteen hundred to two thousand pounds, and not long ago a very fine specimen, bred and fattened in this country, was slaughtered for the express purpose of testing its epicurean qualities;

ELAND —(BOSELAPHUS OREAS.)

royalty, both on this and the other side of the Channel, pronouncing the venison to be of first-rate quality. Sir Cornwallis Harris, in his charming folio volume, entitled "Game and Wild Animals of Southern Africa," speaks of this food in the following amusing and laudatory terms:—"Both in grain and colour it resembles beef, but it is far better tasted and more delicate, possessing a pure game flavour, and exhibiting the most tempting-looking layers of fat and lean, the surprising quantity of the former ingredient with which it is interlarded exceeding that of any other game quadruped with which I am acquainted. The venison fairly melts in the mouth, and as for the brisket, that is absolutely a cut for a monarch! With what satisfaction would not King Jamie of hunting memory have drawn his good blade adown the breast of a plump Eland, to be rewarded with five full inches of 'prime white fat on that ilk,' instead of three, as on the occasion in Greenwich Park, when Nigel assisted at the sylvan ceremony. The vast quantity of tallow yielded by the fat bulls," adds Sir Cornwallis Harris, "furnished us with constant material for manufacturing dips in a candle-mould with which we were provided; and during the greater part of our journey it was to the flesh of this goodly beast that we principally looked for our daily rations."

In regard to the zoological characters by which this gigantic antelope is distinguished, we may observe that the adult male stands fully six feet high at the shoulder, the length of the body being, in some cases, as much as nine feet from the tip of the nose to the root of the tail. The horns are nearly straight, massive, conical, furnished with a strongly developed spiral edge, which gradually disappears at the upper third, where the ends become attenuated and sharply pointed. In the female the horns are longer, slighter, and less markedly furrowed. The forehead of the male is clothed with a thick bundle of stiff, wiry, brownish hairs, the tuft being bordered on either side by a band of yellow orange colour. The ears are comparatively small, the muzzle broad, the neck thick, the dewlap very prominent and fringed with long brown hairs, the legs rather short, the shoulders and hind quarters enormously developed, the fur short and of a rufous-dun or ashy-grey colour generally, the tail being about twenty-six inches long and tufted at the extremity. The female exhibits a bead-like tuft of hair on the under part of the neck, has a more ferruginous colour, and is furnished with four teats.

Respecting the habits of the Eland, it is well known than it frequents only the more open plains of the interior, rejoicing especially "in low belts of shaded hillocks, and in the isolated groves of *Acacia capensis*, which, like islands in the ocean, are scattered over many of the stony and gravelly plains of the interior; large herds of them are also to be seen grazing like droves of oxen on the more verdant meadows, through which some silver rivulet winds in rainbow brightness betwixt fringes of sighing bulrushes. Fat and lethargic groups may be seen scattered up and down the gentle acclivities, some grazing on the hillside, and others lazily basking in the morning sunbeam. Advancing, they appear to move like a regiment of cavalry in single files, the goodliest bulls leading the van; whereas, during a retreat, these it is that uniformly bring up the rear." At one time Elands were abundant in the immediate neighbourhood of Cape Town, but now very few are found within the borders of the colony. Considering the facilities which exist for their destruction, every effort should now be made to follow up the experiments of domestication so successfully commenced by Lord Derby and the Zoological Society; and steps should likewise be taken to secure more specimens from the colony, ere they are driven far up into the interior or become altogether exterminated.

EPYORNIS. A bird much larger than the Ostrich, which at one time must have been indigenous to Madagascar, when that large island was perhaps the remnant of a former vast eastern continent now submerged. From the eggs of this bird, which have been found and brought to Europe, it is clear that they were capacious enough to hold the contents of 240 hen's eggs of the ordinary size. Casts of this egg may now be seen in most of our museums. It is but fair, however, to say that eggs are not always proportioned to the size of the birds which lay them. Many of the aquatic birds, and some of the Gallinacea, lay eggs out of all proportion to the size of the birds.

EUPLECTELLA. A remarkable genus of siliceous sponges. In the third volume of the "Transactions of the Zoological Society of London," the discovery of *Euplectella* is announced in the following manner:—"Mr. Cuming has entrusted to me for description," writes Prof. Owen, "one of the most singular and beautiful as well as the rarest of the marine productions with which his researches in the Philippine Islands have enabled him to enrich the zoological collections of his native country. This production forms part, however, of a member of the lowest class of organised bodies, being the skeleton or framework of a species of sponge belonging to the cylindrical and reticulate, or Alcyonoid family. It is a hollow, sub-circular, slightly conical, and gently curved case or tube, resembling a delicate cornucopia with the apex removed. It measures eight inches in length, two inches across the base, and one inch and a quarter across the apex, which is truncated. The base or wider aperture of the tube is sub-elliptical, and is closed by a cap of coarse and somewhat irregular network, gently convex externally, the circumference of which is divided from the walls of the cylinder, like the

EUPLECTELLA CUCUMER

base of the *Aspergillum*, or water-pot shell, by a thin projecting plate, standing out like a ruff or frill." At the time this description was given, it was probably little thought that only after a period of some fifteen years had elapsed would a second example of this singular sponge come to light, and still less was it imagined that the first described form could be exceeded in beauty; yet, such has been the case; and the second specimen, moreover, turns out to be a distinct species. The first was entitled *Euplectella aspergillum*, and that more recently described *E. cucumer*. In regard to the characters of the latter, which are given in the twenty-second volume of the "Linnæan Transactions," we have only space to observe that the specimen was six inches long two inches wide at the base, expanded to the extent of another half inch more in the centre, and contracted at the apex to about one inch and a quarter. Those who desire a more minute account of these lovely glass-like sponges, should consult the original records, especially the one last mentioned; but we cannot close this brief *résumé* without directing attention to the appropriate reflections which a consideration of these humble structures invariably induces in every rightly constituted mind, and which, in the instance just referred to, have been expressed by the distinguished Professor in the following terms:—"To the question put by almost every one to whom the Euplectella is shown, as to how the threads could have been so regularly, yet intricately, interwoven, I have sometimes replied, that there has been no such thing as interweaving in the case; that no thread, as such, was ever laid across another in the construction of the Euplectella; that the analogy of human textile fabrics does not apply to this beautiful natural object. The artificial lace-work, the several stages of a complex result, must be taken in the succession indicated by painful and exact calculation; in organic lace-work different stages are done at once. Thus it is that the Divine works surpass those of man's utmost ingenuity. The threads of the *Euplectella* were not first spun and then interwoven, but were formed as interwoven, the two processes going on simultaneously, or *pari passu*. Just as in the cancellous structure of bone, the plates of bone are not first formed and then fitted to one another, as in building a house of cards, but the forming and the fitting go on together in the course of molecular growth. I presume also, that in the beautiful object which we call the *Euplectella*, we have but its skeleton; and that, in the living state, the exquisite structure of the flinty framework may be veiled by the delicate gelatinous enveloping organic tissue." Admirable figures (from which the one here given is a reduced copy) accompany the memoirs published in the Transactions of the Linnæan and Zoological Societies. *Euplectella aspergillum* forms part of the magnificent natural history collection of Hugh Cuming, Esq., F.R.S.; whilst *E. cucumer* is the property of Dr. Arthur Farre, F.R.S., a distinguished member of the medical profession.

FAIRY-SHRIMP. A beautiful species of crustacean, found occasionally in fresh-water pools in this country, as at Blackheath. It is about an inch in length, and is nearly perfectly transparent. Dr. Baird has described it in his British Entomostraca: it swims on its back, and in fine warm weather, when the sun is not too strong, it may be seen balancing itself near the surface by means of its branchial feet, which are in constant motion, and form a most beautiful object in the microscope. On the least disturbance the Fairy-shrimp strikes the water rapidly with its tail from right to left, and darts away to conceal itself by diving into the soft mud, or amongst the weeds at the bottom of the pool. When put into a glass of clear water, it forms an attractive sight from its elegant form, easy and graceful motions, and from the delicate and silvery transparency as well as its large black eyes. Dr. Baird remarks, that the male is especially beautiful; the uninterrupted undulatory wavy motion of its graceful branchial feet, slightly tinged with a light reddish hue, the brilliant mixture of transparent bluish green and bright red of its prehensile antennæ, and its bright red tail, with the plumose setæ springing from it, all combine to make it an attractive object. These feet, when moving, keep the animal suspended in the water, serving as oars to the body, which may be likened to a boat. They cause the water also in this boat-like cavity to be compressed, and to mount up, as along a canal, carrying in the current the particles destined for its food towards the mouth. It is the *Chirocephalus diaphanus* of naturalists.

GALECYNUS. A genus of fossil carnivorous mammalia, founded for the reception of a curious skeleton, which was discovered in a quarry at Œningen by one of the workmen, who sold it to Sir Roderick Murchison. That distinguished geologist presented it to the British Museum. When first obtained it was invested with a hard calcareous crust, so that its general outline was alone visible. Dr. Mantell undertook to work out the specimen, and by carefully scraping and chiselling away the enclosing mineral, he succeeded in exposing the skeleton of a fox-like animal; which differs from our reynard in its limbs being much stouter, and in its feet being much more elongated. Sir Roderick Murchison has given a history of the ancient lake of Œningen, in the new series of the "Geological Transactions." In the quarries there many curious fossils have been found. Two of the most interesting of these are deposited in the British Museum, one being the identical specimen of an ancient Salamander, on which Scheuchzer, more learned than scientific, wrote his celebrated treatise "*Homo Diluvii Testis*," the other being the beautifully perfect specimen of the *Galecynus Œningensis*, or Fossil Fox of Œningen, described and figured by Sir Roderick Impey Murchison.

GASTORNIS PARISIENSIS. A large extinct bird, first described by Prof. Hebert, its fossil remains having been discovered in the lower eocene deposits at Meudon in the neighbourhood of Paris. Though as tall, and even more bulky, than our living ostrich, its structural peculiarities point to a closer affinity with the grallatorial or wading birds.

GORILLA. (*Troglodytes Gorilla.*) So slow is the progress of truth among the mass of human minds, that we have daily to deplore a credulity which asserts the existence of things without any other foundation than such as is afforded by the fertility of an erratic imagination; yet, on the other hand, it occasionally happens that the announcement of some grand scientific truth, or the promulgation of a well-authenticated fact, is for a time cast aside as frivolous or unworthy of belief, to be resuscitated after the lapse of many bygone years. Thus the slaves of credulity and incredulity are by turns confounded, whilst the establishment of golden precious truth is slowly and finally secured. The history of the Sea Serpent may be taken to exemplify the rule, whilst that of the Gorilla supplies the more instructive exception.

About five hundred years antecedent to the Christian era, one Hanno, a Carthaginian admiral, set sail from the Mediterranean coast for the purpose of colonising the western shores of Africa. His fleet consisted of no less than sixty vessels, having on board thirty thousand souls. After passing the Pillars of Hercules, and planting colonies along the coast, he at length reaches a spot which is thus described in his "Periplus," or record of the voyage:—"Passing the Streams of Fire," he says, "we came to a bay called the Horn of the South. In the recess there was an island like the first, having a lake, and in this there was another island full of wild men. But much the greater part of them were women with hairy bodies, whom the interpreters termed 'Gorillas.' But,

GORILLA.—(TROGLODYTES GORILLA.)

pursuing them, we were not able to take the men; *they* all escaped, being able to climb the precipices, and defended themselves with pieces of rock. But three women, who bit and scratched those who led them, were not willing to follow. However, having killed them, we flayed them, and conveyed the skins to Carthage; for we did not sail any further, as provisions began to fail."

This extremely brief but interesting narrative of the old Carthaginian navigator, in all human probability, refers to the subject of our present notice. The circumstance of the women being flayed renders it highly probable that they were rather beasts than humankind, whilst their hairy covering and behaviour leave us in little doubt as to their quadrumanous origin. The near approach of these creatures to the human figure is quite sufficient to account for their being described as men and women, for subsequent history has shown that the great anthropoid apes were in old times looked upon as monstrous varieties of our genus, and even the great Linnæus himself, in later times, spoke of a *Homo Troglodytes*, by which, however, he referred to the Orang and Chimpanzee.

In the year 1589 of our era, an English sailor, Andrew Battel by name, accidentally found his way to the shores of Western Tropical Africa. He was, indeed, taken to Angola as a prisoner by the Portuguese, and during his residence there became acquainted with our anthropomorphous ape. He speaks of two kinds of monsters, namely, the Pongo and the *Enjocko*. The latter is evidently identical with the Chimpanzee, whilst the former is unquestionably a synonym of the Gorilla. "The Pongo," he says, "is in all its proportions like a man, except the legs, which have no calves; but he is of gigantic height. The face, hands, and ears of these animals are without hair; their bodies are covered, but not very thickly, with hair of a dunnish colour. When they walk on the ground it is upright, with the hands on the nape of the neck. They sleep on trees, and make a covering to shelter them from the rain. They eat no flesh, but feed on nuts and other fruit; nor have they any understanding beyond instinct. When the people of the country travel through the woods they make fires in the night, and in the morning when they are gone the Pongos will come and sit round them until they go out; for they do not possess sagacity enough to lay on more wood. They go in bodies, and kill many negroes who travel in the woods. When elephants happen to come and feed where they are, they will fall on them, and so beat them with their clubbed fists and sticks, that they are forced to run away roaring. The grown Pongos are never taken alive, owing to their strength, which is so great that ten men cannot hold one of them. The young hang upon their mother's belly with their hands clasped about her; many of them are taken by shooting the mothers with poisoned arrows. When they die among themselves they cover their dead with great heaps of boughs and wood."

Subsequent travellers have confirmed the accuracy of Battel's statements; but nearly all of them have erred in describing the Gorilla and Chimpanzee as one and the same animal. The distinguished natural historian Buffon appears to have entertained a clear persuasion as to the separate existence of the Pongo or Gorilla, which, however, he supposed to be a large species of Orang; but the great anatomist Cuvier contemptuously rejected Buffon's reasonings, regarding them as "nothing more than the imaginary product of his combinations." Thus, for a long season again, the Gorilla was declared to be a myth, and it was not until the year 1847 that science obtained indisputable evidence of the existence of this huge anthropomorphous ape. In April of that year Dr. P. S. Savage, an American missionary, sent a letter to Prof. Owen, enclosing a sketch of the skull, requesting his opinion upon it, and in the interval he returned to America, carrying with him several crania, both young and adult, of either sex. These, together with other portions of the skeleton, were forthwith described by the missionary doctor and Prof. Wyman in the fifth volume of the "Journal of the Natural History Society of Boston." Shortly afterwards Prof. Owen, by the aid of Mr. Stutchbury, of Bristol, procured additional specimens, descriptions of which were communicated to the Zoological Society, and published in the Society's Proceedings and Transactions. In 1849 a preserved Gorilla was conveyed to Paris by Dr. Franquet, a surgeon in the French navy; and this was followed by the accession of entire skeletons within the walls of the British Museum and the Hunterian Collection of the Royal College of Surgeons. Since the last-named date other specimens have arrived, and very recently our national depôt has received (by purchase from an enterprising collector, Mr. Paul B. Du Chaillu) several remarkably fine but ill-preserved specimens, which may now be seen in the Museum together with the original specimen stuffed by Mr. Bartlett. The annexed woodcut is a representation of the individual last mentioned, being the same that for a season figured so conspicuously at the Crystal Palace at Sydenham.

Having said thus much in regard to the singular history of this illustrious ape, we shall in the next place indicate a few of its more important anatomical peculiarities. In doing this we shall in the main reflect the opinions of Prof. Owen, whose examinations and comparisons have been made with that remarkable skill and rare acumen for which he is so distinguished; but at the same time it must be acknowledged that the structural differences (in respect of the brain more particularly) which are said to separate the higher apes from man, are not so great as the learned professor would have us suppose. Anatomically speaking, these differences merely resolve themselves into a question of form and volume; and, therefore, we have always maintained, and still maintain, that the specific and family distinctness of man rests upon higher grounds than upon the degree of those modifications of the material fabric which pervade all nature. Unityof plan bespeaks a corresponding unity of design, and the faith of the enlightened teleologist is in no way disturbed by the puerilities of transmutation hypotheses; it rather gathers strength from a contemplation of the marvels of created life, and the beauteous phenomena of its varied being.

A patient examination of the skins and skeletons of the Gorilla preserved in our National Museum, will enable the visitor to observe all the more important features peculiar to this beast. Directing attention to the head, we find a remarkable shortness of the neck, which, in a front view, seems altogether absent, the chin resting on the upper and anterior part of the chest. This apparent abrogation of the neck is in part due to the extended development backwards of the spines of the lower cervical or neck vertebræ, whose tips form a continuous line or level with the crown of the skull. The brain cavity is thus placed very low down, whilst the face is thrust forward so as to impart to the beast a peculiarly hideous physiognomy. The superciliary ridges are very strongly marked, concealing a pair of comparatively small, green-coloured eyes, in front of which we observe a prominent nose and an astonishingly large brute-like mouth furnished with thick projecting lips; the upper one displays a straight border, looking as if it were trimmed or clipped with a pair of scissors. The chin is inconspicuous; there are no eyebrows, but well-developed eyelashes. We may also note the small ears, the wrinkled and nude face, the large and wide shoulders, the huge and capacious chest, the prominent belly, the prodigious arms, and comparatively short legs; the hand is very broad and thick, long in the palm, the fingers not being separated so low down as in man. This semi-webbed character is still more conspicuous in the foot. The backs of the fingers are callous and devoid of hair, in consequence of the habit which the animal has of walking knuckle downwards. The thumb is short and rather narrow. The legs (or more strictly the posterior arms) are relatively short, and conspicuous for their uniformity of breadth, the so-called calf making scarcely any appearance. The heel of the large foot is prominent, the big toe (or thumb) being particularly stout and widely separated from the adjoining digit; the breadth of the sole and consequent grasp conveying a suggestive notion of its prehensile power.

In regard to the skeletal characters of the Gorilla, these, as will be readily supposed, closely correspond with the external peculiarities above mentioned. The skull displays a still more brutish look than the outside of the face and head. The cranium, indeed, has a decidedly carnivorous aspect, as may be seen by its possession of a remarkably prominent crest, strongly projecting orbital ridges, broad zygomatic arches, powerful jaws, and more particularly two pairs of huge sharply pointed canine teeth. The strength of the jaws with their formidable armature is also attested not only by their bulk, but by the enormously developed temporal muscles, which are primarily concerned in the action of biting and tearing. Among the other osteological features of the Gorilla are the large columnar spines of the lower cervical and upper dorsal vertebræ, the remarkably long and large scapulæ, the broad and capacious ribs (thirteen in number on either side), the short and wide sternum, the great length of the bones proper to the arm and fore arm, the long digital phalanges, the well-developed pelvis, strong femora and short leg-bones, and, lastly, the powerful development of the osseous elements of the foot, in which the calcaneum or heel-bone is more especially significant.

Leaving thus, as briefly as possible, gone over the principal external and internal characteristics of this anthropomorphous ape, which, in a structural point of view, is considered, perhaps rightly, the nearest approach to the human being, we have finally to consider the habits of the Gorilla; and here it is, especially, that we notice how very far this beast is beneath man, as respects intelligence, and the possession of those psychical endowments which pre-eminently characterise our species, despite the brutalising effects of ignorance and evil.

A few years ago science had only acquired the most meagre details of the habits of the Gorilla, and even these were for the most part either stamped with exaggeration or were mixed up with statements altogether fabulous. At the present time, indeed, it is not very easy to present a truthful picture of the behaviour of this beast in its native haunts, notwithstanding that we have travellers' tales in abundance purporting to be the substantial records of personal observation. Although a sincere admirer of that warm generosity which the British public have, rightly or wrongly, extended to a successful and enterprising collector, whose "Adventures in Equatorial Africa" are now so familiar to the general reader, the author of this article, having carefully watched the controversy which Mr. Du Chaillu's volume has provoked, gives preference to other sources of information than are supplied by the work in question. Probably the best, if not the fullest, account of the habits of the Gorilla which has yet been made public is that communicated by Prof. Owen to the Zoological Society of London on the 11th of January, 1859. This record appears to have been culled with great care and judiciousness from the writings and statements of Battel, Savage, Wilson, Laboulaye, Stutchbury, Captain Wagstaff, and others, who have either resided at or visited the neighbourhood of the Gaboon. The subject is introduced as follows:—

"Gorilla-land is a richly wooded extent of the western part of Africa, traversed by the rivers Danger and Gaboon, and extending from the equator to the tenth or fifteenth degree of south latitude. The part where the Gorilla has been most frequently met with presents a succession of hill and dale, the heights crowned with lofty trees, the valleys covered by coarse grass, with partial scrub or scattered shrubs. Fruit trees of various kinds abound both on the hills and in the valleys; some that are crude and uncared for by the negroes are sought out and greedily eaten by the Gorillas, and as different kinds come to maturity at different seasons, they afford the great denizen of the woods a successive and unfailing supply of indigenous fruits. Of these Prof. Owen specified the following sources:—

"The palm-nut (*Elais guiniensis*), of which the Gorillas greatly affect the fruit and upper part of the stipe, called the 'cabbage.' The negroes of the Gaboon have a tradition that their forefathers first learnt to eat the 'cabbage' from seeing the Gorilla eat it, concluding that what was good for him must be good for man.

"The 'ginger-bread tree' (*Parinarium excelsum*), which bears a plum-like fruit

"The papau tree (*Carica papaya*).

"The banana (*Musa Sapientium*), and another species (*M. paradisiaca*).

"The *Amomum Afzelii* and *A. grandiflorum*.

"A tree with a shelled fruit like a walnut, which the Gorilla breaks open with the blow of a stone.

"A tree, also botanically unknown, with a fruit like a cherry.

"Such fruits, and other rich and nutritious productions of the vegetable kingdom, constitute the staple food of the Gorilla, as they do of the Chimpanzee. The molar teeth, which alone truly indicate the diet of an animal, accord with the statements as to the frugivorous character of the Gorilla; but they also sufficiently answer to an omnivorous habit to suggest that the eggs and callow brood of nests discovered in the trees frequented by the Gorilla might not be unacceptable.

"The Gorilla makes a sleeping-place like a hammock, connecting the branches of a sheltered and thickly leaved part of a tree by means of the long, tough, slender stems of parasitic plants, and lining it with the broad dried fronds of palms, or with long grass. This hammock-like abode may be seen at different heights, from ten to forty feet from the ground, but there is never more than one such nest in a tree.

"They avoid the abodes of man, but are most commonly seen in the months of September, October, and November, after the negroes have gathered their outlying rice crops, and have returned from the bush to the village. So observed they are described to be usually in pairs, or, if more, the addition consists of a few young ones, of different ages and apparently of one family. The Gorilla is not gregarious. The parents may be seen sitting on a branch resting the back against the tree trunk—the hair being generally rubbed off the back of the old Gorilla from that habit—perhaps munching fruits, whilst the young Gorillas are at play, leaping and swinging from branch to branch with hoots or harsh cries of boisterous mirth.

"If the old male be seen alone, or when in quest of food, he is usually armed with a stout stick, which the negroes aver to be the weapon with which he attacks his chief enemy the elephant. Not that the elephant directly or intentionally injures the Gorilla, but, deriving its subsistence from the same source, the ape regards the great proboscidian as a hostile intruder. When, therefore, he discerns the elephant pulling down and wrenching off the branches of a favourite tree, the Gorilla, stealing along the bough, strikes the sensitive proboscis of the elephant with a violent blow of his club, and drives off the startled giant trumpeting shrilly with rage and pain.

"In passing from one detached tree to another, the Gorilla is said to walk semi-erect with the aid of his club, but with a waddling awkward gait; when without a stick he has been seen to walk as a biped, with his hands clasped across the back of his head, instinctively so counterpoising its forward projection. If the Gorilla be surprised and approached while on the ground, he drops his stick, betakes himself to all-fours, applying the back part of the bent knuckles of his fore hands to the ground, and makes his way rapidly, with an oblique swinging kind of gallop, to the nearest tree. There he awaits his pursuer, especially if his family be near and requiring his defence. No negro willingly approaches the tree in which the male Gorilla keeps guard. Even with a gun the experienced negro does not make the attack, but reserves his fire in self-defence. The enmity of the Gorilla to the whole negro race, male and female, is uniformly attested.

"The young men of the Gaboon tribe make armed excursions into the forest in quest of ivory. The enemy they most dread on these occasions is the Gorilla. If they have come unawares too near him with his family, he does not, like the lion, sulkily retreat, but comes rapidly to the attack, swinging down to the lower branches, and clutching at the nearest foe. The hideous aspect of the animal, with his green eyes flashing with rage, is heightened by the skin over the prominent roof of the orbits being drawn quickly backward and forward, with the hair erected, causing a horrible and fiendish scowl. If fired at and not mortally hit, the Gorilla closes at once upon his assailant, and inflicts most dangerous, if not deadly wounds, with his sharp and powerful tusks. The commander of a Bristol trader told the author he had seen a negro at the Gaboon frightfully mutilated by the bite of the Gorilla, from which he had recovered. Another negro exhibited to the same voyager a gun-barrel bent and partly flattened by the bite of a wounded Gorilla in its death-struggle. Negroes, when stealing through the gloomy shades of the tropical forest, become sometimes aware of the proximity of one of these frightfully formidable apes by the sudden disappearance of one of their companions, who is hoisted up into the tree, uttering perhaps a short choking cry. In a few minutes he falls to the ground a strangled corpse. The Gorilla, watching his opportunity, has let down his huge hind hand, seized the passing negro by the neck with vice-like grip, has drawn him up to higher branches, and dropped him when his struggles had ceased."

The above account may be taken as a fair record of the habits and prowess of this formidable beast, whose strength is apparently equal to that of a full-grown lion. The statements of Hanno and Battel as to the animal's gregariousness do not appear to be fully borne out by recent testimony. As to their reputed misconduct in obliging negresses to accompany them to their sylvan haunts, that is clearly an unjust accusation. Our anthropomorphous brother, as some would call him, is guiltless in this respect. If the brute deserves consideration, it lies in the notorious circumstance that both parents display a most affectionate regard for their offspring, not unfrequently risking their lives to save the wee four-handed bairns from the clutches of the natural history collector, or rather from the negroes who collect for him. Notwithstanding the expression of these sympathies, it is our sincere desire that the enterprising Mr. Du Chaillu, or some other individual having an equal eye to business and the requirements of science, may yet succeed in transporting hither a live member of the fraternity, whose animated features it shall be our happiness to scrutinise in the delightful grounds of the Zoological Society, Regent's Park.

HECTOCOTYLUS. A name first employed by Cuvier to indicate a supposed genus of animals parasitic on cuttle-fishes. It is now known, however, that these Hectocotyli are neither more nor less than detached arms of the male Argonaut and other allied cephalopods. The arm is specially modified into a copulatory organ, which is capable of being periodically cast off and renewed. [See ARGONAUT, p. 35.]

HYÆNODON. The generic title of an extinct carnivorous animal, whose skeletal remains are to be found in the upper eocene beds of Hordwell in Hampshire. A second species occurs in the miocene strata of Débruge and Alais in France. These quadrupeds, judging by their cutting teeth, appear to have possessed highly destructive propensities; they were about the size of our larger carnivora, such as wolves, leopards, or hyænas.

HYLÆOSAURUS. This name was proposed by Dr. Mantell to characterise a genus of huge dinosaurian reptiles, whose remains were found in the Wealden strata at Tilgate Forest in the year 1832. This gigantic quadrupedal lizard was at least twenty-five feet in length; it was furnished with a thick coat of mail, consisting of dense osseous scales or scutes, like those of our existing crocodiles, and the ridge of the back supported a number of these defensive plates, which, however, in this situation presented a conical outline, being prolonged into most formidable spines. This is not a mere matter of conjecture, because these structures were found in one slab, at least, in the immediate neighbourhood of the vertebral segments, without doubt originally occupying the position so faithfully represented in the restored forms modelled under the superintendence of Mr. Waterhouse Hawkins, and now displayed in the Crystal Palace grounds at Sydenham. In his excellent Report on British Fossils communicated to the meeting of the British Association held at Plymouth in 1841, Prof. Owen refers to the detection of this animal's fossil remains in these terms: "The Hylæosaurus has not been made known, like the Megalosaurus, from detached parts of the skeleton successively discovered and analogically recomposed, but was at once brought into the domain of palæontology by the discovery of the following parts of the skeleton in almost natural juxtaposition, viz., the anterior part of the trunk, including ten of the anterior vertebræ in succession, supporting a small fragment of the base of the skull; the two coracoids, the coracoid extremities of both scapulæ, detached vertebræ, several ribs more or less complete, and some remarkable parts of the dermal skeleton, including, apparently, enormous vertical plates or spines, arranged, as is supposed, in the form of a median dorsal ridge or crest of singular dimensions." The several bony elements have all been minutely described in the "Philosophical Transactions" and elsewhere by the original discoverer, Dr. Mantell, whose patient and honest labours in the cause of science were eventually rewarded by a medal presented to him by the Royal Society. To the original descriptions Prof. Owen has added very largely, especially in the "Report" above referred to.

KIANG. (*Equus Hemionus.*) This comparatively rare quadruped is intermediate in character between the Ass and the Horse. It is an inhabitant of the sandy steppes of Central Asia at a height of sixteen thousand feet above the level of the sea; here they herd together in small numbers. The Kiang or Tschiktei, as it is sometimes called, has a smooth coat and ears of moderate length; the body exhibits a bright rufous-bay tint, the legs being of a pale straw colour. A dark broad streak runs along the central line of the back; but it is not crossed by any similar band over the shoulders. The winter coat is comparatively thick, curly, and darker than in summer. The males are fine animals, standing as much as fourteen hands high at the shoulder; and, moreover, they

KIANG.—(EQUUS HEMIONUS.)

neigh like horses. A living specimen was recently brought over to this country by Major W. E. Hay, F.L.S., by whom it was presented to the Zoological Society. This gentleman has given us a most interesting account of the behaviour of the Kiang during its tedious and perilous journey from Fort Rûdôk, in Little Thibet, down to Kurrachi, where after keeping it a month, Major Hay says: "I took a passage in the barque Sumner, laying in a large quantity of hay, kirbee, and dried lucern, also grain. The latter was worm-eaten, and it was long before the animal could be induced to touch it. Our passage was very long, and, the captain's people having unceremoniously used my provisions to feed their own stock, the Kiang was *twice* reduced to eat the straw with which the sailors' beddings had been stuffed. This proves the hardiness of the animal. At first it refused to drink any tainted water; but before reaching St. Helena, where I had to lay in fresh supplies, it would eat or drink almost anything. The putting it on board ship at Kurrachi was very difficult, and the poor thing struggled so much that it was painful to watch it as it was lowered into the boat to be conveyed to the ship. So anxious were my friends concerning its safety, that a lady and gentleman who had allowed it to stand in their stables, and had given it many a tidbit of lucern, carrots, &c., came off in the boat with it. The sea was rough, and we had some miles to go to the ship; the shaking of the sails frightened it much. However, at last it was hoisted into the ship and placed in a house which had been built for it, and in which it continued until it reached England." This same Kiang may now be seen in the Regent's Park Menagerie. It is exceedingly docile and susceptible of kindness. In its native haunts it appears to have numerous enemies besides ourselves, being especially the victim of a white panther and a large variety of wolf. Major Hay saw them ascending and descending the hills with wonderful rapidity, but they never lost their footing. They prefer, however, the even ground, and more particularly unfrequented spots in the immediate neighbourhood of ponds and lakes.

LABYRINTHODON. A Batrachian reptile of very large size, the remains of which are found in different parts of the British Islands. These large reptiles belonged to an order, the species of which at present existing on the globe are generally of small size, and represented here by frogs, toads, and

CANINE TOOTH OF LABYRINTHODON

the water-newts. In Japan, however, there still exists a gigantic species, named *Sieboldia;* but this, though a giant when compared with the representatives of the order in England, was but a dwarf when contrasted with the Labyrinthodon. In various quarries of the new red sandstone formation in Cheshire, Warwickshire, and Lancashire, the impressions made by the feet of these large Batrachians can be distinctly seen and traced consecutively. These footsteps are partly concave and partly in relief, according as they are on the upper or under surface of the sandstone strata; the impressions in relief are only found on the lower surfaces, and are solid casts of the impressions made on the soft clay by the feet, toes, and claws of the reptiles as they crawled along. It was a considerable time before any remains of bones were found, to justify the naturalists in pronouncing on the cause of these impressions. Dr. Kaup suggested that they were the footmarks of a marsupial animal; he named it *Cheirotherium*, and some geologists were even fanciful enough to imagine that they were the remains of branched seaweeds. It is now ascertained, from various remains of bones and teeth associated with them, that they were formed by huge Batrachians, to which Professor Owen has applied the name of *Labyrinthodon*, in allusion to the peculiar structure of the teeth, the substances composing them being blended together in a very complex labyrinthic manner. The learned Hunterian Professor has shown that the skull of this reptile is attached to the neck-bones by two joints, and that the teeth are situated both on the proper jaw-bones, and on the bone of the roof of the mouth called "vomer;" characters found at the present day only in the frogs and salamanders. The hind foot, as in the frog and toad, was much larger than the fore foot, although they are nearly similar in form. Mr. Waterhouse Hawkins, following the indications of Professor Owen, has built up, on one of the islands in the grounds of the Crystal Palace, a highly graphic restoration of one of the two species of this remarkable genus. It will be thus easily perceived by any eye, how the large footsteps impressed in the sandstone were formed by animals walking on the land, and air-breathers, resembling in most points, except in bulk, the batrachian reptiles at present existing on the globe.

LEIODON ANCEPS. The names given by Professor Owen to a marine Lizard, whose fossil skeletal remains have been found in the chalk of Norfolk. It was closely allied to the great *Mosasaurus* [which see below].

LEPTOPLEURON. [See TELERPETON.] A reptile found in a fossil state in the old red sandstone.

LITHORNIS. The name given by Professor Owen to the fossil remains of a bird preserved in the Hunterian collection. Although the remains of mammalia are abundant in the fresh-water deposits, and in the marine drift of the newer pliocene period, the traces of the existence of birds are very seldom met with. Professor Owen easily accounts for this; he says, that "the light bodies of birds float along on the surface after death; and for one bird that becomes imbedded in the sediment at the bottom, perhaps ninety-nine are devoured before decomposition has sufficiently advanced to allow the skeleton to sink."

The remains of the Lithornis were found in the London clay of Sheppey, that fertile source of supply to the cabinets of collectors. The professor believes, and has given the reasons for his reference, that the remains belong to some species of the vulture family, a family of birds now restricted to the warmer parts of the earth. The fossil fruits described in the work of Mr. Bowerbank, and collected by that able naturalist on the Isle of Sheppey, are mostly referred to families of plants which now inhabit the more tropical portions of the globe; so that the animal and vegetable remains found in a fossil state on this interesting spot seem clearly to show that at the time they were alive, a warmer temperature prevailed.

LOPHIODON. A generic title proposed by Baron Cuvier to indicate an extinct herbivorous quadruped, whose fossil remains were discovered at the beginning of the present century in the tertiary beds near Issel in Languedoc. The Lophiodon was closely allied to our existing Tapirs and Rhinoceroses, as shown by its dental peculiarities, and also still more so to the extinct *Palæotherium* [which see at p. 485].

MACHAIRODUS (μαχαιρα, a sword; οδου, a tooth). A genus of carnivorous mammalia now found only in a fossil state. It derives its name from, and is chiefly characterised by, the long curved compressed canine teeth, the crowns of which have finely serrated margins. Remains of this quadruped have been found in several parts of Europe, and also in this country. Professor Owen says, "In this island, anterior to the deposition of the drift, there was associated with the great extinct tiger, bear, and hyæna of the caves, in the destructive task of controlling the numbers of the richly developed order of the herbivorous mammalia, a feline animal as large as the tiger, and, to judge by its instruments of destruction, of greater ferocity." This creature was the Machairodus, which Cuvier, from the imperfect materials before him, judged to have been a bear. "When we are informed," writes Professor Owen, "that, in some districts of India, entire villages have been depopulated by the destructive incursions of a single species of a large feline animal, the tiger, it is scarcely conceivable that man, in an early and rude condition of society, could have resisted the attacks of the more formidable tiger, bear, and machairodus of the cave epoch. And this consideration may lead us the more readily to receive the negative evidence of the absence of well-authenticated human fossil remains, and to conclude that man did not exist in the land which was ravaged simultaneously by three such formidable Carnivora, aided in their work of destruction by troops of savage hyænas." There are fine specimens of the teeth of this animal, and a cast of its formidable head, in the British Museum. In the British Museum there are specimens of two species found by Dr. Falconer and Major Cautley in the tertiary deposits of the Sewalik Hills in India.

MACRAUCHENIA. This term has been given by Professor Owen to a genus of extinct camuline Ruminants, whose fossil remains have been discovered in South America. These animals were of gigantic proportions as compared with the existing Llamas of that continent, which in other respects, however, they appear to have very closely resembled.

MACROTHERIUM. The generic name employed by Lartet to characterise an extinct edentate animal, whose fossil remains occur in the tertiary beds of Eppelsheim in Hesse-Darmstadt, and also in a lacustrine formation near Auch, Hautes-Pyrénées. This animal formerly represented in Europe the now existing ant-eaters of South America, as well as the Pangolins and *Orycteropus* of the adjacent African continent. Its anatomical peculiarities oscillate principally between *Manis* and the last-named genus; but it also bore some characters common to the sloths.

MEGACEROS. A genus of the Deer family, the bones of which are found in various parts of Ireland, in the shell marl, below the peat or bog earth. The gigantic Irish Deer or Fossil Elk, as it has been called (*Megaceros hibernicus*, Owen), was an immense creature. Specimens have been found, which measured, from the foot to the summit of the antlers, ten feet and four inches; while the measurement from the tip of one antler to another, in full-grown specimens, was from ten to twelve feet. The horns are very wide, and from being flattened out, somewhat as in the Elk or Moose, were regarded by some of the older naturalists as having belonged to a variety of that animal. To support these immense horns, the vertebræ of the neck were of much larger size than in any existing species of deer, while the legs were of stronger proportions. The skull and antlers of a specimen in the Museum of the Royal Dublin Society weigh eighty-seven pounds. It was once thought that the females of this immense creature must have been horned equally with the male, as in the recent reindeer: the late researches of Professor Owen have shown that the female was hornless, as in our Fallow Deer and Red Stag.

Remains of this Deer have been found in the Isle of Man and in different parts of England, such as Norfolk, Essex, and Lancashire; but they have never occurred to the extent that they are met with in Ireland. They seem to have become extinct before the introduction of man on the British Islands. One naturalist, from the appearances exhibited by a perforated rib, thought that the hole must have been produced by a sharp-pointed instrument, which did not penetrate so far as to cause the animal's death, but which remained long enough in the opening to alter the growth of the bone. Professor Owen, however, has shown that an arrow-head, of the dimensions fitted to make such a hole, and sticking in a rib with its point in the chest, must have soon killed the animal, by piercing the contiguous viscera and producing inflammation. He shows that the instrument which pierced the rib must have been instantly withdrawn, and not left to impede the growth of the bone, and concludes that, as male stags are very combative, the injury was most likely produced by the pointed branch of the formidable antler.

The bones of the Megaceros are generally of a dark-brown colour, with patches of the blue phosphate of iron; and in some instances, the hollows of the long bones have contained marrow so fresh as almost to resemble suet.

MICROLESTES. A genus of extinct Marsupials possessing extraordinary palæontological interest from the circumstance that it at present constitutes the earliest record of mammalian life upon our planet. In regard to it, Prof. Owen observes, that the "teeth from German and English trias indicate a very small insectivorous quadruped, to which the above generic name was given by Prof. Plieninger. The German specimens were discovered in 1847, in a bone breccia at Diegerloch, about two miles from Stuttgardt, the geological relations of which are well determined, as between lias and Keuper sandstone." In the year 1858, Mr. Charles Moore, F.G.S., discovered four teeth in this country at Frome, and these having been submitted to the scrutiny of our great palæontologist, were also determined by him to be referable to the same genus. Among living marsupials the Microlestes presents a tolerably near approach to the Australian genus *Myrmecobius*, but it is still nearer in affinity to the extinct *Plagiaulax* [which see below].

MICROTHERIUM. A genus of very small herbivorous mammals, whose fossil remains are abundant in the lacustrine calcareous eocene beds of the Puy-de-Dôme, in France. Their affinities appear to connect them in some degree with existing ruminants, but more closely to the extinct *Anoplotherium.*

MOSASAURUS. A genus of extinct gigantic marine Lizards, whose skeletal remains have been found in the cretaceous strata of Europe and North America. In England a small species (*M. gracilis*) has been discovered in the county of Sussex. The skull of a large European species (*M. Hofmanni*) measures as much as five feet in length, the original cranium being found at St. Peter's Mount, near Maestricht, in the year 1780. A glance at the restored Mosasaurus in the Crystal Palace Gardens at Sydenham will best convey an accurate conception of the vast magnitude of this lacertian monster. The American species has been named *Mosasaurus Maximilliani.*

MYLODON ROBUSTUS. The combined generic and scientific title applied by Prof. Owen to a remarkable form of extinct Sloth, whose fossil remains were discovered in a pleistocene fluviatile deposit not far from the city of Buenos Ayres, South America, in the year 1841. An almost entire skeleton of this animal (which is here figured) may be seen in the Hunterian Museum of the Royal College of Surgeons in a most excellent state of preservation. At the instance of the Council of the College, this splendid relic of past greatness amongst the sloths has been specially described and figured in a magnificent quarto work by their former Hunterian

Professor. The Mylodon, though not so large a the great terrestrial Megatherium, was nevertheless a formidable creature, measuring at least eleven feet from the point of the muzzle to the end of the tail. Structurally

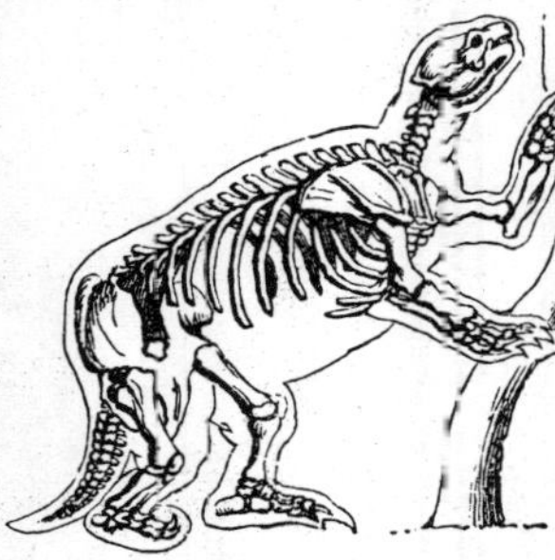

MYLODON ROBUSTUS.

it was closely allied to the sloths of the present day, and like them obtained its living by devouring the leaves of trees, which latter, however, it did not ascend, but rased to the ground by its prodigious strength. [For osteological and other details, see MYLODON, at p. 443.]

NOTHOSAURUS. A genus of gigantic sauropterygian reptiles or fin-lizards, whose skeletal remains are very abundant in the muschelkalk formation of Bayreuth and Luneville. The species of this genus, as in the more familiarly known and allied PLESIOSAURUS, appear to have been very numerous, for at least seven of the former have been indicated by the writings of Von Meyer and Münster, whilst Prof. Owen states that he has described, or otherwise become acquainted with, upwards of twenty distinct forms of Plesiosaur. Considering the formidable prehensile and offensive armature of their jaws, one can readily understand what an enormous destruction of life took place at the commencement of the mesosoic or secondary geological epoch, when multitudes of these creatures preyed upon the co-existing fishes and amphibia of the triassic and liassic period. The Nothosaurs were furnished with extended massive heads, and numerous large, sharp, conical teeth, most of them being provided with long necks like those of the Plesiosaur. In regard to their natatory organs of locomotion, Prof. Owen informs us that "the bones of the limbs, although evidently those of fins or paddle-shaped extremities, are better developed than in Plesiosaurus, and more resemble the corresponding bones in the turtle (*Chelones*). The tuberosities or processes for muscular attachment near the head of the humerus are better marked, especially that on the concave side of the shaft; the distal end is thicker and less expanded. The whole bone is more curved than in any *Plesiosauri*. The femur is relatively longer and less expanded at its distal end. The bones of the forearm, like those of the leg, are longer than in Plesiosaurus. The articular surfaces present the foramina with raised borders, which characterise those in Plesiosauri, and which indicate the fibro-cartilaginous nature of the joints." One species of this genus has been found in the grès-bizarre deposits of the lower trias at Soulz-les-Bains. It is described by Von Meyer under the title of *N. Schimperi*.

NOTORNIS. A curious genus of birds, closely allied to the Gallinules and Water-hens. It was founded by Professor Owen for the reception of some bones sent from New Zealand in a fossil state, and found along with those of the Dinornis. From its osteological characters, the Hunterian Professor inferred that the bird must have been a large modified form of the same natural family of Grallatorial birds as the Porphyrio or Purple Hen; but, from the structure of the sternum, without the power of flight. Three years after, a specimen of the bird was taken alive in a remote part of the South Island of New Zealand; and this specimen, which is still unique, and seems likely to remain so, was sent to Dr. Mantell, and, after his death, passed into the noble collection of birds in the British Museum. The *Notornis Mantellii* was known to the natives of New Zealand by the name of Moho; and even now it is called by the settlers "Mountain-Mr'd." At one time it seems to have been abundant, and to have supplied to the inhabitants, ere Cook visited the islands, a favourite article of food. Partly from this cause, but also in some measure from the persecuting attacks of wild cats and dogs, it seems to be a species on the very verge of being exterminated. The specimen in the Museum was caught by some men engaged in the pursuit of seals, as they were searching the coasts for them and other game. These men observed the traces of a large bird on the snow; and after following the foot-tracks for a considerable distance, came upon the bird, which ran with great speed, and was taken alive. It screamed loud, and fought and struggled violently. The body was eaten by the crew of the schooner, and was said to be delicious. The bird is about two feet high, has a short strong beak, which, with the legs, is of a bright scarlet colour. The neck and body are of a dark purple colour, the wings and back being shot with green and gold. The tail is white beneath, and the wings are feeble in structure and plumage.

NOTOTHERIUM. A genus originally founded by Prof. Owen to indicate the characters of a large extinct herbivorous marsupial mammal (*N. Mitchelli*), and also agreeing with a second species, whose skeletal remains have been found in the pleistocene lacustrine deposits at King's Creek, Darling Downs, Australia. The last-named species was first noticed by Mr. Maclean, under the title of *Zygomaturus trilobus*. Structurally speaking, the Nototheria appear to have resembled the Kangaroos, and at the same time to have approached the singular existing *Phascolarctus* or Koala.

ODONTOSAURUS. The generic title applied by Hermann Von Meyer to an extinct species of labyrinthodont reptile (*O. Voltzii*), whose fossil remains were found in the bunter sandstone of Soulz-les-Bains.

ORNITHICNITES. A term originally employed by Prof. Hitchcock of America, to denote a genus of gigantic birds, the impressions of whose feet occur in the sandstones of Connecticut. [See BRONTOZOUM.] This term, however, is more properly employed in a less restricted sense, to the impression of extinct birds' feet in general, in whatever geological formation they may be found.

OUDENODON BAINII. The generic and specific title given to an extinct cryptodont reptile, whose fossil remains were discovered in an argillaceous limestone in Southern Africa. These organic relics have been added to the vast palæontological treasures of the British Museum, and are described in Prof. Owen's recent Summary of Extinct Animals. The cryptodonts are those reptiles in which the teeth are either concealed or altogether absent.

OXYURIS. (Thread-worms.) A group of nematode Entozoa whose species have pointed tails; hence the generic name given by Deslongchamps. The different species are small, varying from 1-4th to an entire inch in length. The form most commonly known is the little *O. vermicularis*, which infests the lower bowel of children and often proves exceedingly troublesome, giving rise to disagreeable symptoms, if it be not got rid of by suitable aperients. The body is white, filiform, attenuated at either extremity, the head being furnished with three lobes similar to those of Ascaris. The recent researches of Dr. George Walter prove the Oxyurides to be possessed of an extraordinarily developed nervous system. They are unisexual, and are usually found crowded together by thousands, the females being nearly twice as numerous as the males. The history of the development of these creatures is not yet thoroughly understood.

PALAPTERIX. The generic name applied by Prof. Owen to certain extinct struthious birds, whose skeletal remains are found in the caverns, turbaries, and recent pleistocene deposits of New Zealand. The species have been included under the above title in consequence of their close affinity with the existing *Apterix*, inhabiting the same islands. The gigantic Dinornis displayed a very similar anatomical structure.

PALÆOPHIS. A genus of extinct ophidian reptiles, chiefly interesting from the circumstance that its members form the earliest record of this ordinal type upon our planet. Their fossil remains have been found in the eocene beds at Sheppey and Bracklesham, and it appears probable that some of the species attained a length of twenty feet.

PALÆOPHRYNOS. The generic name given to certain extinct tailless Batrachians or Toads, evidences of whose former existence occur in the tertiary strata at Œningen. Fossil remains of their closely allied friends, the frogs, are still more numerous in the eocene and pliocene formations of the European continent, and Prof. Owen has indicated one species of the latter group in the tertiary shales of Bombay.

PALÆOSAURUS. A genus of thecodont reptiles, whose fossil teeth were first described by Messrs. Riley and Stutchbury, and were obtained "from the dolomitic conglomerate at Redland, near Bristol," which deposit seems to be referable to the early triassic period. Associated with these remains were others which Prof. Owen has referred to a distinct genus, under the title of *Thecodontosaurus*. In regard to relations and other points of interest connected with the anatomy of these genera, Prof. Owen observes, that "the following conclusions may be drawn from the knowledge at present possessed of the osteology of the Thecodontosaurus and Palæosaurus; in their thecodont type of dentition, biconcave vertebræ, double-jointed ribs, and proportionate size of the bones of the extremities, they agree with the amphicælian crocodiles; but they combine a dinosaurian femur, a lacertian form of tooth, and a lacertian structure of the pectoral and probably pelvic arch with these crocodilian characters; and they have distinctive modifications, such as the moniliform spinal canal, in which, however, the almost contemporary Rhyncosaur participates. It would be interesting to ascertain whether the caudal vertebræ are characterised, as in the Thuringian Protosaur, by double diverging spinous processes." The fossil remains of the *Rhyncosaurus*, just alluded to, occur in the new red sandstone triassic beds of Shropshire, having been excavated at the Grinsill quarries, near Shrewsbury. The bones were accompanied by footprints probably referable to the same reptile. Respecting the remains of this animal, our distinguished palæontologist also remarks that "the resemblance of the mouth to the compressed beak of certain sea-birds, the bending down of the curved and elongated premaxillaries, so as to be opposed to the deep symphysial extremity of the lower jaw, are further indications that the ancient Rhynchosaur may have had its jaws encased by a bony sheath, as in birds and turtles, the dentinal ends of the premaxillaries projecting from or forming the deflected end of the upper mandible. There are few genera of extinct reptiles of which it is more desirable to obtain the means of determining the precise modifications of the locomotive extremities than the Rhyncosaurus. The fortunate preservation of the skull has brought to light modifications of the lacertine structure leading towards Chelonia, and birds which were before unknown; the vertebræ likewise exhibit very interesting deviations from the lacertian type. The entire reconstruction of the skeleton of the *Rhynchosaurus* may be ultimately accomplished, if due interest is taken in the collection and preservation of the fossils of the Grinsell quarries." The skull of a Rhynchosaur has been found in the so-called sandstones near Elgin.

PARTHENOGENESIS. A term proposed by Prof. Owen to indicate "procreation without the immediate influence of the male," which thus might be understood to embrace within its meaning all the phenomena of dissimilar digenesis or alternate generation, as well as some modes of extension of the species which are not strictly referable to alternate generation. Parthenogenesis includes reproduction by spontaneous division or fission, such as we see taking place in *Actiniadæ*, by budding or gemmation, such as we see in *Hydridæ* and polyps generally, and by germ-cells, as in the sexual viviparous *Aphides*. Both in plants and animals true sexually mature individuals have been produced without the immediate impregnation of the ova or germs from which they have directly originated. [For details of Parthenogenesis, see ALTERNATION OF GENERATION above.]

PHALLUSIA. A name proposed by Savigny to indicate a genus of tunicated mollusca or Ascidians. It is synonymous with the genus *Cynthia*, in which the sessile body is covered by a leathery tunic furnished with quadrangular anal and branchial orifices. [See ASCIDIA, p. 39.]

PHASCOLOMYS GIGAS. At p. 745 will be found a description of the Wombat (*P. wombat*), which had an ancient representative of comparatively gigantic proportions. This old marsupial, which has been described under the generic and specific title above given, appears to have attained the magnitude of our existing tapir, as is evinced by the size of its teeth, procured from the pleistocene deposits of the Australian continent.

PHASCOLOTHERIUM. A highly interesting genus of marsupial quadrupeds, the fossil remains of which have been found in the Stonesfield slate. The species was first described by Mr. Broderip from a specimen which he purchased, when a student at Oxford, from an old stonemason, who brought it fresh from the quarry. He named it *Didelphis Bucklandi*, in compliment to the able author of the Bridgewater Treatise on Mineralogy and Geology. Professor Owen has shown how nearly it was related to some of the living marsupials of Australia, and has indicated the great similarity of, and interesting correspondence between, other existing forms now confined to the Australian continent and the circumjacent seas, and the organic remains of the British oolite. For instance, the bony "palates," as they have been called, which are found in that formation, have been evidently the teeth of gigantic forms of cartilaginous fishes allied to the existing *Cestracion*, a Ray-like shark inhabiting the seas of Australia. *Trigoniæ*, interesting bivalves confined to the Australian seas, are met with in a fossil state in the same beds which contain the Phascolotherium. The plants, too, found in the oolite are closely allied to the Araucaria and different Cycadeous plants which grow in Australia. This curious genus is only known by the jaw, which is preserved in the British Museum. Some naturalists had referred it to the reptile class; but the detailed examination and reasoning of Professor Owen have shown the correctness of its original describer in referring it to the marsupial animals. It is one of the oldest mammiferous remains which have yet occurred on the globe.

PISTOSAURUS LONGÆVUS. The name applied by Von Meyer to indicate a species and genus of sauropterygian reptiles or fin-lizards, whose fossil remains have been discovered in the muschelkalk formation at Bayreuth.

PLACODUS. A genus of extinct vertebrates generally classed with the fishes, but recently pointed out by Prof. Owen to be referable to his sauropterygian order of reptiles. The genus embraces many species, whose chief characteristics lay in the possession of powerful internal palatine and pterygoidean teeth, in addition to the ordinary well-marked dentition of the jaws. The grinding teeth are massive, resembling paving stones, and "the size of the last tooth in *P. laticeps* surpasses that of any of the teeth in the previously discovered species. In proportion to the entire skull, it is the largest grinding tooth in the animal kingdom, the elephant itself not excepted. We cannot contemplate," adds the Professor, "the extreme and peculiar modification of form of the teeth in the genus Placodus, without a recognition of their adaptation to the pounding and crushing of hard substances, and a suspicion that the association of the fossils with shell-clad mollusks in such multitudes as to have suggested special denominations to the strata containing *Placodus* (e. g. muschelkalk, terebratulitenkal, &c.), is indicative of the class whence the Placodi derived their chief subsistence." As will be readily understood, several interesting cranial peculiarities co-ordinated with these astonishing dental modifications.

PLAGIAULAX. In the so-called "dirt-bed" formation, Mr. Beccles has discovered a number of extremely interesting fossils, among the most valuable of which are remains of two extinct mammals, severally described by Dr. Falconer under the titles of *P. Becclesii* and *P. minor*. These vertebrates are supposed by Prof. Owen to have been carnivorous marsupials, having some close affinities to a much larger marsupial (*Thylacoleo*), whose fossil remains occur in the newest tertiary deposits of Australia. The generic name has reference to the oblique grooving on the crowns of the grinding teeth.

PLEUROSTERNON. The generic name applied by Prof. Owen to several extinct tortoise-like reptiles, whose fossil remains occur in the freshwater limestone of Purbeck. The fossils representing the various species of this genus, associated with other allied Chelonians of the Wealden formation, are all described and figured in a monograph, forming part of the Palæontographical Society's publication for the year 1853.

PLIOLOPHUS VULPICEPS, or Fox-headed Plioloph. In the fourteenth volume of the "Journal of the Proceedings of the Geological Society," Prof. Owen has enriched the annals of mammalian palæontology with an elaborate memoir on the above-named animal. In his view, the fossil remains of this small, extinct hoofed herbivore form the best and most instructive record of mammalian life hitherto obtained from the eocene deposits of the tertiary age. The entire skull of the Pliolophus was discovered embedded in an ovoid septarian nodule about a foot in diameter, in the so-called Roman cement bed of the London clay, at Kyson, in Suffolk. The specimen was transmitted by the Rev. R. Bull, of Harwich, to the Professor, by whom we are informed that the cranium presented a peculiarity of "dentition not exhibited by any later or existing species of mammal." The Pliolophus displays close affinities to the extinct Palæotherium and the existing Tapirs.

PLIOPITHECUS ANTIQUUS. The name employed by Gervais to denote an extinct ape (allied to *Hylobates*), whose fossil remains occur in the miocene formation of Southern France.

PLIOSAURUS. The name given by Prof. Owen to a genus of sauropterygian reptiles, or finned lizards, whose fossil remains "are peculiar to the Oxfordian and Kimmeridgian divisions of the upper oolitic system." Some members of the genus have been discovered in similar beds in Russia. The most marked characters of the genus are those which have reference to the short, thick, conical, subtrihedral teeth, and more particularly to the abbreviated neck, in which, as in the allied Ichthyosaur, the vertebræ are much flattened and compressed together. The trunk and extremities resembled those of *Plesiosaurus;* and, notwithstanding the shortness of the neck, it is conjectured that some of the Pliosaurs attained a length of nearly forty feet.

PLUTEUS. This name was originally employed to indicate a genus of minute marine animalcules, the forms of which are now known, chiefly through the researches of Prof. John Müller, of Berlin, to be merely larval conditions of various species of Sea-urchin (*Echinidæ*). *Bipinnaria* and *Actinotrocha* are also larval echinoderms, the former being the young of Asterias. Pluteus larvæ have a quadrangular pyramidal figure, with a central body surmounted by a dome-like eminence. They also usually display four or six arms or tentacles, the margins of which are clothed with cilia, and on this account they bear some resemblance to the ciliograde Medusæ. The young Echinus or the Ophiurus is developed on one side of the body of the Pluteus, becomes detached, and ultimately carries away the digestive organs of the poor larva, which, being thus unceremoniously deprived of its stomach, gradually succumbs to the force of circumstances, and soon perishes. The Bipinnaria is supplied with twelve or fourteen non-ciliated arms; it has a long cylindrical body, forming a

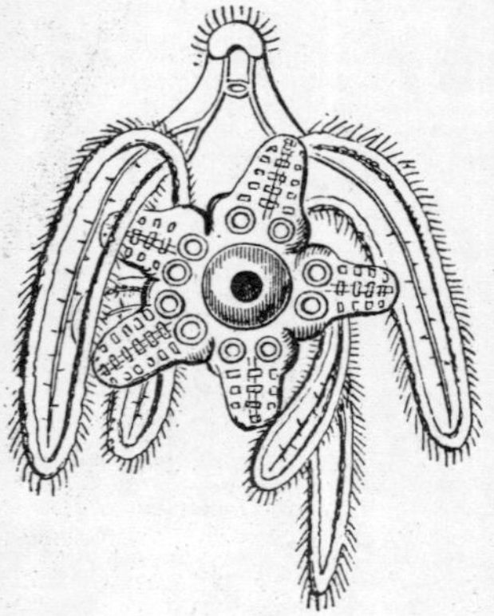

PLUTEUS, OR LARVAL OPHIURUS.

kind of hood at the upper part, on the summit of which the young Star-fish (*Asterias*) is developed, in a similar manner to the Ophiurus and Echinus.

POLYPTYCHODON. A genus of finned lizards closely allied to the above, whose fossil remains "have hitherto been met with only in the cretaceous formations: in the green sand of Kent and Cambridge, also at Kursk, in Russia, and in the chalk of Kent and Sussex." The generic name implies that the conical crowns of the teeth are surrounded by *numerous* longitudinal *ridges*. The *P. interruptus* is supposed to have equalled the Pliosaur in respect of bulk.

POLYZOA. [See BRYOZOA above.]

POTAMOCHŒRUS. Dr. Gray characterises this distinct genus of the family Suidæ by the large bony protuberance with which the male is furnished on each side of his face, about half-way between the end of the nose and the eyes; a feature whence the various species have derived their expressive name of "Wart-hog." The genus is also known by the ears in both sexes being elongated and rapidly attenuated, while they end in a pencil of long hairs. The tail is thick, long, and placed high up the back. The species are confined to Africa. The Wart-hog of South Africa (*P. larvatus*) is black, washed with grey, the sides of the face being white, and a large spot under the eyes being black. Like the rest of its family, it turns up the ground with its dilated muzzle, in search of the roots on which it principally feeds. It is a very active animal, and, from the great size of its head and the sharpness of its teeth, is a most formidable animal. Dr. Gray has lately described a fine new species, from the Camaroon River in West Africa, which is now in the Zoological Gardens in the Regent's Park. It is a very beautiful animal, for a pig, being of a rather bright reddish-brown; the face, forehead, ears, and some large blotches on the legs black. These colours are delicately contrasted with the pure white of the edges of the eyes, of the whiskers, of a streak over and under the eyes, and with the white continuous crested streak along the middle of the back. Dr. Gray has named it *Potamochœrus pictus*, the Camaroon Wart-pig. It is the prince of its family, and seems to thrive well in its new home.

PROTOROSAURUS. A name applied by Von Meyer to a genus of saurians, whose fossil remains (*P. Speneri*) were originally discovered in a copper mine near Eisenach, in Thuringia, at a depth of 100 feet below the surface of the ground. The specimen first described by a Berlin physician, named Spener, in the year 1710, may now be seen in the Hunterian Collection of the Royal College of Surgeons. In respect to the anatomical organisation of the Protorosaur, Prof. Owen remarks as follows:—"It may be concluded, from the length and strength of the tail, and the peculiar provision for muscular attachments in that part, and from the proportions of the hind limbs, that the Protosaurus was of aquatic habits, and that the strength of its head and neck, and the sharpness of its teeth, enabled it to seize and overcome the struggles of the active fishes of the waters which deposited the old Thuringian copper slates."

PUFF-ADDER. (*Clotho arietans.*) One of the most venomous of all the serpent tribe. It is a native of Africa, and seems to be a widely distributed inhabitant, the specimens from the Cape of Good Hope and from Egypt all evidently belonging to the same species. The name Puff-adder is derived from a habit which this deadly serpent has of inflating itself and puffing when provoked, a habit which may be easily seen in the Reptile House of the Zoological Gardens. The colour of this creature so closely resembles that of the sandy and stony places which it frequents, that it is difficult to notice it, especially as it flattens itself against the ground when not in an excited state. Dr. Burchell says, that the venom of this species takes effect so rapidly as to leave the person who is bitten no chance of saving his life but by instantly cutting out the flesh surrounding the wound. The same traveller informs us of a peculiarity which renders it more dangerous, and which it would be well for those to know, who are likely to fall in with it. Unlike most snakes, which dart forwards when irritated, this species, it is said, throws itself backwards, so that those ignorant of its habits would place themselves in the very direction of death, when they thought they were escaping the danger. The natives, Dr. Burchell says, by keeping always in front, are enabled to destroy the Puff-adder without much risk.

PYROSOMA. A singular genus of tunicated molluscs allied to the *Salpæ*. The individuals are very small, but are closely aggregated, and so combined as to form transparent cylinders three or four inches in length, and hollowed out at one extremity. Such a colony may be compared to the de-

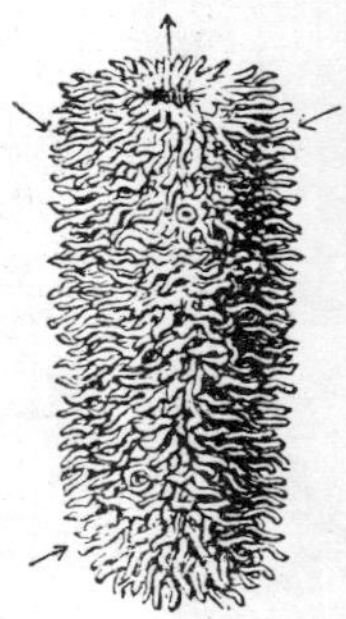

PYROSOMA.

tached finger of a glove, the external surface being, however, covered with minute tubercles. Each of these tubercles represents the anterior end of the individual (ascidiozooid), and below the conical prominent tubercles, which Prof. Huxley (in his recent valuable memoir in the twenty-third volume of the "Linnæan Transactions") calls the lip, lies the mouth, which leads to the digestive cavity within, terminating by an anal orifice into a general cavity (atrium), which latter opens by a minute orifice into the hollow of the cylinder, or, as it is more scientifically called, the *ascidiarium*. These beautiful creatures are inhabitants of tropical seas, and are extremely phosphorescent.

RAMPHORHYNCHUS. The name employed by Hermann von Meyer to designate a genus of Pterodactyle lizards, whose fossil remains abound in the Bavarian lithographic slates of the middle oolite. The species were furnished with long slender tails; and it is even thought that the anterior part of the muzzle was covered by a horny beak.

RHIZOPODA. Another name for the *Foraminifera*, constituting a class of minute animalcules belonging to the Protozoa, or lowermost division of the animal kingdom. Most of the Rhizopods are encased in a horny or calcareous envelope, which frequently assumes the form of a many-chambered shell, and on this account they were formerly thought to be molluscous animals. For similar reasons they are sometimes denominated *Polythalamia*. Anatomically their structure is exceedingly simple, for they only consist of a gelatinous mass, portions of which, during life, are thrust out, in the form of root-like processes or false feet (pseudopods). These so-called feet are prehensile, and by contraction serve to convey small particles of organic matter into the general mass for the purposes of nourishment and growth. They have no special sensory organs or digestive apparatus. The simplest form of Rhizopod is the *Amœba diffluens*, which will be found described under the title of PROTEUS, at p. 548. Among the more complicated forms may be mentioned the *Nummulites*, whose fossil shells were extraordinarily abundant during the tertiary epoch. The pyramids of Egypt are entirely composed of nummulitic limestone.

RHYNCHOSAURUS. [See PALÆOSAURUS above.]

SALPA. A singular and interesting genus of covered or tunicated molluscs. The species live in open seas ; each individual exhibits an elongated form, and is covered by an external transparent tunic, which is highly elastic and open at both ends. This tunic is cartilaginous in texture, and has been found to contain cellulose, a substance chemically identical with that forming the cellular basis of plants. Two kinds of Salpæ have been described, namely, the *solitary* and *aggregated;* but these forms are only varying conditions of the same species. The

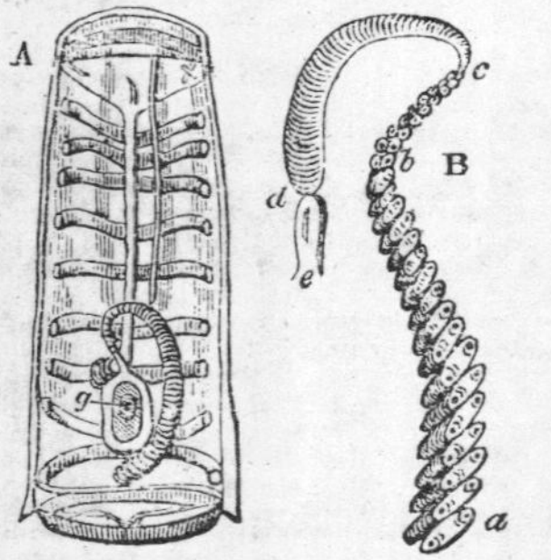

SALPA : A, SOLITARY ; B, AGGREGATED.

anatomical structure, though differing in details, conforms to the molluscan type ; but some of the life phenomena exhibited by these singular creatures are altogether unique: thus the circulation of the blood, instead of proceeding in one uniform direction, becomes periodically reversed, the alternating currents occurring at regular intervals. Here, therefore, we have a physiological phenomenon utterly at variance with our ordinary experience, and one which, while it curiously illustrates the varying results capable of being produced by a simple mechanical contrivance, at the same time suggests a caution as to the conclusions which we seem entitled to draw from known anatomical facts.

SCELIDOSAURUS. A genus of gigantic dinosaurian reptiles, whose skeletal remains have been discovered in the liassic deposits at Charmouth in Dorsetshire.

SCELIDOTHERIUM. The name given by Prof. Owen to a genus of extinct terrestrial Sloths, whose species attained gigantic proportions and are closely allied to the Megatherium.

SIEBOLDIA MAXIMA. Under this combined generic and specific title the gigantic Salamander of Japan is now very generally described, although it has hitherto been noticed under other names. It is by far the largest of all living naked Amphibia, and is said to be capable of attaining a length of upwards of three feet. It is only equalled by the closely allied extinct reptile of Scheuchzer, found in the tertiary freshwater beds at Œningen, and which acquired so much celebrity under the title of *Homo Diluvii Testis*. A living example of the Sieboldia may now be seen in the Zoological Society's Gardens, Regent's Park. It was originally purchased in the market at Nagasaki by Captain Charles Taylor, of the ship Tung Yu, and was brought to England in a wooden tub placed on deck. This reptile is only found in the island of Niphon, inhabiting the lakes and streams of the volcanic range of mountains of the interior. In the full-grown state there is no gill aperture present, the branchiæ disappearing at an early period. As in our common Salamanders, the fore feet are furnished with four toes, the posterior extremities being five-toed or pentadactylous. The eyes are remarkably small, and at first almost escape observation amongst the tubercles and dark spots by which the plicated skin is marked. The cuticle is frequently shed, and when microscopically examined is found to consist of numerous hexagonal flattened cells, very similar to those of the Water-newt. It is a curious habit of the Salamanders to assist in the shedding of their outer skin by seizing loose portions with the mouth and dragging it from before backwards over the tail. Frogs occasionally do the same thing.

SIVATHERIUM. A genus of extinct ruminating quadrupeds, the remains of which were discovered by Dr. Falconer and Major Cautley, on the Sewalik Hills, in India. These naturalists named the genus after *Siva*, an Indian deity. The specimens found by them were presented, with the rest of their collection, to the British Museum. The creature must have resembled a gigantic antelope in the shape of its body, while its head must have borne some likeness to that of an elk. The head was short and thick, and had two pairs of horns ; the front pair small, the hind pair much larger, and, in all probability, palmated and set behind. The eyes were small, and placed on the sides of the head. The lips must have been great ; and, from the large projecting bone over the nasal cavity, it must have had a very long proboscis, an organ not found in any existing ruminating animal. The skull in bulk nearly approaches that of an elephant ; the neck was shorter than in the giraffe, and much stronger, and well suited to support the weight of the heavy head with its two pairs of horns, the hinder of which were most probably massive. There seems to have been two species of this singular genus. The larger has given the two naturalists above cited materials for a fine memoir on its peculiarities. They named it *Sivatherium giganteum*; the second species is named *Sivatherium Perimense*. They are found associated with the remains of extinct and extraordinary species of Elephants, Rhinoceros, Camels, and other quadrupeds; the specimens of which, carefully worked out of the solid stone, are among the most interesting monuments of epochs, compared with which in remoteness, Indian antiquity is but of yesterday.

SPALACOTHERIUM. A genus of extinct mole-like insectivorous mammals founded by Prof. Owen from the examination of fossil remains discovered by Mr. W. R. Brodie in the so-called dirt-bed at Purbeck in Dorsetshire. The Spalacothere is of especial interest in a palæontological point of view, from the circumstance that its skeletal records point to the occurrence of mammalian life "about midway between the older oolitic and the oldest tertiary periods."

STEREOGNATHUS. The generic title of a small extinct quadruped, whose fossil remains occur in a fragmentary condition in the Stonesfield oolite. The teeth and other fossils of this little mammal seem to indicate that it was a hoofed herbivore, although the molars displayed a marked approximation to the insectivorous type.

STROBILA. A term applied to a group of animal forms associated together so as to constitute a colony of individuals arranged in linear series. Thus when the term is used in the works of those who have studied the Entozoa or internal parasites, then it is applicable to the full-grown tape-worm, which is in reality a colony of several hundred individuals in single file. The term is also given to a similarly constituted chain of larval jelly-fishes, in which, however, none of the individuals attain sexual maturity whilst they remain attached. [For a detailed account of the singular metamorphoses of the Strobila, see *Art.* ALTERNATION OF GENERATION, in the present SUPPLEMENT.]

TELERPETON. The name assigned by the late Dr. Mantell to a curious reptile, remains of which were found in a fossil state by Captain Brickenden and Patrick Duff, Esq., at Elgin, in the old red sandstone. Dr. Mantell, from the remains, was unable to make out whether they should be referred to the lizards or the aquatic salamanders, as the cranium was in a mutilated condition, and it was impossible to ascertain the mode in which the teeth had been implanted. From the length and character of the ribs, the situation of the pelvis, and the well-developed limbs, the fossil exhibits affinities to the lizards. The most remarkable character is the great slenderness of the ribs, from which circumstance Professor Owen applied to the genus the name of *Leptopleuron*. There are twenty-four pairs of them, and they appear to have been attached by a simple head, as in the lizards. The great interest of these remains consists in their having afforded the first certain proofs of the existence of reptiles at the time the strata of the old red sandstone were deposited. From the recent discovery of vegetable remains and other interesting observations, it is to be hoped that Mr. Miller will add some other chapters to his interesting work on the Old Red Sandstone. Dr. Mantell has called this reptile *Telerpeton Elginense*, from the locality where it was found.

THECODONTOSAURUS. [See PALÆOSAURUS, above.]

THYLACOLEO. In the "Philosophical Transactions" for 1859, Prof. Owen has described a remarkable extinct carnivorous marsupial, whose bulk and proportions appear to have been equalled only by our existing African lion. The fossil remains of this formidable quadruped (*T. carnifex*) are found imbedded in the pliocene strata of the Australian continent.

THYLACOTHERIUM. A name synonymous with AMPHITHERIUM (which see). Professor Owen has described a second species from the curious beds of Stonesfield slate, which he has named after Mr. Broderip. These remains are particularly interesting to the naturalist, as they prove the great antiquity of the mammalian type of organisation. Professor Owen has compared these remains with the bones of the *Myrmecobius*, a recent insect-eating quadruped of Australia. In the same beds in which the so-called Thylacotheria occur, the wing covers of coleopterous insects are met with. Like the shrews and moles, these quadrupeds seem to have been particularly adapted for living on insects, their molar teeth being covered with sharp points, which crush the hard cases of the beetles and other insects.

TIGER-WOLF. A rare marsupial quadruped, indigenous to Van Diemen's Land, where it is now becoming scarce. Since the year 1850, there has been a pair of these curious animals in the Zoological Gardens, where they attract great attention from the striped tiger-like markings on their backs. From the observations of Mr. Gunn, it would appear, that before the introduction of civilisation into their native island, these carnivorous creatures must have lived chiefly on phalangers and kangaroos, as they obstinately refuse the flesh of the wombat, which is one of the most common inhabitants of their district. Since the introduction of sheep into Van Diemen's Land, the Tiger-wolf has found a favourite food; and so extensive have been their ravages among the flocks, that the shepherds have been obliged to destroy them by every possible means. They are extremely active animals, and, from the ruthless persecution they encounter, are likely soon to be extirpated. The species belongs to the genus Thylacinus; and from its head, resembling that of a wolf or dog, it has been named *T. cynocephalus*. Professor Owen has found traces of what appears from the fossil bones to have been a second species of the genus. These remains were found by Sir Thomas Mitchell in the caverns of Wellington Valley, in New South Wales.

TROGONOTHERIUM. This generic title refers to a species of extinct rodent, closely allied to our existing beaver, but possessing comparatively gigantic proportions. Its fossil remains occur in caverns and recent tertiary deposits of Europe, associated with the skeletal débris of several allied genera.

XYPHODON. Under this name two or more species of extinct mammalia have been described (*X. gracile* and *X. Geylensis*), whose affinities are so strongly connected with the *Anoplotherium* that one of the forms was originally described by Cuvier under the last-named genus. The *Xyphodon* is also closely related to the *Dichodon* [above described].

ZEUGLODON. The generic name of a gigantic species of extinct whale, whose fossil remains occur in the miocene strata of Europe and North America. The above title—equivalent to yoke-toothed animal—has been given in consequence of a remarkable character affecting the posterior teeth, the crowns of which are curiously contracted, so as to exhibit the peculiar form here illustrated. Under the name of *Basilosaurus*, Dr. Harlan of America has described a cetacean answering to the same genus, and very probably a distinct species from that commonly known as *Zeuglodon cetoïdes;* whilst a third form has been indicated by M. Grateloup, who gave it the name of

TEETH OF ZEUGLODON

Squalodon. An almost entire skeleton of Zeuglodon has been discovered in Alabama, from which it is conjectured that the creature attained the length of about seventy feet. Its habits were evidently carnivorous.

ZYGOSAURUS. This name has been given by Eichwald to characterise a genus of Labyrinthodont reptiles, whose skeletal remains occur in the copper-bearing Permian strata at Orenburg.